# The Environmental Resource Handbook

2011/12

Sixth Edition

# The Environmental Resource Handbook

| | |
|---|---|
| Associations | Environmental Statistics |
| Research Centers | Green City Rankings |
| Environmental Health | Grants |
| Publications | Government Agencies |
| Educational Programs | Consultants |
| Environmental Law | Green Product Catalogs |
| Trade Shows | Web Sites |

A Universal Reference Book

Grey House Publishing

PUBLISHER: Leslie Mackenzie
EDITORIAL DIRECTOR: Laura Mars
STATISTICS EDITOR: David Garoogian

PRODUCTION MANAGER: Kristen Thatcher
PRODUCTION ASSISTANTS: Diana Delgado, Erica Schneider, Marko Udovicic

MARKETING DIRECTOR: Jessica Moody

A Universal Reference Book
Grey House Publishing, Inc.
4919 Route 22
Amenia, NY 12501
518.789.8700
FAX 845.373.6390
www.greyhouse.com
e-mail: books@greyhouse.com



First edition published 2001
Sixth edition published 2011
Printed in Canada

The environmental resource handbook. – 6th
Ed. (2011/12)
1014 p. 27.5 cm.
Annual
Spine title: ERH

1. Environmental protection – United States – Directories. 2. Environmental agencies – United States – Directories. 3. Conservation of natural resources – Societies, etc. – Directories. 4. Nature conservation – Societies, etc. – Directories. 5. Environmental protection – United States – Bibliography. 6. Conservation of natural resources – Bibliography. 7. Nature conservation – Bibliography. I. Grey House Publishing, Inc. II. Title: ERH.

GE20.E586
363'7 – dc21
ISBN: 978-1-59237-739-8 softcover

# Table of Contents

## SECTION ONE: RESOURCES

## SECTION TWO: STATISTICS & RANKINGS

# Introduction

This sixth edition of *The Environmental Resource Handbook* offers immediate access to a unique combination of 6,433 resources and close to 200 statistical and ranking charts, tables and maps, all revised with the most current information available.

Comprehensive and thoughtfully arranged for a variety of users, including environmentalists, educators, researchers and students of environmental studies, this latest edition includes thousands of ways to reach exactly the person or place you need, including 3,807 emails, 5,297 web sites, and 6,566 key contact names.

**Praise for previous editions:**

*". . . includes the most current data necessary to understand the changes in energy consumption, solid-waste production, and carbon emissions. . . . Some charts date back to 1940 and will prove valuable to researchers seeking to identify broader trends."*

Library Journal

*". . . This handbook will be valued by anyone working in an environmental area, or just interested in where to get more information on just about any environmental topic. It should be contained in the reference section of all public and university libraries."*

American Reference Book Annual

## Section One: Resources

Includes 15 major chapters from **Associations** to **Green Product Catalogs**, outlined below, which are further arranged into 68 subchapters that organize the wealth of information in this directory by specific environmental issues from **Air & Climate** to **Water Resources.**

1. **Associations & Organizations** disseminate information, host seminars, provide educational literature and promote studies. These listings are defined by both national categories and state listings.

2. **Awards & Honors** list organizations that recognize education and business professionals for excellence in environmental sciences.

3. **Conferences & Trade Shows** include large conventions, as well as small, specialized conferences, and offer information, strategies and solutions for dozens of environmental issues.

4. **Consultants** offer environmental consulting services, such as hazardous material screening, construction requirements, and habitat conservation.

5. **Environmental Health** addresses health issues caused by the environment. These listings are divided into two sections of Associations – both general and those that

focus on environmental issues affecting **Pediatric Health** – plus a third section on pediatric health publications.

6. **Environmental Law** offers resources for legal solutions and advocacy that are needed when environmental issues turn into legal ones.

7. **Financial Resources** include grants, foundations and scholarships to finance educational programs, research and environmental clean-up programs.

8. **Government Agencies & Programs** list Federal and State agencies, and includes a separate state-by-state list of **National Forests, Parks, and Refuges**.

9. **Publications** include books and periodicals that focus on the environment as a business, course of study, subject of activism, and scientific research. In addition to individual publications, this section includes **Environmental Library Collections** and **Publishers**.

10. **Research Centers** helps those looking to research environmental issues. The centers listed here are divided into two categories – those that operate within universities, and those that are run by commercial corporations.

11. **Educational Programs** are divided into two sections – public and private educational institutions, and workshops and camps. These listings offer environmental educational experiences in a wide variety of specific topics for beginning students, professionals new to the work force, and seasoned environmentalists.

12. **Industry Web Sites, Online Databases, and Videos** offer ways to connect electronically to environmental resources.

13. **Green Product Catalogs** lists the best catalogs for environmentally friendly products for home, office and everything in between, from tree-free paper to cruelty-free aromatherapy products.

## Section Two: Statistics & Rankings

The hundreds of tables, charts and maps in this section comprise 17 main topics from **Agriculture** to **UV Index,** plus valuable **Green Metro Area Rankings.** They have been updated with the most current data available, and include a variety of maps, graphs, and pie charts to make the information as accessible as possible.

You will find a great number of state statistics and rankings, making it easy to compile environmental snapshots by state, as well as to compare individual states, counties, and regions of the country. This section is designed to show which communities are taking an active role in protecting our environment and preserving our natural resources. Using the most current data available from more than 40 sources, this section helps complete the picture for those doing research, conducting business, or providing education and consulting services.

## Section Three: Appendices

- **Abbreviations & Acronyms**, provided by the Environmental Protection Agency, contains over 1,000 ways to identify and define the political and educational language of the industry.
- **Glossary of Terms**, also from the EPA, defines nearly 2,000 commonly used environmental terms in non-technical language.

## Section Four: Indexes

- **Entry**: Lists all entries alphabetically, identified by record number.
- **Geographic**: Lists entries alphabetically by state.
- **Subject**: Facilitates a fine-tuned search of resources and statistical tables by more than 250 specific environmental categories.

In addition to this print directory, *The Environmental Resource Handbook* is available for subscription on G.O.L.D, Grey House OnLine Databases. This gives you immediate access to the most valuable US environmental contacts, plus offers easy-to-use keyword searches, organization type and subject searches, hotlinks to web sites and emails, and so much more. Call 800-562-2139 or visit http://gold.greyhouse.com for more information.

## User Guide

Descriptive listings in *The Environmental Resource Handbook (ERH)* are organized into 15 chapters and 68 subchapters. You will find the following types of listings throughout the book:

- Associations & Organizations
- Conferences & Trade Shows
- Print & Electronic Media
- Foundations
- Government Agencies
- Research Centers
- Educational Programs
- Catalogs

Below is a sample listing illustrating the kind of information that is or might be included in an Association entry. Each numbered item of information is described in the paragraphs on the following page.

---

**12345**

**Water Environment Association of South Central US**
1762 South Major Drive
Suite 200
New Orleans, LA 98087

800-000-0000
058-884-0709
058-884-0568

info@wenvi.com
www.wenvi.com

Barbara Pierce, Executive Director
Diane Watkins, Marketing Director
Robert Goldfarb, Administrative Assistant
Ann Klein, Wastewater Consultant

The mission of the Association is to develop and disseminate information concerning waste quality management and the nature, collection, treatment, and disposal of wastewater. The Association publishes information, including a monthly magazine, manages a web site, and offers workshops and consultation on health and legal issues. A variety of educational programs are offered throughout the year on the history, ecology and culture of local rivers and streams, both on site, and in community schools.

Founded 1964

*18 pages*

*Monthly*

## User Key

**Record Number:** Entries are listed alphabetically within each category and numbered ,sequentially. Entry numbers, rather than page number, are used in the indexes to refer to listings.

**Title:** Formal name of company or association. Where association names are completely capitalized, the listing will appear at the beginning of the alphabetized section.

**Address:** Location or permanent address of the association.

**Toll-Free Number:** This is listed when provided by the association.

**Phone Number:** The listed phone number is usually for the main office of the association, but may also be for the sales, marketing, or public relations office as provided.

**Fax Number:** This is listed when provided by the association.

**E-Mail:** This is listed when provided, and is generally the main office e-mail.

**Web Site:** This is listed when provided, and is also referred to as an URL address. These web sites are accessed through the Internet by typing http://before the URL address.

**Key Personnel:** Names and titles of department heads of the association.

**Association Description:** This paragraph contains a brief description of the association and their services.

**Year Founded:** The year in which the association was established or founded. If there has been a name change, the founding date is usually for the earliest name under which it was known.

**Number of Pages:** Number of pages if the listing is a publication.

**Frequency:** How often it is published if it is a publication.

# Associations & Organizations

## National: Air & Climate

1 **Action Environment**
38 Tobin Crescent
A1A-2J3
709-722-1925
To take an active role in protecting, restoring and enhancing the environment; committed to taking an advocacy and activist role in our community.
*Majorie Evans, Contact*

2 **Action on Smoking & Health (ASH)**
1000 10080 Jasper Avenue
Edmonton, AB T5J 1-1V9
780-426-7867
Fax: 780-426-7872
E-mail: website@ash.ca
http://www.ash.ca
To reduce and prevent tobacco use.

3 **Air and Waste Management Association**
One Gateway Center, 3rd Floor
420 Fort Duquesne Boulevard
Pittsburgh, PA 15222
412-232-3444
800-270-3444
Fax: 412-232-3450
E-mail: info@awma.org
http://www.awma.org
A nonprofit, nonpartisan professional organization that enhances knowledge and expertise by providing a neutral forum for information exchange, professional development, networking opportunities, public education, and outreach to morethan 8000 environmental professionals in 65 countries.
*Founded: 1907*
*Jeffry Muffat, President*
*Merlyn Hough, President Elect*

4 **American Meteorological Society**
45 Beacon Street
Boston, MA 2108-3693
617-227-2425
Fax: 617-742-8718
E-mail: kseitter@ametsoc.org
http://www.ametsoc.org
Promotes the development and dissemination of information and education on the atmospheric and related oceanic and hydrologic sciences and the advancement of their professional applications.
*Founded: 1919 14,000 Members*
*Jon Malay, President*
*Louis Uccellini, President Elect*

5 **Center for Clean Air Policy**
750 First Street NE
Suite 940
Washington, DC 20002
202-408-9260
Fax: 202-408-8896
http://www.ccap.org
Significantly advances cost-effective and pragmatic air quality and climate policy through analysis, dialogue and education to reach a broad range of policy-makers and stakeholders worlwide.
*Founded: 1985*
*Tony Earl, Chairman*
*Robert Shinn, Vice Chairman*

6 **Clean Air Council**
135 S 19th Street
Suite 300
Philadelphia, PA 19103
215-567-4004
Fax: 215-567-5791
E-mail: joe_minott@cleanair.org
http://www.cleanair.org
A member-supported, non-profit environmental organization dedicated to protecting everyone's right to breathe clean air. Works through public education, community advocacy, and government oversight to ensure enforcement ofenvironmental laws.
*Founded: 1967*
*Russ Allen, President*
*Patrick J Feeley, Vice President*

7 **Climate Institute**
900 17th Street NW
Suite 700
Washington, DC 20006
202-552-4723
Fax: 202-737-6410
E-mail: info@climate.org
http://www.climate.org
Catalyze innovative and practical policy solutions to protect the balance between climate and life on Earth.
*Founded: 1986*
*John Topping, President*
*Mark Goldberg, Chairman*

8 **Conservation International**
2011 Crystal Drive
Suite 500
Arlington, VA 22202
703-341-2400
800-429-5660
http://www.conservation.org
Building a strong foundation of science, partnership and field demonstration, CI empowers societies to responsibly and sustainably care for nature, our global biodiversity, for the well-being of humanity.
*Peter Seligmann, Chairman/CEO*
*Russell Mittermeier PhD, President*

9 **Institute for Global Environmental Strategies**
1600 Wilson Boulevard
Suite 600
Arlington, VA 22209
703-312-0823
Fax: 703-312-8657
E-mail: info@strategies.org
http://www.strategies.org
A leader in Earth and space science education, communication and outreach, and in fostering national and international cooperation in global Earth observations.
*Nancy Colleton, President*
*Adrian Ash, Office Manager and Project A*

10 **Institute of Clean Air Companies**
1220 N. Fillmore Street
Suite 410
Arlington, VA 22201-4535
703-812-4811
Fax: 202-331-1388
E-mail: icacinfo@icac.com
http://www.icac.com
The national association of companies that supply air pollution monitoring and control systems, equipment, and services for stationery sources. Members are leading manufacturers of equipment to monitor and control emissions ofparticulate, VOC, SO2, NOx, air toxics and greenhouse gases.
*Founded: 1960*
*David C Foerter, Executive Director*

11 **Institute of Global Environment and Society**
4041 Powder Mill Road
Suite 302
Calverton, MD 20705-3106
301-595-7000
Fax: 301-595-9793
E-mail: www@cola.iges.org
http://www.iges.org
Established to improve understanding and predicition of the variations of the Earth's climate through scientific research and the tools necessary to carry out this research with society as a whole.
*Jagadish Shukla, President*
*Anastasia Shukla, Business Manager*

12 **International Center for Arid and Semiarid Land Studies: Texas Tech University**
Texas Tech University
601 Indiana Avenue
PO Box 45004
Lubbock, TX 79409-1036
806-742-3667
Fax: 806-742-1286
http://www.iaff.ttu.edu/home/icasals
The purpose is to stimulate, coordinate and implement teaching, research, and public service activities concerning all aspects of the world's arid and semiarid regions, their people and their problems.
*Founded: 1966*
*Aderbal C. Correa, Director*
*Heather Bradley, Administrative Business Asst*

13 **International Research Institute for Climate and Society**
Columbia University/Lamont Campus
Monell Building, 61 Route 9W
Palisades, NY 10964
845-680-4468
http://iri.columbia.edu
Enhances the society's capability to understand, anticipate and manage the impacts of seasonal climate fluctuations in order to improve human welfare and the environment, especially in developing countries.
*Ann Binder, Manager*

**14 National Association of Clean Air Agencies**
444 North Capitol Street NW 202-624-7864
Suite 307 Fax: 202-624-7863
Washington, DC 20001 E-mail: 4cleanair@4cleanair.org
http://www.4cleanair.org
State and local air pollution control officials formed NACAA (formerly STAPPA/ALAPCO) to improve their effectiveness as managers of air quality programs. It encourages the exchange of information among officials, enhances communicationand cooperation among federal, state and local regulatory agencies, and promotes good management of air resources.
*Dave Shaw, Co-President*
*Lynne Liddington, Co-President*

## National: Business & Education

**15 20/20 Vision**
1463 E. Republican St. #182 206-686-2777
Seattle, WA 98112 Fax: 301-587-1848
E-mail: vision@2020vision.org
http://www.2020vision.org
A national grassroots nonprofit organization that works to increase citizen participation in public policy related to peace and the environment. Each month 20/20 Vision produces an action-alert postcard so members can quickly andeasily contact policymakers and weigh in on timely issues. Works with government and corporations and collaborates with dozens of groups and experts.
*Founded: 1994*
*Charles Hamilton, Co-Owner*
*Martin Potter, Co-Owner*

**16 American Association of Zoo Keepers**
3601 SW 29th Street 785-273-9149
Suite 133 Fax: 785-273-1980
Topeka, KS 66614-2054 E-mail: ed.hansen@aazk.org
http://www.aazk.org
AAZK is a nonprofit 501(c)(3) volunteer organization made up of professional zoo keepers and other interested persons dedicated to professional anmical care and conservation.
*Shane Good, President*
*Bob Cisneros, Vice President*

**17 American Chemical Society**
1155 16th Street Northwest 202-872-4600
Washington, DC 20036 800-227-5558
Fax: 202-872-4615
E-mail: help@acs.org
http://www.chemistry.org
Promotes the public perception and understanding of chemistry and the chemical sciences through public outreach programs and public awareness campaigns.
*Nancy B. Jackson, President*
*Bassam Z. Shakhashiri, President Elect*

**18 American Chemistry Council**
700 Second St., NE 20- 2-9 70
Washington, DC 20002 Fax: 20- 2-9 61
http://www.americanchemistry.com
Represents the companies that make the products that make a modern life possible, while working to protecet the environment, public health, and the security of our nation.
*Founded: 1872*
*Calvin M Dooley, President/CEO*
*Lisa Harrison, VP Communications*

**19 American Federation of Teachers**
555 New Jersey Avenue NW 202-879-4400
Washington, DC 20001 Fax: 202-879-4597
http://www.aft.org
Founded to represent the economic, social and professional interests of classroom teachers.
*Founded: 1916 1.4 Million Members*
*Randi Weingarten, President*
*Loretta Johnson, Executive VP*

**20 American Forest & Paper Association**
1111 Nineteenth St NW 202-463-2700
Suite 800 800-878-8878
Washington, DC 20036 E-mail: info@afandpa.org
http://www.afandpa.org
The national trade association of the forest, paper, and wood products industry and advances public policies that promote a strong and sustainable US forest products industry in the global marketplace.
*Founded: 1993*
*James B. Hannan, Chairman*
*Doyle R. Simons, First Vice Chairman*

**21 American National Standards Institute**
1899 L Street NW 202-293-8020
11th Floor Fax: 202-293-9287
Washington, DC 20036 http://www.ansi.org
The Institute's mission is to enhance both the global competitivness of US business and the US quality of life by promoting and facilitating voluntary consensus standards and conformity assessment systems, and safeguarding theirintegrity.
*Founded: 1918*
*Arthur E. Cote, Chairman*
*Joe Bhatia, President*

**22 American Public Works Association**
2345 Grand Boulevard 816-472-6100
Suite 700 800-848-2792
Kansas City, MO 64108-2625 Fax: 816-472-1610
http://www.apwa.net
An international educational and professional association of public agencies, private sector companies, and individuals dedicated to providing high quality public works goods and services.
*Founded: 1937*
*George Cromble, President*

**23 American Society for Testing and Materials International**
100 Barr Harbor Drive 610-832-9500
PO Box C700 http://www.astm.org
West Conshohocken, PA 19428-2959
One of the largest voluntary standards development organization in the world-a trusted source for technical standards for materials, products, systems, and services.
*Founded: 1898*
*James Thomas, President*
*Kenneth Pearson, Senior Vice President*

**24 Association for Educational Communications**
P.O. Box 2447 812-335-7675
Bloomington, IN 47402 877-677-AECT
E-mail: aect@aect.org
http://www.aect.org
The mission of the Association for Educational Communications and Technology is to provide leadership in educational communications and technology by linking professionals holding a common interest in the use of educational technologyand its application to the learning process.
*Barbara Lockee, President*
*Ana Donaldson, President Elect*

**25 Association of American Geographers**
1710 16th Street NW 202-234-1450
Washington, DC 20009 Fax: 202-234-2744
E-mail: gaia@aag.org
http://www.aag.org
The Association of American Geographers (AAG) is a scientific and educational society whose 6,500 members share interests in the theory, methods, and practice of geography. These interests are cultivated through Annual Meetings, twoscholarly journals (the Annals of the Association of American Geographers and The Professional Geographer), a monthly AAG Newsletter, and the

activities of its two affinity groups, nine regional divisions and 53 specialty groups.
*Founded: 1904*
*Kenneth E. Foote, President*
*Audrey L. Koboyashi, Vice President*

**26 Association of Environmental and Resource Economists (AERE)**
1616 P Street NW
Suite 600
Washington, DC 20036
202-328-5125
Fax: 202-939-3460
E-mail: voigt@rff.org
http://http://www.aere.org/

Offers a way to exhange ideas, stimulate research, and promote graduate training in resource and environmental economics. Members come from academic institutions, the public sector, and private industry.
*Founded: 1979*
*Catherine L. Kling, President*
*Wiktor L. Adamowicz, Vice President*

**27 Association of State Wetland Managers**
2 Basin Road
Windham, ME 4062
208-892-3399
Fax: 207-892-3089
E-mail: aswm@aswm.org
http://www.aswm.org

Nonprofit organization dedicated to the protection and restoration of the nations wetlands. Our goal is to help public and private wetland decision-makers utilize the best possible scientific information and techniques in wetlanddelineation, assessment, mapping, planning, regulation, acquisition, restoration, and other management.
*Founded: 1983*
*Jeanne Christie, Executive Director*
*Jon Kusler, Associate Director*

**28 Association of Zoos and Aquariums**
8403 Colesville Road
Suite 710
Silver Springs, MD 20910
301-562-0777
Fax: 301-562-0888
E-mail: generalinquiry@aza.org
http://www.aza.org

Non profit organization dedicated to the advancement of accredited zoos and aquariums in the areas of animal care, wildlife conservation, education and science.
*Founded: 1924*
*Jeffrey P. Bonner, President/CEO*
*L. Patricia Smmons, President Elect*

**29 Bank Information Center**
1100 H Street NW
Suite 650
Washington, DC 20005
202-737-7752
Fax: 202-737-1155
E-mail: info@bicusa.org
http://www.bicusa.org

The Bank Information Center partners with civil society to help countries influence the World Bank and other international financial institutions to promote social and economic justice and ecological sustainability.
*Athena Ballesteros, Director*
*Mamadou Goita, Executive Director*

**30 Biotechnology Industry Organization**
1201 Maryland Avenue NW
Suite 900
Washington, DC 20024
202-962-9200
Fax: 202-488-6301
E-mail: info@bio.org
http://www.bio.org

The mission of BIO is to be the champion of biotechnology and the advocate for its member organizations-both large and small.
*Founded: 1993*
*James C Greenwood, President/CEO*

**31 Bureau of Land Management**
Environmental Education Program
1849 C Street NW
Rm. 5665
Washington, DC 20240
202-208-3801
Fax: 202-208-5242
E-mail: woinfo@blm.gov
http://http://www.blm.gov/

Provides public education about public lands resources and management issues. Identifies educational needs and resource gaps and collaborates with partner groups, volunteers, schools, and other agencies. This group also makes productsand programs available to BLM field offices, communities and schools.
*Bob Abbey, Director*
*Mike Pool, Deputy Director*

**32 Business for Social Responsibility**
111 Sutter Street
12th Floor
San Francisco, CA 94104
415-984-3200
Fax: 415-984-3201
http://www.bsr.org

Works with its global network of more than 250 member companies to develop sustainable business strategies and solutions through consulting, research and cross-sector collaboration.
*Founded: 1992*
*Aron Cramer, President/CEO*
*Pamela Passman, Vice President*

**33 CERES (Coalition for Environmentally Responsible Economies)**
99 Chauncy Street
6th Floor
Boston, MA 2111
617-247-0700
Fax: 617-267-5400
E-mail: info@ceres.org
http://www.ceres.org

A national network of investors, environmental organizations and other public interest groups working with companies and investors to address sustainability challenges such as global climate change.
*Founded: 1989*
*Majora Carter, President*
*Julie Fox Gorte, Senior Vice President*

**34 CO2 Science**
PO Box 25697
Tempe, AZ 85285
480-966-3719
E-mail: staff@co2science.org
http://www.co2science.org

Reviews articles, books, and other educational materials, attempting to separate reality from rhetoric in the debate that surrounding the subject of carbon dioxide and global change. The Center maintains on-line intructions on how toconduct CO2 enrichment and depletion experiments in its Global Change Laboratory.
*Craig D. Idso, Founder*
*Sherwood D. Idso, President*

**35 Center for Policy Alternatives**
1875 Connecticut Avenue NW
Suite 710
Washington, DC 20009
202-387-6030
Fax: 202-387-8529
E-mail: info@cfpa.org
http://www.stateaction.org

The nation's leading nonpartisan progressive public policy and leadership development center serving state legislators, state policy organizations, and state grassroots leaders. CPA is a 501 (c)(3) nonprofit corporation with thesethree unique programs: Leadership Development; Policy Tools; and Network Building.
*Founded: 1976*
*Nan Grogan-Orrock, Co-Chair*
*Tim McFeeley, Executive Director*

**36 Chelonian Research Institute**
402 South Central Avenue
Oviedo, FL 32765
407-365-6347
Fax: 407-977-5142
E-mail: chelonianri@aol.com
http://http://www.chelonian.org

Dedicated to the study and conservation of turtles and tortoises worldwide
*Founded: 1992*
*Peter Pritchard, President*
*Russell A. Mittermeier, President*

**37 Chemical Producers and Distributors Association**
1730 Rhode Island Ave NW
Suite 812
Washington, DC 20036
202-386-7407
Fax: 202-386-7409
E-mail: cpda@cpda.com
http://www.cpda.com

The preeminent US based trade association representing the interests of generic pesticide registrants, with a membership that includes manufacturers, formulators, and distributors of pesticide

products. Membership also includesmanufacturers and suppliers of inert ingredients used to enhance the delivery and efficacy of pesticide products.
*Dr Susan Ferenc, President*
*Diane Schute, Communications/Programs Dir*

**38 Chlorine Institute**
1300 Wilson Boulevard
Suite 525
Arlington, VA 22209
703-894-4140
Fax: 703-894-4130
http://www.chlorineinstitute.org
Exists to support the chlor-alkali industry and serve the public by fostering continuous improvements to safety and the protection of human health and the environment connected with the production, distribution and use of chlorine,sodium and potassium hydoroxides, and sodium hypochlorite; and the distribution and use of hydrogen chloride.
*Founded: 1924 220 Members*
*Arthur E Dungan, President*

**39 Consumer Specialty Products Association**
900 17th Street NW
Suite 300
Washington, DC 20006
202-872-8110
Fax: 202-872-8114
E-mail: info@cspa.org
http://www.cspa.org
To foster high standards for the industry; concern for the health, safety and environmental impacts of its products; address legislative and regulatory challenges at the federal, state and local level; meet the needs of industry fortechnical and legal guidance; provide a forum to share ideas for scientific and marketing excellence.
*Founded: 1914*
*Christopher Cathcart, President/CEO*
*Keith Fulk, Senior Vice President*

**40 Corps Network**
1100 G Street NW
Suite 1000
Washington, DC 20005
202-737-6272
Fax: 202-737-6277
E-mail: sprouty@corpsnetwork.org
http://www.nascc.org
A proud advocate and representative of the nation's Service and Conservation Corps. The number one goal is to sustain and grow the Corps movement. The Corps Network's member Service and Conservation Corps operate in 42 state and theDistrict of Columbia. Over 26,000 Corpsmembers, ages 16-25, contribute and generate more than 16 million hours of service annually.
*Founded: 1985*
*Sally T Prouty, President/CEO*
*James Jones, Senior Vice President*

**41 Council of State Governments**
2760 Research Park Drive
PO Box 11910
Lexington, KY 40578-1910
859-244-8000
Fax: 859-244-8001
E-mail: membership@csg.org
http://www.csg.org
A region-based forum that fosters the exchange of insights and ideas to help state officials shape public policy. CSG serves the exeecutive, judicial and legislative branches of state government through leadership education, researchand information services.
*Founded: 1933*
*Carol Juet, Director*
*David Akins, Executive Director*

**42 Earth First!**
PO Box 964
Lake Worth, FL 33460
561-249-2071
E-mail: collective@earthfirstjournal.org
http://www.earthfirstjournal.org
Earth First! was founded in response to a compromising and increasingly corporate environmental community.
*Founded: 1979*

**43 Earth Share**
7735 Old Georgetown Road
Suite 900
Bethesda, MD 20814
240-333-0300
800-875-3863
Fax: 240-333-0301
E-mail: info@earthshare.org
http://www.earthshare.org
Mission is to engage individuals and organization in creating a healthy and sustainable environment.
*Founded: 1988*
*Kalman Stein, President/CEO*

**44 Ecological Society of America**
1990 M Street NW
Suite 700
Washington, DC 20036
202-833-8773
Fax: 202-833-8775
E-mail: esahq@esa.org
http://www.esa.org
To promote ecological science by improving communication among ecologists; raise the public's level of awareness of the importance of ecological science; increase the resources available for the conduct of ecological science; andensure the appropriate use of ecological science in environmental decision making by enhancing communication between the ecological community and policy-makers.
*Founded: 1915*
*Alison Power, President*
*Katherine S McCarter, Executive Director*

**45 Environmental Council of the States**
50 F Street NW
Suite 350
Washington, DC 20001
202-266-4920
Fax: 202-266-4937
E-mail: ecos@sso.org
http://www.ecos.org
National non profit and non partisan association of state and territorial environmental agency leaders. Its purpose is to improve the capability of state environmental agencies and their leaders to proteect and improve humanhealth andthe environment of the United States of America.
*Founded: 1993*
*R Steven Brown, Executive Director*
*Carolyn Hanson, Deputy Executive Director*

**46 Environmental Industry Associations**
4301 Connecticut Avenue NW
Suite 300
Washington, DC 20008
202-244-4700
800-424-2869
Fax: 202-966-4824
http://www.envasns.org
The Environmental Industry Associations (EIA) is the parent organization for the National Solid Waste Management Association and the Waste Equipment Technology Association. Through these two associations, EIA represents companies andindividuals who manage solid and medical wastes; manufacture and distribute waste equipment; and to provide environmental managment, consulting and pollution-prevention-related services.

**47 Environmental Media Association**
5979 W. 3rd Street
Suite 204
Los Angles, CA 90036
323-556-2790
Fax: 323-556-2791
E-mail: ema@ema-online.org
http://www.ema-online.org
Dedicated to the broadcast of balanced news about the environment.
*Founded: 1989*
*Debbie Levin, President*
*Greg Baldwin, Executive Director*

**48 Federal Wildlife Association**
FWOA Secretary
PO Box 144
Washington, KS 66968
E-mail: fwoasecretary@aol.com
http://www.fwoa.org
Dedicated to the protection of wildlife, the enforcement of federal wildlife law, the fostering of cooperation and communication among federal wildlife officers, and the perpetuation, enhancement and defense of the wildlife enforcementprofession.
*Dave Hubbard, President*
*John Brooks, Vice President*

**49 Federation of Environmental Technologists**
W175 N11081 Stonewood Drive
Suite 203
Germantown, WI 53022
262-437-1700
Fax: 262-437-1702
E-mail: info@fetinc.org
http://www.fetinc.org

A nonprofit organization formed to assist industry in interpretation of and compliance with environmental regulations.
*Founded: 1982 600-700 Members*
*Barbara Hurula, Executive Director*
*Julie Jansett, Executive Assistant*

**50 Florida Center for Environmental Studies**
5353 Parkside Drive 561-799-8554
Building SR Fax: 56- 7-9 85
Jupiter, FL 33458 E-mail: jjolley@ces.fau.edu
http://www.ces.fau.edu/
Represents the ten state universities and the major private universities in regard to environmental studies and research.
*Founded: 1995*
*Leonard Berry Phd, Director*
*Jo Ann Jolley, Associate Director*

**51 Forestry Conservation Communications Association**
National Office
122 Baltimore Street 717-338-1505
Gettysburg, PA 17325 Fax: 717-334-5656
E-mail: ed@fcca-usa.org
http://www.fcca-usa.org
A national organization; some of whose functions, such as that of frequency coordination, are conducted on a regional level. A certified frequency coordinator for all public safety frequencies, including those in the 700 MHz and 800MHz bands
*Lloyd M. Mitchell, President*
*Roy Mott, Vice President*

**52 Get America Working!**
1700 N Moore Street
Arlington, VA 22209 http://www.getamericaworking.org
A non-profit national organization whose mission is to create 70 million jobs through structural changes in the US economy.
*Marca Bristo, President*
*Sandra Nathan, Vice President*

**53 Global Environmental Management Initiative**
1155 15th Street NW 202-296-7449
Suite 500 Fax: 202-296-7442
Washington, DC 20005 E-mail: info@gemi.org
http://www.gemi.org
Business helping business improve EHS performance, shareholder value and corporate citizenship.
*Founded: 1990*
*Keith Miller, Chairman*
*Bill Gill, Vice Chairman*

**54 Green America**
1612 K Street NW
Suite 600 800-584-7336
Washington, DC 20006 E-mail: ccafaq.htm
http://www.coopamerica.org
Formerly known as Co-op America, whose mission is to harness economic power-the strength of consumers, investors, businesses, and the marketplace-to create a socially just and environmentally sustainable society.
*Founded: 1982*
*Melissa Bradley, President*
*Eric Henry, Vice President*

**55 Green Media Toolshed**
1915 I Street NW 20- 6-9 77
Suite 700 Fax: 86- 8-5 52
Washington, DC 20006 E-mail: info@greenmediatoolshed.org
http://www.greenmediatoolshed.org
Green Media Toolshed is a nonprofit organization that provides the environmental community with access to a host of high quality communications tools for an affordable cost.
*Founded: 2000*
*Martin Kearns, Executive Director*
*Bobbi Russell, Associate Director*

**56 Green Seal**
1001 Connecticut Avenue NW 202-872-6400
Suite 827 Fax: 202-872-4324
Washington, DC 20036-5525 E-mail: greenseal@greenseal.org
http://www.greenseal.org
An independent non-profit organization dedicated to safeguearding the environment and transforming the marketplace by promoting the manufacture, purchase, and use of environmentally responsible products and services.
*Founded: 1989*
*Arthur B. Weissman, President/CEO*
*Gary Petersen, Chairman*

**57 Greenpeace USA**
702 H Street NW 202-462-1177
Suite 300 1 8-0 7-2 69
Washington, DC 20001 E-mail: info@wdc.greenpeace.org
http://www.greenpeace.org
Greenpeace is the leading independent campaigning organization that uses non-violent direct action and creative communication to expose global environmental problems and to promote solutions that are essential to a green and peacefulfuture.
*Founded: 1971*
*Phil Radford, Executive Director*

**58 H John Heinz III Center for Science**
900 17th Street NW 202-737-6307
Suite 700 Fax: 202-737-6410
Washington, DC 20006 E-mail: info@heinzctr.org
http://www.heinzctr.org
The Center is a nonpartisan, nonprofit institute dedicated to improving the scientific and economic foundation for environmental policy through nmultisectoral collaboration. The Heinz Center fosters collaboration among industry,environmental organizations, academia, and all levels of government in each of its program areas.
*Founded: 1995*
*Deb Callahan, President*
*Ralph Ashton, Project Director*

**59 Halogenated Solvents Industry Alliance**
1530 Wilson Boulevard 703-875-0683
Suite 690 Fax: 703-875-0675
Arlington, VA 22209 E-mail: info@hsia.org
http://www.hsia.org
The mission is to present the interests of users and producers of chlorinated solvents. To promote the continued safe use of these products and to promote the use of sound science in assessing their potential health effects.
*Founded: 1980*

**60 Honor the Earth**
2104 Stevens Avenue South 612-879-7529
Minneapolis, MN 55404 http://www.honorearth.org
Honor the Earth is a Native-led organization, established by Winona LaDuke and Indigo Girls Amy Ray and Emily Saliers, in 1993 to address the two primary needs of the Native environmental movement: the need to break the geographic andpolitical isolation of Native communities and the need to increase financial resources for organizing and change.
*Founded: 1993*
*Winona LaDuke, Founder/Author/Activist*
*Amy Ray, Founder*

**61 Institute for Earth Education**
Cedar Cove 304-832-6404
Greenville, WV 24945 Fax: 304-832-6077
E-mail: info@ieetree.org
http://www.eartheducation.org
The Institute for Earth Education is the world's alternative agency-and industry-sponsored supplemental environmental education. IEE develops and disseminates instructional programs aimed at helping people live more lightly,harmoniously, and joyously with the natural world.
*Fran McKeever, Branch Coordinator*

**62 Institute of Clean Air Companies**
1220 N. Fillmore Street
Suite 410
Arlington, VA 22201-4535
703-812-4811
Fax: 202-331-1388
E-mail: icacinfo@icac.com
http://www.icac.com
The nonprofit national association of companies working in the stationary source air pollution control and monitoring sector. Members include leading system and component suppliers of monitoring and control technologies for PM, NOx,SO2, Hg, VOCs and other HAPs.
*Vince Albanese, President*
*Michael Durham, Vice President*

**63 Institute of Environmental Sciences and Technology**
2340 S Arlington Heights Rd
Suite 100
Arlington Heights, IL 60005
84- 9-1 01
Fax: 84- 9-1 41
E-mail: information@iest.org
http://www.iest.org
An international professional society that serves the environmental sciences in the areas of contamination control in electronics manufacturing and pharmaceutical processes, design, test and evaluation of commercial and militaryequipment and product reliability issues.
*R. Vijayakumar, President*
*Christine Peterson, President Elect*

**64 Institute of Hazardous Materials Management**
11900 Parklawn Drive
Suite 450
Rockville, MD 20852-2624
301-984-8969
Fax: 301-984-1516
E-mail: ihmminfo@ihmm.org
http://www.ihmm.org
A nonprofit organization that protects the environment and the public's health and safety through the administration of credentials recognizing professionals who have demonstrated a high level of knowledge, expertise, and excellence inthe management of hazardous materials.
*Founded: 1984*
*Michael S. Davidson, Vice President*
*Robbie Tidwell, Operations Specialist*

**65 International Center for the Solution of Environmental Problems**
5120 Woodway Drive
Suite 8009
Houston, TX 77056
713-527-8711
Fax: 713-961-5157
E-mail: icsep@airmail.net
http://http://icsep.com
To anticipate and detect environmental problems and either solve the problems or design and demonstrate their solutions using scientific methods in concert with nature. To provide environmental information.
*Founded: 1975*
*Leonel Castillo, Immigration Expert*

**66 Interstate Mining Compact Commission**
445A Carlisle Drive
Herndon, VA 20170
703-709-8654
Fax: 703-709-8655
E-mail: bbotsis@imcc.isa.us.
http://www.imcc.isa.us
A multi-state governmental agency/organization that represents the natural resource interests of its member states.
*Founded: 1970*
*Gregory E Conrad, Executive Director*
*Beth A Botsis, Diretor of Programs*

**67 Jane Goodall Institute for Wildlife Research, Education and Conservation**
4245 North Fairfax Drive
Suite 600
Arlington, VA 22203
703-682-9220
Fax: 703-682-9312
E-mail: webmaster@janegoodall.org
http://www.janegoodall.org
A tax-exempt, nonprofit corporation, founded to concentrate on research, education and conservation of wildlife, pursuant to the life work of Jane Goodall.
*William Johnston, President/CEO*
*Rich Hays, Executive Vice President*

**68 Land Improvement Contractors of America**
3080 Ogden Avenue
Suite 300
Lisle, IL 60532
630-548-1984
Fax: 630-548-9189
E-mail: nlica@aol.com
http://www.licanational.com
Organization composed primarily of small contractors whose activities related to the conservation, use and improvement of land and water resources ranging from grading, excavating, paving, landscaping, wetland development, drainage,and site preparation. Strives to improve the climate within which members conduct their businesses by working for better legislation and regulations.
*Founded: 1951*
*Dirk Riniker, Chairman*
*Steve Gerten, President*

**69 National Association for Environmental Management**
1612 K Street NW
Suite 1102
Washington, DC 20006
202-986-6616
800-391-6236
Fax: 202-530-4408
E-mail: programs@naem.org
http://www.naem.org
Provides peer-to-peer networking for EHS managers; develop EHS professionals as leaders; advance the integration of EHS into business as a value driver; and promote the growth and implementation of EHS management systems worldwide; so,as to offer tangible benefits to the regulated entity and other stakeholders.
*Founded: 1990*
*Pat Perry, President*
*Stephen Evanoff, Vice President*

**70 National Association of Biology Teachers**
1313 Dolley Madison Blvd
Suite 402
McLean, VA 22101
703-264-9696
800-406-0775
Fax: 703-790-7672
E-mail: office@nabt.org
http://www.nabt.org
Includes more than 9,000 educators who share experience and expertise with colleagues from around the globe; keep up with trends and developments; and grow professionally. The NABT empowers educators to provide the best biology andlife science education for all students.
*Daniel Ward, President*
*Donald French, President Elect*

**71 National Association of Environmental Professionals**
PO Box 460
Collingswood, NJ 8108
856-283-7816
Fax: 856-210-1619
E-mail: naep@bowermanagementservices.com
http://www.naep.org
NAEP is a multidisciplinary association dedicated to the advancement of environmental professionals in the US and abroad, and a forum for state-of-the-art information on environmental planning, research and management. A network ofprofessional contacts and a resource for structured career development, this organization is a strong proponent of ethics and the highest standards of practice in the environmental professions.
*Ron Deverman, President*
*Paul Looney, Vice President*

**72 National Conference of State Legislatures**
7700 E First Place
Denver, CO 80230
303-364-7700
Fax: 303-364-7800
http://www.ncsl.org
The National Conference of State Legislatures serves the legislators and staffs of the nation's 50 states and its commonwealths and territories. NCSL is a bipartisan organization with three objectives: to improve the quality andeffectiveness of state legislatures; to foster interstate communications and cooperation; and to ensure states a strong cohesive voice in the federal system.
*Founded: 1975*
*William T Pound, Executive Director*

**73 National Council for Science and the Environment**
1101 17th Street NW
Suite 250
Washington, DC 20036
202-530-5810
Fax: 202-628-4311
E-mail: info@ncseonline.org
http://www.ncseonline.org

National nonprofit organization working to improve the scientific basis for making decisions on environmental issues. The Council promotes a new crosscutting approach to environmental science that integrates interdisciplinary research,scientific assessment, environmental education and communication of science-based information to decision makers and the public.
*Karim Ahmed, President*
*James Buizer, Executive Director*

**74 National Energy Foundation**
4516 Sout 700 East
Suite 100
Salt Lake City, UT 84107
800-616-8326
800-616-8326
Fax: 801-908-5400
E-mail: info@nefl.org
http://www.nefl.org

A nonprofit educational organization dedicated to the development, dissemination and implementation of supplemental educational materials and programs primarily related to the environment, conservation, science, energy, water, andnatural resources.
*Founded: 1976*
*Robert Poulson, President*
*Dari Scott, Vice President*

**75 National Environmental Education Foundation**
4301 Connecticut Avenue NW
Suite 160
Washington, DC 20008
202-833-2933
Fax: 202-261-6464
http://www.neefusa.org

Provides objective environmental information and education to help Americans live better. Environmental education programs include those tailored to the adult public, for health professionals, and in the schools.
*Diane W Wood, President*
*Deborah A Sliter, VP Programs*

**76 National Environmental Trust**
901 E St. NW
Washington, DC 20004
202-552-2000
Fax: 202-552-2299
E-mail: info@pewtrusts.org
http://www.net.org

Informs citizens about environmental problems and how they affect human health and quality of life. The Trust uses public education campaigns and modern communication techniques to localize the impacts of national problems.
*Founded: 1994*
*Rebecca W. Rimel, President*
*Rebecca A. Cornejo, Senior Officer*

**77 National FFA Organization**
PO Box 68960
6060 FFA Drive
Indianapolis, IN 46268
317-802-6060
Fax: 317-802-6051
http://www.ffa.org

The organization (formerly Future Farmers of America) is dedicated to making a positive difference in the lives of young people by developing their potential for premier leadership, personal growth and career success throughagricultural education.
*Founded: 1928*
*Ray Nash, President*
*Greg Curlin, President Elect*

**78 National Geographic Society**
1145 17th Street NW
Washington, DC 20036
202-857-7000
800-647-5463
Fax: 202-775-6141
http://www.nationalgeographic.com

Its mission, in a nutshell, is to inspire people to care about the planet. Programs support scientific fieldwork and expeditions; encourage geography education for students; and promote natural and cultural conservation.
*Founded: 1888*
*John Fahey, President/CEO*
*Terrence B Adamson, Executive VP*

**79 National Governors Association**
Hall of States
444 N Capitol St, Suite 267
Washington, DC 20001
202-624-5300
Fax: 202-624-5313
E-mail: webmaster@nga.org
http://www.nga.org

The Association works closely with the administration and Congress on state and federal policy issues, serves as a vehicle for sharing knowledge of innovative programs among states, and provides technical assistance and consultantservices to governors on a wide range of management and policy issues. Part of the organization is the Center for Best Practices which undertakes demonstration projects and provides anticipatory research on important policy issues.
*Founded: 1908*
*Raymond Scheppach, Executive Director*
*Jay Hyde, Director Communications*

**80 National Institute for Global Environmental Change**
University of California, Davis
2850 Spafford Street
Suite A
Davis, CA 95618
530-757-3350
Fax: 530-756-6499
E-mail: nigec@ucdavis.edu
http://nigec.ucdavis.edu

Sponsored by the Office of Science at the US Department of Energy, the institute is operated by the University of California under a cooperative agreement. Its overall mission is to assist the nation in its response to human-inducedclimate and environmental change.
*Dr Lawrence B Coleman, Interim National Director*
*Thanos Toulopoulos, Program & Info Systems*

**81 National Network of Forest Practitioners**
8 North Court Street
Suite 411
Athens, OH 45701
740-593-8733
Fax: 401-273-6508
E-mail: info@nnfp.org
http://www.nnfp.org

Promotes the mutual well-being of workers, rural communities and forests by supporting individuals and groups that build sustainable relationships between forests and people.
*Colin Donahue, Executive Director*
*Scott Bagley, Program Director*

**82 National Parent Teachers Association**
1250 N. Pitt Street
Alexandria, VA 22314
70- 5-8 12
800-307-4782
Fax: 70- 8-6 09
E-mail: info@pta.org
http://www.pta.org

Part of the mission of the National PTA is to support and speak on behalf of children and youth in the school, in the community and before governmental bodies and other organizations. School Construction and Environmental Health is oneof the program focuses where the National PTA aims to ensure that facilities are free from health and environmental hazards.
*Charles J. Saylors, President*
*Betsy Landers, President Elect*

**83 National Religious Partnership for the Environment**
49 South Pleasant Street
Suite 301
Amherst, MA 1002
413-253-1515
Fax: 413-253-1414
E-mail: nrpe@nrpe.org
http://www.nrpe.org

The mission of the National Religious Partnership for the Environment is to permanently integrate issues of environmental sustainability and justice across all aspects of organized religious life.
*Matthew A. Stembridge, Executive Director*
*Joan B. Campbell, Director*

**84 National Solid Waste Management Association**
Environmental Industry Associations
4301 Connecticut Avenue NW
Suite 300
Washington, DC 20008
202-244-4700
800-424-2869
Fax: 202-966-4824
E-mail: membership@envasns.org
http://www.nswma.org; www.envasns.org

The trade association that represents the private scetor companies in North America that provide solid, hazardous and medical

waste collection, recycling and disposal services, and companies that provide professionals and consultingservices to the waste services industry.
*Founded: 1962*
*Bruce J Parker, President/CEO*
*Chaz Miller, Director*

**85 Nature's Classroom**
19 Harrington Rd 508-248-2741
Charlton, MA 01507 800-433-8375
E-mail: info@naturesclassroom.org
http://www.naturesclassroom.org
Nature's Classroom is a residential environmental education program, at fifteen wonderful sites in New York and New England. We give students, and teachers, the chance to experience education from another perspective, outside the wallsof the classroom. After spending time in Nature's Classroom, living and learning together, students develop a sense of community, a confidence in themselves and an appreciation for others that carries over to the school community.
*Dr. John Santos, Director*

**86 North American Association for Environmental Education**
2000 P Street NW 202-419-0412
Suite 540 Fax: 202-419-0415
Washington, DC 20036 E-mail: brian@naaee.org
http://www.naaee.org
A professional association for environmental education that promotes excellence in environmental education and serves environmental educators for the purpose of achieving environmental literacy in order for present and futuregenerations to benefit from a safe and health environment and a better quality of life.
*Founded: 1971*
*Linda Rhoads, Executive Director*
*Sue Bumpous, Program/Communications Mgr*

**87 Office of Environmental Affairs**
Environmental Education
100 Cambidge Street 617-626-1000
Suite 900 Fax: 617-626-1181
Boston, MA 2114 E-mail: env.internet@state.ma.us
http://www.state.ma.us/envir
A resource for environmental literacy, whose goal is to reconnect people to the natural world, and to inspire a sense of public responsibility. Learn how to link to our collective health, making a difference through our decisions andchoices.
*Martha Coakley, Attorney General*

**88 Orion Society**
187 Main Street 413-528-4422
Great Barrington, MA 1230 888-909-6568
E-mail: orion@orionsociety.org
http://www.orionsociety.org
A nonprofit organization supported by donations from individuals, families and foundations, and corporate and government grants. Mission is to inform, inspire, and engage individuals and grassroots organizations in becoming asignificant cultural force for healing nature and community.
*Founded: 1982*
*M G H Gilliam, President*
*H Emerson Blake, Executive Director*

**89 Public Citizen**
1600 20th Street NW 202-588-1000
Washington, DC 20009 http://www.citizen.org
A consumer advocacy organization founded to represent consumer interests in Congress, the executive branch, and the courts.
*Founded: 1971*
*Robert C. Fellmeth, Chairman*
*Jim Bildner, Managing Director*

**90 Public Employees for Environmental Responsibility (PEER)**
2000 P Street NW 202-265-7337
Suite 240 Fax: 202-265-4192
Washington, DC 20036 E-mail: info@peer.org
http://www.peer.org
A national non-profit alliance of local, state and federal scientists, law enfonrcement officers, land managers and other professionals dedicated to upholding environmental laws and values.
*Howard Wilshire, Chairman*
*Carol Goldberg, Associate Director*

**91 Renew America**
1200 18th Street NW 202-721-1545
Suite 1100 Fax: 202-467-5780
Washington, DC 20036 http://www.solstice.crest.org
Renew America is a nonprofit organization founded in 1989. They coordinate a network of community and environment groups, businesses, government leaders and civic activists to exchange ideas and expertise for improving the enviroment.By finding and promoting programs that work, Renew America helps inspire communities and businesses to meet today's environmental challenges.
*Founded: 1989*
*Kenneth Brown, Executive Director*
*L Hunter Lovins, President*

**92 Renewable Fuels Association**
425 Third Street, SW 202-289-3835
Suite 1150 Fax: 202-289-7519
Washington, DC 20024 E-mail: info@ethanolrfa.org
http://www.ethanolrfa.org
The national trade association for the US ethanol industry, promotes policies, regulations and research and development initiative that will lead to the increased production and use of fuel ethanol.
*Founded: 1981*
*Bob Dinneen, President/CEO*
*Mary Giglio, Director*

**93 Roger Tory Peterson Institute of Natural History**
311 Curtis Street 716-665-2473
Jamestown, NY 14701 800-758-6841
Fax: 716-665-3794
E-mail: mail@rtpi.org
http://www.rtpi.org
Nature education organization that works on the national level, with an audience of primarily adults who are interested in gaining skills for educating young people about the natural world. RTPI also houses the life's work of RogerTory Peterson.
*Founded: 1984*
*Jim Berry, President*
*Mark Baldwin, Director of Education*

**94 Silicones Environmental, Health and Safety Council of North America**
2325 Dulles Corner Boulevard 703-788-6570
Suite 500 Fax: 703-788-6545
Herndon, VA 20171 E-mail: sehsc@sehsc.com
http://www.sehsc.com
An organization of North America silicone chemical producers and importers. It promotes the safe use of silicones through product stewardship and environmental, health and safety research. It focuses on coordinating research andsubmitting it for peer review through independent advisory boards and publication of peer-reviewed literature.
*Karluss Thomas, Executive Director*

**95 Smithsonian Institution**
Smithsonian Information
PO Box 37012 202-633-1000
SI Building Room 153 E-mail: info@si.edu
Washington, DC 20013 http://www.si.edu
Independent trust instrumental of the United States holding more than 140 million artifacts and specimens in its trust for the public interest and knowledge. Also a center for research dedicated to

public education, national service,and scholarship in the arts, sciences, and history.
*Founded: 1846*
*Cristian Samper, Secretary*

**96 Society of American Foresters**
5400 Grosvenor Lane
Bethesda, MD 20814
301-897-8720
866-897-8720
E-mail: safweb@safnet.org
http://www.safnet.org
To advance the science, education, technology, and practice of forestry; to enhance the competency of its members; to establish professional excellence; and to use the knowledge, skills, and conservation ethic of the profession toensure the continued health and use of forest ecosystems present and future of availability of forest resources to benefit society.
*Founded: 1900*
*Michael B. Lester, President*
*Michael T. Goergen Jr., Executive Vice President*

**97 Student Pugwash USA**
1015 18th Street NW
Suite 704
Washington, DC 20036
202-429-8900
800-969-2784
Fax: 202-429-8905
E-mail: spusa@spusa.org
http://www.spusa.org
Promotes social responsibility in science and technology. Prepare science, technology and policy students to make social responsibility a guiding focus of their academic and professional endeavors.
*Ben Austin, Vice President*
*Cameron D. Bess, President*

**98 Synthetic Organic Chemical Manufacturers Association**
1850 M Street NW
Suite 700
Washington, DC 20036
202-721-4100
Fax: 202-296-8120
E-mail: info@socma.com
http://www.socma.com
The leading international trade association serving the small and mid-sized batch chemical manufacturers. Advocates flexible policies grounded in sound science and works to ensure that Congress and the regulatory agencies do not adopta one-size fits-all approach to the industry.
*Founded: 1921*
*Larry Brotherton, Chairman*
*Nicholas Shackley, Vice Chairman*

**99 U.S. Global Change Research Information Office**
1717 Pennsylvania Avenue NW
Suite 250
Washington, DC 20006
202-223-6262
Fax: 202-223-3065
E-mail: information@gcrio.org
http://www.gcrio.org
The U.S. Global Change Research Information Office (GCRIO) provides access to data and information on climate change research, adaptation/mitigation strategies and technologies and global change-related educational resources on behalfof the various US federal agencies that are involved in the US Global Change Research Program.
*Founded: 1993*

**100 United Nations Environment Programme New York Office**
2 UN Plaza
Room DC2-803
New York, NY 10017
212-963-8210
Fax: 212-963-7341
E-mail: info@nyo.unep.org
http://www.nyo.unep.org
Provides leadership and to encourage partnership in caring for the environment by inspiring, informing, and enabling nationa and peoples to improve their quality of life without compromising that of future generations.
*Founded: 1972*
*Juanita Castano, Director*
*Zehra Aydin, Senior Programme Officer*

**101 University of North Texas**
Center for Environmental Philosophy
1155 Union Circle # 310980
Denton, TX 76203
940-565-2727
Fax: 940-565-4439
E-mail: cep@unt.edu.
http://www.cep.unt.edu
Conducts environmental research and offers environmental education as is relates to decisions we make when dealing with environmental issues.
*Jan Dickson, Executive Director*

**102 Welder Wildlife Foundation**
PO Box 1400
Sinton, TX 78387
361-364-2643
Fax: 361-364-2650
E-mail: welderfoundation@welderwildlife.org
http://www.welderwildlife.org
A private non profit organization that has gained international recognition through its research program. The mission of the Foundation is to conduct research and education in the fields of wildlife management and conservation andother closely related fields.
*Founded: 1954*
*Terry Blankenship, Director*
*Selma Glasscock, Assistant Director*

## National: Design & Architecture

**103 ABS Group**
American Bureau of Shipping
16855 Northchase Drive
Houston, TX 77060
281-673-2800
1 8-0 7-9 11
Fax: 281-673-2950
http://www.abs-group.com
A subsidiary of American Bureau of Shipping offering risk management, safety, quality and environmental consulting and certification services to a wide range of industries and companies throughout the world.
*Founded: 1990*
*Tony Nassif, President/CEO*
*Gary Graham, Vice President*

**104 American Society of Landscape Architects**
636 Eye Street NW
Washington, DC 20001
202-898-2444
888-999-2752
Fax: 202-898-1185
E-mail: info@asla.org
http://www.asla.org
To lead, to educate and to participate in the careful stewardship, wise planning and artful design of our cultural and natural environments.
*Founded: 1899*
*Johnathan Mueller, President*
*Susan M. Hatchell, President Elect*

**105 Environmental Action Foundation Associated General Contractors of America**
2300 Wilson Boulevard
Suite 400
Arlington, VA 22201
703-548-3118
800-242-1767
Fax: 703-548-3119
E-mail: info@agc.org
http://www.agc.org
The mission is demonstrating the construction industry's commitment to improving the environment and quality of the life through: improving environmental education, supporting sensible application of environmental laws, promotingenvironmental awareness campaigns and assisting in environmental litigation.
*Founded: 1999*
*Ted Aadland, President*
*Kristine L. Young, Senior Vice President*

**106 Environmental Design Research Association**
1760 Old Meadow Road
Suite 500
McLean, VA 22102
703-506-2895
Fax: 703-506-3266
E-mail: edra@edra.org
http://www.edra.org

Advances and disseminates behavior and design research toward improving understanding of the relationships between people and their environments.
*Founded: 1968*
*Lynne Dearborn, Chairman*
*Nick Watkins, Chairman Elect*

**107 Land Improvement Contractors of America**
3080 Ogden Avenue
Suite 300
Lisle, IL 60532
630-548-1984
Fax: 630-548-9189
E-mail: nlica@aol.com
http://www.licanational.com
Organization composed primarily of small contractors whose activities related to the conservation, use and improvement of land and water resources ranging from grading, excavating, paving, landscaping, wetland development, drainage,and site preparation. Strives to improve the climate within which members conduct their businesses by working for better legislation and regulations.
*Founded: 1951*
*Dirk Riniker, Chairman*
*Steve Gerten, President*

**108 Rocky Mountain Institute**
2317 Snowmass Creek Road
Snowmass, CO 81654
970-927-3851
http://www.rmi.org
An independent, entrepreneurial, nonprofit think and do tank. Envisage a world thriving, verdant, and secure, for all.
*Michael Potts, President*
*Carrie Sullivan, Executive Assistant*

## National: Disaster Preparedness & Response

**109 Global Response**
215 Prospect Street
Cambridge, MA 2139
617-441-5400
Fax: 617-441-5417
E-mail: culturalsurvival@cs.org
http://www.globalresponse.org
Provides a worldwide connection for a better global environment through effective letter-writing campaigns. Empowers people of all ages, cultures, and nationalities to protect the environment by creating partnerships for effeectivecitizen action.
*Sarah Fuller, President*
*Richard A. Grounds, Vice Chairman*

**110 International Association of Wildland Fire**
1418 Washburn Street
Missoula, MT 59801
406-531-8264
888-440-4293
E-mail: iawf@iawfonline.org
http://www.iawfonline.org
A professional association representing members of the global wildland fire community. It facilitates communication and provides leadership for the wildland fire community.
*Chuck Bushey, President*
*Kris Johnson, Vice President*

**111 National Association of Flood and Stormwater Management Agencies**
1333 H Street NW W Tower
10th Floor
Washington, DC 20005
202-289-8625
Fax: 202-530-3389
E-mail: info@nafsma.org
http://www.nafsma.org
They are an organziation of public agencies whose function is the protection of lives, property and economic activity from the adverse impacts of storm and flood waters. The mission of the association is to advocate public policy,encourage technologies and conduct education programs which facilitate and enhance the achievement of the public service functions of its members.
*Founded: 1978*
*James Fiedler, President*
*Dusty Williams, Vice President*

**112 National Council on Radiation Protection and Measurements**
7910 Woodmont Avenue
Suite 400
Bethesda, MD 20814
301-657-2652
Fax: 301-907-8768
E-mail: ncrp@ncrponline.org
http://www.ncrponline.org
The National Council on Radiation Protection and Measurements (NCRP) seeks to formulate and widely disseminate information, guidance and recommendations on radiation protection and measurements which represent the consensus of leadingscientific thinking.
*Founded: 1964*
*D A Schaer, Executive Director*

**113 National Fire Protection Association**
1 Batterymarch Park
Quincy, MA 2169
617-770-3000
800-344-3555
Fax: 617-770-0700
http://www.nfpa.org
Reduces the worldwide burden of fire and other hazards on the quality of life by providing and advocating consensus codes and standards, research, training and education.
*Founded: 1896 81,000 members*
*James M Shannon, President*
*Thomas W. Jaeger, President*

**114 Natural Hazards Center**
University of Colorado
482 UCB
Boulder, CO 80309
303-492-6818
Fax: 303-492-2151
E-mail: hazctr@colorado.edu
http://www.colorado.edu/hazards
Advances and communicates knowledge on hazards mitigation and disaster preparedness, response, and recovery. Using an all-hazards and interdisciplinary framework, the Center fosters information sharing and integration of activitiesamong researchers, practitioners, and policy makers from around the world; supports and conducts research; and provides educational opportunities for the next generation of hazards scholars and professionals.
*Founded: 1976*
*Kathleen Tierney, Director*
*Dennis S. Mileti, Senior Research Scientist*

**115 Office of Response and Restoration-NOAA**
National Oceanic and Atmospheric Administration
National Ocean Service
1305 East West Highway
Silver Spring, MD 20910
301-713-4248
Fax: 301-713-4389
E-mail: orr.webmaster@noaa.gov
http://response.restoration.noaa.gov/
Protects coastal and marine resources, mitigates threats, reduces harm, and restores ecological function. Provides comprehensive solutions to environmental hazards caused by oil, chemicals, and marine debris.

**116 Safety Equipment Institute**
1307 Dolley Madison Boulevard
Suite 3A
McLean, VA 22101
703-442-5732
Fax: 703-442-5756
E-mail: info@seinet.org
http://www.seinet.org
A private, nonprofit organization established to administer the first non-governmental, third-party certification program to test and certify a broad range of safety equipment products.
*Founded: 1981*
*Bradley M. Sant, Chairman*
*Ronald N. Herring Jr., Vice Chairman*

## National: Energy & Transportation

**117 Alliance to Save Energy**
1850 M Street NW
Suite 600
Washington, DC 20036
202-857-0666
Fax: 202-331-9588
E-mail: info@ase.org
http://www.ase.org

A non-profit coalition of business, government, environmental and consumer leaders. Promotes energy efficient worldwide to achieve a healthier economy, a cleaner environment and greater energy security.
*Founded: 1977*
*Kateri Callahan, President*
*Jeff Harris, Senior Vice President for Pr*

**118 Alternative Energy Resources Organization**
432 N Last Chance Gulch
Helena, MT 59601
406-443-7272
Fax: 406-442-9120
E-mail: aero@aeromt.org
http://www.aeromt.org

A grassroots nonprofit organization dedicated to solutions that promote resource conservation and local economic vitality. Nutures individuals and community self reliance through programs that support sustainable agriculture, renewableenergy and environmental quality.
*Founded: 1974*
*Jonda Crosby, Executive Director*

**119 American Coal Ash Association**
15200 East Girard Avenue
Suite 3050
Aurora, CO 80014
720-870-7897
Fax: 720-870-7889
E-mail: info@acaa-usa.org
http://www.acca-usa.org

ACAA's mission is to advance the management and use of coal combustion products in ways that are environmentally responsible, technically sound, commercially competitive and more supportive of a sustainable global community.
*Thomas H Adams, Executive Director*

**120 American Council for an Energy-Efficient Economy**
529 14th Street NW
Suite 600
Washington, DC 20045
202-507-4000
Fax: 202-429-2248
E-mail: info@aceee.org
http://www.aceee.org

A nonprofit organization dedicated to advancing energy efficiency as a means of promoting both economic prosperity and environmental protection. Fulfills its mission by: conducting technical and policy assessments; advising governmentsand utilities; publishing books, conference proceedings and research reports and informing consumers.
*Founded: 1980*
*Scott Bernstein, President*
*Carl Blumstein, Chairman*

**121 American Gas Association**
400 North Capitol Street NW
Washington, DC 20001
202-824-7000
Fax: 202-824-7115
http://www.aga.org

The Association represents 202 local energy companies that deliver natural gas throughout the United States.
*Founded: 1918*
*John W. Somerhalder, President/CEO*
*Dave McCurdy, President/CEO*

**122 American Petroleum Institute**
1220 L Street NW
Washington, DC 20005
202-682-8000
http://www.api.org
API is the major national trade association representing all aspects of the nation's oil and natural gas industry. There 400 corporate members that include producers, refiners, pipeline operators, and service companies. Theassociation's broad range of programs include: advocacy, research and statistics, certification, standards, and education.
*Founded: 1911*
*Jack N. Gerard, President*

**123 American Public Power Association**
1875 Connecticut Ave NW
Suite 1200
Washington, DC 20009-5715
202-467-2900
800-515-2772
Fax: 202-467-2910
E-mail: mrufe@appanet.org
http://www.appanet.org

The American Public Power Association (APPA) is the service organization for the nation's more than 2,000 community-owned electric utilities that serve more than 43 million Americans. It was created as a non-profit, non-partisanorganization. Its purpose is to advance the public policy interests of its members and their consumers, and provide member services to ensure adequate, reliable electricity at a reasonable price with the proper protection of the environment.
*Founded: 1940*
*Lonnie Carter, Chairman*
*William Carroll, Chairman Elect*

**124 American Solar Energy Society**
2400 Central Avenue
Suite A
Boulder, CO 80301
303-443-3130
Fax: 303-443-3212
E-mail: ases@ases.org
http://www.ases.org

Nonprofit association of solar professionals and grassroots advocates. The mission is to speed the transition to a sustainable energy economy.
*Margot McDonald, Chairman*
*Jeff Lyng, Vice Chairman*

**125 Association of Energy Engineers**
4025 Pleasantdale Road
Suite 420
Atlanta, GA 30340
770-447-5083
Fax: 770-446-3969
E-mail: info@aeecenter.org
http://www.aeecenter.org

A nonprofit professional society that promotes the scientific and educational interests of those engaged in the energy industry and fosters action for sustainable development.
*Founded: 1977*
*Albert Thumann, Executive Director*
*Ruth Whitlock, Executive Administrator*

**126 Civil Engineering Forum for Innovation (CEFI)**
American Society of Civil Engineers
1801 Alexander Bell Drive
Reston, VA 20191
703-295-6314
Fax: 202-833-6315
E-mail: sskemp@asce.org
http://www.content.asce.org/cefi

Established by the ASCE to strenthen the profession and industry through public policy and technical innovation. It is focused on advancing the engineering profession.
*Susan Skemp, Executive VP*
*Laurie Hanson, Executive Assistant*

**127 Clean Fuels Development Coalition**
4641 Montgomery Avenue
Suite 350
Bethesda, MD 20814
301-718-0077
Fax: 301-718-0606
E-mail: cfdcinc@aol.com
http://www.cleanfuelsdc.org

Actively supportsthe increased production and use of fuels that can reduce air pollution and oil imports.
*Douglas A Durante, Executive Director*

**128 Compressed Gas Association**
4221 Walney Road
5th Floor
Chantilly, VA 20151
703-788-2700
Fax: 703-961-1831
E-mail: cga@cganet.com
http://www.cganet.com

The mission of the CGA is to promote the safe manufacture, transportation, storage, transfilling, and disposal of industrial and medical gases and their containers.
*Founded: 1913*

**129 Consumer Energy Council of America**
2737 Devonshire Place NW
Suite 102
Washington, DC 20008
202-468-8440
Fax: 703-590-5920
E-mail: info@cecarf.org
http://www.cecarf.org

A senior public interest organization in the US focusing on energy, telecommunications and other network industries that provide essential services to consumers.
*Founded: 1973*
*Ellen Berman, CEO*
*Richard Aiken, President*

**130 Environmental Coalition on Nuclear Power**
433 Orlando Avenue 814-237-3900
State College, PA 16803 Fax: 814-237-3900
E-mail: johnstrud@uplink.net
Groups and individuals concerned with the nuclear power and energy policies. Maintains speakers bureau and conducts research and educational programs.
*Dr Judith Johnsrud, Executive Officer*

**131 Institute for Energy and Environmental Research (IEER)**
6935 Laurel Ave, Suite 201 301-270-5500
Takoma Park, MD 20912 Fax: 301-270-3029
E-mail: info@ieer.org
http://www.ieer.org
IEER's aim is to provide people with literature which has a quality equal to that in scientific journals, but which doesn't require you to go back to college to get a degree in science to understand it. Our focus has been mainly on twoareas: ozone layer depletion and energy-related climate issues; and environmental and security aspects of nuclear weapons production and nuclear technology.
*Founded: 1987*
*Arjun Makhijani, President*

**132 International Association for Energy Economics**
28790 Chagrin Boulevard 216-464-5365
Suite 350 E-mail: iaee@iaee.org
Cleveland, OH 44122 http://www.iaee.org
A non-profit, professoinal organization that provides an interdisciplinary forum for the exchage of ideas, experiences and issues among professionals interested in the field of energy economics.
*Founded: 1977*
*Mine Yucel, President*
*Lars Bergman, President Elect*

**133 Interstate Oil and Gas Compact Commission**
900 NE 23rd Street 405-525-3556
Oklahoma City, OK 73105 Fax: 405-525-3592
E-mail: iogcc@iogcc.state.ok.us
http://www.iogcc.state.ok.us
A multi-state government agency that is passionate about advancing the quality of life for all Americans. Works to ensure our nation's oil and natural gas resources are conserved and maximized while protecting health, safety and theenvironment.
*Mary Fallin, Chairman*
*Erica Carr, Communications Manager*

**134 NW Energy Coalition**
811 1st Avenue 206-621-0094
Suite 305 Fax: 206-621-0097
Seattle, WA 98104 E-mail: nwec@nwenergy.org
http://www.nwenergy.org
The Coalition is an alliance of over 100 environmental, civic and human service organizations, progressive utilities and businesses in Oregon, Washington, Idaho, Montana, Alaska and British Columbia. They promote renewable energy andenergy conservation, consumer protection, low-income energy assistance, and fish and wildlife restoration on the Columbia and Snake Rivers.
*1981 pages*
*Ken Miller, Chairman*
*Pat Ford, Vice Chairman*

**135 National Energy Foundation**
4516 Sout 700 East 801-908-5800
Suite 100 800-616-8326
Salt Lake City, UT 84107 Fax: 801-908-5400
E-mail: info@nef1.org
http://www.nef1.org
A nonprofit educational organization dedicated to the development, dissemination and implementation of supplemental educational materials and programs primarily related to the environment, conservation, science, energy, water, andnatural resources.
*Founded: 1976*
*Robert Poulson, President*
*Dari Scott, Vice President*

**136 Northwest Power and Conservation Council**
851 SW 6th Avenue 503-222-5161
Suite 1100 800-452-5161
Portland, OR 97204 Fax: 503-820-2370
E-mail: info@nwcouncil.org
http://www.nwcouncil.org
Develops and maintains a regional power plan and a fish and wildlife program to balance the Northwest's environment and energy needs.
*Bruce A. Measure, Chair*
*Dick Wallace, Vice Chair*

**137 Renewable Fuels Association**
425 Third Street, SW 202-289-3835
Suite 1150 Fax: 202-289-7519
Washington, DC 20024 E-mail: info@ethanolrfa.org
http://www.ethanolrfa.org
The national trade association for the US ethanol industry, promotes policies, regulations and research and development initiatives that will lead to the increased production and use of fuel ethanol.
*Founded: 1981*
*Bob Dinneen, President/CEO*
*Mary Giglio, Director*

**138 Rocky Mountain Institute**
2317 Snowmass Creek Road 970-927-3851
Snowmass, CO 81654 http://www.rmi.org
An independent, entrepreneurial, nonprofit think-and-do tank that envisages a world thriving, verdant, and secure, for all.
*Founded: 1982*
*Michael Potts, President*
*Carrie Sullivan, Executive Assistant*

## National: Environmental Engineering

**139 American Academy of Environmental Engineers**
130 Holiday Court 410-266-3311
Suite 100 Fax: 410-266-7653
Annapolis, MD 21401 E-mail: info@aaee.net
http://www.aaee.net
Dedicated to excellence in the practice of environmental engineering to ensure the public health, safety, and welfare to enable humankind to co-exist in harmony with nature.
*Founded: 1955*
*Joseph S Cavaretta, Executive Director*
*J. Sammi Olmo, Manager*

**140 American Institute of Chemical Engineers**
3 Park Avenue 203-702-7660
New York, NY 10016-5991 800-242-4363
Fax: 203-775-5177
http://www.aiche.org
Organization for chemical engineering professionals. AIChE has the breadth of resources and expertise you need whether you are in core process industries or emerging areas, such as nanobiotechnology.
*Founded: 1908 40,000 Members*
*Maria K. Burka, President*
*David a. Rosenthal, President Elect*

**141 American Institute of Chemists**
315 Chestnut Street 215-873-8224
Philadelphia, PA 19106 Fax: 215-629-5224
E-mail: info@theaic.org
http://www.theaic.org
To advance the chemical sciences by establishing high professional standards of practice and to emphasize the professional, ethical, economic, and social status of its members for the benefit of society as a whole.
*Founded: 1923*
*David Manuta, Chairman*
*David W. Riley, Vice Chairman*

**142 American Society of Agricultural and Biological Engineers**
2950 Niles Road
St Joseph, MI 49085
269-429-0300
800-371-2723
Fax: 269-429-3852
E-mail: hq@asabe.org
http://www.asabe.org

An educational and scientific organization dedicated to the advancement of engineering applicable to agricultural, food, and biological systems.
*Founded: 1907 9,000 Members*
*Ronald L. McAllister, President*
*Sonia M. Maassel Jacobsen, President Elect*

**143 American Society of Civil Engineers**
1801 Alexander Bell Road
Reston, VA 20191
703-295-6000
800-548-2723
Fax: 703-295-6333
http://www.asce.org

Represents members of the civil engineering profession worldwide, and is the oldest national engineering society.
*Founded: 1852 133,000 Members*
*D Wayne Klotz, President*
*Patrick J Natale, Executive Director*

**144 American Society of Safety Engineers**
1800 E Oakton Street
Des Plaines, IL 60018
847-699-2929
Fax: 847-768-3434
E-mail: customerservice@asse.org
http://www.asse.org

The oldest and largest professional safety organization whose members manage, supervise and condult on safety, health, and environmental issues in industry, insurance, government and education.
*Founded: 1911 32,000 Members*
*Darryl C. Hill, President*
*Terrie S. Norris, President Elect*

**145 American Society of Sanitary Engineering**
901 Canterbury
Suite A
Westlake, OH 44145
440-835-3040
Fax: 440-835-3488
E-mail: info@asse-plumbing.org
http://www.asse-plumbing.org

A nonprofit organization that is comprised of individuals and sustaining members who represent all disciplines of the Plumbing Industry.
*Founded: 1906*
*James Bickford, President*
*Donald R. Summers Jr., First Vice President*

**146 Association for Environmental Health and Sciences**
150 Fearing Street
Amherst, MA 1002
413-549-5170
Fax: 413-549-0579
E-mail: paul@aehs.com
http://http://www.aehsfoundation.org/

The AEHS was created to facilitate communication and foster cooperation among professionals concerned with the challenge of soil protection and cleanup. Members represent the many disciplines involved in making decisions and solvingproblems affecting soils.
*Founded: 2009*
*Paul Kostecki, Executive Director*
*Ed Calabrese, Editor in Chief*

**147 Association of Energy Engineers**
4025 Pleasantdale Road
Suite 420
Atlanta, GA 30340
770-447-5083
Fax: 770-446-3969
E-mail: info@aeecenter.org
http://www.aeecenter.org

A non profit professional society that promotes the scientific and educational interests of those engaged in the energy industry and fosters action for sustainable development.
*Founded: 1977 9,500 Members*
*Eric A. Woodroof, President*
*Gary Hogsett, President Elect*

**148 Association of Environmental Engineering and Science Professors**
Business Office
2303 Naples Court
Champaign, IL 61822
217-398-6969
Fax: 217-355-9232
E-mail: joanne@aeesp.org
http://www.aeesp.org

The association is made up of professors of worldwide academic programs who provide education in the sciences and technologies of environmental protection. It has more than 700 members.
*Founded: 1963*
*Nancy G. Love, President*
*Joel G. Burken, President Elect*

**149 Association of Ground Water Scientists and Engineers**
601 Dempsey Road
Westerville, OH 43081
614-898-7791
800-551-7379
Fax: 614-898-7786
E-mail: ngwa@ngwa.org
http://www.ngwa.org/agwse

The Scientists and Engineers membership division of the National Ground Water Association (NGWA).
*Founded: 1948*
*Art Becker, President*
*John Pitz, President Elect*

**150 Environmental and Engineering Geophysical Society**
1720 South Bellaire Street
Suite 110
Denver, CO 80222
303-531-7517
Fax: 303-820-3844
E-mail: staff@eegs.org
http://www.eegs.org

An applied scientific organization with 700 members. Among its goals are the promotion of the science of geophysics especially as it is applied to environmental and engineering problems; and to foster common scientific interests ofgeophysicists and their colleagues in other related sciences and engineering.
*Founded: 1992 650 Members*
*John Stowell, President*
*Mark Dunscomb, President Elect*

**151 Institute for Alternative Futures**
100 North Pitt Street
Suite 235
Alexandria, VA 22314
703-684-5880
Fax: 703-684-0640
E-mail: futurist@altfutures.com
http://www.altfutures.com

A nonprofit research and educational organization that specializes in aiding organization and individuals to more wisely choose and create their preferred futures.
*Founded: 1977*
*Jonathan Peck, President*
*Clement Bezold, Chairman*

**152 Institute of Noise Control Engineering**
9100 Purdue Road
Suite 200
Indianapolis, IN 46268
317-735-4063
Fax: 317-280-8527
E-mail: ibo@inceusa.org
http://www.inceusa.org

A non-profit professional organization incorporated with the primary purpose to promote engineering solutions to environmental, product, machinery, industrial and other noise problems.
*James K. Thompson, President*
*Eric Wood, President Elect*

**153 Inter-American Association of Sanitary and Environmental Engineering**
USAIDIS
PO Box 7737
McLean, VA 22106
703-247-8730
Fax: 703-243-9004
E-mail: turnerje@cdm.com
http://www.aidis-usa.org

To further the goals of AIDIS Interamericana through programs and services that promote sound environmental practices, policies, management, and education to improve the quality of life throughout the Americas.
*Quincalee Brown, President*
*Jacqueline Rose, Senior Vice President*

**154 National Registry of Environmental Professionals (NREP)**
PO Box 2099
Glenview, IL 60025
847-724-6631
Fax: 847-724-4223
E-mail: nrep@nrep.org
http://www.nrep.org
The NREP are dedicated to professionally and legally enhancing the recognition of those individuals who possess the education, training, and experience as qualified environmental engineers, technologists, managers, technicians, andscientists. We aim to consolidate such recognition into a single, viable source so that the public at large, the government, insurers, and employers can easily see, understand, and justify the importance of these individuals.
*Scott Spear, Chairperson*
*Richard Young, Executive Director*

## National: Environmental Health

**155 Greenguard Environmental Institute**
2211 Newmarket Parkway, Suite 110
Marietta, GA 30067
770-984-9903
800-427-9681
Fax: 770-980-0072
E-mail: info@greenguard.org
http://www.greenguard.org
The Greenguard Environmental Institute aims to protect human health and improve quality of life by enhancing indoor air quality and reducing people's exposure to chemicals and other pollutants.
*Founded: 2001*
*Henning Bloech, Executive Director*
*Dr. Marilyn Black, Founder*

## National: Gaming & Hunting

**156 Action Volunteers for Animals (AVA)**
PO Box 64578
Unionville, ON L3R-0M9
416-439-8770
E-mail: actionvolunteers@yahoo.com
http://www.actionvolunteersforanimals.com
An organization of people who are active on behalf of non-human anmicals, and who give their time and energy without monetary reward.
*Founded: 1972*

**157 American Bass Association**
402 N Prospect Avenue
Redondo Beach, CA 90277
310-376-1026
Fax: 310-376-5072
E-mail: feedback@americanbass.com
http://www.americanbass.com
American Bass develops and institutes programs that will protect, enhance and improve the environment and natural resources while providing high value to sponsors and advertisers through quality run events and promotions. They areinvolved in such programs as: stocking bass in public lakes, planting trees, and protection of natural spawning and juvenile fish.
*David Plotnik, President*

**158 American Birding Association**
4945 N Street
Suite 200
Colorado Springs, CO 80919
719-578-9703
800-850-2473
Fax: 719-578-1480
E-mail: member@aba.org
http://www.aba.org
The Association represents the interests of birdwatchers in various arenas, and helps birders increase their knowledge, skills, and enjoyment of birding. ABA also contributes to bird conservation by linking the skills of its members toon-the-ground projects. ABA promotes field-birding skills through meetings, workshops, equipment, and guided involvement in birding, promoting national and international birders networks and publications.
*Founded: 1969*
*Jeffrey A. Gordon, President/CEO*
*Jane Alexander, Director*

**159 American Eagle Foundation**
PO Box 333
Pigeon Forge, TN 37868
865-429-0157
800-232-4537
Fax: 865-429-4743
E-mail: eaglemail@eagles.org
http://www.eagles.org
Not-for-profit organization of concerned citizens and professionals to develop and conduct bald eagle and environmental recovery programs in the United States and to assist private, state and federal projects that do the same.
*Founded: 1985*
*Al Louis Cecere, President/CEO*
*Bobby Halliburton, Vice President*

**160 American Fisheries Society**
5410 Grosvenor Lane
Bethesda, MD 20814
301-897-8616
Fax: 301-897-8096
E-mail: main@fisheries.org
http://www.fisheries.org
The American Fisheries Society (AFS) is an international professional and scientific organization of nearly 9,000 fisheries managers and aquatic scientists. Founded in 1870, AFS is the worlds oldest and largest organization dedicatedto strengthening the fisheries profession, advancing fisheries science and conserving fisheries resource. AFS chapters exist worldwide and throughout North America.
*Ghassan N. Rassam, Executive Director*
*Myja Merritt, Office Administration Mgr*

**161 American Humane Association**
63 Inverness Drive East
Englewood, CO 80112
303-792-9900
800-227-4645
Fax: 303-792-5333
http://www.americanhumane.org
The Associations mission, as a network of individuals and organizations, is to prevent cruelty, abuse, neglect and exploitation of children and animals and to assure that their interests and well-being are fully, effectively, andhumanely guaranteed by an aware and caring society.
*Founded: 1877*
*Robin R. Ganzert, President/CEO*
*Dale Austin, Chief Operating Officer*

**162 American Livestock Breeds Conservancy**
PO Box 477
Pittsboro, NC 27312
919-542-5704
Fax: 919-545-0022
E-mail: albc@albc-usa.org
http://www.albc-usa.org
Protects genetic diversity in livestock and poultry species through the conservation and promotion of endangered breeds. These rare breeds are a part of our national heritage and represent a unique piece of the earth's bio-diversity.
*Founded: 1977*
*Charles Bassett, Executive Director*
*Marjorie E F Bender, Research/Technical Prog Dir*

**163 American Medical Fly Fishing Association**
PO Box 760
Lock Haven, PA 17745
570-769-7375
E-mail: amffa@cub.kcnet.org
http://www.amffa.org
The Association began with the idea of combining professional medical interests with that of then interest in fly fishing and to promote conservation of the natural resources as pertains to the sport and support those causes andefforts oriented towards these latter goals. The organization has grown steadily over the years and presently is a strong and viable organization.
*Founded: 1969*
*Veryl Frye MD, Secretary/Treasurer*

**164 American Pheasant and Waterfowl Society**
6220 Bullbeggar Road
Withams, VA 23488
E-mail: alpat@apexotics.com
http://www.apws.org

To promote the rights and interest of the members to keep pheasants, waterfowl, and other upland aquatic and ornamental birds. To collect and distribute pertinent and scientific data and information relating to these rights. Thesociety advocates and encourages public appreciation and understanding of wildlife conservation and publishes a monthly magazine.

*$25/Year*
*Al Novosad, President*
*Terry Smith, Secretary*

**165 American Society of Ichthyologists and Herpetologists**
Florida Intl Univ-Bio Sciences
11200 SW 8th Street
Miami, FL 33199
305-348-1235
Fax: 305-348-4172
E-mail: asih@fiu.edu
http://www.asih.org

The Society is dedicated to the scientific study of fishes, amphibians and reptiles. Its mission is to increase knowledge about these organisms, to disseminate that knowledge through publications, conferences, symposia, and othermeans, and to encourage and support young scientists who will make future advances in these fields.

*Founded: 1913*
*Michael E. Douglas, President*
*Steven J. Beaupre, Vice President*

**166 American Society of Mammalogists**
PO Box 7060
Lawrence, KS 66044
800-627-0629
Fax: 785-843-1274
E-mail: dodell@cfl.rr.com
http://www.mammalsociety.org

Established for the purpose of promoting interest in the study of mammals. There are over 4,500 members and they provide information for public policy, resources management, conservation and education.

*Founded: 1919*
*Robert M Timm, President*
*Nancy Solomon, VP*

**167 Animal Protection Institute**
1122 S Street
Sacramento, CA 95811
916-447-3085
Fax: 916-447-3070
E-mail: info@api4animals.org
http://www.api4animals.org

A national animal advocacy nonprofit organization established to advocate for the protection of animals from cruelty and exploitation. Its primary campaign areas include: farmed animals, companion animals, compassionate consumerism,wildlife protection, and animals used in entertainment.

*Founded: 1968*
*Ronnie Wilkie, President*
*Susan Lock, Secretary/Treasurer*

**168 Birds of Prey Foundation**
2290 S 104th Street
Broomfield, CO 80020
303-460-0674
E-mail: raptor@birds-of-prey.org
http://www.birds-of-prey.org

Treats injured and orphaned wildlife, primarily raptors, such as eagles, hawks, falcons and owls and return healthy members of the breeding population to the natural habitat; fosters compassion for wildlife in distress and teach newgenerations through mentorships, internships, lectures, and volunteer programs; seeks protecton for raptors in the wild through education, invention and interventions.

*Sigrid Ueblacker, President/Executive Director*
*Brenda Leap, Secretary*

**169 Cetacean Society International**
PO Box 953
Georgetown, CT 6829
203-770-8615
Fax: 860-561-0187
E-mail: rossiter@csiwhalesalive.org
http://www.csiwhalesalive.org

All volunteer, nonprofit conservation, educational and research organization to benefit whales, dolphins, porpoises and the marine environment. Promotes education and conservation programs, including whale and dolphin watching, andnoninvasive, benign research. Advocates for laws and treaties to prevent commercial whaling, habitat destruction and other harmful or destructive human interactions. CSI's world goal is to minimize cetacean killing and captures and to enhance publicawareness.

*William Rossiter, President*

**170 Federal Wildlife Officers Association**
PO Box 206
Amesbury, MA 1913
http://www.fwoa.org

An organization dedicated to the protection of wildlife, the enforcement of federal wildlife law, the fostering of cooperation and communication among federal wildlife officers, and the perpetuation, enhancement and defense of thewildlife enforcement profession.

*Dave Hubbard, President*
*John Brooks, Vice President*

**171 Foundation for North American Wild Sheep**
720 Allen Avenue
Cody, WY 82414
307-527-6261
Fax: 307-527-7117
E-mail: fnaws@fnaws.org
http://www.fnaws.org

The Foundation's mission is to promote and enhance populations of indigenous wild sheep on the North American continent, and to fund programs for the professional management of these populations, while keeping administrative costs to aminimum.

*Founded: 1974*
*Gray Thornton, President/CEO*
*Dan Boone, Chairman*

**172 Friends of the Australian Koala Foundation**
214 W 29th Street
Suite 1004
New York, NY 10001
212-967-8200
Fax: 212-967-7292
E-mail: akf@savethekoala.com
http://www.savethekoala.com

*Founded: 1986*
*Debbie Tabart, CEO*
*Lorraine O'Keefe, Administration & Finance*

**173 Friends of the Earth**
1100 15th Street NW
11th Floor
Washington, DC 20005
202-783-7400
877-843-8687
Fax: 202-783-0444
E-mail: foe@foe.org
http://www.foe.org

National nonprofit advocacy organization dedicated to protecting the planet from environmental degradation; preserving biological, cultural and ethnic diversity, and empowering citizens to have an influential voice in decisionsaffecting the quality of their environment and their lives.

*Founded: 1969*
*Erich Pica, President*
*Nick Berning, Director of Communications*

**174 Fund for Animals**
200 West 57th Street
New York, NY 10019
212-246-2096
Fax: 212-246-2633
E-mail: hdquarters@fund.org
http://www.fundforanimals.org/about/

The Fund for Animals was founded in 1967 by author/humanitarian Cleveland Amory to speak for those who can't. In 2005, The Fund for Animals merged with The Humane Society of the United States, to avoid duplication of program andincrease strength and coordination in the areas of legislation, litigation, humane education and disaster relief.

*Founded: 1967*
*Marian Probst, Chair*

**175 Game Conservancy USA**
2 Soundview Drive
Suite 100
Greenwich, CT 6830
203-661-7900
Fax: 203-661-7997
E-mail: gbignell@gcusa.org
http://www.gcusa.org

Seeks to increase support for research of conservation of game and wildlife habitats in the US and UK through an expanding membership base.
*Bruce D. Sargent, President*
*John M.B. O'connor, Treasurer*

**176 Hawk Migration Association of North America**
P.O. Box 721
Plymouth, NH 3264
E-mail: info@hmana.org
http://www.hmana.org
The mission is to conserve raptor populations through the scientific study, enjoyment and appreciation of raptor migration.
*Founded: 1974*
*Gil Randell, Chair*
*Allen Hale, Vice Chair*

**177 Hawkwatch International**
2240 South 900 East
Salt Lake City, UT 84106
801-484-6808
800-726-4295
Fax: 801-484-6810
E-mail: hwi@hawkwatch.org
http://www.hawkwatch.org
Thier mission is to monitor and protect hawks, eagles, other birds of prey and their environment through research, education and conservation.
*Founded: 1986*
*John Witmer, Chair*
*Cathy Lambourne, Vice Chairman*

**178 Humane Society of the United States**
2100 L Street NW
Washington, DC 20037
202-452-1100
Fax: 202-778-6132
E-mail: corprelations@hsus.org
http://www.hsus.org
HSUS is the nation's largest animal-protection organization with more than 5 million constituents. The HSUS was founded to promote the humane treatment of animals, to foster respect, understanding, and compassion for all creatures.Today our message of care and protection embraces not only the animal kingdom but also the Earth and its environment. To achieve our goals we work through legal, educational, legislative and investigative means.
*Founded: 1954*
*Wayne Pacelle, President/CEO*
*Michael Markarian, VP/COO*

**179 International Association for Bear Research and Management**
1300 College Road
Fairbanks, AK 99071
907-459-7238
Fax: 907-451-9723
http://www.bearbiology.com
A Volunteer organization open to professional biologists, wildlife managers and others dedicated to the conservation of all species of bears. Consists of several hundred members from over 20 countries. Supports the scientificmanagement of bears through research and distribution of information.
*Frank Van Manen, President*
*Harry Reynolds, Vice President*

**180 International Society for Endangered Cats**
3070 Riverside Drive
Suite 160
Columbus, OH 43221
614-487-8760
Fax: 614-487-8769
http://www.isec.org/isec-left-index.htm
A not-for-profit organization dedicated to the conservation of wild cats throughout the world. Most species of wild cats are threatened or endangered in all or parts of their native ranges and, unless action is taken in the nearfuture, many will be lost to extinction. The society's goals are to raise public understanding and knowledge of wild cats and support and facilitate research on ecology, captive breeding and reintroduction of cats to their native habitats.
*Bill Simpson, President*
*Patricia Currie, Executive Director*

**181 International Wild Waterfowl Association**
10114 54th Place NE
Everett, WA 98205
425-334-8223
Fax: 425-397-8136
E-mail: dye@greatnorthern.net
http://www.wildwaterfowl.org
IWWA is committed to protecting and enhancing wild waterflow habitats. Supports the captive breeding and restoration of endangered species, and supports the establishment and maintenance of genetically diverse and disease-free captivepopulations of endangered waterfowl.
*Walter Sturgeon, President*

**182 International Wildlife Coalition**
Whale Adoption Program
70 East Falmouth Highway
PO Box 388
East Falmouth, MA 2536
508-548-8328
Fax: 508-548-8542
http://www.iwc.org
Adoption fees are used to help purchase rescue and research vessels, hire crews, and rescue whales and other marine mammals from fishing nets, conduct crucial research, and help enforce wildlife protection laws and treaties.
*Daniel J Morast, President*

**183 International Wildlife Conservation Society**
New York Zoological Society
Bronx, NY 10460
718-364-4275
Fax: 212-220-7114
*John G Robinson, Director*

**184 International Wolf Center**
3410 Winnetka Avenue North
Minneapolis, MN 55427
763-560-7374
Fax: 763-560-7368
E-mail: mortiz@wolf.org
http://www.wolf.org
The International Wolf Center advances the survival of the wolf populations by teaching about wolves, their relationship to wildlands and the human role in their future.
*Founded: 1985*
*Nancy Jo Tubbs, Chair*
*L. David Mech, Vice Chair*

**185 Last Chance Forever**
PO Box 460993
San Antonio, TX 78246
210-499-4080
Fax: 210-499-4305
E-mail: raptor@ddc.net
http://www.lastchanceforever.org
Dedicated to the rehabilitation of sick, injured and orphaned birds of prey, scientific investigation, and also just as importantly, the education of the public.
*Founded: 1978*
*John A Karger, Executive Director*

**186 Muskies**
PO Box 120870
New Brighton, MN 58112
701-239-9540
E-mail: info@muskiesinc.org
http://www.muskiesinc.org
Supports growth and interest in the sport of muskie fishing. We are reaching out to protect our existing fisheries and to develop new fisheries.
*Founded: 1966*
*Greg Wells, President*

**187 National Bison Association**
8690 Wolff Ct. #200
Westminster, CO 80031
303-292-2833
Fax: 303-845-081
E-mail: info@bisoncentral.com
http://www.bisoncentral.com
Created from a merger of the American Bison Association and the National Buffalo Association, the NBA has over 2400 members in all 50 states and 16 foreign countries. It is a nonprofit association which promotes the preservation,production and marketing of bison. NBA activities and services serve to better inform and educate our members and the general public about bison.
*Founded: 1995*
*John Flocchini, President*
*Peter Cook, Vice President*

**188 National Endangered Species Act Reform Coalition**
1050 Thomas Jefferson Street NW
6th Floor
Washington, DC 20007
202-333-7481
Fax: 202-338-2461
E-mail: nesarc@vnf.com
http://www.nesarc.org

A broad based coalition of roughly 150 member organziations, representing millions of individuals across the United States, that is dedicated to bringing balance back to Endangered Species Act. Our membership includes rural irrigators,municipalities, farmers, electric utilities and many other individuals and organizations that are directly affected by the ESA.

*Nancy Macan McNally, Executive Director*
*Jordan Smith, Associate Director*

**189 National Hunters Association**
PO Box 820
Knightdale, NC 27545
919-365-7157
Fax: 919-366-2142
E-mail: nhadvs@worldnet.att.net
http://www.nationalhunters.com

Our organization is dedicated to preserving the rights of hunters, promoting hunter safety among the youth and all hunters by demanding game controls and hunting laws, maintaining the rights to use and own firearms and protecting theenvironment and maintaining a healthy habitat.

*D Smith, President*
*Faye Smith, Secretary/Treasurer*

**190 National Marine Fisheries Service Office of Protective Resources**
1315 East-West Highway
Silver Spring, MD 20910
301-713-2322
Fax: 301-713-0376
E-mail: jim.lecky@noaa.gov
http://www.nmfs.noaa.gov

*James H Lecky, Director*

**191 National Rifle Association of America**
11250 Waples Mill Road
Fairfax, VA 22030
703-267-1000
800-672-3888
Fax: 703-267-3909
http://www.nra.org

*Founded: 1871*
*Chris Cox, Executive Director*

**192 National Shooting Sports Foundation**
11 Mile Hill Road
Newtown, CT 6470
203-426-1320
Fax: 203-426-1087
E-mail: info@nssf.org
http://www.nssf.org

Leading trade association of the firearms and recreational shooting sports industry. A nonprofit communications and marketing organization, the NSSF manages a variety of programs designed to promote a better understanding of and a moreactive participaqtion in the shooting sports.

*Founded: 1961*
*Robert L. Scott, Chairman*
*Stephen Hornady, Co-Vice Chairman*

**193 National Trappers Association**
524 5th Street
Bedford, IN 47421
812-277-9670
Fax: 812-277-9672
E-mail: ntaheadquarters@nationaltrappers.com
http://www.nationaltrapper.com

To promote conservation, legislation and administrative procedures; to save and faithfully defend from waste the natural resources of the United States; to promote environmental education programs; and to promote a continued annual furharvest using the best tools presently available for that purpose.

*Steve Fitzwater, President*

**194 National Walking Horse Association**
4059 Iron Works Parkway
Suite 4
Lexington, KY 40511
859-252-6942
Fax: 859-252-0640
http://www.nwha.com

An alliance of people committed to preserving and fostering the natural abilities and welfare of the Walking Horse. Improves the lives of horses and people by encouraging responsibility and sportsmanship. Promotes educational andrecreational activities, while preserving the unique qualities of the Walking Horse.

*Lori Snyder-Lowe, President*
*Margaret Hershberger, Vice President*

**195 National Wildlife Federation**
11100 Wildlife Center Drive
Reston, VA 20190
202-797-6800
800-822-9919
Fax: 202-797-6646
http://www.nwf.org

The National Wildlife Federation is the largest member-supported conservation group, uniting individuals, organizations, businesses and government to protect wildlife, wild places and the environment.

*Larry J Schweiger, President/CEO*
*Jaime Matyas, Executive Vice President*

**196 National Wildlife Federation: Office of Federal and International Affairs**
1400 16th Street NW
Washington, DC 20036
202-797-6800
Fax: 202-797-6646

A field office of the NWF staffed by people experienced on policy issues, grassroots outreach, law, government affairs, and media. The Office educates, mobilizes and advocates to preserve and strengthen protection for wildlife and wildplaces.

**197 Native American Fish and Wildlife Society**
From The Eagle's Nest
8333 Greenwood BLVD
Suite 260
Denver, CO 80221
303-466-1725
866-890-7258
Fax: 303-466-5414

National tribal organization to develop a national communications network for the exchange of information and management techniques related to self-determined tribal fish and wildlife management.

*Founded: 1983*
*D Fred Matt, Executive Director*

**198 North American Bear Society**
4061 E Hartford Avenue
PO Box 555774
Phoenix, AZ 85078
602-971-2338
Fax: 602-971-2100
E-mail: bearsociety@nonline.com
http://www.nonprofitnet.com/nabs

To support the conservation and management of bear populations throughout North America for the benefit of the general public and future generations. Dedicated to the conservation and management of the indigenous bears of NorthAmerica.

*William T Smaltz, Founder*

**199 North American Native Fishes Association**
123 W Mount Airy Avenue
Philadelphia, PA 19119
E-mail: nanfa@att.net
http://www.nanfa.org

NANFA is dedicated to the enjoyment, study and conservation of the continent's native fishes. The objectives are: to increase and disseminate knowledge about fishes and their habitats amoung aquarium hobbists, biologists, fish andwildlife officals, anglers, educators, students and others, through publications, electronic media, regional and national meetings, and other means; to promote the conservation and the protection/restoration of natural habitats; to advance thecaptive husbandry.

**200 Organization of Wildlife Planners**
500 Lafayette Road
PO Box 25
Saint Paul, MN 55155
http://www.owpweb.org

A professional organization of creative, committed people concerned with the management and future of government agencies that manage fish and wildlife populations and habitat. The purpose of OWP is to improve fish and wildliferesource management capabilities through informed decision-making.

*Founded: 1978*
*Verdie J. Abel, President*
*Michele Beucler, President Elect*

201 **POWS Wildlife Rehabilitation Center**
15305 44th Avenue West 425-743-1884
PO Box 1037 Fax: 425-742-5711
Lynnwood, WA 98046 E-mail: kparker@paws.org
http://www.paws.org/wildlife
Each wildlife patient requires individualized, specialized care. But for every animal the goal is identical: rehabilitation and release into the wild.
*Kip Parker, Director*

202 **Pacific Whale Foundation**
300 Maalaea Road 808-249-8811
Suite 211 800-942-5311
Wailuku, HI 96793 Fax: 808-243-9021
E-mail: info@pacificwhale.org
http://www.pacificwhale.org
Mission is to promote appreciation, understanding and protection of whales, dolphins, coral reefs, and our planet's oceans. We accomplish this by educating the public from a scientific perspective about the marine environment.
*Founded: 1980*
*Gregory Kaufman, President*

203 **People for the Ethical Treatment of Animals**
501 Front Street 757-622-7382
Norfolk, VA 23510 Fax: 757-622-0457
E-mail: info@peta.org
http://www.peta.org
With more than 850,000 members, PETA is the largest animal rights organization in the world. Also dedicated to establishing and protecting the rights of all animals. Operates under the simple principal that animals are not ours to eat,wear, experiment on, or use for entertainment.
*Founded: 1980*
*Ingrid Newkirk, President*

204 **Pheasants Forever**
1783 Buerkle Circle 651-773-2000
St. Paul, MN 55110 877-773-2070
Fax: 651-773-5500
E-mail: contact@pheasantsforever.org
http://www.pheasantsforever.org
A non profit organization dedicated to the protection and enhancement of pheasant and other wildlife populations in North America. This mission is carried out through habitat improvement, land management, public awareness, andeducation.
*Founded: 1983*
*Howard K Vincent, President/CEO*
*Dave Nomsen, Vice President*

205 **Pope and Young Club**
273 Mill Creek Road 507-867-4144
PO Box 548 Fax: 507-867-4144
Chatfield, MN 55923 E-mail: pyclub@isl.net
http://www.pope-young.org
The Pope and Young Club is one of North America's leading bowhunting and conservation organizations. Founded in 1961 as a nonprofit scientific organization, the Club is patterned after the prestigious Boone and Crockett Club. The Clubadvocates and encourages responsible bowhunting by promoting quality, fair chase hunting, and sound conservation practices. Today it fosters and nourishes bowhunting excellence and acts in the best interest of our bowhunting heritage everywhere.
*Founded: 1961*
*Donald Ace Morgan, President*

206 **Purple Martin Conservation Association**
Edinboro University of Pennsylvania 814-734-4420
219 Meadville Street Fax: 814-734-5803
Edinboro, PA 16444 E-mail: pmca@edinboro.edu
http://www.purplemartin.org
The only organization devoted exclusively to the scientific study of purple martins, their biology, and habitat requirements.
*Founded: 1986*
*James Hill, Founder*
*John Tautin, Executive Director*

207 **Safari Club International**
4800 W Gates Pass Road 520-620-1220
Tucson, AZ 85745 888-486-8724
Fax: 520-622-1205
http://www.safariclub.org/
A leader in wildlife conservation, hunter education, and protecting the freedom to hunt.
*Dennis Anderson, President*
*John W Nelson, Vice President*

208 **Scientists Center for Animal Welfare**
7833 Walker Drive 301-345-3500
Suite 410 Fax: 301-345-3503
Greenbelt, MD 20770 E-mail: INFO@SCAW.COM
http://WWW.SCAW.COM
A nonprofit educational association of individuals whose mission is to promote humane care, use, and management of animals involved in research, testing or education in laboratory, agricultural, wildlife or other settings.
*Founded: 1978*
*Joanne Zurlo, President*
*Ernest D Prentice, Vice President*

209 **Sea Shepherd Conservation Society**
PO Box 2616 360-370-5650
Friday Harbor, WA 98250 Fax: 360-370-5651
E-mail: info@seasheperd.org
http://www.seashepherd.org
The mandate of this organization was mammal protection and conservation with an immediate goal of shutting down illegal whaling and sealing operations. We are committed to the eradication of private whaling, poaching, shark finning,unlawful habitat destruction, and violations of established laws in the world's oceans.
*Founded: 1977*
*Paul Watson, Founder*

210 **Sea Turtle Restoration Project**
PO Box 400 415-488-0370
Forest Knolls, CA 94933 Fax: 415-488-0372
E-mail: info@seaturtles.org
http://www.seaturtles.org
Fights to protect endangered sea turtle populations in ways that meet the ecological needs of the sea turtles and the oceans and the needs of the locla communities who share the beaches and waters with these gentle, beautifulcreatures.
*Founded: 1989*
*Erica Heimberg, Managing Director*

211 **Society for Animal Protective Legislation**
Georgetown Station 202-337-2334
PO Box 3719 Fax: 202-338-9478
Washington, DC 20007 E-mail: sapl@saplonline.org
http://www.saplonline.org
Prepares information for use by Members of Congress and their staffs to protect animals. Also sends action alerts to individuals and organizations interested in animal protective legislation, informing them of ways in which they mayhelp.
*Founded: 1955*
*Madeleine Bemelmans, President*

212 **Society of Tympanuchus Cupido Pinnatus**
College of Natural Resources 715-346-3859
University of Wisconsin-Stevens Point Fax: 715-346-3624
Stevens Point, WI 54481 E-mail: canderson@uwsp.edu
http://www.uwsp.edu/wildlife/programs
*Russell Schallert, President*

213 **Sportsmans Network**
111 S Main Street 859-824-6526
PO Box 257 800-680-8058
Dry Ridge, KY 41035 Fax: 606-824-0556
E-mail: sportsman@sportsmansnetwork.org
http://www.sportsmansnetwork.org
To raise awareness of animal wildlife conservation issues through various forms of public media and to address the subject

in all appropriate ways including hunting, fishing, trapping, and related activities.
*William Krebs, Treasurer*

**214 Trout Unlimited**
1300 North 17th Street
Suite 500
Arlington, VA 22209
703-522-0200
800-834-2419
Fax: 703-284-9400
E-mail: trout@tu.org
http://www.tu.org

Mission: To conserve. protect and restore North America's trout and salmon fisheries and their watersheds. We accomplish this mission on local, state and national levels with an extensive and dedicated volunteer network.
*Founded: 1959*
*Charles Gauvin, President/CEO*
*Steve Moyer, Vice President*

**215 Trumpeter Swan Society**
3800 County Road 24
Maple Plain, MN 55359
763-694-7851
Fax: 763-476-1514
E-mail: ttss@hennepinparks.org
http://www.trumpeterswansociety.org

Private, nonprofit organization dedicated to assuring the vitality and welfare of wild Trumpeter Swan populations, and to restoring the species to as much of its former range as possible.
*Founded: 1968*
*Madeleine Linck, Editor*
*Ruth E Shea, Executive Director*

**216 US Sportsmen's Alliance**
801 Kingsmill Parkway
Columbus, OH 43229
614-888-4868
Fax: 614-888-0326
E-mail: info@USSPORTSMEN.org
http://www.ussportsmen.org

The US Sportsmen's Alliance is a national non-profit association that protects hunting, fishing and trapping and scientific wildlife management programs.
*Founded: 1978*
*Bud Pidgeon Jr, President/CEO*

**217 Waterfowl USA**
Waterfowl Building
Box 50
Edgefield, SC 29824
803-637-5767
Fax: 803-637-6983
E-mail: president@waterfowlusa.org
http://www.waterfowlusa.org

National non profit, conservation organization, dedicated to using funds in the areas in which they were raised for local and state waterfowl projects. Founded by biologists for the purpose of preserving and improving wintering andbreeding habitat within the United States.
*Founded: 1983*
*Roger L White, President/CEO*

**218 Whitetails Unlimited**
2100 Michigan Street
PO Box 720
Strugeon Bay, WI 54235
920-743-6777
800-274-5471
Fax: 920-743-4658
http://www.whitetailsunlimited.com

A national conservation organization that has remained true to its mission and has made great strides in the field of conservation. They have gained the reputation of being the nation's premier organization dedicating resources to thebetterment of the white-tail deer and its environment.
*Founded: 1982*
*Jeffrey Schinkten, President*
*Peter Gerl, Executive Director*

**219 Wilderness Society**
1615 M Street NW
Washington, DC 20036
800-843-9453
E-mail: member@tws.org
http://www.wilderness.org

Protects the wilderness and inspires Americans to care for the wild places.
*Founded: 1935*
*Leslie Jones, General Counsel*

**220 Wildlife Trust**
460 West 34th Street
17th Floor
New York, NY 10001
212-380-4460
Fax: 212-380-4465
E-mail: homeoffice@wildlifetrust.org
http://www.wildlifetrust.org

Works in the United States and worldwide to save threatened species from extinction, protect habitat, and link nature protection with health through collaborative projects with local scientists. They also communicate our results toeducators, health professionals, policy experts and civic leaders.
*Founded: 1971*
*Mary C Pearl PhD, President*
*Anthony M Ramos, Marketing/Communications*

**221 Xerces Society**
4828 SE Hawthorne Boulevard
Portland, OR 97215
503-232-6639
Fax: 503-233-6794
E-mail: info@xerces.org
http://www.xerces.org

An international non profit organization dedicated to protecting biological diversity through inverterate conservation. The Society advocates for invertebrates and their habitats by working with scientists, land managers, educators,and citizens on conservation and education projects. Core programs focus on endangered species, natice pollinators, and watershed health.
*Founded: 1971*
*Scott Hoffman Black, Executive Director*
*Mace Vaughan, Conservation Director*

## National: Habitat Preservation & Land Use

**222 Abundant Life Seeds**
PO Box 279
Cottage Grove, OR 97424
541-767-9606
Fax: 866-514-7333
E-mail: als@abundantlifeseeds.com
http://www.abundantlifeseeds.com

Focuses on the genetic diversity of rare and endangered food crops. Offers varieties grown using only certified organic or biodynamic farming methods.
*Founded: 1975*

**223 Agricultural Resources Center-Pesticide Education Project**
206 New Bern Place
Raleigh, NC 27601
919-833-5333
877-667-7729
E-mail: info@PESTed.org
http://www.ibiblio.org

*Fawn Pattison, Executive Director*
*Billie Karel, Program Coordinator*

**224 Aldo Leopold Foundation**
E13701 Levee Road
PO Box 77
Baraboo, WI 53913
608-355-0279
Fax: 608-356-7309
E-mail: mail@aldoleopold.org
http://www.aldoleopold.org

The Aldo Leopold Foundation is a 501 (c)3 nonprofit organization that works to promote the philosophy of Aldo Leopold and the land ethic he so eloquently defined in his writing. The foundation actively integrates programs on landstewardship, environmental education and scientific research to promote care of natural resources and have an ethical relationship between people and land.
*Founded: 1982*
*Wellington B Huffaker IV, Executive Director*

**225 America the Beautiful Fund**
725 15th Street NW
Suite 605
Washington, DC 20005
202-638-1649
Fax: 202-638-2175
E-mail: info@america-the-beautiful.org
http://www.america-the-beautiful.org

America the Beautiful Fund encourages volunteer citizen efforts to protect the natural and historic beauty of America. Programs

include Operation Green Plant:Free Seeds! which provides free seeds and bulbs for environmental educationand preservation, community gardens and hunger relief.
*Founded: 1965*
*Nanine Bilski, President/CEO*
*Kay Lautman, VP*

**226 American Cave Conservation Association**
119 E Main Street 270-786-1466
PO Box 409 Fax: 270-786-1467
Horse Cave, KY 42749 E-mail: acca@caven.org
http://www.cavern.org/acca
A national nonprofit association dedicated to the protection of caves, karstlands and groundwater. The ACCA operates the American Cave Museum and Karst Center, an educational center that includes the American Cave Museum and HiddenRiver Cave.
*Founded: 1978*
*David G Foster, Executive Director*
*Peggy A Nims, Community Outreach*

**227 American Conservation Association**
1200 New York Avenue NW 202-289-2431
Suite 400 Fax: 202-289-1396
Washington, DC 20005
Operative private organization dedicated to promote the knowledge and understanding of conservation, to preserve the live beauty of the landscape and the natural resources and organisms in areas of the United States and others, and toeducate to the public in the adapted use of these areas.
*Charles Clusen, Executive Director*
*R Greathead, Secretary*

**228 American Council on the Environment**
1301 20th Street NW 202-659-1900
Suite 113
Washington, DC 20036
*John H Gullett, Executive Officer*

**229 American Geological Institute**
4220 King Street 703-379-2480
Alexandria, VA 22302 Fax: 703-379-7563
E-mail: asm@agiweb.org
http://www.agiweb.org
Comprised of more than 50 scientific and professional associations that represent more than 120,000 geologists, geophysicists, and other earth scientists. Provides information services to geoscientists, serves as a voice of sharedinterests in the profession, plays a major role in strengthening geoscience education, and strives to increase public awareness of the vital role the geosciences play in society's use of resources and interaction with the environment.
*Founded: 1948*
*P. Patrick Leahy, Executive Director*
*Patrick Burks, Controller*

**230 American Institute of Fishery Research Biologists**
1315 E West Highway
Silver Spring, MD 20910
*Gary Sakagawa, President*
*Barbara Warkentine, Secretary*

**231 American Land Conservancy**
1388 Sutter Street 415-749-3010
Suite 810 Fax: 415-749-3011
San Francisco, CA 94109 E-mail: mail@alcnet.org
http://www.alcnet.org
ALC is dedicated to the preservation of land and water as enduring public resources, to protect and enhance our nation's natural, ecological historical, recreational, and scenic heritage.
*Harriet Burgess, President*

**232 American Lands**
230 N. Highland Street 703-216-8624
Arlington, VA 22201 E-mail: info@americanlands.org
http://www.americanlands.org
America's wildlife and wildlands continue to be threatened by logging, roadbuilding, grazing, off road vechicles and mining. A broad coalition of activists and organizations is dedicated to protecting our forest heritage and restoringecological integrity to the landscape.
*Jim Jontz, Executive Director*
*Randi Spivak, President*

**233 American Littoral Society**
468 S Green Street 609-294-3111
PO Box 1306 Fax: 609-294-8044
Tuckerton, NJ 08087 E-mail: alssj@earthlink.net
http://www.littoralsociety.org
The American Littoral Society is an environmental organization concerned about issues that affect the littoral zone: that area on the beach between low and high tide. The American Littoral Society is a national, nonprofit,public-interest organization comprised of over 6,000 professional and amateur naturalists.
*Michael Hubertt, President*
*Angela Cristini, Secretary*

**234 American Public Gardens Association**
100 West 10th Street 302-655-7100
Suite 614 Fax: 302-655-8100
Wilmington, DE 19801 E-mail: mquigley@publicgardens.org
http://www.publicgardens.org
Formerly the American Association of Botanical Gardens and Arboreta. APGA is committed to increasing the knowledge of public garden professionals through information sharing, professional development, education, research and plantconservation.
*Founded: 1940*
*Daniel J Stark, Executive Director*
*Madeline Quigley, Communications/Marketing*

**235 American Shore and Beach Preservation Association**
5460 Beaujolais Lane 239-489-2616
Fort Myers, FL 33919 Fax: 239-489-9917
E-mail: exdir@asbpa.org
http://www.asbpa.org
Dedicated to preserving, protecting and enhancing the beaches, shores and other coastal resources of America, recognizing their important quality-of-life assets.
*Founded: 1926*
*Harry Simmons, President*
*Kate & Ken Gooderham, Executive Directors*

**236 Antarctica Project**
1630 Connecticut Avenue NW 202-234-2480
3rd Floor Fax: 202-387-4823
Washington, DC 20009 E-mail: antarctica@igc.org
http://www.asoc.org
The only conservation organization in the world that works exclusively for Antarctica. Leads domestic and international campaigns to protect Antarctica's pristine wilderness and environment, as the international secretariat for theAntarctic and Southern Ocean Coalition (ASOC). ASOC is a global coalition with 214 member organizations working in 44 countries on six continents.
*Beth Clark, Director*
*Josh Stevens, Campaign Associate*

**237 Arctic Network**
PO Box 102252 907-272-2452
Anchorage, AK 99510 Fax: 907-272-2453
Works to the conserve Arctic ecosystem. Focus includes indigenous cultures.

**238 Association for Conservation Information**
1000 Assembly Street 843-762-5032
PO Box 167 Fax: 843-734-3951
Columbia, SC 29202
ACI, the Association for Conservation Information, is a nonprofit association of information and education professionals representing state, federal and Canadian agencies and private conservation organizations.
*Bob Campbell, President*

**239 Association for Environmental Health and Sciences**
150 Fearing Street 413-549-5170
Amherst, MA 01002 Fax: 413-549-0579
E-mail: paul@aehs.com
http://www.aehs.com
The AEHS was created to facilitate communication and foster co-operation among professionals concerned with the challenge of soil protection and cleanup. Members represent the many disciplines involved in making decisions and solvingproblems affecting soils. AEHS recognizes that widely acceptable solutions to the problem can be found only through the integration of scientific and technological discovery, social and political judgement and hands on practice.
*Paul Kostecki PhD, Executive Director*
*Cindy Langlois, Managing Director*

**240 Association for Natural Resources Enforcement Training**
Missouri Department of Conservation 573-751-4115
Box 180 E-mail: yamnil@mail.conservation.state.mo.us
Jefferson City, MO 65102 http://www.dirdid.com\anret
*Dave Windsor, VP*

**241 Association of Consulting Foresters of America, Inc**
312 Montgomery Street 703-548-0990
Suite 208 888-540-8733
Alexandria, VA 22314 Fax: 703-548-6395
E-mail: director@acf-foresters.org
http://www.acf-foresters.org
To protect the public welfare and property in the practice of forestry. To raise the professional standards and work of ACF Consultants and all other consulting foresters. To develop and expand the services of ACF Consultants.
*Lynn C Wilson, Executive Director*

**242 Association of Field Ornithologists**
University of Maine at Machias
9 O'Brian Avenue
Machias, ME 04654 http://www.afonet.org
Society of professional and amateur ornithologists dedicated to the scientific study and dissemination of information about birds in their natural habitats. Especially active in bird-banding and development of field techniques.Encourages participation of amateurs in research, and emphasizes conservation biology of birds.
*Founded: 1922*
*Charles Duncan, President*

**243 Association of Great Lakes Outdoor Writers**
301 Cross Street
Sullivan, IN 47882 877-472-4569
E-mail: edir@aglow.info
http://www.aglow.info
Dedicated to communicating the outdoor experience in word and image.
*Bob Whitehead, President*
*Curt Hicken, Executive Director*

**244 Association of Partners for Public Lands**
8375 Jumpers Hole Road 410-647-9001
Suite 104 Fax: 410-647-9003
Millersville, MD 21108 E-mail: appl@appl.org
http://www.appl.org
To provide the highest levels of program and service to public agencies entrusted with the care of America's natural and cultural heritage.
*Donna Asbury, Executive Director*

**245 Association of State Wetland Managers**
2 Basin Road 208-892-3399
Windham, ME 04062 Fax: 207-892-3089
E-mail: aswm@aswm.org
http://www.aswm.org
Nonprofit organization dedicated to the protection and restoration of the nations wetlands. Our goal is to help public and private wetland decision-makers utilize the best possible scientific information and techniques in wetlanddelineation, assessment, mapping, planning, regulation, acquisition, restoration, and other management.
*Founded: 1983*
*Jeanne Christie, Executive Director*
*Jon Kusler, Associate Director*

**246 Bat Conservation International**
PO Box 162603 512-327-9721
Austin, TX 78716 Fax: 512-327-9724
E-mail: batinfo@batcon.org
http://www.batcon.org
Dedicated to preserving and restoring bat populations and habitats around the world. Uses a non-confrontational approach to educate the public about the ecological and economic value of bats, advance scientific knowledge about bats andthe ecosystems that rely on them and preserve critical bat habitats through win-win solutions that benefit both humans and bats.
*Founded: 1982 $35/Yr*
*Dave Walden, Ex-Director*
*Linda Moore, Director Administration*

**247 Big Thicket Association**
PO Box 198 936-274-5000
Saratoga, TX 77585 E-mail: director@bigthicket.org
http://www.btatx.org
Formed to save remnants of the once extensive historic Big Thicket forests with its remarkable diversity.
*Founded: 1964*
*Dr Bruce Drury, President*
*Elaine Allums, Secretary*

**248 Birds of Prey Foundation**
2290 S 104th Street 303-460-0674
Broomfield, CO 80020 E-mail: raptor@birds-of-prey.org
http://www.birds-of-prey.org
The foundation seeks protection for raptors in the wild through education, invention and intervention; treats injured and orphaned wildlife; and works continuously to improve the quality of care and housing of captive raptorseverywhere.
*Sigrid Ueblacker, Executive Officer*
*Brenda Leap, Secretary*

**249 Boone and Crockett Club**
250 Station Drive 406-542-1888
Missoula, MT 59801 Fax: 406-542-0784
E-mail: bcclub@boone-crockett.org
http://www.boone-crockett.org
Organization founded by Theodore Roosevelt with the vision of establishing a coalition of dedicated conservationists and sportsmen to provide leadership in the issues affecting hunting, wildlife and wildlife habitat.
*Founded: 1887*
*Lowell E Baier, Executive VP*
*Keith Balfourd, Director Marketing*

**250 Brotherhood of the Jungle Cock**
PO Box 576
Glen Burnie, MD 21061 E-mail: bosleywright@hotmail.com
http://www.flyfishersofvirginia.org/bojc.htm
*Founded: 1940*
*Bosley Wright, Executive VP*

**251 Camp Fire Conservation Fund**
230 Campfire Road 914-769-5508
Chappaqua, NY 10514 Fax: 914-923-0977
E-mail: lvallen491@aol.com
The fund is a wildlife and habitat conservation organization that has been around for over 100 years.
*Gordon Whiting, President*

**252 Canyonlands Field Institute**
1320 South Highway 191 435-259-7750
Moab, UT 84532 800-860-5262
Fax: 435-259-2335
E-mail: info@canyonlandsfieldinst.org
http://www.canyonlandsfieldinst.org

To increase understanding of, connection to and care for the Colorado Plateau, expand perception of and appreciation for the beauty and integrity of the natural world, improve the quality of field oriented, experiential teaching andlearning for students and adults and to encourage individuals to be involved in the care of their own home places.
*Founded: 1984*
*Karla VanderZanden, President/Director*

**253 Caribbean Conservation Corporation & Sea Turtle Survival League**
4424 NW 13th Street 352-373-6441
Suite B-11 800-678-7853
Gainesville, FL 32609 Fax: 352-375-2449
E-mail: ccc@cccturtle.org
http://www.cccturtle.org
Strives to ensure the survival of sea turtles within the Wider Caribbean basin abd Atlantic through research, education, training, advocacy and protection of habitats. CCC was the first marine turtle conservation organization in theworld, and has more than 40 years of experience in national and international sea turtle conservation, research and educational endeavors.
*Founded: 1959*
*David Godfrey, Executive Director*
*Dan Evans, Outreach/Field Programs*

**254 Carrying Capacity Network**
2000 P Street NW 202-296-4548
Suite 240 Fax: 202-296-4609
Washington, DC 20036 http://www.carryingcapacity.org
An information exchange network. Interests include resource conservation, population stabilization and environmental protection.

**255 Center for Clean Air Policy**
750 1st Street Northeast 202-408-9260
Suite 940 Fax: 202-408-8896
Washington, DC 20002 E-mail: communications@ccap.org
http://www.ccap.org
A leader in climate and air quality policy, it advances cost-effective and pragmatic policies. International and domestic programs focus on these four major areas: GHG Emission and Mitigation Economics; Emerging Technologies andTechnology Investment; Transportation and Land Use; and Adaptation.
*Founded: 1985*
*Ned Helme, President*
*Stacey Davis, Director Domestic Programs*

**256 Center for Plant Conservation**
PO Box 299 314-577-9450
St. Louis, MO 63166 Fax: 314-577-9465
http://www.mobot.org/CPC/
The CPC is a consortium of 28 American botanical gardens and arboreta whose mission is to conserve and restore the rare native plants of the US To meet this end, they are involved in plant conservation, research and education. Thissite includes information about the National Collection of Endangered Plants, which is maintained by the group.

**257 Center for Wildlife Information**
PO Box 8289 406-721-8985
Missoula, MT 59807 E-mail: bearinfo@cfwi.org
http://www.centerforwildlifeinformation.org
*Chuck Bartlebaugh, Contact*

**258 Center for the Study of Tropical Birds**
218 Conway 210-828-5306
San Antonio, TX 78209 Fax: 210-828-9732
E-mail: office@cstBbnc.org
http://www.cstbinc.org
Devoted to the conservation of neotropical birdlife through collaborative programs of research and education.
*Jack Eitniear, Director*

**259 Clean Sites**
1199 N Fairfax Street 703-519-2135
Suite 400 Fax: 703-548-8733
Alexandria, VA 22314 E-mail: cses@cleansites.com
http://www.cleansites.com
We apply sound project management principles, real-world experience, and cost control measures to find creative solutions to environmental remediation and land reuse problems.
*Douglas Ammon, Contact*

**260 Committee for Conservation and Care of Chimpanzees**
3819 48th Street NW 202-362-1993
Washington, DC 20016 Fax: 202-686-3402
*Dr. Geza Teleki, Chairman*

**261 Committee for the Preservation of the Tule Elk**
PO Box 3696
San Diego, CA 92103
*Jolene W Steigerwalt, Secretary*

**262 Community Greens Shared Parks in Urban Blocks**
1700 N Moore Street 703-527-8300
Suite 2000 Fax: 703-527-8383
Arlington, VA 22209 E-mail: kherrod@ashoka.org
http://www.communitygreens.org
Community Greens aims to transform the interiors of urban blocks across the US into resplendent shared parks and gardens that are owned and managed by the residents who live around them. These community greens would provide a greenoasis for hassled city dwellers, offer a safe place for children to interact, increase home values and improve a sense of community. They would also help the environment by providing microhabitats for birds and small wildlife, clean the air andreduce urban sprawl.
*Kate Herrod, Director*

**263 Conservation Education Association**
3914 Foxdale Road 573-751-4115
New Bloomfield, MO 65063
Dedicated to the education of individuals from all walks of life who are interested in making a difference to the environment. The association offers workshops, advocacy, literature and hands-on events.

**264 Conservation Education Center**
2473 160th Road 641-747-8383
Guthrie Center, IA 50115 Fax: 641-747-3951
E-mail: ajay.winter@dnr.state.ia.us
We are an educational facility serving 20,000 participants each year. We focus on hands-on natural resource topics.
*AJay Winter, Specialist*

**265 Conservation International**
2011 Crystal Drive 703-341-2400
Suite 500 800-429-5660
Arlington, VA 22202 Fax: 202-887-5188
http://www.conservation.org
Our mission is to conserve the Earth's living natural heritage, our global biodiversity, and to demonstrate that human societies are able to live harmoniously with nature.
*Peter A Seligmann, Chairman of the Board*

**266 Conservation Management Institute**
School of Natural Resources
203 W Roanoke Street 540-231-7348
Blackburg, VA 24061 Fax: 540-231-8825
E-mail: sklopfer@ut.edu
http://www.cmi.vt.edu
To better address multi-disciplinary research questions that affect conservation management effectiveness in Virginia, North America and the world. Faculty from research institutions work collaboratively on projects ranging fromendangered species propagation to natural resource-based satellite imagery interpretation.
*Founded: 2000*
*Scott Kloper, Executive Director*

**267 Conservation Treaty Support Fund**
3705 Cardiff Road
Chevy Chase, MD 20815
301-654-3150
800-654-3150
Fax: 301-652-6390
E-mail: ctsf@conservationtreaty.org
http://www.conservationtreaty.org

CTSF's mission is to support the major inter-governmental treaties which conserve wild natural resources and habitat for their own sake and the benefit of people. These include the Endangered Species Convention and the InternationalWetlands Convention. CTSF raises support for treaty projects from individuals, corporations, foundations and government agencies. It also develops educational and informational materials including videos and the CITES Endangered Species Book.
*Founded: 1986*
*George A Furness Jr, President*

**268 Counterpart International**
1200 18th Street NW
Suite 1100
Washington, DC 20036
888-296-9676
Fax: 888-296-9676
http://www.counterpart.org

Promotes socioeconomics, health care, biodiversity, and natural resource management in over 60 countries.

**269 Defenders of Wildlife**
1130 17th Street NW
Washington, DC 20036
202-682-9400
Fax: 202-682-1331
E-mail: defenders@mail.defenders.org
http://www.defenders.org

Dedicated to the protection of all native wild animals and plants in their natural communities. Focus is placed on what scientists consider two of the most serious environmental threats to the planet: the accelerating rate ofextinction of species and the associated loss of biological diversity, and habitat alteration and destruction. Long known for leadership on endangered species issues.
*Roger Schlickeisen, President*

**270 Delta Wildlife**
PO Box 276
Stoneville, MS 38776
601-686-3370
Fax: 601-686-4780
E-mail: teycoo@yahoo.com
http://www.nawf.org/dwf/

Recognizing the need for an agressive, but reasonable effort to develop wildlife, in 1990, one-hundred agri-business leaders, representing every country in Mississippi Delta, had the vision and dedication to form Delta Wildlife so theyand others could do more for conservation.
*Bill Kennedy, Chairman*

**271 Department of Fisheries, Wildlife and Conservation Biology**
University of Minnesota
University of Minnesota/Hodson Hall
1980 Folwell Avenue
St Paul, MN 55108
612-624-3600
Fax: 612-625-5299
E-mail: cuthb001@umn.edu
http://fwcb.cfans.umn.edu

Fosters ahigh quality natural environment by contributing to the management, protection, and sustainable use of fisheries and wildlife resources through teaching, research and outreach.
*Founded: 1972*
*Francesca Cuthbert, Interim Department Head*
*Betsy Chastain, Program Coordinator*

**272 Desert Fishes Council**
PO Box 337
Bishop, CA 93515
760-872-8751
Fax: 760-872-8751
E-mail: phil@desertfishes.org
http://www.desertfishes.org

The mission of the Desert Fishes Council is to preserve the biological integrity of desert aquatic ecosystems and their associated life forms, to hold symposia to report related research and management endeavors, and to effect rapiddissemination of information concerning activities of the Council and its members.
*Nathan Allan, President*
*Phil Pister, Executive Secretary*

**273 Desert Protective Council**
PO Box 3635
San Diego, CA 92163
619-342-5524
E-mail: terryweiner@dpcinc.org
http://www.dpcinc.org

To safeguard for wise and reverent use by this and succeeding generations those desert areas of unique scenic, scientific, historical, spiritual or recreational value and to educate children and adults to a better understanding of thedeserts.
*Founded: 1954*
*Nick Ervin, President*

**274 Desert Tortoise Council**
PO Box 3273
Beaumont, CA 92223
909-884-9700
E-mail: mojotort@yahoo.com
http://www.deserttortoise.org

The Council is a private, nonprofit organization made up of hundreds of professionals, and lay-persons from all walks of life, from across the United States, and several continents. We share a common fascination with wild deserttortoises and environment they depend upon.
*Founded: 1976*
*Peter Woodman, Senior Co-Chair*
*Doug Duncan, Junior Co-Chair*

**275 Dragonfly Society of the Americas**
Bulletin of American Odonatology
2091 Partridge Lane
Bingahamton, NY 13903
607-722-4939
E-mail: tdonnel@binghamton.edu
http://www.afn.org/~iori/dsaintro.html

The Dragonfly Society of the Americas was organized in 1988. It is a nonprofit society whose purpose is to encourage scientific research, habitat preservation and aesthetic enjoyment of Odonata (dragon flies).
*TW Donnelly, Editor*

**276 Ducks Unlimited**
One Waterfowl Way
Memphis, TN 38120
901-758-3825
800-453-8257
Fax: 847-438-9236
http://www.ducks.org

The mission is to fulfill the annual life cycle needs of North American waterfowl by protecting, enhancing, restoring and managing important wetlands and associated uplands.
*Don A Young, Executive VP*
*Ken Babcock, Senior Director Conservation*

**277 Eagle Nature Foundation**
300 E Hickory Street
Apple River, IL 61001
815-594-2306
Fax: 815-594-2305
E-mail: eaglenature.tni@juno.com
http://www.eaglenature.com

A nonprofit organization dedicated to the preservation of the bald eagle, our national symbol, and other endangered species from extinction and to increase public awareness of unique endangered plants and animals. They monitor baldeagle and other endangered species populations, strive to preserve habitat essential to their survival, and develops materials for schools to inform students about the needs of the bald eagle and how we can help preserve and protect their naturalenvironment.
*Founded: 1995*
*Terrence N Ingram, Founder/President*
*Eugene Small, Vice President*

**278 Earth Ecology Foundation**
612 N 2nd Street
Fresno, CA 93702
559-442-3034
*Erik Wunstell, Director*

**279 Earth Island Institute**
2150 Allston Way
Suite 460
San Francisco, CA 94704-1375
415-788-3666
Fax: 415-788-7324
E-mail: johnknox@earthisland.org
http://www.earthisland.org

A non-profit, public interest, membership organization that supoprts people who are creating solutions to protect our shared planet.
*Founded: 1982*
*John A Knox, Executive Director*
*Dave Phillips, Executive Director*

**280 Elephant Interest Group**
106 E Hickory Grove 313-540-3947
Bloomfield Hills, MI 48304 http://www.webdirectory.com
*Hezy Shoshani, Contact*

**281 Elsa Wild Animal Appeal USA**
PO Box 675 630-833-8896
Elmhurst, IL 60126 http://www.2ca.com/fullefx/2king4.htm
*Donald A Rolla, President*

**282 Endangered Species Coalition**
1101 14th Street NW 202-756-2804
Suite 1400 Fax: 202-429-3958
Washington, DC 20005 E-mail: esc@stopextinction.org
The Coalition is one of the most unique organizations in the United States. An organization of ogranizations that supports endangered species issues for over 430 environmental, religious, scientific, humane, and business groups aroundthe country. The vast majority of our member groups are small, local grassroots organizations, who struggle to protect species and habitats in their region.
*Michael Bean*

**283 Ford Foundation**
320 E 43rd Street 212-573-5000
New York, NY 10017 Fax: 212-599-4584
http://www.fordfound.org
Concerned with natural resource preservation. Provides funding for environmental projects throughout the world.
*Janet Maughan, Rural Poverty and Resources*

**284 Forest Guild**
PO Box 519 505-983-8992
Santa Fe, NM 87504 Fax: 505-986-0798
E-mail: info@forestguild.org
http://www.forestguild.org
The Forest Trust merged with Forest Stewards Guild to form Forest Guild. The practice of conservation forestry and the promotion of stewardship are central to the guild's mission.
*Founded: 1984*
*Howard Gross, Director*

**285 Forest History Society**
701 Vickers Avenue 919-682-9319
Durham, NC 27701 E-mail: stevena@duke.edu
http://www.lib.duke.edu/forest/
Brings lessons of forest and conservation history to bear on current issues in natural resource management. Identifies, collects, preserves, interprets and disseminates information on forest and conservation history with the goals ofimproving the public's understanding of the forest industry and forest products; providing original resource material to researchers writing forest history, to assure accuracy on facts and interpretation; and facilitating development of naturalresource policy.
*Steven Anderson, President*

**286 Friends of Acadia**
43 Cottage Street 207-288-3340
PO Box 45 800-625-0321
Bar Harbor, ME 04609 Fax: 207-288-8938
E-mail: info@friendsofacadia.org
http://www.friendsofacadia.org
The mission of Friends of Acadia is to preserve and protect the outstanding natural beauty, ecological vitality and cultural distinctiveness of Acadia National Park and the surrounding communities. Their methods are: raises and donatesprivate funds to park and communities; advocates before legislatures and agencies; counters threats to park; and represents users in betterment of its operations.
*Founded: 1986*
*Marla Stellpflug, President*
*Theresa Begley, Projects/Events Coordinator*

**287 Friends of the Sea Lion Marine Mammal Center**
20612 Laguna Canyon Road 949-494-3050
Laguna Beach, CA 92651 Fax: 949-494-2802
E-mail: info@fslmmc.org
http://www.fslmmc.org
Friends of the Sea Lion Marine Mammal Center is a nonprofit organization staffed by dedicated volunteers and funded by donations. Its mission is to rescue, medically treat and rehabilitate seals and sea lions that are stranded alongOrange County, California beaches due to injury and illness; release healthly animals back to their natural habitat; and increase public awareness of the marine environment through education and research.
*WH Ford, Executive Officer*

**288 Friends of the Sea Otter**
2150 Garden Road 831-373-2747
Suite A-3 Fax: 831-373-2749
Monterey, CA 93922 E-mail: info@seaotters.org
http://www.seaotters.org
Friends of the Sea Otter is a nonprofit organization founded in 1968 dedicated to the protection of a threatened species, the Southern Sea Otter, as well as Sea Otters throughout their north pacific range, and all sea otter habitat.
*Matt Rutishauser, Science Director*
*Mailee Flower, Education Director*

**289 Friends of the Trees Society**
PO Box 4469 509-486-4726
Bellingham, WA 98229 Fax: 509-486-4726
E-mail: tern@geocities.com
http://www.geocities.com/RainForest/4663
Friends of the Trees Society is dedicated to doubling the world's forest. This involves doubling the area covered by trees, doubling the number of trees and most importantly, doubling the weight of the world's forest biomass.
*Michael Pilarski, Executive Officer*

**290 Garden Club of America**
14 E 60th Street 212-753-8287
New York, NY 10022 Fax: 212-753-0134
E-mail: info@gcamerica.org
http://www.gcamerica.org
To stimulate the knowledge and love of gardening, to share the advantages of association by means of educational meetings, conferences, correspondence and publications, and to restore, improve, and protect the quality of theenvironment through educational programs and action in the fields of conservation and civic improvement.
*Frederik Hansen, President*

**291 George Wright Society**
PO Box 65 906-487-9722
Hancock, MI 49930 Fax: 906-487-9405
http://www.georgewright.org
Concerned with the preservation of natural and cultural parks in the United States.

**292 Global Coral Reef Alliance**
324 Bedford Road 914-238-8788
Chappaqua, NY 10514 Fax: 914-238-8768
E-mail: goreau@bestweb.net
http://www.people.fas.harvard.edu/~goreau
A nonprofit organization for the protection and sustainable management of coral reef.
*Dr. Thomas J Goreau, President*

**293 Gopher Tortoise Council**
Florida Museum of Natural History
PO Box 117800 904-362-1721
University of Florida http://www.gohertortoisecouncil.org
Gainesville, FL 32611

Formed by a group of biologists and others concerned about the range-wide decline of the gopher tortoise.
*Founded: 1978*
*Christian Newman, Co-Chair*
*Terry Norton, Co-Chair*

**294 Grand Canyon Trust**
2601 N Fort Valley
Flagstaff, AZ 86001
928-774-7488
888-428-5550
E-mail: info@grandcanyontrust.org
http://www.grandcanyontrust.org
The mission of The Grand Canyon Trust is to protect and restore the canyon country of the Colorado Plateau. Our vision for this unique region 100 years from now is of a landscape still characterized by vast open spaces and dominated bywildness and of healthy and restored natural ecosystems.
*Founded: 1985*
*Bill Hedden, Executive Director*
*Darcy Allen, Director of Administration*

**295 Grassland Heritage Foundation**
PO Box 394
785-748-0955
Shawnee Mission, KS 66201
http://www.grasslandheritage.org
Grassland Heritage Foundation is a nonprofit membership organization to prairie preservation and education.
*Founded: 1976*
*Rex Powell, Contact*

**296 Great Bear Foundation**
802 E Front St
PO Box 9383
Missoula, MT 59807
406-721-5420
Fax: 406-721-9917
E-mail: awr@wildrockies.org
http://www.greatbear.org
The Foundation was established to promote conservation of wild bears and their natural habitat worldwide.
*Liz Sedler, President*
*Brian Horejsi, Vice President*

**297 Hawkwatch International**
1800 South West Temple
Suite 226
Salt Lake City, UT 84115
801-484-6808
800-726-4295
Fax: 801-484-6810
E-mail: hwi@hawkwatch.org
http://www.hawkwatch.org
Thier mission is to monitor and protect hawks, eagles, other birds of prey and their environment through research, education and conservation.
*Founded: 1986*
*Jeff Smith PhD, Conservation Science Dir*
*Jen Hajj, Education Director*

**298 Henry A Wallace Institute for Alternative Agriculture**
9200 Edmonston Road
Suite 117
Greenbelt, MD 20770
301-441-8777
Fax: 301-220-0164
E-mail: hawiaa@access.digex.net
http://www.hawiaa.org
The Institute is a nonprofit, tax-exempt, research and education organization established in 1983 to encourage and facilitate the adoption of low-cost, resource-conserving, and environmentally sound farming systems. Providesleadership, policy research and analysis to influence national agricultural educational and research institutions, producer groups, farmers, scientists, advocates, and other organizations that provide agricultural research, education, and informationservices.
*Founded: 1983*
*I Garth Youngberg, Executive Director*

**299 Holy Land Conservation Fund**
969 Park Avenue
New York, NY 10028
718-965-1057
*Bertel Bruun, President*

**300 Humane Society of the United States**
2100 L Street NW
Washington, DC 20037
202-452-1100
Fax: 202-778-6132
E-mail: corprelations@hsus.org
http://www.hsus.org
HSUS is the nation's largest animal-protection organization with more than 5 million constituents. The HSUS was founded in 1954 to promote the humane treatment of animals, to foster respect, understanding, and compassion for allcreatures. Today our message of care and protection embraces not only the animal kingdom but also the Earth and its environment. To achieve our goals we work through legal, educational, legislative and investigative means.
*Founded: 1954*
*Wayne Pacelle, President/CEO*
*Michael Markarian, Executive VP/COO*

**301 Hummingbird Society**
6560 Highway 179
Suite 204
Sedona, AZ 86351
928-284-2251
800-529-3699
Fax: 928-284-2251
E-mail: info@hummingbirdsociety.org
http://www.hummingbirdsociety.org
Nonprofit corporation organized for the purpose of encouraging international understanding and conservation of hummingbirds by publishing and disseminating information, promoting and supporting scientific study and protecting habitat.
*Founded: 1996*
*Dr Robert L Gell, President*
*Dr H Ross Hawkins, Executive Director/Secretary*

**302 Inland Bird Banding Association**
1409 Childs Road East
Bellevue, NE 68005
419-447-0005
E-mail: jingold@pilot.lsus.edu
http://www.aves.net/inlandbba/
Inland Bird Banding Association was organized in 1922, and now supports the largest membership of any bird banding association in America. Inland Bird Banding Association is an organization for all individuals interested in the seriousstudy of birds, their life-history, ecology, and conservation.
*Founded: 1922*
*James Ingold, President*
*Vernon Kleen, Vice President*

**303 Institute for Conservation Leadership**
2000 P Street NW
Suite 413
Washington, DC 20036
202-466-3330
Fax: 202-659-3897
http://www.icl.org
The mission of this Institute is to train and empower volunteer leaders and to build volunteer institutions that protect and conserve the Earth's environment. We lend a helping hand to our dedicated friends in pursuit of a better worldfor everyone.

**304 Institute for the Human Environment**
PO Box 552
Vineburg, CA 95487
707-935-9335
Community planning and land use.
*Norman Gilroy, President*
*Shelley Arrowsmith, Vice President*

**305 International Association for Bear Research and Management**
2841 Forest Avenue
Berkley, CA 94705
510-549-3116
E-mail: ucumari@aol.com
http://www.bearbiology.com
The International Association for Bear Research and Management (IBA)is a nonprofit tax-exempt volunteer organization open to professional biologists, wildlife managers and others dedicated to the conservation of all species of bears.The organization consists of several hundred members from over 20 countries. It supports the scientific management of the bears through research and distrubition of information.
*Bernie Peyton, Secretary*

**306 International Association of Theoretical and Applied Limnology**
University of Alabama
Department of Biology
Tuscaloosa, AL 35487
205-348-1793
http://www.limnology.org
To further the study and understanding of all aspects of limnology. Promotes and communicates new and emerging knowledge

among limnologists to advance the understanding of inland aquatic ecosystems and their management.
*Dr. Robert G Wetzel, Gen. Secretary/Treasurer*

**307 International Bird Rescue Research Center**
4369 Cordelia Road 707-207-0380
Suisun City, CA 94585 818-222-9453
Fax: 707-207-0395
E-mail: jlewis@ibrrc.org
http://www.ibrrc.org
Dedicated to mitigating the human impact on aquatic birds and other wildlife, worldwide. This is achieved through emergency response, education, research and planning.
*Founded: 1972*
*Jay Holcomb, Executive Director*

**308 International Council for Bird Preservation, US Section**
World Wildlife Fund
1250 24th Street NW 202-467-8348
Washington, DC 20037 Fax: 202-293-9342
E-mail: abc@mnsinc.com
http://www.worldwildlife.org
*Founded: 1994*
*William K Reilly, Chairman*
*Edward P Bass, Vice Chairman*

**309 International Crane Foundation**
E-11376 Shady Lane Road 608-356-9462
PO Box 447 Fax: 608-356-9465
Baraboo, WI 53913 E-mail: cranes@savingcranes.org
http://www.savingcranes.org
Works worldwide to conserve cranes and the wetland and grasslands communities on which they depend. Dedicated to providing experience, knowledge, and inspiration to involve people in resolving threats to these ecosystems.
*Founded: 1973*
*James H Hook, President/CEO*

**310 International Ecology Society**
R Kramer
1471 Barclay Street 651-774-4971
St Paul, MN 55106 http://www.businessfinance.com
*Richard Kramer, President*
*Corey Pierce, Founder/COO*

**311 International Erosion Control Association**
3001 S Lincoln Ave 970-879-3010
Suite A Fax: 970-879-8563
Steamboat Springs, CO 80487 E-mail: ecinfo@ieca.org
http://www.ieca.org
An organization dedicated to minimizing accelerated soil erosion. IECA offers an annual conference, chapter events, training courses as well as a variety of topic specific publications and a quarterly news letter.
*Founded: 1972*
*Doug Wimble, President*
*Becky Gauthier, Secretary*

**312 International Fund for Animal Welfare**
411 Main Street 508-744-2000
PO Box 193 800-932-4329
Yarmouth Port, MA 02675 Fax: 508-744-2009
E-mail: info@ifaw.org
http://www.ifaw.org
IFAW's mission is to improve the welfare of the wild and domestic animals throughout the world by reducing commerical exploitation of animals, protecting wildlife habitats, and assisting animals in distress. We seek to motivate thepublic to prevent cruelty to animals and to promote animal welfare and conservation policies that advance the well-being of both animals and people.
*Brian Davies, Founder*

**313 International Osprey Foundation**
PO Box 250 941-472-1862
Sanibel Island, FL 33957 http://www.sancap.com
Nonprofit corporation dedicated to the continuing recovery and preservation of the osprey, others in the raptor family, wildlife and the environment as a whole. Conducts monitoring activities and accumulates data specific to thebreeding activities of the osprey population. Publishes newsletter, issues grants for researchers whose studies involve environmental concerns. Directs and participates in all areas of wildlife and habitat maintenance and restoration.
*Tim Gardner, President*
*Anne Mitchell, VP*

**314 International Snow Leopard Trust**
4649 Sunnyside Avenue N 206-632-2421
Suite 325 Fax: 206-632-3967
Seattle, WA 98103 E-mail: info@snowleopard.org
http://www.snowleopard.org
The trust is dedicated to the conservation of the endangered Snow Leopard and its mountain ecosystem. Since its founding by Helen Freeman, ISLT has worked on more than 100 projects with local people throughout Central Asia. Focus is onsmall, creative and sustainable programs to make conservation happen now and in the future. Has hosted snow leopard symposia and developed the Snow Leopard Information Management System. This allows for range-wide comparison and sharing ofinformation.
*Founded: 1981*
*Brad Rutherford, Executive Director*
*Tom McCarthy, Conservation Director*

**315 International Society for the Preservation of the Tropical Rainforest**
3302 N Burton Avenue 626-572-0233
Rosemead, CA 91770 800-932-4329
Fax: 818-990-3333
E-mail: forest@nwc.net
http://www.isptr-pard.org/orderlist.html
ISPTR is mandated to publish, inform policy makers and the public, as well as conduct programs toward the conservation of tropical forests, fauna and flora and tribal peoples globaly.
*Founded: 1994*
*Arnold Newmann, Co-Director*
*Roxanne Kremer, Co-Director*

**316 International Society for the Protection of Mustangs and Burros**
PO Box 14194 602-502-7900
Scottsdale, AZ 85267 Fax: 602-502-2205
E-mail: 103053.1112@compuserve.com
http://www.ispmb.com
Dedicated to preserving wild horses/burros and their habitats. We also run a rescue program and purchase slaughter bound wild horses and burros, placing them in permanent loving homes, or transporting entire herds of horses.
*Karen Sussman, President*

**317 International Union for Conservation of Nature and Natural Resources**
World Conservation Union
1630 Connecticut Avenue NW 202-387-4826
Washington, DC 20009 Fax: 202-387-4823
E-mail: postmaster@iucnus.org
http://www.iucn.org
*Founded: 1948*
*Russell Mittermeier, Chairman*
*Gary Allport, Senior Conservation Policy A*

**318 International Union for the Conservation of Nature's Primate Specialist Group**
1015 18th Street NW 202-429-5660
Suite 1000 Fax: 202-887-5188
Washington, DC 20036 http://www.conservation.org
The mission is to conserve the Earth's living natural heritage, our global biodiversity, and to demostrate that human societies are able to live harmoniously with nature.
*Russell A Mittermeier*

**319 International Wildlife Rehabilitation Council**
PO Box 8187
San Jose, CA 95155
408-271-2685
Fax: 408-271-9285
E-mail: office@iwrc-online.org
http://www.iwrc-online.org

A nonprofit international membership organization. Founded in 1972, the IWRC works to enhance the integrity of native wildlife systems and conserve biological diversity worldwide, through rehabilitation of wildlife, support ofrehabilitators, public education and advocacy.
*Founded: 1974*
*Dody Wyman, President*
*Susan Heckly, Secretary*

**320 Island Conservation Effort**
90 Edgewater Drive
Suite 901
Coral Gables, FL 33133
305-666-5381
Fax: 305-663-9941
E-mail: tropbird@unspoledqueen.com
*Martha Walsh-McGehee, President*
*Rosemarie Gnam, Secretary/Tresurer*

**321 Izaak Walton League of America**
707 Conservation Lane
Gaithersburg, MD 20878
301-548-0150
800-453-5463
Fax: 301-548-0146
E-mail: general@iwla.org
http://www.iwla.org

The mission is to conserve, maintain, protect and restore the soil, forest water and other natural resources of the United States and other lands; to promote means and opportunities for the education of the public with respect to suchresources and their enjoyment and wholesome utilization.
*Founded: 1922*
*David W Hoskins, Executive Director*
*William Grant, Associate Executive Director*

**322 Jackson Hole Preserve**
30 Rockefeller Plaza
Room 5600
New York, NY 10112
212-649-5819
Fax: 212-649-5729
http://www.undueinfluence.com
*Founded: 1940*
*Laurance Rockefeller, Chairman of the Board*
*Clayton Frye, President*

**323 Life of the Land**
76 N King Street
Suite 203
Honolulu, HI 96817
808-533-3454
Fax: 808-533-0993
E-mail: henry@lifeoftheland.net
http://www.lifeofthelandhawaii.org

Hawaii's own environmental and community action group protecting our fragile natural and cultural resources through research, education, advocacy and litigation.
*Founded: 1970*
*Kapua Sproat, President*
*Henry Curtis, Executive Director*

**324 Lighthawk**
PO Box 29231
The Presidio, Building 1007
San Francisco, CA 94129
415-561-6250
Fax: 415-561-6251
E-mail: sfo@lighthawk.org
http://www.lighthawk.org

Nonprofit organization founded in 1979, addresses critical environmental issues by providing an aerial perspective on areas of concern in the US, Canada and Central America. Using small aircraft, we fly partner organizations, electedofficials, industry and media representatives, activists, and indigenous groups over protected and threatened regions. Our program flights give passengers both intellectual and visceral understanding of what is at stake.
*Founded: 1979*
*Rick Durden, President*

**325 Marin Conservation League**
55 Mitchell Boulevard
Suite 21
San Rafael, CA 94903
415-472-6170
Fax: 415-472-1404
E-mail: mcl@conservationleague.org
http://www.conservationleague.org

The Marin Conservation League is an a nonprofit organization founded in 1934 to preserve, protect and enhance the natural assets of Marin County for all people. MCL is Marin's oldest, locally based environmental organization,championing a sound balance between the needs of Marin's citizens and its beautiful and fragile environment.
*Susan Stompe, President*
*Robert Berner, Secretary*

**326 Mineral Policy Center**
1612 K Street NW
Suite 808
Washington, DC 20006
202-887-1872
Fax: 202-887-1875
E-mail: info@earthworksaction.org
http://www.mineralpolicy.org

Mining causes serious environmental problems for local communities across the United States and throughout the world. From the perpetual water pollution caused by mine drainage to cyanide spills and heavy metals contamination; from thedesecration of scared sities to the creation of toxic waste rock- mining creates devastating environmental consequences. Mineral Policy Center carries out research and publishes comprehensive reports on the environmental impacts of mining.
*Stephen D'Esposito, President/CEO*
*Cathy Carlson, Policy Advisor*

**327 Monitor Consortium of Conservation Groups**
1506 19th Street NW
Washington, DC 20036
202-234-6576
*Craig VanNote, Executive VP*

**328 Mountain Institute**
3000 Connecticut Avenue NW
Suite 138
Washington, DC 20008
202-234-4050
Fax: 202-234-4051
E-mail: summit@mountain.org
http://www.mountain.org

The Mountain Institute's objectives are to: conserve high priority mountain eco-systems; increase environmentally and culturally sustainable livelihoods for mountain communities; and promote support for mountain cultures and issuesthrough advocacy, education and outreach.
*Founded: 1972*
*Bill Carmean, Chairman/CEO*
*Elsie Walker, Director Special Projects*

**329 National Association of Conservation Districts**
509 Capitol Court NE
Washington, DC 20002
202-547-6223
Fax: 202-547-6450
E-mail: krysta-harden@nacdnet.org
http://www.nacdnet.org

A nongovernmental, nonprofit organization representing 3000 local soil and water conservation districts as well as the 17,000 that serve on their governing boards. Among its goals are: provide useful information to the districts andtheir state associations; represent them as a national unified voice; analyze programs and policy issues that have an impact on local districts; and offer needed and cost-effective services.
*Founded: 1946*
*Krysta Harden, CEO*
*Lisa Lerwick, Director Communications*

**330 National Audubon Society**
700 Broadway
New York, NY 10003
212-979-3000
800-274-4201
Fax: 212-979-3188
E-mail: AudubonAtHome@Audubon.org
http://www.audubon.org

The National Audubon Society is one of the oldest, largest, and most powerful nature appreciation and conservation organizations in the country. NAS works on a broad range of concerns related to the protection of the world'secosystems; preserving wetlands, population planning, eliminating acid rain and reducing air pollution, promoting environmental justice, and protecting water quality.
*Founded: 1905*
*John Flicker, President*
*Monique Quinn, Chief Finance Officer*

**331 National Audubon Society: Project Puffin**
159 Sapsucker Woods Road
Ithaca, NY 14850
607-257-7308
Fax: 607-257-6231
E-mail: puffin@audubon.org
http://www.projectpuffin.org
Established in 1973 in an effort to learn how to restore puffins to historic nesting islands in the Gulf of Maine. Although puffins are abundant in Newfoundland, Iceland and Britain, they are rare in Maine. Project Puffin has a yearround staff of six which increases to include more than 50 biologists and researchers during the seabird breeding season in spring and summer. The project is based in Ithaca, New York and the Todd Wildlife Sanctuary in mid-coast Maine.
*Founded: 1973*
*Stephen Kress, Director*

**332 National Conservation Foundation**
509 Capitol Court NE
Washington, DC 20002
202-547-6223
Fax: 202-547-6450
E-mail: libbysimon@conservationfoundation.com
http://www.nacdnet.org
*Founded: 1982*
*Donald Spickler, Chairman*
*Clear Spring, Managing Director*

**333 National Council for Environmental Balance**
4169 Westport Road
PO Box 7732
Louisville, KY 40257
502-896-8731
Fax: 502-339-1745
http://www.exxonsecrets.org
*Irwin Tucker, President*

**334 National Fish and Wildlife Foundation**
1133 15th Street, NW
Suite 1100
Washington, DC 20005
202-857-0166
Fax: 202-857-0162
E-mail: decarolis@nfwf.org
http://www.nfwf.org
The National Fish and Wildlife Foundation is a nonprofit organization dedicated to the conservation of fish, wildlife, and plants and the habitats on which they depend. Among its goals are species habitat protection, environmentaleducation, public policy development, natural resource management, habitat and ecosystem rehabilitation and restoration, and leadership training for conservation professionals.
*Founded: 1984*
*John Berry, Executive Director*
*Peter Bower, CFO*

**335 National Forest Foundation**
Fort Missoula Road
Building 27, Suite 3
Missoula, MT 59804
202-298-6740
Fax: 406-542-2810
E-mail: bpossiel@nationalforests.org
http://www.natlforests.org
The official nonprofit partner of the USDA Forest Service. To accept and administer private contributions, undertaking activities that further the purposes for which the National Forest System was established and conductingeducational, technical, and other activities that support the multiple use, research, and forestry programs administered by the Forest Service.
*Bill Possiel, President*
*David Bell, Chief Operating Officer*

**336 National Garden Clubs**
4401 Magnolia Avenue
St Louis, MO 63110
314-776-7574
800-550-6007
Fax: 314-776-5108
E-mail: headquarters@gardenclub.org
http://www.gardenclub.org
National Garden Clubs is a nonprofit educational organization with its headquarters in St. Louis, Missouri, US. It is composed of 50 State Garden Clubs and the National Capital Area, 8,858 member garden clubs and 235,316 members. Inaddition, NGC proudly recognizes 200 Internationl Affiliates from Canada to Mexico and South America.
*Founded: 1891*
*Barbara May, President*
*Barbara Baker, Recording Secretary*

**337 National Grange**
1616 H Street NW
Washington, DC 20006
202-628-3507
888-447-2643
Fax: 202-347-1091
E-mail: rfrederick@nationalgrange.org
http://www.nationalgrange.org
The Grange is a family based community organization with a special interest in agriculture and rural America, as well as in legislative efforts regarding these issues.
*Founded: 1867*
*William Steel, President*
*Shirley Lawson, Secretary*

**338 National Institute for Urban Wildlife**
10921 Trotting Ridge Way
Columbia, MD 21044
301-596-3311
Promotes the preservation of wildlife in urban settings, providing support to individuals and organizations invloved in maintaining a place for wildlife in expanding American cities and suburbs. The Institute conducts researchexploring the relationship between humans and wildlife in these habitats, publicizes urban wildlife management methods, and raises public awareness of the value of wildlife in city settings. The Institute also provides consulting services.
*Founded: 1973*

**339 National Military Fish and Wildlife Association**
12428 Pinecrest Lane
Newburg, MD 20664
540-663-4186
Fax: 540-663-4016
E-mail: nmfwa@nmfwa.org
http://www.nmfwa.org
*Thomas Wray II, President*

**340 National Park Foundation**
1201 Eye Street NW
Suite 550B
Washington, DC 20005
202-354-6460
Fax: 202-371-2066
E-mail: ask-npf@nationalparks.org
http://www.nationalparks.org
Mission is to strengthen the connection between the American people and their National Parks by raising private funds, making strategic grants, creating innovative parnerships and increasing public awareness.
*Ken Salazar, Chair*

**341 National Park Trust**
401 E Jefferson Street
Suite 102
Rockville, MD 20850
301-279-7275
866-281-5971
Fax: 301-279-7211
E-mail: npt@parktrust.org
http://www.parktrust.org
*Founded: 1983*
*Grace K Lei, Executive Director*
*William Brownell, Chairman*

**342 National Parks Conservation Association**
1300 19th Street NW
Suite 300
Washington, DC 20036
800-628-7275
Fax: 202-659-0650
E-mail: npca@npca.org
http://www.npca.org
Since 1919, the National Parks Conservation Association has been the voice of the American people in the fight to safeguard the scenic beauty, wildlife, and historical and cultural treasures of the largest and most diverse park systemin the world.
*Founded: 1916*
*Thomas Kiernan, President*
*Tom Martin, Executive Vice President*

**343 National Prairie Grouse Technical Council**
Wildlife Research Center
317 W Prospect Road
Fort Collins, CO 80526
970-484-2836
Fax: 970-490-2621
*Kenneth M Giesen, Executive Officer*

**344 National Recreation and Park Association**
Parks and Recreation Magazine

22377 Belmont Ridge Road
Ashburn, VA 20148
703-858-0784
Fax: 703-858-0794
E-mail: info@nrpa.org
http://www.nrpa.org

The NRPA, headquartered in Ashburn Virginia, is a national non-profit organization devoted to advancing park, recreation and conservation efforts that enhance the quality of life for all Americans. The Association works to extendsocial, health, cultural and economic benefits of parks and recreation, through its network of 23,000 recreation and park professionals and civic leaders. NRPA encourages recreation initiatives for youth in high-risk environments.

*Founded: 1965*
*John A Thorner, Exectutive Director*
*M Lauren Yost, Human Resources Manager*

**345 National Speleological Society**
2813 Cave Avenue
Huntsville, AL 35810
256-852-1300
Fax: 256-851-9241
E-mail: nss@caves.org
http://www.caves.org

Founded for the purpose of advancing the study, conservation, exploration, and knowledge of caves. More than 12,000 members in 200 grottos conduct regular meetings to bring cavers trogether within their general area to coordinateactivities which may include mapping, cleaning and investigating sensitive caves.

*William Tozer, President*
*Stephanie Searles, Operations Manager*

**346 National Tree Society**
PO Box 10808
Bakersfield, CA 93389
805-589-6912
Fax: 775-248-6035
E-mail: tnc@natural-connection.com
http://www.natural-connection.com

Organization to preserve the earth's biosphere by planting and caring for trees. Seeks to raise public understanding of the need for trees and the role they play in maintaining a healthy environment; works to acquire forest and otherlands to ensure the continued growth of trees on such lands; establishes nurseries to supply the trees.

*Gregory W Davis, Contact*

**347 National Trust for Historic Preservation**
1785 Massachusetts Avenue NW
Washington, DC 20036
202-588-6000
800-944-6847
Fax: 202-588-6038
E-mail: feedback@nthp.org
http://www.nthp.org

Provides leadership, education, resources and advocacy to save America's diverse historic places and revitalize our communities.

*Founded: 1949*
*Richard Moe, President*
*David J Brown, Executive Vice President*

**348 National Wildflower Research Center**
Lady Bird Johnson Wildflower Center
4801 Lacross Avenue
Austin, TX 78739
512-471-1525
Fax: 512-471-1551
E-mail: facilityrentals@wildflower.org
http://www.wildflower.org

Combines native plants with local culture, reflecting the specifics and peculiarities of Central Texas Hill Country ecosystems. Walking through the center, you'll find native plants in gardens and natural areas, an unparalleledrainwater collection and storage system, recycled building materials, American folk art, environmentally conscious construction and engaging educational faciltities— all designed to learn to live more gently on the land.

*Founded: 1897*
*Robert.A. Wooster, President*
*Larry McNeill, Vice President*

**349 National Wildlife Federation**
11100 Wildlife Center Drive
Reston, VA 20190
800-822-9919
http://www.nwf.org

Gives voice to the wildlife conservation values that are part of our country's heritage.

*Founded: 1936*
*Larry J Schweiger, President/CEO*

**350 National Wildlife Refuge Association**
1010 Wisconsin Avenue NW
Suite 200
Washington, DC 20007
202-333-9075
877-396-NWRA
Fax: 202-333-9077
E-mail: nwra@refugenet.org
http://www.refugenet.org

Aims to protect, enhance and expand the National Wildlife Refuge System lands, set aside by the American public to protect our country's diverse wildlife heritage.

*Founded: 1975*
*Charlie Estes, Executive Director*
*William H Meadows, President*

**351 National Wildlife Rehabilitators Association**
2625 Clearwater Rd
Suite 110
St. Cloud, MN 56301
320-230-9920
Fax: 320-230-3077
E-mail: nwra@nwrawildlife.org
http://www.nwrawildlife.org

The National Wildlife Rehabilitators Association is a nonprofit international membership organization committed to promoting and improving the integrity and professionalism of wildlife rehabilitation and contributing to thepreservation of natural ecosystems.

*Founded: 1982*
*Wendy Fox, President*
*Di Conger, Vice President*

**352 National Woodland Owners Association**
374 Maple Avenue E
Suite 310
Vienna, VA 22180
800-476-8733
Fax: 703-281-9200
E-mail: info@woodlandowners.org
http://www.woodlandowners.org/

NWOA is independent of the forest products industry and forestry agencies. They independently work with all organizations to promote non-industrial forestry and the best interests of woodland owners.

*Founded: 1983*
*Keith A Argow, President*
*Trish Hugg, Member Services Coordinator*

**353 Natural Area Council**
725 15th street NW
Suite 605
Washington, DC 20005
202-638-1649
Fax: 202-638-2175
http://www.america-the-beautiful.org

*Founded: 1965*
*Nanine Bilski, President*

**354 Natural Areas Association**
PO Box 1504
Bend, OR 97709
541-317-0199
E-mail: mail@naturalarea.org
http://www.naturalarea.org/

To advance the preservation of natural diversity. To inform, unite and support persons engaged in identifying, protecting, managing, and studying natural areas and biological diversity across landscapes and ecosystems.

*Founded: 1980*
*Deb Kraus, Executive Director*

**355 Natural Resources Defense Council**
40 West 20th Street
New York, NY 10011
212-727-2700
Fax: 212-727-1773
E-mail: nrdcinfo@nrdc.org
http://www.nrdc.org

Mission: To safeguard the Earth: its people, its plants and animals, and natural systems on which all life depends.

*Founded: 1990*
*Frances Beinecke, President*
*Peter Lehner, Executive Director*

**356 Neighborhood Parks Council**
451 Hayes Street, 2nd Floor 415-621-3260
San Francisco, CA 94102 Fax: 415-703-0889
E-mail: council@sfnpc.org
http://www.sfnpc.org
Neighborhood Parks Council (NPC) advocates for a superior, equitable and sustainable park and recreation system. NPC provides leadership and support to park users through community-driven stewardship, education, planning and research.
*Founded: 1996*
*Meredith Thomas, Executive Director*
*Victoria Bell, Deputy Director*

**357 Nicodemus Wilderness Project**
PO Box 40712
Albuquerque, NM 87196 E-mail: mail@wildernessproject.org
http://www.wildernessproject.org
The Nicodemus Wilderness Project was founded because of the need for environmental restoration, stewardship, and projection of neglected public lands.
*Robert Dudley, President/Executive Director*

**358 North American Bluebird Society (NABS)**
P.O. Box 244 330-359-5511
Wilmot, OH 44689 Fax: 330-359-5455
E-mail: info@nabluebirdsociety.org
http://www.nabluebirdsociety.org
Nonprofit conservation, education and research organization, promotes the recovery of the bluebirds and other native cavity-nesting bird species. NABS supports conservation through such continent wide programs as the TranscontinentalBluebird Trail and the NABS Nestbox Approval Process. NABS also produces award-winning educational materials.
*Founded: 1977*
*Steve Garr, President*
*Julie Kutruff, Vice President*

**359 North American Crane Working Group**
PO Box 566
Gambier, OH 43022 http://www.nacwg.org
NACWG is an organization of professional biologists, aviculturists, land managers and other interested individuals dedicated to the conservation of cranes and their habitats in North America. They sponsor a North American CraneWorkshop every 3-4 years, promulgates technical information including a published Proceedings of a North American Workshop and a semi-annual newsletter, address conservation issues affecting cranes and their habitat, promote appropriate research oncrane conservation.
*Tom Hoffman, Contact*

**360 North American Falconers Association**
3509 Whippoorwill Cove
White Hall, AR 71602 E-mail: nafaprez@yahoo.com
http://www.n-a-f-a.org
Provides communication among and disseminates relevant information to interested members; scientific study of raptorial species, their care, welfare and training; promotes conservation of the birds of prey and an appreciation of theirvalue in nature and in wildlife conservation programs; urges recognition of falconry as a legal field sport; and establishs traditions which will aid, perpetuate and further the welfare of falconry and raptors it employs.
*Founded: 1961*
*Darryl A Perkins, President*
*Rusty Scarborough, Secretary*

**361 North American Loon Fund**
PO Box 329 603-528-4711
Holderness, NH 03245 800-462-5666
http://www.facstaff.uww.edu
The North American Loon Fund's mission is to promote the preservation of loons and their lake habitats through research, public education, and the involvement of people who share their lakes with loons.
*Linda O'Bara, Director*

**362 North American Wildlife Park Foundation Wolf Park**
4008 East 800 North 765-567-2265
Battle Ground, IN 47920 Fax: 765-567-4299
E-mail: wolfpark@wolfpark.org
http://www.wolfpark.org/
*Founded: 1972*
*Erich Klinghammer, President (Founder)*

**363 North American Wolf Society**
PO Box 3243
Boulder, CO 80307

**364 Open Space Institute**
1350 Broadway 212-290-8200
Room 201 Fax: 212-290-3441
New York, NY 10018 E-mail: info@osiny.org
http://www.osiny.org
Protects land for public benefit and supports the efforts of citizen activists working to improve environmental regulations in their communities.
*Meera Subramanian, Communications Manager*
*Joseph Martens, President*

**365 Openlands Project**
25 E Washington Street 312-427-4256
Suite 1650 Fax: 312-427-6251
Chicago, IL 60602 E-mail: info@openlands.org
http://www.openlands.org
Nonprofit project, founded in 1963, is an independent, nonprofit organization dedicated to preserving and enhancing public open space in northeastern Illinois. Openlands bridges political boundaries and build consensus on open spacegoals and regional growth strategies.
*Founded: 1963*
*Gerald Adelmann, Executive Director*
*Joyce O'Kese, Deputy Director*

**366 Organization for Bat Conservation**
39221 Woodward Ave 248-645-3232
P.O. Box 801 Fax: 517-339-5618
Bloomfield Hills, MI 48303 E-mail: obcbats@aol.com
http://www.batconservation.org
A nonprofit organization, our mission is to preserve bats and their habitats through education, collaboration, and research. We also work with local health departments and government agencies to aid in public health issues associatedwith bats, and we have trained field biologists to research endangered bats.
*Kim Williams, Director*
*Rob Mies, Director*

**367 Ozark Society**
PO Box 2914 479-587-8757
Little Rock, AR 72203 E-mail: racross@uark.edu
http://www.ozarksociety.net
The Ozark Society has remained a strong regional organization because is has not allowed itself to be diverted from its principal purpose: the preservation of wild and scenic rivers, wilderness and unique natural areas. It's primaryfocus is the Ozark-Ouachita region and its associated bottom land habitat.
*Founded: 1962*
*Bob Cross, President*

**368 Ozarks Resource Center**
PO Box 3 417-679-4773
Brixey, MO 65618 Fax: 417-679-4773
E-mail: jlorrain@goin.missouri.org
http://www.ic.org/resources
The purpose of the organization is to provide research, education, technical assistance and dissemination of information on: renewable resource-based appropriate technology, environmentally responsible practices, sustainableagriculture, community economic development and self-reliance for the family, farm community, the Ozarks and their bioregions.
*Janice Lorrain, Executive Director*

**369 Partners in Parks**
Department of Parks and Recreation

City-County Building
Room 452
Pittsburgh, PA 15219
202-364-7244
Fax: 202-255-2364
E-mail: partpark@cqi.com
http://www.partnersinparks.org

Organize and direct volunteers for work in city parks and trails. Outdoor work can be as simple as litter pick-up to trail cleaning, tree planting, etc.

*Founded: 1988*

*David Kekil, Secretary/Treasurer*

**370 Peregrine Fund**

5668 W Flying Hawk Lane
Boise, ID 83709
208-362-3716
Fax: 208-362-2376
E-mail: tpf@peregrinefund.org
http://www.peregrinefund.org

The Peregrine Fund was founded to restore the Peregrine Falcon, which was removed from the U.S. Endangered Species List in 1999. That success has encouraged the organization to expand its focus and apply its experience andunderstanding to raptor conservation efforts on behalf of 87 species in 61 countries worldwide, including the California Condor and Aplomado Falcon in the United States. The organization is non-political, solution-oriented, and hands-on.

*Founded: 1970*

*J Peter Penny, President*

*Susan Whaley, Public Relations Coordinator*

**371 Public Lands Foundation**

PO Box 7226
Arlington, VA 22207
703-790-1988
Fax: 703-893-1500
E-mail: leaplf@erols.com
http://www.publicland.org

Founded in 1987, the Public Lands Foundation is a private nonprofit organization dedicated to the proper use and protection of the public lands administrated by the Bureau of Land Management, implementation of the Federal Land Policyand Management Act, and professional land management by professional employees.

*George Lea, President*

*Carl Enix, Secretary*

**372 Quail Unlimited**

PO Box 610
Edgefield, SC 29824
803-637-5731
Fax: 803-637-0037
http://www.qu.org

Quail Unlimited was established in 1981 to battle the problem of dwindling quail and wildlife habitat. Quail Unlimited is the only national conservation organization dedicated to the wise management of America's wild quail as avaluable and renewable resource.

*Rocky Evans, Executive VP*

*Roger Wells, National Habitat Coordinator*

**373 RARE Conservation**

1840 Wilson Boulevard
Suite 204
Arlington, VA 22201
703-522-5070
Fax: 703-522-5027
E-mail: rare@rareconservation.org
http://www.rareconservation.org

They work globally to equip people in the most threatened natural areas with tools and the motivation needed to care for their natural resources.

*Founded: 1973*

*Brett Jenks, President/CEO*

*Laurie Wilkison, VP Communication/Development*

**374 Rainforest Alliance**

665 Broadway
Suite 500
New York, NY 10012
212-677-1900
Fax: 212-677-2187
E-mail: canopy@ra.org
http://www.rainforest-alliance.org

Dedicated to tropical forest conservation for the benefit of the global community.

*Founded: 1987*

*Daniel Katz, President*

*Tensie Whelan, Executive Director*

**375 Raptor Education Foundation**

PO Box 200 400
Denver, CO 80220
303-680-8500
Fax: 303-680-8502
E-mail: raptor2@usaref.org
http://www.usaref.org

One of the most important challenges facing the world today is the preservation, protection and appropriate use of natural resources. The environmental decisions made today will have a monumental impact on wildlife on a global scaleand effect the overall quality of life for mankind. We must strive to restore and maintain a level of dynamic balance in nature and minimize the rate at which species of animals, plants and other natural resources are declining.

*Founded: 1980*

*Peter Reshetniak, Chairman*

*Ann Price, Secretary*

**376 Ruffed Grouse Society**

451 McCormick Road
Coraopolis, PA 15108
412-262-4044
888-564-6747
Fax: 412-262-9207
E-mail: RGS@ruffedgrousesociety.org
http://www.ruffedgrousesociety.org

The Ruffed Grouse Society's role in conservation of wildlife habitat is to enhance the environment for the Ruffed Grouse, American Woodcock, and other forest wildlife that require or utilize thick, young forests. Since forests aredynamic and constantly changing and man has virually eliminated the fires that shaped much of the forested land we know today, forests must be managed.

*Founded: 1961*

*D Wayne Jacobson Jr, President*

*Michael Zagata, Executive Director/CEO*

**377 Save the Dunes Council**

444 Barker Road
Michigan City, IN 46360
219-879-3937
Fax: 219-872-4875
E-mail: tom@savedunes.org
http://www.savedunes.org

The Save the Dunes Council of northwest Indiana was founded in 1952, one of the oldest grassroots conservation organizations in the country. Its objectives are to maintain and restore the integrity and quality of the naturalenvironment of the Indiana Dunes region. The hard work of their members led to the establishment of the Indiana Dunes National Lakeshore in 1966; the group continues to work on a wide variety of issues concerning the Dunes and the environmentalquality of the area.

*Founded: 1952*

*Deborah Chubb, President*

*Charlotte Read, Vice President*

**378 Save the Manatee Club**

500 N Maitland Avenue
Maitland, FL 32751
407-539-0990
800-432-5646
Fax: 407-539-0871
E-mail: education@savethemanatee.org
http://www.savethemanatee.org

Save the Manatee Club (SMC) is a nonprofit organization, established in 1981 by US Senator Bob Graham and singer/songwriter Jimmy Buffett so the general public could participate in conservation efforts to save endangered manatees fromextinction. The purpose of SMC is to promote public awareness and education; fund manatee research, rescue, and rehabilitation efforts; lobby for the protection of manatees and their habitat, and take appropriate legal action.

*Founded: 1981*

*Judith Vallee, Executive Director*

*Patti Thompson, Director of Science*

**379 Save the Whales**

Animal Welfare Institute
Georgetown Station
PO Box 3650
Washington, DC 20027
703-836-4300
Fax: 703-836-0400
E-mail: awi@awionline.org
http://www.awionline.org

Save the Whales' purpose is to educate children and adults about marine mammals, their environment and their preservation.

Founded in 1977, Save the Whale is a 801(c)(3) Educational Nonprofit Corporation.
*Founded: 1951*
*Cathy Liss, President*

**380 Scenic America**
1250 Eye Street NW 202-638-1839
Suite 750 Fax: 202-638-3171
Washington, DC 20005 E-mail: tracy@scenic.org
http://www.scenic.org
A national nonprofit organization dedicated solely to protecting natural beauty and distinctive community character. They provide technnical assistance across the nation and through our state affiliates on scenic byways, billboard andsign control, context sensitive highway design, wireless telecommunications tower location, transportation enhancements, and other scenic conservation. We advance our number one goal, to build a citizen movement for scenic conservation througheducation.
*Mary Tracy, President*
*Brad Cownover, Scenic Conservation Svcs*

**381 Sierra Club**
85 Second Street 415-977-5500
2nd Floor Fax: 415-977-5799
San Francisco, CA 94105 E-mail: information@sierraclub.org
http://www.sierraclub.org
To advance the preservation and protection of the natural environment by empowering the citizenry, especially democratically-based grassroots organizations, with charitable resources to further the cause of environmental protection.The vehicle through which The Sierra Club Foundation generally fulfills its charitable mission.
*Founded: 1892*
*Allison Chin, President*
*Carl Pope, Executive Director*

**382 Smithsonian Institution**
Independence Avenue & 6th Street SW 202-357-2700
Washington, DC 20560 Fax: 202-357-2426
E-mail: info@si.edu
http://www.si.edu
*Cristian Samper, Acting Secretary*

**383 Society for Marine Mammalogy**
7600 San Point Way NE PO Box 206-526-4016
Seattle, WA 98115 Fax: 206-526-6615
http://www.marinemammalogy.org/
To evaluate and promote the educational, scientific and managerial advancement of marine mammal science. Gather and disseminate to members of the Society, the public and private institutions, scientific, technical and managementinformation through publications and meetings. Provide scientific information, as required, on matters related to the conservation and management of marine mammal resources.
*Founded: 1981*
*Daniel P Costa, Secretary*
*Kit M Kovacs, President*

**384 Sonoran Institute**
7650 E Broadway 520-290-0828
Suite 203 Fax: 520-290-0969
Tucson, AZ 85710 E-mail: sonoran@sonoraninstitute.org
http://www.sonoraninstitute.org
Nonprofit organization that works collaboratively with local people and interests to conserve and restore important natural landscapes in western North America, engaging partners such as landowners, public land managers, local leaders,commuunity residents and nongovernmental organizations. Community Stewardship is an innovative approach to conservation.
*Founded: 1990*
*Luther Propst, Executive Director*
*Don Chatfield, PhD, Deputy Director*

**385 Tall Timbers Research Station**
Tall Timbers
13093 Henry Beadel Drive 850-893-4153
Tallahassee, FL 32312 Fax: 850-668-7781
E-mail: rose@ttrs.org
http://www.talltimbers.org
Dedicated to protecting wildlands and preserving natural habitats. Promotes public education on the importance of natural disturbances to the environment and the subsequent need for wildlife and land management. Conducts fire ecologyresearch and other biological research programs through the Tall Timbers Research Station. Operates museum.
*Founded: 1958*
*Lane Green, Executive Director*
*Rose Rodriguez, Information Resources Mgr*

**386 The Wildlife Society**
5410 Grosvenor Lane 301-897-9770
Suite 200 Fax: 301-530-2471
Bethesda, MD 20814-2144 E-mail: yanin@wildlife.org
http://http://joomla.wildlife.org
TWS is a professional international non-profit scientific and educational association dedicated to excellence in wildlife stewardship through science and education. Their mission is to enhance the ability of wildlife professionals toconserve diversity, sustain productivity, and ensure the responsible use of wildlife resources for the benefit of society.
*Founded: 1937*

**387 Theodore Roosevelt Conservation Alliance**
27 Fort Missola Road 406-549-0101
Suite 4K 877-770-8722
Missoula, MT 59804 Fax: 406-549-7402
E-mail: info@trca.org
http://www.trca.org
To inform and engage Americans to foster our conservation legacy while working to nurture, enhance and protect our fish, wildlife and habitat resources in our National Forest System.
*Founded: 1991*
*Kristen Munson, Executive Assistant*
*Robert Munson, Executive Director*

**388 Tread Lightly!**
298 24th Street 801-627-0077
Suite 325 800-966-9900
Ogden, UT 84401 Fax: 801-621-8633
E-mail: treadlightly@treadlightly.org
http://www.treadlightly.org
National nonprofit organization dedicated to proactively protecting recreation access and opportunities through education and stewardship.
*Founded: 1990*
*Lori McCullough, Executive Director*
*Jill Scott, Assistant Director*

**389 TreePeople**
12601 Mulholland Drive 818-753-4600
Beverly Hills, CA 90210 Fax: 818-753-4635
E-mail: info@treepeople.org
http://www.treepeople.org
To inspire the people of Los Angeles to take personal responsibility for their environment, training and supporting them as they plant and care for trees and improve the neighborhoods in which they live, work and play. Througheducation, planting projects, policy development and research, the organization is helping lead the promotion of integrated urban watershed management.
*Founded: 1973*
*Andy Lipkis, President/Founder*
*Tom Hansen, Executive Director*

**390 Trees for Life**
3006 W St Louis 316-945-6929
Wichita, KS 67203 Fax: 316-945-0909
E-mail: info@treesforlife.org
http://www.treesforlife.org
Empowers people by demonstrating that in helping each other, they can unleash extraordinary power that impacts lives. By planting fruit trees in developing countries, we protect the environment and provide a low-cost, self-renewingsource of food for

a large number of people. Activities include three elements: education, health and environment.
*Founded: 1984*
*Balbir Mathur, President*
*David Kimble, Executive Director*

**391 Trust for Public Land**
116 New Montgomery Street 415-495-4014
4th Floor 800-714-5263
San Francisco, CA 94105 Fax: 415-495-4103
E-mail: info@tpl.org
http://www.tpl.org
A national nonprofit land conservation organization working to protect land for human enjoyment and well-being. Initiatives include: Parks for People; Working Lands; Natural Lands; Heritage Lands; and Land & Water.
*Founded: 1972*
*William Rogers, President/CEO*
*Matthew Shaffer, Media Contact*

**392 UNEP:United Nations Environment Programme/ Regional Office for North America**
900 17h St, NW 202-785-0465
Suite 506 Fax: 202-785-2096
Washington, DC 20006 http://www.unep.org
It's mission is to provide leadership and encourage partnership in caring for the environment by inspiring, informing, and enabling nations and peoples to improve their quality of life without compromising that of future generations.
*Achim Steiner, Executive Director*
*Angela Cropper, Deputy Executive Director*

**393 Unexpected Wildlife Refuge: New Beaver Defenders**
PO Box 765 856-697-3541
Newfield, NJ 08344 Fax: 856-697-5081
http://users.snip.net/~qdi/
The refuge includes 540 acres of swmps, bogs, forests and lakes and is home to seven active beaver lodges.
*Sarah Summerville, Director*

**394 Way of Nature Fellowship**
PO Box 3388 877-818-1881
Tucson, AZ 85722 E-mail: info@sacredpassage.com
*John P Milton, President*

**395 Western Hemisphere Shorebird Reserve Network**
Manomet Center 508-224-6521
81 Stage Point Road, PO Box 1770 Fax: 508-224-9220
Manomet, MA 02345 E-mail: info@whsrn.org
http://www.whsrn.org
WHSRN is a voluntary, community-based coalation of over 185 organizations across the US and other counries in the Western Hemisphere that have joined together to protect, restore and manage critical wetland habitats for migratorybirds.
*John Cecil, Contact*

**396 Western Society of Naturalists**
California State University
Department of Biology 818-677-3256
Northridge, CA 91330 Fax: 818-677-2034
E-mail: jolene.koester@csun.edu
http://www.csun.edu
Members include researchers, educators, academics and others with an interest in the area's biology, particularly its marine life. Membership is $15.00 per year for individuals and $7.00 for students.
*Jolene Koester, President*
*Claire Cavallaro, Chief of Staff*

**397 Whooping Crane Conservation Association**
715 Earl Drive 337-234-6339
Lawrenceburg, TN 38464E-mail: webadmin@whoopingcrane.com
http://www.whoopingcrane.com
The mission is to advance conservation, protection and propagation of the Whooping Crane population, to prevent its extinction, to establish and maintain a captive management program for the perpetuation of the species. We collect anddisseminate knowledge of this species; and advocate and encourage public appreciation and understanding of the Whooping Crane's educational, scientific and economic values.

**398 Wild Canid Survival and Research Center**
PO Box 760 636-938-5900
Eureka, MO 63025 Fax: 636-938-6490
E-mail: wildcanidcenter@onemain.com
http://www.wolfsanctuary.org
A private nonprofit conservation organization dedicated to the preservation of the wolf and other endangered canids through education, research and captive breeding.
*Founded: 1971*
*Susan Lindsey PhD, Executive Director*
*Kim Scott, Assistant Director*

**399 Wild Horse Organized Assistance**
PO Box 555 702-851-4817
Reno, NV 89504 http://www.wildhorseorganizedassistance.org
The mission is to save diminishing herds of wild horses in the Western United States.
*Founded: 1971*
*Dawn Y Lappin, Executive Director*

**400 Wild Horses of America Registry**
6212 E Sweetwater 602-991-0273
Scottsdale, AZ 85254 Fax: 602-991-2920
E-mail: 103053.1112@campuserve.com
http://www.ispmb.com
Recognizes wild horses and burros of America that have been removed from public lands.
*Karen Sussman, Registrar*

**401 Wilderness Society**
1615 M Street NW
Washington, DC 20036 800-846-9453
E-mail: member@tws.org
http://www.wilderness.org
Protects the wilderness and inspires Americans to care for the wild places.
*Founded: 1935*
*Leslie Jones, General Counsel*

**402 Wilderness Watch**
PO Box 9175 406-542-2048
Missoula, MT 59807 Fax: 406-542-7714
E-mail: wild@wildernesswatch.org
http://www.wildernesswatch.org
Keep It Wild! we are Ameriica's leading citizens' voice for protecting our nation's designated wilderness and wild rivers.
*Founded: 1989*
*Kevin Proescholdt, President*
*Jon Dettmann, Vice President*

**403 Wildlife Action**
PO Box 866 843-464-8473
Mullins, SC 29574 800-753-2264
Fax: 843-464-8859
E-mail: info@wildlifeaction.com
http://www.wildlifeaction.com
Raises public awareness about wildlife habitat, security, protection and management; protects the rivers and wetlands from unnecessary destruction and development and works to reduce poaching, trespassing and other illegal outdooractivities.
*Founded: 1977*
*M Gault-Beeson, President*
*Ian Beeson, Secretary*

**404 Wildlife Conservation Society**
2300 Southern Boulevard 718-220-5100
Bronx, NY 10460 Fax: 718-220-2685
E-mail: feedback@wcs.org
http://www.wcs.org
WCS is at work in 53 nations across Africa, Latin America and North America, protecting wild landscapes that are home to a variety of species from butterflies to tigers. We uniquely combine the resources of wildlife parks in New Yorkwith field projects

around the globe to inspire care for nature, provide leadership in environmental eduation, and help sustain our planet's biological diversity.
*Founded: 1895*
*Steven Sanderson, President*
*Richard Lattios, Vice President*

**405 Wildlife Disease Association**
PO Box 1897 785-843-1235
Lawrence, KS 66044 800-627-0629
Fax: 785-843-1274
E-mail: charles_van_riper@usgs.gov
http://www.wildlifedisease.org
Our mission is to acquire, disseminate and apply knowledge of the health and diseases of wild animals in relation to their biology, conservation and interactions with human and domestic animals.
*Founded: 1952*
*Charles van Riper III, President*
*Lynn Creekmore, VP*

**406 Wildlife Forever**
2700 Freeway Boulevard 763-253-0222
Minneapolis, MN 55430 Fax: 763-560-9961
E-mail: info@wildlifeforever.com
http://www.wildlifeforever.com
Wildlife Forever conserves America's wildlife heritage through preservation of habitat, conservation education and management of fish and wildlife.
*Founded: 1987*
*Douglas Grann, President/CEO*

**407 Wildlife Habitat Enhancement Council**
8737 Colesville Road 301-588-8994
Suite 800 Fax: 301-588-4629
Silver Spring, MD 20910 E-mail: Whc@wildlifehc.org
http://www.wildlifehc.org
The Wildlife Habitat Council is a nonprofit groups of corporations, conservations, and individuals dedicated to protecting and enhensing wildlife habitat.
*Founded: 1998*
*Bill Howard, President*
*Emer O'Broin, Vice President*

**408 Wildlife Management Institute**
1440 Upper Bermudian Road 717-677-4480
Gardners, PA 17324 E-mail: info@wildlifemgt.org
http://www.wildlifemgt.org
WMI is a private, nonprofit, scientific and educational organization. It is committed to the conservation, enhancement and professional management of North America's wildlife and other natural resources.
*Founded: 1946*
*Steven A Williams, President*

**409 Wildlife Society**
5410 Grosvenor Lane 301-897-9700
Suite 200 Fax: 301-530-2471
Bethesda, MD 20814 E-mail: tws@wildlife.org
http://www.wildlife.org
A nonprofit scientific and educational organization that serves professionals such as government agencies, academia, industry, and non-government organizations in all areas related to the conservation of wildlife and natural resourcesmanagement.
*Founded: 1937*
*Thomas Franklin, Executive Director/CEO*
*Darryl Walker, Membership/Mktg/Conf Dir*

**410 Wilson Ornithological Society**
OSNA 254-399-9636
5400 Bosque Boulevard, Ste 680E-mail: business@osnabirds.org
Waco, TX 76710http://www.ummz.umich.edu/birds/wos/index.html
World-wide organization of approximately 2,500 people who share a curiosity about birds.
*Founded: 1888*
*Dr Doris J Watt, President*
*John A Smallwood, Secretary*

**411 Windstar Foundation**
2317 Snowmass Creek Road 303-927-4777
Snowmass, CO 81654 Fax: 970-927-4779
E-mail: webhelp@wstar.org
http://www.wstar.org
Windstar is a nonprofit environmental education organization which promotes a holistic approach to addressing environmental concerns. Founded in 1976 for singer/songwriter and environmentalist John Denver along with Aikido Master TomCrum.
*Founded: 1976*
*Ron Deutschendorf, President*
*Pam Peterson, Secretary*

**412 Wolf Education and Research Center**
3909 NE MLK Blvd
Suite 202 888-422-1110
Portland, OR 97212 E-mail: info@wolfcenter.org
http://www.wolfcenter.org
Dedicated to providing public education concerning the gray wolf anf its habitat in the Northern Rocky Mountains. Provides the public with the rare opportunity to observe and learn about wolves in their natural habitat.
*Randy Stewart, Education Coordintor*
*Chris Anderson, Executive Director*

**413 Wolf Fund**
PO Box 471 307-733-0740
Moose, WY 83012 Fax: 307-733-0962
*Renee Askins, Executive Officer*

**414 Wolf Haven International**
3111 Offut Lake Road 360-264-4695
Tenino, WA 98589 800-448-9653
Fax: 360-264-4639
E-mail: director@wolfhaven.org
http://www.wolfhaven.org
Wolf Haven International's objectives include protection of the remaining wild wolves and their habitat, promotion of wolf re-establishment in historic ranges, provision of a sanctuary for captive wolves, and public education on thevalue of all wildlife.
*Founded: 1982*
*John Blakenship, Executive Director*
*Trudy Soucoup, Director Development*

**415 World Bird Sanctuary**
PO Box 270270 636-938-6193
St. Louis, MO 63127 Fax: 636-938-9464
E-mail: info@worldbirdsanctuary.org
http://www.worldbirdsantuary.org
The World Bird Sanctuary's mission is to preserve the earth's biological diversity and to secure the future of threatened bird species in their natural environment. We work to fulfill that mission through education, propagation, fieldstudies and rehabilitation.
*Dennis Breite, Treasurer*

**416 World Forestry Center**
4033 SW Canyon Road 503-228-1367
Portland, OR 97221 Fax: 503-228-4608
E-mail: mail@worlforestry.org
http://www.worldforestry.org
They educate and inform people about the world's forests and trees and their importance to all life, in order to promote a balanced and sustainable future. The WFC also operates a museum in Portland, OR, with local, national, andinternational programs and three demonstration forests.
*Founded: 1964*
*Gary Hartshorn PhD, President/CEO*
*Mark Reed, Operations Director*

**417 World Nature Association**
PO Box 673 301-593-2522
Silver Spring, MD 20901 Fax: 301-593-2522
*Donald H Messersmith, President*

**418 World Wildlife Fund**
1250 24th Street NW
PO Box 97180
Washington, DC 20090
202-293-4800
800-225-5993
Fax: 202-293-9211
http://www.worldwildlife.org

Dedicated to protecting endangered species and their habitats through field work, advocacy, policy engagement, pioneering work, and education. The largest multinational conservation organization in the world. WWF works in 100 countriesand is supported by 1.2 million members in the US and almost 5 million members globally.
*Founded: 1960*
*Bruce Babbitt, Chairman*
*Carter S Roberts, President*

## National: Recycling & Pollution Prevention

**419 Abandoned Mined Lands Reclamation Council**
One Natural Resources Way
Springfield, IL 62701
217-782-0588
Fax: 217-524-4819
http://dnr.state.il.us

Currently reclaiming abandoned mine lands in Illinois
*Founded: 1974*
*Joe Angleton, Executive Director*

**420 Acoustical Society of America**
Suite 1NO1
2 Huntington Quadrangle
Melville, NY 11747
516-576-2360
Fax: 516-576-2377
E-mail: asa@aip.org
http://asa.aip.org/

For years the society has been involved in studies of noise as far as measurements, effects, and ways of reducing noise to improve the human environment.
*Founded: 1929*
*Gilles A Daigle, President*
*Charles E Schmid, Executive Director*

**421 Air and Waste Management Association**
One Gateway Center, 3rd Floor
420 Fort Duquesne Boulevard
Pittsburgh, PA 15222
412-232-3444
800-270-3444
Fax: 412-232-3450
E-mail: info@awma.org
http://www.awma.org

The Air and Waste Management Association (A&WMA) is a nonprofit, nonpartisan professional organization that provides information, networking opportunities, public educational and professional development to more than 9,000environmental professionals in 65 countries.
*Founded: 1907*
*Antoon van der Vooren PhD, President*
*Adrianne Carolla, Executive Director*

**422 Association of Battery Recyclers**
PO Box 290286
Tampa, FL 33687
813-626-6151
Fax: 813-622-8388
E-mail: info@batteryrecyclers.com
http://www.batteryrecyclers.com

To keep members abreast of environmental, health, and safety requirements that affect our industry. The association meets two times a year and membership consists of recyclers of lead-acid batteries and their components, manufacturersand environmental consulting services.
*Founded: 1984*
*Joyce Morales, Sectrary Treasurer*
*Earl Cornette, Chairman*

**423 Association of Energy Engineers**
4025 Pleasantdale Road
Suite 420
Atlanta, GA 30340
770-447-5083
Fax: 770-446-3969
E-mail: info@aeecenter.org
http://www.aeecenter.org

A non profit professional society that promotes the scientific and educational interests of those engaged in the energy industry and fosters action for sustainable development.
*Albert Thumann, Executive Director*
*Ruth Whitlock, Executive Administrator*

**424 Association of State and Territorial Solid Waste Management Officials**
444 N Capitol Street NW
Suite 315
Washington, DC 20001
202-624-5828
Fax: 202-624-7875
E-mail: swmtrina@sso.org
http://www.astswmo.org

To enhance and promote effective state and territorial waste management programs, and affect national waste management policies.
*Founded: 1974*
*Mary Zdanowicz, Executive Director*
*Dania Rodriguez, Deputy Director*

**425 Container Recycling Institute**
1776 Massachusetts Avenue NW
Suite 800
Washington, DC 20036
202-263-0999
E-mail: info@container-recycling.org
http://www.container-recycling.org

A nonprofit organization that studies and promotes policies and programs that increase recovery and recycling of beverage containers, and shift the social and environmental costs associated with manufacturing, recycling and disposal ofcontainer and packaging waste from government and taxpayers to producers and consumers.
*Founded: 1991*
*Jenny Gitlitz, Research Director*

**426 Earth Regeneration Society**
1442A Walnut Street
Number 57
Berkeley, CA 94709
510-849-4155
Fax: 510-849-0183
E-mail: csiri@igc.apc.org
http://www.imaja.com

The Earth Regeneration Society does research and education on climate change, ozone, and pollution, and calls for full employment and full social support based on surival programs and national and international networking.
*Alden Bryant, President*

**427 Environmental Industry Associations**
4301 Connecticut Avenue NW
Suite 300
Washington, DC 20008
202-244-4700
800-424-2869
Fax: 202-966-4818
E-mail: membership@envasns.org
http://www.envasns.org

The Environmental Industry Associations (EIA) is the parent organization for the National Solid Waste Management Association and the Waste Equipment Technology Association. It supports these through research and administrative, legal,federal affairs and public relations resources.
*Bruce Parker, President*

**428 Get Oil Out**
914 Anacapa Street
Santa Barbara, CA 93102
805-965-1519
E-mail: getoilout.goo@verizon.net
http://www.getoilout.org

A Santa Barbara public group dedicated to the protection of the Santa Barbara Channel and coastline from the environmental, economic and esthetic impact of oil development. The organization was formed in response to a 1969 accidentthat blackened the beaches, poisoned the ocean's water and killed many creatures that depend on clean water for survival.
*Henry Feniger, President*

**429 GrassRoots Recycling Network**
PO Box 282
Cotati, CA 94931
707-321-7883
E-mail: linda@grrn.org
http://www.grrn.org

A national network of waste reduction activists and recycling professionals. The voice calling for Zero Waste in the United States by promoting the message that individuals must go beyond

recycling and go upstream to the headwaters ofthe waste stream which is the industrial designer's desk.
*Eric Lombardi, Board President*
*Linda Christopher, Executive Director*

**430 Hazardous Waste Resource Center**
Environmental Technology Council
1112 16th Street NW 202-783-0870
Suite 420 Fax: 202-737-2038
Washington, DC 20036 E-mail: mail@etc.org
http://www.etc.org
The Environmental Technology Council (ETC) is a trade association of commercial environmental firms the recycle, treat and dispose of industrial and hazardous wastes; and firms involved in cleanup of contaminated sites.
*Founded: 1982*
*David R Case, Executive Director*
*Scott Slesunger, VP Goverment Affairs*

**431 Institute of Clean Air Companies**
1730 M Street NW 202-457-0911
Suite 206 Fax: 202-331-1388
Washington, DC 20036 E-mail: icacinfo@icac.com
http://www.icac.com
To promote the air pollution control industry and encourage improvement of engineering and technical systems. Members are leading manufacturers of equipment to monitor and control emissions of particulate, VOC, SO2, NOX, and airtoxics.
*Richard A Hovan, President*
*David C Foerter, Executive Director*

**432 Institute of Scrap Recycling Industries**
1615 L Street NW 202-662-8500
Suite 600 Fax: 202-626-0900
Washington, DC 20036 E-mail: robinweiner@isri.org
http://www.isri.org
Trade association that represents more than 1,200 companies of the scrap processing and recycling industry.
*Robin Weiner, President*
*David Krohne, Director Communications*

**433 Kids Against Pollution**
311 Main Street 315-266-0185
3rd Floor Fax: 315-266-0186
Utica, NY 13501 E-mail: christine@kidsagainstpollution.org
http://www.kidsagainstpollution.org
Nonprofit organization of active youth dedicated to solving and preventing pollution problems through educational projects and events in order to protect children's health and the planet.
*Founded: 1987*
*Christine Shahin, Director*

**434 Manufacturers of Emission Controls Association**
1730 M Street NW 202-296-4797
Suite 206 Fax: 202-331-1388
Washington, DC 20036 E-mail: info@meca.org
http://www.meca.org
Offers current and relevant technical information on emission control technology thereby facilitating strong state, federal and local air quality programs that promote public health, environmental quality and industrial progress.
*Founded: 1976*
*Bruce I Bertelsen, Executive Director*

**435 Municipal Waste Management Association**
1620 Eye Street NW 202-293-7330
Suite 300 Fax: 202-293-2352
Washington, DC 20006 E-mail: info@usmayors.org
http://www.usmayors.org/USCM/mwma
MWMA promotes operational efficiencies, facilitates information, fosters innovation and promotes legislation advocacy around Superfund, brownfields redevlopment, clean air and water, and waste energy regulations.
*Founded: 1982*
*Susan Jarvis, Senior Program Manager*
*Ted Fischer, Director Membership Services*

**436 NORA An Association of Responsible Recyclers**
5965 Amber Ridge Road 703-753-4277
Haymarket, VA 20169 Fax: 703-753-2445
E-mail: sparker@noranews.org
http://www.noranews.org
A trade association that represents almost 200 leading companies in the liquid recycling industry. It defends and promotes the liquid recycling industry and business.
*Scott D Parker, Executive Director*

**437 National Association for PET Container Resources**
PO Box 1327 707-996-4207
Sonoma, CA 95476 Fax: 707-935-1998
E-mail: information@napcor.com
http://www.napcor.com
A trade association formed in 1987, that helps communities establish recycling programs and conducts promotional and educational activities to promote PET plastic container recycling. The members of NAPCOR are manufacturers ofpolyester resins and bottles.
*Founded: 1987*
*Dennis Sabourin, Executive Director*
*Kate Eagles, Communications*

**438 National Association of Chemical Recyclers**
1900 Main Street NW 202-296-1725
Suite 750 Fax: 202-296-2530
Washington, DC 20036 E-mail: info@nacr-r2.org
The National Association of Chemical Recyclers is comprised of companies that recycle solvents and other chemicals for reuse by industry. Its members include both large conglomerates and smaller companies. The association's responsiblerecycling program ensures that all recyclers adhere to the same standards and regulations members in the assocation pledge to meet the ten principles of responsible recycling.
*Brenda Pulley, Executive Director*
*Christopher Goebel, Director*

**439 National Association of Clean Air Agencies**
444 North Capitol Street NW 202-624-7864
Suite 307 Fax: 202-624-7863
Washington, DC 20001 E-mail: 4cleanair@4cleanair.org
http://www.4cleanair.org
State and local air pollution control officials formed NACAA (formerly STAPPA/ALAPCO) to improve their effectiveness as managers of air quality programs. It encourages the exchange of information among officials, enhances communicationand cooperation among federal, state and local regulatory agencies, and promotes good management of air resources.
*Bill Becker, Executive Director*

**440 National Council for Air and Stream Improvements**
4815 Emperor Boulevard 919-941-6400
Canterbury Hall, Ste 110 Fax: 919-941-6401
Durham, NC 27703 E-mail: ryeske@ncasi.org
http://www.ncasi.org
The Council is a technical organization devoted to finding solutions to environmental protection problems in the manufacture of pulp, paper, and wood products in industrial forestry. It now has about 75 member companies.
*Founded: 1943*
*Ronald Yeske, President*

**441 National Recycling Coalition**
805 15th Street NW 202-789-1430
Suite 425 Fax: 202-789-1431
Washington, DC 20005 E-mail: inf0@nrc-recycle.org
http://www.nrc-recycle.org
Nonprofit advocacy group with members from all aspects of the waste reduction, reuse and recycling industries. It is dedicated to the advancement and improvement of recycling, waste prevention, composting and reuse.
*Founded: 1978*
*Kate Krebs, Executive Director*
*Anjia Nicolaidis, Deputy Director*

**442 Noise Pollution Clearinghouse**
PO Box 1137
Montpelier, VT 05601
888-200-8332
E-mail: freenpc@nonoise.org
http://www.nonoise.org

A national non-profit organization with extensive online noise related resources. Creates more civil cities and more natural rural and wilderness areas by reducing noise pollution at the source.

**443 Public Citizen**
1600 20th Street NW
Washington, DC 20009
202-588-1000
Fax: 202-588-7796
E-mail: pcmail@citizen.org
http://www.citizen.org

Founded as a consumer advocacy organization to represent consumer interests in Congress, the executive branch, and the courts. Among other things it fights for clean, safe and sustainable energy sources and strong health, safety andenvironmental protections.
*Founded: 1971*
*Joan Claybrook, President*

**444 Secondary Materials and Recycled Textiles Association**
131 E Broad Street
Suite 206
Falls Church, VA 22046
703-358-1000
Fax: 703-538-6305
E-mail: smartasn@erols.com
http://www.smartasn.org

Since 1932, SMART has represented the interests of companies dealing with pre-consumer and post-consumer recycable textile materials. This material includes fibers, remnants, recycled clothing and shoes, and other related materials.SMART members also manufacture and distribute industrial and commercial wipers.
*Founded: 1932*
*Bill Schapiro, President*

**445 Solid Waste Association of North America**
1100 Wayne Ave
Suite 700
Silver Spring, MD 20910
301-585-2898
800-467-9262
Fax: 301-589-7068
E-mail: info@swana.org
http://www.swana.org

SWANA is dedicated to education and training of its members by advancing the practice of environmentally and economically sound management of solid waste in North America.
*Founded: 1965*
*Mike Wegner, Director*
*Marc Rogoff, Vice Director*

**446 Steel Recycling Institute**
680 Andersen Drive
Pittsburgh, PA 15220
412-922-2772
800-937-1226
Fax: 412-922-3213
E-mail: jimw@recycle-steel.org
http://www.recycle-steel.org

The Steel Recycling Institute, a unit of the American Iron and Steel Institute, is an industry association that promotes and sustains the recycling of all steel products. The SRI educates the solid waste industry, government, bussinesand ultimately the consumer about the benefit of steel's infinite recycling cycle.
*Founded: 1988*
*Bill Heenan, President*
*Gregory L Crawford, VP Operations*

**447 US Conference of Mayors: Environment Projects**
1620 I Street NW
Washington, DC 20006
202-293-7330
Fax: 202-293-2352
E-mail: info@usmayors.org
http://www.usmayors.org/uscm/uscm_project_services/environment

Since implementing the National Office Paper Recycling Strategy, a coalition of office products manufacturers, waste management companies, civic and environmental nonprofit organizations has been developed. Recycling at Work is anotherproject that focues on toner and cartridge recycling.
*Founded: 1990*
*Douglas H Palmer, President*
*Tom Cochran, Executive Director*

## National: Sustainable Development

**448 Alliance for Sustainability**
1521 University Avenue SE
Minneapolis, MN 55414
612-331-1099
Fax: 612-379-1527
E-mail: iasa@mtn.org
http://www.mtn.org/iasa

The Mission of the Alliance is to bring about personal, organizational and planetary sustainability through support of projects that are ecologically sound, economically viable, socially just and humane. The Alliance for Sustanabilityis a Minnesota-based, tax-deductible nonprofit supporting model sustanability projects on the local, national and international levels.
*Founded: 1983*
*Terry Gips, President*

**449 American Crop Protection Association**
1156 15th Street NW
Suite 400
Washington, DC 20005
202-296-1585
Fax: 202-463-0474
http://www.acpa.org

ACPA promotes the environmentally sound use of crop protection products for the economical production of safe, high quality, abundant food, fiber and other crops.
*Founded: 1933*
*Allan Noe, Director*

**450 American Forest Foundation**
1111 19th Street NW
Suite 780
Washington, DC 20036
202-463-2462
888-889-4466
Fax: 202-463-2461
E-mail: info@forestfoundation.org
http://www.affoundation.org

A nonprofit 501 (C)(3) conservation and education organization that strives to ensure the sustainability of America's family forests for present and future generations. The vision is to create a future where North American forests aresustained by the public which understands and values the social, economic, and environmental benefits they provide to our communities, our nation, and our world.
*Founded: 1982*
*Tom Martin, President/CEO*
*Jim Wyerman, VP Communications*

**451 American Forests**
734 15th Street NW
Suite 800
Washington, DC 20005
202-737-1944
800-368-5748
Fax: 202-737-2457
E-mail: info@amfor.org
http://www.americanforests.org

Works to protect, restore and enhance the naural capital of trees and forests. Healthy forest filter water, remove air pollution, catch carbon, and provide homes for wildlife.
*Founded: 1875*
*Gerald Gray, VP Forest Policy Center*
*Deborah Gangloff, Executive Director*

**452 American Planning Association**
122 S Michigan Avenue
Suite 1600
Chicago, IL 60603
312-431-9100
Fax: 312-431-9985
http://www.planning.org

The American Planning Association is a nonprofit public interest and research organization representing 30,000 practicing officials, and citizens involved with urban and rural planning issues. Sixty-five percent of APA's members areemployed by state and local government agencies. These members are involved, on a day-to-day basis, in formulating planning policies and preparing land use regulations.
*Frank So, Executive Director*

**453 American Society of Agronomy**
677 S Segoe Road 608-273-8080
Madison, WI 53711 Fax: 608-273-2021
E-mail: headquarters@agronomy.org
http://www.agronomy.org

ASA is dedicated to development of agricultural interests in harmony with environmental and human values. The Society supports scientific, educational and professional activities that enhance communication and technology transfer amongagronomists and those in related disciplines on topics of local, regional, national and international significance.

*Jerry L Hatfield, President*
*Ellen Bergfeld, CEO*

**454 Ancient Forest International**
PO Box 1850 707-923-4475
Redway, CA 95560 Fax: 707-923-4475
E-mail: afi@ancientforests.org
http://www.ancientforests.org

Ancient Forest International has been instrumental in the protection of primary forests around the world. With the help of its international ancient forest network, AFL develops opportunities for wildlands philanthropists andcommunities to work together to acquire and protect strategic and invaluable forestlands. AFL has helped coordinate the purchase of nearly a million acres of ecologically critical forested land, primarily along the Pacific coast of North and SouthAmerica.

*Founded: 1989*

**455 Association of Fish and Wildlife Agencies**
444 N Capitol Street NW 202-624-7890
Suite 725 Fax: 202-624-7891
Washington, DC 20001 E-mail: info@fishwildlife.org
http://www.fishwildlife.org

Mission is to promote the sustainable use of natural resources; encourage cooperation and coordination of fish and wildlife management at all levels of government; develop coalitions among conservation organizations or promote fish andwildlife interests; encourage the professional management of fish and wildlife; foster public understanding of the need for conservation.

*Founded: 1902*
*Matt Hogan, Executive Director*
*Laura MacLean, Communications/Marketing*

**456 Atlantic Center for the Environment Quebec-Labrador Foundation**
QLF-Atlantic Center for the Environment
55 S Main Street 978-356-0038
Ipswich, MA 01938 Fax: 978-356-7322
E-mail: atlantic@qlf.org
http://www.qlf.org

A not-for-profit corporation in the United States and a registered Charity in Canada. QLF exissts to support the rural communities and environment of eastern Canada and New England and to create models for stewardship of naturalresources and cultural heritage that can be applied worldwide.

*Founded: 1963*
*Lawrence B Morris, President*

**457 CONCERN**
1794 Columbia Road NW 202-328-8160
Suite 6 Fax: 202-387-3378
Washington, DC 20009 E-mail: concern@concern.org
http://www.sustainable.org

CONCERN is a national nonprofit environmental education organization with a focus on sustainable communities. CONCERN disseminates examples of successful initiatives, offers numerous resources and guidelines for action, serves as aclearinghouse for information and collaborates with others to carry out its programs. Through its Sustainable Communities Network CONCERN seeks to increase public understanding of and participation in initatives that are environmentally and sociallysound.

*Founded: 1970*
*Susan Boyd, Director*

**458 Center for Ecoliteracy**
2528 San Pablo Avenue 510-845-4595
Berkeley, CA 94702 E-mail: info@ecoliteracy.org
http://www.ecoliteracy.org

The Center for Ecoliteracy is dedicated to fostering a profound understanding of the natural world, grounded in direct experience that leads to sustainable patterns of living.

*Janet Brown, Program Officer*
*Jim Koulias, Managing Director*

**459 Conservation Fund**
1655 N Fort Myer Drive 703-525-6300
Suite 1300 Fax: 703-525-4610
Arlington, VA 22209-3199 E-mail: postmaster@conservationfund.org
http://www.conservationfund.org

Forges partnerships to protect America's legacy of land and water resources. Through land acquisition, community initiatives and leadership training, the Fund and its partners demonstrate sustainable conservation solutions emphasizingthe integration of economic and environmental goals.

*Founded: 1985*
*J Rutherford Seydel II, Chairman*
*Lawrence A Selzer, President*

**460 Earth Island Institute**
2150 Allston Way 415-788-3666
Suite 460 Fax: 415-788-7324
Berkeley, CA 94704-1375 E-mail: johnknox@earthisland.org
http://www.earthisland.org

A non-profit, public interest, membership organization that supports people who are creating solutions to protect our shared planet.

*Founded: 1982*
*John A Knox, Executive Director*
*Dave Phillips, Executive Director*

**461 Environmental Policy Center Global Cities Project**
2962 Fillmore Street 415-775-0791
San Francisco, CA 94123 Fax: 415-775-4159
E-mail: epc@globalcities.org
http://www.globalcities.org

Provides assistance and information on sustainable development and environmental conservation to communities within North America.

**462 Environmental and Energy Study Institute**
122 C Street NW 202-628-1400
Suite 630 888-788-3378
Washington, DC 20001 Fax: 202-628-1825
E-mail: eesi@eesi.org
http://www.eesi.org

They are a nonprofit organization dedicated to promoting environmentally sustainable societies. EESI believes meeting this goal requires transitions to social and economic patterns that sustain people, the environment and the naturalresources upon which present and future generations depend.

*Founded: 1984*
*Richard Ottinger, Chairman*
*John Sheehan, Vice Chairman*

**463 Forest History Society**
Forest History Society
701 Vickers Avenue 919-682-9319
Durham, NC 27701 Fax: 919-682-2349
E-mail: recluce2@duke.edu
http://www.foresthistory.org

A non-profit educational institution that links the past to the future by identifying, collecting, preserving, interpreting, and disseminating information on the history of people, forests, and their related resources.

*Founded: 1946*
*Steven Anderson, President*
*Cheryl Oakes, Librarian*

**464 Friends of the Earth**
1717 Massachusetts Avenue NW
Suite 600
Washington, DC 20036
202-783-7400
877-843-8687
Fax: 202-783-0444
E-mail: foe@foe.org
http://www.foe.org

National nonprofit advocacy organization dedicated to protecting the planet from environmental degradation; preserving biological, cultural and ethnic diversity, and empowering citizens to have an influential voice in decisionsaffecting the quality of their environment and their lives.
*Founded: 1969*
*Brent Blackwelder, President*

**465 Global Action Network**
50 California Street
Suite 3325
San Francisco, CA 94111
415-477-2303
Fax: 415-477-2334
http://www.globalactionnetwork.org

A new model of relationship development and community building for the reproductive health field that employs an Internet-based model of networking proven successful in other sectors. Our goal is to provide resources and opportunitiesfor Network members to share knowledge, collaborate with other individuals working in the field, engage in mentoring relationships, and utilize online contacts and information to build networks with communities around the world.
*Founded: 1998*
*Andrea Johnston, Co-Director*
*Jessica Klein, Program Associate*

**466 Global Committee of Parliamentarians on Population and Development**
345 E 45th Street
12th Floor
New York, NY 10017
212-953-7947
Fax: 212-557-2061

**467 Global Tomorrow Coalition**
1325 G Street NW
Suite 1010
Washington, DC 20005
202-628-4016
Fax: 202-628-4018

Global Tomorrow Coalition - a national leadership alliance on sustainable development, with membership in business and industry, conservation and environment, education and community planning, and social issues and development.
*Donald R Lesh, President*

**468 Institute for Agriculture and Trade Policy**
2105 First Avenue South
Minneapolis, MN 55404
612-870-0453
Fax: 612-870-4846
http://www.iatp.org

IATP works locally and globally at the intersection of policy and practice to ensure fair and sustainable food, farm and trade systems.
*Founded: 1980*
*Jim Harkness, President*

**469 Institute for Sustainable Communities**
535 Stone Cutters Way
Montpelier, VT 05602
802-229-2900
Fax: 802-229-2919
E-mail: isc@iscvt.org
http://www.iscvt.org

Promotes sustainable environmental practices in the US and around the world. It leads transformative communtiy-driven projects for environmental problem solving and other challenges.
*Founded: 1991*
*George Hamilton, President*
*Elizabeth Coleman, VP Communication/Development*

**470 Institute for the Human Environment**
PO Box 552
Vineburg, CA 95487
707-935-9335

Community planning and land use.
*Norman Gilroy, President*
*Shelley Arrowsmith, Vice President*

**471 International Mountain Society**
613-523-8891
E-mail: jackives@pidgeon.carleton.ca

An association registered in Berne, Switzerland, for the purpose of advancing knowledge and disseminating information about mountain research and development throughout the world. Aims to promote sustainable mountain developmentthrough improved communication among institutions and individuals, with a particular focus on mountain eco-regions in the developing world. Collaborates with like-minded institutions, and is a joint publisher of the journal Mountain Research andDevelopment.
*Dr Jack Ives, President*

**472 International Society of Arboriculture**
PO Box 3129
Champaign, IL 61826
217-355-9411
888-472-8733
Fax: 217-355-9516
E-mail: isa@isa-arbor.com
http://www.isa-arbor.com

Through research, technology, and education promote the professional practice of arboriculture and foster a greater public awareness of the benefits of trees.
*Founded: 1924*
*Jim Skiera, Executive Director*

**473 International Society of Tropical Foresters**
5400 Grosvenor Lane
Bethesda, MD 20814
301-897-8720
Fax: 301-897-3690
E-mail: goergenm@safnet.org
http://www.safnet.org

The International Society of Tropical Foresters, Inc. (ISTF) is a nonprofit organization committed to the protection, wise management and rational use of the world's tropical forests. Established in 1950, ISTF has about 1500 members inmore than 100 countries. Financial support comes from membership dues, donations and grants. ISTF sponsors meetings, promotes chapters in other countries, maintains a web site and has chapters at universities.
*Founded: 1900 24 pages*
*Michael Goergen, Chief Executive*
*Larry Burner, CFO*

**474 Interstate Mining Compact Commission**
445A Carlisle Drive
Herndon, VA 20170
703-709-8654
Fax: 703-709-8655
http://www.imcc.isa.us

A multi-state governmental organization which represents its 24 member states on issues of mining and environmental regulations. It works closely with several federal agencies such as Office of Surface Mining Reclamation andEnforcement, US EPA, and US Bureau of Land Management.
*Founded: 1970*
*Gregory E Conrad, Executive Director*
*Beth A Botsis, Diretor of Programs*

**475 Island Resources Foundation**
1718 P Street NW
Suite T-4
Washington, DC 20036
202-265-9712
Fax: 202-232-0748
E-mail: irf@irf.org
http://www.irf.org

Island Resources Foundation is a private, nonprofit research and education organization based at Red Hook in St. Thomas, US Virgin Islands, dedicated to solving the environmental problems of developing in small tropical island.
*Founded: 1972*
*Bruce Potter, President*
*Charles Consolvo, Secretary*

**476 Kids for Saving Earth Worldwide**
PO Box 421118
Minneapolis, MN 55442
763-559-1234
Fax: 651-674-5005
E-mail: kseww@aol.com
http://www.kidsforsavingearth.org/

To help protect the Earth through kids and adults. To educate and inspire them to participate in Earth-saving actions.
*Tessa Hill, President*

**477 Kids for Saving the Earth Worldwide**
PO Box 421118
Minneapolis, MN 55442
763-559-1234
Fax: 651-674-5005
E-mail: ksеww@aol.com
http://www.kidsforsavingearth.org/
To help protect the Earth through kids and adults. To educate and inspire them to participate in Earth-saving actions.
*Tessa Hill, President/Director*

**478 Land Institute**
2440 E Water Well Road
Salina, KS 67401
785-823-5376
Fax: 785-823-8728
E-mail: theland@landinstitute.org
http://www.landinstitute.org
This nonprofit research and education organization is engaged in a 25-year research program that marries ecology and agronomy to produce a Natural Systems Agriculture. The Land Institute is developing perennial grain plants to be grownin fields of mixed species patterned after native prairies. These domestic prairies will be plowed rarely, need few manufactured inputs because they will provide their own fertility and manage pests and diseases. Year-round roots will hold soils fromerosion.
*Founded: 2000*
*Wes Jackson, President*
*Ken Warren, Managing Director*

**479 Land Trust Alliance**
1660 L Street NW
Suite 1100
Washington, DC 20036
202-638-4725
Fax: 202-638-4730
E-mail: lta@lta.org
http://www.lta.org
National leader of the private land conservation movement, promoting voluntary land conservation across the country and providing resources, leadership and training to the nation's 1200 plus nonprofit, grassroots land trusts, helpingthem to protect important open spaces. Provides an array of pograms, including direct grants to land trusts, training programs, answers to more than 3,000 inquiries for technical assistance each year, and one-on-one mentoring.
*Founded: 1983*
*Rand Wentworth, President*
*Mary Pope Hutson, Vice President*

**480 Manomet Center for Conservation Sciences**
81 Stage Point Road
PO Box 1770
Manomet, MA 02345
508-224-6521
Fax: 508-224-9220
http://www.manomet.org
Manomet mission is to conserve natural resources for the benefit of wildlife and human populations. Through research and collaboration. Builds science-based, cooperative solutions to environmental problems.
*Founded: 1970*
*Jeptha H Wade, Chairperson*
*D Reid Weedon, Jr., Vice Chairman*

**481 Manufacturers of Emission Controls Association**
1730 M Street NW
Suite 206
Washington, DC 20036
202-296-4797
Fax: 202-331-1388
E-mail: info@meca.org
http://www.meca.org
Offers current and relevant technical information on emissions control technology thus helping to strengthen state, federal and local air quality programs in promoting public health, environmental quality, and industrial progress.
*Founded: 1976*
*Joe Kubsh, Executive Director*

**482 National Association of State Departments of Agriculture**
1156 15th Street NW
Suite 1020
Washington, DC 20005
202-296-9680
Fax: 202-296-9686
E-mail: nasda@nasda.org
http://www.nasda.org
This organization's mission is to support and promote the American agriculture industry, while protecting consumers and the environment, through the development, implementation, and communication of sound public policy and programs.There are twenty national organizations affiliated with NASDA. They are made up of persons of similar responsibilities with the state departments of agriculture and other agencies of state government.
*Founded: 1915*
*Richard Kirchhoff, Chief Executive*
*Steve Cox, Chief Operating Officer*

**483 National Association of State Land Reclamationists**
Southern Illinois University
Coal Research Center
Carbondale, IL 62901
618-536-5521
Fax: 618-453-7346
E-mail: President@notes.siu.edu
http://www.siu.edu
As a nationally recognized authority on the reclamation of mined lands, the National Association of State Land Reclamationists (NASLR) advocates the use of research, innovative technology and professional dicource to foster therestoration of lands and waters affected by mining related activities.
*Founded: 1869*
*Anna Caswell, Secretary/Treasurer*

**484 National Audubon Society: Everglades Campaign**
444 Brickell Avenue
Suite 850
Miami, FL 33131
305-371-6399
Fax: 305-371-6398
E-mail: danderson@audubon.org
http://www.audubonofflorida.org
The mission of the NAS Everglades Conservation Office is to ensure the restoration and conservation of the Greater Everglades Ecosystem in order to achieve an ecologically and economically sustainable South Florida. Our Miami-basedoffice has a five-part program including science, education, advocacy, outreach and grassroots action.
*Founded: 1886*
*David Anderson, Executive Director*
*John Flicker, President*

**485 National Environmental Development Association**
1440 New York Avenue NW
Suite 300
Washington, DC 20005
202-638-1230
Fax: 202-639-8685
Companies and others concerned with balancing environmental and economic interests to obtain both a clean environment and a strong economy.
*Andrew McElwaine, Director*

**486 National FFA Organization**
6060 FFA Drive
PO Box 68960
Indianapolis, IN 46268
317-802-6060
Fax: 317-802-6061
http://www.ffa.org
The organization (formerly Future Farmers of America) is dedicated to making a positive difference in the lives of young people by developing their potential for premier leadership, personal growth and career success throughagricultural education.
*Founded: 1928*
*Larry Case, CEO/National Advisor*
*C Coleman Harris, Executive Secretary*

**487 National Forestry Association**
374 Maple Avenue East
Suite 310
Vienna, VA 22180
703-255-2700
800-476-8733
Fax: 703-281-9200
E-mail: info@woodlandowners.org
http://www.woodlandowners.org
Nation's largest referral program to link up private forest owners with professional foresters.To be supplemented with a new innovative Forest Practices Certification program for landowners. Landowners who complete the review processand follow designated practices will be certified by the National Forestry Association.
*Founded: 1981*
*Keith Argow, President*
*Bob Playfair, Vice President*

**488 National Gardening Association**
1100 Dorset Street
South Burlington, VT 05403
802-863-5251
Fax: 802-864-6889
E-mail: alisonw@garden.org
http://www.garden.org

The mission of the National Gardening Association is to sustain and renew the fundamental links between people, plants and the earth. NGA achieves its mission through youth and community gardening programs, industry research, freegardening information and memberships.
*Founded: 1973*
*Mike Metallo, President*
*Tony Vargo, VP*

**489 National Mining Association**
101 Constitution Avenue NW
Suite 500E
Washington, DC 20001
202-463-2600
Fax: 202-463-2666
E-mail: craulston@nma.org
http://www.nma.org
The mission of the National Mining Association is to create and maintain a broad base of political support in Congress, the administration and the media for the mining industry of the US. In doing so, a secondary goal is to help ournation and the world realize the full promise and potential of the natural resources derived from America's mining industry.
*Founded: 1917*
*Hai Quinn, President/CEO*
*Carol Raulston, Communications*

**490 Native Forest Council**
PO Box 2190
Eugene, OR 97402
541-688-2600
Fax: 541-461-2156
E-mail: info@forestcouncil.org
http://www.forestcouncil.org
To provide visionary leadership and to ensure the integrity of public land ecosystems without compromising people or forests.
*Founded: 1987*
*Timothy Hermach, President/Founder*

**491 Native Seeds/SEARCH**
526 N 4th Avenue
Tucson, AZ 85705
520-622-5561
866-622-5561
Fax: 520-622-5591
E-mail: info@nativeseeds.org
http://www.nativeseeds.org
Promote the use of ancient crops and their wild relatives by gathering, safeguarding, and distributing their seeds, while sharing benefits with tradtional communities. Work to preserve knowledge about their uses. Through research,training, and community education, works to protect biodiversity and to celebrate cultural diversity.
*Founded: 1983*
*Suzanne Nelson, Conservation*
*Julie Evans, Marketing/Operations*

**492 Natural Land Institute**
320 South 3rd Street
Rockford, IL 61104
815-964-6666
Fax: 815-964-6661
E-mail: nli@aol.com
http://www.naturalland.org
Dedicated to preserving land and natural diversity for future generations. Since 1958, NLI has protected, managed, and restored thousands of acres throughout Illinois and southern Wisconsin. These include prairies, forests, wetlands,and river corridors.
*Founded: 1958*
*Jerry Paulson, Executive Director*
*Jill Kennay, Assisstant Director*

**493 Natural Resources Council of America**
11100 Wiodlise Centre Drive
Reston, VA 20190
703-438-6000
Fax: 703-438-3570
E-mail: nrca@naturalresourcescouncil.org
http://www.naturalresourcescouncil.org
The Council dedicated to strengthing the conservation movement as a whole. For more than 50 years the Council has been the Crossroads of Conservation, keeping conservationists connected, informed and prepared to face the challenges ofthe future. The Council provides their membership- more than 85 conservation groups and nearly 100 individual supporters- with unique networking opportunities, valuable leadership training and cost-saving services.
*Founded: 1946*
*Andrea Yank, Executive Director*
*Carlton Gleed, Program Coordinator*

**494 Negative Population Growth**
2861 Duke St
Suite 36
Alexandria, VA 22314
703-370-9510
Fax: 703-370-9514
E-mail: npg@npg.org
http://www.npg.org
Leader in the movement for a sound population policy and advocates a smaller and truly sustainable population through voluntary incentives for smaller families and reduced immigration levels.
*Founded: 1972*
*Donald Mann, President*

**495 New England Coalition for Sustainable Population**
PO Box 2375
Acton, MA 01720
603-283-6686
E-mail: info@necsp.org
http://www.necsp.org
Provides progressive leadership and programs to educate the public about population stabilization and its beneficial relationship to environmental conservation, human rights and sustainable economic development.
*Joseph Bish, Executive Director*

**496 Pacific Institute for Studies in Development, Environment and Security**
654 13th Street
Oakland, CA 94612
510-251-1600
Fax: 510-251-2203
E-mail: info@pacinst.org
http://www.pacinst.org
Nonprofit policy research group, bringing knowledge to power on issues of environmental, economical development, and international peace and security.
*Founded: 1987*
*Nancy Ross, Communications Director*
*Gigi Coe, Chairperson*

**497 Panos Institute**
1322 18th Street,NW
Suite 26
Washington, DC 20036
202-429-0730
Fax: 202-223-7947
E-mail: panoswashington@aol.com
http://www.panosinst.org
Founded in 1986, the Panos Institute is an international, nonprofit, nongovernmental organization with offices in Budapest, London, Paris, and Washington DC, working to raise public understanding of sustainable development issues.
*Founded: 1986*
*Melanie Oliviero, Executive Director*

**498 Pinchot Institute for Conservation**
1616 P Street NW
Suite 100
Washington, DC 20036
202-797-6580
Fax: 202-797-6583
E-mail: pinchot@pinchot.org
http://www.pinchot.org
Strives to advance conservation and sustainable natural resource management by developing innovative, practical, and broadly-supported solutions to conservation challenges and opportunities. This is accomplished through nonpartisanresearch, education and technical assistance.
*Founded: 1963*
*V Alaric Sample, President*
*J Robert Hicks Jr, Chairman*

**499 Population Communications International**
777 United Nations Plaza
5th Floor
New York, NY 10017
212-687-3366
Fax: 212-661-4188
E-mail: info@population.org
http://www.population.org
PCI's mission is to work creatively with the media and other organizations to motivate individuals and communities to make

choices that influence population trends encouraging sustainable development and environmental protection.
*Founded: 1985*
*Victoria A Staebler, Senior Financial Advisor*

**500 Population Connection**
2120 L Street NW
Suite 500
Washington, DC 20037
202-332-2200
800-767-1956
Fax: 202-332-2302
E-mail: info@populationconnection.org
http://www.populationconnection.org/
A national nonprofit organization working to slow population growth and achieve a sustainable balance between the Earth's people and its resources. They seek to protect the environment and ensure a high quality of life for present andfuture generations.
*Founded: 1965*
*John Seager, President/CEO*
*Marian Starkey, Communications Manager*

**501 Population Crisis Committee**
1120 19th Street NW
Suite 550
Washington, DC 20036
202-659-1833
Fax: 202-293-1795
*J Joseph Speidel, President*

**502 Population Institute**
107 2nd Street NE
Washington, DC 20002
202-544-3300
888-787-0038
Fax: 202-544-0068
E-mail: web@populationinstitute.org
http://www.populationinstitute.org
The Population Institute is the World's largest independent non-profit, educational organization dedicated exclusively to achieving a more equitable balance between the worlds population, environment, and resources. Established in 1969,the Institute, with members in 172 countries, is headquartered on Capitol Hill in Washington DC. The Institute uses a variety of resources and programs to bring its concerns about the consequences of rapid poulation growth to the forefront of thenational agenda.
*Founded: 1969*
*Werner Fornos, President*
*Victor Morgan, Executive Director*

**503 Population Reference Bureau**
1875 Connecticut Avenue NW
Suite 520
Washington, DC 20009
202-483-1100
800-877-9881
Fax: 202-328-3937
E-mail: popref@prb.org
http://www.prb.org
PRB informs policymakers, educators, the media, and concerned citizens working in the public interest around the world through a broad range of activities, including publications, information services, seminars and workshops, andtechnical support. They work with both public-sector and private-sector partners.
*Founded: 1929*
*William Butz, President/CEO*
*Ellen Carnevale, Director Communications*

**504 Population Resource Center**
15 Roszel Road
Princeton, NJ 08540
609-452-2822
Fax: 609-452-0010
E-mail: prc@prcnj.org
http://www.prcdc.org
The mission of the Population Resource Center is to promote the use of accurate population data and sound, objective analysis of these data in the making of public policy.
*Founded: 1985*
*Jane S De Lung, President*
*Linda Rosen, Director of Policy Analysis*

**505 Population: Environment Balance**
2000 P Street Northwest
Suite 600
Washington, DC 20036
202-955-5700
800-866-8269
Fax: 202-955-6161
E-mail: uspop@us.net
http://www.balance.org
Population Environment Balance is a national, nonprofit membership organization dedicated to maintaing the quality of the United States population stabilization.
*Founded: 1973*
*Aaron Beckwith, Vice President*

**506 Population: Environmental Council**
2000 P Street NW 600
Washington, DC 20036
202-955-5700
Fax: 202-955-6161
E-mail: uspop@balance.org
http://www.balance.org
Population-Environment Balance is dedicated to public education regarding the adverse effects of population growth on the environment. Founded in 1973, Population-Environment Balance has 8,800 members. It advocates measures that wouldencourage population stabilization.
*Founded: 1973*
*Aaron Beckwith, Vise President*

**507 Rainforest Relief**
PO Box 150566
Brooklyn, NY 11215
718-398-3760
Fax: 718-398-3760
E-mail: relief@igc.org
http://www.rainforestrelief.org
Rainforest relief works to end the loss of the world's tropical and temperate rainforests by reducing the demand for materials for which rainforests are destroyed. These include rainforest woods such as mohogany, lauan and cedar,agricultural products such as bananas, chocolate, coffee and cut flowers, and mining products such as petroleum, gold, aluminum and copper. Rainforest relief works through research, education and non-violent direct action campaigns.
*Founded: 1989*
*Tim Keating, Executive Director*
*Jeff Lockwood, West Coast Chapter Director*

**508 Rural Advancement Foundation International USA**
PO Box 640
Pittsboro, NC 27312
919-542-1396
Fax: 919-542-0069
E-mail: regina@raiusa.org
http://www.rafiusa.org
RAFI USA is a nonprofit organization promoting community, equity and sustainability for family farmers and rural communities. Our headquarters serves as a model for green building. In addition to daylighting, solar and energyconservation features, the building showcases the use of salvaged materials from the deconstruction of an 1830s farmhouse.
*Founded: 1990*
*Betty B Bailey, Executive Director*
*Regina Bridgman, Communications Manager*

**509 Save America's Forests**
4 Library Court SE
Washington, DC 20003
202-544-9219
Fax: 202-544-7462
E-mail: info@saveamericasforests.org
http://www.saveamericasforests.org
A nationwide campaign to end clearcutting and protect and restore our America's wild and natural forests. A coalition of groups throught America working together to protect each other's local forests and to protect our nation's forestsand forests throughout the world. A network of individual citizens from the country, the cities, and the suburbs who love forests and want to save them.
*Founded: 1994*
*Carl Ross, Executive Director*

**510 Society for Ecological Restoration**
285 W. 18th Street
Suite 1
Tucson, AZ 85701
520-622-5485
Fax: 520-622-5491
E-mail: info@ser.org
http://www.ser.org
To serve the growing field of Ecological Restoration through facilitating dialogue among restorationists; encouraging research, promoting awareness and public support for restoration and restorative management; contributing to publicpolicy discussions; recognizing those who have made outstanding contributions to

the field of restoration; and promoting ecological restoration around the globe.
*Founded: 1987*
*M K LeSeeour, Executive Director*

**511 Southface Energy Institute**
241 Pine Street NE
Atlanta, GA 30308
404-872-3549
Fax: 404-872-5009
E-mail: info@southface.org
http://www.southface.org

An environmental nonprofit working to promote sustainable homes, workplaces and communities through education, research, advocacy and technical assistance.
*Founded: 1978*
*Dennis Creech, President*

**512 Synthetic Organic Chemical Manufacturers Association**
1850 M Street NW
Suite 700
Washington, DC 20036
202-721-4100
Fax: 202-296-8120
E-mail: info@socma.com
http://www.socma.com

A trade association that serves the specialty, batch and custom chemical industry. SOCMA member companies make the products and refine the raw materials that make our standard of living possible; from pharmaceuticals to cosmetics,soaps to plastics, and all manner of industrial and construction products. SOCMA promotes innovative, safe and environmentally responsible operations, which are internationally competitive and contribute to a healthy, productive economy.
*Founded: 1921*
*Joseph Acker, President*
*Diane McMahon, Director Business Operations*

**513 World Environment Center**
734 15th Street NW
Suite 720
Washington, DC 20005
202-312-1370
Fax: 202-637-2411
E-mail: info@wec.org
http://www.wec.org

The World Environment Center is an independent, global non-profit, non-advocacy organization that advances sustainable development worldwide through the business practices of member companies and in partnership with governments,mulit-lateral organizations, private sector organizations, universities and other stake holders.
*Founded: 1974*
*Terry F Yosie, President/CEO*
*Gwen Davidow, Dir Global Corporate Program*

**514 Worldwatch Institute**
1776 Massachusetts Avenue NW
Suite 800
Washington, DC 20036
202-452-1999
Fax: 202-296-7365
http://www.worldwatch.org

The Worldwatch Institute is an independent, nonprofit environmental research organization in Washington DC. Its mission is to foster a sustainable society in which human needs are met in ways that do not threaten the health of thenatural environment or future generations. To this end, this Institute conducts interdisciplinary research on emerging global issues, the results of which are published and disseminated to decision-makers and the media.
*Founded: 1974*
*Christopher Flavin, President*
*Leanne Mitchell, Director Communications*

## National: Travel & Tourism

**515 American Association for Physical Activity and Recreation**
1900 Association Drive
Reston, VA 20191
703-476-3400
800-213-7193
Fax: 703-476-9527
E-mail: aapar@aahperd.org
http://www.aahperd.org/aapar

The AALR and the AAALF (american Association for Active Lifestyles and Fitness) have merged are in the process of reorganizing. The American Association for Leisure and Recreation (AALR) serves recreation professionals, practitoners,educators, and students who advance the profession and enhance the quality of life of all Americans through creative and meaningful leisure and recreation experiences.
*Founded: 2005*
*Mariah Burton Nelson, Executive Director*

**516 American Hiking Society**
1422 Fenwick Lane
Silver Spring, MD 20910
301-565-6704
Fax: 301-565-6714
E-mail: info@americanhiking.org
http://www.americanhiking.org

American Hiking Society is a recreation based conservation organization working to cultivate a nation of hikers dedicated to establishing, protecting, and maintaning foot trails in America. The more than 10,000 individual members andhiking club members contribute to this national effort.
*Founded: 1976*
*Gregory A Miller, President*
*Celina Montorfano, VP Programs*

**517 American Recreation Coalition**
1225 New York Avenue NW
Suite 450
Washington, DC 20005
202-682-9530
Fax: 202-682-9529
E-mail: arc@funoutdoors.com
http://www.funoutdoors.com

Washington based nonprofit that partners to enhance and protect outdoor recreational opportunities and resources. ARC also monitors legislative and regulatory proposals that influence recreation.
*Founded: 1979*
*Derrick Crandall, President*
*Catherine Ahern, Vice President*

**518 American Whitewater**
PO Box 1540
Cullowhee, NC 28723
828-586-1930
866-262-8429
Fax: 828-586-2840
E-mail: info@amwhitewater.org
http://www.americanwhitewater.org/

Restores rivers adversely affected by hydropower dams, eliminates water degradation, improves public land management and protects public access to rivers for responsible recreational use.
*Founded: 1954*
*Mark Singleton, Executive Director*
*Ben VanCamp, Outreach Director*

**519 Association of Zoos and Aquariums**
8403 Colesville Road
Suite 710
Silver Spring, MD 20910
301-562-0777
Fax: 301-562-0888
E-mail: generalinquiry@aza.org
http://www.aza.org

As the leading accrediting organization for zoos and aquariums, it is dedicated to the advancement of accredited zoos and aquariums in the area of animal care, wildlife conservation, education and science.
*Founded: 1924*
*Jim Maddy, President/CEO*
*Kris Vehrs, Executive Director*

**520 Federation of Western Outdoor Clubs**
PO Box 129
Selma, OR 97538
E-mail: jack.jan.indiancreek@mailbug.com
http://www.federationofoutdoorwesternclubs.org

The Federation is composed of organizations that engage in hiking, camping, birding and other similar activities that rely on an outdoor environment where natural conditions predominate. Organizations in the West that have suchprograms, and that have an active interest in protecting the natural environment, are invited to affiliate.
*Founded: 1932*
*Raelene Gold, President*
*Jack Walker, Treasurer*

**521 Green Hotels Association**
PO Box 420212
Houston, TX 77242
713-789-8889
Fax: 713-789-9786
E-mail: green@greenhotels.com
http://www.greenhotels.com
Helping hotels save water, energy, solid and money world wide; to become environmentally-friendly properties.
*Founded: 1993*
*Patricia Griffin, President*

**522 Lighthawk**
PO Box 653
Lander, WY 82520
307-332-3242
Fax: 307-332-1641
E-mail: info@lighthawk.org
http://www.lighthawk.org
Nonprofit organization founded in 1979, addresses critical environmental issues by providing an aerial perspective on areas of concern in the US, Canada and Central America. Using small aircraft, they fly partner organizations, electedofficials, industry and media representatives, activists, and indigenous groups over protected and threatened regions. Program flights give passengers both intellectual and visceral understanding of what is at stake.
*Founded: 1979*
*Rudy Engholm, Executive Director*

**523 National Association of Recreation Resource Planners**
2001 Jefferson Davis Highway
Suite 1004
Arlington, VA 22202
703-416-0060
Fax: 703-416-0014
E-mail: info@narrp.org
http://www.narrp.org
NARRP, is an organization comprised of outdoor recreation professionals and others interested in recreation resource planning. It is a nationwide organization with members in nearly every state representing federal and state agencies,land managers, consultants, and academic institutions. The mission of NARRP is to Advance the Art, the Science and the Profession of Recreation Resource Planning: and enhance the provision of recreation opportunities for all Americans.
*Glenn Haas, President*
*Bob McLean, Association Manager*

**524 National Association of State OutdoorRecreation Liaison Officers**
105H ABNR University of Missouri
Columbia, MO 65211-7230
573-353-2702
Fax: 573-882-9521
E-mail: nasorlo@embavqmail.com
An association of gubernatorial appointed state and territorial officials that work across the nation to provide places for outdoor recreation through the use of the Land and Water Conservation Fund.
*Dr Douglas Eiken, Executive Director*
*Tim Hagsett, President*

**525 National Council of State Tourism Directors**
TIA
1100 New York Avenue NW
Suite 450
Washington, DC 20005
202-408-8422
Fax: 202-408-1255
E-mail: feedback@tia.org
http://www.tia.org/councils/NCSTD/
The first council of the TIA it brings together tourism directors from all 50 states, DC, and territories to serve as a: unified voice, catalyst for developing programs for the benefit of all, and harmonizer in the diversity of needs,priorities and values.
*Founded: 1969*
*Todd Davidson, Chairman*

**526 National Wildlife Federation: Expeditions**
11100 Wildlife Center Drive
Reston, VA 20190
202-797-6800
800-606-9563
Fax: 202-797-6646
E-mail: expeditions@nwf.org
http://www.nwf.org
Travel with the NWF with expert guides, in an environmentally-sensitive way, and into natural ecosystems. The National Wildlife Federation is the largest member-supported conservation group, uniting individuals, organizations,businesses and government to protect wildlife, wild places and the environment.
*Larry J Schweiger, President/CEO*
*Dan Gifford, Manager Travel Programs*

**527 Rails-to-Trails Conservancy**
The Duke Ellington Building
2121 Ward Ct NW
5th Floor
Washington, DC 20037
202-331-9696
Fax: 202-331-9680
E-mail: railtrails@railtrails.org
http://www.railtrails.org
A nonprofit organization working with communities to preserve unused rail corridors by transforming them into trails, enhancing the health of American's environment, economy, neighborhoods and people.
*Founded: 1986*
*Keith Laughlin, President*
*Jeff Ciabotti, VP Trail Development*

**528 Safari Club International**
4800 W Gates Pass Road
Tucson, AZ 85745
520-620-1220
888-486-8724
Fax: 520-622-1205
http://www.safariclub.org/
A leader in wildlife conservation, hunter education, and protecting the freedom to hunt.
*Dennis Anderson, President*
*John W Nelson, Vice President*

**529 Wilderness Education Association**
900 E 7th Street
Bloomington, IN 47405
812-855-4095
Fax: 812-855-8697
http://www.weainfo.org/
WEA provides professional instruction, leadership training and wilderness travel. Promoting safe, ethical and professional wilderness leaders.
*Founded: 1977*
*Scott Jordan, President*
*Mary Williams, National Office Contact*

## National: Water Resources

**530 Adopt-A-Stream Foundation**
Northwest Stream Center
600-128th Street SE
Everett, WA 98208
425-316-8592
Fax: 425-338-1423
E-mail: aasf@streamkeeper.org
http://www.streamkeeper.org
The mission of the AASF is to teach people how to become stewards of their watershed. That mission is carried out by conducting classes, producing environmental education materials, and providing local communitites stram and wetlandrestoration technical assistance.
*Founded: 1985*
*Tom Murdoch, Executive Director*

**531 American Canal Society**
117 Main Street
Freemansburg, PA 18017
610-691-0956
http://www.americancanals.org
Dedicated to Historic Canal Research, Preservation, Restoration, and Parks. Promotes the wise use of America's many historic canal resources through research, preservation, restoriation, recreation, and parks. Acts as a nationalclearing house of canal information and co-operates with local, state, and international canal societies, groups, and individuals to identify historic canal resources, to publicize canal history, activities, and problems, and to take action onthreatened canals and sites
*Founded: 1972*
*David G Barber, President*
*Charles W Derr, Secretary/Treasurer*

**532 American Ground Water Trust**
50 Pleasant Street Ste 2
Concord, NH 03301
603-228-5444
Fax: 603-228-6557
E-mail: trustinfo@agwt.org
http://www.agwt.org

A not-for-profit education organization. Promotes opportunity, cooperation and action among individuals, groups and organizations in order to educate the public, and further its mission: to protect ground water, promote publicawareness of the environment and economic importance of groundwater and provide accurate information to assist public participation in water resources decisions and management.
*Founded: 1986*
*Andrew Stone, Executive Director*

### 533 American Rivers
1101 14th Street NW
Suite 1400
Washington, DC 20005
202-347-7550
Fax: 202-347-9240
E-mail: outreach@americanrivers.org
http://www.americanrivers.org

The only national organization that is dedicated to protecting and restoring rivers nationwide. Founded over 30 years ago it now has over 65,000 members and supports nationwide and two regional and six field offices.
*Founded: 1973*
*Rebecca R Wodder, President*
*Katherine Baer, Director Healthy Waters*

### 534 American Shore and Beach Preservation Association
5460 Beaujolais Lane
Fort Myers, FL 33919
239-489-2616
Fax: 239-489-9917
E-mail: exdir@asbpa.org
http://www.asbpa.org

This Association is formed in recognition of the fact that shores of our oceans, lakes and rivers constitute important assets for promoting the health and physical well-being of the people of this nation, and that their contiguity togo out great centers of population affords an opportunity for wholesome and necessary rest and recreation not equally available in any other form. The purpose of the Association is to bring together for cooperation and mutural helpfulness.
*Founded: 1926*
*Harry Simmons, President*
*Kate & Ken Gooderham, Executive Directors*

### 535 American Water Resources Association
4 West Federal Street
PO Box 1626
Middleburg, VA 20118
540-687-8390
Fax: 540-687-8395
E-mail: terry@awra.org
http://www.awra.org

Founded in 1964, the American Water Resources Association is a non-profit professional association dedicated to the advancement of men and women in water resources management, research, and education. AWRA's membership ismultidisciplinary; its diversity is its hallmark. It is the professional home of a wide variety of water resources experts including engineers, educators, foresters, biologists, ecologists, geographers, managers, regulators, hydrologists andattorneys.
*Kenneth D Reid, Executive VP*
*Terry Meyer, Marketing Director*

### 536 American Water Works Association
6666 W Quincy Avenue
Denver, CO 80235
303-794-7711
800-926-7337
Fax: 303-347-0804
http://www.awwa.org

Dedicated to the promotion of public health and welfare in the provision of drinking water of unquestionable quality and sufficient quantity. AWWA must be proactive and effective in advancing the technology, science, management andgovernment policies relative to the stewardship of water.
*Founded: 1881*
*Darcy Burke, Executive Director*
*JoAnn Taniguchi, Secretary*

### 537 Association of Ground Water Scientists and Engineers
601 Dempsey Road
Westerville, OH 43081
614-898-7791
800-551-7379
Fax: 614-898-7786
E-mail: smasters@ngwa.org
http://www.ngwa.org/agwse

The Scientists and Engineers membership division of the NGWA. The mission of the National Ground Water Association is to enhance the skills and credibility of all ground water professionals, develop and exchange industry knowledge, andpromote the ground water industry and understanding of ground water resources.
*Sandy Masters, Liaison*

### 538 Association of Metropolitan Sewerage Agencies
1816 Jefferson Place NW
Washington, DC 20036
202-833-2672
Fax: 202-833-4657
E-mail: info@nacwa.org
http://www.amsa-cleanwater.org

Represents the interests of the country's wastewater treatment agencies, true environmental practitioners that serve the majority of the sewered population in the US, and collectively treat and reclaim more than 17 billion gallons ofwastewater each day. Maintains a key role in the development of environmental legislation, and works closely with federal regulatory agencies in the implementation of environmental programs.
*Ken Kirk, Executive Director*

### 539 Association of Metropolitan Water Agencies
1620 I Street NW
Suite 500
Washington, DC 20006
202-331-2820
Fax: 202-785-1845
E-mail: info@amwa.net
http://www.amwa.net

An organization of the largest publicly owned drinking water systems in the nation.
*Founded: 1981*
*Diane VanDe Hei, Executive Director*
*Michael Arceneaux, Deputy Executive Director*

### 540 Association of State Floodplain Managers
2809 Fish Hatchery Road
Suite 204
Madison, WI 53713
608-274-0123
Fax: 608-274-0696
E-mail: asfpm@floods.org
http://www.floods.org

The Association is an organization of professionals involved in floodplain management, flood hazard mitigation, the National Flood Insurance Program, and flood preparedness, warning and recovery.
*Founded: 1977*
*Larry Larson, Executive Director*
*Diane Brown, Administrative Manager*

### 541 Association of State Wetland Managers
2 Basin Road
Windham, ME 04062
208-892-3399
Fax: 207-892-3089
E-mail: aswm@aswm.org
http://www.aswm.org

Nonprofit organization dedicated to the protection and restoration of the nations wetlands. Our goal is to help public and private wetland decision-makers utilize the best possible scientific information and techniques in wetlanddelineation, assessment, mapping, planning, regulation, acquisition, restoration, and other management.
*Founded: 1983*
*Jeanne Christie, Executive Director*
*Jon Kusler, Associate Director*

### 542 Association of State and Interstate Water Pollution Control Administrators
750 1st Street NE
Suite 1010
Washington, DC 20002
202-898-0905
Fax: 202-898-0929
E-mail: admin1@asiwpca.org
http://www.asiwpca.org

Maintain and enhance the quality of the nation's water resources and protect the public health through improving the State's capability to develop and implement effective Federal and State water management programs.
*Roberta Savage, Executive Director*
*Linda Eichmiller, Deputy Director*

### 543 CEDAM International
1 Fox Road
Croton on Hudson, NY 10520
914-271-5365
Fax: 914-271-4723
E-mail: cedamint@aol.com
http://www.cedam.org

Conservation, Education, Diving, Awareness and Marine research International is a nonprofit organization dedicated to the understanding, protection and preservation of the world's marine resources. Through our expeditions, CEDAMInternational volunteer divers actively participate in scientific research and conservation-oriented education projects. The results of our findings and efforts are disseminated to both the scientific and lay communities. The CEDA also publishes anannual newsletter.
*Susan Sammon, Director*

**544 Center for Coastal Studies**
59 Commercial Street
Provincetown, MA 02657
508-487-3623
Fax: 508-487-4695
E-mail: ccs@coastalstudies.org
http://www.coastalstudies.org
Private nonprofit organization for research, conservation and education in the coastal and marine environments.
*Roslyn Garfield, President*

**545 Center for Great Lakes Environmental Education**
PO Box 56
Buffalo, NY 14205
E-mail: info@greatlakesed.org
http://www.greatlakesed.org
Mission leads to awareness of and access to information about Great Lakes environmental subjects by promoting learning links for teachers, students and other stakeholders in the international Great Lakes-St Lawrence basin ecosystem.

**546 Center for Marine Conservation**
2029 K Street, NW
Washington, DC 20006
202-429-5609
800-519-1541
Fax: 202-872-0619
E-mail: info@oceanconservancy.org
http://www.cmc-ocean.org
The mission of the CMC is to protect ocean ecosystems and conserve the global abundance and diversity of marine wildlife. Through sciencebased advocacy, research and public education, CMC informs, inspires and empowers people to speakand act for the oceans.
*Roger Rufe Jr, President*
*Thomas J Tepper, Senior VP Operations*

**547 Center for Watershed Protection**
8390 Main Street
Second Floor
Ellicott City, MD 21043
410-461-8323
Fax: 410-461-8324
E-mail: center@cwp.org
http://www.cwp.org
The center is a nonprofit 501(c)3 organization dedicated to finding new ways to protect and restore our nation's streams, lakes, rivers and estuaries. The center publishes numerous technical publications on all aspects of watershedprotection, including stormwater management, watershed planning and better site design. Publications are available online.
*Founded: 1992*
*Hye Yeong Kwon, Executive Director*
*Karen Cappiella, Program Manager Research*

**548 Center for the Great Lakes**
435 N Michigan Avenue
Suite 1408
Chicago, IL 60611
312-263-0785
Fax: 312-201-0683
*Daniel K Ray, Director*

**549 Clean Harbors Cooperative**
4601 Tremley Point Road
Linden, NJ 07036
908-862-7500
Fax: 908-862-7560
E-mail: chcllc@aol.com
http://www.apicom.org/members.html
*Edward M Wirkowski, Manager*
*Dennis J McCarthy, Director*

**550 Clean Water Action**
1010 Vermont Avenue NW
Suite 400
Washington, DC 20005
410-235-8808
Fax: 410-235-8816
E-mail: cwa@cleanwater.org
http://www.cleanwateraction.org
National citizens' organization working for clean, safe and affordable water, prevention of health-threatening pollution, creation of environmentally-safe jobs and businesses, and empowerment of people to make democracy work.Organizes strong grassroots groups, coalitions and campaigns to protect our environment, health, economic well-being and community quality of life.
*Founded: 1972*
*Bob Wendelgass, Executive Director*

**551 Clean Water Fund**
1010 Vermont Avenue, NW
Suite 1100
Washington, DC 20005
202-895-0432
Fax: 202-895-0438
E-mail: cwf@cleanwater.org
http://www.cleanwaterfund.org
Brings diverse communities together to work for changes that improve our lives, promoting sensible solutions for people and the environment.
*Founded: 1972*
*Peter Lockwood, President*
*David Zwick, Executive VP*

**552 Clean Water Network**
1200 New York Avenue NW
Suite 400
Washington, DC 20005
202-298-2421
Fax: 202-289-1060
E-mail: info@cwn.org
http://www.cleanwaternetwork.org
A nonprofit network of over 1,000 organizations that deal with clean water issues covered by the Clean Water Act. Our member organizations consist of a variety of organizations representing environmentalists, family farmers, recreationanglers, commercial fishermen, surfers, boaters, faith communities, labor unions and civic associations. We publish a monthly newsletter and various reports.
*Natalie Roy, Executive Director*

**553 Coastal Conservation Association**
6919 Portwest
Suite 100
Houston, TX 77024
713-626-4234
800-201-3474
Fax: 713-951-3801
E-mail: ccantl@joincca.org
http://www.joincca.org/
A national nonprofit organization of 17 coastal state chapters dedicated to the conservation and preservation of marine resources.
*Founded: 1977*
*Pat Murray, VP/Director Conservation*
*Ted Venker, Director Communications*

**554 Coastal Society**
PO Box 3590
Williamsburg, VA 23187
757-565-0999
Fax: 757-565-0299
E-mail: coastalsoc@aol.com
http://www.thecoastalsociety.org
Organization of private sector, academic, government professionals and students dedicated to actively addressing emerging coastal issues by fostering dialogue, forging partnerships and promoting communication and education.
*Founded: 1975*
*Kristen Fletcher, President*
*Judy Tucker, Executive Director*

**555 Cook Inletkeeper**
PO Box 3269
3734 Ben Walters Lane
Homer, AK 99603
907-235-4068
Fax: 907-235-4069
E-mail: keeper@inletkeeper.org
http://www.inletkeeper.org
Dedicated to protecting the vast Cook Inlet watershed and the life it sustains; it is a community-based nonprofit organization that combines education, advocacy and science to reach this goal.
*Founded: 1995*
*Rob Ernst, President*
*Bob Shavelson, Executive Director*

**556 Coral Reef Alliance**
351 Caliornia Street
Suite 650
San Francisco, CA 94104
415-834-0900
Fax: 415-834-0999
E-mail: info@coral.org
http://www.coral.org

Unites and empowers communities to save coral reefs. Provides tools, education, and inspiration to residents of coral reef destinations to support local projects that benefit both reefs and people. Originally founded in 1994 togalvanize te dive community for conservation, CORAL has grown from a small grassroots organization into the only international nonprofit organization that works exclusively to protect our planet's coral reefs.
*Founded: 1994*
*Brian Huse, Executive Director*
*Diana Williams, Development Director*

**557 Earth Island Institute**
2150 Allston Way
Suite 4600
Berkeley, CA 94704-1375
415-788-3666
Fax: 415-788-7324
E-mail: johnknox@earthisland.org
http://www.earthisland.org

A non-profit, public interest, membership organization that supports people who are creating solutions to protect our shared planet.
*Founded: 1982*
*John A Knox, Executive Director*
*Dave Phillips, Executive Director*

**558 Ecological Society of America**
1990 M St, NW
Suite 700
Washington, DC 20036
202-833-8773
Fax: 202-833-8775
E-mail: esahq@esa.org
http://www.esa.org

To promote ecological science by improving communication among ecologists; raise the public's level of awareness of the importance of ecological science; increase the resources available for the conduct of ecological science; andensure the appropriate use of ecological science in environmental decision making by enhancing communication between the ecological community and policy-makers.
*Founded: 1915*
*Alison Power, President*
*Katherine S McCarter, Executive Director*

**559 Emergency Committee to Save America's Marine Resources**
1552 Osprey Court
Manasquan Park, NJ 08736
732-223-5729
Fax: 732-528-1056
E-mail: cristori@aol.com

*Founded: 1955*
*Allan J Ristori, Chairman*

**560 Foresta Institute for Ocean and Mountain Studies**
2400 E Speedway
Suite 118-293
Tucson, AZ 85716
520-881-6174
Fax: 520-323-2751

Works to educate teachers and youth about environmental conservation.

**561 Freshwater Society**
2500 Shadywood Road
Excelsior, MN 55331
952-471-9773
888-471-9773
Fax: 952-471-7685
E-mail: freshwater@freshwater.org
http://www.freshwater.org

*Founded: 1968*

**562 Future Fisherman Foundation**
225 Reinekers Lane
Suite 420
Alexandria, VA 22314
703-519-9691
Fax: 703-519-1872
E-mail: info@futurefisherman.org
http://www.asafishing.org

The Future Fisherman Foundation supports groups that offers training in fishing and aquatic resource stewardship by developing curriculum materials for environmental and angler education. Working through state fish and wildlifeagencies, the foundation strives to increase aquatic resource education by using available federal funds.
*Anne Glick, Executive Director*
*John Bryan, Chief Philanthropy Officer*

**563 Gaia Institute**
400 City Island Avenue
Bronx, NY 10464
718-885-1906
Fax: 718-885-0882
E-mail: Gaia@gaia-inst.org
http://www.gaia-inst.org

The purpose of the Gaia Institute is to test through demostration the means by which ecological components of backyards, communities, towns and cities, as well as watersheds and estuaries, can be enhanced through integratedwastes-into-resources technologies.
*Founded: 1972*
*Paul S Mankiewicz, Director*
*Julie A Mankiewicz, PhD, Director*

**564 Global Water Policy Project**
107 Larkspur Drive
Amherst, MA 01002
413-256-4808
E-mail: info@globalwaterpolicy.org
http://www.globalwaterpolicy.org

Global Water Policy Projects aims to promote the preservation and sustainable use of Earth's fresh water through research, writing, outreach, and public speaking.
*Founded: 1994*
*Sandra Postel, Founder/Director*

**565 Great Lakes United**
4525 Rue DeRouen
Montreal Quebec, QC H1V 1
514-396-3333
Fax: 514-396-0297
E-mail: glu@glu.org
http://www.glu.org

Great Lakes United is an international coalition dedicated to preserving and restoring the Great Lakes-St. Lawrence River ecosystem. Great Lakes United is made up of member organizations representing environmentalists,conservationists, hunters and anglers, labor unions, community groups, and programs, and promotes citizen action and grassroots leadership to assure.
*Founded: 1982*
*Derek Stack, Executive Director*
*Patty O'Donnell, President*

**566 International Association for Environmental Hydrology**
PO Box 35324
San Antonio, TX 78235
210-344-5418
Fax: 210-344-9941
E-mail: hydroweb@mail.org
http://www.hydroweb.com

**567 International Desalination Association**
PO Box 387
Topsfield, MA 01983
978-887-0410
Fax: 978-887-0411
E-mail: info@idadesal.org
http://www.idadesal.org

IDA is committed to the development and promotion of the appropriate use of desalination and desalination technology worldwide. We endeavor to carry out these goals by encouraging research and development, exchanging, promotingcommunication and disseminating information.
*Founded: 1972*
*Lisa Henthorne, President*
*Michel Canet, First Vice President*

**568 International Oceanographic Foundation**
4600 Rickenbacker Causeway
PO Box 499900
Miami, FL 33149
305-361-4000
Fax: 305-361-4711
http://www.rsmas.miami.edu

International Oceanographic Foundation (IOF) is a nonprofit supporting organization to the University of Miami's Rosenstiel School of Marine and Atmospheric Science. IOF was chartered in 1953 to encourage scientific investigation ofthe sea, to provide the public with current, accurate and unbiased information pertaining to marine environments and to promote awareness of the importance of Earth's oceans to humankind. IOF is a 501(c)(3) nonprofit organization, contributions aretax deductable.
*Founded: 1953*
*Otis B Brown, Dean*

**569 International Rivers**
2150 Allston Way
Suite 300
Berkeley, CA 94704
510-848-1155
Fax: 510-848-1008
E-mail: info@internationalrivers.org
http://www.internationalrivers.org
Mission is to protect rivers and defend the rights of communities that depend on them. We oppose destructive dams and the development model they advance, and encourage better was of meeting people's needs for water, energy andprotection from damaging floods. To achieve this mission, we collaborate with a network of local communities, social movements, non-governmental organizations and other partners.
*Founded: 1985*
*Patrick McCully, Executive Director*
*Anne Carey, Office Manager*

**570 International Water Resources Association**
Southern Illinois University Carbondale
4535 Faner Hall
Carbondale, IL 62901
618-453-5138
Fax: 618-453-6465
E-mail: iwra@siu.edu
http://www.iwra.siu.edu
IWRA strives to improve water management worldwide through dialogue, education, and research. It seeks to improve water resources outcomes by improving our collective understanding of the physical, biological, chemical, institutonal,and socioeconomic aspects of water.
*Founded: 1972 160 pages*
*ISSN: 0250-8060*
*Benedykt Dziegielewski, Executive Director*
*John W Nicklow, Treasurer*

**571 Interstate Council on Water Policy**
955 L'Enfant Plaza SW
6th Floor
Washington, DC 20024
202-466-7287
Fax: 202-646-6210
http://www.icwp.org
The Interstate Council on Water Policy is the national organization of state and regional water resource management agencies. It provides a means to exchange information, ideas, and experience and to work with federal agencies whichshare water management responsibilities. In paricular, ICWP focuses on water quality and water quantity issues, and on the dynamic interface state and federal roles.
*Founded: 1959*
*Joe Hoffman, Executive Director*

**572 Marine Technology Society**
5565 Sterrett Place
Suite 108
Columbia, MD 21044
410-884-5330
Fax: 410-884-9060
E-mail: mtsdir@aeros.com
http://www.mtsociety.org
Addresses coastal zone management, marine mineral and energy resources, marine environmental protection, and ocean engineering issues.
*Founded: 1963*
*Jerry Streeter, President*
*Judith Krauthamer, Executive Director*

**573 National Association for State and Local River Conservation Programs**
8630 Fenton Street
Suite 910
Silver Spring, MD 20910
301-589-9455
Fax: 301-589-6121
*Barry Beasley, President*

**574 National Association of Clean Water Agencies**
1816 Jefferson Place NW
Washington, DC 20039
202-833-2672
Fax: 202-833-4657
E-mail: info@nacwa.org
http://www.nacwa.org
The NACWA is a recognized leader in environmental policy and a technical resource on water quality and ecosystem protection issues. It represents the interests of the country's wastewater treatment agencies and true environmentalpractioners.
*Ken Kirk, Executive Director*
*Paula Dannenfeldt, Deputy Executive Director*

**575 National Association of Flood and Stormwater Management Agencies**
1301 K Street NW
Suite 800 East
Washington, DC 20005
202-218-4122
Fax: 202-478-1734
http://www.nafsma.org
The National Association of Flood and Stormwater Management Agencies is an organization of public agencies whose function is the protection of lives, property and economic activity from the adverse impacts of storm and flood waters.The mission of the association is to advocate public policy, encourage technologies and conduct education programs which facilitate and enhance the adchievement of the public service functions of its members.
*Founded: 1978*
*Susan Gilson, Executive Director*
*Kerry Wilson, Membership Services*

**576 National Audubon Society: Living Oceans Program**
550 S Bay Avenue
Islip, NY 11751
516-859-3032
Fax: 516-581-5268
E-mail: mlee@audubon.org
http://www.audubon.org/campaign/10
Living Oceans is the marine conservation program of National Audubon Society. Audubon's Living Oceans uses science-based policy analysis, education and grassroots advocacy to put science to work on behalf of marine fish and oceanecosystems.
*Carl Safina, Director*
*Mercedes Lee, Assistant Director*

**577 National Boating Federation**
PO Box 4111
Annapolis, MD 21403
Fax: 360-297-3505
E-mail: rpdavid@capecod.net
http://www.n-b-f.org
Nonprofit, all volunteer organization that represents over 2 million of America's boaters. It is an alliance of yacht and boating clubs, recreational boating organizations and includes associate and individual members.
*Founded: 1966*
*Marlene Barrington, President*
*Margot J Brown, Executive Director*

**578 National Coalition for Marine Conservation**
4 Royal Street SE
Leesburg, VA 20175
703-777-0037
Fax: 703-777-1107
E-mail: christine@savethefish.org
http://www.savethefish.org
National Coalition for Marine Conservation is dedicated exclusively to conserving ocean fish, preventing overfishing, reducing bycatch, and protecting marine habitat.
*Founded: 1973*
*Ken Hinman, Chairman*
*Christine Snovell, Communications/Development*

**579 National Council for Air and Stream Improvements**
4815 Emperor Boulevard
Canterbury Hall, Ste 110
Durham, NC 27703
919-941-6400
Fax: 919-941-6401
E-mail: ryeske@ncasi.org
http://www.ncasi.org
The Council is a technical organization devoted to finding solutions to environmental protection problems in the manufacture of pulp, paper, and wood products in industrial forestry. It now has about 75 member companies.
*Founded: 1943*
*Ronald Yeske, President*

**580 National Ground Water Association**
601 Dempsey Road
Westerville, OH 43081
614-898-7791
800-551-7379
Fax: 614-898-7786
E-mail: ngwa@ngwa.org
http://www.ngwa.org
The mission of the National Ground Water Association is to provide professional and technical leadership for the advancement of

the ground water industry and for the protection, promotion, and responsible development and use of groundwater resources.
*Founded: 1948*
*Kevin B McCray, Executive Director*
*Paul Humes, VP Operations/CFO*

**581 National Institutes for Water Resources**
University of Massachusetts Water Resources Center
47 Harkness Road 413-253-5686
Pelham, MA 01002 Fax: 413-253-1309
E-mail: godfrey@tei.umass.edu
http://www.snr.unl.edu/niwr

The National Institutes for Water Resources is a network of Research Institutes in every state. They conduct basic and applied research to solve water problems unique to their area. The programmatic responsibilities stipulated by theWater resources Research Act provide a unified focus for the federal and non-federal components of the Institute Program.
*Founded: 1970*
*Paul Godfrey, Executive Secretary*

**582 National Water Center**
5473 Highway 23N
Eureka Springs, AR 72631 E-mail: peace@ipa.net
http://www.nationalwatercenter.org

Strives to look at water with the broadest ideas of ecological balance and harmony. Support appropriate technology, bioregionalism, composting toilets, dowsing, ecology, flow forms, stream monitoring, vibrational water, watershedplanning.
*Barbara Helen Harmony, President*

**583 National Water Resources Association**
3800 North Fairfax Drive 703-524-1544
Suite 4 Fax: 703-524-1548
Arlington, VA 22203 E-mail: nwra@nwra.org
http://www.nwra.org

The National Water Resources Association consists of individuals or groups, such as irrigation districts, canal companies and conservancy districts, municipalities and the public in general, who are interested in water resourcedevelopment projects. It was started in 1932 and now has about 5,000 members.
*Founded: 1932*
*W E West Jr, President*

**584 National Watershed Coalition**
1023 Manvel, Suite D 405-627-0670
PO Box 556 E-mail: nwchdqtrs@sbcglobal.net
Chandler, OK 74834 http://www.watershedcoalition.org

A nonprofit Coalition made up of national, regional, state, and local organizations, associations, and individuals, that advocate dealing with natural resource problems and issues using watersheds as the planning and implementationunit.
*Founded: 1989*
*Dr Dan Sebert PhD, CEO/Executive Director*

**585 National Waterways Conference**
4650 Washington Blvd 703-243-4090
# 608 Fax: 866-371-1390
Arlington, VA 22201 E-mail: info@waterways.org
http://www.waterways.org

Coalition of trade and regional associations that have an interest in national waterways policy issues.
*Connie Waterman, Director Internal Operations*

**586 National Xeriscape Council**
PO Box 163172 512-392-6225
Austin, TX 78716 http://www.xeriscape.org

Systematic concept for saving water in landscaped areas.

**587 North American Benthological Society**
PO Box 7065
Lawrence, KS 66044 E-mail: amorin@outtawa.ca
http://www.benthos.org

An international scientific organization that promotes better understanding of biotic communities of lake and stream bottoms and their role in aquatic ecosystems.
*Founded: 1953*
*N LeRoy Poff, President*
*Lucinda B Johnson, Secretary*

**588 North American Lake Management Society**
PO Box 5443 608-233-2836
Madison, WI 53705 Fax: 608-233-3186
E-mail: info@nalms.org
http://www.nalms.org

Their focus is on lake management for a wide variety of uses, but they also get involved with other issues on a watershed level: land, streams, wetlands, and estuaries. The society forges partnerships among citizens, scientists andprofessionals to foster the management and protection of lakes and reservoirs.
*Founded: 1980*
*Ken Wagner, President*
*S Sue Robertson, Treasurer*

**589 Ocean Conservancy**
2029 K Street NW 202-429-5609
Washington, DC 20006 800-519-1541
Fax: 202-872-0619
E-mail: info@oceanconservancy.org
http://www.oceanconservancy.org

To conserve and protect the oceans. Advocate for the oceans, with an emphasis on conserving and protecting significant parts of our oceans.
*Founded: 1972*
*Vikki N Spruill, President/CEO*
*Thomas F Tepper Jr, Senior VP Operations*

**590 Oceana**
2501 M Street NW 202-833-3900
Suite 300 877-762-3262
Washington, DC 20037 Fax: 202-833-2070
E-mail: info@oceana.org
http://www.oceana.org

Oceana is a nonprofit international advocacy organization dedicated to protecting and restoring the world's oceans through policy advocacy, science, law and public education. Oceana's constituency includes members and activists frommore than 190 countries and territories who are committed to saving the world's marine environment. In 2002, American Oceans Campaign became part of Oceana's international effort to protect ocean eco-systems and sustain the circle of life.
*Founded: 2001*
*Mike Hirschfield, Senior VP/Chief Scientist*
*Romanus Berg, VP Operations/CIO*

**591 Oceanic Society**
Fort Mason Center 415-441-1106
Quarters 35N 800-326-7491
San Francisco, CA 94123 Fax: 415-474-3395
E-mail: office@oceanic-society.org
http://www.oceanic-society.org

Founded for the protection of the marine environment, environmental education and conservation-based field research.
*Founded: 1969*

**592 River Network**
520 SW 6th Avenue 503-241-3506
Suite 1130 800-423-6747
Portland, OR 97204 Fax: 503-241-9256
E-mail: info@rivernetwork.org
http://www.rivernetowrk.org

River Network is a national nonprofit organization for citizen groups working for river and watershed protection. Thier mission is to help people understand, protect and restore rivers and their watersheds. They provide conservationpartners with with information training, consultation, grants, referrals to other service organizations and networking opportunities.
*Founded: 1988*
*Don Elder, President/CEO*
*Susan Schwartz, CAO*

593 **Scientific Committee on Oceanic Research: Department of Earth and Planetary Science**
Johns Hopkins University
Baltimore, MD 21218
410-516-4070
Fax: 410-516-4019
E-mail: scor@jhu.edu
http://www.jhu.edu

The Scientific Committee on Oceanic Research (SCOR), is an international nonprofit organization whose purpose is to encourage international cooperation in a branch of ocean research.
*Founded: 1957*
*Edward Urban, Executive Director*
*Elizabeth Gross, Finance Officer*

594 **Seacoast Anti-Pollution League**
163 Court Street
PO Box 1136
Portsmouth, NH 03802
603-431-5089
E-mail: info@sapl.org
http://www.sapl.org
A nonprofit environmental group that makes it our business to monitor threats to public health and safety, wildlife, and ecosystems in our-and your-community.
*Founded: 1969*
*David Hills, President*

595 **Soil and Water Conservation Society**
945 SW Ankeny Road
Ankeny, IA 50023
515-289-2331
800-843-7645
Fax: 515-289-1227
E-mail: memberservices@swcs.org
http://www.swcs.org
Their mission is to foster the science and art of natural resource conservation. Their work targets the conservation of soil, water, and related natural resources.
*Founded: 1943*
*Craig A Cox, Executive Director*
*Lindey Krug, Member Services*

596 **Steamboaters**
1665 Evergreen Drive
Eugene, OR 97404
541-688-4980
Fax: 541-607-3763
E-mail: steamboaters@jeffnet.org
http://www.steamboaters.org
Nonprofit organization dedicated to work to restore the North Umpqua river system's wild fish stocks, particularly steelhead, to a sustainable level that is consistent with optimum natural population levels. Protect, preserve andrestore fish habitat, including adequate and consistent flows of high quality water in the North Umpqua and its tributaries.
*Founded: 1975*
*Joe Ferguson, President*

597 **The Groundwater Foundation**
PO Box 22558
Lincoln, NE 68542
402-434-2740
800-858-4844
Fax: 402-434-2742
E-mail: info@groundwater.org
http://www.groundwater.org
The Groundwater Foundation is a nonprofit organization that is dedicated to informing the public about one of our greatest hidden resources, groundwater. Since 1985, our programs and publications present the benefits everyone receivesfrom groundwater and the risks that threaten groundwater quality. We make learning about groundwater fun and understandable for kids and adults alike.
*Founded: 1985*
*Jane Griffin, President*
*Cindy Kreifels, Executive VP*

598 **United Citizens Coastal Protection League**
PO Box 46
Cardiff By The Sea, CA 92007
760-753-7477
*Founded: 1982*
*Robert Bonde, Executive Director*

599 **Water Environment Federation**
601 Wythe Street
Alexandria, VA 22314
703-684-2400
800-666-0206
Fax: 703-684-2492
E-mail: webfeedback@wef.org
http://www.wef.org
A not-for-profit technical and educational organization with members from varied disciplines who work toward the WEF vision of preservation and enhancement of the global water environment.
*Founded: 1928*
*William Bertera, Executive Director*
*Lori Harrison, Media Relations*

600 **Water Quality Association**
4151 Naperville Road
Lisle, IL 60532
630-505-0160
Fax: 630-505-9637
E-mail: info@mail.wqa.org
http://www.wqa.org
Nonprofit international trade association representing the water treatment industry.
*Founded: 1974*
*Peter J Censky, Executive Director*
*James Abston, Product Cert Coordinator*

601 **Water Resources Congress**
2300 Claredon Boulevard
Suite 404
Arlington, VA 22201
703-525-4881
Fax: 703-527-1693
*Kathleen A Phelps, Executive Director*

602 **Western Hemisphere Shorebird Reserve Network**
Manomet Center
81 Stage Point Road, PO Box 1770
Manomet, MA 02345
508-224-6521
Fax: 508-224-9220
E-mail: info@whsrn.org
http://www.whsrn.org
WHSRN is a voluntary, community-based coalation of over 185 organizations across the US and other counries in the Western Hemisphere that have joined together to protect, restore and manage critical wetland habitats for migratorybirds.
*John Cecil, Contact*

603 **Wildlands Conservancy**
3701 Orchid Place
Emmaus, PA 18049
610-965-4397
Fax: 610-965-7223
E-mail: info@wildlandspa.org
http://www.wildlandspa.org
Wildlands Conservancy is a nonprofit organization dedicated to preserving precious land, river restoration, keeping our waterways healthy, teaching the community about nature and caring for orphaned or injured wildlife.
*Founded: 1973*
*Christopher Kocher, Chairman*
*Daniel Giannelli, COO*

604 **World Aquaculture Society**
143 JM Parker Coliseum
Louisiana State University
Baton Rouge, LA 70803
225-578-3137
Fax: 225-578-3493
E-mail: carolm@was.org
http://www.was.org
The World Aquaculture Society is an international nonprofit society, whose commitment to excellence in science, technology, education and information exchange, will contribute to progressive and sustainable development of aquaculturethroughout the world.
*Founded: 1970*
*Carol Mendoza, Home Office Director*

## Alabama

**605 Alabama Association of Soil & Water Conservation Districts**
PO Box 304800 334-242-2620
Montgomery, AL 36130 Fax: 334-242-0551
E-mail: vpayne@swcc.state.al.us
http://www.swcc.state.al.us
The function of the NRCS at the local level is to provide technical leadership, delivery of special programs, and overall leadership of each office.
*Founded: 1937*
*George Robertson, President*
*Stephen M Cauthen, Executive Director*

**606 Alabama Environmental Council**
City Council Office 205-322-3126
2717 7th Avenue S 800-982-4364
Birmingham, AL 35203 Fax: 205-254-2603
E-mail: stateoffice@aeconline.ws
http://www.aeconline.ws
Statewide grassroots. The oldest environmental advocacy and education organization dedicated to the preservation and protection of Alabama's natural heritage.
*Founded: 1967*
*Ouida Fritschi, President*
*Larry Crenshaw, Secretary*

**607 Alabama National Safety Council: Birmingham**
242 W. Valley Avenue 205-328-7233
Suite 106 800-457-7233
Birmingham, AL 35209 Fax: 205-328-1467
E-mail: alabama@nsc.org
http://www.alabama.nsc.org
The National Safety Council's network of chapters couducts safety, health and environmental efforts at the community level, providing training, conferences, workshops, consultation, newsletters, updates and safety support materials, aswell as valuable networking avenues. Our network extends the Council's visibility and provides a local voice for advocating issues that can educate, inform, protect, and save lives.
*Donny Ward, Executive Director*

**608 Alabama Solar Association**
PO Box 143 256-650-5120
Huntsville, AL 35804 Fax: 256-650-5119
E-mail: isimon1027@aol.com
A non-profit organization leading the US to a sustainable energy economy.
*Erwin H Simon, Contact*

**609 Alabama Waterfowl Association**
1346 Country Road #11 205-259-2509
Scottsboro, AL 35768 E-mail: awa@alabamawaterfowl.org
http://www.alabamawaterfowl.org
A state voice for over 12,000 Alabama waterfowl hunters, concerning the waterfowl hunting season and the federal migratory bird regulations in Alabama. AWA networks with other state waterfowl associations in the north americanwaterfowl federation for wetlands conservation.
*Jerry Davis, CEO*
*Gary Benefield, Executive Director*

**610 Alabama Wildlife Federation**
3050 lanark road 334-285-4550
Millbrook, AL 36054 800-822-9453
Fax: 334-285-4959
E-mail: awf@alabamawildlife.org
http://www.alabamawildlife.org
Mission is to promote the conservation of Alabama's wildlife and related natural resources, as a basis for the social and economic prosperity of present and future generations, through wise use and responsible stewardship of ourwildlife, forests, fish, soils, water and air.
*Founded: 1935*
*Tim Gothard, Executive Director*
*Rebecca Prichett, Representative*

**611 American Lung Association of Alabama**
American Lung Association
3125 Independence Drive 205-933-8821
Suite 325 Fax: 205-930-1717
Birmingham, AL 35209 E-mail: aodom@alabamalung.org
http://www.lungusa2.org
Mission is to prevent lung disease and promote lung health through research, education, community service and advocacy. Our goals include reducing tobacco use, especially among young people, preventing and controlling air pollution,and providing education and funding research to make life more comfortable for people with asthma or other lung disease.
*Founded: 1904*
*John Kirkwood, President/CEO*
*Norman Edelman, Executive Vice President*

**612 American Society of Landscape Architects: Alabama Chapter**
E-mail: lchadwatkins@yahoo.com
The national professional association representing landscape architects.
*Lester C Watkins, President*
*Sharon Nelson-Deep, President-Elect*

**613 BAMA Backpaddlers Association**
307 Madison Place 205-951-0320
Trussville, AL 35173 E-mail: backpaddlers@aol.com
http://www.members.aol.com/backpaddlers
To promote recreation, conservation and education of Alabama's rivers.
*Founded: 1978*
*Pam Belrose, President*

**614 BASS Anglers Sportsman Society**
P.O. Box 17900 334-272-9530
Montgomery, AL 36141 Fax: 334-270-7148
E-mail: bassmail@mindspring.com
http://www.bassmaster.com
Goal is to create a credible and honorable tournament trail, to improve our environment by uniting and amplifying the voices of anglers and to secure a future for our youth.
*Founded: 1967*
*Gary Jones, Federation Director*

**615 National Safety Council: Tennessee Valley Office)**
2042 Beltline Road SW 256-308-1133
Building A, Suite 110 Fax: 256-308-1161
Decatur, AL 35601 E-mail: alabama@nsc.org
Network of chapters conducts safety, health and environmental efforts at the community level, providing training, conferences, workshops, consultation, newsletters, updates and safety support materials, as well as valuable networkingavenues.
*Donny Ward, Executive Director*

**616 Nature Conservancy: Colorado Field Office**
2424 Spruce Street 303-444-2950
Boulder, CO 80302 Fax: 303-444-2986
E-mail: colorado@tnc.org
http://www.nature.org/wherewework/
The mission of the Nature Conservancy is to preserve plants, animals and natural communities that represent the diversity of life on earth by protecting the lands and waters they need to survive. To date, the conservation and its morethan one million members have been responsible for the protection of more than 12 million acres in 50 states.
*Founded: 1951*

**617 Sierra Club**
1330 21st Way S
Suite 110
Birmingham, AL 35205
251-599-8699
E-mail: maggie@campmcdowell.com
http://alabama.sierraclub.org
To advance the preservation and protection of the natural environment by empowering the citizenry, especially democratically-based grassroots organizations, with charitable resources to further the cause of environmental protection.The vehicle through which The Sierra Club Foundation generally fulfills its charitable mission.
*Founded: 1908*
*Margaret Wade Johnston, Chair*
*David Underhill, Secretary*

## Alaska

**618 Alaska Conservation Alliance**
PO Box 100660
Anchorage, AK 99510
907-258-6171
Fax: 907-258-6177
E-mail: info@akvoice.org
http://www.akvoice.org
Statewide non-profit organization whose primary mission is to protect Alaska's natural environment through voter education and engagement, and advocacy.
*Founded: 1997*
*Caitlin Higgins, Executive Director*
*Sonya Wellman, Development Director*

**619 Alaska Natural Resource & Outdoor Education**
200 W. 34th Street
Suite 1007
Anchorage, AK 99503
907-292-1772
Fax: 907-207-1795
E-mail: info@anroe.org
http://www.anroe.wordpress.com
Mission is to promote and implement excellence in natural resource, outdoor, and environmental education for all Alaskans
*Founded: 1986*
*Courtney Sullivan, President*
*Kristen Romanoff, Secretary*

**620 Alaska Wildlife Alliance**
PO Box 202022
Anchorage, AK 99520
907-277-0897
Fax: 907-277-7423
E-mail: info@akwildlife.org
http://www.akwildlife.org
Mission is the protection of Alaska's natural wildlife for its intrinsic value as well as for the benefit of present and future generations.
*Founded: 1978*
*Tina M Brown, President*
*Art Greenwalt, Vice President*

**621 American Lung Association of Alaska**
500 West International Airport Road
Suite A
Anchorage, AK 99518
907-276-5864
800-586-4872
Fax: 907-565-5587
E-mail: marge@aklung.org
http://www.aklung.org
Promoting lung health and preventing lung disease in Alaska.
*Founded: 1934*
*Marge Larson, COO*

**622 American Society of Landscape Architects: Alaska Chapter**
Municipality of Anchorage
PO Box 196650
Anchorage, AK 99519
907-343-8368
888-999-2752
Fax: 907-343-8088
E-mail: info@asla.org
http://www.asla.org
National professional association representing landscape architects.
*Founded: 1899*
*Peter Briggs, President*
*Nancy Casey, President-Elect*

**623 Audubon Alaska**
Nation Audubon Society
441 West Fifth Ave
Suite 300
Anchorage, AK 99501
907-76 -069
Fax: 907-276-5069
http://www.ak.audubon.org
Founded in 1977, this office has played a crucial role in protecting the Great Land and its extraordinary ecosystems through a combination of science and stewardship, public policy, and education. The mission is to conserve Alaska'snatural ecosystems, for the benefit and enjoyment of current and future generations.
*Founded: 1977*
*Stanley E Senner, Executive Director*
*Kristi Bailey, General Manager*

**624 National Wildlife Federation Alaska Regional Center**
750 W 2nd Avenue
Suite 200
Anchorage, AK 99501
907-339-3900
Fax: 907-339-3980
E-mail: mcgeeh@nwf.org
http://http://nwf.org/en/Regional-Centers/Pacific-Region-Alaska.asp
It is committed to protecting Alaska's wildlife and wild places by working in partnership with concerned citizens, grassroots groups, and communities. They also offer the Alaska Youth for Environmental Action program which is designedto help youth become stronger and more effective environmental leaders.
*Founded: 1988*
*Jim Adams, Regional Executive Director*
*Shannon Kuhn, AYEA Program Coordinator*

**625 Northern Alaska Environmental Center**
830 College Road
Fairbanks, AK 99701
907-452-5021
Fax: 907-452-3100
E-mail: info@northern.org
http://www.northern.org
Promotes conservation of the environment in Interior and Arctic Alaska through advocacy, education, and sustainable resource stewardship.
*Founded: 1971*
*Karen Max Kelly, Executive Director*
*Laenne Thompson, Communications & Development*

**626 Sierra Club**
333 W 4th Avenue
Suite 307
Anchorage, AK 99501
907-276-4048
Fax: 907-258-6807
E-mail: dan.ritzman@sierraclub.org
http://www.alaska.sierraclub.org
The Alaska Chapter of the Sierra Club works to protect the last frontier from the damage of logging, drilling and the like. It strives for renewable energy, sustainable use of resources, preservation of wild places, and the smartgrowth of urban areas.
*Founded: 1892*
*Pam Brodie, Chair*
*Richard Hellard, Conservation Committee Chair*

**627 Trustees for Alaska**
1026 W 4th Avenue
Suite 201
Anchorage, AK 99501
907-276-4244
Fax: 907-276-7110
E-mail: ecolaw@trustees.org
http://www.trustees.org
A public interest law firm whose mission is to provide legal counsel to sustain and protect Alaska's natural environment. We represent local and national environmental groups, Alaska Native villages, nonprofit organizations, communitygroups, hunters, fishers and others where the outcome of our advocacy could benefit Alaska's environment.
*Founded: 1974*
*Trish Rolfe, Executive Director*
*Tracy Lohman, Development Director*

**628 Wildlife Society: Alaska Chapter**
1271 Lowbush Lane
Fairbanks, AK 99709
907-456-8682
Fax: 907-257-2774
E-mail: www.membership@wildlife.org
http://www.wildlife.org/chapters/ak
A nonprofit scientific and educational organization that serves professionals such as government agencies, academia, industry,

and non-government organizations in all areas related to the conservation of wildlife and natural resourcesmanagement.
*Founded: 1971*
*Kris Hundertmark, President*
*Michelle Davis, Branch Manager*

## Arizona

**629 American Environmental Health Foundation**
8345 Walnut Hill Lane
Suite 225
Dallas, TX 75231
214-361-9515
800-428-2343
Fax: 214-361-2534
E-mail: aehf@aehf.com
http://www.aehf.com
The Environmental Health Foundation's mission is two-fold: to fund scientific and/or medical research into the causes of environmentally linked disease; and to educate the public about environmentally linked illness and how to preventexposure through lifestyle changes.
*Founded: 1975*
*William Rea, Founder*
*David Hicks, Director*

**630 American Lung Association: Arizona**
The American Lung Association
2819 East Broadway
Tucson, AZ 85716
520-323-1812
800-586-4872
Fax: 520-323-1816
http://www.lungusa.org/south/
The mission of the American Lung Association is to prevent lung disease and promote lung health.
*Founded: 1912*
*Heidi Miller, Vice Chair*
*Bonnie Light, Executive Director*

**631 Arizona ASLA: American Society of LandscapeArchitects**
30 N Third Avenue
Suite 300
Phoenix, AZ 85003
602-258-8668
Fax: 602-273-6814
http://www.azasla.org
The Arizona state affiliate of the American Society of Landscape Architects.
*Michael Buschbacher II, President*
*Deanna Anderson, Executive Director*

**632 Arizona Automotive Recyclers Association**
1030 Baseline Road
#105-1025
Tempe, AZ 85283
480-609-3999
E-mail: admin@aara.com
http://www.aara.com
A group of automotive recyclers that know quality, value and service go hand in hand.
*Founded: 1910*
*Mike Pierson Jr, President*
*Layla Ressler, VP*

**633 Arizona BASS Chapter Federation**
PO Box 84006
Phoenix, AZ 85071
623-434-3520
http://www.azbassfederation.com
Bass fishing tournament organization.
*Founded: 1972*
*Mike Johnson, President*
*Kip Pollay, Executive Director*

**634 Arizona Chapter, National Safety Council**
1606 W Indian School Road
Phoenix, AZ 85015
602-264-2394
Fax: 602-277-5485
E-mail: main@acnsc.org
http://www.acnsc.org
Purpose is to educate and motivate people to live safer and healthier lives whether at home, work, school, play or on the highway.
*Founded: 1949*
*John Keeler, President*
*Margaret Cather, Executive Director*

**635 Arizona Solar Energy Industries Association**
Solar Energy Industries Association
3008N Civic Center Plaza
Scottsdale, AZ 85251
602-253-8180
Fax: 602-258-3422
E-mail: Solar-guy@msn.com
http://www.arizonasolarindustry.org
A non-profit trade association representing local, national and international solar companies in the Arizona market.
*Michael Neary, Executive Director*

**636 Arizona Water Well Association**
1030 E Baseline Road
#105-1025
Tempe, AZ 85283
480-609-3999
Fax: 480-609-3939
E-mail: admin@azwwa.org
http://www.azwwa.org
Promotes protection and wise development of underground water resources.
*Founded: 1957*
*Gary Hix, President*
*Larry Coffelt, Secretary/Treasurer*

**637 Arizona-Sonora Desert Museum**
2021 N Kinney Road
Tucson, AZ 85743
520-883-2702
Fax: 520-883-2500
E-mail: info@desertmuseum.org
http://www.desertmuseum.org
A museum, zoo and botanical garden center. The Sonora Desert of New Mexico and North America is the primary focus.
*Founded: 1952*
*Patricia Engels, Chair*
*Francis Boyle, Secretary*

**638 Bureau of Land Management**
222 W 7th Avenue
#13
Anchorage, AK 99513
907-271-5960
Fax: 907-271-3684
E-mail: ak_akso_public_room@blm.gov
http://www.blm.gov
A nonprofit scientific and educational organization that serves professionals such as government agencies, academia, industry, and non-government organizations in all areas related to the conservation of wildlife and natural resourcesmanagement.
*Bud Cribley, State Director*
*Julia Dougan, Associate State Director*

**639 Center for Biological Diversity**
PO Box 710
Tucson, AZ 85702-0710
866-357-3349
Fax: 520-623-9797
http://www.biologicaldiversity.org
Works through science, law and creative media to secure a future for all specied, great or small, hovering on the brink of extinction.
*Kieran Suckling, Executive Director*
*Linda Wells, Finance Director*

**640 Sierra Club**
202 E McDowell Road
Suite 277
Phoenix, AZ 85004
602-253-8633
E-mail: sierraclubinfo@gmail.com
http://www.arizona.sierraclub.org
The Grand Canyon Chapter is made up of six groups that utilize conservation, political and legislative activism, administrative work, and outings to protect the state.
*Jim Vaaler, Chair*
*Sandy Bahr, Outreach Director*

## Arkansas

**641 American Society of Landscape Architects: Arkansas Chapter**
Dept of Landscape Architecture
230 Memorial Hall
Fayetteville, AR 72701
202-898-2444
800-999-2752
Fax: 202-898-1185
E-mail: fbeatty@uark.edu
http://www.asla.org

Founded in 1899, the American Society of Landscape Architects is the national professional association representing landscape architects.
*Founded: 1899*
*Fran Beatty, President*
*Travis G Brooks, President-Elect*

**642 Arkansas Association of Conservation Districts**
101 E Capitol
Suite 350
Little Rock, AR 72201
501-682-2915
Fax: 501-682-3991
E-mail: Randy.Young@mail.state.ar.us
http://www.aracd.org
Affiliated with the National Association of Conservation Districts. Membership is $25.00 per year for individuals and $1,200.00 per year for organizations and companies.
*Founded: 1937*
*Troy Odom, President*
*Charles Glover, First Vice President*

**643 Arkansas Environmental Federation**
1400 W Markham Street
Suite 302
Little Rock, AR 72201
501-374-0263
Fax: 501-374-8752
http://www.environmentark.org
Formerly the Arkansas Federation of Water and Air Users. Membership is $25.00 per year for individuals and $200.00 to $2,070.02 per year for organizations.
*Founded: 1967*
*Randy Thurman, Executive Director*
*Elyse Cullen, Communication Director*

**644 Arkansas Respiratory Health Association**
211 Natural Resources Drive
Little Rock, AR 72205
501-224-5864
Fax: 501-224-5645
E-mail: lcollers@lungark.org
http://www.lungusa.org
Mission of the Arkansas Respiratory Health Association is to prevent lung disease and promote lung health through education, advocacy and research.

**645 Northern Arkansas: Safety Council of the Ozarks**
1111 South Glenstone
Springfield, MO 65804
417-869-2121
800-334-1349
Fax: 417-869-2133
E-mail: dbiggs@nscoazarks.org
http://www.nscozarks.org
Dedicated to interpreting current safety & health issues and developing practical, cost effective, methodologies based on best practices for protecting human life, property and the environment.
*Debora S Biggs, Executive Director*
*Jane Whillock, Office Manager*

**646 Sierra Club**
1308 W 2nd Street
Little Rock, AR 72201
501-301-8280
E-mail: sierraclubinfo@gmail.com
http://arkansas.sierraclub.org
To practice and promote responsible use of the earth's ecosystems and resources; educate and enlist humanity to protect and restore the quality of the natural environment.
*Rob Leflar, Chair*
*David Lyon, Vice Chair*

## California

**647 American Fisheries Society:Fish Health Section**
California State University
California State University
Department of Biological Sciences
Hayward, CA 94542
510-885-3471
Fax: 510-885-4747
E-mail: eric.cheatham@csueastbay.com
http://www.csuhayward.edu
Goals and missions are to maintain an association of persons involved in safe-guarding the health of fish and other aquatic animals.
*Michael Kent, President*

**648 American Lung Association of California**
American Lung Association
424 Pendleton Way
Oakland, CA 94621
510-638-5864
Fax: 510-638-8984
E-mail: rmcginnis@alac.org
http://www.californialung.org
Fights lung disease through education, community service, advocacy and research.
*Founded: 1904*
*Janet Warner, Chief Executive Officer*
*Ben Abate, President*

**649 American Lung Association of California: Redwood Empire Branch**
115 Talbot Avenue
Santa Rosa, CA 95404
707-527-5864
Fax: 707-542-6111
E-mail: tjohnson@ala.org
http://www.californialung.org
Fighting lung disease through education, community service, advocacy and research.
*Founded: 1953*
*Karen Fulton, Vice President*
*Terrie Johnson, Manager*

**650 American Lung Association of California: East Bay Branch**
American Lung Association
1900 Powell Street
Suite 800
Emeryville, CA 94608
510-893-5474
Fax: 510-893-9008
E-mail: eastbaylung@alaebay.org
http://www.alaebay.org
The American Lung Association of the East Bay, serves the Greater Bay Area. Serves the people that live and work in the 7 Bay Area counties.
*Founded: 1904*
*karen Fulton, President*
*Anita Lee, Vice President*

**651 American Lung Association of California: Superior Branch**
10 Landing Circle
Suite 1
Chico, CA 95973
530-345-5864
800-586-4872
Fax: 530-345-6035
E-mail: sbrantley@alac.org
http://www.lungusa2.org
The American Lung Association of California and its offices statewide work to prevent lung disease and promote lung health.
*Founded: 1904*
*Stephen R O'Kane, Chairman*
*Don Brunson, Vice Chairman*

**652 American Lung Association of Central California**
American Lung Association
4948 North Arthur
Fresno, CA 93705
559-222-4800
800-586-4872
Fax: 559-221-2081
E-mail: myang@alac.org
http://www.lungusa.org/associations/states/california/
The American Lung Association of California and its offices statewide work to prevent lung disease and to promote lung health.
*Founded: 1904*
*Michael Peterson MD, Chairman of the Board*
*Josette Merce Bello, CEO*

**653 American Lung Association of Sacramento: Emigrant Trails**
921 11th Street
Suite 700
Sacramento, CA 95814
916-442-4446
800-586-4872
Fax: 916-442-8585
E-mail: pknepprath@alac.org
http://www.californialung.org
The American Lung Association of California and its offices statewide work to prevent lung disease and promote lung health.
*Founded: 1917*
*Earl Wisthycombe, President*
*Jane Hagerdorn, CFO*

**654 American Lung Association of San Diego & Imperial Counties**
American Lung Association
2750 Fourth Avenue
San Diego, CA 92103
619-297-3901
800-586-4872
Fax: 619-297-8402
E-mail: info@lungsandiego.org
http://www.lungsandiego.org

The mission of the American Lung Association is to prevent lung disease and promote lung health.
*Founded: 1904*
*Janie Davis, Chief Executive Officer*
*Yolanda Orres, Manager*

**655 American Lung Association of Santa Barbara and Ventura Counties**
American Lung Association
1510 San Andres Street
Santa Barbara, CA 93101
805-963-1426
800-586-4872
Fax: 805-456-0892
E-mail: jayne@lungsbvc.org
http://www.californialung.org

The American Lung Association of California and its offices statewide work to prevent lung disease and to promote lung health.
*Founded: 1904*
*John Kirkwood, President*
*Charles A Heinrich, Chairman*

**656 American Lung Association of the Central Coast**
550 Camino El Estero
Suite 100
Monterey, CA 93940
831-373-7306
800-586-4872
Fax: 831-373-5530
E-mail: admin@alaccoast.org
http://www.californialung.org

The American Lung Association of California and its offices statewide work to prevent lung disease and promote lung health.
*Founded: 1953*
*Karen Fulton, Chief Executive Officer*
*John Morrison, Vice President*

**657 American Rivers: California Region:Fairfax**
2150 Allston Way
Suite 320
Berkeley, CA 94704
510-809-8010
E-mail: outreach@americanrivers.org
http://www.americanrivers.org

American Rivers is the only national organization standing up for healthy rivers so our communities can thrive.
*Founded: 1973*
*Rebecca Wodder, President*
*Fay Augustyn, Conservation Associate*

**658 American Rivers: California Region:Nevada City**
PO Box 559
Nevada City, CA 95959
530-478-5672
Fax: 530-478-5849
E-mail: outreach@americanrivers.org
http://www.americanrivers.org

The conservation organization standing up for healthy rivers so communities can thrive. American Rivers protects and restores America's rivers for the benefit of people, wildlife and nature.
*Founded: 1973 65,000 Members*
*Rebecca Wodder, President*
*Fay Augustyn, Conservation Associate*

**659 American Society of Landscape Architects: Northern California Chapter**
5 Third Street
Suite 724
San Francisco, CA 94103
415-974-5430
Fax: 415-543-2112
E-mail: staff@asla-ncc.admn.org
http://host.asla.org/chapters/norcal

Founded in 1899, the American Society of Landscape Architects is the national professional association representing landscape architects.
*Founded: 1899*
*Jeffrey George, President*
*Joe Owen, Executive Director*

**660 American Society of Landscape Architects: San Diego Chapter**
PO Box 81521
San Diego, CA 92138
619-225-8155
Fax: 619-225-8151
E-mail: aslasd@sbcglobal.net
http://www.asla-sandiego.org

Founded in 1899, the American Society of Landscape Architects is the national professional association representing landscape architects.
*Tracy Morgan Hollingworth, Executive Director*

**661 American Society of Landscape Architects: Sierra Chapter**
PO Box 188232
Sacramento, CA 95818
916-447-7400
Fax: 916-447-8270
E-mail: asla-sierra@sbcglobal.net
http://www.asla-sierra.org

Founded in 1899, the American Society of Landscape Architects is the national professional association representing landscape architects.
*Mike Heacox, President*
*Kenneth Tatarka, President-Elect*

**662 American Society of Landscape Architects: Southern California Chapter**
1100 Irvine Boulevard
Suite 371
Tustin, CA 92780
714-838-3615
Fax: 714-730-6296
E-mail: sccasla@aol.com
http://host.asla.org/chapters/southca

Founded in 1899, the American Society of Landscape Architects is the national professional association representing landscape architects.
*Andrew Bowden, President-Elect*
*Vicki Phillipy, Executive Director*

**663 Asian Pacific Environmental Network**
310 8th Street
Suite 309
Oakland, CA 94607
510-834-8920
Fax: 510-834-8926
E-mail: apen@apen4ej.org
http://www.apen4ej.org

The Asian Pacific Environmental Network (APEN) empowers low-income Asian Pacific Islander (API) communities to take action on environmental and social justice issues. APEN builds organizations in dis-empowered API communities todevelop lasting capacity of the community to achieve solutions to problems affecting people's lives.
*Founded: 1991*
*Vivian Chang, Executive Director*
*Lisa DeCastro, Development Associate*

**664 Bio Integral Resource Center**
PO Box 7414
Berkeley, CA 94707
510-524-2567
Fax: 510-524-1758
E-mail: birc@igc.org
http://www.birc.org

The goal of the Bio Integral Resources Center is to reduce pesticide use by educating the public about effective, less toxic alternatives for pest problems.
*Founded: 1979*
*Dr. William Quarles, Executive Director*

**665 Breathe California Golden Gate Public Health Partnership**
2171 Junipero Serra Boulevard
Suite 720
Daly City, CA 94014
650-994-5864
Fax: 650-994-4601
E-mail: info@ggbreathe.org
http://www.ggbreathe.org

Through grassroots education, advocacy and services, Breathe California fights lung disease, advocates for clean air, and advances public health.
*Founded: 1908*
*Linda Civitello-Joy, President/CEO*
*Elis Treveno, Vice President Operations*

**666 Breathe California Of Los Angeles County**
5858 Wilshire Boulevard
Suite 300
Los Angeles, CA 90036
323-935-8050
Fax: 323-935-1873
http://www.breathela.org

Formerly the American Lung Association of Los Angeles County, is dedicated to providing communities in Los Angeles County with the resources needed to promote clean air initiatives through its continued commitment to researcheducation, and advocacy.
*Founded: 1903*
*Enrique M Chiock MD, President/CEO*
*Dan Witzling, Vice President*

**667 Breathe California of the Bay Area: Alameda, Santa Clara & San Benito Counties**
1469 Park Avenue
San Jose, CA 95126
408-998-5865
Fax: 408-998-0578
E-mail: info@lungsrus.org
http://www.lungsrus.org

Fights lung disease in all its forms and works with its communities to promote lung health.
*Founded: 1904*
*Margo Leathers Sidener, Executive Director*
*Steve French, Development Director*

**668 California Academy of Sciences Library**
55 Music Concourse Drive
Golden Gate Park
San Francisco, CA 94118
415-321-8000
Fax: 415-321-8633
E-mail: info@calacademy.org
http://www.calacademy.org

Founded in 1853 to survey and study the vast resources of California and beyond, the California Academy of Sciences is the oldest scientific institution in the West. The Academy has grown to become the 4th largest natural historymuseum in the country. The academy's mission is to explore, explain, and protect the natural world.
*Founded: 1853*
*Patrick Kociolek, Executive Director*
*Allison Brown, CFO*

**669 California Air Resources Board**
State of California
PO Box 2815
Sacremento, CA 95812
916-322-2990
800-242-4450
Fax: 916-445-5025
E-mail: helpline@arb.ca.gov
http://www.arb.ca.gov

The California ARB is the state agency charged with coordinating efforts to attain and maintain ambient air quality standards, conduct research into the causes of and solutions to air pollution and its adverse health impacts, andattack systematically the serious problem caused by motor vehicles, which are the major source of air pollution in many areas of the state. The California ARB's mission is to promote and protect public health, welfare and ecological resources.
*Founded: 1967*
*Mary Nichols, Chairman*
*James Goldstene, Executive Officer*

**670 California Association of Environmental Health**
3700 Chaney Court
Carmichael, CA 95608
530-676-0715
Fax: 530-676-0515
E-mail: justin@ccdeh.com
http://www.ccdeh.com

Environmental Health Departments provide the delivery of Local Environmental Health Programs.
*Justin Malan, Executive Director*
*Sheryl Baldwin, CCDEH Manager*

**671 California Association of Resource Conservation Districts**
801 K Street
Suite 1318
Sacramento, CA 95814
916-457-7904
Fax: 916-457-7934
E-mail: info@carcd.org
http://www.carcd.org

CARCD is a voluntary association whose primary purpose is to provide a unified means for California Resource Conservation Districts (RCDs) to meet major conservation goals.
*Karen Buhr, Executive Director*
*Ron Harben, Project Manager*

**672 California BASS Chapter Federation**
13350 Racquet Court
Poway, CA 92064
858-748-9459
E-mail: info@californiabass.org
http://www.californiabass.org

The Bass Federation Conservation Dept is dedicated to the preservation, restoration, enhancement and public access to their resources.
*Founded: 1969*
*Larry Wilson, President*
*George Fedor, Conservation Director*

**673 California Birth Defects Monitoring Program**
1947 Center Street
Berkeley, CA 94704
510-981-5300
Fax: 510-981-5395
E-mail: info@cbdmp.org
http://www.cbdmp.org

The California Birth Defects Monitoring Program is a public health program devoted to finding the causes of birth defects so they can be prevented. The program is funded through the California Department of Health Services and isjointly operated with the March of Dimes Birth Defects Foundation.
*Founded: 1982*
*Janet Berreman, Health Officer*
*Kate Clayton, Health Promotion, Chief*

**674 California Certified Organic Farmers**
2155 Delaware Avenue
Suite 150
Santa Cruz, CA 95060
831-423-2263
Fax: 831-423-4528
E-mail: ccof@ccof.org
http://www.ccof.org

Premier organic certification since 1973, for growers, processors, handlers, packers, retailers. Public education of organic, and advocacy for public policy to support organic. Newsletter subscriptions available through a supportingmembership program.
*Founded: 1973*
*Jake Lewin, Chief Certification Officer*
*Peggy Miars, Executive Director*

**675 California Council for Environmental and Economic Balance**
100 Spear Street
Suite 805
San Francisco, CA 94105
415-512-7890
Fax: 415-512-7897
E-mail: cceeb@cceeb.org
http://www.cceeb.org

A coalition of California business, labor and public leaders which strive to advance collaborative strategies for a sound California.
*Founded: 1976*
*Victor Weisser, President*
*Bill Quinn, Executive Director*

**676 California Renewable Fuels Council**
1516 Ninth Street
MS-29
Sacramento, CA 95814
916-654-4058
800-555-7794
E-mail: renewable@energy.state.ca.us
http://www.energy.ca.gov/renewables

Lobbying organization
*Founded: 2002*
*Robert Weisenmiller, Chairman*
*James Boyd, Vice Chair*

**677 California Solar Energy Industries Association**
P.O. BOX 782
Rio Vista, CA 94571
916-747-6987
Fax: 707-374-4767
E-mail: info@calseia.or
http://www.calseia.org

Promotes the growth of California's solar energy industry. Membership is $325-650 per year.
*Founded: 1977*
*Les Nelson, President*
*Pat Redgate, Vice President*

**678 California Trappers Association (CTA)**
907 Holmes Flat Road 707-222-4259
Attn: Rita Clark, Membership Secretary 888-457-2873
Redcrest, CA 95569 E-mail: chris.adamski@fwmedia.com
http://www.trapperpredatorcaller.com
CTA's primary concern is California's fur bearing mammals, and the category of small predatory mammals. Approximately 20 of the association's 2,400 members do lengthy studies and assist State and Federal Agencies in the field withtheir biologists. Educational forums and talks are hosted at local universities. Farming agencies and timber companies commonly use their services. Members commonly work for numerous environmental public projects, both public and governmental.
*Founded: 1863*
*John Clark, President*
*James C Schmerker Jr, Vice President*

**679 California Waterfowl Association**
4630 Northgate Boulevard 916-648-1406
Suite 150 Fax: 916-648-1665
Sacramento, CA 95834 E-mail: cwa_hq@calwaterfowl.org
http://www.calwaterfowl.org
A statewide nonprofit organization whose principal objectives are the preservation, protection, and enhancement of California's waterfowl resources, wetlands, and associated hunting heritage.
*Founded: 1945*
*Jake Messerli, Vice President, Conservation*
*Dan Loughman, Waterfowl Projects Superviso*

**680 California Wildlife Foundation**
428 13th Street 510-208-4436
10th Floor, Suite A Fax: 510-268-9948
Oakland, CA 94612 E-mail: info@californiawildlifefoundation.org
http://www.californiawildlifefoundation.org
A nonprofit organization dedicated to the conservation, enhancement, scientific management and wise use of all our natural resources. It seeks to accomplish these objectives through education, scientific research, charitable donationsand support projects and other organizations having compatible purposes.
*Janet Cobb, Principal*
*Amy Larson, Marketing and Project Manage*

**681 Californians for Population Stabilization (CAPS)**
1129 State Street 805-564-6626
Suite 3-D Fax: 805-564-6636
Santa Barbara, CA 93101 E-mail: info@capsweb.org
http://www.capsweb.org
A nonprofit, public interest organization that works to protect California's environment and quality of life by turning the tide of population growth.
*Founded: 1986*
*Diana Hull PhD, President*
*Ben Zuckerman, Vice President*

**682 Colorado River Board of California**
770 Fairmont Avenue 818-500-1625
Suite 100 Fax: 818-543-4685
Glendale, CA 91203 E-mail: crb@crb.ca.gov
http://www.crb.ca.gov
Established to protect California's rights and interests in the resources provided by the Colorado River and to represent California in discussions and negotiations regarding the Colorado River and its management.
*Founded: 1937*
*Christopher Harris, Executive Director*

**683 Communities for a Better Environment**
1940 Franklin Street 510-302-0430
Suite 600 Fax: 510-302-0437
Oakland, CA 94612 E-mail: cbeca@mail.com
http://www.cbecal.org
Nonprofit statewide, multiracial, urban environmental health and justice organization that works with urban communities and grassroots organizations, using science based research, legal tactics and organizing strategies to prevent airand water pollution, eliminate toxic hazards and improve public health. Long-term goals are to develop an environmentally sustainable manufacturing base, minimize the use of toxins, expand pollution prevention strategies and involve people most atrisk.
*Alina Bokde, Deputy Executive Director*
*Adrienne Bloch, Attorney*

**684 Concerned Citizens of South Central Los Angeles**
4111 S Central Avenue 213-846-2505
Suite 101 Fax: 213-846-2508
Los Angelos, CA 90011 E-mail: info@ccscla.org
http://www.ccscla.org/
The mission of the Concerned Citizens of South Central Los Angeles is to fight for social, economic and environmental justice and to encourage resident participation in the process.
*Founded: 1985*
*Robin Cannon, President*
*Noreen McClendon, Vice President/Exectuive Dir*

**685 Desert Tortoise Preserve Committee**
Tortoise T-R-A-C-K-S
4067 Mission Inn Avenue 909-683-3872
Riverside, CA 92501 Fax: 909-683-6949
E-mail: dtpc@pacbell.net
http://www.tortise-tracks.org
A non-profit organization formed to promote the welfare of the desert tortoise in its native wild state.
*Mary Kotschwar, Preserve Manager*

**686 Earth Island Institute**
2150 Allston Way 510-859-9100
Suite 460 Fax: 510-859-9091
Berkeley, CA 94704-1375 http://www.earthisland.org
A non-profit, public interest, membership organization that supports people who are creating solutions to protect our shared planet.
*John Knox, Executive Director*
*Dave Phillips, Executive Director*

**687 Ecology Center**
2530 San Pablo Avenue 510-548-2220
Berkeley, CA 94702 Fax: 510-548-2240
E-mail: info@ecologycenter.org
http://www.ecologycenter.org
The Ecology Center promotes environmentally and socially responsible practices through programs that educate, demonstrate and provide direct services. The Ecology Center currently runs Ecology Center Bookstore, 3 Berkeley farmers'markets, a curbside recycling program and an Environmental Resource Center/Library.
*Founded: 1969*
*Martin Bourque, Executive Director*
*Martin Bourque, Executive Director*

**688 Environmental Defense Center**
906 Garden Street 805-708-0127
Santa Barbara, CA 93101 Fax: 805-962-3152
E-mail: edc@rain.org
http://www.rain.org
The Environmental Defense Center is a nonprofit, public interest organization that provides legal, educational and advocacy support to advance environmental quality. EDC primarily serves community groups on California's South CentralCoast. EDC selects cases and projects that preserve the environment for future generations, protect human health, promote appropriate management and use of natural resources, and enhance the character of the community.
*Linda Krop, Executive Director*

**689 Environmental Health Coalition**
2727 Hoover Avenue 619-474-0220
Suite 202 Fax: 619-474-1210
National City, CA 91950 E-mail: ehc@environmentalhealth.org
http://www.environmentalhealth.org
One of the oldest and most effective grassroots organizations in the US, using social change strategies to achieve environmental and social justice. We believe that justice is accomplished by empowered communities acting together tomake social change. We organize and advocate to protect public health and the environment threatened by toxic pollution. EHC supports broad efforts

that create a just society which foster a healthy and sustainable quality of life.
*Founded: 1985*
*Diane Takvorian, Executive Director*
*Lilia Escalante, Fiscal Manager*

**690 Environmental Health Network**
PO Box 1155 415-541-5075
Larkspur, CA 94977 E-mail: fdadockets@oc.fda.gov
http://www.ehnca.org
Nonprofit, volunteer organization, whose main goal is to promote public awareness of environmental sensitivities and causative factors. EHN's focus is on issues of access and developments relating to the health and welfare of theenvironmentally sensitive.
*Barb Wilkie, President*

**691 Friends of the River**
1418 20th Street 916-442-3155
Suite 100 888-464-2477
Sacramento, CA 95811 Fax: 916-442-3396
E-mail: info@friendsoftheriver.org
http://www.friendsoftheriver.org
Friends of the River is dedicated to preserving and restoring California's rivers, streams, and their watersheds as well as advocating for sustainable water management.
*Founded: 1973*
*Bob Cushman, President*
*Mark DuBois, Director Emeritus*

**692 Institute of International Education**
Institute of International Education, New Yourk
530 Bush Street, Suite 1000 415-362-6520
San Francisco, CA 94108 Fax: 415-392-4667
E-mail: iiesf@iie.org
http://www.iie.org
An independent non-profit organization that is a world leader in the exchange of people and ideas.
*Founded: 1919*
*Allen Goodman, President*
*Tricia Tierney, Director*

**693 Laotian Organizing Project**
220 25th Street 510-236-4616
Richmond, CA 94804 Fax: 510-236-4572
E-mail: apen@apen4ej.org
http://www.apen4ej.org/organize_lop.htm
The Laotian Organization Project (LOP), a project of the Asian Pacific Environmental Network, is a membership-based organization of Laotian residents in Richmond and San Pablo, California. LOP works to bring people from the Laotiancommunity together to identify problems, develop solutions, and take action for a more healthy, safe, and just community.
*Amber Chan, PAO Lead Organizer*
*Roger Kim, Executive Director*

**694 Lead Safe California**
100 Pine Street 415-397-9401
26th Floor Fax: 415-397-5159
San Francisco, CA 94111 E-mail: cehn@cehn.org
http://www.cehn.org
Lead Safe California is a nonprofit, public interest organization dedicated to the prevention of childhood lead poisoning and the preservation of affordable, safe housing.
*Ellen Widess, Executive Director*

**695 League to Save Lake Tahoe**
2608 Lake Tahoe Road 530-541-5388
South Lake Tahoe, CA 96150 Fax: 530-541-5454
E-mail: info@keeptahoeblue.org
http://www.keeptahoeblue.org
Our mission is the preservation and restoration of the magnificent natural attributes of the Tahoe Basin's waters, forests, wildlife and landscape for the enjoyment of present and future generations.
*Founded: 1957*
*Rochelle Nason, Executive Director*
*Carl Young, Program Director*

**696 Marine Mammal Center**
2000 Bunker Road 415-289-7325
Fort Cronkite Fax: 415-289-7333
Sausalito, CA 94965-2619 http://www.marinemammalcenter.org
A nonprofit veterinary research hospital and educational cetner dedicated to the rescue and rehabilitation of ill and injured marine mammals, primarily elephant seals, harbor seals, and California sea lions.
*Founded: 1975*
*Dr Jeff Boehm, Executive Director*
*Marci Davis, CFO/COO*

**697 Mountain Lion Foundation**
PO Box 1896 916-442-2666
Sacramento, CA 95812 800-319-7621
Fax: 916-442-2871
E-mail: mlf@mountainlion.org
http://www.mountainlion.org
A national nonprofit conservation and education organization dedicated to protecting the mountain lion, its wild habitat, and the wildlife that shares that habitat-for present and future generations. The Foundation is dedicated to theproposition that much can be done to preserve the cougar as a viable species of this effort can assure the survival of other species.
*Founded: 1986*
*Tim Dunbar, Executive Director*
*Fred Hall, Operations Manager*

**698 National Institute for Global Environmental Change: Western Regional Center**
University of Califonia-Davis
1 Shields Avenue 530-752-7300
Davis, CA 95616 Fax: 530-752-7302
E-mail: westgec@ucdavis.edu
http://www.ucdavis.edu/
This overview provides insight into the major environmental programs and permitting requirements governing industrial processes and activities.
*Founded: 1990*
*Dr. Susin Ustin, Director*

**699 National Safety Council: California Chapter**
4553 Glencoe Avenue 310-827-9781
Suite 150 800-421-9585
Marina Del Rey, CA 90292 Fax: 310-827-9861
E-mail: california@nsc.org
http://www.nsc.org
Mission is to educate and influence people to prevent accidental injury and death. Vision is to make our world safer.
*Janet Froetscher, President*
*Paulette Moulos, Chief Operating Officer*

**700 Nature Conservancy: Western Division Office**
201 Mission Street 415-777-0487
4th Floor Fax: 415-777-0244
San Francisco, CA 94105 E-mail: comment@tnc.org
http://www.tnc.org
A leading conservation organization working around the world to protect ecologically important lands and waters for nature and people.
*Mark Burgett, President*
*Mick Sweeney, CFO*

**701 NorCal Solar/Northern California Solar Energy Society**
PO Box 3008 510-705-8813
Berkeley, CA 94703 Fax: 510-548-8896
E-mail: info@norcalsolar.org
http://www.norcalsolar.org
A non-profit educational organization whose mission is to accelerate the use of solar energy technology through the exchange of information.
*Founded: 1975*
*Claudia Wentworth, President*
*Erin Middleton, Office Manager*

**702 Northcoast Environmental Center**
575 H Street
Arcata, CA 95521
707-822-6918
Fax: 707-822-0827
E-mail: nec@yournec.org
http://www.yournec.org

A nonprofit educational organization devoted to illuminating people concerning the Biosphere. It has a library, information and referral, radio show, nationally circulated newsletter and national membership. It is focused on theredwood region and northwest California and the Bioregion along the California-Oregon border.
*Founded: 1971*
*Tim McKay, Executive Director*
*Sid Dominitz, Editor*

**703 Outdoor Programs**
University of California San Francisco
500 Parnassus Avenue
PO Box 0234-A
San Francisco, CA 94143
415-476-2078
Fax: 502-620-7415
E-mail: outdoors@cls.ucsf.edu
http://www.outdoors.ucsf.edu

Creates outdoor experiences for students, staff and families in UCSF and the local community.
*Founded: 1970*
*Kirk McLaughlin, Parnassus Programs Superviso*
*Colleen Massey, Mission Bay Programs Supervi*

**704 Pesticide Action Network North America**
49 Powell Street
Suite 500
San Francisco, CA 94102
415-981-1771
Fax: 415-981-1991
E-mail: panna@panna.org
http://www.panna.org

Pesticide Action Network North American advocates the adoption of ecologically sound practices in place of hazardous pesticide use. PANNA works with more than 100 affiliated organizations in Canada, Mexico and US, as well as withPesticide Action Network partners around the world to demand that development agencies and governments redirect support from pesticides to safe alternatives.
*Founded: 1984*
*Kathryn Gilje, Executive Director*
*Stephen Scholl Buckwald, Managing Director*

**705 Pesticide Education Center**
PO Box 225279
San Francisco, CA 94122
415-665-4722
Fax: 415-665-2693
E-mail: pec@igc.org
http://www.pesticides.org

The mission of the Pesticide Education Center is to educate the public about the adverse health effects of exposure to pesticides in the home, community and at work.
*Founded: 1988*
*Marion Moses, President*

**706 Rails-to-Trails Conservancy**
Western Regional Office
235 Montgomery Street
San Francisco, CA 94104
415-814-1100
Fax: 415-989-1255
http://www.railstotrails.org

A nonprofit organization whose mission it is to create a nationwide network of trails from former rail lines and connecting corridors to build healthier places for healthier people.
*Laura Cohen, Regional Director*
*Steve Schweigerdt, Manager Trail Development*

**707 Rainforest Action Network**
221 Pine Street
5th Floor
San Francisco, CA 94104
415-398-4404
Fax: 415-398-2732
E-mail: answers@ran.org
http://www.ran.org

Rainforest Action Network works to protect the Earth's rainforest and support the rights of their inhabitants through education, grassroots organizing, and non-violent direct action.
*Founded: 1985*
*Amanda Starbuck, Energy and Finance Program D*
*Annie Sartor, Global Finance Campaigner*

**708 Redwood Empire Solar Living Association**
c/o Solar Living Institute
13771 S Highway 101
PO Box 836
Hopland, CA 95449
707-744-2017
E-mail: sli@solarliving.org
http://www.solarliving.org

The Solar Living Institute joined the ASES (American Solar Energy Society) as the Redwood Empire Solar Living Association (RESLA). RESLA includes these seven counties in northwest CA: Del Norte, Humboldt, Lake, Mendocino, Siskiyou,Sonoma, Trinity. The Institute promotes sustainable living through environmental education.
*John Schaeffer, Chairman/Founder*
*Bob Gragson, Executive Director*

**709 San Diego Renewable Energy Society (SDRES)**
PO Box 23490
San Diego, CA 92123-3490
E-mail: info@sdres.org
http://sdres.org

A regional chapter of the American Solar Energy Society and membership nonprofit organization dedicated to increasing intelligent use of renewable and sustainable energy technologies in San Diego County. Members are encouraged todevelop and create new technology and applications for alternate energy.
*Bruce Rogow, Chairman*
*Dan Gibbs, Vice Chair*

**710 Save San Francisco Bay Association**
350 Frank Ogawa Plaza
Suite 900
Oakland, CA 94612
510-452-9261
Fax: 510-452-9266
E-mail: savebay@savesfbay.org
http://www.savesfbay.org

Save the Bay has worked for over 40 years to protect the San Francisco Bay-Delta from pollution, fill, shoreline destruction and fresh water diversion. We have launched a century of renewal to restore bay fish and wildlife, reclaimtidal wetlands and make the bay safe and accessible to all.
*Founded: 1961*
*David Lewis, Executive Director*
*Robin Erickson, Director Finance/Administrat*

**711 Scenic California**
2215 5th Street
Berkeley, CA 94710
510-883-0390
Fax: 510-883-0391
E-mail: sceniccalifornia@lsa-assoc.com
http://www.sceniccalifornia.org

Scenic America is the only national nonprofit organization dedicated to protecting natural beauty and distinctive community character. We provide technical assistance across the nation and through affiliates on scenic byways, billboardand sign control, context sensitive highway design, wireless telecommunications tower location, transportation enhancements, and other scenic conservation issues.
*Sheila Brady, Manager*

**712 Sierra Club: Angeles Chapter**
3435 Wilshire Blvd
Suite 320
Los Angeles, CA 90010
213-387-4287
Fax: 213-387-5383
E-mail: contact.us@angeles.sierraclub.org
http://www.angeles.sierraclub.org

The Angeles Chapter has 58,000 members located in Los Angeles and Orange counties.
*Founded: 1920*
*Mike Sappingfield, Chair*
*Ron Silverman, Senior Director*

**713 Sierra Club: Kern Kaweah Chapter**
PO Box 3357
Bakersfield, CA 93385
661-323-5569
E-mail: chair@kernkaweha.sierraclub.org
http://www.kernkaweah.sierraclub.org

To advance the preservation and protection of the natural environment by empowering the citizenry, especially democratically-based grassroots organizations, with charitable resources to further the cause of environmental protection.
*Arthur Unger, Chair*
*Ara Marderosian, Conservation*

**714 Sierra Club: Loma Prieta Chapter**
3921 East Bayshore Road 650-390-8411
Suite 204 Fax: 650-390-8497
Palo Alto, CA 94303 E-mail: melissa.hippard@sierraclub.org
http://www.lomaprieta.sierraclub.org
One of the largest chapters of the Sierra Club.
*Founded: 1933*
*Melissa Hippard, Director*
*Kristen Ohiaeri, Chapter Coordinator*

**715 Sierra Club: Los Padres Chapter**
PO Box 31241 805-965-9719
Santa Barbara, CA 93130 E-mail: motodata@adelphia.net
http://www.lospadres.sierraclub.org
The Chapter represents the members in Ventura and Santa Barbara counties in Southern California.
*Mike Stubblefield, Chair*
*Trevor Smith, Vice Chair*

**716 Sierra Club: Mother Lode Chapter**
801 K Street 916-557-1100
Suite 2700 Fax: 916-557-9669
Sacramento, CA 95814 E-mail: info@mlc.sierraclub.org
http://motherlode.sierraclub.org
The 20,000 members of this chapter come from a large region stretching from Stanislaus County to the Oregon border to the north, the California/Nevada border in the east and to the west of the Central Valley.
*Barbara Williams, Chair*

**717 Sierra Club: Redwood Chapter**
55A Ridgeway Avenue 707-544-7651
PO Box 466 Fax: 707-544-9861
Santa Rosa, CA 95402 E-mail: penningt@sonic.net
http://www.redwood.sierraclub.org
The Club's purpose is to protect and restore wild places, public health and wildlife for future generations.
*Margaret Pennington, Chair*
*Jay Holcomb, Vice Chair/Forest Protection*

**718 Sierra Club: San Diego Chapter**
3820 Ray Street 619-299-1743
San Diego, CA 92104 Fax: 619-299-1742
E-mail: san-diego.chapter@sierraclub.org
http://www.sandiego.sierraclub.org
A nonprofit organization dedicated to preserving, protecting and enjoying the earth.
*Cheryl Reiff, Chapter Coordinator*
*Martha Bertles, Office*

**719 Sierra Club: San Francisco Bay Chapter**
2530 San Pablo Avenue 510-848-0800
Suite I Fax: 510-848-3383
Berkeley, CA 94702 E-mail: info@sfbaysc.org
http://www.sanfranciscobay.sierraclub.org
Represents over 30,000 members from Alameda, Contra Costa, Marin and SF. Nonprofit, member-supported, public interest organization that promotes enjoyment and preservation of the national and local forests, waters, wildlife andwilderness. In addition to a wide variety of environmental and conservation interests and activities, it is active in the areas of pollution prevention and climate change.
*Founded: 1924*
*Michael Bornstein, Senior Director*
*William Walsh, Development Director*

**720 Sierra Club: San Gorgonio Chapter**
4079 Mission Inn Avenue 951-684-6203
Riverside, CA 92501 Fax: 951-684-6172
E-mail: ralphsalisbury@charter.net
http://www.sangorgonio.sierraclub.org
To explore, enjoy and protect the wild places of the earth, to practice and promote the responsible use of the earth's ecosystems and resources; to educate and enlist humanity to protect and restore the quality of the natural and humanenvironment; and to use all lawful means to carry out these objectives.
*Ralph Salisbury, Chair*
*Bill Cunningham, Vice Chair*

**721 Sierra Club: Tehipite Chapter**
PO Box 5396 559-229-4031
Fresno, CA 93755 E-mail: tehipite.chapter@sierraclub.org
http://www.tehipite.sierraclub.org
The Sierra Club is dedicated to protecting the quality of life in Fresno County from unwise land development.
*John Rasmussen, Chair*
*Gary Lasky, Vice Chair*

**722 Sierra Club: Ventana Chapter**
PO Box 5667 831-624-8032
Carmel, CA 93921 Fax: 831-624-3371
E-mail: chapter@ventana.sierraclub.org
http://www.ventana.sierraclub.org
To protect, preserve, and restore the wilderness qualities and biodiversity of the public lands within California's northern Santa Lucia Mountains and Big Sur coast.
*Rita Dalessio, Chair*
*Mary Gale, Secretary*

**723 Society for the Conservation of Bighorn Sheep**
PO Box 94182 310-679-2102
Pasadena, CA 91109 E-mail: info@scbs-desertbighorn.com
http://www.desertbighorn.cjb.net/
Mission and ultimate goal is the full restoration of the California Desert Bighorn to its historic habitat and the establishment of self-sustaining populations throughout those ranges.
*Founded: 1964*
*John Nelson, President*

**724 Southern California Chapter:American Solar Energy Society**
PO Box 931090 323-962-2163
Hollywood, CA 90093 Fax: 323-957-1495
E-mail: HBTeam@earthlink.net
The American Solar Energy Society (ASES) is the United States section of the International Solar Energy Society (ASES) a non profit organization dedicated to the development and adoption of renewable energy in all its forms.
*Sharyn Romano, Contact*

**725 Southwestern Herpetologists Society**
PO Box 7469 818-503-2052
Van Nuys, CA 91409 E-mail: maven@post.com
http://www.swhs.org
To further serve members in central California. Open to anyone interested in the study and conservation of reptiles and amphibians.
*Founded: 1954*
*Bud James, President*
*Sabine Bradley Phillips, Vice President*

**726 Trout Unlimited**
1300 North 17th Street 703-522-0200
Suite 500 800-834-2419
Arlington, VA 22209 Fax: 703-284-9400
E-mail: trout@tu.org
http://www.tu.org
Mission: To conserve, protect and restore North America's trout and salmon fisheries and their watersheds. We accomplish this mission on local, state and national levels with an extensive and dedicated volunteer network.
*Founded: 1959*
*Charles Gauvin, President/CEO*
*Steve Moyer, Vice President*

**727 Urban Habitat Program**
Presido Station
436 14th Street 510-839-9510
# 1205 Fax: 510-839-9610
Oakland, CA 94612 E-mail: contact@urbanhabitatprogram.org
http://www.urbanhabitatprogram.org

Dedicated to building a multicultural majority that provides urban environmental leadership in order to create socially just, ecologically sustainable communities in the Bay Area.
*Founded: 1989*
*Carl Anthony, Director*

**728 Western Occupational and Environmental Medical Association**
575 Market Street 415-927-5736
Suite 2125 Fax: 415-927-5726
San Francisco, CA 94105 E-mail: woema@hp-assoc.com
http://www.woema.org
The mission is to represent and be a resource to members in the profession and practice of occupational and environmental medicine and to enhance their efforts to promote and improve health in the workplace.
*Founded: 1941*
*Robert Orford, President*

**729 Wildlife Society: San Joaquin Valley Chapter**
1234 E Shaw Avenue 559-243-4005
Fresno, CA 93710 Fax: 559-243-4022
E-mail: ekleinfe@dfg.ca.gov
http://www.tws-west.org/sjvc
A nonprofit scientific and educational organization that serves professionals such as government agencies, academia, industry and non-government organizations in all areas related to the conservation of wildlife and natural resourcesmanagement.
*Eric Kleinfelter, President*

## Colorado

**730 American Lung Association of Colorado**
American Lung Association
5600 Greenwood Plaza Boulevard 303-388-4327
Suite 100 Fax: 303-377-1102
Greenwood Village, CO 80111 http://www.lungusa.org
The mission of the American Lung Association is to prevent lung disease and to promote lung health.
*Founded: 1904*
*Curt Hubbert, President*

**731 American Society of Landscape Architects: Colorado Chapter**
PO Box 200822 303-830-6616
Denver, CO 80220 Fax: 303-220-5833
E-mail: info@ccasla.org
http://www.ccasla.org
Serves as the state professional society representing landscape architects in Colorado and Wyoming. Promotes the landscape architecture profession and advances the practice through advocacy, education, communication and networking
*Founded: 1973*
*Kim Douglas, President*
*Ian Anderson, President-Elect*

**732 Arkansas River Compact Administration**
109 SW 9th Street 785-296-3710
First Floor Fax: 785-296-1176
Topeka, KS 66612 E-mail: www.ksda.gov/interstate_water_issues/contact/
http://www.ksda.gov/interstate_water_issues/
Protecting Kansas' interests in interstate rivers.
*David Barfield, Chief Engineer & Director*
*Paul Graves, Assistant Chief Engineer*

**733 Aspen Global Change Institute**
100 East Francis Street 970-925-7376
Aspen, CO 81611 Fax: 970-925-7097
E-mail: agcimail@agci.org
http://www.agci.org
A Colorado nonprofit dedicated to furthering the understanding of Earth systems through interdisciplinary science meetings, publications, and educational programs about global environmental change.
*John Katzenberger, Director*
*Sue Bookhout, Director Operations*

**734 Association of Midwest Fish and Game Law Enforcement Officers**
Division of Wildlife
6060 Broadway 303-291-7223
Denver, CO 80216 http://www.midwestgamewarden.org/organization.phtml
Lead group among wildlife enforcement organizations in the development and maintenance of training for field officers that protects the resource and benefits the citizens of our countries, provinces and states.
*Founded: 1944*
*Garry Bogdan, Executive Secretary*

**735 Bison World**
National Bison Association
8690 Wolff Ct. # 200 303-292-2833
Westminster, CO 80234 Fax: 303-659-3739
E-mail: info@bisoncentral.com
http://www.bisoncentral.com
An organization of bison producers dedicated to awareness of the healthy properties of bison meat and bison production.
*Founded: 1978*
*Dave Carter, Executive Director*
*Merle Maas, Chairman of Board*

**736 Colorado Association of Conservation Districts**
743 Horizon Court 303-232-6242
Suite 322 Fax: 303-232-1624
Grand Junction, CO 81506 http://www.cascd.com
Promotes soil and water conservation. Membership is $15.00 per year for individuals and $250.00 per year for organizations.
*12 pages Quarterly*
*Jerry Schwien, Author*
*Callie Hendrickson, Executive Vice President*

**737 Colorado BASS Chapter Federation**
4485 Enchanted Circle North 719-597-2304
Colorado Springs, CO 80917 E-mail: nozlnut36@comcast.net
http://www.coloradobassfederation.org
Main purpose is to stimulate public awareness of bass fishing as a major sport and to offer the Colorado Department of Wildlife and other organizations moral and political support and encouragement.
*John Bentz, President*
*Dave Lishman, Vice President*

**738 Colorado Forestry Association**
1413 Ash 970-221-1336
Ft. Collins, CO 80521 E-mail: billgheerardi@comcast.net
http://www.coloradoforestry.org/
The original mission was to support conservation, management and renewal of forests through forestry legislation, public education and creation of forest reserves.
*Founded: 1884*
*Bill Gherardi, President*
*Ben Wudtke, Secretary*

**739 Colorado Renewable Energy Society**
PO Box 933 303-806-5317
Golden, CO 80402 E-mail: info@cres-energy.org
http://www.cres-energy.org
Nonprofit membership organization that works for the sensible adoption of cost-effective energy efficient and renewable energy technologies by Colorado businesses and consumers.
*Sheila Townsend, Executive Director*

**740 Colorado Safety Association**
4730 Oakland Street 303-373-1937
Suite 500 800-727-0519
Denver, CO 80239 Fax: 303-373-1955
E-mail: melodye@coloradosafety.org
http://www.coloradosafety.org

Not-for-profit, non-governmental educational organization specializing in occupational safety and health issues.
*Melodye Turek, President*

**741 Colorado Solar Energy Industries Association**
841 Front Street 303-333-7342
Louisville, CO 80027 Fax: 303-604-6988
E-mail: info@coseia.org
http://www.coseia.org
Represents and serves energy professionals and renewable energy users. We promote the use of solar energy and conservation to improve the environment and create a sustainable future.
*Founded: 1989*

**742 Colorado Trappers Association**
PO Box 397 970-268-5554
Empire, CO 80438 E-mail: steve@coloradotrapper.com
http://www.coloradotrapper.com/
An organization dedicated to promoting wildlife education and management, and upholding the ideals of our unique trapping heritage.
*Debra Watts, Secretary*

**743 Colorado Water Congress**
1580 Logan Street 303-837-0812
#400 Fax: 303-837-1607
Denver, CO 80203 E-mail: cwc@cowatercongress.org
http://www.cowatercongress.org
Protects and conserves Colorado's water resources by means of advocacy and education.
*Founded: 1958*
*Doug Kemper, Executive Director*

**744 Colorado Wildlife Federation**
1410 Grant Street 303-987-0400
Suite C-313 Fax: 303-987-0200
Denver, CO 80203 E-mail: cwfed@coloradowildlife.org
http://www.coloradowildlife.org
Mission is to promote the conservation, sound management, and sustainable use of Colorado's wildlife and wildlife habitat through education and advocacy.
*Founded: 1953*
*Suzanne O'Neill, Executive Director*

**745 Keystone Center and Keystone Science School**
Keystone Center & Science School
1628 Sts. John Road 970-513-5800
Keystone, CO 80435 Fax: 970-262-0152
E-mail: info@keystone.org
http://www.keystone.org
Nonprofit public policy and educational organization founded in 1975. Strives to develop creative problem-solving processes that assist diverse parties address issues of importance and to provide qualified science education throughhands-on inquiry of the natural world. Keystone Center pursues this end through its two divisions, the Science and Public Policy Program and the Keystone Science School.
*Founded: 1975*
*Robert W Craig, President*
*Dirk Forrister, Managing Director*

**746 National Wildlife Federation Rocky Mountain Natural Resource Center**
2260 Baseline Road 303-786-8001
Suite 100 800-822-9919
Boulder, CO 80302 Fax: 303-786-8911
http://www.nwf.org
Protects and restores wildlife habitat on tribal lands and is involved with other issues that impact wildlife and wild places of the West and Great Plains regions.
*Larry J Schweiger, President/CEO*
*Julie McGarvey, Senior Field Education*

**747 Nature Conservancy**
1881 9th Street 303-444-2950
Suite 200 800-628-6860
Boulder, CO 80302 Fax: 303-444-2986
E-mail: comment@tnc.org
http://www.nature.org
A leading conservation organization working around the world to protect ecologically important lands and waters for nature and people.
*Founded: 1915*
*Steven J McCormick, Chief Executive*

**748 Sierra Club-Rocky Mountain Chapter**
1536 Wynkoop Street 303-861-8819
4th Floor Fax: 303-449-6520
Denver, CO 80202 E-mail: dan.disner@rmc.sierraclub.org
http://www.rmc.sierraclub.org
To advance the preservation and protection of the natural environment by empowering the citizenry, especially democratically based grassroots organizations with charitable resources to further the cause of environmental protection, thevehicle through which the Sierra Club Foundation generally fulfills its charitable mission.
*Susan Lefever, Director*
*Dan Disner, Chapter Coordinator*

**749 Trout Unlimited**
1300 North 17th Street 703-522-0200
Suite 500 800-834-2419
Arlington, VA 22209 Fax: 703-284-9400
E-mail: trout@tu.org
http://www.tu.org
Mission: To conserve, protect and restore North America's trout and salmon fisheries and their watersheds. We accomplish this mission on local, state and national levels with an extensive and dedicated volunteer network.
*Founded: 1959*
*Charles Gauvin, President/CEO*
*Steve Moyer, Vice President*

**750 Wildlife Society**
317 W Prospect 970-472-4461
Fort Collins, CO 80526 E-mail: rsell@co.blm.gov
http://www.wildlife.org
A nonprofit scientific and educational organization that serves professionals such as government agencies, academia, industry, and non-government organizations in all areas related to the conservation of wildlife and natural resourcesmanagement.
*Robin Sell, Secretary*

**751 Yellowstone Grizzly Foundation**
104 Hillside Court 307-734-8643
Boulder, CO 80302
A non-profit organization dedicated to the conservation of the threatened grizzly bear in the greater Yellowstone Ecosystem. YGF pursues the conservation of the Yellowstone Grizzly by conducting independent research and a wide range ofeducation programs.

## Connecticut

**752 American Association in Support of Ecological Initiatives**
150 Coleman Road 860-364-2967
Middletown, CT 06457 Fax: 860-347-8459
E-mail: wwasch@wesleyan.edu
http://www.wesleyan.edu
AASEI is a US 501 nonprofit organization which supports international environmental initiatives in Russian Nature Reserves. In cooperation with Russia and foreign scientists, students, and universities, AASEI organizes scientificresearch projects, academic internships, work camps, environmental exchanges, and eco-tourism. Our aim is to provide practical support to Russian

Reserves, expand opportunities for international scientific research, and promote internationalunderstanding.
*Founded: 1994*
*Stephanie Hitztaler, Acting Executive Director*
*Brendan Sweeney, President*

**753 American Lung Association of Connecticut**
American Lung Association
45 Ash Street
East Hartford, CT 06108
860-289-5401
800-586-4872
Fax: 860-289-5405
E-mail: bcase@alact.org
http://www.alact.org
The American Lung Association of Connecticut offers a wide variety of lung health services to the people of Connecticut.
*Founded: 1904*
*John Zinn, Chief Executive Officer*
*Margaret LaCroix, Vice President*

**754 American Rivers: Northeast Region**
20 Bayberry Road
Glastonbury, CT 06033
860-652-9911
Fax: 860-652-9922
http://www.americanrivers.org
American Rivers is the only national organization standing up for healthy rivers so our communities can thrive.
*Brian Graber, Associate Director*

**755 American Society of Landscape Architects: Connecticut Chapter**
87 Willow Street
New Haven, CT 06511
203-966-7071
800-878-1474
Fax: 203-972-0770
E-mail: brobinsonla@gmail.com
http://www.ctasla.org
As a chapter of the ASLA, it aims to lead, educate and participate in the careful stewardship, wise planning, and artful design of cultural and natural environments.
*Brian A Robinson, President-Elect*
*Jeff Mills, Executive Director*

**756 Connecticut Audubon Society**
2325 Burr Street
Fairfield, CT 06824
203-259-6305
Fax: 203-254-7673
E-mail: casfairfield@ctaudubon.org
http://www.ctaudubon.org
A statewide, nonprofit, membership organization dedicated to protecting the Connecticut environment by providing citizens of all ages with top-quality education and outdoor experiences. Each year, through school programs, teachertraining workshops, youth activities, adult and family trips, community events and legislative initiatives, the Society reaches more than 175,000 people.
*Robert Martinez, President*
*Barbara Strickland, Chairman*

**757 Connecticut Botanical Society**
PO Box 9004
New Haven, CT 06532
860-633-7557
E-mail: lemmon@snet.net
http://www.ct-botanical-society.org
A group of amateur and professional botanists who share an interest in the plants and habitats of Connecticut and the surrounding region. The goals are to increase knowledge of the state's flora, to accumulate a permanent botanicalrecord, and to promote conservation and public awareness of the state's rich natural heritage.
*Founded: 1903*
*Carol Lemmon, President*

**758 Connecticut Forest and Park Association**
16 Meriden Road
Rockfall, CT 06481
860-346-2372
Fax: 860-347-7463
E-mail: info@ctwoodlands.org
http://www.ctwoodlands.org
An organization for forest and wildlife conservation. Develops outdoor recreation and natural resources. Provides forest management, construction of hiking trails and consultation in the areas of forestry and environment.
*Founded: 1895*
*Richard Whitehouse, President*
*Gordon Anderson, Vice President*

**759 Connecticut Fund for the Environment**
Fact Sheet
205 Whitney Ave
1st floor
New Haven, CT 06511
203-787-0646
Fax: 203-787-0246
E-mail: protect@cfenv.org
http://www.cfenv.org
States non-profit legal champion for the environment. CFE utilizes law, science and education to better air and water quality, control toxic contamination, minimize the adverse impacts of highways and traffic congestion, protect publicwater supplies, and preserve the open space and wetlands so crucial to both the state's citizens and its wildlife.
*Founded: 1978*
*Donald Strait, Executive Director*
*Curt Johnson, Senior Staff Attorney*

**760 Friends of Animals**
777 Post Road
Suite 205
Darien, CT 06820
203-656-1522
Fax: 203-656-0267
E-mail: info@friendsofanimals.org
http://www.friendsofanimals.org
A non-profit, international animal advocacy organization that works to cultivate a respectful view of non-human animals, free-living and domestic.
*Founded: 1957*
*Priscilla Feral, President*

**761 Litchfield Environmental Council: Berkshire**
Oriion Grassroots Network
PO Box 552
Lakeville, CT 06039
860-435-2004
E-mail: wml61@comcast.net
http://www.berklitchfieldenviro.org
To promote an understanding of both the environmental and economic needs, and of the preservation and conservation issues, within the bioregion.
*Ellery Sinclair, Executive Secretary*
*B Blake Levitt, Media Relations*

**762 Save the Sound**
126 E. Putnam Avenue
Cos Cob, CT 06870
203-422-2563
888-728-3547
Fax: 203-967-2677
E-mail: savethesound@snet.net
http://www.savethesound.org
Save the Sound Inc is a nonprofit membership organization dedicated to the restoration, protection and appreciation of Long Island Sound and its watershed through education, research and advocacy.
*Founded: 1972*
*John Atkin, President*

**763 Sierra Club**
645 Farmington Avenue
Hartford, CT 06105
860-236-4405
E-mail: connecticut.chapter@sierraclub.org
http://connecticut.sierraclub.org
To advance the preservation and protection of the natural environment by empowering the citizenry, especially democratically based grassroots organizations with charitable resources to further the cause of environmental protection, thevehicle through which the Sierra Club Foundation generally fulfills its charitable mission.
*John Blake, Chair*
*John Calandrelli, Chapter Coordinator*

**764 Trout Unlimited**
1300 North 17th Street
Suite 500
Arlington, VA 22209
703-522-0200
800-834-2419
Fax: 702-284-9400
E-mail: trout@tu.org
http://www.tu.org
Mission: To conserve, protect and restore North America's trout and salmon fisheries and their watersheds. We accomplish this

mission on local, state and national levels with an extensive and dedicated volunteer network.
*Charles Gauvin, President/CEO*
*Steve Moyer, Vice President*

## District of Columbia

**765 African American Environmentalists Association**
1629 K Street NW 443-569-5102
Suite 300 http://www.aaenvironment.com
Washington, DC 20036
A national, nonprofit environmental organization dedicated to protecting the environment, enhancing human, animal and plant ecologies, promoting the efficient use of natural resources and increasing Afrian American participation in theenvironmental movement.
*Founded: 1985*
*Norris McDonald*

**766 Alliance for Climate Protection**
901 E Street NW 202-628-1999
Washington, DC 20004-2037 http://www.climateproject.org
Seeks to uncover the complete truth about the climate crisis in a way that ignites the moral courage in each of us.
*Maggie L Fox, President/CEO*

**767 Alliance to Save Energy**
1850 M Street NW 202-857-0666
Suite 600 http://www.ase.org
Washington, DC 20036
A nonprofit organization that promotes efficiency worldwide through research, education and advocacy. Strives to be the world's premier organization promoting energy efficiency to achieve a healthier economy, a cleaner environment andgreater energy security.
*Kateri Callahan, President*

**768 American Lung Association of the District of Columbia**
475 H Street Northwest 202-682-5864
Washington, DC 20001 Fax: 202-682-5874
E-mail: info@aladc.org
http://www.aladc.org
The core of the American Lung Association's mission is to help promote lung health and prevent lung disease
*Katrina Jones, Project Cordinator*

**769 American Society of Landscape Architects: Potomac Chapter**
PO Box 18184 703-838-5095
Washington, DC 20036 E-mail: info@potomacasla.org
http://www.potomacasla.org
Serves the metro DC area. It advocates responsible design and use of land and advances the professional success of its members.
*Ron Kagawa, President*

**770 Environmental Working Group**
1436 U Street NW 202-667-6982
Suite 100 Fax: 202-232-2592
Washington, DC 20009 http://www.ewg.org
The Environmental Working Group is a small, computer powered research organization dedicated to improving environmental protection through the analysis of federal and state regulatory policies and performance and through technicalassistance and education.
*Founded: 1993*
*Ken Cook, Presudent*
*Richard Wiles, Senior Vice President*

**771 Human Environment Center**
1930 18th Street NW 202-588-8036
Suite 24 Fax: 202-588-9422
Washington, DC 20009
*Hector Eriksen-Mendoza, Executive Director*

**772 Resources for the Future**
1616 P Street NW 202-328-5000
Suite 600 Fax: 202-939-3460
Washington, DC 20036 http://www.rff.org
A nonprofit and nonpartisan organization that conducts independent research, rooted primarily in economics and other social sciences on environmental, energy, natural resource and environmental health issues.
*Founded: 1952*
*Philip Sharp, President*

**773 Sierra Club**
PO Box 6093 202-363-4366
Washington, DC 20005 Fax: 202-244-4438
E-mail: amanda.brinton@sierraclub.org
http://dc.sierraclub.org
The more than 3,200 members of the DC Chapter are actively involved in local conservation and political efforts in conjunction with the national Sierra Club mission.
*Amanda Brinton, Chapter Organizer*
*Gwyn Jones, Executive Committee Chair*

**774 Washington, DC: Chesapeake Region Safety Council**
Rutherford Business Center 410-298-4770
17 Governor's Court 800-875-4770
Baltimore, MD 21244 Fax: 410-281-1350
E-mail: safety@chesapeakesc.org
http://www.chesapeakesc.org
A private, non-profit, non-governmental public service organization whose mission is to provide the safety training and education that will reduce disabling injuries and save lives.
*Dave Minford, CEO*
*Connie Schultheis, Vice President*

## Delaware

**775 American Lung Association of Delaware**
1021 Gilpin Avenue 302-655-7258
Suite 202 800-586-4872
Wilmington, DE 19806 Fax: 302-655-8546
E-mail: dbrown@lunginfo.org
http://www.alade.org/main.htm
Since the turn of the century, we have been fighting lung disease through education, community services, advocacy and research. Lung disease, including asthma, emphysema, and lung cancer, is the third leading cause of death in America.
*Founded: 1904*
*Deborah Brown, Director Programs*
*Susan DeNardo, Director Development*

**776 Atlantic Waterfowl Council**
Division of Fish and Wildlife
89 Kings Highway 302-739-5295
Dover, DE 19901 Fax: 302-739-6157
http://www.flyways.us/flyways/atlantic
Committed to the preservation and advancement of waterfowl.
*Founded: 1948*
*William C Wagner II, Chairman*

**777 Delaware Association of Conservation Districts**
PO Box 242 302-789-4411
Dover, DE 19903 Fax: 302-739-6724
E-mail: terry.pepper@state.de.us
http://www.nacdnet.org/about/districts/directory/de.phtml
Mandated to preserve and protect our state's soil, water and coastal resources.
*Terry Pepper, President*
*William Vanderwende, First Vice President*

**778 Delaware BASS Chapter Federation**
2453 S State Street
Camden, DE 19901
302-698-9257
800-463-6062
Fax: 720-302-1230
E-mail: pr@deltbf.org
http://www.deltbf.com

An organization striving to represent the interests of local Bass chapters on a state wide basis. We offer the Delaware Department of Natural Resources & Environmental Control our organized and political support and encouragement. TheDFB promotes full adherence to all conservation codes of existing regulatory standards.
*Founded: 1975*
*Ron Horton, President*
*Gary Brandt, Vice President*

**779 Delaware Greenways**
100 W. 10th St., Suite 1001
P.O Box 2095
Wilmington, DE 19899
302-655-7275
Fax: 302-655-7274
E-mail: greenways@dca.net
http://www.delawaregreenways.org/

Committed to the preservation and advancement of Delaware's natural, scenic, historic, cultural, and recreational resources. Works to accomplish this in preserving and connecting open space greenways, increasing opportunities forwalking and biking and creating more livable communities.
*Founded: 1990*
*Robert J Valihura, Jr, President*
*Tim Plemmons, Executive Director*

**780 Delaware Nature Society**
Delaware Nature Society
PO Box 700
Hockessin, DE 19707
302-239-2334
Fax: 302-239-2473
E-mail: dnsinfo@delawarenaturesociety.org
http://www.delawarenaturesociety.org

DNS members benefit from discounts and free admissions, previews and special programs, priority registration, guest passes, and more. Memberships begin at $30 for the college student, through $1000 as a member of the Director's Circle.to many species of amphibians, birds, mammals, fish, reptiles, and native plants.
*Michael Riska, Executive Director*

**781 Delaware: Chesapeake Region Safety Council**
Rutherford Business Center
17 Governor's Court
Baltimore, MD 21244
410-298-4770
800-875-4770
Fax: 410-281-1350
E-mail: safety@chesapeakesc.org
http://www.chesapeakesc.org

A private, non-profit non-governmental public service organization whose mission is to provide the safety training and education that will reduce disabling injuries and save lives.

**782 Nature Conservancy**
260 Chapman Road
Suite 210D
Newark, DE 19702
303-369-4144
Fax: 302-369-4143
E-mail: comment@tnc.org
http://www.nature.org

The Nature Conservancy is a leading conservation organization working around the world to protect ecologically important lands and waters for nature and people. The mission of the Nature Conservancy is to preserve the plants, animalsand natural communities that represent the diversity of life on Earth by protecting the lands and waters they need to survive.
*Founded: 1984*
*Roger Jones, State Director*
*Terrence Scanlon, President*

**783 Save Wetlands and Bays**
41 Beaver Circle
Lewes, DE 19958
302-945-8578
*Henry Glowiak, President*

**784 Sierra Club**
100 West 10th Street
Suite 106
Wilmington, DE 19801
302-351-2776
E-mail: delaware.chapter@sierraclub.org
http://delaware.sierraclub.org
*Nancy Carig, Vice Chair*

## Florida

**785 American Fisheries Society: Agriculture Economics Section**
University of California
PO Box 240
Gainsville, FL 32611
352-392-4991
Fax: 352-392-3646
E-mail: adams@fred.ifas.ufl.edu

Committed to the preservation and advancement of Florida's natural resources.
*Charles Adams, President*

**786 American Lung Association of Florida**
American Lung Association
5526 Arlington Road
Jacksonville, FL 32211
904-743-2933
800-940-2933
Fax: 904-743-2916
E-mail: alaf@lungfla.org
http://www.lungfla.org

The mission of the American Lung Association is to prevent lung disease and promote lung health.
*Founded: 1904*
*Pablo Mila, Director*

**787 American Lung Association: Central Area Office**
American Lung Association
1333 West Colonial Drive
Orlando, FL 32804
407-425-5864
800-586-4872
Fax: 407-425-2876
E-mail: alafcentral@lungfla.org
http://www.lungfla.org

The mission of the American Lung Association is to prevent lung disease and promote lung health.
*Founded: 1904*
*Martha Bogdan, Chief Executive*

**788 American Lung Association: Gulfcoast Area**
American Lung Association
110 Carillon Parkway
Saint Petersburg, FL 33716
727-347-6133
Fax: 727-345-0287
E-mail: alagf@alagf.org
http://www.gulflung.org

The mission of the American Lung Association is to prevent lung disease and to promote lung health.
*Founded: 1904*
*Shirley Westrate, Area Executive Director*

**789 American Lung Association: Gulfcoast Area: Southwest Office**
American Lung Association
12734 Kenwood Lane
Suite 25
Fort Meyers, FL 33907
239-275-7577
800-586-4872
Fax: 239-275-6739
E-mail: alagf@alagf.org
http://www.gulflung.org

The mission of the American Lung Association is to prevent lung disease and to promote lung health
*Founded: 1904*
*Darius Joseph, President*
*Shirley M Westrate, Chief Operating Officer*

**790 American Lung Association: Gulfcoast Area: South Bay Office**
3333 Clark Road
Suite 100
Sarasota, FL 34231
941-377-5864
Fax: 941-342-6099

The mission of the American Lung Association is to prevent lung disease and to promote lung health.

**791 American Lung Association: North Area: Northwest Office**
American Lung Association
4300 Bayou Boulevard
Suite 2
Pensacola, FL 32503
850-478-5864
800-586-4872
Fax: 850-474-6354
E-mail: alafnw@networktel.net
http://www.lungfla.org
The mission of the American Lung Association is to prevent lung disease and to promote lung health.
*Founded: 1904*
*John L Kirkwood, President*

**792 American Lung Association: North Area: Big Bend Office**
539 Silver Slipper Lane
Suite A
Tallahassee, FL 32303
850-386-2065
Fax: 850-422-1894
E-mail: alaf@earthlink.net
http://www.lungfla.org
The mission of the American Lung Association is to prevent lung disease and to promote lung health.
*Brenda Olsen, Director of Governmental Aff*
*Kelsey Ryan, President*

**793 American Lung Association: North Area: Daytona Office**
American Lung Association
412 S. Palmetto Avenue
Daytona Beach, FL 32114
386-255-6447
800-LUN- USA
Fax: 386-253-2410
E-mail: alafspaceport@cfl.rr.com
http://www.lungusa.org/
The mission of the American Lung Association is to prevent lung disease and to promote lung health
*Founded: 1904*
*Charles A Heinrich, Chairman*
*Robert A Green, Vice Chairman*

**794 American Lung Association: South Area Office**
2020 South Andrews Avenue
Fort Lauderdale, FL 33316
954-524-4657
800-524-8010
Fax: 954-524-3162
E-mail: info@sflung.org
http://www.sflung.org
The mission of the American Lung Association is to prevent lung disease and to promote lung health.
*Founded: 1904*
*Denise Grimsley, President*

**795 American Lung Association: Southeast Area Office**
American Lung Association: Florida
2701 North Australian Avenue
Suite 100
West Palm Beach, FL 33409
561-659-7644
800-LUN-GUSA
Fax: 561-835-8967
E-mail: alafse@lungfla.org
http://www.lungfla.orge.com
The mission of the American Lung Association is to save lives by improving lung health and preventing lung disease.
*Founded: 1937*
*Carol A Ruggeri, Executive Director*

**796 American Lung Association: Southeast Area: Belle Glade Office**
American Lung Association of Florida
136 South Main Street
Belle Glade, FL 33430
561-993-3632
800-586-4872
Fax: 561-993-3433
E-mail: amlungself@enhaleexhale.org
http://www.lungusa.org
The mission of the American Lung Association is to prevent lung disease and to promote lung health.
*Founded: 1904*
*Paul Polisena, President*
*James Sugarman, Executive Director*

**797 American Lung Association:Gulfcoast Area: Nature Coast Office**
American Lung Association
PO Box 1445
Inverness, FL 34451
352-860-0616
800-586-4872
Fax: 352-860-0336
http://www.lungusa.org
The mission of the American Lung Association is to prevent lung disease and to promote lung health
*Founded: 1904*
*Heinrich, Chairman*

**798 American Lung Association:Gulfcoast Area: East Bay Office**
110 Carolon Parkway
St. Petersburg, FL 33716
813-962-4448
Fax: 727-345-0287
The mission of the American Lung Association is to prevent lung disease and to promote lung health.

**799 Association of Battery Recyclers**
PO Box 290286
Tampa, FL 33687
813-626-6151
Fax: 813-622-8388
E-mail: info@batteryrecyclers.com
To keep members abreast of environmental, health, and safety requirements that affect our industry.
*Founded: 1984*
*Earl Cornette, President*
*Joyce Morales Caramella, Secretary*

**800 Audubon of Florida**
Travenier Science Center
115 Indian Mound Trail
Tavernier, FL 33070
305-852-5318
Fax: 305-852-8012
E-mail: wmones@audubon.org
http://www.audubonofflorida.org
The science center is the research arm of the Everglades Campaign. The mission of Audubon of Florida is to ensure the restoration and conservation of the Greater Everglades Ecosystem in order to achieve an ecologically and economicallysustainable South Florida. Our Everglades Conservation office has a five-part program including science, education, advocacy, outreach and grassroots action.
*Founded: 1938*
*David Anderson, Executiv Director*
*Mark Krawse, Deputy Director*

**801 Central and North Florida Chapter: National Safety Council**
7001 Lake Ellenor Drive
Suite 120
Orlando, FL 32809
407-370-4098
800-427-2713
Fax: 407-370-9902
E-mail: floridacn@nsc.org
http://www.floridacn.nsc.org
Goal and mission is to educate and influence people to prevent accidental injury and death
*Bob Wilson, Executive Director*

**802 Citizens for a Scenic Florida**
4401 Emerson Street
Suite 10
Jacksonville, FL 32207
904-396-0037
Fax: 904-398-4647
E-mail: scenicfl@scenicflorida.org
http://www.scenicflorida.org
Promotes and carries out programs that protect natural beauty in the environment, protect historical and cultural resources; promotes education of the public about such issues; and coordinates local, regional and state efforts topreserve and enhance visual resources.
*William C Jonson, President*
*William D Brinton, Secretary*

**803 Florida Chapter: American Society of Landscape Architects**
5123 Kernwood Court
Suite 100
Palm Harbor, FL 34685
727-938-6752
Fax: 727-942-4570
E-mail: info@flasla.org
http://www.flasla.org

A non-profit association, operating under the national professional society, that represents the landscape architecture profession throughout the state of Florida.
*Dana K Worthington, President*
*Sue Fern, Association Manager*

**804 Florida Defenders of the Environment**
4424 NW 13th Street 352-378-8465
Suite C-8 Fax: 352-377-0869
Gainesville, FL 32609 E-mail: fde@fladefenders.org
http://www.fladefenders.org
One of the oldest and most accomplished conservation organizations in Florida with a network of scientists, economists and other professionals dedicated to preserving and protecting the state's natural resources. FDE's top priority iscurrently the restoration of a 16-mile stretch of the Ocklawaha River and its 9,000-acre floodplain forest by removal of Rodman Dam- the last vestige of the Cross-Florida Barge Canal.
*Founded: 1969*
*Nick Williams, Executive Director*

**805 Florida Environmental Health Association**
PO Box 271823 863-499-2550
Tampa, FL 33688 Fax: 863-499-2654
E-mail: fehaweb@feha.org
http://www.feha.org
Promotes public health by means of advanced environmental control.
*Founded: 1967*
*Timothy Mayer, President*
*Jennifer Williams, Executive Director*

**806 Florida Forestry Association**
PO Box 1696 850-222-5646
Tallahassee, FL 32302 Fax: 850-222-6179
E-mail: info@forestfla.org
http://www.floridaforest.org
Mission is to promote the responsible use of Florida's forests. This is accomplished through a variety of programsand services designed to keep the state of Florida green while it grows.
*Founded: 1923*
*Jeff Doran, Executive Vice President*
*Debbie Bryant, Director of Member Services*

**807 Florida Keys Wild Bird Rehabilitation Center**
93600 Overseas Highway 305-852-4486
Tavernier, FL 33070 Fax: 305-852-3186
E-mail: fkwbc@terrnova.net
http://www.florida-keys.fl.us
The mission is to reduce the suffering of sick and injured wild birds, to reduce the incidents of their injury and environmental hazards that place these birds at the risk through education.
*Horn Bruce, President*
*Laura Quinn, Executive Director*

**808 Florida Ornithological Society**
143 Beacon Lane 850-942-2489
Jupiter, FL 33469 E-mail: necox@nettally.com
http://www.fosbirds.org
The purpose of the organization is to promote field ornithology and to facilitate contact between those persons interested in birds.
*Founded: 1972*
*Jack Hailman, President*
*Susan B Whiting, Vice President*

**809 Florida Public Interest Research Group**
926 E. Park Ave. 850-224-3321
Tallahassee, FL 32301 Fax: 850-224-1310
E-mail: info@floridapirg.org
http://www.floridapirg.org
State wide, nonprofit, public interest advocacy organization that focuses primarily on environmental and consumer protection.
*Mark Ferrulo, Executive Director*
*Holly Binns, Field Director*

**810 Florida Renewable Energy Association**
510 Hermits Trail 352-241-4733
Altamonte Springs, FL 3270E-mail: info@cleanenergyflorida.org
http://www.cleanenergyflorida.org
Dedicated to expanding the use of clean, renewable energy technolociges through public awareness, political advocacy, and individual initiative.
*Robert Stonerock, President*

**811 Florida Solar Energy Industries Association**
231 W Bay Avenue 407-339-2010
Longwood, FL 32750 800-426-5899
Fax: 407-260-1582
E-mail: bruce@flaseia.org
http://www.flaseia.org
A nonprofit professional association of companies involved in the solar energy industry.
*Founded: 1977*
*R Bruce Kershner, Executive Director*

**812 Florida Trail Association**
5415 SW 13th Street 352-378-8823
Gainesville, FL 32608 Fax: 352-378-4550
E-mail: fta@floridatrail.org
http://www.florida-trail.org
Builds, maintains and preserves the Florida Trail, a 1300 mile trail from Big Cypress Preserve to Gulf Islands National Seashore for hikers and backpackers. Sponsors hikes and canoe trips.
*Founded: 1964*
*Deborah Stewart-Kent, Executive Director*
*Diane Wilkins, Office Manager*

**813 International Association for Hydrogen Energy**
5783 SW 40 Street 305-284-4666
303 Fax: 305-284-4792
Miami, FL 33155 E-mail: info@iahe.org
http://www.iahe.org
Advances the day when hydrogen energy will become the principal means by which the world will achieve its long-sought goal of abundant clean energy. Toward this end, the Association endeavors to inform scientists and the public of theimportant role of hydrogen energy in the planning of an inexhaustible and clean energy system through its publications (International Journal of Hydrogen Energy) and conferences.
*Founded: 1974*
*T Nejat Veziroglu, President*
*Tokio Ohta, Vice President*

**814 International Game Fish Association**
300 Gulf Stream Way 954-927-2628
Dania Beach, FL 33004 Fax: 954-924-4299
E-mail: hq@igfa.org
http://www.igfa.org
Founded as record-keeper and to maintain fishing rules. Today, emphasis is on conservation and education. Encourages youngsters to enter the sport and maintains a huge library on the subject of fishing. Has a network of well over 300representatives around the world, many of whom are conservation leaders in their communities.
*Founded: 1939*
*Rob Kramer, President*
*Phil Hott, Finance Manager*

**815 Keep Florida Beautiful**
201 Park Avenue 850-385-1528
Tallahassee, FL 32301 Fax: 850-385-4020
Non-profit organization dedicated to litter prevention, beautification and community improvement, and minimization of the impacts of waste on communities.
*Jeff Koons, Chairman*

**816 Legal Environmental Assistance Foundation (LEAF)**
1114 Thomasville Road 850-681-2591
Suite E Fax: 850-224-1275
Tallahassee, FL 32303 E-mail: cvalencic@leaflaw.org
http://www.leaflaw.org

To protect human health and life-sustaining natural resources from pollution in Florida, Georgia and Alabama.
*Founded: 1979*
*David Lupder, President*
*Cynthia Valencic, Vice President*

**817 National Wildlife Federation: Everglades Project**
2590 Golden Gate Parkway 239-643-4111
Suite 109 Fax: 239-643-5130
Naples, FL 34105 E-mail: adamsk@nwf.org
http://www.nwf.org
To advocate and support restoration of the greater Everglades ecosystem and protection of the western Everglades through planning, education and management activities. We seek to re-create a more natural hydrologic flow through thegreater Everglades that ensures the long-term viability of native habitats, threatened and endangered species and associated wildlife.
*Founded: 1934*
*Larry Schweiger, President*
*Dulce Zormelo, Chief Finance Officer*

**818 Pelican Man's Bird Sanctuary**
1708 Ken Thompson Parkway 941-388-4444
Sarasota, FL 34236 Fax: 941-388-3258
E-mail: mail@pelicanman.org
http://www.pelicanman.org
Southwest Florida's largest rescue and rehabilitation center for wildlife, emphasizing, but not limited to, birds.
*Mona Schonbrunn PhD, President*
*Tonya Clauss, Director of Veterinary Medic*

**819 Rails-to-Trails Conservancy**
Florida Field Office 850-942-2379
2623 Blair Stone Road http://www.railstotrails.org
Tallahassee, FL 32301
A non-profit organization working with communities to preserve unused rail corridors by transforming them into trails, enhancing the health of America's environment, economy, neighborhoods and people.
*Ken Bryan, State Director*

**820 Reef Relief**
Reef Relief Environmental Center 305-294-3100
PO Box 430 Fax: 305-293-9515
Key West, FL 33041 E-mail: info@relier.org
http://www.reefrelief.org
Reef Relief is a global nonprofit membership organization dedicated to preserve and protect living coral reef ecosystems.
*Founded: 1986*
*Michael Blades, Project Director*
*DeeVon Quirolo, Executive Director*

**821 Sanibel-Captiva Conservation Foundation**
3333 Sanibel-Captiva Road 239-472-2329
PO Box 839 Fax: 239-472-6421
Sanibel, FL 33957 E-mail: sccf@sccf.org
http://www.sccf.org
Land aquisition, native plant nursery, environmental education, habitat management, marine laboratory. Dedicated to the preservation of natural resources and wildlife habitat on and around the barrier islands.
*Founded: 1967*
*Erick Lindblad, Executive Director*
*Jean Laswell, Business Manager*

**822 Sierra Club**
Regional Office 727-824-8813
111 Second Avenue NE, Ste 1001 Fax: 727-824-0936
St Petersburg, FL 33701 E-mail: frank.jackalone@sierraclub.org
http://florida.sierraclub.org
To advance the preservation and protection of the natural environment by empowering the citizenry, especially democratically-based grassroots organizations, with charitable resources to further the cause of environmental protection.
*Betsy Roberts, Chair*
*John Glenn, Conservation Chair*

**823 Society of Environmental Toxicology and Chemistry**
SETAC N America Office
1010 N 12th Avenue 850-469-1500
Pensacola, FL 32501 Fax: 850-469-9778
E-mail: setac@setac.org
http://www.setac.org
Provides a forum for individuals and institutions engaged in study of environmental issues, management and conservation of natural resources, environmental education and environmental research and development.
*Founded: 1979*
*Mimi Meredith, Sr Manager*

**824 South Florida Chapter, National Safety Council**
4171 West Hillsborro Boulevard 954-422-5757
Suite 5 800-392-5101
Coconut Creek, FL 33073 Fax: 954-418-9290
E-mail: occupational@safetycouncil.com
http://www.safetycouncil.com
Not-for-profit, non-governmental, public service organization dedicated to the safety and health of the Broward, Dade, Palm Beach and surrounding communities.
*Michael Walters, VP Operations*

**825 Southeastern Association of Fish and Wildlife Agencies**
8005 Freshwater Farms Road 850-893-1204
Tallahassee, FL 32308 Fax: 850-893-6204
E-mail: seafwa@aol.com
http://www.seafwa.org
An organization whos members are the state agencies with primary responsibility for management and protection of the fish and wildlife resources in 16 states.
*John Frampton, Director*
*Robert Cook, Executive Director*

**826 Suncoast Seabird Sanctuary**
18328 Gulf Blvd 727-391-6211
Indian Shores, FL 33785 800-406-3400
Fax: 727-399-2923
E-mail: seabird@seabirdsanctuary.com
http://www.seabirdsanctuary.com
The sanctuary is the largest wild bird hospital in the United States dedicated to the rescue, repair, rehabilitation and hopeful release of sick and injured native birds. Over 600 birds permanently reside at our beachfront sanctuary.We are open free of charge to the public 365 days a year. Tours and educational programs are available.
*Founded: 1971*
*Ralph Heath Jr, Founder/Director*
*Michelle Simoneau, Marketing/PR Manager*

**827 Tallahassee Museum of History and Natural Science**
3945 Museaum Drive 850-575-8684
Tallassee, FL 32310 Fax: 850-574-8243
E-mail: rdaws@tallasseemuseum.org
http://www.tallahasseemuseum.org
One of the few museums in the nation combining a natural habitat zoo of indigenous wildlife, a collection of historical buildings and artifacts and an environmental science center on a beautiful 52 acre lake side setting.
*Founded: 1957*
*Russell Daws, Director/CEO*
*Jennifer Golden, Director Education*

**828 Wildlife Foundation of Florida**
PO Box 11010 850-922-1066
Tallahassee, FL 32302 Fax: 850-921-5786
E-mail: foundation@myfwc.com
http://www.wildlifeflorida.org
Goal is to ensure that Florida's wildlife survives and thrives for future generations of Florida residents and visitors.
*Founded: 1994*
*Daphne Wood, Chairman*
*Brett Boston, Executive Director*

## Georgia

**829 Agency for Toxic Substances and Disease Registry**
Centre for Disease Control
1600 Clifton Road NE 404-639-3311
Mail Stop F-32 800-311-3435
Atlanta, GA 30333 Fax: 770-488-4178
E-mail: atsdric@cdc.gov
http://www.atsdr.cdc.gov

To prevent exposure and adverse human health effects and diminished quality of life associated with exposure to hazardous substances from waste sites, unplanned releases, and other sources of pollution present in the enviroment. ATSDRis an operating division of the US Department of Health and Human Services. It divides its activities between those related to a particular site and those related to a specific hazardous substance.
*Founded: 1987*
*Julie Gerberding, Director*
*Henry Falk, Associate Director*

**830 American Academy of Sanitarians**
1568 Le Grande Circle 678-407-1051
Lawrenceville, GA 30043 Fax: 678-407-1051
http://www.sanitarians.org

A nonprofit corporation governed to carry out the programs and to meet the objectives stated in its constitution by laws.
*Gary Noonan, Executive Secretary/Treas.*

**831 American Lung Association of Georgia**
American Lung Association
2452 Spring Road 770-434-5864
Smyrna, GA 30080 800-586-4872
Fax: 770-319-0349
E-mail: mail@alaga.org
http://www.lungusa2.org/georgia/index.html

The mission of the American Lung Association is to prevent lung disease and promote lung health.
*Founded: 1904*
*Charles A Heinrich, Chairman*
*Robert A Green, Vice-Chairman*

**832 American Society of Landscape Architects: Georgia Chapter**
PO Box 9803 404-969-1400
Savannah, GA 31412 E-mail: info@gaasla.org
http://www.gaasla.org

A local source of professional support to practicing professionals.
*450 Members*
*Jamie Csizmadia, President*

**833 Center for a Sustainable Coast**
221 Mallory Street 912-638-3612
Suite B Fax: 912-638-3615
St. Simons Island, GA 31522 E-mail: susdev@gate.net
http://www.sustainablecoast.org

The purpose of the non-profit membership organization is to improve the responsible use, protection, and conservation of coastal Georgia's resources- natural historic, and economic.
*Founded: 1997*
*David Kyler, Executive Director*
*Helen Alexander, Administrative Assistant*

**834 Centers for Disease Control and Prevention**
United States of health and human services
1600 Clifton Road NE 404-639-3311
Atlanta, GA 30333 800-311-3435
Fax: 404-639-7111
http://www.cdc.gov

Protects the public health of the nation by providing leadership and direction in the prevention and control of diseases and other preventable conditions and responding to public health emergencies.
*Founded: 1946*
*Julie Louise Gerberding, Director*
*William Gimson, Chief Operating Officer*

**835 Coastal Conservation Association of Georgia**
Coastal Conservation Association
515 Demark St 912-764-6222
Suite 300 800-266-0693
Statesboro, GA 30458 Fax: 912-764-6497
E-mail: info@ccaga.org
http://www.ccaga.org

A non-profit organization dedicated to promoting the preservation, conservation, restoration and protection of the marine fisheries and habitats of the Georgia coast both in shore and offshore, for the benefit and responsibleutilization by the general public.
*Founded: 1986*
*Jonathan Cartwright, Executive Director*

**836 Coosa River Basin Initiative**
408 Broad Street 706-232-2724
Rome, GA 30161 Fax: 706-235-9066
E-mail: info@coosa.org
http://www.coosa.org

We are a nonprofit environmental advocacy organization dedicated to creating a cleaner, healthier, more economically viable Coosa River Basin.
*Carolyn Turner, Executive Director*
*Ben Harrison, President*

**837 Council of State and Territorial Epidemiologists (CSTE)**
2872 Woodcock Boulevard 770-458-3811
Suite 303 Fax: 770-458-8516
Atlanta, GA 30341 E-mail: fellowship@cste.org
http://www.cste.org

The Council of State and Territorial Epidemiologists is an organization of epidemiologists working together to establish effective relationships amoung state and other epidemiologists, to consult and advise with appropriate disciplinesin other health agencies, and to provide technical assistance to the Association of State and Territorial Health Officials.
*Founded: 1951*
*Melvin Kohn, President*
*Perry Smith, Vice President*

**838 Earth Share of Georgia**
1447 Peachtree Street 404-873-3173
Suite 214 Fax: 404-873-3135
Atlanta, GA 30309 E-mail: info@earthsharega.org
http://www.earthsharega.org

Nonprofit federation of local, national and global environmental groups addressing the critical environmental issues. ESGA raises funds for these groups through workplace giving campaigns, special events and individual contributions.
*Founded: 1992*
*Madeline Reamy, Executive Director*
*Cheryl Kortemeier, President, Board of Director*

**839 Environmental Justice Resource Center**
223 James Brawley Drive SW 404-880-6911
Atlanta, GA 30314 Fax: 404-880-6909
E-mail: ejrc@cau.edu
http://www.ejrc.cau.edu

Since 1994, a research, policy and information clearinghouse on issues related to environmental justice, race and the environment, civil rights, facility siting, land use planning, brownfields, transportation equity, suburban sprawland Smart Growth. The overall goal of the center is to assist, support, train and educate people of color, students, professionals and grassroots community leaders with the goal of facilitating their inclusion into the mainstream of environmentaldecision-making.
*Robert Bullard, Director*
*Michelle Dawkins, Program Manager*

**840 Georgia Association of Conservation District Supervisors**
PO Box 111 706-542-3065
Athens, GA 30603 E-mail: info@gacds.org
http://www.gacds.org

Dedicated to the protection and conservation of the state's natual resources
*Founded: 1943 370 Members*
*Christa Carrell, Executive Director*

**841 Georgia Chapter, National Safety Council**
3300 NE Expressway
Suite 5A
Atlanta, GA 30341
770-457-5100
800-441-5103
Fax: 770-457-6189
E-mail: georgia@nsc.org
http://georgia.nsc.org
Mission is to educate and influence people to prevent accidental injury and death. Raising awareness about safety issues that affect all of us regardless of industry, helps make our world a safer place.
*Bob Wilson, Executive Director*

**842 Georgia Conservancy**
817 West Peachtree Street
Suite 200
Atlanta, GA 30338
404-876-2900
Fax: 404-872-9229
E-mail: mail@gaconservancy.org
http://www.georgiaconservancy.org
A catalyst for the stewardship of our natural environment through education, principled advocacy, and inclusive decision-making in order to make Georgia a premier environmental state.
*Founded: 1967*
*Allison Wall, Senior VP*
*Piere Howard, President*

**843 Georgia Environmental Health Association**
Golden Isles Parkway
Route 2, Box 1140
Hawkinsville, GA 31036
706-595-5478
E-mail: clerk@gehaorg.net
http://www.geha-online.org
A non-profit, professional organization, dedicated to promoting, supporting, training, and registering individuals working in environmental health fields throughout government, academia, industry and business.
*Krissa Jones, President*
*Dwain Butler, VP*

**844 Georgia Environmental Organization**
3185 Center Street
Smyma, GA 30080
404-605-0000
Fax: 404-350-9997
E-mail: geoco@geoco.org
Devoted to the preservation of the natural diversity of the plant and animal species, and their habitats, through the prevention of environmental degredation and destruction.
*Olin Ivey, Executive Director*

**845 Georgia Federation of Forest Owners**
2402 Manchester Drive
Waycross, GA 31501
912-283-0871
E-mail: info@woodlandowners.org
http://www.woodlandowners.org
The most active, independent landowners group in the country.
*William Hubbard, South Regional Editor*

**846 Georgia Solar Energy Association**
PO Box 2788
Macon, GA 31203
646-431-4476
E-mail: info@gasolar.org
http://www.gasolar.org
The Georgia Chapter of the American Solar Energy Society. A nonprofit organization working to promote renewable energy in the state of Georgia through education, research and advocacy.
*Kevin Formby, Contact*

**847 Georgia Trappers Association**
PO Box 1033
Pine Mountain, GA 31822
912-782-5417
E-mail: info@gatrappersassoc.com
http://www.gatrappersassoc.com
Non profit organization interested in promoting the education and growth of hunting, fishing, and trapping while maintaining the highest standards of sportsmanship.
*Gene Pritchett, President*
*Chris Johnson, Executive Director*

**848 Georgia Water and Pollution Control Association**
PO Box 6129
Marietta, GA 30065
770-618-8690
Fax: 770-618-8695
E-mail: info@gwpca.org
http://www.gwpca.org
The GW+PCA is dedicated to education, dissemination of technical and scientific information, increased public understanding and promotion of sound public laws and programs in the water resources and related environmental fields.Founded in 1932.
*Jack C Dozier, Executive Director*
*Terry Cole, President*

**849 Georgia Wildlife Federation**
11600 Hazelbrand Road
Covington, GA 30094
770-787-7887
Fax: 770-787-9229
E-mail: gwf@gwf.org
http://www.gwf.org
A member supported, not-for-profit conservation organization and the state affiliate of the NationaL Wildlife Federation.
*Founded: 1936*
*Jerry McCollum, President/CEO*
*Glenn Dowling, Executive Vice President*

**850 Human Ecology Action League (HEAL)**
PO Box 29629
Atlanta, GA 30359
404-248-1898
Fax: 404-248-0162
E-mail: healnatnl@aol.com
http://http://members.aol.com/healnatnl
The Human Ecology Action League Inc (HEAL) is a nonprofit organization founded in 1977 to serve those whose health has been adversely affected by environment exposures; to provide information to those who are concerned about the healtheffects of chemicals; and to alert the general public about the potential dangers of chemicals. Referrals to local HEAL chapters and other support groups are available from the League.
*Founded: 1977*
*Katherine P Collier, Manager*

**851 Mountain Conservation Trust of Georgia**
104 N Main Street
Suite B3
Jasper, GA 30143
706-253-4077
Fax: 706-253-4078
E-mail: info@mctga.org
http://www.mctga.org
Dedicated to the permanent conservation of the scenic beauty and natural resources of the mountains and foothills of North Georgia through land protection, education and collaborative partnerships.
*Founded: 1994*
*Dan Pool, Vice President*
*Clay Johnston, President*

**852 National Wildlife Federation Southeastern Natural Resource Center**
730 Peachtree Street NE
Suite 1000
Atlanta, GA 30308
404-876-8733
Fax: 404-892-1744
E-mail: online.nwf.org/southeastern
http://www.nwf.org
Works to protect the ecosystems of the Southeastern US including the coastal plain estuaries, the everglades, the Appalachian highlands, and public lands.
*Larry J Schweiger, President/CEO*

**853 Nature Conservancy: Georgia Chapter**
1330 W Peachtree Street
Suite 410
Atlanta, GA 30309
404-873-6946
Fax: 404-873-6984
E-mail: comment@tnc.org
http://www.nature.org
A leading conservation organization working around the world to protect ecologically important lands and waters for nature and people.
*Founded: 1951*
*Tavia McCuean, VP*
*Allen Harrison, Director of Operation*

**854 Sierra Club: Georgia Chapter**
1401 Peachtree Street NE
Suite 345
Atlanta, GA 30309
404-607-1262
Fax: 404-876-5260
E-mail: georgia.chapter@sierraclub.org
http://georgia.sierraclub.org

A grassroots organization dedicated to the preservation, protection and enjoyment of our environment. We work towards those goals through public education, political advocacy, an active outings program and litigation when necessary.
*Genie Strickland, Administrative Coordinator*

**855 Trees Atlanta**
96 Poplar Street NW
Atlanta, GA 30303
404-522-4097
Fax: 404-522-6855
E-mail: info@treesatlanta.org
http://www.treesatlanta.org

A non-profit citizen's group dedicated to protecting and improving our urban environment by planting and conserving trees.
*Founded: 1985*
*Marcia Bansley, Executive Director*

**856 Upper Chattahoochee Riverkeeper**
3 Puritan Mill
916 Joseph Lowery Boulevard
Atlanta, GA 30318
404-352-9828
Fax: 404-352-8676
E-mail: sbethea@ucriverkeeper.org
http://www.ucriverkeeper.org

Mission is to advocate and secure the protection and stewardship of the chattahoochee River, its tributaries and watershed, in order to restore and preserve their ecological health for the people, fish and wildlife that depend on theRiver system.
*Founded: 1994*
*Sally Bethea, Executive Director*
*Birgit Bolton, Programs Coordinator*

**857 Wildlife Society**
5410 Grosvenor Lane
Suite 200
Bethesda, MD 20814
301-897-9770
Fax: 301-530-2471
E-mail: tws@Wildlife.org
http://www.wildlife.org

A nonprofit scientific and educational organization that serves professionals such as government agencies, academia, industry, and non-government organizations in all areas related to the conservation of wildlife and natural resourcesmanagement.
*Founded: 1937*
*William R. Rooney, President*
*Jane Pelky, Financial Coordinator/Office*

## Hawaii

**858 American Lung Association in Hawaii**
680 Iwilei Road
Suite 575
Honolulu, HI 96817
808-537-5966
Fax: 808-537-5971
E-mail: lung@ala-hawaii.org
http://www.ala-hawaii.org

Mission is to prevent lung disease and promote lung health. We strive to reach this goal by delivering customer driven quality programs to fight lung disease, developing the financial base to support these activities, and engaging thecommitment of our board staff, volunteers and customers.
*Founded: 1929*
*Jean Evans, Executive Director*
*Karen Lee, President*

**859 American Lung Association of Hawaii: East Hawaii Office**
39 Ululani Street
Hilo, HI 96720
808-935-1206
Fax: 808-935-7474
E-mail: alahbi@ala-hawaii.org
http://www.ala-hawaii.org

The mission of the American Lung Association is to prevent lung disease and to promote lung health.
*Founded: 1929*
*Mary Miller, CEO*
*Malcolm Koga, President*

**860 American Lung Association of Hawaii: Kauai Office**
29992 Umi Street
Lihue, HI 96766
808-245-4142
Fax: 808-245-8488
E-mail: alahkaui@pixi.com
http://www.ala-hawaii.org/

The mission of the American Lung Association is to prevent lung disease and to promote lung health.
*Malcolm Koga, President*
*Sterling Yee, Vice President*

**861 American Lung Association of Hawaii: Maui Office**
American Lung Association of Hawaii
95 Mahalani Street
Cameron Center
Wailuku, HI 96793
808-244-5110
Fax: 808-242-9041
E-mail: alahmaui@ala-hawaii.org
http://www.ala-hawaii.org

The mission of the American Lung association is to prevent lung disease and to promote lung health.
*Founded: 1929*
*Mary Miller, President*
*Didier Decler, VP Finance*

**862 American Lung Association of Hawaii: West Hawaii Office**
American Lung Association
74-5588 Pawai Place
Building P
Kailua-Kona, HI 96740
808-326-4755
800-LUN- USA
Fax: 808-326-9149
E-mail: alahkona@pixi.com
http://www.lungusa.org/site/

The mission of the American Lung Association is to prevent lung disease and to promote lung health.
*Founded: 1904*
*James M Anderson, Secretory*

**863 American Society of Landscape Architects: Hawaii Chapter**
1164 Bishop Street, Suite 124
Box 246
Honolulu, HI 96813
808-521-5631
E-mail: info@hawaiiasla.org
http://www.hawaiiasla.org

The purpose of the society is the advancement of knowledge, education, and skill in the art and science of landscape architecture as an instrument of service for the public welfare.
*Founded: 1969*
*Chris Dacus, President*

**864 Big Island Rain Forest Action Group**
PO Box 341
Kurtistown, HI 96760
808-966-7622
E-mail: ja@interpac.net
*Founded: 1989*
*Jim Albertini, Co-ordinator*

**865 EarthTrust**
Windward Environmental Center
1118 Maunawili Road
Kailua, HI 96734
808-261-5339
Fax: 206-202-3893
E-mail: sue@flipperfund.com
http://www.earthtrust.org

EarthTrust is the impossible missions team for wildlife and the environment. Its low-overhead high-tech campaigns are always positive and effective. Dedicated to saving marine mammals, reforming unsustainable fisheries and ending thetrade in endangered species around the world. EarthTrust is a relatively small organization which may have directly saved more marine wildlife biomass than any other organization in history.
*Founded: 1976*
*Don White, President*

**866 Flipper Foundation**
Windward Environmental Center
1118 Maunawili Road
Kailua, HI 96734
808-261-5339
Fax: 815-333-1158
E-mail: sue@flipperfund.com
http://www.earthtrust.org

The Flipper Foundation exists to save dolphins and to revolutionize consumer control over environmental destruction by world

fisheries. Its mission is to establish and maintain the highest world standard of dolphin safety and fisheriessustainability; to educate consumers worldwide while directly granting funds to save marine mammals and their habitats. Its primary way of accomplishing this is to engage fishery firms in voluntary partnerships to phase out destructive fishingtechnologies.
*Founded: 1992*
*Don White, President*

**867 Greenpeace Foundation**
Windward Environmental Center
1118 Maunawili Road
Kailua, HI 96734
808-263-4388
Fax: 630-604-6129
E-mail: sw@gpfdn.com
http://www.greenpeacefoundation.com
The oldest and original greenpeace organization in the US Greenpeace Foundation is dedicated to peaceful no-nonsense environmental advocacy. Greenpeace Foundation seeks to preserve biodiversity on a green and peaceful planet. Proudlyunaffiliated with Greenpeace International we make no apologies for standing up for the earth; a human voice for the majority of earth's life which has none so that citizens may have a voice in what sort of planet we will leave to our children andtheirs.
*Founded: 1976*
*Sharon White, President*

**868 Hawaii Association of Conservation Districts**
PO Box 404
Hana, HI 96713
808-248-7725
Fax: 808-248-7725
http://www.hi.nrcs.usda.gov/partnerships/hacd.html
HACD coordinates and facilitates partners and governmental agenciesin identifying and implimenting projects and practiceswith cultural sensitivity to assure the protection of Hawaii's environment.
*David Nobriga, President*
*Mike Tulang, Executive Director*

**869 Hawaii Nature Center**
2131 Makiki Heights Drive
Honolulu, HI 96822
808-955-0100
Fax: 808-955-0116
E-mail: hawaiinaturecenter@hawaii.rr.com
http://www.hawaiinaturecenter.org
The Hawaii Nature Center is a private nonprofit organization specializing in environmental education field program for children, adults and families. Its mission is to foster awareness appreciation and understanding of Hawaii andencourage wise stewardship of the Islands. The Nature Center provides full day field trips for 20,000 students on two islands each year and features an interactive nature museum at its field site on Maui.
*Founded: 1981*
*Meredith Ching, President*
*Jeff Case, First Vice Chairman*

**870 Hawaiian Botanical Society**
University of Hawaii
3190 Maile Way
Honolulu, HI 96822
808-956-8072
Fax: 808-956-3923
http://www.botany.hawaii.edu
Dedicated to the understanding and preservation of hawaii's fauna and wildlife.
*Founded: 1924*
*Eileen Helmstetter, President*
*Vickie Caraway, Vice-President:*

**871 Nature Conservancy: Hawaii Chapter**
Nature Conservancy
923 Nuuanu Avenue
Honolulu, HI 96817
808-537-4508
E-mail: egoldstein@tnc.org
http://www.nature.org
The mission is to preserve the plants, animals, and natural communities that represent the diversity of life on earth by protecting the lands and waters they need to survive.
*Founded: 1903*
*Suzanne Case, Executive Director*
*Steven McCormick, President/CEO*

**872 Sierra Club**
PO Box 2577
Honolulu, HI 96803
808-538-6616
Fax: 808-537-9019
E-mail: mikulina@lava.net
http://www.hi.sierraclub.org
To explore, enjoy and protect wild places and the environment, the club uses a multi-pronged approach to protecting and restoring Hawaii's environmental quality. Through the volunteer efforts of group leaders they conduct interpretiveand educational outings; lead fun and challenging hikes; conduct service projects involving fencing, cleaning streams, trail building and noxious plant control; and advocate and lobby for environmental protection.
*Jeff Mikulina, Director*

## Idaho

**873 American Lung Association of Idaho/Nevada: Boise Office**
2405 Vassar Street
Suite B
Reno, NV 89502
775-829-5864
Fax: 775-829-5850
http://www.lungs.org
The mission of the American Heart Association is to prevent lung disease and to promote lung health.
*Founded: 1904*

**874 American Society of Landscape Architects: Idaho/Montana Chapter**
c/o Hatchmueller PC
611 Sherman Avenue
Coeur d'Alene, ID 83814
E-mail: keithd@architectswest.com
http://www.imasla.org
Works to increase public's awareness of and appreciation for the profession of landscape architecture.
*Keith Dixon, President*
*Jolene Rieck, President-Elect*

**875 Energy Products of Idaho**
4006 Industrial Avenue
Coeur d' Alene, ID 83815-8928
208-765-1611
Fax: 208-765-0503
E-mail: epi2@energyproducts.com
http://www.energyproducts.com
EPI is a world leader in the development and implementation of proprietary and patented technologies used to convert biomass and other waste fuels into usable forms of energy. Since 1973, EPI has pioneered and perfected fluidized bedand related technologies for utility, industrial and commercial uses. Although our primary focus remains our world renowned fluidized bed technologies, our vast experience has lead to a stable of superior auxiliary and related proprietarytechnologies.
*Founded: 1973*
*Kent Pope, Director, Business Dev.*
*Patrick Travis, Business Development Mgr*

**876 Idaho Association of Soil Conservation Districts**
39 W Pine Ave
Suite B20
Meridian, ID 83642
208-338-5900
Fax: 208-338-9537
E-mail: kent.foster@agri.idaho.gov
http://www.iascd.state.id.us
Provides action at the local level for promoting wise and beneficial conservation of natural resources with emphasis on soil and water.
*Founded: 1944*
*J Kent Foster, Executive Director*

**877 Idaho Conservation League**
PO Box 844
710 North Sixth Street
Boise, ID 83701
208-345-6933
877-345-6933
Fax: 208-344-0344
E-mail: icl@wildidaho.org
http://www.wildidaho.org

The mission of the Leage is to preserve Idaho's clean water, wilderness and quality of life through citizen action, public education, and professional advocacy.
*Founded: 1973*
*Justin Hayes, Program Director*
*Suki Molina, Deputy Director*

**878 Idaho Forest Owners Association**
233 E. Palouse River Drive 208-883-4488
PO Box 9748 Fax: 208-883-1098
Moscow, ID 83843 E-mail: NWManage@consulting-foresters.com
http://www.consulting-foresters.com
A full service forestry consulting firm. Provides the best forest planning and management practices for forest landowners throughout the Inland Northwest area.
*Founded: 1983*
*Vincent Corrao, President*
*Tom Richards, Vice President*

**879 Northwest Power and Conservation Council**
450 West State Street 3rd Floor 208-334-6970
PO Box 83702 Fax: 208-334-2112
Boise, ID 83702 E-mail: kdunn@nwcouncil.org
http://www.nwcouncil.org/contact/id.asp
The council develops and maintains a regional power plan and a fish and wildlife program to balance the Northwest's environment and energy needs.
*Founded: 1980*
*Karen Dunn, Administrator*
*Shirley Lindstrom, Power Policy Analyst*

**880 Rocky Mountain Elk Foundation**
5705 Grant Creek 406-523-4500
Missoula, MT 59808 800-225-5355
Fax: 208-775-3263
E-mail: info@rmef.org
http://www.rmef.org
Mission is to ensure the future of elk, other wildlife and thier habitat.
*Founded: 1984*
*Buddy Smith, Interim President & Ceo*

**881 Sierra Club-Northern Rockies Chapter**
PO Box 522 208-384-1023
Boise, ID 83701 E-mail: jessica.ruehrwein@sierraclub.org
http://idaho.sierraclub.org
The Northern Rockies Chapter represents over 4,000 members in Idaho and eastern Washington.
*Don Crawford, Conservation Chair*
*Jessica Ruehrwein, Regional Representative*

**882 Wildlife Society**
The Wild Life Society
5410 Grosvenor Lane 301-897-9770
Suite 200 Fax: 301-530-2471
Bethesda, MD 20814 E-mail: aowsiak@idfg.state.id.us
http://www.wildlife.org
A nonprofit scientific and educational organization that serves professionals such as government agencies, academia, industry, and non-government organizations in all areas related to the conservation of wildlife and natural resourcesmanagement.
*Anna Owsiak, Secretary*

## Illinois

**883 American College of Occupational and Environmental Medicine**
25 Northwest Point Blvd 847-818-1800
Suite 700 Fax: 847-818-9266
Elk Grove Village, IL 60007 http://www.acoem.org
Made up of physicians in industry, government, academia, private practice and the military, who promote the health of workers through preventive medicine, clinical care, research and education.
*Founded: 1916*
*Pamela Hymel, President*
*Barry Eisenberg, Executive Director*

**884 American Lung Association of Illinois/Iowa**
American Lung Association
3000 Kelly Lane 217-787-5864
Springfield, IL 62707 800-586-4872
Fax: 217-787-5916
E-mail: info@lungil.org
http://www.lungusa.org/site
*Founded: 1904*
*Harold Wimmer, CEO*
*Lori Younker, Manager*

**885 American Lung Association: Chicagoland Collar Counties**
American Lung Association
1749 South Naperville Road 630-260-9600
Suite 202 Fax: 630-260-1111
Wheaton, IL 60187 E-mail: info@lungil.org
http://www.lungil.org
*Founded: 1904*
*Herald Wimmer, Chief Executive Officer*
*Ted Schlake, Senior Manager*

**886 American Lung Association: Northern Illinois**
1330 East State Street 815-962-6412
Rockford, IL 61104 Fax: 815-962-6413
E-mail: info@lungil.org
http://www.lungusa.org
*Founded: 1904*
*James M Anderson, Secretary*

**887 American Lung Association: Southwestern Illinois**
American Lung Association
1600 Golfview Drive 618-344-8891
Suite 260 800-586-4872
Colinsville, IL 62234 Fax: 618-344-8933
E-mail: info@lungil.org
http://www.lungfla.org
*Founded: 1917*
*Tina Barnard, President*
*Harold Wimmer, Executive Director*

**888 American Medical Association**
515 N State Street
Chicago, IL 60610 800-621-8335
http://www.ama-assn.org
Medical doctors concerned with environmentally related health issues.
*Founded: 1847*
*John C Nelson, President*

**889 American Society of Landscape Architects: Illinois Chapter**
PO Box 4566 630-833-4516
Oak Brook, IL 60522 Fax: 630-833-4030
E-mail: info@il-asla.org
http://www.il-asla.org
The Illinois Chapter is one of the larger ASLA chapters with over 500 members.
*Founded: 1899*
*Ann Viger, President*

**890 Audubon Council of Illinois**
National Audubon Society
700 Broadway 212-979-3000
New York, NY 10003 Fax: 212-979-3188
E-mail: ltennefoss@audbon.org
http://www.audubon.org
*Jess Morton, Board of Directors*

**891 Chicago Chapter: National Safety Council**
1121 Spring Lake Drive
Suite 100
Itasca, IL 60143
630-775-2213
800-621-2855
Fax: 630-775-2136
E-mail: chicago@nsc.org
http://www.chicago.nsc.org
*Alan McMillian, President*

**892 Chicago Zoological Society**
Brookfield Zoo
3300 Golf Road
Brookfield, IL 60513
708-688-8000
800-201-0784
E-mail: bzadmin@brookfieldzoo.org
http://www.brookfieldzoo.org
To inspire conservation leadership by connecting people with wildlife and nature.
*Founded: 1934*
*Stuart Strahl, President/CEO*

**893 Eagle Nature Foundation**
Eagle Nature Foundation, LTD
300 E. Hickory St.
Apple River, IL 61001
815-594-2306
Fax: 815-594-2305
E-mail: eaglenature.tni@juno.com
http://www.eaglenature.com
A nonprofit organization dedicated to the preservation of the bald eagle, our national symbol, and other endangered species from extinction and to increase public awareness of unique endangered plants and animals. We monitor bald eagleand other endangered species populations and strive to preserve habitiat essential to their survival. We develop materials for schools to inform students about the needs of the bald eagle and how we can help preserve and protect their naturalenvironment.
*Founded: 1995*
*Terrence N Ingram, Chief Executive Officer*
*Eugene Small, Vice President*

**894 Environmental Education Association of Illinois**
2112 Behan Road
Crystal Lake, IL 60014
815-479-5779
Fax: 815-479-5766
E-mail: dchap19265@aol.com
http://www.web.stclair.k12.il.us/eeai
*Karen Zuckerman, President*

**895 Great Lakes Sport Fishing Council**
PO Box 297
Elmhurst, IL 60126
630-941-1351
Fax: 630-941-1196
E-mail: hdqtrs@great-lakes.org
http://www.great-lakes.org
Our mission is to inform and educate the outdoor recreational community (sport fishing, boating and general public) through educational outreach programs about natural resource conservation and enhancement, wise conservation andboating policies, and the spread of unintentional introductions on nonindigenous aquatic nuisance species (exotics).
*Founded: 1971*
*Dan Thomas, President*
*Robert Mitchell, Vice President*

**896 Illinois Association of Conservation Districts**
9313 Bull Valley Road
Woodstock, IL 60098
815-338-7664
Fax: 815-338-2773
E-mail: conserveone@aol.com
*Founded: 1972*
*John Todt, President*
*Ken Fiske, Assistant Secretary*

**897 Illinois Association of Environmental Professionals**
PO Box 81551
Chicago, IL 60681
773-325-2771
E-mail: kkopija@cbbel.com
http://www.iaepnetwork.org
Enhnacing environmental awareness for the businesses, communities and citizens of Illinois.
*Nathan Quaglia, President*
*Gregory Merritt, VP*

**898 Illinois Audubon Society**
425 B N Gilbert Street
Danville, IL 61832
217-446-5085
Fax: 217-446-6375
E-mail: director@pdnt.com
http://www.illinoisaudubon.org
A membership organization dedicated to the preservation of Illinois Wildlife and the habitats which support them. Has sanctuaries, conservation education and land acquisition programs and publishes quarterly magazines and newsletters.
*Founded: 1897*
*Marilyn F Campbell, Executive Director*

**899 Illinois Environmental Council**
Po Box 81551
Chicago, IL 60681
217-544-5954
Fax: 217-544-5958
E-mail: kellytzoumis@ameritech.net
http://www.ilenviro.org
Coalition of over 70 environmental, conservation and health groups.
*Founded: 1975*
*Jonathan Goldman, Executive Director*
*Rand , Sparling, President*

**900 Illinois Prairie Path**
P.O. Box 1086
Wheaton, IL 60189
630-752-0120
http://www.ipp.org
*Founded: 1905*
*Ray Bartels, President*

**901 Illinois Recycling Association**
PO Box 3717
Oak Park, IL 60303
708-358-0050
Fax: 708-358-0051
E-mail: executivedirector@illinoisrecycles.org
http://www.illinoisrecycles.org
The association's mission is to encourage the responsible use of resources and protecting the environment by promoting effective programs and practices regarding waste reduction, re-use of materials, and recycling.
*Founded: 1980*
*Michael Mitchell, Executive Director*

**902 Illinois Solar Energy Association**
PO Box 634
Wheaton, IL 60189
630-260-0424
Fax: 630-420-1517
http://www.illinoissolar.org
Its mission is to provide energy education to the Illinois public and promote the widespread application of solar, renewable and sustainable energy methods and technologies.
*Founded: 1975*
*Mark Burger, President*
*Ted Lowe, Secretary*

**903 Lake Michigan Federation**
17 North State Street
Suite 1390
Chicago, IL 60602
312-939-0838
Fax: 313-939-2708
E-mail: chicago@greatlakes.org
http://www.lakemichigan.org
Works to restore fish and wildlife habitat, conserve land and water and eliminate pollution in the watershed of America's largest lake. We achieve these through education, research, law, science, economics and strategic partnerships.
*Founded: 1971*

**904 Nature Conservancy: Illinois Chapter**
8 S Michigan Avenue
Suite 900
Chicago, IL 60603
312-580-2100
E-mail: illinois@tnc.org
http://www.nature.org
*Founded: 1915*
*Steven J McCormick, Chief Executive Officer*

**905 Outside Chicagoland: Iowa/Illinois Safety Council**
8013 Douglas Avenue
Urbandale, IA 50322
515-276-4724
800-568-2495
Fax: 515-276-8038
E-mail: iiscadmin@iisc.org
http://www.iisc.org

Our mission is to educate society to adopt safety, health, and environmental practices and to provide high quality, value added training and services.
*Laura Johnson, Executive Director*

**906 Prairie Rivers Network**
Prairie Rivers Network
809 S 5th Street 217-344-2371
Champaign, IL 61820 Fax: 217-344-2381
E-mail: info@prairierivers.org
http://www.prairierivers.org

The only statewide river conservation organization in Illinois. They strive to protect rivers and streams of Illinois and to promote the lasting health and beauty of watershed communities by providing information, sound science andhands-on assistance. They also help individuals and community groups become effective river conservation leaders.
*Founded: 1967*
*Brad Walker, Program Coordinator*

**907 Respiratory Health Association of Metropolitan Chicago**
1440 West Washington Boulevard 312-628-0245
Chicago, IL 60607 888-880-5864
Fax: 312-243-3954
E-mail: burbaszewski@lungchicago.org
http://www.lungchicago.org

*Founded: 1906*
*Brian Urbaszewski, Director, Environmental Prgm*
*Ashley Collins, Environmental Health Assoc*

**908 Safer Pest Control Project**
Safer Pest Control Project
4611 N Ravenswood Ave. 773-878-7378
Suite 107 Fax: 773-878-8250
Chicago, IL 60640 E-mail: general@spcpweb.org
http://www.spcpweb.org

Dedicated to reducing the health risks and environmental impacts of pesticides and promoting safer alternatives in Illinois.
*Founded: 1994*
*Rachel Lerner Rosenberg, Executive Director*
*Kim Stone, Associate Director*

**909 Sierra Club: Illinois Chapter**
70 East Lake Street 312-251-1680
Suite 1500 Fax: 312-251-1780
Chicago, IL 60601 E-mail: illinois.chapter@sierraclub.org
http://illinois.sierraclub.org

The Illinois Chapter is a statewide organization representing more than 26,000 individuals committed to protecting the Illinois environment.
*Jack Darin, Director*
*Jennifer Hensley, Grassroots Coordinator*

**910 Upper Mississippi River Conservation Committee**
555 Lester Avenue 608-783-8432
Onalaska, WI 54650 E-mail: UpperMississippiRiver@fws.gov
http://www.mississippi-river.com

An organization of natural resource managers from IL, IA, MN, MO, and WI, created to promote a continuing cooperation between conservation agencies on the Upper Mississippi River. This is accomplished through workshops, publicationsand annual meetings.
*Founded: 1943*
*Scott Yess, UMRCC Coordinator*

## Indiana

**911 Acres Land Trust**
1802 Chapman Road 260-637-2273
Huntertown, IN 46748 Fax: 260-637-2273
E-mail: acres@acreslandtrust.org
http://www.acreslandtrust.org

Exists to collect and protect the lasdst of the natural habitats in northeast Indiana and to teach Hoosiers the value of keeping natural tracts intact.
*Founded: 1960*
*Jason Kissel, Executive Director*

**912 American Fisheries Society: Equal Opportunities**
5410 Grosvenor Lane 301-897-8616
Bethesda, MD 20814 Fax: 301-897-8096
E-mail: feuker@fisheries.org
http://www.fisheries.org

The section of the American Fisheries Society promotes the representation and involvement of diverse ethnic, racial and cultural groups and women in the fisheries profession. The group fosters mentoring of under-represented groups,administers awards for travel and academic achievement and provides information on social and professional diversity in fisheries.
*Gus Rassam, Executive Director*

**913 American Lung Association of Indiana: Northern Office**
American Lung Association
115 West Washington Street 317-819-1181
Suite 1180 South Fax: 317-819-1187
Indianapolis, IN 46204 E-mail: info@lungin.org
http://www.lungin.org

*Audrey Ferguson, Master Trainer*

**914 American Lung Association of Indiana: State Office & Support Office**
American Lung Association
9445 Delegates Row 317-573-3900
Indianapolis, IN 46240 800-586-4872
Fax: 317-573-3909
E-mail: info@lungin.org
http://www.lungin.org

*Founded: 1904*
*Audrey Ferguson, Master Trainer*

**915 American Society of Landscape Architects: Indiana Chapter**
PO Box 441195
Indianapolis, IN 46244-1195 E-mail: indiana.asla@gmail.com
http://www.inasla.org

Members include private sector landscape architects, public practitioners, scholars, government officials and other professionals representing landscape architecture, planning, preservation and ecology.
*200 Members*
*Katie Clark, President*

**916 Central/Southern Indiana: National Safety Council: Kentucky Office**
1121 Spring Lake Drive 630-285-1121
Itasca, IL 60143 800-621-7619
Fax: 630-285-1315
E-mail: info@nsc.org
http://www.nsc.org

**917 Conservation Technology Information Center**
1220 Potter Drive 765-494-9555
West Lafayette, IN 47906 Fax: 765-494-5969
E-mail: ctic@conservationinformation.org
http://www.ctic.purdue.edu

A nonprofit organization dedicated to environmentally responsible and economically viable agricultural decision-making.
*Founded: 1982*
*Karen Scanlon, Executive Director*
*Kyle Nickel, Communications Director*

**918 Dj Case And Associates Wildlife Society**
Dj Case And Associates
317 East Jefferson Blvd. 574-258-0100
Mishawaka, IN 46545 Fax: 574-258-0189
E-mail: info@djcase.com
http://www.djcase.com

a nonprofit scientific and educational organization that serves professionals such as government agencies, academia, industry,and non-governmentorganizations. in all areas related to the conservation of wildlife and naturalresources managment.
*Dave Case, President*
*Phil Seng, Vice President*

**919 Indiana Audubon Society**
Indiana Audubon Society Inc.
3497 S Bird Sanctuary Road
Connersville, IN 47331
765-827-0908
Fax: 765-825-9788
E-mail: indianaaudubon@yahoo.com
http://www.indianaudubon.org
*Founded: 1898*
*Dan Leach, President*
*Tom Goldsmith, Director*

**920 Indiana Forestry and Woodland Owners Association**
Purdue University
West Lafayette, IN 47907
765-494-4600
E-mail: steward@inwoodlands.org
http://www.inwoodlands.org
*Founded: 1977*
*Morgan R Olsen, Executive Vice President*

**921 Indiana State Trappers Association**
6420 Street Road 47 N
Crawfordsville, IN 47933
812-939-3215
http://www.geocities.com
*Ken Brosman, President*

**922 Indiana Water Environment Association Purdue University**
7439 Woodland Drive
Indianapolis, IN 46278
E-mail: hcheslek@greeley-hansen.com
http://www.indianawea.org
*Ricky Dodd, President*

**923 Northwestern Indiana: National Safety Council: Chicago Chapter**
1121 Spring Lake Drive
Suite 100
Itasca, IL 60143
630-775-2213
800-621-2855
Fax: 630-775-2136
E-mail: chicago@nsc.org
http://www.chicago.nsc.org

**924 Sierra Club**
1915 W 18th Street
Suite D
Indianapolis, IN 46202
317-822-3750
E-mail: sierra@netdirect.net
http://www.hoosier.sierraclub.org
Indiana/Hoosier Chapter of the Sierra Club
*Founded: 1975*
*Christine Fiordalis, Co-Chair*
*Steve Francis, Co-Chair*

**925 Wildlife Society**
1010 Yeardley Lane
Mishawaka, IN 46544
219-258-0100
Fax: 219-258-0189
E-mail: phil@djcase.com
A nonprofit scientific and educational organization that serves professionals such as government agencies, academia, industry, and non-government organizations in all areas related to the conservation of wildlife and natural resourcesmanagement.
*Phil Seng, President*

## Iowa

**926 American Lung Association of Iowa**
5601 Douglas Avenue
Des Moines, IA 50310
515-309-9507
Fax: 515-334-9564
E-mail: amlung@lungia.org
http://www.lungia.org
*Founded: 1904*

**927 American Society of Landscape Architects: Iowa Chapter**
8345 University Boulevard
Suite F-1
Des Moines, IA 50325
515-225-2323
Fax: 515-225-6363
E-mail: ia-asla@assoc-serv.com
http://www.iaasla.org
The chapter organizes educational and social activities for its more than 155 members.
*Founded: 1899*
*Emily Lawson, President*
*Kim Wagner, Association Manager*

**928 Asla Iowa Chapter**
8345 University Blvd
Suite F-1
Des Moines, IA 50325
515-225-2323
Fax: 515-225-6363
E-mail: ia-asla@assoc-serv.com
http://www.iaasla.org

**929 Indian Creek Nature Center**
6665 Otis Road SE
Cedar Rapids, IA 52403
319-362-0664
Fax: 319-362-2876
E-mail: naturecenter@indiancreeknaturecenter.org
http://www.indiancreeknaturecenter.org
A private nonprofit organization open to the public. The Nature Center provides about 300 acres of natural land and provides an array of educational programs.
*Founded: 1973*
*Rich Patterson, Director*
*Dana Wood, Office Manager*

**930 Iowa Academy of Science**
Iowa Academy of Science
UNI - 175 Baker Hall
2607 Campus Street
Cedar Falls, IA 50614-0508
319-273-2581
Fax: 319-273-2807
E-mail: craig.johnson@uni.edu
http://www.iacad.org
Iowa Academy of Science is a professional scientific organization.
*Founded: 1875*
*Craig Johnson, Executive Director*

**931 Iowa Association of Soil and Water Conservation District Commissioners**
38995 Honeysuckle Road
Oakland, IA 52560
641-774-4461
Fax: 712-482-3386
*Bernie Bolton, Secretary*

**932 Iowa BASS Chapter Federation**
3282 Midway
Marion, IA 52302
319-393-1481
E-mail: mail@iabass.com
http://www.iabass.com
*Tom Bowler, President*

**933 Iowa Native Plant Society**
Iowa State University
Botany Department, 341A Bessey Hall
Iowa State University
Ames, IA 50011
515-294-9499
Fax: 515-294-1337
E-mail: mottll@grinnell.edu
http://www.public.iastate.edu
An organization of amateur and professional botanists and native plant enthusiasts who are interested in the scientific, educational and cultural aspects, as well as the preservation and conservation of the native plants of Iowa. TheSociety was organized in 1995 to create a forum where plant enthusiasts, gardners and professional botanists could exchange ideas and coordinate activities such as field trips, work shops, and restoration of natural areas.
*Larissa Mottl, President*
*Connie Mutel, Vice President*

**934 Iowa Renewable Fuels Association**
5505 NW 88th Street
# 100
Johnston, IA 50131
515-252-6249
Fax: 515-225-0781
E-mail: info@iowarfa.org
http://www.iowarfa.org

Brings together ethanol and biodiesel producers to promote the development and growth of the state's renewable fuels industry through education and infrastructure development.
*Walter Wendland, President*
*Monte Shaw, Executive Director*

**935 Iowa Trappers Association**
Gene Purdy
122 2nd Street
Fontanelle, IA 50846
641-682-3937
Fax: 641-682-9092
E-mail: cegrillo@fbcom.net
http://www.iowatrappers.com
*Spencer Hill, President*
*Chris Grillot, Secretary*

**936 Iowa Wildlife Rehabilitators Association**
1005 Harken Hill Drive
PO Box 217
Osceola, IA 50213
515-342-2783
http://www.earthweshare.org
*Marlene Ehresman, President*
*Wendy DeWalle, Secretary*

**937 Iowa-Illinois Safety Council**
8013 Douglas Avenue
Urbandale, IA 50322
515-276-4724
800-568-2495
Fax: 515-276-8038
E-mail: iiscadmin@iisc.org
http://www.iisc.org
Our mission is to educate society to adopt safety, health and environmental practices and to provide high quality, value added training and services.
*1200 Members*
*Laura Johnson, Executive Director*

**938 Macbride Raptor Project**
6301 Kirkwood Boulevard SW
Cedar Rapids, IA 52406
319-398-5495
Fax: 319-398-5495
E-mail: jcancil@kirkwood.edu
http://www.macbrideraptorproject.org
A nonprofit organization jointly sponsored by the University of Iowa and Kirkwood Community College. The project has two main facilities, the educational display facility and rehabilitation flight cage at the Macbride NatureRecreational Area and the medical clinic on the Kirkwood Campus.
*Founded: 1985*
*Jodeane Cancilla, Director*
*Luke Hart, Assistant Director*

**939 Nature Conservancy: Iowa Chapter**
303 Locust Street
Des Moines, IA 50309
515-244-5044
800-628-6860
Fax: 515-244-8890
E-mail: iowa@tnc.org
http://http://nature.org
*Founded: 1951*
*Margaret Collison, State Director*

**940 Practical Farmers of Iowa**
2035 190th Street
Boone, IA 50036
515-432-1560
http://www.practicalfarmers.org

**941 Sierra Club**
3839 Merle Hay Road
Suite 280
Des Moines, IA 50310
515-277-8868
E-mail: iowa.chapter@sierraclub.org
http://iowa.sierraclub.org
With approximately 6,000 members, the Iowa Chapter has been working together to protect the community and the planet.
*Founded: 1972*
*Pam Mackey-Taylor, Chair*
*Jane Clark, Vice Chair*

**942 Soil and Water Conservation Society**
945 SW Ankeny Road
Ankeny, Io 50023
515-289-2331
Fax: 515-289-1227
E-mail: swcs@swcs.org
http://www.swcs.org
Fosters the science and the art of soil, water and related natural resource management to achieve sustainability. Promote and practice an ethic recognizing the interdependence of people and the environment.
*Founded: 1943*
*Craig Cox, Executive Director*
*Sue Ann Lynes, Executive Assistant*

**943 State of Iowa Woodlands Associations**
204 Park Rd
Iowa City, IA 52246
515-233-1161
Fax: 515-233-1131
http://www.iowawoodlandsowners.org
*Al Manning, President*

## Kansas

**944 American Lung Association of Kansas**
4300 Southwest Drury Lane
Topeka, KS 66604
785-272-9296
Fax: 785-272-9297
E-mail: jkeller@kslung.org
http://www.kslung.org
*Judy Keller, Executive Director*
*Kris Scothorn, Office Manager*

**945 Audubon of Kansas**
210 Southwind Place
Manhattan, KS 66503
785-537-4385
Fax: 785-537-4395
E-mail: aok@audubonofkansas.org
http://www.audubonofkansas.org
*Founded: 1999*
*Ryan Klataske, Special Projects Coordinator*
*Ron Klataske, Executive Director*

**946 Conservation and Research Foundation**
PO Box 909
Shelburne, VT 05482
913-268-0076
Fax: 913-268-0076
*Founded: 1953*
*Mary Wetzel, President*

**947 Heartland Renewable Energy Society**
8214 W 75th Street
Overland Park, KS 66204
816-224-5550
E-mail: sharla@hathmore.com
http://www.heartland-res.org
The Missouri/Kansas Chapter of the American Solar Energy Society.
*Sharla Riead, President*
*Davis Roberts, Treasurer*

**948 Kansas Academy of Science**
1700 SW College Avenue
Topeka, KS 66621
785-231-1010
E-mail: webmaster@washburn.edu
http://www.washburn.edu/kas
*Founded: 1868*
*Jerry B. Farley, President*

**949 Kansas Association for Conservation and Environmental Education**
2610 Claflin Road
Manhatten, KS 66502
785-532-3322
Fax: 785-532-3305
E-mail: ldowney@kacee.edu
http://www.kacee.org
Statewide no-profit dedicated to promoting quality sound, non-brased environmental education in Kansas through professional development and technical assistance.
*Laura Downey, Executive Director*

**950 Kansas Natural Resources Council**
PO Box 2635
Topeka, KS 66601
316-265-0767
E-mail: lerick@ksu.edu
http://www.knrc.ws
Protect the quality and supplies of Kansas' water. Support sustainable family farming practices that respect and restore the land and the community. Ensure a competitive energy market where renewable resources and conservation canflourish. Reduce the

exposure to hazardous and nuclear wastes. Encourage environmentally sound industrial practices.
*Founded: 1980*
*Larry Erickson, President*

**951 Kansas Rural Center**
PO Box 133
Whiting, KS 66552
913-873-3431
Fax: 913-873-3432
E-mail: ksrc@rainbowtel.net
http://www.kansasruralcenter.org
A non-profit promoting the long-term health of the land and its people through research, education and advocacy. KRC is committed to economically viable, environmentally sound, and socially sustainable rural culture.
*Founded: 1979*
*Dan Nagengast, Executive Director*
*Mary Fund, Communications Director*

**952 Kansas Wildflower Society**
2045 Constant Avenue
Lawrence, KS 66047
785-864-3453
Fax: 785-864-5093
http://www.naturalkansas.org
*Dwight Platt, President*
*Cynthia Ford, Secretary*

**953 Kansas Wildscape Foundation**
Riverfront Plaza
Suite 311
Lawrence, KS 66044
785-843-9453
Fax: 785-843-6379
E-mail: wildscape@sunflower.com
http://www.kansaswildscape.org
*Founded: 1991*
*Jim Huntington, President*
*Harland Priddle, Executive Director*

**954 North Dakota Natural Science Society**
Department of Biological Sciences
600 Park Street
Hays, KS 67601
785-628-4214
Fax: 785-628-4156
E-mail: efinck@fhsu.edu
http://www.fhsu.edu/biology/pn/prarienat.htm
Regional organization with interests in the natural history of grasslands and the Great Plains.
*Founded: 1967*
*Chris Deperno, President*

**955 Safety & Health Council of Western Missouri & Kansas**
5829 Troost Avenue
Kansas City, MO 64110
816-842-5223
Fax: 816-842-6226
E-mail: shc@safetycouncilmoks.com
http://www.safetycouncilmoks.com
Is a private not-for-profit community service organization which has been helping to make our community a safer place to live, work and play. We are dedicated to preventing unintentional injuries where ever they occur.
*Kathy Zents, Executive Director*

**956 Sierra Club**
9844 Georgia
Kansas City, KS 66109
913-707-3296
E-mail: info@kansas.sierraclub.org
http://kansas.sierraclub.org
Protection of the Kansas environment through education, and the practice and promotion of responsible ecosystem and resource use.
*Yvonne Cather, Chair*
*Craig Lubow, Vice Chair*

**957 Wildlife Society**
Kansas State University
205 Ackert
Manhattan, KS 66506
758-532-0978
E-mail: wildlife@ksu.edu
A nonprofit scientific and educational organization that serves professionals such as government agencies, academia, industry, and non-government organizations in all areas related to the conservation of wildlife and natural resourcesmanagement.
*Jason Tarwater, President*

## Kentucky

**958 American Lung Association of Kentucky**
American Lung Association
Po Box 9067
Louisville, KY 40209
502-363-2652
800-586-4872
E-mail: info@kylung.org
http://www.kylung.org
*Founded: 1905*
*Jim Sugarman, Executive Director*

**959 American Society of Landscape Architects: Kentucky Chapter**
c/o Egbers Land Design
200 Aristocrat Drive
Florence, KY 41042
859-371-2555
E-mail: phays@egberslanddesign.com
http://www.kyasla.com
*Pamela Hays, President*
*Sara Moser, President-Elect*

**960 Kentucky Association for Environmental Education**
PO Box 146
Kirkland
Kirkland, WA 98083
425-814-5095
http://www.kall.org

**961 Kentucky Audubon Council**
Kentucky Audubon Council
306 Hoover Hill Road
Hartford, KY 42347
859-277-1711
E-mail: kac@kentuckyaudubon.org
http://www.kentuckyaudubon.org
Serves as an effective support and coordinating organizatoin to the Commonwealth's Audubon Society Chapters and members of the National Audubon Society.
*Brenda Little, President*

**962 Kentucky Resources Council**
PO Box 1070
Frankfort, KY 40602
502-875-2428
800-372-7181
Fax: 502-875-2845
E-mail: fitzKRC@aol.com
http://www.kyrc.org
*Tom Fitzserald, Director*

**963 Land Between the Lakes Association**
345 Maintenance Road
Golden Pond, KY 42211
800-455-5897
E-mail: information@friendsoflbl.org
http://www.friendsoflbl.org
Assists with the education, improvement, promotion, conservation, and wise use of the USDA Forest Service's Land Between The Lakes National Recreation Area.
*John Rufli, Executive Director*

**964 National Safety Council, Kentucky Office: Central/Southern Indiana & Cincinnati**
1121 Spring Lake Drive
Itasca, IL 60143
800-621-7615
E-mail: customerservice@nsc.org
http://www.nsc.org

**965 Nature Conservancy: Kentucky Chapter**
642 W Main Street
Lexington, KY 40508
859-259-9655
E-mail: kentucky@tnc.org
http://www.nature.org
*Founded: 1951*
*James R Aldrich, State Director*
*Diane Davis, Director Of Philanthropy*

**966 Scenic Kentucky**
Scenic Kentucky
PO Box 2646
Louisville, KY 40201
502-459-9497
Fax: 502-459-5278
E-mail: keitheiken@msn.com
http://www.scenickentucky.org
Scenic America is the only national nonprofit organization dedicated to protecting natural beauty and distinctive community character. We provide technical assistance across the nation and

through affiliates on scenic byways, billboardand sign control, context sensitive highway design, wireless telecommunications tower location, transportation enhancements, and other scenic conservation issues.
*Keith P Eiken, Executive Director*
*Frederic H Davis, President*

**967 Sierra Club**
PO Box 1368
Lexington, KY 40588
859-296-4335
E-mail: staff@kentucky.sierraclub.org
http://kentucky.sierraclub.org
Advances the preservation and protection of the natural environment by empowering the citizenry, especially democratically-based grassroots organizations, with charitable resources to further the cause of environmental protection. Thevehicle through which The Sierra Club Foundation generally fulfills its charitable mission.
*Founded: 1968*
*Ray Barry, Chair*
*Sherry Otto, Chapter Coordinator*

**968 Southeastern Association of Fish and Wildlife Agencies**
1 Sportsman's Lane
Frankfort, KY 40601
800-858-1549
E-mail: info.center@ky.gov
http://http://fw.ky.gov/
*Dr. Johathan W. Gassett, Commissioner*

## Louisiana

**969 American Lung Association of Louisiana**
2325 Severn Avenue
Suite 8
Metairie, LA 70001
504-828-5864
800-586- 872
Fax: 504-828-5867
E-mail: aline@bellsouth.net
http://www.louisianalung.org
A resource for information and data on lung diseases with a focus on education and prevention.
*Founded: 1904*
*Thomas P Lotz, Executive Director*
*Steven M Lee, Program Director*

**970 American Society of Landscape Architects: Louisiana Chapter**
Brown+Danos Landdesign Inc
601 Laurel Street
Baton Rouge, LA 70801
225-571-9534
E-mail: dbrown@browndanos.com
http://www.lcasla.org
*Dana Nunez Brown, President*
*Shannon Blakeman, President-Elect*

**971 Calcasieu Parish Animal Control and Protection Department**
Department of Animal Services
5500A Swift Plant Road
Lake Charles, LA 70615
337-721-3730
Fax: 337-437-3343
E-mail: dmorales@cppj.net
http://cpac.cppj.net
*David Marcantel, Operations Supervisor*

**972 Louisiana Association of Conservation Districts**
663 Holmes Road
Keatchie, LA 71046
318-933-5375
Fax: 318-872-3178

**973 Louisiana BASS Chapter Federation**
603 Terri Drive
Luling, LA 70070
504-785-9069
E-mail: kevgobear@home.com
http://www.louisanabass.org
*Kevin Gaubert, President*
*Elvis Jeanminette, Vice President*

**974 Louisiana Wildlife Federation**
PO Box 65239
Audubon Station
Baton Rouge, LA 70896
225-344-6707
Fax: 225-344-6707
E-mail: lwf@lawildlifefed.org
http://www.lawildlifefed.org
*Founded: 1940*
*Ken Dancak, President*
*Keith Saucier, VP*

**975 National Safety Council: Ark-La-Tex Chapter**
8101 Kingston Road
#107
Shreveport, LA 71108
318-687-7550
Fax: 318-687-7298
E-mail: altasafetycouncil@wnonline.net
http://www.nscarklatexsafetycouncil.com
Specialists in professional safety training.
*Sally Head, Executive Director*

**976 Nature Conservancy: Louisiana Chapter**
PO Box 4125
Baton Rouge, LA 70821
225-338-1040
Fax: 225-338-0103
E-mail: lafo@tnc.org
http://http://www.nature.org
*Founded: 1957*
*Steve McCormack, President*

**977 Sierra Club: Delta Chapter**
PO Box 19469
New Orleans, LA 70179
504-891-9642
E-mail: chair@louisiana.sierraclub.org
http://louisiana.sierraclub.org
Advances the cause of protecting Louisiana's environment in a variety of ways including lobbying the state legislature in Baton Rouge, raising public awareness about climate change, working to keep the Atchafalaya Basin river swampalive and wild, and sponsoring a Mercury Public Education Campaign.
*Leslie March, Chair*
*Aaron Viles, Secretary*

**978 Tulane Environment Law Clinic**
Tulane University
6823 St. Charles Avenue
New Orleans, LA 70118
504-865-5794
800-873-9283
E-mail: pr@tulane.edu
http://www.tulane.edu
Since 1989, the Tulane Law School, through its Environmental Law Clinic, has provided free legal assistance on wide varitey of environmental issues. In addition, the Clinic assists community groups with scientific and organizationalissues.
*Founded: 1989*
*Scott S. Cowen, President*

**979 Wildlife Society**
200 Quail Drive
Baton Rouge, LA 70808
225-765-2800
http://www.wlf.state.la.us
A nonprofit scientific and educational organization that serves professionals such as government agencies, academia, industry and non-government organizations in all areas related to the conservation of wildlife and natural resourcesmanagement.
*Kathleen Babineaux Blanco, Govenor*

## Maine

**980 American Lung Association of Maine**
122 State Street
Augusta, ME 04330
207-622-6394
800-499-5864
Fax: 639-426-2919
E-mail: info@mainelung.org
http://www.mainelung.org
Leads the lung health promotion and lung disease prevention for Maine.
*Founded: 1911*
*Edward Miller, Senior VP*
*Norman Anderson, Regional Director EHR*

**981 Atlantic Salmon Federation**
PO Box 807
Calais, ME 04619
506-529-1033
Fax: 506-529-4438
E-mail: asfweb@nbnet.nb.ca
http://www.asf.ca

An international non-profit organization that promotes the conservation and wise management of wild Atlantic salmon and their environment.
*Bill Taylor, President/CEO*

**982 Maine Association of Conservation Commissions**
PO Box 702
Bath, ME 04530
207-443-2925
Fax: 207-443-6913
E-mail: macc@clinic.net
http://www.clinic.net/usa/macc

*Founded: 1973*
*Robert C Cummings, Executive Director*

**983 Maine Association of Conservation Districts**
97A Exchange St
Suite 305
Portland, ME 04101
207-752-0392
E-mail: info@maineswcds.org
http://www.maineswcds.org

The statewide voice of Maine's 16 local conservation districts. By working with landowners, organizations and government, districts have helped to protect our soil, water, forestry, wildlife and other natural resources for over 60years.

**984 Maine Audubon**
20 Gilsland Farm Road
Falmouth, ME 04105
207-781-2330
Fax: 207-781-0974
E-mail: info@maineaudubon.org
http://www.maineaudubon.org

Maine Audubon works to conserve Maine's wildlife and wildlife habitat by engaging people of all ages in education, conservation and action.
*Founded: 1843*
*Elyse Tipton, Communications Director*
*Kevin Karley, Executive Director*

**985 Maine Coast Heritage Trust**
1 Bowdoin Mill Island
Suite 201
Topsham, ME 04086
207-729-7366
Fax: 207-729-6863
E-mail: info@mcht.org
http://www.mcht.org

Conserves and stewards Maine's coastal lands and islands for their renowned scenic beauty, outdoor recreational opportunities, ecological diversity and working landscapes.
*Founded: 1970*
*Tim Glidden, President*
*Karin Marchetti Ponte, General Counsel*

**986 National Association of School Nurses**
PO Box 1300
Scarbough, ME 04070
207-883-2117
877-627-6476
Fax: 207-883-2683
E-mail: gdurgin@nasn.org
http://www.nasn.org

The mission of The National Association of School Nurses is to advance the practice of school nursing and provide leadership in the delivery of quality health programs to school communities.
*Founded: 1979*
*Wanda Miller, Executive Director*
*Gloria Durgin, Administrator*

**987 Sierra Club**
44 Oak Street
Suite 301
Portland, ME 04101
207-761-5616
Fax: 207-773-6690
E-mail: maine.chapter@sierraclub.org
http://maine.sierraclub.org

*Jim Frick, Chair*
*David Mokler, Treasurer*

**988 Small Woodland Owners Association of Maine**
153 Hospital Street
PO Box 836
Augusta, ME 04332
207-626-0005
877-467-9626
E-mail: info@swoam.com
http://www.swoam.com

Promoting sound forest management and strengthening long-term woodland stewardship.
*Founded: 1975*
*Tom Doak, Executive Director*

## Massachusetts

**989 Alternatives for Community and Environment**
2181 Washington St
Suite 301
Boston, MA 02119
617-442-3343
Fax: 617-442-2425
E-mail: info@ace-ej.org
http://www.ace-ej.org

Builds the power of communities of color and lower income communities in New England to eradicate environmental racism and classism and achieve environmental justice. We believe that everyone has the right to a healthy environment andto be decision-makers in issues affecting our communities.
*Founded: 1994*
*Kalila Barnett, Executive Director*
*Eugene B Benson, Program Director/Legal Couns*

**990 Association for Environmental Health and Sciences**
150 Fearing Street
Amherst, MA 01002
413-549-5170
Fax: 413-549-0579
E-mail: paul@aehs.com
http://www.aehs.com

The AEHS was created to facilitate communication and foster cooperation among professionals concerned with the challenge of soil protection and cleanup. Members represent the many disciplines involved in making decisions and solvingproblems affecting soils. AEHS recognizes that widely acceptable solutions to the problem can be found only through the integration of scientific and technological discovery, social and political judgement and hands on practice.
*Paul Kostecki PhD, Executive Director*
*Cindy Langlois, Managing Director*

**991 Boston Society of Landscape Architects**
19 Harrison Street
Framingham, MA 01702
508-620-5018
Fax: 508-879-4892
E-mail: info@bslaweb.org
http://www.bslaweb.org

The Boston Chapter of the American Society of Landscape Architects (ASLA). It consists of landscape architects in the states of Massachusetts, New Hampshire and Maine.
*Bob Corning, President*
*Vicki Carr, Administrator*

**992 Earthwatch Institute**
3 Clock Tower Place
Suite 100, Box 75
Maynard, MA 01754
978-461-0081
800-776-0188
Fax: 978-461-2332
E-mail: info@earthwatch.org
http://www.earthwatch.org

Earthwatch is a diverse community of scientists, educators, students, businesspeople, and resolute explorers who work together to get the fullest benefit from scientific expeditions. In addition to 150 dedicated staff in the UnitedStates, England, Australia, and Japan, Earthwatch supports more than 130 scientists each year and builds networks of hundreds of students and teachers.
*Founded: 1971*
*Whitney L Johnson, Chairman*
*Ruth C Scheer, Vice Chairman*

**993 Environmental League of Massachusetts**
14 Beacon
Suite 714
Boston, MA 02108
617-742-2553
Fax: 617-742-9656
E-mail: elm@environmentalleague.org
http://www.environmentalleague.org

Dedicated to protecting the air, water, and land for the people of the commonwealth. We do this by voicing citizens' concerns, educating the public, advocating for strong environmental laws, and ensuring that our laws are implementedand enforced.
*Founded: 1898*
*James Gomes, President*

**994 Ethnobotany Specialist Group**
Oxford Street 617-495-2326
Cambridge, MA 02138 Fax: 617-495-5667

**995 Genesis Fund/National Birth Defects Center**
40 2nd Avenue 781-466-9555
Suite 520 800-322-5014
Waltham, MA 02451 Fax: 781-487-2361
E-mail: nbdc@thegenesisfund.org
http://www.thegenesisfund.org
The National Birth Defects Center provides diagnosis and treatment to children born with birth defects, genetic diseases and mental retardation. The Center consists of physicians and consultants in pediatrics, genetics, orthopedics,cardiology, neurology, ophthalmology, endocrinology, cranialfacial surgery, plastic surgery, and other specialties.
*Founded: 1984*
*Jane E O'Brien, Managing Director*
*Murray Feingold, Founder*

**996 MASSPIRG**
44 Winter Street, 4th Floor 617-292-4800
Boston, MA 02108 Fax: 617-292-8057
E-mail: www.masspirg/about-us/contact-us
http://www.masspirg.org
MASSPIRG is an advocate for the public interest. When consumers are cheated or the voices of ordinary citizens are drowned out by special interest lobbyists, MASSPIRG speaks up and takes action. We uncover threats to public health andwell-being and fight to end them, using time-tested tool of investigative research, media exposes, grassroots organizing, advocacy and litigations. Their mission is to deliver persistent, result-oriented public interest activism.

**997 Massachusetts Association of Conservation Districts**
319 Littleton Road 978-692-9395
Suite 205 Fax: 978-392-1305
Westford, MA 01886 http://www.middlesexconservation.org
*Founded: 1947*
*David Williams, Chairman*
*Elizabeth McGuire, Administrator*

**998 Massachusetts Association of ConservationCommissions**
10 Juniper Road 617-489-3930
Belmont, MA 02478 Fax: 617-489-3935
E-mail: staff@maccweb.org
http://www.maccweb.org
We educate and advocate on behalf of all 351 Conservation Commissions in Massachusetts. We also host the MACC Annual Environment Conference, the largest such event in New England, with over 40 workshops and nearly 50 exhibitors on thefirst Saturday of March at the College of the Holy Cross in Worcester.
*Founded: 1961*
*Patrick Garner, Board President*
*Linda Orel, Executive Director*

**999 Massachusetts Audubon Society**
208 S Great Road 781-259-9500
Lincoln, MA 01773 800-AUD-UBON
E-mail: webmaster@massaudubon.org
http://www.massaudubon.org
Works to protect the nature of Massachusetts for people and wildlife. Together with more than 100,000 members, we care for 32,000 acres of conservation land, provide educational programs for 200,000 children and adults annually, andadvocate for sound environmental policies at local, state, and federal levels.
*Founded: 1896*
*Laura A Johnson, President*

**1000 Massachusetts Environmental Education Society**
290 Turnpike Road 508-792-7270
PO Box 105 Fax: 508-792-7275
Westboro, MA 01581 E-mail: admin@mees.org
http://www.mees.org
Dedicated to the promotion, preservation and improvement of environmental education in the State of Massachusetts. A non-profit organization whose members include classroom teachers, environmental educators, outdoor leaders,naturalists and administrators committed to encouraging education and awareness of the inter-relationship of the natural world, and re-establishing the balance between nature and people.
*Founded: 1977*
*Amy Nelson, President*

**1001 Massachusetts Forest Landowners Association**
PO Box 623 413-549-5900
Leverett, MA 10504 Fax: 413-339-5526
E-mail: massforests@verizon.net
http://www.massforests.org
MFLA's mission is to be exemplary stewards of our forest resources, and help others understand, respect, care for, and use this renewable resource. The only statewide, non-profit organization with an exclusive focus on the forests andtrees of Massachusetts. Focuses on positive, constructive ways of improving and ensuring the health, care, and use of the trees, forests and associated resources of the state for generations to come.
*Founded: 1970*
*Elisa Campbell, Executive Director*
*Cinda Jones, President*

**1002 Massachusetts Trapper's Association**
277 Main Street 508-868-8896
Spencer, MA 01562 E-mail: flat.tail@verizon.net
http://www.masstrappers.org
An organization founded for the purpose of preserving the tradition of fur harvesting, while promoting education and preserverence into the future. Today, the Association's focus is to maintain our trapping priviledges, regain thosethat have been lost, and to encourage and promote education of trappers, both seasoned and inexperienced, as well as those unfamiliar with our sport.
*Malcolm Spencer, President*

**1003 Massachusetts Water Pollution Control Association**
PO Box 221 971-939-0918
Groveland, MA 01834 Fax: 781-939-0907
E-mail: mwpca1965@verizon.net
http://www.mwpca.org
*John Connor, Secretary/Treasurer*

**1004 Mount Grace Land Conservation Trust**
1461 Old Keene Road 978-248-2043
Athol, MA 01331 Fax: 978-248-2053
E-mail: landtrust@mountgrace.org
http://www.mountgrace.org
Protects significant natural, agricultural, and scenic areas and encourages land stewardship in N Central Massachusetts for the benefit of the environment, the economy, and future generations. Mount Grace has protected over 11,000acres of land. We currently own 1,270 acres of land and we hold conservation restrictions on 2,164 acres.
*Founded: 1986*
*Leigh Youngblood, Executive Director*
*Lisa Cormier, Office Manager*

**1005 New England Water Environment Association**
10 Tower Office Park 781-939-0908
Suite 601 Fax: 781-939-0907
Woburn, MA 01801 E-mail: mail@newea.org
http://www.newea.org
A not-for-profit organization whose objective is the advancement of fundamental knowledge and technology of design, construction, operation and management of waste treatment works and other water pollution contral activities anddedication to the preservation of water quality and water resources.
*Founded: 1929*
*Elizabeth Cutone, Executive Director*

**1006 Northeast Sustainable Energy Association**
50 Miles Street
Greenfield, MA 01301
413-774-6051
Fax: 413-774-6053
E-mail: nesea@nesea.org
http://www.nesea.org

NESEA is a leading regional membership organization focused on promoting the understanding, development, and adoption of energy conservation and non-polluting renewable energy technologies. It works to bring clean electricity, greentransportation, and healthy, efficient buildings into everyday use.

*Founded: 1974*
*David Barclay, Executive Director*
*Arianna Alexsandra Grindrod, Education Director*

**1007 Save the Harbor/Save the Bay**
Boston Fish Pier
212 Northern Ave, Suite 304 W
Boston, MA 02210
617-451-2860
Fax: 617-451-0496
http://www.savetheharbor.org

Mission is to restore and protect the harbor and the bay, and to reconnect them with Bostonians from every neighborhood, regional residents and visitors alike, so that we can all enjoy the benefits of the enormous public and privateinvestment in our revitalized harbo and waterfront.

*Founded: 1986*
*Patricia A Folly, President*
*Matt Wolfe, VIce President*

**1008 Sierra Club**
100 Boylston Street
Boston, MA 02116
617-423-5775
Fax: 617-423-5858
E-mail: director@sierraclubmass.org
http://www.sierraclubmass.org

*James Bryan McCaffrey, Director*
*Alexander Oster, Administrator*

**1009 Walden Pond Advisory Committee**
Page Road
Lincoln, MA 01773
781-259-9544
http://www.concordnet.org

*Founded: 1975*

**1010 Walden Woods Project**
44 Baker Farm Rd
Lincoln, MA 01773
781-259-4700
800-554-3569
Fax: 781-259-4710
E-mail: wwproject@walden.org
http://www.walden.org

*Founded: 1990*

## Maryland

**1011 Alliance for the Chesapeake Bay: Baltimore Office**
6600 York Road
Suite 100
Baltimore, MD 21212
410-377-6270
Fax: 410-377-7144
E-mail: mail@acb-online.org
http://www.acb-online.org

A regional nonprofit organization that builds and fosters partnerships to protect and to restore the Bay and its rivers. The Alliance develops methods and tools for restoration activities and trains citizens to use them. Also mobilizesdecision-makers, stakeholders, and other citizens to learn about Bay issues and participate in resolving them. Provides analysis, information and evaluation of Bay policies, proposals, and institutions.

*Founded: 1971*
*David Bancroft, President*
*Darlin Hicks, Finance Director*

**1012 American Bass Association of Maryland**
622 Powhattan Beach Road
Pasadena, MD 21122
410-255-0499

*Clancy Thorn, Presidentt*

**1013 American Lung Association of Maryland**
Executive Plaza 1, Suite 600
11350 McCormick Road
Hunt Valley, MD 21031
410-560-2120
800-642-1184
Fax: 410-560-0829
E-mail: info@marylandlung.org
http://www.marylandlung.org

Founded as the Maryland Tuberculosis Association, and although the focus of the Association has changed since a cure for TB was found, we remain constant in working toward the mission - to prevent lung disease and promote lung health.

*Founded: 1919*
*Stephen Peregoy, President/CEO*

**1014 American Society of Landscape Architects: Maryland Chapter**
PO Box 4825
Baltimore, MD 21211
http://www.mdasla.org

The professoinal assocation ofr landscape architects in Maryland. Promotes the landscape architecture profession and advances the practice through advocacy, education, communication, and fellowship.

*Founded: 1972 370 Members*
*Colleen Bathon, President*
*Naomi Reetz, Executive Director*

**1015 Audubon Naturalist Society of the Central Atlantic States**
8940 Jones Mill Road
Chevy Chase, MD 20815
301-652-9188
Fax: 301-951-7179
E-mail: contact@audubonnaturalist.org
http://www.audubonnaturalist.org

Fosters stewardship of the region's environment by educating citizens about the natural world, promoting conservation of biodiversity, and protecting wildlife habitat. The independent nonprofit society focuses its efforts in themid-Atlantic region.

*Founded: 1897*
*Neal Fitzpatrick, Executive Director*
*Kathy Rushing, Board President*

**1016 Center for Chesapeake Communities**
229 Hanover Street
Suite 101
Annapolis, MD 21401
410-267-8595
Fax: 410-267-8597
http://www.chesapeakecommunities.org

Technical assistance on environmental, land use, energy and water quality issues for local government in Chesapeake Bay Watershed.

*Gary Allen, Executive Director*

**1017 Chesapeake Bay Foundation**
Save The Bay Maryland Office
Philip Merrill Environmental Center
6 Herndon Avenue
Annapolis, MD 21403
410-268-8816
Fax: 410-268-6687
http://www.cbf.org

Fights for stong and effective laws and regulations. CBF also works cooperatively with government, business, and citizens in partnerships to protect and restore the Bay. Their mission is: Save the Bay, defined as achieving a HealthIndex for the Bay of 70 by the year 2050.

*Founded: 1966*
*William C Baker, President*

**1018 Chesapeake Wildlife Heritage**
46 Pennsylvania Avenue
PO Box 1745
Easton, MD 21601
410-822-5100
Fax: 410-822-4016
E-mail: info@cheswildlife.org
http://www.cheswildlife.org

Dedicated to creating, restoring and protecting wildlife habitat and establishing a more sustainable agriculture through direct action, education and research in partnership with private landowners.

*Founded: 1980*
*John E Gerber, Executive Director*
*Chris Pupke, Development Director*

**1019 Conservation Federation of Maryland**
Keep It Country

300 Lenora St., PMB-b156
Seattle, WA 98121
206-441-3137
Fax: 206-374-0858
E-mail: f.a.r.m@erols.com
http://www.darnet.com

*Dolores Milmoe, President*

**1020 Eastern Shore Land Conservancy**
PO Box 169
Queenstown, MD 21658
410-827-9756
Fax: 410-827-5765
E-mail: info@eslc.org
http://www.eslc.org

Mission is to sustain the Eastern Shore's rich landscapes through strategic land conservation and cound land use planning.
*Founded: 1990*
*Robert J Etgen, Executive Director*

**1021 Environmental Health Education Center**
University of Maryland, School of Nursing
655 W Lombard Street
Room 665
Baltimore, MD 21201
410-706-1849
Fax: 410-706-0295
E-mail: cehn@cehn.org
http://www.cehn.org/cehn/resourceguide/ehec.html

The overall mission of the Center is to engage in research and provide training and education programs on topics related to occupational and environmental health and safety. Our focus is broad and the workplace, community and home areall included in our defination of environment. The audiences for our training and education programs include professionals, labor, industry and community members. Through our efforts we hope to prevent occupation and/or environment related injuriesand illnesses.
*Lynn Goldman, Chairman*
*Dick J Batchelor, Vice Chairman*

**1022 Institute of Hazardous Materials Management**
11900 Parklawn Drive
Suite 450
Rockville, MD 20852
301-984-8969
Fax: 301-984-1516
E-mail: ihmminfo@ihmm.org
http://www.ihmm.org

Mission is to provide recognition for professionals engaged in the management and engineering control of hazardous materials who have attained the required level of education, experience and competence; foster continued professionaldevelopment of Certified Hazardous Materials Managers (CHMM).
*Founded: 1984*
*John H Frick PhD, CHMM, Executive Director*
*Betty Fishman, Assistant Executive Director*

**1023 Izaak Walton League of America**
707 Conservation Lane
Gaithersburg, MD 20878
301-548-0150
E-mail: executivedirector@iwla.org
http://www.iwla.org

One of the nation's oldest and most respected conservation organizations. With a powerful grassroots network of nearly 300 local chapters nationwide, the League takes a common-sense approach toward protecting our country's nautralheritage and improving outdoor recreation opportunities for all Americans.
*Founded: 1922*
*David W Hoskins, Executive Director*

**1024 Maryland Association of Soil Conservation Districts**
53 Slama Road
Edgewater, MD 21037
410-956-5771
Fax: 410-956-0161
E-mail: lynnehoot@aol.com
http://www.mascd.net

*Founded: 1956*
*Lynne Hoot, Executive Director*

**1025 Maryland BASS Chapter Federation**
PO Box 3620
Baltimore, MD 21214
301-842-3200
E-mail: teamroger@aol.com
http://www.mdbass.com

*Roger Trageser, President*
*Bill Bennett Jr, 1st Vice President*

**1026 Maryland Native Plant Society**
PO Box 4877
Silver Spring, MD 20914
E-mail: info@mdflora.org
http://www.mdflora.org

Promote awareness, appreciation, and conservation of Maryland's native plants and their habitats.
*Founded: 1992*
*Kirsten Johnson, President*
*Carolyn Fulton, Secretary*

**1027 Maryland Recyclers Coalition**
c/o Mariner Management
PO Box 1046
Laurel, MD 20725
888-496-3196
Fax: 301-238-4579
E-mail: info@marylandrecyclers.org
http://www.marylandrecyclers.org

Mission is to promote sustainable reduction, reuse and recycling of materials otherwise destined for disposal and promote and increase buying products made with recycled matieral content. Seeks to accomplish this mission through acombination of education programs, advocacy activities to affect public policy, technical assistance efforts, and the development of markets to purchase recycled materials and manufacture products with recycled content.
*Virginia Lipscomb, Vice President*
*Brian Ryerson, President*

**1028 Multiple Chemical Sensitivity Referral and Resources**
508 Westgate Road
Baltimore, MD 21229
410-362-6400
Fax: 410-448-3317
E-mail: adonnay@mcsrr.org
http://www.mcsrr.org

A non-profit organization engaged in professional outreach, patient support and public advocacy devoted to the diagnosis, treatment, accommodation and prevention of Multiple Chemical Sensitivity disorders.
*Founded: 1994*
*Albert Donnay MHS, Co-Founder/Executive Directo*

**1029 Nature Conservancy: Maryland/DC Chapter**
5410 Grosvenor Lane
Suite 100
Bethesda, MD 20814
301-897-8570
800-628-6860
Fax: 301-897-0858
E-mail: comment@tnc.org
http://www.tnc.org

To preserve the plants, animals and natural communities that represent the diversity of life on Earth by protecting the lands and waters they need to survive.
*Founded: 1950*
*Steven J McCormick, President/CEO*

**1030 Potomac Region Solar Energy Association**
PO Box 809
Pasadena, MD 21123-0809
866-477-5369
E-mail: info@prsea.org
http://www.prsea.org

A non-profit organization whose purposes are to further the development, use of, and support for solar energy and related arts, sciences, and technologies with concern for the economic, environmental, and social fabric of the region.
*Jim Crowley, Chair*

**1031 Rachel Carson Council**
PO Box 10779
Silver Spring Marlyand, MD 20914
301-593-7507
E-mail: rccouncil@aol.com
http://www.rachelcarsoncouncil.com

An association for the integrity of the environment, seeks to inform and advise the public about the effects of pesticides that threaten the health, welfare, and survival of living organisms and biological systems.
*Founded: 1965*
*Diana Post, Executive Director*

**1032 Sierra Club**
7338 Baltimore Avenue
Suite 101A
College Park, MD 20740
301-277-7111
Fax: 301-277-6699
http://maryland.sierraclub.org

A grassroots environmental organization which promotes appreciation of nature with hikes and outtings. We work to protect the environment in Maryland through legislative and grassroots organizing efforts.
*Founded: 1865*
*Michael Martin, Chair*
*Ron Henry, Vice Chair*

**1033 Spill Control Association of America**
2105 Laurel Bush Road 443-640-1085
Suite 200 Fax: 443-640-1086
Bel Air, MD 21015 http://www.scaa-spill.org
Organized to actively promote the interests of all groups within the spill response community. The organization represents spill response contractors, manufacturers, distributors, consultants, instructors, government & traininginstitutions and corporation working in the industry.
*Founded: 1973*
*John Parker, President*
*Jackie King, Executive Director*

**1034 Trout Unlimited**
1300 North 17th Street 703-522-0200
Suite 500 800-834-2419
Arlington, VA 22209 Fax: 703-284-9400
E-mail: trout@tu.org
http://www.tu.org
Mission: To conserve, protect and restore North America's trout and salmon fisheries and their watersheds. We accomplish this mission on local, state and national levels with an extensive and dedicated volunteer network.
*Founded: 1989*
*Charles Gauvin, President/CEO*
*Steve Moyer, Vice President*

**1035 White Lung Association**
PO Box 1483 410-243-5864
Baltimore, MD 21203 Fax: 410-254-4602
E-mail: jfite@whitelung.org
http://www.whitelung.org
National nonprofit organization dedicated to the education of the public to the hazards of asbestos exposure. Has developed programs of public education and consults with victims of asbestos exposure, school boards, building owners,government agencies and others interested in identifying asbestos hazards and developing control programs.
*James Fite, Executive Director*

**1036 Wildfowl Trust of North America**
600 Discovery Lane 410-827-6694
PO Box 519 E-mail: info@bayrestoration.org
Grasonville, MD 21638 http://www.bayrestoration.org
Mission is to be responsible and protactive environmentally. We strive to improve the health of the Chesapeake Bay. Specifically, we promote environmental stewardship at the 510-acre site, the Chesapeake Bay Environmental Center,through education, restoration and conservation.
*Founded: 1979*
*Judy Wink, Executive Director*
*Beth Poulsen, Executive Assistant*

**1037 Wildlife Society**
5410 Grosvenor Lane 301-897-9770
Suite 200 Fax: 301-530-2471
Bethesda, MD 20814 E-mail: tws@wildlife.org
http://www.wildlife.org
An international, non-profit scientific and educational organization serving and representing wildlife professionals in all areas of wildlife conservation and resource management. Our goal is to promote excellence in wildlifestewardship through science and education.
*Founded: 1937*
*Michael Hutchins, President/CEO*

## Michigan

**1038 American Lung Association**
12751 S Saginaw Street 810-953-3950
Suite 503 Fax: 810-953-3940
Grand Blanc, MI 48439 http://www.midlandlung.org
Serves the Counties of: Arenac, Bay, Genessee, Gladwin, Huron, Lapeer, Livingston, Midland, Saginaw, Sanliac, Shiawassee and Tuscola.
*Founded: 1906*
*Tracy Ross, CEO*

**1039 American Lung Association of Michigan**
American Lung Association
25900 Greenfield 248-784-2000
Suite 401 800-543-5864
Oak Park, MI 48237 Fax: 248-784-2008
E-mail: www.alam.org
http://www.alam.org
*Founded: 1904*
*Ray Maloni, Interim CEO*

**1040 American Lung Association of Michigan: Capital Region Office**
American Lung Association
403 Seymour Avenue 517-484-4541
Lansing, MI 48933 800-678-5864
Fax: 517-484-2118
E-mail: alam@alam.org
http://www.alam.org
Serves the Counties of: Clare, Clinton, Eaton, Gratiot, Hillsdale, Ingham, Ionia, Isabella, Jackson, Lenawee, Mecosta, Montcalm and Osceola.
*Founded: 1904*
*Kevin M Chan, Managing Director*
*Stephen D Moore, President*

**1041 American Lung Association of Michigan: Grand Valley Region**
c/o Wege Center for Health and Learning
300 Lafayette Street SE 616-752-5051
Suite 3400 Fax: 616-752-6972
Grand Rapids, MI 49503
Serves the Counties of: Allegan, Barry, Berrien, Branch, Calhoun, Cass, Kalamazoo, Kent, Muskegon, Newago, Oceana, Ottowa, St. Joseph and Van Buren.

**1042 American Society of Landscape Architects: Michigan Chapter**
1000 W St Joseph Highway 517-485-4116
Suite 200 Fax: 517-485-9408
Lansing, MI 48915 E-mail: chapterinfo@michiganasla.org
http://www.michiganasla.org
*Founded: 1899*
*Norman Cox, President*
*Derek Dalling, Associate Manager*

**1043 Association of Midwest Fish and Wildlife Agencies**
PO Box 30028 517-373-1263
Lansing, MI 48909 Fax: 517-373-6705
http://www.mafwa.iafwa.org
*Founded: 1934*
*Steve Gray, President*
*Becky Humphries, Director-at-Large*

**1044 Ecology Center of Ann Arbor**
117 North Division Street 734-761-3186
Ann Arbor, MI 48104 Fax: 734-663-2414
E-mail: info@ecocenter.org
http://www.ecocenter.org
A member-based, nonprofit environmental organization that is now a regional leader that works for a safe and healthly environment where people live, work, and play.
*Founded: 1970*
*Mike Garfield, Director*

**1045 Great Lakes Commission**
Eisenhower Corporate Park
2805 S Industrial Highway
Suite 100
Ann Arbor, MI 48104
734-971-9135
Fax: 734-971-9150
E-mail: landrews@gic.org
http://www.glc.org

A binational public agency dedicated to the use, management and protection of the water, land and other natural resources of the Great Lakes-St Lawrence system.
*Founded: 1955*
*Tim A Eder, Executive Director*

**1046 Great Lakes Renewable Energy Association**
257 South Bridge Street
PO Box 346
Dimondale, MI 48821
517-646-6269
Fax: 517-646-8584
E-mail: info@glrea.org
http://www.glrea.org

A non-profit organization that educates, advocates, promotes, and publicly demonstrates renewable energy technologies.
*Jennifer Alvarado, Executive Director*

**1047 Home Chemical Awareness Coalition**
Michigab State University
Natural Resource Building
E Lansing, MI 48824
517-355-9578
Fax: 517-353-8994
*Cynthia Frigden, Chairperson*

**1048 Michigan Association of Conservation Districts**
3001 Coolidge Road
Suite 250
East Lansing, MI 48823
517-324-4421
Fax: 517-324-4435
E-mail: lori.phalen@macd.org
http://www.macd.org

A non-governmental, non-profit organization, established to represent and provide services to Michigan's 80 Conservation Districts.
*Tom Middleton, President*
*Lori Phalen, Executive Director*

**1049 Michigan BASS Chapter Federation**
1010 S W Avenue
Jackson, MI 49203
517-789-1008
Fax: 517-789-5603
E-mail: psacks@michiganbass.net
http://www.michiganbass.org

*Paul Sacks, President*

**1050 Michigan Forest Association**
6120 S Clinton Trail
Eaton Rapids, MI 48827
517-663-3423
E-mail: miforest@acd.net
http://www.michiganforests.com

To promote good management on all forest land, to educate our members about good forest practices and stewardship of the land, and to inform the general public about forestry issues and the benefits of good forest management.

**1051 Michigan Natural Areas Council**
c/o Matthaei Botanical Gardens
1800 N Dixboro Road
Ann Arbor, MI 48109
E-mail: mnac@cyberspace.org
http://www.cyberspace.org/~mnac
*Founded: 1946*
*Phyllis Higman, Chair*

**1052 Michigan United Conservation Clubs**
2101 Wood Street
PO Box 30235
Lansing, MI 48912
517-371-1041
Fax: 517-371-1505
http://www.mucc.org

The largest statewide conservation organization with nearly 100,000 members and more than 500 affiliated clubs. MUCC works to conserve Michigan's wildlife, fisheries, waters, forests, air, and soils by providing information, educationand advocacy.
*Founded: 1937*
*Dennis Muchmore, Executive Director*

**1053 National Wildlife Federation Great Lakes Natural Resource Center**
213 W Liberty
Suite 200
Ann Arbor, MI 48104
734-769-3351
Fax: 734-769-1449
E-mail: greatlakes@nwf.org
http://www.nwf.org/greatlakes

Responsible for the National Wildlife Federation's eight state Great Lakes region. The Great Lakes are of global importance but have been used as a garbage dump. The federation has scientists, lawyers, organizers and educators allcontributing their skills to make a change on the health of these freshwater seas.
*Larry J Schweiger, President/CEO*
*Polly Carr, Program Manager*

**1054 Nature Conservancy: Michigan Chapter**
101 E Grand River
Lansing, MI 48906
517-316-0300
Fax: 517-316-9886
E-mail: michigan@tnc.org
http://www.nature.org/michigan

The mission of The Nature Conservancy is to preserve the plants, animals and natural communities that represent the diversity of life on Earth by protecting the lands and waters they need to survive.
*Founded: 1952*
*Helen Taylor, State Director*

**1055 Scenic Michigan**
445 E Mitchell Street
Petoskey, MI 49770
231-347-1171
Fax: 231-347-1185
E-mail: info@scenicmichigan.org
http://www.scenicmichigan.org

An affiliate of the national non-profit organization Scenic America. Work to enhance the scenic beauty of Michigan's communities and roadsides. The principal activity is informing the public of the economic, social and culturalbenefits of highway beautification. Promotes and sponsors programs to encourage natural beauty in the environment, enhance landscapes, protect historical and cultural resources, and improve community appeareance.
*Founded: 1996*
*Abby Dart, Executive Director*
*Jim Lagowski, President*

**1056 Sierra Club: Mackinac Chapter**
109 E Grand River Avenue
Lansing, MI 48906
517-484-2372
Fax: 517-484-3108
E-mail: mackinac.chapter@sierraclub.org
http://michigan.sierraclub.org

To advance the preservation and protection of the natural environment by empowering the citizenry, especially democratically-based grassroots organizations, with charitable resources to further the cause of environmental protection.
*Founded: 1967*
*Anne Woiwode, Director*
*Gayle Miller, Conservation Coordinator*

**1057 Trout Unlimited**
1300 North 17th Street
Suite 500
Arlington, VA 22209
703-522-0200
800-834-2419
Fax: 703-284-9400
E-mail: trout@tu.org
http://www.tu.org

Mission: To conserve, protect and restore North America's trout and salmon fisheries and their watersheds. We accomplish this mission on local, state and national levels with an extensive and dedicated volunteer network.
*Charles Gauvin, President/CEO*
*Steve Moyer, Vice President*

**1058 Wildflower Association of Michigan**
3853 Farrell Road
Hastings, MI 49058
269-948-2496
700-333-6459
Fax: 269-948-2957
E-mail: wam@iserv.net
http://www.wildflowersmich.org

A nonprofit organization whose mission is to promote, coordinate, and participate in education, enjoyment, science, and stewardship of native wildflowers and their habitats - including

promoting public education of proper principles,ethics, and methods of landscaping with native wildflowers and associated habitats.
*Founded: 1986*
*Cheryl Smith Tolley, President*
*Esther Durnwald, Vice President*

**1059 Wildlife Society: Michigan Chapter**
Michigan State University
Department of Natural Resources 517-353-2042
Room 13 Fax: 517-432-1699
East Lansing, MI 48824 E-mail: campa@msu.edu
http://www.wildlife.org/chapters/mi/index.cfm?tname=officers
A nonprofit scientific and educational organization that serves professionals such as government agencies, academia, industry, and non-government organizations in all areas related to the conservation of wildlife and natural resourcesmanagement.
*Founded: 1982*
*Brent Rudolph, President-Elect*
*Scott Winterstein, President*

## Minnesota

**1060 American Lung Association in Minnesota**
490 Concordia Avenue 651-227-8014
Saint Paul, MN 55103 800-586-4872
Fax: 651-227-5459
E-mail: info@lungmn.org
http://www.lungmn.org
To save lives, improve lung health, and prevent lung disease.
*Founded: 1903*
*Penny Fena, Executive Director*
*Harold Wimmer, CEO*

**1061 American Lung Association of Minnesota: Greater Minnesota Branch Office**
424 West Superior Street 218-726-4721
Suite 203 800-548-8252
Duluth, MN 55802 Fax: 218-726-4722
E-mail: info@alamn.org
http://www.alamn.org/
To prevent lung disease and promote lung health.
*Founded: 1903*
*Jerry Orr, Chief Executive Officer*

**1062 American Society of Landscape Architects: Minnesota Chapter**
International Market Square 612-339-0797
275 Market Street, Suite 54 Fax: 612-338-7981
Minneapolis, MN 55405 E-mail: info@masla.org
http://www.masla.org/
*Ellen Stewart, President*
*Tom Moua, Association Manager*

**1063 Institute for Agriculture and Trade Policy**
2105 First Avenue S 612-870-0453
Minneapolis, MN 55404 Fax: 612-870-4846
E-mail: iatp@iatp.org
http://www.iatp.org
Promotes resilient family farms, rural communitites and ecosystems around the world through research and education, science and technology, and advocacy.
*Founded: 1987*
*Jim Harkness, President*

**1064 Minnesota Association of Soil and Water Conservation Districts**
Soil and Water Conservation Districts
790 Cleveland Avenue S 651-690-9028
Suite 201 Fax: 651-690-9065
St. Paul, MN 55116 http://www.maswcd.org
MASWCD is a nonprofit organization which exists to provide leadership and a common voice for Minnesota's soil and water conservation districts and to maintain a positive, results-oriented relationship with rule making agencies,partners and legislators; expanding education opportunities for the districts so they may carry out effective conservation programs.
*Founded: 1952*
*Ken Pederson, President*
*Steve Sunderland, Vice President*

**1065 Minnesota BASS Chapter Federation**
PO Box 225 612-339-5609
Howard Lake, MN 55349 E-mail: jbarnett@mnbfn.org
http://www.mnbfn.org
Mission is to stimulate public awareness of bass fishing as a major sport; to offer our State Conservation Department, our organized and moral and political supports and encouragement, to promote full adherence to, and enforcement ofexisting conservation regulations; to promote and encourage youth fishing and teach youth the importance of Catch and Release; to improve our skills as bass anglers through a friendly exchange of ideas and techniques used in tournament fishing.
*Joe B Barnett, President*

**1066 Minnesota Conservation Federation**
542 Snelling Avenue S 651-690-3077
#104 800-531-3077
Saint Paul, MN 55116 Fax: 651-690-2208
E-mail: mncf@mtn.org
http://www.mncf.org
A common sense conservation organization made up of hunters, anglers and others who are dedicated to the enjoyment, education and ethical use of our natural resources.
*Founded: 1936*
*Steve Maurice, President*

**1067 Minnesota Ground Water Association**
4779 126th Street North 651-296-7822
White Bear Lake, MN 55110 E-mail: office@mgwa.org
http://www.mgwa.org
A non-profit, volunteer organization which promotes public policy and scientific education about ground water.
*Founded: 2000*
*Scott Alexander, President*
*Jon Pollock, Secretary*

**1068 Minnesota Renewable Energy Society**
2928 Fifth Avenue S 612-308-4757
Minneapolis, MN 55408 E-mail: info@mnrenewables.org
http://www.mnrenewables.org
A member-run, non-profit organzation founded to promote the use of, and to engage in advocacy for, renewable energies in Minnesota through education and through the demonstration of practical applications.
*Founded: 1978*
*David Boyce, Chairman*

**1069 Minnesota Wings Society**
Bobwhite Quail Society of Minnesota
PO Box 11323 612-588-2966
Minneapolis, MN 55411 E-mail: wtcn.nature@att.net
http://www.nmu.edu/sbp/us_off.html
*Founded: 1975*
*Thurman Tucker, President*
*Martin Hanson, Secretary*

**1070 National Flyway Council: Mississippi Office Section of Wildlife Natural Resources**
North American Flyways
PO Box 30444 517-373-1263
Lansing, MI 48909 Fax: 517-373-6705
E-mail: humphrir@state.mi.us
http://www.npwrc.usgs.gov/info/flyway/flychair.htm
*Roger Holmes, Chairman*
*Joshua L Sandt, Deputy Director*

**1071 Nature Conservancy: Minnesota Chapter**
1101 W River Park Way
Suite 200
Minneapolis, MN 55415
612-331-0750
Fax: 612-331-0770
E-mail: minnesota@tnc.org
http://www.nature.org

Aims to preserve plants, animals, and natural communities that represent the diversity of life on Earth by protecting the lands and waters they need to survive.
*Founded: 1951*
*Steven McCormick, President/CEO*

**1072 Parks and Trails Council of Minnesota**
275 E 4th Street
Suite 250
Saint Paul, MN 55101
651-726-2457
800-944-0707
Fax: 651-726-2458
E-mail: info@parksandtrails.org
http://www.parksandtrails.org

Mission: To acquire, protect and enhance critical lands for the public's enjoyment now and in the future.
*Founded: 1954*
*Judith Erickson, Government Relations*
*Beth Coleman, Executive Director*

**1073 Raptor Center**
The College of Veterinary Sciences
University of Minnesota
1920 Fitch Avenue
St.Paul, MN 55108
612-624-4745
Fax: 612-624-8740
E-mail: raptor@umn.edu
http://www.raptor.cvm.umn.edu

Specializes in the medical care, rehabilitation, and conservation of eagles, hawks, owls and falcons. In addition to treating approxminately 800 birds a year, the internationally known program reaches more than 240,000 people each yearthrough public education programs and events, provides training in avian medicine and surgery for veterinarians from around the world, and identifies emerging issues related to raptor health and populations.
*Founded: 1974*
*Dr Julia Ponder, Executive Director*

**1074 Sierra Club-North Star Chapter**
2327 E Franklin Avenue
Suite 1
Minneapolis, MN 55406
612-659-9124
Fax: 612-659-9129
http://www.northstar.sierraclub.org

To advance the preservation and protection of the natural environment by empowering the citizenry, especially democratically based grassroots organizations, with charitable resources to further the cause of environmental protection,the vehicle through which The Sierra Club Foundation generally fulfills its charitable mission.
*Scott Elkins, State Director*
*Heather Cusick, Conservation Director*

## Mississippi

**1075 American Lung Association of Mississippi**
PO Box 2178
Ridgeland, MS 39158
601-206-5810
800-586-4872
Fax: 601-206-5813
E-mail: jcofer@alams.org
http://www.alams.org

To prevent lung disease and promote lung health through direct assistance, education programs, advocacy and research.
*Founded: 1914*
*Sandra Holman, President*
*Robin Robinson, Vice President*

**1076 American Society of Landscape Architects: Mississippi Chapter**
PO Box 55726
Jackson, MS 39296
601-898-0775
Fax: 601-898-9112
http://www.msasla.org

*George Ewing III, President*
*Robert Mercier, President-Elect*

**1077 Crosby Arboretum**
Mississippi State University
PO Box 1639
Picayune, MS 39466
601-799-2311
Fax: 601-799-2372
E-mail: crosbyar@datastar.net
http://www.crosbyarboretum.msstate.edu

Dedicated to educating the public about their environment. THis mission is carried out by preserving, protecting, and displaying plants native to the Pearl River Drainage Basin ecosystem, providing environmental and botanical researchopportunities, and offering culutral, scientific, and recreational programs.
*Patricia Knight, Director/Head*
*Melinda Lyman, Senior Curator/On-Site Dir*

**1078 Mississippi Solar Energy Society**
PO Box 141
Columbia, MS 39429
504-319-4701
*Sammy C Germany, Contact*

**1079 Mississippi Wildlife Federation**
855 South Pear Orchard Rd
Suite 500
Ridgeland, MS 39157
601-206-5703
Fax: 601-206-5705
E-mail: cshropshire@mswf.org
http://www.mswildlife.org

Established to advance the protectin of wildlife in Mississippi
*Founded: 1946*
*Dr Jon Jackson, President*
*Dr Cathy Shropshire, Executive Director*

**1080 Sierra Club**
PO Box 4335
Jackson, MS 39296
601-352-1026
Fax: 601-355-1506
E-mail: sierrams@bellsouth.net
http://mississippi.sierraclub.org

The Chapter is currently involved with the air pollution concern that local residents were exposed to after Hurricane Katrina. The general mission is to advance the preservation and protection of the natural environment by empoweringthe citizenry, especially democratically-based grassroots organizations, with charitable resources to further the cause of environmental protection.
*Howard Page, Chairman*
*Becky Gillette, Vice Chair*

**1081 Wildlife Society**
PO Box 820161
Vicksburg, MS 39182
601-631-7133
Fax: 601-631-7133
E-mail: julie.b.marcy@usaace.army.mil
http://www.wildlife.org

A nonprofit scientific and educational organization that serves professionals such as government agencies, academia, industry, and non-government organizations in all areas related to the conservation of wildlife and natural resourcesmanagement.
*Founded: 1936*
*Jane Pelkey, Finance Coordinator*

## Missouri

**1082 American Fisheries Society: North Central Division**
420 New Haven Road
Columbia, MO 65201
573-875-5399
Fax: 573-876-1896
E-mail: pamela_haverland@usgs.gov

*Pamela Haverland, President*

**1083 American Lung Association of Missouri**
1118 Hampton Avenue
Saint Louis, MO 63139
314-645-5505
800-586-4872
Fax: 314-645-7128
http://www.lungusa.org

*Founded: 1904*

**1084 American Lung Association of Missouri: Southeast Missouri Office**
PO Box 482
Cape Girardeau, MO 63702
573-651-3313
Fax: 573-651-1883
http://www.lungusa2.org/missouri/
*Founded: 1907*

**1085 American Lung Association of Missouri: Kansas City Office**
2400 Troost Ave
Suite 4300
Kansas City, MO 64108
816-842-5242
Fax: 816-842-5470
E-mail: kcmo@alawmo.com
http://www.lungusa.org
*Founded: 1904*

**1086 American Lung Association of Missouri: Southwest Missouri Office**
2053-D South Waverly
Springfield, MO 65804
417-883-7177
Fax: 417-883-7026
http://www.lungusa2.org/missouri
*Founded: 1907*

**1087 American Society of Landscape Architects: Prairie Gateway Chapter**
104 W 9th Street
Suite 101
Kansas City, MO 64105
816-421-1054
http://www.pgasla.org
Represents memberhip from the states of Kansas and Missour. The purpose is to promote the profession of landscape architecture and advancement of the practice through advocacy, education, communicatio, and fellowship.
*Joe Daly, President*
*Mike McGrew, President-Elect*

**1088 American Society of Landscape Architects: St Louis Chapter**
1831 Chestnut Street
Suite 700
St Louis, MO 63125
314-206-4313
E-mail: millerl@pbworld.com
http://www.stlouisasla.org
Represents the eastern half of Missouri and is responsible for the promotion and legislation of the landscape architect profession in the St Louis region.
*Lenn Miller, President*

**1089 Kansas BASS Chapter Federation**
9712 Juniper Lane
Overland Park, KS 66207
913-385-2277
http://www.kbcf.com
The purpose is to stimulate public awareness of bass fishing as a major sport.
*Eric Strong, President*

**1090 Missouri Audubon Council**
2620 Forum Boulevard
Suite C-1
Columbia, MO 65201
573-447-2249
Fax: 573-447-2428
E-mail: missouri@audubon.org
http://http://mo.audubon.org
To conserve and restore natural ecosystems, focusing on birds and other wildlife, and their habitats for the benfit of humanity and the earth's biological diversity.
*Founded: 1990*
*Bruce Carr, Executive Director/VP*

**1091 Missouri Forest Products Association**
611 E Capitol Ave
Jefferson City, MO 65101
573-634-3252
Fax: 573-636-2591
E-mail: moforest@moforest.org
http://www.moforest.org
Brings together timberland owners, forest products companies, state agencies, and professional foresters to promote good forest stewardship throughout the state. A membership organization that works together to educate and assistprimary and secondary wood processors, and timberland owners through programs, educational opportunities, publications, and membership benefits.
*Brian Brookshire, Executive Director*

**1092 Missouri Prairie Foundation**
PO Box 200
Columbia, MO 65205
579-356-7828
888-843-6739
Fax: 573-442-0260
E-mail: missouriprairie@yahoo.com
http://www.moprairie.org
Works with public and private partners to protect and restore our prairie and native grassland communities through land acquisition, management, education and research.
*Founded: 1966*
*Steve Mowry, President*
*Justin Johnson, Executive Director*

**1093 Missouri Public Interest Research Group**
310A North Euclid
Saint Louis, MO 63108
314-454-9560
Fax: 314-454-0787
E-mail: info@mopirg.org
http://www.mopirg.org
To deliver persistent, result-oriented public interest activism that protects our environment, encourages a fair, sustainable economy, and fosters responsive, democratic government.
*Founded: 1972*

**1094 Missouri Stream Team: Missouri Department of Conservation**
PO Box 180
Jefferson City, MO 65102
800-781-1989
E-mail: streamteam@mdc.mo.gov
http://www.mostreamteams.org
A working partnership of citizens who are concerned about Missouri Stream. The Stream Team Program Provides an opportunity for all interested to get involved in river conservation.

**1095 Rocky Mountain Elk Foundation**
5705 Grant Creek
Missoula, MT 59808
406-523-4500
800-225-5355
http://www.rmef.org
Committed to conserving, restoring and enhancing natural habitats; promoting the sound management of wild, free-ranging elk, which may be hunted orotherwise enjoyed; fostering cooperation among federal, state, tribal and privateorganization and individuals in wildlife management and habitat conservation; and educating members and the public about habitat conservation, the value of hunting, hunting ethics and wildlife management.
*Founded: 1984 155,000 Members*
*M David Allen, President/CEO*

**1096 Scenic Missouri**
3963 Wyoming Street
St Louis, MO 63116
314-265-5328
E-mail: info@scenicmissouri.org
http://www.scenicmissouri.org
To preserve and enhance the scenic beauty of Missouri
*John Regenbogen, Executive Director*

**1097 Sierra Club**
7164 Manchester Avenue
Maplewood, MO 63143
314-644-1011
800-628-5333
E-mail: missouri.chapter@sierraclub.org
http://missouri.sierraclub.org
*Ginger Harria, Chair*

**1098 Society for Environmental Geochemistry and Health**
1870 Miner Circle
Rolla, MO 65409
573-341-4831
Fax: 303-556-4822
http://www.segh.net
Established to provide a forum for scientists from various disciplines to work together in understanding the interaction between the geochemical environment and the health of plants, animals, and humans.
*Founded: 1971*
*Andrew Hunt, President*

**1099 Trout Unlimited**
1300 North 17th Street
Suite 500
Arlington, VA 22209
703-522-0200
800-834-2419
Fax: 703-284-9400
E-mail: trout@tu.org
http://www.tu.org

Mission: To conserve, protect and restore North America's trout and salmon fisheries and their watersheds. We accomplish this mission on local, state and national levels with an extensive and dedicated volunteer network.
*Founded: 1959*
*Charles Gauvin, President/CEO*
*Steve Moyer, Vice President*

## Montana

**1100 American Lung Association of the Northern Rockies**
825 Helena Avenue
Helena, MT 59601
406-442-6556
Fax: 406-442-2346
E-mail: ala-nr@ala-nr.org
http://www.ala-nr.org

**1101 Chemical Injury Information Network**
PO Box 301
White Sulphur Springs, MT 59645
406-547-2255
Fax: 406-547-2455
http://www.ciin.org

A support and advocacy organization dealing with Multiple Chemical Sensitivities. It is run by the chemically injured for the benefit of the chemically injured, crediable research into MCS, and the empowerment of the chemicallyinjured.
*Founded: 1990*
*Cinthia Wilson, Executive Director*
*John Wilson, President*

**1102 Craighead Environmental Research Institute**
201 S Wallace Avenue
Bozeman, MT 59715
405-585-8705
Fax: 406-587-5951
E-mail: info@craigheadresearch.org
http://craigheadresearch.org

A network of biologists dedicated to providing reliable information, through innovative research and state-of-the-art conservation planning, to foster ecologically sound management of wildlife and their habitats. CERI focuses itsefforts in the Yellowstone-to-Yukon and Coastal Rainforest regions of North America
*Lance Craighead, Executive Director*

**1103 Craighead Wildlife: Wetlands Institute**
5200 Upper Miller Creek Road
Missoula, MT 59803
406-251-3867
Fax: 406-251-5069
http://www.grizzlybear.org

*John A Mitchell PhD, Director*

**1104 Foundation for Research on Economics and the Environment (FREE)**
662 Ferguson Avenue
Bozeman, MT 59718
406-585-1776
Fax: 406-585-3000
http://www.free-eco.org

FREE mission is to advance conservation and environmental values consistent with individuals freedom and responsibility. The Foundation's intellectual entrepreneurs develop environmental policies featuring private property rights,market incentives, and voluntary organizations. FREE achieves its mission by working with leaders in universities, businesses, environmental groups, government, the media, and think tanks.
*John Baden, Chairman*

**1105 Greater Yellowstone Coalition**
PO Box 1874
Bozeman, MT 59771
406-586-1593
800-775-1834
Fax: 406-556-2839
E-mail: gyc@greateryellowstone.org
http://www.greateryellowstone.org

People protecting the lands, waters, and wildlife of the Greater Yellowstone Ecosystem, now and for future generations.
*Founded: 1983*
*Mike Clark, Executive Director*

**1106 Montana Association of Conservation Districts**
501 N Sanders
Helena, MT 59601
406-443-5711
Fax: 406-443-0174
E-mail: mail@macdnet.org
http://www.macdnet.org

Montana's 58 Conservation Districts utilize locally-led and largely non-regulatory approaches to successfully address general natural resource issues. CD's have a decades-long history of conserving our state's resources by helpinglocal people match their needs with technical and financial resources, thereby getting good conservation practices on the ground to benefit all of Montanans.
*Sarah Carlson, Executive Director*

**1107 Montana Audubon**
PO Box 595
Helena, MT 59624
406-443-3949
Fax: 406-443-7144
E-mail: myaudubon@montana.com
http://www.mtaudubon.org

Promotes appreciation, knowledge and conservation of native birds, other wildlife, and their habitats.
*Founded: 1976*
*Steve Hoffman, Executive Director*

**1108 Montana Environmental Information Center**
PO Box 1184
Helena, MT 59624
406-443-2520
Fax: 406-443-2507
E-mail: meic@meic.org
http://www.meic.org

MEIC's purpose is to protect and restore Montana's natural environment. It works to do this by: monitoring and influencing the decisions and activities of the state, local and federal governments; educating individuals and by assistingindividuals and other nonprofit organizations.
*Jim Jensen, Executive Director*
*Anne Hedges, Program Director*

**1109 Montana Land Reliance**
324 Fuller Avenue
PO Box 355
Helena, MT 59624
406-443-7027
Fax: 406-443-7061
E-mail: info@mtlandreliance.org
http://www.mtlandreliance.org

Montana's only private, statewide land trust, an apolitical, nonprofit corporation. Our mission is to provide permanent protection for private lands that are ecologically significant for agricultural production, fish and wildlifehabitat and scenic open space.
*Bill Long, Managing Director*
*Rock Ringling, Managing Director*

**1110 Montana Water Environment Association**
516 N Park Street
Suite A
Helena, MT 59601
406-449-7913
Fax: 406-449-6350

**1111 Montana Wildlife Federation**
PO Box 1175
Helena, MT 59624
404-458-0227
800-517-7256
Fax: 403-458-0373
E-mail: mwf@mtwf.org
http://www.montanawildlife.com

An organization of conservation minded people who share a mission to protect and enhance Montana's piblic wildlife, lands, waters and fair chase hunting and fishing heritage.
*Craig Sharpe, Executive Director*
*Jan Cronin, Development Director*

**1112 National Wildlife Federation Northern Rockies Natural Resource Center**
240 N Higgins
Suite 2
Missoula, MT 59802
406-721-6705
Fax: 406-721-6714
E-mail: scaggs@nwf.org
http://www.nwf.org

Protects the national treasure of the northern Rockies, uniting people throughout Montana, North Dakota, South Dakota and Idaho to protect and enhancethe habitat, and fish and wildlife populations.
*S Scaggs, Contact*

**1113 Northwest Power and Conservation Council**
Capitol Station
1301 Lockey
Helena, MT 59620
406-444-3952
Fax: 406-444-4339
E-mail: ptyree@nwcouncil.org
http://www.nwcouncil.org/contact/mt.asp

The council develops and maintains a regional power plan and a fish and wildlife program to balance the Northwest's environment and energy needs.
*Bruce Measure, Council Vice Chair*
*Rhonda Whiting, Council Member*

**1114 Rocky Mountain Elk Foundation**
5705 Grant Creek
Missoula, MT 59808
406-523-4500
800-225-5355
http://www.rmef.org

Committed to conserving, restoring and enhancing natural habitats; promoting the sound management of wild, free-ranging elk, which may be hunted or otherwise enjoyed; fostering cooperation among federal, state, tribal and privateorganizations and individuals in wildlife management and habitat conservation; and educating members and the public about habitat conservation, the value of hunting, hunting ethics and wildlife management.
*155,000 Members*
*M David Allen, President/CEO*

**1115 Sierra Club**
PO Box 1290
Bozeman, MT 59771
406-582-8365
Fax: 408-582-9417
http://montana.sierraclub.org

To advance the preservation and protection of the natural environment by empowering the citizenry, especially democratically-based grassroots organizations, with charitable resources to further the cause of environmental protection.The vehicle through which The Sierra Club Foundation generally fulfills its charitable mission.
*Jeff van den Noort, Chair*
*Ron Mueller, Vice Chair*

**1116 Trout Unlimited**
1300 North 17th Street
Suite 500
Arlington, VA 22209
703-522-0200
800-834-2419
Fax: 703-284-9400
E-mail: trout@tu.org
http://www.tu.org

Mission: To conserve, protect and restore North America's trout and salmon fisheries and their watersheds. We accomplish this mission on local, state and national levels with an extensive and dedicated volunteer network.
*Charles Gauvin, President/CEO*
*Steve Moyer, Vice President*

**1117 Wildlife Society**
3630 Columbus
Butte, MT 59701
406-533-3445
Fax: 406-533-3600
E-mail: montanatws@montanatws.org
http://www.montanatws.org

A nonprofit scientific and educational organization that serves professionals such as government agencies, academia, industry, and non-government organizations in all areas related to the conservation of wildlife and natural resourcesmanagement.
*Founded: 1937*
*Tom Carlsen, President*
*Barb Pitman, Secretary*

## Nebraska

**1118 American Lung Association of Nebraska**
7101 Newport Avenue
Suite 303
Omaha, NE 68152
402-572-3030
Fax: 402-572-3028
E-mail: ala@lungnebraska.org
http://www.lungnebraska.org
*Founded: 1904*
*Sara Dreiling, CEO*

**1119 American Society of Landscape Architects: Great Plains Chapter**
c/o HDR Engineering
8404 Indian Hills Drive
Omaha, NE 68114
402-399-1399
Fax: 402-392-6713
E-mail: jay.gordon@HDRINC
http://www.asla.org/chapters/greatplains
*Jay Gordon, President*
*Bradley Young, President-Elect*

**1120 Iowa Prairie Network**
6736 Laurel
Omaha, NE 68104
402-571-6230
Fax: 402-571-6230
E-mail: pollockg@top.net
http://www.iowaprairienetwork.org
*Glenn Pollock, President*
*David Hansen, Director*

**1121 Nature Conservancy: Nebraska Chapter**
1025 Leavenworth Street
Suite 100
Omaha, NE 68102
402-342-0282
Fax: 402-342-0474
E-mail: nebraska@tnc.org
http://www.nature.org

The mission of the Nature Conservancy is to preserve the plants, animals and natural communities that represent the diversity of life on Earth by protecting the lands and waters they need to survive.
*Founded: 1915*
*Steven J McCormick, President*
*Henry M Paulson Jr, Chairman*

**1122 Nebraska Association of Resource Districts**
601 S 12th Street
Suite 201
Lincoln, NE 68508
402-471-7670
Fax: 402-471-7677
E-mail: nard@nrdnet.org
http://www.nrdnet.org

The mission is to assist NRDs in a coordinated effort to accomplish collectively what mat not be accomplished individually to conserve, sustain, and improve our natural resources and environment.
*Dean E Edson, Executive Director*
*Jeanne Dryburgh, Office Manager*

**1123 Nebraska BASS Chapter Federation**
National B.A.S. S. Chapter Federation
1518 Kozy Drive
Columbus, NE 68601
402-563-2297
E-mail: admin@nebraskabass.com
http://www.nebraskabass.com
*Joe Citta, President*
*Dave Knuth, Vice President*

**1124 Nebraska Wildlife Federation**
PO Box 81437
Lincoln, NE 68501
402-477-1008
Fax: 402-994-2021
E-mail: nebraskawildlife@altel.net
http://www.nebraskawildlife.org

A state-wide, non-profit membership organization dedicated to fish and wildlife conservation through environmental education, fish and wildlife conservation, and common sense public policy.
*Founded: 1970*
*Dan Stahr, Executive Director*

**1125 Sierra Club**
PO Box 4664
Omaha, NE 68104
http://nebraska.sierraclub.org

The Nebraska Chapter is divided into four groups and is active in statewide programs in conservation; legislative involvement; outings and programs.
*Dick Boyd, Chair*
*JoEllen Polzien, Vice Chair*

**1126 Wildlife Society**
The Wildlife Society
45090 Elm Island Road
Gibbon, NE 68840
308-865-5308
Fax: 308-865-5309
E-mail: mhumpert@lycosmail.com
http://www.wildlifeconsult.com/netws/

A nonprofit scientific and educational organization that serves professionals such as government agencies, academia, industry, and non-government organizations in all areas related to the conservation of wildlife and natural resourcesmanagement.
*Founded: 1937*
*Chris Helzer, President*
*Renae Held, Secretary*

## Nevada

**1127 American Lung Association of Nevada**
10615 Double R Bldv
Reno, NV 89521
775-829-5864
Fax: 775-829-5850
E-mail: dszabo@lungnevada.org
*Founded: 1904*
*Dorothy Szabo, Office Manager*

**1128 American Society of Landscape Architects: Nevada Chapter**
c/o Stone Peak Services
PO Box 12507
Las Vegas, NV 89112
702-454-3057
Fax: 702-454-3097
E-mail: nvasla@earthlink.net
http://host.asla.org/chapters/snasla
*Tammi A Gaudet, President*
*Helen Stone, Executive Director*

**1129 Nevada Wildlife Federation**
PO Box 71238
Reno, NV 89570
775-885-0405
Fax: 775-885-0405
E-mail: nvwf@nvwf.org
http://www.nvwf.org

The federation represents the views of hunters, fisherman and anyone who deeply cares about our wildlife and wild lands.
*Kevin Cobble, President*

**1130 Sierra Club: Toiyabe Chapter**
PO Box 8096
Reno, NV 89507
702-323-3162
http://nevada.sierraclub.org
The Toiyabe Chapter (Eastern Sierra) serves Nevada and Eastern California. Advances the preservation and protection of the natural environment by empowering the citizenry, especially democratically-based grassroots organizations, withcharitable resources to further the cause of environmental protection.
*David Hornbeck, Chair*
*Dennis Ghiglieri, Conservation*

**1131 Solar NV**
10624 S Eastern Ave
Suite A-609
Henderson, NV 89052
702-507-0093
Fax: 702-507-0093
E-mail: info@solarnv.org
http://www.solarnv.org
The Southern Nevada Chapter of the American Solar Energy Society. Educates Southern Nevadans about the benefits of renewable energy and to encourage and promote the use of sustainable energy technology.

**1132 Sunrise Sustainable Resources Group**
PO Box 19074
Reno, NV 89511
775-348-7192
E-mail: president@sunrisenevada.org
http://www.sunrisenevada.org

Empowers Nevadans to use resources responsibly through education, advocacy and community.
*Founded: 1996*
*Philip Moore, President*
*Brian Bass, VP*

**1133 Tahoe Regional Planning Agency**
PO Box 5310
Stateline, NV 89449
702-588-4547
Fax: 702-588-4527
E-mail: trpa@trpa.org
http://www.trpa.org
To cooperatively lead the effort to preserve, restore and enhance the unique natural and human environment of the Lake Tahoe region now and in the future.
*John Singlaub, Executive Director*

## New Hampshire

**1134 American Bass Association of New Hampshire**
235 Ridgeview Road
Weare, NH 03281
603-529-2642
*John Cowan, President*

**1135 Audubon Society of New Hampshire**
84 Silk Farm Road
Concord, NH 03301
603-224-9909
Fax: 603-226-0902
E-mail: asnh@nhaudubon.org
http://www.nhaudubon.org
A nonprofit state wide membership organization that is dedicated to the conservation of wildlife and habitat throughout the state. The mission is to protect New Hampshire's natural environment for wildlife anf for people.
*Founded: 1914*
*Michael J Bartlett, President/CEO*
*Carol Foss, Director, Conservation*

**1136 Breathe New Hampshire**
9 Cedarwood Drive
Unit 12
Bedford, NH 03110
603-669-2411
800-835-8647
Fax: 603-645-6220
E-mail: info@breathenh.org
http://www.breathnh.org
Formerly the American Lung Association of New Hampshire, is committed to eliminating lung disease and improving the quality of life for those with lung disease in New Hampshire.
*Founded: 1916*
*Daniel Fortin, President/CEO*

**1137 Nature Conservancy: New Hampshire Chapter**
22 Bridge Street
4th Floor
Concord, NH 03301
603-224-5853
Fax: 603-228-2459
E-mail: naturenewhampshire@tnc.org
http://www.nature.org
The mission of the Nature Conservancy is to preserve the plants, animals, and nature communities that represent the diversity of life on earth by protecting the lands and waters they need to survive.
*Founded: 1987*
*Daryl Burtnett, State Director*

**1138 New Hampshire Association of Conservation Commissions**
54 Portsmouth Street
Concord, NH 03301
603-224-7867
Fax: 603-228-0423
E-mail: info@nhacc.org
http://www.nhacc.org

The New Hampshire Association of Conservation Commissions is a private, non-profit association of municipal conservation commissions. Its purpose is to foster conservation and appropriate use of New Hampshire's natural resources byproviding assistance to conservation commissions, facilitating communication

and cooperation among commissions, and helping to create a climate in which commissions can be successful.
*Founded: 1970*
*Carol K Andrews, Executive Director*

**1139 New Hampshire Association of Conservation Districts**
PO Box 2311
Concord, NH 03302
603-796-2615
Fax: 603-796-2600
E-mail: director@nhacd.org
http://www.nhacd.org
Provides statewide coordination, representation, and leadership for Conservation Districts to conserve, protect, and promote responsible use of New Hampshire's natural resources.
*Founded: 1946*
*Michele L Tremblay, Executive Director*

**1140 New Hampshire Lakes Association (NH Lakes)**
84 Silk Farm Road
Concord, NH 03301
603-226-0299
Fax: 603-224-9442
E-mail: info@nhlakes.org
http://www.nhlakes.org
A nonprofit, tax-exempt volunteer organization established by the merger of two citizens' groups, each with a history of accomplishments in the protection of lakes. Works on issues concerning shoreland and watershed protection; waterquality improvement; boating safety; lake environment education; and fisheries and wildlife preservation.
*Founded: 1992*
*Jared A Twutsch, President*

**1141 New Hampshire Wildlife Federation**
54 Portsmouth Street
Concord, NH 03301
603-224-5953
Fax: 603-228-0423
E-mail: info@nhwf.org
http://www.nhwf.org
A non-profit member organization promoting conservation, environmental education, sportsmanship, and outdoor activities such as hunting, fishing, trapping, camping and photography. The mission is to be the leading advocate for thepromotion and protection of hunting, fishinf and trapping as well as the conservation of fish and wildlife habitat.
*Founded: 1933*
*Janice Boynton, President*

**1142 Northeast Resource Recovery Association**
2101 Dover Road
Epsom, NH 03234
603-736-4401
Fax: 603-736-4402
E-mail: info@nrra.net
http://www.recyclewithus.org
The Northeast Resource Recovery Association is a pro-active nonprofit working with its membership to make their recycling programs strong, efficient, and financially successful by providing cooperative marketing, cooperativepurchasing, education and networking opportunities; developing innovative recycling programs; creating sustainable alternatives to reduce the volume and toxicity of the waste, and educating and informing local officials about recycling and solidwaste issues.
*Founded: 1981*
*Rick Cooper, President*
*Paula Dow, Executive Director*

**1143 Sierra Club: New Hampshire Chapter**
40 North Main Street
2nd Floor
Concord, NH 03301
603-224-8222
Fax: 603-224-4719
E-mail: feedback@nh.sierraclub.org
http://portal.nhsierraclub.org
Nonprofit member-supported public interest organization that promotes conservation of the natural environment through public policy decisions.
*Founded: 1992*
*Aline Lotter, Acting Chair*

**1144 Trout Unlimited**
1300 North 17th Street
Suite 500
Arlington, VA 22209
703-522-0200
800-834-2419
Fax: 703-284-9400
E-mail: trout@tu.org
http://www.tu.org
Mission: To conserve, protect and restore North America's trout and salmon fisheries and their watersheds. We accomplish this mission on local, state and national levels with an extensive and dedicated volunteer network.
*Founded: 1959*
*Charles Gauvin, President/CEO*
*Steve Moyer, Vice President*

## New Jersey

**1145 American Bass Association of Eastern Pennsylvania/New Jersey**
7 Logan Drive
Somerville, NJ 08876
908-526-7721
Fax: 908-685-0970
E-mail: ehargraves@sdamechanical.com
http://www.aba-of-eastern-pa-nj.com
Organized with three primary purposes for: to ensure the future of fishing through the protection and enhancement of the fishery resource; to promote bass fishing across America as a major sport; and to introduce youngsters to the joyof fishing, sportsmanship and instill in them an appreciation of the life-giving waters of America.
*Founded: 1974*
*Ed Hargraves, President*
*James Norton, Secretary*

**1146 American Lung Association of the Mid-Atlantic**
3001 Old Gettysburg Road
Camp Hill, PA 07083
717-541-5864
Fax: 717-541-8828
E-mail: info@lunginfo.org
http://www.lunginfo.org
The mission of the American Lung Association is to save lives through the prevention of lung disease and the promotion of lung health. Covers PA, WV, & NJ areas.
*Founded: 1904*
*Kenneth G Hysdock, Board Chair*

**1147 American Society of Landscape Architects: New Jersey Chapter**
414 River View Plaza
Trenton, NJ 08611
609-393-7500
Fax: 609-393-9891
E-mail: jsimonetta@publicstrategiesimpact.com
http://www.njasla.net
*Founded: 1901*
*Jeffrey A Tandul, President*
*Joseph Simonetta, Executive Director*

**1148 Association of New Jersey Environmental Commissions**
PO Box 157
Mendham, NJ 07945
973-539-7547
Fax: 973-539-7713
E-mail: info@anjec.org
http://www.anjec.org
To promote the public interest in natural resource protection, sustainable development and reclamation and to support environmenal commissions and open space committees working with citizens and other non-profit organizations.
*Sandy Batty, Executive Director*

**1149 Edison Facilities**
2890 Woodbridge Avenue
Ms 100
Edison, NJ 08837
732-321-6754
Fax: 732-321-4381

**1150 Environmental and Occupational Health Science Institute**
Rutgers University

170 Frelinghuysen Road 732-445-0200
PO Box 1179 Fax: 732-445-0131
Piscataway, NJ 08854 http://eohsi.rutgers.edu
Sponsors research, education and service programs in a setting that fosters interaction among experts in environmental health, toxicology, occupational health, exposure assessment, public policy and health education. The Institute alsoserves as an unbiased source of expertise about environmental problems for communities, employers and government in all areas of occupational and environmental health, toxicology and risk assessment.
*Dr Deborah Cory-Slechta, Director*

**1151 New Jersey BASS Chapter Federation**
77 Kenvil Avenue 201-584-9387
Succasunna, NJ 07876 E-mail: amgoing@verizon.net
http://www.njbassfed.org
An organization of chapters dedicated to the sport of bass fishing.
*Founded: 1917*
*Tony Going, President*
*Forest R Honeywell, Executive Director*

**1152 New Jersey Department of Health and Senior Services**
PO Box 360 609-292-7837
Trenton, NJ 08625 800-367-6543
http://www.state.nj.us/health
The mission of the Child and Adolescent Health Program is to promote optimum health and development of the children of New Jersey through the promotion of preventive services, linkages with primary medical care and healthy physcial andpsychosocial environments.
*Heather Howard, Commissionerector*
*Jon S Corzine, Governor*

**1153 New Jersey Environmental Lobby**
204 W State Street 609-396-3774
Trenton, NJ 08608 Fax: 609-396-4521
http://www.njenvironment.org
Nonprofit organization devoted to lobbying for legislation and/or regulations that will preserve and protect New Jersey's natural resources and environment — both natural and built — and protect the public health.
*Founded: 1969*
*Anne Poole, President*

**1154 New Jersey Public Interest Research Group**
143 E State Street 609-394-8155
Suite 6 http://www.njpirg.org
Trenton, NJ 08608
Delivers persistent, result-oriented public interest activism that protects consumers, encourages a fai, sustainable economy, and fosters responsive, democratic government.
*Allison Cairo, Executive Director*

**1155 New York/New Jersey Trail Conference**
156 Ramapo Valley Road 201-512-9348
Mahwah, NJ 07430 Fax: 201-512-9012
E-mail: info@nynjtc.org
http://www.nynjtc.org
A federation of 104 hiking clubs and environmental organizations and 10,000 individuals dedicated to building and maintaining marked hiking trails and protecting related open spaces in the bi-state region.

**1156 Passaic River Coalition**
94 Mt Bethel Road 908-222-0315
Warren, NJ 07059 Fax: 908-222-0357
E-mail: prc@passaicriver.org
http://www.passaicriver.org
Interested in preserving, maintaining and/or enhancing the water quality and quantity in the Passaic River Basin. Advocates on related issues, carries out projects that further its goals, and participates in land acquisition activitiesin order to provide open space.
*Founded: 1969*
*Ella F Filippone, Executive Director*

**1157 Sierra Club: NJ Chapter**
145 West Hanover Street 609-656-7612
Trenton, NJ 08618 Fax: 609-656-7618
E-mail: njsierra1@verizon.net
http://newjersey.sierraclub.org
Mission is to explore, enjoy, and protect the wild places of the earth; to practice and promote the responsible use of the earth's ecosystems and resources; to educate and enlist humanity to protect and restore the quality of thenatural and human environments.
*Ken Johnson, Chair*

## New Mexico

**1158 American Society of Landscape Architects: New Mexico**
School of Architecture & Planning
2414 Central Ave SE http://host.asla.org/chapters/newmexico/
Albuquerque, NM 87131
Statewide professional organization that is open to landscape architects and their associates.
*C Patricia Westbrook, President*
*Laurie Firor, President-Elect*

**1159 Holistic Management International**
1010 Tijeras Ave NW 505-842-5252
Albuquerque, NM 87102 Fax: 505-843-7900
E-mail: hmi@holisticmanagement.org
http://www.holisticmanagement.org
Enhance the efficiency, natural health, productivity and profitability of their land; increase natural annual profits; provide a framework for family, owners, managers, foreman, communal agriculturalists and other ranch/farmstakeholders to work together toward a common future; and enable development agencies working with marginalized farmers or pastoral people to break the cycle of food and water insecurity.
*Founded: 1984*
*Peter Holter, CEO*
*Tracy Favre, Sr Director Contract Sales*

**1160 National Parks Conservation Association**
1300 19th Street, NW, Suite 300
Washington, DC 20036 800-628-7275
Fax: 202-659-0650
E-mail: npca@npca.org
http://www.npca.org
To protect and enhance America's National Parks for presents and future generations.
*Founded: 1919*
*Thomas C Kiernan, President*

**1161 Nature Conservancy: New Mexico Chapter**
Nature Conservancy
212 E Marcy Street 505-988-3867
Suite 200 800-628-6860
Santa Fe, NM 87501 Fax: 505-988-4095
E-mail: nm@tnc.org
http://www.nature.org
*Founded: 1951*
*Terry Sullivan, State Director*

**1162 New Mexico Association of Conservation Districts**
163 Trail Canyon Road 505-981-2400
Carlsbad, NM 88220 Fax: 505-981-2400
http://www.nm.nacdnet.org
To facilitate conservation of natural resources in New Mexico by providing opportunities & quality support to local conservation districts through representative & leadership.
*Founded: 1946*
*Eddie Vigil, President*
*Debbie Hughes, Executive Director*

**1163 New Mexico Association of Soil and Water Conservation**
163 Trail Canyon Road 505-981-2400
Carlsbad, NM 88220 Fax: 505-981-2422
*Debbie Hughes, Executive Director*

**1164 New Mexico Center for Wildlife Law**
University of New Mexico School of Law
Albuquerque, NM 87131
E-mail: cbyers@unm.edu
http://ipl.unm.edu
Provides expertise in wildlife and biodiversity law and policy, including training, youth education, facilitation, legislation, research, teaching and publication.
*Founded: 1990*
*Carolyn Byers, Director*

**1165 New Mexico Rural Water Association**
3413 Carlisle Boulevard NE 505-884-1031
Albuquerque, NM 87110 800-819-9893
Fax: 505-884-1032
E-mail: nmrwa@nmrwa.org
http://www.nmrwa.org
A non-profit membership organization that provides free training and technical assistance to water and wastewater systems.
*Matthew Holmes, Executive Director*

**1166 New Mexico Solar Energy Association**
1009 Bradbury SE 505-246-0400
#35 Fax: 505-246-2251
Albuquerque, NM 87106 E-mail: info@nmsea.org
http://www.nmsea.org
An educational nonprofit organization dedicated to the promotion of solar energy and related sustainable practices. Membership includes building contractors, architects, planners, educators and others who support renewable energy.
*Founded: 1972*
*Mary McArthur, Executive Director*

**1167 Sierra Club: Rio Grande Chapter**
142 Truman NE 505-243-7767
Albuquerque, NM 87108 Fax: 505-243-7771
E-mail: daniel.lorimier@sierraclub.org
http://riogrande.sierraclub.org
*Founded: 1892*
*Susan Martin, Chair*
*Dan Lorimier, Conservation Coordinator*

**1168 Wildlife Society**
PO Box 35936 505-992-8651
Albuquerque, NM 87176 800-299-0196
E-mail: triley@trcp.org
http://www.leopold.nmsu.edu
A nonprofit scientific and educational organization that serves professionals such as government agencies, academia, industry, and non-government organizations in all areas related to the conservation of wildlife and natural resourcesmanagement.
*Founded: 1937*
*Terry Z Riley, President*
*Valerie A Williams, Secretary*

## New York

**1169 Adirondack Council**
103 Hand Avenue 518-873-2240
Suite 3 877-873-2240
Elizabethtown, NY 12932 Fax: 518-873-6675
E-mail: info@adirondackcouncil.org
http://www.adirondackcouncil.org
A not-for-profit environmental group that has been working since 1975 to protect the open-space resources of New York State's six million acre Adirondack Park and to help sustain the natural and human communities of the region. Basedin the Adirondacks with a second office in Albany, the Adirondack Council has a staff of 15 and a strog and vocal membership in all 50 states.
*Founded: 1975*
*Brian L Houseal, Executive Director*
*John F Sheehan, Communications Director*

**1170 Adirondack Land Trust**
PO Box 65 518-576-2082
Keene Valley, NY 12943 E-mail: adirondacks@tnc.org
http://www.nature.org
Dedicated to protecting open space, working landscapes such as farmlands and managed forests, as well as other lands contributing to the quality of life of Adirondack residents.
*Founded: 1984*

**1171 Adirondack Mountain Club**
310 Hamilton Street 518-668-4447
Albany, NY 12210-1738 Fax: 518-449-3875
E-mail: info@adk.org
http://www.adk.org
Dedicated to the protection and responsible recreational use of the New York State Forest Preserve, and other parks, wild lands, and waters vital to the members and chapters.
*Founded: 1922*
*James Bird, President*
*Neil Woodworth, Executive Director*

**1172 American Council on Science and Health**
1995 Broadway 212-362-7044
2nd Floor 866-905-2694
New York, NY 10023 Fax: 212-362-4919
E-mail: acsh@acsh.org
http://www.acsh.org
A consumer education organization based in New York City that promotes scientifically balanced evaluations of food, chemicals and the environment, and their relationship to human health.
*Founded: 1978*
*Elizabeth M Whelan, President*
*Gilbert Ross MD, Medical/Executive Director*

**1173 American Lung Association of New York State: Albany Office**
155 Washington Avenue 518-465-2013
Suite 210 Fax: 518-465-2926
Albany, NY 12210 http://www.lungusa2.org

**1174 American Lung Association of New York State: Long Island Office**
700 Veterans Memorial Highway 631-265-3848
Hauppauge, NY 11788 Fax: 631-265-6123

**1175 American Lung Association of the City of New York**
116 John Street 212-889-3370
30th Floor Fax: 212-889-3375
New York, NY 10038 E-mail: infonyc@alany.org
http://www.alany.org
Mission is to prevent lung disease and promote lung health - is realized through a broad variety of community education programs, research projects and advocacy initiatives.
*Deborah Carlto, CEO*

**1176 American Rivers: Mid-Atlantic Region**
1 Danker Avenue 518-482-2631
Albany, NY 12206 Fax: 518-482-2632
http://www.americanrivers.org
*Stephanie Lindloff, Director*

**1177 American Rivers: Southeast Region**
2231 Devine Street 803-771-7114
Suite 100 Fax: 803-771-7580
Columbia, SC 29205 http://www.americanrivers.org
*Gerrit Jobsis, Director Conservation*

**1178 American Society of Landscape Architects: New York Upstate Chapter**
c/o R. Rivers, The Rivers Organizat
312 W Commercial Street
East Rochester, NY 14445
585-586-6906
Fax: 585-385-6053
E-mail: rick@riversorg.com
http://www.nyuasla.org

Landscape Architecture is a comprehencive discipline of land analysis, planning, design, management, preservation, and rehabilitation. The New York Updstate Chapter of ASLA promotes the landscape architecture profession and advancesthe practice through advocacy, education, communication, and feloowship amongst Lanscape Architects and allied professionals.
*Doug McCord, President*
*Rick Rivers, Executive Director*

**1179 American Society of Landscape Architects: New York Chapter**
42 Broadway
Suite 1827-35
New York, NY 10004
212-269-2987
E-mail: info@nyasla.org
http://www.nyasla.org
*Susannah Drake, President*
*Jane Cooke, Executive Director*

**1180 Catskill Forest Association**
PO Box 336
Arkville, NY 12406
845-586-3054
Fax: 845-586-4071
E-mail: cfa@catskill.net
http://www.catskillforest.org

A non-profit organization dedicated to enhancing all aspects of the forest in New York's Catskill region.
*Founded: 1982*
*Keith Laurier, President*
*Jim Waters, Executive Director*

**1181 Clean Ocean Action**
18 Hartshorne Drive, Suite 2
Sandy Hook, NJ 07732
732-872-0111
Fax: 732-872-8041
E-mail: sandyhook@cleanoceanaction.org
http://www.cleanoceanaction.org

A borad-based coalition of over 150 groups and individuals that work to clean up and protect the waters of the New York and New Jersey coasts.
*Founded: 1984*
*Cindy Zipf, Executive Director*
*Mary-Beth Thompson, Operations Director*

**1182 Cornell Lab of Ornithology**
159 Sapsucker Woods Road
Ithaca, NY 14850
607-254-2473
800-843-2473
E-mail: cornellbirds@cornell.edu
http://www.birds.cornell.edu

The Lab uses the best science and technology, and inspires the widest range of people and organizations-to solve critical problems facing wildlife. The mission is to interpret and conserve the earth's biological diversity throughresearch, education, and citizen science focused on birds.
*Founded: 1950*

**1183 Environmental Action Coalition**
625 Broadway
9th Floor
New York, NY 10012
212-677-1601
Fax: 212-505-8613
E-mail: eac@eacnyc.org
http://www.eacnyc.org

A network of concerned citizens who devote time and money to spreading information on ecological and environmental clean-up efforts. Offers environmental education and sponsors programs for professionals in the field, citizenactivists, volunteers, teachers, students and labor leaders.
*Paul C Berizzi, Executive Director*

**1184 Federation of New York State Bird Clubs**
New York Birders
PO Box 440
Loch Sheldrake, NY 12759
E-mail: mkoeneke@a-znet.com
http://www.fnysbc.com

The objectives are to document the ornithology of New York State; to foster interest in and appreciation of birds; and to protect birds and their habitats.
*Sue Adadair, Treasurer*

**1185 Great Lakes United**
State University College at Buffalo
Cassety Hall
1300 Elmwood Avenue
Buffalo, NY 14222
716-886-0142
Fax: 716-204-9521
E-mail: glu@glu.org
http://www.glu.org

*Founded: 1982*
*Derek Stack, Executive Director*

**1186 Hudsonia Limited**
30 Campus Road
PO Box 5000
Annandale, NY 12504
845-758-7053
Fax: 845-758-7033
http://www.hudsonia.org

A not-for-profit institute for research, education, and technical assistance in the environmental sciences.
*Founded: 1981*
*Erik Kiviat, Executive Director*

**1187 INFORM**
5 Hanover Square
19th Floor
New York, NY 10004
212-361-2400
E-mail: ramsey@informinc.org
http://www.informinc.org

INFORM prides itself on more than three decades of identifying innovative technologies, practices and products that provide practical solutions to complex environmental and health-related problems.
*Founded: 1974*
*Virginia Ramsey, President*

**1188 Montefiore Medical Center Lead Poisoning Prevention Program**
111 E 210th Street
401 Moses
Bronx, NY 10467
718-920-5016
Fax: 718-920-4377

The Montefiore Lead Poisoning Prevention Program addresses all aspects of childhood lead poisoning from diagnosis and treatment to education and research. Their mission is to treat lead-poisoned children and their families and toeducate families at risk, other medical providers and local, state and national legislators and policy makers.
*Nancy Redkey, Project Coordinator*

**1189 Nature Conservancy: New York Long Island Chapter**
The Nature Conservancy
250 Lawerence Hill Road
Cold Spring Harbor, NY 11724
631-367-3225
800-628-6860
Fax: 516-367-4715
E-mail: comment@tnc.org
http://www.nature.org

*Founded: 1951*
*Henry M Paulson Jr, Chairman*
*Steven J McCormick, President/CEO*

**1190 New York Association of Conservation Districts**
335E HRC, HVCC
80 Vandenburgh Avenue
Troy, NY 12180
518-629-7645
Fax: 518-629-7646
E-mail: nyacd@nycap.rr.com
http://www.nyacd.org

Provides leadership in the wise use of soil, water, and related natural resources.
*Founded: 1958*
*Linda Coffin, President*
*Gregory Bell, Executive Director*

**1191 New York Forest Owners Association**
124 E 4th Street
PO Box 210
Watkins Glen, NY 14891
607-535-9790
Fax: 607-535-9794
E-mail: mjpacker@nyfoa.org
http://www.nyfoa.org

Promotes sustainable forestry practices and improved stewardship on provately owned woodlands in New York State. A not-for-profit group of people who care about NYS trees and for-

ests and are interested in the thoughtful management ofprivate forests for the benefit of current and future generations.
*Founded: 1905*
*Mary Jeanne Packer, Executive Director*

**1192 New York Healthy Schools Network**
773 Madison Ave
Albany, NY 12208
518-462-0632
Fax: 518-462-0433
E-mail: info@healthyschools.org
http://www.healthyschools.org
A national environmental health organization that does research, information, education, coalition-building, and advocacy to ensure that every child has a healthy learning environment that is clean and in good repair.
*Founded: 1995*
*Claire L Barnett, Executive Director*

**1193 New York State Council of Landscape Architects**
235 Lark Street
Albany, NY 12210
212-431-3609
E-mail: kmatthews@mnlandscape.com
http://www.nyscla.org
*W Dean Gomolka, President*
*Alexander F Kurnicki, President-Elect*

**1194 New York State Department of Environmental Conservation**
625 Broadway
Albany, NY 12233
518-402-8540
Fax: 518-402-9016
http://www.dec.ny.gov
DEC protects, improves and conserves the state's land, water, air, fish, wildlife and other resources to enhance the health, safety and welfare of the people and their overall economic and social well-being.
*Joe Martens, Commissioner*

**1195 New York State Ornithology Society**
PO Box 95
Durhamville, NY 13054
E-mail: president1@nybirds.org
http://www.nybirds.org
The objectives of the Federation are to document the ornithology of New York State; to foster interest in and appreciation of birds; and to protect birds and their habitats.
*Founded: 1947*
*Andy Mason, President*
*John Confer, Conservation*

**1196 New York Turtle and Tortoise Society**
PO Box 878
Orange, NJ 07051
212-459-4803
http://www.nytts.org
A nonprofit organization dedicated to the conservation, preservation of habitat, and the promotion of proper husbandry and captive propagation of turtles and tortoises. The Society emphasizes the education of its members and the publicin all areas relevant to the appreciation of these unique animals.

**1197 New York Water Environment Association**
525 Plum Street
Suite 102
Syracuse, NY 13204
315-422-7811
Fax: 315-422-3851
E-mail: pcr@nywea.org
http://www.nywea.org
The New York Water Environment Association is a nonprofit educational association dedicated to the development and dissemination of information concerning water quality management and the nature, collection, treatment, and disposal ofwastewater. Founded in 1929, the Association has over 2,100 members. The NYWEA is a member association of the Water Environment Federation.
*Founded: 1929*
*Patricia Cerro-Reehil, Executive Director*

**1198 Parks and Trails New York**
29 Elk Street
Albany, NY 12207
518-434-1583
Fax: 518-427-0067
E-mail: ptny@ptny.org
http://www.ptny.org
The only organization working statewide to protect New York's parks and help communities create new parks. A non-profit organization whose mission is to expand, protect and promote a network of parks, trails, and open spaces throughoutour state for use and enjoyment by all.
*Founded: 1985*
*Jeffrey P Swain, Chairman*
*Robin Dropkin, Executive Director*

**1199 Rene Dubos Center for Human Environments**
81 Pondfield Road
Suite 387
Bronxville, NY 10708
914-337-1636
Fax: 914-771-5206
http://www.dubos.org
The Rene Dubos Center for Human Environments is a non-profit education and research organization focused on the social and humanistic aspects of environmental problems.
*Noel Brown, President*

**1200 Riverkeeper**
20 Secor Road
Ossining, NY 10562
800-217-4837
E-mail: info@riverkeeper.org
http://www.riverkeeper.org
A member supported watchdog organization dedicated to defending Hudson River and its tributaries and protecting the drinking water supply of nine million New York City and Hudson Valley residents.
*Founded: 1966*
*Paul Gallay, Executive Director*

**1201 Sagamore Institute**
PO Box 40
Raquette Lake, NY 13436
315-354-5311
Fax: 315-354-5851
E-mail: info@greatcampsagamore.org
http://www.greatcampsagamore.org
A non-profit corporation dedicated to the stewardship of Great Camp Sagamore, in Raquette Lake, NY, and to its use for educational and interpretive purposes.
*Founded: 1973*
*Beverly Bridger, Executive Director*
*Dr Michael Wilson, Associate Director*

**1202 Scenic Hudson**
One Civic Center Plaza
Suite 200
Poughkeepsie, NY 12601
845-473-4440
Fax: 845-473-2648
E-mail: info@scenichudson.org
http://www.scenichudson.org
Scenic Hudson works to protect and restore the Hudson River and its majestic landscape as an irreplaceable national treasure and a vital resource for residents and visitors.
*Founded: 1963*
*Ned Sullivan, President*

**1203 Selikoff Clinical Center for Occupational & Environmental Medicine**
Mount Sinai School of Medicine
Department of Community Medicine
1 Gustave Levy Place, Box 1043
New York, NY 10029
212-241-8689
Fax: 212-360-6965
http://www.mssm.edu/cpm
Internationally respected diagnostic referral center and an important interface between the research programs of the Division of Environmental Health Science and populations exposed to environmental hazards.

**1204 Sierra Club: Atlantic Chapter**
353 Hamilton Street
Albany, NY 12210
518-426-9144
Fax: 518-427-0381
E-mail: chaptercoord@newyork.sierraclub.org
http://newyork.sierraclub.org
The Atlantic Chapter applies the principles of the national Sierra Club to the environmental issues facing New York State.
*Ken Baer, Chair*
*Bobbie Josepher, Chapter Coordinator*

**1205 Sierra Club: New York City Office**
116 John Street
Suite 3100
New York, NY 10038
212-791-3600
Fax: 212-791-0839
E-mail: ne.field@sierraclub.org
http://www.sierraclub.org

Part of the national Sierra Club with over 700,000 members, headquarters in San Francisco, offices in Washington DC, and staff in state capitals around the nation. Our mission is to enjoy and protect the earth's ecosystems andresources, and to enlist others to do the same. The Atlantic Chapter consists of 32,000 members in 11 local groups throughout New York State. We use the media, grassroots education and personal contact to bring our issues to our communities and ourpublic officials.
*Founded: 1892 Quarterly*

**1206 Sierra Club: Northeast Office**
85 Washington Street 518-587-9166
Saratoga Springs, NY 12866 Fax: 518-583-9062
E-mail: ne.field@sierraclub.org
http://newyork.sierraclub.org

To advance the preservation and protection of the natural environment by empowering the citizenry, especially democratically based grassroots organizations with charitable resources to further the cause of environmental protection, thevehicle through which The Sierra Club Foundation generally fulfills its charitable mission.
*Founded: 1892*

**1207 Trout Unlimited**
1300 North 17th Street 703-522-0200
Suite 500 800-834-2419
Arlington, VA 22209 Fax: 703-284-9400
E-mail: trout@tu.org
http://www.tu.org

Mission: To conserve, protect and restore North America's trout and salmon fisheries and their watersheds. We accomplish this mission on local, state and national levels with an extensive and dedicated volunteer network.
*Charles Gauvin, President/CEO*
*Steve Moyer, Vice President*

**1208 Tug Hill Tomorrow Land Trust**
PO Box 6063 315-779-8240
Watertown, NY 13601 Fax: 315-782-6192
E-mail: thtomorr@northnet.org
http://www.tughilltomorrowlandtrust.org

A regional, private, nonprofit founded by a group of Tug Hill residents to serve the region of 2,100 square miles serving portions of Jefferson, Lewis, Oneida & Oswego Counties. The mission is two-fold: increase awareness andappreciation of the Tug Hill Rgion through education; and to help retain the forest, farm, recreation, and wild land of the region through voluntary, private land protection efforts.
*Founded: 1990*
*Linda Garrett, Executive Director*

**1209 Waterkeeper Alliance**
17 Battery Place 212-747-0622
Suite 1329 Fax: 212-747-0611
New York, NY 10004 E-mail: info1@waterkeeper.org
http://www.waterkeeper.org

Provides a way for communities to stand up for their right to clean water and for the wise and equitable use of water resources, both locally and globally. The vision of the Waterkeeper movement is for fishable, swimmable and drinkablewaterways worldwide.
*Marc A Yaggi, Interim Executive Director*

**1210 West Harlem Environmental Action**
271 W 125th Street 212-961-1000
Suite 308 Fax: 212-961-1015
New York, NY 10027 E-mail: peggy@weact.org
http://www.weact.org

A non-profit, community-based, environmental justice organization dedicated to building community power to fight environmental racism and improve environmental health, protection and policy in communities of color.
*Peggy M Shepard, Executive Director*

**1211 Women's Environment and Development Organization**
355 Lexington Avenue 212-973-0325
3rd Floor Fax: 212-973-0335
New York, NY 10017 E-mail: wedo@wedo.org
http://www.wedo.org

An international organization that advocates for women's equality in global policy. It seeks to empower women as decisionmakers to achieve economic, social and gender justice, a healthy, peaceful planet and human rights for all.
*Founded: 1990*
*June Zeitlin, Executive Director*

## North Carolina

**1212 Acid Rain Foundation**
1410 Varsity Drive 919-828-9443
Raleigh, NC 27606 Fax: 919-515-3593
*Dr. Harriett S Stubbs, Executive Director*

**1213 American Lung Association of North Carolina**
3801 Lake Boone Trail 919-832-8326
Suite 190 800-892-5650
Raleigh, NC 27607 Fax: 919-856-8530
E-mail: info@lungnc.org
http://www.lungnc.org

The voluntary health organization dedicated to eliminating lung disease and fostering healthy breathing for all people through prevention, outreach, education, research and advocacy.
*Founded: 1904*
*Deborah C Bryan, President/CEO*

**1214 American Society of Landscape Architects: North Carolina Chapter**
1829 East Franklin Street 919-215-3117
Suite 600 888-999-2752
Chapel Hill, NC 27514 Fax: 919-278-2647
E-mail: manager@ncasla.org
http://www.ncasla.org

*Founded: 1899*
*Luther Smith, President*
*Kelly Langston, Association Manager*

**1215 Carolina Bird Club**
1809 Lakepark Drive 910-791-9034
Raleigh, NC 27612 Fax: 910-791-7228
E-mail: hq@carolinabirdclub.org
http://www.carolinabirdclub.org

A nonprofit educational and scientific association, open to anyone interested in the study and conservation of wildlife, particularly birds. Meets each winter, spring and fall. Meeting sites are selected to give participants anopportunity to see many different kinds of birds. Guided field trips, informative programs and business sessions are combined for an exciting weekend of meeting with people who share an enthusiasm and concern for birds.
*Founded: 1937*
*Taylor Piephoff, President*

**1216 Carolina Recycling Association**
274 Pittsboro Elementary School Rd 919-545-9050
Suite #6, PO Box 1578 Fax: 919-545-9060
Pittsboro, NC 27312 E-mail: cra@cra-recycle.org
http://www.cra-recycle.org

Conserves resources by advancing waste reducation and recycling throughout the Carolinas
*Founded: 1989*
*Kerry Krumsiek, Executive Director*

**1217 Environmental Educators of North Carolina**
PO Box 4904 919-250-1050
Chapel Hill, NC 27515 Fax: 919-250-1058
E-mail: eenc@rtpnet.org
http://www.eenc.org

To promote excellence in professional development and facilitate networking opportunities, inspiring educators to create an environmentally literate citizenry.
*Lois Nixon, President*

**1218 Forest History Society**
701 Vickers Avenue
Durham, NC 27701
919-682-9319
Fax: 919-682-2349
E-mail: oakes@duke.edu
http://www.foresthistory.org
The Forest History is a non-profit educational institution that links the past to the future by indentifying, collecting, preserving, interpreting, and disseminating information on the history of people, forests, and their relatedresources.
*Founded: 1946*
*Steven Anderson, President*
*Cheryl Oakes, Secretary*

**1219 North Carolina Association of Soil & Water Conservation Districts**
PO Box 27943
Raleigh, NC 27611
919-733-2302
http://www.ncaswcd.org
An independent nonpartisan conservation organization created to represent the interests of 96 local soil and water conservation districts and the 492 district supervisors who direct their local district's conservation programs.
*Founded: 1944*
*William F Pickett Jr, President*

**1220 North Carolina Chapter of the Wildlife Society**
PO Box 37742
Raleigh, NC 27627
704-732-1391
http://www.nctws.org
Seeks to provide a forum for wildlife professionals and others to interact to improve wildlife conservation and management while fostering high professional standards and ethics for its members
*Founded: 1983*

**1221 North Carolina Coastal Federation**
3609 Highway 24
Newport, NC 28570
252-393-8185
800-232-6210
Fax: 252-393-7508
http://www.nccoast.org
Works with citizens to safeguard the state's coastal rivers, creeks, sounds and beaches. The state's only non-profit organization focused exclusivley on protecting and restoring the coast of Nort Carolina through education, advocacy,and habitat preservation and restoration.
*Founded: 1982*
*Todd Miller, Executive Director*
*Melvin Shepard, Jr, President*

**1222 North Carolina Museum of Natural Sciences**
11 W Jones Street
Raleigh, NC 27601
919-733-7450
Fax: 919-733-1573
E-mail: museum@naturalsciences.org
http://www.naturalsciences.org
*Founded: 1985*
*Angela B Baker-James, Executive Director*

**1223 North Carolina Native Plant Society**
c/o North Carolina Botanical Garden
CB# 3375, Totten Center
UNC-Chapel Hill
Chapel Hill, NC 27599
E-mail: tom@ncwildflower.org
http://www.ncwildflower.org
The purpose of the Society is to promote enjoyment and conservation of native plants and their habitats through education, protection, and propagation.

**1224 North Carolina Sustainable Energy Association**
PO Box 6465
Raleigh, NC 27628
919-832-7601
E-mail: info@energync.org
http://www.ncsustainableenergy.org
A non-profit membership organization of individuals and businesses interested in sustainable energy. Works to ensure a sustainable future by promoting renewable energy and energy efficiency in North Carolina through education, publicpolicy and economic development.
*Founded: 1978*
*Ivan Urlaub, Executive Director*
*Van Crandall, Development Director*

**1225 Sierra Club: North Carolina Chapter**
112 S Blount Street
Raleigh, NC 27601
919-833-8467
Fax: 919-833-8460
E-mail: nc.sierraclub.org
http://info@nc.sierraclub.org
The North Carolina Chapter currently has 19,000 members and 13 local statewide groups.
*Founded: 1970*
*Molly Diggins, State Director*
*Tom Jensen, Conservation Organizer*

**1226 Trout Unlimited**
1300 North 17th Street
Suite 500
Arlington, VA 22209
703-522-0200
800-834-2419
Fax: 703-284-9400
E-mail: trout@tu.org
http://www.tu.org
Mission: To conserve, protect and restore North America's trout and salmon fisheries and their watersheds. We accomplish this mission on local, state and national levels with an extensive and dedicated volunteer network.
*Founded: 1959*
*Charles Gauvin, President/CEO*
*Steve Moyer, Vice President*

## North Dakota

**1227 American Lung Association in North Dakota**
212 North 2nd Street
Bismarck, ND 58501-3819
701-223-5613
800-252-6325
Fax: 701-223-5727
E-mail: jmourhess@lungnd.org
http://www.lungnd.org
Offers a wide array of lung health services to the people of North Dakota. The mission is to save lives, improve lung hea;th and prevent lung disease. ALAND advocates for clean in-/outdoor air quality; provides asthma managementresources; provices prevention and cessation programs and services; monitors lung health; and are
*Judy Mourhess, Associate, Program Services*

**1228 International Association for Impact Assessment**
1330 23rd Street S
Suite C
Fargo, ND 58103
701-297-7908
Fax: 701-297-7917
E-mail: info@iaia.org
http://www.iaia.org
The International Association for Impact Assessment is an interdisciplinary society dedicated to developing international capacity to anticipate, plan and manage the consequences of development. The Association has over 2,500 membersin over 100 nations. IAIA seeks to ensure that political, environmental, social and technological dimensions of decisions are understood by those making them.
*Founded: 1980*
*Rita Hamm, CEO*

**1229 North Dakota Association of Soil Conservation Districts**
3310 University Drive
Bismarck, ND 58504
701-223-8518
Fax: 701-223-1291
E-mail: ndascd@btinet.net
http://ndascd.org
The purpose is to further the widespread application of sound and practical soil and water conservation practices in North Dakota. The goal is to provide quality membership services and nursery products to carry out the soilconservation program of the soil conservation districts of North Dakota.
*Thomas Hanson, Executive Director*

## Ohio

**1230 American Lung Association of Ohio**
1950 Arlingate Lane
Columbus, OH 43228
800-586-4872
Fax: 614-279-4940
E-mail: tracy@ohiolung.org
http://www.ohiolung.org

Helps Ohioans breathe easier. We lead the fight to prevent lung disease through our mission to promote lung health through research, education, community service, and advocacy.
*Founded: 1901*
*Tracy Ross, President/CEO*

**1231 American Lung Association of Ohio: Northeast Region**
6100 Rockside Woods Boulevard
#260
Independence, OH 44131
216-524-5864
800-586-4872
Fax: 216-524-7647
E-mail: northeast@ohiolung.org
http://www.ohiolung.org

*Founded: 1901*

**1232 American Lung Association of Ohio: Northwest Region**
226 State Route 61
Norwalk, OH 44857
419-663-5864
Fax: 419-668-2575
E-mail: northwest@ohiolung.org
http://www.ohiolung.org

**1233 American Lung Association of Ohio: Southwest Region**
4050 Executive Park Drive
#402
Cincinatti, OH 45241
513-985-3990
Fax: 513-985-3995
E-mail: southwest@ohiolung.org
http://www.ohiolung.org

**1234 American Society of Landscape Architects: Ohio Chapter**
579 High Street
Worthington, OH 43085
614-436-4431
Fax: 614-436-4451
E-mail: ocasla@ocasla.org
http://www.ocasla.org

*Richard Espe, President*
*Beth Adamson, Association Manager*

**1235 Association for Facilities Engineering**
12801 Worldgate Drive
Suite 500
Herndon, VA 20170
571-203-7171
Fax: 571-766-2142
E-mail: info@afe.org
http://www.afe.org

The premier organization for facility engineers and maintenance personnel. Unites a large community of likeminded professionals for networking opportunities, knowledge-sharing, and support, as well as offers members a world variety ofeducational opportunities, training, and certification for career advancement.
*Laurence Gration, CEO*

**1236 Central Ohio Anglers and Hunters Club**
1773 Huy Road
PO Box 28224
Columbus, OH 43224
614-447-0116
http://www.dnr.state.oh.us
*Doug Eakens, President*

**1237 Cincinnati Nature Center**
4949 Tealtown Road
Milford, OH 45150
513-831-1711
Fax: 513-831-8052
E-mail: cnc@cincynature.org
http://www.cincynature.org

To inspire passion for nature and promote environmentally responsible choices through experience and education.
*Founded: 1965*
*Bill Hopple, Executive Director*

**1238 Great Lakes Tomorrow**
9315 Glenwood Trail
Brecksville, OH 44141
440-838-4176
Fax: 440-838-4176
E-mail: jcowdeni@ibm.net

*James W Cowden, Director*

**1239 Green Energy Ohio**
7870 Olentangy River Road
#209
Columbus, OH 43235
614-985-6131
Fax: 614-888-9716
E-mail: geo@greenenergyohio.org
http://www.greenenergyohio.org

Green Energy Ohio promotes renewable energy statewide by acting as a clearinghouse for information for Ohioans on sustainable energy.
*Michelle Greenfield, President*
*William A Spratley, Executive Director*

**1240 Holden Arboretum**
9500 Sperry Road
Kirtland, OH 44094
440-946-4400
E-mail: holden@holdenarb.org
http://www.holdenarb.org

Connects people with nature for inspiration and enjoyment, fosters learning and promotes conservation.
*Founded: 1931*
*Clem Hamilton, President*

**1241 League of Ohio Sportsmen**
642 West Broad Street
Columbus, OH 43215
614-224-8970
Fax: 614-224-8971
E-mail: president@leagueofohiosportsmen.org
http://www.leagueofohiosportsmen.org

Dedicated to supporting conservation, restoration, and education that promotes the wise use and enjoyment of our natural resources including wildlife management.
*Founded: 1908*
*Larry Mitchell Sr, President*
*John Hobbs, Vice President*

**1242 Native Plant Society of Northeastern Ohio**
10761 Pekin Road
Newbury, OH 44065
440-286-9504
E-mail: npsohio@hotmail.com

Mission is to promote conservatoin of all native plants an dnative plant communities through habitat protection and other means; to encourage public education and appreciation of native plants; to support proper ethics and methods ofnatural landscaping; to encourage surveys and research on native plants and publication of the information; to promote cooperation with other programs and organizations concerned with the conservation of nautral resources.
*Judy Barnhart, President*

**1243 Nature Conservancy: Ohio Chapter**
6375 Riverside Drive
Suite 50
Dublin, OH 43017
614-717-2770
Fax: 614-717-2777
E-mail: ohio@tnc.org
http://www.nature.org

A global conservation organization dedicated to preserving plants, animals and natural communities that represent the diversity of life on Earth by protecting the lands and water they need to survive. Since its inception in 1951, TheNature Conservancy has protected more than 12 million acres in the US and helped through partnerships preserve more tan 80 million acres in Latin America, the Caribbean, Canada, Asia and the Pacific.
*Founded: 1915*
*Dr Richard L Shank, State Director*

**1244 Ohio Alliance for the Environment**
14 Beck St
Canal Winchester, OH 43110
614-833-4223
Fax: 614-833-4223
E-mail: probasco@ohioalliance.org
http://www.ohioalliance.org

*Founded: 1978*
*Peggy Smith, Executive Director*
*Mike Parkes, President*

**1245 Ohio BASS Chapter Federation**
43 Portsmouth Rd
Gallipolis, OH 45631
740-446-9810
Fax: 740-446-9819
E-mail: jdoss@zoomnet.net
http://www.ohiobass.org

*Jim Doss, President*

**1246 Ohio Energy Project**
670 Enterprise Drive
Suite A
Lewis Center, OH 43035
614-785-1717
Fax: 614-785-1731
E-mail: rsmith@ohioenergy.org
http://www.ohioenergy.org

An organization providing energy and energy efficiency education using current complete and unbrased information, as well as hands-on, engaging and innovative techniques. OEP's kids teching kids approach also helps develop leadershipteam work and presentation skills. An affiliate of NEED, OEP has been named a National Energy Champion and one of the Top 12 Environmental Education Programs in Ohio.
*Debby Yerkes, Executive Director*
*Rich Smith, Director*

**1247 Ohio Environmental Council**
1207 Grandview Avenue Suite 201
Columbus, OH 43212
614-487-7506
Fax: 614-487-7510
E-mail: oec@theOEC.org
http://www.theoec.org

The state's premier advocate for our air, land and water.
*Keith Dimoff, Executive Director*

**1248 Ohio Federation of Soil and Water Conservation Districts**
PO Box 24518
Columbus, OH 43224
614-784-1900
Fax: 614-784-9181
E-mail: laurahollingsworth@ofswcd.org
http://www.ofswcd.org

To provide leadership and support to the board supervisors, soil and water conservation districts, and their partners through grassroots programs that promote natural resource stewardship.
*Founded: 1943*
*Lawrence Burdell, President*
*Laura Hollingsworth, Administrative Assistant*

**1249 Ohio Parks and Recreation Association**
1069A W Main Street
Westerville, OH 43081
614-895-2222
800-238-1108
Fax: 614-895-3050
E-mail: opra@opraonline.org
http://www.opraonline.org

A non-profit, public interest organization representing over 1200 professionals and citizen board members involved in providing leisure facilities and opportunities to all Ohioans as well as the tourists who visit the state each year.Dedicated to the promotion of parks and recreation services for all Ohioans and the sound stewardship of Ohio's natural resources.
*Founded: 1938*
*Woody Woodward, Executive Director*
*Mindy McInturf, Business Manager*

**1250 Rails-to-Trails Conservancy**
Midwest Regional Office
10 S High Street Suite A
Canal Winchester, OH 43110
614-837-6782
Fax: 614-837-6783
http://www.railstotrails.org
*Rhonda Border-Boose, Regional Director*
*Eric Oberg, Manager Trail Development*

**1251 Sierra Club**
131 N High Street
Suite 605
Columbus, OH 43215
614-461-0734
Fax: 614-461-0710
E-mail: Enid.Nagel@thomson.com
http://ohio.sierraclub.org

To advance the preservation and protection of the natural environment by empowering the citizenry, especially democratically-based grassroots organizations, with charitable resources to further the cause of environmental protection.The vehicle through which The Sierra Club Foundation generally fulfills its charitable mission.
*Enid Nagel, Chairman*
*Ellen Hawkeye Carmichael, Conservation Program Mgr*

**1252 Wildlife Society**
952 Lima Avenue
Box A
Findlay, OH 45840
419-424-5000
Fax: 419-422-4875

A nonprofit scientific and educational organization that serves professionals such as government agencies, academia, industry, and non-government organizations in all areas related to the conservation of wildlife and natural resourcesmanagement.
*J Butterworth, President*

## Oklahoma

**1253 American Fisheries Society: Fisheries Management Section**
OK Fish RS Laboratory
500 E Consellation
Norman, OK 73072
405-325-7288
Fax: 405-325-7631
E-mail: main@fisheries.org
http://www.fisheries.org

*Founded: 1870*
*Kurk Kuklinski, President*
*Brent Gordon, Seretary*

**1254 American Lung Association of Oklahoma**
1010 E 8th Street
Tulsa, OK 74120
918-747-3441
Fax: 918-747-4629
E-mail: mcrump@oklung.org
http://www.oklung.org

*Margaret Crump, Contact*

**1255 American Lung Association: Oklahoma City Office**
11212 N May Ave #405
Oklahoma City, OK 73120
405-748-4674
Fax: 405-748-6274
E-mail: jwilliams@breathehealthy.org
http://www.breathehealthy.oth

*Founded: 1904*
*Heather Griswold, Contact*

**1256 American Society of Landscape Architects: Oklahoma Chapter**
E-mail: b.tartar@lpci.com

*Barbara Tartar, President*
*Joan Lindley, President-Elect*

**1257 Nature Conservancy: Oklahoma Chapter**
2727 E 21st Street
Suite 102
Tulsa, OK 74114
918-585-1117
Fax: 918-585-2383
E-mail: mfuhr@tnc.org
http://www.tnc-oklahoma.org

*Mike Fuhr, State Director*

**1258 Oklahoma Association of Conservation Districts**
PO Box 107
Chelsea, OK 74016
405-340-8884
Fax: 405-842-8744
E-mail: claypope@pldi.net
http://www.oacd.us

Provides leadership, resources, and partnership opportunities for conservation districts and those who manage the land to enhance out natural resources for a better Oklahoma.
*Clay Pope, Exeucitve Director*

**1259 Oklahoma BASS Chapter Federation**
17316 E 110 Street N
Owasso, OK 74055
E-mail: okgwgl@cox.net
http://www.okbass.org

*Founded: 1972*
*Gary Gunter, President*

**1260 Oklahoma Ornithological Society**
PO Box 2931
Claremore, OK 74018
918-343-7701
Fax: 918-343-7563
E-mail: info@okbirds.org
http://www.okbirds.org

*Founded: 1982*
*Rebecca Renfro, President*

**1261 Sierra Club**
PO Box 60644
Oklahoma City, OK 73146
405-366-5694
E-mail: c.wesner@sbcglobal.net
http://oklahoma.sierraclub.org

To explore, enjoy and protect the planet; to practice the responsible use of the earth's ecosystem and resources; to educate and enlist humanity; to protect and restore the quality of the natural and human environmenta; and to use alllawfule means to carry out those objectives.
*Charles Wesner, Chair*
*Rick Wicker, Treasurer*

**1262 Trout Unlimited**
1300 North 17th Street
Suite 500
Arlington, VA 22209
703-522-0200
800-834-2419
Fax: 703-284-9400
E-mail: trout@tu.org
http://www.tu.org

Mission: To conserve, protect and restore North America's trout and salmon fisheries and their watersheds. We accomplish this mission on local, state and national levels with an extensive and dedicated volunteer network.
*Charles Gauvin, President/CEO*
*Steve Moyer, Vice President*

## Oregon

**1263 American Fisheries Society: Water Quality Section**
324 25th Street
Ogden, UT 84401
801-625-5358
Fax: 801-625-5756
E-mail: glampman@fs.fed.us

Section objectives are to: maintain an association of persons involved in the protection of watersheds, water quality, and aquatic habitat, and the abatement of water pollution and aquatic habitat and water deterioration.
*Gina Lampman, President*

**1264 American Lung Association of Oregon**
7420 Southwest Bridgeport Road
Suite 200
Tigard, OR 97224
503-924-4094
Fax: 503-924-4120
E-mail: dana@lungoregon.org
http://www.lungoregon.org

The oldest, nationwide, non-profit, voluntary public health organization in Oregon. Governed by a voluntary Board of Directors, we are the only community health agency dedicated solely to fighting lung disease and promoting lung healthin Oregon.
*Founded: 1972*
*Dana Kaye, Executive Director*

**1265 American Society of Landscape Architects: Oregon Chapter**
503-887-3439
E-mail: tarabyler@hotmail.com
http://www.aslaoregon.org

*Tara Byler, President*
*P Annie Kirk, VP Chapter Services*

**1266 Columbia Basin Fish and Wildlife Authority**
851 SW Sixth Avenue, Suite 260
Pacific First Building
Portland, OR 97204
503-229-0191
Fax: 503-229-0443
http://www.cbfwa.org

A non-profit corporation to provide an opportunity for the Agencies and Tribes of the Pacific Northwest to become directly involved in the fiscal, administrative and managerial aspects of jointly funded activities.
*Founded: 1993*
*Brian Lipscomb, President/Executive Director*

**1267 Ecotrust**
721 NW Ninth Avenue
Suite 200
Portland, OR 97209
503-227-6225
Fax: 503-222-1517
E-mail: contact@ecotrust.org
http://www.ecotrust.org

To inspire fresh thinking that creates economic opportunity, social equity and environmental well-being.
*Spencer B Beebe, President*
*Adam Lane, CFO/COO*

**1268 Natural Resources Information Council**
StreamNet Library
Dean Walton
520 University of Oregon
Eugene, OR 97403
541-346-2871
Fax: 541-346-3485
E-mail: dpwalton@uoregon.edu
http://www.nric.info

The main purpose of NRIC is to facilitate the exchange of information among librarians specializing in natural resource libraries and collections in both public (government, academic), and private (NGO, consulting) organizations.
*Founded: 1993*
*Dean Walton, President*

**1269 Northwest Coalition for Alternatives to Pesticides**
PO Box 1393
Eugene, OR 97440
541-344-5044
Fax: 541-344-6923
E-mail: info@pesticide.org
http://www.pesticide.org

Protects the health of people and the environment by advancing alternatives to pesticides.
*Founded: 1977*
*Kim Leval, Executive Director*

**1270 Northwest Power and Conservation Council**
1642 Franklin Avenue
Astoria, OR 97103
503-229-5171
Fax: 503-229-5173
E-mail: cmoreland@nwcouncil.org
http://www.nwcouncil.org/contact/or.asp

The council develops and maintains a regional power plan and a fish and wildlife program to balance the Northwest's environment and energy needs.
*Joan Dukes, Council Member*

**1271 Oregon Refuse and Recycling Association**
PO Box 2186
Salem, OR 97308
800-527-7624
Fax: 503-399-7784
E-mail: orrainfo@orra.net
http://www.orra.net

A 200 member voluntary association of solid waste management companies and businesses which specialize in offering equipment and services important to the industry. ORRA provides legislative advocacy, education, group insurance,meeting facilities and advice on regulatory matters to its memebers.
*Founded: 1965*
*Kristen Mitchell, Executive Director*

**1272 Oregon State Public Interest Research Group**
1536 SE 11th Avenue
Portland, OR 97214
503-231-4181
Fax: 503-231-4007
E-mail: info@ospirg.org
http://www.ospirg.org

OSPIRG's mission is to deliver persistent, result-oriented public interest activism that protects consumers, encourages a fair, sustainable economy, and fosters responsive democratic government.
*Founded: 1983*
*Maureen Kirk, Executive Director*

**1273 Oregon Trout**
65 SW Yamhill Parkway 503-222-9091
Portland, OR 97204 Fax: 503-222-9187
E-mail: info@ortrout.org
http://www.ortrout.org

A statewide non-profit organization headquartered in Portland, Oregon with satellite offices in Bandon, Bend, Corvallis and Medford. Oregon Trout works to restore freshwater health through innovation and education.
*Founded: 1983*
*Joe Whitworth, Executive Director*

**1274 Oregon Water Resources Congress**
1201 Court Street NE 503-363-0121
Suite 303 Fax: 503-371-4926
Salem, OR 97301 E-mail: owrc_info@yahoo.com
http://www.owrc.org

To promote the protection and use of water rights and the wise stewarship of water resources.
*Founded: 1912*
*Anita Winkler, Executive Director*
*Lea Rasura, Office Administrator*

**1275 Oregon Wild**
5825 N Greeley 503-283-6343
Portland, OR 97217 Fax: 503-283-0756
E-mail: rm@oregonwild.org
http://www.oregonwild.org

Formerly Oregon Natural Resources Council, Oregon Wild works to protect and restore Oregon's wildlands, wildlife and waters as an enduring leagacy for all Oregonians.
*Founded: 1974*
*Regna Merritt, Executive Director*

**1276 Pacific Rivers Council**
PO Box 10798 541-345-0119
Eugene, OR 97440 Fax: 541-345-0710
E-mail: info@pacrivers.org
http://www.pacrivers.org

One of the most influential river conservation organizations in the United States. The mission is to protect and restore rivers, their watersheds, and native aquatic species.
*Founded: 1987*
*Holly Spencer, Acting Executive Director*

**1277 Rising Tide North America**
PO Box 1588
Hood River, OR 97031 http://www.risingtidenorthamerica.org

An international, all-volunteer, grassroots network of groups and individuals who organize locally, promote community-based solutions to the climate crisis and take direct action to confront the root causes of climate change.

**1278 Sierra Club-Oregon Chapter**
1821 SE Ankeny St 503-238-0442
Portland, OR 97214 E-mail: oregon.chapter@sierraclub.org
http://oregon.sierraclub.org

Non profit member supported organization that promotes conservation of the state's natural environment for the public interest by influencing public policy decisions.
*Ivan Maluski, State Grassroots Organizer*
*Brian Pasko, Chapter Director*

**1279 Solar Oregon**
205 SE Grand Ave 503-231-5662
Suite 205 E-mail: info@solaror.org
Portland, OR 97214 http://www.solaror.org

A non-profit membership organization providing public education and community outreach to encourage Oregonians to choose solar energy
*Christopher Luttkus, President*
*Michael VanDerwater, Executive Director*

**1280 University of Oregon Environmental Studies Program**
5223 University of Oregon 541-346-5000
Eugene, OR 97403 Fax: 541-346-5954
E-mail: ecostudy@uoregon.edu
http://envs.uoregon.edu

Environmental Studies crosses the boundaries of traditional disciplines, challenging faculty and students to look at the relationship between humans and their environment from a new perspective. They are dedicated to gaining greaterunderstanding of the natural world from an ecological perspective; devising policy and behavior that address contemporary environmental problems; and promoting a rethinking of basic cultural premises, ways of structuring knowledge and the rootmetaphors of society.
*Founded: 1983*
*Alan Dickman, Program Director*
*RaDonna Aymong, Office Manager*

**1281 Wildlife Society**
PO Box 2378 541-937-2131
Corvallis, OR 97339 Fax: 541-937-3401

To promote wise conservation and management of wildlife resources in Oregon by serving and representing wildlife professionals.
*Mark Penninger, President*

## Pennsylvania

**1282 Air and Waste Management Association**
One Gateway Center, 3rd Floor 412-232-3444
420 Fort Duquesne Boulevard 800-270-3444
Pittsburgh, PA 15222 Fax: 412-232-3450
E-mail: info@awma.org
http://www.awma.org

A nonprofit, nonpartisan professional organization that enhances knowledge and expertise by providing a neutral forum for information exchange, professional development, networking opportunities, public education, and outreach to morethan 9000 environmental professionals in 65 countries. A&WMA also promotes global enviromnetal responsibility and increases the effectivenessof organization to make critical decisions that benefit society.
*Founded: 1907*
*Adrianne Carolla, Executive Director*

**1283 Alliance for the Chesapeake Bay**
3310 Market Street 717-737-8622
Suite A Fax: 717-737-8650
Camp Hill, PA 17011 E-mail: acbpa@acb-online.org
http://www.acb-online.org

*Founded: 1971*
*David Bancroft, President*

**1284 American Rivers: Mid-Atlantic Region**
355 N 21st Street 717-763-0741
Suite 309 Fax: 717-763-0743
Camp Hill, PA 17011 http://www.americanrivers.org
*Sara Deuling, Field Associate*

**1285 American Society of Landscape Architects: Pennsylvania/Delaware Chapter**
908 North Second Street 717-441-6041
Harrisburg, PA 17102 http://www.padeasla.org
*Founded: 1899*
*Richard Rauso, President*
*John D Wanner, Executive Director*

**1286 Appalachian States Low-Level Radioactive Waste Commission**
Pennsylvania DEP/BRP 410-537-3345
400 Market Street, 13th Floor Fax: 410-537-4133
Harrisburg, PA 17101 E-mail: kmcginty@state.pa.us
http://www.dep.state.pa.us

The commission was ratified by Maryland, Delaware, Pennsylvania and West Virginia to assure intertstate cooperation for the

proper packaging and transportation of low-level radioactive waste. Pennsylvania is the host state and handlesthe administrative duties of the commission at this time.
*Founded: 1986*
*Kathleen A McGinty, Chair/Executive Director*
*Richard R Janati, Administrator*

**1287 Audubon Society of Western Pennsylvania at the Beechwood Farms Nature Reserve**
614 Dorseyville Road
Pittsburgh, PA 15238
412-963-6100
Fax: 412-963-6761
E-mail: aswp@aswp.org
http://www.aswp.org/
To inspire and educate the people of southwesterm Pennsylvania to be respectful and responsible stewards of the natural world.
*Founded: 1916*
*Carolyn Sanford, President*

**1288 Brandywine Conservancy**
PO Box 141
Chadds Ford, PA 19317
610-388-2700
Fax: 610-388-1197
E-mail: emc@brandywine.org
http://www.brandywineconservancy.org
The Conservancy is a nonprofit land and water conservation organization protecting natural resources in southeastern PA and northern DE. It provides conservation services to landowners, farmers and municipalities through acomprehensive approach to cutting-edge environmental planning and management. Through conservation easements, historic preservation, and water protection efforts, the Conservancy has been instrumental in permanently protecting more than 43,000 acresof land.
*Founded: 1967*
*James H. Duff, Executive Director*
*Sherri Evans-Stantos, Director, EMC*

**1289 Global Education Motivators**
9601 Germantown Avenue
Philadelphia, PA 19128
215-248-1150
877-451-7925
Fax: 215-248-7056
http://www.gem-ngo.org
A non-profit organization to help schools meet the complex challenges of living in a global society.
*Founded: 1981*
*Wayne Jacoby, President*
*Sabrina Cusimano, Educational Outreach Dir*

**1290 Hawk Mountain Sanctuary Association**
1700 Hawl Mountain Road
Kempton, PA 19529
610-756-6961
Fax: 601-756-4468
http://www.hawkmountain.org
To conserve birds of prey worldwide by providing leadership in raptor conservation science and education, and by maintaining Hawk Mountain Sanctuary as a model observation, research and education facility.
*Lee Schisler Jr, President*

**1291 Nature Conservancy: Pennsylvania Chapter**
Nature Conservancy
15 East Ridge Pike
Suite 500
Conshohocken, PA 19428
610-834-1323
800-628-6860
Fax: 610-834-6533
E-mail: pa_chapter@tnc.org
http://http://nature.org/wherewework/northamerica/states/pennsylvan
*Founded: 1951*
*Steven J McCormick, President*
*Philip J James, Senior Vice President*

**1292 Penn State Institutes of Energy and the Environment**
Land and Water Research Building
University Park, PA 16802
814-863-0291
Fax: 814-865-3378
E-mail: bad5@psn.edu
http://www.environment.psu.edu
The mission is to expand Penn State's capacity to pursue the newest frontiers in energy and environmental research by encouraging cooperation across disciplines and the participation of local, state, federal and internationalstakeholders.
*Brian Dempsey, Interim Director*

**1293 Pennsylvania Association of Accredited Environmental Laboratories**
316 Roosevelt Street
Sayre, PA 18840
570-888-4768
Fax: 570-882-8538
E-mail: judygraves@paael.org
http://www.paael.org
A non-profit association of PA DEP accredited laboratories and related industry representatives which takes a leadership role in promoting the advancement of environmental laboratories by: providing educational opportunities,professional development, and a forum for information exchange; providing an arena for memebers to effectively interact with state and national regulatory agencies; and encouraging ethical conduct of environmental laboratories.
*Founded: 1987*
*Judy Graves, Executive Director*

**1294 Pennsylvania Association of Conservation Districts**
25 N Front Street
Harrisburg, PA 17101
717-238-7223
Fax: 717-238-7201
E-mail: pacd@pacd.org
http://www.pacd.org
A nonprofit organization that supports, enhances and promotes Pennsylvania's Conservation Districts and their programs. PACd provides districts with education and information to help them in their work in land and water conservation.
*Founded: 1950*
*Larry Kehl, President*
*Susan Fox Marquart, Executive Director*

**1295 Pennsylvania BASS Chapter Federation**
769 N Cottage Road
Mercer, PA 16137
http://www.pabass.com
A non-profit service organization comprised of seven geographically divided districts across the state.
*Mark Heckaman, President*

**1296 Pennsylvania Environmental Council**
130 Locust Street
Suite 200
Harrisburg, PA 17101
717-230-8044
Fax: 717-230-8045
E-mail: bhill@pecpa.org
http://www.pecpa.org
Protects and restores the natural and built environments through innovation, collaboration, education and advocacy.
*Brian Hill, President/CEO*
*Michael Hudson, VP/COO*

**1297 Pennsylvania Forestry Association**
56 E Main Street
Mechanicsburg, PA 17055
717-766-5371
800-835-8065
E-mail: thepfa@verizon.net
http://pfa.cas.psu.edu
A broad-based citizens organization, provides leadership in sound forest management advice and education and promotes wise stewardship to private land owners, resulting in benefits for the resident of the Commonwealth.
*Founded: 1886*
*Linda Finley, President*
*Marc Lewis, Vice President*

**1298 Pennsylvania Resources Council**
3606 Providence Road
Newtown Square, PA 19073
610-353-1555
Fax: 610-353-6257
E-mail: vanclief@prc.org
http://www.prc.org
To promote conservation of our natural resources and protection of scenic beauty through public education and outreach in a collaborative effort with government agencies, business, charitable foundations and other nonprofitorganizations.
*Founded: 1939*
*Larry Myers, Executive Director*
*Barley Van Clief, Regional Director*

**1299 Pocono Environmental Education Center**
RR2 Box 1010 570-828-2319
Dingsman Ferry, PA 18328 Fax: 570-828-9695
E-mail: peec@peec.org
http://www.peec.org
PEEC enhances environmental awareness, knowledge, and appreciation through hands-on experience in a natural outdoor classroom. Located in the Delaware Water Gap Nat'l Rec Area, PEEC is open year-round and welcomes school groupsfamilies, retreats, and volunteers.
*Founded: 1972*

**1300 Rails-to-Trails Conservancy**
Northeast Regional Office 717-238-1717
2133 Market Street, Ste 222 Fax: 717-238-7566
Camp Hill, PA 17011 http://www.railstotrails.org
*Tom Sexton, Regional Director*
*Carl Knoch, Manager Trail Development*

**1301 Rodale Institute**
611 Siegfriedale Road 610-683-1400
Kutztown, PA 19530 Fax: 610-683-8548
E-mail: info@rodaleinst.org
http://www.rodaleinstitute.org
Works with people worldwide to achieve a regenerative food system thaty renews environmental and human health working with the philosophy that healthy soil = healthy food = healthy people.
*Tim LaSalle, President*

**1302 Sierra Club: Pennsylvania Chapter**
PO Box 663 717-232-0101
Harrisburg, PA 17108 Fax: 717-238-6330
E-mail: pennsylvania.chapter@sierraclub.org
http://pennsylvania.sierraclub.org
Includes 10 local Sierra Club groups. Emphasis is on state environmental policy advocacy, outings, education and local environmental protection efforts.
*Jeff Schmidt, Senior Chapter Director*
*Joan Wilson, Chapter Coordinator*

**1303 Western Pennsylvania Conservancy**
209 Fourth Avenue 412-288-2777
Pittsburgh, PA 15222 866-586-2390
Fax: 412-281-1792
E-mail: webmaster@paconserve.org
http://www.paconserve.org
Protects, conserves and restores land and water for the diversity of the region's plants, animals and their ecosystems. Through science-based strategies, collaboration, leadership and recognition of the relationship between humankindand nature, WPC achieves tangible conservation outcomes for present and future generations.
*Thomas D Saunders, President/CEO*

## Rhode Island

**1304 American Lung Association of Rhode Island**
298 West Exchange Street 401-421-6487
Providence, RI 02903 Fax: 401-331-5266
E-mail: alari@lungri.org
http://www.lungusa.org/rhodeisland
Offers a wide variety of lung health services to the people of Rhode Island

**1305 American Society of Landscape Architects: Rhode Island Chapter**
http://www.riasla.org
The purpose shall be the advancement of knowledge, education, and skill in the art and science of landscape architecture as an instrument of service in the public welfare.
*Mike Dowhan, President*
*Art Eddy, President-Elect*

**1306 Audubon Society of Rhode Island**
12 Sanderson Road 401-949-5454
Smithfield, RI 02917 Fax: 401-949-5788
E-mail: audubon@asri.org
http://www.asri.org
An independent, nonprofit, state organization dedicated to the conservation of wildlife habitat, the education of young and old about natural ecosystems and the need to preserve them, and advocacy in order to promote continued effortsat preserving our natural heritage.
*Founded: 1897*
*Lawrence Taft, Executive Director*

**1307 Nature Conservancy: Rhode Island Chapter**
159 Waterman Street 401-331-7110
Providence, RI 02906 Fax: 401-273-4902
E-mail: ri@tnc.org
http://www.nature.org
An international nonprofit organization dedicated to preserving the plants, animals and natural communities that represent the diversity of life on Earth by protecting the lands and waters they need to survive.
*John Cook, Regional Managing Director*

**1308 Sierra Club: Rhode Island Chapter**
17 Gordon Avenue 401-521-4734
Suite 208 E-mail: chris.wilhite@sierraclub.org
Providence, RI 02905 http://rhodeisland.sierraclub.org
Represents one of 64 chapters across the U.S. and Canada.
*2500 Members*
*Chris Wilhite, Chapter Director*

## South Carolina

**1309 American Lung Association of South Carolina**
1817 Gadsden Street 803-779-5864
Columbia, SC 29201 800-849-5864
Fax: 803-254-2711
E-mail: alasc@lungsc.org
http://www.lungsc.org
*Gabrielle Steele, Regional Director*

**1310 American Lung Association of South Carolina: Coastal Region**
1941 Savage Road 843-556-8451
Suite 200-A Fax: 843-556-3332
Charleston, SC 29407 E-mail: scatlin@lungsc.org
*Sally Catlin, Regional Director*

**1311 American Lung Association of South Carolina: Upstate Region**
11 Brendan Way 864-233-0517
B-2 Fax: 864-233-2124
Greenville, SC 29615 E-mail: altompkins@lungsc.org
*Al Tompkins, Regional Director*

**1312 American Society of Landscape Architects: South Carolina Chapter**
7 Lafayette Place 843-681-6618
Hilton Head Island, SC 29926 Fax: 843-681-7086
http://www.scasla.org
*Andrea Almond, President*
*Barrett Anderson, President-Elect*

**1313 Friends of the Reedy River**
PO Box 9351 864-255-8946
Greensville, SC 29604 http://www.friendsofthereedyriver.org
The Reedy River is an economic and social resource for the community that impacts our quality of life. Maintaining and protecting its health, above and below the surface, is pivotal in maintaining the natural, social and economicalhealth of the Upstate of South Carolina.
*Charles Chamberlain, President*

**1314 Nature Conservancy: South Carolina Chapter**
PO Box 5475 803-254-9049
Columbia, SC 29250 Fax: 803-252-7134
E-mail: southcarolina@tnc.org
http://www.nature.org

Nonprofit conservation organization dedicated to preserving the plants, animals and natural communities that represent the diversity of life on Earth by protecting the lands and waters they need to survive. Buys significant tracts ofland in its project areas and later re-sells the tracts to a public agency partner such as US Fish and Wildlife Service, US Forest Service and the SC Department of Natural Resources. Also supports and encourages conservation easements.
*Mark Robertson, Executive Director*

**1315 Sierra Club: South Carolina Chapter**
1314 Lincoln Street 803-256-8487
Suite 211 Fax: 803-256-8448
Columbia, SC 29202 http://southcarolina.sierraclub.org
*Susan Corbett, Chair*

**1316 South Atlantic Fishery Management Council**
4055 Faber Place Drive 843-571-4366
Suite 201 866-723-6210
North Charleston, SC 29405 Fax: 843-769-4520
E-mail: safmc@safmc.net
http://www.safmc.net

Responsible for the conservation and management of fish stocks within the federal 200-mile limit of the Atlantic off the coasts of North Carolina, South Carolina, Georgia and east Florida to Key West.
*George J Geiger, Chairman*

**1317 South Carolina BASS Chapter Federation**
1469 Schurlknight Road 803-567-4680
St. Stephen, SC 29479
*Tony Bennett, President*

**1318 South Carolina Native Plant Society**
PO Box 491
Norris, SC 29667 E-mail: bill.stringer@scnps.org
http://www.scnps.org

A non-profit ogranization committed to the preservation and protection of native plant communities in South Carolina.
*Bill Stringer, President*

**1319 South Carolina Solar Council**
1201 Main Street 803-737-9852
Suite 430 Fax: 803-737-9846
Columbia, SC 29201 E-mail: sdubose@energy.sc.gov
*Sonny Dubose, Project Coordinator*

**1320 Southern Appalachian Botanical Society**
Newberry College
2100 College Street 803-321-5257
Newberry, SC 29108 Fax: 803-321-5636
E-mail: charles.horn.@newberry.edu
http://www.sabs.appstate.edu

This is a professional organization for those interested in botanical research, especially in the areas of ecology, floristics and systematics. To this end, we publish a journal, CASTANEA, and a newsletter, CHINQUAPIN.
*Founded: 1936*
*Conley McMullen, President*
*Charles Horn, Treasurer*

## South Dakota

**1321 American Lung Association of South Dakota**
PO Box 1524 605-336-7222
Sioux Falls, SD 57104 800-873-5864
Fax: 605-336-7227
E-mail: lung@americanlungsd.org
http://www.lungusa.org

*Founded: 1904*
*Kathleen Sweere, Executive Director*

**1322 Great Plains Native Plant Society**
PO Box 321 605-745-3397
Hermosa, SD 57747 Fax: 605-745-3397
E-mail: info@gpnps.org
http://www.gpnps.org

Mission is to engage in scientific research regarding plants of the Great Plains of North America; to disseminate this knowledge through the creation of one or more educational botanic gardens of plants of the Great Plains, featuringbut not limited to Barr's discoveries; and to engage in any educational activities which may further public familiarity with plants of the Great Plains, their uses and enjoyment.
*Founded: 1984*

**1323 Nature Conservancy: South Dakota Chapter**
The Nature Conservancy
2601 South Minnesota Avenue 612-331-0700
Suite 105-319 Fax: 612-331-0770
Sioux Falls, SD 57105 E-mail: comment@tnc.org
http://nature.org

*Founded: 1951*
*Bob Paulson, Black Hills Program Director*
*Mary Miller, Manager*

**1324 Sierra Club: South Dakota Chapter**
PO Box 1624 605-348-1345
Rapid City, SD 57709 Fax: 605-348-1344
http://southdakota.sierraclub.org

*Jim Margadant, Chair*
*Gerry Bloomer, Secretary*

**1325 South Dakota Association of Conservation Districts**
PO Box 275 605-895-4099
Pierre, SD 57501 Fax: 605-895-9424
http://www.sdconservation.org

Mission is to lead, represent and assist South Dakota's conservation districts in promoting a healthy environment. Specific areas of concern include wind and water erosion, water quality and quantity including preservation of theMissouri main stem dams, air quality, forestry, rangeland, wildlife and recreation.
*Founded: 1942*
*Angela Ehlers, Executive Director*

**1326 South Dakota Ornithologists Union**
3108 South Holly Avenue 605-677-6175
Sioux Falls, SD 57105 Fax: 605-677-6557
E-mail: nancy.drilling@rmbo.org
http://www.sdou.org

To encourage the study of birds in South Dakota and to promote the study of orinthology by more closely uniting the students of this branch of natural science.
*Founded: 1949*
*Nancy Drilling, President*
*Ricky Olson, Vice President*

**1327 South Dakota Wildlife Federation**
PO Box 7075 605-224-7524
Pierre, SD 57501 Fax: 605-224-7524
E-mail: sdwf@mncomm.com
http://www.sdwf.org

Represents the interests of all South Dakotans in wildelife, outdoor recreation, natural resources, and a quality environment.
*Founded: 1945*
*Chris Hesla, Executive Director*

**1328 Wildlife Society**
Box 218
DeSmet, SD 57231
605-854-9105
E-mail: paul.coughlin@state.sd.us
A nonprofit scientific and educational organization that serves professionals such as government agencies, academia, industry, and non-government organizations in all areas related to the conservation of wildlife and natural resourcesmanagement.
*Founded: 1937*
*Will Morlock, South Dakota State President*

## Tennessee

**1329 American Lung Association of Tennessee: Southeast Office**
1466 Riverside Drive
Suite D
Chattanooga, TN 37406
423-629-1098
800-432-5864
Fax: 423-629-0054
E-mail: scudabacalatn@comcast.net
http://www.lungtn.org
*Founded: 1910*
*Shirley Cudabac, Development Director*

**1330 American Lung Association of Tennessee:Middle Region**
State Office
One Vantage Way, Ste B130
Nashville, TN 37228
615-329-1151
800-432-5864
Fax: 615-329-1723
E-mail: alastaff@alatn.org
http://www.alatn.org
*Founded: 1910*

**1331 American Society of Landscape Architects: Tennessee Chapter**
E-mail: hollie@tnasla.org
http://www.tnasla.org
*Trey Benefield, President*
*Hollie Cummings, Executive Director*

**1332 Kentucky-Tennessee Society of American Foresters**
, KY
http://www.ktsaf.org
The mission of the Society of American Foresters is to advance the science, education, technology, and practice of forestry; to enhance the competency of its members; to establish professional excellence; and, to use the knowledge,skills, and conservation ethic of the profession to ensure the continued health and use of forest ecosystems and the present and future availability of forest resources to benefit society.
*428 Members*
*Tamara Cushing, Chair*
*Heather M Slayton, Secretary*

**1333 Kids for a Clean Environment**
PO Box 158254
Nashville, TN 37215
615-331-7381
Fax: 615-333-9879
E-mail: kidsface@mindspring.com
http://www.kidsface.org
Established to help children who wanted to learn more about the world in which they live, provide a way for children to be involved in the protection of nature and connect children with other children who share their concerns aboutglobal environmental issues.
*Founded: 1989*
*Melissa Poe, Founder*

**1334 Nature Conservancy: Tennessee Chapter**
2021 21st Avenue South
Suite C-400
Nashville, TN 37212
615-383-9909
800-628-6860
Fax: 615-383-9717
E-mail: tennessee@tnc.org
http://www.nature.org
The Tennessee Chapter has protected more than 220,000 acres in the state.
*Founded: 1951*
*Scott Davis, State Director*
*Gabby Call, Associate State Director*

**1335 Scenic Tennessee**
PO Box 12174
Murfreesboro, TN 37129
931-962-1813
E-mail: margedavis@comcast.net
http://www.scenictennessee.org
The only organization in the state devoted exclusively to issues of scenic beauty.
*Leslee Dodd Karl PhD, President*
*Pamela Glaser, Secretary*

**1336 Sierra Club: Tennessee Chapter**
2021 21st Avenue S
Suite 436
Nashville, TN 37212
615-386-3640
http://tennessee.sierraclub.org
*Katherine Pendleton, Chair*
*Gloria Griffith, Vice Chair*

**1337 Tennessee Association of Conservation Districts**
PO Box 107
Hickory Valley, TN 38042
731-764-2909
Fax: 731-658-6726
E-mail: barry.lake@tnacd.org
http://tnacd.org
*Founded: 1982*
*Barry Lake, President*
*Danny Sells, Executive Director*

**1338 Tennessee Citizens for Wilderness Planning**
130 Tabor Road
Oak Ridge, TN 37830
865-481-0286
E-mail: groton87@att.net
http://www.tcwp.org
Dedicated to achieving and perpetuating protection of natural lands and waters by means of public ownership, legislation, or cooperation of the private sector.
*Founded: 1966*
*Jimmy Groton, President*
*Frank Hensley, Vice President*

**1339 Tennessee Environmental Council**
One Vantage Way
Suite D 105
Nashville, TN 37228
615-248-6500
Fax: 615-248-6545
E-mail: tec@tectn.org
http://www.tectn.org
The mission of the Tennessee Environmental Council is to educate and advocate for the conservation and improvement of Tennessee's environment, communities, and public health.
*Founded: 1970*
*John McFadden, Executive Director*
*Claudia Schenck, Office Director*

**1340 Tennessee Woodland Owners Association**
PO Box 1400
Crossville, TN 38557
615-484-5535
Fax: 915-484-1924
E-mail: reharrison@multipro.com
*Robert Harrison, Secretary/Treasurer*

**1341 Toxicology Information Response Center**
1060 Commerce Park
MS 6480
Oak Ridge, TN 37830
865-576-1746
Fax: 865-574-9888
E-mail: slusherkg@ornl.gov
http://www.ornl.gov/TechResources/tirc/hmepg.html
TIRC provides customer search services to both scientific and public communities as a convenient and efficient way to obtain comprehensive scientific information on any subject of interest.
*Founded: 1971*
*Kim Slusher, Administrator*

**1342 Trout Unlimited**
1300 North 17th Street
Suite 500
Arlington, VA 22209
703-522-0200
800-834-2419
Fax: 703-284-9400
E-mail: trout@tu.org
http://www.tu.org

Mission: To conserve, protect and restore North America's trout and salmon fisheries and their watersheds. We accomplish this mission on local, state and national levels with an extensive and dedicated volunteer network.
*Founded: 1960*
*Charles Gauvin, President/CEO*
*Steve Moyer, Vice President*

**1343 Wildlife Society**
Ellington Agricultural Center
PO Box 40747
Nashville, TN 37204
423-253-8416
E-mail: mdodson@fs.fed.us
http://www.utm.edu/TN-TWS
A nonprofit scientific and educational organization that serves professionals such as government agencies, academia, industry, and non-government organizations in all areas related to the conservation of wildlife and natural resourcesmanagement.
*Founded: 1968 $10/Year*
*Alan Peterson, President*
*Mary Dodson, Secretary*

## Texas

**1344 American Lung Association of Texas: Central Region**
PO Box 26460
Austin, TX 78755
512-467-6753
800-252-5864
Fax: 512-467-7621
E-mail: info@texaslung.org
http://www.texaslung.org
*Laura Chapman, Senior Program Director*

**1345 American Lung Association of Texas:Alamo and Southern Region**
8207 Callaghan Road
Suite 140
San Antonio, TX 78230
210-308-8978
Fax: 210-308-8992
E-mail: alasoutx@texaslung.org
http://www.texaslung.org
*Jerilyn Miller, Senior Program Director*
*Linda Nichols, Regional VP/Advocacy*

**1346 American Lung Association of Texas:Dallas/Ft Worth Region**
8150 Brookriver Drive
S-102, LB-151
Dallas, TX 75247
214-631-5864
800-LUN- USA
Fax: 214-630-8092
E-mail: dallas@texaslung.org
http://www.texaslung.org
*Sara Dreiling, CEO*
*Yolanda Sims, Program Director*

**1347 American Lung Association of Texas:Houston and Southeast Region**
2030 North Loop West
Suite 250
Houston, TX 77018
713-629-5864
800-586-4872
Fax: 713-629-5828
E-mail: batkins@texaslung.org
http://www.texaslung.org
*Bob Atkins, Regional VP Development*
*Chantel L Henderson, Program Coordinator*

**1348 American Lung Association of Texas:Western Region**
4141 Pinnacle Street
Suite 212
El Paso, TX 79902
915-532-6776
800-252-5864
Fax: 915-532-7231
E-mail: terrazas@texaslung.org
http://www.texaslung.org
*Miguel Escobedo, Program Coordinator*

**1349 American Society of Landscape Architects: Texas Chapter**
E-mail: jean.kavanagh@ttu.edu
http://www.texasasla.org
*Jean Kavanagh, President*
*Chad St John, President-Elect*

**1350 Association of Texas Soil and Water Conservation Districts**
4311 South 31st Street
Suite 125
Temple, TX 76502
254-778-8741
800-792-3485
Fax: 254-773-3311
E-mail: bwhite@tsswcb.state.tx.us
http://www.tsswcb.state.tx.us/swcds/atswcd
The nonprofit organization attempts to make owners and operators of agricultural land aware of the need to conserve and protect the soil and water resources in Texas. It promotes SWCDs (soil and water conservation districts) througheducational, scientific, charitable, and religious activities.
*Scott Buckles, President*

**1351 Big Bend Natural History Association**
PO Box 196
Big Bend National Park, TX 79834
432-477-2236
Fax: 432-477-2234
E-mail: info@bigbendbookstore.org
http://www.bigbendbookstore.org
The association's goal is to educate the public and increase their appreciation of the Big Bend Area. It conducts seminars, publishes and supplies books, maps and other materials.
*Founded: 1956*
*Mike Boren, Executive Director*

**1352 Center for Environmental Philosophy**
University of North Texas
PO Box 310980
Denton, TX 76203
940-565-2727
Fax: 940-565-4439
E-mail: cep@unt.edu.
http://www.cep.unt.edu
Publishes the journal Environmental Ethics, maintains a reprint book series in environmental philosophy, promotes education in the field of environmental philosophy, sponsors conference, workshops.
*Founded: 1979*
*Eugene C Hargrove, President*
*Jan Dickson, Executive Director*

**1353 National Wildlife Federation Gulf States Natural Resource Center**
44 East Avenue
Suite 200
Austin, TX 78701
512-476-9805
Fax: 512-476-9810
E-mail: kaderka@nwf.org
http://online.nwf.org/gulfstates
The focus of the four state region (TX, LA, OK, MO) is to restore clean rivers and estuaries, conserve wetlands and natural river systems, protect wildlife populations, promote sustainable land and water use, and educate the public onthese issues.
*Susan Kaderka, Director*
*Lacey McCormick, Communications Manager*

**1354 North Plains Groundwater Conservation District**
603 E 1st Street
PO Box 795
Dumas, TX 79029
806-935-6401
Fax: 806-935-6633
http://www.npwd.org
*Founded: 1949*
*Steven Walthour, General Manager*
*Mikki Wittie, Public Relations/Education*

**1355 Scenic Texas**
3015 Richmond Ave
Suite 220
Houston, TX 77098
713-533-9149
Fax: 713-629-0485
E-mail: scenic@scenictexas.org
http://www.scenictexas.org
Scenic Texas is a nonprofit organization dedicated to the preservation and enhancement of the tstate's visual environemnt. It seeks and supports public policies which promote scenic conservation and beautification and limits harmfulactions to the visual environment.
*Founded: 1967*
*Margaret Lloyd, Policy Director*
*Anne Culver, Executive Director, Houston*

**1356 Sierra Club: Lone Star Chapter**
PO Box 1931
Austin, TX 78767-1931
512-477-1729
Fax: 512-477-8526
E-mail: lonestar.chapter@sierraclub.org
http://texas.sierraclub.org
The Lone Star Chapter consists of over 25,000 members and serves as the grassroots communications center. The chapter also represents memebers as they fight at the state level to protect and conserve Texas' diverse natural heritage.
*Founded: 1965*
*Ken Kramer, Director*
*Donna Hoffman, Communications Coordinator*

**1357 Texas Conservation Alliance**
3532 Bee Caves Road
Suite 110
Austin, TX 78746
512-441-1122
Fax: 512-327-2115
E-mail: bezanson@texas.net
http://tconr.org
Formerly the Texas Committee on Natural Resources. A statewide conservation organization protecting native wildlife habitat and urging the wise and efficient use of natural resources.
*Founded: 1970*
*Janice Bezanson, Executive Director*

**1358 Texas Environmental Health Association**
PO Box 860099
Plano, TX 75086
972-461-9644
Fax: 972-429-9066
E-mail: steve.berry@myteha.org
http://www.myteha.org
A professional nonprofit educational organization that was originally founded as the Texas Association of Sanitarians and then merged with the National Environmental Health Association and changed its name in 1971. For professionals inall program areas of the environmental health field.
*Founded: 1956*
*Betty Richardson, President*
*Steve Berry, Executive Secretary*

**1359 Texas Solar Energy Society**
PO Box 1447
Austin, TX 78767
512-326-3391
800-465-5049
Fax: 512-444-0333
E-mail: info@txses.org
http://www.txses.org
Their mission is to increase the awareness of the potential of solar and other renewable energy applications and promote the wise use of these sustainable and non-polluting resources.
*Natalie Marquis, Executive Director*

**1360 Texas State Soil and Water Conservation Board (TSSWCB)**
4311 South 31st Street
Suite 125
Temple, TX 76502
254-773-2250
800-792-3485
Fax: 254-773-3311
http://www.tsswcb.state.tx.us
The state agency that administers Texas' soil and water conservation laws and coordinates conservation and nonpoint source pollution abatement programs through the State. The Board is composed of 7 members, 2 Governor appointed and 5landowners, from across Texas, and is the lead state agency for planning, management, and abatement of agricultural and silvicultural (forestry) nonpoint source pollution, and administers the Texas Brush Control Program. There are regional officesthroughout Texas.
*Founded: 1939*
*Jose Dodier Jr, Chairman*
*Rex Isom, Executive Director*

**1361 Texas Water Conservation Association**
221 E 9th Street
Suite 206
Austin, TX 78701
512-472-7216
Fax: 512-472-0537
E-mail: robbins@twca.org
http://www.twca.org
Devoted to conserving, developing, protecting, and using water resources in the state of Texas for all beneficial purposes.
*Founded: 1944*
*Greg Rothe, President*
*Leroy Goodson, General Manager*

## Utah

**1362 American Lung Association in Utah**
1930 South 1100 East
Salt Lake City, UT 84106
801-484-4456
Fax: 801-484-5461
E-mail: info@lungutah.org
http://www.lungutah.org
The American Lung Association is committed to preventing lung disease and promoting lung health, through education, research, and advocacy.
*Founded: 1904*
*Craig Cutright, Executive Director*

**1363 American Society of Landscape Architects: Utah Chapter**
PO Box 511125
Salt Lake City, UT 84151
E-mail: kelly@crsarchitects.com
http://host.asla.org/chapters/utahasla
*John Gillman, President*
*Rodney Sylvester, President-Elect*

**1364 Grand Canyon Trust: Utah Office**
HC 64
PO Box 1801
Maob, UT 84532
435-259-5284
Fax: 435-259-5348
http://www.grandcanyontrust.org
A regional, non-profit conservation organization that advocates collaborative, common sense solutions to the significant problems affecting the region's natural resources. Our work is focused in the greater Grand Canyon region ofnorthern Arizona, and in the forests and red rock country of central and southern Utah.
*Bill Hedden, Executive Director*

**1365 Jack H Berryman Institute**
Utah State University
5230 Old Main Hill NR 206
Logan, UT 84322-5230
435-797-2436
Fax: 435-797-1871
http://www.berrymaninstitute.org
The Berryman Institute is a national organization based in the Department of Wildland Resources at Utah State University and the Department of Wildlife & Fisheries at Mississippi State University. The Berryman Instituite is dedicatedto improving human-wildlife relationships and resolving human-wildlife conflicts through teaching, research, and extension.
*Dr Johan du Toit, Co-Director*
*Dr Bruce D Leopold, Co-Director*

**1366 Nature Conservancy: Utah Chapter**
559 E South Temple
Salt Lake City, UT 84102
801-531-0999
800-628-6860
Fax: 801-531-1003
E-mail: utah@tnc.org
http://www.tnc.org/utah
*Founded: 1951*
*Scott Groene, Executive Director*
*Heidi McIntosh, Conservation Director*

**1367 Sierra Club**
2159 S 700 E
Suite 210
Salt Lake City, UT 84106
801-467-9297
Fax: 801-467-9296
E-mail: utah.chapter@sierraclub.org
http://utah.sierraclub.org
To advance the preservation and protection of the natural environment by empowering the citizenry, especially democratically-based grassroots organizations, with charitable resources to further the cause of environmental protection.The vehicle through which The Sierra Club Foundation generally fulfills its charitable mission.
*Founded: 1892*
*Al Herring, Chair*
*Mary Herring, Secretary*

**1368 Southern Utah Wilderness Alliance**
425 East 100 South
Salt Lake City, UT 84111
801-486-3161
E-mail: info@suwa.org
http://www.suwa.org

The mission is the preservation of the outstanding wilderness at the heartof the Colorado Plateau, and the mangement of these lands in their natural state for the benefit of all Americans.

*Founded: 1983*
*Scott Groene, Executive Director*

**1369 Utah Association of Conservation Districts**
1860 N 100 E
Logan, UT 84341
435-753-6029
Fax: 435-755-2117
http://www.uacd.org

A nonprofit corporation representing Utah's 38 soil conservation districts. Provides technicians and planners to design conservation projects for private landowners, staff to coordinate watershed and conservation district projects andconservation education outreach.

*Gordon L Younker, Executive VP*

**1370 Utah Association of Soil Conservation Districts**
1860 N 100 E
Logan, UT 84341
435-753-6029
Fax: 435-753-4037
*William Rigby, Executive Board Member*

**1371 Utah Division of Wildlife Resources**
1594 West North Temple
Suite 2110 Box 146301
Salt Lake City, UT 84114-6301
801-538-4700
Fax: 801-538-4709
E-mail: dwrcomment@utah.gov
http://wildlife.utah.gov

Serve people of Utah as trustee and guardian of the state's wildlife.

**1372 Utah Solar Energy Association**
PO Box 25263
Salt Lake City, UT 84125
800-671-0169
http://utsolar.org

Organized to promote the usage of renewable energy, with a focus on solar energy, through education, public outreach, participation in policy development and other activities to accomplish the goals of the organization.

## Vermont

**1373 American Lung Association of Vermont**
372 Hurricane Lane
Suite 101
Williston, VT 05495
802-876-6500
Fax: 802-876-6505
E-mail: info@vtlung.org
http://www.lungusa2.org/vermont/

Offers a wide variety of lung health services to the people of Vermont.

*Danielle Pinders, Contact*

**1374 American Society of Landscape Architects: Vermont Chapter**
PO Box 1263
Montpelier, VT 05601
802-886-2267
Fax: 877-270-3655
E-mail: patrick.mclean@stantec.com
http://host.asla.org/chapters/vermont/index.cfm

The chapter holds monthly meetings to encourage dialogue among practioners and members, collaborate with those in related fields, and embrace professional guidelines.

*James Palmer, President*
*Gail King, President-Elect*

**1375 Bluebirds Across Vermont Project**
The Birdhouse Network
255 Sherman Hollow Road
Green Mountain Abdubon Society
Huntington, VT 05462
802-434-3068
Fax: 802-434-4686
E-mail: bluebirdhousing@ellijay.com
http://www.cornell.edu

*Jim Shallow, Executive Director*

**1376 National Wildlife Federation Northeastern Natural Resource Center**
58 State Street
Montpelier, VT 05602
802-229-0650
800-822-9919
Fax: 802-229-4532
http://www.nwf.org

The Northeastern Field Office works with state-based affiliates and like-minded organizations to protect valuable woods, water and wildlife resources across New England through education, advocacy and research.

*Larry J Schweiger, President/CEO*

**1377 Noise Pollution Clearinghouse**
PO Box 1137
Montpelier, VT 05601
888-200-8332
http://www.nonoise.org

A national non-profit organization with extensive online noise related resources. The mission of the Clearinghouse is to create more civil cities and more natural rural and wilderness areas by reducing noise pollutions at the source.

**1378 Northeast Recycling Council**
139 Main Street
Suite 401
Brattleboro, VT 05301
802-254-3636
Fax: 802-254-5870
E-mail: info@nerc.org
http://www.nerc.org

To advance an environmentally sustainable economy by promoting source and toxicity reduction, recycling, and the purchasing of environmentally preferable products and services.

*Lynn Rubinstein, Executive Director*

**1379 Sierra Club: Vermont Chapter**
73 Main Street, Room 28
PO Box 611
Montpelier, VT 05601
802-229-5151
Fax: 802-229-2255
E-mail: ne.field@sierraclub.org
http://vermont.sierraclub.org

**1380 Trout Unlimited**
1300 North 17th Street
Suite 500
Arlington, VA 22209
703-522-0200
800-834-2419
Fax: 703-284-9400
E-mail: trout@tu.org
http://www.tu.org

Mission: To conserve, protect and restore North America's trout and salmon fisheries and their watersheds. We accomplish this mission on local, state and national levels with an extensive and dedicated volunteer network.

*Charles Gauvin, President/CEO*
*Steve Moyer, Vice President*

**1381 Vermont Association of Conservation Districts**
487 Rowell Hill Road
Berlin, VT 05602
802-229-9250
Fax: 802-229-6920
E-mail: mdomi15978@aol.com
http://www.vacd.org

A non-profit organization of Vermont's 14 Conservation Districts whose mission is to help the Districts carry out natural resource oriented programs at the local level.

*Michael Domingue, President*

**1382 Vermont BASS Chapter Federation**
Bassin' USA
19 Pinewood Road
Montpelier, VT 05602
802-223-7793
E-mail: nsk@together.net
http://www.bassinusa.com

*Founded: 1968*
*Brendan Cucinello, President*
*Bob Crino, Secretary*

**1383 Vermont Haulers and Recyclers Association**
PO Box 976
Williston, VT 05495
802-864-3615
Fax: 802-660-8553
http://www.zella.com

**1384 Vermont Land Trust**
8 Bailey Avenue
Montpelier, VT 05602
802-223-5234
800-639-1709
Fax: 802-223-4223
E-mail: info@vlt.org
http://www.vlt.org
One of the most effective land trusts in the country. Its primary focus is on permanently conserving productive, recreational, and scenic lands vital to Vermont's and rural economy and environment.
*Founded: 1977*
*Gil Livingston, President*
*Dawn Lee Minter, Executive Assistant*

**1385 Vermont Public Interest Research Group**
141 Main Street
Suite 6
Montpelier, VT 05602
802-223-5221
Fax: 802-223-6855
E-mail: vpirg@vpirg.org
http://www.vpirg.org
The largest nonprofit consumer and environmental advocacy organization in the state, with approximately 20,000 members and supporters. VPIRG's mission is to promote and protect the health of Vermont's people, environment andlocally-based economy by informing and mobilizing citizens statewide.
*Founded: 1972*
*Paul Burns, Executive Director*
*Colleen Thomas, Associate Director*

**1386 Vermont State-Wide Environmental Education Programs**
9 Bailey Avenue
Montpelier, VT 05602
802-985-8686
*Susan Clark, Chair*

**1387 Wildlife Society**
RD1 Box 2161
Pittsford, VT 05763
E-mail: scott.darling@anr.state.vt.us
A nonprofit scientific and educational organization that serves professionals such as government agencies, academia, industry, and non-government organizations in all areas related to the conservation of wildlife and natural resourcesmanagement.
*Scott Darling, President*

## Virginia

**1388 American Bird Conservancy**
4249 Loudoun Ave
PO Box 249
The Plains, VA 20198-2237
540-253-5780
888-247-3625
Fax: 540-253-5782
http://www.abcbirds.org
A not-for-profit organization whose mission is to conserve native birds and their habitats throughout the Americas.
*George H Fenwick, President*

**1389 American Lung Association of Virginia**
9221 Forest Hill Avenue
Richmond, VA 23235
804-267-1900
800-586-4872
Fax: 804-267-5634
E-mail: resourcecenter@lungva.org
http://www.lungusa.org
To promote lung health and prevent lung disease.
*Founded: 1909*
*Catherine Hamm, President/CEO*
*Richard Pierson, Finance Director*

**1390 American Society of Landscape Architects: Virginia Chapter**
11712C Jefferson Avenue
# 249
Newport News, VA 23606
757-412-2664
Fax: 757-412-4637
E-mail: info@vaasla.org
http://www.vaasla.org
*Founded: 1899*
*Lynn Crump, President*
*Kevin Barnes, President-Elect*

**1391 Arlington Outdoor Education Association**
Phoebe Hall Knipling Outdoor Laboratory
PO Box 5646
Arlington, VA 22205
703-228-7650
Fax: 540-349-3336
http://www.outdoorlab.org
Founded to own and operate the Phoebe Hall Knipling Outdoor Laboratory. Its primary purpose is to provide a facility and support a school program designed to give urban school children who live in Arlington, Virginia, an opportunity tolearn science, outdoor skills, arts and humanities in a natural setting.
*Founded: 1967*
*Neil Heinekamp, Director, Outdoor Lab*

**1392 Ashoka**
1700 N Moore Street
Suite 2000, 20th Floor
Arlington, VA 22209
703-527-8300
Fax: 703-527-8383
E-mail: info@ashoka.org
http://www.ashoka.org
Strives to shape a global, entrepreneurial, competitive citizen sector: one that allows social entrepreneurs to thrive and enables the world's citizens to think and act as changemakers.
*Bill Drayton, CEO/Chair*

**1393 Audubon Society of Northern Virginia**
4022 Hummer Road
Annandale, VA 22003
703-256-6895
Fax: 703-256-2060
E-mail: info@audubonva.org
http://www.audubonva.org
To conserve and restore natural ecosystems. ASNV carries out conservation, education and advocacy programs throughout the region from Alexandria to Manassas in Fairfax, Prince William, Loudoun and Arlington counties, and beyond.
*Founded: 1980*
*Bruce Johnson, President*
*Jill Miller, Organization Administrator*

**1394 Center for Health, Environment and Justice**
PO Box 6806
Falls Chruch, VA 22040
703-237-2249
E-mail: chej@chej.org
http://www.chej.org
Works to build healthy communities, with social justice, economic well-being, and democratic governance. Through training, coalition-building and one-on-one technical and organizing assistance, the Center works to level the playingfield so that people can have a say in the environmental policies and decisions that affect their health and well-being.
*Founded: 1981*
*Lois Marie Gibbs, Executive Director*
*Maryll Kleibrink, Director of Development*

**1395 Center for a New American Dream**
455 Second Street SE
Suite 101
Charlottesville, VA 22902
301-891-3683
E-mail: newdream@newdream.org
http://www.newdream.org
Seeks to cultivate a new American dream - one that emphasizes community, ecological sustainability, and a celebration of non-material values, while upholding the spirit of the traditional American dream of life, liberty, and thepursuit of happiness.
*Founded: 1997*
*Wendy Philleo, Executive Director*

**1396 Center for the Evaluation of Risks to Human Reproduction**
NIEHS EC-32
PO Box 12233
Research Triangle Park, NC 27709
919-541-3455
Fax: 919-316-4511
E-mail: shelby@niehs.nih.gov
http://cerhr.niehs.nih.gov
Established to serve as an environmental health resource to the public and to regulatory and health agencies. The Center provides scientifically-based, uniform assessments of the potential for adverse effects on reproduction anddevelopment caused by agents to which humans may be exposed.
*Founded: 1998*
*Dr Michael D Shelby, Director*

**1397 Chesapeake Bay Foundation**
Capitol Place
1108 E Main Street
Suite 1600
Richmond, VA 23219
804-780-1392
Fax: 804-648-4011
E-mail: chesapeake@cbf.org
http://www.cbf.org
*Founded: 1964*
*Ann Jennings, Executive Director*

**1398 Citizens Clearinghouse for Hazardous Waste**
PO Box 6806
Falls Church, VA 22040-6806
703-237-2249
Fax: 703-237-8389
E-mail: info@chej.org
http://www.chej.org
Nonprofit organization serves citizens' groups, individuals and small municipalities working to solve hazardous and solid waste problems. Supplies information needed to understand, prevent, reduce or eliminate exposure to toxicchemicals through customized assistance, both in-house and on referral, a research library and service, publications and newsletters.
*Founded: 1981*
*Louis Marie Gibbs, Executive Director*
*Stephen Lester, Science Director*

**1399 Institute of Scrap Recycling Industries**
1615 L Street NW
Suite 6000
Washington, DC 20036
202-622-8500
Fax: 202-626-0900
E-mail: frankc@cozzigroup.com
http://www.isri.org
*Founded: 1987*
*Frank Cozzi, President*
*John Sacco, Vice President*

**1400 Nature Conservancy**
490 Westfield Road
Charlottesville, VA 22901
434-295-6106
800-628-6860
E-mail: dwhite@tnc.orgg
http://www.tnc.org
To preserve the plants, animals and natural communities that represent the diversity of life on Earth by protecting the lands and waters they need to survive.

**1401 Potomac Appalachian Trail Club**
118 Park Street SE
Vienna, VA 22180
703-242-0693
Fax: 703-242-0968
E-mail: info@patc.net
http://www.patc.net
A volunteer-based organization, founded by the men and women who planned and built the Appalachian Trail. The Club now manages over 1000 miles of hiking trails in the Mid-Atlantic region, along with cabins, shelters, and hundreds ofacres of conserved land.
*Founded: 1927*
*Lee Sheaffer, President*
*Wilson Riley, Director Administration*

**1402 Scenic Virginia**
4 East Main Street
Suite 2A
Richmond, VA 23219
804-643-8439
Fax: 804-643-8438
E-mail: email@scenicvirginia.org
http://www.scenicva.org
The sole statewide organization in the Commonwealth dedicated to the preservation, protection and enhancement of Virginia's scenic beauty and community character. Promotes and sponsors programs that enhance landscapes, promote tourismand economic development, encourage natural beauty in the environment, preserve historical and cultural resources, and improve community appearance.
*Founded: 1998*
*Hylah Boyd, Chairman*
*Leighton Powell, Executive Director*

**1403 Sierra Club: Virginia Chapter**
422 East Franklin Street
Suite 302
Richmond, VA 23219
804-225-9113
Fax: 804-225-9114
E-mail: michael.town@sierraclub.org
http://virginia.sierraclub.org
To advance the preservation and protection of the natural environment by empowering the citizenry, especially democratically-based grassroots organizations, with charitable resources to further the cause of environmental protection.The vehicle through which The Sierra Club Foundation generally fulfills its charitable mission.
*Michael Town, Director*

**1404 Society for Occupational and Environmental Health**
6728 Old McLean Village Drive
McLean, VA 22101
703-556-9222
Fax: 703-556-8729
E-mail: soeh@degnon.org
http://www.soeh.org
Provides a neutral forum where occupational safety and health and environmental issues can be discussed and resolved. Actively seeks to improve the quality of both working and living places.
*George K Degnon, Executive Director*

**1405 Student Conservation Association**
1800 N Kent Street
Suite 102
Arlington, VA 22209
703-524-2441
Fax: 703-524-2451
E-mail: dcinfo@thesca.org
http://www.thesca.org
To build the next generation of conservation leaders and inspire lifelong stewardship of our environment and communities by engaging young people in hands-on service to the land.
*Founded: 1957*
*Flip Hagood, Sr VP Business Development*

**1406 Teratology Society**
1821 Michael Faraday Drive
Suite 300
Reston, VA 20190
703-438-3104
Fax: 703-438-3113
E-mail: tshq@teratology.org
http://teratology.org/
A multidisciplinary scientific society founded in 1960, the members of which study the causes and biological processes leading to abnormal development and birth defects at the fundamental and clinical level, and appropriate measuresfor prevention.

**1407 Trout Unlimited**
1300 North 17th Street
Suite 500
Arlington, VA 22209
703-522-0200
800-834-2419
Fax: 703-284-9400
E-mail: trout@tu.org
http://www.tu.org
Mission: To conserve, protect and restore North America's trout and salmon fisheries and their watersheds. We accomplish this mission on local, state and national levels with an extensive and dedicated volunteer network.
*Founded: 1959*
*Charles Gauvin, President/CEO*
*Steve Moyer, Vice President*

**1408 Virginia Association of Soil and Water Conservation Districts**
7308 Hanover Green Drive
Suite 100
Mechanicsville, VA 23111
804-559-0324
Fax: 804-559-0325
E-mail: info@vaswcd.org
http://www.vaswcd.org
The Virginia Association of Soil and Water Conservation Districts (VASWCD) is a private nonprofit association of 47 soil and water conservation districts in Virginia. It is a voluntary, nongovernmental association of Virginia'sdistricts that provides and promotes leadership in the conservation of natural resources through stewardship and education programs.
*Founded: 1930*
*Wilkie W Chaffin PhD, President*

**1409 Virginia Conservation Network**
422 E Franklin Street
Suite 303
Richmond, VA 23219
804-644-0283
Fax: 804-644-0286
E-mail: vcn@vcna.org
http://www.vcnva.org
Devoted to advancing a common, environmentally sound vision for Virginia. The network's membership is comprised of more

than 115 member organizations committed to protecting Virginia's natural resources.
*Founded: 1990*
*Nathan Lott, Executive Director*

**1410 Virginia Forestry Association**
3808 Augusta Avenue
Richmond, VA 23230
804-278-8733
E-mail: vfa@vaforestry.org
http://www.vaforestry.org
Promotes stewardship and wise use of the Commonwealth's forest resources for the economic and environmental benefits of all Virginians.
*Founded: 1943*

**1411 Virginia Native Plant Society**
400 Blandy Farm Lane
Unit 2
Boyce, VA 22620
540-837-1600
Fax: 540-837-1523
E-mail: vnpsofa@shentel.net
http://www.vnps.org
A statewide organization with approximately 2000 members supported primarily by dues and contributions. The Society's programs emphasize public education, protection of endangered species, habitat preservation, and encouragement ofappropriate landscape use of natice plants.
*Founded: 1982*
*Sally Anderson, President*

**1412 Virginia Waste Industries Association**
508 Somerset Avenue
Richmond, VA 23226
757-686-5960
Fax: 757-686-0010
E-mail: mdobson@envasns.org
http://www.vwia.com
To promote the management of waste in a manner that is environmentally responsible, efficient, profitable, and ethical while benefiting the public and protecting the employees.
*Mike Dobson, Manager*

**1413 Water Environment Federation**
601 Wythe Street
Alexandria, VA 22314
703-684-2400
800-666-0206
Fax: 703-684-2492
E-mail: csc@wef.org
http://www.wef.org
A not-for-profit technical and educational organization with 32,000 individual members and 80 affiliated Member Associations representing and additional 50,000 water quality professionals throughout the world. WEF and its memberassociations proudly work to achieve our mission of preserving and enhancing the global water environment.
*Founded: 1928*
*William Bertera, Executive Director*

## Washington

**1414 American Lung Association of Washington: Spokane Branch**
American Lung Association
1817 East Springfield
Suite E
Spokane, WA 99202
509-325-6516
800-732-9339
Fax: 509-323-5380
E-mail: alaw@alaw.org
http://www.alaw.org
The mission of the American Lung Association is to prevent lung disease and to promote lung health.
*Founded: 1904*
*Marina Cofer-Wildsmith, Chief Executive Officer*
*Leanne Noren, Chief Operations Officer*

**1415 American Lung Association of Washington: Eastern Region**
110 South 9th Avenue
Yakima, WA 98902
509-248-4384
Fax: 509-248-4943
E-mail: alaw@alaw.org
http://www.alaw.org
The mission of the American Lung Association is to eliminate lung disease and to promote lung health.
*Founded: 1906*
*Marina Cofer-Wildsmith, Chief Execuive Officer*

**1416 American Lung Association of Washington:Western Region**
American Lung Association
223 Tacoma Avenue South
Tacoma, WA 98402
253-272-8777
Fax: 253-593-8827
E-mail: lnoren@alaw.org
http://www.ala.org
The mission of the American Lung Association is to prevent lung disease and to promote lung health.
*Founded: 1905*
*marina Cofer-Wildsmith, Chief Executive Director*
*leanne Noren, Operations Officer*

**1417 American Lung Association- Mountain Pacific**
2625 Third Avenue
Seattle, WA 98121
206-441-5100
800-586-4872
Fax: 206-441-3277
E-mail: alaw@alaw.org
http://www.alaw.org
Mission is to save lives by improving lung health and preventing lung disease.
*Founded: 1906*
*Renee Klein, President/CEO*

**1418 American Rivers: Northwest Region Seattle**
4005 20th Ave West
Suite 221
Seattle, WA 98199
206-213-0330
Fax: 206-213-0334
E-mail: arnw@amrivers.org
http://www.americanrivers.org
American Rivers is the only national organization standing up for healthy rivers so our community can thrive. Through national advocacy, innovative solutions and our growing network of strategic partners, we protect and promote ourrivers as valuable assests that are vital to our health, safety and quality of life.
*Founded: 1992*
*Ross Freeman, Associate Director*
*Amy Souers Kober, Communications Director*

**1419 American Rivers: Northwest Region:Portland**
320 SW Stark Street
Suite 412
Portland, OR 97204
503-827-8648
Fax: 503-827-8654
http://www.americanrivers.org
American Rivers is the only national organization standing up for healthy rivers so our communities can thrive. Through national advocacy, innovative solutions and our growing network of strategic partners, we protect and promote ourrivers as valuable assests that are vital to our health, safety and quality of life.
*Founded: 1992*
*J David Moryc, Associate Director Programs*

**1420 American Society of Landscape Architects: Washington Chapter**
603 Stewart Street
Suite 610
Seattle, WA 98101
206-443-9484
Fax: 425-450-9077
E-mail: curtis.lapierre@otak.com
http://www.wasla.org
Mission is to lead, to educate and to participate in the careful stewardship, wise planning and artful design of our cultural and natural environments.
*Founded: 1899*
*Curtis LaPierre, President*
*Christopher Overdorf, President-Elect*

**1421 Conservation Northwest**
1208 Bay Street
Bellingham, WA 98225
360-671-9950
Fax: 360-671-8429
E-mail: wild@conservationnw.org
http://www.conservationnw.org

A non-profit organization with 4 offices and 23 staff around the state that are supported by 5,000 families and hundreds of volunteers who together provide 70 percent of our funding.
*Founded: 1988*
*Alex Loeb, President*
*Bill Donnelly, Vice President*

**1422 Environmental Education Association of Washington**
Environmental Education Association of Washington
EEAW 360-943-6643
P.O. Box 6277 Fax: 360-497-7132
Olympia, WA 98507 E-mail: eeaw@eeaw.org
http://www.eeaw.org

Dedicated to increasing the awareness of and support for environmental education in the state of Washington.
*Founded: 1991*
*Beverly Walker, President*
*Martin Fortin, Treasurer*

**1423 Friends of Discovery Park**
PO Box 99662 206-285-6862
Seattle, WA 98199 888-291-6104
Fax: 253-872-6668
E-mail: info@discoveryparkfriends.org
http://www.discoveryparkfriends.org

An all-volunteer group formed to defend the integrity of Discovery Park and to create and protect there an open space of quiet and tranquility, a sanctuary where the words of man are minimized.
*Founded: 1970*
*Valerie Cholvin, President*

**1424 Friends of the San Juans**
PO Box 1344 360-378-2319
Friday Harbor, WA 98250 877-757-3629
Fax: 360-378-2324
E-mail: friends@sanjuans.org
http://www.sanjuans.org

Mission is to protect the land, water, sea and livability of the San Juan islands through science, education, law and citizen action.
*Founded: 1979*
*Stephanie Buffum, Executive Director*
*Tina Whitman, Science Director*

**1425 Great Peninsula Conservancy**
3721, Kitsap Way 360-373-3500
Suite 5 866-373-3504
Bremerton, WA 98312 Fax: 360-377-0239
E-mail: info@greatpeninsula.org
http://www.greatpeninsula.org

The Great Peninsula Conservancy is a private nonprofit land trust dedicated to forever protecting the rural landscapes, natural habitat and open spaces of our region.
*Founded: 2000*
*Ann D Haines, Executive Director*
*Kate Kuhlman, Director of Development*

**1426 International Bicycle Fund**
4887 Columbia Drive South 206-767-0848
Seattle, WA 98108 Fax: 206-767-0848
E-mail: ibike@ibike.org
http://www.ibike.org

A non-governmental, nonprofit, advocacy organization, promoting sustainable transport and international understanding. Major areas of activity are non-motorized urban planning, economic development, bike safety education, responsibletravel and bicycle tourism, and cross-cultural, educational programs.
*Founded: 1983*
*David Mozer, President*
*John Dowlin, Executive Director*

**1427 Issaquah Alps Trails Club**
PO Box 351 425-392-4432
Issaquah, WA 98027 E-mail: wilbs@worldnet.att.net
http://www.issaquahalps.org

Mission is to act as custodian of the trails and the lush, open tree-covered mountaintops known as the Issaquah Alps. Offers free guided hikes and a voice for protection of our open spaces, trails, and quality of life.
*Founded: 1979*
*Steve Williams, President*

**1428 Mountaineers Conservation Division**
The Mountaineers
300 3rd Avenue West 206-284-6310
Seattle, WA 98119 800-573-8484
Fax: 206-284-4977
E-mail: clubmail@mountaineers.org
http://www.mountaineers.org

Mission is to be the premier outdoor recreation club, dedicated to the responsible employment and protection of natural areas.
*Founded: 1906*
*Ron Eng, President*
*Steve Costie, Executive Director*

**1429 National Wildlife Federation Western Natural Resource Center**
6 Nickerson Street 206-285-8707
Suite 200 Fax: 206-285-8698
Seattle, WA 98109 E-mail: nwnrc@nwf.org
http://online.nwf.org/western

The NWF, through its western regional center works hard to protect the landscapes along the Pacific Coast and the habitats of Washington, Oregon, California, and Hawaii.
*Joelle Robinson, Contact*

**1430 Nature Conservancy: Washington Chapter**
217 Pine Street 206-343-4344
Suite 1100 800-628-6860
Seattle, WA 98101 Fax: 206-343-5608
E-mail: washington@tnc.org
http://www.tnc-washington.org

A leading conservation organization working around the world to protect ecologically important lands and waters for nature and people.
*Founded: 1951*
*John Rose, Chairman*

**1431 North Cascades Conservation Council**
PO Box 95980 206-282-1644
Seattle, WA 98145 Fax: 206-684-1379
E-mail: steveb@premier1.net
http://www.northcascades.org

Mission is to protect and preserve the North Cascade's scenic, scientific, recreational, educational, and wilderness values.
*Founded: 1957*
*Marc Bardsley, President*

**1432 Northwest Power and Conservation Council**
110 Y Street 360-693-6951
Vancouver, WA 98661 Fax: 360-693-6079
E-mail: jblack@nwcouncil.org
http://www.nwcouncil.org/contact/wa.asp

The council develops and maintains a regional power plan and a fish and wildlife program to balance the Northwest's environment and energy needs.
*Frank L Cassidy, Council Member*

**1433 Olympic Park Associates**
2433 Del Campo Drive 206-364-3933
Everett, WA 98208 Fax: 206-364-6379
E-mail: pollytdyer@juno.com
http://www.drizzle.com/~rdpayne/opa.html/

An organization working to preseve Olympic Park's wilderness, beauty and spelndor.
*Founded: 1948*
*Polly Dyer, President*
*Bruce Babbit, Secretary*

**1434 Olympic Region Clean Air Agency**
2940-B Limited Lane, NW 360-586-1044
Olympia, WA 98502 800-422-5623
Fax: 360-491-5308
E-mail: info@orcaa.org
http://www.orcaa.org

A local government agency charged with regulatory and enforcement authority for air quality issues. It is one of the seven such regional air pollution control agencies in Washington State. The agency also administers laws andregulations regarding such programs as solid fuel burning devices, asbestos abatement, and open burning. ORCAA's jurisdiction: Clallam, Grays Harbor, Jefferson, Mason, Pacific and Thurston Counties.
*Founded: 1968*
*Richard A Stedman, Executive Director*
*Dan Nelson, Public Information Officer*

**1435 People for Puget Sound**
911,Western Avenue 206-382-7007
Suite 580 Fax: 206-382-7006
Seattle, WA 98104 E-mail: people@pugetsound.org
http://www.pugetsound.org

People for Puget Sound is a regional citizen's organization founded in 1991 to educate and involve ordinary - and extraordinary - people in protecting and restoring the land and waters of Puget Sound. People for Puget Sound's programsare based on partnership and collaborations, scientific credibility, creative use of communications and technology, and a hands-on-style. People for Puget Sound publishes a quarterly newsletter, and many scientific publications.
*Founded: 1991*
*Kathy Fletcher, Executive Director*
*Stacey Jurgensen, Communications Director*

**1436 Rivers Council of Washington**
509 10th Avenue E 206-568-1380
Suite 200 Fax: 206-568-1381
Seattle, WA 98102 E-mail: RIVERSWA@BRIadoon.com

The Pacific Northwest's oldest river conservation non-profit advocating the protection of free-flowing rivers for recreation, fisheries, and responsible water use.
*Doug North, President*

**1437 Sea Shepherd**
PO Box 2616 360-370-5650
Friday Harbor, WA 98250 Fax: 360-370-5651
http://www.seashepherd.org

An international non-profit, marine wildlife conservation organization whose mission is to end the destruction of habitat and slaughter of wildlife in the world's oceans in order to conserve and protect ecosystems and species.
*Founded: 1977*
*Captain Paul Watson, Founder/President*

**1438 Sierra Club: Cascade Chapter**
180 Nickerson Street 206-378-0114
Suite 202 Fax: 206-378-0034
Seattle, WA 98109 E-mail: cascade.chapter@sierraclub.org
http://cascade.sierraclub.org

The Sierra Club is the nation's oldest, largest, and most influential grassroots environmental organization. The Cascade Chapter is its voice for most of Washington State. Inspired by nature, we work together to protect our communitiesand the planet.
*Trevor Kaul, Director*
*Terri M Morgan, Development Officer*

**1439 Solar Washington**
6848 23rd Avenue NE 206-333-5191
Seattle, WA 98115 E-mail: info@solarwashington.org
http://www.solarwashington.org

A private not-for-profit 501(c)3 association of solar energy equipment manufacturers, system integrators, distributors, dealers, designers, consultants, students, and interested people
.
*Jason Keyes, President*

**1440 Student Conservation Association Northwest**
1265 S Main Street 206-324-4649
Suite 210 Fax: 206-324-4998
Seattle, WA 98144 E-mail: webmaster@thesca.org
http://www.thesca.org/

SCA is a national organization with regional offices in Seattle, Oakland, Pittsburg, Washington DC and headquartered in Charlestown NH. Our mission is to build the next generation of conservation leaders and inspire lifelongstewardship of our environment and communities by engaging young people in hands-on service to the land. We offer a wide range of internships and crew based programs for ages 16 years and up.
*Founded: 1957*
*Dale M Penny, President/CEO*
*Mark Bodin, Executive VP/COO*

**1441 Trout Unlimited**
1300 North 17th Street 703-522-0200
Suite 500 800-834-2419
Arlington, VA 22209 Fax: 703-284-9400
E-mail: trout@tu.org
http://www.tu.org

Mission: To conserve, protect and restore North America's trout and salmon fisheries and their watersheds. We accomplish this mission on local, state and national levels with an extensive and dedicated volunteer network.
*Charles Gauvin, President/CEO*
*Steve Moyer, Vice President*

**1442 Washington Association of Conservation Districts**
185 Beebe Road 509-773-5065
Goldendale, WA 98620 Fax: 509-773-5600
E-mail: wacd@ncia.com
http://www.wacd.org

A non-profit organization representing Washington's 48 Conservation Districts, whos mission is to advance the purposes of Conservation Districts and their constituents by providing leadership, information, and representation.
*Mike Bailey, Finance Director*

**1443 Washington Environmental Council**
615 2nd Avenue 206-622-8103
Suite 380 800-561-8294
Seattle, WA 98104 Fax: 206-622-8113
E-mail: wec@wecprotects.org
http://www.wecprotects.org

Protects what Washingtonians care about- our land and water, fish, and wildlife, and our special way of life. We engage the public and decision makers to improve and enforce protections for the health and well-being of our communities.
*Founded: 1967*
*Joan Crooks, Executive Director*

**1444 Washington Public Interest Research Group**
3240 Eastlake Avenue E 206-568-2850
Suite 100 800-213-7383
Seattle, WA 98102 Fax: 206-568-2858
E-mail: washpirg@pirg.org
http://www.washpirg.org

When consumers are cheated or the voices of ordinary citizens are drowned out by special interest lobbyists, WashPIRG speaks up and takes action.
*Robert Pregulman, Executive Director*

**1445 Washington Recreation and Park Association**
4405 7th Avenue SE 360-459-9396
Suite 202 888-459-0009
Lacey, WA 98503 Fax: 360-459-4160
E-mail: wrpa@seanet.com
http://www.wrpatoday.org

A not-for-profit professional and public interest organization which is dedicated to enhancing and promoting parks, recreation and arts pursuits in Washington State.
*Tracy Thomas, Secretary*

**1446 Washington Refuse and Recycling Association**
4160 6th Avenue SE
Suite 205
Lacey, WA 98503
360-943-8859
866-788-9772
Fax: 360-357-6958
E-mail: office@wrra.org
http://www.wrra.org
Represents Washington's diverse and multifaceted solid waste handling industry, providing its members with general legal support, educational seminars, workshops, and representation before regulatory agencies and the Legislature.
*Steve Wheatley, President*
*Jay Alexander, Vice-President*

**1447 Washington Toxics Coalition**
4649 Sunnyside Avenue N
Suite 540
Seattle, WA 98103
206-632-1545
800-844-7233
Fax: 206-632-8661
E-mail: info@watoxics.org
http://www.watoxics.org
Washington Toxics Coalition protects public health and the environment by eliminating toxic pollution. WTC promotes alternatives, advocates policies, empowers communities, and educates people to create a healthy environment.
*Founded: 1981*
*Jennifer Dold, President*
*Maureen A Judge, Executive Director*

**1448 Washington Wilderness Coalition**
4649 Sunnyside Avenue N
Suite 520
Seattle, WA 98103
206-633-1992
800-627-0062
Fax: 206-633-1996
E-mail: info@wawild.org
http://www.wawild.org
Mission is to preserve and restore wild areas of Washington State throuh citizen empowerment, support for grassroots community groups and advocacy and public education.
*Founded: 1979*
*Tom Geiger, President*

**1449 Washington Wildlife Federation**
PO Box 1656
Bellevue, WA 98118
206-769-5627
E-mail: info@washingtonwildlife.org
http://www.washingtonwildlife.org
To preserve, enhance, and perpetuate Washington's fish, wildlife and habitat through education and conservation programs.
*Kyle Smith, Executive Director*

**1450 Wildlife Society**
3526 103rd Place SE
Everett, WA 98208
509-997-2131
Fax: 509-997-9770
E-mail: tillzinger@communitynet.org
http://http://www.washingtonwildlifesoc.org
A nonprofit scientific and educational organization that serves professionals such as government agencies, academia, industry, and non-government organizations in all areas related to the conservation of wildlife and natural resourcesmanagement.
*Bill Vogel, President*
*Peter Singleton, President-Elect*

## West Virginia

**1451 American Lung Association of West Virginia**
American Lung Association
PO Box 3980
Charleston, WV 25339
304-342-6600
800-LUN- USA
Fax: 304-342-6096
E-mail: tatty@alawv.org
http://www.alawv.org
The mission of the American Lung Association is to prevent lung disease and promote lung health.
*Founded: 1904*
*Sara Crickenberger, Executive Director*
*Chantal Fields, Assistant Executive Director*

**1452 American Society of Landscape Architects: West Virginia Chapter**
5088 Washington Street West
Cross Lanes, WV 25313
304-776-7473
Fax: 304-776-6426
E-mail: tschoolcraft@elrobinson.com
http://www.wvasla.org
A professional society representing members of the landscape architecture profession throughout the state of West Virginia.
*Peter J Williams, President*

**1453 Sierra Club: West Virginia Chapter**
PO Box 4142
Morgantown, WV 26504
E-mail: jim_scon@yahoo.com
http://westvirginia.sierraclub.org
*Jim Sconyers, Chair*

**1454 West Virginia Bureau for Public Health**
West Virginia Departmwent of Health and Human Reso
350 Capitol Street
Suite 702
Charleston, WV 25301
304-558-2971
Fax: 304-558-1035
E-mail: ASOsupport@wvdhhr.org
http://www.wvdhhr.org
Organizational activities not directed specifically toward children are education, regulation and research.
*Chris Curtis, Commisioner*
*Ronald Forren, Deputy Commisioner*

**1455 West Virginia Forestry Association**
PO Box 718
Ripley, WV 25271
304-372-1955
888-372-9663
Fax: 304-372-1957
E-mail: wvfa@wvadventures.net
http://www.wvfa.org
The West Virginia Forestry Association is a non-profit organization funded by its membership. Our members include individuals and businesses involved in forest management, timber production and wood product manufacturing. Our membersare concerned with protecting the environment, as well as enhancing the future of West Virginia's forests through multiple-use management.
*Richard Waybright, Executive Director*

**1456 West Virginia Highlands Conservancy**
HC 64
Box 281
Hillsboro, WV 24946
304-342-8989
E-mail: blittle@citynet.net
http://www.wvhighlands.org
One of the state's oldest environmental activist organizations. A coalition of recreational users of the West Virginia Highlands came together to address a whole host of environmental threats to our state.
*Founded: 1967*
*Hugh Rogers, President*
*Peter Shoenfeld, Senior Vice President*

**1457 West Virginia Woodland Owners Association**
PO Box 13695
Sissonville, WV 25360
304-532-4351
Fax: 304-594-3648
A nonprofit membership organization started and continues to be operated exclusively by independent West Virginia woodland owners.
*Russ Richardson, VP*

## Wisconsin

**1458 American Society of Landscape Architects: Wisconsin Chapter**
PO Box 851
Madison, WI 53701
E-mail: wiasla@wiasla.com
http://www.wiasla.com
The Wisconsin state chapter supports professional development, visibility and network working opportunities.
*Pam Linn, President*
*Rebecca Flood, President-Elect*

**1459 Botanical Club of Wisconsin**
Wisconsin Academy of Science, Arts, & Letters
1922 University Avenue 608-262-5489
Madison, WI 53705 Fax: 608-265-2993
The Botanical Club serves the interests of amateurs and professionals, toward the common goal of learning more about our state's diverse vegitation.
*Emmet Judziewicz, President*

**1460 Central Wisconsin Environmental Station (CWES)**
10186 County Road MM 715-824-2428
Amherst Junction, WI 54407 Fax: 715-824-3201
E-mail: sjohnson@uwsp.edu
http://www.uwsp.edu/cnr/cwes/
Mission is to foster in adults and youth the appreciation, understanding, skill development, and motivation needed to help them build a sustainable balance between the environment, economy, and community.
*Founded: 1975*
*Scott Johnson, Director*

**1461 Citizens for Animals: Resources and Environment**
PO Box 18772 414-466-1250
Milwaukee, WI 53218
*Debi Zweifel, Director*

**1462 Midwest Renewable Energy Association**
7558 Deer Road 715-592-6595
Custer, WI 54423 Fax: 715-592-6596
E-mail: info@the-mrea.org
http://www.the-mrea.org
A non profit organization promoting renewable energy, energy efficiency, and sustainable living through education and demonstration. There are over 3200 active international members representing 39 states and 3 foreign countries.
*Founded: 1990*
*Tehri Parker, Executive Director*
*Amy Heart, Events & Outreach*

**1463 River Alliance of Wisconsin**
306 E Wilson 608-257-2424
Suite 2W Fax: 608-260-9799
Madison, WI 53703 E-mail: wisrivers@wisconinrivers.org
http://www.wisconsinrivers.org
Mission is to advocate for the protection, enhancement and restoration of Wisconsin's rivers and watersheds.
*Jake Barnes, Treasurer*

**1464 Sierra Club: John Muir Chapter**
222 S Hamilton Street 608-256-0565
Suite 1 Fax: 608-256-4562
Madison, WI 53703 E-mail: john.muir.chapter@sierraclub.org
http://wisconsin.sierraclub.org
Preserve and protect the natural environment by empowering citizens, especially democratically based grassroots organizations with charitable resources to further the cause of environmental protection, the vehicle through which TheSierra Club Foundation generally fulfills its charitable mission.
*Jim Steffens, Chair*
*Patrea Wilson, Administrator*

**1465 Sixteenth Street Community Health Center**
1032 S 16th Street 414-672-1353
S Cesar E Chavez Drive Fax: 414-672-9190
Milwaukee, WI 53204 http://www.sschc.org
The mission of the Sixteenth Street Community Health Center is to improve the health and well-being of Milwaukee's Near South Side residents by providing quality, family-based health care, health education and social services, freefrom linguistic, cultural and economic barriers.
*Founded: 1969*
*John Bartkowski, Chief Executive Officer*

**1466 Trees for Tomorrow Natural Resources Educational Center**
519 Sheridan Street 715-479-6456
PO Box 609 Fax: 715-479-2318
Eagle River, WI 54521 E-mail: learning@treesfortomorrow.com
http://www.treesfortomorrow.com
Accredited natural resource specialty school. Hosts workshops for middle/high school students during the school year. Workshops emphasize conservation, proper land management and environmental basics.
*Founded: 1944*
*Maggie Bishop, Executive Director*
*Sheri Buller, Assistant Director*

**1467 Trout Unlimited**
1300 North 17th Street 703-522-0200
Suite 500 800-834-2419
Arlington, VA 22209 Fax: 703-284-9400
E-mail: trout@tu.org
http://www.tu.org
Mission: To conserve, protect and restore North America's trout and salmon fisheries and their watersheds. We accomplish this mission on local, state and national levels with an extensive and dedicated volunteer network.
*Charles Gauvin, President/CEO*
*Steve Moyer, Vice President*

**1468 Wildlife Society**
6315 Clovernock Road 608-221-6344
Middletown, WI 53562 Fax: 608-221-6353
E-mail: barteg@dnr.state.wi.us
A nonprofit scientific and educational organization that serves professionals such as government agencies, academia, industry, and non-government organizations in all areas related to the conservation of wildlife and natural resourcesmanagement.
*Gerald Adrian Bartelt Wydeven, President*

**1469 Wisconsin Association for Environmental Education**
8 Nelson Hall 715-346-2796
University of Wisconsin Fax: 715-346-3835
Stevens Point, WI 54481 E-mail: waee@uwsp.edu
http://www.uwsp.edu/cnr/waee/
A non-profit organization that sponsors conferences, workshops, and gatherings to promote professional growth and networking opportunities.
*Founded: 1975*
*Abbie Enlund, Administrative Assistant*

**1470 Wisconsin Association of Lakes**
4513 Vernon Boulevard 608-661-4313
Suite 101 800-542-5253
Madison, WI 53705 Fax: 715-346-3624
E-mail: wal@coredcs.com
http://www.wisconsinlakes.org
The Wisconsin Association of Lakes is the only statewide nonprofit organization working exclusively to protect and enhance the quality of Wisconsin's 15,000 lakes.
*Jim Burgess, President*

**1471 Wisconsin Land and Water Conservation Association**
702 East Johnson Street 608-441-2677
Madison, WI 53703 Fax: 608-441-2676
E-mail: julian@wlwca.org
http://www.wlwca.org
Mission: To assist county Land Conservation Committees and Departments with the protection, enhancement and sustainable use of Wisconsin's natural resources and to represent them through education and governmental interaction.
*Julian Zelazny, Executive Director*

**1472 Wisconsin Society for Ornithology**
2022 Sherryl Lane 262-547-6128
Waukesha, WI 53188 E-mail: wso1939@hotmail.com
http://www.wsobirds.org

Emphasizes all of the many aspects of birding and to support the research and habitat protection necessary to preserve Wisconsin's birdlife.
*Founded: 1939*
*Jesse Peterson, President*
*Thomas R Schultz, VP*

**1473 Wisconsin Wildlife Federation**
242 Keoller Avenue
Oshkosh, WI 54901
412-235-9136
Fax: 414-235-6030
E-mail: wiwf@execpc.com
http://www.wiwf.org
Made up of hunters, fishers, trappers, and others that are actively engaged in the outdoors. Recognizes the importance of protecting fish and wildlife habitat.
*Russell Hitz, Representative*

**1474 Wisconsin Woodland Owners Association**
PO Box 285
Stevens Point, WI 54481
715-346-4798
800-838-9472
Fax: 715-346-4821
E-mail: mrz@mwwb.net
http://www.wisconsinwoodlands.org
The Wisconsin Woodland Owners Association, a nonprofit educational organization, was established in 1979 to advance the interests of woodland owners and the cause of forestry; develop public appreciation for the value of Wisconsin'swoodlands and their importance in the economy and overall welfare of the state; foster and encourage wise use and management of Wisconsin's woodlands for timber production, wildlife habitat and recreation; and to educate those interested in managingthe woodlands.
*Founded: 1979*
*Dale Zaug, President*
*Merlin Becker, President Elect*

## Wyoming

**1475 Jackson Hole Conservation Alliance**
685 S Cache
PO Box 2728
Jackson, WY 83001
307-733-9417
Fax: 307-733-9008
E-mail: info@jhalliance.org
http://www.jhalliance.org
An organization dedicated to responsible land stewardship in Jackson Hole, Wyoming to ensure that human activities are in harmony with the area's irreplaceable wildlife, scenery and other natural resources.
*Founded: 1979*
*Trevor Stevenson, Executive Director*
*Cindy Harger, Managing Director*

**1476 Nature Conservancy: Wyoming Chapter**
Nature Conservancy
258 Main Street
Suite 200
Lander, WY 82520
307-332-2971
Fax: 307-332-2974
http://www.nature.org
The leading conservation organization working around the world to protect ecologically important lands and waters for nature and people.
*Founded: 1950*
*Andrea Erickson, State Director*
*Paula Hunker, Associate State Director*

**1477 Powder River Basin Resource Council**
934 North Main
Sheridan, WY 82801
307-672-5809
Fax: 307-672-5800
E-mail: resources@powderriverbasin.org
http://www.powderriverbasin.org
Committed to the preservation and enrichment of Wyoming's agricultural heritage and rural lifestyle; the conservation of Wyomings unique land, mineral, water and clean air resources, consistent with responsible use of those resourcesto sustain the vitality of present and future generations; the education and empowerment of Wyoming's citizens to raise a coherent voice in decisions. They are the only group in Wyoming that addresses both agricultural and conservation issues.
*Bernie Barlow, Chairman*
*Clay Rowley, Vice Chairman*

**1478 Sierra Club: Wyoming Chapter**
Po Box 12047
Jackson, WY 83002
307-733-4557
Fax: 307-733-4558
E-mail: wychaptersc@vcn.com
http://www.wyoming.sierraclub.org
To advance the preservation and protection of the natural environment by empowering the citizenry, especially democratically-based grassroots organizations, with charitable resources to further the cause of environmental protection.The vehicle through which The Sierra Club Foundation generally fulfills its charitable mission.
*Steve Thomas, Regional Director*

**1479 Trout Unlimited**
1300 North 17th Street
Suite 500
Arlington, VA 22209
703-522-0200
800-834-2419
Fax: 703-284-9400
E-mail: trout@tu.org
http://www.tu.org
Mission: To conserve, protect and restore North America's trout and salmon fisheries and their watersheds. We accomplish this mission on local, state and national levels with an extensive and dedicated volunteer network.
*Founded: 1959*
*Charles Gauvin, President/CEO*
*Steve Moyer, Vice President*

**1480 Western Association of Fish and Wildlife Agencies**
5400 Bishop Boulevard
Cheyenne, WY 82006
307-777-4569
Fax: 307-777-4699
E-mail: ikruck@state.wy.us
http://www.wafwa.org
WAFWA is a strong advocate of the rights of states and provinces to manage fish and wildlife within their borders. The association has been a key organization in promoting the principles of sound resource management and the building ofpartnerships at the regional, national, and international levels in order to enhance wildlife conservation efforts and the protection of associated habitats and the public interest.
*Ken Ambrock, President*

**1481 Wyoming Association of Conservation Districts**
517 E 19th Street
Cheyenne, WY 82001
307-632-5716
Fax: 307-638-4099
E-mail: waocd@tribcsp.com
http://www.conservewy.com
Mission is to provide leadership for the conservation of Wyoming's soil and water resources, promotes the controll of soil erosion, promotes and protects the quality of Wyomings's waters, reduce siltation of stream channels andreservoirs, promote wise use of Wyoming's water, and all other natural resources, preserve and enhance wildlife habitat, protect the tax base and promote the health, safety and general welfare of the citzens of the state through a responsibleconservation ethic.
*Founded: 1945*
*Ralph Brokaw, President*
*Shawn Sims, Vice President*

**1482 Wyoming Native Plant Society**
PO Box 2500
Laramie, WY 82073
307-766-3020
E-mail: lmflora@alluretech.net
http://uwadmnweb.uwyo.edu/wyndd/wnps/info.asp?p=3182
Goals are to encourage the appreciation and conservation of the native flora and plant communities of Wyoming through education, research, and communication.
*Founded: 1981*
*Lynn Moore, President*
*Brian Elliott, Vice President*

**1483 Wyoming Wildlife Federation**

PO Box 106 307-637-5433
Cheyenne, WY 82003 800-786-5434
Fax: 307-637-6629
E-mail: admin@wyomingwildlife.org
http://www.wyomingwildlife.org

The Wyoming Wildlife Federation, established in 1937, is Wyomings oldest and largest conservation group advocating sportsmen and sportswomen. The Federation's mission is to work for hunters, anglers and other wildlife enthusiasts toprotect and enhance habitat; propetuate quality hunting and fishing; protect citizens rights to use public lands and waters; and promote ethical hunting and fishing.
*Founded: 1937*
*Dave Gowdey, Executive Director*
*Mark Winland, President of the Board*

# Awards & Honors

## Environmental

**1484 Adirondack Council Conservationist of the Year**
103 Hand Avenue 518-873-2240
Suite 3 877-873-2240
Elizabethtown, NY 12932 Fax: 518-873-6675
E-mail: info@adirondackcouncil.org
http://www.adirondackcouncil.org

This is awarded to the individual or organization who has provided the greatest contribution towards protecting the health of Adirondack Park. The award is presented each year at the Council's Forever WildÆDay, and winners receive aspecailly commissioned, museum-quality, hand-carved common loon in recognition of their achievements.

*John F Sheehan, Communications Director*
*Brian Houseal, Executive Director*

**1485 Aerospace Medical Association**
320 S Henry Street 703-739-2240
Alexandria, VA 22314 Fax: 703-739-9652
E-mail: rrayman@asma.org
http://www.asma.org

AsMA is dedicated to uniting the world's professionals in aviation, space and environmental medicine: advancing the frontiers of aerospace medicine by dissemination of knowledge throughout industry, the general public, and governmentalagencies worldwide. Ensuring the highest level of safety, health and performance of those involved in aerospace, AsMA is recognized as the international authority in aerospace medicine.
*Russell Rayman, Executive Director*

**1486 Air Force Association**
1501 Lee Highway 703-247-5800
Arlington, VA 22209 800-727-3337
Fax: 703-247-5853
E-mail: service@afa.org
http://www.afa.org

The Air Force Association's mission is to advocate aerospace power and a strong national defense; to support the United States Air Force and Air Force Family; and to promote aerospace education to the American people.

*Kathy Hartness, Industry Relations*

**1487 Air and Waste Management Association**
1 Gateway Center 412-232-3444
3rd Floor Fax: 412-232-3450
Pittsburgh, PA 15222 E-mail: info@awma.org
http://www.awma.org/

The Air and Waste Management Association is a nonprofit, nonpartisan professional organization that provides training, information and networking opportunites to thousands of environmental professionals in 65 countries.
*Adrianne Carolla, Secretary*

**1488 American Association of Engineering Societies**
6522 Meadowridge Rd 202-296-2237
Suite 101 Fax: 202-296-1151
Elkridge, MD 21075 E-mail: info@aaes.org
http://www.aaes.org

Multidisciplinary organization dedicated to advancing the knowledge, understanding and practice of engineering in the public interest. Its members represent the mainstream of US engineering-affecting over 1,000,000 engineers inindustry, government and education. Through its councils, commissions, committees and task forces, the AAES addresses questions relating to the engineering profession.

*William Koffel, Executive Director*

**1489 American Chemical Society**
American Chemical Society
1155 16th Street NW
Washington, DC 20036 800-227-5558
Fax: 202-776-8258
E-mail: webmaster@acs.org
http://www.chemistry.org

The American Chemical Society is a self-governed individual membership organization that consists of more than 158,000 members in the field of chemistry. The organizations provides a broad range of opportunities for peer intereactionand career development, regardless of professional or scientific interests.
*Catherine T. Hunt, President*

**1490 American Conference of Governmental Industrial Hygienists**
1330 Kemper Meadow Drive 513-742-6163
Cincinnati, OH 45240 Fax: 513-742-3355
E-mail: mail@acgih.org
http://www.acgih.org

The American Conference of Governmental Industrial Hygeienists (ACGIH) is a member-based organization and community of professionals that advances worker health and safety through education and the development and dissemination ofscientific and technical knowledge.
*Beverly S Cohen, Chair*
*Lawrence M Gibbs, Vice Chair*

**1491 American Forest and Paper Association**
111 Nineteenth Street, NW 202-463-2700
Suite 800 800-878-8878
Washington, DC 20036 E-mail: info@afandpa.org
http://http://afandpa.org

The American Forest and Paper Association (AF&PA) is the national trade association of the forest, pulp, paperboard wood products industry. We represent member companies engaged i ngrowning, harvesting and processing wood and woodfiber, manufacturing pulp, paper and paperboard products from both virgin and recycled fiber, and producing engineered and traditional wood products.

**1492 American Institute of Chemical Engineers**
3 Park Avenue 212-591-8100
New York, NY 10016 800-242-4363
Fax: 212-591-8888
E-mail: xpress@aiche.org
http://www.aiche.org

Founded in 1908, a professional association of more than 50,000 members that provides leadership in advancing the chemical engineering profession. Fosters and disseminates chemical engineering knowledge, supports the professional andpersonal growth of its members, and applies the expertise of its members to address societal needs through the world.
*Scott Berger, Director*
*Bette Lawler, Sr. Director, Operations*

**1493 American Institute of Mining, Metallurgical and Petroleum Engineers**
3 Park Avenue 212-419-7676
New York, NY 10016 Fax: 212-419-7671
E-mail: aimeny@aimeny.org
http://www.idis.com/aime

Organized and operated exclusively to advance, record and disseminate significant knowledge of engineering and the arts and sciences involved in the production and use of minerals, metals, energy sources and materials for the benefitof humankind, both directly as AIME and through Member Services.
*Nellie Guernsey, Executive Director*

**1494 American Nuclear Society**
555 North Kensington Ave. 708-352-6611
La Grange Park, IL 60526 800-323-3044
Fax: 708-352-0499
E-mail: nucleus@ans.org
http://www.ans.org

The American Nuclear Society is a not-for-profit, international, scientific and educational organization. It was established by a group of individuals who recognized the need to unify the professional activities within the diversefields of nuclear science and technology.

*Harry Bradley, Executive Direcotr*

**1495 American Society of Civil Engineers**
1801 Alexander Bell Drive
Reston, VA 20191
703-295-6300
800-548-2723
Fax: 703-295-6222
http://www.asce.org
Founded in 1852, the American Society of Civil Engineers represents more than 137,500 members of the civil engineering profession worldwide, and is America's oldest national engineering society. ASCE's vision is to position engineersas global leaders building a better quality of life.
*William Marcuson, President*

**1496 American Society of Heating, Refrigerating and Air-Conditioning (ASHRAE)**
1791 Tullie Circle NE
Atlanta, GA 30329
404-636-8400
Fax: 404-321-5478
E-mail: ashrae@ashrae.org
http://www.ASHRAE.org
ASHRAE, founded in 1894, is an international organization of some 50,000 persons. ASHRAE fulfills its mission of advancing heating, ventilation, air conditioning and refrigeration to serve humanity and promote a sustainable worldthrough reserach, standards writing, publishing and continuing education.
*Jeff Littleton, Executive Vice President*

**1497 American Sportfishing Association**
225 Reinekers Lane
Suite 420
Alexandria, VA 22314
703-519-9691
Fax: 703-519-1872
E-mail: info@asafishing.org
http://www.asafishing.org
The American Sportfishing Association is the sportfishing industry's trade association, committed to looking out for the interests of the entire sportfishing community. We give the industry a unified voice, speaking out on behalf ofsportfishing and boating industries, state and federal natural resource agencies, conservation organizations, angler advocacy groups and outdoor journalists when emerging laws and policies could significantly affect sportfishing business orsportfishing itself.
*Mike Nussman, President/CEO*
*Gordon Robertson, VP*

**1498 American Water Resources Association**
4 West Federal Street
PO Box 1626
Middleburg, VA 20118
540-687-8390
Fax: 540-687-8395
E-mail: terry@awra.org
http://www.awra.org
Founded in 1964, the American Water Resources Association is a non-profit professional association dedicated to the advancement of men and women in water resources management, research, and education. AWRA's membership ismultidisciplinary; its diversity is its hallmark. It is the professional home of a wide variety of water resources experts including engineers, educators, foresters, biologists, ecologists, geographers, managers, regulators, hydrologists andattorneys.
*Kenneth D Reid, Executive VP*
*Terry Meyer, Marketing Director*

**1499 Association for Conservation Information**
Montana Department of Fish, Wildlife and Parks
1420 E 6th Street
PO Box 200701
Helena, MT 59620
406-444-4038
Fax: 406-444-4952
E-mail: raasheim@mt.gov
http://http://fwp.mt.gov
ACI, the Association for Conservation Information, is a non-profit association of information and education professionals representing state, federal and Canadian agencies and private conservation organizations. ACI memberprofessionals play a major role in providing natural resource, environmental, wildlife and other information and education to the public through a variety of means, many of which are continental in scope.
*Ron Aasheim, Administrator*

**1500 Association of Conservation Engineers**
2901 w. Truman Boulevard
PO Box 180
Jefferson City, MO 65109
573-751-4115
Fax: 573-751-4467
http://www.conservation.state.mo.us
Organization of conservation engineers and technicians who are working to conserve and improve our nation's natural heritage. Brings together engineers and allied personnel employed by conservation and recreation agencies andconsultants who have a community of specialized interests in the areas of fish, wildlife, parks, forests and related conservation/recreation fields. Members pool experience and information pertaining to conservation engineering to make naturalresources more accessible.
*Anita B Gorman, Chairman*
*Lowell Mohler, Vice-Chairman*

**1501 Association of Consulting Foresters of America**
312 Montgomery Street
Suite 208
Alexandria, VA 22314
703-548-0990
Fax: 703-548-6395
E-mail: director@acf-foresters.org
http://www.acf-foresters.org
To protect the public welfare and property in the practice of forestry, to raise the professional standards and work of ACF Consultants and all other consulting foresters. To develop and expand the services of ACF Consultants.
*Lynn C Wilson, Executive Director*

**1502 Audubon Naturalist Society of Central Atlantic**
8940 Jones Mill Road
Chevy Chase, MD 20815
301-652-9188
Fax: 301-951-7179
E-mail: contact@audubonnaturalist.com
http://www.audubonnaturalist.com
The Audubon Naturalist Society is an independent environmental education and conservation organization with over 10,000 members in the Washington DC area. The society offers a wide variety of natural history classes and campaigns forthe protection and renewal of the Mid-Atlantic regions natural resources.

*Neal Fitzpatrick, Executive Director*
*Anne Cottingham, Board President*

**1503 Audubon Society of New Hampshire**
3 Silk Farm Road
Concord, NH 03301
603-224-9909
Fax: 603-226-0902
E-mail: asnh@nhaudubon.org
http://www.nhaudubon.org
The Audubon Society of New Hampshire, a nonprofit statewide membership organization, is dedicated to the conservation of wildlife and habitat throughout the state. Independent of the National Audubon Society, ASNH has offered programsin wildlife conservation, land protection, environmental policy and environmental education since 1914.

*Partricia Casey, Human Resource Manager*

**1504 Audubon of Florida: Center for Birds of Prey**
1101 Audubon Way
Maitland, FL 32751
407-644-0190
Fax: 407-644-8940
http://www.audubonofflorida.org
To conserve, protect and restore Florida's natural resources and to create a conservation ethic among all Floridians. The mission is to conserve and restore natural ecosystems, focusing on birds and other wildlife for the benefit ofhumanity and the earth's biological diversity. The Center for Birds of Prey is dedicated to promoting a stewardship ethic towards Florida's birds of prey and their habitats through medical rehabilitation, interactive education and practical research.
*Katie Warner, Center Administrator*

**1505 Big Thicket Conservation Association**
PO Box 198
Saratoga, TX 77585
409-892-8976
http://www.btatx.org
*Dr. Bruce Drury, President*

**1506 Botanical Society of America**
PO Box 299 314-577-9566
St. Louis, MO 63166 Fax: 314-577-9515
E-mail: wdahl@botany.org
http://www.botany.org

Promote botany, the field of basic science dealing with the study and inquiry into the form, function, diversity, reproduction, evolution, and uses of plants and their interactions within the biosphere.
*Bill Dahl, Executive Director*

**1507 Chicago Community Trust**
111 East Wacker Drive 312-616-8000
Suite 1400 Fax: 312-616-7955
Chicago, IL 60601 E-mail: info@cct.org
http://www.cct.org

*Terry Mazany, President/CEO*
*Greg White, Strategy & Operations VP*

**1508 Connecticut River Watershed Council**
15 Bank Row 413-772-2020
Greenfield, MA 01301 Fax: 413-772-2090
E-mail: cgwyther@ctriver.org
http://www.ctriver.org

The Connecticut River Watershed Council (CRWC) is the only broad-based citizen advocate for the environmental well-being of the entire Connecticut River. Our primary mission is to promote improvement of water quality and therestoration, conservation, wise development and use of the natural resources of the Connecticut River watershed.
*Chelsea Gwyther, Executive Director*

**1509 Ecological Society of America**
1990 M St NW 202-833-8773
Suite 700 Fax: 202-833-8775
Washington, DC 20036 E-mail: esahq@esa.org
http://www.esa.org

To promote ecological science by improving communication among ecologists; raise the public's level of awareness of the importance of ecological science; increase the resources available for the conduct of ecological science; andensure the appropriate use of ecological science in environmental decision making by enhancing communication between the ecological community and policy-makers.

*Alison Power, President*
*Katherine S McCarter, Executive Director*

**1510 Federal Aviation Administration**
Office of Public Affairs
800 Independence Avenue SW 202-267-3883
Washington, DC 20591 Fax: 202-267-5047
http://www.faa.gov

The major roles of the Federal Aviation Administration (a part of the Department of Transportation) include regulation, development, and research in the areas of civil aviation, civil aeronautics, and U.S. commercial spacetransportation.
*Marion C Blakey, Administrator*
*Robert A Sturgell, Deputy Administrator*

**1511 Federation of Fly Fishers**
215 E. Lewis 406-222-9369
Livingston, MT 59047 Fax: 406-222-5823
E-mail: van@fedflyfishers.org
http://www.fedflyfishers.org

The Federation of Fly Fishers seeks to cultivate and advance the art science and sport of flyfishing as the most sporting and enjoyable method of angling and the way of fishing most consistent with the preservation and use of game fishresources; to be the voice for organized fly fishing; to promote conservation of recreational resources; to facilitate and improve the knowledge of fly fishing; and to elevate the standard of integrity, honor and courtesy of anglers.
*RP van Gytenbeek, Chief Executive Officer*
*Bob Wiltshire, Chief Operating Officer*

**1512 Frank A Chambers Award**
Air and Waste Management Association
One Gateway Center, 3rd Floor 412-232-3444
420 Fort Duquesne Boulevard 800-270-3444
Pittsburgh, PA 15222 Fax: 412-232-3450
http://www.awma.org

Award for outstanding achievement in the science and art of air pollution control. It requires accomplishment of a technical nature on the part of the recipient which is considered to be a major contribution to the science and art ofair pollution control, the merit of which has been widely recognized by persons in the field.
*Steve Wafalosky*

**1513 German Marshall Fund of the United States**
1744 R Street NW 202-745-3950
Washington, DC 20009 Fax: 202-265-1662
E-mail: info@gmfus.org
http://www.gmfus.org

To stimulate the exchange of ideas and promote cooperation between the United States and Europe in the spirit of the postwar Marshall Plan. GMF was created in 1972 by a gift from the German people as a permanent memorial to MarshallPlan aid.
*Craig Kennedy, President*

**1514 Global Tomorrow Coalition**
Capital Research Center
1513 16th Street NW 202-483-6900
Washington, DC 20036 Fax: 202-483-6990
E-mail: contact@capitalresearch.org
http://www.capitalresearch.org/gw

Green Watch is an online database and information clearinghouse providing factual information on over 500 nonprofit environmental groups. This free service identifies the location, leadership and membership of each profiled group.Green Watch also produces timely news reports and analyses of the environmental movement.
*Terrence Scanlon, President*

**1515 Golden Gate Audubon Society**
2530 San Pablo Avenue 510-843-2222
Suite G Fax: 510-843-5351
Berkeley, CA 94702 E-mail: ggas@goldengateaudubon.org
http://www.goldengateaudubon.org

A conservation and education organization that has birds as its key component. We seek to protect and enjoy wildlife and their natural habitat in San Francisco and East Bay through interaction between our members and the community.

*Mark Welther, Executive Director*

**1516 Goldman Environmental Foundation**
211 Lincoln Blvd 415-345-6330
San Francisco, CA 94129 Fax: 415-345-9686
E-mail: info@goldmanprize.org
http://www.goldmanprize.org

Goldman Environmental Prizes are awarded for sustained and important efforts to preserve the natural environment, including, but not limited to:Æprotecting endangered ecosystems and species, combatting destructive development projects,promoting sustainability, influencing environmental policies and striving for environmental justice.

**1517 Great Lakes Commission**
2805 S Industrial Highway 734-971-9135
Suite 100 Fax: 734-971-9150
Ann Arbor, MI 48104 E-mail: eschmidt@glc.org
http://www.glc.org

Binational agency that promotes the orderly, integrated and comprehensive development, use and conservation of the water and related natural resources of the Great Lakes basin and St Lawrence River.
*Tim A. Eder, Executive Director*

**1518 Honorary Membership**
Air and Waste Management

1 Gateway Center, 3rd Floor
420 Fort Duquesne Blvd
Pittsburgh, PA 15222
412-232-3444
800-270-3444
Fax: 412-232-3450
E-mail: info@awma.org
http://www.awma.org

May be conferred upon persons who have attained eminence in some field related to the mission and objectives of the Association who have rendered valuable service to the Association.
*Steve Wafalosky, Business Contact*

**1519 Institute of Environmental Sciences and Technology**
Arlington Place One
2340 S Arlington Heights Rd, Suite 100
Arlington Heights, IL 60005
847-255-1561
Fax: 847-255-1699
E-mail: iest@iest.org
http://www.iest.org

An international professional society that serves the environmental sciences in the areas of contamination control in electronics manufacturing and pharmaceutical processes, design, test and evaluation of commercial and militaryequipment and product reliability issues.
*Julie Kendrick, Executive Director*
*Kristin Thryselius, Publication Sales Coor*

**1520 International Desalination Association**
PO Box 387
Topsfield, MA 01983
978-887-0410
Fax: 978-887-0411
E-mail: info@idadesal.org
http://www.idadesal.org

IDA is committed to the development and promotion of the appropriate use of desalination and desalination technology worldwide. We endeavor to carry out these goals by encouraging research and development, exchanging, promotingcommunication and disseminating information.
*Jose Antonio Medina, President*
*Lisa R Henthorne, First Vice President*

**1521 International Studies Association**
University of South Carolina
817 Henderson Street
Columbia, SC 29208
803-777-3109
Fax: 803-777-8255
E-mail: mgross@sc.edu
http://www.cas.sc.edu/poli

*Melissa Gross, Business Manager*
*Pamela Mauldin, Faculty Coordinator*

**1522 International Wildlife Film Festival: Media Center**
718 S Higgins Avenue
Missoula, MT 59801
406-728-9380
Fax: 406-728-2881
E-mail: iwff@wildlifefilms.org
http://www.wildlifefilms.org

Goal is to be the preeminent wildlife film, television and media organization, showcasing the world's best wildlife films and television programs, providing educational resources and events seminars, workshops, field classes, filmtours and many hands-on activities, that emphasize the most up-to-date, factual and ethical scienced based information. Our mission—-to promote awareness, knowledge and understanding of wildlife, habitat, people and nature through excellent film,television and media.
*Janet Rose, Executive Director*

**1523 Irrigation Association**
6540 Arlington Boulevard
Falls Church, VA 22042
703-536-7080
Fax: 703-536-7019
E-mail: webmaster@irrigation.org
http://www.irrigation.org

To improve the products and practices used to manage water resources and to held shape the worldwide business environment of the irrigation industry.
*Karen Koenig, Certification Manager*

**1524 John Burroughs Association**
15 West 77th Street
New York, NY 10024
212-769-5169
E-mail: breslof@amnh.org
http://http://research.amnh.org/burroughs

Each year a medal is awarded to the author of a distinguished book of natural history, a list of exceptional national history books for young readers is selected. and an outstanding nature essay is identified.
*Robert Abrams, Director*
*Lisa Breslof, Secretary*

**1525 Keep America Beautiful**
1010 Washington Boulevard
Stamford, CT 06901
203-323-8987
Fax: 203-325-9199
E-mail: info@kab.org
http://www.kab.org

Nonprofit organization whose network of local, statewide and international affiliate progams educates individuals about litter prevention and ways to reduce, reuse, recycle and properly manage wase materials.
*G Raymon Empson, President*

**1526 Keep North Carolina Beautiful**
1503 Mail Service Center
Raleigh, NC 27699
919-733-3109
877-dot-4you
Fax: 919-733-9980
http://www.ncdot.org

North Carolina Keep America Beautiful is a nonprofit public education organization dedicated to enhancing the natural beauty of North Carolina communities, improving waste handling practices and empowering individuals to take greaterresponsibility for improving community environments.

**1527 Lawrence K Cecil Award**
American Institute of Chemical Engineers
3 Park Avenue
New York, NY 10016
212-591-8100
800-242-4363
Fax: 212-591-8888
E-mail: awards@aiche.org
http://www.aiche.org

Recognizes an individual's outstanding chemical engineering contribution and achievement in the preservation or improvement of the environment.
*Larry Evans, President*

**1528 Lyman A Ripperton Award**
Air and Waste Management Association
One Gateway Center, 3rd Floor
420 Fort Duquesne Boulevard
Pittsburgh, PA 15222
412-232-3444
Fax: 412-232-3450
E-mail: info@awma.org
http://www.awma.org

Awarded for distinguished achievement as an educator in some field of air pollution control. Awarded to an individual, who by precept and example, has inspired students to achieve excellence in all their professional and socialendeavors.
*Antoon Van Der Vooren, President*

**1529 NSF International**
789 N Disboro Road
PO Box 130140
Ann Arbor, MI 48113
734-769-8010
800-nsf-mark
Fax: 734-769-0109
E-mail: info@nsf.org
http://www.nsf.org

NSF International, The Public Health and Safety Company, is an independent, not-for-profit organization providing a wide range of services around the world. For more than 55 years, NSF has been committed to public health, safety andprotection of the enviroment.
*Robert Ferguson, Vice President*
*Tom Bruursema, General Manager*

**1530 National Association for Environmental Education**
PO Box 400
Troy, OH 45373
937-698-6493
Fax: 937-335-5623

The National Association for Environmental Education is a network of professionals, students and volunteers working in the field of environmental education throughout North America and in over 55 countries around the world. NAAEE takesa cooperative, nonconfrontational, scientifically-based approach to promoting education about environmental issues.
*Joseph Baust, President*
*Martha Monroe, President Elect*

**1531 National Association of Conservation Districts**
Service Center
612 West Main Street
League City, TX 77574
281-332-3402
Fax: 281-332-5259
E-mail: beth-mason@nacdnet.org
http://www.nacdnet.org/news/awards/index.phtml

The association's annual Awards Program recognizes individuals and organizations for outstanding work and leadership in soil and water conservation. Awards include: NACD Friend of Conservation; Distinguished Service; President's;Excellence in Communications; District Excellence; and Collaborative Conservation.
*Krysta Harden, CEO*
*Beth Mason, Awards Contact*

**1532 National Audubon Society**
National Audubon Society
700 Broadway
New York, NY 10003
212-979-3000
Fax: 212-979-3188
E-mail: webmaster@audubon.org
http://www.audubon.org

The mission of the National Audubon Society is to conserve and restore natural ecosystems, focusing on birds and other wildlife for the benefit of humanity and the earth's biological diversity. Founded in 1905, the National AudubonSociety is named for John James Audubon, famed orithologist, exployer, and wildlife artist.
*Jess Morton, Contact*

**1533 National Bison Association**
The National Bison Association
8690 Wolff Ct
Suite 200
Westminster, CO 80031
303-292-2833
Fax: 303-659-3739
http://www.bisoncentral.com
*Dave Carter, Executive Director*

**1534 National Environmental Training Association**
5320 N 16th Street
Suite 114
Phoenix, AZ 85018
602-956-6099
Fax: 602-956-0399
http://ehs-training.org

The National Environmental Training Assoication, is a nonprofit international organization of enviromental, health and safety, other technical training professionals. Activities centeral to NETA's educational services include itssupport for trainer networking, professional development and competency certification for its members, EH&S training information and programs for industry, and development of training competency standards.
*Charles L Richardson, Executive Director*

**1535 National Ocean Industries Association**
National Ocean Industries Association
1120 G Street NW
Suite 900
Washington, DC 20005
202-347-6900
Fax: 202-347-8650
E-mail: mkearns@noia.org
http://www.noia.org

The National Ocean Industries Assoication, founded in 1972 with 35 members, represents all facets of the domestic offshore and related industries. Today, our more than 300 member companies are dedicated to the development of offshoreoil and natural gas for the coninued growth and secrity of the United States. NOIA members are engaged in many business activities, in addition to those listed below, including enviromental safeguards, equipment supply, gas transmission,naviogation,ect.

**1536 National Press Club**
National Press Club
529 14th Street NW
Washington, DC 20045
202-662-7500
E-mail: info@press.org
http://www.press.org

Professional organization of reporters, writers and newspeople employed by newspapers, wire services, magazines, radio and television stations, and other forms of news media; and former newspeople and associates of newspeople. Sponsorsprofessional, sports, travel and cultural events; book rap sessions with news figures and authors; and newsmaker and luncheon speaker sessions. Houses reference library and archives. Offers computer training. Publishes a weekly newsletter.
*Jerry Zremski, President*

**1537 National Recreation and Park Association**
National Recreation And Park Association
22377 Belmont Ridge Road
Ashburn, VA 20148
703-858-0784
Fax: 703-858-0794
E-mail: info@nrpa.org
http://www.nrpa.org

The mission of the National Recreation and Park Association is to a advance parks, recreation and environmental conservation efforts that enhance the quality of life for all people.
*Craig Baker, Exposition Manager*
*Krista Barnes, Senior Director*

**1538 National Recycling Coalition**
National Recycling Coalition
805 15th Street Nw
Suite 425
Washington, DC 20005
202-789-1430
Fax: 202-789-1431
E-mail: info@nrc-recycle.org
http://www.nrc-recycle.org

NRC is a not-for-profit organization dedicated to the advancement and improvement of recycling, source reduction, composting, and reuse by providing technical information, education, training, outreach, and advocacy services to itsmembers in order to conserve resources and benefits the environment.
*Kate Krebs, Executive Director*
*Anjian Nicolaidis, Deputy Director*

**1539 National Water Resources Association**
National Water Resources Association
3800 N Fairfax Drive
Suite 4
Arlington, VA 22203
703-524-1544
Fax: 703-524-1548
E-mail: nwra@nwra.org
http://www.nwra.org/

The National Water Resources Association is a nonprofit federation of state organizations whose membership includes rural water districts, municipal water entities, commerical companies and individuals. As an Association we areconcerned with the appropriate management, conservation, and use of water and land resources on a national scope.

**1540 National Wild Turkey Federation**
770 Augusta Road
PO Box 530
Edgefield, SC 29824
803-637-3106
800-THE-NWTF
Fax: 803-637-0034
E-mail: nwtf@nwtf.net
http://www.nwtf.org/

The NWTF, an international nonprofit conservation and education organization dedicated to conserving wild turkeys and preserving hunting traditions. Growth and progress define the NWTF as it has expanded from 1,300 members in 1973 tonearly a half million today.
*Tammy Bristow Sapp, President Communications*

**1541 National Wildlife Federation**
11100 Wildlife Center Drive
Reston, VA 20190
800-822-9919
http://www.nwf.org

The National Wildlife Federation inspires Americans to protect wildlife for our children's future. They represent the power and commitment of over five million members and supporters joined by affiliated wildlife organizationsthroughout the states and territories. They channel the energy of thousands of volunteers from all walks of life to take action because they care about wildlife.
*Larry J Schweiger, President/CEO*

**1542 Natural Resources Defense Council**
40 West 20th Street
New York, NY 10011
212-727-2700
Fax: 212-727-1773
E-mail: nrdcinfo@nrdc.org
http://www.nrdc.org

Mission: To safeguard the Earth: its people, its plants and animals, and natural systems on which all life depends.
*Frances Beinecke, President*
*Peter Lehner, Executive Director*

**1543 Nature Conservancy**
4245 N Fairfax Drive
Suite 100 800-628-6860
Arlington, VA 22203 E-mail: webmaster@tnc.org
http://www.nature.org
The Nature Conservancy is a leading international, nonprofit organization dedicated to rpeserving the diversity of life on Earth. The mission of The Nature Conservancy is to preserve teh plants, animals and natural communities thatrepresent the diversity of life on Earth by protecting the lands and waters they need to survive.

**1544 New England Wildflower Society**
180 Hemenway Road 508-877-7630
Framingham, MA 01701 Fax: 508-877-3658
E-mail: information@newenglandwild.org
http://www.newenglandwild.org
New England Wild Flower Society is a recognized leader in native plant conservation. Founded in 1900, the Society is the oldest plant conservation organization in the US. Its purpose is to promote the conservation of temperate NorthAmerican plants through key programs-conservation and research, education, horticulture and habitat preservation. They publish two magazines annually, a seed catalog, nursey catalog, brochures and pamphlets about native plant conservation andhorticulture

*Debbi Edelstein, Executive Director*
*Francis H Clark, Chair, Board of Trustees*

**1545 New York Botanical Garden**
200th Street & Kazimiroff Boulevard 718-817-8700
Bronx, NY 10458 Fax: 718-562-8474
http://www.nybg.org
Founded in 1891, the Garden is one of the world's great collections of plants, the region's leading educational center for gardening and horticulture, and an international center for plant research. The New York Botanical Garden is anadvocate for teh plant kingdom.
*Wilson Nolen, Chairman*
*Gregory Long, President*

**1546 Outdoor Writers Association of America**
121 Hickory Street 406-728-7434
Suite 1 800-692-2477
Missoula, MT 59801 Fax: 406-728-7445
E-mail: krhoades@owaa.org
http://www.owaa.org
The mission of Outdoor Writers Association of America is to improve the professional skills of our members, set the highest ethical and communications standards, encourage public enjoyment and conservation of natural resources, and bementors for the next generation of professional outdoor communicators.
*Kevin Rhoades, Executive Director*

**1547 Ozark Society**
63 Robinwood Drive 501-219-4293
Little Rock, AR 72227 E-mail: alice209ok@yahoo.com
http://www.ozarksociety.net
The Ozark Society, was founded in 1962 by Dr. Neil Compton of BEntonville, an Ozark native, and group of associates for the immediate purpose of saving the Buffalo River from dams proposed by the US Army Corps of Engineers. Societyfounders, working with Sen. JW Fullbright, helped get the National Park Service to survey the Buffalo River area and then began to campiagn for the creation of the Buffalo National River as an alternative to the dams.
*Alice Andrews, President*

**1548 Pennsylvania Association of Environmental Professionals**
174 Crestview Drive 814-355-2467
Bellefonte, PA 16823 Fax: 814-355-2452
E-mail: info@paep.org
http://www.paep.org
A nonpolitical interdisciplinary organization of individuals working in environmental management, planning, impact assessment, environmental protection, compliance, research, engineering, design and education.
*Jeff Prawdzik, President*
*Jason Minnich, Vice President*

**1549 Power**
McGraw-Hill
11 W 19th Street 212-337-4060
New York, NY 10011 Fax: 212-627-3811

**1550 Sea Grant Association**
University Of Maine
5784 York Complex 207-581-1435
Orono, ME 04469 Fax: 207-581-1426
E-mail: panderson@maine.edu
http://www.sga.seagrant.org
The Sea Grant Association (SGA) is a non-profit organization dedicated to furthering the Sea Grant program concept. SGA provides the mechanism for academic institutions to coordinate their activities, to set program priorities at boththe regional and national level, and to proved a unified voice for the institutions on issues of importance to the oceans and coasts.
*Paul S Anderson, President*

**1551 Sierra Club**
85 Second Street 415-977-5500
2nd Floor Fax: 415-977-5799
San Francisco, CA 94105 E-mail: information@sierraclub.org
http://www.sierraclub.org
To advance the preservation and protection of the natural environment by empowering the citizenry, especially democratically-based grassroots organizations, with charitable resources to further the cause of environmental protection.The vehicle through which The Sierra Club Foundation generally fulfills its charitable mission.
*Lisa Renstrom, President*

**1552 Society of American Foresters**
5400 Grosvenor Lane 301-897-8720
Bethesda, MD 20814 866-897-8720
Fax: 301-897-3690
E-mail: safweb@safnet.org
http://www.safnet.org
The mission of the Society of American Foresters is to advance the science, education, technology, and practice of forestry; to enhance the competency of its members; to establish professional excellence; and, to use the knowledge,skills, and conservation ethic of the profession to ensure the continued health and use of forest ecosystems and the present and future availability of forest resources to benefit society.

*Michael T Goergen, Jr, Executive VP/CEO*
*Carol McKernon, Member Services Manager*

**1553 Society of American Travel Writers**
7044 South 13th Street 414-908-4949
Oak Creek, WI 53154 Fax: 414-768-8001
E-mail: satw@satw.org
http://www.satw.org
SATW is a tax-exempt professional association whose purpose is to promote responsible journalism, provide professional support and development for our members. and encourage the conservation and preservation of travel resourcesworldwide.

**1554 Society of Petroleum Engineers**
PO Box 833836 972-952-9393
Richardson, TX 75083 Fax: 972-952-9434
E-mail: spedal@spe.org
http://www.spe.org
*Eve Sprunt, President*

**1555 Soil and Water Conservation Society**
945 Sw Ankeny Road 515-289-2331
Ankeny, IA 50023 Fax: 515-289-1227
E-mail: swcs@swcs.org
http://www.swcs.org/
Foster the science and the art of soil, water and related natural resource management to achieve sustainability. To promote and

practice an ethic recognizing the interdependence of the people in the environment.
*Deborah Cavanaugh-Grant, President*

**1556 Solar Energy Industries Association**
805 15th Street NW
Suite 510
Washington, DC 20005
202-682-0556
Fax: 202-628-7779
E-mail: info@seia.org
http://www.seia.org/
SEIA's primary mission is to expand the use of solar technologies in the global marketplace. National members combined with chapter members in 22 states exceed 500 compines providing solar thermal and solar electric products andservices.
*Chris O'Brien, Chairman*
*Jeffrey D Wolfe, Division Chair*

**1557 TWS Awards**
5410 Grosvenor Lane
Suite 200
Bethesda, MD 20814-2144
301-897-9770
Fax: 301-530-2471
E-mail: yanin@wildlife.org
http://http://joomla.wildlife.org
The Wildlife Society's Awards Program honors individuals and groups who have made notable contributions to wildlife conservation. With more than a dozen awards in all, visit thieir website for full details and nomination information.

**1558 Trout Unlimited**
1300 North 17th Street
Suite 500
Arlington, VA 22209
703-522-0200
800-834-2419
Fax: 703-284-9400
E-mail: trout@tu.org
http://www.tu.org
Mission: To conserve, protect and restore North America's trout and salmon fisheries and their watersheds. We accomplish this mission on local, state and national levels with an extensive and dedicated volunteer network.
*Charles Gauvin, President/CEO*
*Steve Moyer, Vice President*

**1559 US Army Corps of Engineers**
20 Massachusetts Avenue NW
Washington, DC 20314
202-761-0011
E-mail: webmaster@usace.army.mil
http://www.usace.army.mil/
Our mission is to provide quality, responsive engineering services to the nation including: planning, desiging, building and operating water resources and other civila works projects.

**1560 US Department of Energy**
1000 Independence Avenue SW
Washington, DC 20585
800-dia-ldoe
Fax: 202-586-4403
http://www.energy.gov
The Department of ENergy's mission is to advance the national, economic and energy security of the US; to promote scientific and technological innovation; and to ensure the environmental cleanup of the national nuclear weapons complex.
*Samuel W Bodman, Secretary Of Energy*

**1561 US Department of the Interior**
1849 C Street NW
Washington, DC 20240
202-208-3100
E-mail: webteam@ios.doi.gov
http://www.doi.gov

**1562 US Environmental Protection Agency**
Ariel Rios Building
1200 Pennsylvania Ave Nw
Washington, DC 20460
800-438-2474
http://www.epa.gov
The mission of the EPA is to protect human health and the environment.

**1563 Underwater Society of America**
PO Box 628
Daly City, CA 94017
650-583-8492
Fax: 408-294-3496
http://www.underwater-society.org/
The Underwater Society of America was founded in 1959 by the existing skin-diving councils; it was composed of and represented all divers in North America. It is the public diving organization of the United States. It is controlled byits Executive committee, board of directors and delegates of the member councils and clubs meeting annually.
*Carol Rose, President*

**1564 Washington Journalism Center**
Po Box 15239
Washington, DC 20003
202-296-8455
Fax: 808-588-365
E-mail: terrymichael@wcpj.org
http://www.wcpj.org
*Terry Michael, Executive Director*

**1565 Water Environment Federation**
601 Wythe Street
Alexandria, VA 22314
800-666-0206
Fax: 703-684-2492
E-mail: webfeedback@wef.org
http://www.wef.org
Founded in 1928, the Water Environment Federation (WEF) is a not-for-profit technical and educational organization with members from varied disciplines who work toward the WEF vision of preservation and enhancement of the global waterenvironment. The WEF network includes more than 100,000 water quality professionals from 77 member associations in 31 countries.
*William J Bertera, Executive Director*

**1566 Western Forestry and Conservation Association**
4033 SW Canyon Road
Portland, OR 97221
503-226-4562
Fax: 503-226-2515
E-mail: richard@westernforestry.org
http://www.westernforestry.org
Offers continuing education workshops and seminars for professional foresters.

**1567 Whooping Crane Conservation Association**
715 Earl Drive
Lawrenceburg, TN 38464
337-234-6339
E-mail: webadmin@whoopingcrane.com
http://www.whoopingcrane.com
The mission is to advance conservation, protection and propagation of the Whooping Crane population, to prevent its extinction, to establish and maintain a captive management program for the perpetuation of the species. We collect anddisseminate knowledge of this species; and advocate and encourage public appreciation and understanding of the Whooping Crane's educational, scientific and economic values.

**1568 Wilderness Society**
1615 M Street NW
Washington, DC 20036
202-833-2300
800-THE-WILD
E-mail: member@tws.org
http://www.wilderness.org
Deliver to future generations an unspoiled legacy of wild places, with all the precious values they hold: Biological diversity; clean air and water; towering forests, rushing rivers, and sage-sweet, silent deserts.
*Brenda Davis, Chair*
*Doug Walker, Vice Chair*

**1569 Willowbrook Wildlife Haven Preservation**
National Wildlife Rehabilitation Association
2625 Clearwater Road
Suite 110
St. Cloud, MN 56301
320-230-9920
Fax: 320-230-3077
E-mail: nwra@nwrawildlife.org
http://www.nwrawildlife.org
The National Wildlife Rehabilitators Association is a nonprofit international membership organization committed to promoting and improving the integrity and professionalism of wildlife rehabilitation and contributing to thepreservation of natural ecosystems.
*Lessie Davis, President*

**1570 World Environment Center**
734 15th Street NW
Suite 720
Washington, DC 20005
202-312-1370
Fax: 202-682-1682
E-mail: info@wec.org
http://www.wec.org
The World Environment Center is an independent, global non-prfot, non-advocacy organization that advances sustainable

development worldwide through the business practices of member companies, and in partnerships with governments,multi-lateral and private sector organizations, universities, and other stake holders.

*Terry F Yosie, President & CEO*
*Gwen Davidow, Dir Global Corporate Program*

**1571 World Wildlife Fund**

1250 24th Street NW
Po Box 97180
Washington, DC 20090
202-293-4800
E-mail: membership@wwfus.org
http://www.worldwildlife.org/

World Wildlife Fund is dedicated to protecting the world's wildlife and wildlands. The largest privately supported international conservation organization in the world, WWF has more than 1 million members in the US alone. Since itsinception in 1961, WWF has invested in over 13,100 projects in 157 countries.

*Carter S Roberts, President*

# Conferences & Trade Shows

## Environmental

**1572 Air and Waste Management Association Annual Conference and Exhibition**
1 Gateway Center, 3rd Floor
420 Fort Duquesne Boulevard
Pittsburgh, PA 15222
412-232-3444
800-270-3444
Fax: 412-232-3450
E-mail: info@awma.org
http://www.awma.org
Environmental professionals from all sectors of the economy including colleges, universities, natural resource manufacturing and process industries, consultants, local state, provincial, regional and federal governments, construction,utilities industries.
*Edith M Ardiente, President*
*Richard C Scherr, Secretary*

**1573 American Academy of Environmental Medicine Conference**
7701 East Kellogg
Suite 625
Wichita, KS 67207
316-684-5500
Fax: 316-684-5709
E-mail: administrator@aaem.com
http://www.aaem.com
Aims to support physicians and other professionals in serving the public through education about the interaction between humans and their environment, and to promote optimal health through prevention and safe, effective treatment ofthe causes, not the illness.

*35+ booths with 200 attendees and 35+ exhibits*
*James W Willoughby II, President*
*James F Coy, President Elect*

**1574 American Board of Industrial Hygiene Professional Conference**
American Board of Industrial Hygiene
6015 West St Joseph
Suite 102
Lansing, MI 48917
517-321-2638
Fax: 517-321-4624
E-mail: abih@abih.org
http://www.abih.org
Industrial hygiene certification organization. Certified industrial hygienist is offered based on education, experience and examination.
*Lynn O'Donnell CIH, Executive Director*

**1575 American Conference of Governmental Industrial Hygienists**
1330 Kemper Meadow Drive
Cincinnati, OH 45240
513-742-6163
Fax: 513-742-3355
E-mail: mail@acgih.org
http://www.acgih.org
Advances occupational and environmental health.
*3,000 Members*
*90 booths*
*Jimmy L Perkins, Chair*
*A Anthony Rizzuto, Executive Director*

**1576 American Industrial Hygiene Association Conference and Exposition**
2700 Prosperity Avenue
Suite 250
Fairfax, VA 22031
703-849-8888
Fax: 703-207-3561
E-mail: infonet@aiha.org
http://www.aiha.org
AIHA promotes, protects and enhances industrial hygienists and other occupational health, safety and environmental professionals in their efforts to improve the health and well-being of workers, the community and the environment.
*Kim Bacon, Assistant Manager*
*Carol Tobin, Director*

**1577 American Solar Energy Society Conference**
American Solar Energy Society
2400 Central Avenue
Suite A
Boulder, CO 80301
303-443-3130
Fax: 303-443-3212
E-mail: ases@ases.org
http://www.ases.org
the american solar energy society conference (ases) is the united states section of the international solar energy society. ASES is a non profit organization dedicated to the development and adoption of renewal energy in all forms.
*Bradley D. Collins, Executive Director*

**1578 American Water Resources Association Conference**
4 West Federal Street
PO Box 1626
Middleburg, VA 20118
540-687-8390
Fax: 540-687-8395
E-mail: terry@awra.org
http://www.awra.org
Founded in 1964, the American Water Resources Association is a non-profit professional association dedicated to the advancement of men and women in water resources management, research and education.
*Kenneth D Reid, Executive VP*
*Terry Meyer, Marketing Director*

**1579 Arkansas Association of Conservation Districts Annual Conference**
101 East Capitol
Suite 350
Little Rock, AR 72201
501-682-2915
Fax: 501-682-3991
E-mail: debbiepinreal@aol.com
http://www.aracd.org
Affiliated with the National Association of Conservation Districts. Membership is $25.00 per year for individuals and $1,200.00 per year for organizations and companies.

*November*
*Debbie Moreland, Program Administrator*

**1580 Atlantic States Marine Fisheries Commission Annual Meeting**
1444 I Street NW
6th Floor
Washington, DC 20005
202-289-6400
Fax: 202-289-6051
E-mail: info@asmfc.org
http://www.asmfc.org
The Atlantic States Marine Fisheries Commission was formed by the 15 Atlantic coast states in 1942 in recognition that fish do not adhere to political boundaries. The Commission serves as a deliberative body, coordinating theconservation and management of the states shared near shore fishery resources-marine, shell, and anadromous-for sustainable use.
*John V O'Shea, Executive Director*
*George D Lapointe, Chair*

**1581 Children's Environmental Health: Research, Practice, Prevention and Policy**
110 Maryland Avenue Northeast
Suite 505
Washington, DC 20002
202-543-4033
Fax: 202-543-8797
E-mail: cehn@cehn.org
http://www.cehn.org
Children's Environmental Health Network is a national non profit organization focused on environmental health. The work of the work focuses on promoting pediatric research, prevention and practice
*Nsedu Obot Witherspoon, Executive Director*
*Joanne Perodin, Program Coordinator*

**1582 Coastal Society Conference**
PO Box 3590
Williamsburg, VA 23187
757-565-0999
Fax: 703-933-1596
E-mail: coastalsoc@aol.com
http://www.thecoastalsociety.org
The Coastal Society is an organization of private sector, academic, and government professionals and students dedicated to actively addressing emerging coastal issues by fostering dialogue, forging partnerships, and promotingcommunication and education.
*Judy Tucker, Executive Director*

**1583 Colorado Water Congress Annual Meeting**
1580 Logan Street
Suite 400
Denver, CO 80203
303-837-0812
E-mail: cwc@cowatercongress.org
http://www.cowatercongress.org
Protects and conserves Colorado's water resources by means of advocacy and education.

*January*
*350 attendees and 9 exhibits*
*Doug Kemper, Executive Director*

**1584 Connecticut Forest and Park Association Annual Meeting**
16 Meriden Road
Rockfall, CT 06481
860-346-2372
Fax: 860-347-7463
E-mail: info@ctwoodlands.org
http://www.ctwoodlands.org
An organization for forest and wildlife conservation. Develops outdoor recreation and natural resources. Provides forest management, construction of hiking trails and consultation in the areas of forestry and environment.

*Spring*
*Richard Whitehouse, President*
*Gordon Anderson, VP*

**1585 ESTECH**
Institute of Environmental Sciences & Technology
Arlington Place One, 2340 S.
Arlington Heights Rd. Suite 100
Arlington Heights, IL 60005
847-981-0100
Fax: 847-981-4130
E-mail: iest@iest.org
http://www.iest.org
An international professional society that serves the environmental sciences in the areas of contamination control in electronics manufacturing and pharmaceutical processes, design, test and evaluation of commercial and militaryequipment and product reliability issues.

*April*
*50 booths with 500 attendees*
*Corrie Roesslein, Managing Director*
*Roberta Burrows, Deputy Executive Director*

**1586 Environmental Technology Expo**
Association of Energy Engineers
4025 Pleasantdale Road
Suite 420
Atlanta, GA 30340
770-447-5083
Fax: 770-446-3969
E-mail: whit@aeecenter.org
http://www.aeecenter.org
AEE is a source of information in the field of energy efficiency, utility deregulation, facility management, plant engineering, and environmental compliance. Outreach programs include technical seminars, conferences, books, joblistings and certification programs.
*Ruth Whitlock, Executive Administrator*

**1587 Federation of Environmental Technologists**
9451 N 107th Street
Milwaukee, WI 53224
414-354-0070
Fax: 414-354-0073
E-mail: info@fetinc.org
http://www.fetinc.org
A nonprofit organization formed to assist industry in interpretation of and compliance with environmental regulations. Membership is open to all industries, municipalities, organizations and individuals concerned about environmentalregulations. Currently there are approximately 1000 members and 125 patron companies.

*March*
*200 attendees and 70 exhibits*
*Barbara Hurula, Executive Director*

**1588 Forestry Conservation Communications Association Annual Meeting**
Fcca
PO Box 3217
Gettysburg, PA 17325
717-338-1505
Fax: 717-334-5656
E-mail: nfc@fcca-usa.org
http://www.fcca-usa.org
The FCCA is a national organization. Its main function is to assist federal, state and local governments in public safety two-way radio operations by locating suitable frequencies within specified operating areas, recommending theirassignment to the FCC for licensing, and protecting them once licensed.
*Ralph Haller, Executive Director*

**1589 Global Warming International Conference and Expo**
Po Box 50303
Palo Alto, CA 94303
630-910-1551
Fax: 630-910-1561
http://www.globalwarming.net
The GWIC is the international body disseminating information on global warming science and policy, serving both governmental,and non-governmental organizations and industries in more than 145 countries. It sponsors unbiased researchsupporting the understanding of global warming and its mitigation.
*Sinyan Shen*

**1590 GlobalCon**
Association of Energy Engineers
4025 Pleasantdale Road
Suite 420
Atlanta, GA 30340
404-761-0509
Fax: 770-446-3969
E-mail: info@aeecenter.org
http://www.globalconevent.com
Energy/environmental technological equipment.
*Ruth Bennett, Information Services Dir.*

**1591 Greenprints: Sustainable Communities by Design**
Southface Energy Institute
241 Pine Street Northeast
Atlanta, GA 30308
404-872-3549
Fax: 404-872-5009
E-mail: info@southface.org
http://www.southface.org
Southface promotes sustainable homes, workplaces and communities through education, research, advocacy and technical assistance.Greenprints is a conference and trade show produced by the Southface Energy Institute.

*March*
*100 booths with 1200 attendees*
*Dave Boles, Controller*

**1592 HydroVision**
HCI Publications
410 Archibald Street
Kansas City, MO 64111
816-931-1311
Fax: 816-931-2015
E-mail: info@hcipub.com
http://www.hcipub.com
Serves the hydroelectric industry.

**1593 Institute of Scrap Recycling Industries Convention**
1615 L Street Northwest
Suite 600
Washington, DC 20036
202-662-8500
Fax: 202-626-0900
E-mail: robinwiener@isri.org
http://www.isri.org
Equipment for the recycling industries.
*Robin Weiner, President*
*Marion White, Mailroom Supervisor*

**1594 International Association for Energy Economics Conference**
International Association for Energy Economics
28790 Chagrin Boulevard
Suite 350
Cleveland, OH 44122
216-464-5365
Fax: 216-464-2737
E-mail: iaee@iaee.org
http://www.iaee.org
The IAEE is a nonprofit professional organization that provides a forum for the exchange of ideas and experiences among energy professionals. The conference attracts delegates governmental, corporate and academic energydecision-makers.
*David L Williams, Executive Director*

**1595 International Conference on Solid Waste**
Widener University, Civil Engineering
One University Place 610-499-4042
Chester, PA 19013 Fax: 610-499-4461
E-mail: solid.waste@widener.edu
http://www.widener.edu/solid.waste

An annual conference on solid waste technology and management. Over 150 speakers from 40 countries present their work. Proceedings available.

*March*
*Ronald L Mersky, Chair*

**1596 Maryland Recyclers Coalition Annual Conference**
Maryland Recyclers Coalition
Po Box 1046 888-496-3196
Laurel, MD 20725 Fax: 301-238-4579
E-mail: recycle@marylandrecyclers.org
http://www.marylandrecyclers.org

MRC's mission is to promote sustainable reduction, reuse and recycling of materials otherwise destined for disposal and promote and increase buying products made with recycled material content.

*June*
*Peter Houstle, Executive Director*

**1597 Massachusetts Association of Conservation Commissions Conference**
10 Juniper Road 617-489-3930
Belmont, MA 02478 Fax: 617-489-3935
E-mail: staff@maccweb.org
http://www.maccweb.org

We host the MACC Annual Environmental Conference, the largest such event in New England, with over 40 workshops and nearly 50 exhibitors.

*March*
*1100 attendees and 50 exhibits*
*Sally Zielinski PhD, Board President*
*Linda Mack, Executive Director*

**1598 Massachusetts Water Pollution Control Association Annual Conference**
PO Box 221 978-374-0170
Groveland, MA 01834 Fax: 978-374-0170
E-mail: mwpca1965@verizon.net
http://www.mwpca.org

*September*
*John Connor, Secretary/Treasurer*

**1599 Michigan Association of Conservation Districts Annual Meeting**
Po Box 99 231-876-0328
Cadillac, MI 49601 Fax: 231-876-0372
E-mail: macd@macd.org
http://www.macd.org

The Michigan Association of Conservation Districts is a non-governmental, non-profit organization, established to represent and provide services to Michigan's 80 Conservation Districts. The Association represents its members at thestate level by working with legislators, cooperating agencies, and special interest groups whose programs affect the care and management of Michigan's natural resources, especially on private lands.

*November*
*Teresa Salveta, Michigan Coordinator*

**1600 Michigan Forest Association Annual Meeting**
6120 South Clinton Trail 517-663-3423
Eaton Rapids, MI 48827 Fax: 517-663-3423
http://www.michiganforests.com

Mission: To promote good management on all forest land, to educate our members about good forest practices and stewardship of the land, and to inform the general public about forestry issues and the benefits of good forest management.

*Summer*
*William Botti, Executive Director*

**1601 Minnesota Association of Soil and Water Conservation Districts Annual Meeting**
790 Cleveland Avenue S 651-690-9028
Suite 201 Fax: 651-690-9065
St. Paul, MN 55116 http://www.maswcd.org

MASWCD is a nonprofit organization which exists to provide leadership and a common voice for Minnesota's soil and water conservation districts and to maintain a positive, results-oriented relationship with rule making agencies,partners and legislators; expanding education opportunities for the districts so they may carry out effective conservation programs.

*December*
*Ken Pederson, President*
*Steve Sunderland, Vice President*

**1602 Montana Association of Conservation Districts Annual Meeting**
501 North Sanders 406-443-5711
Helena, MT 59601 Fax: 406-443-0174
E-mail: mail@macdnet.org
http://www.macdnet.org

Montana's 58 Conservation Districts utilize locally-led and largely non-regulatory approaches to successfully address general natural resource issues. CD's have a decades-long history of conserving our state's resources by helpinglocal people match their needs with technical and financial resources, thereby getting good conservation practices on the ground to benefit all of Montanans.

*November*
*Sarah Carlson, Executive Director*

**1603 Montana Water Environment Association Annual Meeting**
516 N Park Street 406-449-7913
Suite A Fax: 406-449-6350
Helena, MT 59601

*Spring*

**1604 NEHA Annual Education Conference and Exhibition**
National Environmental Health Association
720 South Colorado Boulevard 303-756-9090
Suite 1000 N Fax: 303-691-9490
Denver, CO 80246 E-mail: staff@neha.org
http://www.neha.org

A revealing look at how the Environmental Health Profession is Evolving

*June*
*100 exhibits*
*Nelson Fabian, Executive Director*
*Larry Marcum, Manager*

**1605 NESEA BuildingEnergy Conference**
50 Miles Street 413-774-6051
Greenfield, MA 01301 Fax: 413-774-6053
E-mail: nesea@nesea.org
http://www.nesea.org

Held in Boston every March, this conference is the oldest and largest regional buuilding energy and renewable energy conference and trade show for practitioners in the Northeast. Over 4,000 people passed through the doors of the 2009event to take advantage of the high-level educational sessions and top quality exhibits.

*March*
*4000 attendees and 150 exhibits*
*Arianna Alexsandra Grindrod, Education Director*
*Jenny Spencer, Trade Show Manager, BE Conf.*

**1606 National Association Civilian Conservation Corps Alumni**
16 Hancock Avenue 314-487-8666
Saint Louis, MO 63125 E-mail: naccca@aol.com
http://www.cccalumni.org
The NACCCA was established as a non-profit organization in 1977 in California. The NACCCA offers annual national reunions, and a scholarship to a descendent of a NACCCA member.

*10 booths*
*Gene Morris, Civilian Records*

**1607 National Association of Environmental Professionals**
100 North 20th Street 215-564-3484
4th Floor Fax: 215-564-2175
Philadelphia, PA 19103 http://www.naep.org
NAEP is a multidisciplinary association dedicated to the advancement of environmental professionals in the US and abroad, and a forum for state-of-the-art information on environmental planning, research and management. A network ofprofessional contacts and a resource for structured career development, this organization is a strong proponent of ethics and the highest standards of practice in the environmental professions.
*Ron Deverman, President*
*Paul Looney, Vice President*

**1608 National Conference of Local Environmental Health Administrators**
University of Washington 206-616-2097
Dept Enviro Health, Box 357234 Fax: 206-543-8123
Seattle, WA 98195 E-mail: ctreser@u.washington.edu
http://www.ncleha.org
The NCLEHA's purpose is to provide a forum for local administrators to share common concerns and solutions to mutual problems, and to provide a professional organization for environmental health administrators, focused on the issuesand problems of local environmental health programs.
*Dave Riggs, Secretary*

**1609 National Environmental Balancing Bureau Meeting**
National Environmental Balancing Bureau
8575 Grovemont Circle 301-977-3698
Gaithersburg, MD 20877 Fax: 301-977-9589
E-mail: barry@neb.org
http://www.nebb.org
The NEBB is a nonprofit organization founded by contractors in the heating, ventilating and air conditioning (HVAC) industry. NEBB exists to help architects, engineers, building owners and contractors produce great buildings with HVACsystems that perform in ways they have visualized and designed.

**1610 National Environmental Health Association Annual Education Conference**
National Environmental Health Association
720 South Colorado Boulevard 303-756-9090
Suite 1000 N Fax: 303-691-9490
Denver, CO 80246 E-mail: staff@neha.org
http://www.neha.org
The NEHA AEC and Exhibition is a six-day educational event consisting of nine different environmental health and protection conferences and highlighting a two-day exhibition. It is the only conference that emcompasses all areas ofenvironmental health and protection, including, but not limited to: food protection, onsite wastewater, chemical and bioterrorism preparedness, indoor air quality, hazardous waste, and drinking water.

*Late June-Early July*
*120 booths with 1300 attendees*
*Larry Marcum, Manager*
*Nelson Fabian, Executive Director*

**1611 National Environmental, Safety and Health Training Association**
PO Box 10321 602-956-6099
Phoenix, AZ 85064-0321 Fax: 602-956-6399
E-mail: neshta@neshta.org
http://http://neshta.org
The National Environmental, Safety and Health Training Association is a non-profit international society for environmental, safety, health and other technical training and adult education professionals. NESHTA promotes trainercompetency through training and education standards, voluntary certification, and peer networking.

*Charles L. Richardson, Executive Director*
*Suzanne M. Lanctot, Membership Services*

**1612 National Real Estate Environmental Conference**
National Society of Environmental Consultants
PO Box 12528 210-225-2897
San Antonio, TX 78212 800-486-3676
Fax: 956-225-8450
Environmentally responsible management of real estate.

**1613 National Recycling Congress Show**
805 15th Street Nw 202-789-1430
Suite 425 Fax: 202-789-1431
Washington, DC 20005 E-mail: info@nrc-recycle.org
http://www.nrc-recycle.org
Founded in 1978, the National Recycling Coalition, Inc. provides technical education, disseminates public information on selected recycling issues, shapes public and private policy on recycling and operates programs that encouragerecycling markets and economic development.
*Kate Krebs, Executive Director*

**1614 National Solar Energy Conference**
2400 Central Avenue 303-443-4308
Suite A Fax: 303-442-3212
Boulder, CO 80301 E-mail: pmcfadden@ases.org
http://www.ases.org
The American Solar Energy Society (ASES) presents the Conference along with Green Energy Ohio. The event combines a premiere technical conference, plenary and forum sessions, a Renewable Energy Products and Services exhibit, workshops,tours and special events of interest to professionals and consumers.

*750 attendees and 75-100 exhibits*
*Pam McFadden, Registration/Sales*

**1615 National Water Resources Association Annual Conference**
3800 North Fairfax Drive 703-524-1544
Suite 4 Fax: 703-524-1548
Arlington, VA 22203 E-mail: nwra@nwra.org
http://www.nwra.org
Conservation of water resources in the 17 western reclamation states.

**1616 Nebraska Association of Resources Districts Annual Meeting**
601 South 12th Street 402-471-7670
Suite 201 Fax: 402-471-7677
Lincoln, NE 68508 E-mail: nard@nrdnet.org
http://www.nrdnet.org
Our mission is to assist NRDs in a coordinated effort to accomplish collectively what may not be accomplished individually to conserve, sustain, and improve our natural resources and environment.

*September*
*Dean E Edson, Executive Director*
*Jeanne Dryburgh, Office Manager*

**1617 New England Enviro Expo**
Zweigwhite

330 North Wabash
Suite 3201
Chicago, IL 60611
312-628-5870
Fax: 312-628-5878
E-mail: info@zweigwhite.com
http://www.enviroexpo.com

Environmental products/services for industrial, municipal, and government uses.

*May*
*400 booths with 5000 attendees*
*Dick Ryan, President*
*Fred White, Executive Vice President*

**1618 New England Water Environment Association Annual Meeting**
NEWEA
10 Tower Office Park
Suite 601
Woburn, MA 01801
781-939-0908
Fax: 781-939-0907
E-mail: mail@newea.org
http://www.newea.org

We are a regional member association of the Water Environmental Federation. We provide technical and education for the waste water industry.

*150+ booths with 1500 attendees*
*Erin Mosley, President*

**1619 New Hampshire Association of Conservation Commissions Annual Meeting**
54 Portsmouth Street
Concord, NH 03301
603-224-7867
Fax: 603-228-0423
E-mail: info@nhacc.org
http://www.nhacc.org

The New Hampshire Association of Conservation Commissions is a private, non-profit association of municipal conservation commissions. Its purpose is to foster conservation and appropriate use of New Hampshire's natural resources byproviding assistance to conservation commissions, facilitating communication and cooperation among commissions, and helping to create a climate in which commissions can be successful.

*November*
*Carol K Andrews, Executive Director*

**1620 New Jersey Society for Environmental Economic Development Annual Conference**
222 West State Street
Trenton, NJ 08608
609-695-3481
Fax: 609-695-0151
http://www.njslom.org

*October*
*William G. Dressel, Executive Director*

**1621 New Jersey Water Environment Association Conference**
PO Box 1212
Fair Lawn, NJ 07410
201-296-0021
Fax: 201-296-0031
http://www.njwea.org

The New Jersey Water Environment Association is a nonprofit educational organization dedicated to preserving and enhancing the water environment.
*Joseph Bonaccorso, President*

**1622 New Mexico Association of Soil and Water Conservation Annual Conference**
New Mexico Association Of Conservation Districts
163 Trail Canyon Road
Carlsbad, NM 88220
505-981-2400
Fax: 505-981-2422
E-mail: conserve@hughes.net
http://http://www.nm.nacdnet.org

The mission of NMACD is to facilitate conservation of the natural resources in New Mexico by providing opportunities and quality support to local conservation districts through representation and leadership.

*Fall*
*Debbie Hughes, Executive Director*

**1623 New York Water Environment Association Semi-Annual Conferences**
Nywea
525 Plum Street
Suite 102
Syracuse, NY 13204
315-422-7811
Fax: 315-422-3851
E-mail: pcr@nywea.org
http://www.nywea.org

The New York Water Environment Association is a nonprofit educational association dedicated to the development and dissemination of information concerning water quality management and the nature, collection, treatment, and disposal ofwastewater. Founded in 1929, the Association has over 2,500 members. The NYWEA is a member association of the Water Environment Federation.

*Winter and Summer*
*Patricia Cerro-Reehil, Executive Director*

**1624 North American Lake Management Society International Symposium**
4513 Vernon Boulevard, Suite 103
PO Box 5443
Madison, WI 53703
608-233-2836
Fax: 608-233-3186
E-mail: dbrown@nalms.org
http://www.nalms.org

The North American Lake Management Society's mission is to forge partnerships among citizens, scientists and professionals to foster the management and protection of lakes and reservoirs for today and tomorrow.

*October and Novembe*
*100 booths with 850 attendees and 50 exhibits*
*Darcy Brown, Administrative Assistant*

**1625 North Carolina Association of Soil and Water Conservation Districts Annual Conference**
Po Box 27943
Raleigh, NC 27611
919-733-2302
E-mail: bridget.munger@ncmail.net
http://www.ncaswcd.org

The association is an indepented, nonpartisan conservation organization created in 1944 to represent the interests of the 96 local soil and water conservation districts and the 492 district supervisors who direct their local district'sconservation programs.

*January*
*Don Rawls, President*

**1626 North Dakota Association of Soil Conservation Districts Annual Conference**
Lincoln Oaks Nursurey
3310 University Drive
Bismarck, ND 58504
701-223-8575
Fax: 701-223-1291
E-mail: lincolnoaks@btinet.net
http://www.lincolnoakes.com

*November*

**1627 Northeast Recycling Council Conference**
139 Main Street
Suite 401
Brattleboro, VT 05301
802-254-3636
Fax: 802-254-5870
E-mail: patty@nerc.org
http://www.nerc.org

NERC's mission is to leverage the strengths and resources of its member states to advance an environmentally stable economy in the Northeast by promoting source reduction, recycling, and the purchasing of environmentally preferableproducts and services.

*March and October*
*Moon Morgan, Office Manager*
*Patty Dillon, Manager Of Toxics*

**1628 Northeast Resource Recovery Association Annual Conference**
2101 Dover Road
Epsom, NH 03234
603-736-4401
Fax: 603-736-4402
E-mail: info@nrra.net
http://www.recyclewithus.org

The Northeast Resource Recovery Association is a pro-active nonprofit working with its membership to make their recycling programs strong, efficient, and financially successful by providing cooperative marketing, cooperativepurchasing, education and networking opportunities; developing innovative recycling programs; creating sustainable alternatives to reduce the volume and toxicity of the waste, and educating and informing local officials about recycling and solidwaste issues.

*June*
*Rick Cooper, President*
*Paula Dow, Executive Director*

**1629 Pacific Fishery Management Council Conferences**
7700 Northeast Ambassador Place
Suite 101
Portland, OR 97220
503-820-2280
866-806-7204
Fax: 503-820-2299
E-mail: john.coon@noaa.gov
http://www.pcouncil.org

The Pacific Council has developed fishery management plans for salmon, groundfish and coastal species in the US Exclusive Economic Zone off the coast of Washington, Oregon and California, and recommends Pacific halibut harvestregulations to the International Pacific Halibut Commission.

*5x year*
*Donald McIsaac, Executive Director*
*John Coon, Deputy Director*

**1630 Parks and Trails Council of Minnesota Annual Meeting**
275 E 4th Street
Suite 250
Saint Paul, MN 55101
651-726-2457
800-944-0707
Fax: 651-726-2458
E-mail: info@parksandtrails.org
http://www.parksandtrails.org

Mission: To acquire, protect and enhance critical lands for the public's enjoyment now and in the future.

*March*
*Judith Erickson, Government Relations*
*Beth Coleman, Executive Director*

**1631 Plant and Facilities Expo**
Association of Energy Engineers
4025 Pleasantdale Road
Suite 420
Atlanta, GA 30340
404-761-0509
Fax: 770-446-3969
E-mail: info@aeecenter.org
http://www.aeecenter.org

Occupational health/safety systems.
*Ruth Bennett, Information Services Dir.*

**1632 Renewable Energy Roundup & Green Living Fair**
PO Box 9507
Austin, TX 78766
512-345-5446
800-465-5049
E-mail: info@txses.org
http://www.txses.org; www.theroundup.org

Organized by the Texas Solar Energy Society and Texas Renewable Energy Industries Association featuring solutions to global warming such as rainwater harvesting, green and sustainable building, alternative transporation, and energyconservation.

*September*
*Natalie Marquis, Executive Director*
*Russel Smith, Event Coordinator*

**1633 Solar Cookers International World Conference**
1919 21st Street
Suite 101
Sacramento, CA 95814
916-444-6616
Fax: 916-444-5379
E-mail: sbci@igc.apc.org

Equipment for solar cooking and pasteurization of drinking water.

**1634 South Dakota Association of Conservation Districts Conference**
PO Box 275
Pierre, SD 57501
605-895-4099
Fax: 605-895-9424
http://www.sdconservation.org

*September*
*Angela Ehlers, Executive Director*

**1635 South Dakota Environmental Health Association Annual Conference**
State Department of Health
600 East Capitol Avenue
Pierre, SD 57501
605-773-3361
800-738-2301
E-mail: DOH.INFO@state.sd.us
http://http://doh.sd.gov

*April*
*Doneen Hollingsworth, Secretary of Health*

**1636 Southeastern Association of Fish and Wildlife Agencies Annual Meeting**
8005 Freshwater Farms Road
Tallahassee, FL 32308
850-893-1204
Fax: 850-893-6204
E-mail: seafwa@aol.com
http://www.seafwa.org

The SEAFWA conducts an annual conference each fall to provide a forum for presentation of information and exchange of ideas regarding the management and protection of fish and wildlife resources throughout the nation but with emphasison the southeast.

*October/November*
*Robert M. Brantly, Executive Secretary*

**1637 Take it Back**
Raymond Communications
5111 Berwyn Road
Suite 115
College Park, MD 20740
301-345-4237
Fax: 301-345-4768
E-mail: bruce@raymond.com
http://www.raymond.com

Top recycling experts and practical sessions.

*250 attendees*
*Bruce Popka, Vice President*

**1638 Texas Environmental Health Association Annual Education Conference**
PO Box 10
Leesburg, TX 75451
903-572-7278
Fax: 903-572-4193
E-mail: ginger.shaffer@myteha.org
http://www.myteha.org

The mission of TEHA is to work for the betterment of the health and welfare of people through the improvement of the environment.

*March*
*Margie N Earl, President*
*Ginger Shaffer, Executive Secretary*

**1639 Texas Water Conservation Association Annual Conference**
221 E 9th Street
Suite 206
Austin, TX 78701
512-472-7216
Fax: 512-472-0537
E-mail: goodson@twca.org
http://www.twca.org

*March*
*Leroy Goodson, Contact*
*Dean Robbins, Contact*

**1640 Utah Association of Conservation Districts Annual Conference**
1860 N 100 East
Logan, UT 84341
435-753-6029
Fax: 435-755-2117
E-mail: amber.beck@ut.nacdnet.net
http://http://uacd.org

The Utah Association of Conservation Districts is a nonprofit corporation representing Utah's 38 soil conservation districts. By working with landowners, organizations and government, the conservation districts work through voluntary,incentive-based programs to protect soil, water quality and other natural resources.

*November*
*Gordon Younker, Executive Vice President*
*Susan Stillion, Executive Assistant*

**1641 Virginia Association of Soil and Water Conservation Districts Annual Conference**
7308 Hanover Green Drive
Suite 100
Mechanicsville, VA 23111
804-559-0324
Fax: 804-559-0325
E-mail: info@vaswcd.org
http://www.vaswcd.org

The Virginia Association of Soil and Water Conservation Districts (VASWCD) is a private nonprofit association of 47 soil and water conservation districts in Virginia. It is a voluntary, nongovernmental association of Virginia'sdistricts that provides and promotes leadership in the conservation of natural resources through stewardship and education programs.

*December*
*Wilkie WO Chaffin PhD, President*

**1642 Virginia Forestry Association Annual Conference**
3308 Augusta Avenue
Richmond, VA 23230
804-278-8733
E-mail: vafa@verizon.net
http://www.vaforestry.org

VFA promotes stewardship and wise use of the Commonwealth's forest resources for the economic and environmental benefits of all Virginians. Membership consists of forest landowners, forest product businesses, forestry professionals,and a variety of individuals and groups who are concerned about the future and well-being of Virginia's forest resources.

*Late Spring*
*Patrick Gottschalk, Secretary of Commerce*

**1643 WEFTEC Show**
Water Environment Federation
601 Wythe Street
Alexandria, VA 22314
800-666-0206
Fax: 703-684-2492
E-mail: csc@wef.org
http://www.wef.org

North America's largest annual water quality conference and exposition. Covers a wide spectrum of critical water quality issues.

*Fall*
*16,000 attendees and 800+ exhibits*
*Tom Wolfe, Director of Advertising*

**1644 Waste Expo**
Environmental Industries Association
4301 Connecticut Avenue NW
Suite 300
Washington, DC 20008
202-244-4700
Fax: 202-966-4818
http://www.envasns.org
Waste/recycling equipment and technology.
*Bruce Parker, President*

**1645 WasteExpo**
11 Riverbend Drive S
Stamford, CT 06907
203-358-9900
800-559-0620
Fax: 203-358-5816
E-mail: laura.magliola@penton.com
http://www.wasteexpo.com

WasteExpo is the largest tradeshow in North America serving the $43 billion solid waste and recycling industries.

*11,500 attendees and 450 exhibits*
*Laura Magliola, Marketing Manager*

**1646 West Virginia Forestry Association**
PO Box 718
Ripley, WV 25271
304-372-1955
888-372-9663
Fax: 304-372-1957
E-mail: wvfa@wvadventures.net
http://www.wvfa.org

The West Virginia Forestry Association is a non-profit organization funded by its membership. Our members include individuals and businesses involved in forest management, timber production and wood product manufacturing. Our membersare concerned with protecting the environment, as well as enhancing the future of West Virginia's forests through multiple-use management.

*Summer*
*Richard Waybright, Executive Director*

**1647 Western Association of Fish and Wildlife Agencies Annual Meeting**
5400 Bishop Blvd
Cheyenne, WY 82006
307-777-4569
Fax: 307-777-4699
http://www.wafwa.org

*July*
*Larry Kruckenberg, Secretary*

**1648 Western Forestry and Conservation Association Conference**
4033 Canyon Road SW
Portland, OR 97221
503-226-4562
Fax: 503-226-2515
E-mail: richard@westernforestry.org
http://www.westernforestry.org

Offers continuing education workshops and seminars for professional foresters throughout the west.

*January/February*

**1649 Western Society of Naturalists Annual Meeting**
San Diego State University Department Of Biology
5500 Campanile Drive
San Diego, CA 92182
818-677-3256
Fax: 818-677-2034
http://www.wsn-online.org

Members include researchers, educators, academics and others with an interest in the area's biology, particularly its marine life. Membership is $15.00 per year for individuals and $7.00 for students.
*Mark Carr, President*

**1650 Wildlife Habitat Council Annual Symposium**
Wildlife Habitat Council
8737 Colesville Road
Suite 800
Silver Spring, MD 20910
301-588-8994
Fax: 301-588-4629
E-mail: whc@wildlifehc.org
http://www.wildlifehc.org

The Wildlife Habitat Council is a nonprofit, group of corporations, conservation organizations, and individuals dedicated to restoring and enhancing wildlife habitat.
*Emer OBroin, Chairman*
*Lawrence A Selzer, Vice Chairman*

**1651 Wildlife Society Annual Conference**
Wildlife Society
5410 Grosvenor Lane
Suite 200
Bethesda, MD 20814
301-897-9770
Fax: 301-530-2471
E-mail: tws@wildlife.org
http://www.wildlife.org

Annual conference of wildlife professionals, organized by the Wildlife Society.

*September*
*50 booths with 1400 attendees*
*Michael Hutchins, Executive Director/CEO*
*Darryl Walter, Dir Membership/Mktg/Conf*

**1652 Wisconsin Association for Environmental Education Annual Conference**

8 Nelson Hall
University of Wisconsin
Stevens Point, WI 54481
715-346-2796
Fax: 715-346-3835
E-mail: waee@uwsp.edu
http://www.uwsp.edu/waee

WAEE is a statewide non-profit organization composed of people interested in learning about and helping others learn about environmental issues. Our goal is to promote responsible environmental action through education in the classroomand in the community.

*Fall*
*Cassie Bauer, Student Representitive*

**1653 Wisconsin Land and Water Conservation Association Annual Conference**

702 East Johnson Street
Madison, WI 53703
608-441-2677
Fax: 608-441-2676
E-mail: julian@wlwca.org
http://www.wlwca.org

Mission: To assist county Land Conservation Committees and Departments with the protection, enhancement and sustainable use of Wisconsin's natural resources and to represent them through education and governmental interaction.

*December*
*Julian Zelazny, Executive Director*

**1654 Wisconsin Woodland Owners Association Annual Conference**

PO Box 285
Stevens Point, WI 54481
715-346-4798
E-mail: nbozek@uwsp.edu
http://www.wisconsinwoodlands.org

The Wisconsin Woodland Owners Association, a nonprofit educational organization, was established in 1979 to advance the interests of woodland owners and the cause of forestry; develop public appreciation for the value of Wisconsin'swoodlands and their importance in the economy and overall welfare of the state; foster and encourage wise use and management of Wisconsin's woodlands for timber production, wildlife habitat and recreation; and to educate those interested in managingthe woodlands.

*October*
*William J. Horvath, Director*

**1655 World Energy Engineering Congress**

Association of Energy Engineers
4025 Pleasantdale Road
Suite 420
Atlanta, GA 30340
770-447-5083
Fax: 770-446-3969
E-mail: info@aeecenter.org
http://www.aeecenter.org

Equipment and services.
*Ruth Bennett, Information Services Dir.*

# Consultants

## Environmental

**1656 3D/International**
1900 West Loop South
Suite 400
Houston, TX 77027
713-871-7000
Fax: 713-871-7171
E-mail: contact@3di.com
http://www.3di.com

Enrvironmental compliance and consulting services

*John Murph, PE, President/CEO*
*Gary Boyd, AIA, Executive VP*

**1657 AAA Lead Consultants and Inspections**
1307 West 6th Street
Suite 134
Corona, CA 92882
951-582-9071
Fax: 951-582-9073
E-mail: aaalead@sbcglobal.net
http://http://aaalead.net/index.html/

Offers quality consulting, inspections, monitoring and project design for lead based paint.

*Michael Cohn, CEO*

**1658 AB2MT Consultants**
9400 South Dadeland Boulevard
Suite 370
Miami, FL 33156
305-670-1011
Fax: 305-670-1016
E-mail: ab2mt@aol.com

Environmental and engineering consulting.
*Paula H Church, President*

**1659 ABS Consulting**
16800 Greenspoint Park Drive
Suire 300 South
Houston, TX 77060
281-673-2800
Fax: 281-673-2812
E-mail: info@absconsulting.com
http://www.plg.com

ABS Consulting provides rational engineering, science and technology-based solutions that blend effective management controls, state-of-the-art engineering analyses, practical loss-control measures and innovative risk-transfer options.
*Frank Iarossi, Chairman/CEO*

**1660 ACC Environmental Consultants**
7977 Capwell Drive
Suite 100
Oakland, CA 94621
510-638-8400
800-525-8838
Fax: 510-638-8404
E-mail: general@accenv.com
http://www.accenv.com

An employee owned environmental and energy consulting firm. Helps companies and public agencies throughout California identify and manage environmental hazards, comply with their OSHA and EPA requirements.
*Kenneth R Churchill, CEO*
*James Wilson, President*

**1661 ACRT Environmental Specialists**
1333 Home Avenue
Akron, OH 44310
330-945-7500
800-622-2562
Fax: 330-945-7200
E-mail: askacrt@acrtinc.com
http://www.acrt.com

Appraisal, Research and Training is an international consulting service and training organization in the utility and urban forestry, arboricultural, environmental, natural resource and horticultural services.

**1662 ADS LLC**
4940 Research Drive
Huntsville, AL 35805
800-633-7246
Fax: 256-430-6633
E-mail: adssales@idexcorp.com
http://www.adsenv.com

ADS LLC develops and provides technology-based hardware and software products and services for the water, wastewater, gas, and hydroelectric industries through three divisions, ADS environmental services, hydra-stop and Accusonictechnologies. ADS pioneered the industry's first flow monitoring hardware and software products over 32 years ago, and today continues to provide the highest quality products and services to its clients.

*Karl Boone, President*
*Joseph Goustin, Cheif Financial Officer*

**1663 AECOS**
45-939 Kamehameha Highway
Room 104
Kaneohe, HI 96744
808-234-7770
Fax: 808-234-7775
E-mail: aecos@aecos.com
http://www.aecos.com

Environmental counseling firm providing the services of scientists and facilities in the environmental sciences to clients throughout the Pacific area. Specializes in aquatic (both fresh water and marine) biology and water quality,with practiced expertise in analytical chemistry, oceanography, water pollution, and marine and fresh water ecology.
*Eric Guinther, President*
*Michele , Office Manager*

**1664 AF Meyer and Associates**
9060 Meadowood Street
Baton Rouge, LA 70815
225-925-0630
Fax: 225-928-7848
E-mail: afmal@webtv.net
http://www.erols.com/afma

Environmental consulting firm.
*AF Meyer, President*

**1665 AKT Peerless Environmental Services**
22725 Orchard Lake Road
Farmington, MI 48336
248-615-1333
Fax: 248-615-1334
http://www.akt.com

Providing environmental services to facilitate real estate transfer, development, and redevelopment. Services include phase I ESA, subsurface investigation, remediation, Brownfield's redevelopment, Brownfield's financial incentives.

**1666 AM Kinney**
2900 Vernon Place
Cincinnati, OH 45219
513-421-2265
800-AMK-3682
Fax: 513-281-1123
E-mail: nielseng@amkinney.com
http://www.amkinney.com

Provides creative and cost effective solutions in the planning, design and delivery of clients' projects.
*George Finch, President*

**1667 ANA-Lab Corporation**
2600 Dudley Road
PO Box 9000
Kilgore, TX 75663
903-984-0551
Fax: 903-984-5914
E-mail: corp@ana-lab.com
http://www.ana-lab.com

Environmental laboratory. Offers ICP-MS which allows Ana-Lab to offer improved turn around time, reduce costs, and achieve better quantitation of regulated parameters. Tests are performed by methods specified by the EPA. Specializes inenvironmental chemistry.

*C H Whiteside, President*
*Bill Peery, Jr., SVP/COO*

**1668 APEC-AM Environmental Consultants**
2525 Northwest Expressway
Suite 301D
Oklahoma City, OK 73112
405-840-9327
Fax: 405-840-9328
E-mail: apecapec@msn.com

Environmental assessment
*Charlie Bowlin, Principal*
*Saleem Nizami, Principal*

**1669 ARCADIS**
630 Plaza Drive
Suite 200
Highlands Ranch, CO 80129
720-344-3500
Fax: 720-344-3535
http://www.arcadis-us.com

A leading, global, knowledge-driven service provider. Active in the fields of infrastructure, environment and buildings. Feasibil-

ity studies, design, engineering, project management, implementation and facility management, plusrelated legal and financial services.
*John Boyette, President*
*Steven B Blake, CEO*

**1670 ATC Associates**
600 W. Cummings Park 781-932-9400
Suite 5500 Fax: 781-952-6211
Woburn, MA 01801 http://www.atcassociates.com
Environmental consulting firm with 1,600 experts in 65 offices throughout the United States, including engineers, scientists, technicians, and regulatory specialists.
*Pam O'Deen, Business Development*
*Bobby Toups, CEO*

**1671 ATC Associates: Omaha**
3712 S 132nd Street 402-697-9747
Omaha, NE 68144 Fax: 402-697-9170
http://www.atcassociates.com
Environmental and worker exposure consulting firm for EPA and OSHA compliance.

**1672 ATS-Chester Engineers**
260 Airside Drive 412-809-6600
Moon Township, PA 15108 Fax: 412-809-6611
http://www.atsengineers.com
Provides services in waste water treatment and air pollution control.
*Robert Agbede, President*

**1673 Aarcher**
910 Commerce Road 410-897-9100
Annapolis, MD 21401 Fax: 410-897-9104
E-mail: cschwartz@aarcherinc.com
http://www.aarcherinc.com
Aarcher is a small business providing environmental management, assessment and planning services nationwide from its headquarters and regional offices. Aarcher provides environmental compliance audits; NEPA analysis and documentation;natural and cultural resource management planning; site assessment and investigation; plans and permits; and environmental liability assessment and control. Our consulting services are guided by a comprehensive understanding of current environmentalregulations.
*Craig J Schwartz, President*

**1674 Abacus Environmental**
123 Pinney Street 860-871-6216
PO Box 365 800-343-9970
Ellington, CT 06029 Fax: 860-872-8044
E-mail: donweek@jx.netcom.com
http://www.abacusenvironmental.com
Full service environmental and occupational safety and health consulting firm. Services include project management, indoor air quality and industrial hygiene, asbestos and lead project management, expert witness testimony and allaspects of workplace safety. Goals are to help reduce client's operating costs, minimize liability for environmental and occupational safety reguations, guidelines, and requirements.
*Donald M Weekes Jr, President*

**1675 Abco Engineering Corporation**
6901 South Yosemite Street 303-220-8220
Suite 205 Fax: 303-796-0810
Englewood, CO 80112 E-mail: info@abco-corp.com
http://www.abco-corp.com
Provides a full spectrum of engineering and environmental services pertaining to both new construction and existing buildings, including Property Condition Assessment Reports, Phase I Environmental Assessments, Quality Control Reportsand other technical support services related to buildings and building systems including feasibility reports, construction observation and cost eliminating.
*Joe Johnson, Director*
*Michael R Dannecker, Director*

**1676 Abonmarche Environmental**
95 West Main Street 269-927-2295
PO Box 1088 Fax: 269-927-1017
Benton Harbor, MI 49023 E-mail: aci@abonmarche.com
http://www.abonmarche.com
Full-service architectural, engineering, land surveying and planning firm.

**1677 AccuTech Environmental Services**
43 West Front Street 732-739-6444
Rear Suite 800-644-ISRA
Keyport, NJ 07735 Fax: 732-739-0451
E-mail: info@accutechenvironmental.com
http://www.accutechenvironmental.com
Aims to meet the environmental consulting needs generated by New Jersey's Environmental Cleanup Responsibility Act by preparing and managing complete environmental sampling and cleanup programs.
*Harry Moscatello, President*

**1678 Accutest Laboratories**
2235 Route 130 732-329-0200
Dayton, NJ 08810 Fax: 732-329-3499
E-mail: infonj@accutest.com
http://www.accutest.com
Privately held, independent testing laboratory delivering legally defensible data, providing a full range of environmental analytical services to industrial, engineering/consulting and government clients throughout the United States.Operating from coordinated laboratories in New Jersey, Massachusetts, Florida and Texas, resources include a staff of over 200, five million dollars worth of laboratory instrumentation and equipment, and more than 80,000 square feet of laboratoryspace.

**1679 Acheron Engineering Services**
147 Main Street 207-368-5700
Newport, ME 04953 Fax: 207-368-5120
E-mail: WBall@AcheronEngineering.com
http://www.acheronengineering.com
Provides solutions to the most challenging engineering, environmental and geologic issues.
*William B Ball, President*
*Kirk Ball, Engineering Field Technician*

**1680 Activated Carbon Services**
409 Meade Drive 724-457-6576
Coraopolis, PA 15108 800-367-2587
Fax: 724-457-1214
E-mail: hnpacs@aol.com
http://www.pacslabs.com
Training courses and conferences. Provides short courses in spectrocopy, chomatography, quality, safety, environmental, and management. Provides professional manuals and software products. Provides laboratory testing and consultingservices. Company also goes by the following names: Activated Carbon Services, PACS Testing and Consulting, PACS Courses and Conferences. PACS provides: Testing, Training, R & D Conferences, and software for activated carbon users.

*Henry G Nowicki PhD/MBA, President*
*Barbara Sherman, Manager of Operations*

**1681 Acumen Industrial Hygiene**
1032 Irving Street 415-242-6060
#922 Fax: 415-242-6006
San Francisco, CA 94122 E-mail: info@acumen-ih.com
http://www.acumen-ih.com
Industrial hygiene consultation.
*Michael Connor, Principal*

**1682 Advanced Chemistry Labs**
3039 Amwiler Road 770-409-1444
Suite 100 Fax: 770-409-1844
Atlanta, GA 30360 E-mail: acl@acl-labs.net
http://www.advancedchemistrylabs.com

Environmental teststing laboratory.

*John Andros, Technical Director*

**1683 Advanced Resources International**
4501 Fairfax Drive
Suite 910
Arlington, VA 22203
703-528-8420
Fax: 703-528-0439
E-mail: ari-info@adv-res.com
http://www.adv-res.com

Independent consulting firm focused on providing technical services to the international energy industry.
*Vello A Kuuskraa, President*
*Jonathan Kelafant, Senior Vice President*

**1684 Advanced Waste Management Systems**
6430 Hixson Pike
Hixson, TN 37343
423-843-2206
Fax: 423-843-2310
E-mail: info@awm.net
http://www.awm.net/

Provides a wide range of environmental and engineering services to domestic and international clients, including governments, corporations, and provate citizens.
*Richard Ellis PhD, CEO*
*James Mullican, PE, President*

**1685 Aerosol Monitoring and Analysis**
4475 Forbes Boulevard
Lanham, MD 20706
301-459-2640
Fax: 301-459-2643
E-mail: amalab@aol.com

Aerosol Monitoring & Analysis, provides Industrial Hygiene, Environmental and Health & Safety to government agencies, institutions, building owners, property managers, architects and engineers.

**1686 Aguirre Engineers**
13276 E Fremont Place
PO Box 3814
Englewood, CO 80112
303-799-8378
Fax: 303-799-8392
E-mail: infoteam@aquirre1.com

Aguirre Engineers is an environmental engineering firm based in Englewwod, CO. Over the past 20 years we have augmented our service offering, from a commercial geotechnical leader, into providing full-scale government contractingservices for radioactive and hazardous waste remediation. The continued growth or our company throughout the years is a solid indicator of high-quality performance and client satisfaction.

**1687 Air Consulting and Engineering**
2106 NW 67th Place
Suite 4
Gainesville, FL 32653
352-335-1889
Fax: 352-335-1891
http://http://www/airconsulting.us

Air Consulting and Engineers, provides air pollution testing services utilizing United States Environmental Protection Agency. The company was founded in April 1984 to provide prefessional source emission testing and engineering, andair permitting to industries located in Florida and throughout the world.
*Charles Simon, Sr. Scientist*

**1688 Air Sciences**
1301 Washington Avenue
Suite 200
Golden, CO 80401
303-988-2960
Fax: 303-988-2968
E-mail: air@airsci.com
http://www.airsci.com

Air Sciences was founded in the Denver-metro area in 1980 with the purpose of providing superior air pollution consulting services. Air Sciences attained this goal and presently enjoys a unique reputation as a firm that provides bothindustry and government a high quality service in air quality consulting. Our future is focused on emerging disciplines in the air quality arena driven by new air quality standards and regional haze regulations. Air Sciences is an employee ownedfirm.

**1689 Aires Consulting Group**
1550 Hubbard Avenue
Batavia, IL 60510
630-879-3006
800-247-3799
Fax: 630-879-3014
E-mail: info@airesconsulting.com
http://www.airesconsulting.com

National full-service industrial hygiene, environmental and occupational health consulting firm. Assists clients in the control of liability through the application of risk management principals.
*Rich Rapacki, VP*
*Kevin Bannon, Marketing Director*

**1690 Airtek Environmental Corporation**
39 West 38th Street
12th Floor
New York, NY 10018
212-768-0516
Fax: 212-768-0759
E-mail: info@airtekenv.com
http://www.airtekenv.com

Environmental investigation and mangement professionals specializing in multi-jurisdictional regulatory climates.
*Mike S Zouak, President*

**1691 Alan Plummer Associates, Inc.**
1320 South University Drive
Suite 300
Fort Worth, TX 76107
817-806-1700
Fax: 817-870-2536
E-mail: aplummer@apaienv.com
http://www.apaienv.com

Civil and environmental engineering consulting.

*Alan R. Tucker, President*
*Julie Lippe, Marketing Coordinator*

**1692 All 4 Inc**
2393 Kimberton Road
PO Box 299
Kimberton, PA 19442
610-933-5246
Fax: 610-933-5127
E-mail: jegan@all4inc.com
http://www.all4inc.biz

An environmental consulting company specializing in air quality services, primarily assisting clients with complex air permitting, modeling, continuous monitoring, and regulation compliance.
*John Egan, Principal Consultant*
*Dan Holland, Principal Consultant*

**1693 Allee, King, Rosen and Fleming**
440 Park Avenue South
7th Floor
New York, NY 10016
212-696-0670
Fax: 212-779-9721
E-mail: nycinfo@akrf.com
http://www.akrf.com

Environmental consulting firm.
*Debra C Allee, AICP, Principal*
*Edward A Applebome, ITE, Principal*

**1694 Allied Engineers**
PO Box 2760
San Ramon, CA 94583
925-867-4646
Fax: 925-867-0736
http://www.alliedengineersinc.com

Consulting services in wastewater and industrial waste, including emissions testing.
*Robert Dawyat, President*

**1695 Allstate Power Vac**
928 East Hazelwood Avenue
Rahway, NJ 07065
732-815-0220
Fax: 732-815-9892
http://www.aspvac.com

Industrial and environmental waste management.

**1696 Allwest Environmental**
530 Howard Street
Suite 300
San Francisco, CA 94105
415-391-2510
Fax: 415-391-2008
E-mail: info@allwest1.com
http://www.allwest1.com

Practical, business-oriented consulting firm specializing in Environmental and Engineering Due Diligence offering expertise to the real estate industry. Helps clients to understand and manage potential environmental and buildingliabilites, and to advocate their interests through the discovery and mitigation process.
*Marc Cunningham, President*
*Chris Marinescu, Vice President*

**1697 Alpha-Omega Environmental Services**
933 Northwest 31 Avenue 954-969-5906
Pompano Beach, FL 33069 866-969-6653
Fax: 954-969-5232
E-mail: dave@aomegagroup.com
http://www.aomegagroup.com
Environmental engineering consulting

**1698 Alternative Resources**
1732 Main Street 978-371-2054
Concord, MA 01742 Fax: 978-371-7269
E-mail: info@alt-res.com
http://www.alt-res.com
Alternative Resources is an independent consulting firm providing management, engineering, environmental, economic and financial advisory in the fields of water and wastewater treatment, solid waste management, residuals management,environmental compliance, and energy production.
*Gretchen Karlson, Personnel*

**1699 Ambient Engineering**
100 Main Street
Suite 330 888-262-6232
Concord, MA 01742 Fax: 978-369-8380
E-mail: info@ambient-engineering.com
http://ambient-engineering.com
An environmental engineering and consulting firm incorporated in 1994. We provide environmental, site civil engineering and regulatory compliance services with a focus on the six New England states.

*T J Stevenson, PhD, President*
*Kenneth Pyzocha, PE, Director of Engineering*

**1700 American Archaeology Group LLC**
208 West Second Street 512-556-4100
Lampasas, TX 76550 Fax: 512-556-3373
E-mail: info@american-archaeology.com
http://www.american-archaeology.com
Professional archeological firm that hand;es state and federal permitting on projects.

*Michael R. Bradle, President*

**1701 American Engineering Testing**
550 Cleveland Avenue North 651-659-1308
St Paul, MN 55114 800-972-6364
Fax: 651-659-1379
E-mail: rkaiser@amengtest.com
http://www.amengtest.com
America's people, technology, innovation and quality commited to fulfilling client tequirements.
*Robert A Kaiser, VP*

**1702 American Services Associates**
18154 41st Place SE 425-641-5130
Issaquah, WA 98027 Fax: 425-641-5138
E-mail: airsampler@aol.com
http://www.asaearth.com
Consultants in emission testing, permitting, emission control system design, training and continuous emission moniters (CEMs). Producer of video training programs on E{A emission sampling methods in CD and VHS formats. Offices are inIssaquah, Washington.
*Wes Snowden, President*
*John Vareski, Vice President*

**1703 Andco Environmental Processes**
415 Commerce Drive 716-691-2100
Amherst, NY 14228 Fax: 716-691-2880
E-mail: Andco@Localnet.com
Manufacturers of waste disposal treatment systems.
*Jack I Reich, Sales Manager*

**1704 Andersen 2000 Inc/Crown Andersen**
1015 Tyrone Road 770-486-2000
Suite 410 800-241-5424
Tyrone, GA 30290 Fax: 770-487-5066
E-mail: tom.vanremmen@and2k.com
http://www.crownandersen.com
Supplying World Industry with Incineration and Air Pollution Control Systems

**1705 Anderson Consulting Group**
PO Box 407 610-918-7461
Downingtown, PA 19335 Fax: 610-918-9469
E-mail: info@andersonconsultinggroup.com
http://www.andersonconsultinggroup.com
Anderson Consulting Group has helped companies and publics agencies manage their project development risk, drive down construction cost, and improve schedules. Anderson Consulting Group's environmental and geotechnical services areuniquely designed to address client objectives. Our engineering solutions have earned engineering leadeship and innovation awards.

**1706 Apollo Energy Systems**
4100 North Powerline Road 954-969-7755
Building D3 Fax: 954-969-7788
Pompano Beach, FL 33073E-mail: www.apolloenergysystems.com
Products and services includes lead cobalt batteries, alkaline fuel cells, power plants and EV programs.
*Robert Aronsson, Director/CEO*
*Sonny Spoden, Director/CFO*

**1707 Applied Ecological Services, Inc.**
17921 Smith Road 608-897-8641
PO Box 256 Fax: 608-897-8486
Brodhead, WI 53520 E-mail: Info@AppliedEco.com
http://www.appliedeco.com
AES is a broad-based ecological consulting, contracting and restoration firm providing services to foundations, government units, corporations, and commercial/residential developers nationwide. Our staff consultants consisting of amultidisciplinary team of geologists, hydrologists, ecologists, botanists, wildlife biologists, wetland scientists, landscape architects, and prairie and ecosystem restoration specialists manage over 100 projects a year.

*Steve Apfelbaum, President*
*Carl V Korfmacher, MLA, Vice President*

**1708 Applied Geoscience and Engineering**
1300 New Holland Road 610-777-5027
Reading, PA 19607 Fax: 610-777-4276
E-mail: office@appliedgeoscience.com
http://www.appliedgeoscience.com
Environmental engineering, site assessments, and testing services through subcontractors.

**1709 Applied Marine Ecology**
658 NE 70th St. 305-757-0018
Miami, FL 33145 Fax: 305-759-3999
Marine ecology research firm.
*Anitra Thorhaug, President*
*Andrew Oerke, CEO*

**1710 Applied Science Associates**
55 Village Square Drive 401-789-6224
South Kingston, RI 02879 Fax: 401-789-1932
E-mail: asa@asascience.com
http://www.asascience.com
A global science and technology solutions company. Through consulting, environmental modeling, and application development, ASA helps a diverse range of clients in government, industry, and academia investigate their issues of concernand obtain functional answers.

*Eoin Howlett, CEO*
*J Craig Swanson, Senior Principal*

**1711 Aqua Sierra**
8350 South Mariposa Drive
Morrison, CO 80465
303-697-5486
800-524-FISH
Fax: 303-697-5069
E-mail: info@aqua-sierra.com
http://www.aqua-sierra.com

Aqua Sierra is a complete company servicing fisheries, aquaculture, water quality, wastewater, and database management interest. Aqua Sierra can assure an efficient, effective cost-conscious approach to managing aquatic resources bycombining a broad base of experience in all aspects of fisheries, aquatic ecology, and water quality management.

*William J Logan, President*

**1712 Aqua Survey**
469 Point Breeze Road
Flemington, NJ 08822
908-788-8700
Fax: 908-788-9165
E-mail: Mail@AquaSurvey.com
http://www.aquasurvey.com

Aqua Survey is a full service ecotoxicology company founded in 1975. Aqua Survey provides laboratory testing, field sampling and consulting services to a wide variety of clients throughout the world including many of the largest UScorporations, internationally reconized environmental consulting firms, and the public sector.

*Kenneth R Hayes, President*

**1713 Aqualogic**
30 Devine Street
North Haven, CT 06473
203-248-8959
800-989-8959
Fax: 203-288-4308
E-mail: rheller@aqualogic.com
http://www.aqualogic.com

Industrial Wastewater Treatment Systems and Chemical Recovery Systems.

*Dick Heller, Sales Engineer*

**1714 Arcadis**
630 Plaza Drive
Suite 200
Highlands Ranch, CO 80129
720-344-3500
800-225-8419
Fax: 720-344-3535
E-mail: arcadisgm@arcadis-us.com
http://www.arcadis-us.com

Complete environmental services and remediation.

*Alan Hurley, Area Manager*

**1715 Architectural Energy Corporation**
2540 Frontier Avenue
Suite 201
Boulder, CO 80301
303-444-4149
800-450-4454
Fax: 303-444-4304
E-mail: aecinfo@archenergy.com
http://www.archenergy.com

*Michael J Holtz, FAIA, President*

**1716 Arctech**
14100 Park Meadow Drive
Chantilly, VA 20151
703-222-0280
Fax: 703-222-0299
E-mail: dwalia@arctech.com
http://www.arctech.com

Arctech a diverse American Corporation is providing cost-effective solutions for energy, environmental, and agriculture market sectors. Arctech group through 25 years of experience in energy, energetics, environment and agriculture,has created holistic solutions in these interrelated market sectors. The enterpreneurial scientist and engineers at Arctech have pioneered the use of our vast resources of coal to make coal-derived humic acid products.

*Daman S Walia, President/CEO*
*Madhu Walia, Administration Director*

**1717 Ardea Consulting**
10 1st Street
Woodland, CA 95695
530-669-1645
Fax: 530-669-1674
E-mail: birdtox1@ardeacon.com
http://www.ardeacon.com

Ardea Consulting provides avian and wildlife toxicology guidance to engineering and environmental firms, government agencies, business and non-governmental organizations.

*Joseph P Sullivan PhD, Sr. Consultant*

**1718 Argus/King Environmental Limited**
7271 Wurzbach
Suite 202
San Antonio, TX 78240
210-493-2560
800-698-6018
Fax: 210-342-9027

Industrial hygiene and indoor air quality management. Mold and bacteria sampling, asbestos and lead testing and forensic.

*Robert W Miller, CIH*
*Henry King, Consultant*

**1719 Arro Consulting**
270 Granite Run Drive
Lancaster, PA 17601
717-569-7021
800-229-6009
Fax: 717-560-0577
E-mail: info@thearrogroup.com
http://www.thearrogroup.com

Environmental engineering.

*GM Brown, President*
*Darrell L. Becker, P.E., Vice President*

**1720 Arro Laboratory**
PO Box 686
Caton Farm Road
Joliet, IL 60434
815-727-5436
Fax: 815-740-3238

Testing and analysis laboratory specializing in resource conservation and recovery act sampling. Research results published in confidential reports to clients.

**1721 Artemel and Associates**
218 North Lee Street
Suite 316
Alexandria, VA 22314
703-683-3838
Fax: 703-836-1370
E-mail: aiusa@artemel.com
http://www.artemel.com

Artemel & Associates is the technical arm of the Artemel Group of companies, with planning, engineering and analytical capabilities. The firm's ares of professinal expertise directly complement Artemel International's areas ofspecialization. Artemel & Associates has been serving clients since 1984, and has benn credited with a variety of successful technical accomplishments both in the United States and around the globe.

*Engin Artemel, President*

**1722 Ascension Technology**
107 Catamount Drive
Milton, VT 05468
802-893-6657
800-321-6596
Fax: 802-893-6659
E-mail: ascension@ascension-tech.com
http://www.ascension-tech.com

Manufactures motion tracking equipment and provides a full year's warranty, free telephonic support, and on-site support.

*Edward C Kern Jr, President*

**1723 Associates in Rural Development**
159 Bank Street
Suite 300
Burlington, VT 05401
802-658-3890
Fax: 802-658-4247
E-mail: ard@ardinc.com
http://www.ardinc.com

ARD was founded in 1977 as a Vermont corporation. Vermont's reputation for leadership in environmental affairs and its heritage of local participatory government embody ARD's ideals. Services include: watershed management,resource/sector assessments, EIA, urban environmental magagement, policy and action planning, natural resource assessment and evaluation, NR-based enterprise development, biodiversity conservation and finance, integrated water resource planning andmanagement.

*George Burrill, President*
*Jim Talbot, Sr. VP*

**1724 Astbury Environmental Engineering**
5645 West 79th Street 317-472-0999
Indianapolis, IN 46278 Fax: 317-472-0993
E-mail: info@aeeindy.com
http://www.aeeindy.com
Astbury Environmental Engineering is a privately owned Indianapolis company that provides a full range of environmental services that include invironemntal management, site investigation and corrective action, health and safety, aircompliance, solid and hazardous waste management, and wastewater.
*Steve Wilcox, President/CEO*
*Fred Nichols, VP, Business Development*

**1725 Astorino Branch Environmental**
227 For Pitt Boulevard 412-765-1700
Pittsburg, PA 15222 800-518-0464
Fax: 412-765-1711
E-mail: marketing@ldastorino.com
Astorino Companies is an architectural, engineering and environmental consulting firm headquartered in downtown Pittsburg, PA.
*Louis Astorino, President & CEO*

**1726 Athena Environmental Sciences**
1450 S Rolling Road 410-455-6319
Baltimore, MD 21227 888-892-8408
Fax: 410-455-1155
E-mail: athenaes@athenaenvironmental.com
http://www.athenaenvironmental.com
Designs and develops novel products that represent environmentally responsible and economically sound solutions to environmental problems. Contract services are provided to clients for product development. Products developed includethe company's own Spill Pill (TM), a proprietary cleaning agent for petroleum contamination on concrete and other building surfaces, Bilge Tech, Inc.'s Bilge Pill (TM), a cleaning agent for removing oil and dirt buildup in boat bilges, and expertisein biotechnology.
*Sheldon Broedel PhD, CEO*

**1727 Atkins Environmental HELP**
PO Box 222320 661-260-2260
Santa Clarita, CA 91322 800-750-0622
Fax: 661-253-3555
E-mail: info@atkinsenvironmental.com
http://www.atkinsenvironmental.com
Environmental, health and safety compliance. Support services.
*BJ Atkins, President*

**1728 Atlantic Testing Laboratories**
6431 US Highway 11 315-386-4578
Canton, NY 13617 Fax: 315-386-1012
E-mail: info@AtlanticTesting.com
http://www.atlantictest.com
ATL is a full-service engineering support firm offering environmental services, subsurface investigations, geoprobe services, water-based investigations, geotechnical engineering, construction materials testing and engineering, specialinspection services, pavement engineering, nondestructive testing, and surveying from our ten offices. The firm currently has extensive capabilities in the areas of underground and aboveground storage tank testing and management and other relatedareas.

*Marijean Remington, President*
*Thomas Cronin, Vice President*

**1729 Atlas Environmental Engineering**
15701 Chemical Lane 714-890-7129
Huntington Beach, CA 92649 Fax: 714-890-7149
E-mail: info@aeei.com
http://www.aeei.com
Atlas Environmental Engineering provides very cost effective site assessments, investigations, corrective and remedial action plans and risk-based corrective action for low risk sites, along with groundwater monitoring, sampling, freeproduct removal activities and reporting. We also provide complete groundwater and soil remediation and cleanup activities, including all necessary equipment. Our goal is to provide clients with site closure in a minimal time period.
*Karl H Kerner, VP*

**1730 Ayres Associates**
3433 Oakwood Hills Parkway 715-834-3161
PO Box 1590 800-666-3103
Eau Claire, WI 54701 Fax: 715-831-7500
E-mail: LeithS@AyresAssociates.com
http://www.ayresassociates.com
Ayres Associates is a multi-specialty architectural/engineering consulting firm that has assisted public and private clients since 1959. Our staff of approximately 330 people provides services in transportation, civil, structural,water resources, levee and river engineering; planning; architecture; environmental science; energy corridors; surveying; geospatial services; and geographic information systems (GIS).

*Thomas Pulse, President*
*Sue Leith, PG, Manager, Marketing*

**1731 BBS Corporation**
1103 Schrock Road 614-888-3100
Suite 400 Fax: 614-888-0043
Columbus, OH 43229 E-mail: email@bbsengineers.com
http://www.bbsengineers.com
A full service multi-disciplinary engineering firm specializing in the planning, design and construction administration of water and wastewater treatment, distribution and collection systems. Other services include data conversion anddatabase design for geographical information systems projects.
*Edward Vance, Chairman*

**1732 BCI Engineers and Scientists**
2000 East Edgewood Drive, Suite 215 863-667-2345
Lakeland, FL 33803 Fax: 877-550-4224
E-mail: info@bcieng.com
http://www.bcieng.com
BCI was founded in the early 1970s by former MIT professor, Dr. L.G. Bromwell. At that time, the majority of our clients were in the phosphate and Geotechnical industry. Since its beginning, BCI has grown into a multidisciplinary firmwith over 100 employees. We are proud to offer our valued clients a diverse team of professionals with a unique and complimentary blend of expertise. Our unusual blend of experience allows us to develop solutions to complex engineering andenvironmental challenges.
*Richard M Powers, President*

**1733 BE and K/Terranext**
155 South Madison Street 303-399-6148
Suite 311 Fax: 303-399-6146
Denver, CO 80209 E-mail: kmartin@terranext.net
http://www.terranext/index.html
We have been nationally recognized for excellence in environmental services since 1985. The goal of management is to develop and execute appropriate solutions to complex issues and act as strong advocates for our clients.
*Kim Martin, President*

**1734 BHE Environmental**
11733 Chesterdale Road 513-326-1500
Cincinnati, OH 45246 Fax: 513-326-1550
http://www.bheenv.com
BHE's mission is to provide a full range of environmental consulting and remediation services that set the standard for quality and responsibility. We strive to serve the total needs of clients and to create a challenging andsupportive work environment for our employees.
*John M. Bruck, PE, President*

**1735 BRC Acoustics and Technology Consulting**
1741 First Avenue S 206-523-3350
Suite 401 800-843-4524
Seattle, WA 98105 Fax: 206-270-8690
E-mail: brc@brcacoustics.com
http://www.brcacoustics.com
A Seattle-based acoustical and technology consulting firm providing diverse services to public and private clients throughout

the United States. Services include architectural and mechanical acoustics, vibration measurement andanalysis, multimedia system design, noise monitoring, acoustical modeling, and noise contour mapping for environmental noise projects.
*Daniel C Bruck PhD, President*
*Roger Andrews, General Manager*

**1736 Bac-Ground**
3216 Georgetown
Houston, TX 77005
713-664-8452
Fax: 713-664-2629
E-mail: ebaca@bac-ground.com
http://www.bac-ground.com
Environmental consultant
*Ernesto Baca*

**1737 Badger Laboratories and Engineering Company**
501 West Bell Street
Neenah, WI 54956
920-729-1100
800-776-7196
Fax: 920-729-4945
E-mail: rlarson@badgerlabs.com
http://www.badgerlabs.com
Badger Laboratories and Engineering provides customers with analytical, engineering and technical services focusing on the environmental field.

*Richard Larson, President*
*Jeff Wagner, Laboratory Services*

**1738 Baltec Associates**
69 Fields Lane
Brewster, NY 10509
845-279-7448
Fax: 845-279-7467
E-mail: info@baltecusa.com
http://www.baltecusa.com
Baltec Associates is an international environmental consulting firm specializing in groundwater and soil remediation. We offers professional expertise and technical services for environmental management and planning, assessment,engineering, and remediation projects around the world.

**1739 Barco Enterprises**
11200 Pulaski Highway
White Marsh, MD 21162
800-832-7538
Fax: 410-335-0790
E-mail: barco.enterprises@verizon.net
http://www.barcoenterprises.com
Hazardous materials handling

**1740 Barer Engineering**
199 Main Street
Suite 600
Burlington, VT 05401
518-236-7070
800-878-2806
Fax: 518-236-5796
E-mail: info@barer.com
http://www.barer.com
Environmental engineering; pollution control

**1741 Baron Consulting Company**
273 Pepes Farm Road
Milford, CT 06460
203-874-5678
Fax: 203-874-7863
E-mail: analyze@baronconsulting.com
http://www.baronconsulting.com
Chemical, environmental and biological testing firm. Analytical also.

*Harry Agahigian, Technical Director*
*Barbara Obert, Lab Manager*

**1742 Barr Engineering Company**
450 South Wagner Road
Ann Arbor, MI 48103
734-922-4400
800-270-5017
Fax: 734-922-4401
E-mail: askbarr@barr.com
http://www.barr.com
Barr provides engineering, environmental, and information technology services to clients across the nation and around the world. We were incorporated as an employee-owned firm in 1966 and trace our orgins back to the early 1900s.Today, our more than 300 engineers, scientists, and technical support staff in Minnesota, Michigan, and Missouri work with clients in numerous industries, as well as at all levels of government.
*Karin Clemon, Contact*

**1743 Batta Environmental Associates**
6 Garfield Way
Newark, DE 19713
302-737-3376
800-543-4807
Fax: 302-737-5764
E-mail: bcbatta@battaenv.com
http://www.battaenv.com
BATTA was establihed in 1982 and is a Deleware Corporation registered to conduct work in the states of the Mid-Atlantic Region. BATTA has the in-house expertise in the scientific disciplines of geology, hydrogeology, civil andenvironmental engineering, chemistry, toxicology, health and safety, project design and construction management to adequately perform work without the use of outside consultants.
*Naresh C Batta*

**1744 Baxter and Woodman**
8678 Ridgefield Road
Crystal Lake, IL 60012
815-459-1260
Fax: 815-455-0450
E-mail: info@baxterwoodman.com
http://www.baxterwoodman.com
Municipal waste, water, transportation, control systems, and mapping services. Our mission statement is: we will be the leader in consulting engineering based on our reputation for trust, integrity, and client service.
*Darrel R Gavle, PE, DEE, President/CEO*
*Steve A Larson, PE, DEE, VP*

**1745 Baystate Environmental Consultants, Inc**
296 North Main Street
East Longmeadow, MA 01028
413-525-3822
Fax: 413-525-8348
E-mail: ccarranza@b-e-c.com
http://www.b-e-c.com
BEC offers a wide range of civil engineering, water resources and environmental expertise. BEC was incorporated in 1972 and specializes in lake and pond restoration services, environmental assessment under MEPA/NEPA and wetlandscience. The staff at BEC is exceptionally diverse, having had formal training and long-term experience in multi-disiplinary projects involving site development options, conceptual layout planning, enivronmental permitting and civil engineeringservices.
*Carlos Carranza, President*

**1746 Beak Consultants**
4600 Northgate Boulevard
Suite 215
Sacramento, CA 95834
916-565-7929
Fax: 916-565-7900
Beak provides a fully integrated approach to environmental planning, assessment and problem solving. Our professional and technical specialists include ecologists, environmental auditors, risk assessors, geochemists, ecotoxicologists,contaminant hydrogeologists, modellers and environmental engineers. We integrate these specialties to provide our clients with a broad range of services.

*Rick Swift*
*Amy Stuhr*

**1747 Beals and Thomas**
144 Turnpike Road
Route 9
Southborough, MA 01772
508-366-0560
Fax: 508-366-4391
E-mail: info@btiweb.com
http://www.btiweb.com
Beals and Thomas is a multidisciplinary consulting firm providing services to support the development and conservation of land and water resources throughout New England and the northeastern United States. Founded in 1984, BTI islocated in Southborough MA. Our mission is to advocate and assist in the attainment of our clients' project goals. We strive to provide creative and solution-oriented land eplanning and design services that are balanced with an environmental ethic.
*John E Thomas, President*
*John E Bensley, Principal*

**1748 Bear West Company**
145 S 400 E
Salt Lake City, UT 84111
801-355-8816
Fax: 801-355-2090
E-mail: bearwest@bearwest.com
Consultants on envionmental issues
*Ralph Becker, President*

**1749 Beaumont Environmental Systems**
108 Lintel Drive
McMurray, PA 15317
724-941-1743
Fax: 561-382-6455
http://www.besmp.com
Beaumont company provides Particulate and Gaseous Air Pollution Control Equipment, Systems and Services for Power Generation, Waste Incineration, Utility, Steel, Mining, Cement, Foundry and Pulp and Paper Industries. They design andfurnish systems that include Fabric Filters, Electrostatic Precipitators, Wet Scubbers, Semi Dry Scrubbers and Evaporative Cooloers along with the other necessary systems components.
*Will Goss, President*

**1750 Becher-Hoppe Associates**
330 Fourth Street
Wausau, WI 54403
715-845-8000
800-845-8009
Fax: 715-845-8008
E-mail: mailbox@becherhoppe.com
http://www.bhassoc.com
Becher-Hoppe Associates is a firm of consulting engineers, architects, scientists, real estate specialists and surveyors. We provide a spectrum of professional services to governmental, industry and the private sector for airport,highway, municipal, facilities maintainance, water/wastewater, solid waste and environmental projects. From our location in central Wisconsin, we provide upper Midwest clients with neighborly promptness and efficiency.
*Randy W Van Natta, PE, President*
*Phil Valitchka, Business Development*

**1751 Benchmark Environmental Consultants**
6116 North Central Expressway
Suite 808
Dallas, TX 75206
214-363-5996
Fax: 214-363-5994
E-mail: info@benchmarkenviro.com
http://www.benchmarkenviro.com/index2.htm
A progressive environmental consulting firm which specialized in solving environmental problems by using a practical business and technical approach.

**1752 Bendix Environmental Research**
1950 Addison Street
Suite 202
Berkeley, CA 94704
415-861-8484
Fax: 510-845-8484
http://home.earthlink.net/~bendix/
Specializes in toxicology, hazardous materials management, and preparation of environmental documents, or appropriate parts of environmental documents dealing with hazardous materials. Provides expert witness serrvices and litigationresearch and support for toxic tort cases, including workplace and environmental exposures and chemical cancer causation.

**1753 Beta Associates**
518 Fearrington Post
Pittsboro, NC 27312
919-545-0481
Fax: 919-545-0481
E-mail: BetaBob@BetaAssociates.com
http://www.betaassociates.com/
A professional project management organization which provides complete problem analysis, feasibility study and design services. These include specification and construction management for control and monitoring systems.

**1754 Better Management Corporation of Ohio**
41738 Esterly Drive
PO Box 130
Columbiana, OH 44408
330-482-9028
800-445-7887
Fax: 330-482-9242
E-mail: bmc@bmcohio.com
http://www.bmcohio.com
A focus of providing transportation and disposal services of baled, compacted and loose municipal solid waste as well as C&D material to both public and private waste transfer station companies located mainly along the East Coast fromMaine to Florida.
*Lee Stoneburner, President*
*Tim Waltemire, Vice President*

**1755 Beyaz and Patel**
800 South Broadway
Suite 200
Walnut Creek, CA 94596
925-934-0707
888-431-0707
Fax: 925-934-0318
E-mail: info@beyazpatel.com
http://www.beyazpatel.com
Structural engineering firm specializing in the structural design and construction management of public works infrastructure projects.
*Yogesh B Patel, President*
*Subhash Patel, VP*

**1756 Bhate Associates**
1608 13th Avenue South
Suite 300
Birmingham, AL 35205
205-918-4000
800-806-4001
Fax: 205-918-4050
E-mail: kgallant@bhate.com
http://www.bhate.com
Consulting environmental engineers
*Kathleen Gallant, CPA, Human Resources Director*

**1757 Bioengineering Group**
18 Commercial Street
Salem, MA 01970
978-740-0096
Fax: 978-740-0097
E-mail: mail@bioengineering.com
http://www.bioengineering.com
Provides a full range of consulting services in the field of bioengineering for erosion control, water quality, habitat restoration and stormwater management.

**1758 Bioenvironmental Associates**
3212 Wetterhorn Drive
Suite A
Fort Collins, CO 80525
Fax: 970-481-8386
http://www.toolcity.net/~richreen/Bio.htm
Management plans, environmental permitting and compliance monitoring. We specialize in biological inventories for threated and endangered species.
*Rex E Thomas PhD, Principal*

**1759 Biological Frontiers Institute**
PO Box 313
Sonoma, CA 95476
707-996-2863
http://www.zoogenetics.com
Genetics, preservation of endangered species
*Fred T Shultz, President*

**1760 Biological Monitoring**
1800 Kraft Drive
Suite 101
Blacksburg, VA 24060
540-953-2821
877-953-2821
Fax: 540-951-1481
E-mail: bmi@biomon.com
http://www.biomon.com/
Environmental consulting group

**1761 Biological Research Associates**
3910 US Highway 301 N
Suite 180
Tampa, FL 33619
813-664-4500
Fax: 813-664-0440
E-mail: callahan@BiologicalResearch.com
http://www.biolresearch.com/
BRA professionals act as proponents of our clients' interest, helping them through the maze of environmental regulations.
*Richard Callahan, CEO*
*J Steve Godley, President*

**1762 Biospec Products**
PO Box 788
Bartlesville, OK 74005
918-336-3363
800-617-3363
Fax: 918-336-6060
E-mail: info@biospec.com
http://www.biospec.com

Laboratory scientific equipment.

*Tim Hopkins, President*
*Jeff Anderson, Manager*

**1763 Bison Engineering**
1400 11th Avenue 406-442-5768
Helena, MT 59601 Fax: 406-449-6653
E-mail: hrobbins@bison-eng.com
http://www.bison-eng.com

Bison Engineering, Inc. is a full-service environmental consulting firm with extensive experience in air quality permitting, air emissions testing and ambient air monitoring. Our knowledge of the water quality regulatory environmentis substantial and expanding. We also provide an array of other environmental services through our staff and associates to complement our air and water quality expertise.

*Hal Robbins, President*

**1764 Bjaam Environmental**
455 Beverly Avenue 330-854-5300
PO Box 523 800-666-5331
Canal Futon, OH 44614 Fax: 330-854-5340
E-mail: info@realtimeboss.com
http://www.riskassessment.com/

Provide one source for reliable, affordable environmental consulting and contracting services, as well as industrial wastewater pre-treatment systems and service

**1765 Black and Veatch Engineers: Architects**
11401 Lamar Avenue 913-458-2000
Overland Park, KS 66211 Fax: 913-458-2934
E-mail: info@bv.com
http://www.bv.com

*Linda Heil, Communications Specialist*
*Corrine Smith, Vice President*

**1766 Blasland, Bouck and Lee**
6723 Towpath Road 315-446-9120
PO Box 66 Fax: 315-449-0017
Syracuse, NY 13214 E-mail: info@bbl-inc.com
http://www.bbl-inc.com/bblinc/

Hazardous waste, environmental compliance, air quality; engineering, solid waste, water, wastewater engineering
*Robert K Goldman, PE, CEO/Chairman*

**1767 Block Environmental Services**
2451 Estand Way 925-682-7200
Pleasant Hill, CA 94523 800-682-7255
Fax: 925-686-0399
E-mail: dblock@blockenviron.com
http://www.blockenviron.com

A environmental consulting firm specializing in indoor air quality and toxicology. We are a Certified Aquatic Bioassay Laboratory.

*Ronald Block, President*
*David Block, Vice President*

**1768 Boelter and Yates**
1300 Higgins Road 847-692-4700
Suite 301 Fax: 847-692-3127
Park Ridge, IL 60068 E-mail: info@boelter-yates.com
http://www.boelter-yates.com/

Provide environmental engineering, occupational health and safety management, design engineering, and consulting services.

*Fred Boelter, Chairperson*
*Thomas Kowalski, President/CEO*

**1769 Bollyky Associates Inc.**
31 Strawberry Hill Avenue 203-967-4223
Stamford, CT 06902 Fax: 203-967-4845
E-mail: ljbbai@bai-ozone.com
http://www.bai-ozone.com

Engineering firm specializing in Ozone technology, water and wastewater treatment, treatability studies.

*L Joseph Bollyky, President*
*Thomas Kleiber, Office Manager*

**1770 Bottom Line Consulting**
27248 Twin Pond Road 847-381-0597
Lake Barrington, IL 06010 Fax: 847-381-0598
Plastics recycling.
*John Fearncombe, President*

**1771 Braun Intertec Corporation**
11001 Hampshire Avenue South 952-995-2000
Minneapolis, MN 55438 800-279-6100
Fax: 952-995-2020
E-mail: info@braunintertec.com
http://www.braunintertec.com

An engineering firm providing consulting, management and testing services to clients in the commercial, industrial and residential real estate, institutional, retail, financial and government markets.
*George D Kluempke, PE, President/CEO*

**1772 Bregman and Company**
5272 River Road 301-652-4818
Suite 550 Fax: 301-652-4819
Bethesda, MD 20816 E-mail: bob@bregmanandcompany.com
http://www.bregmanandcompany.com

Environmental consulting firm.
*Robert Edell, President/CEO*

**1773 Brinkerhoff Environmental Services**
1913 Atlantic Avenue 732-223-2225
Suite R 5 800-246-7358
Manasquan, NJ 08736 Fax: 732-223-3666
E-mail: lbrinkerhoff@brinkenv.com
http://www.brinkenv.com

Groundwater remediation, environmental site assessments and sensitive area mapping; hazardous material management.
*Laura A Brinkerhoff, President/CEO*
*Eileen Della Volle, Consultant*

**1774 Brooks Laboratories**
9 Issac Street 203-853-9792
Norwalk, CT 06850 800-843-1631
Fax: 203-853-0273
E-mail: brooklabs@aol.com
http://www.brookslabs.com

Consulting and testing air, soil and water for contamination. Accident and disease prevention.
*Michael Zubarev, President*
*Kalonji Diyoka, VP*

**1775 Brown, Vence and Associates**
115 Sansome Street 415-434-0900
Suite 800 Fax: 415-956-6220
San Francisco, CA 94104 http://www.brownvence.com
Waste management energy consulting firm.

**1776 Buck, Seifert and Jost**
65 Oak Street 201-767-3111
PO Box 415 Fax: 201-767-3178
Norwood, NJ 07648 E-mail: bsjinc@bsjinc.com
http://www.bsjinc.com

Consultancy for the water and wastewater industries.
*Ronald von Autenried, PE, President*
*Guido von Autenried, PE, Director/Chief Engineer*

**1777 Burk-Kleinpeter**
4176 Canal Street 504-486-5901
New Orleans, LA 70119 Fax: 504-488-1714
E-mail: mjackson@bkiusa.com
http://www.bkiusa.com

A full service firm bringing together resources from our Engineering, Architecture, Planning and Environmental Science Divisions. Our Divisions, which may function independently also

work as a team, providing our clients withassistance from the first conceptual idea through final construction. We also provide services through our professional support groups which include landscape architecture, construction management and inspection, graphic design, aerial photographyand marketing.
*Michael G Jackson, PE, Executive VP*

**1778 Burns and McDonnell**
9400 Ward Parkway 816-333-9400
Kansas City, MO 64114 Fax: 816-333-3690
E-mail: busdev@burnsmcd.com
http://www.burnsmcd.com/index.html
A multidisciplinary engineering, architectural, construction and environmental service firm. More than 2,900 engineers, architects, scientists and other specialists plan, design and build quality projects around the world.

**1779 C&H Environmental**
224 Stiger Street 908-852-4855
PO Box 188 Fax: 908-852-5275
Hackettstown, NJ 07840 http://www.candhenvironmental.com
Environmental consulting firm.

*John H Crow, PhD*
*Timir B Hore, PhD*

**1780 CA Rich**
17 DuPont Street 516-576-8844
Plainview, NY 11803 Fax: 516-576-0093
E-mail: info@carichinc.com
http://www.carichinc.com
An independently owned, private consulting firm providing targeted, solution oriented hydrogeologic and environmental engineering services. Assists in the conception, development, design, implementation, documentation and defense ofsite evaluations and remedial action.

*Charles A Rich, Founder/President*
*Richard J Izzo, Human Resources Director*

**1781 CBA Environmental Services**
57 Park Lane 570-682-8742
Hegins, PA 17938 Fax: 570-682-8915
E-mail: info@cbaenvironmental.com
http://www.cbaenvironmental.com
Provides environmental solutions from general plant maintenance and cleaning to large-scale soil remediation projects.
*Bruce L Bruso, Principal*

**1782 CDS Laboratories**
75 Suttle Street 303-247-4220
PO Box 2605 800-553-6266
Durango, CO 81302 Fax: 303-247-4227
Specializes in analytical analysis and testing, consulting, QA, environmental, and dyes.

**1783 CEDA**
3519 Old Red Trail 701-663-0307
PO Box 787 Fax: 701-667-2090
Mandan, ND 58554
Provides services in hazardous waste, spill response, and asbestos abatement.

*WF Mowatt, President*

**1784 CIH Environmental**
1044 Victory Circle 610-372-6692
Reading, PA 19605 Fax: 610-372-0862
E-mail: cihenv@fast.net
http://www.cihenv.com
Provides services in indoor air quality and industrial hygiene and mold/bacterial contaminations.
*James E Detwiler, President*

**1785 CIH Services**
7148 Creekside Lane 317-797-7768
Indianapolis, IN 46250 Fax: 317-913-1895
E-mail: cihservices@juno.com
http://www.cih-services.com
Services in indoor air quality.
*John Beltz*

**1786 CII Engineered Systems**
6767 Forrest Hill Avenue 804-320-1405
Richmond, VA 23225 800-768-2545
Fax: 804-320-9625
Services in energy conservation.
*Jack Thacker*

**1787 CK Environmental**
1020 Turnpike Street 781-828-5200
#8 888-253-0303
Canton, MA 02021 Fax: 781-828-5380
E-mail: info@ckenvironmental.com
http://www.eco-web.com
Serivces in regulatory compliance.

**1788 CRB Geological and Environmental Services**
4573 Ponce De Leon Boulevard 305-447-9777
Coral Gables, FL 33146 Fax: 305-567-2853
E-mail: blivieri@crbgeo.net
http://www.crbgeo.net/
Environmental consulting.
*Frederick R Baddour, President*

**1789 CTE Engineers**
303 East Wacker Drive 312-938-0300
Suite 600 Fax: 312-938-1109
Chicago, IL 60601 E-mail: tony.bouchard@cte.aecom.com
http://www.cte-eng.com
Consulting services in the use of computers and electronics to solve pollution problems.

*Tony Bouchard, Environmental Services*
*Carl Mahr, Environmental Services*

**1790 CTI and Associates, Inc**
12482 Emerson Drive 248-486-5100
Brighton, MI 48116 800-284-8633
Fax: 248-486-5050
http://www.cticompanies.com
Environmental engineering firm.

*Matthew Jerue, Environmental Services*

**1791 CTL Environmental Services**
24404 South Vermont Avenue
Suite 307 800-777-0605
Harbor City, CA 90710 Fax: 310-530-0792
E-mail: info@ctles.com
http://www.ctles.com
Industrial hygiene and safety, asbestos/lead based paint surveys, environmental site assessments, risk mangement, indoor air quality, radon testing and mold investigation.

**1792 CZR**
2151 Alternate A1A South 561-747-7455
Suite 2000 Fax: 561-747-7576
Jupiter, FL 33477 E-mail: czrinc@czr-inc.com
http://www.czr-inc.com/czrwilm
Environmental impact studies, wetlands delineation, threatened species surveys, environmental resource permitting.

*Samuel E Wiley, Vice President*

**1793 Cabe Associates**
144 South Governors Avenue 302-674-9280
PO Box 877 800-542-7979
Dover, DE 19904 Fax: 302-674-1099
E-mail: jpj@cabe.com
http://www.cabe.com

Environmental and pollution control.
*Robert Kerr, Contact*

**1794 California Environmental**
1001 Street
P.O. Box 2815
Sacramento, CA 95812
818-991-1542
Fax: 818-991-0793
Industrial safety and hygiene.
*Michael R Tiffany, CIH*

**1795 California Geo-Systems**
1545 Victory Boulevard
2nd Floor
Glendale, CA 91201
818-500-9533
Fax: 818-500-0134
E-mail: geosys@pacebell.net
http://www.geosys1.com
Geotechnical environmental services.
*Vince Carnegie, President*
*Rachel Fischer, Sr. Environmental Geologist*

**1796 Cambridge Environmental**
58 Charles Street
Cambridge, MA 02141
617-225-0810
Fax: 607-225-0813
E-mail: info@cambridgeenvironmental.com
http://www.cambridgeenvironmental.com
Consulting and research firm that assesses and helps to minimize risks to health and the environment.
*Laura Green, Senior Scientist*
*Michael Ames, Senior Engineer*

**1797 Camiros Limited**
411 South Wells
Suite 400
Chicago, IL 60607
312-922-9211
Fax: 312-922-9689
E-mail: sbland@camiros.com
http://www.camiros.com
Camiros is an active proponent of Sustainable Growth as well as other environmentally sensitive aspects of urban planning and design. In particular, the firm has drafted land use plans and zoning ordinances that pay careful attentionto environmental issues.

*Shirelle Brand, Administrative*

**1798 Camo Pollution Control**
1610 State Route 376
Wappingers Falls, NY 12590
845-463-7310
Environmental consultants.
*Michael Tremper, Vice President*

**1799 Camtech**
4550 McKnight Road
Suite 210
Pittsburg, PA 15237
412-931-1210
Fax: 412-931-1304
Environmental consulting.

**1800 Canin Associates**
500 Delaney Avenue
Suite 404
Orlando, FL 32801
407-422-4040
Fax: 407-425-7427
E-mail: trhenderson@canin.com
http://www.canin.com

Environmental services.
*Elizabeth Doran, Business Development*
*Myrna Canin, VP*

**1801 Cape Environmental Management**
5545 W. Raymond St.
Suite E
Atlanta, GA 30345
770-908-7200
800-488-4372
Fax: 770-908-7219
http://www.capeenv.com

Environmental consulting firm.
*Fernando Rios, President*

**1802 Capital Environmental Enterprises**
2244 Profit Drive
Indianapolis, IN 46241
317-240-8085
888-376-4315
Fax: 317-241-4180
http://www.capitalenvironmental.net

Environmental consulting.

**1803 Cardinal Environmental**
3303 Paine Avenue
Sheboygan, WI 53081
920-459-2500
800-413-7225
Fax: 920-459-2503
E-mail: info@cardinalenvironmental.com
http://www.cardinalenvironmental.com
Environmental consulting/analytical laboratory.

*Scott A Hanson, President*

**1804 Carpenter Environmental Associates**
PO Box 656
Monroe, NY 10949
845-781-4844
Fax: 845-782-5591
E-mail: b.bell@cea-enviro.com
http://www.ceaenviro.com/
Environmental engineering and assessment services, including wastewater and storm water management, wetlands and ecological investigations, site assessments, environmental compliance and contingency planning, permitting services, andlitigation support.

*Bruce A Bell, PhD, PE, DEE, President*

**1805 Carr Research Laboratory**
17 Waban Street
Wellesley, MA 02482
508-651-7027
Fax: 508-647-4737
E-mail: cann@carr-research.lab.com
http://www.carr-research.lab.com
Environmental consulting research laboratory providing these services: hydrology, wetlands, lakes, oceanography, water pollution, geology, and environmental engineering.

*Jerome B Carr, PhD, President*
*Deseng Wang, PhD, PE, Dir Engineering Services*

**1806 Catlin Engineers and Scientists**
220 Old Dairy Road
PO Box 10279
Wilmington, NC 28404
910-452-5861
Fax: 910-452-7563
E-mail: info@catlinusa.com
http://www.catlinusa.com
Specializes in providing quality service in the fields of environmental, civil, and geotechnical engineering. Services include soil and ground water remediation, wastewater treatment system design, public infrastructure,environmentally secure landfills, and safe, clean, water supplies.
*Richard Catlin, President*

**1807 Center for Energy and Environmental Analysis Oak Ridge Laboratory**
Energy and Environmental Analysis
PO Box 2008
Oak Ridge, TN 37831
865-576-4160
Fax: 865-574-8884
E-mail: sdb@ornl.gov
http://www.ornl.gov/ceea/
In the Center for Energy and Environmental Analysis, a part of the Energy Division at Oak Ridge National Laboratory, we provide our customers with analysis of energy and environmental issues of local, regional, national, and globalimportance so as to provide decision makers with information on which to base major policy, program, and project decisions.
*Dr. Michael O Lerner, President*

**1808 Central States Environmental Services**
1079 Copple Road
Centralia, IL 62801
618-532-4784
Fax: 618-532-5615
Environmental clean-up contractor.
*Elvin Copple, President*

**1809 Challenge Environmental Systems Inc**
2270 Worth Lane
Suite D
Springdale, AR 72764
479-927-1008
Fax: 479-927-1000
E-mail: kent@challenge-sys.com
http://www.challenge-sys.com
Challenge Environmental Systems Inc manufactures environmental equipment for respiration monitoring on wastewater processes, soil remediation, and composting.
*Mark L Kuss, Founder*

**1810 Chapman Environmental Control**
PO Box 288
Osceola, IN 46561 800-675-8706
Air pollution control.
*Frank X Chapman, President*

**1811 Chelsea Group**
89 Awawa Rd
PO Box 68 800-626-6722
Maunaloa, HI 96770 Fax: 630-729-3189
E-mail: info@chelsea-grp.com
http://www.chelsea-grp.com
Specializes in strategic, technical, and marketing consulting to major corporations for enhanced positioning of products and services relating to the indoor environment.
*George Benda, Senior Princpial*
*David Munn, PE, Principal*

**1812 Chemical Data Management Systems**
6515 Trinity Court 925-551-7300
Suite 201 800-735-1761
Dublin, CA 94568 Fax: 925-829-3886
E-mail: info@cdms.com
http://www.cdms.com
Provides a full range of hazardous material and OSHA regulatory compliance services to industries using hazardous materials, including implementing compliance programs, submitting necessary reports to all regulatory agencies, andproviding a full range of training services to industry, agencies, and industrial groups.

**1813 Chicago Chem Consultants Corporation**
14 North Peoria Street 312-226-2436
Suite 2C Fax: 312-226-8886
Chicago, IL 60607 E-mail: info@chichem.com
http://www.chichem.com
Provides innovative, technologically sophisticated, cost-effective, and risk protective environmental and engineering services.
*Jeffrey P Perl, President*
*Stanley Yoslov, Senior Associate*

**1814 Cigna Loss Control Services**
Oklahoma City, OK 73150 405-524-2127
http://www.cigna.com
Provides services in the field of industrial hygiene.
*H Edward Hanway, Chairman/CEO*
*Michael W Bell, Executive VP/CFO*

**1815 Clayton Group Services**
45525 Grand River Avenue 248-344-2661
Suite 200 Fax: 248-344-2656
Novi, MI 48374 E-mail: info@claytongrp.com
http://www.claytongrp.com
A full service environmental, occupational health and safety, and laboratory services consulting firm serving both public and private clients.
*Lisa Barnes, PE CIH, Sr VP/COO*
*MJ Haught, Marketing Director*

**1816 Clean Air Engineering**
500 West Wood Street 847-991-3300
Palatine, IL 60067 800-627-0033
Fax: 847-991-3385
E-mail: contact@cleanair.com
http://www.cleanair.com
Environmentally consulting and permitting, process engineering, equipment rental and manufacture, measurement and analytical services.

*Allen Kephart, Vice President*
*Jim Pollack, Director of Sales*

**1817 Clean Environments**
10803 Gulfdale 210-349-7242
Suite 210 Fax: 210-349-1132
San Antonio, TX 78216 E-mail: cei@cleanenvironments.com
http://www.cleanenvironments.com
Environmental consulting firm.

**1818 Clean Technologies**
2700 Kirkwood Hwy 302-999-0924
Newark, DE 19711 Fax: 302-999-0925
Environmental engineering firm.
*Deborah A Buniski, President*

**1819 Clean World Engineering**
1737 S Naperville Road 630-260-0200
Suite 200 800-761-9603
Wheaton, IL 60187 Fax: 630-260-0797
E-mail: cwe@clean-world.com
http://www.clean-world.com
A woman-owned environmental engineering firm established in 1985. They specialize in meeting the needs of small businesses to FORTUNE 500 companies to large government agencies. They develop practical and cost-efficient soulutions toincreasingly stringent and complex environmental reguations.
*Rita Kapur, President/CEO*

**1820 Coastal Lawyer**
173 E Blithedale Avenue 415-383-3715
Suite 3 Fax: 415-383-3718
Mill Valley, CA 94941 E-mail: del@greendogcampaigns.com
Consultation for environmental causes and initiatives. Legal representation, public relations and campaign consulting.
*Dotty E LeMieux, Principal*

**1821 Coastal Planning and Engineering**
2481 Northwest Boca Raton Boulevard 561-391-8102
Boca Raton, FL 33431 Fax: 561-391-9116
E-mail: mail@coastalplanning.net
http://www.coastalplanning.net
Environmental consulting firm providing services in coastal engineering, coastal planning, coastal surveying, environmental science, and regulatory permitting.

**1822 Cohen Group**
3 Waters Park Drive 650-349-9737
Suite 226 Fax: 650-349-3378
San Mateo, CA 94403 E-mail: jcohen@thecohengroup.com
http://www.thecohengroup.com
Provides a complete range of environmental health and safety services to business and government including indoor air quality, asbestos, respiratory protection, and industrial hygiene safety.
*Joel M Cohen, President*

**1823 Cohrssen Environmental**
3450 Sacremento St 415-775-1105
San Francisco, CA 94123
Industrial hygiene services.
*Barbara Cohrssen, Principal Executive*

**1824 Columbia Analytical Services**
1317 South 13th Avenue 360-577-7222
Kelso, WA 98626 800-695-7222
Fax: 360-425-9096
http://www.caslab.com
Areas of expertise and services include environmental testing of air, water, soil, hazardous waste, sediments and tissues; process and quality control testing; analytical method development; sampling and mobile laboratory services; andconsulting and data management services.

*Stephen W Vincent, President/CEO*
*Ed Wilson, Employee Representative*

**1825 Combustion Unlimited**
PO Box 8856 215-537-0871
Philadelphia, PA 19117 Fax: 215-884-3074
Engineering consulting services in air pollution control and combustion.

*John F Straitz III, President*

**1826 Committee for Environmentally Effective Packaging**
601 13th Street NW 202-783-5594
Suite 900S Fax: 203-783-5595
Washington, DC 20005 http://www.epa.gov/epaoswer
Monitors legislation and regulations affecting packaging in the food service industry and educates decision makers on packaging.

**1827 Commonwealth Engineering and Technology CET Engineering Services**
1240 North Mountain Road 717-541-0622
Harrisburg, PA 17112 Fax: 717-541-8004
E-mail: contact@cet-inc.com
http://www.cetinc.com
Pollution control utilizing the ability of Geographic Informaiton Systems as an analytical tool to be used to analyze water and wastewater systems.

*Jeffrey Wendle, President*

**1828 Community Conservation Consultants Howlers Forever**
50542 One Quiet Lane 608-735-4717
Gays Mills, WI 54631 Fax: 512-519-8494
E-mail: communityconservation@mwt.net
http://www.communityconservation.org
Works together with local rural people to aid in the protection of their wildlife and forests. Projects undertaken have mainly been in India, Belize and Wisconsin with an emphasis on primates and other species. Additional projects areevolving in Madagascar, Costa Rica and other countries.

*Robert Horwich, Director*

**1829 Compass Environmental**
1751 McCollum Parkway 770-499-7127
Kennesaw, GA 30144 Fax: 770-423-7402
E-mail: staff@compassenv.com
http://www.compassenv.com
Asbestos indoor air quality and industrial hygiene services. Established as an alternative to large companuies that are typically structured to provide routine testing and consulting services.
*Eva M Ewing, Vice President*
*William M Ewing, President*

**1830 Comprehensive Environmental**
21 Depot Street 603-424-8444
Merrimack, NH 03054 800-725-2550
Fax: 800-331-0892
E-mail: webmaster@ceiengineers.com
http://www.ceiengineers.com
Provides water, wastewater and hazardous waste services with a mission to protect the client, public health and the environment, to be client advocates, to build trust, and to provide objectivity.
*Eileen Pannetier, President*

**1831 Comprehensive Environmental Strategies**
11950 Rocky Brook Court 703-791-7700
Manassas, VA 20112 Fax: 703-368-6821
Asbestos, indoor air quality, industriel hygiene services.
*Reginald B Simmons, Principal Executive*

**1832 Conestoga-Rovers and Associates**
2055 Niagra Falls Boulevard 716-297-6150
Suite 3 Fax: 716-297-2265
Niagra Falls, NY 14304 E-mail: info@cra.com
http://www.craworld.com
Family of companies that provide a full-service engineering, environmental, construction and information technology services worldwide.

*Frank A Rovers, Principal Executive*

**1833 Conservtech Group**
5885 Rickenbacker Road 323-867-9044
Commerce, CA 90040 Fax: 323-867-9045
E-mail: bob@conservtechgroup.com
http://www.conservtechgroup.com
Pollution control systems, site assessment, waste problems, water and waste systems.

*Robert J MacDonald, President/CEO*

**1834 Consoer Townsend Envirodyne Engineers**
303 East Wacker Drive 312-938-0300
Suite 600 Fax: 312-938-1109
Chicago, IL 60601 E-mail: tony.bouchard@cte.aecom.com
http://www.cte-eng.com
Environmental engineering, hazardous and toxic waste, environmental and energy products, cost-effective soulutions to modern, urban infrastructure problems.

*Tony Bouchard, Environmental Services*
*Carl Mahr, Environmental Services*

**1835 Consultox**
PO Box 51210 504-529-7500
New Orleans, LA 70125 Fax: 504-926-0638
E-mail: info@consultox.com
http://www.consultox.com
Toxicology consulting firm.

*Richard A Parent, President*

**1836 Continental Shelf Associates**
759 Parkway Street 561-746-7946
Jupiter, FL 33477 Fax: 561-747-2954
E-mail: csa@gate.net
Environmental consulting firm.
*Dr David A Gettleson, President*
*Robert C Stevens Jr, CEO*

**1837 Converse Consultants**
222 East Huntington Drive 626-930-1200
Suite 211 Fax: 626-930-1212
Monrovia, CA 91016 E-mail: converse@converseconsultants.com
http://www.converseconsultants.com
Environmental consulting and engineering firm with offices throughout the United States.

**1838 Cook Flatt and Strobel Engineers**
2930 SW Woodside Drive 785-272-4706
Topeka, KS 66614 Fax: 785-272-4736
E-mail: cfsengr@cfse.com
http://www.cfse.com
Environmental engineering.

**1839 Cornerstone Environmental, Health and Safety**
880 Lennox Court 317-733-2637
Zionsville, IN 46077 800-285-2568
Fax: 317-577-2481
E-mail: info@corner-enviro.com
http://www.corner-enviro.com
Environmental, health, and safety services.
*Jill Para, Contact*

**1840 Corporate Environmental Advisors**
127 Hartwell Street 508-835-8822
West Boylston, MA 01583 Fax: 508-835-8812
Environmental engineering and consulting firm.

**1841 Cox Environmental Engineering**
82 Dresser Hill Road 508-248-5185
Charlton, MA 01507 Fax: 508-248-5003
Environmental engineering consulting.

**1842 Crouse & Company**
912 Greengate North Plaza
Greensburgh, PA 15601
724-832-3114
Fax: 724-832-3627
http://www.crouse.com
Natural resources management company dedicated to ushering traditional engineering and natural sciences consulting services into the age of modern information technology.
*Jeffery P Evers, Director*

**1843 Cultural Resource Consultants International Archaeology & Ecology**
7400 Jones Drive
Suite 313
Galveston, TX 77551
832-592-9549
Fax: 713-468-4263
E-mail: postmaster@culturalresource.com
http://www.culturalresource.com
Historic land use and environmentally sensitive projects.
*Robert P d'Aigle, Owner*
*Nataliya Hryshechko, Laboratory Services Director*

**1844 Curt B Beck Consulting**
408 W Kingsmill Street
PO Box 2442
Pampa, TX 79065
806-665-9281
Fax: 806-665-1965
E-mail: curtbbeck@cableone.net
Pollution control services.
*Curt B Beck, Owner*

**1845 Custom Environmental Services**
233 Forest Drive
Santa Barbara, CA 93117
805-968-2112
Fax: 805-968-2137
E-mail: info@Custom-env.com
http://www.custom-env.com
Small environmental consulting business offering services that help individuals and companies of all sizes comply with environmental laws, including the Clean Air Act, Clean Water Act, and Medical Waste Management Act.
*Rosalie A Skefich, Founder*

**1846 D'Appolonia**
275 Center Road
Monroeville, PA 15146
412-856-9440
Fax: 412-856-9535
E-mail: info@dappolonia.com
http://www.dappolonia.com
Provides engineering, scientific and construction management services for projects involving large civil works and special earth/structure interaction issues.

**1847 D/E3**
18234 S Miles Road
Suite 44
Cleveland, OH 44128
216-663-1500
Fax: 216-663-1501
Environmental impact statements, corrective process, pollution abatement, hazardous waste, radon and asbestos hazards.
*Harold N Danto, President*

**1848 DPRA**
200 Research Drive
Manhattan, KS 66503
785-539-3565
Fax: 785-539-5353
E-mail: info@dpra.com
http://www.dpra.com
Environmental, economic, regulatory and technical research company. Research results published by information services.
*Richard Seltzer, President*

**1849 DW Ryckman and Associates: REACT Environmental Engineers**
1120 South 6th Street
St. Louis, MO 63104
314-678-1398
800-325-1398
Fax: 314-678-6610
E-mail: stewart-ryckman@react-env.com
http://www.react-env.com
D.W. Ryckman & Associates, Inc dba REACT Environmental Engineers, was founded in 1975 to provide rapid response and remediation services for environmental and hazardous contamination problems.
*SE Ryckman, President*

**1850 Datanet Engineering**
11416 Reisterstown Road
Owings Mills, MD 21117
410-654-1800
888-896-7133
Fax: 410-654-3711
E-mail: info@datanetengineering.com
http://www.datanetengineering.com
Fuel systems repairs maintenance and installation are also provided
*John V Cignatta PhD PE, President*

**1851 DeVany Industrial Consultants**
14507 NW 19th Avenue
Vancouver, WA 98685
360-546-0999
Fax: 360-546-0777
E-mail: mdevany@earthlink.net
Strive to provide a full range of safety and industrial hygiene services customized for your particular environment.

**1852 Dennis Breedlove and Associates**
330 West Canton Avenue
Winter Park, FL 32789
407-677-1882
800-304-1882
Fax: 407-657-7008
Environmental and natural resources consulting firm.

**1853 Detail Associates: Environmental Engineering Consultants**
300 Grand Avenue
Englewood, NJ 07631
201-569-6708
Fax: 201-569-4378
http://www.daienviro.com
Asbestos management programs. Indoor air quality. Analytical services. Lead surveys. Phase I, II and III environmental audits.

**1854 Donald Friedlander**
1091 Willowbrook Road
Staten Island, NY 10314
718-698-7545

**1855 Dunn Corporation**
151 Seventh Avenue
2nd Floor
Brooklyn, NY 11215
718-388-9407
Fax: 718-388-0638
E-mail: info@dunndev.com
http://www.dunndev.com

**1856 ENSR Consulting and Engineering**
2 Technology Park Drive
Westford, MA 01886
978-589-3000
800-722-2440
Fax: 978-589-3100
http://www.ensr.com
Provides consulting, engineering, remediation, and related services to industrial and commercial companies, municipalities, and regulated government agencies throughout the United States, Europe, Latina America, and Asia.
*Robert C Weber, President/CEO*

**1857 ENTRIX**
5252 Westchester
Suite 250
Houston, TX 77005
800-368-7511
Fax: 713-666-5227
E-mail: webmaster@entrix.com
http://www.entrix.com
Provides environmental and natural resource management consulting
*Todd Williams, President*
*Richard Firth, Executive VP*

**1858 ENVIRON Corporation**
4350 N Fairfax Drive
Arlington, VA 22203
703-516-2300
Fax: 703-516-2345
Health and environmental sciences consultants.
*Grover Wrenn, President*
*Cindy Holloman, Contact*

**1859 ESS Group**
401 Wampanoag Trail
Suite 400
East Providence, RI 02915
401-434-5560
Fax: 401-434-8158
E-mail: questions@essgroup.com
http://www.essgroup.com

Environmental Science Services is an multi-disciplinary environmental consulting and engineering firm with offices located in Wellesley, Massachusetts and Providence, Rhode Island. ESS was established in 1979, and has experiencedsteady growth and market diversification to become a recognized leader in the environmental consulting and engineering services business.

*Charles J Natale, Jr., President/CEO*
*Christopher G Rein, Sr VP*

**1860 ETS**
1401 Municipal Road Northwest
Roanoke, VA 24012
540-265-0004
Fax: 540-265-0131
E-mail: jmck@esti-inc.com
http://www.esti-inc.com

ETS is a full-service environmental consulting and training firm specializing in air emissions control, measurement, engineering and consulting services.
*John McKenna, Contact*
*Jack Mycock, Contact*

**1861 Earth Science Associates**
4300 Long Beach Blvd
Suite 310
Long Beach, CA 90802
562-428-3181
Fax: 562-428-3186
E-mail: contactESA@earthsci.com
http://www.earthsci.com

Earth Science Associates is a consultancy serving the international oil and gas industry. ESA specializes in resource assessment, economic evaluation and risk studies and the development of custom geographic information systems. Oiland gas companies use our assessment studies in evaluating the geologic potential of prospects, plays and basins.

**1862 Earth Science Associates (ESA Consultants)**
PO Box 12067
Knoxville, TN 37912
800-467-6380
E-mail: esa@halos.com
http://www.halos.com

**1863 EcoLogic Systems**
7977 Capwell Drive
Suite 150
Oakland, CA 94621
510-635-7400
800-223-0609
Fax: 510-634-7402
E-mail: gjames@accenv.com
http://www.ecologicsystems.com

The developer of leading suite of hazardous material management and environmental health and safety compliance software.
*Geoff James, Business Development*

**1864 Ecology and Environment**
368 Pleasant View Drive
Lancaster, NY 14086
716-684-8060
Fax: 716-684-0844
E-mail: ckarpowicz@ene.com
http://www.ene.com

Ecology and Environment is a multidisciplinary environmental science and engineering company with more than 25 offices in the US and offices and partners in more than 35 countries. We are a world leader in providing environmentalconsulting services and litigation support.
*Cheryl Karpowicz, VP*
*Janet Steinbruckner, Director of Human Resources*

**1865 Ed Caicedo Engineers & Consultants**
PO Box 22256
Lexington, KY 40522
859-259-0042
E-mail: info@eciengineers.com
http://www.eciengineers.com

We are a Kentucky-based Consulting Engineering firm which provides professional and technical services to the public and private sectors. These services cover all facets of the construction process, including the preparation offeasibility studies, conceptual designs, environmental impact studies, cost-effectiveness analysis, final project design, construction administration, operation and maintenance management.
*Eduardo Caicedo, President*
*William H Meadows, Principal Engineer*

**1866 Elinor Schwartz**
318 South Abingdon Street
Arlington, VA 22204
703-920-5389
Fax: 703-920-5402
E-mail: es@elinorschwartz.com

Representing state agencies and providing research on natural resources, energy and environmental issues.
*Elinor Schwartz, Washington Representative*

**1867 EnSafe**
5724 Summer Trees Drive
Memphis, TN 38134
901-372-7962
800-588-7962
Fax: 901-372-2454
http://www.ensafe.com

Services include implementation of occupational health and safety programs, worker's compensation programs, and environmental management systems for clients ranging from large corporations to small businesses to decrease accidentfrequency and accident costs.
*Phillip G Coop, President/CEO*
*Michael Wood, Vice President/CFO*

**1868 Energy Technology Consultants**
2020 E 1st Street
Santa Ana, CA 92705
714-835-6886
Fax: 714-667-7147

**1869 EnviroTest Laboratories**
315 Fillerton Avenue
Newbrugh, NY 12550
845-562-0890
Fax: 845-562-0841
E-mail: info@envirotestlab.com

Test soil and water.
*Scott Morris, President*

**1870 Envirocorp**
7020 Portwest Drive
Suite 100
Houston, TX 77024
713-880-4640
800-535-4105
Fax: 713-880-3248
E-mail: pfh@subsurfacegroup.com
http://www.envirocorpinc.com

Envirocorp, Inc. is a consultant firm with over 25 years of experoence around the world involving site assessments and remediations. Envirocorp's expertise includes UST removals and remediations, hazardous waste reporting,environmental audits, oil & gas property assessments, and asbestos inspection. Envirocorp has offices in South Bend, Indiana; Baton Rouge, Louisiana and its's corporate office in Houston, Texas.

**1871 Environmental Compliance Consulting**
PO Box 11417
Green Bay, WI 54307
920-434-6380
888-ECC-INOW
Environmental assessments.

**1872 Environmental Consultants**
391 Newman Avenue
Clarksville, IN 47129
812-282-8481
Fax: 812-282-8554
Environmental consulting firm.
*Robert E Fuchs, President*

**1873 Environmental Resource Associates**
6000 West 54th Avenue
Arvada, CO 80002
303-431-8454
800-372-0122
Fax: 303-421-0159
E-mail: info@eraqc.com
http://www.eraqc.com

**1874 Environmental Resources Management**
3352 128th Avenue
Holland, MI 49424
616-399-3500
Fax: 616-399-3777
http://www.erm.com

*Robin Bidwell, Executive Chair*

**1875 Environmental Risk Limited**
120 Mountain Avenue 860-242-9933
Bloomfield, CT 06002 Fax: 860-243-9055
E-mail: info@erl.com
http://www.erl.com
Environmental consulting and engineering firm offers environmental permitting and compliance assistance, site investigation and remediation services, air quality impact analyses, pollution prevention planning, aquatic toxicitylaboratory, hazardous waste management and chemical accident prevention program assistance.
*Mitchell M. Wurmbrand, C.C.M., Principal*

**1876 Environmental Risk Management**
3109 N McColl Road 956-686-6569
PO Box 3213 800-880-9582
McAllen, TX 78502 Fax: 956-668-7227
E-mail: office@enrisk.com
http://www.enrisk.com
Provides experienced environmental consulting to South Texas. Services includes Phase I, II and III environmental site assessments, leaking petroleum storage tank assessments and project management, asbestos inspections and managementplans, remedial services and non process waste management.
*Mark Barron, President*

**1877 Environmental Science Associates**
225 Bush Street 415-896-5900
Suite 1700 Fax: 415-896-0332
San Francisco, CA 94104 E-mail: mabell@esassoc.com
http://www.esassoc.com
Environmental Science Associates is an environmental consulting firm committed to helping clients meet the environmental challenges of tommorrow today.
*Marty Abell, VP*

**1878 Environmental Strategies Corporation**
11911 Freedom Drive 703-709-6500
Suite 900 Fax: 703-709-8505
Reston, VA 20190 E-mail: jchizzonite@escva.com
http://www.environmental-strategies.com
ECS is a complete environmental consulting, management, and engineering firm specializing in identifying potential or actual environmental liabilities and preventing or remediating them.
*John Simon, Executive Partner*
*Jan Chizzonite, Managing Executive Partner*

**1879 Environmental Testing and Consulting**
2790 Whitten Road 901-213-2400
Memphis, TN 38133 Fax: 901-213-2440
E-mail: nathan.pera@etcmemphis.com
http://www.etcmemphis.com
Environmental laboratory
*Nathan Pera, President*

**1880 Enviroplan Consulting**
81 Two Bridges Road 973-575-2555
Fairfield, NJ 07004 Fax: 973-575-6617
E-mail: contact@enviroplan.com
http://www.enviroplan.com/
An air pollution company with 14 offices throughout the United States. Specialize in three areas: air pollution consulting including greenhouse gas emmissions, inventory development and mitigation; ambient air quality and meterologicalmonitoring programs; and wind resource analyses
*Howard Ellis, Contact*

**1881 Epcon Industrial Systems NV, Ltd**
17777 I-45 South 936-273-3300
Conroe, TX 77385 800-447-7872
Fax: 936-273-4600
E-mail: epcon@epconinc.com
http://www.epconind.com
Provides a broad line of technology advanced, yet user friendly air pollution control products, finishing systems and heat processing equipment.

**1882 Foothill Engineering Consultants**
18590 Hwy 49 303-278-0622
Plymouth, CA 95669 Fax: 303-278-0624
E-mail: sales@foothillmc.com
http://www.foothillmc.com
Offers services in environmental studies and design, decontamination and decommissioning support, mining services, radiological engineering, waste management, civil engineering, waste management, water resources planning andengineering, and cultural, natural, and physical resources evaluation.
*Darrin Punceles*

**1883 Franklin D Aldrich MD, PhD**
1094 Quince Avenue 303-443-2316
Boulder, CO 80304 Fax: 303-938-9420
E-mail: w1@fa@hotmail.com
Environmental/ clinical toxicology and consulting.

**1884 GBMC & Associates**
219 Brown Lane 501-847-7077
Bryant, AK 72022 Fax: 501-847-7943
E-mail: vblubaugh@gbmcassoc.com
http://www.gbmcassoc.com
GBMC & Associates are a consulting firm providing strategic environmental services to industrial clients and air permitting support, water quality and toxicity studies, storm water management, environmental program development andreporting.
*Vince Blubaugh, Principal/Sr Project Mgr*
*Chuck Campbell, Senior Engineer*

**1885 GEO/Plan Associates**
30 Mann Street 617-740-1340
Hingham, MA 02043 Fax: 617-740-1340
*Michu Tcheng, Partner*
*Peter Rosen, Partner*

**1886 Gabbard Environmental Services**
7611 Hope Farm Road 260-493-2982
Fort Wayne, IN 46815 Fax: 219-493-4043
Consulting services for environmental affairs such as permitting, compliance, plans and programs.
*William D Gabbard, President*

**1887 Galson Corporation**
6601 Kirkville Road 315-432-0506
East Syracuse, NY 13057 800-950-0506
Fax: 315-437-0509
Environmental consulting and engineering and analytical services, specializing in the air management of indoor and outdoor environments.
*Cindy Kuiper*

**1888 Geo-Marine Technology**
4226 Lincoln Road 406-721-1599
Missoula, MT 59802 E-mail: jr@geomarinetech.com
http://www.geomarinetech.com
Provides geological, geophysical, and hydrographic survey consultancy services to offshore oil and gas industries, offshore survey industries, and governments.

**1889 GeoResearch**
7806 MacArthur Boulevard 301-229-8111
Cabin John, MD 20818 Fax: 301-229-7980
http://www.georesearch.com
Provides grography-related services from forestry to telecommunications, real-time mobile interactive geographic technologies and databases.

**1890 Geomet Technologies**
20251 Century Boulevard 301-428-9898
Suite 300 800-296-9898
Germantown, MD 20874 Fax: 301-428-9482
E-mail: marketing@geonet.com
http://www.geomet.com
Provides consultant, technical, and material evaluation services in the areas of personal protective systems, indooor and ambient air quality, energy, chemical testing, and environmental services to government agencies, private andcommercial clients.

**1891 Geospec**
17912 Sotile Drive 225-753-8811
Baton Rouge, LA 70809 877-503-5618
Fax: 225-753-8877
E-mail: info@geospec-llc.com
http://www.geospec-llc.com
Geophysical services. Gound penetrating radar, EM, conductivity and resistivity surveys to detect and identify potential environmental hazards or contamination. Borehole and excavation utility clearance.

**1892 Gradient Corporation**
20 University Road 617-395-5000
Cambridge, MA 02138 Fax: 617-395-5001
E-mail: info@gradientcorp.com
http://www.gradientcorp.com
A consulting firm with nationally recognized specialities in risk and environmental sciences.

**1893 Granville Composite Products Corporation**
600 Round House Road 717-247-2879
Lewistown, PA 17044 800-350-4660
Fax: 412-291-3291
E-mail: infosales@granville.cc
http://www.granville.cc
Plastic recycling. Molder of recycled pastics.

**1894 Great Lakes Educational Consultants**
4109 Apple Bluff Drive 269-382-2314
Kalamazoo, MI 49006 Fax: 616-382-6495
E-mail: rjonaiti@kresanet.org
A consulting firm which develops safety/security/emergency plans to protect the educational/business environment including buildings, grounds, personnel and students. We conduct a hazard analysis to determine planning requirements andprovide a proposal for your consideration.
*Robert Jonaitis, President*

**1895 Greeley-Polhemus Group**
105 South High Street 215-692-2224
West Chester, PA 19382 Fax: 215-692-4052
Specializes in providing consulting services to the United States Army Corps of Engineers and to non-Federal local sponsors of proposed Federal projects. Provides services in the areas of project planning, economics, finance,institutional strategy development, and environmental studies related to flood control, land uses, recreation, water supply, and navigation.

**1896 Groundwater Technology**
100 River Ridge Drive 781-769-7600
Suite 300 800-635-0053
Norwood, MA 02062 Fax: 781-769-7992
Groundwater Technology has been a leader in the development and application of advanced technologies for environmental restoration of contaminated sites. One of the largest environmental consulting, engineering and remediation firms.GTI is widely recognized for its innovative, bioremedial technology for rapid cleanup of soil and groundwater, both above-ground and in situ.

**1897 HC Nutting, A Terracon Company**
611 Lunken Park Drive 513-321-5816
Cincinnati, OH 45226 Fax: 513-321-0294
E-mail: cincinnati@hcnutting.com
http://www.hcnutting.com
Materials tesing company, geotechnical and environmental engineering firm.

*Jack Scott, President*
*Ron Ebelhar, Sr Principal, Envir Srvs*

**1898 HE Cramer Company**
8249 Shangrila Circle 801-561-4964
Sandy, UT 84094 Fax: 801-561-4964
E-mail: checo1@qwest.net
H.E. Cramer Company does air pollution consulting, computer software development, and environmental consulting. Research results are published in project reports and professional journals.

*Jay R Bjorklund, President*

**1899 HYGIENETICS Environmental Services**
432 Columbia Street 617-621-0363
Suite 16A Fax: 617-621-1609
Cambridge, MA 02141 http://www.hygienetics.com
Hygienetics provides comprehensive analysis, design, and program management services to a diverse group of private sector customers. Primary areas of expertise include environmental site assessments for property transaction, soil andgroundwater investigation and remediation, air resource management, industrial hygiene and asbestos/lead management.
*Carmen Pombiero, General Information Contact*

**1900 Harold I Zeliger PhD**
1270 Sacandaga Road 518-882-6800
West Charlton, NY 12010 Fax: 518-882-6926
E-mail: hiz@zeliger.com
http://www.zeliger.com
Areas of expertise include occupational and environmental exposure to toxic chemicals, hazard communication, chemical formulating and processing impact and toxic waste.
*Dr Harold I Zeliger, Principal*

**1901 Hart Crowser, Inc.**
1700 Westlake Avenue North 206-324-9530
Seattle, WA 98109 Fax: 206-328-5581
E-mail: mike.ehlebracht@hartcrowser.com
http://www.hartcrowser.com
Hart Crowser, Inc. provides a full range of services from initial site studies through regulatory permitting design, and construction. They integrate these services as required for each project, They know what kind of information isimportant, how to collect it and apply it to te selection of viable solutions, and how actions are perceived by regulatory agencies and the public. Consequently, they design an approach that is practical, cost-effective, and client-oriented.

*Mike Ehlebracht, Principal, Env. Svce. Mgr.*

**1902 Hasbrouck Geophysics**
2473 North Leah Lane 928-778-6320
Prescott, AZ 86301 Fax: 928-778-6320
E-mail: jim@hasgeo.com
http://www.hasgeo.com
Over 30 years experience in all major surface, airbone, and borehole geophisical methods plus strong geological background.
*Jim Hasbrouck, Principal*

**1903 HazMat Environmental Group**
60 Commerce Drive 716-827-7200
Buffalo, NY 14218 Fax: 716-827-7217
E-mail: rwickham@hazmatinc.com
http://www.hazmatinc.com
Transportation services - specializing in hazardous materials and hazardous waste transportation.

*Ricky Wickham, General Manager of Operation*

**1904 Heritage Environmental Services**
7901 West Morris Street
Indianpolis, IN 46231
317-243-7475
877-436-8778
Fax: 317-486-5095
E-mail: webmaster@heritage-enviro.com
http://www.heritage-enviro.com
Provides environmental management, integrated environment remediation services, product recovery and recycling services, waste services, analytical services, consulting and engineering services, and plant and industrial services.
*Mike Karpinski, Quality Manager*

**1905 Hermann Associates**
117 Church Road
Winnetka, IL 60093
Fax: 847-446-7640

**1906 Huff and Huff**
915 Harger Road
Suite 330
Oak Brooke, IL 60523
630-684-9100
Fax: 630-684-9120
E-mail: jhuff@huffnhuff.com
http://www.huffnhuff.com
Multi-diciplined firm providing environmental, civil, and chemical engineering and consulting services.
*James Huff, PE*
*Richard Trzupek*

**1907 Hydrogeologic**
11107 Sunset Hills Road
Suite 400
Reston, VA 20190
703-478-5186
Fax: 703-471-4180
*Jack Robertson, President*

**1908 In-Flight Radiation Protection Services**
211 E 70th Street
Suite 12G
New York, NY 10021
212-288-7201
E-mail: robbarish@aol.com
http://robbarish.tripod.com
Our mission is to educate flight crew members and business frequent flyers about the risks of cosmic radiation exposure during air travel.

**1909 Integrated Chemistries Inc.**
Po Box 10558
White Bear Lake, MN 55110
651-426-3224
Fax: 651-426-3114
E-mail: info@integratedchemistries.com
http://www.capsuleinc.com
Environmental management.
*James E Nash, President*

**1910 Integrated Environmental Management**
9040 Executive Park Drive
Suite 205
Knoxville, TN 37923
865-531-9140
Fax: 865-531-9130
IEM is a women-owned small business that provides strategic consulting and services in the areas of radiation, radioactivity, and the environment.

**1911 International Certification Accreditation Board**
PO Box 2099
Glenview, IL 60025
847-724-6631
Fax: 847-724-4223
E-mail: icab@icab.cc
http://www.icab.cc
A legally recognized, nonprofit, accreditation organization. Established in 2000, ICAB accredits credentialing programs of environmental, safety, medical, pharaceutical, information technology, educational, industrial and trainingorganizations.
*Christopher Young, Executive Director*

**1912 Interpoll Laboratories**
4500 Ball Road NE
Circle Pines, MN 55014
763-786-6020
Fax: 763-786-7854
E-mail: interpoll@interpoll-labs.com
http://www.interpoll-labs.com
*Dan Despen, President*
*Timothy MacDonald, Manager Field Services*

**1913 JJ Keller and Associates**
3003 Breezewood Lane
Neenah, WI 54956
920-722-2848
Fax: 920-727-7455
E-mail: sales@jjkeller.com
http://www.jjkeller.com
Regulatory compliance, best practices, and training for environmental, safety, and transportation issues.

**1914 Jack J Bulloff**
8140 Township Line Road
Indianapolis, IN 46260
317-824-0014
E-mail: jbulloff@ind.net
Environmental consultant
*Jack J Bulloff*

**1915 James Anderson and Associates**
2123 University Park Drive
Suite 130
Okemos, MI 48864
517-349-8066
Fax: 517-349-7870
E-mail: info@jaa-hlp.cpm
http://www.safe-at-work.com
A leading worldwide provider of noise control, sound exposure, and hearing loss prevention services for the industry.
*Lee D Hagev, Executive VP*

**1916 James W Sewall Company**
136 Center Street
PO Box 433
Old Town, ME 04468
207-827-4456
Fax: 207-827-3641
E-mail: info@jws.com
http://www.jws.com
Founded in 1880, Sewall provides comphrehensive services in forestry appraisal and inventory, aerial imagery, GIS consulting and engineering. Sewall's expertise in GIS project implementation is supported by 50 years' experience inaerial photography and photogrammetry and 30 years' experience in data conversion, database design and application development.
*Scott E Graham, PE, Vice President*
*Aaron Shaw, PE, Project Manager*

**1917 John Zink Company**
11920 East Apache
Tulsa, OK 74116
918-234-1800
Fax: 918-234-2700
E-mail: info@johnzink.com
http://www.johnzink.com
John Zink offers technologically advanced equipment and systems for the clean and efficient combustion of fossil fuels and for the removal of contaminants from process affluents entering the atmosphere.
*Bill Hermann, Global Marketing Director*

**1918 Kemstar Corporation**
3456 Wade Street
Los Angeles, CA 90066
310-390-0180
Fax: 310-391-8143

**1919 Kimre Inc**
16201 Southwest 95 Avenue
Suite 303
Miami, FL 33157
305-233-4249
888-315-1673
Fax: 305-233-8687
E-mail: sales@kimre.com
http://www.kimre.com
Technology provides superior air pollution control particulate/mist elimination: Mist Eliminators, Phase Separation, Scrubbers, Mass Transfer, Engineered clog-resistant interlaced mesh (large surface area/void spaces) providesoptimized efficiency using selected filaments/arrangements.
*George C Pedersen, CEO*
*Frederick H Mueller, Sales Manager*

**1920 LA Weaver Company**
308 E Jones Street
Releigh, NC 27601
919-832-6242
Fax: 919-831-1130
E-mail: aweaver1@bellsouth.net
http://www.laweaverco.com
Occupational and Environmental safety consulting services.
*Al Weaver, President*

**1921 LSI Adapt Engineering**
615 8th Avenue South
Seattle, WA 98104
206-654-7045
800-643-9932
Fax: 360-674-7048
E-mail: seattle@lsi-industries.com
http://www.lsiadapt.com

Consulting and engineering firm specializing in petroleum engineering, geotechnical and environmental issues. Team of professional engineers, licensed environmental site assessors, hydrogeologiststs and geotechnical engineers offer avariety of strengths and services to clients.

**1922 Landau Associates**
10 N Post Street
Suite 218
Spokane, WA 99201
509-327-9737
Fax: 509-327-9691
E-mail: information@landauinc.com
http://www.landauinc.com

Provided environmental and geotechnical services on nearly 2,000 projects for more than 500 private and public clients in the Pacific Northwest and western US since we opened our doors in 1982. A valued resource in waterfrontdevelopment for public and private ports, also expanded beyond the waterfront to serve clients in some of the best known industries, municipal government, and site development.

*Steve Johnston, Principal and CEO*
*Dennis Stettler, Principal*

**1923 Law Environmental**
3200 Town Point Drive NW
#100
Kennesaw, GA 30144
770-421-3400
Fax: 770-421-3486

**1924 Lawler, Matusky and Skelly Engineers**
1 Blue Hill Plaza
Pearl River, NY 10965
845-735-8300
Fax: 845-735-7466
E-mail: cnevel@lmseng.com
http://www.lmseng.com

Environmental engineering and consulting firm. Research results published in client reports and professional journals.
*Christy Nevel, Director Marketing*

**1925 Lenox Institute of Water Technology**
101 Yokun Avenue
PO Box 1639
Lenox, MA 01240
413-637-3025
Fax: 413-637-3362

Provides services in the area of municipal and industrial water and wastewater treatment systems.
*Charles L Smith, Executive VP*

**1926 Les A Cartier and Associates**
191 Main Street
PO Box 559
Candia, NH 03034
603-483-2180
800-639-7703
Fax: 603-483-8986

Environmental service company; Health and Safety Course in their Hazardous Materials Management Series
*Leslie A Cartier, President*

**1927 Louis Berger Group**
100 Halsted Street
East Orange, NJ 07018
973-678-1960
Fax: 973-672-4284
E-mail: ctompkins@louisberger.com
http://www.louisberger.com

Offers professional services in the areas of civil, structural, mechanical, electrical and environmental engineering; program management; planning; environmental sciences; cultural resources; information services; economics; policy andmanagement analysis; and construction management and support.
*Nicholas J Masucci, President*
*Cindy Tompkins, Programs Coordinator*

**1928 Louis Defilippi**
208 Edgewood Lane
Palatine, IL 60067
847-925-8524
Fax: 847-303-1731
E-mail: defilip1@flash.net
http://www.flash.net/~defilip1/Default.htm

We offer consulting services in three broad areas: industrial biotechnology, bioprocessing, and proteomics; environmental and regulatory compliance; biotechnology and applied engineering for the microbiological treatment of hazardouswaste. We are especially valuable when you don't really need a full time in-house expert, but need someone to take an important load off your shoulders, to review your processes and procedures, or just to get things moving faster, on an as-neededbasis.

*Dr. Louis DeFilippi, President*

**1929 Marc Boogay Consulting Engineer**
326 Main Street
Vista, CA 92084
760-407-4000
Fax: 760-407-4004
E-mail: boogay@sdnc.quik.com
http://www.boogay.com

Marc Boogay and staff have completed more than 600 site assessments. This work has include all varieties of developed and undeveloped properties. Projects have range from Phase I investigations through sampling surveys, remediationdesigns, abatement monitoring, risk assessment, and expert witness testimony. Projects have benn conducted in several states, meeting standards of government agencies as well as many leading/ investment institutions.
*Marc Boogay, PE, Principal*
*Todd Jacquay, Soils Engineer*

**1930 McVehil-Monnett Associates**
44 Inverness Drive East
Building C
Englewood, CO 80112
303-790-1332
Fax: 303-790-7820
http://www.mcvehil-monnett.com

MMA is a experienced consulting firm of atmospheric scientists, engineers and environmental specialists providing air quality and environmental management system (EMS) services worldwide. Serves the mining, oil and gas, electric powerand manufacturin industries, as well as government agencies and engineering and law firms. Leader in air quality permitting, modeling, monitoring and litigation supoort services as well as environmental management and planning services.

*William R Monnett, President*
*George McVehil, Principal*

**1931 Mercury Technology Services**
23014 Lutheran Church Road
Tomball, TX 77377
281-255-3775
Fax: 281-357-0721
E-mail: smw@htech.com
http://www.hgtech.com

Specialists in solving problems related to mercury pollution and contamination. MTS provides technical services to companies having metals contamination and expert testimony on mercury pollution and remediation.
*Mark Wilhelm, President*

**1932 Meteorological Evaluation Services Company**
165 Broadway
Amityville, NY 11701
516-691-3395
Fax: 516-691-3550
http://www.mesamity.com

Consultants in Applied Meteorology, Air Quality and the Environment.

**1933 Miceli Kulik Williams and Associates**
39 Park Avenue
Rutherford, NJ 07070
201-933-7809
Fax: 201-933-8702
E-mail: jwilliams@mkwla.com
http://www.mkwla.com

Offers complete services covering the various aspects of landscape architecture, site planning, and urban design. Present scope of work includes neighborhood rehabilitation, housing and community development, park, recreational andopen space planning, landscape architecture, impact assessment, educational, municipal, commercial and industrial commissions. Project involvement extends throughout the Eastern States.

**1934 Michael Baker Corporation**
Airside Business Park
100 Airside Drive 412-269-6300
Moon Township, PA 15108 800-553-1153
Fax: 412-375-3983
E-mail: CorpCom@mbakercorp.com
http://www.mbakercorp.com
Michael Baker Corporation provides engineering and operations and maintenance services for its clients' most complex challenges worldwide. The firm's primary practice areas are aviation, environmental, facilities, geospatialinformation technologies, linear utilities, transportation, water/wastewater, and oil & gas. With approximately 4,500 employees in over 400 offices across the U.S. and internationally, Baker is focused on providing services that span the completelife cycle of infrastucture.

*David Higie, VP Corporate Communications*

**1935 Michael Brandman Associates**
220 Commerce 714-508-4100
Suite 200 Fax: 714-508-4110
Irvine, CA 92602 http://www.brandman.com
Michael Brandman Associates is a comprehensive environmental planning services firm specializing in environmental documentation, planning, and natural resources management.
*Erika Bennett, Marketing Manager*
*Michael Brandman, President/CEO*

**1936 Micro-Bac**
3200 North Interstate Highway 35 512-310-9000
Round Rock, TX 78681 877-559-1800
Fax: 512-310-8800
E-mail: mail@micro-bac.com
http://www.micro-bac.com
Delvelops and manufactures biological products for remediation of contaminated substances; reduction of waste and odor in food processing, agriculture, and sewage; and control of paraffin in oil production.

**1937 Midstream Farm**
20004 Sterling Creek Lane 804-749-8720
Rockville, VA 23146 Fax: 804-749-8719
http://www.usaclem.com

*Clement Mesavage Jr, Proprietor*

**1938 Mostardi Platt Environmental**
1520 Kensington Road 630-993-2100
Suite 204 Fax: 630-993-9017
Oak Brook, IL 60523 http://www.mostardiplattenv.com
Mostardi Platt Environmental-your full service environmental management partner. Offers innovative solutions and strategies to assist our clients comply with environmental, health and safety regulations and develop environmentalprograms that save long-term costs. We understand our clients need the best possible compliance options. We evaluate a wide variety of technical and economic concerns and work with our clients to establish the best path towards compliance.
*Joseph J Macak III, President*
*Robert A Gere, Engineering Consultant*

**1939 NTH Consultants**
38955 Hills Tech Drive 248-553-6300
PO Box 9173 Fax: 248-324-5179
Farmington Hills, MI 48333 http://www.nthconsultants.com
NTH Consultants has provided consulting engineering services to clients throughout the United States since 1968. Headquartered in Farmington Hills, MI, NTH has maintained an office in downtown Detroit since 1980, a regional,full-service office in Exton, PA and offices in Lansing and Grand Rapids, MI since 1992.
*Jerome C Neyer, Chairman*

**1940 National Environment Management Group**
PO Box 5131 708-771-7350
River Forest, IL 60305 Fax: 312-733-2478
Environmental consulting.
*Jack Hughes, Chairman*

**1941 National Environmental**
1019 W Manchester Boulevard 310-645-4516
Suite 102 800-870-1719
Inglewood, CA 90301 Fax: 310-645-0148
E-mail: customerservice@natlenviro.com
Training school. Training in use of lead, asbestos and hzardous materials.
*James McFarland, President*
*David P Fuller, VP*

**1942 National Institute for Urban Wildlife**
10921 Trotting Ridge Way 301-596-3311
Columbia, MD 21044
Promotes the preservation of wildlife in urban settings, providing support to individuals and organizations invloved in maintaining a place for wildlife in expanding American cities and suburbs. The Institute conducts researchexploring the relationship between humans and wildlife in these habitats, publicizes urban wildlife management methods, and raises public awareness of the value of wildlife in city settings. The Institute also provides consulting services.

**1943 National Sanitation Foundation**
789 North Dixboro Road 734-769-8010
PO Box 130140 800-NSF-MARK
Ann Arbor, MI 48113 Fax: 734-769-0109
E-mail: info@nsf.org
http://www.nsf.org
NSF International, The Public Health and Safety Company, is an independent, not for profit organization providing a wide range of services around the world. For more than 55 years, NSF has been committes to public health, safety andprotection of the environment.
*Robert Ferguson, Vice President*
*Tim Bruursema, General Manager*

**1944 National Society of Environmental Consultants**
303 West Cypress Street 210-271-0781
PO Box 12528 800-486-3676
San Antonio, TX 78212 Fax: 210-225-8450
E-mail: lincolncenter@worldnet.att.net
http://nsec.lincoln-grad.org/
The mission is to encourage an awareness of environmental risk and the regulations regarding their impact on real property value, to advocate reponsible use and development of real estate resources in harmony with the environment, toelevate the competency of the membership through information and education and to promote the development of ethics and standards of professional practice for the speciality of environmental consultants
*Gary T Deane, Executive Director*

**1945 Natural Resources Consulting Engineers**
131 Lincoln Avenue 970-224-1851
Suite 300 Fax: 970-224-1885
Fort Collins, CO 80524 E-mail: office@nrce.com
http://www.nrce.com
Water supply investigations. Native American water rights expert witness testimony.

**1946 Network Environmental Systems**
1141 Sibley Street 916-353-2360
Folsom, CA 95630 800-637-2384
Fax: 916-353-2375
E-mail: office@nesglobal.net
http://www.nesglobal.net
Network Environmental Systems was incorporated in 1988 to privide high quality professional industrial hygiene and environmental management services through customer service excellence.
*Jerry Bucklin, President/CEO*
*Donald Rothenbaum, Senior Vice President*

**1947 Ninyo and Moore**
5710 Ruffin Road 858-576-1000
San Diego, CA 92123 Fax: 858-576-9600
E-mail: nminquiries@ninyoandmoore.com
http://www.ninyoandmoore.com
As a leading geotechnical and environmental scieces engineering and consulting firm, Ninyo & Moore provides specialized services to clients in both public and private sectors.
*Avram Ninyo, Principal Engineer*

**1948 Nordlund and Associates**
813 East Ludington Avenue 231-843-3485
Ludington, MI 49431 Fax: 231-843-7676
E-mail: Nordlund@T-one.net
Water systems, wastewater treatment, sanitary landfills and hydrogeological studies.
*James T Nordlund Sr, President*

**1949 Normandeau Associates**
25 Nashua Road 603-472-5191
Bedford, NH 03110 Fax: 603-472-7052
E-mail: nai@normandeau.com or pkinner@normandeau.com
http://www.normandeau.com
Normandeau Associates is an employee owned natural resources management consulting and testing services firm that provides: permit assistance, water quality studies, aquatic and terrestrial ecology, environmental impact assessments,property transfer site assessments, wetlands services, contamination studies and biological laboratory services.
*Pamela Hall, President*
*Peter Kinner, Senior VP*

**1950 Norton Associates**
46 Leland Road 508-528-3357
Norfolk, MA 02056 E-mail: norton@designofmachinery.com
http://www.designofmachinery.com
Professor Norton and his associates have been providing engineering consulting services since 1970. Areas of expertise include: cam design and analysis, linkage design and analysis, street analysis, vibrations in machinery, dynamicsignal analysis, machinery monitoring, and machine dynamic analysis. We also can provide short courses and seminars on site in cam design, dynamic signal analysis and machinery vibrations.
*Robert L Norton, President*

**1951 NuChemCo**
5765-F Burke Centre Parkway 703-548-3200
#149 800-682-4362
Burke, VA 22015 Fax: 703-978-0642
E-mail: info@nuchemco.com
http://www.nuchemco.com
*Neil B Jurinski*
*Joseph B Jurinski*

**1952 OCCU-TECH**
6501 East Commerce Avenue 816-231-5580
Suite 230 800-950-1953
Kansas City, MO 64120 Fax: 816-231-5641
E-mail: service@occutec.com
http://www.occutec.com
OCCU-TECH is a leading safety, health and environmental services company. From OSHA to EPA issues, safety assessments to program development, asbestos inspections to environmental management, our expertise has been relied on for over16 years.

**1953 Oak Creek**
60 Oak Creek 207-929-6375
Buxton, ME 04093 Fax: 207-929-6374
E-mail: jssmith@oak-creek.net
http://www.oak-creek.net
*James S Smith Jr, PhD, President/Toxicologist*
*Brad House, Senior Scientist*

**1954 Occupational Health and Safety Management**
117 La Farge 303-665-8528
Louisville, CO 80027 Fax: 303-673-0785
Industrial hygiene/safety consulting.
*Mary Ann Heaney*

**1955 Occupational Safety and Health Consultants**
12000 6th Street East 727-345-1552
Saint Petersburg, FL 33706 Fax: 727-363-8151
E-mail: oshc@oshc.com
http://www.oshc.com
Air pollution control/industrial hygiene.

**1956 Occupational and Environmental Health Consultiing Services**
635 Harding Road 630-325-2083
Hinsdale, IL 60521 Fax: 630-325-2098
E-mail: bobb@safety-epa.com
http://www.safety-epa.com
A full service regulatory, safety, industrial hygiene, and environmental engineering consulting firm. Specialize in assisting all sizes of companies and corporations. Clients include very small businesses up to Fortune 100corporations.
*Bob Brandys PhD,MPH,PE,CIH, President*

**1957 Occusafe**
135 Mountain View Drive 413-323-1036
Belchertown, MA 01007 Fax: 413-323-1039
E-mail: occusafe@map.com
http://www.occusafe.net
OCUSAFE is a full service consulting firm specializing in assistance to management in the areas of occupational safety, industrial hygiene, and environment.

**1958 Ocean City Research**
50 Tennessee Avenue 609-399-2417
Ocean City, NJ 08226 Fax: 609-399-5233
E-mail: jrepp@corrpro.com
http://www.corrpro.com
Ocean City Research Corporation, incorporated in 1963, is a wholly owned subsidiary of Corrpro Companies, Collectively, the Corrpro affiliated companies represent the largest, independent consulting corrosion engineering organizationin the world.
*J Peter Ault, PE*

**1959 Omega Waste Management**
957 Colusa Street 530-824-1890
PO Box 495
Corning, CA 96021
A consulting firm, whose unique and innovative approach to waste removal and recycling has made it one of the largest volume purchasers of waste services in the nation.

**1960 Owen Engineering and Management Consultants**
5353 West Dartmouth Avenue 303-969-9393
Suite 402 Fax: 303-969-9394
Denver, CO 80227
Water/Wastewater design systems.
*Webster J Owen, President*

**1961 PACE Analytical Services**
1700 Elm Street 612-607-1700
Minneapolis, MN 55414 Fax: 612-607-6444
E-mail: info@pacelabs.com
http://www.pacelabs.com
Provider of air, water, soil and environmental testing services.
*Bruce E Warden, General Manager*
*Jeff Smith, Marketing Contact*

**1962 PAR Environmental**
1906 21st Street 916-739-8356
PO Box 160756 Fax: 916-739-0626
Sacramento, CA 95816 E-mail: mlmaniery@aol.com
http://www.parenvironmental.com
PAR Environmantal Services mission is to provide technical reports on time, within budget, and with meticulous attention to detail.
*Mary L Maniery, President*

**1963 PBR HAWAII**
1001 Bishop Street, ASB Tower 808-521-5631
Suite 650 Fax: 808-523-1402
Honolulu, HI 96813 E-mail: sysadmin@pbrhawaii.com
http://www.pbrhawaii.com
Consulting services in environmental studies,permitting land planning and landscape architecture and graphic design.

*Thomas S Witten ASLA, President*
*Frank Brandt Falsa, Chairman*

**1964 PBS Environmental Building Consultants**
4412 SW Corbett Ave 503-248-1939
Portland, OR 97239 888-248-1939
Fax: 503-248-0223
http://www.pbsenv.com
PBS specializes in program development, identification, assessment, testing and corrective action consultation in the areas of: Environmental Engineering, Geotechnical Engineering, Hazardous Materials Management, Industrial HygieneServices, Natural Resources Studies, Training and Laboratory.
*Stephen Smiley, President*

**1965 PE LaMoreaux and Associates**
PO Box 2310 205-752-5543
Tuscaloosa, AL 35403 Fax: 205-752-4043
E-mail: info@pela.com
http://www.pela.com
For over three decades, PELA's integration of qualified personnel, up-to-date technology, and sound management has established PELA as an international leader in the environmental consulting field. PELA's expertise in hydrogeology,geotechnical analysis, design and construction management, remediation, computer graphics and models, and permitting can get your project on two feet quicker than you might think.
*James W Moreaux, Chairman Of Board*
*James M Lee, President*

**1966 PEER Consultants**
12300 Twinbrook Parkway 301-816-0700
Suite 410 Fax: 301-816-9291
Rockville, MD 20852 E-mail: peercpc@peercpc.com
http://www.peercpc.com
For nearly a quarter of a century, PEER Consultants has provided civil, sanitary, and environmental engineering consulting services for public and private sector clients nationwide.
*Lilia Abron, President*

**1967 Pacific Soils Engineering**
10653 Progress Way 714-220-0770
Cypress, CA 90630 Fax: 714-220-9589
Services include: Geotechnical Services, Laboratory Testing, Field Observation and Testing, Consultation and Review of Geotechnical Reports.

*Daniel Martinez, President*

**1968 Parish and Weiner Inc**
297 Knollwood Road 914-997-7200
Suite 315 Fax: 914-997-7201
White Plains, NY 10607 E-mail: pwm@verizon.net
Consulting firm which prepares environmental impact studies, traffic studies, zoning and site plan studies for private developers, non-profit organizations, governmental entities. Also provide expert consultation to lawyers forlitigation and hearings.

*Nat Parish, President*

**1969 Pavia-Byrne Engineering Corporation**
7443 Obyx St. 504-288-8406
New Orleans, LA 70184 Fax: 504-283-4090
Provides services for environmental control and water treatment including definition, process development, and start up services.
*Edgar H Pavia, President*

**1970 Perry-Carrington Engineering Corporation**
214 West Second Street 715-384-2133
Marshfield, WI 54449 Fax: 715-384-9797
E-mail: 2perryear@temet.com
Water pollution control systems.
*David L LaFontaine, President*

**1971 Petra Environmental**
10550 North 6th Avenue 715-536-7870
Merrill, WI 54452 800-458-3772
Fax: 715-536-7890
E-mail: info@petraenvironmental.net
http://www.petraenvironmental.net
PETRA Environmental Consultants, is an environmental engineering firm specializing in environmental compliance, hydrogeological investigations, and environmental assessments.
*Anthony M Ungerer, PG, Principal Hydrogeologist*
*Christopher J Rog, PG, Senior Hydrogeologist*

**1972 Phase One**
2680 Walnut Avenue 714-669-8055
Suite B 800-524-8877
Tustin, CA 92780 Fax: 714-669-8025
E-mail: info@phasei.com
http://www.phasei.com
A focused environmental consulting practice that specializes in real property assessments for any type of property transfer, leasing development, special uses, and/or financing purposes. Founded in response to the business community'sneed for affordable, standardized and consistently high quality assessment reports that provide recommendations for sound real estate decisions.
*Eric D Kieselbach, President/CEO*

**1973 Planning Resources**
402 W Liberty Drive 630-668-3788
Wheaton, IL 60187 Fax: 630-668-4125
http://www.planres.com
Land use and environmental planning.
*Keven Graham, Managing Principal*

**1974 Post, Buckley, Schuh and Jernigan**
2001 Northwest 107th Avenue 305-592-7275
Miami, FL 33172 800-597-7275
Fax: 305-599-3809
http://www.pbsj.com
PBS&J was founded in 1960 by four respected engineers who joined forces to help develop Florida's first planned community. Their tenacity in meeting production schedules, commitment to client service, and ability to provide innovativesolutions to difficult challenges quickly earned our firm a reputation for excellence and laid the foundation for future grouth.
*Todd J Kenner, President*

**1975 Presnell Associates**
815 West Market 502-585-2222
Suite 300 Fax: 502-581-0406
Louisville, KY 40202 E-mail: presnell@thepoint.net
http://www.qk4.com
The professional practice of Prenell encompasses a variety of services directly related to preserving the environment, including potable water system planning and design, municipal and industrial wastewater treatment, solid wastemanagement, landfill siting, asbestos management, contamination screening assessments, indoos air quality, and lead paint abatement.
*Wendell Wright, President*

**1976 Priester and Associates**
1345 Garner Lane 803-798-4377
Suite 105 877-798-4377
Columbia, SC 29210 Fax: 803-798-4378
E-mail: priester@conterra.com
Provides personalized environmental services ranging from short-term consulting to extensive remediation and management activities.
*LE Priester, President*

**1977 Process Applications**
2627 Redwing Road
Suite 340
Fort Collins, CO 80526
970-223-5787
Fax: 970-223-5786
Environmental engineering consultants.
*Bob A Hegg, President*

**1978 Professional Analytical and Consulting Services (PACS)**
409 Meade Drive
Coraopolis, PA 15108
724-457-6576
800-367-2587
Fax: 724-457-1214
E-mail: hnpacs@aol.com
http://www.pacslabs.com
Training courses and conferences. Provides short courses in spectrocopy, chomatography, quality, safety, environmental, and management. Provides professional manuals and software products. Provides laboratory testing and consultingservices. Company also goes by the following names: Activated Carbon Services, PACS Testing and Consulting, PACS Courses and Conferences. PACS provides: Testing, Training, R & D Conferences, and software for activated carbon users.

*Henry G Nowicki PhD/MBA, President*
*Barbara Sherman, Manager of Operations*

**1979 Psomas and Associates**
4540 California Avenue
Suite 210
Bakersfield, CA 93309
661-631-2311
866-9PS-OMAS
Fax: 661-631-2782
E-mail: info@psomas.com
http://www.psomas.com
Psomas is a leading consulting engineering firm offering services in land development, water and natural resources, transportaion, public works, survey and information systems to public and private sector clients.
*George Psomas, Chairman*

**1980 QORE**
4201 Pleasant Hill Road Northwest
Suite A
Duluth, GA 30096
770-232-0235
877-767-3462
Fax: 770-232-0238
E-mail: corporate@qore.net
http://www.qore.net
Consultants in property science, in fields of geology, geotechnical and environmental engineering.
*Richard D Heckel, PE, President*
*Ed Heustess, Chief Financial Officer*

**1981 RDG Geoscience and Engineering**
10360 Sapp Brothers Drive
Omaha, NE 68138
402-894-2678
888-260-0893
Fax: 402-894-9043
E-mail: info@rdgge.com
http://www.rdgge.com
Is an earth science and engineering consulting firm that has completed over 1200 projects throughout the mid-west and mountain west of US.
*Jon Gross, President*
*Robert Kalinski, Vice President*

**1982 RGA Environmental**
1466 66th Street
Emeryville, CA 94608
510-547-7771
800-776-5696
Fax: 510-547-1983
E-mail: rga@rgaenv.com
http://www.rgaenv.com
Founded in 1985, RGA Environmental is a specialty consultant in the environmental sciences. Our mission is to provide high-quality environmental engineering, health & safety consulting services to meet the special needs of our clients.
*Steven C Rosas, COO, Director of Business*

**1983 RMT**
744 Heartland Trail
PO Box 8923
Madison, WI 53717
608-831-4444
800-283-3443
Fax: 608-831-3334
E-mail: info@rmtinc.com
http://www.rmtinc.com
Serves industrial compaines throughout the world who value environmental and engineering solutions that improve productivity and profitability. RMT's diversified staff of over 550 engineers, scientists and technicians takesresponsibility for managing environmental issues so clients can concentrate on their core business.
*Jodi Burmester, Corporate Communications*

**1984 RMT Inc.**
744 Heartland Trail
Madison, WI 53717
608-831-4444
800-283-3443
Fax: 608-831-3334
E-mail: info@rmtinc.com
http://www.rmtinc.com
RMT delivers environmental engineering health and safety and construction solutions that help industrial companies solve complex problems while improving their bottom line.

**1985 Raterman Group**
75 East Wacker Drive
Suite 500
Chicago, IL 60601
312-345-0111
Fax: 312-345-9950
E-mail: susan@ratermangroup.com
http://www.ratermangroup.com
Industrial hygiene and environmental assessments.
*Susan M Raterman, President*

**1986 Reclamation Services Unlimited**
701 Temple Street
Central City, KY 42330
270-754-3976
Fax: 270-754-4374
Environmental consulting services.
*Sue Poole Cardwell, President*

**1987 Redniss and Mead**
22 1st Street
Stamford, CT 06905
203-327-0500
800-404-2060
Fax: 203-357-1118
E-mail: a.mead@rednissmead.com
http://www.rednissmead.com
Redniss & Mead, Inc. provides land surveying, civil engineering and land planning services.
*Aubrey E Mead, Jr., PE, VP*
*Raymond L Redniss, PLS, Senior Vice President*

**1988 Refuse Management Systems**
99 Tulip Avenue
#303
Floral Park, NY 11001
516-354-1212
800-346-5926
Fax: 516-354-2434
E-mail: enviroeq@ix.netcom.com
Environmental consultants.
*Harvey Podolsky, President*

**1989 Regional Services Corporation**
3200 Sycamore Court
Suite 2B
Columbus, IN 47203
812-372-9511
Fax: 812-372-9520
Solid waste disposal.
*Mark Richards, President*

**1990 Regulatory Management**
6190 Lehman Drive
Suite 106
Colorado Springs, CO 80918
719-531-6883
Fax: 719-599-4410
E-mail: maxlab@usa.net
Environmental consulting group.
*James T Egan, President*

**1991 Resource Applications**
9291 Old Keene Mill Road
Burke, VA 22015
703-644-0404
Fax: 703-644-7143
Hazardous waste management, pollution prevention/site remediation.
*Paul Singh, Director, Corp. Development*

**1992 Resource Concepts**
340 North Minnesota Street
Carson City, NV 89703
775-883-1600
Fax: 775-883-1656
E-mail: john@rci-nv.com
http://www.rci-nv.com
RCI has years of experience and demonstrated accomplisjment working with environmentally sensitive projects. Combining technical abilities and excellent working relationships with regulatory agencies results in highly effective projectplanning and permitting services.

*Bruce R Scott, Principal*
*John McLain, Principal*

**1993 Resource Decisions**
934 Diamond Street
San Francisco, CA 94114
415-282-5330
E-mail: marvin@resourcedecisions.net
http://www.resourcedecisions.net
Assisting clients to evaluate trade-offs which foster the wise allocation of resources is primary mission of Resource Decisions. To accomplish this mission we apply a wide range of economic and decision-making tools.
*Marvin Feldman, PhD, Principal*

**1994 Resource Management**
625 Chapin Road
Chapin, SC 29036
803-345-0200
Fax: 803-345-6520
E-mail: resourc9@winusa.com
Hazardous waste management.
*Don Dicus, President*

**1995 Resource Technology Corporation (RTC)**
2931 Soldier Springs Road
PO Box 1346
Laramie, WY 82070
800-576-5690
Fax: 307-745-7936
E-mail: RTC@RT-corp.com
http://www.rt-corp.com
They offer Laboratory Proficiency Testing for drinking water, waste water and USEPA RCCRA Program. Certified analytical standards and Certified Reference Materials.

**1996 Respec Engineering**
3824 Jet Drive
Rapid City, SD 57703
605-394-6400
877-4RE-SPEC
http://www.respec.com
Since our founding in 1969, RESPEC has remained committed to its original purpose of providing clients with high-quality technical and advisory services.

**1997 Reston Consulting Group (RCG)**
462 Herndon Parkway
Suite 203
Herndon, VA 20170
703-834-1155
Fax: 703-834-3086
E-mail: information@rcg.com
http://www.rcg.com

**1998 Rich Tech**
2410 Devonshire Drive
Rockford, IL 61107
815-229-1122
Fax: 815-229-1525
Water pollution control.
*Gail Rivitts, President*
*Rich Rivitts, Vice President*

**1999 Rizzo Associates**
1 Grant Street
Framingham, MA 01701
508-903-2000
Fax: 508-903-2001
E-mail: rmoore@rizzo.com
http://www.rizzo.com
A leading engineering, transportation, and environmental engineering firm. We work with you throughout the development process to reslove the challenges that arise in planning, permitting, design, and construction phases of complexprojects.
*Rick Moore, President*

**2000 Robert B Balter Company**
18 Music Fair Road
Owings Mills, MD 21117
410-363-1555
Fax: 410-363-8073
Environmental consultation.
*Sharon Ames-Burgess, Marketing Director*

**2001 Rockwood Environmental Services Corporation**
50 Kearney Road
Needham, MA 02494
781-449-8740
Fax: 781-449-8741
E-mail: bwhite@rockwood-enviro.com
http://www.rockwood-enviro.com
Rockwood specializes in solving the problems of hazardous waste management and disposal for New England generators. By shipping wastes directly to ultimate disposal sites on a regular basis, Rockwood reduces current disposal costs andreduces long-term liability exposure.
*William A White III, President*

**2002 Rodriguez, Villacorta and Weiss**
8765 Springs Cypress
Suite L#177
Spring, TX 77379
281-447-1726
Fax: 281-447-2299
E-mail: sweiss@rvw.net
http://rvw.net/
Our mission is to provide cost-effective and thorough work product for claims services and loss control. Maximum integration of all in-house and affiliated expertise will guarantee prompt service, nurturing strong client relationshipsbased on dependability, trust and competence.
*Steve Weiss, Principal*

**2003 Roux Associates**
209 Shafter Street
Islandia, NY 11749
631-232-2600
Fax: 631-232-9898
E-mail: sisadiker@rouxinc.com
http://www.rouxinc.com
Environmental Consulting and Management.

*Steve Sadiker, Vice President*

**2004 SLC Consultants/Constructors**
295 Mill Street
Lockport, NY 14094
716-433-0776
800-932-0157
Fax: 716-433-0802

**2005 Safina**
953 N Plum Grove Road
Suite A-1
Schaumburg, IL 60173
847-956-8617
Fax: 847-956-8619
Environmental due diligence.

*Sanjiv Pillai, General Manager*

**2006 Schneider Instrument Company**
8115 Camargo Road
Cincinnati, OH 45243
513-561-6803
Fax: 513-527-4375
E-mail: schneidxcompany@aol.com
*Gary Schneider, Vice President*

**2007 Schoell and Madson**
15050 23rd Ave North
Plymouth, MN 55447
765-746-1600
Fax: 765-746-1699
E-mail: mail@schoellmadson.com
http://www.schoellmadson.com
We are dedicated to creatively serving our clients by meeting or exceeding their needs in a responsive and cost-effective manner while providing an interesting and rewarding experience for our employees.

**2008 SciComm**
7735 Old Georgetown Road
12th Floor
Bethesda, MD 20814
301-652-1900
E-mail: info@scicomm.com
http://www.scicomm.com
A professional services firm specializing in communications, engineering, environmental, and information management services. Organized to carry out the interest, expertise, and vision of co-founder Laura Chen and Dan Lewis.

*Laura Chen, Founder*

**2009 SevernTrent Laboratories**
4101 Shuffel Drive NW
North Canton, OH 44720
330-497-9396
Fax: 330-497-0772
http://www.stl-inc.com

The two compaines merged as Wadsworth/Alert Laboratories in early 1980's and the core business focused on environmental testing, with a specialization in on-site and emergency response projects. Mobile Labs were placed as far north asMichigan, and south to Florida, east to New York, and west to Missouri.
*Rachel Jannetta, President*

**2010 Shaw Environmental**
4171 Essen Lane
Baton Rouge, NJ 70809
225-952-2500
E-mail: general@shawgrp.com
http://www.shawgrp.com

Hazardous waste remediation.
*Ron Prann, Division Manager*

**2011 Shell Engineering and Associates**
2403 West Ash Street
Columbia, MO 65203
573-445-0106
Fax: 573-445-0137
E-mail: Charles@shellengr.com
http://www.shellengr.com

Shell Engineering provides services firm specializing in communications, engineering, environmental monitoring and engineering. Shell Engineering has completed hundreds of projects since 1975 throughout the United States, Canada.Centeral America, South America, Asia and Africa.
*Harvey D Shell, CEO/Chairman*
*Charles A Shell, President*

**2012 Sierra Geological and Environmental Consultants**
91 South Main Street
PO Box 136
Kent City, MI 49330
616-678-5157
E-mail: info@sierraconsultants.net
http://www.sierraconsultants.net

A full service environmental consulting firm providing assessment, investigation, and cleanup services throughout Michigan and the Great Lakes States.

**2013 Slakey and Associates**
375 Village Square
PO Box 944
Orinda, CA 94563
925-254-4164
Fax: 925-254-0679

Consulting, civil, mechanical, environmental engineers with 40 years experience in indoor air quality, air pollution control. Design of systems and equipment for collection abatement of fugitive and source missions of dusts, odor andfumes. Industrial clients only.
*Philip Slakey, President*

**2014 Slosky & Company**
303 E 17th Avenue
Suite 1080
Denver, CO 80203
303-825-1911
Fax: 303-892-3882
E-mail: Lslosky@slosky.com
http://www.slosky.com

Full service environmental consulting firm.

*Leonard Slosky, President*

**2015 Snyder Research Company**
111 West Saint John
Suite 200
San Jose, CA 95113
408-414-5950
Fax: 408-275-6219
E-mail: info@sdforum.org
http://www.sdforum.org

**2016 Staunton-Chow Engineers**
100 W 32nd Street
7th Floor
New York, NY 10001
212-695-9390
Fax: 212-779-2092
http://www.stauntonchow.com

Known widely as a small premiere multidisciplined engineering/architectural consulting firm providing professional services for new construction, repair, alterations, and maintenance for nearly 50 years.
*Kin Chow, President*

**2017 Strata Environmental Services**
110 Perimeter Park
Suite E
Knoxville, TN 37922
865-539-2077
Fax: 865-539-3970
E-mail: info@strataenv.com
http://www.strataenv.com

Founded to provide consulting services in geosciences, engineering, air quality, water quality, regulatory compliance, and environmental due diligence.

**2018 TECHRAD Environmental Services**
4619 North Santa Fe Avenue
Oklahoma City, OK 73118
405-528-7016
800-375-7016
Fax: 405-528-3346

Analytical laboratory, environmental site assessments, underground storage tank management and remediation, industrial hygiene, stormwater and hazardous waste management, asbestos consulting and analysis and regulatory compliance.
*Edward M Wall, President/CEO*

**2019 THP**
40 Brunswick Woods Drive
East Brunswick, NJ 08816
732-257-4040
Fax: 732-257-7953

Engineering traffic and engineering planning consulting firm.
*Lester J Nebenzahl, President*

**2020 Technos**
10430 Northwest 31st Terrace
Miami, FL 33172
305-718-9594
Fax: 305-718-9621
E-mail: info@technos-inc.com
http://www.technos-inc.com

A geologic and geophysical consulting firm specializing in subsurface site characterization for geotechnical, environmental, and groundwater projects.

*Lynn Yuhr, President*
*Ron Kaufmann, VP*

**2021 Terryn Barill**
301 N Harrison
Suite 484
Princeton, NJ 08540
800-718-6690
Fax: 609-243-8703
E-mail: terryn1@mail.com
http://www.terryn.com

Audits/assessments, training, implementation and facilitation.

**2022 Tetra Tech**
3475 East Foothill Boulevard
Pasadena, CA 91107
626-351-4664
Fax: 626-351-5291
E-mail: info@tetratech.com
http://www.tetratech.com

We provide comprehensive resource management , infraestructure and communications services, including, research and development, applied science, management consulting, engineering and architectural design, construction management, andoperation and maintenance.
*Sam W Box, President*
*Mark A Walsh, Senior VP*

**2023 Theil Consulting**
1136 South Fort Thomas Avenue
Fort Thomas, KY 41075
859-781-2651
Fax: 859-781-2356
E-mail: larry@theilair.com
http://www.theilair.com

Experts in industrial process exhausts—especially submicron particles created by heat or other high energy in a process.
*Greg Theil, Technical Director*
*Larry Olson, Sales Manager*

**2024 Titan Corp. Ship and Aviation Engineering Group**
11955 Freedom Drive
Reston, VA 20190
703-434-4000
Fax: 703-434-5075
E-mail: corpcomm@titan.com
http://www.titan.com

TITAN provides a wide range of engineering and environmental services. Experience includes ISO 14001 and ISO 9000 series and

its implementation, pollution prevention planning, hazardous materials/waste management, database management.
*Gene W Ray, Chairman of the Board*
*Lawrence J Delaney, VP of Operations*

**2025 Tradet Laboratories**
Rr 2 304-547-9094
Box 227 Fax: 304-547-9097
Triadelphia, WV 26059
Coal, analytical and environmental services.
*G William Kald, President*

**2026 Transviron**
1624 York Road 410-321-6961
Lutherville, MD 21093 Fax: 410-494-9321

**2027 Trinity Consultants**
12770 Merit Drive 972-661-8100
Suite 900 800-229-6655
Dallas, TX 75251 Fax: 972-385-9203
E-mail: information@trinityconsultants.com
http://www.trinityconsultants.com
An environmental consulting company that assists industrial facilities with issues related to regulatory compliance and environmental management. Founded in 1974, this nationwide firm has particular expertise in air quality issues.Trinity also sells environmental software and professional education. T3, a Trinity Consultants Company, provides EH&S management information systems (EMIS) implementation and integration services.

*Jay Hofmann, President*
*Richard H Schulze, CEO*

**2028 Troppe Environmental Consulting**
24 N. High Street 330-375-1900
Akron, OH 44308 Fax: 330-375-1904
Provides level I and level II assessments, water and oil testing, and amtm standards.
*Fred Troppe, President*

**2029 Versar**
6850 Versar Center 703-750-3000
Springfield, VA 22151 800-283-7727
Fax: 703-642-6807
E-mail: info@versar.com
http://www.versar.com
Versar is a public-held, international professional services firm that applies technology, science, and management skills to enhance its customers' performance.
*Dennis Rankin, VP*

**2030 Water and Air Research**
6821 SW Archer Road 352-372-1500
Gainesville, FL 32608 800-242-4927
Fax: 352-378-1500
E-mail: services@waterandair.com
http://www.waterandair.com
Mission is to be an international environmental consulting firm that achieves extraordinary results by partnering with clients that to make informed and responsible decisions regarding the environment.

*William C Zegel, President*
*William Kinser, Director/Manager*

**2031 Weavertown Group Optimal Technologies**
2 Dorrington Road 724-746-4850
Carnegie, PA 15106 800-746-4850
Fax: 724-746-9024
E-mail: optimal@optimaltech.com
http://www.weavertown.com
We are an environmental engineering and consulting firm.
*Dawn Fuchs, President*

**2032 Wenck Associates**
1800 Pioneer Creek Center 763-479-4200
PO Box 249 800-472-2232
Maple Plain, MN 55359 Fax: 763-479-4242
E-mail: wenckmp@wenck.com
http://www.wenck.com
Our mission is to provide our customers strategic advice and technical excellence.

**2033 Weston Solutions, Inc**
1400 Weston Way 610-701-3000
Box 2653 800-7WE-STON
West Chester, PA 19380 Fax: 610-701-3186
E-mail: info@westonsolutions.com
http://www.westonsolutions.com
Weston is a leading infrastructure redevelopment services firm delivering integrated environmental engineering solutions to industry and government worldwide. With an emphasis on creating lasting economic value for its clients, thecompany provides services in site remediation, redevelopment, infrastructure operations and knowledge management.
*Stacey Bamihos, Marketing Dev/Advertising*
*Patrick G McCann, CEO/President*

**2034 Zapata Engineering, Blackhawk Division**
301 Commercial Road 303-278-8700
Suite B Fax: 303-278-0789
Golden, CO 80401 E-mail: geoinfo@blackhawkgeo.com
http://www.blackhawkgeo.com
High quality geophysical contracting and consulting services over the full spectrum of geophysical technologies, and to apply the geophysical technologies to several cross-cutting areas of engineering and exploration.
*Jim Hild, Manager/Senior Geophysicist*

# Environmental Health

## Associations

**2035 Acadia Environmental Society**
626 Old Students' Union Building 902-585-2149
Wolfville, Nova Scotia B4P-2R6 Fax: 902-542-3901
E-mail: aes@acadiau.ca
Provides resources on environmental issues. The Society's goal is to encourage and help the Acadia community to adopt and maintain environmentally sound and sustainable practices.
*Hillary Barter, Coordinator*

**2036 Acid Rain Foundation**
1410 Varsity Drive 919-828-9443
Raleigh, NC 27606 Fax: 919-515-3593
*Dr. Harriett S Stubbs, Executive Director*

**2037 Action on Smoking and Health**
2013 H Street NW 202-659-4310
Washington, DC 20006 http://ash.org
Organized to use the power of the law to protect the rights of nonsmokers. Emphasis is placed on legal efforts to protect nonsmokers and to get courts to support the rights of nonsmokers. Also conducts educational and awarenesscampaigns regarding the problem of smoking and the rights of nonsmokers.
*John Banzhaf, Executive Director*

**2038 Advanced Foods & Materials Network**
150 Research Lane 519-822-6253
Suite 215 Fax: 519-824-8453
Guelph ON N1G-4T2 http://www.afmnet.ca
Canada's front line of research and development in the area of advanced foods and bio-materials, including new, low-cost antibiotics, improved frozen food quality, and fast healing wound dressings.
*Ron Woznow, CEO*
*Rickey Yada, Chief Research Officer*

**2039 Agency for Toxic Substances and Disease Registry**
1825 Century Blvd 800-232-4636
Atlanta, GA 30345 E-mail: cdcinfo@cdc.gov
http://www.atsdr.cdc.gov
The mission of the agency is to prevent exposure and adverse human health effects and diminished quality of life associated with exposure to hazardous substances from waste sites, unplanned releases, and other sources of pollutionpresent in the enviroment. ATSDR is an operating division of the US Department of Health and Human Services. It divids its activities between those related to a particular site and those related to a specific hazardous substance.
*Julie L. Gerberding, MD, MPH, Administrator*
*Howard Frumkin, MD, DrPH, Director*

**2040 Air and Waste Management Association**
1 Gateway Center, 3rd Floor 412-232-3444
420 Fort Duquesne 800-270-3444
Pittsburgh, PA 15222 Fax: 412-232-3450
E-mail: info@awma.org
http://www.awma.org
The Air & Waste Management Association (A&WMA) is a nonprofit, nonpartisan professional organization that provides training, information, and networking opportunities to thousands of environmental professionals in 65 countries.
*Antoon van der Vooren, PhD, PE, President*

**2041 Alliance for Acid Rain Control and Energy Policy**
444 N Capitol Street 202-624-5475
Suite 602 Fax: 202-508-3829
Washington, DC 20001

**2042 Alternatives for Community and Environment**
2181 Washington Street 617-442-3343
Roxbury, MA 02119 Fax: 617-442-2425
E-mail: info@ace-ej.org
http://www.ace-ej.org
ACE is a community-based, nonprofit, environmental justice, law and education center. ACE works in partnership with community groups from low income communities and communities of color to help them address their environmental andenvironmental heath issues by providing free legal, educational and organizing services.
*Kalila Barnett, Executive Director*
*Eugene B Benson, Program Director*

**2043 American Academy of Environmental Medicine**
7701 E Kellog Street 316-684-5500
Suite 625 Fax: 316-684-5709
Wichita, KS 67207 E-mail: administrator@aaem.com
http://www.aaem.com
The Academy is interested in expanding the knowledge of interactions between human individuals and their environment, as these may be demonstrated to be reflected in their total health. The Academy is comprised primarily of medicalprofessionals who sponsor publications, seminars and courses. A newsletter and journal are among the organization's publications.
*Gerald D Natzke, President*

**2044 American Association for the Support of Ecological Initiatives**
150 Coleman Road 860-346-2967
Middletown, CT 06457 Fax: 860-347-8459
E-mail: Wwasch@wesleyan.edu
http://http://www.wesleyan.edu/aasei
AASEI is a US 501 nonprofit organization which suports international environmental initiatives in Russian Nature Reserves. In cooperation sith Russia and foreign scientists, students, and universities, AASEI organizes scientificresearch projects, academic internships, work camps, environmental exchanges, and eco-tourism. Our aim is to provide practical support to Russina Reserves, expand opportunities for international scientific research, and promote internationalunderstanding.
*Brendan Sweeney, President/Founder*

**2045 American Association of Poison Control Centers**
515 King Street 703-894-1858
Suite 510 Fax: 703-683-2812
Alexandria, VA 22314 E-mail: info@aapcc.org
http://www.aapcc.org
A non-profit national organization that represents the poison control centers of the United States and the interests of poison prevention and treatment of poisoning.

*Jim Hirt, Executive Director*

**2046 American Board of Environmental Medicine**
65 Wehrle Drive 716-833-2213
Buffalo, NY 14225 Fax: 716-833-2244
http://www.americanboardofenvironmentalmedicine.org
To establish and maintain the educational and testing criteria for board certification to ensure optimal standard and quality of the environmental physician.
*Dr K Patel, President*

**2047 American Board of Industrial Hygiene**
6015 W St Joseph Highway 517-321-2638
Suite 102 Fax: 517-321-4624
Lansing, MI 48917 E-mail: abih@abih.org
http://www.abih.org
Premier organization for certifying professionals in the practice of industrial hygiene. Responsible for ensuring high-quality certification application and examination processes, certifcation maintenance and ethics governance andenforcement.
*Lynn O'Donnell, CIH, Executive Director*

**2048 American Cancer Society**
1599 Clifton Road NW
Atlanta, GA 30329
404-329-7686
Fax: 404-321-4669
http://www.cancer.org

The American Cancer Society is the nationwide, community-based, voluntary health organization dedicated to eliminating cancer as a major health problem by preventing cancer, saving lives, and diminishing suffering from cancer throughresearch, education, advocacy and service.

*John R Seffrin PhD, CEO*

**2049 American College of Occupational and Environmental Medicine**
25 NW Point Blvd
Suite 700
Elk Grove Village, IL 60007
847-818-1800
Fax: 847-818-9266
E-mail: mdreger@acoem.org
http://www.acoem.org

Made up of physicians in industry, government, academia, private practice and the military, who promote the health of workers through preventive medicine, clinical care, research and education.

*Robert K McLellan, President*

**2050 American Conference of Governmental Industrial Hygienists**
1330 Kemper Meadow Drive
Cincinnati, OH 45240
513-742-6163
Fax: 513-742-3355
E-mail: mail@acgih.org
http://www.acgih.org

The American Conference of Governmental Industrial Hygeienists (ACGIH) is a member-based organization and community of professionals that advances worker health and saftey through education and the development and dissemination ofscientific and technical knowledge.

*90 pages Magazine*
*Beverly S Cohen, Chair*
*Lawrence M Gibbs, Vice Chair*

**2051 American Council on Science and Health**
1995 Broadway
2nd Floor
New York, NY 10023
212-362-7044
866-905-2694
Fax: 212-362-4919
E-mail: acsh@acsh.org
http://www.acsh.org

A consumer education consortium concerned with issues related to food, nutrition, chemicals, pharmaceuticals, lifestyle, the environment and health.

*Elizabeth M Whelan, President*
*Gilbert Ross MD, Medical/Executive Director*

**2052 American Indian Environmental Office**
1200 Pennsylvania Avenue NW
Washington, DC 20460
202-564-0303
Fax: 202-564-0298
http://www.epa.gov/indian

Coordinates the US environmental Protection Agency-wide effort to strengthen public health and environmental protection in Indian Country, with a special emphasis on building Tribal capacity to administer their own environmentalprograms.

*Carol Jorgensen, Director*
*Christopher Hoff, Acting Deputy Director*

**2053 American Industrial Hygiene Association**
2700 Prosperity Avenue
Suite 250
Fairfax, VA 22031
703-849-8888
Fax: 703-207-3561
E-mail: infonet@aiha.org
http://www.aiha.org

To promote the highest quality of occupational and environmental health and safety within the workplace and the community through advocacy and the provision of services and tools to enhance the professional practice of our members.

*Peter J O'Neil, Executive Director*
*Cathy L Cole CIH CSP, President*

**2054 American Institute of Biological Sciences**
1444 I Street NW
Suite 200
Washington, DC 20005
202-628-1500
Fax: 202-628-1509
E-mail: rogrady@aibs.org
http://www.aibs.org

AIBS facilities communication and interactions among biologists, biological societies, and biological disciplines in order to serve and advance the interests of organismal and integrative biology in the broader scientific community andother components of society on issues related to research, education, and public policy.

*Dr. Richard O'Grady, Executive Director*

**2055 American Lung Association**
61 Broadway
6th Floor
New York, NY 10006
212-315-8700
http://www.lungusa.org

The American Lung Association has been fighting lung disease in all its forms with emphasis on environmental health, asthma, and tobacco control. The work continues as they strive to make breathing easier for everyone througheducation, community service, advocacy and research programs.

*Stephen J Nolasn Esq, President*

**2056 American Medical Association**
515 N State Street
Chicago, IL 60610
312-464-5000
800-621-8335
Fax: 312-464-4184
http://www.ama-assn.org

Mission: To promote the art and science of medicine and the betterment of public health.

*Ronald M Davis, President*

**2057 American Public Health Association**
800 I Street NW
Washington, DC 20001
202-777-2742
Fax: 202-777-2534
E-mail: comments@apha.org
http://www.apha.org

Aims to protect all Americans and their communities from preventable, serious health threats and strives to assure community-based health promotion and disease prevention activities and preventive health services are universallyaccessible in the United States.

*Georges C Benjamin, Executive Director*

**2058 American Society for Microbiology**
1752 N Street NW
Washington, DC 20036
202-737-3600
Fax: 202-942-9333
E-mail: oed@asmusa.org
http://www.asm.org

A scientific society of individuals interested inthe microbiological sciences. The mission is to advance microbiological sciences through the pursuit of scientific knowledge and dissemination of the results of fundamental and appliedresearch.

*43,000 Members*
*Alison O'Brien, President*

**2059 American Society of Safety Engineers**
1800 East Oakton Street
Des Plaines, IL 60018
847-699-2929
Fax: 847-768-3434
E-mail: customerservice@asse.org
http://www.asse.org

ASSE is a global association providing professional development and representation for those engaged in the practice of safety, health and environmental issues. Provides services to the private and public sectors to protect people,property and the environment.

*Michael W Thompson, President*
*Fred J Fortman, Executive Director*

**2060 Appalachian States Low-Level Radioactive Waste Commission**
Pennsylvania DEP/BRP
400 Market Street, 13th Floor
Harrisburg, PA 17101
410-537-3345
Fax: 410-537-4133
E-mail: kmcginty@state.pa.us
http://www.dep.state.pa.us

The commission was ratified by Maryland, Delaware, Pennsylvania and West Virginia to assure intertstate cooperation for the proper packaging and transportation of low-level radioactive waste. Pennsylvania is the host state and handlesthe administrative duties of the commission at this time.

*Kathleen A McGinty, Chair/Executive Director*
*Richard R Janati, Administrator*

**2061 Asbestos Information Association of North America**
PO Box 2227 703-560-2980
Arlington, VA 22202 Fax: 703-560-2981
E-mail: aiabjpigg@aol.com
The Asbestos Information Association/North America was founded in 1970 to represent the interest of the asbestos industry and to collect and disseminate information about asbestos and asbestos products, with emphasis on safety, health,and environmental issues. The Association appears before Federal regulatory bodies and works with Government agencies to develop and implement standards for worker protection.

*Bob Pigg, President*

**2062 Asian Pacific Environmental Network**
310 8th Street 510-834-8920
Suite 309 Fax: 510-834-8926
Oakland, CA 94607 E-mail: apen@apen4ej.org
http://www.apen4ej.org
The Asian Pacific Environmental Network (APEN) empowers low-income Asian Pacific Islander (API) communities to take action on environmental and social justice issues. APEN builds organizations in dis-empowered API communities todevelop lasting capacity of the community to achieve solutions to problems affecting people's lives.
*Vivian Chang, Executive Director*

**2063 Association for Environmental Health and Sciences**
150 Fearing Street 413-549-5170
Amherst, MA 01002 Fax: 413-549-0579
http://www.aehs.com
The Association for Environmental Health and Sciences (AEHS) was created to facilitate communication and foster cooperation among professionals concerned with the challenge of soil protection and cleanup.
*Paul Kostecki, Executive Director*
*Marc A Nascarella, Managing Ed/Conference Coor*

**2064 Association of American Pesticide Control Officials**
PO Box 466 302-422-8152
Milford, DE 19963 Fax: 302-422-2435
E-mail: aapco-sfireg@comcast.net
http://www.aapco.org
Organization formed to provide a rational forum and representation for state pesticide control officials in the development, implementation, and communication of parties and programs related to the sale, transport, application anddisposal of pesticide.

*Chuck Andrews, President*
*Gena Davis, President-Elect*

**2065 Association of Battery Recyclers**
PO Box 290286 813-626-6151
Tampa, FL 33687 Fax: 813-622-8388
E-mail: info@batteryrecyclers.com
http://www.batteryrecyclers.com
The Association of Battery Recyclers is a non-profit trade association. ABR strives to keep its members abreast on environmental and health matters and also provides a means for communication with government officials on issuesaffecting the lead recycling industry.
*Joyce Morales, Contact*

**2066 Association of State and Territorial Health Officials**
2231 Crystal Drive 202-371-9090
Suite 450 Fax: 202-527-3189
Arlington, VA 22202 E-mail: pjarris@astho.org
http://www.astho.org
Dedicated to formulating and influencing sound public health policy, and to assuring excellence in state-based public health practice.
*Paul E Jarris MD, Executive Director*
*Gino D Marinucci, Sr Dir Enviro Health Policy*

**2067 Asthma and Allergy Foundation of America**
1233 20th Street, NW 202-466-7643
Suite 402 Fax: 202-466-8940
Washington, DC 20036 E-mail: info@aafa.org
http://www.aafa.org
AAFA provides practical information, community based services and support through a national network of chapters and support groups. AAFA develops health education, organizes state and national advocacy efforts and funds research tofind better treatments and cures.
*Chris Ward, President*
*Mary Brasler, Programs & Services Director*

**2068 Beyond Pesticides**
701 E Street SE 202-543-5450
Suite 200 Fax: 202-543-4791
Washington, DC 20003 E-mail: info@beyondpesticides.org
http://www.beyondpesticides.org
Beyond Pesticides works with allies in protecting public health and the environment to lead the transition to a world free of toxic pesticides.
*Robina Suwol, President*
*Jay Feldman, Executive Director*

**2069 Bio Integral Resource Center**
PO Box 7414 510-524-2567
Berkeley, CA 94707 Fax: 510-524-1758
E-mail: birc@igc.org
http://www.birc.org
The goal of the Bio Integral Resources Center is to reduce pesticide use by educating the public about effective, less toxic alternatives for pest problems.
*Dr. William Quarles, Executive Director*

**2070 Bison World**
National Bison Association
8690 Wolff Ct 303-292-2833
Suite 200 Fax: 303-659-3739
Westminster, CO 80031 E-mail: info@bisoncentral.com
http://www.bisoncentral.com
An organization of bison producers dedicated to awareness of the healthy properties of bison meat and bison production.
*Dave Carter, Executive Director*

**2071 Center for Health, Environment and Justice Library**
150 S Washington Street, Ste 300 703-237-2249
PO Box 6806 Fax: 703-237-8389
Falls Church, VA 22040-6806 E-mail: chej@chej.org
http://www.chej.org
The Center for Health, Environment and Justice works to build healthy communities, with social justice, economic well-being, and democratic governance. We believe this can happen when individuals from communities have the power to playan integral role in promoting human health and environmental integrity. Our role is to provide the tools to build strong, healthy communities where people can live, work, learn, play and pray.

*Lois Marie Gibbs, Executive Director*

**2072 Center for Science in the Public Interest**
1875 Connecticut Avenue NW 202-332-9110
Suite 300 Fax: 202-265-4954
Washington, DC 20009 E-mail: cspi@cspinet.org
http://www.cspinet.org
Mission: To provide useful, objective information to the public and policymakers and to conduct research on food, alcohol, health, the environment, and other issues related to science and technology; to represent the citizen'sinterests before regulatory, judicial and legislative bodies on food, alcohol, health, the environment, and other issues; and to ensure that science and technol-

ogy are used for the public good and to encourage scientists to engage in public-interestactivities.
*Kathleen O'Reilly, President*
*Michael F Jacobson, Treasurer*

### 2073 Center for the Evaluation of Risks to Human Reproduction

Po Box 12233
Research Triangle Park, NC 27709
919-541-3455
Fax: 919-316-4511
E-mail: shelby@niehs.nih.gov
http://http://cerhr.niehs.nih.gov

The Center's mission includes the following: to provide timely and unbiased, scientifically sound assessments of reproductive health hazards associated with human exposure to naturally occurring and man-made chemicals; to make theseassessments readily available to the public, to state and federal agencies and to the scientific community; and to build an electronic resource for providing, or directing one to, information of public interest concerning human reproductive health.
*Dr. Michael D Shelby*

### 2074 Centers for Disease Control & Prevention

National Center for Environmental Health
1600 Clifton Road
Atlanta, GA 30333
404-639-3311
800-311-3435
http://www.cdc.gov/nceh

To provide national leadership, through science and service, to promote health and quality of life by preventing and controlling disease and death resulting from interactions between people and their environment.

### 2075 Chemical Injury Information Network

PO Box 301
White Sulphur Springs, MT 59645
406-547-2255
Fax: 406-547-2455
E-mail: chemicalinjury@ciin.org
http://www.ciin.org

Nonprofit tax-exempt support and advocacy organization run by the chemically injured for the benefit of the chemically injured. CIIN serves an international membership, and focuses primarily on eductaion, credible multiple sensitivityresearch and the empowerment of the chemically injured.

*Cinthia Wilson, Executive Director*
*John Wilson, President*

### 2076 Chlorine Institute

1300 Wilson Boulevard
Arlington, VA 22209
703-741-5760
Fax: 703-741-6068
http://www.chlorineinstitute.org

Exists to support the chlor-alkali industry and serve the public by fostering continuous improvements to safety and the protection of human health and the environment connected with the production, distribution and use of chlorine,sodium and potassium hydoroxides, and sodium hypochlorite; and the distribution and use of hydrogen chloride.

*Arthur E Dungan, President*

### 2077 Columbia Analytical Services

1317 South 13th Avenue
Kelso, WA 98626
360-577-7222
800-695-7222
Fax: 360-425-9096
http://www.caslab.com

Areas of expertise and services include environmental testing of air, water, soil, hazardous waste, sediments and tissues; process and quality control testing; analytical method development; sampling and mobile laboratory services; andconsulting and data management services.

*Stephen W Vincent, President/CEO*
*Ed Wilson, Employee Representative*

### 2078 Commonweal

PO Box 316
Bolinas, CA 94924
415-868-0970
Fax: 415-868-2230
E-mail: commonweal@commonweal.org
http://www.commonweal.org

A health and environment research institute that conducts programs that contribute to human and ecosystem health. The Commonweal Health and Environment Program focuses on environmental contaminants.

*Michael Lerner, President*
*Waz Thomas, General Manager*

### 2079 Communities for a Better Environment

1440 Broadway
Suite 701
Oakland, CA 94612
510-302-0430
Fax: 510-302-0437
E-mail: cbeca@mail.com
http://www.cbecal.org

Mission: To achieve environmental health and justice by building grassroots power in and with communities of color and working-class communities.
*Richard Toshiyuki Drury, President*
*Lizette Ruiz, Vice President*

### 2080 Community-Based Hazard Management Program

George Perkins Marsh Institute
Clark University
950 Main Street
Worcester, MA 01610
508-793-7711
E-mail: otaylor@clarku.edu
http://www.clarku.edu

The Community-Based Hazardous Management Program (formerly the Childhood Cancer Research Institute) is engaged in capacity building in communities affected by nuclear weapons production and testing and also specializes in radiationhealth risk assessment and management.
*Octavia Taylor, Program Manager*

### 2081 Corporate Accountability International

10 Milk Street
Suite 610
Boston, MA 02108
617-695-2525
800-688-8797
Fax: 617-695-2626
E-mail: info@stopcorporateabuse.org
http://www.stopcorporateabuse.org

For more than 30 years, Corporate Accountability International has successfully challenged corporations like GE, NestlŠ, and Philip Morris to halt abusive practices that threaten public health, the environment and our democracy. Todayour campaigns challenge the dangerous practices of some of the world's most powerful industries.

*Kelle Louaillier, Executive Director*
*Leslie Samuelrich, Chief of Staff*

### 2082 Council of State and Territorial Epidemiologists (CSTE)

2872 Woodcock Boulevard
Suite 303
Atlanta, GA 30341
770-458-3811
Fax: 770-458-8516
http://www.cste.org

CSTE promotes the effective use of epidemiologic data to guide public health practice and improve health. CSTE accomplishes this by supporting the use of effective public health surveillance and good epidemiologic practice throughtraining, capacity development, and peer consultation, developing standards for practice, and advocating for resources and scientifically based policy.
*Melvin Kohn, President*
*Perry Smith, Executive Director*

### 2083 Dangerous Goods Advisory Council

1100 H Street NW
Suite 740
Washington, DC 20005
202-289-4550
Fax: 202-289-4074
E-mail: info@dgac.org
http://www.hmac.org

HMAC promotes improvement in the safe transportation of hazardous materials/dangerous goods globally by: providing education, assistance, and information to the private and public sectors; through our unique status with regulatorybodies; and the adversity and technical strengths of our membership.
*Mike Morrissette, President*
*Alan I Roberts, Vice President*

**2084 Earth Regeneration Society**
1442A Walnut Street 510-849-4155
# 57 Fax: 510-849-0183
Berkeley, CA 94709 E-mail: csiri@igc.apc.org
http://www.imaja.com/as/environment/ers
The Earth Regeneration Society does research and education on climate change, ozone, and pollution, and calls for full employment and full social support based on surival programs and national and international networking.
*Alden Bryant, President*

**2085 EarthSave International**
PO Box 96 718-459-7503
New York, NY 10108 800-362-3648
Fax: 718-228-2491
E-mail: information@earthsave.org
http://www.earthsave.org
Educates people on the powerful effects that our food choices have on the environment, our health, and all life on Earth, and supports people in moving toward a plant-based diet. Founded by John Robbins, author of Diet for a NewAmerica.
*Patricia Carney, Executive Director*

**2086 Environmental Defense Fund**
257 Park Avenue South 212-505-2100
17th Floor Fax: 212-505-2375
New York, NY 10010 E-mail: members@edf.org
http://www.edf.org
Dedicated to protecting the environmental rights of all people, including future generations. Among these rights are clean air, clean water, healthy, nourishing food and a flourishing ecosystem. Advocates solutions based on science,even when it leads in unfamiliar directions. Works to create solutions that win lasting political, economic and social support because they are bipartisan, efficient and fair.
*Fred Krupp, President*
*David Yarnold, Executive Director*

**2087 Environmental Hazards Management Institute**
10 New Market Road 603-868-1496
Durham, NH 03821 800-558-3464
Fax: 603-868-1547
E-mail: info@ehmi.org
http://www.ehmi.org
An independent, nonprofit organization dedicated to understanding enhancement and preservation of our environment. A catalyst for informed environmental decision making by gathering, refining, and disseminating objective information toall stakeholders with emphasis on the role played by individuals and communities of individuals.
*Alan J Borner, President*

**2088 Environmental Health Coalition**
401 Mile of Cars Way 619-474-0220
Suite 310 Fax: 619-474-1210
National City, CA 91950 E-mail: ehc@environmentalhealth.org
http://www.environmentalhealth.org
One of the oldest and most effective grassroots organizations in the US, using social change strategies to achive environmental and social justice. We believe that justice is accomplished by empowered communities acting together tomake social change. We organize and advocate to protect public health and the environment threatened by toxic pollution. EHC supports broad efforts that create a just society which foster a healthy and sustainable quality of life.
*Diane Takvorian, Executive Director*
*Sonya Holmquist, Associate Director*

**2089 Environmental Health Education Center**
655 West Lombard Street 410-706-1849
Room 665 Fax: 410-706-0295
Baltimore, MD 21201 http://www.envirn.umaryland.edu
Mission: Supporting nursing professionals seeking accurate, timely and credible scientific information on environmental health and nursing. The ultimate goal is to prevent environmental disease by increasing the numbers of nursingprofessionals who can recognize environmental etiologies and risk factors of disease, promote health through risk reduction and control strategies and empower individuals, families and communities through partnering, advocacy and education.
*Barbara Sattler RN DrPH*

**2090 Environmental Health Network**
PO Box 1155 415-541-5075
Larkspur, CA 94977 http://www.chnca.org
Nonprofit, volunteer organization, whose main goal is to promote public awareness of environmental sensitivities and causative factors. EHN's focus is on issues of access and developments relating to the health and welfare of theenvironmentally sensitive.

**2091 Environmental Health Strategy Center**
PO Box 2217 207-827-6331
Bangor, ME 04402 Fax: 207-827-5755
E-mail: info@preventharm.org
http://www.preventharm.org
The Environmental Health Strategy Center works to protect human health by reducing exposure to toxic chemicals, expanding the use of safer alternatives, and building partnerships that focus on the environment as a public healthpriority.
*Michael Belliveau, Executive Director*

**2092 Environmental Information Association**
6935 Wisconsin Avenue 301-961-4999
Suite 306 888-343-4342
Chevy Chase, MD 20815 Fax: 301-961-3094
E-mail: info@eia-usa.org
http://www.eia-usa.org
A nonprofit organization dedicated to providing environmental information to individuals, members, and industry. They specialize in the dissemination of information about the abatement of asbestos and lead based paint, and about safetyand health issues, analytical issues and environmental site assessments.
*Brent Kynoch, Managing Director*
*Kelly Rutt, Development/Communications*

**2093 Environmental Justice Resource Center**
223 James Brawley Drive SW 404-880-6911
Atlanta, GA 30314 Fax: 404-880-6909
E-mail: ejrc@cau.edu
http://www.ejrc.cau.edu
Since 1994, a research, policy and information clearinghouse on issues related to environmental justice, race and the environment, civil rights, facility siting, land use planning, brownfields, transportation equity, suburban sprawland Smart Growth. The overall goal of the center is to assist, support, train and educate people of color, students, professionals and grassroots community leaders with the goal of facilitating their inclusion into the mainstream of environmentaldecision-making.
*Robert D Bullard PhD, Director*

**2094 Environmental Mutagen Society**
1821 Michael Faraday Drive 703-438-8220
Suite 300 Fax: 703-438-3113
Reston, VA 20190 E-mail: emshq@ems-us.org
http://www.ems-us.org
The Environmental Mutagen Society is the primary scientific society fostering research on the basic mechanisms of mutagensis as well as on the application of this knowledge in the field of genetic toxicology. EMS has seven corescientific content areas.
*Martina Veigl, President*

**2095 Environmental Resource Center**
471 Washington Ave. N 208-726-4333
PO Box 819 Fax: 208-726-1531
Ketchum, ID 83340 E-mail: jennifer@ercsv.org
http://www.ercsv.org
An oraganization offering environmental education for the community.
*Jennifer Colson, Program Director*

**2096 Environmental Resource Management (ERM)**
1001 Connecticut Avenue NW 202-466-9090
Suite 1115 Fax: 202-466-9191
Washington, DC 20036 http://www.erm.com
ERM works around the world with the private sector assessing how their business is likely to be impacted by environmental and social issues, new regulations, consumer concerns and supply chain issues and help companies developappropriate policies and management systems to manage these business risks.
*Robin Bidwell, Executive Chair*
*Peter Regan, CEO*

**2097 Environmental Safety**
1700 North Moore Street 703-527-8300
Arlington, VA 22209 Fax: 703-527-8383
E-mail: usprogram@ashoka.org
http://www.ashoka.org
Ashoka's mission is to shape a citizen sector that is entrepreneurial, productive and globally integrated, and to develop the profession of social entrepreneurship around the world. Ashoka identifies and invests in leading socialentrepreneurs-extraordinary individuals and unprecedented ideas for change in their communities-supporting the individual, idea and institution through all phases of their career. Once elected to Ashoka, Fellows benefit from being part of the globalfellowship for life.
*Barbara Kazdan, Director*

**2098 Environmental Working Group**
1436 U Street Northwest 202-667-6982
Suite 100 Fax: 202-232-2592
Washington, DC 20009 http://www.ewg.org
Mission: To use the power of public information to protect public health and the environment
*Ken Cook, President*
*Richard Wiles, Executive Director*

**2099 Environmental and Occupational Health Science Institute**
Rutgers University
170 Frelinghuysen Road 732-445-0200
Piscataway, NJ 08854 Fax: 732-445-0131
E-mail: webmaster@eohsi.rutgers.edu
http://www.eohsi.rutgers.edu
Environmental and Occupational Health Sciences Institute sponsors research, education and service programs in a setting that fosters interaction among experts in environmental health, toxicology, occupational health, exposureassessment, public policy and health education. The Institute also serves as an unbiased source of expertise about environmental problems for communities, employers and government in all areas of occupational and environmental health, toxicology andrisk assessment.
*Kenneth Reuhl PhD, Associate Director*

**2100 Food Safety and Inspection Service**
Food Safety Education Office
1400 Independence Avenue SW 301-504-9605
Washington, DC 20250 Fax: 301-504-0203
E-mail: fsis.webmaster@usda.gov
http://www.fsis.usda.gov
The Food Safety and Inspection Services (FSIS) is the public health agency in the U.S. Department of Agriculture responsible for ensuring that the nation's commercial supply of meat, poultry, and egg products is safe, wholesome, andcorrectly labeled and packaged.

**2101 Food and Drug Administration**
US Department of Health and Human Services
Office of Public Affairs 301-827-3666
5600 Fishers Lane 888-463-6332
Rockville, MD 20857 Fax: 301-443-0017
http://www.fda.gov
The FDA is responsible for protecting the public health by assuring the safety, efficacy, and security of human and veterinary drugs, biological products, medical devices, our nation's food supply, cosmetics, and products that emitradiation.
*Lester M Crawford, DVM, PhD, Commissioner*

**2102 Friends of the River**
915 20th Street 916-442-3155
Sacramento, CA 95814 Fax: 916-442-3396
E-mail: info@friendsoftheriver.org
http://www.friendsoftheriver.org
Friends of the River educates, organizes, and advocates to protect and restore California rivers, streams, and watersheds.
*Pete Ferenbach, Executive Director*

**2103 Halogenated Solvents Industry Alliance**
1300 Wilson Boulevard 703-741-5780
12th Floor Fax: 703-741-6077
Arlington, VA 22209 E-mail: info@hsia.org
http://www.hsia.org
The mission is to present the interests of users and producers of chlorinated solvents. To promote the continued safe use of these products and to promote the use of sound science in assessing their potential health effects.

*Steven Risotto, Executive Director*

**2104 Hazardous Waste Resource Center Environmental Technology Council**
1112 16th Street NW 202-783-0870
Suite 420 Fax: 202-737-2038
Washington, DC 20036 E-mail: mail@etc.org
http://www.etc.org
The Environmental Technology Council (ETC) is a trade association of commercial firms that recycle, treat and dispose of industrial and hazardous wastes; and firms involved in cleanup of contaminated sites.
*David R Case, Executive Director*
*Scott Slesunger, VP Goverment Affairs*

**2105 Holistic Management International**
1010 Tijeras Ave NW 505-842-5252
Albuquerque, NM 87102 Fax: 505-843-7900
E-mail: hmi@holisticmanagement.org
http://www.holisticmanagement.org
Enhance the efficiency, natural health, productivity and profitability of their land; increase natural annual profits; provide a framework for family, owners, managers, foreman, communal agriculturalists and other ranch/farmstakeholders to work together toward a common future; and enable development agencies working with marginalized farmers or pastoral people to break the cycle of food and water insecurity.

*Peter Holter, CEO*
*Tracy Favre, Sr Director Contract Sales*

**2106 Human Ecology Action League (HEAL)**
PO Box 29629 404-248-1898
Atlanta, GA 30359 Fax: 404-248-0162
E-mail: HEALNatnl@aol.com
http://members.aol.com/HEALNatnl/index/html
The Human Ecology Action League Inc (HEAL) is a nonprofit organization founded in 1977 to serve those whose health has been adversely affected by environment exposures; to provide information to those who are concerned about the healtheffects of chemicals; and to alert the general public about the potential dangers of chemicals. Referrals to local HEAL chapters and other support groups are available from the League.
*Katherine P Collier, Contact*

**2107 INFORM**
318 W 39th Street 212-361-2400
5th Floor Fax: 212-361-2412
New York, NY 10018 E-mail: ramsey@informinc.org
http://www.informinc.org
Dedicated to educating the public about the effects of human activity on the environment and public health. The goal is to empower citizens, businesses and government to adopt practices and policies that will sustain our planet forfuture generations.
*Virginia Ramsey, President*
*Yon Lam, Communications*

**2108 Institute for Agriculture and Trade Policy**
2105 First Avenue South 612-870-0453
Minneapolis, MN 55404 Fax: 612-870-4846
http://www.sustain.org
The mission of the Institute for Agriculture and Trade Policy is to foster socially, economically and environmentally sustainable rural communities and regions.
*Jim Harkness, President*

**2109 Institute of Hazardous Materials Management**
11900 Parklawn Drive 301-984-8969
Suite 450 Fax: 301-984-1516
Rockville, MD 20852 E-mail: ihmminfo@ihmm.org
http://www.ihmm.org
Mission is to provide recognition for professionals engaged in the managment and engineering control of hazardous materials who have attained the required level of education, experience and competence; foster continued professionaldevelopment of Certified Hazardous Materials Managers (CHMM).
*John H Frick, PhD, CHMM, Executive Director*
*Betty Fishman, Assistant Executive Director*

**2110 Laotian Organizing Project**
310 8th Street 510-834-8920
Suite 309 Fax: 510-834-8926
Oakland, CA 94607 E-mail: apen@apen4ej.org
http://www.apen4ej.org
The Laotian Organization Project (LOP), a project of the Asian Pacific Environmental Network, is a membership-based organization of Laotian residents in Richmond and San Pablo, California. LOP works to bring people from the Laotiancommunity together to identify problems, develop solutions, and take action for a more healthy, safe, and just community.
*Vivian Chang, Executive Director*
*Elaine Shen, Chair*

**2111 MCS Referral and Resources**
618 Wyndhurst Avenue #2 410-889-6666
Baltimore, MD 21210 Fax: 410-889-4944
E-mail: adonnay@mcsrr.org
http://www.mcsrr.org
The mission of MCS Refferal and Resources is to further the diagnosis, treatment, accomodation and prevention of multiple chemical sensitivity (MCS) disorders.
*Dr. Anne McCampbell*

**2112 Midwest Center for Environmental Science and Public Policy**
One East Hazelwood Drive
Champaign, IL 61820 800-407-0261
E-mail: glrppr@istc.illinois.edu
http://www.glrppr.org
A professional organization dedicated to promoting information exchange and networking to P2 professionals in the Great Lakes regions of the United States and Canada
*Bob Iverson, Contact*

**2113 Mount Sinai School of Medicine: Division of Environmental Health Science**
Department of Community and Preventive Medicine
1 Gustave Levy Place 212-241-8689
Box 1043 Fax: 212-360-6965
New York, NY 10029 http://www.mssm.edu/cpm
The Division's ultimate goal is the protection of the public's health by understanding, elucidating and preventing diseases that arise from environmental exposures.
*Philip J Landrigan M.D.*

**2114 National Alliance for Hispanic Health**
1501 16th Street NW 202-387-5000
Washington, DC 20036 Fax: 202-797-4353
E-mail: alliance@hispanichealth.org
http://www.hispanichealth.org
The mission of the National Alliance for Hispanic Health is to improve the health and well-being of Hispanics. Issues covered include the full range of health and human services issues, including environmental health.
*Jane L Delgado PhD, MS, President/CEO*

**2115 National Association of City and County Health Officials**
1100 17th Street NW 202-783-5550
2nd Floor Fax: 202-783-1583
Washington, DC 20036 http://www.naccho.org
The National Assoication of County and City Health Officials is a nonprofit, membership organization serving all 3,000 local health departments nationwide. NACCHO is dedicated to improving the health of people and communities byassuring an effective local public health system. As the Voice of local public health officials at the national level, NACCHO is able to promote the local perspective on national health programs and policies.
*Robert M Pestronk MPH, Executive Director*

**2116 National Association of Noise Control Officials**
53 Cubberly Road 609-586-2684
West Windsor, NJ 08550 Fax: 609-799-2616
http://www.arcat.com
The association consists of employees of the federal and state governments, consultants, scientists, and students concerned with acoustical control in the environment. It now has about 70 members.

*Edward J DiPolzere, Executive Director*

**2117 National Association of Physicians for the Environment**
6410 Rockledge Drive 307-571-9790
Suite 412 Fax: 301-530-8910
Bethesda, MD 20817 E-mail: nape@napenet.org
The National Association of Physicians for the Environment works to involve physicians and other health care professionals, particularly through their geographic and medical specialty organizations, to deal with the impact ofpollutants on organs and systems of the human body.
*Betty Farley, Executive Assistant*

**2118 National Association of School Nurses**
8484 Georgia Avenue 240-821-1130
Suite 420 866-627-6767
Silver Spring, MD 20910 Fax: 301-585-1791
E-mail: nasn@nasn.org
http://www.nasn.org
The mission of The National Association of School Nurses is to advance the practice of school nursing and provide leadership in the delivery of quality health programs to school communities.
*Amy Garcia, Executive Director*

**2119 National Cancer Institute**
National Institutes of Health
Building 31 Room 10A19 301-496-6641
31 Center Drive MSC 2580 Fax: 301-496-0846
Bethesda, MD 20892 E-mail: ncipressofficers@mail.nih.gov
http://www.cancer.gov
Leads the Nation's fight against cancer by supporting and conducting ground-breaking research in cancer biology, causation, prevention, detection, treatment and survivorship.
*John E Niederhuber, Director*

**2120 National Capital Poison Center**
3201 New Mexico Avenue NW
Suite 310 800-222-1222
Washington, DC 20016 Fax: 202-362-8377
E-mail: pc@poison.org
http://www.poison.org
This mission of the Poison Center is to prevent poisonings, save lives, and limit injury from poisoning. In addition, the Center decreases health care costs of poisoning cases. The Center provides 24-hour telephone guidance, teachingmaterials, and professional education.

**2121 National Center for Disease Control and Prevention**
1600 Clifton Road 404-639-3311
Atlanta, GA 30333 http://www.cdc.gov
Mission: To promote health and quality of life by preventing and controlling disease, injury, and disability.
*Lynn Austin, Chief of Staff*

**2122 National Center for Environmental Health Strategies**
1100 Rural Avenue 856-429-5358
Voorhees, NJ 08043 Fax: 856-429-5358
E-mail: mary@ncehs.org
http://www.ncehs.org/
Fosters the development of creative solutions to environmental health problems with a focus on indoor air quality, chemical sensitivites and environmental disabilities.
*Mary Lamielle, Executive Director*

**2123 National Center for Healthy Housing**
10320 Little Patuxent Parkway 410-992-0712
Suite 500 877-312-3046
Columbia, MD 21044 Fax: 443-539-4150
E-mail: rmorley@centerforhealthyhousing.org
http://www.centerforhealthyhousing.org
Formerly known as the National Center for Lead-Safe Housing, it deveops and promotes practical methods to protect children fron environmental health hazards in homes while preserving affordable housing.

*Rebecca L Morley, Executive Director*
*Jonathan W Wilson, Deputy Director*

**2124 National Conference of Local Environmental Health Administrators**
University of Washington
Department of Environmental Health 206-616-2097
Box 357234 Fax: 206-543-8123
Seattle, WA 98195 E-mail: webmaster@ncleha.org
http://www.ncleha.org
The NCLEHA's purpose is to provide a forum for local administrators to share common concerns and solutions to mutual problems, and to provide a professional organization for environmental health administrators, focused on the issuesand problems of local environmental health programs.
*Keith L Krinn, Chair*

**2125 National Conference of State Legislatures**
7700 E First Place 303-364-7700
Denver, CO 80230 Fax: 303-364-7800
http://www.ncsl.org
The National Conference of State Legislatures serves the legislators and staffs of the nation's 50 states and its commonwealths and territories. NCSL is a bipartisan organization with three objectives: to improve the quality andeffectiveness of state legislatures; to foster interstate communications and cooperation; and to ensure states a strong cohesive voice in the federal system.

*William T Pound, Executive Director*

**2126 National Education Association Health Information Network**
1201 16th Street NW 202-822-7570
Suite 216 800-718-8387
Washington, DC 20036 Fax: 202-822-7775
E-mail: info@neahin.org
http://www.neahin.org
The National Education Association Health Information Network believes that sound public education must begin with school employees and students who are healthy and free of preventable diseases and supported with information, materialsand training opportunities that reaffirm these values.
*Jerald Newberry, Executive Director*

**2127 National Environmental Coalition of Native Americans**
PO Box 988 918-342-3041
Claremore, OK 74018 E-mail: noteno_84@hotmail.com
http://www.alphacdc.com/necona/
Nonprofit organization formed to educate Indians and Non-Indians about the health dangers of radioactivity and the transport of nuclear waste on America's rails and roads. Networks with environmentalists to develop grassrootscounter-movement to the efforts of the nuclear industry and develop Tribal nuclear free zones across the nation.
*Grace Thorpe, President*

**2128 National Environmental Health Association (NEHA)**
720 S Colorado Boulevard 303-756-9090
Suite 1000-N Fax: 303-691-9490
Denver, CO 80246 E-mail: staff@NEHA.org
http://www.neha.org
NEHA is the only national association that represents all of environmental health and protection from terrorism and all-hazards preparedness, to food safety and protection and on site wastewater systems. Over 4500 members and theprofession are served by the association through its Journal of Environmental Health, Annual Education Conference and Exhibition credentialing programs, research and development activities and other services.

*Nelson Fabian, Executive Director*

**2129 National Environmental Health Science and Protection Accreditation Council**
2632 Southeast 25th Avenue 503-235-6047
Suite D Fax: 503-235-7300
Portland, OR 97202 E-mail: ehacinfo@aehap.org
http://www.ehacoffice.org
The National Environmental Health Science and Protection Accreditation Council promotes a high quality education for persons studying environmental health science and protection; promotes commonality in coverage of basic concepts ofenvironmental health science and protection education; and promotes undergraduate curricula of a quality and content compatible with admission prerequisites of graduate programs in environmental health science and protection.
*Randall K Bentley, Chair*

**2130 National Environmental Trust**
1200 18th Street NW 202-887-8800
5th Floor Fax: 202-887-8877
Washington, DC 20036 E-mail: cdelany@net.org
http://www.environet.org
Manages comprehensive media and public policy campaigns around national environmental issues.
*Kymberly Escobar, Director*

**2131 National Institute for Global Environmental Change**
University of California, Davis
2850 Spafford Street 530-757-3350
Suite A Fax: 530-756-6499
Davis, CA 95618 E-mail: nigec@ucdavis.edu
http://nigec.ucdavis.edu
Mission: To assist the nation in its response to human-induced climate and environmental change.
*Lawrence B Coleman, Interim National Director*

**2132 National Institute of Environmental Health Sciences**
Po Box 12233 919-541-3345
Research Triangle Park, MD 27709 Fax: 919-541-2260
http://www.niehs.nih.gov
The mission of the National Institute of Environmental Health Sciences is to reduce the burden of environmentally associated diseases and dysfunctions.
*David A Schwartz, Director*

**2133 National Oceanic & Atmospheric Administration**
14th Street/Constitution Avenue NW 202-482-6090
Room 6217 Fax: 202-482-3154
Washington, DC 20230 E-mail: d.james.baker@noaa.gov
http://www.noaa.gov
Describes and predicts changes in the Earth's environment and conserves and wisely manages the nation's coastal and marine re-

sources. Goals and objectives include advance short-term warning and forecast services.
*Conrad C Lautenbacher Jr, NOAA Administrator*

**2134 National Pesticide Information Center Oregon State University**
310 Weniger Hall
Oregon State University 800-858-7378
Corvallis, OR 97331 Fax: 541-737-0761
E-mail: npic@ace.orst.edu
http://http://npic.orst.edu
Provides objective, science-based information about a wide variety of pesticide-related topics, including: pesticide product information, information on the recognition and management of pesticide poisonings, toxicology andenvironmental chemistry. Highly trained specialists can also provide referrals for the following: investigation of pesticide incidents, emergency treatment information, safety information, health and environmental effects, and clean-up and disposalprocedures.
*Dave Stone, Director*

**2135 National Religious Partnership for the Environment**
49 South Pleasant Street 413-253-1515
Suite 301 Fax: 413-253-1414
Amherst, MA 01002 E-mail: nrpe@nrpe.org
http://www.nrpe.org
The mission of the National Religious Partnership for the Environment is to permanently integrate issues of environmental sustainability and justice across all aspects of organized religious life.
*Paul Gorman, Executive Director*

**2136 National Safety Council**
Environmental Health Center
1025 Connecticut Avenue NW 202-293-2270
Suite 1200 Fax: 202-293-0032
Washington, DC 20036 E-mail: info@nsc.org
http://www.nsc.org
The Mission of the Environmental Health Center is to foster improved public understanding of significant health risk and challenges facing modern society. This goal reinforces the National Safety Council's commitment to increased andmore effective citizen involvement in safety, health and environmental decision-making.
*Jeffrey Shavelson, Policy Analyst*

**2137 Natural Resources Defense Council**
40 West 20th Street 212-727-2700
New York, NY 10011 Fax: 212-727-1773
E-mail: nrdcinfo@nrdc.org
http://www.nrdc.org
Mission: To safeguard the Earth: its people, its plants and animals, and natural systems on which all life depends.
*Frances Beinecke, President*
*Peter Lehner, Executive Director*

**2138 Navy Environmental Health Center**
620 John Paul Jones Circle 757-953-0700
Suite 1100 Fax: 757-953-0999
Portsmouth, VA 23708 http://www.-nehc.med.navy.mil
Ensures Navy and Marine Corps readiness through leadership in prevention of disease and promotion of health.
*Captain David Hiland, Commanding Officer*

**2139 Navy and Marine Corps Public Health Center**
620 John Paul Jones Circle 757-953-0700
Suite 1100 E-mail: ask-nmcphc@med.navy.mil
Portsmouth, VA 23708 http://www-nehc.med.navy.mil
The Navy and Marine Corps center for public health services that provides leadership and expertise to ensure mission readiness through disease prevention and health promotion in support of the National Military Strategy.
*CAPT Bruce A Cohen MC USN, Commanding Officer*
*CAPT Mike Henderson MSC USN, Executive Officer*

**2140 Noise Pollution Clearinghouse**
PO Box 1137
Montpelier, VT 05601 888-200-8332
E-mail: webmaster@nonoise.org
http://www.nonoise.org
The Noise Pollution Clearinghouse is a nonprofit organization with extensive online noise related resources. The mission is to create more civil cities and more natural rural and wilderness areas by reducing noise pollution and itssources.

**2141 North American Association for Environmental Education**
2000 P Street NW 202-419-0412
Suite 540 Fax: 202-419-0415
Washington, DC 20036 E-mail: brian@naaee.org
http://www.naaee.org
A professional association for environmental education that promotes excellence in environmental education and serves environmental educators for the purpose of achieving environmental literacy in order for present and futuregenerations to benefit from a safe and health environment and a better quality of life.
*Brian A Day, Executive Director*
*Sue Bumpous, Program/Communications Mgr*

**2142 Northwest Coalition for Alternatives to Pesticides**
PO Box 1393 541-344-5044
Eugene, OR 97440 Fax: 541-344-6923
E-mail: info@pesticide.org
http://www.pesticide.org
The Northwest Coalition for Alternatives to Pesticides protects the health of people and the environment by advancing alternatives to pesticides.
*Jean Cameron, President*
*Norma Grier, Executive Director*

**2143 Novozymes North America Inc**
77 Perry Chapel Church Road 919-494-3000
Franklinton, NC 27525 Fax: 919-494-3450
E-mail: enzymesna@novozymes.com
http://www.novozymes.com/en
Novozymes is the biotech bases world leader in enzymes and microorganisms. Using nature's own technologies, they continuously expand the frontiers of biological solutions to improve industrial performance everywhere.
*Steen Riisgaard, President/CEO*

**2144 Occupational Safety and Health Administration: US Department of Labor**
Office of Administrative Services
200 Constitution Avenue NW 202-693-1999
Room N-310 800-321-6742
Washington, DC 20210 http://www.osha.gov
Mission: To assure the safety and health of America's workers by setting and enforcing standards; providing training, outreach, and education; establishing partnerships; and encouraging continual improvement in workplace safety andhealth.

**2145 Pesticide Action Network North America**
49 Powell Street 415-981-1771
Sutie 500 Fax: 415-981-1991
San Francisco, CA 94102 E-mail: panna@panna.org
http://www.panna.org
Pesticide Action Network North American advocates the adoption of ecologically sound practices in place of hazardous pestices in place of pesticide use. PANNA works with more than 100 affiliated oragnizations in Canada, Mexico and US,as well as with Pesticide Action Network partners around the world to demand that development agencies and governments redirect support from pesticides to safe alternatives.

*Kathryn Gilje, Executive Director*
*Stephen Scholl Buckwald, Managing Director*

**2146 Physicians for Social Responsibility**
1875 Connecticut Avenue, Northwest 202-667-4260
Suite 1012 Fax: 202-667-4201
Washington, DC 20009 E-mail: psrnatl@psr.org
http://www.psr.org
A non-profit advocacy organization that is the medical and public health voice for policies to prevent nuclear war and proliferation and to slow, stop and reverse global warming and toxic degradation of the environment.
*Peter Wilk MD, Executive Director*

**2147 Public Citizen**
1600 20th Street NW 202-588-1000
Washington, DC 20009 http://www.citizen.org
Founded by Ralph Nader in 1971, Public Citizen is the consumer's eyes in Washington. With the support of more than 15,000 people like you, we fight for safer drugs and medical devices, cleaner and safer energy sources, a cleanerenvironment, fair trade and a more open and democratic government.
*Joan Claybrook, President*

**2148 Rachel Carson Center for Natural Resources**
Churchill High School 541-687-3421
1850 Bailey Hill Road E-mail: haberman@4j.lane.edu
Eugene, OR 97405 http://schools.4j.lane.edu/carson/
Offers an alternative to the traditional high school curriculum, providing students with experience, knowledge, and skills that relate to the natural environment.
*Helen Haberman, Environmental Instructor*
*Tim Whitley, Director*

**2149 Rachel Carson Council**
PO Box 10779 301-593-7507
Silver Spring, MD 20914 E-mail: rccouncil@aol.com
http://www.members.aol.com/rccouncil/ourpage
Independent nonprofit scientific organization dedicated to protecting the environment against toxic and chemical threats, particularly those of pesticides.
*Dr. Diana Post, President*

**2150 Rene Dubos Center**
The Rene Dubos Center 914-337-1636
Suite 387 Fax: 914-771-5206
Bronxville, NY 10708 E-mail: dubos@mindspring.com
http://www.dubos.org
The Rene Dubos Center for Human Environments is a non-profit education and research organization focused on the social and humanistic aspects of environmental problems.
*Ruth A Eblen, President*

**2151 Rodale Institute**
611 Siegfriedale Road 610-683-1400
Kutztown, PA 19530 Fax: 610-683-8548
E-mail: info@rodaleinst.org
http://www.rodaleinstitute.org
The Institute offers creative opportunities and solutions that contribute to regenerating environmental and human health worldwide. Their mission statement is clear: The Rodale Institute works worldwide to achieve a regenerative foodsystem that improves environmental and human health.
*Tim LaSalle, President*

**2152 Safer Pest Control Project**
4611 North Ravenswood Avenue 773-878-7378
Suite 107 Fax: 773-878-8250
Chicago, IL 60640 E-mail: general@spcpweb.org
http://spcpweb.org
Dedicated to reducing the health risks and environmental impacts of pesticides and promoting safer alternatives in Illinois.
*John Johnson, President*
*Rachel Rosenberg, Executive Director*

**2153 Second Nature Inc.**
Consortium for Environmental Education in Medicine
18 Tremont Street 617-224-1610
Suite 1120 E-mail: acortese@secondnature.org
Boston, MA 02108 http://www.secondnature.org
Dedicated to advancing our quality of life by demostrating the close links between human health and the environment. The center's goal is to make the relationship of environment to human health an integral part of medical education.
*Dr. Anthony Cortese, President*

**2154 Silicones Environmental, Health and Safety Council of North America**
2325 Dulles Corner Boulevard 703-788-6570
Suite 500 Fax: 703-788-6545
Herndon, VA 20171 E-mail: sehsc@sehsc.com
http://www.sehsc.com
An organization of North America silicone chemical producers and importers. It promotes the safe use of silicones through product stewardship and environmental, health and safety research. It focuses on coordinating research andsubmitting it for peer review through independent advisory boards and publication of peer-reviewed literature.
*Karluss Thomas, Executive Director*

**2155 Society for Occupational and Environmental Health**
6728 Old McLean Village Drive 703-556-9222
McLean, VA 22101 Fax: 703-556-8729
E-mail: soeh@degnon.org
http://www.soeh.org
Provides a neutral forum where occupational safety and health and environmental issues can be discussed and resolved. Actively seeks to improve the quality of both working and living places.
*George K Degnon, Executive Director*

**2156 Society of Environmental Toxicology and Chemistry**
SETAC N America Office
1010 N 12th Avenue 850-469-1500
Pensacola, FL 32501 Fax: 850-469-9778
E-mail: setac@setac.org
http://www.setac.org
Mission: To support the development of principles and practices for protection, enhancement and management of sustainable environmental quality and ecosystem integrity.
*Greg Schiefer, Executive Director*
*Jason Anderson, IT Manager*

**2157 Synthetic Organic Chemical Manufacturers Association**
1850 M Street NW 202-721-4100
Suite 700 Fax: 202-296-8120
Washington, DC 20036 E-mail: info@socma.com
http://www.socma.com
A trade association that serves the specialty, batch and custom chemical industry. SOCMA member companies make the products and refine the raw materials that make our standard of living possible; from pharmaceuticals to cosmetics,soaps to plastics, and all manner of industrial and construction products. SOCMA promotes innovative, safe and environmentally responsible operations, which are internationally competitive and contribute to a healthy, productive economy.

*Joseph Acker, President*
*Diane McMahon, Director Business Operations*

**2158 Teratology Society**
1821 Michael Faraday Drive 703-438-3104
Suite 300 Fax: 703-483-3113
Reston, VA 20190 E-mail: tshq@teratology.org
http://www.teratology.org
A multidisciplinary scientific society founded in 1960, the members of which study the causes and biological processes leading to abnormal development and birth defects at the fundamental and clinical level, and appropriate measuresfor prevention.
*Elaine M Faustman, President*
*Thomas B Knudson, Vice President*

**2159 Toxicology Information Response Center**
1060 Commerce Park 865-576-1746
MS 6480 Fax: 865-574-9888
Oak Ridge, TN 37830 E-mail: slusherkg@ornl.gov
http://www.ornl.gov/TechResources/tirc/hmepg.html

TIRC provides customer search services to both scientific and public communities as a convenient and efficient way to obtain comprehensive scientific information on any subject of interest.

*Kim Slusher, Administrator*

**2160 US Consumer Product Safety Commission**
4330 East West Highway 301-504-7923
Bethesda, MD 20814 800-638-2772
Fax: 301-504-0124
E-mail: info@cpsc.gov
http://www.cpsc.gov

An independent federal regulatory agency. Helps keep American families safe by reducing the risk of injury or death from consumer products.
*Patsy Semple, Executive Director*
*John Horner, Director*

**2161 US Nuclear Regulatory Commission**
One White Flint North 301-415-7000
11555 Rockville Pike 800-368-5642
Rockville, MD 20852 Fax: 301-415-8200
E-mail: pdr@nrc.gov
http://www.nrc.gov

Ensures adequate protection of the public health and safety, the common defense and security, and the environment in the use of nuclear materials in the United States.
*Dale E Klein, Chairman*
*Peter B Lyons, Commissioner*

**2162 US Public Interest Research Group**
218 D Street SE 202-546-9707
Washington, DC 20003 Fax: 202-546-2461
http://www.pirg.org

Mission: To deliver persistent, result-oriented public interest activism that protects our environment, encourages a fair, sustainable economy, and fosters responsive, democratic government.
*Christy Leavitt, Clean Water Advocate*

**2163 US-Mexico Border Health Association**
5400 Suncrest Dr 915-833-6450
Suite C-5 Fax: 915-833-7840
El Paso, TX 79912 http://www.usmbha.org

Promotes public and individual health along the United States-Mexico border through reciprocal technical cooperation.
*Rebeca Ramos, Interim Executive Director*

**2164 Water Environment Federation**
601 Wythe Street 703-684-2400
Alexandria, VA 22314 800-666-0206
Fax: 703-684-2492
http://www.wef.org

Nonprofit international membership organization that develops and disseminates technical information on the nature, collection, treatment and disposal of domestic and industrial wastewater.
*Janet Blatt, Director*
*Mincaiee Brown, Executive Director*

**2165 Western Occupational and Environmental Medical Association**
575 Market Street 415-927-5736
Suite 2125 Fax: 415-927-5726
San Francisco, CA 94105 E-mail: woema@hp-assoc.com
http://www.woema.org

The mission is to represent and be a resource to members in the profession and practice of occupational and environmental medicine and to enhance their efforts to promote and improve health in the workplace and the community.
*Kerry Parker, Executive Director*
*Michael Barack, Administrative Manager*

**2166 White Lung Association**
PO Box 1483 410-243-5864
Baltimore, MD 21203 Fax: 410-243-5234
E-mail: jfite@whitelung.org
http://www.whitelung.org

National nonprofit organization dedicated to the education of the public to the hazards of asbestos exposure. Has developed programs of public education and consults with victims of asbestos exposure, school boards, building owners,government agencies and others interested in identifying asbestos hazards and developing control programs.
*James Fite, Executive Director*

## Pediatric Health: Associations

**2167 Academic Pediatric Association**
6728 Old McLean Village Drive 703-556-9222
McLean, VA 22101 Fax: 703-556-8729
E-mail: info@academicpeds.org
http://www.academicpeds.org

The mission of the APA is to foster the health and well-being of children and their families by: promoting health services, education and research in general pediatrics; affecting public and governmental policies regarding issues vitalto child health and to education and research in general pediatrics; and supporting the professional growth and development of faculty in general pediatrics.

*Danielle Laraque, President*
*Janet Serwint MD, President Elect*

**2168 Allergy and Asthma Network: Mothers of Asthmatics**
2751 Prosperity Avenue
Suite 150 800-878-4403
Fairfax, VA 22031 Fax: 703-573-7794
http://www.aanma.org

Our mission is to eliminate suffering and death due to asthma and allergies through education, advocacy, community outreach and research.
*Nancy Sander, President*
*Mary McGowan, Executive Director*

**2169 American Academy of Pediatrics: Committee on Environment Health**
141 NW Point Boulevard 847-434-4000
Elk Grove Villiage, IL 60007 Fax: 847-434-8000
http://www.aap.org

The AAP is commited to the attainment of optimal physical, mental and social health for all infants, children, and young adults. This mission will be accomplished by engaging in the following activities: professional education,advocacy for children and youth, advocacy for pediatricians, public education, membership service and research.
*Jay E Berkelhamer, President*

**2170 American Federation of Teachers**
555 New Jersey Avenue NW 202-879-4463
Washington, DC 20001 Fax: 202-879-4597
http://www.aft.org

The American Federation of Teachers is a union that represents K-12 teachers and other school employees, health care professionals and public employees. The union considers itself an advocacy organization for children and the public.
*Edward J McElroy, President*

**2171 Association of Maternal and Child Health Program**
1220 19th Street N.W. 202-775-0436
Suite 801 Fax: 202-775-0061
Washington, DC 20036 http://www.amchp.org

AMCHP accomplishes its mission through the active participation of its members and vital partnerships with government agencies, families and advocates, health care purchasers and providers, academic and research professionals, andothers at the national, state and local levels.
*Barbara Laur, Interim CEO*
*Nora Lam, Executive Assistant*

**2172 Childhood Lead Poisoning Prevention Program**
Ohio Department of Health

246 N High Street
Columbus, OH 43215
614-728-9454
Fax: 614-728-6793
E-mail: BCFHS@odh.ohio.gov
http://www.odh.state.oh.us

The mission of Ohio's Childhood Lead Poisoning Prevention Program is to eliminate childhood lead poisoning through screening, environmental inspection, abatement, education and case management.

*Alvin D Jackson, Director*

**2173 Children's Defense Fund**

25 E Street NW
Washington, DC 20001
202-628-8787
800-233-1200
E-mail: cdfinfo@childrensdefense.org
http://www.childrensdefense.org

Mission: To ensure every child a healthy start, a head start, a fair start, a safe start, and a moral start in life and successful passage to adulthood with the help of caring families and communities.

*Marian Wright Edelman, President*

**2174 Children's Environmental Health Network**

110 Maryland Avenue NE
Suite 505
Washington, DC 20002
202-543-4033
Fax: 202-543-8797
E-mail: cehn@cehn.org
http://www.cehn.org/

Mission is to promote a healthy environment and to protect the fetus and child from environmental hazards. Three areas of concentration for the Network are education, research and policy. Network's goals are: to promote the developmentof sound public health and child-focused national policy; to stimulate prevention-oriented research; to educate health professionals, policymakers and community members in preventive strategies; and to elevate public awareness of environmentalhazards to children.

*Nsedu Obot Witherspoon, Executive Director*
*E Ramona Trovato, Chair*

**2175 Coalition for Clean Air**

811 West 7th Street
Suite 1100
Los Angelos, CA 90017
213-630-1192
Fax: 213-630-1158
E-mail: air@coalitionforcleanair.org
http://www.coalitionforcleanair.org

The Coalition for Clean Air is committed to restoring clean, healthy air to all of California and strengthening the environmental movement by promoting broad-based community involvement, advocating responsible public policy andproviding technical expertise.

*Alberto B Mendoza, President & CEO*
*Tim Carmichael, Senior Director of Policy*

**2176 Genesis Fund/National Birth Defects Center: Pregnancy Environmental Hotline**

40 2nd Avenue
Suite 510
Waltham, MA 02451
781-890-4282
800-322-5014
Fax: 781-487-2361
E-mail: info@thegenesisfund.org
http://www.thegenesisfund.org/peh.htm

General information service that provides information regarding exposure to environmental factors during pregnancy and the effects on the developing fetus.

**2177 Healthy Child Healthy World**

12300 Wilshire Blvd
Suite 320
Los Angeles, CA 90025
310-820-2030
Fax: 310-820-2070
E-mail: info@healthychild.org
http://www.healthychild.org

Healthy Child Healthy World is dedicated to protecting the health and well being of children from harmful environmental exposures. We educate parents, support protective policies, and engage communities to make responsible decisions,simple everyday choices, and well-informed lifestyle improvements to create healthy environments where children and families can flourish.

*Rachel Lincoln Sarnoff, Executive Director*
*Mandy Geisler, Project Manager*

**2178 Healthy Mothers, Healthy Babies**

2000 North Beauregard Street
6th Floor
Alexandria, VA 22311
703-837-4792
Fax: 703-684-5968
E-mail: info@hmhb.org
http://www.hmhb.org

Healthy Mothers, Healthy Babies is a coalition of national, state and local providers, advocates and administrators concerned about health of pregnant women, infants and families. The coalition serves as a forum for informationexchange and as a catalyst to encourage collaborative partnerships amoung its members and colleagues.

*George Guido, Chairperson*
*Hampton Shaddock, Vice Chair*

**2179 Healthy Schools Network**

773 Madison Avenue
Albany, NY 12208
518-462-0632
Fax: 518-462-0433
E-mail: info@fhealthyschools.org
http://www.healthyschools.org

HSN is a nationally recognized state-based advocate for the protection of children's environmental health in schools. Engages in research, education, outreach, technical assistance and coalition building to create schols that areenvironmentally responsible to children, and to their communities. Publishes a quarterly newsletter and maintains an Information Clearinghouse and Referral Service.

*John Shaw, President*
*Beatriz Barraza, Vice-President*

**2180 Institute of Medicine: Board on Children, Youth and Families**

500 Fifth Street, NW
Washington, DC 20001
202-334-1935
Fax: 202-334-3829
E-mail: bocyf@nas.edu
http://www.bocyf.org

The Board on Children, Youth and Families addresses a variety of policy-relevant issues related to the health and development, of children, youth and families. It does so by convening experts to weigh in on matters from the perspectiveof the behavioral, social, and health sciences. The Board operates under the National Research Council and the Institute of Medicine of the National Academies.

*Rosemary Chalk, Director*

**2181 Kids for Saving Earth Worldwide**

PO Box 421118
Minneapolis, MN 55442
763-559-1234
Fax: 651-674-5005
E-mail: kseww@aol.com
http://www.kidsforsavingearth.org/

To help protect the Earth through kids and adults. To educate and inspire them to participate in Earth-saving actions.

*Tessa Hill, President and Director*

**2182 Kids for a Clean Environment**

PO Box 158254
Nashville, TN 37215
615-331-7381
Fax: 615-333-9879
E-mail: kidsface@mindspring.com
http://www.kidsface.org

Mission: To provide information on environmental issues to children, to encourage and facilitate youth's involvement with effective environmental action and to recognize those efforts which result in the improvement of nature.

*Melissa Poe, Founder*

**2183 March of Dimes Birth Defects Foundation**

1275 Mamaroneck Avenue
White Plains, NY 10605
914-997-4488
http://www.modimes.org

The mission of the March of Dimes Birth Defects Foundation is to improve the health of babies by preventing birth defects and reducing infant mortality. The March of Dimes carries out the mission through research, community service,education and advocacy.

*Ann Umemoto, Associate Director*

**2184 Montefiore Medical Center Lead Poisoning Prevention Program**
111 E 210th Street 718-920-5016
Bronx, NY 10467 Fax: 718-920-4377
The Montefiore Lead Poisoning Prevention Program addresses all aspects of childhood lead poisoning from diagnosis and treatment to education and research. Their mission is to treat lead-poisoned children and their families and toeducate families at risk, other medical providers and local, state and national legislators and policy makers.
*Nancy Redkey, Project Coordinator*

**2185 National Institute of Child Health and Human Development**
Po Box 3006
Rockville, MD 20847 800-370-2943
Fax: 301-984-1473
E-mail: NICHDinformationresourcecenter@mail.nih.gov
http://www.nichd.nih.gov
National Institute of Child Health and Human Development supports and conducts basic, clinical and epidemiological research on the reproductive, neurobiological, developmental and behavioral processes that determine and maintain thehealth of children, adults, families and populations.
*Duane Alexander, Director*

**2186 National Parent Teachers Association**
541 North Fairbanks Court 312-670-6782
Suite 1300 800-307-4782
Chicago, IL 60611 Fax: 312-670-6783
E-mail: info@pta.org
http://www.pta.org
The mission of the National PTA is to support and speak on behalf of children and youth in the school, in the community and before governmental bodies and other organizations; to assist parents in developing the skills they need toraise and protect their children and to encourage parent and public involvement in the public schools.
*Warlene Gary, CEO*

**2187 Office of Children's Health Protection**
US Environmental Protection Agency
Ariel Rios Building 202-564-2188
1200 Pennsylvania Avenue NW Fax: 202-564-2733
Washington, DC 20460 http://www.epa.gov
The mission of the Office of Children's Health Protection is to make the protection of children's health a fundamental goal of public health and environmental protection in the United States.
*Joanne Rodman, Associate Director*

**2188 Oklahoma Childhood Lead Poisioning PreventionProgram**
Oklahoma Department of Health
1000 Northeast 10th Street 405-271-6617
Oklahoma, OK 73117 Fax: 405-271-4971
E-mail: oklppp@health.ok.gov
http://lpp.health.ok.gov
Mission: To reduce blood lead levels to below a level of concern in all Oklahomans.

## Pediatric Health: Publications

**2189 Child Health and the Environment**
Oxford University Press
198 Madison Avenue 212-726-6000
New York, NY 10016 Fax: 919-677-1303
http://www.oup-usa.org
Focus on environmental threats to child health. The first three chapters provide overviews of key children's environmental health issues as well as the role of environmental epidemiology and risk assessment in child health protection.Later chapters address the health affects of metal, PCBs, dioxins, pesticides, hormonally active agents, radiation, indoor and outdoor air pollution, and water contaminants.
*416 pages*
*ISBN: 0-195135-59-8*
*Donald T Wigle, Author*

**2190 Children's Defense Fund**
25 E Street NW 202-628-8787
Washington, DC 20001 800-233-1200
E-mail: cdfinfo@childrensdefense.org
http://www.childrensdefense.org
Mission: To ensure every child a healthy start, a head start, a fair start, a safe start, and a moral start in life and successful passage to adulthood with the help of caring families and communities.
*Marian Wright Edelman, President*

**2191 Handbook of Pediatric Environmental Health**
American Academy of Pediatrics
141 NW Point Boulevard 847-434-4000
Elk Grove Village, IL 60007 Fax: 847-434-8000
http://www.aap.org
The AAP is committed to the attainment of optimal physical, mental and social health for all infants, children, and young adults. This mission will be accomplished by engaging in the following activities: professional education,advocacy for children and youth, advocacy for pediatricians, public education, membership service and research.
*723 pages*
*ISBN: 1-581100-29-9*
*Jay E Berkelhamer, President*

**2192 Handle with Care: Children and Environmental Carcinogens**
Natural Resources Defense Council
40 West 20th Street 212-727-2700
New York, NY 10011 Fax: 212-727-1773
E-mail: nrdcinfo@nrdc.org
http://www.nrdc.org
Mission: To safeguard the Earth: its people, its plants and animals, and the natural systems on which all life depends.
*50 pages*
*Francis Beinecke, Executive Director*

**2193 Kids Count Data Book: State Profiles of Child Well-Being**
701 St. Paul Street 410-547-6600
Baltimore, MD 21202 Fax: 410-547-3610
E-mail: webmail@aecf.org
http://www.aecf.org/kidscount
Kids Count is national and state-by-state effort to provide makers and citizens with benchmarks of child well-being. It includes variables such as percentage of low birth-weight babies, child death rates, percentage of childern inpoverty, and percentage of childern without health insurance.

**2194 Pesticides and the Immune System: Public Health Risks**
10 G Street NE 202-729-7600
Suite 800 Fax: 202-729-7610
Washington, DC 20002 E-mail: lauralee@wri.org
http://www.wri.org
Brings together for the frist time an extensive body of experimental and epidemiological research from around the world documenting the the effects of widely used pesticides on the immune system and the attendent health risks. In sodoing, it documents that pesticide-related health risks are much more serious than genrally known, especially in developing countries where exposure is widespread and infectious diseases take a heavy toll.
*100 pages*
*ISBN: 1-569730-87-3*
*Robert Repetto, Author*
*Sanjay S Baliga, Author*

**2195 Resource Guide on Children's Environmental Health**

600 Grant Street
Suite 800
Denver, CO 80203
303-861-5165
Fax: 303-861-5315
http://www.cchn.org

The Children's Environmental Health Network has developed the Resource Guide on Childern's Environmental Health to assist community leaders, policy makers, health and environmental specialists, members of the advocacy community andmedia, and the general public in identifying and accessing key resources in childern's environmental health.

*Annette Kowal, CEO*
*Kitty Bailey, COO*

# Environmental Law

## Associations

**2196 Atlantic States Legal Foundation**
658 West Onondaga Street 315-475-1170
Syracuse, NY 13204 Fax: 315-475-6719
E-mail: atlantic.states@aslf.org
http://www.aslf.org
Atlantic States Legal Foundation was established in 1982 to provide legal, technical, and organizational assistance on environment issues to citizen organizations (NGOs), individuals, local governments, and others.
*Samuel H Sage, President*
*Daisy Hollis, Chair*

**2197 Business & Legals Reports**
141 Mill Rock Road East 860-510-0100
Old Saybrook, CT 06475 800-727-5257
Fax: 860-510-7220
E-mail: service@blr.com
http://www.blr.com
Provides essential tools for safety and environmental compliance and training needs.
*Brian Gurnham, Chief Operating Officer*
*Peggy Carter-Ward, Editor-in-Chief*

**2198 Center for Community Action and Environmental Justice**
PO Box 33124 951-360-8451
Riverside, CA 92519 Fax: 951-360-5950
E-mail: admin@ccaej.org
http://www.ccaej.org
The Center for Community Action and Environmental Justice serves as a resource center for community groups working on environmental justice issues.
*Renae Bryant, President*
*Wendy Eads, President-Elect*

**2199 Center for Health, Environment and Justice Library**
150 S Washington Street, Ste 300 703-237-2249
PO Box 6806 Fax: 703-237-8389
Falls Church, VA 22040-6806 E-mail: chej@chej.org
http://www.chej.org
The Center for Health, Environment and Justice works to build healthy communities, with social justice, economic well-being, and democratic governance. We believe this can happen when individuals from communities have the power to playan integral role in promoting human health and environmental integrity. Our role is to provide the tools to build strong, healthy communities where people can live, work, learn, play and pray.
*Lois Marie Gibbs, Executive Director*

**2200 Center for International Environmental Law**
1350 Connecticut Avenue NW 202-785-8700
Suite 1100 Fax: 202-785-8701
Washington, DC 20036 E-mail: info@ciel.org
http://www.ciel.org
The Center for International Environmental Law (CIEL) is a nonprofit organization working to use international law and institutions to protect the environment, promote human health, and ensure a just and sustainable society. We providea wide range of services including legal counsel, policy research, analysis, advocacy, education, training, and capacity building.
*Daniel B Magraw Jr, President*
*Cameron Aishton, Administrator*

**2201 Center for Investigative Reporting**
2927 Newbury Street 510-809-3160
Suite A Fax: 510-849-1813
Berkeley, CA 94703 E-mail: center@cironline.org
http://www.muckraker.org
The only independent, nonprofit organization in the country dedicated to investigative reporting in the public interest on a broad range of issues.
*Tom Goldstein, President*
*Christa Scharfenberg, Acting Executive Director*

**2202 Communities for a Better Environment**
1400 Broadway 510-302-0430
Suite 701 Fax: 510-302-0437
Oakland, CA 94612 E-mail: cbeca@mail.com
http://www.cbecal.org
Mission: To achieve environmental health and justice by building grassroots power in and with communities of color and working-class communities.
*Richard Toshiyuki Drury, President*
*Lizette Ruiz, Vice President*

**2203 Community Environmental Council**
26 West Anapamu Street 805-963-0583
2nd Floor Fax: 805-962-9080
Santa Barbara, CA 93101 E-mail: admin@cecmail.org
http://www.communityenvironmentalcouncil.org
The Community Environmental Council is a nonprofit environmental organization headquartered in Santa Barbara, California. Our community involvement includes managing two recycling centers, a household hazardous waste facility, anurban farm, and three community gardens as well as the environmental education program Art From Scrap. In addition CEC provides research, technical assistance and education on local and statewide land use planning, and solid waste and integrated pestmanagement.
*Marilyn Parke, CFO*
*Dave Davis, Executive Director*

**2204 Community Rights Counsel**
1301 Connecticut Avenue NW 202-296-6889
Suite 502 Fax: 202-296-6895
Washington, DC 20036 E-mail: crc@communityrights.org
http://www.communityrights.org
A nonprofit public interest law firm that was formed in 1997 to assist communities in protecting their health and welfare by regulating permissible land uses, and that provides strategic assistance to state and local governmentattorneys in defending land use laws.
*Douglas T Kendall, Founder/Executive Director*
*Timothy J Dowling, Chief Counsel*

**2205 Conservation Law Foundation**
62 Summer Street 617-350-0990
Boston, MA 02110 Fax: 617-350-4030
E-mail: ihartclf.org
http://www.clf.org
A nonprofit, member-supported organization that works to solve the environmental problems that threaten the people, natural resources and communities of New England. CLF's advocates use law, economics and science to design andimplement strategies that conserve natural resources, protect public health and promote vital communities in our region.
*Philip Warburg, President*
*Peter Shelley, Vice President*

**2206 Earthjustice**
426 17th Street 510-550-6700
6th Floor 800-584-6460
Oakland, CA 94612 Fax: 510-550-6740
E-mail: info@earthjustice.org
http://www.earthjustice.org
Nonprofit public interest law firm dedicated to protecting the magnificent places, natural resources, and wildlife of this earth and to defending the right of all people to a healthy environment. It enforces and strengthensenvironmental laws on behalf of hundreds of organizations and communities.
*Vawter (Buck) Parker, Executive Director*
*Melinda Carmack, VP of Development*

**2207 Environmental Defense Fund**
257 Park Avenue South
17th Floor
New York, NY 10010
212-505-2100
Fax: 212-505-2375
E-mail: members@edf.org
http://www.edf.org

Dedicated to protecting the environmental rights of all people, including future generations. Among these rights are clean air, clean water, healthy, nourishing food and a flourishing ecosystem. Advocates solutions based on science,even when it leads in unfamiliar directions. Works to create solutions that win lasting political, economic and social support because they are bipartisan, efficient and fair

*Fred Krupp, President*
*David Yarnold, Executive Director*

**2208 Environmental Law Alliance Worldwide**
1877 Garden Avenue
Eugene, OR 97403
541-687-8454
Fax: 541-687-0535
E-mail: elawus@elaw.org
http://www.elaw.org

E-LAW advocates serve low income communities around the world, helping citizens strengthen and enforce laws to protect communities from toxic pollution and environmental degradation.

*Bern Johnson, Executive Director*
*Lori Maddox, Associate Director*

**2209 Environmental Law Institute**
2000 L Street NW
Suite 620
Washington, DC 20036
202-939-3800
Fax: 202-939-3868
E-mail: law@eli.org
http://www.eli.org

Community Education and Training Program provides citizens and grassroots groups with information on environmental law and policy that can help them participate effectively in the decisions that impact public health and the environmentin their communities. Program's activities have included training courses on right-to-know laws and a series of workshops in demystifying the law, which focus on using the tools of public participation to address issues ranging from hazardous wasteto land use.

*Leslie Carothers, President*
*William Eichbaum, Chairman*

**2210 Environmental Law and Policy Center of the Midwest**
35 E Wacker Drive
Suite 1600
Chicago, IL 60601
312-673-6500
Fax: 312-795-3730
E-mail: elpc@elpc.org
http://www.elpc.org

*Howard Drucker, President/Executive Director*
*Richard Day, Chair*

**2211 Environmental Support Center**
1500 Massachusetts Avenue NW
Suite 25
Washington, DC 20005
202-331-9700
Fax: 202-331-8592
E-mail: general@envsc.org
http://www.envsc.org/

The mission of the Environmental Support Center is to promote the quality of the natural environment, human health, and community sustainability by increasing the organizational effectiveness of local, state, and regional organizationsworking on environmental issues and for environmental justice. To be eligible for assistance, your organization must be a local, state or regional nonprofit organization with a portion of its resources devoted to environmental issues.

*Judy Hatcher, Interim Co-Director*

**2212 Federation of Environmental Technologists**
9451 N 107th Street
Milwaukee, WI 53224
414-354-0070
Fax: 414-354-0073
E-mail: info@fetinc.org
http://www.fetinc.org

A nonprofit organization formed to assist industry in interpretation of and compliance with environmental regulations. Membership is open to all industries, municipalities, organizations and individuals concerned about environmentalregulations. Currently there are approximately 1000 members and 125 patron companies.

*Barbara Hurula, Executive Director*

**2213 Greenpeace**
702 H Street NW
Suite 300
Washington, DC 20001
202-462-1177
800-326-0959
E-mail: info@WDC.greenpeace.org
http://www.Greenpeaceusa.org

Greenpeace is an independent campaigning organization which uses non-violent creative confrontation to expose global environmental problems, and to force solutions that are essential to a green and peaceful future.

*John Passacantando, Executive Director*

**2214 Harvard Environmental Law Society**
Harvard Law School
Pound Hall
1563 Massachusetts Avenue
Cambridge, MA 02138
617-495-3125
E-mail: els@law.harvard.edu
http://www.law.harvard.edu/students/orgs/els

The Harvard Environmental Law Society was founded by three Harvard Law students who perceived a pressing need for the Law School, and the law in general, to respond more effectively to the nation's environmental problems. To this end, they created an organization that was committed to preparing students to creatively and intelligently use the law in the service of the environment.

*Myra Blake, President*
*Nigel Barella, Secretary/Treasurer*

**2215 Humane Society Legislative Fund**
519 C Street Northeast
Washington, DC 20002
202-676-2314
E-mail: humanesociety@hslf.org
http://www.fund.org

Humane Society Legislative Fund (HSLF) is a social welfare organization. HSLF works to pass animal protection laws at the state and federal level, to educate the public about animal protection issues, and to support humane candidatesfor office.

*Lisa Gallo, Public Relations*

**2216 LandWatch Monterey County**
PO Box 1876
Salinas, CA 93902
831-422-9390
Fax: 831-422-9391
E-mail: landwatch@mclw.org
http://www.landwatch.org

LandWatch is a nonprofit membership organization, founded in 1997. LandWatch works to promote and inspire sound land use legislation at the city, country, and regional lands, through grassroots community action.

*Michael DeLapa, President*
*Chris Fitz, Executive Director*

**2217 League of Conservation Voters**
1920 L Street NW
Suite 800
Washington, DC 20036
202-785-8683
Fax: 202-835-0491
http://www.lcv.org

Works to create a Congress more responsive to your environmental concerns. As the nonpartisan political voice for over nine million members of environmental and conservation groups, LCV is the only national environmental organizationdedicated full-time to educating citizens about the environmental voting records of Members of Congress.

*Gene Karpinski, President*
*Bill Roberts, Chair*

**2218 League of Women Voters of the United States**
1730 M Street NW
Suite 1000
Washington, DC 20036
202-429-1965
Fax: 202-429-0854
E-mail: lwv@lwv.org
http://www.lwv.org

The League of Women Voters, a nonpartisan political organization, encourages the informed and active participation of citizens in government, works to increase understanding of major public policy issues and influences public policythrough education and advocacy.

*Mary G Wilson, President*
*Nancy Tate, Executive Director*

**2219 Legacy International**
1020 Legacy Drive 540-297-5982
Bedford, VA 24523 Fax: 540-297-1860
E-mail: mail@legacyintl.org
http://www.legacyintl.org

Creates environments where people can address personal, community, and global needs while developing skills and effective responses to change.Whether working with youths, corporate leaders, educational professionals, entrepreneurs, orindividuals on opposing sides of a conflict, our goal is the same. Programs provide experiences, skills, and strategies that enable people to build better lives for themselves and others around them.
*J E Rash, President*
*Shanti Thompson, VP/Director of Training*

**2220 National Association of Conservation Districts League City Office**
509 Capitol Ct NE 202-547-6223
Washington, DC 20002 Fax: 202-547-6450
http://www.nacdnet.org

To serve conservation districts by providing national leadership and a unified voice for natural resource conservation.
*Olin Sims, President*
*Steve Robinson, First Vice President*

**2221 Natural Resources Defense Council**
40 West 20th Street 212-727-2700
New York, NY 10011 Fax: 212-727-1773
E-mail: nrdcinfo@nrdc.org
http://www.nrdc.org

Mission: To safeguard the Earth: its people, its plants and animals, and natural systems on which all life depends.
*Frances Beinecke, President*
*Peter Lehner, Executive Director*

**2222 Natural Resources Law Center**
University of Colorado
School of Law 303-492-1288
Box 401 E-mail: lawadmin@colorado.edu
Boulder, CO 80309 http://www.colorado.edu/Law/centers/nrlc/index.htm

Mission: Promote sustainability in the rapidly changing American West by informing and influencing natural resources policies, and decisions.
*Mark Squillace, Director*
*Heidi Horten, Special Asst. to the Dir.*

**2223 New Mexico Environmental Law Center**
1405 Luisa Street 505-989-9022
Suite 5 Fax: 505-989-3769
Santa Fe, NM 87505 E-mail: nmelc@nmelc.org
http://www.nmenvirolaw.org

The New Mexico Environmental Law Center works to protect New Mexico's communities and their environments through public education, legislative initiatives, administrative negotiations and litigation.
*Douglas Meiklejohn, Executive Director*
*Sebia Hawkins, Development Director*

**2224 Southern Environmental Law Center**
201 W Main Street 434-977-4090
Suite 14 Fax: 434-977-1483
Charlottesville, VA 22902 E-mail: selcva@selcva.org
http://www.southernenvironment.org

Dedicated to protecting the natural resources of Alabama, Georgia, North Carolina, South Carolina, Tennessee and Virginia. Works with more than 100 partner groups to safeguard southern forests, wetlands, coastal resources, rivers, airand water quality, wildlife habitat and rural landscapes through policy reform, public education, and direct legal action.
*Frederick S Middleton III, President*
*Rick Middleton, Executive Director*

**2225 Stanford Environmental Law Society**
559 Nathan Abbott Way 650-723-4421
Stanford, CA 94305 E-mail: sranchod@stanford.edu
http://www.elj.stanford.edu

Provides students with a unique set of opportunities to tap into structured programs or to create and pursue their own projects. Both organizations complement the Stanford Environmental and Natural Reources Law and Policy Program.
*Zachary Fabish, Co-President*
*Craig Segall, Co-President*

**2226 Student Environmental Action Coalition**
PO Box 31909 215-222-4711
Philadelphia, PA 19104 E-mail: webteam@seac.org
http://www.seac.org

Student and youth run national network of progressive organizations and individuals whose aim is to uproot environmental injustices through action and education. Works to create progressive social change on both the local and globallevels.
*Matt Reitmann, Working Comte Coordinator*

**2227 US Public Interest Research Group**
218 D Street SE 202-546-9707
Washington, DC 20003 Fax: 202-546-2461
E-mail: webmaster@pirg.org
http://www.uspirg.org

Mission: To deliver persistent, result-oriented public interest activism that protects our environment, encourages a fair, sustainable economy, and fosters responsive, democratic government.
*Douglas H Phelps, President*
*Margie Alt, Executive Director*

**2228 Western Environmental Law Center**
1216 Lincoln Street 541-485-2471
Eugene, OR 97401 Fax: 541-485-2457
E-mail: eugene@westernlaw.org
http://www.westernlaw.org

The Western Environmental Law Center is a non-profit public interest law firm that works to protect and restore western wildlands and advocates for healthy environments on behalf of communities throughout the West.
*Corrie Yackulic, President*
*Lori Maddox, Vice President*

## Publications

**2229 A Guide to Environmental Law in Washington DC**
Environmental Law Institute
2000 L Street NW 202-939-3800
Suite 620 Fax: 202-939-3868
Washington, DC 20036 E-mail: law@eli.org
http://www.eli.org

Community Education and Training Program provides citizens and grassroots groups with information on environmental law and policy that can help them participate effectively in the decisions that impact public health and the environmentin their communities. Program's activities have included training courses on right-to-know laws and a series of workshops in demystifying the law, which focus on using the tools of public participation to address issues ranging from hazardous wasteto land use.
*Leslie Carothers, President*
*William M Eichbaum, Chairman*

**2230 Buying Green: Federal Purchasing Practices and the Environment**
Government Printing Office
732 North Capitol Street NW 202-512-0000
Washington, DC 20401 E-mail: www.gpo.gov
*William H Turri, COO*
*Robert C Tapella, Cheif of Staff*

**2231 Clean Water Act Twenty Years Later**
Island Press

1718 Connecticut Avenue NW
Suite 300
Washington, DC 20009
202-232-7933
Fax: 202-234-1328
E-mail: info@islandpress.org
http://www.islandpress.org

*333 pages*
*ISBN: 1-559632-65-8*
*Richard W Alder; Jessica C Landman; Diane Cameron, Author*

**2232 Comparative Environmental Law and Regulation**
Oceana Publications, Inc
198 Madison Avenue
New York, NY 10016
800-334-4249
Fax: 212-726-6476
http://www.oceanalaw.com

Key environmental laws, regulations and implementation systems and agencies of 37 countries from around the world.
*2 vol pages Semi-Annual*
*ISBN: 0-379012-51-0*
*Nicholas A Robinson, Editor*

**2233 Environmental Defense Fund**
257 Park Avenue South
17th Floor
New York, NY 10010
212-505-2100
Fax: 212-505-2375
E-mail: members@edf.org
http://www.edf.org

Dedicated to protecting the environmental rights of all people, including future generations. Among these rights are clean air, clean water, healthy nourishing food and a flourishing ecosystem. Advocates solutions based on science,even when it leads in unfamiliar directions. Works to create solutions that win lasting political, economic and social support because they are bipartisan, efficient and fair.

*Fred Krupp, President*
*David Yarnold, Executive Director*

**2234 Environmental Defense Newsletter**
257 Park Avenue South
17th Floor
New York, NY 10010
212-505-2100
Fax: 212-505-2375
E-mail: members@edf.org
http://www.edf.org

Dedicated to protecting the environmental rights of all people, including future generations. Among these rights are clean air, clean water, healthy, nourishing food and a flourishing ecosystem. The solutions we advocate will be basedon science, even when it leads in unfamiliar directions. We will work to create solutions that win lasting political, economic and social support because they are bipartisan, efficient and fair.

*Fred Krupp, President*
*David Yarnold, Executive Director*

**2235 Environmental Law and Compliance Methods**
Oceana Publications, Inc
198 Madison Avenue
New York, NY 10016
800-334-4249
Fax: 212-726-6476
http://www.oceanalaw.com

Presents practical information tailored to professionals responsible for day-to-day compliance with the environmental laws of the US.
*678 pages One Time*
*ISBN: 0-379214-26-1*
*Edward E Shea, Author*

**2236 Environmental Politics and Policy**
Congressional Quarterly Press
1255 22nd Street NW
Suite 400
Washington, DC 20037
202-729-1800
*366 pages*
*ISBN: 1-568028-78-4*
*Walker A Rosenbaum, Author*

**2237 Environmental Regulatory Glossary**
Government Institutes
4 Research Place
Suite 200
Rockville, MD 20850
301-921-2300
Fax: 301-921-0373
*623 pages*
*Thomas F P Sullivan, Author*

**2238 How Wet is a Wetland?: The Impacts of the Proposed Revisions to the Federal Wetlands Manual**
Environmental Defense Fund
257 Park Avenue South
17th Floor
New York, NY 10010
212-505-2100
Fax: 212-505-2375
E-mail: members@edf.org
http://www.edf.org

To prevent environmentally induced harm to human populations.

*Fred Krupp, President*
*David Yarnold, Executive Director*

**2239 Insider's Guide to Environmental Negotiation**
Lewis Publishers
2719 Dover Lane
Albany, GA 31721
229-438-1080
E-mail: tlewis@lewispub.com
http://www.lewispub.com

*242 pages*
*ISBN: 0-873715-09-8*
*Dale M Gorczynski, Author*

**2240 International Environmental Policy: From the Twentieth to the Twenty-First Century**
Duke University Press
905 W Main Street, Ste 18-B
PO Box 90660
Durham, NC 27708
919-687-3600
Fax: 919-688-4574
E-mail: orders@dukepress.edu
http://www.dukepress.com

*496 pages*
*Ken Wissoker, Editor in Chief*
*Courtney Berger, Assistant Editor*

**2241 Making Development Sustainable: Redefining Institutions, Policy, and Economics**
Island Press
1718 Connecticut Avenue NW
Suite 300
Washington, DC 20009
202-232-7933
Fax: 202-234-1328
E-mail: info@islandpress.org
http://www.islandpress.org

*362 pages*
*ISBN: 1-559632-13-5*
*Johan Holmberg, Author*

**2242 Managing Planet Earth: Perspectives on Population, Ecology and the Law**
Greenwood Publishing Group
88 Post Road W
Westport, CT 06881
203-226-3571
800-225-5800
http://www.greenwood.com

*184 pages*
*ISBN: 0-897892-16-X*
*Miguel A Santos, Author*

**2243 Natural Resources Policy and Law: Trends and Directions**
Island Press; Natural Resources Law Center
1718 Connecticut Avenue NW
Suite 300
Washington, DC 20009
202-232-7933
Fax: 202-234-1328
E-mail: info@islandpress.org
http://www.islandpress.org

*255 pages*
*ISBN: 1-559632-46-1*
*Lawrence J MacDonnell; Sarah F Bates, Author*

**2244 New Mexico Environmental Law Center: Green Fire Report**

1405 Luisa Street
Suite 5
Santa Fe, NM 87505
505-989-9022
Fax: 505-989-3769
E-mail: nmelc@nmelc.org
http://www.nmenvirolaw.org

A publication from the organization dedicated to protecting New Mexico's natural environment and communities from pollution and degradation. Over 80 percent of our clients are indigenous Native American or Hispanic and low income.Cases often include mining issues, growth impacts, water protection, air pollution, public lands protection or indigenous land claims. The organization is supported by grants from foundations, contributions from individuals and fees.

*12 pages Quarterly*
*Douglas Meiklejohn, Executive Director*
*Sebia Hawkins, Development Director*

**2245 Oversight of Implementation of the Clean Air Act Amendments of 1990**

Government Printing Office
732 North Capitol Street NW
Washington, DC 20401
202-512-0000
Fax: 202-512-2104
E-mail: contactcenter@gpo.gov
http://www.gpo.gov

*ISBN: 0-160388-26-0*

**2246 People for the Ethical Treatment of Animals**

501 Front Street
Norfolk, VA 23510
757-622-7382
E-mail: info@peta.org
http://www.peta.org

People for the Ethical Treatment of Animals, with more than seven hundred members, is the largest animal rights organization in the world. Founded in 1980, PETA is dedicated to establishing and protecting the rights of all animals.PETA operates under the simple principle that animals are not ours to eat, wear, experiment on, or use for entertainment.

**2247 Renewable Resource Policy: The Legal-Institutional Foundation**

Island Press
1718 Connecticut Avenue NW
Suite 300
Washington, DC 20009
202-232-7933
Fax: 202-234-1328
E-mail: info@islandpress.org
http://www.islandpress.org

*572 pages*
*ISBN: 1-559632-25-9*
*David A Adams, Editor*

**2248 Saving All the Parts: Reconciling Economics and the Endangered Species Act**

Island Press
1718 Connecticut Avenue NW
Suite 300
Washington, DC 20009
202-232-7933
Fax: 202-234-1328
E-mail: info@islandpress.org
http://www.islandpress.org

*280 pages*
*ISBN: 1-559632-02-X*
*Rocky Barker, Author*

**2249 Searching Out the Headwaters: Change and Rediscovery in Western Policy**

Island Press
1718 Connecticut Avenue NW
Suite 300
Washington, DC 20009
202-232-7933
Fax: 202-234-1328
E-mail: info@islandpress.org
http://www.islandpress.org

*253 pages*
*ISBN: 1-559632-17-8*
*Sarah F Bates, et al, Author*

**2250 Setting National Priorities: Policy for the Nineties**

Brookings Institution
1775 Massachusetts Ave., NW
Washington, DC 20036
202-797-6000
800-275-1447
Fax: 202-797-6004
http://www.brookings.edu

**2251 Trade and the Environment: Law, Economics and Policy**

Island Press
1718 Connecticut Avenue NW
Suite 300
Washington, DC 20009
202-232-7933
Fax: 202-234-1328
E-mail: info@islandpress.org
http://www.islandpress.org

*333 pages*
*ISBN: 1-559632-67-4*
*Durwood Zaelke et al, Editors*

**2252 Understanding Environmental Administration and Law**

Island Press
1718 Connecticut Avenue NW
Suite 300
Washington, DC 20009
202-232-7933
Fax: 202-234-1328
E-mail: info@islandpress.org
http://www.islandpress.org

*239 pages*
*ISBN: 1-559634-74-X*
*Susan J Buck, Author*

# Financial Resources

## Foundations & Charities

**2253 AMETEK Foundation**
37 N Valley Road
PO Box 1764
Paoli, PA 19301
610-647-2121
800-473-1286
Fax: 215-323-9337
E-mail: www.ametek.com
http://webmaster@ametek.com
The AMETEK Foundation-the charitable arm of AMETEK Inc., a global manufacturer of electronic insturments and electric motors, has long supported efforts to improve early childhood literacy in schools and libraries near its plants andfacilities.
*Robert W Yannarell, Assistant Secretary*

**2254 ARCO Foundation**
515 South Flower Street
Los Angeles, CA 90071
213-486-3342
The foundation awards education grants both on the national and regional level. Education programs that are national in scope are funded through the headquarters located in Los Angeles. Regional grants are made to nonprofitorganizations in states where ARCO has facilities and personnel.
*Russell Sakaguchi, Program Officer*

**2255 Abelard Foundation**
2530 San Pable Avenue
Suite B
Berkeley, CA 94702
510-644-1904
A family foundation with a 40 year history of progressive funding. The foundation is committed to supporting grassroots social change organizations which engage in community organizing.
*Leah Brumer, Executive Director*

**2256 Acid Rain Foundation**
1410 Varsity Drive
Raleigh, NC 27606
919-828-9443
Fax: 919-515-3593
Designed to significantly reduce emissions responsible for acid deposition.
*Dr. Harriet S Stubbs, Executive Director*

**2257 Acorn Foundation**
2530 San Pablo Avenue
Suite B
Berkeley, CA 94702
510-644-1904
Environmental issues, bio-diversity, health issues related to environmental hazards.
*Leah Brummer, Executive Director*

**2258 African Wildlife Foundation**
1717 Massachusetts Avenue NW
Suite 602
Washington, DC 20036
202-939-3333
Fax: 202-939-3332
The African Wildlife Foundation, together with the people of Africa, work to ensure the wildlife and wild lands of Africa will endure forever. The AFW is the leading international conservation organization focused soley on Africa. Webelieve that protecting Africa's wildlife and wild landscapes is the key to the future prosperity of Africa and its people.
*Paul T Schindler, President*

**2259 Amax Foundation**
200 Park Avenue
New York, NY 10166
212-856-4250
*Sonja Michaud, President*

**2260 American Association of Petroleum Geologists Foundation**
Po Box 979
Tulsa, OK 74101
918-584-2555
Fax: 918-560-2636
E-mail: info@aapg.org
http://www.aapg.org
News for explorationists of oil, gas and minerals as well as for geologists with environmental and water well concerns.
*Richard D Fritz, Executive Director*

**2261 American Electric Power**
1 Riverside Plaza
Columbus, OH 43215
614-223-1000
E-mail: corpcomm@aep.com
http://www.aep.com
One of the largest electric utilities in the United States, delivering electricity to more then 5 million customers in 11 states. AEP ranks among the nations largest generators of electricity, owning more then 38,000 megawatts ofgenerating capacity in the US.
*Richard M McMorrow, Assistant VP*

**2262 American Rivers**
1101 14th Street NW
Suite 1400
Washington, DC 20005
202-347-7550
Fax: 202-347-9240
E-mail: wnelson@americanrivers.org
http://www.americanrivers.org
Support and donations for the protection and restoration of America's rivers for the benefit of people, fish and wildlife.
*Wilke Nelson, VP Resource Development*

**2263 Amoco Foundation**
200 East Randolph Drive
Chicago, IL 60601
312-856-6306
E-mail: foundation@amoco.com
http://www.amoco.com
The BP Amoco awards grants for education, primarily in the field of science and engineering, as well as community organizations in BP Amoco communities.
*Patricia Wright, Executive Director*

**2264 Andrew W. Mellon Foundation**
140 East 62nd Street
New York, NY 10021
212-838-8400
Fax: 212-223-2778
The foundation concentrates most of its grantmaking in a few areas. Institutions and programs receiving support are often leaders in fields of Foundation activity, but they may also be promising newcomers, or in a position todemonstrate new ways of overcoming obstacles to achieve program goals.
*William Robertson IV, Cons. and Envir. Pro. Dir.*

**2265 Asthma and Allergy Foundation of America**
1233 20th Street, NW
Suite 402
Washington, DC 20036
202-466-7643
Fax: 202-466-8940
E-mail: info@aafa.org
http://www.aafa.org
Dedicated to improving the quality of life for people with asthma and allergies through education, advocacy and research.

**2266 Atherton Family Foundation**
1164 Bishop Street
Suite 800
Honolulu, HI 96813
808-566-5524
Fax: 808-521-6286
E-mail: foundations@hcf-hawaii.org
http://www.athertonfamilyfoundation.org
The Atherton Family Foundation is now one of the largest endowed grantmaking private resource in the State of Hawaii, devoted exclusively to the support of charutabe activities. It perpetuates the philanthropic commitment expressedduring the lifetime of Juliette M. Atherton and Frank C. Atherton, and of the family who have followed them.
*Lisa Schiff, Priv Found Services Officer*

**2267 Audubon Naturalist Society of the Central Atlantic States**
8940 Jones Mill Road
Chevy Chase, MD 20815
301-652-9188
Fax: 301-951-7179
The Audubon Naturalist Society is an independent environmental education and conservation organization with over 10,000 members in the Washington DC area. The society offers a wide variety of natural history classes and campaigns forprotection and renewal of the Mid-Atlantic regions natural resources.
*Neal Fitzpatrick, Executive Director*
*Anne Cottingham, Board President*

**2268 BP America**
200 Public Square 36-A
Cleveland, OH 44114
216-586-8625
http://www.bp.com
The purpose is to provide products that satisfy human needs, fuel progress and economic growth and to maintain and invest in a sustainable environment.
*Lance C Buhl, Dir. Corporate Contributions*

**2269 Baltimore Gas & Electric Foundation**
PO Box 1475
Baltimore, MD 21203
410-685-0123
800-685-0123
E-mail: corporate.communications@bge.com
http://www.bge.com
The mission is to safely, economically, reliably, and profitably deliver gas and electricity to our customers. The vision is to be a recognized leader in energy delivery by enhancing our customer's quality of life, our shareholdersvalue, and our team's well being.
*Gary R Fuhrman, Asst. Secretary/Treasurer*

**2270 Bauman Foundation**
1731 Connecticut Avenue
Suite 400
Washington, DC 20009
202-234-8547
Fax: 202-234-8584
The Bauman Foundation was funded by the estate of Lionel R. Bauman, a New York City lawyer and businessman. He was a partner in the real estate development firm of Eugene M. Grant & Co. The foundation is managed by Lionel's daughter.Patricia Bauman.
*Patricia Bauman, Co-Director*

**2271 Bay and Paul Foundations, The**
17 West 94th Street
New York, NY 10025
212-663-1115
E-mail: info@bayandpaulfoundations.org
http://www.bayandpaulfoundations.org
The Bay and Paul Foundations Inc. was formed in January 2005 by the merger of 2 foundations. The Bay Foundation and the Josephine Bay Paul and C. Michael Paul Foundation.
*Robert W Ashton, Executive Director*

**2272 Blandin Foundation**
100 North Pokegama Avenue
Grand Rapids, MI 55744
218-326-0523
877-882-2257
Fax: 218-327-1949
Blandin Foundation is focused on the economic viability of rural Minnesota communities, as part of our mission to help strengthen rural Minnesota and the Grand Rapids area, our home.
*Paul M Olson, President*

**2273 Boise Cascade Corporation**
1111 West Jefferson Street
Po Box 50
Boise, ID 83728
208-384-6161
Fax: 208-384-7189
E-mail: bcweb@bc.com
http://www.bc.com
As we focus on delivering the best return for our investors, we can be trusted to do what we say and take responsibility for our actions, which we base on values and principles.
*Connie E Weaver, Contributions Administrator*

**2274 Cape Branch Foundation**
5 Independence Way
Princeton, NJ 08540
609-987-0300
Fax: 609-452-1024
A private foundation which provides grant support for higher education, museums, and land conservation in the New Jersey area. There are no grants to individuals.
*Dorothy Frank, Partner*

**2275 Cargill Foundation**
15407 McGinty Road
PA-50
West Wayzata, MN 55391
952-742-2546
Fax: 952-742-7224
http://www.cargill.com
Cargill is an international provider of food, agricultural and risk management products and services. With 158,000 employees in 66 countries, the company is committed to using its knoweldge and experience to collaborate with customersto help them succeed.
*James S Hield, Secretary, Contributions*

**2276 Caribbean Conservation Corporation**
4424 NW 13th Street
Suite A1
Gainesville, FL 32609
352-373-6441
800-678-7853
Fax: 352-375-2449
E-mail: ccc@cccturtle.org
http://cccturtle.org
Caribbean Conservation Corporation is a nonprofit membership organization based in Gainesville. CCC was the first marine turtle conservation organization in the world, and has more than 40 years of experience in national andinternational sea turtle conservation, research and educational endeavors.
*David Godfrey, Executive Director*

**2277 Carolyn Foundation**
818 W 46th Street
Suite 203
Minneapolis, MN 55419
612-596-3266
Fax: 612-339-1951
E-mail: berdahl@carolynfoundation.org
http://www.carolynfoundation.org
The Carolyn Foundation is a small general foundation. Please check out our website www.carolynfoundation.org for the most up-to-date information regarding funding priorities, guidelines and application process.
*Rebecca Erdahl, Executive Director*
*Kristen Cullen, Foundation Administrator*

**2278 Caterpillar Foundation**
100 Northeast Adams Street
Peoria, IL 61629
309-675-4464
http://www.caterpillar.com
Provides funding and support from a corporate perspective. Formed in 1952, the Foundation has distributed almost $200 million to support education, health and human services, and civic, cultural, and environmental causes.
*Edward W Siebert, Vice President/Manager*

**2279 Charles Engelhard Foundation**
645 5th Avenue
7th Floor
New York, NY 10022
212-935-2430
Provides funding to a wide range of causes including education, medical research, cultural institutions, and wildlife conservation.
*Elaine Catterall, Secretary*

**2280 Chesapeake Bay Foundation**
6 Herndon Avenue
Annapolis, MD 21403
410-268-8816
Fax: 410-268-6687
http://www.cbf.org
The only independent organization dedicated soley to restoring and protecting the Chesapeake Bay and its tributary rivers.
*William C Baker, President*

**2281 Chevron Corporation**
575 Market Street
San Francisco, CA 94105
415-894-7700
E-mail: chevweb@chevron.com
http://www.chevron.com
As a global enterprise that is highly competitive across all energy sectors, Chevron brings together a wealth of talent, shared values and a strong commitment to developing vital energy resources worldwide.
*Skip Rhodes, Mgr. Corporate Contributions*

**2282 Clean Water Action**
1010 Vermont Avenue NW
Suite 400
Washington, DC 20005
202-895-0420
Fax: 202-895-0438
E-mail: cwa@cleanwater.org
http://www.cleanwateraction.org
Clean Water Action is a national organization of diverse people and groups working together for clean water, protecting health, creating jobs and making democracy work.
*Bob Wendelgass, Executive Director*

**2283 Collins Foundation**
1618 SW 1st Avenue
Suite 305
Portland, OR 97201
503-227-7171
http://www.collinsfoundation.org

An independent private foundation, exists to improve, enrich and give greater expression to the religious, educational, cultural, and scientific endeavors in the state of Oregon and to assist in improving the quality of life in thestate.
*Brian Coakley, President*

**2284 Compton Foundation**
255 Shorline Drive 650-508-1181
Suite 540 Fax: 650-508-1191
Redwood City, CA 94065 http://www.comptonfoundation.org
Seeks to foster human and ecological security by addressing contemporary threats to these inalienable rights. We support responsible stewardship that respects the rights of future generations to a balanced and healthy ecology, bothpersonal and global, allowing for the full richness of human experience.
*Edith T Eddy, Executive Director*

**2285 Conservation International**
2011 Crystal Drive 703-341-2400
Suite 500 800-429-5660
Arlington, VA 22202 Fax: 202-887-5188
http://www.conservation.org
Our mission is to conserve the Earth's living natural heritage, our global biodiversity, and to demonstrate that human societies are able to live harmoniously with nature.
*Peter A Seligmann, Chairman of the Board*

**2286 Conservation Treaty Support Fund**
3705 Cardiff Road 301-654-3150
Chevy Chase, MD 20815 800-654-3150
Fax: 301-652-6390
E-mail: ctsf@conservationtreaty.org
http://www.conservationtreaty.org
CTSF's mission is to support the major inter-governmental treaties which conserve wild natural resrouces and habitat for their own sake and the benefit of the people. These includ the Endangered Species Convention and the InternationalWetlands Convention. CTSF raises support for treaty projects from indivudals, corporations, foundations and government agencies. It also develops educational and informational materials including videos and the CITES Endangered Species Book.
*1986 pages*
*George A Furness Jr, President*

**2287 Conservation and Research Foundation**
PO Box 909 913-268-0076
Shelburne, VT 05482 Fax: 913-268-0076
*Dr Mary Wetzel, President*

**2288 Cooper Industries Foundation**
PO Box 4446 713-209-8800
Houston, TX 77210 Fax: 713-209-8982
E-mail: info@cooperindustries.com
http://www.cooperindustries.com
Cooper was primarily a one-market company, manufactguring power and compression equipment for the transmission of natural gas. Eventually broadening its product lines to include petroleum and industrial equipment, electrical powerequipment, automotive products tools and hardware.
*Victoria B Guennewig, President*

**2289 Cricket Foundation**
Exchange Place 617-570-1130
Suite 2200 Fax: 617-523-1231
Boston, MA 02109
Dedicated to improving the quality of life
*George W Butterworth III, Counsel*

**2290 Curtis and Edith Munson Foundation**
321 N Clark Street 312-527-5545
Suite 950 Fax: 312-527-9064
Chicago, IL 60610
Over the past 15 years, we have emphasized partnerships, collaborations, and seed funding for new projects and organizations within the framework of our programs as defined by our guidelines.
*C Wolcott Henry III, President*

**2291 Deer Creek Foundation**
720 Olive Street 314-241-3228
Suite 1975
St. Louis, MO 63101
Projects should focus on the preservation and advancement of majority rule in our society, including the protection of basic rights
*Mary Stake Hawker, Administrator*

**2292 Defenders of Wildlife**
1130 17th Street NW 202-682-9400
Washington, DC 20036 Fax: 202-682-1331
E-mail: defenders@mail.defenders.org
http://www.defenders.org
Dedicated to the protection of all native wild animals and plants in their natural communities. Focus is placed on what scientists consider two of the most serious environmental threats to the planet: the accelerating rate ofextinction of species and the associated loss of biological diversity, and habitat alteration and destruction. Long known for leadership on endangered species issues.
*Roger Schlickeisen, President*

**2293 Digital Equipment Corporation**
111 Powder Mill Road 508-493-5111
Unit B14 Fax: 508-493-8780
Maynard, MA 01754 http://www.digitalcentury.com
A leading worldwide supplier of networked computer systems, software and services. Its products serve a variety of applications, such as scientific analysis, industrial control, time-sharing, commercial data processing, graphic arts,word processing, office automation, health care, instrumentation, engineering, and simulation.
*Jane Hamel, Mgr. Corporate Contributions*

**2294 Dunspaugh-Dalton Foundation**
1533 Sunset Dr. 305-668-4192
Suite 150 Fax: 305-668-4247
Coral Gables, FL 33143
Supports educational, social, medical and cultural institutions in Florida, California and North Carolina.
*William A Lane Jr, President*

**2295 Earth Share**
7735 Old Georgetown Road 240-333-0300
Suite 900 800-875-3863
Bethesda, MD 20814 http://www.earthshare.org
A nationwide network of America's leading non-profit environmental and conservation organizations, works to promote environmental education and charitable giving through workplace giving campaigns.

**2296 Earth Society Foundation**
238 E 58th Street 212-832-3659
Suite 2400 Fax: 212-826-6213
New York, NY 10022 E-mail: trusteeone@aol.com
http://www.earthsite.org
Started the original Earth Day, which is devoted to peace, justice and the care of earth. It invites everyone to think and act as trustees of earth.
*Mary Carlin, Secretary*
*John McConnell, Chairman Emeritus*

**2297 Echoing Green**
494 Eighth Ave 212-689-1165
2nd Floor Fax: 212-689-9010
New York, NY 10001 E-mail: info@echoinggreen.org
http://www.echoinggreen.org
Echoing Green is a global science venture fund that provides seed funding and support to visionary leaders with bold new ideas for social change.
*Cheryl Dorsey, President*

### 2298 Edward John Noble Foundation
Po Box 954
Ridgefield, CT 06877
203-438-5690
*EJ Noble Smith, Executive Director*

### 2299 Energy Foundation
1012 Torney Avenue #1
San Francisco, CA 94129
415-561-6700
Fax: 415-561-6709
http://www.ef.org

The Energy Foundation is a partnership of major donors interested in solving the world's energy problems. Our mission is to advance energy efficiency and renewable energy-new components of a clean energy future.
*Hal Harvey, Executive Director*

### 2300 Environmental Law Institute
2000 L Street NW
Suite 620
Washington, DC 20036
202-939-3800
Fax: 202-939-3868
E-mail: law@eli.org
http://www.eli.org

Provides information services, advice, publications, training courses, seminars, research programs and policy recommendations to engage and empower environmental leaders the world over.
*Leslie Carothers, President*
*Kenneth Berlin, Chairman*

### 2301 Exxon Education Foundation
225 E John W Carpenter Freeway
Room 1429
Irving, TX 75062
972-444-1000
http://www.exxon.com

It is ExxonMobil's longstanding belief that education is the key to progress, development and economic growth, and we are committed to being a responsible partner in the communities where we operate.
*Leonard Fleischer, Mgr. Corporate Contributions*

### 2302 First Hawaiian Foundation
999 S Bishop Street
29th Floor
Honolulu, HI 96813
808-525-7000
http://www.fhb.com

First Hawaiian Foundation is the charitable arm of First Hawaiian Bank. The foundation funds educational opportunities, access to health care, services for children and youth, human service needs, and the many ways that the arts enrichour lives.
*Herbert E Wolff, Secretary*

### 2303 First Interstate Bank of Nevada Foundation
PO Box 11007
Reno, NV 89520
775-784-3844
*Kevin Day, President*

### 2304 FishAmerica Foundation
6101 E Apache Street
Tulsa, OK 74115
918-836-5581
800-444-5581
Fax: 918-836-3542
http://www.fishamerica.org

Unites the sportfishing industry with conservation groups, government natural resource agencies, corporations, and charitable foundations to invest in fish and habitat conservation and research across the country.

### 2305 FishAmerica Foundation.
Grant Guidelines
225 Reinekers Lane
Suite 420
Alexandria, VA 22314
703-519-9691
Fax: 703-519-1872
E-mail: fishamerica@asafishing.org
http://www.fishamerica.org

The FishAmerica Foundation provides funding for local, hands-on projects to enhance fish populations, restore fisheries habitat, improve water quality, and advancing fisheries research in North America; thereby increasing theopportunity for sportfishing success.

*Johanna Laderman, Managing Director*
*Jeff Bloem, Grants Coordinator*

### 2306 Frank Weeden Foundation
747 Third Avenue
34th Floor
New York, NY 10017
212-888-1672
Fax: 212-888-1354
E-mail: weedenfdn@weedenfdn.org

From its inception in 1963, the Foundation embraced the protection of biodiversity as its main priority.

*Alan N Weeden, President*

### 2307 Friends of the Earth Foundation
1717 Massachusetts Avenue NW 600
Washington, DC 20036
877-843-8687
Fax: 202-783-0444
E-mail: foe@foe.org
http://www.foe.org

Defends the environment and champions a healthy and just world.
*Brent Blackwelder, President*

### 2308 Frost Foundation
511 Armijo
Suite A
Santa Fe, NM 87501
505-986-0208
E-mail: info@frostfound.org
http://www.frostfound.org

Created to be operated exclusively for educational, charitable and religious purposes. The foundation possesses all powers, rights, privileges, capacities and immunities which non profit corporations are authorized to possess.
*Mary Amelia Whited-Howell, President*
*Philip B. Howell, Executive Vice President*

### 2309 Fund for Animals
200 West 57th Street
New York, NY 10019
212-246-2096
Fax: 212-246-2633
E-mail: hdquarters@fund.org
http://www.fundforanimals.org/about/

The Fund for Animals was founded in 1967 by author/humanitarian Cleveland Amory to speak for those who can't. In 2005, The Fund for Animals merged with The Humane Society of the United States, to avoid duplication of program andincrease strength and coordination in the areas of legislation, litigation, humane education and disaster relief.

*Marian Probst, Chair*

### 2310 Fund for Preservation of Wildlife and Natural Areas
Boston Safe Deposit and Trust Company
1 Boston Place
Boston, MA 02108
617-722-7340
Fax: 617-722-7129

Accounting services for mutual fund companies.
*Sylvia Salas, Director*

### 2311 George B Storer Foundation
PO Box 1270
Saratoga, WY 82331
307-326-8308

Provides support for higher education and social services, especially for the blind, youth organizations, conservation, hospitals, and cultural programs.
*Peter Storer, President*

### 2312 Georgia Pacific Foundation
133 Peachtree Street NE
Atlanta, GA 30303
404-652-4000
http://www.gp.com

Invests in educational efforts that empower youth, and provide workers with job readiness training. We also invest in scholarships and technical programs that give workers the skills necessary for today's workplace.
*Wayne Tamblyn, President*

### 2313 Geraldine R. Dodge Foundation
163 Madison Avenue
PO Box 1239
Morristown, NJ 07962
973-540-8442
Fax: 973-540-1211
E-mail: info@grdodge.org
http://www.grdodge.org/foundation

The mission of the Geraldine R. Dodge Foundation is to support and encourage those educational, cultural, social and environ-

mental values that contribute to making our society more humaine and our world more liveable.
*David Grant, Executive Director*
*Robert Perry, Program Director*

**2314 Greensward Foundation**
Po Box 610 Lenox Hill Station
New York, NY 10021 E-mail: info@greenswardsparks.org
http://www.greenswardparks.org
The Greensward Foundation, through its local branches, celebrates and suppports our communities's public parks. We are nonprofit. We receive no public funding, subsisting entirely on private grants and member contributions.
*Robert M Makla, Director*

**2315 HKH Foundation**
121 W 27th St. 212-682-7522
10th Floor
New York, NY 10001
Gives a major portion of its funding to the Adirondack Historical Association. Additional funding is distributed to the disarmament and prevention of war, civil liberties and human rights, and environmental protection.
*Harriet Barlow, Adv.*

**2316 Helen Clay Frick Foundation**
7227 Reynolds Street 412-371-0600
PO Box 86190
Pittsburgh, PA 15208
Devoted to the interpretation of the life and times of Henry Clay Frick.
*DeCoursey E McIntosh, Executive Director*

**2317 Henry L and Consuelo S Wenger Foundation**
100 Renaissance Center 313-567-1212
Detroit, MI 48226
*Shelly Raines, Principal Manager*

**2318 Hoffman-La Roche Foundation**
PO Box 278 973-235-3797
Nutley, NJ 07110
Hoffman-La Roche is the US prescription drug unit of the Roche Group, one of the world's leading research-oriented health care groups with core businesses in pharmaceuticals and diagnostics.
*Rosemary Bruner, Administrative Director*

**2319 INFORM**
120 Wall Street 212-361-2400
14th Floor Fax: 212-361-2412
New York, NY 10005 http://www.informinc.org
INFORM is an independent research organization that examines the effects of business practices on the environment and on human health. Our goal is to identify ways of doing business that ensure environmentally sustainable economicgrowth. Our reports are used by government, industry, and environmental leaders around the world.
*Joanna Underwood, President*

**2320 International Primate Protection League**
PO Box 776 843-871-2280
Summerville, SC 29484 Fax: 843-871-7988
E-mail: info@ippl.org
http://www.ippl.org
An organization that works worldwide for the conservation and protection of apes and monkeys.

*Dr Shirley McGreal, Executive Director*
*Barbara Allison, Office Manager*

**2321 International Wildlife Coalition**
634 N Falmouth Highway 508-457-1898
Box 388 Fax: 508-457-1898
North Falmouth, MA 02556 E-mail: iwchq@iwc.org
http://www.iwc.org
A federally recognized, non-profit tax-exempt charitable organization. The Coalition is dedicated to public education, research, resuce, rehabilitation, litigation, legislation and international treaty negotiations concerning globalwildlife and natural habitat protection issues.
*Daniel J Morast, President*

**2322 International Wildlife Conservation Society**
Grants Management Association
2300 Southern Boulevard 718-220-5100
Bronx, NY 10460 http://www.wcs.org
Saves wildlife and wetlands. We do so through careful science, international conservation, education, and the management of the world's largest system of urban wildlife parks, led by the flagship Bronx Zoo.
*Ala H Reid, Administrator*

**2323 Jessie Smith Noyes Foundation**
6 E 39th Street 212-684-6577
12th Floor Fax: 212-689-6549
New York, NY 10016 E-mail: noyes@noyes.org
http://www.noyes.org
Promotes a sustainable and just social and natural system by supporting grassroots organizations and movements committed to this goal.

*Victor DeLuca, President*

**2324 John D and Catherine T MacArthur Foundation**
140 S Dearborn Street 312-726-8000
Suite 1100 Fax: 312-920-6258
Chicago, IL 60603 E-mail: 4answers@macfound.org
http://www.macfound.org
A private independent grantmaking institution dedicated to helping groups and individuals foster lasting improvement in the human condition.

**2325 Jules and Doris Stein Foundation**
PO Box 30 323-276-2101
Beverly Hills, CA 90213
Founded the Jules Stein Eyes Institute at UCLA in the 1960's. Founded as a multidisciplinary center for vision science.
*Linda L Valliant, Secretary*

**2326 Kangaroo Protection Foundation**
1900 L Street NW 202-452-1100
Suite 526
Washington, DC 20036
*Marian Newman, Program Director*

**2327 Keep America Beautiful**
1010 Washington Boulevard 203-323-8987
Stamford, CT 06901 Fax: 203-325-9199
E-mail: blyons@kab.org
http://www.kab.org
Nonprofit organization whose network of local, statewide and international affiliate programs educates individuals about litter prevention and ways to reduce, reuse, recycle and properly manage waste materials. Through partnershipsand strategic alliances with citizens, businesses and government, Keep America Beautiful's programs motivate millions of volunteers annually to clean up, beautify and improve their neighborhoods, thereby creating healthier and safer communityenvironments.
*Becky Lyons, VP Training/Affiliate*

**2328 Kraft General Foods Foundation**
Kraft Court 847-998-7031
Unit 2W 800-543-5335
Glenview, IL 60025 http://www.kraftfoods.com
Based on the values of innovation, quality, safety, respect, integrtity and openness. These values are what we stand for, the standard of conduct we hold ourselves to and our commitment to the people who work with us, invest in us andpurchase our products.
*Pamela Hollie, Dir. Corporate Contributions*

**2329 Kroger Company Foundation**
1014 Vine Street 513-762-4443
PO Box 1199 866-221-4141
Cincinnati, OH 45202 http://www.kroger.com
Spans many states with store formats that include grocery and multi-department store, convenience stores and mall jewelry stores. We operate under nearly 2 dozen banners, all of which share the same belief in building strong local tiesand brand loyalty with our customers.
*Paul Bernish, VP/Secretary*

**2330 Liz Claiborne Foundation**
1441 Broadway Avenue
New York, NY 10018 E-mail: corporate.secretary@liz.com
http://http://www.lizclaiborneinc.com/foundation/default.asp
Established to serve as the Company's center for charitable activiteis. Works to meet the needs of the communities where the major facilities of Liz Claiborne, Inc. are located. Projects focus primarily on helping disadvantaged womengain their self-sufficiency through job training and microenterprise development. The Foundation also provides ongoing support to many artistic and cultural institutions which enhance the livability of our communities.

*Paul R Charron, Chairman of the Board/CEO*
*Angela J Ahrendts, Executive Vice President*

**2331 Liz Claiborne and Art Ortenberg Foundation**
650 5th Avenue 212-333-2536
15th Floor Fax: 212-956-3531
New York, NY 10019 E-mail: lcaof@fcc.net
The Foundation has 2 primary program interests: mitigation of conflict between the land and resources needs of local communities and conservation of biological diversity in rural landscapes outside of parks and reserves; implementationof relevant, field based scientific, technical and practical training programs for local people. The Foundation typically funds modest, carefully designed field activities-primarily in developing countries and the Northern Rockies.
*James Murtaugh, Director*
*Jeffery T Olson, Director*

**2332 Louis and Anne Abrons Foundation**
437 Madison Avenue 212-756-3376
New York, NY 10017
*Richard Abrons, President and Director*

**2333 Louisiana Land and Exploration Company**
PO Box 60350 504-566-6500
New Orleans, LA 70160
One of the largest independent oil and gas exploration compaines in the United States. It operates a crude oil refinery and conducts exploration and production operations in the United States and selected foreign countries.
*Karen A Overson, Contributions Coordinator*

**2334 MNC Financial Foundation**
10 Light Street 301-244-5000
PO Box 987-MS251001
Baltimore, MD 21203
*Geeorge BP Ward Jr, Secretary/Treasurer*
*Alfred Lerner, Chairman*

**2335 Mark and Catherine Winkler Foundation**
4900 Seminary Road 703-998-0400
Alexandria, VA 22311 Fax: 703-578-7899
*Lynne Ball, Executive Director*

**2336 Mars Foundation**
6885 Elm Street 703-821-4900
McLean, VA 22101 E-mail: ifoundmars@aol.com
http://www.multiple-sclerosis-mf.org
The mission of the MARS foundation is to be committed to finding a cure for multiple sclerosis by funding medical research.
*Roger G Best, Secretary*

**2337 Marshall and Ilsley Foundation**
770 North Water Street 414-765-7835
Milwaukee, WI 53201 http://www.micorp.com
Provides comprehensive financial products and services and unparalleled customer service to personal, business, corporate and institutional customers nationwide.
*Diana L Sebion, Secretary*

**2338 Mary Reynolds Babcock Foundation**
2920 Reynolda Village 336-748-9222
Winston-Salem, NC 27106 Fax: 336-777-0095
E-mail: info@mrbr.org
http://www.mrbf.org
Our mission is to help people and places to move out of poverty and achieve greater social and economic justice.
*Gayle Williams, Executive Director*
*Sandra Mikush, Assistant Director*

**2339 Max McGraw Wildlife Foundation**
PO Box 9 847-741-8000
Dundee, IL 60118 Fax: 847-741-8157
E-mail: mcgrawwild@AOL.COM
http://www.mcgrawwildlife.org
The foundation's mission: education, research, and land management. Currently, the Foundation is invovled in over 15 research and land management projects through the Chicago region, including participation in the Chicago Wildernessinitiative. Situated on 1,225 acres, the Foundation property is managed by professional land management staff.
*Stanley W Koenig, Executive Director*
*John Thompson, Director Research*

**2340 Max and Victoria Dreyfus Foundation**
50 Main Street 914-682-2008
White Plains, NY 10606
A leading figure in the music publishing business, the Foundation's grantmaking supports organizations in the arts, education, health care, hospitals, social services, civic affairs, and religion.
*Lucy Gioia, Administrative Assistant*

**2341 May Stores Foundation**
611 Olive Street 314-342-6300
St. Louis, MO 63101 Fax: 314-342-4461
http://www.maycompany.com
*James Abrams, VP Corporate Communications*

**2342 McIntosh Foundation**
15840 Meadows Wood Drive 202-338-8055
Wellington, FL 33414 Fax: 202-234-0745
E-mail: mcf@aol.com
The McIntosh foundation began in 1949 who founded the Great Atlantic & Pacific Tea Company...later renamed A&P
*Michael A McIntosh, President*

**2343 Nathan Cummings Foundation**
475 Tenth Avenue 212-787-7300
Fourteenth Floor Fax: 212-787-7377
New York, NY 10018 E-mail: info@nathancummings.org
http://http://www.nathancummings.org
The Nathan Cummings Foundation is rooted in the Jewish tradition and committed to democratic values and social justice, including fairness, diversity, and community. They seek to build a socially and economically just society thatvalues nature and protects the ecological balance for future generations; promotes humane health care; and fosters arts and cultures that enriches communities.
*, Environment Program Director*

**2344 National Arbor Day Foundation**
100 Arbor Avenue 402-474-5655
Nebraska City, NE 68410 888-448-7337
Fax: 402-474-0820
E-mail: info@arborday.org
http://www.arborday.org
A nonprofit educational, environmental organization that helps people plant and care for trees. We are committed to tree-planting

and environmental stewardsip. Newsletter free with $10.00 annual membership.
*8 pages Bi-Monthly*
*John Rosenow, President*
*Gary Brienzo, Info. Coordinator*

**2345 National Audubon Society**
700 Broadway
New York, NY 10003
212-979-3000
Fax: 212-979-3188
E-mail: jbianchi@audubon.org
http://www.audubon.org
The mission of the National Audubon Society is to conserve and restore natural ecosystems, focusing on birds and other wildlife for the benefit of humanity and the earth's biological diversity. Founded in 1905, the National AudubonSociety is named for John James Audubon, famed orinthologist, exployer, and wildlife artist.
*John Flicker, President*

**2346 National Fish and Wildlife Foundation**
1120 Connecticut Avenue NW
Suite 900
Washington, DC 20036
202-857-0166
Fax: 202-857-0162
http://www.nfwf.org
Sustains, restores and enhances the Nation's fish, wildlife, plants, and habitats.
*John Berry, Executive Director*
*Karen Sprecher-Keating, General Counsel*

**2347 National Forest Foundation**
Fort Missoula Road
Building 27, Suite 3
Missoula, MT 59804
202-298-6740
Fax: 406-542-2810
E-mail: bpossiel@nationalforests.org
http://www.natlforests.org
The official nonprofit partner of the USDA Forest Service. To accept and administer private contributions, undertaking activities that further the purposes for which the National Forest System was established and conductingeducational, technical, and other activities that support the multiple use, research, and forestry programs administered by the Forest Service.
*Bill Possiel, President*
*David Bell, Chief Operating Officer*

**2348 National Geographic Society Education Foundation**
1145 17th Street NW
Washington, DC 20036
202-828-6672
Fax: 202-775-6141
The mission is to motivate and enable each new generation to become geographically literate.
*Dori Jacobson, Program Development Officer*

**2349 National Parks Conservation Association**
1300 19th Street
#300
Washington, DC 20036
202-223-6722
Fax: 202-659-0650
http://www.npca.org
Mission is to protect and enhance America's National Parks for present and future generations.
*Thomas Kiernan, President*

**2350 National Wildlife Federation**
11100 Wildlife Center Drive
Reston, VA 20190
202-797-6800
800-822-9919
Fax: 202-797-6646
http://www.nwf.org
The National Wildlife Federation is the largest member-supported conservation group, uniting individuals, organizations, businesses and government to protect wildlife, wild places and the environment.
*Larry J Schweiger, President/CEO*

**2351 Nature Conservancy**
1815 North Lynn Street
Arlington, VA 22209
703-841-5300
Fax: 703-841-1283
http://www.nature.org
The mission is to preserve plants, animals and natural communities that represent the diversity of life on Earth by protecting the lands and waters they need to survive.
*John C Sawhill, President*

**2352 New England Biolabs Foundation**
8 Enon st #2B
Beverly, MA 01915
978-927-2404
Fax: 978-998-6837
E-mail: fosters@nebf.org
http://www.nebf.org
NEBF funds grass roots organizations in developing countries that focus on environmental issues and education.

**2353 New York Times Company Foundation**
229 W 43rd Street
New York, NY 10036
212-556-1234
Fax: 212-556-3690
http://www.nytco.com
Strongly committed to protecting the environment in all of the many communities in which it operates.
*Arthur Gelb, President*

**2354 New-Land Foundation**
1114 Avenue of the Americans
46th Floor
New York, NY 10036
212-479-6162
Fax: 212-841-6275
Seeks to foster positive change throughout the global community through its grant making.

*Robert Wolf, President*

**2355 Norcross Wildlife Foundation**
250 W 88th Street
New York, NY 10024
212-362-4831
Fax: 212-812-4299
E-mail: norcross_wf_po@prodigy.net
http://www.norcrossws.org
A place of refuge where all wildlife is encouraged not just to survive but also to proliferate naturally, and where certain species, now threatned with extinction, may again attain more normal distribution and benefit the public bytheir survival.
*Richard Reagan, President*

**2356 Northwest Area Foundation**
332 Minnesota Street
Suite E-1201
St. Paul, MN 55101
651-224-9635
Fax: 651-225-3881
Committed to helping communities reduce poverty for the long term.
*Terry Tinson Saario, President*

**2357 Oliver S and Jennie R Donaldson Charitable Trust**
US Trust Company of New York
114 W 47th Street
New York, NY 10036
212-852-3683
Fax: 212-852-3377
Philantropic organization working to promote social change that contributes to a more just, sustainable and peaceful world.
*Anne L Smith-Ganey, Secretary*

**2358 Overbrook Foundation**
122 East 42nd Street
Suite 2500
New York, NY 10168
212-661-8710
Fax: 212-661-8664
E-mail: contact@overbrookfoundation.org
http://www.overbrookfoundation.org
The Overbrook Foundation strives to improve the lives of people by supporting projects that protect human and civil rights, advance the self-sufficiencey and well being of individuals and their communities, and conserve the naturalenvironment.
*Ms Sheila McGoldrick*

**2359 Pacific Whale Foundation**
101 N Kihei Road
Suite 25
Kihei, HI 96753
808-879-8811
800-942-5311
Fax: 808-879-2615
http://www.pacificwhale.org
Mission is to promote appreciation, understanding and protection of whales, dolphins, coral reefs and our plantet's oceans. We accomplish this by educating the public from a scientific perspective about the marine environment. Wesupport and conduct responsible marine research and address marine conservation issues in Hawaii and the Pacific. Through educational ecotours, we

model and promote sound ecotourism practices and responsible wildlife watching.
*Gregory D Kaufman, President*

**2360 Patrick and Anna Cudahy Fund**
1609 Sherman Avenue 708-866-0760
#207
Evanston, IL 60201
Types of support: general/operating support; continuing support; annual campaigns; building/renovation; equipment; program development; seed money; technical assistance; matching funds.
*Judith Borchers, OSB, Executive Director*

**2361 Pew Charitable Trusts**
One Commerce Square 215-575-9050
2005 Market Street Suite 1700 Fax: 215-575-4939
Philadelphia, PA 19103 http://www.pewtrusts.org
Driven by the power of knowledge to solve today's most challenging problems. Pew applies a rigorous, analytical approach to improve public policy, inform the public and stimulate civic life.
*Joshua S Reichert, Dir. Environmental Programs*

**2362 Providence Journal Charitable Foundation**
75 Fountain Street 401-277-7000
Providence, RI 02902 http://www.projo.com
Focuses on offering local and regional news, information, advertising and interactive opportunities for our audience.
*Phil Kukielski, Managing Editor*
*John Kostrzewa, Business Editor*

**2363 RARE Center for Tropical Bird Conservation**
1529 Walnut Street 215-568-0420
Philadelphia, PA 19102
Works globally to equip people in the world's most threatened natural areas with the tools and motivation they need to care for their natural resources.
*John Guarnaccia, Executive Director*

**2364 Rainforest Action Network**
221 Pine Street 415-398-4404
Suite 500 Fax: 415-398-2732
San Francisco, CA 94104 E-mail: answers@ran.org
http://www.ran.org
Rainforest Action Network works to protect the Earth's rainforests and support the rights of their inhabitants through education, grassroots organizing and non-violent direct action.
*Michael Brune, Executive Director*

**2365 Rainforest Alliance**
665 Broadway 212-677-1900
Suite 500 888-693-2784
New York, NY 10012 Fax: 212-677-2187
E-mail: info@ra.org
Works to conserve biodiversity and ensure sustainable livelihoods by transforming land-use practices, business practices and consumer behavior.
*Daniel R Katz, President/Executive Director*

**2366 Raytheon Company**
870 Winter Street 781-552-3000
Waltham, MA 02451 http://www.raytheon.com
Raytheon is a technology leader specializing in defense, homeland security, and other government markets throughout the world.
*Janet Taylor, Admin. Corp. Contributions*

**2367 Richard King Mellon Foundation**
PO Box 2930 412-392-2800
Pittsburg, PA 15230 Fax: 412-392-2837
http://www.foundationcenter.org/grantmaker/rkmellon
Regional focus for Pennsylvania are conservation, land preservation, watershed protection and restoration, sustainable environments, regional economic development, children, youth, and young adults and education.
*George H Taber, Vice President and Director*

**2368 Richard Lounsberry Foundation**
1020 19th Street NW 202-872-8080
Suite LL60 Fax: 202-872-9292
Washington, DC 20036 http://http://rlounsbery.org
Aims to enhance national strengths in science and technology through support programs in the areas of science and technology components of key US policy issues, elementary and secondary science and math education, historical studiesand contemporary assessments of key trends in the physical and biomedical sciences and start up assistance for establishing the infrastructure of research projects.
*Alan F McHenry, President*

**2369 Rockefeller Brothers Foundation**
437 Madison Avenue 212-812-4200
37th Floor Fax: 212-812-4299
New York, NY 10022 E-mail: info@rfb.org
http://www.rbf.org
A philantrophic organization working to promote social change that contributes to a more just, sustainable and peaceful world. The Fund's programs are intended to develop leaders, strenghten institutions, engage citizens, buildcommunity, and foster partnerships that include government, business, and civil society.
*Benjamin R Shute Jr, Secretary*

**2370 Rockefeller Family Fund**
437 Madison Avenue 212-812-4252
37th Floor Fax: 212-812-4299
New York, NY 10022 E-mail: mmccarthy@rffund.org
http://www.rffund.org
For thirty years, the Rockefeller Family Fund has worked at the cutting edge of advocacy in such areas as environmental protection, advancing the economic rights of women, and holding public and private institutions accountable fortheir actions.
*, Director*

**2371 Safari Club International Foundation**
4800 West Gates Pass Road 602-620-1220
Tucson, AZ 85745 800-377-5399
Fax: 602-622-1205
Safari Club Internation Foundation is a charitable organization that funds and manages worldwide programs dedicated to wildlife conservation, outdoor education and humanitarian services.
*Doug Yajko, President*

**2372 Samuel Roberts Noble Foundation**
2510 Sam Noble Parkway 508-223-5810
Ardmore, OK 73401 Fax: 508-224-6217
http://www.noble.org
One of the largest international offshore drilling contractors in the world.

*Michael Cawley, President/CEO*
*Patrick Jones, VP/CFO/Treasurer*

**2373 Save the Redwoods League**
114 Sansome Street 415-362-2352
Room 1200 888-836-0005
San Francisco, CA 94104 Fax: 415-362-7017
E-mail: info@savetheredwoods.org
http://www.savetheredwoods.org
Guided by their science-based master plan to save redwoods throughout their natural ranges, the Leagues purchases priority pieces of land and donates or sells the property to government agencies for protection as parks and reserves.The league funds restoration, supports research to expand knowledge of redwood forest dynamics, and educates the public about redwoods and their ecosystems, in order to reconnect people with the peace and beauty of these wonders of the natural world.

*Ruskin K Hartley, Executive Director*

**2374 Scherman Foundation**
16 E 52nd Street 212-832-3086
Suite #601 Fax: 212-838-0154
New York, NY 10022 http://www.scherman.org

The giving program of the Foundation emphasizes long-term general support, reflecting the director's commitment to sustained support for current grantees, and the belief that strong nonprofit leaders who are closest to the issues canbest decide on the most effective use of grant funds.
*Mr Mike Pratt, Program Officer/Treasurer*

**2375 Sequoia Foundation**
820 A Street 206-627-1634
Suite 345
Tacoma, WA 98402
A private non-profit organization dedicated to the identification and reduction of environmental threats to public health. We seek to support the efforts of local, state, national-and international-public health agencies in promotingand implimenting effective public health policy. This mission is achieved through research collaborations with local, state, federal, and international agencies, community-based organizations, and hospitals and universities.
*John S. Petterson Ph.D., Executive Director*
*Pam Petree, Contracts Manager*

**2376 Sierra Club Foundation**
220 Sansome Street 415-291-1800
Suite 1100 Fax: 415-291-1791
San Francisco, CA 94104
To advance the preservation and protection of the natural environment by empowering the citizenry, especially democratically-based grassroots organizations, with charitable resources to further the cause of environmental protection.The vehicle through which The Sierra Club Foundation generally fulfills its charitable mission.
*Stephen M Stevick, Executive Director*

**2377 Social Justice Fund NW**
603 Stewart Street 206-624-4081
Suite 1007 Fax: 206-382-2640
Seattle, WA 98101 E-mail: office@sodialjusticefund.org
http://www.socialjusticefund.org
Progressive foundation dedicated to creating a more just society. Funds grass-roots community-based organizations in Idoho, Montana, Wyoming, Washington, and Oregon.
*Stephanie Austin, Administrative Assistant*

**2378 Switzer Foundation New Hampshire Charitable Foundation**
Po Box 293 207-338-5654
Belfast, ME 04915 800-464-6641
Fax: 603-225-1700
E-mail: info@nhcf.org
Identifies and nurtures environmental leaders who have the ability and determination to make a significant impact, and supports initiatives that will have direct and measurable results to improve environmental quality.
*Judith Burrows, Director Student Aid*

**2379 Texaco Foundation**
2000 Westchester Avenue 914-701-0320
White Plains, NY 10650 http://www.texaco.com
The foundation focuses on early childhood education in math and science through its Early Notes (music) program and its Touch Science program, which supports scientific discovery through hands-on learning.
*Maria Mike-Mayer, Secretary*

**2380 Threshold Foundation**
Po Box 29903 415-561-6400
San Francisco, CA 94109 Fax: 415-561-6401
E-mail: info@thresholdfoundation.org
http://www.thresholdfoundation.org
A progressive foundation and a community of individuals united through wealth, mobilizing money, people and power to create a more just, joyful and sustainable world.
*Drummond Pike, Foundation Manager*

**2381 Times Mirror Foundation**
202 W First Street 213-237-3945
Los Angeles, CA 90012 Fax: 213-237-2116
http://www.timesmirrorfoundation.org
The Times Mirror Foundation, an affiliate of Tribune Company, is dedicated to supporting nonprofit organizations that measurably improve the quality of life in communities we serve. The Foundation focuses its support on programs thatimprove the quality of journalism, education and literacy, strengthen the fabric of the community, and enhance cultural appreciation and understanding.
*Cassandra Malry, Treasurer*

**2382 Tinker Foundation Inc**
55 E 59th Street 212-421-6858
New York, NY 10022 Fax: 212-223-3326
E-mail: tinker@tinker.org
http://www.foundationcenter.org/grantmaker/tinker
Endeavors to promote better understanding among the peoples of the US, Latin America, and Iberia. In the environmental policy program area, grants are awarded to to 501(c)(3) or equivalent organizations for projects addressingresource-based economic activities and for improving the formulation of effective environmental governance.

*Meg Cushing, Program Officer*

**2383 Town Creek Foundation**
121 N West Street 410-763-8171
Easton, MD 21601 Fax: 410-763-8172
E-mail: info@towncreekfdn.org
http://www.towncreekfdn.org
A private philanthropic foundation dedicated to a sustainable environment

*Stuart A Clark, Executive Director*

**2384 TreePeople**
12601 Mulholland Drive 818-753-4600
Beverly Hills, CA 90210 Fax: 818-753-4625
http://www.treepeople.org
A nonprofit organization that has been serving the Los Angeles area for over three decades. Simply put, our work is about helping nature heal our cities.
*Andy Lipkis, Executive Director*

**2385 Trout Unlimited**
1300 North 17th Street 703-522-0200
Suite 500 800-834-2419
Arlington, VA 22209 Fax: 703-284-9400
E-mail: trout@tu.org
http://www.tu.org
Mission: To conserve, protect and restore North America's trout and salmon fisheries and their watersheds. We accomplish this mission on local, state and national levels with an extensive and dedicated volunteer network.
*Charles Gauvin, President/CEO*
*Steve Moyer, Vice President*

**2386 True North Foundation**
508 Westwood Drive 970-223-5285
Fort Collins, CO 80524 Fax: 970-495-0892
Committed to preventing damage to the natural systems, water, air, and land on which all life depends.
*Kerry K Anderson, President*

**2387 Turner Foundation**
133 Luckie Street NW 404-681-9900
2nd Floor Fax: 404-681-0172
Atlanta, GA 30303 http://www.turnerfoundation.org
This Foundation is committed to preventing damage to the natural systems, water, air and land, on which all life depends.

*Micheal Finley, President*

**2388 US-Japan Foundation**
145 East 32nd Street
New York, NY 10016
212-481-8753
Fax: 212-481-8762
E-mail: info@us-jf.org
http://www.us-jf.org
Committed to promoting stronger ties between Americans and Japanese by supporting projects that foster mutual knowledge and education, deepen understanding, create effective channels of communication, and address common concerns in anincreasingly interdependent world.
*Tom Foran, Program Officer*

**2389 USF and G Foundation**
100 Light Street
Baltimore, MD 21202
410-685-3047

**2390 Union of Concerned Scientists**
2 Brattle Square
Cambridge, MA 02238
617-547-5552
Fax: 617-864-9405
E-mail: ucs@ucsusa.org
http://www.ucsusa.org
The Union of Scientists is the leading science-based nonprofit organization working for a healthy environment and a safer world. Since 1969. we've used rigorous scientific analysis, innovative policy development, and tenacious citizenadvocacy to advance practical solutions for the environment.

*Kevin Knobloch, President*
*Kathleen Rest, Executive Director*

**2391 Unitarian Universalist Veatch Program at Shelter Rock**
48 Shelter Rock Road
Manhasset, NY 11030
516-627-6576
Fax: 516-627-6596
E-mail: jan@veatch.org
Supports Unitarian Universalist organizations that foster the growth and development of the denomination and that increase the involvement of Unitarian Universalists in social action and non-denominational organizations whose goalsreflect UU principles.
*Ned Wight, Executive Director*

**2392 Victoria Foundation**
946 Bloomfield Avenue
Glen Ridge, NJ 07028
973-748-5300
Fax: 973-748-0016
E-mail: info@victoriafoundation.org
http://www.victoriafoundation.org
A private grantmaking institution. Since the early 1960's the Foundation's trustees have targeted giving to programs that impact the cycle of poverty in Newark, New Jersey.

*Catherine McFarland, Executive Officer*

**2393 Vidda Foundation**
250 West 57th Street
Suite 916
New York, NY 10016
212-696-4052
Fax: 212-889-7791
The Vidda Foundation is a private non-operating foundation interested in supporting programs that will have lasting impact in the areas of conservation, education, healthcare, human services, and the arts.
*Gerald E Rupp, Manager*

**2394 Virginia Environmental Endowment**
Three James Center
1051 East Cary Street
PO Box 790
Richmond, VA 23206
804-644-5000
E-mail: info@vee.org
http://www.freenet.vcu.edu/vee
Mission is to improve the quality of the environment by using its capital to encourage all sectors to work together to prevent pollution, conserve natural resources, and promote environmental literacy.
*Gerald P McCarthy, Executive Director*

**2395 W Alton Jones Foundation**
232 East High Street
Charlottesville, VA 22902
804-295-2134
Fax: 804-295-1648
E-mail: earth@wajones.org
http://www.wajones.org
Helps to fund hundreds of environmental groups.
*Dr. JP Meyers, Director*

**2396 Wallace Genetic Foundation**
4801 Massachusetts Avenue NW
Suite 400
Washington, DC 20016
202-966-2932
http://www.voideinternational.org
Committed to funding a variety of interests including agricultural research, preservation of farmland, ecology, conservation and sustainable development.

**2397 Wilderness Society**
1615 M Street NW
Washington, DC 20036
202-833-2300
800-THE-WILD
E-mail: member@tws.org
http://www.wilderness.org
Deliver to future generations an unspoiled legacy of wild places, with all the precious values they hold: Biological diversity; clean air and water; towering forests, rushing rivers, and sage-sweet, silent deserts.
*Brenda Davis, Chair*
*Doug Walker, Vice Chair*

**2398 Wildlife Preservation Trust International**
3400 West Girard Avenue
Philadelphia, PA 19104
215-222-3636
Fax: 215-222-2191
Empowers local conservation scientists worldwide to protect nature and safeguard ecosystem and human health.

*Dr. Mary Pearl*

**2399 William Bingham Foundation**
20305 Center Ridge Road
Shite 629
Rocky River, OH 44116
440-331-6350
E-mail: info@wbinghamfoundation.org
Supports organizations in in education, science, health and human servces and the arts.
*Ms Laura C Hitchcox, Director*

**2400 William H Donner Foundation**
60 East 42nd Street
Suite 1560
New York, NY 10165
212-949-0404
Fax: 212-949-6022
When we build let us...build forever. Let it not be for present delight nor for present alone. Let it be such work as our descendants will thank us for.
*Joseph W Donner, Jr., President*
*Deborah Donner, Vice President*

**2401 William Penn Foundation**
Two Logan Square 11th Floor
100 North 18th Street
Philadelphia, PA 19103
215-988-1830
Fax: 215-988-1823
E-mail: moreinfo@williampennfoundation.org
http://www.williampennfoundation.org
To improve the quality of life in the greater Philadelphia region through efforts that foster rich cultural expression, strenghten children's futures, and deepen connections to nature and community. In partnerships with others, we workto advance a vital, just and caring community.
*Feather O Houstoun, President*
*Olive Mosier, Director*

**2402 William and Flora Hewlett Foundation**
2121 Sand Hill Road
Menlo Park, CA 94025
650-234-4500
Fax: 650-234-4501
http://www.hewlett.org
The Foundation concentrates its resources on activities in education, environment, global development, performing arts and population. In addition, the Foundation has programs that make grants to advance the field of philanthropy, andto support disadvantaged communities in the San Francisco Bay Area.
*Stephen Toben, Program Officer*

**2403 Winston Foundation for World Peace**
2040 S Street, NW 202-483-4215
Washington, DC 20009 Fax: 202-483-4219
E-mail: winstonfoun@igc.apc.org
*John H. Adams, Director*

**2404 Wisconsin Energy Corporation Foundation**
231 W. Michigan St. 414-221-2345
Milwaukee, WI 53203 Fax: 414-221-2554
http://www.wisconsinenergy.com
Wisconsin Energy's principal business is providng electric and natural gas service to customers across Wisconsin and the Upper Peninsula of Michigan.
*Jerry G Remmel, Treasurer*

**2405 World Parks Endowment**
1616 Place Street NW 202-939-3808
Suite 200 Fax: 202-939-3868
Washington, DC 20036 E-mail: worldparks@worldparks.org
http://www.worldlandtrust.org
The World Parks Endowment provides the opportunity to buy rainforest land and establish new protected areas that conserve rainforests and other sites of high biodiversity value. Our projects target lands that conserve rare orendangered species, and are low price, so the minimum amount of the funds protect high priority areas.
*Daniel Katz, President*

**2406 World Research Foundation**
41 Bell Rock Plaza 928-284-3300
Sedona, AZ 86351 Fax: 928-284-3530
E-mail: info@wrf.org
http://www.wrf.org
The purpose of the foundation is to locate, gather, codify, evaluate, classify and disseminate information dealing with health and the environment. All countries are contacted to collect the best information in an unbiased, neutral andindependent manner.
*LaVerne Boeckmann, Vice President/Founder*

**2407 World Resources Institute**
10 G Street, NW 202-729-7600
Suite 800 Fax: 202-729-7610
Washington, DC 20002 http://www.wri.org
An independent nonprofit organization with a staff of more than 100 scientists, economists, policy experts, business analysts, statistic analysists, mapmakers and communicators working to protect the Earth and improve people's lives.Our four goals are: protect the Earth's living systems, increase access to information, create sustainable enterprise and opportunity and reverse global warming.
*Jonathan Lash, President*

**2408 World Society for the Protection of Animals**
89 South Street 617-896-9214
Suite 201 800-883-9772
Boston, MA 02111 Fax: 617-737-4404
E-mail: wspa@wspausa.com
http://www.wspa-usa.org or www.wspa-internationa.org
The world's largest alliance of animal welfare organization whose vision is a world where animal welfare matters and animal cruelty ends. We strive to bring about change from grassroots to government levels to benefit animals. WSPAsupports and develops high-profile campaigns, scientifically-backed projects and innovative education initiatives. Its work is recognized by the United Nations and Council of Europe.

*Cecily West, USA Executive Director*
*Dena Jones, USA Programs Director*

**2409 World Wildlife Fund**
1250 24th Street NW 202-293-4800
Washington, DC 20037 Fax: 202-293-9211
http://www.worldwildlife.org
The largest multinational conservation organization in the world, WWF works in 100 countries and is supported by 1.2 million members in the United States and close to 5 million globally.
*Jennifer A Zadwick, Program Information Coord.*

**2410 Xerces Society**
4828 Se Hawthorne Blvd. 503-232-6639
Portland, OR 97215 Fax: 503-222-2763
Works with farmers, land managers, golf course staff, public agencies, and gardners to promote the conservation and recovery of native pollinator insects and their habitat.
*Melody Allen, Executive Director*

## Scholarships

**2411 AGI Minority Geoscience Scholarship**
American Geological Institute
4220 King Street 703-379-2480
Alexandria, VA 22302 800-336-4764
Fax: 703-379-7563
Provides information services, serves as a voice of shared interests in our profession, plays a major role in strengthening geoscience education, and strives to increase public understanding of the vital role in geosciences play insociety's use of resources and interaction with the environment.
*Patrick Leahy, Executive Director*
*Ann E. Benbow, Education, Outreach Director*

**2412 Abundant Life Seed Foundation**
930 Lawrence Street 425-385-5660
PO Box 772 Fax: 360-385-7455
Port Townsend, WA 98368 E-mail: abundant@olypen.com
http://www.abundantlifeseed.org
A nonprofit organization dedicated to the preservation of rare, heirloom and native seeds. ALSF grows and distributes open-pollinated seeds and offers them for sale in an annual catalog. Seeds are also sent to people in need throughthe World Seed Fund. ALSF teaches seed saving through workshops, appreticeships and school programs.
*Matthew Dillon, Executive Director*
*Elsa Golts, Board President*

**2413 Alexander Hollaender Distinguished Postdoctoral Fellowships**
PO Box 117 423-576-9975
Oak Ridge, TN 37831 E-mail: holmesl@orau.gov
http://www.orau.gov/orise/contacts.htm
Prepares and distributes program literature to universities and laboratories across the country, accepts application, convenes a panel to make award recommendation, and issues stipend checks.
*Barbara Dorsey*
*Linda Holmes*

**2414 American Association for the Advancement of Science**
1333 H Street NW 202-326-6600
Washington, DC 20005 Fax: 202-289-4950
http://www.aaas.org
An international non-profit organization dedicated to advancing science around the world by serving as an educator, leader, spokesperson and professional association.

**2415 American Geophysical Union Member Programs Division**
2000 Florida Avenue NW 202-462-6900
Washington, DC 20009 Fax: 202-328-0566
Organized to represent the US in the International Research Council's International Union Of Geodesy and Geophysics and to serve as the National Research Council Committee on Geophysics.

**2416 American Indian Science and Engineering Society**
2305 Renard SE 505-765-1052
PO Box 9828 Fax: 505-765-5608
Albuquerque, NM 87119 E-mail: info@aises.org
http://www.aises.org

The mission is to increase substantially the representation of American Indian and Alaskan Natives in engineering, science and other related technology disciplines.

*Pamala Silas, CEO*
*Shirley LaCourse, Deputy CEO*

**2417 American Museum of Natural History**
Central Park West & 79th Street
New York, NY 10024

Mission is to discover, interpret, and disseminate-through scientific research and education, knowledge about human cultures, the natural world, and the universe.

**2418 American Nuclear Society**
555 North Kensington Avenue 708-352-6611
La Grange Park, IL 60525 800-323-3044
Fax: 708-352-0499

Not-for-profit, international, scientific and educational organization. Established by a group of individuals who recognized the need to unify the professional activities within the diverse fields of nuclear science and technology.

**2419 American Society of Naturalists**
Queens College - CUNY
Department of Biology 718-997-3426
Flushing, NY 11367

Purpose is to advance and to diffuse knowledge of organic evolution and other broad biological principals so as to enhance the conceptual unification of the biological sciences.

**2420 American Sport Fishing Association**
225 Reinekers Lane 847-381-9490
Suite 420 Fax: 847-381-9518
Alexandria, VA 22314 E-mail: info@asafishing.org
http://www.asafishing.org

Unites more then 650 members of the sportfishing and boating industries with state fish and wildlife agencies, federal land and water management agencies, conservation organizations, angler advocacy groups and outdoor journalists. Wesafeguard and promote the enduring social, economic and conservation values of sportfishing.

*Mike Nussman, President And CEO*
*Gordon Robertson, Vice President*

**2421 Apple Computer Earth Grants: Community Affairs Department**
1 Infinite Loop 408-996-1010
Cupertino, CA 95014 800-692-7753
http://www.apple.com

*Beverly Long, Program Manager*

**2422 Beldon Fund**
99 Madison Avenue 212-616-5600
8th Floor 800-591-9595
New York, NY 10016 Fax: 212-616-5656
E-mail: info@beldon.org
http://www.beldon.org

Mission is by supporting effective, nonprofit, advocacy organizations, the Beldon Fund seeks to build a national consensus to achieve and sustain a healthy planet.

*Judy Donald, Executive Director*

**2423 Beldon II Fund: Old Kent Bank and Trust Company**
Old Kent Bank 616-771-5326
300 Old Kent Bank Building
Grand Rapids, MI 49503

*John R Hunting, President and Director*

**2424 Charles A. and Anne Morrow Lindbergh Foundation**
2150 Third Avenue North 763-576-1596
Suite 310 Fax: 763-576-1664
Anoka, MN 55303 E-mail: info@lindberghfoundation.org
http://www.lindberghfoundation.org/grants/

Each year, the Charles A. and Anne Morrow Lindbergh Foundation provides grants to men and women whose individual initiative and work in a wide spectrum of disciplines furthers the Lindbergh's vision of a balance between the advance oftechnology and the perservation of the natural/human environment.

*Knox Bridges, President*
*Shelley Nehl, Grants Administrator*

**2425 Cousteau Society**
710 Settler's Landing Road 757-523-9335
Hampton, VA 23669 800-441-4395
Fax: 757-722-8185
E-mail: cousteau@cousteausociety.org
http://www.cousteausociety.org

Mission is to educate people to understand, to love and to protect the water systems of the planet, marine and fresh water, for the wellbeing of future generations.

*Francine Cousteau, President*

**2426 DRB Communications**
1234 Summer Street
Stamford, CT 06905 800-323-1550
Fax: 203-324-7175

*Robyn DeWolf*

**2427 Delmar Publishers Scholarship**
National FAA Foundation
6060 FFA Drive 317-802-6060
PO Box 68960 Fax: 317-802-6061
Indianapolis, IN 46268

*Carrie Powers, Contact*

**2428 Du Pont de Nemours and Company**
1007 Market Street 302-774-2036
Room 8065 800-441-7515
Wilmington, DE 19898 E-mail: info@dupont.com
http://www.dupont.com

Creating sustainable solutions essential to a better, safer, healthier, life for people everywhere.

*Peter C Morrow, Mgr. Corporate Contributions*

**2429 Earth Island Institute-Brower Youth Awards**
300 Broadway 415-788-3666
Suite 28 Fax: 415-788-7324
San Francisco, CA 94133 E-mail: bya@earthisland.org
http://www.earthisland.org

Incubates and supports over 30 projects working on environmental issues worldwide. Publishes quarterly Earth Island Journal. Project support programs help aspiring and veteran activists alike put ideas into action. Youth programincreases the visibility, effectiveness and influence of youth leadership in the environmental movement, inspiring other young people to work for the Earth.

*John A Knox, Executive Dir Operations*
*Sharon Smith, Program Director*

**2430 Environmental Defense Fund**
257 Park Avenue South 212-505-2100
17th Floor 800-684-3322
New York, NY 10010 Fax: 212-505-2375
E-mail: members@edf.org
http://www.edf.org

Dedicated to protecting the environmental rights of all people, including future generations. Among these rights are clean air, clean water, healthy, nourishing food and a flourishing ecosystem. Advocates solutions based on science,even when it leads in unfamiliar directions. Works to create solutions that win lasting political, economic and social support because they are bipartisan, efficient and fair.

*Fred Krupp, President*
*David Yarnold, Executive Director*

**2431 Environmental Grantmakers Association**
55 Exchange Place 646-747-2655
Suite 405 Fax: 646-747-2656
New York, NY 10005
Mission is to help member organizations become more effective environmental grantmakers through information sharing, collaboration and networking.

**2432 Environmental Protection Agency: Grants Administration Division**
Grants Operation Branch 202-260-5260
401 M Street SW
Washington, DC 20460
Programs include air and water pollution controll, toxic substances, pesticides, and drinking water regulation, wetlands protection, hazardous waste management, hazardous waste site cleanup and some regulation of radioactive materials.

**2433 Environmental and Engineering Fellowship**
American Association for the Advancement
1333 H Street NW 202-326-6600
Washington, DC 20005 Fax: 202-289-4950
Aimed at postdoctoral to midcareer professionals from any discipline of science, engineering or any relevant interdisciplinary fields.

**2434 Financial Support for Graduate Work**
Women's Seamen's Friend Society of Connecticut
291 Whitney Avenue 203-777-2165
New Haven, CT 06511
Restricted to Connecticut residents who are students at state maritime schools, or Connecticut residents majoring in Marine Sciences at any college or university or residents of any state majoring in Marine Sciences at a Connecticutcollege or university.

**2435 Ford Motor Company Fund**
American Road 313-845-8711
PO Box 6248 800-392-3673
Dearborn, MI 48126 http://www.fordfound.com
Ford Motor Company Fund is a not-for-profit corporation organized in 1949. Made possible by Ford Motor Company profits, Ford Motor Company Fund supports initiatives and institutions that enhance and improve opportunities for those wholive in the communities where Ford Motor Company operates.
*Leo J Brennan Jr, Executive Director*

**2436 Forest History Society**
701 William Vickers Avenue 919-682-9319
Durham, NC 27701 Fax: 919-682-2349
E-mail: coakes@duke.edu
http://www.foresthistory.org
The Forest History Society is a non-profit educational institution that links the past to the future by identifying, collecting, preserving, interpreting, and disseminating information on the history of people, forests, and theirrelated resources.

*Cheryl Oakes, Librarian*
*Steven Anderson, President*

**2437 Garden Club of America**
14 East 60th Street 212-753-8287
3rd Floor Fax: 212-753-0134
New York, NY 10022 http://www.gcamerica.org
Purpose is to stimulate the knowledge and love of gardening, to share the advantages of association by means of educational meetings, conferences, correspondence and publications, and to restore, improve, and protect the quality of theenvironment through educational programs and action in the fields of conservation and civic improvement.
*Sellers Thomas Jr, President*

**2438 Georgia M. Hellberg Memorial Scholarships**
National Future Federation of America
6060 FFA Drive 317-802-6060
PO Box 68960 Fax: 317-802-6061
Indianapolis, IN 46268 http://www.ffa.org
*Carrie Powers, Contact*

**2439 German Marshall Fund of the United States**
1744 R Street NW 202-745-3950
Suite 750 Fax: 202-265-1662
Washington, DC 20009 http://www.gmfus.org
A nonpartisan American public policy and grantmaking institution dedicated to promoting greater cooperation and understanding between the United States and Europe.
*Marianne L Gindburg, Program Officer*

**2440 Great Lakes Protection Fund**
1560 Sherman Avenue 847-425-8150
Suite 880 Fax: 847-424-9832
Evanston, IL 60201 http://www.glpf.org
A private non profit organization formed by the Govenors of the Great Lakes States. It is a permanent environmental endowdment that supports collaborative actions to improve the health of the Great Lakes ecosystem.

*Russell Van Herik, Executive Director*

**2441 Hawk Mountain Sanctuary Association**
1700 Hawk Mountain Road 215-756-6961
Kempton, PA 19529 Fax: 215-756-4468
Mission is to conserve birds of prey worldwide by providing leadership in raptor conservation science and education, and my maintaining Hawk Mountain Sanctuary as a model observation, research and education facility.
*Cynthia R Lenhart, Executive Director*

**2442 Hazardous Waste Reduction Loan Program**
California Department of Commerce
1001 I Street 916-341-6181
PO Box 4025 E-mail: grants@ciwmb.ca.gov
Sacramento, CA 95812 http://www.ciwmb.ca.gov
Loans assist small business to redude waste generation or to reduce the hazardous properties of waste generated. Proceeds can only be used to finance hazardous waste equipment acquistion, installation and processes.
*Merri Stevenson*

**2443 Heller Charitable and Educational Fund**
244 California Street 415-434-3160
San Francisco, CA 94111 Fax: 415-434-3807
Mission is to protect and improve the quality of life through support of programs in the environment, human health, education, and the arts.
*Ruth B Heller, Correspondence Secretary*

**2444 JM Kaplan Fund**
261 Madison Avenue 212-767-0630
19th Floor Fax: 212-767-0639
New York, NY 10016 E-mail: info@jmkfund.org
Support for the arts, the environment, human rights, and a robust civil society. New interests emerged in programs to support New York City neighborhoods parks and libraries as well as historic preservation and municipal design work inLower Manhattan.
*Anthony C Wood, Program Officer*

**2445 Jessie Ball duPont Religious, Charitable and Educational Fund**
1 Independent Drive 904-353-0890
Suite 1400 800-252-3452
Jacksonville, FL 32202 Fax: 904-353-3870
E-mail: contactus@dupontfund.org
http://www.dupontfund.org
A private grantmaking foundation limited in its giving to approximately 330 eligible organizations to which Mrs. duPont personnally contributed to in a five year period 1960-1964. The

duPont fund accomplishes its mission by workingcreatively with these organizations and their partners.

*Sherry Magill, PhD, President*
*Davena Sawyer, Exec. Asst to President*

**2446 Johnson's Wax Fund**
1525 Howe Street
Racine, WI 53403
262-260-2000
http://www.scjohnsonwax.com
Through the SC Johnson Fund, in the US, we donate, on average, 5% pre-tax profits every year to increase local and global well-being. Our contributions are targeted to advancing the three legs of sustainability: economic vitality,social progress, and a healthy environmnet.

**2447 Joint Oceanographic Institutions**
1201 New York Avenue NW
Suite 400
Washington, DC 20005
202-232-3900
Fax: 202-265-4409
E-mail: info@joiscience.org
http://www.joiscience.org
A consortium of 20 premier oceanographic research institutions that serves the US scientific community through management of large scale, global research programs in the fields of marine geology and geophysics and oceanography.
*Steven Bohlen, President*
*Amy Castner, Executive Program Associate*

**2448 LSB Leakey Foundation**
P.O. Box 29346
1002A O'Reilly Aveanue
San Francisco, CA 94129
415-561-4646
Fax: 415-561-4647
E-mail: info@leakeyfoundation.org
http://www.leakeyfoundation.org/foundation.f4jsp
The mission of the Leakey Foundation is to increase scientific knowledge and public understanding of human origins and evolution.

**2449 MJ Murdock Charitable Trust**
703 Broadway
Suite 710
Vancouver, WA 98660
360-694-8415
Fax: 360-649-1819
http://www.murdock-trust.org
The mission is to enrich the quality of life in the Pacific Northwest by providing grants organizations that seek to strenghten the region's educational and cultural base in creative and sustainable ways.
*Neal O Thorpe, Executive Director*

**2450 Mary Flagler Cary Charitable Trust**
122 East 42nd Street
Room 3505
New York, NY 10168
212-953-7700
Fax: 212-953-7720
http://www.carytrust.org/
The Trust was established as a testamentary, charitable trust by the will of the late Mary Flagler Cary. The trustees have worked to use the assets of the Trust to carry forward Mrs. Cary's interests, and to elaborate on them in lightof new circumstances and needs. A major part of the Trust's assets continue to be devoted to special commitments relating to the origins of the Trust, especially the Institute of Ecosystem Studies at the Mary Flagler Cary Arboretum in Millbrook, NewYork.

**2451 Maryland Sea Grant**
University of Maryland
4321 Hardwick Road
Suite 300
College Park, MD 20740
301-405-7500
Fax: 301-314-5870
E-mail: mdsg@mdsg.umd.edu
http://www.mdsg.umd.edu
Supports innovative marine research and education, with a special focus on the Chesapeake Bay.
*Dr Jonathan Kramer, Director*

**2452 National Academy of Sciences**
2001 Constitution Avenue NW
Washington, DC 20418
202-334-2000
Fax: 202-334-2158
http://www.nasonline.org
The National Academy of Sciences (NAS) is an honorific society of distinguished scholars engaged in scientific and engineering research, dedicated to the furtherance of science and technology and to their use for the general welfare.

**2453 National Center for Atmospheric Research**
PO Box 3000
Boulder, CO 80307
303-497-1601
Fax: 303-497-1314
NCAR provides the university research and teaching community with tools such as aircraft and radar to observe the atmosphere and with the technology and assistance to interpret and use these observations, including supercomputeraccess, computer models, and user support.

**2454 National Environmental Health Association NEHA/AAS Scholarships**
720 S Colorado Boulevard
Suite 970-S
Denver, CO 80246
303-756-9090
Fax: 303-691-9490
E-mail: staff@neha.org
http://www.neha.org
Mission is to advance the environmental health and protection professional for the purpose of providing a healthful environment for all.
*V Potter, Member Liaison*

**2455 Needmor Fund**
42 South Saint Clair Street
Toledo, OH 43604
419-255-5560
http://www.fdncenter.org
Mission is to work with others to bring about social justice. The Needmor Fund supposrt people who work together to change the social, economic or politcal conditions which bar their access to participation in a democratic society.
*Lynn Gisi, Coordinator*

**2456 Nixon Griffis Fund for Zoological Research: New York Zoological Society**
Bronx Zoo
185th Street & Southern Boulevard
Bronx, NY 10460
212-220-5152
Fax: 212-220-7114
Supports research in zoology, conservation, and marine science. Grants lilmited to $3,000. Grants made four times a year. Applications reviewed by selected US zoo personnel.
*John Behler, Contact*

**2457 North American Loon Fund Grants**
PO Box 329
Holderness, NH 03245
603-528-4711
800-462-5666
THe North American Loon Fund's mission is to promote the preservation of loons and their lake habitats through research, public education, and the involvement of people who share their lakes with loons.
*Linda O'Bara, Director*

**2458 Oak Ridge Institute Science & Engineering Education Division**
MS 36 120 Badger Avenue
PO Box 117
Oak Ridge, TN 37831
865-576-3424
Fax: 865-241-2923
E-mail: westm@orau.gov
http://www.orau.org
Strive to advance scientific research and education by creating mutually beneficial collaborative partnerships involving academe, government, and industry.

**2459 Oklahoma State University**
003 Life Sciences E
Stillwater, OK 74078
405-744-9995
Fax: 405-744-7673
The major fields of study include biological science, botany, cell and molecular biology, conservation sciences, medical technology, microbiology, physiology, and zoology.
*Dr. Norman N Durham, Director*

**2460 Resources for the Future**
1616 P Street NW
Washington, DC 20036
202-328-5000
Fax: 202-939-3460
E-mail: info@rff.org
http://www.rff.org
RFF is a nonprofit and nonpartisan organization, or think tank, that conducts independent research-rooted primarily in economics and other social sciences-on environmental, energy, climate

change and natural resource issues. Itsresearch scope comprises programs in nations around the world.

*Phil Sharp, President*
*Peter Nelson, Director of Communications*

**2461 The Center for Environmental Biotechnology at the University of Tennessee at Knoxville**

1414 Circle Drive, 676 Dabney Hall 865-974-8080
Knoxville, TN 37996 Fax: 865-974-8086
E-mail: ceb@utk.edu
http://www.ceb.utk.edu

One of the nations oldest and largest university-based multidisciplinary research units devoted to environmental analysis. The CEB is a leader in the development of whole cell bioluminescent bioreporters for the detection of organicand inorganic pollutants including environmental endocrine disruptors and toxicants.

*Gary S Sayler, Professor & Director*
*John Sanseverino, Managing Director*

**2462 University of Colorado: Boulder**

Campus Box 216 303-492-1143
Boulder, CO 80309 Fax: 303-492-1149

Recognized as one of the outstanding public universities in the United States. The Boulder campus has 5 colleges and 4 schools, offering 3,400 courses in about 150 areas of study.

*Dr. Robert Sievers, Director*

**2463 WERC Undergraduate Fellowships**

New Mexico State University 505-646-2038
Box 30001, MSC WERC 800-523-5996
Las Cruces, NM 88003 Fax: 505-646-4149
E-mail: bdelrio@nmsu.edu
http://www.werc.net

WERC a consortium for environmental education and technology development. The consortium's mission is to develop the human resources and technologies needed to address environmental issues.

*Barbara Valdez, Contact*

**2464 Water Environment Federation**

601 Wythe Street 703-684-2400
Alexandria, VA 22314 800-666-0206
Fax: 703-684-2492
http://www.wef.org

Founded in 1928, the Water Environment Federation (WEF) is a not-for-profit technical and educational organization with members from varied disciplines who work toward the WEF vision of preservation and enhancement of the global waterenvironment. The WEF network includes more than 100,000 water quality professionals from 77 member associations in 31 countries.

*Janet Blatt, Director*
*Mincaiee Brown, Executive Director*

**2465 Weston Institute**

1400 Weston Way 610-701-3000
Po Box 2653 800-779-7866
West Chester, PA 19380 Fax: 610-701-3186
http://www.westonsolutions.com

A leading environmental and redevelopment firm focused on restoring efficiency to your essential resources: air, land, water, people, and facilities. We can help you develop solutions that maximize resource value and turn environmentalresponsibility into economic growth.

*William Gaither, Director*

**2466 Wildlife Conservation Society**

Bronx Zoo 718-220-5100
2300 Southern Blvd Fax: 718-365-3694
Bronx, NY 10460 http://www.wcs.org

The Wildlife Conservation Society saves wildlife and wild lands. We do so through careful science, internation conservation, education, and the management of the world's largest system of urban wildlife parks, led by the flagship BronxZoo. Together, these activities change individual attitudes toward nature and help people imagine wildlife and humans living in sustainable interaction on both a local and a global scale.

**2467 Women's Seamen's Friend Society of Connecticut**

291 Whitney Avenue 203-467-3887
New Haven, CT 06511

**2468 Yale Institute for Biospheric Studies (YIBS)**

21 Sachem Street 203-432-9856
PO Box 208105 Fax: 203-432-9927
New Haven, CT 06520 E-mail: roserita.riccitelli@yale.edu
http://www.yale.edu/yibs

The Yale Institute for Biospheric Studies (YIBS) serves as a principal focus for Yale University's research and training efforts in the environmental sciences, and is committed to the teaching of environmental studies to futuregenerations. It provides physical and intellectual centers for research and education that address fundamental questions that will inform the ability to generate solutions to the biosphere's most critical environmental solutions.

*Rose Rita Riccitelli, Assistant Director*
*Jeffrey Park, Director*

# Government Agencies & Programs

## Federal

**2469 Advisory Committee on Nuclear Waste**
US Nuclear Regulatory Commission
Office Of Public Affairs
Washington, DC 20555
301-415-8200
800-368-5642
Fax: 301-415-5575
E-mail: ram2@nrc.gov
http://www.nrc.gov

*Dr John T Larkins, Acting Executive Director*

**2470 Advisory Council on Historic Preservation**
1100 Pennsylvania Avenue NW
Suite 803, Old Post Office Building
Washington, DC 20004
202-606-8503
E-mail: achp@achp.gov
http://www.achp.gov
Mission: To promote the preservation, enhancement, and productive use of our Nation's historic resources, and advise the President and Congress on national historic preservation policy.
*John Nau, Chairman*

**2471 Advisory Panel for Ecology**
National Science Foundation
4201 Wilson Boulevard
Arlington, VA 22230
703-292-5111
E-mail: info@nsf.gov
http://www.nsf.gov
To promote the progress of science; to advance the national health, prosperity, and welfare; to secure the national defense.
*Arden L Bement, Director*
*Kathie Olsen, Deputy Director*

**2472 Agency for Toxic Substances and Disease Registry**
Centers for Disease Control and Prevention
1600 Clifton Road
Atlanta, GA 30333
404-639-3311
800-311-3435
E-mail: ATSDRIC@cdc.gov
http://www.cdc.gov
This agency provides leadership and direction to programs and activities designed to protect both the public and workers from exposure or adverse health effects of hazardous substances in storage sites or released in fires, explosions,or transportation accidents. The agency also collects, maintains, analyzes and disseminates information relating to serious diseases, mortality and human exposure to toxic or hazardous substances.

**2473 Air and Radiation Research Committee**
Environmental Protection Agency
1200 Pennsylvania Avenue NW
Washington, DC 20460
202-564-7400
Fax: 202-501-0826
http://www.epa.gov/air

*Stephen L Johnson, Administrator*
*Marcus Peacock, Deputy Administrator*

**2474 American Farmland Trust**
1200 18th Street NW
Washington, DC 20036
202-331-7300
Fax: 202-659-8339
E-mail: info@farmland.org
http://www.farmland.org
Mission: To stop the loss of productive farmland and to promote farming practices that lead to a healthy environment.
*Ralph Grossi, President*
*Vicki Edwards, CFO*

**2475 American Indian Environmental Office**
1200 Pennsylvania Avenue NW (4104M)
Washington, DC 20460
202-564-0303
Fax: 202-264-0298
http://www.epa.gov/indian/
Coordinates the US environmental Protection Agency-wide effort to strengthen public health and environmental protection in Indian Country, with a special emphasis on building Tribal capacity to administer their own environmentalprograms.
*Carol Jorgensen, Director*
*Christopher Hoff, Acting Deputy Director*

**2476 Animal and Plant Health Inspection Service Protection Quarantine**
1400 Independence Avenue SW
Whitten Building, Room 302-E
Washington, DC 20250
202-720-5601
Fax: 202-690-0472
E-mail: aelder@aphis.usda.gov
http://www.aphis.usda.gov
Mission: To protect the health and value of American agriculture and natural resources.
*Alfred Elder, Deputy Administrator*

**2477 Antarctica Project and Southern Ocean Coalition**
1630 Connecticut Avenue NW
3rd Floor
Washington, DC 20009
202-234-2480
Fax: 202-387-4823
E-mail: info@asoc.org
http://www.asoc.org
Concerned with educating the public about environmental problems in the arctic regions. Conducts research pertaining to Antarctica.
*Jim Barnes, Executive Director*
*Deidre Zoll, Assistant to the Director*

**2478 Aquatic Nuisance Species Task Force**
4401 N Fairfax Drive
Suite 840
Arlington, VA 22203
703-358-1796
Fax: 703-358-1800
E-mail: scott_newsham@fws.gov
http://www.anstaskforce.gov/taskforce.php
An intergovernmental organization dedicated to preventing and controlling aquatic nuisance species. The task forec consists of 10 federal agency reps and 12 ex-officio members
*Mamie Parker, Co-Chairman*
*Scott Newsham, Executive Secretary*

**2479 Argonne National Laboratory**
9700 S Cass Avenue
Argonne, IL 60439
630-252-2000
http://www.anl.gov/
One of the US Department of Energy's largest research centers.

**2480 Army Corps of Engineers**
441 G Street NW
Washington, DC 20314
202-761-0660
http://www.usace.army.mil
The Army Corps of Engineers serves as the Army's real property manager; manages and executes civil works programs, including research and development, planning, design, construction, operation and maintenance and real estate activitiesrelated to rivers, harbors and waterways; administers laws for protection and preservation of navigable waters and related resources such as wetlands, and assists in recovery from natural disasters.
*Robert L Van Antwerp, Commander/Chief Engineers*

**2481 Aspen Institute**
One Dupont Circle NW
Suite 700
Washington, DC 20036
202-736-5800
Fax: 202-467-0790
http://www.aspeninst.org
A forum that addresses critical environmental issues. Interests include energy, the environment and economics. Partner institutes located in Japan, Italy, Germany and France.
*Walter Isaacson, President/CEO*
*Katie Loughary, Vice President*

**2482 Atlantic States Marine Fisheries Commission**
1444 I Street NW
6th Floor
Washington, DC 20005
202-289-6400
Fax: 202-289-6051
E-mail: info@asmfc.org
http://www.asmfc.org
The Atlantic States Marine Fisheries Commission was formed by the 15 Atlantic coast states in 1942 in recognition that fish do not adhere to political boundaries. The Commission serves as a deliberative body, coordinating theconservation and management of the states shared near shore fishery resources-marine, shell, and anadromous-for sustainable use.
*John V O'Shea, Executive Director*
*Tina L Berger, Public Affairs/Resource Spcl*

**2483 Blue Mountain Natural Resource Institute Advisory Board**
US Department of Agriculture
1400 Independence Avenue SW 202-205-8333
Room 240W http://www.fs.fed.us
Washington, DC 20250

*Charles R Hilty, Management Office*

**2484 Bureau of Economic Analysis**
Economics and Statistics Administration
1441 L Street NW 202-606-9900
Room 6006 Fax: 202-606-5311
Washington, DC 20230 E-mail: john.landefeld@bea.doc.gov
http://www.bea.gov
Mission: To promote a better understanding of the US economy by providing the most timely, relevant, and accurate economic accounts data in an objective and cost-effective manner.
*John Landefeld, Director*

**2485 Bureau of Land Management, Land & Renewable Resources**
1849 C Street NW Room 406LS 202-452-5125
Washington, DC 20240 Fax: 202-452-5124
E-mail: woinfo@blm.gov
http://www.blm.gov
The mission of the Bureau of Land Management is to sustain the health, diversity, and productivity of the public lands for the use and enjoyment of present and future generations. Offers environmental education, news about theactivities of the Bureau, events, and regulations. In addition, there is information about ALMRS (Automated Land and Management Record System). This is an information system that contains more than one billion land and mineral records.
*James M Hughes, Acting Director*
*Luke Johnson, Chief of Staff*

**2486 Bureau of Oceans International Environmental & Scientific Affairs**
US Department of State
2201 C Street NW 202-647-4000
Washington, DC 20520 http://www.state.gov
OES coordinates US international oceans, environmental and health policy, integrating US domestic concerns with geopolitical concerns. OES promotes the full range of US interests in the oceans to advance our national security,facilitate commerce, manage fish resources, foster scientific understanding and protect the marine environment through bilateral, regional and multilateral fora.
*John D Negroponte, Deputy Secretary*

**2487 Center for Disease Control and Prevention**
National Center for Environmental Health
1600 Clifton Road 404-639-3311
Atlanta, GA 30333 800-311-3435
http://www.cdc.gov
To provide national leadership, through science and service, to promote health and quality of life by preventing and controlling disease and death resulting from interactions between people and their environment.

**2488 Center for Environmental Finance: Environmental Finance Center Network (EFC)**
Environmental Protection Agency
1200 Pennsylvania Ave NW (2731R) 202-564-4994
Washington, DC 20460 Fax: 202-565-2587
E-mail: efin@epa.gov
http://www.epa.gov/efinpage/
Part of the EPA's EFP, the EFC is a network of 9 university-based programs in eight EPA regions.
*Vanessa Bowie, Staff Director*

**2489 Centers for Disease Control and Prevention**
1600 Clifton Road NE 404-639-3311
Atlanta, GA 30333 800-311-3435
http://www.cdc.gov
The Centers for Disease Control and Prevention protect the public health of the nation by providing leadership and direction in the prevention and control of diseases and other preventable conditions, and responding to public healthemergencies.
*Julie Gerberding, Director*
*William Gimson, Chief Operating Officer*

**2490 Chemical, Bioengineering, Environmental & Transport Systems**
National Science Foundation
4201 Wilson Boulevard 703-292-5111
Arlington, VA 22230 http://www.eng.nsf.gov/bes/
CBET supports research and education in the rapidly evolving fields of bioengineering and environmental engineering and in areas that involve the transformation and/or transport of matter and energy by chemical, thermal, or mechanicalmeans.
*Judy A Raper, Division Director*
*Robert M Wellek, Deputy Division Director*

**2491 Chesapeake Bay Critical Areas Commission**
1804 W Street 410-260-3460
Suite 100 http://www.dnr.state.md.us/criticalarea/
Annapolis, MD 21401
Develops criteria used by local jurisdictions to develop individual Critical Area programs and amend local comprehensive plans, zoning ordinances and subdivision regulations. Programs are designed to address the unique characteristicsand needs of each county and municipality and together they represent a comprehensive land use strategy for preserving and protecting Maryland's most important natural resource, the Chesapeake Bay.
*Kerrie Gallo, Natural Resources Planner*
*LeeAnne Chandler, Science Advisor*

**2492 Chief of Engineers Environmental Advisory Board**
441 G Street NW 202-761-0008
Washington, DC 20314 Fax: 202-761-1683
http://www.hq.usace.army.mil/hqhome/
Serves the Armed Forces and the Nation by providing vital engineering services and capabilities, as a public service, across the full spectrum of operations; from peace to war; in support of national interests.
*Robert L Van Antwerp, Commander/Chief of Engineers*

**2493 Civil Division: Consumer Litigation Office**
950 Pennsylvania Avenue NW 202-514-6786
Washington, DC 20530 http://www.usdoj.gov/civil/home.html
The Civil Division's Office of Consumer Litigation is responsible for criminal and civil litigation and related matters arising under a variety of federal statutes administered by its client agencies that protect public health andsafety.
*Eugene M Thirolf, Director*

**2494 Clean Air Scientific Advisory Committee**
US Environmental Protection Agency
Ariel Rios Building 202-564-2188
1200 Pennsylvania Avenue NW http://www.epa.gov
Washington, DC 20460
The Clean Air Scientific Advisory Committee (CASAC) has a statutorily mandated responsibility to review and offer scientific advice on the air quality criteria and regulatory documents which form the basis for the National Ambient AirQuality Standards (NAAQS), which are currently lead, particulate matter (PM), ozone and other photochemical oxidants ($O_3$), carbon monoxide (CO), nitrogen oxides ($NO_x$) and sulfur oxides ($SO_x$).
*Joanne Rodman, Associate Director*

**2495 Coast Guard**
2100 2nd Street SW 202-267-2229
Washington, DC 20593 http://www.uscg.mil
Its core roles are to protect the public, the environment, and US ecomonic and security interests in any maritime region in which those interests may be at risk, including international waters and America's coasts, ports, and inlandwaterways.

**2496 Coastal States Organization**
444 N Capitol Street NW 202-508-3860
Hall of the States, Suite 322 Fax: 202-508-3843
Washington, DC 20001 E-mail: cso@sso.org
http://www.coastalstate.org

Mission: To support the shared vision of the coastal states, commonwealths and territories for the protection, conservation, responsible use and sustainable economic development of the nation's coastal and ocean resources.
*Kacky Andrews, Executive Director*

**2497 Committee of State Foresters**
NASF-444 N Capitol Street NW 202-624-5415
Suite 540 Fax: 202-624-5407
Washington, DC 20001 E-mail: nasf@stateforesters.org
http://www.stateforesters.org

*James J Farrell Jr, Executive Director*
*LouAnn B Gilmer, Dir Finance, Grants & Admin*

**2498 Committee on Agriculture, Nutrition, and Forestry**
Russell Senate Office Building 202-224-2035
Room SR-328A Fax: 202-224-1725
Washington, DC 20510 http://agriculture.senate.gov
*Tom Harkin, Chair*

**2499 Committee on Appropriations**
The Capitol Building 202-224-7363
Room S-131 http://appropriations.senate.gov
Washington, DC 20510
*Robert C Byrd, Chair*
*Thad Cochran, Ranking Member*

**2500 Committee on Commerce**
Committee on Energy and Commerce
2125 Rayburn House Office Building 202-225-2927
Washington, DC 20515 E-mail: commerce@mail.house.gov
http://www.house.gov/commerce

*John D Dingell, Chair*

**2501 Committee on Commerce, Science, and Transportation**
508 Dirksen Senate Office Building 202-224-5115
Washington, DC 20510E-mail: webmaster@commerce.senate.gov
http://commerce.senate.gov

*Daniel K Inouye, Chair*
*Ted Stevens, Vice Chair*

**2502 Committee on Energy and Natural Resources**
304 Dirksen Senate Building 202-224-4971
Washington, DC 20510 E-mail: committee@energy.senate.gov
http://energy.senate.gov

*Jeff Bingaman, Chair*
*Pete V Domenici, Ranking Member*

**2503 Committee on Environment and Public Works Republicans**
Senate Dirksen Office Building 202-224-8832
Room 410 http://epw.senate.gov
Washington, DC 20510
*Barbara Boxer, Chair*

**2504 Committee on Government Reform and Oversight**
2157 Rayburn House Office Building 202-225-5051
Washington, DC 20515 http://www.house.gov/reform/
*Henry A Waxman, Chair*

**2505 Committee on Natural Resources**
US House of Representatives
1324 Longworth House Office Bldg. 202-225-6065
Washington, DC 20515 Fax: 202-225-1931
http://resourcescommittee.house.gov

*Nick J Rahall II, Chair*

**2506 Committee on Science and Technology**
2320 Rayburn Building 202-225-6375
Washington, DC 20515 Fax: 202-225-3895
http://science.house.gov
*Bart Gordon, Chair*

**2507 Committee on Small Business and Entrepreneurship: US Senate**
428A Russell Senate Office Building 202-224-5175
Washington, DC 20510 Fax: 202-224-5619
http://sbc.senate.gov

*John Kerry, Chair*

**2508 Committee on Small Business: House of Representatives**
Small Business Committee
2361 Rayburn House Office Bldg. 202-225-4038
Washington, DC 20515 Fax: 202-225-7209
E-mail: smbiz@mail.house.gov
http://www.house.gov/smbiz/

*Nydia M Velazquez, Chair*

**2509 Committee on Transportation and Infrastructure**
2165 Rayburn House Office Building 202-225-4472
Washington, DC 20515 Fax: 202-226-1270
http://transportation.house.gov

*James Oberstar, Chair*

**2510 Community Greens**
1700 N Moore Street 703-527-8300
Suite 2000 Fax: 703-527-8383
Arlington, VA 22209 E-mail: info@communitygreens.org
http://www.communitygreens.org

Mission: To catalyze the development of shared green spaces inside residential blocks in cities across the United States.
*Kate Herrod, Director*

**2511 Cooperative Forestry Research Advisory Council**
Department of Agriculture
Cooperative State Research Service 202-720-4318
Washington, DC 20250 http://www.reeusda.gov
*Peter A Muscato*

**2512 Council on Environmental Quality**
722 Jackson Place, Northwest 202-395-5750
Washington, DC 20503 Fax: 202-456-6546
http://www.whitehouse.gov/ceq

Formulates and recommends national environmental policies.

*James L Connaughton, Chair*

**2513 Dangerous Goods Advisory Council**
1100 H Street, Northwest 202-289-4550
Suite 740 Fax: 202-289-4074
Washington, DC 20005 E-mail: info@dgac.org
http://www.dgac.org

DGAC is an international, nonprofit, educational organization that promotes safety in the transportation of hazardous materials and dangerous goods, including hazardous substances and hazardous wastes.

*Mike Morrissette, President*
*Alan I Roberts, Vice President*

**2514 Department of Agriculture**
USDA Forest Service
PO Box 96090 202-205-1657
Washington, DC 20090 E-mail: zbowden@fs.fed.us
http://www.fs.fed.us

**2515 Department of Agriculture: Research Department, Forest Environment Research**
USDA Forest Service, Research 202-205-1657
PO Box 96090 E-mail: zbowden@fs.fed.us
Washington, DC 20090 http://www.fs.fed.us/research
*Ann M Bartuska, Deputy Chief*
*James Reaves, Associate Deputy Chief*

**2516 Department of Agriculture: Forest Inventory, Economics**
Research Department
USDA Forest Service, Research 202-205-1657
PO Box 96090 E-mail: zbowden@fs.fed.us
Washington, DC 20090 http://www.fs.fed.us/research
*Ann M Bartuska, Deputy Chief*
*James Reaves, Associate Deputy Chief*

**2517 Department of Agriculture: Forest Service Public Affairs**
Sidney R Yates Federal Building 202-205-8333
1400 Independence Avenue SW E-mail: webmaster@fs.fed.us
Washington, DC 20250 http://www.fs.fed.us

**2518 Department of Agriculture: National Forest Watershed and Soil Resource**
USDA Forest Service 202-205-1657
PO Box 96090 E-mail: zbowden@fs.fed.us
Washington, DC 20090 http://www.fs.fed.us/spf/

**2519 Department of Agriculture: Natural Resources State and Private Forestry Division**
USDA Forest Service 202-205-1657
PO Box 96090 E-mail: zbowden@fs.fed.us
Washington, DC 20090 http://www.fs.fed.us/spf
Links forestry and conservation with people from the inner city to the rural countryside. Connects people to resources, ideas and to one another so we can all care for the forests and sustain our communities.
*Jim Hubbard, Deputy Chief*
*Robin Thompson, Associate Deputy Chief*

**2520 Department of Agriculture: Research Department**
USDA Forest Service, Research 202-205-1657
PO Box 96090 E-mail: zbowden@fs.fed.us
Washington, DC 20090 http://www.fs.fed.us/research
*Ann M Bartuska, Deputy Chief*
*James Reaves, Associate Deputy Chief*

**2521 Department of Agriculture: Research Department Fire Sciences Program**
USDA Forest Service 201 202-205-1657
PO Box 96090 http://www.fs.fed.us/recreation
Washington, DC 20090

**2522 Department of Agriculture: State & Private Forestry**
USDA Forest Service 201, 4th Floor 202-205-1657
PO Box 96090 E-mail: zbowden@fs.fed.us
Washington, DC 20090 http://www.fs.fed.us/spf/

**2523 Department of Commerce: National Oceanic & Atmospheric Administration**
14th Street & Constitution Ave NW 202-482-6090
Room 6217 Fax: 202-482-3154
Washington, DC 20230 http://www.noaa.gov
*Maureen Wylie, CFO*
*William Broglie, Chief Administrative Officer*

**2524 Department of Commerce: National Marine**
1315 E West Highway 301-713-2379
9th Floor Fax: 301-713-2385
Silver Spring, MD 20910 http://www.nmfs.noaa.gov
Mission: Stewardship of living marine resources through science-based conservation and management and the promotion of healthy ecosystems.
*Christopher M Moore, Division Chief*

**2525 Department of Commerce: National Ocean Service**
Office of Ocean Resources Conservation/Assessment
1305 East-West Highway 301-713-2989
Silver Spring, MD 20910 Fax: 301-713-4389
http://www.oceanservice.noaa.gov
The National Ocean Service works to observe, understand, and manage our nation's coastal and marine resources.
*John H Dunnigan, Assistant Administrator*

**2526 Department of Energy: Office of NEPA Policy and Compliance**
1000 Independence Avenue SW
EH-42 800-472-2756
Washington, DC 20585 Fax: 202-586-7031
E-mail: denise.freeman@eh.doe.gov
http://www.eh.doe.gov/nepa
This office serves as the contact point for NEPA matters for the US Department of Energy.
*Carol M Borgstrom, Director*

**2527 Department of Energy: Transportation and Alternative Fuels**
1000 Independence Avenue SW 202-586-1723
Washington, DC 20585 Fax: 202-586-9811
E-mail: john.garbak@hq.doe.gov
http://www.ott.doe.gov/oaat/afv.html
*John Bgarbak*

**2528 Department of Justice: Environment and Natural Resources Division**
PO Box 4390 202-514-2000
Ben Franklin Station http://www.usdoj.gov/enrd
Washington, DC 20044
*Robert Bruffy, Executive Officer*

**2529 Department of Justice: Environment and Resources, Environmental Defense**
PO Box 4390 202-514-2219
Ben Franklin Station Fax: 202-514-0557
Washington, DC 20044

**2530 Department of State: Bureau of Economic and Business**
2201 C Street NW Room 3529 202-647-1498
Washington, DC 20520 Fax: 202-647-8758
*Earl Anthony Wayne, Assistant Secretary*

**2531 Department of State: Bureau of Oceans and International Environmental and Scientific Affair**
2201 C Street NW Room 7831 202-647-2232
Washington, DC 20520 Fax: 202-647-0217
E-mail: www.state.gov/g/oes
Mission: We advance sustainable development internationally through leadership in oceans, environment, science and health.
*Claudia McMurray, Assistant Secretary*

**2532 Department of State: Ocean and Fisheries Affairs**
US Department of State
Office of Marine Conservation 202-647-2335
2201 C Street NW, Room 5806 Fax: 202-736-7350
Washington, DC 20520 http://www.foia.state.gov/records.asp
*David Hogan, Deputy Director*

**2533 Department of State: Office of Ecology, Health, and Conservation**
2201 C Street NW 202-647-2418
OES/ETC, Room 4333 Fax: 202-736-7351
Washington, DC 20520
*Paul Blakeburn, Director*

**2534 Department of State: Office of Global Change**
Office of Global Change 202-647-4069
OES/EGC, US Department State Fax: 202-647-0191
Washington, DC 20520 http://www.state.gov/g/oes/climate
*Paula Dobriansky*

**2535 Department of Transportation: Office of Marine Safety, Security & Environmental**
U.S. Coast Guard
2100 2nd Street SW Room 2408
Washington, DC 20593
202-267-2200
Fax: 202-267-4839
http://www.dot.gov/safety.html

**2536 Department of Transportation: Office of Pipeline Safety**
1200 New Jersey Avenue
SE East Building 2nd Floor
Washington, DC 20590
202-366-4595
Fax: 202-366-4566
http://ops.dot.gov/request.htm
*Alex P Alvarado, Chief*
*Darline Coleman, Contact*

**2537 Department of the Interior**
1849 C Street Northwest
Washington, DC 20240
202-208-3100
E-mail: webteam@ios.doi.gov
http://www.doi.gov
Mission: To protect and provide access to our Nation's natural and cultural heritage and honor our trust responsibilities to Indian Tribes and our commitments to island communities.
*Dirk Kempthorne, Secretary*
*Lynn Scarlett, Deputy Secretary*

**2538 Department of the Interior, U.S. Fish & Wildlife Service**
1849 C Street NW Room 3242
Washington, DC 20240
202-208-4646
Fax: 202-208-6916
http://www.fws.gov
Mission: Working with others, to conserve, protect and enhance fish, wildlife, and plants and their habitats for the continuing benefit of the American people.

**2539 Department of the Interior: National Parks Service**
1849 C Street Northwest
Washington, DC 20240
202-208-6843
http://www.nps.gov
Mission: The National Park Service preserves unimpaired the natural and cultural resources and values of the national park system for the enjoyment, education, and inspiration of this and future generations. The Park Service cooperateswith partners to extend the benefits of natural and cultural resource conservation and outdoor recreation throughout this country and the world.
*Mary A Bomar, Director*

**2540 Department of the Interior: Bureau of Land Management**
1849 C Street Northwest
Room 406-LS
Washington, DC 20240
202-452-5125
Fax: 202-452-5124
E-mail: woinfo@blm.gov
http://www.blm.gov

*James M Hughes, Acting Director*
*Luke Johnson, Chief of Staff*

**2541 Department of the Interior: Division of Parks and Wildlife**
Office of the Solicitor
1849 C Street NW Room 6557
Washington, DC 20240
202-208-4344
Fax: 202-208-3877

**2542 Department of the Interior: National Resources Department**
National Park Service
Interior Main Interior Building
1849 C Street NW, Room 3127
Washington, DC 20240
202-208-5391
Fax: 202-208-4620
E-mail: gmachlis@uidaho.edu
*F Eugene Hester, Associatate Director*

**2543 Department of the Interior: Office of the Solicitor**
Division of Land and Water
1849 C Street NW, MS 6412
Washington, DC 20240
202-208-5757
Fax: 202-219-1792
http://www.doi.gov/solicitor

*David Longly Bernhardt, Solicitor*

**2544 Department of the Interior: Soil, Water & Air**
Land and Renewable Rangeland Resources Division
1620 L Street NW Room 204
Washington, DC 20036
202-208-4621
Fax: 202-208-7520
*Donald D White, Chief*

**2545 Department of the Interior: Water and Science, Water Resources Division**
Water Resources Division
12201 Sunrise Valley Drive
Reston, VA 20192
703-648-4000
E-mail: rhirsch@usgs.gov
http://water.usgs.gov
Provides reliable, impartial, timely information to understand the water resources of the United States.
*Robert M Hirsch, Associate Director*
*Matt Larsen, Chief Hydrologist*

**2546 Department of the Interior: Water and Science Bureau of Reclamation**
1849 C Street NW
Washington, DC 20240
202-513-0501
Fax: 202-513-0309
http://www.usbr.gov
To manage, develop, and protect water and related resources in an environmentally and economically sound manner in the interest of the American public.

*Michael L Connor, Commissioner*
*Kerry Rae, Chief of Staff*

**2547 Department of the Interior: Wild Horses and Burros**
Land and Renewable Resources
1620 L Street NW Room 204
Washington, DC 20036
202-653-5258
*Bob Bainbridge, Advisor*

**2548 Dept. of Agriculture: National Forest Watershed and Hydrology**
14th Street SW Building 201
3rd Floor
Washington, DC 20250
202-205-0886
*Keith McLaughlin, Water Quality/Hydrology Prog*

**2549 Dialogue Committee on Phosphoric Acid Product Consensus and Dispute Resolution**
Environmental Protection Agency
400 M Street SW
Washington, DC 20460
202-260-5495
*Deborah Dalton*

**2550 EPA: Office of Solid Waste, Municipal & Industrial Solid Waste**
401 M Street SW Room M2105
Washington, DC 20460
703-308-7267
Fax: 703-308-8686
E-mail: levy.steve@epamail.epa.gov
*Bruce R Weddle, Director*

**2551 Emission Standards Office**
National Air Pollution Control Techniques
Office of Air Quality
Mail Drop 13
Research Triangle Park, NC 27711
919-541-5571
Fax: 919-541-0072
The Emission Standards Division (ESD) is responsible for establishing emission standards and managing federal programs for nationawide control of hazardous and criteria pollutant emissions from stationary sources. The Division developsand implements emission standards for hazardous and criteria air pollutants, new source performance standards, control technique guidelines, hazardous waste standards, alternative control techniques documents, and guidance for implementing standards.
*Bruce C Jordon, Designated Federal Officer*
*Sally Shaver, Director*

**2552 Environment and Natural Resources: Environmental Crimes Section**
Department of Justice

PO Box 4390 202-514-0424
Ben Franklin Station Fax: 202-514-4231
Washington, DC 20044 E-mail: webcontentmgr.enrd@usdoj.gov
http://www.usdoj.gov/enrd
*Robert L Bruffy, Executive Officer*

**2553 Environmental Change and Security Program: Woodrow Wilson International Center for Scholars**
1300 Pennsylvania Avenue NW 202-691-4000
Washington, DC 20004 E-mail: ecsp@wwic.si.edu
http://www.wilsoncenter.org/ecsp
The ECSP provides specialists and interested individuals with a road-map to the myriad conceptions, activities and policy initiatives related to environment, population and security. The project pursues three basic activities:gathering information on related international academic and policy initiatives; organizing meetings of experts and public seminars; and publishing the ECSP Report, The China Environment Series and related papers.ECSP explores a wide range ofenvironment related issues.

*Geoffrey D Dabelko, Director*
*Meaghan Parker, Writer/Editor*

**2554 Environmental Financial Advisory Board (EFAB)**
US EPA, Office of Enterprise Technology & Innovat.
Environmental Finance Program (Mail 202-564-4994
Code) 2731R 1200 Pennsylvania Ave. NW Fax: 202-565-2587
Washington, DC 20460 E-mail: efin@epa.gov
http://www.epa.gov/efinpage/efab.htm
The EFAB provides advice to the Environmental Protection Agency's Administrator and Program Offices around the financial aspects of environmental protection. They are a federally chartered advisory committee operating under the FederalAdvisory Committee Act.

*Vanessa Bowie, Staff Director*

**2555 Environmental Health Sciences Review Committee**
Division of Extramural Research and Training
PO Box 12233 919-541-7508
Research Triangle Park, NC 27709 Fax: 919-541-2503
E-mail: malone@niehs.nih.gov
http://www.niehs.nih.gov/dert/home.htm
*Dennis R Lang, Director*
*David Balshaw, Health Science Administrator*

**2556 Environmental Management**
US Department of Energy 202-586-5000
1000 Independence Ave SW, Room 5A-014 800-342-5363
Washington, DC 20585 Fax: 202-586-4403
E-mail: em.webcontentmanager@em.doe.gov
http://www.em.doe.gov
The Assistant Secretary for EM provides program policy guidance, manages the assessment and cleanup of departmental inactive waste sites and facilities in compliance with federal, state legal and regulatory requirements. The AS alsodirects a program of safe and effective waste management operations, develops and implements an aggressive applied waste research and development program to provide innovative environmental technologies to yield permanent and cost-effective disposalsolutions.
*James A Rispoli, Assistant Secretary*
*Ines R Triay, COO*

**2557 Environmental Management Advisory Board**
1000 Independence Avenue SW 202-586-7709
Room 180671 Fax: 202-586-0293
Washington, DC 20585 http://www.em.doe.gov
The mission of the Environmental Management Advisory Board is to provide advice, information and recommendations to the Assistant Secretary for Environmental Management regarding environmental restoration and waste management issues.
*James Ajello, Chair*
*Dennis P Ferrigno, Vice Chair*

**2558 Environmental Protection Agency**
Ariel Rios Building 202-272-0167
1200 Pennsylvania Avenue NW http://www.epa.gov
Washington, DC 20460
EPA's mission is to protect human health and to safeguard the natural environment- air, awter and land- upon which life depends. For 30 years, EPA has been working for a cleaner, healthier environment for the American people.
*Stephen L Johnson, Administrator*
*Marcus Peacock, Deputy Administrator*

**2559 Environmental Protection Agency Air & Radiation**
Office of Radiation
Ariel Rios Building 202-564-7404
1200 Pennsylvania Avenue NW Fax: 202-501-0986
Washington, DC 20460 http://www.epa.gov/air
Mission: Lead the US and the world in protecting human health and the environment.
*Robert J Meyers, Assistant Administrator*
*David Bloomgren, Director of Communications*

**2560 Environmental Protection Agency Climate Change Division**
Ariel Rios Building 202-343-9990
1200 Pennsylvania Avenue E-mail: climatechange@epa.gov
Washington, DC 20460 http://www.epa.gov
EPA's Climate Change Division works to assess and address global climate change and the associated risks to human health and the environment.

**2561 Environmental Protection Agency Ground Water and Drinking Water**
Ariel Rios Building 202-564-3750
1200 Pennsylvania Avenue Fax: 202-564-3753
Washington, DC 20460 http://www.epa.gov/safewater
Mission: To protect public health by ensuring safe drinking water and protecting ground water.
*Cynthia Dougherty, Director*
*Nanci Gelb, Deputy Director*

**2562 Environmental Protection Agency Resource Conservation and Recovery Act**
Ariel Rios Building 202-260-4808
1200 Pennsylvania Avenue Fax: 202-260-1400
Washington, DC 20460 http://www.epa.gov
RCRA gave EPA the authority to control hazardous waste from the cradle-to-grave. This includes the generation, transportation, treatment, storage, and disposal of hazardous waste.

**2563 Environmental Protection Agency: Indoor Air Division**
Office of Radiation
Ariel Rios Building 202-233-9315
1200 Pennsylvania Avenue NW http://www.epa.gov
Washington, DC 20460
Mission: To protect the public and the environment from the risks of radiation and indoor air pollution.
*Elizabeth Cotsworth, Director*

**2564 Environmental Protection Agency: Office of Pollution Prevention & Toxics**
Pollution Prevention Division
Ariel Rios Building 202-564-3810
1200 Pennsylvania Avenue NW http://www.epa.gov/oppt
Washington, DC 20460
*Charles M Auer, Director*
*Wendy Cleland-Hamnett, Deputy Director*

**2565 Environmental Protection Agency: Water**
Environmental Protection Agency
Ariel Rios Building 202-272-0167
1200 Pennsylvania Avenue NWhttp://www.epa.gov/ow/index.html
Washington, DC 20004
*Benjamin Grumbles, Assistant Administror*
*Michael H Shapiro, Deputy Asst Administrator*

**2566 Federal Aviation Administration**
800 Independence Avenue, SW
Washington, DC 20591 866-835-5322
http://www.faa.gov

Mission: To provide the safest, most efficient aerospace system in the world.
*Marion C Blakey, Administrator*
*Ramesh K Punwani, CFO*

**2567 Federal Energy Regulatory Commission**
888 1st Street NE 202-502-6088
Room 11A-1 http://www.ferc.gov
Washington, DC 20426
FERC is an independent commission within the department which has retained many of the functions of the Federal Power Commission, such as setting rates and charges for the transporation and scale of natural gas and for the transmissionand sale of electricity and the licensing of hydroelectric power projects. In addition, the commission establishes rates or charges for the transportation of oil by pipeline, as well as the valuation of such pipelines.
*Joseph T Kelliher, Chair*
*Daniel L Larcamp, Chief of Staff*

**2568 Federal Highway Administration**
1200 New Jersey Avenue SE 202-366-0660
Washington, DC 20590 E-mail: douwes@fhws.dot.gov
http://www.fhwa.dot.gov
Mission: Enhancing mobility through innovation, leadership, and public service.
*Richard Capka, Adiministrator*
*Frederick G Wright Jr, Executive Director*

**2569 Federal Railroad Administration**
1120 Vermont Avenue NW 202-493-6000
Mail Stop 35 E-mail: webmaster@fra.dot.gov
Washington, DC 20005 http://www.fra.dot.gov
Office of Acquisition and Grants Services is a centralized procurement Office that negotiates, awards, and administers contracts, purchases grants, and cooperative agreemnts in support of the Federal Railroad Administration. The Officeprocures supplies, services, research development, architecture-engineering, information technology and services, and other requirements related to FRA's mission.
*Joseph H Boardman, Administrator*
*Clifford C Eby, Deputy Administrator*

**2570 Federal Task Force on Environmental Education**
Office of Environmental Education Code 111
1849 C Street NW 202-452-5078
Washington, DC 20240 Fax: 202-452-5199
http://www.epa.gov/enviroed/FTFmemws.html
The Federal Task Force on Environmental Education facilities communication and collaboration among federal agencies and departments that have common interests in supporting and implementing EE programs. The task force places emphasison supporting joint interagency EE projects that leverage both federal and non-federal dollars.
*Kathleen MacKinnon*

**2571 Federal Transit Administration**
US Department of Transportation
1200 New Jersey Avenue SE 202-366-4043
4th & 5th Floor, East Building E-mail: fta.webmaster@dot.gov
Washington, DC 20590 http://www.fta.dot.gov
*James S Simpson, Administrator*
*Sherry E Little, Deputy Administrator*

**2572 Food Safety and Inspection Service**
Technical Service Center
1299 Farnam Street 402-344-5000
Landmark Center, Suite 300 800-233-3935
Omaha, NE 68102 Fax: 402-344-5005
E-mail: TechCenter@fsis.usda.gov
http://www.fsis.usda.gov
The Food Safety and Inspection Service (FSIS) is thepublic health agency in the U.S. Department of Agriculture responsible for ensuring that the nation's commercial supply of meat, poultry, and egg products is safe, wholesome, andcorrectly labeled and packaged.
*James Kile, Director*
*Alfred V Almanza, Administrator*

**2573 Food and Drug Administration**
US Department of Health and Human Services
Office of Public Affairs 301-827-3666
5600 Fishers Lane 888-463-6332
Rockville, MD 20857 Fax: 301-443-4915
http://www.cfsan.fda.gov
Mission: Responsible for protecting the public health by assuring the safety, efficacy, and security of human and veterinary drugs, biological products, medical devices, our nation's food supply, cosmetics, and products that emitradiation. The FDA is also responsible for advancing the public health by helping to speed innovations that make medicines and foods more effective, safer, and more affordable.

**2574 General Services Administration**
1800 F Street NW 202-501-1231
GSA Building Fax: 202-501-1300
Washington, DC 20405 E-mail: sharon.lighton@gsa.gov
http://www.gsa.gov
Our mission is to provide other federal agencies the workspace, products, services, technology, and policy they need to accomplish their missions. The mission statement contained in our Strategic Plan reflects our recognition that wemust provide Federal agencies with the highest quality service at a competitive cost.
*Lurita Alexis Doan, Administrator*

**2575 Global Learning and Observations to Benefit the Environment**
UCAR-The GLOBE Program
PO Box 3000 800-858-9947
Boulder, CO 80307 Fax: 970-491-8768
E-mail: help@globe.gov
http://www.globe.gov
Mission: To promote the teaching and learning of science, enhance environmental literacy and stewardship, and promote scientific discovery.

*Edward Geary, Director*
*Teresa Kennedy, Deputy Director*

**2576 House Committee on Agriculture**
1301 Longworth House Office Bldg 202-225-2171
Washington, DC 20515 Fax: 202-225-8510
E-mail: agriculture@mail.house.gov
http://agriculture.house.gov

*Collin C. Peterson, Chairman*
*Tim Holden, Vice Chairman*

**2577 House Committee on Foreign Affairs**
2170 Rayburn House Office Building 202-225-5021
Washington, DC 20515
Our Committee is charged with overseeing US foreign policy programs and agencies. We manage legislation regarding foreign policy, State Department management, foreign assistance, trade promotion, export controls, foreign arms sales,student exchanges, international broadcasting and many other issues.
*Lynne Weil, Contact*

**2578 House Committee on Transportation and Infrastructure**
2165 Rayburn House Office Bldg 202-225-4472
Washington, DC 20515 Fax: 202-226-1270
http://transportation.house.gov
*James Oberstar, Chair*

**2579 Installation Management**
600 Army Pentagon 703-693-3233
Room 1E668 Fax: 703-693-3507
Washington, DC 20310 http://www.army.mil
This office is responsible for policy and oversight of construction, utilization, improvement, alteration, maintenance, repair and disposal of real estate and facilities.

**2580 Inter-American Foundation**
901 N Stuart Street 703-306-4301
10th Floor Fax: 703-306-4365
Arlington, VA 22203 E-mail: info@iaf.gov
http://www.iaf.gov
The Inter-American Foundation (IAF) is an independent foreign assistance agency of the United States government, working to promote equitable, responsive, and participatory self-help development in Latin America and the Caribbean.
*Larry Palmer, President*
*Linda Borst-Kolko, VP for Operations*

**2581 International Joint Commission**
1250 23rd Street NW 202-736-9024
Suite 100 Fax: 202-467-0746
Washington, DC 20440 http://www.ijc.org
Mission: Prevents and resolves disputes between the United States and Canada, and pursues the common good of both countries as an independent and objective advisor to the two governments.
*Frank Bevacqua, Public Information Officer*
*Greg McGillis, Public Affairs Advisor*

**2582 Land and Minerals Management**
1849 C Street NW 202-208-5676
Washington, DC 20240 Fax: 202-208-3144
E-mail: bob.armstrong@mms.gov
http://www.mms.gov
*Randall B Luthi, Director*

**2583 Land and Minerals Office of Surface Mining Reclamation & Enforcement**
Department of the Interior
1951 Constitution Avenue NW 202-208-2719
Washington, DC 20240 E-mail: getinfo@osmre.gov
http://www.osmre.gov
Our mission is to carry out the requirments of the Surface Mining Control and Reclamation Act in cooperation with States and Tribes. Our primary objectives are to ensure that coal mines are operated in a manner that protects citizensand the environment during mining and assures that the land is restored to beneficial use following mining, and to mitigate the effects of past mining byaggressively pursuing reclamation of abandoned coal mines.

**2584 Management and Budget Office: Natural Resources, Energy and Science**
Old Executive Office Building 202-395-4561
Washington, DC 20503 Fax: 202-395-4639
http://omb.gov
This division of the Management and Budget Office assists the President by clearing and coordinating advice on proposed legislation and by making recommendations as to presidential action on legislative enactments related to naturalresources, energy and science.
*Rob Portman, Director*
*Stephen McMillin, Deputy Director*

**2585 Manpower, Reserve Affairs, Installations and Environment**
Environment, Safety and Occupational Health
1665 Air Force Pentagon 703-697-9297
Washington, DC 20330 Fax: 703-614-2884
http://www.af.mil
This office provides guidance, direction, and oversight for the department on all matters pertaining to the environment, safety, and occupational health.

**2586 Marine Mammal Commission**
4340 East West Highway 301-504-0087
Suite 700 Fax: 301-504-0099
Bethesda, MD 20814 E-mail: mmc@mmc.gov
http://www.mmc.gov
Developing, reviewing and making recommendations on domestic and international actions and policies with respect to marine mammal protection, conservation and with carrying out a research program. Primary objective is to ensure thatfederal programs are being administered in ways that maintain the health and stability of marine ecosystems and do not disadvantage marine mammal populations or species.
*Timothy J Ragen, Executive Director*
*Robert C Gisiner, Scientific Program Director*

**2587 Maritime Administration**
Maritime Administration (MARAD)
Southeast Federal Center 202-366-5807
1200 New Jersey Avenue SE 800-996-2723
Washington, DC 20590 E-mail: pao.marad@dot.gov
http://www.marad.dot.gov
Mission: To improve and strengthen the US marine transportation system-including infrastructure, industry and labor-to meet the economic and security needs of the Nation.
*Sean T Connaughton, Administrator*
*Julie Nelson, Deputy Administrator*

**2588 Migratory Bird Conservation Commission**
4401 N Fairfax Drive 703-358-1716
Suite 622 Fax: 703-358-2223
Arlington, VA 22203 E-mail: mbcc@fws.gov
http://www.fws.gov/refuges/realty/mbcc.html
The Migratory Bird Conservation Commission (MBCC) is responsible for considering and approving for acquistion areas of migratory bird habitat (other than waterfowl production areas) that have been submitted by regional offices andrecommended by the Secretary. The MBCC is composed of representatives from the Legislative and Executive Branches of government, fixes the price at which such areas may be purchased or rented, and meets three times a year.
*Ken Salazar, Chair*

**2589 Migratory Bird Regulations Committee Office of Migratory Bird Management**
4401 N Fairfax Drive 703-358-1714
MBSP 4107 Fax: 703-358-2217
Arlington, VA 22203 E-mail: brian_a_millsap@fws.gov
http://policy.fws.gov/723fw2.html
The Service Regulations Committee will review information provided to it each year on regulatory issues and submit recommendations to the Director. In this regard, the committe receives guidance fromt he office of Migratory BirdManagement, the Division of Law Enforcement, and from the Regional migratory Bird Coordinators. In addition, Flyway Consultants, from the four Flyway Councils, may provide technical data and certain advice as limited by memoranda of understanding.
*Krista Holloway, Contact*

**2590 Mine Safety and Health Administration**
1100 Wilson Boulevard 202-693-9500
21st Floor Fax: 202-693-9501
Arlington, VA 22209 http://www.msha.gov
The mission of the Mine Safety and Health Administration is to administer the provisions of the Federal Mine Safety and Health Act of 1977 and to enforce compliance with mandatory safety and health standards as means to eliminate fatalaccidents; to reduce the frequency and severity of nonfatal accidents; to minimize health hazards; and to promote improved safety and health conditions in the Nation's mines. MSHA carries out the mandates of the Mine Act at all mining and mineralprocessing areas.
*Richard E Stickler, Assistant Secretary*

**2591 NOAA Sanctuaries and Reserves Management Divisions**
Department of Commerce
NOAA's National Marine Sanctuaries 301-713-3125
1305 E West Highway, 11th Floor Fax: 301-713-0404
Silver Spring, MD 20910 E-mail: sanctuaries@noaa.gov
http://www.sanctuaries.nos.noaa.gov
Mission: To serve as the trustee for the nation's system of marine protected areas, to conserve, protect and enhance their biodiversity, ecological integrity and cultural legacy.

**2592 National Aeronautics and Space Administration**
NASA

NASA Headquarters
Suite 5K39
Washington, DC 20546
202-358-0001
Fax: 202-358-3469
E-mail: public-inquiries@hq.nasa.gov
http://www.nasa.gov

Research includes a variety of global environmental conditions.
*Michael D Griffin, Administrator*
*Shana Dale, Deputy Administrator*

**2593 National Cancer Institute: Cancer Epidemiology and Genetics Division**
6116 Executive Boulevard
Room 3036A
Bethesda, MD 20892
800-422-6237
http://www.cancer.gov

The National Cancer Institute expands existing scientific knowledge on cancer cause and prevention as well as on the diagnosis, treatment, and rehabilitation of cancer patients. This division conducts research on cancer epdiemiologyand genetics.
*John E Niederhuber, Director*

**2594 National Center for Health Statistics**
Nat. Cen. for Health Stats. Division of Data Serv.
3311 Toledo Road
Hyattsville, MD 20782
800-232-4636
E-mail: nchsquery@cdc.gov
http://www.cdc.gov/nchs/

The mission of the National Center for Health Statistics (NCHS) is to provide statistical information that will guide actions and policies to improve the health of the American people. As the Nation's principal health statisticsagency, NCHS leads the way with accurate, relevant, and timely data.
*Edward Sondik, Director*

**2595 National Climatic Data Center**
Federal Building
151 Patton Avenue
Asheville, NC 28801
828-271-4800
Fax: 828-271-4876
http://www.ncdc.noaa.gov

Mission: To manage the Nation's resource of global climatological in-situ and remotely sensed data and information to promote global environmental stewardship; to describe, monitor and assess the climate; and to support efforts topredict changes in the Earth's environment.
*Thomas R Karl, Director*

**2596 National Council on Radiation Protection and Measurements**
7910 Woodmont Avenue
Suite 400
Bethesda, MD 20814
301-657-2652
800-229-2652
Fax: 301-907-8768
E-mail: ncrp@ncrponline.org
http://www.ncrponline.org

The National Council on Radiation Protection and Measurements (NCRP) seeks to formulate and widely disseminate information, guidance and recommendations on radiation protection and measurements which represent the consensus of leadingscientific thinking.
*Thomas S Tenforde, President*
*David Schauer, Executive Director*

**2597 National Environmental Justice Advisory Council**
401 M Street Southwest (MC 2201A)
Washington, DC 20460
202-564-2515
800-962-6215
Fax: 202-501-0740
E-mail: environmental-justice-epa@epamail.epa.gov
http://www.epa.gov/compliance/environmentaljustice/nejac/index.html

A federal advisory committee established by charter to provide independent advice, consultation and recommendations to the Administrator of the US Environmental Protection Agency on matters related to environmental justice.
*Richard Moore, Chair*

**2598 National Environmental Satellite Data & Information Service**
National Oceanic and Atmospheric Administration
Suitland & Silver Hill Roads
Federal Office Building, Room 2069
Suitland, MD 20233
757-824-3446
Fax: 757-824-7300
E-mail: rwinokur@nesdis.noaa.gov
http://www.nesdis.noaa.gov

The National Environmental Satellite, Data and Information Service operates a national environmental satellite system. It acquires, stores and disseminates worldwide environmental data through its data centers.
*Mary Ellen Kicza, Assistant Administrator*

**2599 National Health Information Center (NHIC)**
US Department of Health and Human Services
PO Box 1133
Washington, DC 20013
301-565-4167
800-336-4797
Fax: 301-984-4256
E-mail: info@nhic.org
http://www.health.gov/nhic

A health information referral line. NHIC links consumers and health professionals who have health questions and organizations best able to provide reliable health information. Maintains an online directory of more than than 1600 healthorganizations that can provide provide information. They include Federal and State agenices, voluntary a professional associations and universities. The database is accessible to the public through the healthfinder web site.
*Rachel Langston, Information Services Manager*

**2600 National Institute for Occupational Safety and Health**
395 E Street SW, Suite 9200
Patriots Plaza Building
Washington, DC 20201
202-245-0625
800-356-4674
Fax: 513-533-8573
http://www.cdc.gov/niosh

To ensure safe and healthful working conditions for all working people, occupational safety and health standards are developed, and research and other activities are carried out, through the National Institute for Occupational Safetyand Health.
*John Howard, Director*
*Frank J Hearl, Chief of Staff*

**2601 National Institute of Environmental Health Sciences**
PO Box 12233
Research Triangle Park, NC 27709
919-541-3201
Fax: 919-541-2260
E-mail: webcenter@niehs.nih.gov
http://www.niehs.nih.gov

Environmental research.
*David A Schwartz, Director*
*Samuel Wilson, Deputy Director*

**2602 National Institutes of Health**
9000 Rockville Pike
Bethesda, MD 20892
301-496-4000
E-mail: nihinfo@od.nih.gov
http://www.nih.gov

The National Institutes of Health conducts, supports and promotes biomedical research to improve the health of the American people by increasing the understanding of processes underlying human health, disability and disease. Theinstitutes advance knowledge concerning the health effects of interactions between humans and the environment.
*Dr Elias A Zerhouni, Director*
*Raynard Kington, Deputy Director*

**2603 National Lead Information Center**
422 South Clinton Avenue
Rochester, NY 14620
800-424-LEAD
Fax: 585-232-3111
E-mail: hotline.lead@epamail.epa.gov
http://www.epa.gov/lead/pubs/nlic.htm

Provides the general public and professionals with information about lead hazards and their prevention.

**2604 National Marine Fisheries Service**
National Oceanic & Atmospheric Administration
14th Street & Constitution Ave NW
Room 6217
Washington, DC 20230
202-482-6090
Fax: 202-482-3154
E-mail: roland.schmitten@noaa.gov
http://www.noaa.gov

NMFS conducts an integrated program of management, research and services for the protection and rational use of living marine resources. It also is responsible for the protection of marine mammals.

**2605 National Oceanic & Atmospheric Administration**
14th Street & Constitution Ave NW 202-482-6090
Room 6013 Fax: 202-482-3154
Washington, DC 20230 http://www.rdc.noaa.gov
The Administration's mission is to explore, map, and chart the global ocean and its living resources and to manage, use, and conserve those resources; to describe, monitor, and predict conditions in the atmosphere, ocean, sun, andspace environment; to issue warnings against impending destructive natural events; to assess the consequences of inadvertent environmental modification over several scales of time, and to manage and disseminate long-term environmental information.
*Maureen Wylie, CFO*

**2606 National Organic Standards Board Agricultural Marketing Service**
USDA/AMS
1400 Independence Avenue SW 202-720-3252
Washington, DC 20250 Fax: 202-205-7808
E-mail: NOSB.Materials@usda.gov
http://www.ams.usda.gov/nop
*Andrea M Caroe, Chair*
*Julie S Weisman, Vice Chair*

**2607 National Park Service: Fish, Wildlife and Parks**
National Park Service
1849 C Street NW 202-208-4621
PO Box 37127 Fax: 202-208-7889
Washington, DC 20240 E-mail: nps_director@nps.gov
http://www.nps.gov
The National Park Service preserves unimpaired the natural and cultural resources and values of the national park system for the enjoyment, education, and inspiration of this and future generations. The Park Service cooperates withpartners to extend the benefits of natural and cultural resource conservation and outdoor recreation throughout this country and the world.

**2608 National Petroleum Council**
1625 K Street NW 202-393-6100
Suite 600 Fax: 202-331-8539
Washington, DC 20006 E-mail: info@npc.org
http://www.npc.org
The purpose of the NPC is solely to represent the views of the oil and natural gas industries in advising, informing, and making recommendations to the Secretary of Energy with respect to any matter relating to oil and natural gas, orto the oil and gas industries submitted to it or approved by the Secretary. The NPC does not concern itself with trade practices, nor does it engage in any of the usual trade association activities.
*Samuel W Bodman, Secretary*
*Carla Scali Byrd, Information Coordinator*

**2609 National Science Foundation**
4201 Wilson Boulevard 703-292-5111
Arlington, VA 22230 800-877-8339
E-mail: info@nsf.gov
http://www.nsf.gov
It is the National Science Foundation's mission to promote the progress of science; to advance the national health, prosperity, and welfare; and to secure the national defense.
*Arden L Bement Jr, Director*
*Kathie Olsen, Deputy Director*

**2610 National Science Foundation Office of Polar Programs**
4201 Wilson Boulevard 703-292-8030
Room 755 S Fax: 703-292-9081
Arlington, VA 22230 http://www.nsf.gov/od/opp/
OPP shares in the vision and goals expressed in NSF's strategic plan: enable world leadership in science and engineering; promote discovery, dissemination and employment of new knowledge; and support excellence in science mathematical,engineering and technology education. Polar research and the associated logistics activities make a recognized and visible contribution to these goals.
*Karl A Erb, Director*
*Michael L Van Woert, Executive Officer*

**2611 National Water Supply Improvement Association**
PO Box 102 301-855-1173
St. Leonard, MD 20685 Fax: 410-586-2844
*Jack C Jorgensen, Executive Director*

**2612 Natural Resources Conservation Service**
PO Box 2890 202-720-3210
Washington, DC 20013 Fax: 202-720-1564
http://www.nrcs.usda.gov
Provides leadership in a partnership effort to help people conserve, maintain and improve our natural resources and environment. NRCS is the technical delivery arm of USDA and provides conservation information, incentive programs andtechnical assistance at the state and county levels. Contact the Conservation Communications Staff.
*Jack Carlson, Chief Information Officer*

**2613 Naval Sea Systems Command**
1333 Isaac Hull Avenue SE 202-781-4124
Washington Navy Yard, DC 20376 Fax: 202-781-4713
http://www.navsea.navy.mil
The Naval Sea Systems Command provides material support to the Navy and Marine Corps, and for mobilization purposes to the Department of Defense and Department of Transportation, for ships, submarines, and other sea platforms,shipboard combat systems and components, other surface and undersea warfare and weapons systems, and ordinance expendables not specifically assigned to other system commands.

**2614 Navy Environmental Health Center**
620 John Paul Jones Circle 757-953-0700
Suite 1100 Fax: 757-953-0999
Portsmouth, VA 23708 http://www-nehc.med.navy.mil
Mission: The Navy and Marine Corps center for public health services. We provide leadership and expertise to ensure mission readiness through disease prevention and health promotion in support of the National Military Strategy.
*Captain William Stover, Commanding Officer*
*Captain John B Burgess, Executive Officer*

**2615 New Forests Project**
1025 Vermont Avenue, NW 202-547-3800
7th Floor Fax: 202-546-4784
Washington, DC 20005 E-mail: etoledo@newforests.org
http://www.newforestsproject.com
The New Forest Project strives to protect, conserve and enhance the health of the Earth's ecosystems along with the people depending on them, by supporting integrated grassroots efforts in agroforestry, reforestation, protection ofwatersheds, water and sanitation and renewable energy initiatives.
*Erick Toledo, Director*

**2616 Nuclear Materials, Safety, Safeguards & Operations**
Nuclear Regulatory Commission
One White Flint N Building 301-415-7000
11555 Rockville Pike 800-368-5642
Rockville, MD 20852 http://www.nrc.gov
The Division of Industrial and medical Nuclear safety within the Office of Nuclear Materials Safety and safeguards at the NRC has the responsibility for NRC's principal rulemaking and guidance development, licensing, inspection, eventresponse and regulatory activities for material licensed under the Atomic Energy Act of 1954, as amended, to ensure safety and quality associated with the possession, processing, and handling of nuclear material.
*Peter B Lyons, Commissioner*
*Dale E Klein, Chair*

**2617 Occupational Safety and Health Administration: US Department of Labor**
Office of Administrative Services

200 Constitution Avenue NW 202-693-1999
Room N-310 800-321-6742
Washington, DC 20210 http://www.osha.gov
OSHA's mission is to send every worker home whole and healthy every day. Since the agency was created in 1971, workplace fatalities have been cut in half and occupational injury and illness rates have declined 40 percent. At the sametime, US employment has nearly doubled from 56 million workers at 3.5 million worksites to 105 milion workers at nearly 6.9 million sites.

**2618 Oceanic and Atmospheric Research Office**
National Oceanic and Atmospheric Administration
1315 E West Highway 301-713-2458
Room 11627 http://www.oar.noaa.gov
Silver Spring, MD 20910
The Office of Oceanic and Atmospheric Research is where much of the work is done that results in better weather forecasts, longer warning lead times for natural disasters, new products from the sea, and greater understanding of ourclimate, atmosphere, and oceans.
*Richard W Spinrad, Assistant Administrator*
*Craig McLean, Deputy Asst Administrator*

**2619 Office of Civil Water Enforcement Division**
Environmental Protection Agency
US Env. Protect. Agency, Water Enf. 202-564-2240
1200 Pennsylvania Avenue Fax: 202-564-0018
Washington, DC 20460 http://www.epa.gov

**2620 Office of Research & Engineering Hazardous Materials**
National Transportation Safety Board
490 L'Enfant Plaza East SW 202-382-6585
Room 5131 http://www.ntsb.gov
Washington, DC 20594

**2621 Office of Solid Waste Management & Emergency Response**
1200 Pennsylvania Avenue NW 202-382-7486
Washington, DC 20460 http://www.epa.gov/swerrims
Develops guidelines and standards for the land disposal of hazardous wastes and for underground storage tanks. Furnishes techincal assistance in the development, management and operation of solids waste activities and analyzes therecovery of useful energy from solid waste. The Office has undertaken the development and implementatin of a program to respond to abandoned and active hazardous waste sites and accidental release as well as the encouragement of new technology.

**2622 Office of Surface Mining**
1951 Constitution Avenue NW 202-208-2719
Washington, DC 20240 E-mail: getinfo@osmre.gov
http://www.osmre.gov
Aids in maintaining proper safety precautions during coal reclamation.

**2623 Office of Surface Mining Reclamation & Enforcement**
US Department of the Interior
1951 Constitution Avenue NW 202-208-2719
Washington, DC 20240 E-mail: getinfo@osmre.gov
http://www.osmre.gov
The Office of Surface Mining is the bureau of the US Department of the Interior with responsability, in cooperation with the state and indian Tribes, to protect citizens and the environment during coal mining and reclamation, and toreclaim mines abandoned before 1977.

**2624 Office of the Chief Economist**
Jamie L Whitten Federal Building 202-720-5447
Room 112-A http://www.usda.gov/oce
Washington, DC 20250
The Office of the Chief Economist advises the Secretary on the economic implications of policies and programs affecting the US food and fiber systems and rural areas. The Chief Economist coordinates, reviews, and approves theDepartment's commodity and farm sector forecast.
*Keith Collins, Chief Economist*
*Brenda Chapin, Information Officer*

**2625 Office of the Executive Clerk**
17th Street & Pennsylvania Ave NW 202-456-2226
Washington, DC 20500 Fax: 202-456-2569
http://www.whitehouse.gov
This office provides information on when a bill was signed or vetoed, the dates of presidential messages, executive orders, and dates of other presidential actions.
*George T Saunders, Executive Clerk*

**2626 Office of the General Counsel**
14th & Constitution Avenue NW 202-482-4772
Washington, DC 20230 Fax: 202-482-0042
E-mail: generalcounsel@doc.gov
http://www.ogc.doc.gov
The Office of the General Counsel provides legal services for all programs, operations and activities of the department.
*Cameron F Kerry, General Counsel*
*Geovette E Washington, Deputy General Counsel*

**2627 Office of the Secretary of Energy**
1000 Independence Avenue SW 202-586-6210
Forrestal Building, Room 7A-257 Fax: 202-586-4403
Washington, DC 20585 http://www.doe.gov
The Secretary of Energy provides the framework for a comprehensive and balanced national energy plan through the coordination and administration of the energy functions of the federal government.

**2628 Office of the Secretary of Health and Human Services**
200 Independence Avenue SW 202-619-0257
Hubert H Humphery Building http://www.os.dhhs.gov
Washington, DC 20201
The Secretary of Health and Human Services advises the President on health, welfare, and income security plans, policies, and programs of the federal government.
*Mike Leavitt, Sect'y Health/Human Service*

**2629 Office of the Secretary of the Interior**
1849 C Street NW 202-208-3100
Interior Building, Room 6156 E-mail: webteam@ios.doi.gov
Washington, DC 20240 http://www.doi.gov
The Secretary of the Interior is responsible for the administration of over 500 million acres of federal land, and holds in trust approximately 50 million acres of land, mostly Indian reservations; the conservation and development ofmineral and water resources; the conservation, development, and utilization of fish and wildlife resources; the coordination of federal and state recreational programs; the preservation and administration of the nation's scenic and historic areas.
*Dirk Kempthorne, Secretary of the Interior*
*P Lynn Scarlett, Deputy Secretary*

**2630 Office of the Solicitor**
1849 C Street NW 202-208-3100
Interior Building, Room 6352 http://www.doi.gov/solicitor
Washington, DC 20240
The Solicitor performs all of the legal work of the department and is the principal legal adviser to the secretary and the chief law officer of the department.
*David Bernhardt, Solicitor*

**2631 Office of the US Trade Representative**
600 17th Street NW 202-395-7360
Washington, DC 20508 http://www.ustr.gov
This division of the Office of the US Trade Representative is responsible for the direction of all trade negotiations and the formulation of trade policy for the United States as related to the environment and natural resources.
*Susan C Schwab, Ambassador*

**2632 Peace Corps**
1111 20th Street NW 202-692-2100
Washington, DC 20526 800-424-8580
http://www.peacecorps.gov
The goal of the Peace Corps is to help people of interested countries in meeting their need for trained men and women, to promote a better understanding of Americans on the part of peolple served and to help promote a betterunderstanding of other peoples on the part of Americans.
*Ronald A Tschetter, Director*
*Josephine Olsen, Deputy Director*

**2633 Research, Education and Economics**
1400 Independence Avenue SW 202-720-5923
Room 216W Whitten Building Fax: 202-690-2842
Washington, DC 20250 E-mail: carolyne.foster@osec.usda.gov
http://www.ree.usda.gov
Mission: Federal leadership responsibility for the discovery of knowledge spanning the biological, physical, and social sciences, and involving agricultural research, economic analysis, statistics, outreach, and higher education.
*Raj Shah Chief Scientist USDA, & Under Secretary of REE*

**2634 Research, Education, and Economics National Agricultural Statistics Service**
1400 Independence Avenue SW
S Building, Room 4117 800-727-9540
Washington, DC 20250 http://www.nass.usda.gov
Mission: Provides timely, accurate, and useful statistics in service to US agriculture.
*R Ronald Bosecker, Administrator*

**2635 Risk Assessment and Cost Benefit Analysis Office**
Office of the Chief Economist
1400 Independence Avenue SW 202-720-8022
S Building, Room 4032 E-mail: jcallahan@oce.usda.gov
Washington, DC 20250 http://www.usda.gov/oce/risk_assessment
ORACBA's primary role is to ensure that major regulations proposed by USDA are based on sound scientific and economic analysis.
*Michael McElvaine, Deputy Director*

**2636 Rural Utilities Service**
1400 Independence Avenue SW 202-720-9540
S Building, Room 5151 E-mail: webmaster@rus.usda.gov
Washington, DC 20250 http://www.usda.gov/rus
Modern utilities came to rural America through some of the most successful goverment initiatives in American history, carried out through the USDA working with rural cooperatives, nonprofit associations, public bodies, and for-profitutilities. Today, they carry on this tradition helping rural utilities expand and keep their technology up to date, helping establish new and vital services such as distance learning and telemedicine.
*James M Andrew, Administrator*
*Curtis M Anderson, Deputy Administrator*

**2637 Saint Lawrence Seaway Development Corporation**
US Department of Transportation
1200 New Jersey Avenue SE 202-366-0091
Suite W32-300 800-785-2779
Washington, DC 20590 Fax: 202-366-7147
E-mail: research@sls.dot.gov
http://www.seaway.dot.gov
Saint Lawrence Seaway Development Corporation operates and maintains the Great Lakes/St. Lawrence System, which encompasses the St. Lawrence River and the five Great lakes.
*Collister Johnson Jr, Administrator*
*Craig H Middlebrook, Deputy Administrator*

**2638 Science Advisory Board Environmental Protection Agency**
401 M Street SW 202-382-4126
Room 1145 E-mail: thomas.patrical@epa.gov
Washington, DC 20460 http://www.epa.gov/sab
The SAB was established by Congress to provide independent scientific and engineering advice to the EPA Administrator on the technical basis for EPA regulations. Expressed in terms of the current parlance of the risk assessment/riskmanagement paradigm of decision making, they deal with risk assessment issues and only that portion of risk management that deals strictly with the technical issues associated with various contorol options.

**2639 Science and Technology Policy Office**
Old Executive Office Bldg, Rm 424 202-456-7116
725th 17th Street Room 5228 Fax: 202-456-6021
Washington, DC 20502 E-mail: info@ostp.gov
http://www.ostp.gov
The Science and Technology Policy Office serves as a source of scientific, engineering, and technological analysis and judgement for the President with respect to major policies, plans, and programs of the federal government. Incarrying out this mission, the office advises the President of scientific and technological considerations involved in areas of national concern, including the economy, national security, health, foreign relations, and the environment.
*John H Marburger III, Director*
*Sharon Hays, Deputy Director*

**2640 Senate Committee on Appropriations**
The Capitol S-131 202-224-7363
Washington, DC 20510 http://appropriations.senate.gov
*Robert C Byrd, Chair*
*Thad Cochran, Ranking Member*

**2641 Senate Committee on Energy and Natural Resources**
304 Dirksen Senate Building 202-224-4971
Washington, DC 20510 Fax: 202-224-6163
http://energy.senate.gov
*Jeff Bingaman, Chair*
*Pete V Domenici, Ranking Member*

**2642 Senate Committee on Foreign Relations**
Dirksen Senate Office Builing 202-224-6797
Washington, DC 20510 Fax: 202-224-0836
E-mail: senator@biden.senate.gov
http://foreign.senate.gov
No foreign policy can be sustained without the informed consent of the American people. Our government works best when citizens care enough to become involved.
*Joseph R Biden Jr, Chairman*

**2643 Smithsonian Tropical Research Institute**
Smithsonian Institution Research Department
1100 Jefferson Drive 202-633-4014
Suite 3123 Fax: 202-786-2557
Washington, DC 20560 http://www.stri.org
*Ira Rubinoff, Director*
*Georgina de Alba, Associate Director*

**2644 Take Pride in America Advisory Board Department of the Interior**
1849 C Street NW 202-208-3100
Washington, DC 20240 E-mail: webteam@ios.doi.gov
http://www.doi.gov
The purposes of the program include the following: 1-to establish and maintain a public awareness campaign in cooperation with public and private organizations and individuals; 2- To conduct a national awards program to honor thoseindividuals and entities which, in the opinion of the Secretary of the Interior, have distinguished themselves in the above mentioned activities.

**2645 Technology Administration: National Institute of Standards & Technology**
Technology Policy Office
1000 Bureau Drive 301-975-4500
Gaithersburg, MD 20899 E-mail: inquiries@nist.gov
http://www.nist.gov
The National Institute of Standards and Technology assists industry in the development of technology needed to improve product quality, modernize manufacturing processes, ensure product reliability, and facilitate rapidcommercialization of products based on new scientific discoveries. NIST's primary mission is to pro-

mote US economic growth by working with industry to develop and apply technology, measurements, and standards.

*William A Jeffrey, Director*
*James Turner, Deputy Director*

**2646 US Agency for International Development Information Center**
Ronald Reagan Building
Washington, DC 20523
202-712-4320
Fax: 202-216-3524
http://www.usaid.gov

Aids in the development of urban environmental programs, forest conservation teams, watershed management and promotes improved pollution control.

*James R Kunder, Acting Deputy Administrator*
*Alonzo Fulgham, COO*

**2647 US Consumer Product Safety Commission**
4340 E West Highway
Bethesda, MD 20814
800-638-2772
Fax: 301-504-0124
http://www.cpsc.gov

An independent federal regulatory agency. Helps keep American families safe by reducing the risk of injury or death from consumer products.

*Nancy Nord, Acting Chair*
*Thomas Moore, Commissioner*

**2648 US Customs & Border Protection**
1300 Pennsylvania Avenue NW
Washington, DC 20229
202-927-1000
http://www.customs.ustreas.gov

Mission: Safeguard the American homeland at and beyond our borders.

*W Ralph Basham, Commissioner*
*Deborah J Spero, Deputy Commissioner*

**2649 US Department of Agriculture**
1400 Independence Avenue SW
Washington, DC 20250
202-720-8732
http://www.usda.gov

USDA Mission: Enhance the quality of life for the American people by supporting production of agriculture: ensuring a safe, affordable, nutritious, and accessible food supply; caring for agricultural, forest, and range lands;supporting sound development of rural communities; providing economic opportunities for farm and rural residents; expanding global markets for agricultural and forest products and services; and working to reduce hunger in America and throughout theworld.

*Mike Johanns, Secretary*

**2650 US Department of Education**
400 Maryland Avenue SW
Washington, DC 20202
800-872-5327
Fax: 202-401-0689
http://www.ed.gov

Mission: To strengthen the Federal commitment to assuring access to equal educational opportunity for every individual; improve the coordination of Federal education programs; improve the management of Federal education activities; andincrease the accountability of Federal education programs to the President, the Congress, and the public

**2651 US Department of Housing and Urban Development**
451 7th Street SW
Washington, DC 20410
202-708-1112
http://www.hud.gov

*Alphonso Jackson, Secretary*

**2652 US Department of Housing and Urban Development**
Office of Lead Hazard Control
451 7th Street SW
Washington, DC 20410
202-755-1785
Fax: 202-755-1000
http://www.hud.gov/lea/leahome.html

The office works to ensue that hazard controls are conducted in the safest, most cost-effective and efficient way possible to preserve our nation's stock of affordable housing while still ensuring that our children are properlyprotected.

**2653 US Department of Labor**
200 Consititution Avenue NW
Washington, DC 20210
866-487-2365
http://www.dol.gov

The US Department of Labor is charged with preparing the American workforce for new and better jobs, and ensuring the adequacy of America's workplaces. It is responsible for the administration and enforcement of over 180 federalstatutes. These legislative mandates and the regulations produced to implement them cover a wide variety of workplace activities including protecting workers' wages, health and safety, employment and pension rights, promoting equal employmentopportunity.

*Elaine L Chao, Secretary*

**2654 US Department of Treasury**
1500 Pennsylvania Avenue NW
Washington, DC 20220
202-622-2000
Fax: 202-622-6415
http://www.ustreas.gov

*Henry M Paulson Jr, Secretary*

**2655 US Department of the Army: Office of Public Affairs**
1500 Army Pentagon
Washington, DC 20310
703-693-0677
http://www.army.mil

Mission: Public Affairs fulfills the Army's obligation to keep the American people and the Army informed, and helps to establish the conditions that lead to confidence in America's Army and its readiness to conduct operations inpeacetime, conflict and war.

*Anthony A Cucolo III, Chief of Public Affairs*

**2656 US Environmental Protection Agency**
Ariel Rios Building
1200 Pennsylvania Avenue NW
Washington, DC 20460
202-272-0167
http://www.epa.gov

The mission of the US Environmental Protection Agency is to protect human health and to safeguard the natural environment; air, water and land upon which life depends.

*Stephen L Johnson, Administrator*
*Marcus Peacock, Deputy Administrator*

**2657 US Environmental Protection Agency Office of Children's Health Protection**
1200 Pennsylvania Avenue NW
Mail Code 1107A, Room 2512 Ariel Rios N
Washington, DC 20004
202-564-2188
Fax: 202-564-2733
http://www.epa.gov/children

The mission of this office is to make the protection of children's environmental health a fundamental goal of public health and environmental protection in the US.

*Melanie Marty, Chair*

**2658 US Environmental Protection Agency: Clean Air Markets Division**
1200 Pennsylvania Avenue NW
Mail Code 6204J
Washington, DC 20460
202-233-9150
http://www.epa.gov/airmarkets

Mission: To improve human health and the natural environment through the skillful design, operation, and evaluation of cap and trade and other innovative programs that cost-effectively lower harmful air emissions and their deposition.

*Sam Napolitano, Director*
*Kevin Culligan, Chief*

**2659 US Environmental Protection Agency: Office of Air and Radiation**
1200 Pennsylvania Avenue NW
Washington, DC 20760
866-411-4372
E-mail: oar_comments@epa.gov
http://www.epa.gov/oar/

Protects human health and the environment by preventing air pollution and exposure to radiation through effective management of public and private resources.

**2660 US Environmental Protection Agency: Office of Environmental Justice**
1200 Pennsylvania Avenue NW 202-564-2515
Washington, DC 20460 800-962-6215
Fax: 202-501-0740
E-mail: environmental-justice-epa@epamail.epa.org
http://www.epa.gov/compliance/environmentaljustice/index.html
Serves as a focal point for ensuring that communities comprised mainly of people of color or low income receive protection under environmental laws.

**2661 US Forest Service**
U.S. Dept. of Agriculture
1400 Independence Avenue SW 202-205-8333
Washington, DC 20250 http://www.fs.fed.us.
Manages National Forests and Grasslands. Provides assistance to private forest operators.

**2662 US Geological Survey: National Wetlands Research Center**
700 Cajundome Boulevard 337-266-8501
Lafayette, LA 70506 Fax: 337-266-8513
E-mail: nwrcdirector@usgs.gov
http://www.nwrc.usgs.gov
To develop and disseminate scientific information needed for understanding the ecology and values of our nation's wetlands and for managing and restoring wetland habitats and associated plant and animal communtities.

*Greg Smith, Director*

**2663 US Nuclear Regulatory Commission**
Reference Librarian 301-415-8200
Public Document Room (01F-13) 800-368-5642
Washington, DC 20555 E-mail: pdr@nrc.gov
http://www.nrc.gov
Mission: To regulate the nation's civilian use of byproduct, source, and special nuclear materials to ensure adequate protection of public health and safety, to promote the common defense and security, and to protect the environment.
*Peter B Lyons, Commissioner*
*Dale E Klein, Chair*

## Alabama

**2664 Agriculture and Industries Department**
Pesticide Laboratory
1445 Federal Drive 334-240-7171
Montgomery, AL 36107 800-642-7761
http://www.agi.stste.al.us/
Mission: To provide timely, fair and expert regulatory control over product, business entities, movement, and application of goods and services for which applicable state and federal law exists and strive to protect and provide serviceto Alabama consumers.
*Ron Sparks, Commissioner*
*Douglas Rigney, Deputy Commissioner*

**2665 Alabama Cooperative Extension System**
224 Duncan Hall Annex 334-844-5270
Auburn University, AL 36849 Fax: 334-844-5276
E-mail: webmaster@aces.edu
http://www.aces.edu
Operates as the primary outreach organization for the land-grant function of Alabama A&M University and Auburn University. Identifies statewide educational needs, audiences, and optimal educational programs that are delivered through anetwork of public and private partners supported by county, state, and federal governments.
*Paul Waddy, Extension Coordinator*

**2666 Alabama Department of Environmental Management**
1400 Coliseum Boulevard 334-271-7700
PO Box 301463 Fax: 334-271-7950
Montgomery, AL 36130 E-mail: webmaster@adem.state.al.us
http://www.adem.state.al.us
Alabama Department of Environmental Management is the state agency responsible for the adoption and fair enforcement of rules and regulations set to protect and improve the quality of Alabama's environment and the health of all itscitizens. Monitor environmental conditions in Alabama and recommend changes in state law or revise regulations as needed to respond appropriately to changing environmental conditions.

*Lance R Lefleur, Director*

**2667 Alabama Forestry Commission**
513 Madison Avenue 334-240-9300
Montgomery, AL 36104 Fax: 334-240-9390
http://www.forestry.state.al.us
Mission: To serve Alabama by protecting and sustaining our forest resources using professionally applied stewardship principles and education. We will ensure Alabama's forests contribute to abundant timber and wildlife, clean air andwater, and a healthy economy.
*Jerry Dwyer, Director*
*Linda Casey, State Forester*

**2668 Conservation and Natural Resources Department**
64 North Union Street 334-242-3486
Montgomery, AL 36130 Fax: 334-242-0999
http://www.dcnr.state.al.us/Lands.htm
*M Barnett Lawley, Commissioner*
*Hobbie L Sealy, Assistant Commissioner*

**2669 EPA: National Air and Radiation Environmental Laboratory**
540 S Morris Avenue 334-270-3400
Montgomery, AL 36115 Fax: 334-270-3454
http://www.epa.gov/narel/
The National Air and Radiation Environmental Laboratory is a comprehensive environmental laboratory, and provides services to a wide range of clients, including other EPA offices, Federal agencies, and, in somes cases, the privatesector. The mission is the commitment to developing and applying the most advanced methods for measuring environmental radioactivity and evaluating its risk to the public.

**2670 Geological Survey of Alabama, Agency of the State of Alabama**
University of Alabama
420 Hackberry Lane 205-349-2852
Tuscaloosa, AL 35486 Fax: 205-349-2861
E-mail: info@gsa.state.al.us
http://www.gsa.state.al.us
To survey and investigate the mineral, energy, water, and biological resources of the state, to maintain adequate geological, topographic, hydrologic, and biologic databases, and to prepare maps and reports on the state's naturalresources to encourage the safe and prudent development of Alabama's natural resources while providing for the safety, health and well-being of all Americans.

*Berry H (Nick) Tew Jr, Director*

## Alaska

**2671 Alaska Cooperative Fish and Wildlife Research Unit**
University of Alaska
209 Irving 1 907-474-7661
PO Box 757020 Fax: 907-474-7872
Fairbanks, AK 99775 http://www.akcfwru.uaf.edu
The Alaska Unit is a part of a nationwide program created to foster college-level research and graduate student training in support of science-based management of fish and wildlife, and their habitats. The Unit exists by cooperativeagreement between the AK Department of Fish and Game, University of Alaska Fair-

banks, US Geological Survey, Wildlife Management Institute, and US Fish and Wildlife Service. The Unit is staffed by 5 USGS scientists, who are also research faculty.

*Kathleen Pearse, Administrative Assistant*

**2672 Alaska Department of Fish and Game**
PO Box 115526 907-465-4100
Juneau, AK 99802 http://www.adfg.state.ak.us
Aims to manage, protect, maintain and improve the fish, game and aquatic plant resources of Alaska. The primary goals are to ensure that Alaska's renwable fish and wildlife resources and their habitats are conserved and managed on thesustained yield prinicpal, and the use of development of these resources are in the best interest of the economy and well-being of the people of the state.
*Denby Lloyd, Commissioner*
*David Bedford, Deputy Commissioner*

**2673 Alaska Department of Public Safety**
5700 E Tudor Road 907-269-5511
Anchorage, AK 99507 http://www.dps.state.ak.us
Provides functions relative to the protection of life, property and wildlife resources.
*Walt Monegan, Commissioner*
*John D Glass, Deputy Commissioner*

**2674 Alaska Division of Forestry: Central Office**
550 W 7th Avenue 907-269-8463
Suite 1450 Fax: 907-269-8931
Anchorage, AK 99501 http://www.dnr.state.ak.us/forestry
Mission: To develop, conserve, and enhance Alaska's forests to provide a sustainable supply of forest resources for Alaskans.
*Chris Maisch, State Forester*
*Dean Brown, Deputy State Forester*

**2675 Alaska Division of Forestry: Coastal Region Office**
400 Willoughby Avenue 907-465-2491
3rd Floor Fax: 907-586-3113
Juneau, AK 99801 http://www.dnr.state.ak.us/forestry
*Mike Curran, Regional Forester*

**2676 Alaska Division of Forestry: Delta Area Office**
Mi. 267.5 Richardson Highway 907-895-4225
PO Box 1149 Fax: 907-895-4934
Delta Junction, AK 99737 http://www.dnr.state.ak.us/forestry
*Al Edgren, Area Forester*

**2677 Alaska Division of Forestry: Fairbanks Area Office**
3700 Airport Way 907-451-2600
Fairbanks, AK 99709 Fax: 907-458-6895
http://www.dnr.state.ak.us/forestry
*Marc Lee, Area Forester*

**2678 Alaska Division of Forestry: Kenai/Kodiak Area Office**
42499 Sterling Highway 907-260-4200
Soldotna, AK 99669 Fax: 907-260-4205
http://www.dnr.state.ak.us/forestry
*Ric Plate, Area Forester*

**2679 Alaska Division of Forestry: Mat-Su/Southwest Area Office**
101 Airport Road 907-761-6300
Palmer, AK 99645 Fax: 907-761-6319
http://www.dnr.state.ak.us/forestry
*Ken Bullman, Area Forester*

**2680 Alaska Division of Forestry: Northern Region Office**
3700 Airport Way 907-451-2670
Fairbanks, AK 99709 Fax: 907-451-2690
http://www.dnr.state.ak.us/forestry
*Mark Eliot, Regional Forester*

**2681 Alaska Division of Forestry: State Forester's Office**
3700 Airport Way 907-451-2666
Fairbanks, AK 99709 Fax: 907-451-2690
http://www.dnr.state.ak.us/forestry
*Chris Maisch, State Forester*

**2682 Alaska Division of Forestry: Tok Area Office**
Mile 123 Glenn Highway 907-883-5134
Box 10 Fax: 907-883-5135
Tok, AK 99780
*Jeff Hermanns, Area Forester*

**2683 Alaska Division of Forestry: Valdez/Copper River Area Office**
Mile 110 Richardson Highway 907-822-5534
Box 185 Fax: 907-822-8600
Glennallen, AK 99588 http://www.dnr.state.ak.us/forestry
*Gary Mullen, Fire Management Officer*

**2684 Alaska Health Project**
218 E 4th Avenue 907-276-2864
Anchorage, AK 99517 Fax: 907-279-3089

**2685 Alaska Oil and Gas Conservation Commission**
333 W 7th Avenue 907-279-1433
Suite 100 Fax: 907-276-7542
Anchorage, AK 99501 E-mail: aogcc.customer.svc@alaska.gov
http://www. state.ak us./local/ak pages/ ADMIN/ogc/homeogc.html
Protecting the oil and gas of Alaska.
*John K Norman, Commissioner*

**2686 Alaska Resource Advisory Council**
Bureau of Land Management
222 W 7th Avenue 907-271-5555
Suite 13 Fax: 907-271-3684
Anchorage, AK 99513 http://www.blm.gov/ak/advisory.html
A statewide resource advisory council that advises BLM on land management issues for 80 million acres of federal public lands in Alaska. Membership is comprised of representatives from industry, conservation, recreation, Alaska Nativeorganizations, an elected offical, and the public at large. The council meets three times a year.

*Thomas P Lonnie, BLM/AK State Director*
*Sharon Wilson, Alaska RAC Coordinator*

**2687 Anchorage Office: Alaska Department of Environmental Conservation**
555 Cordova Street 907-269-7634
Anchorage, AK 99501 Fax: 907-269-3098
E-mail: Tom_Chapple@dec.state.ak.us
http://www.dec.state.ak.us
The people and industries that operate in our state have both the corporate conscience and the technical ability to work with us on constuctive solutions to basic environmental management and public health issues. We anticipate,collaborate, negotiate, educate and communicate to address the most important environmental and public health risks to Alaska and Alaskans. Investigation, legislation, regulation and litigation are available tools, but not the first tools of choice.
*Larry Hartig, Commissioner*

**2688 Cooperative Extension Service: University of Alaska Fairbanks**
PO Box 756180 907-474-7246
Fairbanks, AK 99775 Fax: 907-474-5139
E-mail: dffes@uaa.alaska.edu
http://www.uaf.edu/coop-ext
Mission: To interpret and extend relevant research-based knowledge in an understandable and usable form; and to encourage the application of this knowledge to solve the problems and meet the challenges that face the people of Alaska;and, to bring the concerns of the community back to the university.
*Pete Pinney, Interim Director*

**2689 Fairbanks Office: Alaska Department of Environmental Conservation**
610 University Avenue
Fairbanks, AK 99709
907-451-2143
Fax: 907-451-2155
http://www.dec.state.ak.us

**2690 Juneau Office: Alaska Department of Environmental Conservation**
410 Willoughby Avenue
Suite 303
Juneau, AK 99811
907-465-5100
Fax: 907-465-5129
http://www.dec.state.ak.us

**2691 Kenai Office: Alaska Department of Environmental Conservation**
43335 Kalifornsky Beach Road
Suite 11, Red Diamond Center
Soldotna, AK 99669
907-262-5210
Fax: 907-262-2294
http://www.dec.state.ak.us

**2692 Kodiak Office: Alaska Department of Environmental Conservation**
PO Box 515
Kodiak, AK 99615
907-486-3350
Fax: 907-486-5032
http://www.dec.state.ak.us

**2693 Natural Resources Department Public Affairs Information Office**
550 West 7th Avenue
Suite 1450
Anchorage, AK 99501
907-269-8463
Fax: 907-269-8931
http://www.dnr.state.ak.us/pic/dnrdirectory.htm
*Tom Irwin, Commissioner*

**2694 Palmer Office: Alaska Department of Environmental Conservation**
500 S Alaska Street
Suite A
Palmer, AK 99645
907-747-3236
http://www.dec.state.ak.us

**2695 Sitka Office: Alaska Department of Environmental Conservation**
901 Halibut Point Road
Suite 3
Sitka, AK 99835
907-747-8614
Fax: 907-747-7419
http://www.dec.state.ak.us

**2696 Subsistance Resource Commission Cape Krusenstern National Monument**
National Park Service
Box 1029
Kotzebue, AK 99752
907-442-3890
Fax: 907-442-8316
http://www.nps.gov/cakr

**2697 Subsistence Resource Gates of the Artic National Park**
National Park Service
National Park Service-Fairbanks HQ
4175 Geist Road
Fairbanks, AK 99709
907-457-5752
Fax: 907-455-0601
http://www.nps.gov/gaar

**2698 United States Department of the Army: US Army Corps of Engineers**
PO Box 6898
Elmendorf AFB, AK 99506
907-753-2522
http://www.usace.army.mil
Design and constructs military projects for the Army, Air Force, civil works and water resources development projects for coastal communities. Conducts military Real Estate transactions, is responsible for Emergency Operationsinvolving national emergency and natural disaster, and regulates development in navigable waters, and placement of fill material in waters and wetlands.

**2699 Valdez Office: Alaska Department of Environmental Conservation**
213 Meals Avenue, Room 17
Po Box 1709
Valdez, AK 99686
907-835-8012

## Arizona

**2700 Arizona Department of Agriculture: Animal Services Division**
1688 West Adams Street
Phoenix, AZ 85007
602-542-4373
http://www.azda.gov/ASD/asd.htm
Mission: Protect consumers from contagious and infectious disease in livestock, poultry, commercially raised fish, meat, milk, and eggs; enforce laws concerning the movement, sale, importation, transport, slaughter, and theft oflivestock.
*Donald Butler, Director*

**2701 Arizona Department of Environmental Quality**
1110 West Washington Street
Phoenix, AZ 85007
602-771-2300
800-234-5677
http://www.azdeq.gov/index.html
The Arizona Department of Environmental Quality was established in 1987 to preserve, protect and enhance the environmental and public health through the maintenance of air, land and water resources. The department oversees compliancewith state and federal environmental regulations and works with industry and local governments.
*Steve Owens, Director*

**2702 Arizona Game & Fish Department**
2221 West Greenway Road
Phoenix, AZ 85023
602-942-3000
http://www.gf.state.az.us
Aims to conserve, enhance and restore Arizona's diverse wildlife resources and habitats through aggressive protection and management programs, and to provide wildlife resources and safe watercraft and off-highway vehicle recreation forthe enjoyment, appreciation, and use by present and future generations.

**2703 Arizona Game & Fish Department: Region I**
2878 East White Mountain Blvd
Pinetop, AZ 85935
928-367-4281

**2704 Arizona Game & Fish Department: Region II**
3500 South Lake Mary Road
Flagstaff, AZ 86001
928-774-5045

**2705 Arizona Game & Fish Department: Region III**
5325 North Stockton Hill Road
Kingman, AZ 86409
928-692-7700

**2706 Arizona Game & Fish Department: Region IV**
9140 East 28th Street
Yuma, AZ 85365
928-342-0091

**2707 Arizona Game & Fish Department: Region V**
555 North Greasewood Road
Tucson, AZ 85745
520-628-5376

**2708 Arizona Game & Fish Department: Region VI**
7200 East University
Mesa, AZ 85207
480-981-9400

**2709 Arizona Geological Survey**
416 W Congress Street
Suite 100
Tucson, AZ 85701
520-770-3500
Fax: 520-770-3505
http://www.azgs.state.az.us
Our mission is to inform and advise the public about the geologic character of Arizona in order to foster understanding and prudent development of the State's land, water, mineral and energy resources.
*M Lee Allison, Director*

**2710 Arizona State Parks**
1300 West Washington
Phoenix, AZ 85007
602-542-4174
800-285-3703
E-mail: feedback@pr.state.az.us
http://www.pr.state.az.us
*William C Cordasco, Chair*

**2711 Arizona Strip District-US Department of Interior Bureau of Land Management**
345 E Riverside Drive 435-688-3200
St George, UT 84790 Fax: 435-688-3528
http://www.blm.gov/az/st/en/fo/arizona_strip_field.html
Manages nearly 2 million acres in northwestern Arizona, including the Vermilion Cliffs National Monument.
*Scott Florence, District Manager*
*Becky Hammond, Acting Field Manager*

**2712 Environmental and Analytical Chemistry Laboratory**
Arizona Department of Health 602-542-1188
250 North 17th Avenue Fax: 602-542-0760
Phoenix, AZ 85007 http://www.hs.state.az.us
State public health laboratory both in chemistry and microbiology. Supports investigations into environmental contamination by analyzing water, soil, air, hazardous materials, food and miscellaneous items for the presence of hazardousand toxic chemicals. Microbiology tests for pathogens and/or indicator organisms.
*Joseph J Soltis, Hazardous Material Manager*

**2713 Gila Box Riparian National Conservation Area BLM Safford District Office**
711 14th Avenue 928-348-4400
Safford, AZ 85546 Fax: 928-348-4450
http://www.az.blm.gov
There are more than 14 million acres of public lands in Arizona that people have put in our trust. It's an awesome responsiblity, and one that we take very seriously. We don't try to do it alone. Every day, we work with people to helpmake sure we are doing what is right for Arizona's envrionment, wildlife, culture, and history... for the people who rely upon the land to earn a living or to manufacture the things which make our lives a little easier...and most importantly, forArizona's future.
*Tom Schnell, Manager*

**2714 Phoenix District Advisory Council: BLM**
21605 North 7th Avenue 623-580-5500
Phoenix, AZ 85027 Fax: 623-580-5580
http://www.az.blm.gov
Manages public lands and resources in central Arizona, and supports related statewide initiatives and functions to sustain their health, diversity and productivity while providing for customer service and meeting public demandresulting from the expanding Phoenix metropolitan area and growth of adjoining communities.
*Teri A Raml, District Manager*

## Arkansas

**2715 Arkansas Department of Parks and Tourism**
One Capitol Mall 501-682-7777
Little Rock, AR 72201 http://www.arkansas.com

**2716 Arkansas Fish and Game Commission**
2 Natural Resources Drive 501-223-6300
Little Rock, AR 72205 800-364-4263
http://www.agfc.com
The Arkansas Game and Fish Commission plays an important role in keeping The Natural State true to its name.
*Scott Henderson, Director*
*Sonny Varnell, Chair*

**2717 Arkansas Natural Heritage Commission**
1500 Tower Building 501-324-9619
323 Center Street Fax: 501-324-9618
Little Rock, AR 72201 E-mail: arkansas@naturalheritage.org
http://www.naturalheritage.org
Mission: To identify and protect remaining high-quality natural communities and maintain information on the distribution and status of rare species that live within the state.
*Karen Smith, Director*
*Chris Colclasure, Deputy Director*

**2718 Arkansas State Plant Board**
1 Natural Resources Drive 501-225-1598
Little Rock, AR 72205 http://www.plantboard.org
Mission: To protect and serve the citizens of Arkansas and the agricultural and business communities by providing information and unbiased enforcement of laws and regulations thus ensuring quality products and services.
*Darryl Little, Director*

**2719 Department of Environmental Quality**
5301 Northshore Drive 501-682-0744
North Little Rock, AR 72118 http://www.adeq.state.ar.us
Mission: To protect, enhance and restore the natural environment for the well-being of Arkansas.
*Teresa Marks, Director*

## California

**2720 American Cetacean Society**
PO Box 1391 310-548-6279
San Pedro, CA 90733 Fax: 310-548-6950
E-mail: info@acsonline.org
http://www.acsonline.org
A non profit organization that is the oldest whale conservation group in the world. Founded to protect whales, dolphins, porpoises, and their habitats through public education, research grants, and conservation actions.

**2721 California Department of Education Office of Environmental Education**
1430 North Street 916-319-0800
Suite 5602 E-mail: www.cde.ca.gov
Sacramento, CA 95814
Mission: Guiding principles, goals, and objectives of the California Department of Education.
*Kenneth Noonan, President*
*Jack O'Connell, Superintendent*

**2722 California Department of Fish and Game**
1416 9th Street 916-445-0411
Sacramento, CA 95814 http://www.dfg.ca.gov
Manages California's diverse fish, wildlife, and plant resources, and the habitats upon which they depend, for their ecological values and for their use and enjoyment by the public.
*Ryan Broddrick, Director*
*John McCamman, Chief Deputy Director*

**2723 California Department of Water Resources**
1416-9th Street 916-653-5791
Room 1104-1 Fax: 916-653-3310
Sacramento, CA 95814 http://www.dwr.water.ca.gov
Mission: To manage the water resources of California in cooperation with other agencies, to benefit the State's people, and to protect, restore, and enhance the natural and human environments.
*Lester A Snow, Director*
*Mark Cowin, Deputy Director*

**2724 California Desert District Advisory Council Bureau of Land Management**
California State Office
22835 Calle San Juan De Los Lagos 951-697-5200
Moreno Valley, CA 82553 Fax: 951-697-5299
http://www.ca.blm.gov
Mission: To protect the natural, historic, recreation and economic riches, and scenic beauty of the California Desert.

**2725 California Environmental Protection Agency**
PO Box 2815 916-323-2514
Sacramento, CA 95812 E-mail: cepacomm@calepa.ca.gov
http://www.calepa.ca.gov

Mission: To restore, protect and enhance the environment, to ensure the public health, environmental quality and economic vitality.
*Linda S Adams, Secretary*

**2726 California Institute of Public Affairs**
PO Box 189040
Sacramento, CA 95818
916-442-2472
E-mail: info@cipahq.org
http://www.cipahq.org
A forum for policy dialogue and research on California and international environmental issues. Publishes the online World Directory of Environmental Organizations.
*Thaddeus C Trzyna, President*
*John Davidson, Senior Associate*

**2727 California Pollution Control Financing Authority**
915 Capitol Mall, Room 457
Sacramento, CA 95814
916-654-5610
Fax: 916-657-4821
E-mail: cpcfa@treasurer.ca.gov
http://www.treasurer.ca.gov/cpcfa

*Michael Paparian, Executive Director*
*Sherri Wahl, Deputy Executive Director*

**2728 Department of Agriculture: Forest Service, Pacific Southwest Region**
1323 Club Drive
Vallejo, CA 94592
707-562-8737
http://www.fs.fed.us/r5

**2729 Department of the Interior: National Parks Pacific West Region**
One Jackson Center
1111 Jackson Street Suite 700
Oakland, CA 94607
510-817-1300
http://www.nps.gov/legacy/regions.html

**2730 Energy Commission**
1516 9th Street
MS-29
Sacramento, CA 95814
916-654-4287
http://www.energy.ca.gov
*B B Blevins, Executive Director*

**2731 Environmental Protection Agency Region IX**
75 Hawthorne Street
San Francisco, CA 94105
415-947-8000
http://www.epa.gov/region09/
Region 9 covers Arizona, California, Hawaii, Nevada, the Pacific Islands subject to US law, and approximately 140 Tribal Nations. We work together with state, local, and tribal governments in the region to carry out the nationsenvironmental laws.
*Wayne Nastri, Regional Administrator*

**2732 Environmental Protection Office: Hazard Identification**
Environmental Health Hazard Assessment Office
301 Capitol Mall 2nd Floor
Room 205
Sacramento, CA 95814
916-445-6900
Fax: 916-327-1097
The mission of the Office of Environmental Health Hazard Assessment (OEHHA) is to protect and enhance public health and the environment by objective scientific evaluation of risks posed by hazardous substances.

**2733 Environmental Protection Office: Toxic Substance Control Department**
400 P Street 4th Floor
PO Box 806
Sacramento, CA 95812
916-323-9723
Fax: 916-323-3215
http://www.calepa.ca.gov
The Department's mission is to restore, protect and enhance the environment, to ensure public health, environmental quality and economic vitality by regulating hazardous waste, conducting and overseeing cleanups, and developing andpromoting pollution prevention.

**2734 Golden Gate National Recreation Area**
Fort Mason
Building 201
San Francisco, CA 94123
415-561-4700
http://www.nps.gov/goga
The Golden Gate National Recreation Area (GGNRA) is the largest urban national park in the world. The total park area is 74,000 acres of land and water. Approximately 28 miles of coastline line within its boundaries. It is nearly twoand one-half times the size of San Francisco.
*Brian O'Neill, General Superintendent*
*Mai-Liis Bartling, Deputy Superintendent*

**2735 Inter-American Tropical Tuna Commission**
8604 La Jolla Shores Drive
La Jolla, CA 92037
858-546-7100
Fax: 858-546-7133
E-mail: webmaster@iattc.org
http://www.iattc.org/homeeng.htm
The IATTC, established by international convention in 1950, is responsible for the conservation and management of fisheries for tunas and other species taken by tuna-fishing vessels in the eastern Pacific Ocean. The IATTC also hassignificant responsibilies for the implementation of the International Dolphin Conservation Program (IDCP), and provides the Secretariat for that program.
*Guillermo Compean, Director*

**2736 Klamath Fishery Management Council US Fish & Wildlife Service**
1829 South Oregon Street
Yreka, CA 96097
530-842-5763
Fax: 530-842-4517
http://www.fws.gov/yreka/kfmc.htm
*Phil Detrich, Supervisor*

**2737 Native American Heritage Commission**
915 Capitol Mall
Room 364
Sacramento, CA 95814
916-653-4082
Fax: 916-657-5390
E-mail: nahc@pacbell.net
http://www.nahc.ca.gov
The mission of the Native American Heritage Comm. is to provide protection to Native American burials from vandalism and inadvertent destruction, provide a procedure for the notification of most likely descendents regarding thediscovery of Native American human remains and associated grave goods, bring legal action to prevent severe and irreparable damage to sacred shrines, ceremonial sites, sanctified cemeteries and place of worship on pub. property, and maintain aninventory of sacred places.

*Larry Myers, Executive Secretary*

**2738 Pesticide Regulation, Environmental Monitoring and Pesticide Management**
1001 I Street
PO Box 4015
Sacramento, CA 95812
916-445-4300
Fax: 916-324-1452
http://www.cdpr.ca.gov
Mission: Protect human health and the environment by regulating pesticide sales and use, and by fostering reduced-risk pest management.
*Chris Reardon, Chief Deputy Director*

**2739 Resources Agency: California Coastal Commission**
45 Fremont Street
Suite 2000
San Francisco, CA 94105
415-904-5250
Fax: 415-904-5400
http://www.coastal.ca.govs
Mission: To protect, conserve, restore, and enhance environmental and human-based resources of the California coast and ocean for environmentally sustainable and prudent use by current and future generations.

**2740 Resources Agency: California Conservation Corps**
1719 24th Street
Sacramento, CA 95816
916-341-3100
800-952-5627
Fax: 916-323-8922
http://www.ccc.ca.gov
Engages young men and women in meaningful work, public service educational activities that assist them in becoming more responsible citizens, while protecting and enhancing California's environment, human resources and communities.
*David Muraki, Director*
*Lucia Becerra, Chief Deputy Director*

**2741 Resources Agency: State Coastal Conservancy**
1330 Broadway
13th Floor
Oakland, CA 94612
510-286-1015
Fax: 510-286-0470
E-mail: dwayman@scc.ca.gov
http://www.coastalconservancy.ca.gov
The Coastal Conservancy acts with others to preserve, protect and restore the resources of the California Coast.
*Quarterly Magazine*
*Samuel Schuchat, Executive Officer*
*Dick Wayman, Communications Director*

**2742 Southwestern Low-Level Radioactive Waste Commission**
1731 Howe Ave #611
Sacramento, CA 95825
916-448-2390
Fax: 916-720-0144
E-mail: swllrwcc@swllrwcc.org
http://www.swllrwcc.org
The Southwestern Low-Level Radioactive Waste Commission is the governing body for the Southwestern Low-Level Radioactive Waste Disposal Compact, consisting of Arizona, California, North Dakota, and South Dakota. Created by public law100-712 in 1988, its key duties include controlling the importation and exportation of low-level waste into and out of the region. The Commission has no authority over disposal facility siting, but can make recommendations and comments to ensure safedisposal.
*Kathy A David, Executive Director*

**2743 United States Department of Agriculture Research Education and Economics**
800 Buchanan Street
Albany, CA 94710
510-559-6060
Fax: 510-559-5779
http://www.ree.usda.gov
Mission: Dedicated to the creation of a safe, sustainable, competitive US food and fiber system and strong, healthy communities, families, and youth through integrated research, analysis and education.
*Dwayne R Buxton, Area Director*
*Gale A Buchanan, Undersecretary*

**2744 United States Department of the Army US Army Corps of Engineers**
1455 Market Street
San Francisco, CA 94103
415-503-6800
http://www.usace.army.mil
*Lt Col Craig W Kiley, Commander*

## Colorado

**2745 Bureau of Land Management**
Department of the Interior
2815 H Road
Grand Junction, CO 81506
970-244-3000
Fax: 970-244-3083
http://www.blm.gov/co/st/en.html
Manages 8.3 million acres of public lands in Colorado. These lands are managed for a multitude of uses including, but not limited to, recreation, mining, wildlife habitat and grazing. Along with these 8.3 million acres, BLM oversees27.3 million subsurface acres for mineral development.
*Catherine Robertson, Field Manager*

**2746 Bureau of Land Management: Little Snake Field Office**
Little Snake Field Office
455 Emerson Street
Craig, CO 81625
970-826-5000
Fax: 970-526-5002
http://www.blm.gov
Encompasses 4.2 million acres of federal, state and private lands in Moffat, Routt, and Rio Blanco counties.
*John Husband, Director*

**2747 Canon City District Advisory Council**
3170 E Main Street
Canon City, CO 81212
719-269-8500
Fax: 719-269-8599
http://www.co.blm.gov/ccdo/canon.htm

**2748 Cheyenne Mountain Zoological Park**
4250 Cheyenne Mountain Zoo Road
Colorado Springs, CO 80906
719-633-9925
Fax: 719-633-2254
E-mail: info@cmzoo.org
http://www.cmzoo.org
Mission: To foster an appreciation and respect for all living things. Actifely provide survival assistance for species in peril. Provide a high quality recreational experience. Be source of pride and economic strength.

**2749 Colorado Department of Agriculture**
700 Kipling Street
Suite 4000
Lakewood, CO 80215
303-239-4100
Fax: 303-239-4125
http://www.ag.state.co.us
Mission: To strengthen and advance Colorado's agriculture industry; ensure a safe, high quality, and sustainable food supply; and protect consumers, the environment, and natural resources.
*John R Stulp, Commisioner*

**2750 Colorado Department of Natural Resources**
1313 Sherman Street
Room 718
Denver, CO 80203
303-866-3311
800-536-5308
Fax: 303-866-2115
E-mail: feedback.dnr@state.co.us
http://www.dnr.state.co.us
*Harris D Sherman, Executive Director*
*Mike King, Deputy Director*

**2751 Colorado Department of Natural Resources: Division of Water Resources**
1313 Sherman Street
Room 818
Denver, CO 80203
303-866-3581
Fax: 303-866-3589
E-mail: askdwr@state.co.us
http://www.water.state.co.us
The Colorado Division of Water Resources is an agency of the State of Colorado, Department of Natural Resources, operating under the direction of specific state stautes, court decrees, and interstate compacts. The DWR is empowered toadminister all surface and ground water rights throughout the state and ensure that the doctrine of prior appropiation is enforced.

**2752 Colorado Department of Public Health Environment Consumer Protection Division**
4300 Cherry Creek Drive South
Denver, CO 80246
303-692-3620
Fax: 303-753-6809
http://www.cdphe.state.co.us/cp
The Consumer Protection Division assumes the responsiblity for protecting Colorado residents and visitors by prevention of a wide array of health hazards.

**2753 Colorado Department of Public Health and Environment**
4300 Cherry Creek Drive South
Denver, CO 80246
303-692-2000
800-886-7689
http://www.cdphe.state.co.us
Mission: Committed to protecting and preserving the health and environment of the people of Colorado.
*James B Martin, Executive Director*

**2754 Colorado State Forest Service**
Colorado State University
5060 Campus Delivery
Fort Collins, CO 80523-5060
970-491-6303
Fax: 970-491-7736
http://www.csfs.colostate.edu
Mission: To provide for the stewardship of forest resources and to reduce related risks to life, property and the environment for the benefit of present and future generations.

**2755 Environmental Protection Agency Region VIII (CO, MT, ND, SD, UT, WY)**
1595 Wynkoop Street
Denver, CO 80202
303-312-6312
800-227-8917
http://www.epa.gov/region8/
To restore and protect the ecological integrity of the mountains, plains and deserts and to protect the health of their inhabitants.

**2756 Governors Office of Energy, Management and Conservation: Colorado**
225 E 16th Avenue 303-866-2100
Suite 650 800-632-6662
Denver, CO 80203 Fax: 303-866-2930
E-mail: geo@state.co.us
http://www.state.co.us/oemc
Supports cost-effective programs, grants and partnerships that benefit Colorado's economic and natural environment.

**2757 Minerals Management Service/Minerals Revenue Management**
PO Box 25165 303-231-3162
Denver, CO 80225 http://www.mms.gov
Mission: To manage the ocean energy and mineral resources on the Outer Continental Shelf and Federal and Indian mineral revenues to enhance public and trust benefits, promote responsible use, and realize fair value.
*Randall B Luthi, Director*

**2758 Natural Resources Department: Air Quality Division**
Department of the Interior
PO Box 25287 303-969-2070
Denver, CO 80225 Fax: 303-969-2822
E-mail: christine_shaver@nps.gov

**2759 Natural Resources Department: Oil & Gas Conservation Commission**
1120 Lincoln Street Suite 801 303-894-2100
Denver, CO 80203 Fax: 303-894-2109
E-mail: dnr.ogcc@state.co.us
http://www.colorado.gov/cogcc
*David Neslin, Director*

**2760 Natural Resources Department: Wildlife Division**
6060 Broadway 303-291-7227
Denver, CO 80216 http://wildlife.state.co.us
Manages the state's 960 wildlife species. Regulates hunting and fishing activities by issuing licenses and enforcing regulations. Conducts research to improve wildlife management activities, provides technical assistance to private andother land owners concerning wildlife and habitat management and develops programs to protect and recover threatened and endangered species.
*Mark B Konishi, Acting Director*

**2761 Office of Surface Mining Reclamation & Enforcement**
1999 Broadway 303-844-1400
Suite 3320 Fax: 303-844-1546
Denver, CO 80202 http://www.wrcc.osmre.gov/
*Al Klein, Regional Director*

**2762 Rocky Mountain Low-Level Radioactive Waste Board**
1675 Broadway 303-825-1912
Suite 1400 Fax: 303-892-3882
Denver, CO 80202 E-mail: board@rmllwb.us
http://www.rmllwb.us
*Leonard Slosky, Executive Director*
*Sheri Reynolds, Administrator*

**2763 United States Forest Service: United States Department of Agriculture**
740 Simms Street 303-275-5350
Golden, CO 80401 http://www.fs.fed.us/r2

## Connecticut

**2764 Connecticut Department of Agriculture**
165 Capitol Avenue 860-713-2500
Hartford, CT 06106 Fax: 860-713-2514
E-mail: ctdeptag@po.state.ct.us
http://www.state.ct.us/doag
*Steven K Reviczky, Commissioner*

**2765 Connecticut Department of Environmental Protection**
Department of Environmental Protection
79 Elm Street 860-424-3000
Hartford, CT 06106 E-mail: dep.webmaster@po.state.ct.us
http://dep.state.ct.us
Mission: To conserve, improve and protect the natural resources and environment of the State of Connecticut in such a manner as to encourage the social and economic development of Connecticut while preserving the natural environmentand the life forms it supports in a delicate, interrelated and complex balance, to the end that the state may fulfill its responsibility as trustee of the environment for present and future generations.

**2766 Connecticut Department of Public Health**
410 Capitol Avenue 860-509-8000
PO Box 340308 http://www.dph.state.ct.us
Hartford, CT 06134
Has long recognized the adverse public health impact of environmental sources of lead in many of Connecticut's childern. Established dedicated staff to evaluate these environmental sources and began funding local programs in the1970's. The Childhood Lead Posioning Prevention Program has continued to be active in addressing this issue by implementing additional state and community programs, especially in towns that have been identified as high risk.
*J Robert Galvin, Commissioner*
*Norma Gyle, Deputy Commissioner*

## District of Columbia

**2767 District of Columbia State Extension Services**
4200 Connecticut Avenue 202-274-7115
Building 352, Suite 322 Fax: 202-274-7130
Washington, DC 20008

## Delaware

**2768 Delaware Association of Conservation Districts**
PO Box 242 302-739-9921
Dover, DE 19903 Fax: 302-739-6724
http://www.nacdnet.org/about/districts/directory/de.phtml
Coordinates the three state conservation districts.
*Martha Pileggi, Staff Assistant*

**2769 Delaware Cooperative Extension**
University of Delaware
Townsend Hall 302-831-2501
Newark, DE 19716 Fax: 302-831-6758
E-mail: jseitz@udel.edu
http://ag.udel.edu/extension/
Contact person Janice A Seitz's title is Associate Dean for Extension and Outreach Director of Extension College of Agriculture and Natural Resources.
*Janice A Seitz, Director*

**2770 Delaware Department of Agriculture**
2320 S DuPont Highway 302-698-4500
Dover, DE 19901 http://dda.delaware.gov
As part of the state government, the department's mission is to sustain and promote the viability of food, fiber and agricultural industries in Delaware through quality services that protect and enhance the environment, health andwelfare of the general public.
*Michael T Scuse, Secretary*
*Harry D Shockley, Deputy Secretary*

**2771 Delaware Department of Natural Resources and Environmental Control**
DNREC
89 Kings Highway 302-739-9902
Dover, DE 19901 Fax: 302-739-6242
http://www.dnrec.state.de.us

Protects and manages the state's vital natural resources, protects public health and safety, provides quality outdoor recreation, and serves and educates the citizens of the First State about the wise use, conservation and enhancementof Delaware's environment.

**2772 Delaware Sea Grant Program**
University of Delaware 302-831-8083
Newark, DE 19716 E-mail: marinecom@udel.edu
http://www.ocean.udel.edu/seagrant

**2773 Mid-Atlantic Fishery Management Council**
300 S New Street 302-674-2331
Room 2115 Fax: 302-674-5399
Dover, DE 19904 E-mail: info@mafmc.org
http://www.mafmc.org
The Mid-Atlantic Fishery Management Council is responsible for management of fisheries in federal waters which occur predominantly off the mid-Atlantic coast.

*Daniel T Furlong, Executive Director*

**2774 United States Department of the Interior United States Fish and Wildlife Service**
Delaware Bay Estuary Project
2610 Whitehall Neck Road 302-653-9152
Smyrna, DE 19977 Fax: 302-653-9421
http://www.fws.gov/delawarebay/
Delaware Bay Estate Project is a field office of the US Fish & Wildlife service's coastal program.
*Gregory Breese, Project Leader*

## Florida

**2775 Department of Commerce National Oceanic & Atlantic Oceanographic & Meteorological Laboratory**
4301 Rickenbacker Causeway 305-361-4450
Miami, FL 33149 E-mail: webmaster@aoml.noaa.gov
http://www.aoml.noaa.gov
*Robert M Atlas, Director*

**2776 Fish & Wildlife Conservation Commission**
620 South Meridian Street 850-488-4676
Tallahassee, FL 32399 http://www.ststc.fl.us/fwc
Mission: Managing fish and wildlife resources for their long-term well-being and the benefit of the people.
*Rodney Barreto, Chair*

**2777 Florida Department of Agriculture & Consumer Service**
The Capitol, Pl 10 850-488-3022
400 South Monroe Street Fax: 850-488-7585
Tallahassee, FL 32399 http://www.doacs.state.fl.us
Mission: To safeguard the public and support Florida's agriculture economy by: ensuring the safety and wholesomeness of food and other consumer products through inspection and testing programs; protecting consumers from unfair anddeceptive business practices and providing consumer information; assisting Florida's farmers and agriculture industries with the production and promotion of agriculture products; and conserving and protecting the state's agriculture and naturalresources.
*Charles H Bronson, Commissioner*

**2778 Florida Department of Environmental Protection**
3900 Commonwealth Boulevard 850-245-2118
M.S. 49 Fax: 850-245-2128
Tallahassee, FL 32399 E-mail: citizensservices@dep.state.flu.us
http://www.dep.state.fl.us
Mission: To promote the efficient and effective operation of the Agency consistent with its Administratice and statutory responsibilities.
*Cynthia Kelly, Director, Div of Admin Svcs*
*Herschel Vinyard, Jr., Secretary*

**2779 Florida State Department of Health**
2585 Merchants Row Blvd 850-245-4444
Tallahassee, FL 32399 http://www.doh.state.fl.us
Mission: To promote and protect the health and safety of all people in Florida through the delivery of quality public health services and the promotion of health care standards.
*Ana M Viamonte Ros, Secretary*

**2780 Gulf of Mexico Fishery Management Council**
2203 North Lois Avenue 813-348-1630
Suite 1100 888-833-1844
Tampa, FL 33607 Fax: 813-348-1711
E-mail: gulfcouncil@gulfcouncil.org
http://www.gulfcouncil.org
The Gulf of Mexico Fishery management Council is one of eight regional Fishery Management Councils which were established by the Fishery conservation and Management Act in 1976 (now called the Magnuson-Stevens Fishery Conservation andMagnuson Act). The Council prepares fishery plans which are designed to manage fishery resources from where state waters end to the 200 mile limit of the Gulf of mexico.
*Wayne E Swingle, Executive Director*
*Rick Leard, Deputy Executive Director*

**2781 Lee County Parks & Recreation**
3410 Palm Beach Boulevard 239-461-7400
Fort Myers, FL 33916 Fax: 239-461-7450
http://www.leeparks.org/
Our mission is to provide safe, clean and functional Parks & Recreation facilities; to provide programs and services that add to the quality of life for all Lee County residents and visitors; to enhance tourism through special eventsand attractions. We are committed to fulfilling this mission through visionary leadership, individual dedication and the trustworthy use of available resources.
*John Yarbrough, Director*
*Barbara Manzo, Deputy Director*

**2782 Natural Resources Department: Recreation & Parks Division**
3900 Commonwealth Boulevard 850-245-2157
Tallahassee, FL 32399 http://www.dep.state.fl.us/parks
Mission: To provide resource-based recreation while preserving, interpreting and restoring natural and cultural resources. Our goal is to help create a sense of place by showing park visitors the best of Florida's diverse natural andcultural heritage sites.
*Mike Bullock, Director*

**2783 Southwest Florida Water Management District**
2379 Broad Street 352-796-7211
Brookville, FL 34604 Fax: 352-754-6885
http://www.watermatters.org
Manages the water and water-related resources within its boundaries. Maintains balance between the water needs of current and future users while protecting and maintaining the natural systems that provide the District with its existingand future water supply. The Conservation Projects Section, in the Resource Conservation and Development Department, is reponsible for managing water conservation, reclaimed water and other alternative source projects, and estimating future waterdemands.

## Georgia

**2784 Blackbeard Island National Wildlife Refuge**
1000 Business Center Drive 912-652-4415
Suite 10 Fax: 912-652-4385
Savannah, GA 31405 http://www.fws.gov/blackbeardisland
This Georgia barrier island's 5,618 acres includes maritime forest, saltmarsh, freshwater marsh, and beach habitat, 3,000 of which has been set aside as National wilderness of variety of recreational activities are availableyear-round, including wildlife observation, birdwatching, hiking and beachcombing.

**2785 Board of Scientific Counselors: Agency for Toxic Substance and Disease Registry**
61 Forsyth Street SW 404-562-1788
7th Floor Room 7T90 Fax: 404-562-1790
Atlanta, GA 30303 E-mail: ATSDRIC@cdc.gov
http://www.atsdr.cdc.gov

*Beverly Samuels, Secretary*
*Ben Moore, Regional Representative*

**2786 Georgia Department of Agriculture**
19 Martin Luther King Jr Drive SW 404-656-3645
Atlanta, GA 30334 E-mail: tirvin@agr.state.ga.us
http://www.agr.georgia.gov
Mission: To provide excellence in services and regulatory functions, to protect and promote agriculture and consumer interests, and to ensure an abundance of safe food and fiber for Georgia, America and the world by usingstate-of-the-art technology and a professional workforce.
*Tommy Irvin, Commissioner*

**2787 Georgia Department of Education**
2054 Twin Towers East 404-656-2800
Atlanta, GA 30334 Fax: 404-651-6867
http://www.doe.k12.ga.us
To function as a service oriented and policy driven agency that meets the needs of local school systems as they go about the business of preparing all students for college or a career in a safe and drug free environment where we ensurethat no chils is left behind.
*Kathy Cox, CEO*

**2788 Georgia Department of Natural Resources: Historic Preservation Division**
34 Peachtree Street NW 404-656-2840
Suite 1600 Fax: 404-651-8739
Atlanta, GA 30303 E-mail: ray luce@dnr.state.ga.us
http://www.gashpo.org
To promote the preservation and use of historic places for a better Georgia.
*Ray Luce, Director*

**2789 Georgia Department of Natural Resources: Pollution Prevention Assistance Division**
7 Martin Luther King Jr Drive 404-651-5120
Suite 450 800-685-2443
Atlantic, GA 30334 Fax: 404-651-5130
E-mail: info@p2ad.org
http://www.p2ad.org

*Bob Donaghue, Director*
*David Gipson, Assistant Director*

**2790 Georgia Sea Grant College Program**
University of Georgia
220 Marine Science Building 706-542-6009
Athens, GA 30602 Fax: 706-542-3652
http://alpha.marsci.uga.edu/gaseagrant.html
Goal is to better understand the complex interactions between the physical, chemical, biological and geological processes that are manifested in the area where land and sea come together, and to make that knowledge available and usefulto Georgia's citizens. Sea Grant strives to deepen our understanding of coastal and estuarine ecology, the critical role of fresh water interaction and to expand our knowledge of action beyond the marshes and estuaries and into the life of the riversand streams.
*Keith Gates, Marine Advisory Serv Leader*

**2791 National Center for Environmental Health**
4770 Buford Highway NE 770-488-7030
Mail Stop F-29 888-232-6789
Atlanta, GA 30341 Fax: 770-488-7042
E-mail: ncehinfo@cehod1.em.cdc.gov
http://www.cdc.gov/nceh
Mission: Plans, directs, and coordinates a national program to maintain and improve the health of the American people by promoting a healthy environment and by preventing premature death and avoidable illness and disability caused bynon-infectious, non-occupational and related factors.

**2792 Natural Resource Department**
2 Martin Luther King Jr Drive SE 404-656-3500
Suite 1252 East Tower http://www.state.ga.us/dnr/
Atlanta, GA 30334
The mission of the Department of Natural Resources is to sustain, enhance, protect and conserve Georgia's natural, historic and cultural resources for present and future generations, while recognizing the importance of promoting thedevelopment of commerce and industry that utilize sound environmental practices.
*Noel Holcomb, Commissioner*

**2793 Natural Resources Department: Air Protection**
4244 International Parkway 404-363-7000
Suite 120 Fax: 404-363-7100
Atlanta, GA 30354 E-mail: james.capp@dnr.state.ga.us
http://www.gaepd.org
The Air Protection Branch helps provide Georgia's citizens with clean air and works closely with other branches of Georgia's Environmental Protection Division to assure compliance with environmental laws so that, in addition to clearair, we have clean water, healthy lives and productive land.
*James A Capp, Branch Chief*

**2794 Natural Resources Department: Coastal Resources Division**
1 Conservation Way 912-264-7218
Brunswick, GA 31520 Fax: 912-262-3143
http://crd.dnr.state.ga.us

*Susan Shipman, Director*
*Kathy Herrin, Administrative Assistant*

**2795 Natural Resources Department: Environmental Protection Division**
2 Martin Luther King Jr Drive 404-657-5947
Suite 1152 East Tower 888-373-5947
Atlanta, GA 30334 http://gaepd.org
Provides Georgia's citizens with clean air, clean water, healthy lives and productive land by assuring compliance with environmental laws and by assisting others to do their part for a better environment.
*Carol Couch, Director*

**2796 United States Department of the Army US Army Corps of Engineers**
US Army Engineer Distric 912-652-5822
PO Box 889 http://www.sas.usace.army.mil
Savannah, GA 31402
Mission: To provide quality, responsive engineering services to the nation including: planning, designing, building and operating water resources and other civil works projects; designing and managing the construction of militaryfacilities for the Army and Air Force; and providing design and construction management support for other defense and federal agencies.
*Robert L Van Antwerp, Commander/Chief of Engineers*
*Ronald L Johnson, Deputy Commander*

**2797 Wassaw National Wildlife Refuge**
1000 Business Center Drive 912-652-4415
Suite 10 http://www.fws.gov/wassaw
Savannah, GA 31405
The most primitive of Georgia's barrier islands, the 10,053 acre refuge, includes beaches with rolling dunes, live oak and slash pine woodlands, and vast salt marshes. The island supports rookeries for egrets and herons, and a varietyof leading birds are abundant in the summer months. Wassaw also provides prime nesting habitat for the loggerhead seaturtles. Refuge visitors may enjoy recreational activities such as birdwatching, beachocombing, hiking, and general nature studies.

## Hawaii

**2798 Agriculture Department**
1428 South King Street
Honolulu, HI 96814
808-973-9560
E-mail: hdoainfo@exec.state.hi.vs
http://www.hawaiiag.gov/hdoa/

*Sandra Lee Kunimoto, Chair*

**2799 College of Tropical Agriculture and Human Resources**
University of Hawaii
3050 Maile Way
Gilmore 202
Honolulu, HI 96822
808-956-8131
Fax: 808-956-9105
http://www2.ctahr.hawaii.edu/extout.asp

Mission: Committed to the preparation of students and all citizens of Hawaii for life in the global community through research and educational programs supporting tropical agriculture systems that foster viable communities, adiversified economy, and a healthy environment.

*Andrew G Hashimoto, Dean/Director*

**2800 Department of Land and Natural Resources Division of Water Resource Management**
1151 Punchbowl Street
Room 227
Honolulu, HI 96813
808-587-0214
Fax: 808-587-0219
E-mail: dlnr.cwrm@hawaii.gov
http://www.hawaii.gov/dlnr/cwrm

*Laura H Thielen, Interim Chairperson*

**2801 Environmental Center**
University of Hawaii
2500 Dole Street
Krauss Annex 19
Honolulu, HI 96822
808-956-7361
Fax: 808-956-3980
E-mail: envctr@hawaii.edu
http://www.hawaii.edu/envctr

The Center's three areas of focus are education, research and service. The education function of the Center includes the administration of the Environmental Studies Major Equivalent and Certificate program. It fulfills its researchfunction by identifying and addressing environmentally related research needs, particularly those pertinent to Hawaii. The service function primarily involves the coordination and transfer of technical information from the University community togovernment agencies.

*James Moncur, Director*

**2802 Hawaii Department of Agriculture**
1428 South King Street
Honolulu, HI 96814
808-973-9560
http://www.hawaii.gov/hdoa

Contains devisions such as: Administrative; Animal Industry; Marketing; Measurement Standards; and Plant Industry. Carries out programs to conserve, develop and utilize the agricultural resources of the state. Enforces laws, andformulates and enforces rules and regulation to further control the management of these resources.

*Sandra Lee Kunimoto, Chairperson*

**2803 Hawaii Institute of Marine Biology University of Hawaii**
PO Box 1346
Kane'ohe, HI 96744
808-236-7401
Fax: 808-236-7443
http://www.hawaii.edu/HIMB

*Jo-Ann C Leong, Director*

**2804 Health Department: Environmental Quality Control**
235 S Beretania Street
Room 702
Honolulu, HI 96813
808-586-4185
Fax: 808-586-4186
E-mail: oeqc@doh.hawaii.gov
http://hawaii.gov/health/environmental/oeqc/index/html

The office is tasked to implement Chapter 343, Hawaii Revised Statues and Title 11, Chapter 200. This is a systematic process to ensure consideration is given to the environmental consequences of actions proposed within our state. Thereview process offers many opportunities to prevent environmental degradation and protect human communities through decreased citizen involvement and informed decision making.

*Gary L. Hooser, Director*

**2805 State of Hawaii: Department of Land andNatural Resources**
Kalanimoku Building
1151 Punchbowl Street
Honolulu, HI 96813
808-587-0400
Fax: 808-587-0390
E-mail: dlnr@hawaii.gov
http://www.hawaii.gov/dlnr

State agency.

*William J Aila Jr, Chairperson*

**2806 Water Resources Research Center University of Hawaii**
2540 Dole Street
Holmes Hall 283
Honolulu, HI 96822
808-956-7847
Fax: 808-956-5044
E-mail: morav@hawaii.edu
http://www.wrrc.hawaii.edu

The Water Resources Research Center was organized under the federal Water Resources Research Act of 1964. The Center is supported by university funds, external grants, contracts and a small annual federal grant. WRRC faculty coverthe areas of engineering, hydrology, microbiology, ecology, economics and zoology. Cooperating faculty come from numerous other disciplines. WRRC is open to consideration of any question related to water supply or water quality.

*James Moncur, Director*

## Idaho

**2807 Idaho Association of Soil Conservation Districts**
39 W Pine Ave
Ste B20
Meridian, ID 83642
208-338-5900
Fax: 208-338-9537
E-mail: kent.foster@agri.idaho.gov
http://www.iascd.state.id.us

Provides action at the local level for promoting wise and beneficial conservation of natural resources with emphasis on soil and water.

*J Kent Foster, Executive Director*

**2808 Idaho Cooperative Extension**
1000 West Hubbard
Suite 145
Coeur D'Alene, ID 83814
208-292-2522
Fax: 208-292-2535
http://www.uidaho.edu/extension/

**2809 Idaho Department of Environmental Quality: Pocatello Regional Office**
444 Hospital Way #300
Pocatello, ID 83201
208-236-6160
888-655-6160
Fax: 208-236-6168
http://www.state.id.us/deq

Mission: To protect human health and preserve the quality of Idaho's air, land, and water for use and enjoyment today and in the future.

*Toni Hardesty, Director*

**2810 Idaho Department of Environmental Quality: State Office**
1410 North Hilton
Boise, ID 83706
208-373-0502
Fax: 208-373-0417
http://www.deq.idaho.gov

Mission: To protect human health and preserve the quality of Idaho's air, land, and water for use and enjoyment today and in the future.

*Toni Hardesty, Director*

**2811 Idaho Department of Environmental Quality: Idaho Falls Regional Office**
900 N Skyline
Suite B
Idaho Falls, ID 83402
208-528-2650
800-232-4635
Fax: 208-528-2695
http://www.state.id.us/deq

Mission: To protect human health and preserve the quality of Idaho's air, land, and water for use and enjoyment today and in the future.
*Toni Hardesty, Director*

**2812 Idaho Department of Fish & Game: Clearwater Region**
3316 16th Street 208-799-5010
Lewiston, ID 83501 Fax: 208-799-5012
http://wwwfishandgame.idaho.gov
Mission: All wildlife, including all wild animals, wild birds, and fish, within the state of Idaho, is hereby declared to be the property of the state of Idaho.

**2813 Idaho Department of Fish & Game: Headquarters**
600 S Walnut Street 208-334-3700
PO Box 25 Fax: 208-334-2114
Boise, ID 83707 http://www.fishandgame.idaho.gov
Mission: All wildlife, including all wild animals, wild birds, and fish, within the state of Idaho, is hereby declared to be the property of the state of Idaho.

**2814 Idaho Department of Fish & Game: Magic Valley Region**
319 South 417 East 208-324-4359
Jerome, ID 83338 Fax: 208-324-1160
http://www.fishandgame.idaho.gov
Mission: All wildlife including all wild animals, wild birds, and fish, within the state of Idaho, is hereby declared to be the property of the state of Idaho.

**2815 Idaho Department of Fish & Game: McCall**
555 Deinhard Lane 208-634-8137
McCall, ID 83638 Fax: 208-634-4320
http://www.fishandgame.idaho.gov
Mission: All wildlife, including all wild animals, wild birds, and fish, within the state of Idaho, is hereby declared to be the property of the state of Idaho.

**2816 Idaho Department of Fish & Game: Panhandle Region**
2885 West Kathleen Avenue 208-769-1414
Coeur d'Alene, ID 83815 Fax: 208-769-1418
http://www.fishandgame.idaho.gov
Mission: All wildlife, including all wild animals, wild birds, and fish, within the state of Idaho, is hereby declared to be the property of the state of Idaho.

**2817 Idaho Department of Fish & Game: Salmon Region**
99 Highway 93 N 208-756-2271
PO Box 1336 Fax: 208-756-6274
Salmon, ID 83467 http://www.fishandgame.idaho.gov
Mission: All wildlife, including all wild animals, wild birds, and fish, within the state of Idaho, is hereby declared to be the property of the state of Idaho.

**2818 Idaho Department of Fish & Game: Southeast Region**
1345 Barton Road 208-232-4703
Pocatello, ID 83204 Fax: 208-233-6430
http://www.fishandgame.idaho.gov
Mission: All wildlife, including all wild animals, wild birds, and fish, within the state of Idaho, is hereby declared to be the property of the state of Idaho.

**2819 Idaho Department of Fish & Game: Southwest Region**
3101 S Powerline Road 208-465-8465
Nampa, ID 83686 Fax: 208-465-8467
http://www.fishandgame.idaho.gov
Mission: All wildlife, including all wild animals, wild birds, and fish, within the state of Idaho, is hereby declared to be the property of the state of Idaho.

**2820 Idaho Department of Fish & Game: Upper Snake Region**
4279 Commerce Circle 208-525-7290
Idaho Falls, ID 83401 Fax: 208-523-7604
http://www.fishandgame.idaho.gov
Mission: All wildlife, including all wild animals, wild birds, and fish, within the state of Idaho, is hereby decalred to be the property of the state of Idaho.

**2821 Idaho Department of Lands**
300 N 6th Street, Suite 103 208-334-0200
Po Box 83720 Fax: 208-334-3698
Boise, ID 83720-0050 E-mail: public_records_request@idl.idaho.gov
http://www.idl.idaho.gov
Mission: To manage endowment trust lands to maximize long-term financial returns to the benficiary institutions and provide protection to Idaho's natural resources.
*George Bacon, Director*
*Kathy Opp, Deputy Director*

**2822 Idaho Department of State Parks and Recreation**
PO Box 83720 208-334-4199
Boise, ID 83720 Fax: 208-334-5232
http://www.idahoparks.org
Manages 27 state parks. We also run the registration program for snowmobiles, boats and off-highway vehicles. Money from registrations and other sources goes to develop and maintain trails, facilities and programs statewide for thepeople who use those vehicles.
*Robert L Meinen, Director*

**2823 Idaho Department of Water Resources**
322 E Front Street 208-287-4800
Boise, ID 83720 Fax: 208-287-6700
E-mail: IDWRInfo@idwr.idaho.gov
http://www.idwr.idaho.gov/about/
Working for a controlled development and wise management of Idaho's resources. Documents and reports on topics of public interest such as drought, salmon, wilderness and the Snake River Basin.

*David R Tuthill Jr, Director*

**2824 Idaho Geological Survey**
University of Idaho
Morrill Hall 208-885-7991
3rd Floor Fax: 208-885-5826
Moscow, ID 83844 E-mail: igs@uidaho.edu
http://www.idahogeology.org
*Roy M Breckenridge, Director*
*Kurt L Othberg, Director*

**2825 Idaho State Department of Agriculture**
2270 Old Penitentiary Road 208-332-8500
Boise, ID 83712 Fax: 208-334-2170
E-mail: info@agri.idaho.gov
http://www.agri.idaho.gov

*Celia R Gould, Director*
*Brian Oakey, Deputy Director*

**2826 Lands Department: Soil Conservation Commission**
Po Box 790 208-332-8650
Boise, ID 83701 http://www.scc.state.id.us/scc_facts.htm
Responsibilities of the Commission are: organize Districts and provide assistance, coordination, information and training to Disrict supervisors; ensure that Districts function legally and properly as local subdivisions of stategovernment; administer general funds appropriated by the Idaho Legislature to Districts so they can install resource conservation practices and provide technical assistance personnel to Districts administering water quality projects and conductingsoil surveys.
*Jerry Nicolescu, Administrator*

**2827 United States Department of the Interior Bureau of Land Management**
3948 Development Avenue 208-384-3300
Boise, ID 83705 Fax: 208-384-3493

## Illinois

### 2828 Association of Illinois Soil and Water Conservation Districts

4285 North Walnut Street 217-744-3414
Springfield, IL 62707 Fax: 217-744-3420
E-mail: sherry.finn@aiswcd.org
http://www.aiswcd.org

To foster and promote charitable and educational purposes designed to further the principles of soil conservation and stewardship, water conservation and energy conservation. Provides, conducts and sponsors programs to aidindividuals, groups, organizations, government bodies, association and all entities in combating soil erosion and energy water waste.

*Richards W Nichols, Executive Director*

### 2829 Construction Engineering Research Laboratory

US Army Engineer Research and Development Center
PO Box 9005 217-352-6511
Champaign, IL 61826 800-USA-CERL
E-mail: Dana.L.Finney@erdc.usace.mil
http://www.cecer.army.mil

CERL conducts research to support sustainable military installations. Research is directed toward increasing the Army's ability to more efficiently construct, operate and maintain its installations and ensure environmental quality andsafety at a reduced life-cycle cost. Excellent facilities support the Army's training, readiness, mobilization and sustainability missions.

*Ilker Adiguzel, Director*

### 2830 Department of Natural Resources: Division of Education

Illinois Department of Natural Resources
One Natural Resources Way 217-524-4126
Springfield, IL 62702 E-mail: dnr.teachkids@illinois.gov
http://www.dnr.state.il.us/education/index.htm

Responsible for the development and dissemination of educational programs and materials and for training in their use. The website provides contests for students, loan materials, education materials, and grant information, in additionto graduate program information, podcasts and workshops. Overall, an excellent resource for education professionals, parents, and students.

*Valerie Keener, Administrator*

### 2831 Environmental Protection Agency Bureau of Water

Water Bureau
4500 South Sixth Street Road 217-786-6892
Springfield, IL 62706 http://www.epa.state.il.us/water

Mission: To ensure that Illinois' rivers, streams and lakes will support all uses for which they are designated including protection of aquatic life, recreation and drinking water supplies; ensure that every illinois Public Watersystem will provide water that is consistently safe to drink; and protect Illinois' groundwater resource for designated drinking water and other beneficial uses.

### 2832 Environmental Protection Agency: Region 5

77 West Jackson Boulevard 312-353-2000
Chicago, IL 60604 800-621-8431
http://www.epa.gov/region5

*Phillippa Cannon, Media Relations Leader*

### 2833 Illinois Conservation Foundation

One Natural Resources Way 217-785-2003
Springfield, IL 62702 Fax: 217-785-8405
E-mail: www.ilcf.org/contact_us/
http://www.ilcf.org

Established by law, the ILCF is a volunteer group with a 13-member Board, chaired by the Director of the IL Dept of Natural Resources. The role of the ILCF and its partners is to preserve and enhance our precious natural resources bysupporting and fostering ecological, educational, and recreational programs for the benefit of all citizens of Illinois and for future generations.

*Marc Miller, Director*

### 2834 Illinois Department of Agriculture Bureau of Land and Water Resources

PO Box 19281 217-782-2172
Springfield, IL 62794 Fax: 217-785-4505
http://www.agr.state.il.us

*Charles A Hartke, Director*

### 2835 Illinois Department of Transportation

2300 S Dirksen Parkway 217-782-7820
Springfield, IL 62764 E-mail: Monseurmj@nt.dot.state.il.us
http://www.dot.state.il.us

Provides cost-effective, safe and efficient transportation for the people who live, work, visit and do business in Illinois, and ensures that the system supports the state's economic growth.

*Ann L Schneider, Director*
*Vincent E Rangel, Deputy Director*

### 2836 Illinois Nature Preserves Commission

One Natural Resources Way 217-785-8686
Springfield, IL 62702 Fax: 217-785-2438
E-mail: kelly.neal@illinois.gov
http://www.dnr.state.il.us/inpc

To assist private and public landowners in protecting high quality natral areas and habitats of endangered and threatened species in perpetuity, through voluntary dedication or registration of such lands into the Illinois NaturePreserves Systems. Offering programs in Defence, Protection, and Stewardship.

*Deborah Stone, Director*
*Kelly Neal, Stewardship Project Manager*

### 2837 United States Department of the Army US Army Corps of Engineers

US Army Engineer District
Clock Tower Building 309-794-4200
PO Box 2004 Fax: 309-794-5793
Rock Island, IL 61204 http://www.usace.army.mil

## Indiana

### 2838 Indiana Department of Natural Resources

402 West Washington Street 317-232-4020
Indianapolis, IN 46204 Fax: 317-232-8036
http://www.ai.org/dnr

Protects, enhances, preserves, and wisely uses natural, cultural, and recreational resources for the benefit of Indiana's citizens through professional leadership, management and education.

*Robert E Carter, Director*
*Todd Tande, Deputy Director*

### 2839 Indiana State Department of Agriculture, Soil Conservation

101 West Ohio 317-232-8770
Suite 1200 Fax: 317-232-1362
Indianapolis, IN 46204 http://www.in.gov/isda

Mission: To facilitate the protection and enhancement of Indiana's land and water.

*Jerod Chew, Director*

### 2840 Indiana State Department of Health

2 N Meridan Street 317-233-1325
Indianapolis, IN 46204 E-mail: gwilson@isdh.state.in.us
http://www.state.in.us/isdh

Mission: To support Indiana's economic prosperity and quality of life by promoting, protecting and providing for the health of Hoosiers in their communities.

*Judith A Monroe, State Health Commissioner*
*Mary L Hill, Dep State Health Comm*

**2841 Natural Resources Department: Fish & Wildlife**
402 West Washington Street RMW273 317-232-4080
Indianapolis, IN 46204 Fax: 317-232-8150
E-mail: dfw@dnr.in.gov
http://www.in.gov/dnr/fishwild
Mission: To professionally manage Indiana's fish and wildlife for present and future generations, balancing ecological, recreational, and economic benefits.

## Iowa

**2842 Iowa Association of County Conservation Boards**
405 SW 3rd Street 515-963-9582
Suite 1 Fax: 515-963-9582
Ankeny, IA 50023 E-mail: iaccb@ecity.net
http://www.ecity.net/iaccb
IACCB is a nonprofit organization assisting member county conservation boards in areas of board member education, public relations and legislation. The association's main purposes are to promote the objectives and supplement theactivities of conservation boards, exchange information, assist boards and members in program development and provide a unified voice in the legislature. IACCB is governed by a nine-member board elected by member counties.

**2843 Iowa Department of Agriculture, and Land Stewardship Division of Soil Conservation**
502 E 9th 515-281-5851
Wallace State Office Building Fax: 515-281-6170
Des Moines, IA 50319 http://www.iowaagriculture.org
The Division of Soil Conservation is responsible for state leadership in the protection and management of soil, water and mineral resources, assisting soil and water conservation districts and private landowners to meet theiragricultural and environmental protection needs.
*Chuck Gipp, Director*
*Karen Fynaardt, Administrative Assistant*

**2844 Iowa Department of Natural Resources Administrative Services Division**
502 E 9th Street 515-281-5918
Wallace Office Building Fax: 515-281-8895
Des Moines, IA 50319 http://www.iowadnr.com
Aims to manage, protect, conserve and develop Iowa's natural resources in cooperation with other public and private organizations and individuals, so that the quality of life for Iowans is significantly enhanced by the use, enjoymentand understanding of those resources.

**2845 Iowa State Extension Services**
1032 Wallace Road Office Building 515-294-6192
Ames, IA 55011 Fax: 515-294-4715
E-mail: jlpease@iastate.edu
http://www.extension.iastate.edu/
*Dr James L Pease, Specialist*

**2846 Natural Resource Department**
502 E 9th Street 515-281-5385
Des Moines, IA 50319 Fax: 515-281-8895
http://www.iowadnr.com
Mission: To conserve and enhance our natural resources in cooperation with individuals and organizations to improve the quality of life for Iowans and ensure a legacy for future generations.
*Richard Leopold, Director*
*Liz Christiansen, Deputy Director*

## Kansas

**2847 Emporia Research and Survey Office Kansas Department of Wildlife & Parks**
1830 Merchant
PO Box 1525 620-342-0658
Emporia, KS 66801 E-mail: randys@wp.state.ks.us
http://www.kdwp.state.ks.us

**2848 Environmental Protection Agency: Region 7, Air & Toxics Division**
901 North 5th Street 913-551-7003
Kansas City, KS 66101 Fax: 913-551-7066
http://www.epa.gov/region7/
Responsible for management of programs for air, hazardous waste, toxic substances, radiation and pollution prevention in Iowa, Kansa, Missouri and Nebraska as required by the following legistlation: The Clean Air Act, The ResourceConservation and Recovery Act, the Toxic Substances Control Act and the Emergency planning and Community Right-to Know Act.
*Patrick Bustos, Director*
*Hattie Thomas, Deputy Director*

**2849 Health & Environment Department: Air & Radiation**
100 SW Jackson 785-296-1593
Suite 310 Fax: 785-296-8464
Topeka, KS 66612 E-mail: jmitchell@kdheks.gov
http://www.kdheks.gov/bar
Mission: To protect the public from the harmful effects of radiation and air pollution and conserve the natural resources of the state by preventing damage to the environment from releases of radioactive materials or air contaminants.
*John Mitchell, Director*

**2850 Health & Environment Department: EnvironmentDivision**
1000 SW Jackson Street 785-296-1535
Suite 400 Fax: 785-296-8464
Topeka, KS 66612 E-mail: jmitchell@kdheks.gov
http://www.kdheks.gov/environment
The mission of the Division of Environment is the protection of the public health and environment. The division conducts regulatory programs involving public water supplies, industrial discharges, wastewater treatment systems, solidswaste landfills, hazardous waste, air emissions, radioactive materials, asbestos removal, refined petroleum storage tanks and other sources which impact the environment.
*John Michtell, Director*

**2851 Health & Environment Department: Waste Management**
1000 SW Jackson 785-296-1600
Suite 320 Fax: 785-296-8909
Topeka, KS 66612 http://www.kdheks.gov/waste
Regulates landfills, HHW, Hazardous Waste Permitting, Solid Waste Permitting, Public Outreach, Illegal Dumps.
*William Bider, Uirector*
*F JJ, 06262001*

**2852 Kansas Cooperative Fish & Wildlife Research Unit**
Kansas State University
205 Leasure Hall 785-532-6070
Manhattan, KS 66506 Fax: 785-532-7159
E-mail: kscfwru@ksu.edu
http://www.k-state.edu/kscfwru
Unit Research contributes to understanding ecological systems within the Great Plains. Unit staff, collaborators, and graduate students conduct research with both natural and altered systems, particularly those impacted by agriculture.Unit projects investigate ways to maintain a rich diversity of endemic wild animals and habitats while meeting the needs of people. The Unit focuses on projects that involve graduate students, and the research needs of cooperators are given priority.
*Philip Gipson, Leader*
*Jack Cully Jr, Assistant Leader*

**2853 Kansas Corporation Commission Conservation Division**
Finney State Office Building 316-337-6200
130 S Market, Room 2078 Fax: 316-337-6211
Wichita, KS 67202 http://www.kcc.state.ks.us

State of Kansas oilfield regulatory agency. The KCC is responsible for the preservation of Kansas' hydro and carbo resources, protection of corrullative right and the prevention and remediation of oil field pollution.
*Doug Louis, Director*

**2854 Kansas Department of Health & Environment**
1000 SW Jackson Street
Topeka, KS 66612
785-296-1500
Fax: 785-368-6368
E-mail: info@kdhe.state.ks.us
http://www.kdhe.state.ks.us
An organization dedicated to optimizing the promotion and protection of the health of Kansas through efficient and effective public health programs and services and through preservation, protection and remediation of natural resourcesof the environment.
*Roderick L Bremby, Secretary*
*Aaron Dunkel, Deputy Secretary*

**2855 Kansas Department of Wildlife & Parks Region 2**
3300 SW 29th
Topeka, KS 66614
785-273-6740
Fax: 785-273-6757
http://www.kdwp.state.ks.us
Manages and promotes the wildlife and natural resources of Kansas. Administered by a secetary of Wildlife and Parks and is advised by a seven-member Wildlife and Parks Commission.

**2856 Kansas Department of Wildlife & Parks Region 3**
1001 McArtor Drive
Dodge City, KS 67801
620-227-8609
Fax: 620-227-8600
http://www.kdwp.state.ks.us

**2857 Kansas Department of Wildlife & Parks Region 4**
6232 E 29th Street N
Wichita, KS 67220
316-683-8069
http://www.kdwp.state.ks.us

**2858 Kansas Department of Wildlife & Parks Region 5**
1500 W 7th Street
Po Box 777
Chanute, KS 66720
620-431-0380
Fax: 620-431-0381
http://www.kdwp.state.ks.us
This region is made up of 18 counties in the southeastern corner of the state. This area is dominated by the Osage Questas physiographic region, which is characterized by rolling grasslands, limestone bluffs, and heavily timberedbottomlands.

**2859 Kansas Department of Wildlife and Parks**
512 SE 25th Avenue
Pratt, KS 67124
620-672-5911
Fax: 620-672-2972
http://www.kdwp.state.ks.us

**2860 Kansas Geological Survey**
University of Kansas
1930 Constant Avenue
Lawrence, KS 66047
785-864-3965
Fax: 785-864-5317
http://www.kgs.ku.edu
Conducts geological studies and research and collects, correlates, preserves and disseminates information leading to a better understanding of the geology of Kansas, with special emphasis on nautral resources of economic value, waterquality and quantity and geologic hazards. This information is published in books and maps both technical and educational and also provides computer programs and data bases derived from geologic investigations.
*William E Harrison, Director & State Geologist*

**2861 Kansas Health & Environmental Laboratories**
Forbes Field
Building 740
Topeka, KS 66620
785-291-3162
Fax: 785-296-1641
http://www.kdhe.state.ks.us/labs
Provides timely and accurate analytical information for public health benefit in Kansas and assures the quality of statewide laboratory sevices though certification and improvement programs.
*Dennis L Dobson, Acting Director*

**2862 Kansas Water Office**
901 S Kansas Avenue
Topeka, KS 66612
785-296-3185
888-KAN-WATE
E-mail: kfreed@kwo.state.ks.us
http://www.kwo.org
Works to achieve proactive solutions for resource issues of the state and to ensure good quality water to meet the needs of the people and the environment of Kansas. Evaluates and develops public policies, coordinating the waterresource operations of agencies at all levels of government.
*Tracy Streeter, Director*
*Ken Grotewiel, Assistant Director*

**2863 Pratt Operations Office Kansas Department of Wildlife & Parks**
512 SE 25th Avenue
Pratt, KS 67124
620-672-5911
E-mail: kenb@wp.state.ks.us
http://www.kdwp.state.ks.us

## Kentucky

**2864 Attorney General's Office Civil and Environmental Law Division**
700 Capitol Avenue
Capitol Building, Suite 118
Frankfort, KY 40601
502-696-5300
http://ag.ky.gov/civil/uninsured.htm
*Jack Conway, Attorney General*
*Margaret Everson, Asst Deputy Attorney General*

**2865 Department for Energy Development & Independence**
500 Mero Street
12th Floor Capital Plaza Tower
Frankfort, KY 40601
502-564-7192
Fax: 502-564-7484
E-mail: amanda.cook@ky.gov
http://www.energy.ky.gov
Provides leadership to maximize the benefits of energy effeciency and alternate energy through awareness, technology development, energy preparedness and new partnerships and resources.
*John Davies, Deputy Commissioner*

**2866 Department for Environmental Protection**
300 Fair Oaks Lane
Frankfort, KY 40601
502-564-2150
Fax: 502-564-4245
http://www.dep.ky.gov
Mission: To protect and enhance Kentucky's environment. This mission is important because it has a direct impact on Kentucky's public health, our citizens' safety and the quality of Kentucky's valuable natural resources-ourenvironment.
*Robert W Logan, Commisioner*

**2867 Division of Mine Reclamation and Enforcement**
Two Hudson Hollow Road
Frankfort, KY 40601
502-564-2340
Fax: 502-564-5848
E-mail: j.hamon@ky.gov
http://www.dmre.ky.gov
The Division of Mine Reclamation and Enforcement is responsible for inspecting all surface and underground coal mining permits in the state to assure compliance with the 1977 Federal Surface Mining Control Act. The DMRE is alsoresponsible for regulating and enforcing the surface mining reclamation laws for non-coal mining sites in the state, including limestone, sand, gravel, clay, shale and the surface effects of dredging river sand and gravel.

**2868 Economic Development Cabinet: Community Development Department Brokerage Division**
Old Capitol Annex
300 West Broadway
Frankfort, KY 40601
502-564-7140
800-626-2930
Fax: 502-564-3256
http://www.thinkkentucky.com
*John Hindman, Secretary*

**2869 Environmental Protection Department: Waste Management Division**
Fort Boone Plaza 502-564-6716
14 Reilly Road Fax: 502-564-4049
Frankfort, KY 40601 E-mail: waste@ky.gov
http://www.waste.ky.gov
Mission: To protect human health and the environment by minimizing adverse impacts on all citizens of the commonwealth through the development and implementation of fair, equitable and effective waste management programs.
*Bruce Scott, Director*
*Tony Hatton, Assistant Director*

**2870 Environmental Protection Department: Water Division**
Fort Boone Plaza 502-564-3410
14 Reilly Road Fax: 502-564-0111
Frankfort, KY 40601 E-mail: water@ky.gov
http://www.water.ky.gov
*David Morgan, Director*
*Sandy Gruzesky, Assistant Director*

**2871 Fish and Wildlife Resources Department: Fisheries Division**
1 Sportsman's Lane 502-564-3596
Frankfort, KY 40601 800-858-1549
Fax: 502-564-6501
E-mail: info.center@ky.gov
http://www.fw.ky.gov
Mission: To conserve and enhance fish and wildlife resources and provide opportunity for hunting, fishing, trapping, boating and other wildlife related activities.
*John Gassett, Commissioner*
*Benjy T Kinman, Fisheries Director*

**2872 Kentucky Department for Public Health**
275 E Main Street 502-564-3970
Franfort, KY 40621 http://chfs.ky.gov/dph
*William D Hacker, Commissioner*

**2873 Kentucky Environmental and Public Protection Cabinet**
500 Metro Street 502-564-3350
Capital Plaza Tower, 5th Floor Fax: 502-564-3354
Frankfort, KY 40601 E-mail: environment@ky.gov
http://www.environment.ky.gov
Provides a safe, clean environment in the Commonwealth, while working with business and industry to help ensure adequate jobs and a strong economy.
*Teresa J Hill, Secretary*

**2874 Kentucky State Cooperative Extension Services**
University of Kentucky
S-107 Agricultural Science Bldg N 859-257-4302
Lexington, KY 40546 Fax: 859-257-3501
http://www.ca.uky.edu/ces

**2875 Kentucky State Nature Preserves Commission**
801 Schenkel Lane 502-573-2886
Frankfort, KY 40601 Fax: 502-573-2355
E-mail: naturepreserves@ky.gov
http://www.naturepreserves.ky.gov
Aims to protect Kentucky's natural heritage by identifying, acquiring and managing natural areas that represent the best known occurrences of rare native species and natural communities and working together to protect biologicaldiversity.
*Don Dott, Director*

**2876 Natural Resources Department: Conservation Division**
375 Versailles Road 502-573-3080
Frankfort, KY 40601 Fax: 502-573-1692
E-mail: steve.coleman@ky.gov
http://www.conservation.ky.gov
Assists Kentucky's local conservation districts in the development and implementation of sound soil and water conservation programs to manage, enhance, and promote the wise use of the Commonwealth's natural resources.
*Stephen A Coleman, Director*

**2877 Natural Resources Department: Division of Forestry**
627 Comanche Trail 502-564-4496
Frankfort, KY 40601 Fax: 502-564-6553
E-mail: gwen.holt@ky.gov
http://www.forestry.ky.gov
*Gwen Holt, Contact*

**2878 Natural Resources and Environment Protection Cabinet: Environmental Quality Commission**
14 Reilly Road 505-564-3410
Frankfort, KY 40601 E-mail: eqc@ky.gov
http://www.eqc.ky.gov

**2879 Tourism Cabinet: Parks Department**
2400 Capital Plaza Tower 22nd Floor 502-564-4930
500 Metro Street http://www.kentuckytourism.com/stateparks
Frankfort, KY 40601

## Louisiana

**2880 Agriculture & Forestry: Soil & Water Conservation**
Louisiana Department of Agriculture and Forestry
PO Box 3554 225-922-1269
Baton Rouge, LA 70821 Fax: 225-922-2577
http://www.ldaf.state.la.us
*Bob Odom, Commissioner*

**2881 Central Interstate Low-Level Radioactive Waste Commission**
1033 "O" Street 402-476-8247
Suite 636 Fax: 402-476-8205
Lincoln, NE 68508 E-mail: rita@cillrwcc.org
http://www.cillrwcc.org
*Rita Houskie, Administrator*

**2882 Culture, Recreation and Tourism**
PO Box 94361 225-342-8115
Baton Rouge, LA 70802 Fax: 225-342-3207
http://www.crt.state.la.us
*Angele Davis, Secretary*
*Dawn Romero Watson, Undersecretary*

**2883 Department of Natural Resources: Office of Mineral Resources**
PO Box 2827 225-342-4615
Baton Rouge, LA 70821 Fax: 225-342-4527
E-mail: OMR@dnr.state.la.us.us
http://www.dnr.louisiana.gov/MIN/
Provides staff support to the State Mineral Board in granting and administering leases on state-owned lands and waterbottoms for the production and development of minerals, primarily oil and gas, for the purpose of optimizing revenueto the state from the royalties, bonuses and rentals generated therefrom.
*Stacey Talley, Deputy Assistant Secretary*

**2884 Louisana Department of Natural Resources**
PO Box 94396 225-342-2710
PO Box 94396 Fax: 225-342-5861
Baton Rouge, LA 70804 http://www.dnr.state.la.us
*Scott Angelle, Secretary*
*Lori LeBlanc, Deputy Secretary*

**2885 Louisiana Cooperative Extension Services**
PO Box 25100 225-578-6083
Baton Rouge, LA 70894 Fax: 225-578-4225
http://www.lsu.agcenter.com
*Paul Coreil, Director*

**2886 Louisiana Department of Natural Resources Office of Coastal Restoration and Management**
625 N 4th Street
Baton Rouge, LA 70802
225-342-7308
Fax: 225-342-9417
E-mail: philp@dnr.state.la.us
http://dnr.louisiana.gov/crm

Develops, implements and monitors costal vegetated wetland restoration, creation and conservation measures. Preforms engineering, planning and monitoring functions essential to successful development and implementation of wetlandconservation and restoration plans and projects as directed by the Costal Wetlands Conservation and Restoration Plan.
*Gerald M Duszynski, Acting Assistant Secretary*

**2887 Natural Resources: Conservation Office**
617 N 3rd Street
PO Box 94275
Baton Rouge, LA 70804
225-342-5540
Fax: 225-342-3705
http://www.dnr.louisiana.gov

Regulatory oil and gas agency, State of Louisiana.
*James H Welsh, Commissioner of Conservation*
*Gary P Ross, Asst. Commissioner, Conserva*

**2888 Natural Resources: Injection & Mining Division**
PO Box 94275
Baton Rouge, LA 70804
225-342-5515
Fax: 225-342-3441

Has the responsibility for implementation of major environmental programs statutorily charged to the Office of Conservation. Administers a regulatory and permit program to protect underground sources of drinking water fromendangerment; is responsible for regulating exploration, development and surface mining operations for coal and lignite; and protection of state and private lands.
*Joseph S Ball Jr, Director*

## Maine

**2889 Maine Cooperative Fish & Wildlife Research Unit**
University of Maine
USGS Biological ResourcesDiscipline
5755 Nutting Hall
Orono, ME 04469
207-581-2862
Fax: 207-581-2858
http://www.wle.umaine.edu

Mission: To facilitate and strengthen professional education and training of fisheries and wildlife scientists; carry out research programs of aquatic, mammalian, and avian organisms and their habitats; and disseminate research resultsthrough the appropriate media, especially peer-review scientific articles.

*Dr Cyndy Loftin, Leader*

**2890 Maine Department of Environmental Protection: Augusta**
17 State House Station
Augusta, ME 04333
207-287-7688
800-452-1942
http://www.maine.gov/dep

Responsible for environmental protection and regulation in the state of Maine. Engages in a wide range of activities, makes reccomendations to the Legistlature regarding measures to minimize and eliminate environmental pollution,grants licenses, initiates enforcement actions, and provides information and technical assistance.
*David P Littell, Commissioner*

**2891 Maine Department of Conservation**
22 State House Station
Augusta, ME 04333
207-287-2211
Fax: 207-287-2400
http://www.maine.gov/doc

*Bill Beardsley, Commissioner*

**2892 Maine Department of Conservation: Ashland Regional Office**
45 Radar Road
Ashland, ME 04732
207-435-7963
Fax: 207-435-7184
http://www.maine.gov/doc

**2893 Maine Department of Conservation: Bangor Regional Office**
106 Hogan Road
BMHI Complex
Bangor, ME 04401
207-941-4014
Fax: 207-941-4222
http://www.maine.gov/doc

**2894 Maine Department of Conservation: Bolton Hill Regional Office**
2870 North Belfast Avenue
Augusta, ME 04330
207-624-3700
Fax: 207-287-8534
http://www.maine.gov/doc

**2895 Maine Department of Conservation: Bureau of Parks & Lands**
22 State House Station
2nd & 1st Floors
Augusta, ME 04333
207-287-3821
Fax: 207-287-3823
http://www.maine.gov/doc

**2896 Maine Department of Conservation: Entomology Laboratory**
50 Hospital Street
Augusta, ME 04330
207-287-2431
Fax: 207-287-2432
http://www.maine.gov/doc

**2897 Maine Department of Conservation: Farmington Regional Office**
25 Maine Street
PO Box 327
Farmington, ME 04938
207-778-8231
Fax: 207-778-5932
http://www.maine.gov/doc

**2898 Maine Department of Conservation: Greenville Regional Office**
Lake View Drive
PO Box 1107
Greenville, ME 04441
207-695-3721
Fax: 207-695-2380
http://www.maine.gov/doc

**2899 Maine Department of Conservation: Hallowell Regional Office**
1 Beech Street
Stevens Complex
Hallowell, ME 04330
207-624-6080
Fax: 207-624-6081
http://www.maine.gov/doc

**2900 Maine Department of Conservation: Jonesboro Regional Office**
Route 1A
PO Box 130
Jonesboro, ME 04648
207-434-2627
Fax: 207-434-2624
http://www.maine.gov/doc

**2901 Maine Department of Conservation: Land Use Regulation Commission**
22 State House Station
4th Floor
Augusta, ME 04333
207-287-2631
Fax: 207-287-7439
http://www.maine.gov/doc

The Maine Land Use Regulation Commission meets monthly in various locations throughout the state to discuss jurisdiction-related issues and to act upon pending cases.

**2902 Maine Department of Conservation: Millinocket Regional Office**
191 Main Street
East Millinocket, ME 04430
207-746-2244
Fax: 207-746-2243
http://www.maine.gov/doc

**2903 Maine Department of Conservation: Old Town Regional Office**
Airport Road
PO Box 415
Old Town, ME 04468
207-827-1818
Fax: 207-827-6295
http://www.maine.gov/doc

**2904 Maine Department of Conservation: Rangeley Regional Office**
Route 4
PO Box 887
Rangeley, ME 04970
207-864-5064
Fax: 207-864-5252
http://www.maine.gov/doc

**2905 Maine Department of Environmental Protection: Presque Isle**
1235 Central Drive
Skyway Park
Presque Isle, ME 04769
207-764-0477
888-769-1053
Fax: 207-760-3143
http://www.maine.gov/dep

**2906 Maine Department of Environmental Protection: Portland**
312 Canco Road
Portland, ME 04103
207-822-6300
888-769-1036
Fax: 207-822-6303
http://www.maine.gov/dep

**2907 Maine Inland Fisheries & Wildlife Department**
284 State Street Station 41SHS
41 State House Station
Augusta, ME 04333
207-287-8000
Fax: 207-287-6395
E-mail: ifw.webmaster@maine.gov
http://www.maine.gov/ifw
The Department of Inland Fisheries & Wildlife was established to ensure that all species of wildlife and aquatic resources in the State of Maine are maintained and perpetuated for their intrinsic and ecological values, for theireconomic contribution, and for their recreational, scientific and educational use by the people of the State. The Department is also responsible for the establishment and enforcement of rules and regulations.
*Roland Martin, Commissioner*
*Paul F Jacques, Deputy Commissioner*

**2908 Maine Natural Areas Program**
93 State House Station
Augusta, ME 04333
207-287-8044
Fax: 207-287-8040
E-mail: maine.nap@maine.gov
http://www.mainenaturalareas.org
Mission: To ensure the maintenance of Maine's natural heritage for the benefit of present and future generations. MNAP facilitates informed decision-making in development planning, conservation, and natural resources management. TheProgram's success relies upon using consistent and objective methods to collect, organize, and interpret information.
*Molly Docherty, Director*

**2909 Maine Sea Grant College Program**
University of Maine
5784 York Complex
The University of Maine
Orono, ME 04469
207-581-1435
Fax: 207-581-1426
E-mail: umseagrant@maine.edu
http://www.seagrant.umaine.edu
*Paul Anderson, Director*

**2910 University of Maine Cooperative Extension Forestry & Wildlife Office**
5755 Nutting Hall
Room 105
Orono, ME 04469
207-581-2892
Fax: 207-581-3466
http://www.umext.maine.edu/
*Catherine Elliott, Wildlife Specialist*
*Les Hyde, Extension Educator*

## Massachusetts

**2911 Connecticut River Salmon Association**
103 E Plumtree Road
Sunderland, MA 01375
413-548-8002
Fax: 413-548-9746
E-mail: info@ctriversalmon.org
http://www.ctriversalmon.org
To support the effort to restore Atlantic salmon in the Connecticut River basin. An invitation is extended to explore the site to learn more about our organization and the work being done to reestablish a species that had been extinctin this region since about 1800.

**2912 Department of Agricultural Resources**
251 Causeway Street
Suite 500
Boston, MA 02114
617-626-1700
Fax: 617-626-1850
E-mail: dwebber@state.ma.us
http://www.mass.gov/agr/animalhealth/index.htm
The Bureau of Animal Health focuses its efforts on ensuring the health and safety of the Commonwealth's domestic animals. Through diligent inspection, examination, licensing, quarantine, and enforcement of laws, regulations and ordersand the provision of technical assistance the Bureau promotes the welfare of companion and food-producing animals in Massachusetts.

**2913 Department of Environmental Protection**
One Winter Street
Boston, MA 02108
617-292-5500
Fax: 617-556-1049
http://www.mass.gov/dep
The Department of Environmental Protection is the state agency responsible for ensuring clean air and water, the safe management of toxics and hazards, the recycling of solid and hazardous wastes, the timely cleanup of hazardous wastesites and spills, and the preservation of wetlands and coastal resources.
*Arleen O'Donnell, Acting Commissioner*

**2914 Department of Fish & Game**
251 Causeway Street
Suite 400
Boston, MA 02114
617-626-1500
http://www.mass.gov/dfwele
The Department of Fish & Game works to preserve the state's natural resources and people's right to conservation of those resources, as protected by Article 97 of the Massachusetts Constitution. To carry out this mission, theDepartment exercises responsibility over the Commonwealth's marine and freshwater fisheries, wildlife species, plants, and natural communities, as well as the habitats that support them.
*Mary B Griffin, Commissioner*

**2915 EPA: Region 1, Air Management Division**
1 Congress Street
Suite 1100
Boston, MA 02114
617-565-3800
http://www.epa.gov/region01
The region's Air Enforcement program is responsible for conducting various compliance monitoring and enforcement activities under the Clean Air Act that govern the operation of stationary pollution sources. In addition, we are chargedwith the responsibility of overseeing the performance of state air pollution enforcement program.

**2916 Environmental Affairs Bureau of Markets**
251 Causeway Street
Suite 500
Boston, MA 02114
617-626-1750
Fax: 617-626-1850
http://www.massdfa.org/agricult.html

**2917 Environmental Affairs: Hazardous Waste Facilities Site Safety Council**
100 Cambridge Street
Boston, MA 02202
617-727-6629

**2918 Environmental Protection Agency Region 1 (CT, ME, MA, NH, RI, VT)**
1 Congress Street
Suite 1100
Boston, MA 02114
617-918-1111
Fax: 617-918-1112
http://www.epa.gov/region1
The mission of the US Environmental Protection Agency is to protect human health and to safeguard the natural environment- air, water and land- upon which life depends.

**2919 Executive Office of Energy & Environmental Affairs**
Executive Office of Environmental Affairs
100 Cambridge Street
9th Floor
Boston, MA 02114
617-626-1000
Fax: 617-626-1181
E-mail: env.internet@state.ma.us
http://www.mass.gov/envir/eoea.htm
*Ian Bowles, Secretary*
*Kathleen Baskin, Director of Water Policy*

**2920 Massachusetts Highway Department**
10 Park Plaza
Suite 3170
Boston, MA 02116
617-973-7800
Fax: 617-973-8040
E-mail: feedback@mhd.state.ma.us
http://www.mhd.state.ma.us

*Luisa Paiewonsky, Commissioner*

**2921 New England Interstate Water Pollution Control Commission**
Boott Mills South
116 John Street
Lowell, MA 01852
978-323-7929
Fax: 978-323-7919
E-mail: mail@neiwpcc.org
http://www.neiwpcc.org

The New England Interstate Water Pollution Control Commission, a nonprofit interstate agency established by an Act of Congress, serves and assist its members states individually and collectively by providing coordination, publiceducation, training and leadership in the management and protection of water quality in the New England Region and New York.
*Ronald F Poltak, Executive Director*
*Susan Sullivan, Deputy Director*

**2922 United States Department of the Army US Army Corps of Engineers**
696 Virginia Road
Concord, MA 01742
978-318-8220
Fax: 978-318-8821
http://www.usace.army.mil

**2923 Waquoit Bay National Estuarine Research Reserve**
149 Waquoit Highway
PO Box 3092
Waquoit, MA 02536
508-457-0495
Fax: 617-727-5537
http://www.waquoitbayreserve.org/employ.htm

In 1979, The Commonwealth of Massachusetts designated Waquoit Bay as an Area of Critical Environment Concern (ACEC) in recognition of its significant natural resources. The designation provides a state-wide umbrella of protection andoversight under the exisiting regulations of different state agencies which include higher standards of protection for ACEC's. The Waquoit Bay ACEC includes parts of the Bay proper. Although the ACEC covers 2522 acres, including Washburn Island andSouth Cape Beach.
*Brendan Annett, Reserve Manager*

## Maryland

**2924 Chesapeake Bay Executive Council**
Chesapeake Bay Program Office
410 Severn Avenue, Suite 109
Annapolis, MD 21403
410-267-5700
800-908-7229
Fax: 410-267-5777
http://www.chesapeakebay.net/exec.htm

The Chesapeake Bay Executive Council establishes the policy direction for the restoration and protection of the Chesapeake Bay and its living resources. A series of Directives, Agreements, and Amendments signed by the ExecutiveCouncil set goals and guide policy for the Bay restoration. The Council meets annually.

*Edward Rendell, Chairman*

**2925 Interstate Commission on the Potomac River Basin**
51 Monroe Street
Suite PE-08
Rockville, MD 20850
301-984-1908
Fax: 301-984-5841
E-mail: info@icprb.org
http://www.potomacriver.org

The Interstate Commission on the Potomac River Basin is an interstate compact agency established to help protect the Potomac River and its 14,670-square-mile watershed. Its mission is to enhance, protect, and conserve the water andassociated land resources of the Potomac River and its tributaries through regional and interstate cooperation.

*Joseph Hoffman, Executive Director*
*Curtis Dalpra, Communications Manager*

**2926 Maryland Department of Agriculture**
50 Harry S Truman Parkway
Annapolis, MD 21401
410-841-5700
800-492-5590
Fax: 410-841-5914
http://www.mda.state.md.us

Established on the basis of agriculture's growing importance and impact to the economy of the state. Many activities are regulatory in nature, others are assigned to a category of public service and some are educational or promotionalin scope. All are intended to provide the maximum protection possible for the consumer as well as promote the economic well-being of farmers, food and fiber processors and businesses engage in agricultural related operations.
*Earl F Hance, Secretary*

**2927 Maryland Department of Agriculture: State Soil Conservation Committee**
50 Harry S Truman Parkway
Annapolis, MD 21401
410-841-5863
Fax: 410-841-5736
http://www.mda.state.md.us

*Bruce Yerkes, Chair*

**2928 Maryland Department of Health and Mental Hygiene**
201 W Preston Street
Baltimore, MD 21201
410-767-6860
877-463-3464
Fax: 410-767-6489
http://www.dhmh.state.md.us/

The Maryland Department of Health and Mental Hygiene's mission is to protect and promote health and prevent disease and injury. This is accomplished through the provision of population-based health services and core public health;assessment, assurance and policy development.
*John M Colmers, Secretary*

**2929 Maryland Department of the Environment: Water Management Field Office**
160 South Water Street
Frostburg, MD 21532
301-689-1486
Fax: 301-689-6543
E-mail: mdecambr@intercom.net
http://www.mde.state.md.us

To restore and maintain the quality of the State's ground and surface waters, protect wetland habitats throughout the State, and manage the utilization of Maryland's mineral resources.
*Scott Boylan, Program Chief*

**2930 Maryland Department of the Environment/Water Management: Nontidal Wetlands & Waterways**
District Court/Multi Service Bldg
201 Baptist Street, Suite 22
Salisbury, MD 21801
410-713-3682
Fax: 410-713-3683
E-mail: salisb@shore.intercom.net
http://www.mde.state.md.us

**2931 Maryland Department of the Environment: Air & Radiation Management Field Office**
201 Baptist Street
Suite 15
Salisbury, MD 21801
410-713-3680
Fax: 410-713-3681
E-mail: awilliams@mde.state.md.us
http://www.mde.state.md.us

*Jay Bozman, Manager*

**2932 Maryland Department of the Environment: Air and Radiation Management Main Field Office**
160 South Water Street
Frostburg, MD 21532
301-689-5756
Fax: 301-689-6544
E-mail: frostbur@hereintown.net
http://www.mde.state.md.us

**2933 Maryland Department of the Environment: Field Operations Office**
416 Chinquapin Round Road
Annapolis, MD 21401
443-482-2700
http://www.mde.state.md.us

*John Steinfort, Manager*

**2934 Maryland Department of the Environment: Main Office**
1800 Washington Blvd
Baltimore, MD 21230
410-537-3000
800-633-6101
E-mail: mdeprf@olg.com
http://www.mde.state.md.us
To protect and restore the quality of Maryland's air, land, and water resources, while fostering economic development, healthy and safe communities, and quality environmental education for the benefit of the environment, public health,and future generations.

**2935 Maryland National Capital Park & Planning Commission**
6611 Kenilworth Avenue
Riverdale, MD 20737
301-454-1740
Fax: 301-454-1750
E-mail: webmanager@mncppc.org
http://www.mncppc.org
Is a bi-county agency empowered by the State of Maryland to acquire, develop, maintain and administer a regional system of parks withing Montgomery and Prince George's Counties, and to prepare and administer a general plan for thephysical development of the two counties.

**2936 NOAA Chesapeake Bay Office**
410 Severn Avenue
Chesapeake Bay Office
Annapolis, MD 21403
410-267-5660
Fax: 410-267-5666
http://noaa.chesapeakebay.net
NOAA Chesapeake Bay Office works to help protect and restore the Chesapeake Bay through its programs in fisheries management, habitat restoration, coastal observations, and education, and represents NOAA in the Chesapeake Bay Program.

## Michigan

**2937 Great Lakes Environmental Research Laboratory**
2205 Commonwealth Boulevard
Ann Arbor, MI 48105
734-741-2235
Fax: 734-741-2055
http://www.glerl.noaa.gov
Conducts integrated, interdisciplinary environmental research in support of resource management and environmental services in coastal and estuarine waters with a special emphasis on the Great Lakes. Laboratory performs field,analytical, and laboratory investigations to improve understanding and prediction of coastal and estaurine processes and the interdependencies with the atmosphere and sediments.
*Michael Quigley, Public Affairs Officer*

**2938 Great Lakes Fishery Commission**
2100 Commonwealth Boulevard
Suite 100
Ann Arbor, MI 48105
734-662-3209
Fax: 734-741-2010
E-mail: info@glfc.org
http://www.glfc.org
The Great Lakes Fishery Commision was established by the Convention on Great Lakes Fisheries between Canada and the US in 1955. It has two major responsibilities: to develop coordinated programs of research on the Great Lakes and, onthe basis of the findings, to recommend measures which will permit the maximum sustained productivity in stocks of fish of common concern; and to formulate and implement a program to eradicate or minimize sea lampry populations in the Great Lakes.
*Charles Krueger, Science Director*

**2939 Michigan Department of Community Health**
Capitol View Building
201 Townsend Street
Lansing, MI 48913
517-373-3740
800-649-3777
Fax: 517-335-8509
http://www.michigan.gov/MDCH
A state agency which continually and diligently endeavors to prevent disease, prolong life and promote the public health.
*Janet Olszewski, Director*

**2940 Michigan Department of Environmental Quality**
525 W Allegan Street, 6th Floor
PO Box 30473
Lansing, MI 48909
517-373-7917
http://www.michigan.gov/DEQ
*Steven*

**2941 Michigan Department of Natural Resources**
Box 30444
Lansing, MI 48909
517-241-7427
Fax: 517-241-7428
E-mail: dnr-wld-webpages@state.mi.us
http://www.michigan.gov/dnr

**2942 Michigan State University Extension**
Agriculture Hall
Room 108
East Lansing, MI 48824
517-355-2308
Fax: 517-355-6473
http://www.msue.msu.edu/msue/
Michigan State University Extension helps people improve their lives through an educational process that applies knowledge to critical issues, needs and opportunities.
*Thomas G Coon, Director*

**2943 Natural Resources: Wildlife Division**
PO Box 30028
Lansing, MI 48909
517-373-1263
Fax: 517-373-1547
http://www.dnr.state.mi.us

## Minnesota

**2944 Minnesota Board of Water & Soil Resources**
520 Lafayette Road North
St Paul, MN 55155
651-296-3767
Fax: 651-297-5615
http://www.bwsr.state.mn.us
The Minnesota Board of Water and Soil Resources assists local governments to manage and conserve their irreplaceable water and soil resources.
*Brian Napstad, Chairman of the Board*
*John Jaschke, Executive Director*

**2945 Minnesota Department of Agriculture**
625 Robert Street North
St Paul, MN 55155
651-201-6000
800-967-2474
Fax: 651-297-5522
http://www.mda.state.mn.us
The MDA's mission is to work toward a diverse ag industry that is profitable as well as environmentally sound; to protect the public health safety regarding food and ag products; and to ensure orderly commerce in agricultural and foodproducts. We have two major branches of the department to accomplish this mission: regulatory divisions and non-regulatory divisions.
*Gene Hugoson, Commissioner*

**2946 Minnesota Department of Natural Resources**
500 Lafayette Road
Box 21
St. Paul, MN 55155
651-259-5483
888-646-6367
Fax: 651-297-4946
E-mail: info@dnr.state.mn.us
http://www.dnr.state.mn.us
The DNR vision hinges on the concept of sustainability. To DNR, sustainability means protecting and restoring the natural environment while enhancing economic opportunity and community well-being. DNR endorsed ecosystem-basedmanagement as its method to achieve sustainability goals. Sustainability addresses three related elements: the environment, the economy and the community. The goal is to maintain all three elements in a healthy state indefinitely.

**2947 Minnesota Environmental Quality Board**
658 Cedar Street
Centennial Building
St Paul, MN 55155
651-296-9535
E-mail: eqb@mnplan.state.mn.us
http://www.eqb.state.mn.us
The mission of the Environmental Quality Board is to lead Minnesota environmental policy by responding to key issues, providing appropriate review and coordination, serving as a public forum and developing long-range strategies toenhance Minnesota's environmental quality. The Environmental Quality Board

consists of 10 state agency commissioners or directors and five citizen members. It was established by the Minnesota Legislature in 1973.
*Michael Sullivan, Executive Director*
*Gene Hugoson, Chair*

**2948 Minnesota Pollution Control Agency**
520 Lafayette Road North 651-296-6300
St. Paul, MN 55155 800-657-3864
Fax: 651-296-7923
http://www.pca.state.mn.us
Established in 1967 to protect Minnesota's environment through monitoring environmental quality and enforcing environmental regulations.
*Brad Moore, Commissioner*

**2949 Minnesota Pollution Control Agency: Duluth**
525 S Lake Avenue 218-723-4660
Suite 400 800-657-3864
Duluth, MN 55802 Fax: 218-723-4727
http://www.pca.state.mn.us
MPCA staff at Duluth and the other six agency offices: identify environmental problems through testing, monitoring, inspections and research; develop environmental priorities; set standards and propose rules to protect people and theenvironment; develop permits; provide technical assistance; respond to emergencies and encourage pollution prevention and sustainability.
*Suzanne Hanson, Manager*

**2950 Minnesota Sea Grant College Program**
University of Minnesota
144 Chester Park 218-726-8106
31 W College Street Fax: 218-726-6556
Duluth, MN 55812 E-mail: seagr@d.umn.edu
http://www.seagrant.umn.edu
Minnesota Sea Grant is dedicated to providing the tools and technology for responsible management and policy decisions to maintain and enhance Lake Superior and Minnesota's inland aquatic economies and resources. We involveuniversities, federal and state agencies, the public and industry in a partnership to understand the complex nature of the multidisciplinary problems facing us, and then help in the development of the infrastructure necessary for innovativesolutions.

*Jeffrey Gunderson, Acting Director*
*Judy Zomerfelt, Executive Secretary*

**2951 United States Department of the Army US Army Corps of Engineers**
190 5th Street East 651-290-5200
Suite 401 Fax: 651-290-5752
St. Paul, MN 55101 http://www.usace.army.mil

## Mississippi

**2952 Gulf Coast Research Laboratory**
703 E Beach Drive 228-872-4200
Ocean Springs, MS 39564 Fax: 228-872-4204
http://www.usm.edu/gcr

*Dr William C Hawkins, Executive Director*
*Dr Jeffrey M Lotz, Chair*

**2953 Mississippi Alabama Sea Grant Consortium**
703 East Beach Drive 228-818-8836
Ocean Springs, MS 39564 Fax: 228-818-8841
E-mail: swanndl@auburn.edu
http://masgc.org
A federal state partnership that is dedicated to activities that foster the conservation and sustainable development of coastal and marine resources in Mississippi and Alabama.

*Dr La Don Swann, Director*
*Martha Saunders, President*

**2954 Mississippi Department Agriculture & Commerce**
PO Box 1609 601-359-1100
Jackson, MS 39215 Fax: 601-354-6290
E-mail: Spell@mdac.state.ms.us
http://www.mdac.state.ms.us/Index.asp
The mission of the Mississippi Department of Agriculture and Commerce is to regulate and promote agricultural-related businesses within the state and to promote Mississippi's products throughout both the state and the rest of the worldfor the benefit of all Mississippi citizens.
*Lester Spell Jr, Commissioner*

**2955 Mississippi Department of Environmental Quality**
PO Box 20305 601-961-5171
Jackson, MS 39289 E-mail: charles-chisolm@deq.state.ms.us
http://www.deq.state.ms.us
Mission: To safeguard the health, safety, and welfare of present and future generations of Mississippians by conserving and improving our environment and fostering wise economic growth through focused research and responsibleregulation.
*Trudy D Fisher, Executive Director*

**2956 Mississippi Department of Wildlife, Fisheries and Parks**
1505 Eastover Drive 601-432-2400
Jackson, MS 39211 http://www.mdwfp.com
It is the mission of the Mississippi Department of Wildlife, Fisheries and Parks to conserve and enhance Mississippi's natural resources, to provide continuing outdoor recreational opportunities, to maintain the ecological integrityand aesthetic quality of the resources and to ensure socioeconomic and educational opportunities for present and future generations.

**2957 Mississippi Forestry Commission**
301 North Lamar Street Suite 300 601-359-1386
Jackson, MS 39201 Fax: 601-359-1349
http://www.mfc.state.ms.us
Mission: To provide active leadership in forest protection, forest management, forest inventory and effective forest information distribution, necessary for Mississippi's sustainable forset-based economy.
*Charlie W Morgan, State Forester*
*Robert E Cox, Chair*

**2958 Mississippi State Department of Health Bureau of Child/Adolescent Health**
PO Box 1700 601-960-7634
Jackson, MS 39215 http://www.msdh.state.ms.us
The Mississippi Department of Health's mission is to promote and protect the health of the citizens of Missippi. Geographic focus: Missippi other organizational activies, not directed specifically toward children; advocacy, directservice delivery, education, organizing, regulation, social services
*Sam Valentine, Director*

**2959 United States Department of the Army US Army Corps of Engineers**
4155 Clay Street 601-631-5053
Vicksburg, MS 39180 http://www.usace.army.mil

## Missouri

**2960 Missouri Conservation Department**
1907 Hillcrest Drive 573-884-6861
Columbia, MO 65201 http://mdc.mo.gov

**2961 Missouri Department of Natural Resources**
PO Box 176 573-751-3443
Jefferson City, MO 65102 800-361-4827
Fax: 573-751-7627
E-mail: contact@dnr.mo.gov
http://www.dnr.mo.gov
The Department of Natural Resources preserves, protects and enhances Missouri's natural, cultural and energy resources and

works to inspire their enjoyment and responsible use for present and future generations. Our staff work toensure that our state enjoys clean air to breathe, clean water for drinking and recreation and land that sustains a diversity of life.

**2962 Natural Resources Department: Air Pollution Control**
PO Box 176
Jefferson City, MO 65102
800-334-6946
Fax: 573-751-8656
E-mail: cleanair@mail.dnr.state.mo.us
http://www.dnr.mo.gov
Mission: To maintain the purity of Missouri's air to protect the health, general welfare and property of the people.

**2963 Natural Resources Department: Energy Center**
PO Box 176
Jefferson City, MO 65102
573-751-2254
800-361-4827
Fax: 573-751-6860
E-mail: contact@dnr.mo.gov
http://www.dnr.mo.gov
A nonregulatory state agency that works to protect the environment and stimulate the economy through energy efficiency and renewable energy resources and technologies.

**2964 Natural Resources Department: Environmental Improvement and Energy Resources Authority**
PO Box 744
Jefferson City, MO 65102
573-751-4919
Fax: 573-635-3486
E-mail: eiera@dnr.mo.gov
http://www.dnr.mo.gov
A quasi-governmental agency that serves as the financing arm for the Missouri Department of Natural Resources.

**2965 United States Department of the Army US Army Corps of Engineers**
US Army Engineer District
601 E 12th Street
Room 736
Kansas City, MO 64106
816-389-3486
http://www.usace.army.mil

## Montana

**2966 Butte District Advisory Council**
106 N Parkmont
Butte, MT 59702
406-533-7600
Fax: 406-533-7660
E-mail: mt_butte_fo@blm.gov
http://www.mt.blm.gov/bdo/

*Rick Hotaling, Field Manager*

**2967 Crown of the Continent Research Learning Center - Glacier National Park**
PO Box 128
West Glacier, MT 59936
406-888-7800
Fax: 406-888-7808
E-mail: tara_carolin@nps.gov
http://www.nps.gov/glac/naturescience/ccrlc/htm
Designed to increase the effectiveness and communication of research and science results in national parks through facilitating the use of parks for scientific inquiry, supporting science-informed decision making, communicating therelevance of and providing access to knowledge gained through scientific research, and promoting science literacy and resource stewardship
*Tara Carolin, Director*
*Melissa Sladek, Science Comm Specialist*

**2968 Environmental Quality Council**
State Capitol, Room 171
PO Box 201704
Helena, MT 59620
406-444-3742
Fax: 406-444-3971
E-mail: teverts@mt.gov
The EQC is a state legislative committee create by the 1971 Montana Environmental Policy Act. As outlined in MEPA, the EQC'S purpose is to encourage conditions under which people can coexist with nature in productive harmony. TheCouncil fulfills this purpose by assisting the Legislature in the development of natural resource and environmental policy, by conducting studies on related issues and by serving in an advisory capacity to the state's natural resource programs.

**2969 Lewiston District Advisory Council Bureau of Land Management**
920 Northeast Main Street
PO Box 1160
Lewistown, MT 59457
406-538-1900
Fax: 406-538-1904
http://www.blm.gov.mt

**2970 Montana Department of Agriculture**
PO Box 200201
Helena, MT 59620
406-444-3144
Fax: 406-444-5409
E-mail: agr@mt.us
http://www.agr.state.mt.us
Mission: To protect producers and consumers and to enhance and develop agriculture and allied industries.
*Joel A Clairmont, Acting Director*

**2971 Montana Natural Heritage Program**
1515 East 6th Avenue
Helena, MT 59620
406-444-5354
Fax: 406-444-0581
E-mail: mtnhp@mt.gov
http://mtnhp.org
The Montana Natural Heritage Program is the state's source for information on the status and distribution of our native animals and plants, emphasizing species of concern and high quality habitats such as wetlands.
*Susan Crispin, Director*
*Darlene Patzer, Finance/Grants Administrator*

**2972 Natural Resources & Conservation Department**
1625 11th Avenue
Helena, MT 59620
406-444-2074
Fax: 406-444-2684
http://dnrc.mt.gov
*Mary Sexton, Director*

## Nebraska

**2973 Department of Agriculture: Natural Resources Conservation Service**
National Soil Survey Center
100 Centennial Mall N Room 152
Lincoln, NE 68508
402-437-5499
Fax: 402-437-5336
http://soils.usda.gov

**2974 Department of Natural Resources**
301 Centennial Mall S
PO Box 94676
Lincoln, NE 68509
402-471-2363
Fax: 402-471-2900
http://www.dnr.ne.gov
State Natural Resources Agency
*Ann Saloman Bleed, Director*
*Brian P Dunnigan, Deputy Director*

**2975 Nebraska Department of Agriculture**
301 Centennial Mall S
PO Box 94947
Lincoln, NE 68509
402-471-2341
Fax: 402-471-6876
http://www.agr.ne.gov
Regulatory state agency.

*Greg Ibach, Director*
*Bobbie Kriz-Wickham, Assistant Director*

**2976 Nebraska Department of Environmental Quality**
1200 N Street Suite 400
PO Box 98922
Lincoln, NE 68509
402-471-2186
Fax: 402-471-2909
E-mail: MoreInfo@NDEQ.state.NE.US
http://www.deq.state.ne.us
Mission: To protect the quality of Nebraska's environment-our air, land, and water resources. Enforce regulations and provide assistance.

**2977 Nebraska Ethanol Board**
301 Centennial Mall S
PO Box 94922
Lincoln, NE 68509
402-471-2941
Fax: 402-471-2470
E-mail: todd.sneller@nebraska.gov
http://www.ne-ethanol.org

The Nebraska Ethanol Board assists ethanol producers with programs and strategies for marketing ethanol and related co-products. The Board supports organizations and policies that advocate the increased use of ethanol fuels, andadministers public information, education and ethanol research projects. The Board also assists companies and organizations in the development of ethanol production facilities in Nebraska.

*Todd C Sneller, Administrator*

**2978 Nebraska Game & Parks Commission: Fisheries Division**
2200 N 33rd Street
PO Box 30370
Lincoln, NE 68503
402-471-5515
Fax: 402-471-4992
E-mail: don.gabelhouse@nebraska.gov
http://www.outdoornebraska.org

**2979 Nebraska Games & Parks Commission**
2200 North 33rd Street
Lincoln, NE 68503
402-471-0641
E-mail: webmaster@ngpc.ne.gov
http://www.ngpc.state.ne.us

*Rex Amack, Director*
*Kirk Nelson, Assistant Director*

**2980 Nebraska Games & Parks: Wildlife Division**
2200 N 33rd Street
PO Box 30370
Lincoln, NE 68503
402-471-5411
Fax: 402-471-4992
E-mail: jim.douglas@nebraska.gov
http://www.outdoornebraska.gov

**2981 United States Department of the Army US Army Corps of Engineers**
US Army Engineer District
215 N 17th Street
Omaha, NE 68102
402-221-3902
Fax: 402-221-4626
http://www.usace.army.mil

*David C Press, Commander*
*Quay B Jones, Deputy Commander*

## Nevada

**2982 Bureau of Land Management**
Department of the Interior
HC 33 Box 33500
Ely, NV 89301
775-289-1800
Fax: 775-289-1910
http://www.nv.blm.gov/Ely

**2983 Carson City Field Office Advisory Council**
Bureau of Land Management
5665 Morgan Mill Road
Carson City, NV 89701
775-885-6000
Fax: 775-885-6147
http://www.nv.blm.gov/carson/default.htm

A Federal Land Management Agency.

**2984 Colorado River Basin Salinity Control Program**
US Bureau of Land Reclamation
125 S State Street
Room 7311
Salt Lake City, UT 84138-1147
801-524-3753
Fax: 801-524-3847
E-mail: kjacobson@usbr.gov
http://www.usbr.gov

The Colorado River and its tributaries provide municipal and industrial water to about 27 million people and irrigation to nearly four million acres of land in the US. The river also serves about 2.3 million people and 500,000 acres inMexico. The threat of salinity is a major concern in both the US and Mexico. Salinity affects agricultural, municipal and industrial water users. We work to control the salinity of the Colorado river and thereby to protect the land and people.

*Kib Jacobson, Program Manager*
*Brad Parry, Program Coordinator*

**2985 Conservation and Natural Resources Department**
Wildlife Division, Conservation Education
901 South Stewart Street
Suite 5001
Carson City, NV 89701
775-684-2700
Fax: 775-684-2715
E-mail: ndowinfo@govmail.state.nv.us
http://dcnr.nv.gov

The Department of Conservation and Natural Resources (DCNR) is responsible for the establishment and administration of goals, objectives and priorities for the preservation of the State's natural resources.

*Allen Biaggi, Director*
*Kay Scherer, Deputy Director*

**2986 Conservation and Natural Resources: Water Resources Division**
901 South Stewart Street
Suite 2002
Carson City, NV 89701
775-684-2800
Fax: 775-684-2811
E-mail: hricci@wr.state.nv.us
http://ndwr.state.nv.us

*Tracy Taylor, State Engineer*

**2987 Department of the Interior: Bureau of Reclamation**
Lower Colorado Regional Office
PO Box 61470
Boulder City, NV 89006
702-293-8000
Fax: 702-293-8418
http://www.usbr.gov/lc

Manage a number of environmental managment programs related to the lower Colorado River.

*Jayne Harkins, Deputy Regional Director*

**2988 Las Vegas Bureau of Land Management**
4701 N Torrey Pines Drive
Las Vegas, NV 89130
702-515-5000
http://www.nv.blm.gov

The Bureau of Land Management's mission is to help sustain the health, diversity and productivity of public lands so they can be used and enjoyed by both present and future generations.

**2989 Nevada Bureau of Mines & Geology**
University of Nevada
Mail Stop 178
Reno, NV 89557
775-784-6691
Fax: 775-784-1709
http://www.nbmg.unr.edu

*Jonathan Price, Director/State Geologist*

**2990 Nevada Department of Wildlife**
1100 Valley Road
Reno, NV 89512
775-688-1500
Fax: 775-688-1595
http://www.ndow.org

*Kenneth Mayer, Director*
*Doug Hunt, Deputy Director*

**2991 Nevada Natural Heritage Program**
901 South Stewart Street
Suite 5002
Carson City, NV 89701
775-684-2900
Fax: 775-684-2909
http://heritage.nv.gov

The mission is to develop and maintain a cost-effective central information source and inventory of locations, biology, and status of all threatened, endangered, rare and at-risk plants and animals in Nevada.

*Jennifer Newmark, Administrator*

**2992 Tahoe Regional Planning Agency (TRPA) Advisory Planning Commission**
128 Market Street
Stateline, NV 89449
775-588-4547
Fax: 775-588-4527
E-mail: trpa@trpa.org
http://www.trpa.org

The TRPA leads the cooperative effort to preserve, restore and enhance the natural and human environment of the Lake Tahoe Region. The Code of Ordinances regulates, among other things, land use, density, rate of growth, land coverage,excavation and scenic impacts. These regulations are designed to bring the region into compliance with the threshold standards established for water quality, air quality, soil conservation, wildlife habitat, vegetation, noise, recreation and scenicresources.

## New Hampshire

**2993 New Hampshire Department of Environmental Services**
29 Hazen Drive
PO Box 95
Concord, NH 03302
603-271-3503
Fax: 603-271-2867
http://www.des.state.nh.us
Mission: To help sustain a high quality of life for all citizens by protecting and restoring the environment and public health in New Hampshire.

**2994 New Hampshire Fish and Game Department**
11 Hazen Drive
Concord, NH 03301
603-271-3511
Fax: 603-271-1438
E-mail: info@wildlife.state.nh.us.gov
http://www.wildnh.com
As the guardian of the states fish, wildlife and marine resources, the department works with the public to: conserve, manage and protect these resources and their habitats; inform and advise the public about these resources; providethe public with opportunities to use and appreciate these resources.

*Gelnn Normandeau, Executive Director*
*Tanya L Haskell, Administrative Assistant*

**2995 New Hampshire State Conservation Committee**
PO Box 3907
Concord, NH 03302
603-271-1092
http://www.nh.gov/scc
Coordinates the work of the ten county conservation districts in the state of New Hampshire.
*Stephen H Taylor, Commissioner*

**2996 Northeastern Forest Fire Protection Compact**
P.O. Box 6192
21 Parmenter Terrace
China Village, ME 04926
207-968-3782
Fax: 207-968-3782
E-mail: info@nffpc.org
http://www.nffpc.org
The mandate of the Northeastern Forest Fire Protection Commission is to provide the means for its member states and provinces to cope with fires that might be beyond the capabilities of a singler member through infromation, technologyand resources sharing activities.

**2997 Resources & Development Council: State Planning**
Department of Resources & Economic Development
172 Pembroke Road
P.O. Box 1856
Concord, NH 03302
603-271-2411
Fax: 603-271-2629
E-mail: sboucher@dred.state.nh.us
http://www.dred.state.nh.us

*George Bald, Commissioner*

**2998 University of New Hampshire Cooperative Extension**
Taylor Hall
59 College Road
Durham, NH 03824
603-862-1520
Fax: 603-862-1585
http://extension.unh.edu
The University of New Hampshire Cooperative Extension provides New Hampshire citizens with research-based education and information, enhancing their ability to make informed decisions that strengthen youth, families and communities,sustain natural resources, and improve the economy.
*John Pike, Dean/Director*

## New Jersey

**2999 New Jersey Department of Agriculture**
PO Box 330
Trenton, NJ 08625
609-292-3976
Fax: 609-292-3978
http://www.nj.gov/agriculture
The New Jersey Department of Agriculture is an agency which oversees programs that serve virutally all New Jersey citizens. A major priority of the NJDA is to promote, protect and serve the Garden State's diverse agriculture andagribusiness industries.
*Louis A Bruni, Chief of Operations*

**3000 New Jersey Department of Environmental Protection**
401 E State Street
7th Floor, E Wing
Trenton, NJ 08625
609-292-2885
Fax: 609-292-7695
http://www.state.nj.us/dep
A state department dedicated to protecting New Jersey's air, land, water and natural resources.

*Bob Martin, Commissioner*
*John Hazen, Director of Leg Affairs*

**3001 New Jersey Department of Environmental Protection: Site Remediation Program**
401 E State Street 6th Floor E Wing
PO Box 028
Trenton, NJ 08625
609-292-1250
Fax: 609-777-1914
http://www.state.nj.us/edt
*I Kropp, Assistant Commissioner*

**3002 New Jersey Division of Fish & Wildlife**
501 E State Street, 3rd Fl
PO Box 400
Trenton, NJ 08625
609-292-2965
http://www.njfishandwildlife.com
Our mission is to protect and manage the state's fish and wildlife to maximize their long-term biological, recreation and economic values for all New Jerseyans.

**3003 New Jersey Geological Survey**
PO Box 427
Trenton, NJ 08625
609-292-1185
Fax: 609-633-1004
E-mail: njgsweb@dep.state.nj.us
http://www.njgeology.org
State agency that maps, interprets and provides geoscience information to the public on geology and ground water resources.

*Karl Muessig, State Geologist*

**3004 New Jersey Pinelands Commission**
PO Box 7
15 Springfield Road
New Lisbon, NJ 08064
609-894-7300
Fax: 609-894-7330
E-mail: info@njpines.state.nj.us
http://www.state.nj.us/pinelands/
Mission: To preserve, protect, and enhance the natural and cultural resourcse of the Pinelands National Reserve, and to encourage compatible economic and other human activities consistent with that purpose.
*John Stokes, Executive Director*
*Nadine Young, Executive Assistant*

## New Mexico

**3005 Albuquerque Bureau of Land Management**
435 Montano Road NE
Albuquerque, NM 87107
505-761-8700
Fax: 505-761-8911
http://www.nm.blm.gov

**3006 Attorney General**
Environmental Enforcement
PO Drawer 1508
Santa Fe, NM 87504
505-827-6000
Fax: 505-827-5826
http://www.ago.state.nm.us

*Gary King, Attorney General*

**3007 Energy, Minerals & Natural Resources: Energy Conservation & Management Division**
1220 South Street 505-476-3310
Francis Drive http://www.emnrd.state.nm.us/ecmd/
Santa Fe, NM 87505
Encourages efficient energy use in New Mexico by offerin programs and information for state agencies, companies and induviduals.
*Fernando Martinez, Director*

**3008 Energy, Minerals and Natural Resources Department**
1220 S St. Francis Drive 505-476-3230
Santa Fe, NM 87505 Fax: 505-476-3234
http://www.emnrd.state.nm.us
Mission is to provide leadership in the protection, conservation, management and responsible use of New Mexico's natural resources.
*Sandra Haug, Director*

**3009 New Mexico Bureau of Geology & Mineral Resources**
801 Leroy Place 505-835-5420
Socorro, NM 87801 Fax: 505-835-6333
E-mail: scholle1@nmt.edu
http://www.geoinfo.nmt.edu
A service and research division of the New Mexico Institute of Mining and Technology. Acts as the geological survey of New Mexico.

*Peter Scholle, Director & State Geologist*
*Bruce Allen, Field Geologist*

**3010 New Mexico Cooperative Fish & Wildlife Research Unit**
New Mexico University MSC 4901 505-646-6053
PO Box 30003 Fax: 505-646-1281
Las Cruces, NM 88003 http://fws-nmcfwru.nmsu.edu
The New Mexico Research Unit conducts, research on problems of mutual concern to cooperators; graduate academic training; technical assistance in fish and wildlife management; and conservation education through publications, lectures,and demonstrations.
*Colleen Caldwell, Leader*
*Louis Bender, Assisstant Unit Leader*

**3011 New Mexico Department of Game & Fish**
PO Box 25112 505-476-8000
Santa Fe, NM 87504 http://www.wildlife.state.nm.us
*Bruce Thompson, Director*

**3012 New Mexico Environment Department**
1190 St. Francis Drive
Suite N4050 800-219-6157
Santa Fe, NM 87505 http://www.nmenv.state.nm.us
The New Mexico Environment Department's mission is to provide the highest quality of life throughout the state by promoting a safe, clean, and productive environment.
*Ron Curry, Cabinet Secretary*
*Cindy Padilla, Deputy Secretary*

**3013 New Mexico Soil & Water Conservation Commission**
MSC APR PO Box 30005 505-646-2642
Las Cruces, NM 88003 Fax: 505-646-1540
E-mail: acoleman@nmda.nmsu.edu

**3014 Roswell District Advisory Council: Bureau of Land Management**
2909 W 2nd Street 505-627-0272
Roswell, NM 88201 Fax: 505-627-0276
http://www.nm.blm.gov/rfo/index.htm
*Larry Ashley, Engine Module Leader*

**3015 United States Department of the Interior: United States Fish and Wildlife Service**
500 Gold Avenue SW 505-248-6911
Albuquerque, NM 87102 E-mail: diane_knudson@fws.gov
http://southwest.fws.gov
Mission: To work with others, to conserve, protect and enhance fish, wildlife, and plants and their habitats for the continuing benefit of the American people.
*Benjamin Tuggle, Director*

## New York

**3016 Adirondack Park Agency**
1133 NYS Route 86 518-891-4050
PO Box 99 Fax: 518-891-3938
Ray Brook, NY 12977 http://www.apa.state.ny.us
Mission: To protect the public and private resources of the Park through the exercise of the powers and duties provided by law.

**3017 Department of Environmental Conservation**
625 Broadway 518-402-8540
Albany, NY 12233 Fax: 518-402-8541
http://www.dec.ny.gov
Mission: To conserve, improve, and protect New York State's natural resources and environment, and control water, land and air pollution, in order to enhance the health, safety and welfare of the people of the state and their overalleconomic and social well being.
*Pete Grannis, Commissioner*
*Stuart Gruskin, Executive Dep Commissioner*

**3018 Department of Environmental Conservation: Division of Air Resources**
625 Broadway 518-402-8452
Albany, NY 12233 http://www.dec.ny.gov/about/644.html
Maintains and improves New York State air quality through research, permitting and enforcement.
*Jared Snyder, Assistant Commissioner*

**3019 Department of Environmental Conservation: Division of Mineral Resources**
625 Broadway 518-402-8076
3rd Floor Fax: 518-402-8060
Albany, NY 12233 http://www.dec.ny.gov/about/636.html
The Division of Mineral Resources is responsible for ensuring the environmentally sound, economic development of New York's non-renewable energy and mineral resources for the benefit of current and future generations. To carry out thismission, we regulate the extraction of oil and gas, and require the reclamation of land after mining.
*Val Washington, Deputy Commissioner*

**3020 New York Cooperative Fish & Wildlife Research Unit**
Cornell University, Fernow Hall 607-255-2151
Natural Resources Department Fax: 607-255-1895
Ithaca, NY 14853 E-mail: dnrcru-mailbox@cornell.edu
http://www.dnr.cornell.edu/f&wres/nycf&wru.htm

*Dr. Milo Richmond, Unit Leader*
*Richard A Malecki, Assistant Leader*

**3021 New York Department of Health**
Flanigan Square 518-402-7500
547 River Street Fax: 518-402-7509
Troy, NY 12180 http://www.health.state.ny.us
Working together and committed to excellence, we protect and promote the health of New Yorkers through prevention, science and the assurance of quality health care delivery.
*Richard F Daines, Commissioner*

**3022 New York State Office of Parks, Recreation and Historic Preservation**
Empire State Plaza 518-474-0456
Agency Building 1 http://www.nysparks.com
Albany, NY 12238
The agency operates 168 parks offering a wide variety of recreational, cultural and education activities, and 35 state historic sites; sponsors boating and snowmobiling, nature study and outreach programs; manages grant programs forboating and snowmobiling enforcement and aid to zoos, botanical gardens, aquariums; and administers funds for federal historical preservation and parks programs, the Environmental Protection Fund and the 1996 Clean Water/Clean Air Bond act.
*Carol Ash, Commissioner*

**3023 New York State Soil and Water Conservation Committee**
10B Airline Drive 518-457-3738
Albany, NY 12235 Fax: 518-457-3412
E-mail: lauren.hoeffner@agmkt.state.ny.us
http://www.nys-soilandwater.org
The New York State Soil and Water Conservation Committee is composed of voting and advisory members who represent a wide range of agricultural, environmental and other interests. The Committee operates through a network of partnershipsbetween state, federal and local agencies, as well as citizen interests and the private sector. The mission of the Committee is to develop and oversee and agriculatural nonpoint source water quality program for New York State.
*Ronald Kaplewicz, Director*
*Michael Latham, Assistant Director*

**3024 Tug Hill Tomorrow Land Trust**
PO Box 6063 315-779-8240
Watertown, NY 13601 Fax: 315-782-6192
E-mail: thtomorr@northnet.org
http://www.tughilltomorrowlandtrust.org
A regional, private, nonprofit founded by a group of Tug Hill residents to serve the region of 2,100 square miles serving portions of Jefferson, Lewis, Oneida & Oswego Counties. The mission is two-fold: increase awareness andappreciation of the Tug Hill Rgion through education; and to help retain the forest, farm, recreation, and wild land of the region through voluntary, private land protection efforts.

*Linda Garrett, Executive Director*

**3025 United States Department of the Army US Army Corps of Engineers**
1776 Niagara Street 716-879-4104
Buffalo, NY 14207 http://www.lrb.usace.army.mil

## North Carolina

**3026 Carnivore Preservation Trust**
1940 Hanks Chapel Road 919-542-4684
Pittsboro, NC 27312 Fax: 919-542-4454
E-mail: info@cptigers.org
http://www.cptigers.org
Is a wildlife sanctuary, offering unique opportunities to learn about these animals and their critical importance to our quality of life on Earth.
*Pam Fulk, Executive Director*
*Amanda Byrne, IT Administrator*

**3027 Department of Agriculture & Consumer Services**
2 West Edenton Street 919-733-7125
Raleigh, NC 27601 http://www.ncagr.com
Mission: To improve the state of agriculture in North Carolina by providing services to farmers and agribusinesses, and to serve the citizens of North Carolina by providing services and enforcing laws to protect consumers.
*Steve Troxler, Commissioner*

**3028 North Carolina Board of Science and Technology**
301 North Wilmington Street 919-733-6500
1326 Mail Service Center Fax: 919-733-8356
Raleigh, NC 27699 E-mail: ncbst@nccommerce.com
http://www.ncscienceandtechnology.com
Encourages, promotes, and supports scientific, engineering, and industrial research applications in North Carolina.
*Robert McMahan, PhD, Executive Director*
*Margaret Dardess, Chair*

**3029 North Carolina Department of Environment and Natural Resources**
1601 Mail Service Center 919-733-4984
Raleigh, NC 27699 Fax: 919-715-3060
http://www.enr.state.nc.us
*William G Ross Jr, Secretary*
*Mary Penny Thompson, General Counsel*

**3030 United States Department of the Army US Army Corps of Engineers**
69 Darlington Avenue 910-251-4625
PO Box 1890 910-251-4185
Wilmington, NC 28402 http://www.saw.usace.army.mil
Provides quality planning, design, construction, and operations products and services to meet the needs of civilian and military customers.

## North Dakota

**3031 Dakotas Resource Advisory Council: Department of the Interior**
Bureau of Land Management
99 23rd Avenue West 701-227-7700
Suite A Fax: 701-227-7701
Dickinson, ND 58601 E-mail: corrine_walter@blm.gov
http://www.blm.gov
The Dakotas Council currently has 15 members. It is structured to provide a balance of membership by area of expertise, training, and experience. It consists of five individuals in each of three categories.

**3032 ND Game and Fish Department**
100 N Bismarck Expressway 701-328-6300
Bismarck, ND 58501 Fax: 701-328-6352
E-mail: ndgf@nd.gov
http://gf.nd.gov
To protect, conserve and enhance fish and wildlife populations and their habitats for sustained public consumptive and nonconsumptive use.
*Terry Steinwood, Director*
*Roger Rostvet, Deputy Director*

**3033 North Dakota Forest Service**
307 First Street E 701-228-5422
Bottineau, ND 58318 Fax: 701-228-5448
E-mail: forest@nd.gov
http://www.nd.gov/forest
The ND Forest Service administers forestry programs statewide. The agency operates a nursery at Towner specializing in the production of conifer tree stock. The nursery is the sole supplier of evergreen seedlings in North Dakota.Technial assistance relating to the management of private forest lands, state forest lands, urban and community forests, tree planting and wildland fire protection is provided by the agency. The ND Forest Service also owns and manages app. 13,278acres of state lands.

*Larry Kotchman, State Forester*

**3034 North Dakota Parks and Recreation Department**
1600 E Century Ave. 701-328-5357
Suite 3 Fax: 701-328-5363
Bismarck, ND 58503 E-mail: parkrec@state.nd.us
http://www.ndparks.com

The state government agency charged with managing North Dakota's state parks and recreation areas; the state's nature preserves and natural area programs; motorized and non-motorized trail programs; recreational grants and state-widerecreation planning; and state scenic byways program.

*Douglass A Prchal, Director*
*Dorothy Streyle, Coordinator*

## Ohio

**3035 Division of Environmental Services**
Ohio EPA
8955 East Main Street 614-644-4247
Reynoldsburg, OH 43068 Fax: 614-644-4272
http://www.epa.state.oh.us/des
The Division of Environmental Services provides biological and chemical data and technical assistance to other divisions within Ohio EPA, state and local agencies, and private entities in order to help monitor and protect human healthand the environment to ensure a high quality of life in Ohio.
*Steve Roberts, Quality Assurance Supervisor*

**3036 Environmental Protection Agency: Ohio Division of Surface Water**
50 West Town Street 614-644-2001
Suite 700 Fax: 614-644-2745
Columbus, OH 43215 http://www.epa.state.oh.us/dsw
To protect, enhance and restore all waters of the state for the health, safety and welfare of present and future generations. We accomplish this mission by monitoring the aquatic environment, permitting, enforcing environmental laws,using and refining scientifically sound methods and regulations, planning, coordinating, educating, providing technical assistance and encouraging pollution prevention practices.

**3037 Lead Poisoning Prevention Program**
Ohio Department of Health
246 N High Street 614-466-1450
Columbus, OH 43266 877-668-5323
Fax: 614-752-4157
E-mail: lead@odh.ohio.gov
http://www.odh.state.oh.us/data
The Lead Poisoning Prevention Program ensures the public receives safe and proper lead abatement, detection, and analytical services by requiring those services be conducted according to federal and state regulations, and by trainedand licensed personnel.

**3038 Ohio Department of Natural Resources Division of Geological Survey**
2045 Morse Rd 614-265-6576
Bldg C Fax: 614-447-1918
Columbus, OH 43229 E-mail: geo.survey@dnr.state.oh.us
http://www.ohiodnr.com/geosurvey
Provides geologic information and services for responsible management of Ohio's natural resources. Geologic maps, reports and data files developed by the division can be used by individuals, educators, industry, business andgovernment.

*Larry Wickstrom, State Geologist*

**3039 Ohio Environmental Protection Agency**
50 West Town Street 614-644-2782
Suite 700 Fax: 614-644-2329
Columbus, OH 43215 E-mail: request@www.epa.state.oh.us
http://www.epa.state.oh.us
Mission: To protect the environment and public health by ensuring compliance with environmental laws and demonstrating leadership in environmental stewardship.
*Christopher Korleski, Director*

**3040 Ohio River Valley Water Sanitation Commission**
5735 Kellogg Avenue 513-231-7719
Cincinnati, OH 45228 Fax: 513-231-7761
E-mail: info@orsanco.org
http://www.orsanco.org
ORSANCO operates programs to improve water quality in the Ohio River and its tributaries, including: setting waste water discharge standards; performing biological assessment; monitoring for the chemical and physical properties of thewaterways; and conducting special surveys and studies. Also coordinates emergency response activities for spills or accidental discharges to the river and promotes public participating programs.
*Alan Vicory, Executive Director*
*Peter Tennant, Deputy Executive Director*

**3041 Ohio Water Development Authority**
480 South High Street 614-466-5822
Columbus, OH 43215 Fax: 614-644-9964
http://owda.org
Provides financial assistance for environmental infrastructure from the sale of municipal revenue bonds through loans to local governments in Ohio and issuing Industrial Revenue Bonds for qualified projects. The vision of OWDA is tocontinue to provide assistance for environmental infrastructure by being responsive to the needs of local government agencies, enhancing the provision of financial and technical assistance and developing new financial assistance products for theprivate sector.

*Steven J Grossman, Executive Director*
*Scott Campbell, COO*

## Oklahoma

**3042 Oklahoma Department of Environmental Quality**
PO Box 1677 405-702-1000
Oklahoma City, OK 73101 Fax: 405-702-1001
http://www.deq.state.ok.us
Administers environmental laws considering both the economy of today and the environment of tomorrow.

**3043 Oklahoma Department of Health**
1000 Northeast 10th Street 405-271-4471
Oklahoma, OK 73117 Fax: 405-271-6199
http://www.health.state.ok.us
The Oklahoma State Department of Health is to protect and promote the health of citizens of Oklahoma and to prevent disease and injury, and to assure the conditions by which our citizens can be healthy.
*James Michael Crutcher, Commissioner*

**3044 Salt Plains National Wildlife Refuge**
Route 1 Box 76 580-626-4794
Jet, OK 73749 Fax: 580-626-4793
E-mail: debbie_pike@fws.gov
http://saltplains.fws.gov
*Debbie Pike, Park Ranger*

**3045 Water Quality Division**
5225 N Shurtel 405-702-8100
Oklahoma City, OK 73118 Fax: 405-810-1046
http://www.deq.state.ok.us/WQDNew
*Larry Edmison, Director*

## Oregon

**3046 Burns District: Bureau of Land Management**
Department of the Interior

28910 Highway 20 West
Hines, OR 97738
541-573-4400
E-mail: OR_Burns_Mail@blm.gov
http://www.blm.gov/or/districts/burns

*Dana Shuford, Manager*
*Melissa Towers, Administrative Officer*

**3047 Department of Transportation**
355 Capitol St. NE
Salem, OR 97301
888-275-6368
Fax: 503-986-3432
http://www.oregon.gov/odot
Mission: To provide a safe, efficient transportation system that supports economic opportunity and livable communities for Oregonians.
*Matthew Garrett, Director*
*Margi Lifsey, Sustainability Coordinator*

**3048 Eugene District: Bureau of Land Management**
Department of the Interior
PO Box 10226
Eugene, OR 97440
541-683-6600
888-442-3061
Fax: 541-683-6981
E-mail: OR_Eugene_Mail@blm.gov
http://www.blm.gov/or/districts/eugene
The Eugene District manages several ecosystems ranging from coastal inlands to dense Douglas-fir, hemlock, and cedar forests. The wide variation in the lands managed by the District offers the perfect compromise between the urban parksin the cities and the high elevation recreation opportunities in the adjacent Willamette, Siuslaw and Umpqua National Forest.
*Ginnie Grilley, Manager*

**3049 Klamath River Compact Commission**
280 Main Street
Klamath Falls, OR 97601
541-882-4436
Created by the Klamath River Compact in 1957, KRCC is a three member commission whose purpose, with respect to the water of the Klamath River Basin, is to faciliate and promote the orderly, integrated, and comprehensive developement,use, conservation and control of water for development of lands by irrigation, protection of fish and wildlife, domestic and industrial use, hydropower, navigation, and flood protection.

**3050 Lakeview District: Bureau of Land Management**
Department of the Interior
1301 South "G" Street
Lakeview, OR 97630
541-947-2177
Fax: 541-947-6399
E-mail: OR_Lakeview_Mail@blm.gov
http://www.blm.gov/or/districts/lakeview

**3051 Medford District: Bureau of Land Management**
3040 Biddle Road
Medford, OR 97504
541-618-2200
http://www.blm.gov/or/districts/medford
The Bureau of Land Management's Medford District oversees approximately 862,000 acres of scattered public lands between the Cascade and Siskiyou mountain ranges and from the Oregon/California border to Canyon Creek and southern DouglasCounty. This large land base is divided into four Resource Areas: Ashland, Butte Falls, Grants Pass and Glendale.

**3052 Oregon Department of Environmental Quality**
811 SW Sixth Avenue
Portland, OR 97204
503-229-5696
800-452-4011
Fax: 503-229-6124
E-mail: deq.info@deq.state.or.us
http://www.oregon.gov/deq
Mission: To be a leader in restoring, maintaining and enhancing the quality of Oregon's air, land and water.
*Stephanie Hallock, Director*
*Dick Pedersen, Deputy Director*

**3053 Oregon Department of Fish and Wildlife**
3406 Cherry Avenue NE
Salem, OR 97303
503-947-6000
800-720-6339
E-mail: odfw.info@state.or.us
http://www.dfw.state.or.us
Commissioners formulate general state programs and policies concerning management and conservation of fish wildlife resources and establishes seasons, methods and bag limits for recreational and commercial take.
*Ed Bowles, Fish Administrator*
*Ron Anglin, Wildlife Administrator*

**3054 Oregon Department of Forestry**
2600 State Street
Salem, OR 97310
503-945-7200
800-437-4490
Fax: 503-945-7212
http://www.odf.state.or.us
Mission: To serve the people of Oregon by protecting, managing, and promoting stewardship of Oregon's forests to enhance environmental, economic, and community sustainability.
*Doug Decker, State Forester*
*Dan Postrel, Agency Affairs Director*

**3055 Oregon Department of Land Conservation and Development**
635 Capitol Street NE
Suite 150
Salem, OR 97301
503-373-0050
Fax: 503-378-5518
http://www.lcd.state.or.us
Mission: Support all of our partners in creating and implementing comprehensive plans that reflect and balance the statewide planning goals, the vision of citizens, and the interests of local, state, federal and tribal governments.
*Lane Shetterly, Director*

**3056 Oregon Water Resource Department**
725 Summer Street Northeast
Suite A
Salem, OR 97301
503-986-0900
Fax: 503-986-0904
http://www.wrd.state.or.us
Mission: To serve the public by practicing and promoting responsible water management through two key goals: to directly address Oregon's water supply needs; and to restore and protect streamflows and watersheds in order to ensure thelong-term sustainability of Oregon's ecosystems, economy, and quality of life.
*Phillip C Ward, Director*
*Brenda Bateman, Public Information Officer*

**3057 Prineville District: Bureau of Land Management**
Department of the Interior
3050 NE Third Street
Prineville, OR 97754
541-416-6700
Fax: 541-416-6798
E-mail: OR_Prineville_Mail@blm.gov
http://www.blm.gov/or/districts/prineville
*Debbie Henderson-Norton, Manager*
*Steve Robertson, Associate Manager*

**3058 Roseburg District: Bureau of Land Management**
Department of the Interior
777 NW Garden Valley Boulevard
Roseburg, OR 97470
541-440-4930
Fax: 541-440-4948
E-mail: or100mb@blm.gov
http://www.blm.gov/or/districts/roseburg
Public lands of the Roseburg District, located in southwestern Oregon, contain some of the most productive forests in the world. An important mainstay of the local economy, which acquires timber from both private and federal lands inthe region. The district is criss-crossed with streams and rivers that support sport fishing. With Interstate 5 running through the middle of the district, and east-west state highways connecting Crater Lake to the Pacfic coast, the district drawsmany tourists.

**3059 Salem District: Bureau of Land Management**
Department of the Interior
1717 Fabry Road Southeast
Salem, OR 97306
503-375-5646
Fax: 503-375-5622
E-mail: OR_Salem_Mail@blm.gov
http://www.blm.gov/or/districts/salem
BLM Mission: to sustain the health, diversity and productivity of the public lands for the use and enjoyment of present and future generations. Salem District manages 400,000 acres scattered across 13 counties. Seventy three percent ofOregon's population live within the boundries of this district. Their major focus is an ecosystem management approach involving many different disci-

plines. Salem employs 200 full-time employees working in forestry, land surveying, wildlife biology,hydrology, etc.
*Aaron Horton, Manager*
*Don Hollenkamp, Associate Manager*

**3060 United States Department of the Army: US Army Corps of Engineers**
US Army Engineer Division
PO Box 2946 503-808-5150
Portland, OR 97208

**3061 Vale District: Bureau of Land Management**
Department of the Interior
100 Oregon Street 541-473-3144
Vale, OR 97918 Fax: 541-473-6213
E-mail: OR_Vale_Mail@blm.gov
http://www.blm.gov/or/districts/vale
The Vale District of the Bureau of Land Management manages 4.9 million acres of public land in eastern Oregon. The mission of the BLM is to sustain the health, diversity, and productivity of the public lands for the use and enjoymentof present and future generations.
*Dave Henderson, Manager*
*Larry Frazier, Associate Manager*

## Pennsylvania

**3062 Allegheny National Forest**
US Forest Service
4 Farm Colony Road 814-728-6100
Warren, PA 16365 Fax: 814-726-1465
E-mail: anf/r9_allegheny@fs.fed.us
http://www.fs.fed.us/r9/allegheny
An organization dedicated to providing advice for development of the corridor management plan for the northern section of the Allegheny River that has been designated as a National Wild and Scenic River.
*Mark Conn, Forest Supervisor*

**3063 Childhood Lead Poisoning Prevention Program**
Pennsylvania Department of Health
PO Box 90 717-783-8451
Harrisburg, PA 17108 Fax: 717-772-0323
The mission of the Pennsylvania Department of Health, Childhood Lead Poisoning Prevention Program is to make the citizens of the Commonwealth aware of the dangers of lead poisoning and to reduce the number of children who becomelead-poisoned.

**3064 Citizens Advisory Council**
Pennsylvania Department Environmental Protection
13th Floor, RCSOB 717-787-4527
PO Box 8459 Fax: 717-787-2878
Harrisburg, PA 17105 E-mail: suswilson@state.pa.us
http://www.depweb.state.pa.us/cac
Advises the Department of Environmental Protection, the Governor and the General Assembly on environmental issues and the work of the Department.

*Susan Wilson, Executive Director*
*Joyce Hatala, Chair*

**3065 Department of the Interior: National Parks**
200 Chestnut Street 215-597-7013
5th Floor Fax: 215-597-8015
Philadelphia, PA 19106

**3066 Environmental Protection Agency: Region III**
1650 Arch Street 215-814-5000
Philadelphia, PA 19103 http://www.epa.gov/region03
Region III is responsible for federal environmental programs in Delaware, Maryland, Pennsylvania, Virginia, West Virginia and District of Columbia. Programs include air and water pollution control; toxic substances, pesticides, anddrinking water regulation; wetlands protection; hazardous waste management, hazardous waste dump site cleanup; and some aspects of radioactive materials regulation.

**3067 Lacawac Sanctuary Foundation**
94 Sanctuary Road 570-689-9494
Lake Ariel, PA 18436 Fax: 570-689-2017
E-mail: info@lacawac.org
http://www.lacawac.org
A nature preserve and historic site in northeastern PA's Pocono Mountains. The main function is the protection of its natural areas and historic buildings, including the pristine glacial Lake Lacawac. Research space is available forvisiting researchers, and participants in the Pocono Comparative Lakes Program.
*Jen Naugle, Chairperson*

**3068 Pennsylvania Department of Conservation and Natural Resources**
Rachel Carson State Office Building 717-787-2869
400 Market Street, PO Box 8767 Fax: 717-772-9106
Harrisburg, PA 17105 http://www.dcnr.state.pa.us
Mission: To maintain, improve and preserve state parks; to manage state forest lands to assure their long-term health, sustainability and economic use; to provide information on Pennsylvania's ecological and geological resources; andto administer grant and technical assistance programs that will benefit rivers conservation, trails and greenways, local recreation, regional heritage conservation and environmental education programs across Pennsylvania.
*Michael DiBerardinis, Secretary*
*Jennifer Stepulitis, Executive Secretary*

**3069 Pennsylvania Fish & Boat Commission: Northeast Region**
5566 Main Road 570-477-5717
Sweet Valley, PA 18656 Fax: 570-477-3221
E-mail: wdietz@state.pa.us
http://www.fishandboat.com
Mission: To protect, conserve, and enhance the Commonwealth's aquatic resources and provide fishing and boating opportunities.
*John Arway, Executive Director*

**3070 Pennsylvania Forest Stewardship Program**
PO Box 8552 717-787-2106
Harrisburg, PA 17105 http://vip.cas.psu.edu/PAprogram.html
Forest stewardship is a US Forest Service program with the goal of helping private landowners manage their lands for various objectives. Landowners participating in the Forest Stewardship Program work with a private forestry consultantto depelop a customized plan for their land and objetives. Studies show that landowners that work with professionals and follow their customized plan are more likely to engage in practices that sustain forest values.

**3071 Pennsylvania Game Commission**
2001 Elmerton Avenue 717-787-4250
Harrisburg, PA 17110 Fax: 717-772-0542
http://www.pgc.state.pa.us
The Pennsylvania Game Commission has the specific responsibility of acting as steward of the Commonwealth's wild birds and wild animals for the benefit of present and future generations. In carrying out this state constitutionalmandate, the Pennsylvania Game Commission will: Protect, conserve and manage the diversity of wildlife and their habitats; Provide wildlife related education, services, and recreational opportunities for both consumptive and non-consumptive uses ofwildlife.

*Thomas Boop, President*
*Roxane Palone, Vice President*

**3072 Susquehanna River Basin Commission**
1721 N Front Street 717-238-0423
Harrisburg, PA 17102 Fax: 717-238-2436
E-mail: srbc@srbc.net
http://www.srbc.net
The responsibility of SRBC is to enhance public welfare through comprehensive planning, water supply allocation & management

of the water resources of the Susquehanna River Basin. The SRBC works to reduce damages caused by floods;provide for the reasonable & sustained development & use of surface & ground water for municipal, agricultural, recreational, commercial & industrial purposes; protect & restore fisheries, wetlands & aquatic habitat; protect water quality & instreamuses.
*Paul O Swartz, Executive Director*

**3073 United States Department of the Army US Army Corps of Engineers**
US Army Engineer District
1000 Liberty Avenue 412-395-7500
Pittsburgh, PA 15222 Fax: 412-644-2811
http://www.usace.army.mil

*Michael P Crall, District Engineer*
*Peter A Steinig, Deputy District Engineer*

## Rhode Island

**3074 Division of Parks and Recreation**
2321 Hartford Avenue 401-222-2632
Johnston, RI 02919 Fax: 401-934-6010
E-mail: riparks@earthlink.net
http://www.riparks.com
The objective of the Division of Parks & Recreation is to provide all Rhode Island Residents and Visitors the opportunity to enjoy a diverse mix of well-maintained, scenic, safe, accessible areas and facilities within our park systemand to offer a variety of outdoor recreational opportunities and programming which may benefit and enhance our Quality of Life.

**3075 Environmental Management: Division of Fish and Wildlife**
4808 Tower Hill Road 401-789-3094
Wakefield, RI 02879 Fax: 401-783-4460
http://www.state.ri.us/dems/programs/
Agency manages the fish and wildlife resources of the State of Rhode Island inluding marine fisheries. The division has 60 employees and is located in 4 stations statwide.

*Michael Lapisky, Chief*

**3076 Rhode Island Department of Environmental Management**
235 Promenade Street 401-222-6800
Providence, RI 02908 http://www.dem.ri.gov
We are committed to preserving the quality of Rhode Island's environment, maintaining and safety of its residents and protecting the natural systems upon which life depends. Together with many partners, we offer assistance toindividuals, business and municipalities, conduct research, find solutions, and enforce laws created to protect the environment.
*W Michael Sullivan, Director*

**3077 Rhode Island Department of Evironmental Management: Forest Environment**
1037 Hartford Pike 401-647-4389
North Scituate, RI 02857 Fax: 401-647-3590
http://www.dem.ri.gov/programs.bnaters/forest/index.htm
Coordinates a statewide forest fire protection plan, provides forest fire protection on state lands, assists rural volunteer fire departments, and develops forest and wildlife management plans for private landowners who choose tomanage their property in ways that will protect these resources on their land. The program promotes public understanding of environmental conservation, enforces Department rules and regulations on DEM lands.
*Catherine Sparks, Chief*

**3078 Rhode Island Water Resources Board**
1 Capitol Hill 401-574-8400
3rd Floor Fax: 401-574-8401
Providence, RI 02908 http://www.wrb.state.ri.us
An executive agency of state government charged with managingther proper development, utilization and conservation of water resources. The primary responsibility is to ensure that sufficient water supply is available for present andfuture generations, apportioning available water to all areas of the state.
*Daniel W Varin, Chairman*
*Juan Mariscal, General Manager*

## South Carolina

**3079 Department of Interior: South Carolina Fish and Wildlife**
Clemson University
261 Lehotsky Hall 864-656-2432
Clemson, SC 29634 Fax: 864-656-1350
E-mail: southeast@fws.gov
http://www.fws.gov/southeast

**3080 Department of Parks, Recreation and Tourism**
1205 Pendleton Street 803-734-1700
Edgar A Brown Building 866-224-9339
Columbia, SC 29201 http://www.travelsc.com
Mission: To learn more about our purpose, mission, and vision.
*Pam Benjamin, Director of Human Resources*

**3081 Office of Environmental Laboratory Certification**
PO Box 72 803-896-0970
State Park, SC 29147 Fax: 803-896-0850
http://www.scdhec.net/envserv/html
We offer certification to any environmental laboratory wishing to analyze samples for South Carolina's Department of Health and Environmental Control [DHEC]. This scope of certification covers the Safe Drinking Water Act (SDWA), theClean Water Act (NPDES), and solid & hazardous wastes including RCRA and CERCLA requirements (SW846 methodologies).

**3082 South Atlantic Fishery Management Council**
4055 Faber Place Drive 843-571-4366
Suite 201 Fax: 843-769-4520
North Charleston, SC 29405 E-mail: safmc@safmc.net
http://www.safmc.net
The South Atlantic Fishery Management Council is headquartered in Charleston, SC, and is responsible for the conservation and management of fish stocks within the 200-mile limit of the Atlantic off the coasts of North Carolina, SouthCarolina, Georgia, and east florida to Key West.
*Robert Mahood, Executive Director*
*Gregg Waugh, Deputy Executive Director*

**3083 South Carolina Department of Health and Environmental Control**
2600 Bull Street 803-898-3432
Columbia, SC 29201 http://www.scdhec.net
Mission: We promote and protect the health of the public and the environment.

**3084 South Carolina Department of Natural Resources**
1000 Assembly Street 803-734-4007
Rembert C Dennis Building Fax: 803-734-4300
Columbia, SC 29201 http://www.dnr.state.sc.us
Mission: To serve as the principal advocate for and steward of South Carolina's natural resources.

*John E Frampton, Director*
*Hank Stallworth, Chief of Staff*

**3085 South Carolina Forestry Commission**
5500 Broad River Road 803-896-8800
Columbia, SC 29212 Fax: 803-798-8097
http://www.trees.sc.gov
The mission of the Forestry Commission is to protect, promote, enhance and nurture the forest lands of South Carolina in a manner consistent with achieving the greatest good for its citizens. Responsibilities extend to all forestlands, both rural and urban, and to all associated forest values and amenities including, but

not limited to: timber, wildlife, water quality, air quality, soil protection, recreation and aesthetics.

*Henry E Kodama ter, State Forester*
*Joel T Felder, Deputy State Forester*

**3086 United States Department of the Army US Army Corps of Engineers**
69A Hagood Avenue 843-329-8000
Charleston, SC 29403 Fax: 843-329-2332
http://www.sac.usace.army.mil

The Charleston District (USACE), South Atlantic Division serves the citizens of South Carolina, the Region, and the Nation by providing quality water resources, value engineering/value management, environmental, and international andinteragency projects and services.
*Lt. Col. Trey Jordan, Commander*

## South Dakota

**3087 Attorney General's Office**
1302 East Highway 14 605-773-3215
Suite 1 Fax: 605-773-4106
Pierre, SD 57501 E-mail: atghelp@state.sd.us
http://www.state.sd.us/attorney/

The Natural Resources Division of the South Dakota Attorney General's Office provides specialized legal counsel to state agencies in environmental, agricultural, financial, Indian law and natural resource matters. It focuses on (1)state boards and agencies which issue environmental, water, and agricultural permits and the lease of state mineral lands; (2) environmental litigation before boards and agencies and in the courts; and (3) jurisdictional disputes.
*Larry Long, Attorney General*

**3088 Department of Environment & Natural Resources**
523 E Capitol Avenue 605-773-3151
Pierre, SD 57501 Fax: 605-773-6035
E-mail: denrinternet@state.sd.us
http://www.state.sd.us/denr/denr.html

Our mission is to provide environmental and natural resources assessment, financial assistance, and regulation in a customer service manner that protects the public health, conserves natural resources, preserves the environment andpromotes economic development.
*Steve Pirner, Secretary of the Department*

**3089 Department of Wildlife and Fisheries Sciences**
South Dakota State University
Box 2140b 605-688-6121
Brookings, SD 57007 Fax: 605-688-4515
E-mail: charles.scalet@sdstate.edu
http://wfs.sdstate.edu/wfsci.htm

*Charles Berry Jr, Leader*

**3090 South Dakota Department of Game, Fish & Parks**
523 E Capitol Avenue 605-773-3485
Pierre, SD 57501 Fax: 605-773-5842
E-mail: wildinfo@state.sd.us
http://sdgfp.info/index.htm

The purpose of the Department of Game, Fish and Parks is to perpetuate, conserve, manage, protect and enhance South Dakota's wildlife resources, parks and outdoor recreational opportunities for the use, benefit and enjoyment of thepeople of this state and its visitors, and to give the highest priority to the welfare of this states's wildlife and parks, and their environment, in planning and decisions.
*Jeff Vonk, Secretary*

**3091 South Dakota Department of Health**
600 E Capitol Avenue 605-773-3361
Pierre, SD 57501 800-738-2301
http://www.state.sd.us/doh

Mission: To prevent disease and promote health, ensure access to needed, high-quality health care, and to efficiently manage public health resources.
*Doneen Hollingsworth, Secretary*

**3092 South Dakota State Extension Services**
South Dakota State University
Box 2207D 605-688-4792
Brookings, SD 57007 Fax: 605-688-6347
http://sdces.sdstate.edu

Extension serves the people of South Dakota by helping them apply unbiased, scientific knowledge to improve their lives. Extension also offers educational information, programs, and services in response to local issues and needs.
*Gregg Carlson, Interim Extension Director*

## Tennessee

**3093 Carbon Dioxide Information Analysis Center**
Oak Ridge National Laboratory
Building 1509 865-574-0390
Bethel Valley Road, PO Box 2008 Fax: 865-574-2232
Oak Ridge, TN 37831 E-mail: cdiac@ornl.gov
http://cdiac.esd.ornl.gov/

The primary global-change data and information analysis center of the US Department of Energy. Responds to data and information requests from users from all over the world who are concerned with the greenhouse effect and global climatechange.
*Tom Boden, Director*

**3094 Obed Wild & Scenic River**
PO Box 429 423-346-6294
Wartburg, TN 37887 Fax: 423-346-3362
E-mail: rebecca_schapansky@nps.gov
http://www.nps.gov/obed/

Approximately 45 miles of wild and scenic river are comprised of the Obed River, Clear Creek, Daddy's Creek and Emory River. These water courses have cut rugged gorges leaving exciting whitewater gorges with bluffs as high as 500 feetabove the water.
*Position Vacant, Director*

**3095 Tennessee Department of Agriculture**
Ellington Agricultural Center 615-837-5103
Melrose Station, PO Box 40627 Fax: 615-837-5333
Nashville, TN 37204 E-mail: TN.Agriculture@state.tn.us
http://www.tennessee.gov/agriculture

Mission: To serve the citizens of Tennessee by promoting wise uses of our agriculture and forest resources, developing economic opportunities, and ensuring safe and dependable food and fiber.
*Ken Givens, Commissioner*

**3096 Tennessee Department of Environment and Conservation**
401 Church Street 615-532-0109
L&C Annex, 1st Floor 888-891-8332
Nashville, TN 37243 http://state.tn.us/environment
*Jim Fyke, Commissioner*
*Paul Sloan, Deputy Commissioner*

**3097 Tennessee Valley Authority**
400 W Summit Hill Drive 865-632-2101
Knoxville, TN 37902 E-mail: tvainfo@tva.com
http://www.tva.gov

TVA generates prosperity in the Tennessee Valley by promoting economic development, supplying low-cost, reliable power and supporting a thriving river system.
*Tom Kilgore, President/CEO*
*William R McCollum Jr, COO*

**3098 United States Army Engineer District: Memphis**
167 N Main Street 901-544-3221
Memphis, TN 38002 800-317-4156
http://www.mvm.usace.army.mil

Provides flood control, navigation, environmental stewardship, emergency operations, and other authorized civil works to benefit the region and the Nation.
*Robert L Van Antwerp, Commander/Chief of Engineers*
*Ronald L Johnson, Deputy Commander*

**3099 University of Tennessee Extension**
2621 Morgan Circle
121 Morgan Hall
Knoxville, TN 37996
865-974-7114
Fax: 865-974-1068
E-mail: tlcross@utk.edu
http://www.utextension.tennessee.edu
Statewide educational organization that brings research-based information about agriculture, family and consumer sciences, and resource development to the people of Tennessee where they live and work.

**3100 Wildlife Resources Agency**
PO Box 40747
Nashville, TN 37204
615-781-6500
Fax: 615-741-4606
http://state.tn.us/twra
The Tennessee Wildlife Resources Agency develops, manages and maintains sound programs of hunting, fishing, trapping, boating, and other wildlife related outdoor recreational activities.

*Ed Carter, Executive Director*

**3101 Wildlife Resources Agency: Fisheries Management Division**
PO Box 40747
Nashville, TN 37204
615-781-6575
Fax: 615-781-6667
http://state.tn.us/twra

## Texas

**3102 Attorney General of Texas Natural Resources Division (NRD)**
300 W 15th Street
PO Box 12548
Austin, TX 78711
512-463-2100
Fax: 512-475-2994
http://www.oag.state.tx.us
The NRD represents the enviromental and energy agencies of the State of Texas in court. NRD's primary activity is the prosecution of lawsuits, referred by state agencies, that involve violations of the state's enviromental and naturalresources protection laws. NRD also defends permits issued by agencies uder those laws and defends challenges to the statues and regulations themselves.NRD also has primary enforcement responsibility for protecting the public's access to Texasbeaches.
*Greg Abbott, Attorney General*

**3103 Bureau of Economic Geology**
University of Texas at Austin
University Station
Box X
Austin, TX 78713-8924
512-471-1534
888-839-4365
Fax: 512-471-0140
E-mail: beg@utexas.edu
http://www.beg.utexas.edu
The Bureau provides wide-ranging advisory, technical, informational, and research-based services to industries, nonprofit organizations, and Federal, State, and local agencies.

*Scott W Tinker, Director*
*Ian Duncan, Associate Director*

**3104 Chihuahuan Desert Research Institute**
PO Box 905
Fort Davis, TX 79734
432-364-2499
Fax: 432-364-2686
http://www.cdri.org
Conducts research on the Chihuahuan Desert.
*Cathryn A Hoyt, Executive Director*
*Jan Hill, Education/Volunteer Coord*

**3105 Environmental Protection Agency: Region VI**
1445 Ross Avenue
Suite 1200
Dallas, TX 75202
214-665-6444
http://www.epa.gov/region06
Region 6 encompasses the ecologically, demographically and economically diverse of states of Arkansas, Louisiana, New Mexico, Oklahoma and Texas. The regional vision is to meet the environmental needs of a changing world.
*Richard Greene, Administrator*
*Larry Starfield, Deputy Administrator*

**3106 Guadalupe: Blanco River Authority**
933 E Court Street
Seguin, TX 78155
830-379-5822
Fax: 830-379-9718
E-mail: comments@gbra.org
http://www.gbra.org
Aims to conserve and protect the water resources of the Guadalupe River basin and make them available for beneficial use. Services include water and wastewater treatment, water quality testing, the management of water rights anddelivery of stored water, the production of electricity from seven hydroelectric plants and engineering design support.

**3107 Parks & Wildlife: Public Lands Division**
4200 Smith School Road
Austin, TX 78744
512-389-4866
Fax: 512-389-4960
http://www.tpwd.state.tx.us
In 1963 the Parks Board merged with the game and Fish Commission to form the Texas Parks and Wildlife Department. The merger created the Parks Division, currently the Public Lands Division. In 1967 park acquisition and developmentincreased with the passage of a $75 million parks bond authorization and the dedication of a portion of the state's cigarette tax to the development of state and local parks.

**3108 Parks & Wildlife: Resource Protection Division**
4200 Smith School Road
Austin, TX 78744
512-389-4864
http://www.tpwd.state.tx.us
The Resource Protection Division protects Tezas fich, wildlife, plant and mineral resources from degradation or depletion. The division investigates any environmental contamination that may cause loss of fish or wildlife. It providesinformation and recommendations to other government agencies and participates in administrative and judicial proceedings concerning pollution incidents, development, development projects and other actions that may affect fish and wildlife.

**3109 Pecos River Commission**
PO Box 340
Monahans, TX 79756
432-940-1753
Fax: 432-943-3267
http://www.tceq.state.tx.us
The Pecos River Compact Commission administers the Pecos River Compact to ensure that Texas receives its equitable share of quality water from the Pecos River and its tributaries as appointed by the Compact. The Compact includes thestates of New Mexico and Texas.
*Julian W Thrasher Jr, Commissioner*

**3110 Rio Grande Compact Commission**
PO Box 1917
El Paso, TX 79950
915-834-7075
Fax: 915-834-7080
http://www.tceq.state.tx.us
The Rio Grande Compact Commission administers the Rio Grande Compact to ensure that Texas receives its equitable share of quality water from the Rio Grande and its tributaries as appointed by the Compact. The Compact includes thestates of Colorado, New Mexico, and Texas.
*Patrick R Gordon, Commisioner*

**3111 Sabine River Compact Commission**
PO Box 556
Mauriceville, TX 77626
409-745-3135
E-mail: webmaster@www.state.tx.us
http://www.tceq.state.tx.us
The Sabine River Compact Commission administers the Sabine River Compact to ensure that Texas receives its equitable share of quality water from the Sabine River and its tributaries as appor-

tioned by the Compact. The Compact includesthe states of Texas and Louisiana.
*Gary E Gagnon, Commissioner*
*Frank E Parker, Commissioner*

**3112 Texas Animal Health Commission**
2105 Kramer Lane 512-719-0700
Austin, TX 78758 Fax: 512-719-0719
http://www.tahc.state.tx.us
TAHC works to keep pests from reoccurring as major livestock health hazards. Ultimately, the TAHC mission and role is the assurance of marketability and mobility of Texas livestock. TAHC works to sustain and continue to make a vitalcontribution to a wholesome and abundant supply of meat, eggs, and dairy products at affordable costs.
*Dee Ellis, Executive Director*
*Gene Snelson, General Counsel*

**3113 Texas Cooperative Extension**
Jack K Williams Administration Bldg 979-845-7800
Room 112, 7101 TAMU Fax: 979-845-9542
College Station, TX 77843 E-mail: tce@tamu.edu
http://texasextension.tamu.edu
*Edward G Smith, Director*

**3114 Texas Department of Agriculture**
PO Box 12847 512-463-7476
Austin, TX 78711 Fax: 888-223-8861
http://www.agr.state.tx.us
*Todd Staples, Commissioner*

**3115 Texas Department of Health**
1100 W 49th Street 512-458-7375
Austin, TX 78756 888-983-7111
Fax: 512-458-7686
E-mail: web.master@dshs.state.tx.us
http://www.tdh.state.tx.us
The Texas Department of Health is the state government agency charged with protecting and promoting the health of the public.

**3116 Texas Forest Service**
301 Tarrow 979-458-6606
Suite 364 Fax: 979-458-6610
College Station, TX 77840 http://texasforestservice.tamu.edu
The mission is to provide statewide leadership and professional assistance to assure the states's forest, tree and related natural resources are wisely used, nurtured, protected and perpetuated for the benefit of all Texans.
*James Hull, Director*

**3117 Texas Natural Resource Conservation Commission**
12100 Park 35 Circle 512-239-1000
Austin, TX 78753 Fax: 512-239-4430
http://www.tnrcc.state.tx.us
The Texas Natural Resource Conservation Commission strives to keep our state's human and natural resources consistent with sustainable economic development. Our goal is clean air, clean water and the safe management of waste.

**3118 Texas Parks & Wildlife Department**
4200 Smith School Road 512-389-4800
Austin, TX 78744 800-792-1112
Fax: 512-389-4814
E-mail: webcomments@tpwd.state.tx.us
http://www.tpwd.state.tx.us
Mission: To manage and conserve the natural and cultural resources of Texas and to provide hunting, fishing and outdoor recreation opportunities for the use and enjoyment of present and future generations.
*Carter Smith, Executive Director*

**3119 Texas State Soil and Water Conservation Board (TSSWCB)**
4311 South 31st Street 254-773-2250
Suite 125 800-792-3485
Temple, TX 76501 Fax: 254-773-3311
http://www.tsswcb.state.tx.us
The state agency that administers Texas' soil and water conservation laws and coordinates conservation and nonpoint source pollution abatement programs through the State. The Board is composed of 7 members, 2 Governor appointed and 5landowners, from across Texas, and is the lead state agency for planning, management, and abatement of agricultural and silvicultural (forestry) nonpoint source pollution, and administers the Texas Brush Control Program. There are regional officesthroughout Texas.
*Hose Dodier Jr, Chairman*
*Barry Mahler, Executive Director*

**3120 United States Department of the Army: US Army Corps of Engineers**
PO Box 17300 817-886-1326
Fort Worth, TX 76102 http://www.usace.army.mil
*Christopher W Martin, Commander*

## Utah

**3121 Cedar City District: Bureau of Land Management**
Department of the Interior
176 East D.L. Sargent Drive 435-586-2401
Cedar City, UT 84720 Fax: 435-865-3058
http://www.ut.blm.gov/cedarcity_fo/index.htm
*Todd S Christiansen, Field Manager*

**3122 Colorado River Basin Salinity Control Advisory Council: Upper Colorado Region**
US Bureau of Reclamation
125 S State Street 801-524-3774
Room 6107 Fax: 801-524-3856
Salt Lake City, UT 84138 http://www.usbr.gov/uc
*Rick Gold, Director*

**3123 Moab District: Bureau of Land Management**
Department of the Interior
82 East Dogwood 435-259-2100
Moab, UT 84532 Fax: 435-259-2106
http://www.ut.blm.gov/moab/index.html
*Maggie Wyatt, Manager*

**3124 Richfield Field Office: Bureau of Land Management**
Department of the Interior
150 E 900 N 435-896-1500
Richfield, UT 84701 Fax: 435-896-1550
http://www.ut.blm.gov/richfield/index.htm
*Cornell Christensen, Manager*

**3125 Salt Lake District: Bureau of Land Management**
Department of the Interior
2370 South 2300 West 801-977-4300
Salt Lake City, UT 84119 http://www.ut.blm.gov/
The BLM administers public lands within a framework of numerous laws. The most comprehensive of these is the Federal Land Policy and Management Act of 1976. All Bureau policies, procedures and management actions must be consistent withFLPMA and the other laws that govern use of the public lands. It is their mission to sustain the health, diversity and productivity of the public lands for the use and enjoyment of present and future generations.
*Glenn Carpenter, Manager*

**3126 Upper Colorado River Commission**
355 S 400 E Street 801-531-1150
Salt Lake City, UT 84111 Fax: 801-789-4883
The Upper Colorado River Commission is an interstate compact administration agency created by the Upper Colorado River Basin Compact of 1948. Since its inception, the Commission (made up of Commissioners appointed by the Governor ofeach Upper Division State and one appointed by the President of the United States) has actively participated in the development, utilization

and conservation of the water resources of the Colorado River Basin.

**3127 Utah Department of Agriculture and Food**
350 N Redwood Road 801-538-7100
Salt Lake City, UT 84114 Fax: 801-538-7126
http://ag.utah.gov

**3128 Utah Geological Survey**
1594 W North Temple, Suite 3110 801-537-3300
PO Box 146100 Fax: 801-537-3400
Salt Lake City, UT 84114 http://geology.utah.gov
The Utah Geological Survey is an applied scientific agency that creates, interprets and provides information about Utah's geologic environment, resources and hazards to promote safe, beneficial and wise use of land.
*Richard Allis, Director*
*Kimm Harty, Deputy Director*

**3129 Utah Natural Resources: Water Resources Section**
1594 W North Temple, Room 310 801-538-7230
Salt Lake City, UT 84116 Fax: 801-538-7279
http://www.water.utah.gov
Mission: Plan, conserve, develop and protect Utah's water resources.
*Dennis J Strong, Director*

**3130 Utah Natural Resources: Wildlife Resource Division**
1594 W N Temple, Suite 2110 801-538-4700
Salt Lake City, UT 84116 Fax: 801-538-4745
http://www.wildlife.utah.gov
Mission: To serve the people of Utah as trustee and guardian of the state's wildlife and to ensure its future and values through management, protection, conservation and education.

**3131 Utah State Department of Natural Resources: Division of Forestry, Fire, & State Lands**
1594 W North Temple, Suite 3520 801-538-5555
PO Box 145703 http://www.ffsl.utah.gov
Salt Lake City, UT 84114
*Dick Buehler, State Forester/Director*

**3132 Vernal District: Bureau of Land Management**
Department of the Interior
170 South 500 East 435-781-4400
Vernal, UT 84078 Fax: 435-781-4410
http://www.ut.blm.gov/vernal/index.html
*Bill Stringer, Manager*

## Vermont

**3133 Department of Forests, Parks, and Recreation**
103 South Main Street 802-241-3670
Waterbury, VT 05671 Fax: 802-244-1481
http://www.vtfpr.org
Mission: To practice and encourage high quality stewardship of Vermont's environment by: monitoring and maintaining the health, integrity and diversity of important species, natural communities, and ecological processes; managingforests for sustainable use; providing and promoting opportunities for compatible outdoor recreation; and furnishing related information, education, and service.
*Jonathan Wood, Commissioner*

**3134 Vermont Agency of Agriculture, Food and Markets**
116 State Street 802-828-2416
Montpelier, VT 05620 E-mail: jwa@agr.state.vt.us
http://www.vermontagriculture.com
*Jon Anderson, Executive Secretary*

**3135 Vermont Agency of Natural Resources**
103 S Main Street 802-241-3600
Waterbury, VT 05671 Fax: 802-244-1102
http://www.anr.state.vt.us
*George Crombie, Secretary*

**3136 Vermont Department of Health**
108 Cherry Street 802-863-7200
PO Box 70 800-464-4343
Burlington, VT 05402 Fax: 802-951-1275
http://healthvermont.gov
The Vermont Department of Health, the state's public health agency, works to protect and improve the health of our population through core public health functions. Core public health functions are those activities that lay thegroundwork for healthy communities.
*Henry Chen MD, Commissioner*

## Virginia

**3137 Commerce and Trade: Mines, Minerals and Energy Department**
Depart. Mines, Minerals & Energy 804-692-3200
9th St Office Bldg, 202 North 9th Street Fax: 804-692-3200
Richmond, VA 23219 http://www.mme.state.va.us/
Mission: To enhance the development and conservation of energy and mineral resources in a safe and environmentally sound manner in order to support a more productive economy in Virginia.
*George P. Willis, Director*

**3138 Conservation & Development of Public Beaches Board**
Virginia Department of Conservation & Recreation
203 Governor Street 804-786-1712
Suite 213 http://www.dcr.state.va.us/sw/pubbeach.htm
Richmond, VA 23219

**3139 Department of Conservation & Recreation: Division of Dam Safety**
203 Governor Street 804-371-6095
Suite 210 Fax: 804-786-0536
Richmond, VA 23219 E-mail: dam@dcr.virginia.gov
http://www.dcr.virginia.gov
The program's purpose is to provide for safe design, construction, operation and maintenance of dams to protect public safety.

**3140 Division of Mineral Resources**
PO Box 3667 804-951-6340
Charlottesville, VA 22903 Fax: 804-951-6365
http://www.dmme.virginia.gov
Mission: To enhance the development and conservation of energy and mineral resources in a safe and environmentally sound manner to support a more productive economy in Virginia. DMR generates, collects, complies, and evaluates geologicdata, creates and publishes geologic maps and report, works cooperatively with other state and federal agencies, and is the primary source of information on geology, minerla and energy resources, and geologic hazards for both the mineral and energyindustries.
*Ed Erb, Director*

**3141 Division of State Parks**
Virginia Department of Conservation & Recreation
203 Governor Street 804-692-0403
Suite 306 800-933-7275
Richmond, VA 23219 Fax: 804-786-9294
E-mail: resvs@dcr.virginia.gov
http://www.dcr.virginia.gov

**3142 Secretary of Commerce and Trade**
PO Box 1475 804-786-7831
Richmond, VA 23218 Fax: 804-371-0250
E-mail: mdd@mme.state.va.us
http://www.commerce.virginia.gov

The secretayr of Commerce and Trade oversees the economic, community, and workforce development of the Commonwealth. Each of the 13 Commerce and Trade agencies actively contributes to the Commonwealth's economic strength and highquality of life.
*Patrick O Gottschalk, Sec of Commerce & Trade*

**3143 US Geological Survey**
12201 Sunrise Valley Drive 703-648-4000
Reston, VA 20192 http://va.water.usgs.gov
Mission: The Unites States Geological Survey serves the Nation by providing reliable scientific information to describe and understand the Earth; minimize loss of life and property from natural disasters; manage water, biological,energy, and mineral resources; and enhance and protect our quality of life.
*Mark D Myers, Director*
*Bob Doyle, Deputy Director*

**3144 Virginia Cooperative Fish & Wildlife Research Unit**
Virginia Polytechnic Institute & State Unversity
106 Cheatham Hall 540-231-4934
Blacksburg, VA 24061 E-mail: vacfwru@listserve.vt.edu
http://www.coopunits.org
A field station of the US Geological Survey, dedicated to research and management of fish and wildlife resources in Virginia and surrounding states. Expertise includes freshwater fish and mollusks and large game mammals. The unitincludes 3 research scientists.
*Dr. Richard Neves, Unit Leader*
*Paul Angermeier, Assistant Unit Leader*

**3145 Virginia Department of Environmental Quality**
629 E Main Street 804-698-4000
PO Box 1105 800-592-5482
Richmond, VA 23218 Fax: 804-698-4500
E-mail: vanaturally@deq.virginia.gov
http://www.deq.virginia.gov
Virginia's regulatory state agency for air, water waste management and coastal resources. The department is also the coordinating clearinghouse for environmental education and information; and maintains the state's gateway.
*David K Paylor, Director*

**3146 Virginia Department of Game & Inland Fisheries: Wildlife Division**
4010 West Broad Street 804-367-9588
Richmond, VA 23230 Fax: 804-367-9147
http://www.dgif.state.va.us

**3147 Virginia Department of Game & Inland Fisheries Fisheries Division**
4010 West Broad Street 804-367-8704
Richmond, VA 23230 Fax: 804-367-9147
http://www.dgif.state.va.us

**3148 Virginia Department of Game and Inland Fisheries**
4010 West Broad Street 804-367-1000
PO Box 11104 Fax: 804-364-9147
Richmond, VA 23230 E-mail: dgifweb@dgif.virginia.gov
http://www.dgif.state.va.us
To manage Virginia's wildlife and inland fish to maintain optimum population of all species to serve the needs of the commonwealth; to provide opportunity for all to enjoy wildlife, inland fish, boating and related outdoor recreation;to promote safety for persons and property in connection with boating, hunting and fishing.
*J Carlton Courter III, Director*

**3149 Virginia Department of Health Commissioners Office**
1500 East Main Street Suite 214 804-786-3561
Richmond, VA 23219 Fax: 804-786-4616

**3150 Virginia Department of Mines, Minerals & Energy: Division of Mined Land Reclamation**
3405 Mountain Empire Road 276-523-8100
Big Stone Gap, VA 24219 Fax: 276-523-8148
E-mail: dmlrinfo@dmme.virginia.gov
http://www.mme.state.va.us

**3151 Virginia Department of Mines, Minerals and Energy: Division of Mineral Resources**
PO Box 3667 804-951-6340
Charlottesville, VA 22903 Fax: 804-951-6365
http://www.geology.state.va.us
Mission: To enhance the development and conservation of energy and mineral resources in a safe and environementally sound manner to support a more productive economy in Virginia.

**3152 Virginia Museum of Natural History**
21 Starling Avenue 276-634-4141
Martinsville, VA 24112 Fax: 276-634-4199
http://www.vmnh.net
We are a state museum of natural history with research scientists in marine biology, vertebrate and invertebrate paleontology, archaeology, earth sciences, entomology, and mammalogy. Creates education programs, exhibits, and fieldtrips focused on natural history and environmental issues. Its publishing division specializes in works by natural scientists and environmental educators in the US and abroad. Writing, editorial, and design services available for books, reports, textbooks, etc..
*Ryan Barber, Director of Marketing*
*Zach Ryder, Marketing Associate*

**3153 Virginia Sea Grant Program**
1208 Greate Road 804-684-7530
PO Box 1346 E-mail: rickards@virginia.edu
Gloucester Point, VA 23062 http://www.virginia.edu
Virginia Sea Grant facilitates research, educational, and outreach activities promoting sustainable management of marine resources.
*William DuPaul, Interim Director*
*Cynthia L Suchman, Assistant Director*

## Washington

**3154 Department of Commerce: Pacific Marine Environmental Laboratory**
7600 Sand Point Way NE 206-526-6239
Seattle, WA 98115 Fax: 206-526-6815
http://www.pmel.noaa.gov
PMEL carries out interdisciplinary scientific investigations in oceanography and atmospheric science.
*Eddie N Bernard, Director*

**3155 Department of Fish and Wildlife**
1111 Washington Street SE 360-902-2200
Olympia, WA 98501 Fax: 360-902-2947
http://wdfw.wa.gov
*Jeffrey P Koenings, Director*

**3156 Environmental Protection Agency: Region 10 Environmental Services**
1200 6th Avenue 206-553-1200
Seattle, WA 98101 800-424-4372
Fax: 206-553-1809
E-mail: philip.jeff@epa.gov
http://www.epa.gov/r10earth

**3157 Julia Butler Hansen Refuge for the Columbian White-Tailed Deer**
PO Box 566 360-795-3915
Cathlamet, WA 98612 Fax: 360-795-0803
http://pacific.fws.gov/refuges/field/WA_julia.htm
Offers critical habitat for the endangered Columbian white-tailed deer. The refuge also provides a wintering area for tundra swans, Canada geese, mallards, American wigeon and pintails. Deer and

elk are easily observed and photographedfrom the country road that circles the mainland portion of the refuge. Evenings and mornings are the best time to spot animals. Open year-round. No fees charged.

**3158 Washington Cooperative Fish & Wildlife Research Unit**
University of Washington
Box 355020 206-543-6475
Seattle, WA 98195 E-mail: washcoop@u.washington.edu
http://www.coopunits.org
*Christian Grue, Unit Leader*
*David Beauchamp, Assistant Unit Leader*

**3159 Washington Department of Ecology**
PO Box 47600 360-407-6000
Olympia, WA 98504 Fax: 360-459-6007
http://www.ecy.wa.gov
The mission is to protect, preserve and enhance Washington's environment and to promote the wise management of our air, land and water.

**3160 Washington Department of Fish & Wildlife: Fish and Wildlife Commission**
600 Capitol Way N 360-902-2267
Olympia, WA 98501 Fax: 360-902-2448
E-mail: commission@dfw.wa.gov
http://wdfw.wa.gov/commission
The Fish and Wildlife Commission's primary role is to establish policy and direction for fish and wildlife species and their habitats in Washington and to monitor the Department's implementation of the goals, policies and objectivesestablished by the Commission. The Commission also classifies wildlife and establishes the basic rules and regulations governing the time, place, manner, and methods used to harvest or enjoy fish and wildlife.
*Miranda Wecker, Chair*
*Gary Douvia, Vice Chair*

**3161 Washington Department of Fish & Wildlife: Habitat Program**
600 Capitol Way N 360-902-2534
Olympia, WA 98501 Fax: 360-902-2946
E-mail: habitatprogram@dfw.wa.gov
http://wdfw.wa.gov
Includes Maps and Digital Info Requests.

**3162 Washington Department of Natural Resources: Southeast Region**
713 Bowers Road 509-925-8510
Ellensburg, WA 98926 E-mail: southeast.region@dnr.wa.gov
http://www.dnr.wa.gov

**3163 Washington Dept. of Natural Resources: Northwest Division**
919 North Township Street 360-856-3500
Sedro Woolley, WA 98284 Fax: 360-856-2150
E-mail: northwest.region@dnr.wa.gov
http://www.dnr.wa.gov

**3164 Washington Dept. of Natural Resources: South Puget Sound Region**
950 Farman Avenue North 360-825-1631
Enumclaw, WA 98022 E-mail: southpuget.region@dnr.wa.gov
http://www.dnr.wa.gov

**3165 Washington Sea Grant Program**
University of Washington
3716 Brooklyn Avenue NE 206-543-6600
Seattle, WA 98105 Fax: 206-685-0380
E-mail: seagrant@u.washington.edu
http://www.wsg.washington.edu/
Mission: Washington Sea Grant serves communities, industries and the people of Washington state, the Pacific Northwest and the nation through research, education and outreach by: identifying and addressing important marine issues;providing better tools for management of the marine environment and use of its resources; and initiating and supporting strategic partnerships within the marine community.
*Penelope D Dalton, Director*
*Raechel Waters, Associate Director*

**3166 Washington State Parks & Recreation Commission: Eastern Region**
7150 Cleanwater Drive SW 360-902-8844
PO Box 42650 http://www.parks.wa.gov
Olympia, WA 98504
The Washington State Parks and Recreation Commission aquires, operates, enhances and protects a diverse system of recreational, cultural, historical, and natural sites. The Commission fosters outdoor recreation and education statewideto provide enjoyment and enrichment for all and a valued legacy to future generations.
*Rex Derr, Director*
*Judy Johnson, Deputy Director*

## West Virginia

**3167 Capitol Conservation District**
418 Goff Mountain Road 304-759-0736
Suite 102 Fax: 304-776-5326
Cross Lanes, WV 25313 E-mail: ccd@wvca.us
http://www.wvca.us
*Clyde Bailey, Chair*

**3168 Gauley River National Recreation Area Advisory National Park Service**
PO Box 246 304-465-0508
Glen Jean, WV 25846 Fax: 304-465-0591
E-mail: katy_miller@nps.gov
http://www.nps.gov/gar www.nps.gov/neri www.nps.gov/blue
Located in the southern West Virginia, New River Gorge National River was established in 1978 to conserve and protect 53 miles of the New River as a free-flowing waterway. This unit of the National Park System encompasses over 70,000acres of land along the New River between the towns of Hinton and Fayetteville. New River Gorge National River and Bluestone National Scenic River are both managed by our same office in Glen Jean, WV.
*Lorrie Sprague, Public Information Officer*

**3169 West Virginia Cooperative Fish & Wildlife Research Unit USGS**
W Virginia University 304-293-3794
PO Box 6125 Fax: 304-293-4826
Morgantown, WV 26506 E-mail: wvcoop@wvu.edu
http://www.coopunits.org
*Patricia Mazik, Unit Leader Fisheries*
*Petra Wood, Assistant Leader Wildlife*

**3170 West Virginia Department of Environmental Protection**
WV Dpt of Environmental Protection 304-926-3647
1356 Hansford Street Fax: 304-926-3637
Charleston, WV 25301 http://www.wvdep.org

**3171 West Virginia Division of Natural Resources**
324 4th Avenue 304-558-2754
South Charleston, WV 25303 Fax: 304-558-2768
http://www.wvdnr.gov
*Frank Jezioro, Director*
*Harry F Price, Executive Secretary*

**3172 West Virginia Geological & Economic Survey**
1 Mont Chateau Road 304-594-2331
Morgantown, WV 26508 Fax: 304-594-2575
E-mail: info@geosrv.wvnet.edu
http://www.wvgs.wvnet.edu
*Michael Hohn, Director/State Geologist*

## Wisconsin

**3173 Great Lakes Indian Fish and Wildlife Commission**
PO Box 9 — 715-682-6619
Odanah, WI 54861 — Fax: 715-682-9294
http://www.glifwc.org
Mission: To help ensure significant, off-reservation harvests while protecting the resources for generations to come.

*James Zorn, Executive Administrator*
*Gerald DePerry, Deputy Administrator*

**3174 Natural Resources Department**
PO Box 7921 — 608-266-2621
Madison, WI 53707 — Fax: 608-261-4380
http://www.dnr.state.wi.us
Mission: To protect and enhance our natural resources; to provide a healthy, sustainable environment; to ensure the right of all people; to work with people; and to consider the future and generations to follow.
*P Scott Hassett, Secretary*

**3175 Wisconsin Cooperative Fishery Research Unit**
University of Wisconsin
College of Natural Resources — 715-346-2178
Stevens Point, WI 54481 — Fax: 715-346-3624
E-mail: coopfish@uwsp.edu
http://www.uwsp.edu/cnr/wicfru

**3176 Wisconsin Department of Agriculture Trade & Cosumer Protection: Land & Water Resources Bureau**
2811 Agriculture Drive — 608-224-4622
PO Box 8911 — Fax: 608-224-4615
Madison, WI 53708 — http://www.datcp.state.wi.us

**3177 Wisconsin Geological & Natural History Survey**
University of Wisconsin Extension
3817 Mineral Point Road — 608-262-1705
Madison, WI 53705 — Fax: 608-262-8086
http://www.uwex.edu/wgnhs
Mission: The survey conducts earth-science surveys, field studies, and research. We provide objective scientific information about the geology, mineral resources, water resources, soil, and biology of Wisconsin. We collect, interpret,disseminate, and archive natural resource information. We communicate the results of our activities through publications, technical talks, and responses to inquiries from the public.
*James Robertson, State Geologist & Director*

**3178 Wisconsin State Extension Services Community Natural Resources & Economic Development**
University of Wisconsin Extension
432 N Lake Street — 608-263-2781
Madison, WI 53706 — Fax: 606-262-9166
http://www.uwex.edu/ces/

*Rick Klemme, Dean/Director Coop Ext*

## Wyoming

**3179 Casper District: Bureau of Land Management**
Department of the Interior
2987 Prospector Drive — 307-261-7600
Casper, WY 82604 — Fax: 307-261-7587
E-mail: casper_wymail@blm.gov
http://www.blm.gov/Director/fo_map/casper_fo.lhtml

**3180 Environmental Quality Department**
122 W 25th Street — 307-777-7937
Herschler Building — Fax: 307-777-7682
Cheyenne, WY 82002 — E-mail: deqwyo@state.wy.us
http://deq.state.wy.us
DEQ contributes to Wyoming's quality of life through a combination of monitoring, permitting, inspection, enforcement and restoration/remediation activities which protect, conserve and enhance the environment while supportingresponsible stewarship of our state's resources.
*John Corra, Director*

**3181 Rock Springs Field Office: Bureau of Land Management**
Department of the Interior
280 Highway 191 N — 307-352-0256
Rock Springs, WY 82901 — Fax: 307-352-0329
http://www.wy.blm.gov
BLM's Rock Springs Field Office is a federal agency in the USA that manages over 3.6 million acres of public land surface and 3.5 million acres of public sub-surface minerals in the southwestern part of the great State of Wyoming. Forthese public lands, BLM administers a variety of programs including mineral exploration and development, wildlife habitat, outdoor recreation, wild horses, lifestock grazing and historic trails.

**3182 Wyoming Board of Land Commissioners**
Herschler Building 3W — 307-777-7331
122 West 25th Street — Fax: 307-777-5400
Cheyenne, WY 82002 — E-mail: slfmail@state.wy.us
http://slf-web.state.wy.us/admin/sblc.aspx

**3183 Wyoming Cooperative Fish and Wildlife Research Unit**
University of Wyoming
Biological Sciences Building — 307-766-5415
Box 3166 — Fax: 307-766-5400
Laramie, WY 82071 — http://uwadmnweb.uwyo.edu
Unit conducts fish and wildlife research for the state of Wyoming conservation, fish department and federal agencies.

**3184 Wyoming State Forestry Division**
1100 W 22nd Street — 307-777-7586
Cheyenne, WY 82002 — Fax: 307-777-5986
E-mail: forestry@state.wy.us
http://slf-web.state.wy.us/forestry.aspx
The Forestry Division's general reposnsibility and objectives are to promote and assist the multiple use management and protection of Wyoming's 270,000 acres of state and 1.9 million acres of private forest lands; to provide forestryassistance and information to landowners, industry, communities and public agencies; and to help provide rural, range, and forest land fire protection, equipment and training.

**3185 Wyoming State Geological Survey**
PO Box 1347 — 307-766-2286
Laramie, WY 82073 — Fax: 307-766-2605
E-mail: wsgs-info@uwyo.edu
http://www.wsgs.uwyo.edu
The Wyoming State Geological Survey's mission is to promote the beneficial and environmentally sound use of Wyoming's vast geologic, mineral, and energy resources while helping to protect the public from geologic hazards.

*Ronald C Surdam, Director*

## Alabama: US Forests, Parks, Refuges

**3186 Bon Secour National Wildlife Refuge**
12295 State Highway 180 — 251-540-7720
Gulf Shores, AL 36542 — Fax: 251-540-7301
E-mail: bonsecour@fws.gov
http://bonsecour.fws.gov/index_files/slide0001.htm
Gulf Shores offer nature enthusiasts much to explore. The Bon Secour NWR, which lies just 6 miles west of Gulf Shores, caters equally to the angler, the hiker and the birder. The refuge encourages guests to enjoy a leisurely hikethrough the grounds or a fishing excursion on the 40-acre fresh water Gator Lake. Pack a picnic lunch and your binoculars, park your blanket on one of the

many secluded beaches and savor the scenery. Call 1-866-SEA TURTLE to report sea turtleactivity.

**3187 Choctaw National Wildlife Refuge**
PO Box 150 251-843-5238
Gilbertown, AL 36908 Fax: 251-843-2568
E-mail: choctaw@fws.gov
http://www.fws.gov/choctaw

The objectives of the Refuge are: to manage habitat for wintering waterfowl, maintain habitat and provide protection for threatened and endangered species, manage wood duck nest boxes and brood rearing habitat, maintain wildlifediversity, manage forest to be productive bottomland hardwoods, and to provide wildlife dependent recreation.

*Robert Dailey, Manager*

**3188 Little River Canyon National Preserve**
2141 Gault Avenue N 256-845-9605
Fort Payne, AL 35967 Fax: 256-997-9129
E-mail: LIRI_Superintendent@nps.gov
http://www.nps.gov/liri

Little River flows for most of its length atop Lookout Mountain in northeast Alabama. The river and canyon systems are spectacular Appalachian Plateau landscapes any season of the year. Forested uplands, waterfalls, canyon rims andbluffs, stream riffles and pools, boulders and sandstone cliffs offer settings for a variety of recreational activities. Natural resources and cultural heritage come together to tell the story of the preserve, a special place in the SouthernAppalachians.

**3189 Wheeler National Wildlife Refuge Complex**
2700 Refuge Headquarters Road 256-353-7243
Decatur, AL 35603 Fax: 256-340-9728
E-mail: wheeler@fws.gov

The 35,000 acre wildlife refuge was established in 1938 The refuge is located between Decatur and Huntsville in the Tennessee River Valley of northern Alabama.

**3190 William B Bankhead National Forest**
Bankhead Ranger District 205-489-5111
PO Box 278 Fax: 205-489-3427
Double Springs, AL 35553

The William B Bankhead national Forest covers 180,000 acres in Franklin, Winston and Lawrence counties. Within the forest are the 26,000 acre Sipsey Wilkderness and the Sipsey Wild and Scenic River, offering 61.4 miles of seasonalcanoeing.

## Alaska: US Forests, Parks, Refuges

**3191 Alagnak Wild River Katmai National Park**
PO Box 245 907-246-3305
King Salmon, AK 99613 Fax: 907-246-2116
http://www.nps.gov/alag

The Alagnak river offers 69 miles of outstanding white-water floating. The river is also noted for abundant wildlife and sport fishing for five species of salmon.

**3192 Alaska Maritime National Wildlife Refuge**
95 Sterling Highway 907-235-6546
Suite 1 Fax: 907-235-7783
Homer, AK 99603 E-mail: alaskamaritime@fws.gov
http://alaska.fws.gov/nwr/akmar/index.htm

To administer a national network of lands and waters for the conservation management and where appropiate, restoration of the fish, wildlife, and plant resources and their habitats within the US for the benefit of present and futuregeneration of Americans.

**3193 Alaska Peninsula National Wildlife Refuge**
PO Box 277 907-246-3339
King Salmon, AK 99613 Fax: 907-246-6696
http://www.gorp.com/gorp/resource/us_nwr/ak_ak_pe.htm

**3194 Becharof National Wildlife Refuge**
PO Box 277 907-246-3339
MS 545 Fax: 907-246-6696
King Salmon, AK 99613 E-mail: becharof@fws.gov
http://becharof.fws.gov

*Daryle Lons, Manager*

**3195 Chugach National Forest**
3301 C Street 907-743-9500
Suite 300 Fax: 907-743-9476
Anchorage, AK 99503 http://www.fs.fed.us/r10/chugach

**3196 Denali National Park and Preserve**
PO Box 9 907-683-2294
Denali Park, AK 99755 Fax: 907-683-9612
E-mail: denali_info@nps.gov
http://www.nps.gov/dena

**3197 Innoko National Wildlife Refuge**
PO Box 69 907-524-3251
MS 549 Fax: 907-524-3141
McGrath, AK 99627 E-mail: innoko@fws.gov
http://innoko.fws.gov

The Innoko National Wildlife Refuge was established December 2, 1980, with the passage of the Alaska National Interest Lands Conservation Act. This 3.85 million acre refuge supports a large nesting waterfowl population, and is wellpopulated with moose, bear, and other animals, as well as a variety of game birds and neotropical bird species and is a relatively flat plain covering much of the drainage area of the Innoko and Iditarod rivers. The vegetation of the reguse is atransition zone.
*William H Schaff, Manager*

**3198 Izembek National Wildlife Refuge**
Box 127 907-532-2445
MS 515 877-837-6332
Cold Bay, AK 99571 Fax: 907-532-2549
E-mail: izembek@fws.gov
http://izembec.fws.gov

Established to conserve fish, wildlife and habitats in their natural diversity including, waterfowl, shorebirds, other migratory birds, brown bears and salmon; to fulfill treaty obligations; to provide the opportunity for continuedsubsistence uses by local residents consistent with the purposes previously mentioned; and to ensure necessary water quality and quantity.

*Nancy Hoffman, Wildlife Refuge Manager*

**3199 Kanuti National Wildlife Refuge**
101 12th Avenue 907-456-0329
MS 555 Room 262 877-220-1853
Fairbanks, AK 99701 Fax: 907-456-0428
http://kanuti.fws.gov

*Mike Spindler, Manager*

**3200 Katmai National Park and Preserve**
PO Box 7 907-246-3305
King Salmon, AK 99613 Fax: 907-246-2116
http://www.nps.gov/katm

**3201 Kenai Fjords National Park**
PO Box 1727 907-224-7500
Seward, AK 99664 Fax: 907-224-7505
http://www.nps.gov/kefj

This park encompasses over 600,000 acres of wild coastal Alaska. The Harding Icefield dominates most of the park. This 300-square mile bowl of ice spills out into numerous glaciers at its edges. Tidewater glaciers and the amazingmarine wildlife of the park can be viewed from boat or air. Humpback whales, orca, many species of sea birds, Steller sea lions and other marine wildlife come here because of the rich variety of foods. Sea birds and sea lions also raise their youngon the rocky sites.

**3202 Kenai National Wildlife Refuge**
PO Box 2139 907-262-7021
Soldotna, AK 99669 Fax: 907-487-2144
http://www.gorp.com/gorp/resource/us_nwr/ak_kenai.htm

**3203 Kobuk Valley National Park**
PO Box 1029 907-442-3890
Kotzebue, AK 99752 Fax: 907-442-8316
E-mail: NWAK_superintendant@nps.gov
http://www.nps.gov/kova
Mission: Cooperative stewardship for the conservation and understanding of natural and cultural resources in Northwest Alaska.

**3204 Kodiak National Wildlife Refuge**
1390 Buskin River Road 907-487-2600
MS 559 888-408-3514
Kodiak, AK 99615 Fax: 907-487-2144
E-mail: kodiak@fws.gov
http://kodiak.fws.gov
Kodiak National Wildlife Refuge was established to conserve Kodiak brown bears, salmon, sea otters, sea lions, other marine mammals, and migratory birds; to fulfill treaty obligations; to provide for continued subsistence uses; and toensure necessary water quanlity and quantity.
*Gary Wheeler, Manager*

**3205 Koyukuk and Nowitna National Wildlife Refuge**
101 Front Street 907-656-1231
PO Box 287 800-656-1231
Galena, AK 99741 Fax: 907-656-1708
E-mail: r7kynwr@fws.gov
http://koyukuk.fws.gov
Approximately 200 miles west of Fairbanks, the refuge lies within a solar basin encircled by rolling hills capped by alpine tundra. The Nowitna River, a nationally designated Wild River, bisects the refuge and forms a broad meanderingfloodplain. The river passes through a scenic 15 mile canyon with peaks up to 2,100 feet.
*Kenton Moos, Manager*

**3206 Lake Clark National Park and Preserve**
240 West 5th Avenue 907-781-2218
Suite 236 E-mail: jeelan_eastlack@nmps.gov
Anchorage, AK 99501 http://www.nps.gov/lacl
Lake Clark National Park and Preserve was created to protect scenic beauty, populations of fish and wildlife, watersheds essential for red salmon, and the traditional lifestyle of local residents.
*Dick Proenneke, Wilderness Steward*

**3207 Selawik National Wildlife Refuge**
160 2nd Avenue 907-442-3799
PO Box 270 800-492-8848
Kotzebue, AK 99752 Fax: 907-442-3124
E-mail: selawik@fws.gov
http://selawik.fws.gov
Selawik National Wildlife Refuge was established to conserve the Western Arctic caribou herd, waterfowl, shorebirds, other migratory birds, salmon, and sheefish; to fulfill treaty obligations; to provide for continued subsistence uses;and to ensure necessary water quality and quantity.
*LeeAnne Ayres, Manager*

**3208 Tetlin National Wildlife Refuge**
PO Box 779 907-883-5312
Tok, AK 99780 Fax: 907-883-5747
E-mail: tetlin@fws.gov
http://tetlin.fws.gov

*Tony Booth, Refuge Manager*

**3209 Togiak National Wildlife Refuge**
PO Box 270 907-842-1063
MS 569 800-817-2538
Dillingham, AK 99627 Fax: 907-842-5402
E-mail: togiak@fws.gov
http://togiak.fws.gov
Established to conserve fish and wildlife populations and habitats in their natural diversity including salmon, marine birds, mammals, migrating birds and large mammals, to fulfill international treaty obligations; to provide forcontinued subsistence uses; and to ensure necessary water quality and quantity.
*Paul Liedberg, Manager*

**3210 Tongass National Forest: Chatham Area**
204 Siginaka Way 907-747-6671
Sitka, AK 99835 Fax: 907-747-4331
http://www.fs.fed.us/r10/tongass
The Tongass National Forest is a forest of islands and trees and rain. It also abounds in animals and birds and fish, with unsurpassed scenery. It's a place where eagles are commonplace, most every road is a deer crossing, and bearsuse the trails too. The spirituality and scenery demands respect.
*Fred Salinas, Forest Supervisor*

**3211 Tongass National Forest: Ketchikan Area**
Federal Building 907-225-3101
648 Mission Street Fax: 907-228-6215
Ketchikan, AK 99901 http://www.fs.fed.us/r10/tongass
The Tongass National Forest is a forest of islands and trees and rain. It also abounds in animals and birds and fish, with unsurpassed scenery. It's a place where eagles are commonplace, most every road is a deer crossing, and bearsuse the trails too. The Tongass is a wild place, where the natural world is a strong presence that nurtures spirituality and materially demands respect.

**3212 Tongass National Forest: Stikine Area**
15 N 12th Street 907-772-3841
PO Box 309 Fax: 907-772-5895
Petersburg, AK 99833 http://www.fs.fed.us/r10/tongass
The Tongass National Forest is a forest of islands and trees and rain. It also abounds in animals and birds and fish, with unsurpassed scenery. It's a place where eagles are commonplace, most every road is a deer crossing, and bearsuse the trails too. The Tongass is a wild place, where the natural world is a strong presence that nurtures spirituality and materially demands respect.

**3213 Yukon Delta National Wildlife Refuge**
807 Chief Eddie Hoffman Road 907-543-3151
PO Box 346 MS 535 Fax: 907-543-4413
Bethel, AK 99559 E-mail: yukondelta@fws.gov
http://yukondelta.fws.gov
Yukon Delta National Wildlife Refuge was established to conserce shorebirds, seabirds, whistling swans, emperor, white-fronted and Canada geese, black brant and other migratory birds, salmon, muskox, and marine mammals; to fulfilltreaty obligations; to provide for continued subsistence uses; and to ensure necessary water quality and quantity.
*Michael Rearden, Manager*

**3214 Yukon Flats National Wildlife Refuge**
101 12th Avenue 907-456-0440
Room 264 MS 575 800-531-0676
Fairbanks, AK 99701 Fax: 907-456-0447
E-mail: yukonflats@fws.gov
http://yukonflats.fws.gov
Located about 100 air miles north of Fairbanks, encompassing about 12 million acres along the Yukon River. In the spring, millions of migrating birds converge on the refuge. With its 40,000 lakes and other wetlands, it has one of thehighest waterfowl nesting densities in North America for ducks, geese, sandhill cranes, loons, grebes and songbirds. Each year, the Yukon Flats is a major contributor to the migrations that occur along the North American flyways.
*Robert Jess, Refuge Manager*
*Wennona Brown, Deputy Refuge Manager*

**3215 Yukon-Charley Rivers National Preserve**
PO Box 167 907-547-2233
Eagle, AK 99738 Fax: 907-547-2247
E-mail: yuch_eagle_cheifofoperation
http://www.nps.gov/yuch

## Arizona: US Forests, Parks, Refuges

**3216 Apache-Sitgreaves National Forest**
PO Box 640 928-333-4301
Springerville, AZ 85938 Fax: 928-333-5966
http://www.fs.fed.us/r3/asnf
Taking care of the land while making the forest resources available to all shareholders. Resources include: high quality water, wilderness, and outdoor recreation; quality habitat for many plants and animals; wood for paper, homes, andhundreds of other uses; forage for wildlife and livestock; a source of minerals.
*Chris Knopp, Forests Supervisor*

**3217 Bill Williams River National Wildlife Refuge**
60911 Highway 95 928-667-4144
Parker, AZ 85344 Fax: 928-667-3402
E-mail: dick_gilbert@fws.gov
http://www.billwilliamsriver.org
Bill Williams River National Wildlife Refuge is located along the Bill Williams River in La Paz and Mojave Counties, Arizona, with the river as the dividing line between the two counties. The refuge was established in 1941 as part ofHavasu NWR as mitigation for the Boulder and Parker Dam projects. In 1993, the two refuges were seperated and the Bill W Unit became the Bill Williams River NWR.
*Richard A Gilbert, Manager*

**3218 Buenos Aires National Wildlife Refuge**
PO Box 109 520-823-4251
Sasabe, AZ 85633 Fax: 520-823-4247
http://southwest.fws.gov/refuges/arizona/buenosaires

**3219 Cabeza Prieta National Wildlife Refuge**
1611 N 2nd Avenue 520-387-6483
Ajo, AZ 85321 Fax: 520-387-5359
E-mail: cabezaprieta@fws.gov
http://www.fws.gov/southwest/refuges/arizona/cabeza.htm
*Roger Di Rosa, Manager*

**3220 Chiricahua National Monument**
12856 East Rhyolite Creek Road 520-824-3560
Wilcox, AZ 85643 Fax: 520-824-3421
http://www.nps.gov/chir
The monument is mecca for hikers and birders. At the intersection of the Chiricahuan and Sonoran deserts, and the southern Rocky Mountains and northern Sierra Madre in Mexico, Chiricahua plants and animals represents one of the premierareas for biological diversity in the northern hemisphere.

**3221 Coconino National Forest**
1824 South Thompson Street 928-527-3600
Flagstaff, AZ 86001 Fax: 928-527-3620
http://www.fs.fed.us/r3/coconino/

**3222 Coronado National Forest**
300 W Congress Street 520-388-8300
Tucson, AZ 85701 http://www.fs.fed.us/r3/coronado
The Coronado National Forest covers 1,780,000 acres of southeastern Arizona and southwestern New Mexico. Elevations range from 3,000 feet to 10,720 feet in 12 widely scattered mountain ranges or sky islands that rise dramatically fromthe desert floor, supporting plant communities as biologically diverse as those encountered on a trip from Mexico to Canada. The views are spectacular from these mountains, and visitors may experience all four seasons during a single day's journey.

**3223 Glen Canyon National Recreation Area**
Glen Canyon NRA 928-608-6200
PO Box 1507 E-mail: GLCA_CHVC@nps.gov
Page, AZ 86040 http://www.nps.gov/glca
Glen Canyon National Recreation Area offers unparalleled opportunities for water-based & backcountry recreation. The recreation area stretches for hundreds of miles from Lees Ferry in Arizona to the Orange Cliffs of southern Utah,encompassing scenic vistas, geologic wonders, and a panorama of human history. Additionally, the controversy surrounding the construction of Glen Cayon Dam and the creation of Lake Powell contributed to the birth of modern day environmental movement.

**3224 Grand Canyon National Park**
PO Box 129 928-638-7888
Grand Canyon, AZ 86023 Fax: 928-638-7797
http://www.nps.gov/grca

**3225 Imperial National Wildlife Refuge**
PO Box 72217 928-783-3371
Yuma, AZ 85365 Fax: 928-783-0652
E-mail: FW2_RW_Imperial@fws.gov
http://www.fws.gov/southwest/refuges/arizona/imperial/index
Imperial National Wildlife Refuge protects wildlife habitat along 30 miles of the lower Colorado River in Arizona and California, including the last unchannelized section before the river enters Mexico.
*Elaine Johnson, Manager*

**3226 Kaibab National Forest**
800 South Sixth Street 928-635-8200
Williams, AZ 86046 Fax: 928-635-8208
http://www.fs.fed.us/r3/kai/

**3227 Kofa National Wildlife Refuge**
356 West 1st Street 928-783-7861
Yuma, AZ 85364 E-mail: FW2_RW_Kofa@fws.gov
http://gorp.away.com/gorp/resource/us_nwr/az_kofa.htm

**3228 Organ Pipe Cactus National Monument**
10 Organ Pipe Drive 520-387-6849
Ajo, AZ 85321 E-mail: orpi_information@nps.gov
http://www.nps.gov/orpi
Organ Pipe Cactus National Monument celebrates the life and landscape of the Sonoran Desert. Here, in this desert wilderness of plants and animals and dramatic mountains and plains scenery, you can drive a lonely road, hike abackcountry trail, camp beneath a clear desert sky, or just soak in the warmth and beauty of Southwest.

**3229 Petrified Forest National Park**
PO Box 2217 928-524-6228
Petrified Forest, AZ 86028 Fax: 928-524-3567
E-mail: PEFO_superintendant@nps.gov
http://www.nps.gov/pefo
Petrified Forest is a surprising land of scenic wonders and fascinating science. The park is located in northeast Arizona and features one of the world's largest and most colorful concentrations of petrified wood. Also included in thepark's 93,533 acres are the multihued badlands of the Chinle Formation known as the Painted Desert, historic structures, archeological sites and displays of 225 millio-year-old fossils.

**3230 Prescott National Forest**
344 S Cortez Street 928-443-8000
Prescott, AZ 86303 http://www.fs.fed.us/r3/prescott
This involves taking care of the land while making the forest resources available to all shareholders. Resources include high quality water, wilderness and outdoor recreation; quality habitat for many plants and animals; wood forpaper, homes and hundreds of other uses; forage for wildlife and livestock; and minerals.

**3231 Saguaro National Park**
3693 S Old Spanish Trail 520-733-5153
Tuscon, AZ 85730 Fax: 520-733-5183
E-mail: mailto:sagu_information@nps.gov
http://www.nps.gov/sagu
This unique desert is home to the most recognizable cactus in the world, the majestic saguaro. Visitors of all ages are fascinated and enchanted by these desert gaints, especially their many interesting and complex interrelationshipswith other desert life. With the average life span of 150 years, a mature saguaro may grow to the height of 50 feet and weigh over 10 tons.

**3232 San Bernardino/Leslie Canyon National Refuge**
PO Box 3509 520-364-2104
Douglas, AZ 85607 Fax: 520-364-2130
http://www.fws.gov/southwest/refuges/arizona/sanbernardino.htm
*Bill Radke, Manager*

**3233 Sunset Crater Volcano National Monument**
6082 Sunset Crater Road 928-526-0502
Flagstaff, AZ 86004 Fax: 928-714-0565
http://www.nps.gov/sucr
Welcome to the Flagstaff Area National Monuments! There is something for everyone: prehistoic cliff dwellings at Walnut Canyon, the mountain scenery and geology of Sunset Crater Volcano, and the painted desert landscape and masonrypueblos of Wupatki National Monument. Here at Sunset Crater Volcano, amid lava and cinders, one can imagine a landscape still hot to the touch. Imagine the thoughts of the prehistoric people who lived here when the eruption occured.

**3234 Tonto National Forest**
2324 East McDowell Road 602-225-5200
Phoenix, AZ 85006 Fax: 602-225-5295
http://www.fs.fed.us/r3/tonto
The Tonto National Forest occupies about 2.8 million acres which generally lie northeast of Phoenix, Ariz., to the Mogollon Rim and east to the San Carlos and Fort Apache Indain Reservations. The west side approximately interstate 17which stretches north of Phoenix to Flagstaff. The lower elevations are of the Sonoran Desert type while the northern portion of the Forest is generally Pinon, Juniper, and Ponderosa Pine types.

**3235 Walnut Canyon National Monument**
3 Walnut Canyon Road 928-526-3367
Flagstaff, AZ 86004 Fax: 928-527-0246
http://www.nps.gov/waca
Hike down into Walnut Canyon and walk in the footsteps of people who lived here over 900 years ago. Built under limestone overhangs, these dwellings were occupied from about 1100 to 1250. Look down into the canyon and imagine the creekrunning through. Visualize a woman hiking up from the bottom with a pot of water on her back. Imagine the men on the rim farming corn or hunting deer. Think of a cold winter night with your family huddled around the fire.

**3236 Wupatki National Monument**
25137 N Wupatki Loop 928-679-2365
Flagstaff, AZ 86004 Fax: 928-679-2349
http://www.nps.gov
Wupatki, comprised of big skies, open grassland, and desert scrub, with Painted Desert to the east and San Francisco Peaks to the west. A visit to this beautiful landscape will remind you what life was like in this region 900 yearsago.
*Chuck Sypher, District Ranger*

## Arkansas: US Forests, Parks, Refuges

**3237 Bald Knob National Wildlife Refuge**
1439 Coal Chute Road 501-724-2458
Bald Knob, AR 72010 Fax: 501-724-2460
E-mail: baldknob@fws.gov
http://www.fws.gov/baldknob
The refuge facts established in 1993 has 14,800 acres and the location is in White County, Ar approximately two miles south of Bald Knob, AR on Coal Chute Road. It provides habitat for migratory waterfowl and other birds. And also forendangered species recreational and environmental education opportunities.
*Robert Alexander, Manager*

**3238 Buffalo National River**
402 N Walnut 870-741-5443
Suite 136 Fax: 870-741-7286
Harrison, AR 72601 E-mail: buff_information@nps.gov
http://www.nps.gov/buff
One of the few remaining unpolluted, free-flowing rivers in the lower 48 states offering both swift-running and placid stretches. The river encompasses 135 miles of the 150 mile long river. It begins as a trickle in the BostonMountains 15 miles above the park boundary. Following what is likely an ancient riverbed, the Buffalo cuts its way through massive limestone bluffs traveling eastward throug the Ozarks and into the Whtie River.

**3239 Felsenthal National Wildlife Refuge**
5531 Highway 82 West 870-364-3167
Crossett, AR 71635 Fax: 870-364-3757
E-mail: felsenthal@fws.gov
http://www.fws.gov/felsenthal
65,000 acre national wildlife refuge with abuntant water resources - 15,000 acres, and vast bottomland hardwood forest that rises to the pine uplands.
*Bernard J Petersen, Project Leader*

**3240 Greers Ferry National Fish Hatchery**
349 Hatchery Road 501-362-3615
Heber Springs, AR 72543 Fax: 501-362-4007
E-mail: greersferry@fws.gov
http://www.fws.gov/greersferry
Through self-guided tours at the hatchery, visitors can observe techniques of trout production, view information exhibits in the aquarium, and see trout in the outdoor raceways. Adjacent to the hatchery, visitors can camp at JFK Parkand trout fish the Little Red River. Nearby, the US Army Corps of Engineers has a visitor center, two mini-hiking trails and an overlook of Greers Ferry Dam. Greers Ferry Lake on the other side of the dam offers camping, swimming, fishing and otherwater sports.
*Sherri Shoults, Hatchery Manager*

**3241 Holla Bend National Wildlife Refuge**
10448 Holla Bend Road 479-229-4300
Dardanelle, AR 72834 Fax: 479-229-4302
E-mail: hollabend@fws.gov
http://www.fws.gov/HollaBend
Part of a system of over 475 national wildlife refuges located across the country. Administered by US Fish and Wildlife Service, this system of refuges, the finest in the world, protects important habitat needed to provide a home for awide variety of wildlife. These refuges also provide the public with valuable opportunities to see and learn about wildlife and to enjoy outdoor activities such as hunting and fishing. Holla Bend's main purpose is to provide a winter home for ducksand geese.

*Durwin Carter, Manager*

**3242 Hot Springs National Park**
101 Reserve Street 501-623-2701
Hot Springs, AR 71901 Fax: 501-624-3458
E-mail: HOSP_Interpretation@nps.gov
http://www.nps.gov/hosp
The park protects eight historic bathhouses with the former luxurious Fordyce Bathhouse housing the park visitor center. The entire Bathhouse Row area is national Historic Landmark District that contains the grandest collection ofbathhouses of its kind in North America. By protecting the 47 hot springs and their watershed, the National Park Service continues to provide visitors with historic leisure activities such as hiking, picnicking and scenic drives.

*Josie Fernandez, Superintendent*
*Mardi Arce, Deputy Superintendent*

**3243 Ouachita National Forest**
Federal Building 501-321-5202
PO Box 1270 Fax: 501-321-5353
Hot Springs, AR 71902 http://www.fs.fed.us/r8/ouachita
Mission: To sustain the ecological health and productivity of lands and waters entrusted to our care and provide for human uses compatible with that goal. We understand that our greatest asset is the land, our greatest strength is ourworkforce, and our greatest challenge is achieving public understanding, trust, and confidence in all that we do.

**3244 Overflow National Wildlife Refuge**
3858 Highway 8 East 870-473-2869
Parkdale, AR 71661 Fax: 870-473-5191
http://southeast.fws.gov/Overflow/index.html
Refuge objectives are to: provide a diversity of habitat types for migratory waterfowl and other birds; provide habitat and protection for the delisted bald eagle; provide opportunities for environmental and ecological research;provide a variety of recreational opportunities consistent with primary wildlife objectives; and expand the public's understanding of and appreciation for the environment with special emphasis on natural resources.

*Lake Lewis, Manager*

**3245 Ozark-St. Francis National Forest**
605 W Main Street 479-964-7200
Russleville, AR 72801 http://www.fs.fed.us/oonf/ozark
The Ozark-St. Francis National Forests are really two separate Forests with many differences. They are distinct in their own topographical, geological, biological, cultural and social differences, yet each makes up a part of theoverall National Forest system.

**3246 Pond Creek National Wildlife Refuge**
1958 Central Road 870-289-2126
Lockesburg, AR 71846 Fax: 870-289-2127
http://www.fws.gov/southeast/PondCreek
Pond Creek National Wildlife Refuge plans to: protect the area's wetland and bottomland hardwood hbitat for natural diversity of wildlife; provide habitat for neo-tropical migratory birds; provide wintering habitat for migratorywaterfowl; provide breeding and nesting habitat for wood ducks; and to provide opportunities for compatible public outdoor recreation.

*Paul Gideon, Manager*

**3247 Wapanocca National Wildlife Refuge**
178 Hammond Avenue Highway 42 East 870-343-2595
PO Box 279 Fax: 870-343-2416
Turrell, AR 72384 E-mail: glen_miller@fws.gov
The 5,484 acre refuge is an important stopover for waterfowl traveling the Mississippi Flyway and for songbirds as they migrate to and from Central and South America. The refuge is open to limited small and big game hunting. Auto tourroutes offers excellent wildlife observation, photogrophy and hiking opportunities. An observation platform is located on the east side of the 600 acre Wapanocca Lake.

**3248 White River National Wildlife Refuge**
57 South CC Camp Road 870-282-8200
PO Box 205 Fax: 870-282-8234
St Charles, AR 72140 E-mail: whiteriver@fws.gov
http://www.fws.gov/whiteriver
Refuge objectives are to provide: optimum habitat for migratory birds; habitat and protection for endangered species: a natural diversity of wildlife common to the White River bottoms; opportunities and facilities for wildlife orientedrecreation and environmental education; cooperation with other water and land managing agencies and private interests to foster proper management of the White River Basin's resources; and preservation of appropriate wooded areas in their naturalcondition.

## California: US Forests, Parks, Refuges

**3249 Angeles National Forest**
701 North Santa Anita Avenue 626-574-1613
Arcadia, CA 91006 Fax: 626-574-5233
http://www.fs.fed.us/r5/angeles
The Angeles National Forest covers 650,000 acres and is the backyard playground to the huge metropolitan area of Los Angeles. The Los Angeles National Forest manages the watersheds eithin its boundaries to provide valuable water tosouthern California and to protect surrounding communities from catastrophics floods.

**3250 Bear Valley National Wildlife Refuge**
Klamath Basin NWR Complex 916-667-2231
Route 1 Box 74 Fax: 916-667-3299
Tulelake, CA 96134 http://www.fws.gov
Refuge was established to protect a major winter night roost site for bald egales. The acquisition program was completed in 1991. Klamath Basin hosts the largest wintering popluation of blad eagles in the contiguous United States, withnumbers some years approaching 1,000. Refuge serves as one of serveral eagle roots in the Basin. It consists of large stands of old-growth tinmber, which protects the birds at night from the harsh winter weather.

**3251 Bitter Creek National Wildlife Refuge**
Hopper Mountain NWR 805-644-5185
Po Box 5839 Fax: 805-644-1732
Ventura, CA 93005 http://hoppermountain.fws.gov/bitterck
The primary wildlife traditional feeding and roosting habitat for the California condor. Also provides habitat for the San Joaquin kit fox, golden eagle, Southern bald eagle and American peregrine falcon. 14,000 contiguous acres,mostly annual grasslands with some juniper and scrub oak with grass understory. Public use is severely limited because of the sensitive situation of the California condor. The refuge can be viewed from the Cerro Noroeste Road.

**3252 Blue Ridge National Wildlife Refuge**
Kern NWR 661-725-2767
Po Box 670 http://hoppermountain.fws.gov/blueridge
Delano, CA 93216
Primary wildlife area is a traditional summer roosting site for the endangered Califorina condor. The habitat includes 897 acres of rugged mountains, rock outcroppings, chaparral and coniferous trees. The refuge is closed to publicaccess due to the sensitivity of California condors and its isolation and difficulty in access.

**3253 Castle Rock National Wildlife Refuge**
Humboldt Bay NWR 707-733-5406
PO Box 576 Fax: 707-733-1946
Loleta, CA 95551 http://www.gorp.com
The Refuge is a 14-acre offshore rock with steep cliffs and sparse vegetation. It was established in 1981 to protect an important migration staging area of the threatened Aleutian Canada goose. Over 21,000 of these roost on the island,which contains the second largest seabird breeding colony in California. Haul-out for a variety of marine mammals, including California sea lion, Stellar sea lion and northern elephant seal. Not open to the public, but wildlife can be observed fromshore.

**3254 Channel Islands National Park**
1901 Spinnaker Drive 805-658-5730
Ventura, CA 93001 Fax: 805-658-5799
E-mail: chis_interpretation@nps.gov
http://www.nps.gov/chis/siteindex.htm
Encompasses five of the eight California Channel Islands and their ocean environment, preserving and protecting a wealth of natural and cultural resources. Marine life ranges from microscopic plankton to the blue whale, the largestanimal to live on Earth. Archeological and cultural resources span a period of more than 10,000 years of human habitation.

**3255 Clear Lake National Wildlife Refuge**
Klamath Basin NWR Complex 916-667-2231
Route 1 Box 74 Fax: 916-667-3299
Tulelake, CA 96134 http://klamathbasinrefuges.fws.gov
Clear Lake National Wildlife Refuge plans to: maintain habitat for endangered, threatened and sensitive species; provide and enhance habitat for fall and spring migrant waterfowl; protect native habitats and wildlife representativeofthe natural biological diversity of the Klamath Basin; integrate the maintenance of productive wetland habitats and sustainable agriculture; and to provide high quality wildlife-dependent visitor services.

**3256 Cleveland National Forest**
10845 Rancho Bernardo Road 858-673-6180
Suite 200 Fax: 858-673-6192
San Diego, CA 92127 http://www.fs.fed.us/r5/cleveland

**3257 Coachella Valley National Wildlife Refuge**
906 W Siclair Road 760-348-5278
PO Box 120 E-mail: clark_bloom@fws.gov
Calipatria, CA 92233 http://www.fws.gov
Contains 13,000 acres consisting of palm oasis woodlands, perennial desert pools and blow-sand habitat. This habitat is critical for the Coachella Valley fringe-toed lizard (Uma inornata) and flat-tailed horned lizard. These threatenedspecies are restricted to the refuge dune system and a few other small areas. Also has the state's second largest grove of native fan palms and the Coachella milk-vetch, a species of special concern.

**3258 Death Valley National Park**
PO Box 579 760-786-3200
Death Valley, CA 92328 Fax: 760-786-3283
E-mail: deva_superintendant@nps.gov
http://pacific.fws.gov/refuges/field/CA_delevan.htm
Death Valley National Park has more than 3.3 million acres of spectacular desert scenery, interesting and rare desert wildlife, complex geology, undisturbed wilderness and sites of historical and cultural interest. The National ParkService is dedicated to the protection and preservation of this park's unique resources for everyone to enjoy now and for future generations.

**3259 Delevan National Wildlife Refuge**
Sacramento NWR Complex 530-934-2801
752 County Road 99W http://sacramentovalleyrefuges@fws.gov
Willows, CA 95988
Delevan NWR is part of the Sacramento NWR Complex and is located in the Sacramento Valley of north-central California. The refuge consists of nearly 5,800 acres comprised of seasonal marsh, permanent ponds, watergrass and uplands inColusa County.

**3260 Devil's Postpile National Monument**
PO Box 3999 760-934-2289
Mammoth Lakes, CA 93546 Fax: 760-934-4780
http://www.nps.gov/depo
The geologic formation that is the Postpile is the world's finest example of unusual columnar basalt. Its columns of lava, with their four to seven sides, display of honeycomb pattern of order and harmony. Another jewel in the Monumentis the San Joaquin River.

**3261 Eldorado National Forest**
100 Forni Road 530-622-5061
Placeville, CA 95667 Fax: 530-621-5297
http://www.fs.fed.us/r5/eldorado
Situated near the California gold discovery site on the American River at Coloma, this forest still boast numerous gold-bearing rivers and streams.Fishing opportunities are abundant. Only 34 miles of waterways are stocked, theremainder contain resident trout. Winter sports are cross country ski, sonowmobile and snowshoe. Backcountry exploration takes place year round in the Desolation and Mokelumme wildernesses.

**3262 Ellicott Slough National Wildlife Refuge**
San Francisco Bay NWR Complex 510-792-0222
Po Box 524 http://www.fws.gov
Newark, CA 94560

**3263 Havasu National Wildlife Refuge**
317 Mesquite Ave 760-326-3853
Needles, CA 92363 Fax: 760-326-5745
E-mail: carol_berry@fws.gov
http://www.fws.gov/southwest/refuges/arizona/havasu

*Carol Berry, Refuge Manager*

**3264 Hopper Mountain National Wildlife Refuge**
PO Box 5839 805-644-5185
Ventura, CA 93005 Fax: 805-644-1732
http://www.fws.gov/hoppermountain
The area is a traditional feeding site for the endangered California condor. Condors use the area frequently from October through May. A variety of other birds occur during migration and year round. The habitat includes 2,471 acres ofgrassland, chaparral and coastal sage scrub. There is a small, 350 acre area of intact California black walnut groves, some of the last remaining in southern California.

**3265 Inyo National Forest**
351 Pacu Lane 760-873-2400
Suite 200 http://www.r5.fs.fed.us/inyo/
Bishop, CA 93514
The Inyo National Forest is a unique and special area of public land located along the aestern edge of California and Sierra Nevada. Extending 165 miles along the California/Nevada border between Los Angeles and Reno, the Inyo Nationalforest includes 1.9 million acres of pristine lakes, fragile meadows, winding streams, rugged Sierra Nevada peaks, and arid Great Baisn Mountains. Elevations range frome 4,000 to 14,495 feet, providing diverse habitats that support vegetationpatterns ranging.

**3266 Joshua Tree National Park**
74485 National Park Drive 760-367-5500
Twentynine Palms, CA 92277 Fax: 760-367-6392
E-mail: JOTR_info@nps.gov
http://www.nps.gov/jotr
Joshua Tree National Park's 794,000 acres span the transition between the Mojave and Colorado deserts of Southern California. Proclaimed a National Monument in 1936 and a Biosphere Reserve in 1984, Joshua Tree was designated a NationalPark In 1994. The area possesses a rich human history and a pristine natural environment.

**3267 Kern National Wildlife Refuge**
PO Box 670 661-725-2767
Delano, CA 93216 Fax: 661-725-6041
http://natureali.org/KNWR.htm

*David Hardt, Manager*

**3268 Kings Canyon National Park**
Sequoia & Kings Canyon National Park
47050 Generals Highway 559-565-3341
Three Rivers, CA 93271 Fax: 559-565-3730
http://kingscanyon.areaparks.com
Kings Canyon National Park, located in California's Sierra Nevada Mountains, is a park most famous for its pristine stands of Giant Sequoia trees. Visitors explore Grant Grove along a network of easy trails. For the more adventurous,longer trails including those in the remote Kings Canyon Backcountry provide a greater challenge.

**3269 Klamath Basin National Wildlife Refuges**
4009 Hill Road 530-667-2231
Tulelake, CA 96134 Fax: 530-667-8337
E-mail: michele_nuss@fws.gov
http://http://klamathbasinrefuges.fws.gov
The Klamath Basin National Wildlife Refuges Complex consists of six refuges in Northern California and Southern Oregon. The refuges support the largest concentration of migratory water fowl on the west coast and the largest winteringnumbers of bald eagles in the lower 48 states.
*Ron Cole, Refuge Manager*
*Greg Austin, Assistant Refuge Manager*

**3270 Klamath National Forest**
1312 Fairlane Road 530-842-6131
Yreka, CA 96097 http://www.fs.fed.us/r5/klamath
The Klamath National Forest covers an area of 1,700,000 acres located in Siskiyou County, Northern California and Jackson County, Oregon. The forest comprises some five wilderness areas, Marble Mountain, Russian Wilderness Area,Trinity Alps, Red Buttes Wilderness Area and Siskiyou Wilderness Area.
*Peg Boland, Forest Supervisor*
*Patricia A Grantham, Deputy Forest Supervisor*

**3271 Lake Tahoe Basin Management Unit**
35 College Drive 530-543-2600
S Lake Tahoe, CA 96150 Fax: 530-573-2693
http://www.fs.fed.us/r5/ltbmu/
Majestic sceenery and diverse recreation oppotunities draw millions of visitors to the Lake Tahoe Basin annually. Changing colors throughout the year afford a brilliant backdrop to the many available activities in all seasons. TheBasin is home to a rich diversity of plants and animals that can be viewed during walks at interpertive sites and on many forest trails.

**3272 Lassen National Forest**
2550 Riverside Drive 530-257-2151
Susanville, CA 96130 http://www.fs.fed.us/r5/lassen
Lassen National Forest lies at the heart of a fascinating part of California, a crossroads of people and nature. This is where the Sierra Nevada, the Cascades, the Modoc Plateau and the Great Basin meet. Within Lassen National Forest,you can explore a lava tube or the land of Ishi, the last survivor of the Yahi Tana Native American tribe: watch prong-horn antelope glide across sage flats; drive four-wheel trails into granite country appointed with sapphire lakes or discoverwildflowers on foot.

**3273 Lava Beds National Monument**
1 Indian Well Headquarters 530-667-8100
Tulelake, CA 96134 Fax: 530-667-2737
E-mail: LABE_SUperintendent@nps.gov
http://www.nps.gov/labe
Volcanic eruptions on the Medicine Lake shield volcano have created an incredibly rugged landscape punctuated by cinder cones, lava flows, spatter cones, lava tube caves and pit craters. During the Modoc War of 1872-1873, the ModocIndains used these tortuous lava flows to their advantage. Under the leadership of Captain Jack, the Modocs took refuge in Captain Jack's Stronghold, a natural lava fortress.

**3274 Los Padres National Forest**
6755 Hollister Avenue 805-968-6640
Suite 150 http://www.fs.fed.us/r5/lospadres
Goleta, CA 93117
Los Padres National Forest encompasses nearly 2 million acres of the central coastal mountains of California.
*Ken Heffner, Forest Supervisor*
*Ann Garland, Deputy Forest Supervisor*

**3275 Lower Klamath National Wildlife Refuge**
Hill Road, Route 1 530-667-2231
Box 74 http://klamathbasinrefuges.fws.gov
Tulelake, CA 96134
The objectives of the refuge are to: maintain habitat for endangered, threatened and sensitive species; provide and enhance habitat for fall and spring migrant waterfowl; protect native habitats and wildlife representative of thenatural biological diversity of the Klamath Basin; integrate the maintenance of productive wetland habitats and sustainable agriculture; provide high quality wildlife-dependent visitor services.

**3276 Mendocino National Forest**
825 North Humboldt Avenue 530-934-3316
Willows, CA 95988 E-mail: mailroom_r5_mendocino@fs.fed.us
http://www.fs.fed.us/r5/mendocino
The Mendocino national Forest was set aside by President Roosevelt in 1907. It was frist named the Stony Creek Reserve and then the Stony Creek National Forest. It was later named the California National Forest and in 1932 became theMendocino National Forest. The MNF straddles the eastern spur of the Coastal Mountain Range in northernwestern Califonia, just a three hour drive north of San Francisco and Sacramento.
*Tom Contreras, Forest Supervisor*

**3277 Modoc National Forest**
800 West 12th Street 530-233-5811
Alturas, CA 96101 http://www.fs.fed.us/r5/modoc
A land of contrasts and unspoiled vaction-hideaway settings. Nestled in the extreme northeastern corner of California, The Modoc National Forest is 140 miles east of Redding on Highway 299, and 169 miles north of Reno, Nevada, viahighway 395. The Modoc National Forest features several mountain areas. The Warner Mountains, on its east, are the western edge of Great Basin Province, the Medicine Lake Highlands, to northwest, are a couthern spur of the Cascade Range.
*Stanley G Sylva, Forest Supervisor*

**3278 Modoc National Wildlife Refuge**
PO Box 1610 530-233-3572
Alturas, CA 96101 Fax: 530-233-4143
E-mail: modoc@fws.gov
http://modoc.fws.gov
A 7,000+ acre refuge established to manage and protect migratory waterfowl.

*Steve Clay, Refuge Manager*
*Sean Cross, Assistant Refuge Manager*

**3279 Pinnacles National Monument**
5000 Highway 146 831-389-4485
Paicines, CA 95043 Fax: 831-389-4489
http://www.nps.gov/pinn
Rising out of the chaparral-covered Gabilan Mountains, east of central California's Salinas Valley, are the spectacular remains of an ancient volcano. Massive monoliths, spires, sheer-walled canyons and talus passages define millionsof years of erosion, faulting and tectonic plate movement is reowned for the beauty and variety of its spring wildflowers. Hiking, rock climbing, picnicing and sildlife observation can be enjoyed throughout the year.

**3280 Pixley National Wildlife Refuge**
Kern NWR 661-725-2767
Po Box 670http://www.fws.gov/pacific/refuges/field/CA_Pixley.htm
Delano, CA 93216
Pixley national Wildlife Refuge provides for the endangered San Joaquin kit fox, Tipton kangaroo rat, and blunt-nosed leopard lizard.

**3281 Plumas National Forest**
159 Lawrence Street 530-283-2050
Quincy, CA 95971 http://www.fs.fed.us/r5/plumas
The Plumas National Forest has fresh conifer forests, rugged cayons, crystal clear lakes, grassy meadows, trout filled streams and brilliant star-filled skies. Located where the Sierra Nevada and Cascade Mountain ranges meet, thisforest has more than 100 lakes, 1,000 miles of rivers and streams, and over a million acres of National Forest.
*Alice Carlton, Forest Supervisor*
*Maria Garcia, Deputy Supervisor*

**3282 Redwood National Park**
1111 2nd Street 707-464-6101
Crescent City, CA 95531 Fax: 707-464-1812
http://www.nps.gov/redw
The world's tallest living trees can found along the northern California coast. Of the coast redwood forests still around today, almost one half of them can be found within the projected boundaries of Redwood National and State Parks.In 1994, the National Park Service and the California State Parks joined forces to manage four parks: Redwood National, Jebediah Smith, Del Norte Coast, and Prairie Creek Redwoods State Parks collectively known as Redwood National and State Parks.

**3283 Sacramento National Wildlife Refuge Complex**
752 County Road 99 West 530-934-2801
Willows, CA 95988 Fax: 530-934-7814
E-mail: sacramentovalleyrefuges@fws.gov
http://sacramentovalleyrefuges.fws.gov
The Complex consists of five national wildlife refuges and three wildlife management areas that comprise over 35,000 acres of wetlands and uplands in the Sacramento Valley of California. In addition there are over 30,000 acres ofconservation easements in the Complex. The refuges and easements serve as resting and feeding areas for nearly half the migratory birds on the Pacific Flyway.

**3284 Salinas River National Wildlife Refuge**
PO Box 524 510-792-0222
Newark, CA 94560 http://www.fws.gov
367 acres of diverse habitats including ocean, beach, dunes, grassland, river, lagoon, and salt marsh.

**3285 San Bernardino National Forest**
602 South Tippecanoe Avenue 909-382-2600
San Bernardino, CA 92408 http://www.fs.fed.us/r5/sanbernardino
In the San Bernardino Mountains, the forest service has developed an extensive network of campgrounds and dozens of picnic areas for families and groups who want to enjoy a day in the mountains. The forest offers camping, picnicking,fishing, boating, swimming, hiking, horseback riding and more. During the winter, visitors come to the forest to cross-contry and down ski, snowboard and snowmobile.
*Jeanne Wade Evans, Forest Supervisor*
*Max Copenhagen, Deputy Forest Supervisor*

**3286 San Francisco Bay National Wildlife Refuge Complex**
9500 Thornton Avenue 510-792-0222
Newark, CA 94560 E-mail: sfbaynwrc@fws.gov
http://www.fws.gov/sfbayrefuges
The San Francisco Bay National Refuge Complex is a collection of seven National Wildlife Refuges administered by the US Fish and Wildlife Service-Antioch Dunes National Wildlife Refuge, Don Edwards San Francisco Bay National Wildliferefuge, Ellicott Slough National Wildlife Refuge, Farallon National Wildlife Refuge, Marin Islands National Wildlife Refuge, Salinas River National Wildlife Refuge, and San Pablo National Wildlife Refuge.

**3287 San Luis National Wildlife Refuge**
PO Box 2176 209-826-3508
Los Banos, CA 93635 Fax: 209-826-1445
http://http://www.fws.gov/sanluis
Mission: Working with others, to conserve, protect and enhance fish, wildlife, and plants and their habitats for the continuing benefit of the American people.

**3288 San Pablo Bay National Wildlife Refuge**
7715 Lakeville Highway 707-769-4200
Petaluma, CA 94954 http://www.pickleweed.org
San Pablo Bay National Wildlife Refuge protects and preserves habitat critical to the survival of the endangered California clapper rail and salt marsh harvest mouse. The Refuge and surrounding San Pablo Bay area also provide winteringhabitat for millions of shorebirds and thousands of waterfowl, including the largest wintering population of canvasbacks on the west coast.
*Jerry Karr, President*
*B K Cooper, Vice President*

**3289 Santa Monica Mountains National Recreation**
401 West Hillcrest Drive 805-370-2301
Thousand Oaks, CA 91360 Fax: 805-370-2351
http://www.nps.gov/samo
Santa Monica Mountains rise above Los Angeles, widen to meet the curve of Santa Monica Bay and reach their highest peaks facing the ocean, forming a beautiful and multi-faceted landscape. Santa Monica Mountains National Recreation Areais a cooperative effort that joins federal, state and local park agencies with private preserves and landowners to protect the natural and cultural resources of this transverse mountain range and seashore.

**3290 Sequoia National Forest**
1839 South Newcomb Street 559-784-1500
Porterville, CA 93257 Fax: 559-781-4744
http://www.fs.fed.us/r5/sequoia
The Sequoia National Forest is at the southern tip of the Sierra Nevada range. Its highest point is 12,432 foot Florence Peak in the Golden Trout Wilderness. The forest has five wildernesses, a scenic byway and four wild and scenicrivers. About 10,000 cows graze on the forest land. Camping, water sports, hiking, downhill and cross-country skiing and horseback riding are amoung the forest's many recreational activities.

**3291 Shasta-Trinity National Forest**
3644 Avtech Parkway 530-226-2500
Redding, CA 96002 Fax: 530-226-2470
http://www.fs.fed.us/r5/shastatrinity
*J Sharon Heywood, Forest Supervisor*

**3292 Sierra National Forest**
1600 Tollhouse Road 559-297-0706
Clovis, CA 93611 E-mail: dkohut@fs.fed.us
http://www.fs.fed.us/r5/sierra
*Edward C Cole, Forest Supervisor*

**3293 Six Rivers National Forest**
1330 Bayshore Way 707-442-1721
Eureka, CA 95501 Fax: 707-442-9242
http://www.fs.fed.us/r5/sixrivers
Six Rivers National Forest lies east of Redwood State and National Parks in northwestern California, and stretches southward from the Oregon border for about 140 miles.
*Tyrone Kelley, Forest Supervisor*
*Will Metz, Deputy Forest Supervisor*

**3294 Sonny Bono Salton Sea National Wildlife Refuge Complex**
906 W Sinclair Road 760-348-5278
Calipatria, CA 92233 http://saltonsea.fws.gov
The Refuge is composed of two disjunctive units, separated by 18 miles of private lands. Each unit contains managed wetland habitat, agricultural fields, and tree rows. The courses of the New and Alamo rivers run through the Refuge,providing freshwater inflow to the Salton Sea.

**3295 Stanislaus National Forest**
19777 Greenley Road 209-532-3671
Sonora, CA 95370 Fax: 209-533-1890
http://www.fs.fed.us/r5/stanislaus
The Stanislaus National Forest, created on February 22, 1897, is among the oldest of the National Forests. It is named for the Stanislaus River whose headwaters rise within Forest boundaries. The Spanish explorer Gabriel Moraga namedthe river Our Lady of Guadalupe during an 1806 expedition. Later, the river was renamed in honor of Estanislao, an Indian leader.

**3296 Sutter National Wildlife Refuge**
Sacramento NWR Complex 530-934-2801
752 County Road 99W Fax: 530-934-7814
Willows, CA 95988 http://www.fws.gov

**3297 Tahoe National Forest**
631 Coyote Street 530-265-4531
Nevada City, CA 95959 http://www.fs.fed.us/r5/tahoe/
The Tahoe National Forest straddles the crest of the Sierra Nevada mountains in northern California, and encompasses a vast territory, from the golden foothills on the western slope to the high peaks of the Sierra crest.

**3298 Tijuana Slough National Wildlife Refuge**
Tijuana River NERR 619-575-2704
301 Caspian Way Fax: 619-575-6913
Imperial Beach, CA 91932 http://www.fws.gov
Established in 1980 to conserve and protect endangered and threatened fish, wildlife and plant species. Conservation of the light-footed clapper rail was the primary impetus for the creation of the refuge. The refuge is part of alarger unit called the Tijuana River National Estuarine Research Reserve, which is administered by the National Oceanographic and Atmospheric Administration.

**3299 Tule Lake National Wildlife Refuge**
4009 Hill Road 530-667-2231
Tulelake, CA 96134 Fax: 530-667-3299
http://www.fws.gov/klamathbasinrefuges/tulelake/tulelake.html
The objectives of Tule Lake are to: maintain habitat for endangered, threatened and sensitive species; provide and enhance habitat for fall and spring migrant waterfowl; protect native habitats and wildlife representative of thenatural biological diversity of the Klamath Basin; integrate the maintenance of productive wet-

land habitats and sustainable agriculture; and ensure that the refuge agricultural practices confirm to the principles of integrated pest management.

**3300 Yosemite National Park**
PO Box 577 209-372-0200
Yosemite National Park, CA 95389 http://www.nps.gov
Yosemite National Park embraces a spectacular tract of mountain- and-valley scenery in the Sierra Nevada, which was set aside as a national park in 1890. The park harbors a grand collection of waterfalls, meadows, and forests thatinclude groves of giant sequoias, the world's largest living things. Highlights of the park include Yosemite Valley, and its high cliffs and waterfalls;Wawona's history center and historic hotel; the Mariposa Grove, which contains hundreds of ancientgiant sequoias.

## Colorado: US Forests, Parks, Refuges

**3301 Alamosa/Monte Vista/ Baca National Wildlife Refuge Complex**
9383 El Rancho Lane 719-589-4021
Alamosa, CO 81101 Fax: 719-587-0595
E-mail: alamosa@fws.gov
http://alamosa.fws.gov
The Valley extends over 100 miles from north to south and 50 miles from east to west, with dwarfing mountains in three directions. The surrounding mountains feed the arid valley with precious surface water, as well as replenish anexpansive underground reservoir. This liquid wealth has made two National Wildlife Refuges possible in the San Luis Valley: Alamosa and Monte Vista. These wetland gems near the heart and on the edge of the Valley are places for wildlife andpeople.

**3302 Arapaho National Wildlife Refuge**
953 JC Rd 32 970-723-8202
Walden, CO 80480 Fax: 970-723-8528
E-mail: arapaho@fws.gov
http://arapaho.fws.gov
Arapaho National Wildlife Refuge supports diverse wildlife habitats including sagebrush grassland uplands, grassland meadows, willow riparian areas, wetlands and mixed conifer and aspen woodland. This refuge is one in a system of over500 National Wildlife Refuges, a network of lands set aside and managed specifically for wildlife. It is administered by the US Fish and Wildlife Service.

**3303 Browns Park National Wildlife Refuge**
1318 Highway 318 970-365-3613
Maybell, CO 81640 Fax: 970-365-3614
E-mail: brownspark@fws.gov
http://www.brownspark.fws.gov
The primary purpose of Browns Park Refuge is to provide high quality nesting and migration habitat for the Great Basin Canada Goose, ducks and other migratory birds. Before Flaming Gorge Dam was constructed in 1962, the Green Riverflooded annually, creating excellent waterfowl nesting, feeding and resting marshes in the backwater sloughs and old stream meanders. The dam stopped the flooding, eliminating much of this waterfowl habitat.

**3304 Colorado National Monument**
Fruita, CO 81521 970-858-3617
Fax: 970-858-0372
E-mail: COLM_Info@nps.gov
http://www.nps.gov/colm
Colorado National Monument consists of geological features including: towering red sandstone monoliths, deep sheer-walled canyons and a variety of wildlife.

**3305 Curecanti National Recreation Area**
102 Elk Creek 970-641-2337
Gunnison, CO 81230 Fax: 970-641-3127
E-mail: CURE_Vis_Mail@nps.gov
http://www.nps.gov/cure
Three reservoirs, named for corresponding dams on the Gunnison River, form the heart of Curecanti National Recreation Area; Blue Mesa Reservoir, Morrow Point Reservoir and the Crystal Reservoir.

**3306 Dinosaur National Monument**
4545 East Highway 40 970-374-3000
Dinosaur, CO 81610 Fax: 970-374-3003
http://www.nps.gov/dino
Dinosaur Monument protects a large deposit of fossil dinosaur bones that lived millons of years ago.

**3307 Florissant Fossil Beds National Monument**
PO Box 185 719-748-3253
Florissant, CO 80816 Fax: 719-748-3164
E-mail: FLFO_Information@nps.gov
http://www.nps.gov/flfo
Huge petrified redwoods and incredibly detailed fossils of ancient insects and plants reveal a very different Colorado of long ago. A lake formed in the valley and the fine-grained sediments at its bottom became the final resting-placefor thousands of insects and plants. These sediments compacted into layers of shale and preserved the delicated details of these organisms as fossils.

**3308 Great Sand Dunes National Park & Preserve**
11999 Highway 150 719-378-6399
Mosca, CO 81146 Fax: 719-378-6310
http://www.nps.gov/grsa
These dunes are the tallest in North America, rising 750 feet from the valley floor. The dunes are home to some unique and spectacular species of flora and fauna. Besides a large variety of birds, there are quite a few species ofmammals that visit or reside within the dunes. Few reptiles are found here due to the high altitude.

**3309 Rio Grande National Forest**
1803 W US Highway 160 719-852-5941
Monte Vista, CO 81144 http://www.fs.fed.us/r2/riogrande
The Rio Grande National Forest is 1.86 million acres located in southwestern Colorado and remains one of the true undiscovered jewels of Colorado. The Continental Divide runs for 236 miles along most of the western border of theForest. The Forest present myraid ecosystems; from 7600-ft alpine desert to over 14,300-ft in the majestic Sangre de Cristo Wilderness on the eastern side.

**3310 Rocky Mountain National Park**
1000 Highway 36 970-586-1206
Estes Park, CO 80517 Fax: 970-586-1256
E-mail: ROMO_informatin@nps.gov
http://www.nps.gov/romo

**3311 San Juan National Forest**
15 Burnett Court 970-247-4874
Durango, CO 81301 Fax: 970-385-1243
http://www.fs.fed.us/r2/sanjuan
San Jaun National Forest, a region of forested mountains, 14,000-foot peaks, scenic roads, geological wonders, hisoric and prehistoric communities, and a narrow-gauge railroad.
*Mark Stiles, Forest Supervisor*
*Howard Sargent, Deputy Forest Supervisor*

**3312 White River National Forest**
900 Grand Avenue 970-945-2521
PO Box 948 Fax: 970-945-3266
Glenwood Springs, CO 81602 http://www.fs.fed.us/r2/whiteriver
The two and one quarter million acre White River National Forest is located in the heart of the Colorado Rocky Mountains, approximately two to four hours west of Denver on Interstate 70. The scenic beauty of the area, along with ampledeveloped and undeveloped recreation opportunities on the forest, accounts for the fact the White River consistently ranks as one of the top five Forests nationwide for total recreation use.

## Connecticut: US Forests, Parks, Refuges

**3313 Stewart B McKinney National Wildlife Refuge**
733 Old Clinton Road 860-399-2513
Westbrook, CT 06498 Fax: 860-399-2515
E-mail: r5rw_sbmnwr@fws.mail.gov
http://www.fws.gov/northeast/mckinney

## Delaware: US Forests, Parks, Refuges

**3314 Bombay Hook National Wildlife Refuge**
2591 Whitehall Neck Road 302-653-6872
Smyrna, DE 19977 E-mail: fw5rw_bhnwr@fws.gov
http://bombayhook.fws.gov
Stretching about eight miles along Delaware Bay and covering nearly 16,000 acres, Bombay Hook NWR was established as a refuge for migratory waterfowl. Today, the Refuge provides habitat for a diversity of wildlife. The refuge offersauto tours, walking trails, observation towers, and interpretive displays for the visiting public.

**3315 Prime Hook National Wildlife Refuge**
11978 Turkle Pond Road 302-684-8419
Milton, DE 19968 Fax: 302-684-8504
E-mail: FW5RW_PHNWR@fws.gov
http://primehook.fws.gov
The Prime Hook National Wlidlife Refuge spans about 10,000 acres along the western Delaware Bay. The marshes of the refuge are ideal habitat for thousands of migrating ducks, geese, and shorebirds. The refuge is also home to woodlandand grassland birds, reptiles, amphibians, and mammals, including the endangered Delmarva Peninsula Fox Squirrel. Avid photographers can enjoy the beauty of wildlife from a photography blind and wheel-chair accessible observation platform.

## District of Columbia: US Forests, Parks, Refuges

**3316 Battleground National Cemetery**
6625 Georgia Avenue, NW 202-426-6924
Washington, DC 20240 Fax: 202-426-1845
http://www.nps.gov/cwdw/btcemet.htm
Battleground National Cemetery, located at 6625 Georgia Avenue NW, was established shortly after the Battle of Fort Stevens in the summer of 1864. The battle, which lasted two days (July 11-12, 1864) marked the defeat of General JubalA Early's Confederate campaign to launch an offensive action against the poorly defended Nation's Capital. Near the entrance are monuments commemorating those units which fought at Fort Stevens.

## Florida: US Forests, Parks, Refuges

**3317 Arthur R Marshall Loxahatchee National Wildlife Refuge**
10216 Lee Road 561-734-8303
Boynton Beach, FL 33437 http://www.fws.gov/loxahatchee
Welcome to the Arthur R. Marshall Loxahatchee National Wildlife Refuge, the last northernmost portion of the unique Everglades. With over 221 square miles of Everglades habitat, A.R.M. Loxahatchee National Wildlife Refuge is home tothe American alligator and the endangered Everglades snail kite. In any given year, as many as 257 species of birds may use the refuge's diverse wetland habitats.

**3318 Big Cypress National Preserve**
33100 Tamiami Trail East 239-695-1201
Ochopee, FL 34141 Fax: 239-695-3901
E-mail: bob_degross@nps.gov
http://www.nps.gov/bicy
The 729,000 acre Big Cypress National Preserve was set aside to ensure the preservation, conservation, and protection of the natural scenic, floral and faunal, and recreational values of the Big Cypress Watershed. The importance ofthis watershed to the Everglades National Park was a major consideration for its establishment. The name Big Cypress refers to the large size of this area. Vast expanses of cypress strands span this unique landscape.

**3319 Biscayne National Park**
9700 SW 328 Street 305-230-7275
Homestead, FL 33033 Fax: 305-230-1190
E-mail: BISC_Information@nps.gov
http://www.nps.gov/bisc
Turquoise waters, emeral islands and fish-bejeweled reefs make Biscayne National Park a paradise for wildlife-watching, snorkeling, diving, boating, fishing and other activities. Within the park boundaries are the longest stretch ofmangrove forest left on Florida's east coast, the clear shallow waters of Biscayne Bay, over 40 of the northernmost Florida Keys, and a spectacular living coral reef. Superimposed on all of this natural beauty is 10,000 years of human history.

**3320 Canaveral National Seashore**
212 South Washington Avenue 321-267-1110
Titusville, FL 32796 Fax: 321-264-2906
E-mail: cana_resource_management@nps.gov
http://www.nbbd.com/godo/cns/
Canaveral National Seashore is a step into the past, protection for the present and a doorway into the future. The 100 Native American Archeological sites that are within our boundaries are evidence of past generations of people thatlived here. Canaveral National Seashore covers 57,000 acres and is the longest stretch (24 miles) of undeveloped beach on Florida's east coast. Fourteen endangered species make their home within Canaveral's boundaries.
*Carol Clark, Superintendent*

**3321 Chassahowitzka National Wildlife Refuge Complex**
1502 SE Kings Bay Drive 352-563-2088
Crystal River, FL 34429 Fax: 352-795-7961
E-mail: chassahowitzka@fws.gov
http://chassahowitzka.fws.gov
A complex of 5 National Wildlife Refuges on the Gulf Coast of Florida including Crystal River NWR. Established in 1983 for the protection of the endangered West Indian manatee. Office hours are from 7:30 AM to 4:00 PM Monday thruFriday. The office is also open Saturdays and Sundays during the winter months (November 15 - March 31) please call the office at 352-563-2088 for more information.
*James Kraus, Manager*

**3322 Dry Torgus National Park & Everglades National Park**
40001 State Road 9336 305-242-7700
Homestead, FL 33034 Fax: 305-242-7711
E-mail: drto_information@nps.gov
http://www.nps.gov/drto
Recognized for its near-pristine natural resources including sea grass beds, fisheries, and sea turtle and nesting habit. The area also lays claim to a rich cultural heritage with a diverse array of themes. Fort Jefferson, on GardenKey, is the park's central cultural feature and one of the largest 19th century American masonry coastal forts.
*Dan Kimball, Superintendent*
*Keith Whisenant, Deputy Superintendent*

**3323 Egmont Key National Wildlife Refuge**
1502 SE Kings Bay Drive 352-563-2088
Crystal River, FL 34429 Fax: 352-795-7961
http://www.fws.gov/egmontkey

This barrier island refuge is approximately 350 acres and was established to provide nesting, feeding and resting habitat for brown pelicans and other migratory birds. The combined resources of the US Fish and Wildlife Service andFlorida Park Service provide protection for Egmont Key and its wildlife, as well as an enjoyable experience for the visitor.
*James Kraus, Manager*

**3324 Everglades National Park**
40001 State Road 9336 — 305-242-7700
Homestead, FL 33034 — Fax: 305-242-7711
E-mail: EVER_Information@nps.gov
http://www.nps.gov/ever
The largest subtropical wilderness in the United States, boasts rare and endangered species. It has been designated a World Heritage Site, international Biosphere Reserve, and Wetland of International Importance, significant to allpeople of the world.
*Dan Kimball, Superintendent*
*Keith Whisenant, Deputy Superintendent*

**3325 Florida Panther National Wildlife Refuge**
3860 Tollgate Boulevard — 239-353-8442
Suite 300 — Fax: 239-353-8640
Naples, FL 34114 — E-mail: floridapanther@fws.gov
http://www.fws.gov/floridapanther
Mission: To conserve and manage lands and waters in concert with other agency efforts within the Big Cypress Watershed, primarily for the Florida Panther, other endangered and threatened species, natural diversity, and culturalresources for the benefit of the American people.
*Layne Hamilton, Manager*

**3326 Hobe Sound National Wildlife Refuge and Nature Center**
PO Box 645 — 772-546-6141
Hobe Sound, FL 33475 — Fax: 772-545-7572
E-mail: hobesound@fws.gov
http://www.fws.gov/hobesound
Refuge objectives are to: maintain and restore diverse habitats designed to achieve refuge purposes and wildlife population objectives; maintain viable diverse populations of native flora and fauna consistent with sound biologicalprinciples; manage natural and cultural resources through land protection and partnership; and develop and implement wildlife dependent recreation and environmental education that leads to enjoyable recreation experiences and a greater understandingof resources.

*Margo Stahl, Refuge Manager*
*Debbie Fritz-Quincy, Nature Center Director*

**3327 JN Darling National Wildlife Refuge**
1 Wildlife Drive — 239-472-1100
Sanibel, FL 33957 — Fax: 239-472-4061
E-mail: dingdarling@fws.gov
http://www.fws.gov/dingdarling
The refuge on Sanibel Island is a subtropical barrier island in the Gulf of Mexico hemmed by mangrove trees, shallow bays and white sandy beaches. The 6,300-acre refuge is connected to the mainland by a three-mile causeway. Named in1967 for Jay Norwood (Ding) Darling, an editorial cartoonist, pioneer conservationist and originator of the federal Duck Stamp Program. Darling, who was the first director of what is now the US Fish and Wildlife Service, wintered on neighboringCaptiva Island.
*Robert Jess, Manager*

**3328 Lake Woodruff National Wildlife Refuge**
2045 Mud Lake Road — 904-985-4673
DeLeon Springs, FL 32130 — Fax: 904-985-0926
E-mail: lakewoodruff@fws.gov
http://www.fws.gov/lakewoodruff
Encompasses two large lakes offering sights of diverse habitats and a variety of wildlife. Fishing and boating are the primary recreational activities.
*Harold Morrow, Project Leader*

**3329 National Key Deer Refuge**
179 Key Deer Blvd — 305-872-2239
Big Pine Key Plaza — Fax: 305-872-2154
Big Pine Key, FL 33043 — E-mail: keydeer@fws.gov
http://nationalkeydeer.fws.gov
Refuge objectives are: to protect and preserve Key deer and other wildlife resources in the Florida Keys; to conserve endangered and threatened fish, wildlife and plants; to provide habitat and protection for migratory birds; and toprovide opportunities for environmental education and public viewing of refuge wildlife and habitats.
*Anne Morkill, Manager*

**3330 Pelican Island National Wildlife Refuge**
1339 20th Street — 772-562-3909
Vero Beach, FL 32960 — Fax: 772-299-3101
E-mail: pelicanisland@fws.gov
http://www.fws.gov/pelicanisland

*Charlie Pelizza, Refuge Manager*
*Nick Wirwa, Refuge Manager*

**3331 St. Marks National Wildlife Refuge**
1255 Lighthouse Road — 850-925-6121
St Marks, FL 32355 — E-mail: saintmarks@fws.gov
http://www.fws.gov/saintmarks
Saint Marks National Wildlife Refuge is in Wakulla, Jefferson and Taylor counties along the Gulf coast of north Florida. The refuge is approximately 25 miles south of Tallahassee. The refuge encompasses 65,000 acres of divided tidalflats; and freshwater impoundments harbor a large variety of wildlife, including 434 verebrate species, excluding fish. Over quarter million vistors enjoy a variety of outdoor recreation opportunities annually.

**3332 St. Vincent National Wildlife Refuge**
PO Box 447 — 850-653-8808
Apalachicola, FL 32329 — Fax: 850-653-9893
E-mail: saintvincent@fws.gov
http://www.fws.gov/saintvincent
The historic St Vincent National Wildlife Refuge is a large barrier island, four miles wide and nine miles long. It was inhabited as early as 240A.D. and is known to have been visited by Franciscan friars in the early 1600s. Over itshistory, private landowners developed the island into a preserve housing Asian and African wildlife and an assortment in between. The US Fish and Wildlife Service purchased the island in 1968 bringing an end to the exotic jungle.
*Monica Harris, Manager*

**3333 Timucuan Ecological & Historic Preserve**
12713 Fort Caroline Road — 904-221-5568
Jacksonville, FL 32225 — Fax: 904-221-5248
http://www.nps.gov/timu
The 46,000 acre Timucuan Ecological and Historic Preserve was established to protect one of the last unspoiled coastal wetlands on the Atlantic Coast and to preserve historic and prehistoric sites within the area. The estuarineecosystem includes aslt marsh, coastal dunes, hardwood hammock, as well as salt, fresh, and brackish waters, all rich in native vegetation and animal life.

*Barbara Goodman, Superintendent*

## Georgia: US Forests, Parks, Refuges

**3334 Chattahoochee-Oconee National Forest**
1755 Cleveland Highway — 770-297-3000
Gainesville, GA 30501 — Fax: 770-297-3011
http://www.fs.fed.us/conf
Together, these forest offer more than ten wilderness areas, six beaches, and thousands of acres of lakes and stream. The forest is real draw for history buffs, who can get absorbed in tracing events such as the Trail of Tears and manya Civil War battle.

**3335 Cumberland Island National Seashore**
PO Box 806 912-882-4336
Saint Marys, GA 31558 Fax: 912-882-6284
http://www.nps.gov/cuis
Cumberland Island is 17.5 miles long and totals 36,415 acres of which 16,850 are marsh, mud flats, and todal crreks. Well know for its sea turtles, abundant shore bierds, dune fields, maritime forest, salt marshes, and historicstructures.

**3336 Okefenokee National Wildlife Refuge**
Route 2 Box 3330 912-496-7836
Folkston, GA 31537 http://www.fws.gov/okefenokee
Okefenokee NWR was established in 1937 to preserve the 438,000 acre Okefenokee Swamp. Presently the refuge encompasses approximately 396,000 acres of which 353,000 are designated as Wilderness. Habitats include open wet prairies,cypress forests, scrub-shrub, oak hammocks and longleaf pine forests. The prosperity and survival of the swamp, and the species dependent on it, is directly tied with maintaining the integrity of complex ecological processes, including hydrology andfire.

**3337 Piedmont National Wildlife Refuge**
718 Juliette Road 478-986-5441
Hillsboro, GA 31038 Fax: 478-986-9646
E-mail: piedmont@fws.gov
http://www.fws.gov/piedmont
35,000 acre national wildlife refuge with hiking trails, gravel roads throughout, wildlife drive, hunting and fishing opportunties, bird watching.

**3338 Pinckney Island National Wild Refuge**
Savannah Coastal Refuges
1000 Business Center Drive 912-652-4415
Suite 10 Fax: 912-652-4385
Savannah, GA 31405 http://www.fws.gov/pinckneyisland
A group of islands and small hammocks, the 4,053-acre refuge includes a variety of land types: saltmarsh, forestland, brushland, fallow fields and freshwater ponds. Pinckney, the largest of the refuge islands, is 3.8 miles long and1.75 miles across at its greatest width. Boaters who navigate the refuge's estuarine waters may view shore and wading birds, including the endangered wood stork, that feeds on mudflats, oysterbeds and shores.

**3339 Savannah National Wildlife Refuge**
1000 Business Center Drive 912-652-4415
Suite 10 Fax: 912-652-4385
Savannah, GA 31405 E-mail: savannacoastal@fws.gov
http://www.fws.gov/savannah
Refuge objectives are to provide: a refuge and feeding ground for native birds and wild animals; habitat and protection for threatened and endangered plants and animals; habitat and sanctuary for migratory birds consistent with theobjectives of the Atlantic Flyway; habitats for other species of indigenous wildlife and fishery resources; management of furbearers, deer and other upland animals; and opportunities for environmental education, interpretation and recreation for thevisiting public.

## Hawaii: US Forests, Parks, Refuges

**3340 Haleakala National Park**
PO Box 369 808-572-4400
Makawao, HI 96768 http://www.nps.gov/hale
The Park preserves the outstanding volcanic landscape of the upper slopes of Haleakala on the island of Maui and protects the unique and fragile ecosystems of Kipahulu Valley, the scenic pools along Oheo gulch, and many rare andendangered species. Haleakala National Park was designated an International Biosphere Reserve.

**3341 Hawaii Volcanoes National Park**
PO Box 52 808-985-6000
Hawaii National Park, HI 96718 Fax: 808-985-6004
http://www.nps.gov/havo
Established in 1916, this 333,000 acre National Park encompasses coastal lava plains, rain forests and deserts. It preserves and protects active volcanoes, rare and endangered plants and animals, and Hawaiian archeological sites.

**3342 Huleia National Wildlife Refuge: Kauai**
PO Box 1128 808-828-1413
Kilauea Fax: 808-828-6634
Kauai, HI 96754 E-mail: shannon_smith@fws.gov
http://www.gorp.com
Located on the southeast side of Kauai, lies adjacent to the famous Menehune Fish Pond, a registered National Historic Landmark. The Huleia Refuge is approximately 241 acres of wetlands which provides habitat for five endangeredHawaiian waterbirds.
*Shannon Smith, Refuge Manager*
*Mike Mitchell, Deputy Refuge Manager*

**3343 James Campbell National Wildlife Refuge**
Oahu Refuge Complex 808-637-6330
66-590 Kam Highway, Rm 2C Fax: 808-637-3578
Haleiwa, HI 96712 E-mail: sylvia_pelissa@fws.gov
http://www.fws.gov/pacificislands/wnwr/ojamesnwr.html
James Campbell NWR lies at the northernmost tip of Oahu near the community of Kahuku and serves as a strategic landfall for native and migratory birds coming from as far away as Alaska, Siberia, and Asia. The specific purpose of theRefuge is to provide habitat for Hawaii's four endemic, endangered waterbirds and other native wildlife, as well as migratory waterfowl and shorebirds. A total of 102 bird species have been documented on the Refuge since its creation.

*Sylvia R Pelizza, Manager*

**3344 Kauai National Wildlife Refuge Complex**
PO Box 1128 808-828-1413
Kilauea, HI 96754 Fax: 808-828-1414
http://www.fws.gov/pacificislands/wnwr/kauainwrindex.html
Kauai is also known as the Garden Isle for its lush vegetation and spectacular waterfalls. It is the oldest of the Hawaiian islands chain and is approximately 553 square miles in size. The Kauai National Wildlife Refuge Complexconsists of Kilauea Point, Hanalei, and Huleia National Wildlife Refuges.

**3345 Kealia Pond National Wildlife Refuge**
Maui Refuge Complex 808-875-1582
PO Box 1042 Fax: 808-875-2945
Kihei, HI 96753 http://www.fws.gov/pacificislands
*Glynnis Nakai, Manager*

**3346 Kilauea Point National Wildlife Refuge**
PO Box 1128 808-828-1413
Kilauea, HI 96754 Fax: 808-828-1414
http://www.fws.gov/pacificislands/wnwr/kkilaueanwr.html
This National Wildlife Refuge provides nesting habitat for seabirds; notably red-footed boobies, Laysan albatross and wedge-tailed shearwaters. The Refuge is also home to the historic Kilauea Point Lighthouse and native marine coastalplant communities. In winter, the 180 foot high precipice provides an ideal site for viewing humpback whales in offshore waters.
*Mike Hawkes, Refuge Manager*

**3347 Oahu National Wildlife Refuge Complex**
66-590 Kamehameha Highway 808-637-6330
Room 2C Fax: 808-637-3578
Haleiwa, HI 96712 http://www.fws.gov/pacificislands
*Sylvia Pelizza, Manager*

**3348 Pearl Harbor National Wildlife Refuge**
66-590 Kamehameha Highway 808-637-6330
Room 2C Fax: 808-637-3578
Haleiwa, HI 96712 http://www.fws.gov/pacificislands
*Sylvia Pelizza, Manager*

## Idaho: US Forests, Parks, Refuges

**3349 Bear Lake National Wildlife Refuge**
PO Box 9 208-847-1757
Montpelier, ID 83254 Fax: 208-847-1319
E-mail: annette_deknijf@fws.gov
http://www.fws.gov/pacific/refuges/field/ID_Bearlk.htm
18,000 acres of marsh and uplands north of Bear Lake proper. Approx 7 miles south of Montpelier. White-faced ibis, Franklin's gulls, sandhill cranes, lots of ducks, Canada geese, Trumpeter swans.

*Annette deKnijf, Refuge Manager*

**3350 Boise National Forest**
1249 S Vinnell Way 208-373-4100
Boise, ID 83709 E-mail: r4_boise_info@fs.fed.us
http://wwwfs.fed.us/r4/boise
The predominantly Ponderosa pine and Douglas fir ecosystem provides homes for fish and wildlife; fiber for wood and paper products; forage for cows and sheep; precious metals for industrial and personal use; and an unlimited menu ofyear round opportunities. The Boise National Forest also contains a number of unique sites, including the Experimental Forest, the Lucky Peak Nursery and Bogus Basin Ski Area.

**3351 Camas National Wildlife Refuge Southeast Idaho Refuge Complex**
2150 E 2350 N 208-662-5423
Hamer, ID 83425 http://www.fws.gov/pacific/refuges
Cama NWR is 36 miles north of Idaho Falls in southeast Idaho, in the Cama Creek floodplain. Elevation is 4,800 feet. About half of its acreage consists of lakes, ponds and marshlands; the remainder is grass sagebrush uplands, meadowsand farm fields. Camas Creek flows for 8 miles through the length of the refuge and is the source of water for many of the lakes and ponds. Tall cottonwood trees along the creek attract a wide variety of songbirds.

**3352 Caribou-Targhee National Forest**
1405 Hollipark Drive 208-524-7500
Idaho Falls, ID 83401 E-mail: jjbennett@fs.fed.us
http://www.fs.fed.us/r4/caribou-targhee
The Caribou National Forest was created to help preserve wilderness land in an area marked by mining activity and westward migration. The forest now covers more than 1 million acres in southeast Idaho, with small portions in Utah andWyoming. Several north-south mountain ranges of the Overthrust Belt dominate the landscape. Caribou National Park offers a wide variety of outdoor activities, including camping, hiking, fishing, climbing, skiing and horseback riding.
*Larry Timchak, Supervisor*
*Lynn Ballard, Public Affairs*

**3353 Clearwater National Forest**
12730 Highway 12 208-476-4541
Orofino, ID 83544 Fax: 208-476-8329
E-mail: elozar@fs.fed.us.com
http://www.fs.fed.us/r1/clearwater/
The Clearwater National Forest covers 1.8 million acres from the jagged peaks of the Bitterroot Mountains in the east to the river canyons and rolling hills of the Palouse Prairie in the west. The North Fork of the Clearwater & theLochsa rivers provide miles of tumbling white water interspersed with quiet pools for migratory and resident fish. The mountains provide habitat for elk, moose, whitetail & mule deer, gray wolf, cougar, mountain goats, and many smaller mammals.
*Tom Reilly, Forest Supervisor*
*Kimberly Nelson, Public Affairs*

**3354 Craters of the Moon National Monument & Preserve**
PO Box 29 208-527-3257
Arco, ID 83213 Fax: 208-527-3073
http://www.nps.gov/crmo
A sea of lava flows with scattered islands of cinder cones and sagebrush characterizes this wierd and scenic landscape known as Craters of the Moon. Craters of the Moon National Monument and Preserve contains three young lava fieldscovering almost half a million acres. These remarkably well preserved volcanic features resulted from geologic events that appear to have happened yesterday and will likely continue tomorrow.

**3355 Deer Flat National Wildlife Refuge**
13751 Upper Embankment Road 208-467-9278
Nampa, ID 83686 Fax: 208-467-1019
E-mail: deerflat@fws.gov
http://deerflat.fws.gov
One of the nation's oldest refuges. It includes the Lake Lowell sector and the Snake River Islands sector. The refuge provides a mix of wildlife habitats, including open waters, wetland edges around the lake, sagebrush uplands,grasslands and riparian forests. More than 250 birds and 30 mammals have been seen on the refuge, providing excellent wildlife observation and photography opportunities.

**3356 Grays Lake National Wildlife Refuge**
74 Grays Lake Road 208-574-2755
Wayan, ID 83285 http://www.fws.gov/pacific/refuges
Twenty-seven miles north of Soda Springs in southeast Idaho, the refuge lies in a high mountain valley at 6,400 feet. Grays Lake is actually a large, shallow marsh with dense vegetation and little open water. Most of the marshvegetation is bulrush and cattail. Adjacent lands are primarily wet meadows and grasslands.

**3357 Idaho Panhandle National Forest**
3815 Schreiber Way 208-765-7223
Coeur d'Alene, ID 83815 Fax: 208-765-7307
http://www.fs.fed.us/ipnf

**3358 Kootenai National Wildlife Refuge**
287 Westside Rd 208-267-3888
Bonners Ferry, ID 83805 Fax: 208-267-5570
E-mail: tammie_coon@fws.gov
http://kootenai.fws.gov/
The 2,774 acre refuge is located in the northern panhandle of Idaho, and serves as a resting and feeding area for migratory birds.

**3359 Minidoka National Wildlife Refuge**
961 E Minidoka Dam 208-436-3589
Rupert, ID 83350 http://www.fws.gov/pacific/refuges
Lying 12 miles northeast of Rupert in the Snake River Valley in south-central Idaho, the refuge extends upstream about 25 miles from the Minidoka Dam along both shores of the Snake River and includes all of Lake Walcott. Over half therefuge is open water, with some small marsh area.

**3360 Nez Perce National Forest**
1005 Highway 13 208-983-1950
Grangeville, ID 83530 Fax: 208-983-4099
http://www.fs.fed.us/r1/nezperce
The Forest is best known for its wild character. Nearly half of the Forest is designated wilderness. It also sports two rivers popular with thrill-seeking floaters-the Selway and the Salmon.
*Jane Cottrell, Forest Supervisor*

**3361 Payette National Forest**
800 W Lakeside Avenue 208-634-0700
McCall, ID 83638 E-mail: webmaster@fs.fed.us
http://www.fs.fed.us/r4/payette
Payette National Forest spans over 2.3 million acres of some of west-central Idaho's most beautiful and diverse country. In one day you can travel from hot desert grasslands through cool conifer forests to snow-capped peaks. Thespecacular land is bordered by two of the deepest canyons in North America— the Salmon River Canyon On the north and Hells Canyon of the Sake River on the west. To the east lies 2.4 million-acre the largest Congressionally designated wilderness inthe lower 48 states.
*Suzanne Rainville, Forest Supervisor*

**3362 Sawtooth National Forest**
2647 Kimberly Road E 208-737-3200
Twin Falls, ID 83301 http://www.fs.fed.us/r4/sawtooth
The Sawtooth National Forest encompasses 2.1 million acres of some of the nation's most magnificent country. Managed and protected by the US Department of Agriculture's Forest Service, the Sawtooth National Forest is working, producingforest that has been providing goods and services to the american people since its establishment in 1905.
*Jane Kollmeyer, Forest Supervisor*

**3363 Southeast Idaho National Wildlife Refuge Complex**
4425 Burley Drive 208-237-6615
Suite A http://www.fws.gov/pacific/refuges/field/ID_SEID.htm
Chubbuck, ID 83202

## Illinois: US Forests, Parks, Refuges

**3364 Mark Twain National Wildlife Refuge Complex**
1704 N 24th Street 217-224-8580
Quincy, IL 62301 E-mail: Dick_Steinbach@fws.gov
http://midwest.fws.gov/marktwain
Administrative office over five refuges; Port Louisa NWR, Great River NWR, Clarence Common NWR, Two Rivers NWR, and Middle Mississippi River NWR.
*Richard Steinbach, Manager*

**3365 Shawnee National Forest**
50 Highway 145 South 618-253-7114
Harrisburg, IL 62946 800-699-6637
Fax: 618-253-1060
http://www.fs.fed.us/r9/forests/shawnee
The Shawnee National Forest lies in the rough, unglaciated areas known as the Illinois Ozark and Shawnee Hills. The geology is spectacular and divergent, with numerous stone bluffs and overlooks transcending to lowland areas.
*Rebecca Banker, Public Affairs*

**3366 Upper Mississippi River National Wildlife & Fish Refuge: Savanna District**
Riverview Road 815-273-2732
Thomson, IL 61285 Fax: 815-273-2960
http://www.fws.gov/midwest/uppermississippiriver
The Upper Mississippi River National Wildlife and Fish Refuge covers nearly 240,000 acres and extends along 281 miles of the Mississippi River. This Refuge is home to a diverse collection of wildlife, including bald eagles, great blueherons, sandhill cranes and spectacular concentrations of waterfowl. Local residents and visitors enjoy a wide array of opportunties throughout the year such as fishing, hunting, wildlife observation, photography, interpretation and environmentaleducation.

*Ed Britton, District Manager*
*Pam Steinhaus, Visitor Service Manager*

## Indiana: US Forests, Parks, Refuges

**3367 Indiana Dunes National Lakeshore**
1100 N Mineral Springs Road 219-926-7561
Porter, IN 46304 http://www.nps.gov/indu
Indiana Dunes National Lakeshore, authorized by Congress in 1996, is located approximately 50 miles southeast of Chicago, Illinois in the counties of Lake, Porter and LaPorte in Northwest Indiana. The national lakeshore runs for nearly25 miles along southern Lake Michigan, bordered by Michigan City, Indiana on the east and Gary on the west. The park contains approximately 15,000 acres, 2,182 of which are located in Indiana Dunes State Park managed by the Indiana Department ofNatural Resources.

## Iowa: US Forests, Parks, Refuges

**3368 DeSoto National Wildlife Refuge**
1434 316th Lane 712-642-4121
Missouri Valley, IA 51555 Fax: 712-642-2877
E-mail: Larry_Klimek@fws.gov
http://www.fws.gov/midwest/desoto
DeSoto Refuge is located along the Missouri River, 25 miles north of Omaha, NE. Popular activities include fishing, picnicking, mushroom-picking, hiking, boating, and wildlife observation. Peak viewing of 500,000 waterfowl inmid-November. The DeSoto Visitor Center houses over 200,000 artifacts from the 1860's steamboat Bertrand.

*Larry Klimek, Refuge Manager*
*Mindy Sheets, Assistant Refuge Manager*

**3369 McGregor District Upper: Mississippi River National Wildlife & Fish Refuge**
PO Box 460 563-873-3423
McGregor, IA 52157 Fax: 563-873-3803
E-mail: uppermississippiriver@fws.gov
http://www.fws.gov/midwest/uppermississippiriver

*T Yager, Manager*

## Kansas: US Forests, Parks, Refuges

**3370 Kirwin National Wildlife Refuge**
702 East Xavier Road 785-543-6673
Kirwin, KS 67644 http://www.fws.gov/kirwin
*Craig Mowry, Manager*
*Diane Stockman, Administrative Assistant*

## Kentucky: US Forests, Parks, Refuges

**3371 Daniel Boone National Forest**
1700 Bypass Road 859-745-3100
Winchester, KY 40391 Fax: 859-744-1568
http://www.fs.fed.us/r8/boone
The mission is to achive quality land management under the sustainable multiple-use management concept to meer the diverse need of people.

**3372 Mammoth Cave National Park**
1 Mammoth Cave Parkway 270-758-2180
PO Box 7 E-mail: MACA_Park_information@nps.gov
Mammoth Cave, KY 42259 http://www.nps.gov/maca
Established to preserve the cave system, including Mammoth Cave, the scenic river valleys of the Green and Nolin Rivers, and a section of the hilly country of south central Kentucky. This is the longest recorded cave system in theworld with more than 360 miles explored and mapped. Established July 1, 1941. Designated a World Heritage Site October 27, 1981. Designated a Biosphere Reserve in 1990.
*Patrick Reed, Superintendent*
*Bruce Powell, Deputy Superintendent*

## Louisiana: US Forests, Parks, Refuges

**3373 Atchafalaya National Wildlife Refuge**
PO Box 127 318-566-2251
Krotz Springs, LA 70750 http://www.fws.gov/atchafalaya
The State of Louisiana's Sherburne Wildlife Management Area is located in the upper third of the Atchafalaya River Basin between Interstate Highway 10 and US Highway 190. It covers approximately 11,780 acres and was established in 1983by the Louisiana Department of Wildlife and Fisheries. The area supervisor's

headquaters is located east of Krotz Springs, Louisiana, on LA 975 approximately three miles south of US Highway 190.

**3374 Cameron Prairie National Wildlife Refuge**
1428 Highway 27 337-598-2216
Bell City, LA 70630 Fax: 337-598-2492
E-mail: cameronprairie@fws.gov
http://www.fws.gov/cameronprairie/index.html
The refuge contains 9,621 acres of fresh marsh, coastal prairie and old rice fields. East Cove Unit of the refuge contains 14,927 acres of brackish and salt marsh. Seasonal visitors include the Peregrine falcon, alligators, whitetailed deer, wading and shorebirds, ducks and geese and various migratory birds.
*Glen Harris, Manager*

**3375 Catahoula National Wildlife Refuge**
PO Drawer Z 318-992-5261
Rhinehart, LA 71363 Fax: 318-992-6023
E-mail: catahoula@fws.gov
http://catahoula.fws.gov
The objectives of the refuge are: to provide wintering habitat for migratory waterfowl consistent with Mississippi Flyway objectives; to provide habitat and protection for endangered species; preserve bottomland hardwoods and providehabitat necessary for wildlife diversity; provide opportunities for environmental education, interpretation and wildlife oriented recreation.
*Greg Harper, Manager*

**3376 D'Arbonne National Wildlife Refuge**
11372 Highway 143 318-726-4400
Farmerville, LA 71363 Fax: 318-726-4667
E-mail: northlarefuges@fws.gov
http://www.fws.gov/darbonne
The Refuge provides habitat for a diversity of migratory birds and resident wildlife species, provides habitat and protection for endangered species such as the bald eagle, wood stork and red-cockaded woodpecker and providesopportunities for wildlife-oriented recreation, environmental education and interpretation.

*Kelby Ouchley, Manager*

**3377 Kisatchie National Forest**
2500 Shreveport Highway 318-473-7160
Pineville, LA 71360 E-mail: www.fs.fed.us/r8/kisatchie
The Kisatchie National Forest has a lot to offer vistors, such as 355 miles of trails for hiking, camping, mountain biking, horseback riding, ORV riding. Other recreational opportunites include four lakes, an 8700 acre Wilderness anddozens of caming sites. The forest also provides opportunities to hunt and fish.

**3378 Lacassine National Wildlife Refuge**
209 Nature Road 337-774-5923
Lake Arthur, LA 70549 Fax: 337-774-9913
E-mail: lacassine@fws.gov
http://www.fws.gov/swlarefugecomplex/lacassine
The nearly 35,000 acre refuge is mostly freshwater marsh habitat. It preserves one of the major wintering grounds for waterfowl in the US. Wintering populations of ducks and geese at Lacassine are among the largest in the NationalWildlife Refuge System. Portions of the refuge are open year-round from one hour before sunrise until one hour after sunset. Please consult refuge brochures or contact the refuge office for more details. A vicinity map and refuge map are available.
*Larry Narcisse, Manager*

**3379 Sabine National Wildlife Refuge**
3000 Holly Beach Highway 337-762-3816
Hackberry, LA 70645 Fax: 337-762-3780
E-mail: sabine@fws.gov
http://www.fws.gov/sabine/
The objectives of the refuge is to provide habitat for migratory waterfowl and other birds, to preserve and enhance coastal marshes for fish and wildlife, and to provide outdoor recreation and environmental education for the public.

*Don Voros, Project Manager*
*Terence Delaine, Refuge Manager*

## Maine: US Forests, Parks, Refuges

**3380 Acadia National Park**
PO Box 177 207-288-8800
Bar Harbor, ME 04609 Fax: 208-288-8813
E-mail: Acadia_Information@nps.gov
http://www.nps.gov/acad/faqs.htm
The purpose of the commission is to consult with the Secretary of the Interior, or his designee, on matters relating to the management and development of the park, including but not liited to the acquisition of lands and interests inlands and termination of rights of use and occupancy.

**3381 Sunkhaze Meadows National Wildlife Refuge**
1168 Main Street 207-827-6138
Old Town, ME 04468 Fax: 207-827-6099
E-mail: info@sunkhaze.org
http://www.sunkhaze.org
Sunkhaze Meadows NWR was established to protect a large peat bog and its associated wildlife. There are three divisions of the refuge, totalling 11,772 acres. The areas are open to wildlife-dependent recreation.

## Maryland: US Forests, Parks, Refuges

**3382 Antietam National Battlefield**
PO Box 158 301-432-5124
Sharpsburg, MD 21782 Fax: 301-432-4590
http://www.nps.gov/anti

**3383 Assateague Island National Seashore**
7206 National Seashore Lane 410-641-1441
Berlin, MD 21811 Fax: 410-641-1099
http://www.nps.gov/asis/home.htm
Natural resource management and environmental education related to coastal resources in general and Assateague Island in particular.
*Trish Kicklighter, Superintendent*

**3384 Patuxent Research Refuge**
10901 Scarlet Tanager Loop 301-497-5580
Laurel, MD 20708 Fax: 301-497-5515
http://www.fws.gov/northeast/patuxent/index.htm
One of over 500 refuges in the National Wildlife Refuge System, a network of lands and waters specifically for the protection of wildlife and wildlife habitat.
*Brad Knudsen, Project Leader/Manager*

## Massachusetts: US Forests, Parks, Refuges

**3385 Boston National Historical Park**
Charleston Navy Yard 617-242-5642
Boston, MA 02129 Fax: 617-242-6006
http://www.nps.gov/bost
Boston National Historical Park tells the story of the events that led to the American Revolution and the Navy that kept the nation strong.

**3386 Cape Cod National Seashore**
99 Marconi Station Site Road 508-349-3785
Wellfleet, MA 02667 Fax: 508-349-9052
E-mail: CACO_Superintendent@nps.gov
http://www.nps.gov/caco/home.html
Cape Cod National Seashore comprises 43,604 acres of shoreline and upland landscape features, including a forty-mile long stretch of pristine sandy beach, dozens of clear, deep, freshwater kettle ponds, and upland scenes that depictevidence of how peo-

ple have used the land. A variety of historic structures are within the boundary of the Seashore, including lighthouses, a lifesaving station, and numerous Cape Cod style houses.

**3387 Great Meadows National Wildlife Refuge**
73 Weir Hill Road 978-443-4661
Sudbury, MA 01776 Fax: 978-443-2898
http://www.fws.gov/northeast/greatmeadows
The Sudbury office serves as the headquarters for the eight refuge Eastern Massachusetts National Wildlife Refuge Complex. Public use opportunities are available at a number of the complex's refuges.
*Libby Herland, Complex Manager*

**3388 Silvio O Conte National Fish & Wildlife Refuge**
52 Avenue A 413-863-0209
Turners Falls, MA 01376 Fax: 413-863-3070
E-mail: boardman@k12.oit.umass.edu
http://www.fws.gov/r5soc
The Connecticut River Watershed, 7.2 million acres in four states, is larger and more heavily populated than areas usually considered when creating a refuge. The refuge's purposes are also much broader than usual. The new scientificand social challenge of protecting natural diversity cannot be met by land acquistion alone. The refuge's primary action is to involve people of the watershed, especially landowners and land managers, in environmental education programs andcooperative management.

## Michigan: US Forests, Parks, Refuges

**3389 Huron-Manistee National Forest**
1755 S Mitchell Street 231-775-2421
Cadillac, MI 49601 800-821-6263
Fax: 231-775-5551
http://www.fs.fed.us/r9/hmnf/
The Huron-Manistee National Forest comprise almost a million acres of public lands extending across the northern lower peninsula of Michigan. The Huron-Manistee national Forest provide recreation opportunities for visitors, habitat forfish and wildlife, and resources for local industry.

**3390 Ottawa National Forest**
E6248 US Highway 2 906-932-1330
Ironwood, MI 49938 Fax: 906-932-0122
http://www.fs.fed.us/r9/ottawa
The almost one million acres of the Ottawa National Forest are located in the western Upper Peninsula of Michigan. It extends from the south shore of Lake Superior down to Wisconsin and the Nicolet National Forest. The area is rich inwildlife viewing opportunities; topography in the northern portion is the most dramatic with breathtaking views of rolling hills dotted with lakes, rivers and spectacular waterfalls.

**3391 Pictured Rocks National Lakeshore**
N8391 Sand Point Road 906-387-2607
PO Box 40 Fax: 906-387-4025
Munising, MI 49862 http://www.nps.gov/piro/home.htm
Multicolored sandstone cliffs, beaches, and dunes, waterfalls, inland lakes, wildlife and the forest of Lake Superior shoreline beckon visitors to explore this 73,000+ acre park. Attractions include a lighthouse and former Coast Guardlife-saving stations along with old farmsteads and orchards. The park is a four season recreational destination where hiking, camping, hunting, nature study, and winter activities abound.

**3392 Seney National Wildlife Refuge**
1674 Refuge Entrance Road 906-586-9851
Seney, MI 49883 Fax: 906-586-3800
E-mail: seney@fws.gov
http://www.fws.gov/midwest/seney
Seney National Wildlife Refuge was established as a refuge and breeding ground for migratory birds and other wildlife. Today, Seney supports a variety of wildlife, including protected and reintroduced species. Bald eagles, commonloons, and trumpeter swans are regularly seen during the summer months, especially June and July, when they are raising their young.

*Mark Vaniman, Manager*
*Greg McClellan, Deputy Manager*

**3393 Shiawassee National Wildlife Refuge**
6975 Mower Road 989-777-5930
Saginaw, MI 48601 Fax: 989-777-9200
E-mail: shiawassee@fws.gov
http://www.fws.gov/midwest/shiawassee
The refuge is comprised of over 9400 acres of wetlands, uplands, and bottomland hardwood forests. Four rivers flow through. Over 12 miles of hiking trails, bank fishing sites, and two photography blinds are available. Environmentaleducation progrms are offered at Green Point Environmental Learning Center in Saginaw.

*Steve Kahl, Manager*
*Ed DeVries, Assistant Manager*

**3394 Sleeping Bear Dunes National Lakeshore**
9922 Front Street 231-326-5134
Empire, MI 49630 Fax: 231-326-5382
http://www.nps.gov/slbe
Sleeping Bear Dunes National Lakeshore encompasses a 60 km stretch of Lake Michigan's eastern coastline, as well as North and South Manitou Islands. The park was established primarily for its outstanding natural features, includingforests, beaches, dune formations and ancient glacial phenomena. The Lakeshore also contains many cultural features including an 1871 lighthouse, three former Life-Saving Service (Coast Guard) Stations and an extensive rural historic farm district.

## Minnesota: US Forests, Parks, Refuges

**3395 Agassiz National Wildlife Refuge**
22996 290th Street Northeast 218-449-4115
Middle River, MN 56737 Fax: 218-449-3241
E-mail: agassiz@fws.gov
http://midwest.fws.gov/agassiz/
The National Refuge provides resting, nesting and feeding habitat for waterfowl and other migratory birds. Agassiz NWR is designated a Globally Important Bird Area by the American Bird Conservancy. It protects endangered and threatenedspecies. It also provides for biodiversity, public opportunities for outdoor recreation and environmental education.
*Margaret Anderson, Manager*

**3396 Big Stone National Wildlife Refuge**
44843 County Road 19 320-273-2191
Odessa, MN 56276 Fax: 320-273-2231
E-mail: BigStone@fws.gov
http://midwest.fws.gov/bigstone
Fig Stone NWR is one of more than 545 National Wildlife Refuges administered as part of the National Wildlife Refuge System by the US Fish and Wildlife Service. The Refuge now overlays 11,585.8 acres of the Minnesota River Valley inwestern Minnesota. A unique visual and geological feature of the Refuge is the red, lichen covered granite outcrops for which the Refuge was named. The Refuge offers an auto tour route, nature trails, wildlife observation, hunting and fishingopportunities.

*Alice Hanley, Manager*

**3397 Chippewa National Forest**
200 Ash Avenue Northwest 218-335-8600
Cass Lake, MN 56633 http://www.fs.fed.us/r9/chippewa/
The Chippewa was the first National Forest established east of the Mississippi. The Forest boundary encompasses 1.6 millon, of which over 666,618 acres are managed by the USDA Forest Service. Aspen, birch, pines, balsam fir and maplesblanket the uplands. Water is abundant, with over 1300 lakes, 923 miles of rivers and streams, and 400,000 acres of wetlands.
*Robert M Harper, Forest Supervisor*

**3398 Crane Meadows National Wildlife Refuge**
19502 Iris Road 320-632-1575
Little Falls, MN 56345 Fax: 320-632-5471
E-mail: cranemeadows@fws.gov
http://www.fws.gov/midwest/CraneMeadows
Crane Meadows National Wildlife Refuge was established in 1992 to preserve a large, natural wetland complex. The refuge is located in central Minnesota and serves as an important stop for many species of migrating birds.

**3399 Detroit Lakes Wetland Management District**
26624 N Tower Road 218-847-4431
Detroit Lakes, MN 56501 Fax: 218-847-4156
E-mail: DetroitLakes@fws.gov
http://www.fws.gov/Midwest.detroitlakes/
is divided into three general landscape areas covering approximately 6000 square miles. From east to west, these are: the Red River Valley floodplain, the glacial moraine/prairie pothole region, and the hardwood/coniferous forest.Land acquisition and management efforts are focused in the prairie pothole region of the WMD, with a goal of providing habitat for nesting waterfowl.

*Scott Kahan, Manager*

**3400 Hiawatha National Forest**
2727 North Lincoln Road 906-786-4062
Escanaba, MI 49829 Fax: 906-789-3311
http://www.fs.fed.us/r9/forests/hiawatha
Located in the central and easter Upper Peninsula of Michigan, the firest affords visitors access to white sand, scenic beaches and relatively undeveloped shorelines along three of Americas's Inland Seas, Lake Superior, Michigan andHuron.

**3401 Mille Lacs National Wildlife Refuge**
36289 State Highway 65 218-768-2402
McGregor, MN 55760 Fax: 218-768-3040
E-mail: millelacs@fws.gov
http://midwest.fws.gov/millelacs
Comprised of two small islands in Mille Lacs Lake in central Minnesota. The islands are boulder and gravel outcrops for colonial nesting birds including common terns and ring-billed gulls.

*Walt Ford, Manager*

**3402 Minnesota Valley National Wildlife Refuge**
3815 American Blvd East 612-854-5900
Bloomington, MN 55425 Fax: 612-725-3279
E-mail: minnesotavalley@fws.gov
http://www.fws.gov/midwest/MinnesotaValley
Mission: To restore and manage the ecological communities of the Lower Minnesota River Valley and its watershed while providing environmental education and wildlife dependent recreation.
*Thomas Larson, Manager*

**3403 Rice Lake National Wildlife Refuge**
36289 State Highway 65 218-768-2402
McGregor, MN 55760 Fax: 218-768-3040
E-mail: ricelakes@fws.gov
http://www.fws.gov/midwest/RiceLake
*Walt Ford, Manager*

**3404 Rydell National Wildlife Refuge**
17788 349th Street Southeast 218-687-2229
Erskine, MN 56535 Fax: 218-687-2225
E-mail: dave_bennett@fws.gov
http://www.fws.gov/midwesty/rydell

*Dave Bennett, Manager*

**3405 Sherburne National Wildlife Refuge**
17076 293rd Avenue 763-389-3323
Zimmerman, MN 55398 Fax: 763-389-3493
E-mail: sherburne@fws.gov
http://www.fws.gov/midwest/sherburne/index.htm
Mission: To represent a diverse biological community characteristic of the transition zone between tallgrass prairie and forest.

**3406 Superior National Forest**
8901 Grand Avenue Place 218-626-4300
Duluth, MN 55808 Fax: 218-626-4398
E-mail: r9 superior NF@fs.fed.us
http://www.fs.fed.us/r9/forests/superior
Located in northeastern tip of Minnesota, the Superior National Forest stretches 150 miles along the US-Canadian border, encompassing 3.85 million acres. This hilly deep pine forest is home to moose, wolves, black bears, loons andmigratory birds. More than 2,250 miles of stream flow within the forest, including the renowned Boundary Waters Canoe Area, where you can canoe, portage and camp in the spirit of the French Canadian voyagers of 200 years ago.

*Jim Sanders, Forest Supervisor*

**3407 Tamarac National Wildlife Refuge**
35704 County Highway 26 218-847-2641
Rochert, MN 56578 Fax: 218-847-9141
http://www.fws.gov/midwest/tamarac
Established in 1938, Tamarac Refuge is dedicated to providing a breeding ground and sanctuary for migratory birds and other wildlife. Situated at a unique transitional zone where hardwood forest, boreal forest and tallgrass prairiemeets. Tamarac provides boundless opportunities for visitors to observe wildlife in their natural surroundings. Spring and fall migrations of songbirds and waterfowl can be spectacular. Their visitor center offers interpretive displays and programs.

**3408 Voyageurs National Park**
3131 Highway 53 218-283-9821
International Falls, MN 56649 Fax: 218-285-7407
http://www.nps.gov/voya
The park lies in the southern part of the Canadian Shield, representing some of the oldest exposed rock formations in the world. This bedrock has been shaped and carved by at least four periods of glaciation. The topography of the parkis rugged and varied; rolling hills are interspersed between bogs, beaver ponds, swamps, islands, small lakes and four large lakes.

**3409 Winona District National Wildlife Refuge Upper Mississippi River National Wildlife and Fish**
51 E 4th Street 507-452-4232
Winona, MN 55987 Fax: 507-452-0851
http://www.fws.gov/midwest/uppermississippiriver
Refuge objectives are to: protect and preserve one of America's premier fish and wildlife areas; provide habitat for migratory birds, fish, plants and resident wildlife; protect and enhance habitat for endangered species; provideinterpretation, environmental education and wildlife-oriented recreational public use opportunities; and conserve a diversity of plant life.
*Don Hultman, Manager*

## Mississippi: US Forests, Parks, Refuges

**3410 Bienville National Forest**
3473 Highway 35 S 601-469-3811
Forest, MS 39074 http://www.fs.fed.us/r8/mississippi/bienville
The forest offers camping, picnicking, swimming, hiking, fishing and historic sites. Bienville boasts the largest known cluster of old growth pine forest in Mississippi in the 180-acre Bienville Pines Scenic Area. Here visitors canwander among towering loblolly and shortleaf pines, many more than two centuries old. The 23-mile Shockaloe Horse Trail, a national recreation trail, starts near the town of Forest.

**3411 Mississippi Sandhill Crane National Wildlife Refuge**
7200 Crane Lane 228-497-6322
Gautier, MS 39553 Fax: 228-497-5407
E-mail: mississippisandhillcrane@fws.gov
http://www.fws.gov/mississippisandhillcrane
The Refuge is located in southeast Mississippi in Jackson County a few miles north of the Gulf of Mexico. This 20,000 acre refuge was established in 1975 to protect the endangered Mississippi

Sandhill Cranes and the wet pine savannahabitat they prefer. Visitor Center hours Tues-Sat 9-3, featuring exhibits, informational videos, and a walking trail - where you can view carnivorous plants. Free entry.

*Ted Rentmeister, Refuge Manager*
*Douglas Hunt, Refuge Manager*

**3412 Noxubee National Wildlife Refuge**
2970 Bluff Lake Road 662-323-5548
Brooksville, MS 39739 Fax: 662-323-6390
E-mail: noxubee@fws.gov
http://noxubee.fws.gov

Established in 1940 to protect and enhance habitat for the conservation of migratory birds, endangered species and other wildlife. The recreational and educational opportunities provided on the refuge help the public experience natureand learn how sound management ensures that future generations continue to enjoy fish and wildlife and their habitats.

**3413 Panther Swamp National Wildlife Refuge**
13695 River Road 662-746-5060
Yazoo City, MS 39194 Fax: 662-746-5055
E-mail: Bo_Sloan@fws.gov
http://pantherswamp.fws.gov/index.html

Refuge objectives are: to provide resting, nesting and feeding habitat for waterfowl and other migratory birds; to provide habitat for resident wildlife; to protect endangered and threatened species; and to provide public useopportunities for outdoor recreation and environmental education.
*Bo Sloan, Refuge Manager*

## Missouri: US Forests, Parks, Refuges

**3414 Mark Twain National Forest**
401 Fairgrounds Road 573-364-4621
Rollo, MO 65401 Fax: 573-364-6844
http://www.fs.fed.us/r9/forests/marktwain

The Mark Twain National forest is located in southern and centeral Missouri, and extends from the St Francois Mountains in the southeast to glades in the southwest, from the southwest, from the prairie lands the Missouri River to thenation's most ancient mountains in the south.

**3415 Mingo National Wildlife Refuge**
24279 State Highway 51 573-222-3589
Puxico, MO 63960 Fax: 573-222-6343
E-mail: mingo@fws.gov
http://www.fws.gov/midwest/mingo

Established as a resting and wintering area for migratory waterfowl and other birds. The 21,592-acre Refuge contains approximately 15,000 acres of bottomland hardwood forest, 3,500 acres of marsh and water, 506 acres of cropland, 704acres of seasonally flooded impoundments, and 474 acres of grassy openings.

**3416 Ozark National Scenic Riverways**
404 Watercress Drive 573-323-4236
PO Box 490 Fax: 573-323-4140
Van Buren, MO 63965 E-mail: ozar_superintendent@nps.gov
http://www.nps.gov/ozar

Missouri's largest National Park and America's first to preserve a free flowing river in its wild state. Covers some 80,000 acres along 134 miles of the Current and Jacks Fork Rivers. Staff provide elementary level environmentaleducation programs on natural history, with an emphasis on karst and water issues. Publishes More Than Skin Deep, a Teacher's Guide to Caves and Groundwater, a curriculum guide suitable for grades K-12.

**3417 Swan Lake National Wildlife Refuge**
16194 Swan Lake Avenue 660-856-3323
Sumner, MO 64681 Fax: 660-856-3687
E-mail: john_benson@fws.gov
http://www.fws.gov/midwest/swanlake

*Steve Whitson, Project Director*

## Montana: US Forests, Parks, Refuges

**3418 Beaverhead-Deerlodge National Forest**
420 Barrett Street 406-683-3900
Dillon, MT 59725 http://www.fs.fed.us/r1/bdnf
*Bruce Ramsey, Forest Supervisor*
*Thomas D Osen, District Ranger*

**3419 Benton Lake National Wildlife Refuge**
922 Bootlegger Trail 406-727-7400
Great Falls, MT 59404 Fax: 406-727-7432
http://www.fws.gov/bentonlake
*Kathleen Burchett, Project Leader*
*Robert F Johnson Jr, Deputy Manager*

**3420 Bighorn Canyon National Recreation Area**
5 Avenue B 406-666-2412
PO Box 7458 Fax: 406-666-2415
Fort Smith, MT 59035 http://www.nps.gov/bica

This dam, named after the famous Crow chairman Robert Yellowtail, harnessed the waters of the Bighorn River and turned this variable stream into a magnificent lake. The Afterbay Lake below the Yellowtail Dam is a good spot for troutfishing and wildlife viewing for ducks, geese and other animals. The Bighorn River below the Afterbay Dam is a world class trout fishing area. Bighorn Canyon National Recreation Area boasts breath-takign scenery, countless varieties of wildlife.

**3421 Bitterroot National Forest**
1801 N First Street 406-363-7100
Hamilton, MT 59840 Fax: 406-363-7106
E-mail: rl_bitterroot_comments@fs.fed.us
http://www.fs.fed.us/rl/bitterroot/

The 1.6 million acre Bitterroot National Forest, in west central Montana and east central Idaho, is part of the Norther Rocky Mountains. National Forest land begins above the foothills of the Bitterroot River Valley in two mountainranges—the Bitterroot Mountains on the west and the Sapphire Mountains on the east side of the valley.

**3422 Bowdoin National Wildlife Refuge: Refuge Manager**
194 Bowdoin Auto Tour Road 406-654-2863
Malta, MT 59538 Fax: 406-654-2866
E-mail: bowdoin@fws.gov
http://www.fws.gov/bowdoin

Bowdoin National Wildlife Refuge was established in 1936 as a migratory bird refuge. It is located in the short and mixed greens prairie region of North-central Montana and encompasses 15,551 acres.

**3423 Custer National Forest**
1310 Main Street 406-657-6200
Billings, MT 59105 Fax: 406-657-6222
http://www.fs.fed.us/r1/custer/

The Custer national Forest is made up of 1.2 million acres of high alpine mountain country, and small pockets of timbered buttes and grasslands scattered across two states, Montana and South Dakota.
*Mary C Erickson, Acting Forest Supervisor*
*Chris C Worth, Deputy Forest Supervisor*

**3424 Flathead National Forest**
1935 3rd Avenue E 406-758-5200
Kalispell, MT 59901 Fax: 406-758-5363
http://www.fs.fed.us/r1/flathead/

The 2.3 million-acre Flathead national Forest is bordered by Canada to the north, Glacier National Park to the north and east and Clark National Forest to the east, the Lolo national Forest to the south, and the Kootenai NationalForest to the west.

**3425 Gallatin National Forest**
PO Box 130 406-587-6701
Bozeman, MT 59771 http://www.fs.fed.us/r1/gallatin

**3426 Helena National Forest**
2880 Skyway Drive 406-449-5201
Helena, MT 59602 Fax: 406-449-5436
http://www.fs.fed.us/r1/helena/
The Helena National Forest offers close to one million acres of diverse landscapes and wildland opportunities. Located in west central Montana, the Helena National Forest boasts some of the most vivid glimpses into the past of thishistorically rich area.

**3427 Kootenai National Forest**
31374 Highway 2 West 406-293-6211
Libby, MT 59923 Fax: 406-283-7709
http://www.fs.fed.us/r1/kootenai
The Kootenai National Forest, conataining 2.2 million acres, is located in the extreme northwest corner of Montana, Bordered on the north by Canada and on the west by Idaho. Of the total acres, 50,384 are in the state of Idaho. Accessinto the forest is available from US highways 2 and 93, and Montana State Highways 37, 56, 200, and 508.
*Paul Bradford, Forest Supervisor*

**3428 Lee Metcalf National Wildlife Refuge**
4567 Wildfowl Lane 406-777-5552
Stevensville, MT 59870 Fax: 406-777-2489
http://www.fws.gov/leemetcalf
Mission is to manage habitat for a diversity of wildlife species with emphasis on migratory birds and endangered and threatened species.

**3429 Lewis & Clark National Forest**
1101 15th Street N 406-791-7700
Great Falls, MT 59401 http://www.fs.fed.us/rl/lewisclark/
The 1.8 million acres of the Lewis and clark National Forest are scatteered into seven separate mountain ranges. The Forest is situated i west central Montana. The boundaries spread eastward from the rugged, mountainous ContinentalDivide onto the plains. When looking at a map, the National Forest System lands appear as islands of forest within oceans of prairie. Because of its wide-ranging land pattern, the forest is separated into two divisions: the Rocky Mountain and theJefferson.

**3430 Lolo National Forest**
Building 24-A Fort Missoula 406-329-3750
Missoula, MT 59804 http://www.fs.fed.us/r1/lolo
Of the 15 National Forests in the Northern Region of the USDA Forest Service, the Lolo National Forest is estimated to be the third largest. It is located in western Montana. Several major tributaries to the Clark Fork River of theColumbia River Basin flow through the Forest. Its 2.1 million acres of diverse and spectacular mountainous country extend into seven counties.

**3431 Medicine Lake National Wildlife Refuge Complex**
223 North Shore Road 406-789-2305
Medicine Lake, MT 59247 Fax: 406-789-2350
http://www.fws.gov/medicinelake
Medicine Lake National Wildlife Refuge is located on the heavily glaciated rolling plains of northeaster Montana, between Missouri River and the Canadian Border.

**3432 National Bison Range National Wildlife Refuge**
58355 Bisons Range Road 406-644-2211
Moiese, MT 59824 Fax: 406-644-2661
http://www.fws.gov/bisonrange/nbr
*Jeff King, Project Leader*

**3433 Red Rocks Lakes National Wildlife Refuge**
27820 Southside Centennial Road 406-276-3536
Lima, MT 59739 Fax: 406-276-3538
E-mail: redrocks@fws.gov
http://www.fws.gov/redrocks
Primarily a high elevation mountain wetland-riparian area. Red Rock Creek flows through the upper end of the Centennial Valley, within which the Refuge lies, creating the impressive Upper Red Rock Lake, River Marsh and Lower Red RockLake marshlands. The rugged Centennial Mountains border the Refuge on the south, catching the snows of winter that replenish the Refuge's lakes and marshes.

## Nebraska: US Forests, Parks, Refuges

**3434 Crescent Lake National Wildlife Refuge**
10630 Road 181 308-762-4893
Ellsworth, NE 69340 Fax: 308-762-7606
E-mail: crescentlake@fws.gov
http://crescentlake.fws.gov
Located in the panhandle of Western Nebraska, Crescent Lake consists of 45,818 acres of rolling sandhills interspersed with numberous shallow wetlands and lakes. Plant and animal species call Crescent Lake home, while visitors canparticipate in a variety of public use activities.

**3435 Fort Niobrara National Wildlife Refuge**
Fort Niobrara/Valentine NWR Complex
HC 14, Box 67 402-376-3789
Valentine, NE 69201 Fax: 402-376-3217
E-mail: fortniobrara@fws.gov
http://www.fws.gov/fortniobrara
National Wildlife Refuge Complex includes Fort Niobrara NWR, Valentine NWR and Seier NWR. Complex lands include riverine, riparian, marshland, hand prairie, and sandhills prairie habitats.
*Steven A Hicks, Project Leader*

**3436 Nebraska & Samuel R McKelvie National Forest**
125 N Main Street 308-432-0300
Chadron, NE 69337 http://www.fs.fed.us/r2/nebraska

## Nevada: US Forests, Parks, Refuges

**3437 Anaho Island National Wildlife Refuge**
Stillwater Wildlife Management
9604 Auction Road 702-423-5128
PO Box 1236 http://www.fws.gov/stillwater/anaho.html
Fallon, NV 89406
Primary wildlife Anaho Island has one of the largest white pelican nesting colonies in North America, as well as cormorant, great blue heron and gull nesting colonies. An island in Pyramid Lake in Great Basin. Closed to public entry toprotect wildlife. Contact refuge manager for information.

**3438 Ash Meadows National Wildlife Refuge**
HCR 70 775-372-5435
Box 610-Z Fax: 775-372-5436
Amargosa Valley, NV 89020 http://www.fws.gov/desertcomplex/ashmeadows
Refuge staff are responsible for managing the 22,117 acre refuge, most of which is spring-fed wetland and alkaline desert upland The refuge area habitat for at least 24 plants and animals found nowhere else in the world. Four fishesand one plant are currently listed as endangered. Species found on the refuge include numerous endemic species, the greatest concentration in the US and the second greatest in all of North America.
*Sharon McKelvey, Manager*

**3439 Desert National Wildlife Refuge Complex**
1500 N Decatur Boulevard 702-646-3401
Las Vegas, NV 89108 http://www.fws.gov/desertcomplex
Established in 1936 for perpetuating the desert bighorn sheep. Two threatened, and twenty-nine species of concern can be found at the refuge. Wildlife observation is one of the most popular refuge activities. Big game hunting is verylimited, but also very popular. Bird watching is another popular activity. A growing program provides additional opportunities and students are able to earn college credits through an internship at the refuge.

**3440 Great Basin National Park**
100 Great Basin National Park 775-234-7331
Baker, NV 89311 Fax: 775-234-7269
http://www.nps.gov/grba
Great Basin National Park includes streams, lakes, alpine plants, abundant wildlife, a variety of forest types including groves of ancient bristlecone pines, and numerous limestone caverns, including beatiful Lehman Caves.

**3441 Humboldt-Toiyabe National Forest**
1200 Franklin Way 775-331-6444
Sparks, NV 89431 http://www.fs.fed.us/r4/htnf
The Humboldt-Toiyabe National Forest encompasses all of Nevada and the far Eastern edge of California. The Humboldt-Toiyabe is the largest forest in the lower 48 states.
*Edward Monnig, Forest Supervisor*

**3442 Lake Mead National Recreation Area (NRA)**
601 Nevada Way 702-293-8990
Boulder City, NV 89005 Fax: 702-293-8936
E-mail: LAME_Interpretation@nps.gov
http://www.nps.gov/lame
Lake Mead NRA, which includes Lake Mohave, offers a wealth of things to do and places to go year-round. Its huge lakes cater to boaters, swimmers, sunbathers and fishermen while its desert rewards hikers, wildlife photographers androadside sightseers. Three of America's four desert ecosystems: the Mojave, the Great Basin and the Sonoran Desert meet in here. As a result, this seemingly barren area contains a surprising variety of plants animals, some of which may be foundnowhere else.

**3443 Moapa Valley National Wildlife Refuge**
4701 N Torrey Pines Drive 702-515-5450
Las Vegas, NV 89130 Fax: 702-515-5460
http://www.fws.gov/desertcomplex/moapavalley

*Amy Lavoie, Refuge Manager*

**3444 Pahranagat National Wildlife Refuge**
PO Box 510 775-725-3417
Alamo, NV 89001 Fax: 775-725-3389
E-mail: merry-maxwell@fws.gov
http://www.fws.gov/desertcomplex/pahranagat
Refuge staff are responsible for managing the 5,380 acre refuge, a mixture of desert, open water, native grass meadows, cropland and marsh. The refuge provides habitat for migratory birds of the Pacific Flyway and several speciesabound at this desert oasis. Wildlife observation is one of the most popular refuge activities. Waterfowl and small game hunting are also very popular, as is bird watching. A growing program provides additional opportunities, including internshipsfor college students.

*Merry Maxwell, Manager*

## New Hampshire: US Forests, Parks, Refuges

**3445 Lake Umbagog National Wildlife Refuge**
PO Box 240 603-482-3415
Route 16 North Fax: 603-482-3308
Errol, NH 03579 E-mail: lakeumbagog@fws.gov
http://lakeumbagog.fws.gov/
Northern Forest refuge of New Hampshire and Maine provides long-term conservation of important wetland/upland habitats for wildlife, migratory birds and protected species.

**3446 White Mountain National Forest**
719 N Main Street 603-528-8721
Laconia, NH 03246 Fax: 603-528-8783
http://www.fs.fed.us/r9/forests/white_mountain
The White Mountain National Forest is located in northern New Hampshire and southwestern Maine, and lies within Carroll, Coos, and Grafton Counties in New Hampshire, and Oxford County in Maine.
*Tom Wagner, Forest Supervisor*

## New Jersey: US Forests, Parks, Refuges

**3447 Cape May National Wildlife Refuge**
24 Kimbles Beach Road 609-463-0994
Cape May Courthouse, NJ 08210 Fax: 609-463-1667
E-mail: capemay@fws.gov
http://www.fws.gov/northeast/capemay/

**3448 Great Swamp National Wildlife Refuge**
241 Pleasant Plains Road 973-425-1222
Basking Ridge, NJ 07920 Fax: 973-425-7309
E-mail: greatswamp@fws.gov
http://www.fws.gov/northeast/greatswamp
Swamp woodland, hardwood ridges, cattail marsh and grassland are typical of this approximately 7,800 acre refuge. The Swamp contains many large old oak and beach trees, stands of mountain laurel, mosses, ferns and species of many otherplants of both Northern and Southern botanical zones.
*William Koch, Refuge Manager*

## New Mexico: US Forests, Parks, Refuges

**3449 Bitter Lake National Wildlife Refuge**
4067 Bitter Lake Road 505-622-6755
Roswell, NM 88201 Fax: 505-623-9039
http://southwest.fws.gov/refuges/newmex/bitter.html
Native grasses, sand dunes, brushy bottomlands, seven lakes and a red-rimmed plateau make up Bitter Lake National Wildlife Refuge, winter home for thousands of migratory birds. The Lakes on the refuge were formed within the ancientriver beds of the Pecos River. These lakes store about 1,000 acres of water at their highest levels, while nearby marshland, mudflats and the Pecos Rriver provide an additional 24,500 acres of habitat.

**3450 Bosque del Apache National Wildlife Refuge**
PO Box 1246 505-835-1828
Socorro, NM 87801 Fax: 505-835-0314
http://southwest.fws.gov/refuges/newmex/bosque
Bosque del Apache NWR is located in south-central New Mexico, along the Rio Grande in the northern reach of the Chihuahuan Desert. Habitats include cottonwood forests, seasonally managed wetlands, farm fields, saltgrass meadows, anddesert uplands. The refuge features large concentrations of sandhill cranes, light geese, and migrating waterfowl in fall and winter, shorebirds and songbirds travel through in spring and fall, and hummingbirds are abundant in summer.

**3451 Capulin Volcano National Monument National Park Service**
PO Box 40 505-278-2201
Capulin, NM 88414 Fax: 505-278-2211
http://www.nps.gov/cavo
Capulin Volcano is long extinct, and today the forested slopes provide habitat for mule deer, wild turkey, black bear and other wildlife. Abundant displays of wildflowers bloom on the mountain each summer. A two mile paved roadspiraling to the volcano rim makes Capulin Volcano one of the most accessible volcanos

in the world. Trails leading around the rim allow exploration of this classic cinder cone.
*, Superintendent, Park Ranger*

**3452 Carlsbad Caverns National Park**
3225 National Parks Highway 505-785-2232
Carlsbad, NM 88220 Fax: 505-785-2133
E-mail: cave_park_information@nps.gov
http://www.nps.gov/cave
Established to preserve Carlsbad Cavern and numerous other caves within a Permian-age fossil reef, the park contains over 100 known caves, including Lechuguilla Cave-the nation's deepest limestone cave and third longest. CarlsbadCavern, with one of the world's largest underground chambers and countless formations, is highly accessible, with a variety of tours offered year-round.
*John Benjamin, Superintendent*
*Chuck Barat, Deputy Superintendent*

**3453 Carson National Park**
208 Cruz Alta Road 505-758-6200
Taos, NM 87571 Fax: 505-758-6213
Some of the finest mountain scenery in the southwest is found in the 1.5 million acres covered by the Carson National Forest. Elevations rise from 6,000 feet to 13,161 feet. The scenic Sangre de Cristo Mountains include Wheeler Peak,the highest peak in New Mexico.

**3454 Cibola National Forest**
2113 Osuna Road Northeast Suite A 505-346-3900
Albuquerque, NM 87113 Fax: 505-346-3901
http://www.fs.fed.us/r3/cibola
Cibola, pronounced See'-bo-lah, is thought to be the original Zuni Indian name for their group of pueblos or tribal lands. Later, the Spanish interpreted the word to mean, buffalo. Valued for its recreation opportunities, naturalbeauty, timber, watersheds, water, forage, and wilderness resources, the forest is managed to give the American people the greatest benefits that can be produced on a permanent basis.

**3455 El Malpais National Monument**
123 E Roosevelt Avenue 505-876-2783
201 E Roosevelt Fax: 505-285-5661
Grants, NM 87020 E-mail: Leslie_DeLong@nps.gov
This monument preserves 114,277 acres of which 109,260 acres are federal and 5,017 acres are private. Volcanic features such as lava flows, cinder cones, pressure ridges and complex lava tube systems dominate the landscape. Sandstonebluffs and mesas border the eastern side, providing to vast wilderness.
*Douglas E Eury*

**3456 Las Vegas National Wildlife Refuge**
Route 1 Box 399 505-425-3581
Las Vegas, NM 87701 http://www.gorp.com
*Joe B Rodriguez, Manager*

**3457 Lincoln National Forest**
LNF Federal Building 505-437-6030
11th and New York http://www.fs.fed.us/r3/lincoln/
Alamogordo, NM 88310

**3458 Maxwell National Wildlife Refuge**
PO Box 276 505-375-2331
Maxwell, NM 87728 Fax: 505-375-2332
E-mail: fw2_rw_maxwell@fws.gov
http://southwest.fws.gov/refuges/newmex/maxwell.html
At an altitude of 6,050 feet, the refuge is made up of more than 3,000 acres of gently rolling prairie, playa lakes and farmland for waterfowl. Rangeland and reclaimed farmland on the refuge are made up of a variety of grassesincluding blue grama, galleta, sand dropseed, threeawn and buffalo grass, as well as fourwing saltbush and cactus. Several lakes provide approximately 700 acres of roosting and feeding habitat for waterfowl. Supports waterfowl nesting and is alsobeneficial to shore birds.
*Patty Hoban, Refuge Manager*

**3459 San Andres National Wildlife Refuge**
PO Box 756 505-382-5047
Las Cruces, NM 88004 Fax: 505-382-5454
E-mail: fw2_rw_sanandres@fws.gov
http://www.fws.gov/southwest/refuges/newmex/sanandres/index.html
Refuge not open to the public due to its location within the boundaries of U.S. Department of Army, White Sands Missile Range. Primary emphasis has been focused on restoring a remnant population of desert bighorn sheet (OvisCanadensis Mexicana)

*Kevin Cobble, Refuge Manager*

**3460 Sevilleta National Wildlife Refuge**
PO Box 1248 505-864-4021
Socorro, NM 87801 Fax: 505-864-7761
E-mail: fw2_rw_sevilleta@fws.gov
http://southwest.fws.gov/refuges/newmex/sevill.html
Home to over 1200 species of plants, 89 species of mammals, 225 species of birds and 15 species of amphibians. More commonly seen species include mule deer, coyotes, pronghorns, red-tailed hawks, northern harriers, western diamondbackrattlesnakes, roadrunners, sandhill cranes and many different types of waterfowl and migrating shorebirds. Bobcats, elk, bighorn sheep and an occasional mountain lion also roam the hillsides.

*Kathy Granillo, Manager*

**3461 White Sands National Monument**
PO Box 1086 505-479-6124
Holloman Air Force Base, NM 88330 Fax: 505-479-4333
White Sands National Mounment preserves a major portion of this gypsum dune field, along with the plants and animals that have successfully adapted to this constantly changing environment.
*Dennis L Ditmanson*

## New York: US Forests, Parks, Refuges

**3462 Fire Island National Seashore**
120 Laurel Street 631-687-4750
Patchogue, NY 11772 Fax: 631-289-4898
E-mail: fiis_interpretation@nps.gov
http://www.nps.gov/fiis/home.htm
There are 32 miles of sandy beaches and saltwater marshes, a sunken forest of 300 year old holly trees, hiking trails, a wilderness area and many other sites on the Fire Island National Seashore.

*Michael T Reynolds, Superintendent*
*K Christopher Soller, Superintendent*

**3463 Gateway National Recreation Area**
Building 69 Floyd Bennett Field 718-338-3575
Brooklyn, NY 11234 E-mail: carole_silano@nps.gov
Gateway NRA is a 26,000 acre recreation area located in the heart of the New York metropolitan area. The park extends through three New York City boroughs and into northern New Jersey. Parks sites offer a variety of recreationopportunities, along with a chance to explore many significant cultural resources.
*Kevin Buckley, Supt*

**3464 Iroquois National Wildlife Refuge**
1101 Casey Road 585-948-5445
Basom, NY 14013 Fax: 585-948-9538
http://http://iroquoisnwr.fws.gov
Iroquois National Wildlife Refuge lies within the rural township of Alabama, New York, midway between Buffalo and Rochester. Part of what the locals call the Alabama Swamps, its 10,818 acres of freshwater marshes, and hardwood swampsbounded by woods, forest, pastures and wet meadows, serve the habitat needs of many animals as a major stopover for migrating birds and as a year-round residence.

*Robert Lamayr, Refuge Manager*

**3465 Montezuma National Wildlife Refuge**
3395 Routes 5 & 20 East 315-568-5987
Seneca Falls, NY 13148 Fax: 315-568-8835
E-mail: andrea_vanbeusichem@fws.gov
http://www.fws.gov/r5mnwr
Montezuma is the premiere refuge in New York State. View bald eagles year-round. Spring and fall bring tens of thousands of migrating ducks and geese. View Shaebirds in early spring and late summer. May and June are great for warblerwatching. Volunteer opportunities available Spring- Fall.

*Tom Jasikoff, Manager*
*Andrea VanBeusichem, Visitor Services Manager*

**3466 Seatuck National Wildlife Refuge: Long Island National Wildlife Refuge Complex**
Long Island NWR Complex 516-581-1538
PO Box 21 Fax: 516-581-2003
Shirley, NY 11967 E-mail: R5RW_STKNWR@fws.gov
Located along the southern shore of Long Island, the Refuge consists of half salt marsh and half freshwater wetlands, ponds and sparsely wooded areas. It is part of the larger Great South Bay, which is a significant coastal habitat formigrating birds. Limited recreation includes viewing wildlife and bird watching.
*Charles Stenvall, Manager*

**3467 Target Rock National Wildlife Refuge Long Island National Wildlife Refuge Complex**
PO Box 21 516-286-0485
Shirley, NY 11967 Fax: 516-286-4003
The refuge was established in 1967. It consists of mixed upland forest, a half mile of rocky beach, a brackish and several vernal ponds. The offshore, beach and pond habitats provide foraging areas for piping plover, winteringwaterfowl and fish species.

## North Carolina: US Forests, Parks, Refuges

**3468 Alligator River National Wildlife Refuge**
PO Box 1969 252-473-1131
Manteo, NC 27954 Fax: 252-473-1668
E-mail: alligatorriver@fws.gov
http://http://www.fws.gov/alligatorriver

*Bonnie Strawser, Wildlife Interpretive Spec*

**3469 Cape Hatteras National Seashore**
1401 National Park Drive 252-473-2111
Manteo, NC 27954 Fax: 252-473-2595
E-mail: CAHA_Information@nps.gov
http://www.nps.gov/caha/
A thin broken strand of islands curves out into the Atlantic Ocean and then back again in a sheltering embrace of North Carolina's mainland coast and its offshore sounds. These are the Outer Banks of North Carolina. Today their longstretches of beach, sand dunes, marshes and woodlands are set aside as Cape Hatteras National Seashore.
*Lawrence A Belli, Superintendent*

**3470 Cape Lookout National Seashore**
131 Charles Street 252-728-2250
Harkers Island, NC 28531 Fax: 252-728-2160
E-mail: CALO_information@nps.gov
http://www.nps.gov/calo
The seashore is a 56 mile long section of the Outer Banks of North Carolina running from Ocracoke Inlet on the northeast to Beafort Inlet on the southeast. The four undeveloped barrier island, make up the seashore- North Core Banks,South Core Banks, Middle Core Banks and Shackleford Banks- may seem barren and isolated but they offer many natural and historical features that can make a visit very rewarding.
*Robert A Vogel, Superintendent*
*Donna O Tiptor, Administrative Officer*

**3471 Cedar Island National Wildlife Refuge**
38 Mattamuskeet Road 252-926-4021
Swan Quarter, NC 27885 Fax: 252-926-1743

**3472 Currituck National Wildlife Refuge**
Mackay Island NWR 919-429-3100
PO Box 39 Fax: 919-429-3185
Knotts Island, NC 27950 http://www.gorp.com

**3473 Mattamuskeet National Wildlife Refuge**
Route 1 919-926-4021
Box N-2 E-mail: r4rw_nc.mtk@fws.gov
Swan Quarter, NC 27885 http://www.gorp.com
Located in eastern North Carolina in Hyde County, the Mattamuskeet Refuge consists of more than 50,000 acres of water, marsh, timber and crop lands. The refuge's most significant feature is lake Mattamuskeet, the largest natural lakein North Carolina. The lake is 18 miles long and five to 6 miles wide, encompassing approximately 40,000 acres, but averages 2 feet in depth.
*Don Temple, Manager*

**3474 Nantahala National Forest**
160-A Zillicoa Street 828-257-4200
Asheville, NC 28802 Fax: 828-257-4263
The National Forests of North Carolina include four national forests covering 1.2 million acres from the mountains to the sea. The Nantahala is located in the Appalachians of southwest North Carolina. The Nantahala is the largest ofthe four forests, totaling 528,541 acres. The Nantahala sits adjacent to Great Smokey Mountains National Park.

**3475 Pea Island River National Wildlife Refuge**
PO Box 1969 252-473-1131
Manteo, NC 27954 Fax: 252-473-1668
E-mail: alligatorriver@fws.gov
http://http://www.fws.gov/peaisland
*Bonnie Strawser, Wildlife Interprtive Spec*

**3476 Pee Dee National Wildlife Refuge**
Route 1, Box 92 704-694-4424
Wadesboro, NC 28170 Fax: 704-694-6570
E-mail: fw4_rw_pee_dee@fws.gov
http://peedee.fws.gov//index.html
Refuge objectives are to: provide habitat for migratory waterfowl and song birds; to provide habitat and protection for an endangered species, the red-cockaded woodpecker; to provide recreation, environmental education andinterpretation for the public; to engage in dynamic partnering.
*Dan Frisk, Refuge Manager*

**3477 Pocosin Lakes National Wildlife Refuge**
205 S Ludington Drive 252-796-3004
PO Box 329 Fax: 252-796-3010
Columbia, NC 27925 E-mail: pocosinlakes@fws.gov
http://www.fws.gov/pocosinlakes/
The 112,000 acre refuge was established to protect and ehnhance a unique habitat called a pocosin and contains a variety of wildlife including endangered species such as the red wolf, bald eagle, peregrine falcon and red-cockadedwoodpecker as well as natural vegetation and scenic areas.

*Howard Philips, Refuge Manager*
*David Kitts, Deputy Manager*

**3478 Roanoke River National Wildlife Refuge**
114 W Water Street 252-794-3808
PO Box 430 Fax: 252-794-3780
Windsor, NC 27983 E-mail: roanokeriver@fws.gov
http://roanokeriver.fws.gov/
Refuge objectives are: to provide habitat for migratory waterfowl, neo-tropical migrants and other birds; to provide migrating, spawning and nursery habitat for anadromous fish; (i.e. blueback herring, alewife, hickory shad and stripedbass); to enhance and protect forested wetlands consisting of bottomland hardwoods and swamps; to protect and manage for endangered and threat-

ened wildlife; and to provide recreation and environmental education for the public

*Harvey Hill, Refuge Manager*

**3479 Swanquarter National Wildlife Refuge**
Route 1, Box N-2 252-926-4021
Swan Quarter, NC 27885 Fax: 252-926-1743

## North Dakota: US Forests, Parks, Refuges

**3480 Crosby Wetland Management District**
70700 Highway 42 NW 701-965-6488
Crosbye, ND 58730 E-mail: crosbywetlands@fws.gov
Previously a part of the Des Lacs NWR Complex, the Crosby WMD includes of 17,000 acres of Waterfowl Production Areas (WPAs), numerous grassland and wetland easment contracts, and the 3,219 acre Lake Zahl NWR.
*Tim K Kessler, WMD Manager*

**3481 Des Lacs National Wildlife Refuge**
County Road 1A, 1 Mile West of Town 701-385-4046
Kenmare, ND 58746 Fax: 701-385-3214
E-mail: deslacs@fws.gov
http://http://mountain-prairie.fws.gov/dslcomplex
Previously a part of the old Des Lacs NWR complex, the Des Lacs NWR is a smaller parcel that is home to wildlife and water fowl. The Lakes no longer encompass the the wetlands that are part of Crosby Wetlands district.

**3482 Devils Lake Wetland Management District**
Devils Lake WMD 701-662-8611
PO Box 908 Fax: 701-662-8612
Devils Lake, ND 58301
Located in the heart of the Prairie Pothole Region of the US. The northeastern North Dakota counties of Towner, Cavalier, Pembina, Benson, Ramsey, Walsh, Nelson and Grand Forks are included in the District. Managed by the US Fish andWildlife Service, the district provides wetland areas needed by waterfowl in the spring and summer for nesting and feeding. Hundreds of thousands of waterfowl also use these wetlands in the spring and fall for feeding and resting during longmigratory flights.

**3483 Lake Ilo National Wildlife Refuge**
489 102 Avenue SW 701-548-8110
Dunn Center, ND 58626 Fax: 701-548-8108
E-mail: lakeilo@fws.gov
http://www.fws.gov/lakeilo
Located near the center of Dunn County in west central North Dakota, the refuge habitat is made up of native prairie, planted grasslands and wetlands. The uplands are characterized by gently sloping hills and terraces with creeks andan occasional slough. The average rainfall of 16.8 inches supports a prairie environment with a climate of hot dry summers, occasional thunderstorms and cold winters.

*Kory Richardson, Refuge Manager*

**3484 Long Lake National Wildlife Refuge**
12000 353rd Street 56 701-387-4397
Moffit, ND 58560 Fax: 701-387-4767
E-mail: r6rw-llk@fws.gov
http://www.r6.fws.gov/refuge
The Refuge is about 18 miles long and contains 22,300 acres. The Refuge attracts a diversity and abundance of animals and waterfowl, both resident and migratory. Over 200 species of birds use the Refuge for breeding, rearing theiryoung and as a migratory stop. Long Lake Refuge is open for birdwatching, fishing, photography, boating, hiking and regulated hunting.

**3485 Lostwood National Wildlife Refuge**
8315 Highway 8 701-848-2722
Kenmare, ND 58746 Fax: 701-848-2702
Lies in the highly productive pothole region that produces more ducks than any other region in the lower 48 states. The refuge is a land of rolling hills mantled in short-grass and mixed with grass prairie interspersed with numerouswetlands. Established to preserve a unique wildlife habitat, Lostwood is an important link in our nation's system of more than 410 wildlife refuges.

**3486 Theodore Roosevelt National Park**
Box 7 701-623-4466
Medora, ND 58645 Fax: 701-623-4840
E-mail: susan_recce@nps.gov
http://www.nps.gov
Here in the North Dakota badlands, where many of his personal concerns first gave rise to his later environmental efforts, Roosevelt is remembered with a national park that bears his name and honors the memory of this greatconservationist. Theodore Roosevelt NAtional Park is colorful North Dakota badlands and is home to a variety of plants and animals, including bison, prairie dogs, and elk.

## Ohio: US Forests, Parks, Refuges

**3487 Cuyahoga Valley National Park**
15610 Vaughn Road 216-524-1497
Brecksville, OH 44141 800-445-9667
E-mail: cuva_canal_visitor_center@nps.gov
http://www.nps.gov/cuva or www.dayinthevalley.com
Cuyahoga Valley National Park protects 33,000 acres along the Cuyahoga River between Cleveland and Akron, Ohio. Managed by the National Park Service, CVNP combines cultural, historical, recreational and natural activities in onesetting. Visitors can hike, bike, birdwatch, golf, fish, ski, ride Cuyahoga Valley Scenic Railroad, explore the history of the Ohio and Erie Canal on a 20 mile section of the Towpath Trail, and attend national park ranger-guided programs, concerts,art exhibits and more.
*Colleen Brown, Visitor Use Assistant*

**3488 Ottawa National Wildlife Refuge**
14000 W State Route 2 419-898-0014
Oak Harbor, OH 43449 Fax: 419-898-7895
E-mail: dan_frisk@fws.gov
http://midwest.fws.gov/ottawa.html
Refuge objectives are: to restore optimum acreage to a natural floodplain condition; to improve and restore wetland habitat, to improve fishery and wildlife resources, to provide for biodiversity; and to provide public opportunitiesfor outdoor recreation and environmental education.
*Dan Frisk, Refuge Manager*

## Oklahoma: US Forests, Parks, Refuges

**3489 Deep Fork National Wildlife Refuge**
PO Box 816 918-756-0815
Okmulgee, OK 74447 Fax: 918-756-0275
E-mail: fw2_rw_deepfork@fws.gov
http://southwest.fws.gov/refuges/oklahoma/deepfrk.html
Protecting important wetlands along the Deep Fork River, Deep Fork National Wildlife Refuge in eastern Oklahoma is a newcomer to the National Wildlife Refuge System. Established in 1993, the 9,000 acre refuge is subject to flooding atleast once a year. This flooding results in excellent conditions for waterfowl, including mallard, blue-winged teal, shoveler, pintail and wood ducks.
*Darrin B Unruh, Manager*

**3490 Little River National Wildlife Refuge**
PO Box 340 405-584-6211
Broken Bow, OK 74728 Fax: 405-584-2034
http://www.gorp.com

*Berlin A Heck, Manager*

**3491 Oklahoma Bat Caves National Wildlife Refuge**
Route 1 Box 18-A
Vian, OK 74962
918-773-5251
Fax: 918-773-5252
http://www.gorp.com

**3492 Optima National Wildlife Refuge**
20834 east 940 Road
Butler, OK 73625
580-664-2205
Fax: 580-664-2206
E-mail: washita@fws.gov
http://southwest.fws.gov/refuges/oklahoma/optima.html
Located in the middle of the Oklahoma panhandle, the 4,333-acre refuge is made up of grasslands and wooded bottomland on the Coldwater Creek arm of the Army Corps of Engineers Optima Reservoir Project.

*Daniel Moss, Refuge Manager*

**3493 Sequoyah National Wildlife Refuge**
Route 1 Box 18 A
Vian, OK 74962
918-773-5251
Fax: 918-773-5598
E-mail: FW2_RW_Seqouyah@fws.gov
http://http://southwest.fws.gov/refuges/oklahoma/sequoy.html
*Jeff Haas, Refuge Manager*
*Scott Gilje, Assistant Refuge Manager*

**3494 US Fish & Wildlife Service Tishomingo National Wildlife Refuge**
12000 S Refuge Road
Tishomingo, OK 73460
580-371-2402
Fax: 580-371-9312
E-mail: fw2_rw_tishomingo@fws.gov
http://http://southwest.fws.gov/refuges/oklahoma/tishomingo
The 16,464-acre Refuge lies at the upper Washita arm of Lake Texoma and is administered for the benefit of migratory waterfowl in the Central Flyway. It offers a variety of aquatic habitats for wildlife. The murky water of theCumberland Pool provides abundant nutrients for innumerable microscopic plants and animals. Seasonally flooded flats and willow shallows lying at the Pool's edge also provide excellent wildlife habitat.
*Kris Patton, Refuge Manager*

**3495 Washita National Wildlife Refuge**
Route 1 Box 685
Butler, OK 73625
405-664-2205
E-mail: r2rw_wa@fws.gov
http://www.gorp.com

*Jon M Brock, Manager*

## Oregon: US Forests, Parks, Refuges

**3496 Ankeny National Wildlife Refuge**
Western Oregon NWR Complex
20301 Wintel Road
Jefferson, OR 97352
503-588-2071
Fax: 541-757-4450
http://www.gorp.com
Refuge's primary management goal is to provide vital wintering habitat for dusky Canada geese. The refuge includes flat to gently rolling land near the confluence of the Willamette and Sanitiam rivers. The refuge's fertile farmedfields, hedgerows, forests, and wetlands provide a variety of wildlife habitats. the refuge is open to limited opportunites for wildlife-oriented education and recreation. Ducks, geese, and swans are commonly seen in refuge fields and ponds throughthe fall and winter.

**3497 Bandon Marsh National Wildlife Refuge**
Western Oregon NWR Complex
26208 Finley Refuge Road
Corvallis, OR 97333
503-757-7236
http://www.gorp.com

**3498 Baskett Slough National Wildlife Refuge**
Western Oregon NWR Complex
26208 Finley Refuge Road
Corvallis, OR 97333
503-757-7236
http://www.gorp.com
Includes 2,492 acres typical of Willamette Vallley's irrigated hillsides, oak-covered knolls and grass fields. Wetlands include Morgan Lake and Baskett Slough. The refuge's objective is the protection and management of winteringhabitat for dusky Canada geese. Several species of waterfowl, herons, hawks, quail , shorebirds, mourning doves, woodpeckers and a variety of songbirds frequent the area, as well as mammals, amphibians and reptiles. Recreation includes observation,study and photography.
*Richard Guadagno, Manager*

**3499 Cape Meares National Wildlife Refuge**
Western Oregon NWR Complex
26208 Finley Refuge Road
Corvallis, OR 97333
503-757-7236
http://www.gorp.com

**3500 Cold Springs National Wildlife Refuge**
PO Box 700
Umatilla, OR 97882
541-992-3232
E-mail: gary_hagedorn@fws.gov
http://www.gorp.com
Cold Springs NWR lies in sharp contrast with the arid desert surroundings of northeastern Oregon. The refuge, a tree-lined reservoir, lies 7 miles east of the agricultural community of Hermiston. The variety of refuge habitats attractsan abundance of wildlife. Cold Springs supports peak populations of over 45,000 winter waterfowl comprised mainly of mallards and Canada geese.

**3501 Columbia River Gorge National Scenic Area**
902 Wasco Avenue
Waucoma Center, Suite 200
Hood River, OR 97031
541-386-2333
http://www.fs.fed.us/r6/columbia/
The Columbia River Gorge is a espaectacular river canyon cutting the only sea-level route through the Cascade Mountain Range. It's 80 miles long and to 4,00 feet deep with the north canyon walls in Washington State and the south canyonwalls in Oregon State.

**3502 Crater Lake National Park**
Highway 62
PO Box 7
Crater Lake, OR 97604
541-594-2211
Fax: 541-594-2261
http://www.nps.gov/crla/home.html
During the summer, visitors may navigate the Rim Drive around the lake, enjoy boat tours, stay in the historic Crater Lake Lodge Camp or hike some of the park's various trails. The winter brings some of the heaviest snowfall in thecountry, averaging 533 inches per year. Although park facilities mostly close for the snow season, visitors may view the lake during fair weather, enjoy cross-country skiing, and participate in weekend snowshoe hikes.
*Dave Morris, Supt*

**3503 Deschutes National Forest**
1645 US Highway 20 NE
Bend, OR 97701
541-388-2715
Scenic backdrop of volcanic attraction, evergreen forest, mountain lakes, caves, desert areas and alpine meadows.

**3504 Fremont Winema National Forest**
1300 S G Street
Lakeview, OR 97630
541-947-2151
Fax: 541-947-6399
http://www.fs.fed.us/r6/fremont/
Located in Oregon's Outback, the forest provides the self reliant recreationist the opportunity to discover nature in a rustic environment.

**3505 Malheur National Forest**
431 Patterson Bridge Road
PO Box 909
John Day, OR 97845
541-575-1731
The 1,460,000 acre Malheur National Forest is located in the blue Mountains of Eastern Oregon. The diverse and beautiful scenery of the forest includes high desert grasslands, sage and juniper, pine, fir and other tree species, and thehidden gem of alpine lakes and meadows. Elevations vary from about 4000feet (1200 meters) to the 9038 foot (2754 meters) top Strawberry Mountain. The Strawberry Mountain range extends east to west through the center of the forest.
*Bonnie J Wood, Forest Supervisor*

**3506 Malheur National Wildlife Refuge**
36391 Sodhouse Lane 541-493-2612
Princeton, OR 97721 Fax: 541-493-2405
E-mail: tim_bodeen@fws.gov
http://www.fws.gov/malheur

*Tim Bodeen, Manager*

**3507 McKay Creek National Wildlife Refuge**
Umatilla NWR Complex 503-922-3232
PO Box 239 http://www.gorp.com
Umatilla, OR 97882

**3508 Mount Hood National Forest**
16400 Champion Way 503-668-1700
Sandy, OR 97055 Fax: 503-668-1641
Located 20 miles east of the city of Portland and the northern Willamette River valley, Mt Hood National Forest extends south from the strikingly beautiful Columbia River Gorge across more than sixty miles of forested mountains, lakesand streams to OlAllie Scienic Area, a high lake basin under the slopes of Mt Jefferson. Our many visitors enjoy fishing, caming, boating and hiking in the summer, hunting in the fall, skiing and other snow sports in the winter.

**3509 National Park Service: John Day Fossil Beds National Monument**
32651 Highway 19 541-987-2333
Kimberly, OR 97845 Fax: 541-987-2336
E-mail: joda_interpretation@nps.gov
http://www.nps.gov/joda
Within the heavily eroded volcanic deposits of the scenic John Day Fossil Basin is a great diversity of well-preserved plant and animal fossils. This remarkably complete record spans more than 40 of the 65 million years of the CenozoicEra (the Age of Mammals). The monument was established in 1975.
*John Fiedor, Chief Visitor Services*

**3510 Ochoco National Forest**
3160 NE 3rd Street 541-416-6500
Prineville, OR 97754 Fax: 541-416-6695
http://www.fs.fed.us/r6/centraloregon
With a total of almost 1,500 square miles, the Ochoco National Forest is endowed with vast natural resources, scenic grandeur and tremendous recreation opportunities. People are drawn to the Ochoco for its majestic ponderosa pinestands, picturesque rimrock vantage points, deep canyons, unique geologic formations, abundant wildlife and plentiful sunshine.

**3511 Oregon Caves National Monument**
19000 Caves Highway 541-592-2100
Cave Junction, OR 97523 Fax: 541-592-3981
http://www.nps.gov/orca
Oregan Caves National Monument is small in size, 480 acres, but rich in diversity. Above ground, the monument encompasses a remnant old-growth coniferous forest. It harbors a fantastic array of plants, and a Douglas-fir tree with thewildest known girth in Oregon. Three hiking trails access this forest. Below ground is an active marble cave created by natural forces over hundreds of thousands of years in one of the world's most diverse geologic realms.

*Vicki Snitzler, Superintendent*

**3512 Oregon Coastal Refuges**
2127 SE OSU Drive 541-867-4550
Newport, OR 97365 Fax: 541-867-4551
*Nancy Morrissey, Manager*

**3513 Oregon Islands National Wildlife Refuge**
Western Oregon NWR Complex 503-757-7236
26208 Finley Refuge Road http://www.gorp.com
Corvallis, OR 97333

**3514 Rogue River National Forest**
333 W 8th Street 541-858-2200
PO Box 520 Fax: 541-858-2220
Medford, OR 97501 http://www.fs.fed.us/r6/rogue
The Rogue River National Forest encompasses roghly 630,000 acres of Southern Oregon's most beautiful territory. Staddling the Siskiyou and Cascade mountain ranges, the forest provides vistors and local residents with an array ofresources and recreation opportunities.

**3515 Sheldon National Wildlife Refuge**
US Fish and Wildlife Service- Pacific Region
PO Box 111 503-947-3315
18 South G Fax: 503-947-4414
Lakeview, OR 97630 http://www.fws.gov
*Brian Day, Manager*

**3516 Siskiyou National Forest**
200 NE Greenfield Road 541-471-6500
Po Box 440 Fax: 541-471-6514
Grants Pass, OR 97528
The Siskiyou National Fforest embodies the most complex soils, geology, landscape, and plant communities in the Pacific Northwest. World-class rivers, biological diversity, fisheries, and complex watersheds rank the Siskiyou high inthe Nation as an outstanding resource.

**3517 Siuslaw National Forest**
4077 SE Research Way 541-750-7000
PO Box 1148 Fax: 541-750-7234
Corville, OR 97339
The Siuslaw National Forest encompasses one of the most productive and diverse landscapes in the world from fertile soils, which support tall stands of Douglas fir, western hemlock and Sitka spruce forests laced with miles of riversand streams, to miles of open sand dunes. These rich settings from habitats for a broad array of plants and animals and provide endless opportunities for learning.

**3518 Three Arch Rocks National Wildlife Refuge**
Western Oregon NWR Complex 503-757-7236
26208 Finley Refuge Road http://www.gorp.com
Corvallis, OR 97333

**3519 Umatilla National Forest**
2517 SW Hailey Avenue 541-278-3716
Pendleton, OR 97801
The Umatilla National Forest, located in the Blue Mountains of southeast Wasington and northeast Oregon, covers 1.4 million acres of diverse landscapes and plant communities. The forest has some mountainous terrain, but most of theforest consists of v-shaped valleys separated by narrow ridges or plateaus.

**3520 Umatilla National Wildlife Refuge**
Mid-Columbian River Refuges
2805 St. Andrews Loop 509-545-8588
PO Box 2527 Fax: 509-545-8670
Pasco, WA 99301 http://http://midcolumbiariver.fws.gov
*Morris C LeFever, Manager*

**3521 Umpqua National Forest**
2900 NW Stewart Parkway 541-672-6601
PO Box 1008 Fax: 541-957-3495
Roseburg, OR 97470 E-mail: jcaplan@fs.fed.us
The Umpqua National Forest covers nearly one million acres and is located in the western slopes of the Cascades in Southwest Oregon. The forest encommpasses a diverse area of rugged peaks, high rolling meadows, sparkling rivers andlakes and deep canyons producing a wealth of water resources, timber, forage, minerals, wildlife and outdoor recreation opportunities.

**3522 Wallowa-Whitman National Forest**
1500 Dewey Avenue 541-523-6391
PO Bos 907 Fax: 541-523-1315
Baker, OR 97814
The Wallowa-Whitman National Forest contains 2.3 million acres ranging in elevation from 875 feet in Hells Canyon, to 9845

feet in the Eagle Cap Wilderness. Our varied forests are managed as sustainable ecosystems providing cleanwater, wildlife habitat and valuable forest products. For things to do and places to be, the Wallowa-Whitman is the setting for a variety of year-round recreation. You are welcome at the Wallowa-Whitman National Forest.

**3523 William L Finley National Wildlife Refuge**
26208 Finley Refuge Road 503-757-7236
Corvallis, OR 97333 http://www.gorp.com

**3524 Winema National Forest**
2819 Dahlia Street 541-883-6714
Klamath Falls, OR 97601 Fax: 541-883-6709
The 1.1 million acre Winema National Forest lies on the eastern slopes of the Cascade Mountain Range in South Central Oregon, an area noted for its year-round sunshine. The Forest borders Crater Lake National Park near the crest of theCascades and stretches eastward into the Klamath River Basin. Near the floor of the Basin the Forest gives way to vast marshes and meadows assoicated with Upper Klamath Lake and the Williamson River.

## Pennsylvania: US Forests, Parks, Refuges

**3525 Allegheny National Forest**
4 Farm Colony Road 814-723-5150
Warren, PA 16365 Fax: 814-726-1465
E-mail: anf/r9_allegheny@fs.fed.us
http://www.fs.fed.us/r9/allegheny
An organization dedicated to providing advice for development of the corridor management plan for the northern section of the Allegheny River that has been designated as a National Wild and Scenic River.
*Mark Conn, Staff Contact*

**3526 Delaware National Scenic River: Delaware Water Gap National Recreation Area**
Delaware Water Gap National Recreation Area
HQ River Road - Route 209 570-588-2435
Bushkill, PA 18324 Fax: 570-588-2780
E-mail: dewa_interpretation@nps
http://www.nationalparksgallery.com/

**3527 Erie National Wildlife Refuge**
11296 Wood Duck Lane 814-789-3585
Guys Mills, PA 16327 Fax: 814-789-2909
http://www.fws.gov/northwest/erie
A haven for migratory birds consisting of two divisions: the Sugar Lake Division and the Seneca Division. Refuge management objectives include: providing waterfowl and other migratory birds with nesting, feeding, brooding, and restinghabitat; providing habitat to support a diversity of other wildlife species; and enhancing opportunities for wildlife-oriented public recreation and environmental education.

*Patty Nagel, Deputy Refuge Manager*

**3528 Gettysburg National Military Park**
1195 Baltimore Pike, Suite 100 717-334-1124
Gettysburg, PA 17325 Fax: 717-334-1891
E-mail: gett_superintendant@nps.gov
http://www.nps.gov
A unit of the national park service preserving 6000 acres of Gettysburg battlefield, and the Soldiers' National Cemetery, site of Lincoln's Gettysburg Address.

**3529 John Heinz National Wildlife Refuge at Tinicum**
86th Street & Lindbergh Boulevard 215-365-3118
Philadelphia, PA 19153 Fax: 215-365-2846
E-mail: FW5RW_JHTNWR@fws.gov
http://http://Heinz.fws.gov
The John Heinz National Wildlife Refuge at Tinicum is administered by the Department of Interior's U.S. Fish and Wildlife Service and is located in Philadelphia and Delaware Counties, Pennsylvania. The refuge protects the last 200acres of freshwater tidal marsh in Pennsylvania. The refuge has become a resting and feeding area for more than 20 species of birds, 80 of which nest here. Fox, deer, muskrat, turtles, fish, frogs and a wide variety of wildflowers and plants callthe refuge home.

**3530 Upper Delaware Scenic & Recreational River**
Rural Route 2 570-729-7134
Box 2428 Fax: 570-729-7918
Beach Lake, PA 18405 http://www.npa.gov/upde
As a part of the National Wild and Scenic Rivers System, upper Delaware Scenic and Recreational River streches 73.4 miles (118.3 km) along the New York/Pennsylvania border. The longest free flowing river in the Northeast, it includesriffles and Class I and II rapids between placid pools eddies. Public fishing and boating accesses are provided, although most land along the river is privately owned. Wintering bald eagles are among the wildlife that may be seen here.
*Dave Forney, Superintendent*
*Michael Reubet, Chief Resource Management*

## Rhode Island: US Forests, Parks, Refuges

**3531 Rhode Island National Wildlife Refuge Complex**
3769 D Old Post Road 401-364-0170
PO Box 307 Fax: 401-364-0170
Charlestown, RI 02813 E-mail: fw5rw_rinwr@fws.gov
http://www.northeast.fws./ri.htm
*Charles Vandemoer, Complex Refuge Manager*
*Gary M Andres, Deputy Refuge Manager*

## South Carolina: US Forests, Parks, Refuges

**3532 Ace Basin National Wildlife Refuge**
PO Box 848 843-889-3084
Hollywood, SC 29449 Fax: 843-889-3282
http://acebasin.fws.gov
The Ace Basin National Wildlife Refuge was established in 1990 to assist in preserving the nationally significant wildlife and related habitats within the 350,000-acre Ashepoo, Combahee and South Edisto (ACE) rivers basin. The wetlandshabitat of the area has been preserved during the last several centuries through careful management by private landowners. An antebellum mansion that survived the Civil War now serves in part as office space for the refuge.
*Jane Griess, Refuge Manager*

**3533 Cape Romain National Wildlife Refuge**
5801 Highway 17 North 843-928-3264
Awendaw, SC 29429 Fax: 843-928-3803
http://caperomain.fws.gov/DonnyBrowning
Refuge objectives are to: provide habitat for waterfowl, shorebirds, wading birds and resident species; provide habitat and management of endangered and threatened species; provide protection of Class I Wilderness Area; and provideenvironmental education and recreation for the public.
*James "Donny" Browning, Refuge Manager*

**3534 Carolina Sandhills National Wildlife Refuge**
Route 2 Box 330 803-335-8401
McBee, SC 29101 Fax: 803-335-8406
http://www.gorp.com
*Richard P Ingram, Manager*

**3535 Francis Marion-Sumter National Forest**
4931 Broad River Road 803-561-4000
Columbia, SC 29212 http://www.fs.fed.us/r8/fms/
Headquaters in the capital city of Colombia, both forests are managed for many uses; including timber and wood production, watershed protection and improvement, habitat for wildlife and fish

species, wilderness area management,minerals leasing and outdoor recreation.

**3536 Santee National Wildlife Refuge**
2125 Fort Watson Road 803-478-2217
Summerton, SC 29148 Fax: 803-478-2314
E-mail: santee@fws.gov
http://www.fws.gov/santee

Wildlife viewing opportunities available at the refuge on hiking, biking, canoeing/kayaking, and driving trails. A refuge visitor center is open Tuesday thru Saturday from 8-4. Located off of I-95 at exit 102 in Summerton, SC, SanteeNational Wildlife offers something for everybody.

*Marc Epstein, Refuge Manager*
*Susan Heisey, Park Ranger*

## South Dakota: US Forests, Parks, Refuges

**3537 Badlands National Park**
PO Box 6 605-433-5361
Interior, SD 57750 Fax: 605-433-5404
E-mail: badl_information@nps.gov
http://www.nps.gov/badl/

Consists of acres of sharply eroded buttes, pinnacles and spires blended with the largest protected mixed grass prarie in the US. The Badlands Wilderness Area covers 64,000 acres and is the site of the reintroduction of theblack-footed ferret, the most endangered land mammal in North America. The Stronghold Unit is co-managed with the Oglala Sioux Tribe and includes site of 1890s Glost Dances. Over 11,000 years of human history pale to the ages old paleontologicalresources.

*William R Supernaugh, Supt*

**3538 Black Hills National Forest**
25041 N Highway 16 605-673-9200
Custer, SD 57730 Fax: 605-673-9350
http://www.fs.fed.us/r2/blackhills/

Eleven reservoirs, 30 campgrounds, 2 scenic byways, 1300 miles of streams, 13,000 acres of wilderness, 353 miles of trails, and much more. The forest is managed for multiple use so don't be surprised to see mining, logging, cattlegrazing, and summer homes on your travel.

**3539 Huron Wetland Management District**
200 4th Street SW 605-352-5894
Federal Building, Room 309 Fax: 605-352-6709
Huron, SD 57350 E-mail: huronwetlands@fws.gov
http://huron wetlands.fws.gov

The public lands of the HWMD, called Waterfowl Production Areas, are part of the National Wildlife Refuge System. The refuges and WPAs are vitally important to wildlife and people. They provide food, water, cover and space for hundredsof species of birds, mammals, reptiles, amphibians, fish and plants. Managed to benefit endangered species, migratory birds and other wildlife and provide places to learn about and enjoy wildlife. HWMD's mission is to preserve wetlands and managehabitat.
*harris Hoisted, Project Leader*

**3540 Jewel Cave National Monument**
Rural Route 1 Box 60-AA 605-673-2288
Custer, SD 57730 Fax: 605-673-3294
E-mail: mailto:JECA_Interpretation@nps.gov
http://www.dgif.state.va.us

With more than 125 miles sureyed, jewel cave is recognized as the third longest cave in the world. Airflow within its passages indicates a vast area yet to be explored. Cave tours provide opportunities for viewing this pristine cavesystem and its wide varitey of speleothems including stalactites, stalagmites, draperies, frostwork, flowstone, boxwork and hydromagnesite balloons. The cave is an important hibernaculum for several species of bats.
*Kate Cannon, Supt*

**3541 Sand Lake National Wildlife Refuge**
39650 Sand Lake Drive 605-885-6320
Columbia, SD 57433 Fax: 605-885-6333
E-mail: sandlake@fws.gov
http://sandlake.fws.gov

Sand Lake Refuge is haven for wildlife and those who enjoy it. Home to more than 266 species of birds, 40 mammal species and a variety of fish, reptiles and amphibians, this 22,000 acre refuge is a mosaic of wildlife and the wildplaces they need. Sand Lake is also a very popular recreation spot. Wildlife observation, fishing, hunting, photography, interpretation and environmental education are all popular activities at the refuge.

*Refuge Manager*

**3542 Wind Cave National Park**
26611 US Highway 385 605-745-4600
Hot Springs, SD 57747 Fax: 605-745-4207
http://www.npa.gov/wica

One of the world's longest and most complex caves lies beneath the 28,295 acres of rolling, mixed grass prairie ecosystems of Wind Cave National Park. The park is home to a large variety of prairie wildlife such as bison, pronghornantelope and prairie dogs. The cave is famous for the rare cave formation called boxwork.

*Vidal Davila, Park Superintendent*

## Tennessee: US Forests, Parks, Refuges

**3543 Big South Fork National River Recreation Area**
4564 Leatherwood Road 931-879-3625
Onieda, TN 37841 Fax: 423-569-5505

The free-flowing Big South Fork of the Cumberland River and its tributaries pass through 90 miles of scenic gorges and valleys containing a wide range of natural and historic features. The area offers a broad range of recreationalopportunities including camping, whitewater rafting, kayaking, canoeing, hiking, horseback riding, mountain biking, hunting and fishing, The US Army Corps of Engineers, with its experience in managing river basins, was charged with land acquisition,planning and deve
*William K Dickinson, Supt*

**3544 Cherokee National Forest**
2800 N Ocoee Street NW 423-476-9700
Po Box 2010 Fax: 423-476-9721
Cleveland, TN 37320

The Cherokee is steeped in colorful history and rich in the grandeur of the Appalachian Mountains. The forest is separated into two sections by Great Smoky Mountains Park and shares other boundaries with national forest in Georgia,North Carolina and Virginia.

**3545 Chickasaw National Wildlife Refuge**
1505 Sand Bluff Road 731-635-7621
Ripley, TN 38063 Fax: 731-635-0178
E-mail: fw4_rw_chickasaw.fws.gov
http://www.fws.gov

Established to provide essentail habitat for migratory birds in the Lower Mississippi Valley. The refuge supports a variety of wildlife. Visitors can see large numbers of migratory wasterfowl in the winter. Neotropical migratory birdsand shorebirds are a common site yearround. The refuge is open to hunting and fishing-special regulations apply. Please contact the refuge manager for current regulations.
*Curt McMurl, Refuge Manager*

**3546 Cross Creeks National Wildlife Refuge**
643 Wildlife Road 931-232-7477
Dover, TN 37058 Fax: 931-232-5958
E-mail: fw4_web_manager@fws.gov
http://www.fws.gov/crosscreeks/

*Vicki C Grafe, Manager*

**3547 Great Smokey Mountains National Park**
107 Park Headquarters Road 865-436-1200
Gatlinburg, TN 37738 Fax: 865-436-1220
E-mail: grsm_smokies_information@nps.gov
http://www.nps.gov
The national park, in the state of North Carolina is world renowned for thr diversity of its plant and animal resources, the beaty of its ancient mountains, the quality of its remmants of Southern Appalachin mountain culture, and thedepth and integrity of the wildernees sanctuary within its boundaries, it is one the largest protected areas in the east.
*Randall R Pope, Supt*

**3548 Hatchie National Wildlife Refuge**
4172 Highway 76 South 901-772-0501
Brownsville, TN 38012 Fax: 901-772-7839
E-mail: r4rw_tn.htc@fws.gov
http://www.gorp.com
*Marvin L Nichols, Manager*

**3549 Lower Hatchie National Wildlife Refuge**
1505 Sand Bluff Road 901-635-7621
Ripley, TN 38063 Fax: 901-635-7621
E-mail: r4rw_tn.rlf@fws.gov
http://www.gorp.com

**3550 Reelfoot National Wildlife Refuge**
4343 Highway 157 901-538-2481
Union City, TN 38261 Fax: 901-538-9760
E-mail: r4rw_tn.rlf@fws.gov
http://www.gorp.com
*Randy Cook, Manager*

**3551 Tennessee National Wildlife Refuge**
PO Box 849 901-642-2091
Paris, TN 38242 Fax: 901-644-3351
E-mail: r4rw_tn.tns@fws.gov
http://www.gorp.com
*John Taylor, Manager*

## Texas: US Forests, Parks, Refuges

**3552 Alibates Flint Quarries National Monument: Lake Meredith National Recreation Area**
PO Box 1460 806-857-3151
Fritch, TX 79036 Fax: 806-857-2319
http://www.nps.fov/alfl
ALIBATES: The only national monument in Texas. Preserves over 700 archeological sites. The monument can only be viewed by ranger-led guided tours. LAKE MEREDITH: A 45,000 acre recreation area that includes a 10,000 acre reservoirwhere visitors can enjoy water and land recreational activities such as hunting, fishing, boating, horseback riding, off-road vehicles, jetskies and the like.
*Karren Brown, Supt*

**3553 Amistad National Recreation Area**
PO Box 420367 830-775-7491
Del Rio, TX 78840 Fax: 830-775-7299
E-mail: interpretation@nps.gov
Situated on the United States-Mexico border, is know primarily for excellent year round, water-based recreation including: fishing , boating, swimming, suba diving. Also provides opportunities for picnicking, camping and hinting. Thereservoir, at the confluence of the Rio Grande, Devils and Pecos rivers, was created by Amistad Dam in 1969, This area is reach in technology and rock art, and contains a wide variety of plant and animal life.
*Robert Reyes, Supt*

**3554 Angelina National Forest**
701 N 1st Street 409-639-8620
Room 100 Fax: 409-639-8624
Lufkin, TX 75901
The Angelina National Forest is located in the heart of Texas. The reservoir, a 114,500 acre lake on the Angelina River is noted for its fishing, boating and water skiing.

**3555 Big Bend National Park**
PO Box 129 915-477-2251
Big Bend National Park, TX 79834 Fax: 915-477-1175
The Big Bend National Park is situated on the boundry with Mexico along the Rio Grande. It is a place where countries and cultures meet, also a place that merges natural environments, from desert to mountains. It's a place where southmeets north and east meets west, creating a great diversity of plants and animals. The park covers over 801,000 acres of west Texas in the place where the Rio Grande makes a sharp turn - the Big Bend.
*Robert Arnberger, Supt*

**3556 Big Thicket National Preserve**
3785 Milam Street 409-951-6800
Beaumont, TX 77701 Fax: 409-951-6868
E-mail: BITH_Administration@nps.gov
http://www.nps.gov/bith/default.htm
The Preserve consists of nine land units and six water corridors encompassing more than 97,000 acres. Big Thicket was the first Preserve in the National Park System and protects and area of rich biological diversity. A convergence ofecosystems occured here during the last Ice Age. It brought together, in one geographical location, the eastern hardwood forests, the Gulf coastal plains and the midwest praries.
*Ronald Switzer, Supt*

**3557 Grulla National Wildlife Refuge**
Muleshoe NWR 806-946-3341
PO Box 549 Fax: 806-946-3317
Muleshoe, TX 79347 http://fws.gov/southwest
Located in Roosevelt County, New Mexico, near the small town of Arch, approximately 25 miles northwest of Muleshoe National Wildlife Refuge. Grulla NWR, which is managed by the staff at Muleshoe NWR, has 3,236 acres, more than 2,000 ofwhich make up the saline lake bed of Salt Lake. The rest of the refuge is grassland. When the lake holds sufficient water, Grulla NWR is a wintering area for lesser sandhill cranes. Ring-necked pheasant, scaled quail and lesser prairie chickens maybe seen.
*Jude Smith, Refuge Manager*

**3558 Guadalupe Mountains National Park**
400 Pine Canyon Drive 915-828-3251
Salt Flat, TX 79847 Fax: 915-828-3269
E-mail: GUMO_Superintendent@nps.gov
http://www.nps.gov/gumo
This mountain mass contains portions of the world's most extensive and significant Permian limestone fossil reef, earth fault peaks, unusual flora and fauna. Guadalupe Peak, highest point in Texas at 8,749 feet.
*John Lujan, Superintendent*
*Fred Armstrong, Chief of Resource Management*

**3559 Hagerman National Wildlife Refuge**
6465 Refuge Road 903-786-2826
Sherman, TX 75092 Fax: 903-786-3327
E-mail: fw2_rw_hagerman@fws.gov
http://southwest.fws.gov/refuges/texas/hagermn.html
Hagerman NWR lies on the Big Mineral Arm of Lake Texoma, on the Red River between Oklahoma and Texas. Established in 1946, the refuge includes 3,000 acres of marsh and water and 8,000 acres of upland and farmland. During fall, winterand spring, the marshes and waters are in constant use by migrating and wintering waterfowl.
*Kathy Whaley, Manager*

**3560 Padre Island National Seashore**
PO Box 181300 361-949-8173
Corpus Christi, TX 78480 Fax: 361-949-8023
http://www.nps.gov/pais

Encompassing 130,434 acres, the longest remaining undeveloped stretch of barrier island in the world, and offers a wide variety of flora and fauna as well as recreation.

**3561 Santa Ana National Wildlife Refuge**
Route 2
Box 202A
Alamo, TX 78516
956-784-7500
Fax: 956-787-8338
E-mail: r2rw_sta@fws.gov
http://southwest.fws.gov/refuges/texas/santana.html
The 2,088 acre refuge along the banks of the lower Rio Grande was established in 1943 for the protection of migratory birds. Considered the jewel of the refuge system, this essential island of thorn forest habitat is host or home tonearly 400 different types of birds and a myriad of other species, including the indigo snake, malachite butterfly and the endangered ocelot. Provides habitat for thousands of migrating birds and about one half of all butterfly species found in NorthAmerica.
*Jeff Howland, Refuge Manager*

## Utah: US Forests, Parks, Refuges

**3562 Arches National Park**
PO Box 907
Moab, UT 84532
435-719-2299
Fax: 435-719-2300
E-mail: archinfo@nps.gov
Arches National Park preserves over two thousand natural sandstone arches and a variety of other unique geological resources. The extraordinary features of the park are highlighted by a striking environment of contrasting colors,landforms and textures. Administered by Canyonlands National Park
*] pages*
*Laura Jess, Supt*

**3563 Ashley National Forest**
PO Box 279
Manila, UT 84046
435-784-3445
Fax: 435-781-5295

**3564 Bear River Migratory Bird Refuge**
58 S
Brigham City, UT 84302
801-723-5887
E-mail: R6RW_BRR@fws.gov
To date, close to 1 million cubic yards of earth has been moved to restore and enhance the refuge. Forty-seven primary water control structures have been restored along with over forty-seven miles of dikes. Through volunteer efforts,debris has been removed from the old headquaters site and a new pavilion, restroom, demonstration pond, and kiosk have been built on the site. The 12-mile auto tour route has been reopened to the public.
*Alan K Trout, Manager*

**3565 Bryce Canyon National Park**
PO Box 170001
Bryce Canyon, UT 84717
435-834-5322
Fax: 435-834-4102
http://www.nps.gov/brca
Consists of 37,277 acres of scenic colorful rock formations and desert wonderland. Bryce Canyon National Park is named for one of a series of horseshoe-shaped amphitheaters carved from the eastern edge of the Paunsaugunt Plateau insouthern Utah. Erosion has shaped colorful Claron limestones, sandstones and mudstones into thousands of spires, fins, pinnacles and mazes. Collectively called hoodoos, these unique formations are whimsically arranged and tinted with colors toonumerous to name.
*Craig C Axtell*

**3566 Canyonlands National Park**
2282 S West Resource Boulevard
Moab, UT 84532
435-719-2313
Fax: 435-719-2300
E-mail: canyinfo@nps.gov
http://www.nps.gov/cany/home.htm
Canyonlands National Park preserves a stunning landscape of sedimentary sandstones eroded into countless canyons, mesas and buttes by the Colorado River and its tributaries. Largely undeveloped, the park is a popular backcountrydestination and scientific research site.

**3567 Capitol Reef National Park**
HC 70 Box 15
Torrey, UT 84775
435-425-3791
Fax: 435-425-3026
E-mail: care_administration@nps.gov
http://www.capitol.reef.national-park.com
The Waterpocket Fold, a 100 mile long wrinkle in the earth's know as a monoclide, extends from nearby Thousand Lakes Mountain to the Colorado River. Capitol Reef National Park was established to protect this grand and colorful geologicfeature, as well as the unique historical and cultural history found in the area.
*Albert J Hendricks, Supt*

**3568 Cedar Breaks National Monument**
2390 W Highway 56
Suite 11
Cedar City, UT 84720
435-586-9451
Fax: 435-586-3813
Millons of years of sedimentation, uplift and erosion continue to create a deep canyon of rock walls, fins, spires and columms, that spans some three miles, and over 2,000 feet deep. The rim of the canyon is over 10,000 feet above sealevel, and is forested with islands of Englemann spruce, subalpine fir and aspen: separated by broad meadows of brillant summertime wild flowers.
*Denny Davies, Supt*
*Ateve Robinson, Chief Ranger*

**3569 Dixie National Forest**
82 N 100 E
Cedar City, UT 84720
435-865-3700
Fax: 435-865-3791
The Dixie is located adjacent to three National Parks, Bryce Canyon, Zion and Capitol Reef.The red sandstone formations of Red Canyon rival those of Bryce Canyon National park. From the top of Powell Point, it is possible to see farinto three different states. Boulder Mountain and the many different lakes provide opportunities for hiking, fishing and viewing outstanding scenery.
*Mary Wagner, Forest Supervisor*

**3570 Fish Springs National Wildlife Refuge**
PO Box 568
Dugway, UT 84022
435-831-5353
Fax: 435-831-5354
Located at the southern end of the Great Salt Lake Desert in western Utah, Fish Springs National Wildlife Refuge encompasses 17,992 acres between two small mountain ranges. Five major springs and several lesser springs and seeps flowfrom a faultline at the base of the eastern front of the Fish Springs Mountain Range, These warm, saline springs provide virtually all of the water for the Refuge's 10,000-acre marsh system.
*Jerry Bana, Manager*

**3571 Fishlake National Forest**
115 E 900 N
Richfield, UT 84701
435-896-9233
Fax: 435-896-9347
http://fs.fed.us/r4/fishlake
The Fishlake National Forest in central Utah features majestic stands of aspen encircling open mountain meadows that are lush with a diverse community of forbs and grasses. The mountains of the Fishlake are a source of water for manyof the neighboring communities and agricultural valleys in the region. Hunting, fishing and OHV use are among the most popular forms of recreation enjoyed by forest visitors.
*Allen Rowley, Forest Supervisor*

**3572 Manti-LaSai National Forest**
599 W Price River Drive
Price, UT 84501
435-637-2817
Fax: 435-637-4940
*Elaine Zieroth, Forest Supervisor*

**3573 Natural Bridges National Monument**
HC 60
Box 1
Lake Powell, UT 84533
435-692-1234
Fax: 435-692-1111
E-mail: nabrinfo@nps.gov
http://www.nps.gov/nabr
Natural Bridges protects some of the finest examples of ancient stone architecture in the southwest. The monument is located in the southeast Utah on a pinyon-juniper covered mesa bisected by deep canyons of Permian age Ceder MesaSandstone. Where me-

andering streams cut through the cayon walls, three natural bridges formed: Kachina, Owachomo and Sipapu.

*Coralee S Hays, Superintendent*

**3574 Ouray National Wildlife Refuge**
266 West 100 North #2
Vernal, UT 84078
801-789-0351
Fax: 801-789-4805
http://www.gorp.com
*Gary Montoya, Manager*

**3575 Timpanogos Cave National Monument**
Rural Route 3 Box 200
American Fork, UT 84003
801-756-5239
Fax: 801-756-5661
http://www.nps.gov/tica
Timpanogos Cave Natioanl Monument sits high in the Wasatch Mountains. The cave system consists of three spectacularly decorated caverns. Each cavern has unique colors and formations. Helicitites and anthodites are just a few of themany dazzling formations to be found in the many chambers. As visitors climb to the cavern entrance, on a hike gaining over 1,000 feet in elevation, they are offered incredible views of American Fork Canyon.

*Dennis Davis, Supt*

**3576 Uinta National Forest**
88 W 100 N
Provo, UT 84601
801-342-5100
Fax: 801-342-5144
The Uinta National Forest ranges from high western desert at Vernon to lofty mountain peaks such as Mount Nebo (elevation 11,877 feet, the highest peak in the Wasatch Range) and Mount Timpanogos (elevation 11,750 feet). The forestcontains three wilderness areas: the Lone Peak, the Mount Timpanogos and the Mount Nebo Wildernesses. The Forest surrounds the Timpanogos Cave National Mounment.
*Peter Karp, Forest Supervisor*

**3577 Wasatch-Cache National Forest**
8236 Federal Building
125 S State Street
Salt Lake City, UT 84138
801-524-3900
Fax: 801-524-3172
Wasatch-Cache National Forest lands are located in three major areas: the northern and western slopes of the Uinta Mountains. The Wasatch Front from Lone Peak north to the Idaho border including the Wasatch, Monte cristo, and BearRiver Ranges. The Stansbury Range, in the Great Basin.
*Tom Tidwell, Forest Supervisor*

**3578 Zion National Park**
Star Route 9
Springdale, UT 84767
435-772-3256
Fax: 435-772-3426
E-mail: zion_park_information@nps.gov
http://www.ups.gov/zion
Protected within Zion National Park's 229 square miles (593.1 km) is a spectacular cliff-and-cayon landscape and wilderness full of the unexpected including the world's largest arch -Kolob Arch- with a span that measures 310 feet(94.5m). Wildlife such as mule deer, golden eagles, and mountain lions also inhabit the park. Mukuntuweap National Monument proclaimed July 31, 1909; incorporated in Zion National Monument March 18, 1918; established as national park Nov. 19, 1919.
*Ron Terry, Public Information Officer*

## Vermont: US Forests, Parks, Refuges

**3579 Green Mountains & Finger Lakes National Forest**
231 North Main Street
Rutland, VT 05701
802-747-0300
Fax: 802-747-6766
http://www.fs.fed.us/r9/gmfl/Green Mountain National Forest

**3580 Missisquoi National Wildlife Refuge**
29 Tabor Road
Swanton, VT 05488
802-868-4781
Fax: 802-868-2379
E-mail: missisquoi@fws.gov
http://http://missisquoi.fws.gov
The 6,592-acre refuge includes most of the Missisquoi River delta where it flows into Missisquoi Bay. The refuge consists of quiet waters and wetlands which attract large flocks of migratory birds.

*Mark Sweeny, Manager*

## Virginia: US Forests, Parks, Refuges

**3581 George Washington National Forest**
5162 Valleypointe Parkway
PO Box 233
Roanoke, VA 24019
540-265-5100
Fax: 540-265-5145
Outstanding hiking trails, campsites, fishing and canoeing are the hallmarks of George Washington Forest in Virginia and West Virginia, part of the George Washington and Jefferson National Forest.

**3582 Jefferson National Forest**
5162 Valleypointe Parkway
Roanoke, VA 24019
540-265-5100
888-265-0019
The Jefferson National Forest is prize Appalachia country: tumbling waterfalls, rare wildflowers, vividly colored hills and Virginia's highest peak. Jefferson National Forest spreads 690,000 acres of hardwood and conifer forest acrosswest-central Virginia, West Virgina and Kentucky, including the ridge province of the Blue Ridge mountains.

**3583 Mason Neck National Wildlife Refuge**
14344 Jefferson Davis Highway
Woodbridge, VA 22191
703-690-1297
The refuge, the Mason Neck State Park, the Northern Virginia Park Authority, the Gunston Hall Plantation and the Virginia Department of Game and Inland Fisheries are cooperating in the management of their combined 5,000+ acres on theMason Neck peninsula. This cooperation provides a wide variety of recreational activities while protecting natural resources. The primary objective of the refuge is to protect essential nesting, feeding and roosting habitat for bald eagles.
*J Frederick Milton, Manager*

**3584 Shenandoah National Park**
3655 US Highway 211 East
Luray, VA 22835
540-999-2243
Fax: 540-999-3500
Shenandoah National park lies astride a beautiful section of the blue rige mountains, which from the eastern rampart of the Appalachian Mountains between Pennsylvania and Georgia. The Shenandoah River flows through the valley to thewest, with Massanutten Mountain, 40 miles long, standing between the river's north and south forks. The rolling Piedmont country lies to the east of the park. Skyline Drive, a 105- mile road that winds along the crest of the mountainsthrough thelength of the park.
*William Wade, Supt*

## Washington: US Forests, Parks, Refuges

**3585 Colville National Forest**
765 South Main Street
Coleville, WA 99114
509-684-3711
Fax: 509-684-7280
http://www.fed.us/cvnf
The Colville National Forest encompasses over one million acres in northeastern Washington. The Sherman Pass National Scenic Byway leads through a portion of the Forest, with camping, fishing, hiking, picnicking, mountain biking,cross-country skiing, sonowmobiling and other recreational activities. 49 Degrees North, a full service ski resort, is located east of Chewelah, about one hour north of Spokane. The Salmo-Prist Wilderness Area sits in the northeast corner of theforest.

**3586 Conboy Lake National Wildlife Refuge**
PO Box 5 509-364-3410
Glenwood, WA 98619 Fax: 509-364-3667
E-mail: harold_cole@fws.gov
Located in the northwest corner of Klickitat County, Washington, the refuge was established primarily for waterfowl. The broad range of habitat diversity provides for a broad diversity of resident wildlife species.
*Harold E Cole, Manager*

**3587 Gifford Pinchot National Forest**
6926 E 4th Plain Boulevard 360-891-5000
PO Box 8944 Fax: 360-891-5045
Vancouver, WA 98668 http://www.fs.fed.us/gpnf/
The Gifford Pinchot National Forst is one of the oldest National Forests in the United States. Include as part of The Mount Rainier Forest Reserve in 1897, this area was set aside as the columbia National Forest in 1908, and renamedthe Gifford Pinchot National Forest in 1949. The Forest, located in southwest Washington State, now contains 1,312,000 acreas and includes the 110,000- acre Mount St. Helens National Volcanic Mounument established by congress in 1982.

**3588 Lewis & Clark National Wildlife Refuge**
Julia Butler Hansen NWR 206-795-3915
PO Box 566 http://www.gorp.com
Cathlamet, WA 98612

**3589 McNary National Wildlife Refuge**
64 Maple Road 509-547-4942
PO Box 544
Burbank, WA 99723
A resting and feeding area for up to 100,000 migrating waterfowl. It includes 3,629 acres of water and marsh, croplands, grasslands, trees and shrubs.
*David Linehan, Manager*

**3590 Mount Baker-Snoqualmie National Forests**
21905 64th Avenue West 425-744-3200
Mountlake Terrace, WA 98043 800-627-0062

**3591 Mount Rainier National Park**
Tahoma Woods, Star Route 253-569-2211
Ashford, WA 98304 Fax: 360-569-2170
E-mail: MORAInfo@nps.gov
Established in 1899. 235,625 acres (97% is designated Wilderness). Includes mount Rainier (14,410'), an active volcano encased in over 35 square miles of snow and ice. The park contains outstanding examples of old growth forests andsubalpine meadows. Designated a National Historic landmark Distract in 1997 as a showcase for the NPS Rustic style architecture of the 1920s and 1930s.
*William Briggle, Supt*

**3592 North Cascades National Park Service Complex**
810 State Route 20 360-856-5700
Sedro-Woolley, WA 98284 Fax: 360-856-1934
E-mail: NOCA_Interpretation@nps.gov
http://www.nps.gov/noca

**3593 Okanogan National Forest**
1240 S 2nd Avenue 509-826-3068
PO Box 950
Okanogan, WA 98840
There is a variety of country from craggy peaks to rolling meadows, to rich old growth forest and classic groves of ponderosa pine. We're called the Sunny Okanogan and for good reason: summers here are hot and dry, and our winters arefamous for brilliant clear skies and plenty of snow.

**3594 Okanogan and Wenatchee National Forests Headquarters**
215 Melody Lane 509-664-9200
Wenatchee, WA 98801 Fax: 509-664-9280
http://www.fs.fed.us/r6/wenatchee/forest/formain.htm
The 2.2 million acre Wenatchee National Forest extends about 135 miles along the east side of the crest of the Cascade Mountains in Washington State. This National Forest is most noted for its wide range of recreation opportunities.There truly is something for everyone who likes to have fun in the outdoors.

**3595 Olympic National Forest**
1835 Black Lake Boulevard SW 360-956-2402
Olympia, WA 98512 http://www.fs.fed.us/r6/olympic
The National Forests are part of America's great outdoors and are public lands. They are managed for the multiple uses of recreation, wildlife, timber, gazing, mining, oil and gas, watershed and wilderness. The Olympic National Forestis over 632,000 acres in size and is divided into two Ranger Districts: Hood Canal and Pacific.

**3596 Olympic National Park**
600 East Park Avenue 360-565-3000
Port Angeles, WA 98362 Fax: 360-565-3015
E-mail: olym_visitor_center@hps.gov
Often referred to as three parks in one, Olympic National Park encompasses three distinctly different ecosystems-rugged glacier capped mountains, over 73 miles of wild Pacific coast and magnificent mountains are still largely pristinein character and are Olympic's gift to you.

*Karen Gustin, Superintendent*

**3597 Ross Lake National Recreation Area North Cascades National Park**
810 State Route 20 360-856-5700
Sedro Woolley, WA 98284 Fax: 360-856-1934
E-mail: mailto:NOCA_Interp@nps.gov
http://www.nps.gov/rola
The most accessible part of the North Cascades National Park Service Complex. Is also the corridor for scenic Washington State Route 20, the North Cascades Highway, and includes three reserviors.

**3598 Toppenish National Wildlife Refuge**
21 Pumphouse Road
Toppenish, WA 98948 800-344-9453
E-mail: refuges@fws.gov
http://www.fws.gov/refuges/
An important migration and wintering area for waterfowl in the Yakima Valley of eastern Washington. Wetland impoundments along Toppenish Creek provide natural foods for wintering mallards and other ducks. Ducks and other water birdsbreed in the wetland impoundments during the summer. Native shrub-steppe communities and riparian areas along Toppenish and Snake creeks provide habitat for many other species of birds. The refuge has active hunting and wildlife-viewing programs.
*George J Fenn, Manager*

**3599 Willapa National Wildlife Refuge**
HC 01 206-484-3482
Box 910
Llwaco, WA 98624
Located on Willapa Bay in Pacific County, the southernmost coastal county in Washington. The upland forest varies in successional stages from recently logged areas to a unique remnant of virgin, coastal cedar-hemlock forest home todeer, bear, elk, grouse, beaver and numerous songbirds and small mammals.
*James A Hidy, Manager*

## West Virginia: US Forests, Parks, Refuges

**3600 Monongahela National Forest**
200 Sycamore Street 304-636-1800
Elkins, WV 26241 Fax: 304-636-1875
The Monongahela National Forest was established following passage of the 1911 Weeks Act. This act authorized the purchase of land for long-term watershed protection and natural resource management following massive cutting of theEastern forests in the late 1800's and at the turn of the century.

**3601 Ohio River Islands National Wildlife Refuge**
PO Box 1811 304-422-0752
Parkersburg, WV 26102 Fax: 304-422-0754
E-mail: fw5rw_ohrinwr@fws.gov
The refuge extends 362 river miles from Shippingport, Pennsylvania to Manchester, Ohio along one of the nation's busiest waterways. Ohio River Islands and their back channels have long been recognized for high quality fish andwildlife, recreation, scientific and natural heritage values.
*Jerry L Wilson, Manager*

## Wisconsin: US Forests, Parks, Refuges

**3602 Apostle Islands National Lakeshore**
415 Washington Street 715-779-3397
Route 1, Box 4 Fax: 715-779-3049
Bayfield, WI 54814 E-mail: APIS_Webmaster@nps.gov
The national lakeshore includes 21 islands and 12 miles of mainland Lake Superior shoreline, featuring pristine stretches of sand beach, spectacular sea caves, remnant old growth forests, resident bald eagles and black bears and thelargest collection of lighthouses anywhere in the National Park System.
*Robert J Krumenaker, Superintendant*

**3603 Chequamegon National Forest**
1170 4th Avenue S 715-762-2461
Park Falls, WI 54552 Fax: 715-762-5179
Shaped principally by glacial action some 10,000 years ago, the forest offers a variety of hiking, ATV, and cross-country ski trails at different levels of difficulty. These campgrounds are located on either a lake or a river and offerfishing and boating.

**3604 Ice Age National Scientific Reserve**
PO Box 7921 608-266-2183
Madison, WI 53707 Fax: 608-267-7474
E-mail: brigit.brown@dnr.state.wi.us
This first national scientific reserve contains nationally significant features of continental glaciation. State parks in the area are open to the public.
*Tom Gilbert, Supt*

**3605 Nicolet National Forest**
68 S Stevens Street 715-362-1300
Rhinelander, WI 54501 Fax: 715-362-1359
Located in Wisconsin's Northwoods where towering pine and hardwood forests are interspersed with hundreds of crystal clear lakes and streams, the Nicolet offers you many opportunities to enjoy the outdoors. Within a day's drive of theChicago, Milwaukee, St. Paul and Minneapolis metropolitan areas, the forest is a place where urban dwellers can truly get away from it all in the scenic beauty of the northwoods.

**3606 St Croix National Scenic Riverway**
401 N Hamilton Street 715-483-3284
St Croix Falls, WI 54024 Fax: 715-483-3288
http://www.nps.gov
The St. Croix National Scenic Riverway is home to the endangered Higgins Eye and Winged Mapleleaf mussels, bald eagles, gray wolves, and the prehistoric paddlefish. The 252 miles of Riverway provide numerous recreational opportunitiesfor boaters, canoeists, kayakers and others.

*Tom Bradley, Superintendent*
*Ron Erickson, Education Team Manager*

## Wyoming: US Forests, Parks, Refuges

**3607 Bighorn National Forest**
2013 Eastside 2nd Street 307-674-2600
Sheridan, WY 82801 Fax: 307-674-2668
The forest has 32 campgrounds, 14 picnic areas, 2 visitor centers, 2 ski areas, 7 lodges, 2 recreation lakes, 3 scenic byways and over 1500 miles of trails. The Bighorn National Forest is 80 miles long and 30 miles wide. The mostcommon tree is the lodgepole pine. The Bighorn River, flowing along the west side of the forest was first named by American Indians due to the great herds of bighorn sheep at its mouth.

**3608 Bridger-Teton National Forest**
340 N Cache-Forest Service Bldg 307-739-5500
PO Box 1888 Fax: 307-739-5010
Jackson, WY 83001 E-mail: r4_b-t_info@fs.fed.us
http://www.fs.fed.us/btnf/
With it's 3.4 million acres, it is the second largest National Forest outside Alaska. Included are more than 1.2 million acres of wilderness in the Bridger, Gros Ventre, and Teton Wildernesses. The Bridger-Teton is a land of variedrecreational opportunities, beautiful vistas, and abundant wildlife. Its crystal blue skies are puctuated by awesome mountain ranges which include the Gros Ventre, Teton, Salt River, Wind River, and Wyoming Mountain Ranges.
*Carole Hamilton, Forest Supervisor*
*Pamela Bode, Resources Staff Officer*

**3609 Devils Tower National Monument**
PO Box 10 307-467-5283
Devils Tower, WY 82714 Fax: 307-467-5350
E-mail: deto_interpretation@nps.gov
http://www.nps.gov/deto/home.html
This unit of the National Park Service protects the nearly vertical monolith known as Devil's Tower. The rolling hills of this 1347 acre park are covered with pine forests, deciduous woodlands, and prairie grasslands. Known byseveral northern plains tribes as Bear Lodge, it is sacred to many American Indians. Devil's Tower was proclaimed in September, 1906 as the nation's first national monument by President Theodore Roosevelt.

*Dorothy FireCloud, Superintendent*

**3610 Fossil Butte National Monument**
PO Box 592 307-877-4455
Kemmerer, WY 83101 Fax: 307-877-4457
E-mail: FOBU_Superintendent@nps.gov
http://www.nps.gov/fobu
Located in southwest Wyoming, Fossil Butte National Monument represents one of the richest fossil localities in the world. Fifty million-year-old fish, insects, birds, reptiles, and plants are nearly perfectly preserved in limestone.
*David E McGinnis, Superintendent*
*Marcia Fagnant, Park Ranger*

**3611 Grand Teton National Park**
PO Box 170 307-739-3300
Moose, WY 83012 Fax: 307-739-3438
E-mail: GRTE_info@nps.gov
http://www.nps.gov/grte
Established in 1929 and enlarged in 1950 to protect a rugged, awe-inspiring mountain range with numerous piedmont lakes nestled amoung its flanks, and a wide sagebrush-covered valley called Jackson Hole. Administered by the NationalPark Service under the Department of the Interior, Grand Teton is one of 384 units within the national park system. It encompasses approximately 310,000 acres or 485 square miles of northwestern Wyoming, just south of Yellowstone National Park.

*Mary Gibson Scott, Supt*

**3612 Medicine Bow National Forest**
2468 Jackson Street 307-745-2300
Laramie, WY 82070 Fax: 307-745-2398
http://www.fs.fed.us/r2/mbr/about/districts/laramie.shtml/
The Medicine Bow National Forest dates back to May 22, 1902, with the establishment of the Medicine Bow Forest Reserve by President Theodore Roosevelt. In 1929, the former Hayden National Forest along the Continental Divide wasformerly a War Department target and maneuver reservation under joint administration by the Forest Service and the War Department. In 1959, the area formerly used by the military was added to the Medicine Bow National Forest.

**3613 National Elk Refuge**
PO Box 510 307-733-9212
Jackson, WY 83001 Fax: 307-733-9729
E-mail: don_delong@fws.gov
http://nationalelkrefuge.fws.gov/
More than 7,500 elk make the winter range of National Elk Refuge their home from October until May. Adjacent to the north side of Jackson, Wyoming, the 25,000-acre refuge includes nearly 1600 acres of open water and marsh lands, 47different mammals and nearly 175 species of birds.
*Mike Hedrick, Manager*

**3614 Seedskadee National Wildlife Refuge**
PO Box 700 307-875-2187
Green River, WY 82935 Fax: 307-875-4425
E-mail: Seedskadee@fws.gov
http://http://seedskadee.fws.gov/
Fishery resource is managed cooperatively with the state G&F and includes a special regulations area to promote catch and release fishing for trophy trout (brown, Snake River cutthroat and rainbow trout). Refuge lands are rich inhistorical and cultural resources as the area was utilized by nomadic Indian tribes, fur trappers, early pioneers and travelers heading for the better life of California and Oregon. Many of the old campsites, river crossings and early structuresstill exist.

**3615 Shoshone National Forest**
808 Meadow Lane 307-527-6241
Cody, WY 82414 Fax: 307-578-1212
The Shoshone consists of 2.4 million acres of varied terrain ranging from sagebrush flats to rugged mountain peaks and includes portions of the Absaroka, Wind River, and Beartooth Mountain Ranges. Elevations on the Shoshone range from4,600 feet at the mouth of the spectacular Clarks Fork Canyon to 13,804 feet on ganneett Peak, Wyoming's highest point. Geologists delightedly call the Shoshone's varied topography an open book.

**3616 Yellowstone National Park**
PO Box 168 307-344-7381
Yellowstone National Park, WY 82190 Fax: 307-344-2005
E-mail: yell_visitor_services@nps.gov
http://www.nps.gov
Established on March 1, 1872, Yellowstone National Park is the first and oldest national park in the world. Preserved within Yellowstone are Old Faithful Geyser and some 10,000 hot springs and geysers, the majority of the plant'stotal. These geothermal wonders are evidence of one of the world's largest active volcanoes; its last eruption created a crater or caldera that spans almost half of the park.
*Suzanne Lewis, Supt*

# Publications

## Directories & Handbooks: Air & Climate

**3617 Acid Rain**
Watts, Franklin
90 Sherman Turnpike
Danbury, CT 06816
203-797-3500
800-621-1115
Fax: 203-797-3657

Lists over 4,000 citations, with abstracts, to the worldwide literature on the sources of acid rain and its effects on the environment.

**3618 Weather America, 3rd Edition**
Grey House Publishing
4919 Route 22
Amenia, NY 12501
518-789-8700
800-562-2139
Fax: 845-373-6390
E-mail: books@greyhouse.com
http://www.greyhouse.com

Provides extensive climatological data for over 4,000 national and cooperative weather stations throughout the US. Includes a new major storms section and a nationwide ranking section that provides rankings for maximum and minimumtemperatures, precipitation, snowfall, fog, humidity and wind speed. Each of 50 state sections contains a city index for locating the nearest weather station to the city/county being researched and a narrative description of the state's climaticconditions.
*Publication Date: 2010 2,013 pages*
*ISBN: 1-891482-29-7*
*Leslie Mackenzie, Publisher*
*David Garoogian, Editor*

## Directories & Handbooks: Business

**3619 American Caves**
American Caves Conservation Association
119 E Main Street
PO Box 409
Horse Cave, KY 42749
270-786-1466
Fax: 270-786-1467
E-mail: debraheavers@caven.org
http://www.cavern.org

A bi-annual membership publication. Published by the American Caves Conservation Association and available by subscription to nonmembers.
*David Foster, Executive Director/Author*
*Debra Heavers, Editor*

**3620 Associations Canada**
Grey House Publishing Canada
555 Richmond Street West
2nd Floor
Toronto, Ontario M5V 3B1
416-644-6479
866-433-4739
Fax: 416-644-1904
E-mail: info@greyhouse.ca
http://www.greyhouse.ca

Annual directory of Canadian associations and environmental groups including industry, commerical and professional organizations.
*Publication Date: 2011 1200 pages*
*Bryon Moore, General Manager*

**3621 Business and the Environment: A Resource Guide**
Island Press
1718 Connecticut Avenue NW
Suite 300
Washington, DC 20009
202-232-7933
800-828-1302
Fax: 202-234-1328
E-mail: info@islandpress.org
http://www.islandpress.org

Includes more than 1,000 references to material from scholarly journals, government agencies, case clearing-houses, research organizations, trade magazines and the popular press. It was the most current (1992) listing of research onself-monitoring and compliance programs and environmental performance strategies for corporate competitiveness.
*Publication Date: 1992 382 pages*
*ISBN: 1-559631-59-7*
*Barbara Dean, Executive Editor*
*Jonathan Cobb, Executive Editor*

**3622 California Certified Organic Farmers: Membership Directory**
2155 Delaware Ave.
Suite 150
Santa Cruz, CA 95060
831-423-2263
Fax: 831-423-4528
E-mail: ccof@ccof.org
http://www.ccof.org

*Annual*
*Brian Leahy, Executive Director*
*Helge Hellberg, Marketing Director*

**3623 Directory of Environmental Information Sources**
Government Institutes
4 Research Place
Suite 200A
Rockville, MD 20850
301-907-1000
Fax: 301-921-2362

Over 1,400 federal and state government agencies, professional and scientific organizations and trade associations are profiled.
*322 pages*

**3624 Directory of New York City Environmental Organizations**
Council on the Environment of New York City
51 Chambers Street
Room 228
New York, NY 10007
212-788-7900
Fax: 212-788-7913
E-mail: conyc@cenyc.org
http://www.cenyc.org

Promotes environmental awareness and solutions to environmental problems.

**3625 Directory of Professional Services**
Professional Services Institute
1730 Rhode Island Avenue NW
Suite 1000
Washington, DC 20036
202-659-4613
800-424-2869
Fax: 202-775-5917

**3626 Directory of Socially Responsible Investments**
Funding Exchange
666 Broadway
Suite 500
New York, NY 10012
212-529-5300
Fax: 212-982-9272
E-mail: mailto:%20FEXEXC@aol.com
http://www.udc.edu/index-b.htm

Network of 15 community foundations around the country with a national office in New York City. Staff at the national office are responsible for three main program areas: grantmaking, donor programs and member fund saervices.

**3627 EPA Information Resources Directory**
National Technical Information Service
5285 Port Royal Road
Springfield, VA 22161
703-605-6000
http://www.ntis.gov

Supports the nation's economic growth and job creation by providing access to information that stimulates innovation and discovery. NTIS accomplishes this mission through two major programs: information collection and dissemination tothe public and production and other services to federal agencies.

**3628 EnviroSafety Directory**
IEI Publishing Division
1635 W Alabama St
Houston, TX 77006
713-529-1616
800-654-1480
Fax: 281-529-0936
E-mail: iei@mail.infohwy.com
http://www.oilonline.com

Approximately 6,000 environmental services, state agencies and EPA/Superfund sites within the EPA regions 4, 6 and 9.
*James W Self, Editor*
*Janis Johnson, Managing Editor*

**3629 Environment: Books by Small Presses of the General Society of Mechanics & Tradesmen**
Small Press Center
20 West 44th Street
New York, NY 10036
212-764-7021
Fax: 212-840-2046
E-mail: smallpress@aol.com
http://www.smallpress.org

*Publication Date: 1992  250 pages*
*ISBN: 0-962276-93-6*
*Paula Matta, Author*

**3630 Environmental Address Book: The Environment's Greatest Champions and Worst Offenders**
Perigee Books
200 Madison Avenue
New York, NY 10016
212-951-8400
http://www.penguinputnam.com

**3631 Fibre Market News: Paper Recycling Markets Directory**
Recycling Media Group GIE Publishers
4012 Bridge Avenue
Cleveland, OH 44113
216-961-4130
800-456-0707
Fax: 216-961-0364

A list of over 2,000 dealers, brokers, packers and graders of paper stock in the US and Canada.
*Dan Sandoval, Internet/Senior Editor*
*Jim Keefe, Group Publisher*

**3632 Greenpeace Guide to Anti-Environmental Organizations**
Odonian Press
PO Box 776
Berkeley, CA 94701
510-486-0313
800-326-0959
Fax: 415-512-8699

Corporations, foundations and public relations firms determined to be anti-environmental despite their attempts to project the green image.
*Publication Date: 1993  112 pages*
*ISBN: 1-878825-05-3*
*Carl Deal, Author*

**3633 Guide to Curriculum Planning in Environmental Education**
125 S Webster Street
PO Box 7841
Madison, WI 53707
608-266-2188
800-441-4563
Fax: 608-267-9110
E-mail: sandi.mcnamer@dpi.state.wi.us
http://www.dpi.state.wi.us/pubsales

Provides a direction in planning a comprehensive environmental education program based on perceptual awareness knowledge, environmental ethics, citizen action skills and citizen action experience.
*Publication Date: 1994  167 pages  Book*
*Sandi McNamer, Publications Director*

**3634 Handbook on Air Filtration**
IEST
2340 S. Arlington Heights Road
Suite 100
Arlington Heights, IL 60005
847-981-0100
Fax: 847-981-4130
E-mail: iest@iest.org
http://www.iest.org

Covers a broad range of applications for users who require removal of airborn particulate contamination for maximum air cleanliness.

*ISBN: 1-877862-60-6*
*Julie Kendrick, Executive Director*

**3635 Harbinger File**
Harbinger Communications, Inc
5 N Union Street
Elgin, IL 60123
847-622-0905
800-320-7206
Fax: 847-622-0830
E-mail: info@harbingeronline.com
http://www.harbingeronline.com

**3636 National Environmental Data Referral Service**
US National Environmental Data Referral Service
1825 Connecticut Avenue NW
Washington, DC 20235
202-606-4089

More than 22,200 data resources that have available data on climatology and meteorology, ecology and pollution, geography, geophysics and geology, hydrology and limnology, oceanography and transmissions from remote sensing satellites.

**3637 National Environmental Organizations**
US Environmental Directories
PO Box 65156
St Paul, MN 55165

**3638 New Jersey Environmental Directory**
Youth Environmental Society
PO Box 441
Cranbury, NJ 08512
609-655-8030

Environmental education and leadership programs for high school students in New Jersey.

**3639 Opportunities in Environmental Careers**
VGM Career Books
4255 W Touphy Avenue
Lincolnwood, IL 60646
847-679-5500
800-323-4900
Fax: 847-679-2494

*Odom Fanning, Author*

**3640 Research Services Directory**
Grey House Publishing
4919 Route 22
Amenia, NY 12501
518-789-8700
800-562-2139
Fax: 845-373-6390
E-mail: books@greyhouse.com
http://www.greyhouse.com

This Ninth Edition provides access to well over 8,000 corporate and independent commercial research firms and laboratories offering contract services for hands-on, basic or applied research in environmental and other areas. Providesthe company's name and addresses, as well as a company description and research and technical fields served.
*Publication Date: 2003  1,200 pages*
*ISBN: 1-891482-30-0*
*Leslie Mackenzie, Publisher*
*Richard Gottlieb, Editor*

**3641 State Environmental Agencies on the Internet**
Government Institutes
4 Research Place
Suite 200A
Rockville, MD 20850
301-907-1000
Fax: 301-921-2362
http://www.govinst.com

Provides a concise profile of each state agency's requirements and resources-including hard-to-find online laws, rules, and regulations-in one quicl-reference guide.

**3642 Water Environment & Technology Buyer's Guide and Yearbook**
Water Environment Federation
601 Wythe Street
Alexandria, VA 22314
800-666-0206
Fax: 703-684-2492
E-mail: webfeedback@wef.org
http://www.wef.org

Founded in 1928, the Water Environment Federation (WEF) is a not for profit technical and educational organization with members from varied disciplines who work toward the WEF vision of preservation and enhancement of the global waterenvironment. The WEF network includes water quality professionals from 76 Member Associations in 30 countries.
*William J Bertera, Executive Director*

**3643 World Directory of Environmental Organizations Online**
California Institute of Public Affairs

PO Box 189040 916-442-2472
Sacramento, CA 95818 http://www.interenvironment.org
A guide to governmental and nongovernmental organizations and programs concerned with protecting the earth's resources. It also covers national and international organizations throughout the world. Only available online.

## Directories & Handbooks: Design & Architecture

**3644 Directory of International Periodicals and Newsletters on Built Environments**
Van Nostrand Reinhold

More than 1,400 international periodicals and newsletters that cover architectural design and the building industry and the aspects of the environment that deal with the industry are covered.
*Publication Date: 1992 175 pages*
*ISBN: 0-442230-03-6*
*Frances C Gretes, Author*

## Directories & Handbooks: Disaster Peparedness & Response

**3645 Association of State Floodplain Managers**
Association of State Floodplain Managers
2809 Fish Hatchery Road 608-274-0123
Suite 204 Fax: 608-274-0696
Madison, WI 53713 E-mail: asfpm@floods.org
http://www.floods.org
A complete name/address/phone listing for all key floodplain managers in the nation, comprehensive summary of ASFPM's activities of past year and planned future directions, key federal agency programs, much more. Free to currentmembers.
*Diane Watson, Editor*

**3646 El Environmental Services Directory Online**
Environmental Information Limited
PO Box 390266 952-831-2473
Edina, MN 55439 Fax: 952-831-6550
E-mail: ei@enviro-information.com
http://www.envirobiz.com
The most comprehensive, largest directory of environmental services in the United States. Coverage includes asbestos & lead abatement, consulting, laboratories, transportation, industrial cleaning, municipal solid waste facilities,hazardous waste facilities, indsutrial waste facilities, well drilling, soil boring, drum reconditioning, spill response, and remediation services.
*Cary Perket*

**3647 Floodplain Management: State & Local Programs**
Association of State Floodplain Managers
2809 Fish Hatchery Road 608-274-0123
Madison, WI 53713 Fax: 608-274-0696
E-mail: asfpm@floods.org
http://www.floods.org
The most comprehensive source assembled to date, this report summarizes and analyzes various state and local programs and activities.
*Publication Date: 2005*

**3648 Hazardous Materials Regulations Guide**
JJ Keller
3003 West Breezewood Lane 920-722-2848
PO Box 368 877-564-2333
Neenah, WI 54957 Fax: 800-727-7516
E-mail: sales@jjkeller.com
http://www.jjkeller.com
A complete reference guide of hazardous materials regulations.
*May/Novemeber*
*ISBN: 0-934674-94-9*
*Tom Ziebell, Editor*

**3649 Institute of Chemical Waste Management Directory of Hazardous Waste Treatment**
National Solid Wastes Management Assn.
4301 Connecticut Ave. 202-244-4700
Suite 1000 800-424-2869
Washington, DC 20008 Fax: 202-966-4824
http://www.nswma.org

**3650 Pesticide Directory: A Guide to Producers and Products, Regulators, and Researchers**
Thomson Publications
Box 9335 559-266-2964
Fresno, CA 93791 Fax: 559-266-0189
http://www.agbook.com
This directory is for the person who needs to know anything about the US pesticide industry. It includes basic manufacturers and formulators along with their products, key personnel, managers, district/regional offices and otherpertinent information. Other sections include Universities, State Extension Centers, USDA, EPA, National Organizations, US Forest Service, Poison Control Centers and much more.
*Publication Date: 1987 153 pages Biannual*
*ISBN: 0-913702-45-5*
*WT Thomson, Author*
*Susan Heflin, President/Owner*

**3651 SEEK**
520 Lafayette Rd N. 651-215-0256
St. Paul, MN 55155 888-668-3224
E-mail: seek@moea.state.mn.us
http://www.seek.state.mn.us/comment.cfm
Minnesotas's interactive directory of environmetal education resources.

**3652 The Grey House Homeland Security DirectoryGrey House Publishing**
Grey House Publishing
4919 Route 22 518-789-8700
Amenia, NY 12510 800-562-2139
Fax: 518-789-0556
E-mail: books@greyhouse.com
http://www.greyhouse.com
A comprehensive, annual resource for national, state and local officials responsible for homeland security along with manufacturers of homeland security products and services.
*Publication Date: 2011 900 pages*
*Leslie Mackenzie, Publisher*
*Jessica Moody, Marketing Director*

**3653 Tracking Toxic Wastes in CA: A Guide to Federal and State Government Information Sources**
INFORM
120 Wall Street 212-361-2400
14th Floor Fax: 212-361-2412
New York, NY 10005 http://www.informinc.org/INFORM.html

## Directories & Handbooks: Energy & Transportation

**3654 Alternative Energy Network Online**
Environmental Information Networks
119 South Fairfax Street 703-683-0774
Alexandria, VA 22314 Fax: 703-683-3893
E-mail: sales@eintoday.com
http://www.eintoday.com
Reports on news of all energy sources designed as alternatives to conventional fossil fuels, including wind, solar and alcohol fuels.

**3655 Current Alternative Energy Research and Development in Illinois**
Department of Energy & Natural Resources
325 W Adams 217-785-2800
Room 300 800-252-8955
Springfield, IL 62704 Fax: 217-785-2618

**3656 Department of Energy Annual Procurement and Financial Assistance Report**
US Department of Energy
Mail Stop 142
Washington, DC 20585 800-342-5303
Fax: 202-586-4403
Offers a list of universities, research centers and laboratories that represent the Department of Energy.

**3657 Directory of Solar-Terrestrial Physics Monitoring Stations**
Air Force Geophysics Laboratory
Department of Defense 781-377-3977
Hanscom Air Force Base, MA 01731 Fax: 781-377-4498

**3658 Energy Science and Technology**
US Department of Energy
1 Science Gov Way 865-576-1188
Oak Ridge, TN 37830 Fax: 865-576-2865
E-mail: ISTIWebmaster@osti.gov
http://www.osti.gov/resource.html
To collect, preserve, disseminate, and leverage the scientific and technical information (STI) resources of the Department of Energy to provide access to national and global STI for use by DOE, the scientific research community,academia, US industry, and the public to expand the knowledge base of science and technology.

**3659 Energy Statistics Spreadsheets**
Institute of Gas Technology
3424 S State St 312-842-4100
Chicago, IL 60616 Fax: 773-567-5209
The coverage of this database encompasses worldwide energy industry statistics, including production, consumption, reserves, imports and prices.

**3660 Interstate Oil Compact Commission and State Oil and Gas Agencies Directory**
Interstate Oil & Gas Compact Commission
2101 N. Lincoln Blvd. 405-521-2302
Oklahoma City, OK 73105 800-822-4015
Fax: 405-521-3099
E-mail: iogcc@iogcc.state.ok.us
http://www.iogcc.oklaosf.state.ok.us
A directory of members and in the back is a list of state oil and gas agencies

**3661 Women's Council on Energy and the Environment Membership Directory**
PO Box 33211 703-351-7850
Washington, DC 20033 Fax: 202-478-2098
E-mail: info@wcee.org
http://www.wcee.org
Offers valuable information on over 800 members representing consulting firms, private industry and the environmental community.
*Publication Date: 1980*
*Clare Piercy, Executive Director*
*JoAnne Scribner, Deputy Executive Director*

## Directories & Handbooks: Environmental Engineering

**3662 Association of Conservation Engineers: Membership Directory**
Engineering Section Alabama Dept. of Conservation
573-522-2323
Fax: 573-522-2324
E-mail: mihalg@mail.conservation.state.mo.us
http://www.conservation.state.mo.us/engineering/ace

**3663 Energy Engineering: Directory of Software for Energy Managers and Engineers**
Fairmont Press
700 Indian Trail 770-925-9388
Liburn, GA 30047 Fax: 770-381-9865
E-mail: linda@fairmontpress.com
http://www.fairmontpress.com
Directory of services and supplies to the industry.
*Publication Date: 1904 80 pages Bimonthly*
*ISSN: 0199-8895*
*Wayne C Turner, Author*
*Wayne C Turner, Editor*

**3664 NEPA Lessons Learned**
Office of NEPA Policy & Compliance
1000 Independence Avenue SW 202-586-4600
EH-42 800-472-2756
Washington, DC 20585 Fax: 202-586-7031
E-mail: denise.freeman@eh.doe.gov
http://www.eh.doe.gov/nepa
*Publication Date: 1994 Quarterly*
*Carol M Borgstrom, Director*

**3665 New York State Conservationist**
NYS Department of Environmental Conservation
625 Broadway 518-402-8047
2nd Floor 800-678-6399
Albany, NY 12233 Fax: 518-402-9036
E-mail: dinnelson@gw.dec.state.ny.us
http://www.dec.state.ny.us/
An informative and entertaining full-color bi-monthly magazine featuring New York State's natural resources and peoples' enjoyment of those resources.
*Bi-Monthly*
*ISSN: 0010-650X*
*David Nelson, Editor*

**3666 Pollution Abstracts**
Cambridge Scientific Abstracts
7200 Wisconsin Avenue 301-961-6700
Suite 601 800-843-7751
Bethesda, MD 20814 Fax: 301-961-6720
E-mail: sales@csa.com
http://www.csa.com
This database provides fast access to the environmental information necessary to resolve day to day problems, ensure ongoing compliance, and handle emergency situations more effectively.
*James P McGinty, President*
*Ted Caris, Publisher*

## Directories & Handbooks: Environmental Health

**3667 American Academy of Environmental Medicine Directory**
American Academy of Environmental Medicine
7701 E Kellogg 316-684-5500
Suite 625 Fax: 316-684-5709
Wichita, KS 67207 E-mail: administrator@aaem.com
http://www.aaem.com
To suppoprt physicians and other professionals in serving the publi through education about the interaction between humans and their environment. Also to promote optimal health through prevention, and safe and effective treatment ofthe causes, not the illness.
*Dee Rogers, Contact*

**3668 Canadian Environmental Resource Guide**
Grey House Publishing Canada

555 Richmond Street West
2nd Floor
Toronto, Ontario M5V 3B1
416-644-6479
866-433-4739
Fax: 416-644-1904
E-mail: info@greyhouse.ca
http://www.greyhouse.ca

Annual directory — Canada's most complete reference of environmental associations and organizations, government regulators and purchasing groups, product and service companies and special libraries.
*Publication Date: 2011 1200 pages*
*Bryon Moore, General Manager*

**3669 Directory of NEHA Credentialed Professionals**
720 S Colorado Boulevard
Suite 1000-N
Denver, CO 80246
303-756-9090
Fax: 303-691-9490
E-mail: staff@neha.org
http://www.neha.org

This is a directory of all NEHA credentialed professionals. It is available to NEHA credentialed professionals only.
*Catalog 569*

**3670 Ecosystem Change and Public Health: A Global Perspective**
Johns Hopkins University Press
2715 N Charles Street
Baltimore, MD 21218
410-516-6900
800-537-5487
Fax: 410-516-6968
http://www.press.jhu.edu/books

The strength of the John Hopkins University Press' publications in medicine is in part a reflection of the university and medical institution's excellence and long term tradition of exceptional research and clinical care. Joan Aron'sEcosystem Change and Public Health is the first textbook devoted to this emerging field. The book covers such topics as global climate change, stratospheric ozone depletion, water resources management, ecology and infectious disease. Paperback
*Publication Date: 2001 526 pages*
*ISBN: 0-801865-82-4*
*Joan Aron, Author*
*Joan Aron, Editor*
*Jonathan Pratz, Editor*

**3671 Environmental Encyclopedia**
Thomson Gale
27500 Drake Road
Farmington Hills, MI 48331
248-699-4253
800-877-4253
Fax: 800-414-5043
http://www.galegroup.com

Consisting of nearly 1,300 signed articles and term definitions. The encyclopedia provides in-depth, worldwide coverage of environmental issues. Each article written in a nontechnical style and provides current status, analysis andsuggests solutions whenever possible.
*Publication Date: 2002 2000 pages*
*ISBN: 0-787654-86-8*
*Virginia Regish, Contact*

**3672 Environmental Guidebook: A Selective Guide to Environmental Organizations and Related Entities**
Environmental Frontlines
PO Box 43
Menlo Park, CA 94026
650-323-8452
E-mail: info@envirofront.org
http://www.envirofront.org

Designed to serve as an essential reference book profiling nearly 500 national organizations and other entities actively engaged in environmental issues in the US and beyond.
*312 pages*
*ISBN: 0-972068-50-3*
*Jeff Staudinger, Author*

**3673 Environmental Key Contacts and Information Sources**
Government Institutes
4 Research Place
Suite 200A
Rockville, MD 20850
301-907-9100
Fax: 301-921-2362
http://www.govinst.com

An updated and revised compilation of Government Institutes' two previous directories, this reference contains more than 400 pages of contact information for more than 2,700 federal, state, and local environmental agencies andorganizations. This directory also includes contacts for information concerning environmental protection, hazardous waste materials, clean water and air, environmental assessment and management, pesticides, pollution control, recycling, naturalresources and conservation.
*Publication Date: 1998 424 pages*
*ISBN: 0-865876-39-8*
*Charlene Ikonomou and Diane Pacchione, Author*

**3674 Pesticide Directory: A Guide to Producers and Products, Regulators, and Researchers**
Thomson Publications
Box 9335
Fresno, CA 93791
559-266-2964
Fax: 559-266-0189
http://www.agbook.com

This directory is for the person who needs to know anything about the US pesticide industry. It includes basic manufacturers and formulators along with their products, key personnel, managers, district/regional offices and otherpertinent information. Other sections include Universities, State Extension Centers, USDA, EPA, National Organizations, US Forest Service, Poison Control Centers and much more.
*Publication Date: 1987 153 pages Biannual*
*ISBN: 0-913702-45-5*
*WT Thomson, Author*
*Susan Heflin, President/Owner*

## Directories & Handbooks: Habitat Preservation & Land Use

**3675 Alliance for Wildlife Rehabilitation and Education Wildlife Care Directory**
1912 Harbor Boulevard
Costa Mesa, CA 92627
949-722-0606

**3676 Biodiversity Action Network**
1630 Connecticut Avenue
3rd Floor
Washington, DC 20009
202-547-8902
Fax: 202-265-0222
http://www.bionet-us.org

An information exchange network launched by the Center for International Environmental Law.

**3677 Conservation Directory 2004: The Guide to Worldwide Environmental Organizations**
National Wildlife Federation
11100 Wild Life Center Drive
Reston, VA 20190
800-822-9919
http://www.nwf.org/conservationDirectory/

Your guide to thousands of environmental non-profit, education, commercial, and government groups operating across the planet. As of July 07, 2009, there are 4,248 groups listed. Visit and search for free online.
*Robin Assa, Sales Assistant*

**3678 County Conservation Board Outdoor Adventure Guide**
Iowa Association of County Conservation Boards
405 SW 3rd Street
Suite 1
Ankeny, IA 50021
515-963-9582
Fax: 515-963-9582
E-mail: iaccb@ecity.net
http://http://george.ecity.net/iaccb/guide.htm

The 2002 issue has information on 1,614 areas covering approximately 159,899 acres managed by County Conservation Boards. This guide also includes a map of each county with the area to be shaded in or a pinpoint of the location, andhas information on cabin rentals, camping, shelters, playgrounds, swimming, fishing, boating, boat rental, sports and fields, hunting, nature centers, praires, historic sites, wildlife exhibits and more.
*184 pages*
*Don Brazelton, Contact*

**3679 DOCKET**
US Environmental Protection Agency
US EPA Region 3 215-814-2993
1650 Arch Street (3PM52) Fax: 215-814-5102
Philadelphia, PA 19103 E-mail: teller.lawrence@epa.gov
http://www.epa.gov
This database offers the complete text of summaries of all justice cases filed by the US Department of Justice on behalf of the US Environmental Protection Agency.

**3680 Directory of Resource Recovery Projects and Services**
Institute of Resource Recovery
1730 Rhode Island Avenue NW 202-659-4613
Suite 1000 Fax: 202-775-5917
Washington, DC 20036

**3681 Ecology Abstracts**
Cambridge Scientific Abstracts
7200 Wisconsin Avenue 301-961-6700
Suite 601 800-843-7751
Bethesda, MD 20814 Fax: 301-961-6720
E-mail: sales@csa.com
http://www.csa.com
This large database updated continuously, offers over 150,000 citations, with abstracts, to the worldwide literature available on ecology and the environment.
*James P McGinty, President*
*Theodore Caris, Publisher*

**3682 Environmental Bibliography**
International Academy at Santa Barbara
5385 Hollister Avenue 805-683-8889
#210 Fax: 805-965-6071
Santa Barbara, CA 93111 E-mail: info@iasb.org
http://www.iasb.org
Over 615,000 citations are offered in this database, aimed at scientific, technical and popular periodical literature dealing with the environment.

*ISSN: 1053-1440*

**3683 Environmental Concerns: Directory of the Environmental Industry in Colorado**
Business Research Division-Univ. of Colorado
420 UCB 303-492-8227
Boulder, CO 80309-0420 Fax: 303-492-3620
http://leeds.colorado.edu/brd
Approximately 1,300 private businesses, government organizations and corporations in Colorado that contribute to environmental protection and rehabilitation.

*ISBN: 1-883226-02-3*
*Gin Hayden, Editor*
*Sean Shepherd, Editor*

**3684 Environmental Guide to the Internet**
Government Institutes
4 Research Place 301-907-9100
Suite 200A Fax: 301-921-2362
Rockville, MD 20850 http://www.govinst.com
Provides information for the best sites in the internet dealing with the preservation and protection of the environment, ecology, and conservation and offers over 320 new listings and addresses. Writin for environmental consultants,industry professionals, researchers, lawyers, educators, and students, contains the top 1,200 environmental internet resources, including, newletters and journals, and world wide web sites
*Publication Date: 1997 384 pages*
*ISBN: 0-865875-78-2*
*Carol Briggs-Erickson and Toni Murphy, Author*

**3685 Environmental Guidebook: A Selective Guide to Environmental Organizations and Related Entities**
Environmental Frontlines
PO Box 43 650-323-8452
Menlo Park, CA 94026 E-mail: info@envirofront.org
http://www.envirofront.org
Designed to serve as an essential reference book profiling nearly 500 national organizations and other entities actively engaged in environmental issues in the US and beyond.
*312 pages*
*ISBN: 0-972068-50-3*
*Jeff Staudinger, Author*

**3686 Helping Out in the Outdoors: A Directory of Volunteer Opportunities on Public Lands**
American Hiking Society
1422 Fenwick Lane 301-565-6704
Silver Spring, MD 20910 Fax: 301-565-6714
E-mail: info@americanhiking.org
http://www.americanhiking.org
*Mary Margaret Sloan, President*

**3687 Hospitality Directory**
Human Ecology Action League (HEAL)
PO Box 29629 404-248-1898
Atlanta, GA 30359 Fax: 404-248-0162
E-mail: HEALNatnl@aol.com
http://members.aol.com/HEALNatnl/index.html
Nonprofit organization founded in 1977 to serve those whose health has been adversely affected by environment exposures; to provide information to those who are concerned about the health effects of chemicals; and to alert the generalpublic about the potential dangers of chemicals.
*Katherine P Collier, Contact*

**3688 Human Ecology Action League Directory**
Human Ecology Action League
PO Box 29629 404-248-1898
Atlanta, GA 30359 Fax: 404-248-0162
E-mail: HEALNatnl@aol.com
http://members.aol.com/HEALNatnl/index.html
The Human Ecology Action League Inc (HEAL) is a nonprofit organization founded in 1977 to serve those whose health has been affected by environmental exposures; to provide information to those who are concerned about the health effectsof chemicals; and to alert the general public about the potential dangers of chemicals. Referrals to local HEAL chapters and other support groups are available from the League.
*Katherine P Collier, Contact*

**3689 Hummingbird Connection**
6560 Highway 179
Suite 204 928-284-2251
Sedona, AZ 86351 800-529-3699
E-mail: info@hummingbird.org
http://www.hummingbirdsociety.org
Published by the Hummingbird Society.
*Publication Date: 1992 16 pages Quarterly*
*ISSN: 1097-3427*
*H Ross Hawkins, Author/Editor*

**3690 International Society of Tropical Foresters: Membership Directory**
5400 Grosvenor Lane 301-897-8720
Bethesda, MD 20814 Fax: 301-897-3690
E-mail: istfi@igc.apc.org
The International Society of Tropical Foresters, Inc. (ISTF) is a nonprofit organization committed to the protection, wise management and rational use of the world's tropical forests. Established in 1950, ISTF was reactivated in 1979.It has about 1500 members in more than 100 countries. Financial support comes from membership dues, donations and grants. ISTF sponsors meetings, promotes chapters in other countries, maintains a web site and has chapters at universities.
*Warren T Doolittle, President*

**3691 Journal of Wildlife Rehabilitation**
International Wildlife Rehabilitation Council

PO Box 8187 408-271-2685
San Jose, CA 95155 Fax: 408-271-9285
E-mail: office@iwrc-online.org
http://www.iwrc-online.org

A peer reviewed scientific journal that has served as a primary reference for wildlife rehabilitators and others involved in the care and conservation of wildlife. Features articles, columns and reviews, with topics ranging from allaspects of wildlife care to administration, fundraising, education programs, case studies, environmental issues, legalities, ethics and more. And is also a benefit of membership to IWRC.

*Publication Date: 1978 40 pages Quarterly*
*Jennifer Gursu, Executive Director*

**3692 LEXIS Environmental Law Library**
Lexis Nexis Group
PO Box 933 937-865-6800
Dayton, OH 45401 800-227-9597
Fax: 937-865-6909
http://www.lexis-nexis.com

This database contains decisions related to environmental law from the Supreme Court and other legislative bodies.

**3693 Learning About Our Place**
311 Curtis Street 716-665-2473
Jamestown, NY 14701 800-758-6841
Fax: 716-665-3794
E-mail: mail@rtpi.org
http://www.rtpi.org

47 lesson plans that connect learning to nature and the outdoors

*15-30 pages Quarterly*
*Jim Berry, President*

**3694 Managed Area Basic Record**
The Nature Conservancy
4245 N. Fairfax Drive 804-295-6106
Suite 100 Fax: 804-979-0370
Arlington, VA 22203 E-mail: cmullen@tnc.org
http://http://nature.org

**3695 Minienvironments**
IEST
2340 S. Arlington Heights Road 847-981-0100
Suite 100 Fax: 847-981-4130
Arlington Heights, IL 60005 E-mail: iest@iest.org
http://www.iest.org

The purpose of this document is to provide a framework for describing minienvironments for microelectronics and similar applications.

*Publication Date: 2002 28 pages*
*ISBN: 1-877862-83-5*
*Julie Kendrick, Executive Director*

**3696 Morrison Environmental Directory**
PO Box 2312 316-262-0100
Wichita, KS 67201

*ISSN: 1060-488*

**3697 National Directory of Conservation Land Trusts**
Land Trust Alliance
1660 L St. NW 202-638-4725
Suite 1100 Fax: 202-638-4730
Washington, DC 20036 E-mail: info@lta.org
http://www.lta.org

More than 1,200 nonprofit land conservation organizations at the local and regional levels are profiled.

*210 pages*

**3698 New York State Department of Environmental Conservation Personnel Directory: Internet Only**
NYS Department of Environmental Conservation
625 Broadway 518-402-8013
Albany, NY 12233 Fax: 518-402-9036
E-mail: dinnelson@gw.dec.state.ny.us
http://www.dec.state.ny.us

Internet only, this directory includes DEC's executive management and division directors. Executive managers are appointed by the Governor to carry out the policies of the state. Division directors have direct management responsibilityfor the department's programs.

*Mary A Kadlecek, Chief Internet Publications*

**3699 Nonprofit Sample and Core Repositories Open to the Public in the United States**
Branch of Sedimentary Processes
MS 939 Federal Center 303-236-5760
Denver, CO 80225 Fax: 303-236-0459
*Walter E Dean, Contact*

**3700 Range and Land Management Handbook**
Wyoming Association of Conservation Districts
517 E 19th Street 307-632-5716
Cheyenne, WY 82001 Fax: 307-638-4099
http://www.conservewy.com

This publication is intended for people from all walks of life who want to gain an appreciation of rangelands. This publication is also an introduction to the various fields of range management.

*Annual*

**3701 Takings Litigation Handbook: Defending Takings Challenges to Land Use Regulations**
American Legal Publishing Corporation
432 Walnut Street
Suite 1200 800-445-5588
Cincinnati, OH 45202 Fax: 513-763-3562
E-mail: customerservice@amlegal.com
http://www.amlegal.com

No government attorney, land use planner or other local official can effectively protect their community from harmful land use without a working knowledge of takings law. Developers and other landowners increasingly are attempting touse takings litigation, or the mere threat of takings litigation, to convince government agencies to relax or abandon vital protections for our neighborhoods and natural environment.

*Publication Date: 2000 404 pages*
*Kendall, Dowling and Schwartz, Author*
*Douglas Kendall, Executive Director*

**3702 Trout Unlimited Chapter and Council Handbook**
Trout Unlimited
1300 North 17th Street 703-522-0200
Suite 500 Fax: 703-284-9400
Arlington, VA 22209 E-mail: trout@tu.org
http://www.tu.org

*Charles Gauvin, President/CEO*
*Kenneth Mendez, Executive VP/COO*

**3703 Turtle Help Network**
New York Turtle and Tortoise Society
PO Box 878 212-459-4803
Orange, NJ 07051

**3704 Wilson Journal of Ornithology**
OSNA 254-399-9636
5400 Bosque Boulevard, Ste 680E-mail: business@osnabirds.org
Waco, TX 7671(http://www.ummz.umich.edu/birds/wos/index.html

Scholarly journal consisting of articles on bird studies, orinthological news, reviews of new bird books and related subjeects.

*Quarterly*
*Dr Doris J Watt, President*
*John A Smallwood, Secretary*

**3705 Wisconsin Department of Public Instruction**
125 S Webster Street 608-266-2188
PO Box 7841 800-441-4563
Madison, WI 53707 Fax: 608-267-9110
E-mail: sandi.mcnamer@dpi.state.wi.us
http://www.dpi.state.wi.us/pubsales

State education department publisher of K-12 curriculum planning guides in 25 subject areas including environmental education and science.
*Sandi McNamer, Publications Director*

## Directories & Handbooks: Recycling & Pollution Prevention

**3706 A Glossary of Terms and Definitions Relating to Contamination Control**
IEST
5005 Newport Drive 847-255-1561
Suite 506 Fax: 847-255-1699
Rolling Meadows, IL 60008 E-mail: iest@iest.org
http://www.iest.org
A publication from the Institute of Environmental Science and Technology.
*Publication Date: 1995 32 pages*
*ISBN: 1-877862-28-2*
*Julie Kendrick, Executive Director*

**3707 American Recycling Market Directory: Reference Manual**
Recycling Data Management Corp.
PO Box 577 315-471-0707
Ogdensburg, NY 13669 800-267-0707
Fax: 613-471-3258
Comprehensive directory/reference manual to materials recycling markets. Helps individuals locate buyers and sellers of recyclable materials on a regional basis throughout North America. Contains 20,000 cross-referenced company andagency listings. Sections include: scrap metals, waste paper, paper mills, auto dismantlers, demolition, glass, oil, rubber and textiles recyclers, recycling centers, MRF's composting, equipment and consulting services, industry references UBC specsand more.

**3708 Analysis of the Stockholm Convention on Persistent Organic Pollutants**
Oceana Publications, Inc
198 Madison Ave
New York, NY 10016 800-334-4249
Fax: 212-726-6476
E-mail: oxfordonline@oup.com
http://www.oceanalaw.com
This book analyzes the Stockholm Convention on Persistent Organic Pollutants. Prepared under the auspices of the UN Environment Programme Chemical Division.
*Publication Date: 2003 200 pages One Time*
*ISBN: 0-379215-06-3*
*Mario Antonio Olsen, Author*

**3709 Criteria Pollutant Point Source Directory**
North American Water Office
3394 Lake Elmo Ave 651-770-3861
Lake Elmo, MN 55042 Fax: 651-770-3976
E-mail: gwillc@mtn.org
http://http://www.nawo.org

**3710 Directory of Municipal Solid Waste Management Facilities**
The Institute of Solid Waste Disposal
1730 Rhode Island Avenue NW 202-659-4613
Suite 1000 Fax: 202-296-7915
Washington, DC 20036

**3711 EI Environmental Services Directory**
Environmental Information Networks
8525 Arjons Drive 858-695-0050
Suite H Fax: 952-831-6550
San Diego, CA 92126 E-mail: ei@mr.net
http://www.envirobiz.com
Waste-handling facilities, transportation and spill response firms, laboratories and the broad scope of environmental services. Online versions are also available.
*ISSN: 1053-475N*
*Cary Perket*

**3712 Environmental Encyclopedia**
Thomson Gale
27500 Drake Road 248-699-4253
Farmington Hills, MI 48331 800-877-4253
Fax: 800-414-5043
http://www.galegroup.com
Consisting of nearly 1,300 signed articles and definitions. The encyclopedia provides in depth, worldwide coverage of environmental issues. Each article written in a non technical style and provides current status, analysis andsuggests solutions whenever possible.
*Publication Date: 2002 2000 pages*
*ISBN: 0-787654-86-8*
*Virginia Regish, Contact*

**3713 Environmental Guide to the Internet**
Government Institutes
4 Research Place 301-907-9100
Suite 200A Fax: 301-921-2362
Rockville, MD 20850 http://www.govinst.com
Provides information for the best sites in the internet dealing with the preservation and protection of the environment, ecology, and conservation and offers over 320 new listings and addresses. Writin for environmental consultants,industry professionals, researchers, lawyers, educators, and students, contains the top 1,200 environmental internet resources, including, newletters and journals, and world wide web sites
*Publication Date: 1997 384 pages*
*ISBN: 0-865875-78-2*
*Carol Briggs-Erickson and Toni Murphy, Author*

**3714 Environmental Guidebook: A Selective Guide to Environmental Organizations and Related Entities**
Environmental Frontlines
PO Box 43 650-323-8452
Menlo Park, CA 94026 E-mail: info@envirofront.org
http://www.envirofront.org
Designed to serve as an essential reference book profiling nearly 500 national organizations and other entities actively engaged in environmental issues in the US and beyond.
*312 pages*
*ISBN: 0-972068-50-3*
*Jeff Staudinger, Author*

**3715 Fibre Market News: Paper Recycling Markets Directory**
Recycling Media Group GIE Publishers
4012 Bridge Avenue 216-961-4130
Cleveland, OH 44113 800-456-0707
Fax: 216-961-0364
A list of over 2,000 dealers, brokers, packers and graders of paper stock in the US and Canada.
*Dan Sandoval, Editor*

**3716 Hazardous Materials Regulations Guide**
JJ Keller
3003 West Breezewood Lane
PO Box 368 877-564-2333
Neenah, WI 54957 Fax: 800-727-7516
E-mail: sales@jjkeller.com
http://www.jjkeller.com
A complete reference guide of hazardous materials regulations.
*May/November*
*ISBN: 0-934674-94-9*
*Tom Ziebell, Editor*

**3717 How-To: 1,400 Best Books on Doing Almost Everything**
R.R. Bowker Company
630 Central Ave. 908-286-1090
New Providence, NJ 07974 888-269-5372
E-mail: info@bowker.com
http://www.bowker.com/bowkerweb/

**3718 Institute of Chemical Waste Management Directory of Hazardous Waste Treatment and Disposal**
National Solid Wastes Management Assn.
4301 Connecticut Avenue NW 202-244-4700
Suite 300 Fax: 202-966-4824
Washington, DC 20008 http://www.nswma.org

**3719 International Handbook of Pollution Control**
Greenwood Publishing Group
88 Post Road W 203-226-3571
Westport, CT 06881 E-mail: webmaster@greenwood.com
http://www.greenwood.com

*Publication Date: 1989 482 pages*
*ISBN: 0-313240-17-5*
*Edward J Kormondy, Author*

**3720 KIND News**
National Assn for Humane & Environmental Education
67 Norwich Essex Turnpike 860-434-8666
East Haddam, CT 06423 Fax: 860-434-6282
E-mail: nahee@nahee.org
http://www.nahee.org

Classroom newspaper for kids in grades K-6. It features articles, puzzles and celebrity interviews that teach children the value of showing kindness and respect to animals, the environment, and one another.
*Publication Date: 1985 4 pages 9 per year*
*William DeRosa, Executive Director*
*Dorothy Weller, Director Outreach & Fulf.*

**3721 List of Water Pollution Control Administrators**
Assn. of State and Interstate Water Pollution Con.
1221 Connecticut Ave. NW 202-898-0905
2nd Floor Fax: 202-898-0929
Washington, DC 20036 E-mail: admin1@aswipca.org
http://www.asiwpca.org

*Roberta Savage, Executive Director*
*Linda Eichmiller, Deputy Director*

**3722 Nebraska Recycling Resource Directory**
Nebraska Dept of Environmental Quality
1200 N Street Suite 400 402-471-2186
PO Box 98922 877-253-2603
Lincoln, NE 68509 Fax: 402-471-2909
E-mail: MoreInfo@nebraska.gov
http://www.deq.state.ne.us

*Publication Date: 1986 139 pages Bi-Annually*
*Steve Danahy, Unit Supervisor*

**3723 Pesticide Directory: A Guide to Producers and Products, Regulators, and Researchers**
Thomson Publications
Box 9335 559-266-2964
Fresno, CA 93791 Fax: 559-266-0189
http://www.agbook.com

This directory is for the person who needs to know anything about the US pesticide industry. It includes basic manufacturers and formulators along with their products, key personnel, managers, district/regional offices and otherpertinent information. Other sections include Universities, State Extension Centers, USDA, EPA, National Organizations, US Forest Service, Poison Control Centers and much more.
*Publication Date: 1987 153 pages Biannual*
*ISBN: 0-913702-45-5*
*WT Thomson, Author*
*Susan Heflin, President/Owner*

**3724 Pollution Abstracts**
Cambridge Scientific Abstracts
7200 Wisconsin Avenue 301-961-6700
Suite 601 800-843-7751
Bethesda, MD 20814 Fax: 301-961-6720
E-mail: sales@csa.com
http://www.csa.com

This database provides fast access to the environmental information necessary to resolve day to day problems, ensure ongoing compliance, and handle emergency situations more effectively.
*James P McGinty, President*
*Ted Caris, Publisher*

**3725 Product Cleanliness Levels and Contamination Control Program**
IEST
5005 Newport Drive 847-255-1561
Suite 506 Fax: 847-255-1699
Rolling Meadows, IL 60008 E-mail: iest@iest.org
http://www.iest.org

Intended to provide a basis for specifying product cleanliness levels and contamination control program requirments with emphasis on contaminants that affect product performance.
*Publication Date: 2002 20 pages*
*ISBN: 1-877862-82-7*
*Julie Kendrick, Executive Director*

**3726 Recycling Related Newsletters, Publications And Periodicals**
Continnuus
PO Box 416 303-575-5676
Denver, CO 80201 Fax: 970-292-2136

**3727 Recycling Today: Recycling Products & Services Buyers Guide**
Recycling Today GIE Publishers
4020 Kinross Lakes Parkway 216-961-4130
#201 800-456-0707
Richfield, OH 44286 Fax: 216-961-0364
E-mail: jkeefe@gie.net
http://www.recyclingtoday.com

Directory of services and supplies to the industry.
*Dan Sandoval, Internet/Senior Editor*
*James Keefe, Group Publisher*

**3728 Scholastic Environmental Atlas of the United States**
Scholastic
730 Broadway 212-505-3000
New York, NY 10003 E-mail: uwpress@washinton.edu
http://www.washington.edu/uwpress/

**3729 Tracking Toxic Wastes in CA: A Guide to Federal and State Government Information Sources**
INFORM
5 Hanover Square 212-361-2400
Floor 19 Fax: 212-361-2412
New York, NY 10004 http://www.informinc.org

**3730 Waste Age: Resource Recovery Acitivities Update Issue**
National Solid Wastes Management Assn.
1730 Rhode Island Avenue NW 202-659-4613
Suite 1000
Washington, DC 20036

**3731 Waste Age: Waste Industry Buyer Guide**
National Solid Wastes Management
1730 Rhode Island Avenue NW 202-659-4613
Suite 1000 800-424-2869
Washington, DC 20036 Fax: 202-659-0925

**3732 Wastes to Resources: Appropriate Technologies for Sewage Treatment and Conversion**
National Center for Appropriate Techology

3040 Continental Drive 406-494-4572
Butte, MT 59701 800-275-6228
Fax: 406-494-2905
E-mail: info@ncat.org
http://www.ncat.org
*Kathy Hadley, Executive Director*

## Directories & Handbooks: Sustainable Development

**3733 Solar Energy Resource Guide/SERG**
NorCal Solar
PO Box 3008 530-852-0354
Berkeley, CA 94703 E-mail: info@norcalsolar.org
http://www.norcalsolar.org
Articles and resources on solar electric, solar thermal, financial analysis, etc. Also a guidebook for education on the workings and intstallation of solar technology.
*96 pages*
*Claudia Wentworth, President*
*Liz Merry, Program Manager*

## Directories & Handbooks: Travel & Tourism

**3734 Access America: An Atlas and Guide to the National Parks for Visitors with Disabilities**
Northern Cartographic
4050 Williston Road 802-860-2886
South Burlington, VT 05403 Fax: 802-865-4912
E-mail: info@ncarto.com
http://www.ncarto.com
*Publication Date: 1988*
*ISBN: 0-944187-00-5*

**3735 Audubon Society Field Guide to the Natural Places of the Northeast**
National Audubon Society
700 Broadway 212-979-3000
New York, NY 10003 Fax: 212-979-3188
http://www.audubon.org

**3736 Complete Guide to America's National Parks: The Official Visitor's Guide**
National Park Foundation
11 Dupont Circle NW 202-238-4200
Suite 600 800-285-2448
Washington, DC 20036 Fax: 202-234-3103
E-mail: ask-npf@nationalparks.org

**3737 Field Guide to American Windmills**
University of Oklahoma Press
2800 Venture Drive 405-325-2000
Norman, OK 73069 800-627-7377
Fax: 405-364-5798
This guide to America's windmills is both a complete general history of turbine wheel mills and an identification guide to the 112 most common models, which still dot landscapes today.
*Publication Date: 1985 528 pages*
*T Lindsay Baker, Author*

**3738 Guide to the National Wildlife Refuges**
Macmillan Publishing Company
National Wildlife Guide 212-832-2101
590 Madison Avenue http://www.nationalwildlifeguide.com
New York, NY 10022
More than 500 National Wildlife Refuges and satellite refuges are listed.
*684 pages*

**3739 National Parks Visitor Facilities and Services**
Conference of National Park Concessioners
PO Box 29041 480-967-6006
Phoenix, AZ 85038 http://www.nps.gov/legacy/business.html
Within the parks, private businesses provide accommodations and services for visitors under concession contracts.
*Rex G Maughan, Chairman*

**3740 National Parks: National Park Campgrounds Issue**
National Parks Conservation Association
1300 19th Street Northwest
Suite 300 800-628-7275
Washington, DC 20036 Fax: 202-659-0650
E-mail: npca@npca.org
http://www.npca.org
To safeguard the scenic beauty, wildlife, and historical and cultural treasures of the largest and most diverse park system in the world.
*Thomas C Kiernan, President*
*Tom Martin, Executive Vice President*

**3741 National Wildlife Refuges: A Visitor's Guide**
Fish and Wildlife Services, Interior Department
1849 C Street NW 703-358-2043
Washington, DC 20242 E-mail: webteam@ios.doi.gov
http://www.fws.gov
Contains a map showing national wildlife refuges that provide recrational and educational opportunities. Provides tips for visiting national wildlife refuges. Also list refuges in all 50 States, Puerto Rico and the Virgin Islans, withthe best wildlife viewing season and the features of each refuge.

*ISBN: 0-160617-00-6*

**3742 Nature Center Directory**
Wisconsin Association for Environmental Education
8 Nelson Hall 715-346-2796
University of Wisconsin-Stevens Point Fax: 715-346-3835
Stevens Point, WI 54481 E-mail: waee@uwsp.edu
http://www.uwsp.edu/waee
*Annual*

**3743 Rails-to-Trails Magazine**
1100 17th Street NW 202-331-9696
10th Floor Fax: 202-331-9680
Washington, DC 20036 E-mail: railtrails@railtrails.org
http://www.railtotrails.org
Official magazine of the Rails-to-Trails Conservancy (RTC). The RTC is a national nonprofit organization dedicated to creating a nationwide network of trails from former rail lines and connecting corridors. It does not own or manageany rail trails.
*Keith Laughlin, President*
*Jeff Ciabotti, VP Trail Development*

**3744 Recreation Sites in Southwestern National Forests**
USDA Forest Service
Public Affairs Office 505-842-3292
333 Broadway Blvd SE Fax: 505-842-3106
Albuquerque, NM 87102 http://www.fs.fed.us/r3
Listings for all recreation sites for Arizona and New Mexico.
*72 pages*
*Corbin Newman, Regional Forester*

**3745 Sierra Club Guide to the Natural Areas of California**
Sierra Club
85 2nd Street 415-977-5500
2nd Floor Fax: 415-977-5799
San Francisco, CA 94105 E-mail: information@sierraclub.org
http://www.sierraclub.org
Revised and updated, this comprehensive guide makes more than 200 wilderness areas in California, including many lesser known natural areas, accessible to the outdoor enthusiast.
*Publication Date: 1997 352 pages*
*ISBN: 0-871568-50-0*
*John Perry and Jane Greverus Perry, Author*

**3746 Thermal Springs of Wyoming**
Wyoming State Geological Survey
PO Box 1347
Laramie, WY 82073
307-766-2286
Fax: 307-766-2605
E-mail: wsgs-info@uwyo.edu
http://www.wsgs.uwyo.edu

*Ronald C Surdam, Agency Director*
*Richard W Jones, Editor*

**3747 Traveler's Guide to the Smoky Mountains Region**
Harvard Common Press
535 Albany Street
Boston, MA 02118
617-423-5803
Fax: 619-695-9794
E-mail: orders@harvardcommonpress.com
http://www.harvardcommonpress.com

Features museums, events of the South Appalachians of Tennessee, North Carolilna, Virginia and Georgia
*Publication Date: 1985 288 pages*
*ISBN: 0-916782-64-6*
*Valerie Cimino, Executive Editor*
*Christine Alaimo, Associate Publisher*

**3748 Wild Places & Open Spaces Map**
Division of Fish and Wildlife
PO Box 400
Trenton, NJ 08625
609-292-9450
Fax: 609-984-1414
http://www.njfishandwildlife.com

Designed similar to a road map, offers the outdoors person a welath of information on locating and exploring New Jersey's open spaces in compact and easy to read format. Showcasing a full color map of New Jersey, with more than 700,000acres of public open space.
*Carol Nash, Customer Service*

## Directories & Handbooks: Water Resources

**3749 Citizen's Directory for Water Quality Abuses**
Izaak Walton League of America
707 Conservation Lane
Gaithersburg, MD 20878
301-548-0150
800-453-5463
Fax: 301-548-0146
E-mail: general@iwla.org
http://www.iwla.org

*Paul Hansen, Executive Director*

**3750 Coordination Directory of State and Federal Agency Water Resources Officials: Missouri Basin**
Department of Water Resources
5231 South 19th Street
Lincoln, NE 68512
402-471-2363
http://ne.water.usgs.gov

**3751 How Wet is a Wetland?: The Impacts of the Proposed Revisions to the Federal Wetlands Manual**
Environmental Defense Fund
257 Park Avenue South
New York, NY 10010
212-505-2100
800-684-3322
Fax: 212-505-0892
E-mail: media@environmentaldefense.org
http://www.environmentaldefense.org

*Publication Date: 1992*
*Fred Krupp, President*

**3752 Hydro Review: Industry Source Book Issue**
HCI Publications
410 Archibald Street
Kansas City, MO 64111
816-931-1311
Fax: 816-931-2015
E-mail: info@hcipub.com
http://www.hcipub.com

List of over 800 manufacturers and suppliers of products and services to the hydroelectric industry in the US and Canada.
*January*
*Carl Vansant, Editor-In-Chief*

**3753 List of Water Pollution Control Administrators**
Assn. of State and Interstate Water Pollution Con.
1221 Connecticut Ave. NW
2nd Floor
Washington, DC 20036
202-898-0905
Fax: 202-898-0929
E-mail: admin1@aswipca.org
http://www.asiwpca.org

*Roberta Savage, Executive Director*
*Linda Eichmiller, Deputy Director*

**3754 Water Environment & Technology Buyer's Guide and Yearbook**
Water Environment Federation
601 Wythe Street
Alexandria, VA 22314
703-684-2400
800-666-0206
Fax: 703-684-2492
E-mail: webfeedback@wef.org
http://www.wef.org

Founded in 1928, the Water Environment Foundation (WEF) is a not for profit technical and educational organization with members from varied disiplines who work toward the WEF vision of preservation and enhancement of the globalwaterenvironment. The WEF network includes water quality professionals from 76 Member Associations in 30 countries.
*William J Bertera, Executive Director*

## Periodicals: Air & Climate

**3755 Air/Water Pollution Report**
Business Publishers
PO Box 17592
Baltimore, MD 21297
301-589-5103
800-274-6737
Fax: 301-589-8493
E-mail: custserv@bpines.com
http://www.bpinews.com

Regulatory activities and governmental legislation and litigation are covered in this pulication.
*Publication Date: 1963 Monthly*
*Leonard Eiserer, Publisher*

**3756 Bulletin of the American Meteorological Society**
45 Beacon Street
Boston, MA 02108
617-227-2425
Fax: 617-742-8718
E-mail: amsinfo@ametsoc.org
http://www.ametsoc.org

The American Meteorological Society promotes the development and dissemination of information and education on the atmospheric and related oceanic and hydrologic sciences and the advancement of their professional applications.
*Publication Date: 1919 Monthly*
*Ronald D McPherson, Executive Director*

**3757 Climate Institute: Climate Alert**
Coping with Climate Change
1785 Massachusetts Avenue NW
Washington, DC 20036
202-547-0104
Fax: 202-547-0111
E-mail: info@climate.org
http://www.climate.org

The Climate Institute works to protect the balance between climate and life on earth by facilitating dialogue among scientists, policy makers, business executives and citizens. In all its efforts, the institute strives to be a sourceof objective, reliable information.
*Publication Date: 1988 8-12 pages Quarterly*
*ISSN: 1071-3271*
*John Topping, President*

**3758 Earth Share of Georgia Newsletter**
1447 Peachtree Street
Suite 214
Atlanta, GA 30309
404-873-3173
Fax: 404-873-3135
E-mail: info@earthsharega.org
http://www.earthsharega.org

Nonprofit federation of local, national and global environmental groups addressing the critical environmental issues. ESGA raises

funds for these groups through workplace giving campaigns, special events and individual contributions.
*Publication Date: 1992 Bi-Monthly*
*Madeline Reamy, Executive Director*

**3759 Environmental Policy Alert**
Inside Washington Publishers
1225 South Clark Street
Suite 1400
Arlington, VA 22202
703-416-8500
800-424-9068
Fax: 703-416-8543
E-mail: iwp@iwpnews.com
http://www.iwpnews.com
Addresses the legislative news and provides reports on the federal environmental policy process.
*Publication Date: 1980*
*Paul Finger, Publisher*

**3760 Journal of the Air Pollution Control Association**
Air Pollution Control Association
420 Fort Duquesne
Boulevard #3
Pittsburgh, PA 15222
412-232-3444
A comprehensive journal offering information to the environment and conservation industry.

**3761 Journal of the Air and Waste Management Association**
Air and Waste Management Association
420 Fort Duquesne Blvd.
Pittsburgh, PA 15222
412-232-3444
800-270-3444
Fax: 412-232-3450
E-mail: info@awma.org
http://www.awma.org
Published for the working environmental professional and carries peer-reviewed technical papers on a variety of topics form control technology to science.
*Publication Date: 1951 Monthly*
*Maura Moktar, Managing Editor*

**3762 Population Reference Bureau: Household Transportation Use and Urban Pollution**
1875 Connecticut Avenue NW
Suite 520
Washington, DC 20009
202-483-1100
800-877-9881
Fax: 202-328-3937
E-mail: popref@prb.org
http://www.prb.org
*Publication Date: 1929*
*Peter Donaldson, President*
*Mary Mederios Kent, Editor*

**3763 Population Reference Bureau: Population & Environment Dynamics**
1875 Connecticut Avenue NW
Suite 520
Washington, DC 20009
202-483-1100
800-877-9881
Fax: 202-328-3937
E-mail: popref@prb.org
http://www.prb.org
PRB publishes the quarterly Population Bulletin, the annual World Population Data Sheet, and PRB Reports on America, as well as specialized publications covering population and public policy issues in the U.S. and abroad, particularlyin developing countries.
*Publication Date: 1929*
*Peter Donaldson, President*
*Mary Mederios Kent, Editor*

**3764 Population Reference Bureau: Water**
1875 Connecticut Avenue NW
Suite 520
Washington, DC 20009
202-483-1100
800-877-9887
Fax: 202-328-3937
E-mail: popref@prb.org
http://www.prb.org
*Publication Date: 1929*
*Peter Donaldson, President*
*Mary Mederios Kent, Editor*

**3765 Trinity Consultants Air Issues Review**
12770 Merit Drive
Suite 900
Dallas, TX 75251
972-661-8100
800-229-6655
Fax: 972-385-9203
E-mail: information@trinityconsultants.com
http://www.trinityconsultants.com
An environmental consulting company that assists industrial facilities with issues related to regulatory compliance and environmental management. Founded in 1974, this nationwide firm has particular expertise in air quality issues.Trinity also sells environmental software and professional education. T3, a Trinity Consultants Company, provides EH&S management information systems (EMIS) implementation and integration services.
*Publication Date: 1990 8 pages Quarterly*
*John Hofmann, VP*
*Patrick Delamater, VP*

**3766 Weather & Climate Report**
Nautilus Press
1054 National Press Building
Washington, DC 20045
202-347-6643
Reports on federal actions which impact weather, climate research and global changes in climate.
*Monthly*
*ISSN: 0730-8256*
*John R Botzum, Editor*

**3767 World Resource Review**
SUPCON International
PO Box 50303
Palo Alto, CA 94303
630-910-1551
Fax: 630-910-1561
E-mail: syshen@megsinet.net
http://www.globalwarming.net
For business and government readers, provides expert worldwide reviews of global warming and extreme events in relation to the management of natural, mineral and material resources. Subjects include global warming impacts onagriculture, energy, and infrastructure, monitoring of changes in resources using remote sensing, actions of national and international bodies, global carbon budget, greenhouse budget and more.
*Publication Date: 1990 Quarterly*
*ISSN: 1042-8011*
*Dr. Sinya Shen, Production Manager*

**3768 World Watch**
Worldwatch Institute
1776 Massachusetts Avenue NW
Washington, DC 20036
202-452-1999
Fax: 202-296-7365
E-mail: worldwatch@worldwatch.org
http://www.worldwatch.org
Magazine on global environmental issues.
*Publication Date: 1975 40 pages*
*ISSN: 0896-0615*
*Ed Ayres, Author*
*Lester Brown, Founding Publisher*
*Lisa Mastny, Senior Editor*

## Periodicals: Business

**3769 AFE Journal**
8160 Corporate Park Drive
Suite 125
Cincinnati, OH 45242
513-489-2473
Fax: 513-247-7422
E-mail: mail@afe.org
http://www.afe.org
AFE Journal is a bimonthly publication from the Association for Facilities Engineering.
*48 pages Bimonthly*
*ISSN: 1088-5900*
*Gabriella Jacobs, Author*
*Gabriella Jacobs, Editor*

**3770 ALBC News**
American Livestock Breeds Conservancy

PO Box 477
Pittsboro, NC 27312
919-542-5704
Fax: 919-545-0022
E-mail: albc@albc-usa.org
http://www.albc-usa.org

ALBC News is a bi-monthly newsletter published by the American Livestock Breeds Conservancy.
*Publication Date: 1987 20 pages Bi-Monthly*
*ISSN: 1064-1599*
*Cindy Rubel, Author*
*Cindy Rubel, Editor*

## 3771 Abstracts of Presentations

Wildlife Society
5410 Grosvenor Lane
Suite 200
Bethesda, MD 20814
301-897-9770
Fax: 301-530-2471
E-mail: tws@wildlife.org
http://www.wildlife.org

A yearly publication of the Wildlife Society.
*Publication Date: 1994 300 pages Yearly*
*Gene Pozniak, Production Editor*

## 3772 Advisor

Great Lakes Commission
Eisenhower Corporate Park
2805 South Industrial Hwy, Suite 100
Ann Arbor, MI 48104
734-665-9135
Fax: 734-971-9150
E-mail: glc@great-lakes.net
http://www.glc.org

Covers economic and environmental issues of the Great Lakes region with a special focus on activities of the Great Lakes Commission.
*Publication Date: 1955 12 pages Bi-Monthly*
*Julie Wagemakers, Production Manager*

## 3773 Agribusiness Fieldman

Western Agricultural Publishing Company
4969 E Clinton Way
Suite 104
Fresno, CA 93727
559-252-7000
Fax: 559-252-7387

Aimed at keeping the professional agriculture consultant posted on changes in the agricultural-chemical industry. Provides news about pests and control measures for all segments of the agricultural-chemical industry.
*Paul Baltimore, Publisher*
*Margi Katz, Editor*

## 3774 American Environmental Laboratory

International Scientific Communications
PO Box 870
Shelton, CT 06484
203-926-9300
Fax: 203-926-9310
E-mail: iscpubs@iscpubs.com
http://www.iscpubs.com

Laboratory activities, new equipment, and analysis and collection of samples are the main topics.
*Bi-Monthly*
*Brian Howard, Publisher/Editor-in-Chief*

## 3775 Annual Newsletter and Report

The Peregrine Fund
5668 W Flying Hawk Lane
Boise, ID 83709
208-362-3716
Fax: 208-362-2376
E-mail: tpf@peregrinefund.org
http://www.peregrinefund.org

Yearly publication from The Peregrine Fund. Free with $25 membership fee.
*Publication Date: 1970 Yearly*
*William Burnham, Author*
*Dr William Burnham, President*

## 3776 Bison World

National Bison Association
8690 Wolff Ct.
Suite 200
Westminster, CO 80031
303-292-2833
Fax: 303-659-3739
E-mail: info@bisoncentral.com
http://www.bisoncentral.com

Published by the NBA, an organization of bison producers dedicated to awareness of the healthy properties of bison meat and bison production.
*Publication Date: 1975 100 pages Quarterly*
*Sam Albrecht, Publisher*
*Laurie Dineen, Editor*

## 3777 Business and the Environment

Cutter Information Corporation
37 Broadway
Suite 1
Arlington, MA 02474
781-648-8700
800-888-8939
Fax: 781-648-1950
E-mail: service@cutter.com
http://www.cutter.com

Environmental investment trends, deals and market developments.
*Karen Fine Coburn, Publisher*
*Kathleen Victory, Editor*

## 3778 CAC Annual Reports

Citizens Advisory Council
13th Floor, RCSOB
PO Box 8459
Harrisburg, PA 17105
717-787-4527
Fax: 717-787-2878
E-mail: mioff.stephanie@state.pa.us
http://www.cacdep.state.pa.us

Publisher by the Citizens Advisory Council.
*Publication Date: 1977 20-40 pages Annual*
*Susan Wilson, Executive Director*
*Stephanie Mioff, Administrative Assistant*

## 3779 Chemosphere

Pergamon Press
660 White Plains Road
Tarrytown, NY 10591
914-524-9200
Fax: 914-592-3625

Related to environmental affairs. Accepts advertising.
*100 pages*
*T Stephen, Editor*
*Rosemarie Fazzolari, Advertising*

## 3780 Connecticut Sea Grant

1080 Shennecossett Road
Groton, CT 06340
860-405-9128
Fax: 860-405-9109
http://www.seagrant.uconn.edu

Based at the University of Connecticut, CT Sea Grant is part of the National Sea Grant network, whose mission is the conservation and wise use of coastal and marine resources through research, education and outreach.
*Peg Van Patten, Communications Director*

## 3781 Earth First! Journal

PO Box 3023
Tucson, AZ 85702
520-620-6900
Fax: 413-254-0057
E-mail: collective@earthfirstjournal.org
http://www.earthfirstjournal.org/

Earth First! Journal was founded in 1979 in response to a lethargic, compromising and increasingly corporate environmental community. Earth First! takes a decidedly different tack toward environmental issues. We believe in using allthe tools in the toolbox, ranging from grassroots organizing and involvement in the legal process to civil disobedience and monkeywrenching.
*Publication Date: 1979 64 pages Bimonthly*
*ISSN: 1055-8411*

## 3782 Earth Island Journal

Earth Island Institute
2150 Allston Way
Suite 460
Berkeley, CA 94704-1375
415-788-3666
Fax: 415-788-7324
E-mail: editor@earthisland.org
http://www.earthisland.org

Publication from the Earth Island Institute - cutting-edge news, analysis and commentary on vital international environmental news.
*Publication Date: 1987 64 pages Quarterly*
*ISSN: 1041-0406*
*Audrey Webb, Editor*
*Jason Mark, Editor*

**3783 Economic Opportunity Report**
Business Publishers
PO Box 17592
Baltimore, MD 21297
301-587-6300
800-274-6737
Fax: 301-587-4530
E-mail: custserv@bpinews.com
http://www.bpinews.com

Antipoverty news coverage and analysis which gives insight into developments that affect social programs.
*Publication Date: 1963*
*Leonard A Eiserer, Publisher*
*Beth Early, Operations Director*

**3784 Environmental Business Journal**
ZweigWhite
321 Commonwealth Road
Suite 101
Wayland, MA 01778
508-651-1559
800-466-6275
Fax: 800-842-1560
E-mail: info@zweigwhite.com
http://www.environmentalbusinessjournal.com

EBJ is the leading business newsletter for the environmental industry, providing competive strategies, new business opportunities, and up-to-date market trends and data. Now published by ZweigWhite, the EBJ comes out every month.
*Publication Date: 1988 16+ pages Monthly*
*, President*

**3785 Environmental Compliance Update**
High Tech Publishing Company
PO Box 1275
Amherst, MA 01004
413-534-4500

Identifies and analyzes the issues and business and economic impact of environmental compliance laws and regulations. Monitors the relevant changes due to legislation, court decisions, private rulings and technology.
*Lori Reilly, Editor*

**3786 Environmental News**
CA Business Publications
PO Box 3359
Fort Worth, TX 76113
817-924-5301
Fax: 817-922-8893
E-mail: txenv@aol.com

Follows the progress of public environmental stock companies, provides updates on environmental contract opportunities, news of international environmental opportunities, and profiles innovative new companies.
*Carolyn Ashford, Publisher/Editor*

**3787 Environmental Packaging**
Thompson Publishing Group
1725 K Street NW
Suite 700
Washington, DC 20006
202-872-4000
800-677-3789
Fax: 202-739-9578
http://www.thompson.com

A newletter aimed at the environmental regulatory specialist, product development managers, purchasing managers, legal counsel and package designers covering state-by-state regulations and the FCA guidelines and enforcement.
*Publication Date: 1972*
*Daphne Musselwhite, Publisher*

**3788 Florida Forests Magazine**
Florida Forestry Association
PO Box 1696
Tallahassee, FL 32302
850-222-5646
Fax: 850-222-6179
E-mail: info@forestfla.org
http://www.floridaforest.org

A publication of the Florida Forestry Association.
*Publication Date: 1997 26-32 pages Quarterly*
*J Doran, Executive VP*

**3789 George Miksch Sutton Avian Research Center Sutton Newsletter**
PO Box 2007
Bartlesville, OK 74005
918-336-7778
Fax: 918-336-7783
E-mail: gmsarc@aol.com
http://www.suttoncenter.org

Newsletter published by George Miksch Sutton Avian Research Center.
*Publication Date: 1990 8-10 pages Semiannual*
*Steve Sherrod, Executive Director*
*Alan Jenkins, Assistant Director*

**3790 Green Business Letter**
Tilden Press
6 Hillwood Place
Oakland, CA 94610
202-332-1700
Fax: 202-332-3028
E-mail: gbl@greenbiz.com
http://www.greenbiz.com

Hands-on journal for environmentally conscious companies, covering management strategies, facilities management, personnel policies and procurement with environmental consciousness. Emphasis on products, resources and how-toinformation.
*8 pages*
*ISSN: 1056-490X*
*Joel Makower, Editor*

**3791 In Business: The Magazine for Sustainable Enterprises and Communities**
JG Press, Inc
419 State Avenue
Emmaus, PA 18049
610-967-4135
E-mail: advert@jgpress.com
http://www.jgpress.com
*Jerome Goldstein, Editor*

**3792 International Environment Reporter**
Bureau of National Affairs
1231 25th Street NW
Washington, DC 20037
202-452-4200
800-372-1033
Fax: 202-822-8092
http://www.bna.com

A four-binder information and reference service covering international environmental law and developing policy in the major industrial nations.
*William A Beltz, Publisher*

**3793 International Environmental Systems Update**
BSI Management Systems
12110 Sunset Hills Road
Suite 200
Reston, VA 20190
703-437-9000
800-862-4977
Fax: 703-435-7979
E-mail: solutions@bsiamencas.com
http://www.bsiamericas.com

Provides accurate, up-to-date and useful information for environmental professionals around the globe. Monthly publication brings current environmental events into the limelight, dissecting complex issues, helping hundreds oforganizations improve their environmental and business preformance.
*Publication Date: 1994 24 pages Monthly*
*ISSN: 1079-0837*
*Marcus Darby, Publisher*

**3794 McCoy's RCRA Unraveled**
McCoy & Associates
12596 West Bayaud Avenue
Suite 210
Lakewood, CO 80228
303-526-2674
Fax: 303-526-5471
E-mail: info@mccoyseminars.com
http://www.understandRCRA.com

This book addresses the most troublesome areas in 40 CFR Parts 261 and 262 of the federal regulations. Our engineers have researched every scrap of guidance EPA has ever issued on these troublesome topics, studied the Federal Registerpreamble language, and talked to thousands of people who attented our RCRA seminars and shared their real-world experiences. It includes a keyword index with more than 1,300 entries, 200 probing examples from EPA's own guidance documents and ahelpful acronym list.
*Publication Date: 2005 828 pages Yearly*
*Paul Gallagher, President*
*Nancy Pribble, Marketing Manager*

**3795 NFPA Journal**
One Batterymarch Park
Quincy, MA 02169
617-770-3000
Fax: 617-770-0700
E-mail: nfpa@nfpa.org
http://www.nfpa.org

A bi-monthly journal published by the National Fire Protection Association.
*Bi-Monthly*

**3796 NSS News**
National Speleological Society
2813 Cave Avenue
Huntsville, AL 35810
256-852-1300
Fax: 256-851-9241
E-mail: nss@caves.org
http://www.caves.org

Published by the National Speleological Society.
*Publication Date: 1942 Monthly*
*ISSN: 0027-7010*
*Dave Bunnell, Editor*

**3797 Newsleaf**
4949 Tealtown Road
Milford, OH 45150
513-831-1711
Fax: 513-831-8052
E-mail: cnc@cincynature.org
http://www.cincynature.org

Newsleaf is a quarterly publication for Cincinatti Nature Center members. This publication provides informative articles that teach readers about native flora and fauna.
*Publication Date: 1965 20-24 pages Quarterly*
*Rhonda Barnes-Kloth, Communications Manager*

**3798 Proceedings of the Desert Fishes Council**
Desert Fishes Council
PO Box 337
Bishop, CA 93515
760-872-8751
http://www.desertfishes.org

Yearly publication from Desert Fishes Council.
*Publication Date: 1969 Yearly*
*ISSN: 1068-0381*
*E P Pister, Executive Secretary*

**3799 Proceedings of the Southeastern Association of Fish and Wildlife Agencies**
8005 Freshwater Farms Road
Tallahassee, FL 32308
850-893-1204
Fax: 850-893-6204
E-mail: SEAFWA@aol.com
http://www.seafwa.org

Proceedings of the SEAFWA - an annual publication.
*Publication Date: 1947 4-900 pages Yearly*
*Robert M Brantly, Executive Secretary*

**3800 Pumper**
COLE Publishing
1720 Maple Lake Dam Road
PO Box 220
Three Lakes, WI 54562
715-546-3346
800-257-7222
Fax: 715-546-3786
E-mail: info@pumper.com
http://www.pumper.com

Emphasis on companies, individuals and industry events while focusing on customer service, environmental issues and employment trends.
*Publication Date: 1947*
*Ken Lowther, Editor*

**3801 Regulatory Update**
Arkansas Environmental Federation
1400 W Markham Street
Suite 302
Little Rock, AR 72201
501-374-0263
Fax: 501-374-8752
http://www.environmentark.org

The AEF focuses on practical, common-sense laws and regulations based on sound science; a teamwork approach to compliance; waste minimization and pollution prevention. The AEF enables information to be exchanged on a daily basisbetween its members, government regulators, and policy makers.
*Publication Date: 1967 Semi-Annual*
*Randy Thurman, Executive Director*

**3802 Risk Policy Report**
Inside Washington Publishers
1225 South Clark Street
Suite 1400
Arlington, VA 22202
703-416-8500
800-424-9068
E-mail: iwp@iwpnews.com
http://www.iwnews.com

Contains analysis, great perspectives, industry news, policymaking profiles and a calendar of events.
*Monthly*
*David P Clarke, Publisher/Editor*

**3803 Semillero**
731 8th Street SE
Washington, DC 20003
202-547-3800
Fax: 202-546-4784
E-mail: etoledo@newforestsproject.com
http://www.newforestsproject.com

Our electronic publication has more than 2500 subscribers in the US and Latin America. It provides useful information and references to specifice resources regarding agro-forestry, rural development and grant information.
*Publication Date: 1982*
*Erick Toledo and Catalina Serna, Author*
*Erick Toledo, Director*

## Periodicals: Design & Architecture

**3804 LICA News**
Land Improvement Contractors of America
3080 Ogden Avenue
Suite 300
Lisle, IL 60532
630-548-1984
Fax: 630-548-9189
E-mail: nlica@aol.com
http://www.licanational.com

Official publication of Land Improvement Contractors of America.
*Gerald J Biuso Sr, Executive VP*
*Eileen Levy, Publisher*

**3805 MSW Management**
Forester Communications
PO Box 3100
Santa Barbara, CA 93130
805-682-1300
Fax: 805-682-0200
E-mail: erosion@ix.netcom.com
http://www.mswmanagement.com

Provides general news on facility construction, financing, new equipment and revenue issues.
*Publication Date: 1991 7 Times Yearly*
*Daniel Waldman, Publisher*

## Periodicals: Disaster Preparedness & Response

**3806 Hazard Technology**
EIS International
555 Herndon Parkway
Herndon, VA 20170
703-478-9808
Fax: 703-787-6720

Application of technology to the field of emergency and environmental management to save lives and protect property.
*James W Morentz PhD, Publisher*
*Leslie Atkin, Managing Editor*

**3807 Hazardous Materials Newsletter**
Hazardous Materials Publishing
243 West Main Street
Kutztown, PA 19530
610-683-6721
Fax: 610-683-3171
E-mail: lheffner@hazmat-tsp.com
http://www.hazmatpublishing.com

Focuses on response to and control of hazardous materials emergencies. Particularly appropriate tools, equipment, materials, methods, procedures, strategies and lessons learned. Addresses leak, fore and spill control for incidentcommanders and experienced responders, including incident clauses, prevention and re-

medial actions; decisionmaking; scene management; control and containment; response teams; and product identification and hazards.
*12 pages*

**3808 Natural Hazards Observer**
Natural Hazards Center
University of Colorado 303-492-6818
482 UCB Fax: 303-492-2151
Boulder, CO 80309 E-mail: hazctr@colorado.edu
http://www.colorado.edu/hazards
A periodical of the Natural Hazards Center that covers current disaster issues; new international, national, and local disaster management, mitigation, and education programs; hazards research; political and policy developments; newinformation sources and Web sites; upcoming conference; and recent publications
*Bi Monthly*
*Dan Whipple, Editor*
*Kathleen Tierney, Director*

## Periodicals: Energy & Transportation

**3809 AERO SunTimes**
Alternative Energy Resources Organization
432 N Last Chance Gulch 406-443-7272
Helena, MT 59601 Fax: 406-442-9120
E-mail: aero@aeromt.org
The SunTimes is the newsletter for AERO. The organization is a statewide grassroots group whose members work together to strengthen communities through promoting sustainable agriculture, local food production and citizen-based SmartGrowth community planning.
*Publication Date: 1978 4-24 pages Quarterly*
*ISSN: 1046-0993*
*Kiki Hubbard, Editor*
*Jean Duncan, Editor*

**3810 Butane-Propane News**
Butane-Propane News, Inc
PO Box 660698 626-357-2168
Arcadia, CA 91066 800-214-4386
Fax: 626-303-2854
E-mail: arey@bpnews.com
http://www.bpnews.com
Offers information to professionals that are involved in the distribution, production, shipping and sales of butane and propane in the US and internationally. $32 for US one year subscription; $60 for international one yearsubscription.
*Publication Date: 1939 48-96 pages Monthly*
*ISSN: 0007-7259*
*Ann Rey, Editor/Director*

**3811 Cars of Tomorrow**
Northeast Sustainable Energy Association
50 Miles Street 413-774-6051
Greenfield, MA 01301 Fax: 413-774-6053
E-mail: nesea@nesea.org
http://www.nesea.org
Offers a look at the energy options of vehicles of the future.
*Curriculum*
*David Barclay, Executive Director*
*Arianna Alexsandra Grindrod, Education Director*

**3812 E&P Environment**
Pasha Publications
1600 Wilson Boulevard 703-528-1244
Suite 600 800-424-2908
Arlington, VA 22209 Fax: 703-528-1253
Reports on environmental regulations, advances in technology and litigation aimed specifically at the exploration and production segments of the oil and gas industry.
*Harry Baisden, Group Publisher*
*Michael Hopps, Editor*

**3813 Energy Engineering**
Association of Energy Engineers
4025 Pleasantdale Road 770-447-5083
Suite 420 Fax: 770-446-3969
Atlanta, GA 30340 E-mail: webmaster@aeecenter.org
http://www.aeecenter.org
Engineering solutions to cost efficiency problems and mechanical contractors who design, specify, install, maintain, and purchase non-residential heating, ventilating, air conditioning and refrigeration equipment and components.
*Wayne Turner, Editor-in-Chief*

**3814 Energy Journal**
International Association for Energy Economics
28790 Chagrin Boulevard 216-464-5365
Suite 350 Fax: 216-464-2737
Cleveland, OH 44122 E-mail: iaee@iaee.org
http://www.iaee.org
Promotes the advancement and dissemination of new knowledge on energy matters and related topics. Topics include: transportation, electricity markets, environmental issues, natural gas topics, and carbon emissions reduction.
*Publication Date: 1980 200 pages Quarterly*
*David L Williams, Executive Director*

**3815 Getting Around Clean & Green**
Northeast Sustainable Energy Association
50 Miles Street 413-774-6051
Greenfield, MA 01301 Fax: 413-774-6053
E-mail: nesea@nesea.org
http://www.nesea.org
This interdisciplinary science/social studies curriculum allows students to explore the transportation and environmental issues in their own lives. Activities cover: transportation systems, health impacts, environmental andtransportation histories, carpooling, and mass transit.
*90 pages Curriculum*
*David Barclay, Executive Director*
*Arianna Alexsandra Grindrod, Education Director*

**3816 Getting Around Without Gasoline**
Northeast Sustainable Energy Association
50 Miles Street 413-774-6051
Greenfield, MA 01301 Fax: 413-774-6053
E-mail: nesea@nesea.org
http://www.nesea.org
An interdisciplinary unit that explores the feasibility of getting around without using gasoline. Students can conduct various activities to compare powering vehicles with gasoline versus electricity.
*60 pages S & H Only*
*David Barclay, Executive Director*
*Arianna Alexsandra Grindrod, Education Director*

**3817 Heliographs**
Illinois Solar Energy Association
PO Box 634 630-260-0424
Wheaton, IL 60189 E-mail: info@illinoissolar.org
http://www.illinoissolar.org
*Publication Date: 1975 12 pages Quarterly*

**3818 IAEE Membership Directory**
International Association for Energy Economics
28790 Chagrin Boulevard 216-464-5365
Suite 350 Fax: 216-464-2737
Cleveland, OH 44122 E-mail: iaee@iaee.org
http://www.iaee.org
One of the three periodicals put out by the International Association for Energy Economics (IAEE). It lists members contact information and affiliation and general association information. Also available online.
*Annual*
*David L Williams, Executive Director*

**3819 IAEE Newsletter**
International Association for Energy Economics

28790 Chagrin Boulevard
Suite 350
Cleveland, OH 44122
216-464-5365
Fax: 216-464-2737
E-mail: iaee@iaee.org
http://www.iaee.org

Association information including: upcoming events, conferences, special reports, affiliate activities, and chapter news.

*Quarterly*

*David L Williams, Executive Director*

**3820 International Journal of Hydrogen Energy**

5783 SW 40 Street
#303
Miami, FL 33155
305-284-4666
Fax: 305-284-4792
E-mail: info@iahe.org
http://www.iahe.org

A monthly publication serving to inform scientists and the public of advances made in hydrogen energy research and development.

*Publication Date: 1976 120 pages Monthly*

*ISSN: 0360-3199*

*T Nejat Veziroglu, Author*

*T Neja Veziroglu, Editor-in-Chief*

**3821 Midwest Renewable Energy Association Newsletter**

7558 Deer Road
Custer, WI 54423
715-592-6595
Fax: 715-592-6596
E-mail: info@the-mrea.org
http://www.the-mrea.org

ReNews includes articles on energy issues, book reviews, case studies, and other general information about renewable energy.

*Quarterly*

*Tehri Parker, Executive Director*

**3822 Northeast Sun**

Northeast Sustainable Energy Association
50 Miles Street
Greenfield, MA 01301
413-774-6051
Fax: 413-774-6053
E-mail: nesea@nesea.org
http://www.nesea.org

Promotes responsible use of energy for a stronger economy and cleaner environment. Northeast Sun is a bi-annual Spring and Fall publication that includes articles by leading authorities on sustainable energy practices, energyefficiency and renewable energy, and each Fall issue includes the Sustainable Green Pages Directory of engery professionals in the Northeast. Subscription is free with membership.

*Bi-Annual*

*David Barclay, Executive Director*

*Arianna Alexsandra Grindrod, Education Director*

**3823 Nuclear Monitor**

Nuclear Information & Resource Services
1424 16th Street NW
Suite 404
Washington, DC 20036
202-328-0002
Fax: 202-462-2183
E-mail: nirsnet@nirs.org
http://www.nirs.org

Nuclear power, radioactive waste and sustainable energy news for environmental activities, state and local officials and investment communities.

*12 pages Monthly*

*Michael Mariotte, Editor*

**3824 Nuclear Waste News**

Business Publishers
8737 Colesville Road
PO Box 17592
Baltimore, MD 21297
301-589-5103
800-274-6737
Fax: 301-589-8493
E-mail: custserv@bpinews.com
http://www.bpinews.com

Worldwide coverage of the nuclear waste management industry, including waste generation, radiological environmental remediation, packaging, transport, processing and disposal.

*Weekly*

*Leonard A Eiserer, Publisher*

*Beth Early, Associate Publisher*

**3825 Radwaste Magazine**

American Nuclear Society
555 North Kensington Avenue
LaGrange Park, IL 60526
708-352-6611
800-323-3044
Fax: 708-352-0499
E-mail: radwaste@ans.org
http://www.ans.org

Addresses issues in all fields of radioactive waste management, removal, handling, disposal, treatment, cleanup and environmental restoration.

*6x per Year*

*Nancy Zacha, Publisher*

**3826 ReNews**

Midwest Renewable Energy Association
PO Box 249
Amherst, WI 54406
715-592-6595
Fax: 715-592-6596
E-mail: info@the-mrea.org
http://www.the-mrea.org

ReNews is a quarterly newsletter that includes articles on energy issues, book reviews, case studies, and other general information about renewable energy.

*Quarterly*

**3827 Solar Energy**

Elsevier Science
360 Park Avenue South
11th Floor
New York, NY 10010
212-989-5800
Fax: 212-633-3990
E-mail: usinfo-f@elsevier.com
http://www.elsevier.com

*John A Duffie, Editor*

**3828 Solar Energy Report**

PO Box 782
Rio Vista, CA 94571
949-837-7430
Fax: 949-709-8043
http://www.calseia.org

*Bi-Monthly*

**3829 Solar Reflector**

Texas Solar Energy Society
PO Box 1447
Austin, TX 78767
512-326-3391
800-465-5049
Fax: 512-444-0333
E-mail: info@txses.org
http://www.txses.org

A Texas Solar Energy Society publication. Promotes the wise use of sustainable and non-polluting resources.

*Quarterly*

*Natalie Marquis, Executive Director*

**3830 Solar Today**

American Solar Energy Society
2400 Central Avenue
Suite A
Boulder, CO 80301
303-443-3130
Fax: 303-443-3212
E-mail: ases@ases.org
http://www.ases.org

Provides information and case histories and reviews of a variety of renewable energy technologies, including solar, wind, biomass and geothermal.

*Donna McClane, Publications*

**3831 Sustainable Green Pages Directory**

Northeast Sustainable Energy Association
50 Miles Street
Greenfield, MA 01301
413-774-6051
Fax: 413-774-6053
E-mail: nesea@nesea.org
http://www.nesea.org

The SPG Directory lists over 30 categories of sustainable energy professionals working throughout the Northeast, including architects, engineers, builders, energy auditors, consultants and renewable energy installers and manufacturers.It is the largest directory of its kind and the only one that targets the Northeastern USA. Published in the Northeast Sun magazine and onlines.

*September Annual*

*David Barclay, Executive Director*

*Arianna Alexsandra Grindrod, Education Director*

## Periodicals: Environmental Engineering

**3832 Air and Waste Management Association's Magazine for Environmental Managers**
Air and Waste Management Association
One Gateway Center 412-232-3444
420 Fort Duquesne Blvd., 3rd Floor 800-270-3444
Pittsburgh, PA 15222 Fax: 412-232-3450
E-mail: info@awma.org
http://www.awma.org
A magazine that contains sections of Washington and Canadian reports, a calendar of events, government affairs, news focus, campus research, business briefs, district control news, porfessional development programs, professionalservices and other issues facing the environmental professionals.
*Todd Zahniser, Publisher/Editor*

**3833 Asbestos & Lead Abatement Report**
Business Publishers
PO Box 17592 301-589-5103
Baltimore, MD 21297 800-274-6737
Fax: 301-589-8493
E-mail: bpinews@bpinews.com
http://www.bpinews.com
Tracks the major legal, legislative, regulatory, business and technological developments in the asbestos and lead abatement industries.
*Weekly*
*Leonard A Eiserer, Publisher*
*Beth Early, Associate Publisher*

**3834 Curt B Beck Consulting Engineer: Newsletter**
408 W Kingsmill Street 806-665-9281
PO Box 2442 888-665-9281
Pampa, TX 79006 Fax: 806-665-1965
E-mail: curtbbeck@cableone.net
http://www.pan-tex.net/usr/b/beck
Pollution control services.
*Publication Date: 1984 Monthly*
*Curt B Beck, Owner*

**3835 Defense Cleanup**
Business Publishers
PO Box 17592 301-587-6300
Baltimore, MD 21297 800-274-6737
Fax: 301-587-4530
E-mail: bpinews@bpinews.com
http://www.bpinews.com
Covers the latest news and analysis of defense cleanup activity. Including base remediation and closure, contract awards and site cleanups.
*Weekly*
*Leonard A Eiserer, Publisher*
*Beth Early, Associate Publisher*

**3836 EI Digest**
Environmental Information
PO Box 390266 952-831-2473
Edina, MN 55439 Fax: 952-831-6550
E-mail: ei@enviro-information.com
http://www.envirobiz.com
Contains market studies of commercial hazardous waste management companies with in-depth analysis of trends in policy, regulations, technology and business.
*Publication Date: 1983*
*ISSN: 1042-251X*
*Cary Perket, Editor*

**3837 Environment**
Helen Dwight Reid Educational Foundation
1319 18th Street NW 202-296-6267
Washington, DC 20036 800-365-9753
Fax: 202-296-5149
E-mail: subscribe@heldref.org
http://www.heldref.org
Provides environment professionals and concerned citizens with authoritative yet accessible articles that provide critical analysis of environmental science and policy issues, book recommendations, commentaries, news briefs and reviewson environmental websites and major governmental and institutional reports.
*Publication Date: 1958 48 pages Monthly*
*ISSN: 0013-9157*
*Douglas Kirkpatrick, Executive Director*
*Barbara T Richman, Managing Editor*

**3838 Environment 21**
Florida Environments Publishing
4010 Newberry Road 352-373-1401
#F Fax: 352-373-1405
Gainesville, FL 32607 E-mail: info@enviroworld.com
http://www.enviroworld.com
Regulations, wildlife, hazard waste/materials, ground/surface/drinking water and other issues concerning Florida's environment are emphasized in this publication for the environmental management team.
*Dave Newport, Publisher*

**3839 Environmental Engineer Magazine**
American Academy of Environmental Engineers
130 Holiday Court 410-266-3311
Suite 100 Fax: 410-266-7653
Annapolis, MD 21401 E-mail: info@aaee.net
http://www.aaee.net
Official magazine of the American Academy of Environmental Engineers. It addresses issues and practice with: updates on legal developments affecting environmental engineering, documentation of the profession's heritage, articles onenvironmental policy, and profiles of leading environmental engineers.
*Quarterly*
*Lawrence C Pencak, Executive Director*
*Yolanda Y Moulden, Production Manager*

**3840 Environmental Engineering Science**
Mary Ann Liebert
140 Huguenot Street 914-740-2100
3rd Floor 800-MLI-EBER
New Rochelle, NY 10801 Fax: 914-740-2101
E-mail: info@liebertpub.com
http://www.liebertpub.com
The focus is on pollution control of the suface, ground, and drinking water, and highlight research news and product developments that aid in the fight against pollution.
*Bi-Monthly*
*ISSN: 1092-8758*
*Mary Ann Liebert, Publisher*
*Dumpnico Grosso PhD, Editor-in-Chief*

**3841 Environmental Manager**
Air and Waste Management Association
420 Fort Duquesne Blvd 412-232-3444
3rd Floor 800-270-3444
Pittsburgh, PA 15222 Fax: 412-232-3450
E-mail: info@awma.org
http://www.awma.org
Features timely articles on business, regulatory, and technical issues of interest to the environmental industry.
*Tim Keener, Editor*
*Lisa Bucher, Managing Editor*

**3842 Environmental and Energy Study Institute**
122 C Street NW 202-628-1400
Suite 630 Fax: 202-628-1825
Washington, DC 20001 E-mail: eesi@eesi.org
http://www.eesi.org
A nonprofit organization dedicated to promoting environmentally sustainable societies. EESI believes meeting this goal requires transitions to social and economic patterns that sustain people, the environment and the natural resourcesupon which present and future generations depend. EESI produces credible, timely information and innovative public policy initiatives that

lead to these transitions. These products take the form of publications, briefings, work shops and taskforces.
*Publication Date: 1984*
*Carol Werner, Executive Director*

**3843 Federation of Environmental Technologists**
9451 N 107th Street 414-354-0070
Milwaukee, WI 53224 Fax: 414-354-0073
E-mail: info@fetinc.org
http://www.fetinc.org
A nonprofit organization formed to assist industry in interpretation of and compliance with environmental regulations. Membership is open to all industries, municipalities, organizations and individuals concerned about environmentalregulations. Currently there are approximately 1000 members and 125 patron companies.
*Publication Date: 1982 Monthly*
*Barbara Hurula, Executive Director*

**3844 Food Protection Trends**
International Association for Food Protection
6200 Aurora Avenue 515-276-3344
Suite 200W 800-369-6337
Des Moines, IA 50322 Fax: 515-276-8655
E-mail: info@foodprotection.org
http://www.foodprotection.org
Published as the general membership publication by the International Association for Food Protection, each issue contains refereed articles on applied research, applications of current technology and general interest subjects for foodsafety professionals. Regular features include industry and association news, an industry related product section and a calendar of meetings, seminars and workshops.Updates of government regulations and sanitary design is also featured. All membersreceive FPT.
*Publication Date: 1981 80+ pages Monthly*
*ISSN: 1043-3546*
*David W Tharp, Executive Director*
*Lisa K Hovey, Managing Editor*

**3845 Hazardous Materials Intelligence Report**
World Information Systems
129 Mount Auburn 617-491-5100
Cambridge, MA 02238 Fax: 617-492-3312
Provides news analysis on environmental business, hazardous materials, waste management, pollution prevention and control. Covers regulations, legislation and court decisions, new technology, contract opportunities and awards andconference notices.
*Richard S Golob, Publisher*
*Roger B Wilson Jr, Editor*

**3846 Hazmat Transport News**
Business Publishers
PO Box 17592 301-587-6300
Baltimore, MD 21297 800-274-6737
Fax: 301-587-4530
E-mail: bpinews@bpinews.com
http://www.bpinews.com
Reports on the regulatory, enforcement, legislative and litigation developments affecting hazardous materials transportation.
*Monthly*
*Leonard A Eiserer, Publisher*
*Beth Early, Associate Publisher*

**3847 Integrated Waste Management**
McGraw Hill
PO Box 182604 877-833-5524
Columbus, OH 43272 Fax: 614-759-3749
http://www.mcgraw-hill.com
Articles geared toward integration of solid waste management.
*8 pages*
*Kevin Hamilton, Publisher*

**3848 International Journal of Phytoremediation**
Taylor & Francis Inc
325 Chestnut Street 413-549-5170
Suite 800 Fax: 413-549-0579
Philadelphia, PA 19106 http://www.aehs.com
An official journal of the Association for Environmental Health and Sciences (AEHS). Dedicated to current laboratory and field research on how to use plant systems to remediate contaminated environments.
*6 Issues Yr*
*ISSN: 1522-6514*
*Jason White, Managing Editor*

**3849 Iowa Academy of Science Journal**
Iowa Academy of Science
UNI - 175 Baker Hall 319-273-2021
2607 Campus Street Fax: 319-273-2807
Cedar Falls, IA 50614-0508 E-mail: iascience@uni.edu
http://www.iacad.org
*Quarterly*
*Craig Johnson, Executive Director*

**3850 Journal of Environmental Engineering**
American Society of Civil Engineers
1801 Alexander Bell Drive 703-295-6300
Reston, VA 20191 Fax: 703-295-6211
E-mail: onlinejls@asce.org
http://http://pubs.asce.org/journals/environmental
The journal of Environmental Engineering presents a collection of broad interdisciplinary information on the practice and status of research in environmental engineering science, systems engineering, and sanitation.
*Publication Date: 1875*
*Raymond A Ferrara, Editor*
*Melissa Junior, Director, Journals*

**3851 Journal of Environmental Quality**
American Society of Agronomy
677 S Segoe Road 608-273-8080
Madison, WI 53711 Fax: 608-273-2021
http://www.agronomy.org
Written for university, government and industry scientists interested in the impacts of environmental perturbations on the biological and physical sciences. Domestic member price: $50 (Int'l $103); Domestic non-member price: $650(Int'l $703).
*Bi-Monthly*
*Dennis Corwin, Editor*
*Susan Ernst, Managing Editor*

**3852 Journal of Environmental and Engineering Geophysics: JEEG**
1720 South Bellaire Street 303-531-7517
Suite 110 Fax: 303-820-3844
Denver, CO 80222 E-mail: staff@eegs.org
http://www.eegs.org
A peer reviewed journal of the EEGS made available to members and a variety of libraries.
*Publication Date: 1992 Quarterly*
*ISSN: 1083-1363*
*Kathie Barstnar, Executive Director*
*Janet Simms, Editor*

**3853 Journal of IEST**
Institute of Environmental Sciences & Technology
2340 S. Arlington Heights Rd 847-981-0100
Suite 100 Fax: 847-981-4130
Arlingotn Heights, IL 60005 E-mail: iest@iest.org
http://www.iest.org
An annual journal published by the Institute of Environmental Sciences & Technology.
*Publication Date: 1957 185 pages Annual*
*ISSN: 1098-4321*
*Charles W Berndt, VP Communications*

**3854 Kennedy-Jenks Consultants: Alert Newsletter**
303 Second Street 415-243-2150
Suite 300 S Fax: 415-896-0999
San Francisco, CA 94107 http://www.kennedyjenks.com

Environmental engineering consulting company.
*Publication Date: 1990 6 pages 3 times/year*
*Gordon Morris, Graphics Services Manager*
*Rena Chin, Editor*

**3855 Kennedy-Jenks Consultants: Spotlights**
303 Second Street 415-243-2150
Suite 300 S Fax: 415-896-0999
San Francisco, CA 94107 http://www.kennedyjenks.com
Environmental engineering consulting company.
*Publication Date: 1981 12 pages 3 times/year*
*Gordon Morris, Graphics Services Manager*
*Rena Chin, Editor*

**3856 Lead Detection & Abatement Contractor**
IAQ Publications
7920 Norfolk Avenue 301-913-0115
Suite 900 Fax: 301-913-0119
Bethesda, MD 20814 E-mail: iaqpubs@aol.com
http://www.iaqpubs.com
Feature articles include new on legislation, operational and safety issues that affect the removal of lead and lead by-products from paint, water, soil, and air.
*Susan Valenti, Editor*

**3857 Leading Edge**
Society of Exploration Geophysicists
8801 South Yale Avenue 918-497-5500
Tulsa, OK 74137 Fax: 918-497-5557
E-mail: jlawnick@seg.org
http://www.seg.org
Addresses a broad spectrum of topics related to applied geophysics. Material immediately accessible to a broad audience.
*Publication Date: 1930 116 pages*
*ISSN: 1070-485X*
*Dean Clark, Editor*
*Linda Holeman, Associate Editor*

**3858 McCoy's CAA Unraveled**
McCoy & Associates
12595 W Bayaud AvenueRoad 303-526-2674
Suite 210 Fax: 303-526-5471
Lakewood, CO 80228 E-mail: info@mccoyseminars.com
http://www.understandRCRA.com
*414 pages*

**3859 McCoy's RCRA Reference**
McCoy & Associates
12595 W Bayaud AvenueRoad 303-526-2674
Suite 210 Fax: 303-526-5471
Lakewood, CO 80228 E-mail: info@mccoyseminars.com
http://www.understandRCRA.com
*1190 pages*

**3860 Medical Waste News**
Business Publishers
PO Box 17592 301-589-5103
Baltimore, MD 21297 800-274-6737
Fax: 301-589-8493
E-mail: custserv@bpinews.com
http://www.bpinews.com
Reports on the rapidly evolving legislative and regulatory actions in medical waste management. Includes coverage of incineration, laboratory wastes, infection control, liability and legal issues and waste transport.
*Publication Date: 1963*
*Leonard A Eiserer, Publisher*
*Beth Early, Operations Director*

**3861 Noise Control Engineering Journal**
Institute of Noise Control Engineering
PO Box 17592 515-294-6142
Baltimore, MD 21297 Fax: 515-294-3528
E-mail: ibo@inceusa.org
http://www.inceusa.org
The technical publication of the Institute of Noise Control Engineering. It contains technical articles on all aspects of noise control engineering.
*Bi-Monthly*
*ISSN: 0736-2501*
*Joseph M Cuschieri, Executive Director*
*Courtney Burroughs, Editor in Chief*

**3862 Noise Regulation Report**
Business Publishers
PO Box 17592 301-589-5103
Baltimore, MD 21297 800-274-6737
Fax: 301-589-8493
E-mail: custserv@bpinews.com
http://www.bpinews.com
Exclusive coverage of airport, highway, occupational and open space noise, noise control and mitigation issues.
*Publication Date: 1963*
*Leonard A Eiserer, Publisher*
*Beth Early, Operations Director*

**3863 Plumbing Standards**
American Society of Sanitary Engineering
901 Canterbury 440-835-3040
Suite A Fax: 440-835-3488
Westlake, OH 44145 E-mail: info@asse-plumbing.org
http://www.asse-plumbing.org
Disseminates industry-wide technical information on standards, updates, water, wastewater, plumbing design, and other topics related to the water industry.
*Shannon M Corcoran, Executive Director*
*Megan Bryant, Managing Editor*

**3864 Pollution Engineering**
Cahners Business Information
2000 Clearwater Drive 630-320-7000
Oak Brook, IL 60523 Fax: 630-288-8282
http://www.pollutionengineering.com
Serves the field of pollution control in manufacturing industries, utilities, consulting engineers and constructors. Also serves government agencies including administration of federal, state and local environmental programs.

*ISSN: 0032-3640*
*Barbara Olsen, Publisher*
*Roy Bigham, Managing Editor*

**3865 RMT Newsletter**
744 Heartland Trail 608-831-4444
PO Box 8923 800-283-3443
Madison, WI 53717 Fax: 608-831-3334
E-mail: info@rmtinc.com
http://www.rmtinc.com
Global engineering and management consulting firm that develops environmental solutions for industry. With a 600 person staff and 20 offices throughout the US and Europe helping clients sustain the environment while meeting theirbusiness objectives. Engineers, scientists and construction managers can take a project from conception through successful completion. Expertise includes air, water and waste permitting, remediation, hazardous/solid waste management, air pollutioncontrol and more.
*8 pages Quarterly*
*Jodi Burmester, Marketing Communications*

**3866 SPAC Newsletter**
Soil and Plant Analysis Council
300 Speedway Cirlce 402-476-0300
Suite 2 Fax: 402-476-0302
Lincoln, NE 68502 E-mail: bvaughan@ogsource.com
http://www.spcouncil.com

Quarterly newsletter.
*Quarterly*
*Mark Flock, President*
*Bryon Vaughan, Secretary/Treasurer*

**3867 Sludge**
Business Publishers

PO Box 17592
Baltimore, MD 21297
301-589-5103
800-274-6737
Fax: 301-589-8493
E-mail: custserv@bpinews.com
http://www.bpinews.com

Premier insider guide to the biosolids industry. Follows developments in and management of beneficial use and wastewater residuals, with practical information about industrial sludge, incineration, special wastes, permits andlandfills.

*BiWeekly*
*Leonard A Eiserer, Publisher*
*Beth Early, Operations Director*

**3868 Soil & Sediment Contamination:An International Journal**
Taylor & Francis Inc
325 Chestnut Street
Suite 800
Philadelphia, PA 19106
413-549-5170
Fax: 413-549-0579
http://www.aehs.com

An official journal of the Association for Environmental Health and Sciences (AEHS). An internationally peer-reviewed publication that focuses on sediment and soil contamination.

*Bi Monthly*
*ISSN: 1532-0383*
*James Dragun, Editor-in-Chief*
*Paul Kostecki, Managing Editor*

**3869 Solid Waste Report**
Business Publishers
PO Box 17592
Baltimore, MD 21297
301-589-5103
800-274-6737
Fax: 301-589-8493
E-mail: custserv@bpinews.com
http://www.bpinews.com

Comprehensive news and analysis of legislation, regulation and litigation in solid waste management. Regularly features federal rules, congressional actions, state updates and business trends.

*Weekly*
*Leonard A Eiserer, Publisher*
*Beth Early, Associate Publisher*

**3870 Waste News**
Crain Communications
1725 Merriman Road
#300
Akron, OH 44313
330-836-9180
Fax: 330-836-1692
E-mail: editorial@wastenews.com
http://www.wastenews.com

Editorial content focuses on waste management and recycling issues, primarily how businesses deal with the waste they generate. Covers waste management service providers, legistlative and regulatory environmental issues, emergingtechnologies, municipal recycling and waste issues, commodity market price, mergers, aquisitions and expansions.

*Publication Date: 1995 Bi-Weekly*
*ISSN: 1091-699*
*Allan Gerlat, Editor*
*Brennan Lafferty, Managing Editor*

**3871 Widener University: International Conference on Solid Waste Proceedings**
One University Place
Chester, PA 19013
610-499-4042
Fax: 610-499-4461
E-mail: solid.waste@widener.edu
http://www.widener.edu/solid.waste

Publication of the annual conference on solid waste technology and management. Over 100 speakers from 35 countries present their work. Proceedings available.

*Publication Date: 1983 Quarterly*
*ISSN: 1088-1697*
*Ronald L Mersky, Editor*
*Iraj Zandi, Founder*

## Periodicals: Environmental Health

**3872 American College of Toxicology**
9650 Rockville Pike
Bethesda, MD 20814
301-571-1840
Fax: 301-571-1852
E-mail: ekagan@actox.org
http://www.actox.org

The American College of Toxicology is a 501-0-3 nonprofit organization. It is not a degree-granting organization. The American College of Toxicology is dedicated to providing an interactive forum for the advancement and exchange oftoxicologic information between industry, goverment, and academia. There is an annual meeting in November each year. The ACT publiches a journal, International Journal of Toxicology on a bi-monthly basis.

*ISSN: 1091-5818*
*Carol L Lemire, Executive Director*
*Eve Gamzu Kagan, Assistant Executive Director*

**3873 American Journal of Public Health**
American Public Health Association
800 I Street NW
Washington, DC 20001
202-777-2742
888-320-2742
Fax: 202-777-2533
E-mail: ellen.meyer@apha.org
http://www.apha.org

Peer-reviewed journal of the American Public Health Association (APHA) for public health workers and academics. Its emphasis is on research and practioners experiences.

*12 Issues/Yr*
*Ellen Meyer, Director Publications*
*Nancy Johnson, Executive Editor*

**3874 Applied and Environmental Microbiology**
American Society for Microbiology
1752 N Street NW
Washington, DC 20036
202-737-3600
Fax: 202-942-9333
E-mail: oed@asmusa.org
http://www.asm.org

Contains current significant research in industrial microbiology, microbial ecology, biotechnology, public health microbiology and food microbiology.

*Samuel Kaplan, Chairman Publications Board*
*Linda M Illig, Director Journals*

**3875 Asbestos & Lead Abatement Report**
Business Publishers
PO Box 17592
Baltimore, MD 21297
301-589-5103
800-274-6737
Fax: 301-589-8493
E-mail: custserv@bpinews.com
http://www.bpinews.com

Contains articles on regulation compliance, environmental trends and business opportunities.

*Leonard A Eiserer, Publisher*

**3876 Aviation, Space and Environmental Medicine**
Aerospace Medical Association
320 S Henry Street
Alexandria, VA 22314
703-739-2240
Fax: 703-739-9652
E-mail: pday@asma.org
http://www.asma.org

Provides contact with physicians, life scientists, bioengineers, and medical specialists working in both basic medical research and in its clinical applications.

*Publication Date: 1929 962 pages Monthly*
*ISSN: 0095-6562*
*Sarah A Nunneley MD, Author*
*Sarah A Nunneley, Editor-in-Chief*
*Sarah A Pierce-Rubio, Editor Assistant*

**3877 Bio Integral Resource Center: Common Sense Pest Control**
PO Box 7414
Berkeley, CA 94707
510-524-2567
Fax: 510-524-1758
E-mail: birc@igc.org
http://www.birc.org
Features least toxic solutions to pest problems of the home and garden. Those who are chemically sensitive and looking for alternatives may find what they need in the Quarterly.
*Publication Date: 1984 24 pages Quarterly*
*ISSN: 8756-7881*
*Dr. William Quarles, Executive Director*

**3878 Bio Integral Resource Center: IPM Practitioner**
PO Box 7414
Berkeley, CA 94707
510-524-2567
Fax: 510-524-1758
E-mail: birc@igc.org
http://www.birc.org
Focuses on management alternatives for pests such as insects, mites, ticks, vertebrates, weeds and plant pathogens.
*Publication Date: 1979 24 pages 10 Times a Year*
*ISSN: 0738-969x*
*Dr. William Quarles, Executive Director*

**3879 Center for Statistical Ecology & Environmental Statistics: Environmental & Ecological Statistics**
Kluwer Academic Publishers
Pennsylvania State University
421 Thomas Building, Dept of Statistics
University Park, PA 16802
814-865-9442
Fax: 814-865-1278
E-mail: gpp@stat.psu.edu
http://www.stat.psu.edu/~gpp
The Center is the first of its kind in the world and enjoys national and international reputation. They have an ongoing program of research that integrates statistics, ecology and the environment. The emphasis is on the environment andcollaborative research, training and exposition on improving the quantification and communication of man's impact on the environment. Major interest also lies in statistical investigations of the impact of the environment on man. Contact them forfull listings.
*Publication Date: 1984 110 pages Quarterly*
*ISSN: 1352-8505*
*Ganapati P Patil, Editor-in-Chief*

**3880 EH&S Software News Online**
Donley Technology
220 Garfield Avenue
PO Box 152
Colonial Beach, VA 22443
804-224-9427
Fax: 804-224-7958
E-mail: donleytech@donleytech.com
http://www.donleytech.com
Reports on news and upgraded software products, database, and on-line systems from commercial developers and government resources.
*Quarterly*
*John Donley, Editor*
*Elizabeth Donley, Managing Editor/Publisher*

**3881 Enterprise Software: Essential EH&S**
Essential Technologies
1401 Rockville Pike
#500
Rockville, MD 20852
301-284-3000
Fax: 301-284-3001
E-mail: info@essentech.com
http://www.essential-technologies.com
Integrated solutions for emissions management, hazard communication, compliance management, occupational health and safety and contingency management.
*James Morentz, Publisher*

**3882 Environmental Connections**
Connecticut College
270 Mohegan Avenue
Box 5293
New London, CT 06320
860-439-5417
Fax: 860-439-2418
E-mail: ccbes@conncoll.edu
http://www.ccbes.conncoll.edu
*Publication Date: 1998 10 pages 2 times per year*
*Robert Askins, Director*
*Diane Whitelaw, Assistant Director*

**3883 Environmental Dimensions**
Trine Publishers
28 Kilbarry Crescent
Ottawa, Ontario K1K
613-749-3735
Fax: 613-749-6807
E-mail: trine@istar.ca
http://www.envirodim.com
Provides the environmental professionals with news and information on current environmental health issues, solution, and hazards, as well as examining governmental policies, legal news and Canadian's environmental preformance.
*22 times per year*
*Roland Blassnig, Editor*

**3884 Environmental Health Letter**
Business Publishers
PO Box 17592
Baltimore, MD 21297
301-589-5103
800-274-6737
Fax: 301-589-8493
E-mail: custserv@bpinews.com
http://www.bpinews.com
Comprehensive coverage of the latest policies and ground-breaking research that explores the potential links between environmental factors and human health.
*8 pages*
*Leonard A Eiserer, Publisher*
*Beth Early, Operations Director*

**3885 Environmental Health perspectives**
National Inst. of Environmental Health Sciences
C/O Blogar & Partners
14600 Weston Parkway, Suite 300
Cary, NC 27513
919-653-2581
866-541-3841
Fax: 919-678-8696
E-mail: ehponline@niehs.nih.gov
http://www.ehponline.org
Annual subscription.
*224 pages Monthly*
*Hugh Tilson, Editor-in-Chief*
*Karen A Warren, Business Manager*

**3886 Florida Journal of Environmental Health**
Florida Environmental Health Association
5101 Ortega Blvd
Jacksonville, FL 32210
904-384-0838
E-mail: Scexedir@aol.com
http://www.feha.org
Promotes public health by means of advanced environmental control.
*Quarterly*
*ISSN: 0897-1823*
*Lu Grimm, Editor*
*Jennifer M Willimas, Executive Director*

**3887 Healthy Schools Network Newsletter**
773 Madison Avenue
Albany, NY 12208
518-462-0632
Fax: 518-462-0433
E-mail: info@healthyschools.rog
http://www.healthyschools.org
HSN is a nationally recognized state-based advocate for the protection of children's environmental health in schools. Engages in research, education, outreach, technical assistance and coalition building to create schols that areenvironmentally responsible to children, and to their communities. Publishes a quarterly newsletter and maintains an Information Clearinghouse and Referral Service.
*Quarterly*
*Claire L Barnett, Executive Director*

**3888 Human Ecology Action League Magazine**
2250 N Druid Hills Road NE
Atlanta, GA 30329
404-248-1898
Fax: 404-248-0162
E-mail: HEALNatnl@aol.com
http://members.aol.com/HEALNatnl/index/html
The Human Ecology Action League Inc (HEAL) is a nonprofit organization founded in 1977 to serve those whose health has been adversely affected by environment exposures; to provide information to those who are concerned about the healtheffects of chemicals; and to alert the general public about the potential dan-

gers of chemicals. Referrals to local HEAL chapters and other support groups are available from the League.
*Publication Date: 1977 35 pages Quarterly*
*ISSN: 8755-7878*
*Diane Thomas, Editor*

**3889 Indoor Environment Review**
IAQ Publications
7920 Norfolk Avenue 301-913-0115
Suite 900 Fax: 301-913-0119
Bethesda, MD 20814 E-mail: iaqpubs@aol.com
http://www.iaqpubs.com
New technology, research and legislation concerning all indoor air and water quality issues.
*Robert Morrow, Editor*

**3890 Industrial Health and Hazards Update**
InfoTeam
PO Box 15640 954-473-9560
Plantation, FL 33318 Fax: 954-473-0544
E-mail: infoteamma@aol.com
Covers occupational safety, health hazards, and disease; mitigation and control of hazardous situations; waste recycling and treatment.
*20 pages*
*Dr. David Allen, Associate Editor*

**3891 Journal of Environmental Health**
National Environmental Health Association
720 S Colorado Boulevard 303-756-9090
Suite 1000-N Fax: 303-691-9490
Denver, CO 80246 E-mail: staff@neha.org
http://www.neha.org
A practical journal containing information on a variety of environmental health issues.
*Publication Date: 1937 70-76 pages 10 Times a Year*
*ISSN: 0022-0892*
*Nelson E Fabain, Executive Director*
*Julie Collins, Content Editor*

**3892 Journal of Medical Entomology**
Journal of Entomology
10001 Derekwood Lane 301-731-4535
Suite 100 Fax: 301-731-4538
Lanham, MD 20706 E-mail: esa@entsoc.org
http://www.entsoc.org
Contributions report on all phases of medical entomology and medical acarology, including the systematics and biology of insects, acarines, and other arthropods of public health and veterinary significance.
*Bi=Monthly*
*John Edman, Editor-in-Chief*

**3893 Journal of Pesticide Reform**
PO Box 1393 541-344-5044
Eugene, OR 97440 Fax: 541-344-6923
E-mail: info@pesticide.org
http://www.pesticide.org
Pesticide factsheets, alternatives factsheets for common pest problems, and helpful information on how to take action for change are featured in this journal. Each issue also includes updates on NCAP's work, news on pesticide issues,and reviews of books and videos.
*Publication Date: 1984 24 pages Quarterly*
*Caroline Cox, Editor*
*Norma Grier, Executive Director*

**3894 Nation's Health, The**
American Public Health Association
800 I Street NW 202-777-2742
Washington, DC 20001 888-320-2742
Fax: 202-777-2533
E-mail: ellen.meyer@apha.org
http://www.apha.org
Monthly newspaper of the American Public Health Association (APHA). It focuses on the latest public health news that public health professionals need to know such as food safety, racial and ethnic disparities, patients' rights,environmental issues, and health screening.
*10 Issues/Yr*
*Ellen Meyer, Director Publications*
*Michele Late, Executive Editor*

**3895 National Institute of Environmental Health Sciences Journal**
111 T.W. Alexander Drive 919-496-2433
Research Triangle Park, NC 27709 Fax: 919-496-8276
E-mail: olden@NIEHS.nih.gov
http://www.niehs.nih.gov
The National Institute of Environmental Health Sciences is the principal federal agency for basic biomedical research on the health effects of environmental agents. It is the headquarters for the National Toxicology Program whichcoordinates toxicology studies within the Department of Health and Human Services.
*Publication Date: 1972 150 pages Monthly*
*ISSN: 0091-6765*
*Kenneth Olden PhD, Director*

**3896 Natural Resources Council of America: Environmental Resource Handbook**
Universal Reference Publications
1616 P Street 202-232-6631
NW Suite 340 Fax: 240-465-0467
Washington, DC 20036 http://www.NaturalResourcesCouncil.org
Environmental Resource Handbook updates.
*4 pages*
*Laura Seal, Membership Coordinator*

**3897 Natural Resources Council of America: Conservation Voice**
1616 P Street 202-232-6631
NW Suite 340 Fax: 240-465-0467
Washington, DC 20036 http://www.naturalresourcescouncil.org
Charts the news, events, and personnel that shape the face of the conservation movement and includes the quarterly supplemental publication NEPA news.
*Publication Date: 1958 6-8 pages Bi-Monthly*
*Andrea Yank, Executive Director*

**3898 Natural Resources Council of America: NEPA News**
1616 P Street 202-232-6631
NW Suite 340 Fax: 240-465-0467
Washington, DC 20036 http://www.naturalresourcescouncil.org
*Publication Date: 1994 8 pages Quarterly*
*Andrea Yank, Executive Director*

## Periodicals: Gaming & Hunting

**3899 American Bass Association Newsletter**
402 N Prospect Avenue 310-376-1026
Redondo Beach, CA 90277 Fax: 310-376-5072
E-mail: feedback@americanbass.com
http://www.americanbass.com
*Publication Date: 1989 24 pages Quarterly*
*Craig Sutherland, Editor*

**3900 Caller**
770 Augusta Road 803-637-3106
PO Box 530 800-THE-NWTF
Edgefield, SC 29824 Fax: 803-637-0034
E-mail: nwtf@nwtf.net
http://www.nwtf.org
News magazine is the NWTF's member publication that allows NWTF volunteers to share information about their successful conservation projects, Wild Turkey Super Fund Banquets and educational events for youth, women and disabledindividuals. Free with membership.
*Quarterly*
*Jason Gilbertson, Editor*

3901 **IPPL News**
PO Box 776 843-871-2280
Summerville, SC 29484 Fax: 843-871-7988
E-mail: info@ippl.org
http://www.ippl.org
Educates readers in more than 50 countries about action that can be taken to protect primates.
*Publication Date: 1974 32 pages 3 Times a Year*
*ISSN: 1040-3027*
*Dr Shirley McGreal, Chairperson*
*Marjorie Doggett, Secretary*

3902 **International Game Fish Association Newsletter**
300 Gulf Stream Way 954-927-2628
Dania Beach, FL 33004 Fax: 954-924-4299
E-mail: hq@igfa.org
http://www.igfa.org
Founded as record-keeper and to maintain fishing rules. Today, emphasis is on conservation and education. Newsletters published are World Record Game Fish, annually and The International Angler, bi-monthly.
*Bi-Monhtly*
*Rob Kramer, President*

3903 **JAKES Magazine**
Wild Turkey Center 803-637-3106
PO Box 530 800-THE-NWTF
Edgefield, SC 29824 Fax: 803-637-0034
E-mail: nwtf@nwtf.net
http://www.nwtf.org
JAKES (Juniors Acquiring Knowledge, Ethics and Sportsmanship) is a magazine which provides fun and educational articles focusing on young hunters, outdoor activities, the environment and other items of interest to readers 17 years oldand younger. Free with membership.
*Quarterly*
*Matt Lindler, Editor*

3904 **Mid-Atlantic Fishery Management Council Newsletter**
300 S New Street 302-674-2331
Federal Building, Room 2115 877-446-2362
Dover, DE 19904 Fax: 302-674-5399
http://www.mafmc.org
*Publication Date: 1998 8 pages Quarterly*
*Daniel T Furlong, Executive Director*

3905 **Turkey Call**
770 Augusta Road 803-637-3106
PO Box 530 800-THE-NWTF
Edgefield, SC 29824 Fax: 803-637-0034
E-mail: nwtf@nwtf.net
http://www.nwtf.org
A magazine for turkey hunting enthusiasts, provides articles to help you improve your hunting skills and learn how to enhance your land for wildlife. Free with membership.
*Bi-Monthly*
*Doug Howlett, Editor*

3906 **Wheelin' Sportsmen**
770 Augusta Road 803-637-3106
PO Box 530 800-THE-NWTF
Edgefield, SC 29824 Fax: 803-637-0034
E-mail: nwtf@nwtf.net
http://www.nwtf.org
Magazine for all disabled people and their able-bodied partners who are interested in the outdoors, especially recreational shooting, hunting and fishing. Free with membership.
*Quarterly*
*Karen Roop, Editor*

3907 **Women in the Outdoors**
770 Augusta Road 803-637-3106
PO Box 530 800-THE-NWTF
Edgefield, SC 29824 Fax: 803-637-0034
E-mail: nwtf@nwtf.net
http://www.nwtf.org
Magazine that delivers features on a variety of outdoor topics of interest to the novice and experienced outdoorswoman. Free with membership.
*Quarterly*
*Karen Roop, Editor*

## Periodicals: Habitat Preservation & Land Use

3908 **ANJEC Report**
Association/New Jersey Environmental Commissions
PO Box 157 973-539-7547
Mendham, NJ 07945 Fax: 973-539-7713
E-mail: info@anjec.org
http://www.anjec.org
Nonprofit organization promoting public interest in natural resorce protection and supporting municipal environmental commissions throughout New Jersey.
*Publication Date: 1970 32 pages Quarterly*
*Sandy Batty, Executive Director*

3909 **Afield**
4705 University Drive 919-403-8558
Suite 290 Fax: 919-403-0379
Durham, NC 27707 E-mail: northcarolina@tnc.org
http://www.nature.org/northcarolina
Published by The Nature Conservancy, saving the last great places of North Carolina.
*Quarterly Newsletter*
*Katherine Skinner, Executive Director*

3910 **Aldo Leopold Foundation: The Leopold Outdoor**
E13701 Levee Road 608-355-0279
Baraboo, WI 53913 Fax: 608-356-7309
E-mail: mail@aldoleopold.org
http://www.aldoleopold.org
A nonprofit organization founded in 1982, works to promote the philosophy of Aldo Leopold and the land ethic he so eloquently defined in his writing. The foundation actively integrates programs on land stewardship, environmentaleducation and scientific research to promote care of natural resources and have an ethical relationship between people and land.
*Publication Date: 1999 8 pages Quarterly*
*Steve Swenson, Ecologist*
*Buddy Huffaker, Executive Director*

3911 **American Association for Advancement of Science: Animal Keeper's Forum**
1200 New York Avenue Northwest 202-326-6400
Washington, DC 20005 Fax: 202-289-4985
E-mail: webster@aaas.org
http://www.aaas.org
The American Association for the Advancement of Science is the world's largest general science and publisher of the peer-reviewed journal. With more than 138,000 members and 275, AAAS serves as an authoritative source for informationon the latest developments in science and bridges gaps among scientists, policy- makers and the public to advance science and science education.
*Publication Date: 1947 54 pages Monthly*
*ISSN: 0164-9531*
*Alan Leshner, Executive Director*

3912 **American Birding Association: Birding**
4945 N Street 719-578-9703
Suite 200 800-850-2473
Colorado Springs, CO 80919 Fax: 719-578-1480
E-mail: member@aba.org
http://www.aba.org
The American Birding Association represents the interests of birdwatchers in various arenas, and helps birders increase their knowledge, skills, and enjoyment of birding. ABA also contributes to bird conservation by linking the skillsof its members to on-the-ground projects. ABA promotes field-birding skills

through meetings, workshops, equipment, and guided involvement in birding, promoting national and international birders networks and publications.
*Publication Date: 1972 8 pages Bi Monthly*
*Rob Robinson, Executive Director*

**3913 American Birding Association: Winging It**
4945 N. 30th St. 719-578-9703
Suite 200 800-850-2473
Colorado Springs, CO 80919 Fax: 719-578-1480
E-mail: member@aba.org
http://www.americanbirding.org
The American Birding Association represents the interests of birdwatchers in various arenas, and helps birders increase their knowledge, skills, and enjoyment of birding. ABA also contributes to bird conservation by linking the skillsof its members to on-the-ground projects. ABA promotes field-birding skills through meetings, workshops, equipment, and guided involvement in birding, promoting national and international birders networks and publications.
*Publication Date: 1989 24 pages Bi Monthly*
*Steve Runnels, Executive Director*
*Rick Wright, Editor*

**3914 American Entomologist**
Journal of Entomology
10001 Derekwood Lane 301-731-4535
Suite 100 Fax: 301-731-4538
Lanham, MD 20706 E-mail: esa@entsoc.org
http://www.entsoc.org
American Entomologist is a quarterly, general interest entomology magazine written for both scientists and nonscientists. It publishes colorful, illustrated feature articles, peer-reviewed scientific reports, provocative and humorouscolumns, letters, book reviews, and obituaries.
*Publication Date: 1955 64 pages Quarterly*
*ISSN: 1046-2821*
*Gene Kristky, Editor-in-Chief*

**3915 American Forests Magazine**
PO Box 2000 202-955-4500
Washington, DC 20013 800-368-5748
Fax: 202-955-4588
E-mail: info@amfor.org
http://www.americanforests.org
Published by American Forests, the oldest national citizens conservation organization in the US.
*Quarterly*
*Gerald Gray, VP Forest Policy Center*
*Deborah Gangloff, Executive Director*

**3916 Annals of the Entomological Society of America**
Journal of Entomology
10001 Derekwood Lane 301-731-4535
Suite 100 Fax: 301-731-4538
Lanham, MD 20706 E-mail: esa@entsoc.org
http://www.entsoc.org
Contributions report on the basic aspects of the biology of anthropods and are divided into categories by subject matter; systematics; ecology and population biology; arthropods in relation to plant disease; conservation biology andbiodiversity; physiology, biochemistry, and toxicology; morphology, history, and fine sructure; genetics and behavior.
*Bi-Monthly*
*Joe B Keiper, Editor-in-Chief*

**3917 Annual Review of Entomology**
Annual Reviews
10001 Derekwood Lane 301-731-4535
Suite 100 Fax: 301-731-4538
Lanham, MD 20706 E-mail: esa@entsoc.org
http://www.entsoc.org
This is published in Januray and made available through ESA on a regular subscription basis. The series occupies a special place within the field of entomology. Authoritative critical reviews by eminent scientists provide a valuableresource for students, teachers, and researchers: specialists and nonspecialists.
*Yearly*
*Alan Kahan, Director Communications*

**3918 Appalachian Mountain Club**
Appalachian Mountain Club Books
5 Joy Street 617-523-0655
Boston, MA 02108 800-262-4455
Fax: 617-523-0722
E-mail: information@outdoors.org
http://www.outdoors.org
The AMC, founded in 1876, promotes the protection, enjoyment, and wise use of the mountains, rivers and trails of the Northeast. We encourage people to enjoy and protect the natural world because we believe that successful conservationdepends on this experience. The AMC publishes an award-winning magazine and more than 60 guide books to the Northeast.
*Andrew Falender, Executive Director*
*Clair O'Connell, Director Development*

**3919 Arthropod Management Tests**
Journal of Entomology
10001 Derekwood Lane 301-731-4535
Suite 100 Fax: 301-731-4538
Lanham, MD 20706 E-mail: esa@entsoc.org
http://www.entsoc.org
This is published in late spring. The purpose is to promote timely dissemination of information on preliminary and routine screening tests on management of arthropods, both beneficial and harmful. Pest management methods tested andreported may include the use of chemical pesticides as well as other materials or agents, such as insect growth regulators, pheromones, natural enemies for biological control, or pest-resistant plants/animals. Reports are based on tests conducted byreseachers.
*Yearly*
*David L Kerns, Editor-in-Chief*

**3920 Biodiversity Institute Newsletter**
University of Kansas
Dyche Hall 785-864-4540
Lawrence, KS 66045 Fax: 785-864-5335
A comprehensive research, graduate education and public service institution dedicated to biodiversity science and collections. Collections of more than 7 million plant and animal specimens, with particular strengths in neotropicalamphibians, great plains, flora, bees and antarctic plant fossils.
*Publication Date: 1978 6 pages Quarterly*
*Dr Leonard Krishtalka, Director*

**3921 Blowhole**
3625 Brigantine Boulevard 609-266-0538
PO Box 773 Fax: 609-266-6300
Brigantine, NJ 08203 E-mail: mmsc@verizon.net
http://marinemammalstrandingcenter.org
A quarterly newsletter published by the Marine Mammal Stranding Center.
*Publication Date: 1978 8 pages Quarterly*
*Robert C Schoelkopf, Director*
*Sheila M Dean, Co-Director*

**3922 Carolina Bird Club**
6325 Falls of the Neuse Road 910-791-5726
STE 9 PMB 150 Fax: 910-791-7228
Raleigh, NC 27615 E-mail: hq@carolinabirdclub.org
http://www.carolinabirdclub.org
A nonprofit educational and scientific association, open to anyone interested in the study and conservation of wildlife, particularly birds. Meets each winter, spring and fall. Meeting sites are selected to give participants anopportunity to see many different kinds of birds. Guided field trips, informative programs and business sessions are combined for an exciting weekend of meeting with people who share an enthusiasm and concern for birds.
*Publication Date: 1937 Quarterly*
*ISSN: 0009-1987*
*Dana Harris, HQ Secretary*
*Ken Fiala, Editor*

**3923 Coalition for Education in the Outdoors**
2217 Professional Studies Building 607-753-4971
SUNY Cortland Fax: 607-753-5982
Cortland, NY 13045 E-mail: info@outdooredcoalition.org
http://www.outdooredcoalition.org
A network of organizations, business, institutions, centers, agencies, and associations linked and communicating in support of the broad purpose of education in, for, and about the outdoors. Takes a board view of outdoor education andseeks not to duplicate or compete with the work of organizations, but to provide services not easily performed by other groups.
*Publication Date: 1987 48-52 pages Bi-Annual*
*ISSN: 1065-5204*
*Charles Yaple, Executive Director*

**3924 College of Tropical Agriculture and Human Resources: Impact Report**
University of Hawaii
3050 Maile Way 808-956-7056
Gilmore 119 Fax: 808-956-5966
Honolulu, HI 96822 E-mail: ocs@ctahr.hawaii.edu
http://www.ctahr.hawaii.edu
The vision of the college is to be the premier resource for tropical agricultural systems and resource management in the Asia-Pacific region. Its mission outlines a commitment to the preparation of students and all citizens of Hawaiifor life in a global community through research and education programs supporting tropical agricultural systems that foster viable communities, a diversified economy and a healthy environment.
*Annual*
*Andrew G Hashimoto, Dean/Director*

**3925 Connecticut Woodlands**
Middlefield 860-346-2372
16 Meriden Road Fax: 860-347-7463
Rockfall, CT 06481 E-mail: conn.forest.assoc@snet.net
http://www.ctwoodlands.org
Quarterly publication of the Connecticut Forest and Park Association, an organization for forest and wildlife conservation. Develops outdoor recreation and natural resources. Provides forest management, construction of hiking trailsand consultation in the areas of forestry and environment.
*Quarterly*
*Chris Woodside, Editor*

**3926 Conservancy of Southwest Florida**
Eye On The Issues
1450 Merrihue Drive 232-262-0304
Naples, FL 34102 Fax: 232-262-0672
E-mail: info@conservancy.org
http://www.conservancy.org
*8 pages Quarterly*
*Kathy Prosser, President/CEO*
*Sheila Etalamaki, Director Finance*

**3927 Conservation & Natural Resources: Water Resources Division, Nevada Wildlife Almanac**
901 S. Stewart Street 775-681-2800
Suite 2002 Fax: 775-684-2811
Carson City, NV 89701 E-mail: hricci@wr.state.nv.us
http://ndwr.state.nv.us
*Publication Date: 1996 12 pages Twice/year*
*Hugh Ricci, State Engineer*

**3928 Conservation Commission News**
New Hampshire Association of Conservation Comm.
54 Portsmouth Street 603-224-7867
Concord, NH 03301 E-mail: info@nhacc.org
http://www.nhacc.org
Encourage conservation and appropriate use of New Hampshire's natural resources by providing assistance to New Hampshire's municipal conservation commissions and by facilitating communication among commissions and between commissionsand other public and private agencies involved in conservation.
*8 pages*
*Marjory Swope, Publisher*

**3929 Conservation Communique**
Wyoming Association of Conservation Districts
517 East 19th Street 307-632-5716
Cheyenne, WY 82001 Fax: 307-638-4099
http://www.conservewy.com
*Quarterly*
*Kelly Brown, Editor*

**3930 Conservation Conversation**
Montana Association of Conservation Districts
PO Box 99 231-876-0328
Cadillac, MI 49601 Fax: 231-876-0372
E-mail: macd@macd.org
http://www.macd.org
*Quarterly*

**3931 Conservation Leader**
Utah Association of Conservation Districts
1860 North 100 East 435-753-6029
Logan, UT 84341 Fax: 435-755-2117
*Quarterly*

**3932 Conservation Notes: New England Wildflower Society**
180 Hemenway Road 508-877-7630
Framingham, MA 01701 508-877-6553
Fax: 508-877-3658
E-mail: information@newenglandwild.org
http://www.newenglandwild.org
*36 pages Yearly*
*Debbi Edelstein, Executive Director*
*Frances Clark, Chair, Board of Trustees*

**3933 Conservation Partner**
Arkansas Association of Conservation Districts
101 E Capitol 501-682-2915
Suite 350 Fax: 501-682-3991
Little Rock, AR 72201 http://www.aracd.org
*Quarterly*
*Debbie Moreland, Program Administrator*

**3934 Conservation Visions**
Nebraska Association of Resources Districts
601 S 12th Street 402-471-7670
Suite 201 Fax: 402-471-7677
Lincoln, NE 68508 E-mail: nard@nrdnet.org
http://www.nrdnet.org
*Bi-Monthly*
*Dean E Edson, Executive Director*

**3935 Conservogram**
Soil and Water Conservation Society
945 Southwest Ankeny Road 515-289-2331
Ankeny, IA 50023 Fax: 515-289-1227
E-mail: pubs@swcs.org
http://www.swcs.org
Published for the professionals in the natural resource fields, and contains highlights on the news and ideas in the preservation of natural resources.
*Monthly*
*Suzi Case, Editor*

**3936 Consultant**
Assoc of Consutling Foresters of America, Inc
312 Montgomery Street 703-548-0990
Suite 208 888-540-8733
Alexandria, VA 22314 Fax: 703-548-6395
E-mail: director@acf-foresters.org
http://www.acf-foresters.org
Publication from Association of Consulting Foresters of America, Inc.
*Publication Date: 1948*
*Lynn C Wilson, Executive Director*

**3937 Cornell Lab of Ornithology: Birdscope**
Birdscope

159 Sapsucker Woods Road
Ithaca, NY 14850
607-254-2451
Fax: 607-254-2415
E-mail: cornellbirds@cornell.edu
http://www.birds.cornell.edu

*Quarterly,Newsletter*
*Miyoko Chu, Communications Director*

### 3938 Cornell Lab of Ornithology: Living Bird

Birdscope
159 Sapsucker Woods Road
Ithaca, NY 14850
607-254-2475
Fax: 607-254-2415
E-mail: cornellbirds@cornell.edu
http://www.birds.cornell.edu

*Quarterly, Magazine*
*Miyoko Chu, Communications Director*

### 3939 Department of Natural Resources

301 Centennial Mall South
PO Box 94676
Lincoln, NE 68509
402-471-2363
Fax: 402-471-2900
E-mail: webmaster@dnr.state.ne.us
http://www.dnr.state.ne

*Quarterly*
*Roger K Patterson, Director*

### 3940 District Connection Newsletter

North Carolina Assoc. of Soil/Water Cons. Dist.
1614 Mail Service Center
Raleigh, NC 27699
919-733-2302
Fax: 919-715-3559
http://www.enr.state.nc.us\DSWC

*Monthly*
*David Williams, Acting Director*

### 3941 E Magazine

Doug Moss
28 Knight St
Norwalk, CT 06851
203-854-5559
Fax: 203-866-0602
E-mail: jessica@emagazine.com
http://www.emagazine.com

A comprehensive magazine dealing with environmental issues and national conservation concerns.
*Publication Date: 1990 64 pages Bimonthly*
*Doug Moss, Publisher/Executive Director*

### 3942 ESA Newsletter

Journal of Entomology
10001 Derekwood Lane
Suite 100
Lanham, MD 20706
301-731-4535
Fax: 301-731-4538
E-mail: esa@entsoc.org
http://www.entsoc.org

ESA Newsletter is a monthly publication presenting timely information of interest to ESA members. In addition to feature articles, it contains meeting announcements, society business, listing of employment oppurtunities, notice ofgrants and awards, member profiles, and branch and section news.
*Monthly*
*Lisa Spurlock, Editor*

### 3943 Eagle Nature Foundation Ltd: Bald Eagle News

Bald Eagle News
300 E Hickory
Apple River, IL 61001
815-594-2306
Fax: 815-594-2305
E-mail: eaglenature.tni@juno.com
http://eaglenature.com

A quarterly publication from the Eagle Nature Foundation.
*Publication Date: 1992 20 pages Quarterly*
*Terrence N Ingram, Executive Director*

### 3944 Earth Steward

Michigan Association of Conservation Districts
3001 Coolidge Road
Suite 250
East Lansing, MI 48823-6362
517-324-4421
Fax: 517-324-4435
E-mail: macd@macd.org
http://www.macd.org

Quarterly membership newsletter
*Quarterly*
*Mike Lawless, President MACD State Council*
*Lori Phalen, Executive Director*

### 3945 Ecosphere

Forum International
91 Gregory Lane
Suite 21
Pleasant Hill, CA 94523
925-997-1864
800-252-4475
Fax: 925-946-1500
E-mail: fti@foruminternational.com
http://www.foruminternational.com

Accepts advertising. The first ever environmental/ecological magazine. It is dedicated to the interrelations of man in nature and a balanced approach of its biological, economic, socio-political and spiritual components. Since 1965.
*Publication Date: 1965 16-48 pages Bi Monthly*
*Dr. Nicolas Hetzer, Production Manager*
*J McCormack, Circulation Director*

### 3946 Elm Leaves

Elm Research Institute
11 Kit Street
Keene, NH 03431
603-358-6198
800-367-3567
Fax: 603-358-6305
E-mail: libertyelm@webryders.com
http://www.libertyelm.com

A semi annual publication published by the Elm Research Institute. Free with membership.
*Publication Date: 1967 4-9 pages Semi-Annually*
*John Hansel, Editor/Executive Director*
*Yvonne Spalthoff, Assistant Director*

### 3947 Environmental Concern: The Wonders of Wetlands, an Educators Guide

POW-The Planning of Wetlands
201 Boundary Lane
PO Box P
St Michaels, MD 21663
410-745-9620
Fax: 410-745-3517
E-mail: order@wetland.org
http://www.wetland.org

Since its founding in 1972, EC has been specializing in consulting, planning design, education services, construction services and research related to all aspects of wetlands. As wetlands and contiguous upland forests and meadows areinteracting ecosystems EC specializes in consulting, planning, design, and project supervision services for such upland ecosystem constructions and restorations for the purpose of wetland buffers, reforestation, wildlife habitat and critical areas ofpreservation.
*Publication Date: 1995 Quarterly*
*ISSN: 1095-2063*
*Suzanne Pittengar Slear, President*

### 3948 Environmental Entomology

Journal of Entomology
10001 Derekwood Lane
Suite 100
Lanham, MD 20706
301-731-4535
Fax: 301-731-4538
E-mail: esa@entsoc.org
http://www.entsoc.org

Contributions report on the interaction of insects with biological, chemical, and physiological and chemical ecology (abiotic effects, pheromonea, effects of miscellaneous pollutants), community/ecosystem ecology (trophic-levelsstudies, associations), population ecology (mating, reproduction, movement, behavior, parasitism, predation, microbial ecology, insect-plant relations), pest management and sampling (integrated pest management, sampling, distribution), and biologicalcontrol.
*Bi-Monthly*
*E Alan Cameron, Editor-in-Chief*

### 3949 Everglades Reporter

Friends of the Everglades
7800 Red Road
Suite 215K
South Miami, FL 33143
305-669-0858
Fax: 305-669-4108
E-mail: info@everglades.org
http://www.everglades.org

Protecting the Everglades. A bi-annual publication from Friends of the Everglades.
*Publication Date: 1971 8 pages Quarterly*
*David Reiner, President*

**3950 Fisheries**
American Fisheries Society
5410 Grosvenor Lane
Bethesda, MD 20814
301-897-8616
Fax: 307-897-8096
E-mail: main@fisheries.org
http://www.fisheries.org
Peer-reviewed articles that address contemporary issues and problems, techniques, philosophies and other areas of interest to the general fisheries profession. Monthly features include letters, meeting notices, book listings andreviews, environmental essays and organization profiles.
*Kristin Merriman-Clarke, Editor*

**3951 Fisheries Focus: Atlantic**
Atlantic States Marine Fisheries Commission
1444 I Street Northwest
6th Floor
Washington, DC 20005
202-289-6400
Fax: 202-289-6051
E-mail: comments@asmfc.org
http://www.asmfc.org
*12 pages Monthly*
*John V O'Shea, Executive Director*
*Tina L Berger, Public Affairs/Resource Spec*

**3952 Forest History Society**
701 William Vickers Avenue
Durham, NC 27701
919-682-9319
Fax: 919-682-2349
E-mail: coakes@duke.edu
http://www.foresthistory.org
The Forest History Society is a non-profit educational institution that links the past to the future by identifying, collecting, preserving, interpreting, and disseminating information on the history of people, forests, and theirrelated resources.
*Publication Date: 1946*
*Cheryl Oakes, Librarian*
*Steven Anderson, President*

**3953 Forest Voice**
Native Forest Council
PO Box 2190
Eugene, OR 97402
541-688-2600
Fax: 541-461-2156
E-mail: info@forestcouncil.org
http://www.forestcouncil.org
Quarterly publication from Native Forest Council.
*Publication Date: 1989 16 pages Quarterly*
*ISSN: 1069-2002*
*Timothy Hermach, President*

**3954 Forestry Source**
Society of American Foresters
5400 Grosvernor Lane
Bethesada, MD 20814
301-897-8720
866-897-8720
Fax: 301-897-3690
E-mail: safweb@safnet.org
http://www.safnet.org
Tabloid newsletter covering important information regarding critical issues in forestry research and technology, legislative updates and news about SAF programs and activities on a national and local level.
*20 pages Monthly*
*Matt Walls, Editor*

**3955 Game & Fish Commission Wildlife Management Division Newsletter**
2 Natural Resources Drive
Little Rock, AR 72205
501-223-6300
800-364-4263
Fax: 501-223-6452
http://www.agfc.com
Dedicated to managing wildlife in the state of Arkansas.
*Publication Date: 1920 33-35 pages 5x year*
*Doyle Shook, Chief*

**3956 Golden Gate Audubon Society**
2530 San Pablo Avenue
Suite G
Berkeley, CA 94702
510-843-2222
Fax: 510-843-5351
E-mail: ggas@goldengateaudubon.org
http://www.goldengateaudubon.org
Monthly publication from Golden Gate Audubon Society. Published 10 times yearly.
*Publication Date: 1917 12 pages Monthly*
*ISSN: 0164-971x*
*Mark Welther, Executive Director*

**3957 Great Plains Native Plant Society Newsletter**
PO Box 641
Hot Springs, SD 57747
605-745-3397
Fax: 605-745-3397
E-mail: cascade@gwtc.net
The Society's mission is to engage in scientific research regarding plants of the Great Plains of North America; to disseminate this knowledge through the creation of one or more educational botanical garders of such flora, featuringbut not limited to Barr's discoveries; and to engage any educational activities which may further public familiarity with the plants of the Great Plains, their uses and enjoyment.
*Publication Date: 1984 4-8 pages Intermittant*
*Cynthia Reed, President*

**3958 Green Space**
New York Parks and Conservation Association
29 Elk Street
Albany, NY 12207
518-434-1583
Fax: 518-427-0067
E-mail: ptny@ptny.org
http://www.nypca.org
*Semi-Annual*
*Robin Dropkin, Executive Director*

**3959 Habitat Hotline**
Atlantic States Marine Fisheries Commission
1444 I Street Northwest
6th Floor
Washington, DC 20005
202-289-6400
Fax: 202-289-6051
E-mail: info@asmfc.org
http://www.asmfc.org
*6 pages Quarterly*
*John V O'Shea, Executive Director*
*Tina L Berger, Public Affairs/Resource Spec*

**3960 Illinois Audubon Society**
PO Box 2547
Springfield, IL 62708
217-544-BIRD
Fax: 217-544-7433
E-mail: ias@illinoisaudubon.org
http://www.illinoisaudubon.org
A membership organization dedicated to the preservation of Illinois Wildlife and the habitats which support them. Has sanctuaries, conservation education and land acquisition programs and publishes quarterly magazines and newsletters.
*Publication Date: 1916 28 pages Quarterly*
*ISSN: 1061-9801*

**3961 Illinois Environmental Council: IEC Bulletin**
107 W Cook Street
Suite E
Springfield, IL 62704
217-544-5954
Fax: 217-544-5958
E-mail: iec@ilenviro.org
http://www.ilenviro.org
The IEC is a coalition of over 70 environmental, conservation and health groups.
*Bi-Monthly*
*Jonathan Goldman, Executive Director*
*Jennifer Sublett, Outreach Coordinator*

**3962 In Brief**
223 S King Street
Suite 400
Honolulu, HI 96813
808-599-2436
Fax: 808-521-6841
E-mail: eajushi@earthjustice.org
http://www.earthjustice.org
Newsletter of Earthjustice, a nonprofit public interest law firm dedicated to protecting the magnificent places, natural resources and wildlife of this earth and to defending the right of all people to a healthy environment. We bringabout far-reaching change by enforcing and strengthening environmental laws on behalf of hundreds of organizations and communities.
*Quarterly*
*Douglas Hannold, Managing Attorney*

**3963 Iowa Cooperative Fish & Wildlife Research Unit: Annual Report**
Iowa State University
Science Hall II
Ames, IA 50011
515-294-3056
Fax: 515-294-5468
*Publication Date: 1932 50 pages Annual*

**3964 Iowa Native Plant Society Newsletter**
Iowa State University
Botany Department
Ames, IA 50011
515-294-9499
Fax: 515-294-1337
E-mail: dlewis@iastate.edu
An organization of amateur and professional botanists and native plant enthusiasts who are interested in the scientific, educational and cultural aspects, as well as the preservation and conservation of the native plants of Iowa. TheSociety was organized in 1995 to create a forum where plant enthusiasts, gardners and professional botanists could exchange ideas and coordinate activities such as field trips, work shops, and restoration of natural areas.
*Publication Date: 1995 12 pages 3/4 times X year*
*Tom Rosburg, President*
*Deb Lewis, Contact Person*

**3965 Journal of Caves & Karst Studies**
National Speleological Society
2813 Cave Avenue
Huntsville, AL 35810
256-852-1300
Fax: 256-851-9241
E-mail: nss@caves.org
http://www.caves.org
A quarterly journal published by the National Speleological Society.
*Quarterly*
*ISSN: 1090-6924*
*Stephanie Searles, Operations Manager*
*Dave Bunnell, Editor*

**3966 Journal of Economic Entomology**
Journal of Entomology
10001 Derekwood Lane
Suite 100
Lanham, MD 20706
301-731-4535
Fax: 301-731-4538
E-mail: esa@entsoc.org
http://www.entsoc.org
Contributions report on the economic significance of insects and are divided into categories by subject matter: apiculture and social insects; arthropods in relation to plant disease; biological and microbial disease; ecology andbehavior; ecotoxicology; extension; field and forage crops; forest entomology; horticultural entomology; household and structural insects; insecticide resistance and resistance management; medical entomology; plant resistance; sampling andbiostatistics.
*Bi-Monthly*
*John T Trumble, Editor-in-Chief*

**3967 Land Use Law Report**
Business Publishers
PO Box 17592
Baltimore, MD 21297
301-587-6300
800-274-6737
Fax: 301-587-4530
E-mail: custserv@bpinews.com
http://www.bpinews.com
Provides timely news on court decisions, legislation and regulations that impact today's most pressing land-use policy planning and legal issues.
*Biweekly*
*James D Lawlor, Author*
*Adam Goldstein, Publisher*
*James D Lawlor, Editor*

**3968 Land and Water Magazine**
Land and Water
320 A Street
Fort Dodge, IA 50501
515-576-3191
Fax: 515-576-2606
E-mail: landandwater@frontiernet.net
http://www.landandwater.com
Edited for contractors, engineers, architects, government officials and those working in the field of natural resource management and restoration from idea stage through project completion and maintenance.
*Publication Date: 1974 72 pages Bimonthly*
*ISSN: 0192-9453*
*Amy Dencklau, Publisher/Editor*

**3969 Leaves Newsletter**
Michigan Forest Association
6120 S. Clinton Trail
Eaton Rapids, MI 48827
517-663-3423
E-mail: mfa@i-star.com
http://www.michiganforests.com
A monthly publication from the Michigan Forest Association.
*Monthly*
*McClain B Smith Jr, Executive Director*

**3970 MACC Newsletter**
Alba Press
10 Juniper Road
Belmont, MA 02478
617-489-3930
Fax: 617-489-3935
E-mail: staff@maccweb.org
http://www.maccweb.org
Published six times a year, each issue features carefully chosen technical and interpreative articles, updates on government actions and policies, notices of workshops and meetings, publications, listings and a professional directory.
*16 pages Bi-Monthly*
*Lindsay Martucci, Editor*

**3971 Michigan Forests Magazine**
Michigan Forest Association
1558 Barrington
Ann Arbor, MI 48103
734-665-8279
Fax: 734-913-9167
E-mail: mfa@i-star.com
http://www.michiganforests.com
A quarterly magazine published by the Michigan Forest Association.
*Quarterly*
*McClain B Smith Jr, Executive Director*

**3972 Minnesota Plant Press**
Minnesota Native Plant Society
220 Biological Science Center
1445 Gortner Avenue
Saint Paul, MN 55108
E-mail: president@mnnps.org
http://www.mnnps.org
A nonprofit organization dedicated to the conservation of the native plants of Minnesota through public education and advocacy. Offered are monthly meetings, field trips, symposia and a regular newsletter.
*4 per year*
*Jason Husveth, President*
*Gerry Drewry, Editor*

**3973 Monitor**
Florida Defenders of the Environment
4424 NW 13 Street
Suite C-8
Gainesville, FL 32609
352-378-8465
Fax: 352-377-0869
E-mail: fde@fladefenders.org
http://www.fladefenders.org
Newsletter of Florida Defenders of the Environment, one of the oldest and most accomplished conservation organizations in Florida with a network of scientists, economists and other professionals dedicated to preserving and protectingthe state's natural resources. FDE's top priority is currently the restoration of a 16-mile stretch of the Ocklawaha River and its 9,000-acre floodplain forest by removal of Rodman Dam- the last vestige of the Cross-Florida Barge Canal.
*Publication Date: 1982 6-8 pages Bi-Annually*
*Nick Williams, Executive Director*

**3974 Montana Land Reliance Newsletter**
324 Fuller Avenue
PO Box 355
Helena, MT 59624
406-443-7027
Fax: 406-443-7061
E-mail: info@mtlandreliance.org
http://www.mtlandreliance.org
Montana's only private, statewide land trust, an apolitical, nonprofit corporation. Our mission is to provide permanent protection for private lands that are ecologically significant for

agricultural production, fish and wildlifehabitat and scenic open space. We publish a newsletter twice per year.
*8 pages Spring/Fall*
*Jay Erickson, Managing Director*
*Rock Ringling, Managing Director*

**3975 NACD News & Views**
National Association of Conservation Districts
509 Capitol Court, NE 202-547-6223
Washington, DC 20002 Fax: 202-547-6450
http://www.nacdnet.org
Newsletter of the nonprofit organization that represents the nation's 3,000 conservation districts and 17,000 men and women who serve on their governing boards. Conservation districts, local units of government established under statelaw to carry out natural resource management programs at the local level, work with more than 2.5 million cooperating landowners and operators to help them amange and protect land and water resources on nearly 98% of the private lands in the UnitedStates.
*Publication Date: 1952 8 pages Bi-Monthly*
*Maxine Mathis, Production Manager*

**3976 National Gardener Magazine**
National Garden Clubs, Inc
4401 Magnolia Avenue 314-776-7574
St Louis, MO 63110 800-550-6007
Fax: 314-776-5108
E-mail: headquarters@gardenclub.org
http://www.gardenclub.org
National Garden Clubs, Inc publishes The National Gardener Magazinequarterly.
*Publication Date: 1970 48 pages Quarterly*
*ISSN: 0027-9331*
*Susan Davidson, Author*
*Susan Davidson, Editor*

**3977 National Grange Newsletter**
1616 H Street Northwest 202-628-3507
Washington, DC 20006 888-447-2643
Fax: 202-347-1091
E-mail: rweiss@nationalgrange.org
http://www.nationalgrange.org
The Grange is a family based community organization with a special interest in agriculture and rural america as well as in legislative efforts regarding these issues.
*6 pages Bi-Monthly*
*William Steele, President*
*Richard Weiss, COO*

**3978 National Recreation and Park Association, (NRPA): Parks & Recreation Magazine**
Parks and Recreation Magazine
22377 Belmont Ridge Road 703-858-0784
Ashburn, VA 20148 Fax: 703-858-0794
E-mail: info@nrpa.org
http://www.nrpa.org
The NRPA, headquartered in Ashburn Virginia, is a national nonprofit organization devoted to advancing park, recreation and conservation efforts that enhance the quality of life for all Americans. The Association works to extendsocial, health, cultural and economic benefits of parks and recreation, through its network of 23,000 recreation and park professionals and civic leaders. NRPA encourages recreation initiatives for youth in high-risk environments.
*100 pages Monthly*
*ISSN: 0031-2215*
*John Thorner, Executive Director*
*Rachel Roberts, Editor*

**3979 National Wildlife Magazine**
National Wildlife Federation
11100 Wildlife Center Drive
Reston, VA 20190 800-822-9919
http://www.nwf.org
The official member magazine of the National Wildlife Federation.
*6 Issues/Yr*
*Larry J Schweiger, President/CEO*
*Mark Wexler, Editorial Director*

**3980 National Woodlands Magazine**
National Woodland Owners Association
374 Maple Avenue E 703-255-2700
Suite 310 800-476-8733
Vienna, VA 22180 Fax: 703-281-9200
E-mail: argow@nwoa.net
http://www.nationalwoodlands.org
Provides timely information about forestry and forest practices with news from Washington, DC and state capitals. Written for non-industrial, private woodland owners. Includes state landowner association news.
*28 pages Quarterly*
*Keith A Argow, Publisher*
*Eric Johnson, Editor*

**3981 Natural Resources Department: Fish & Wildlife Newsletter**
402 West Washington Street RMW273 317-232-4080
Indianapolis, IN 46204 Fax: 317-232-8150
http://www.state.in.us/dnr/fishwild/index.htm
*Publication Date: 1985 12 pages Quarterly*

**3982 Nature Conservancy: Nebraska Chapter Newsletter**
1025 Leavenworth Street 402-342-0282
Omaha, NE 68102 Fax: 402-342-0474
E-mail: nebraska@tnc.org
http://nature.org
The mission of the Nature Conservancy is to preserve the plants, animals and natural communities that represent the diversity of life on Earth by protecting the lands and waters they need to survive.
*Publication Date: 1987 12 pages Quarterly*
*Jill Jeffrey, Donor Relations Manager*

**3983 Nature's Voice, The DNS Online Newsletter**
Delaware Nature Society
PO Box 700 302-239-2334
Hockessin, DE 19707 Fax: 302-239-2473
E-mail: dnsinfo@delawarenaturesociety.org
http://www.delawarenaturesociety.org
Now only available online, the newsletter offers trail information, nature center updates and programs, volunteer opportunities and more about what's happening in the outdoors of Delaware.
*Michael Riska, Executive Director*

**3984 New England WildFlower: New England Wildflower Society**
180 Hemenway Road 508-877-7630
Framingham, MA 01701 Fax: 508-877-3658
E-mail: information@newenglandwild.org
http://www.newenglandwild.org
*36 pages Twice a Year*
*Debbi Edelstein, Executive Director*
*Frances Clark, Chair, Board of Trustees*

**3985 New Jersey Environmental Lobby News**
204 W State Street 609-396-3774
Trenton, NJ 08608 Fax: 609-396-4521
E-mail: njelcurtis@aol.org
http://www.njenvironment.org
Quarterly publication from New Jersey Environmental Lobby.
*Publication Date: 1971 8 pages Quarterly*
*ISSN: 1535-2021*
*Anne Poole, President*
*Marie A Curtis, Executive Director*

**3986 North Dakota Association of Soil Conservation Districts Newsletter**

3310 University Drive
Bismarck, ND 58504
701-223-8518
Fax: 701-223-1291
E-mail: gpuppe@tic.bisman.com
http://www.ndascd.org

*Quarterly*
*Gary Puppe, Executive VP*

**3987 Outdoors Unlimited**

121 Hickory Street
Suite 1
Missoula, MT 59801
406-728-7434
Fax: 406-728-7445
E-mail: owaa@montana.com
http://www.owaa.org

Magazine of the Outdoor Writers Association of America. Membership fees are $175.00 (individual), $325.00 (supporting), $40.00 (student)
*Publication Date: 1962 35 pages Monthly*
*Kevin Rhoades, Executive Director*

**3988 Pacific Fishery Management Council Newsletter**

7700 NE Ambassador Place
Suite 200
Portland, OR 97220
503-820-2280
866-806-7204
Fax: 503-820-2299
E-mail: Donald.McIsaac@noaa.gov
http://www.pcouncil.org

*24 pages 5x Year*
*Donald McIsaac, Executive Director*
*John Coon, Deputy Director*

**3989 Parks and Trails Council of Minnesota: Newsletter**

275 E 4th Street
Suite 250
Saint Paul, MN 55101
651-726-2457
800-944-0707
Fax: 651-726-2458
E-mail: info@parksandtrails.org
http://www.parksandtrails.org

Mission: To acquire, protect and enhance critical lands for the public's enjoyment now and in the future.
*Quarterly*
*Judith Erickson, Government Relations*
*Beth Coleman, Executive Director*

**3990 Powder River Basin Resource Council: Powder River Breaks Newsletter**

934 North Main
Sheridan, WY 82801
307-672-5809
Fax: 307-672-5800
E-mail: resources@powderriverbasin.org
http://www.powderriverbasin.org

Committed to the preservation and enrichment of Wyoming's agricultural heritage and rural lifestyle; the conservation of Wyomings unique land, mineral, water and clean air resources, consistent with responsible use of those resourcesto sustain the vitality of present and future generations; the education and empowerment of Wyoming's citizens to raise a coherent voice in decisions. They are the only group in Wyoming that addresses both agricultural and conservation issues.
*Publication Date: 1973 8-12 pages 6x year*
*Jillian Malone, Editor*
*Stephanie Avey, Assistant Editor*

**3991 Prairie Naturalist Magazine**

600 Park Street
Department of Biological Sciences
Hays, KS 67601
785-628-4214
Fax: 316-341-5607
E-mail: efinck@fhsu.edu
http://www.fhsu.edu

Published by the North Dakota Natural Science Society, a regional organization with interests in the natural history of grasslands and the Great Plains.
*Publication Date: 1968 260 pages Quarterly*
*ISSN: 0091-0376*
*Elmer J Finck, Editor*

**3992 Reef Line**

Reef Relief Environmental Center
PO Box 430
Key West, FL 33041
305-294-3100
Fax: 305-293-9515
E-mail: info@reefrelief.org
http://www.reefrelief.org

Reef Line is a quarterly publication from Reef Relief.
*Publication Date: 1986 16 pages Quarterly*
*Michael Blades, Project Director*
*DeeVon Quirdo, Executive Director*

**3993 SWOAM News**

Small Woodland Owners Association of Maine
153 Hospital Street
PO Box 836
Augusta, ME 04332
207-626-0005
877-467-9626
Fax: 207-626-7992
E-mail: info@swoam.com
http://www.swoam.com

*Monthly*
*Tom Doak, Executive Director*

**3994 Save San Francisco Bay Association: Watershed Newsletter**

350 Frank H Ogawa Plaza
Suite 900
Oakland, CA 94612
510-452-9261
Fax: 510-452-9266
E-mail: SAVEBAY@savesfbay.org
http://www.savesfbay.org

Save the Bay has worked for over 40 years to protect the San Francisco Bay-Delta from pollution, fill, shoreline destruction and fresh water diversion. We have launched a century of renewal to restore bay fish and wildlife, reclaimtidal wetlands and make the bay safe and accessible to all.
*8-10 pages 3-4 times/year*
*David Lewis, Executive Director*

**3995 Scenic America Newsletter**

Scenic America
1250 Eye Street NW
Suite 750
Washington, DC 20005
202-638-1839
Fax: 202-638-3171
E-mail: tracy@scenic.org
http://www.scenic.org

*12 pages 3 Times Per Year*
*Mary Tracy, President*
*Peggy Lint, Office Manager*

**3996 Shore and Beach**

5460 Beaujolais Lane
Fort Myers, FL 33919
239-489-2616
Fax: 239-489-9917
E-mail: ExDir@asbpa.org
http://www.asbpa.org

A quarterly publication from the American Shore and Beach Preservation Association.
*24 pages Quarterly*
*ISSN: 0037-4237*
*Ken Gooderham, Executive Director/Editor*
*Kate Gooderham, Executive Director/Editor*

**3997 Sierra Club, NJ Chapter: The Jersey Sierran**

139 W Hanover Street
Trenton, NJ 08618
609-656-7612
Fax: 609-656-7618
E-mail: webmaster@sierraactivist.org
http://www.sierraactivist.org

The Sierra Club is our country's oldest and most effective grassroots environmental organization. Hikes and outings are scheduled throughout the year. We are dedicated to fighting sprawl and over-development.
*Publication Date: 1992 14 pages Quarterly*
*Jeff Tittle, Director*
*Dennis Schvejda, Conservation Coordinator*

**3998 Sierra Club: Pennsylvania Chapter Newsletter**

600 North 2nd Street
Suite 300A
Harrisburg, PA 17101
717-232-0101
Fax: 717-238-6330
E-mail: sierraclub.pa@paonline.com
http://www.sierraclub.org/chapter/pa/

The Pennsylvania chapter includes 11 local Sierra Club groups. Emphasis is on state environmental policy advocacy, outings, education and local environmental protection efforts.
*18 pages Quarterly*
*Jeff Schmidt, Sr. Chapter Director*

**3999 Southern Appalachian Botanical Society: Gastanea**
Newberry College
2100 College Street 803-321-5257
Newberry, SC 29108 Fax: 803-321-5636
E-mail: chorn@newberry.edu
http://www.newberrynet.com/sabs/
This is a professional organization for those interested in botanical research, especially in the areas of ecology, floristics and systematics. To this end, we publish a journal, CASTANEA, and a newsletter, CHINQUAPIN.
*Publication Date: 1936 350 pages Quarterly*
*ISSN: 0008-7475*
*Michael E Held, PhD, President*
*Charles Horn, Treasurer*

**4000 Terrain Magazine**
2530 San Pablo Avenue 510-548-2220
Berkeley, CA 94702 Fax: 510-548-2240
E-mail: info@ecologycenter.org
http://www.ecologycenter.org
A quarterly magazine published by the Ecology Center of Berkeley, CA.
*Publication Date: 1971 39 pages Quarterly*
*ISSN: 1526-8322*
*Linnea Due, Editor-in-Chief*

**4001 Tidbits Newsletter**
Minnesota Association/Soil and Water Cons. Dist.
790 Cleveland Avenue S 651-690-9028
Suite 216 Fax: 651-690-9065
St. Paul, MN 55075 E-mail: leann.buck@maswcd.org
http://www.maswcd.org
*Quarterly*
*Le Ann Buck, Executive Director*
*Sheila Vanney, Editor*

**4002 Tide**
Coastal Conservation Association
4801 Woodway Drive 713-626-4234
Houston, TX 77056 800-201-FISH
E-mail: ccantl@joincca.org
TIDE is the official bimonthly magazine of the Coastal Conservation Association. It has received local, state and national acclaim for writing, photography and layout and currently boasts a circulation of more than 70,000. TIDE isavailable only to members of the Coastal Conservation Association.

**4003 Upper Mississippi River Conservation Committee Newsletter**
4469 48th Avenue Court 309-793-5800
Rock Island, IL 61201 Fax: 309-793-5804
E-mail: umrcc@mississippi-river.com/umrcc
http://mississippi-river.com/umrcc
A bimonthly newsletter published by the Upper Mississippi River Conservation Committee.
*10 pages Bi monthly*
*Mike McGhee, Chairman*

**4004 Urban Land Magazine**
Urban Land Institute
1025 Thomas Jefferson Street NW 202-624-7000
Suite 500 W 800-321-5011
Washington, DC 20007 Fax: 202-624-7140
http://www.uli.org
Nonprofit research and education organization dedicated to improving land use policy and development practice. Publishes a monthly magazine, several quarterly publications and books. Topics relate to real estate development includinggovernment sensitive development, smart growth, sustainable development and city parks.
*Monthly*
*Kristina Kessler, Chief Editor*

**4005 Utah Geological Survey: Survey Notes**
1594 W N Temple Suite 3110 801-537-3300
PO Box 146100 Fax: 801-537-3400
Salt Lake City, UT 84114 http://www.ugs.state.ut.us
The Utah Geological Survey is an applied scientific agency that creates, interprets and provides information about Utah's geologic environment, resources and hazards to promote safe, beneficial and wise use of land. This is theirpublication, which is issued three times yearly.
*Publication Date: 1964 3 Times Yearly*
*ISSN: 1061-7930*
*Richard Allis, Director*

**4006 Virginia Forests Magazine**
Virginia Forestry Association
3808 Augusta Avenue 804-278-8733
Richmond, VA 23255 Fax: 804-320-1447
E-mail: vafa@erols.com
http://www.vaforestry.org
Quarterly magazine published by the Virginia Forestry Association.
*Quarterly*
*Paul Howe, VFA Executive Vice President*

**4007 WAEE Bulletin**
Wisconsin Association for Environmental Education
8 Nelson Hall 715-346-2796
University of Wisconsin E-mail: waee@uwsp.edu
Stevens Point, WI 54481
*Quarterly*
*Carol Weston, Administrative Assistant*

**4008 West Virginia Forestry Association Newsletter**
PO Box 718 304-372-1955
Ripley, WV 25271 888-372-9663
Fax: 304-372-1957
E-mail: wvfa@wvadventures.net
http://www.wvfa.org
*Monthly*
*Richard Waybright, Executive Director*

**4009 Western Pennsylvania Conservancy Newsletter**
209 4th Avenue 412-288-2777
Pittsburgh, PA 15222 866-JOI-NWPC
Fax: 412-281-1792
E-mail: wpc@paconserve.org
http://www.paconserve.org
WPC, working together to save the places we care about, protects natural lands, promotes healthy and attractive communities, and preserves Fallingwater, Frank Lloyd Wright's masterwork in Mill Run, which was entrusted to theConservancy in 1963. Since its inception in 1932, the Conservancy has protected more than 280,000 acres of natural lands in Pennnsylvania. We continue to work to secure lands of ecological significance that frequently offer recreational and scenicvalues.
*16 pages Quarterly*
*Larry Schweiger, President*

**4010 Western Proceedings Newsletter**
Western Association of Fish and Wildlife Agencies
5400 Bishop Boulevard 307-777-4569
Cheyenne, WY 82006 Fax: 307-777-4699
E-mail: ikruck@state.wy.us
*Annual*

**4011 Western Society of Naturalists: Newsletter**
San Diego State University Biology Department
5500 Campanile Street 818-677-3256
San Diego, CA 92182 Fax: 818-677-2034
http://www.wsn-online.org
*Mark Carr, President*
*Ralph Larson, President Elect*

**4012 Wetlands in the United States**
American Ground Water Trust
50 Pleasant Street Ste 2
Concord, NH 03301
603-228-5444
Fax: 603-228-6557
E-mail: trustinfo@agwt.org
http://www.agwt.org

*15 pages Quarterly*
*Andrew Stone, Director*

**4013 Whalewatcher**
PO Box 1391
San Pedro, CA 90733
310-548-6279
Fax: 310-548-6950
E-mail: info@acsonline.org
http://www.acsonline.org

A bi-annual publication from the American Cetacean Society. Cost included with membership fees.
*Publication Date: 1967 30 pages Bi-Annual*
*Diane Alps, Administrative Assistant*

**4014 Wilderness Education Association**
900 E 7th Street
Bloomington, IN 47405
812-855-4095
Fax: 812-855-8697
E-mail: wea@indiana.edu
http://www.weainfo.org/

*Publication Date: 1976 4-6 pages 3 Times Per Year*

**4015 Wildfowl Trust of North America: Newsletter**
Wildfowl Trust of North America
600 Discovery Lane
PO Box 519
Grasonville, MD 21638
410-827-6694
Fax: 410-827-6713
E-mail: cbec@cbec-wtna.org
http://www.cbec-wtna.org

Published by the Wildfowl Trust of North America.
*Publication Date: 1995 8 pages Qaurterly*
*Judy Wink, Executive Director*
*Sharyn B Harlow, Executive Admin. Assistant*

**4016 Wildlife Law News Quarterly and Weekly Alerts**
University of New Mexico School of Law
MSC 11 6060rd NE
One University of New Mexico
Albuquerque, NM 87131
505-277-5006
Fax: 505-277-7064
E-mail: musgrave@unm.edu
http://wildlifenews.unm.edu

A quarterly publication from the New Mexico Center for Wildlife Law.
*Publication Date: 1993 16 pages Quarterly*
*ISSN: 1085-7338*
*R Musgrave, Editor*
*D Macke, Editor*

**4017 Wildlife Society Bulletin**
Wildlife Society
5410 Grosvenor Lane
Suite 200
Bethesda, MD 20814
301-897-9770
Fax: 301-530-2471
E-mail: tws@wildlife.org
http://www.wildlife.org

A quarterly publication from the Wildlife Society.
*Publication Date: 1973 Quarterly*
*ISSN: 0091-7648*
*Warren Ballard, Editor*

**4018 Woodland Management Newsletter**
Wisconsin Woodland Owners Association
PO Box 285
Stevens Point, WI 54481
715-346-4798
Fax: 715-346-4821
E-mail: nbozek@uwsp.edu
http://www.wisconsinwoodlands.org

*Quarterly*
*Tim Eisele, Editor*

**4019 Woodland Report**
National Woodland Owners Association
374 Maple Avenue E
Suite 310
Vienna, VA 22180
703-255-2700
800-476-8733
Fax: 703-281-9200
E-mail: argow@nwoa.net
http://www.nationalwoodlands.org

Provides timely information about forestry and forest practices with news from Washington, DC and state capitals. Written for non-industrial, private woodland owners. Includes state landowner association news.
*2 pages*
*Keith A Argow, Publisher*
*Eric Johnson, Editor*

**4020 World Wildlife Fund: US Focus**
1250 24th Street NW
Suite 500
Washington, DC 20037
202-293-4800
http://www.worldwildlife.org

WWF projects.
*8 pages*
*Pat Sullivan, Publisher*

## Periodicals: Recycling & Pollution Prevention

**4021 AARA Newsletter**
Arizona Automotive Recyclers Association
1030 E Baseline Rd
#105-1025
Tempe, AZ 85283
480-609-3999
E-mail: admin@aara.com
http://www.aara.com

Quarterly newsletter of the AARA, a select group of recyclers providing quality recycled parts for the benefit of our customers, communities and environment. There are 90 member companies. AARA is affiliated with the AutomotiveDismantlers and Recyclers Association.
*Quarterly*
*Mike Pierson Jr, President*
*Layla Ressler, Vice President*

**4022 Air/Water Pollution Report**
Business Publishers
8737 Colesville Road
10th Floor
Silver Spring, MD 20910
301-589-5103
800-274-6737
Fax: 301-589-8493
E-mail: custserv@bpinews.com
http://www.bpinews.com

Provides comprehensive coverage of economic, political, legislative, regulatory and domestic and international implications of air and water pollution.
*Weekly*
*Leonard A Eiserer, Publisher*
*Beth Early, Associate Publisher*

**4023 American Waste Digest**
Carasue Moody
226 King Street
Pottstown, PA 19464
610-326-9480
800-442-4215
Fax: 610-326-9752
E-mail: awd@americanwastedigest.com
http://www.americanwastedigest.com

Provides reviews on new products, profiles on sucessful waste removal businesses, and provides discussion on legislation on municipal regulations on recycling.
*100 pages Monthly*
*Carasue Moody, Publisher/Editor*

**4024 BNA's Environmental Compliance Bulletin**
Bureau of National Affairs
1231 25th Street NW
Washington, DC 20037
202-452-4200
800-372-1033
Fax: 202-452-5331
http://www.bna.com

Cover the water and air pollution, waste management and regulatory updates, as well as a summary of selected regulatory actions and a list of key environmental compliance dates.
*Kevin Fepherston, Managing Editor*

**4025 Bio-Integral Resource Center: IPM Practitioner**
PO Box 7414 510-524-2567
Berkeley, CA 94707 Fax: 510-524-1758
E-mail: birc@igc.org
http://www.birc.org
The goal of the Bio Integral Resources Center is to reduce pesticide use by educating the public about effective, least-toxic alternatives for pest problems.
*Publication Date: 1979 6-12 pages*
*ISSN: 0738-968x*
*Dr. William Quarles, Executive Director*

**4026 C&D Recycler**
Gie Publishing
4012 Bridge Avenue 216-961-4130
Cleveland, OH 44113 800-456-0707
Fax: 216-961-0364
E-mail: btaylor@gie.net
http://www.cdrecycler.com
*Brian Taylor, Editor*

**4027 Common Sense Pest Control Quarterly**
PO Box 7414 510-524-2567
Berkeley, CA 94707 Fax: 510-524-1758
E-mail: birc@igc.org
http://www.birc.org
A quarterly publication published by the Bio Integral Resource Center
*Publication Date: 1984 24 pages Quarterly*
*ISSN: 8756-7881*
*Dr. William Quarles, Executive Director*

**4028 Composting News**
McEntee Media Corporation
9815 Hazelwood Avenue 440-238-6603
Cleveland, OH 44149 Fax: 440-238-6712
E-mail: ken@recycle.cc
http://www.recycle.cc
New composting projects, research, regulations and legislation, as well as the latest news in the composting industry.
*Publication Date: 1992 Monthly*
*Ken McEntee, Publisher*

**4029 Daily Environment Report**
Bureau of National Affairs
1231 25th Street NW 202-452-4200
Washington, DC 20037 800-372-1033
Fax: 202-822-8092
http://www.bna.com
A 40-page daily report providing comprehensive, in-depth coverage of national and international environmental news. Each issue contains summaries of the top news stories, articles, and in-brief items, and a journal of meetings, agencyactivities, hearings and legal proceedings. Coverage includes air and water pollution, hazardous substances, and hazardous waste, solid waste, oil spills, gas drilling, pollution prevention, impact statements and budget matters.
*40 pages*
*ISSN: 1060-2976*
*William A Beltz, Publisher*

**4030 E-Scrap News**
Resource Recycling
PO Box 42270 503-233-1305
Portland, OR 97242 Fax: 503-233-1356
E-mail: info@resource-recycling.com
http://www.resource-recycling.com
*64 pages*
*ISSN: 0744-4710*
*Justin Gast, Editor*

**4031 Earth Preservers**
PO Box 6 908-654-9293
Westfield, NJ 07091 E-mail: earthpreservers@att.net
http://www.earthpreserves.com
Award winning monthly environmental newspaper for school children aged 7 to 15.
*4 pages*
*Bill Paul, Publisher*

**4032 Environment Reporter**
Bureau of National Affairs
1231 25th Street NW 202-452-4200
Washington, DC 20037 800-372-1033
Fax: 202-822-8092
http://www.bna.com
A weekly notification and reference service covering the full-spectrum of legislative, administrative, judicial, industrial and technological developments affecting pollution control and environmental protection.
*ISSN: 0013-9211*
*William A Beltz, Publisher*
*Patricia Spencer, Managing Editor*

**4033 Environmental Engineering Science**
Mary Ann Liebert
140 Huguenot Street 914-740-2100
3rd Floor 800-MLI-EBER
New Rochelle, NY 10801 Fax: 914-740-2101
E-mail: info@liebertpub.com
http://www.liebertpub.com
The focus is on pollution control of the suface, ground, and drinking water, and highlight research news and product developments that aid in the fight against pollution.
*Bi-Monthly*
*ISSN: 1092-8758*
*Mary Ann Liebert, Publisher*
*Dumpnico Grosso PhD, Editor-in-Chief*

**4034 Environmental Notice**
235 S Beretania Street 808-586-4185
Room 702 Fax: 808-586-4186
Honolulu, HI 96813 E-mail: oeqc@doh.hawaii.gov
http://hawaii.gov/health/environemtnal/oeqc/index/html
A bi-monthly publication from the Health Department Environmental Quality Control division.
*Publication Date: 1978 24 pages Bi monthly*
*Kathy Kealoha, Director*

**4035 Environmental Regulation**
State Capitals Newsletters
PO Box 7376 703-768-9600
Alexandria, VA 22307 Fax: 703-768-9690
E-mail: newsletters@statecapitals.com
http://www.statecapitals.com
Weekly news from the state capitals keeps you informed on state programs, recycling, wetlands, ground water protection, beach renourishment, land management, greenspace laws, brownfields, livestock regulation, wilderness preservation,urban sprawl and solid waste.
*Publication Date: 1946 4-10 pages Newsletter 48x/Yr*
*ISSN: 1061-9682*
*Ellen Klein, Editor*

**4036 Environmental Regulatory Advisor**
JJ Keller
3003 W Breezewood Lane
PO Box 368 877-564-2333
Neenah, WI 54957 Fax: 800-727-7516
E-mail: sales@jjkeller.com
http://www.jjkeller.com
Covers developments at the EPA.
*12 pages*
*ISSN: 1056-3164*
*Tom Ziebell, Editor*

**4037 Environmental Science and Technology**
American Chemical Society
1155 16th Street NW
Washington, DC 20036
800-221-5558
Fax: 202-872-4615
E-mail: help@acs.org
http://www.acs.org

Articles on pollution control, waste treatment, climate changes and various other environmental interests.

*110 pages Semi-Monthly*
*Bruce Poorman, Ad Manager*
*Steve Cole, Managing Editor*

**4038 Environmental Systems Corporation Newsletter**
200 Tech Center Drive
Knoxville, TN 37912
865-688-7900
Fax: 865-687-8977
E-mail: esccorp@envirosys.com
http://www.envirosys.com

Data acquisition and reporting systems for electric power producers and industrial sources, ESC is the leading supplier of CEM and ambient data systems in the US Newsletter is free.

*Publication Date: 1994 4 pages Quarterly*
*Steve Drevik, Sr. Marketing Manager*

**4039 Environmental Times**
Environmental Assessment Association
1224 N Nokomis NE
Alexandria, MN 56308
320-763-4320
Fax: 320-763-9290
E-mail: eaa@iami.org
http://www.iami.org/eaa.cfm

This publication contains environmental conferences and expos, industry trends, federal regulations related to the environment and industry assessments.

*Robert Johnson, Publisher/Editor*

**4040 From the Ground Up**
Ecology Center
117 Division Street
Ann Arbor, MI 48104
734-761-3186
Fax: 734-663-2414
E-mail: info@crocenter.org
http://www.crocenter.org

Progressive environmental news from southeast Michigan.

*32 pages*
*Michael Garfield, Editor*

**4041 Full Circle**
Northeast Resource Recovery Association
PO Box 721
Concord, NH 03302
603-798-5777
Fax: 603-798-5744
E-mail: nrra@tds.net
http://www.recyclewithus.org

*Bi-Monthly*
*Elizabeth Bedard, Executive Director*

**4042 Hauler**
Hauler Magazine
166 South Main Street
PO Box 508
New Hope, PA 18938
215-997-3622
Fax: 215-997-3623
E-mail: mag@thehauler.com
http://www.thehauler.com

This magazine serves as an advertising guide to new products in the waste management, recycling, and environmental industries.

*Publication Date: 1978 Monthly*
*Thomas N Smith, Publisher/Editor*

**4043 HazMat Management**
Business Information Group
1450 Don Mills Road
Don Mills, Ontario M3B
416-442-2223
888-702-1111
Fax: 416-442-2917
E-mail: sales@hazmatmag.com
http://www.hazmatmag.com

Solutions for the environment.

*Publication Date: 1989 Bi-Annual*
*ISSN: 0843-9303*
*Thea Papadakis, Publisher*
*Connie Vitello, Editor*

**4044 Hazardous Waste News**
Business Publishers
PO Box 17592
Baltimore, MD 21297
301-589-5103
800-274-6737
Fax: 301-589-8493
E-mail: custserv@bpinews.com
http://www.bpinews.com

Comprehensive federal, state and local coverage of legislation and regulation affecting all aspects of the hazardous waste industry including Superfund, Resource Conservation and Recovery Act, US EPA, incineration, land disposal andmore.

*8 pages*
*Leonard A Eiserer, Publisher*
*Beth Early, Operations Director*

**4045 Hazardous Waste/Superfund Week**
Business Publishers
PO Box 17592lle Road
Baltimore, MD 21297
301-589-5103
800-274-6737
Fax: 301-589-8493
E-mail: custserv@bpinews.com
http://www.bpinews.com

Provides comprehensive coverage on hazardous waste disposal and cleanup, behind-the-scenes coverage of congressional action, EPA initiatives, Superfund sites, regulatory changes, court cases, enforcement news, contract opportunities,new technologies, research findings and business developments.

*Weekly*
*Leonard A Eiserer, Publisher*
*Beth Early, Associate Publisher*

**4046 Hazmat Transportation News**
Bureau of National Affairs
1801 S. Bell Street
Arlington, VA 22202
800-372-1033
Fax: 202-822-8092
http://www.bna.com

A two-binder service containing the full-text of rules and regulations governing shipment of hazardous material by rail, air, ship, highway and pipeline, including DOT's Hazardous Materials Tables and EPA's rules for its hazardouswaste tracking system.

*Stan Pond, Managing Editor*

**4047 Inside EPA**
Inside Washington Publishers
1225 South Clark Street
Suite 1400
Arlington, VA 22202
703-416-8500
Fax: 703-415-8543
E-mail: iwp@iwpnews.com
http://www.iwpnews.com

Gives timely information on all facets of waste, water, air, and other environmental regulatory programs.

*Publication Date: 1980 Weekly*
*Al Sosenko, Publisher*

**4048 Institute of Scrap Recycling Industries**
1615 L Street NW
Suite 600
Washington, DC 20036
202-662-8500
Fax: 202-626-0900
E-mail: kentkiser@scrap.org
http://www.isri.org

*Publication Date: 1988 148 pages Bi monthly*
*ISSN: 0036-9527*
*Frank Cozzi, President*
*Kent Kiser, Publisher/Editor-in-Chief*

**4049 Journal of Environmental Education**
Heldref Publications
1319 18th Street NW
Washington, DC 20036
202-296-6267
Fax: 202-296-5149
E-mail: webmaster@heldref.org
http://www.heldref.org

The issues featured are case studies, environmental philosophy and policy discussions, new research evaluations and information on environmental education.

*Douglas Kirkpatrick, Publisher/Executive Director*

**4050 Legislative Bulletin**
Arkansas Environmental Federation

1400 W Markham Street
Suite 302
Little Rock, AR 72201
501-374-0263
Fax: 501-374-8752
http://www.environmentark.org
*Publication Date: 1967*
*Randy Thurman, Executive Director*

**4051 Minnesota Pollution Control Agency Minnesota Environment Magazine**
520 Lafayette Road North
St. Paul, MN 55155
651-296-6300
800-657-3864
Fax: 651-296-7923
E-mail: vicki.schindeldecker@pca.state.mn.us
http://www.pca.state.mn.us
Established in 1967 to protect Minnesota's environment through monitoring environmental quality and enforcing environmental regulations.
*Publication Date: 2000 16-20 pages Quarterly*
*Paul Eger, Comm./Chair Citizens' Board*

**4052 Northeast Recycling Council Bulletin**
139 Main Street
Suite 401
Brattleboro, VT 05301
802-254-3636
Fax: 802-254-5870
E-mail: info@nerc.org
http://www.nerc.org
*Monthly*
*Lynn Rubinstein, Executive Director*

**4053 Northeast Recycling Council News**
139 Main Street
Suite 401
Brattleboro, VT 05301
802-254-3636
Fax: 802-254-5870
E-mail: info@nerc.org
http://www.nerc.org
*3x Year*
*Lynn Rubenstein, Executive Director*

**4054 Oregon Refuse and Recycling Association Newsletter**
PO Box 2186
Salem, OR 97308
503-588-1837
800-527-7624
Fax: 503-399-7784
E-mail: orrainfo@orra.net
http://www.orra.net
*Monthly*
*Max Brittingham, Executive Director*
*Kristin Mitchell, Editor*

**4055 Plastics Recycling Update**
Resource Recycling
PO Box 42270
Portland, OR 97242
503-233-1305
Fax: 503-233-1356
E-mail: subscriptions@resource-recycling.com
http://www.resource-recycling.com
*Monthly*
*Justin Gast, Managing Editor*
*Mary Lynch, Circulation*

**4056 Pollution Equipment News**
Rimbach Publishing
8650 Babcock Boulevard
Pittsburgh, PA 15237
412-364-5366
800-245-3182
Fax: 412-369-9720
E-mail: info@rimbach.com
http://www.rimbach.com
Provides information to those responsible for selecting products and services for air, water, wastewater and hazardous waste pollution abatement.
*Publication Date: 1967 64 pages Bi-Monthly*
*ISSN: 0032-3659*
*Raquel Rimbach, Editor*
*Norberta Rimbach, Publisher/President*

**4057 Pollution Prevention News**
US EPA
1200 Pennsylvania Avenue, NW
Washington, DC 20460
202-272-0167
Fax: 202-260-2219
http://www.epa.gov
Articles include recent information on source reduction and sustainable technologies in industry, transportation, consumer, agriculture, energy, and the international sector.
*Maureen Eichelberger, Editor*

**4058 Recharger Magazine**
Recharger Magazine
1050 East Flamingo Road
Suite N237
Las Vegas, NV 89119
702-438-5557
Fax: 702-873-9671
E-mail: info@rechargermag.com
http://www.rechargermag.com
Information including articles that cover business and marketing, technical updates, association and industry news, and company profiles. On the remanufactured imaging supplies industry, related features focus on the importance ofrecycling, government legislation, and product comparisons. Annual trade event in Las Vegas.
*Publication Date: 1989 300+ pages Monthly*
*ISSN: 1053-7503*
*Julie Kerrane, Author*
*Phyllis Gurgevich, Publisher*
*Amy Turner, Managing Editor*

**4059 Recycled News**
Maryland Recyclers Coalition
PO Box 1046
Laurel, MD 20725
888-496-3196
Fax: 301-238-4579
E-mail: info@marylandrecyclers.org
http://www.marylandrecyclers.org
*2 pages Bi-Monthly*
*Jackie King, Executive Director*

**4060 Recycling Laws International**
Raymond Communications
5111 Berwin Road
Suite 115
College Park, MD 20740
301-345-4237
Fax: 301-345-4768
E-mail: circulation@raymond.com
http://www.raymond.com
Covers recycling, takeback and green labeling policy for business in 38 countries. Available online.
*Publication Date: 1995 Bi-Monthly*
*Bruce Popka, Vice President*

**4061 Recycling Markets**
NV Business Publishers Corporation
43 Main Street
Avon by the Sea, NJ 07717
732-502-0500
Fax: 732-502-9606
E-mail: jcurley@nvpublications.com
http://www.nvpublications.com
Contains profiles on recycling mills, as well as large users and generators of recycled materials for the broker, dealers and processors of paper stock, scrap metal, plastics and glass.
*Jim Curley, Editor*
*Anna Dutko Rowley, Managing Editor*

**4062 Recycling Product News**
Baum Publications
201-2323 Boundary Road
Vancouver, Can, BC V5M
604-291-9900
Fax: 604-291-1906
E-mail: webadmin@baumpub.com
http://www.baumpub.com
Published for the recycling center operators and other waste mangers, articles discuss technology and new products.
*Engelbert J Baum, Publisher*
*Keith Barker, Editor*

**4063 Recycling Today**
GIE Media
4020 Kinross Lakes Parkway
#201
Richfield, OH 44286
800-456-0707
Fax: 216-925-5022
E-mail: dtoto@gie.net
http://www.recyclingtoday.com
Published for the secondary commodity processing/recycling market.
*James R Keefe, Group Publisher*
*Brian Taylor, Editor*

**4064 Resource Recovery Report**
5313 38th Street NW
PO Box 3356
Warrenton, VA 20188
540-347-4500
800-627-8913
Fax: 540-348-4540
E-mail: rwill@coordgrp.com
http://www.coordgrp.com

Covers all alternatives to landfills, i.e., recycling, energy recovery, composting in North America, Government, industry, associations, universities, etc. are included.
*12 pages*
*Richard Will, Production Manager*

**4065 Resource Recycling Magazine**
Resource Recycling
PO Box 42270
Portland, OR 97242
503-233-1305
Fax: 503-233-1356
E-mail: info@resource-recycling.com
http://www.resource-recycling.com

*Monthly*
*Justin Gast, Managing Editor*
*Mary Lynch, Circulation*

**4066 Reuse/Recycle Newsletter**
Technomic Publishing Company
PO Box 3535
Lancaster, PA 17601
717-291-5609
800-233-9936
Fax: 717-295-4538
E-mail: aflannery@techpub.com
http://www.techpub.com

Provides news and information on important developments in both industrial and municipal recycling, and focuses on large-scale post-consumer, post-commercial, and post-industrial waste recycling.
*8 pages*
*ISSN: 0048-7457*
*Susan E Selke, Author*
*Amy Flannery, Marketing*

**4067 Scrap**
Institute of Scrap Recycling Industries
1615 L Street NW
Suite 600
Washington, DC 20036
202-662-8500
Fax: 202-626-0945
E-mail: ellenross@scrap.org
http://www.scrap.org

Serves the scrap processing and recycling industry. Subscription: $32.95 (US), $38.95 (Canada/Mexico) & $104.95 (all other international)
*Bi-Monthly*
*Kent Kiser, Publisher/Editor-in-Chief*
*Ellen Ross, Production Director*

**4068 Solid Waste & Recycling**
Southam Environment Group
1450 Don Mills Road
Don Mills, Ontario M3B
416-442-5600
800-387-0273
Fax: 416-510-5130
E-mail: bobrien@solidwastemag.com
http://www.solidwastemag.com

Published to emphasize on municipal and commercial aspects of collection, handling, transportation, hauling, disposal and treatment of solid waste , including incineration, recycling and landfill technology.
*Brad O'Brien, Publisher*
*Guy Crittenden, Editor-in-Chief*

**4069 Solid Waste Report**
Business Publishers
PO Box 17592
Baltimore, MD 21297
301-589-5103
800-274-6737
Fax: 301-589-8493
E-mail: custserv@bpinews.com
http://www.bpinews.com

Comprehensive news and analysis of legislation, regulation and litigation in solid waste management including resource recovery, recycling, collection and disposal. Regularly features international news, state updates and businesstrends.
*Bi-Weekly*
*Leonard A Eiserer, Publisher*
*Beth Early, Operations Director*

**4070 State Recycling Laws Update**
Raymond Communications
5111 Berwin Road
Suite 115
College Park, MD 20740
301-345-4237
Fax: 301-345-4768
E-mail: michele@raymond.com
http://www.raymond.com

Provides coverage of recycling legislation affecting business, as well as the outlook on future legislation across the US and Canada.
*Bruce Popka, Vice President*

**4071 Waste Age**
Environmental Industry Associations
4301 Connecticut Avenue NW
#300
Washington, DC 20008
202-244-4700
800-424-2869
Fax: 202-966-4868
E-mail: ptom@primediabusiness.com
http://www.wasteage.com

Contents focus on new system technologies, recycling, resource recovery and sanitary landfills with regular features on updates in the status of government regulations, new products, guides, company profiles, exclusive surveyinformation, legislative implications and news.
*Patricia-Anne Tom, Editor*
*Stephen Ursery, Managing Editor*

**4072 Waste Age's Recycling Times**
Environmental Industry Associations
4301 Connecticut Avenue NW
#300
Washington, DC 20008
202-244-4700
800-424-2869
Fax: 202-966-4868
E-mail: ptom@primediabusiness.com
http://www.wasteage.com

Features municipalities, recycling goals and rates, program innovations, waste habits, and new materials being recycled.
*Patricia-Anne Tom, Editor*
*Stephen Ursery, Managing Editor*

**4073 Waste Handling Equipment News**
Lee Publications
6113 State Highway 5
PO Box 121
Palatine Bridge, NY 13428
518-673-3237
800-218-5586
Fax: 518-673-2381
E-mail: rbrown@leepub.com
http://www.leepub.com

Dicusses the latest developments in woodwaste, C&D, scrapmetal, concrete, asphalt, recycling and composting with the emphasis on equipment.
*Publication Date: 1993 50 pages Monthly*
*Coyle Rockwell, Author*
*Holly Reiser, Editor*
*Richard Brown, Production Coordinator*

**4074 Waste Recovery Report**
Icon/Information Concepts
211 S 45th Street
Philadelphia, PA 19104
215-349-6500
Fax: 215-349-6502
E-mail: wasterec@aol.com
http://www.icodat.com

Contains information on waste-to-energy, recycling, composting and other technologies.
*Publication Date: 1975 6 pages Monthly*
*ISSN: 0889-0072*
*Alan Krigman, Publisher/Editor*

## Periodicals: Sustainable Development

**4075 AERO SunTimes**
Alternative Energy Resources Organization

432 N Last Chance Gulch 406-443-7272
Helena, MT 59601 Fax: 406-442-9120
E-mail: aero@aeromt.org

The SunTimes is the newsletter for AERO. The organization is a statewide grassroots group whose members work together to strengthen communities through promoting sustainable agriculture, local food production and citizen-based SmartGrowth community planning.

*Publication Date: 1978 4-24 pages Quarterly*
*ISSN: 1046-0993*
*Kiki Hubbard, Editor*
*Jean Duncan, Editor*

**4076 California Association of Resource Conservation Districts- CCP News**

801 K Street 916-457-7094
Suite 1415 Fax: 916-457-7934
Sacramento, CA 93101 http://www.carcd.org

*Quarterly*
*Patrick Truman, President*
*Brian Leahy, Executive Director*

**4077 Californians for Population Stabilization: CAPS News**

1129 State Street 805-564-6626
Suite 3-D Fax: 805-564-6636
Santa Barbara, CA 93101 E-mail: info@capsweb.org
http://www.capsweb.org

A nonprofit, public interest organization that works to protect California's environment and quality of life by turning the tide of population growth.

*Publication Date: 1986 8 pages 3x year*
*Diana Hull PhD, President*
*Ben Zuckerman PhD, Vice President*

**4078 Cultivar**

Center for Agoecology
1156 High Street 831-459-2506
Santa Cruz, CA 95064 Fax: 831-459-2867
http://www.agroecology.org

*Publication Date: 1985 9-12 pages Bi-Yearly*
*Steven Gliessman, Professor Agroecology*
*Martha Brown, Editor*

**4079 Ecosphere**

Forum International
91 Gregory Lane
Suite 21 800-252-4475
Pleasant Hill, CA 94523 Fax: 925-946-1500
E-mail: fti@foruminternational.com
http://www.foruminternational.com

Accepts advertising. The first ever environmental/ecological magazine. It is dedicated to the interrelations of man in nature and a balanced approach of its biological, economic, socio-political and spiritual components. Since 1965.

*Publication Date: 1965 16-48 pages Bi-monthly*
*Dr. Nicolas Hetzer, Production Manager*
*J McCormack, Circulation Director*

**4080 EnviroNews**

600 Forbes Avenue 412-396-6000
331 Fisher Hall 800-456-0590
Pittsburgh, PA 15282 Fax: 412-396-4092
E-mail: bembic@duq.edu
http://www.science.duq.edu

Educating environmental professionals for the twenty-first century is the focus of the Duquesne University Environmental Science and Management (ESM) Masters Degree Program. The program grew out of the perceived need to combine depthof knowledge in environmental science with a comprehensive understanding of the business, legal and policy implications surrounding environmental issues.

*4 pages Semester Newsletter*
*Sonia Bembic, Program Advisor*

**4081 Environmental News**

Arkansas Environmental Federation
1400 W Markham Street 501-374-0263
Suite 302 Fax: 501-374-8752
Little Rock, AR 72201 E-mail: rthurman@environmentark.org
http://www.environmentark.org

*Randy Thurman, Executive Director*

**4082 Forest Magazine**

Forest Service Employees for Environmental Ethics
PO Box 11615 541-484-2692
Eugene, OR 97440 Fax: 541-484-3004
E-mail: fseee@fseee.org
http://www.fseee.org

FSEEE is the largest forest watchdog organization in the nation. Since 1989, FSEEE has defended the rights and responsibilities of brave scientists and resource professionals working to assure the long-term health and vitality of ournational forests. FSEEE publishes Forest Magazine quarterly to educate the public on forest issues.

*Publication Date: 1989 50 pages Bi-Monthly*
*ISSN: 1534-9284*
*Andy Stahl, Executive Director*
*Patricia Marshall, Editor*

**4083 Forest Service Employees for Environmental Ethics**

PO Box 11615 541-484-2692
Eugene, OR 97440 Fax: 541-484-3004
E-mail: fseee@fseee.org
http://www.fseee.org

FSEEE is the largest forest watchdog organization in the nation. Since 1989, FSEEE has defended the rights and responsibilities of brave scientists and resource professionals working to assure the long-term health and vitality of ournational forests. FSEEE publishes Forest Magazine quarterly to educate the public on forest issues.

*Andy Stahl, Executive Director*

**4084 Greens/Green Party USA Green Politics Green Politics**

PO Box 1134 978-682-4353
Lawrence, MA 01842 866-GRE-ENS2
Fax: 978-682-4318
E-mail: gpusa@greens.org
http://www.greenparty.org

Is a national non-profit membership organization dedicated to advancing the Green Ten Key Values as a guiding force in American society and politics.

*12 pages Quarterly*
*Don Fitz, Editor*

**4085 International Boreal Forest Newsletter**

Institute for World Resource Research
PO Box 50303 630-910-1551
Palo Alto, CA 94303 Fax: 630-910-1561
http://www.globalwarming.net

Covers all phases of developments in forestry and reforestation of northern nations including the US, Canada, Russia, Sweden, Finland, Norway, China, Japan and others. Its goal is to increase the worldwide understanding of theecological and economic roles of the northern forest regions of the world.

*Dr. Yuan Lee, Editor-in-Chief*
*BJ Jefferson, Advertising/Sales*

**4086 International Society of Tropical Foresters: ISTF Notices**

5400 Grosvenor Lane 301-897-8720
Bethesda, MD 20814 Fax: 301-897-3690
E-mail: istfi@igc.apc.org

The International Society of Tropical Foresters, Inc. (ISTF) is a nonprofit organization committed to the protection, wise management and rational use of the world's tropical forests. Established in 1950, ISTF was reactivated in 1979.It has about 1500 members in more than 100 countries. Financial support comes from membership dues, donations and grants. ISTF sponsors meetings, promotes chapters in other countries, maintains a web site and has chapters at universities.

*Warren T Doolittle, President*

**4087 Jackson Hole Conservation Alliance: Alliance News**
PO Box 2728 307-733-9417
Jackson, WY 83001 Fax: 307-733-9008
E-mail: info@jhalliance.org
http://www.jhalliance.org
An organization dedicated to responsible land stewardship in Jackson Hole, Wyoming to ensure that human activities are in harmony with the area's irreplaceable wildlife, scenery and other natural resources.
*20 pages Quarterly*
*Cindy Harger, Managing Director*

**4088 Leopold Letter**
Leopold Center for Sustainable Agriculture
209 Curtiss Hall 515-294-3711
Ames, IA 50011 Fax: 515-294-9696
E-mail: leocenter@iastate.edu
http://www.leopold.iastate.edu
To inform diverse audiences about Leopold Center programs and activities; to encourage increased interest in and use of sustainable farming practicies; and to stimulate public discussion about sustainable agriculture in Iowa.
*Publication Date: 1987 12 pages Quarterly*
*ISSN: 1065-2116*
*Jerry DeWitt, Director*
*Laura Miller, Editor*

**4089 Minnesota Department of Agriculture: MDA Quarterly**
625 Robert St. N. 651-201-6000
St Paul, MN 55155 800-967-2474
Fax: 651-297-5522
E-mail: webinfo@mda.state.mn.us
http://www.mda.state.mn.us
The MDA's mission is to work toward a diverse ag industry that is profitable as well as environmentally sound; to protect the public health safety regarding food and ag products; and to ensure orderly commerce in agricultural and foodproducts. We have two major branches of the department to accomplish this mission: regulatory divisions and non-regulatory divisions.
*Publication Date: 2000*
*Gene Hugoson, Commissioner*
*Michael Schommer, Editor*

**4090 Mountain Research and Development**
PO Box 1978 530-752-8330
Davis, CA 95617 http://www.mrd-journal.org/about_mrd.htm
The leading journal specifically devoted to the world's mountains. It has been published since 1981 and has established itself as a renowned international publication containing well-researched, peer-reviewed scientific articles byauthors from around the world.
*Professor Hans Hurni, Editor-in-Chief*

**4091 Northeast Sun**
Northeast Sustainable Energy Association
50 Miles Street 413-774-6051
Greenfield, MA 01301 Fax: 413-774-6053
E-mail: nesea@nesea.org
http://www.nesea.org
Promotes responsible use of energy for a stronger economy and cleaner environment. Northeast Sun is a bi-annual Spring and Fall publication that includes articles by leading authorities on sustainable energy practices, energyefficiency and renewable energy, and each Fall issue includes the Sustainable Green Pages Directory of engery professionals in the Northeast. Subscription is free with membership.
*Bi-Annual*
*David Barclay, Executive Director*
*Arianna Alexsandra Grindrod, Education Director*

**4092 Pinchot Letter**
1616 P Street NW 202-797-6580
Suite 100 Fax: 202-797-6583
Washington, DC 20036 E-mail: pinchot@pinchot.org
http://www.pinchot.org
A tri-annual newsletter published by the Pinchot Institute for Conservation, an independent nonprofit organization that works collaboratively with all Americans-from federal and state policymakers to citizens in rural communities-tostrengthen forest conservation by advancing sustainable forest management, developing conservation leaders and providing science-based solutions to natural resource issues.
*Publication Date: 1995 20 pages Tri-Annual*
*Dr V Alaric Sample, President*

**4093 Population Institute Newsletter**
107 2nd Street NE 202-544-3300
Suite 207 188-787-0038
Washington, DC 20002 Fax: 202-544-0068
E-mail: web@populationinstitute.org
http://www.populationinstitute.org
The Population Institute is the World's largest independent nonprofit, educational organization dedicated exclusively to achieving a more equitable balance between the worlds population, environment, and resources. Established in 1969,the Institute, with members in 172 countries, is headquartered on Capitol Hill in Washington DC. The Institute uses a variety of resources and programs to bring its concerns about the consequences of rapid poulation growth to the forefront of thenational agenda.
*Publication Date: 1988 8 pages Bi-Monthly*
*Werner Fornos, President*
*Hal Burdett, Executive Editor*

**4094 Population Reference Bureau: World Population Data Sheet**
1875 Connecticut Avenue NW 202-483-1100
Suite 520 800-877-9881
Washington, DC 20009 Fax: 202-328-3937
E-mail: popref@prb.org
http://www.prb.org
Up-to-date demographic data and estimates for all the countries and major regions of the world.
*William P Butz, President/CEO*

**4095 Reporter**
Population Connection
2120 L Street NW 202-332-2200
Suite 500 Fax: 202-332-2302
Washington, DC 20037 E-mail: info@popconnect.org
http://www.popconnect.org
Looks at the connections between overpopulation and the environment around the world and features reports from our activists on Capitol Hill Days 2005. This publication is included in your $25.00 memberhsip fee.
*Publication Date: 1972 24 pages Quarterly*
*ISSN: 0199-0071*
*John Seager, President/CEO*
*Mara Nelson Grynavinski, Editor*

**4096 Resource Development Newsletter**
University of Tennessee
PO Box 1071 865-974-7448
Knoxville, TN 37901 Fax: 423-974-7448
Community development information.
*4 pages*
*Dr Alan Barefield, Publisher*

**4097 Restoration Ecology Magazine**
Blackwell Science
350 Main Street 781-388-8200
Malden, MA 02148 Fax: 781-388-8210
http://www.blackwellpublishing.com
Provides the most recent developments in the ecological and biological restoration field for both the fundamental and practical implications of restorations.
*Richard Hobbs, Editor*

**4098 Society of American Foresters Information Center Newsletter**
5400 Grosvenor Lane 301-897-8720
Bethesda, MD 20814 Fax: 301-897-3690
E-mail: safweb@safnet.org
http://www.safnet.org
An organization that represents the forestry profession in the United States. Its mission is to advance the science, education, technology and practice of forestry.
*Publication Date: 1996 24 pages Monthly*
*Michael T Goergen, Jr, EVP/CEO*

**4099 Solar Energy Magazine**
Elsevier Science
360 Park Avenue S 212-989-5800
11th Floor Fax: 212-633-3680
New York, NY 10010 E-mail: usinfo-f@elseview.com
http://www.elseview.com
Devoted exclusively to the science and technology of solar energy applications.
*Publication Date: 1957*
*ISSN: 0380-92X*
*D Yogi Goswami, Editor-in-Chief*

**4100 Solar Energy Report**
California Solar Energy Industries Association
PO Box 782 916-747-6987
Rio Vista, CA 94571 Fax: 707-374-4767
E-mail: info@calseia.org
http://www.calseia.org
*Bi-Monthly*
*Les Nelson, President*

**4101 Southface Journal of Sustainable Building**
Southface Energy Institute
241 Pine Street NE 404-872-3549
Atlanta, GA 30308 Fax: 404-872-5009
E-mail: info@southface.org
http://www.southface.org
Contains articles on numerous sustainable building topics. Free to members and available online.
*Publication Date: 1978 24 pages Quarterly*
*Dennis Creech, Executive Director/Editor*

**4102 Sustainable Green Pages Directory**
Northeast Sustainable Energy Association
50 Miles Street 413-774-6051
Greenfield, MA 01301 Fax: 413-774-6053
E-mail: nesea@nesea.org
http://www.nesea.org
The SPG Directory lists over 30 categories of sustainable energy professionals working throughout the Northeast, including architects, engineers, builders, energy auditors, consultants and renewable energy installers and manufacturers.It is the largest directory of its kind and the only one that targets the Northeastern USA. Published in the Northeast Sun magazine and online.
*September Annual*
*David Barclay, Executive Director*
*Arianna Alexsandra Grindrod, Education Director*

**4103 Tall Timbers Research Station: Bulletin Series**
13093 Henry Beadel Drive 850-893-4153
Tallahassee, FL 32312 Fax: 850-668-7781
E-mail: rose@ttrs.org
http://www.talltimbers.org
Dedicated to protecting wildlands and preserving natural habitats. Promotes public education on the importance of natural disturbances to the environment and the subsequent need for wildlife and land management. Conducts fire ecologyresearch and other biological research programs through the Tall Timbers Research Station. Operates museum.
*Publication Date: 1962*
*ISSN: 0496-7631*
*Lane Green, Executive Director*
*R Todd Engstrom, Editor*

**4104 Tall Timbers Research Station: Fire Ecology Conference Proceedings**
13093 Henry Beadel Drive 850-893-4153
Tallahassee, FL 32312 Fax: 850-668-7781
E-mail: rose@ttrs.org
http://www.talltimbers.org
Dedicated to protecting wildlands and preserving natural habitats. Promotes public education on the importance of natural disturbances to the environment and the subsequent need for wildlife and land management. Conducts fire ecologyresearch and other biological research programs through the Tall Timbers Research Station. Operates museum.
*Publication Date: 1962*
*ISSN: 0082-1527*
*Lane Green, Executive Director*
*R Todd Engstrom, Editor*

**4105 Tall Timbers Research Station: Game Bird Seminar Proceedings**
13093 Henry Beadel Drive 850-893-4153
Tallahassee, FL 32312 Fax: 850-668-7781
E-mail: rose@ttrs.org
http://www.talltimbers.org
Dedicated to protecting wildlands and preserving natural habitats. Promotes public education on the importance of natural disturbances to the environment and the subsequent need for wildlife and land management. Conducts fire ecologyresearch and other biological research programs through the Tall Timbers Research Station. Operates museum.
*Publication Date: 1980*
*ISSN: 1087-4372*
*Lane Green, Executive Director*
*R Todd Engstrom, Editor*

**4106 Tall Timbers Research Station: Miscellaneous Series**
13093 Henry Beadel Drive 850-893-4153
Tallahassee, FL 32312 Fax: 850-668-7781
E-mail: rose@ttrs.org
http://www.talltimbers.org
Dedicated to protecting wildlands and preserving natural habitats. Promotes public education on the importance of natural disturbances to the environment and the subsequent need for wildlife and land management. Conducts fire ecologyresearch and other biological research programs through the Tall Timbers Research Station. Operates museum.
*Publication Date: 1961*
*ISSN: 0494-764x*
*Lane Green, Executive Director*
*R Todd Engstrom, Editor*

**4107 Totally Tree-Mendous Activities**
Northeast Sustainable Energy Association
50 Miles Street 413-774-6051
Greenfield, MA 01301 Fax: 413-774-6053
E-mail: nesea@nesea.org
http://www.nesea.org
Resource for teachers and parents that offers creative and fun tree-based projects for students.
*40 pages*
*David Barclay, Executive Director*
*Arianna Alexsandra Grindrod, Education Director*

**4108 Woodland Steward**
Massachusetts Forestry Association
270 Jackson Street 413-323-7326
Belchertown, MA 01007 Fax: 413-323-9594
This publication is full of information about Massachusett's forest and ways that landowners can manage their woodlands to achieve their goals in an environmentally sustainable manner. Free with membership.
*Bi-Monthly*
*Gregory Cox, Executive Director*

**4109 Worldwatch Institute: State of the World**
1776 Massachusetts Avenue NW 202-452-1999
Washington, DC 20036 Fax: 202-296-7365
E-mail: worldwatch@worldwatch.org
http://www.worldwatch.org
The most authorative go-to resource for those who understand the importance of nuturing a safe, sane and healthy global environment through both policy and action.
*Annual*
*ISBN: 0-393326-66-7*
*Christopher Flavin, President*
*Tom Prugh, Editor*

**4110 Worldwatch Institute: Vital Signs**
1776 Massachusetts Avenue NW 202-452-1999
Suite 800 Fax: 202-296-7365
Washington, DC 20036 http://www.worldwatch.org
Provides comprehensive, user-friendly information on key trends and includes tables and graphs that help readers access the developments that are changing their lives for better or for worse.
*Annual*
*ISBN: 0-393326-89-6*
*Christopher Flavin, President*
*Tom Prugh, Editor*

**4111 Worldwatch Institute: World Watch**
1776 Massachusetts Avenue NW 202-452-1999
Suite 800 Fax: 202-296-7365
Washington, DC 20036 http://www.worldwatch.org
The Worldwatch Institute is an independent, nonprofit environmental research organization in Washington DC. Its mission is to foster a sustainable society in which human needs are met in ways that do not threaten the health of thenatural environment or future generations. To this end, this Institute conducts interdisciplinary research on emerging global issues, the results of which are published and disseminated to decision-makers and the media.
*Bi-Monthly*
*Christopher Flavin, President*
*Tom Prugh, Editor*

**4112 Worldwatch Institute: Worldwatch Papers**
1776 Massachusetts Avenue NW 202-452-1999
Suite 800 Fax: 202-296-7365
Washington, DC 20036 http://www.worldwatch.org
Provides cutting-edge analysis on an environmental topic that is making - or is about to make - headlines worldwide.
*50-70 pages 5x times year*
*Christopher Flavin, President*
*Tom Prugh, Editor*

## Periodicals: Travel & Tourism

**4113 New York State Parks, Recreation and Historic Preservation**
Empire State Plaza 518-474-0456
Agency Building 1 Fax: 518-486-2924
Albany, NY 12238 http://www.nysparks.com
Publishes New York State Boat Launching Guide, Camping/Cabin Reservation Info, Snowmobiling Guide and Preservation Magazine.
*Bernadette Castro, Commissioner*

**4114 Noxubee National Wildlife Refuge Newsletter**
2970 Bluff Lake Road 662-323-5548
Brooksville, MS 39739 Fax: 662-323-6390
E-mail: noxubee@fws.gov
http://http://www.fws.gov/noxubee/
Noxubee National Wildlife Refuge was established in 1940 to protect and enhance habitat for the conservation of migratory birds, endangered species and other wildlife. The recreational and educational opportunities provided on therefuge help the public experience nature and learn how sound management ensures that future generations continue to enjoy fish and wildlife and their habitats.
*2 pages Bi-Annual*
*Andrea Duncan, Editor*

**4115 Parks and Recreation Magazine**
National Recreation and Park Association
22377 Belmont Ridge Road 703-858-0784
Ashburn, VA 20148 Fax: 703-858-0794
E-mail: info@nrpa.org
http://www.activeparks.comornrpa.org
Informs, motivates and inspires professionals, civic leaders and citizens to elevate the value of parks and recreation as a public service.
*Monthly*
*John Thorner, Exectutive Director*
*Rachel Roberts, Editor*

**4116 Potomac Appalachian**
118 Park Street SE 703-242-0693
Vienna, VA 22180 Fax: 703-242-0968
E-mail: info@patc.net
http://www.patc.net
Published by the Potomac Appalachian Trail Club, which through volunteer efforts, education and advocacy, acquires, maintains and protects the trail and lands of the Appalachian Trail, other trails and related facilities in theMid-Atlantic Region for the enjoyment of present and future hikers. PATC publishes hiking guides, maps and history books of the Appalachian Trail and other trails in our area of responsibility. The monthly newsletter is sent to members and uponrequest. Free to members.
*16-20 pages Monthly*
*ISSN: 098 -8154*
*Thomas R Johnson, President*

## Periodicals: Water Resources

**4117 Air Water Pollution Report's Environment Week**
Business Publishers
PO Box 17592 301-589-5103
Baltimore, MD 21297 800-274-6737
Fax: 301-589-8493
E-mail: custserv@bpinews.com
http://www.bpinews.com
Provides a balanced, insightful update on the week's most important environmental news from Washington, DC.
*Leonard A Eiserer, Publisher*
*David Goeller, Editor*

**4118 Air/Water Pollution Report**
Business Publishers
PO Box 17592 301-589-5103
Baltimore, MD 21297 800-274-6737
Fax: 301-589-8493
E-mail: custserv@bpinews.com
http://www.bpinews.com
Regulatory activities and governmental legislation, in addition to litigation are covered in this pulication.
*Leonard A Eiserer, Publisher*
*David Goeller, Editor*

**4119 America's Priceless Groundwater**
American Ground Water Trust
50 Pleasant Street Ste 2 603-228-5444
Concord, NH 03301 Fax: 603-228-6557
E-mail: trustinfo@agwt.org
http://www.agwt.org

*15 pages Quarterly*
*Andrew Stone, Director*

**4120 American Fisheries Society: Water Quality Matters**
324 25th Street
Ogden, UT 84401
801-625-5358
Fax: 801-625-5756
E-mail: glampman@fs.fed.us

*8 pages 1-2 per year*
*Georgina Lampman, President*
*Gregg Lomincky, Editor*

**4121 American Water Resources Association: Journal of the American Water Resources Association**
PO Box 1626
Middleburg, VA 20118
540-687-8390
Fax: 540-687-8395
E-mail: terry@awra.org
http://www.awra.org
AWRA is a nonprofit, scientific educational association for individuals and organizations involved in all aspects of water resources. Its goal is to advance multidisciplinary water resources management and research through itsconferences, publications, technical commettees, state sections and student chapters.
*Publication Date: 1964 Bi-Monthly*
*ISSN: 1093-474X*
*Kenneth D Reid, Executive VP*
*Terry Meyer, Marketing Director*

**4122 American Water Resources Association: Water Resources IMPACT**
PO Box 1626
Middleburg, VA 20118
540-687-8390
Fax: 540-687-8395
E-mail: terry@awra.org
http://www.awra.org
AWRA is a nonprofit, scientific educational association for individuals and organizations involved in all aspects of water resources. Its goal is to advance multidisciplinary water resources management and research through itsconferences, publications, technical commettees, state sections and student chapters.
*Bi-Monthly*
*ISSN: 1093-474X*
*Kenneth D Reid, Executive VP*
*Terry Meyer, Marketing Director*

**4123 Arsenic and Groud Water Home**
American Ground Water Trust
50 Pleasant Street Ste 2
Concord, NH 03301
603-228-5444
Fax: 603-228-6557
E-mail: trustinfo@agwt.org
http://www.agwt.org

*15 pages Quarterly*
*Andrew Stone, Director*

**4124 Bacteria and Water Wells**
American Ground Water Trust
50 Pleasant Street Ste2
Concord, NH 03301
603-228-5444
Fax: 603-228-6557
E-mail: trustinfo@agwt.org
http://www.agwt.org

*15 pages Quarterly*
*Andrew Stone, Director*

**4125 Blue Planet Magazine**
The Ocean Conservancy
2029 K Street
Washington, DC 20006
202-429-5609
800-519-1541
Fax: 202-429-0056
E-mail: info@oceanconservancy.org
http://www.oceanconservancy.org
To educate peoeple about ocean issues; inspire readers with the beauty and wonder of oceans; encourage dedication to appreciating and protecting marine resources; and enlist new volunteers in the ocean community. Free with membershipfee of $25.00
*46 pages Quarterly*
*Roger Rufe,Jr, President/CEO*
*Sara Bennington, Editor*

**4126 Clean Water Network: CWN Status Water Report**
Spills and Kills
1200 New York Avenue, NW
Suite 400
Washington, DC 20005
202-289-2421
Fax: 202-289-1060
E-mail: info@cwn.org
http://www.cwn.org
A nonprofit network of over 1,000 organizations that deal with clean water issues covered by the Clean Water Act. Our member organizations consist of a variety of organizations representing environmentalists, family farmers, recreationanglers, commercial fishermen, surfers, boaters, faith communities, labor unions and civic associates. We publish a monthly newsletter and various reports.
*8-12 pages Monthly*
*Katherine Smitherman, Executive Director*

**4127 Clean Water Report Newsletter**
Business Publishers
PO Box 17592
Baltimore, MD 21297
301-589-5103
800-274-6737
Fax: 301-589-8493
E-mail: custserv@bpinews.com
http://www.bpinews.com
Follows the latest news from the EPA, Congress, the states, the courts, and private industry. A key information source for environmental professionals, covering the important issues of ground and drinking water, wastewater treatment,wetlands, drought, coastal protection, non-point source pollution, agrichemical contamination and more.
*8 pages Bi-Weekly*
*ISSN: 0009-8620*
*Leonard A Eiserer, Publisher*
*Louise Harris, Editor*

**4128 Clearwaters Magazine**
New York Water Environment Association
525 Plum Street
Suite 102
Syracuse, NY 13204
315-422-7811
Fax: 315-422-3851
E-mail: pcr@nywea.org
http://www.nywea.org
Published by The New York Water Environment Association, a nonprofit educational association dedicated to the development and dissemination of information concerning water quality management and the nature, collection, treatment, anddisposal of wastewater. Founded in 1929, the Association has over 2,500 members. The NYWEA is a member association of the Water Environment Federation.
*Quarterly*
*Patricia Cerro-Reehil, Executive Director*
*Robert D Hennigan, Executive Editor*

**4129 Colorado Department of Natural Resources: Division of Water Resources: StreamLines**
1313 Sherman Street
Room 818
Denver, CO 80203
303-866-3581
Fax: 303-866-3589
http://www.water.state.co.us
The Colorado Division of Water Resources is an agency of the State of Colorado, Department of Natural Resources, operating under the direction of specific state stautes, court decrees, and interstate compacts. The DWR is empowered toadminister all surface and ground water rights throughout the state and ensure that the doctrine of prior appropiation is enforced.
*Publication Date: 1988 4-8 pages Quearterly*
*Hal D Simpson, Director Water Resources*
*Russell George, Executive Director*

**4130 Colorado Water Rights**
1580 Logan Street
#400
Denver, CO 80203
303-837-0812
Fax: 303-837-1607
E-mail: cwc@cowatercongress.org
http://www.cowatercongress.org
This newsletter helps the Colorado Water Congress protect and conserve Colorado's water resouces by educating its readers.
*Publication Date: 1982 4-16 pages Quarterly*

*Doug Kemper, Executive Director*

**4131 Confluence**
Texas Water Conservation Association

221 E 9th Street
Suite 206 http://www.twca.org
Austin, TX 78701
The official newsletter of the Texas Water Conservation Association. For those interested in water issues from river authorities to industrial concerns.
*Quarterly*

**4132 Domestic Water Treatment for Homeowners**
American Ground Water Trust
50 Pleasant Street Ste 2 603-228-5444
Concord, NH 03301 Fax: 603-228-6557
E-mail: trustinfo@agwt.org
http://www.agwt.org

*15 pages Quarterly*
*Andrew Stone, Director*

**4133 Environmental Policy Alert**
Inside Washington Publishers
1225 South Clark Street 703-416-8500
Suite 1400 800-424-9068
Arlington, VA 22202 Fax: 703-416-8543
E-mail: iwp@iwpnews.com
http://www.iwpnews.com
Is a reliable resource for all regulatory, congressional and litigation developments in air quality, waste cleanup, clean water and other environmental quality issues. Also provides a special focus on efforts to reinvent environmentalpolicies.
*Publication Date: 1984 Bi-Weekly*
*Jeremy Bernstein, Editor*

**4134 Georgia Water and Pollution Control Association: Operator**
2121 New Market Parkway 770-618-8690
Suite 144 Fax: 770-618-8695
Marietta, GA 30067 E-mail: info@gwpca.org
http://www.gawponline.org
The GW+PCA is dedicated to education, dissemination of technical and scientific information, increased public understanding and promotion of sound public laws and programs in the water resources and related environmental fields.Founded in 1932.
*Publication Date: 1970 56-68 pages Quarterly*
*Jack C Dozier, PE, Executive Director*

**4135 Georgia Water and Pollution Control Association: News & Notes**
2121 New Market Parkway 770-618-8690
Suite 144 Fax: 770-618-8695
Marietta, GA 30067 E-mail: info@gwpca.org
http://www.gawponline.org
The GW+PCA is dedicated to education, dissemination of technical and scientific information, increased public understanding and promotion of sound public laws and programs in the water resources and related environmental fields.Founded in 1932.
*Publication Date: 1970 20-28 pages Monthly*
*Jack C Dozier, PE, Executive Director*

**4136 Groundwater: A Course of Wonder**
American Ground Water Trust
50 Pleasant Street Ste 2 603-228-5444
Concord, NH 03301 Fax: 603-228-6557
E-mail: trustinfo@agwt.org
http://www.agwt.org

*15 pages Quarterly*
*Andrew Stone, Director*

**4137 Gulf of Mexico Science**
Dauphin Island Sea Lab
101 Bienville Boulevard 251-861-2141
Dauphin Island, AL 36528 Fax: 251-861-4646
http://www.disl.org
Journal devoted to disemminating knowledge of the Gulf of Mexico and adjacent areas. Appropriate topics of consideration for publication include all areas of marine science.
*2x Year*
*ISSN: 1087-688X*
*Carolyn Wood, Assistant Editor*

**4138 International Desalination and Water Reuse Quarterly**
Lineal Publishing Company
306 Eagle Dr 561-451-9429
Jupiter, FL 33477 Fax: 561-451-9435
Disseminates technical information, reviews and analyzes regional developments in the field, as well as new products and processes. The publication provides, on a continuing basis, a major vehicle in which to promote desalination andwater reuse technologies, equipment, and design to potential users.
*Irv Lineal, Publisher*

**4139 Journal of Soil and Water Conservation**
Soil & Water Conservation Society
945 SW Ankeny Road 515-289-2331
Ankeny, IA 50023 Fax: 515-289-1227
E-mail: swcs@swcs.org
http://www.swcs.org
Publication includes a variety of conservation subjects, as well as international conservation issues.
*Craig Cox, Executive Director*
*Deb Happe, Editor/Communications Dir*

**4140 Journal of the American Shore and Beach Preservation Association**
American Shore & Beach Preservation Association
5460 Beaujolais Lane 239-489-2616
Fort Myers, FL 33919 Fax: 239-489-9917
E-mail: exdir@asbpa.org
http://www.asbpa.org
Peer-reviewed journal of the ASBPA. It provides sound, interesting technical information concerning shores and beaches of the nation and worldwide.
*24 pages Quarterly*
*ISSN: 0037-4237*
*Kate & Ken Gooderham, Editors*
*Dr Beth Sciaudone, Managing Editor*

**4141 Journal of the New England Water Environment**
New England Water Environment Association
10 Tower Office Park 781-939-0908
Suite 601 Fax: 781-939-0907
Woburn, MA 01801 E-mail: mail@newea.org
http://www.newea.org
Bi-annual publication from New England Water Environment Association.
*Publication Date: 1929 150 pages Bi-Annual*
*ISSN: 1077-3002*
*Susan Landon, Journal/Publications Dir*

**4142 Journal of the North American Benthological Society**
North American Benthological Society
PO Box 7065
Lawrence, KS 66044 E-mail: amorin@outtawa.ca
http://www.benthos.org
The society is an international scientific organization that promotes better understanding of biotic communities of lake and stream bottoms and their role in aquatic ecosystems. The journal includes articles that promote the furtherunderstanding of benthic communities and helps members to keep current on interests.
*Quarterly*
*ISSN: 0887-3593*
*N LeRoy Poff, President*

**4143 Mass Waters**
Massachusetts Water Pollution Control Association
PO Box 221 978-374-0170
Groveland, MA 01834 Fax: 978-521-4083
E-mail: mwpca1965@verizon.net
http://www.mwpca.org

*Quarterly*
*John Connor, Secretary/Treasurer*

**4144 Mono Lake Committee Newsletter**
Corner of Hwy 395 & 3rd Street 760-647-6595
PO Box 29 Fax: 760-647-6377
Lee Vining, CA 93541 E-mail: info@monolake.org
http://www.monolake.org
Nonprofit citizen's group dedicated to: protecting and restoring the Mono Basin ecosystem; educating the public about Mono Lake and the impacts on the environment of excessive water use; promoting cooperative solutions that protectMono Lake and meet real water needs without transferring environmental problems to other areas.
*Publication Date: 1978 28 pages Quarterly*
*Geoffrey McQuilkin, Executive Director*
*Arya Degenhardt, Editor*

**4145 Montana Environmental Training Center Newsletter**
Hagener Science Center
Rm #110 406-265-3763
MSU-Northern Fax: 406-265-3750
Havre, MT 59501 E-mail: boylej@msun.edu
http://www.msun.edu/grants/metc/gary.asp
METC is a cooperative effort between Montana State University-Northern and the Montana Department of Environmental Quality. Basic, advance training, and continuing education in the areas of water and wastewater operation, maintenance,safety, process control, cross connection and backflow prevention along with courses in basic water science and watershed awareness define the training activities of METC. A newsletter and Training Announcement are published quarterly.
*Quarterly*
*Jan Boyle, Director*

**4146 Montana Environmental Training Center Training Announcement**
Hagener Science Center
Rm #110 406-265-3763
MSU-Northern Fax: 406-265-3750
Havre, MT 59501 E-mail: boylej@msun.edu
http://www.msun.edu/grants/metc/gary.asp
METC is a cooperative effort between Montana State University-Northern and the Montana Department of Environmental Quality. Basic, advance training, and continuing education in the areas of water and wastewater operation, maintenance,safety, process control, cross connection and backflow prevention along with courses in basic water science and watershed awareness define the training activities of METC. A newsletter and Training Announcement are published quarterly.
*Quarterly*
*Jan Boyle, Director*

**4147 Montana Water Environment Association: Newsletter**
516 N Park Street 406-449-7913
Suite A Fax: 406-449-6350
Helena, MT 59601
*Semi-Annual*
*Carl Anderson, President*

**4148 New Mexico Rural Water Association Newsletter**
3413 Carlisle Boulevard NE 505-884-1031
Albuquerque, NM 87110 800-819-9893
Fax: 505-884-1032
E-mail: contact @nmrwa.org
http://www.nmrwa.org
To provide top quality, responsive technical assistance and training for rural water and wastewater systems in New Mexico.
*Quarterly*
*Matthew Holmes, Executive Director*
*Robert Matthews, Co-Editor*

**4149 New York Water Environment Association Clearwaters**
525 Plum Street 315-422-7811
Suite 102 Fax: 315-422-3851
Syracuse, NY 13204 E-mail: pcr@nywea.org
http://www.nywea.org
Contains articles on environmental issues, regulatory changes, technological advances as well as, updates on members and activities.
*50 pages Quarterly*
*Patricia Cerro-Reehil, Executive Director*
*Hope Dodge, Editor*

**4150 Oregon Water Resources Congress Newsletter**
1201 Court Street NE 503-363-0121
Salem, OR 97301 Fax: 503-371-4926
E-mail: owrc@owrc.org
http://www.owrc.org
Is to promote the protection and use of water rights and the wise stewardship of water resources.
*Quarterly*
*Anita Winkler, Executive Director*
*Carol Zielinski, Editor*

**4151 Ozark National Scenic Riverways**
Ozark National Scenic Riverways
PO Box 490 573-323-4236
Van Buren, MO 63965 Fax: 573-323-4140
E-mail: ozar_superintendent@nps.gov
http://www.nps.gov/ozar
Missouri's largest National Park and America's first to preserve a free flowing river in its wild state. Covers some 80,000 acres along 134 miles of the Current and Jacks Fork Rivers. Staff provide elementary level environmentaleducation programs on natural history, with an emphasis on karst and water issues. Publishes More Than Skin Deep, a Teacher's Guide to Caves and Groundwater, a curriculum guide suitable for grades K-12.
*Publication Date: 1964 Annual*
*Noel Poe, Superintendent*

**4152 Pacific Rivers Council: Freeflow**
PO Box 10798 541-345-0119
Eugene, OR 97440 Fax: 541-345-0710
E-mail: info@pacrivers.org
http://www.pacrivers.org
Promoting the protection and restoration of rivers, their watersheds, and native aquatic species.
*Quarterly*
*David Bayles, Executive Director*
*Holly Spencer, Editor*

**4153 Pipeline**
National Evironmental Services Center
NRCCE Building, Evandale Drive 304-293-4191
PO Box 6064 800-624-8301
Morgantown, WV 26506 Fax: 304-293-3161
E-mail: nsfc_orders@mail.nesc.wvu.edu
http://www.nsfc.wvu.edu
Newsletter of the National Small Flows Clearinghouse, a nonprofit national source of information about small flows technologies-those systems that have fewer than one million gallons of wastewater flowing through them per day-rangingfrom individual septic systems to small sewage treatment plants. Free to US residents.
*Publication Date: 1990 8 pages Quarterly*
*ISSN: 1060-0043*
*Dr Gerald Iwan, Director*
*Jen Hause, Engineering Scientist*

**4154 Puerto Rico Water Resources and Environmental Research Institute Newsletter**
University of Puerto Rico
College of Engineering 787-833-0300
PO Box 9040 Fax: 787-832-0119
Mayaguez, PR 00681 E-mail: PRWRERI@uprm.edu
http://www.ece.uprm.edu/rumhp/prwrri
Its objectives are to: conduct research aimed at resolving local and national water resources problems; train scientists and engineers through hands-on participation in research; and to facilitate the incorporation of research resultsin the knowledge base of water resources professionals.
*Publication Date: 1990 4 pages Quarterly/thru Email*
*Jose R Cedeno, Associate Director*

**4155 Runoff Rundown**
Center for Watershed Protection
8390 Main Street 410-461-8323
Second Floor Fax: 410-461-8324
Ellicott City, MD 21043 E-mail: center@cwp.org
http://www.cwp.org or www.stormwatercenter.net
Electronic newsletter published by the Center for Watershed Protection, a nonprofit 501(c)3 organization dedicated to finding new ways to protect and restore our nation's streams, lakes, rivers and estuaries. The center publishesnumerous technical publications on all aspects of watershed protection, including stormwater management, watershed planning and better site design. All of our publications are available oneline at www.cwp.org.
*Publication Date: 2000 Quarterly*
*Hye Yeong Kwon, Executive Director*
*Lauren Lasher, Editor*

**4156 Save San Francisco Bay Association: Watershed Newsletter**
350 Frank H Ogawa Plaza 510-452-9261
Suite 900 Fax: 510-452-9266
Oakland, CA 94612 E-mail: SAVEBAY@savesfbay.org
http://www.savesfbay.org
Save the Bay has worked for over 40 years to protect the San Francisco Bay-Delta from pollution, fill, shoreline destruction and fresh water diversion. We have launched a century of renewal to restore bay fish and wildlife, reclaimtidal wetlands and make the bay safe and accessible to all.
*8-10 pages 3-4 times/year*
*David Lewis, Executive Director*
*Paul Revier, Editor*

**4157 Small Flow Quarterly**
NRCCE Building, Evandale Drive 304-293-4191
PO Box 6064 800-624-8301
Morgantown, WV 26506 Fax: 304-293-3161
E-mail: nsfc_orders@mail.nesc.wvu.edu
http://www.nsfc.wvu.edu
Magazine of the National Small Flows Clearinghouse, a nonprofit national source of information about small flows technologies-those systems that have fewer than one million gallons of wastewater flowing through them per day-rangingfrom individual septic systems to small sewage treatment plants. Free to US residents.
*Publication Date: 2000 50 pages Quarterly*
*ISSN: 1528-6827*
*Dr Gerald Iwan, Director*
*Jen Hause, Engineering Scientist*

**4158 South Carolina Sea Grant Consortium**
287 Meeting Street 843-727-2078
Charleston, SC 29401 Fax: 843-727-2080
http://www.scseagrant.org
A state agency that supports coastal and marine research, education, outreach, and one technical assistance program that fosters sustainable economic development and resource conservation. The consortium represents eight university andstate research organizations and induces a number of information products on coastal and marine resource topics.
*Publication Date: 1982 16 pages*
*M Richard DeVoe, Executive Director*

**4159 TCS Bulletin**
PO Box 25408 703-768-1599
Alexandria, VA 22313 Fax: 703-768-1596
E-mail: coastalsoc@aol.com
http://www.thecoastalsociety.org
Organization of private sector, academic, government professionals and students dedicated to actively addressing emerging coastal issues by fostering dialogue, forging partnerships and promoting communication and education. Thispublication covers issues of aquaculture-related law and coastal management research.
*Publication Date: 1975 24 pages Yearly*
*Paul Ticco, President*
*John Duff, Editor*

**4160 Tide**
Coastal Conservation Association
6919 Portwest 713-626-4234
Suite 100 800-201-FISH
Houston, TX 77024 E-mail: ccantl@joincca.org
http://www.joincca.org
TIDE is the official bimonthly magazine of the Coastal Conservation Association. It has received local, state and national acclaim for writing, photography and layout and currently boasts a circulation of more than 70,000. TIDE isavailable only to members of the Coastal Conservation Association.
*Bi-Monthly*
*Pat Murray, Executive Director*
*Ted Venker, Editor*

**4161 Utah Watershed Review**
Utah Association of Conservation Districts
1860 North 100 East 435-753-6029
Logan, UT 84341 Fax: 435-755-2117
http://www.uacd.org
Provides information about what's new in Utah and watershed volunteer work and management.
*Bi-Monthly*
*Gordon Younker, EVP*
*Jack Wilbur, Editor*

**4162 Water & Wastes Digest**
Scranton Gillette Communications
3030 W. Salt Creek Lane 847-391-1000
Suite 201 Fax: 847-390-0408
Arlington Heights, IL 60005 http://www.scrantongillette.com
This serves readers in the water and/or wastewater industries. These people work for municipalities, in industry, or as engineers. They design, specify, buy, operate and maintain equipment, chemicals, software and wastewater treatmentservices.
*128 pages*
*ISSN: 0043-1181*
*Dennis Martyka, Publisher*
*Tim Gregorski, Editorial Director*

**4163 Water Conservation in Your Home**
American Ground Water Trust
50 Pleasant Street Ste 2 603-228-5444
Concord, NH 03301 Fax: 603-228-6557
E-mail: trustinfo@agwt.org
http://www.agwt.org
*15 pages Quarterly*
*Andrew Stone, Director*

**4164 Water Quality Products**
Scranton Gillette Communications
3030 W. Salt Creek Lane 847-391-1000
Suite 201 Fax: 847-390-0408
Arlington Heights, IL 60005 http://www.scrantongillette.com
Provides balanced editorial content including developments in water conditioning, filtration and disinfection for residential, commercial and industrial systems.
*68 pages*
*ISSN: 1092-0978*
*Dennis Martyka, Publisher*
*Tracy Fabre, Editor*

**4165 Water Resource Center: Minnegram**
University of Minnesota
173 McNeal 1985 Buford Avenue 612-624-9282
St Paul, MN 55108 Fax: 612-625-1263
Fax: '
E-mail: ander045@umn.edu
http://www.wrc.coafes.umn.edu
Four University water programs, Extension Water Quality Program, Center for Hydrocultural Impacts on Water Quality, Water Resources Research Center and Water Resources Science Graduate Program make up the Water Resources Center. Thecenter sponsors and coordinates programs in research, graduate educa-

tion, outreach and service to address water resource management issues.
*Publication Date: 1986 Quarterly*
*Jim Anderson, Co-Director*
*Debra Swackhamer, Co-Director*

**4166 WaterMatters**
Southwest Florida Water Management District
2379 Broad Street 352-796-7211
Brookville, FL 34604 Fax: 352-754-6885
http://www.watermatters.org
Newletter of the Southwest Florida Water Management District, which manages the water and water-related resources within its boundaries. Maintains balance between the water needs of current and future users while protecting andmaintaining the natural systems that provide the District with its existing and future water supply. The Conservation Projects Section, is reponsible for managing water conservation, reclaimed water, other alternative source projects, and estimatingfuture water demands.
*2 pages Monthly*
*Dave Moore, Executive Director*
*Rebecca Bray, Editor*

## Books: Air & Climate

**4167 Air Pollution Control and the German Experience: Lessons for the United States**
Center for Clean Air Policy
750 1st Street NE 202-408-9260
Suite 940 Fax: 202-408-8896
Washington, DC 20002 E-mail: communications@ccap.org
http://www.ccap.org

**4168 Caring for Our Air**
Enslow Publishers, Inc
40 Industrial Road, Dept. F61 908-771-9400
PO Box 398 800-398-2504
Berkeley Heights, NJ 07922 Fax: 908-771-0925
E-mail: customerservice@enslow.com
http://www.enslow.com
*Publication Date: 1976*

**4169 Center for Resource Economics**
1718 Connecticut Avenue NW 202-232-7933
Suite 300 Fax: 202-234-1328
Wasington, DC 20009 E-mail: info@islandpress.org
http://www.islandpress.org
Works to educate the public about global environmental issues. Methods include publishing literature on environmental concerns.

**4170 Confronting Climate Change: Strategies for Energy Research and Development**
National Academy Press
500 5th Street NW 202-334-3313
Lockbox 285 888-624-8373
Washington, DC 20055 Fax: 202-334-2451
http://www.nap.edu
*Publication Date: 1990 144 pages*
*ISBN: 0-309043-47-6*

**4171 Fight Global Warming: 29 Things You Can Do**
Environmental Defense Fund
257 Park Avenue South 212-505-2100
17th Floor 800-684-3322
New York, NY 10010 Fax: 212-505-2375
E-mail: members@edf.org
http://www.edf.org
*Fred Krupp, President*
*David Yarnold, Executive Director*

**4172 Fundamentals of Stack Gas Dispersion**
Milton R. Beychok Consulting
1126 Colony Plaza 949-718-1360
Newport Beach, CA 92660 Fax: 949-718-1360
E-mail: mbeychok@air-dispersion.com
http://www.air-dispertion.com
The most comprehensive single-source reference book on dispertion modeling of continuous buoyant pollution plumes.
*Milton R Beychok, Principal*

**4173 Healing the Planet: Strategies for Resolving the Environmental Crisis**
Addison-Wesley Publishing Company
75 Arlington Street 617-848-7500
Suite 300 http://www.aw.com
Boston, MA 02116

**4174 Indoor Air Quality: Design Guide Book**
Fairmont Press
700 Indian Trail 770-925-9388
Liburn, GA 30047 Fax: 770-381-9865
E-mail: linda@fairmountpress.com
http://www.fairmontpress.com

**4175 Ozone Depletion and Climate Change: Constructing a Global Response**
SUNY Press
194 Washington Ave 518-472-5000
Suite 305 800-666-2211
Albany, NY 12210 Fax: 518-472-5038
E-mail: info@sunypress.edu
http://www.sunypress.edu
Available in both soft and hardcover, this book offers solutions that address climate change from a global viewpoint.
*Publication Date: 2005 276 pages*
*Matthew J Hoffman, Author*

**4176 Politics of Air Pollution: Urban Growth, Ecological Modernization, and Symbolic Inclusion**
SUNY Press
194 Washington Ave 518-472-5000
Suite 305 800-666-2211
Albany, NY 12210 Fax: 518-472-5038
E-mail: info@sunypress.edu
http://www.sunypress.edu
Available in both soft and hardcover, this title addresses the relationship between urban growth and pollution.
*Publication Date: 2005 152 pages*
*George A Gonzalez, Author*

**4177 To Breath Free: Eastern Europe's Environmental Crisis**
John Hopkins University Press
3400 N Charles Street 410-516-6900
Baltimore, MD 21218 800-537-5487
http://www.jhubookis.com
*Adam Glazer, Promotions Manager*

## Books: Business

**4178 Environmental Career Guide: Job Opportunities with the Earth in Mind**
J Wiley & Sons
605 3rd Avenue 212-850-6000
6th Floor Fax: 212-850-6088
New York, NY 10158 http://www.wiley.com
*Publication Date: 1991 208 pages*
*ISBN: 0-471534-13-7*
*Nicholas Basta, Author*

**4179 Environmental Disputes: Community Involvement in Conflict Resolution**
Island Press

PO Box 7
Covelo, CA 95428
707-983-6432
800-828-1302
Fax: 707-983-6414

A book published by Island Press which helps citizen groups, business and government understand how Envionmental Dispute Settlement-a set of procedures for settling disputes over environmental policies without litigation-can work forthem.

*Publication Date: 1990 295 pages*
*ISBN: 0-933280-74-2*
*James E Crowfoot, Julia Wondolleck, Author*

**4180 Globalization and the Environment: Greening Global Political Economy**
SUNY Press
194 Washington Ave
Suite 305
Albany, NY 12210
518-472-5000
800-666-2211
Fax: 518-472-5038
E-mail: info@sunypress.edu
http://www.sunypress.edu

Also in hardcover, 40.00.

*Publication Date: 2004 175 pages*
*Gabriela Kutting, Author*

**4181 Shopping for a Better Environment: Brand Name Guide to Environmentally Responsible Shopping**
Meadowbrook Press
5451 Smetana Drive
Minnetonka, MN 55343
800-338-2232
Fax: 952-930-1940
E-mail: info@meadowbrookpress.com
http://www.meadowbrookpress.com

## Books: Design & Architecture

**4182 Designing Healthy Cities**
Krieger Publishing Co.
PO Box 9542
Melbourne, FL 32902
321-724-9542
800-724-0025
Fax: 321-951-3671
E-mail: info@krieger-publishing.com
http://www.krieger-publishing.com

Krieger Publishing Company produces quality books in various fields of interest. We have an extensive Natural Science listing.

*Publication Date: 1998 158 pages*
*ISBN: 0-894649-27-2*
*Cheryl Stanton, Advertising*

**4183 Indoor Air Quality: Design Guide Book**
Fairmont Press
700 Indian Trail
Liburn, GA 30047
770-925-9388
Fax: 770-381-9865
E-mail: linda@fairmountpress.com
http://www.fairmontpress.com

## Books: Disaster Preparedness & Response

**4184 Acceptable Risk?: Making Decisions in a Toxic Environment**
University of California Press
2120 Berkeley Way
Berkeley, CA 94704
510-642-4247
Fax: 510-643-7127
E-mail: askucp@ucpress.edu
http://www.ucpress.edu

**4185 Borrowed Earth, Borrowed Time: Healing America's Chemical Wounds**
Plenum Publishers
233 Spring Street
New York, NY 10013
212-460-1500
Fax: 212-460-1575
http://www.plenum.com

*Publication Date: 1991*

## Books: Energy & Transportation

**4186 Coming Clean: Breaking America's Addiction to Oil And Coal**
Sierra Club Books
85 Second Street
2nd Floor
San Francisco, CA 94105
415-977-5500
Fax: 415-977-5799
E-mail: books.publishing@sierraclub.org
http://www.sieraaclub.org/books/

As Americans awaken to their addiction to oil and coal, we want to take action towards a cleaner path. This title provides the road map, showing how we can promote real solutions, and collectively pressure government and corporationsto change their energy priorities.

*256 pages*
*Michael Brune, Author*

**4187 Confronting Climate Change: Strategies for Energy Research and Development**
National Academy Press
500 5th Street NW
Washington, DC 20055
202-334-3313
888-624-8373
Fax: 202-334-2451
http://www.nap.edu

*Publication Date: 1990 144 pages*
*ISBN: 0-309043-47-6*

**4188 Energy & Environmental Strategies for the 1990's**
Fairmont Press
700 Indian Tr.
Lilburn, GA 30047
770-925-9388
Fax: 770-381-9865
E-mail: linda@fairmontpress.com
http://www.fairmontpress.com

**4189 Energy Management and Conservation**
National Conference of State Legislatures
7700 E First Place
Denver, CO 80230
303-364-7700
Fax: 303-364-7800
http://www.ncsl.org

**4190 Getting Around Clean & Green**
Northeast Sustainable Energy Association
50 Miles Street
Greenfield, MA 01301
413-774-6051
Fax: 413-774-6053
E-mail: nesea@nesea.org
http://www.nesea.org

This interdisciplinary science/social studies curriculum allows students to explore the transportation and environmental issues in their own lives. Activities cover: transportation systems, health impacts, environmental andtransportation histories, carpooling, and mass transit.

*90 pages Curriculum*
*David Barclay, Executive Director*
*Arianna Alexsandra Grindrod, Education Director*

**4191 Getting Around Without Gasoline**
Northeast Sustainable Energy Association
50 Miles Street
Greenfield, MA 01301
413-774-6051
Fax: 413-774-6053
E-mail: nesea@nesea.org
http://www.nesea.org

An interdisciplinary unit that explores the feasibility of getting around without using gasoline. Students can conduct various activities to compare powering vehicles with gasoline versus electricity.

*60 pages S & H Only*
*David Barclay, Executive Director*
*Arianna Alexsander Gridrod, Education Director*

**4192 Global Science: Energy, Resources, Environment**
Kendall-Hunt Publishing Company
4050 Westmark Drive 319-589-1000
PO Box 1840 800-772-9165
Dubuque, IA 52004 http://www.kendallhunt.com

**4193 Oil, Globalization, and the War for the Arctic Refuge**
SUNY Press
194 Washington Ave 518-472-5000
Suite 305 800-666-2211
Albany, NY 12210 Fax: 518-472-5038
E-mail: info@sunypress.edu
http://www.sunypress.edu

Also in hardcover, 71.50.
*Publication Date: 2006 227 pages*
*David M Stanlea, Author*

**4194 Transporting Atlanta: The Mode of Mobility Under Construction**
SUNY Press
194 Washington Ave 518-472-5000
Suite 305 800-666-2211
Albany, NY 12210 Fax: 518-472-5038
E-mail: info@sunypress.edu
http://www.sunypress.edu

*Publication Date: 2009 220 pages*
*Miriam Konrad, Author*

## Books: Environmental Engineering

**4195 Principles of Environmental Science and Technology**
Elsevier Science Publishers
360 Park Avenue South 212-989-5800
New York, NY 10010 Fax: 212-633-3990
http://www.elsevier.com

## Books: Environmental Health

**4196 Ecologue: The Environmental Catalogue and Consumer's Guide for a Safe Earth**
Prentice Hall Press (Simon & Schuster Division)
1 Gulf & Western Plaza 212-373-8500
New York, NY 10023 800-223-1360
http://www.prenhall.com

## Books: Gaming & Hunting

**4197 Better Trout Habitat: A Guide to Stream Restoration**
Island Press
1718 Connecticut Avenue NW 202-232-7933
Suite 300 Fax: 202-234-1328
Washington, DC 20009 E-mail: info@islandpress.org
http://www.islandpress.org

## Books: Habitat Preservation & Land Use

**4198 50 Simple Things Kids Can Do to Save the Earth**
Andrews and McMeel
4520 Main Street 816-932-6700
Suite 700 Fax: 816-932-6706
Kansas City, MO 64111

**4199 Access EPA: Clearinghouses and Hotlines**
National Technical Information Service
5285 Port Royal Road 703-487-4650
Springfield, VA 22161 E-mail: info@ntis.gov
http://www.ntis.gov
*Publication Date: 1991 57 pages*

**4200 Access EPA: Library and Information Services**
National Technical Information Service
5285 Port Royal Road 703-487-4650
Springfield, VA 22161 E-mail: info@ntis.gov
http://www.ntis.gov
*Publication Date: 1990 110 pages*

**4201 After Earth Day: Continuing the Conservation Effort**
University of North Texas Press
PO Box 311336 940-565-2142
Denton, TX 76203 800-826-8911
Fax: 940-565-4590
E-mail: rchrisman@unt.edu
http://www.unt.edu/untpress
*Publication Date: 1992 241 pages*
*ISBN: 1-574414-44-0*
*Karen DeVinney, Managing Editor*

**4202 Agatha's Feather Bed: Not Just Another Wild Goose Story**
Peachtree Publishers
1700 Chattahoochee Avenue 404-876-8761
Atlanta, GA 30318 Fax: 404-875-2578
E-mail: hello@peachtree-online.com
http://www.peachtree-online.com

*32 pages*
*ISBN: 1-561450-08-1*

**4203 America in the 21st Century: Environmental Concerns**
Population Reference Bureau
1707 H Street NW 202-530-5810
Suite 200 800-877-9881
Washington, DC 20006 Fax: 202-328-3937
E-mail: popref@prb.org
http://www.prb.org; www.popplanet.org

**4204 Ancient Ones: The World of the Old-Growth Douglas Fir**
Sierra Club Books
85 2nd Street 415-977-5500
2nd Floor Fax: 415-977-5799
San Francisco, CA 94105 http://www.sierraclub.org/books/
A children's book that offers insight on one of the oldest species of trees.
*32 pages*
*ISBN: 0-871566-82-6*
*Barbara Bash, Author*
*Suzanne Head, Editor*
*Robert Heinzman, Editor*

**4205 Association of State Wetland Managers Symposium**
Association of State Wetland Managers
2 Basin Road 207-892-3399
Windham, ME 04062 Fax: 207-892-3089
E-mail: aswm@aswm.org
http://www.aswm.org

*Jeanne Christie, Executive Director*
*Jon Kusler, Associate Director*

**4206 At Odds with Progress: Americans and Conservation**
University of Arizona Press
355 S Euclid Avenue 520-621-1441
Suite 103 Fax: 520-621-8899
Tucson, AZ 85719 http://www.uapress.arizona.edu
*Publication Date: 1991 255 pages*
*ISBN: 0-816509-17-4*
*Bret Wallach, Author*

**4207 Balancing on the Brink of Extinction**
Island Press

1718 Connecticut Avenue NW
Suite 300
Washington, DC 20009
202-232-7933
Fax: 202-232-1328
E-mail: info@islandpress.org
http://www.islandpress.org

*Publication Date: 1991 329 pages*
*Kathryn A Kohm, Author*

**4208 Beyond the Beauty Strip: Saving What's Left of Our Forests**
Tilbury House Publishers
2 Mechanic Street
Suite 3
Gardiner, ME 04345
207-582-1899
Fax: 202-582-8227
E-mail: tilbury@tilburyhouse.com
http://www.tilburyhouse.com

**4209 Biodiversity and Ecosystem Function**
Springer-Verlag
233 Spring Street
New York, NY 10013
212-460-1500
800-777-4643
Fax: 212-460-1575
E-mail: service-ny@springer-sbm.com
http://www.springeronline.com

*Publication Date: 1994 528 pages*

**4210 Bioemediation**
McGraw-Hill
1221 Avenue of the Americas
New York, NY 10020
212-512-2000
800-722-4726
http://www.magraw-hill.com

**4211 Bluebird Bibliography**
North American Bluebird Society
PO Box 43
Miamiville, OH 45147
812-988-1876
Fax: 330-359-5455
E-mail: info@nabluebirdsociety.org
http://www.nabluebirdsociety.org

**4212 Butterflies of Delmarva**
Delware Nature Society and Tidewater Publishers
PO Box 700
Hockessin, DE 19707
301-239-2334
Fax: 302-239-2473
E-mail: e-mail@dnashland.org
http://www.delawarenaturesociety.org

The result of the author's lifelong interest in the 61 adult butterfly species that naturally occur on the Delmarva Peninsula, this field guide clearly identifies the adult, larva and pulpa stages, discusses the differences in colorand wing patterns between sexes, as well as the habitat, range, and food sources of each species. 132 full-color photographs illustrate the text. Includes general butterfly information, and how to attrract them to your garden. Paperback

*Publication Date: 1998 138 pages 2nd Edition*
*ISBN: 0-870334-53-0*
*Dr. Elton N Woodbury, Author*

**4213 Clean Sites Annual Report**
Clean Sites
46161 West Lake Drive
Suite 230-B
Potomac Falls, VA 20165
703-519-2140
Fax: 703-519-2141
E-mail: cses@cleansites.com
http://www.cleansites.com

We apply sound project management principles, real-world experience, and cost control measures to find creative solutions to environmental remediation and land reuse problems.

*Douglas Ammon, Contact*

**4214 Connections: Linking Population and the Environment Teaching Kit**
Population Reference Bureau
1875 Connecticut Avenue NW
Suite 520
Washington, DC 20009
202-483-1100
800-877-9881
Fax: 202-328-3937
E-mail: popref@prb.org
http://www.prb.org

**4215 Conservation and Research Foundation Five Year Report**
Conservation and Research Foundation
PO Box 909
Shelburne, VT 05482
913-268-0076
Fax: 913-268-0076

Publication from Conservation and Research Foundation which is published every five years and distributed to contributors. This publication is also available upon request. Next edition to be published in Fall 2003. Please, call beforefaxing!

*Publication Date: 1998 43 pages*
*Dr Mary Wetzel, President*

**4216 Decade of Destruction: The Crusade to Save the Amazon Rain Forest**
Henry Holt and Company
175 Fifth Avenue
New York, NY 10010
646-307-5095
Fax: 212-633-0748
E-mail: publicity@hholt.com
http://www.henryholt.com

*215 pages*
*Adrian Cowell, Author*

**4217 Discordant Harmonies: A New Ecology for the Twenty-first Century**
Oxford University Press
198 Madison Avenue
New York, NY 10016
212-679-7300
Fax: 212-725-2972
http://www.oup.co.uk

*Publication Date: 1992 254 pages*
*ISBN: 0-195074-69-6*
*Daniel B Botkin, Author*

**4218 Earth Keeping**
Zondervan Publishing House
5300 Patterson Avenue SE
Grand Rapids, MI 49530
616-698-6900
Fax: 616-698-3439
http://www.zondervan.com

**4219 Earthright**
Prima Publishing & Communications
PO Box 1260BK
Rocklin, CA 95677
916-786-0426
800-632-8676
Fax: 916-632-4405
http://www.primapublishing.com

**4220 Ecology of Greenways: Design and Function of Linear Conservation**
University of Minnesota Press
111 Third Avenue South
Suite 290
Minneapolis, MN 55401
612-627-1970
800-388-3863
Fax: 612-627-1980
E-mail: lfreeman@epx.cis.umn.edu
http://www.upress.umn.edu

*Publication Date: 1994 238 pages*
*ISBN: 0-816621-57-8*
*Daniel S Smith, Paul Cawood Hellmund, Author*

**4221 Eli's Songs**
MacMillan Publishing Company
866 3rd Avenue
New York, NY 10022
212-702-2000
800-257-5755
http://www.macmillian.com

**4222 Endangered Kingdom: The Struggle to Save America's Wildlife**
John Wiley & Sons
605 3rd Avenue
New York, NY 10158
212-850-6890
800-825-7550
Fax: 212-850-8800
http://www.wiley.co.uk

*Publication Date: 1991 241 pages*
*ISBN: 0-471528-22-6*
*Roger L DiSilvestro, Author*

**4223 Environment in Peril**
Smithsonian Institution Press
SI Building, Room 153, MRC 010
PO Box 37012
Washington, DC 20013
202-633-1000
800-782-4612
Fax: 202-633-5285
E-mail: info@si.edu
http://www.si.edu

**4224 Environmental Concern in Florida and the Nation**
University of Florida Press
15 NW 15th Street
Gainesville, FL 32611
352-392-1351
800-226-3822
Fax: 352-392-7302
http://www.upf.com
*Publication Date: 1997 144 pages*
*ISBN: 0-813010-56-X*
*Lance Dehaven-Smith, Author*

**4225 Environmental Concern: A Comprehensive Review of Wetlands Assessment Producers**
POW-The Planning of Wetlands
201 Boundary Lane
PO Box P
St Michaels, MD 21663
410-745-9620
Fax: 410-745-3517
E-mail: order@wetland.org
http://www.wetland.org
Since its founding in 1972, EC has been specializing in consulting, planning design, education services, construction services and research related to all aspects of wetlands. As wetlands and contiguous upland forests and meadows areinteracting ecosystems EC specializes in consulting, planning, design, and project supervision services for such upland ecosystem constructions and restorations for the purpose of wetland buffers, reforestation, wildlife habitat and critical areas ofpreservation.
*Publication Date: 1999 Quarterly*
*ISBN: 1-883226-04-x*
*Suzanne Pitenger Slear, President*

**4226 Environmental Concern: Evaluation for Planned Wetlands**
POW-The Planning of Wetlands
201 Boundary Lane
PO Box P
St Michaels, MD 21663
410-745-9620
Fax: 410-745-3517
E-mail: order@wetland.org
http://www.wetland.org
Since its founding in 1972, EC has been specializing in consulting, planning design, education services, construction services and research related to all aspects of wetlands. As wetlands and contiguous upland forests and meadows areinteracting ecosystems EC specializes in consulting, planning, design, and project supervision services for such upland ecosystem constructions and restorations for the purpose of wetland buffers, reforestation, wildlife habitat and critical areas ofpreservation.
*Publication Date: 1994 Quarterly*
*ISBN: 1-883226-03-1*
*Suzanne Pittenger Slear, President*

**4227 Environmental Concern: The Wonders of Wetlands**
POW-The Planning of Wetlands
201 Boundary Lane
PO Box P
St Michaels, MD 21663
410-745-9620
Fax: 410-745-3517
E-mail: order@wetland.org
http://www.wetland.org
Since its founding in 1972, EC has been specializing in consulting, planning design, education services, construction services and research related to all aspects of wetlands. As wetlands and contiguous upland forests and meadows areinteracting ecosystems EC specializes in consulting, planning, design, and project supervision services for such upland ecosystem constructions and restorations for the purpose of wetland buffers, reforestation, wildlife habitat and critical areas ofpreservation.
*Publication Date: 1995*
*ISBN: 1-888631-00-7*
*Suzanne Pittenger Slear, President*

**4228 Environmental Crisis: Opposing Viewpoints**
Greenhaven Press
PO Box 289009
San Diego, CA 92128
858-485-9549
800-231-5163
Fax: 800-550-5448
E-mail: info@grennhaven.com
http://www.greenhaven.com

**4229 Environmental Profiles: A Global Guide to Projects and People**
Garland Publishing
717 5th Avenue
25th Floor
New York, NY 10022
212-751-7447
Fax: 212-308-9399
http://www.garlandpub.com
*Publication Date: 1993 1112 pages*
*ISBN: 0-815300-63-8*

**4230 Friends of the Earth Foundation Annual Report**
Friends of the Earth Found.
218 D Street SE
Washington, DC 20003
202-544-2600
Fax: 202-543-4710
http://www.oceanic-society.org

**4231 Future Primitive**
University of North Texas Press
PO Box 311336
Denton, TX 76203
940-565-2142
800-826-8911
Fax: 940-565-4590
E-mail: rchrisman@unt.edu
http://www.unt.edu/untpress
*Publication Date: 1996 223 pages*
*ISBN: 1-574410-07-5*
*Ronald Chrisman, Director*
*Karen DeVinney, Managing Editor*

**4232 Going Green: A Kid's Handbook to Saving the Planet**
Puffin Books
375 Hudson Street
New York, NY 10014
212-366-2403
http://www.puffin.co.uk
Out of print—limited availability.
*Publication Date: 1990*
*ISBN: 0-140345-97-3*

**4233 Guide to Spring Wildflower Areas**
Minnesota Native Plant Society
1520 St. Olaf Avenue
Northfield, MN 55057
507-786-2222
E-mail: MNPS@HotPOP.com
http://www.stolaf.edu
Updated its guide to over 40 wildflower sites in the Twin Cities area. The guide contains a description and location for each.
*4 per year*

**4234 Guide to Urban Wildlife Management**
National Institute for Urban Wildlife
10921 Trotting Ridge Way
Columbia, MD 21044
301-596-3311

**4235 Information Please Environmental Almanac**
Houghton Mifflin Company
222 Berkeley Street
30th Floor
Boston, MA 02116
617-725-5000
http://www.hmco.com
*Publication Date: 1992 704 pages*
*ISBN: 0-395637-67-8*

**4236 International Protection of the Environment**
Oceana Publications, Inc
198 Madison Avenue
New York, NY 10016
800-334-4249
Fax: 212-726-6476
E-mail: info@oceanalaw.com
http://www.oceanalaw.com

This set provides the documents which form the framework of softlaw administrative instruments for the implementation of international environment treaties under Agenda 21.
*Publication Date: 1995 7 vol pages Bi-Monthly*
*ISBN: 0-379102-95-1*
*Nicholas A Robinson & Wolfgang Burhenne, Author*

**4237 International Society for Endangered Cats**
3070 Riverside Drive Suite 160 614-487-8760
Columbus, OH 43221 Fax: 614-487-8769
*Publication Date: 1990 237 pages*
*ISBN: 0-816019-44-4*
*Bill Simpson, President*
*Patricia Currie, Executive Director*

**4238 Just A Dream**
Houghton Mifflin Company
Beacon Street 617-725-5000
30th Floor
Boston, MA 02108

**4239 Krieger Publishing Company: Wildlife Habitat Management of Wetlands**
Krieger Publishing Co.
PO Box 9542 321-724-9542
Melbourne, FL 32902 800-724-0025
Fax: 321-951-3671
E-mail: info@krieger-publishing.com
http://www.krieger-publishing.com
Krieger Publishing Company produces quality books in various fields of interest. We have an extensive Natural Science listing.
*Publication Date: 1992 572 pages*
*ISBN: 1-575240-89-0*
*Cheryl Stanton, Advertising*

**4240 Krieger Publishing Company: Wildlife Habitat Management of Forestlands/Rangelands/Farmlands**
Krieger Publishing Co.
PO Box 9542 321-724-9542
Melbourne, FL 32902 800-724-0025
Fax: 321-951-3671
E-mail: info@krieger-publishing.com
http://www.krieger-publishing.com
Krieger Publishing Company produces quality books in various fields of interest. We have an extensive Natural Science listing.
*Publication Date: 1994 868 pages*
*ISBN: 1-575240-93-9*
*Cheryl Stanton, Advertising*

**4241 Last Extinction**
MIT Press
5 Cambridge Center 617-253-5646
Cambridge, MA 02142 Fax: 617-258-6779
Today there is a new and more widespread awareness of what some consider to be the great tragedy of our time - organisms which took many thousands or even millions of years to evolve are being snuffed out permanently owing to humanactivity.
*Publication Date: 1993*
*Les Kaufman, Kenneth Mallory, Author*

**4242 Mastering Nepa: A Step-By-Step Approach**
Solano Press
PO Box 773
Point Arena, CA 95468 800-931-9373
Fax: 707-884-4109
http://www.solano.com
*Publication Date: 1993 250 pages*
*ISBN: 0-923956-14-x*
*Ronald E Bass, Albert I Herson, Author*

**4243 National Wildlife Rehabilitators Association Annual Report**
National Wildlife Rehabilitators Association
14 N 7th Avenue 320-259-4086
St Cloud, MN 56303 Fax: 320-259-4086
E-mail: nwra@nwrawildlife.org
http://www.nwrawildlife.org
Please, call before you fax!

**4244 Nature and the American: Three Centuries of Changing Attitudes**
Unviersity of Nebraska Press
1111 Lincoln Mall 402-472-3581
Lincoln, NE 68588 http://www.nebraskapress.unl.edu

**4245 Ordinance Information Packet**
Scenic America
1634 I Street NW 202-638-0550
Suite 510 Fax: 202-638-3171
Washington, DC 20006 http://www.scenic.org

**4246 Ozone Diplomacy: New Directions in Safeguarding the Planet**
Harvard University Press
79 Garden Street 617-495-2600
Cambridge, MA 02138 Fax: 617-495-5898
E-mail: botref@oeb.harvard.edu
http://www.hup.harvard.edu
Offers an insider's view of the politics, economics, science and diplomacy involved in creating the precedent-setting treaty to protect the Earth: the 1987 Montreal Protocol on Substances That Deplete the Ozone Layer.
*Richard Elliot Benedick, Author*

**4247 Practical Guide to Environmental Management**
1616 P Street NW 202-939-3800
Suite 200 Fax: 202-939-3868
Washington, DC 20036

**4248 Preserving the World Ecology**
H W Wilson Company
950 University Avenue 718-588-8400
Bronx, NY 10452 Fax: 718-588-6365
http://www.hwwilson.com

**4249 Protecting Our Environment: Lessons from the European Union**
SUNY Press
194 Washington Ave 518-472-5000
Suite 305 800-666-2211
Albany, NY 12210 Fax: 518-472-5038
E-mail: info@sunypress.edu
http://www.sunypress.edu
Available in both hard and soft cover, this title deals with the environment as a global issue.
*Publication Date: 2005 204 pages*
*Janet R Hunter, Zachary A Smith, Author*

**4250 Quill's Adventures in Grozzieland**
John Muir Publications
PO Box 613 505-982-4078
Santa Fe, NM 87504 800-888-7504
Fax: 505-988-1680

**4251 RARE Center for Tropical Conservation Annual Report**
Rare Center for Tropical Conservation
1616 Walnut Street 215-735-3510
Suite 911 Fax: 215-735-3615
Philadelphia, PA 19103 http://www.rarecenter.org

**4252 Resource Conservation and Management**
Wadsworth Publishing Company
10 Davis Drive 415-595-2350
Belmont, CA 94002 Fax: 415-637-7544
http://www.wadsworth.com

**4253 Revolution for Nature: From the Environment to the Connatural World**
University of North Texas Press
PO Box 311336
Denton, TX 76203
940-565-2142
800-826-8911
Fax: 940-565-4590
E-mail: rchrisman@unt.edu
http://www.unt.edu/untpress

*145 pages*
*ISBN: 1-574417-0X-*
*Ronald Chrisman, Director*
*Karen DeVinney, Managing Editor*

**4254 Saving Sterling Forest: The Epic Struggle to Preserve New York's Highlands**
SUNY Press
194 Washington Ave
Suite 305
Albany, NY 12210
518-472-5000
800-666-2211
Fax: 518-472-5038
E-mail: info@sunypress.edu
http://www.sunypress.edu

Also in hardcover, 59.50.
*Publication Date: 2007 216 pages*
*Ann Botshon, Author*

**4255 Seed Listing**
Native Seeds/SEARCH
526 N. Fourth Avenue
Tucson, AZ 85705
520-622-5561
Fax: 520-622-5591
http://www.nativeseeds.org

**4256 Statement of Policy and Practices forProtection of Wetlands**
National Wildlife Fed. Corporate Conservation Coun
1400 16th Street NW
Washington, DC 20036
202-797-6870
Fax: 202-797-6871

**4257 Student Conservation Association Northwest: Lightly on the Land**
1265 S Main Street
Suite 210
Seattle, WA 98144
206-324-4649
Fax: 206-324-4998
http://www.sca-inc.org

SCA is a national organization with regional offices in Seattle, Oakland, Pittsburg, Washington DC and headquartered in Charlestown NH. Our mission is to build the next generation of conservation leaders and inspire lifelongstewardship of our environment and communities by engaging young people in hands-on service to the land. We offer a wide range of internships and crew based programs for ages 16 years and up.
*Publication Date: 1996 267 pages*
*ISBN: 0-898869-91-7*
*Su Thieds, Director of Regional Progams*

**4258 Transactions of Annual North American Wildlife and Natural Resources Conference**
Wildlife Management
1101 14th Street NW
Suite 801
Washington, DC 20005
202-371-1808
Fax: 202-408-5059

**4259 Urban Wildlife Manager's Notebook**
National Institute for Urban Wildlife
10921 Trotting Ridge Way
Columbia, MD 21044
301-596-3311

**4260 Wetlands Protection: The Role of Economics**
Environmental Law Institute
1616 P Street NW
Suite 200
Washington, DC 20036
202-939-3800
Fax: 202-939-3868

**4261 Wilderness Society Annual Report**
1615 M Street NW
Washington, DC 20036
202-833-2300
800-THE-WILD
E-mail: member@tws.org
http://www.wilderness.org

Deliver to future generations an unspoiled legacy of wild places, with all the precious values they hold: Biological diversity; clean air and water; towering forests, rushing rivers, and sage-sweet, silent deserts.
*Brenda Davis, Chair*
*Doug Walker, Vice Chair*

**4262 Wildlife Conservation in Metropolitan Environments**
National Institute for Urban Wildlife
10921 Trotting Ridge Way
Columbia, MD 21044
301-596-3311

**4263 Wildlife Habitat Relationships in Forested Ecosystems**
Timber Press
133 SW 2nd Avenue
Suite 450
Portland, OR 97204
503-227-2878
Fax: 503-227-3070
http://www.timber-press.com

Available by special order.
*Publication Date: 1997*
*David R Patton, Author*

**4264 Wildlife Research and Management in the National Parks**
University of Illinois Press
1325 S Oak Street
Champaign, IL 61820
217-333-0950
Fax: 217-244-8082
http://www.press.uillinois.edu

*Publication Date: 1992 240 pages*
*ISBN: 0-252018-24-9*
*Gerald R Wright, Author*

**4265 Wildlife Reserves and Corridors in the Urban Environment: A Guide to Ecological Landscape**
National Institute for Urban Wildlife
10921 Trotting Ridge Way
Columbia, MD 21044
301-596-3311

Out of print—limited availability.
*Publication Date: 1989 91 pages*
*ISBN: 0-942015-02-9*
*Lowell W Adams, Louise E Dove, Author*

**4266 Wildlife-Habitat Relationships: Concepts and Applications**
University of Wisconsin Press
114 N Murray Street
Madison, WI 53715
608-262-8782

Anyone working with wildlife must be concerned with its habitat identification, measurement and analysis. Wildlife-Habitat Relationships goes beyond introductory wildlife biology texts and specialized studies of single species toprovide a broad but advanced understanding of habitat relationships applicable to all terrestrial species.
*Publication Date: 1998 416 pages*
*ISBN: 0-299156-40-0*
*Michael L Morrison, Bruce G Marcot, Author*

## Books: Recycling & Pollution Prevention

**4267 An Ontology of Trash: The Disposable and Its Problematic Nature**
SUNY Press
194 Washington Ave
Suite 305
Albany, NY 12210
518-472-5000
800-666-2211
Fax: 518-472-5038
E-mail: info@sunypress.edu
http://www.sunypress.edu

Also in hardcover, 65.00.
*Publication Date: 2008 238 pages*
*Greg Kennedy, Author*

**4268 Aunt Ipp's Museum of Junk**
HarperCollins
10 E 53rd Street
New York, NY 10022
212-207-7000
800-424-6234
Fax: 212-207-7433
http://www.harpercollins.com

**4269 Beyond 40 Percent: Record-Setting Recycling and Composting Programs**
Island Press
1718 Connecticut Avenue NW
Suite 300
Washington, DC 20009
202-232-7933
Fax: 202-234-1328
E-mail: info@islandpress.org
http://www.islandpress.org
*Publication Date: 1991 280 pages*
*ISBN: 1-559630-73-6*

**4270 Borrowed Earth, Borrowed Time: Healing America's Chemical Wounds**
Plenum Publishers
233 Spring Street
New York, NY 10013
212-460-1500
Fax: 212-460-1575
http://www.plenum.com
*Publication Date: 1991*

**4271 Caring for Our Air**
Enslow Publishers
40 Industrial Road, Dept. F61
PO Box 398
Berkeley Heights, NJ 07922
800-398-2504
Fax: 908-964-4116
http://www.enslow.com

**4272 Community Recycling: System Design to Management**
Prentice Hall
Route 9W
Englewood Cliffs, NJ 07632
201-592-2000
800-947-7700
E-mail: orders@prenhall.com
http://www.prenhall.com
A guide for getting into the growing business of community recycling, for those with little or no previous experience with the technical details of recycling. Discusses marketing, management, equipment, profit comparisons of variousprocessing methods and legal considerations.
*Publication Date: 1992 240 pages*
*ISBN: 0-131557-89-0*
*Nyles V Reinfeld, Carl M Layman, Author*

**4273 Garbage and Recycling**
Kingfisher Publications
http://www.kingfisherpub.com
*Publication Date: 1995 32 pages*
*ISBN: 1-856976-15-7*
*Rosie Harlow, Sally Morgan, Author*

**4274 Hey Mr. Green: Sierra Magazine's Answer Guy Tackles Your Toughest Green Living Questions**
Sierra Club Books
85 Second Street
2nd Floor
San Francisco, CA 94105
415-977-5500
Fax: 415-977-5799
http://www.sierraclub.org/books
When is the right time to replace an old refridgerator? Is it more environmentally correct to buy your beer in bottles or cans? And is it okay to knit a sweater with acrylic (pertoleum-based) yarn? Bob Schildgen has been Mr. Green inSierra Magazine for several years now, providing fact-backed replies to reader's questions. Well organized, funny, and supremely useful, this title offers green-living tips for everyday.
*Publication Date: 2009 224 pages*
*ISBN: 1-578051-43-4*
*Bob Schildgen, Author*

**4275 How On Earth Do We Recycle Glass?**
Millbrook Press
2 Old New Milford Road
PO Box 335
Brookfield, CT 06804
203-740-2220
800-462-4703
Fax: 203-740-2526
http://www.millbrookpress.com

**4276 Let's Talk Trash: The Kids' Book About Recycling**
Waterfront Books
85 Crescent Road
Burlington, VT 05401
802-658-7477
http://www.waterfrontsbooks.com

**4277 Plastic: America's Packaging Dilemma**
Island Press
1718 Connecticut Avenue NW
Suite 300
Washington, DC 20009
202-232-7933
Fax: 202-234-1328
E-mail: info@islandpress.org
http://www.islandpress.org

**4278 Pollution Knows No Frontiers**
Paragon House of Publishers
1925 Oakcrest Avenue
Suite 7
St. Paul, MN 55113
651-644-3087
Fax: 651-644-0997
E-mail: paragon@paragonhouse.com
http://www.paragonhouse.com

**4279 Recycle!: A Handbook for Kids**
Little, Brown & Company
1271 Avenue of the Americas
New York, NY 10020
800-759-0190
Fax: 212-522-0885
http://www.twbookmark.com
*Publication Date: 1996 32 pages*
*Gail Gibbons, Editor*

**4280 Recycling Paper: From Fiber to Finished Product**
TAPPI Press
15 Technology Parkway South
Norcross, GA 30092
770-446-1400
Fax: 770-446-6947
http://www.tappi.org

**4281 Reducing Toxics**
Island Press
1718 Connecticut Avenue NW
Suite 300
Washington, DC 20009
202-232-7933
Fax: 202-234-1328
E-mail: info@islandpress.org
http://www.islandpress.org
*Publication Date: 1995 460 pages*
*Robert Gottlieb, Editor*

## Books: Sustainable Development

**4282 Biodiversity Prospecting: Using Genetic Resources for Sustainable Development**
World Resources Institute
10 G Street NE
Suite 800
Washington, DC 20002
202-729-7600
Fax: 202-729-7610
http://www.wri.org

**4283 Building Sustainable Communities: An Environmental Guide for Local Government**
Global Cities Project
2926 Philmore Street
San Francisco, CA 94123
415-775-0791
http://www.globalcities.org

**4284 Center for Ecoliteracy**
2528 San Peblo Avenue
Berkeley, CA 94702
510-845-4595
Fax: 510-845-1439
E-mail: info@ecoliteracy.org
http://www.ecoliteracy.org

The Center for Ecoliteracy is dedicated to fostering a profound understanding of the natural world, grounded in direct experience that leads to sustainable patterns of living.
*Publication Date: 2000 90 pages Paperback*
*ISBN: 0-967565-23-5*
*Zenobia Barlow, Executive Director*

**4285 Constructing Sustainable Development**
SUNY Press
194 Washington Avenue
Suite 305
Albany, NY 12210
518-472-5000
800-666-2211
Fax: 518-472-5038
*Publication Date: 2000 188 pages*
*Neil E Harrison, Author*

**4286 Ecological Literacy: Educating Our Children for a Sustainable World**
Sierra Club Books
85 Second Street
2nd Floor
San Francisco, CA 94105
415-977-5500
Fax: 415-977-5799
http://www.sierraclub.org/books
*256 pages*
*ISBN: 1-578051-53-3*
*Michael K Stone and Zenobia Barlow, Author*

**4287 Environmental Defense Annual Report**
Environmental Defense
257 Park Avenue South
17th Floor
New York, NY 10010
212-505-2100
800-684-3322
Fax: 212-505-2375
E-mail: members@edf.org
http://www.edf.org
Environmental Defense believes that a sustainable environment will require economic and social systems that are equitable and just.
*Fred Krupp, President*
*David Yarnold, Executive Director*

**4288 Environmental Integration: Our Common Challenge**
SUNY Press
194 Washington Ave
Suite 305
Albany, NY 12210
518-472-5000
800-666-2211
Fax: 518-472-5038
E-mail: info@sunypress.edu
http://www.sunypress.edu
Also in hardcover, 85.00.
*Publication Date: 2009 290 pages*
*Ton Buhrs, Author*

**4289 Environmental Policy Making: Assessing the Use of Alternative Policy Instruments**
SUNY Press
194 Washington Ave
Suite 305
Albany, NY 12210
518-472-5000
800-666-2211
Fax: 518-472-5038
E-mail: info@sunypress.edu
http://www.sunypress.edu
Also in hardcover, 85.00.
*Publication Date: 2005 276 pages*
*Michael T Hatch, Author*

**4290 Environmental Profiles: A Global Guide to Projects and People**
Garland Publishing
717 5th Avenue
25th Floor
New York, NY 10022
212-751-7447
Fax: 212-308-9399
http://www.garlandpub.com
*Publication Date: 1993 1112 pages*
*ISBN: 0-815300-63-8*

**4291 Global Environment**
Jones and Bartlett Publishers
40 Tall Pine Drive
Sudbury, MA 01776
800-832-0034
Fax: 978-443-8000
E-mail: info@jbpub.com
http://www.jbpub.com

**4292 Gnat is Older than Man: Global Environment and Human Agenda**
Princeton University Press
41 William Street
Princeton, NJ 08540
609-258-4900
800-777-4726
Fax: 609-258-6305
http://www.pup.princeton.edu

**4293 Implementation of Environmental Policies in Developing Countries**
SUNY Press
194 Washington Ave
Suite 305
Albany, NY 12210
518-472-5000
800-666-2211
Fax: 518-472-5038
E-mail: info@sunypress.edu
http://www.sunypress.edu
A Case of Protected Areas and Tourism in Brazil. Also in hardcover, 50.00.
*Publication Date: 2008 150 pages*
*Jose Antonio Puppim de Oliveira, Author*

**4294 Managing Sustainable Development**
Earthscan Publications
8-12 Camden High Street
London
207-387-8558
Fax: 207-387-8998
*Publication Date: 2001 304 pages*
*Michael Carley, Editor*

**4295 Practice of Sustainable Development**
Urban Land Institute
1025 Thomas Jefferson Street NW
Suite 500 W
Washington, DC 20007
202-624-7000
Fax: 202-624-7140
E-mail: customerservice@uli.org
http://www.uli.org
*Publication Date: 2000 160 pages*
*ISBN: 0-874208-31-9*
*Douglas R Porter, Author*

**4296 Sustainable Planning and Development**
WIT Press
978-667-5841
Fax: 978-667-7582
E-mail: salesUSA@witpress.com
http://www.witpress.com
*Publication Date: 2003 1048 pages*
*ISBN: 1-853129-85-2*
*Linda Ouellette, Customer Service Manager*

**4297 The Incompleat Eco-Philosopher:Essay from the Edges of Environmental Ethics**
SUNY Press
194 Washington Ave
Suite 305
Albany, NY 12210
518-472-5000
800-666-2211
Fax: 518-472-5038
E-mail: info@sunypress.edu
http://www.sunypress.edu
Also in hardcover for 65.50
*Publication Date: 2009 210 pages*
*Anthony Weston, Author*

**4298 Urban Sprawl, Global Warming, and the Empire of Capital**
SUNY Press
194 Washington Ave
Suite 305
Albany, NY 12210
518-472-5000
800-666-2211
Fax: 518-472-5038
E-mail: info@sunypress.edu
http://www.sunypress.edu

Also in hardcover, 60.00.
*Publication Date: 2009 170 pages*
*George A Gonzalez, Author*

**4299 Who Gets What? Domestic Influences on International Negotiations Allocating Shared Resources**
SUNY Press
194 Washington Ave 518-472-5000
Suite 305 800-666-2211
Albany, NY 12210 Fax: 518-472-5038
E-mail: info@sunypress.edu
http://www.sunypress.edu
Also in hardcover, 60.00.
*Publication Date: 2008 192 pages*
*Aslaug Asgeirsdottir, Author*

**4300 Worldwatch Paper 101: Discarding the Throwaway Society**
Worldwatch Intitutes
1776 Massachusetts Ave. NW 202-452-1999
Washington, DC 20036 Fax: 202-296-7365
http://www.worldwatch.org

## Books: Travel & Tourism

**4301 Appalachian Mountain Club**
Appalachian Mountain Club Books
5 Joy Street 617-523-0636
Boston, MA 02108 800-262-4455
Fax: 617-523-0722
E-mail: information@outdoors.org
http://www.outdoors.org
The AMC, founded in 1876, promotes the protection, enjoyment, and wise use of the mountains, rivers and trails of the Northeast. We encourage people to enjoy and protect the natural world because we believe that successful conservationdepends on this experience. The AMC publishes an award-winning magazine and more than 60 guide books to the Northeast.
*Andrew Falender, Executive Director*
*Chase O'Connell, Director Development*

**4302 Prospect Park Handbook**
Greensward Found
Lenox Hill Station 212-473-6283
PO Box 610 http://www.greenswardparks.org
New York, NY 10021

## Books: Water Resources

**4303 And Two if By Sea: Fighting the Attack on America's Coasts**
Coast Alliance
202-546-9609
This book is the benchmark in the effort to save the coasts.
*Publication Date: 1986*

**4304 Comparative Health Effects Assessment of Drinking Water Treatment Technologies**
Government Institutes Division
16855 Northchase Drive 281-673-2800
Houston, TX 77060 http://www.govinst.com
The report evaluates the public health impact of the most widespread drinking water treatment technologies, with particular emphasis on disinfection.
*Publication Date: 1988 20 pages*

**4305 Dying Oceans**
Gareth Stevens, Inc
PO Box 360140 414-332-3520
Strongsville, OH 44136 800-542-2595
Fax: 414-332-3567
http://www.garethstevens.com
*Publication Date: 1991*
*ISBN: 0-836804-76-7*

**4306 Freshwater Resources and Interstate Cooperation: Strategies to Mitigate an Environmental Risk**
SUNY Press
194 Washington Ave 518-472-5000
Suite 305 800-666-2211
Albany, NY 12210 Fax: 518-472-5038
E-mail: info@sunypress.edu
http://www.sunypress.edu
Also in hardcover, 60.00.
*Publication Date: 2008 184 pages*
*Frederick D Gordon, Author*

**4307 Living Waters: Reading the Rivers of the Lower Great Lakes**
SUNY Press
194 Washington Ave 518-472-5000
Suite 305 800-666-2211
Albany, NY 12210 Fax: 518-472-5038
E-mail: info@sunypress.edu
http://www.sunypress.edu
Also in hardcover for 45.00
*Publication Date: 2009 213 pages*
*Margaret Wooster, Author*

**4308 Managing Troubled Water: The Role of Marine Environmental Monitoring**
Duke University Press
905 W Main Street 919-687-3600
Suite 18B Fax: 919-688-4574
Durham, NC 27701 http://www.dukepress.edu
*Publication Date: 1990*

**4309 Turning the Tide: Saving the Chesapeake Bay**
Island Press
1718 Connecticut Avenue NW 707-983-6432
Suite 300 800-828-1302
Washington, DC 20009 E-mail: info@islandpress.org
http://www.islandpress.org
The Chesapeake Bay is one of the most productive and important ecosystems on earth, and as such is a model for other estuaries facing the demands of commerce, tourism, transportation, recreation and other uses. Turning the Tidepresents a comprehensive look at two decades of efforts to save the bay, outlining which methods have worked and which have not.
*Publication Date: 2003 352 pages*
*ISBN: 1-559635-48-7*
*Tom Horton, Author*

**4310 Using Common Sense to Protect the Coasts: The Need to Expand Coastal Barrier Resources**
Coast Alliance
PO Box 505 732-872-0111
Sandy Hook, NJ 07732 E-mail: coast@coastalliance.org
http://www.coastalliance.org
This report gives a common sense approach to protecting coastal areas from unwise development that would benefit American taxpayers.
*Publication Date: 1990*

## Library Collections

**4311 3M: 201 Technical Library**
3M Center 651-575-1300
St. Paul, MN 55133 Fax: 651-736-3940
http://www.3M.com
High-tech library that manages its collection with 3M digital identification.

**4312 Acres International Library**
100 Sylvan Parkway 716-689-3737
Amherst, NY 14228 Fax: 716-689-3749
E-mail: amherst@acres.com
http://www.acres.com
Serves clients in the hydroelectric power, highways and bridges, mining, heavy industrial, civil/geotechnical and environmental and hazardous waste sectors.
*Marion D'Amboise, Librarian*

**4313 Alaska Department of Fish and Game Habitat Library**
333 Raspberry Road 907-267-2314
Anchorage, AK 99518 Fax: 907-349-1723
*Celia Rozen, Contact*

**4314 Alaska Resources Library and Information Services**
ARLIS Suite 111 Library Building 907-272-7547
3211 Providence Drive Fax: 907-786-7652
Anchorage, AK 99508 E-mail: reference@arlis.org
http://www.arlis.org
ARLIS is the mother lode of Alaska resources information. ARLIS has served as the central library for rresource information supporting management of 235 million acres of federal and 100 million acres of state and water resourcesthroughout Alaska.
*Publication Date: 1997*
*Carrie Holba, Reference Services Coord*

**4315 American Academy of Pediatrics**
141 NW Point Boulevard 847-434-4000
PO Box 747 Fax: 847-434-8000
Elk Grove Village, IL 60007 http://www.aap.org
Dedicated to the health of all children.

**4316 American Water Works Association**
6666 W Quincy Avenue 303-794-7711
Denver, CO 80235 Fax: 303-347-0804
http://www.awwa.org
Dedicated to the promotion of public health and welfare in the provision of drinking water of unquestionable quality and sufficient quantity. AWWA must be proactive and effective in advancing the technology, science, management andgovernment policies relative to the stewardship of water.
*Jack W Hoffbuhr, Executive Director*

**4317 Aquatic Research Institute: Aquatic Sciences and Technology Archive**
2242 Davis Court 510-782-4058
Hayward, CA 94545 Fax: 510-784-0945
Library and data base in aquatic sciences and technologies also research faculty in aquatic sciences.
*V Parker, Archv*

**4318 Arizona State Energy Office Information Center**
3800 N Central 602-280-1402
Suite 1200 Fax: 602-280-1445
Phoenix, AZ 85012 E-mail: energy@azcommerce.com
*Maxine Robertson, Assistant Director*

**4319 Arizona State University Architecture and Environmental Design Library**
College of Architecture and Environmental Design
4300 480-965-6400
Tempe, AZ 85287 Fax: 480-727-6965
E-mail: deborah.koshinsky@asu.edu
http://www.asu.edu/caed/AEDlibrary
*Deborah H Koshinsky, Director*

**4320 Arkansas Energy Office Library**
1 State Capitol Mall 501-682-1370
Little Rock, AR 72201 Fax: 501-682-2703
E-mail: cbenson@1800arkansas.com

**4321 Atmospheric Sciences Model Division Library**
US Environmental Protection Agency
79 TW Alexander Drive 919-541-4536
4201 Building, Room 308 Fax: 919-541-1379
Research Triangle Park, NC 27711 http://www.epa.gov
Serves the NOAA Division assigned to support the EPA National Exposure Laboratory and Office of Air Quality Planning and Standards. The major field of interest is the meteorological aspects of air pollution, including numerical andphysical model development and application.
*Evelyn M Poole-Kober, Tech. Pubns.*

**4322 Belle W Baruch Institute for Marine Biology and Coastal Research Library**
607 EWS Building 803-777-5288
Columbia, SC 29208 Fax: 803-777-3935

**4323 Bickelhaupt Arboretum Education Center**
340 S 14th Street 319-242-4771
Clinton, IA 52742

**4324 Brown University Center for Environmental Studies Library**
135 Angel Street 401-863-3449
Box 1943 Fax: 401-863-3503
Providence, RI 02912 E-mail: envstudies@brown.edu
http://www.envstudies.brown.edu
The Center for Environmental Studies at Brown University was established with the primary aim of educating individuals to solve challenging environmental problems both at the local and global levels. It also works directly to improvehuman well-being and environmental quality through community, city, and state partnerships in service and research.
*J Timmons Roberts, Program Director*

**4325 Burroughs Audubon Center and Library**
Burroughs Audubon Society
21905 SW Woods Chapel Road 816-795-8177
Independence, MO 64050

**4326 CH2M Hill**
Corvallis Regional Office Library
80112 541-752-4271
http://www.ch2m.com
The firm's solutions keep sustainability always in mind, along with government regulations, environmental concerns, maintenance requirements, and public perceptions. A team of experts brings the knowledge gained from a wide range ofprojects around the world, rigourous attention to detail, and a capability to create innovative solutions that are also models for the industry.
*Shirley Fisher, COO*

**4327 California Energy Commission Library**
1516 9th Street 916-654-4292
MS 10 Fax: 916-654-4046
Sacramento, CA 95814 E-mail: library@energy.state.ca.us
http://www.energy.ca.gov/library.index.html
The state's central repository for information on all forms of energy. The collection consists of more than 22,000 titles on energy policy, energy conservation, energy consumption, electric utilities, environmental issues, petroleum,natural gas, solar, wind, biomass, nuclear power and related subjects. The Library serves Energy Commission staff, California state government agencies, the Legislature and its staff, and members of the public.
*Karen Kasuba n, Librarian*

**4328 California State Resources Agency Library**
1416 9th Street 916-653-2225
Room 117 Fax: 916-653-1856
Sacramento, CA 95814

Contains books, documents and subscriptions on topics including: flood control; natural resources (in California); endangered species (in California); soil conservation; water; water pollution; water quality; water resources;conservation and water supply.

**4329 Center for Coastal Fisheries and Habitat Research: Rice Library**
101 Pivers Island Road 252-728-8713
Beaufort, NC 28516 Fax: 252-838-0809
E-mail: patti.marraro@noaa.gov
http://www8.nos.noaa.gov/ricelibrary
Ensures the delivery of scientific, technical, and legistlative information to library users including NOAA staff, general public, academia, industry, and governmental agencies. Houses comprehensive coverage of marine fisheries,fisheries statistics, habitat restoration, mapping and remote sensing, marine chemistry, pollution and toxicology, living marine resources, protected species, and oceanography.
*Patti M Marraro, Technical Info Specialist*

**4330 Center for Health, Environment and Justice Library**
150 S Washington Street, Ste 300 703-237-2249
PO Box 6806 Fax: 703-237-8389
Falls Church, VA 22040-6806 E-mail: chej@chej.org
http://www.chej.org
Works to build healthy communities, with social justice, economic well-being, and democratic governance. We believe this can happen when individuals from communities have the power to play an integral role in promoting human health andenvironmental integrity. Our role is to provide the tools to build strong, healthy communities where people can live, work, learn, play and pray.
*Lois Marie Gibbs, Executive Director*

**4331 Clinton River Watershed Council Library**
101 Main Street 248-601-0606
Suite 100 Fax: 248-601-1280
Rochester, MI 48307 http://www.crwc.org

**4332 Colorado River Board of California**
770 Fairmont Avenue 818-500-1625
Suite 100 Fax: 818-543-4685
Glendale, CA 91203 E-mail: crb@crb.ca.gov
http://www.crb.ca.gov

**4333 Columbia River Inter-Tribal Fish Commission**
StreamNet Library
729 NE Oregon Street 503-736-3581
Suite 190 Fax: 503-731-1260
Portland, OR 97232 E-mail: fishlib@critfc.org
http://www.fishlib.org
Serving the scientific and environmental community of the Pacific Northwest, The StreamNet Library works in cooperation with the region's fish and wildlife recovery efforts. The library provides access to technical information on theColumbia Basin fisheries, ecosystem and other relevant subjects for states in the Pacfic Northwest. The library collections emphasize less commonly available grey literature, such as consultant's reports, state documents and nonprofit organizations'reports.
*Lenora Oftedahl, Librarian*
*Todd Hannon, Assistant Librarian*

**4334 DER Research Library**
Pennsylvania Department of Environmental Resources
Box 8458 717-787-9647
Harrisburg, PA 17105 Fax: 717-772-0288

**4335 Dawes Arboretum Library**
7770 Jacksontown Road SE 740-323-2355
Newark, OH 43056 800-44D-AWES
Fax: 740-323-4058
http://www.dawesarb.org

**4336 Delaware River Basin Commission Library**
25 State Police Drive 609-883-9500
Box 7360 Fax: 609-883-9522
West Trenton, NJ 08628 http://www.drbc.net
The Commission is a federal/interstate agency responsible for managing the water resources at the Delaware River Basin.
*Carol R Collier, Executive Director*
*Clarke Rupert, Communications Director*

**4337 Division of Water Resources Library**
Kansas Department of Agriculture
109 SW 9th Street 785-296-3717
2nd Floor Fax: 785-296-1176
Topeka, KS 66612

**4338 Duke University Biology: Forestry Library**
Duke University
Perkins Library 919-660-5880
Durham, NC 27708 Fax: 919-684-2855
http://www.lib.duke.edu
*David M Talbert*

**4339 Earthworm Recycling Information Center**
35 Medford Street 617-628-1844
Somerville, MA 02143 Fax: 617-628-2773
*John Perkins, Contact*

**4340 Eastern States Office Library**
US Bureau of Land Management
7450 Boston Boulevard 703-440-1561
Springfield, VA 22153 Fax: 703-440-1599
*Terry Lewis, Contact*

**4341 Eastern Technical Associates Library**
PO Box 1009 919-878-3188
Garner, NC 27529 Fax: 919-872-5199
E-mail: tomrose@eta-is-opacity.com
http://www.eta-is-opacity.com
Environmental consulting firm. Research results published in government reports.
*Publication Date: 1979*
*Thomas H Rose, President*

**4342 Ecology Center Library**
2530 San Pablo Avenue 510-548-2220
Berkeley, CA 94702 Fax: 510-548-2240
E-mail: info@ecologycenter.org
http://www.ecologycenter.org.

**4343 Environment and Natural Resources Branch Library**
US Department of Justice
One Congress Street 617-918-1807
Suite 1100 Fax: 617-918-1810
Boston, MA 02114 E-mail: friedman.fred@epa.gov
http://www.eoa.gov
Research library for Solid Wasteto conduct research and answer questions in the subject fields of nonhazarodus solid waste and recycling.
*Leola Decker, Librarian*

**4344 Environmental Action Coalition Library: Resource Center**
625 Broadway 212-677-1601
2nd Floor Fax: 212-505-8613
New York, NY 10012 http://www.enviro-action.org
*Paul Berizzi, Executive Director*

**4345 Environmental Coalition on Nuclear Power Library**
433 Orlando Avenue 814-237-3900
State College, PA 16803 Fax: 814-237-3900
*Dr Judith Johnsrud, Executive Officer*

**4346 Environmental Contracting Center Library**
ENSR Consulting and Engineering
Box 2105 970-493-8878
Fort Collins, CO 80522 800-722-2440
E-mail: faq/default.asp
http://www.ensr.com
*Beth Mullan, Librarian*

**4347 Environmental Research Associates Library**
PO Box 219 610-449-7400
Villanova, PA 19085 Fax: 610-449-7404
Research and consulting ecologists and testing firm. Research results published in professional journals. Research results for private clients.
*Publication Date: 1970*
*M H Levin PhD, Director*

**4348 Federated Conservationists of Westchester County (FCWC) Office Resource Library**
78 N Broadway 914-422-4053
White Plains, NY 10603 Fax: 914-289-0539
E-mail: info@fcwc.org
http://www.fcwc.org

**4349 Fish and Wildlife Reference Service**
5430 Grosvenor Lane 301-492-6403
Suite 110 800-582-3421
Bethesda, MD 20814 Fax: 301-564-4059
To provide policy guidance regarding the operation and use of the Fish and Wildlife Reference Service.
*Paul E Wilson, Project Manager*

**4350 Florida Conservation Foundation**
1191 Orange Avenue 407-644-5377
Winter Park, FL 32789

**4351 Forest History Society Library and Archives**
701 William Vickers Avenue 919-682-9319
Durham, NC 27701 Fax: 919-682-2349
E-mail: coakes@duke.edu
http://www.foresthistory.org
The Forest History Society is a non-profit educational institution that links the past to the future by identifying, collecting, preserving, interpreting, and disseminating information on the history of people, forests, and theirrelated resources.
*Publication Date: 1946*
*Cheryl Oakes, Librarian*
*Steven Anderson, President*

**4352 Galveston District Library**
US Army Corps of Engineers
Box 1229 409-766-3196
Galveston, TX 77553 Fax: 409-766-3905
E-mail: clark.bartee@usace.army.mil
http://www.swg.usace.army.mil/library.htm
*Clark Bartee*

**4353 Georgia State Forestry Commission Library**
PO Box 819 912-751-3480
Macon, GA 31202 Fax: 912-751-3465
*Fred Allen, Director*

**4354 Glen Helen Association Library**
405 Corry Street 937-767-7375
Yellow Springs, OH 45387

**4355 Great Lakes Environmental Research Laboratory**
2205 Commonwealth Boulevard 734-741-2235
Ann Arbor, MI 48105 Fax: 734-741-2055
http://www.glerl.com
Conducts integrated interdiciplinary environmental research in support of resource management and environmental services in costal and esturine water with special emphasis on the Great Lakes.

**4356 Huxley College of Environmental Studies**
Western Washington University 360-650-3000
Bellingham, WA 98225 E-mail: huxley@cc.wwu.edu
One of the oldest environmental studies colleges in the nation. Innovative and indisciplinary academic programs reflect a broad view of the physical, biological, social and cultural world.
*Hailey Outzs, Coordinater*

**4357 Illinois State Water Survey Library**
208 Water Survey Research Center 217-244-5459
2204 Griffith Drive Fax: 217-333-6540
Champaign, IL 61820 E-mail: library@sws.uiuc.edu
http://www.sws.uiuc.edu/chief
The Illinois State Water Survey, a division of the office of Scientific Research and Analysis of the Illinois Department of Natural Resources and affiliated with the University of Illinois, is the primary agency in Illinois concernedwith water and atmosheric resources.
*Patricia G Morse, Librarian*

**4358 Institute of Ecosystem Studies**
65 Sharon Turnpike 845-677-5343
Box AB Fax: 845-677-5976
Millbrook, NY 12545 E-mail: Cadwalladerj@ecostudies.org
http://www.ecostudies.org
Ecology research and education institution; independent; international.
*Jill Cadwallader, Public Information Officer*

**4359 International Academy at Santa Barbara Library**
800 Garden Street 805-965-5010
Suite D Fax: 805-965-6071
Santa Barbara, CA 93101
*Susan J Shaffer, Office Manager*

**4360 International Game Fish Association**
300 Gulf Stream Way 954-927-2628
Dania Beach, FL 33004 Fax: 954-924-4299
E-mail: hq@igfa.org
http://www.igfa.org
Founded as record-keeper and to maintain fishing rules. Today, emphasis is on conservation and education. Encourages youngsters to enter the sport and maintains a huge library on the subject of fishing. Has a network of well over 300representatives around the world, many of whom are conservation leaders in their communities.
*Rob Kramer, President*

**4361 Interstate Oil and Gas Compact Commission Library**
900 NE 23rd Street 405-525-3556
Box 53127 Fax: 405-525-3592
Oklahoma City, OK 73152 E-mail: iogcc@iogcc.state.ok.us
http://www.iogcc.oklaosf.state.ok.us
*W Timothy Dowd, Executive Director*

**4362 Lake Michigan Federation**
17 N State Street 312-939-0838
Suite 1390 Fax: 312-939-2708
Chicago, IL 60602 E-mail: chicago@greatlakes.org
http://www.lakemichigan.org
Works to restore fish and wildlife habitat, conserve land and water, and eliminate pollution in the watershed of America's largest lake. We achieve these through education, research, law, science, economics and strategic partnerships.

**4363 Lionael A Walford Library**
74 Magruder Road 732-872-3035
Highlands, NJ 07732 Fax: 732-872-3088
*Claire L Steimle, Librarian*

**4364 Los Angeles County Sanitation District Technical Library**
PO Box 4998 562-699-7411
Whittier, CA 90607 Fax: 562-699-5422
http://www.lacsd.org

**4365 Louisiana Department of Environmental Quality Information Resource Center**
7290 Bluebonnet Boulevard 225-765-0169
2nd Floor Fax: 225-765-0222
Baton Rouge, LA 70810 E-mail: pattyb@deq.state.la.us
To promote a healthy environment by providing a specialized environmental library to meet the informational and educational needs of the DEQ employees and the citizens of Louisiana.
*Patty Birkett, Tech. Librarian*

**4366 Marine Environmental Sciences Consortium**
Dauphin Island Sea Lab
101 Bienville Boulevard 251-861-2141
Dauphin Island, AL 36528 Fax: 251-861-4646
http://www.disl.org
*Carolyn Wood, Assistant Editor*

**4367 Massachusetts Audubon Society's Berkshire Wildelife Sanctuaries**
Pleasant Valley Wildlife Sanctuary
472 W Mountain Road 413-637-0320
Lenox, MA 01240 Fax: 413-637-0499
E-mail: berkshires@massaudubon.org
http://www.masssaudubon.org
The Massachusetts Audubon Society is an environmental organization with emphases in conservation, advocacy and education. The advocacy effort is statewide and features a legislative team in Boston.
*Publication Date: 1896*
*Rene Laubach, Sanctuary Director*

**4368 Minneapolis Public Library and Information Center**
Technology and Science Department
300 Nicolet Mall 612-372-6570
Minneapolis, MN 55401 Fax: 312-372-6546
The varied collection in the Technology/Science/Government Documents department runs from agriculture to zoology, computers to cooking, engineering to handicrafts, medicine to motorcycle repair to military science. Special resourcesinclude a complete US Patent and Trademark collection and the CASSIS Patent Trademark Databases, a collection of US industrial standards, including publications from ANSI (American National Standards Institutes).

**4369 Minnesota Department of Natural Resources DNR Library**
500 Lafayette Road 651-297-4929
Box 21 Fax: 651-297-4946
St. Paul, MN 55155 E-mail: dnr.library@state.mn.us
15,000 titles on natural resource subjects available on interlibrary loan.
*Jo Ann Musumeci, Librarian*

**4370 Minnesota Department of Trade and Economic Development Library**
500 Metro Square 651-296-8902
121 7th Place E Fax: 651-296-1290
St. Paul, MN 55101
*Pat Fenton, Sr. Librarian*

**4371 Minot State University Bottineau Library**
105 Simrall Boulevard 701-228-5454
Bottineau, ND 58318 Fax: 701-228-5468
http://www.misu-b.nodak.edu
*Jan Wysocki, Library Director*

**4372 Mississippi Department of Environmental Quality Library**
PO Box 20307 601-961-5024
Jackson, MS 39289 Fax: 601-354-6965
E-mail: ronnie_sanders@deq.state.ms.us
http://www.deq.state.ms.ud
Geology and environmental reference library. Holdings in geosciences, hydrology, pollution control, paleontology, petroleum geology, land and water resources. Special collections; Topographic maps, United States, State andInternational Geoloical survey publications.
*Ronnie Sanders, Librarian*

**4373 Missouri Department of Natural Resources Geological Survey & Resource Assessment Division**
Box 250 573-368-2101
Rolla, MO 65401 Fax: 573-368-2111
*Mimi Garstang, Director/State Geologist*

**4374 National Audubon Society: Aullwood Audubon Center and Farm Library**
1000 Aullwood Road 937-890-7360
Dayton, OH 45414 Fax: 937-890-2382
E-mail: aullwood@gemair.com
Known as the Miami Valley's first educational farm, here visitors will discover a variety of native grasses and flowers, 300 year old oak trees and threatened bird species.

**4375 National Institute for Urban Wildlife Library**
10921 Trotting Ridge Way 301-596-3311
Columbia, MD 21004 http://www.webdirectory.com/wildlife/
*Louise E Dove, Wildlife Biology*

**4376 Native Americans for a Clean Environment Resource Office**
Box 1671 918-458-4322
Tahlequah, OK 74465 Fax: 918-458-0322
NACE is to raise the consciousness of Indian people and the general public about environment hazards, with an emphasis on the nuclear industry.
*Lance Hughes, Executive Director*

**4377 Nature Conservancy Long Island Chapter**
Uplands Farm Environmental Center
250 Lawrence Hill Road 516-367-3225
Cold Spring Harbor, NY 11724 Fax: 516-367-4715

**4378 Nebraska Natural Resources Commission Planning Library**
301 Centennial Mall S 402-471-2081
Box 94876 Fax: 402-471-3132
Lincoln, NE 68509 E-mail: mosaic@nrcdec.nrc.state.ne.us
http://www.nrc.state.ne.us/

**4379 New England Coalition on Nuclear Pollution Library**
PO Box 545 802-257-0336
Brattleboro, VT 05302 E-mail: energy@necnp.org
http://www.necnp.org

**4380 New England Governors' Conference Reference Library**
76 Summer Street 617-423-6900
Boston, MA 02110 Fax: 617-423-7327
E-mail: info@negc.org
http://www.negc.org

**4381 Occupational Safety and Health Library**
1111 3rd Avenue 206-553-5930
Suite 715 Fax: 206-553-6499
Seattle, WA 98101

**4382 Ohio Environmental Protection Agency Library**
122 South Front Street 614-644-3024
Columbus, OH 43215 Fax: 614-728-9500
http://www.epa.state.oh.us
*Ruth Ann Evans, Librarian*

**4383 Peninsula Conservation Foundation Library of the Environment**
3921 E Bayshore Road 650-962-9876
Palo Alto, CA 94303 Fax: 650-962-8234

**4384 People, Food and Land Foundation Library**
35751 Oak Springs 559-855-3710
Tollhouse, CA 93667 E-mail: sunmt@sunmt.org
http://www.sunmt.org

**4385 Rainforest Action Network Library**
221 Pine Street 415-398-4404
Suite 500 Fax: 415-398-2732
San Francisco, CA 94104 E-mail: rainforest @ran.org
Rainforest Action Network works to protect the Earth's rainforests and support the rights of their inhabitants through education, grassroots organizing and non-violent direct action.
*Michael Brune, Executive Director*

**4386 Region 2 Library**
US Environmental Protection Agency
290 Broadway 212-637-3185
16th Floor Fax: 212-637-3086
New York, NY 10007 http://www.epa.gov/region02/library
Is a research and reference library for use by EPA staff, EPA contractors, other government agencies, and the public. The library contains or has access to scientific and technical materials in paper and electronic media related to awide variety of environmental issues, with and emphasis on EPA's Region 2.
*Eveline M Goodman, Head Librarian*

**4387 Region 9 Library**
US Environmental Protection Agency
75 Hawthorne Street 13th Floor 415-774-1510
San Francisco, CA 94105 Fax: 415-744-1474
E-mail: libaray-reg9@epa.gov
http://www.epa.gov/region9/library
*Deborra Samuels, Hd. Libn./Coord.*

**4388 Rob and Bessie Welder Wildlife Foundation Library**
Walker Wildlife Foundation 361-364-2643
PO Box 1400 Fax: 361-364-2650
Sinton, TX 78387 E-mail: welderwf@aol.com
http://www.members.aol.com/welderwf/welderhome
Private, nonprofit operation foundation which conducts research and education in wildlife management and related fields. Funds graduate fellowships and conducts its reserach and education program on its 7,800 acre wildlife refuge inthe surrounding South Texas region and throughout the United States.
*Dr. D Lynn Drawe, Director*
*Vandra Davis, Librarian*

**4389 Schuylkill Center for Environmental Education**
8480 Hagy's Mill Road 215-482-7300
Philadelphia, PA 19128 Fax: 215-482-8158
http://www.schuylkillcenter.org
*Karin James, Resource Librarian*

**4390 Society of American Foresters Information Center**
5400 Grosvenor Lane 301-897-8720
Bethesda, MD 20814 Fax: 301-897-3690
http://www.safnet.org
An organization that represents the forestry profession in the United States. Its mission is to advance the science, education, technology and practice of forestry.
*Jeff Ghannam, Director Media Relations*

**4391 Solartherm Library**
1315 Apple Avenue 301-587-8686
Silver Spring, MD 20910 Fax: 301-587-8688
http://www.solartherm.com

**4392 Solid Waste Association of North America**
1100 Wayne Avenue
Suite 700 800-467-9262
Silver Spring, MD 20910 Fax: 301-589-7068
E-mail: info@swana.org
http://www.swana.org
Nonprofit trade association designed to serve the municipal solid waste industry in cutting-edge informational and technilogical practices.
*Dr John Skinner, Executive Director*

**4393 Southeast Fisheries Laboratory Library**
75 Virginia Beach Drive 305-361-4229
Miami, FL 33149 Fax: 305-361-4499

**4394 Southwest Research and Information Center**
105 Stanford SE 505-262-1862
PO Box 4524 Fax: 505-262-1864
Albuquerque, NM 87106 E-mail: sricdon@earthlink.net
http://www.sric.org
SRIC exists to provide timely, accurate information to the public on matters that affect the environment, human health, and communities in order to protect natural resources, promote citizen participation, and ensure environmental andsocial justice now and for future generations.
*Dan Hancock, Administrator*
*Annette Aguayo, Information Specialist*

**4395 St Paul Plant Pathology Library**
395 Borlaug Hall 612-625-9777
St Paul Campus
St Paul, MN 55108
Subject oriented library, specializing in plant diseases, plant virology, mycology, mycotoxicology and the effects of air pollution on vegetation. The collection contains approximately 8000 volumes of books, periodicals and PlantPathology theses. Over 50 current periodicals are recieved.

**4396 St. Paul Forestry Library**
University of Minnesota
B-50 Skok Hall 612-624-3222
2003 Upper Buford Circle Fax: 612-624-3733
St. Paul, MN 55108 E-mail: heroL228@umn.edu
http://http://forestry.lib.umn.edu
Houses a general collection of books, journals, government documents, maps, and pamphlets relating to the subjects of forestry, forest products, outdoor recreation, range management, and remote sensing. There is also a small generalrefernce section. Also compiles and maintains four databases focused on aspects of forestry: Social Sciences in Forestry, Urban Forestry, Tropical Conservation and Development, Trails Planning Construction and Maintenance Planning.
*Philip Herold, Librarian*

**4397 State University of New York**
College of Environmental Science and Forestry
Environamental Science and Forestry 315-470-6715
Syracuse, NY 13210 Fax: 315-470-6512
E-mail: spweiter@esf.edu
http://www.esf.edu/moonlib
Moon Library supports the SUNY College of Environmental Science and Forestry where students major in Engineering, Chemistry, Biology, Landscape Architecture, Forest Resources Management and Environmental Studies.
*Stephen Weiter, Director/College Libraries*

**4398 Staten Island Institute of Arts and Sciences**
William T Davis Education Center
75 Stuyvesant Place 718-987-6233
State Island, NY 10301 Fax: 718-273-5683
*Patricia Salmon, Curator of History*

**4399 Texas Water Commission Library**
PO Box 13087 512-463-7834
Austin, TX 78711

**4400 Turner, Collie and Braden Library**
Box 130089 713-267-2826
Houston, TX 77219 Fax: 713-780-0838
E-mail: rushbrookd@tcbhou.com
http://www.tcbhou.com
*David Rushbrook, Librarian*
*Renee Miller, Library Assistant*

**4401 US Bureau of Land Management**
California State Office
2135 Butano Drive 916-978-4400
Sacramento, CA 95825 Fax: 916-978-4305
It is the mission of the Bureau of Land management to sustain the health, diversity and productivity of the public lands for the use an employment of present and future generations.

**4402 US Bureau of Land Management Library**
Denver Federal Center Building 50 303-236-6648
Box 25047 Fax: 303-236-4810
Denver, CO 80225 E-mail: blm_library@blm.gov
http://www.blm.gov/nstc/library/library.html
The BLM Library serves the information and research needs of BLM personnel. The library also serves as the point of contract for bureau publications and information with other federal agen-

cies and the public. The collection covers allaspects of land management and natural resources.
*Barbara Campbell, Director*

**4403 US Department of Agriculture: National Agricultural Library, Water Quality Info Center**
10301 Baltimore Boulevard 301-504-6077
Beltsville, MD 20705 Fax: 301-504-7098
E-mail: wqic@nal.usda.gov/wqic
http://www.nal.usda.gov/wqic
Collects, organizes and communicates the scientific findings, educational methologies and public policy issues related to water and agriculture.
*Joseph R Makuch, Coord. WQIC*

**4404 US Geological Survey: Great Lakes Science Center**
1451 Green Road 734-994-3331
Ann Arbor, MI 48105 Fax: 734-994-8780
E-mail: GS-B-GLSC-Webmaster@usgs.gov
http://www.glsc.usgs.gov
The USGS Great Lakes Science Center exists to meet the Nation's need for scientific information for restoring, enhancing, managing, and protecting living resources and their habitats in the Great Lakes. The center is headquartered inAnn Arbor, Michigan, and has biological research stations and vessels located throughout the Great Lakes basin.
*Russell Strach, Center Director*
*Jacqueline F Savino, Deputy Center Director*

**4405 US Geological Survey: National Wetlands Research Center**
700 Cajundome Boulevard 337-266-8692
Lafayette, LA 70506 Fax: 337-266-8841
E-mail: nwrclibrary@usgs.gov
http://www.nwrc.usgs.gov/library
The National Wetlands Research Center is a source and clearinghouse of science information about wetlands in the United States and the world for fellow agencies, private entities, academia, and the public at large. Staff members obtainand provide this information by performing original scientific research and developing research results into literature and technological tools. They then disseminate that information through a variety of means.

**4406 US Geological Survey: Upper Midwest Environmental Sciences Center Library**
2630 Fanta Reed Road 608-781-6215
La Crosse, WI 54603 Fax: 608-783-6066
A federal library with technical holdings mainly in aquatic sciences, bird and amphibean materials.
*Kathy Mannstedt, Librarian*

**4407 US Geological Survey: Water Resources Division Library**
375 S Euclid Avenue 520-670-6201
Tucson, AZ 85719

**4408 Unexpected Wildlife Refuge Library**
110 Unexpected Road 856-697-3541
Newfield, NJ 08344 http://www.animalplace.org

**4409 University of California**
1 Shields Avenue 530-752-1011
Davis, CA 95616

**4410 University of Florida Coastal Engineering Archives**
209 Yon Hall 352-392-2710
Gainesville, FL 32611 Fax: 352-392-2710
E-mail: Twedell@coastal.ufl.edu
*Helen Twedell, Archivist*
*Kimberly Hunt, Sr. Library Technical Asst*

**4411 University of Hawaii at Manoa Water Resources Center**
2540 Dole Street 808-956-7847
Homes Hall 283 Fax: 808-956-5044
Honolulu, HI 96822 E-mail: morav@hawaii.edu
http://www.wrrc.hawaii.edu
Coordinates and conducts research to identify, characterize and quantify water/environmental related problems in the state of Hawaii. Based on the research WRRC makes recommendations to all agencies and organizations withresponsibilities to manage the water/ environmental resources in Hawaii.
*Phillip Morakik, Technology Transfer Spec*
*James Moncur, Director*

**4412 University of Illinois at Chicago**
Energy Resource Center
851 S Morgan Street 312-996-4490
12th Floor Fax: 312-996-5420
Chicago, IL 60607 E-mail: rsanka1@uic.edu
http://h008.erc.uic.edu/welcome.htm
The Energy Resources center is an interdiciplinary public service, research, and special projects organization dedicated to improving energy efficiency and the environment. Conducts studies in the fields of energy and environment andprovides industry, utilities, government agencies and the public with assistance, information, and advice on new technologies, public policy, and professional development training.
*James Hartnett, Director*

**4413 University of Maryland: Center for Environmental Science Chesapeake Biological Lab**
1 Willams Street 410-326-7287
Box 38 Fax: 410-326-7302
Solomons, MD 20688 http://www.cbl.umces.edu
*Kathleen A Heil, Librarian*

**4414 University of Montana Wilderness Institute Library**
Forestry Building 406-243-5361
Room 207 Fax: 406-243-4845
Missoula, MT 59812 E-mail: wi@forestry.umt.edu
http://www.forestry.umt.edu/wi

**4415 Vermont Institute of Natural Sciences Library**
27023 Church Hill Road 802-457-2779
Woodstock, VT 05091 Fax: 802-457-2779

**4416 Voices from the Earth**
Box 4524 505-262-1862
Albuquerque, NM 87106 E-mail: sricdon@earthlink.net
http://www.sric.org
SRIC exists to provide timely, accurate information to the public on matters that affect the environment, human health, and communities in order to protect natural resources, promote citizen participation, and ensure environmental andsocial justice now and for future generations.
*Dan Hancock, Administrator*
*Annette Aguayo, Information Specialist*

**4417 Wasserman Public Affairs Library**
University of Texas at Austin
General Libraries 512-495-4400
Sid Richardson Hall 3243 Fax: 512-495-4347
Austin, TX 78712 E-mail: pal@lib.utexas.edu
http://www.lib.utexas.edu/pal
*Stephen Littrell, Head Librarian*

**4418 Western Ecology Division Library**
US Environmental Protection Agency
200 SW 35th Street 541-754-4731
Corvallis, OR 97333 Fax: 541-754-4799
E-mail: obrien.mary@epa.gov
http://www.epa.gov/libraries/wed/html
*Publication Date: 1966*
*Kathy Martin, Program Analyst*
*Mary O'Brien, Librarian*

**4419 Wildlife Management Institute Library**
1101 14th Street NW 202-371-1808
Suite 725 Fax: 202-408-5059
Washington, DC 20005
*Richard E McCabe, Sec./Dir., Pubns.*

**4420 Wisconsin Department of Natural Resources Library**
Box 7921 608-266-8933
Madison, WI 53707 Fax: 608-266-5226
Contains books, journals, and EPA reports on air pollution, geology, hazardous waste, natural resources management, recycling, soil pollution, solid waste, toxic substances, waste minimization, wastewater, water pollution, andwetlands.
*Erin Matiszik, Librarian*

**4421 Wisconsin's Water Library at UW Madison**
University of Wisconsin
1975 Willow Drive, Floor 2 608-262-3069
Madison, WI 53706 Fax: 608-262-0591
E-mail: AskWater@aqua.wisc.edu
http://www.aqua.wisc.edu/waterlibrary
Wisconsin's Water Library is a collection of materials that cover all major topics in water resources, but is particularly strong in Wisconsin and Great Lakes water issues, groundwater protection, wetlands issues, and the impacts ofagricultural chemicals. The collection consists of over 31,000 hard copy and microfiche documents, over 35 journals and 130 newsletters. The collection may be searched at http://madcat.library.wisc.edu.
*Anne K Moser, Special Librarian*

**4422 Yale University School of Forestry and Environmental Studies Library**
205 Prospect Street 203-432-5132
New Haven, CT 06511 Fax: 203-432-5942
http://www.library.yale.edu/scilib/forestl.html
A part of the Yale University Library System, the library serves the resource needs of the graduate students and faculty of Yale's 100 year old school of Forestry and Environmental Studies.
*Carla Heister, Librarian*

## Publishers

**4423 Academic Press: New York**
Academic Press
15 E 26th Street 212-592-1000
15th Floor E-mail: ap@acad.com
New York, NY 10010 http://customerservice.apnet.com

**4424 Adison Wesley Longman**
Pearson
26 Prince Andrew Place 905-853-7888
Toronto ONT M3C-2T8 800-563-9196
Fax: 800-263-7733
E-mail: webinfo.pubcanada@pearson.com
http://pearson.com
One integrated and diverse company offering learning resources on an extraordinary level. Pearson has an estblished reputation for producing market-leading educational products and services as well as a comprehensive range ofbest-selling consumer, environmental technical and professional titles.

**4425 Blackwell Publishers**
Blackwell Publishers
350 Main Street 781-388-8200
Malden, MA 02148 Fax: 781-388-8210
Blackwell Publishers are dedicated to serving the global academic community. We recognize that publishing is about making connections. Knowledge is not constrained by national or liguistic boundries. Many academics are engaged in bothteaching and research. Our readers are often our authors as well. We develop books for students which take account of the latest research and we aim to make the journals we publish as acessible as possible.

**4426 Boxwood Press**
183 Ocean View Boulevard 408-375-9110
Pacific Grove, CA 93950 Fax: 408-375-0430
E-mail: boxwood@boxwoodpress.com
Publishes significant titles in the areas of Natural History, Area Studies, General Sciences and Local and Special Interest. Founded in 1952, it first published lab manuals, then expanded to include a variety of mainly biologicaltitles.

**4427 CABI Publishing**
CAB International 617-395-4056
875 Massachusetts Ave, 7th Floor 800-552-3083
Cambridge, MA 02139 Fax: 617-354-6875
E-mail: cabi-nao@cabi.org
http://www.cabi.org
CABI publishing is a leading international, nonprofit publisher in applied life sciences, including animal science, nutrition, integrated crop management and forestry. Our products have a global reputation for quality, relevance andauthority, and are used in over 100 countries. Our long-established print publishing activities include a substantial book and reference work list, and an expanding primary and review journal program.

**4428 CRC Press**
CRC Press
2000 NW Corporate Boulevard 561-994-0555
Boca Raton, FL 33431 800-272-7737
Fax: 800-374-3401
http://www.crcpress.com
CRC Press LLC is recognized as a leader in scientific, medical, environmental science, engineering, business, technical, mathamatical, and statistics publishing. CRC Press LLC publishes books, journals, newsletters and databases.Customers have access to publications through individual purchases, bookstores, libraries and on-line acess at www.crcpress.com

**4429 Chelsea Green Publishing Company**
85 N Main Street 802-295-6300
PO Box 428 800-639-4099
White River Junction, VT 05001 Fax: 802-295-6444
E-mail: publicity@chelseagreen.com
http://www.chelseagreen.com
Chelsea Green publishes information that helps us lead pleasurable lives on a planet where human activities are in harmony and balance with nature.Free catolog listing over 250 titles on sustainable living, innovative shelter andorganic gardening.
*Alice Blackmer, Publicity Director*

**4430 DK Publishing**
DK Publishing
95 Madison Avenue 212-213-4800
New York, NY 10016 Fax: 212-213-5240
Dorling Kindersley is an international publishing company specialising in the creation of high quality, illustrated information books, interactive software, TV programs and online resources for childern and adults. Founded in London1974, DK now has offices in the UK, USA, Australia, South Africa, India France, Germany and Russia.
*Publication Date: 1974*

**4431 Elsevier Science**
Elsevier Sciences
655 Avenue of the Americas 212-633-3730
New York, NY 10010 Fax: 212-633-3680
E-mail: usinfo-f@elsevier.com
Our focus will be entirely on scientific, technical and medical publishing. Together we can offer customers choice across our portfolio, with outstanding platforms for the delivery of electronic services and a high level of investmentto ensure the development of leading electronic products.

**4432 Environmental Working Group**
1436 U Street 202-667-6982
Suite 100 http://www.ewg.org
Washington, DC 20009
The Environmental Working Group is a leading content provider for public interest groups and concerned citizens who are campaigning to protect the environment. Offers reports, articles, technical assistance and the development ofcomputer databases and Internet resources.
*Ken Cook, President*

**4433 Global and Environmental Education Resources**
Global Change Research Information

Suite 250 1717
Pennsylvania Ave NW
Washington, DC 20006
202-223-6262
Fax: 202-223-3065
E-mail: information@gcrio.org
http://www.gcrio.org/edu.html

Multidisciplinary and international in scope, this collection of resources was selected for its relevance to global change and environmental education. Included is a wide range of resources in a variety of formats for educators andstudents at all levels (K-12 and higher education), librarians, citizens and community groups.

**4434 Grey House Publishing**
4919 Route 22
Amenia, NY 12501
518-789-8700
800-562-2139
Fax: 845-373-6390
E-mail: books@greyhouse.com
http://www.greyhouse.com

Directories, handbooks and reference works for public, high school and academic libraries and the business and health communities. Publishes environmental directories for US and Canadian markets.
*Leslie Mackenzie, Publisher*
*Richard Gottlieb, Editor*

**4435 Grey House Publishing Canada**
555 Richmond Street West
2nd Floor
Toronto, Ontario M5V 3B1
416-644-6479
866-433-4739
Fax: 416-644-1904
E-mail: info@micromedia.ca
http://www.micromedia.ca

Canada's largest developer, publisher and distributor of value-added reference information for the academic, library, government and corporate markets. Our mission is to be Canada's one stop shop for information products and services.We license content from media, government and other sources and organize, abstract and compile this content into databases. Through a combination of technology expertise and a full service approach, our solutions provide access to a wide range ofinformation.
*Bryon Moore, General Manager*

**4436 Institute for Food and Development Policy**
398 60th Street
Oakland, CA 94618
510-654-4400
Fax: 510-654-4551
E-mail: foodfirst@foodfirst.org
http://www.foodfirst.org

Publishes books on poverty, agriculture and development, also backgrounders, policy briefs and development reports.
*Publication Date: 1975*
*Eric Holt-GimŠNez, Executive Director*

**4437 Island Press**
Distribution Center
PO Box 7
Covelo, CA 95428
707-983-6432
800-828-1302
Fax: 707-983-6414
E-mail: service@islandpress.com
http://www.islandpress.com

Mission-oriented nonprofit publisher organized in 1984 to help meet the need for accessible, solutions-oriented information through a unique approach that addresses the multidisciplinary nature of environmental problems. Our program isdesigned to translate technical information from a range of disciplines into a book format that is accessible and informative to citizen activists, educators, students and professionals involved in the study or management of environmental problems.
*Bernice Hiatt, Customer Service*

**4438 It's Academic**
29 West 35th Street
New York, NY 10001
212-216-7800
Fax: 212-564-7854

A tool for teachers who use Routledge books in their classes. To aid in finding the books best suited for your needs, we offer: pages which highlight books designed specifically for your courses, a list of conferences at which wedisplay our books, journal information, a forum for instructors to send us their comments, supplements available on line and a subject search menu.

**4439 Kluwer Academic Publishers**
101 Philip Drive
Assinippi Park
Norwell, MA 02061
781-871-6600
Fax: 781-871-6528

A sector of the Wolters Kluwer publishing group. Operates world-wide from offices in Dordrecht, Boston, New York and London. All over the world, scientists and professionals hold our publications in high esteem.

**4440 Krieger Publishing Company**
Krieger Publishing Co.
PO Box 9542
Melbourne, FL 32902
321-724-9542
800-724-0025
Fax: 321-951-3671
E-mail: info@krieger-publishing.com
http://www.krieger-publishing.com

Krieger Publishing Company produces quality books in various fields of interest. We have an extensive Natural Science listing.
*Cheryl Stanton, Advertising*

**4441 MIT Press**
55 Hayward Street
Cambridge, MA 02142
617-253-5646
Fax: 617-258-6779
http://mitpress.mit.edu

The only university press in the US whose list is based in science and technology. Our environment list is strong in policy and the social sciences. We are committed to the edges and frontiers of the world - to exploring new fields andnew modes of inquiry. We publish about 200 new books a year and over 40 journals including Global Environmental Politics. We have a long-term commitment to both design excellence and the efficient and creative use of new technologies.
*Clay Morgan, Senior Acquisition Editor*

**4442 McGraw-Hill Education**
The McGraw-Hill Companies
1221 Avenue of the Americas
40th Floor
New York, NY 10020
212-512-2000
Fax: 212-512-6111

A global leader in educational materials and professional information, with offices in more than 30 countries and publications in more than 40 languages, we develop products that influence people's lives from preschool through career.The scope of our operations, the quality of our editorial product and the pace at which we are developing new media to fulfill our customers' information requirements are increasing.

**4443 National Information Service Corporation**
NISC USA, Wyman Towers
3100 St. Paul Street
Baltimore, MD 21218
410-243-0797
Fax: 410-243-0982

Publishes information products for access through BiblioLine, our Web search service, or on CD-ROM. Some of our abstract and index services are available in print. NISC's bibliographic and full-text databases cover a wide range oftopics in the natural and social sciences, arts and humanities. Some titles provide comprehensive coverage of particular geographic regions, such as Latin America, Africa, South-East Asia or the Arctic and Antarctic.

**4444 O'Reilly & Associates**
101 Morris Street
Sebastopol, CA 95472
800-998-9938
Fax: 707-829-0104

Premier information source for leading-edge computer technologies. We offer the knowledge of experts through our books, conferences and web sites. Our books, known for their animals on the covers, occupy a treasured place on theshelves of developers building the next generation of software. Conferences and summits bring innovators together to shape the ideas that spark new industries. From the Internet to the web, Linux, Open Source and peer-to-peer networking, we puttechnologies on the map.

**4445 Random House**
1540 Broadway
New York, NY 10036
212-782-9000
Fax: 212-302-7985

The world's largest English-language general trade book publisher. It is a division of the Bertelsmann Book Group of

Bertelsmann AG, one of the foremost media companies in the world.

**4446 Simon & Schuster**
1230 Avenue of the Americas 212-698-7000
New York, NY 10020 Fax: 212-698-2359
E-mail: ken.riel@simonandschuster.com.
*Ken Riel*

**4447 Springer-Verlag New York**
175 Fifth Avenue 212-460-1500
New York, NY 10010 Fax: 212-473-6272
Founded in 1964 and maintained its position last year as the Springer Group's largest foreign subsidiary. In 1999, 426 new titles were released. In addition, 50 journals were published, most of them available in electronic form as wellas via the Springer information system LINK. The number of license agreements in the North American market has increased fivefold due to the increasing demand for this leading Online Library.

**4448 Virginia Museum of Natural History**
1001 Douglas Avenue 540-666-8600
Martinsville, VA 24112 Fax: 540-632-6487
http://www.vmnh.org
Our publishing division specializes in works by natural scientists and environmental educators in the US and abroad. Writing, editorial, and design services available for books, reports, manuals, text books, presentations, fieldguides, etc.. A catalogue is available on request.
*24 pages Quarterly*
*ISSN: 1085-5084*
*Susan Felker, Managing Editor\Outreach*

**4449 WW Norton & Company**
500 5th Avenue 212-354-5500
New York, NY 10110 Fax: 212-869-0856
The oldest and largest publishing house owned wholly by its employees, strives to carry out the imperative of its founder to publish books of long-term value in the areas of fiction, nonfiction and poetry. The roots of the company dateback to 1923, when William Warder Norton and his wife, M.D. Herter Norton, began publishing lectures delivered at the People's Institute, the adult education division of New York City's Cooper Union.

**4450 Wiley North America**
605 3rd Avenue 212-850-6000
New York, NY 10158 Fax: 212-850-6088
The company was founded in 1807, during the Jefferson presidency. In the early years, Wiley was best known for the works of Washington Irving, Edgar Allen Poe, Herman Melville and other 19th century American literary giants. By theturn of the century, Wiley was established as a leading publisher of scientific and technical information.

# Research Centers

## Corporate & Commercial Centers

**4451 AAA & Associates**
28 West Adams
Suite 1511
Detroit, MI 48226
313-961-4122
Fax: 313-588-6232

*Katherine Banicki, President*

**4452 AB Gurda Company**
6061 Whitnall Way
Hales Corners, WI 53130
414-529-3116
Environmental testing and analysis firm.

**4453 ABC Research Corporation**
3437 SW 24th Avenue
Gainesville, FL 32607
352-372-0436
Fax: 352-378-6483
E-mail: info@abcr.com
http://www.abcr.com
Research and analysis laboratory. Research results published in scientific journals.
*Dr William L Brown, President*
*Dr Peter Bodnaruck, VP*

**4454 ACRES Research**
6621 W Ridgeway Avenue
Waterloo, IA 50701
319-277-6661
Fax: 319-266-7569
E-mail: acresres@aol.com
Environmental research and testing.
*Bert Schou PhD, President*

**4455 ACZ Laboratories, Inc**
2773 Downhill Drive
Steamboat Springs, CO 80487
970-879-6590
800-334-5493
Fax: 970-879-2216
E-mail: sales@acz.com
http://www.acz.com
A full service environmental analytical lab with inorganic, organic and radiochemical capabilities. We perform analysis on a wide variety of matrices including water, wastewater, waste, soil, plant and animal tissue as well as fishtissue.

*Tim VanWyngarden, Manager Business Development*
*Audrey Stover, President/CEO*

**4456 ADA Technologies**
8100 Shaffer Parkway
Suite 130
Littleton, CO 80127
303-792-5615
800-232-0296
Fax: 303-792-5633
E-mail: ada@adatech.com
http://www.adatech.com
Product development and testing of environmental technologies.
*Judith Armstrong PhD, President*

**4457 AECOM**
10 Iverness Center Parkway
Suite 120
Birmingham, AL 35242
205-980-0054
800-722-2440
Fax: 205-980-1509
E-mail: askenvironment@aecom.com
http://www.aecom.com
AECOM is a global provider of environmental and energy development services to industry and government. As a full-service environmental firm, AECOM's professionals provide clients with consulting, engineering, remediation, and relatedservices from over 24 countries.

*John Petraglia, PR/Enviro Service Inquiries*
*Paul Genarro, PR/Enviro Service Inquiries*

**4458 AER**
131 Hartwell Avenue
Lexington, MA 02421
781-761-2288
Fax: 781-761-2299
E-mail: ross@aer.com
http://www.aer.com

**4459 AMA Analytical Services**
4475 Forbes Boulevard
Lanham, MD 20706
301-459-2640
Fax: 301-459-2643
http://www.amalab.com
Environmental research. Asbestos, lead and explosives analysis.
*David P Hood, CEO*

**4460 ANA-Lab Corporation**
2600 Dudley Road
PO Box 9000
Kilgore, TX 75663
903-984-0551
Fax: 903-984-5914
E-mail: corp@ana-lab.com
http://www.ana-lab.com
Environmental laboratory. Offers ICP-MS which allows Ana-Lab to offer improved turn around time, reduce costs, and achieve better quantitation of regulated parameters. Tests are performed by methods specified by the EPA. Specializes inenvironmental chemistry.

*C H Whiteside, President*
*Bill Perry, SVP/COO*

**4461 APC Lab**
13760 Manolia Avenue
Chino, CA 91710
909-590-1828
Fax: 909-590-1498
E-mail: apcl@apclab.com
Environmental and industrial testing laboratory. Research results published in journals and conference reports.
*Irene Huang, Public Relations*

**4462 APS Technology**
7 Laser Lane
Wallingford, CT 06492
860-613-4450
Fax: 203-284-7428
E-mail: info@aps-tech.com
http://www.aps-tech.com
Product development, conceptual design, engineering, prototype manufacture and test analysis.
*William E Turner, President*
*Denis Bigin, VP*

**4463 ARDL**
400 Aviation Drive
Mount Vernon, IL 62864
618-244-3235
Fax: 618-244-1149
Environmental sampling and testing laboratory; Research Development Engineering. Alternate Name: Applied Research and Development Laboratories, Inc.
*Larry Gibbons PhD, President*
*Don Gillespie, Marketing Manager*

**4464 ASW Environmental Consultants**
20 N Plains Industrial Road
PO Box 495
Wallingford, CT 06492
203-265-0509
Fax: 203-265-1476
*Jason J Sarojak, PE*

**4465 ATC Associates**
7988 Centerpoint Drive
Suite 100
Indianapolis, IN 46256
317-849-4990
877-282-4756
Fax: 317-849-4278
http://www.atcassociates.com
Technical engineering research and environmental consulting firm.
*John Mundell, Director/Manager*

**4466 ATC Environmental**
720 E Benson Road
Sioux Falls, SD 57103
605-338-0555
*Donald Beck*

**4467 ATL**
2912 W Clarendon Avenue
Phoenix, AZ 85017
602-241-1097
Fax: 602-277-1306

Technical engineering evaluation firm. Research results published in test summaries and project reports.
*Frank C Rivera, President*
*David P Hayes, VP*

**4468 AW Research Laboratories**
16326 Airport Road
Brainerd, MN 56401
218-829-7974
Fax: 218-829-1316
http://www.awlab.com
A.W. Research Laboratories, Inc. (AWRL) provides environmental consulting services and water quality analysis. AWRL specializes in the use of remote sensing techniques for lake analysis and management.
*Alan W Cibuzar, CEO*

**4469 AZTEC Laboratories**
6402 Stadium Drive
PO Box 7953
Kansas City, MO 64129
816-921-3922
Data collection and analysis, systems design and product development firm.

**4470 Aaron Environmental**
189 Atwater Street
Plantsville, CT 06479
860-276-1201
800-372-1233
Fax: 860-276-1233
E-mail: info@aaronenvironmental.com
http://www.aaronenvironmental.com
*Joyce Kogut, President*
*Mike Bolegh, Business Manager*

**4471 Accurate Engineering Laboratories**
2707 W Chicago Avenue
Chicago, IL 60622
773-384-4522
Fax: 773-384-8681
Environmental engineering laboratory.
*Noel Buczkowski, President*

**4472 Accutest Laboratories**
2235 Route 130 S
Building B
Dayton, NJ 08810
732-329-0200
Fax: 732-329-3499
http://www.accutest.com
Environmental testing firm.
*Vincent Pagrissi, President*

**4473 Acts Testing Labs**
100 Northpointe Parkway
Buffalo, NY 14228
716-505-3300
Fax: 716-505-3301
Global consumer products testing organization providing quality assurance testing, inspections and consulting services.
*Tom Fatta, Contact*

**4474 Adelaide Associates**
7 Holland Avenue
White Plains, NY 10603
914-949-3109
Fax: 914-949-8103
E-mail: adelaide@bestweb.net
Environmental health consulting and testing firm. Additional offices: White Plains, NY, Poughkeepsie, NY and Perth Amboy, NJ.
*Ron Birlinski, CEO*

**4475 Adelaide Environmental Health Associates**
111-115 Court Street
Binghamton, NY 13901
607-722-6839
Fax: 607-771-0752
*Roland E Bielinski, President*

**4476 Adirondack Environmental Services**
314 N Pearl Street
Albany, NY 12207
518-434-4546
Fax: 518-434-0891
E-mail: aes@adirondackenvironmental.com
http://www.adirondackenvironmental.com
Analytical medical laboratory.

**4477 Adirondack Lakes Survey Corporation**
Route 86
PO Box 296
Ray Brook, NY 12977
518-897-1354
Fax: 518-897-1364
E-mail: admin@adirondacklakessurvey.org
http://www.adirondacllakessurvey.org
Determines the extent and magnitude of acidification of lakes and ponds in the Adironack region.

**4478 Advance Pump and Filter Company**
10 Calef Highway
Lee, NH 03824
603-868-3212
800-863-3212
Fax: 603-868-3230
E-mail: advancepump@comcast.net
http://http://advanceh2o.com
Services include water treatment, submersible pumps, jet pumps, water tanks, sewage and sump systems.
*Cathleen Pleadwell, Business Manager*

**4479 Advanced Terra Testing**
833 Parfet Street
Unit A
Lakewood, CO 80215
303-232-8308
888-859-8378
Fax: 303-232-1579
E-mail: terratest@aol.com
http://www.terratesting.com
Geotechnical and geosynthetic testing firm.
*Chris Wienecke, Director/Manager*

**4480 AeroVironment**
181 West Huntington Drive
Suite 202
Monrovia, CA 91016
626-357-9983
Fax: 626-359-9628
E-mail: info@avinc.com
http://www.aerovironment.com
Research, service and consulting firm specializing in the environment, alternative energy and aerodynamic design. Research results published in project reports and technical journals.

**4481 Aerosol Monitoring & Analysis**
PO Box 646
Hanover, MD 21076
410-684-3327
Fax: 410-684-3384
http://www.amatraining.com
Environmental services firm.

**4482 Agvise Laboratories**
604 Highway 15 West
PO Box 510
Northwood, ND 58267
701-587-6010
Fax: 701-587-6013
E-mail: agvise@polarcomm.com
http://www.agvise.com
Applied and product research in environmental applications.
*Robert Wallace, CEO*

**4483 Alan Plummer and Associates**
1320 South University Drive
Suite 300
Fort Worth, TX 76107
817-806-1700
Fax: 817-870-2536
http://www.apaienv.com
Civil and environmental engineering consulting.
*Alan H Plummer Jr, President*

**4484 Alar Engineering Corporation**
9651 West 196th Street
Mokena, IL 60448
708-479-6100
Fax: 708-479-9059
*Alex Doncer, President*

**4485 Alden Research Laboratory**
30 Shrewsbury Street
Holden, MA 01520
508-829-6000
Fax: 508-829-5939
E-mail: arlmail@aldenlab.com
http://www.aldenlab.com
Hydraulic engineering firm solving air and water flow problems using physical and CFD models and field testing, for areas such as fish passage/protection systems, free surface and closed conduit flow, pump/turbine performance,hydraulic structures, environmental hydraulics, fluid equipment, 3D air flow, and flow meter calibration.
*Edward P Taft II, President*

**4486 Allied Laboratories**
716 North Iowa
Villa Park, IL 60181
630-279-0390
Fax: 630-279-3114
*Irving I Domsky, Director*

**4487 Alloway Testing**
1101 N Cole Street
Lima, OH 45805
419-223-1362
Fax: 419-227-3792
http://www.alloway.com
Environmental sampling and analysis laboratory.
*John R Hoffman, President*

**4488 Alpha Manufacturing Company**
100 Old Barnwell Road
PO Box 2809
West Columbia, SC 29171
803-739-4500
Fax: 803-739-0517
E-mail: mail@alphamfg.com
http://www.alphamfg.com
Physical testing of environmental testing and repair services, and instrument design.
*William L Cowley, President*

**4489 Alton Geoscience**
21 Technology Drive
Irvine, CA 92618
949-753-0101
Fax: 949-753-0111
http://www.trcsolutions.com
Environmental remediation and consulting firm.
*Jenny Rue, Public Relations*
*Larry Farrington, Contact*

**4490 Amalgamated Technologies**
13901 N 73rd Street
Suite 208
Scottsdale, AZ 85260
480-991-2901
Firm providing metals and materials development, processing and testing.
*Roy E Beal, President*

**4491 American Environmental Network**
9151 Rumsey Road
Suite 150
Columbia, MD 21045
410-730-8525
Fax: 410-997-2586
*Paul Jackson, Marketing Manager*

**4492 American Testing Laboratory**
255 West Street
South Hackensack, NJ 07606
201-489-8573
Fax: 201-489-9365
E-mail: info@mytestlab.com
http://www.americantestinglaboratory.com
Multidisciplinary testing laboratory offering a wide range of confidential testing services including environmental simulations (humidity, high/low temperature extremes).
*Daniel Narbone, Manager*

**4493 American Waste Processing**
2100 West Madison Street
Maywood, IL 60153
708-681-3999
800-841-6900
Fax: 708-681-3583
E-mail: american@american-waste.com
http://www.american.waste.com
Non-hazardous waste management, disposal/transfer, station/recycling, roll off containers.
*William Vajdik, President*

**4494 Analab**
630 Heron Drive
PO Box 336
Bridgeport, NJ 08014
856-467-4555
800-262-5229
Fax: 856-467-1212
E-mail: info@analab1.com
http://www.analab1.com
Compliance lab services for EMC/EMI, safety and ESD.
*Jason Smith, Director/Manager*

**4495 Analyte Laboratories**
2121 Cedar Circle Drive
Catonsville, MD 21228
410-747-3844
Fax: 410-747-4007

**4496 Analytical Laboratories and Consulting**
361 West 5th Avenue
Eugene, OR 97401
541-485-8404
Fax: 541-484-5995
*Rory E White, Sr. Analyst*

**4497 Analytical Process Laboratories**
8222 W Calumet Road
Milwaukee, WI 53223
414-355-3909
Fax: 414-355-3099
http://www.apl-lab.com
Materials analysis laboratory and environmental engineering.
*Jitendra Shah, President*
*Dr. Taxla Shah, CEO*

**4498 Analytical Resources**
4611 S 134th Place
Tukwila, WA 98168
206-695-6200
Fax: 206-695-6201
http://www.arilabs.com
Environmental testing and analysis laboratory.
*Mark Weidner, President*
*Stephanie Lucas, Project Manager*

**4499 Analytical Services**
110 Technology Parkway
Nocross, GA 30092
770-734-4200
Fax: 770-734-4201
http://www.asi-lab.com
Environmental testing and analysis firm.
*G Wyn Jones, President*

**4500 Anametrix**
1961 Concourse Drive
Suite E
San Jose, CA 95131
408-432-8192
Fax: 408-432-8198
*Doug Robbins, President*

**4501 AndCare**
PO Box 14566 Parkway
Research Triangle Park, NC 27709
919-544-8220
Fax: 919-544-9808
Development and commercialization of low cost, simple to use diagnostic devices and tests for medical, environmental, and laboratory markets.
*Dr. Steven Wagner, PhD, President*

**4502 Anderson Engineering Consultants**
10205 W Rockwood Road
Little Rock, AR 72204
501-455-4545
Fax: 501-455-4552
Firm providing engineering, inspection, and testing services specializing in geotechnology and materials, and environmental sciences. Services include site studies, soil testing, engineering surveys, specification evaluation, andfailure investigation.

**4503 Andrea Aromatics**
PO Box 3091
Princeton, NJ 08543
609-695-7710
Fax: 609-392-8914
E-mail: orders@andreaaromatics.com
Natural essential oils, fragrances, deodorants, odor neutralizers.
*Michael D'Andrea, VP*
*Richard D'Andrea, President*

**4504 Anlab**
1910 S Street
Sacramento, CA 95814
916-447-2946

**4505 Anteon Corporation**
3211 Jermantown Road
Suite 200
Fairfax, VA 22030
703-246-0200
Fax: 703-246-0797
*Don MacDougall, President Services*

**4506 Applied Biomathematics**
100 N Country Road
Setauket, NY 11733
631-751-4350
Fax: 631-751-3435
E-mail: info@ramas.com
http://www.ramas.com
Environmental and ecological software development firm.
*Lev Ginzburg, President*

**4507 Applied Coastal Research & Engineering**
766 Falmouth Road
Building A, Unit 1-C
Mashpee, MA 02649
508-539-3737
Fax: 508-539-3739
E-mail: info@appliedcoastal.com
http://www.appliedcoastal.com

Environmental analysis.
*Mark Bynes PhD, President*

**4508 Applied Technical Services**
1190 Atlanta Industrial Drive
Marietta, GA 30066
770-423-1400
Fax: 770-514-3299
http://www.atslab.com

Environmental, chemical, and mechanical testing and consulting company.
*Jim F Hill, President*

**4509 Aquatec Chemical International**
408 Auburn Avenue
Pontiac, MI 48342
313-334-4747
*Douglas Schwartz, President*

**4510 Architectural Energy Corporation**
2540 Frontier Avenue
Suite 201
Boulder, CO 80301
303-444-4149
Fax: 303-444-4304
E-mail: aecinfo@archenergy.com
http://www.archenergy.com

Energy, daylighting and sustainable design and analysis; LEED certification services; building commissioning, energy auditing and diagnostic testing; home energy rating software (RFOM/Rate); commercial energy analysis software(VisualDOE); and data acquisition equipment.

*Michael J Holtz, FAIA, President*

**4511 Ardaman & Associates**
8008 S Orange Avenue
Orlando, FL 32809
407-855-3860
800-683-SOIL
Fax: 407-859-8121
E-mail: mmongeau@ardaman.com
http://www.ardaman.com

Geotechnical, environmental and materials consultants.

*Mark L Mongeau PE, Vice President*

**4512 Arete Associates**
5000 Van Nuys Boulevard
PO Box 6024
Sherman Oaks, CA 91413
818-501-2880
Fax: 818-501-2905
http://www.arete.com

Environmental research.
*Dr Stephen C Lubard, President*

**4513 Aroostook Testing & Consulting Laboratory**
160 Airport Drive
Presque Isle, ME 04769
207-762-5771
Fax: 207-764-8123
E-mail: atclabs@ainop.com

Toxiological and environmental laboratory.

*G Noel Currie III, President*

**4514 Arro**
Caton Farm Road
PO Box 686
Joliet, IL 60434
815-727-5436
Fax: 815-740-3234

*Robert J Rolih, President*

**4515 Artesian Laboratories**
630 Churchmans Road
Newark, DE 19702
302-266-9121
Fax: 302-454-8720

Environmental sampling and testing laboratory. Research results published in methods reports to clients.

**4516 Association of Ecosystem Research Centers**
730 11th Street NW
Washington, DC 20001
202-628-1500
Fax: 202-628-1509
E-mail: aerc@culter.colorado.edu

Brings together 39 US research programs in universities and private, state and federal laboratories that conduct research, provide training and analyze policy at the ecosystem level of environmental science and natural resourcesmanagement. Although AERC is an association of professional scientists rather than environmental activists, its goals and interest complement those of conservation organizations.
*John E Hobbie, President*

**4517 Astro-Chem Services**
4102 2nd Avenue W
PO Box 972
Williston, ND 58802
701-572-7355
Fax: 701-774-3907

*David Vander, President*

**4518 Astro-Pure Water Purifiers**
1441 SW 1st Way
Deerfield Beach, FL 33441
954-422-8966
Fax: 954-422-8966

Manufacturers complete line of water treatment equipment, purifiers, De Calcifiers, R.O., V.V., iron filters, chemical feed equipment. Sizes for portable, point of use, central commercial and industrial. Manufacturers and privatelabels counter top units.

*RL Stefl, President*

**4519 Atlantic Testing Laboratories**
6431 US Highway 11
PO Box 29
Canton, NY 13617
315-386-4578
Fax: 315-386-1012
E-mail: atl-test@northweb.com
http://www.atlantictest.com

ATL is a full service engineering support firm offering environmental services, subsurface investigations, geoprobe services, water-based investigations, geotechnical engineering, construction materials testing and engineering, specialinspection services, pavement engineering, non-destructive testing, and surveying from ten office sites. The firm currently has extensive capabilities in the areas of underground and above-ground storage tank testing and management, and relatedareas.

*MarikeanF Remington, President*
*Thomas Cronin, Vice President*

**4520 Atlas Weathering Services Group**
45601 N 47th Avenue
Phoenix, AZ 85087
623-465-7356
800-255-3738
Fax: 623-465-9409
E-mail: info@atlaswsg.com
http://www.atlaswsg.com

Technical research firm specializing in environmental testing. Research results published in reports to clients and in archival journals.
*Jack Martin, President*

**4521 Atmospheric & Environmental Research**
131 Hartwell Avenue
Lexington, MA 02421
781-761-2288
Fax: 781-761-2299
E-mail: aer@aer.com
http://www.aer.com

Firm providing research, consulting, and assessment on atmospheric chemistry, meteorology, climate, and air quality.
*Nien Dak Sze, President*
*Cecilia Sze, CEO*

**4522 Axiom Laboratories**
24 Tobey Road
Bloomfield, CT 06002
860-242-6291
Fax: 860-286-0634
E-mail: mackeyw@worldnet.att.net

Environmental and materials analytical testing services.
*William AG Macke, President*

**4523 B&P Laboratories**
5635 Delridge Way SW
Seattle, WA 98106
206-937-3644
Fax: 206-937-1348
E-mail: bplabor@hotmail.com

Environmental testing and chemical laboratory water analyses (ICP);Mercury analyzer sulfates, fluorides, chlorides

(DIONEX); cyandes (CONTES); storm waters; fats, oil & grease (FOG); Karl Fischer corrosion testing; process solutions

*Victor Broto, President*

**4524 BC Analytical**
4100 Atlas Court 661-327-4911
Bakersfield, CA 93308 Fax: 661-327-1918
http://www.bclabs.com/
Lab capabilities include diversified sample matrices for drinkg waters, ground water monitoring and waste acceptance. Diversified analytical methods include general chemistry and field services including field analysis, sampling andcourier service.

**4525 BC Laboratories**
4100 Atlas Court 661-327-4911
Bakersfield, CA 93308 Fax: 661-327-1918
Chemical analysis and environmental monitoring of hazardous waste. Research results published in project reports.

**4526 BC Research**
3650 Westbrook Mall 604-224-4331
Vancouver, BC V6S2L Fax: 604-224-0540
Research results published in scientific journals and trade magazines.
*Hugh Wynne-Edwards, President*
*Marion Webber, Director/Manager*

**4527 BCI Engineers & Scientists**
2000 Edgewood Drive 863-667-2345
Suite 215 877-550-4224
Lakeland, FL 33807 Fax: 863-667-2262
http://http://bcieng.com
Environmental and civil engineering, geotechnical processes, and chemical research and development firm.
*Rick Powers, President/CEO*
*Wendy Lee, Chief Financial Officer*

**4528 BCM Engineers**
920 Germantown Pike 610-313-3100
Suite 200 Fax: 610-313-3151
Plymouth Meeting, PA 19462 http://www.atcassociates.com/default.asp
Services include environmental, geotechnical and materials, remedial design, industrial hygiene and hazardous management planning.
*Bobby Toups, Chief Executive Officer*
*Albert Petersen, Authority Engineer*

**4529 Babcock & Wilcox Company**
1562 Beeson Street 330-753-4511
Alliance, OH 44601 Fax: 330-823-0639

*Robert E Howson, President*

**4530 Badger Laboratories & Engineering Company**
501 West Bell Street 920-729-1100
Neenah, WI 54956 800-776-7196
Fax: 920-729-4945
http://www.badgerlabs.com
*Stephen Taylor, President*

**4531 Baird Scientific**
532 Oak Street 417-358-5567
Carthage, MO 64836
*Gary Baird, President/Owner*

**4532 Baker Environmental**
420 Rouser Road 412-269-6000
Coraopolis, PA 15108 Fax: 412-269-2534
Environmental engineering company.
*Andrew P Paja, President*

**4533 Baker-Shiflett**
5701 East Loop 820 South 817-478-8254
Fort Worth, TX 76119 Fax: 817-478-8874
*Larry Gardner, Administrative Manager*

**4534 Barnebey & Sutcliffe Corporation**
835 North Cassady Avenue 614-258-9501
PO Box 2526 Fax: 614-258-3464
Columbus, OH 43216

*Amanda L Fisher, Marketing Coordinator*

**4535 Barton & Loguidice**
290 Elwood Davis Road 315-457-5200
Box 3107 Fax: 315-451-0052
Syracuse, NY 13220 E-mail: B&L@BartonandLoguidice.com
http://www.BartonandLoguidice.com
Since 1961,Barton and Loguidice, P.C. has assisted a wide variety of clients in meeting their engineering requirements. As a full service, multi-disciplinary firm, B and L has the expertise and capacity to perform a wide array of highquality engineering services including bridge and highway, facilities, water, wastewater, environmental, and solid waste engineering. The firm continues to serve as an engineering services leader in Upstate New York.
*Tom Aiston, President*

**4536 Baxter and Woodman**
8678 Ridgefield Road 815-459-1260
Crystal Lake, IL 60012 Fax: 847-948-2887
Environmental engineering firm.

**4537 Bell Evaluation Laboratory**
17300 Mercury Drive 281-488-3701
Houston, TX 77058 Fax: 281-488-8543
E-mail: bellabs@bellabs.com
http://www.bellabs.com
Coating testing and evaluation firm.
*Robert T Bell, Contact*

**4538 Belle W Baruch Institute for Marine Biology and Coastal Research**
607 EWS Building 803-777-5288
Columbia, SC 29208 Fax: 803-777-3935
Conducts basic and applied research in marine and coastal environments.

**4539 Beltran Associates**
1133 East 35th Street 718-338-3311
Brooklyn, NY 11210 Fax: 718-253-9028
E-mail: info@beltrantechnologies.com
http://www.beltrantechnologies.com
A leader in advanced gas cleaning and air pollution control for a broad spectrum of industrial processes and emission requirements. Our reputation is based on 50 years of successful research, development and problem solving leading tomore than 1000 installations worldwide.

*Mike Beltran, President*
*Swapan Mitra, Sales Manager*

**4540 Benchmark Analytics**
4777 Saucon Creek Road 215-974-8100
Center Valley, PA 18034 Fax: 610-974-8104
E-mail: bma@epix.net
http://www.benchmarkanalyticslabs.com
Benchmark Analytics is an independent analytical testing laboratory. Benchmark analyzes many types of samples including drinking water, wastewater and soil in addition to testing food, mold, sludge, soot, air samples and industrialproducts.
*Stephanie Olexa, President*

**4541 Bendix Environmental Research**
1950 Addison Street 415-861-8484
Suite 202 Fax: 510-845-8484
Berkeley, CA 94704 http://home.earthlink.net/~bendix/#A1
*Selina Bendix, PhD, President*

**4542 Bhate Environmental Associates**
5115 Maryland Way 615-377-0725
Brentwood, TN 37027 Fax: 615-661-4226
Environmental consulting.

**4543 Bio-Chem Analysts Inc**
4940 North Memorial Pkwy 256-859-2161
PO Box 3270 Fax: 256-859-9222
Huntsville, AL 35810
Environmental testing of water, wastewater, air, soil, and hazardous waste.

*Vijay Thakore, President*

**4544 Bio-Science Research Institute**
4813 Cheyenne Way 909-628-3007
Chino, CA 91710 Fax: 909-590-8948
Independent environmental testing laboratory.

**4545 Bio/West**
1063 West 1400 North 435-752-4202
Logan, UT 84321 Fax: 435-752-0507
*Paul Holden, Principal*

**4546 Biological Research Associates**
3910 N US Highway 301 813-664-4500
Suite 180 Fax: 813-664-0440
Tampa, FL 33619
Environmental research and consulting firm.

**4547 Biomarine**
16 E Main Street 978-281-0222
PO Box 1153 Fax: 978-283-6296
Gloucester, MA 01930 E-mail: info@biomarinelab.com
http://www.biomarinelab.com
Provides water and seafood analysis and consulting for the public, private companies and government.

*John Marletta, Laboratory Director*

**4548 Bionetics Corporation Analytical Laboratories**
20 Research Drive 757-865-0880
Hampton, VA 23666 Fax: 757-865-8014
*Joseph A Stern, President*

**4549 Bioscience**
1550 Valley Center Parkway 610-974-9693
Suite 140 800-627-3069
Bethlehem, PA 18017 Fax: 610-691-2170
E-mail: bioscience@bioscienceinc.com
http://www.bioscienceinc.com
Specialized microbes for wastewater and hazardous waste, biological treatment. BOD and COD monitoring instruments and test kits for water and wastewater analysis.

*Thomas G Zitrides, President*
*Richard Bleam, Director of Technical Svc.*

**4550 Biospherics**
12051 Indian Creek Court 301-419-3900
Beltsville, MD 20705 Fax: 301-210-4909

*Gilbert V Levin, PhD, President*

**4551 Black Rock Test Lab**
5 Eastgate Plaza 304-296-8347
Morgantown, WV 26501

**4552 BlazeTech Corporation**
24 Thorndike Street 617-661-0700
Cambridge, MA 02141 Fax: 617-661-9242
E-mail: office@blazetech.com
http://www.blazetech.com
An engineering consulting firm specializing in fire, explosion, environmental safety and homeland defense. They have developed specialized software for the chemical, petroleum, aerospace and power industries: ADORA, BLAZETANK andothers.
*Albert Moussa, President*

**4553 Bollyky Associates Inc**
31 Strawberry Hill Avenue 203-967-4223
Stamford, CT 06902 Fax: 203-967-4845
E-mail: ljbbai@bai-ozone.com
http://www.bai-ozone.com
Engineering firm specializing in Ozone technology, water and wastewater treatment, treatability studies.

*L Joseph Bollyky, President*
*Thomas Kleiber, Office Manager*

**4554 Bolt Technology Corporation**
4 Duke Place 203-853-0700
Norwalk, CT 06854 Fax: 203-854-9601

*Raymond M Soto, President*

**4555 Braun Intertec Corporation**
11001 Hampshire Avenue South 952-995-2000
Minneapolis, MN 55438 800-279-6100
Fax: 952-995-2020
E-mail: info@braunintertec.com
http://www.braunintertec.com
Full-service engineering, environmental and infrastructure consulting and testing organization.

*George D Kluempke, President*
*Scott C Barnard, CFO*

**4556 Braun Intertec Northwest**
11001 Hampshire Avenue S 952-995-2000
Minneapolis, MN 55438 Fax: 952-995-2020
E-mail: info@braunintertec.com
http://www.braunintertec.com
Testing and quality control monitoring laboratory specializing in construction inspections, materials testing, soils engineering and geological services.
*George D Kluempke PE, Chief Executive Officer*
*Robert J Janssen PE, Chief Operations Officer*

**4557 Briggs Associates**
100 Weymouth 781-871-6040
Rockland, MA 02370 Fax: 781-871-7982
Environmental engineering and testing facility.

**4558 Brighton Analytical**
2105 Pless Drive 810-229-7575
Brighton, MI 48116 Fax: 810-229-8650
*J Shawn Letwin, Laboratory Director*

**4559 Brooks Companies**
9 Isaac Street 203-853-9792
Norwalk, CT 06850 Fax: 203-853-0273
*Margaret Y Brooks PhD, President*

**4560 Brooks Laboratories**
9 Isaac Street 203-853-9792
Norwalk, CT 06850 800-843-1631
Fax: 203-853-0273
E-mail: brookslabs@aol.com
http://www.brookslabs.com
Consulting and testing air, soil and water for contamination. Accident and disease prevention.

**4561 Brotcke Engineering Company**
750 Merus Court 636-343-3029
PO Box 1168 800-969-3029
Fenton, MO 63026 Fax: 636-343-3773
E-mail: sheartlein@brotcke.com
http://www.brotcke.com
Professional engineering firm that provides well drilling and pump services.
*Paul Brotcke, President*

**4562 Buchart-Horn**
445 W Philadelphia Street
York, PA 17404
717-852-1400
800-274-2224
E-mail: pace40@aol.com

This company provides environmental engineering, consulting, civil engineering, facility design and planning as well as laboratory and testing services.

*Dennis Miner, VP Business Development*
*Gene Schenck, Public Relations Director*

**4563 Burt Hill Kosar Rittelmann Associates**
400 Morgan Center
Butler, PA 16001
724-285-4761
Fax: 724-285-6815
*P Richard Rittelmann, FAIA, Sr. VP*

**4564 Business Health Environmental Lab**
33 E 7th Street
300 Doctors Building
Covington, KY 41011
859-431-6224
Fax: 859-431-6228
*Dan Moos, President*

**4565 C L Technology Division of Microbac Lab**
280 N Smith Avenue
Corona, CA 92880
909-734-9600
Fax: 909-734-2803
E-mail: info@microbac.com
http://www.microbac.com

Research and development laboratory specializing in analysis and testing. Services include: air quality; environmental testing, consulting and analysis; nutraceutical testing and analysis; consumer products testing; food producttesting; fuel and manufactured products testing.

*J Trevor Boyce, President/CEO*
*A Wayne Boyce, Chairman*

**4566 CDS Laboratories**
RR 2
Box 234
Loganton, PA 17747
570-725-3411

**4567 CENSOL**
582 Hawthorne
L9T-4N8
Milton, ON
416-219-6950
Fax: 905-878-8775
E-mail: info2003@censol.ca
http://www.censol.ca

Environmental consulting and testing firm. Research results published in technical association papers.

**4568 CET Environmental Services**
7032 South Revere Parkway
Englewood, CO 80112
720-875-9115
Fax: 720-875-9114
http://www.cetenvironmental.com

Provides env. consulting, engineering, remediation & construction servies. The Group has three primary segments: Industrial services, environmental remediation & gvmt. programs. The industrial services include water & wastewatertreatment, facility cleaning, operation, construction & emergency response. The environmental rem. services include environmental assessment & remedial investigation studies. The government programs include emergency response, remediation, &water-related services.

*Steven H Davis, President & CEO*
*Dale W Bleck, CFO*

**4569 CONSAD Research Corporation**
121 N Highland Avenue
Pittsburgh, PA 15206
412-363-5500
Fax: 412-363-5509
E-mail: info@consad.com
http://www.consad.com

Social science research and consulting firm. Research results published in journals and project and government reports.

*Wilbur A Steger, President*
*Frederick H Rueter, VP*

**4570 CPAC**
2364 Leicester Road
Leicester, NY 14481
585-382-3223
Fax: 585-382-3031
E-mail: cpacinfo@cpac.com
http://www.cpac-fuller.com

CPAC, Inc. manages holdings in two industries: Cleaning and Personal Care and Imaging. The Fuller Brands segment develops, manufactures, and markets over 2799 branded and private lavel products for commercial cleaning, householdcleaning, and personal care. CPAC Imaging manufactures, packages, and distrubtes branded and private label chemicals for photographic, health care, and graphic arts markets as well as associated imaging equipment and silver refining services.

*Thomas N Hendrickson, President*

**4571 Cascadia Research**
218 1/2 W 4th Avenue
Olympia, WA 98501
206-943-7325
http://www.cascadiaresearch.org

Nonprofit tax-exempt scientific and educational organization founded to conduct research needed to manage and protect threatened marine mammals.

**4572 Cedar Grove Environmental Laboratories**
100 Gallagherville Road
Downingtown, PA 19335
610-269-6977
Fax: 610-269-6965
http://www.cgelab.com

Environmental analysis firm serving agriculture and industry.

**4573 Ceimic Corporation**
10 Dean Knauss Drive
Narragansett, RI 02882
401-782-8900
Fax: 401-782-8905

Environmental testing for water and soil.

*Margaret Marple, Marketing*

**4574 Center for Solid & Hazardous Waste Management**
2207 NW 13th Street
Suite D
Gainesville, FL 32609
352-392-6264
Fax: 352-846-0183
E-mail: center@floridacenter.org
http://www.floridacenter.org

The center serves the citizens of Florida by providing leadership in the field of waste management research and by supporting the Florida Department of Environmental Protection in its mission to preserve and protect the state's naturalresources.

*John D Schert, Executive Director*

**4575 Center for Technology, Policy & Industrial Development**
1 Amherst Street, Room E40-239
Cambridge, MA 02139
617-253-1664
Fax: 617-452-2265
http://web.mit.edu/www/tel/

Environmental research.

*Nicholas Ashferdadeh, Director/Manager*

**4576 Central Virginia Laboratories and Consultants**
3109 Odd Fellows Road
PO Box 10938
Lynchburg, VA 24506
804-847-2852
Fax: 804-847-2830
*Adrian K Mood, President*

**4577 Century West Engineering Corporation**
1444 NW College Way
Bend, OR 97709
541-388-3500

**4578 Chas. T Main: Environmental Division**
Prudential Center
Boston, MA 02199
617-262-3200
Fax: 781-401-2575

Environmental consulting firm specializing in site assessments, surveys, and tests and analysis.

**4579 Chemical Resource Processing**
2525 Battleground Road
PO Box 1914
Deer Park, TX 77536
281-930-2525
Fax: 281-930-2535

Environmental consulting and chemical processing firm.

**4580 Chemical Waste Disposal Corporation**
4214 19th Avenue 718-274-3339
Astoria, NY 11105 Fax: 718-726-7917
Environmental waste disposal and consulting company.

**4581 Chemir Analytical Services**
2672 Metro Boulevard 314-291-6620
Maryland Heights, MO 63043 800-659-7659
Fax: 314-291-6630
E-mail: info@chemir.com
http://www.chemir.com
Provides a wide range of chemical analysis and chemical testing services. Experienced at solving difficult problems including product failure analysis, materials identification, plastic testing or reverse engineering. An independenttesting lab with scientists that can provide litigation support such as expert witness testimony in intellectual property or products liability cases.

*Dr Shri Thanedar, President*

**4582 Chesner Engineering**
38 Park Avenue 516-431-4031
Suite 200 E-mail: mail@chesnerengineering.com
Long Beach, NY 11561 http://www.chesnerengineering.com
Civil, environmental, and waste management firm that provides professional services to industry and government. Specializes in the areas of waste and by-product material recycling and stabliization, marine and dredge environmentalmanagement, risk assessment and environmental modeling, environmental database program development, remedial site investigations and cleanup management, and water and wastewater treatment.
*Warren Chesner, President*

**4583 Chihuahuan Desert Research Institute**
PO Box 1334 915-837-8370
Alpine, TX 79831
Conducts research on the Chihuahuan Desert.

**4584 Chopra-Lee**
1815 Love Road 716-773-7625
Grand Island, NY 14072 800-508-5419
Fax: 716-773-7624
E-mail: pburger@mailexcite.com
Environmental services, assessments, sampling, and data collection, testing, and analysis firm. Industrial hygiene/indoor air quality; materials testing/failure analysis and hazardous waste analysis.

**4585 ChromatoChem**
2837 Fort Missoula Road 406-728-5897
Missoula, MT 59801 Fax: 406-728-5924
Biotechnology firm. Research results published in proposals to the Environmental Protection Agency.

**4586 Chyun Associates**
267 Wall Street 609-924-5151
Princeton, NJ 08540

**4587 Clark Engineering Corporation**
621 Lilac Drive N 763-545-9196
Minneapolis, MN 55422 E-mail: mheller@clark-eng.com
http://www.clark-eng.com
*Stephen E Clark, President*

**4588 Clark's Industrial Hygiene and Environmental Laboratory**
1801 Route 51 S 412-387-1001
Building 9 Fax: 412-387-1027
Jefferson Hills, PA 15025 E-mail: bhunt@clarklabsllc.com
http://www.clarklabsllc.com/IH/index.htm
General industrial hygiene consulting and field services laboratory that provides support to the environmental efforts of industry, both light and heavy, refineries, power industry, aluminum, steel and environmental consulting firms.Maintains AIHA (American Industrial Hygiene Association) accreditation for asbestos (PLM and PCM), metals, organic solvents, diffusive samplers and silica.
*Brando Hunt, Analytical Projects Manager*
*Lee Rogers, Quality Assurance Manager*

**4589 Clean Air Engineering**
500 West Wood Street 847-991-3300
Palatine, IL 60067 800-627-0033
Fax: 847-991-3385
E-mail: contact@cleanair.com
http://www.cleanair.com
Environmentally consulting and permitting, process engineering, equipment rental and manufacture, measurement and analytical services.

*Allen Kephart, Vice President*
*Jim Pollack, Director of Sales*

**4590 Clean Harbors**
1200 Crown Colony Drive 781-849-1800
PO Box 9137 Fax: 781-848-2141
Quincy, MA 02169
Environmental consulting firm specializing in soil analysis, site assessments, and water sample testing.

**4591 Clean Water Systems**
2322 Marina Drive 541-882-9993
Klamath Falls, OR 97601 Fax: 541-882-9994
E-mail: cws@internetcds.com
http://www.cleanwatersysintl.com
Environmental research firm. Designs, develops and manufactures ultra-violet electronic water purification units and systems and electronic measuring systems. The Company has developed lines of proprietary electronic monitoring andcontrol systems and electronic ballast.
*Charles Romary, President*

**4592 Coastal Resources**
25 Old Solomons Island Road 410-956-9000
Annapolis, MD 24101 Fax: 410-956-0566
E-mail: coastal@coastal-resources.net
http://www.coastal-resources.net
Environmental impact assessments and nontidal wetlands identification expert testimony. Also conducts field investigations for a broad range of natural resources including soils, wetlands, streams, water quality, forests, wildlife,habitats and rare, threatened and endangered species.

*Betsy M Weinkam, President/Enviro Biologist*
*Chuck Weinkam, Sr Environmental Scientist*

**4593 Colorado Analytical**
240 S Main Avenue 303-659-2313
PO Box 507 Fax: 303-659-2315
Brighton, CO 80601
Agricultural consulting and testing laboratory. Research results published in project reports and test summaries.

**4594 Colorado Research Associates**
3380 Mitchell Lane 303-415-9701
Boulder, CO 80301 Fax: 303-415-9702
http://www.co-ra.com
Environmental research.
*David C Fritts, VP*

**4595 Columbus Instruments International**
950 N Hague Avenue 614-276-0861
Columbus, OH 43204 Fax: 614-276-0529
E-mail: sales@colinst.com
http://www.colinst.com
Manufacturer of biomedical and environmental research equipment which includes respirometers and gas analysis monitoring systems.
*Jan Czekdjewski, President*
*Ken Kober, Sales Manager*

**4596 Columbus Water and Chemical Testing Laboratory**
4628 Indianola Avenue 614-262-4372
Columbus, OH 43214

**4597 Commonwealth Technology**
2526 Regency Road 859-276-3091
Lexington, KY 40503 800-467-3091
Fax: 859-276-4374
E-mail: fyi@ctienv.com
http://www.ctienv.com
Environmental engineering and analysis firm.

**4598 CompuChem Environmental Corporation**
501 Madison Avenue 919-379-4000
Cary, NC 27513 800-833-5097
Fax: 919-379-4050
http://www.compuchemlabs.com
Environmental testing laboratory.
*Gerard Verkerk, Contact*

**4599 Conjun Laboratories**
9283 Highway 15 606-633-8027
Isom, KY 41824

**4600 Conservation Foundation**
1919 M Street NW 202-912-1000
Suite 600 Fax: 202-912-0765
Washington, DC 20036 http://www.conservation.org
Conducts research and develops knowledge and techniques to improve the quality of the environment.

**4601 Consumer Testing Laboratories**
430 S Congress Avenue 561-330-3081
Suite 1B Fax: 561-330-7712
Delray Beach, FL 33445
Research in textiles, safety wear.
*Stewart Satter, President*

**4602 Container Testing Laboratory**
607 Fayette Avenue 914-381-2600
Mamaroneck, NY 10543 Fax: 914-381-0143
E-mail: contestlab@hotmail.com
http://www.containertechnology.com
Independent third party testing laboratory.
*Anton Cotaj, Laboratory Director/Manager*

**4603 Conti Testing Laboratories**
3190 Industrial Blvd 412-833-7766
PO Box 174 Fax: 412-854-0373
Bethel Park, PA 15102 E-mail: contilab@verizon.net
Analytical commerical laboratory, fuel analysis (coal, coke, alternative fuels), metals, ore barge gauging, customize analysis.
*Patricia A Otroba, President*
*Timothy Otrobe, CEO*

**4604 Continental Systems**
7870 Deering Avenue 818-340-3217
Canoga Park, CA 91304 Fax: 818-340-2405
Environmental laboratory specializing in soil and water analysis.
*Janis Butler, President*

**4605 Controlled Environment Corporation**
29 Sanford Drive 207-854-9126
Gorham, ME 04038 Fax: 207-854-4357
Firm providing research, design, and development services relating to clean rooms and contamination control.
*Matthew F Pec, President*

**4606 Controls for Environmental Pollution**
1925 Rosina Street 505-982-9841
Box 5351 800-545-2188
Santa Fe, NM 87502 Fax: 505-982-9289

*James J Mueller, President*

**4607 Converse Consultants**
3 Century Drive 973-605-5200
Parsippany, NJ 07054 Fax: 973-605-8145
E-mail: convers@mailidt.net
Applied and product research in environmental studies.
*R Brian Ellwood, VP*

**4608 Copper State Analytical Lab**
710 East Evans Boulevard 520-388-4922
Tucson, AZ 85713 Fax: 520-884-5133
E-mail: csalinc@aol.com
http://www.csalinc.com
Hazardous waste characterization, organic and inorganic waste oil characterization, waste water analysis, potable water analysis, microbiology, general waters and soil chemistry.

*DA Shah, President*

**4609 Corning**
1 Riverfront Plaza 607-974-9000
Corning, NY 14831 Fax: 607-974-8091
http://www.corning.com

**4610 Corrosion Testing Laboratories**
60 Blue Hen Drive 302-454-8200
Newark, DE 19711 Fax: 302-454-8204
E-mail: ctl@corrosionlab.com
http://www.corrosionlab.com
Corrosion testing laboratory. Research results published in technical journals and conference proceedings.
*Bradley D Krantz, Corrosion Lab Supervisor*
*Shari Nathanson Rosenbloom PhD, Principal Metallurgical Eng*

**4611 Coshocton Environmental Testing Service**
709 Main Street 740-622-3328
Coshocton, OH 43812 Fax: 740-622-3368
Environmental testing service.

**4612 Crane Environmental**
2600 Eisenhower Avenue 610-631-7700
Trooper, PA 19043 Fax: 610-631-6800
http://www.craneenv.com
Industrial water treatment equipment.

**4613 Crosby & Overton**
1610 W 17th Street 562-432-5445
Long Beach, CA 90813 800-827-6729
Fax: 562-436-7540
http://www.crosbyoverton.com
Fully permitted RCRA Part B TSD facility located in Southern California. The Facility can process both bulk and drummed waste, including lab-packs. Crosby & Overton can process a wide variety of D,F,K,P and U listed RCRA waste.
*Bob Ritter, Sales Manager*
*Michelle Dalot, Sales*

**4614 Curtis & Tompkins**
2323 5th Street 510-486-0900
Berkeley, CA 94710 Fax: 510-486-0532
*C Bruce Godfrey, President*

**4615 Cutter Environment**
37 Broadway 781-678-8702
Arlington, MA 02474 800-964-8702
Fax: 781-648-1950
E-mail: environment@cutter.com
http://www.cutter.com/environment
Environmental research.
*Karen Coburn, President*

**4616 Cyberchron Corporation**
US Route 9 845-265-3700
PO Box 160 Fax: 845-265-3752
Cold Spring, NY 10516
Computer manufacturing firm specializing in the development of computers designed to withstand extreme travel, environmental, and work conditions.

**4617 Cyrus Rice Consulting Group**
200 High Tower Road 412-788-2468
Suite 302 Fax: 412-788-1797
Pittsburgh, PA 15205
*Al Owens, VP*

**4618 DE3**
18234 S Miles Road 216-663-1500
Cleveland, OH 44128 Fax: 216-663-1501
Environmental engineering research and testing firm.
*Harold N Danto, President*

**4619 DLZ Laboratories - Cleveland**
1000 Rockefeller Building 216-771-1090
614 W Superior Avenue 800-340-5227
Cleveland, OH 44113 Fax: 216-771-0334
E-mail: dlzroundabouts.com
http://www.dlz.com OR www.dlzcorporation.com
DLZ, a minority-owned business enterprise, is a full-service, multidisciplinary professional corporation that provides complete architectural, engineering, and environmental services to both the public and private sectors. Researchstudies are published in professional journals and project reports.
*Bill Sampson, Marketing Director*

**4620 DLZ Laboratories - Columbus/Corporate**
6121 Huntley Road 614-848-4333
Columbus, OH 43229 800-340-5227
Fax: 614-841-0818
E-mail: dlzroundabouts.com
http://www.dlz.com OR www.dlzcorporation.com
DLZ, a minority-owned business enterprise, is a full-service, multidisciplinary professional corporation that provides complete architectural, engineering, and environmental services to both the public and private sectors. Researchstudies are published in professional journals and project reports.
*Bill Sampson, Marketing Director*

**4621 DLZ Laboratories - Cuyahoga Falls**
2162 Front Street 330-923-0401
Cuyahoga Falls, OH 44221 800-340-5227
Fax: 330-928-1029
E-mail: dlzroundabouts.com
http://www.dlz.com OR www.dlzcorporation.com
DLZ, a minority-owned business enterprise, is a full-service, multidisciplinary professional corporation that provides complete architectural, engineering, and environmental services to both the public and private sectors. Researchstudies are published in professional journals and project reports.
*Bill Sampson, Marketing Director*

**4622 DOWL HKM**
4041 B Street 907-562-2000
Anchorage, AK 99503 Fax: 907-563-3953
http://www.dowlhkm.com
Serves clients' needs in the areas of environmental planning, National Environmental Policy Act (NEPA) documentation, permitting, engineering and public involvement. Environmental studies and analyses include wetland delineation andfunction and values assessment, vegetation mapping, GIS mapping and analysis, environmental site assessment, air and noise impact analysis. Section 106 consultation, hydrology studies, and secondary and cumulative impact analysis.

*Jennifer Payne, Corporate Development Mgr*
*Kristen J Hansen, Service Environ. Spec.*

**4623 DPRA**
200 Research Drive 785-539-3565
Manhattan, KS 66503 Fax: 785-539-5353
http://www.dpra.com
Environmental, economic, regulatory and technical research company. Research results published by information services.
*Richard Seltzer, President*

**4624 DW Ryckman and Associates REACT Environmental Engineers**
1120 S 6th Street 314-678-1398
St. Louis, MO 63104 800-325-1398
Fax: 314-678-6610
E-mail: stewart_ryckman@react-env.com
http://www.react-env.com/home1.htm
D.W. Ryckman & Associates, Inc., d.b.a. REACT Environmental Engineers, was founded in 1975 to provide rapid response and remediation services for environmental and hazardous contamination problems.

*SE Ryckman, President*

**4625 Daily Analytical Laboratories**
1621 West Candletree Drive 309-692-5252
Peoria, IL 61614 Fax: 309-692-0488
*Kurt Stepping, Chief Chemist*

**4626 Dan Raviv Associates**
57 E Willow Street 973-564-6006
Millburn, NJ 07041 Fax: 973-564-6442
E-mail: ddrai@ix.netcom.com
http://www.danraviv.com
Environmental consulting firm specializing in environmental impact studies, waste management, site assessment and litigation support.
*Dan D Raviv PhD, President*
*John J Trela PhD, Director/Manager*

**4627 Danaher Corporation**
2099 Pennsylvania Avenue NW 202-828-0850
Washington, DC 20006 Fax: 202-828-0860
http://www.danaher.com
Development of process and environmental controls, tools and components.

**4628 Datachem Laboratories**
960 W Levoy Drive 801-266-7700
Salt Lake City, UT 84123 Fax: 801-268-9992
Analytical laboratory provides lab analysis of soil, water, air and asbestos samples.
*Brent Stephens, VP/Laboratory Director*

**4629 Davis Research**
134 Hobart Road 662-332-1943
Avon, MS 38723 Fax: 662-332-0081
Agricultural, food and environmental testing and research firm.
*R G Davis, President*
*Diane Barnham, Director/Manager*

**4630 Dellavalle Laboratory**
1910 W McKinley Avenue 559-233-6129
Suite 110 800-228-9896
Fresno, CA 93728 Fax: 559-268-8174
http://www.dellavallelab.com
Agricultural laboatory analyzes plant, soil, manure and water (ag, domestic, wastewater). Certified Professional Soil Scientits/ Agronomists/Crop Advisors and others provide consultation on nutrient and fertilizer management, cropfeasibility, regulatory compliance, troubleshooting and related areas.

*Hugh A Rathbun, President*

**4631 Douglass Environmental Services**
8649 Bash Street 317-595-9108
Indianapolis, IN 46256 Fax: 317-822-8362
Environmental engineering and testing firm.

**4632 Duke Solutions**
1 Winthrop Square 617-482-8228
Boston, MA 02110 Fax: 617-482-3784

**4633 Dynamac Corporation**
22575 Research Boulevard
Suite 500
Rockville, MD 20850
301-417-9800
Fax: 301-417-9801
E-mail: ibaumel@dynamac.com
http://www.dynamac.com
A scientific research, engineering, and information technology company, conducting state of the art field and laboratory research, and providing scientific and technical support to federal and state environmental programs.
*William J Silver, President*
*Diana Mac Arthur, CEO*

**4634 E&A Environmental Consultants**
11629 Central Street
Stoughton, MA 02072
781-344-6446
Fax: 781-575-8915
E-mail: EAEnviron@aol.com
http://http://members.aol.com/eaenviron/index.html
An environmental consulting firm, E&A provides waste management solutions through the utilization of alternative and innovative management and treatment techniques. E&A has developed expertise in all aspects of composting and organicwaste utilization. Research results are published in presentations, journals, and newsletters.
*Eliot Epstein PhD, Ch. Environmental Scientist*
*Charles M Alix PE, Senior Engineer*

**4635 E&S Environmental Chemistry**
PO Box 609
Corvallis, OR 97339
541-758-5777
Fax: 541-758-4413
Environmental research.
*Tim Sullivan, President*

**4636 EA Engineering Science and Technology**
11019 McCormick Road
Hunt Valley, MD 21031
410-584-7000
Fax: 410-771-1625
E-mail: ea@eaest.com
http://www.eaest.com
*Loren Jensen, President*
*James Gift, R&D*

**4637 EADS Group**
1126 8th Avenue
Altoona, PA 16602
814-944-5035
800-626-0904
Fax: 814-944-4862
E-mail: ibelsel@eadsgroup.com
http://www.eadsgroup.com
The EADS Group has experienced personnel that provide environmental risk assessments and site investigations, wetlands delineation and mitigation, and all related permitting. The scope of services covers terrestrial and aquaticecology, water resources, threatened and endangered species, vegetation and wetlands, soils and geology, air quality, noise, hazardous waste, socioeconomics and land use. The EADS Group has five offices in Pennsylvania and one in Maryland.
*Dennis M Stidinger, President*
*Janet L Helsel, Marketing Director*

**4638 EAI Corporation**
1308 Continental Drive
Suite J
Abingdon, MD 21009
410-676-1449
Fax: 410-671-7241
E-mail: info@eaicorp.com
http://www.eaicorp.com
Environmental engineering and scientific firm. Research results published in private reports to clients.
*Charles Speranzella, President*
*Tom Albro, VP*

**4639 EMCO Testing & Engineering**
PO Box 266
Taftville, CT 06380
860-886-0697
Fax: 860-886-0697
E-mail: emco@99main.com
Water treatment and environmental research and consulting, including: storm water pollution prevention; well water contamination investigation; property contamination investigation; solar energy application to commercial andresidential buildings.
*Dr Ernie Cohen, President/R&D*

**4640 EMCON Alaska**
201 E 56th Avenue
Suite 300
Anchorage, AK 99518
907-562-3452
Fax: 907-563-2814
Environmental engineering firm.

**4641 EMMES Corporation**
401 North Washington Street
Suite 700
Rockville, MD 20850
301-251-1161
Fax: 301-251-1355
E-mail: info@emmes.com
http://www.emmes.com
Firm providing medical data management and statistical support services. Research results published in scientific literature.
*Donald M Stablein, PhD, President*

**4642 EMS Laboratories, Inc.**
117 W Bellevue Drive
Pasadena, CA 91105
626-568-4065
800-675-5777
Fax: 626-796-5282
E-mail: contact@emslabs.com
http://www.emslabs.com
Environmental testing lab services. Asbestos/Lead/,I.H. testing. Fully accreditted AIHA lab for metal waste characterization.
*Bernadine Kolk, President*
*Anthony Kolk, CEO*

**4643 EN-CAS Analytical Laboratories**
2359 Farrington Point Drive
Winston-Salem, NC 27107
336-785-3252
Fax: 336-785-3262
http://www.en-cas.com
Chemical and environmental testing and analysis company.

**4644 ENSR Consulting and Engineering**
1601 Prospect Parkway
Fort Collins, CO 80525
970-493-8878
Fax: 970-493-0213
Environmental engineering and consulting firm.
*Will Wright, Director/Manager*

**4645 ENSR-Anchorage**
1835 S Bragaw Street
Suite 490
Anchorage, AK 99508
907-561-5700
800-722-2440
Fax: 907-273-4555
E-mail: askensr@ensr.aecom.com
http://www.ensr.aecom.com/Office/44/67/index.jsp
ENSR is a global provider of environmental and energy development services to industry and government. As a full-service environmental firm, ENSR's professionals provide clients with consulting, engineering, remediation, and relatedservices from over 15 countries.
*John Petraglia, PR/Enviro Service Inquiries*
*Paul Genarro, PR/Enviro Service Inquiries*

**4646 ENSR-Billings**
2048 Overland Avenue
Suite 101
Billings, MT 59102
406-652-7481
800-722-2440
Fax: 406-652-7485
E-mail: askensr@ensr.aecom.com
http://www.ensr.aecom.com/Office/44/67/index.jsp
ENSR is a global provider of environmental and energy development services to industry and government. As a full-service environmental firm, ENSR's professionals provide clients with consulting, engineering, remediation, and relatedservices from over 15 countries.
*John Petraglia, PR/Enviro Service Inquiries*
*Paul Genarro, PR/Enviro Service Inquiries*

**4647 ENSR-Carmel**
879 W Carmel Drive
Carmel, IN 46032
317-843-0008
800-722-2440
Fax: 317-843-1850
E-mail: askensr@ensr.aecom.com
http://www.ensr.aecom.com/Office/44/67/index.jsp
ENSR is a global provider of environmental and energy development services to industry and government. As a full-service environmental firm, ENSR's professionals provide clients with

consulting, engineering, remediation, and relatedservices from over 15 countries.
*John Petraglia, PR/Enviro Service Inquiries*
*Paul Genarro, PR/Enviro Service Inquiries*

**4648 ENSR-Chicago**
27755 Diehl Road 630-836-1700
Suite 100 800-722-2440
Warrenville, IL 60555 Fax: 630-836-1711
E-mail: askensr@ensr.aecom.com
http://www.ensr.aecom.com/Office/44/67/index.jsp
ENSR is a global provider of environmental and energy development services to industry and government. As a full-service environmental firm, ENSR's professionals provide clients with consulting, engineering, remediation, and relatedservices from over 15 countries.
*John Petraglia, PR/Enviro Service Inquiries*
*Paul Genarro, PR/Enviro Service Inquiries*

**4649 ENSR-Columbia (MD)**
8320 Guilford Road 410-884-9280
Suite L 800-722-2440
Columbia, MD 21046 Fax: 410-884-9271
E-mail: askensr@ensr.aecom.com
http://www.ensr.aecom.com/Office/44/67/index.jsp
ENSR is a global provider of environmental and energy development services to industry and government. As a full-service environmental firm, ENSR's professionals provide clients with consulting, engineering, remediation, and relatedservices from over 15 countries.
*John Petraglia, PR/Enviro Service Inquiries*
*Paul Genarro, PR/Enviro Service Inquiries*

**4650 ENSR-Fort Collins**
1601 Prospect Parkway 970-493-8878
Fort Collins, CO 80525 800-722-2440
Fax: 970-493-0213
E-mail: askensr@ensr.aecom.com
http://www.ensr.aecom.com/Office/44/67/index.jsp
ENSR is a global provider of environmental and energy development services to industry and government. As a full-service environmental firm, ENSR's professionals provide clients with consulting, engineering, remediation, and relatedservices from over 15 countries.
*John Petraglia, PR/Enviro Service Inquiries*
*Paul Genarro, PR/Enviro Service Inquiries*

**4651 ENSR-Harvard**
325 Ayer Road 978-772-2345
Harvard, MA 01451 800-722-2440
Fax: 978-772-4956
E-mail: askensr@ensr.aecom.com
http://www.ensr.aecom.com/Office/44/67/index.jsp
ENSR is a global provider of environmental and energy development services to industry and government. As a full-service environmental firm, ENSR's professionals provide clients with consulting, engineering, remediation, and relatedservices from over 15 countries.
*John Petraglia, PR/Enviro Service Inquiries*
*Paul Genarro, PR/Enviro Service Inquiries*

**4652 ENSR-Kalamazoo**
220 Parkwood Avenue 978-772-2345
Kalamazoo, MI 49001 800-722-2440
Fax: 978-772-4956
E-mail: askensr@ensr.aecom.com
http://www.ensr.aecom.com/Office/44/67/index.jsp
ENSR is a global provider of environmental and energy development services to industry and government. As a full-service environmental firm, ENSR's professionals provide clients with consulting, engineering, remediation, and relatedservices from over 15 countries.
*John Petraglia, PR/Enviro Service Inquiries*
*Paul Genarro, PR/Enviro Service Inquiries*

**4653 ENSR-Minneapolis**
4500 Park Glen Road 952-924-0117
Suite 210 800-722-2440
St Louis Park, MN 55416 Fax: 952-924-0317
E-mail: askensr@ensr.aecom.com
http://www.ensr.aecom.com/Office/44/67/index.jsp
ENSR is a global provider of environmental and energy development services to industry and government. As a full-service environmental firm, ENSR's professionals provide clients with consulting, engineering, remediation, and relatedservices from over 15 countries.
*John Petraglia, PR/Enviro Service Inquiries*
*Paul Genarro, PR/Enviro Service Inquiries*

**4654 ENSR-New Orleans**
1555 Poydras Street 504-592-3559
Suite 1860 800-722-2440
New Orleans, LA 70112 Fax: 504-522-2085
E-mail: askensr@ensr.aecom.com
http://www.ensr.aecom.com/Office/44/67/index.jsp
ENSR is a global provider of environmental and energy development services to industry and government. As a full-service environmental firm, ENSR's professionals provide clients with consulting, engineering, remediation, and relatedservices from over 15 countries.
*John Petraglia, PR/Enviro Service Inquiries*
*Paul Genarro, PR/Enviro Service Inquiries*

**4655 ENSR-Norcross**
4155 Shackleford Road 770-381-1836
Suite 245 800-722-2440
Norcross, GA 30093 Fax: 770-381-9583
E-mail: askensr@ensr.aecom.com
http://www.ensr.aecom.com/Office/44/67/index.jsp
ENSR is a global provider of environmental and energy development services to industry and government. As a full-service environmental firm, ENSR's professionals provide clients with consulting, engineering, remediation, and relatedservices from over 15 countries.
*John Petraglia, PR/Enviro Service Inquiries*
*Paul Genarro, PR/Enviro Service Inquiries*

**4656 ENSR-Portland (ME)**
42 Market Street 207-773-9501
Portland, ME 04101 800-722-2440
Fax: 207-773-9637
E-mail: askensr@ensr.aecom.com
http://www.ensr.aecom.com/Office/44/67/index.jsp
ENSR is a global provider of environmental and energy development services to industry and government. As a full-service environmental firm, ENSR's professionals provide clients with consulting, engineering, remediation, and relatedservices from over 15 countries.
*John Petraglia, PR/Enviro Service Inquiries*
*Paul Genarro, PR/Enviro Service Inquiries*

**4657 ENSR-Sacramento**
10461 Old Placerville Road 916-362-7100
Suite 170 800-722-2440
Sacramento, CA 95827 Fax: 916-362-8100
E-mail: askensr@ensr.aecom.com
http://www.ensr.aecom.com/Office/44/67/index.jsp
ENSR is a global provider of environmental and energy development services to industry and government. As a full-service environmental firm, ENSR's professionals provide clients with consulting, engineering, remediation, and relatedservices from over 15 countries.
*John Petraglia, PR/Enviro Service Inquiries*
*Paul Genarro, PR/Enviro Service Inquiries*

**4658 ENSR-Shawnee Mission**
6400 Glenwood Street 913-362-8444
Suite 105 800-722-2440
Shawnee Mission, KS 66202 Fax: 913-362-1044
E-mail: askensr@ensr.aecom.com
http://www.ensr.aecom.com/Office/44/67/index.jsp
ENSR is a global provider of environmental and energy development services to industry and government. As a full-service envi-

ronmental firm, ENSR's professionals provide clients with consulting, engineering, remediation, and relatedservices from over 15 countries.
*John Petraglia, PR/Enviro Service Inquiries*
*Paul Genarro, PR/Enviro Service Inquiries*

**4659 ENSR-St Petersburg**
9700 16th Street N 727-577-5430
St Petersburg, FL 33716 800-722-2440
Fax: 727-577-5892
E-mail: askensr@ensr.aecom.com
http://www.ensr.aecom.com/Office/44/67/index.jsp
ENSR is a global provider of environmental and energy development services to industry and government. As a full-service environmental firm, ENSR's professionals provide clients with consulting, engineering, remediation, and relatedservices from over 15 countries.
*John Petraglia, PR/Enviro Service Inquiries*
*Paul Genarro, PR/Enviro Service Inquiries*

**4660 ENSR-Stamford**
100 Toms Road 203-323-6620
Stamford, CT 06906 800-722-2440
Fax: 203-348-0809
E-mail: askensr@ensr.aecom.com
http://www.ensr.aecom.com/Office/44/67/index.jsp
ENSR is a global provider of environmental and energy development services to industry and government. As a full-service environmental firm, ENSR's professionals provide clients with consulting, engineering, remediation, and relatedservices from over 15 countries.
*John Petraglia, PR/Enviro Service Inquiries*
*Paul Genarro, PR/Enviro Service Inquiries*

**4661 ENTRIX**
5252 Westchester Street 800-368-7511
Suite 250 Fax: 713-666-5227
Houston, TX 77005 http://www.entrix.com
Provides environmental and natural resource management engineering
*Todd Williams, CEO/President*
*Richard Firth, Executive VP*

**4662 ENVIRO Tech Services**
361 N Ohio Street 785-827-1682
Salina, KS 67401 Fax: 785-827-8765
*Jack Esidck, President*

**4663 ESA Laboratories**
Laboratories, 22 Alpha Road 978-250-7150
Chelmsford, MA 01824 Fax: 978-250-7171
Environmental and biological testing laboratory.

**4664 ESS Group, Inc.**
888 Worcester Street 781-431-0500
Suite 240 Fax: 781-431-7434
Wellesley, MA 02482 E-mail: jhay@essgroup.com
http://www.essgroup.com
The ESS team of scientists, engineers, and regulatory specialists provides a comprehensive range of services related to energy facility development, land and waterfront development, water resource management and ecology, and industrialpermitting and compliance.

*Charles Natale, President/CEO*
*Christopher Rein, Senior Vice President*

**4665 ETS**
1401 Municipal Road NW 540-265-0004
Roanoke, VA 24012 Fax: 540-265-0131
E-mail: jmck@etsi-inc.com
http://www.etsi-inc.com
Research results published by The Environmental Protection Agency, Department of Energy, and Air Pollution Control Association in papers and government reports. ETS is a full-service environmental consulting and training firmspecializing in air emissions control, measurement, engineering and consulting services.
*John McKenna, CEO*

**4666 ETTI Engineers and Consultants**
1000 Rand Road 847-526-1606
Unit 210 Fax: 847-826-7443
Wauconda, IL 60084 E-mail: ettinc@ettinc.com
http://www.ettinc.com
Engineering research and testing laboratory specializing in construction services and environmental needs. Research results published in project reports and test summaries.

**4667 Eaglebrook Environmental Laboratories**
1152 Junction Avenue 219-322-0450
Schererville, IN 46375 Fax: 219-322-0440
Environmental testing service.

**4668 Earth Dimensions**
1091 Jamison Road 716-655-1717
Elma, NY 14059 Fax: 716-655-2915
Geotechnical soil investigations and wetland delineations.

*Don Owens, President*
*Brian Bartron, Geologist/Drilling Manager*

**4669 Earth Regeneration Society**
1442A Walnut Street 510-849-4155
Number 57 Fax: 510-849-0183
Berkeley, CA 94709 E-mail: csiri@igc.apc.org
http://www.imaja.com
The Earth Regeneration Society does research and education on climate change, ozone, and pollution, and calls for full employment and full social support based on surival programs and national and international networking.
*Alden Bryant, President*

**4670 Earth Tech**
4135 Technology Parkway 920-458-8711
PO Box 1067 Fax: 920-458-0537
Sheboygan, WI 53083 E-mail: earthtech.com
http://www.earthtech.com
Specializes in the planning, design, and construction management and observation of environmental and infrastructure projects including water/wastewater; solid, hazardous and process waste facilities; environmental restoration;transportation; and architecture. The company's staff size, multiple office locations and comprehensive mix of expertise and experience combined iwht an in-depth knowledge of technical and regulatory issues, provide our clients with a valuableresource for solutions.
*Diane Creel, President*

**4671 Earth Technology Corporation**
300 Oceangate 562-951-2000
Suite 700 Fax: 562-951-2100
Long Beach, CA 90802 http://www.earthtech.com/
*Alan P. Krusi, President*

**4672 Earthwatch International**
3 Clock Tower Place, Suite 100 978-461-0081
Box 75 800-776-0188
Maynard, MA 01754 Fax: 978-461-2332
E-mail: info@earthwatch.org
http://www.earthwatch.org
Engages people worldwide in scientific research and education to promote the understanding and action necessary for a sustainable environment.
*Ed Wilson, Acting President*

**4673 East Texas Testing Laboratory**
1717 E Erwin Street 903-595-4421
Tyler, TX 75702 Fax: 903-595-6113
Engineering research and testing laboratory specializing in construction services and environmental needs. Research results published in project reports and test summaries.

**4674 Eastern Technical Associates**
3302 Anvil Place 919-878-3188
Raleigh, NC 27603 Fax: 919-872-5199
E-mail: tomrose@eta-is-opacity.com
http://www.eta-is-opacity.com
Environmental consulting firm services of which include visible emissions training & certification. Research results published in government reports.

*Thomas H Rose, President/Co-Founder*
*Willie S Lee, Co-Founder*

**4675 Eberline Analytical, Lionville Laboratory**
364 Welsh Road 610-280-3000
Exton, PA 19341 800-841-5487
Fax: 610-280-3041
E-mail: info@eberlineservices.com
http://www.eberlineservices.com
Analytical, consulting, and field services offer broad capabilities in: radiological characterization and analyses; environmental chemical analyses; hazardous, radioactive, and mixed waste management; and environmental, safety, andhealth management.

*Carter Nulton, Laboratory Manager*
*William F Niemeyer, Business Development Manager*

**4676 Eberline Services - Albuquerque**
7021 Pan American Freeway NE 505-262-2694
Albuquerque, NM 87109 877-477-8989
Fax: 505-262-2698
E-mail: info@eberlineservices.com
http://www.eberlineservices.com
Provides analytical, consulting and field services, offering broad capabilities in radiological characterizaion and analysis; hazaradous, radioactive, and mixed waste management; and environmental, safety, and health management. Inaddition, Eberline Services provides onsite staff and services for site characterization and remediation, decontamination and decommissioning, waste management, and facility operations.
*Leva Jensen, Laboratory Manager*
*William F Niemeyer, Project Manager*

**4677 Eberline Services - Los Alamos**
PO Box 80 505-667-0104
AE-VO2 877-477-8989
Los Almos, NM 87544 Fax: 505-667-7316
E-mail: info@eberlineservices.com
http://www.eberlineservices.com
Provides analytical, consulting and field services, offering broad capabilities in radiological characterizaion and analysis; hazaradous, radioactive, and mixed waste management; and environmental, safety, and health management. Inaddition, Eberline Services provides onsite staff and services for site characterization and remediation, decontamination and decommissioning, waste management, and facility operations.
*Leva Jensen, Laboratory Manager*
*William F Niemeyer, Project Manager*

**4678 Eberline Services - Oak Ridge**
601 Scarboro Road 865-481-0683
Oak Ridge, TN 57830 877-477-8989
Fax: 865-483-4621
E-mail: info@eberlineservices.com
http://www.eberlineservices.com
Provides analytical, consulting and field services, offering broad capabilities in radiological characterizaion and analysis; hazaradous, radioactive, and mixed waste management; and environmental, safety, and health management. Inaddition, Eberline Services provides onsite staff and services for site characterization and remediation, decontamination and decommissioning, waste management, and facility operations.
*Leva Jensen, Laboratory Manager*
*William F Niemeyer, Project Manager*

**4679 Eberline Services - Richland**
3200 George Washington Way 509-371-1506
Richland, WA 99352 877-477-8989
Fax: 509-371-1415
E-mail: info@eberlineservices.com
http://www.eberlineservices.com
Provides analytical, consulting and field services, offering broad capabilities in radiological characterizaion and analysis; hazaradous, radioactive, and mixed waste management; and environmental, safety, and health management. Inaddition, Eberline Services provides onsite staff and services for site characterization and remediation, decontamination and decommissioning, waste management, and facility operations.
*Leva Jensen, Laboratory Manager*
*William F Niemeyer, Project Manager*

**4680 Eberline Services - Richmond**
2030 Wright Avenue 510-235-2633
Richmond, CA 94804 800-841-5487
Fax: 510-235-0438
E-mail: info@eberlineservices.com
http://www.eberlineservices.com
Analytical, consulting, and field services offer broad capabilities in: radiological characterization and analysis; hazardous, radioactive, and mixed waste management; and environmental, safety, and health management.

*Leva Jensen, Laboratory Manager*
*William F Niemeyer, Program Manager*

**4681 Eco-Analysts**
105 E 2nd Street 208-882-2588
Suite 1 Fax: 208-883-4288
Moscow, ID 83843 E-mail: eco@ecoanalysts.com
http://http://www.ecoanalysts.com/contact.html
Specializes in Aquatic Taxonomy and Bioassessment. An independent environmental consulting firm located in Moscow, Idaho. Experienced in the identification of freshwater organisms; macroinvertebrates, periphyton, plankton, and fish.Offer aquatic bioassessment and biological monitoring services.

*Gary Lester*

**4682 EcoTest Laboratories**
377 Sheffield Avenue 631-422-5777
North Babylon, NY 11703 Fax: 631-422-5770
Environmental testing laboratory.

**4683 Ecological Engineering Associates**
13 Marconi Lane 508-748-3224
Marion, MA 02738 Fax: 508-748-9740
*Bruce Strong, Operations Manager*

**4684 Ecology and Environment**
Buffalo Corporate Center 716-684-8060
368 Pleasant View Drive Fax: 716-684-0844
Lancaster, NY 14086 E-mail: nsiekmann@ene.com
http://www.ene.com
E and E is a multidisciplinary environmental science and engineering company with more than 25 offices in the US and offices and partners in more than 35 countries. We are a world leader in providing environmental consulting servicesand litigation support.

*Ronald J Skare, Sr. VP Marketing/Sales*

**4685 Economists**
1200 New Hampshire Avenue NW 202-223-4700
Suite 400 Fax: 202-296-7138
Washington, DC 20036 http://www.ei.com
Firm providing economic analysis and public policy evaluation, with emphasis on private antitrust litigation, communications regulation, and the Environment.
*Bruce M Owen, President*

**4686 Ecotope**
4056 9th Avenue NE 206-322-3753
Seattle, WA 98105 Fax: 206-325-7270
http://www.ecotope.com
Energy efficiency research, architecture, and engineering.
*David Baylon, President*

**4687 Eder Associates**
480 Forest Avenue 516-671-8440
Locust Valley, NY 11560 Fax: 516-671-3349
Environmental engineering and consulting firm.
*Leonard J Eden, President*

**4688 Eichleay Corporation of Illinois**
11919 S Avenue O 773-731-7010
Chicago, IL 60617
Environmental consulting firm.

**4689 El Dorado Engineering**
2964 W 4700 South 801-966-8288
Suite 109 Fax: 801-966-8499
Salt Lake City, UT 84118 E-mail: eldorado50@aol.com
Environmental applications.
*Ralph W Haye, President*

**4690 Electron Microprobe Laboratory Bilby Research Center**
Northern Arizona University 928-523-9565
PO Box 6013 Fax: 928-523-7290
Flagstaff, AZ 86011 E-mail: james.wittke@nau.edu
http://http://www4.nau.edu/microanalysis/Microprobe/Probe.html
The primary goal at the Bilby Research Center is to promote research across the Northern Arizona University campus. Research support services include professional editing; imaging services include illustration, photography, andvideography; and website design.
*James H Wittke, Director*
*Marcelle Coder, Project Director*

**4691 Elm Research Institute**
11 Kit Street 603-358-6198
Keene, NH 03431 800-367-3567
Fax: 603-358-6305
E-mail: info@elmresearch.org
http://www.elmresearch.org
A nonprofit organization dedicated to the restoration and preservation of the American Elm. Provides disease-resistant American Liberty Elms to municipalities, colleges and volunteer nonprofit groups for public planting. Distributionof Elm Fungicide for treatment of Dutch Elm Disease.

*John Hansel, Executive Director*
*Yvonne Spalthoff, Assistant Director*

**4692 Emcon Baker-Shiflett**
5701 E Loop S 817-478-8254
Fort Worth, TX 76119 Fax: 817-478-8874
Consulting engineers providing research services to the construction industry.
*Larry Gardner, Contact*

**4693 Endyne Labs**
342 River Street 802-223-7088
Montpelier, VT 05602 Fax: 802-223-1013
E-mail: info@endynelabs.com
http://www.endynelabs.com
Endyne Labs is a full-service environmental testing laboratory that specializes in the analysis of organic, inorganic, metals and microbiological contaminants in a variety of matrices including drinking water, wastewater, soil,hazardous waste and air.
*Harry B Locker PhD, President*

**4694 Energetics**
7164 Columbia Gateway Drive 410-290-0370
Columbia, MD 21046 Fax: 410-290-0377
Environmental engineering firm.

**4695 Energy & Environmental Technology**
110 Daventry Lane 502-458-0600
Louisville, KY 40223
Environmental research.
*Shirish Phulgaonkar, President*

**4696 Energy Conversion Devices**
2956 Waterview Drive 248-293-0440
Rochester Hills, MI 48309 800-528-0617
Fax: 248-844-1214
http://www.ovonic.com
Maintained a strong core competence in materials research and advanced product development throughout its forty plus year history. The company protects the results of these efforts through an extensive patent collection.

*Stanford Ovshinsky, President/CTO*
*Iris Ovshinsky, Vice President*

**4697 Energy Laboratories**
2393 Salt Creek Highway 307-235-0515
PO Box 3258 888-235-0515
Casper, WY 82602 Fax: 307-234-1639
E-mail: casper@energylab.com
http://www.energylab.com
Environmental data collection, testing and analysis firm.

**4698 Energy and Environmental Analysis**
1655 Fort Myer Drive 703-528-1900
Suite 600 Fax: 703-528-5106
Arlington, VA 22209 http://www.eea-imc.com
Consulting firm offering technical, analytical, and managerial services in the energy/environmental field.
*Michael O Lerner, President*

**4699 Engineering & Environmental Management Group**
11251 Roger Bacon Drive 703-318-4522
Reston, VA 20910 Fax: 703-318-4729
Environmental applications.

**4700 Engineering Analysis**
715 Arcadia Circle 256-533-9391
Huntsville, AL 35801 Fax: 256-533-9325
E-mail: eai@mindspring.com
http://eai.home.mindspring.com
Environmental and safety research and analysis organization. Research results published in client and technical reports and professional journals.
*Frank B Tato, President*

**4701 Entek Environmental & Technical Services**
1724 5th Avenue 518-271-2000
Troy, NY 12180 800-888-9200
Fax: 518-273-6595
E-mail: McDonough@entek-env.com
http://www.entek-env.com/
Environmental consulting and engineering firm.
*Patrick J McDonough, Director/Manager*

**4702 Entropy**
PO Box 90067 919-781-3550
Raleigh, NC 27675 800-486-3550
Fax: 919-787-8442
E-mail: sales@entropyinc.com
http://www.entropyinc.com
Provides air emission testing services.
*Robert Drew, President*

**4703 Enviro Dynamics**
1340 Old Chain Bridge Road 703-760-0023
Suite 300 Fax: 703-760-9382
Mc Lean, VA 22101 E-mail: ian@2edi.com
http://www.2edi.com
Occupational and environmental health analysis and consulting firm.
*William J Keanet, President*

**4704 Enviro Systems**
1 Lafayette Road 603-926-3345
PO Box 778 Fax: 603-926-3521
Hampton, NH 03842 E-mail: pkarbe@envirosystems.com
http://www.envirosystems.com
Environmental compliance testing services, specializing in analytical chemistry and environmental toxicity testing with fresh and salt water, soil and sediment.

*Petra Karbe, VP Marketing*

**4705 Enviro-Bio-Tech**
140 Sharsville Road 610-488-7664
Bernville, PA 19506 Fax: 610-488-9185
http://www.harpi@aol.com

Chemical analysis firm.
*Harpal Singh, President*

**4706 Enviro-Lab**
45-10 Court Square 718-392-0185
Long Island City, NY 11101 Fax: 718-392-8654
E-mail: info@envirolab.com
http://http://www.envirolab.com
Environmental Toxicology Laboratory (ETL) is a research, development and testing laboratory; concentrating its efforts on new approaches to toxicity testing. The mission of ETL is the further advancement of the Tetramitis Assay aswell as the promotion and commercialization of the test.

*Dr. Robert L Jaffe, Ph.D, Director of Lab. Science*

**4707 Enviro-Sciences**
111 Howard Boulevard 973-398-8183
Suite 108 Fax: 973-398-8037
Mount Arlington, NJ 07856 http://www.enviro-sciences.com
Environmental sciences firm.
*Irving D Cohen, CEO*
*Glenn Lechner, Marketing Manager*

**4708 EnviroAnalytical**
627 Main Street 203-459-1800
Monroe, CT 06468 800-459-5060
Fax: 203-459-1466
E-mail: dsethi@enviroanalytical.com
Environmental compliance analysis, R&D, personnel training and analytical method development.
*Dr S Sethi, Job Director*

**4709 Enviroclean Technology**
13015 SW 89th Place 305-232-8249
Suite 212 Fax: 305-232-1011
Miami, FL 33176 E-mail: bill lorenz@erm.com
http://www.erm.com
Environmental laboratory service company.

**4710 Envirodyne Engineers**
303 E Wacker Drive 312-938-0300
Suite 600 Fax: 312-938-1109
Chicago, IL 60601
Environmental science research firm.

**4711 Environ Laboratories**
9725 Girard Avenue S 952-888-7795
Minneapolis, MN 55431 800-826-3710
Fax: 952-888-6345
http://www.environlab.com
Laboratory providing environmental and physical testing of products and materials to commercial and military specifications.
*Alan G Thompson, President*

**4712 Environment Associates**
9618 Variel Avenue 818-709-0568
Chatsworth, CA 91311 800-354-1522
Fax: 818-709-8914
E-mail: info@eatest.com
http://www.eatest.com
Provides a full spectrum of environmental test services to Aerospace, Military and commercial manufacturers including temperature, humidity, altitude, thermal vacuum, shock, vibration, corrosive atmosphere, a DSCC approved connectortest lab, hydraulic and pneumatic test capabilities, flow testing firewall testing, EMMI/EMC testing and more. Services also include test procedure development formal test reports, certification and test fixture modifications and adaptations.

*William Spaulding, President*
*Andrew Spaulding, Sales*

**4713 Environment/One Corporation**
2773 Balltown Road 518-346-6161
Niskayuna, NY 12309 Fax: 518-346-6188
Environmental analysis and instrumentation firm. Research results published in technical journals.

**4714 Environmental Acoustical Research**
PO Box 2146 303-447-2619
Boulder, CO 80306 866-327-7584
Fax: 303-447-2637
E-mail: info@earinc.com
http://www.earinc.com
Specialized hearing protection and enhancement systems.
*G. Gordon, CEO*
*A. Gordon, President*

**4715 Environmental Analysis**
3278 N Highway 67 314-921-4488
Florissant, MO 63033 Fax: 314-921-4494
Environmental analytical laboratory.
*R M Ferris, President*

**4716 Environmental Analytical Laboratory**
95 Beaver Street 781-893-3124
Waltham, MA 02453 Fax: 781-893-4414
E-mail: sboyle@hubtesting.net
http://www.hubtesting.net
Environmental testing services company specializing in asbestos consulting and analysis, waste water, soils and surveys. Also mold remediation and screening.

*Frederick Boyle, President*

**4717 Environmental Audits**
120 Bishops Way 262-785-9322
Suite 130 Fax: 262-785-9323
Brookfield, WI 53005 E-mail: info@environmentalaudits.net
http://www.environmentalaudits.net
Consulting firm specializing in environmental science.

*John R Ruetz, President*

**4718 Environmental Chemical**
6954 Cornell Road 513-489-2001
Suite 300 http://www.ecc.net/Offices/about_office.html
Cincinnati, OH 45242
Research and testing firm.

**4719 Environmental Consultants**
391 Newman Avenue 812-282-8481
Clarksville, IN 47129 Fax: 812-282-8554
Environmental consulting firm.
*Robert E Fuchs, President*

**4720 Environmental Consulting Laboratories**
1005 Boston Post Road 203-245-0568
Madison, CT 06443 800-246-9624
E-mail: eclinc@aol.com
Environmental testing and consulting firm.

**4721 Environmental Consulting Laboratories, Inc.**
1005 Boston Post Road 203-245-0568
Madison, CT 06443 800-246-9624
Fax: 203-318-0830
E-mail: eclinc@aol.com
Environmental testing and consulting firm. Specializing in Micro Biology and aquatic tixicity.

*David Barris, President/Lab Director*

**4722 Environmental Control**
5 Baca Lane 505-473-0982
Santa Fe, NM 87507 Fax: 505-438-3801
Environmental consulting and analytical laboratory. Research results published in project reports and test summaries.
*James J Meuller, President*
*Lisa Ann Wilburn, Marketing Manager*

**4723 Environmental Control Laboratories**
38818 Talyor Industrial Parkway 440-353-3700
North Ridgeville, OH 44039 800-962-0118
Fax: 440-353-3773
E-mail: eclabs@compuserve.com
E.C. Labs does environmental testing, such as: waste water, soil, asbestos, remediation and constrution projects.
*Ron Schiedel, Marketing Manager*
*Phyllis Conley, Lab Manager*

**4724 Environmental Data Resources**
440 Wheelers Farms Road 203-255-6606
Milford, CT 06460 800-352-0050
Fax: 800-231-6802
http://www.edrnet.com
Applied and product research in environmental applications.
*Robert D Barber, CEO*
*Mark Cerino, COO*

**4725 Environmental Elements Corporation**
7380 Coca-Cola Drive 410-368-7000
Handover, MD 21076 800-333-4331
Fax: 410-368-7252
http://www.EEC1.com
Environmental Elements Corporation, the leading supplier of air pollution control systems for over 50 years, designs, installs, and maintains electrostatic precipitators, fabric filters, gas, and particulate scrubbing andAmmonia-on-Demand (AOD) Systems. EEC technologies enable customers in a broad range of power and generation, pulp, and paper, waste-to-energy, rock products, metals and petrochemical industries to operate their facilities in compliance withparticulate and gaseous emissions.
*John L Sans, President/CEO*
*Neil R Davis, Sr. VP Operations*

**4726 Environmental Health Sciences Research Laboratory**
127 New Market Street 504-394-2233
PO Box 379 Fax: 504-394-7982
Belle Chasse, LA 70037

**4727 Environmental Innovations**
9600 West Flag Avenue 414-358-7760
Milwaukee, WI 53225 Fax: 414-358-7770
Environmental engineering and consulting firm.

**4728 Environmental Laboratories**
142 Temple Street 203-789-1260
New Haven, CT 06510 Fax: 203-789-8261
*Ray Macaluso, VP/General Manager*

**4729 Environmental Management: Guthrie**
PO Box 700 405-282-8510
Guthrie, OK 73044 Fax: 405-282-8533
http://www.emiok.com
Full service environmental firm provides emergency response remediation routine waste management for hazardous and non hazardous materials and consulting. The firm owns the transportation and remediation equipment along with providinga full technical staff.

*Terry Bobo, President*

**4730 Environmental Management: Waltham**
95 Beaver Street 781-891-4750
Waltham, MA 02453 Fax: 781-893-4414
E-mail: sboyle@hubtesting.net
http://www.hubtesting.net
Environmental testing services, consulting services, environmental abatement services, water quality/chemical testing, asbestos remediation monitoring and inspections, mold testing/mold remediation, industrial hygiene services.
*Susan Boyle, Vice President*

**4731 Environmental Measurements**
2660 California Street 415-567-8089
San Francisco, CA 94115 Fax: 415-398-7664
E-mail: sales@langan.net
http://http://lpi.langan.net

**4732 Environmental Monitoring Laboratory**
59 N Plains Industrial Road 203-284-0555
Suite A Fax: 203-284-2064
Wallingford, CT 06492
Laboratory providing environmental chemistry services including analysis, bioassays, product efficacy and research and development in the areas of water and wastewater, agricultural chemicals, protective coatings, petroleum products,metals and chemicals.
*Jan D Dunn PhD, Director/Manager*

**4733 Environmental Quality Protection Systems Company**
5150 Keele Street 601-961-5650
Jackson, MS 39206 Fax: 601-354-6612
Environmental engineering and science firm.

**4734 Environmental Research Associates**
414 Mill Road 610-449-7400
Havertown, PA 19083 Fax: 610-449-7404
Research and consulting ecologists and testing firm. Research results published in professional journals. Research results for private clients.

*M H Levin PhD, Director*

**4735 Environmental Resource Associates**
5540 Marshall Street 800-372-0122
Arvada, CO 80002 Fax: 303-421-0159
Engineering consultant.

**4736 Environmental Risk Limited**
120 Mountain Avenue 860-242-9933
Bloomfield, CT 06002 Fax: 860-243-9055
E-mail: info@erl.com
http://www.erl.com
Environmental consulting and engineering firm offers environmental permitting and compliance assistance, site investigation and remediation services, air quality impact analyses, pollution prevention planning, aquatic toxicitylaboratory, hazardous waste management and chemical accident prevention program assistance.
*Gordon T Brookman, President*

**4737 Environmental Risk: Clifton Division**
1373 Broad Street 973-773-8322
Suite 301 Fax: 973-243-9055
Clifton, NJ 07013
Environmental engineering and consulting services.

**4738 Environmental Science & Engineering**
8901 N Industrial Road 309-692-4422
Peoria, IL 61615 Fax: 309-692-9364
Comprehensive environmental and engineering consulting firm.
*Richard Holm, Director/Manager*

**4739 Environmental Services International**
6404 Maccorkle Avenue 304-768-2233
Saint Albans, WV 25177 Fax: 304-768-9988
E-mail: esi@citynet.net
Consulting, engineering, and analytical firm. Research results published in reports to clients.

**4740 Environmental Systems Corporation**
200 Tech Center Drive 865-688-7900
Knoxville, TN 37912 Fax: 865-687-8977
E-mail: esccorp@envirosys.com
http://www.envirosys.com
Data acquisition and reporting systems for electric power producers and industrial sources, ESC is the leading supplier of CEM and ambient data systems in the US.
*Steve Drevik, Senior Marketing Manager*

**4741 Environmental Technical Services**
834 Castle Ridge Road 512-327-6672
Austin, TX 78746 Fax: 512-327-1974
http://www.wetlands.com/
Firm conducts sewer rehabilitation and tank testing.

**4742 Environmental Testing Services**
95 Beaver Street 781-893-8339
Waltham, MA 02453 Fax: 781-893-4414
E-mail: sboyle@hubtesting.net
http://www.hubtesting.net
Environmental testing services company specializing in asbestos consulting and analysis, waste water, soils and surveys. Also mold remediation and screening.
*Frederick Boyle, President*

**4743 Environmental Testing and Consulting**
2924 Walnut Grove Road 901-327-2750
Memphis, TN 38111 Fax: 901-327-6334
Environmental testing service. Research results published in proprietary reports.

**4744 Environmental Working Group**
1718 Connecticut Avenue NW 202-667-6982
Suite 600 Fax: 202-232-2592
Washington, DC 20009 http://www.ewg.org
Cutting-edge research on health and the environment.

**4745 Enviropro**
9765 Eton Avenue 818-998-7197
Chatsworth, CA 91311 Fax: 818-998-7258
Environmental engineering services. Spcialize in: Site investigation/remediation; Real Estate transfers: Phase I and II assessments; feasibility studies; clean-up of contaminated property, soil and groundwater; Methane gasinvestigations.
*Zvia Uziel, President*
*Dr Michael Uziel, Director/Manager*

**4746 Enviroscan Inc**
1051 Columbia Avenue 717-396-8922
Lancaster, PA 17603 Fax: 717-396-8746
E-mail: email@enviroscan.com
http://www.enviroscan.com
Specializes in non-intrusive, non-deestructive land marine and borhole geophysics for engineers, environmental consultants, architects, industry, government and others. Geophysics is the earth science equivalent of medical radiology,and is used to locate subsurface objects such as utilities, underground storage tanks, drums, bedrock depths, sinkholes, contaminant plumes, fractures, graves, downed aircraft(in oceans, lakes) and submerged items.
*Mary Christie, Technical Accounts Manager*

**4747 Envisage Environmental**
PO Box 152 440-526-0990
Richfield, OH 44286 Fax: 440-526-8555
Environmental engineering firm.

**4748 Eppley Laboratory**
12 Sheffield Avenue 401-847-1020
PO Box 419 Fax: 401-847-1031
Newport, RI 02840 E-mail: eplab@mail.bbsnet.com
Produces radiometer, pyranometers, pyrheliometers and pyrgeometers that measure solar and terrestrial radiation.
*George L Kirk, President*

**4749 Era Laboratories**
4730 Oneota Street 218-727-6380
Duluth, MN 55807 Fax: 218-727-3049
http://www.eralabs.com
Environmental laboratory serving the agricultural industry through chemical analysis and sampling.
*Robert D Manuson, President*

**4750 Ernaco**
3740 Capulet Terr. 301-598-5025
Silver Spring, MD 20906
Firm offers biomedical, health and environmental research services.
*Dr Muriel M Lippman, President*

**4751 Eureka Laboratories**
6794 Florin Perkins Road 916-381-7953
Sacramento, CA 95828
*Shao-Pin Yo, Laboratory Director*

**4752 Eustis Engineering Services, LLC**
3011 28th Street 504-834-0157
Metairie, LA 70002 800-966-0157
Fax: 504-834-0354
E-mail: info@eustiseng.com
http://www.eustiseng.com
Geotechnical firm performing complete investigations, dynamic pile testing, cone penetrometer testing, CQC and materials testing and environmental services.
*William W Gwyn, President*
*John R Eustis, Executive Vice President*

**4753 Evans Cooling Systems**
255 Route 41 North 860-364-5130
Sharon, CT 06069 Fax: 860-364-0888
Environmental applications.

**4754 Everglades Laboratories**
1602 Clare Avenue 561-833-4200
West Palm Beach, FL 33401 Fax: 561-833-7280
E-mail: info@evergladeslabs.com
http://www.evergladeslabs.com/
*Dr. Ben Martin, Director*

**4755 Excel Environmental Resources**
825 Georges Road 732-545-9525
2nd Floor Fax: 732-545-9425
New Brunswick, NJ 08902 http://www.excelenv.com
Environmental research.
*Laura Dodge, President*

**4756 First Coast Environmental Laboratory**
8818 Arlington Expressway 904-725-4847
Jacksonville, FL 32211 Fax: 904-725-2215
Analytical laboratory.
*Adolph W Wollitz, Director/Manager*

**4757 Fishbeck, Thompson, Carr & Huber, Inc.**
1515 Arboretum Drive, SE 616-575-3824
Grand Rapids, MI 49546 Fax: 616-464-3993
E-mail: info@ftch.com
http://www.ftch.com
Environmental consulting and engineering firm.
*James A. Susan, P.E., President*
*Kenneth G. Wiley, CPG, Vice President*

**4758 Flowers Chemical Laboratories**
481 Newburyport Avenue 407-339-5984
PO Box 150597 800-669-5227
Altamonte Springs, FL 32715 Fax: 407-260-6110
E-mail: jeff@flowerslabs.com
http://www.flowerslabs.com
Analytical consulting firm. Research results published in reports to clients. Displays report in a pdf or html format.

*Dr Jefferson Flowers, President*

**4759 Forensic Engineering**
PO Box 102 775-359-4692
Sparks, NV 89432 Fax: 775-358-6090

*Joe M Beard, President*

**4760 Fredericktowne Labs**
3020 Ventrie Court 301-293-3340
PO Box 245 800-332-3340
Myersville, MD 21773 Fax: 301-293-2366
E-mail: info@Fredericktownelabs.com
http://www.Fredericktownelabs.com
Environmental testing lab performing analyses on drinking water, waste water and natural waters for microbiological, inorganic, metal and organic contaminants. State certified laboratory. State certified sample collectors.Consulting services.

*Mary Miller, PhD, Laboratory Director*
*Kathy Ryan, Special Projects Coordinator*

**4761 Free-Col Laboratories: A Division of Modern Industries**
11618 Cotton Road 814-724-6242
Meadville, PA 16335 Fax: 814-333-1466
E-mail: johnp@modernind.com
http://www.modernind.com
Full service environmental laboratory - drinking water, waste water, solid waste, industrial hygiene testing; materials testing & engineering; non-destructive testing; mechanical testing; chemical analysis; failure analysis;consulting.
*John Paraska, Director*

**4762 Froehling & Robertson**
3015 Dumbarton Road 804-264-2701
Richmond, VA 23228 Fax: 804-264-1202
Environmental and construction materials testing lab.

**4763 FuelCell Energy**
3 Great Pasture Road 203-825-6000
Danbury, CT 06813 E-mail: dferenz@fce.com
http://www.fce.com
Developer and manufacturer of clean and efficient electric power generators. Products are designed for distributed generation users including schools, data centers, hospitals, buildings, waste water treatment plants and othercommercial and industrial applications.

**4764 Fugro McClelland**
1107 W Gibson Street 512-443-6551
Austin, TX 78704 Fax: 512-444-3996
http://www.fmmg.fugro.com/
Environmental, geotechnical, marine geoscience and environmental engineering firm.
*Frank Marshall, President*

**4765 G&E Engineering**
5601 NW 72nd Street 405-840-0301
Suite 290 Fax: 405-840-4307
Oklahoma City, OK 73132
Environmental impact assessment firm.
*Richard Adams, President*

**4766 GE Osmonics: GE Water Technologies**
5951 Clearwater Drive 952-933-2277
Minnetonka, MN 55343 Fax: 952-933-0141
E-mail: anthony.kobilnyk@ge.com
http://www.gewater.com
Water and process technologies from GE Osmonics provide water, wastewater and process systems solutions.

*Jeffrey R Immelt, Chairman/CEO*
*Anthony Kobilnyk, Media Relations*

**4767 GEO Plan Associates**
30 Mann Street 781-740-1340
Hingham, MA 02043-1316 Fax: 781-740-1340
E-mail: geoplanassoc@gmail.com
Scientific and technical consulting, environmental planning and analysis firm. Experienced in coastal and shallow marine geology, land-use planning, and transportation.

*Peter S Rosen PhD, Partner*
*Michu Tcheng, Partner*

**4768 GEO-CENTERS**
7 Wells Avenue 617-964-7070
Newton, MA 02459 Fax: 617-527-7592
http://www.geo-centers.com
Provider of WMD homeland security preparedness services and products with major strengths in chemical and biological research.
*Edward P Marram, CEO*

**4769 GKY and Associates**
4229 Lafayette Ctr Dr 703-870-7000
Suite 1850 Fax: 703-870-7039
Chantilly, VA 21051 E-mail: sstein@gky.com
http://www.gky.com
Civil and environmental systems engineering consulting organization. Research results published in project reports, government publications, and professional journals.

*Stuart Stein, President*
*Brett Martin, VP*

**4770 GL Applied Research**
142 Hawley Street 847-223-2220
PO Box 187 Fax: 847-223-2287
Grayslake, IL 60030 E-mail: glapplied@aol.com
Analytical and process control instrumentation development and manufacture; photometric analyzers.

*Edgar Watson Jr, President*

**4771 GSEE**
599 Waldron Road 615-793-7547
La Vergne, TN 37086 Fax: 615-793-5070
E-mail: gsee@gseeinc.com
http://http://gseeinc.com
GSEE provides environmental engineering and technical services for municipal, industrial and governmental clientele as well as other engineering consultants and manufacturers throughout the United States and abroad. GSEE offersexisting plant evaluations, process engineering, detail engineering, and operations training.
*Wendy Ingram, Lab Director*

**4772 GZA GeoEnvironmental**
One Edgewater Drive 781-278-3700
Norwood, MA 02062 Fax: 781-278-5701
E-mail: info@gza.com
http://www.gza.com
Geotechnical and geohydrological testing and analysis firm.

*William R Beloff, President/CEO*
*Joseph P Hehir, CFO*

**4773 Gabriel Laboratories**
1421 North Elston 773-486-2123
Chicago, IL 60622 Fax: 773-486-0004
*Donna Panek, Laboratory Director*

**4774 Galson Laboratories**
6601 Kirkville Road 315-432-5227
PO Box 369 888-432-5227
East Syracuse, NY 13057 Fax: 315-437-0509
http://www.galsonlabs.com
Environmental industrial hygiene and biological testing service.

**4775 Gas Technology Institute**
1700 S Mount Prospect Road 847-768-0500
Des Plaines, IL 60018 Fax: 847-768-0501
E-mail: info@gastechnology.org
http://www.gastechnology.org
Energy and environmental research.
*John F Riordan, President*
*Robert A Stokes, R&D*

**4776 Gaynes Labs**
9708 Industrial Drive 708-223-6655
Bridgeview, IL 60455 Fax: 708-233-6985
E-mail: Gayneslabs@aol.com
http://www.nrinc.com/gaynes/
Research laboratories. Research results published in confidential test summaries and project reports. Services include environmental testing, atmospheric simulations, frequency vibration testing, temperature and humidity simulationtesting.
*Philip Ross, Material Testing Manager*

**4777 General Engineering Labs**
2040 Savage Road 843-556-8171
Charleston, SC 29407 Fax: 843-766-1178
http://www.gel.com
Environmental testing lab.

**4778 General Oil Company/GOC-Waste Compliance Services**
31478 Industrial Road 734-266-6500
Suite 100 800-323-9905
Livonia, MI 48150 Fax: 734-266-6400
E-mail: twesterdale@generaloilco.com
http://www.generaloilco.com
Waste Compliance Services (WCS) is an analytical laboratory, serving the industrial community. WCS services a broad client base that includes manufacturing industries, environmental consultants, independent contractors, andindividuals. Services include effluent management programs (permit negotiation and management, sampling, self monitoring reports and waste treatment assistance) and analytical testing services.
*Timothy A Westerdale, President/CEO*
*Adam Westerdale, VP/Chief Operating Officer*

**4779 General Sciences Corporation**
4600 Powder Mill Road 301-931-2900
Suite 400 Fax: 301-931-3797
Beltsville, MD 20705 http://www.saic-gsc.com
Consulting and research firm specializing in environmental sciences.
*Jeffrey Chen, President*

**4780 General Systems Division**
1025 West Nursery Road 410-636-8700
Suite 120 Fax: 410-636-8708
Linthicum Heights, MD 21090 http://www.nct-active.com
Environmental studies.
*Michael Parella, President*

**4781 Geo Environmental Technologies**
140 Broadway 401-421-4140
Providence, RI 02903 Fax: 401-751-8613
Environmental testing and consulting company.

**4782 Geo-Con**
4075 Monroeville Boulevard 412-856-7700
Suite 400 Fax: 412-373-3357
Monroeville, PA 15146
Environmental services firm provides soil remediation by mixing soil with chemicals designed to eliminate the contaminants.

**4783 GeoPotential**
22323 E Wild Fern Lane 503-622-0154
Brightwood, OR 97011 Fax: 503-492-4404
E-mail: geopotential@aol.com
http://www.members.aol.com/resiii/geomain.htm
GeoPotential provides subsurface mapping surveys to locate underground objects such as underground storage tanks, utilities, geology, etc. They use geophysical methods consisting of ground penetrating radar, magnetics,electromagnetics and gravity.

*Ralph Soule, President*

**4784 Geological Sciences & Laboratories**
3133 N Main Street 606-439-3373
Hazard, KY 41701

**4785 Geomatrix**
2443 Sidney Avenue # A 410-752-5388
Baltimore, MD 21230
Environmental engineering and consulting firm.

**4786 Geomet Technologies**
20251 Century Boulevard 301-428-9898
Germantown, MD 20874 Fax: 301-428-9482

*Robert L Durfee, President*

**4787 Geophex**
605 Mercury Street 919-839-8515
Raleigh, NC 27603 Fax: 919-839-8528
E-mail: geophex@geophex.com
http://www.geophex.com
Environmental services firm.
*IJ Won, President*

**4788 George Miksch Sutton Avian Research Center**
PO Box 2007 918-336-7778
Bartlesville, OK 74005 Fax: 918-336-7783
E-mail: gmsarc@aol.com
http://www.suttoncenter.org
Finding cooperative conservation solutions for birds and the natural world through science and education.

*Steve Sherrod, Executive Director*
*Alan Jenkins, Assistant Director*

**4789 Geotechnical and Materials Testing**
22446 Davis Drive 703-406-8702
Suite 127 Fax: 703-406-8708
Sterling, VA 20164
Environmental engineering and consulting firm.
*Ahmed N Elrefai, President*

**4790 Gerhart Laboratories**
Route 219 814-634-0820
Garrett, PA 15542
Environmental testing laboratory.
*Michael Gerhart, President*

**4791 Giblin Associates**
PO Box 6172 707-528-3078
Santa Rosa, CA 95406 Fax: 707-528-2837
Environmental and geotechnical engineering firm.
*Jere A Giblin, President*

**4792 Global Geochemistry Corporation**
6919 Eton Avenue 818-992-4103
Canoga Park, CA 91303 Fax: 818-992-8940

Consulting firm in the fields of geochemistry and environmental sciences. Research results published by the firm's scientists in journals.
*Isaac Kaplan, President*

**4793 Globetrotters Engineering Corporation**
300 S Wacker Drive 312-922-6400
Suite 200 Fax: 312-922-2953
Chicago, IL 60606
*Niranjan S Shah, Chair*

**4794 GoodKind & O'Dea**
31 Saint James Avenue 617-695-3400
Suite 1601 Fax: 617-695-3310
Boston, MA 02116
Architectural and engineering consulting firm.
*David K Blake, Contact*

**4795 Gordon & Associates**
6975 North County Road 765-478-4801
550 West POÆBox 25 Fax: 765-478-9073
Benstonville, IN 47331
Environmental consulting firm specializing in waste residuals.
*Paul W Gordon, President*

**4796 Gordon Piatt Energy Group**
Box 650 316-221-4770
Winfield, KS 67156 Fax: 316-221-6289
*Jim Salomon, President*

**4797 Grand Junction Laboratories**
435 North Avenue 970-242-7618
Grand Junction, CO 81501 Fax: 970-243-7235
*Brian Bauer, Director*

**4798 Greeley-Polhemus Group**
1310 Birmingham Road 610-793-9440
West Chester, PA 19382
Environmental engineering and economic analysis firm.

**4799 Ground Technology**
14227 Fern Drive 281-597-8866
Houston, TX 77079 Fax: 281-597-8308
E-mail: ground@groundtechinc.com
http://http://www.groundtechinc.com
A multi-disciplinary engineering firm specializing in environmental services, geotechnical engineering, and construction materials and inspection services. GTI is a woman owned business enterprise as well as minority/disadvantagedenterprise certified by TxDOT, METRO, the State of Texas, and City of Houston, HISD, and the Houston Minority Business Council.
*Ruma Acharya, President*

**4800 Groundwater Specialists**
3806 Telluride Place 303-494-8122
Boulder, CO 80305 Fax: 303-494-5443
E-mail: gws@qwest.net
Groundwater exploration; dewatering; waterwell design; mitigation of high groundwater problems.
*William H Bellis, Hydrologist*

**4801 Gruen, Gruen & Associates**
564 Howard Street 415-433-7598
San Francisco, CA 94105
*Nina J Gruen, Principle Sociologist*

**4802 Guanterra Environmental Services**
1721 S Grand Avenue 714-258-8610
Santa Ana, CA 92705 Fax: 714-258-0921
Chemical analysis technical and consulting research firm. Research results published in project reports and professional journals.

**4803 Guardian Systems**
1108 Ashville Road NE 205-699-6647
PO Box 190 866-729-7211
Leeds, AL 35094 Fax: 205-699-3882
E-mail: gsilab@gsilab.com
http://www.gsilab.com
Laboratory division provides a full range of analysis for inorganic, organicand physical testing of drinking water, wastewater, groundwater sediments, sludge, waste materials and soils. Industrial Hygiene division provides equipmentand analysis to meet OSHA requirements. Bio-Assay division can accommodate Aquatic Toxicity monitoring requirement.
*Gerald Miller, President*
*Linda Miller, Executive Vice President*

**4804 Gulf Coast Analytical Laboratories**
7979 GSRI Avenue 225-769-4900
Baton Rouge, LA 70820 Fax: 225-767-5717
E-mail: edg@gcal.com
http://www.gcal.com
Environmental and industrial testing laboratory.
*Ed Gallagher, Sales Director*

**4805 Gutierrez, Smouse, Wilmut and Associates**
11117 Shady Trl 972-620-1255
Dallas, TX 75229 Fax: 972-620-8028
Environmental engineering consulting firm. Research results published in project reports and technical journals.
*Charles G Wilmut, President*

**4806 H John Heinz III Center for Science**
900 17th Street NW 202-737-6307
Suite 700 Fax: 202-737-6410
Washington, DC 20006 E-mail: info@heinzctr.org
http://www.heinzctr.org
The Center is a nonpartisan, nonprofit institute dedicated to improving the scientific and economic foundation for environmental policy through nmultisectoral collaboration. The Heinz Center fosters collaboration among industry,environmental organizations, academia, and all levels of government in each of its program areas.
*Deb Callahan, President*

**4807 H2M Group: Holzmacherm McLendon & Murrell**
575 Broad Hollow Road 631-756-8000
Melville, NY 11747 Fax: 631-694-4122
E-mail: labs@h2m.com
http://www.h2mlabs.com
Engineers, hydrogeologists, geologists and scientists strive to balance society's dynamic industrial and commercial growth with appropriate development and conservation of natural and man-made resources.
*John J Molloy, President*

**4808 HC Nutting, A Terracon Company**
611 Lunken Park Drive 513-321-5816
Cincinnati, OH 45226 Fax: 513-321-0294
E-mail: cincinnati@hcnutting.com
http://www.hcnutting.com
A materials testing company, geotechnical and environmental engineering firm.
*Jack Scott, President*
*Jim Cahill, CEO*

**4809 HTS**
416 Pickering Street 713-692-8373
Houston, TX 77091
*Ron Langston, President*

**4810 HWS Consulting Group**
PO Box 80358 — 402-479-2200
Lincoln, NE 68501 — Fax: 402-479-2276
*James Linderholm, President*

**4811 Hach Company**
Box 389 — 970-669-3050
Loveland, CO 80539 — Fax: 970-669-2932

*Bruce Hach, President*

**4812 Haley & Aldrich**
9040 Friars Road, Suite 220 — 619-285-7119
San Diego, CA 92108 — Fax: 619-285-7169
E-mail: info@haleyaldrich.com
http://www.haleyaldrich.com
Haley & Aldrich provides leading edge underground engineering and environmental consulting services, nationally and internationally. The staff encompass a wide range of disciplines, offering their clients integrated solutions.Environmental services include corrective action; environmental management, health and safety consulting; and environmental site assessment/due diligence.

*Bruce E Beverly, CEO*

**4813 Halliburton Company**
500 North Akard — 214-978-2600
Dallas, TX 75201 — Fax: 214-978-2611
One of the world's largest providers of products and services to the oil and gas industries.

*Thomas H Cruikshank, Chairman/CEO*

**4814 Hamilton Research, Ltd.**
80 Grove Street — 914-631-9194
Tarrytown, NY 10591 — Fax: 914-631-6134
E-mail: rwh@rwhamilton.com
Hamilton Research is a consulting firm specializing in environmental physiology. Our focus is mainly on exposure of people to pressures less and greater than atmospheric, and involves dealing with different breathing gases, especiallyhigh and low levels of oxygen, and the consequences of changes in pressure, especially decompression. Another important area of interest is hyperbaric oxygen therapy.

*R W Hamilton, President*

**4815 Hampton Roads Testing Laboratories**
611 Howmet Drive — 757-826-5310
Hampton, VA 23661
Independent third party testing laboratory. Performs sampling and analysis in accordance with the required ASTM or ISO Standards.

**4816 Handex Environmental Recovery**
500 Campus Drive — 732-536-8500
PO Box 451 — Fax: 732-536-7751
Morganville, NJ 07751
Environmental management and analysis firm.
*CL Smith, CEO*

**4817 Hart Crowser**
1910 Fairview Avenue E — 206-324-9530
Suite 100 — Fax: 206-328-5581
Seattle, WA 98102 — E-mail: rick.moore@hartcrowser.com
http://www.hartcrowser.com
Hart Crowser, Inc. provides a full range of services from initial site studies through regulatory permitting, design, and construction. They integrate thses services as required by each project. They know what kind of informaiton isimportant, how to collect it and apply it to the selection of viable solutions, and how actions are perceived by regulatory agencies and the public. Consequently, they design an appraoch that is practical, cost-effective, and client-oriented.

*Rick Moore, Principal, Env. Svce. Mgr.*

**4818 Hatch Mott MacDonald**
PO Box 1008 — 973-379-3400
27 Bleeker Street — 800-832-3272
Millburn, NJ 07041 — Fax: 973-376-1072
E-mail: info@hatchmott.com
http://www.hatchmott.com
Hatch Mott MacDonald is a client-focused consulting firm providing planning, investigation, design and management capabilities in engineering disciplines and environmental sciences. Areas of expertise include industrial wastewater,site utilities engineering, hazardous and solid waste management and environmental site assessments.

*Russell Shallieu, Associate*
*Dennis Suler, Executive VP*

**4819 Hatcher-Sayre**
905 Southlake Boulevard — 804-794-0216
Richmond, VA 23236 — Fax: 804-379-8934
Environmental consulting and engineering services firm.

**4820 Havens & Emerson**
700 Bond Court Building — 216-621-2407
1300 E 9th Street — Fax: 216-621-4972
Cleveland, OH 44114
Environmental engineering firm.
*Gary Siegel, President/CEO*

**4821 Hayden Environmental Group**
6015 Manning Road — 937-439-3764
Dayton, OH — Fax: 937-439-3767
Testing, sampling, and analysis service.

**4822 Hayes, Seay, Mattern & Mattern**
PO Box 13446 — 540-857-3100
Roanoke, VA 24034 — Fax: 540-857-3296
*Troy S Kincer, PE, Principal Associate*
*Guy E Slagle, PE, LS, Vice President*

**4823 HazMat Environmental Group**
60 Commerce Drive — 716-827-7200
Buffalo, NY 14218 — Fax: 716-827-7217
http://www.hazmatinc.com
Transportation services - specializing in hazardous materials and hazardous waste transportation.

*Ricky F Wickham, General Manager of Operation*

**4824 Henry Souther Laboratories**
24 Tobey Road — 860-242-6291
Bloomfield, CT 06002 — Fax: 860-286-0634

*Richard J Lombardi, VP*

**4825 Heritage Laboratories**
7901 W Morris Street — 317-243-8304
Indianapolis, IN 46231 — Fax: 317-486-5085
Environmental testing laboratory.

**4826 Heritage Remediation Engineering**
4925 Heller Street — 502-473-0638
Louisville, KY 40218 — Fax: 502-459-4988
Environemtal management and remediation company.

**4827 Hess Environmental Services**
6057 Executive Centre Drive — 901-377-9139
Suite 6 — Fax: 901-377-9150
Memphis, TN 38134 — E-mail: HES@hessenv.com
http://www.hessenv.com
Hess Environmental Services, Inc. (HES) is an environmental consulting/engineering firm. Their primary activities are: Indoor air quality (IAQ) (Mold and Bacterial) Investigations; Title V Air and Other Permit Applications; Phase I,II, III Property Assessments; Remedial Investigations, Audits, Enviromental

Health and Safety, Storm Water, Wastewater, Air Monitoring, Asbestos Inspection and Sampling.

*Connie Hess, President*
*Gary Siebenschuh, VP*

**4828 Hidell-Eyster Technical Services**
PO Box 325 781-749-8040
Accord, MA 02018 Fax: 781-749-2304
E-mail: hidell@hidelleyster.com
Environmental and bottled water assessment and consulting company.
*Henry R Hidell, President*

**4829 Hillmann Environmental Company**
1080 Cedar Avenue 908-686-3335
Union, NJ 07083 Fax: 908-686-2636
*Joseph Hillmann, Executive VP*

**4830 Honeywell Technology Center**
3660 Technology Drive 612-951-1000
Minneapolis, MN 55418 800-328-5111
Fax: 612-951-7438
E-mail: info@htc.honeywell.com
http://www.htc.honeywell.com
Parent holding company with numerous high-tech units involved in environmental, energy, computer hardware and industrial automation research and development.

**4831 Hoosier Microbiological Laboratory**
912 West McGalliard 765-288-1124
Muncie, IN 47303 Fax: 765-288-8378
*Donald A Hendrickson, Owner*

**4832 Horner & Shifrin**
5200 Oakland Avenue 314-531-4321
Saint Louis, MO 63110 Fax: 314-531-6966
http://www.hornershifrin.com
Civil, structural, and environmental engineering firm.

*Jim McCleish, Associate VP*

**4833 Houston Advanced Research Center**
4800 Research Forest Drive 281-367-1348
The Woodlands, TX 77381 Fax: 281-363-7914
http://www.harc.edu

Environmental studies

*Robert Harriss, CEO/President*
*James Lester, Vice President*

**4834 Humphrey Energy Enterprises**
216 7th Avenue South 406-538-3132
Lewistown, MT 59457
*Dr. John P Humphrey, Consultant*

**4835 Huntingdon Engineering & Environmental**
1940 Orange Tree Lane 909-793-2691
Redlands, CA 92374 Fax: 909-793-1704
http://www.dell.com/outlet
Offers the following service(s): Environmental remediation, engineering services, environmental consultant, environmental research, petroleum, mining, and chemical engineers and sanitary engineers.

**4836 Hydro Science Laboratories**
320 West Water Street 732-349-9692
P.O. Box 4978 800-624-3100
Toms River, NJ 08753 Fax: 609-693-4682
E-mail: info@HydroscienceInc.com
http://www.hydroscienceinc.com
Environmental testing and analysis laboratory.
*Robert Salt, Director of Business Dvlpmt*

**4837 Hydro-logic**
1927 North 1275 Road 785-542-2518
Eudora, KS 66025 Fax: 785-542-3971
E-mail: Logic913@aol.com
http://www.hydro-logic.com
Offers professional environmental services, specializing in hydraulic soil and groundwater sampling. Maintains a multidisiplinary team of geologists, hydrologists, chemists, and regulatory compliance specialists.
*Thomas Barr, President*

**4838 Hydrocomp**
2386 Branner Drive 650-561-9030
Menlo Park, CA 94025 Fax: 650-561-9031
*Dr. N Crawford, President*

**4839 Hydrologic**
122 Lyman Street 828-258-3973
Asheville, NC 28801 Fax: 828-258-3973
Environmental laboratory services firm.
*Thomas Barr, President*

**4840 IAS Laboratories Inter Ag Services**
2515 E University Drive 602-273-7248
Phoenix, AZ 85034 Fax: 602-275-3836
E-mail: ias_at_iaslabs.com
http://www.iaslabs.com
IAS is an international agricultural laboratory and research facility serving that provides a variety of environmental services including: soil fertility and water suitability testing; ASTM, AASHTO, ADOT & CDOT testing; quality controltesting for the fertilizer industry; and plant tissue, petiole and soil analysis for agriculture and farmers.
*Paul J Eberhardt PhD, President*
*Sheri K McLane, Laboratory Manager*

**4841 IC Laboratories**
PO Box 721 914-962-2477
Amawalk, NY 10501 Fax: 914-962-5564
Firm providing qualitative and quantitative materials analysis through X-ray diffraction. Studies focus on powders, metals, fibers, and clays, including analysis of crystallinity, thin films, environmental dusts, geological materials,and fabrics. Also provides limited research and development and consulting.

**4842 ICS Radiation Technologies**
8416 Florence Avenue 562-923-1837
Suite 207 Fax: 562-923-3609
Downey, CA 90240 E-mail: mike@icsrad.com
http://www.icsrad.com
Testing, engineering and consulting firm specializing in radiation effects in semiconductor devices.
*Dr Michael K Gauthier, President*

**4843 IHI Environmental**
4527 N 16th Street 602-776-0300
Suite 105 Fax: 602-776-0301
Phoenix, AZ 85016 E-mail: phoenix@ihi-env.com
http://www.ihi-env.com/

**4844 INFORM**
5 Hanover Street 212-361-2400
14th Floor Fax: 212-361-2412
New York, NY 10004 E-mail: inform@informinc.org
http://www.informinc.org
INFORM is an independent research organization that examines the effects of business practices on the environment and on human health. Our goal is to identify ways of doing business that ensure environmentally sustainable economicgrowth. Our reports are used by government, industry, and environmental leaders around the world.
*Joanna Underwood, President*

**4845 Ike Yen Associates**
867 Marymount Lane 714-621-2302
Claremont, CA 91711

**4846 Image**
4525 Kingston Street 303-371-3338
Denver, CO 80239 Fax: 303-371-3299
Biochemistry and environmental research firm. Research results published in professional journals.

**4847 ImmuCell Corporation**
56 Evergreen Drive 207-878-2770
Portland, ME 04103 Fax: 207-878-2117
http://www.immucell.com
Biotechnology testing kits, animal health products and environment water testing.
*Michael F Brigham, President*

**4848 Industrial Laboratories**
4046 Youngfield Street 303-287-9691
Wheat Ridge, CO 80033 800-456-5288
Fax: 303-287-0964
http://www.industriallabs.net
Provides quality laboratory analysis and consultation. ICP Mineral Analysis.

*Larisa Moore, Business Development Manager*

**4849 Informatics Division of Bio-Rad**
3316 Spring Garden Street 215-382-7800
Philadelphia, PA 19104 Fax: 215-382-7800
*Richard Shaps, Division Manager*

**4850 Innovative Biotechnologies International**
335 Lang Boulevard 716-773-4232
Grand Island, NY 14702 Fax: 716-773-4257
E-mail: info@ibi.cc
http://www.ibi.cc
Manufacturing technology of biosensing technology.

**4851 Inprimis**
500 West Cypress Creek Road 954-556-4020
Suite 1 Fax: 954-556-4031
Fort Lauderdale, FL 33309 E-mail: info@ener1.com
http://www.inprimis.com/
Provides hardware and software technology, communications solutions that enbale data transmission, connectivity of devices, and access to applications and information via the Internet, personal computers, and/or server-basedenvironments. Also designs, manufactures, markets, and supports quality, innovative products that have a cost, performance, and time-to-market advantage.
*Kevin P. Fitzgerald, Chairman/CEO*
*Ronald N. Stewart, Executive VP*

**4852 Institute for Alternative Agriculture**
9200 Edmonston Road 301-441-8777
Suite 117 Fax: 301-220-0164
Greenbelt, MD 20770 E-mail: hawiaa@access.digex.net
The Wallace Institute advances this goal by providing the leadership, and policy research and analysis necessary to influence national agriculturalpolicy. It is a contributing member of a growing national alternaative agriculturenetwork, and works directly with government agencies, educational and research institutions, producer groups, farmers, scientists, advocates, and other organizations that provide agricultural research, education, and information services.
*Dr. I Garth Youngberg, Executive Director*

**4853 Institute for Applied Research**
840 La Goleta Way 916-482-3120
Sacramento, CA 95864

**4854 Institute for Environmental Education**
16 Upton Drive 978-658-5272
Wilmington, MA 01887 800-823-6239
Fax: 978-658-5435
E-mail: sales@ieetrains.com
http://www.ieetrains.com
IEE is New England's largest environmental training provider with over 57 classes in Asbestos, OSHA, Lead-paint and Environmental Health and Safety.

*Martin Wood, President*
*Roy Teresky, VP Marketing*

**4855 Integral System**
5000 Philadelphia Way 301-731-4233
Suite A Fax: 301-731-9606
Lanham, MD 20706 http://www.integ.com
Custom computer systems for satellite control; environmental monitoring.

**4856 Inter-Mountain Laboratories**
1673 Terra Avenue 307-672-8945
Sheridan, WY 82801 Fax: 307-672-6053
Provides high-quality analytical, engineering and field services to industry and governmental agencies.
*Duane Madsen, President*

**4857 International Asbestos Testing Laboratories**
16000 Horizon Way 856-231-9449
Unit 100 Fax: 856-231-9818
Mount Laurel, NJ 08054 E-mail: info@iatl.com
An environmental laboratory specializing in asbestos, lead and mold analysis. Provides environmental laboratory services to environmental consultants, engineers, building owners and govt. agencies throughout the US, Canada and othercountries. Accredited by numerous agencies including the National Voluntary Laboratory Accreditation Program (NVLAP) and the American Industrial Hygiene Association (AIHA).

*Emil M Ondra, President*
*Shirley Clark, Business Development*

**4858 International Maritime, Inc**
110 Pine Avenue 562-624-4343
Suite 1070
Long Beach, CA 90802

**4859 International Science and Technology Institute**
1820 North Fort Myer Drive 703-807-2080
Suite 600 Fax: 703-807-1126
Arlington, VA 22209 E-mail: isti@istiinc.com
http://www.istiinc.com
Provides technical assistance in project design, implementation, and evaluation; database development and maintenance; institutional and human resource development; policy and economic analysis, methodological research and analysis;strategic planning; and workshop and conference design and organization.
*BK Wesley Copeland, Vice Chair*

**4860 International Society of Chemical Ecology**
University of California
Department of Entomology 909-787-5821
Riverside, CA 92521 Fax: 909-787-3086
E-mail: jocelyn.millar@ucr.edu
http://www.isce.ucr.edu/Society/
ISCE is organized specifically to promote the understanding of interactions between organisms and their environment. Research areas include the chemistry, biochemistry and function of natural products, their importance at all levels ofecological organization, their evolutionary origin and their practical application.
*John Hildebrand, President*

**4861 Interpoll Laboratories**
4500 Ball Road NE 763-786-6020
Circle Pines, MN 55014 Fax: 763-786-7854
E-mail: interpoll@interpoll-labs.com
http://www.interpoll-labs.com
Interpoll is a full service environmental laboratory with a multidisciplinary staff. They provide their clients with responsive and accurate solutions to their environmental needs. Interpoll offers a full range of environmentaltesting services including stationary source testing, laboratory analysis, groundwa-

ter monitoring, ambient air monitoring and pharmaceutical analysis.

*Dan Despen, President*
*Timothy MacDonald, Manager Field Services*

**4862 Invensys Climate Controls**
191 East North Avneue 630-260-3402
Carol Stream, IL 60188 http://www.icca.invensys.com
Formely the Robertshaw Controls Company. Founded after a successfully designing and manufacturing a line of top quality smoke alarms for the residential smoke alarm market.

**4863 J Dallon and Associates**
16 Fox Hollow Road 201-825-4574
Ramsey, NJ 07446
Research and consulting firm specializing in hortoculture.
*Dr Joseph Dallon Jr, President*

**4864 J Phillip Keathley: Agricultural Services Division**
25330 Ruess Avenue 209-599-2800
Ripon, CA 95366
*Dr. J Phillip Keathley, President*

**4865 JABA**
2766 North Country Club Road 520-327-7440
Tucson, AZ 85716 Fax: 520-327-7450
E-mail: jbriscoe@jaba.com
http://www.jaba.com
Mining exploration and environmental analysis firm.
*James A Briscoe, President*

**4866 JH Kleinfelder & Associates**
7133 Koll Center Parkway 925-484-1700
Suite 100 Fax: 925-484-5838
Pleasanton, CA 94566
Geotechnical and environmental Engineering firm.

**4867 JH Stuard Associates**
22 Tanglewood Drive 802-878-5171
Woodstock, VT 05091
Environmental Consulting firm.
*Joe Shockcor, President*

**4868 JK Research Associates**
86 Gold Hill Road 970-453-1760
Breckenridge, CO 80424

**4869 JL Rogers & Callcott Engineers**
PO Box 5655 864-232-1556
Greenville, SC 29606 Fax: 864-233-9058
Environmental engineering research firm.

**4870 JM Best**
119 S College Street 724-222-2102
Washington, PA 15301
Performs geologic, economic and engineering evaluations for oil and gas well drilling. Also provides completion operations, environmental studies, map preparations and investigative studies.

**4871 JR Henderson Labs**
123 Seaman Avenue 732-341-1211
Beachwood, NJ 08722 Fax: 732-505-1658
Environmental laboratory.
*Elmer Hemphill, President*

**4872 JWS Delavau Company**
2140 Germantown Avenue 215-235-1100
Philadelphia, PA 19122 Fax: 215-671-1401
International environmental and technical company.
*David L Sokol, President*

**4873 James R Reed and Associates**
770 Pilot House Drive 757-873-4703
Newport News, VA 23606 800-873-4703
Fax: 757-873-1498
E-mail: claiborne@jrreed.com
http://www.jrreed.com
Full service environmental testing facility offering quality analysis and reliable technical services to industry, local and federal government, engineers and private citizens. Areas of expertise include organic and inorganic chemicalanalyses, microbiological testing, and aquatic toxicity monitoring. Certificationto include NELAC Certification for the State of Virginia.

*Han Ping Huang, President*
*Elaine Claiborne, Laboratory Director*

**4874 James W Bunger and Associates**
PO Box 520037 801-975-1456
Salt Lake City, UT 84152 Fax: 801-975-1530
Energy research and development firm specializing in environmental and oil remediation.

**4875 Jane Goodall Institute for Wildlife Research, Education and Conservation**
8700 Georgia Avenue 240-645-4000
Suite 500 800-99C-HIMP
Silver Spring, MD 20910 Fax: 301-565-3188
http://www.janegoodall.org
A tax-exempt, nonprofit corporation, founded in 1977 focusing on Jane Goodall.
*William Johnson, President & CEO*

**4876 John D MacArthur Agro Ecology Research Center**
300 Buck Island Ranch Road 727-669-0242
Lake Placid, FL 33852 Fax: 863-699-2217
E-mail: maerc@archbold-station.org
http://www.archbold-station.org/maerc/maerc.htm
The MacArthur Agro Ecology Research Center at Buck Island Ranch is dedicated to a mission of long-term research, education and outreach related to the ecological and social value of subtropical grazing lands. The Center is at a 10,300acre cattle ranch on a long-term lease to Archbold Biological Station from the John D and Catherine T MacArthur Foundation. Provides researchers the opportunity to evaluate the relationship between economic and ecological factors and how these changeover time.

**4877 Johnson Company**
100 State Street 802-229-4600
Suite 600 Fax: 802-229-5876
Montpelier, VT 05602 E-mail: info@jcomail.com
http://www.johnsonco.com
Environmental science and engineering consulting

*Chris M Crandell, President*
*Michael B Moore, VP*

**4878 Johnson Controls**
5757 N Green Bay Avenue 414-228-1200
Milwaukee, WI 53209 Fax: 414-228-2446
Research in environmental controls.
*James Keyes, President*

**4879 Johnson Research Center**
University of Alabama at Huntsville 256-890-6343
Huntsville, AL 35899 Fax: 256-890-6848
Environmental research.
*Dr. Michael Eley, CEO*

**4880 Jones & Henry Laboratories**
2567 Tracy Road 419-666-0411
Northwood, OH 43619 Fax: 419-666-1657
E-mail: jhlabs@glasscity.net
Environmental sampling and testing laboratory.
*Fred W Doering, President*
*David Collins, Marketing Manager*

**4881 Joyce Environmental Consultants**
5051 North Lane 407-297-7980
Orlando, FL 32808 Fax: 407-290-0388

*Connie Morrison, VP*

**4882 KAI Technologies**
16 Marin Way 603-778-1888
Stratham, NH 03885 Fax: 603-778-0700
E-mail: cliff@kaitech.com
http://www.kaitech.com
Applied and product research in the environment.
*Bruce L. Cliff, Director*

**4883 KCM**
1917 1st Avenue 206-443-5300
Seattle, WA 98101 Fax: 206-443-5372
http://www.tetratech.com
Applied and product research in the environment.
*Stephen Wagner, President*

**4884 KE Sorrells Research Associates**
8100 National Drive 501-562-8139
Little Rock, AR 72209-4839 Fax: 501-562-7025
E-mail: sorrells@comcast.net
http://sorrellsresearch.com
Analytical chemistry and applied research company providing consulting services in water technology and stream ecology.

*KE Sorrells, President*
*Cecil Sorrells, CEO*

**4885 KLM Engineering**
3394 Lake Elmo Avenue N 651-773-5111
PO Box 897 888-959-5111
Lake Elmo, MN 55042 Fax: 651-773-5222
E-mail: jkollmer@klmengineering.com
Structural engineering and inspection firm specializing in the industry of steel and concrete plate structures.

*Jack R Kollmer, President/Principal*
*Shawn A Mulhern, Vice President-Sales/Mktg.*

**4886 Kag Laboratories International**
2323 Jackson Street 920-426-2222
Oshkosh, WI 54903 800-356-6045
Fax: 920-426-2664
E-mail: info@kaglab.com
http://www.kaglab.com/
An independent agricultural testing and consulting laboratory. Professional scientific services for agriculture, soil, feed, plant, water and other fields. Total farm management services including high value crops such as cranberry,stevia, blueberry, ginseng, strawberry, herbs, etc. Consultation and recommendation to increase net yield. Available for contractual applied research for all agribusiness industries in Wisconsin, North America and world-wide.
*Dr. Akhtar Khwaja, President*
*Ruma Roy, Vice President/Chemist*

**4887 Kansas City Testing Laboratory**
2012 W 104th Street 913-648-2303
Shawnee Mission, KS 66206 Fax: 913-321-8181
Consulting engineering firm employed in geotechnical, materials, and environmental engineering. Research results published in Project reports.
*Donald Cesso, President*

**4888 Kar Laboratories**
4425 Manchester Road 269-381-9666
Kalamazoo, MI 49001 Fax: 269-381-9698
E-mail: info@karlabs.com
http://www.karlabs.com
Environmental testing laboratory, wastewater, drinking water, hay waste, soil and air.
*William Rauch, President*
*Jayne Rauch, Marketing Manager*

**4889 Kemper Research Foundation**
122 Main Street 513-249-2489
Milford, OH 45150
*Richard Kemper, Director*

**4890 Kemron Environmental Services**
109 Starlite Park 740-373-4071
Marietta, OH 45750 Fax: 740-373-4835
Environmental testing and analysis firm.
*Cindy Arnold, Contact*

**4891 Kennedy-Jenks Consultants**
303 Second Street 415-243-2150
Suite 300 S Fax: 415-896-0999
San Francisco, CA 94107 http://www.kennedyjenks.com
Environmental engineering consulting company.
*Gordon Morris, Graphics Services Manager*
*Rena Chin, Editor*

**4892 Kentucky Resource Laboratory**
Highway 421 606-598-2605
Manchester, KY 40962 Fax: 606-598-1544
Environmental testing firm.
*Roy Rice, President*

**4893 Kenvirons**
452 Versailles Road 502-698-4357
Frankfort, KY 40601 Fax: 502-695-4363
E-mail: rrussell@kenvirons.com
http://kenvirons.com
A multi-disciplined environmental and civil engineering firm. Offers engineering services in a range of areas to include water and wastewater related studies and system design, dam design, hydrological studies, environmentalassessments, air and water quality studies, urban and industrial planning, solid waste management, energy-environment interface, computer technology and laboratory services.

*Randall Russell, President*
*Douglas Griffin, Chair*

**4894 Keystone Labs**
600 East 17th Street South 800-858-5227
Newton, IA 50208 Fax: 641-792-7989
http://http://www.keystonelabs.com
Keystone Laboroatories, Inc. is a full service environmental laboratory committed to providing the highest quality services at competitive prives.
*, President*

**4895 Kinnetic Laboratories**
307 Washington Street 831-457-3950
Santa Cruz, CA 95060 Fax: 831-426-0405
Environmental marine, physical, toxicological, water quality, biological research and scientific consulting and services.
*Mary Lee Kinney, President*
*Mark Savoie, VP*

**4896 Kleinfelder**
981 Garcia Avenue 925-427-6477
Suite A Fax: 925-427-6478
Pittsburg, CA 94565
Laboratories testing.
*Gerry Salontai, President/CEO*

**4897 Konheim & Ketcham**
175 Pacific Street 718-330-0550
Brooklyn, NY 11201 Fax: 718-330-0582
E-mail: csk@konheimketcham.com
http://www.konheimketcham.com
Environmental and transportation planning.
*Carolyn S Konheim, President*

**4898 Kraim Environmental Engineering Services**
11437 Etiwanda Avenue 818-363-0952
Northridge, CA 91326 Fax: 818-363-0492
E-mail: luftmench@msn.com
Environmental engineering firm.

*Jerry Kraim, President*

**4899 Kramer & Associates**
4501 Bogan Avenue NE 505-881-0243
Suite A1 Fax: 505-881-7738
Albuquerque, NM 87109
Environmental monitoring firm. Research results published in conference proceedings.
*Gary Kramer, Contact*

**4900 Ktech Corporation**
10800 Gibson S E 505-998-5830
Albuquerque, NM 87123 Fax: 505-998-5848
E-mail: rswanson@ktech.com
http://www.ktech.com/corporate/history.cfm
Ktech Corporation, an employee-owned company based in Albuquerque, New Mexico, is dedicated to providing outstanding technical support services, sound scientific and engineering work, and proven management expertise to a wide varietyof government and industry clients.

*Steven E Downie, President*
*Robert E Swanson, VP/Chief Operations Officer*

**4901 LaBella Associates P.C.**
300 State Street 585-454-6110
Suite 201 Fax: 585-454-3066
Rochester, NY 14614 E-mail: info@labellapc.com
http://www.labellapc.com
Civil and environmental engineering firm.

*Robert Healy, President*
*Sergio Esteban, Chief Executive Officer*

**4902 LaQue Center for Corrosion Technology**
521 Fort Fisher Blvd, North 910-256-2271
Kure Beach, NC 28449 Fax: 910-256-9816
E-mail: info@laque.com
http://www.laque.com
Corrosion technology firm. Research results published in trade journals and presented at technical association meetings.
*W T Raines, President*
*D G Melton, VP*

**4903 Laboratory Corporation of America Holdings**
1904 Alexander Drive 919-572-6900
Research Triangle Park, NC 27709 800-533-0567

**4904 Laboratory Services Division of Consumers Energy**
135 W Trail Street 517-788-2238
Jackson, MI 49201 800-736-4147
Fax: 517-788-1104
E-mail: naserafin@cmsenergy.com
http://www.laboratoryservices.com
Laboratory Services is a full-service testing laboratory. Services include: calibration, nondestructive testing, metallurgy, materials testing and chemistry. They are A2LA accredited (ISO/IEC 17025)-request scope-and 10CFR50 AppendixB authorized.

*Nick Serafin, Marketing Manager*

**4905 Lancaster Laboratories**
2425 New Holland Pike 717-656-2300
Lancaster, PA 17601 Fax: 717-656-2681
http://www.lancasterlabs.com
Premier contract testing laboratory serving environmental, pharmaceutical and biophamaceutical clients worldwide. Offers a broad range of high quality analytical services in full compliance with EPA and FDA regulations and clientrequirements.

*J Wilson Hershey PhD, President*
*Anne Osborn, Dir/Marketing/Communications*

**4906 Lancy Environmental**
181 Thorn Hill Road 724-772-0044
Warrendale, PA 15086 Fax: 724-772-1360
*Gerald Rogers, President*

**4907 Land Management Decisions**
3048 Research Drive 814-231-1248
State College, PA 16801 Fax: 814-231-1253
*Dr. Dale E Baker, President*

**4908 Land Management Group Inc**
3805 Wrightsville Avenue 910-452-0001
Suite 15 Fax: 910-452-0060
Wilmington, NC 28403 E-mail: rmoul@lmgroup.net
http://www.lmgroup.net
LMG provides environmental servicesin the following discplines: wetland delineations and permitting; soil mapping and waste water suitability studies; phase I & II environmental site assiessments; EA and EIS land use and ecologicalstudies; wetland mitigations; and coastal management (CAMA) permitting assistance.

*Robert L Moul, President*

**4909 Land Research Management**
1300 N Congress Avenue 561-686-2481
Suite C Fax: 561-684-8709
West Palm Beach, FL 33409 E-mail: lrmi@bellsouth.net
Land planning and zoning, environmental assessments and market analysis firm. Research results published in reports.

*Kevin McGinley, President*

**4910 Lark Enterprises**
16 Sunset Drive 508-949-2672
Dudley, MA 01571 Fax: 508-943-8833
E-mail: rjlark@aol.com

Environmental research.
*Lother Frank, President*

**4911 Laticrete International**
1 Laticrete Park N 203-393-0010
Bethany, CT 06524 Fax: 203-393-1684
http://www.laticrete.com
Firm providing chemical, mechanical, and environmental simulation testing of concrete and aggregrate building materials.

**4912 Law & Company of Wilmington**
1711 Castle Street 910-762-7082
Wilmington, NC 28403 Fax: 910-762-8785

*Richard W Spivey, President*

**4913 Lawler, Matusky and Skelly Engineers**
1 Blue Hill Plaza 845-735-8300
Pearl River, NY 10965 Fax: 845-735-7466
E-mail: cnevel@lmseng.com
http://www.lmseng.com
We anticipate the environmental and engineering needs of our clients and contribute to their success by providing creative solutions.
*Christy Nevel, Director Marketing*

**4914 Lawrence Berkeley Laboratory: Structural Biology Division**
1 Cyclotron Road 510-486-4311
Mail Stop 3-0226 Fax: 510-486-6059
Berkeley, CA 94720 http://www.lbl.gov/sbdiv/

**4915 Lawrence G Spielvogel**
21506 Valley Forge Circle — 610-783-6350
King of Prussia, PA 19406 — Fax: 610-783-6349
A consulting engineer who specializes in energy management and procurement and problem solving in buildings.
*Lawrence G Spielvogel, President*

**4916 Ledoux and Company**
359 Alfred Avenue — 201-837-7160
Teaneck, NJ 07666 — Fax: 201-837-1235

*LA Ledoux, President*

**4917 Lee Wilson and Associates**
105 Cienega Street — 505-988-9811
Santa Fe, NM 87501 — Fax: 505-986-0092
Environmental consulting firm. Research results published in project reports.
*Lee Wilson, President*

**4918 Leighton & Associates**
17781 Cowan Street — 949-250-1421
Irvine, CA 92614 — Fax: 949-250-1114
http://www.leightongeo.com
Geotechnical and environmental engineering firm.
*Bruce Clark, Contact*

**4919 Life Science Resources**
2 Fremontia Street — 650-851-0225
Portola Valley, CA 94028
Biomedical and environmental sciences research firm.

**4920 Life Systems**
24755 Highpoint Road — 216-464-3291
Suite 1 — Fax: 216-464-8146
Cleveland, OH 44122
Environmental engineering research and consulting organization. Research results published In project reports and in technical journals.
*R Wynveen, President*

**4921 Los Alamos Technical Associates**
2400 Louisiana Boulevard NE — 505-665-8616
Building 1, Suite 400 — 800-952-5282
Albuquerque, NM 87110 — Fax: 505-880-3560
http://www.lata.com
Environmental studies.
*LP Reinig, CEO*

**4922 Louisville Testing Laboratory**
1401 West Chestnut Street — 502-584-5914
Louisville, KY 40203 — Fax: 502-584-5914
*Kenneth Smith Jr, President*

**4923 Lowry Systems**
146 South Street
Blue Hill, ME 04614 — 800-434-9080
Fax: 207-374-3503
E-mail: info@lowryh2o.com
http://www.lowryh2o.com
Environmental research.
*Sylvia Lowry, President*

**4924 Lycott Environmental Inc**
600 Charlton Street — 508-765-0101
Southbridge, MA 01550 — 800-462-8211
Fax: 508-765-1352
E-mail: lycottinc@aol.com
http://www.lycott.com
Environmental science and ecological planning consultant and research firm. Research results published in project reports.
*Lee D Lyman, President*

**4925 Lyle Environmental Management**
1507 Chambers Road — 614-488-1022
Columbus, OH 43212 — Fax: 614-488-1198
Chemical research and consulting service.

**4926 Lyle Laboratories**
1507 Chambers Road — 614-488-1022
Columbus, OH 43212 — Fax: 614-488-1198
*Dr. Thomas Eggers, Director*

**4927 Lynntech**
7610 Eastmark Drive — 979-693-0017
Suite 105 — Fax: 979-764-7479
College Station, TX 77840
*Oliver J Murphy, President*

**4928 MBA Labs**
340 South 66th Street — 713-928-2701
Houston, TX 77261 — 800-472-1485
Fax: 281-292-7492
E-mail: mbalabs@mbalabs.com
http://www.mbalabs.com
Independently owned and operated since 1968, mba Labs serves industry, government agencies and private citizens in Houston, the continental US and even across the globe. Conform to standards established by the EPA, the TNRCC and meetthe equivalent of ISO 9000 requirements for laboratories through their accreditation by NELAC.

*Herman J Kresse*

**4929 MBA Polymers**
500 West Ohio Avenue — 521-231-9031
Richmond, CA 94804 — Fax: 521-231-0320
E-mail: info@mbapolymers.com
http://www.mbapolymers.com/
Environmental research.
*Mike Biddle, President*

**4930 MBC Applied Environmental Sciences**
3000 Redhill Avenue — 714-850-4830
Costa Mesa, CA 92626 — Fax: 714-850-4840
E-mail: info@mbcnet.net
http://www.mbcnet.net
Environmenatl consultants since 1969. Specializing in marine biology and ecology, oceanography, EIR, EIS, EA, toxicity testing, technical meetings, expert witnesses. MBE/DBE certified.

**4931 MWH Global**
380 Interlocken Crescent — 303-533-1900
Suite 200 — Fax: 303-533-1901
Broomfield, CO 80021 — E-mail: webinfo@mwhglobal.com
http://www.mwhglobal.com
MWH, globally driving the wet infrastructure sector, is leading the world in results-oriented management services, technical engineering, construction services and solutions to create a better world. The wet infrastructure sectorencompasses a full range of water related projects and programs from water supply, treatment and storage, dams, water management for the natural resources industry and coastal restoration to renewable power and environmental services.

*Robert Uhler, President/CEO*
*Alan Krause, President/COO*

**4932 MWH Laboratories**
750 Royal Oaks Drive — 626-386-1100
#100 — 800-566-5227
Monrovia, CA 91016 — Fax: 626-386-1101
E-mail: mwhlabs@mwhglobal.com
http://www.mwhlabs.com
Environmental testing laboratory that provides water and wastewater analyses including: drinking water synthetic organic and volatile organic tests; recycled water tests; organic disinfection byproduct and precursor analyses; inorganicdisinfection byproducts and precursors; inorganic tests including a complete

suite of metals to low-reporting levels; microbiological analyses ; and radiochemical analyses.

*Andrew Eaton Ph.D, Technical Director*
*Ed Wilson, Laboratory Director*

**4933 Mabbett & Associates: Environmental Consultants and Engineers**
5 Alfred Circle 781-275-6050
Bedford, MA 01730 800-877-6050
Fax: 781-275-5651
E-mail: info@mabbett.com
http://www.mabbett.com
Mabbett & Associates (M&A) provides multi-disciplinary environmental, health and safety services to manufacturing and commercial industry, institutions and public agencies. M&A's services include pollution prevention and wasteminimization, site assessment and remediation, environmental pollution control, environmental management systems and auditing, training and occupational safety and health.

*Arthur N Mabbett, President*
*Paul D Steinberg, VP/General Manager*

**4934 Mack Laboratories**
2199 Dartmore Street 412-885-2900
Pittsburgh, PA 15210

**4935 Magma-Seal**
10116 Aspen Street 512-836-4936
Austin, TX 78758 Fax: 512-836-4936
E-mail: tkdw39a@prodigy.com
Develops materials (plastic and rubber) to withstand severe environmental conditions.
*Earl Dumitro, President*

**4936 Malcolm Pirnie**
104 Corporate Park Drive 914-694-2100
White Plains, NY 10602 800-478-6870
Fax: 914-694-9286
E-mail: webmaster@pirnie.com
http://www.pirnie.com
Provides environmental engineering, science and consulting services to over 3,000 public and private clients.
*Paul L Busch, PhD, President*

**4937 Maryland Spectral Services**
1500 Caton Center Drive 410-247-7600
Suite G Fax: 410-247-7602
Baltimore, MD 21227
*Samuel Hamner, VP*

**4938 Massachusetts Technological Lab**
330 Pleasant Street 617-484-7314
Belmont, MA 02178 E-mail: masstechlab@juno.com
Applies research in the following areas: telecommunications and Internet.
*Dr Ta-Ming Fang, President*

**4939 Mateson Chemical Corporation**
1025 E Montgomery Avenue 215-423-3200
Philadelphia, PA 19125 Fax: 215-423-1164
Environmental, toxic, materials, hazerdous waste research.

**4940 Mayhew Environmental Training Associates (META)**
PO Box 786 785-842-6382
Lawrence, KS 66044 800-444-6382
Fax: 785-842-6993
E-mail: salesmeta@cs.com
http://www.metaworld.org
Environmental testing lab offering site assessments.
*Thomas Bradford Mayhew, President*

**4941 McCoy & McCoy Laboratories**
1800 Kentucky Avenue 270-444-6547
Paducah, KY 42003 Fax: 270-444-6572
Environmental assessment laboratory.

**4942 McIlvaine Company**
191 Waukegan Road 847-784-0012
Suite 208 Fax: 847-784-0061
Northfield, IL 60093 E-mail: editor@mcilvainecompany.com
http://www.mcilvainecompany.com
Environmental research and consulting firm. Research results published in manuals updated by newsletters and abstracts.
*Robert W McIlvaine, President*
*Marilyn McIlvaine, Marketing Manager*

**4943 McLaren-Hart**
3039 Kilgore Road 916-638-3696
Rancho Cordova, CA 95670 Fax: 916-638-6840
http://www.mclaren-hart.com
Environmental research.

**4944 McNamee Advanced Technology**
3135 S State Street 734-665-5553
Suite 301 Fax: 734-665-2570
Ann Arbor, MI 48108
Environmental engineering firm, offering environmental consulting and environmental testing services.

**4945 McVehil-Monnett Associates**
44 Inverness Drive E 303-790-1332
Suite C Fax: 303-790-7820
Englewood, CO 80112 http://www.mcvehil-monnett.com
Experienced consulting firm of atmospheric scientsits, engineers and environmental specialists providing air quality and environmental management system (EMS) services worldwide. Serves the mining, oil and gas, electric power andmanufacturing industries as well as government agencies and engineering and law firms.

*William R Monnett, President/CEO*

**4946 McWhorter and Associates**
33 Bull Street 912-234-8891
Box 9419 Fax: 912-234-8892
Savannah, GA 31412

*Thomas McWhorter, President*

**4947 Mega Engineering**
10800 Lockwood Drive 301-681-4778
Silver Spring, MD 20901 Fax: 301-681-5683

*Richard E Dame, PE*

**4948 Membrane Technology & Research Corporate Headquarters**
1360 Willow Road 650-328-2228
Suite 103 Fax: 650-328-6580
Menlo Park, CA 94025 E-mail: sales@mtrinc.com
http://www.mtrinc.com
Supplier of membrane-based hydrocarbon recovery systems natural gas treatment systems and hydrogen recovery systems. Company capabilities include membrane and module manufacturing, process and system design, project engineering andcommissioning services.

*Dr. Hans Wijmans, President*

**4949 Merck & Company**
126 East Lincoln Avenue 732-574-4000
Rahway, NJ 07065 Fax: 732-594-3810

*Bill Hamilton, Dir/NJ Environmental Affairs*

**4950 Merrimack Engineering Services**
66 Park Street 978-475-3555
Andover, MA 01810 Fax: 978-475-1448
E-mail: merreng@aol.com
Research of all forms of environmental studies.
*Stephen Stapinski, President*

**4951 Metro Services Laboratories**
6309 Fern Valley Pass 502-964-0865
Louisville, KY 40228 Fax: 502-241-4347
Environmental testing laboratory offering air, water and soil testing services.

**4952 Michael Baker Jr: Civil and Water Division**
4301 Dutch Ridge Road 724-495-7711
Beaver, PA 15009 Fax: 724-495-4017
E-mail: hchakrav@mbakercorp.com
http://www.mbakercorp.com

**4953 Michael Baker Jr: Environmental Division**
420 Rouser Road 412-269-6000
Coraopolis, PA 15108 Fax: 412-269-6097
*Andrew P Pajak, President*

**4954 Mickle, Wagner & Coleman**
3434 Country Club Avenue 479-649-8484
PO Box 1507 Fax: 479-649-8486
Fort Smith, AR 72903 E-mail: info@mwc-engr.com
http://www.mwc-engr.com/
Provides civil and environmental engineering services, offering clients a broad range of plan designs and development capabilities from water, sewer, and drainage to streets, bridges, and dams, airports, recreational facilities, andresidential subdivisions.
*Patrick J Mickle, PE, Chief Engineer*

**4955 Microseeps, Inc**
220 William Pitt Way 412-826-5245
Pittsburgh, PA 15238 800-659-2887
Fax: 412-826-3433
E-mail: rpirkle@microseeps.com
http://www.microseeps.com
A full service, NELAP certified environmental laboratory which specializes in the evaluation of groundwater geochemistry for use in in-situ remediation processes.

*Robert J Pirkle, President*
*Frank Phillips, VP Sales*

**4956 Microspec Analytical**
3352 128th Avenue 616-399-6070
Holland, MI 49424 Fax: 616-399-6185
E-mail: info@mspec.com
http://www.mspec.com
Environmental research and resting firm. Research results published in journals and client reports.
*Tom Beamish, President*

**4957 Midwest Environmental Assistance Center**
6561 N Seeley Avenue 773-973-4850
Chicago, IL 60645 Fax: 773-973-4851
E-mail: meac2@aol.com
Noise pollution research firm.
*Howard R Schechter, President*

**4958 Midwest Laboratories, Inc.**
13611 B Street 402-334-7770
Omaha, NE 68144 Fax: 402-334-9121
E-mail: pohlman@midwestlabs.com
http://www.midwestlabs.com
Midwest Laboratories, Inc. offers analytical services to agriculture, industry and municipal entities throughout the US and Canada. Using wet chemistry methods, they have the capability of testing soil, water, feed, food, plants,fertilizers and residues. Their quality assurance program (QA/QC) provides consistent production of reliable data with high accuracy and precision.

*Ken Pohlman, President*
*John DeBoer, VP*

**4959 Midwest Research Institute**
425 Volker Boulevard 816-753-7600
Kansas City, MO 64110 Fax: 816-753-8420
E-mail: bduncan@mriresearch.org
http://www.mriresearch.org
Midwest Research Institute is an independent, not-for-profit organization that performs contract research for clients in business, industry and government. MRI conducts programs in the areas of environment, health, engineering,technology development and energy research.
*James Spigarelli, President*

**4960 MikroPul Environmental Systems Division of Beacon Industrial Group**
17 Wachung Avenue 973-635-1115
Chatham, NJ 07928 Fax: 973-635-0678
E-mail: info@mikropul.com
http://www.mikropul.com
Established in 1929, MikroPul is a manufacturer of dust control and product recovery products, from small unit collectors to complete engineered systems, for industrial applications worldwide.
*Lacy Hayes, President/Beacon Ind Group*
*Richard Bearse, Chairman/Beacon Ind Group*

**4961 Miller Engineers**
5308 S 12th Street 920-458-6164
Sheboygan, WI 53081 Fax: 920-458-0369
Civil and environmental engineering firm.
*Roger G Miller, President*

**4962 Minnesota Valley Testing Laboratories**
1126 N Front Street 507-354-8517
New Ulm, MN 56073 Fax: 507-359-1231
Independent bacteriological and chemical analysis firm, with services in environmental, agricultural, and energy fields. Research results published in project reports.
*Henry Nupson, President*

**4963 Mirage Systems**
PO Box 820 386-740-9222
DeLand, FL 32721 Fax: 386-740-9444
Environmental research.
*Robert S Ziernicki, President*

**4964 Montgomery Watson Mining Group**
1475 Pine Grove Road 970-879-6260
Suite 109 Fax: 970-879-9048
Steamboat Springs, CO 80477 http://www.mw.com
Mine engineering and environmental services firm.
*Alan Krause, SVP/COO*

**4965 Mycotech**
630 S Utah 406-782-2386
PO Box 4109 Fax: 406-782-9912
Butte, MT 59702
*Clifford Bradley, Director R&D*

**4966 Myra L Frank & Associates**
811 W 7th Street 213-627-5376
Suite 800 Fax: 213-627-6853
Los Angeles, CA 90017 E-mail: fwilliams@myrafrank.com
http://www.myrafrank.com
Environmental impact analysis firm. Architectural historic surveys.
*Florence Williams*

**4967 Mystic Air Quality Consultants**
1204 North Road 860-449-8903
Route 117 800-247-7746
Groton, CT 06340 Fax: 860-449-8860
E-mail: maqc2@aol.com
http://www.mysticair.com
Indoor air quality and industrial hygiene services.

*Chris Eident, CEO*

**4968 NET Pacific**
11135 Rush Street 626-350-4241
Suite Q
South El Monte, CA 91733

**4969 National Institute for Urban Wildlife**
10921 Trotting Ridge Way 301-596-3311
Columbia, MD 21044
Promotes the preservation of wildlife in urban settings, providing support to individuals and organizations invloved in maintaining a place for wildlife in expanding American cities and suburbs. The Institute conducts researchexploring the relationship between humans and wildlife in these habitats, publicizes urban wildlife management methods, and raises public awareness of the value of wildlife in city settings. The Institute also provides consulting services.

**4970 National Loss Control Service Corporation**
1 Kemper Drive 847-320-2488
Long Grove, IL 60049 Fax: 847-320-4331
Environmental science laboratory.
*Joan Wronski, Laboratory Manager*

**4971 National Oceanic & Atmospheric Administration**
325 Broadway 303-497-3000
Boulder, CO 80305 http://www.noaa.gov
Earth system research including climate, weather, and atmospheric chemistry, space weather research and forecasts.

**4972 National Renewable Energy Laboratory/NREL**
1617 Cole Boulevard 303-275-3000
Golden, CO 80401 Fax: 303-275-4053
E-mail: public_affairs@nrel.gov
http://www.nrel.gov
The National Renewable Energy Laboratory/NREL began operating in 1977 as the Solar Energy Research Institute. It was designated a national laboratory of the U.S. Department of Energy (DOE) in September 1991 and its name changed toNREL. NREL develops renewable energy and energy efficiency technologies and practices, advances related science and engineering, and transfers knowledge and innovations to address the nation's energy and environmental goals.
*Dan Arvizu, Director*
*William Glover, Deputy Lab Director*

**4973 Neilson Research Corporation**
245 South Grape Street 541-770-5678
Medford, OR 97501 Fax: 541-770-2901
E-mail: clientservices@nrclabs.com
http://www.nrclabs.com
Provides analytical services to support environmental projects including testing of drinking water, wastewater, ground and surface water, foods soils, sediments, sludges, filters, air, and hazardous waste samples.
*John WT Neils, CEO*

**4974 Neponset Valley Engineering Company**
378 Page Street 781-297-7040
Suite 10 Fax: 781-297-7050
Stoughton, MA 02072
Environmental engineering analysis and consulting firm.

**4975 New England Testing Laboratory**
1254 Douglas Avenue 401-353-3420
North Providence, RI 02904 Fax: 401-354-8951
*Mark Bishop, VP Operations*

**4976 New York Testing Laboratories**
143-05 Emery Avenue 718-658-7300
Jamaica, NY 11432 800-281-3329
Fax: 718-657-3902
http://www.nytesting.com
Consulting on a range of disciplines including environmental.

*Charles Realmuto, Director Marketing*

**4977 Newport Electronics**
2229 South Yale Street 714-540-4914
Santa Ana, CA 92704 800-639-7678
Fax: 203-968-7311
E-mail: info@newportus.com
http://www.newportus.com
Manufacturer of industrial and environmental instrumentation including signal conditioners, digital panel meters, PID controllers and temperature sensors.

*Milton Hollander, President*

**4978 Nobis Engineering**
18 Chenell Drive 603-224-4182
Concord, NH 03301 Fax: 603-224-2507
Environmental engineering consulting firm.

**4979 Normandeau Associates**
102 South Boundary 803-652-2206
New Ellenton, SC 29809 Fax: 803-652-7428
*Jean Eidson*

**4980 North American Environmental Services**
PO Box 26521 512-264-2828
Austin, TX 78755
Environmental science research firm.
*D Craig Kissock, President*

**4981 Northeast Test Consultants**
587 Spring Street 207-854-3939
Westbrook, ME 04092 Fax: 207-854-3658
Asbestos and lead testing/industrial hygiene.

*Stephen Broadhead, Laboratory Manager*

**4982 Northern Lights Institute**
210 N Higgins #326 406-721-7415
PO Box 8084 Fax: 406-721-7415
Missoula, MT 59807
*Donald Snow, Program Director*

**4983 Nuclear Consulting Services**
7000 Huntley Road 614-846-5710
Columbus, OH 43229 Fax: 614-431-0858
*Joseph C Enneking, VP*

**4984 O'Brien & Gere Engineers**
Box 4873 315-437-6100
Syracuse, NY 13221 Fax: 315-463-7554

*Cornelius B Murphy, President*

**4985 OA Laboratories and Research, Inc.**
1430 N Stadium Drive 317-639-2626
Indianapolis, IN 46202 Fax: 317-636-6760
E-mail: oalabs@dajanigroup.com
OA Laboratories and Research, Inc. serves customers in Indiana and throughout the United States by meeting their Analytical needs.

*Bela Jones, Senior Chemist*

**4986 Oak Ridge Institute for Science and Education**
210 Badger Avenue 865-576-3424
PO Box 117, MS 36 Fax: 865-241-2923
Oak Ridge, TN 37831 E-mail: westm@orau.gov
http://www.orau.org

*Dr. Nathaniel W Revis, Director*

**4987 Occupational Health Conservation**
5118 N 56th Street 813-626-8156
Tampa, FL 33610 800-229-8156
Fax: 813-623-6702

Environmental impact assessment firm.
*James F Rizk, President*

**4988 Occusafe**
240 East Lake Street
Addison, IL 60101
630-941-3001
800-323-7597
Fax: 630-941-3865
E-mail: info@occusafe-inc.com
http://www.occusafe-inc.com
Employee safety, industrial hygiene and environmental consulting firm.
*Bob McKinley, President*

**4989 Ogden Environment & Energy Services Company**
4455 Brookfield Corporate Drive
Suite 100
Chantilly, VA 20151
703-488-3700
Fax: 703-488-3701
Environmental engineering and consulting company.
*J Mark Elliot, President*

**4990 Ogden Environment & Energy Services Company**
5510 Morehouse Drive
San Diego, CA 92121
858-458-9044
Fax: 858-458-0943
Scientific and environmental engineering; analytical chemistry.
*Mike Nienberg, Executive VP*

**4991 Oil-Dri Corporation of America**
410 N Michigan Avenue # 400
Chicago, IL 60611
312-321-1515
Fax: 312-321-1271
Absorbents for consumers, industrial, agricultural, environmental and fluid purification.

**4992 Olver**
1116 S Main Street
Blacksburg, VA 24060
540-552-5548
Fax: 540-552-5577
E-mail: info@olver.com
http://www.olver.com
Engineering research and consulting firm specializing in environmental design and analysis. Research results published in project reports.

**4993 Omega Thermal Technologies**
21 Elbo Lane
Mount Laurel, NJ 08054
856-232-1399
Fax: 856-232-1772
E-mail: contact@ottusa.com
http://www.ottusa.com
Technology consultants, designers and constructors offering technical expertise and hardware design for thermal processing and environmental studies.
*Kenneth W Hladun, President*

**4994 Oneil M Banks**
336 S Main Street
Suite 2D
Bel Air, MD 21014
410-879-4676
Fax: 410-836-8685
Industrial and environmental hygiene and toxicology consulting company.

**4995 Online Environs**
201 Broadway
Suite 7
Cambridge, MA 02139
617-577-0202
Fax: 617-577-0772
http://www.environs.com
Telecommunications and Internet research.
*Anrew Yu, President*

**4996 Operational Technology Corporation**
4100 NW Loop 410 Street
Suite 230
San Antonio, TX 78229
210-731-0000
800-677-8072
Fax: 210-731-0008
E-mail: webmaster@otcorp.com
http://www.otcorp.com
Employment research firm providing information technologies, computer sales and service and environmental services.
*John Fernandez, CEO*

**4997 Orlando Laboratories**
820 Humphries Avenue
Orlando, FL 32814
407-896-6645
Fax: 407-898-6588
Independent environmental testing and analysis laboratory.

**4998 Ostergaard Acoustical Associates**
200 Executive Drive
West Orange, NJ 07052
973-731-7002
Fax: 973-731-6680
E-mail: kherbert@acousticalconsultant.com
http://acousticalconsultant.com
Environmental, acoustic and noise control testing and analysis firm. Research results published in project reports.
*R Kring Herbert, Principal*
*Edward M. Clark, Principal*

**4999 Ozark Environmental Laboratories**
PO Box 806
Rolla, MO 65402
573-364-8900
Fax: 573-341-2040
Firm providing construction materials testing on soils, aggregates, and asphaltic and portland cement concrete; water and wastewater physical and chemical analysis; and quality control studies encompassing physical measurements andchemical analysis.

**5000 P&P Laboratories**
2025 Woodlynne Avenue
Oaklyn, NJ 08107
856-962-6188
Environmental testing and chemical toxicology laboratory.

**5001 PACE**
100 Marshall Drive
Warrendale, PA 15086
724-772-0610
Fax: 724-772-1686
Environmental testing and analysis firm.

**5002 PACE Analytical Services**
1700 Elm Street
Suite 200
Minneapolis, MN 55414
612-607-1700
E-mail: info@pacelabs.com
http://www.pacelabs.com
Provider of air, water, soil and environmental testing services.
*Steve A Vanderboom, CEO*
*Gabe LeBrun, Director/Manager*

**5003 PACE Environmental Products**
5240 W Coplay Road
Whitehall, PA 18052
610-262-3818
800-303-4532
Fax: 610-262-4445
E-mail: sales@pacecems.com
http://www.pacecems.com
Manufacturer and Integrator of continuous emissions monitoring systems (EMS). Regulatory, process, and certification stack testing. In-shop analyzer repair, CEMS field service. Parts, sales, rentals, repairs and service.
*Damian Gaiotti, Sales Manager*

**5004 PACE Resources, Incorporated**
40 S Richland Avenue
York, PA 17404
717-852-1300
800-711-8075
Fax: 717-852-1301
E-mail: pace40@aol.com
This company is the parent of units involved in environmental engineering and consulting, civil engineering, architectural planning, data processing, printing and other services.
*Russell E Horn, Jr, President*

**5005 PARS Environmental**
6A S Gold Drive
Robbinsville, NJ 08691
609-890-7277
Fax: 609-890-9116
E-mail: hgill@parsenviro.com
http://www.parsenviro.com
Environmental consulting company.
*HS Gill, President*

**5006 PCCI**
300 N Lee Street
Suite 201
Alexandria, VA 22314
703-684-2060
Fax: 703-684-5343
E-mail: use form on website
http://www.pccii.com
Provides sensible solutions to difficult engineering and environmental problems in coastal, ocean and inland environments. Spe-

cialties include: environmental compliance; all hazards emergency response planning, trainings, drills andexercises; and marine engineering.
*Robert W Urban, President*
*Alan R Becker, VP*

**5007 PDC Laboratories**
2231 West Altorser Drive 309-692-9688
Peoria, IL 61615 Fax: 309-692-9689
Environmental laboratory performs air sample analysis, soil analysis, and potential toxic waste analysis.

**5008 PE LaMoreaux & Associates**
PO Box 2310 205-752-5543
Tuscaloosa, AL 35403 Fax: 205-752-4043
Consulting hydrologists, geologists, engineers, and environmental scientists. Research results published in brochures, pamphlets, news releases, speeches, seminars, studies, and reports.
*James W Lamoreaux, President*

**5009 PEI Associates**
11499 Chester Road 513-782-4700
Suite 200 Fax: 513-782-4807
Cincinnati, OH 45246
Environmental consulting firm. Research results published in government publications.

**5010 PELA**
PO Box 2310 205-752-5543
Tuscaloosa, AL 35403 Fax: 205-752-4043
E-mail: pela@dbtech.net
http://www.pela.com
For over three decades, PELA's integration of qualified personnel, up to date technology, and sound management has established PELA as an international leader in the environmental consulting field. PELA's expertise in hydrology,geotechnical analysis, design and construction management, remediation, computer graphics and models, and permitting can get your project on the two feet quicker than you might think.

*James W LaMoreaux, President*

**5011 PRC Environmental Management**
233 N Michigan Avenue 312-938-0300
Suite 1621 Fax: 312-931-1109
Chicago, IL 60601
*Robert Banosten, VP*

**5012 PRD Tech**
1776 Mentor Avenue 513-731-1800
Suite 400-A Fax: 513-984-5710
Cincinnati, OH 45212 E-mail: rsmprdt@aol.com
http://www.prdtechinc.com
Biological and chemical research and commercial technology development firm, serving primarily the baking, brewing, and other food industry segments with their environmental control needs - odor and volatile organic compound (VOC)control applications.
*Ramesh Melarkode, President*

**5013 PSC Environmental Services**
550 Pinetown Road 215-643-5466
Suite 166 800-292-2510
Fort Washington, PA 19034 Fax: 215-643-2772
http://pscnow.com
Environmental services
*Ed Boner, Location Manager*

**5014 PSI**
1901 South Meyers Road 630-691-1587
Suite 400 800-548-7901
Oakbrook Terrace, IL 60181 Fax: 630-691-1587
E-mail: info@psiusa.com
http://www.psiusa.com
Distinguished as a leader in environmental consulting, geotechnical engineering, and construction testing services, PSI is nationally recognized in several disciplines, including: construction services, materials testing, roofconsulting and asbestos management.

**5015 Pace**
2400 Cumberland Drive 219-464-2389
Valparaiso, IN 46383 Fax: 219-462-2953
Environmental testing laboratory.
*Les Arnold, President*

**5016 Pace Laboratory**
9893 Brewers Court 301-490-9860
Laurel, MD 20723
Environmental testing laboratory.

**5017 Pace New Jersey**
284 Raritan Center Parkway 973-257-9300
Edison, NJ 08837 Fax: 973-257-0777
Environmental analytical laboratory and data management firm.

**5018 Pacific Gamefish Research Foundation**
47-381 Kealakehe Parkway 808-329-6105
PO Box 4800 Fax: 808-329-1148
Kailua Kona, HI 96740

**5019 Pacific Northwest National Lab**
902 Battelle Boulevard 509-375-2121
PO Box 999 888-375-7665
Richland, WA 99352 Fax: 509-375-2491
E-mail: inquiry@pnl.gov
http://www.pnl.gov
Contract research and development for the government environmental restoration, energy, national security and health.

**5020 Pacific Northwest Research Institute**
720 Broadway 206-726-1200
Seattle, WA 98122 Fax: 206-726-1217
http://www.pnri.org
Established as Pacific Northwest Research Foundation in 1956 by Dr. William B Hutchinson, Sr. as the first private nonprofit biomedical and clinical research institute in the Northwest. As founder and first director, Dr. Hutchinson'sprimary objective was to provide a facility for basic and clinical research dedicated to the improvement ofpatient care. Sponsors basic science efforts in biochemistry, molecular biology and immunology as they pertain to the clinical areas of cancerand diabetes.
*R. Paul Robertson, MD, CEO/Scientific Director*

**5021 Pacific Nuclear**
1010 South 336th Street 253-874-2235
Federal Way, WA 98003 Fax: 253-874-2401

**5022 Package Research Laboratory**
41 Pine Street 973-627-4405
Rockaway, NJ 07866 Fax: 973-627-4407
E-mail: info@package-testing.com
http://www.package-testing.com
Packaged product testing facility. Research results published in reports, videos and pictures. Custom tests designed. DOT/UN certification on hazardous materials. Extreme environment testing. Pallet load and pallet merchandizingtesting. Design and packaging development, consulting, package analysis, project management, and vendor audits.
*David Dixon, VP*
*Brian Berg, R&D*

**5023 Pan American Laboratories**
4099 Highway 190 985-893-4097
Covington, LA 70433 Fax: 985-893-6195
E-mail: pamlab@pamlab.com
http://www.pamlab.com/
Pharmaceutical manufacturer.
*Mary L Lipps, President*

**5024 Pan Earth Designs**
16525 103rd Street SE 360-458-9173
Suite A Fax: 360-458-9123
Yelm, WA 98597

Environmental research firm.

**5025 Par Environmental Services**
1906 21st Street 916-739-8356
PO Box 160756 Fax: 916-739-8356
Sacramento, CA 95816 http://www.parenvironmental.com
Environmental research firm.

**5026 Parsons Engineering Science**
100 W Walnut Street 626-440-2000
Pasadena, CA 91124 Fax: 626-440-2630
E-mail: erin.kuhlman@parsons.com
http://www.parsons.com
Environmental engineering testing and consulting company with expertise in advanced wastewater treatment.

*Frank A DeMartino, President*
*Erin Kuhlman, VP Corporate Relations*

**5027 Penniman & Browne**
6252 Falls Road 410-825-4131
PO Box 65309 Fax: 410-321-7384
Baltimore, MD 21209 E-mail: clientservices@pandbinc.com
http://www.pandbinc.com
Independent testing laboratory whose mission is to provide excellent client service with its scope of both engineering and chemical services.

*Hans V Steer, Client Services Manager*

**5028 Peoria Disposal Company**
4700 N Sterling Avenue 309-688-0760
Suite 2 Fax: 309-688-0881
Peoria, IL 61615 http://www.pdclab.com
Environmental services firm, especially hazardous waste testing.

**5029 Pharmaco LSR**
Mettlers Road 732-873-2550
Box 2360 Fax: 732-873-3992
East Millstone, NJ 08875

*Dr. Geoffrey K Hogan, President*

**5030 Philip Environmental Services**
210 W Sand Bank Road 618-281-7173
PO Box 230 Fax: 618-281-5120
Columbia, IL 62236 http://www.philipinc.com
Environmental research and analysis firm.
*Jenny Penland, President*

**5031 Physical Sciences**
20 New England Business Center 978-689-0003
Andover, MA 01810 Fax: 978-689-3232
http://www.psicorp.com
PSI focuses on providing contract research and development services in a variety of technical areas to both government and commercial customers. Our interests range from basic research to technology development, with an amphasis onapplied research.
*George Caledonia, President & CEO*
*David Green, President, R & D Operations*

**5032 Pittsburgh Mineral & Environmental Technology**
700 5th Avenue 724-843-5000
New Brighton, PA 15066 Fax: 724-843-5353
E-mail: pmet@pmet-inc.com
http://www.pmet-inc.com
A full service company specializing in metals and mineral processing, coal ash utilization, waste stream management, and precision analysis. Also develops technologies dedicated to waste minimibation, treatment, and conversion tosafe,usable, profitable products.

*Thomas E Weyand, President*
*William F Sutton, EVP*

**5033 Planning Concepts**
309 Commercial Street 530-265-8068
Nevada City, CA 95959 Fax: 916-265-5042
Environmental impact assessment firm.

**5034 Planning Design & Research Engineers**
2000 Lindell Avenue 615-298-2065
Nashville, TN 37203 Fax: 615-269-4119
E-mail: ttichenor@pdre.net
http://PDRE.net
Environmental engineers, asbestos, lead paint design, testing underground tanks, hazardous waste projects, Phase I and II site assessments.

*Teresa Tichenor, Office Manager*

**5035 Planning Resources**
402 W Liberty Drive 630-668-3788
Wheaton, IL 60187 Fax: 630-668-4125
E-mail: dri@sprintmail.com
Land use and environmental planning.
*Lan R Richart, President*
*Pamela J Richart, VP*

**5036 Plant Research Technologies**
525 Del Rey Avenue Unit C 408-245-4423
PO Box 6008 Fax: 408-245-8043
Sunnyvale, CA 94086
Contact research organization which provides agricultural and analytical applied services.
*Basil Burke PhD, President*

**5037 Plasma Science & Fusion Center**
167 Albany Street 617-253-8100
Cambridge, MA 02139 Fax: 617-253-0570
E-mail: info@psfc.mit.edu
http://www.psfc.mit.edu
Plasma science and technology and plasma fusion energy research.
*Miklos Porkolab, Director/Manager*
*Paul Rivenberg, Communications Manager*

**5038 Polaroid Corporation**
549 Technology Square 617-577-2000
Cambridge, MA 02139 Fax: 617-577-5618
http://www.polaroid.com

**5039 Polyengineering**
1935 Headland Avenue 334-793-4700
Dothan, AL 36303 http://www.polyengineering.com
Offers a broad range of professional engineering and architectural services as well as financial services and administrative support.

*AE Parsons, President*

**5040 Polytechnic**
3740 W Morse Avenue 847-677-0450
Lincolnwood, IL 60712 Fax: 847-677-0480

**5041 Porter Consultants**
4400 Old William Penn Highway 412-380-7500
Suite 200 Fax: 214-689-9
Monroeville, PA 15146 http://www.porter-consulting.com
Executive recruiting firm specializing in national and international placement of Sales, Marketing, Management, Executive-level, and Technical Support professionals within a wide rang of industries including High Tech, Exhibit.Telecommunications, Medical, and Pharmaceutical.
*SW Porter Jr, COO*

**5042 Powell Labs Limited**
1915 Aliceanna Street 410-558-3540
Baltimore, MD 21231
Provides services in the specialty fields of metallurgical investigations, failure analysis, metal overheating and corrosion failures, remaining life assessments of high temperature

components, identification of casting andmanufacturing defects, microbiological investigations, alloy identification, cycle water, cooling water, drinking water, high purity water, industrial process water, waste water, water and stream formed deposits, field examinations and training.

**5043 Precision Environmental**
180 Canada Larga Road 805-641-9333
Ventura, CA 93001 800-375-7786
Fax: 805-648-6999
http://http://www.precisionenv.com
Precision Environmental, Inc. was founded at Stanford University with the purpose of providing quality environmental contracting services to clients with asbestos contamination problems. State licensed and registered.

**5044 Princeton Energy Resources International**
1700 Rockville Pike 301-881-0650
Suite 550 Fax: 301-230-1232
Rockville, MD 20852 http://www.perihq.com
Engineering and consulting firm: engineering and environmental technology, environmental management and global climate change issues, economic research, aviation economics, and human factors
*Adolfo Menendez, President*

**5045 Priorities Institute**
3233 Vallejo Street 303-477-3792
#3B Fax: 303-838-8105
Denver, CO 80211 E-mail: mail@priorities.org
http://www.priorities.org
Nonprofit, educational research organization that explores issues of critical importance that are not adequately researched by existing educational, media, research, governmental or other organizations.
*Logan Perkins, Director/Founder*

**5046 Professional Service Industries**
1211 W. Cambridge Circle Drive 913-310-1600
Kansas City, KS 66103 Fax: 913-310-1601
*Elizabeth Noakes, Department Manager*

**5047 Professional Service Industries Laboratory**
4106 NW Riverside Drive 816-741-9466
Riverside, MO 64150 Fax: 816-587-2996
Engineering test laboratory.
*Stephen Fitzer, President*

**5048 Professional Service Industries/Jammal & Associates Division**
1675 Lee Road 407-645-5560
Winter Park, FL 32789 Fax: 407-645-1320
*William N Phillips, Executive VP*

**5049 Q-Lab**
1005 SW 18th Avenue 305-245-5600
PO Box 349490 Fax: 305-245-5656
Homestead, FL 33034 E-mail: mcrewdson@q-panel.com
http://www.q-panel.com
Firm providing environmental simulation testing.
*Michael J Crewdson, Director/Manager*

**5050 QC**
1205 Industrial Highway 215-355-3900
Southampton, PA 18966 Fax: 215-355-7231
Environmental testing lab.

**5051 Quantum Environmental**
167 Little Lake Drive 734-930-2600
Ann Arbor, MI 48104 Fax: 734-930-2798
Environmental remediation firm.

**5052 R&R Visual**
1828 W Olson Road 219-223-5426
Rochester, IN 46975 800-656-4225
Fax: 219-223-7953
E-mail: info@rapidview.com
http://www.rapidview.com
Developing and providing unique inspection solutions to the nuclear, petrochemical, industrial and municipal sewer industries.
*Rex Robinson, President*

**5053 RE/SPEC**
Box 725 605-394-6400
Rapid City, SD 57709 Fax: 605-394-6456
*Tom Zeller, VP Finance*

**5054 RETEC Group/ENSR-Seattle**
1011 SW Klickitat Way 206-624-9349
Suite 207 Fax: 206-624-2839
Seattle, WA 98134 E-mail: askensr@ensr.aecom.com
http://www.retec.com/index.html OR www.ensr.aecom.com
The RETEC Group/ENSR is an environmental management consulting and engineering firm that solves complex problems throughout the three main stages of the business life cycle-from new asset development to ongoing operations to finalasset disposition and restoration. RETEC develops integrated solutions logically aligned to optimize these business needs-financially, operationally, environmentally-ultimately, creating healthier businesses. RETEC merged with ENSR in February 2007.
*Steve McInerney, Systems Operations*
*Randy Kabrick, Engineering Services*

**5055 RMC Corporation Laboratories**
214 W Main Plaza 417-256-1101
West Plains, MO 65775 Fax: 417-256-1103
Environmental waste studies. Research results published in journals.
*Joseph Cooke, President*
*Dr R Soundararajan, Director R&D*

**5056 RMT**
744 Heartland Trail 608-831-4444
PO Box 8923 800-283-3443
Madison, WI 53717 Fax: 608-831-3334
E-mail: info@rmtinc.com
http://www.rmtinc.com
Global engineering and management consulting firm that develops environmental solutions for industry. With a 600 person staff and 20 offices throughout the US and Europe helping clients sustain the environment while meeting theirbusiness objectives. Engineers, scientists and construction managers take a project from conception through successful completion. Expertise includes air, water and waste permitting, remediation, hazardous/solid waste management, air pollutioncontrol and more.
*Jodi Burmester, Corporate Communications*

**5057 RV Fitzsimmons & Associates**
1860 Arthur Road 630-231-0680
West Chicago, IL 60185 Fax: 630-231-0811
Environmental testing and consulting firm.
*Robert Fitzsimmons, President*

**5058 Radian Corporation**
PO Box 201088 512-244-0100
Austin, TX 78720 Fax: 512-388-0966
Environmental science and industrial safety research and consulting firm. Research results published in project reports and in professional journals.

**5059 Ralph Stone and Company**
10954 Santa Monica Boulevard 310-478-1501
Los Angeles, CA 90025 800-813-9613
Fax: 310-478-7359
E-mail: rstoneco@aol.com
Environmental - Phase 1&2; Remediation; Geology
*Richard Kahle, President*

**5060 Ramco**
6362 Ferris Square 858-452-5963
Suite C Fax: 858-453-0625
San Diego, CA 92121
*Richard A McCormack, President*

**5061 Raytheon Company**
870 Winter Street 781-522-3000
Waltham, MA 02141 Fax: 781-522-3001
E-mail: cjkovalsky@raytheon.com
http://www.raytheon.com
An environmental testing firm, one of Raytheon's unique testing resources is the Andover Environmental Test Laboratory (ETL), a full-service, state-of-the-art facility. ETL specializes in performing static, dynamic (vibration, shock,and acceleration) and climatic test procedures, as well as comprehensive failure analysis studies.
*William H Swanson, Chairman/CEO*
*Connie Kovalsky, Media Relations*

**5062 Recon Environmental Corporation**
5 Johnson Drive 908-526-1000
PO Box 130 Fax: 908-526-7886
Raritan, NJ 08869
Environmental engineering, consulting, and laboratory services. Research results published in project reports and government publications.
*Norman J Weinstein, President*

**5063 Recon Systems**
5 Johnson Drive 908-526-1000
PO Box 130 Fax: 908-526-7886
Raritan, NJ 08869

*Dr. Norman J Weinstein, President*

**5064 Recra Environmental**
10 Hazelwood Drive
Suite 110 800-527-3272
Amherst, NY 14228 Fax: 716-691-2617
http://www.clu-in.org/products/site/complete/rcraenvi.htm
Research and development chemical and environmental measurement information.

*Kenneth Kinecki, Technology Developer Contact*

**5065 Reed and Associates**
2430 South Arlington Heights Road 847-718-0101
Arlington Heights, IL 60005 Fax: 847-718-0202
Environmental testing laboratory.

**5066 Reid, Quebe, Allison, Wilcox & Associates**
4755 Kingsway Drive 317-255-6060
Suite 400 Fax: 317-255-8354
Indianapolis, IN 46205
Architectural and environmental engineering research firm.
*J Edward Doyle, President*

**5067 Reliance Laboratories**
Benedum Airport Industrial Park 304-842-5285
PO Box 625 Fax: 304-842-5351
Bridgeport, WV 26330
*William F Kirk Jr, President*

**5068 Remtech**
110 12th Street NW 205-682-7900
Suite E 106 Fax: 205-682-7953
Birmingham, AL 35203
Systems design and engineering firm specializing in energy and environmental control applications. Research results published in project reports and are presented in papers at conferences.
*Gene Fuller, President*

**5069 Research Planning**
1121 Park Street 803-256-7322
Columbia, SC 29201 Fax: 803-254-6445
Scientific consulting firm specializing in the environment and natural resource assessment. Extensive experience in field surveys, EIS, spatial data analysis, and international work in Central America, West Africa and the Middle East.Research results published in professional journals, proceedings, and project reports. Woman-owned, small business concern.
*Jacqueline Michel, President*

**5070 Resource Technologies Corporation**
248 E Calder Way 814-237-4009
Suite 300 Fax: 814-237-1769
State College, PA 16801 E-mail: clients@resourcetec.com
http://www.resourcetec.com
An independent research, development and technical services firm located in central Pennsylvania. Specializes in appraisal and assessment services, information system development, assessment appeals and digitalmapping, web basedapplications, geotechnical services, environmental and ecological analysis and planning and management services.
*Jeffrey R Stern, President*
*Ronald W Stingelin, Contact*

**5071 Resources for the Future**
1616 P Street NW 202-328-5000
Washington, DC 20036 Fax: 202-939-3460
E-mail: info@rff.org
http://www.rff.org
RFF is a nonprofit and nonpartisan think tank located in Washington DC that conducts independent research-rooted primarily in economics and other social sciences on environmental and natural resource issues. RFF was founded in 1952.

*Lesli Creedon, Vice Pres., External Aff.*
*Stan Wellborn, Diretor of Communications*

**5072 Responsive Management**
130 Fraklin Street 540-432-1888
Harrisonburg, VA 22801 Fax: 540-432-1892
E-mail: mdduda@rica.net
http://www.responsivemanagement.com
Responsive Management is a Virginia-based public opinion polling and survey research firm specializing in fisheries, wildlife, natural resource, outdoor recreation and environmental issues.
*Mark Duda, Executive Director*

**5073 Revet Environmental and Analytical Lab**
181 Cedar Hill Street 508-460-7600
Marlborough, MA 01752 Fax: 508-460-7777
Environmental analysis and consulting laboratory.
*V Taylor, President*

**5074 Ricerca Biosciences LLC**
7528 Auburn Road 440-357-3300
Concord, OH 44077 888-742-3722
Fax: 440-354-6276
E-mail: info@ricerca.com
http://www.ricerca.com
Ricerca, a premier solution provider, offers expertise in both biology and chemistry to enable life sciences companies to fully leverage integrated, cost-effective, best practices approach to lead optimization and drug development.Services include in-vitro/in-vivo ADME, pharmacology, toxicology, medicinal, process, analytical chemistry, cGMP API scale-up production, regulatory support.

*Mark Crane, VP Business Dev/Marketing*

**5075 Rich Technology**
2410 Devonshire Drive 815-229-1122
Rockford, IL 61107 Fax: 815-229-1525
Environmental engineering research firm.

**5076 Riviana Foods: RVR Package Testing Center**
1702 Taylor Street 713-861-8221
Houston, TX 77007 Fax: 713-861-9939
*Lejo C Brana, Director Packaging*

**5077 Robert Bosch Corporation**
32104 State Road 2 574-237-2100
New Carlisle, IN 46552 Fax: 219-654-8755
Controlled-road environmental testing of automotive components for passenger cars, trucks, buses, tractor-trailers and off-road vehicles; certification to federal brake commission and fuel economy requirements.

**5078 Robert D Niehaus**
140 E Carrillo Street 805-962-0611
Santa Barbara, CA 93117 Fax: 805-962-0097
http://www.rdniehaus.com
Socioeconomic and environmental planning organization. Research results published in reports.

**5079 Rone Engineers**
8908 Ambassador Row 214-630-9745
Dallas, TX 75247 Fax: 214-630-9819
http://www.roneengineers.com
Provider of Geotechnical, Construction Materials Testing and Environmental Consulting services throughout Texas and the Southwest.
*Mark Gray, PE, Geotechnical & Engineering*

**5080 Roux Associates**
209 Shafter Street 631-232-2600
Islandia, NY 11749 Fax: 631-232-9898
E-mail: sisadiker@rouxinc.com
http://www.rouxinc.com
Environmental Consulting and Management.

*Steve Sadiker, Vice President*

**5081 Rummel, Klepper & Kahl**
81 Mosher Street 410-728-2900
Baltimore, MD 21217 800-787-3755
Fax: 410-728-2992
http://www.rkkengineers.com
Civil, site, transpotation, environmental, structural engineering services.

**5082 S-F Analytical Laboratories**
6125 W National Avenue 414-475-6700
PO Box 14513 800-300-6700
Milwaukee, WI 53214 Fax: 414-475-7216
E-mail: dkliber@sflabs.com
http://www.sflabs.com
Environmental and materials testing laboratory.

*David L Kliber, President/CEO*

**5083 SCS Engineers**
3900kilroy Airport Way 562-426-9544
Suite 100 Fax: 562-427-0805
Long Beach, CA 90806 E-mail: service@scsengineers.com
http://www.scsengineers.com
Delivers economically and environmentally sound solutions for solid waste management and site remediation projects throughout the world. Provides engineering, construction, and contract operations services to private and public sectorclients through a network of more than 40 offices and 500 professional staff working in the US and abroad.

**5084 SGI International**
1200 Prospect Street 858-551-1090
Suite 325 Fax: 858-551-0247
La Jolla, CA 92037 E-mail: info@sgiinternational.com
http://www.sgiinternational.com
Environmental applications.
*Michael L Rose, President*

**5085 SGS Environmental Services Inc**
200 W Potter Drive 907-562-2343
Anchorage, AK 99518 Fax: 907-562-0119
E-mail: julie.shumway@sgs.com
http://www.us.sgs.com
Environmental laboratory services.

*Chuck Homestead, General Manager*
*Julie Shumway, Business Development*

**5086 SHB AGRA**
3232 W Virginia Avenue 602-995-3916
Phoenix, AZ 85009 Fax: 602-995-3921
Geotechnical and environmental research firm.

**5087 SP Engineering**
45 Congress Street, Building 4 978-745-4569
PO Box 848, Shetland Park Fax: 978-745-4881
Salem, MA 01970 E-mail: brucepoolesp@aol.com
http://www.spengineeringinc.com
SP Engineering specializes in all areas of environmental compliance. Services include testing for chemical or bacterial contamination of well water and assessments for the presence of petroleum products or hazardous waste, in additionto performing environmental audits insuring owners that all the tenants in their industrial complexes are in compliance with government regulations relative to hazardous waste disposal.

*Bruce Poole, Executive Director*

**5088 SPECTROGRAM Corporation**
287 Boston Post Road 203-318-0535
Madison, CT 06443 Fax: 203-318-0535
E-mail: spectrogram@msn.com
http://www.spectrogram.com
Research, development and manufacturing firm which produces analytical instrumentation and systems in the fields of analytical chemistry (environmental) and elastomeric physical testing (rubber and plastics). Also offers a line ofproducts, each of which is involved in on-line environmental monitoring for the detection of an accidental release of petroleum products (oil spills).

*HR Gram, President/CEO*

**5089 STL Denver**
4955 Yarrow Street 303-736-0100
Arvada, CO 80002 800-572-8958
Fax: 303-431-7171
Testing and analysis services.

**5090 STS Consultants**
750 Corporate Woods Parkway 847-279-2500
Vernon Hills, IL 60061 800-859-7871
Fax: 847-279-2510
http://stsltd.com
Consulting engineering firm offering an integrated package of services in geotechnical engineering, waste management, environmental management, and construction technology.
*Thomas W Wolf, CEO*

**5091 STS Consultants**
111 Pfingsten Road 630-272-6520
Northbrook, IL 60062 Fax: 847-498-2721

*Mike Russell, President*

**5092 Saint Louis Testing Laboratories**
2810 Clark Avenue 314-531-8080
Saint Louis, MO 63103 Fax: 314-531-8085
E-mail: testlab@labinc.com
http://www.labinc.com/
Research and testing laboratory specializing in chemical, metallurgical, nondestructive and environmental testing and field services. Research results published in project reports.

*W Trowbridge, President*

**5093 Samtest**
3730 James Savage Road 989-496-3610
Midland, MI 48642 Fax: 989-496-3190
Geotechnical and environmental services firm.

**5094 Sari Sommarstrom**
PO Box 219 530-467-5783
Etna, CA 96027 Fax: 530-467-3623
E-mail: sari@sisqtel.net

**5095 Savannah Laboratories**
PO Box 13548 912-354-7854
Savannah, GA 31416 Fax: 912-352-0165
Environmental and biological research and testing laboratory with expertise in fish farming technology.

**5096 Scitest**
Route 66 Professional Center 802-728-6313
PO Box 339 Fax: 802-728-6044
Randolph, VT 05060
Environmental testing and analysis laboratory.
*Roderick J Lamothe, President*

**5097 Separation Systems Technology**
4901 Morena Boulevard 858-581-3765
Suite 809 Fax: 858-581-1211
San Diego, CA 92117 E-mail: riley1034@aol.com
Environmental research.
*Robert L Riley, President*

**5098 Shannon & Wilson**
PO Box 300303 206-632-8020
Seattle, WA 98103 Fax: 206-695-6777
http://www.shannonwilson.com
Environmental research.

**5099 Sheladia Associates**
15825 Shady Grove Road 301-590-3939
Suite 100 Fax: 301-948-7174
Rockville, MD 20850
Consulting firm specializing in environmental studies. Research results published in research reports for the government.
*A Moytayek, President*

**5100 Shell Engineering and Associates**
2403 West Ash 573-445-0106
Columbia, MO 65203 Fax: 573-445-0137
*Harvey D Shell, COO*

**5101 Sherry Laboratories**
2417 West Pinhook Road 337-235-0483
Lafayette, LA 70508 Fax: 337-233-6540
Analytical environmental laboratory.
*Mel Burnell, President*

**5102 Shive-Hattery Engineers & Architects**
800 1st Street Northwest 319-364-0227
PO Box 1803 Fax: 319-364-4251
Cedar Rapids, IA 52406
*Donald P Hattery, Chairman*

**5103 Siebe Appliance Controls**
2809 Emerywood Parkway 804-756-6500
Richmond, VA 23294 Fax: 804-756-6563
Automatic temperature, environmental, electronic appliance, heating, cooling and gas safety controls and valves; thermostats and oven burners.

**5104 Siemens Water Technologies**
1239 Willow Lake Boulevard 651-766-2700
Vadnais Heights, MN 55110 800-224-9474
Fax: 651-766-2701
E-mail: controlsystems.water@siemens.com
http://www.water.siemens.com/en/Pages/default.aspx
Products and services includes: environmental devices and controls; system troubleshooting/diagnostics; system startup; instrumentation calibration and commissioning, and radio topographic path analysis.
*Phil Williams, General Manager*
*David L Lee, Marketing/Sales Director*

**5105 Simpson Electric Company**
853 Dundee Avenue # 859 847-697-2260
Elgin, IL 60120 Fax: 847-697-2272
Analog and digital panel meters, meter relays, controllers, volt-ohm-milliammeters, scopes and industrial and environmental test instruments.

**5106 Skinner and Sherman Laboratories**
1st Avenue 781-890-7200
Waltham, MA 02451 Fax: 781-890-7003

**5107 Smith & Mahoney**
540 Broadway 518-463-4107
PO 22047 Fax: 518-463-3823
Albany, NY 12201
*Michael W McNarney, President*

**5108 Snell Environmental Group**
1425 Keystone Avenue 517-393-6800
PO Box 22127 Fax: 517-272-7390
Lansing, MI 48909 E-mail: seg-adm@ix.netcom.com
http://www.dlzcorp.com
Consulting structural engineers.
*John O'Mallia, President*

**5109 Soil Engineering Testing/SET**
9301 Bryant Avenue S 952-884-6833
Suite 107 Fax: 952-884-6923
Bloomington, MN 55420 E-mail: labinfo@soilengineeringtesting.com
http://www.soilengineeringtesting.com/
A comprehensive soil mechanics laboratory facility for engineering disciplines, environmental and hydrological applications. Scope of services includes: water content; unit mass; liquid limit; sieve analysis; specific gravity; pH;organic content; unconfined compression; and expansion index.
*Gordon R Eischens, President*

**5110 Solar Testing Laboratories**
1125 Valley Belt Road 216-741-7007
Brooklyn Heights, OH 44131 Fax: 216-741-7011
E-mail: stl@solartestinglabs.com
http://www.solartestinglabs.com/html/
Geotechnical, environmental engineering, materials testing, and construction inspection laboratory. Services include environmental site assessments, assisting in the selection and coordination of the work of remediation contractors,asbestos inspection and abatement supervision, micro purgegroundwater sampling, sediment control inspection, landfill closure quality assurance, radiological assessments, U.S.T. closures and RCRA closures and facility investigation.
*George J Ata PE, President*

**5111 Southeastern Engineering & Testing Laboratories**
4761 SW 51st Street 954-584-4322
Davie, FL 33314 Fax: 954-584-4338
E-mail: jack@seetl.com
http://www.seetl.com
Geotechnical and environmental engineering consulting firm and construction materials engineering laboratory.
*Jack Krouskroup, Director*

**5112 Southern Petroleum Laboratory/SPL**
8880 Interchange Drive 713-660-0901
PO Box 20807 800-969-6773
Houston, TX 77054 Fax: 713-660-8975
E-mail: HRBrown@spl-inc.com
http://www.spl-inc.com
SPL provides technical and analytical services to the oil and gas industry including environmental and hydrocarbon analytical services as well as field (gas & liquid measurement) services.
*HR Brown, President*

**5113 Southern Research Institute COBRA Training Facility Center for Domestic Preparedness**
61 Responder Drive 256-847-2515
PO Box 5129 Fax: 256-847-2525
Anniston, AL 35211 E-mail: secrist@southernresearch.org
http://www.southernresearch.org
Southern Research Institute is an independent research corporation with established capabilities in pharmaceutical discovery and development, engineering, chemical and biological defense, environmental and energy-related sciences.Research is conducted through contracts and grants with government and commerical clients.
*John A Secrist III PhD, President/CEO*
*Tommy Hurn, Facilities Director*

**5114 Southern Research Institute Corporate Office: Life Sciences/Environment/Energy**
2000 Ninth Avenue S 205-581-2000
PO Box 55305 800-967-6774
Birmingham, AL 35205 Fax: 205-581-2726
E-mail: secrist@southernresearch.org
http://www.southernresearch.org
Southern Research Institute is an independent research corporation with established capabilities in pharmaceutical discovery and development, engineering, chemical and biological defense, environmental and energy-related sciences.Research is conducted through contracts and grants with government and commerical clients.
*John A Secrist III PhD, President/CEO*
*David A Rutledge, Chief Financial Officer*

**5115 Southern Research Institute: Carbon To LiquidsDevelopment Center**
5201 International Drive 919-806-3456
Durham, NC 27712 Fax: 919-806-2306
E-mail: secrist@southernresearch.org
http://www.southernresearch.org
Southern Research Institute is an independent research corporation with established capabilities in pharmaceutical discovery and development, engineering, chemical and biological defense, environmental and energy-related sciences.Research is conducted through contracts and grants with government and commerical clients.
*John A Secrist III PhD, President/CEO*
*Tommy Hurn, Facilities Director*

**5116 Southern Research Institute: Chemical Defense Training Facility-Missouri**
CDTF-Building 5101 573-596-0131
Fort Leonard Wood, MO 65473 Fax: 573-596-0722
E-mail: secrist@southernresearch.org
http://www.southernresearch.org
Southern Research Institute is an independent research corporation with established capabilities in pharmaceutical discovery and development, engineering, chemical and biological defense, environmental and energy-related sciences.Research is conducted through contracts and grants with government and commerical clients.
*John A Secrist III PhD, President/CEO*
*Tommy Hurn, Facilities Director*

**5117 Southern Research Institute: Engineering Research Center**
757 Tom Martin Drive 205-581-2000
Birmingham, AL 35211 800-967-6774
Fax: 205-581-2726
E-mail: secrist@southernresearch.org
http://www.southernresearch.org
Southern Research Institute is an independent research corporation with established capabilities in pharmaceutical discovery and development, engineering, chemical and biological defense, environmental and energy-related sciences.Research is conducted through contracts and grants with government and commerical clients.
*John A Secrist III PhD, President/CEO*
*Michael D Johns, VP Engineering Division*

**5118 Southern Research Institute: Environment & Energy Research**
3000 Aerial Center Parkway 919-806-3456
Suite 160 Fax: 919-806-2306
Morrisville, NC 27560 E-mail: secrist@southernresearch.org
http://www.southernresearch.org
Southern Research Institute is an independent research corporation with established capabilities in pharmaceutical discovery and development, engineering, chemical and biological defense, environmental and energy-related sciences.Research is conducted through contracts and grants with government and commerical clients.
*John A Secrist III PhD, President/CEO*
*Stephen D Piccot, Environment/Energy Director*

**5119 Southern Research Institute: Infectious Disease Research Facility**
431 Aviation Way 301-694-3232
Frederick, MD 21701 Fax: 301-694-7223
E-mail: secrist@southernresearch.org
http://www.southernresearch.org
Southern Research Institute is an independent research corporation with established capabilities in pharmaceutical discovery and development, engineering, chemical and biological defense, environmental and energy-related sciences.Research is conducted through contracts and grants with government and commerical clients.
*John A Secrist III PhD, President/CEO*
*Michael G Murray PhD, Infectious Disease Director*

**5120 Southern Research Institute: Power Systems Development Facility**
PO Box 1069 205-670-5068
Highway 25 N Fax: 205-670-5843
Wilsonville, AL 35186 E-mail: secrist@southernresearch.org
http://www.southernresearch.org
Southern Research Institute is an independent research corporation with established capabilities in pharmaceutical discovery and development, engineering, chemical and biological defense, environmental and energy-related sciences.Research is conducted through contracts and grants with government and commerical clients.
*John A Secrist III PhD, President/CEO*
*Tommy Hurn, Facilities Director*

**5121 Southern Testing & Research Laboratories**
3809 Airport Drive 252-237-4175
Wilson, NC 27896 Fax: 252-237-9341
http://www.southerntesting.com
Full-service laboratory with over 75 chemists, microbiologists and support personnel that provides personalized service to clients. Capabilities include pharmaceutical, foods and feeds, environmental, industrial hygiene, agriculturaland microbiological sciences. Laboratory is FDA-inspected GLP/cGMP laboratory utilizing AOAC, USP, EPA, USDA, AACC, AOCS, ISO, client and in-house validated methods.
*Robert Dermer, Managing Director*
*Walter Hogg, Business Development*

**5122 Spears Professional Environmental & Archeological Research Services**
13858 S Highway 170 479-839-3663
West Fork, AR 72774 Fax: 479-839-2575
E-mail: SPEARSC@aol.com
Archeological research service. The company conducts cultural resources studies including background studies for Environmental Impact Studies, archeological surveys, significance testing, and data recovery/excavation. Large multi-yearprojects have included cultural resources surveys for timber sales and studies for proposed interstates and utilities.
*Carol S Spears, President/Owner*

**5123 Spectrochem Laboratories**
545 Commerce Street 201-337-4774
Franklin Lakes, NJ 07417 Fax: 201-337-1255

Research and development firm specializing in environmental sciences and inorganic chemistry. Research results published in proceedings at technical conferences.
*Irene Van Dren, President*

**5124 Spectrum Sciences & Software**
91 Hill Avenue NW 850-796-0909
Fort Walton Beach, FL 32548 Fax: 850-244-9560
E-mail: lcars@specsci.com OR ir@spectrumholdingscorp.com
http://www.specsci.com
An environmental research firm, Spectrum provides diversified capabilities of a large business in a number of advanced technologies. Services include all the disciplines and technologies relevant to operations and maintenance; computerand system sciences; manufacturing; comprehensive planning and environmental assessment technology; and system design testing and evaluation.
*Jeremy Maines, Information Technology*
*Dwight Howard, VP Business Development*

**5125 Spotts, Stevens and McCoy**
1047 North Park Road 610-621-2000
Reading, PA 19610 Fax: 610-621-2001
E-mail: information@ssmgroup.com
http://www.ssmgroup.com
An engineering and consulting firm, serving business, industry, and government, SSM provides consulting services in the areas of environmental health and safety, regulatory compliance and training. In addition, SSM provides costeffective, well-engineered solutions to environmental health and safety issues facing facility owners in industry, education, healthcare and local government.
*Eileen Kaley, Marketing Director*

**5126 Standard Testing and Engineering - Corporate**
3400 Lincoln Boulevard 405-528-0541
Oklahoma City, OK 73105 800-725-0541
Fax: 405-528-0559
E-mail: stantech@stantest.com OR bburris@stantest.com
http://www.stantest.com
Standard Testing and Engineering Company was founded in 1951 as a professional engineering firm specializing in materials testing and engineering for the construction and manufacturing industries. Standard Testing also providesenvironmental services such as groundwater studies. Standard Testing's capabilities include a wide variety of specialties such as environmental engineering, and industrial hygiene.
*Charles B Burris PE, Engineer/Director*

**5127 Standard Testing and Engineering - Enid**
902 Trails West Loop 580-237-3130
Enid, OK 73703 800-725-3130
Fax: 580-237-3211
E-mail: stantech@stantest.com OR gbeckham@stantest.com
http://www.stantest.com
Standard Testing and Engineering Company was founded in 1951 as a professional engineering firm specializing in materials testing and engineering for the construction and manufacturing industries. Standard Testing also providesenvironmental services such as groundwater studies. Standard Testing's capabilities include a wide variety of specialties such as environmental engineering, and industrial hygiene.
*G Beckman, Engineer*

**5128 Standard Testing and Engineering - Lawton**
900 Southeast Second 580-353-0872
Lawton, OK 73501 800-725-0541
Fax: 580-353-1263
E-mail: stantech@stantest.com OR kwayman@stantest.com
http://www.stantest.com
Standard Testing and Engineering Company was founded in 1951 as a professional engineering firm specializing in materials testing and engineering for the construction and manufacturing industries. Standard Testing also providesenvironmental services such as groundwater studies. Standard Testing's capabilities include a wide variety of specialties such as environmental engineering, and industrial hygiene.
*K Wayman, Engineer*

**5129 Standard Testing and Engineering - Oklahoma City Environmental Services Division**
4300 N Lincoln Boulevard 405-424-8378
Oklahoma, OK 73105 800-725-8378
Fax: 405-424-8129
E-mail: stantech@stantest.com OR sbaber@stantest.com
http://www.stantest.com
Standard Testing and Engineering Company was founded in 1951 as a professional engineering firm specializing in materials testing and engineering for the construction and manufacturing industries. Standard Testing also providesenvironmental services such as groundwater studies. Standard Testing's capabilities include a wide variety of specialties such as environmental engineering, and industrial hygiene.
*S Baber, Engineer*

**5130 Standard Testing and Engineering - Tulsa**
5358 S 125th East Avenue 918-459-2700
Tulsa, OK 74146 800-725-4592
Fax: 918-459-2715
E-mail: stantech@stantest.com OR mhunter@stantest.com
http://www.stantest.com
Standard Testing and Engineering Company was founded in 1951 as a professional engineering firm specializing in materials testing and engineering for the construction and manufacturing industries. Standard Testing also providesenvironmental services such as groundwater studies. Standard Testing's capabilities include a wide variety of specialties such as environmental engineering, and industrial hygiene.
*M Hunter, Engineer*

**5131 Stanford Technology Corporation**
57 Poplar Street 203-348-4080
PO Box 2100D Fax: 203-327-5225
Glenbrook, CT 06906 E-mail: stctestlab@aol.com
http://www.stctestlab
High technology research firm. Research results published in confidential project reports.
*Charles C Cullari, President*
*Gerald T Ciccone, VP*

**5132 Stantec Consulting Services**
4875 Riverside Drive 478-474-6100
Macon, GA 31210 Fax: 478-474-8933
E-mail: media@stantec.com
http://www.stantec.com
Stantec, founded in 1954, provides professional design and consulting services in planning, engineering, architecture, surveying, economics, and project management. Stantec supports public and private sector clients in a diverse rangeof markets in the infrastructure and facilities sector at every stage, from initial concept and financial feasibility to project completion and beyond.
*Tony Franceschini, President/CEO*
*Jay Averill, Communications/Media*

**5133 Steven Winter Associates - New York NY**
307 7th Avenue 212-564-5800
Suite 1201 Fax: 212-645-9931
New York, NY 10001 E-mail: swa@swinter.com
http://www.swinter.com
New York client base includes City, State, and Federal agencies, and owners of a wide array of buildings from small residential to well-known sustainable buildings, such as Battery Park City, 4 Times Square, AOL/Time Warner and HearstHeadquarters. SWA is working with many developers, architects, engineers, and building scientists to help deliver higher performance buildings throughout the NYC metropolitan area and in the surrounding region.
*Steven Winter, President*

**5134 Steven Winter Associates - Norwalk CT**
50 Washington 203-857-0200
Norwalk, CT 06854 Fax: 203-852-0741
E-mail: swa@swinter.com
http://www.swinter.com
Founded in 1972, Steven Winter Associates, Inc. (SWA) provides a variety of services including: building system assessment,

green materials and product specifications, LEED assessments and certification, green building commissioning,accessibility conformance, energy auditing capabilities, builder/operator training, preparation of green guidelines, HVAC troubleshooting, indooor air quality analysis and testing, solar and PV design engineering.

*Steven Winter, President*

**5135 Steven Winter Associates - Washington DC**
1112 16th Street NW 202-628-6100
Suite 240 Fax: 202-393-5043
Washington, DC 20036 E-mail: swadc@swinter.com
http://www.swinter.com
Located just a few blocks from the White House in the historic downtown section, SWA/DC focuses on technology transfer, buildings-related policy analysis, information dissemination, media outreach & publishing, association management,classroom and web-based training, and buildings research work on behalf of U.S. DOE, HUD, EPA, and the national energy laboratories. SWA/DC also provides logistical support to HUD's Office of Native American Programs (ONAP).
*Steven Winter, President*

**5136 Stone Environmental**
535 Stone Cutters Way 802-229-4541
Montpelier, VT 05602 800-959-9987
Fax: 802-229-5417
E-mail: sei@stone-env.com
http://www.stone-env.com
Environmental consulting services and technologies that include: environmental planning and documentation; environmental compliance; waste management; environmental spatial analysis; remediation. Scientific disciplines include: civilengineering; environmental engineering and chemistry; forest biology; hydrogeology; geology; soil science; and geographic information systems.
*Christopher Stone, President*
*David Healy, Vice President*

**5137 Stork Heron Testing Laboratories**
1200 Westinghouse Boulevard 704-588-1131
Suite A 888-786-7555
Charlotte, NC 28273 Fax: 704-588-5412
E-mail: info.herron@stork.com
http://www.storksmt.com
Analysis and testing laboratory offering environmental control services including mechanical, metallurgical, and chemical analysis in addition to materials testing, nondestructive testing, failure analysis and product evaluation.
*J Doehring, Chief Operations Officer*
*Elizabeth Huber, Failure Analyst*

**5138 Stork Southwestern Laboratories**
222 Cavalcade Street 713-692-9151
PO Box 8768 Fax: 713-696-6307
Houston, TX 77249 E-mail: info.smt@stork.com
http://www.storksmt.com
Analysis and testing laboratory offering environmental control services including materials testing; nondestructive testing; polymer testing and polymeric materials testing; electrical and thermal testing; failure analysis;construction materials testing and engineering; product evaluation; surface testing, and air emissions.
*J Doehring, Chief Operations Officer*
*Elizabeth Huber, Failure Analyst*

**5139 Suburban Laboratories**
4140 Litt Drive 708-544-3260
Hillside, IL 60162 Fax: 708-544-8587
E-mail: Info@suburbanlabs.com
http://www.suburbanlabs.com
Environmental laboratory providing chemical, chromatographic, and spectrographic analysis of biological materials, including water and groundwater, soil, and hazardous materials for priority pollutants, metals, and pesticide residues.
*Andy Groeper, Marketing Director*
*Jarrett Thomas, Manager*

**5140 Sunsearch**
PO Box 590 203-453-6591
393A Soundview Road 800-338-0258
Guilford, CT 06437 Fax: 203-458-9011
E-mail: ebarber@sunsearchinc.com
http://www.sunsearchinc.com
Designs, installs and services solar energy systems. Additional services includes: feasibility studies; inspections or assessments of existing systems; repair and redesign of existing systems; service agreements and evaluation ofcomplex systems.
*Everett M Barber Jr, President*

**5141 Systech Environmental Corporation**
3085 Woodman Drive 937-643-1240
Suite 300 800-888-8011
Dayton, OH 45420 Fax: 937-643-1203
E-mail: Erica.Hawk@lafarge-na.com
http://www.sysenv.com
Provider of alternative fuels to cement kilns.
*Erica Hawk, Corporate Mktg Specialist*

**5142 TAKA Asbestos Analytical Services**
PO Box 208 631-261-2117
Greenlawn, NY 11740 Fax: 631-261-2120
TAKA provides environmental consultation, testing and analytical services for the assessment and detection of onsite asbestos materials. The president and owner of TAKA, Dr. Thomas A. Kubic, has extensive experience with forensicmicroscopy and advanced techniques in sampling and evaluation of airborne asbestos particles using Polarized Light Microscopy.
*Thomas A Kubic PhD/MS/JD/FABC, President/Owner*

**5143 TRAC Laboratories**
113 Cedar Street 940-566-3359
PO Box 215 Fax: 940-566-2698
Denton, TX 76201 E-mail: sjunot@traclaboratories.com
http://www.traclaboratories.com
Provide multi-disciplinary problem-solving approaches to environmental and public health issues.
*Barney J Venables PhD, Research Director*
*Stephen B Junot, Laboratory Director*

**5144 TRC Environmental Corporation-Alexandria**
4615 Parliament Drive 318-445-8544
Suite 101 Fax: 318-448-4453
Alexandria, LA 71303 E-mail: cobrien@trcsolutions.com
http://www.trcsolutions.com
A provider of engineering, financial, risk management and construction services to large industrial and government customers throughout the United States, TRC provides customer focused solutions in three primary markets: environmental,energy and infrastructure. Environmental services include project development, resolving legacy environmental issues, ensuring compliance for continuing operations, and identifing and mitigating future environmental risks.

*Christopher P Vincze, CEO/Corporate Office*
*Caren O'Brien, TRC Service Inquiries Mgr*

**5145 TRC Environmental Corporation-Atlanta**
PO Box 80310 404-932-9343
Atlanta, GA 30366 Fax: 770-451-4745
E-mail: cobrien@trcsolutions.com
http://www.trcsolutions.com
A provider of engineering, financial, risk management and construction services to large industrial and government customers throughout the United States, TRC provides customer focused solutions in three primary markets: environmental,energy and infrastructure. Environmental services include project development, resolving legacy environmental issues, ensuring compliance for continuing operations, and identifing and mitigating future environmental risks.

*Christopher P Vincze, CEO/Corporate Office*
*Caren O'Brien, TRC Service Inquiries Mgr*

**5146 TRC Environmental Corporation-Augusta**
249 Western Avenue 207-621-7000
Augusta, ME 04330 Fax: 207-621-7001
E-mail: cobrien@trcsolutions.com
http://www.trcsolutions.com
A provider of engineering, financial, risk management and construction services to large industrial and government customers throughout the United States, TRC provides customer focused solutions in three primary markets: environmental,energy and infrastructure. Environmental services include project development, resolving legacy environmental issues, ensuring compliance for continuing operations, and identifing and mitigating future environmental risks.

*Christopher P Vincze, CEO/Corporate Office*
*Caren O'Brien, TRC Service Inquiries Mgr*

**5147 TRC Environmental Corporation-Boston**
31 Milk Street 617-350-3444
11th Floor Fax: 617-350-3443
Boston, MA 02109 E-mail: cobrien@trcsolutions.com
http://www.trcsolutions.com
A provider of engineering, financial, risk management and construction services to large industrial and government customers throughout the United States, TRC provides customer focused solutions in three primary markets: environmental,energy and infrastructure. Environmental services include project development, resolving legacy environmental issues, ensuring compliance for continuing operations, and identifing and mitigating future environmental risks.

*Christopher P Vincze, CEO/Corporate Office*
*Caren O'Brien, TRC Service Inquiries Mgr*

**5148 TRC Environmental Corporation-Bridgeport**
10 Middle Street 203-335-3500
6th Floor Fax: 203-335-3550
Bridgeport, CT 06604 E-mail: cobrien@trcsolutions.com
http://www.trcsolutions.com
A provider of engineering, financial, risk management and construction services to large industrial and government customers throughout the United States, TRC provides customer focused solutions in three primary markets: environmental,energy and infrastructure. Environmental services include project development, resolving legacy environmental issues, ensuring compliance for continuing operations, and identifing and mitigating future environmental risks.

*Christopher P Vincze, CEO/Corporate Office*
*Caren O'Brien, TRC Service Inquiries Mgr*

**5149 TRC Environmental Corporation-Chicago**
10 S Riverside Plaza 312-879-0191
Suite 1770 Fax: 312-879-0081
Chicago, IL 60606 E-mail: cobrien@trcsolutions.com
http://www.trcsolutions.com
A provider of engineering, financial, risk management and construction services to large industrial and government customers throughout the United States, TRC provides customer focused solutions in three primary markets: environmental,energy and infrastructure. Environmental services include project development, resolving legacy environmental issues, ensuring compliance for continuing operations, and identifing and mitigating future environmental risks.

*Christopher P Vincze, CEO/Corporate Office*
*Caren O'Brien, TRC Service Inquiries Mgr*

**5150 TRC Environmental Corporation-Ellicott City**
9056 Chevrolet Drive 410-465-7927
Ellicott City, MD 21042 Fax: 410-465-7535
E-mail: cobrien@trcsolutions.com
http://www.trcsolutions.com
A provider of engineering, financial, risk management and construction services to large industrial and government customers throughout the United States, TRC provides customer focused solutions in three primary markets: environmental,energy and infrastructure. Environmental services include project development, resolving legacy environmental issues, ensuring compliance for continuing operations, and identifing and mitigating future environmental risks.

*Christopher P Vincze, CEO/Corporate Office*
*Caren O'Brien, TRC Service Inquiries Mgr*

**5151 TRC Environmental Corporation-Henderson**
1009 Whitney Ranch Drive 702-248-6415
Henderson, NV 89014 Fax: 702-248-0626
E-mail: cobrien@trcsolutions.com
http://www.trcsolutions.com
A provider of engineering, financial, risk management and construction services to large industrial and government customers throughout the United States, TRC provides customer focused solutions in three primary markets: environmental,energy and infrastructure. Environmental services include project development, resolving legacy environmental issues, ensuring compliance for continuing operations, and identifing and mitigating future environmental risks.

*Christopher P Vincze, CEO/Corporate Office*
*Caren O'Brien, TRC Service Inquiries Mgr*

**5152 TRC Environmental Corporation-Honolulu**
677 Ala Moana Boulevard 808-728-4111
Suite 920 Fax: 808-638-5649
Honolulu, GA 30366 E-mail: cobrien@trcsolutions.com
http://www.trcsolutions.com
A provider of engineering, financial, risk management and construction services to large industrial and government customers throughout the United States, TRC provides customer focused solutions in three primary markets: environmental,energy and infrastructure. Environmental services include project development, resolving legacy environmental issues, ensuring compliance for continuing operations, and identifing and mitigating future environmental risks.

*Christopher P Vincze, CEO/Corporate Office*
*Caren O'Brien, TRC Service Inquiries Mgr*

**5153 TRC Environmental Corporation-Indianapolis**
50 E 91st Street 317-819-1300
Suite 201 Fax: 317-819-1301
Indianapolis, IN 46240 E-mail: cobrien@trcsolutions.com
http://www.trcsolutions.com
A provider of engineering, financial, risk management and construction services to large industrial and government customers throughout the United States, TRC provides customer focused solutions in three primary markets: environmental,energy and infrastructure. Environmental services include project development, resolving legacy environmental issues, ensuring compliance for continuing operations, and identifing and mitigating future environmental risks.

*Christopher P Vincze, CEO/Corporate Office*
*Caren O'Brien, TRC Service Inquiries Mgr*

**5154 TRC Environmental Corporation-Irvine**
21 Technology Drive 949-727-7376
Irvine, CA 92618 Fax: 949-727-7312
E-mail: dzarider@trcsolutions.com
http://www.trcsolutions.com
A provider of engineering, financial, risk management and construction services to large industrial and government customers throughout the United States, TRC provides customer focused solutions in three primary markets: environmental,energy and infrastructure. Environmental services include project development, resolving legacy environmental issues, ensuring compliance for continuing operations, and identifing and mitigating future environmental risks.

*Christopher P Vincze, CEO/Corporate Office*
*David Zarider, Environmental Services Mgr*

**5155 TRC Environmental Corporation-Jackson**
5591 Morrill Road
Jackson, MI 49201
517-782-5531
Fax: 517-782-5679
E-mail: cobrien@trcsolutions.com
http://www.trcsolutions.com

A provider of engineering, financial, risk management and construction services to large industrial and government customers throughout the United States, TRC provides customer focused solutions in three primary markets: environmental,energy and infrastructure. Environmental services include project development, resolving legacy environmental issues, ensuring compliance for continuing operations, and identifing and mitigating future environmental risks.

*Christopher P Vincze, CEO/Corporate Office*
*Caren O'Brien, TRC Service Inquiries Mgr*

**5156 TRC Environmental Corporation-Kansas City**
Livestock Exchange Building
1600 Genessee Street, Suite 416
Kansas City, MO 64102
816-474-1500
Fax: 816-474-1853
E-mail: cobrien@trcsolutions.com
http://www.trcsolutions.com

A provider of engineering, financial, risk management and construction services to large industrial and government customers throughout the United States, TRC provides customer focused solutions in three primary markets: environmental,energy and infrastructure. Environmental services include project development, resolving legacy environmental issues, ensuring compliance for continuing operations, and identifing and mitigating future environmental risks.

*Christopher P Vincze, CEO/Corporate Office*
*Caren O'Brien, TRC Service Inquiries Mgr*

**5157 TRC Environmental Corporation-Lexington**
4468 Walnut Creek Drive
Lexington, KY 40509
859-420-6223
Fax: 859-271-9902
E-mail: cobrien@trcsolutions.com
http://www.trcsolutions.com

A provider of engineering, financial, risk management and construction services to large industrial and government customers throughout the United States, TRC provides customer focused solutions in three primary markets: environmental,energy and infrastructure. Environmental services include project development, resolving legacy environmental issues, ensuring compliance for continuing operations, and identifing and mitigating future environmental risks.

*Christopher P Vincze, CEO/Corporate Office*
*Caren O'Brien, TRC Service Inquiries Mgr*

**5158 TRC Environmental Corporation-Littleton**
7761 Shaffer Parkway
Suite 100
Littleton, CO 80127
303-792-5555
Fax: 303-792-0122
E-mail: cobrien@trcsolutions.com
http://www.trcsolutions.com

A provider of engineering, financial, risk management and construction services to large industrial and government customers throughout the United States, TRC provides customer focused solutions in three primary markets: environmental,energy and infrastructure. Environmental services include project development, resolving legacy environmental issues, ensuring compliance for continuing operations, and identifing and mitigating future environmental risks.

*Christopher P Vincze, CEO/Corporate Office*
*Caren O'Brien, TRC Service Inquiries Mgr*

**5159 TRC Environmental Corporation-Lowell**
Wannalancit Mills
630 Suffolk Street
Lowell, MA 01854
978-970-5600
Fax: 978-453-1995
E-mail: gharkness@trcsolutions.com
http://www.trcsolutions.com

A provider of engineering, financial, risk management and construction services to large industrial and government customers throughout the United States, TRC provides customer focused solutions in three primary markets: environmental,energy and infrastructure. Environmental services include project development, resolving legacy environmental issues, ensuring compliance for continuing operations, and identifing and mitigating future environmental risks.

*Christopher P Vincze, CEO/Corporate Office*
*Glenn Harkenss, Energy Services*

**5160 TRC Environmental Corporation-Phoenix**
Wannalancit Mills
630 Suffolk Street
Lowell, MA 01854
978-970-5600
Fax: 978-453-1995
E-mail: gharkness@trcsolutions.com
http://www.trcsolutions.com

A provider of engineering, financial, risk management and construction services to large industrial and government customers throughout the United States, TRC provides customer focused solutions in three primary markets: environmental,energy and infrastructure. Environmental services include project development, resolving legacy environmental issues, ensuring compliance for continuing operations, and identifing and mitigating future environmental risks.

*Christopher P Vincze, CEO/Corporate Office*
*Glenn Harkenss, Energy Services*

**5161 TRC Environmental Corporation-Princeton**
Research Park
322 Wall Street
Princeton, NJ 08540
609-497-1379
Fax: 609-497-1879
E-mail: cobrien@trcsolutions.com
http://www.trcsolutions.com

A provider of engineering, financial, risk management and construction services to large industrial and government customers throughout the United States, TRC provides customer focused solutions in three primary markets: environmental,energy and infrastructure. Environmental services include project development, resolving legacy environmental issues, ensuring compliance for continuing operations, and identifing and mitigating future environmental risks.

*Christopher P Vincze, CEO/Corporate Office*
*Caren O'Brien, TRC Service Inquiries Mgr*

**5162 TRC Environmental Corporation-San Francisco**
55 2nd Street
Suite 575
San Francisco, CA 94105
415-644-3000
Fax: 415-541-9378
E-mail: cobrien@trcsolutions.com
http://www.trcsolutions.com

A provider of engineering, financial, risk management and construction services to large industrial and government customers throughout the United States, TRC provides customer focused solutions in three primary markets: environmental,energy and infrastructure. Environmental services include project development, resolving legacy environmental issues, ensuring compliance for continuing operations, and identifing and mitigating future environmental risks.

*Christopher P Vincze, CEO/Corporate Office*
*Caren O'Brien, TRC Service Inquiries Mgr*

**5163 TRC Environmental Corporation-West Palm Beach**
1665 Palm Beach Lakes Boulevard
Suite 720
West Palm Beach, FL 33401
561-681-3494
Fax: 561-681-3494
E-mail: cobrien@trcsolutions.com
http://www.trcsolutions.com

A provider of engineering, financial, risk management and construction services to large industrial and government customers throughout the United States, TRC provides customer focused solutions in three primary markets: environmental,energy and infrastructure. Environmental services include project development, resolving legacy environmental issues, ensuring compliance for continuing operations, and identifing and mitigating future environmental risks.

*Christopher P Vincze, CEO/Corporate Office*
*Caren O'Brien, TRC Service Inquiries Mgr*

**5164 TRC Environmental Corporation-Windsor**
21 Griffin Road North 860-298-9692
Windsor, CT 06095 Fax: 860-298-6399
E-mail: czoephel@tresolutions.com
http://www.trcsolutions.com
A provider of engineering, financial, risk management and construction services to large industrial and government customers throughout the United States, TRC provides customer focused solutions in three primary markets: environmental,energy and infrastructure. Environmental services include project development, resolving legacy environmental issues, ensuring compliance for continuing operations, and identifing and mitigating future environmental risks.

*Christopher P Vincze, CEO/Corporate Office*
*Carl Zoephel, Business Development Manager*

**5165 TRC Garrow Associates**
3772 Pleasantdale Road 770-270-1192
Suite 200 Fax: 770-270-1392
Atlanta, GA 30340 E-mail: bgarrow@trcgarrow.com
http://www.trcgarrow.com
TRC Garrow Associates provides business consulting services focusing on environmental analysis, planning and development.
*Barbara Garrow, President*

**5166 Talos Technology Consulting**
3336 Fern Hollow Place 703-715-3500
Suite 100 Fax: 703-715-0189
Herndon, VA 20171 E-mail: information@talos.com
http://www.talos.com
An environmental computer company, Talos works with businesses and government organizations to help them capitalize upon emerging technologies and achieve their organizational objectives. Talos professionals perform strategic planning;requirements analysis; design documentation; software trade surveying; system security planning; cost-benefit analysis; and surveys of technology markets.
*Scott Little, Strategic Development*
*Rob Smith, Chief Technology Officer*

**5167 Taylor Engineering**
9000 Cypress Green Drive 904-731-7040
Suite 200 Fax: 904-731-9847
Jacksonville, FL 32256 http://www.taylorengineering.com
Services in coastal engineering consulting, dredging and dredged material management, hydrology and hydraulics, environmental services, and construction support services.

*Terrence Hall P.E., President*

**5168 Tellus Institute**
11 Arlington Street 617-266-5400
Boston, MA 02116 Fax: 617-266-8303
E-mail: info@tellus.org
http://www.tellus.org
Environmental research and strategic development firm. Services include: analyzing energy systems and environmental impacts; evaluating policies for transition to efficient and renewable energy technology; formulating strategies formitigating and adapting to climate change; evaluating long-term solutions that balance competing freshwater needs for basic services; and developing methods to support comprehensive river basin assessment.
*Paul D Raskin PhD, President*
*David McAnulty, Administrative Director*

**5169 TestAmerica-Austin**
14050 Summit Drive 512-244-0855
Suite A100 Fax: 512-244-0160
Austin, TX 78728 E-mail: info@stl-inc.com
http://http://stl-inc.com/contacts/lab_list.htm
Environmental engineering research and consulting firm. Testing capabilities include chemical, physical and biological analyses of a variety of matrices, including aqueous, solid, drinking water, waste, tissue, air and saline/estuarinesamples. Specialty capabilities include air toxics testing, mixed waste testing, tissue preparation and analysis, aquatic toxicology, dioxin/furan testing and microscopy.
*Carl Skelley, Lab Director-Austin*
*Chip Meador, Regional General Manager*

**5170 TestAmerica-Buffalo**
10 Hazelwood Drive 716-691-2600
Amherst, NY 14228 Fax: 716-691-7991
E-mail: info@stl-inc.com
http://http://stl-inc.com/contacts/lab_list.htm
Environmental engineering research and consulting firm. Testing capabilities include chemical, physical and biological analyses of a variety of matrices, including aqueous, solid, drinking water, waste, tissue, air and saline/estuarinesamples. Specialty capabilities include air toxics testing, mixed waste testing, tissue preparation and analysis, aquatic toxicology, dioxin/furan testing and microscopy.
*Chris Spencer, Lab Director-Buffalo*
*Chris Oprandi, Regional General Manager*

**5171 TestAmerica-Burlington**
208 South Park Drive 802-655-1203
Suite 1 Fax: 802-655-1248
Colchester, VT 05446 E-mail: info@stl-inc.com
http://http://stl-inc.com/contacts/lab_list.htm
Environmental engineering research and consulting firm. Testing capabilities include chemical, physical and biological analyses of a variety of matrices, including aqueous, solid, drinking water, waste, tissue, air and saline/estuarinesamples. Specialty capabilities include air toxics testing, mixed waste testing, tissue preparation and analysis, aquatic toxicology, dioxin/furan testing and microscopy.
*William Cicero, Lab Director-Burlington*
*Scott Morris, Regional General Manager*

**5172 TestAmerica-Chicago**
2417 Bond Street 708-534-5200
University Park, IL 60466 Fax: 708-534-5211
E-mail: info@stl-inc.com
http://http://stl-inc.com/contacts/lab_list.htm
Environmental engineering research and consulting firm. Testing capabilities include chemical, physical and biological analyses of a variety of matrices, including aqueous, solid, drinking water, waste, tissue, air and saline/estuarinesamples. Specialty capabilities include air toxics testing, mixed waste testing, tissue preparation and analysis, aquatic toxicology, dioxin/furan testing and microscopy.
*Michael Healy, Lab Director-Chicago*
*Chris Oprandi, Regional General Manager*

**5173 TestAmerica-Connecticut**
128 Long Hill Cross Road 203-929-8140
Shelton, CT 06484 Fax: 203-929-8142
E-mail: info@stl-inc.com
http://http://stl-inc.com/contacts/lab_list.htm
Environmental engineering research and consulting firm. Testing capabilities include chemical, physical and biological analyses of a variety of matrices, including aqueous, solid, drinking water, waste, tissue, air and saline/estuarinesamples. Specialty capabilities include air toxics testing, mixed waste testing, tissue preparation and analysis, aquatic toxicology, dioxin/furan testing and microscopy.
*Peter Frick, Lab Director-Connecticut*
*Scott Morris, Regional General Manager*

**5174 TestAmerica-Corpus Christi**
1733 N Padre Island Drive 361-289-2673
Corpus Christi, TX 78408 Fax: 361-289-2471
E-mail: info@stl-inc.com
http://http://stl-inc.com/contacts/lab_list.htm
Environmental engineering research and consulting firm. Testing capabilities include chemical, physical and biological analyses of a variety of matrices, including aqueous, solid, drinking water, waste, tissue, air and saline/estuarinesamples. Specialty capabilities include air toxics testing, mixed waste testing, tissue prepa-

ration and analysis, aquatic toxicology, dioxin/furan testing and microscopy.
*Amelia Kennedy, Lab Director-Corpus Christi*
*Chip Meador, Regional General Manager*

**5175 TestAmerica-Denver**
4955 Yarrow Street 303-736-0100
Arvada, CO 80002 Fax: 303-431-7171
E-mail: info@stl-inc.com
http://http://stl-inc.com/contacts/lab_list.htm
Environmental engineering research and consulting firm. Testing capabilities include chemical, physical and biological analyses of a variety of matrices, including aqueous, solid, drinking water, waste, tissue, air and saline/estuarinesamples. Specialty capabilities include air toxics testing, mixed waste testing, tissue preparation and analysis, aquatic toxicology, dioxin/furan testing and microscopy.
*Robert Hanisch, Lab Director-Denver*
*Chip Meador, Regional General Manager*

**5176 TestAmerica-Edison**
777 New Durham Road 732-549-3900
Edison, NJ 08817 Fax: 732-549-3679
E-mail: info@stl-inc.com
http://http://stl-inc.com/contacts/lab_list.htm
Environmental engineering research and consulting firm. Testing capabilities include chemical, physical and biological analyses of a variety of matrices, including aqueous, solid, drinking water, waste, tissue, air and saline/estuarinesamples. Specialty capabilities include air toxics testing, mixed waste testing, tissue preparation and analysis, aquatic toxicology, dioxin/furan testing and microscopy.
*Ann Gladwell, Lab Director-Edison*
*Tim O'Shields, Regional General Manager*

**5177 TestAmerica-Houston**
6310 Rothway Street 713-690-4444
Houston, TX 77040 Fax: 713-690-5646
E-mail: info@stl-inc.com
http://http://stl-inc.com/contacts/lab_list.htm
Environmental engineering research and consulting firm. Testing capabilities include chemical, physical and biological analyses of a variety of matrices, including aqueous, solid, drinking water, waste, tissue, air and saline/estuarinesamples. Specialty capabilities include air toxics testing, mixed waste testing, tissue preparation and analysis, aquatic toxicology, dioxin/furan testing and microscopy.
*Norman Flynn, Lab Director-Houston*
*Chip Meador, Regional General Manager*

**5178 TestAmerica-Knoxville**
5815 Middlebrook Pike 865-291-3008
Knoxville, TN 37921 Fax: 865-584-4315
E-mail: info@stl-inc.com
http://http://stl-inc.com/contacts/lab_list.htm
Environmental engineering research and consulting firm. Testing capabilities include chemical, physical and biological analyses of a variety of matrices, including aqueous, solid, drinking water, waste, tissue, air and saline/estuarinesamples. Specialty capabilities include air toxics testing, mixed waste testing, tissue preparation and analysis, aquatic toxicology, dioxin/furan testing and microscopy.
*J Thomas Yoder, Lab Director-Knoxville*
*Rusty Vicinie, Regional General Manager*

**5179 TestAmerica-Los Angeles**
1721 South Grand Avenue 714-258-8610
Santa Ana, CA 92705 Fax: 714-258-0921
E-mail: info@stl-inc.com
http://http://stl-inc.com/contacts/lab_list.htm
Environmental engineering research and consulting firm. Testing capabilities include chemical, physical and biological analyses of a variety of matrices, including aqueous, solid, drinking water, waste, tissue, air and saline/estuarinesamples. Specialty capabilities include air toxics testing, mixed waste testing, tissue preparation and analysis, aquatic toxicology, dioxin/furan testing and microscopy.
*Elizabeth Winger, Lab Director-Los Angeles*
*Mark Weiner, Regional General Manager*

**5180 TestAmerica-Mobile**
900 Lakeside Drive 251-666-6633
Mobile, AZ 36693 Fax: 251-666-6696
E-mail: info@stl-inc.com
http://http://stl-inc.com/contacts/lab_list.htm
Environmental engineering research and consulting firm. Testing capabilities include chemical, physical and biological analyses of a variety of matrices, including aqueous, solid, drinking water, waste, tissue, air and saline/estuarinesamples. Specialty capabilities include air toxics testing, mixed waste testing, tissue preparation and analysis, aquatic toxicology, dioxin/furan testing and microscopy.
*Jesse Smith, Lab Director-Mobile*
*Jack Tuschall PhD, Regional General Manager*

**5181 TestAmerica-New Orleans**
2501 Lexington Avenue 504-469-3685
Kenner, LA 60466 Fax: 504-469-0140
E-mail: info@stl-inc.com
http://http://stl-inc.com/contacts/lab_list.htm
Environmental engineering research and consulting firm. Testing capabilities include chemical, physical and biological analyses of a variety of matrices, including aqueous, solid, drinking water, waste, tissue, air and saline/estuarinesamples. Specialty capabilities include air toxics testing, mixed waste testing, tissue preparation and analysis, aquatic toxicology, dioxin/furan testing and microscopy.
*Michael Antoine, Lab Director-New Orleans*
*Jack Tuschall PhD, Regional General Manager*

**5182 TestAmerica-North Canton**
4101 Shuffel Drive NW 330-497-7058
North Canton, OH 44720 Fax: 330-497-0772
E-mail: info@stl-inc.com
http://http://stl-inc.com/contacts/lab_list.htm
Environmental engineering research and consulting firm. Testing capabilities include chemical, physical and biological analyses of a variety of matrices, including aqueous, solid, drinking water, waste, tissue, air and saline/estuarinesamples. Specialty capabilities include air toxics testing, mixed waste testing, tissue preparation and analysis, aquatic toxicology, dioxin/furan testing and microscopy.
*Opal Davis Johnson, Lab Director-North Canton*
*Rusty Vicinie, Regional General Manager*

**5183 TestAmerica-Orlando**
3400 S Conway Road 407-423-0014
Orlando, FL 32812 Fax: 407-856-0886
E-mail: info@stl-inc.com
http://http://stl-inc.com/contacts/lab_list.htm
Environmental engineering research and consulting firm. Testing capabilities include chemical, physical and biological analyses of a variety of matrices, including aqueous, solid, drinking water, waste, tissue, air and saline/estuarinesamples. Specialty capabilities include air toxics testing, mixed waste testing, tissue preparation and analysis, aquatic toxicology, dioxin/furan testing and microscopy.
*Keith Blanchard, Lab Director-Orlando*
*Scott Morris, Regional General Manager*

**5184 TestAmerica-Pensacola**
3355 McLemore Drive 850-474-1001
Pensacola, FL 32514 Fax: 850-478-2671
E-mail: info@stl-inc.com
http://http://stl-inc.com/contacts/lab_list.htm
Environmental engineering research and consulting firm. Testing capabilities include chemical, physical and biological analyses of a variety of matrices, including aqueous, solid, drinking water, waste, tissue, air and saline/estuarinesamples. Specialty capabilities include air toxics testing, mixed waste testing, tissue prepa-

ration and analysis, aquatic toxicology, dioxin/furan testing and microscopy.
*Susan Rembert, Lab Director-Pensacola*
*Jack Tuschall PhD, Regional General Manager*

**5185 TestAmerica-Phoenix / Aerotech Environmental Laboratories**
4645 East Cotton Center Blvd 602-437-3340
Building 3, Suite 189 866-772-5227
Phoenix, AZ 85040 Fax: 623-445-6192
E-mail: info@stl-inc.com
http://http://stl-inc.com/contacts/lab_list.htm
Environmental engineering research and consulting firm. Testing capabilities include chemical, physical and biological analyses of a variety of matrices, including aqueous, solid, drinking water, waste, tissue, air and saline/estuarinesamples. Specialty capabilities include air toxics testing, mixed waste testing, tissue preparation and analysis, aquatic toxicology, dioxin/furan testing and microscopy.
*Elizabeth Wueschner, Lab Director-Phoenix*
*Mark Weiner, Regional General Manager*

**5186 TestAmerica-Pittsburgh**
301 Alpha Drive 412-963-7058
RIDC Park Fax: 412-963-2468
Pittsburgh, PA 15238 E-mail: info@stl-inc.com
http://http://stl-inc.com/contacts/lab_list.htm
Environmental engineering research and consulting firm. Testing capabilities include chemical, physical and biological analyses of a variety of matrices, including aqueous, solid, drinking water, waste, tissue, air and saline/estuarinesamples. Specialty capabilities include air toxics testing, mixed waste testing, tissue preparation and analysis, aquatic toxicology, dioxin/furan testing and microscopy.
*Larry Matko, Lab Director-Pittsburgh*
*Rusty Vicinie, Regional General Manager*

**5187 TestAmerica-Richland**
2800 George Washington Way 509-375-3131
Richland, WA 99354 Fax: 509-375-5590
E-mail: info@stl-inc.com
http://http://stl-inc.com/contacts/lab_list.htm
Environmental engineering research and consulting firm. Testing capabilities include chemical, physical and biological analyses of a variety of matrices, including aqueous, solid, drinking water, waste, tissue, air and saline/estuarinesamples. Specialty capabilities include air toxics testing, mixed waste testing, tissue preparation and analysis, aquatic toxicology, dioxin/furan testing and microscopy.
*Gregory Jungclaus PhD, Lab Director-Richland*
*Chip Meador, Regional General Manager*

**5188 TestAmerica-San Francisco**
1220 Quarry Lane 925-484-1919
Pleasanton, CA 94566 Fax: 925-484-1096
E-mail: info@stl-inc.com
http://http://stl-inc.com/contacts/lab_list.htm
Environmental engineering research and consulting firm. Testing capabilities include chemical, physical and biological analyses of a variety of matrices, including aqueous, solid, drinking water, waste, tissue, air and saline/estuarinesamples. Specialty capabilities include air toxics testing, mixed waste testing, tissue preparation and analysis, aquatic toxicology, dioxin/furan testing and microscopy.
*Peter Moretan, Lab Director-San Francisco*
*Roger Freize, Regional General Manager*

**5189 TestAmerica-Savannah**
5102 LaRoche Avenue 912-354-7858
Savannah, GA 31404 Fax: 912-351-3673
E-mail: info@stl-inc.com
http://http://stl-inc.com/contacts/lab_list.htm
Environmental engineering research and consulting firm. Testing capabilities include chemical, physical and biological analyses of a variety of matrices, including aqueous, solid, drinking water, waste, tissue, air and saline/estuarinesamples. Specialty capabilities include air toxics testing, mixed waste testing, tissue preparation and analysis, aquatic toxicology, dioxin/furan testing and microscopy.
*Benjamin Gulizia, Lab Director-Savannah*
*Scott Morris, Regional General Manager*

**5190 TestAmerica-St Louis**
13715 Rider Trail North 314-298-8566
Earth City, MO 63045 Fax: 314-298-8757
E-mail: info@stl-inc.com
http://http://stl-inc.com/contacts/lab_list.htm
Environmental engineering research and consulting firm. Testing capabilities include chemical, physical and biological analyses of a variety of matrices, including aqueous, solid, drinking water, waste, tissue, air and saline/estuarinesamples. Specialty capabilities include air toxics testing, mixed waste testing, tissue preparation and analysis, aquatic toxicology, dioxin/furan testing and microscopy.
*Marty Cahill, Lab Director-St Louis*
*Chip Meador, Regional General Manager*

**5191 TestAmerica-Tacoma**
5755 8th Street E 253-922-2310
Tacoma, WA 98424 Fax: 253-922-5047
E-mail: info@stl-inc.com
http://http://stl-inc.com/contacts/lab_list.htm
Environmental engineering research and consulting firm. Testing capabilities include chemical, physical and biological analyses of a variety of matrices, including aqueous, solid, drinking water, waste, tissue, air and saline/estuarinesamples. Specialty capabilities include air toxics testing, mixed waste testing, tissue preparation and analysis, aquatic toxicology, dioxin/furan testing and microscopy.
*Kathy Kreps, Lab Director-Tacoma*
*Roger Freize, Regional General Manager*

**5192 TestAmerica-Tallahassee**
2846 Industrial Plaza Drive 850-878-3994
Suite 100 Fax: 850-878-9504
Tallahassee, FL 32301 E-mail: info@stl-inc.com
http://http://stl-inc.com/contacts/lab_list.htm
Environmental engineering research and consulting firm. Testing capabilities include chemical, physical and biological analyses of a variety of matrices, including aqueous, solid, drinking water, waste, tissue, air and saline/estuarinesamples. Specialty capabilities include air toxics testing, mixed waste testing, tissue preparation and analysis, aquatic toxicology, dioxin/furan testing and microscopy.
*Todd Baumgartner, Lab Director-Tallahassee*
*Scott Morris, Regional General Manager*

**5193 TestAmerica-Tampa**
6712 Benjamin Road 813-885-7427
Suite 100 Fax: 813-885-7049
Tampa, FL 33634 E-mail: info@stl-inc.com
http://http://stl-inc.com/contacts/lab_list.htm
Environmental engineering research and consulting firm. Testing capabilities include chemical, physical and biological analyses of a variety of matrices, including aqueous, solid, drinking water, waste, tissue, air and saline/estuarinesamples. Specialty capabilities include air toxics testing, mixed waste testing, tissue preparation and analysis, aquatic toxicology, dioxin/furan testing and microscopy.
*Shibu Paul PhD/MBA, Lab Director-Tampa*
*Scott Morris, Regional General Manager*

**5194 TestAmerica-Valparaiso**
2400 Cumberland Drive 219-464-2389
Valparaiso, IN 46383 800-688-6522
Fax: 219-462-2953
E-mail: info@stl-inc.com
http://www.stl-inc.com/labs/Valpariaso/Valparaiso_index.htm
Environmental engineering research and consulting firm. Testing capabilities include chemical, physical and biological analyses of a variety of matrices, including aqueous, solid, drinking water, waste, tissue, air and saline/estuarinesamples. Specialty capabilities include air toxics testing, mixed waste testing, tissue prepa-

ration and analysis, aquatic toxicology, dioxin/furan testing and microscopy.
*Kurt Ill, Lab Director-Valparaiso*
*Chris Oprandi, Regional General Manager*

**5195 TestAmerica-West Sacramento**
880 Riverside Parkway 916-373-5600
West Sacramento, CA 95605 Fax: 916-373-5600
E-mail: info@stl-inc.com
http://http://stl-inc.com/contacts/lab_list.htm
Environmental engineering research and consulting firm. Testing capabilities include chemical, physical and biological analyses of a variety of matrices, including aqueous, solid, drinking water, waste, tissue, air and saline/estuarinesamples. Specialty capabilities include air toxics testing, mixed waste testing, tissue preparation and analysis, aquatic toxicology, dioxin/furan testing and microscopy.
*Karia Bluecher, Lab Director-W Sacramento*
*Roger Freize, Regional General Manager*

**5196 TestAmerica-Westfield**
53 Southampton Road 413-572-4000
Westfield, MA 01085 Fax: 413-572-3707
E-mail: info@stl-inc.com
http://http://stl-inc.com/contacts/lab_list.htm
Environmental engineering research and consulting firm. Testing capabilities include chemical, physical and biological analyses of a variety of matrices, including aqueous, solid, drinking water, waste, tissue, air and saline/estuarinesamples. Specialty capabilities include air toxics testing, mixed waste testing, tissue preparation and analysis, aquatic toxicology, dioxin/furan testing and microscopy.
*Steven Hartmann, Lab Director-Westfield*
*Scott Morris, Regional General Manager*

**5197 Testing & Inspection Services, Inc.**
Thornton Laboratories
1145 E Cass Street 813-223-9702
PO Box 2880 Fax: 813-223-9332
Tampa, FL 33601 E-mail: steve.fickett@thorntonlab.com
http://www.thorntonlab.com
Environmental & fertilizer sampling and testing laboratory and general analytical testing lab.

*Steve Fickett, Project Manager*
*Hugh Rodriques, Lab Manager*

**5198 Testing Engineers & Consultants (TEC) - AnnArbor**
3985 Varsity Drive 734-971-0030
Ann Arbor, MI 48108 Fax: 734-971-3721
E-mail: tec@tectest.com OR egalczynski@tectest.com
http://www.tectest.com
TEC specializes in environmental and geotechnical engineering, materials testing, roof systems management, facility asset management, and indoor air quality. Environmental services include: baseline assessments; contaminationassessments; expert testimony; feasibility studies; hazardous materials surveys; and hyrogeological/groundwater investigations.

*Donald Kaylor, Environmental Assessment Mgr*
*Edward Galczynski, Ann Arbor Office Manager*

**5199 Testing Engineers & Consultants (TEC) - Detroit**
601 W Fort Street 313-837-8464
Suite 440 800-835-2654
Detroit, MI 48226 Fax: 313-837-1305
E-mail: tec@tectest.com
http://www.testingengineers.com
TEC specializes in environmental and geotechnical engineering, materials testing, roof systems management, facility asset management, and indoor air quality. Environmental services include: baseline assessments; contaminationassessments; expert testimony; feasibility studies; hazardous materials surveys; and hyrogeological/groundwater investigations.

*Donald Kaylor, Environmental Assessment Mgr*
*Nanette Rose, Senior Marketing Specialist*

**5200 Testing Engineers & Consultants (TEC) - Troy**
1343 Rochester Road 248-588-6200
Troy, MI 48083 Fax: 248-588-6232
E-mail: tec@tectest.com
http://www.tectest.com
TEC specializes in environmental and geotechnical engineering, materials testing, roof systems management, facility asset management, and indoor air quality. Environmental services include: baseline assessments; contaminationassessments; expert testimony; feasibility studies; hazardous materials surveys; and hyrogeological/groundwater investigations.

*Donald Kaylor, Environmental Assessment Mgr*
*Carey J Suhan PE, VP/Mgr Geotechnical/Enviro*

**5201 Tetra Tech - Christiana DE**
240 Continental Drive 302-738-7551
Suite 200 800-462-0910
Newark, DE 19713 Fax: 302-454-5980
http://www.tetratech.com
Tetra Tech is a provider of specialized management consulting and technical services in resource management, infrastructure and communication. The company's clients include a diverse base of public and private sector organizationsserviced through more than 330 offices located in the US and internationally. Tetra Tech's services include research and development, applied science and technology, engineering design, program management, construction management, and operations andmaintenance.

*Harish Mital, Delaware Operations Manager*
*Dan L Batrack, CEO/COO*

**5202 Tetra Tech - Pasadena CA/Corporate**
3475 E Foothill Boulevard 626-351-4664
Pasadena, CA 91107 Fax: 626-351-5291
http://www.tetratech.com
Tetra Tech is a provider of specialized management consulting and technical services in resource management, infrastructure and communication. The company's clients include a diverse base of public and private sector organizationsserviced through more than 200 offices located in the US and internationally. Tetra Tech's services include research and development, applied science and technology, engineering design, program management, construction management, and operations andmaintenance.

*Sam W Box, President*
*Dan L Batrack, CEO/COO*

**5203 Thermo Fisher Scientific**
81 Wyman Street 781-622-1000
Waltham, MA 02254 800-678-5599
Fax: 781-622-1207
E-mail: lori.gorski@thermo.com
http://www.thermofisher.com
Provides a wide range of products, services and solutions for research, analysis, discovery and diagnostics using advanced technologies ranging from mass spectrometry and elemental analysis to chromatography, molecular spectroscopy,and microanalysis. Additional services includes automated systems and technologies from standalone robots to complete liquid handling systems.

*Marijin Dekkers, President/CEO*
*Lori Gorski, Media Relations*

**5204 ThermoEnergy Corporation**
323 Center Street 501-376-6477
Suite 1300, Tower Building Fax: 501-244-9203
Little Rock, AR 72201 E-mail: technology@thermoenergy.com
http://www.thermoenergy.com
ThermoEnergy Corporation is an integrated technologies company seeking to develop and commercialize patented water treatment and clean energy technologies. Products and services solutions include removing nitrogen from wastewaterstreams, converting sewage sludge to a renewable high-energy fuel, and enabling the conversion of coal and other hydrocarbon fuels into energy with zero air emissions.
*Dennis C Cossey, Chairman/CEO*
*Andrew T Melton, EVP/Chief Financial Officer*

**5205 Thermotron Industries**
836 Brooks Avenue
Holland, MI 49423
616-392-1491
Fax: 616-392-5643
E-mail: info@thermotron.com
http://www.thermotron.com
Manufacturers and suppliers of environmental testing, test system integration, screening, simulation equipment, and vibration equipment for transportation and screening test requirements. Additional services includes integrated testingsolutions with Research & Development and Total Quality Control to help insure product reliability and performance.
*Mark Lamers, Technical Manager*
*Kevin Ewing, Marketing/Sales Manager*

**5206 Thompson Engineering**
2970 Cottage Hill Road
Suite 190
Mobile, AL 36606
251-666-2443
Fax: 251-666-6422
E-mail: info@thompsonengineering.com
http://www.thompsonengineering.com
A multi disciplined engineering design, environmental consulting, construction management, construction inspection and materials testing firm. The Environmental Division is comprised of a diverse team of professionals with significantknowledge and experience in environmental compliance and permitting, audits and assessments, engineering design, and monitoring and supervision of remedial activities.

*John H Baker III, President*
*James H Shumock CPA, CEO/Chief Financial Officer*

**5207 Tighe & Bond**
53 Southampton Road
Suite 3
Westfield, MA 01085
413-562-1600
Fax: 413-562-5317
E-mail: info@tighebond.com
http://www.tighebond.com
Tighe & Bond provides engineering and consulting services to a wide variety of clients, from some of the largest municipalities in the country to small, privately-held businesses. Areas of expertise includes water supply, wastewatermanagement, buildings, roadways, environmental permitting, remediation, health and safety training.
*David Pinsky, President*
*Jeffrey P Bibeau, Environmental Manager*

**5208 Timber Products Inspection - Conyers**
1641 Sigman Road
Conyers, GA 30012
770-922-8000
Fax: 770-922-1290
E-mail: info@tpinspection.com
http://www.tpinspection.com
Timber Products Inspection, Inc. (TP) is an independent inspection, testing and consulting company with expertise in all phases of the wood products industry. TP provides quality auditing services in the areas of sawmilling, drying,component fabrication as well as value added processes such as pressure treating and gluing.
*Patrick Edwards, Engineering Director*
*Greg Pittman, Analytical Laboratory Mgr*

**5209 Timber Products Inspection - Vancouver**
105 SE 124th Avenue
Vancouver, WA 98684
360-449-3840
Fax: 360-449-3953
E-mail: info@tpinspection.com
http://www.tpinspection.com
Timber Products Inspection, Inc. (TP) is an independent inspection, testing and consulting company with expertise in all phases of the wood products industry. TP provides quality auditing services in the areas of sawmilling, drying,component fabrication as well as value added processes such as pressure treating and gluing.

*Casey Dean, Director*
*Cliff Eddington, Utility Products*

**5210 Tox Scan**
42 Hangar Way
Watsonville, CA 95076
831-724-4522
Fax: 831-761-5449
E-mail: dlewis@toxscan.com
http://www.toxscan.com
Environmental bioassay and bioacoumulation testing.

*David B. Lewis, Director*

**5211 Transviron**
1624 York Road
Lutherville, MD 21093
410-321-6961
Fax: 410-494-9321
E-mail: Transviron@comcast.net
Civil and environmental engineering firm. Technical consulting services includes: water supply and distribution; storm water management; highways and bridges; hazardous waste management; water and wastewater treatment plant operationsand construction management; wastewater collection and treatment.
*Charles Bao, President*

**5212 Tri-State Laboratories**
2870 Salt Springs Road
Youngstown, OH 44509
330-797-8844
800-523-0347
Fax: 330-797-3264
E-mail: trislabs@aol.com
http://www.tristatelabs.net
Environmental testing laboratory services of which include: asbestos testing; field services; forensic analysis and court testimony; hazardous waste analysis; inorganics/wet chemistry, metals, and organic analysis.

*A Bari Lateef PhD, CEO*
*Wendy Hanna, COO*

**5213 Turner Laboratories**
2445 North Coyote Drive
Suite 104
Tucson, AZ 85745
520-882-5880
Fax: 520-882-9788
E-mail: nturner@turnerlabs.com
http://www.turnerlabs.com
Turner Laboratories is an advanced, full-service environmental testing laboratory specializing in providing a wide range of analytical services including: inorganic, organic, wet chemistry and microbiological testing on soils/solids,drinking water, wastewater and groundwater.

*Nancy D Turner, President*
*Terri Garcia, Technical Director*

**5214 URS**
600 Montgomery Street
26th Floor
San Francisco, CA 94111
415-774-2700
Fax: 415-398-1905
E-mail: media_contact@urscorp.com
http://www.urscorp.com/About_URS/index.php
An environmental analysis and comprehensive engineering service firm, URS provides a full range of planning, design, program and construction management services to a wide variety of private and public sector clients. URS hasapproximately 30,000 employees in a network of more than 370 offices and contract-specific job sites in 20 countries.
*Martin M Koffel, Chairman/CEO*
*Thomas W Biship, VP/Strategic Development*

**5215 US Public Interest Research Group**
44 Winter Street
4th Floor
Boston, MA 02108
617-747-4370
Fax: 617-292-8057
E-mail: info@uspirg.org
http://www.uspirg.org
US PIRG is an advocate for the public interest. We uncover threats to public health and well-being and fight to end them, using the time-tested tools of investigative research, media exposes, grassroots organizing, advocacy andlitigation.
*Andre Delattre, Executive Director*
*Phineas , Senior Policy Analyst*

**5216 US Public Interest Research Group - Washington**
218 D Street SE 202-546-9707
Washington, DC 20003 Fax: 202-546-2461
E-mail: uspirg@pirg.org
http://www.pirg.org
US PIRG is an advocate for the public interest. We uncover threats to public health and well-being and fight to end them, using the time-tested tools of investigative research, media exposes, grassroots organizing, advocacy andlitigation.
*Douglas H Phelps, President/Chairman*
*Anne Aurillo, Legislative Director*

**5217 USDA Forest Service: Pacific Southwest Research Station**
800 Buchanan Street 510-559-6300
West Annex Building Fax: 510-559-6440
Albany, CA 94710 E-mail: psw_webmaster@fs.fed.us
http://www.fs.fed.us/psw/
A Governmental Research Organization specializing in research on forest ecosystems, including fire, watersheds, forest genetics and diversity, wildlife, forest diseases, and urban forestry.
*Jim Baldwin, Project Manager*
*Marilyn Hartley, Communications Director*

**5218 Umpqua Research Company**
125 Volunteer Way 541-863-7770
PO Box 609 Fax: 541-863-7775
Myrtle Creek, OR 97457 E-mail: info@urcmail.net
http://http://www.urc.cc/ OR http://umpquatesting.com/
UMPQUA Research Company (URC), founded in 1973 by David F. Putnam and Gerald V. Colombo, offers technical services in four primary areas: Drinking Water and Environmental Analysis; Air and Water Purification Related EngineeringServices (including NASA Flight Hardware); Research & Development; and Materials Testing. The staff includes chemical, electrical, and mechanical engineers, chemists, physicists, and biological scientists.
*William F Michalek PE, Manager*
*Vern McDonald, Lead Technician*

**5219 United Environmental Services**
2025 Delsea Drive 856-227-5477
Sewell, NJ 08080 Fax: 856-227-6578
E-mail: Mamrakv ues@aol.com
http://www.transtechindustries.com/ues.html
An environmental testing and analysis firm providing a full range of construction, remedial and maintenance services at landfills, commercial and industrial sites, including Brownfield re-development projects.
*John A Mamrak Jr, General Manager*

**5220 Universal Environmental Technologies**
87 Technology Way 603-883-9312
Nashua, NH 03060 Fax: 603-883-9314
E-mail: info@uetenvirosys.com
http://www.uetenvirosys.com/
An environmental research firm, Universal Environmental Technologies specializes in the design, fabrication and installation of integrated groundwater and soil remediation systems that are used on retail petroleum sites, industrialmanufacturing sites, EPA Superfund sites and U.S. military bases.
*Sharon McMillin, VP/Remedial Services*

**5221 Upstate Laboratories**
6034 Corporate Drive 315-437-0255
East Syracuse, NY 13057 Fax: 315-437-1209
E-mail: AScala@Upstatelabs.com
http://www.upstatelabs.com/
Testing laboratory specializing in environmental and organic/synthetic analysis. Services include certification and air quality (mycology).

*Anthony J Scala, President/CEO/Chemist*
*Corey Niland, Quality Assurance & Control*

**5222 Vara International: Division of Calgon Corporation**
1201 19th Place 407-567-1320
Suite 400 Fax: 407-567-4108
Vero Beach, FL 32960 E-mail: customerrelations@calgoncarbon.com
http://www.calgoncarbon.com
An environmental and industrial process research firm, Vara International is a global manufacturer and supplier of granular activated carbon, innovative treatment systems, value added technologies and services for optimizing productionprocesses and safely purifying the environment.
*John S Stanik, Chairman/President/CEO*
*Gail A Gerono, VP/Communications*

**5223 Versar**
6850 Versar Center 703-750-3000
PO Box 1549 800-283-7727
Springfield, VA 22151 Fax: 703-642-6807
E-mail: info@versar.com
http://www.versar.com
Engineering and environmental research organization. Research results published in project reports, government publications, books, articles, and technical reports.
*Theodore Prosciv PhD, President/CEO*
*Lawrence W Sinnot, EVP/Chief Operating Officer*

**5224 Vista Leak Detection**
755 N Mary Avenue 408-830-3300
Sunnyvale, CA 94085 Fax: 408-830-3399
E-mail: info@VistaLD.com
http://www.vistaleakdetection.com
Vista Research provides leak detection products and services to airport, oil industry and military clients for ensuring the integrity of underground/aboveground pipeline and tank systems.
*William W Pickett, VP/Operations*
*Cody Freeman, Contracts Administrator*

**5225 Volumetric Techniques, Ltd. / VTEQE**
317 Bernice Drive 631-472-4848
Bayport, NY 11705 Fax: 631-472-4991
E-mail: vteqe@msn.com
http://http://vteqeltd.com/137.html
Full service environmental engineering organization, VTEQE specializes in all aspects of the environmental services industry, including assessments (ESA Phase 1,2), site engineering, investigations/reports, remediation/cleanupstrategies, Phase 3, and bottled water facility licensing dealing with contaminated water, groundwater, soil, also engineering design, construction management, and full revitalization management.

*Sander Sternig, President/CEO/Chairman*
*Benito San Pedro, Professional Engineer*

**5226 WERC: Consortium for Environmental Education & Technology Development**
New Mexico State University
1080 Frenger Mall 505-646-2038
EC III, 3rd floor, Suite 300 S 800-523-5996
Las Cruces, NM 88003 Fax: 505-646-5474
E-mail: werc@nmsu.edu
http://www.werc.net/about/index.htm
A consortium focusing on environmental education and technology development. The consortium's mission is to develop the human resources and technologies needed to address environmental issues. WERC's program aims to achieveenvironmental excellence through education, public outreach and technology development and deployment.
*Abbas Ghassemi PhD, Executive Director*
*Jack B Tillman, Associate Director*

**5227 Waid & Associates - Austin/Corporate**
14205 N Mopac Expressway 512-255-9999
Suite 600 Fax: 512-255-8780
Austin, TX 78728 E-mail: information@waid.com
http://www.waid.com
Waid & Associates is an engineering and environmental services firm that specializes in air quality services, particularly emissions control, permits, and compliance. Additional services in-

clude wastewater/waste management andenvironmental information management systems.
*Jay R Hoover PE, President/Principal Engineer*
*Glen H Colby III, Senior Systems Analyst*

**5228 Waid & Associates - Houston**
1120 NASA Road 1 281-333-9990
Suite 630 Fax: 281-333-9992
Houston, TX 77058 E-mail: information@waid.com
http://www.waid.com
Waid & Associates is an engineering and environmental services firm that specializes in air quality services, particularly emissions control, permits, and compliance. Additional services include wastewater/waste management andenvironmental information management systems.
*Jay R Hoover PE, President/Principal Engineer*
*Glen H Colby III, Senior Systems Analyst*

**5229 Waid & Associates - Permian Basin**
3000 N Garfield Street 432-682-9999
Suite 212 Fax: 432-682-7774
Midland, TX 79705 E-mail: information@waid.com
http://www.waid.com
Waid & Associates is an engineering and environmental services firm that specializes in air quality services, particularly emissions control, permits, and compliance. Additional services include wastewater/waste management andenvironmental information management systems.
*Jay R Hoover PE, President/Principal Engineer*
*Glen H Colby III, Senior Systems Analyst*

**5230 Waste Water Engineers**
5751 Old Hickory Boulevard 615-883-7100
Suite 207 Fax: 615-889-6101
Hermitage, TN 37076
Environmental science research consultant. Research results published in project reports. Environmental civil engineering consultant.
*Thomas H Patton Jr, President*

**5231 Water and Air Research**
6821 SW Archer Road 352-372-1500
Gainesville, FL 32608 Fax: 352-378-1500
E-mail: info@waterandair.com
http://www.waterandair.com
Environmental research and consulting firm. Research results published in client reports.
*William C Zegel, President*
*Connie Bieber, Director/Manager*

**5232 Watkins Environmental Sciences**
7000 E Genesee Street 315-446-4763
Suite A7 Fax: 315-446-4764
Fayetteville, NY 13066
Environmental assessments, septic system designs, residential water sampling and testing services, home inspections, radon, foundation designs & inspections.

*Andrew A Watkins PE, President*

**5233 Weather Services Corporation**
131A Great Road 781-275-8860
Bedford, MA 01730 Fax: 781-271-0178
*Michael Leavitt, President*

**5234 West Coast Analytical Service**
9840 Alburtis Avenue 562-948-2225
Santa Fe Springs, CA 90670 Fax: 562-948-5850
*DJ Northington, PhD, President*

**5235 West Michigan Testing**
815 E Ludington Avenue 231-843-3353
Ludington, MI 49431 Fax: 231-843-7676
We provide soil borings, geotechnical services, environmental assessments, construction materials testing and asbestos inspection.

*James T Nordlund Jr, Vice President*

**5236 West More Mechanical Testing and Research**
PO Box 388 724-537-8686
Youngstown, PA 15696 Fax: 724-537-3151
*James Dague, Laboratory Manager*

**5237 Western Environmental Services**
913 N Foster Road 307-234-5511
Casper, WY 82601 800-545-5711
Fax: 307-234-8324
E-mail: aroylance@testair.com
http://www.testair.com
Air emission testing.

*Alan Roylance, President*
*J Scott Mortimer, Vice President*

**5238 Western Michigan Environmental Services**
3552 128th Avenue 616-399-6070
Holland, MI 49424 Fax: 616-399-6185

*Cheryl A Dell, President*

**5239 Westinghouse Electric Company**
11 Stanwix Street 412-244-2000
Pittsburgh, PA 15222 Fax: 412-642-4985

**5240 Westinghouse Remediation Services**
675 Park N Boulevard 404-298-7101
Suite F-100 Fax: 404-296-9752
Clarkston, GA 30021
Environmental remediation firm.

**5241 Weston Solutions, Inc**
1400 Weston Way 610-701-3000
Box 2653 800-7WE-STON
West Chester, PA 19380 Fax: 610-701-3124
E-mail: info@westonsolutions.com
http://www.westonsolutions.com
Weston is a leading infrastructure redevelopment services firm delivering integrated environmental engineering solutions to industry and government worldwide. With an emphasis on creating lasting economic value for its clients, thecompany provides services in site remediation, redevelopment, infrastructure operations and knowledge management.
*Patrick McCann, President/Chief Executive*
*William Robertson, Chairman*

**5242 Whibco**
River Road 856-455-9200
PO Box 259 Fax: 856-455-8884
Leesburg, NJ 08327 http://www.whibco.com
*Andrew R Strelczyk, Director Quality Control*

**5243 Wik Associates**
PO Box 230 302-322-2558
New Castle, DE 19720 Fax: 302-322-8921
Environmental testing and analysis firm.

**5244 William T Lorenz & Company**
3541 Norwegian Hollow Road 608-935-9285
Dodgeville, WI 53533 Fax: 608-935-2010
Environmental and water resources marketing, consulting, and product research firm.

**5245 William W Walker Jr**
1127 Lowell Road 978-369-8061
Concord, MA 01742 Fax: 978-369-4230
E-mail: wwwalker@wwwalker.net
http://www.wwwalker.net
*William W Walker, Jr, Environmental Engineer*

**5246 Woods End Research Laboratory**
290 Belgrade Rd 207-293-2457
PO Box 297 Fax: 207-293-2488
Mount Vernon, ME 04352 E-mail: info@woodsend.org
http://www.woodsend.org
Compost analysis; bioremediation design; solvita test kits for soil and compost. Quality Seal of Approval Program for Compost Products.

*William Brinton, President*

**5247 World Resources Company**
1600 Anderson Road 703-734-9800
Suite 200 Fax: 703-790-7245
Mc Lean, VA 22102 http://www.worldresourcescompany.com
World Resources Company (WRC) is a highly specialized environmental risk management company that designs, implements and manages recycling activities and provides environmental services for non-ferrous metal industries nationally andinternationally. This support includes regulatory, environmental, transportation, production, and all other aspects of business inherent to recycling services.
*Peter T Halpin, CEO*

**5248 Yellowstone Environmental Science**
320 South Wilson Avenue 406-586-3905
Bozeman, MT 59715 Fax: 406-587-5109

*Mary M Hunter, Managing Partner*

**5249 Yes Technologies**
320 S Willson Avenue 406-586-2002
Bozeman, MT 59715 Fax: 406-586-8818
E-mail: yes@yestech.com
http://www.yestech.com
Environmental and public health research and development. Patent consulting.
*Mary M Hunter, President*

**5250 Zimpro Environmental**
301 West Military Road 715-359-7211
Rothschild, WI 54474 Fax: 715-355-3219
*William Copa, VP Technical Services*

**5251 Zurn Industries**
5900 Elwin Buchanan Drive 919-775-2255
Sanford, NC 27330 800-997-3876
Fax: 919-775-3541
E-mail: sean.martin@zurn.com
http://www.zurn.com
Environmental systems including air, land, thermal and water; energy systems including steam and heat; mechanical systems.
*Sean Martin, VP Marketing & Sales*
*Michael Boone, General Manager*

## University Centers

**5252 Adirondack Ecological Center**
SUNY College of Environmental Science & Forestry
6312 State Route 28N 518-582-4551
Science & Forestry, Huntington Forest Fax: 518-582-2181
Newcomb, NY 12852 E-mail: aechwf@esf.edu
http://www.esf.edu/aec/
Provides the organizational framework for research, instructional, and public service activities thoughout the Adriondack region.
*Dr. William F Porter, Director*

**5253 Agricultural Research and Development Center**
University of Nebraska
1071 County Road G 402-624-8000
Ithaca, NE 68033 Fax: 402-624-8010
http://ardc.unl.edu
Serves as the primary site for field based reseach with 5,000 acres of row crops and 5,000 domestic farm animals used for teaching and research.
*Daniel J Duncan, Director*

**5254 Akron Center for Environmental Studies**
University of Akron
215 Crouse Hall 330-972-5389
Akron, OH 44325-4102 Fax: 330-972-7611
E-mail: ids@uakron.edu
http://www.uakron.edu/envstudies/
*Ira D Sasowsky, Director*

**5255 Albrook Hydraulics Laboratory**
Washington State University
PO Box 642910 509-335-2576
Pullman, WA 99164 Fax: 509-335-7632
E-mail: rhh@wsu.edu
http://www.wsu.edu
Research laboratory capable of performing projects with physically scaled hydraulic models. 15,000 square feet of floor space, discharge capacity up to 70 cubic feet per second, modern instrumentaion and shop facilities.
*Rollin H Hotchkiss, Director*

**5256 Alternative Energy Institute**
Texas A&M University 806-656-2296
PO Box 248 WT Fax: 806-656-2733
Canyon, TX 79016
*Dr. Vaughn Nelson, Director*

**5257 American Petroleum Institute University**
1220 L Street NW 202-682-8000
Washington, DC 20005 Fax: 202-682-8232
E-mail: training@api.org
http://www.api-u.org
API, through its university, provides training materials to help those in the oil and natural gas business meet regulatory requirements and industry standards. It works with the National Science Teachers Association and othereducational groups to impart scientific literacy and develop critical thinking skills in the classroom.
*Red Cavaney, President/CEO*

**5258 American Society of Primatologists**
University of Washington
PO Box 357330 206-543-0440
Seattle, WA 98195 Fax: 206-685-0305
http://www.asp.org
Conducts research on primates.

**5259 Applied Energy Research Laboratory**
North Carolina State University 919-515-5236
Raleigh, NC 27695 http://www.mae.ncsu.edu/centers/aerl/
*Dr. John A Edwards, Director*

**5260 Aquatic Research Laboratory**
Lake Superior State University
650 W Easterday Avenue 906-635-1949
Sault Sainte Marie, MI 49783 888-800-LSSU
Fax: 906-635-2266
Administered through the college of Arts and Sciences.
*Prof Alex Litvinov, Director*

**5261 Architecture Research Laboratory**
University of Arizona
College of Architecture 520-621-6751
Tucson, AZ 85721 Fax: 520-621-8700
E-mail: Whampton@ccit.arizona.edu
Provides assistance in the areas of education, applied research and public service.

**5262 Biological Reserve**
Denison University 740-587-6261
Granville, OH 43023 E-mail: stocker@denison.edu
http://webby.cc.denison.edu/biology/bioreserve/DUBR.shml

Enhances the education of students in Biology and the Environmental Sciences by providing opportunities for field studies.
*Dr. John E Fauth, Contact*

**5263 Caesar Kleberg Wildlife Research Institute**
Texas A&M University
700 University Boulevard 361-595-3922
MSC 218
Kingsville, TX 78363
Facilitates complex wildlife-related research studies. Includes modern high-tech facilities, specially designed wildlife pens, and rangeland tracts.
*Dr. Sam L Beasom, Director*

**5264 California Sea Grant College Program**
University of California
9500 Gilman Drive 858-534-4440
Deptartment 0232 Fax: 858-534-2231
La Jolla, CA 92093 http://www.csgc.ucsd.edu
*Russell Moll, Director*
*Marsha Gear, Communications Director*

**5265 Cedar Creek Natural History Area**
University of Minnesota
2660 Fawn Lake Drive NE 763-434-5131
Bethel, MN 55005 Fax: 763-434-7361
Twenty-two hundred hectare experimental ecological reserve.

**5266 Center for Applied Energy Research**
University of Kentucky
2540 Research Park Drive 859-257-0305
Lexington, KY 40511 Fax: 859-257-0220
http://www.gaer.uky.edu
An applied research and development center with an international reputation, focusing on the optimal use of Kentucky's energy resources for the benefit of its people.
*Ari Geertsema, Director*

**5267 Center for Applied Environmental Research**
University of Michigan
432 N Saginaw Street 810-767-7373
Suite 805 Fax: 810-767-7183
Flint, MI 48502 E-mail: hblecker@umich.edu
http://www.umf-outreach.edu/caer/index.htm
*Harry S Blecker, Interim Director*

**5268 Center for Aquatic Research and Resource Management**
Florida State University
Conradi Building, 136-B 850-644-4887
Tallahassee, FL 32306 Fax: 850-644-9829
E-mail: livingston@bio.fsu.edu
http://www.bio.fsu.edu/carrma.htm
Conducts research designed to answer aquatic resource-management questions posed by government agencies and private concerns. Research is conducted in lakes, rivers, and near-shore coastal systems throughout the southeastern UnitedStates with a multi-disciplinary approach to topics such as light, nutrients, primary productivity, fate and effects of storm water pollutants, sediment-water interactions, community assemblages of fish and invertebrates in various habitats andtrophic dynamics.
*Dr. Robert J Livingston, Director*

**5269 Center for Cave and Karst Studies**
Western Kentucky University
1906 College Heights Blvd. #31066 270-745-3252
Bowling Green, KY 42101 Fax: 270-745-3961
E-mail: caveandkarst@wku.edu
http://caveandkarst.wku.edu
Promotes research on all aspects of cave and karst studies with emphasis upon solving environmental problems associated with karst.
*Dr. Nicholas Crawford, Director*

**5270 Center for Crops Utilization Research**
Iowa State University of Science & Technology
1041 Food Sciences Building 515-294-0160
Iowa State University Fax: 515-294-6261
Ames, IA 50011 E-mail: ljohnson@iastate.edu
http://www.ag.iastate.edu/centers/ccur
Incorporates various aspects of new product and product research, applications development, and technology transfer. Activities focus on developing technologies for producing food and industrial products from agricultural materials,developing agricultural substitutes for petrochemicals, and exploring and modifying the functional properties of crop-derived materials.

*Dr. Lawrence A Johnson, Director*

**5271 Center for Earth & Environmental Science**
SUNY Plattsburgh
101 Broad Street 518-564-2028
Hudson Hal 102 877-554-1041
Plattsburgh, NY 12901 Fax: 518-564-5267
E-mail: cees@plattsburgh.edu
http://www.plattsburgh.edu/cees
Undergraduate degree programs in environmental science, geology, planning and geography, with special emphasis on watershed science, remote sensing and geographic information systems, and aquatic and terrestrial ecology. The center isone of the oldest and largest environmental programs in the US, with 16 full-time interdisciplinary faculty and diverse field sites.
*Dr Robert D Fuller, Director*

**5272 Center for Environmental Communications (CEC)**
Rutgers University
31 Pine Street 732-932-1966
New Brunswick, NJ 08901 Fax: 732-932-9544
E-mail: cec@aesop.rutgers.edu
The CEC, located on the Cook College Campus, brings together university investigators to provide a social science perspective to environmental problem solving. CEC has gained international recognition for responding to environmentalcommunication dilemmas with research, training, and public service. Established in 1986, CEC is now jointly sponsored by the New Jersey Agricultural Experiment Station and the Edward J. Bloustein School of Planning and Public Policy.

*Caron Chess, Director*

**5273 Center for Environmental Health Sciences**
Massachusetts Institute of Technology
77 Massachusetts Avenue 617-253-6220
Building 16-743 Fax: 617-258-5424
Cambridge, MA 02139 http://www.mit.edu
*William G Thilly, Director*

**5274 Center for Environmental Medicine Asthma & Lung Biology**
University of North Carolina
104 Mason Farm Road, CB #7310 919-962-0126
Room 552 EPA Human Studies Facility Fax: 919-966-9863
Chapel Hill, NC 27599 http://www.med.unc.edu/envlung/
The CEMALB are a group of investigators with diverse research interests that include cardiopulmonary medicine, immunology, lung physiology, cell biology, cell and molecular immunology, molecular toxicology and epidemiology. We conductresearch studies involving human volunteers that are aimed at understanding the negative health effects of air pollution on the lung and heart.
*Philip A Bromberg, MD, Scientific Director*
*David B Peden, MD, MS, Center Director*

**5275 Center for Environmental Research Education**
SUNY Buffalo
1300 Elmwood Avenue 716-878-4329
Upton Hall, Room 314 Fax: 716-878-6644
Buffalo, NY 14222 E-mail: zolnowsa@buffalostate.edu
http://www.buffalostate.edu/~glc/
*Dr. Charles Beasley, Contact*

**5276 Center for Environmental Studies**
Williams College

Kellogg House 413-597-2346
PO Box 632 Fax: 413-597-3489
Williamstown, MA 01267 http://www.williams.edu/CES
Provides students with the opportunity to learn how environmental issues are interconnected with many traditional fields of study. Offered as a concentration, the program encourages students to become well grounded in a single field bypersuing a major in a traditional discipline or department, while focusing several of their elective courses on the interdisciplinary study of the environment.

*Karen Merrill, Director*
*Sarah Gardner, Associate Director*

**5277 Center for Environmental Toxicology and Technology**
Colorado State University 970-491-8522
Foothills Campus Fax: 970-491-8304
Fort Collins, CO 80521 http://www.cvmbs.colostate.edu
*Raymond Yang, Director*

**5278 Center for Field Biology**
Austin Pay State University
PO Box 4718 931-221-7019
Clarksville, TN 37044 Fax: 931-221-6372
E-mail: fieldbiology@apsu.edu
http://www.apsu.edu/field_biology
The Center of Excellence for Field Biology at Austin Peay State University brings together scholars and students from various biological disciplines to conduct research on topics in field biology and ecology, including toxicology,population and community ecology, and the ecology and biology of rare, threatened and endangered species. Major research efforts have focused on the ecology and biology of the flora and fauna of the Land Between the Lakes.
*Dr. Andrew N Barrass, Director*

**5279 Center for Global & Regional Environmental Research**
University of Iowa
424 IATL 319-355-3333
Iowa City, IA 52242 Fax: 319-335-3337
E-mail: jfrank@cgrer.uiowa.edu
http://www.cgrer.uiowa.edu

*Greg Carmichael, Co-Director*
*Jerry Schnoor, Co-Director*

**5280 Center for Global Change Science (MIT)**
Massachusetts Institute of Technology
77 Massachusetts Avenue 617-253-4902
Room 54-1312 Fax: 617-253-0354
Cambridge, MA 02139 E-mail: cgcs@mit.edu
http://mit.edu/cgcs/
Addresses long-standing scientific problems whose solution is necessary for accurate prediction of changes in the global environment. The CGCS is interdisciplinary and interdepartmental, and builds on research and educational programsin earth sciences and engineering. The Center is also involved in substantial cooperative efforts focused on climate modeling, and on climate-policy research.

*Dr. Ronald Prinn, Director*
*Anne Slinn, Administrator*

**5281 Center for Groundwater Research (CGR)**
Oregon Health & Science University
20000 NW Walker Road 503-690-1193
Beaverton, OR 97006 Fax: 503-690-1273
E-mail: rjohnson@ese.ogi.edu
http://http://cgr.ebs.ogi.edu
The CGR coordinates a range of projects relating to the transport and fate of contaminants in soils and groundwater. The scope of the Center includes: the development of new sampling and site characterization techniques; thedevelopment of new analytical techniques; and other improved groundwater remediation techniques.
*Richard Johnson, Director*

**5282 Center for Hazardous Substance Research**
Kansas State University 785-532-6519
Ward Hall 104 Fax: 785-532-5985
Manhattan, KS 66506 E-mail: hsrc@ksu.edu

*Dr. Larry Erickson, Director*

**5283 Center for International Development Research**
Duke University
Sanford Institute of Public Policy 919-684-8894
PO Box 90237 Fax: 919-684-2861
Durham, NC 27708

**5284 Center for International Food and Agricultural Policy**
University of Minnesota
1994 Buford Avenue 612-625-9208
332k Classroom Office Fax: 612-625-2729
St. Paul, MN 55108 E-mail: frunge@apec.umn.edu
http://www.apec.umn.edu/cifap
With its interdisciplinary approach, CIFAP uses its research and education activities to increase international understanding about food, agriculture, nutrition, natural and human resources, and the environment, and to positivelyaffect the policies of both developed and developing countries.

*C Ford Runge, Director*

**5285 Center for Lake Superior Ecosystem Research**
Michigan Technological University
1400 Townsend Drive 906-487-2769
Houghton, MI 49931 E-mail: wkerfoot@mtu.edu
http://www.mtu.edu/level3/centers.html
An interdisiplinary center with goals to promote and strengthen ecological research and graduate programs at MTU through developing and applying technological advances to ecological problems, to advocate an ecosystem perspective forstudying aquatic and terrestrial portions of the Lake Superior watershed and to become a resource center for basic information on watershed and lake properties.

**5286 Center for Marine Biology**
University of New Hampshire
85 Adams Point Road 603-862-2175
Durham, NH 03824 Fax: 603-862-1101
E-mail: ray.grizzle@unh.edu
http://http://marine.unh.edu/jacksonlab.htm
The Center for Marine Biology (CMB) fosters excellence in marine biological research and education. Its primary goals are to strengthen and focus research and graduate education in modern marine biology and to encourage the developmentof high-quality undergraduate programs in all aspects of marine biology. The center helps faculty members compete for external grant funds and fosters coordination of marine research efforts, both with the life sciences and in other disciplines.
*Ray Grizzle, Research Scientist*

**5287 Center for Population Biology**
University of California
One Shields Avenue 530-752-1274
Davis, CA 95616 Fax: 530-752-1449
Founded in 1989, the Center for Population Biology unites UC Davis' population biologists. The center's membership comprises graduate students enrolled in the http://www-eve.ucdavis.edu/popbio.htm, graduate students interested inpopulation biology who are earning their degrees in graduate programs such as ecology or entomology, postdoctoral researchers, from nine academic departments and sections, 17 of whom have faculty appointments in the division.

*Dr H Bradley Shaffer, Director*

**5288 Center for Resource Policy Studies**
University of Wisconsin
1450 Linden Drive 608-262-8254
Room 240
Madison, WI 53706
The Center for Resource Policy Studies and Programs uses interdisciplinary research, teaching and extension efforts to analyze

resource policies and development programs. This center gives particular emphasis to the social scienceaspects of natural resource policy issues.

**5289 Center for Statistical Ecology & Environmental Statistics**
Pennsylvania State University
Dept of Statistics 814-865-9442
421 Thomas Building Fax: 814-865-1278
University Park, PA 16802 E-mail: gpp@stat.psu.edu
http://www.stat.psu.edu/~gpp/aims_scope.htm
The Center is the first of its kind in the world and enjoys national and international reputation. They have an ongoing program of research that integrates statistics, ecology and the environment. The emphasis is on the environment andcollaborative research, training and exposition on improving the quantification and communication of man's impact on the environment. Major interest also lies in statistical investigations of the impact of the environment on man. Contact them forfull listings.
*Ganapati P Patil, Director*

**5290 Center for Streamside Studies**
University of Washington
Box 352100 206-543-6920
Seattle, WA 98195 Fax: 206-543-3254
http://dept.washington.edu/cssuw/
The mission of the Center for Streamside Studies is to provide scientific information necessary for the resolution of management issues related to the production and protection of forest, fish, wildlife, and water resources associatedwith the streams and rivers in the Pacific Northwest.
*Robert J Naiman, Director*

**5291 Center for Tropical Agriculture**
University of Florida
3081 McCarty Hall 352-392-2643
Box 110286 Fax: 352-846-0816
Gainesville, FL 32611 E-mail: hlp@ufl.edu
Enhances research and education on tropical agriculture between University of Florida and tropical countries.
*Dr. Hugh L Popenoe, Director*

**5292 Center for Water Resources and Environmental Research (CWRER)**
The City College of New York
138th Street and Convent Avenue 212-650-7000
New York, NY 10031 E-mail: rk@ce.eng.ccny.cuny.edu
http://www.ccny.cuny.edu
The Center investigates pollution movement, surface water and groundwater cleanup, wetland preservation, watershed management, hydraulics and hydrology of natural flow systems, ecology preservation and the technical and sociopoliticaloutcomes.
*Dr. Reza Khanbilvardi, Director*

**5293 Center for the Management, Utilization and Protection of Water Resources**
Tennessee Technological University
Box 5033 931-372-3507
Cookeville, TN 38505 Fax: 931-372-6346
E-mail: dgeorge@tntech.edu
http://www.tntech.edu/wrc
The Center for the Management, Utilization and Protection of Water Resources at Tennessee Technological University is dedicated to the vision of enhancing environmental education through research. Using interdisciplinary teams ofresearchers, the Center focuses its work in the core areas of environmental resource management and protection, environmental hazards, and environmental information.
*Dr. Dennis George, Director*

**5294 Clean Energy Research Institute**
University of Miami
219 McArthur Building 305-284-4666
Coral Gables, FL 33124 Fax: 305-284-4792
E-mail: veziroglu@miami.edu
Acts as the focal point of energy and environmental related activities in the College of Engineering. Its goals are to conduct research and to generate research proposals to investigate energy and environmental problems; to organizeseminars, workshops and conferences using researchers within and without the University; to assemble, compile, publish and disseminate information on every aspect of energy and environmental problems; and to cooperate with other organs of theUniversity.
*Dr T Nejat Veziroglu, Director*

**5295 Cobbs Creek Community Environment Educational Center (CCCEEC)**
Penn State Cooperative Extension 215-471-2223
4601 Market Street, 2nd Floor Fax: 215-471-2231
Philadelphia, PA 19139 E-mail: ccceec@cobbscreek.org
CCCEEC is designated to institutionalize the practice of Urban Environmental Education. Their mission is to preserve the quality for residents living in the Cobbs Creek area of Philadelphia through the establishment of a center foreducating and informing people about the issues affecting their environment.
*Alan G Fastman, Executive Director*
*Jasa M Porciello, Programing Coordinator*

**5296 College of Forest Resources**
University of Washington
College of Forest Resources 206-543-2730
107 Anderson Hall Fax: 206-685-0790
Seattle, WA 98195 E-mail: cfruw@u.washington.edu
http://www.cfr.washington.edu
The University of Washington College of Forest Resources is dedicated to generating and disseminating knowledge for the stewardship of natural and managed environments and the sustainable use of their products and services throughteaching, research and outreach.

**5297 Colorado Cooperative Fish & Wildlife Research Unit**
Colorado State University
1484 Campus Delivery 970-491-5396
Fort Collins, CO 80523 Fax: 970-491-1413
http://www.colostate.edu/depts/coopunit
The Colorado Cooperative Wildlife Research Unit was founded in 1947, and the Colorado Cooperative Fishery Research Unit was established in 1963. The two Units were combined in 1984 into the Colorado Cooperative Fish and WildlifeResearch Unit. This unit is staffed, supported, and coordinated by the Colorado Division of Wildlife, Colorado State University, the United States Geological Survey , and the Wildlife Management Insistute.
*Dana L Winkelman, Unit Leader*

**5298 Cooperative Fish & Wildlife Research Unit**
University of Missouri
112 Stephens Hall 573-882-3634
Columbia, MO 65211 Fax: 573-884-5070
http://www.conserv.missouri.edu/research/coop.html
*Dr Rose Marie Muzika, Program Director*

**5299 Cornell Waste Management Institute**
Cornell University 607-255-1187
101 Rice Hall Hall, Envrn'l Rsch. Center Fax: 607-255-8207
Ithaca, NY 14853 E-mail: cwmi@cornell.edu
http://cwmi.css.cornell.edu
Conduct applied research and outreach focused on composting and land application of sewage sludges.

*Lauri Wellin, Administrative Assistant*

**5300 ERI Earth Research Institute**
University of California
6831 Ellison Hall 805-893-8231
Santa Barbara, CA 93106 Fax: 805-893-2578
http://www.eri.ucsb.edu
Purpose is to increase our understanding of the geological processes and evolution of the earth's crust and lithosphere, and the impact these processes have on society.
*Douglas W Burbank, Director*

**5301 Eagle Lake Biological Field Station**
California State University

Department of Biology Sciences 530-898-4490
Chico, CA 95929 E-mail: rbogiatto@oavax.csuchico.edu
http://www.csuchico.edu/biol/eaglelakehtml

The Eagle Lake Biological Field Station, located 26 miles northwest of Susanville in Lassen County, California is a ten building facility on the eastern shore of Eagle Lake. The field station is administered by California StateUniversity, Chico and the CSUC Foundation with support from the University of California Natural Reserve System and UC Davis. The ELBFS is open to any individual or group whose purpose is primarily academic and whose activities are consistent withthe isolation.

*Raymond J Bogiatto, Director*

## 5302 Earth Science & Observation Center

University of Colorado Boulder
Main Campus 303-492-5086
318 CIRES Building Fax: 303-492-5070
Boulder, CO 80309 E-mail: lornay.hansen@colorado.edu
http://cires.colorado.edu/esoc

We advance scientific and societal understanding of the Earth System based on innovative remote sensing research. Through our research, we provide fundamental insights into how the Earth system functions, how it is changing, and whatthose changes mean for life on earth, for the benefit of human kind.

*Dr Waleed Abdalati, Director*

## 5303 Ecology Center

Utah State University 435-797-2555
UMC 5205 E-mail: ecol@cc.usu.edu
Logan, UT 84322

The Utah State University Ecology Center is an administrative structure in the University that supports and coordinates ecological research and graduate education in the science of ecology, and provides professional information andadvice for decision makers considering actions that affect the environment.

*Frederic H Wagner, Director*

## 5304 Energy Resources Center

Univeristy of Illinois at Chicago
851 S Morgan Street 312-996-4490
Chicago, IL 60607 Fax: 312-996-5620
E-mail: erc@uic.edu
http://www.erc.uic.edu

The Center is a University of Illinois at Chicago interdisciplinary research and public service organization. It was established in 1973 by the University's Board of Trustees to conduct studies in the field of energy and to providelocal, state and federal governments and the public with current information on energy technology and policy.

*William M Worek, Director*

## 5305 Energy, Environment & Resource Center

University of Tennesse at Knoxville
1414 Circle Drive, 676 Dabney Hall 865-974-8080
Knoxville, TN 37996 Fax: 865-974-8086
E-mail: ceb@utk.edu
http://www.ceb.utk.edu

*Dr. Jack Barkenbus, Director*

## 5306 Environmental & Water Resources Engineering Area

Texas A&M University
Civil Engineering Department 979-845-3011
College Station, TX 77843 Fax: 979-862-1542

*Bill Batchelor, Area Leader*

## 5307 Environmental Center

University of Hawaii
2500 Dole Street 808-956-7361
Krauss Annex 19 Fax: 808-956-3980
Honolulu, HI 96822 E-mail: envctr@hawaii.edu
http://www2.hawaii.edu/~envctr/

The Center's three areas of focus are education, research and service. The education function of the Center includes the administration of the Environmental Studies Major Equivalent and Certificate program. It fulfills its researchfunction by identifying and addressing environmentally related research needs, particularly those pertinent to Hawaii. The service function primarily involves the coordination and transfer of technical information from the University community togovernment agencies.

*Dr John T Harrison, Coordinator*

## 5308 Environmental Chemistry and Technology Program

University of Wisconsin at Madison
660 N Park Street 608-263-3264
Room 122 Fax: 608-262-0454
Madison, WI 53706 http://www.engr.wisc.edu/interd/wcp

*James J Schauer, Professor and Director*

## 5309 Environmental Exposure Laboratory

University of California
1000 Veterans Avenue 310-825-2739
Rehabilitation Center, Room A163
Los Angeles, CA 90024

*Dr. Henry Gong Jr, Director*

## 5310 Environmental Human Toxicology

University of Florida
471 Mowry Road 352-392-2243
PO Box 110885 Fax: 352-392-4707
Gainesville, FL 32611 http://www.floridatox.org

The Center serves as an interface between basic research and its applications for evaluation of human health and environmental risk. The research and teaching activities of the Center provide a resource to identify and reduce riskassociated with environmental pollution, food contamination, and workplace hazards. The center provides a forum for the discussion of specific and general problems concerning the potential adverse human health effects associated with chemicalexposure.

*Dr. Stephen Roberts, Director*
*Lauren Curry, Program Assistant*

## 5311 Environmental Institue and Water Resources Research Institute

Auburn University
101 Comer Hall 334-844-4132
Auburn, AL 36849 Fax: 334-844-4462
E-mail: hatchlu@auburn.edu
http://www.ave.auburn.edu

*Dr Upton Hatch, Director*

## 5312 Environmental Institute of Houston

University ofg Houston
Environmental Institute of Houston 281-283-3950
Box 540 Fax: 281-283-3044
Houston, TX 77058 E-mail: eih@cl.uh.edu

The mission of EIH is to help people in the Houston region participate more effectively in environmental improvement. Information and technology will be obtained and disseminated from research supported by EIH in critical areasincluding pollution prevention, natural resource conservation, public policy and societal issues. EIH will seek to expand balanced environmental education based on objective scholarship to empower the community to make sound decisions onenvironmental issues.

*Candy Allison, Senior Secretary*
*Heather Biggs, GIS Analyst*

## 5313 Environmental Remote Sensing Center

University of Wisconsin
1225 W Dayton Street 608-263-3251
Madison, WI 53706 Fax: 608-262-5964
http://www.ersc.wisc.edu

A university research center focused on application of remote sensing and attending geospatial technologies in government, business and science. Particular heritage in the application of remote sensing in natural resource managementand environmental monitoring. A NASA-sponsored Affiliated Research Center.

*Prof Thomas Lillesand, Director*

## 5314 Environmental Research Institute

University of Idaho
Food Research Center 103 208-885-6580
Moscow, ID 83844 Fax: 208-885-5741
E-mail: crawford@uidaho.edu
http://image.fs.uidaho.edu/

The faculty, associated with the institute, perform multidisciplinary research in environmental molecular ecology, restoration of contaminated soils and waters, and microbial genomics related to environmental processes.
*Dr Ronald L Crawford, Director*

**5315 Environmental Resource Center**
San Jose Southern University
One Washington Square #1 408-924-5467
San Jose, CA 95192 E-mail: ncowan@email.sjsu.edu
The Environmental Resource Center is a nonprofit information and outreach organization within the Environmental Studes department at San Jose Southern University, serving the San Jose community since 1971.
*Annemarie Vallesteros, Executive Director*

**5316 Environmental Science & Engineering Program**
Clarkson University
210 Clarkson Hall 315-268-3786
Potsdam, NY 13699 Fax: 315-268-4266
E-mail: feitelsb@clarkson.edu
*Samuel B Feitelberg PT, MA, FAPTA, Professor*

**5317 Environmental Studies Institute**
University of Pennsylvania
256 Smith Walkity 215-573-3164
Philadelphia, PA 19104 E-mail: ies_penn@sas.upenn.edu
The Institute for Environmental Studies is dedicated to improving the understanding of key scientific, economic, and political issues that underlie environmental problems and their management. The mission of the Institute is to bringscholars together from across the University in order to promote collaborations in education and research endeavors in the area of environmental issues. These collaborative endeavors span basic and applied sciences, engineering and the social andhuman sciences.
*Dr. Bernard Hamel, Director*

**5318 Environmental Systems Application Center**
Indiana University
107 S Indiana Avenue 812-855-4848
Bloomington, IN 47405
The goals of the Center are to promote excellence in environmental science research and to foster increased interdisciplinary collaboration among environmental science faculty on the Indiana University-Bloomington campus. The Centerhas no degree programs. The Center can be listed as an affiliation of the associated faculty in publications and in correspondence.

**5319 Environmental Systems Engineering Institute**
University of Central Florida
4000 Central Florida Boulevard 407-823-2000
Orlando, FL 32816

**5320 Environmental Toxicology Center**
University of Wisconsin
1710 University Avenue 608-263-4580
Room 290 Fax: 608-262-5245
Madison, WI 53706 http://www.wisc.edu/etc/
Environmental Toxicology is the study of the adverse effects on individual life forms and ecosystems of environmental agents (chemical, physcial, biological) whether of natural origin or released through human activity, and origins andcontrol of these harmful agents.
*Prof. Colin R Jefcoate, Director*

**5321 Environmental and Occupational Health Science Institute**
Rutgers University
170 Frelinghuysen Road 732-445-0200
PO Box 1179 Fax: 732-445-0131
Piscataway, NJ 08855 http://www.cohsi-rutgers.edu
The major objectives of the institute are to: improve understanding of the impact of environmental chemicals on human health; to find ways to quantify and prevent exposure to hazardous substances; and develop methods to identify andtreat people adversely affected by environmental agents. Devises approaches for educating the public about the relative risks from chemical exposure. Trains professionals to accomplish these tasks.
*Dr. Cory Slechta, Director*

**5322 Feed and Fertilizer Laboratory**
Louisiana State University
Department Agriculture and Forestry 225-388-2755
Baton Rouge, LA 78021
*David Wall, Contact*

**5323 Field Station & Ecological Reserves**
University of Kansas
c/O Kansas Biological Survey 785-864-1505
2101 Constant Avenue Fax: 785-864-5093
Lawrence, KS 66047 http://www.ksr.ku.edu
The KSR is dedicated to field-based environmental reseach and education. KSR is located within the transition zone (ecotone) between the eastern deciduous forest and tallgrass prairie biomes. The 3,000 acres of diverse native andmanaged habitats, experimental systems, support facilities, and longterm databases are used to undertake an outstanding array of scholarly activities. Environmental stewardship is a stong emphasis as high-quality natural areas are preserved for thefuture.
*Edward Martinko, KSR Director*

**5324 Fitch Natural History Reservation**
University of Kansas
2060 E 1600 Road 785-843-3612
Lawrence, KS 66044

**5325 Florida Cooperative Fish and Wildlife Research Unit**
University of Florida
Box 110450 352-392-1861
Gainesville, FL 32611 Fax: 352-846-0841
http://biology.usgs.gov/coop/unitpages/fl_cfwru.html
The Cooperative Research Unit has three facets to its mission: education—Cooperative Unit scientists teach university courses at the graduate level, provide academic guidence to graduate students, and serve on academic committees;research—Cooperative Unit scientists conduct research that is designed to meet the information needs expressed by unit cooperators; technical Assistance— unit provides technical assistance and training to State and federal personnel and othernatural resources.
*Dr. Wiley M Kitchens, Leader*

**5326 Florida Museum of Natural History**
University of Florida
Museum Drive 352-392-1721
PO Box 117800 Fax: 352-392-8783
Gainesville, FL 32611 E-mail: gdshaak@flmnh.ufl.edu
http://www.flmnh.ufl.edu
The Florida Museum of Natural History, on the University of Florida Campus, is one of the leading university natural history museums in the nation. With over 30 million specimens and artifacts in its permanent collections, it is thelargest collection-based museum in the southeastern US. The museum was established by the Legislature in 1917 at the University of Florida where it functions in a dual capacity as the official state museum of Florida and the University Museum.

*Dr Douglas S Jones, Director*
*Dr Beverly Sensbach, Associate Director*

**5327 Formaldehyde Institute**
1330 Connecticut Avenue NW 202-833-2131
Washington, DC 20036 Fax: 202-659-1699
http://www.ainc.org

*John F Murray, Executive Director*

**5328 Gannett Energy Laboratory**
Florida Institute of Technology
150 W University Boulevard 321-768-8000
Melbourne, FL 32901

**5329 Global Change & Environmental Quality Program**
University of Colorado
Campus Box 214 303-492-7943
Boulder, CO 80309 Fax: 303-492-1414

In addition to addressing CU's overrall objectives, the Global Change and Environmental Quality Program is pursuing three main goals; studying environmental issues at the local level, including the cleanup and restoration of toxicsites, such as Rocky Flats, the Rocky Mountain Arsenal, and mine tailing sites; waste treatment and water quality; and land use.
*Dr. Robert Sievers, Director*

**5330 Graduate Program in Community and Regional Planning**
University of Texas
Main Building 512-471-0134
Austin, TX 78701 Fax: 512-471-0716
The CRP provides its graduates with the theoretical foundations, specific skills and practical experience to succeed in professional planning and related policy careers. They strive to create a diverse student body and program and arecommitted to building a professional planning community that rese,bles those where graduates will work. The program has a strong focus on sustainable development processes and practices. Finding paths that balance growth with improved environmentalperformance.
*Jane Shaughness, Graduate Admissions Coor.*

**5331 Great Lakes Coastal Research Laboratory**
Purdue University
School of Civil Engineering 765-494-4600
West Lafayette, IN 47907

**5332 Great Lakes/Mid-Atlantic Hazardous Substance Research Center**
University of Michigan
EWRE Building 734-763-2274
Suite 181
Ann Arbor, MI 48109
The mission of the Great Lakes Mid- Atlanic Center for Hazardous Substances Research is to foster and support integrated, intersdisciplinary, and collaborative efforts that advance the science and technology of hazardous substancemanagement to benefit human and environmental health and well-being.
*Dr. Walter J Weber Jr, Director*

**5333 Great Plains: Rocky Mountain Hazardous Substance Research Center**
Kansas State University
Ward Hall 101 785-532-6519
Manhattan, KS 66506 Fax: 785-532-6952
E-mail: hsrc@ksu.edu
http://www.engg.ksu.edu/HSRC
Conducts research and transfers technology on hazardous substance management, and remediation of contaminated soil and water.
*Dr. Larry Erickson, Director*

**5334 Greenley Memorial Research Center**
University of Missouri
Greenley Memorial Center 660-739-4410
Box 126 Fax: 660-739-4500
Novelty, MO 63460 E-mail: smootr@missouri.edu
*Randall Smoot, Supervisor*

**5335 HT Peters Aquatic Biology Laboratory**
Bemidji State University 218-755-2877
Bemidji, MN 56601 E-mail: dcloutman@bemidjistate.edu

**5336 Harry Reid Center for Environmental Studies**
University of Las Vegas
4505 Maryland Parkway 702-895-3382
Box 454009 Fax: 702-895-3094
Las Vegas, NV 89154
The HRC was started in 1981 under UNLV's Marjorie Barrick Museum of Natural History. HRC currently includes 65 staff members and a 65,000 square foot building with four laboratories.

**5337 Hawaii Cooperative Fishery Research Unit**
University of Hawaii at Manoa
2538 The Mall 808-956-8350
Honolulu, HI 96822 Fax: 808-956-4238
E-mail: parrishj@hawaii.edu
*Dr. James D Parrish, Unit Leader*

**5338 Hawaii Undersea Research Laboratory**
University of Hawaii at Manoa
1000 Pope Road 808-956-6335
MSB 303
Honolulu, HI 96822
One of six research centers funded by NOAA's National Undersea Research Program. HURL operates two 2000-meter Pisces submersibles and a remotely operated vehicle. Research projects include fisheries research, geology and biology of thedeepsea around the Hawaiian Islands.
*Dr. Alexander Malahoff, Director*

**5339 Henry S Conrad Environmental Research Area**
1210 Grinnell College 641-269-4457
Grinnell, IA 50112 Fax: 641-269-4984

**5340 Highlands Biological Station**
University of North Carolina
265 Sixth Street 828-526-2602
Highlands, NC 28741 Fax: 828-526-2797
E-mail: hbs@email.wcu.edu
http://www.wcu.edu/hbs
The Station is an interinstitutional center of the University of North Carolina and includes the Highlands Nature Center and Botanical Gardens, as well as the Biological Laboratory. Our mission, for more than 75 years has been tofoster education and research focused on the rich natural heritage of the Highlands Plateau.
*Tom Martin, Interim Director*
*Peggy Cowart, Office Assistant IV*

**5341 Hudsonia**
Bard College Field Station 845-758-7053
PO Box 5000 Fax: 845-758-7033
Annandale, NY 12504 E-mail: kiviat@bard.edu
http://www.hudsonia.org
Since 1981 Hudsonia has conducted environmental research, education, training and technical assistance to protect the Hudson River Valley's natural heritage. Nonpartisan and non-ideological, Hudsonia serves as a neutral voice in thechallenging process of land conservation.
*Erik Kiviat, Executive Director*

**5342 Huntsman Environmental Research Center**
Utah State University 435-750-1418
UMC 4105 Fax: 435-797-1248
Logan, UT 84322 E-mail: herc@usu.edu
The establishement of the Hunts man Environmental Research Center recognized the fundamental interdependence of the health of man and the health of the environment. The HERC's mission is to engage in research in the key areas ofrecycling, degradability, improvement of air and water quality and conservation of trees. The center purpose is to solve environmental problems and to provide realistic and comprehensive research solutions for our environment.

*Ron Sims, Director*

**5343 INFORM**
120 Wall Street 212-361-2400
14th Floor Fax: 212-361-2412
New York, NY 10005 http://www.informinc.org
INFORM is an independent research organization that examines the effects of business practices on the environment and on human health. Our goal is to identify ways of doing business that ensure environmentally sustainable economicgrowth. Our reports are used by government, industry, and environmental leaders around the world.
*Joanna Underwood, President*

**5344 Idaho Cooperative Fish & Wildlife Research Unit**
University of Idaho

PO Box 44-1141 208-885-2750
Moscow, ID 83844 Fax: 208-885-9080
E-mail: sarahm@uidaho.edu
http://www.cnrhome.uidaho.edu/coop

Personnel of the Idaho Cooperative Fish and Wildlife Research Unit will: (1) conduct research on problems of state, regional and national interest; (2) train graduate students for careers in the fish and wildlife professions;(3)provide technical assistance to state and federal managers and researchers.

*Dr J Michael Scott, Unit Leader*
*Sarah Martinez, Administrative Asst. II*

**5345 Institute for Biopsychological Studies of Color, Light, Radiation, Health**
San Jose State University
One Washington Square 408-924-1000
Psychology Department Fax: 408-924-1018
San Jose, CA 95192 http://www.sjsu.edu

**5346 Institute for Ecological Infrastructure Engineering**
Losuisiana State University
College of Engineering 225-578-1399
102 ELAB Fax: 225-578-8662
Baton Rouge, LA 70803 E-mail: eielab@eiel.lsu.edu
http://www.eiel.lsu.edu

Institute for Ecological Infrastructure Engineering integrates engineering with science (physical, chemical, life & social) for the co-development of society and nature (ecosystems).
*Lily A Rusch, Director*

**5347 Institute for Environmental Science**
University of Texas at Dallas
PO Box 830688 972-794-4000
Richardson, TX 75083
*Dr. John Ward, Director*

**5348 Institute for Lake Superior Research**
University of Minnesota
135 Medical Building 218-726-8000
Duluth, MN 55812 http://www.uwsuper.edu
*Dr. Carlson, Director*

**5349 Institute for Regional and Community Studies**
Western Illinois University 309-298-1566
Tillman Hall 413B
Macomb, IL 61455

**5350 Institute for Urban Ports and Harbors**
School of Marine and Atmospheric Sciences
Stony Brook University 631-632-8700
Endeavor Hall, Room 145 Fax: 631-632-8820
Stony Brook, NY 11794-5000 E-mail: somas@stonybrook.edu
http://www.somas.stonybrook.edu
*David Conover, Dean/Director*

**5351 Institute of Analytical and Environmental Chemistry**
University of New Haven
300 Orange Avenue 203-932-7171
West Haven, CT 06515 800-342-5864
http://www.newhaven.edu
*Prof. George Wheeler, Director*

**5352 Institute of Chemical Toxicology**
Wayne State University
2727 2nd Avenue 313-577-0100
Room 4000 Fax: 313-577-0082
Detroit, MI 48201 http://www.wayne.edu

**5353 Institute of Ecology**
University of California
1 Shields Avenue 530-752-3026
Davis, CA 95616 Fax: 530-752-3350
E-mail: aking@ucdavis.edu
http://www.ucdavis.edu

**5354 Interdisciplinary Center for Aeronomy & Other Atmospheric Sciences**
University of Florida
311 Space Sciences Research Bldg 352-392-2001
Gainesville, FL 32611 Fax: 352-392-2003
*Prof. Alex ES Green, Director*

**5355 Iowa Cooperative Fish & Wildlife Research Unit**
Iowa State University
NREM-ICFWRU 515-294-3056
339 Science II Fax: 515-294-5468
Ames, IA 50011 E-mail: coppunit@iastate.edu
http://www.cfwru.iastate.edu

The Iowa landscape and economy are dominated by production agriculture. Game and non-game wildlife species inhabinting the state are influenced by the destruction, degradation and frgamentation of wetland, prairie, and forest habitatscaused by intensified agricultural practices. This Unit is designed to identify, and emaphsize these effects through research and education programs.
*VACANT , Leader*

**5356 Iowa Waste Reduction Center**
University of Northern Iowa
113, BCS Building 319-273-8905
Cedar Falls, IA 50614 800-422-3109
Fax: 319-273-6582
E-mail: publicrelations@iwrc.org
http://www.iwrc.org

A service of the University of Northern Iowa, it provides free and confidential environmental regulatory assistance to Iowa small businesses. The IWRC has also developed two products available to the painting and coating industry:LaserPaint and VirtualPaint.
*John Konefes, Director*
*James Olson, Assistant Director*

**5357 James H Barrow Field Station**
Hiram College Biological Station 330-527-2141
Garrettsville, OH 44231 Fax: 330-527-3187

The James H Barrow Field Station was established in 1967 to provide Hiram College students the opportunity to supplement classroom activities with hands-on learning experiences. Over the Last 32 years the Station has grown anddeveloped into an active research and educational facility that not only echances the College's science and environmental studies programs, but also provides a means for the general public to increase their understanding and appreciation of Ohio'snatural history.

**5358 John F Kennedy School of Government Environmental and Natural Resources Program**
Harvard University
79 John F Kennedy Street 617-495-1351
Cambridge, MA 02138 Fax: 617-495-1635
E-mail: enrp@ksg.harvard.edu
http://http://bcsia.ksg.harvard.edu/?program+ENRP
*Henry Lee, Director*

**5359 Juneau Center School of Fisheries & Ocean Sciences**
University of Alaska Fairbanks
1120 Glacier Highway 907-465-6441
Juneau, AK 99801 Fax: 907-465-6447
E-mail: fsfosj@uaf.edu
http://www.sfos.uaf.edu/

JCSFOS has the primary responsibility within the University for education, research and public service in support of fisheries related areas of oceanography, marine biology and limnology with emphasis on Alaskan waters and the Arctic.The school's goal is to maintain and develop the broad expertise among its faculty and students needed to contribute to the wise use of Alaska's natural resources.

**5360 Kresge Center for Environmental Health**
Harvard University
665 Huntington Avenue 617-732-1272
Boston, MA 02115 E-mail: brain@hsph.harvard.edu
http://www.hsph.harvard.edu/kresge

The Kresge Center serves as the focus for research and training activities in environmental health at the Harvard School of Public Health and elsewhere in the University. The Center was established in 1958 to promote interactions amongbiological scientists, physical scientists and engineers working on environmental problems of human health concern.
*Joseph D Brain, Director*

## 5361 Laboratory for Energy and the Environment

Massachusetts Institute of Technology
Muckley Building E40, 4th Floor 617-253-1341
Room E40-455 Fax: 617-253-8013
Cambridge, MA 02139 E-mail: thill@mit.edu
http://www.lfee.mit.edu/metadot/index.pl

The LFEE at the Massachusetts Institute of Technology brings together collaborating faculty and staff in 13 departments to address the complex interrelationships between energy and the environment, and other global environmentalchallenges.
*Ernest J Moniz, Director*
*Theresa Hill PhD, Publications & Programs*

## 5362 Leopold Center for Sustainable Agriculture

Iowa State University of Science & 515-294-3711
Technology, 209 Curtiss Hall Fax: 515-294-9696
Ames, IA 50011 E-mail: leocenter@iastate.edu
http://www.leopold.iastate.edu

The center was created by the 1987 Iowa Ground Water Protection Act with a three fold mission: 1) to identify and reduce adverse environmental impacts of farming practices, 2) develop profitable farming systems that conserve naturalresources, and 3) create educational programs with the ISU Extension Service. The center opertes a competitive grant program and supports several muti-desciplinary research teams and initatives. It is named after internationally acclaimed and Iowaborn Aldo Leopold.
*Frederick L Kirschenmann, Distinguished Fellow*

## 5363 Living Marine Resources Institute

Stony Brook University 631-632-8656
Stony Brook, NY 11794 Fax: 631-632-9441
E-mail: wwise@notes.cc.sunysb.edu
http://www.sunysb.edu

LIMRI is one of several specialized institutes subsumed within the Marine Sciences Research Center of Stony Brook University. LIMRI's program of research includes investigations on marine fisheries, harmful algal blooms, marine law &policy, and aquaculture. The Institute operates the Flax Pond Marine Laboratory, a seaside, seawater-equipped facility for experimental work, located 5 miles north of the main campus on a tidal pond adjacent to Long Island Sound.
*William Wise, Director*

## 5364 Long-Term Ecological Research Project

University of Colorado
1560 30th Street 303-492-3302
Campus Box 450 Fax: 303-492-0434
Boulder, CO 80309 E-mail: tims@culter.colorado.edu
*Dr T R Seastedt, Director*

## 5365 Louisiana Sea Grant College Program

Louisiana State University
Sea Grant Building 225-578-6564
Baton Rouge, LA 70803 Fax: 225-578-6331
http://www.laseagrant.org

Works to promote stewardship of the state's coastal resources through a combination of research, education and outreach programs critical to the cultural, economic, and environmental health of Louisiana's coastal zone. Part of theNational Sea Grant Program, it is one of 32 programs located in coastal, Great Lakes, and Puerto Rican coast areas.

*Charles (Chuck) Wilson PhD, Executive Director*
*Roy Kron, Director of Outreach & Comm*

## 5366 MIT Sea Grant College Program

Massachusetts Institute of Technology
77 Massachusettes Ave. 617-253-7131
E 38-300 Fax: 617-258-5730
Cambridge, MA 02139 E-mail: chrys@mit.edu
http://web.mit.edu/seagrant/
*Chrys Chryssostomidis, Director*

## 5367 Marine Science Institute

University of Texas
750 Channel View Drive 361-749-6711
Port Aransas, TX 78373 Fax: 361-749-6777
http://www.utmsi.utexas.edu

The Marine Institute is an organized research unit of The University of Texas at Austin. Institute scientists are engaged in both multi-investigator, multi-disciplinary studies and individual research projects in the local area andthroughout the world. Many of these projects are combinations of field and laboratory investigations. The Institute receives an operating budget annually that is based on a two-year advanced budget approval by the state legislature.

## 5368 Marine and Freshwater Biomedical Sciences Center

University of Miami
4600 Rickenbacker Cswy 305-361-4736
Rosenstiel School 888-232-8635
Miami, FL 33149 Fax: 305-421-4001
E-mail: toxmaster@rsmas.miami.edu
http://www.rsmas.miami.edu/groups/niehs

The Environmental Health Science Center is an integral part of the University, with 20 faculty postdoctoral fellows and outreach personnel. Supported by the US government our research focus is on human health applications for diseaseprevention. Research includes neurotoxicology, potent marine metabolites present in seafood, environmental intoxicants, fisheries models for hepatic metabolism and development of sentinel species for xenobiotic evaluation. Courses, conferences,seminars, outreach.
*Dr Pat Walsh, Center Director*
*Lora Fleming, Associate Director*

## 5369 Massachusetts Cooperative Fish & Wildlife Unit

University of Massachusetts
Holdsworth Natural Resources Center 413-545-0080
Amherst, MA 01003 Fax: 413-545-4358

## 5370 Masschusetts Water Resources Research Center

University of Massachusetts
Blaisdell House 413-253-5686
Amherst, MA 01003 Fax: 413-253-1309
E-mail: godfrey@tei.umass.edu
http://www.umass.edu/tei/wrrc

The Center has three objectives: 1) to develop, through research, new technology and more efficient methods for resolving local, state and national water resources problems; 2) to train water scientists and engineers through on-the-jobparticipation in water resources research and outreach; 3) to facilitate water research coordination and the application of research results by means of information dissemination, technology transfer and outreach.
*Paul Godfrey, Director*

## 5371 Millar Wilson Laboratory for Chemical Research

Jacksonville University
2800 University Boulevard N 904-744-3950
Jacksonville, FL 32211 Fax: 904-744-0101
http://dept.ju.edu/mwl/

## 5372 Mining and Mineral Resources Research Center

MTU 512 M&ME Building 906-487-2630
1400 Townsend Drive Fax: 906-487-2934
Houghton, MI 49931 http://www.mtu.edu

## 5373 Mississippi Cooperative Fish & Wildlife Research Unit

Mississippi State University
Thompson Hall, Room 271 662-325-3174
Box 9690 Fax: 662-325-8750
Mississippi State, MS 39762

**5374 Mississippi State Chemical Laboratory**
Mississippi State University
PO Box CR 662-325-3584
Mississippi State, MS 39762 Fax: 662-325-1618
http://www.msstate.edu
*Dr. Earl G Allen, State Chemist*

**5375 Montana Cooperative Fishery Research Unit**
Montana St University Dept Ecology 406-994-3491
Po Box 173460 Fax: 406-994-7479
Bozeman, MT 59717 E-mail: bobwhite@montana.edu
*Dr. Robert G White, Leader*

**5376 Monterey Bay Watershed Project**
California State University
100 Campus Center 831-582-4120
Seaside, CA 93955 Fax: 831-582-4122
E-mail: essp_comments@csumb.edu
http://watershed.csumb.edu
The Watershed Institute is a direct action, community based coalition of researchers, restoration ecologists, educators, students, planners and area volunteers dedicated to restoring the watersheds of the Monterey Bay region throughrestoration, education and research. Their policy is to work with state and federal agencies, private landowners and local planners to gain access to critical lands. Institute staff are involved in local land and water planning.
*Robert Curry, Adjunxt Professor*

**5377 Museum of Zoology**
University of Massachusetts
Zoology Department 413-545-2287
Amherst, MA 01003 http://www.umass.edu
*Dr. DJ Klingener, Director*

**5378 National Center for Ground Water Research**
University of Oklahoma
660 Parrington Oval 405-325-0311
Norman, OK 73019 Fax: 405-325-7596
E-mail: canter@ou.edu
*Dr Larry Canter, Director*

**5379 National Center for Vehicle Emissions Control & Safety**
Colorado State University
1584 Campus Delivery 970-491-7240
Fort Collins, CO 80523 Fax: 970-491-7801
E-mail: ncvecs@cahs.colostate.edu
http://www.NCVECS.colostate.edu
NCVECS is a nationally and internationally recognized university based research and training center devoted to motor vehicle emission issues. NCVECS primarily assists states with research and training related to their local vehicleinspection program. In addition, research is conducted on OBDII systems, alternative fuels and diesel vehicle issues.
*Dr Lenora Bohren, Director*

**5380 National Institute for Global Environmental Change: South Central Regional Center**
Tulane University
605 Lindy Boogs Center 504-865-5250
New Orleans, LA 70118 Fax: 504-865-6745
E-mail: nigec@tulane.edu
*Dr Stathis Michaelides, Director*

**5381 National Mine Land Reclamation Center: Eastern Region**
State University of Pennsylvania
106 Land & Water Resources Building 814-863-0291
University Park, PA 16802 Fax: 814-865-3378
E-mail: ajm2@psu.edu

**5382 National Mine Land Reclamation Center: Midwest Region**
Southern Illinois University
1201 W Gregory 618-453-2496
Carbondale, IL 62901 Fax: 217-333-8816

**5383 National Mine Land Reclamation Center: Western Region**
Highway 6 S 701-777-5217
Mandan, ND 58554 E-mail: jsolc@eerc.und.nodak.edu

**5384 National Park Service Cooperative Unit: Athens**
University of Georgia
Institute of Ecology 706-542-8301
Athens, GA 30602
*Dr. Stephen Cover-Shabica, Contact*

**5385 National Research Center for Coal and Energy (NRCCE)**
West Virginia University 304-293-2867
PO Box 6064 Fax: 304-293-3749
Morgantown, WV 26506 E-mail: richard.bajura@mail.wvu.edu
http://www.wvwri.org
A research and training center at West Virginia University, advances innovations for energy and the environment by working with research faculty across WVU and with other university, government, and private sector researchersnationwide. The center is organized into a variety of multidisciplinary programs, centers, and institutes focusing on topics such as clean energy production, energy distribution, energy efficiency, alternative fuels, watershed restorationandpreservation.
*Richard A Bajura, Director*
*Trina K Wafle, Deputy Director*

**5386 Natural Energy Laboratory of Hawaii Authority**
73-4460 Queen Kachumanu Highway 808-329-7341
101 Keahole Point Fax: 808-326-3262
Kailua Kona, HI 96740 E-mail: inquires@nelha.org
http://nelha.org
NELHA, an agency of the State of Hawaii, operates facilities at Keahole Point on Hawaii Island that pump ashore cold deep and warm surface seawater for commercial and research tenants from the private and public sectors. Tenantsutilize the seawater and NELHA's high sunlight and consistant temperatures in a wide range of aquaculture and energy projects.
*Ron Baird, CEO*

**5387 New Hampshire Sea Grant College Program**
University of New Hampshire
Kingman Farm Unit 603-749-1565
Durham, NH 03824 Fax: 603-743-3997
E-mail: steve.adams@unh.edu
http://www.seagrant.unh.edu
A component of the National Sea Grant College Program, NH Sea Grant works toward the conservation, wise use and development of marine resources in the state and region.
*Jonathon Pennock, Director*

**5388 New York Cooperative Fish & Wildlife Research Unit**
Cornell University
Natural Resource Department 607-255-2151
Fernow Hall Fax: 607-255-1895
Ithaca, NY 14853 E-mail: dnrcru-mailbox@cornell.edu
http://www.dnr.cornell.edu/f@wres/nycf@wru.htm
*Dr. Milo Richmond, Unit Leader*

**5389 New York State Water Resources Institute**
Cornell University
204A Rice Hall 607-255-5941
Ithaca, NY 14853 Fax: 607-255-5945
E-mail: nyswri@cornell.edu
http://www.cfe.cornell.edu/wri/
*Keith S Porter, Director*

**5390 Northwoods Field Station**
Hiram College

PO Box 123
Hiram, OH 44234
216-569-3211
http://www.hiram.edu
*Dr WR Knight, Director*

**5391 Occupational & Environmental Health Laboratory**
University of North Alabama
Box 5049
Florence, AL 35632
256-765-4622
Fax: 256-765-4958
http://www2.una.edu/chemdept/

**5392 Ocean & Coastal Policy Center**
University of California
Woolley-5134
Santa Barbara, CA 93106
805-893-8393
Fax: 805-893-8062

**5393 Ocean Engineering Center**
33 College Road
Kingsbury Hall
Durham, NH 03824
603-862-1898
Fax: 603-862-0241
E-mail: kcb@cisunix.unh.edu
*Kenneth Baldwin, Director*

**5394 Oceanic Institute**
Makapuu Point
41-202 Kalanianaole Highway
Waimanalo, HI 96795
808-259-3102
Fax: 808-259-5971
E-mail: oi@oceanicinstitute.org
http://www.oceanicinstitute.org

Oceanic Institute is a not-for-profit organization dedicated to research, development and transfer of oceanographic, marine Environmental, and aquaculture technologies. Oceanic Institute is a world leader in conducting appliedresearch in aquaculture production and marine resource conservation. Its mission is to develop and transfer environmentally responsible technologies to increase aquatic food production while promoting the sustainable use of ocean resources.

*Anthony Ostrowski, President*
*Chisa Woodley, Controller*

**5395 Oregon Cooperative Fishery Research Unit**
Oregon State University
104 Nash Hall
Corvallis, OR 97331
541-737-1938
Fax: 541-737-3590
E-mail: or_cfwru@orst.edu
http://www.orst.edu/

**5396 Oregon Cooperative Park Studies Unit**
Oregon State University
3200 Jefferson Way
Corvallis, OR 97331
541-737-2056
E-mail: starkeye@ccmail.orst.edu
http://www.cof.orst.edu/
*Dr Edward E Starkey, Codirector*

**5397 Oregon Sea Grant College Program**
Oregon State University
322 Kerr Administrative Building
Corvallis, OR 97331
541-737-2714
Fax: 541-737-7958
http://seagrant.oregonstate.edu/
*Stephen B Brandt, Director*

**5398 Pennsylvania Cooperative Fish & Wildlife Research Unit**
Pennsylvania State University
Forest Resources Bldg
University Park, PA 16802
814-865-4511
Fax: 814-863-4710
E-mail: klc2@psu.edu
http://pacfwru.cas.psu.edu

*Tyler Wagner, Assistant Leader*

**5399 Permaculture Gap Mountain**
9 Old County Road
Jaffrey, NH 03452
603-532-6877

**5400 Pesticide Research Center**
Michigan State University
107 Pesticide Research Center
East Lansing, MI 48824
517-353-9430
E-mail: whalon@pilot.msu.edu
http://www.msu.edu

**5401 Planning Institute**
University of Southern California
Von KleinSmid Center 351
Los Angeles, CA 90089
213-740-6842
Fax: 213-740-1801
E-mail: sppd@usc.edu
http://www.usc.edu

*Michael Horst, Contact*

**5402 Program for International Collaboration in Agroecology**
University of California Santa Cruz
MS: PICA
1156 High Street
Santa Cruz, CA 95064
831-459-4051
Fax: 831-459-2867
E-mail: gliess@ucsc.edu
http://www.agroecology.org

Researches, develops, and advances sustainable food and agricultural systems that are environmentally sound, economically viable, socially responsible, nonexploitive, and that serve as a foundation for future generations. A specialfocus on promoting an international network of training programs in agroecology.
*Stephen Gliessman, Professor of Agroecology*
*Vivan Vadakan, PICA Program Manager*

**5403 Program in Freshwater Biology**
University of Mississippi
214 Shoemaker Hall
Biology Department
University, MS 38677
662-232-7203
Fax: 662-915-5144
E-mail: biology@olemiss.edu
http://www.olemiss.edu
*Dr. James Kushlan, Chairman*

**5404 Randolph G Pack Environmental Institute**
One Forestry Drive
107 Marshall Hall
Syracuse, NY 13210
315-470-6636
Fax: 315-470-6915
E-mail: envsty@esf.edu
http://www.esf.edu/ed/pack

The Institute seeks to advance scholoarly and popular knowledge of key contemporary issues related to environmental policy and regulartion. It focuses on how democratic public decisions affecting the natural environment are made,concentrating on such topics as public participation, environmental equity, and sustainable development.

*David A Sonnenfeld, Director*

**5405 Rare and/or Endangered Species Research Center**
215 Mitchell Street
Florence, AL 35630
256-760-4429

**5406 Red Butte Garden and Arboretum**
300 Wakara Way
Salt Lake City, UT 84108
801-581-4747
http://www.redbuttegarden.org

**5407 Remote Sensing/Geographic Information Systems Facility**
Indiana State University
Science Building, Room 159
Terre Haute, IN 47809
812-237-2444
Fax: 812-237-8029
E-mail: gga@baby.indstate.edu
http://http://baby.indstate.edu/geo/ggafolder/ggafacil.html
*Dr Susan Berta, Interim Chairperson*

**5408 Renew America**
1200 18th Street Northwest
Suite 1100
Washington, DC 20036
202-721-1545
Fax: 202-467-5780
http://www.solstice.crest.org
*Tina Hobson, President*

**5409 Research Triangle Institute**
3040 Cornwallis Road
PO Box 12194
Research Triangle Park, NC 27709
919-541-6000
Fax: 919-541-7155
E-mail: listen@rti.org
http://www.rti.org

Clients around the world rely on RTI to conduct innovative, multidisciplinary research to meet their R and D challenges. RTI's staff of more then 1,850 people represents a diverse set of technical capabilities in health and medicine,environmental protection, technology commercialization, decision support systems and education and training.
*Victoria F Haynes, President*
*Dennis F Naugle, VP Enviro and Engineering*

**5410 Resources for the Future**
1616 P Street Northwest
Washington, DC 20036
202-328-5000
Fax: 202-939-3460
E-mail: info@rff.org
http://www.rff.org
RFF is a nonprofit and nonpartisan think tank located in Washington DC that conducts independent research-rooted primarily in economics and other social sciences on environmental and natural resource issues. RFF was founded in 1952.

*Lesli Creedon, Vice Pres., External Affairs*
*Stan Wellborn, Director of Communications*

**5411 Resources for the Future: Energy & Natural Resources Division**
1616 P Street, Northwest
Washington, DC 20036
202-328-5000
Fax: 202-939-3460
http://www.rff.org

*Douglas Bohi, Director*

**5412 Resources for the Future: Quality of the Environment Division**
1616 P Street, Northwest
Washington, DC 20036
202-328-5000
Fax: 202-939-3460
http://www.rff.org

*Raymond J Kopp, Director*

**5413 River Studies Center**
University of Wisconsin
Department of Microbiology
La Crosse, WI 54601
608-785-8238
Fax: 608-785-6460

**5414 Robert J Bernard Biological Field Station**
1400 N Amherst Avenue
Claremont, CA 91711
909-624-6661
E-mail: Stephen.Dreher@cgu.edu
http://www.bfs.claremont.edu/

**5415 Rocky Mountain Biological Laboratory**
PO Box 519
Crested Butte, CO 81224
970-349-7231
Fax: 970-349-7481
E-mail: admin@rmbl.org
http://www.rmbl.org
High-altitude field station whose principal purpose is to provide quality research and teaching facilities for biologists and biology students of all diciplines who can benefit personally and intellectually from studying at thislocation. An important further purpose of the Laboratory is to promote the understanding and protection of the high altitude ecosystems of Colorado and the watershed of the Gunnison River through through the professional activities of its members.

*Ian Billick, Director*
*Dave Larson, Assistant Director*

**5416 Rocky Mountain Mineral Law Foundation**
9191 Sheridan Boulevard #203
Westminster, CO 80031
303-321-8100
Fax: 303-321-7657
E-mail: info@rmmlf.org
http://www.rmmlf.org

*David P Phillips, Executive Director*
*Mark Holland, Associate Director*

**5417 Romberg Tiburon Centers**
San Francisco State University
3150 Paradise Drive
PO Box 855
Tiburon, CA 94920
415-435-7100
Fax: 415-435-7120

**5418 Roosevelt Wildlife Station**
1 Forestry Drive
Syracuse, NY 13210
315-470-6798
Fax: 315-470-6934
E-mail: wfporter@esf.edu

**5419 Salt Institute**
700 N Fairfax Street
Suite 600
Alexandria, VA 22314
703-549-4648
Fax: 703-548-2194
E-mail: info@saltinstitute.org
http://www.saltinstitute.org
The Salt Institute is an international trade association of salt producers. It has information about the environmental impacts of salt production and use.

*Richard L Hanneman, President*

**5420 School for Field Studies**
10 Federal Street
Suite 24
Salem, MA 01970
978-741-3544
800-989-4435
Fax: 978-741-3551
E-mail: admissions@fieldstudies.org
http://www.fieldstudies.org
Students conduct hands-on, community-focused environmental field work around the world. Addresses critical environmental issues including preserving entire ecosystems or individual species, balancing economic development andconservation, and finding ways to manage and maintain wildlife, marine and agricultural resources.
*Bonnie Clendenning, President*

**5421 School of Marine Affairs (SMA)**
University of Washington
3707 Brooklyn Avenue NE
Seattle, WA 98105
206-543-7004
Fax: 206-543-1417
SMA is a masters-level, professional school within the University of Washington specializing in the interdisciplinary teaching and research on contemporary coastal and ocean resources, environmental and developmental problems.
*Marc J Hershman, Professor*

**5422 Science and Public Policy**
Rockefeller University
1230 York Avenue, Box 234
New York, NY 10021
212-327-7917
Fax: 212-327-7519
http://www.rockefeller.edu

**5423 Seatuck Foundation: Seatuck Research Program**
500 Saint Marks Lane
Islip, NY 11751
631-581-6908

**5424 Society for Ecological Restoration**
1955 West Grant Road
Suite 150
Tucsonn, AZ 85745
520-622-5485
Fax: 520-622-5491
E-mail: info@ser.org
http://www.ser.org

*John Rieger, President*

**5425 Society for the Application of Free Energy**
1315 Apple Avenue
Silver Spring, MD 20910
301-587-8686
Fax: 301-587-8688
E-mail: uv@uvbi.com
http://www.solarthem.com

**5426 Soil and Water Research**
4115 Gourrier Avenue
Baton Rouge, LA 70808
225-757-7726
Fax: 225-757-7728
http://mse.ars.usda.gov/la/btn/swr/
Mission is to characterize and quantify the transport and fate of agrochemicals in high water table soils, develop integrated soil, water, and agrochemical management systems that provide profitable yields and improve water table soilsin the humid, warm temperature areas of the US and develop improved soil and water management systems and operational procedures that enhance crop production conditions and increase the efficiency of conducting farming operationsin a timely manner.
*Dr. James Fouss, Research Leader*

**5427 Solar Energy Group**
University of Chicago
5640 S Ellis Avenue 773-702-7756
Chicago, IL 60637 Fax: 773-702-6317
E-mail: winston@rainbow.uchicago.edu

**5428 Solar Energy and Energy Conversion Laboratory**
University of Florida
Mechanical Engineering Department 352-392-0812
Gainesville, FL 32611 Fax: 352-392-1071
E-mail: solar@mae.ufl.edu
http://http://seecl.mae.ufl.edu/solar/

Has uniquely influenced the development of solar energy and renewable energy conversion systems all over the world through its research, education and training. Has pioneered research in many areas of solar energy, energy conversionand conservation. The lab has been designated as an ASME National Landmark.

*Dr. D Yogi Goswami, Professor & Director*

**5429 South Carolina Agromedicine Program**
295 Calhoun Street 843-792-2281
PO Box 250192 Room 121 800-922-5250
Charleston, SC 29425 Fax: 843-792-4702
E-mail: Simpsowm@musc.edu
http://www.musc.edu/oem/

Information, consultation, referral service for professional and lay persons involved in or in contact with agriculture or agricultural products.

*Dr. William M Simpson, Jr., Medical Director*
*JoAnn Stukes, Administative Assistant*

**5430 South Carolina Sea Grant Consortium**
287 Meeting Street 843-727-2078
Charleston, SC 29401 Fax: 843-727-2080
http://www.scseagrant.org

A state agency that supports coastal and marine research, education, outreach, one technical assistance program that foster sustainable economic development and resource conservation. The consortium represents eight university andstate research organizations and induces a number of information products on coastal and marine resource topics.
*16 pages*
*M Richard DeVoie, Executive Director*

**5431 Southern California Pacific Coastal Water Research Project**
7171 Fenwick Lane 714-894-2222
Westminster, CA 92683 Fax: 714-894-9699
http://www.sccwrp.org/#

SCCWRP is a joint powers agency focusing on marine environmental research. A joint powers agency is one that is formed when several government agencies have a common mission that can be better addressed by pooling resources andknowledge. In our case, the common mission is to gather the necessary scientific information so that our member agencies can effectively and cost-efficiently protect the Southern California marine environment.

**5432 Southwest Consortium on Plant Genetics & Water Resources**
New Mexico State University 505-646-6553
Box 3GL http://www.nmsu.edu
Las Cruces, NM 88003
*Dr. John D Kemp, Chairman*

**5433 Strom Thurmond Institute of Government & Public Affairs, Regional Development Group**
200 Excelsior Mill Road 864-646-4700
PO Box 158
Pendleton, SC 29670

**5434 Stroud Water Research Center**
970 Spencer Road 610-268-2153
Avondale, PA 19311 Fax: 610-268-0490
E-mail: webmaster@stroudcenter.org
http://www.stroudcenter.org

The Stroud Water Research Center seeks to understand streams and rivers and to use the knowledge gained from its research to promote environmental stewardship and resolve freshwater challenges throughout the world.
*B W Sweeney, Laboratory Director*

**5435 Sustainable Agriculture Research & Education Program**
ASI at UC Davis
1 Shields Avenue 530-752-7556
Davis, CA 95616 Fax: 530-754-8550
E-mail: sarep@ucdavis.edu
http://www.sarep.ucdavis.edu

*Tom Tomich, Director*
*Bev Ransom, Program Manager*

**5436 Tennessee Cooperative Fishery Research Unit**
Tennessee Technological University
Biology Department 931-372-3094
PO Box 5114 Fax: 931-372-6257
Cookeville, TN 38505 E-mail: jim_layzer@tntech.edu

*James Layzer, Leader*

**5437 Texas Center for Policy Studies**
44 East Avenue 512-474-0811
Suite 306 Fax: 512-474-7846
Austin, TX 78768 http://www.texascenter.org
*Mary Kelly, Executive Director*

**5438 Texas Water Resources Institute**
Texas A&M University 979-845-1851
College Station, TX 77843 Fax: 979-845-8554
E-mail: twri@tamu.edu
http://twri.tamu.edu

*C Allan Jones, Director*

**5439 The Natural History Museum & Biodiversity Research Center**
University of Kansas
Dyche Hall 785-864-4540
1345 Jayhawk Boulevard Fax: 785-864-5335
Lawrence, KS 66045 E-mail: kunhm@ku.edu
http://www.nhm.ku.edu/index.html

At the UK Natural History Museum & Biodiversity Research Center, they study the life of the planet for the benefit of the earth and its inhabitants. They document the diverse life of the earth, uncover its patters and document theresearch in order to better understand the natural environments, enhance power to predict environmental phenomena, and provide knowledge for natural resource management. Their ability to discover, document and diesseminate their research leads thenation.
*Dr. Leanord Krishtalka, Directo, Biodiversity Inst.*

**5440 Throckmorton-Purdue Agricultural Center**
8343 US231 Southy 765-538-3422
Lafayette, IN 47909 Fax: 765-538-3423
E-mail: jjf@aes.purdue.edu

*John Trott, Farm Director*

**5441 Toxic Chemicals Laboratory**
Cornell University
Cornell University, Tower Road 607-255-4538
New York State College of Agriculture
Ithaca, NY 14853

**5442 US Forest Service: Wildlife Habitat & Silviculture Laboratory**
506 Hayter Street 936-569-7981
Nacogdoches, TX 75965 Fax: 936-569-9681
http://www.srs.fs.fed.us/wildlife

We assess impacts of forest management practices on wildlife populations and their habitats and provide guide lines to land managers for improving their management to accommodate wildlife.
*Ronald E Thill, Project Leader*

**5443 USDA Forest Service: Northern Research Station**
1407 South Harrison Road 517-355-7740
Suite 220 Fax: 517-355-5121
East Lansing, MI 48823 E-mail: rhaack@fs.fed.us
*Dr. Robert Haack, Research Entonologist*

**5444 USDA Forest Service: Rocky Mountain Research Station**
2150 Centre Avenue 970-295-5926
Building A Fax: 970-295-5927
Fort Collins, CO 80526
We seek to be an unbiased source of scientific information; provide tools that consider the multidisciplinary nature of natural resource decisions and recognize that resource managers need scientific information that is integrated anddeveloped for application. Research results are made available through a variety of technical reports, seminars, demonstrations, exhibits and personal consultations. These help resource managers and planners balance economic and environmental demandsworldwide.
*Marcia Patton-Mallory, Station Director*

**5445 UVA Institute for Environmental Negotiation**
University of Virginia
104 Emmet Street 434-924-1970
Charlottesville, VA 22903 Fax: 434-924-0231
E-mail: envneg@virgina.edu/
http://www.virginia.edu/ien
The Institute for Environmental Negotiation is committed to building a sustainable future for Virginia's communities and beyond by: bringing people together to develop sustainable solutions; providing people with learning opportunitiesto be creative and collaborative leaders; and building understanding of best collaborative practices.
*Dr E Franklin Dukes, Director*

**5446 University Forest**
University of Missouri
46 University Forest Dr 573-222-8373
Wappapello, MO 63966 Fax: 573-222-8829
*Marie Oboryn, Site Manager*
*Hank Stetzer, Site Supervisor of Research*

**5447 VT Forest Resource Center and Arboretum**
University of Tennessee
901 S Illinois Avenue 865-483-3571
Oak Ridge, TN 37830 Fax: 865-483-3572
E-mail: utforest@utk.edu
http://forestry.tennessee.edu
The UT Forestry Experiment Station mission is to:(1) provide the land and supporting resources necessary for conducting modern and effective forestry, wildlife, and associated social, biological and ecological research programs;(2)demonstrate the application of optimal forest and wildlife management technologies; and (3) assist with transfer of new technology to forest land owners and industries.
*Richard M Evans, Center Director*

**5448 Vantuna Research Group**
Moore Laboratory of Zoology
Occidental College 323-259-2675
1600 Campus Road http://www.oxy.edu
Los Angeles, CA 90041
*John S Stephens, Director*

**5449 Virginia Center for Coal & Energy Research**
Virginia Tech 540-231-5038
Mail Code 0411 Fax: 540-231-4078
Blacksburg, VA 24061 E-mail: vccer@vt.edu
http://www.energy.vt.edu
Created by an Act of the VA General Assembly in 1977 as a study, research, information and resource facility for the commonwealth of VA, and is located at VA Tech. The mission involves four primary functions: research in energy andcoal related issues of interest to the Commonweatlth; coordination of coal and energy research at VA Tech; dissemination of coal and energy data to users in the Commonwealth; examination of socio-economic implications and environmental impacts ofcoal and energy.

*Dr Michael Karmis, Director*
*Margaret Radcliffe, Asst Director for Operations*

**5450 Washington Cooperative Fishery Research Unit**
University of Washington
Box 355020 206-543-6475
Seattle, WA 98195

**5451 Waste Management Education & Research Consortium**
New Mexico State University
Box 30001, MSC WERC 505-646-2038
Las Cruces, NM 88003 800-523-5996
Fax: 505-646-4149
E-mail: bdelrio@nmsu.edu
http://www.werc.net
A key component of WERC is higher education degree programs. To support this component, WERC administers a Fellowship Program at each academic partner institution. The primary objective for the WERC Fellowship program is to helpstudents develop a program which will lead to environmental related career opportunities upon graduation.

*Barbara Valdez, Program Facilitator*

**5452 Waste Management Research & Education Institute**
University of Tennessee
676 Dabney Hall 865-974-8080
Knoxville, TN 37996 Fax: 865-974-8086
E-mail: ceb@utk.edu
http://www.web.utk.edu
*Gary S Sayler, Acting Director*

**5453 Water Quality Laboratory**
Western Wyoming Community College 307-382-1662
PO Box 428 http://www.wwcc.cc.wy.us/
Rock Springs, WY 82901
*Craig Thompson, Director*

**5454 Water Resource Center**
University of Minnesota
1985 Buford Avenue 612-625-2282
St Paul, MN 55108 Fax: 612-625-1263
E-mail: ander045@umn.edu
http://wrc.coafes.umn.edu
The center coordinatoes research, education and extension programs on water resource issues. Administrative responsibility for Water Resource Sciences Graduate Program. Is the Water Resources Institute for Minnesota.
*Jim Anderson, Co-Director*
*Deb Swackhamer, Co-Director*

**5455 Water Resources Institute**
University of Wisconsin
1975 Willow Drive 608-262-0905
Floor 2 Fax: 608-262-0591
Madison, WI 53706 http://www.wri.wisc.edu
The University of Wisconsin Water Resources Institute's primary mission is to plan, develop and coordinate research programs that address present and emerging water-and land-related issues. It has developed a broadly based statewideprogram of basic and applied research that has effectively confronted a spectrum of societal concerns. It is one of 54 institutes or centers located at the Land Grant College in each state.
*Anders W Andren, Director*

**5456 Water Resources Research Institute at Kent University**
Kent State University
230 Research One Bldg 330-672-2529
PO Box 5190 Fax: 330-672-4834
Kent, OH 44242 E-mail: rheath@kent.edu
http://http://dept.kent.edu/wrri

The institute fosters a broad-based approach to the evaluation and analysis of environmental problems related to water use. WRRI is a resouce for citizens, governmental agencies and policy makers, providing reliable scientificinformation on which to base decisions related to the wise use and management of water and land management, water policy decisions and environmental conservation.

*Dr. Robert T Heath, Director*
*Dr. Joseph Otiz, Assistant Director*

**5457 Water Resources Research of the University of North Carolina**
North Carolina State University
1131 Jordan Hall, Box 7912
Raleigh, NC 27695
919-515-2815
Fax: 919-515-2839
E-mail: water_resources@nesu.edu
http://www.ncsu.edu/wrri/

One of 54 state water institutes authorized to administer and promote federal/state partnerships in research and information transfer on water-related issues. Identifies and supports research needed to help solve water quality andwater resources problems in NC. Publishes peer-reviewed reports on completed research projects. Sponsors educational seminars and conferences and provides public information on water issues through publication of a newsletter.

*Dr David H Moreau, Director*
*Kelly Porter, Env. Ed. & Comm. Coordinator*

**5458 Water Testing Laboratory**
Morehead State University
Box 804
Morehead, KY 40351
606-783-2961
http://www.morehead-st.edu
*Rita Wright, Contact*

**5459 Weather Analysis Center**
University of Michigan
Space Research Building
2455 Hayward Street
Ann Arbor, MI 48109
734-936-0482
Fax: 734-763-0437
E-mail: dbaker@umich.edu
http://www.aoss.engin.umich.edu

Atmospheric, planetary and space science engineering.
*Dr Dennis Baker, Director*

**5460 West Virginia Water Research Institute**
West Virginia University
Room 202 NRCCE
Box 6064
Morgantown, WV 26506
304-293-2867
Fax: 304-293-7822
E-mail: pziemkie@wvu.edu
http://www.wvwri.org

The West Virginia Water Research Institute (WVWRI) has served as a statewide vehicle for performing research related to water issues. WVWRI serves as the coordinating body for the following programs: the National Mine Land ReclamationCenter, Appalachian Clean Streams Initiatve, Acid Drainage Technology Initiative, Northern WV Brownfields Assistant Center, Hydrogeology Research Center, State Water Institutes, and more.

*Paul Ziemkiewicz, PhD, Director*

**5461 Western Region Hazardous Substance Research Center**
Oregon State University
204 Merry Field
Corvallis, OR 97331
541-737-2751
Fax: 541-737-3099
E-mail: wrhsrc@engr.orst.edu
http://http://wrhsrc.oregonstate.edu

Funding for the center ended in 2006 and it is no longer in opertian, but information is still available on the website.
*Lew Semprini, Director*

**5462 Western Research Farm**
36515 Highway 34 E
Castana, IA 51010
712-885-2802
E-mail: wroush@iastate.edu
*Wayne B Roush, Ag Specialist*

**5463 Wetland Biogeochemistry Institute**
Louisiana State University
Baton Rouge, LA 70803
225-388-8806
Fax: 225-388-6423
*William H Patrick Jr, Director*

**5464 Wilderness Institute: University of Montana**
School of Forestry
Missola, MT 59812
406-243-5361
800-462-8636
Fax: 406-243-4845
E-mail: wi@forestry.umt.edu
http://www.forestry.umt.edu.si

Mission is to further understand wilderness and its stewardship through education, outreach, and scholarship. Activity is guided by the philosophy that wildlands are increasingly significant, ecologically and socially, and educateddialogue about the role of wild places in our nation's future should be promoted. Engaged in undergraduate education, graduate student research, the dissemination of wilderness information and the promotion of scholarship on wilderness issues.
*Wayne Freimund, Director*
*Laurie Ashley, Outreach Coordinator*

**5465 Wilderness Research Center**
University of Idaho
College of Natural Resources
Room 18a
Moscow, ID 83844
208-885-7911
Fax: 208-885-6226
E-mail: rrt@uidaho.edu
http://http://www.cnr.uidaho.edu/wrc/

The mission of the WRC is to study the human dimensions of wilderness ecosystems. The WRC conducts research and teaches courses on the use of wilderness for personal growth, therapy, education, and leadership development.
*Steve Hollenhorst, Ph.D, Director*
*Lilly Steinhorst, Administrative Assistant*

**5466 Wisconsin Applied Water Pollution Research Consortium: University of Wisconsin-Madison**
University of Wisconsin
3232 Engineering Hall
Madison, WI 53706
608-263-7773
Fax: 608-262-5199
E-mail: harringt@engr.wisc.edu
http://www.engr.wisc.edu/consortial/cawper

This consortium seeks effective and economical solutions to water supply problems and pollution control in Wisconsin. It conducts innovative practical research that cannot be carried out effectively by individual organizations.
*Greg Harrington, Director*

**5467 Wisconsin Rural Development Center**
USDA Rural Development-WI
4949 Kirschling Ct.
Stevens Point, WI 54481
715-345-7615
Fax: 715-345-7614
E-mail: RD.Webmaster@wi.usda.gov
http://www.rurdev.usda.gov/wi/index.htm

*Frank Frassetto, State Director*

**5468 Wisconsin Sea Grant Institute**
University of Wisconsin
1975 Williw Drive
Floor 2
Madison, WI 53706
608-262-0905
Fax: 608-262-0591
http://www.seagrant.wisc.edu

The University of Wisconsin Sea Grant Institute is a statewide program of basic and applied research, education, and technology transfer dedicated to the wise stewardship and sustainable use of Great Lakes and ocean resources. It ispart of a national network of 30 university-based programs.

*Dr. Anders W Andren, Director*

**5469 Yale Institute for Biospheric Studies (YIBS)**
21 Sachem Street
PO Box 208105
New Haven, CT 06520
203-432-9856
Fax: 203-432-9927
E-mail: roserita.riccitelli@yale.edu
http://www.yale.edu/yibs

*Rose Rita Riccitelli, Assistant Director*
*Jeffrey Park, Director*

# Educational Resources & Programs

## Universities

**5470 Academy for Educational Development**
Center for Environmental Strategies
1825 Connecticut Avenue NW
Washington, DC 20009
202-884-8000
Fax: 202-884-8400
E-mail: web@aed.org
http://www.aed.org

Develops sustainable solutions to global environmental protection and natural resource management problems through individual and institutional behavior change, education, training and communication strategies. Efforts are driven by astrong commitment to improve or maintain environmental quality as well as the quality of life for diverse communities and groups through the provision of technical assistance, guided practice and capacity building support.

*Edward Russell, Chair*

**5471 Allegheny College**
Environmental Science/Studies
520 North Main Street
Meadville, PA 16335
814-332-3100
E-mail: info@allegheny.edu
http://www.allegheny.edu

Offers the study of interrelationships between human activities and the environment. Two major programs: 1) Environmental Science. Core courses include biology, chemistry, geology, and mathematics. Upper level courses synthesize,integrate and apply basic sciences toward solving real environmental problems; 2) Environmental Studies. Objective is to study the concept of sustainability in an integrated way.

*Richard J Cook, President*

**5472 Antioch College**
Glen Helen Outoor Education Center
1075 State Route 343
Yellow Springs, OH 45387
937-767-7648
Fax: 937-767-6655
E-mail: rjaramillo@antioch-college.edu
http://www.antioch-college.edu

Training in residential naturalist instruction for upper elementary aged students. Classes and field experience in outdoor education methods and natural history. Care for hawk or owl in Raptor Center.

*Rebecca Jaramillo, Assistant Director*

**5473 Antioch University/New England**
Environmental Studies
40 Avon Street
Keene, NH 03431
800-553-8920
Fax: 603-357-0718
E-mail: admissions@antiochne.edu
http://www.antiochne.edu

For those committed to scholarly excellence and wish to design, implement and evaluate research regarding crucial environmental issues. The PhD program cultivates a dynamic learning community of environmental practioners who addresscomplex regional, national, and global issues responsibly, creatively, and compassionately.

*Joy Ackerman, Chair*
*Laura Andrews, Admissions Co Director*

**5474 Antioch University/Seattle**
Center for Creative Change
2326 Sixth Avenue
Seattle, WA 98121
206-441-5352
E-mail: osmythe@antiochsea.edu/
http://http://www.antiochsea.edu/

Approches environmental concerns by emphasizing social science perspectives and natural science literacy. The program is part of the Center for Creative Change, an integrated professional studies center.

*Ormond Smythe, Academic Dean*

**5475 Arkansas Tech University**
Wildlife Conservation Program
1605 Coliseum Drive
Suite 141
Russellville, AR 72801
800-582-6953
E-mail: sdonnell@atu.edu
http://www.atu.edu

A two-year preparatory program in Wildlife Conservation with an outlined Wildlife Curriculum was developed at Arkansas Tech University in 1956. Two years later, plans were made to elevate this program to a four-year program. During the1959-1960 academic year, a full slate of courses was developed that provided the foudation for degree that specialized in fisheries and wildlife management.

*Shauna Donnell, Vice President*

**5476 Auburn University**
Environmental Institute
101 Comer Hall
Auburn, AL 36849
334-844-4132
Fax: 334-844-4462
E-mail: blockdh@auburn.edu
http://http://auei.auburn.edu/

Serves faculty, governments, and the general public in a coordinating role to bring together teams to develop acceptable and economically feasible means of enhancing the environmental quality of the state and nation.

*Dennis Block, Interim Director*
*M. Kay Stone, Academic Program Assistant*

**5477 Ball State University**
Natural Resources and Environmental Management
2000 West University
Muncie, IN 47306
765-285-5780
Fax: 765-285-2606
E-mail: nrem@bsu.edu
http://www.bsu.edu/nrem

The Natural Resources and Environmental Management Department enhances scientific competence and prepares students for a variety of environmental careers. Programs focus on air, energy, land, parks, recreation, soil, waste management,and water and emergency management.

*James Eflin, Chairperson*

**5478 Bard College**
Environmental and Urban Studies
Annandale-on-Hudson, NY
845-758-6822
http://www.bard.edu

This program focuses on both the lived and built environments. Its goal is to involve students in empirically-based studies that bridge the divisions between natural and artificial, given and created. This approach is designed to buildon the transformations within a range of social and natual sciences, ranging from systems theory to enviornmental toxicology.

*Sanjib Baruah, Director*
*Felicia Keesing, Professor*

**5479 Bemidji State University**
Center for Environmental, Earth & Space Studies
1500 Birchmont Drive NE
Bemidji, MN 56601
218-755-4104
888-345-1721
Fax: 218-755-4107
E-mail: fchang@bemidjistate.edu
http://www.bemidjistate.edu/

The Center for Environmental, Earth and Space Studies (CEESS) offer a unique variety of interdisciplinary degree programs. Degrees in Environmental Studies include both B.S. and M.S., and a B.S. or B.A. with geology minor is alsoavailable. Students in the CEESS program are concerned with both the technological problems and social apects of enviromental issues.

*Fu-Hsian Chang Ph.D, Director & Professor*

**5480 Bradley University**
Geological Sciences Program
1501 West Bradley Avenue
Peoria, IL 61625
800-447-6460
E-mail: admissions@bradley.edu
http://www.admissions.bradley.edu

Aims to develop an awareness of the Earth as a dynamic and unified system in time and space. Curriculum is preparatory for careers in geology, engineering geology, geophysics, hydrogeology, oceanography or secondary Earth scienceteaching.

**5481 Brooklyn College**
Environmental Studies
2900 Bedford Avenue 718-951-4159
Brooklyn, NY 11210 Fax: 718-951-4546
E-mail: yklein@brooklyn.cuny.edu
http://http://www.envirolink.org
Program is aimed at educating students to be fluent in social and physical sciences as related to the environment. In addition, the program draws from other courses in humanities, social sciences, mathematics, and sciences. Thisinterdisciplanary approach is designed to introduce the field of environmental studies and to apply this knowledge to various careers.
*Yehuda Klein, Deputy Director*

**5482 Brown University**
Center for Environmental Studies
135 Angel Street 401-863-3449
Box 1943 Fax: 401-863-3503
Providence, RI 02912 E-mail: envstudies@brown.edu
http://www.envstudies.brown.edu
The Center for Environmental Studies at Brown University was established with the primary aim of educating individuals to solve challenging environmental problems both at the local and global levels. It also works directly to improvehuman well-being and environmental quality through community, city, and state partnerships in service and research.
*J Timmons Roberts, Program Director*

**5483 California Polytechnic State University**
Institute for City & Regional Planning
Cal Poly 805-756-1315
San Luis Obispo, CA 93407 Fax: 805-756-1340
E-mail: crp@calopoly.edu
http://www.planning.calpoly.edu
Developed to coordinate interdisciplinary projects and research relating to the management of watersheds, urban areas, marine environments and related natural and human resources. The Institute offers specialists in various areas suchas biological science, business administation, city and regional planning, civil and environmental engineering, economics, geology, landscape architecture, natural resources management, political science and soil science.
*William Siembieda, Department Head*
*W. David Conn, Vice Provost*

**5484 California State University/Fullerton**
Environmental Studies
PO Box 34080 714-278-2011
Fullerton, CA 92834 E-mail: arwebmaster@fullerton.edu
http://www.fullerton.edu
The Environmental Studies program is an interdisciplinary program that broadens environmental knowledge and awareness. It's designed to prepare students as professionals in the environmental field by providing an opportunity to learnapplicable skills and to develop an appropriate body of knowledge.
*Milton Gordon, President*

**5485 California State University/Seaside**
Capstone Project Program
100 Campus Center 831-582-3689
Seaside, CA 93955 E-mail: laura_lienk@csumb.edu
http://http://watershed.csumb.edu/
Capstone Projects encompass a broad array of student interests, primarily within the programs of Earth Systems Science and Policy. A Capstone Project is similar to a senior thesis project at other universites, and showcases mastery ofESSP skills. They follow a set of outcome based, interdisciplinary criteria used to measure the competence of participants.
*Laura Lee Lienk*

**5486 California University of Pennsylvania**
Biological & Environmental Sciences
250 University Avenue 724-938-4000
California, PA 15419 E-mail: pavtis@cup.edu
http://www.cup.edu
Department includes intensive scientific curricula that prepare students for graduate work in the biological and environmental sciences and career work in related areas.
*Jenifer Siqado, Director*

**5487 Carnegie Mellon University**
Civil and Environmental Engineering Program
5000 Forbes Avenue 412-268-2940
Pittsburgh, PA 12513 Fax: 412-268-7813
E-mail: webmaster@ce.cmu.edu
http://www.ce.cmu.edu
A major function of the Environmental Institute is to enable Carnegie Mellon to play a leadership role in developing educational programs on environmental issues. These include initiatives at both undergraduate and graduate levels.
*Michael Balderson, Administrative Coordinator*
*Deborah Lange, Executive Director*

**5488 Clark University**
Environmental Science & Policy
950 Main Street 508-793-7711
Worcester, MA 01610 E-mail: idce@clarku.edu
http://www.clarku.edu/
An interdisciplinary approach that emphasizes policy questions involving the environment and use and misuse of science and technology. Its goal is to enable individuals to deal with technical and environmental issues in social andpolitical areas. Topics addresses deal with urgent and important issues, including assessment and management of environmental risks to humans and ecosystems, capacity for sustainable development in third world countriie, and integrated watershedmanagement.

**5489 Clemson University**
Environmental Engineering and Earth Sciences
342 Comupter Court 864-656-3276
Anderson, SC 29625 http://www.clemson.edu
Programs in the environmental field focus on environmental process engineering, hydrogeology, environmental health physics and radiochemistry, environmental chemistry and sustainable systems.
*Esin Gulari, Dean*

**5490 Colby-Sawyer College**
Community & Environmental Studies
541 Main Street 603-526-3000
New London, NH 03257 E-mail: kslover@colby-sawyer.edu
http://http://www.colby-sawyer.edu
Bachelor of Science degree in Community and Environmental Studies. A minor in CES is also available.
*Thomas Galligan, President*

**5491 College of Natural Resources**
Conservation Management Institute
1900 Kraft Drive 540-231-8851
Suite 250 Fax: 540-231-7019
Blacksburg, VA 24061 E-mail: vemrick@vt.edu
http://www.cmi.vt.edu
Offers multi-disciplinary research that addresses conservation management effectiveness throughout the world. Faculty from far reaching research institutions work collaboratively on projects ranging from endangered species propagationto natural resource-based satellite imagery interpretation.
*Verl Emerick, Millitary Lands Coordinator*

**5492 College of William and Mary**
Center for Conservation Biology
PO Box 1346 804-684-7000
Rt 1208 Fax: 804-684-7097
Gloucester Point, VA 23062 E-mail: wmaster@vims.edu
http://www.vims.edu/
The Center for Conservation Biology is an organization dedicated to discovering innovative solutions to environmental problems that are both scientifically sound and practical within today's social context. It has been a leader inconservation issues throughout the mid-Atlantic region with a philosophy that uses a

general systems approach to locate critical information needs and to plot a deliberate course of action to reach goals.

*John T. Wells, Dean And Director*
*Roger L. Mann, Director For Researech*

**5493 Colorado Mountain College**
Natural Resource Management Program
831 Grand Avenue
Glenwood Springs, CO 81601
800-621-8559
Fax: 970-947-8324
E-mail: joinus@coloradomtn.edu
http://www.coloradomtn.edu

The Natural Resource Management program grew out of the Environmental Technology, which was one of the most well established programs of its kind in the country. It specializes in helping students graduate with entry-level skills in avariety of environmental fields, while combining aquatic and terrestrial resource management. Students are trained in career fields of environmental site assessment, hydrology, soil science, environmental law and others.

**5494 Colorado School of Mines**
Environmental Engineering & Applied Science
1500 Illinois Street
Golden, CO 80401
303-273-3000
http://www.mines.edu

This public research university devoted to engineering and applied science, offers a curriculum and research program that is geared toward responsible stewardship of the earth and its resources. It has broad expertise in resourceexploration, extration, production and utilization. The programs at Mines are central to balancing resource availability with environmental protection.
*William Scoggine, President*

**5495 Colorado State University**
College of Natural Resouses
101 Natural Resources Building
Campus Delivery 1401
Fort Collins, CO 80523
970-491-6675
Fax: 970-491-0279
E-mail: ask_us_dean@cnr.colostate.edu
http://www.cnr.colostate.edu

The College of Natural Resources is one of the most comprehensive environment and natural resources programs in the nation. With four departments and eight undergraduate majors, 9 minors and 13 concentrations, students address the mostcurrent issues in environment and natural resources, including endangered species, water quality, biological diversity, parks forests and wildlife management, recreation and environmental and ecosystem sciences.
*Joseph T. O'Leary, Dean*

**5496 Columbia University**
Public Admin in Environmental Science & Policy
2960 Broadway
New York, NY 10027
212-854-1754
http://www.columbia.edu

The program is designed to train sophisticated public managers and policymakers who apply innovative, system-based thinking to environmental issues. It emphasizes practical skills and is enriched by ecological and plantatary science.

**5497 Connecticut College**
Goodwin Niering Center for Conservation Biology
270 Mohegan Avenue
Box 5293
New London, CT
860-439-5417
Fax: 860-439-2418
E-mail: goodwin-nieringcenter@conncoll.edu
http://www.conncoll.edu

A comprehensive, interdisciplinary program aimed at understanding contemporary ecological challenges. Its Certificate Program offers students the opportunity to blend thier interest in the environment with a non-science major and is ofparticular interest to those planning careers in environmental policy, law, economies or education.
*Glenn Dreyer, Executive Director*
*Douglas Thompson, Director*

**5498 Conservation Leadership School**
301B Agricultural Admin Bldg
University Park, PA 16802
814-865-8301
877-778-2937
Fax: 814-865-7050
E-mail: csco@psu.edu
http://www.conferences.cas.psu.edu/cls

A one-week residential program for high school students to learn about the world around them through exploration and hands-on activities. The classroom includes over 700 acres of forest, fields, wetlands, and streams, and learningabout the environment includes having fun, meeting new friends and learning leadership skills.

**5499 Cornell University**
Center for the Environment
201 Rice Hall
Ithaca, NY 14853
607-255-7535
Fax: 215-701-1844
E-mail: environment@cornell.edu
http://www.environment.cornell.edu/

Offers opportunities for graduate study in the ecology, management, and policy of fishery, forest, wetland, wildlife, and other environmental resources. There also are opportunities to focus on conservation biology, agroforestry,environmetnal change, and conservation and sustainable development.
*Anthony G. Hay, Director*
*Susan Henry, Dean*

**5500 Dartmouth College**
Department of Earth Sciences
Hanover, NH 03755
603-646-1110
E-mail: cantact@dartmouth.edu
http://www.dartmouth.edu/

Offers opportunities for learning and research in all major disciplines devoted to the study of the earth, including its structure and development, the oceans and atmosphere, weather and climate. Teaching and research at a moreadvanced level emphasize watershed processes, environmental biogeochemistry, geophysics and mechanics, sedimentology, paleontology, economic geology, end remote sensing of the earth from aircraft and satellites.
*James Wright, President*

**5501 Delaware Valley College**
Agronomy & Environmental Science
700 East Butler Avenue
Doylestown, PA 18901
215-345-1500
E-mail: webmaster@delval.edu
http://www.delval.edu

The Department of Agronomy and Environmental Science offers courses designed to give a broad, workable background in the plant, soil, turf or environmental sciences. Focusing on the environmental issues facing society today, thesecourses provide the knowledge and training necessary to be successful in the field or to move on to the graduate level.
*Dorothy Prisco, Vice President*

**5502 Drake University**
Environmental Science & Policy Program
2507 University Avenue
Olin Hall
Des Moines, IA 50311
515-271-2920
800-443-7253
Fax: 515-271-3702
E-mail: thomas.rosburg@drake.edu
http://www.drake.edu/artsci/env

Environmental Science and Policy Program is an interdisciplinary program that awards BS and BA degrees in both Environmental Science and Environmental Policy. There are 50 to 60 students in either of these majors.

*Thomas Rosburg, Director*

**5503 Duke University/Marine Laboratory**
Nicholas School of the Environment
Marine Laboratory
135 Duke Marine Lab Road
Beaufort, NC 28516
252-504-7502
Fax: 252-504-7648
E-mail: hnearing@duke.edu
http://www.nicholas.duke.edu/marinelab

The Laboratory is a campus of Duke University and a unit within the Nicholas School of the Environment. The mission is education, research, and service to understand marine systems, including the human component, and to developapproaches for marine conservation and restoration.

*Helen Nearing*

**5504 Duquesne University**
Environmental Science & Management Program
600 Forbes Avenue 412-396-6632
Pittsburgh, PA 15282 E-mail: admissions@dug.edu
http://http://www.dug.edu
Educating environmental professionals for the twenty-first century is the focus of this program, which grew out of the need to combine depth of knowledge in environmental science with a comprehensive understanding of the business,legal and policy implications surrounding environmental issues.
*Charles J. Dougherty, President*

**5505 Eastern Illinois University**
Aquatic and Fisheries Program
600 Lincoln Avenue 217-581-5000
Charleston, IL 61920 E-mail: csmhp@eiu.edu
http://www.eiu.edu
Offers aquatic ecology, fisheries biology, and physiological ecology. Specific areas of concentration include community analysis of stream fishes, life history and demographics of fish, bioenergetics of development and life historyphenomena, and lipid storage and utilization patterns of fish.
*Mary Herrington-Perry, Assistant Vice President*

**5506 Eastern Michigan University**
Kresge Environmental Education Center
202 Welch Hall 734-487-1849
Ypsilanti, MI 48197 800-468-6368
Fax: 734-487-6559
E-mail: tkasper@emich.edu
http://www.emich.edu
The Kresge Environmental Education Center is located in Mayfield, about six miles north east of the city of Lapeer. The main buildings are located on Fish Lake Road in the middle of the center's 240 acres. These 240 acres sit next to7,000 acres of state land.
*Tom Kasper, Admissions*

**5507 Eastern Nazarene College**
Eastern Environmental Program
23 East Elm Avenue 617-745-3546
Quincy, MA 02170 800-88E-NC88
Fax: 617-745-3907
E-mail: jonathan.e.twining@enc.edu
http://www.enc.edu
Cross-disciplinary program which provides for students strong pereparation in the several scientific disciplines involved in the study of environmental issues. The program is jointly sponsored by the Department of Biology andChemistry in order to provide the appropriate basis in all the sciences for students wishing to pursue environmental careers or graduate school.
*Jonathan E Twining, Instructor, Env. Science*

**5508 Fairleigh Dickinson University**
Environmental Studies/System Science
College at Florham
285 Madison Avenue
Madison, NJ 07940
This program offers students a wide variety of 18 concentrations, including envionmental chemistry, environmental risk assessment, water treatment, environmental planning, groundwater hydrology, environmental remediation, soil science,land-use planning and air pollution.

**5509 Ferrum College**
Environmental Science
PO Box 1000 540-365-2121
Ferrum, VA 24088 800-868-9797
Fax: 540-365-4203
E-mail: webmaster@ferrum.edu
http://www.ferrum.edu
Offers programs in environmental science that includes informationon air pollutant deposition in the Great Lakes and the formulation of membrances for ion-selective electrodes. The program also includes participation in a water qualitymonitoring project on Smith Mountain Lake that uses a geographical information system to model soil loss in its watershed.
*Sandra Prillaman, Executive Assistant*

**5510 Field Station & Ecological Reserves**
University of Kansas
c/O Kansas Biological Survey 785-864-1505
2101 Constant Avenue Fax: 785-864-5093
Lawrence, KS 66047 http://www.ksr.ku.edu
The KSR is dedicated to field-based environmental reseach and education. KSR is located within the transition zone (ecotone) between the eastern deciduous forest and tallgrass prairie biomes. The 3,000 acres of diverse native andmanaged habitats, experimental systems, support facilities, and longterm databases are used to undertake an outstanding array of scholarly activities. Environmental stewardship is a stong emphasis as high-quality natural areas are preserved for thefuture.
*Edward Martinko, KSR Director*

**5511 Florida State University**
Environmental Studies
Tallahassee, FL 32306 850-644-2525
Fax: 850-644-9936
E-mail: admissions@admin.fsu.edu
http://www.fsu.edu
Offers the study of environmental issues as they relate to geological phenomena, which include volcanic and earthquake hazards, resource and land- use planning, air and water pollution, waste disposal, glaciation and sea-level change,landslides, flooding, shoreline erosion, and global change issues.
*Thomas Wetherell, Professor*

**5512 George Washington University**
International Environmental Policy & Management
2121 1st Street NW 202-994-1000
Washington, DC 20052 E-mail: hmerchnt@awu.edu
http://www.gwu.edu/
Offers a program on International Environmental Policy and Management and Marketing Management, held in various locations around the world
*Laurel Price Jones, Vice President*

**5513 Georgetown University**
Environmental Studies
37th O Streets NW 202-687-0100
Washington, DC 20057 E-mail: webmaster@georgetown.edu
http://http://www.georgetown.edu
Environmental Studies is an interdisciplinary program designed to allow an undergraduate of the college majoring in any discipline to focus on environmental issues. Environmental Studies provides a framework for the study offundamental mechanisms of ecosystems and human interaction with the Earth. Environmental studies encompasses the humanities, social sciences and natural sciences as they relate to environmental questions.
*Wayne A. Davis, President*

**5514 Hocking College**
Environmental Restoration Technology Program
3301 Hocking Parkway
Nelsonville, OH 45764 877-462-5464
Fax: 740-753-7065
E-mail: admissions@hocking.edu
http://www.hocking.edu
Growing concern for the environment has increased the need for technicians qualified in the restoration of environmentally unstable land, water, and air. Hocking College's Environmental Restoration technology prepares students for thatchallenge.
*Tammy Andrews*

**5515 Idaho State University**
Geochemistry & Hydrogeology Program
921 South Eighth Avenue 208-282-2362
Pocatello, ID 83209 E-mail: webmaster@isu.edu
http://www.isu.edu/
Emphasizes environmental geochemistry and hydrogeology. This specialty is ideal in southern Idaho where problems of nu-

clear and toxic waste clean-up at the Idaho National Environmental Engineering Laboratory will require study andgenerate research monies for years to come.
*Robert A. Wharton, Provost And Vice President*

**5516 Indiana State University**
Department of Ecology & Organismal Biology
Terre Haute, IN 47809
812-237-3993
888-824-3920
Fax: 812-237-8525
E-mail: rlfaq@isugw.indstate.edu
http://www.indstate.edu/
This department conducts research in the areas of ecology, evolution, and conservation. The M.S. and Ph.D degrees garnered in state-of-the-art laboratories and local field stations enable students to conduct innovative research andplay significant roles as well-trained evnironmentists.
*Jack Maynard, Vp Academic Affairs*

**5517 Iowa State University**
Environmental Studies Program
100 Alumni Hall
Ames, IA 50011
515-294-5836
800-262-3810
Fax: 515-294-2592
E-mail: admissions@iastate.edu
http://www.iastate.edu
The Environmental Studies Program deals with the relationship between humans and nature, or between humans and natural systems. The curriculum is designed to give students an understanding of regional and global environmental issuesand an appreciation of different perspectives regarding these issues. Courses are provided for both students pursuing careers related to the environment and those with an interest in environmental issues.

**5518 Johns Hopkins University**
Department of Geography/Environmental Engineering
3400 N Charles Street
Baltimore, MD 21218
410-516-4050
E-mail: engineeringinfo@jhu.edu
http://www.jhu.edu/
Concerned with understanding the nature and dynamics of ecosystems, engineered systems, and societies. Offers a broad range of graduate programs including the natural, social and engineering sciences.
*Nick Jones, Dean*

**5519 Kansas State University**
Department of Agricultural Economics
342 Waters Hall
Manhattan, KS 66506
785-532-6151
Fax: 785-532-6897
E-mail: kstateag@k-state.edu
http://www.agecom.ksu.edu
The Department of Agricultural Economics has a rich tradition of services to agriculture and related fields. The department has a history of succes in its land-grant mission, teaching, research, and extension outreach, maintaininglarge and diverse programs in undergraduate and graduate instruction, as well as research and extension outreach.
*Dr David Lambert, Department Head*
*Cherie Hodgson, Academic Program Coordinator*

**5520 Keene State College**
Environmental Studies
229 Main Street
Keene, NH 03435
800-572-1909
E-mail: admissions@keene.edu
http://www.keene.edu
Environmental Studies is an interdisciplinary program comprised of courses in Biology, Chemistry, Economics, Geography, Geology, and Political Science. Two concentration options are Environmental Policy and Enviromental Science, bothof which will prepare students for a wide range of environment-related career opportunities.
*Tim Allen, Associate Professor*

**5521 Lake Erie College**
Environmental Management Program
391 West Washington Street
Painesville, OH 44077
440-296-1856
800-533-4996
Fax: 440-375-7005
E-mail: webmaster@lec.edu
http://http://www.lakeerie.edu
Interdisciplinary major, grounded in the sciences and liberal arts,designed for those who want to pursue career paths utilizing environmental science in decision making. Courses include environmental management, biology, chemistry,mathamatics and business. Program is designed to help students build a solid knolwledge base in regional and global environmental issues.
*Jerome T. Osborne, President*
*Michael E. Bee, Executive Vice President*

**5522 Louisiana State University**
Environmental Sciences Program
1285 Energy, Coast and Environment Building
Baton Rouge, LA 70803
225-578-8521
Fax: 225-578-4286
E-mail: edlaws@lsu.edu
http://http://info.envs.lsu.edu
Environmental Sciences program is designed to provide a broad-based graduate education to prepare students for careers in industrial, government, and academia. The program builds on a strong undergraduate background in the sciences.
*Edward Laws, Interim Chair*
*Charolette Romaine, Academic Coordinator*

**5523 Louisiana Tech University**
Wildlife Conservation Program
305 Wisteria Street
PO Box 10197
Ruston, LA 71272
318-257-4287
E-mail: ANSmail@ans.latech.edu
http://www.ans.latech.edu/
The Wildlife Conservation degree program meets the certification requirements of the Wildlife Society, and graduates may apply for certification as an Associate Wildlife Biologist.
*James Liberatos, Dean*

**5524 Miami University**
Institute of Environmental Sciences
501 East High Street
Oxford, OH 45056
513-529-1809
E-mail: secretary@muohio.edu
http://www.muohio.edu/
Offers a masters degree in Environmental Science. This interdisciplinary program stresses problem solving and community service, and provides practical experience in many potential areas of concentration, preparing students for avariety of practical careers in public and private sector jobs.

*David C. Hodge, President*

**5525 Michigan Technological University**
Applied Technology & Environmental Science
1400 Townsend Drive
Houghton, MI 49913
906-487-2454
800-966-3764
Fax: 906-487-2915
E-mail: forest@mtu.edu
http://forestry.mtu.edu/
A degree in Applied Ecology and Environmental Sciences prepares students to address complex environmental problems posed by the use of natural resources. Students learn how to protect the integrity of ecosystems and help assure thatnatural resources will be managed wisely for generations of sustainable use.
*Margaret Gale, Dean*

**5526 Middlebury College**
Environmental Studies
121a South Main Street
Middlebury, VT 05753
802-443-5418
Fax: 802-443-2060
E-mail: blse@breadnet.middlebury.edu
http://www.middlebury.edu
Explores the relationship between humans and their environment. Students pursuing the ES major work in a variety of disciplines, including biology, chemistry, economics, geography, geology, literature, the performing arts, philosophy,political science, religion and sociology.
*James Maddox, Director*

**5527 Montana State University**
Montana Environmental Training Center
2100 16th Avenue South
Great Falls, MT 59405
800-662-6132
E-mail: webmaster@msun.edu
http://www.msun.edu/

A cooperative effort between Montana State University/Northern and the Montana Department of Environmental Quality. Offers basic, advance training, and continuing education in the areas of water and wastewater operation, maintenance,safety, process control, cross connection and backflow prevention along with courses in basic water science and watershed awareness. A newsletter and Training Announcement are published quarterly.
*Alex Capdeville, Chancellor*

**5528 New Mexico State University**
Department of Geological Sciences
Box 30001
MSC 3AB
Las Cruces, NM 88003
505-646-2708
Fax: 505-646-1056
E-mail: geology@nmsu.edu
http://www.nmsu.edu/~geology

Offers both undergraduate and graduate study leading to advanced degrees in geological science. Advanced training qualifies students for employment in such branches of geological science as mining, petroleum, environmental andengineering geology, government service or for further graduate study. The education experience may include sedimentology, geochemistry, volcanology, stratigraphy, geotectonics and paleontology.

*Dr. Nancy S. McMillan, Department Head*

**5529 North Dakota State University**
Department of Biological Sciences
Fargo, ND 58105
701-231-7087
Fax: 701-231-7149
http://www.biology.ndsu.nodak.edu/

Department offers undergraduate and graduate degrees in biological disciplines, including environmental and conservation sciences.

*Martinus Otte, Department Head*

**5530 Northeastern Illinois University**
International Center for Tropical Ecology
5500 N St. Louis
Chicago, IL 60625
773-583-4050
Fax: 773-442-4900
E-mail: admrec@neiu.edu
http://www.neiu.edu

The International Center for Tropical Ecology provides a focal point for interdisciplinary research and graduate education in all aspects of the conservation of tropical ecosystems. The Center, formed in collaboration with the MissouriBotanical Garden, supports a network in the United States of students, scientists, and conservationists from tropical countries to study issues related to biodiversity conservation.

*T. Sonia Arvanitis, Director Of Development*

**5531 Northern Arizona University**
Environmental Sciences
South San Francisco Street
Flagstaff, AZ 86011
928-523-5511
Fax: 928-523-6023
E-mail: distance.programs@nau.edu
http://www.nau.edu/

Designed to offer students a technically rigorous foundation and broad exposure to the environmental science. The core courses in environmental sciences are interdisciplinary, and team taught by scientists with different backgroundsand specialties, providing multiple perspectives and rich learning experiences.
*John D. Haeger, President*

**5532 Northern Michigan University**
Environmental Studies
1401 Presque Isle Avenue
Marquette, MI 49855
906-227-2242
Fax: 906-227-2249
http://www.nmu.edu

Research focuses on the Upper Peninsula environment, ethnic groups, economy, politics, folklore and literature.
*Darlene Walch, Dean*
*Leslie Wong, President*

**5533 Northland College**
Environmental Studies
1411 Ellis Avenue
Ashland, WI 54806
715-682-1699
800-753-1840
Fax: 715-682-1308
E-mail: info@northland.edu
http://www.northland.edu

Offers a comprehensive range of environmental programs that integrate traditional study with a keen eye toward problem-solving and environmental impact.
*Karen Halbersleben, President*
*Rick Fairbanks, Provost*

**5534 Ohio State University**
School of Environment and Natural Resources
2021 Coffey Road
210 Kottman Hall
Columbus, OH 43210
614-292-2265
Fax: 614-292-7432
E-mail: geise.1@osu.edu
http://www.senr.osu.edu

Focuses on the science and management of natural resources and the environment. A variety of integrated undergraduate programs of study provide the foundation to a variety of career paths dealing with natural resources and theenvironment. Graduates are employed as environmental and ecosystem scientists; forest, wildlife and fisheries reserchers and biologists; environmental educators, communicators and naturalists; and park, forest, and wildlife managers.
*Dr Jerry M Bigham, Program Director*

**5535 Oklahoma State University**
Environmental Science Program
Stillwater, OK 74078
405-744-4357
E-mail: osu-it@okstate.edu
http://www.it.okstate.edu

Environmental Science Program is designed to broaden the scope of scientific and technological study through a multidisciplinary approach encompassing social and legal aspects of environmental concerns and based on ecologicalfoundations.
*Michael Shuttic, Compliance Officer*

**5536 Oregon State University**
Environmental Sciences
Corvallis, OR 97331
541-737-1000
Fax: 541-737-3590
http://www.osu.orst.edu

Offers programs that are central to the mission of the university, which includes wise use of natural resources. Recognized as a Land, Sea, and Space Grant institution, OSU has exceptional strength in many of the disciplines that arerequired to provide a high-quality interdisciplinary education for future environmental scientists.
*Micheal Unsworth, Director*

**5537 Pennsylvania State University**
Center for Statistical Ecology
326 Thomas Building
University Park, PA 16802
814-865-1348
Fax: 814-863-7114
E-mail: b2a@stat.psu.edu
http://www.stat.psu.edu

This ground-breaking program enjoys national and international recognition. With an ongoing program of research that integrates statistics, ecology and the environment, the emphasis is on the environment and collaborative research,training and exposition on improving the quantification and communication of human impact on the environment. Studies also include statistical investigations of the impact of the environment on man.
*Bonnie Cain, Financial Secretary*
*Barbara Freed, Administrative Assistant*

**5538 Portland State University**
Environmental Sciences & Resources Program

PO Box 751
Portland, OR 97207
503-725-3194
Fax: 503-725-9040
E-mail: psuinfo@pdx.edu
http://www.pdx.edu

Environmental studies are central to the mission of Portland State University, which serves the state's major urban center. The Environmental Sciences and Resources program offers both undergraduate and graduate degrees.
*John Reuter, Program Director*

**5539 Prescott College**
Department of Environmental Studies
220 Grove Avenue
Prescott, AZ 86301
800-628-6364
Fax: 928-776-5242
E-mail: admissions@prescott.edu
http://www.prescott.edu/

Prescott College is a private liberal arts collage offering a resident BA and limited residency BA, MA, and PhD. Small groups of students work actively on real-world projects with faculty who are leaders in the field of environmentalstudies. Offers dynamic and active laboratories for students and gives them the opportunity to be on the cutting edge of environmental and sustainabilty research.
*Michelle Tissot, RDP Director of Admissions*
*Ted Bouras, ADGP Director of Admissions*

**5540 Purdue University**
Natural Resources & Environmental Science
3440 Lilly Hall
West Lafayette, IN 47907
765-494-8060
Fax: 765-496-2926
E-mail: jgraveel@purdue.edu
http://www.agry.purdue.edu/nres/nres.htm

An interdisciplinary program at the Purdue School of Agriculture designed to prepare students to work with environmental problems which impact our basic natural resources, specifically land, air and water. Faculty from all departmentsin the school contribute to the curriculum. NRES is a flexible program which allows students, working closely with an academic advisor, to develop their personal curriculum to meet individual career goals.
*Dr John Graveel, Program Director*

**5541 Rensselaer Polytechnic Institute**
Lally School of Management and Technology
110 8th Street
Pittsburgh Building
Troy, NY 12180
518-276-6565
E-mail: lally-dean-l@lists.rpi.edu
http://http://lallyschool.rpi.edu

Committed to integrating green business strategy throughout all management curriculum. An MBA with an Environmental Management and Policy Concentration is truly designed as an interdisciplinary degree, enabling graduates to work intraditional business settings with the knowledge and skill to help these businesses realize environmental, health and safety strategy.
*Dr. David A. Gautschi, Dean*

**5542 Rice University**
Urban & Environmental Policy Program
6100 Main
PO Box 1892
Huston, TX 77005
713-348-7423
800-527-owls
E-mail: admi@rice.edu
http://www.rice.edu

This program is designed to introduce students to how environmental policies are developed and how science and engineering issues are included in effective policy.

**5543 Roger Williams University**
Center for Environmental Development
One Old Ferry Road
Bristol, RI 02809
401-253-1040
Fax: 401-254-3310
E-mail: mdg@alpha.rwu.edu
http://www.rwu.edu/

Undergraduate program in marine biology combining chemistry, biology, physics, and mathmatics. Designed to keep and develop interest in the sciences by using field research and laboratory experimentation.

**5544 SUNY/Cortland**
Coalition for Education in the Outdoors
Park Center
PO Box 2000
Cortland, NY 13045
607-753-4971
Fax: 607-753-5982
E-mail: info@outdooredcoalition.org
http://www.outdooredcoalition.org

A nonprofit network of outdoor and environmental education centers, nature centers, conservation and recreation organizations, outdoor education and experimential education associations, public and private schools and fish and wildlifeagencies. All those involved in the coalition share the desire to support and encourage environmental and outdoor education and its goals.

**5545 SUNY/Fredonia**
Environmental Sciences
280 Central Avenue
Fredonia, NY 14063
716-673-3251
800-252-1212
Fax: 716-673-3249
E-mail: admissions@fredonia.edu
http://www.fredonia.edu/

Rigorous, interdisciplinary program in environmental science with 68 semester hours of core courses in mathematics, biology, chemistry, environmental sciences, and geosciences. Students are prepared to pursue graduate studies,professional certifications, or employment in the private or public sector.

*Christopher Dearth, Director Of Admissions*

**5546 SUNY/Plattsburgh**
Environmental Science
101 Broad Street
Plattsburgh, NY 12901
518-564-2000
E-mail: franzida@plattsburg.edu
http://http://web.plattsburg.edu

One of the largest and most established environmental science programs in the US, with 20 interdisciplinary faculty and nearly 300 majors among five degree programs. Opportunities for hands-on work and practical experience are providedby close proximity to the Adirondack Mountains State Forest Preserve, Plattsburgh's location on the banks of Lake Champlain, and affiliations with the Miner Agricultural Research Institute and more.
*David Franzi, Professor of Geology*

**5547 SUNY/Syracuse**
College of Environmental Science & Forestry
One Forestry Drive
106 Bray Hall
Syracuse, NY 13210
315-470-6600
800-777-7373
Fax: 315-470-6933
E-mail: esfinfo@esf.edu
http://www.esf.edu

As part of its education mission, SUNY offers an accredited engineering undergraduate program in Forest Engineering and graduate programs at both the masters and doctoral levels. The Faculty also conducts research and public serviceprograms that study how a variety of events affect our environment.
*Susan Sanford, Director of Admissions*
*Tom Fletcher, Associate of Admissions*

**5548 School for Field Studies**
Environmental Field Studies
10 Federal Street
Suite 24
Salem, MA 01970
978-741-3567
Fax: 978-741-3551
E-mail: admissions@fieldstudies.org
http://www.fieldstudies.org

Teaches students to address critical environmental problems using an interdisciplinary experimental approach to education.
*Diane Robinson, Academic Liaisons*

**5549 Slippery Rock University**
Environmental Geosciences
1 Morrow Way
Slippery Rock, PA 16057
724-738-2495
800-SRU-9111
Fax: 724-738-4807
E-mail: webmaster@sru.edu
http://www.sru.edu/

Prepares students for ocupations with industrial laboratories concerned with air, water and soil pollution control, engineering firms that study industrial pollution and prepare environmental impact statements, and state and federalagencies charged with monitoring the environment.
*James Hathaway, Chair*

**5550 Sonoma State University**
Environmental Studies & Planning Program
1801 E Cotati Avenue
Rohnert Park, CA 94928
707-664-2880
Fax: 707-644-4060
E-mail: cynthia.jowers@sonoma.edu
http://www.sonoma.edu

Founded as an interdisciplinary program during a period of growing environmental concern. The department has evolved and matured, now stressing the development of a global prespective by synthesizing knowledge from a variety ofscientific and academic disciplines, the acquisition of specific professional skills through a focused course of study, and the application of knowledge and skills through effective strategies for environmental management.

*Eduardo M. Ochoa, Provost*
*Cynthia Jowers, Executive Assistant*

**5551 Southern Connecticut State University**
Environmental Education Program
501 Crescent Avenue
Jennings Hall, Room 342
New Haven, CT 06515
203-392-6600
888-500-SCSU
Fax: 203-392-6614
E-mail: cusatos7@southernct.edu
http://southernct.edu

This program focuses on practicality and application of theory bringing about environmental change through educational processes. The objective is to prepare well informed people who are dedicated to improving environmental conditions.
*Cheryl J Norton, President*
*Susan Cusato, Chair Department of Science*

**5552 Southern Oregon University**
Environmental Education Program
1250 Siskiyou Boulevard
Siskiyou Environmental Center
Ashland, OR 97520
541-552-6411
800-482-7672
Fax: 541-552-6337
E-mail: presidentsoffice@sou.edu
http://www.sou.edu/

Designed to promote a better understanding of the environment and environmental issues, including an awarenesss and knowledge of biodiversity and ecosystem complexity. Seeks to prepare students for active roles in education and socialchange related to resolution of environmental problems and conflicts affecting present and future generations.
*Mary Cullinan, President*

**5553 St. Lawrence University**
Environmental Studies
23 Romoda Drive
Canton, NY 13617
315-229-5011
800-285-1856
E-mail: icania@stlawu.edu
http://www.stlawu.edu

Programs offer ten options for combining environmental studies with traditional disciplines (eg. biology, economics) plus B.A. program in Environmental Studies.
*Lisa M. Cania*

**5554 Stanford University**
Center for Environmental Studies
616 Serra Street
Suite 400
Stanford, CA 94305
650-725-6851
Fax: 650-725-1992
E-mail: iiswebmaster@lists.stanford.edu
http://http://cesp.stanford.edu

Focuses on significant environmental problems and draws methods and analyses from multiple diciplines.
*Marshall Burke*

**5555 Sterling College**
Center for Northern Studies
PO Box 72
Craftsbury Common, VT 05827
802-586-7711
Fax: 802-586-2596
E-mail: north@sterlingcollege.edu
http://www.sterlingcollege.edu/CNS

Center for Northern Studies is part of a small, undergraduate teaching and research institution located in Wolcott, Vermont. Its program is interdisciplinary in nature, integrating social and natural sciences, humanities and resourceissues in the Circumpolar North.

*Dr. Pavel Cenkl, Academic Dean*

**5556 Tennessee Technological University**
Bioenvironmental Sciences
Office of Admissions
PO Box 5006
Cookeville, TN 38505
931-372-3888
866-733-8324
E-mail: visit@tntech.edu
http://www.tntech.edu

Prepares graduates for high-level careers in various areas of biology and bioenvironmentl sciences.
*Robert Bell, President*

**5557 Texas A & M University**
Center for Natural Resource Information Technology
113 Administration Building
College Station, TX 77843
409-845-5548
Fax: 409-845-6430
E-mail: j-stuth@tamu.eud
http://cnrit.tamu.eud/cnrit

Serves as a point of contact for external organizations seeking cooperative efforts to assemble and disseminate information, create information technologies, and research critical natural resource concepts. The center strives tofacilitate technology transfer through training of end users and establishing necessary information infrastructures.
*Bob Brown, Director*

**5558 Texas Christian University**
Environmental Science Program
PO Box 298830
Fort Worth, TX 76129
817-257-7506
Fax: 817-257-7789
E-mail: m.slattery@tcu.edu
http://www.ensc.tcu.edu

A program helping students to understand the connection between science and the earth.
*Mike Slattery, Director*

**5559 Treasure Valley Community College**
Biology Department
650 College Boulevard
Ontario, OR 97914
541-881-TVCC
888-987-8822
Fax: 541-881-2721
E-mail: rfindley@tvcc.cc
http://www.tvcc.cc

Offers several courses for those seeking careers in natural resource management including range management, wildland fire management, and forest management.
*James Sorenson, President*

**5560 Tulane University**
Environmental Health & Sciences
6823 St. Charles Avenue
New Orleans, LA 70118
504-865-5368
800-873-9283
Fax: 504-862-8715
E-mail: pr@tulane.edu
http://www.tulane.edu

Environmental Health Sciences offers several graduate degree programs, including Public Health and Science. Graduates will be prepared to meet the needs of public health professionals such as environmental health and health officers,as well as undertake responsible positions in government, industrial facilities, research, or eduational institutions.
*Scott Cowen, President*

**5561 University of Arizona**
Soutywest Environmental Health Sciences Center
1501 Campbell Avenue
PO Box 245018
Tucson, AZ 85724
520-626-4555
Fax: 602-827-2074
E-mail: kchadder@u.arizonia.edu
http://www.ahsc.arizonia.edu

This center serves as a platform to promote the study of health effects of environmental agents. The SWEHSC promotes interdisciplinary research collaborations driven by cutting-edge technologies. Research in the SWEHSC is focused

onmechanisms of action of environmetnal agents in living systems.
*Keith A. Joiner, Vice Provost*

**5562 University of California/Berkeley**
Environmental Management Program
1995 University Avenue 510-642-4111
Suite 110 E-mail: info@unex.berkeley.edu
Berkeley, CA 94704 http://www.unex.berkeley.edu/
The Environmental Management Program prepares students to take on significant leadership roles in the environmental community.
*Diana Wu, Acting Dean*

**5563 University of California/Los Angeles**
Geoscience Engineering
1147 Murphy Hall 310-825-3101
Box 951436 Fax: 310-206-1206
Los Angeles, CA 90095 E-mail: ugadm@saonet.ucla.edu
http://www.ulca.edu
Programs study the principles of, and offer pracitcums in, soil mechanics and foundation engineering in light of geologic conditions, recognition, prediction, and control or abatement of subsidence, landslides, earthquakes, and othergeologic aspects of urban planning and subsurface disposal of liquids and solid wastes.

**5564 University of California/Santa Barbara**
Bren School of Environmental Studies
Bren Hall 805-893-2968
Room 4312 Fax: 805-893-8686
Santa Barbara, CA 93106 E-mail: esprogram@es.ucsb.edu
http://www.es.ucsb.edu
Program provides students with the scholarly background and intellectual skills necessary to understand complex environmental problems and formulate decisions that are environmentally sound. Academic process is interdisciplinary,drawing upon not only environmental science faculty, but also the resources of a variety of related departments and disciplines.

*Jami Nielsen, Program Assistant*
*Eric Zimmerman, Academic Advisor*

**5565 University of California/Santa Cruz**
Environmental Studies
1156 Hight Street 831-459-2634
Santa Cruz, CA 95064 E-mail: env@ucsc.edu
http://http://envs.ucsc.edu
Offers programs that prepare students to make significant contributions in the various fields of environmental study, whether pursing a career with private or non-profit institutions.
*Jeffery Bury, Assistant Professor*

**5566 University of Colorado**
Environmental Engineering Program
Regent Administrative Center 125 303-492-6303
552 UCB E-mail: apply@colorado.edu
Boulder, CO 80309 http://www.colorado.edu
Environmental Engineering Progam in the Department of Civil, Environmental, and Architectural Engineering at the University of Colorado in Boulder welcomes students, alumni, and colleagues in environmental engineering with aninvitation to explore areas of environmental emphasis, B.S., M.S., and Ph.D programs, research, and facilities.
*Bob Sievers, Director*

**5567 University of Florida**
College of Natural Resources & Environment
103 Black Hall 352-846-1634
PO Box 116455 Fax: 352-392-9748
Gainesville, FL 32611 E-mail: kbray@ufl.edu
http://http://snre.ufl.edu
Science based, multidisciplinary and academically rigorous, the College of Natural Resources & Environment has students and a curriculum that includes 200 courses taught in 56 departments of other colleges. The 290 affiliate facultyhave their primary appointments in discipline-centered departments of other colleges.
*Dr Stephen R Humphrey, Director*

**5568 University of Florida/Gainesville**
School of Forest Resources and Conservation
118 Newins-Ziegler Hall 352-846-0850
PO Box 110410 Fax: 352-392-1707
Gainesville, FL 32611
Offers baccalaureate (BSFRC) and graduate (PhD, MS, MFRC, MFAS, incl. a joint JD with the College of Law and a co-major with Dept of Statistics) degree programs; conducts fundamental and applied research; and provides public servicethrough extension programs. Programs include forestry, geomatics, fisheries, and aquatic sciences, natural resource economics, management, and policy, as well as related programs in natural resource education, ecotourism, and agroforestry.
*Timothy L White, Director*
*G Blakeslee T Frazer, Associate Directors*

**5569 University of Georgia**
Savannah River Ecology Laboratory
Drawer E 803-725-2472
Aiken, SC 29803 Fax: 803-725-3309
http://www.uga.edu/srel
Our goal is to attract collaborating scientists from across the DOE complex and the nation for collaborative work at the interface of fundamental and applied environmental research. We strive to improve the management of contaminatedstites. The staff and facilities of AACES are available to researchers in environmental science and engineering and to practitoneers from industry, government, academia and private foundations.
*Rosemary Forrest, Public Relations Coordinator*

**5570 University of Hawaii/Manoa**
Oceanography & Global Environmental Sciences
1000 Pope Road 808-956-9937
MSB 205 Fax: 808-956-9225
Honolulu, HI 96822 E-mail: ges@soest.hawaii.edu
http://www.soest.hawaii.edu/oceanography/ges/index.htm
Bachelor of Science degree in Global Environmental Science offered through the Department of Oceanography, University of Hawaii at Manoa.

*Jane Schoonmaker, Undergraduate Chair*

**5571 University of Idaho**
College of Natural Resources
PO Box 441138 208-885-8981
Moscow, ID 83844 Fax: 208-885-5534
E-mail: cnr@uidaho.edu/cnr
http://www.cnrhome.edu
Consists of 5 departments which together form a comprehensive educational program on the study and management of natural resources. Each department has several degree options to provide students with a flexible curriculum for theirdegree. We educate resource professionals with truly integrated resource management skills using innovative instructional programs. Our education occurs in a residential setting and provides a balance between theoretical and pratical experiences.
*Steven B Daley Laursen, Dean*

**5572 University of Illinois/Springfield**
Environmental Studies
One University Plaza 217-206-4847
MS UHB 1080 888-977-4847
Springfield, IL 62703 E-mail: admissions@uis.edu
http://www.uis.edu/
Goal of the environmental studies program is to enhance society's ability to create an environmentally acceptable future. Program faculty with diverse backgrounds in social and natural sciences and humanities are committed todeveloping interdisciplinary approaches to environmental problem solving. The primary objective is to educate citizens and professionals who are aware of environmental issues and their origins, causes, effects, and resolutions.
*Harry J. Berman, Provost/Vice Chancellor*

**5573 University of Illinois/Urbana**
Department of Natural Resources and Environment

1102 South Goodwin Avenue 217-333-2770
W-503 Turner Hall Fax: 217-244-3219
Urbana, IL 61801 E-mail: nres@uiuc.edu
http://www.nres.uiuc.edu

Establishes and implements research and educational programs that enhance environmental stewardship in the management and use of natural, agricultural, and urban systems in a socially responsible manner.

*Bruce Branham, Associate Prof/Interim Head*
*Jeff Brown, Professor/Head*

**5574 University of Iowa**
MacBride Raptor Project
E216 Field House 319-335-9293
Iowa City, IA 52242 Fax: 319-335-6655
E-mail: rec-services@uiowa.edu
http://www.recserv.uiowa.edu

Jointly sponsored by the University of Iowa and Kirkwood Community College, the project has two main facilities. Classes are held at the educational display facility and rehabilitation flight cage at the Macbride Nature RecreationalArea, and at the Raptor Clinic and educational display at KCC. The project is dedicated to the preservation of birds of prey and their habitats through rehabilitation of injured raptors, public education programs and raptor reseach.

*Jodeane Cancilla, Center Coordinator*
*Wayne Fett, Associate Director*

**5575 University of Maine**
Environmental Studies
23 University Street 207-834-7500
Fort Kent, ME 04743 888-try-umfk
Fax: 207-834-7609
E-mail: umfkadm@maine.edu
http://www.umfk.maine.edu

Offers a broad knowledge of the natural and social sciences, with the ability to focus of an area of personal interest. Students learn to critically identify environmental problems, collect and interpret data, communicate complexenvironmental issues, and explore creative solutions, while working closely with an interdisciplinary group of faculty with expertise in biology, chemistry, forestry, the social sciences and the humanities.

*Steven Selva, Professor*

**5576 University of Maryland**
Environmental Policy Programs
2101 Van Munching Hall 301-405-6330
College Park, MD 20742 Fax: 301-403-4675
E-mail: jbanders@umd.edu
http://www.puaf.umd.edu

This part-time degree program is intented for highly ambitious mid-career professionals who are ready to advance within the field, understand the importance and value of a professional degree and able to attend one or two classes perweek for two years. A minimum of 5 years of policy related work experience is required. For-profit, nonprofit and public sector work will be considered. A minimum udergraduate GPA of 3.0 is required.

*Janet Anderson, Assistant Director*

**5577 University of Maryland/Baltimore**
Environmental Science Studies
1000 Hilltop Circle 410-455-2291
Baltimore, MD 21250 800-UMB-C4US
Fax: 410-455-1094
E-mail: adnmissions@umbc.edu
http://www.umbc.edu

The goal of the MEES program is to train students with career interests in environmental science involving terrestrial, freshwater, marine, or estuarine systems. The program is university-wide and interdisciplinary, allowing studentsto use facilities and interact with all faculty in order to plan a program best suited to their particular interests.

*Lasse Lindahl, Professor*

**5578 University of Maryland/College Park**
Environmental Chemistry
0107 Chemistry Building 301-405-1788
College Park, MD 20742 Fax: 301-314-9121
E-mail: mdoyle3@umd.edu
http://www.chem.umd.edu

The combined Chemistry and Biochemistry Departments offers specialized training at the graduate level in environmental chemistry. In addition to course work in traditional chemistry subjects, students in this specialty take specificenvironmental courses and do research under the guidance of faculty members specializing in this area.

*Michael Doyle, Professor & Chair*

**5579 University of Miami**
Rosenstiel School of Marine & Atmospheric Science
4600 Rickenbacker Causeway 305-421-4000
Miami, FL 33149 Fax: 305-421-4711
E-mail: dean@rsmas.miami.edu
http://www.rsmas.miami.edu/

Established as the Marine Laboratory of the University of Miami. It has grown from its modest beginnings in a boathouse to one of the nations leading institutions for oceanographic research and education.

*Ron Avissar, Dean*

**5580 University of Minnesota/St. Paul**
College of Agriculatural & Environmental Sciences
277 Coffey Hall 612-624-3009
1420 Eckles Avenue Fax: 612-625-1260
St.Paul, MN 55108 E-mail: spccc@umn.edu
http://www.cnr.umn.edu

The programs offered through this college are designed to prepare students for work in a variety of environmental disciplines, specifically those that relate to the agriculture industry

*Allen Levine, Dean*

**5581 University of Montana**
Environmetal Studies
32 Campus Drive 406-243-0211
Missoula, MT 59812 E-mail: david.micus@umontana.edu
http://http://www.umt.edu/evst

Interdisciplinary graduate and undergraduate program in environmental studies.

*David Micus, Registrar*

**5582 University of Nebraska**
Environmental Studies Program
201 Administravitve Building 402-472-7211
UN 1 E-mail: hperlman1@unl.edu
Lincoln, NE 68588 http://www.unl.edu

The Environmental Studies Program is designed to serve a variety of students concerned about environmental issues and change. The program provides a thorough, holistic view of the environment and human-environmental interaction and thetechnical skills for active participation in an environmental career.

*Harvey Perlman, Chancellor*

**5583 University of Nevada/Las Vegas**
Department of Civil & Environmental Engineering
4505 S Maryland Parkway 702-895-3701
Box 454015 Fax: 702-895-3936
Las Vegas, NV 89154 E-mail: ce-info@ce.unlv.edu
http://www.ce.unlv.edu

Department of Civil and Environmental Engineering offers programs leading to a Master of Science in Engineering and Doctor of Philosophy, with concentration in six areas: environmental engineering; fluid mechanics and hydraulics;geotechnical engineering; structural engineering; construction engineering; and transportation systems.

*Allen Sampson, Sr Development Tech*

**5584 University of Nevada/Reno**
Civil & Environmental Engineering

Mail Stop 258
Reno, NV 89557
775-784-1474
Fax: 775-784-4466
E-mail: vdadams@unr.neveda.edu
http://www.coeweb.engr.unr.edu

Offers an educational program in environmental engineering. Environmental engineers have taken an increasingly important role in the application on engineering and scientific principles to protect and preserve human health andenvironment. The curriculum is designed with the goal of providing each student with the necessary fundamentals and background in engineering science and design to address many different challenges.

*Dr Manos Maragakis, Dean*
*Dr Indira Chatterjee, Professor/Assoc. Dean*

**5585 University of New Haven**

Environmental Science
300 Boston Post Road
Westhaven, CT 06516
203-932-7319
E-mail: adminfo@newhaven.edu
http://www.newhaven.edu

The bachelor of science program in environmental science is designed to give students a strong foundation in the fudamental sciences, including biology, chemistry, physics, and geology, and how they relate to our environmentalconcerns.

*Charles Coleman, Buisness Administration*

**5586 University of North Carolina/Chapel Hill**

Environmental Science & Studies
CB #2200
Jackson Hall
Chapel Hill, NC 27599
919-966-3621
Fax: 919-962-3045
E-mail: unchelp@admissions.unc.edu
http://www.admissions.unc.edu

The Environmental Science and Studies program leads to degrees in Environmental Science or Environmental Studies. Students investigate the relationship between the environment and society, focusing on environmental management, law andbusiness. The programs combines traditional classroom teaching with extensive use of interdisciplinary, team-based field projects, internships, study abroad and research.

*Patricia J. Pukkila, Director*
*Martha S. Arnold, Associate Director*

**5587 University of Oregon**

Environmental Studies
5223 University of Oregon
Eugene, OR 97403
541-346-5000
Fax: 541-346-5954
E-mail: ecostudy@uoregon.edu
http://envs.uoregon.edu

Environmental Studies crosses the boundaries of traditional disciplines, challenging faculty and students to look at the relationship between humans and their environment from a new perspective. They are dedicated to gaining greaterunderstanding of the natural world from an ecological perspective; devising policy and behavior that address contemporary environmental problems; and promoting a rethinking of basic cultural premises, ways of structuring knowledge and the rootmetaphors of society.

*Alan Dickman, Program Director*
*RaDonna Aymong, Office Manager*

**5588 University of Pennsylvania**

Natural Resource Conservation Program
3451 Walnut Street
Philadelphia, PA 19104
215-898-5000
E-mail: webmaster@upenn.edu
http://www.upenn.edu

The mission of our department is to bring the time perspective of the Earth scientist/historian to bear on contemporary problems of natural resource conservation and environmental quality.

*Willis Stetson Jr., Dean of Admissions*

**5589 University of Pittsburgh**

Department of Geology & Planetary Science
4107 O'Hara Street
200 SRCC Building
Pittsburgh, PA 15260
412-624-6615
Fax: 412-624-3914
E-mail: mookie@pitt.edu
http://www.geology.pitt.edu

Equips students with an understanding of earth systems and the impact of humans on the biosphere, atmosphere and hydrosphere. Courses in the natural and social sciences, humanities, and schools of law, business, and public healthprovide a comprehensive, interdisciplinary background in environmental issues and public policy.

*Mark Collins, Environ Studies Coordinator*

**5590 University of Redlands**

Environmental Studies
1200 E Colton Avenue
PO Box 3080
Redlands, CA 92373
909-793-2121
Fax: 909-793-2029
E-mail: kerry_robles@redlands.edu
http://www.redlands.edu

Designed to promote a new way of thinking and acting about our relationship to the world, including graduating students who are environmentally literate, sensitive to competing demands and conflicting values of each issue and finally,and have the creativity, confidence and conviction to begin effecting change.

*Kerry Robleton, Department Secretary*

**5591 University of South Carolina**

Belle Baruch Institute of Marine Costalsciences
607 EWS Building
PO Box 1630
Georgetown, SC 29442
803-777-5288
Fax: 803-777-3935
E-mail: bergin.sc.edu
http://www.cas.sc.edu

Environmental research and programs are focused on estuarine systems and their associated watersheds. More than 160 investigators representing 30 academic institutes and agencies are affiliated with over 100 projects. The lab providessupport for undergraduate classes, graduate students, and senior scientists. Long-term environmental monitoring, training programs and outreach activities are sponsored by the North-inlet-Winyah Bay National Estuarine research reserve.

*Margaret Bergin, Buisness Manager*

**5592 University of Southern California**

Environmental Sciences, Policy/Engineering Program
University Park Campus
Hancock Building, Room 232
Los Angeles, CA 90089
213-740-2311
http://www.usc.edu

This multidisciplinary doctoral training program is funded by the National Science Foundation, and prepares students to confront, analyze and resolve the challenges posed by problems of urban sustainablilty. Engineering SustainableCities is high on the list of goals of the program, which allows students to transcend disciplines, and conduct policy-relevant research on major environmental problems.

*Albert A Herrara, Director & Professor*

**5593 University of Southern Mississippi**

Environmental Science Program
118 College Drive
Building #5018
Hattiesburg, MS 39406
601-266-4748
Fax: 601-266-5797
E-mail: admissions@usm.edu
http://www.biology.usm.edu

The Environmental Science concentration focuses on industrial problems related to the working environment, pollution control, and safety. Courses address major industrial issues, including environmental impact statements, industrialhygiene and environmental laws and regulations.

*Frank Moore, Chair*
*Patricia Brewer, Administrative Assistant*

**5594 University of Tennessee**

Geoscience Program
University Street
Martin, TN 38238
901-587-7020
800-829-utm1
Fax: 901-587-7029
E-mail: webmaster@utm.edu
http://www.utm.edu

This program focuses on the application of geology to the interaction between man and the environment. Topics include geohazards, chemical and nuclear contamination of soils and water, remediation of environmental problems andgovernmental environmental agencies and laws.

*Tom Rakes, Chancellor*

**5595 University of Virginia**
Department of Environmental Sciences
291 McCormick Road 434-924-7761
PO Box 400123 Fax: 434-982-2137
Charlottesville, VA 22904 E-mail: ralph@virginia.edu
http://http://www.evsc.virginia.edu
Offers instruction and research opportunities in Ecology, Geosciences, Hydrology, and Atmospheric Sciences. The research endeavors of both faculty and graduate students, whether disciplinary or interdisciplinary, deal largely withproblems of fundamental scientific interest and with applied sciences, management or policy making.
*Allen Ralph, Professor*

**5596 University of Washington**
Environmental Science
18115 Campus Way NE 425-352-5000
Bothell, WA 98011 425.352.5303
E-mail: info@uwb.edu
http://www.uwb.edu
Primary goal of this program is to train a new generation of interdisciplinary scientists who are able to work in both the public and private sectors to address some of the pressing environmental issues that face our society.
*Steve Holland, Director*
*Walt Freytag, Assistant Director*

**5597 University of West Florida**
Environmental Studies
11000 University Parkway 850-474-2200
Pensacola, FL 32514 Fax: 850-474-3131
E-mail: jcavanaugh@uwf.edu
http://www.uwf.edu
The program in Environmental Studies consists of a multi-disciplinary approach that combines natural science and resource management. Students learn to analyze physical and socioeconomic environments and to reach decisions concerningenvironmental use and protection. It offers a core curriculum that is designed to provide the student with a solid foundation in earth and life sciences, as well as in modern methods and techniques.
*John C. Cavanaugh, President*

**5598 University of Wisconsin/Green Bay**
Environmental Science
2420 Nicolet Drive 920-465-2000
Green Bay, WI 54311 E-mail: admissions@uwgb.edu
http://www.uwgb.edu
This program is interdisciplinary, emphasizing an integrated approach to knowledge in the field. Because the study of the environmental science major is grounded in the natural sciences and mathematics, the curriculum includes a socialscience component, enabling students to gain an understanding of environmental economic and policy issues. Field experiences, internships and practicums are emphasized.
*Bruce Shepard, Chancellor*

**5599 University of Wisconsin/Madison**
Environmental Monitoring Program
WI Agricultural Experiment Station 608-261-1432
240 Agricultural Hall Fax: 608-265-9534
Madison, WI 53706 E-mail: research@cals.wisc.edu
http://www.cals.wisc.edu/waes
Remote sensing and geographic information systems offer sophisticated and powerful tools for monitoring the environment on large geographic scales over time. Students in the Environmental Monitoring Program learn to employ thesetechnologies in fields of their choice, from forestry to urban planning to environmental engineering.
*Molly Jahn, Director*
*Angela Seitler, Assistant Director*

**5600 University of Wisconsin/Stevens Point**
Environmental Task Force Progeam
2100 Main Street 715-346-0123
Stevens Point, WI 54481 Fax: 715-346-2561
E-mail: webmaster@uwsp.edu
http://www.uwsp.edu/
This program involves two water chemistry labs which tests for organics and inorganics. This is staffed by five full time workers, plus a part time faculty director, and about 40 students are hired and/or trained each year. Samplingis performed with state-of-the-art field sampling and laboratory analytical equipment nutrients, pesticides, polynuclear aromatic hydrocarbons, polychlorinated biphenyls, and volatile organic compounds.

*Linda Bunnell, Chancellor*

**5601 University of the South**
Environmental Studies
735 University Avenue 931-598-1271
Sewanee, TN 37383 800-522-2234
Fax: 931-598-1667
E-mail: collegeadmission@sewanee.edu
http://www.sewanee.edu/envstudies/
Brings together students, faculty, and staff from thirteen academic departments to study, discuss, and research environmental issues at local, national, and international scales. The program's goal is to expose students to a variety ofviewpoints concerning environmental issues, and to offer the interdisciplinary tools they need to become environmental problem solvers before they graduate from Sewanee.

*Sid Brown, Associate/Professor/Chair*

**5602 Utah State University**
Berryman Institute
Wildland Resources Department 435-797-0242
5230 Old Main Hill Fax: 435-797-3796
Logan, UT 84322 http://http://www.berrymaninstitute.org
The Berryman Institute is a functional component of the Department of Wildland Resources and the College of Natural Resources. Its faculty members hold academics appointments in various departments throughout Utah State University andother universities. This multidisciplinary approach is calculated to speed the discovery and development of innovative methods to solve human wildlife conflict.
*Dr Johan Du Toit, Co-Director*
*Dr Bruce Leopold, Co-Director*

**5603 Vanderbilt University**
Vanderbilt Center for Environmental Management
401 21st Avenue South 615-322-6814
Nashville, TN 37203 Fax: 615-343-7177
E-mail: mark.a.cohen@vanderbilt.edu
http://www.vanderbilt.edu/vcems
VCEMS provides guidance and support for the interdisciplinary study of environmental issues. The Center brings faculty and students together from various disciplines for collaborative study and research on topics such as environmentalrisk assessment, management and communication, policy analysis, civil and criminal liability, environmentally conscious manufacturing and technology management, and global environmental studies.
*Dr. Mark Abkowitz, Co-Director*
*Dr. Mark Cohen, Co-Director*

**5604 Vermont Law School**
Environmental Law Center
Chelsea Street
PO Box 96 800-227-1395
South Royalton, VT 05068 http://www.vermontlaw.edu
The Environmental Law Center administers three different degrees in Environmental Law, each adaptable to career objectives in both public and private sectors. The school's mission is to educate for stewardship, to teach an awareness ofunderlining environmental issues and values, to provide a solid knowledge of environmental law and to develop skills to administer and improve policies.
*J. Scott Cameron, Chair*

**5605 Virginia Polytechnic Institute**
Environmental Science
201 Burruss Hall 540-231-6267
Blackburg, VA 24061 E-mail: vtadmiss@vt.edu
http://www.vt.edu

This program deals with crop production, soil utilization, and environmental stewardship. Its professionals are concerned with helping to feed the world and protect the environment, and include women and men who work to grow crucialcommodities, improve water quality, develop environmentally acceptable methods for protecting crops from pests, and advise municipalities on use of the land resource.
*Steven C Hodges, Department Head*

**5606 Virginia Tech**
Environmental Science
240 Smyth Hall 540-231-6300
W Campus Drive Fax: 540-231-3431
Blacksburg, VA 24061 E-mail: savilel@vt.edu
http://www.ense.vt.edu
Provides a B.S. degree for environmental professionals needed in the private and public sector and by nonprofit organizations. Built on a rigorous interdisciplinary curriculum that stresses the basic sciences, environmentaltechnologies, soils, and analytical skills. Graduates are in high demand in the environmental arena.

*James R McKenna, Interim Department Head*
*Rachel Saville, Administrative Assistant*

**5607 Warren Wilson College**
Natural & Social Sciences
PO Box 9000 828-771-2070
Asheville, NC 28815 E-mail: pfeiffer@warren-wilson.edu
http://www.warren-wilson.edu
Combines rigorous courses in the natural and social sciences with abundant natural resources near the classrooom. Courses and work crews give students a balance of theory, first hand knowledge and field experience. Successful programsmost often result when students, with the help of an advisor, begin planning course work and indentifying goals during their first year.
*William Sanborn Pfeiffer, President*

**5608 Washington State University**
Environmental Science
PO Box 644430 509-335-8538
Pullman, WA 99164 E-mail: esrp@wsu.edu
http://www.esrp.wsu.edu
The students in this diverse program are encouraged to specialize in their specific interest, including agricultural ecology, biological science, environmental education, environmental quality (air and water), natural resourcemanagement, systems, environmental/land use planning or hazardous waste management.
*Elson Floyd, President*

**5609 West Virginia University**
Environmental Geosciences Program
B-33 Stewart Hall 304-293-4006
PO Box 6003 Fax: 304-293-7337
Morgantown, WV 26506 E-mail: cafcs@wvu.edu
http://www.caf.wvu.edu
The Environmental Geosciences program features an interdisciplinary approach to environmental issues. Graduates will be well prepared to face the environmental challenges, whether in government or in the corporate world.
*Dixie Paletta, Assistant Director*
*Brandon Twigg, Buisness Manager*

**5610 Western Montana College**
Environmental Sciences Program
710 S Atlantic Street 406-683-7331
Dillon, MT 59725 866-UMN-MONT
Fax: 406-683-7331
E-mail: admissions@umwestern.edu
http://www.umwestern.edu
The mission of the environmental sciences programs is to provide students with an in-depth understanding of the natural processes which create and shape our environment. Students will become informed, critical thinkers capable ofscientifically evaluating complex issues involving the environment. Student development will occur through interdisciplinary, field-based research projects that have societal relevance.
*Craig Zaspel, Prof Environmental Sciences*

**5611 Williams College**
Center for Environmental Studies
PO Box 518 413-597-2346
Kellogg House Fax: 412-597-3489
Williamstown, MA 01267 E-mail: szepke@williams.edu
http://www.williams.edu
The Environmental Studies program provides students with tools, ideas, and opportunities to engage constructively with the environmental and social issues brought about by changes in population, economic activity, and values. Theenvironmental studies program is interdisciplinary and broad, including the coditions of inner-city poverty as well as the magnificent scenery of wildlands, encompassing the view of planet earth from near space as well as from culturalanthropologists.
*Hank Art, Director*

**5612 Yale University**
Office of Public Affairs
265 Church Street 203-432-1345
Suite 901 Fax: 203-432-1323
New Haven, CT 0651lE-mail: undergraduate.admission@yale.edu
http://www.yale.edu
Our mission is to provide the leadership and knowledge needed to restore and sustain both the health of the biosphere and the well-being of its people. Believing that human enterprise can and must be conducted in harmony with theenvironment, we are committed to using natural resources in ways that sustain both resources and ourselves. Solving environmental problems must incorporate human values and motivations and a deep respect for both human and natural communities.
*Richard Charles Levin, President*

## Workshops & Camps

**5613 A Closer Look at Plant Life**
Educational Images
PO Box 3456 607-732-1090
Westside Station 800-527-4264
Elmira, NY 14905 Fax: 607-732-1183
E-mail: edimages@edimages.com
http://www.educationalimages.com
Access a wealth of information on every major group of vascular and nonvascular plants. Includes details on plant microanatomy; external and internal structures; life cycles; and processes such as growth transpiration andphotosynthesis.

**5614 A Closer Look at Pondlife - CD-ROM**
Educational Images
PO Box 3456 607-732-1090
Westside Station 800-527-4264
Elmira, NY 14905 Fax: 607-732-1183
E-mail: info@edimages.com
http://www.educationalimages.com
Through the wonders of close-up photography, this unique CD-ROM brings students face-to-face with the inner workings of a freshwater pond, the myriad creatures and plants that reside there, and the dynamic interactions that go onbeneath the surface. This disk features a library of reference information, images, illustrations, clip art, video clips and more!

**5615 Abbott's Mill Nature Center**
Delaware Nature Society
15411 Abbott's Pond Road 302-239-2334
Milford, DE 19963 Fax: 302-239-2473
E-mail: dnsinfo@delawarenaturesociety.org
http://www.delawarenaturesociety.org
Abbott's Mill Nature Center features education programs for families, school classes and public groups, walking trails through fields, pine woods and streams, and a historic, fully-operating grstmill.
*Michael Riska, Executive Director*

**5616 Air and Waste Management Association**
One Gateway Center 3rd Floor 412-232-3444
420 Fort Duquesne Blvd. Fax: 412-232-3450
Pittsburg, PA 15222 E-mail: info@awma.org
http://www.awma.org

The Air & Waste Management Association is a non profit, nonpatism professional organization that provides training, information, and networking opportunities to 12,000 environmental professionals in 65 countries. THe Association'sgoals are to stengthen the environmental professionals in critical environmental decision making to benefit society.

**5617 American Museum of Natural History**
Center for Biodiversity and Conservation
Central Park West 212-769-5742
79th Street Fax: 212-769-5292
New York, NY 10024 E-mail: biodiversity@amnh.org
http://www.research.amnh.org

Conducts research and field projects based on information provided by Museum departments.
*Eleanor Sterling, Director*

**5618 Animal Tracks and Signs**
Educational Images
PO Box 3456 607-732-1090
Elmira, NY 14905 800-527-4264
Fax: 607-732-1183
E-mail: edimages@edimages.com
http://www.educationalimages.com

Presents various animal tracks and signs throughout the seasons, and provides useful information about the special characteristics and natural history of the animals that left the signs. Footprints, scratch marks, nesting places,wallows, scats and signs of food gathering are all detailed. Coverage includes deer, fox, porcupine, rabbit, bear, mink, otter, owl, woodpecker, killdeer, wild turkey, sapsucker and grouse.

*Charles R Belinky, Ph.D, CEO*

**5619 Annotated Invertebrate Clipart CD-ROM**
Educational Images
POÆBox 3456 607-732-1090
Elmira, NY 14905 800-527-4264
Fax: 607-732-1183
E-mail: info@edimages.com
http://www.educationalimages.com

780 colorful graphics of invertebrates from protists through urochordates, supported by extensive written annotations in addition to traditional labels. Includes presentation graphics and page after page of supplemental information onclassification, anatomy, evolution, development, reproduction, etc.

*Charles R Belinky, Ph.D, CEO*

**5620 Annotated Vertebrate Clipart CD-ROM**
Educational Images
PO Box 3456 607-732-1090
Elmira, NY 14905 800-527-4264
Fax: 607-732-1183
E-mail: info@edimages.com
http://www.educationalimages.com

792 colorful graphics of vertebrates from urochordates and tunicates through mammals, supported by extensive written annotations in addition to traditional labels. Includes presentation graphics and page after page of supplementalinformation on classification, organ systems, anatomy, evolution, development, reproduction, etc.

*Charles R Belinky, Ph.D, CEO*

**5621 Argonne National Laboratory**
9700 S Cass Avenue 630-252-2000
Argonne, IL 60439 E-mail: dep_webmaster@anl.gov
http://www.anl.gov/

We focus on four broad strategic environmental areas under which specific programs and projects are conducted. Our staff of over 100 multidisciplinary professionals are organized in a matrix fashion to undertake programs and projectswith technical managers and staff from seven sections, each specializing in specific technical disciplines.
*Henry Bienen, President Of Nwu*

**5622 Ashland Nature Center**
Delaware Nature Society
3511 Barley Mill Road 302-239-2334
PO Box 700 Fax: 302-239-2473
Hockessin, DE 19707 E-mail: dnsinfo@delawarenaturesociety.org
http://www.delawarenaturesociety.org

Open year round, seven days a week, Ashland is headquarters of the Delaware Nature Society. Ashland Nature Center offers self-guided nature trails traversing 81 acres of rolling terrain, through meadows, woodlands, and marshes.Programs for all ages are offered, including schools and groups.
*Michael Riska, Executive Director*

**5623 Aspen Global Change Institute**
100 East Francis Street 970-925-7376
Aspen, CO 81611 Fax: 970-925-7097
E-mail: agcimail@agci.org
http://www.agci.org

A Colorado nonprofit dedicated to furthering the understanding of Earth systems through interdisciplinary science meetings, publications, and educational programs about global environmental change.
*John Katzenberger, Director*
*Sue Bookhout, Director Operations*

**5624 Association for Environmental Health and Sciences**
150 Fearing Street 413-549-5170
Amherst, MA 01002 Fax: 413-549-0579
E-mail: info@aehs.com
http://www.aehs.com

Created to facilitate communication and foster cooperation among professionals concerned with the challenge of soil protection and cleanup. Experience over the past decades has revealed the need for a consistent and reliable networkfor the exchange of information derived from multiple sources and disciplines among people who, because of different disciplinary affiliations and interests, may not have easy access to significant portions of the information map.
*Paul Kostecki, Executive Director*

**5625 Audubon Expedition Institute**
29 Everett Street 617-349-8544
Cambridge, MA 02138 800-999-1959
E-mail: info@lesley.edu
http://www.lesley.edu/

Students challenge themselves and their assumptions through experimental learning and direct contact with social, natural, historical and urban environments. Subjects are studied and integrated through real life experiences.
*Donald Perrin, Chair*

**5626 Bio-Integral Resource Center**
PO Box 7414 510-524-2567
Berkeley, CA 94707 Fax: 510-524-1758
E-mail: birc@igc.org
http://www.birc.org

The goal of the Bio Integral Resources Center is to reduce pesticide use by educating the public about effective, least-toxic alternatives for pest problems.
*Dr. William Quarles, Executive Director*

**5627 Biosystems and Agricultural Engineering**
University of Kentucky
128 C.E Barnhart Building 859-257-3000
Lexington, KY 40546 Fax: 859-257-5671
E-mail: gates@bae.uky.edu
http://www.bae.uky.edu/

Biosystems and Agricultural Engineering provides an essential link between the biological sciences and the engineering profession. The linkage is necessary for the development of food and fiber production and processing systems whichpreserves our natural resources base.
*Dr. Richard Gates, Chair*

**5628 Bishop Resource Area**
785 N Main Street 760-872-4881
Suite E E-mail: snelson@ca.blm.gov
Bishop, CA 93514 http://www.wmrs.edu
The Bishop Resource Area has facilitated aerial photo interpretation and remote sensing programs in local schools through corporate and public partnerships. The program incorporates aerial photo interpretation, its relationship tomapping and land use history.
*Steve Nelson, Gis Coordinator*

**5629 Bog Ecology**
Educational Images
PO Box 3456 607-732-1090
Elmira, NY 14905 800-527-4264
Fax: 607-732-1183
E-mail: info@edimages.com
http://www.educationalimages.com
A comprehensive program that explores the origin and formation of bogs, common plants and animals, and compares bogs to other types of wetlands. Bog succession is illustrated by use of diagrams and photographs. 74 frames and guide.
*Charles R Belinky, Ph.D, CEO*

**5630 Camp Fire USA**
1100 Walnut Street 816-285-2010
Suite 1900 800-669-6884
Kansas City, MO 64106 Fax: 816-756-0258
E-mail: info@campfireusa.org
http://www.campfireusa.org
Not-for-profit, youth development organization, Camp Fire USA provides fun, coeducational programs for approximately 650,000 youth from birth to age 21. Helps boys and girls learn and play side by side in comfortable, informalsettings.
*John S Foro, Director Market/Communicate*

**5631 Camp Habitat Northern Alaska Environmental Center**
830 College Road 907-452-5021
Fairbanks, AK 99701 Fax: 907-452-3100
E-mail: camphabitat@northern.org
http://northern.org
Camp Habitat is a nature education program for young people ages 4-17 sponsored by the Northern Alaska Environmental Center, Friends of Creamer's Field, and Alaska Department of Fish & Game. The mission of Camp Habitat is to provideyoung people with guided explorations of their natural surroundings through interactive, hands-on activities. Skilled instructors and resource specialists lead small groups through new outdoor activities focusing on the habitats of Interior Alaska.

**5632 Center for Environmental Research and Conservation**
1200 Amsterdam Avenue 212-854-8179
New York, NY 10027 Fax: 212-854-8188
E-mail: cerc@columbia.edu
http://www.cerc.columbia.edu
CERC, a consortium of five education and research institutions, was created in response to critical environmental concerns facing the Earth. Within the next fifty years human influence will affect every place on the planet. That impactwill almost certainly result in species extinctions, ecosystem degradation and a loss of the benefits those species and ecosystems provide to people.
*Nancy Degnan, Executive Director*

**5633 Center for Geography and Environmental Education**
311 Conference Center Building 865-974-4251
University of Tennessee Fax: 865-974-1838
Knoxville, TN 37996 E-mail: mckeowni@utk.edu
http://eerc.ra.utk.edu/CGEE.html
A research and outreach center at the University of Tennessee. The CGEE focuses on environmental and geography education contributions to education for sustainable development. CGEE is responsible for the Tennessee Solid WasteEducation Project.
*Dr. Rosalyn McKeown-Ice, Director*

**5634 Center for Mathematical Services**
4202 East Fowler Avenue 813-974-9568
Adm 147 Fax: 974-974-2700
Tampa, FL 33620 E-mail: uco@admin.usf.edu
http://www.usf.edu
Mission is to help prepare students of all levels to effectively use mathematics as a tool to analyze situations and resolve problems. In the field of mathematical sciences it serves as an interface for the University with thesecondary schools in the area served by the University of South Florida. By means of this interface special programs in the mathematics, science, and engineering are offered at the University of South Florida for secondary students.
*Nick J Trivunovich, Controller*

**5635 Cetacean Society International**
PO Box 953 203-770-8615
Georgetown, CT 06829 Fax: 860-561-0187
E-mail: rossiter@csiwhalesalive.org
http://www.csiwhalesalive.org
All volunteer, nonprofit conservation, educational and research organization to benefit whales, dolphins, porpoises and the marine environment. Promotes education and conservation programs, including whale and dolphin watching, andnoninvasive, benign research. Advocates for laws and treaties to prevent commercial whaling, habitat destruction and other harmful or destructive human interactions. CSI's world goal is to minimize cetacean killing and captures and to enhance publicawareness.
*William W Rossiter, President*

**5636 Chicago Botanic Garden**
1000 Lake Cook Road 847-835-5440
Glencoe, IL 60022
The Chicago Horticultural Society has been promoting gardens and gardening since 1890. Generations of Chicagoans have been touched by the Society's flower shows, victory gardens, horticultural lectures and more. The mission encompassesthree important components: collections, programs and research. A living museum, the Chicago Botanic Garden serves both a public and a scientific community.

**5637 Clean Ocean Action**
18 Hartshorne Drive 732-872-0111
Sandy Hook Fax: 732-872-8041
, NJ 07732 E-mail: sandyhook@cleanoceanaction.org
http://www.cleanoceanaction.org
Clean Ocean Action is a broad-based coalition of over 150 conservation, community, diving, fishing, environmental, surfing, women's and business groups that works to improve and protect the waters off the New York and New Jersey coast.
*Cindy Zipf, Executive Director*
*Mary-Beth Thompson, Operations Director*

**5638 Cleaner and Greener Environment**
1526 Chandler Street 608-280-0256
Madison, WI 53711 877-977-9277
Fax: 608-255-7202
E-mail: info@cleanerandgreener.org
http://www.cleanerandgreener.org
Cleaner and Greener Environment is a program of Leonardo Academy, a 501 environmental nonprofit organization. Leonardo Academy reports reductions in emissions, and promotes the development of markets for the emission reductions thatresult from energy efficiency, renewable energy, and other emission reduction action.

**5639 Climate Change Program**
Dade County Dept. of Env. Res. Mgm. 305-372-6825
701 NW 1st Court Fax: 305-372-6954
Miami, FL 33136 E-mail: dermp2@itd.metro-dade.com
Receives monies from sources such as fees from pollution prevention events, grants, allocations, appropriations and workshop fees. These funds are then used in developing, promoting and

conducting environmental workshops, expositions,symposia, conferences and other forms of public information for the purpose of educating industry, government and the public about pollution prevention.
*Nichole Hefty, Coordinator*

**5640 Coastal Resources Center**
University of Rhode Island
220 South Ferry Road 401-874-6224
Narragansett, RI 02882 Fax: 401-789-4670
E-mail: info@crc.uri.edu
http://www.crc.uri.edu
Mobilizes governments, business and communities around the world to work together as stewards of coastal ecosystems. With partners we strive to define and achieve the health, equitable allocation of wealth, and sustainable intensitiesof human activity at the transition between the land and sea.
*Stephen Olsen, Director*
*Chip Young, Communications Liasion*

**5641 Comet Halley: Once in a Lifetime!**
Educational Images
PO Box 3456 607-732-1090
Elmira, NY 14905 800-527-4264
Fax: 607-732-1183
E-mail: info@edimages.com
http://www.educationalimages.com
Particularly relevant because of the recent appearance of Hale-Bopp, this program presents the reactions to comets in ancient, historic and relatively modern times, press coverage of the 1910 Halley return, superstitions and beliefs,current scientific knowledge and research, and much more.

*Charles R Belinky, Ph.D, CEO*

**5642 Cooch-Dayett Mills**
Delaware Nature Society
904 Old Baltimore Pike 302-239-2334
Newark, DE 19702 Fax: 302-239-2473
E-mail: dnsinfo@delawarenaturesociety.org
http://www.delawarenaturesociety.org
Programming provided by the delaware nature society features environmental education and natural history for families, classes, groups and the public. The historic roller mills, along with natural features of the site, become theclassroom.
*Michael Riska, Executive Director*

**5643 Cooperative Institute for Research in Environmental Sciences: K-12 and Public Outreach**
University of Colorado
Campus Box 216 303-492-1143
Boulder, CO 80309 Fax: 303-492-1149
E-mail: info@cires.colorado.edu
http://www.cries.colorado.edu
We educate people about Earth and environmental science issues that are relevant to our everyday lives, through outreach to the public and to the K-12 education community.
*Konrad Steffen, Director*

**5644 Coverdale Farm**
Delaware Nature Society
543 Way Road 302-239-2334
Greenville, DE 19807 Fax: 302-239-2473
E-mail: dnsinfo@delawarenaturesociety.org
http://www.delawarenaturesociety.org
School students and guests of all ages participate in seasonal programs, learning about the farm cycle of life and humans' dependence on soil, water, plants and animals for survival.
*Michael Riska, Executive Director*

**5645 Deep Portage Conservation Reserve**
2197 Nature Center Drive NW 218-682-2325
Hackensack, MN 56452 E-mail: portage@uslink.net
http://www.deep-portage.org
Deep Portage serves schools, groups, organizations, research teams, area residents and visitor with resident environmental education programs, weekly classes, interpretive programs, wildflower garden displays, land use demonstrations,summer youth camps and recreation opportunities of birding, hiking, hunting and skiing.
*Molly Malecek, Assistant Director*

**5646 Department of Energy and Geo-Environmental Engineering**
116 Deike Building 814-863-6546
University Park, PA 16802 Fax: 814-863-7708
E-mail: egee@ems.psu.edu
http://www.ems.psu.edu
Through education, research and service, EGEE aspires to insure that socisty is provided with an affordable supply of energy and minerals, concomitant with protecting the environment.
*William Easterling, Dean*

**5647 DuPont Environmental Education Center**
Delaware Nature Society
1400 Delmarva Lane 302-239-2334
Wilmington, DE 19801 Fax: 302-239-2473
E-mail: dnsinfo@delawarenaturesociety.org
http://www.delawarenaturesociety.org
Opening September 2009, the DuPont Environmental Education Center showcases where the city, river, and marsh meet. The Center is on the edge of the 212-acre Russell W Peterson Urban Wildlife Refuge that adjoins the Christina River,which is home to many species of amphibians, birds, mammals, fish, reptiles, and native plants.
*Michael Riska, Executive Director*

**5648 Earth Day Network**
1616 P Street NW 202-518-0044
Suite 340 Fax: 202-518-8794
Washington, DC 20036 E-mail: earthday@earthday.net
http://www.earthday.net
This nonprofit network was created to be a vehicle for increased awareness & responsibility through the promotion of Earth Day. Offers workshops.
*Kathleen Rogers, President*
*Mary Minette, Senior VP Programs*

**5649 Earth Force**
1908 Mount Vernon Avenue 703-579-6867
2nd Floor Fax: 703-299-9485
Alexandria, VA 22301 E-mail: capitol@earthforce.org
http://http://www.earthforce.org
Earth Force offers educators innovatove programs and resources. The young graduates of these programs create lasting solutions to environmental problems in their communities. Earth Force's goal is to help youth become environmentalproblem solvers.

*James Macgregor, Chairman*

**5650 Ecological & Environmental Learning Services**
46 Back Bone Hill Road 732-577-5599
Clarksburg, NJ 08510 800-206-6672
Fax: 732-577-5598
Ecological & Environmental Learning Services provides K-12 education consulting for teacher professional development, curriculum development and education programs and assemblies. EELS has the expertise and experience to providesolutions for enhancing the academic excellence of students in the following areas: Science (particularly in science research); Ecology; Environmental Science; and Environmental Education.

**5651 Ecology and Environmental Sciences**
5782 Winslow Hall 207-581-3198
Room 305 E-mail: mark.anderson@umit.maine.edu
Orono, ME 04469 http://www.umaine.edu
Faculty from five different academic departments, covering biological, physical and social sciences, work together to offer a broad educational experience for our students. Since these faculty have active research programs, studentsnot only get access to the most up-to-date information, but also get employment opportunities in their fields of study during the academic year and the summer months.

**5652 Economic Development/Marketing California Environmental Business Council**
UC Extention 408-748-2170
3120 De La Cruz Fax: 408-748-2189
Santa Clara, CA 95054 E-mail: br1027@aol.com
The CEBC is a nonprofit trade and business assoication that promotes and assists California's environmental technology and services industry at the state, national, and international levels. Founded in 1994, the CEBC currently has morethan 100 member compines and other organizations throughout the state that represent all segments of the environmental industry.

**5653 Energy Thinking for Massachusetts**
Northeast Sustainable Energy Association
50 Miles Street 413-774-6051
Greenfield, MA 01301 Fax: 413-774-6053
E-mail: nesea@nesea.org
http://www.nesea.org
*CD*
*David Barclay, Executive Director*
*Arianna Alexsandra Grindrod, Education Director*

**5654 Energy Thinking for Pennsylvania**
Northeast Sustainable Energy Association
50 Miles Street 413-774-6051
Greenfield, MA 01301 Fax: 413-774-6053
E-mail: nesea@nesea.org
http://www.nesea.org
*CD*
*David Barclay, Executive Director*
*Arianna Alexsandra Grindrod, Education Director*

**5655 Environmental Data Resources**
440 Wheelers Farms Road 203-255-6606
Milford, CT 06461 800-352-0050
Fax: 800-231-6802
E-mail: resinfo@edrnet.com
http://www.edrnet.com
Applied and product research in environmental applications.
*Robert D Barber, CEO*

**5656 Environmental Education Council of Ohio**
1972 Clark Avenue 330-823-3655
Alliance, OH 44601 800-992-6682
Fax: 330-823-8531
E-mail: info@muc.edu
http://www.muc.edu
EECO believes that: we are all learners interacting with others in lifelong process, education is vital for individuals to reach their full potential as members of our global community, a healthy and sustainable environment isessential to the survival of the planet. It is the mission of EECO to lead in facilitating and promoting environmental education which nurtures knowledge, attitudes and behaviors that foster global stewardship.

**5657 Environmental Education K-12**
PO Box 2057 863-465-2571
Lake Placid, FL 33862 Fax: 863-699-1927
E-mail: archbold@archbold-station.org
http://arcbold-station.org
Archbold Biological Station provides environmental education programs to help people af all ages discover and understand the unique and endangered Florida scrib. Several programs for children Grades K-12 are offered each year. Theprogram goals are; promote a sound foundation in ecological principles, nurture a sense of stewardship for Florida scrub habitat, demostrate the value of scientific research, and develop a deeper understanding of the importance of natural habitatsfor investigation.
*Dr. Hilary Swain, Executive Director*

**5658 Environmental Forum of Marin**
PO Box 150459 415-479-7814
San Rafael, CA 94915 E-mail: forum@MarinEFM.org
http://www.marinefm.org
Dedicated to protecting and enhancing the environment by educating its members and the Marin citizenry on environmental issues. In futherance of this goal, the Environmental Forum of Marin conducts annual training programs onenvironmental matters, provides continuing education for its members and public, and supports citizen action to influence environmental decision-making and public policy.
*David McConnell, President*
*Dianne Fruin, Vice President*

**5659 Environmental Law Institute**
1616 P Street NW 202-939-3800
Suite 20036 Fax: 202-939-3868
Washington, DC 20036
*J William Futrell, President*

**5660 Environmental Resources**
W275 N1990 Cabin Creek Ct 262-691-7413
Pewaukee, WI 53072 Fax: 262-691-1579
Owner and co-founder of Moraine Multimedia, and Environmental Resources developes continuing education programs.

**5661 Environmental Sciences**
205 O'Neil Hall 314-977-3131
St. Louis, MO 63108 Fax: 314-977-3117
E-mail: enviro@eas.slu.edu
Environmental Sciences is concerned with the near-surface realm of Earth and the way humans interact with that environment. Environmental scientist are concerned with water availability and equal, waste disposal, the use of Earth'slimited resources, and natural hazards such as earthquakes, landslides, and floods. Environmental scientists use the principles of geology, physics, chemistry, and biology to understand these phenomena and solve environmental problems.
*Cyn Wise, Administrative Secretary*

**5662 Exploring Animal Life - CD-ROM**
Educational Images
PO Box 3456 607-732-1090
Elmira, NY 14905 800-527-4264
Fax: 607-732-1183
E-mail: info@edimages.com
http://www.educationalimages.com
A curriculum oriented presentation and an instant encyclopedia, filled with superb photographs, informative text, exciting video clips, printable diagrams and illustrations, and lab activities. Provides a fascinating survey of themajor divisions of animal life and their characteristics: sponges, molluscs, insects, arthropods, fish, reptiles, birds and mammals are fully presented in the order in which you teach them.
*Charles R Belinky, Ph.D, CEO*

**5663 Exploring Environmental Science Topics**
Educational Images
PO Box 3456 607-732-1090
Elmira, NY 14905 800-527-4264
Fax: 607-732-1183
E-mail: info@edimages.com
http://www.educationalimages.com
Provides a curriculum oriented presentation, an instant encyclopedia, superb photographs, video clips, informative text, printable diagrams & illustrations, & lab activities. This program offers a fascinating survey of environmentaltopics & concerns such as the environmental costs of energy; acid rain; energy flow and the greenhouse effect; oil spills; tundra, chaparral, desert, grassland and forest biomes; the hydrological cycle and water pollution; and the recycling elementsin the biosphere.
*CD-Rom*
*Charles R Belinky, Ph.D, CEO*

**5664 Exploring Freshwater Communities**
Educational Images
PO Box 3456 607-732-1090
Elmira, NY 14905 800-527-4264
Fax: 607-732-1183
E-mail: info@edimages.com
http://www.educationalimages.com
A complete resource for studying freshwater biomes. It provides a fascinating survey of the ecology of swamps, bogs, marshes,

wetlands, streams, ponds, lakes and the Everglades. There is even an introduction to fish restoration andwater pollution.
*CD-Rom*
*Charles R Belinky, Ph.D, CEO*

**5665 Five Winds International**
20 Paoli Pike
Paoli, PA 19301
610-640-2302
Fax: 610-640-2303
E-mail: c.hatfield@fivewinds.com
http://www.fivewinds.com

Five Winds helps companies and organizations understand sustainability to improve their performance and succeed in the marketplace.

*Curt Hatfield, Business Manager*

**5666 Fossil Rim Wildlife Center**
2155 County Road 2008
PO Box 2189
Glen Rose, TX 76043
254-897-2960
888-775-6742
Fax: 254-897-3785
E-mail: vistor-services@fossilrim.org
http://www.fossilrim.org

Fossil Rim Wildlife Center is dedicated to conservation of species in peril, scientific research, training of professionals, creative management of natural resources, and impactful public education. Through these activities we providea diversity of compelling learning experiences which invoke positive change in the way people think, feel and act environmentally. Also provides scenic drives and lodgings for visitors, and is open all seasons.

*Billie Kinnard, Marketing/PR Director*
*Lisa Roberts, Membership Director*

**5667 GLOBE**
Mailstop T28H
Moffett Field, CA 94035
800-858-9947
Fax: 650-604-1913
E-mail: help@globe.gov
http://www.globe.gov

GLOBE is a worldwide hands-on, primary and secondary school based science and education program.

**5668 Glacier Institute**
137 Main Street
PO Box 1887
Kalispell, MT 59903
406-755-1211
Fax: 406-755-7154
E-mail: register@glacierinstitute.org
http://www.glacierinstitute.org

The Glacier Institute serves adults and children as an educational leader in the Crown of the Continent ecosystem with Glacier National Park at its center. Emphasizing field based learning experiences, the Institute provides anobjective and science based understanding of the area's ecology and its interaction with people. Through this non advocacy approach to outdoor education, participants can be better prepared to make informed and constructive decisions which impactthis & other ecosystems.

*Joyce Baltz, Executive Director*

**5669 Global Nest**
Michigan State University
Institute of International Health
B-301 W Fee Hall
East Lansing, MI 48824
517-353-8992
Fax: 517-355-1894
E-mail: secretary@gnest.org
http://www.NESTHOME.com

The Global Nest constitutes an international association of scientists, technologists, engineers and other interested groups involved in all scientific and technological aspects of the environment as well as in application techniquesaiming at the development of sustainable solutions. Its main target is to support and assist the dissemination of information regarding the most contemporary methods for improving quality of life through the development and application oftechnologies.

**5670 Gore Range Natural Science School**
82 East Beaver Creek Blvd Suite 202
PO Box 9469
Avon, CO 81620
970-827-9725
Fax: 970-827-9730
E-mail: science@gorerange.org
http://www.gorerange.org

Offers summer day and overnight programs for students in 3rd grade up. During the academic school year GRNSS provides integrated field science education to local and visiting schools. Its mission is to raise environmental awareness andinspire stewardship of the Eagle River watershed.
*Carolyn Busch, Director Of Marketing*
*Markian Feduschak, Executive Director*

**5671 Groundwater Foundation**
The Groundwater Foundation
Po Box 22558
Lincoln, NE 68542
402-434-2740
800-858-4844
Fax: 402-434-2742
E-mail: info@groundwater.org
http://www.groundwater.org

the groundwater foundation is a non profit organization, that is dedicated to informing the public about one of our greatest hidden resources, groundwater since 1985 our program and publications presents the benefits everyone recievesfrom groundwater,and the risks that threaten groundwater quality. we make learning about ground water fun and understandable for kids and adults alike.

**5672 Hazardous Chemicals: Handle With Care**
Educational Images
PO Box 3456
Elmira, NY 14905
607-732-1090
800-527-4264
Fax: 607-732-1183
E-mail: info@edimages.com
http://www.educationalimages.com

Shows the importance of hazardous chemicals in our daily lives and problems caused by ignorance, mistakes, accidents and occasionally, recklessness in their use. Four case studies show how toxic chemicals were introduced into theenvironment causing serious health and environmental effects. Video, 56 page guide with lesson plans, projects, reproducible handouts.

*Charles R Belinky, CEO*

**5673 Hidden Villa**
26870 Moody Road
Los Altos Hills, CA 94022
650-949-8650
Fax: 650-948-4159
E-mail: info@hiddenvilla.org
http://www.hiddenvilla.org

A non-profit 1600 acre organic farm and wilderness preserve serving approx. 50,000 visitors. Programs and offerings include: Hidden Villa Environmental Education Program, Summer Camp, a resident intern program, Community SupportedAgriculture, a Hostel for domestic and international travelers, meeting/retreat rental space, Community Programs and eight miles of hiking trails and picnic areas.
*Beth Ross, Executive Director*

**5674 Ice Age Relicts: Living Glaciers and Signs of Ancient Ice Sheets**
Educational Images
PO Box 3456
Elmira, NY 14905
607-732-1090
800-527-4264
Fax: 607-732-1183
E-mail: info@edimages.com
http://www.educationalimages.com

Glaciers, living relicts of the ice age, are still important today. They hold much of the earth's fresh water, sculpted much of North America, and promise an early return to finish their work. Provides a coherent picture of howglaciers work, and what they did. 26 page guide. Video or filmstrips.

*Charles R Belinky, Ph.D, CEO*

**5675 International Center for Earth Concerns**
D162 Baldwin Road
Ojai, CA 93023
805-649-3535
Fax: 805-649-1757
E-mail: information@earthconcerns.org
http://www.earthconcerns.org

Dedicated to providing for public use, a world class botanic garden, outdoor learning-ecology center and a 50-passenger all-electric floating classroom on Lake Casitas. These facilities are used to promote a better understanding ofman's place in the

environment, as well as to help develop a sense of respect, responsibility and compassion for animals and nature.
*John Taft, Chairman*

### 5676 Invertebrate Animal Videos

Educational Images
PO Box 3456 607-732-1090
Elmira, NY 14905 800-527-4264
Fax: 607-732-1183
E-mail: info@edimages.com
http://www.educationalimages.com

Four part series. Each a 40-minute multimedia presentation with easy going narration and hundreds of interactive links. Part I: sponges, anemones, corals and flatworms. Part II: molluscs, segmented worms and minor phyla. Part III:the insects. Part IV: noninsect arthropods and echinoderms.

*Charles R Belinky, Ph.D, CEO*

### 5677 Jones & Stokes

11820 Northup Way 425-822-1077
Suite E300 Fax: 425-822-1079
Bellevue, WA 98005

An employee-owned company, Jones & Stokes is the best consulting source for integrated environmental planning and natural resources management services in the western United States.
*Grant Bailey*

### 5678 Killer Whales: Lords of the Sea

Educational Images
PO Boc 3456 607-732-1090
Elmira, NY 14905 800-527-4264
Fax: 607-732-1183
E-mail: info@edimages.com
http://www.educationalimages.com

Separates facts from myth about these majestic, maligned and usually misrepresented scagoing mammals: both wild and captive killer whales, their mental and physical powers, their feeding and reproductive behavior, physiology,sociology and echolocation. Information on other cetaceans is presented for comparison and better understanding. Provides scientific information and reports on ongoing research.

*Charles R Belinky, Ph.D, CEO*

### 5679 Legacy International

1020 Legacy Drive 540-297-5982
Bedford, VA 24523 Fax: 540-297-1860
E-mail: mail@legacyintl.org
http://www.legacyintl.org

Creates environments where people can address personal, community, and global needs while developing skills and effective responses to change.Whether working with youths, corporate leaders, educational professionals, entrepreneurs, orindividuals on opposing sides of a conflict, our goal is the same. Programs provide experiences, skills, and strategies that enable people to build better lives for themselves and others around them.
*J Rash, President*

### 5680 Lesley/Audubon Environmental Education Programs

Lesley University
29 Everett Street 617-349-8320
Cambridge, MA 02138 E-mail: info@lesley.edu
http://www.lesley.edu/gsass/75audubon.html

In partership with Audubon Expedition Institute in Belfast, Lesley University offers a Bachelor of Science degree in Environmental Studies, a Master of Science degree in Environmental Education and a Master of Science in EcologicalTeaching and Learning. Students travel throughout the US earning academic credit and gaining first-hand experience of environmental issues.

### 5681 Let's Grow Houseplants

Educational Images
Po Box 3456 607-732-1090
Elmira, NY 14905 800-527-4264
Fax: 607-732-1183
E-mail: info@edimages.com
http://www.educationalimages.com

Details different kinds of plants and their needs, when to water, selection, how to start from seeds and cuttings, how to make inexpensive pots, etc. Perfect to initiate an elementary classroom gardening project. 74 frames and guide.For elementary and preschool.

*Charles R Belinky, Ph.D, CEO*

### 5682 Lost Valley Educational Center

81868 Lost Valley Lane 541-937-3351
Dexter, OR 97431 E-mail: info@lostvalley.org
http://www.lostvalley.org

Offers a wide variety of programs, including residential interships, educational workshops that emphasize hands on, experiential learning and personal/spiritual growth workshops. The Center also provides a supportive and nourishingplace to hold individuals and organizations who share our vision for an environmentally sound, pollution free world to hold conferences, retreats and workshops.
*Dianne Brause, Co-Founder*

### 5683 MacKenzie Environmental Education Center

Wisconsin Department of Natural Resources
300 Femrite Dr 608-221-2575
Monona, WI 53716 Fax: 608-221-9095
E-mail: info@naturenet.com
http://www.naturenet.com/mackenzie/

Located only 20 miles north of Madison, the MacKenzie Center offers a wide array of outdoor experiences. Five themed nature trails, prairie restorations, picnic area, nature study, three museums and a wildlife exhibit containing liveanimals that are native to Wisconsin, are here to help you gain a better understanding of our natural resources.

*Derek A Duane, Director*

### 5684 Marine Biological Laboratory

7 Mbl Street 508-548-3705
Woods Hole, MA 02543 Fax: 508-457-1924
E-mail: mdonovan@mbl.edu
http://www.mbl.edu

For more than a century, scientists from around the world have been gathering in Woods Hole. The best students from the best universities, the brightest young faculty, the most succesful scientists working at the pinnacle of theprofession, an unmatched collection of researchers and educators congregates every year in the seaside village whose name has become synonimous with science.

### 5685 Microscopic Pond

Educational Images
PO Box 3456 607-732-1090
Elmira, NY 14905 800-527-4264
Fax: 607-732-1183
http://www.educationalimages.com

Introduces students to both the miersccpic plant and animal life of a pond. Various groups of algae are discussed and illustrated, including desmids, Pediastrum, Pithophora, Spyrogyra, Volvox, Nostac, calothrix, Bacillariophyseae,Dinophyseae, and amoebas (includeing Amoeba proteus), Arcella, the testaceans, and many others. With only a few exceptions, all of the organisms in this program were photographed live.

*Charles R Belinky, Ph.D, CEO*

### 5686 Mote Environmental Services

1600 Ken Thompson Parkway 941-388-4441
Sarasota, FL 34236 800-691-6683
http://www.mote.org

Mote Environmental Services offers consulting services focused on marine and coastal issues, where our expertise is strongest. We provide superior, results-oriented investigations and management planning service within our areas oftechnical and policy specialty. MESI is a wholly owned subsidiary of Mote Marine Laboratory, an independent, nonprofit research and public education institution dedicated to excellence in marine and environmental sciences.
*Dr. Kumar Mahadevan, President*

**5687 National Environmental Health Association (NEHA)**
720 South Colorado Boulevard 303-756-9090
Suite 1000-N Fax: 303-691-9490
Denver, CO 80246 E-mail: staff@neha.org
http://www.neha.org

NEHA is the only national association that represents all of environmental health and protection from terrorism and all-hazards preparedness, to food safety and protection and onsite wastewater systems. Over 4500 members and theprofession are served by the association through its Journal of Environmental Health, Annual Educational Conference & Exhibition, credentialing programs, research and development activities and other services.

*Kim Clapper, Executive Assistant*

**5688 National Institute of Environmental Health Sciences**
111 TW Alexander Drive 919-541-3345
PO Box 12233 Fax: 919-541-4395
Research Tri Pk, NC 27709 E-mail: webcenter@niehs.nih.gov
http://www.niehs.nih.gov

The National Institute of Environmental Health Sciences is the principal federal agency for basic biomedical research on the health effects of environmental agents. It is the headquarters for the National Toxicology Program whichcoordinates toxicology studies within the Department of Health and Human Services.

*Dr David A Schwartz, Director*
*Ms Lou Rozler, Public Liason Officer*

**5689 National PTA: Environmental Project**
541 N Faribanks Court 312-670-6782
Suite 1300 800-307-4782
Chicago, IL 60611 Fax: 312-670-6783
E-mail: info@pta.org
http://www.pta.org

The mission of the National PTA is to support and speak on behalf of children and youth in the school, in the community and before governmental bodies and other organizations; to assist parents in developing the skills they need toraise and protect their children and to encourage parent and public involvement in the public schools. Engages in advocacy and education, including workshops and lobbying.
*Jan Harp Domene, President*

**5690 Natural Resources Conservation and Management**
Lexindton Convention And Visitors Bureau
301 East Vine Street 859-244-7706
Lexington, KY 40507 800-845-3959
E-mail: lexadmin@visitlex.com

As a trained professional, you will have a variety of challenging employment opportunities in public agencies and industry to contribute to sustained productivity and equality of all of our natural resources. In addition, some studentsfind that the Natural Resource Conservation and Management program satisfies their desire for a career in environmental education or environmental journalism.
*David Lord, President*

**5691 Nielsen Environmental Field School**
9600 Achenbach Canyon Road 505-532-5535
Las Cruces, NM 88011 Fax: 505-532-5978
E-mail: info@environmentalfieldschool.com
http://www.environmentalfieldschool.com

In 1990, the Nielsen Environmental Field School was created in reponse to a demand from the environmental industry for practically oriented, hands-on environmental field training.
*David Nielsen, Event Coordinator*

**5692 Northwest Environmental Education Council**
Northwest Environmental Education Council
650 S. Orcas Street 206-762-1976
Suite 220 Fax: 206-762-1979
Seattle, WA 98108 E-mail: emcwayne@nweec.org
http://www.nweec.org

The Northwest Environmental Education Council increases environmental awareness, appreciation and stewardship by providing environmental education and science training opportunities for youth and adults.
*Tasya Gray, President*
*Lief Horwitz, Vice President*

**5693 Northwest Interpretive Association**
Northwest Interpretive Association
164 South Jackson Street 206-220-4140
Seattle, WA 98104 877-874-6775
http://www.nwpubliclands.com

Works with public and management agencies to operate educational bookstores. Our mission is to provide visitors with information they need to learn about the nature and natural history of public lands so they can make wise choicesabout the lands use, preservation and protection. We accomplish our mission by selling educational and interpretive materials directly to visitors as well as returning net proceeds to the site where they were guarenteed to help fund other programs.
*Mark Lester, Chair*

**5694 Office of Energy and Environmental Industries**
International Trade Administraction 202-482-5225
1401 Constitution Ave NW Room 4053 800-usa-trad
Washington, DC 20230 Fax: 202-482-5665
http://www.trade.gov/envirotech / export.gov/reee

The Office of Energy and Environmental Industries (OEEI) is the principal resource and key contact point within the US Department of Commerce for American environmental technology companies. OEEI's goal is to facilitate and increaseexports of environmental technologuies-goods and services-by providing support and guidance to US exporters.
*Adam O'Malley, Office Director*
*Man Cho, Team Leader, Energy*

**5695 Perkiomen Watershed Conservancy**
1 Skippack Pike 610-287-9383
Schwenksville, PA 19473 Fax: 610-287-9237
E-mail: pwc@perkiomenwatershed.org
http://www.perkiomenwatershed.org/

A nonprofit organization founded in 1964 by local citizens that works to protect the watershed of the Perkiomen Creek and its tributaries. This is accomplished through environmental education, conservation programs and watershedstewardship activities.

*Crystal Gilchrist, Executive Director*
*Trudy Phillips, Director for Env Education*

**5696 Primary Ecological Succession**
Educational Images
PO Box 3456 607-732-1090
Elmira, NY 14905 800-527-4264
Fax: 607-732-1183
E-mail: info@edimages.com
http://www.educationalimages.com

An illustrated explanation of basic concepts of primary succession: the pioneer community; tolerant vs. intolerant species; stabilization; stratification and the climax community. Concise overview followed by classic, specificexamples of succession - on bare rock, on the sand dunes of Lake Michigan, on the outer banks of North Carolina - all explored in detail. 72 frames and guide.

*Charles R Belinky, Ph.D, CEO*

**5697 Project Oceanology**
University Of Connecticut
Avery Point Campus 860-445-9007
1084 Shennecossett Road 800-364-8472
Groton, CT 06340 Fax: 860-449-8008
E-mail: oceanology@aol.com
http://www.oceanology.org

Project Oceanology is owned and operated by Interdistrict Committee for Project Oceanology and association of 25 educational institutions in Massachusetts, Rodhe Island, Connecticut and New York. Members of this associations includepublic school districts, private schools, states university, public and private colleges, a maritime museum and an aquarium. The project is governed by an assembly of delegates representing the member institutions.

**5698 Project WILD**
5555 Morningside Drive 713-520-1936
Suite 212 Fax: 713-520-8008
Houston, TX 77005 E-mail: info@projectwild.org
Project WILD is one of the most widely-used conservation and environmental education programs among educators of students in kindergarten through high school. Project WILD is based on the premise that young people and educators have avital interest in learning about natural world.
*Cheryl Stanco, Senior Manager*
*Josetta Hawthorne, Executive Director*

**5699 Resource-Use Education Council**
The Virginia Natural Resources Education Guide
PO Box 11104 804-698-4442
Richmond, VA 23230 Fax: 804-698-4522
E-mail: jkcomfort@deg.virginia.gov
http://WWW.virginianaturally.org
In the mid 1950s, representatives from Virginia and federal natural resource agencies, along with professors in the colleges of education and resource management, came together as the Virginia Resource Use Education Council. For 35years, the VRUEC sponsored a summer conservation course for teachers at four of Virginia's colleges.
*Ann Regn, Chairman*

**5700 Risk Management Internet Services**
Managerial Technologies Corporation
2400 East Main Street 630-221-9116
Suite 103- 319 Fax: 312-602-4935
St Charles, IL 60174 E-mail: info@rmis.com
http://rmis.com
The rmFamily of sites and services is dedicated to bringing risk management related professions together with the consultants, developers and providers who service them.

**5701 Ross & Associates Environmental Consulting,Ltd**
1218 Third Avenue 206-447-1805
Suite 1207 Fax: 206-447-0956
Seattle, WA 98101 E-mail: rossmail@ross-assoc.com
http://www.ross-assoc.com
Ross and Associates environmental consulting, is a small group of highly motivated professionals committed to helping environmental and natural resources agencies improve management programs and achieve better environmental results.
*Bill Ross, President And Principal*

**5702 SEEK**
525 South Lake Avenue 218-529-6258
Suite 400 888-668-3224
Duluth, MT 55802
The SEEK directory works as a clearinghouse for all types of environmental education resources, from articles to lesson plans, from performances to displays, and many more. These resources come a variety of organizations throughoutMinnesota, including schools and colleges, government agencies, libraries and businesses.

**5703 Sacramento River Discovery Center**
1000 Sale Lane 530-527-1196
Red Bluff, CA 96080 Fax: 530-527-1312
E-mail: lgreen@tehama.k12.ca.us
http://www.srdc.tehama.k12.ca.us
The mission of the Sacramento River Discovery Center is to educate the public's school programs. Teacher professional development, camping, rafting and tourist events are available.
*Lupe Green, Executive Director*
*Anna Draper, Program Manager*

**5704 Save the Dolphins Project Earth Island Institute**
300 Broadway 415-788-3666
Suite 28 Fax: 415-788-7324
San Francisco, CA 94133 E-mail: johnknox@earthisland.org
http://www.earthisland.org
*John Knox, Executive Director*

**5705 Schlitz Audubon Nature Center**
1111 E Brown Deer Road 414-352-2880
Milwaukee, WI 53217 Fax: 414-352-6091
E-mail: smanning@sanc.org
http://schlitzauduboncenter.com
A unique urban area of green just 15 minutes north of downtown Milwaukee, we are located along the shore of Lake Michigan. Escape from the world of concrete to hike seven miles of trails, walk along the beach and feel far away from thecity or view forests and wildlife from the 60-foot observation tower. Remember to bring your binoculars, you never know what you may want to take a closer look at while visiting the Center. New Sustainable Environmental Learning Center opened in2003!
*Amy Burke, Fun Manager*

**5706 School of Public & Environmental Affairs**
Indiana University
107 S. Indiana Ave. 812-855-4848
Bloomington, IN 47405 800-765-7755
The School of Public Environmental Affairs offer environmental science summer programs to high school students and middle/high school teachers who want answers to environmental questions.
*Michael McRobbie, President*

**5707 Science House**
909 Capability Drive 919-515-6118
Suite 1200 Fax: 919-515-7545
Raleigh, NC 27695 E-mail: science_house@ncsu.edu
http://www.science-house.org
The activities of The Science House is itself a partership of facultu and stuff from science and education departments across the NC State campus, and collaborates with many other k-12 support organization in North Carolina.

**5708 Seacamp Association, Inc**
Newfound Harbor Marine Institute
1300 Big Pine Avenue 305-872-2331
Big Pine Key, FL 33043 Fax: 305-872-2555
E-mail: info@seacamp.org
http://www.seacamp.org
To create awareness of the complex and fragile marine world and to foster critical thinking and informed decision making about man's use of natural resources. One of the few organizations in the US providing experiential education inmarine studies to students aged 8 to 21 years.

*Irene Hooper, Executive Director*
*Grace Upshaw, Director at Seacamp*

**5709 Setting Up a Small Aquarium**
Educational Images
PO Box 3456 607-732-1090
Elmira, NY 14905 800-527-4264
Fax: 607-732-1183
E-mail: info@edimages.com
http://www.educationalimages.com
Details the exact procedure to be followed in setting up either a marine or freshwater aquarium successfully. Methods are scientifically sound, well documented, and up-to-date. A scaled-down version of the methods used in largepublic aquaria, the system works! 90 frames and guide.

*Charles R Belinky, Ph.D, CEO*

**5710 Sierra Club**
85 Second Street 415-977-5500
2nd Floor Fax: 415-977-5799
San Francisco, CA 94105 E-mail: information@sierraclub.org
http://www.sierraclub.org
Aims to explore, enjoy, and protect the wild places of the earth, to practice and promote the responsible use of the earth's ecosystems and resources, to educate and enlist humanity to protect and restore the quality of the naturaland human environment, and to use all lawful means to carry out these objectives.
*Robbie Cox, President*
*Robin Mann, Vice President*

**5711 Smithsonian Environmental Research Center**
Po Box 28 443-482-2200
647 Contees Wharf Road Fax: 443-482-2380
Edgewater, MD 21037 E-mail: jiacintos@si.edu
http://www.serc.si.edu

The Smithsonian Environmental Research Center advances stewardship of the biosphere through interdisciplinary research and educational outreach. SERC's scientists study a variety of interconnected ecosystems at the Center's primaryresearch site here in Maryland, and at affiliated sites around the world.
*Susan Jiacinto, General Admissions Contact*

**5712 Society of Environmental Toxicology and Chemistry**
SETAC N America Office
1010 N 12th Avenue 850-469-1500
Pensacola, FL 32501 Fax: 850-469-9778
E-mail: setac@setac.org
http://www.setac.org

The Society of Environmental Toxicology and Chemistry provides a forum for the examination of environmental issues by environmental professionals from industry, academia, government, and public-interest groups.
*Mimi Meredith, Sr. Manager*

**5713 Southwest Environmental Health Sciences: Community Outreach and Education Program**
University Of Arizona College Of Pharmacy
Room 244 520-626-3692
Po Box 210207 Fax: 520-626-4468
Tucson, AZ 85721 E-mail: swehsc-info@pharmacy.arizonia..edu
http://www.swehsc.pharmacy.arizoniz.edu

The COEP goals are to review, develop, and disseminate quality environmental health science curricula. Develop and host K-12 teacher training workshops, communicate with the general public about local and common environmental healthscience concerns, share research results from SWEHSC investigators with the COEP target audiences.

**5714 Spiders in Perspective: Their Webs, Ways and Worth**
Educational Images
PO Box 3456 607-732-1090
Elmira, NY 14905 800-527-4264
Fax: 607-732-1183

Comprehensive coverage of the nature of spiders, their diversity of structures, and their remarkable behavior patterns. Presents the unique world of creatures you may have ignored before, but probably never will again. Coverslocomotion, the various perceptual senses, silk production, camouflage and mimicry, webs, hunting, predation by wasps, kleptoparasitism, courtship and reproduction, population densities, impact on humans, etc. 2 parts, 76 & 78 frames. Video, slidesor filmstrip.

*Charles R Belinky, Ph.D., CEO*

**5715 Student Conservation Association**
Po Box 550 603-543-1700
689 River Road Fax: 603-543-1828
Charlestown, NH 03603 E-mail: internships@thesca.org
http://www.thesca.org

America's largest and oldest provider of conservation service opportunities, outdoor education and career training for youth. SCA is building the next generation of conservation leaders and inspire lifelong stewardship of ourenvironment and communities.
*Dale Penny, President*
*Shaundrea Kenyon, Operations Director/Recruitm*

**5716 The Groundwater Foundation**
PO Box 22558 402-434-2740
Lincoln, NE 68542 800-858-4844
Fax: 402-434-2742
E-mail: info@groundwater.org
http://www.groundwater.org

The Groundwater Foundation is a nonprofit organization that is dedicated to informing the public about one of our greatest hidden resources, groundwater. Since 1985, our programs and publications present the benefits everyone receivesfrom groundwater and the risks that threaten groundwater quality. We make learning about groundwater fun and understandable for kids and adults alike.

**5717 The Nelson Institute for Environmental Studies**
UW Madison, 550 N Park Street 608-262-7996
70 Science Hall Fax: 608-262-2273
Madison, WI 53706 E-mail: nelson@mailplus.wisc.edu
http://www.nelson.wisc.edu

Few institutions can match the University of Wisconsin-Madison's expertise in environmental studies. Literally hundreds of professors teach and conduct research in environmentally related subjects ranging from agriculture to zoology.Their scholarship and achievement are widely recognized. In dozens of academic fields, the university is consistently rated the nation's best and most prolific.

**5718 Thorne Ecological Institute**
1466 N 63rd Street 303-499-3647
PO Box 19107 Fax: 720-565-3873
Boulder, CO 80308 E-mail: info@thorne-eco.org
http://www.thorne-eco.org

Offers hands-on environmental education for young people along the Front Range of Colorado.
*Kelly Keena, Executive Director*

**5719 Trees for Tomorrow**
519 Sheridan Street 715-479-6456
PO Box 609 800-838-9472
Eagle River, WI 54521 Fax: 715-479-2318
E-mail: learning@treesfortomorrow.com
http://www.treesfortomorrow.com

Private, nonprofit natural resource education school that uses a combination of field studies and classroom presentations to teach conservation values, as well as demonstrate the benefits of modern resource management.

*Maggie Bishop, Executive Director*

**5720 Triumvirate Environmental**
61 Inner Belt Road
Somerville, MA 02143 800-966-9282
Fax: 617-628-8099
E-mail: contactus@triumvirate.com
http://www.triumvirate.com

Triumvirate Environmental is a full-service environmental management firm headquartered in eastern Massachusetts. Serving the environmental and hazardous waste needs of clients throughout the northeast in the areas of biotechnology andpharmaceuticals, education, health care, metal platers and finishers, manufacturing, and utilities, Triumvirate Environmental is the industry leader in personalized service.

**5721 Tropical Forest Foundation**
2121 Eisenhower Avenue 703-518-8834
Suite 200 Fax: 703-518-8974
Alexandria, VA 22314 E-mail: tff@igc.org
http://www.tropicalforestfoundation.org

A non-profit educational institution dedicated to the conservation of tropical forests through sustainable forestry. Its Board of Directors includes representatives from industry, government, science, academia and conservation.

*Keister Evans, President*
*Wendy Baer, Staff Executive*

**5722 US Environmental Protection Agency: Great Lakes National Program Office**
77 West Jackson Boulevard 312-886-4040
Chicago, IL 60604 Fax: 312-353-2018

The focus for the State of the Lakes Ecosystem Conference (SOLEC) 1996 is the nearshore zone of the Great Lakes. Nearshore ecosystems are complex and dynamic with many measurable parameters. The nearshore area is extremely important tooverall ecosystem function. It is the most productive zone within each of the Great Lakes and is the area most affected by human activity. Nearshore zones include embayments, tributaries and tributary mouths, marshes and other wetlands, and dunes.

**5723 Water Resources Management**
550 N Park Street 608-262-7996
70 Science Hall Fax: 608-262-2273
Madison, WI 53706 E-mail: nelsongrad@mailplus.wisc.edu
http://www.nelson.wisc.edu
The program addresses the complex, interdisciplinary aspects of managing resources by helping students integrate the biological and phisical sciences with engineering and law and the social sciences. The workshop provides anopportunity for students to work outside of the textbook environment and tackle a rea-world problem.

**5724 Wilderness Education Association**
Colorado University 970-223-6252
Fort Collins, CO 80523 E-mail: wea@lamar.colostate.edu
http://www.prienet.org/

*David Cockrell, President*

**5725 Windows on the Wild**
World Wildlife Fund
1250 24th Street Northwest 202-293-4800
Po Box 97180 800-225-5993
Washington, DC 20037 Fax: 202-293-9211
Provides educators with interdisciplinary curriculum materials and training programs. By using biodiversity as its organizing theme, WOW provides students with a unique window for exploring a range of topics including science,economics, social studies, language arts, geography and civics.
*Jennifer A Zadwick, Program Information Coord.*

**5726 World Resources Institute**
10 G Street, NE 202-729-7600
Suite 800 Fax: 202-729-7610
Washington, DC 20002 http://www.wri.org
WRI provides information, ideas, and solutions to global environmental problems. Our mission is to move human society to live in ways that protect the environment for current and future generations, with programs that meet globalchallenges by using knowledge to catalyze public and private action. Goals include safeguarding earth's climate from further harm, protecting the ecosystems, and reducing the use of materials and generation of wastes in the production of goods andservices.
*Jonathan Lash, President*

**5727 Young Entomologists Society**
6907 West Grand River Avenue 517-886-0630
Lansing, MI 48906 Fax: 517-886-0630
E-mail: yesbugs@aol.com
http://www.members.@aol.com/YESbugs/mainmenu.html
To provide young people with a combination of programs, publications, and educational materials that enrich their insect and spider studies through dynamic, innovative, and enjoyable learning experience.
*Gary Dunn, Director Of Education*

# Industry Web Sites

## Environmental

### 5728 ABS Consulting Training Services

http://www.absconsulting.com/svc_training.cfm

Government Institutes Division provides continuing education and practical information on government regulatory topics. We recognize that you face unique challenges presented by the ever-increasing number of new regulations and theresulting rapid evolution of new technologies.

### 5729 Academy of Natural Sciences

http://www.acnatsci.org

Our mission is to create the basis for a healthy and sustainable planet through exploration, research and education.

### 5730 ActiveSet.org

http://www.activeset.org

ActiveSet.org was created for the benefit of professionals involved with all aspects of environmental air quality testing, monitoring and management. Environmental Managers can easily research and contact emissions testing firms,services and products either by state, region, or using the site's built-in search engine. Facility owners and managers use ActiveSet.org's free online request for proposals form to reach stack testing firms all over the country for their nextproject.

### 5731 Adirondack Council

http://www.adirondackcouncil.org

The Adirondack Council is a nonprofit environmental group working to protect the open space resources of New York's six million acre Adirondack Park and to help sustain the natural and human communities of the region. It monitorsdevelopment on private lands and ensures the mandated constitutional protection of public lands.

### 5732 Advanced Recovery

http://www.advancedrecovery.com

Advanced Recovery, based in New Jersey, is involved in the recycling industry. They promote the proper disposal of all scrap, but are particularly interested in the disposal of electronic equipment, such as computer monitors consistingof lead.

### 5733 Advanced Technologies And Practices

http://www.advancedbuildings.org

A building professional's guide to more than 90 environmentally-appropriate technologies and practices. Architects, engineers and buildings managers can improve the energy and resource efficiency of commercial, industrial andmulti-unit residential buildings through the use of the technologies and practices described in this web site.

### 5734 Advanced Technology Environmental Education Center

http://www.ateec.org

ATEEC's mission is the advancement of environmental technology education through curriculum development, professional development and program improvement in the nation's community colleges and high schools. ATEEC is funded by theNational Science Foundation and is a partnership of the Hazardous Materials Training and Research Institute, the National Partnership for Environmental Technology Education, and the University of Northern Iowa.

### 5735 African Environmental Research and Consulting Group

http://www.africaenviro.org

The AERCG is a US nonprofit organization with offices in Africa and Western countries. The group focuses on improving the quality of life, mitigating environmental hazards and the protection of human health in Africa. They are involvedin promoting sustainable development in African communities, their site contains information about their efforts.

### 5736 Agency for Toxic Substances and Disease Registry

http://www.atsdr.cdc.gov/

The mission of the Agency for Toxic Substances and Disease Registry, as an agency of the US Department of Health and Human Services, is to prevent exposure and adverse human health effects and diminished quality of life associated withexposure to hazardous substances from waste sites, unplanned releases and other sources of pollution present in the environment.

### 5737 Agriculture Network Information Center

http://www.agnic.org

The Agriculture Network Information Center is a voluntary alliance of the National Agricultural Library, land-grant universities and other agricultural organizations in cooperation with citizen groups and government agencies. AgNICfocuses on providing agricultural information in electronic format over the World Wide Web via the Internet.

### 5738 Air Force Center for Environmental Excellence

http://www.afcee.brooks.af.mil

The Air Force Center for Environmental Excellence provides our customers with a complete range of world class enviromental, architectural and landscape design, planning and construction management services and products.

### 5739 Air and Waste Management Association

http://www.awma.org

The Air and Waste Management Association is a nonprofit, nonpartisan professional organization that provides training, information and networking opportunities to thousands of environmental professionals in 65 countries.

### 5740 Alabama Department of Environmental Management

http://www.adem.state.al.us

Provides environmental stewardship through the implementation of authorized environmental statutes, advocating statutory change as needed.

### 5741 Alaska Chilkat Bald Eagle Preserve

http://www.dnr.state.ak.us/parks/units/eagleprv.htm

The Alaska Chilkat Bald Eagle Preserve was created by the State of Alaska in June 1982. The Preserve was established to protect and perpetuate the world's largest concentration of bald eagles and their critical habitat. It alsosustains and protects the natural salmon runs.

### 5742 Alfred Wegener Institute for Polar and Marine Research

http://www.awi-bremerhaven.de/index-e.html

Polar and Marine research are central themes of Global system and Environmental Science. The Alfred Wegener Institute conducts research in the Arctic, the Antarctic and at temperate latitudes. It coordinates Polar research in Germanyand provides both the necessary equipment and the essential logistic back up for polar expeditions. Website is in the German language.

### 5743 Alliance for Environmental Technology

http://www.aet.org

The Alliance for Environmental Technology is an international association of chemical manufacturers and forest products companies dedicated to improving the environmental performance of the pulp and paper industry. AET supports the useof Elemental Chlorine-Free technology based on chlorine dioxide.

### 5744 American Academy of Environmental Engineers

http://www.aaee.net

The American Academy of Environmental Engineers is dedicated to excellence in the practice of environmental engineering to ensure the public health, safety and welfare to enable humankind to co-exist in harmony with nature.

### 5745 American Chemical Society

http://www.chemistry.org/portal/a/c/s/l/home.html

The mission is to promote the public perception and understanding of chemistry and the chemical sciences through public outreach programs and public awareness campaigns; involve the Society's more than 163,000 member's in improving thepublic's perception of chemistry.

### 5746 American Conference of Governmental Industrial Hygienists

http://www.acgih.org/home.htm

Member-based organization and community of professionals that advances worker health and safety through education and the development and dissemination of scientific and technical knowledge.

### 5747 American Council for an Energy-Efficient Economy

http://www.aceee.org

The American Council for an Energy-Efficient Economy is a nonprofit organization dedicated to advancing energy efficiency as a means of promoting both economic prosperity and environmental protection.

### 5748 American Farmland Trust

http://www.farmland.org

American Farmland Trust is a private nonprofit organization founded in 1980 to protect our nation's farmland. AFT works to stop the loss of productive farmland and to promote farming practices that lead to a healthy environment.

### 5749 American Forests

http://www.americanforests.org

American Forests is a world leader in planting trees for environmental restoration, a pioneer in the science and practice of urban forestry and a primary communicator of the benefits of trees and forests.

### 5750 American Geophysical Union

http://earth.agu.org

AGU's mission is to promote the scientific study of Earth and its environment in space and to disseminate the results to the public, to promote cooperation among scientific organizations involved in geophysics and related disciplinesand to initiate and participate in geophysical research programs.

### 5751 American Hydrogen Association

http://www.clean-air.org

The Mission of AHA is to facilitate achievements of prosperity without pollution and to close the information gap between researchers, industry and the public, drawing on world-wide developments concerning hydrogen, solar, wind, hydro,ocean and biomass resource materials, energy conversion, wealth-addition economics and the environment.

### 5752 American Rivers

http://www.americanrivers.org

American Rivers is a national nonprofit conservation organization dedicated to protecting and restoring America's rivers and to fostering a river stewardship ethic.

### 5753 American Solar Energy Society

http://www.ases.org

The American Solar Energy Society is a national organization dedicated to advancing the use of solar energy for the benefit of US citizens and the global environment. ASES promotes the widespread near-term and long-term use of solarenergy.

### 5754 Ames Laboratory: Environment Technology Department

http://www.etd.ameslab.gov

The Ames Laboratory's Environmental & Protection Sciences Program is playing an important role in the US Department of Energy's initiative to cleanup hazardous waste, responding to remediation problems that need faster, safer, betteror cheaper technological solutions. You'll find information here about those technologies, the scientists behind them and our efforts to move these technologies into the marketplace.

### 5755 Antarctic and Southern Ocean Coalition

http://www.asoc.org

The Antartic Project is the Secretariat of the Antartic and Southern Ocean Coalition which contains nearly 230 organizations in 49 countries and leads the national and international campaigns to protect the biological diversity andpristine wilderness of Antartica, including its oceans and marine life. We work for passage of strong measures which protect the marine ecosystem from the harmful effects of overfishing, and work to ensure that the integrity of the land and animalsis maintained.

### 5756 Argonne National Laboratory

http://www.anl.gov

Argonne National Laboratory is one of the US Department of Energy's largest research centers. It is also the nation's first national laboratory, chartered in 1946.

### 5757 Arizona Geological Survey

http://www.azgs.state.az.us/

To inform and advise the public about the geologic character of Arizona in order to foster understanding and prudent development of the State's land, water, mineral and energy resources.

### 5758 Arkansas Natural Heritage Commission

http://www.naturalheritage.org

The Arkansas Natural Heritage Commission (ANHC) is responsible for maintaining the most up-to-date and comprehensive source of information concerning the rare plant and animal species, and high-quality natural communities of Arkansas.Systematic analysis of this natural heritage data can be used to identify locations that hold exceptional importance for the state's natural diversity, but that lack formal protection.

### 5759 Asia-Pacific Centre for Environmental Law

http://http://law.nus.edu.sg/apcel

APCEL was established in response to the need for capacity-building in environmental legal education and the need for promotion of awareness in environmental issues. It is currently working closely with IUCN's Commission onEnvironmental Law.

### 5760 Association of Energy Engineers

http://www.aeecenter.org

A nonprofit professional society which promotes the scientific and educational interests of those engaged in the energy industry and fosters action for sustainable development.

**5761 Association of State Flood Plain Managers**

http://www.floods.org/home/default.asp

Promotes common interest in flood damage abatement, supports environmental protection for floodplain areas, provides education on floodplain management practices and policy and urges incorporating multi-objective management, approachesto solve local flooding problems.

**5762 Associations of University Leaders for a Sustainable Future**

http://www.ulsf.org

The mission of the Association of University Leaders for Sustainable Future is to make sustainability a major focus of teaching, research, operations and outreach at colleges and universities worldwide. ULSF pursues this missionthrough advocacy, education, research, assessment, membership support and international partnerships to advance education for sustainability.

**5763 Atlantic Salmon Federation**

http://www.asf.ca

The Atlantic Salmon Federation is an international nonprofit organization which promotes the conservation and wise management of the wild Atlantic salmon and its environment.

**5764 Australian Cooperative Research Centres**

http://www.crc.gov.au/Information/default.aspx

The Cooperative Research Centers, generally known as CRCs, bring together researchers from universities, CSIRO and other government laboratories and private industry or public sector agencies in long-term collaborative arrangementswhich support research and development and education activities that achieve real outcomes of national economic and social significance.

**5765 Australian Oceanographic Data Centre**

http://www.aodc.gov.au/about.html

The mission of the Australian Oceanographic Data Centre is to acquire, manage and distribute oceanographic information and provide specialist oceanographic advice to; enable the Australian Defence Force to exploit the above and belowwater physical operating environments for strategic, operational and tactical advantage meet national and international obligations to manage oceanographic information.

**5766 Bat Conservation International**

http://www.batcon.org

The mission of Bat Conservation International is to protect and restore bats and their habitats worldwide.

**5767 Battelle Seattle Research Center**

http://www.seattle.battelle.org/

Battelle Memorial Institute is a multidimensional organization dedicated to making the future better for everyone. Although that may sound a bit grand, it really is Battelle's mission. From 1929 to the present, putting innovation andtechnology to practical use has been our goal.

**5768 Bellona Foundation**

http://www.bellona.no

Bellona Foundation on the web brings you news and background on important environmental issues.

**5769 Benton Foundation**

http://www.benton.org/

Since 1981, the Foundation has worked to articulate a public interest vision for the digital age and to demonstrate the value of communications for solving social problems. Through its projects, the foundation bridges the worlds ofphilanthropy, public policy and community action to promote the use of digital media to engage, equip and connect people for social change.

**5770 Best Manufacturing Practices Center of Excellence**

http://www.bmpcoe.org

The Office of Naval Research's Best Manufacturing Practices Program is a unique, innovative technology transfer effort that improves the competitiveness of the US industrial base both here and abroad. The main goal at BMP is toincrease the quality, reliability and maintainability of goods produced by American firms.

**5771 Biocatalysis/Biodegradation Database**

http://www.umbbd.ahc.umn.edu/

This database contains information on microbial biocatalytic reations and biodegradation pathways for primarily xenobiotic, chemical compounds. The goal of the UM-BBD is to provide information on microbial enzyme catalyzed reactionsthat are important for biotechnology.

**5772 Bioelectromagnetics Society**

http://www.bioelectromagnetics.org/

The Bioeletromagnetics Society was established in 1978 as an independent organization of biological and physical scientists, physicians and engineers interested in the interactions of non-ionizing radiation with biological systems.BEMS is incorporated as a nonprofit organization in the District of Columbia and is registered with the Internal Revenue Service as a educational and training organization.

**5773 Birding on the Web**

http://www.birder.com

Birding is the most extensive section of this site. You will find checklists which span the globe, birding Hot Spots, and rare bird alert phone numbers. In the Backyard Birders section you can also find information on seeds to attractbirds, building bird houses, and links to home pages of other bird watchers.

**5774 British Atmospheric Data Centre**

http://www.badc.rl.ac.uk/

The role of the BADC is to assist UK atmosperic researchers to locate, access and interpret atmospheric data to ensure the long-term integrity of atmospheric data produced by NERC projects. The BADC has substantial data holdings of itsown and also provides information and links to data held by other data centres.

**5775 Brookhaven National Laboratory**

http://www.bnl.gov/world/Default.asp

The department of Energy's Brookhaven National Laboratory conducts research in physical, biomedical and environmental sciences, as well as in energy technologies. Brookhaven also builds and operates major facilities available touniversity, industrial and government scientists.

**5776 Brown is Green**

http://www.brown.edu/Departments/Brown_Is_Green/

The listserv dicussion lists provide for some good conservation, conjecture and occasional facts and figures, but are not reliable sources of information for students, faculty and staff working on campus environmental programs. The WebServer is our attempt at making current information on Brown's and other University campus' environmental programs freely available over the internet.

**5777 Bureau of International Recycling**

http://www.bir.org

BIR is an international trade federation representing the world's recycling industry, covering in particular ferrous and non-ferrous metals, paper and textiles. Plastics, rubber, tires and glass are also studied and traded by some BIRmembers.

**5778 Bureau of Reclamation**

http://www.usbr.gov
The mission of the Bureau of Reclamation is to manage, develop and protect water and related resources in an environmentally and economically sound manner in the interest of the American public.

**5779 Bureau of Transportation Statistics**

http://www.bts.gov
The 1991 Intermodal Surface Transportation Efficiency Act established the Bureau of Transportation Statistics for data collection, analysis and reporting and to ensure the most cost-effective use of transportation-monitoring resources.We strive to increase public awareness of the nation's transportation system and its implications and improve the transportation knowledge base of decision makers.

**5780 Business & Legal Reports**

http://www.blr.com/
Business & Legal Reports has been helping employers avoid legal problems for 25 years. Human Resources, Compensation, Environmental and Safe managers know that they can count on our compliance and training products to keep them out oftrouble. BLR's attorneys are constantly researching federal and state legislation, best practices, industry trends and impending changes that can affect your organization.

**5781 California Conservation Corps**

http://www.ccc.ca.gov/cccweb/index.htm
The CCC is the oldest, largest and longest-running youth conservation corps in the world! Nearly 90,000 young men and women have worked more than 50 million hours to protect and enhance California's environment and communities and haveprovided six million hours of assistance with emergencies like fires, floods and earthquakes. We're proud of our accomplishments and hope you are too!

**5782 California Energy Commission**

http://www.energy.ca.gov
The California Energy Commission is the state's primary energy policy and planning agency. It was created by the Legislature in 1974 and located in Sacramento.

**5783 California Environmental Protection Agency**

http://www.calepa.ca.gov
The mission of the Cailfornia Environmental Protection Agency is to restore, protect and enhance the environment, to ensure public health, environmental quality and economic vitality.

**5784 California Environmental Resources Evaluation System (CERES)**

http://http://resources.ca.gov/
The goal of CERES is to improve environmental analysis and planning by integrating natural and cultural resource information from multiple contributors and by making it available and useful to wide variety of users. CERES collects andintegrates data and information and distributes it via the World Wide Web.

**5785 California League of Conservation Voters**

http://www.ecovote.org/
The California League of Conservation Voters is the nation's largest and oldest state political action organization for the environment. Founded in 1972, the League mobilizes California voters to support environmentally responsiblecandidates and issues and serves as a watchdog to hold elected officials accountable for their environmental votes.

**5786 California Resources Agency**

http://http://resources.ca.gov/
The California Resources Agency is responsible for the conservation, enhancement and management of California's natural and cultural resources, including land, water, wildlife, parks, minerals and historic sites. The Agency is composedof departments, boards, conservations, commissions and programs.

**5787 Canadian Chlorine Chemistry Council**

http://www.cfour.org/cms/
To facilitate dialogue and promote coordinated action in Canada among key stakeholders in order to bring about a balanced view of chlorine chemistry to enable society to make informed, science based decisions on issues involvingchlorine.

**5788 Canadian Council of Ministers of the Environment (CCME)**

http://www.ccme.ca
CCME works to promote cooperation on and coordination of interjurisdictional issues such as waste management, air pollution and toxic chemicals. CCME members propose nationally-consistent environmental standards and objectives so as toachieve a high level of environmental quality across the country.

**5789 Canadian Environmental Assessment Agency**

http://www.ceaa.gc.ca/index_e.htm
To provide Canadians with high quality environmental assessments that contribute to informed decision-making in support of sustainable development.

**5790 Canadian Institute for Environmental Law and Policy (CIELAP)**

http://www.cielap.org
To provide leadership in the research and development of environmental law and policy that promotes the public interest and principles of sustainability.

**5791 Carbon Dioxide Information Analysis Center (CDIAC)**

http://www.cdiac.esd.ornl.gov/home.html
CDIAC responds to data and information requests from users all over the world who are concerned with the greenhouse effect and global climate change. CDIAC's data holdings include records of the concentrations of carbon dioxide andother radiatively active gases in the atmosphere.

**5792 Carnegie Institute of Technology, Department of Civil & Environmental Engineering**

http://www.ce.cmu.edu
Carnegie Mellon's Department of Civil and Environmental Engineering is a part of the engineering college, Carnegie Institute of Technology. The department maintains a commitment to excellence and innovation in education and research.

**5793 Center for Disease Control**

http://www.cdc.gov
To promote health and quality of life by preventing and controlling disease, injury and disability. CDC seeks to accomplish its mission by working with partners throughout the nation and world to monitor health, detect and investigatehealth problems, conduct research to enhance prevention, develop and advocate sound public health policies, implement prevention strategies, promote healthy behaviors, foster safe and healthful environments and provide leadership and training.

**5794 Center for Environmental Biotechnology**

http:////www.ceb.utk.edu/

The Center for Environmental Biotechnology at the University of Tennessee, Knoxville was established in 1986 to foster a multidisciplinary approach for training the next generation of environmental scientists and solving environmentalproblems through biotechnology.

**5795 Center for Environmental Citizenship**

http://www.envirocitizen.org/

The Center for Environmental Citizenship is a national nonpartisan 501 organization. We were founded by young activists in 1992 to encourage college students to be environmental citizens. CEC is dedicated to educating, training andorganizing a diverse national network of young leaders to protect the environment.

**5796 Center for Environmental Design Research College of Environmental Design**

http://www.cedr.berkeley.edu/

The Center for Environmental Design Research is an Organized Research Unit of the University of California at Berkeley. The Center's mission is to encourage research in environmental planning and design, in order to increase thefactual content of design decisions and to promote systematic approaches to design decision making.

**5797 Center for Health Effects of Environmental Contamination**

http:////www.cheec.uiowa.edu/

The University of Iowa Center for Health Effects of Environmental Contamination supports and conducts research to identify, measure and prevent adverse health outcomes related to exposure to environmental toxins. CHEEC organizes andparticipates in educational and outreach programs, provides environmental health expertise to local, state and federal entities and serves as a resource to Iowans in the field of environmental health.

**5798 Center for International Earth Science Information Network (CIESIN)**

http://www.ciesin.org

CIESIN works at the intersection of the social, natural and information services, specializing in on-line data and information management, spatial data integration and training and interdisciplinary research related to humaninteractions in the environment.

**5799 Center for International Environmental Law**

http://www.ciel.org

The Center for International Environmental Law is a public interest nonprofit environmental law firm founded in 1989 to strengthen international and comparative environmental law and policy around the world. CIEL provides a full rangeof environmental legal services in both international and comparative national law.

**5800 Center for Plant Conservation**

http://www.centerforplantconservation.org/

The CPC is a consortium of 28 American botanical gardens and arboreta whose mission is to conserve and restore the rare native plants of the US. To meet this end, they are involved in plant conservation, research and education. Thissite includes information about the National Collection of Endangered Plants which is maintained by the group.

**5801 Center for Renewable Energy and Sustainable Technology**

http://www.crest.org

CREST's goal is to accelerate the use of renewable energy by providing credible information, insightful analysis and innovative strategies amid changing energy markets and mounting environmental needs. The combined CREST organizationboasts a strong platform for research, publication and dissemination of timely information regarding sustainable energy.

**5802 Central European Environmental Data Request Facility**

http://www.cedar.at/sitemap/htm

CEDAR was created to provide computing and Internetworking facilities to support international data exchange with the Central and Eastern European environmental community. Focusing at first on mainly Central and Eastern Europeancountries, CEDAR's activities expanded quickly to an audience all over the world.

**5803 Centre for the Analysis and Dissemination of Demonstrated Energy Technologies**

http://www.caddet.org

A unique source of global information on proven commercial applications covering the full range of energy-saving technologies.

**5804 Cetacean Society International**

http://csiwhalesalive.org/

CSI is all volunteer nonprofit conservation, education and research organization based in the USA, with volunteer representatives in 26 countries around the world. The goal of the Cetacean Society International is to achieve on aglobal basis the optimum utilization of cetacean resources through benign utilization and the elimination of all killing and captive display of whales, dolphins, and porpoises. Our ultimate aim is peaceful coexistence and mutual enrichment for humansand cetaceans.

**5805 Chanslor Wetlands Wildlife Project**

http://www.sonomawetlands.org

Chanslor Wetlands Wildlife Project protects 250 acres of crucial habitat adjoining the historic fishing community of Bodega Bay just 1.25 hours north of San Francisco. Dedicated in 1973, the Chanslor Wetlands encompasses a rarebrackish marsh as well as freshwater marshes, vernal pools and ponds bordered by Salmon Creek.

**5806 Charles Darwin Research Station**

http://www.darwinfoundation.org/

Our mission is to conduct scientific research and environmental education about conservation and natural resource management in the Galapagos archipelago and its surrounding Marine Reserve. Scientific research and monitoring projectsare conducted at the CDRS in conjunction and cooperation with our chief partner, the Galapagos National Park Service, which functions as the principal government authority in charge of conservation and natural resource issues in the Galapagos.

**5807 Chemical Industry Institute of Toxicology**

http://www.ciit.org

Founded in 1974, CIT is a nonprofit toxicology research institute dedicated to providing an improved scientific basis for understanding and assessing the potential adverse effects of chemicals, pharmaceuticals and consumer products onhuman health.

**5808 Chicago Wilderness**

http://www.chiwild.org

The lands stretching south and west from the shores of Lake Michigan hold one of North America's great metropolises. More than nine million people live in northwestern Indiana, northeastern Illinois and southeastern Wisconsin. Livingamong them, on islands of green, are thousands of species of native plants and animals-species that make up some of the rarest natural communities on earth. We call these communities and the lands and waters that are their homes Chicago Wilderness.

**5809 Children of the Green Earth**

707-839-5013
866-983-2784
E-mail: mail@childrenofthegreenearth.com
http://www.childrenofthegreenearth.com

Children of the Green Earth is committed to offering quality, organic, earth friendly products for children of all ages. From grandmothers to babies, we've got something for everyone. All of our products are ethical, bothenvironmentally and socially, meaning only organically grown products and all of the work involved in making the products is sweat shop and child labor free.

**5810 China Council for International Cooperation on Environment and Development**

http://www.iisd.org/trade/cciced/

The China Council for International Cooperation on Environment & Development will continue to act as a bridge between China and other countries in cooperation on environment and development by introducing useful experiences of othercountries to China and communicating to the world the determination and aspiration of the Chinese Government and people for sustainable development.

**5811 Chlorine Chemistry Council**

http://www.c3.org

The Chlorine Chemistry Council, a business council of the American Chemistry Council, is a national trade association based in Arlington, VA representing the manufacturers and users of chlorine and chlorine-related products. Chlorineis widely used as a diease-fighting disinfection agent, as a basic component in pharmaceuticals and myriad other products that are essential to modern life.

**5812 City Farmer**

http://www.cityfarmer.org/

Our nonprofit society promotes urban food production and environmental conservation from a small office in downtown Vancouver, British Columbia and from our demonstration food garden in nearby Kitsilano, a residential neighborhood.

**5813 Climate Change and Human Health**

http://www.jhu.edu/~climate/

The Climate Change and Human Health Integrated Assessment Web provodes recent and relevant information about the potential impacts of climate change through integrated assessment.

**5814 Coastal Conservation Association**

http://www.joincca.org

CCA is a national organization dedicated to the conservation and preservation of marine resources.

**5815 Code of Federal Regulations**

http://www.gpoaccess.gov/cfr/

The Code of Federal Regulations is a codification of the general and permanent rules published in the Federal Register by the Executive departments and agencies of the Federal Government.

**5816 Colorado Department of Natural Resources**

http://www.dnr.state.co.us/index.asp

The Colorado Department of Natural Resources was created to develop, protect and enhance Colorado natural resources for the use and enjoyment of the state's present and future residents, as well as for visitors to the state.

**5817 Colorado School of Mines**

http://www.mines.edu/index_js.shtml

The Colorado School of Mines shall be a specialized baccalaureate and graduate research institution with high admission standards. The Colorado School of Mines shall have a unique mission in energy, mineral and materials science andengineering and associated engineering and science fields.

**5818 Columbia Earth Institute: Columbia University**

http://www.earthinstitute.columbia.edu

The Earth Institute at Columbia University is the world's leading academic center for the integrated study of Earth, its environment and society. The Earth Institute builds upon excellence in the core disciplines-earth sciences,biological sciences, engineering sciences, social sciences and health sciences-and stresses cross-disciplinary approaches to complex problems.

**5819 Connecticut Department of Environmental Protection**

http://http://dep.state.ct.us/

The mission of the Department of Environmental Protection is to conserve, improve and protect the natural resources and environment of the State of Connecticut while preserving the natural environment and the life forms it supports ina delicate, interrelated and complex balance, to the end that the state may fulfill its responsibility as trustee of the environment for present and future generations.

**5820 Conservation Fund**

http://www.conservationfund.org

Works with private and public agencies and organizations to protect wildlife habitats, historic sites and parks.

**5821 Conservation International**

http://www.conservation.org

Our mission is to conserve the Earth's living natural heritage, our global biodiversity, and to demonstrate that human societies are able to live harmoniously with nature.

**5822 Conservation Treaty Support Fund**

http://www.conservationtreaty.org

The unique mission of the Conservation Treaty Support Fund is to support major inter governmental treaties which conserve wild natural resources for their own sake and the benefit of people. The fund believes these undertakings havethe best potential for global conservation, because they stem from the will of the nations of the world, are premised on the goal of sustaining living natural resources and have created a framework for effective conservation supported by manyagencies.

**5823 Consortium on Green Design and Manufacturing (CGDM)**

http://http://cgdm.berkeley.edu/

The Consortium on Green Design and Manfacturing (CGDM) is an interdisciplinary research initiative at the University of California, Berkeley and an industry/government/university partnership to develop linkages between manufacturingand design and their environmental effects and to integrate engineering information, management practices and government policy-making.

**5824 Consultative Group on International Agricultural Research (CGIAR)**

http://www.cgiar.org

To contribute to food security and poverty eradication in developing countries through research, partnerships, capacity building and policy support. Promotes sustainable agricultural development based on the environmentally soundmanagement of natural resources.

**5825 Coral Health and Monitoring Program (CHAMP)**

http://coral.aoml.noaa.gov

The mission of the Coral Health and Monitoring Program is to provide services to help improve and sustain coral reef health throughout the world.

**5826 Coral Reef Alliance**

http://www.coral.org

The Coral Reef Alliance (CORAL) is a member supported, nonprofit organization dedicated to keeping coral reefs alive around the world. Coral reefs are one of nature's most magnificent creations, filled with thousands of unique andvaluable plants and animals. CORAL works with marine park managers, businesses and communities to help increase their capacity to protect their local coral reefs.

**5827 Cornell University Center for the Environment**

http://www.cfe.cornell.edu

The Cornell Center for the Environment is committed to research, teaching and outreach focused on environmental issues, with the goals of enhancing the quality of life, encouraging economic vitality and promoting the conservation ofnatural resources for sustainable future.

**5828 Council for Agricultural Science & Technology (CAST)**

http://www.cast-science.org/

CAST is a nonprofit organization composed of scientific societies and many individual, student, company, nonprofit and associate society members. CAST's Board of Directors is composed of representatives of the scientific societies andindividual members as well as an Executive Committee. CAST assembles, interprets and communicates science based information regionally, nationally and internationally on food, fiber, agricultural, natural resources and related environmental issues tostakeholders.

**5829 Council on Environmental Quality (CEQ)**

http://ceq.eh.doe.gov or www.whitehouse.gov/CEQ/

The Council on Environmental Quality coordinates federal environmental efforts and works closely with agenices and other White House offices in the development of environmental policies and initiatives.

**5830 Coweeta LTER Site**

http://coweeta.ecology.uga.edu/

The program was developed to support research of ecological phenomena that occur on time scales of decades or centuries, periods of time normally investigated with research support from National Science Foundation.

**5831 CropLife Canada**

http://www.cropro.org/

To support sustainable agriculture in Canada, in cooperation with others, by building trust and appreciation for plant life science technologies.

**5832 Declining Amphibian Populations Task Force**

http://www.open.ac.uk/daptf/

Established in 1991, the DAPTF consists of a network of over 3,000 scientists and conservationists belonging to national and regional working groups which now cover more than 90 countries around the world. The mission of the DAPTF isto determine the nature, extent and causes of declines of amphibians throughout the world and to promote means by which declines can be halted or reversed.

**5833 Defenders of Wildlife**

http://www.defenders.org

Dedicated to the protection of all native wild animals and plants in their natural communities. Focus is placed on what scientists consider two of the most serious environmental threats to the planet: the accelerating rate ofextinction of species and the associated loss of biological diversity, and habitat alteration and destruction. Long known for leadership on endangered species issues.

**5834 Defense Technical Information Center (DTIC)**

http://www.dtic.mil

To improve the productivity of those who use scientific and technical information to accomplish a Defense mission objective, DTIC manages 13 Information Analysis Centers staffed by experienced information specialists, scientists andengineers who help customers locate, analyze and use scientific and technical information in a specialized subject area.

**5835 Delaware Department of Natural Resources and Environmental Control**

DNREC

http://www.dnrec.state.de.us

Delaware Department Of Natural Resources and Environmental Control's mission is to protect Delaware's environment for future generations.

**5836 Department of Conservation**

http://www.conservation.state.mo.us

The mission is to protect and manage the fish, forest and wildlife resources of the state; to serve the public and facilitate their participation in resource management activities; to provide opportunity for all citizens to use, enjoyand learn about fish, forest and wildlife resources.

**5837 Department of Energy**

http://www.doe.gov/engine/content.do

The Department of Energy's mission is to foster a secure and reliable energy system that is environmentally and economically sustainable, to be a responsible steward of the nation's nuclear weapons, to cleanup our own facilities and tosupport continued US leadership in science and technology.

**5838 Department of the Interior**

http://www.doi.gov

The Interior Department has had a wide range of responsibilities enstrusted to it: the construction of the national capital's water system, the colonization of freed slaves in Haiti, exploration of western wilderness, oversight of theDistrict of Columbia jail, regulation of territorial governments, management of hospitals and universities, management of public parks, the basic responsibilities for Indians, public lands, patents and pensions.

**5839 DiveWeb**

http://www.sandiegodiving.com/resources/links/items/277.html

DiveWeb is the original comprehensive online resource for information about commerical diving, ROV, marine technology, offshore/telecommunications and inland/coastal underwater industries.

**5840 EDIE: Environmental Data Interactive Exchange**

http://www.edie.net

EDIE is a free, personalized, interactive news, information and communications service for water, waste and environmental professionals around the world. With comprehensive independent coverage, powerful search facilities, e-mailalerts and discussion forums, EDIE provides a one-stop-shop for the exchange of specialized information on the Web.

**5841 EE-Link (Environmental Education-Link)**

http://http://eelink.net/pages/EE-Link+Introduction

EE-Link is a participant in the Environmental Education and Training Partnership of the North American Association for Environmental Education.

**5842 Earth Day Network**

http://www.earthday.net/
Earth Day Network is the nonprofit coordinating body of worldwide Earth Day activities. Our goal is to promote a healthy environment and a peaceful, just, sustainable world by sending environmental awareness through educationalmaterials and publications and by organizing events, activities and annual campaigns. Our network includes more than 5,000 organizations in 184 countries.

**5843 Earth Observing System Amazon Project**

http://boto.ocean.washington.edu/eos/
This project is a NASA Earth Observing System Interdisciplinary Investigation. The purpose of this research project is to understand the biogeochemistry, hydrology and sedimentation of the Amazon River and its drainage basin.

**5844 Earth Preservers**

http://www.earthpreservers.com
The web site for the environmental newspaper-for kids and adults. Features of environmental work being done by kids all over the world. Contains articles and interesting facts.

**5845 Earth Resources Laboratory at MIT**

http://http://eaps.mit.edu/erl/
The Earth Resources Laboratory, formed in 1982, brings together faculty, staff and students dedicated to research in applied geophysics that will further our understanding of the Earth, its resources and the environment.

**5846 EarthVote.com**

http://www.earthvote.com
Twenty-four hour global resource for domestic and global environmental and social issues. Input your own voting topic or vote on current issues.

**5847 Earthlink**

http://www.earthlink.org.au/earthlink/listings.php
Earthlink is dedicated to creating a just sustainable world by harnessing economic power for positive change. Earthlink is designed to educate and encourage people to use their spending and investing power to bring about increasedsocial justice and environmental responsibility.

**5848 Earthwatch Institute**

http://www.earthwatch.org
The mission of the Earthwatch Institute is to promote sustainable conservation of our natural resources and cultural heritage by creating partnerships between scientists, educators and the general public.

**5849 EcoEarth**

http://www.ecoearth.info
Empowering the environmental sustainability movement, EcoEarth provides a landslide of environmental news postings and information, updated daily. Search, blog and network capabilities.

**5850 EcoTradeNet**

http://www.ecosecretariat.org/ECOTradeNet/index.htm
This Site provides trade information in respect of Member States of the Economic Cooperation Organziation. It's objective is not only to facilitate intra-regional trade cooperation but also provide relevant and updated information tothe prospective trader from outside the ECO region.

**5851 Ecologia**

http://www.ecologia.org
Ecologia is a private nonprofit organization providing information, training and technical support for grassroots environmental groups. Ecologia offers technical and humanitarian assistance to individuals and organizations working tosolve ecological problems at the local, regional, national and global levels.

**5852 Ecology Action Centre**

http://www.ecologyaction.ca
The Ecology Action Centre has been an active advocate protecting the environment since 1972. The Centre's earliest projects included recycling, composting and energy conservation and these are now widely recognized environmentalissues.

**5853 Edison Electric Institute**

http://www.eei.org/
Edison Electric Institute is the association of US shareholder owned electric compaines, international affiliates and industry associates worldwide. Our US members serve over 90 percent of all customers served by the shareholder ownedsegment of the industry.

**5854 Edwards Aquifer Research and Data Center**

http://www.eardc.txstate.edu/
The Edwards Aquifer Research and Data Center was established in 1979 by special funding for Southwest Texas State University to provide a public service in the study, understanding and use of the very fragile natural resource known asthe Edwards Aquifer.

**5855 Electric Power Research Institute (EPRI)**

http://http://my.epri.com/portal/server.pt
EPRI is a nonprofit organization committed to providing science and technology-based solutions of indispensable value to our global energy customers. To carry out our mission, we manage a far-reaching program of scientific research,technology development and product implementation.

**5856 Elsevier Science Tables of Contents**

http://www.elsevier.com/wps/find/homepage.cws_home
Elsevier Science has become the undisputed market leader in the publication and dissemination of literature covering the broad spectrum of scientific endeavors.

**5857 Endangered Species Recovery Program**

http://http://esrp.csustan.edu/
The Endangered Species Recovery Program's mission is to facilitate endangered species recovery and resolve conservation conflicts through scientifically based recovery planning and implementation.

**5858 Energy & Environmental Research Center (EERC)**

http://www.eerc.und.nodak.edu/
The EERC is dedicated to moving promising technologies out of the laboratory and into the marketplace to produce energy cleanly and efficiently, minimizing enviromental impacts and conserving precious natural resources.

**5859 Energy Ideas Clearinghouse**

http://www.energyideas.org/
EnergyIdeas is the most comprehensive technical resource that Northwest businesses, industy, government and utilities use to implement energy technologies and practices.

**5860 Energy Technology Data Exchange**

http://www.etde.org/

ETDE through its member countries provides an extensive bibliographic database announcing published energy research and technology information.

**5861 Enviro Village**

http://www.envirovillage.com
Indoor air quality and environment resource

**5862 Enviro$en$e**

http://http://es.epa.gov/
Enviro$en$e is a free public environmental information system resident on the Internet's World Wide Web. This Web provides users with pollution prevention/cleaner production solutions, compliance and enforcement assistance informationand innovative technology and policy options.

**5863 Enviro-Access**

http://www.enviroaccess.ca
Enviro-Access is a business partner investing in the development of environmental technologies by supplying compaines in the environemtal sector with the professional services required during the various steps of bringing theirproducts and services to the market-place.

**5864 EnviroOne.com**

http://enviroone.com
EnviroOne.com is your one stop center for everything environmental. We have created propriety technology that combines a vast array of resources, timely content, innovative tools and e-commerce functionality that closes the loop onsearching for information and having the ability to immediately act upon that information.

**5865 Envirolink Library**

http://library.envirolink.org/start.html
EnviroLink is a nonprofit organization, a grassroots online community that unites hundreds of organizations and volunteers around the world with millions of people in more than 150 countries. EnviroLink is dedicated to providingcomprehensive, up-to-date environmental information and news. We recognize that our technologies are just tools, and that the solutions to our ecological challenges lie within our communities and their connection to the Earth itself.

**5866 Envirolink Network**

http://envirolink.org
EnviroLink is a nonprofit organization, a grassroots online community that unites hundreds of organizations and volunteers around the world with millions of people in more than 150 countries. EnviroLink is dedicated to providingcomprehensive, up-to-date environmental information and news. We recognize that our technologies are just tools, and that the solutions to our ecological challenges lie within our communities and their connection to the Earth itself.

**5867 Environment Council UK**

http://www.the-environment-council.org.uk/
The Environment Council is an independent UK charity which brings together people from all sectors of business, non-governmental organizations, government and the community to develop long term solutions to environmental issues.

**5868 Environment in Asia**

http://www.asianenviro.com/
AsianEnviro is a collaboration between AET Ltd and ERM Japan. This site draws on the established regional networks of environmental professionals in both organizations to deliver targeted environmental business intelligence. Pleasenote that translation is necessary to view this page.

**5869 Environmental Alliance for Senior Involvement**

http://www.easi.org/links.html
The mission of the Environmental Alliance for Senior Involvement is to build, promote and utilize the environmental ethic, expertise and commitment of older persons to expand citizen involvement in protecting and caring for ourenvironment for present and future generations.

**5870 Environmental Assessment Association**

http://www.iami.org/eaa.html
The Environmental Assessment Association is an international organization dedicated to providing members with information and education in the Real Estate Industry in respect to Environmentaal Inspections, Testing and HazardousMaterial Removal.

**5871 Environmental Change Network (ECN)**

http://www.ecn.ac.uk/
To establish and maintain a selected network of sites within the UK from which to obtain comparable long-term datasets through the monitoring of raange of variables identified as being of major environmental importance.

**5872 Environmental Compliance Assistance Center**

http://www.epa.gov/compliance/
The purpose of this site is to bring complicated environmental laws and regulations into everyday language that the normal business person can understand, and to give that person associated information on educational opportunities.

**5873 Environmental Contaminants Encyclopedia**

http://www.nature.nps.gov/hazardssafety/toxic/index.cfm
This product differs from existing databases in that it has an environmental toxicology emphasis and it summarizes information on these issues into a single, easily searchable source.

**5874 Environmental Defense**

http://www.environmentaldefense.org/home.cfm
Environmental Defense is dedicated to protecting the environmental rights of all people, including future generations. Among these rights are clean air, clean water, healthy, nourishing food and flourishing ecosystem.

**5875 Environmental Measurements Laboratory**

http://www.eml.st.dhs.gov/
EML's current mission is to conduct scientific investigations and develop technologies related to environmental restoration, site and facility characterization and environmental surveillance and monitoring.

**5876 Environmental News Network**

http://www.enn.com/
Since 1993, the Environmental News Network has been working to educate the world about environmental issues facing our Earth. We began as a monthly print publication called Environmental News Briefing, and two years later discoveredthe Internet as an effective means of reaching a broader, more diverse audience.

**5877 Environmental Organization Web Directory**

http://www.webdirectory.com/
Our goal is simple: we strive to make it easy for people from around the world to find your web page. We currently have 23 terrific staff members all dedicated to provide a free service to the environmental community.

**5878 Environmental Protection**

http://www.eponline.com

Website dedicated to pollution and waste treatment solutions for environmental professionals.

### 5879 Environmental Protection Agency, US

http://www.epa.gov/

The mission of the United States Environmental Protection Agency is to protect human health and to safeguard the natural environment air, water, and land upon which life depends.

### 5880 Environmental Research Institute of Michigan-Altarum

http://www.altarum.org/

ERIM promotes sustainable societal wellbeing by helping our customers - and through them, society - employ new knowledge and decision support tools to solve complex systems problems in the healthcare, national security and energy,environment and transportation sectors.

### 5881 Environmental Resource Center

http://www.ercweb.com

Environmental Resource Center is a full-service environmental consulting firm that has been serving the needs of private industry and government for over seventeen years with the highest standards of quality at a competitive price.

### 5882 Environmental Resources Information Network/ Environment Australia Online

http://kaos.erin.gov.au

ERIN, the Environmental Resources Information Network, provides environmental information for policy developers and decision makers. ERIN is a National facility, using the latest computing technology to provide access to a vastreservoir of information on the Australian environment, and the analytical tools to interpret it.

### 5883 Environmental Resources Management

http://www.erm.com

Environmental Resources Management has well-established reputation as one of the world's largest providers of environmental management consulting services. We have 25 year's experience working with both the public and private sectorsacross a broad spectrum of industries.

### 5884 Environmental Simulations

http://www.groundwatermodels.com

Our mission is to provide our software clients with superior and technical support. To provide cost-effective in house training courses, state-of-the-art groundwater modeling services and high quality independent hydrogeologicalconsulting.

### 5885 Environmental Treaties and Resource Indicators

http://sedac.ciesin.columbia.edu/entri/

ENTRI, a database of searchable treaties and resource indicators, is provided by the CIESIN organization with the assistance of many other groups. They use nine specific issue areas to search the database. Those areas are: 1) landuse/land cover change and deseration, 2)global climate changr, 3)stratospheric ozone depletion, 4)transboundary air pollution, 5)conservation of biologicaldiversity, 6)deforestation, 7)oceans, 8)trade and environment, and 9) population.

### 5886 Environmental Working Group

http://www.ewg.org

The Environmental Working Group is a leading content provider for public interest groups and concerned citizens who are campaiging to protect the environment.

### 5887 Environmental and Societal Impacts Groups

http://www.isse.ucar.edu/

ESIG studies environmental change and responses to such change inorder to gain insights into how decision makers, from individuals to governements to international coalitions, might better understand and cope with impacts associatedwith the complex relationship of the atmosphere, environment, and society.

### 5888 Essential Information

http://www.essential.org/about.html

Founded by Ralph Nader. A nonprofit, tax-exempt organization. We are involved in a variety of projects to encourage citizens to become active and engaged in their communities. We provide provocative information to the public onimportant topics neglected by the mass media and policy makers. Publishes a monthly magazine, books and reports, sponsors investigative journalism conferences, provides writers with grants to persue investigations and operates information clearinghouses.

### 5889 European Centre for Nature Conservation

http://www.ecnc.nl

ECNC actively promotes, by bringing the gap between science and policy, the conservation of nature and especially of biodiversity in Europe, because of their intrinsic values and their relevance to economy and European culture; therebyECNC seeks the integration of nature conservation considerations into other policies.

### 5890 European Forest Institute (EFI)

http://www.efi.fi

EFI's mission is to promote, conduct and co-operate in research of forests, forestry and forest products at the pan-European level; and to make the results of the research known to all interested parties, notably in the areas of policyformulation and implementation, in order to promote the conservation and sustainable management of forests in Europe.

### 5891 Everglades Digital Library

http://everglades.fiu.edu/

The Everglades Information Network is a program of library and information services in support of research, restoration, and resource management of the south Florida environment. The EIN serves researchers, resource managers,educators, students, researchers, decision makers, and concerned citizens both within south Florida and around the world.

### 5892 Extension Toxicology Network ETOXNET

http://http://extoxnet.orst.edu/

Some of the goals of EXTOXNET are to stimulate dialog on toxicology issues, develop and make available information relevant to extension toxicology, and facilitate the exchange of toxicology-related information in electronic form,accessible to all with access to the Internet.

### 5893 Federal Emergency Management Agency FEMA

http://www.fema.gov/

Advising on building codes and flood plain management, teaching people how to get through a disaster, helping equip local and state emergency preparedness, the range of FEMA's activities is broad indeed.

### 5894 Federal Geographic Data Committee

http://www.fgdc.gov/

The Federal Geographic Data Committee is a 19 member interagency committee composed of representatives from the Executive Office of the President, Cabinet-level and independent agencies. The FGDC is developing the National Spatial DataInfrasturcture (NSDI) in cooperation with organizations from State, local and tribal governments, the academic community, and the private sector.

**5895 Fedworld Information Network**

http://www.fedworld.gov

We here at FedWorld have enjoyed thinking outside the box to offer multiple distribution channels to disseminate information to the public and to the Federal Governement. The modes of access, the variety of documents available, and thetechnological expertise at FedWorld are expanding with technology.

**5896 Finnish Forest Research Institute: METLA**

http://www.metla.fi

The Finnish Forest Research Institute builds the future of the forest sector through research. METLA social task is to promote through research economically, ecologically and socially sustainable management and utilisation of theforests.

**5897 Fish and Wildlife Information Exchange Homepage**

http://www.cmiweb.org/fwie/fwie.html

The FWIE is a technical assistance center and information clearinghouse for fish, wildlife, and land management agencies and organizations. The FWIE also assists with the planning, development, implementation, and maintenance ofinformation management and delivery systems.

**5898 Florida Center for Environmental Studies**

http://www.ces.fau.edu

The Mission of the Center is to collect, analyze, and promote the use of scientifically-sound information concerning tropical and suptropical freshwater ecosystems.

**5899 Florida Cooperative Extension Service**

http://www.wec.ufl.edu/extension

Part of the mission of the Florida Cooperative Extension Service is to disseminate and provide access to science-based information that will contribute to the solution of natural resource problems of concern to the people of Florida. Wildlife Extension specialists in the Department of Wildlife Ecology and Conservation serve , advice, and develop educayional programs for Florida citizens in conjunction with county extension agents and other state , county and local organizations.

**5900 Florida Department of Environmental Protection**

http://www.dep.state.fl.us

The mission of the Department of Environmental Protection is: More Protection, Less Process. The Department accomplishes its mission in a manner that provides stewardship of Florida's ecosystems so that the State's unique quality oflife may be preserved for present and future generations.

**5901 Forest History Society**

http://www.lib.duke.edu/forest

The Forest History Society is a non-profit educational institution that links the past to the future by identifying, collecting, preserving, interpreting, and disseminating information on the history of people, forests, and theirrelated resources.

**5902 Forest Service Employees for Environmental Ethics**

http://www.fseee.org/

Forest Service Employess for Environmental Ethics is a 501 non-profit organization. Our mission is to forge a socially responsible value system for the Forest Service based on a land ethic that ensures ecologically and economicallysustainable resource management.

**5903 Friends of the Earth International**

http://www.foei.org/

Friends of the Earth International is a worldwide federation of national environmental organizations. This federation aims to: protect the earth against futher deterioration and repair damage inflicted upon the environment by humanactivities and negligence; preserve the earth's ecological, cultural and ethic diversity.

**5904 GAP (Gap Analysis Program) National**

http://www.gap.uidaho.edu/

The mission of the GAP Analysis Program is to provide regional assessments of the conservation status of native vertebrate species and natural land cover types and to facilitate the application of this information to land managementactivities.

**5905 GLOBE Program**

http://www.globe.gov

GLOBE is a worldwide hands-on, primary and secondary school based science and education program.

**5906 Galapagos Coalition**

http://www.law.emory.edu/PI/GALAPAGOS/

The Galapagos Coalition is a group of biologists, other scientists, and lawyers with expertise in environemntal and international law, many of whom have done research in the Galapagos and all of whom are interested in the understandingthe relationship between the conservation of the Galapagos and human activities.

**5907 General Accounting Office**

http://www.gao.gov

The General Accounting Office is the audit, evalution and investigative arm of Congress. GAO exists to support the congress in meeting its Constitutional responsibilities and to help improve the performance and ensure accountability ofthe federal government for the American people. GAO examines the use of public funds, evaluates federal programs and activities and provides analysis, options, recommendations and other assistance to help the Congress make effective oversite, policyand funding.

**5908 Geohydrodynamics and Environmental Research**

http://modbg.oce.ulg.ac.be/

The general objective of the MODB is to deliver advanced data products to mediterranean research projects supported by the MAST programme of the European Union. All products are however freely distributed to the whole scientificcommunity.

**5909 Georgia Department of Natural Resources**

http://www.gadnr.org/

The mission of the Department of Natural Resources is to sustain, enhance, protect and conserve Georgia's natural, historic and cultural resources for present and future generations, while recognizing the importance of promoting thedevelopment of commerce and industry that utilize sound environmental practices.

**5910 Geotechnical and Geoenvironmental Software Directory**

http://www.ggsd.com/ggsd/index.cfm

This site contains a summary of links to sites describing software used for seepage/groundwater flow modeling.

**5911 Germinal Project**

http://http://lasig.epfl.ch/projets/germinal/Germinal.html

GERMINAL is an interdissiplinary project established by the institutes of the Department of Rural Engineeing and by the directorship of the Swiss Federal Institute of Technology at Lausanne. Its goal is the development of a global,integrated approach for land use planning and environmental management based on the use of Geographic Information Systems.

### 5912 Global Change Master Directory (GCMD)

http://gcmd.gsfc.nasa.gov/

The mission of the Global Change Master Directory is to assist the scientific community in the discovery of and linkage to Earth science data, as well as to provide data holders a means to advertise their data to the Earth ScienceCommunity.

### 5913 Global Change Research Information Office

http://gcrio.org

The US Global Change Research Information Office provides access to data and information on global change research, adaption/mitigation strategies and technologies, and global change-related educational resources on behalf of thevarious US Federal Agenices and Organizations that are involved in the US Global Change Research Program

### 5914 Global Ecovillage Network

http://http://gen.ecovillage.org/

GEN's main aim is to support and encourage the evolution of sustainable settlements across the world through: Internal and External Communications services; facilitating information exchange and flow about ecovillages anddemonstrations sites.

### 5915 Global Environmental Options (GEO)

http://www.genonetwork.org

GEO was created to bring attention to the impact that buildings and planning have on the environment and biodiversity. To reduce these impacts, GEO develops integrated high performance green building stragies for professionals andother decision makers worldwide, coordinating hands-on building and planningprojects that showcase green design.

### 5916 Global Network of Environment & Technology

http://www.gnet.org/

GNET provides worldwide access to timely information on environmental news, products and services, marketing opportunities, contracts, government programs, policy and law, and business assitance resources via the World Wide Web.

### 5917 Global Research Information Database (GRID)

http://www.grida.no/

GRID-Arendal provides environmental information, communications and capacity buildings services for information management and assessment. Established to strengthen the United Nations through its Environmental Programme, our focus isto make credible, science-based knowledge understandable to the public and to decision making for sustainable development.

### 5918 Great Lakes Fishery Commission

2100 Commonwealth Boulevard 734-662-3209
Ann Arbor, MI 48105 Fax: 734-741-2010
http://www.glfc.org

To develop coordinated programs of research on the Great Lakes, and, on the basis of the findings, to recommend measures which will permit the maximum sustained productivity in stocks of fish of common concern.

### 5919 Great Lakes Information Network (GLIN)

http://www.great-lakes.net

The Great Lakes Information Network is an partnership that provides one place to find information relating to the binational Great Lakes-St Lawrence region of North America. GLIN offers a wealth of data and information about theregion's environment, economy, tourism, education and more.

### 5920 Green Mountain Institute for Environmental Democracy

http://www.gmied.org

The Green Mountain Institute for Environmental Democracy seeks to reinvigorate the essential connections among the public, government and information necessary for effective improvements in environmental quality.

### 5921 Green Seal

http://www.greenseal.org

Green Seal is the independent nonprofit organization dedicated to protecting the environment by promoting the manufacture and sale of enviromentally responsible consumer products. It sets environmental standards and awards a Green Sealof Approval to products that cause less harm to the environment than other similar products.

### 5922 Green University Initiative

http://www.gwu.edu/~greenu/

The George Washington Green University Initiative began as a grassroot movement to implement sustainable practices into all aspects of life at GW. At Green University we work towards ecosystem protection, incorporating environmentaljustice into daily activities and decisions.

### 5923 Greenbelt Alliance

http://www.greenbelt.org/

Our mission is to make the nine-country San Francisco Bay Area a better place to live by protecting the region's Greenbelt and improving the livability of its cities and towns. Scince 1958 we have worked in partnership with diversecoalitions on public policy development, advocacy and education.

### 5924 Greenhouse Gas Technology Information Exchange GREENTIE

http://www.greentie.org

IEA GREENTIE is an international information network that distributes details of suppliers whose technologies help to reduce greenhouse gas emissions. GREENTIE also provides information on leading international organizations and IEAprograms whose R&D and information activities center around clean energy technologies.

### 5925 Ground-Water Remediation Technologies Analysis Center

http://www.gwrtac.org

The Groundwater Remediation Technologies Analysis Center compiles, analyzes and disseminates information on innovative ground-water remediation technologies. GWRTAC prepares reports by technical teams selectively chosen from ConcurrentTechnologies Corporation, the University of Pittsburgh and other supporting institutions, also maintaining an active outreach program.

### 5926 Harbor Branch Oceanographic Institution

http://www.hboi.edu

Harbor Branch Oceanographic Institution is dedicated to exploring the world's oceans, integrating the science and technology of the sea with the needs of humankind.

### 5927 Harvard Forest

http://http://harvardforest.fas.harvard.edu/

Through the years researchers at the Forest have focused on silviculture and forest management, soils and the development of forest site concepts, the biology of temperate and tropical trees, forest ecology and economics and ecosystemdynamics.

**5928 Hawaii Biological Survey**

http://http://hbs.bishopmuseum.org/hbs1.html

It was created to locate, identify and evaluate all native and non-native species of flora and fauna within the State and maintain the reference collections of that flora and fauna for a wide range of uses.

**5929 Hawaiian Ecosystems at Risk (HEAR)**

http://www.hear.org

The mission of the Hawaiian Ecosystem at Risk project is to promote technology, methods and information to decision-makers, resource managers and the general public to aid in the fight against harmful alien species in Hawaii and thePacific Basin.

**5930 Hawk Mountain Sanctuary**

http://www.hawkmountain.org/index.php?pr=The_Sanctuary

Hawk Mountain's mission is to foster the conservation of birds of prey worldwide and to create a better understanding of and further the conservation of the natural environment, particularly the Central Appalachian region.

**5931 Hawkwatch International**

http://www.hawkwatch.org/

Our mission is to protect hawks, eagles, other birds of prey and their environment through research, education, and conservation.

**5932 Hazardous Substance Research Centers**

http://www.hsrc.org

Hazardous Substance Research Cente is a national organization that carries out an active program of basic and applied research, technology transfer and training. Our activities are conducted regionally by five multi-university centers,which focus on different aspects of hazardous substance management.

**5933 Hazardous Waste Clean-Up Information (CLU-IN)**

http://www.clu-in.com

Providing information about innovative treatment and site characterization technologies while acting as a forum for all waste remediation stakeholders.

**5934 Headwaters Science Center**

http://www.hscbemidji.org/

Headwaters Science Center is dedicated to science education and environmental awareness. It features hands-on exhibits, a live animal collection and special events and science-related programs and demonstrations.

**5935 Heartwood**

http://www.heartwood.org/

An association of groups, individuals and businesses dedicated to the health and well being of the native forest of the Central Hardwood region and its interdependent plant, animal and human communities.

**5936 Holland Island Preservation Foundation**

http://www.intercom.net/local/holland/index.html

Our goal is to stabilize and preserve this beautiful island, not only for the people that once lived there, but also for the wildlife that thrives there still. With modern tools and techniques, this fragile ecosystem can be saved fromultimate destrution.

**5937 Horned Lizard Conservation Society (HLCS)**

http://www.hornedlizards.com

The Texas Chapter of the Horned Lizard Conservation Society (HLCS) is devoted to discovering why the Texas Horned Lizard has declined in numbers so dramatically in recent years and what can be done to reverse the process.

**5938 Houston Audubon Society**

http://www.houstonaudubon.org/

The Houston Audubon Society works for the thoughtful conservation of the earth's natural resources by educating people to the value of the natural world; protecting, preserving and enhancing wildlife habitat and encouraging the passageof legislation to protect the environemnt.

**5939 Howl: The PAWS Wildlife Center**

http://www.paws.org/wildlife/index.htm

The PAWS Wildlife Center is a world reowned wildlife rehabiliation facility. Formerly known as HOWL, the PAWS Wildlife Center receives over 5,000 injured or displaced wild animals every year. The center houses and rehabilitateswildanimals, and prepares them for eventual release back into the wild.

**5940 IFAW: International Fund for Animal Welfare**

http://www.ifaw.org

IFAW's mission is to improve the welfare of wild and domestic animals throughout the world by reducing commercial exploitation of animals in distress. We seek to motivate the public to prevent cruelty to animals and promote animalwelfare and conservation policies that advance the well-being of both animals and people.

**5941 IISDnet: International Institute for Sustainable Development**

http://www.iisd.org/

Our mission is to champion innovation, enabling societies to live sustainably.

**5942 Illinois Recycling Association**

http://www.illinoisrecycles.org/

The Illinois Recycling Association's mission is to encourage the responsible use of resources by promoting waste reduction, re-use and recycling.

**5943 Indiana Department of Natural Resources**

http://www.in.gov/dnr/

The mission of the Indiana Department of Natural Resources is to protect, enhance, preserve and wisely use natural, cultural and recreational resources for the benefit of Indiana's citizens through professional leadership, managementand education.

**5944 Information Center for the Environment (ICE)**

http://ice.ucdavis.edu/

The Foundation wishes to participate actively in all activities and efforts to protection, preservation and exploitation of mangrove ecosystem for the prosperity of all people.

**5945 Inland Seas Education Association**

http://www.schoolship.org/

Inland Seas Education Association is a nonprofit organization whose mission is to provide a floating classroom where people of all ages can gain first-hand training and experience in the Great Lakes ecosystem. The knowledge gainedthrough these experiences will provide the leadership, understanding and commitment needed for the long-term stewardship of the Great Lakes.

**5946 International Arid Lands Consortium**

http://http://ag.arizona.edu/OALS/IALC/Home.html

The International Arid Lands Consortium works to achive research and development educational and traning initiahves, demonstration projects, workshops, and other technology-transfer activities applied to the development, management,restoration, and reciamation of and semiand lands in the US the MIddle East, and elsewhere in the world.

**5947 International Association for Energy Economics**

http://www.iaee.org

The International Association for Energy Economics provides a forum for the exchange of ideas, experience and issues among professionals interested in energy economics. Its scope is worldwide, as are its members who come from diversebackgrounds-corporate, academic, scientific and government.

**5948 International Association for Environmental Hydrology**

http://www.hydroweb.com

Worldwide association of environmental hydrologists dedicated to the protection and cleanup of fresh water resources.

**5949 International Canopy Network (ICAN)**

http://www.evergreen.edu/ican/

The International Canopy Network is devoted to facilitating the continuing interaction of people concerned with forest canopies and forest ecosystems around the world. ICAN is a nonprofit organization supported by a global community ofscientists, conservation advocates, canopy educators and environmental professionals. The organization is funded by subscriber dues, donations and grants.

**5950 International Centre for Gas Technology Information**

http://www.gtionline.org/

ICGTI benefits us by: Offering gas technology decisionmakers a competitive information edge; affording immediate connectionn with the world's natural gas technology leaders; and by allowing easy access to provider services andequipment that serve the natural gas industry.

**5951 International Council for the Exploration of the Sea(ICES)**

http://www.ices.dk/ocean/

ICES is a leading forum for the promotion, coordination and gas dissemination of research on the physical, chemical and biological systems in the North Atlantic and advice on human impact on its environment, in particular fisherieseffects in the Northeast Atlantic.

**5952 International Crane Foundation**

http://www.savingcranes.org/

The International Crane Foundation works worldwide to conserve cranes and the wetland and grasslands communities on which they depend. ICF is dedicated to providing experience, knowledge, and inspiration to involve people in resolvingthreats to these ecosystems.

**5953 International Energy Agency Solar Heating and Cooling Programme**

http://www.iea-shc.org

The Solar Heating and Cooling Programme was one of the frist IEA Implementing Agreements to be established. Since 1977, its 21 members have collaborating to advance active solar, passive solar and photovoltaic thecnologies and theirapplication in buildings.

**5954 International Geosphere-Biosphere Programme**

http://www.igbp.kva.se

Our scientific objective is to describe and understand the interactive physical, chemical and biological processes that regulate the total Earth System, the unique environment that it provides for life, the changes that are occurringin this system, and the manner in which they are influenced by human actions.

**5955 International Ground Source Heat Pump Association(IGSHPA)**

http://www.igshpa.okstate.edu/default.htm

As an organization, IGSHPA pursues these goals: Supporting GHP industry research and development; promoting the GHP-related current events internationally. Developing and distuting internationally recognized training materials.

**5956 International Institute for Industrial Environmental Economics**

http://www.iiiee.lu.se/

The mission of the Institute is to the international advancement of sustainable development by conducting at the forefront of issues pretaining to cleaner production, and to educate present and future decision makers within all sectorsof society in the formulation and implemantation of preventive environmental strategies. The Institute is founded on the firm conviction that a preventive approach to environmental problems is necessary for the perpetuation of life on this planet.

**5957 International Marine Mammal Association**

http://www.imma.org

The International Marine Mammal Association is a non-for-profit organization dedicated to promoting the conservation of marine mammals and their habitats worldwide, through research and education.

**5958 International Otter Survival Fund**

http://www.otter.org/

The International Otter Survival Fund is a global organization working to conserve all 13 species of otter by helping to support scientists and other workers in practical conservation, education, research and rescue and rehabilitation.

**5959 International Primate Protection League (IPPL)**

http://www.ippl.org

The International Primate Protection League was funded in 1973, and since this time has been working continuously for the well-being of primates.

**5960 International Research Institute for Climate Prediction**

http://www.ir.columbia.edu/climate/cid/

The vision for the IRI is that of an innovative science institution working to accelerate the ability of societies worldwide to cope with climate fluctuations, especially those that cause devastating impacts on humans and theenvironment, thereby reaping the benefits of decades of research on the predictability of El Nino-Southern Oscillation phenomenon and other climate variations.

**5961 International Rivers Network**

http://www.irn.org/index.html

IRN's mission is to halt and reverse the degradation of river systems; to support local communities in protecting and restoring the well-being of the people, cultures and ecosystems that depend on rivers; to promote sustainable,environmentally sound alternatives to damming and channeling rivers.

**5962 International Satellite Land Surface Climatology Project (ISLSCP)**

http://www.gewex.org

ISLSCP Objective is to demosnstrate the type of surface and near-surface satellite measurements that are relevant to climate and global change studies. Develop and improve algorithms for the interpretation of satellite measurements ofland-surface features.

**5963 International Snow Leopard Trust**

http://www.snowleopard.org

The International Snow Leopard Trust is dedicated to the conservation of the endangered snow leopard and its mountain ecosystem through a balanced approach that considers the needs of the people and the environment.

**5964 International Society for Ecological Modelling (ISEM)**

http://www.isemna.org

The International Society for Ecological Modelling promotes the international exchange of ideas, scientific results, and general knowledge in the area of the application of systems analysis and simulation in ecology and naturalresource management.

**5965 International Society for Environmental Ethics**

http://www.cep.unt.edu

ISEE now maintains this website, which includes the largest bibliography in the world on environmental ethics, over 7,000 entries. Newsletters over the last ten years also available here, and by consulting these a full historicalrecord may be obtained.

**5966 International Society of Arboriculture**

http://www.isa-arbor.com

The mission is through research, technology, and education promote the professional practice of arboriculture and foster a greater public awareness of the benefits of trees.

**5967 International Solar Energy Society**

http://www.ises.org/

The mission is to encourage the use and acceptance of Renewable Energy technologies; to realise a global community of industry, individuals and institutions in support of renewable energy and to create a structure to faciliatecooperation and exchange.

**5968 International Union of Forestry Research Organizations (IUFRO)**

http://www.iufro.org

IUFRO is a nonprofit, non-governmental international network of forest scientists. Its objective are to promote international cooperation in forestry and forest products research. IUFRO's activities are organized primarily through its268 specialized Units in 8 technical Divisions.

**5969 International Wildlife Coalition**

http://www.iwc.org

The International Wildlife Coalition is a federally recognized, nonprofit taxexempt charitable organization. Founded in 1984, the Coalition is dedicated to public education, research, rescue, rebilitation, litigation and internationaltreaty negotiations concerning global wildlife and natural habitat protection issues.

**5970 International Wolf Center**

http://www.wolf.org

The mission of the International Wolf Center is profoundly simple. We support the survival of the wolf around thr world by teaching about its life, its association with other species and dynamic relationships to humans.

**5971 International Year of the Ocean -1998**

http://www.yoto98.noaa.gov

The overall objective is to focus and reinforce the attention of the public, governments and decision makers at large on the importance of the oceans and the marine environment as resources for sustainable development.

**5972 Iowa Department of Natural Resources**

http://www.iowadnr.com

The department's mission is to manage, protect, conserve, and develop Iowa's natural reources in cooperation with other public and private organizations and individuals, so that the quality of life for Iowans is significantly enchancedby the use, enjoyment and understanding of those resources.

**5973 Irish Peatland Conservation Council**

http://www.ipcc.ie

The Irish Peatland Conservation Council is an independent conservation charity. We were established in 1982 to campaign for the conservation of a representative sample of living intact Irish bogs and peatlands and we need your support.

**5974 Island Wildlife Natural Care Centre**

http://www.islandstrust.bc.ca

Its mission is to function in a twofold manner. Frist, by rehabilitating North American wildlife, including marine mammals, with emphasis on alternative, non-toxic, non-invasive treatments. Second, educationally, by furtheringknowledge of treatments available to professionals in the field, and by educationg the public on both rehabilitation and the interaction of man and wild animals.

**5975 Izaak Walton League**

http://www.iwla.org

The mission is to conserve, maintain, protect and restore the soil, forest, water and other natural resources of the United States and other lands; to promote means and opportunities for education of the public with respect to suchresources and their enjoyment and wholesome utilization.

**5976 Jane Goodall Institute**

http://www.janegoodall.org/

The Jane Goodall Institute advances the power of individuals to take informed and compassionate action to improve the environment of all living things.

**5977 Jefferson Land Trust**

http://www.saveland.org

Jefferson Land Trust is a private, nonprofit, grass-roots organization with a mission to conserve property and natural resources. Landowners may work with a Land Trust when they wish to permanently protect the ecological, scenic,historic, or recreational qualities of land they own from inappropriate development.

**5978 John M Judy Environmental Education Consortium**

http://www.utm.edu/departments/ed/cece/john.html

The purpose of the Consortium is to promote and enhance environmental education through systemic change toward infusing a collective environmental consciousness and individual environmental ethic into the learning and teaching process.

**5979 Joint Center for Energy Management (JCEM)**

http://bechtel.colorado.edu/Graduate Programs/Jcem/jcemmain.html

The Joint Center for Energy Management is a research center in the Department of Civil, Environmental, and Architectural Engineering at the University of Colorado at Boulder. It is dedicated to excellence in energy-related research,development, education, and technical assistance.

**5980 Journey North**

http://www.learner.org/jnorth/index.html

Journey North engages students in a global study of wildlife migration and seasonal change. K-12 students share their own field observations with classmates across North America. Widely considered a best-practices model for education,we are the nation's citizen science project for children.

**5981 Kansas Environmental Almanac**

http://www.idir.net/-chsjones/

Kansas is a magnificent place. We owe it to ourselves, to earlier generations and those yet to come, to take good care of our home. With that goal in mind, the Kansas Environmental Almanac will collect and house information pertainingto the Kansas environment and its protection.

**5982 Kentucky Department of Fish and Wildlife Resources**

http://www.kdfwr.state.ky.us/

We are stewards of Kentucky's fish and wildlife resources and their habitats. We manage for the perpetuation of these resources and their use by present and future generations. Through partnerships, we will enhance wildlife diversityand promote sustainable use, including hunting, fishing, boating and other nature-related recreation.

**5983 Kentucky Water Resources Research Institute**

http://www.uky.edu/WaterResources/

The institute mission is to stimulate water resources and water-related environmental research. To assist and stimulate academic units in the conduct of undergraduate and graduate education in water resources and water-relatedenvironmental issues.

**5984 Kola Ecogeochemistry**

http://www.ngu.no/kola

The primary aims are to map the extent of contamination by inorganic elements in various media around industrial centres. To map the content of raionuclides in topsoil throughout the Project area and to shed light on the process anddynamics of trace element cycling in catchments.

**5985 LIFE**

http://life.csu.edu.au/

Founded in 1992, the LIFE Site is Australia's frist information service on the World Wide Web. The main focus is on biological information, especially the environment and biodiversity.

**5986 LTER (US Long-Term Ecological Research)**

http://www.lternet.edu

The mission of the LTER Network is to facilitate and conduct ecological research through: Understanding ecological pheonmena over long temporal and large spatial scales. Creating a legacy of well-designed and documented long-termexperiments and observations for future generations.

**5987 Lake Pontchartrain Basin Foundation**

http://www.saveourlake.org

The Lake Pontchartrain Basin Foundation, a membership-basede citizens organization, is the public's independent voice dedicated to restoring and preserving the Lake Pontchartrain Basin.

**5988 Land Conservancy of San Luis Obispo County**

http://www.special-places.org/

We at the Land Conservancy take pride in our active approach to land conservation. We pursue the protection of open space through land aquisition, conservation easements, restoration, and stewardship.

**5989 Land Trust Alliance**

http://www.lta.org

Founded in 1982, the Land Trust Alliance is the national leader of the private land conservation movement, promoting voluntary land conservation across the country and providing resources, leadership and training to the nation's 1,200plus nonprofit, grassroots land trusts, helping them to protect important open spaces.

**5990 League of Conservation Voters**

http://www.lcv.org

The League of Conservation Voters works to create a Congress more responsive to your environmental concerns. As the nonpartisan political voice for over nine million members of environmental and conservation groups LCV is the onlynational environmental organization dedicated full-time to educating citizens about the environmental voting records of Members of Congress.

**5991 Learning about Backyard Birds**

http://www.birdfeeding.org

Free resource material from the National Birdfeeding Society, including background and guidance for instructors and facilitators as well as project materials for group use.

**5992 Leave No Trace**

http://www.leavenotrace.com

The mission of the Leave No Trace program is to promote and inspire responsible outdoor recreation through education, research, and partnership. The program is managed by LNT, a nonprofit organization located in Boulder, Colorado.

**5993 Living on Earth**

http://www.loe.org

Living on Earth with Steve Curwood is the weekly environmental news and information program distributed by National Public Radio. This Year marks Living on Earth's eighth anniversary.

**5994 Lloyd Center for Environment Studies**

http://www.thelloydcenter.org

The Lloyd Center for Environmental Studies is a nonprofit organization that provides education programs and conducts research to develop a scientific and public understanding of coastal, estaurine, and watershed environments insoutheastern New England.

**5995 Louisiana Department of Agriculture & Forestry**

http://www.ldaf.state.la.us

The Louisiana Department of Agriculture & Forestry was created in accordance with the provisions of Article IV, Section 10 of the Constitution of Louisiana. The commissioner of agriculture and forestry heads the department andexercises all functions of the state relating to the promotion, protection, and advancement of agriculture and forestry, expert research and educational functions expressly allocated by the constitution or by law to other state agencies.

**5996 Louisiana Energy & Environmental Resources & Information Center**

http://www.leeric.lsu.edu/

A primary LEERIC objective is to serve information needs of LSU's faculty, staff, and researchers. LEERIC also provides energy and environmental educational programs for consumers and non-college educators and students.

**5997 Lower Rio Grande Ecosystem Initiative**

http://www.cerc.usgs.gov/lrgrei/lrgrei.htm

The Lower Rio Grande Ecosystem Initiative was established by the Biological Division of the USGS to address research and information needs pertinent to the biotic resources of the river and its adjacent terrestrial habitats.

**5998 Macaw Landing Foundation**

http://www.macawlanding.org/

The foundation is operated solely by volunteers, which allows us to spend all the donated money entirely on Macaws. The Foundation is dedicated to preservation of Macaws.

**5999 Maine Department of Conservation**

http://www.state.me.us/doc/

Created in 1973, the Department of Conservation's Mission is to benefit the citizens, landowners, and users of the state's natural resources by promoting and performing stewardship and ensuring responsible balanced use of Marine'sland, forest, water, and mineral resources.

**6000 Maine Department of Inland Fisheries & Wildlife**

http://www.state.me.us/ifw

The Vision is of an IF&W that: conserves, protects, and enhances the inland fisheries and wildlife resources and promotes efficiency in program management through employee involvement, intitiative, innovation, and teamwork.

**6001 Mangrove Replenishment Initiative**

http://www.mangrove.org

The Mangrove Replenishment Initiative began as a local project along the central east coast of Florida; however, in the last few years it has contributed to wide range of habitat creation and restoration programs that are internationalin scope.

**6002 Manomet Center for Conservation Science**

http://www.manomet.org

Manomet's mission is to conserve natural resources for the benefit of wildlife and human populations. Through research and collaboration, Manomet builds science-based, cooperative solutions to environmental problems.

**6003 Marine Biological Association**

http://www.chm.org.uk

The Marine Biological Association of the United Kingdom is a professional body for marine biologists with some 1200 members world-wide. The current programme encourages fundamental research in marine biology by resident Fellows,interwoven with that of visting scientists.

**6004 Marine Environmental Research Institute**

http://www.downeast.net

The Marine Environmental Research Institute is a 501 nonprofit charitable organization dedicated to scientific research and education on the impacts of pollution on marine life, and to protecting the health and biodiversity of themarine environment for furture generations.

**6005 Marine Mammal Center**

http://www.tmmc.org

We recognize our interdependence with marine mammals, their improtance as sentinels of the ocean environment, and our responsibility to use our awareness, compassion and intelligence to ensure their survival and the conservation oftheir habitat.

**6006 Marine Technology Society**

http://www.mtsociety.org

Our mission is to disseminate marine science and technical knowledge, to promote and support education for marine scientists, engineers and technicians, advance the development of tools and procedures required to explore, study andfurther the responsible and sustainable use of the oceans. Provide servicesthat create a broader understanding of the relevance of marine sciences to other technologies, arts and human affairs.

**6007 Maryland Department of Natural Resources**

http://www.dnr.state.md.us

The Maryland Department of Natural Resources is the state agency which overseas the management and wise use of the living and natural resources of the Chesapeake Bay and its tributaries. The resources of Maryland portion of thewatershed include its state forests and parks, fisheries, wildlife and the recreation of citizens.

**6008 Maryland Forests Association**

http://www.mdforests.org

Incorporated in 1976, the Maryland Forests Association is a nonprofit 501 citizens organization whose membership includes more than 500 individuals and companies from throughout Maryland and the tri-state area.

**6009 Massachusetts Department of Fisheries, Wildlife and Environmental Law Enforcement**

http://www.state.ma.us/dfwele

Massachusetts state agency responsible for the management and conservation of the state's fisheries and wildlife, including rare and endangered species.

**6010 Medomak Valley Land Trust**

http://www.medomakvalley.org

The Medomak Valley Land Trust is a local, private nonprofit oragnization edstablished in 1991 to preserve the natural, recreational, scenic and productive values of the Medomak River watershed. Our goals are to foster a regionalperspective of the watershed and to encourage valley residents to work together to ensure that the resources they value will remain for future generations.

**6011 Messinger Woods Wildlife Care and Education Center**

http://www.wildlifecare.org

The mission is to promote community awareness, education and instruction, involvement, understanding, appre4ciation, and acceptance of our wildlife in order to conserve it. To co-exist and protect each other, our natural surroundings,and all the inhabitants of our earth by education & example.

**6012 Michigan Department of Environment Quality**

http://www.michigan.gov/deq

Our mission is to drive improvements in environmental quality for the protection of public health and natural resources to benefit current and future generations. This will be accomplished through effective administration of agencyprograms, providing for the use of innovative strategies, while helping to foster a strong and sustainable economy.

**6013 Michigan Department of Natural Resources**

http://www.michigan.gov/dnr

The Department of Natural Resources is responsible for the stewardship of Michigan's natural resources and for the provision of outdoor recreational opportunities; a role is has relished sonce creation of the original ConservationDepartment in 1921.

**6014 Michigan Environmental Science Board**

http://www.michigan.gov/mesb

The MESB is an independent state egency established to provide scientific and technical advice to the Governor of Michigan and to state departments, on matters affecting the protection and management of Michigan's environmental andnatural resources.

**6015 Michigan Forest Association**

http://www.michiganforests.com

The mission of the Michigan Forest Association is to promote good management on all forest land, to educate our members about good forest practices and stewardship of the land. To inform the general public about forestry issues and thebenefits of good forest management.

**6016 Michigan United Conservation Clubs**

http://www.mucc.org

MUCC is the largest statewide conservation in the nation. The mission of Uniting Citizens to Conserve Michigan's Natural Resources and Protect our outdoor heritage. MUCC works to conserve Michigan's wildlife, fisheries, waters, forest,air, and soils by providing information, education and advocacy.

**6017 Midwest Renewable Energy Association (MREA)**

http://www.the-mrea.org

The Midwest Renewable Energy Association is a nonprofit network for sharing ideas, resources, and information with individuals, business, and communities to promote a sustainable future through renewable energy and energy efficiency.

**6018 Milton Keynes Wildlife Hospital**

http://mkweb.co.uk/

We are one of the few establishments in the UK to be licensed by the Department of the Environment to care for certain species of birds. A purely voluntary organization and rely solely on donations from the public and local companiesin order to keep the hospital open.

**6019 Mineral Policy Center**

http://www.mineralpolicy.org

Mineral Policy Center is a nonprofit environmental organization dedicated to protecting communities and the environment by preventing the environmental impacts associated with irresponsible mining and mineral development, and bycleaning up pollution caused by past mining.

**6020 Minnesota Department of Natural Resources**

http://www.dnr.state.mn.us

The DNR vision hinges on the concept of sustainability. Protecting and restoring the natural environment while enhancing economic opportunity and community well-being. DNR endorsed ecosystem-based management as its methods to achievesustainability, and uses the concept of ecosystems integrity as a benchmark to measure progress toward sustainability goals. The goal is to maintain environment, econony and the community in a healthy state indefinitely.

**6021 Minnesota Pollution Control Agency**

http://www.pca.state.mn.us

This site includes a great deal of information on air, water, and waste pollution in Minnesota. It also contains data on regulations and permits, clean-up techniques, prevention, publications, and programs to protect Minnesota'senvironment. The MPCA site has a calender of events, information for childern, news releases, training opportunities, and conference information.

**6022 Minnesotans for An Energy-Efficient Economy(ME3)**

http://www.me3.org

ME3 is a coalition working to improve the quality of life, the environment and the economy of Minnesota by promoting energy efficiency and the sound use of renewable energy. Through a program research, public education, and theintervention in the decision making process. ME3 seeks to develop and build consensus for energy vision that will ensure the well being of future generations.

**6023 Missouri Audubon Council**

http://www.audubon.org/chapter/mo

The Audubon Society's mission is to conserve and restore natural ecosystems, focusing on birds and other wildlife for the benefit of humanity and earth's biological diversity. The purpose of the Missouri Audubon Council is to representthe interests of the 14 chapters of the National Audubon Society on a state level.

**6024 Missouri Department of Conservation**

http://www.conservation.state.mo.us

The mission of the Missouri Department of Conservation is to protect and manage the fish, forest, and wildlife resources of the state, to serve the public and facilitate their participation in resources management activities, toprovide opportunity fos all citizens to use, enjoy, and learn about fish, forest and wildlife resources.

**6025 Missouri Prairie Foundation**

http://www.moprairie.org

The Missouri Praire Foundation mission is to preserve the Greater Praire Chicken. To restore the vegetative and faunal balance to the grassland ecosystem, not just to one species, even through the praire chicken is on the verge ofextirpation from Missouri.

**6026 Mmarie**

http://www.kuleuven.ac.be/mmarie

The general objectives of MMARIE are to create an interdisciplinary forum for the exchange of information and experience, related to projects carried out by the participants, and to facilitate collaboration between the partners.

**6027 Monarch Watch**

http://www.monarchwatch.org

Our goals are to further science education, particularly in primary and secondary school systems, to promote the conservation of Monarch butterflies; and to invlove thousands of students and adults in a cooperative study of theMonarch's spectacular fall migration.

**6028 Mountain Lion Foundation**

http://www.mountainlion.org

The Mountain Lion Foudation is a nonprofit conservation and education organization dedicated to protecting the mountain lion, its wild habitat, and the wildlif that shares that habitat. The foundation is dedicated to the propositionthat much can be done to preserve the cougar as a viable species and that the success of this effort can assure the survival of other species.

**6029 NEMO: Oceanographic Data Server**

http://nemo.ucsd.edu

Nemo is a collection of data useful for physical oceanographers here at Scripps Institutions of Oceanography.

**6030 NIREX**

http://www.nda.gov.uk/

NIREX, with the agreement of the Government, to examine safe, environmental and exconomic aspects of deep geological disposal of radioactive waste. We deal with intermediate level waste, which accounts for the majority of radioactivewaste currently in storage, and also with some low-level waste.

**6031 NOAA (National Oceanic and Atmospheric Administration)**

http://www.noaa.gov

NOAA's mission is to describe and predict changes in the Earth's environment, and conserve and wisely manage the Nation's coastal and marine resources. NOAA's strategy consist of seven interrelated Strategic Goals for environmentassessment,, predictions and stewardship.

**6032 Napa County Resource Conservation**

http://www.naparcd.org

Napa County Resource Conservation mission is to encourage and assist acceptance of individual responsibility for watershed management; the goals are enhacement of wildlife habitat, reduction of soil erosion, protection and enhacementof water quality, and promotion of land stewardship and sustainable agriculture.

**6033 National Agricultural Pest Information System**

http://www.ceris.purdue.edu

NAPIS is the database for the Cooperative Agricultural Pest Survey and is maintained by the Center for Environmental and Regulatory Information Systems (CERIS). This site contains pest information, the NAPIS User Guide, the CAPSProgram Guidebook, and the APHIS Environmental Manual. There is also a list of government certified nurseries.

**6034 National Arborist Association**

http://www.natlarb.com

NAA is a trade association of commercial tree care firms that develops that safety and education programs, satndards of tree care practice, and management information for arboriculture firms around the world.

**6035 National Association for Environmental Management**

http://www.naem.org

Dedicated to advancing the profession of environmental management and supports the professional corporate and facility environmental manager.

**6036 National Association for Pet Container Resources**

http://www.napcor.com

The National Association for Pet Container Resources is the trade association for the PET plastic industry in the United States and Canada. Its mission is to facilitate PET plastic recycling and to promote the usage of PET packaging.

**6037 National Association of Conservation Districts**

http://www.nacdnet.org

Thier mission is to serve conservation districts by providing national leadership and a unified voice for natural resource conservation. The association works with landowners, organizations and agency partners in the district helpingto protect the soil, water, forest wildlife and other resources.

**6038 National Association of Environmental Professionals**

http://www.naep.org

NAEP is a multi-disciplinary association dedicated to the advancement of persons in the environmental profession in the US and abroad; a forum for state of the art information on environmental planning, research and management; anetwork of professional contacts and exchange on information among colleagues in industry, government, academic, and the private sector.

**6039 National Association of State Foresters**

http://www.stateforesters.org

State Foresters provide management assistance and protection services for over-two-thirds of the nation's forests.

**6040 National Audubon Society**

http://www.audubon.org

The mission of the National Audubon Society is to conserve and restor natural ecosystems, focusing on birds and other wildlife for the benefit of humanity and the earth's biological diversity.

**6041 National Center for Atmospheric Research**

http://www.ncar.ucar.edu

NCAR's mission to plan, organize, and conduct atmospheric and related research programs in collaboration with the universities and other institutions, to provide state of the art research tools and facilities to the atmosphericsciences community, to support and enhance university atmospheric science education, and to facilitate the transfer of technology to both the public and private sectors.

**6042 National Center for Ecological Analysis and Synthesis**

http://www.nceas.ucsb.edu/

The mission is to advance the state of ecological knowledge through the search for general patterns and principles, to organize and synthesize ecological information in a manner useful to researchers, resource managers, and policymakers addressing important environmental issues.

**6043 National Energy Foundation**

http://www.nef1.org

A nonprofit educational organization dedicated to the development, dissemination and implementation of supplemental educational materials and programs primarily related to energy, water, natural resources, science, conservation and theenvironment.

**6044 National Energy Technology Laboratory**

http://www.alrc.doe.gov

Our mission is to provide stewardship for the Nation's mineral resources by conserving materials produced from minerals.

**6045 National Environmental Health Association(NEHA) , CO**

http://www.neha.org

NEHA is the only national association that represents all of environmental health and protection from terrorism and all-hazards preparedness, to food safety and protection and onsite wastewater systems. Over 4500 members and theprofession are served by the association through it Journal of Environmental Health, Annual Educational Conference and Exhibition, credentialing programs, research and development activities and other services.

**6046 National Estuary Program**

http://www.epa.gov/nep/

The National Estuary Program was established in 1987 by amendments to Clean Water to identify, restore, and protect nationally significant estuaries of the United States. NEP targets a broad range of issues and engages localcommunities in the process. The program focuses not just on improving water quality in an estuary, but on maintaining the integrity of the whole systems-its chemical, physical and biological properties, as well as its economic, recreational, andaesthetic values.

**6047 National Ground Water Association**

http://www.ngwa.org/

Providing and protecting the world's ground water resource. Enhance the skills and credibility of all ground water professionals, develop and exchange industry knowledge and promote the ground water industry and understanding ofground water resources.

**6048 National Institute for Environmental Studies**

http://www.nies.go.jp/index.html

The National Institute for Environmental Studies was established in 1974 at Tsukuba Science City, about 60 Km northeast of Tokyo, as the main research branch of the Environment Agency of the Government of Japan. NIES is the solenational institute for comprehensive research in the environmental sciences.

**6049 National Institute of Environmental Health Science (NIEHS)**

http://www.niehs.nih.gov/

The mission is to reduce the burden of human illness and dysfunction from environmental causes by understanding each of these elements and how they interrelate. The NIEHS achives its mission through multidisciplinary biomedicalresearch programs, prevention and intervention efforts, and communication strategies that encompass training, education, technology transfer, and community outreach.

## 6050 National Library for the Environment

http://www.cnie.org/nle

The National Library for the Environment is a universal, timely, easy to use,single point entry to quality environmental data and information for the use of all participants in the environmental enterprise. This online library icludesdirectories of academic environmental programs, journals, funding sources, meetings, job opportunities, news sources, laws and treaties, reports, reference materials, and more.

## 6051 National Oceanic and Atmospheric Administration

http://www.noaa.gov

NOAA mission is to describe and predict changes in the Earth's environment, and conserve and wisely manage the Nation's coastal and marine resources.

## 6052 National Outdoor Leadership School

http://www.nols.edu/

Over the past 35 years NOLS has become the leader in wilderness education. NOLS is now the largest backcountry permit holder in the United States and runs courses on four continents. NOLS has gone from 100 students in 1965 toapproximately 3,070 students in 1999. As NOLS enters the 21st century, it remainscommited to the quality of courses and programs that it offers, as well as to the wilderness environment that serves as our classroom.

## 6053 National Park and Conservation Association

http://www.npca.org/

The National Parks & Conservation Association has been the sole voice of the American people in the fight to safeguard the scenic beauty, wildlife, and historical and cultural treasures of the largest and most diverse park system inthe world.

## 6054 National Pollutant Inventory

http://www.npi.gov.au

Provides Australians with free access to information on the types and amounts of pollutants being emitted in their community.

## 6055 National Pollution Prevention Center for Higher Education (NPPC)

http://www.css.snre.umich.edu/

The National Pollutant Release Inventory was established under the Canadian Environmental Protection Act, to provide information on the type and quality of pollutantas being released into Canada's environment. The inventory is aNationwide, publicly accessible database of releases and transfers of 178 specified substances to air, water and land.

## 6056 National Renewable Energy Laboratory (NREL)

http://www.nrel.gov/

As the nation's leading center for renewable energy research, NREL is developing new energy technologies to benefit both the environment and the economy.

## 6057 National Sea Grant Library

http://nsgd.gso.uri.edu

The National Sea Grant Library was established as an archive and lending library for Sea Grant funded documents. These documents cover a wide variety of subjects, including oceanography, marine education, aquaculture, fisheries,limonology, coastal zone management, marine recreation and law. NSGL staff lends documents all over the world to aid scientists, teachers, students, fishermen and many other individuals in their reseacrh and studies.

## 6058 National Society for Clean Air

http://www.nsca.org.uk

NCSA is a nonprofit group made up of organizations and individuals who promote clean air through the reduction of air, water, and land pollution, noise and other contaminats, while having due regard for other aspects of theenvironment. The society exmines environmental policy issues from air quality perspective and aims to place them in a broader social and economic context.

## 6059 National Wildlife Health Center

http://www.aphis.usda.gov.ws.nwrc

The National Wildlife Health Center's mission is to improve information, technical assistance, and reseqarch on national and international wildlife health issues. To fulfill the NWHC mission, the Center monitors disease and assessesthe impact od disease on wildlife populations; defines ecological relationships leading to occirrence of disease; transfers technology for disease prevention and control; and provides guidance, training and on site assistance for reducing wildlifelosses.

## 6060 National Wildlife Rehabilitators Association

http://www.nwrawildlife.org

The National Wildlife Rehabilitation Association is a nonprofit international membership organization committed to promoting and improving the integrity and professionalism of wildlife rehabilitation and contributing to thepreservation of natural ecosystems. Please, call before you fax!

## 6061 National Woodland Owners Association

http://www.woodlandowners.org

An independent landowners group with the purpose of developing policy, legislation and representation at the national level, and providing educational and networking opprotunities to landowners throught the country. There are currently35 state partner organizations.

## 6062 Native Americans and the Environment

http://www.cnie.org/NAE

The mission is to educate the public on environmental problems in Native American Communities; to explore the values and historical experiences that Native Americans bring to bear on environmental issues and to promote conservationmeasures that respect Native American land and resource rights.

## 6063 Native Forest Council

http://www.forestcouncil.org

The mission is to provide visionary leadership and to ensure the integrity of public land ecosystems, without compromising people or forests.

## 6064 Native Forest Network

http://www.native forest.org/

The Native Forest Network's mission is to protect the world's remaining native forest by they temperature or otherwise, to ensure they can survive, flourish and maintain their evolutionary potential.

## 6065 Natural Energy Laboratory of Hawaii

E-mail: inquires@nelha.org
http://www.nelha.org

NELHA mangers comprehensive environmental monitoring of seawater. This site describes the participants in this program and the projects in which they are involved. NELHA's goal is to help businesses utilize Hawaii's natural resources.Information available here includes water quality data, NELHA's Seawater Delivery System information, Ocean Thermal Energy Converstion material, and bibliographical information.
*Thomas H Daniel, Scientific Manager*

**6066 Natural Environmental Research Council**

http://www.nerc.ac.uk/

The mission is to promote and support, by any means, high quality basic, strategic and applied research, survey, long-term environmental monitoring and related postgraduate training in terrestrial, marine and freshwater biology andEarth, atmospheric, hydrological, oceanographic and polar sciences and Earth observation. To provide advice on, dieeminate knowledge and promote public understanding of the fields aforesaid.

**6067 Natural Resources Canada**

http://www.nrcan-rncan.gc.ca

Advanced Forest Technologies Program purpose is to develop new approaches in the areas of remote sensing, geographic information systems, artifical intelligence, expert systems and decision support systems to: assist the resource andenvironmental manager with integrated resource planning; provide the land manager with tools for sustainable development.

**6068 Natural Resources Defense Council**

http://www.nrdc.org

Works to safeguard the Earth: its people, its plants and animals, and natural systems on which all life depends.

**6069 Nature Conservancy**

http://www.tnc.org/

The mission of The Nature Conservancy is to preserve the plants, animals and natural communities that represent the diversity of life on Earth by protecting the lands and waters they need to survive.

**6070 Nature Conservancy of Texas**

http://www.tnc.org/texas/

The Nature Conservancy of Texas conserves habitat for native wildlife, using science based research and a cooperative approach to protect the animals and plants that represent Texa's precious natural heritage.

**6071 Nature Node**

http://www.naturenode.com

Online community celebrating the diversity of life on our planet. It is a place to build a virtual community of like-minded individuals.

**6072 Nature Saskatchewan**

http://www.naturesask.ca

Our mission is to promote appreciation and understanding of our natural environment through education, and through conservation and research, to protec and preserve natural ecosystems and their biodiversity.

**6073 NatureNet**

http://naturenet.net/index.html

NatureNet is a voluntary enterprise to provide a good resource for practical nature conservation and countryside management on the Web. Based in UK, and most of the information available on Naturenet relates to the UK, particularyEngland.

**6074 New England Wild Flower Society**

http://www.newfs.org

The New England Wild Flower Society is the oldest plant conservation organization in the United States, promoting the conservation of temperature North American plants through key programs.

**6075 New England Wildlife Center**

http://www.newildlife.com

The New England Wildlife Center is a native wildlife preservation, rehabilitation , animal habitat and environmetal protection, educational organization. We afford humane care to native and naturalized wild animals through our wildlifemedicine hospital. We conduct no research on patients. The center is a nonprofit environmental native habitat protection and preservation organization.

**6076 New Forest Project**

http://www.newforestsproject.com/

To protect conserve, and enhance the health of the Earth's ecosystems by supporting integrated grassrrots efforts to maintain and rebuild the world's forest through the promotion of agroforestry, reforestation, the protection ofwatersheds, and the initiation of renewable energy products.

**6077 New Hampshire Department of Environmental Sciences (DES)**

http://www.des.state.nh.us/

The mission of the Department of Environmental Services is to protect, maintain and anhance environmental quality public health in New Hampshire.

**6078 New Hampshire Fish and Game Department**

http://www.wildlife.state.nh.us/

The mission is to conserve, manage and protect these resources and their habitats; inform and educate the public about these resources and provide the public with opportunities to use and appreciate these resources.

**6079 New Jersey Department of Environmental Protection**

http://www.state.nj.us/dep/

The mission is to assist the residents of New Jersey in preserving, retoring, sustaining, protecting and enhancing the environment to ensure the integration of high environmental quality, public health and economic vitality.

**6080 New Jersey Division of Fish, Game and Wildlife**

http://www.state.nj.us/dep/fgw/

The mission of the Divison of Fish, Game, and Wildlife is to protect and manage the State's fish and wildlife resources to maximize their long term economic, recreational and biological values for the citizens of New Jersey.

**6081 New Mexico Wilderness Alliance**

http://www.nmwild.org

The New Mexico Wilderness Allince is dedicated to the protection, restoration, and rewilding of New Mexico's Wilderness areas. We focus on forward-looking measures to develop an active and educates Wilderness constituency throughoutthe state.

**6082 New Mexico Wildlife Association**

http://www.wildlifewest.org

The New Mexico Wildlife Association is a nonprofit corporation dedicated to yhe preservation of the rich heritage of native New Mexico wildlife and its habitat through education, scientific research, and the sponsorship of wildlifepark.

**6083 New York Association for Reduction, Reuse and Recycling**

http://www.ny.sar3.org/

The mission is to provide state-wide leadership on waste reduction, reuse and recycling issues and practices.

**6084 New York State Department of Environmental Conservation**

http://www.dec.state.ny.us

The mission of the department is to conserve, improve, and protect its natural resources and environment, and control water, land and air ollution, in order to enhance the health, safety and welfare of the peolple of the state andtheir overall economic and social well being.

**6085 North American Commission for Environmental Cooperation**

http://www.cec.org

The Commission for Environmental Cooperation is an international organization created by Canada, Mexico and the US under the North American Agreement on Environmental Cooperation. The CEC was established to address regionalenvironmental concerns, help prevent potential trade and environmental conflicts and to promote the effective enforcement of environmental law.

**6086 North American Lake Management Society**

http://www.nalms.org

Members are academics, lake managers and others interested in furthering the understanding of lake ecology.

**6087 North Carolina Coastal Federation**

http://www.nccoast.org/

NNCF is a nonprofit tax exempt organization which seeks to protect and restore the states's coastal environment, culture and economy through citizen involvement in the management of coastal resources.

**6088 North Carolina Department of Environment and Natural Resources**

http://www.enr.state.nc.us

The mission is to lead stewardship agency for preservation and protection of North Carolina's outstanding natural resources. The organization administers regulatory programs designed to protect air quality, water quality, and thepublic's health.

**6089 North Cascades Conservation Council**

http://www.northcascades.org

The NCCC keeps government officials, environmental organizations, and the general public informed about issues affecting the Greater North Cascade Ecosystem. Action is pursued through legislative, legal, and public participationchannels to protect the lands, waters, plants and wildlife.

**6090 Northeast Advanced Vehicle Consortium**

http://www.navc.org/home.html

The Northeast Advanced Vehicle Consortium is a nonprofit association of private and public sector firms and agencies workimg together to promote advanced vehicle technologies in the Northeast US. NAVC is now the principal multi-state,nonprofit funding mechanism for advanced transportation research, technology development and demostration in the region.

**6091 Northeast Sustainable Energy Association**

http://www.nesea.org

NESEA is a regional membership organization comprised of engineers, educators, builders, students, energy experts, environmental activists, transportation planners, architects, and other citizens interested in responsible energy use.The goal is to bring clean electricity, green transportation, and healthy, efficient buildings into everyday use in order to strengthen the economy and improve the environment.

**6092 Northern Prairie Wildlife Research Center**

http://www.npwrc.usgs.gov/

Our mission is to develop research information on the quantitative requirements for sustainable wildlife populations. To design and conduct studies of numbers and distribution of flora and fauna including identification of changeresulting from habitat loss and modification. To disseminate the latest in technical information and research findings such that interested audiences benefit to the maximum extent possible.

**6093 Ocean Voice International**

http://www.ovi.ca/

Ocean Voice International is a nonprofit membership based marine environmental organization.

**6094 Oceania Project**

http://nornet.nor.com.au/

The Oceania Project mission is to promotr awareness and co-operation to instigate and maintain the process of rehabilitation, preservation and conservation of Cetacea and the Oceans; the promotion and undertaking of scientific researchof Cetacea and the Oceans for the benefit of the community. To provide environmentally sensitive Ocean platforms for non-manipulative research, education and experiential programmes of Cetacea and the Oceans using sensitive vessels.

**6095 Oceanic Resource Foundation**

http://www.orf.org/

The Oceanic Resource Foundation is a nonprofit, scientific researchorganization dedicated to the preservation of the global marine environment and marine biological diversity.

**6096 Office of Energy Efficiency**

http://oee.nrcan.gc.ca/

The office of Energy Efficiency, Canada's centre of excellence for energy efficiency and alternative fuels information, is pplaying a dynamic leadership role in helping Canadians save millions of dollars in energy cost while addressingthe challenges of climate change.

**6097 Office of Protected Resources**

http://www.nmfs.noaa.gov/pr/

The Office of Protected Resources provides program oversight, national policy direction and guidance on the conservation of those marine mammals and endangered species, and their habitats, under the jurisdiction of the Secretary ofCommerce; develops national guidelines and policies for relevant protected resources programs, and provides oversight, advice and guidance on scientific aspects of managing protected species and marine protected areas.

**6098 Ohio Environmental Protection Agency**

http://www.epa.state.oh.us

The Ohio EPA has authority to implement laws and regulations regarding air and water quality standards; solid hazardous and infectious waste disposal standards; quality planning, supervision of sewage treatment and public drinkingwater supplies; and cleanup of unregulated hazardous waste sites. The Ohio EPA cooperates with government and private agencies, manages some federally funded pollution control projects, obtains technical and laboratory services, investigateenvironment problems, etc.

**6099 Ohio Wildlife Center**

http://www.ohiowildlifecenter.org/

The Ohio Wildlife Center is a nonprofit educational organization that promotes increased appreciation and understanding of the natural environment, with particular emphasis on wildlife. OWC is supported by individuals from all walks oflife who wish to improve their own understanding of native wild species and local wildlife issues.

**6100 Oklahoma Department of Wildlife Conservation**

http://www.wildlifedepartment.com

The mission is to manage Oklahoma's wildlife resources and habitat to provide scientific, educational, aesthetic, economic and recreational benefits for present and future generations of hunters, anglers and others who appreciatewildlife.

**6101 Ontario Environment Network**

http://www.oen.ca

The Ontario Environment Network is a nonprofit, nongovernmental network serving Ontario's environmental nonprofit, nongovernmental community. The OENÆseeks to increase awareness of these organizations and encourage discussions aboutmeans to protect the environment.

**6102 Organization of American States: Department of Regional Development and Environment**

http://www.oas.org

Conducts technical cooperation and training programs to assist the member States in their efforts to preserve natural resources. It works with the countries on planning sustainable development, managing the environment and preparinginvestment programs and projects.

**6103 PureZone**

http://www.purezone.com

PureZone is a group effort among manufacturers, suppliers, and installers of Heating and Ventilating systems designed to maximize the quality of aour air in our buildings.

**6104 Rachel Carson Council**

http://www.rachelcarsoncouncil.com

A clearinghouse and library with information at both scientific and layperson levels on pesticides related issues. Rachel Carson Council develops its knowledge from literature searches and conservations with experts. It then providesanswers to the public and also products various publications clarifying pestcide dangers and bringing alternative pest controls to the public's attention.

**6105 Renewable Fuels Association**

http://www.ethanolrfa.org

Members are companies and individuals involved in the production and use of ethanol. Ethanol is sold nationwide as a high-octane fuel that delivers improved vehicle performance while reducing emissions and improving air quality.

**6106 Renewable Natural Resources Foundation**

http://www.rnrf.org

Consortium of professional and scientific organizations with an interest in natural rsources. Established to advance sciences and education in renewable natural resources; promote the application of sound scientific practices inmanaging and conserving renewable natural resources; foster coordination and cooperation among professional, scientific and educational organizations having leadership responsibilities for renewable natural resources; and develop a Renewable NaturalResources Center.

**6107 Society for Conservation Biology (SCB)**

http://www.conbio.org/

The mission of the Center for Conservation Biology Network is to help develop the technical means for the protection, maintenance and restoration of life on this planet-its species, its ecological and evolutionary processes and itsparticular and total environment; to help raise awareness, educate and encourage personal involvement of the public and academics alike.

**6108 Steel Recycling Institute**

http://www.recycle-steel.org

Promotes steel recycling and works to forge a coalition of steelmakers, can manufacturers, legislators, government officials, solid waste managers, business and consumer groups.

**6109 Student Conservation Association**

http://www.sca-inc.org

To build the next generation of conservation leaders and inspire lifelong stewardship of our environment and communities by engaging young people in hands on service to the land.

**6110 Synthetic Organic Chemical Manufacturers Association**

http://www.socma.com

A trade association that serves the specialty, batch and custom chemical industry. SOCMA member companies make the products and refine the raw materials that make our standard of living possible; from pharmaceuticals to cosmetics,soaps to plastics, and all manner of industrial and construction products. SOCMA promotes innovative, safe and environmentally responsible operations, which are internationally competitive and contribute to a healthy, productive economy.

**6111 TechKnow**

http://www.techknow.org

TechKnow is an interactive database that is available over the Internet. The database is a springboard for people interested in Environmentally Sustainable technologies. Originally, TechKnow only contained environmental remediationtechnologies but has now branched off to other forms of sustainable technologies. One of the main goals of the database is to remain current and topical.

**6112 The Environment Directory**

http://www.webdirectory.com

Includes agriculture, animals, arts, business, databases, design, disasters, education, employment, energy, forestry, general environmental interest, government, health, land conservation, parks and recreation, pollution, products andservices, publications, recycling, science, social science, sustainable development, transportation, usenet newsgroups, vegetarianism, water resources, weather and wildlife.

**6113 US Environmental Protection Agency: Environmental Monitoring and Assessment Program**

http://www.epa.gov

A research program to develop the tools necessary to monitor and assess the status and trends of national ecological resources.

**6114 United States Geological Survey: National Earthquake Information Centre**

http://www.earthquake.usgs.gov

Determine location and size of all destructive eartquakes worldwide and immediatly disseminate this information to concerned national and agencies, scientistsm, and the general public

**6115 Vertical Net**

http://www.verticalnet.com

Vertical Net is the Internet's leading bussines e-commerce enabler, providing ent to end e-commerce salutions that are targeted at district business segments through two strategic business units- Vertical Markets and Vertical NetSolutions. While both units focus on a core area of eexpertise, each also leverages the strengths, resources and experience of the other.

**6116 Wisconsin Sea Grant Program**

http://www.seagrant.wisc.edu

Wisconsin Sea grant is a statewide program of basic and applied research, education, and outreach and technology transfer dedicated to the stewardship and sustainable use of the nation's Great Lakes and ocean resources.

**6117 World Data Centre: National Geophysical Data Centre**

http://www.ngdc.noaa.gov

Data management in the broadest sense. Play an integral role in the nation's research into the environment, and at the same tome provide public domain data to a wide group of users.

**6118 World Fish Center**

http://www.worldfishcenter.org/

Our mission is to promote sustainable development and use of living aquatic based on environmentally sound management.

**6119 World Women in Environment and Development Organization**

http://www.wedo.org

International advocacy organization that seeks to increase the power of women worldwide as policymakers at all levels in governments, institutions and forums to achive economic and social justice, healthy and peaceful planet, and humanrights for all.

# Online Databases & Clearinghouses

## Environmental

**6120 Air Risk Information Support Center Hotline**
US Office of Air Quality Planning & Standards
Mail Drop 13
Research Triangle Park, NC 27711
919-541-0888
Fax: 919-541-1818
*Holly Reid, Co-chair*

**6121 Alternative Treatment Technology Information Center**
4 Research Place
Suite 210
Rockville, MD 20850
301-670-6294
Fax: 301-670-3815
*Gary Turner, System Operator*

**6122 Asbestos Ombudsman Clearinghouse Hotline**
401 M Street SW
A-149 C
Washington, DC 20460
703-305-5938
800-368-5888
Fax: 703-305-6462
*Karen V Brown, Ombudsman*

**6123 Bureau of Explosives Hotline**
50 F Street NW
Washington, DC 20001
202-639-2222
Fax: 412-741-0609
http://www.aar.org/aarhome.nsf

**6124 CQS Health and Environmental**
43 Boynton Street
South 2R
Boston, MA 02130
617-522-3466
http://www.cqs.com
*Jonathan Campbell, Health Consultant*

**6125 Carbon Dioxide Information Analysis Center**
Environmental Services Division
Oak Ridge National Lab Bldg 1000
PO Box 2008
Oak Ridge, TN 37831
865-574-0390
Fax: 865-574-2232
*RI Vanhook, Dir. Opns. & User Servs. Mgr*

**6126 Center for Environmental Research Information**
ORD Research Information Unit
26 West Martin Luther King Drive
MS G-72
Cincinnati, OH 45268
513-569-7562
*Dorothy Williams, Executive Officer*

**6127 Center for Environmental and Regulatory Information Systems**
Purdue University
1231 Cumberland Avenue
Suite A
West Layfayette, IN 47906
317-494-7309
Fax: 317-494-9727
http://www.ceris.purdue.edu/ceris
*Eileen Luke, Director*

**6128 Center for Health, Environment and Justice**
PO Box 6806
Falls Church, VA 22040
703-237-2249

The Center for Heath, Environment and Justice trains and assists local people to fight for justice, become empowered to protect their communities from environmental threats and build strong locally controlled organizations. CHEJconnects these strong local groups with each other to build a movement from the bottom up.

*Lois Marie Gibbs, Executive Director*

**6129 Center for Sustainable Systems**
University of Michigan
440 Church Street
Dana Building
Ann Arbor, MI 48109-1041
734-764-1412
Fax: 734-647-5841
E-mail: css.info@umich.edu
http://http://css.snre.umich.edu

CSS advances concepts of sustainability through interdisciplinary research and education. Collaborates with diverse stakeholders to develop and apply life cycle based models and sustainability metrics for systems that meet societalneeds. Promotes tools and knowledge that support the design, evaluation, and improvement of complex systems.

*Jonathan W Bulkley, Co-Director*
*Gregory A Keoleian, Co-Director*

**6130 Chemtrec Center**
1300 Wilson Boulevard
Arlington, VA 22209
800-CMA-8200
Fax: 703-741-6037

**6131 Chemtrec Hotline**
2501 M Street NW
Washington, DC 20037
800-764-9563

**6132 Clean Ocean Action**
18 Hartshorne Drive
Sandy Hook, NJ 07732
732-872-0111
Fax: 732-872-8041
E-mail: SandyHook@cleanoceanaction.org
http://www.cleanoceanaction.org

Clean Ocean Action is a broad-based coalition of over 150 conservation, community, diving, fishing, environmental, surfing, women's and business groups that works to improve and protect the waters off the New York and New Jersey coast.

*Cindy Zipf, Executive Director*
*Mary-Beth Thompson, Operations Director*

**6133 Clean-Up Information Bulletin Board System**
US EPA Technology Innovation Office
401 M Street
OS-10W
Washington, DC 20460
301-589-8368
Fax: 301-589-8487
*Beth Ann Kyle, System Operator*

**6134 Congressional Clearinghouse on the Future**
H2-555 House Annex 2
Washington, DC 20515
202-226-3434

**6135 Conservation Locator**
NYS Department of Environmental Conservation
625 Broadway
2nd Floor
Albany, NY 12233
518-402-8013
Fax: 518-402-9036
http://www.dec.state.ny.us/website/locator

A web-based index of current DEC publications, with listings by subject, links to publications available on the web and directions for obtainig copies of print-only publications.
*Helen Paruolo, EnCon Program Assistant*

**6136 Consumer Energy Council of America**
2000 L Street NW
Suite 802
Washington, DC 20036
202-659-0404
Fax: 202-659-0407
E-mail: outreach@cecarf.org
http://www.cecarf.org

A senior public interest organization in the US focused on the energy, telecommunications and other network industries that provide essential services to consumers.

*Ellen Berman, President*
*Peggy Welsch, Senior VP*

**6137 Consumer Product Safety Commission Hotline**
Office of the Secretary
Washington, DC 20207
800-638-2882

**6138 Control Technology Center**
Emission Standards Division
US EPA, MD 13
Research Triangle Park, NC 27709
919-541-0800
Fax: 919-541-0072
*Bob Blaszczak, ESD/ QAQPS*

**6139 EPA Model Clearinghouse**
Office of Air Quality Planning 919-541-5683
Research Triangle Park, NC 27711 Fax: 919-541-2464
*Dean A Wilson*

**6140 EPA Public Information Center**
US EPA
401 M Street 202-260-7751
PM- 211B
Washington, DC 20460

**6141 EPCRA (SARA Title III) Hotline**
US EPA
401 M Street SW 202-479-2449
OS120 800-535-0202
Washington, DC 20460 Fax: 703-412-3333

**6142 Electronic Bulletin Board System**
US EPA
26 West Martin Luther King Drive 513-569-7358
Cincinnati, OH 800-258-9605
Fax: 513-569-7585
*Charles W Guion*

**6143 Emergency Planning and Community Right-to-Know Information Hotline**
Booz, Allen & Hamilton
1725 Jefferson Davis Highway 703-920-8977
Arlington, VA 22202 800-535-0202
Fax: 703-486-3333
*Dan Kovacs, Contractor*

**6144 Emission Factor Clearinghouse**
US EPA
MD-14 919-541-1000
Research Triangle Park, NC 27709 Fax: 919-541-0684
*Dennis Shipman*

**6145 Environmental Financing Information Network**
US EPA
EFN, WH-547 202-564-4994
401 M Street, East Tower, Room 1117 Fax: 202-565-2694
Washington, DC 20460
*June Lobit*

**6146 Green Committees of Correspondence Clearinghouse**
PO Box 30208 816-931-9366
Kansas City, MO 64112
*Amy Belanger, Coordinator*

**6147 Green Lights Program**
Bruce Company
1850 K Street 290 202-775-6650
Washington, DC 20006 Fax: 202-775-6680
*Maria Theesen*

**6148 Hazardous Waste Ombudsman Program**
US EPA
401 M Street SW 202-260-9361
OS-130, Room SE 315 800-262-7937
Washington, DC 20460 Fax: 202-260-8929
http://www.epa.gov/earth100/records/000154.html
*Bob Knox, Headquarters Contact*

**6149 Indoor Air Quality Information Center**
1200 Pennsylvania Avenue NW 202-343-9370
Washington, DC 20460 800-438-4318
Fax: 202-343-2394
E-mail: iaqinfo@aol.com
http://www.epa.gov/iaq/iaqinfo.html
*Susan Dolgin*

**6150 Inspector General Hotline**
US EPA
1200 Pennsylvania Avenue NW
Washington, DC 20460 800-424-4000
Fax: 202-260-6976
*Ed Maddox*

**6151 International Ground Water Modeling Center**
Colorado School of Mines 303-273-3103
Golden, CO 80401 Fax: 303-273-3278
*Paul van der Hijde, Director*

**6152 Kentucky Partners State Waste Reduction Center**
University of Louisville
Ernst Hall 502-588-7260
Room 312
Louisville, KY 40292
*Joyce St. Clair, Executive Director*

**6153 Methods Information Communications Exchange**
11251 Roger Bacon Drive 703-676-4690
Reston, VA 20190 Fax: 703-318-4646
E-mail: mice@cpmx.saic.com
http://www.epa.gov/sw-846/mice.htm
*Ray Anderson, Contractor*

**6154 Minority Energy Information Clearinghouse**
Office of Minority Economic Impact/US Energy Dept.
100 Independence Ave SW 202-586-8698
Forrestal Building, Room 5B-110 800-543-2325
Washington, DC 20585
*Effie A Young, Officer*

**6155 Montana Natural Resource Information System**
Montana State Library
1515 East 6th Avenue 406-444-3115
Helena, MT 59620 Fax: 406-444-5612
E-mail: msl.state.mt.us
*Alan Cox*

**6156 National Air Toxics Information Clearinghouse**
US EPA
Mail Drop 13 919-541-3586
Office of Air Quality and Standards Fax: 919-541-7674
Research Triangle Park, NC 27709
*Vasu Kilaru, Database Administrator*

**6157 National Capital Poison Center**
Georgetown Univertisy Hospital
3201 New Mexico Avenue 202-362-3867
Suite 310
Washington, DC 20016

**6158 National Center for Biotechnology Information**
National Library of Medicine 301-496-2475
Building 38A, Room 8N805 Fax: 301-480-9241
Bethseda, MD 20894 E-mail: info@ncbi.nlm.nih.gov
http://www.ncbi.nlm.nih.gov
The National center for Biotechnology Information creates public databases, conducts research in computational biology, develops software tools for analyzing genome data, and dissemanates biomediacl information, all for the betterunderstanding of molecular processes affecting human health and disease.

**6159 National Ground Water Information Center**
601 Dempsey Road 800-551-7379
Westerville, OH 43081 800-242-4965
Fax: 614-898-7786
*Kevin McCray, Assistant Executive Director*

**6160 National Pesticide Information Retrieval System**
1231 Cumberland Avenue 317-494-7309
Suite A Fax: 317-494-9727
West Lafayette, IN 4706
*Virginia Walters*

**6161 National Pesticide Telecommunications Network**
Preventive Medicine & Community Health

Texas Tech University
Sciences Center
Lubbock, TX 79430
806-858-7378
800-858-PEST
Fax: 806-743-3094
*Frank L Davido*

**6162 National Radon Hotline**
National Safety Concil
1025 Connecticut Avenue NW
#1200
Washington, DC 20036
202-293-2270
800-767-7236
Fax: 202-293-0032
E-mail: airqual@nsc.org
http://www.nsu.org/issues/radon/index.htm
Provides information on radon and other indoor air quality issues through various toll-free hotlines and publications.
*Kristin Marstiller, Senior Program Manager*

**6163 National Renewable Energy Laboratory**
Technical Inquiry Service
1617 Cole Boulevard
Golden, CO 80401
303-275-4099
*Steve Rubin, Manager*

**6164 National Response Center**
US Coast Guard
2100 2nd Street SW
Room 2611
Washington, DC 20593
202-267-2675
800-424-8802
Fax: 202-267-2181
*Commander David Beach*

**6165 National Small Flows Clearinghouse**
West Virginia University
NRCCE Building, Evandale Drive
PO Box 6064
Morgantown, WV 26506
304-293-4191
800-624-8301
Fax: 304-293-3161
E-mail: nsfc_orders@mail.nesc.wvu.edu
http://www.nsfc.wvu.edu
Nonprofit national source of information about small flows technologies-those systems that have fewer than one million gallons of wastewater flowing through them per day-ranging from individual septic systems to small sewage treatmentplants. Offers more than 450 free and low-cost educational products, a toll-free technical assistance hotline, five computer databases, two free publications and an online discussion group.
*Peter Casey, NSFC Program Coordinator*
*Jen Hause, NSFC Engineering Scientist*

**6166 New York State Department of Environmental Conservation**
625 Broadway
Pollution Prevention Unit
Albany, NY 12233
518-402-8013
*John E Iannotti, PE*

**6167 Northeast Multi-Media Pollution Prevention**
Northeast Waste Management Officials Association
85 Merrimac Street
Boston, MA 02114
617-367-8558
*Terri Goldberg, Program Manager*

**6168 Nuclear Information and Resource Service**
6930 Carroll Avenue
Suite 340
Tacoma Park, MD 20912
301-270-6477
Fax: 301-270-4291
E-mail: nirsnet@nirs.org
http://www.nirs.org
NIRS is the information and networking center for citizens and environmental organizations concerned about nuclear power, radioactive waste, radiation, and sustainable energy issues.
*Michael Mariotte, Executive Director*

**6169 OTS Chemical Assessment Desk**
401 M Street SW
(TS-778)
Washington, DC 20460
202-260-3583
*Terry O'Bryan, Executive Officer*

**6170 Pesticide Action Network North America**
49 Powell Street
Suite 500
San Francisco, CA 94102
415-981-1771
Fax: 415-981-1991
*Kathryn Gilje, Executive Director*

**6171 Powder River Basin Resource Council**
Energy Convervation Education Committee
23 North Scott
Sheridan, WY 82801
307-672-5809
*Jill Morrision, Organizer*

**6172 Public Information Center**
US EPA
401 M Street SW
PM-211B
Washington, DC 20460
202-260-7751
*Alison Cook, Director*

**6173 RACT/BACT/LAER Clearinghouse**
Office of Air Quality Planning
Emissions Standards Division
MD-13
Research Triangle Park, NC 27709
919-541-0800
Fax: 919-541-0072
*Bob Blaszczak, ESD*

**6174 Rachel Carson Council**
8940 Jones Mill Road
Chevy Chase, MD 20815
301-652-1877
*Dr. Diana Post, Executive Director*

**6175 Records of Decision System Hotline**
Computer Sciences Corporation
401 M Street SW
Room L101
Washington, DC 20460
202-260-3770
*Thomas Batts*

**6176 Risk Communication Hotline**
US EPA
401 M Street SW
W Tower, Room 425
Washington, DC 20460
202-260-5606
Fax: 202-260-9757
*Ernestine Thomas*

**6177 Safe Drinking Water Hotline**
LaBat-Anderson, Inc.
800-426-4791

**6178 Small Business Ombudsman Clearinghouse**
1200 Pennsylvania Avenue NW
Washington, DC 20460
202-566-2075
800-368-5888
Fax: 202-566-1505
E-mail: rogers.joanb@epa.gov
http://www.epa.gov/sbo OR www.smallbiz-envirweb.org
Also: Hotline US EPA Office of Small Business Programs, Asbestos and Small Business Ombudsman Program. The EPA Asbestos and Small Business Ombusdman (ASBO) helps the EPA responsibly protect the environment and human health througheconomic and compliance assistance services to small businesses.
*Joan B Rogers, ASBO Ombudsman*

**6179 Solid Waste Information Clearinghouse and Hotline**
1100 Wayne Avenue, Suite 700
PO Box 7219
Silver Spring, MD 20907
800-467-9262
Fax: 301-589-7068
E-mail: tvondeak@swana.org
http://www.swana.org
*Todd von Deak, Dir Marketing/Member Service*

**6180 Stratospheric Ozone Information Hotline**
US Environmental Protection Agency

1200 Pennsylvania Avenue NW
Washington, DC 20460
202-343-9210
800-296-1996
Fax: 202-343-2363
E-mail: miles.louise@epa.gov
http://www.epa.gov/ozone

**6181 Sustainable Buildings Industry Council**
1112 16th Street NW
Suite 240
Washington, DC 20036
202-628-7400
Fax: 202-393-5043
E-mail: SBIC@SBICouncil.org
http://www.sbicouncil.org
*Helen English, Executive Director*

**6182 TNN Bulletin Board System**
Office of Air Quality Planning
Standards Tech Transfer Network
Research Triangle Park, NC 27709
919-541-5616
Fax: 919-541-0824
*Hersch Rorex, System Manager*

**6183 TSCA Assistance Information Service Hotline**
Environmental Assistance Division
1200 Pennsylvania Avenue NW
Mail Code 74080
Washington, DC 20460
202-554-1404
Fax: 202-554-5603
E-mail: tsca-hotline@epamail.epa.gov
The information service furnishes TSCA regulation information to the chemical industry, labor and trade organization, environmental groups and the general public. Technical as well as general information is available.
*John Alter, Primary EPA*

**6184 Toxicology Information Response Center**
Oak Ridge National Laboratory
1060 Commerce Park
MS-6480
Oak Ridge, TN 37831
865-574-4160
Fax: 865-574-0595
http://www.ornl.gov
*Thomas Mason, Director*

**6185 Toxnet**
6707 Democracy Boulevard
2 Democracy Plaza, Suite 510
Bethesda, MD 20892
http://toxnet.nlm.nih.gov
TOXNET offers a high quality database, some scientifically peer-reviewed, in an easy to use interface, and includes links to additional sources of dat related toxicology and environmental health. An array of information to inform thepublic and the scientific community about environmental hazards. TOXNET is a product of the National Library of Medicine's Toxicology and Environmental Infromation Program.

**6186 US Global Change Data and Information System**
61 Route 9w
PO Box 1000
Palisades, NY 10964
914-365-8930
Fax: 914-365-8922
http://globalchange.gov

**6187 Waste Exchange Clearinghouse**
University of Michigan
400 Ann Street NW
Number 201-A
Grand Rapids, MI 49504
616-363-3262
*Jeffery L Duphin*

**6188 Wastewater Treatment Information Exhange**
National Small Flows Clearinghouse
West Virginia University
PO Box 6064
Morgantown, WV 56506
800-624-8301
Fax: 304-293-3161
*Loukis Kissonergis*

**6189 Wetlands Protection Hotline**
Geological Resource Consultants
1555 Wilson Boulevard
Suite 500
Arlington, VA 22209
703-527-5190
800-832-7828
*John Ruffing*

**6190 White Lung Association**
PO Box 1483
Baltimore, MD 21203
410-243-5864
Fax: 410-243-5892
*James Fite, Executive Director*

**6191 Wisconsin Energy Information Clearinghouse**
Wisconsin Division of Energy
101 East Wilson Street
Madison, WI 53702
608-266-8234
Fax: 608-267-6931
E-mail: heat@wisconsin.gov
http://www.doa.state.wi.us/energy/
The Division of Energy Services administers the Wisconsin Home Energy Assistance Program, the Wisconsin Weatherization Program, and the Lead Hazard Reduction Program.
*Judy Ziewacz, State Energy Office Director*

# Videos

## Environmental

**6192 Acid Rain: A North American Challenge**
National Film Board of Canada
1123 Broadway
Suite 307
New York, NY 10010
212-629-8890
Fax: 212-629-8502
E-mail: j.sirabella@nfb.ca
http://www.nfb.ca
Summarizes what we know today about the causes and effect of the menace of acid rain.

**6193 Adventures of the Little Koala & Friends**
Family Home Entertaiment
15400 Sherman Way
PO Box 10124
Van Nuys, CA 91410
818-908-0303
http://www.familyhome ent.com

**6194 Air Pollution: A First Film**
2349 Chaffee Drive
St Louis, MO 63146
314-569-0211

**6195 Alaska: Outrage at Valdez**
Cousteau Collection Volume 1
Facets Media
1517 Fullerton Aveune
Chicago, IL 60614
773-281-4114
http://www.centerstage.net/film/cinemas/

**6196 Black Waters**
Green Mountain Post Films
PO Box 229
Turner Falls, MA 01376
413-863-4754
Fax: 413-863-8248
E-mail: info@gmpfilms.com
http://www.gmpfilms.com

**6197 Bog Ecology**
Educational Images
PO Box 3456
Elmira, NY 14905
607-732-1090
800-527-4264
Fax: 607-732-1183
E-mail: info@edimages.com
http://www.educationalimages.com

*Charles R Belinky PhD, CEO*

**6198 Captain Planet & the Planeteers: Toxic Terror**
Turner Home Entertainment Company
1 CNN N Tower
12th Floor
Atlanta, GA 30348
404-827-1700
http://www.turner.com

**6199 Carnivores**
Walt Disney Home Video
500 S Buena Vista Street
500 South Buena Vista Street
, CA 91521
818-562-3560
http://www.disneyvideos.com

**6200 Chelyabinsk: The Most Contaminated Spot on the Planet**
Filmakers Library
124 E 40th Street
New York, NY 10016
212-808-4980
Fax: 212-808-4983
E-mail: info@filmakers.com
http://www.filmakers.com
The story of the Chelyabinsk atomic weapons complex, including a 1957 explosion, 1967 storm that spread radioactive dust, and the dumping of radioactive waste into a water-supply river.

*Linda Gottesman, Co-President*

**6201 Chemical Kids**
Filmakers Library
124 E 40th Street
New York, NY 10016
212-808-4980
Fax: 212-808-4983
E-mail: info@filmakers.com
http://www.filmakers.com
Alarming facts about man-made chemicals in food and water, especially how children are the most affected.

*Linda Gottesman, Co-President*

**6202 Children of Chernobyl**
Filmakers Library
124 E 40th Street
New York, NY 10016
212-808-4980
Fax: 212-808-4983
E-mail: info@filmakers.com
http://www.filmakers.com
Reveals the tragedy at Chernobyl through exclusive archival film and eyewitness accounts — deception and cover-up on a grand scale.

*Linda Gottesman, Co-President*

**6203 City of the Future**
University of California
2176 Shattuck Avenue
Media Center
Berkeley, CA 94704
510-642-0460
http://www.berkley.edu

**6204 Clouds of Doubt**
Electronics Arts Intermix
536 Braodway
9th Floor
New York, NY 10012
212-966-4605
http://www.eci.org

**6205 Cocos Island: Treasure Island**
ESPN Home Video
ESPN Plaza
Bristol, CT 06010
860-585-2000
http://www.espn.com

**6206 Coral Cities of the Caribbean**
Nancy Sefton/Triton Productions
Earthwise Media
PO Box 1223
Poulsbo, WA 98370
360-271-1584
Fax: 360-394-2168
E-mail: info@earthwisevideos.com
http://www.earthwisevideos.com
Journey the coral reefs of the Caribbean with the fishes and invertebrates that reside in these coral cities.
*Nancy Sefton, Creative Director*
*Wes Nicholson, Technical Director*

**6207 Designing the Environment**
NETCHE
1800 N 33rd Street
Lincoln, NE 68503
402-472-3611
Fax: 402-472-1785
E-mail: netche@unl.edu
http://www.netdb.unl.edu/netchevideo

**6208 Disappearance of the Great Rainforest**
Arthur Mokin Productions
PO Box 1866
Santa Rosa, CA 95402
707-542-4868
800-238-4868

**6209 Dolphin**
Media Guild
PO Box 910534
San Diego, CA 92191
858-755-9191
Fax: 858-755-4931

**6210 Earth Summit: What Next?**
EcuFilm
810 12th Avenue S
Nashville, TN 37203
615-242-6277
800-251-4091
http://www.ecufilm.com

**6211 Earth at Risk Environmental Series**
Library Video Company
PO Box 580
Wunnewood, PA 19096
610-645-4000
800-843-3620
Fax: 610-645-4040
E-mail: comments@libraryvideo.com
http://www.libraryvideo.com

**6212 Earth's Physical Resources**
Media Guild
PO Box 910534
San Diego, CA 92191
858-755-9191
Fax: 858-755-4931

**6213 Earthwise Media**
PO Box 1223
Poulsbo, WA 98370
360-271-1584
Fax: 360-394-2168
E-mail: info@earthwisevideos.com
http://www.earthwisevideos.com
Production company specializing in environmental educational videos, multi-media presentations and computer aided learning tools.

*Wes Nicholson, Technical Director*

**6214 Ecological Realities: Natural Laws at Work**
University of California
2176 Shattuck Avenue
Berkeley, CA 94704
510-642-0460

**6215 Educational Images**
PO Box 3456
Elmira, NY 14905
607-732-1090
800-527-4264
Fax: 607-732-1183
E-mail: info@edimages.com
http://www.educationalimages.com
Produces educational CD ROMs, slide sets and videos on ecology, geology, aquatic life and related biological science and medical topics, plus stock photos for digital and print.
*Charles R Belinky PhD, CEO*

**6216 Effluents of Affluence**
University of Michigan
Film Video Library
919 S University Avenue, Room 207
Ann Arbor, MI 48109
734-764-5360
Fax: 734-764-6849
E-mail: ful.office@umich.edu
http://www.lib.umich.edu

**6217 Elephant Boy**
HBO Home Video
1114 6th Avenue
New York, NY 10036
212-512-7400
http://www.hbohomevideo.com

**6218 Empire of the Red Bear**
Discovery Home Entertainmnet
7700 Wisconsin Avenue
Betseda, MD 20814
301-986-1999
E-mail: letters@discovery.com
http://www.discovery.com

**6219 Endangered Species: Massasauga Rattler and Bog Turtle**
Educational Images
PO Box 3456
Elmira, NY 14905
607-732-1090
800-527-4264
Fax: 607-732-1183
E-mail: info@edimages.com
http://www.educationalimages.com

*Charles R Belinky PhD, CEO*

**6220 Enemies of the Oak**
Carolina Biological Supply Company
2700 York Road
Burlington, NC 27215
919-584-0381
800-334-5551
E-mail: carolina@carolina.com
http://www.carolina.com

**6221 Energetics of Life**
The Media Guild
PO Box 910534
San Diego, CA 92191
858-755-9191
Fax: 858-755-4931

**6222 Energy Now**
Educational Images
PO Box 3456
Elmira, NY 14905
607-732-1090
800-527-4264
Fax: 607-732-1183
E-mail: info@edimages.com
http://www.educationalimages.com

*Charles R Belinky PhD, CEO*

**6223 Energy to Go Around**
The Media Guild
PO Box 910534
San Diego, CA 92191
858-755-9191
Fax: 858-755-4931

**6224 Energy: The Alternatives**
The Media Guild
PO Box 910534
San Diego, CA 92191
858-755-9191
Fax: 858-755-4931

**6225 Everglades Region: An Ecological Study**
The Media Guild
PO Box 910534
San Diego, CA 92191
858-755-9191
Fax: 858-755-4931

**6226 Exploring the Forest**
Alfred Higgins Productions
6350 Laurel Canyon Boulevard
North Hollywood, CA 91606
818-762-3300
Fax: 818-762-8223
http://www.alfredhigginsprod.com

**6227 Fascinating World of Forestry**
Educational Images
PO Box 3456
Elmira, NY 14905
607-732-1090
800-527-4264
Fax: 607-732-1183
E-mail: info@edimages.com
http://www.educationalimages.com

*Charles R Belinky PhD, CEO*

**6228 Flag: The Story of the White White-Tailed Deer**
Educational Images
PO Box 3456
Elmira, NY 14905
607-732-1090
800-527-4264
Fax: 607-732-1183
E-mail: info@edimages.com
http://www.educationalimages.com

*Charles R Belinky PhD, CEO*

**6229 Florida Bay and the Everglades**
Educational Images
PO Box 3456
Elmira, NY 14905
607-732-1090
800-527-4264
Fax: 607-732-1183
E-mail: info@edimages.com
http://www.educationalimages.com

*Charles R Belinky PhD, CEO*

**6230 Food from the Rainforest**
The Media Guild
PO Box 910534
San Diego, CA 92191
858-755-9191
Fax: 858-755-4931

**6231 Forms of Energy**
Educational Images
PO Box 3456
Elmira, NY 14905
607-732-1090
800-527-4264
Fax: 607-732-1183
E-mail: info@edimages.com
http://www.educationalimages.com

*Charles R Belinky PhD, CEO*

**6232 Freshwater and Saltwater Marshes**
Educational Images
PO Box 3456
Elmira, NY 14905
607-732-1090
800-527-4264
Fax: 607-732-1183
E-mail: info@edimages.com
http://www.educationalimages.com

*Charles R Belinky PhD, CEO*

**6233 GPN Educationl Media**
1407 Fleet Street
Baltimore, MD 21231
800-228-4630
Fax: 800-306-2330
E-mail: inquiry@shopgpn.com
http://www.shopgpn.com

Produces educational media - video, CD-ROM, DVD, Internet-for-16. Free previews available on line and on video.

*Jen Haus, VP Marketing*

**6234 Great American Woodlots**
Cornell University
8 Business & Technology Park
Itahaca, NY 14850
607-255-2091
607-255-9946
http://www.cornell.edu

**6235 Green TV**
1125 Hayes Street
San Francisco, CA 94117
415-255-4797
Fax: 415-255-4664
E-mail: fgreen@greentv.org
http://www.greentv.org

Video production company that combines environmental journalism with dramatic wildlife and natural history footage.
*Frank Green, Owner/President*

**6236 Guardians of the Cliff: The Peregrine Falcon Story**
Educational Images
PO Box 3456
Elmira, NY 14905
607-732-1090
800-527-4264
Fax: 607-732-1183
E-mail: info@edimages.com
http://www.educationalimages.com

*Charles R Belinky PhD, CEO*

**6237 Happy Campers with Miss Shirley & Friends**
Kids Express
1106 South Truckee Way
Aurora, CO 80017
888-492-5437
E-mail: missshirleybowers@msn.com
http://www.kids-express.com

**6238 I Walk in the Desert**
Educational Images
PO Box 3456
Elmira, NY 14905
607-732-1090
800-527-4264
Fax: 607-732-1183
E-mail: info@edimages.com
http://www.educationalimages.com

*Charles R Belinky PhD, CEO*

**6239 Joe Albert's Fox Hunt**
Education Development Center
55 Chapel Street
Newton, MA 02160
617-969-7100
800-225-4276
E-mail: www@edc.org
http://www.edc.org

**6240 John Muir**
Educational Images
PO Box 3456
Elmira, NY 14905
607-732-1090
800-527-4264
Fax: 607-732-1183
E-mail: info@edimages.com
http://www.educationalimages.com

*Charles R Belinky PhD, CEO*

**6241 Legacy of an Oil Spill**
Media Guild
PO Box 910534
San Diego, CA 92191
858-755-9191
800-886-9191
Fax: 858-755-4931
E-mail: info@mediaguild.com
http://www.mediaguild.com

Documenting the 1989 oil spill into Alaska's Prince William Sound which damaged 1,000 miles of shoreline and killed hundreds of thousands of wildlife. This video looks at the long-term effects of the spill on several species.
*Ruth Pipitone*

**6242 Life on a Rocky Shore**
Earthwise Media
PO Box 1223
Poulsbo, WA 98370
360-271-1584
Fax: 360-394-2168
E-mail: info@earthwisevideos.com
http://www.earthwisevideos.com

An introduction to the plants and animals of the Pacific Northwest's rich Intertidal Zone. The self-paced presentation uses 100 photos, video clips and graphics to illustrate the story of life in the Zone. Learn about the tides,changing bands of life, and adaptation of life in the Zone. Discusses marine biology and includes teacher's guide for grades 6-12, with review, vocabulary and quizzes.

*Wes Nicholson, Technical Director & Owner*

**6243 Manatees: A Living Resource**
Educational Images
PO Box 3456
Elmira, NY 14905
607-732-1090
800-527-4264
Fax: 607-732-1183
E-mail: info@edimages.com
http://www.educationalimages.com

*Charles R Belinky PhD, CEO*

**6244 Mitzi A Da Si: A Visit to Yellowstone National Park**
Educational Images
PO Box 3456
Elmira, NY 14905
607-732-1090
800-527-4264
Fax: 607-732-1183
E-mail: info@edimages.com
http://www.educationalimages.com

*Charles R Belinky PhD, CEO*

**6245 Modeling Photosynthesis**
Media Guild
PO Box 910534
San Diego, CA 92191
858-755-9191
Fax: 858-755-4931

**6246 Our Precious Environment**
Educational Images
PO Box 3456
Elmira, NY 14905
607-732-1090
800-527-4264
Fax: 607-732-1183
E-mail: info@edimages.com
http://www.educationalimages.com

*Charles R Belinky PhD, CEO*

**6247 RMC Medical Inc**
3019 Darnell Road
Philadephia, PA 19154
215-824-4100
800-332-0672
Fax: 215-824-1371
E-mail: rmcmedical@cs.com
http://www.rmcmedical.com

Manufacturer of decontamination emergency response equipment.

*Lois White, Sales/Marketing Manager*

**6248 Rainbows in the Sea: A Guide to Earth's Coral Reefs**
Earthwise Media
PO Box 1323
Poulsbo, WA 98370
360-271-1584
Fax: 360-394-2168
E-mail: info@earthwisevideos.com
http://www.earthwisevideos.com

Showcases the wide variety of colorful fish and invertebrates living in coral reefs, how they form and what they need to blossom. Includes teacher's guide and recommended by the National Science Teachers Association.
*28 minutes*
*Wes Nicholson, Technical Director*

**6249 Return of the Dragon**
Educational Images
PO Box 3456
Elmira, NY 14905
607-732-1090
800-527-4264
Fax: 607-732-1183
E-mail: info@edimages.com
http://www.educationalimages.com
*Charles R Belinky PhD, CEO*

**6250 Salt Marshes-A Special Resource**
Educational Images
PO Box 3456
Elmira, NY 14905
607-732-1090
800-527-4264
Fax: 607-732-1183
E-mail: info@edimages.com
http://www.educationalimages.com
*Charles R Belinky PhD, CEO*

**6251 Sand Dune Ecology and Formation**
Educational Images
PO Box 3456
Elmira, NY 14905
607-732-1090
800-527-4264
Fax: 607-732-1183
E-mail: info@edimages.com
http://www.educationalimages.com
*Charles R Belinky PhD, CEO*

**6252 Seals**
Media Guild
PO Box 910534
San Diego, CA 92191
858-755-9191
Fax: 858-755-4931

**6253 Song of the Salish Sea: A Natural History of Northwest Waters**
Earthwise Media
PO Box 1323
Poulsbo, WA 98370
360-271-1584
Fax: 360-394-2168
E-mail: info@earthwisevideos.com
http://www.earthwisevideos.com
Examines the fragile habitats that make up the Strait of Georgia, Strait of Juan de Fuca and Puget Sound. Includes teacher's guide, Puget Sound Beach Guide, Kids for Puget Sound Passport, and curriculums for grades 6-12.
*45 minutes*
*Wes Nicholson, Technical Director*

**6254 Survey of Environment**
Educational Images
PO Box 3456
Elmira, NY 14905
607-732-1090
800-527-4264
Fax: 607-732-1183
E-mail: info@edimages.com
http://www.educationalimages.com
*Charles R Belinky PhD, CEO*

**6255 The World Between the Tides: A Guide to Pacific Rocky Shores**
Earthwise Media
PO Box 1323
Poulsbo, WA 98370
360-271-1584
Fax: 360-394-2168
E-mail: info@earthwisevideos.com
http://www.earthwisevideos.com
A narrated journey along the intertidal area that examines harsh conditions, and how animals and plants have adapted. Received the National Communicator Award of Distinction and praise from the NW Aquatic & Marine Educators. Includesteacher's guide.
*23 minutes*
*Wes Nicholson, Technical Director*

**6256 Tropical Rainforest**
The Media Guild
PO Box 910534
San Diego, CA 92191
858-755-9191
Fax: 858-755-4931

**6257 Tropical Rainforests Under Fire**
Educational Images
PO Box 3456
Elmira, NY 14905
607-732-1090
800-527-4264
Fax: 607-732-1183
E-mail: info@edimages.com
http://www.educationalimages.com
*Charles R Belinky PhD, CEO*

**6258 Warm-Blooded Sea Mammals of the Deep**
Cousteau Odyssey Volume 10
Warner Home Video
4000 Warner Boulevard
Burbank, CA 91522
818-954-6000
http://www.store.warnervideo.com

**6259 Warming Warning**
Media Guild
PO Box 910534
San Diego, CA 92191
858-755-9191
Fax: 858-755-4931

**6260 Water Resources Videos**
Educational Images
PO Box 3456
Elmira, NY 14905
607-732-1090
800-527-4264
Fax: 607-732-1183
E-mail: info@edimages.com
http://www.educationalimages.com
*Charles R Belinky PhD, CEO*

**6261 Watershed: Canada's Threatened Rainforest**
Media Guild
PO Box 910534
San Diego, CA 92191
858-755-9191
800-886-9191
Fax: 858-755-4931
E-mail: info@mediaguild.com
http://www.mediaguild.com
Chronicles the expedition to the Pacific Coast's rainforest, including environmental issues, complex ecosystems, and efforts of conservationists.
*Ruth Pipitone*

**6262 Wetlands: Development, Progress, Environmental Protection under Changing Law**
American Law Institute
4025 Chestnut Street
Philadelphia, PA 19104
215-243-1600
800-CLE-NEWS
Fax: 215-243-1664
E-mail: jmendicino@ali.org
http://www.ali.org

**6263 Whales**
Cousteau Undersea World Volume 10
Churchill Media
12210 Nebraska Avenue
Los Angeles, CA 90025
310-207-6600
Fax: 310-207-1330

**6264 Wilderness Video**
New Nature In Motion: Yellowstone, Yosemite, Etc.
PO Box 3150
Ashland, OR 97520
541-488-9363
Fax: 541-488-9363
E-mail: Bob@Wildernessvideo.com
http://www.wildernessvideo.com
Collection of high definition stock footage, including nature, national parks, and cities.
*Bob Glusic, Owner*

**6265 Windrifters: The Bald Eagle Story**
Educational Images

PO Box 3456
Elmira, NY 14905

607-732-1090
800-527-4264
Fax: 607-732-1183
E-mail: info@edimages.com
http://www.educationalimages.com

*Charles R Belinky PhD, CEO*

# Green Product Catalogs

## General

**6266 Acorn Designs**
5066 Mott Evans Road
Trumansburg, NY 14886
800-299-3997
Fax: 607-387-5609
E-mail: info@acorndesigns.org
http://www.acorndesigns.org
Totes and other items with wildlife and nature themes; high post-consumer recycled paper; images from nature as note cards and jouranls; organic cotton tees. Art and images represent the works of over 30 artisits.
*Steve Sierigk, Owner*

**6267 Alexandra Avery Purely Natural Body Care**
4717 SE Belmont Street
Portland, OR 97215
503-236-5926
800-669-1863
Fax: 503-234-7272
E-mail: aavery42@earthlink.net
100% natural and cruelty free aromatherapy products for face and body care.
*Alexandra Avery, President*

**6268 American Resources Group**
374 Maple Avenue
Suite 310
Vienna, VA 22180
703-255-2700
Fax: 703-281-9200
E-mail: info@american-resources.org
http://www.american-resources.org
Provides a listing of resources and links for environmental research centers, associations and consulting engineers.
*Keith A Argow PhD, President*

**6269 Artistic Video**
87 Tyler Avenue
Sound Beach, NY 11789
631-744-5999
888-982-4244
Fax: 631-744-5993
E-mail: bobklien@movementsofmagic.com
http://www.movementsofmagic.com
Instructional videos and DVDs on health and fitness, alternative healing, children's programs about animals, free interactive section with articles, discussion forums, video clips, classes, instructions, Tai-Chi and other natureoriented cultures and practices.

*Bob Klein, President*

**6270 BDM Holdings**
7915 Jones Branch Drive
McLean, VA 22102
703-848-5000
*Bennie Dibona, VP Engineering/Environment*

**6271 Balance of Nature**
Unviersity of Chicago Press
1427 East 60th Street
Chicago, IL 60637
773-702-7700
Fax: 773-702-9756
http://www.press.chicago.edu

**6272 Bio-Sun Systems**
RR 2 Box 134A
Millerton, PA 16936
570-527-2200
800-847-8840
Fax: 570-537-6200
E-mail: biosun@npacc.net
http://www.bio-sun.com
Composting toilets and modular restrooms.
*Donna White, President*
*Al White, VP*

**6273 Cotton Clouds**
5176 S 14th Avenue
Safford, AZ 85546
520-428-7000
Fax: 520-428-6630
E-mail: cottonclouds@az.org
Yarn, patterns, books, video's, kit for weaving crochet, knitting and spinning.

**6274 Earth Options**
Solar Electric Engineering
117 Morris Street
Sebastopol, CA 95472
707-824-4150
882-198-1986
Fax: 707-542-4358
http://882-198-1986
Environmental retail products.

**6275 Earth Science**
PO Box 1925
Corona, CA 91718
909-371-7565
800-222-6720
Fax: 909-371-0509
http://800-222-6720
All-natural, environmentally sound skin and hair care products.
*Kristine Schoenauer, President*

**6276 Eco-Store**
2441 Edgewater Drive
Orlando, FL 32804
407-426-9949
800-556-9949
Fax: 407-649-3148
E-mail: beth@eco-store.com
http://800-556-9949
Environmental home products, gifts, etc.

**6277 Ecology Store**
6928 Queens Boulevard
Flushing, NY 11377
718-446-4444
800-548-9660
Fax: 718-446-9860
http://800-548-9660
Range of environmental products.

**6278 Energy Efficient Environments**
2119 Inverness Lane
Glen View, IL 60025
847-475-3005
800-336-3749
E-mail: info@eeenvironments.com
http://www.eeenvironments.com
Efficient and earth-friendly devices for energy, water and light use.

**6279 Environment Friendly Papers**
Cherry Paper
13520 Liberty Avenue
Jamaica, NY 11419
718-297-3000
Fax: 718-297-2986
E-mail: cherryop@AOL.com
100% recycled gift stationary, with original designs depicting flora and fauna of South Africa and environmental themes.

**6280 Erlander's Natural Products**
Nature's Department Store
2279 Lake Avenue
Altadena, CA 91001
626-797-7004
800-562-8873
Fax: 626-798-2663
E-mail: erlander@webtv.net
http://800-562-8873
Natural olive oil-wine soap; bar and liquid roach killer from herbs, organic red zinfandel wine, 100% organic cotton pillows and mattresses, jewelry - semi-precious and costume, washing compound/non-detergent, sodium sesquicarbonate.
*Leatrice Erlander, Co-Owner*
*Stig Erlander, Co-Owner*

**6281 GAIA Clean Earth Products**
PO Box 1906
York, PA 17405
717-840-1638
800-726-5496
Fax: 800-726-5496
E-mail: gaia@blazenet.net
http://800-726-5496
Environmentally-compatible products.
*Brian N Hartman, President*

**6282 Greenpeace**
564 Mission Street Box 416
San Francisco, CA 94105
510-538-7842
800-326-0959
Fax: 202-462-4517
E-mail: greenpeace@npgear.com
http://800-326-0959

Environmentally and socially responsible apparel, accessories and gifts.

**6283 Jason Natural Cosmetics**
8468 Warern Drive
Culver City, CA 90232
877-JAS-ON01
Fax: 310-838-9274
E-mail: jnp@jason-natural.com
http://877-JAS-ON01

All natural cosmetics.
*Jeffrey Light, President*

**6284 Look Alive!**
Rice Lake Products
100 27th Street NE
Minot, ND 58703
701-857-6357
800-998-7450
Fax: 701-857-6300
E-mail: ricelake@dalotah.com
http://800-998-7450

Environmentally safe movement devices for hunting decoys, as well as bird and pest deterring owls for home garden, boats and businesses.
*Virgil Farstad, President*

**6285 Real Goods**
Real Goods Solar
833 West South Boulder Road
Louisville, CO 80027
800-919-2400
http://www.realgoods.com

Source for simple living products designed to balance with a conscious lifestyle. Offers products that reduce energy consumption, from solar panels and wind turbines to complete solar power systems.

**6286 Real Goods Trading Company**
PO Box 8507
Ukiah, CA 95482
707-744-2100
800-347-0070
Fax: 707-468-9394
http://800-347-0070

Recycled papers, environmental gifts and household goods.

**6287 Second Renaissance Books**
17 George Washington Plaza
Gaylordsville, CT 06755
800-729-6149
Fax: 860-355-7160
E-mail: inquiries@secondrenaissance.com
http://800-729-6149

Editorially selected books and audio tapes. Complete selection of writing and letters by Ayn Rand.

**6288 Sparky Boy Enterprises**
1512 Gold Avenue
Bozeman, MT 59715
406-587-5891
800-289-6656
Fax: 406-587-0223
E-mail: ecostoke@mcn.net
http://800-289-6656

Natural products for the home, lawn and garden.
*Wayne Vinje, President*

**6289 Sunrise Lane Products**
780 Greenwich Street
New York, NY 10014
212-243-4745

Environmentally safe and cruelty-free products for home and personal care.
*Rossella Mocerino, President*

**6290 The Green Catalog**
Advertising Specialty Institute
http://www.asicentral.som

Features products that are friendly to the enviroment, designed to inspire social responsibility.

**6291 The Green Life**
The Green Life Store
2409 Main Street
Santa Monica, CA 90405
310-392-4702
E-mail: info@thegreenlife.com
http://www.thegreenlifecostore.com

Green products that are the most ecologically friendly available, for home & garden, bed & bath, children, cleaning, office, and pets.
*Scott O'Brien, Owner*

**6292 Williams Distributors**
1801 S Cardinal Lane
New Berlin, WI 53151
262-597-9865

All natural products; nutrition, health, home care and personal care, plus water purification systems.
*GL Williams, President*

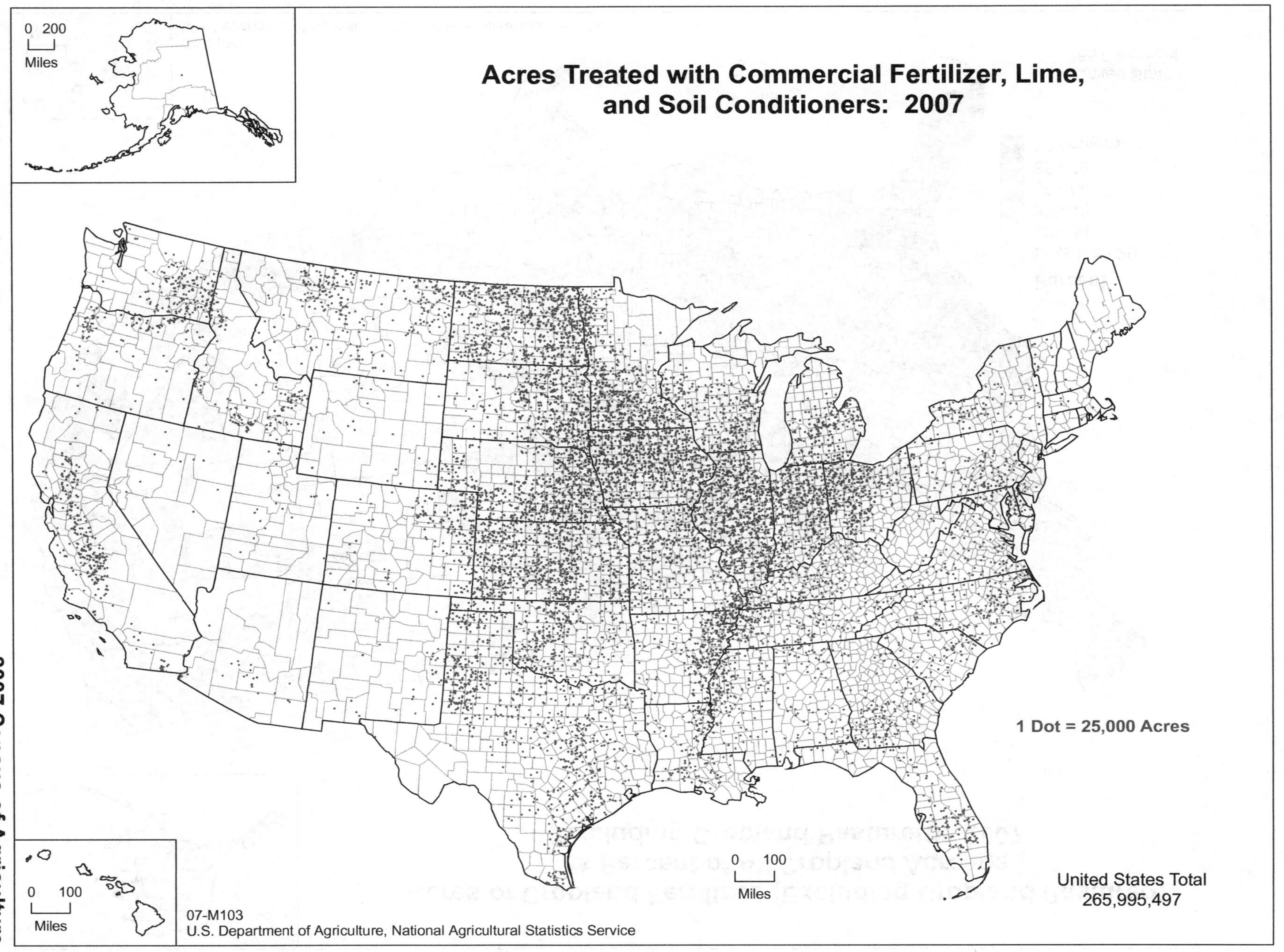


2007 Census of Agriculture

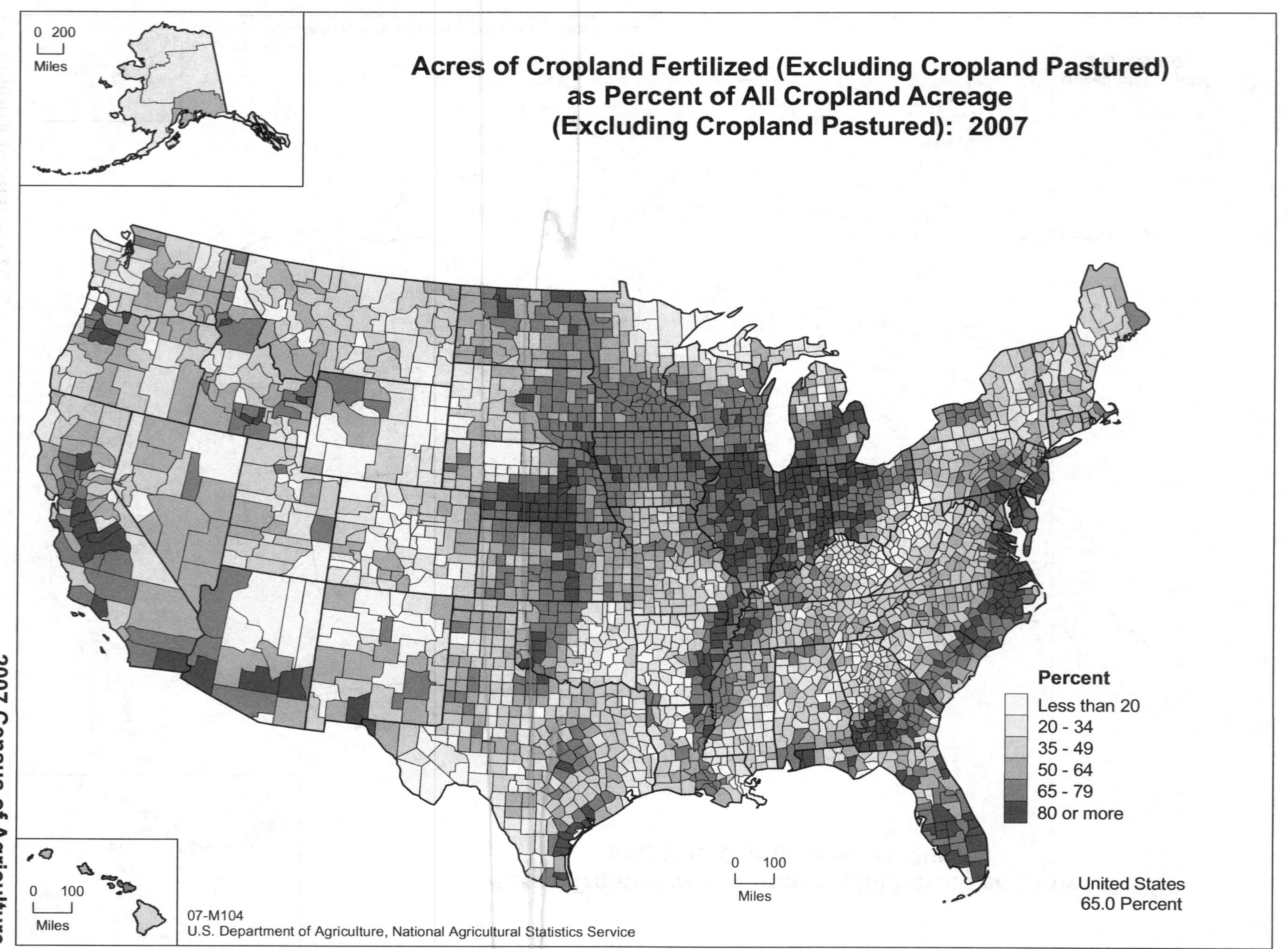
Acres of Cropland Fertilized (Excluding Cropland Pastured)
as Percent of All Cropland Acreage
(Excluding Cropland Pastured): 2007
0 200
Miles
Percent
Less than 20
20 - 34
35 - 49
50 - 64
65 - 79
80 or more
0 100
Miles
United States
65.0 Percent
0 100
Miles
07-M104
U.S. Department of Agriculture, National Agricultural Statistics Service
2007 Census of Agriculture

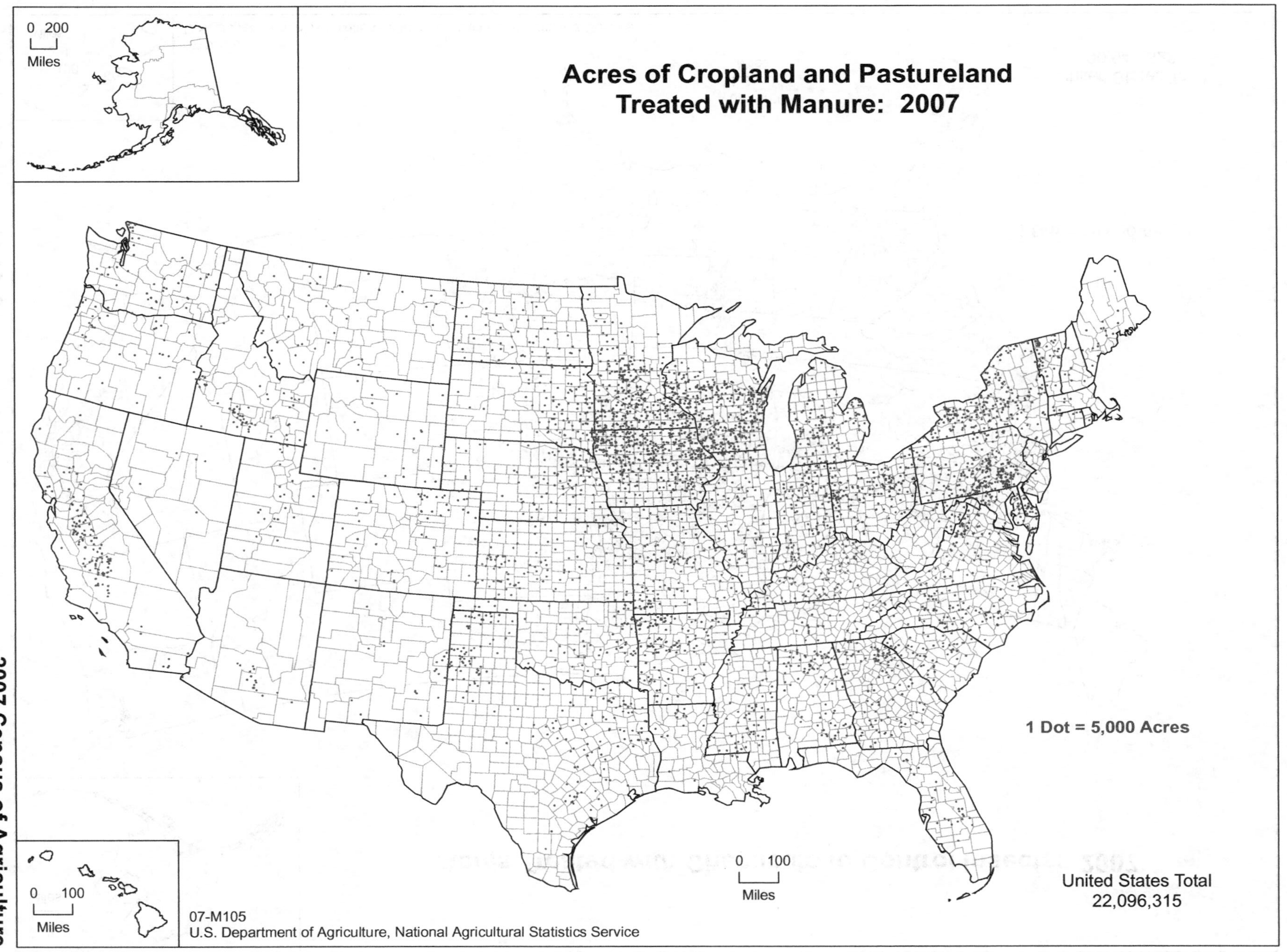


2007 Census of Agriculture

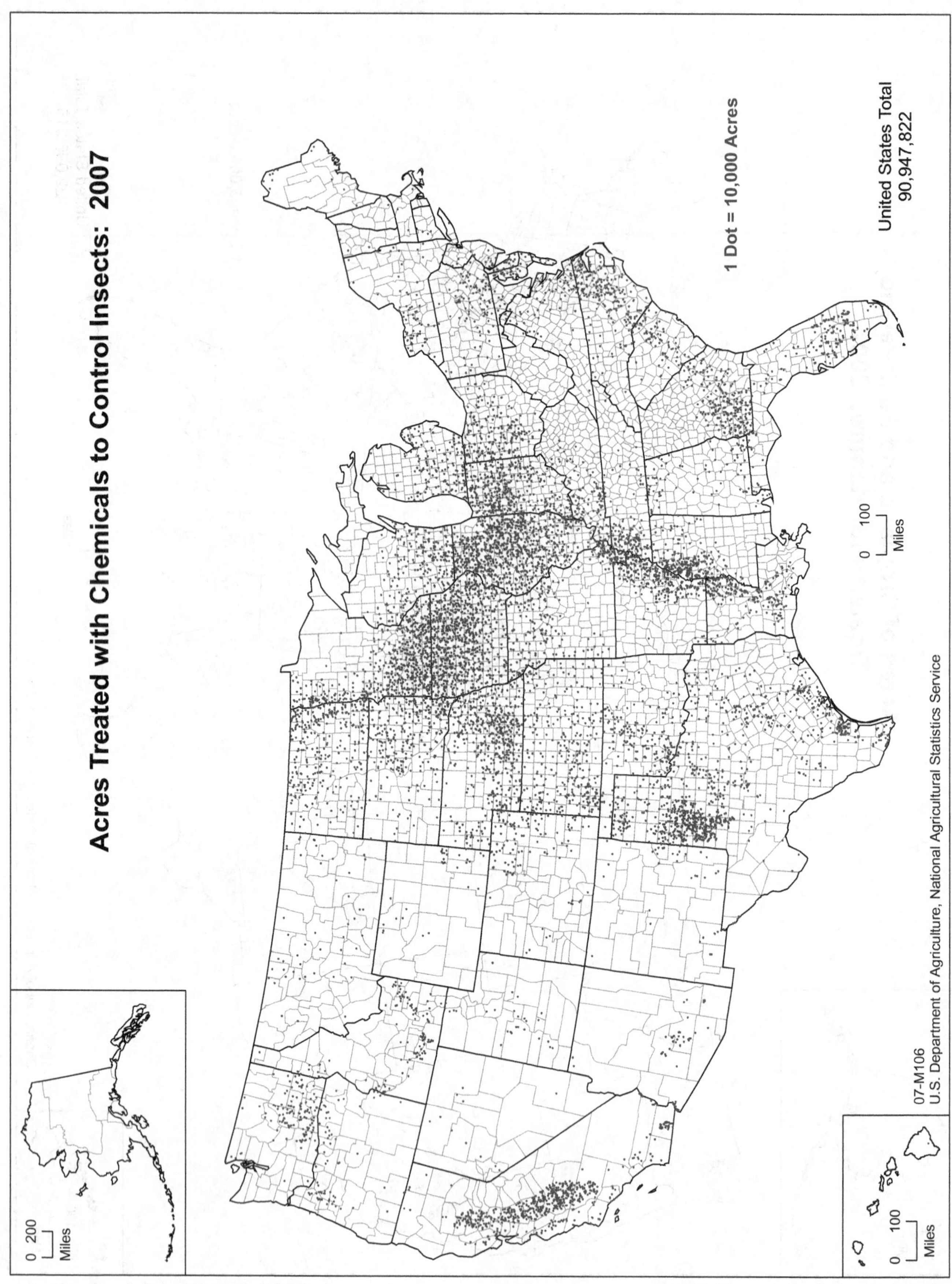
Acres Treated with Chemicals to Control Insects: 2007
1 Dot = 10,000 Acres
United States Total
90,947,822
0 100
Miles
0 200
Miles
0 100
Miles
07-M106
U.S. Department of Agriculture, National Agricultural Statistics Service

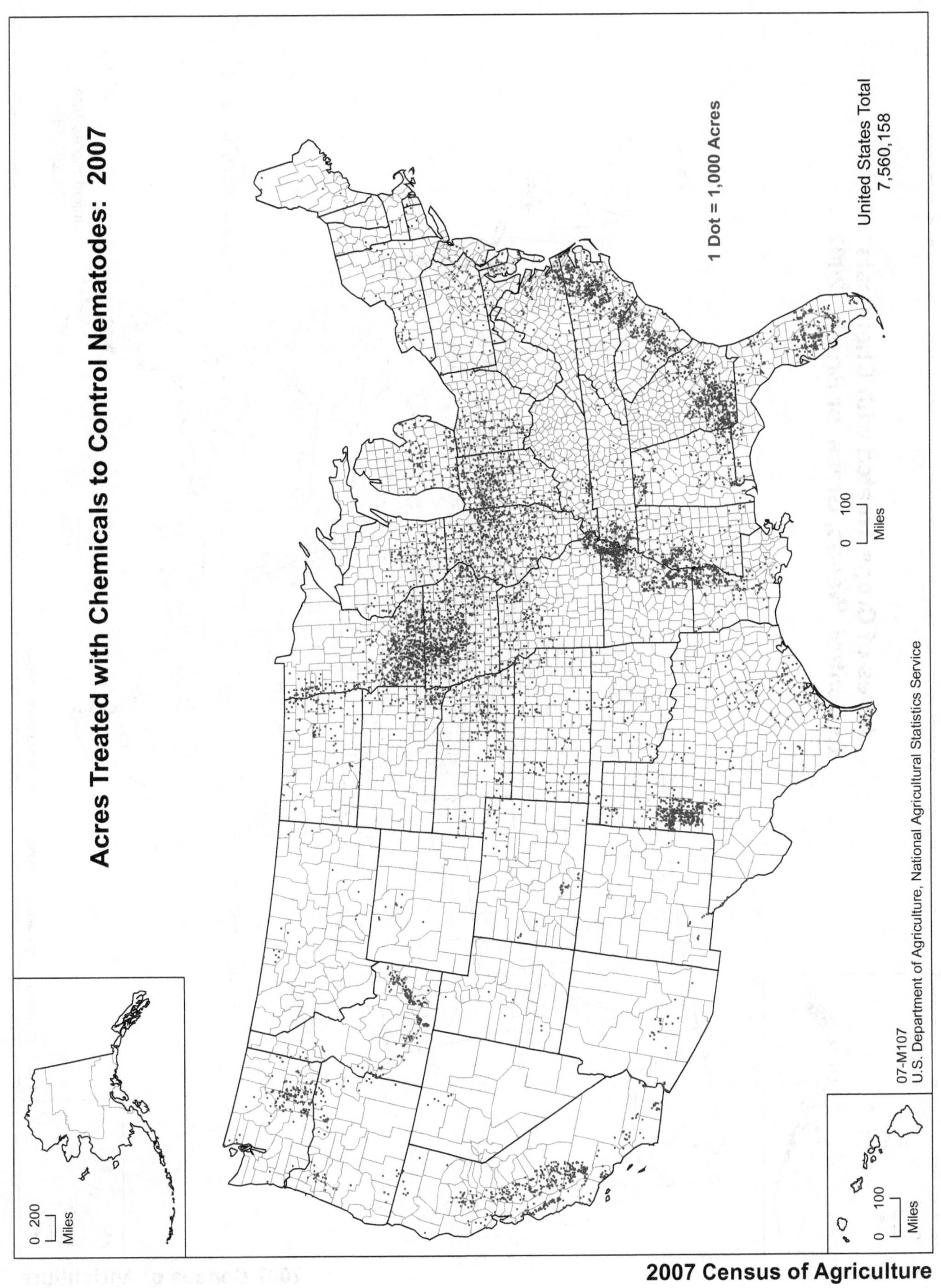


2007 Census of Agriculture

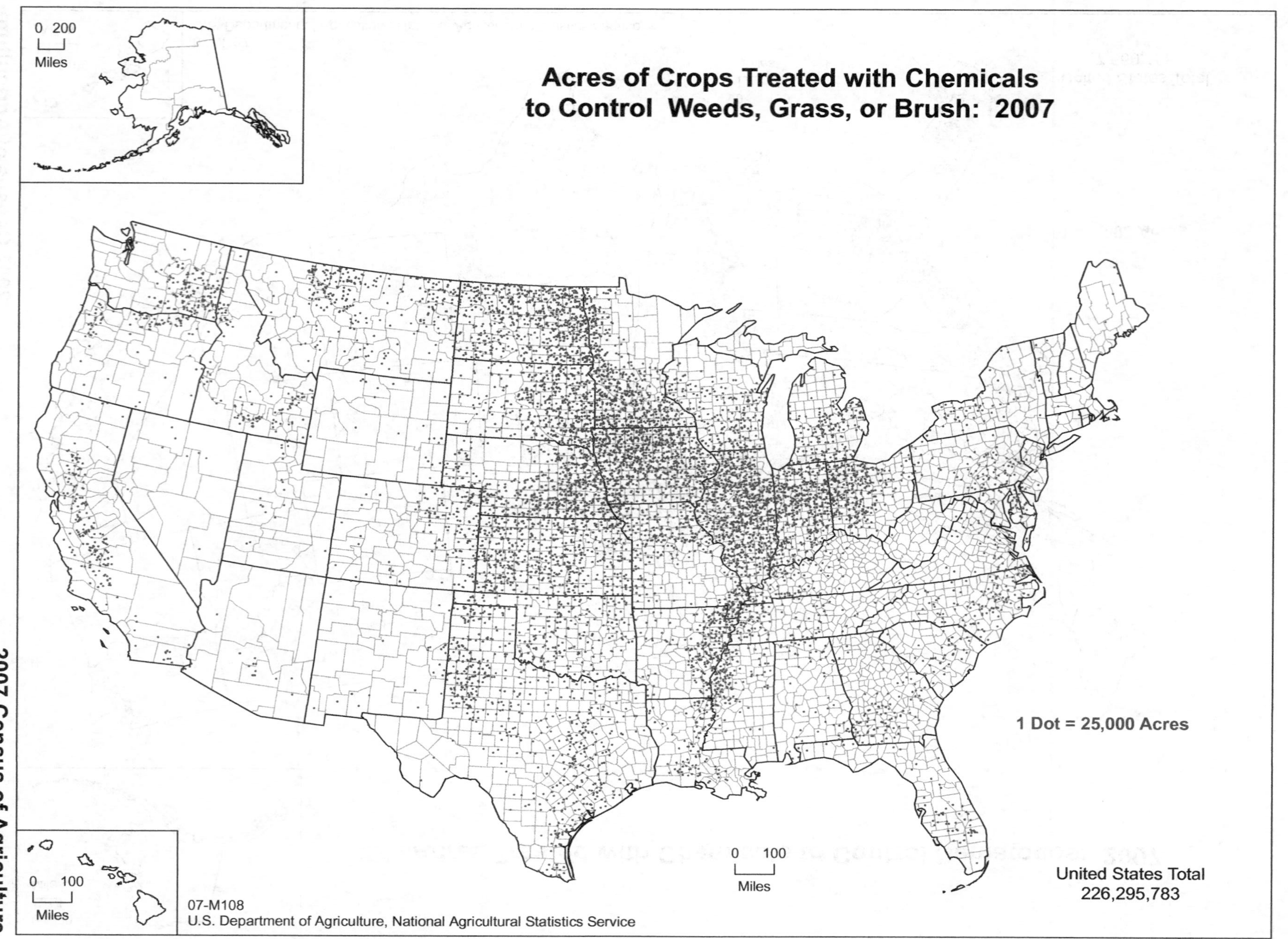


2007 Census of Agriculture

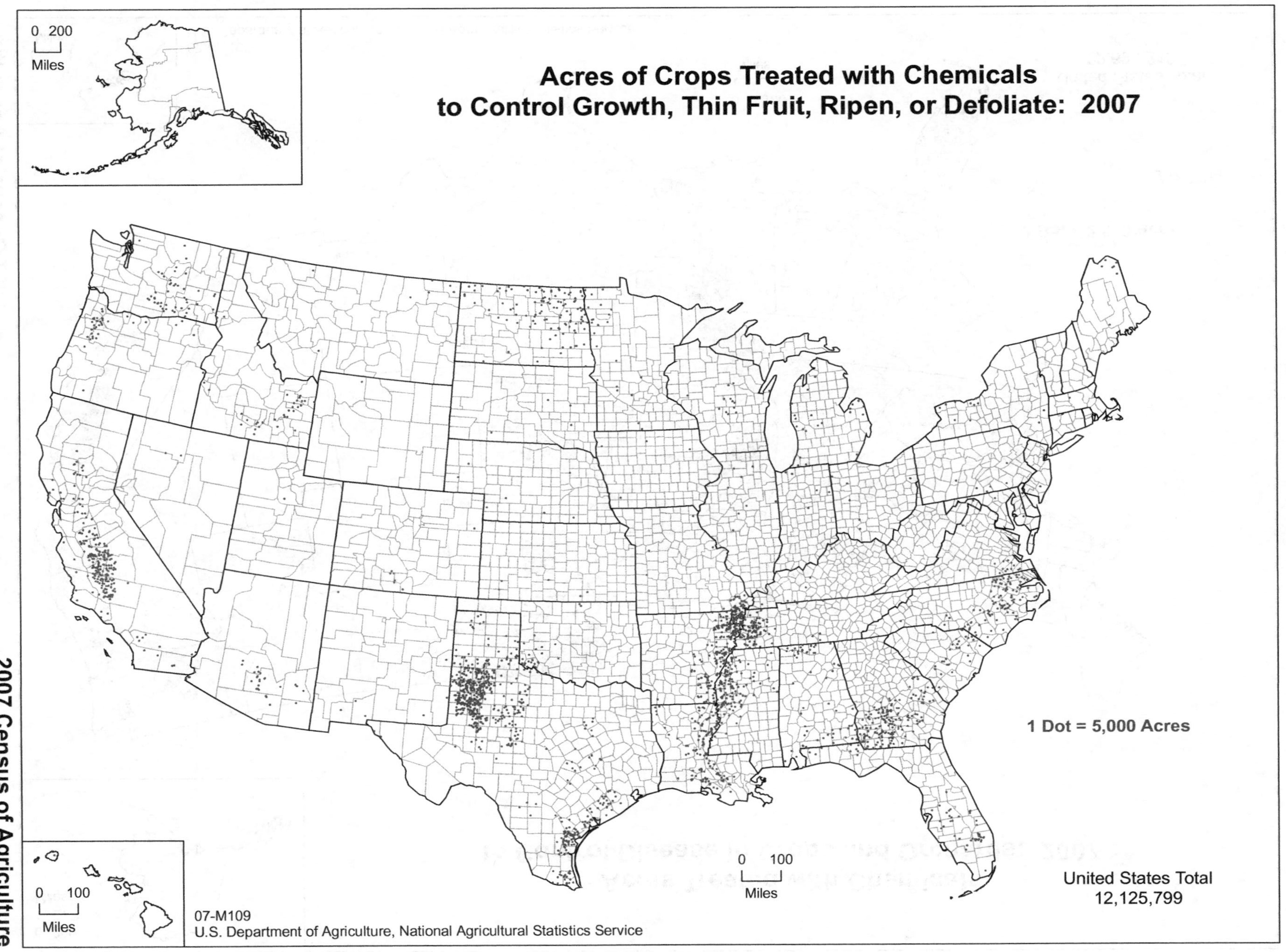
Acres of Crops Treated with Chemicals
to Control Growth, Thin Fruit, Ripen, or Defoliate: 2007
0 200
Miles
1 Dot = 5,000 Acres
0 100
Miles
United States Total
12,125,799
0 100
Miles
07-M109
U.S. Department of Agriculture, National Agricultural Statistics Service
2007 Census of Agriculture

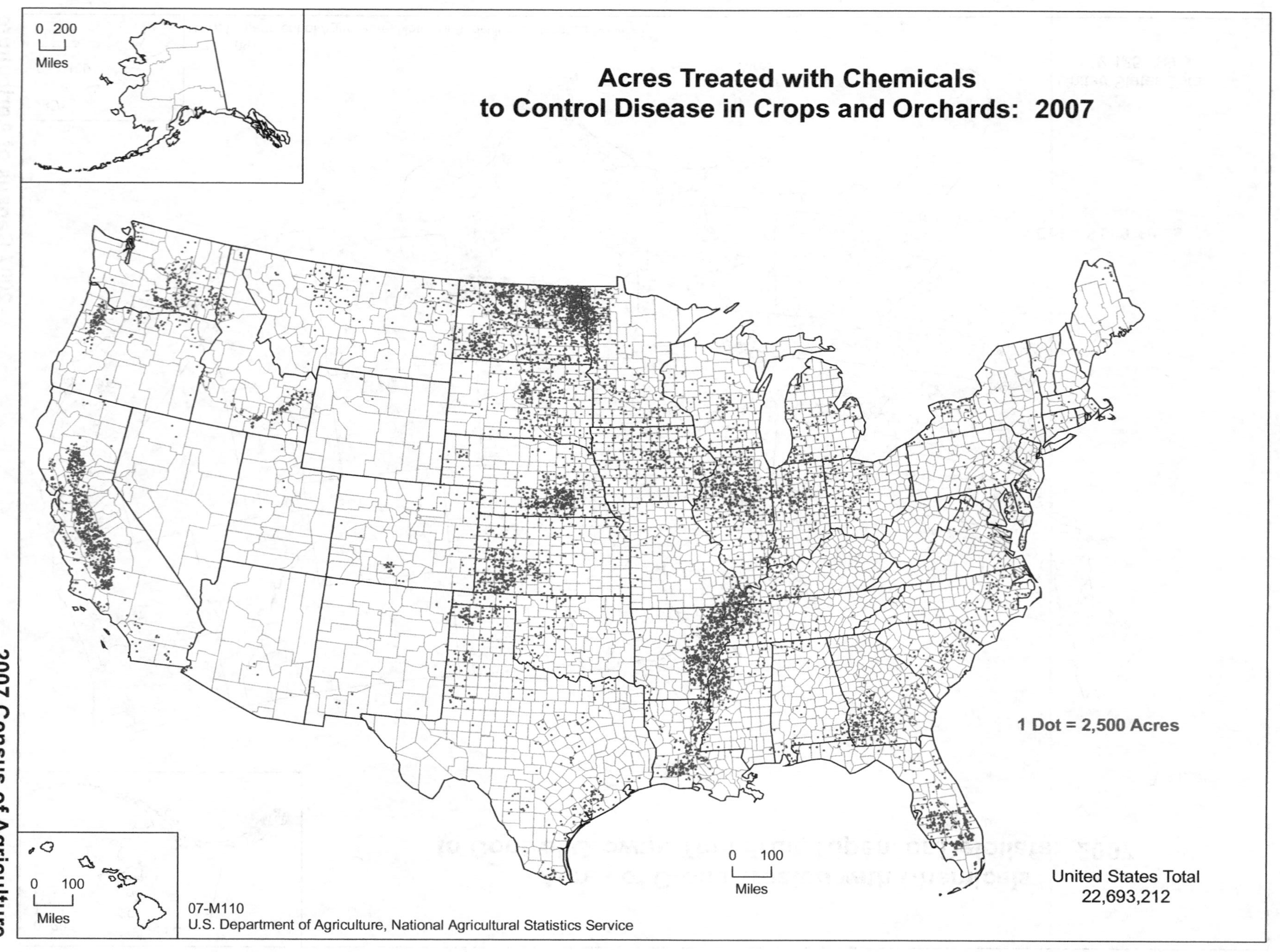


2007 Census of Agriculture

## Air Quality Index Report, 2008

| Metropolitan Statistical Area | Number of Days with AQI Data | Number of Days when Air Quality was... | | | | AQI Statistics | | Number of Days when AQI Pollutant was... | | | | | |
|---|---|---|---|---|---|---|---|---|---|---|---|---|---|
| | | Good | Moderate | Unhealthy for Sensitive Groups | Unhealthy | Max | Median | CO | $NO_2$ | $O_3$ | $SO_2$ | $PM_{2.5}$ | $PM_{10}$ |
| Akron, OH | 306 | 234 | 64 | 8 | 0 | 140 | 39 | 37 | - | 189 | 24 | 56 | - |
| Albany, GA | 148 | 84 | 63 | 1 | 0 | 107 | 47 | - | - | - | - | 139 | 9 |
| Albany-Schenectady-Troy, NY | 336 | 263 | 65 | 8 | 0 | 135 | 38 | 1 | - | 226 | 0 | 109 | - |
| Albuquerque, NM | 275 | 121 | 132 | 15 | 7 | 175 | 53 | 1 | 0 | 143 | - | 94 | 37 |
| Alexandria, LA | 102 | 94 | 8 | 0 | 0 | 64 | 29 | - | - | - | - | 102 | - |
| Allentown-Bethlehem-Easton, PA | 306 | 198 | 99 | 9 | 0 | 147 | 41 | 1 | 0 | 151 | 3 | 131 | 20 |
| Altoona, PA | 305 | 272 | 31 | 2 | 0 | 116 | 31 | 2 | 0 | 206 | 28 | - | 69 |
| Amarillo, TX | 303 | 293 | 10 | 0 | 0 | 79 | 21 | - | - | - | - | 303 | - |
| Anchorage, AK | 214 | 204 | 10 | 0 | 0 | 76 | 18 | 64 | - | - | - | 48 | 102 |
| Ann Arbor, MI | 218 | 170 | 45 | 3 | 0 | 122 | 42 | - | - | 165 | - | 53 | - |
| Appleton-Oshkosh-Neenah, WI | 241 | 211 | 29 | 1 | 0 | 133 | 36 | - | - | 192 | - | 49 | - |
| Asheville, NC | 333 | 254 | 78 | 1 | 0 | 111 | 40 | - | - | 123 | - | 210 | - |
| Athens, GA | 305 | 199 | 101 | 5 | 0 | 116 | 43 | - | - | 168 | - | 137 | - |
| Atlanta, GA | 305 | 128 | 149 | 24 | 4 | 182 | 54 | 0 | 0 | 136 | 0 | 167 | 2 |
| Atlantic-Cape May, NJ | 268 | 215 | 52 | 1 | 0 | 101 | 39 | - | - | 215 | 0 | 45 | 8 |
| Augusta-Aiken, GA | 366 | 238 | 120 | 8 | 0 | 135 | 43 | - | 0 | 236 | - | 130 | 0 |
| Austin-San Marcos, TX | 306 | 235 | 69 | 2 | 0 | 106 | 40 | 1 | 0 | 209 | - | 96 | 0 |
| Bakersfield, CA | 278 | 51 | 134 | 67 | 26 | 203 | 78 | 0 | 0 | 167 | - | 88 | 23 |
| Baltimore, MD | 305 | 175 | 107 | 19 | 4 | 166 | 46 | 0 | 0 | 195 | 0 | 110 | 0 |
| Bangor, ME | 81 | 76 | 5 | 0 | 0 | 78 | 24 | - | - | 0 | - | 72 | 9 |
| Baton Rouge, LA | 366 | 236 | 122 | 8 | 0 | 122 | 44 | 0 | 0 | 214 | 3 | 146 | 3 |
| Beaumont-Port Arthur, TX | 306 | 213 | 89 | 4 | 0 | 137 | 42 | 1 | 0 | 189 | 1 | 115 | - |
| Bellingham, WA | 305 | 300 | 5 | 0 | 0 | 58 | 18 | - | - | - | - | 305 | - |
| Benton Harbor, MI | 209 | 161 | 46 | 2 | 0 | 124 | 41 | - | - | 167 | - | 42 | - |
| Bergen-Passaic, NJ | 275 | 171 | 95 | 9 | 0 | 135 | 43 | 3 | 0 | 147 | 0 | 117 | 8 |
| Billings, MT | 321 | 318 | 3 | 0 | 0 | 87 | 12 | 136 | - | - | 36 | 149 | - |
| Biloxi-Gulfport-Pascagoula, MS | 326 | 232 | 88 | 6 | 0 | 140 | 41 | - | 0 | 122 | 0 | 204 | - |
| Birmingham, AL | 336 | 116 | 204 | 16 | 0 | 150 | 58 | 30 | - | 87 | 1 | 174 | 44 |
| Bismarck, ND | 335 | 330 | 4 | 1 | 0 | 102 | 31 | - | 0 | 303 | 0 | 17 | 15 |
| Bloomington-Normal, IL | 324 | 291 | 33 | 0 | 0 | 97 | 35 | - | - | 272 | - | 52 | - |
| Boise City, ID | 274 | 216 | 56 | 2 | 0 | 128 | 40 | 2 | - | 131 | - | 115 | 26 |
| Boston, MA-NH | 313 | 197 | 108 | 8 | 0 | 137 | 44 | 1 | 0 | 146 | 0 | 159 | 7 |
| Boulder-Longmont, CO | 297 | 221 | 72 | 4 | 0 | 111 | 44 | 5 | - | 257 | - | 32 | 3 |
| Brazoria, TX | 301 | 263 | 32 | 6 | 0 | 127 | 33 | - | 0 | 301 | - | - | - |
| Bremerton, WA | 305 | 260 | 45 | 0 | 0 | 89 | 23 | - | - | - | - | 305 | - |
| Bridgeport, CT | 275 | 229 | 39 | 6 | 1 | 169 | 37 | 43 | 0 | 166 | 21 | 42 | 3 |
| Brockton, MA | 101 | 90 | 11 | 0 | 0 | 80 | 29 | - | 0 | 0 | 0 | 101 | - |
| Brownsville-Harlingen-San Benito, TX | 303 | 227 | 76 | 0 | 0 | 97 | 38 | 1 | - | 102 | - | 198 | 2 |
| Buffalo-Niagara Falls, NY | 336 | 229 | 103 | 4 | 0 | 111 | 42 | 1 | 0 | 157 | 0 | 178 | - |
| Burlington, VT | 295 | 270 | 25 | 0 | 0 | 91 | 19 | 73 | 0 | 0 | - | 220 | 2 |
| Canton-Massillon, OH | 292 | 219 | 62 | 11 | 0 | 116 | 39 | 67 | - | 192 | - | 33 | - |
| Casper, WY | 91 | 91 | 0 | 0 | 0 | 43 | 15 | - | - | - | - | - | 91 |
| Cedar Rapids, IA | 336 | 253 | 82 | 1 | 0 | 110 | 40 | 3 | - | 150 | 6 | 177 | 0 |
| Champaign-Urbana, IL | 333 | 318 | 15 | 0 | 0 | 76 | 30 | - | - | 300 | - | 33 | - |
| Charleston, WV | 305 | 240 | 60 | 5 | 0 | 127 | 39 | - | - | 173 | 10 | 70 | 52 |
| Charleston-North Charleston, SC | 366 | 287 | 78 | 1 | 0 | 119 | 41 | 0 | 0 | 199 | 0 | 167 | 0 |
| Charlotte-Gastonia-Rock Hill, NC-SC | 336 | 177 | 130 | 26 | 3 | 164 | 49 | 1 | 0 | 181 | 0 | 154 | 0 |
| Charlottesville, VA | 204 | 173 | 28 | 3 | 0 | 116 | 42 | - | - | 170 | - | 34 | - |
| Chattanooga, TN-GA | 305 | 166 | 132 | 6 | 1 | 156 | 48 | - | - | 126 | - | 179 | 0 |
| Cheyenne, WY | 91 | 89 | 2 | 0 | 0 | 59 | 17 | - | - | - | - | 38 | 53 |
| Chicago, IL | 336 | 165 | 170 | 1 | 0 | 104 | 51 | 1 | 0 | 99 | 1 | 224 | 11 |
| Chico-Paradise, CA | 259 | 188 | 53 | 12 | 6 | 182 | 43 | 1 | 0 | 229 | - | 23 | 6 |
| Cincinnati, OH-KY-IN | 316 | 183 | 117 | 16 | 0 | 145 | 47 | 1 | 0 | 177 | 0 | 138 | 0 |

| Metropolitan Statistical Area | Number of Days with AQI Data | Number of Days when Air Quality was... | | | | AQI Statistics | | Number of Days when AQI Pollutant was... | | | | | |
|---|---|---|---|---|---|---|---|---|---|---|---|---|---|
| | | Good | Moderate | Unhealthy for Sensitive Groups | Unhealthy | Max | Median | CO | $NO_2$ | $O_3$ | $SO_2$ | $PM_{2.5}$ | $PM_{10}$ |
| Clarksville-Hopkinsville, TN | 315 | 222 | 93 | 0 | 0 | 90 | 42 | - | - | 155 | 8 | 152 | 0 |
| Cleveland-Lorain-Elyria, OH | 306 | 200 | 92 | 14 | 0 | 137 | 44 | 4 | 0 | 165 | 13 | 60 | 64 |
| Colorado Springs, CO | 295 | 227 | 66 | 2 | 0 | 111 | 43 | 1 | - | 270 | - | 16 | 8 |
| Columbia, SC | 366 | 234 | 124 | 8 | 0 | 145 | 45 | 0 | 0 | 217 | 0 | 130 | 19 |
| Columbus, GA-AL | 334 | 205 | 127 | 2 | 0 | 109 | 46 | - | - | 119 | 0 | 213 | 2 |
| Columbus, OH | 306 | 223 | 73 | 10 | 0 | 150 | 42 | 36 | - | 190 | 26 | 49 | 5 |
| Corpus Christi, TX | 302 | 241 | 58 | 3 | 0 | 109 | 38 | - | - | 155 | 0 | 144 | 3 |
| Corvallis, OR | 274 | 264 | 10 | 0 | 0 | 77 | 13 | - | - | - | - | 274 | - |
| Dallas, TX | 306 | 178 | 110 | 18 | 0 | 145 | 48 | 1 | 0 | 208 | 0 | 97 | 0 |
| Danbury, CT | 274 | 227 | 37 | 10 | 0 | 145 | 34 | 0 | 0 | 169 | 62 | 43 | 0 |
| Davenport-Moline-Rock Island, IA-IL | 363 | 230 | 130 | 3 | 0 | 144 | 42 | 28 | 0 | 108 | 0 | 138 | 89 |
| Dayton-Springfield, OH | 366 | 245 | 111 | 10 | 0 | 150 | 44 | 28 | - | 153 | 3 | 177 | 5 |
| Daytona Beach, FL | 274 | 263 | 11 | 0 | 0 | 84 | 33 | - | - | 251 | - | - | 23 |
| Decatur, AL | 276 | 232 | 44 | 0 | 0 | 80 | 40 | - | - | 209 | - | 67 | - |
| Decatur, IL | 335 | 282 | 53 | 0 | 0 | 81 | 37 | - | - | 190 | 5 | 140 | - |
| Denver, CO | 342 | 200 | 129 | 13 | 0 | 150 | 47 | 0 | 0 | 232 | 0 | 52 | 58 |
| Des Moines, IA | 336 | 269 | 67 | 0 | 0 | 90 | 38 | 1 | 0 | 137 | 0 | 194 | 4 |
| Detroit, MI | 311 | 195 | 109 | 6 | 1 | 173 | 42 | 15 | 0 | 119 | 16 | 136 | 25 |
| Dothan, AL | 272 | 247 | 25 | 0 | 0 | 74 | 36 | - | - | 204 | - | 68 | - |
| Dover, DE | 250 | 185 | 57 | 8 | 0 | 122 | 42 | - | - | 198 | - | 52 | - |
| Duluth-Superior, MN-WI | 306 | 268 | 37 | 1 | 0 | 122 | 36 | 1 | - | 200 | - | 103 | 2 |
| Dutchess County, NY | 332 | 305 | 23 | 4 | 0 | 132 | 32 | - | - | 332 | - | - | - |
| El Paso, TX | 306 | 146 | 144 | 16 | 0 | 147 | 52 | 1 | 0 | 135 | 0 | 117 | 53 |
| Elkhart-Goshen, IN | 220 | 167 | 52 | 1 | 0 | 121 | 41 | - | - | 166 | - | 54 | - |
| Elmira, NY | 335 | 318 | 15 | 2 | 0 | 124 | 31 | - | - | 332 | 3 | - | - |
| Erie, PA | 306 | 236 | 67 | 3 | 0 | 149 | 40 | 1 | 0 | 161 | 4 | 134 | 6 |
| Eugene-Springfield, OR | 275 | 217 | 52 | 6 | 0 | 120 | 34 | 1 | - | 132 | - | 142 | 0 |
| Evansville-Henderson, IN-KY | 316 | 201 | 111 | 4 | 0 | 119 | 44 | 0 | 0 | 143 | 2 | 170 | 1 |
| Fargo-Moorhead, ND-MN | 336 | 321 | 14 | 0 | 1 | 161 | 30 | 1 | 0 | 302 | 0 | 32 | 1 |
| Fayetteville, NC | 333 | 225 | 105 | 3 | 0 | 119 | 44 | 0 | - | 146 | - | 186 | 1 |
| Fayetteville-Springdale-Rogers, AR | 366 | 344 | 22 | 0 | 0 | 71 | 32 | - | - | 321 | - | 45 | - |
| Flagstaff, AZ-UT | 305 | 236 | 66 | 3 | 0 | 109 | 45 | - | - | 305 | - | 0 | 0 |
| Flint, MI | 214 | 173 | 39 | 2 | 0 | 111 | 39 | - | - | 174 | - | 40 | - |
| Florence, AL | 275 | 238 | 37 | 0 | 0 | 80 | 39 | - | - | 203 | - | 72 | - |
| Florence, SC | 105 | 74 | 31 | 0 | 0 | 69 | 38 | - | - | - | - | 105 | - |
| Fort Collins-Loveland, CO | 314 | 213 | 94 | 7 | 0 | 114 | 46 | 0 | - | 301 | - | 9 | 4 |
| Fort Lauderdale, FL | 275 | 250 | 24 | 1 | 0 | 145 | 32 | 1 | 0 | 180 | 0 | 79 | 15 |
| Fort Myers-Cape Coral, FL | 274 | 255 | 18 | 1 | 0 | 101 | 32 | - | - | 240 | - | 14 | 20 |
| Fort Pierce-Port St. Lucie, FL | 274 | 264 | 10 | 0 | 0 | 93 | 28 | - | - | 245 | - | 29 | - |
| Fort Smith, AR-OK | 274 | 231 | 43 | 0 | 0 | 77 | 36 | - | 0 | 171 | - | 53 | 50 |
| Fort Wayne, IN | 305 | 178 | 125 | 2 | 0 | 142 | 47 | 1 | - | 81 | - | 206 | 17 |
| Fort Worth-Arlington, TX | 306 | 192 | 87 | 26 | 1 | 156 | 46 | 1 | 0 | 228 | - | 77 | 0 |
| Fresno, CA | 281 | 77 | 133 | 56 | 15 | 207 | 70 | 0 | 0 | 195 | 0 | 77 | 9 |
| Gadsden, AL | 317 | 237 | 80 | 0 | 0 | 94 | 42 | - | - | 110 | - | 207 | - |
| Gainesville, FL | 274 | 258 | 15 | 1 | 0 | 101 | 33 | - | - | 225 | - | 24 | 25 |
| Galveston-Texas City, TX | 303 | 263 | 38 | 1 | 1 | 169 | 33 | - | 0 | 151 | 46 | 100 | 6 |
| Gary, IN | 335 | 223 | 110 | 2 | 0 | 116 | 42 | 1 | 0 | 82 | 0 | 219 | 33 |
| Goldsboro, NC | 283 | 215 | 68 | 0 | 0 | 93 | 36 | - | - | - | - | 283 | - |
| Grand Junction, CO | 296 | 220 | 76 | 0 | 0 | 98 | 44 | 11 | - | 135 | - | 20 | 130 |
| Grand Rapids-Muskegon-Holland, MI | 285 | 178 | 104 | 2 | 1 | 161 | 42 | 1 | - | 141 | 0 | 133 | 10 |
| Great Falls, MT | 320 | 316 | 4 | 0 | 0 | 59 | 9 | 189 | - | - | - | 131 | - |
| Greeley, CO | 297 | 233 | 62 | 2 | 0 | 104 | 42 | 13 | - | 242 | - | 38 | 4 |
| Green Bay, WI | 335 | 271 | 60 | 4 | 0 | 125 | 36 | - | - | 144 | 36 | 155 | - |
| Greensboro-Winston Salem-High Point, NC | 335 | 178 | 145 | 11 | 1 | 154 | 48 | 0 | 0 | 113 | 0 | 222 | 0 |
| Greenville, NC | 235 | 170 | 60 | 4 | 1 | 175 | 42 | - | - | 194 | - | 41 | - |

| Metropolitan Statistical Area | Number of Days with AQI Data | Number of Days when Air Quality was... | | | | AQI Statistics | | Number of Days when AQI Pollutant was... | | | | | |
|---|---|---|---|---|---|---|---|---|---|---|---|---|---|
| | | Good | Moderate | Unhealthy for Sensitive Groups | Unhealthy | Max | Median | CO | $NO_2$ | $O_3$ | $SO_2$ | $PM_{2.5}$ | $PM_{10}$ |
| Greenville-Spartanburg-Anderson, SC | 366 | 198 | 152 | 15 | 1 | 209 | 48 | 0 | 0 | 210 | 0 | 156 | 0 |
| Hagerstown, MD | 245 | 189 | 53 | 3 | 0 | 122 | 42 | - | - | 190 | - | 55 | - |
| Hamilton-Middletown, OH | 305 | 219 | 79 | 7 | 0 | 140 | 41 | - | - | 185 | 60 | 60 | 0 |
| Harrisburg-Lebanon-Carlisle, PA | 306 | 181 | 112 | 13 | 0 | 150 | 45 | 1 | 0 | 150 | 0 | 153 | 2 |
| Hartford, CT | 275 | 220 | 47 | 8 | 0 | 140 | 37 | 20 | 0 | 151 | 0 | 103 | 1 |
| Hattiesburg, MS | 90 | 61 | 29 | 0 | 0 | 89 | 38 | - | - | - | - | 90 | - |
| Hickory-Morganton-Lenoir, NC | 335 | 202 | 129 | 4 | 0 | 122 | 46 | - | - | 116 | - | 219 | 0 |
| Honolulu, HI | 275 | 269 | 6 | 0 | 0 | 89 | 20 | 1 | 0 | 134 | 0 | 28 | 112 |
| Houma, LA | 346 | 287 | 56 | 1 | 2 | 251 | 37 | - | - | 233 | - | 113 | - |
| Houston, TX | 306 | 122 | 162 | 20 | 2 | 177 | 54 | 2 | 0 | 140 | 0 | 136 | 28 |
| Huntington-Ashland, WV-KY-OH | 315 | 203 | 105 | 7 | 0 | 142 | 45 | - | 0 | 178 | 7 | 128 | 2 |
| Huntsville, AL | 366 | 277 | 87 | 2 | 0 | 122 | 41 | - | - | 251 | - | 115 | 0 |
| Indianapolis, IN | 366 | 206 | 155 | 5 | 0 | 129 | 47 | 0 | 0 | 119 | 0 | 246 | 1 |
| Iowa City, IA | 291 | 235 | 55 | 1 | 0 | 106 | 29 | - | - | - | - | 278 | 13 |
| Jackson, MS | 330 | 247 | 83 | 0 | 0 | 90 | 39 | - | - | 81 | - | 249 | - |
| Jackson, TN | 92 | 75 | 17 | 0 | 0 | 74 | 32 | - | - | - | - | 92 | 0 |
| Jacksonville, FL | 275 | 234 | 41 | 0 | 0 | 100 | 39 | 3 | 0 | 182 | 2 | 73 | 15 |
| Jamestown, NY | 335 | 287 | 43 | 5 | 0 | 132 | 38 | - | - | 325 | 2 | 8 | - |
| Janesville-Beloit, WI | 199 | 180 | 19 | 0 | 0 | 77 | 38 | - | - | 199 | - | - | - |
| Jersey City, NJ | 275 | 171 | 92 | 12 | 0 | 137 | 43 | 4 | 0 | 122 | 0 | 145 | 4 |
| Johnson City-Kingsport-Bristol, TN-VA | 314 | 249 | 62 | 3 | 0 | 145 | 42 | 13 | 0 | 213 | 32 | 56 | - |
| Johnstown, PA | 306 | 252 | 54 | 0 | 0 | 92 | 35 | 1 | 0 | 164 | 29 | 71 | 41 |
| Joplin, MO | 273 | 260 | 13 | 0 | 0 | 93 | 19 | - | - | - | - | - | 273 |
| Kalamazoo-Battle Creek, MI | 210 | 164 | 43 | 3 | 0 | 109 | 41 | - | - | 169 | - | 41 | - |
| Kansas City, MO-KS | 305 | 198 | 104 | 3 | 0 | 109 | 45 | 0 | 0 | 161 | 11 | 127 | 6 |
| Kenosha, WI | 243 | 196 | 46 | 1 | 0 | 106 | 38 | - | - | 195 | - | 48 | - |
| Knoxville, TN | 305 | 122 | 160 | 23 | 0 | 150 | 56 | - | - | 202 | 22 | 60 | 21 |
| Kokomo, IN | 88 | 71 | 17 | 0 | 0 | 90 | 33 | - | - | - | - | 88 | - |
| La Crosse, WI-MN | 294 | 262 | 31 | 1 | 0 | 108 | 36 | - | - | 242 | - | 52 | - |
| Lafayette, IN | 320 | 231 | 88 | 1 | 0 | 115 | 36 | - | - | - | - | 320 | - |
| Lafayette, LA | 366 | 315 | 48 | 3 | 0 | 122 | 33 | - | - | 337 | - | 29 | - |
| Lake Charles, LA | 366 | 279 | 85 | 2 | 0 | 101 | 39 | - | 0 | 236 | 0 | 130 | - |
| Lakeland-Winter Haven, FL | 274 | 253 | 18 | 3 | 0 | 106 | 36 | - | - | 237 | - | 18 | 19 |
| Lancaster, PA | 306 | 239 | 56 | 11 | 0 | 119 | 37 | 5 | 0 | 196 | 25 | 48 | 32 |
| Lansing-East Lansing, MI | 214 | 174 | 39 | 1 | 0 | 106 | 39 | - | - | 174 | - | 40 | - |
| Laredo, TX | 306 | 239 | 65 | 2 | 0 | 144 | 37 | 1 | - | 91 | - | 192 | 22 |
| Las Cruces, NM | 274 | 122 | 122 | 19 | 11 | 420 | 52 | - | 0 | 134 | - | 32 | 108 |
| Las Vegas, NV-AZ | 335 | 194 | 128 | 13 | 0 | 135 | 48 | 1 | 0 | 219 | 0 | 31 | 84 |
| Lawrence, MA-NH | 304 | 264 | 37 | 3 | 0 | 145 | 36 | 0 | 0 | 157 | 0 | 147 | 0 |
| Lawton, OK | 332 | 299 | 32 | 1 | 0 | 111 | 37 | - | - | 332 | - | - | - |
| Lewiston Auburn, ME | 72 | 67 | 4 | 1 | 0 | 120 | 27 | - | - | 0 | - | 63 | 9 |
| Lexington, KY | 315 | 226 | 88 | 1 | 0 | 106 | 42 | - | 0 | 180 | 4 | 130 | 1 |
| Lima.Oh | 305 | 280 | 25 | 0 | 0 | 90 | 34 | - | - | 212 | 76 | - | 17 |
| Lincoln, NE | 306 | 300 | 6 | 0 | 0 | 72 | 26 | 64 | - | 191 | - | 51 | - |
| Little Rock-North Little Rock, AR | 366 | 268 | 96 | 2 | 0 | 122 | 40 | 0 | 0 | 201 | 0 | 165 | 0 |
| Longview-Marshall, TX | 305 | 239 | 65 | 1 | 0 | 116 | 39 | - | 0 | 211 | 0 | 94 | 0 |
| Los Angeles-Long Beach, CA | 250 | 71 | 108 | 43 | 28 | 207 | 70 | 3 | 0 | 183 | 0 | 56 | 8 |
| Louisville, KY-IN | 316 | 167 | 140 | 9 | 0 | 140 | 49 | 1 | 0 | 151 | 5 | 159 | 0 |
| Lowell, MA-NH | 305 | 291 | 13 | 1 | 0 | 109 | 31 | 46 | 0 | 238 | 0 | 20 | 1 |
| Lubbock, TX | 31 | 30 | 1 | 0 | 0 | 53 | 26 | - | - | - | - | 31 | - |
| Lynchburg, VA | 108 | 90 | 18 | 0 | 0 | 73 | 30 | - | - | - | - | 108 | - |
| Macon, GA | 305 | 174 | 125 | 6 | 0 | 127 | 48 | - | - | 116 | 0 | 186 | 3 |
| Madison, WI | 330 | 249 | 80 | 1 | 0 | 112 | 38 | - | - | 193 | - | 135 | 2 |
| Manchester, NH | 280 | 270 | 10 | 0 | 0 | 64 | 26 | 35 | 0 | 183 | 44 | 0 | 18 |
| Mcallen-Edinburg-Mission, TX | 303 | 219 | 83 | 1 | 0 | 101 | 40 | - | - | 124 | - | 177 | 2 |
| Medford-Ashland, OR | 275 | 221 | 46 | 7 | 1 | 152 | 36 | 3 | - | 117 | - | 155 | 0 |

| Metropolitan Statistical Area | Number of Days with AQI Data | Number of Days when Air Quality was... | | | | AQI Statistics | | Number of Days when AQI Pollutant was... | | | | | |
|---|---|---|---|---|---|---|---|---|---|---|---|---|---|
| | | Good | Moderate | Unhealthy for Sensitive Groups | Unhealthy | Max | Median | CO | $NO_2$ | $O_3$ | $SO_2$ | $PM_{2.5}$ | $PM_{10}$ |
| Melbourne-Titusville-Palm Bay, FL | 274 | 256 | 17 | 1 | 0 | 104 | 34 | - | - | 242 | 0 | 17 | 15 |
| Memphis, TN-AR-MS | 366 | 201 | 155 | 9 | 1 | 164 | 47 | 0 | 0 | 153 | 0 | 212 | 1 |
| Merced, CA | 276 | 154 | 86 | 27 | 9 | 203 | 48 | - | 0 | 254 | - | 15 | 7 |
| Miami, FL | 275 | 249 | 22 | 4 | 0 | 139 | 33 | 3 | 0 | 184 | 0 | 80 | 8 |
| Middlesex-Somerset-Hunterdon, NJ | 275 | 182 | 74 | 18 | 1 | 501 | 42 | 1 | 0 | 193 | 0 | 81 | - |
| Milwaukee-Waukesha, WI | 335 | 270 | 61 | 3 | 1 | 151 | 36 | - | 0 | 240 | - | 74 | 21 |
| Minneapolis-St. Paul, MN-WI | 335 | 244 | 89 | 2 | 0 | 120 | 42 | 1 | 0 | 173 | 0 | 152 | 9 |
| Missoula, MT | 263 | 247 | 16 | 0 | 0 | 93 | 20 | 22 | - | - | - | 70 | 171 |
| Mobile, AL | 356 | 308 | 44 | 4 | 0 | 124 | 32 | - | - | 190 | - | 51 | 115 |
| Modesto, CA | 281 | 131 | 115 | 28 | 7 | 206 | 53 | 1 | 0 | 190 | - | 86 | 4 |
| Monmouth-Ocean, NJ | 275 | 205 | 52 | 17 | 1 | 161 | 41 | 1 | - | 209 | - | 65 | - |
| Monroe, LA | 364 | 354 | 10 | 0 | 0 | 60 | 28 | - | - | 303 | 0 | 61 | - |
| Montgomery, AL | 320 | 224 | 95 | 1 | 0 | 104 | 42 | - | - | 124 | - | 194 | 2 |
| Muncie, IN | 215 | 182 | 33 | 0 | 0 | 80 | 37 | - | - | 157 | - | 58 | - |
| Naples, FL | 274 | 262 | 12 | 0 | 0 | 100 | 28 | - | - | 274 | - | - | - |
| Nashua, NH | 222 | 202 | 20 | 0 | 0 | 84 | 35 | 0 | 0 | 164 | 0 | 58 | 0 |
| Nashville, TN | 366 | 216 | 139 | 11 | 0 | 137 | 47 | 0 | 0 | 146 | 0 | 220 | 0 |
| Nassau-Suffolk, NY | 335 | 247 | 73 | 14 | 1 | 164 | 42 | 0 | 0 | 223 | 0 | 112 | - |
| New Bedford, MA | 184 | 148 | 32 | 4 | 0 | 132 | 42 | - | - | 184 | 0 | 0 | - |
| New Haven-Meriden, CT | 275 | 208 | 64 | 3 | 0 | 122 | 36 | 5 | 0 | 117 | 0 | 153 | 0 |
| New London-Norwich, CT-RI | 274 | 211 | 57 | 6 | 0 | 147 | 37 | - | - | 158 | - | 116 | - |
| New Orleans, LA | 366 | 252 | 109 | 5 | 0 | 124 | 42 | - | 0 | 206 | 19 | 141 | 0 |
| New York, NY | 336 | 192 | 125 | 18 | 1 | 164 | 47 | 1 | 0 | 127 | 0 | 208 | - |
| Newark, NJ | 275 | 180 | 86 | 9 | 0 | 127 | 43 | 1 | 0 | 150 | 0 | 124 | - |
| Newburgh, NY-PA | 335 | 269 | 58 | 7 | 1 | 179 | 36 | - | - | 193 | - | 142 | - |
| Norfolk-Virginia Beach-Newport News, Va_Nc | 334 | 239 | 80 | 11 | 4 | 163 | 41 | 0 | 0 | 137 | 1 | 195 | 1 |
| Oakland, CA | 275 | 192 | 71 | 11 | 1 | 187 | 42 | 1 | 0 | 151 | 0 | 121 | 2 |
| Ocala, FL | 274 | 258 | 16 | 0 | 0 | 93 | 31 | - | - | 274 | - | - | - |
| Odessa-Midland, TX | 305 | 285 | 19 | 1 | 0 | 103 | 27 | - | - | - | - | 305 | - |
| Oklahoma City, OK | 336 | 266 | 65 | 5 | 0 | 122 | 41 | 1 | 0 | 300 | 0 | 34 | 1 |
| Olympia, WA | 303 | 267 | 33 | 3 | 0 | 123 | 28 | - | - | 144 | - | 159 | - |
| Omaha, NE-IA | 335 | 109 | 222 | 4 | 0 | 125 | 58 | 0 | - | 4 | 14 | 194 | 123 |
| Orange County, CA | 249 | 173 | 61 | 14 | 1 | 172 | 44 | 2 | 0 | 206 | 0 | 37 | 4 |
| Orlando, FL | 275 | 241 | 32 | 2 | 0 | 122 | 37 | 1 | 0 | 229 | 0 | 42 | 3 |
| Owensboro, KY | 306 | 222 | 83 | 1 | 0 | 111 | 41 | - | 0 | 171 | 0 | 135 | - |
| Panama City, FL | 274 | 228 | 43 | 3 | 0 | 124 | 36 | - | - | 234 | - | 34 | 6 |
| Parkersburg-Marietta, WV-OH | 305 | 242 | 59 | 4 | 0 | 150 | 40 | - | - | 188 | 65 | 52 | - |
| Pensacola, FL | 274 | 199 | 69 | 6 | 0 | 129 | 41 | - | 0 | 191 | 1 | 82 | - |
| Peoria-Pekin, IL | 336 | 245 | 91 | 0 | 0 | 85 | 41 | 7 | - | 158 | 9 | 162 | 0 |
| Philadelphia, PA-NJ | 306 | 153 | 123 | 26 | 4 | 260 | 51 | 1 | 0 | 160 | 0 | 143 | 2 |
| Phoenix-Mesa, AZ | 366 | 56 | 121 | 105 | 84 | 501 | 104 | 0 | 0 | 65 | 0 | 10 | 291 |
| Pittsburgh, PA | 306 | 116 | 155 | 34 | 1 | 152 | 56 | 1 | 0 | 79 | 0 | 224 | 2 |
| Pittsfield, MA | 305 | 239 | 63 | 2 | 1 | 161 | 36 | - | - | 121 | - | 184 | - |
| Pocatello, ID | 295 | 284 | 11 | 0 | 0 | 99 | 17 | - | - | - | 88 | 143 | 64 |
| Ponce, PR | 274 | 268 | 6 | 0 | 0 | 79 | 6 | - | - | - | 183 | 35 | 56 |
| Portland, ME | 352 | 330 | 21 | 1 | 0 | 101 | 31 | 2 | 0 | 317 | 0 | 26 | 7 |
| Portland-Vancouver, OR-WA | 305 | 258 | 41 | 6 | 0 | 122 | 32 | 0 | 0 | 186 | 0 | 118 | 1 |
| Portsmouth-Rochester, NH-ME | 284 | 258 | 23 | 3 | 0 | 119 | 32 | - | 0 | 174 | 61 | 49 | 0 |
| Providence-Fall River-Warwick, RI-MA | 306 | 230 | 70 | 6 | 0 | 135 | 38 | 2 | 0 | 152 | 0 | 151 | 1 |
| Provo-Orem, UT | 335 | 211 | 112 | 12 | 0 | 142 | 44 | 3 | 0 | 122 | - | 181 | 29 |
| Pueblo, CO | 129 | 121 | 8 | 0 | 0 | 83 | 25 | - | - | - | - | 52 | 77 |
| Racine, WI | 187 | 173 | 14 | 0 | 0 | 84 | 35 | - | - | 187 | - | - | - |
| Raleigh-Durham-Chapel Hill, NC | 336 | 202 | 122 | 11 | 1 | 170 | 45 | 2 | - | 166 | 0 | 168 | 0 |
| Rapid City, SD | 273 | 237 | 35 | 1 | 0 | 143 | 27 | - | - | - | - | 31 | 242 |
| Reading, PA | 306 | 237 | 54 | 15 | 0 | 132 | 36 | 2 | 0 | 196 | 40 | 46 | 22 |

| Metropolitan Statistical Area | Number of Days with AQI Data | Number of Days when Air Quality was... | | | | AQI Statistics | | Number of Days when AQI Pollutant was... | | | | | |
|---|---|---|---|---|---|---|---|---|---|---|---|---|---|
| | | Good | Moderate | Unhealthy for Sensitive Groups | Unhealthy | Max | Median | CO | $NO_2$ | $O_3$ | $SO_2$ | $PM_{2.5}$ | $PM_{10}$ |
| Redding, CA | 349 | 279 | 51 | 15 | 4 | 250 | 42 | - | - | 340 | - | 9 | 0 |
| Reno, NV | 275 | 208 | 59 | 6 | 2 | 180 | 45 | 2 | 0 | 249 | - | 14 | 10 |
| Richland-Kennewick-Pasco, WA | 305 | 281 | 24 | 0 | 0 | 69 | 22 | - | - | - | - | 201 | 104 |
| Richmond-Petersburg, VA | 337 | 234 | 86 | 14 | 3 | 187 | 40 | 0 | 0 | 148 | 2 | 185 | 2 |
| Riverside-San Bernardino, CA | 366 | 94 | 140 | 79 | 53 | 322 | 73 | 0 | 0 | 231 | 0 | 31 | 104 |
| Roanoke, VA | 336 | 240 | 94 | 2 | 0 | 123 | 42 | 3 | 0 | 117 | 0 | 211 | 5 |
| Rochester, MN | 304 | 204 | 95 | 5 | 0 | 145 | 42 | - | - | 132 | - | 156 | 16 |
| Rochester, NY | 336 | 274 | 58 | 4 | 0 | 127 | 36 | 1 | - | 231 | 0 | 104 | - |
| Rockford, IL | 336 | 311 | 24 | 1 | 0 | 113 | 31 | 6 | - | 277 | - | 53 | - |
| Rocky Mount, NC | 239 | 178 | 58 | 3 | 0 | 116 | 42 | - | - | 196 | - | 43 | - |
| Sacramento, CA | 336 | 195 | 84 | 37 | 20 | 204 | 47 | 4 | 0 | 299 | 0 | 12 | 21 |
| Saginaw-Bay City-Midland, MI | 85 | 71 | 14 | 0 | 0 | 77 | 23 | - | - | - | - | 85 | - |
| Saint Cloud, MN | 306 | 271 | 34 | 1 | 0 | 134 | 33 | 6 | - | 150 | - | 150 | - |
| Saint Joseph, MO | 242 | 102 | 140 | 0 | 0 | 85 | 53 | - | - | - | - | 230 | 12 |
| Saint Louis, MO-IL | 336 | 129 | 197 | 10 | 0 | 135 | 56 | 1 | 0 | 96 | 17 | 173 | 49 |
| Salem, OR | 268 | 248 | 18 | 2 | 0 | 124 | 25 | - | - | 143 | - | 125 | - |
| Salinas, CA | 306 | 293 | 13 | 0 | 0 | 80 | 35 | 1 | 0 | 289 | - | 2 | 14 |
| Salt Lake City-Ogden, UT | 335 | 171 | 142 | 21 | 1 | 153 | 50 | 3 | 0 | 136 | 0 | 126 | 70 |
| San Antonio, TX | 306 | 218 | 78 | 10 | 0 | 116 | 42 | 1 | 0 | 188 | 0 | 116 | 1 |
| San Diego, CA | 308 | 124 | 151 | 29 | 4 | 185 | 58 | 1 | 0 | 206 | 0 | 90 | 11 |
| San Francisco, CA | 275 | 222 | 53 | 0 | 0 | 91 | 37 | 1 | 0 | 109 | 0 | 165 | 0 |
| San Jose, CA | 275 | 192 | 76 | 6 | 1 | 154 | 42 | 1 | 0 | 170 | - | 104 | 0 |
| San Luis Obispo-Atascadero-Paso Robles, CA | 335 | 192 | 100 | 41 | 2 | 154 | 48 | - | 0 | 319 | 0 | 4 | 12 |
| Sanjuan-Bayamon, PR | 275 | 249 | 25 | 1 | 0 | 135 | 31 | 17 | 0 | 61 | 1 | 21 | 175 |
| Santa Barbara-Santa Maria-Lompoc, CA | 308 | 235 | 69 | 4 | 0 | 127 | 44 | 1 | 0 | 283 | 0 | 20 | 4 |
| Santa Cruz-Watsonville, CA | 306 | 293 | 13 | 0 | 0 | 100 | 35 | 1 | 0 | 294 | 0 | 1 | 10 |
| Santa Fe, NM | 274 | 248 | 26 | 0 | 0 | 84 | 42 | - | - | 274 | - | 0 | 0 |
| Santa Rosa, CA | 275 | 268 | 7 | 0 | 0 | 81 | 34 | 3 | 0 | 258 | - | 14 | 0 |
| Sarasota-Bradenton, FL | 274 | 251 | 18 | 5 | 0 | 135 | 35 | 0 | 0 | 242 | 0 | 8 | 24 |
| Savannah, GA | 305 | 229 | 75 | 1 | 0 | 119 | 39 | - | - | 121 | 7 | 171 | 6 |
| Scranton-Wilkes Barre-Hazleton, PA | 306 | 227 | 74 | 4 | 1 | 159 | 38 | 3 | 0 | 166 | 2 | 124 | 11 |
| Seattle-Bellevue-Everett, WA | 306 | 255 | 48 | 3 | 0 | 140 | 35 | 1 | - | 144 | 0 | 161 | - |
| Sharon, PA | 304 | 217 | 80 | 7 | 0 | 124 | 40 | - | - | 174 | 18 | 112 | - |
| Sheboygan, WI | 190 | 165 | 23 | 2 | 0 | 109 | 37 | - | - | 190 | - | - | - |
| Shreveport-Bossier City, LA | 366 | 300 | 64 | 2 | 0 | 114 | 37 | - | - | 252 | 0 | 111 | 3 |
| Sioux City, IA-NE | 103 | 87 | 16 | 0 | 0 | 94 | 30 | - | - | - | - | 82 | 21 |
| Sioux Falls, SD | 273 | 252 | 21 | 0 | 0 | 99 | 31 | - | 0 | 140 | 1 | 55 | 77 |
| South Bend, IN | 224 | 178 | 46 | 0 | 0 | 93 | 39 | - | 0 | 166 | - | 58 | - |
| Spokane, WA | 306 | 272 | 33 | 1 | 0 | 104 | 34 | 8 | - | 131 | - | 131 | 36 |
| Springfield, IL | 336 | 312 | 24 | 0 | 0 | 79 | 32 | 1 | - | 268 | 8 | 59 | - |
| Springfield, MA | 305 | 220 | 74 | 11 | 0 | 150 | 41 | 0 | 0 | 178 | 0 | 127 | 0 |
| Springfield, MO | 305 | 263 | 41 | 1 | 0 | 109 | 37 | 22 | 0 | 192 | 47 | 44 | 0 |
| Stamford-Norwalk, CT | 275 | 205 | 55 | 12 | 3 | 174 | 39 | 5 | 0 | 156 | 11 | 103 | 0 |
| State College, PA | 305 | 225 | 78 | 2 | 0 | 114 | 39 | - | 0 | 155 | 3 | 147 | - |
| Steubenville-Weirton, OH-WV | 310 | 226 | 78 | 6 | 0 | 122 | 42 | 0 | - | 149 | 66 | 78 | 17 |
| Stockton-Lodi, CA | 279 | 184 | 78 | 13 | 4 | 169 | 45 | 0 | 0 | 212 | - | 51 | 16 |
| Syracuse, NY | 336 | 301 | 32 | 3 | 0 | 140 | 35 | 1 | - | 322 | 0 | 13 | - |
| Tacoma, WA | 305 | 255 | 45 | 5 | 0 | 129 | 33 | - | - | 192 | - | 113 | - |
| Tallahassee, FL | 274 | 235 | 36 | 3 | 0 | 109 | 35 | - | - | 222 | - | 52 | - |
| Tampa-St. Petersburg-Clearwater, FL | 275 | 218 | 50 | 6 | 1 | 161 | 42 | 1 | 0 | 191 | 1 | 40 | 42 |
| Terre Haute, IN | 305 | 218 | 87 | 0 | 0 | 100 | 41 | - | - | 100 | 30 | 175 | 0 |
| Texarkana, TX-Texarkana, AR | 46 | 36 | 10 | 0 | 0 | 68 | 33 | - | - | - | - | 46 | - |
| Toledo, OH | 305 | 195 | 106 | 4 | 0 | 109 | 45 | - | - | 158 | - | 126 | 21 |
| Topeka, KS | 305 | 289 | 16 | 0 | 0 | 77 | 36 | - | - | 257 | - | 39 | 9 |

| Metropolitan Statistical Area | Number of Days with AQI Data | Number of Days when Air Quality was... | | | | AQI Statistics | | Number of Days when AQI Pollutant was... | | | | | |
|---|---|---|---|---|---|---|---|---|---|---|---|---|---|
| | | Good | Moderate | Unhealthy for Sensitive Groups | Unhealthy | Max | Median | CO | $NO_2$ | $O_3$ | $SO_2$ | $PM_{2.5}$ | $PM_{10}$ |
| Trenton, NJ | 274 | 216 | 48 | 10 | 0 | 150 | 38 | - | 0 | 225 | - | 47 | 2 |
| Tucson, AZ | 366 | 262 | 102 | 2 | 0 | 111 | 44 | 0 | 0 | 311 | 0 | 2 | 53 |
| Tulsa, OK | 336 | 230 | 98 | 7 | 1 | 161 | 45 | 1 | 0 | 217 | 7 | 107 | 4 |
| Tuscaloosa, AL | 267 | 240 | 27 | 0 | 0 | 87 | 36 | - | - | 195 | - | 72 | - |
| Tyler, TX | 304 | 274 | 29 | 0 | 1 | 164 | 36 | - | 0 | 304 | - | - | - |
| Utica-Rome, NY | 335 | 271 | 62 | 2 | 0 | 119 | 38 | - | - | 221 | 0 | 114 | - |
| Vallejo-Fairfield-Napa, CA | 366 | 262 | 91 | 10 | 3 | 169 | 41 | 0 | 0 | 210 | 0 | 156 | 0 |
| Ventura, CA | 305 | 167 | 107 | 31 | 0 | 150 | 47 | - | 0 | 294 | - | 10 | 1 |
| Victoria, TX | 299 | 287 | 12 | 0 | 0 | 74 | 29 | - | - | 299 | - | - | - |
| Vineland-Millville-Bridgeton, NJ | 272 | 199 | 65 | 8 | 0 | 119 | 41 | - | - | 214 | 0 | 58 | - |
| Visalia-Tulare-Porterville, CA | 305 | 110 | 87 | 84 | 24 | 203 | 77 | - | 0 | 283 | - | 21 | 1 |
| Waco, TX | 303 | 254 | 48 | 1 | 0 | 111 | 39 | 1 | 0 | 202 | 0 | 100 | - |
| Washington, DC-MD-VA-WV | 333 | 196 | 116 | 18 | 3 | 172 | 45 | 0 | 0 | 194 | 1 | 137 | 1 |
| Waterbury, CT | 275 | 249 | 24 | 2 | 0 | 112 | 4 | 141 | 0 | 0 | 42 | 90 | 2 |
| Waterloo-Cedar Falls, IA | 107 | 94 | 12 | 1 | 0 | 102 | 28 | - | - | - | - | 82 | 25 |
| Wausau, WI | 217 | 207 | 10 | 0 | 0 | 93 | 36 | - | - | 217 | - | - | - |
| West Palm Beach-Boca Raton, FL | 274 | 253 | 20 | 1 | 0 | 145 | 31 | 0 | 0 | 206 | 0 | 60 | 8 |
| Wheeling, WV-OH | 305 | 190 | 114 | 1 | 0 | 103 | 44 | - | - | 123 | 4 | 178 | - |
| Wichita, KS | 305 | 274 | 30 | 1 | 0 | 101 | 37 | 0 | 0 | 243 | - | 32 | 30 |
| Williamsport, PA | 305 | 278 | 21 | 6 | 0 | 135 | 29 | - | - | 212 | 78 | - | 15 |
| Wilmington, NC | 333 | 285 | 43 | 4 | 1 | 151 | 31 | - | - | 110 | 64 | 159 | 0 |
| Wilmington-Newark, DE-MD | 306 | 194 | 99 | 12 | 1 | 179 | 44 | 1 | 0 | 189 | 1 | 115 | 0 |
| Worcester, MA | 305 | 243 | 54 | 7 | 1 | 172 | 37 | 9 | 0 | 153 | 0 | 138 | 5 |
| Yakima, WA | 305 | 266 | 38 | 1 | 0 | 110 | 26 | - | - | - | - | 300 | 5 |
| Yolo, CA | 366 | 319 | 38 | 8 | 1 | 156 | 36 | - | 0 | 347 | - | 10 | 9 |
| York, PA | 305 | 234 | 61 | 9 | 1 | 151 | 39 | 4 | 0 | 188 | 1 | 51 | 61 |
| Youngstown-Warren, OH | 310 | 203 | 99 | 8 | 0 | 123 | 43 | - | - | 129 | 5 | 170 | 6 |
| Yuba City, CA | 274 | 202 | 59 | 7 | 6 | 195 | 37 | - | 0 | 207 | - | 58 | 9 |
| Yuma, AZ | 329 | 211 | 104 | 13 | 1 | 245 | 45 | - | - | 132 | - | 9 | 188 |

***Notes:** Dashes indicates data was not available; The Air Quality Index (AQI) is an index for reporting daily air quality. It tells you how clean or polluted your air is, and what associated health concerns you should be aware of. The AQI focuses on health effects that can happen within a few hours or days after breathing polluted air. EPA uses the AQI for six major air pollutants regulated by the Clean Air Act: CO (carbon monoxide), NO (nitrogen dioxide), O (ground-level ozone), SO (sulfur dioxide), PM (particulate matter 10 - particles with diameters of 10 micrometers or less), PM (particulate matter 2.5 - particles with diameters of 2.5 micrometers or less). For each of these pollutants, EPA has established national air quality standards to protect against harmful health effects.*

*The AQI runs from 0 to 500. The higher the AQI value, the greater the level of air pollution and the greater the health danger. For example, an AQI value of 50 represents good air quality and little potential to affect public health, while an AQI value over 300 represents hazardous air quality. An AQI value of 100 generally corresponds to the national air quality standard for the pollutant, which is the level EPA has set to protect public health. So, AQI values below 100 are generally thought of as satisfactory. When AQI values are above 100, air quality is considered to be unhealthy-at first for certain sensitive groups of people, then for everyone as AQI values get higher. Each category corresponds to a different level of health concern. For example, when the AQI for a pollutant is between 51 and 100, the health concern is "Moderate." Here are the six levels of health concern and what they mean:*

*"Good" The AQI value for your community is between 0 and 50. Air quality is considered satisfactory and air pollution poses little or no risk.*
*"Moderate" The AQI for your community is between 51 and 100. Air quality is acceptable; however, for some pollutants there may be a moderate health concern for a very small number of individuals. For example, people who are unusually sensitive to ozone may experience respiratory symptoms.*
*"Unhealthy for Sensitive Groups" Certain groups of people are particularly sensitive to the harmful effects of certain air pollutants. This means they are likely to be affected at lower levels than the general public. For example, children and adults who are active outdoors and people with respiratory disease are at greater risk from exposure to ozone, while people with heart disease are at greater risk from carbon monoxide. Some people may be sensitive to more than one pollutant. When AQI values are between 101 and 150, members of sensitive groups may experience health effects. The general public is not likely to be affected when the AQI is in this range.*
*"Unhealthy" AQI values are between 151 and 200. Everyone may begin to experience health effects. Members of sensitive groups may experience more serious health effects.*
*"Very Unhealthy" AQI values between 201 and 300 trigger a health alert, meaning everyone may experience more serious health effects.*
*"Hazardous" AQI values over 300 trigger health warnings of emergency conditions. The entire population is more likely to be affected.*

***Source:** U.S. Environmental Protection Agency, Office of Air and Radiation, Air Quality Index Report, 2008*

## *Brownfields Prevalence: Number of Sites and Estimated Acreage, 1993/2010*

| City | State | Est. Brownfield Sites in 1993 | Est. Brownfield Sites in 2010 | Est. Avg. Size of Brownfield Sites in 1993 (Acres) | Est. Avg. Size of Brownfield Sites in 2010 (Acres) |
|---|---|---|---|---|---|
| Akron | Ohio | 3 | 6 | 6 | 8 |
| Alameda | California | 250 | 270 | 0.15 | 0.2 |
| Allentown | Pennsylvania | 20 | 10 | 7 | 7 |
| Arlington | Texas | * | 1650 | * | * |
| Arlington Heights | Illinois | * | * | * | * |
| Asheville | North Carolina | 300 | 300 | 2 | 2 |
| Atlanta | Georgia | * | 164 | * | 9.3 |
| Babylon | New York | * | 350 | * | 0.24 |
| Baltimore | Maryland | 1000 | 1000 | * | * |
| Bartlett | Illinois | 4 | 1 | 1 | 0.5 |
| Bessemer | Alabama | 40 | 30 | 10 | 8 |
| Biloxi | Mississippi | 9 | 20 | 5.5 | 5 |
| Binghamton | New York | 23 | 21 | 5 | 5 |
| Boston | Massachusetts | 1250 | 1400 | 0.25 | 0.25 |
| Bridgeport | Connecticut | 500 | 450 | 3 | 3 |
| Camden | New Jersey | 500 | 485 | 3 | 3 |
| Cape Coral | Florida | * | * | * | * |
| Carson | California | * | 217 | * | 4.5 |
| Charleston | South Carolina | 75 | 100 | 10 | 5 |
| Chattanooga | Tennessee | * | * | * | * |
| Chicago Heights | Illinois | 52 | 30 | 1.5 | 2.5 |
| Cincinnati | Ohio | * | 15 | * | 28.116 |
| Clearwater | Florida | * | 244 | * | 1.5 |
| Clovis | New Mexico | 1 | 1 | 1 | 1 |
| Columbia | South Carolina | 11 | 35 | 1 | 1 |
| Columbus | Ohio | 18 | 28 | 30 | 10 |
| Dallas | Texas | * | * | * | * |
| Dayton | Ohio | * | 30 | * | 10 |
| Dearborn | Michigan | * | 20 | * | 50 |
| Denton | Texas | * | * | * | * |
| Dubuque | Iowa | * | 38 | * | 3 |
| Durham | North Carolina | * | * | * | * |
| Elizabeth | New Jersey | * | 160 | * | 1.5 |
| Evansville | Indiana | 350 | 300 | 1.5 | 1 |
| Fayetteville | North Carolina | 100 | 125 | 250 | 300 |
| Florence | Alabama | * | 20 | * | 5 |
| Fort Myers | Florida | * | 50 | * | 0.93 |
| Fresno | California | * | * | * | * |
| Frisco | Texas | * | * | * | * |
| Glendale Heights | Illinois | 2 | 2 | 1 | 1 |
| Gloucester | Massachusetts | * | * | * | * |
| Grand Rapids | Michigan | * | * | * | 1 |
| Green Bay | Wisconsin | 31 | 15 | * | 1.5 |
| Greensboro | North Carolina | * | 30 | * | 2 |
| Gulfport | Mississippi | 2000 | 1709 | * | 6.8 |
| Hagerstown | Maryland | 36 | 33 | * | * |
| Hartford | Connecticut | 20 | 35 | 5 | 3.5 |
| Hoffman Estates | Illinois | * | * | * | * |
| Honolulu | Hawaii | * | * | * | * |
| Houston | Texas | 1000 | 2000 | * | 3 |
| Huntsville | Alabama | * | * | * | * |
| Indianapolis | Indiana | * | 1617 | * | 4.03 |
| Irvine | California | * | * | * | * |
| Isabela | Puerto Rico | 4 | 4 | 15 | 15 |
| Jackson | Mississippi | * | 250 | * | 0.57 |
| Kalamazoo | Michigan | 200 | 340 | 2.5 | 2.5 |
| Kansas City | Kansas | * | * | * | * |
| Lakeland | Florida | 1 | 2 | 85 | * |
| Las Vegas | Nevada | 3 | 2 | 69 | 57.5 |
| Lewiston | Maine | 18 | 11 | * | * |
| Louisville | Kentucky | * | 67 | * | 8 |
| Memphis | Tennessee | 23 | 149 | * | 23 |
| Miami Beach | Florida | * | 1 | * | 2.83 |

| | | | | | |
|---|---|---|---|---|---|
| Milwaukee | Wisconsin | 400 | 300 | 2 | 1 |
| Monroe | Louisiana | 9 | 8 | 3 | 2.5 |
| Naugatuck | Connecticut | 15 | 10 | 50 | 150 |
| New Orleans | Louisiana | 500 | 450 | 1 | 1 |
| Niagara Falls | New York | 1750 | 1750 | 1.5 | 1.5 |
| North Chicago | Illinois | * | * | * | * |
| North Little Rock | Arkansas | 50 | 50 | 1 | 1 |
| North Miami | Florida | * | 0 | * | * |
| Norwalk | Connecticut | * | 261 | * | 0.5 |
| Orland Park | Illinois | * | 1 | * | 20 |
| Palm Beach Gardens | Florida | * | 0 | * | * |
| Pawtucket | Rhode Island | 90 | 88 | 0.5 | 0.5 |
| Pembroke Pines | Florida | * | * | * | * |
| Piscataway | New Jersey | 5 | 8 | 7 | 7 |
| Portland | Oregon | * | 515 | * | 2.8 |
| Racine | Wisconsin | * | 300 | * | * |
| Radnor | Pennsylvania | * | * | * | * |
| Saint Louis | Missouri | * | 304 | * | 0.3 |
| Saint Paul | Minnesota | 225 | 200 | 40 | 10 |
| San Juan | Puerto Rico | * | 130 | * | 5 |
| Sanford | North Carolina | 60 | 28 | 12 | 4.2 |
| Shreveport | Louisiana | * | 460 | * | 5 |
| Somerville | Massachusetts | 750 | 500 | 0.5 | 0.5 |
| Southgate | Michigan | * | 1 | * | * |
| Syracuse | New York | * | 163 | * | 8.2 |
| Tacoma | Washington | * | 322 | * | 0.25 |
| Tallahassee | Florida | * | 1 | * | 3 |
| Toa Baja | Puerto Rico | * | 1 | * | 0.33 |
| Tucson | Arizona | * | 5200 | * | * |
| Vista | California | * | 250 | * | 1 |
| Waco | Texas | * | 100 | * | 3 |
| Warren | Ohio | 5 | 5 | 7 | 7 |
| West Palm Beach | Florida | * | 4236 | * | 1 |
| Wilson | North Carolina | * | 75 | * | 2 |
| York | Pennsylvania | 121 | 70 | 0.526 | 0.573 |
| Zanesville | Ohio | * | * | * | * |

| *99 Respondents* | *Est. Brownfield Sites in 1993* | *Est. Brownfield Sites in 2010* |
|---|---|---|
| *Totals* | 11,824 | 29,624 |

*Source: The U.S. Conference of Mayors, Recycling America's Land: A National Report on Brownfields Redevelopment (1993-2010), November 2010, Volume IX*

### *Brownfields Redeveloped Since 1993: Number of Sites and Estimated Acreage*

| City | State | Since 1993, Sites redeveloped | Since 1993, Acres Developed | In Progress Sites | In Progress Acres |
|---|---|---|---|---|---|
| Akron | Ohio | 12 | 60 | 5 | 50 |
| Alameda | California | 10 | 2 | 2 | 2 |
| Allentown | Pennsylvania | 10 | 62 | 4 | 31 |
| Arlington | Texas | 35 | 380 | 6 | 10 |
| Arlington Heights | Illinois | * | * | * | * |
| Asheville | North Carolina | 1 | 3 | 3 | 14 |
| Atlanta | Georgia | * | * | 15 | 266 |
| Babylon | New York | * | * | 20 | * |
| Baltimore | Maryland | 45 | * | 5 | 40 |
| Bartlett | Illinois | 3 | 10 | 1 | 0.5 |
| Biloxi | Mississippi | * | * | * | * |
| Binghamton | New York | 2 | 35 | * | * |
| Boston | Massachusetts | * | * | 1 | 2.43 |
| Bridgeport | Connecticut | 50 | 150 | 15 | 95 |
| Camden | New Jersey | 15 | 20 | 3 | 37 |
| Cape Coral | Florida | * | * | * | * |
| Carson | California | * | * | 3 | 162 |
| Charleston | South Carolina | 6 | 50 | 200 | 800 |
| Chattanooga | Tennessee | * | * | * | * |
| Chicago Heights | Illinois | 2 | 11 | 1 | 8 |
| Cincinnati | Ohio | 5 | 31 | 5 | 139 |
| Bessemer | Alabama | 10 | 200 | 2 | 25 |
| Clearwater | Florida | 15 | 30 | 10 | 18 |
| Clovis | New Mexico | 1 | 1 | 1 | 1 |
| Columbia | South Carolina | 46 | 46 | 35 | 35 |
| Columbus | Ohio | 19 | 200 | 7 | 100 |
| Dallas | Texas | 53 | 1046 | 5 | 55 |
| Dayton | Ohio | 10 | 150 | 3 | 130 |
| Dearborn | Michigan | 9 | 100 | 2 | 75 |
| Denton | Texas | * | * | 1 | 2 |
| Dubuque | Iowa | 3 | 80 | 4 | 85 |
| Durham | North Carolina | * | * | * | * |
| Elizabeth | New Jersey | 12 | 190 | 9 | 30 |
| Evansville | Indiana | 8 | 75 | 3 | 18 |
| Fayetteville | North Carolina | 1 | 15 | 1 | 100 |
| Florence | Alabama | * | * | 2 | 25 |
| Fort Myers | Florida | 5 | 19 | * | * |
| Fresno | California | * | * | 1 | 3 |
| Frisco | Texas | * | * | * | * |
| Glendale Heights | Illinois | 1 | 1 | * | * |
| Gloucester | Massachusetts | * | * | 2 | 4 |
| Grand Rapids | Michigan | 90 | 280.55 | 10 | 10 |
| Green Bay | Wisconsin | * | * | 3 | 10 |
| Greensboro | North Carolina | * | * | 1 | 10 |
| Gulfport | Mississippi | * | * | * | * |
| Hagerstown | Maryland | 5 | * | * | * |
| Hartford | Connecticut | * | * | 10 | 30 |
| Hoffman Estates | Illinois | * | * | * | * |
| Honolulu | Hawaii | * | * | 1 | 2 |
| Houston | Texas | 68 | 1914.57 | 39 | 682 |

| | | | | | |
|---|---|---|---|---|---|
| Huntsville | Alabama | * | * | * | * |
| Indianapolis | Indiana | 44 | 200 | 35 | 140 |
| Irvine | California | * | * | * | * |
| Isabela | Puerto Rico | * | * | * | * |
| Jackson | Mississippi | * | 45 | 20 | 80 |
| Kalamazoo | Michigan | 30 | 103 | 16 | 133.94 |
| Kansas City | Kansas | * | * | * | * |
| Lakeland | Florida | 1 | 1 | * | * |
| Las Vegas | Nevada | 2 | 11.5 | 2 | 57.5 |
| Lewiston | Maine | 4 | 21 | 4 | 6.5 |
| Louisville | Kentucky | 24 | * | 11 | * |
| Memphis | Tennessee | 20 | * | 4 | * |
| Miami Beach | Florida | 1 | 2.83 | * | * |
| Milwaukee | Wisconsin | 90 | 200 | 40 | 100 |
| Monroe | Louisiana | 1 | 5 | 1 | 5 |
| Naugatuck | Connecticut | 5 | 10 | 3 | 8 |
| New Orleans | Louisiana | 28 | 52 | 5 | 6 |
| Niagara Falls | New York | 3 | 200 | 3 | 100 |
| North Chicago | Illinois | 1 | 2 | 1 | 40 |
| North Little Rock | Arkansas | 10 | 10 | 3 | 180 |
| North Miami | Florida | * | * | * | * |
| Norwalk | Connecticut | * | * | 4 | 10 |
| Orland Park | Illinois | * | * | 1 | 20 |
| Palm Beach Gardens | Florida | * | * | * | * |
| Pawtucket | Rhode Island | 1 | 1 | 3 | 10 |
| Pembroke Pines | Florida | * | * | * | * |
| Piscataway | New Jersey | 1 | 1 | * | * |
| Portland | Oregon | * | * | * | * |
| Racine | Wisconsin | 3 | 11 | 2 | 20 |
| Radnor | Pennsylvania | * | * | * | * |
| Saint Louis | Missouri | 30 | * | 79 | * |
| Saint Paul | Minnesota | 25 | 625 | 8 | 80 |
| San Juan | Puerto Rico | * | * | 130 | 5 |
| Sanford | North Carolina | 32 | 84 | 2 | 6 |
| Shreveport | Louisiana | 42 | 228 | 5 | 28 |
| Somerville | Massachusetts | 25 | 40 | 20 | 100 |
| Southgate | Michigan | * | * | * | * |
| Syracuse | New York | 4 | 160 | 30 | 393 |
| Tacoma | Washington | 3 | 5 | 3 | 6 |
| Tallahassee | Florida | 1 | 2.5 | 2 | 6 |
| Toa Baja | Puerto Rico | * | * | 1 | 0.33 |
| Tucson | Arizona | 25 | * | 12 | * |
| Vista | California | 2 | 2 | * | * |
| Waco | Texas | * | * | 1 | 5 |
| Warren | Ohio | * | * | 2 | 10 |
| West Palm Beach | Florida | * | 25 | 10 | 3.05 |
| Wilson | North Carolina | * | * | * | * |
| York | Pennsylvania | * | * | 2 | 12 |
| Zanesville | Ohio | * | * | 1 | 4.33 |
| *99 Respondents* | | **Sites redeveloped since 1993** | **Acres developed since 1993** | **Sites in progress** | **Acres in progress** |
| | Totals | 1,010.00 | 7,209.95 | 907.00 | 4,682.58 |

*Source: The U.S. Conference of Mayors, Recycling America's Land: A National Report on Brownfields Redevelopment (1993-2010), November 2010, Volume IX*

## *Estimated Annual Tax Revenue Gains from Brownfields Redevelopment*

| City | State | Est. Annual Tax Revenue Gained Conservative | Est. Annual Tax Revenue Gained Optimistic | Est. Annual Tax Revenue Gained Actual Since '93 |
|---|---|---|---|---|
| Akron | Ohio | * | * | * |
| Alameda | California | $550,000.00 | $950,000.00 | * |
| Allentown | Pennsylvania | $150,000.00 | $250,000.00 | $500,000.00 |
| Arlington | Texas | * | * | * |
| Arlington Heights | Illinois | * | * | * |
| Asheville | North Carolina | $1,000,000.00 | $2,000,000.00 | $500,000.00 |
| Atlanta | Georgia | $4,614,000.00 | $23,070,000.00 | * |
| Babylon | New York | $5,000,000.00 | $8,000,000.00 | * |
| Baltimore | Maryland | * | * | * |
| Bartlett | Illinois | $35,000.00 | $50,000.00 | $1,360,000.00 |
| Bessemer | Alabama | $5,000,000.00 | $20,000,000.00 | * |
| Biloxi | Mississippi | * | * | * |
| Binghamton | New York | * | * | * |
| Boston | Massachusetts | $7,000,000.00 | $10,000,000.00 | * |
| Bridgeport | Connecticut | $20,000,000.00 | $50,000,000.00 | * |
| Camden | New Jersey | $54,359,000.00 | $71,253,000.00 | * |
| Cape Coral | Florida | * | * | * |
| Carson | California | * | * | * |
| Charleston | South Carolina | $50,000,000.00 | $25,000,000.00 | * |
| Chattanooga | Tennessee | * | * | * |
| Chicago Heights | Illinois | * | * | * |
| Cincinnati | Ohio | * | * | * |
| Clearwater | Florida | $2,500,000.00 | $5,000,000.00 | * |
| Clovis | New Mexico | $100,000.00 | $100,000.00 | * |
| Columbia | South Carolina | $63,700,000.00 | $100,000,000.00 | $12,500,000.00 |
| Columbus | Ohio | $15,000,000.00 | $15,000,000.00 | $175,000,000.00 |
| Dallas | Texas | $5,000,000.00 | $10,000,000.00 | * |
| Dayton | Ohio | $1,000,000.00 | $3,000,000.00 | * |
| Dearborn | Michigan | $100,000,000.00 | $140,000,000.00 | * |
| Denton | Texas | * | * | * |
| Dubuque | Iowa | $4,000,000.00 | $8,000,000.00 | $4,106,500.00 |
| Durham | North Carolina | * | * | * |
| Elizabeth | New Jersey | $30,000,000.00 | $45,000,000.00 | $6,600,000.00 |
| Evansville | Indiana | * | * | * |
| Fayetteville | North Carolina | $250,000.00 | $1,000,000.00 | * |
| Florence | Alabama | * | * | * |
| Fort Myers | Florida | * | * | * |
| Fresno | California | * | * | * |
| Frisco | Texas | * | * | * |
| Glendale Heights | Illinois | $50,000.00 | $100,000.00 | * |
| Gloucester | Massachusetts | $3,000,000.00 | $12,000,000.00 | * |
| Grand Rapids | Michigan | $2,533,998.00 | $4,500,000.00 | $7,380,170.00 |
| Green Bay | Wisconsin | * | * | * |
| Greensboro | North Carolina | $250,000.00 | $250,000.00 | * |
| Gulfport | Mississippi | $5,000,000.00 | $10,000,000.00 | * |
| Hagerstown | Maryland | * | * | * |
| Hartford | Connecticut | $500,000.00 | $2,000,000.00 | * |
| Hoffman Estates | Illinois | * | * | * |
| Honolulu | Hawaii | * | * | * |
| Houston | Texas | * | * | * |
| Huntsville | Alabama | * | * | * |
| Indianapolis | Indiana | $11,500,000.00 | $20,000,000.00 | * |
| Irvine | California | * | * | * |
| Isabela | Puerto Rico | $5,000,000.00 | $7,000,000.00 | $10,000,000.00 |
| Jackson | Mississippi | $125,000.00 | $200,000.00 | $50,000.00 |
| Kalamazoo | Michigan | $16,000,000.00 | $30,000,000.00 | $4,538,583.00 |

| City | State | Est. Annual Tax Revenue Gained Conservative | Est. Annual Tax Revenue Gained Optimistic | Est. Annual Tax Revenue Gained Actual Since '93 |
|---|---|---|---|---|
| Kansas City | Kansas | * | * | * |
| Lakeland | Florida | $15,000.00 | $35,000.00 | * |
| Las Vegas | Nevada | $16,000,000.00 | $28,182,000.00 | $2,163,250.00 |
| Lewiston | Maine | $252,000,000.00 | $317,000,000.00 | $718,000.00 |
| Louisville | Kentucky | $67,250.00 | $80,700.00 | * |
| Memphis | Tennessee | $8,000,000.00 | $25,000,000.00 | * |
| Miami Beach | Florida | * | * | * |
| Milwaukee | Wisconsin | $50,000,000.00 | $100,000,000.00 | $17,000,000.00 |
| Monroe | Louisiana | * | * | * |
| Naugatuck | Connecticut | $500,000.00 | $1,000,000.00 | * |
| New Orleans | Louisiana | $5,000,000.00 | $20,000,000.00 | $1,500,000.00 |
| Niagara Falls | New York | * | * | * |
| North Chicago | Illinois | $30,000.00 | $50,000.00 | * |
| North Little Rock | Arkansas | * | * | $2,000,000.00 |
| North Miami | Florida | * | * | * |
| Norwalk | Connecticut | * | * | * |
| Orland Park | Illinois | * | * | * |
| Palm Beach Gardens | Florida | * | * | * |
| Pawtucket | Rhode Island | * | * | * |
| Pembroke Pines | Florida | * | * | * |
| Piscataway | New Jersey | * | * | * |
| Portland | Oregon | * | * | * |
| Racine | Wisconsin | * | $1,197,057.00 | * |
| Radnor | Pennsylvania | * | * | * |
| Saint Louis | Missouri | * | * | * |
| Saint Paul | Minnesota | $8,000,000.00 | $10,000,000.00 | $5,900,000.00 |
| San Juan | Puerto Rico | $1,600,000.00 | $2,400,000.00 | * |
| Sanford | North Carolina | $40,000.00 | $400,000.00 | * |
| Shreveport | Louisiana | * | * | * |
| Somerville | Massachusetts | $106,000,000.00 | $120,000,000.00 | * |
| Southgate | Michigan | * | * | * |
| Syracuse | New York | $9,587,119.50 | $38,348,478.00 | $56,224,371.00 |
| Tacoma | Washington | * | * | * |
| Tallahassee | Florida | $1,000,000.00 | $3,000,000.00 | $593,260.00 |
| Toa Baja | Puerto Rico | * | * | * |
| Tucson | Arizona | * | * | * |
| Vista | California | * | * | * |
| Waco | Texas | * | * | * |
| Warren | Ohio | * | * | * |
| West Palm Beach | Florida | * | * | * |
| Wilson | North Carolina | $200,000.00 | $500,000.00 | * |
| York | Pennsylvania | $500,000.00 | $2,000,000.00 | * |
| Zanesville | Ohio | * | * | * |
| Total Respondents: 99 | | *Est. Annual Tax Revenue Gained Conservative* *$871,206,367.50* | *Est. Annual Tax Revenue Gained Optimistic* *$1,291,966,235.00* | *Est. Annual Tax Revenue Gained Actual Since '93* *$308,634,134.00* |

*Source: The U.S. Conference of Mayors, Recycling America's Land: A National Report on Brownfields Redevelopment (1993-2010), November 2010, Volume IX*

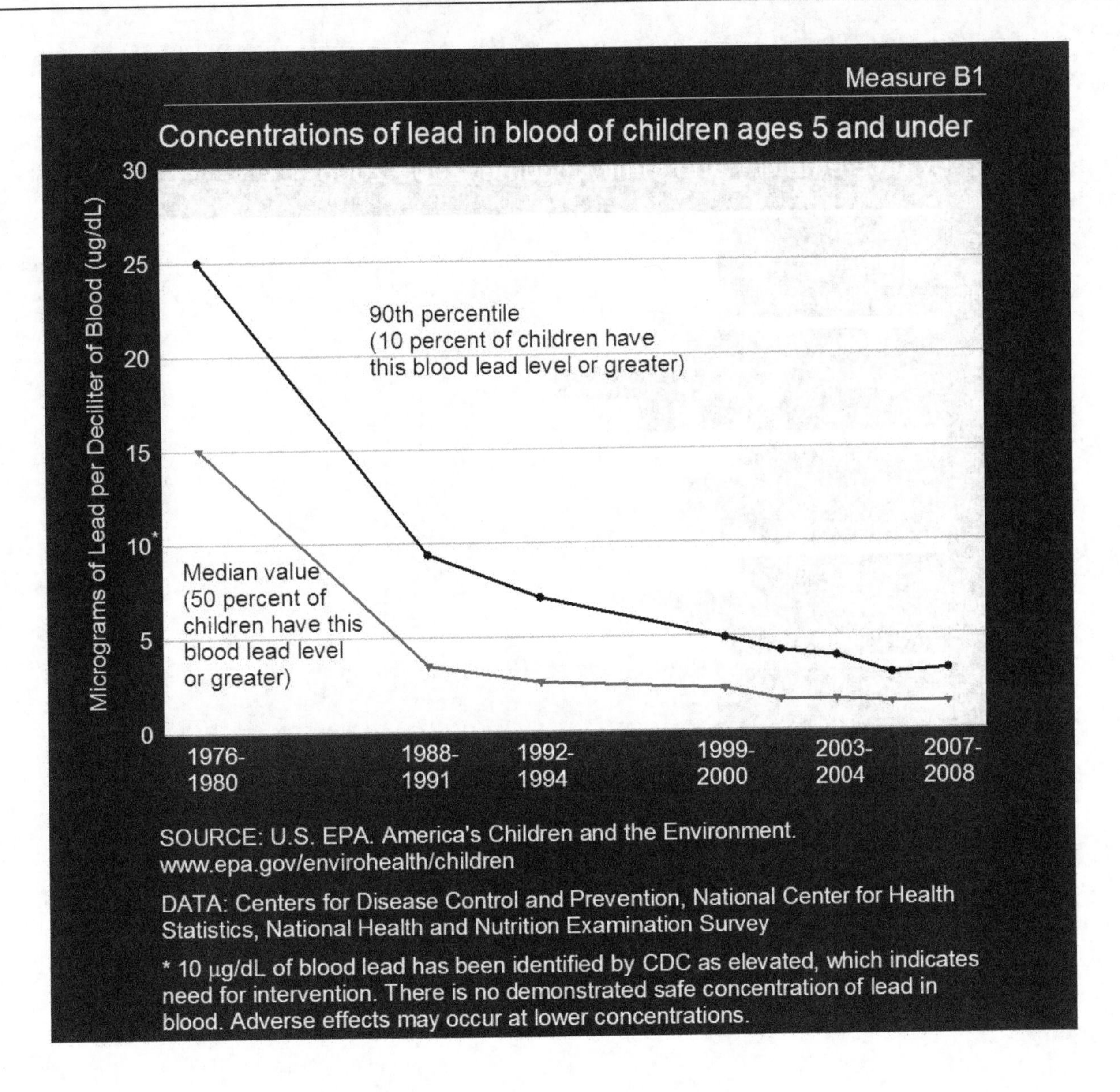
Measure B1
Concentrations of lead in blood of children ages 5 and under
Micrograms of Lead per Deciliter of Blood (ug/dL)
30
25
20
15
10*
5
0
90th percentile
(10 percent of children have
this blood lead level or greater)
Median value
(50 percent of
children have this
blood lead level
or greater)
1976-
1980
1988-
1991
1992-
1994
1999-
2000
2003-
2004
2007-
2008
SOURCE: U.S. EPA. America's Children and the Environment.
www.epa.gov/envirohealth/children
DATA: Centers for Disease Control and Prevention, National Center for Health Statistics, National Health and Nutrition Examination Survey
* 10 µg/dL of blood lead has been identified by CDC as elevated, which indicates need for intervention. There is no demonstrated safe concentration of lead in blood. Adverse effects may occur at lower concentrations.

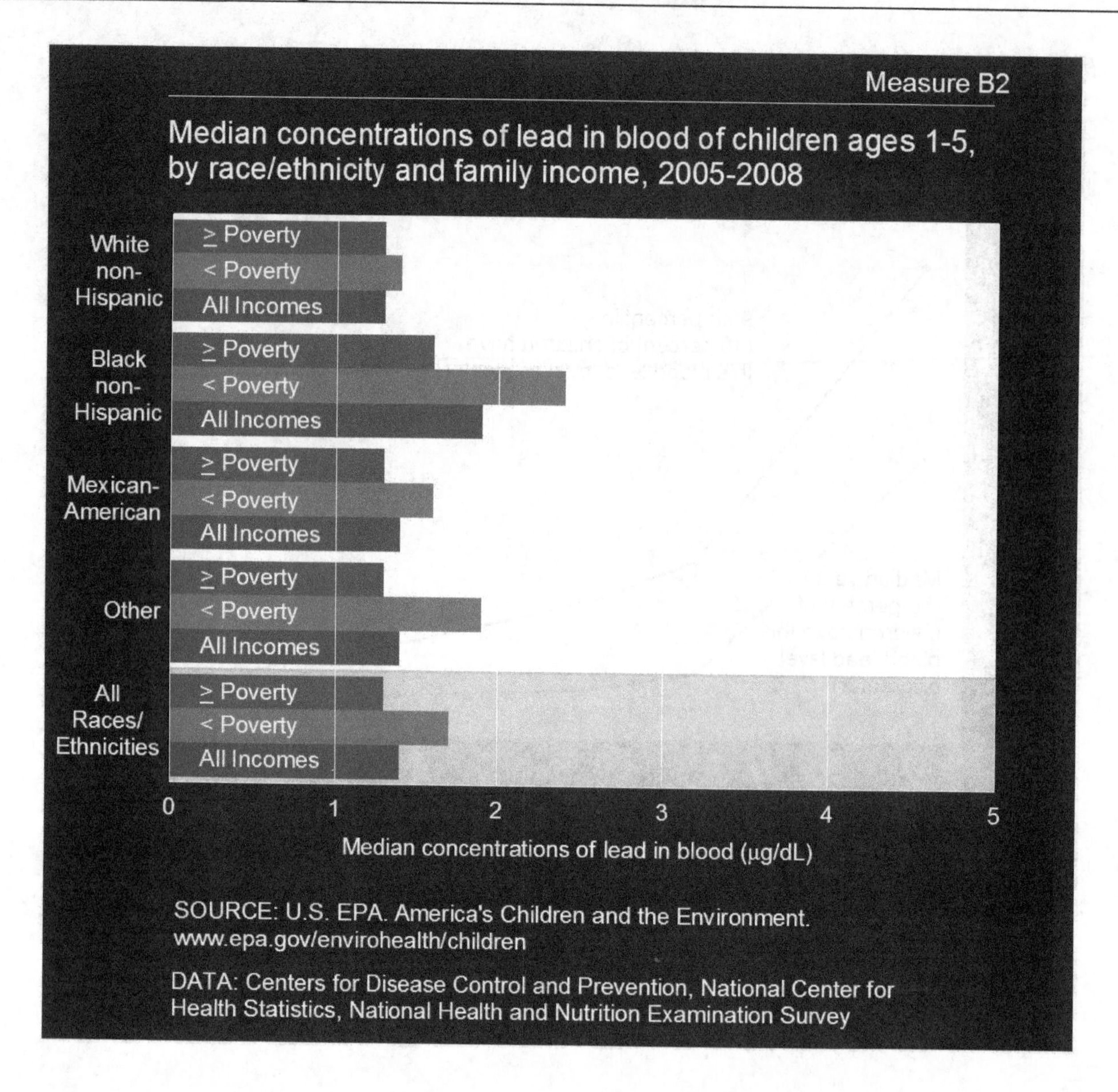
Measure B2
Median concentrations of lead in blood of children ages 1-5, by race/ethnicity and family income, 2005-2008
White non-Hispanic
Black non-Hispanic
Mexican-American
Other
All Races/ Ethnicities
≥ Poverty
< Poverty
All Incomes
0
1
2
3
4
5
Median concentrations of lead in blood (µg/dL)
SOURCE: U.S. EPA. America's Children and the Environment. www.epa.gov/envirohealth/children
DATA: Centers for Disease Control and Prevention, National Center for Health Statistics, National Health and Nutrition Examination Survey

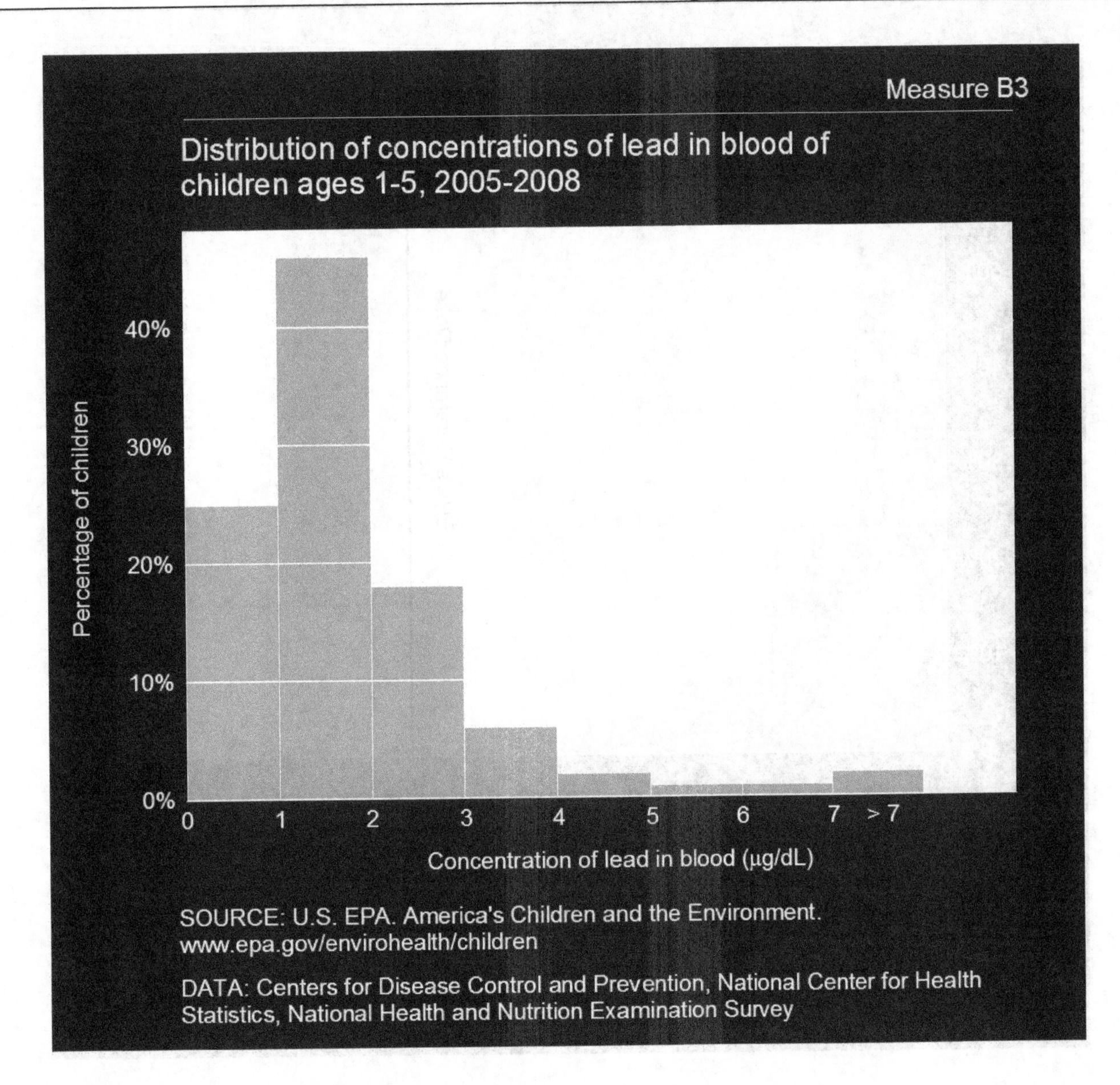
Measure B3
Distribution of concentrations of lead in blood of children ages 1-5, 2005-2008
Percentage of children
40%
30%
20%
10%
0%
0
1
2
3
4
5
6
7
>7
Concentration of lead in blood (µg/dL)
SOURCE: U.S. EPA. America's Children and the Environment. www.epa.gov/envirohealth/children
DATA: Centers for Disease Control and Prevention, National Center for Health Statistics, National Health and Nutrition Examination Survey

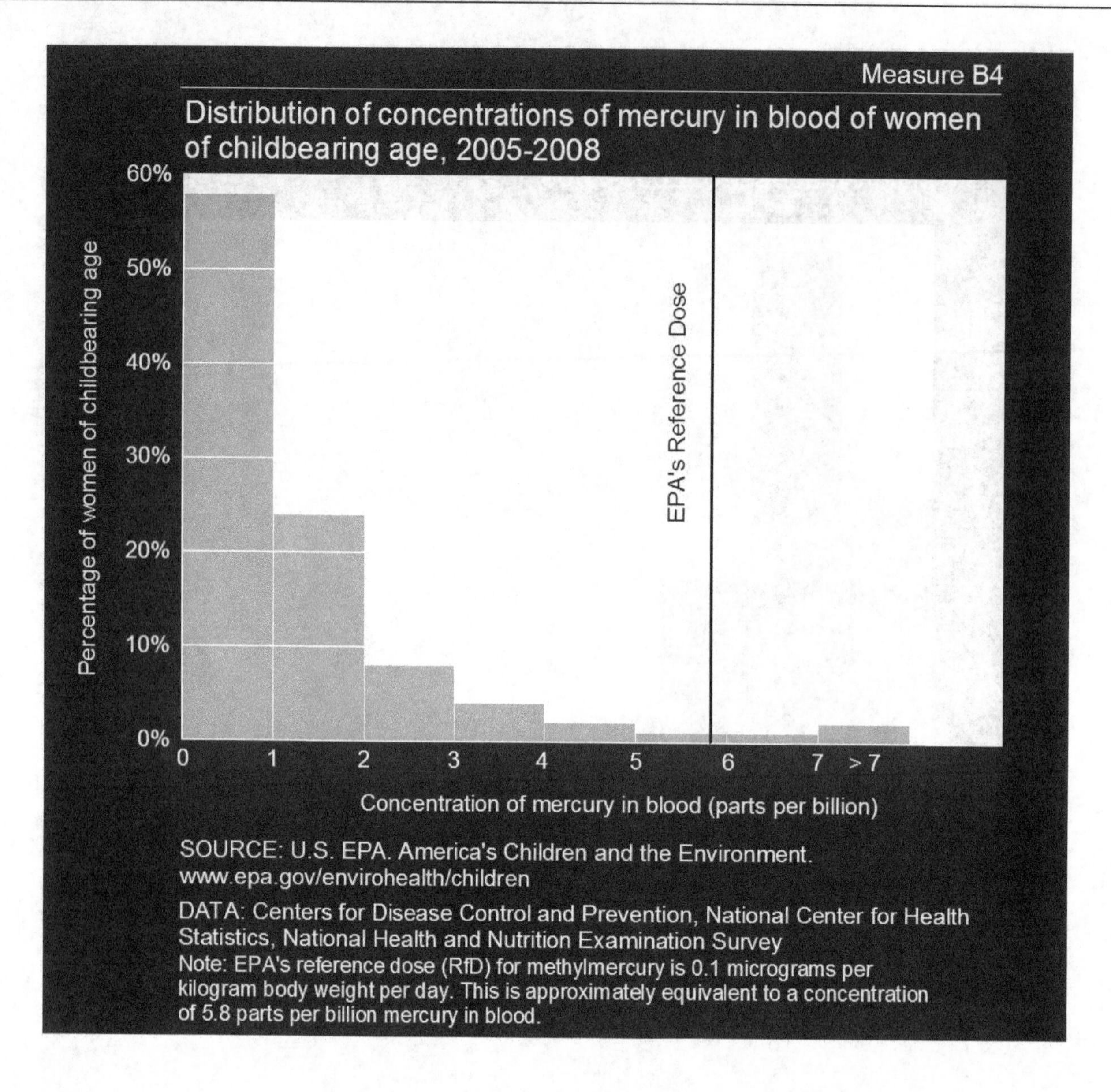
Measure B4
Distribution of concentrations of mercury in blood of women of childbearing age, 2005-2008
60%
50%
40%
30%
20%
10%
0%
Percentage of women of childbearing age
EPA's Reference Dose
0
1
2
3
4
5
6
7
>7
Concentration of mercury in blood (parts per billion)
SOURCE: U.S. EPA. America's Children and the Environment.
www.epa.gov/envirohealth/children
DATA: Centers for Disease Control and Prevention, National Center for Health Statistics, National Health and Nutrition Examination Survey
Note: EPA's reference dose (RfD) for methylmercury is 0.1 micrograms per kilogram body weight per day. This is approximately equivalent to a concentration of 5.8 parts per billion mercury in blood.

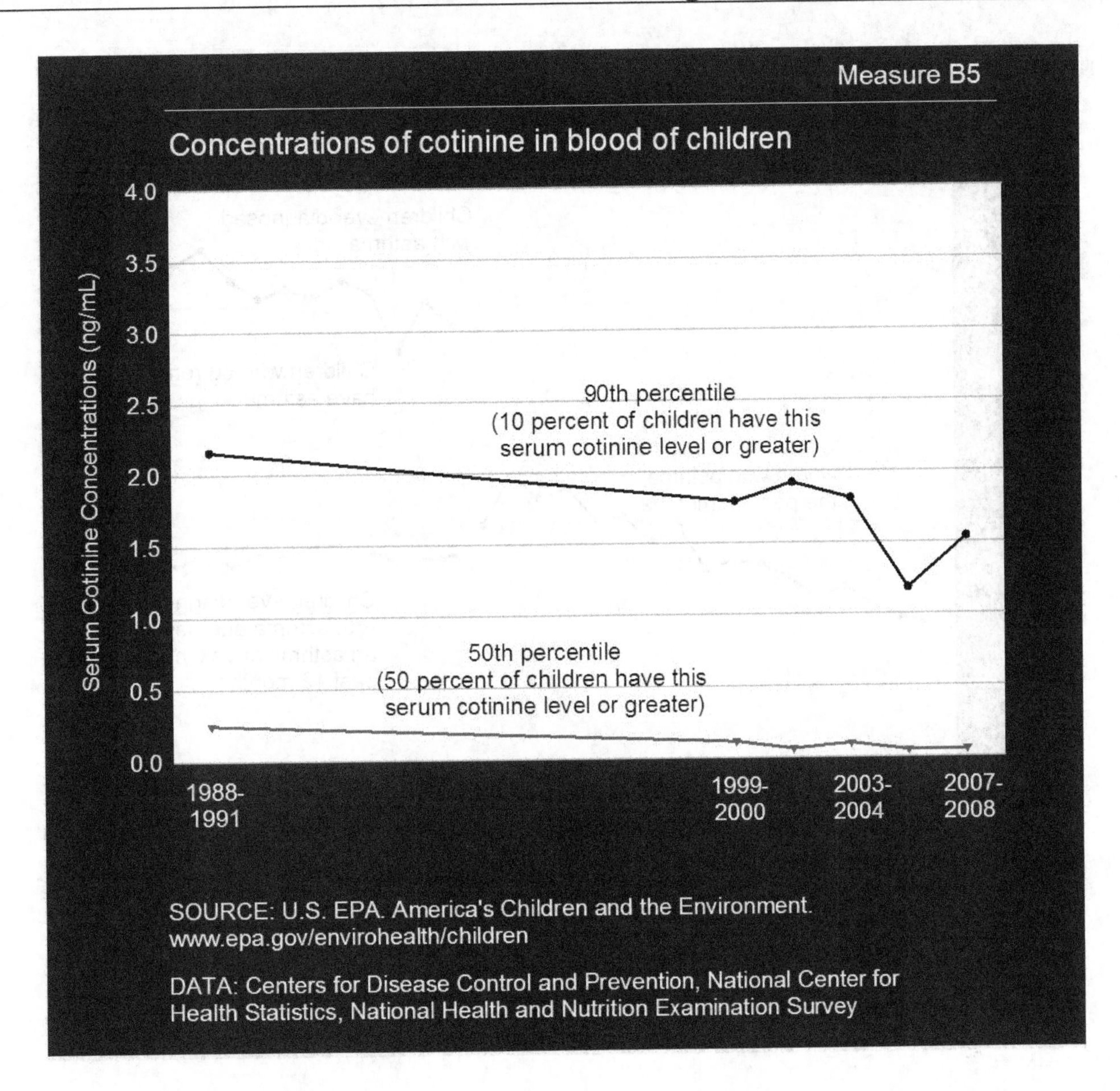
Measure B5
Concentrations of cotinine in blood of children
Serum Cotinine Concentrations (ng/mL)
4.0
3.5
3.0
2.5
2.0
1.5
1.0
0.5
0.0
90th percentile
(10 percent of children have this
serum cotinine level or greater)
50th percentile
(50 percent of children have this
serum cotinine level or greater)
1988-
1991
1999-
2000
2003-
2004
2007-
2008
SOURCE: U.S. EPA. America's Children and the Environment.
www.epa.gov/envirohealth/children
DATA: Centers for Disease Control and Prevention, National Center for
Health Statistics, National Health and Nutrition Examination Survey

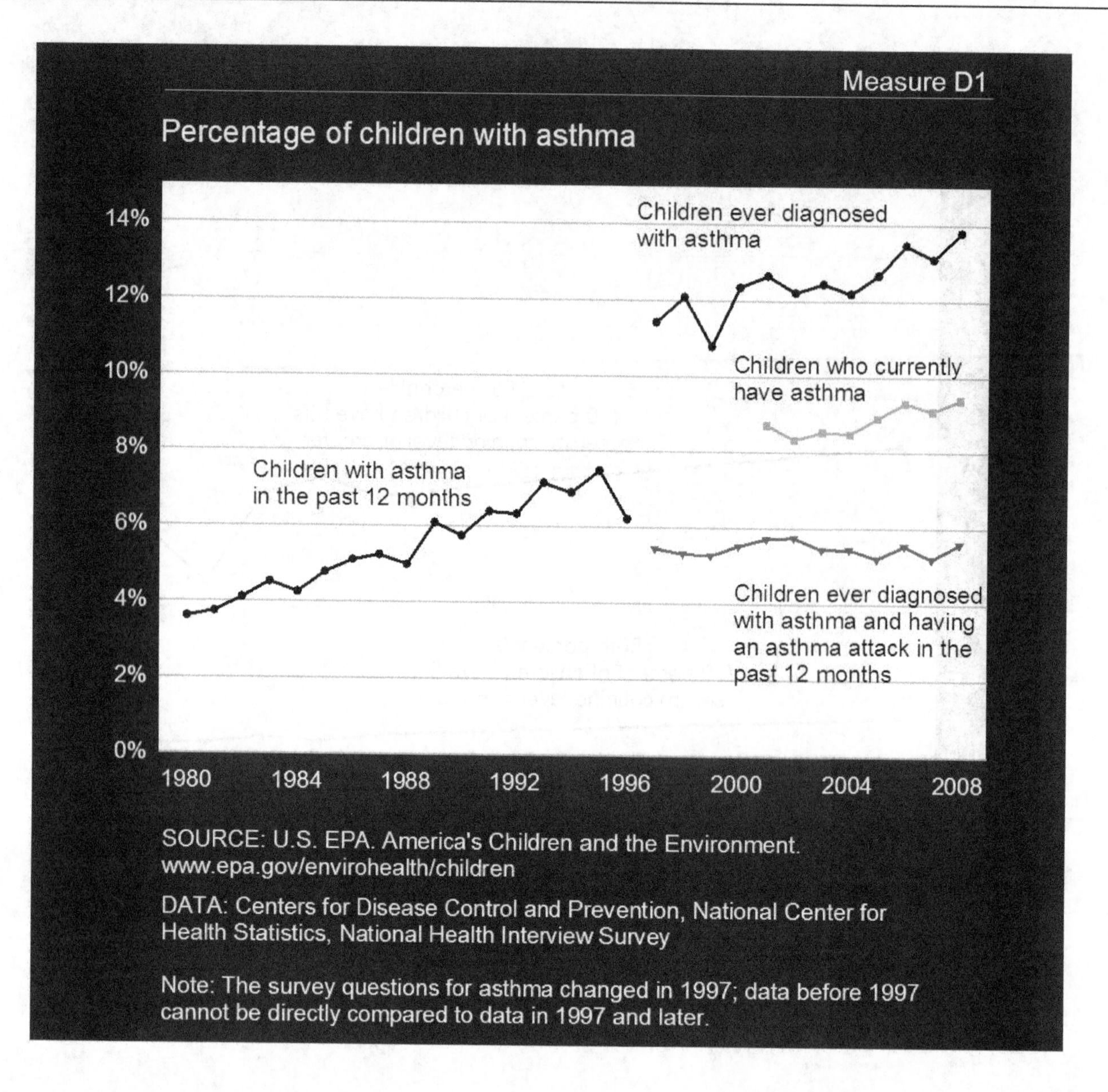
Measure D1
Percentage of children with asthma
14%
12%
10%
8%
6%
4%
2%
0%
1980
1984
1988
1992
1996
2000
2004
2008
Children ever diagnosed with asthma
Children who currently have asthma
Children with asthma in the past 12 months
Children ever diagnosed with asthma and having an asthma attack in the past 12 months
SOURCE: U.S. EPA. America's Children and the Environment. www.epa.gov/envirohealth/children
DATA: Centers for Disease Control and Prevention, National Center for Health Statistics, National Health Interview Survey
Note: The survey questions for asthma changed in 1997; data before 1997 cannot be directly compared to data in 1997 and later.

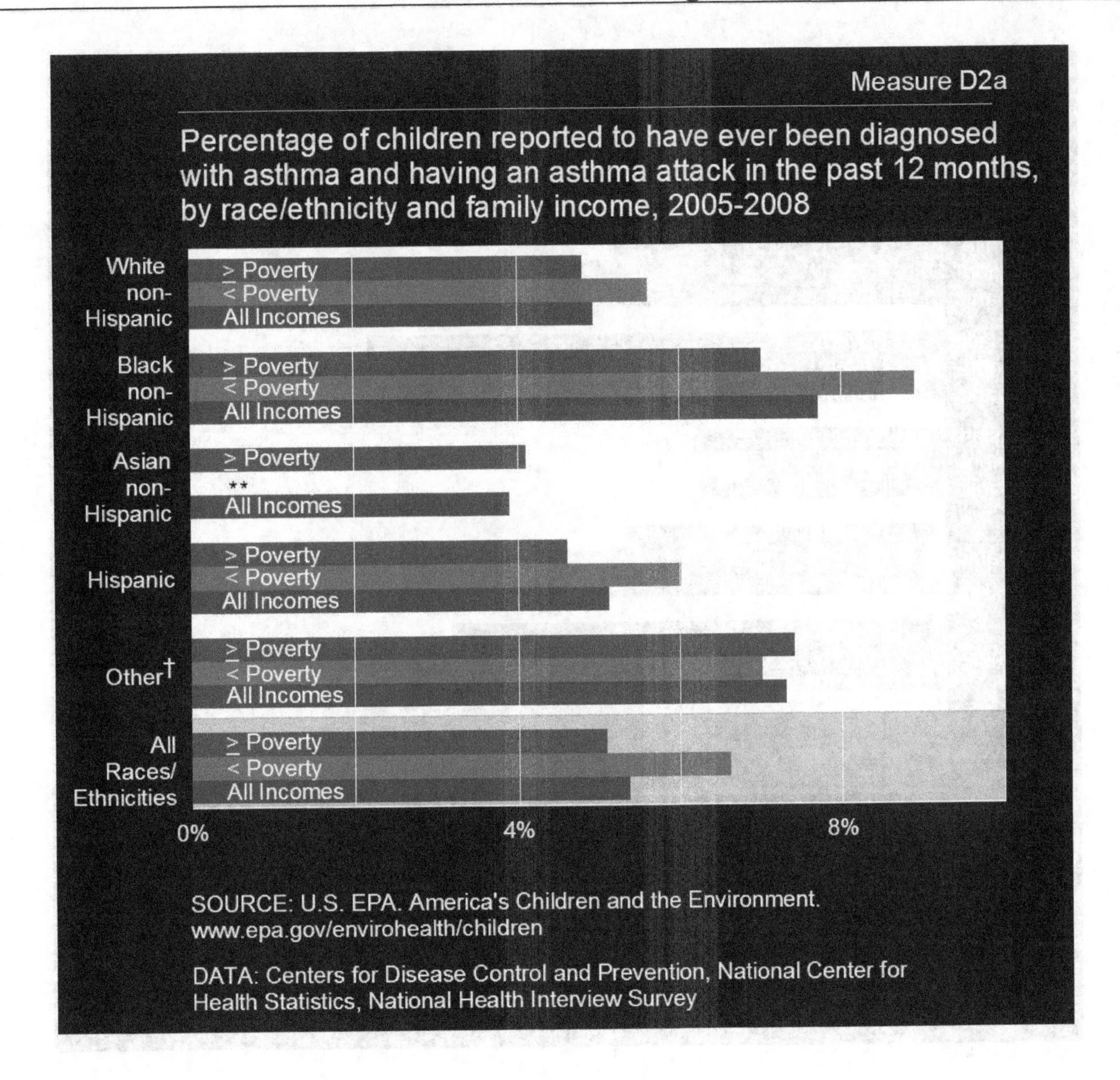
Measure D2a
Percentage of children reported to have ever been diagnosed with asthma and having an asthma attack in the past 12 months, by race/ethnicity and family income, 2005-2008
White non-Hispanic
≥ Poverty
< Poverty
All Incomes
Black non-Hispanic
≥ Poverty
< Poverty
All Incomes
Asian non-Hispanic
≥ Poverty
**
All Incomes
Hispanic
≥ Poverty
< Poverty
All Incomes
Other†
≥ Poverty
< Poverty
All Incomes
All Races/ Ethnicities
≥ Poverty
< Poverty
All Incomes
0%
4%
8%
SOURCE: U.S. EPA. America's Children and the Environment. www.epa.gov/envirohealth/children
DATA: Centers for Disease Control and Prevention, National Center for Health Statistics, National Health Interview Survey

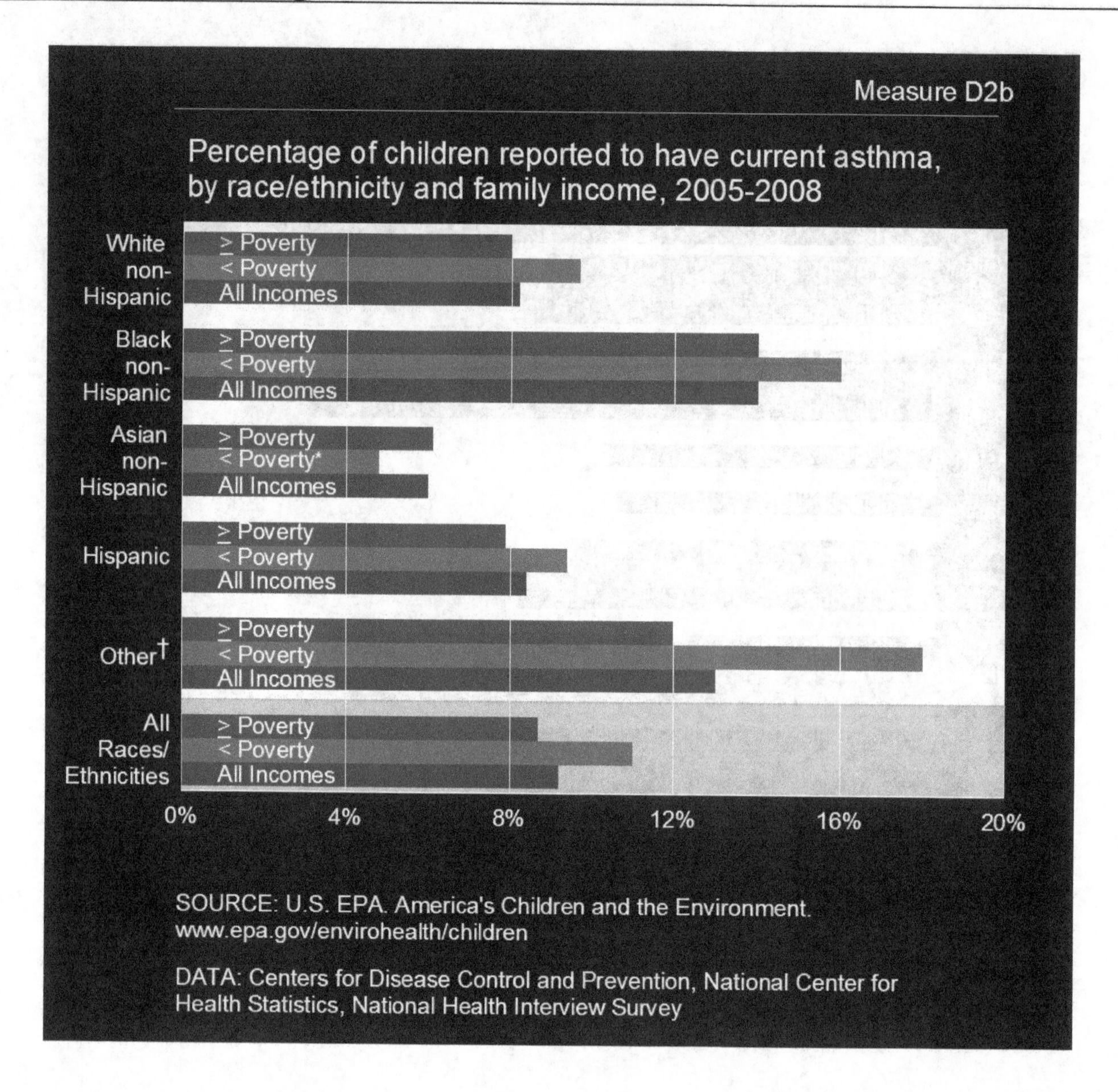
Measure D2b
Percentage of children reported to have current asthma, by race/ethnicity and family income, 2005-2008
White non-Hispanic
≥ Poverty
< Poverty
All Incomes
Black non-Hispanic
≥ Poverty
< Poverty
All Incomes
Asian non-Hispanic
≥ Poverty
< Poverty*
All Incomes
Hispanic
≥ Poverty
< Poverty
All Incomes
Other†
≥ Poverty
< Poverty
All Incomes
All Races/ Ethnicities
≥ Poverty
< Poverty
All Incomes
0%
4%
8%
12%
16%
20%
SOURCE: U.S. EPA. America's Children and the Environment. www.epa.gov/envirohealth/children
DATA: Centers for Disease Control and Prevention, National Center for Health Statistics, National Health Interview Survey

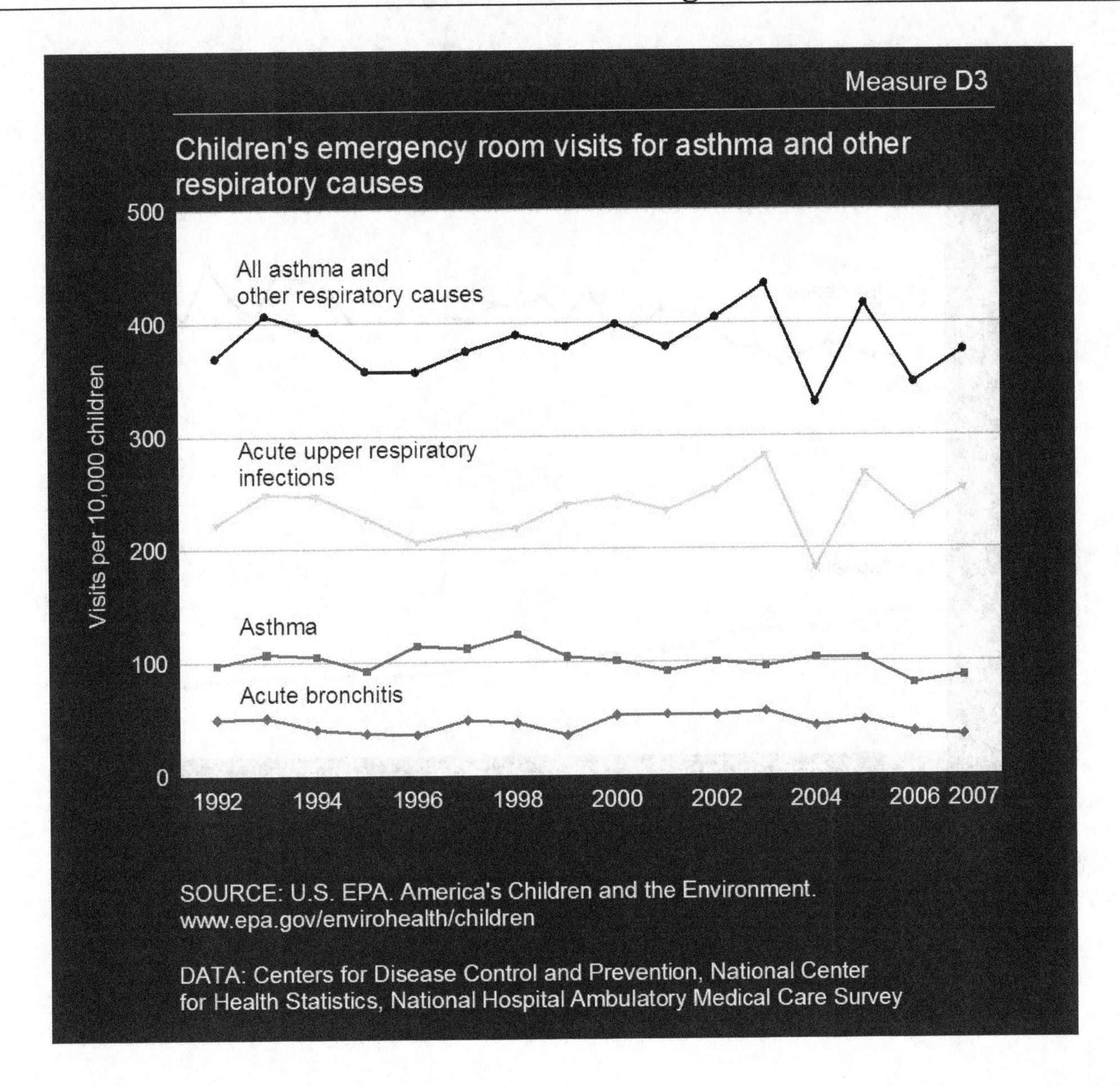
Measure D3
Children's emergency room visits for asthma and other respiratory causes
Visits per 10,000 children
500
400
300
200
100
0
All asthma and other respiratory causes
Acute upper respiratory infections
Asthma
Acute bronchitis
1992
1994
1996
1998
2000
2002
2004
2006
2007
SOURCE: U.S. EPA. America's Children and the Environment. www.epa.gov/envirohealth/children
DATA: Centers for Disease Control and Prevention, National Center for Health Statistics, National Hospital Ambulatory Medical Care Survey

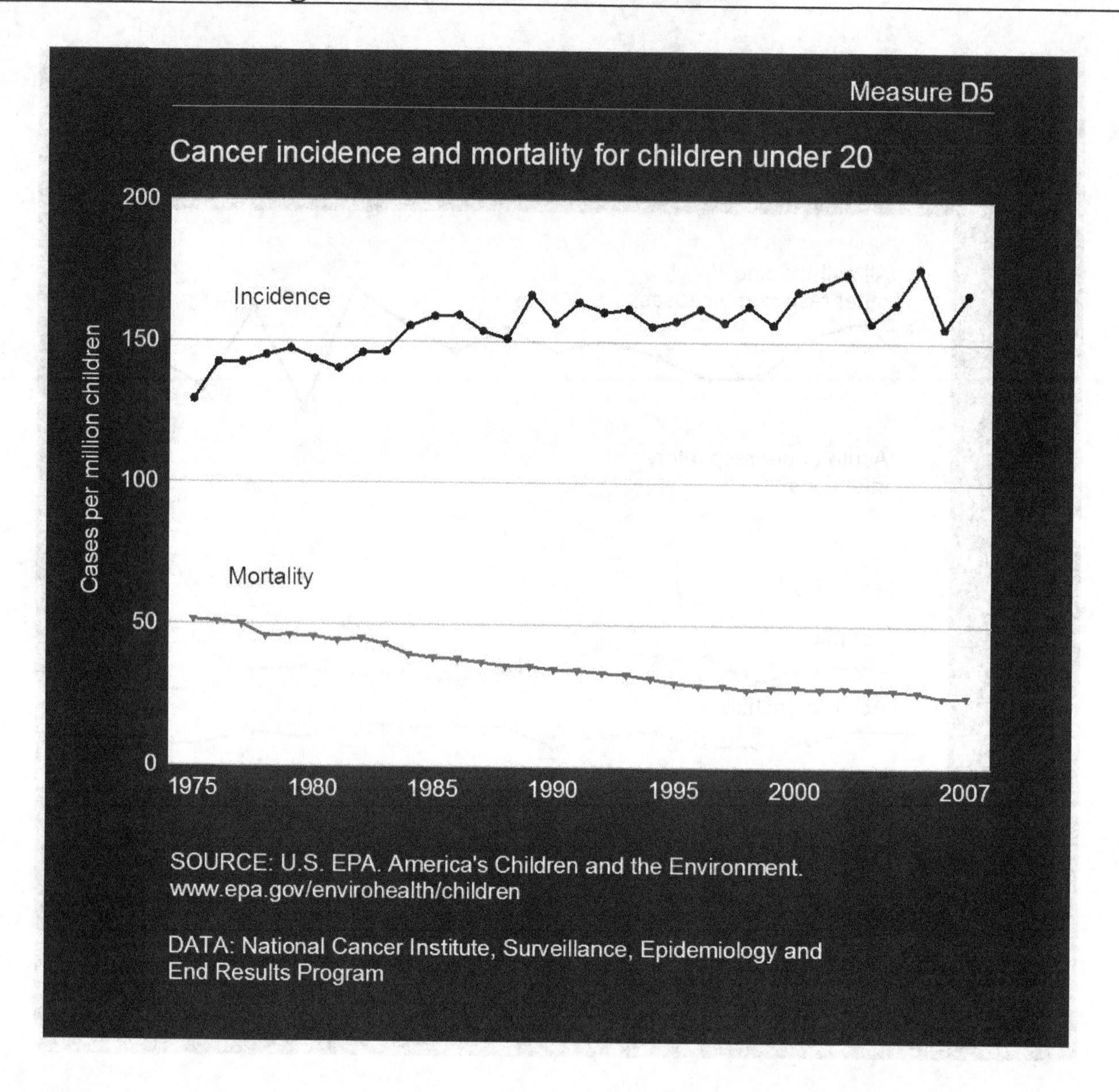
Measure D5
Cancer incidence and mortality for children under 20
Cases per million children
200
150
100
50
0
Incidence
Mortality
1975
1980
1985
1990
1995
2000
2007
SOURCE: U.S. EPA. America's Children and the Environment.
www.epa.gov/envirohealth/children
DATA: National Cancer Institute, Surveillance, Epidemiology and
End Results Program

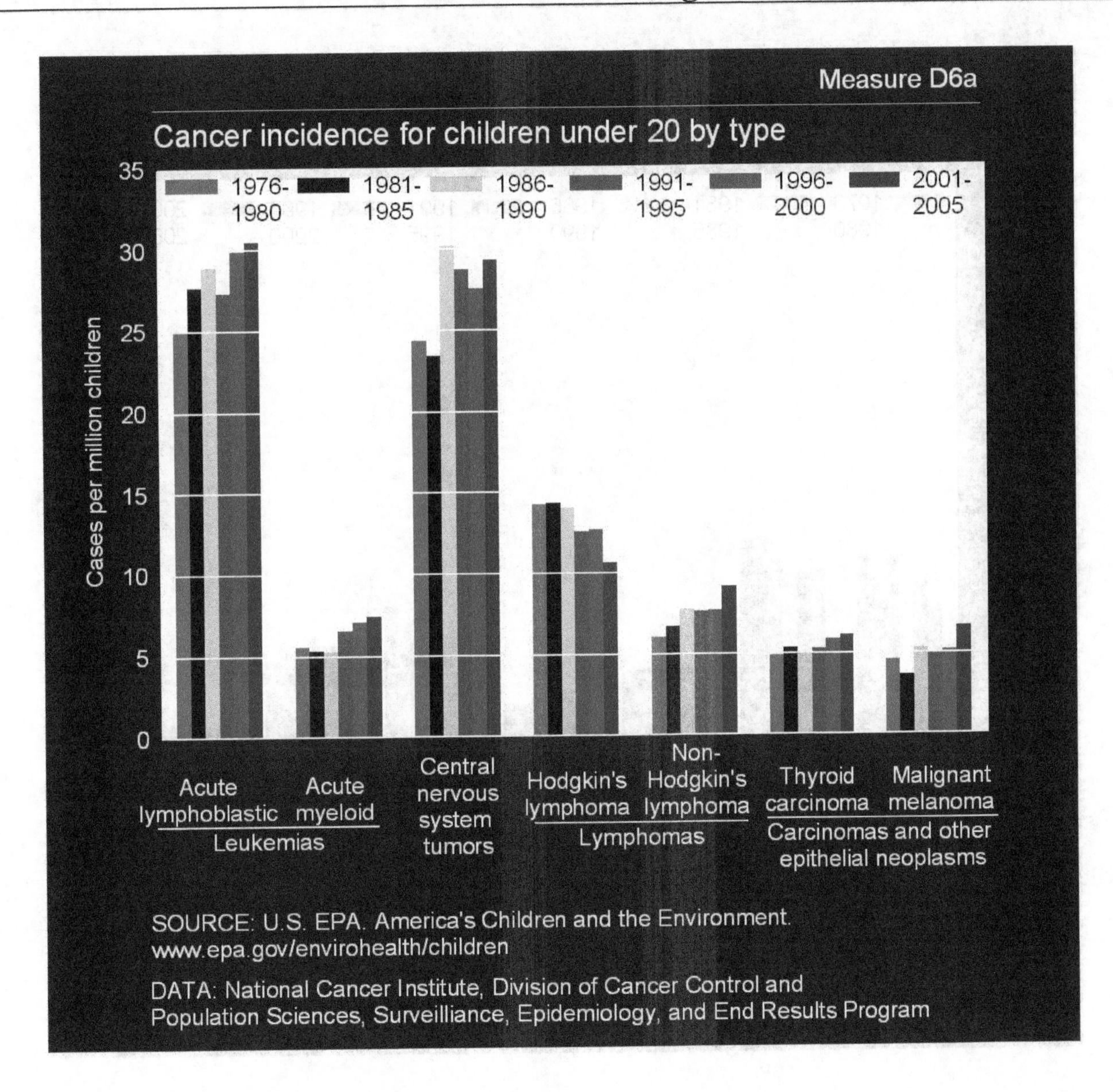
Measure D6a
Cancer incidence for children under 20 by type
1976-1980
1981-1985
1986-1990
1991-1995
1996-2000
2001-2005
Cases per million children
35
30
25
20
15
10
5
0
Acute lymphoblastic
Acute myeloid
Leukemias
Central nervous system tumors
Hodgkin's lymphoma
Non-Hodgkin's lymphoma
Lymphomas
Thyroid carcinoma
Malignant melanoma
Carcinomas and other epithelial neoplasms
SOURCE: U.S. EPA. America's Children and the Environment.
www.epa.gov/envirohealth/children
DATA: National Cancer Institute, Division of Cancer Control and Population Sciences, Surveilliance, Epidemiology, and End Results Program

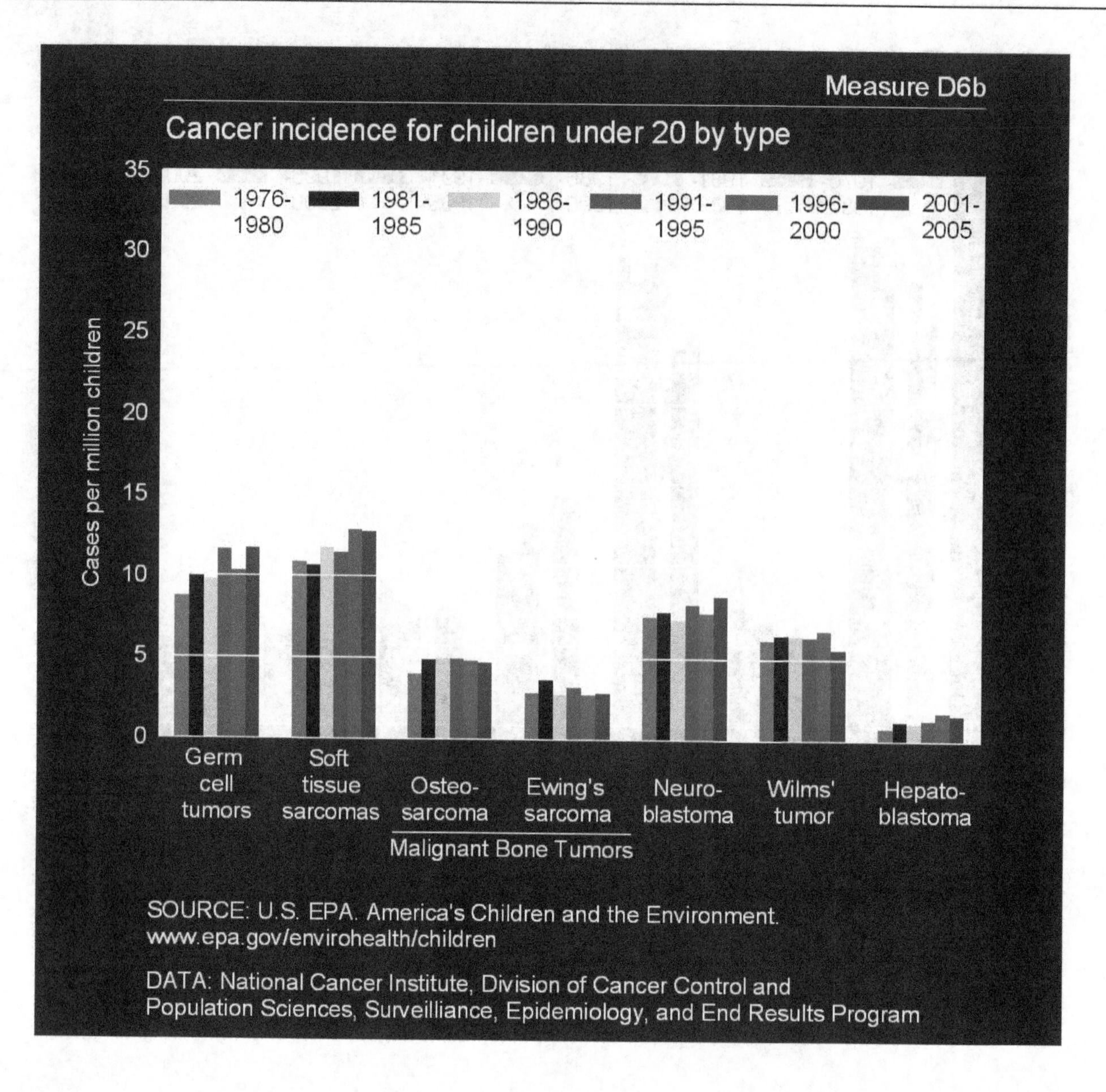
Measure D6b
Cancer incidence for children under 20 by type
1976-1980
1981-1985
1986-1990
1991-1995
1996-2000
2001-2005
Cases per million children
35
30
25
20
15
10
5
0
Germ cell tumors
Soft tissue sarcomas
Osteo-sarcoma
Ewing's sarcoma
Malignant Bone Tumors
Neuro-blastoma
Wilms' tumor
Hepato-blastoma
SOURCE: U.S. EPA. America's Children and the Environment. www.epa.gov/envirohealth/children
DATA: National Cancer Institute, Division of Cancer Control and Population Sciences, Surveilliance, Epidemiology, and End Results Program

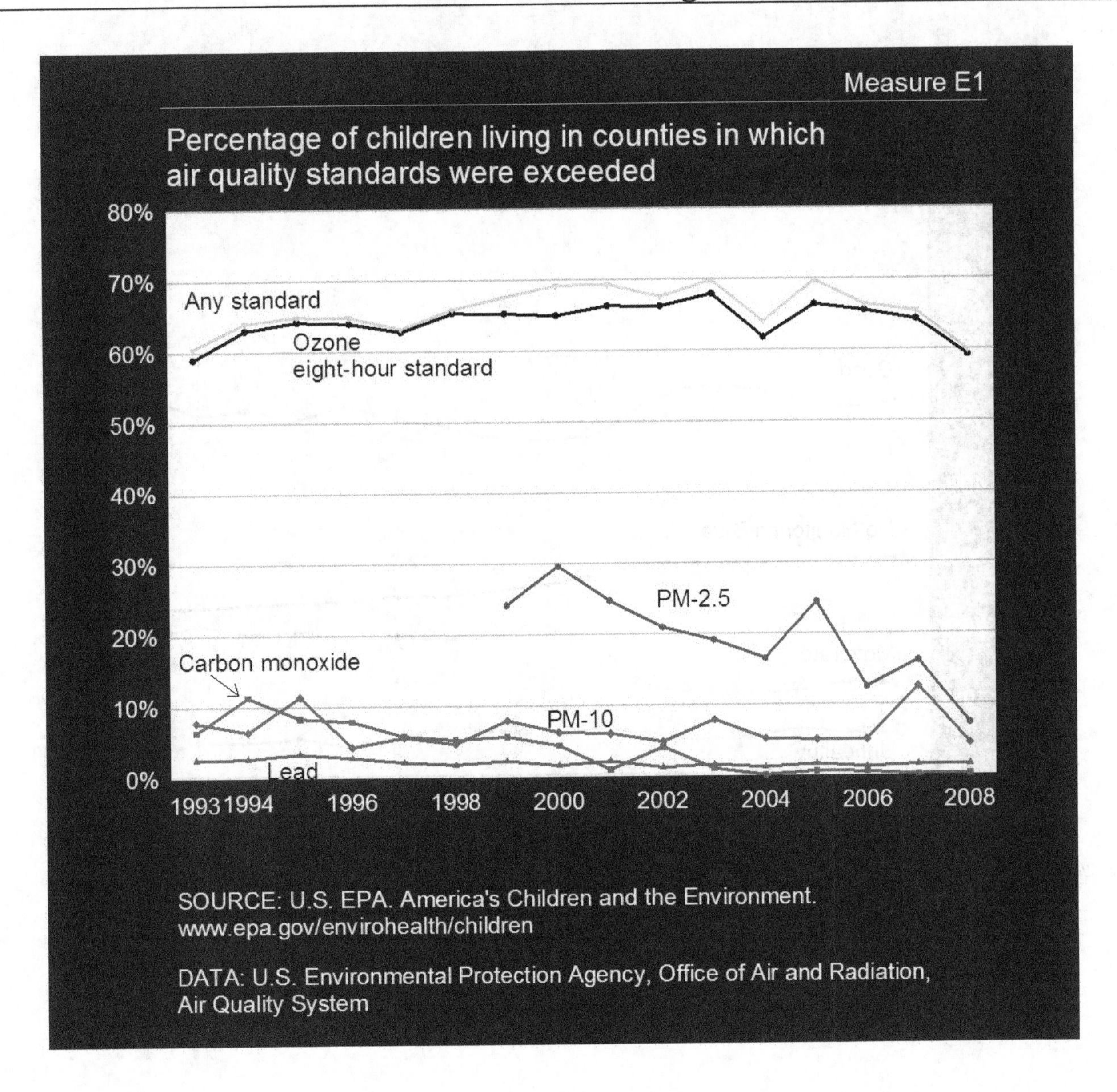
Measure E1
Percentage of children living in counties in which air quality standards were exceeded
80%
70%
60%
50%
40%
30%
20%
10%
0%
Any standard
Ozone eight-hour standard
PM-2.5
Carbon monoxide
PM-10
Lead
1993
1994
1996
1998
2000
2002
2004
2006
2008
SOURCE: U.S. EPA. America's Children and the Environment. www.epa.gov/envirohealth/children
DATA: U.S. Environmental Protection Agency, Office of Air and Radiation, Air Quality System

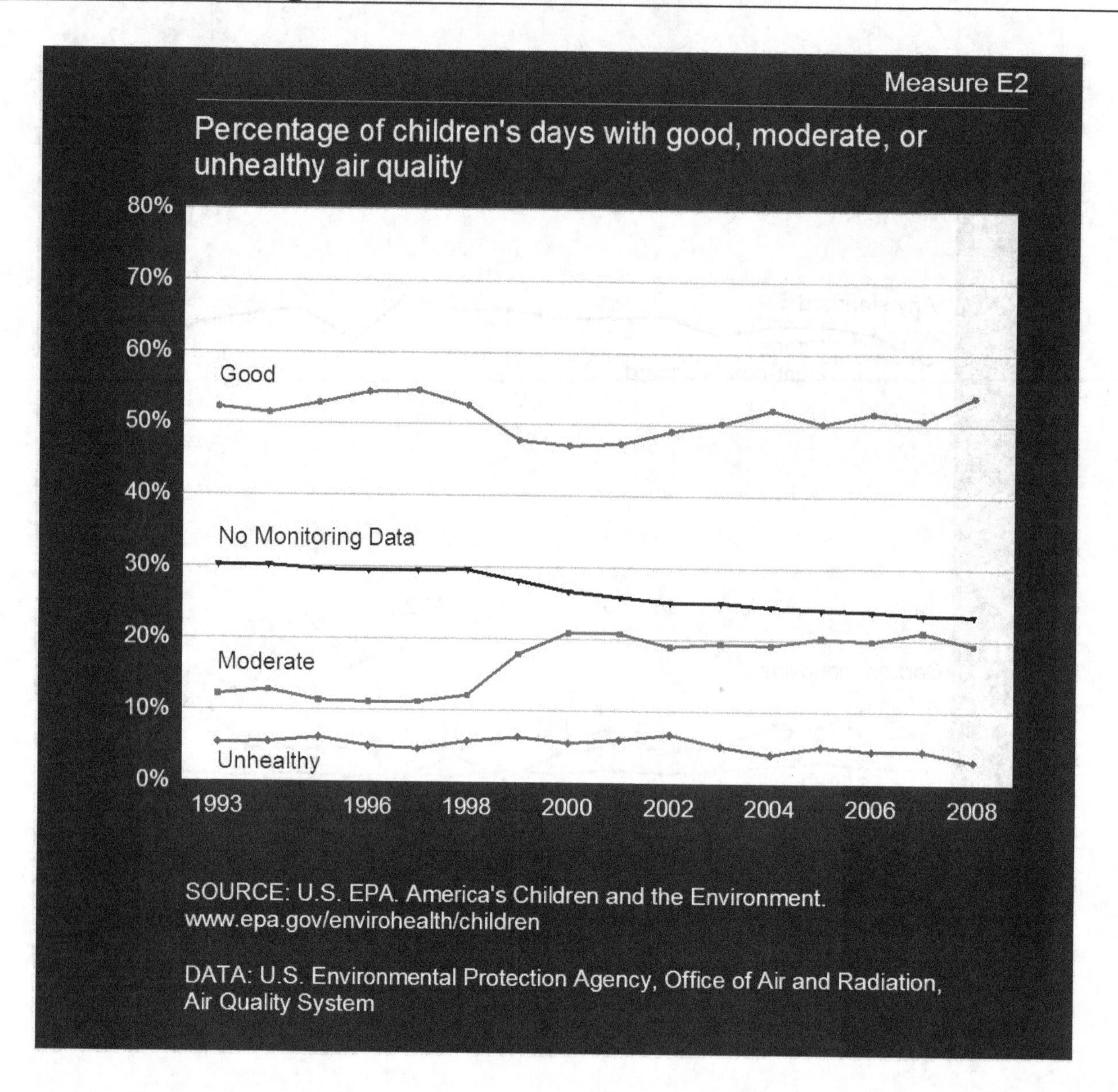
Measure E2
Percentage of children's days with good, moderate, or unhealthy air quality
80%
70%
60%
50%
40%
30%
20%
10%
0%
Good
No Monitoring Data
Moderate
Unhealthy
1993
1996
1998
2000
2002
2004
2006
2008
SOURCE: U.S. EPA. America's Children and the Environment. www.epa.gov/envirohealth/children
DATA: U.S. Environmental Protection Agency, Office of Air and Radiation, Air Quality System

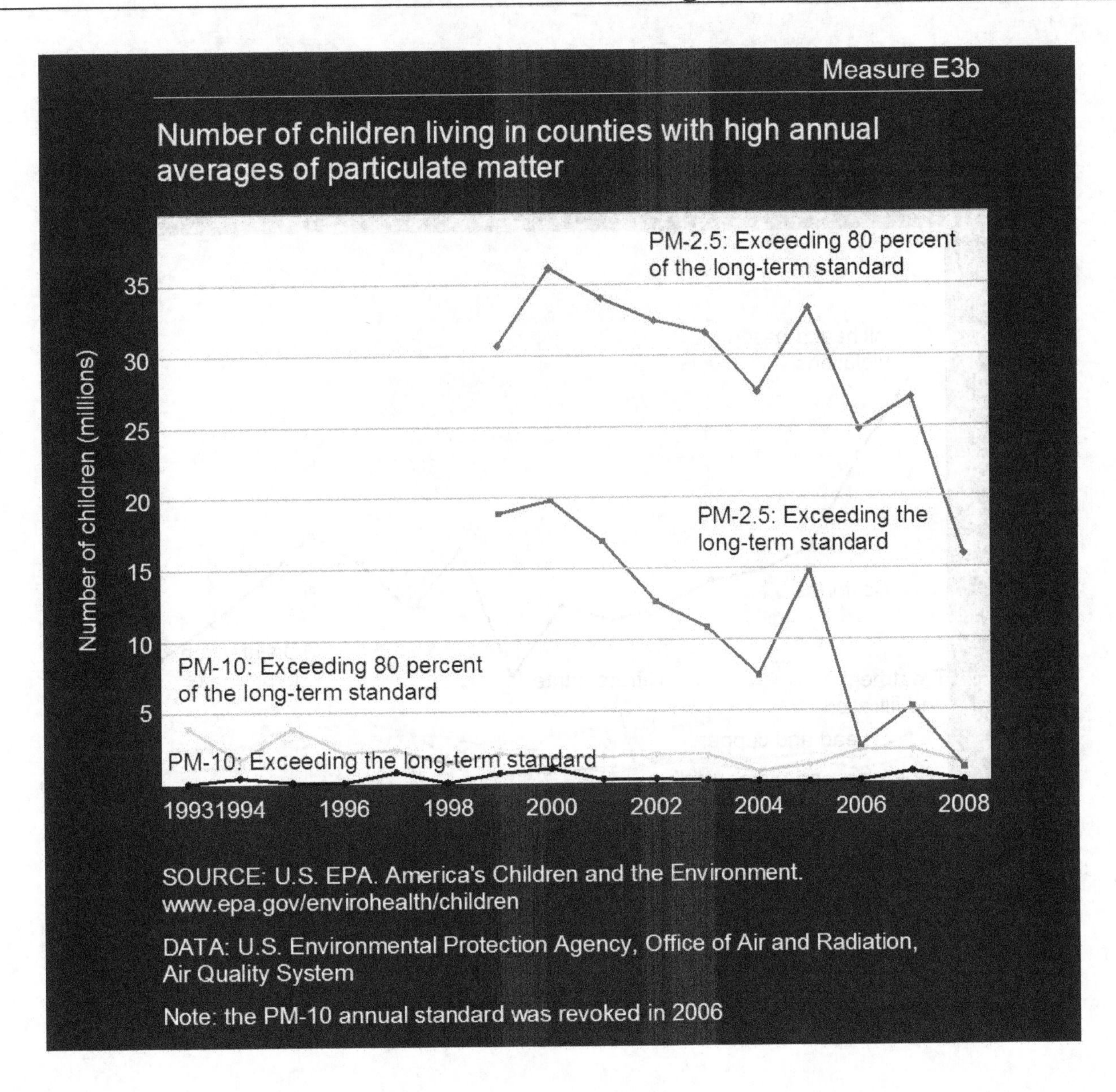
Measure E3b
Number of children living in counties with high annual averages of particulate matter
Number of children (millions)
35
30
25
20
15
10
5
PM-2.5: Exceeding 80 percent of the long-term standard
PM-2.5: Exceeding the long-term standard
PM-10: Exceeding 80 percent of the long-term standard
PM-10: Exceeding the long-term standard
19931994
1996
1998
2000
2002
2004
2006
2008
SOURCE: U.S. EPA. America's Children and the Environment. www.epa.gov/envirohealth/children
DATA: U.S. Environmental Protection Agency, Office of Air and Radiation, Air Quality System
Note: the PM-10 annual standard was revoked in 2006

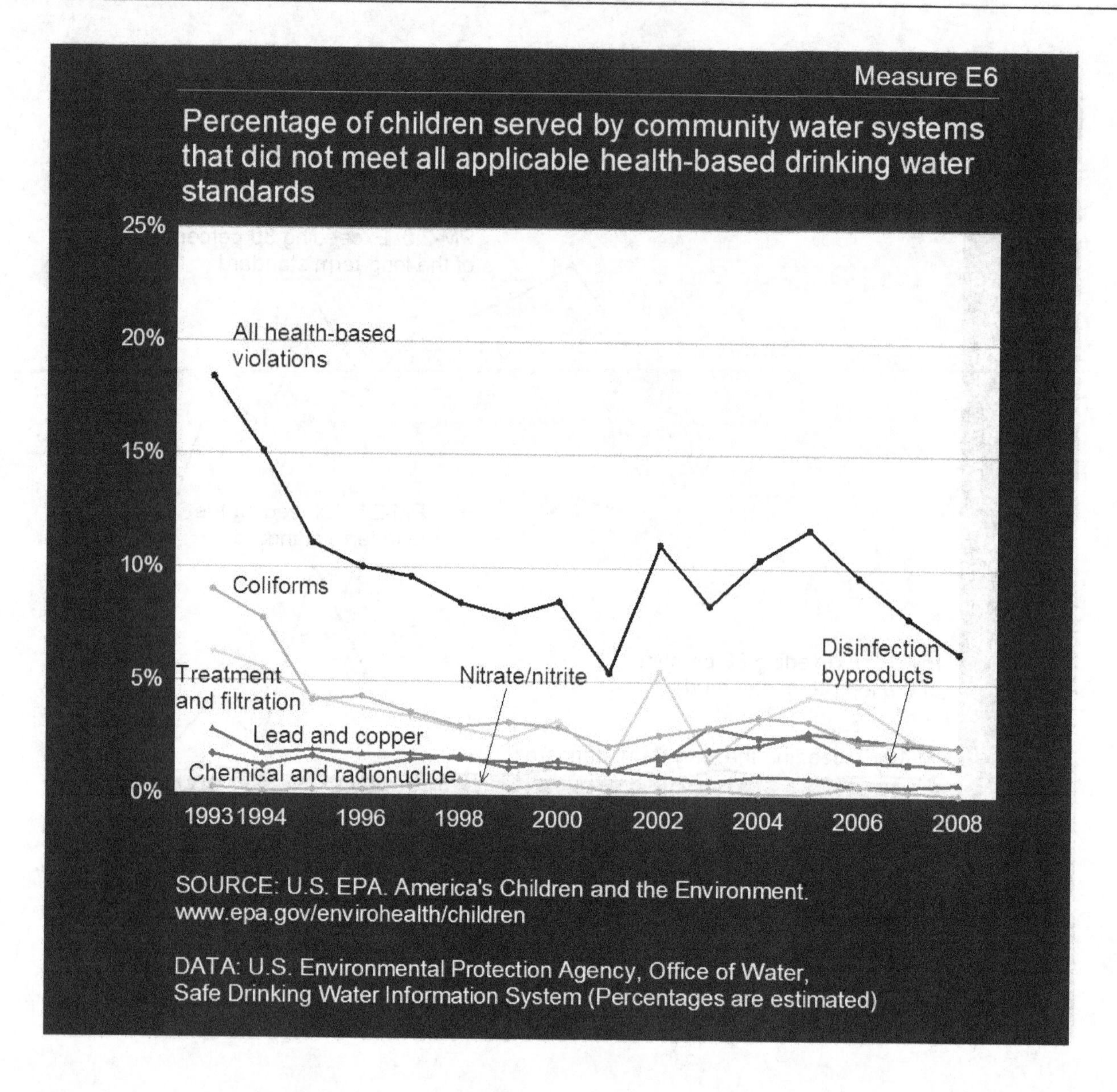
Measure E6
Percentage of children served by community water systems that did not meet all applicable health-based drinking water standards
25%
20%
15%
10%
5%
0%
All health-based violations
Coliforms
Treatment and filtration
Nitrate/nitrite
Disinfection byproducts
Lead and copper
Chemical and radionuclide
1993
1994
1996
1998
2000
2002
2004
2006
2008
SOURCE: U.S. EPA. America's Children and the Environment. www.epa.gov/envirohealth/children
DATA: U.S. Environmental Protection Agency, Office of Water, Safe Drinking Water Information System (Percentages are estimated)

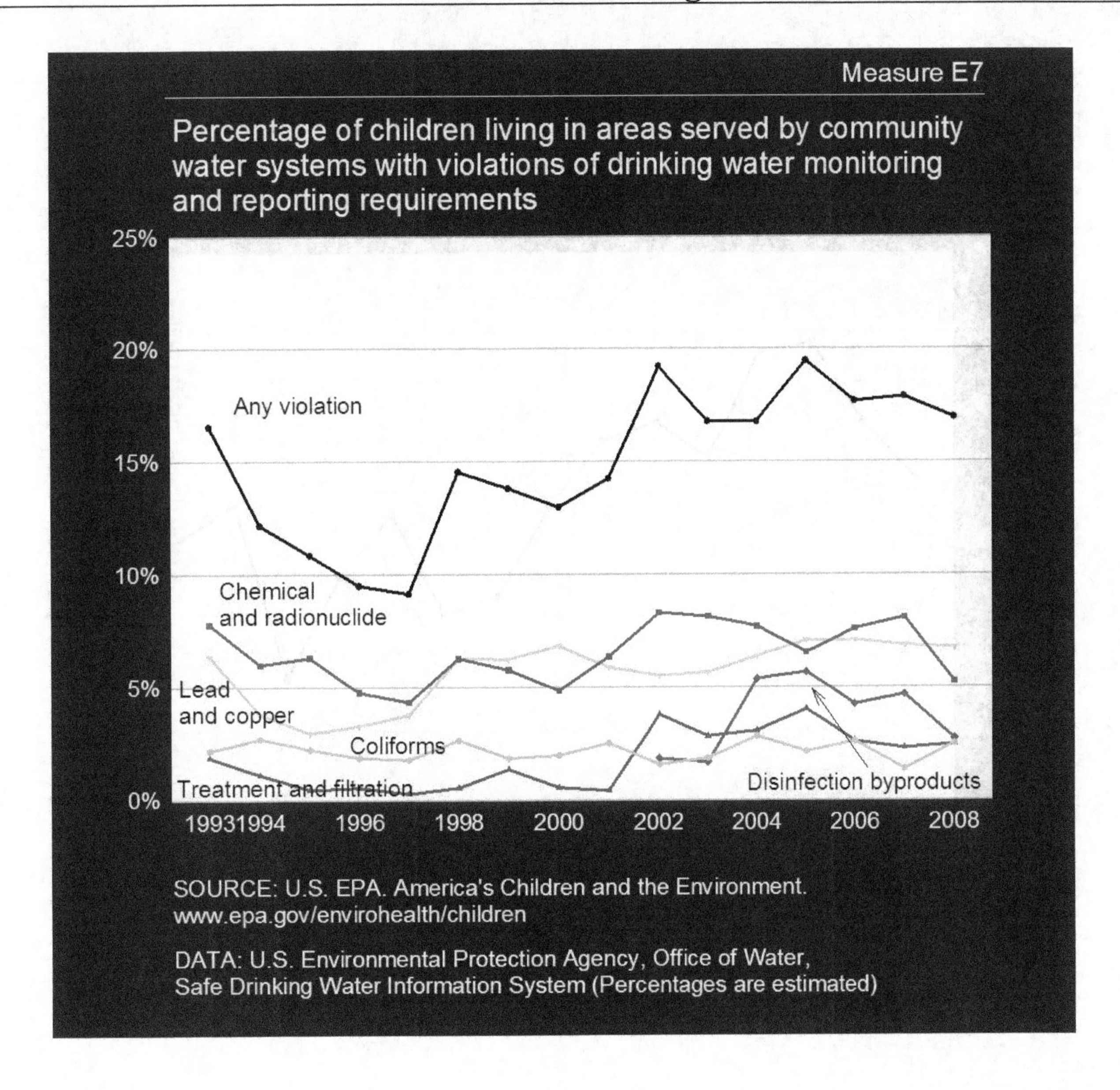
Measure E7
Percentage of children living in areas served by community water systems with violations of drinking water monitoring and reporting requirements
25%
20%
15%
10%
5%
0%
Any violation
Chemical and radionuclide
Lead and copper
Coliforms
Treatment and filtration
Disinfection byproducts
19931994
1996
1998
2000
2002
2004
2006
2008
SOURCE: U.S. EPA. America's Children and the Environment. www.epa.gov/envirohealth/children
DATA: U.S. Environmental Protection Agency, Office of Water, Safe Drinking Water Information System (Percentages are estimated)

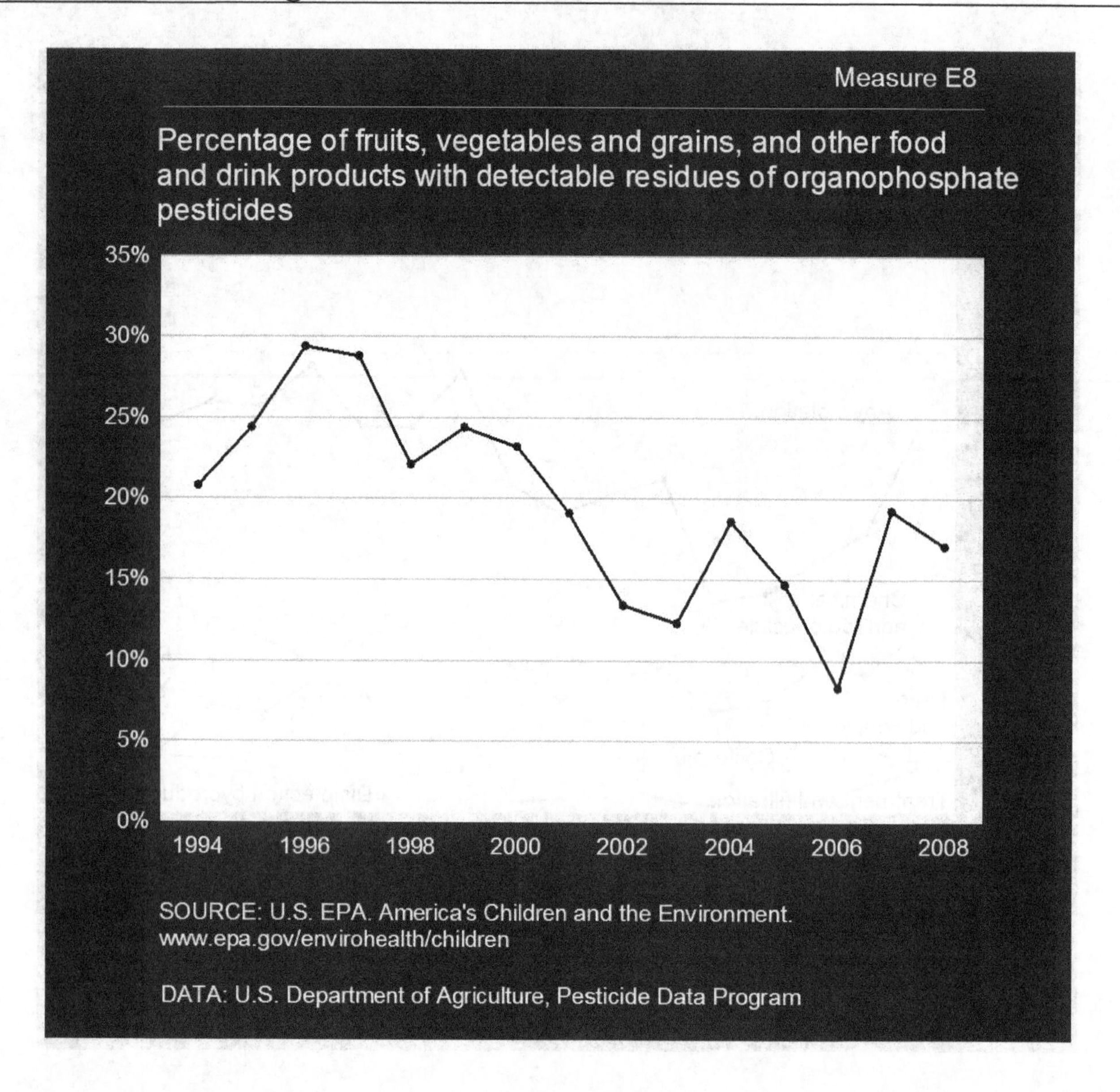
Measure E8
Percentage of fruits, vegetables and grains, and other food and drink products with detectable residues of organophosphate pesticides
35%
30%
25%
20%
15%
10%
5%
0%
1994
1996
1998
2000
2002
2004
2006
2008
SOURCE: U.S. EPA. America's Children and the Environment. www.epa.gov/envirohealth/children
DATA: U.S. Department of Agriculture, Pesticide Data Program

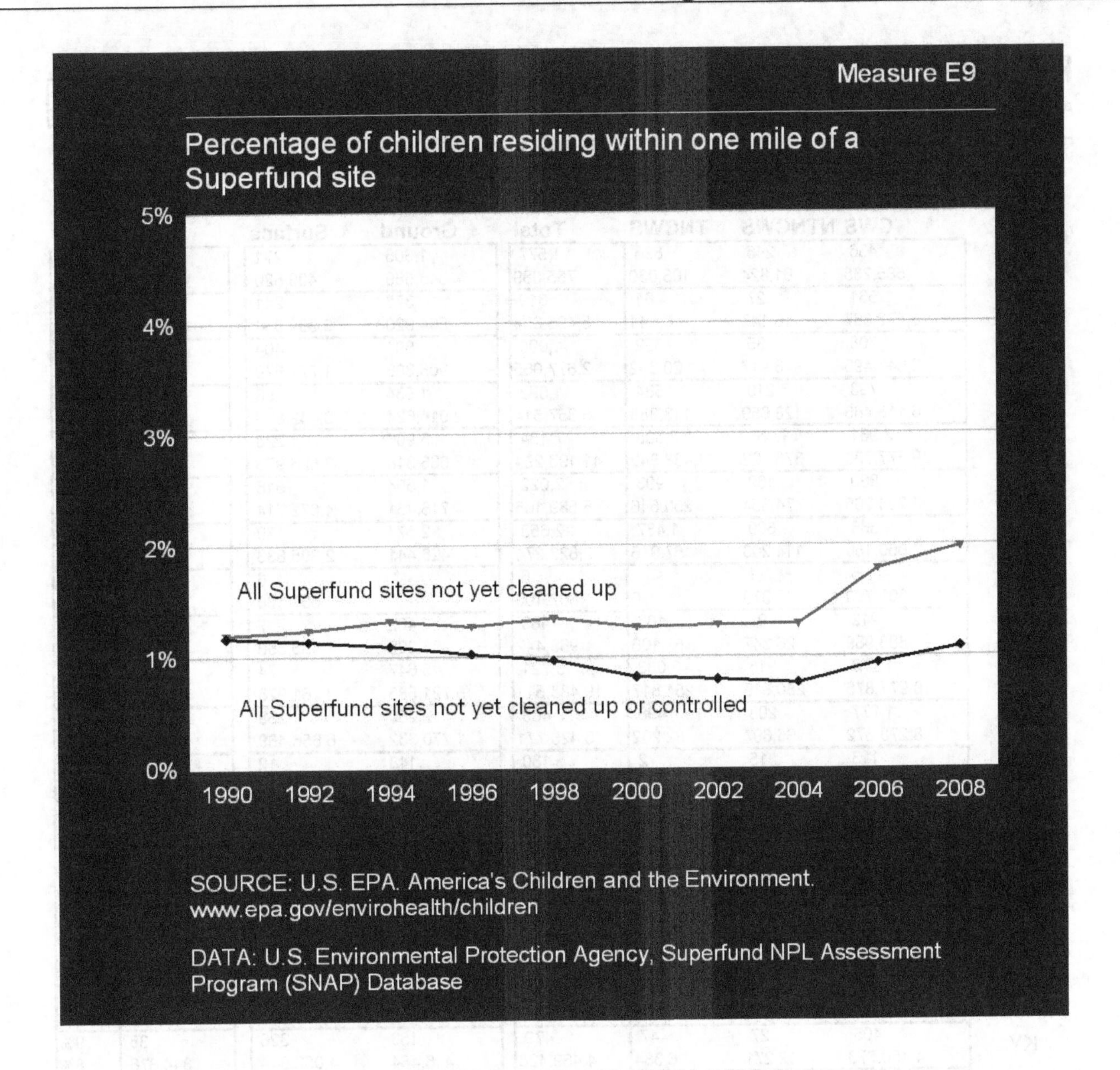
Measure E9
Percentage of children residing within one mile of a Superfund site
5%
4%
3%
2%
1%
0%
All Superfund sites not yet cleaned up
All Superfund sites not yet cleaned up or controlled
1990
1992
1994
1996
1998
2000
2002
2004
2006
2008
SOURCE: U.S. EPA. America's Children and the Environment. www.epa.gov/envirohealth/children
DATA: U.S. Environmental Protection Agency, Superfund NPL Assessment Program (SNAP) Database

# FY2009 Water System Data by State

Active, current systems, from SDWIS/FED 09Q3 frozen inventory table

Health-based (Treatment Technique & Maximum Contaminant Level) violations, from 09Q3 frozen violations table

| # systems |
|---|
| Pop served |

| State | CWS | NTNCWS | TNCWS | Total | Ground | Surface | CWSs w/ reported health-based violations | |
|---|---|---|---|---|---|---|---|---|
| AK | 436 | 245 | 896 | 1,577 | 1,306 | 271 | 76 | 17% |
| | 585,235 | 61,824 | 108,030 | 755,089 | 345,569 | 409,520 | 33,998 | 6% |
| AL | 531 | 27 | 61 | 619 | 378 | 241 | 35 | 7% |
| | 5,472,867 | 16,146 | 6,841 | 5,495,854 | 1,504,920 | 3,990,934 | 227,967 | 4% |
| AR | 708 | 35 | 352 | 1,095 | 691 | 404 | 137 | 19% |
| | 2,647,496 | 8,917 | 20,652 | 2,677,065 | 905,386 | 1,771,679 | 644,578 | 24% |
| AZ | 798 | 210 | 584 | 1,592 | 1,534 | 58 | 106 | 13% |
| | 6,115,466 | 128,659 | 113,389 | 6,357,514 | 2,914,624 | 3,442,890 | 200,764 | 3% |
| CA | 2,961 | 1,365 | 2,808 | 7,134 | 5,906 | 1,228 | 304 | 10% |
| | 39,377,730 | 376,722 | 1,438,842 | 41,193,294 | 7,666,318 | 33,526,976 | 941,301 | 2% |
| CO | 860 | 169 | 993 | 2,022 | 1,504 | 518 | 92 | 11% |
| | 5,264,265 | 74,354 | 250,546 | 5,589,165 | 716,451 | 4,872,714 | 165,381 | 3% |
| CT | 566 | 600 | 1,487 | 2,653 | 2,574 | 79 | 55 | 10% |
| | 2,650,150 | 114,208 | 57,916 | 2,822,274 | 426,441 | 2,395,833 | 70,230 | 3% |
| DC | 5 | 1 | - | 6 | - | 6 | 1 | 20% |
| | 606,730 | 370 | 0 | 607,100 | - | 607,100 | - | 0% |
| DE | 212 | 91 | 186 | 489 | 482 | 7 | 17 | 8% |
| | 888,989 | 26,347 | 53,106 | 968,442 | 485,092 | 483,350 | 24,980 | 3% |
| FL | 1,773 | 915 | 3,033 | 5,721 | 5,647 | 74 | 143 | 8% |
| | 18,977,875 | 250,819 | 254,817 | 19,483,511 | 15,721,585 | 3,761,926 | 842,352 | 4% |
| GA | 1,777 | 208 | 498 | 2,483 | 2,250 | 233 | 88 | 5% |
| | 8,278,872 | 65,697 | 82,202 | 8,426,771 | 1,770,332 | 6,656,439 | 424,043 | 5% |
| HI | 113 | 15 | 2 | 130 | 118 | 12 | 7 | 6% |
| | 1,441,427 | 11,260 | 375 | 1,453,062 | 1,291,355 | 161,707 | 88,715 | 6% |
| IA | 1,134 | 133 | 683 | 1,950 | 1,799 | 151 | 71 | 6% |
| | 2,685,264 | 47,390 | 80,908 | 2,813,562 | 1,510,584 | 1,302,978 | 92,106 | 3% |
| ID | 752 | 225 | 987 | 1,964 | 1,884 | 80 | 133 | 18% |
| | 1,091,272 | 52,116 | 106,193 | 1,249,581 | 990,020 | 259,561 | 181,851 | 17% |
| IL | 1,756 | 402 | 3,573 | 5,731 | 4,996 | 735 | 108 | 6% |
| | 12,049,873 | 129,074 | 359,037 | 12,537,984 | 3,570,467 | 8,967,517 | 529,661 | 4% |
| IN | 820 | 604 | 2,832 | 4,256 | 4,137 | 119 | 64 | 8% |
| | 4,710,557 | 194,504 | 378,003 | 5,283,064 | 2,929,182 | 2,353,882 | 361,196 | 8% |
| KS | 895 | 48 | 90 | 1,033 | 654 | 379 | 110 | 12% |
| | 2,572,953 | 20,645 | 4,195 | 2,597,793 | 708,785 | 1,889,008 | 170,152 | 7% |
| KY | 405 | 27 | 47 | 479 | 153 | 326 | 38 | 9% |
| | 4,450,773 | 12,271 | 6,384 | 4,469,428 | 416,484 | 4,052,944 | 344,876 | 8% |
| LA | 1,069 | 150 | 231 | 1,450 | 1,359 | 91 | 124 | 12% |
| | 4,887,944 | 56,344 | 60,126 | 5,004,414 | 2,994,442 | 2,009,972 | 756,459 | 15% |
| MA | 530 | 271 | 928 | 1,729 | 1,513 | 216 | 71 | 13% |
| | 9,314,153 | 73,462 | 139,444 | 9,527,059 | 1,934,787 | 7,592,272 | 835,037 | 9% |
| MD | 484 | 561 | 2,482 | 3,527 | 3,404 | 123 | 50 | 10% |
| | 5,145,864 | 160,542 | 216,098 | 5,522,504 | 935,549 | 4,586,955 | 1,693,762 | 33% |
| ME | 381 | 359 | 1,160 | 1,900 | 1,818 | 82 | 63 | 17% |
| | 661,560 | 68,358 | 183,582 | 913,500 | 467,894 | 445,606 | 94,128 | 14% |
| MI | 1,404 | 1,464 | 8,686 | 11,554 | 11,247 | 307 | 113 | 8% |
| | 7,614,815 | 336,872 | 1,020,307 | 8,971,994 | 3,068,050 | 5,903,944 | 123,467 | 2% |
| MN | 958 | 523 | 5,781 | 7,262 | 7,149 | 113 | 59 | 6% |
| | 4,191,398 | 78,401 | 536,240 | 4,806,039 | 3,467,539 | 1,338,500 | 205,420 | 5% |
| MO | 1,478 | 250 | 1,057 | 2,785 | 2,538 | 247 | 195 | 13% |
| | 5,175,712 | 76,761 | 116,403 | 5,368,876 | 1,851,384 | 3,517,492 | 276,645 | 5% |
| MS | 1,117 | 84 | 76 | 1,277 | 1,264 | 13 | 100 | 9% |
| | 3,083,388 | 75,091 | 10,216 | 3,168,695 | 2,933,268 | 235,427 | 379,259 | 12% |

# FY2009 Water System Data by State

Active, current systems, from SDWIS/FED 09Q3 frozen inventory table

Health-based (Treatment Technique & Maximum Contaminant Level) violations, from 09Q3 frozen violations table

Legend: # systems / Pop served

| State | CWS | NTNCWS | TNCWS | Total | Ground | Surface | CWSs w/ reported health-based violations | % |
|---|---|---|---|---|---|---|---|---|
| MT | 686 | 254 | 1,157 | 2,097 | 1,881 | 216 | 95 | 14% |
| | 716,606 | 79,028 | 176,244 | 971,878 | 540,052 | 431,826 | 78,998 | 11% |
| NC | 2,129 | 455 | 3,753 | 6,337 | 5,892 | 445 | 322 | 15% |
| | 7,366,427 | 124,936 | 318,186 | 7,809,549 | 1,984,364 | 5,825,185 | 701,741 | 10% |
| ND | 332 | 24 | 152 | 508 | 399 | 109 | 19 | 6% |
| | 568,478 | 3,658 | 13,893 | 586,029 | 265,824 | 320,205 | 13,265 | 2% |
| NE | 596 | 162 | 566 | 1,324 | 1,264 | 60 | 137 | 23% |
| | 1,478,839 | 52,263 | 53,618 | 1,584,720 | 745,620 | 839,100 | 164,688 | 11% |
| NH | 706 | 450 | 1,265 | 2,421 | 2,364 | 57 | 108 | 15% |
| | 854,669 | 96,692 | 318,595 | 1,269,956 | 739,868 | 530,088 | 80,473 | 9% |
| NJ | 614 | 803 | 2,423 | 3,840 | 3,722 | 118 | 44 | 7% |
| | 8,785,575 | 354,170 | 416,911 | 9,556,656 | 3,707,519 | 5,849,137 | 1,672,522 | 19% |
| NM | 623 | 160 | 456 | 1,239 | 1,165 | 74 | 98 | 16% |
| | 1,704,947 | 51,956 | 77,437 | 1,834,340 | 1,085,421 | 748,919 | 232,416 | 14% |
| NV | 215 | 105 | 242 | 562 | 523 | 39 | 37 | 17% |
| | 2,529,692 | 41,609 | 23,171 | 2,594,472 | 294,981 | 2,299,491 | 117,583 | 5% |
| NY | 2,668 | 749 | 5,877 | 9,294 | 7,996 | 1,298 | 281 | 11% |
| | 17,953,613 | 313,262 | 2,845,236 | 21,112,111 | 5,482,338 | 15,629,773 | 1,857,850 | 10% |
| OH | 1,261 | 796 | 2,983 | 5,040 | 4,726 | 314 | 73 | 6% |
| | 10,350,734 | 228,342 | 424,427 | 11,003,503 | 3,461,001 | 7,542,502 | 404,712 | 4% |
| OK | 1,103 | 105 | 363 | 1,571 | 838 | 733 | 213 | 19% |
| | 3,519,562 | 21,230 | 30,432 | 3,571,224 | 660,538 | 2,910,686 | 748,790 | 21% |
| OR | 871 | 336 | 1,423 | 2,630 | 2,324 | 306 | 97 | 11% |
| | 3,199,133 | 71,637 | 212,474 | 3,483,244 | 787,167 | 2,696,077 | 69,210 | 2% |
| PA | 2,073 | 1,227 | 6,109 | 9,409 | 8,816 | 593 | 175 | 8% |
| | 10,757,852 | 521,004 | 779,324 | 12,058,180 | 2,600,087 | 9,458,093 | 586,558 | 5% |
| RI | 87 | 72 | 284 | 443 | 416 | 27 | 10 | 11% |
| | 977,970 | 26,431 | 48,774 | 1,053,175 | 210,121 | 843,054 | 76,574 | 8% |
| SC | 612 | 139 | 736 | 1,487 | 1,143 | 344 | 37 | 6% |
| | 3,819,317 | 42,383 | 40,981 | 3,902,681 | 616,963 | 3,285,718 | 380,537 | 10% |
| SD | 457 | 26 | 173 | 656 | 515 | 141 | 57 | 12% |
| | 686,897 | 8,180 | 22,977 | 718,054 | 312,723 | 405,331 | 35,296 | 5% |
| TN | 483 | 46 | 355 | 884 | 529 | 355 | 38 | 8% |
| | 6,095,241 | 26,463 | 56,565 | 6,178,269 | 1,483,034 | 4,695,235 | 303,020 | 5% |
| TX | 4,649 | 855 | 1,234 | 6,738 | 5,489 | 1,249 | 423 | 9% |
| | 24,630,929 | 511,200 | 249,813 | 25,391,942 | 7,266,177 | 18,125,765 | 1,462,779 | 6% |
| UT | 465 | 68 | 490 | 1,023 | 891 | 132 | 45 | 10% |
| | 2,686,530 | 30,015 | 75,574 | 2,792,119 | 824,644 | 1,967,475 | 134,484 | 5% |
| VA | 1,198 | 565 | 1,116 | 2,879 | 2,492 | 387 | 135 | 11% |
| | 6,553,650 | 307,903 | 171,198 | 7,032,751 | 779,408 | 6,253,343 | 135,054 | 2% |
| VT | 445 | 242 | 679 | 1,366 | 1,248 | 118 | 82 | 18% |
| | 452,470 | 41,814 | 97,807 | 592,091 | 317,127 | 274,964 | 47,372 | 10% |
| WA | 2,253 | 320 | 1,575 | 4,148 | 3,889 | 259 | 215 | 10% |
| | 6,172,424 | 143,150 | 394,578 | 6,710,152 | 3,049,884 | 3,660,268 | 104,501 | 2% |
| WI | 1,074 | 877 | 9,531 | 11,482 | 11,428 | 54 | 93 | 9% |
| | 3,987,846 | 209,247 | 716,891 | 4,913,984 | 3,063,447 | 1,850,537 | 337,714 | 8% |
| WV | 498 | 122 | 456 | 1,076 | 742 | 334 | 37 | 7% |
| | 1,498,069 | 39,313 | 32,911 | 1,570,293 | 306,738 | 1,263,555 | 135,834 | 9% |
| WY | 308 | 89 | 378 | 775 | 637 | 138 | 20 | 6% |
| | 444,979 | 23,369 | 74,672 | 543,020 | 196,626 | 346,394 | 7,576 | 2% |
| | 50,329 | 18,029 | 83,289 | 151,647 | 137,634 | 14,013 | 5,251 | |
| | 287,735,077 | 5,886,409 | 13,276,541 | | 102,308,180 | 204,589,847 | 19,597,875 | |

# CWS Violations Reported by Fiscal Year

FY2009 data from SDWIS/FED 09Q3 frozen tables
FY2008 data from SDWIS/FED 08Q3 frozen tables
FY2007 data from SDWIS/FED 07Q3 frozen tables
FY2006 data from SDWIS/FED 06Q3 frozen tables
FY2005 data from SDWIS/FED 05Q4 frozen tables, except for Chem M/Rs which are from 06Q1
FY2004 data from SDWIS/FED 04Q4 frozen tables, except for Chem M/Rs which are from 05Q1
FY2003 data from SDWIS/FED 03Q4 frozen tables, except for Chem M/Rs which are from 04Q1

### Number of violations

| FY | MCL | MRDL | TT | M/R | Other* | Total |
|---|---|---|---|---|---|---|
| 2009 | 9,971 | 5 | 2,392 | 60,823 | 20,941 | 94,132 |
| 2008 | 9,883 | 5 | 2,647 | 103,064 | 20,774 | 136,373 |
| 2007 | 9,410 | 5 | 3,056 | 61,316 | 23,299 | 97,086 |
| 2006 | 9,076 | 7 | 2,790 | 57,117 | 19,245 | 88,235 |
| 2005 | 9,360 | 4 | 3,036 | 109,167 | 18,414 | 139,981 |
| 2004 | 5,340 | 6 | 2,168 | 87,393 | 14,459 | 109,366 |
| 2003 | 4,688 | 1 | 2,299 | 58,675 | 12,671 | 78,334 |

### Number of systems in violation

| FY | MCL | MRDL | TT | M/R | Other* | Total* |
|---|---|---|---|---|---|---|
| 2009 | 4,451 | 3 | 1,401 | 13,403 | 8,546 | 20,684 |
| 2008 | 4,553 | 5 | 1,506 | 13,084 | 8,631 | 20,797 |
| 2007 | 4,412 | 4 | 1,648 | 12,619 | 10,429 | 21,544 |
| 2006 | 4,371 | 3 | 1,538 | 13,071 | 9,600 | 21,618 |
| 2005 | 4,383 | 4 | 1,625 | 14,291 | 10,258 | 22,772 |
| 2004 | 3,419 | 4 | 1,409 | 13,551 | 8,947 | 21,055 |
| 2003 | 3,042 | 1 | 1,458 | 12,819 | 8,109 | 20,280 |

### Population affected

| FY | MCL | MRDL | TT | M/R | Other* | Total* |
|---|---|---|---|---|---|---|
| 2009 | 15,918,279 | 29,763 | 10,643,904 | 50,360,756 | 20,077,807 | 78,507,217 |
| 2008 | 15,841,822 | 1,021 | 10,086,935 | 57,538,783 | 21,872,736 | 83,385,983 |
| 2007 | 16,105,351 | 263,875 | 10,895,825 | 51,148,667 | 30,434,294 | 87,843,223 |
| 2006 | 16,221,483 | 46,344 | 16,612,762 | 52,289,555 | 23,372,794 | 83,062,903 |
| 2005 | 18,951,548 | 424,859 | 16,879,450 | 61,767,844 | 23,442,800 | 90,890,243 |
| 2004 | 15,018,194 | 3,617 | 16,588,301 | 48,675,006 | 17,833,120 | 76,870,272 |
| 2003 | 15,438,443 | 2,250 | 16,969,087 | 49,497,572 | 12,726,332 | 80,765,799 |

# FY2009 CWS Results by System Size

### Number of violations

| | MCL | MRDL | TT | M/R | Other | Total |
|---|---|---|---|---|---|---|
| Very small | 5,404 | 4 | 1,466 | 37,910 | 15,820 | 60,604 |
| Small | 2,953 | | 584 | 13,441 | 3,505 | 20,483 |
| Medium | 1,023 | | 183 | 4,362 | 1,009 | 6,577 |
| Large | 576 | 1 | 137 | 4,678 | 589 | 5,981 |
| Very large | 15 | | 22 | 432 | 18 | 487 |

### Number of systems in violation

| | MCL | MRDL | TT | M/R | Other | Total* |
|---|---|---|---|---|---|---|
| Very small | 2,491 | 2 | 867 | 8,341 | 5,812 | 12,673 |
| Small | 1,205 | | 333 | 3,488 | 1,858 | 5,299 |
| Medium | 442 | | 106 | 943 | 519 | 1,588 |
| Large | 304 | 1 | 80 | 585 | 341 | 1,051 |
| Very large | 9 | | 15 | 46 | 16 | 73 |

### Population affected

| | MCL | MRDL | TT | M/R | Other | Total* |
|---|---|---|---|---|---|---|
| Very small | 414,466 | 361 | 137,056 | 1,337,033 | 890,270 | 2,031,407 |
| Small | 1,760,191 | | 492,897 | 4,827,782 | 2,551,045 | 7,483,656 |
| Medium | 2,540,925 | | 619,293 | 5,409,412 | 2,952,772 | 9,127,587 |
| Large | 7,425,642 | 29,402 | 2,308,071 | 14,875,363 | 8,757,525 | 27,344,784 |
| Very large | 3,777,055 | | 7,086,587 | 23,911,166 | 4,926,195 | 32,519,783 |

*Totals for the number of systems in violation, and for population affected, should be lower than the sum in each row. This is because some systems will have incurred more than one type of violation.

**Compliance information since FY2003 includes new DBP and IESWTR rules.**

# NTNCWS Violations Reported by Fiscal Year

FY2009 data from SDWIS/FED 09Q3 frozen tables
FY2008 data from SDWIS/FED 08Q3 frozen tables
FY2007 data from SDWIS/FED 07Q3 frozen tables
FY2006 data from SDWIS/FED 06Q3 frozen tables
FY2005 data from SDWIS/FED 05Q4 frozen tables, except for Chem M/Rs which are from 06Q1
FY2004 data from SDWIS/FED 04Q4 frozen tables, except for Chem M/Rs which are from 05Q1
FY2003 data from SDWIS/FED 03Q4 frozen tables, except for Chem M/Rs which are from 04Q1

**Number of violations**

| FY | MCL | MRDL | TT | M/R | Other | Total |
|---|---|---|---|---|---|---|
| 2009 | 3,882 | 1 | 185 | 14,276 | 8,516 | 22,519 |
| 2008 | 4,068 | | 227 | 14,938 | 8,159 | 22,714 |
| 2007 | 4,319 | | 237 | 15,626 | 7,664 | 22,996 |
| 2006 | 4,066 | | 211 | 15,571 | 6,417 | 21,800 |
| 2005 | 3,961 | | 215 | 17,542 | 2,284 | 21,236 |
| 2004 | 3,960 | | 195 | 17,314 | 1,697 | 20,653 |
| 2003 | 4,055 | | 245 | 17,429 | 3,685 | 21,338 |

**Number of systems in violation**

| FY | MCL | MRDL | TT | M/R | Other | Total* |
|---|---|---|---|---|---|---|
| 2009 | 3,882 | 1 | 185 | 14,276 | 8,516 | 22,519 |
| 2008 | 4,068 | | 227 | 14,938 | 8,159 | 22,714 |
| 2007 | 4,319 | | 237 | 15,626 | 7,664 | 22,996 |
| 2006 | 4,066 | | 211 | 15,571 | 6,417 | 21,800 |
| 2005 | 3,961 | | 215 | 17,542 | 2,284 | 21,236 |
| 2004 | 3,960 | | 195 | 17,314 | 1,697 | 20,653 |
| 2003 | 4,055 | | 245 | 17,429 | 3,685 | 21,338 |

**Population affected**

| FY | MCL | MRDL | TT | M/R | Other | Total* |
|---|---|---|---|---|---|---|
| 2009 | 1,289,606 | 300 | 37,459 | 2,434,722 | 826,580 | 3,413,437 |
| 2008 | 548,776 | | 58,580 | 1,959,479 | 859,435 | 2,900,038 |
| 2007 | 538,931 | | 63,160 | 1,999,117 | 731,880 | 2,799,528 |
| 2006 | 1,296,508 | | 50,372 | 2,166,532 | 642,080 | 3,596,226 |
| 2005 | 541,171 | | 86,648 | 2,260,178 | 344,411 | 2,812,854 |
| 2004 | 549,164 | | 51,915 | 2,205,971 | 234,291 | 2,705,723 |
| 2003 | 540,765 | | 40,518 | 2,159,794 | 340,950 | 2,647,611 |

# TNCWS Violations Reported by Fiscal Year

**Number of violations**

| FY | MCL | MRDL | TT | M/R | Other | Total |
|---|---|---|---|---|---|---|
| 2009 | 4,969 | 1 | 259 | 24,397 | 26,422 | 56,048 |
| 2008 | 5,259 | | 384 | 26,589 | 24,437 | 56,669 |
| 2007 | 5,575 | | 435 | 28,599 | 22,078 | 56,687 |
| 2006 | 5,264 | | 474 | 29,848 | 19,010 | 54,596 |
| 2005 | 4,994 | | 502 | 34,880 | 4,800 | 45,176 |
| 2004 | 5,003 | | 415 | 34,295 | 3,028 | 42,741 |
| 2003 | 5,063 | | 573 | 33,666 | 8,640 | 47,942 |

**Number of systems in violation**

| FY | MCL | MRDL | TT | M/R | Other | Total* |
|---|---|---|---|---|---|---|
| 2009 | 3,882 | 1 | 185 | 14,276 | 8,516 | 22,519 |
| 2008 | 4,068 | | 227 | 14,938 | 8,159 | 22,714 |
| 2007 | 4,319 | | 237 | 15,626 | 7,664 | 22,996 |
| 2006 | 4,066 | | 211 | 15,571 | 6,417 | 21,800 |
| 2005 | 3,961 | | 215 | 17,542 | 2,284 | 21,236 |
| 2004 | 3,960 | | 195 | 17,314 | 1,697 | 20,653 |
| 2003 | 4,055 | | 245 | 17,429 | 3,685 | 21,338 |

**Population affected**

| FY | MCL | MRDL | TT | M/R | Other | Total* |
|---|---|---|---|---|---|---|
| 2009 | 1,289,606 | 300 | 37,459 | 2,434,722 | 826,580 | 3,413,437 |
| 2008 | 548,776 | | 58,580 | 1,959,479 | 859,435 | 2,900,038 |
| 2007 | 538,931 | | 63,160 | 1,999,117 | 731,880 | 2,799,528 |
| 2006 | 1,296,508 | | 50,372 | 2,166,532 | 642,080 | 3,596,226 |
| 2005 | 541,171 | | 86,648 | 2,260,178 | 344,411 | 2,812,854 |
| 2004 | 549,164 | | 51,915 | 2,205,971 | 234,291 | 2,705,723 |
| 2003 | 540,765 | | 40,518 | 2,159,794 | 340,950 | 2,647,611 |

*Totals for the number of systems in violation, and for population affected, should be lower than the sum in each row. This is because some systems will have incurred more than one type of violation.
**Compliance information since FY2003 includes new DBP and IESWTR rules.**

# FY2009 MCL, MRDL and TT violations reported

From SDWIS/FED 09Q3 frozen violations table

| # violations |
|---|
| # systems |
| Pop. affected |

| | Very small 25-500 | Small 501-3,300 | Medium 3,301-10,000 | Large 10,001-100,000 | Very Large >100,000 | Total |
|---|---|---|---|---|---|---|
| **Microbials:** | Applies to all water systems | | | | | |
| TCR | 7,265 | 936 | 285 | 262 | 5 | 8,753 |
| | 5,526 | 736 | 240 | 205 | 5 | 6,712 |
| | 596,824 | 1,011,374 | 1,391,789 | 5,039,392 | 2,919,071 | 10,958,450 |
| | Applies to CWS & NTNCWS that disinfect (plus TNCWSs that use ClO2) | | | | | |
| Stage 1 DBP | 1,417 | 1,471 | 522 | 143 | 5 | 3,558 |
| | 590 | 494 | 183 | 69 | 3 | 1,339 |
| | 101,915 | 754,537 | 1,043,531 | 1,387,713 | 532,597 | 3,820,293 |
| | Applies to surface water systems | | | | | |
| SWTR | 733 | 202 | 26 | 15 | 5 | 981 |
| | 403 | 131 | 24 | 13 | 5 | 576 |
| | 64,231 | 175,254 | 157,890 | 380,543 | 3,080,163 | 3,858,081 |
| | Applies to surface water systems | | | | | |
| IE<1SWTR | 162 | 145 | 39 | 68 | 14 | 428 |
| | 63 | 68 | 25 | 36 | 9 | 201 |
| | 12,194 | 105,523 | 145,744 | 1,091,775 | 4,627,993 | 5,983,229 |
| **Organics:** | Applies to CWS and NTNCWS | | | | | |
| VOC | 39 | 7 | 3 | 2 | | 51 |
| | 20 | 5 | 2 | 1 | | 28 |
| | 3,105 | 10,449 | 8,012 | 14,060 | | 35,626 |
| SOC | 26 | 7 | 3 | 1 | | 37 |
| | 12 | 3 | 2 | 1 | | 18 |
| | 1,258 | 2,736 | 10,192 | 17,500 | | 31,686 |
| **Inorganics:** | Applies to CWS and NTNCWS | | | | | |
| Nitrate/Nitrite | 948 | 149 | 18 | 12 | | 1,127 |
| | 551 | 61 | 6 | 6 | | 624 |
| | 62,595 | 73,587 | 29,974 | 110,647 | | 276,803 |
| Arsenic | 1,622 | 515 | 212 | 75 | | 2,424 |
| | 696 | 176 | 49 | 25 | | 946 |
| | 104,370 | 237,810 | 287,016 | 728,314 | | 1,357,510 |
| Other IOC | 289 | 83 | 9 | 12 | 4 | 397 |
| | 101 | 30 | 4 | 7 | 1 | 143 |
| | 16,079 | 38,251 | 27,675 | 283,422 | 100,387 | 465,814 |
| | Applies to CWS | | | | | |
| Radionuclides | 704 | 299 | 81 | 109 | 4 | 1,197 |
| | 232 | 90 | 27 | 24 | 1 | 374 |
| | 39,647 | 119,302 | 150,770 | 643,817 | 155,000 | 1,108,536 |
| | Applies to CWS and NTNCWS | | | | | |
| Lead & Copper | 1,150 | 225 | 40 | 19 | 1 | 1,435 |
| | 836 | 185 | 31 | 16 | 1 | 1,069 |
| | 117,520 | 231,273 | 171,234 | 508,661 | 173,431 | 1,202,119 |

Acronyms:

TCR = Total Coliform Rule
SWTR = Surface Water Treatment Rule
IE<1SWTR = Interim Enhanced and Long Term 1 Surface Water Treatment Rule
DBP = Disinfection By-Products Rule
VOC = Regulated volatile organic contaminants (e.g., Benzene) other than TTHM (which are listed in DBP Rule)
SOC = Synthetic organic chemicals (e.g., Atrazine)
Other IOC = Regulated inorganic contaminants (e.g., Arsenic) other than Copper, Lead, Nitrate, and Nitrite

# FY2009 Monitoring violations reported

From SDWIS/FED 09Q3 frozen violations table

| # violations |
|---|
| # systems |
| Pop. affected |

| | Very small 25-500 | Small 501-3,300 | Medium 3,301-10,000 | Large 10,001-100,000 | Very Large >100,000 | Total |
|---|---|---|---|---|---|---|
| **Microbials:** | Applies to all water systems | | | | | |
| **TCR** | 26,541 | 2,535 | 331 | 186 | 19 | 29,612 |
| | 16,377 | 1,614 | 280 | 150 | 14 | 18,435 |
| | 1,726,892 | 2,015,229 | 1,601,075 | 4,040,901 | 4,546,479 | 13,930,576 |
| | Applies to CWS & NTNCWS that disinfect (plus TNCWSs that use ClO2) | | | | | |
| **Stage 1 DBP** | 3,967 | 1,623 | 456 | 226 | 10 | 6,282 |
| | 2,386 | 1,023 | 274 | 129 | 6 | 3,818 |
| | 405,558 | 1,367,321 | 1,553,060 | 3,332,316 | 3,045,026 | 9,703,281 |
| | Applies to surface water systems | | | | | |
| **SWTR** | 717 | 231 | 76 | 19 | 3 | 1,046 |
| | 229 | 114 | 39 | 11 | 2 | 395 |
| | 39,579 | 138,706 | 233,070 | 206,525 | 1,280,000 | 1,897,880 |
| | Applies to surface water systems | | | | | |
| **IE<1SWTR** | 254 | 201 | 69 | 28 | 5 | 557 |
| | 78 | 82 | 32 | 21 | 3 | 216 |
| | 15,344 | 113,091 | 191,376 | 447,659 | 3,250,636 | 4,018,106 |
| **Organics:** | Applies to CWS and NTNCWS | | | | | |
| **VOC** | 21,049 | 4,053 | 1,014 | 1,666 | 189 | 27,971 |
| | 758 | 165 | 43 | 52 | 7 | 1,025 |
| | 99,705 | 228,800 | 251,563 | 1,511,496 | 2,829,170 | 4,920,734 |
| **SOC** | 9,457 | 1,666 | 643 | 1,119 | 143 | 13,028 |
| | 305 | 85 | 28 | 39 | 3 | 460 |
| | 35,985 | 115,903 | 167,635 | 1,012,757 | 744,787 | 2,077,067 |
| **Inorganics:** | Applies to CWS and NTNCWS | | | | | |
| **Nitrate/Nitrite** | 6,835 | 590 | 99 | 83 | 6 | 7,613 |
| | 5,867 | 463 | 67 | 54 | 5 | 6,456 |
| | 613,949 | 566,609 | 389,129 | 1,546,166 | 1,250,489 | 4,366,342 |
| **Arsenic** | 625 | 209 | 62 | 61 | 9 | 966 |
| | 481 | 132 | 35 | 36 | 5 | 689 |
| | 70,094 | 163,075 | 210,472 | 1,068,506 | 2,556,191 | 4,068,338 |
| **Other IOC** | 1,698 | 456 | 163 | 184 | 13 | 2,514 |
| | 199 | 64 | 11 | 21 | 3 | 298 |
| | 25,788 | 91,819 | 72,617 | 593,839 | 3,367,135 | 4,151,198 |
| | Applies to CWS | | | | | |
| **Radionuclides** | 1,820 | 3,402 | 1,312 | 904 | 9 | 7,447 |
| | 467 | 677 | 158 | 69 | 2 | 1,373 |
| | 93,482 | 1,005,215 | 845,920 | 1,488,266 | 355,710 | 3,788,593 |
| | Applies to CWS and NTNCWS | | | | | |
| **Lead & Copper** | 8,526 | 1,871 | 382 | 247 | 28 | 11,054 |
| | 5,768 | 1,379 | 317 | 202 | 16 | 7,682 |
| | 832,824 | 1,721,331 | 1,813,721 | 4,868,381 | 11,689,885 | 20,926,142 |

Acronyms:
TCR = Total Coliform Rule
SWTR = Surface Water Treatment Rule
IE<1SWTR = Interim Enhanced and Long Term 1 Surface Water Treatment Rule
DBP = Disinfection By-Products Rule
VOC = Regulated volatile organic contaminants (e.g., Benzene) other than TTHM (which are listed in DBP Rule)
SOC = Synthetic organic chemicals (e.g., Atrazine)
Other IOC = Regulated inorganic contaminants (e.g., Arsenic) other than Copper, Lead, Nitrate, and Nitrite

**Waterborne-disease outbreaks (n = six) associated with untreated recreational water, by state — United States, 2006**

| State | Month | Class* | Etiologic agent | Predominant illness † | No. of cases (n = 74) | Type | Setting |
|---|---|---|---|---|---|---|---|
| Florida | May | II | Norovirus G2 | AGI | 50 | Lake | Swimming beach |
| Massachusetts | Aug | III | *Cryptosporidium* | AGI | 6 | Pond | Camp |
| Minnesota | May | II | Norovirus G1 | AGI | 10 | Lake | Private beach |
| Ohio | Aug | IV | Unidentified§ | Skin | 2 | Pond | Pond |
| Tennessee | Jul | IV | *Escherichia coli* O157:H7 | AGI | 3 | Lake | Swimming beach |
| Wisconsin | Jun | IV | *E. coli* O157:H7 | AGI | 3 | Lake | State park |

* On the basis of epidemiologic and water-quality data provided on CDC form 52.12 (available at http://www.cdc.gov/healthyswimming/downloads/cdc_5212_waterborne.pdf) (and Table 1).
† AGI: acute gastrointestinal illness; and Skin: illness, condition, or symptom related to skin.
§ Etiology unidentified: clinical diagnosis of cercarial dermatitis (caused by avian schistosomes).

*Source: Centers for Disease Control and Prevention, Morbidity and Mortality Weekly Report, Surveillance Summaries, September 12, 2008 / Vol. 57 / No. SS-9*

**Waterborne-disease outbreaks (n = 34) associated with treated recreational water, by state — United States, 2006**

| State | Month | Class* | Etiologic agent | Predominant illness† | No. of cases (deaths) (n = 791) | Type | Setting |
|---|---|---|---|---|---|---|---|
| Arkansas | Jul | I | *Legionella pneumophila*§ | ARI | 37 | Pool, spa | Hotel |
| California | Jul | II | *Shigella sonnei* | AGI | 9 | Kiddie pool | Private residence |
| California | Jul | III | *Cryptosporidium*¶ | AGI | 16 | Interactive fountain | Community |
| Colorado | Aug | II | *Cryptosporidium*** | AGI | 12 | Pool | Community |
| Colorado | Oct | II | *L. pneumophila* serogroup 1†† | ARI | 6 | Spa | Private home |
| Florida | Jan | I | *L. pneumophila* serogroup 1 | ARI | 11 (1) | Spa | Hotel |
| Florida | May | III | *Giardia, Cryptosporidium*§§ | AGI | 55 | Interactive fountain | Community |
| Florida | Aug | IV | *Cryptosporidium* | AGI | 3 | Pool | Hotel |
| Georgia | Aug | IV | *Cryptosporidium* | AGI | 19 | Pool | Community |
| Georgia | Feb | IV | Unidentified ¶¶ | Skin | 4 | Spa | Cabin |
| Georgia | Oct | IV | *Cryptosporidium* | AGI | 4 | Pool | Community |
| Illinois | Jan | I | *Legionella**** | ARI | 43 (1) | Pool, spa | Hotel |
| Illinois | Jul | I | *C. hominis*** | AGI | 65 | Pool | Day camp, water park |
| Illinois | Jun | III | Unidentified††† | Skin | 9 | Pool | Community |
| Illinois | Aug | IV | *Cryptosporidium* | AGI | 4 | Pool | Water park |
| Illinois | Aug | IV | *Cryptosporidium* | AGI | 18 | Pool | Water park |
| Indiana | Apr | IV | Unidentified | ARI, ear | 12 | Pool | Membership club |
| Kansas | Dec | IV | *Pseudomonas aeruginosa* | Skin | 8 | Spa | Private residence |
| Louisiana | Jul | II | *Cryptosporidium*** | AGI | 29 | Pool, interactive fountain | Water park |
| Minnesota | Sep | II | *C. hominis* | AGI | 47 | Pool | Schools |
| Missouri | Jul | IV | *Cryptosporidium* | AGI | 6 | Pool | Community |
| Missouri | Jun | III | *Cryptosporidium* | AGI | 116 | Pool, interactive fountain | Water park |
| Montana | Jul | IV | *Cryptosporidium* | AGI | 82 | Pools | Community |
| Nebraska | Dec | III | Unidentified§§§ ¶¶¶ | ARI, eye | 24 | Pool | Hotel |
| New York | Oct | III | *L. pneumophila* serogroup 1 | ARI | 2 | Spa | Water park |
| New York | Mar | IV | Unidentified§§§ | ARI | 9 | Pool | Water park |
| Pennsylvania | Jun | IV | *Cryptosporidium* | AGI | 13 | Pool | Membership club |
| South Carolina | Jul | III | *C. hominis*** | AGI | 12 | Pool | Community |
| Tennessee | Mar | IV | Unidentified¶¶ | Skin | 15 | Pool, spa | Hotel |
| Wisconsin | Feb | III | Unidentified ¶¶ | Skin | 28 | Pool, spa | Hotel |
| Wisconsin | May | I | Norovirus | AGI | 18 | Pool | Hotel |
| Wisconsin | Aug | II | *Cryptosporidium* | AGI | 22 | Pool | Campground |
| Wisconsin | Aug | IV | *Cryptosporidium* | AGI | 4 | Pool | Community |
| Wyoming | Jun | II | *Cryptosporidium*** | AGI | 29 | Pools**** | Community |

* On the basis of epidemiologic and water-quality data provided on CDC form 52.12 (available at http://www.cdc.gov/healthyswimming/downloads/cdc_5212_waterborne.pdf) (see Table 1).

† ARI: acute respiratory illness; AGI: acute gastrointestinal illness; Skin: illness, condition, or symptom related to skin; Ear: illness, condition, or symptom related to ears; and Eye: illness, condition, or symptom related to eyes.

§ Pontiac fever was diagnosed in 34 persons, and Legionnaires' disease was diagnosed in three persons.

¶ Eleven pulsed-field gel electrophoresis-matched cases of *Salmonella stanley* were also detected among persons who visited the fountain during this outbreak; however, the outbreak investigation did not rule out other possible common exposures among those case-patients.

** **Source:** CDC. Cryptosporidiosis outbreaks associated with recreational water use — five states, 2006. MMWR 2007;56:729–32.

†† All cases were diagnosed as Pontiac fever.

§§ Thirty-five persons had stool specimens that tested positive for *Giardia*, seven persons had stool specimens that tested positive for *Cryptosporidium*, and two persons had stool specimens that tested positive for both *Giardia* and *Cryptosporidium*.

¶¶ Etiology unidentified: *Pseudomonas aeruginosa* suspected on the basis of clinical syndrome and setting.

*** Pontiac fever was diagnosed in 40 persons, and Legionnaires' disease was diagnosed in three persons. *L. pneumophila* and *L. maceachernii* were detected in both pool and spa water.

††† Etiology unidentified: low pH suspected on the basis of water testing and symptoms.

§§§ Etiology unidentified: chemical contamination from pool disinfection by-products (e.g., chloramines) suspected.

¶¶¶ **Source:** CDC. Ocular and respiratory illness associated with an indoor swimming pool—Nebraska, 2006. MMWR 2007;56:929–32.

**** Case-patients identified in this outbreak reported exposure to multiple community pools and to an untreated reservoir.

*Source: Centers for Disease Control and Prevention, Morbidity and Mortality Weekly Report, Surveillance Summaries, September 12, 2008 / Vol. 57 / No. SS-9*

Number of waterborne-disease outbreaks (n = 78) associated with recreational water, by etiologic agent(s) and type of water — United States, 2005–2006

| | Type of water | | | | | |
|---|---|---|---|---|---|---|
| | Treated | | Untreated | | Total | |
| Predominant illness | No. of outbreaks | No. of cases | No. of outbreaks | No. of cases | No. of outbreaks (%) | No. of cases (%) |
| Bacteria | 14 | 167 | 8 | 88 | 22 (28.2) | 255 (5.8) |
| *Campylobacter jejuni* | 1 | 6 | 0 | 0 | 1 | 6 |
| *Escherichia coli* spp. | 0 | 0 | 3 | 10 | 3 | 10 |
| *Leptospira* spp. | 0 | 0 | 2 | 46 | 2 | 46 |
| *Legionella* spp.* | 8 | 124 | 0 | 0 | 8 | 124 |
| *Pseudomonas aeruginosa* | 4 | 28 | 0 | 0 | 4 | 28 |
| *Shigella sonnei* | 1 | 9 | 3 | 32 | 4 | 41 |
| Parasites | 31 | 3,784 | 3 | 35 | 34 (43.6) | 3,819 (86.6) |
| *Cryptosporidium* spp. | 29 | 3,718 | 2 | 33 | 31 | 3,751 |
| *Giardia intestinalis* | 1 | 11 | 0 | 0 | 1 | 11 |
| *Naegleria fowleri* | 0 | 0 | 1 | 2 | 1 | 2 |
| *Cryptosporidium* and *Giardia* spp.† | 1 | 55 | 0 | 0 | 1 | 55 |
| Viruses | 1 | 18 | 3 | 68 | 4 (5.1) | 86 (1.9) |
| Norovirus | 1 | 18 | 3 | 68 | 4 | 86 |
| Chemicals/toxins | 1 | 19 | 1 | 3 | 2 (2.6) | 22 (0.5) |
| Copper sulfate | 0 | 0 | 1 | 3 | 1 | 3 |
| Chlorine gas§ | 1 | 19 | 0 | 0 | 1 | 19 |
| Suspected etiology | 9 | 135 | 3 | 17 | 12 (15.4) | 152 (3.4) |
| Suspected chemical exposure¶ | 1 | 9 | 0 | 0 | 1 | 9 |
| Suspected chloramines | 3 | 53 | 0 | 0 | 3 | 53 |
| Suspected norovirus | 0 | 0 | 1 | 13 | 1 | 13 |
| Suspected *P. aeruginosa* | 5 | 73 | 0 | 0 | 5 | 73 |
| Suspected schistosomes | 0 | 0 | 2 | 4 | 2 | 4 |
| Unidentified | 2 | 44 | 2 | 34 | 4 (5.1) | 78 (1.8) |
| Total (%) | 58 (74.4) | 4,167 (94.4) | 20 (25.6) | 245 (5.6) | 78 (100.0) | 4,412 (100.0) |

* Five outbreaks were attributed to *Legionella pneumophila*, two outbreak investigations did not identify a *Legionella* species, and one outbreak investigation detected *L. pneumophila* and *L. maceachernii* in both pool and spa water.

† Thirty-five persons had stool specimens that tested positive for *Giardia*, seven persons had stool specimens that tested positive for *Cryptosporidium*, and two persons had stool specimens that tested positive for both *Giardia* and *Cryptosporidium*.

§ Chlorine gas was released after high levels of liquid chlorine and acid were mixed in the recirculation system and subsequently released into the pool water.

¶ Low pH suspected on the basis of water testing and symptoms.

*Source: Centers for Disease Control and Prevention, Morbidity and Mortality Weekly Report, Surveillance Summaries, September 12, 2008 / Vol. 57 / No. SS-9*

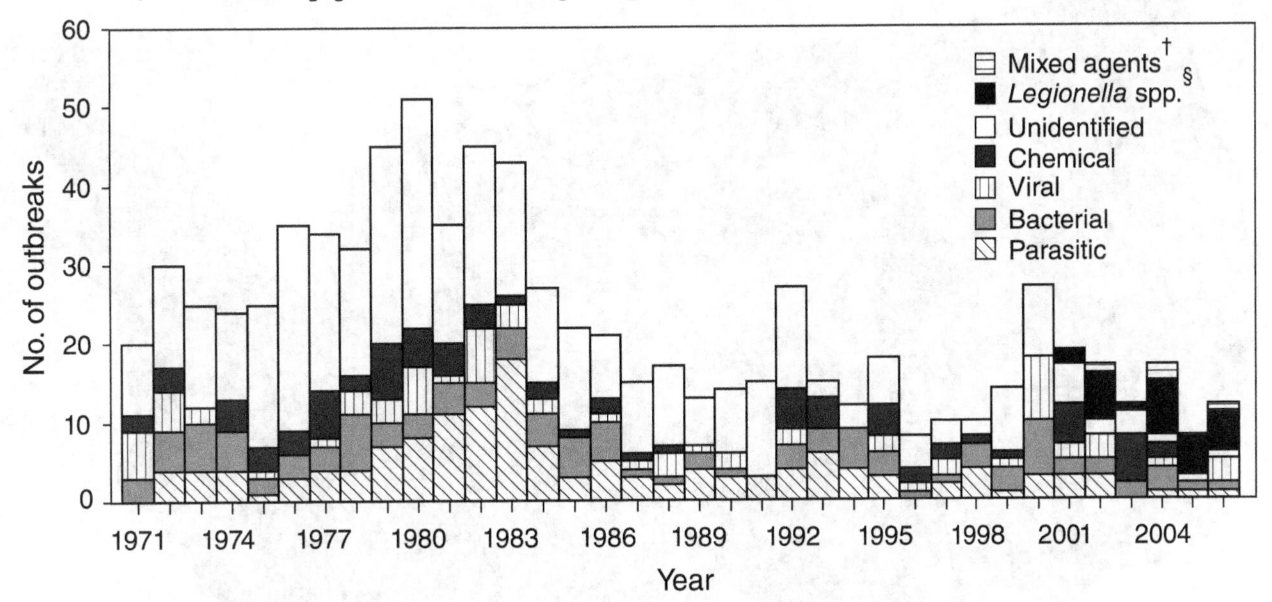


* Single cases of disease related to drinking water (n = 16) have been removed from this figure; therefore, it is not comparable to figures in previous *Surveillance Summaries*.

† Beginning in 2003, mixed agents of more than one etiologic agent type were included in the surveillance system. However, the first observation is a previously unreported outbreak in 2002.

§ Beginning in 2001, Legionnaires' disease was added to the surveillance system, and *Legionella* species were classified separately in this figure.

*Source: Centers for Disease Control and Prevention, Morbidity and Mortality Weekly Report, Surveillance Summaries, September 12, 2008 / Vol. 57 / No. SS-9*

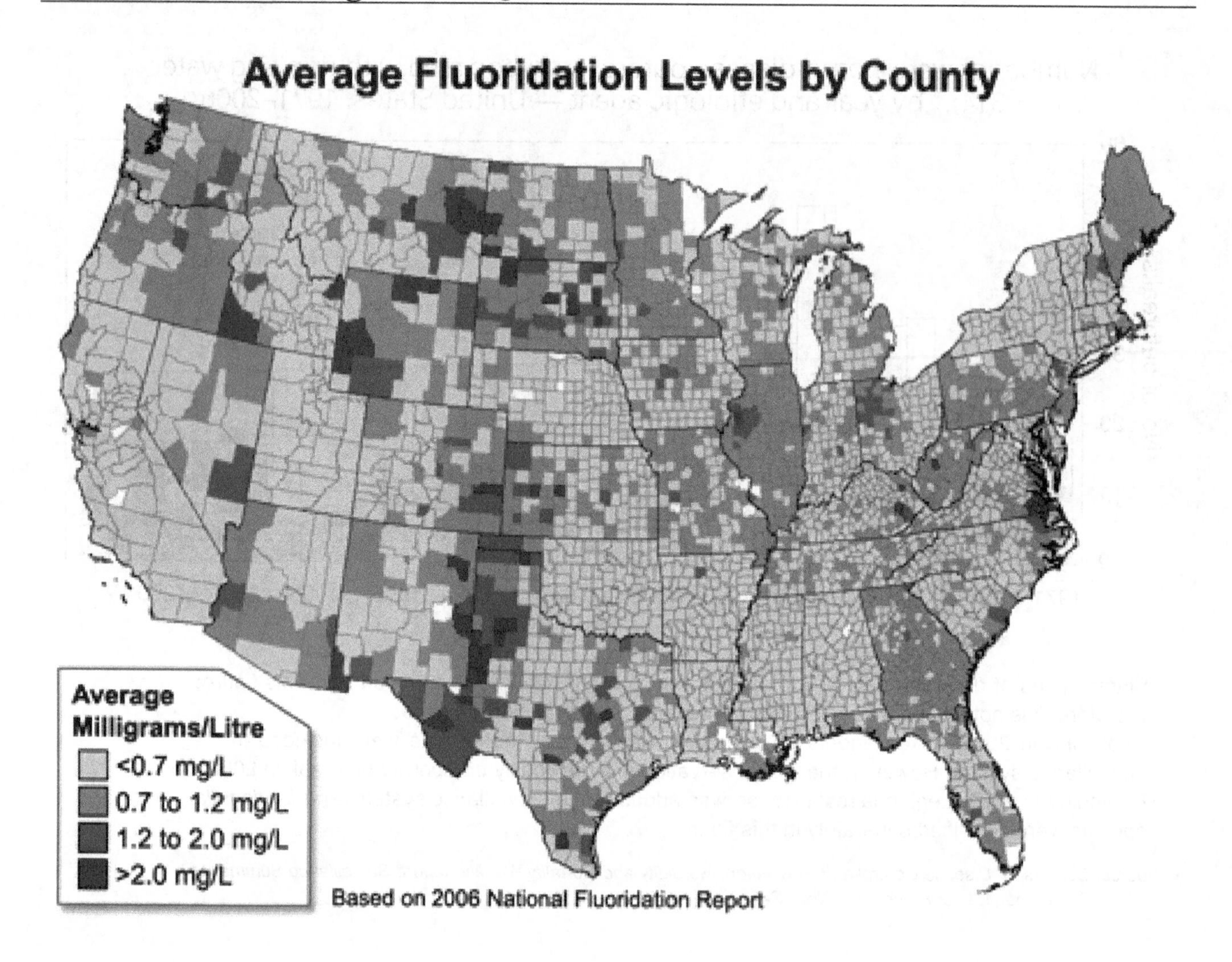


*Note: Page last modified: September 17, 2008*

*Source: Division of Oral Health, National Center for Chronic Disease Prevention and Health Promotion*

## Delisted Species

| Species name | Date species first listed | Date species delisted | Reason delisted |
|---|---|---|---|
| Agave, Arizona (Agave arizonica) | 5/18/1984 | 6/19/2006 | Original Data in Error - Not a listable entity |
| Alligator, American (Alligator mississippiensis) | 7/27/1979 | 6/4/1987 | Recovered |
| Barberry, Truckee (Berberis (=Mahonia) sonnei) | 12/6/1979 | 10/1/2003 | Original Data in Error - Taxonomic revision |
| Bidens, cuneate (Bidens cuneata) | 2/17/1984 | 2/6/1996 | Original Data in Error - Taxonomic revision |
| Broadbill, Guam (Myiagra freycineti) | 8/27/1984 | 2/23/2004 | Extinct |
| Butterfly, Bahama swallowtail (Heraclides andraemon bonhotei) | 4/28/1976 | 8/31/1984 | Original Data in Error - Act amendment |
| Cactus, Lloyd's hedgehog (Echinocereus lloydii) | 11/28/1979 | 6/24/1999 | Original Data in Error - Taxonomic revision |
| Cactus, spineless hedgehog (Echinocereus triglochidiatus var. inermis) | 12/7/1979 | 9/22/1993 | Original Data in Error - Not a listable entity |
| Cinquefoil, Robbins' (Potentilla robbinsiana) | 9/17/1980 | 8/27/2002 | Recovered |
| Cisco, longjaw (Coregonus alpenae) | 3/11/1967 | 9/2/1983 | Extinct |
| Daisy, Maguire (Erigeron maguirei) | 9/5/1985 | 2/18/2011 | Recovered |
| Deer, Columbian white-tailed Douglas County DPS(Odocoileus virginianus leucurus) | 7/24/2003 | 7/24/2003 | Recovered |
| Dove, Palau ground (Gallicolumba canifrons) | 6/2/1970 | 9/12/1985 | Recovered |
| Duck, Mexican U.S.A. only(Anas diazi) | 3/11/1967 | 7/25/1978 | Original Data in Error - Taxonomic revision |
| Eagle, bald lower 48 States(Haliaeetus leucocephalus) | 3/11/1967 | 8/8/2007 | Recovered |
| Falcon, American peregrine (Falco peregrinus anatum) | 6/2/1970 | 8/25/1999 | Recovered |
| Falcon, Arctic peregrine (Falco peregrinus tundrius) | 6/2/1970 | 10/5/1994 | Recovered |
| Flycatcher, Palau fantail (Rhipidura lepida) | 6/2/1970 | 9/12/1985 | Recovered |
| Gambusia, Amistad (Gambusia amistadensis) | 4/30/1980 | 12/4/1987 | Extinct |
| Globeberry, Tumamoc (Tumamoca macdougalii) | 4/29/1986 | 6/18/1993 | Original Data in Error - New information discovered |
| Goose, Aleutian Canada (Branta canadensis leucopareia) | 3/11/1967 | 3/20/2001 | Recovered |
| Hedgehog cactus, purple-spined (Echinocereus engelmannii var. purpureus) | 10/11/1979 | 11/27/1989 | Original Data in Error - Taxonomic revision |
| Kangaroo, eastern gray (Macropus giganteus) | 12/30/1974 | 3/9/1995 | Recovered |
| Kangaroo, red (Macropus rufus) | 12/30/1974 | 3/9/1995 | Recovered |
| Kangaroo, western gray (Macropus fuliginosus) | 12/30/1974 | 3/9/1995 | Recovered |
| Mallard, Mariana (Anas oustaleti) | 12/8/1977 | 2/23/2004 | Extinct |
| Milk-vetch, Rydberg (Astragalus perianus) | 5/27/1978 | 9/14/1989 | Original Data in Error - New information discovered |
| Monarch, Tinian (old world flycatcher) (Monarcha takatsukasae) | 6/2/1970 | 9/21/2004 | Recovered |
| Owl, Palau (Pyrroglaux podargina) | 6/2/1970 | 9/12/1985 | Recovered |
| Pearlymussel, Sampson's (Epioblasma sampsoni) | 6/14/1976 | 1/9/1984 | Extinct |
| Pelican, brown except U.S. Atlantic coast, FL, AL(Pelecanus occidentalis) | 6/2/1970 | 12/17/2009 | Recovered |
| Pelican, brown U.S. Atlantic coast, FL, AL(Pelecanus occidentalis) | 6/2/1970 | 2/4/1985 | Recovered |
| Pennyroyal, Mckittrick (Hedeoma apiculatum) | 7/13/1982 | 9/22/1993 | Original Data in Error - New information discovered |
| Pike, blue (Stizostedion vitreum glaucum) | 3/11/1967 | 9/2/1983 | Extinct |
| Pupfish, Tecopa (Cyprinodon nevadensis calidae) | 10/13/1970 | 1/15/1982 | Extinct |
| Pygmy-owl, cactus ferruginous AZ pop.(Glaucidium brasilianum cactorum) | 3/10/1997 | 4/14/2006 | Original Data in Error - Not a listable entity |
| Seal, Caribbean monk (Monachus tropicalis) | 4/10/1979 | 10/28/2008 | Extinct & Unlist |
| Shrew, Dismal Swamp southeastern (Sorex longirostris fisheri) | 9/26/1986 | 2/28/2000 | Original Data in Error - New information discovered |
| Snail, Utah valvata (Valvata utahensis) | 12/14/1992 | 9/24/2010 | Original Data in Error - New information discovered |
| Sparrow, Santa Barbara song (Melospiza melodia graminea) | 6/4/1973 | 10/12/1983 | Extinct |
| Sparrow, dusky seaside (Ammodramus maritimus nigrescens) | 3/11/1967 | 12/12/1990 | Extinct |
| Springsnail, Idaho (Pyrgulopsis idahoensis) | 12/14/1992 | 9/5/2007 | Original Data in Error - Taxonomic revision |
| Sunflower, Eggert's (Helianthus eggertii) | 5/22/1997 | 8/18/2005 | Recovered |
| Treefrog, pine barrens FL pop.(Hyla andersonii) | 12/18/1977 | 11/22/1983 | Original Data in Error - New information discovered |
| Trout, coastal cutthroat Umpqua R.(Oncorhynchus clarki clarki) | 9/13/1996 | 4/26/2000 | Original Data in Error - Taxonomic revision |

| Species name | Date species first listed | Date species delisted | Reason delisted |
|---|---|---|---|
| Turtle, Indian flap-shelled (Lissemys punctata punctata) | 6/14/1976 | 2/29/1984 | Original Data in Error - Erroneous data |
| Whale, gray except where listed(Eschrichtius robustus) | 6/16/1994 | 6/16/1994 | Recovered |
| Wolf, gray Northern Rocky Mountain DPS(Canis lupus) | 5/5/2010 | 2/27/2008<br>4/2/2009 | Recovered<br>& Recovered |
| Woolly-star, Hoover's (Eriastrum hooveri) | 7/19/1990 | 10/7/2003 | Recovered |

***Notes:*** *Data as of May 20, 2011*
***Source:*** *U.S. Fish & Wildlife Service, Threatened and Endangered Species System (TESS)*

## Summary of Listed Species, Listed Populations[1] and Recovery Plans[2]

| Group | United States[3] | | Foreign | | Total listings (U.S. and foreign) | U.S. listings with active recovery plans[2] |
|---|---|---|---|---|---|---|
| | Endangered | Threatened | Endangered | Threatened | | |
| **Animals** | | | | | | |
| Amphibians | 14 | 10 | 8 | 1 | 33 | 17 |
| Arachnids | 12 | 0 | 0 | 0 | 12 | 12 |
| Birds | 76 | 16 | 198 | 13 | 303 | 85 |
| Clams | 65 | 8 | 2 | 0 | 75 | 70 |
| Corals | 0 | 2 | 0 | 0 | 2 | 0 |
| Crustaceans | 19 | 3 | 0 | 0 | 22 | 18 |
| Fishes | 72 | 67 | 11 | 1 | 151 | 101 |
| Insects | 50 | 10 | 4 | 0 | 64 | 40 |
| Mammals | 70 | 14 | 255 | 20 | 359 | 60 |
| Reptiles | 13 | 24 | 66 | 16 | 119 | 38 |
| Snails | 25 | 11 | 1 | 0 | 37 | 29 |
| Animal Subtotal | 416 | 165 | 545 | 51 | 1,177 | 470 |
| **Plants** | | | | | | |
| Conifers and Cycads | 2 | 1 | 0 | 2 | 5 | 3 |
| Ferns and Allies | 27 | 2 | 0 | 0 | 29 | 26 |
| Flowering Plants | 613 | 145 | 1 | 0 | 759 | 638 |
| Lichens | 2 | 0 | 0 | 0 | 2 | 2 |
| Plant Subtotal | 644 | 148 | 1 | 2 | 795 | 669 |
| Grand Total | 1,060 | 313 | 546 | 53 | 1,972 | 1,139 |

***Notes:*** *Data as of May 20, 2011; (1) A listing has an E or a T in the "status" column of the tables in 50 CFR 17.11(h) or 50 CFR 17.12(h) (the "List of Endangered and Threatened Wildlife and Plants"). 16 animal species (11 in the U.S.3 and 5 Foreign) are counted more than once in the above table, primarily because these animals have distinct population segments (each with its own individual listing status); (2) There are a total of 594 distinct active (Draft and Final) recovery plans. Some recovery plans cover more than one species, and a few species have separate plans covering different parts of their ranges. This count includes only plans generated by the USFWS (or jointly by the USFWS and NMFS), and only listed species that occur in the United States; (3) United States listings include those populations in which the United States shares jurisdiction with another nation.*

***Source:*** *U.S. Fish & Wildlife Service, Threatened and Endangered Species System (TESS)*

# Energy Perspectives

## Overview

**Figure 1. Primary Energy Overview**

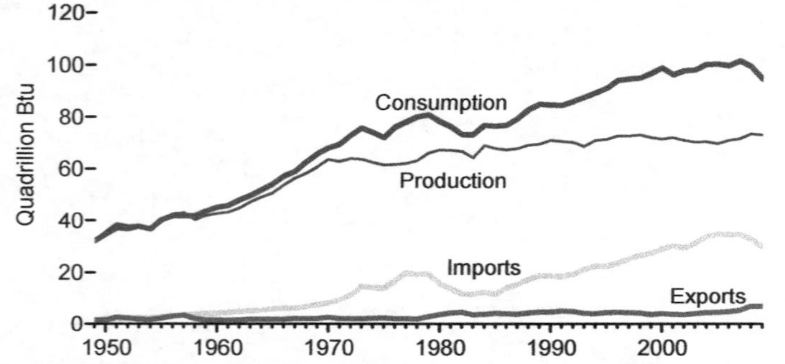

The United States was self-sufficient in energy until the late 1950s when energy consumption began to outpace domestic production. At that point, the Nation began to import more energy to meet its needs. In 2009, net imported energy accounted for 24 percent of all energy consumed.

**Figure 3. Energy Consumption per Real Dollar[1] of Gross Domestic Product**

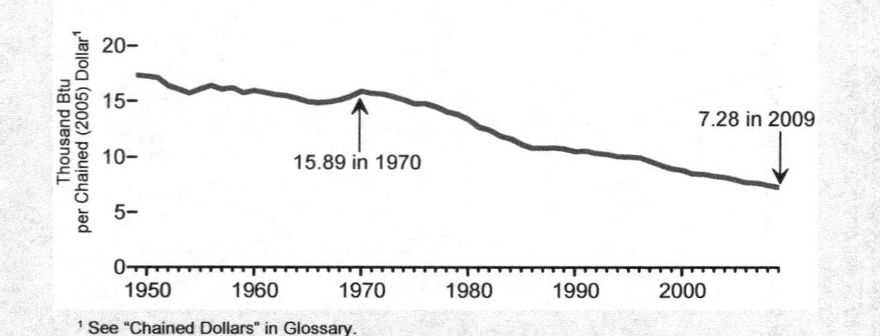


[1] See "Chained Dollars" in Glossary.

After 1970, the amount of energy consumed to produce a dollar's worth of the Nation's output of goods and services trended down. The decline resulted from efficiency improvements and structural changes in the economy. The level in 2009 was 54 percent below that of 1970.

**Figure 2. Energy Consumption per Person**

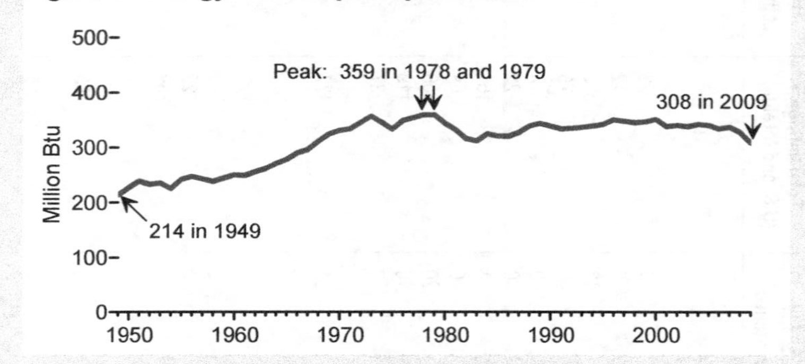


Energy use per person stood at 214 million British thermal units (Btu) in 1949. The rate generally increased until the oil price shocks of the mid-1970s and early 1980s when the trend reversed for a few years. From 1988 on, the rate held fairly steady until the 2008-2009 economic downturn. In 2009, 308 million Btu of energy were consumed per person, 44 percent above the 1949 rate.

**Figure 4. Primary Energy Consumption by Source**

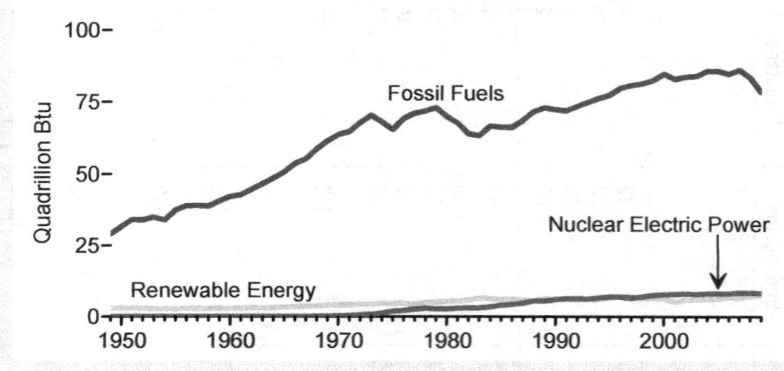


Most energy consumed in the United States comes from fossil fuels. Renewable energy resources supplied a small but growing portion. In the late 1950s, nuclear fuel began to be used to generate electricity. From 1998 through 2009, nuclear electric power surpassed renewable energy.

U.S. Energy Information Administration / Annual Energy Review 2009

# Consumption by Source

**Figure 5. Primary Energy Consumption by Source, 1775-2009**

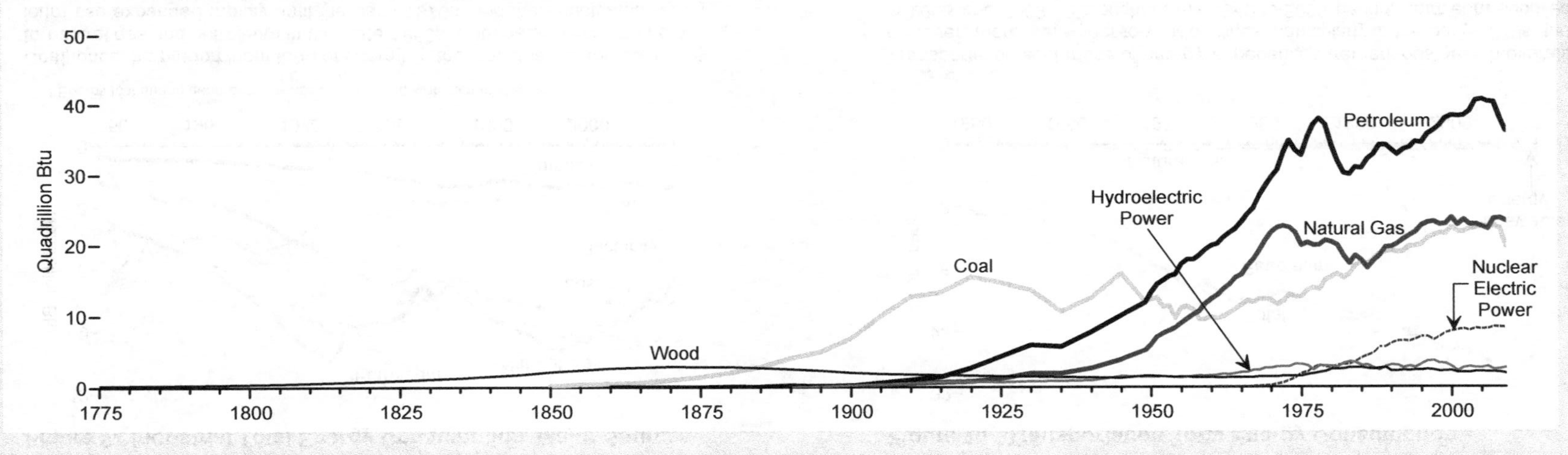


**Figure 6. Energy Consumption Outlook From the Annual Energy Outlook Reference Case, 2010-2035**

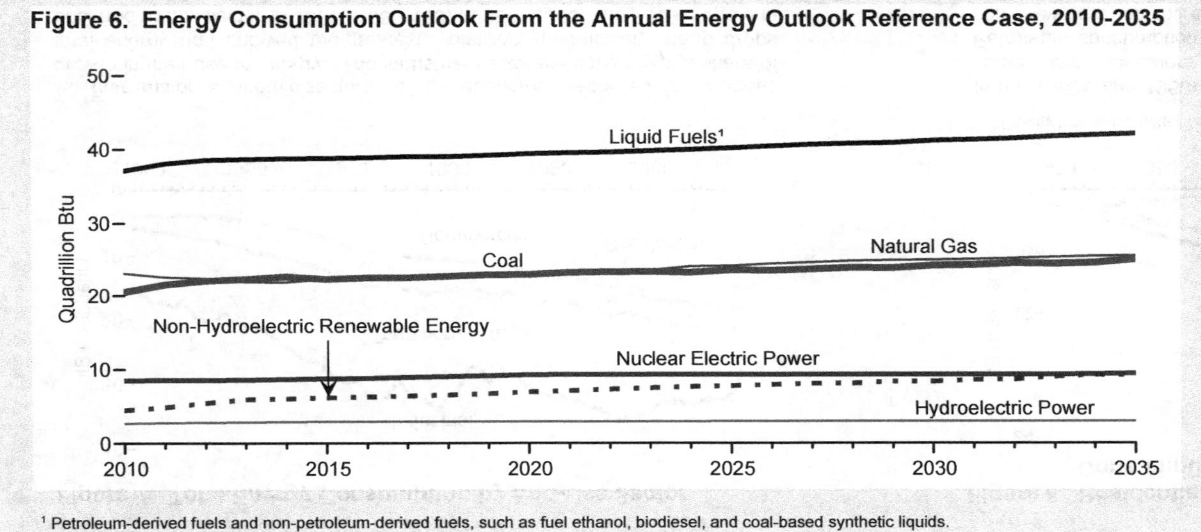


[1] Petroleum-derived fuels and non-petroleum-derived fuels, such as fuel ethanol, biodiesel, and coal-based synthetic liquids.

The Nation's energy history is one of large-scale change as new forms of energy develop. Wood served as the primary form of energy until about 1885, when it was surpassed by coal. Despite its tremendous and rapid expansion, coal was in turn overtaken by petroleum in the middle of the 20th century. In the second half of the 20th century, natural gas experienced rapid development, and coal began to expand again. Late in the century, still other forms of energy–hydroelectric power and nuclear electric power–were developed and supplied significant amounts of energy.

The reference case from the U.S. Energy Information Administration's *Annual Energy Outlook 2010*, which assumes current laws and regulations remain unchanged, projects that fossil fuels continue to provide most of the energy consumed in the United States over the next 25 years. The fossil-fuel share of overall energy use declines, however, as the role of renewable forms of energy grows. Non-hydroelectric renewable energy is projected to double by 2035.

U.S. Energy Information Administration / Annual Energy Review 2009

# Consumption by Sector

**Figure 7. Total Energy Consumption by End-Use Sector**

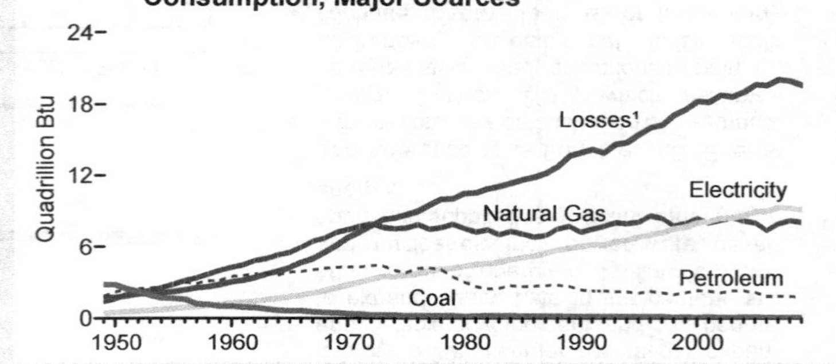


All four major economic sectors of the economy recorded tremendous growth in their use of energy. The industrial sector used the biggest share of total energy and showed the greatest volatility; in particular, steep drops occurred in the sector in 1975, 1980-1982, 2001, 2008, and 2009 largely in response to high oil prices and economic slowdown.

**Figure 8. Residential and Commercial Total Energy Consumption, Major Sources**

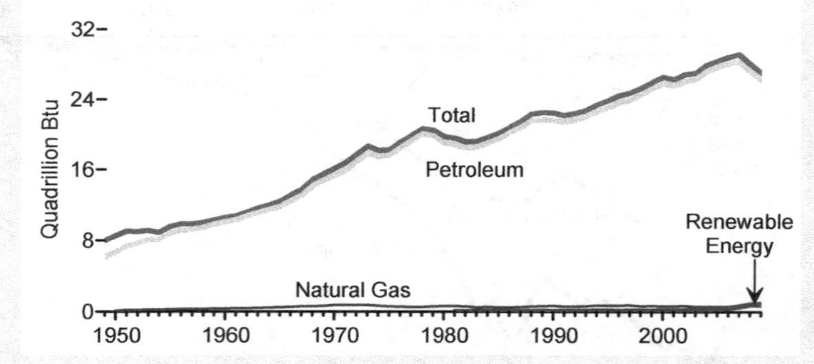


[1] Energy lost during generation, transmission, and distribution of electricity.

In the 1950s and 1960s, coal, which had been important to residential and commercial consumers, was gradually replaced by other forms of energy. Petroleum consumption peaked in the early 1970s. Natural gas consumption grew fast until the early 1970s, and then, with mild fluctuations, held fairly steady in the following years. Meanwhile, electricity use (and related losses) expanded dramatically.

**Figure 9. Industrial Total Energy Consumption, Major Sources**

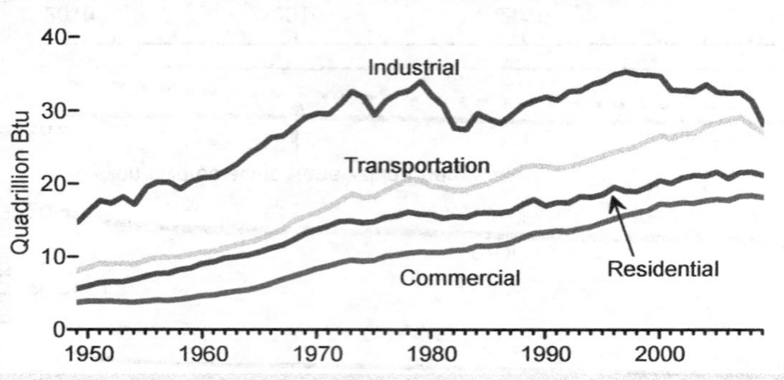


[1] Energy lost during generation, transmission, and distribution of electricity.

Coal, once the predominant form of energy in the industrial sector, gave way to natural gas and petroleum in the late 1950s. Both natural gas and petroleum use expanded rapidly until the early 1970s, and then fluctuated widely over the following decades. Use of electricity and biomass trended upward.

**Figure 10. Transportation Total Energy Consumption**

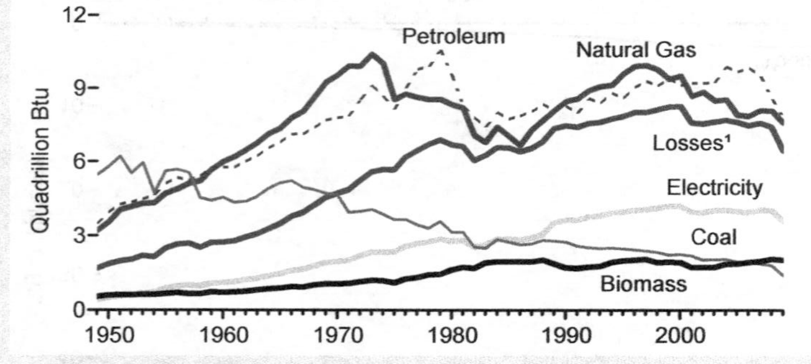


Transportation sector use of energy experienced tremendous growth overall; however, there were year-to-year declines, particularly in the early 1980s and in 2008 and 2009. Throughout the 1949-to-2009 period, petroleum supplied most of the demand for transportation energy; in 2009, petroleum accounted for 94 percent of the transportation sector's total use of energy. Natural gas and renewable energy accounted for the remainder.

U.S. Energy Information Administration / Annual Energy Review 2009

# Production and Trade

**Figure 11. Primary Energy Production by Major Source**

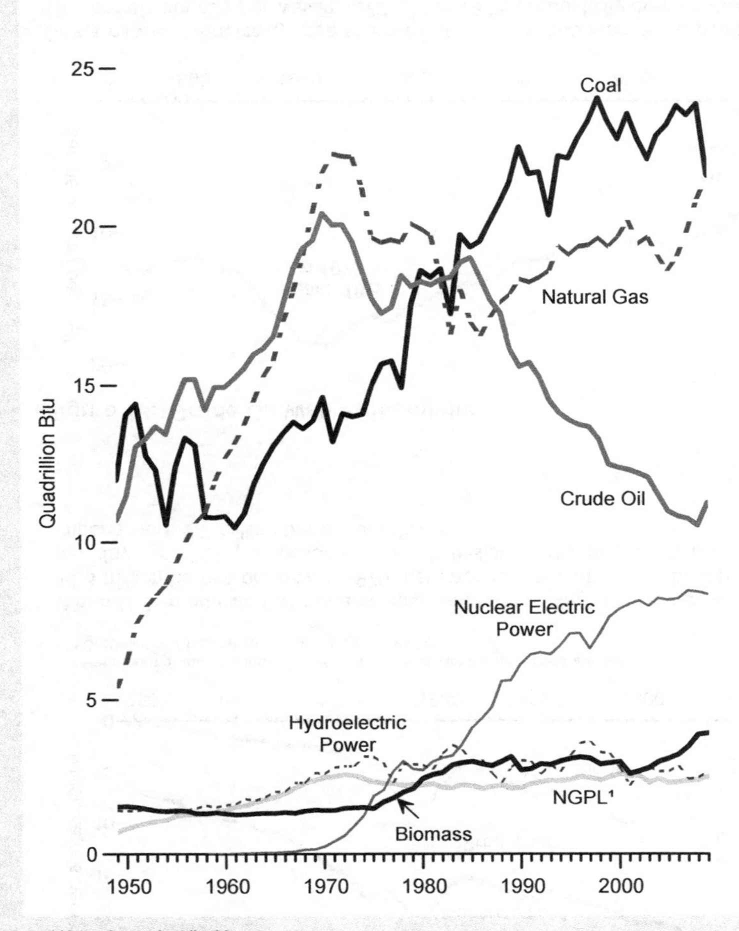


¹ Natural gas plant liquids.

Most energy produced in the United States came from fossil fuels—coal, natural gas, and crude oil. Coal, the leading source at the middle of the 20th century, was surpassed by crude oil and then by natural gas. By the mid-1980s, coal again became the leading energy source produced in the United States, and crude oil declined sharply . In the 1970s, electricity produced from nuclear fuel began to make a significant contribution and expanded rapidly in the following decades.

**Figure 12. Production as Share of Consumption for Coal, Natural Gas, and Petroleum**

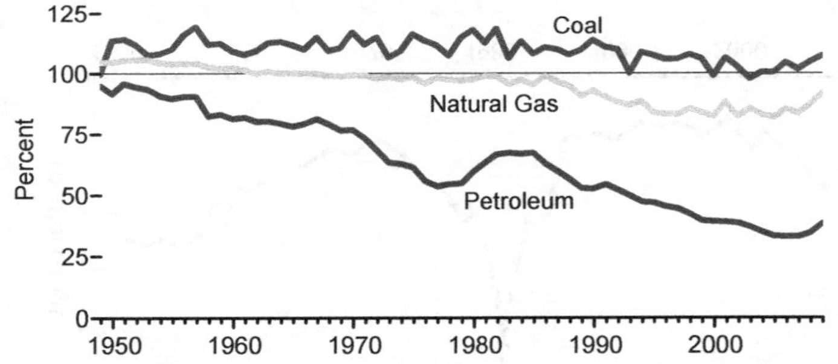

The United States almost always produced more than enough coal for its own requirements. F or many years, the United States w as also self-sufficient in natural gas, but after 1967, it produced less than it consumed each year. Petroleum production fell far short of domestic demands, requiring the Nation to rely on imported supplies.

**Figure 13. Primary Energy Imports and Exports**

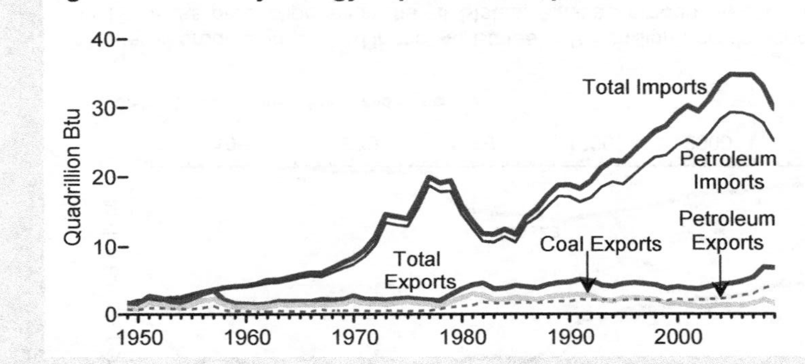


Since the mid-1950s, the Nation imported more energy than it exported. In 2009, the United States imported 30 quadrillion Btu of energy and exported 7 quadrillion Btu. Most imported energy was in the form of petroleum; since 1986, natural gas imports expanded rapidly as well. Through 1992, most exported energy was in the form of coal; after that, petroleum exports often exceeded coal exports.

**U.S. Energy Information Administration / Annual Energy Review 2009**

# Petroleum Overview and Crude Oil Production

**Figure 14. Petroleum Overview**

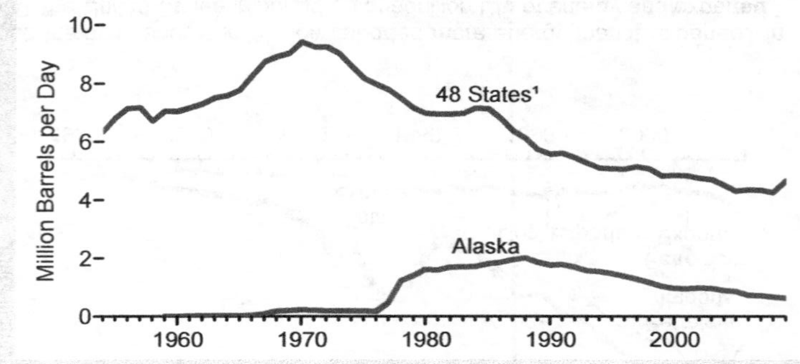


[1] Petroleum products supplied is used as an approximation for consumption.
[2] Crude oil and natural gas plant liquids production.

When U.S. production of crude oil and natural gas plant liquids peaked at 11.3 million barrels per day in 1970, net imports stood at 3.2 million barrels per day. In 2009, production was 7.2 million barrels per day, and net imports were 9.7 million barrels per day.

**Figure 15. 48 States and Alaskan Crude Oil Production**

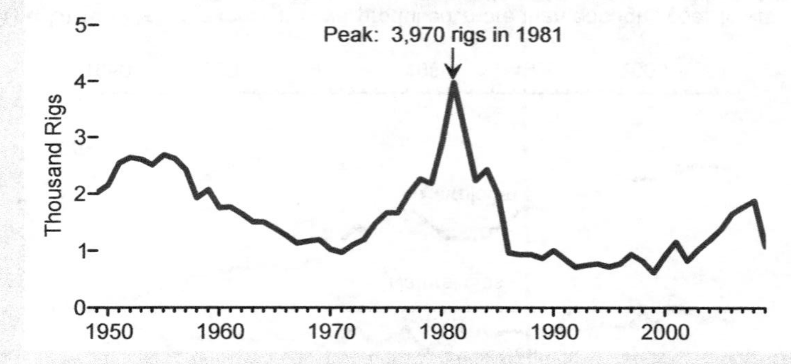


[1] United States excluding Alaska and Hawaii.

Crude oil production peaked in the 48 States at 9.4 million barrels per day in 1970. As production fell in the 48 States, Alaska's production came on-line and helped supply U.S. needs. Alaskan production peaked at 2.0 million barrels per day in 1988; in 2009, Alaska's production stood at 32 percent of its peak level, or 0.6 million barrels per day.

**Figure 16. Crude Oil Well Productivity**

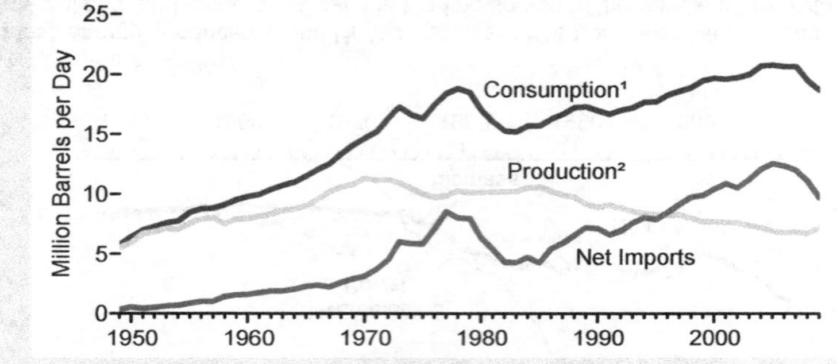


Crude oil well productivity rose sharply in the 1960s and reached a peak of 18.6 barrels per day per well in 1972. After 1972, productivity trended downward to a 55-year low. The 2008 rate of 9.4 barrels per day per well was 51 percent of the 1972 peak. In 2009, productivity rose to 10.1 barrels per day per well.

**Figure 17. Crude Oil and Natural Gas Rotary Rigs in Operation**

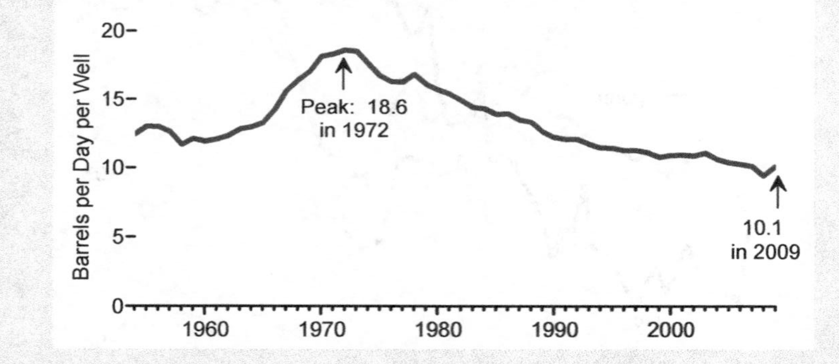


Rotary rig activity declined sharply from 1955 to 1971. After 1971, the number of rigs in operation began to climb again, and a peak of nearly 4 thousand rigs in operation was registered in 1981. In 2009, 1,089 rigs were in operation, a 42 percent drop from 2008 and only 27 percent of the peak level in 1981.

U.S. Energy Information Administration / Annual Energy Review 2009

# Petroleum Consumption and Prices

**Figure 18. Petroleum Consumption[1] by Sector**

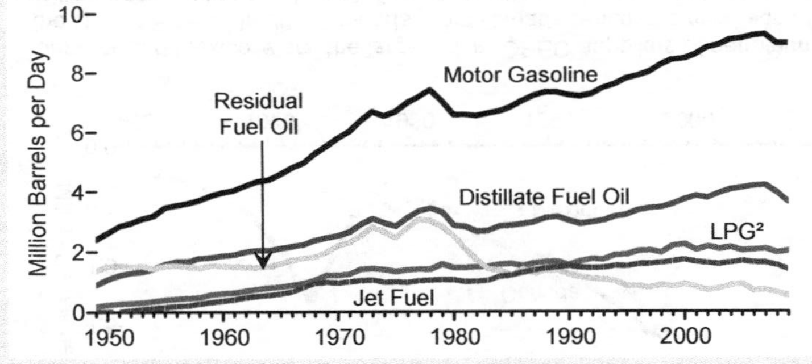


[1] Petroleum products supplied is used as an approximation for consumption.
[2] Through 1988, electric utilities only; after 1988, also includes independent power producers.

Transportation was the largest consuming sector of petroleum and the one showing the greatest expansion. In 2009, 13.3 million barrels per day of petroleum products were consumed for transportation purposes, accounting for 71 percent of all petroleum used.

**Figure 20. Crude Oil Refiner Acquisition Cost[1]**

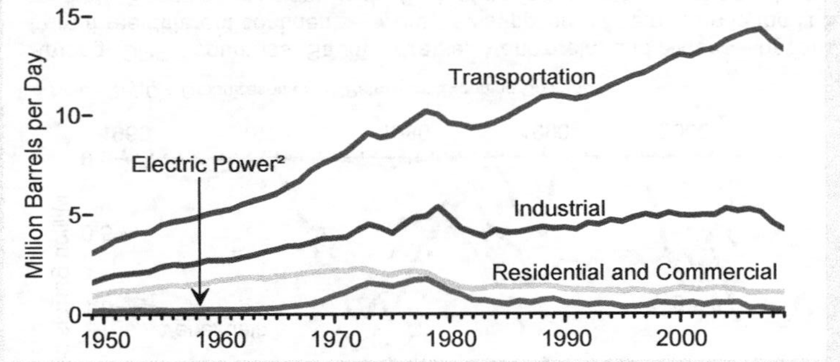


[1] Composite of domestic and imported crude oil.
[2] In chained (2005) dollars, calculated by using gross domestic product implicit price deflator. See "Chained Dollars" in Glossary.

Unadjusted for inflation (nominal dollars), the refiner acquisition composite (domestic and foreign) cost of crude oil reached $35.24 per barrel in 1981. Over the years that followed, the price fell dramatically to a low of $12.52 per barrel in 1998 before rising again. The preliminary nominal price reported for 2009 was $59.27 per barrel, a decrease of 37 percent over the 2008 price.

**Figure 19. Petroleum Consumption[1] by Selected Product**

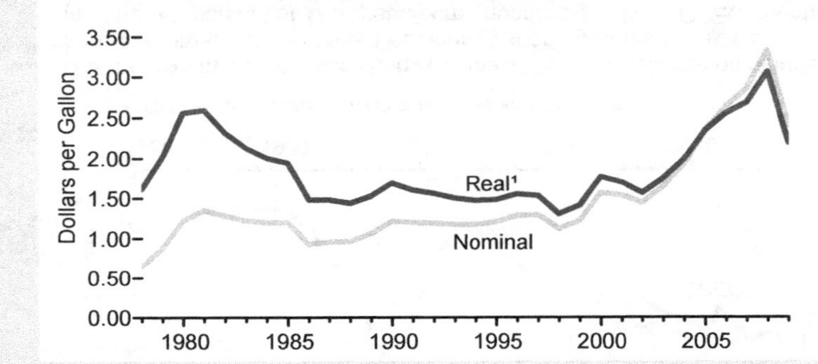


[1] Petroleum products supplied is used as an approximation for consumption.
[2] Liquefied petroleum gases.

Motor gasoline was the single largest petroleum product consumed in the United States. Its consumption stood at 9.0 million barrels per day in 2009, 48 percent of all petroleum consumption. Distillate fuel oil, liquefied petroleum gases (LPG), and jet fuel were other important products. The use of residual fuel oil fell off sharply after 1977.

**Figure 21. Retail Price of Motor Gasoline, All Grades**

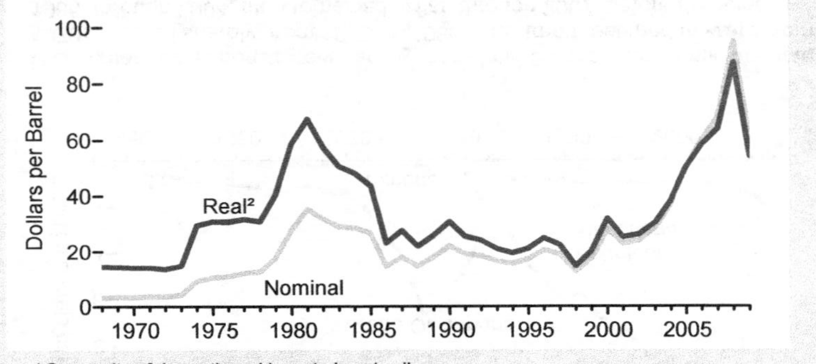


[1] In chained (2005) dollars, calculated by using gross domestic product implicit price deflator. See "Chained Dollars" in Glossary.

In nominal (unadjusted for inflation) dollars, Americans paid an average of 65¢ per gallon for motor gasoline in 1978. The 2009 average price of $2.40 was more than five times the 1978 rate; adjusted for inflation, it was 36 percent higher.

U.S. Energy Information Administration / Annual Energy Review 2009

# Petroleum Trade

**Figure 22. Petroleum Trade**

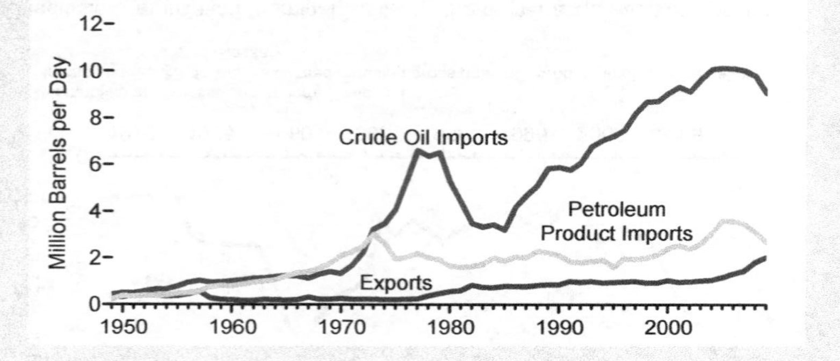


U.S. crude oil imports grew rapidly from mid-20th century until the late 1970s, but fell sharply from 1979 to 1985. The trend resumed upward from 1985 through 2004, then remained flat through 2007, before dropping in 2008 and 2009. In 2009, crude oil imports were 9.1 million barrels per day; petroleum product imports were 2.7 million barrels per day; and, exports were 2.0 million barrels per day, mainly in the form of distillate and residual fuel oils.

**Figure 24. Petroleum Imports From Selected OPEC Countries**

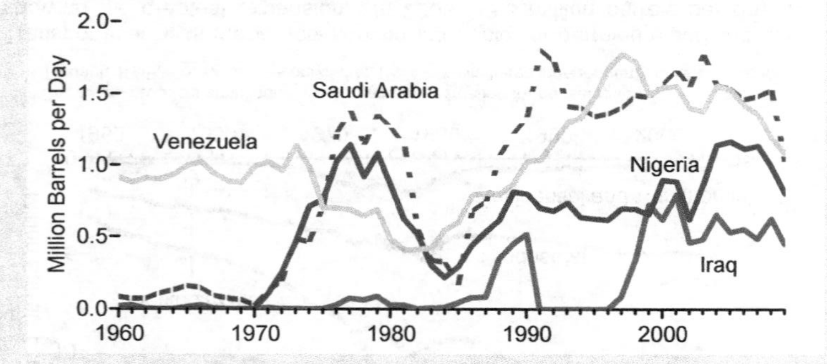


Note: OPEC = Organization of the Petroleum Exporting Countries.

Among OPEC countries, Saudi Arabia, Venezuela, and Nigeria—nations from three different continents—were key suppliers of petroleum to the U.S. market. Each experienced wide fluctuation in the amount of petroleum it sold to the United States over the decades. In 2009, 0.4 million barrels per day of petroleum came into the United States from Iraq.

**Figure 23. Petroleum Imports From OPEC and Non-OPEC**

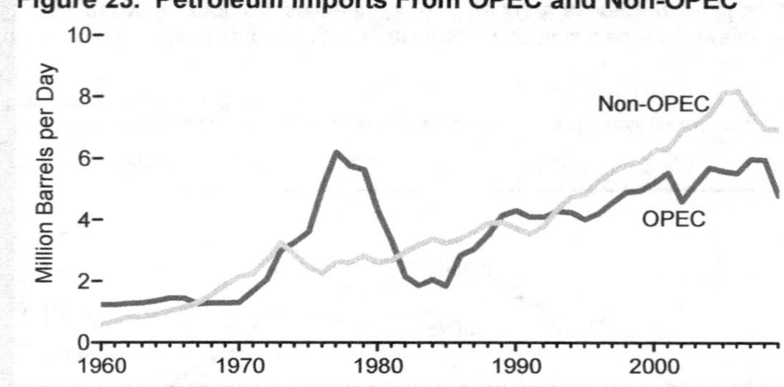


Note: OPEC = Organization of the Petroleum Exporting Countries.

U.S. petroleum imports rose sharply in the 1970s, and reliance on petroleum from the Organization of the Petroleum Exporting Countries (OPEC) grew. In 2009, 41 percent of U.S. petroleum imports came from OPEC countries, down from 70 percent in 1977. After 1992, more petroleum came into the United States from non-OPEC countries than from OPEC countries.

**Figure 25. Petroleum Imports From Canada and Mexico**

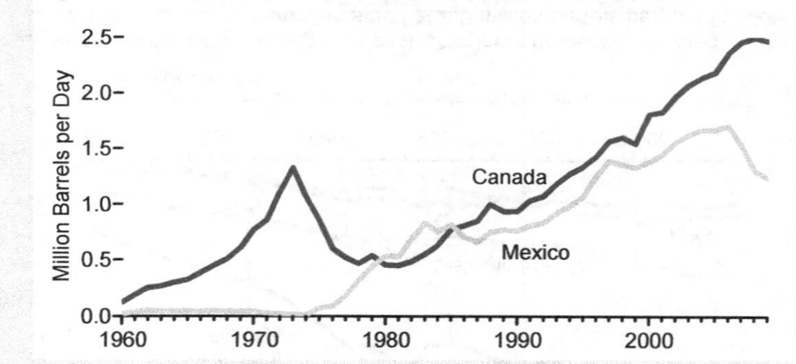


Canada and Mexico were the largest non-OPEC suppliers of petroleum to the United States. In 2009, imports from Canada reached a new high of 2.5 million barrels per day. Imports from Mexico were insignificant until the mid-1970s, when they began to play a key role in U.S. supplies. Canadian and Mexican petroleum together accounted for 32 percent of all U.S. imports in 2009.

# Petroleum Stocks

**Figure 26. Stocks of Crude Oil and Petroleum Products**

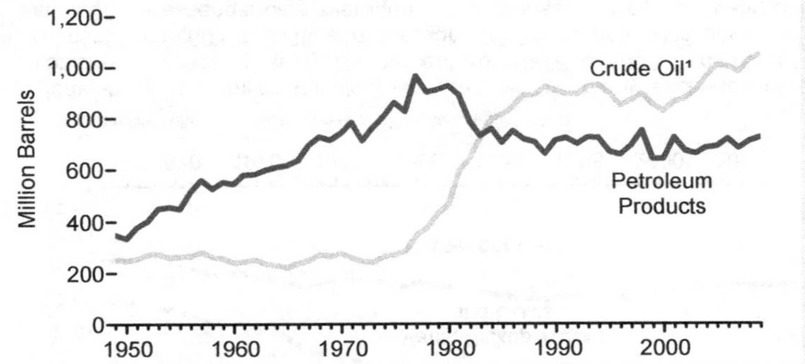

[1] Includes crude oil and lease condensate stored in the Strategic Petroleum Reserve.

Through 1983, the Nation held most of its petroleum storage in the form of products, which were ready for the ma rket. After 1983, most petroleum in storage was in the form of crude oil (including that held by the government in the Strategic Petroleum Reserve) that still needed to be refined into usable end products. At the end of 2009, pet roleum stocks totaled 1.8 billion barrels, 59 percent crude oil and 41 percent products.

**Figure 27. Total Petroleum Stocks and the SPR[1]**

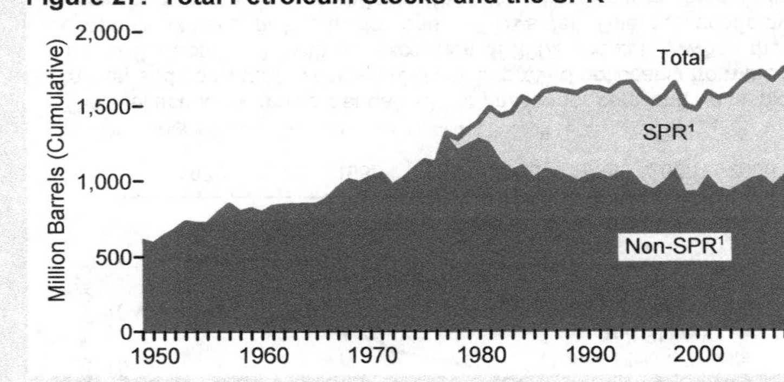


[1] Strategic Petroleum Reserve.

In 1977, the United Stat es began filling the Strat egic Petroleum Reserve (SPR), a national reserve of petroleum stocks in case of emergency. At the end of 2009, the SPR held 727 million barrels of crude oil, 41 percent of all U.S. petroleum stocks.

**Figure 28. Crude Oil Imports for the SPR[1]**

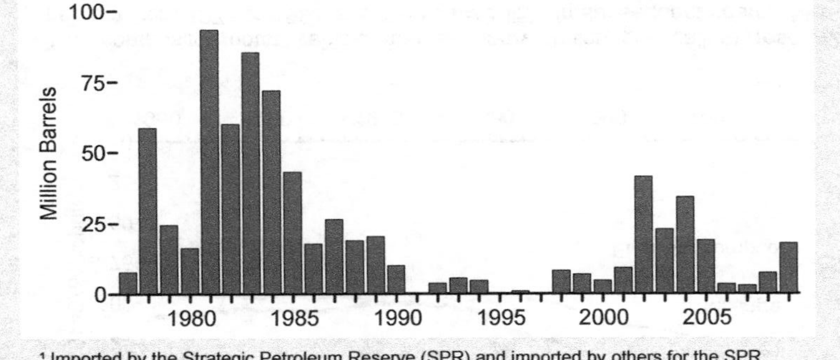


[1] Imported by the Strategic Petroleum Reserve (SPR) and imported by others for the SPR.

Most crude oil in the SPR was imported and came in during the early 1980s. In fact, from 1991 through 1997, only 14 million barrels were imported for the reserve, and in 3 of thos e years, no oil at all w as imported for the reserve. SPR imports picked up again after 1997, and stored another 176 million barrels from 1998 through 2009.

**Figure 29. SPR Stocks as Days of Petroleum Net Imports**

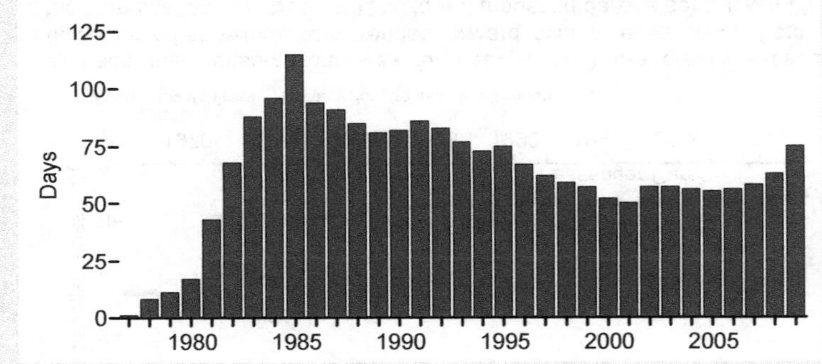


Stocks are often measured by the number of days of total net imports o f petroleum that could be met by the reserve in an emergency. The peak level occurred in 1985 w hen the Strategic Petroleum Reserve (SPR) could have supplied 115 days of petroleum net imports, at the 1985 level. The rate trended down for many years, falling to 50 days in 2001. In 2009, SPR held 75 days of net imports.

# Motor Vehicles

**Figure 30. Motor Vehicle Indicators**

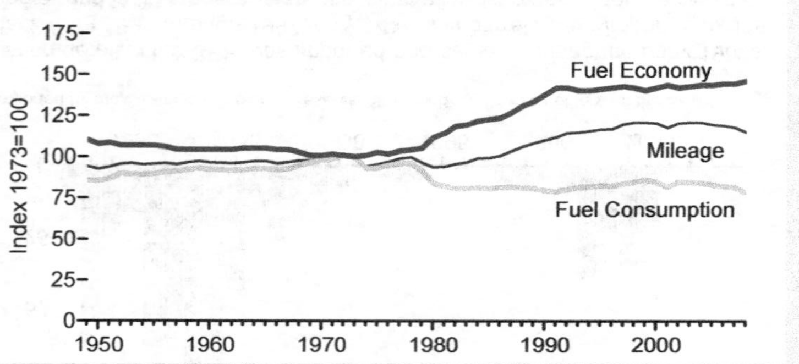


The composite motor vehicle fuel economy (miles per gallon) rose 42 percent from 1973 to 1991 and then varied little in subsequent years. Mileage (miles per vehicle) grew steadily from 1980 to 1998, and then remained near 12 thousand miles per vehicle per year through 2007. Fuel consumption (gallons per vehicle) fell 21 percent from 1973 to 1991, regained 9 percent from 1991 to 1999, and then trended down through 2008.

**Figure 31. Motor Vehicle Fuel Consumption**

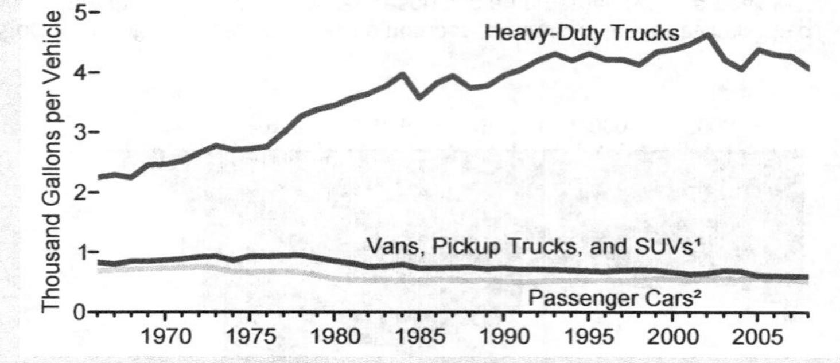


[1] Sport utility vehicles. [2] Motorcycles are included through 1989.

Average fuel consumption rates for heavy-duty trucks greatly exceeded those for other vehicle and trended upw ard over time—doubling from 2.3 thousand gallons per truck in 1966 to 4.6 thousand gallons per truck in 2002. Average fuel consumption rates for passenger cars, and vans, pickup trucks, and sport utility vehicles were much lower and generally trended downward.

**Figure 32. Motor Vehicle Mileage**

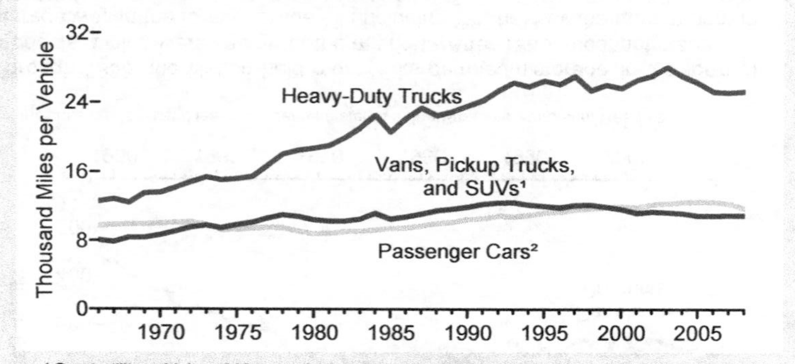


[1] Sport utility vehicles. [2] Motorcycles are included through 1989.

Heavy-duty truck miles traveled per year, which greatly exceeded other vehicle categories, grew by 124 percent from 1966 to 2003, decreased 10 percent from 2003 to 2008, and averaged 25.3 thousand miles per vehicle in 2008. Passenger cars averaged 11.8 thousand miles per vehicle in 2008. Vans, pickup trucks, and sport utility vehicles averaged 11.0 thousand miles per vehicle in 2008.

**Figure 33. Motor Vehicle Fuel Economy**

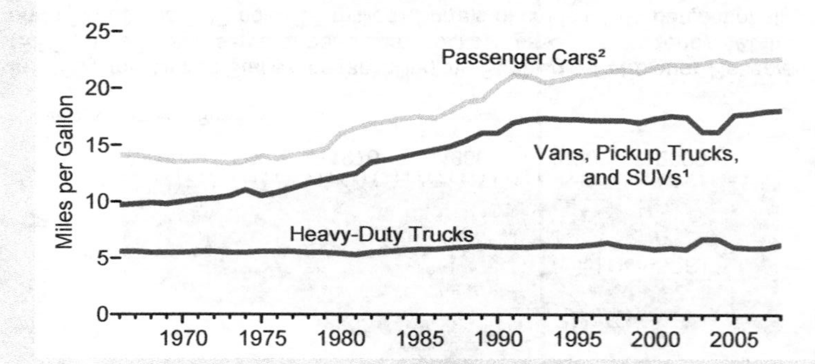


[1] Sport utility vehicles. [2] Motorcycles are included through 1989.

The fuel economy (miles per gallon), of passenger cars and vans, pickup trucks, and sport utility vehicles (SUVs), improved noticeably from the mid-1970s through 2008, with the exception of 2002 and 2003, when the fuel economy of vans, pickup trucks, and SUVs fell. T he fuel economy of heavy-duty trucks was much lower than for other vehicles, largely due to their greater size and weight, and showed far less change over time.

Note: Motor vehicles include passenger cars, motorcycles, vans, pickup trucks, sport utility vehicles, trucks, and buses.

U.S. Energy Information Administration / Annual Energy Review 2009

# Natural Gas

**Figure 34. Natural Gas Overview**

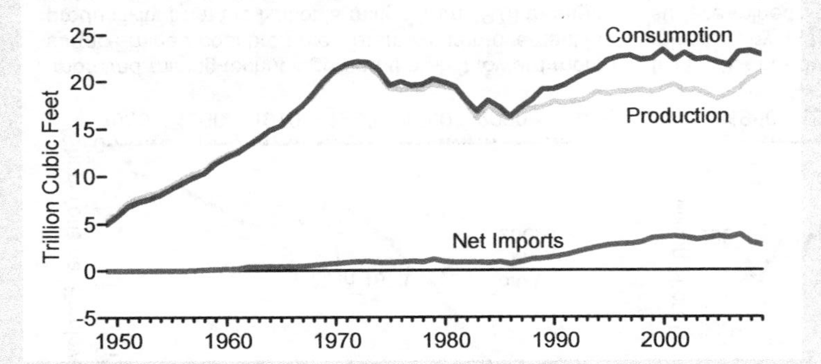


U.S. natural gas production and consumption were nearly in balance through 1986. After that, consumption began to outpace production, and imports of natural gas rose to meet U.S. demand. Production increased from 2006 through 2009. In 2009, production stood at 21.0 trillion cubic feet (Tcf), net imports at 2.7 Tcf, and consumption at 22.8 Tcf.

**Figure 35. Natural Gas Well Average Productivity**

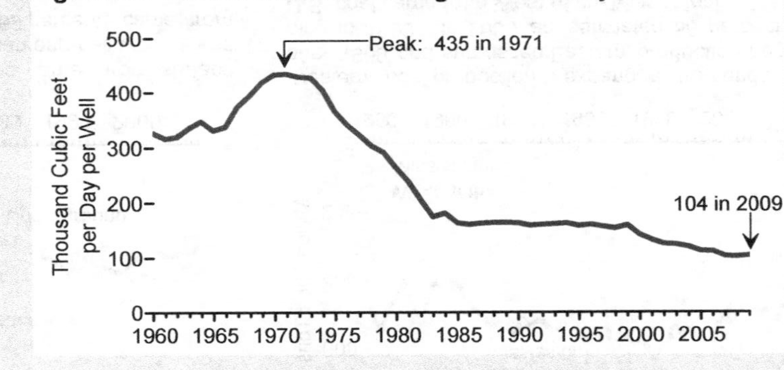


Natural gas well productivity, measured as gross withdrawals per day per well, grew rapidly in the late 1960s, peaked in 1971, and then fell sharply until the mid-1980s. Productivity remained fairly steady from 1985 through 1999, fell annually through 2008, and turned up slightly in 2009.

**Figure 36. Natural Gas Net Imports as Share of Consumption**

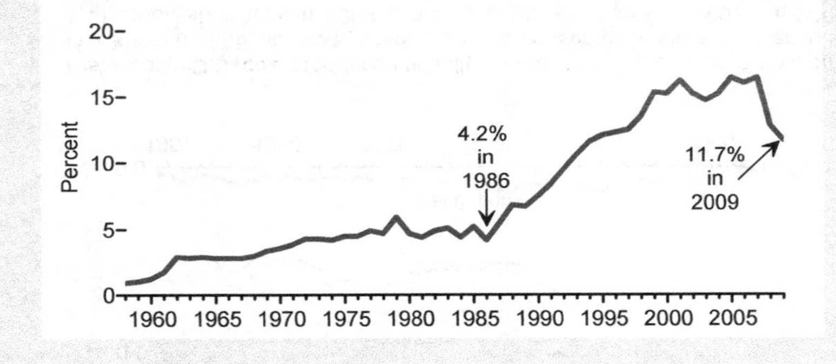


Net imports of natural gas as a share of consumption remained below 6 percent through 1987. Then, during a period when consumption outpaced production, the share rose to a peak of 16.4 percent in 2005 and again in 2007. In 2009, the share was 11.7 percent.

**Figure 37. Natural Gas Consumption by Sector**

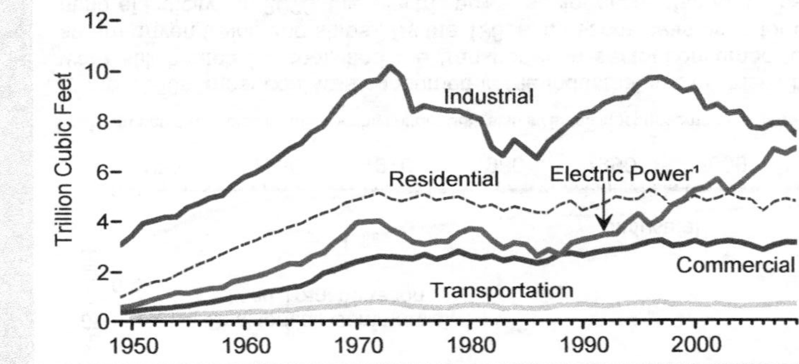


[1] Through 1988, electric utilities only; after 1988, also includes independent power producers.

Throughout the 1949-to-2009 period, the industrial sector consumed more energy than any other sector and accounted for 32 percent of all natural gas consumption in 2009. Big fluctuations in the level of consumption were due to variability in industrial output. Energy consumption by the electric power sector grew substantially over the same period and, in 2009, accounted for 30 percent of all natural gas consumption.

# Coal

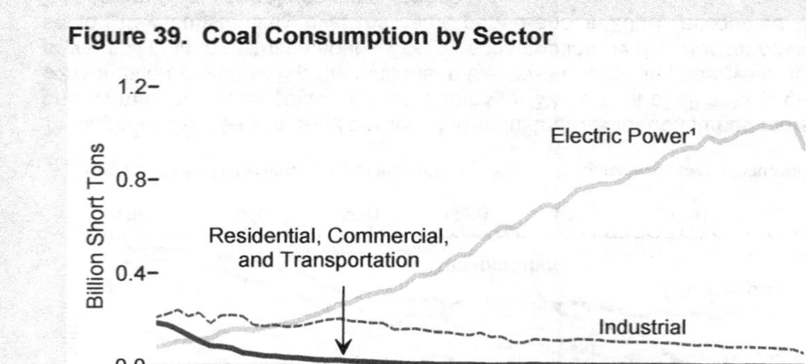


Figure 38. Coal Overview

Historically, U.S. coal production usually surpassed U.S. coal consumption. In 2004 and 2005, however, production and consumption were in balance at 1.11 billion short tons in 2004 and 1.13 billion short tons in 2005. In 2006 through 2009, production again slightly exceeded consumption. Net exports, which peaked at 111 million short tons in 1981, stood at 36 million short tons in 2009.

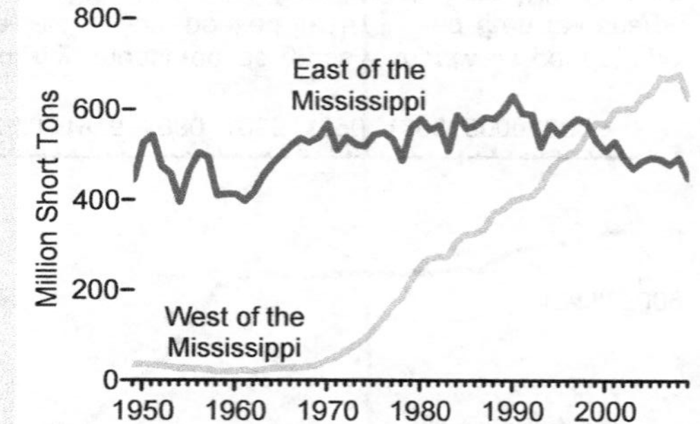


Figure 39. Coal Consumption by Sector

[1] Through 1988, electric utilities only; after 1988, also includes independent power producers.

In the 1950s, most coal was consumed in the industrial sector, many homes were still heated by coal, and the transportation sector consumed coal in steam-driven trains and ships. By the 1960s, most coal was used for generating electricity. In 2009, the electric power sector accounted for 94 percent of all coal consumption, on a tonnage basis.

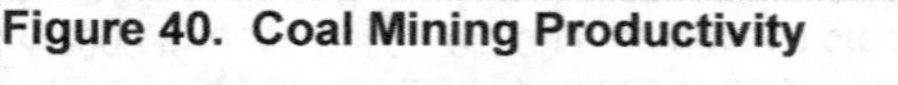

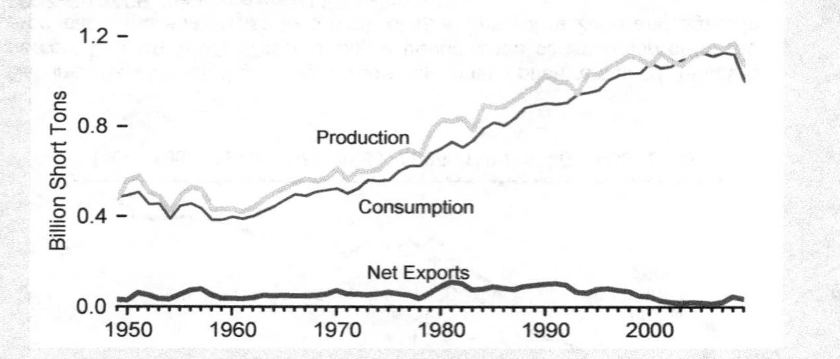


Figure 40. Coal Mining Productivity

Improved mining technology and the shift toward more surface-mined coal promoted dramatic improvement in productivity from the Nation's mines from 1978 through 2000, but productivity declined in most years since then.

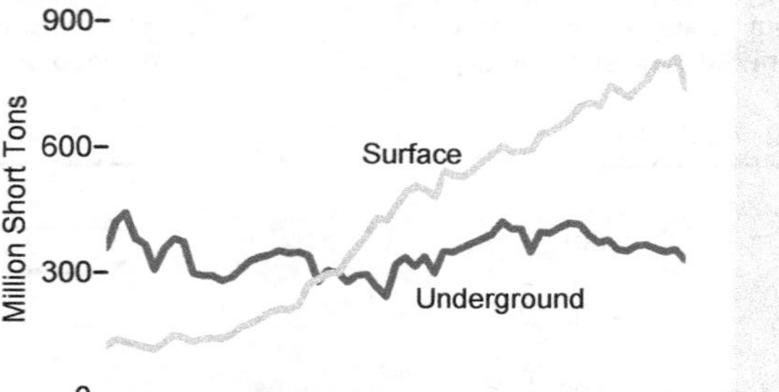


Figure 41. Coal Production by Mining Method

In 1949, one-fourth of U.S. coal came from surface mines; by 1971, more than one-half w as surface-mined; and in 2009, 69 percent came from above-ground mines.

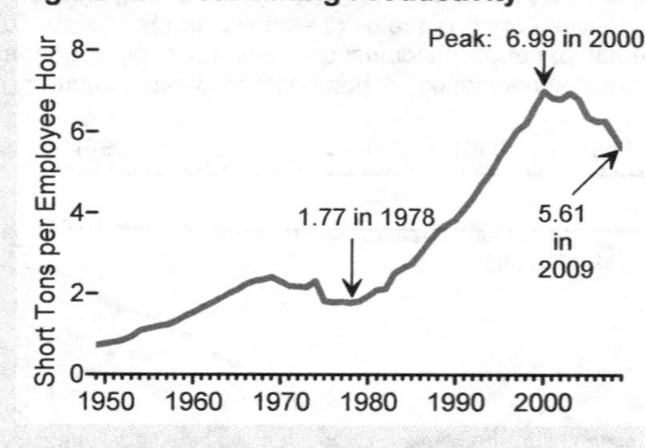


Figure 42. Coal Production by Location

Western coal production expanded tremendously after 1969 and surpassed Eastern production beginning in 1999. In 2009, an estimated 58 percent of U.S. coal came from West of the Mississippi.

# Electricity Net Generation, Electric Net Summer Capacity Change, and Useful Thermal Output

**Figure 43. Electricity Net Generation by Sector**

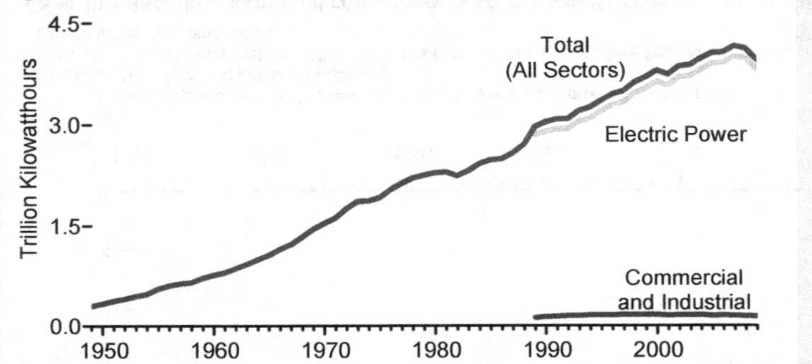


Total electricity net generation in all sectors grew from 0.3 trillion kilowatthours in 1949 to 4.1 trillion kilowatthours in 2009, failing to increase in only 4 years (1982, 2001, 2008, and 2009) over the entire span. Most generation was in the electric power sector, but some occurred in the commercial and industrial sectors.

**Figure 44. Total Electricity Net Generation by Source**

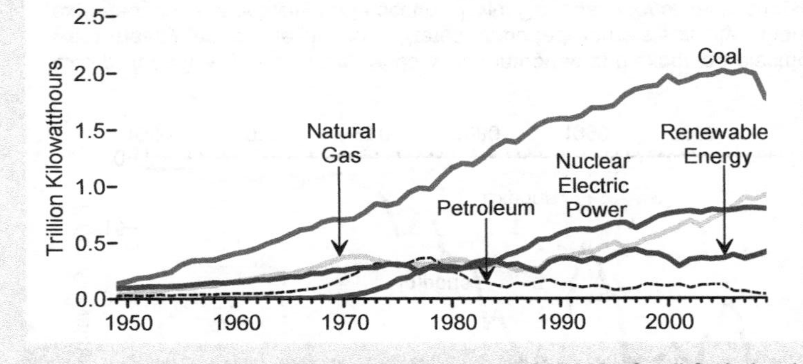


Most electricity net generation came from coal. In 2009, fossil fuels (coal, petroleum, and natural gas) accounted for 69 percent of all net generation, while nuclear electric power contributed 20 percent, and renewable energy resources 10 percent. In 2009, 66 percent of the net generation from renewable energy resources was derived from conventional hydroelectric power.

**Figure 45. Electric Net Summer Capacity Change by Source, 1989-2009**

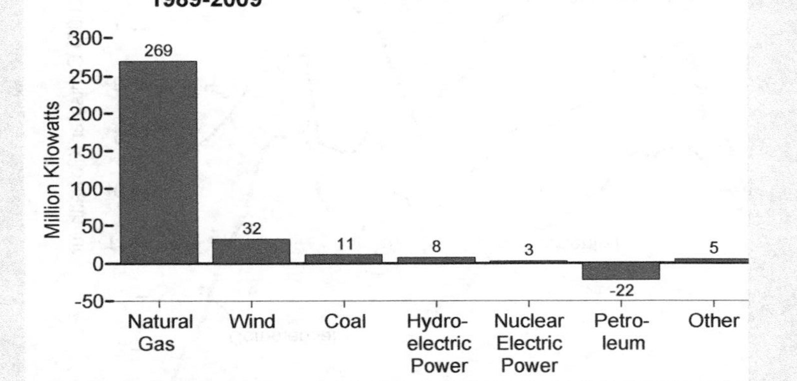


From 1989 through 2009, natural gas-fired electric net summer capacity increased dramatically. Coal, hydroelectric power, and nuclear electric power capacity also increased over the 20-year period. In contrast, petroleum capacity was lower in 2009 than in 1989. Among non-hydroelectric renewable energy sources, wind capacity increased the most. In the "Other" category, wood, waste, and solar capacity registered small increases, whereas geothermal capacity posted a slight decline.

**Figure 46. Useful Thermal Output at Combined-Heat-and-Power Plants by Sector**

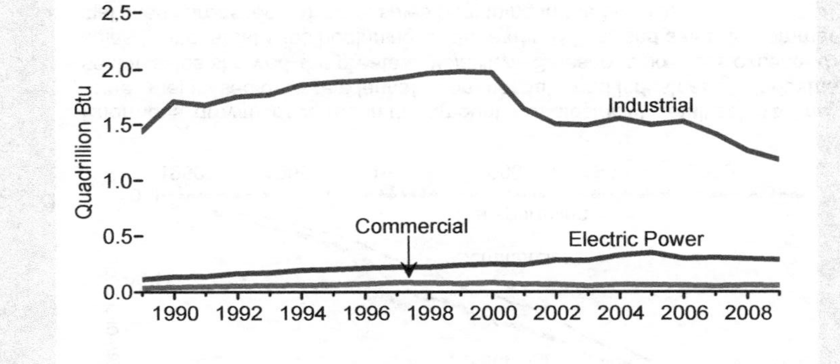


The non-electrical output at a CHP plant is called useful thermal output. Useful thermal output is thermal energy that is available from the plant for use in industrial or commercial processes or heating or cooling applications. In 2009, the industrial sector generated 1.2 quadrillion Btu of useful thermal output; the electric power and commercial sectors generated much smaller quantities.

U.S. Energy Information Administration / Annual Energy Review 2009

# Electricity Prices, Sales, and Trade

### Figure 47. Average Real[1] Retail Prices of Electricity by Sector

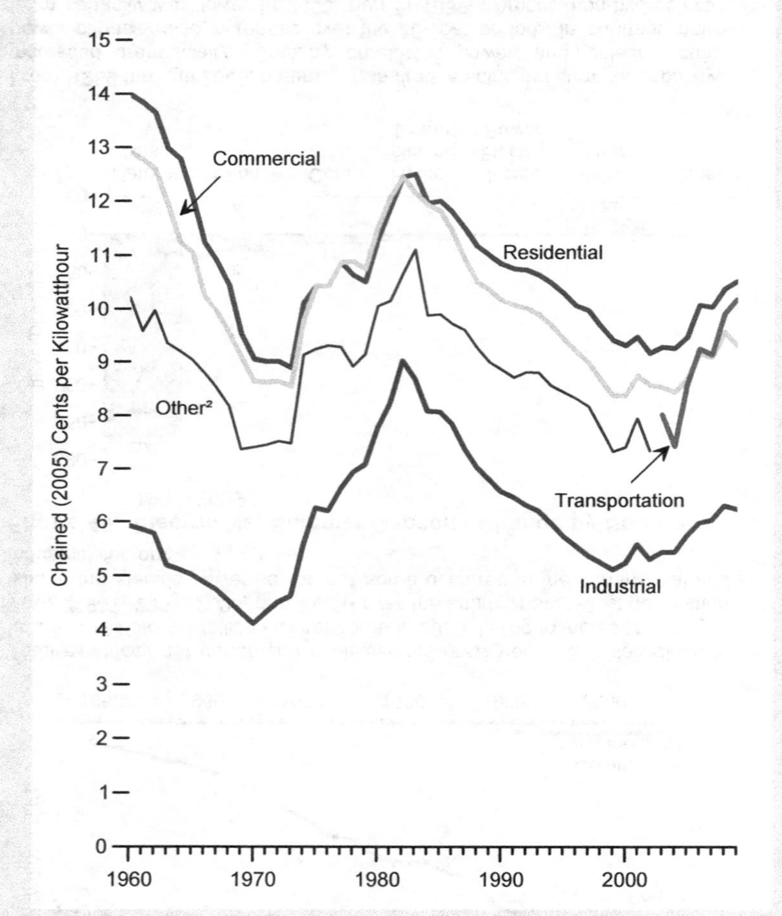


[1] In chained (2005) dollars, calculated by using gross domestic product implicit price deflators. See "Chained Dollars" in Glossary.

[2] In addition to transportation, "Other" includes public street and highway lighting agriculture and irrigation, and other uses.

Over the decades, industrial consumers paid the lowest rates for electricity; residential customers usually paid the highest prices. Inflation-adjusted prices rose in all sectors in 2005, 2006, and 2008 but remained well below the peak price levels of the mid-1980s.

### Figure 48. Electricity Retail Sales by Sector

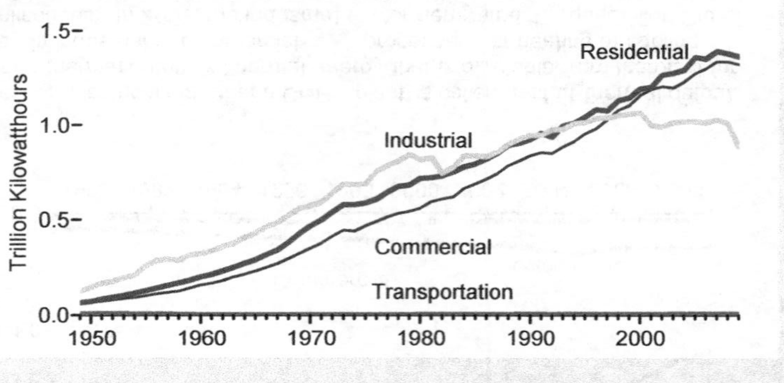


Enormous growth occurred in the amount of electricity retail sales to the three major sectors—residential, commercial, and industrial. Industrial sector sales showed the greatest volatility. Sales to residences exceeded sales to industrial sites beginning in the early 1990s, and sales to commercial sites surpassed industrial sales beginning in the late 1990s.

### Figure 49. Electricity Trade

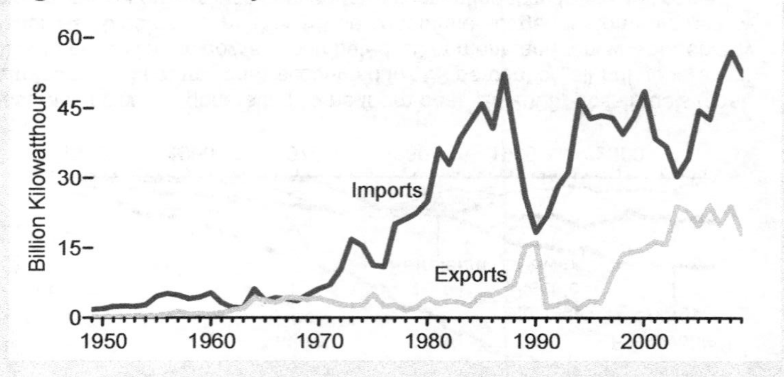


Except for a few years in the 1960s when imported and exported electricity were nearly equal, the United States imported more electricity than it exported. Most electricity trade occurred with Canada; very small exchanges occurred between the United States and Mexico. Nonetheless, in 2009, net imported electricity was less than 0.9 percent of all electricity used in the United States.

U.S. Energy Information Administration / Annual Energy Review 2009

# Nuclear Electric Power

**Figure 50. Nuclear Net Summer Capacity Change, 1950-2009**

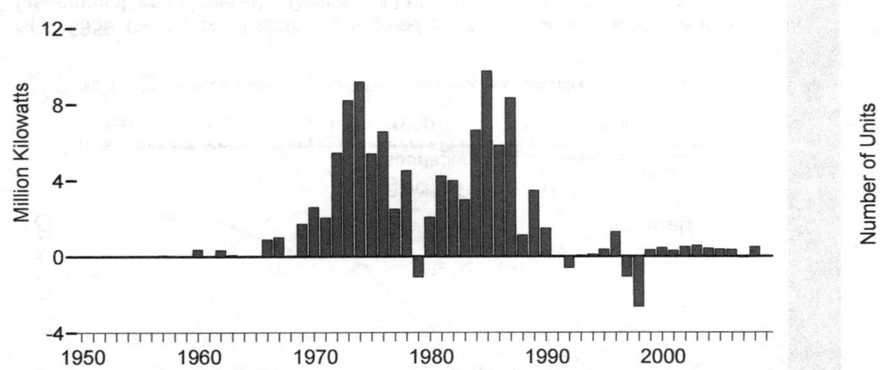


Capacity change reflects capacity additions, retirements, and other changes (such as generator re-ratings). In the nuclear power industry, capacity additions follow the issuing of full-power operating licenses. Year-to-year capacity additions were the greatest in the 1970s and 1980s. In fact, nuclear power capacity was added almost every year from the 1950s through 1990, when growth leveled off.

**Figure 51. Nuclear Operable Units**

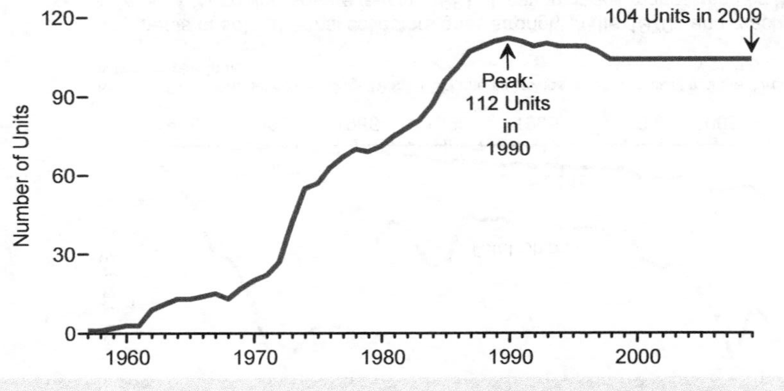


Out of the 132 nuclear units that were granted full-power operating licenses, or equivalent permission, over time, 28 were permanently shut down. The largest number of units ever operable in the United States was 112 in 1990. From 1998 through 2009, 104 units were operable.

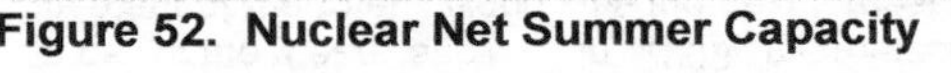

**Figure 52. Nuclear Net Summer Capacity**

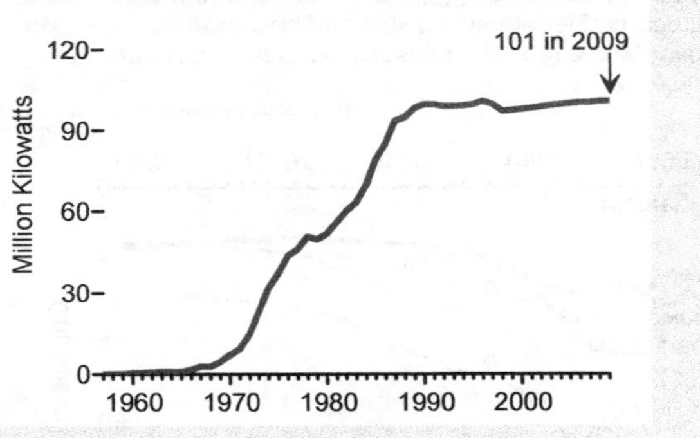


The U.S. nuclear industry's first commercial plant opened in Shippingport, Pennsylvania, in 1957. Nuclear net summer capacity expanded sharply in the 1970s and 1980s. Total net summer capacity stood at 101 million kilowatts in 2009.

**Figure 53. Nuclear Capacity Factor**

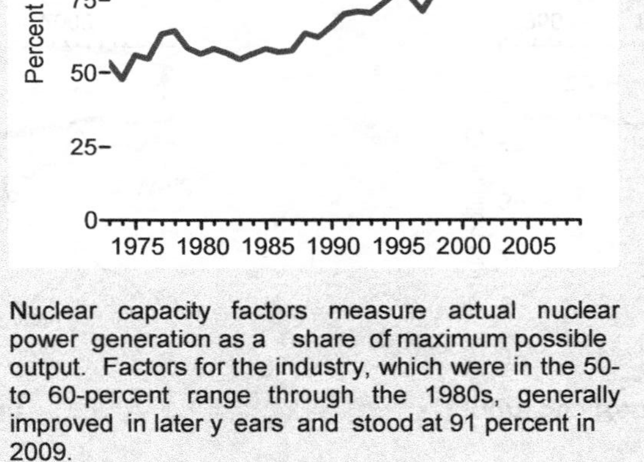


Nuclear capacity factors measure actual nuclear power generation as a share of maximum possible output. Factors for the industry, which were in the 50- to 60-percent range through the 1980s, generally improved in later years and stood at 91 percent in 2009.

**Figure 54. Nuclear Share of Net Generation**

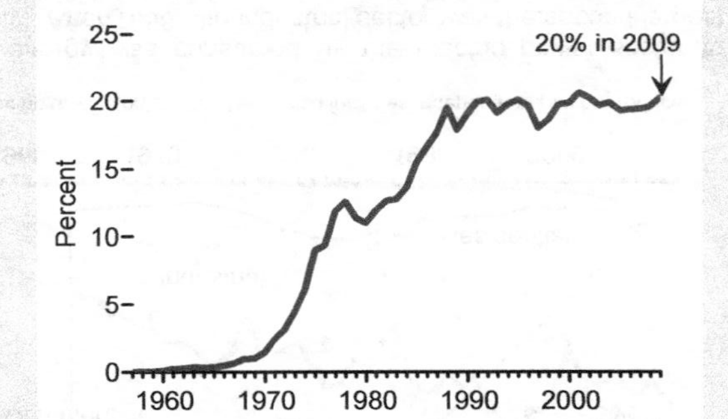


Over the latter part of the last century, nuclear electric power began to play a key role in meeting the Nation's rapidly growing electricity requirements. In 2009, 20 percent of U.S. total electricity net generation came from nuclear electric power.

U.S. Energy Information Administration / Annual Energy Review 2009

# Renewable Energy

**Figure 55. Renewable Energy Total Consumption and Major Sources**

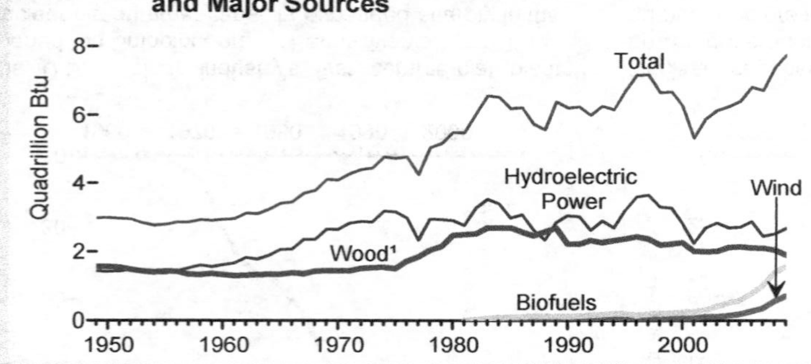


[1] Wood and wood-derived fuels.

Total renewable energy consumption generally followed the pattern of hydroelectric power output, which was the largest component of the total fo r most of the y ears shown. In 2009, hydroelectric power accounted for 35 percent of the total. Wood was the next largest source of renewable energy, followed by biofuels and wind.

**Figure 56. Renewable Energy Consumption by Sector**

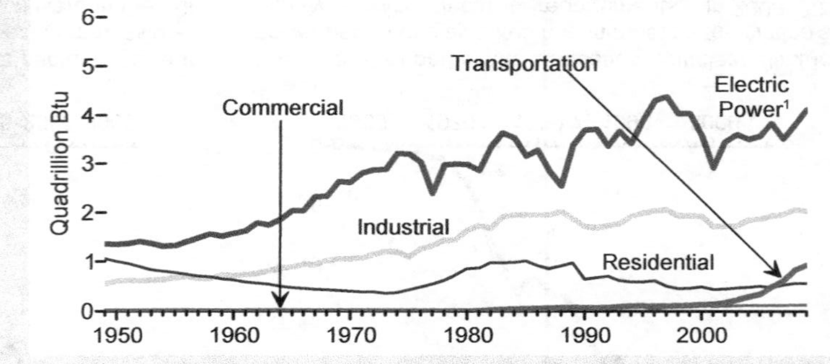


[1] Through 1988, electric utilities only; after 1988, also includes independent power producers.

Most renewable energy was consumed by the electric power sector to generate electricity. After 1958, the industrial sector was the second largest consuming sector of renewable energy; the residential sector was the third largest consuming sector of renewable energy until it w as exceeded by the transportation sector in 2006.

**Figure 57. Biomass Consumption by Sector**

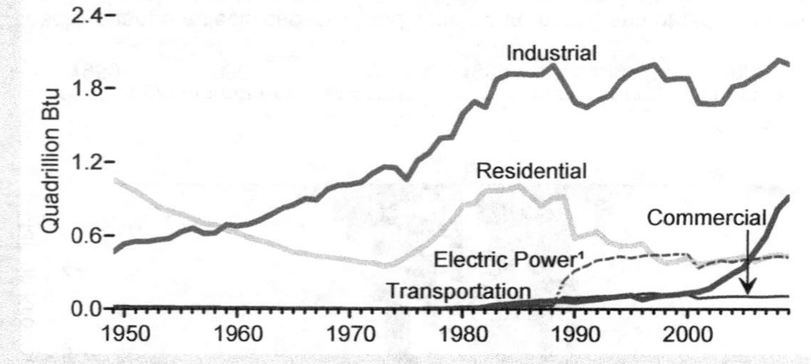


[1] Through 1988, electric utilities only; after 1988, also includes independent power producers.

After 1959, the industrial sector cons umed the most biomass (w ood, waste, fuel ethanol, and biodiesel). Residentia l use of biomass (w ood) fell through 1973, expanded from 1974 through 1985, and then trended downward again. Transportation consumption of biomass (fuel ethanol and biodiesel) expanded after 1996 and, by 2006, exceeded the consumption of biomass (wood and waste) in both the electric power and residential sectors.

**Figure 58. Solar Thermal Collector Shipments and Trade**

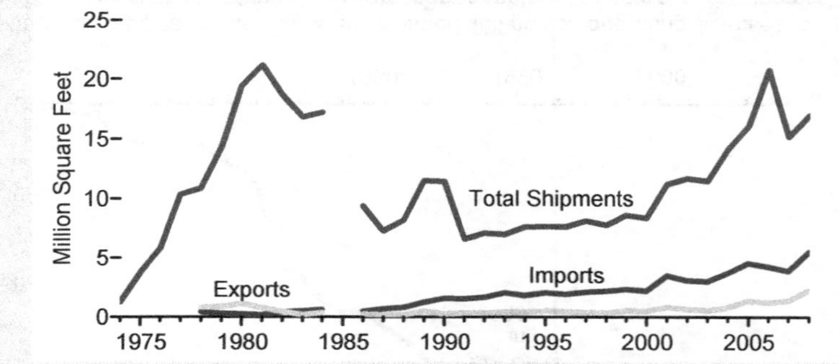


Notes: • Data were not collected for 1985. • Shipments include all domestically manufactured collectors plus imports.

Shipments of solar thermal collectors grew strongly in the 1970s and reached a peak of 21 million square feet in 1981. Uneven perfor mance marked the next decade, followed by a mild upw ard trend during the 1990s. Shipments rose from 2000 to 2002 and 2004 through 2006 before declining in 2007 and rising again in 2008. Imports reached a record level of 5.5 million square feet in 2008.

U.S. Energy Information Administration / Annual Energy Review 2009

# International Energy

**Figure 59. World Primary Energy Production by Source**

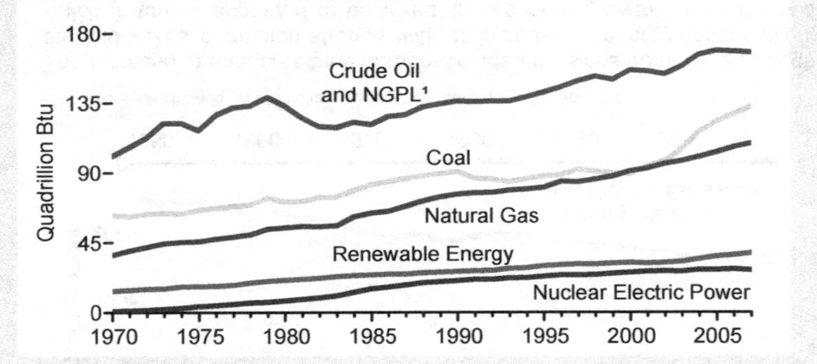


[1] Natural gas plant liquids.

Although crude oil and natural gas plant liquids consistently accounted for the largest share of world primary energy production throughout the 1970-to-2007 period, all major energy sources exhibited growth. In 2007, the fossil fuels (crude oil, natural gas plant liquids, natural gas, and coal) accounted for 86 percent of all energy produced worldwide, renewable energy 8 percent, and nuclear electric power 6 percent.

**Figure 60. World Primary Energy Production by Region**

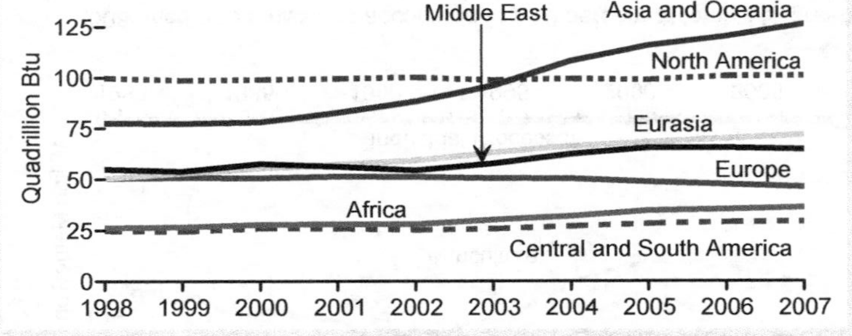


Twenty-one percent of the 475 quadrillion Btu of energy produced worldwide in 2007 came from North America. The largest regional energy producer was Asia and Oceania with 27 percent of the world total in 2007.

**Figure 61. World Crude Oil Production**

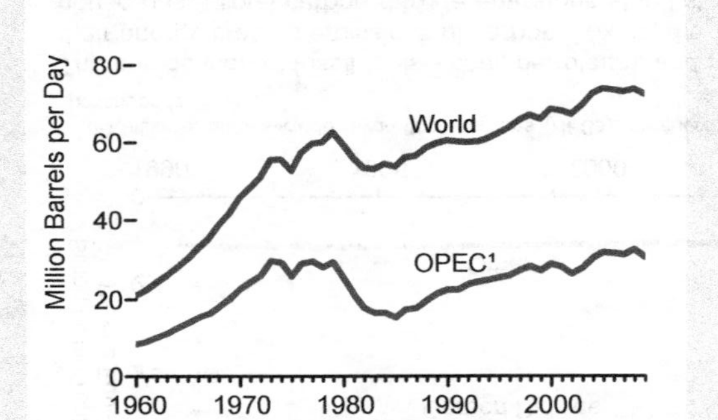


[1] Organization of the Petroleum Exporting Countries.

World crude oil production totaled 72 million barrels per day in 2009, down 2 percent from the level in 2008. OPEC's share of the world total in 2009 was 42 percent, compared to the peak level of 53 percent in 1973.

**Figure 62. Leading Crude Oil Producers**

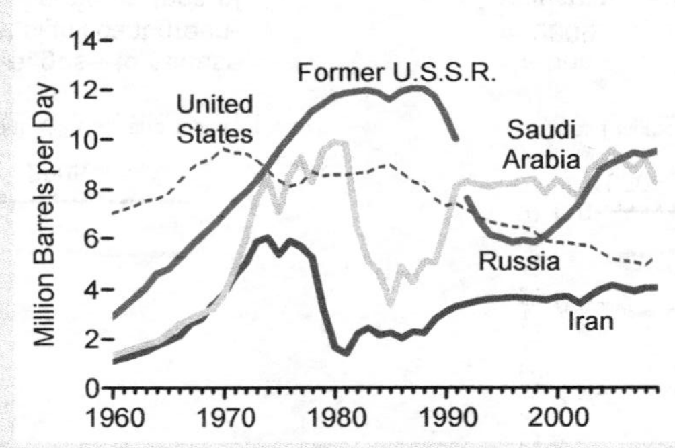


From 1974 through 1991, the former U.S.S.R. was the world's leading crude oil producer. After 1991, Saudi Arabia was the top producer until 2006, when Russia's production exceeded Saudi Arabia's. U.S. production peaked in 1970 but still ranked third in 2009.

**Figure 63. Leading Petroleum Consumers**

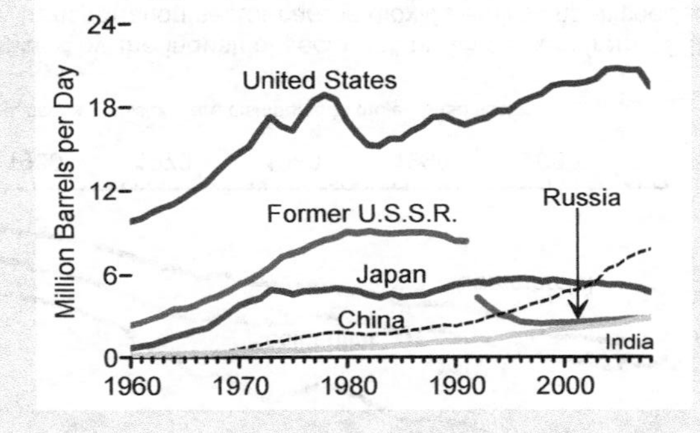


The United States accounted for 23 percent of world petroleum consumption in 2008. China and Japan, the next two leading consumers, together accounted for 15 percent. Russia, Germany, and India were the next largest consumers of petroleum in 2008.

# Emissions

**Figure 64. Greenhouse Gas Emissions, Based on Global Warming Potential**

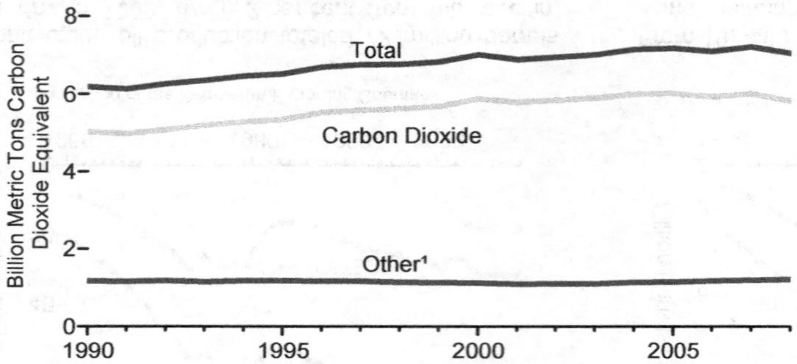

[1] Methane, nitrous oxide, hydrofluorocarbons (HFCs), perfluorocarbons (PFCs), and sulfur hexafluoride ($SF_6$).

The combustion of fossil fuels—coal, petroleum, and natural gas—to release their energy creates emissions of carbon dioxide, the most significant greenhouse gas. Total carbon dioxide emissions stood at 6 billion metric tons of gas in 2008, 16 percent higher than the 1990 level.

**Figure 65. Carbon Dioxide Emissions From Energy Use**

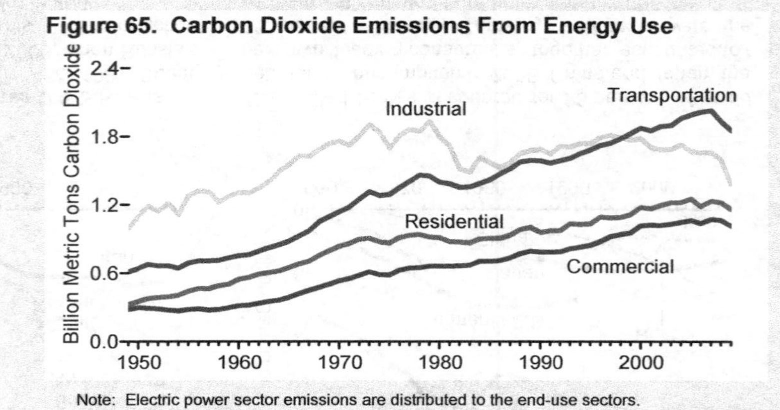


Note: Electric power sector emissions are distributed to the end-use sectors.

Carbon dioxide emitted by the industrial sector fell by 23 percent from 1980 to 2009. By 1999, transportation sector carbon dioxide emissions exceeded industrial sector emissions. Of the major sectors, the commercial secto r generated the least carbon dioxide, but recorded the largest grow th (53 percent) since 1980.

**Figure 66. GDP Growth and Carbon Dioxide Emissions**

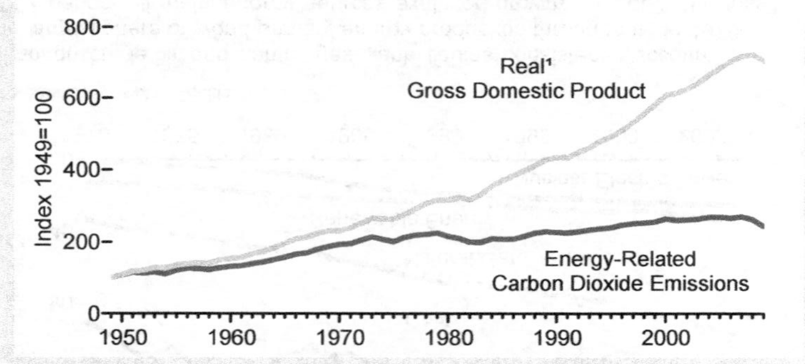


[1] Based on chained (2005) dollars. See "Chained Dollars" in Glossary.

The increase in energy-related carbon dioxide emissions from 1949 through the mid 1970s correlated strongly with an increase in energy consumption, which in turned appeared to be linked to economic growth. After the mid 1970s, however, energy-related carbon dioxide emissions increased more slowly than inflation-adjusted gro ss domestic product. In 2009, during an economic downturn, emissions decreased 7 percent from the 2008 level.

**Figure 67. Methane Emissions by Source**

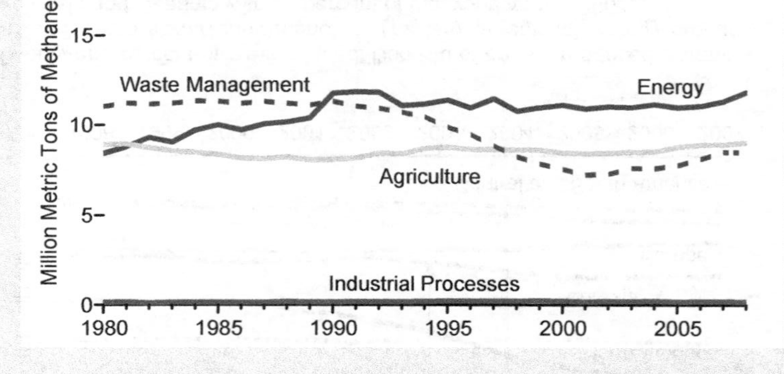


In 2008, methane emissions accounted fo r 10 percent of total U.S. greenhouse gas emissions, w eighted by global warming potential (see "Global Warming Potential" in Glossary ). Most methane emissions came from energy, waste management, and agricultural sources. The production, processing, and distribution of natural gas accounted for 60 percent of the energy-related methane emissions in 2008.

## Energy Flow, 2009 (Quadrillion Btu)

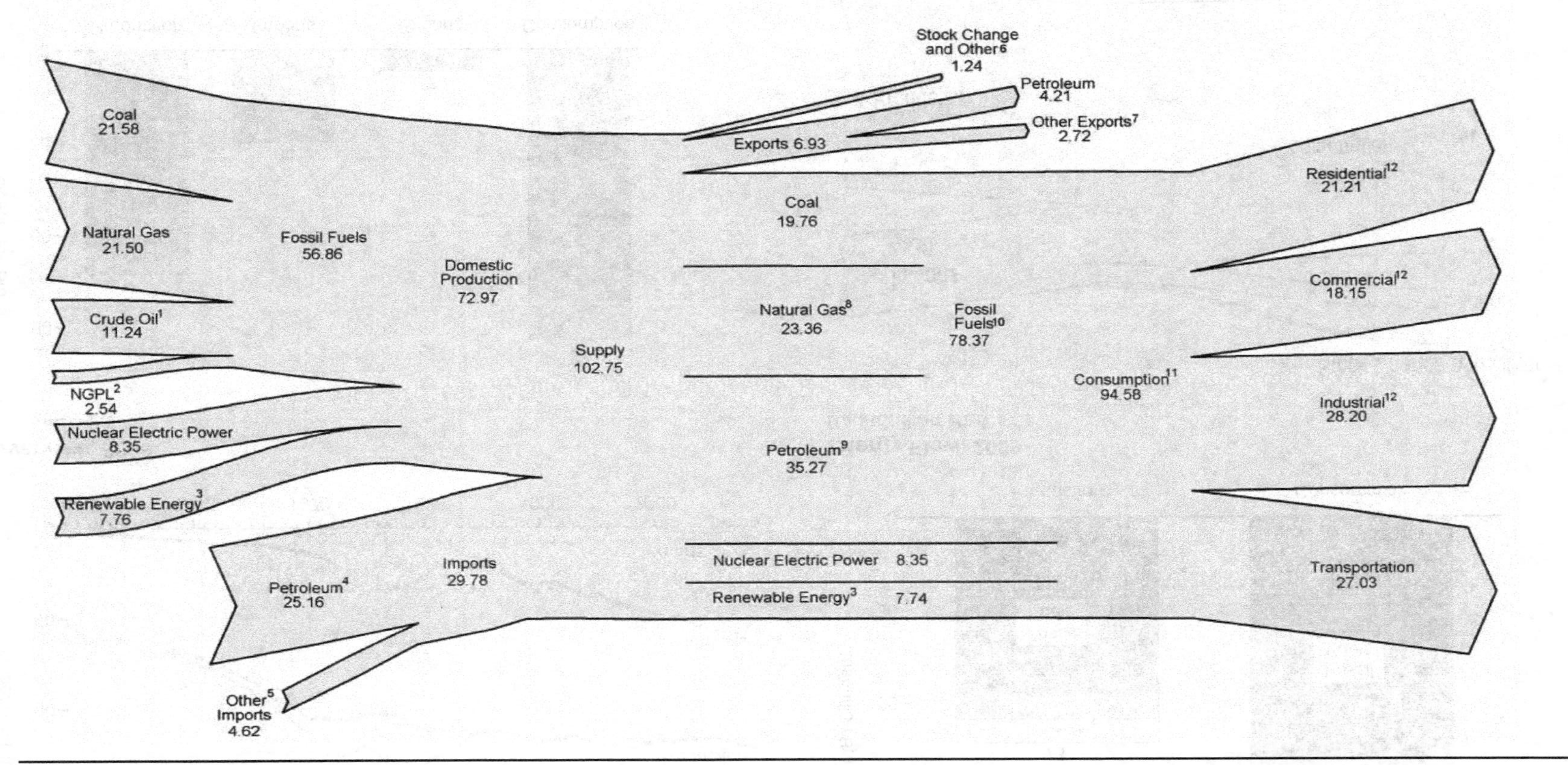


[1] Includes lease condensate.
[2] Natural gas plant liquids.
[3] Conventional hydroelectric power, biomass, geothermal, solar/photovoltaic, and wind.
[4] Crude oil and petroleum products. Includes imports into the Strategic Petroleum Reserve.
[5] Natural gas, coal, coal coke, biofuels, and electricity.
[6] Adjustments, losses, and unaccounted for.
[7] Coal, natural gas, coal coke, electricity, and biofuels.
[8] Natural gas only; excludes supplemental gaseous fuels.
[9] Petroleum products, including natural gas plant liquids, and crude oil burned as fuel.
[10] Includes 0.02 quadrillion Btu of coal coke net exports.
[11] Includes 0.12 quadrillion Btu of electricity net imports.
[12] Total energy consumption, which is the sum of primary energy consumption, electricity retail sales, and electrical system energy losses. Losses are allocated to the end-use sectors in proportion to each sector's share of total electric ity retail sales. See Note, "Electrical Sy stems Energy Losses," at end of Section 2.

Notes: • Data are preliminary. • Values are derived from source data prior to rounding for publication. • Totals may not equal sum of components due to independent rounding.

Sources: Tables 1.1, 1.2, 1.3, 1.4, and 2.1a.

U.S. Energy Information Administration / Annual Energy Review 2009

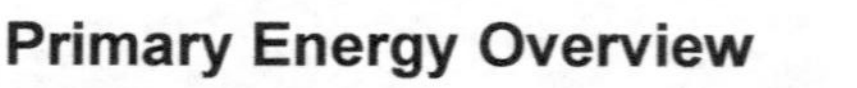

## Primary Energy Overview

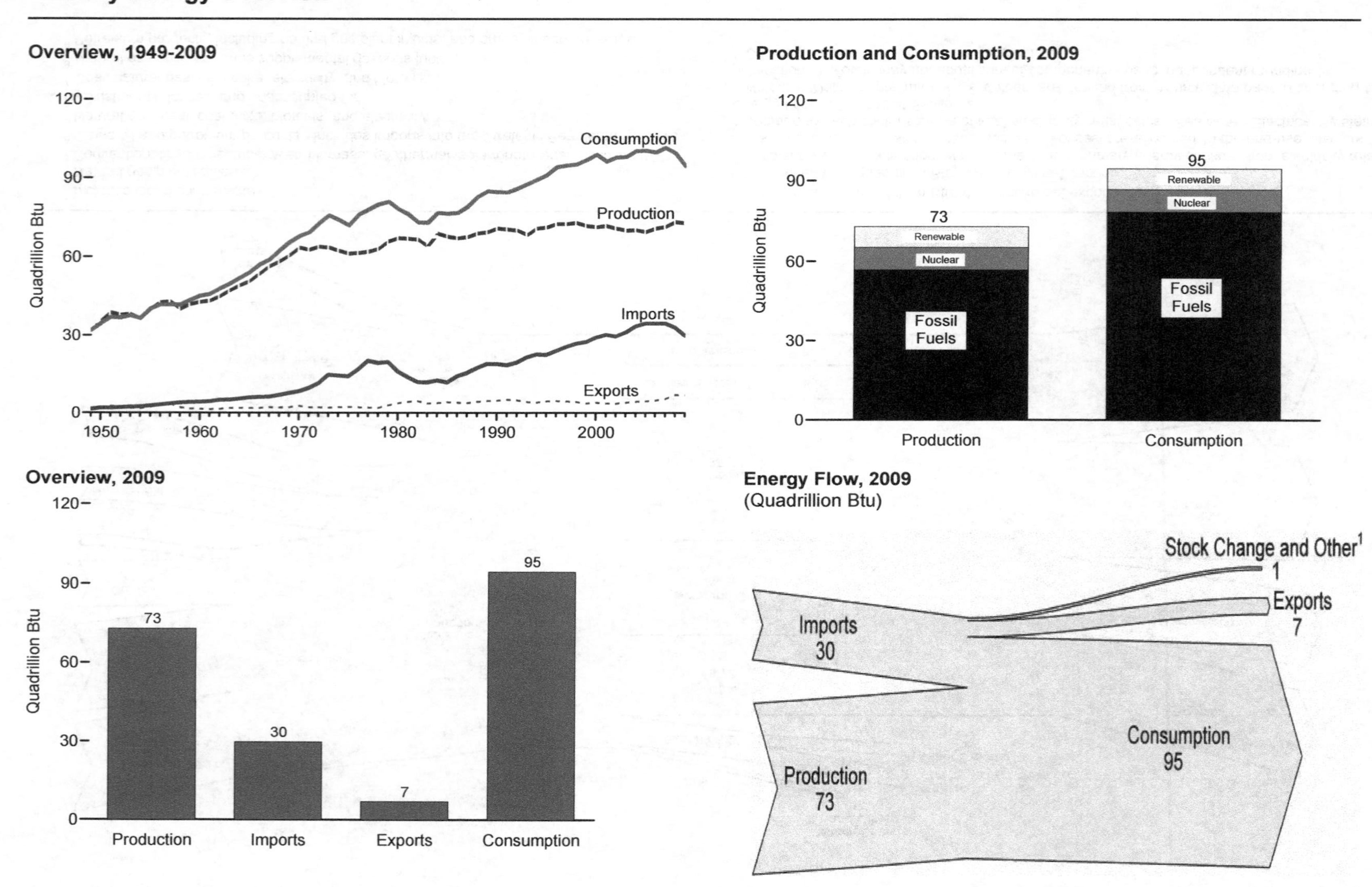


[1] Adjustments, losses, and unaccounted for.

U.S. Energy Information Administration / Annual Energy Review 2009

## Primary Energy Overview, Selected Years, 1949-2009

(Quadrillion Btu)

| Year | Production | | | | Trade | | | | | Stock Change and Other [7] | Consumption | | | |
|---|---|---|---|---|---|---|---|---|---|---|---|---|---|---|
| | Fossil Fuels [2] | Nuclear Electric Power | Renewable Energy [3] | Total | Imports | | Exports | | Net Imports [1] | | Fossil Fuels [8] | Nuclear Electric Power | Renewable Energy [3] | Total [9] |
| | | | | | Petroleum [4] | Total [5] | Coal | Total [6] | Total | | | | | |
| 1949 | 28.748 | 0.000 | 2.974 | 31.722 | 1.427 | 1.448 | 0.877 | 1.592 | -0.144 | 0.403 | 29.002 | 0.000 | 2.974 | 31.982 |
| 1950 | 32.563 | .000 | 2.978 | 35.540 | 1.886 | 1.913 | .786 | 1.465 | .448 | -1.372 | 31.632 | .000 | 2.978 | 34.616 |
| 1955 | 37.364 | .000 | 2.784 | 40.148 | 2.752 | 2.790 | 1.465 | 2.286 | .504 | -.444 | 37.410 | .000 | 2.784 | 40.208 |
| 1960 | 39.869 | .006 | 2.929 | 42.804 | 3.999 | 4.188 | 1.023 | 1.477 | 2.710 | -.427 | 42.137 | .006 | 2.929 | 45.087 |
| 1965 | 47.235 | .043 | 3.398 | 50.676 | 5.402 | 5.892 | 1.376 | 1.829 | 4.063 | -.722 | 50.577 | .043 | 3.398 | 54.017 |
| 1970 | 59.186 | .239 | 4.076 | 63.501 | 7.470 | 8.342 | 1.936 | 2.632 | 5.709 | -1.367 | 63.522 | .239 | 4.076 | 67.844 |
| 1971 | 58.042 | .413 | 4.268 | 62.723 | 8.540 | 9.535 | 1.546 | 2.151 | 7.384 | -.818 | 64.596 | .413 | 4.268 | 69.289 |
| 1972 | 58.938 | .584 | 4.398 | 63.920 | 10.299 | 11.387 | 1.531 | 2.118 | 9.269 | -.485 | 67.696 | .584 | 4.398 | 72.704 |
| 1973 | 58.241 | .910 | 4.433 | 63.585 | 13.466 | 14.613 | 1.425 | 2.033 | 12.580 | -.456 | 70.316 | .910 | 4.433 | 75.708 |
| 1974 | 56.331 | 1.272 | 4.769 | 62.372 | 13.127 | 14.304 | 1.620 | 2.203 | 12.101 | -.482 | 67.906 | 1.272 | 4.769 | 73.991 |
| 1975 | 54.733 | 1.900 | 4.723 | 61.357 | 12.948 | 14.032 | 1.761 | 2.323 | 11.709 | -1.067 | 65.355 | 1.900 | 4.723 | 71.999 |
| 1976 | 54.723 | 2.111 | 4.768 | 61.602 | 15.672 | 16.760 | 1.597 | 2.172 | 14.588 | -.178 | 69.104 | 2.111 | 4.768 | 76.012 |
| 1977 | 55.101 | 2.702 | 4.249 | 62.052 | 18.756 | 19.948 | 1.442 | 2.052 | 17.896 | -1.948 | 70.989 | 2.702 | 4.249 | 78.000 |
| 1978 | 55.074 | 3.024 | 5.039 | 63.137 | 17.824 | 19.106 | 1.078 | 1.920 | 17.186 | -.337 | 71.856 | 3.024 | 5.039 | 79.986 |
| 1979 | 58.006 | 2.776 | 5.166 | 65.948 | 17.933 | 19.460 | 1.753 | 2.855 | 16.605 | -1.649 | 72.892 | 2.776 | 5.166 | 80.903 |
| 1980 | 59.008 | 2.739 | 5.485 | 67.232 | 14.658 | 15.796 | 2.421 | 3.695 | 12.101 | -1.212 | 69.826 | 2.739 | 5.485 | 78.122 |
| 1981 | 58.529 | 3.008 | R5.477 | 67.014 | 12.639 | 13.719 | 2.944 | 4.307 | 9.412 | -.258 | 67.570 | 3.008 | R5.477 | 76.168 |
| 1982 | 57.458 | 3.131 | 6.034 | 66.623 | 10.777 | 11.861 | 2.787 | 4.608 | 7.253 | R-.723 | 63.888 | 3.131 | 6.034 | 73.153 |
| 1983 | 54.416 | 3.203 | R6.561 | R64.180 | 10.647 | 11.752 | 2.045 | 3.693 | 8.059 | .799 | 63.154 | 3.203 | R6.561 | R73.038 |
| 1984 | 58.849 | 3.553 | R6.522 | R68.924 | 11.433 | 12.471 | 2.151 | 3.786 | 8.685 | -.894 | 66.504 | 3.553 | R6.522 | R76.714 |
| 1985 | 57.539 | 4.076 | R6.185 | R67.799 | 10.609 | 11.781 | 2.438 | 4.196 | 7.584 | 1.107 | 66.091 | 4.076 | R6.185 | R76.491 |
| 1986 | 56.575 | 4.380 | R6.223 | R67.178 | 13.201 | 14.151 | 2.248 | 4.021 | 10.130 | -.552 | 66.031 | 4.380 | R6.223 | R76.756 |
| 1987 | 57.167 | 4.754 | R5.739 | R67.659 | 14.162 | 15.398 | 2.093 | 3.812 | 11.586 | -.073 | 68.522 | 4.754 | R5.739 | R79.173 |
| 1988 | 57.875 | 5.587 | R5.568 | R69.030 | 15.747 | 17.296 | 2.499 | 4.366 | 12.929 | .860 | 71.556 | 5.587 | R5.568 | R82.819 |
| 1989 | 57.483 | 5.602 | R6.391 | R69.476 | 17.162 | 18.766 | 2.637 | 4.661 | 14.105 | 1.362 | 72.913 | 5.602 | R6.391 | R84.944 |
| 1990 | 58.560 | 6.104 | R6.206 | R70.870 | 17.117 | 18.817 | 2.772 | 4.752 | 14.065 | -.283 | 72.333 | 6.104 | R6.206 | R84.651 |
| 1991 | 57.872 | 6.422 | R6.237 | R70.531 | 16.348 | 18.335 | 2.854 | 5.141 | 13.194 | .881 | 71.880 | 6.422 | R6.238 | R84.606 |
| 1992 | 57.655 | 6.479 | R5.992 | R70.126 | 16.968 | 19.372 | 2.682 | 4.937 | 14.435 | 1.394 | 73.397 | 6.479 | R5.992 | R85.955 |
| 1993 | 55.822 | 6.410 | R6.261 | R68.494 | 18.510 | 21.273 | 1.962 | 4.258 | 17.014 | R2.093 | R74.835 | 6.410 | R6.261 | R87.601 |
| 1994 | 58.044 | 6.694 | R6.153 | R70.891 | 19.243 | 22.390 | 1.879 | 4.061 | 18.329 | R.037 | R76.257 | 6.694 | R6.153 | R89.257 |
| 1995 | 57.540 | 7.075 | R6.701 | R71.316 | 18.881 | 22.260 | 2.318 | 4.511 | 17.750 | R2.103 | R77.257 | 7.075 | R6.703 | R91.169 |
| 1996 | 58.387 | 7.087 | R7.165 | R72.639 | 20.284 | 23.702 | 2.368 | 4.633 | 19.069 | R2.465 | R79.782 | 7.087 | R7.166 | R94.172 |
| 1997 | 58.857 | 6.597 | R7.177 | R72.631 | 21.740 | 25.215 | 2.193 | 4.514 | 20.701 | R1.429 | 80.874 | 6.597 | R7.175 | R94.761 |
| 1998 | 59.314 | 7.068 | R6.655 | R73.037 | 22.908 | 26.581 | 2.092 | 4.299 | 22.281 | R-.140 | R81.369 | 7.068 | R6.654 | R95.178 |
| 1999 | 57.614 | 7.610 | R6.678 | R71.903 | 23.133 | 27.252 | 1.525 | 3.715 | 23.537 | R1.372 | R82.427 | 7.610 | R6.677 | R96.812 |
| 2000 | 57.366 | 7.862 | R6.257 | R71.485 | 24.531 | 28.973 | 1.528 | 4.006 | 24.967 | R2.517 | R84.732 | 7.862 | R6.260 | R98.970 |
| 2001 | 58.541 | R8.029 | R5.312 | R71.883 | 25.398 | 30.157 | 1.265 | 3.770 | 26.386 | R-1.953 | R82.902 | R8.029 | R5.311 | R96.316 |
| 2002 | 56.894 | R8.145 | R5.892 | R70.931 | 24.673 | 29.407 | 1.032 | 3.668 | 25.739 | R1.183 | R83.749 | R8.145 | R5.888 | R97.853 |
| 2003 | R56.099 | 7.959 | R6.139 | R70.197 | 26.218 | 31.061 | 1.117 | 4.054 | 27.007 | R.927 | R84.010 | 7.959 | R6.141 | R98.131 |
| 2004 | R55.895 | 8.222 | R6.235 | R70.352 | 28.196 | 33.543 | 1.253 | 4.433 | 29.110 | R.851 | R85.805 | 8.222 | R6.247 | R100.313 |
| 2005 | R55.038 | R8.161 | R6.393 | R69.592 | 29.247 | 34.710 | 1.273 | 4.561 | 30.149 | R.704 | R85.793 | R8.161 | R6.406 | R100.445 |
| 2006 | 55.968 | R8.215 | R6.774 | R70.957 | 29.162 | 34.673 | 1.264 | 4.868 | 29.805 | R-.973 | R84.687 | R8.215 | R6.824 | R99.790 |
| 2007 | R56.447 | R8.455 | R6.706 | R71.608 | 28.762 | 34.685 | 1.507 | 5.448 | 29.238 | R.682 | R86.246 | R8.455 | R6.719 | R101.527 |
| 2008 | R57.613 | R8.427 | R7.381 | R73.421 | R27.644 | R32.952 | 2.071 | R7.016 | R25.936 | R.045 | R83.496 | R8.427 | R7.366 | R99.402 |
| 2009P | 56.860 | 8.349 | 7.761 | 72.970 | 25.160 | 29.781 | 1.515 | 6.932 | 22.849 | -1.241 | 78.368 | 8.349 | 7.744 | 94.578 |

[1] Net imports equal imports minus exports. A minus sign indicates exports are greater than imports.
[2] Coal, natural gas (dry), crude oil, and natural gas plant liquids.
[3] See Note "Renewable Energy Production and Consumption" at the end of Section 10.
[4] Crude oil and petroleum products. Includes imports into the Strategic Petroleum Reserve.
[5] Also includes natural gas, coal, coal coke, fuel ethanol, biodiesel, and electricity.
[6] Also includes natural gas, petroleum, coal coke, biodiesel, and electricity.
[7] Calculated as consumption and exports minus production and imports. Includes petroleum stock change and adjustments; natural gas net storage withdrawals and balancing item; coal stock change, losses, and unaccounted for; fuel ethanol stock change; and biodiesel stock change and balancing item.
[8] Coal, coal coke net imports, natural gas, and petroleum.
[9] Also includes electricity net imports.

R=Revised. P=Preliminary.

Notes: • See "Primary Energy," "Primary Energy Production," and "Primary Energy Consumption" in Glossary. • Totals may not equal sum of components due to independent rounding.

Web Page: For all data beginning in 1949, see http://www.eia.gov/emeu/aer/overview.html.

U.S. Energy Information Administration / Annual Energy Review 2009

## Primary Energy Production by Source

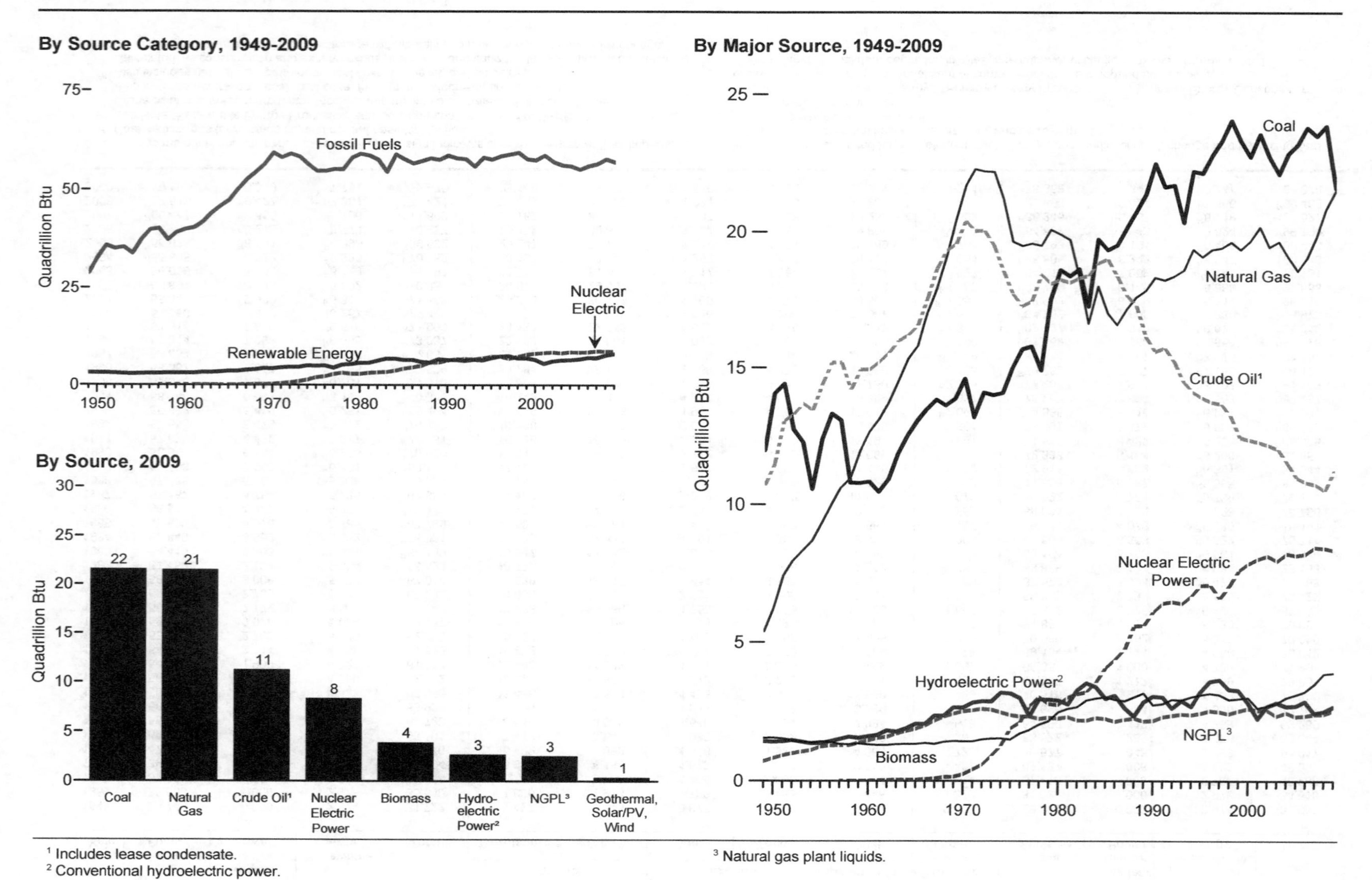


[1] Includes lease condensate.
[2] Conventional hydroelectric power.
[3] Natural gas plant liquids.

U.S. Energy Information Administration / Annual Energy Review 2009

## Primary Energy Production by Source, Selected Years, 1949-2009

(Quadrillion Btu)

| Year | Fossil Fuels | | | | | Nuclear Electric Power | Renewable Energy [1] | | | | | | Total |
|---|---|---|---|---|---|---|---|---|---|---|---|---|---|
| | Coal [2] | Natural Gas (Dry) | Crude Oil [3] | NGPL [4] | Total | | Hydro-electric Power [5] | Geothermal | Solar/PV | Wind | Biomass | Total | |
| 1949 | 11.974 | 5.377 | 10.683 | 0.714 | 28.748 | 0.000 | 1.425 | NA | NA | NA | 1.549 | 2.974 | 31.722 |
| 1950 | 14.060 | 6.233 | 11.447 | .823 | 32.563 | .000 | 1.415 | NA | NA | NA | 1.562 | 2.978 | 35.540 |
| 1955 | 12.370 | 9.345 | 14.410 | 1.240 | 37.364 | .000 | 1.360 | NA | NA | NA | 1.424 | 2.784 | 40.148 |
| 1960 | 10.817 | 12.656 | 14.935 | 1.461 | 39.869 | .006 | 1.608 | .001 | NA | NA | 1.320 | 2.929 | 42.804 |
| 1965 | 13.055 | 15.775 | 16.521 | 1.883 | 47.235 | .043 | 2.059 | .004 | NA | NA | 1.335 | 3.398 | 50.676 |
| 1970 | 14.607 | 21.666 | 20.401 | 2.512 | 59.186 | .239 | 2.634 | .011 | NA | NA | 1.431 | 4.076 | 63.501 |
| 1971 | 13.186 | 22.280 | 20.033 | 2.544 | 58.042 | .413 | 2.824 | .012 | NA | NA | 1.432 | 4.268 | 62.723 |
| 1972 | 14.092 | 22.208 | 20.041 | 2.598 | 58.938 | .584 | 2.864 | .031 | NA | NA | 1.503 | 4.398 | 63.920 |
| 1973 | 13.992 | 22.187 | 19.493 | 2.569 | 58.241 | .910 | 2.861 | .043 | NA | NA | 1.529 | 4.433 | 63.585 |
| 1974 | 14.074 | 21.210 | 18.575 | 2.471 | 56.331 | 1.272 | 3.177 | .053 | NA | NA | 1.540 | 4.769 | 62.372 |
| 1975 | 14.989 | 19.640 | 17.729 | 2.374 | 54.733 | 1.900 | 3.155 | .070 | NA | NA | 1.499 | 4.723 | 61.357 |
| 1976 | 15.654 | 19.480 | 17.262 | 2.327 | 54.723 | 2.111 | 2.976 | .078 | NA | NA | 1.713 | 4.768 | 61.602 |
| 1977 | 15.755 | 19.565 | 17.454 | 2.327 | 55.101 | 2.702 | 2.333 | .077 | NA | NA | 1.838 | 4.249 | 62.052 |
| 1978 | 14.910 | 19.485 | 18.434 | 2.245 | 55.074 | 3.024 | 2.937 | .064 | NA | NA | 2.038 | 5.039 | 63.137 |
| 1979 | 17.540 | 20.076 | 18.104 | 2.286 | 58.006 | 2.776 | 2.931 | .084 | NA | NA | 2.152 | 5.166 | 65.948 |
| 1980 | 18.598 | 19.908 | 18.249 | 2.254 | 59.008 | 2.739 | 2.900 | .110 | NA | NA | 2.476 | 5.485 | 67.232 |
| 1981 | 18.377 | 19.699 | 18.146 | 2.307 | 58.529 | 3.008 | 2.758 | .123 | NA | NA | [R]2.596 | [R]5.477 | 67.014 |
| 1982 | 18.639 | 18.319 | 18.309 | 2.191 | 57.458 | 3.131 | 3.266 | .105 | NA | NA | [R]2.663 | 6.034 | 66.623 |
| 1983 | 17.247 | 16.593 | 18.392 | 2.184 | 54.416 | 3.203 | 3.527 | .129 | NA | (s) | [R]2.904 | [R]6.561 | [R]64.180 |
| 1984 | 19.719 | 18.008 | 18.848 | 2.274 | 58.849 | 3.553 | 3.386 | .165 | (s) | (s) | [R]2.971 | [R]6.522 | [R]68.924 |
| 1985 | 19.325 | 16.980 | 18.992 | 2.241 | 57.539 | 4.076 | 2.970 | .198 | (s) | (s) | [R]3.016 | [R]6.185 | [R]67.799 |
| 1986 | 19.509 | 16.541 | 18.376 | 2.149 | 56.575 | 4.380 | 3.071 | .219 | (s) | (s) | [R]2.932 | [R]6.223 | [R]67.178 |
| 1987 | 20.141 | 17.136 | 17.675 | 2.215 | 57.167 | 4.754 | 2.635 | .229 | (s) | (s) | [R]2.875 | [R]5.739 | [R]67.659 |
| 1988 | 20.738 | 17.599 | 17.279 | 2.260 | 57.875 | 5.587 | 2.334 | .217 | (s) | (s) | [R]3.016 | [R]5.568 | [R]69.030 |
| 1989 | [2]21.360 | 17.847 | 16.117 | 2.158 | 57.483 | 5.602 | 2.837 | .317 | .055 | .022 | [R]3.159 | [R]6.391 | [R]69.476 |
| 1990 | 22.488 | 18.326 | 15.571 | 2.175 | 58.560 | 6.104 | 3.046 | .336 | .060 | .029 | [R]2.735 | [R]6.206 | [R]70.870 |
| 1991 | 21.636 | 18.229 | 15.701 | 2.306 | 57.872 | 6.422 | 3.016 | .346 | .063 | .031 | [R]2.782 | [R]6.237 | [R]70.531 |
| 1992 | 21.694 | 18.375 | 15.223 | 2.363 | 57.655 | 6.479 | 2.617 | .349 | .064 | .030 | [R]2.932 | [R]5.992 | [R]70.126 |
| 1993 | 20.336 | 18.584 | 14.494 | 2.408 | 55.822 | 6.410 | 2.892 | .364 | .066 | .031 | [R]2.908 | [R]6.261 | [R]68.494 |
| 1994 | 22.202 | 19.348 | 14.103 | 2.391 | 58.044 | 6.694 | 2.683 | .338 | .069 | .036 | [R]3.028 | [R]6.153 | [R]70.891 |
| 1995 | 22.130 | 19.082 | 13.887 | 2.442 | 57.540 | 7.075 | 3.205 | .294 | .070 | .033 | [R]3.099 | [R]6.701 | [R]71.316 |
| 1996 | 22.790 | 19.344 | 13.723 | 2.530 | 58.387 | 7.087 | 3.590 | .316 | .071 | .033 | [R]3.155 | [R]7.165 | [R]72.639 |
| 1997 | 23.310 | 19.394 | 13.658 | 2.495 | 58.857 | 6.597 | 3.640 | .325 | .070 | .034 | [R]3.108 | [R]7.177 | [R]72.631 |
| 1998 | 24.045 | 19.613 | 13.235 | 2.420 | 59.314 | 7.068 | 3.297 | .328 | .070 | .031 | [R]2.929 | [R]6.655 | [R]73.037 |
| 1999 | 23.295 | 19.341 | 12.451 | 2.528 | 57.614 | 7.610 | 3.268 | .331 | .069 | .046 | [R]2.965 | [R]6.678 | [R]71.903 |
| 2000 | 22.735 | 19.662 | 12.358 | 2.611 | 57.366 | 7.862 | 2.811 | .317 | .066 | .057 | [R]3.006 | [R]6.257 | [R]71.485 |
| 2001 | [2]23.547 | 20.166 | 12.282 | 2.547 | 58.541 | [R]8.029 | 2.242 | .311 | .065 | .070 | [R]2.624 | [R]5.312 | [R]71.883 |
| 2002 | 22.732 | 19.439 | 12.163 | 2.559 | 56.894 | [R]8.145 | 2.689 | .328 | .064 | .105 | [R]2.705 | [R]5.892 | [R]70.931 |
| 2003 | 22.094 | [R]19.633 | 12.026 | 2.346 | [R]56.099 | 7.959 | 2.825 | .331 | .064 | .115 | [R]2.805 | [R]6.139 | [R]70.197 |
| 2004 | 22.852 | [R]19.074 | 11.503 | 2.466 | [R]55.895 | 8.222 | 2.690 | .341 | .065 | .142 | [R]2.998 | [R]6.235 | [R]70.352 |
| 2005 | 23.185 | [R]18.556 | 10.963 | 2.334 | [R]55.038 | [R]8.161 | 2.703 | .343 | .066 | .178 | [R]3.104 | [R]6.393 | [R]69.592 |
| 2006 | 23.790 | 19.022 | 10.801 | 2.356 | 55.968 | [R]8.215 | 2.869 | .343 | .072 | .264 | [R]3.226 | [R]6.774 | [R]70.957 |
| 2007 | 23.493 | [R]19.825 | 10.721 | 2.409 | [R]56.447 | [R]8.455 | 2.446 | .349 | .081 | .341 | [R]3.489 | [R]6.706 | [R]71.608 |
| 2008 | [R]23.851 | [R]20.834 | [R]10.509 | [R]2.419 | [R]57.613 | [R]8.427 | [R]2.511 | [R].360 | [R].097 | [R].546 | [R]3.867 | [R]7.381 | [R]73.421 |
| 2009[P] | 21.578 | 21.500 | 11.241 | 2.541 | 56.860 | 8.349 | 2.682 | .373 | .109 | .697 | 3.900 | 7.761 | 72.970 |

[1] Most data are estimates. See Tables 10.1-10.2c for notes on series components and estimation.
[2] Beginning in 1989, includes waste coal supplied. Beginning in 2001, also includes a small amount of refuse recovery. See Table 7.1.
[3] Includes lease condensate.
[4] Natural gas plant liquids.
[5] Conventional hydroelectric power.

R=Revised. P=Preliminary. NA=Not available. (s)=Less than 0.0005 quadrillion Btu.
Notes: • See "Primary Energy Production" in Glossary. • Totals may not equal sum of componen ts due to independent rounding.
Web Page: For all data beginning in 1949, see http://www.eia.gov/emeu/aer/overview.html.

U.S. Energy Information Administration / Annual Energy Review 2009

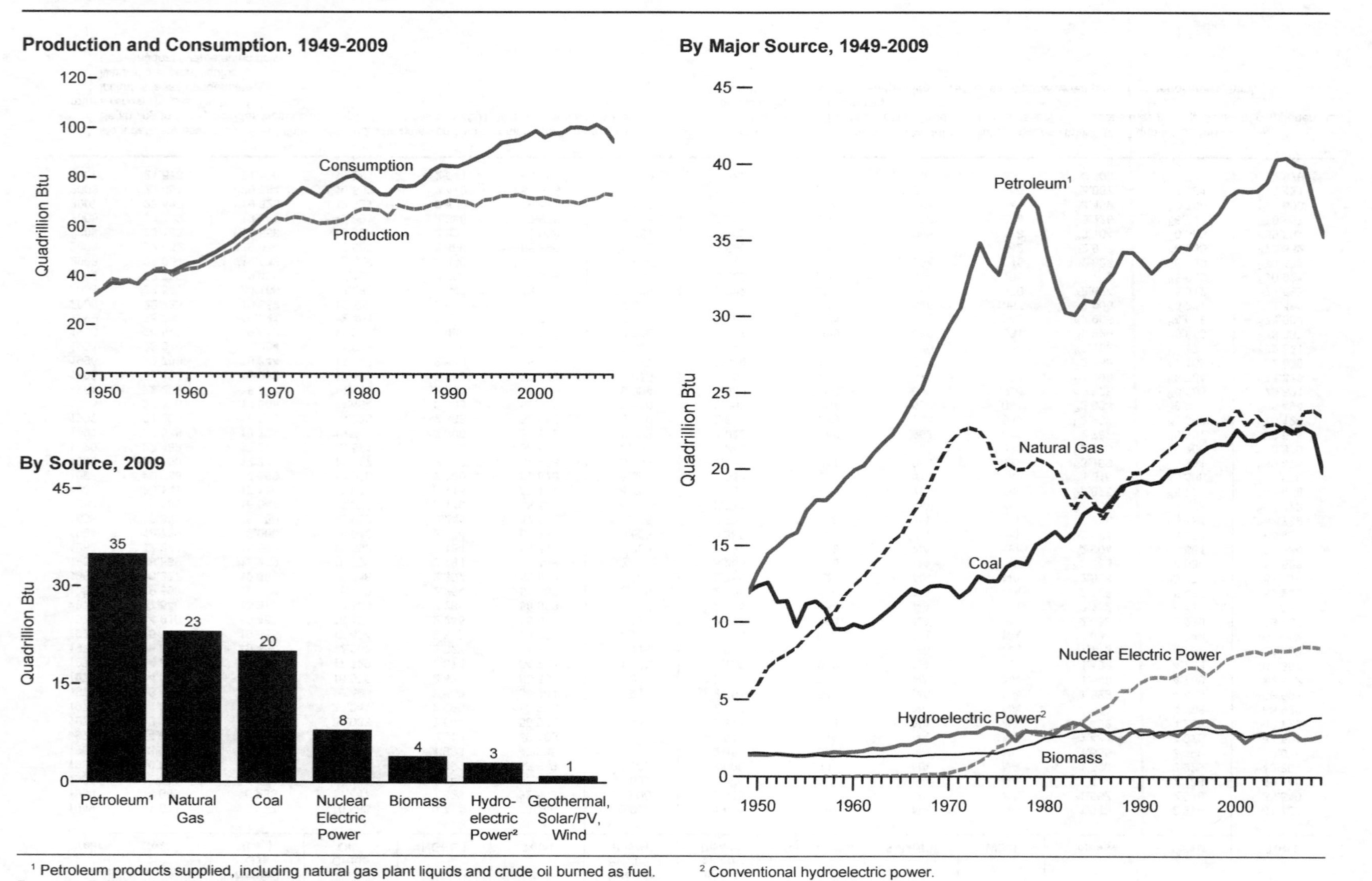


¹ Petroleum products supplied, including natural gas plant liquids and crude oil burned as fuel. Does not include biofuels that have been bl ended with petroleum—biofuels are included in "Biomass."

² Conventional hydroelectric power.

U.S. Energy Information Administration / Annual Energy Review 2009

## Primary Energy Consumption by Source, Selected Years, 1949-2009

(Quadrillion Btu)

| Year | Fossil Fuels | | | | | Nuclear Electric Power | Renewable Energy [1] | | | | | | Electricity Net Imports [2] | Total |
|---|---|---|---|---|---|---|---|---|---|---|---|---|---|---|
| | Coal | Coal Coke Net Imports [2] | Natural Gas [3] | Petroleum [4] | Total | | Hydro-electric Power [5] | Geothermal | Solar/PV | Wind | Biomass | Total | | |
| 1949 | 11.981 | -0.007 | 5.145 | 11.883 | 29.002 | 0.000 | 1.425 | NA | NA | NA | 1.549 | 2.974 | 0.005 | 31.982 |
| 1950 | 12.347 | .001 | 5.968 | 13.315 | 31.632 | .000 | 1.415 | NA | NA | NA | 1.562 | 2.978 | .006 | 34.616 |
| 1955 | 11.167 | -.010 | 8.998 | 17.255 | 37.410 | .000 | 1.360 | NA | NA | NA | 1.424 | 2.784 | .014 | 40.208 |
| 1960 | 9.838 | -.006 | 12.385 | 19.919 | 42.137 | .006 | 1.608 | .001 | NA | NA | 1.320 | 2.929 | .015 | 45.087 |
| 1965 | 11.581 | -.018 | 15.769 | 23.246 | 50.577 | .043 | 2.059 | .004 | NA | NA | 1.335 | 3.398 | (s) | 54.017 |
| 1970 | 12.265 | -.058 | 21.795 | 29.521 | 63.522 | .239 | 2.634 | .011 | NA | NA | 1.431 | 4.076 | .007 | 67.844 |
| 1971 | 11.598 | -.033 | 22.469 | 30.561 | 64.596 | .413 | 2.824 | .012 | NA | NA | 1.432 | 4.268 | .012 | 69.289 |
| 1972 | 12.077 | -.026 | 22.698 | 32.947 | 67.696 | .584 | 2.864 | .031 | NA | NA | 1.503 | 4.398 | .026 | 72.704 |
| 1973 | 12.971 | -.007 | 22.512 | 34.840 | 70.316 | .910 | 2.861 | .043 | NA | NA | 1.529 | 4.433 | .049 | 75.708 |
| 1974 | 12.663 | .056 | 21.732 | 33.455 | 67.906 | 1.272 | 3.177 | .053 | NA | NA | 1.540 | 4.769 | .043 | 73.991 |
| 1975 | 12.663 | .014 | 19.948 | 32.731 | 65.355 | 1.900 | 3.155 | .070 | NA | NA | 1.499 | 4.723 | .021 | 71.999 |
| 1976 | 13.584 | (s) | 20.345 | 35.175 | 69.104 | 2.111 | 2.976 | .078 | NA | NA | 1.713 | 4.768 | .029 | 76.012 |
| 1977 | 13.922 | .015 | 19.931 | 37.122 | 70.989 | 2.702 | 2.333 | .077 | NA | NA | 1.838 | 4.249 | .059 | 78.000 |
| 1978 | 13.766 | .125 | 20.000 | 37.965 | 71.856 | 3.024 | 2.937 | .064 | NA | NA | 2.038 | 5.039 | .067 | 79.986 |
| 1979 | 15.040 | .063 | 20.666 | 37.123 | 72.892 | 2.776 | 2.931 | .084 | NA | NA | 2.152 | 5.166 | .069 | 80.903 |
| 1980 | 15.423 | -.035 | 20.235 | 34.202 | 69.826 | 2.739 | 2.900 | .110 | NA | NA | 2.476 | 5.485 | .071 | 78.122 |
| 1981 | 15.908 | -.016 | 19.747 | 31.931 | 67.570 | 3.008 | 2.758 | .123 | NA | NA | R2.596 | R5.477 | .113 | 76.168 |
| 1982 | 15.322 | -.022 | 18.356 | 30.232 | 63.888 | 3.131 | 3.266 | .105 | NA | NA | R2.663 | 6.034 | .100 | 73.153 |
| 1983 | 15.894 | -.016 | 17.221 | 30.054 | 63.154 | 3.203 | 3.527 | .129 | NA | (s) | R2.904 | R6.561 | .121 | R73.038 |
| 1984 | 17.071 | -.011 | 18.394 | 31.051 | 66.504 | 3.553 | 3.386 | .165 | (s) | (s) | R2.971 | R6.522 | .135 | R76.714 |
| 1985 | 17.478 | -.013 | 17.703 | 30.922 | 66.091 | 4.076 | 2.970 | .198 | (s) | (s) | R3.016 | R6.185 | .140 | R76.491 |
| 1986 | 17.260 | -.017 | 16.591 | 32.196 | 66.031 | 4.380 | 3.071 | .219 | (s) | (s) | R2.932 | R6.223 | .122 | R76.756 |
| 1987 | 18.008 | .009 | 17.640 | 32.865 | 68.522 | 4.754 | 2.635 | .229 | (s) | (s) | R2.875 | R5.739 | .158 | R79.173 |
| 1988 | 18.846 | .040 | 18.448 | 34.222 | 71.556 | 5.587 | 2.334 | .217 | (s) | (s) | R3.016 | R5.568 | .108 | R82.819 |
| 1989 | 19.070 | .030 | 19.602 | 34.211 | 72.913 | 5.602 | 2.837 | .317 | .055 | .022 | R3.159 | R6.391 | .037 | R84.944 |
| 1990 | 19.173 | .005 | 19.603 | 33.553 | 72.333 | 6.104 | 3.046 | .336 | .060 | .029 | R2.735 | R6.206 | .008 | R84.651 |
| 1991 | 18.992 | .010 | 20.033 | 32.845 | 71.880 | 6.422 | 3.016 | .346 | .063 | .031 | R2.782 | R6.238 | .067 | R84.606 |
| 1992 | 19.122 | .035 | 20.714 | 33.527 | 73.397 | 6.479 | 2.617 | .349 | .064 | .030 | R2.932 | R5.992 | .087 | R85.955 |
| 1993 | 19.835 | .027 | 21.229 | 33.744 | R74.835 | 6.410 | 2.892 | .364 | .066 | .031 | R2.908 | R6.261 | .095 | R87.601 |
| 1994 | 19.909 | .058 | 21.728 | R34.561 | R76.257 | 6.694 | 2.683 | .338 | .069 | .036 | R3.028 | R6.153 | .153 | R89.257 |
| 1995 | 20.089 | .061 | 22.671 | R34.436 | R77.257 | 7.075 | 3.205 | .294 | .070 | .033 | R3.101 | R6.703 | .134 | R91.169 |
| 1996 | 21.002 | .023 | 23.085 | 35.673 | R79.782 | 7.087 | 3.590 | .316 | .071 | .033 | R3.157 | R7.166 | .137 | R94.172 |
| 1997 | 21.445 | .046 | 23.223 | R36.159 | 80.874 | 6.597 | 3.640 | .325 | .070 | .034 | R3.105 | R7.175 | .116 | R94.761 |
| 1998 | 21.656 | .067 | 22.830 | R36.816 | R81.369 | 7.068 | 3.297 | .328 | .070 | .031 | R2.928 | R6.654 | .088 | R95.178 |
| 1999 | 21.623 | .058 | 22.909 | R37.837 | R82.427 | 7.610 | 3.268 | .331 | .069 | .046 | R2.963 | R6.677 | .099 | R96.812 |
| 2000 | 22.580 | .065 | 23.824 | R38.263 | R84.732 | 7.862 | 2.811 | .317 | .066 | .057 | R3.008 | R6.260 | .115 | R98.970 |
| 2001 | 21.914 | .029 | 22.773 | R38.185 | R82.902 | R8.029 | 2.242 | .311 | .065 | .070 | R2.622 | R5.311 | .075 | R96.316 |
| 2002 | 21.904 | .061 | 23.558 | R38.225 | R83.749 | R8.145 | 2.689 | .328 | .064 | .105 | R2.701 | R5.888 | .072 | R97.853 |
| 2003 | 22.321 | .051 | R22.831 | R38.808 | R84.010 | 7.959 | 2.825 | .331 | .064 | .115 | R2.807 | R6.141 | .022 | R98.131 |
| 2004 | 22.466 | .138 | R22.909 | R40.292 | R85.805 | 8.222 | 2.690 | .341 | .065 | .142 | R3.010 | R6.247 | .039 | R100.313 |
| 2005 | 22.797 | .044 | R22.561 | R40.391 | R85.793 | R8.161 | 2.703 | .343 | .066 | .178 | R3.117 | R6.406 | .084 | R100.445 |
| 2006 | 22.447 | .061 | 22.224 | R39.955 | R84.687 | R8.215 | 2.869 | .343 | .072 | .264 | R3.277 | R6.824 | .063 | R99.790 |
| 2007 | 22.749 | .025 | R23.702 | R39.769 | R86.246 | R8.455 | 2.446 | .349 | .081 | .341 | R3.503 | R6.719 | .107 | R101.527 |
| 2008 | R22.385 | .041 | R23.791 | R37.279 | R83.496 | R8.427 | R2.511 | R.360 | R.097 | R.546 | R3.852 | R7.366 | .112 | R99.402 |
| 2009[P] | 19.761 | -.024 | 23.362 | 35.268 | 78.368 | 8.349 | 2.682 | .373 | .109 | .697 | 3.883 | 7.744 | .117 | 94.578 |

[1] Most data are estimates. See Tables 10.1-10.2c for notes on series components and estimation.
[2] Net imports equal imports minus exports. A minus sign indicates exports are greater than imports.
[3] Natural gas only; excludes supplemental gaseous fuels. See Note 1, "Supplemental Gaseous Fuels," at end of Section 6.
[4] Petroleum products supplied, including natural gas plant liquids and crude oil burned as fuel. Does not include biofuels that have been blended with petroleum—biofuels are included in "Biomass."
[5] Conventional hydroelectric power.

R=Revised. P=Preliminary. NA=Not available. (s)=Less than 0.0005 and greater than -0.0005 quadrillion Btu.

Notes: • See "Primary Energy Consumption" in Glossary. • See Table E1 for estimated energy consumption for 1635-1945. • See Note 3, "Electricity Imports and Exports," at end of Section 8. • Totals may not equal sum of components due to independent rounding.

Web Page: For all data beginning in 1949, see http://www.eia.gov/emeu/aer/overview.html.

U.S. Energy Information Administration / Annual Energy Review 2009

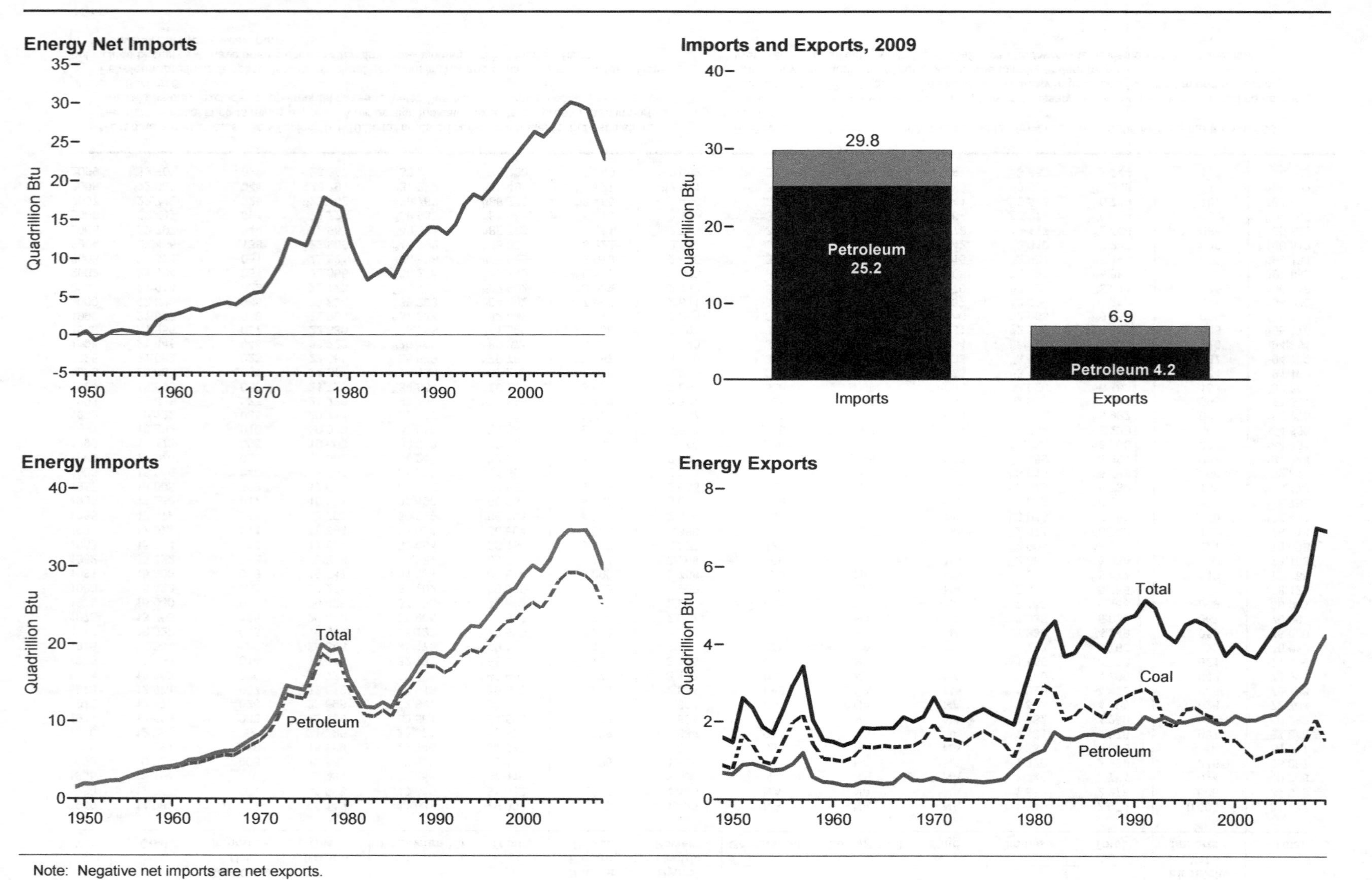
Primary Energy Trade by Source, 1949-2009
Energy Net Imports
Quadrillion Btu
35
30
25
20
15
10
5
0
-5
1950
1960
1970
1980
1990
2000
Imports and Exports, 2009
40
30
20
10
0
29.8
Petroleum
25.2
6.9
Petroleum 4.2
Imports
Exports
Energy Imports
Total
Petroleum
Energy Exports
8
6
4
2
Total
Coal
Petroleum
Note: Negative net imports are net exports.
U.S. Energy Information Administration / Annual Energy Review 2009

## Primary Energy Trade by Source, Selected Years, 1949-2009

(Quadrillion Btu)

| | Imports | | | | | | | | | Exports | | | | | | | | | Net Imports [1] |
|---|---|---|---|---|---|---|---|---|---|---|---|---|---|---|---|---|---|---|---|
| | | | | Petroleum | | | | | | | | | Petroleum | | | | | | |
| Year | Coal | Coal Coke | Natural Gas | Crude Oil [2] | Petroleum Products [3] | Total | Bio-fuels [4] | Elec-tricity | Total | Coal | Coal Coke | Natural Gas | Crude Oil [2] | Petroleum Products [3] | Total | Bio-fuels [5] | Elec-tricity | Total | Total |
| 1949 | 0.008 | 0.007 | 0.000 | 0.915 | 0.513 | 1.427 | NA | 0.006 | 1.448 | 0.877 | 0.014 | 0.021 | 0.192 | 0.488 | 0.680 | NA | 0.001 | 1.592 | -0.144 |
| 1950 | .009 | .011 | .000 | 1.056 | .830 | 1.886 | NA | .007 | 1.913 | .786 | .010 | .027 | .202 | .440 | .642 | NA | .001 | 1.465 | .448 |
| 1955 | .008 | .003 | .011 | 1.691 | 1.061 | 2.752 | NA | .016 | 2.790 | 1.465 | .013 | .032 | .067 | .707 | .774 | NA | .002 | 2.286 | .504 |
| 1960 | .007 | .003 | .161 | 2.196 | 1.802 | 3.999 | NA | .018 | 4.188 | 1.023 | .009 | .012 | .018 | .413 | .431 | NA | .003 | 1.477 | 2.710 |
| 1965 | .005 | .002 | .471 | 2.654 | 2.748 | 5.402 | NA | .012 | 5.892 | 1.376 | .021 | .027 | .006 | .386 | .392 | NA | .013 | 1.829 | 4.063 |
| 1970 | .001 | .004 | .846 | 2.814 | 4.656 | 7.470 | NA | .021 | 8.342 | 1.936 | .061 | .072 | .029 | .520 | .549 | NA | .014 | 2.632 | 5.709 |
| 1971 | .003 | .004 | .964 | 3.573 | 4.968 | 8.540 | NA | .024 | 9.535 | 1.546 | .037 | .083 | .003 | .470 | .473 | NA | .012 | 2.151 | 7.384 |
| 1972 | .001 | .005 | 1.047 | 4.712 | 5.587 | 10.299 | NA | .036 | 11.387 | 1.531 | .031 | .080 | .001 | .466 | .467 | NA | .010 | 2.118 | 9.269 |
| 1973 | .003 | .027 | 1.060 | 6.887 | 6.578 | 13.466 | NA | .057 | 14.613 | 1.425 | .035 | .079 | .004 | .482 | .486 | NA | .009 | 2.033 | 12.580 |
| 1974 | .052 | .088 | .985 | 7.395 | 5.731 | 13.127 | NA | .053 | 14.304 | 1.620 | .032 | .078 | .006 | .458 | .465 | NA | .009 | 2.203 | 12.101 |
| 1975 | .024 | .045 | .978 | 8.721 | 4.227 | 12.948 | NA | .038 | 14.032 | 1.761 | .032 | .074 | .012 | .427 | .439 | NA | .017 | 2.323 | 11.709 |
| 1976 | .030 | .033 | .988 | 11.239 | 4.434 | 15.672 | NA | .037 | 16.760 | 1.597 | .033 | .066 | .017 | .452 | .469 | NA | .008 | 2.172 | 14.588 |
| 1977 | .041 | .045 | 1.037 | 14.027 | 4.728 | 18.756 | NA | .069 | 19.948 | 1.442 | .031 | .056 | .106 | .408 | .514 | NA | .009 | 2.052 | 17.896 |
| 1978 | .074 | .142 | .995 | 13.460 | 4.364 | 17.824 | NA | .072 | 19.106 | 1.078 | .017 | .053 | .335 | .432 | .767 | NA | .005 | 1.920 | 17.186 |
| 1979 | .051 | .099 | 1.300 | 13.825 | 4.108 | 17.933 | NA | .077 | 19.460 | 1.753 | .036 | .056 | .497 | .505 | 1.002 | NA | .007 | 2.855 | 16.605 |
| 1980 | .030 | .016 | 1.006 | 11.195 | 3.463 | 14.658 | NA | .085 | 15.796 | 2.421 | .051 | .049 | .609 | .551 | 1.160 | NA | .014 | 3.695 | 12.101 |
| 1981 | .026 | .013 | .917 | 9.336 | 3.303 | 12.639 | NA | .124 | 13.719 | 2.944 | .029 | .060 | .482 | .781 | 1.264 | NA | .010 | 4.307 | 9.412 |
| 1982 | .019 | .003 | .950 | 7.418 | 3.360 | 10.777 | NA | .112 | 11.861 | 2.787 | .025 | .052 | .500 | 1.231 | 1.732 | NA | .012 | 4.608 | 7.253 |
| 1983 | .032 | .001 | .940 | 7.079 | 3.568 | 10.647 | NA | .132 | 11.752 | 2.045 | .016 | .055 | .348 | 1.217 | 1.565 | NA | .011 | 3.693 | 8.059 |
| 1984 | .032 | .014 | .847 | 7.302 | 4.131 | 11.433 | NA | .144 | 12.471 | 2.151 | .026 | .055 | .384 | 1.161 | 1.545 | NA | .009 | 3.786 | 8.685 |
| 1985 | .049 | .014 | .952 | 6.814 | 3.796 | 10.609 | NA | .157 | 11.781 | 2.438 | .028 | .056 | .432 | 1.225 | 1.657 | NA | .017 | 4.196 | 7.584 |
| 1986 | .055 | .008 | .748 | 9.002 | 4.199 | 13.201 | NA | .139 | 14.151 | 2.248 | .025 | .062 | .326 | 1.344 | 1.670 | NA | .016 | 4.021 | 10.130 |
| 1987 | .044 | .023 | .992 | 10.067 | 4.095 | 14.162 | NA | .178 | 15.398 | 2.093 | .014 | .055 | .319 | 1.311 | 1.630 | NA | .020 | 3.812 | 11.586 |
| 1988 | .053 | .067 | 1.296 | 11.027 | 4.720 | 15.747 | NA | .133 | 17.296 | 2.499 | .027 | .075 | .329 | 1.412 | 1.741 | NA | .024 | 4.366 | 12.929 |
| 1989 | .071 | .057 | 1.387 | 12.596 | 4.565 | 17.162 | NA | .089 | 18.766 | 2.637 | .027 | .109 | .300 | 1.536 | 1.836 | NA | .052 | 4.661 | 14.105 |
| 1990 | .067 | .019 | 1.551 | 12.766 | 4.351 | 17.117 | NA | .063 | 18.817 | 2.772 | .014 | .087 | .230 | 1.594 | 1.824 | NA | .055 | 4.752 | 14.065 |
| 1991 | .085 | .029 | 1.798 | 12.553 | 3.794 | 16.348 | NA | .075 | 18.335 | 2.854 | .020 | .132 | .246 | 1.882 | 2.128 | NA | .008 | 5.141 | 13.194 |
| 1992 | .095 | .052 | 2.161 | 13.253 | 3.714 | 16.968 | NA | .096 | 19.372 | 2.682 | .017 | .220 | .188 | 1.819 | 2.008 | NA | .010 | 4.937 | 14.435 |
| 1993 | .205 | .053 | 2.397 | 14.749 | 3.760 | 18.510 | .001 | .107 | 21.273 | 1.962 | .026 | .142 | .208 | 1.907 | 2.115 | NA | .012 | 4.258 | 17.014 |
| 1994 | .222 | .083 | 2.682 | 15.340 | 3.904 | 19.243 | .001 | .160 | 22.390 | 1.879 | .024 | .164 | .209 | 1.779 | 1.988 | NA | .007 | 4.061 | 18.329 |
| 1995 | .237 | .095 | 2.901 | 15.669 | 3.211 | 18.881 | .001 | .146 | 22.260 | 2.318 | .034 | .156 | .200 | 1.791 | 1.991 | NA | .012 | 4.511 | 17.750 |
| 1996 | .203 | .063 | 3.002 | 16.341 | 3.943 | 20.284 | .001 | .148 | 23.702 | 2.368 | .040 | .155 | .233 | 1.825 | 2.059 | NA | .011 | 4.633 | 19.069 |
| 1997 | .187 | .078 | 3.063 | 17.876 | 3.864 | 21.740 | (s) | .147 | 25.215 | 2.193 | .031 | .159 | .228 | 1.872 | 2.100 | NA | .031 | 4.514 | 20.701 |
| 1998 | .218 | .095 | 3.225 | 18.916 | 3.992 | 22.908 | (s) | .135 | 26.581 | 2.092 | .028 | .161 | .233 | 1.740 | 1.972 | NA | .047 | 4.299 | 22.281 |
| 1999 | .227 | .080 | 3.664 | 18.935 | 4.198 | 23.133 | (s) | .147 | 27.252 | 1.525 | .022 | .164 | .250 | 1.705 | 1.955 | NA | .049 | 3.715 | 23.537 |
| 2000 | .313 | .094 | 3.869 | 19.783 | 4.749 | 24.531 | (s) | .166 | 28.973 | 1.528 | .028 | .245 | .106 | 2.048 | 2.154 | NA | .051 | 4.006 | 24.967 |
| 2001 | .495 | .063 | 4.068 | 20.348 | 5.050 | 25.398 | .002 | .131 | 30.157 | 1.265 | .033 | .377 | .043 | 1.996 | 2.038 | (s) | .056 | 3.770 | 26.386 |
| 2002 | .422 | .080 | 4.104 | 19.920 | 4.753 | 24.673 | .002 | .125 | 29.407 | 1.032 | .020 | .520 | .019 | 2.023 | 2.042 | (s) | .054 | 3.668 | 25.739 |
| 2003 | .626 | .068 | 4.042 | 21.060 | 5.158 | 26.218 | .002 | .104 | 31.061 | 1.117 | .018 | .686 | .026 | 2.124 | 2.150 | .001 | .082 | 4.054 | 27.007 |
| 2004 | .682 | .170 | 4.365 | 22.082 | 6.114 | 28.196 | .013 | .117 | 33.543 | 1.253 | .033 | .862 | .057 | 2.150 | 2.207 | .001 | .078 | 4.433 | 29.110 |
| 2005 | .762 | .088 | 4.450 | 22.091 | 7.156 | 29.247 | .013 | .152 | 34.710 | 1.273 | .043 | .735 | .067 | 2.373 | 2.441 | .001 | .068 | 4.561 | 30.149 |
| 2006 | .906 | .101 | 4.291 | 22.085 | 7.077 | 29.162 | R.068 | .146 | 34.673 | 1.264 | .040 | .730 | .052 | 2.694 | 2.747 | .004 | .083 | 4.868 | 29.805 |
| 2007 | .909 | .061 | 4.723 | 21.914 | 6.849 | 28.762 | .055 | .175 | 34.685 | 1.507 | .036 | .830 | .058 | 2.914 | 2.972 | .035 | .069 | 5.448 | 29.238 |
| 2008 | .855 | .089 | R4.084 | R21.448 | R6.195 | R27.644 | R.085 | .195 | R32.952 | 2.071 | .049 | R1.015 | .061 | R3.653 | R3.713 | .086 | .082 | R7.016 | R25.936 |
| 2009P | .566 | .009 | 3.842 | 19.806 | 5.354 | 25.160 | .026 | .179 | 29.781 | 1.515 | .032 | 1.081 | .093 | 4.115 | 4.208 | .034 | .062 | 6.932 | 22.849 |

[1] Net imports equal imports minus exports. Minus sign indicates exports are greater than imports.
[2] Crude oil and lease condensate. Imports data include imports into the Strategic Petroleum Reserve, which began in 1977.
[3] Petroleum products, unfinished oils, pentanes plus, and gasoline blending components. Does not include biofuels.
[4] Fuel ethanol (including denaturant) and biodiesel.
[5] Biodiesel only.

R=Revised. P=Preliminary. NA=Not available. (s)=Less than 0.0005 quadrillion Btu.
Notes: • Includes trade between the United States (50 States and the District of Columbia) and i ts territories and possessions. • See "Primary Energy" in Glossary. • See Note 3, "Electricity Imports an d Exports," at end of Section 8. • Totals may not equal sum of components due to independent rounding.
Web Page: For all data beginning in 1949, see http://www.eia.gov/emeu/aer/overview.html.

## Energy Consumption and Expenditures Indicators

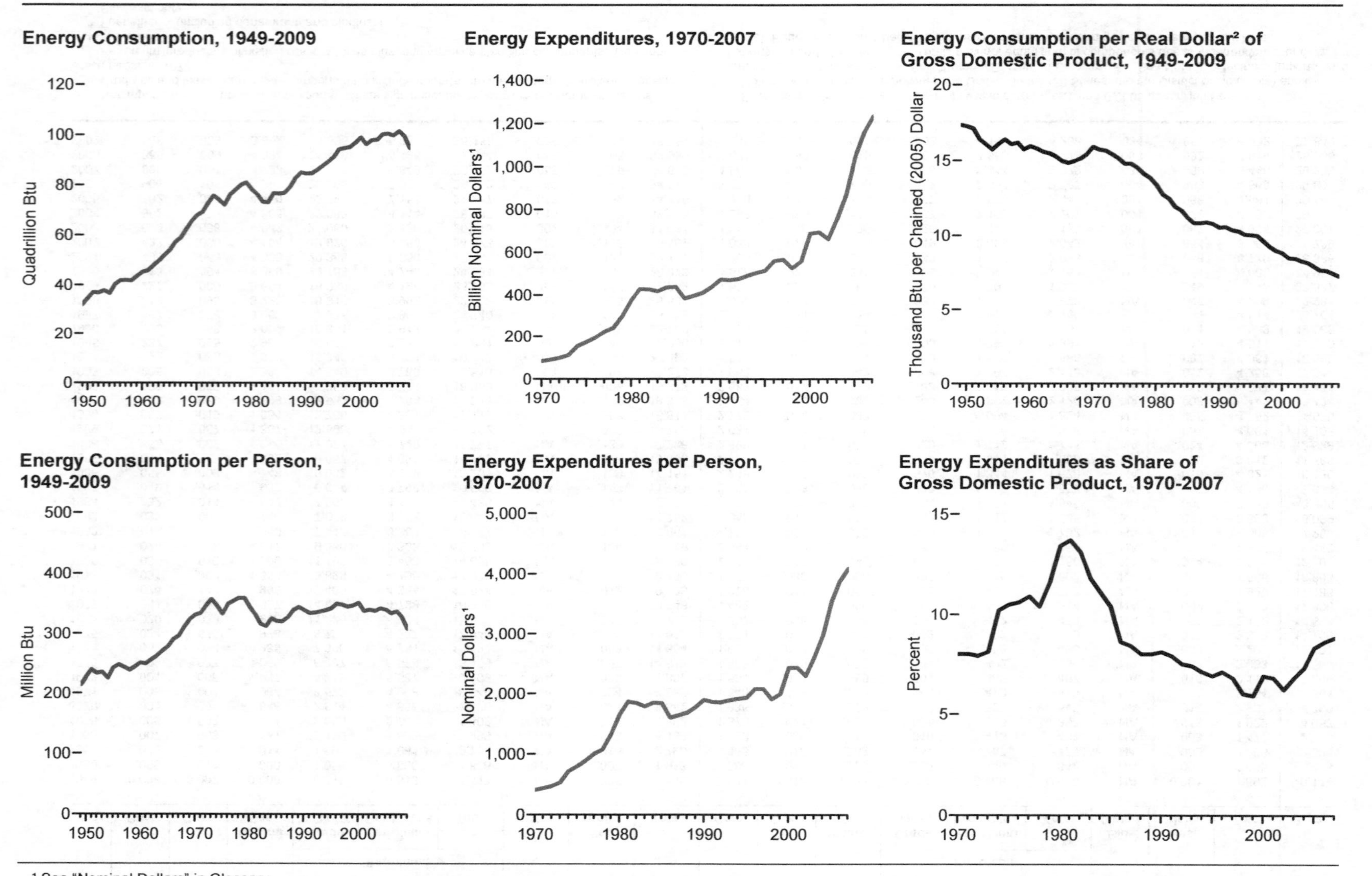


[1] See "Nominal Dollars" in Glossary
[2] In chained (2005) dollars, calculated by using gross domestic product implicit price deflators, See Appendix D1.

U.S. Energy Information Administration / Annual Energy Review 2009

## Energy Consumption, Expenditures, and Emissions Indicators, Selected Years, 1949-2009

| Year | Energy Consumption | Energy Consumption per Person | Energy Expenditures [1] | Energy Expenditures [1] per Person | Gross Domestic Product (GDP) | Energy Expenditures [1] as Share of GDP | Gross Domestic Product (GDP) | Energy Consumption per Real Dollar of GDP | Greenhouse Gas Emissions [2] per Real Dollar of GDP | Carbon Dioxide Emissions [3] per Real Dollar of GDP |
|---|---|---|---|---|---|---|---|---|---|---|
| | Quadrillion Btu | Million Btu | Million Nominal Dollars [4] | Nominal Dollars [4] | Billion Nominal Dollars [4] | Percent | Billion Chained (2005) Dollars [5] | Thousand Btu per Chained (2005) Dollar [5] | Metric Tons Carbon Dioxide Equivalent per Million Chained (2005) Dollars [5] | Metric Tons Carbon Dioxide per Million Chained (2005) Dollars [5] |
| 1949 | 31.982 | 214 | NA | NA | R267.2 | NA | R1,844.7 | R17.34 | NA | R1,196 |
| 1950 | 34.616 | 227 | NA | NA | R293.7 | NA | R2,006.0 | R17.26 | NA | R1,187 |
| 1955 | 40.208 | 242 | NA | NA | R414.7 | NA | R2,500.3 | R16.08 | NA | R1,074 |
| 1960 | 45.087 | 250 | NA | NA | 526.4 | NA | R2,830.9 | R15.93 | NA | R1,029 |
| 1965 | 54.017 | 278 | NA | NA | 719.1 | NA | R3,610.1 | R14.96 | NA | R959 |
| 1970 | 67.844 | 331 | 82,911 | 404 | R1,038.3 | 8.0 | R4,269.9 | R15.89 | NA | R998 |
| 1971 | 69.289 | 334 | 90,071 | 434 | R1,126.8 | 8.0 | R4,413.3 | R15.70 | NA | R977 |
| 1972 | 72.704 | 346 | 98,108 | 467 | R1,237.9 | 7.9 | R4,647.7 | R15.64 | NA | R975 |
| 1973 | 75.708 | 357 | 111,928 | 528 | R1,382.3 | 8.1 | R4,917.0 | R15.40 | NA | R963 |
| 1974 | 73.991 | 346 | 153,370 | 717 | R1,499.5 | 10.2 | R4,889.9 | R15.13 | NA | R935 |
| 1975 | 71.999 | 333 | 171,846 | 796 | R1,637.7 | 10.5 | R4,879.5 | R14.76 | NA | R909 |
| 1976 | 76.012 | 349 | 193,897 | 889 | R1,824.6 | 10.6 | R5,141.3 | R14.78 | NA | R915 |
| 1977 | 78.000 | 354 | 220,461 | 1,001 | R2,030.1 | 10.9 | R5,377.7 | R14.50 | NA | R901 |
| 1978 | 79.986 | 359 | 239,230 | 1,075 | R2,293.8 | 10.4 | R5,677.6 | R14.09 | NA | R862 |
| 1979 | 80.903 | 359 | 297,543 | 1,322 | R2,562.2 | 11.6 | R5,855.0 | R13.82 | NA | R848 |
| 1980 | 78.122 | 344 | R374,346 | R1,647 | R2,788.1 | 13.4 | R5,839.0 | R13.38 | R NA | R817 |
| 1981 | 76.168 | 332 | R427,878 | R1,865 | R3,126.8 | 13.7 | R5,987.2 | R12.72 | R NA | R775 |
| 1982 | 73.153 | 316 | R426,437 | R1,841 | R3,253.2 | 13.1 | R5,870.9 | R12.46 | R NA | R751 |
| 1983 | R73.038 | 312 | R417,418 | R1,785 | R3,534.6 | 11.8 | R6,136.2 | R11.90 | R NA | R714 |
| 1984 | R76.714 | 325 | R435,148 | R1,845 | R3,930.9 | R11.1 | R6,577.1 | R11.66 | R NA | R701 |
| 1985 | R76.491 | R321 | R438,184 | R1,842 | R4,217.5 | 10.4 | R6,849.3 | R11.17 | R NA | R672 |
| 1986 | R76.756 | 320 | R383,409 | R1,597 | R4,460.1 | 8.6 | R7,086.5 | R10.83 | R NA | R650 |
| 1987 | R79.173 | 327 | R396,515 | R1,637 | R4,736.4 | 8.4 | R7,313.3 | R10.83 | R NA | R651 |
| 1988 | R82.819 | 339 | R410,426 | R1,679 | R5,100.4 | 8.0 | R7,613.9 | R10.88 | R NA | R654 |
| 1989 | R84.944 | 344 | R437,611 | R1,773 | R5,482.1 | 8.0 | R7,885.9 | R10.77 | R NA | R643 |
| 1990 | R84.651 | 339 | R472,539 | R1,893 | R5,800.5 | 8.1 | R8,033.9 | R10.54 | R770 | R625 |
| 1991 | R84.606 | 334 | R470,559 | R1,860 | R5,992.1 | R7.9 | R8,015.1 | R10.56 | R766 | R621 |
| 1992 | R85.955 | 335 | R475,587 | R1,854 | R6,342.3 | 7.5 | R8,287.1 | R10.37 | R755 | R612 |
| 1993 | R87.601 | 337 | R491,168 | R1,890 | R6,667.4 | 7.4 | R8,523.4 | R10.28 | R745 | R607 |
| 1994 | R89.257 | 339 | R504,204 | R1,916 | R7,085.2 | 7.1 | R8,870.7 | R10.06 | R729 | R592 |
| 1995 | R91.169 | 342 | R514,049 | R1,930 | R7,414.7 | 6.9 | R9,093.7 | R10.03 | R717 | R583 |
| 1996 | R94.172 | 350 | R559,954 | R2,079 | R7,838.5 | R7.1 | R9,433.9 | R9.98 | R710 | R582 |
| 1997 | R94.761 | 348 | R566,785 | R2,079 | R8,332.4 | 6.8 | R9,854.3 | R9.62 | R686 | R564 |
| 1998 | R95.178 | 345 | R525,738 | R1,906 | R8,793.5 | 6.0 | R10,283.5 | R9.26 | R659 | R545 |
| 1999 | R96.812 | 347 | R556,509 | R1,994 | R9,353.5 | R5.9 | R10,779.8 | R8.98 | R633 | R526 |
| 2000 | R98.970 | 351 | R687,587 | R2,437 | R9,951.5 | R6.9 | R11,226.0 | R8.82 | R624 | R521 |
| 2001 | R96.316 | 338 | R694,515 | R2,436 | R10,286.2 | R6.8 | R11,347.2 | R8.49 | R607 | R506 |
| 2002 | R97.853 | 340 | R661,902 | R2,300 | R10,642.3 | R6.2 | R11,553.0 | R8.47 | R602 | R501 |
| 2003 | R98.131 | 338 | R754,668 | 2,599 | R11,142.1 | R6.8 | R11,840.7 | R8.29 | R592 | R493 |
| 2004 | R100.313 | R342 | R869,112 | 2,966 | R11,867.8 | R7.3 | R12,263.8 | R8.18 | R583 | R485 |
| 2005 | R100.445 | 340 | R1,045,465 | 3,535 | R12,638.4 | R8.3 | R12,638.4 | R7.95 | R568 | R473 |
| 2006 | R99.790 | R334 | R1,158,483 | R3,880 | R13,398.9 | R8.6 | R12,976.2 | R7.69 | R547 | R454 |
| 2007 | R101.527 | 337 | R1,233,058 | R4,089 | R14,077.6 | R8.8 | R13,254.1 | R7.66 | R544 | R452 |
| 2008 | R99.402 | 327 | NA | NA | R14,441.4 | NA | R13,312.2 | R7.47 | R529 | R436 |
| 2009P | 94.578 | 308 | NA | NA | 14,256.3 | NA | 12,987.4 | 7.28 | NA | 416 |

1 Expenditures include taxes where data are available.
2 Greenhouse gas emissions from anthropogenic sources. See Table 12.1.
3 Carbon dioxide emissions from energy consumption. See Table 12.2
4 See "Nominal Dollars" in Glossary.
5 See "Chained Dollars" in Glossary.

R=Revised. P=Preliminary. NA=Not available.
Web Page: For all data beginning in 1949, see http://www.eia.gov/emeu/aer/overview.html.
Sources: **Energy Consumption:** Table 1.3. **Energy Expenditures:** Table 3.5. **Gross Domestic Product:** Table D1. **Population Data:** Table D1. **Greenhouse Gas Emissions:** Table 12.1. **Carbon Dioxide Emissions:** Table 12.2. **Other Columns:** Calculated by U.S. Energy Information Administration.

U.S. Energy Information Administration / Annual Energy Review 2009

## State-Level Energy Consumption and Consumption per Person, 2007

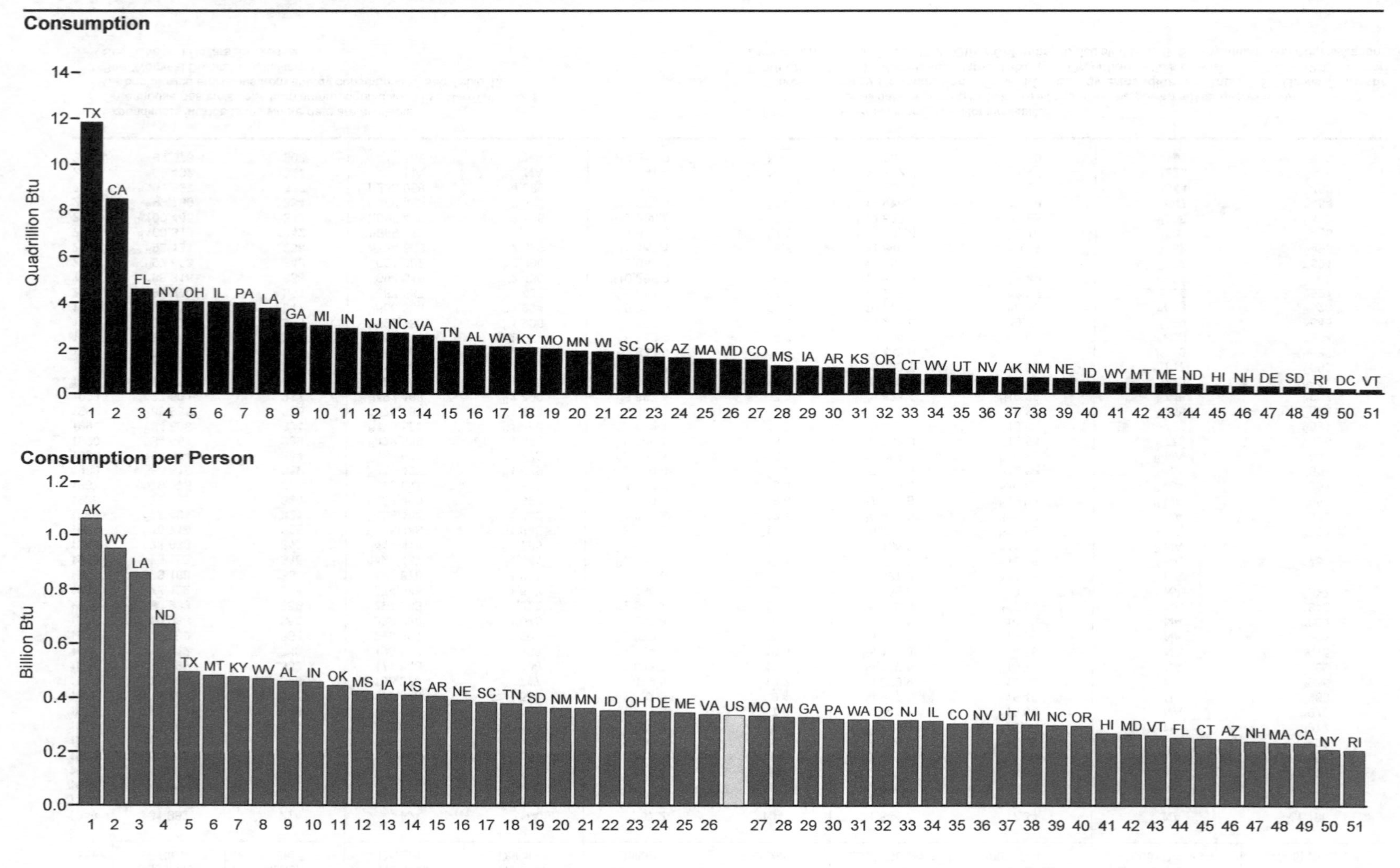


Source: Table 1.6.

U.S. Energy Information Administration / Annual Energy Review 2009

## State-Level Energy Consumption, Expenditures, and Prices, 2007

| Rank | Consumption | | Consumption per Person | | Expenditures [1] | | Expenditures [1] per Person | | Prices [1] | |
|---|---|---|---|---|---|---|---|---|---|---|
| | Trillion Btu | | Million Btu | | Million Dollars [2] | | Dollars [2] | | Dollars [2] per Million Btu | |
| 1 | Texas | 11,834.5 | Alaska | 1,062.3 | Texas | 140,651 | Alaska | 9,191 | Hawaii | 25.20 |
| 2 | California | 8,491.5 | Wyoming | 948.6 | California | 121,829 | Wyoming | 8,687 | Connecticut | 24.93 |
| 3 | Florida | 4,601.9 | Louisiana | 861.2 | New York | 63,642 | Louisiana | 7,688 | District of Columbia | 24.68 |
| 4 | New York | 4,064.3 | North Dakota | 671.1 | Florida | 60,747 | North Dakota | 6,442 | Massachusetts | 23.89 |
| 5 | Ohio | 4,048.9 | Texas | 496.3 | Pennsylvania | 49,301 | Texas | 5,899 | New Hampshire | 23.25 |
| 6 | Illinois | 4,043.2 | Montana | 483.1 | Illinois | 48,297 | Montana | 5,504 | Vermont | 22.90 |
| 7 | Pennsylvania | 4,006.2 | Kentucky | 477.5 | Ohio | 48,190 | Maine | 5,090 | Rhode Island | 22.72 |
| 8 | Louisiana | 3,766.2 | West Virginia | 469.9 | New Jersey | 39,609 | Hawaii | 4,833 | New York | 21.78 |
| 9 | Georgia | 3,133.0 | Alabama | 460.8 | Michigan | 36,882 | Iowa | 4,805 | Maryland | 21.60 |
| 10 | Michigan | 3,026.9 | Indiana | 458.4 | Georgia | 35,678 | Kentucky | 4,796 | Florida | 21.47 |
| 11 | Indiana | 2,904.0 | Oklahoma | 445.8 | Louisiana | 33,624 | Oklahoma | 4,719 | Nevada | 21.12 |
| 12 | New Jersey | 2,743.7 | Mississippi | 424.3 | North Carolina | 32,574 | Alabama | 4,670 | Delaware | 20.82 |
| 13 | North Carolina | 2,700.0 | Iowa | 414.0 | Virginia | 30,509 | West Virginia | 4,624 | Arizona | 20.72 |
| 14 | Virginia | 2,610.9 | Kansas | 409.1 | Indiana | 28,627 | Kansas | 4,610 | California | 20.12 |
| 15 | Tennessee | 2,330.5 | Arkansas | 406.1 | Massachusetts | 25,862 | Mississippi | 4,585 | New Jersey | 19.55 |
| 16 | Alabama | 2,132.0 | Nebraska | 391.6 | Tennessee | 25,462 | New Jersey | 4,577 | Maine | 19.17 |
| 17 | Washington | 2,067.2 | South Carolina | 384.2 | Missouri | 23,342 | Indiana | 4,518 | North Carolina | 19.17 |
| 18 | Kentucky | 2,023.0 | Tennessee | 379.0 | Washington | 23,224 | South Dakota | 4,506 | New Mexico | 19.05 |
| 19 | Missouri | 1,964.1 | South Dakota | 367.2 | Wisconsin | 22,455 | Delaware | 4,465 | Pennsylvania | 18.30 |
| 20 | Minnesota | 1,874.6 | New Mexico | 361.8 | Minnesota | 21,708 | Nebraska | 4,451 | Oregon | 18.23 |
| 21 | Wisconsin | 1,846.3 | Minnesota | 361.7 | Alabama | 21,606 | Arkansas | 4,428 | Alaska | 17.87 |
| 22 | South Carolina | 1,692.3 | Idaho | 354.0 | Maryland | 21,490 | Connecticut | 4,340 | Wisconsin | 17.84 |
| 23 | Oklahoma | 1,608.5 | Ohio | 352.8 | Kentucky | 20,316 | Vermont | 4,329 | Missouri | 17.73 |
| 24 | Arizona | 1,577.8 | Delaware | 350.4 | Arizona | 20,198 | Ohio | 4,199 | Ohio | 17.71 |
| 25 | Massachusetts | 1,514.6 | Maine | 346.3 | South Carolina | 18,130 | Minnesota | 4,189 | Washington | 17.63 |
| 26 | Maryland | 1,488.7 | Virginia | 339.1 | Colorado | 17,033 | Tennessee | 4,141 | Texas | 17.60 |
| 27 | Colorado | 1,479.3 | Missouri | 334.1 | Oklahoma | 17,027 | Nevada | 4,138 | Virginia | 17.58 |
| 28 | Mississippi | 1,239.5 | Wisconsin | 329.8 | Connecticut | 15,146 | South Carolina | 4,116 | South Dakota | 17.45 |
| 29 | Iowa | 1,235.2 | Georgia | 329.0 | Iowa | 14,334 | District of Columbia | 4,069 | Michigan | 17.37 |
| 30 | Arkansas | 1,149.3 | Pennsylvania | 322.6 | Mississippi | 13,392 | New Hampshire | 4,065 | Illinois | 17.27 |
| 31 | Kansas | 1,136.2 | Washington | 320.5 | Oregon | 13,175 | Wisconsin | 4,011 | Montana | 17.26 |
| 32 | Oregon | 1,108.2 | District of Columbia | 318.5 | Kansas | 12,803 | New Mexico | 4,010 | Kansas | 17.23 |
| 33 | Connecticut | 870.7 | New Jersey | 317.1 | Arkansas | 12,533 | Massachusetts | 3,998 | Tennessee | 17.19 |
| 34 | West Virginia | 850.5 | Illinois | 315.2 | Nevada | 10,571 | Missouri | 3,971 | Mississippi | 17.16 |
| 35 | Utah | 805.5 | Colorado | 305.5 | Utah | 8,739 | Pennsylvania | 3,969 | Georgia | 17.10 |
| 36 | Nevada | 777.4 | Nevada | 304.3 | West Virginia | 8,369 | Virginia | 3,963 | South Carolina | 17.04 |
| 37 | Alaska | 723.6 | Utah | 301.8 | New Mexico | 7,877 | Maryland | 3,825 | Colorado | 17.00 |
| 38 | New Mexico | 710.7 | Michigan | 301.2 | Nebraska | 7,877 | Illinois | 3,766 | Minnesota | 17.00 |
| 39 | Nebraska | 692.9 | North Carolina | 298.6 | Maine | 6,696 | Georgia | 3,746 | Oklahoma | 16.76 |
| 40 | Idaho | 529.6 | Oregon | 296.7 | Alaska | 6,260 | Michigan | 3,670 | Nebraska | 16.72 |
| 41 | Wyoming | 496.4 | Hawaii | 269.1 | Hawaii | 6,174 | Idaho | 3,621 | Arkansas | 16.66 |
| 42 | Montana | 462.1 | Maryland | 265.0 | Idaho | 5,418 | North Carolina | 3,603 | Utah | 16.44 |
| 43 | Maine | 455.6 | Vermont | 261.2 | New Hampshire | 5,335 | Washington | 3,601 | Iowa | 16.12 |
| 44 | North Dakota | 428.1 | Florida | 252.9 | Montana | 5,265 | Oregon | 3,527 | Alabama | 16.01 |
| 45 | Hawaii | 343.7 | Connecticut | 249.5 | Wyoming | 4,546 | Colorado | 3,517 | Kentucky | 15.99 |
| 46 | New Hampshire | 314.2 | Arizona | 248.3 | North Dakota | 4,110 | Rhode Island | 3,387 | Idaho | 15.92 |
| 47 | Delaware | 302.0 | New Hampshire | 239.5 | Delaware | 3,849 | California | 3,349 | West Virginia | 15.28 |
| 48 | South Dakota | 292.2 | Massachusetts | 234.2 | South Dakota | 3,585 | Florida | 3,338 | Wyoming | 14.81 |
| 49 | Rhode Island | 217.6 | California | 233.4 | Rhode Island | 3,567 | New York | 3,276 | Indiana | 14.41 |
| 50 | District of Columbia | 187.2 | New York | 209.2 | Vermont | 2,687 | Utah | 3,274 | Louisiana | 14.19 |
| 51 | Vermont | 162.1 | Rhode Island | 206.6 | District of Columbia | 2,392 | Arizona | 3,179 | North Dakota | 13.22 |
| | **United States** | **[3,4]101,468.0** | **United States** | **336.8** | **United States** | **[5]1,233,058** | **United States** | **[6]4,093** | **United States** | **18.23** |

[1] Prices and expenditures include taxes where data are available.
[2] Prices are not adjusted for inflation. See "Nominal Dollars" in Glossary.
[3] Includes 25.2 trillion Btu of coal coke net imports and 378.0 trillion Btu of energy losses and co-products from the production of fuel ethanol that are not allocated to the States.
[4] The U.S. consumption value in this table does not match those in Tables 1.1 and 1.3 because it: 1) does not include biodiesel; 2) does not incorporate the latest data revisions; and 3) is the sum of State values, which use State average heat contents to convert physical units of coal and natural gas to Btu.
[5] Includes $347 million for coal coke net imports, which are not allocated to the States.
[6] Based on population data prior to revisions shown on Table D1.

Note: Rankings based on unrounded data.

Web Page: For related information, see http://www.eia.gov/emeu/states/_seds.html.

Sources: **Consumption:** U.S. Energy Information Administration (EIA), "State Energy Data 2007: Consumption" (August 2009), Tables R1 and R2. **Expenditures and Prices:** EIA, "State Energy Data 2007: Prices and Expenditures" (August 2009), Table R1. "State Energy Data 2007" includes State-level data by end-use sector and type of energy. Consumption estimates are annual 1960 through 2007, and price and expenditure estimates are annual 1970 through 2007.

## Fossil Fuel Production on Federally Administered Lands

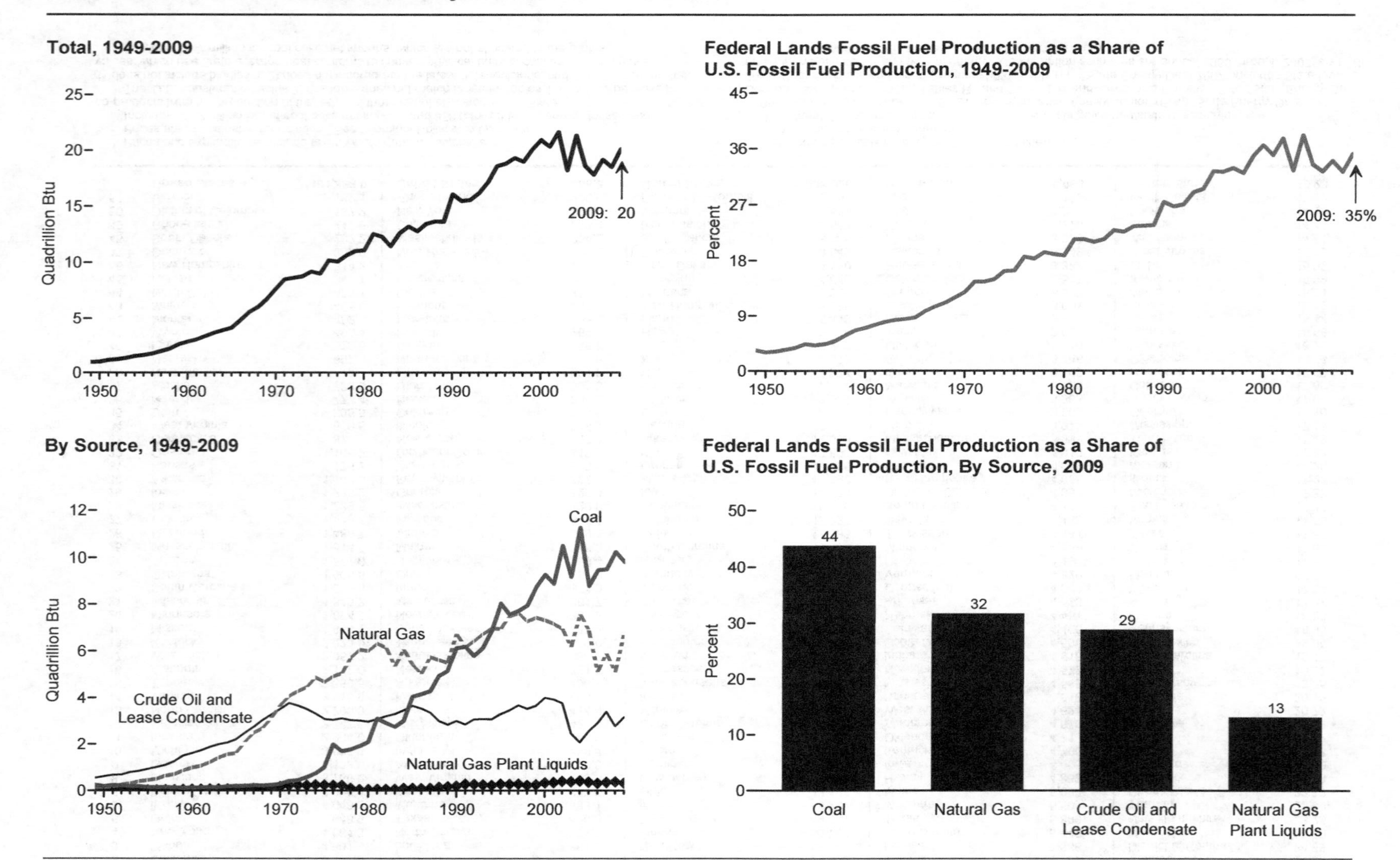


Notes: • Data through 2000 are on a calendar-year basis; data for 2001-2009 are on a fiscal-year basis (October–September). • "Federally Administered Lands" include all classes of land owned by the Federal Government, including acquired military, Outer Continental Shelf, and public lands.

# Fossil Fuel Production on Federally Administered Lands, Selected Years, 1949-2009

| Year | Crude Oil and Lease Condensate | | | Natural Gas Plant Liquids [1] | | | Natural Gas [2] | | | Coal | | | Fossil Fuels | |
|---|---|---|---|---|---|---|---|---|---|---|---|---|---|---|
| | Million Barrels [3] | Quadrillion Btu | Percent of U.S. Total | Million Barrels [3] | Quadrillion Btu | Percent of U.S. Total | Trillion Cubic Feet [3] | Quadrillion Btu | Percent of U.S. Total | Million Short Tons [3] | Quadrillion Btu | Percent of U.S. Total | Quadrillion Btu | Percent of U.S. Total |
| | | | | | | | Calendar-Year Data [4] | | | | | | | |
| 1949 | 95.2 | 0.55 | 5.2 | 4.4 | 0.02 | 2.8 | 0.15 | 0.15 | 2.8 | 9.5 | 0.24 | 2.0 | 0.96 | 3.3 |
| 1950 | 105.9 | .61 | 5.4 | 4.4 | .02 | 2.4 | .14 | .15 | 2.4 | 7.7 | .19 | 1.4 | .98 | 3.0 |
| 1955 | 159.5 | .92 | 6.4 | 6.0 | .03 | 2.1 | .43 | .45 | 4.8 | 5.9 | .15 | 1.2 | 1.55 | 4.1 |
| 1960 | 277.3 | 1.61 | 10.8 | 11.6 | .05 | 3.4 | .95 | .98 | 7.8 | 5.2 | .13 | 1.2 | 2.77 | 6.9 |
| 1965 | 378.6 | 2.20 | 13.3 | 14.3 | .06 | 3.2 | 1.56 | 1.61 | 10.2 | 8.2 | .20 | 1.6 | 4.07 | 8.6 |
| 1970 | 605.6 | 3.51 | 17.2 | 40.6 | .17 | 6.7 | 3.56 | 3.67 | 16.9 | 12.0 | .29 | 2.0 | 7.64 | 12.9 |
| 1971 | 648.9 | 3.76 | 18.8 | 54.0 | .22 | 8.7 | 3.95 | 4.08 | 18.3 | 17.3 | .41 | 3.1 | 8.47 | 14.6 |
| 1972 | 630.5 | 3.66 | 18.2 | 56.7 | .23 | 8.9 | 4.17 | 4.28 | 19.3 | 19.0 | .44 | 3.1 | 8.61 | 14.6 |
| 1973 | 604.3 | 3.51 | 18.0 | 54.9 | .22 | 8.7 | 4.37 | 4.46 | 20.1 | 24.2 | .57 | 4.1 | 8.75 | 15.0 |
| 1974 | 570.2 | 3.31 | 17.8 | 61.9 | .25 | 10.1 | 4.75 | 4.87 | 22.9 | 32.1 | .74 | 5.3 | 9.16 | 16.3 |
| 1975 | 531.5 | 3.08 | 17.4 | 59.7 | .24 | 10.0 | 4.57 | 4.67 | 23.8 | 43.6 | 1.00 | 6.7 | 8.99 | 16.4 |
| 1976 | 525.7 | 3.05 | 17.7 | 57.2 | .23 | 9.7 | 4.81 | 4.91 | 25.2 | 86.4 | 1.98 | 12.6 | 10.16 | 18.6 |
| 1977 | 535.0 | 3.10 | 17.8 | 57.4 | .23 | 9.7 | 4.94 | 5.04 | 25.8 | 74.8 | 1.69 | 10.7 | 10.06 | 18.3 |
| 1978 | 523.6 | 3.04 | 16.5 | 25.9 | .10 | 4.5 | 5.60 | 5.71 | 29.3 | 79.2 | 1.76 | 11.8 | 10.61 | 19.3 |
| 1979 | 519.8 | 3.01 | 16.7 | 11.9 | .05 | 2.1 | 5.93 | 6.05 | 30.1 | 84.9 | 1.91 | 10.9 | 11.02 | 19.0 |
| 1980 | 510.4 | 2.96 | 16.2 | 10.5 | .04 | 1.8 | 5.85 | 6.01 | 30.2 | 92.9 | 2.08 | 11.2 | 11.09 | 18.8 |
| 1981 | 529.3 | 3.07 | 16.9 | 12.3 | .05 | 2.1 | 6.15 | 6.31 | 32.1 | 138.8 | 3.10 | 16.8 | 12.53 | 21.4 |
| 1982 | 552.3 | 3.20 | 17.5 | 15.0 | .06 | 2.7 | 5.97 | 6.14 | 33.5 | 130.0 | 2.89 | 15.5 | 12.29 | 21.4 |
| 1983 | 568.8 | 3.30 | 17.9 | 14.0 | .05 | 2.5 | 5.17 | 5.33 | 32.1 | 124.3 | 2.74 | 15.9 | 11.43 | 21.0 |
| 1984 | 595.8 | 3.46 | 18.3 | 25.4 | .10 | 4.3 | 5.88 | 6.07 | 33.7 | 136.3 | 3.00 | 15.2 | 12.62 | 21.4 |
| 1985 | 628.3 | 3.64 | 19.2 | 26.6 | .10 | 4.5 | 5.24 | 5.41 | 31.8 | 184.6 | 4.04 | 20.9 | 13.19 | 22.9 |
| 1986 | 608.4 | 3.53 | 19.2 | 23.3 | .09 | 4.1 | 4.87 | 5.01 | 30.3 | 189.7 | 4.16 | 21.3 | 12.79 | 22.6 |
| 1987 | 577.3 | 3.35 | 18.9 | 23.7 | .09 | 4.1 | 5.56 | 5.73 | 33.4 | 195.2 | 4.28 | 21.2 | 13.45 | 23.5 |
| 1988 | 516.3 | 2.99 | 17.3 | 37.0 | .14 | 6.2 | 5.45 | 5.61 | 31.9 | 225.4 | 4.92 | 23.7 | 13.67 | 23.6 |
| 1989 | 488.9 | 2.84 | 17.6 | 45.1 | .17 | 8.0 | 5.32 | 5.49 | 30.7 | 236.3 | 5.14 | 24.1 | 13.64 | 23.7 |
| 1990 | 515.9 | 2.99 | 19.2 | 50.9 | .19 | 8.9 | 6.55 | 6.74 | 36.8 | 280.6 | 6.12 | 27.2 | 16.05 | 27.4 |
| 1991 | 491.0 | 2.85 | 18.1 | 72.7 | .28 | 12.0 | 5.99 | 6.17 | 33.8 | 285.1 | 6.18 | 28.5 | 15.47 | 26.7 |
| 1992 | 529.1 | 3.07 | 20.2 | 70.7 | .27 | 11.4 | 6.25 | 6.43 | 35.0 | 266.7 | 5.78 | 26.6 | 15.55 | 27.0 |
| 1993 | 529.3 | 3.07 | 21.2 | 64.4 | .24 | 10.2 | 6.56 | 6.74 | 36.3 | 285.7 | 6.12 | 30.0 | 16.17 | 29.0 |
| 1994 | 527.7 | 3.06 | 21.7 | 60.0 | .23 | 9.5 | 6.78 | 6.97 | 36.0 | 321.4 | 6.88 | 30.9 | 17.14 | 29.5 |
| 1995 | 567.4 | 3.29 | 23.7 | 74.0 | .28 | 11.5 | 6.78 | 6.96 | 36.4 | 376.9 | 8.04 | 36.2 | 18.56 | 32.3 |
| 1996 | 596.5 | 3.46 | 25.2 | 71.2 | .27 | 10.6 | 7.31 | 7.50 | 38.8 | 354.5 | 7.56 | 33.0 | 18.79 | 32.2 |
| 1997 | 632.8 | 3.67 | 26.9 | 74.7 | .28 | 11.3 | 7.43 | 7.62 | 39.3 | 362.6 | 7.72 | 33.0 | 19.29 | 32.8 |
| 1998 | [5]606.3 | [5]3.52 | [5]26.6 | [5]60.3 | [5].23 | [5]9.4 | [5]7.06 | [5]7.27 | [5]37.1 | 371.1 | 7.95 | 33.0 | [5]18.97 | [5]32.0 |
| 1999 | [6]628.9 | [6]3.65 | [6]29.3 | [6]66.5 | [6].25 | [6]9.9 | [6]7.24 | [6]7.44 | [6]38.4 | 414.5 | 8.73 | 37.4 | [6]20.07 | [6]34.8 |
| 2000 | 689.2 | 4.00 | 32.3 | 88.9 | .33 | 12.7 | 7.14 | 7.32 | 37.2 | 440.2 | 9.27 | 40.7 | 20.92 | 36.5 |
| | | | | | | | Fiscal-Year Data [7] | | | | | | | |
| 2001 | 676.5 | 3.92 | 32.0 | 93.0 | 0.35 | 14.0 | 6.98 | 7.17 | 35.7 | 425.4 | 8.87 | 38.1 | 20.31 | 34.9 |
| 2002 | 647.8 | 3.76 | 30.5 | 106.5 | .40 | 15.2 | 6.78 | 6.96 | 35.4 | 507.8 | 10.51 | 45.7 | 21.63 | 37.6 |
| 2003 | [8]422.6 | [8]2.45 | [8]20.4 | 101.0 | .38 | 16.0 | 6.01 | R6.17 | 31.5 | 446.7 | 9.18 | 41.3 | R18.18 | 32.4 |
| 2004 | 356.4 | 2.07 | 17.7 | 110.7 | .41 | 16.8 | 7.38 | R7.58 | 39.4 | 551.1 | 11.27 | 49.7 | R21.32 | 38.1 |
| 2005 | 439.9 | 2.55 | 22.7 | 96.6 | .36 | 14.8 | 6.70 | R6.88 | 36.6 | 431.0 | 8.78 | 37.8 | R18.57 | 33.4 |
| 2006 | 502.1 | 2.91 | 27.4 | 84.1 | .31 | 13.7 | 4.96 | 5.10 | 27.4 | 466.2 | 9.47 | 40.1 | 17.80 | 32.3 |
| 2007 | 584.7 | 3.39 | 31.5 | 94.5 | .35 | 14.7 | 5.73 | R5.90 | R30.1 | 467.5 | 9.51 | 40.4 | R19.15 | R34.0 |
| 2008 | 476.6 | 2.76 | 26.1 | 101.3 | .38 | 15.2 | 4.96 | 5.09 | R24.6 | 506.1 | R10.24 | 43.3 | 18.48 | R32.2 |
| 2009 | 544.3 | 3.16 | 28.8 | 87.3 | .32 | 13.1 | 6.60 | 6.77 | 31.7 | 490.6 | 9.83 | 43.8 | 20.08 | 35.1 |

[1] Includes only those quantities for which the royalties were paid on the basis of the value of the natural gas plant liquids produced. Additional quantities of natural gas plant liquids were produced; however, the royalties paid were based on the value of natural gas processed. These latter quantities are included with natural gas.
[2] Includes some quantities of natural gas processed into liquids at natural gas processing plants and fractionators.
[3] Data from the U.S. Department of the Interior (DOI), U.S. Minerals Management Service (MMS), are for sales volumes.
[4] Through 2000, data are on a calendar-year (January through December) basis. The only exception is in 1949-1974 with production from Naval Petroleum Reserve No. 1, which is on a fiscal-year (July through June) basis.
[5] There is a discontinuity in this time series between 1997 and 1998 due to the sale of "Elk Hills," Naval Petroleum Reserve No. 1.
[6] There is a discontinuity in this time series between 1998 and 1999; beginning in 1999 Naval Petroleum Reserve data have become insignificant and are no longer included.
[7] Beginning in 2001, data are on a fiscal-year (October through September) basis; for example, fiscal-year 2006 data are for October 2005 through September 2006.
[8] A significant amount of Federal offshore crude oil was diverted to the Strategic Petroleum Reserve.
R=Revised.
Note: "Federally Administered Lands" include all classes of land owned by the Federal Government, including acquired military, Outer Continental Shelf, and public lands.
Web Pages: • For all data beginning in 1949, see http://www.eia.gov/emeu/aer/overview.html.
• For related information, see http://www.mrm.mms.gov.
Sources: See end of section.

## Primary Energy Flow by Source and Sector, 2009

(Quadrillion Btu)

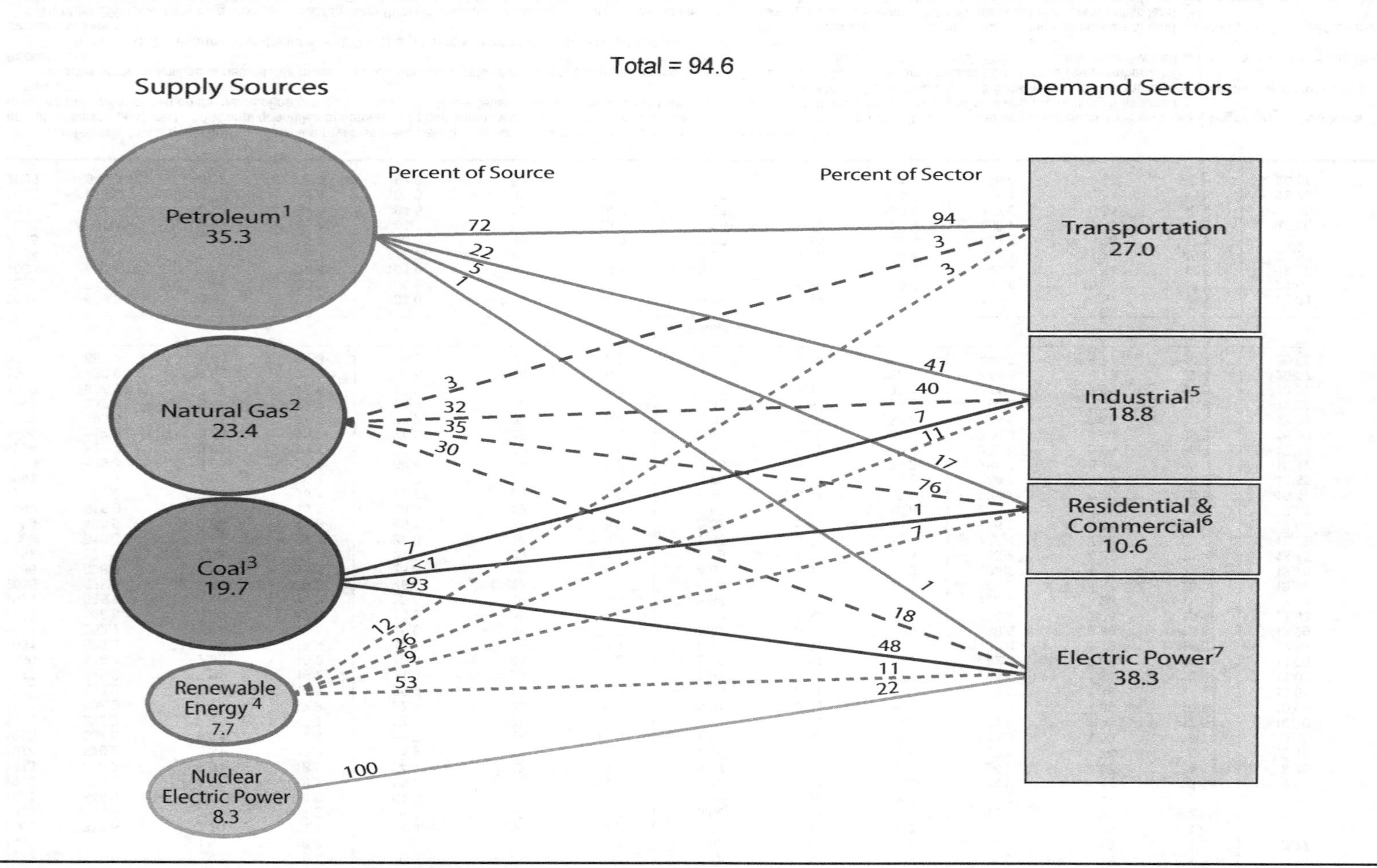


[1] Does not include biofuels that have been blended with petroleum—biofuels are included in "Renewable Energy."

[2] Excludes supplemental gaseous fuels.

[3] Includes less than 0.1 quadrillion Btu of coal coke net exports.

[4] Conventional hydroelectric power, geothermal, solar/PV, wind, and biomass.

[5] Includes industrial combined-heat-and-power (CHP) and industrial electricity-only plants.

[6] Includes commercial combined-heat-and-power (CHP) and commercial electricity-only plants.

[7] Electricity-only and combined-heat-and-power (CHP) plants whose primary business is to sell electricity, or electricity and heat, to the public.

Note: Sum of components may not equal total due to independent rounding.

Sources: U.S. Energy Information Administration, *Annual Energy Review 2009*, Tables 1.3, 2.1b-2.1f , 10.3, and 10.4.

## Energy Consumption by Sector Overview

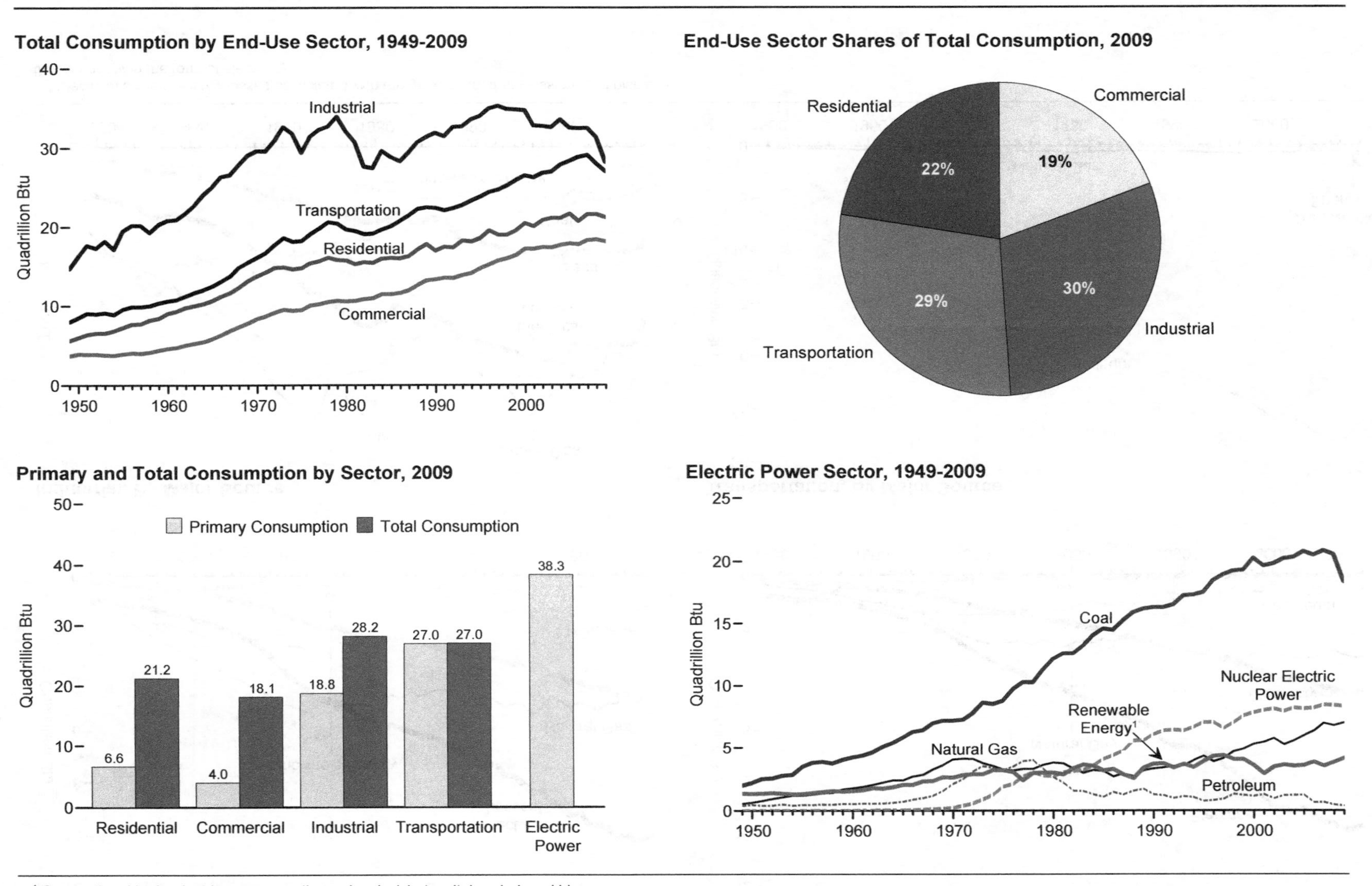


[1] Conventional hydroelectric power, geothermal, solar/photovoltaic, wind, and biomass.
Note: See "Primary Energy Consumption" in Glossary.

U.S. Energy Information Administration / Annual Energy Review 2009

## Energy Consumption by End-Use Sector, 1949-2009

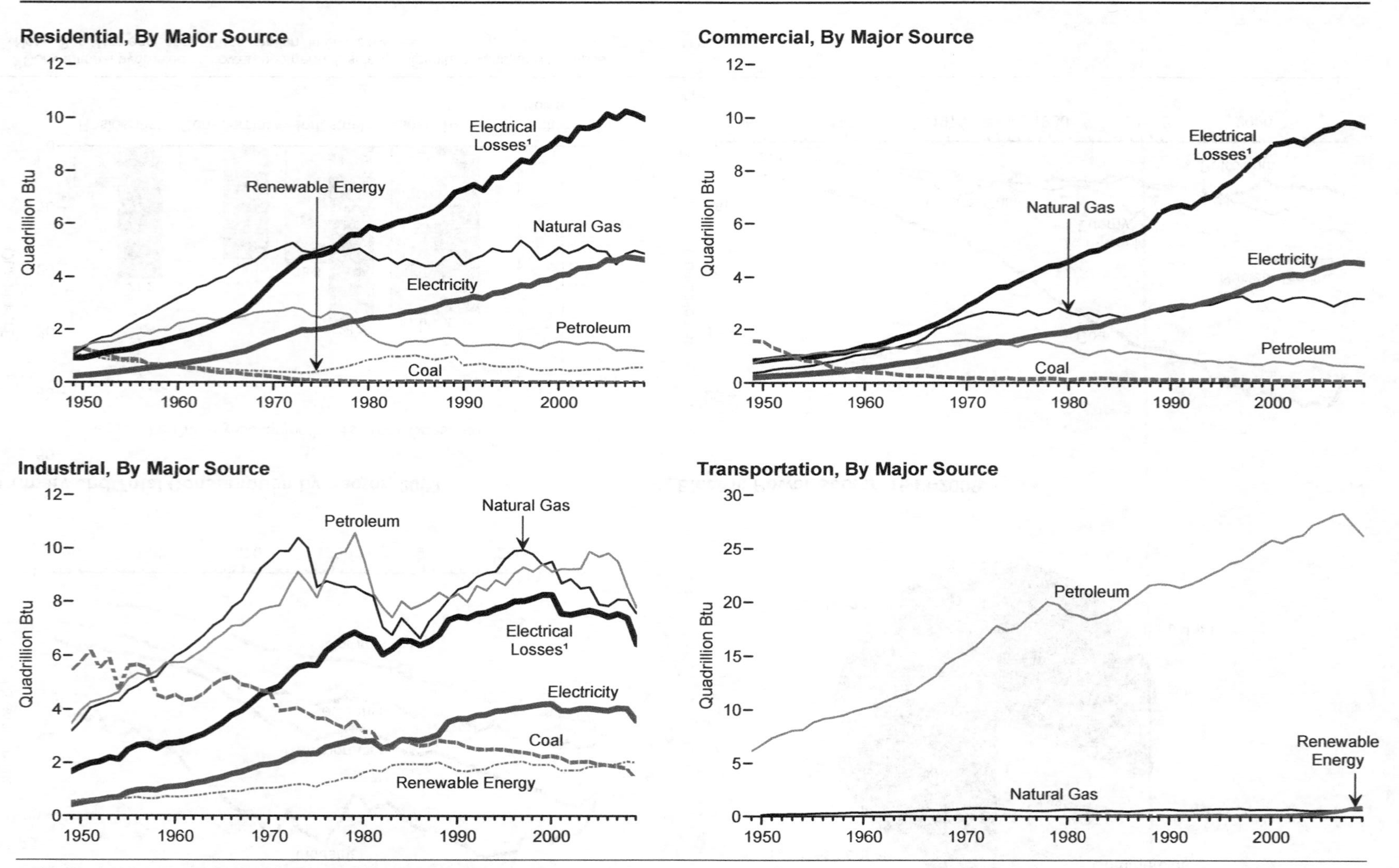


[1] Electrical system energy losses associated with the generation, transmission, and distribution of energy in the form of electricity.

U.S. Energy Information Administration / Annual Energy Review 2009

## Energy Consumption by Sector, Selected Years, 1949-2009

(Trillion Btu)

| | End-Use Sectors | | | | | | | | Electric Power Sector[3,4] | | |
|---|---|---|---|---|---|---|---|---|---|---|---|
| | Residential | | Commercial[1] | | Industrial[2] | | Transportation | | | | |
| Year | Primary[5] | Total[6] | Primary[5] | Total[6] | Primary[5] | Total[6] | Primary[5] | Total[6] | Primary[5] | Balancing Item[7] | Primary Total[8] |
| 1949 | R4,466 | R5,605 | R2,670 | R3,670 | 12,627 | 14,717 | 7,880 | 7,990 | 4,339 | (s) | 31,982 |
| 1950 | R4,836 | R5,995 | R2,836 | R3,895 | 13,881 | 16,233 | 8,384 | 8,493 | 4,679 | (s) | 34,616 |
| 1955 | R5,617 | R7,287 | R2,564 | R3,898 | 16,091 | 19,472 | 9,475 | 9,551 | 6,461 | (s) | 40,208 |
| 1960 | R6,665 | R9,053 | R2,726 | R4,613 | 16,977 | 20,823 | 10,560 | 10,597 | 8,158 | (s) | 45,087 |
| 1965 | R7,297 | R10,658 | R3,182 | R5,851 | 20,124 | 25,075 | 12,400 | 12,434 | 11,014 | (s) | 54,017 |
| 1970 | R8,314 | R13,760 | R4,235 | R8,346 | 22,975 | 29,641 | 16,061 | 16,098 | 16,259 | (s) | 67,844 |
| 1971 | R8,418 | R14,239 | R4,321 | R8,720 | 22,732 | 29,601 | 16,693 | 16,729 | 17,124 | (s) | 69,289 |
| 1972 | R8,615 | R14,851 | R4,409 | R9,185 | 23,532 | 30,953 | 17,681 | 17,716 | 18,466 | (s) | 72,704 |
| 1973 | R8,212 | R14,891 | R4,419 | R9,545 | 24,741 | 32,653 | 18,576 | 18,612 | 19,753 | 7 | 75,708 |
| 1974 | R7,893 | R14,649 | R4,255 | R9,397 | 23,816 | 31,819 | 18,086 | 18,119 | 19,933 | 7 | 73,991 |
| 1975 | R7,973 | R14,810 | R4,055 | R9,498 | 21,454 | 29,447 | 18,209 | 18,244 | 20,307 | 1 | 71,999 |
| 1976 | R8,374 | R15,406 | R4,367 | R10,070 | 22,685 | 31,430 | 19,065 | 19,099 | 21,513 | 8 | 76,012 |
| 1977 | R8,172 | R15,653 | R4,253 | R10,213 | 23,193 | 32,307 | 19,784 | 19,820 | 22,591 | 7 | 78,000 |
| 1978 | R8,238 | R16,122 | R4,303 | R10,515 | 23,276 | 32,733 | 20,580 | 20,615 | 23,587 | 2 | 79,986 |
| 1979 | R7,905 | R15,813 | R4,362 | R10,656 | 24,212 | 33,962 | 20,436 | 20,471 | 23,987 | 2 | 80,903 |
| 1980 | R7,426 | R15,760 | R4,101 | R10,590 | 22,610 | 32,077 | 19,658 | 19,696 | 24,327 | -1 | 78,122 |
| 1981 | R7,030 | R15,268 | R3,833 | R10,629 | 21,338 | 30,756 | 19,476 | R19,512 | 24,488 | 3 | 76,168 |
| 1982 | R7,130 | R15,533 | R3,859 | R10,871 | R19,075 | R27,656 | R19,050 | R19,087 | 24,034 | 4 | 73,153 |
| 1983 | R6,812 | R15,428 | R3,835 | R10,952 | R18,577 | 27,481 | 19,132 | R19,174 | 24,679 | 3 | R73,038 |
| 1984 | R7,194 | R15,971 | R3,996 | R11,463 | R20,197 | R29,624 | R19,606 | R19,653 | 25,719 | 3 | R76,714 |
| 1985 | R7,129 | R16,057 | R3,726 | R11,475 | R19,467 | R28,876 | R20,040 | R20,086 | 26,132 | -4 | R76,491 |
| 1986 | R6,890 | R15,998 | R3,688 | R11,635 | R19,098 | R28,332 | R20,739 | R20,788 | 26,338 | 3 | R76,756 |
| 1987 | R6,908 | R16,288 | R3,769 | R11,977 | R19,976 | R29,442 | R21,418 | R21,468 | 27,104 | -3 | R79,173 |
| 1988 | R7,340 | R17,155 | R3,989 | R12,607 | R20,883 | R30,737 | R22,266 | R22,317 | 28,338 | 3 | R82,819 |
| 1989 | R7,552 | R17,824 | R4,039 | R13,237 | R20,896 | R31,397 | R22,423 | R22,477 | [4]30,025 | 9 | R84,944 |
| 1990 | R6,538 | R16,982 | R3,890 | R13,365 | R21,207 | R31,894 | R22,365 | R22,419 | 30,660 | -9 | R84,651 |
| 1991 | R6,725 | R17,457 | R3,939 | R13,546 | R20,853 | R31,486 | R22,064 | R22,117 | 31,025 | 1 | R84,606 |
| 1992 | R6,930 | R17,393 | R3,985 | R13,487 | R21,785 | R32,660 | R22,362 | R22,414 | 30,893 | (s) | R85,955 |
| 1993 | R7,123 | R18,257 | R3,966 | R13,868 | R21,783 | R32,720 | R22,714 | R22,767 | 32,025 | -10 | R87,601 |
| 1994 | R6,959 | R18,149 | R4,011 | R14,143 | R22,421 | R33,606 | R23,309 | R23,364 | 32,563 | -6 | R89,257 |
| 1995 | R6,915 | R18,547 | R4,094 | R14,729 | R22,746 | R34,045 | R23,790 | R23,846 | 33,621 | 3 | R91,169 |
| 1996 | R7,440 | R19,531 | R4,266 | R15,213 | R23,442 | R34,988 | R24,382 | R24,437 | 34,638 | 4 | R94,172 |
| 1997 | R7,007 | R18,994 | R4,289 | R15,726 | R23,720 | R35,287 | R24,694 | R24,749 | 35,045 | 6 | R94,761 |
| 1998 | R6,390 | R18,986 | R3,998 | R16,014 | R23,209 | R34,926 | R25,200 | R25,255 | 36,385 | -3 | R95,178 |
| 1999 | R6,746 | R19,583 | R4,045 | R16,422 | R22,989 | R34,854 | R25,891 | R25,948 | 37,136 | 6 | R96,812 |
| 2000 | R7,127 | R20,446 | R4,269 | R17,218 | R22,869 | R34,756 | R26,488 | R26,548 | 38,214 | 2 | R98,970 |
| 2001 | R6,839 | R20,065 | R4,076 | R17,180 | R21,833 | R32,803 | R26,212 | R26,275 | R37,362 | -6 | R96,316 |
| 2002 | R6,901 | R20,838 | R4,136 | R17,404 | R21,855 | R32,762 | R26,783 | R26,844 | R38,173 | 5 | R97,853 |
| 2003 | R7,183 | R21,139 | R4,275 | R17,388 | R21,538 | R32,612 | R26,919 | R26,994 | 38,218 | R-1 | R98,131 |
| 2004 | R6,966 | R21,125 | R4,223 | R17,707 | R22,437 | R33,592 | R27,816 | R27,895 | 38,876 | R-6 | R100,313 |
| 2005 | R6,883 | R21,660 | R4,043 | R17,905 | R21,448 | R32,528 | R28,270 | R28,352 | R39,800 | R (s) | R100,445 |
| 2006 | R6,155 | R20,735 | R3,739 | R17,760 | R21,557 | R32,466 | R28,749 | R28,829 | R39,590 | (s) | R99,790 |
| 2007 | R6,607 | R21,600 | R3,923 | R18,314 | R21,430 | R32,499 | R29,030 | R29,118 | R40,540 | R-3 | R101,527 |
| 2008 | R6,765 | R21,606 | R4,043 | R18,411 | R20,503 | R31,358 | R27,944 | R28,027 | R40,147 | R (s) | R99,402 |
| 2009P | 6,606 | 21,207 | 3,974 | 18,148 | 18,751 | 28,199 | 26,951 | 27,033 | 38,304 | -9 | 94,578 |

[1] Commercial sector, including commercial combined-heat-and-power (CHP) and commercial electricity-only plants.
[2] Industrial sector, including industrial combined-heat-and-power (CHP) and industrial electricity-only plants.
[3] Electricity-only and combined-heat-and-power (CHP) plants within the NAICS 22 category whose primary business is to sell electricity, or electricity and heat, to the public.
[4] Through 1988, data are for electric utilities only; beginning in 1989, data are for electric utilities and independent power producers.
[5] See "Primary Energy Consumption" in Glossary.
[6] Total energy consumption in the end-use sectors consists of primary energy consumption, electricity retail sales, and electrical system energy losses. See Note, "Electrical System Energy Losses," at end of section.
[7] A balancing item. The sum of primary consumption in the five energy-use sectors equals the sum of total consumption in the four end-use sectors. However, total energy consumption does not equal the sum of the sectoral components due to the use of sector-specific conversion factors for natural gas and coal.
[8] Primary energy consumption total. See Table 1.3.

R=Revised. P=Preliminary. (s)=Less than 0.5 trillion Btu and greater than -0.5 trillion Btu.

Notes: • See Note 2, "Classification of Power Plants Into Energy-Use Sectors," at end of Section 8.
• Totals may not equal sum of components due to independent rounding.

Web Page: For all data beginning in 1949, see http://www.eia.gov/emeu/aer/consump.html.

## Motor Vehicle Mileage, Fuel Consumption, and Fuel Economy

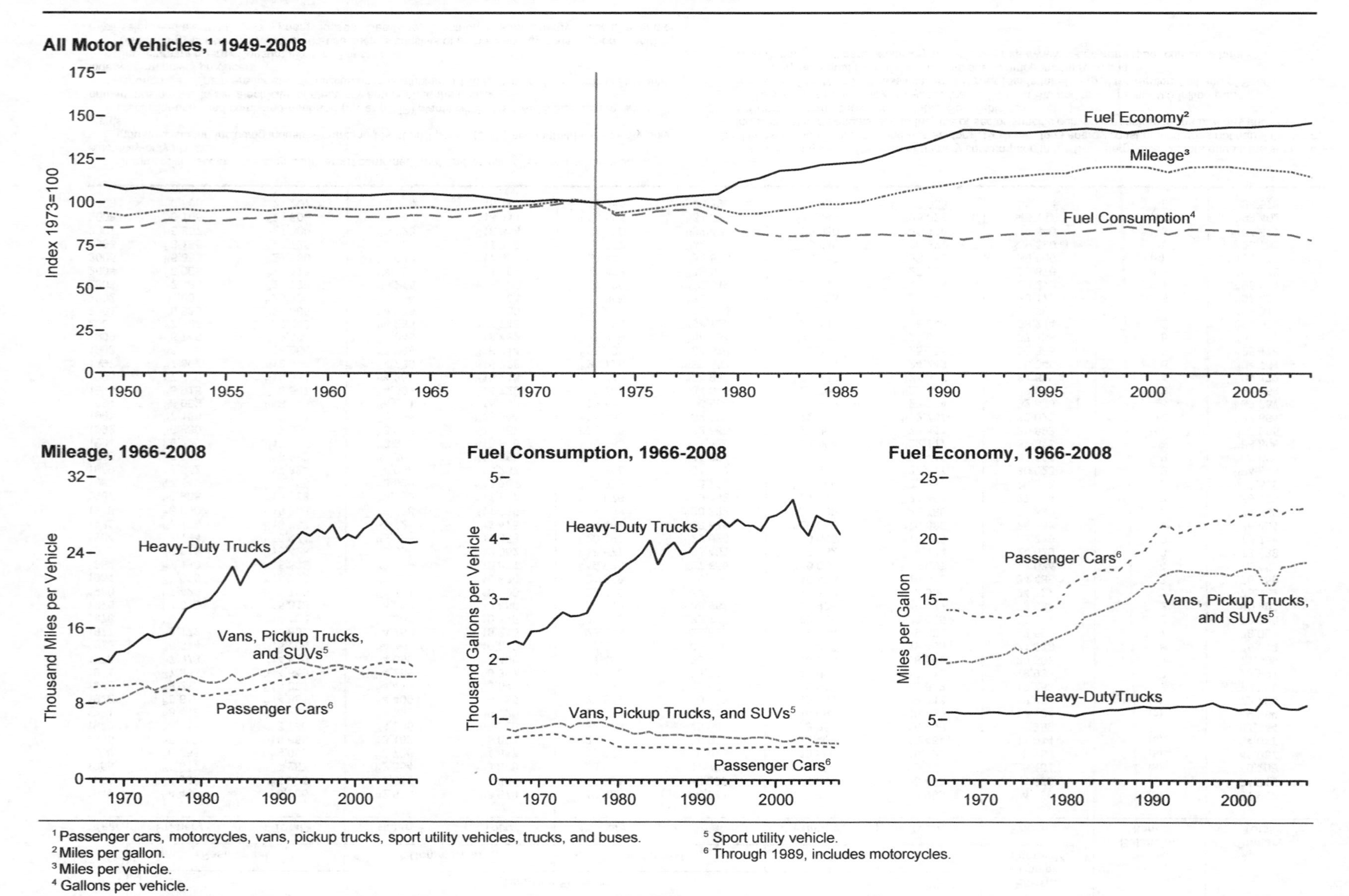


[1] Passenger cars, motorcycles, vans, pickup trucks, sport utility vehicles, trucks, and buses.
[2] Miles per gallon.
[3] Miles per vehicle.
[4] Gallons per vehicle.
[5] Sport utility vehicle.
[6] Through 1989, includes motorcycles.

U.S. Energy Information Administration / Annual Energy Review 2009

## Motor Vehicle Mileage, Fuel Consumption, and Fuel Economy, Selected Years, 1949-2008

| | Passenger Cars [1] | | | Vans, Pickup Trucks, and Sport Utility Vehicles [2] | | | Heavy-Duty Trucks [3] | | | All Motor Vehicles [4] | | |
|---|---|---|---|---|---|---|---|---|---|---|---|---|
| | Mileage | Fuel Consumption | Fuel Economy | Mileage | Fuel Consumption | Fuel Economy | Mileage | Fuel Consumption | Fuel Economy | Mileage | Fuel Consumption | Fuel Economy |
| Year | Miles per Vehicle | Gallons per Vehicle | Miles per Gallon | Miles per Vehicle | Gallons per Vehicle | Miles per Gallon | Miles per vehicle | Gallons per vehicle | Miles per Gallon | Miles per Vehicle | Gallons per Vehicle | Miles per Gallon |
| 1949 | 9,388 | 627 | 15.0 | ([5]) | ([5]) | ([5]) | 9,712 | 1,080 | 9.0 | 9,498 | 726 | 13.1 |
| 1950 | 9,060 | 603 | 15.0 | ([5]) | ([5]) | ([5]) | 10,316 | 1,229 | 8.4 | 9,321 | 725 | 12.8 |
| 1955 | 9,447 | 645 | 14.6 | ([5]) | ([5]) | ([5]) | 10,576 | 1,293 | 8.2 | 9,661 | 761 | 12.7 |
| 1960 | 9,518 | 668 | 14.3 | ([5]) | ([5]) | ([5]) | 10,693 | 1,333 | 8.0 | 9,732 | 784 | 12.4 |
| 1965 | 9,603 | 661 | 14.5 | ([5]) | ([5]) | ([5]) | 10,851 | 1,387 | 7.8 | 9,826 | 787 | 12.5 |
| 1970 | 9,989 | 737 | 13.5 | 8,676 | 866 | 10.0 | 13,565 | 2,467 | 5.5 | 9,976 | 830 | 12.0 |
| 1971 | 10,097 | 743 | 13.6 | 9,082 | 888 | 10.2 | 14,117 | 2,519 | 5.6 | 10,133 | 839 | 12.1 |
| 1972 | 10,171 | 754 | 13.5 | 9,534 | 922 | 10.3 | 14,780 | 2,657 | 5.6 | 10,279 | 857 | 12.0 |
| 1973 | 9,884 | 737 | 13.4 | 9,779 | 931 | 10.5 | 15,370 | 2,775 | 5.5 | 10,099 | 850 | 11.9 |
| 1974 | 9,221 | 677 | 13.6 | 9,452 | 862 | 11.0 | 14,995 | 2,708 | 5.5 | 9,493 | 788 | 12.0 |
| 1975 | 9,309 | 665 | 14.0 | 9,829 | 934 | 10.5 | 15,167 | 2,722 | 5.6 | 9,627 | 790 | 12.2 |
| 1976 | 9,418 | 681 | 13.8 | 10,127 | 934 | 10.8 | 15,438 | 2,764 | 5.6 | 9,774 | 806 | 12.1 |
| 1977 | 9,517 | 676 | 14.1 | 10,607 | 947 | 11.2 | 16,700 | 3,002 | 5.6 | 9,978 | 814 | 12.3 |
| 1978 | 9,500 | 665 | 14.3 | 10,968 | 948 | 11.6 | 18,045 | 3,263 | 5.5 | 10,077 | 816 | 12.4 |
| 1979 | 9,062 | 620 | 14.6 | 10,802 | 905 | 11.9 | 18,502 | 3,380 | 5.5 | 9,722 | 776 | 12.5 |
| 1980 | 8,813 | 551 | 16.0 | 10,437 | 854 | 12.2 | 18,736 | 3,447 | 5.4 | 9,458 | 712 | 13.3 |
| 1981 | 8,873 | 538 | 16.5 | 10,244 | 819 | 12.5 | 19,016 | 3,565 | 5.3 | 9,477 | 697 | 13.6 |
| 1982 | 9,050 | 535 | 16.9 | 10,276 | 762 | 13.5 | 19,931 | 3,647 | 5.5 | 9,644 | 686 | 14.1 |
| 1983 | 9,118 | 534 | 17.1 | 10,497 | 767 | 13.7 | 21,083 | 3,769 | 5.6 | 9,760 | 686 | 14.2 |
| 1984 | 9,248 | 530 | 17.4 | 11,151 | 797 | 14.0 | 22,550 | 3,967 | 5.7 | 10,017 | 691 | 14.5 |
| 1985 | 9,419 | 538 | 17.5 | 10,506 | 735 | 14.3 | 20,597 | 3,570 | 5.8 | 10,020 | 685 | 14.6 |
| 1986 | 9,464 | 543 | 17.4 | 10,764 | 738 | 14.6 | 22,143 | 3,821 | 5.8 | 10,143 | 692 | 14.7 |
| 1987 | 9,720 | 539 | 18.0 | 11,114 | 744 | 14.9 | 23,349 | 3,937 | 5.9 | 10,453 | 694 | 15.1 |
| 1988 | 9,972 | 531 | 18.8 | 11,465 | 745 | 15.4 | 22,485 | 3,736 | 6.0 | 10,721 | 688 | 15.6 |
| 1989 | [1]10,157 | [1]533 | [1]19.0 | 11,676 | 724 | 16.1 | 22,926 | 3,776 | 6.1 | 10,932 | 688 | 15.9 |
| 1990 | 10,504 | 520 | 20.2 | 11,902 | 738 | 16.1 | 23,603 | 3,953 | 6.0 | 11,107 | 677 | 16.4 |
| 1991 | 10,571 | 501 | 21.1 | 12,245 | 721 | 17.0 | 24,229 | 4,047 | 6.0 | 11,294 | 669 | 16.9 |
| 1992 | 10,857 | 517 | 21.0 | 12,381 | 717 | 17.3 | 25,373 | 4,210 | 6.0 | 11,558 | 683 | 16.9 |
| 1993 | 10,804 | 527 | 20.5 | 12,430 | 714 | 17.4 | 26,262 | 4,309 | 6.1 | 11,595 | 693 | 16.7 |
| 1994 | 10,992 | 531 | 20.7 | 12,156 | 701 | 17.3 | 25,838 | 4,202 | 6.1 | 11,683 | 698 | 16.7 |
| 1995 | 11,203 | 530 | 21.1 | 12,018 | 694 | 17.3 | 26,514 | 4,315 | 6.1 | 11,793 | 700 | 16.8 |
| 1996 | 11,330 | 534 | 21.2 | 11,811 | 685 | 17.2 | 26,092 | 4,221 | 6.2 | 11,813 | 700 | 16.9 |
| 1997 | 11,581 | 539 | 21.5 | 12,115 | 703 | 17.2 | 27,032 | 4,218 | 6.4 | 12,107 | 711 | 17.0 |
| 1998 | 11,754 | 544 | 21.6 | 12,173 | 707 | 17.2 | 25,397 | 4,135 | 6.1 | 12,211 | 721 | 16.9 |
| 1999 | 11,848 | 553 | 21.4 | 11,957 | 701 | 17.0 | 26,014 | 4,352 | 6.0 | 12,206 | 732 | 16.7 |
| 2000 | 11,976 | 547 | 21.9 | 11,672 | 669 | 17.4 | 25,617 | 4,391 | 5.8 | 12,164 | 720 | 16.9 |
| 2001 | 11,831 | 534 | 22.1 | 11,204 | 636 | 17.6 | 26,602 | 4,477 | 5.9 | 11,887 | 695 | 17.1 |
| 2002 | 12,202 | 555 | 22.0 | 11,364 | 650 | 17.5 | 27,071 | 4,642 | 5.8 | 12,171 | 719 | 16.9 |
| 2003 | 12,325 | 556 | 22.2 | 11,287 | 697 | 16.2 | 28,093 | 4,215 | 6.7 | 12,208 | 718 | 17.0 |
| 2004 | 12,460 | 553 | 22.5 | 11,184 | 690 | 16.2 | 27,023 | 4,057 | 6.7 | 12,200 | 714 | 17.1 |
| 2005 | 12,510 | 567 | 22.1 | 10,920 | 617 | 17.7 | 26,235 | 4,385 | 6.0 | 12,082 | 706 | 17.1 |
| 2006 | 12,485 | 554 | 22.5 | 10,920 | 612 | 17.8 | 25,231 | 4,304 | 5.9 | 12,017 | 698 | 17.2 |
| 2007 | R12,304 | 547 | 22.5 | R10,962 | 609 | 18.0 | R25,152 | R4,275 | 5.9 | R11,920 | R693 | 17.2 |
| 2008P | 11,788 | 522 | 22.6 | 10,951 | 605 | 18.1 | 25,254 | 4,075 | 6.2 | 11,619 | 667 | 17.4 |

[1] Through 1989, includes motorcycles.
[2] Includes a small number of trucks with 2 axles and 4 tires, such as step vans.
[3] Single-unit trucks with 2 axles and 6 or more tires, and combination trucks.
[4] Includes buses and motorcycles, which are not separately displayed.
[5] Included in "Heavy-Duty Trucks."
R=Revised. P=Preliminary.

Web Pages: • For all data beginning in 1949, see http://www.eia.gov/aer/consump.html. • For related information, see http://www.fhwa.dot.gov/policy/ohpi/hss/index.htm.

Sources: **Passenger Cars, 1990-1994:** U.S. Department of Transportation, Bureau of Transportation Statistics, *National Transportation Statistics 1998,* Table 4-13. **All Other Data:** • 1949-1994—Federal Highway Administration (FHWA), *Highway Statistics Summary to 1995,* Table VM-201A. • 1995 forward—FHWA, *Highway Statistics,* annual reports, Table VM-1.

## Petroleum Flow, 2009 (Million Barrels per Day)

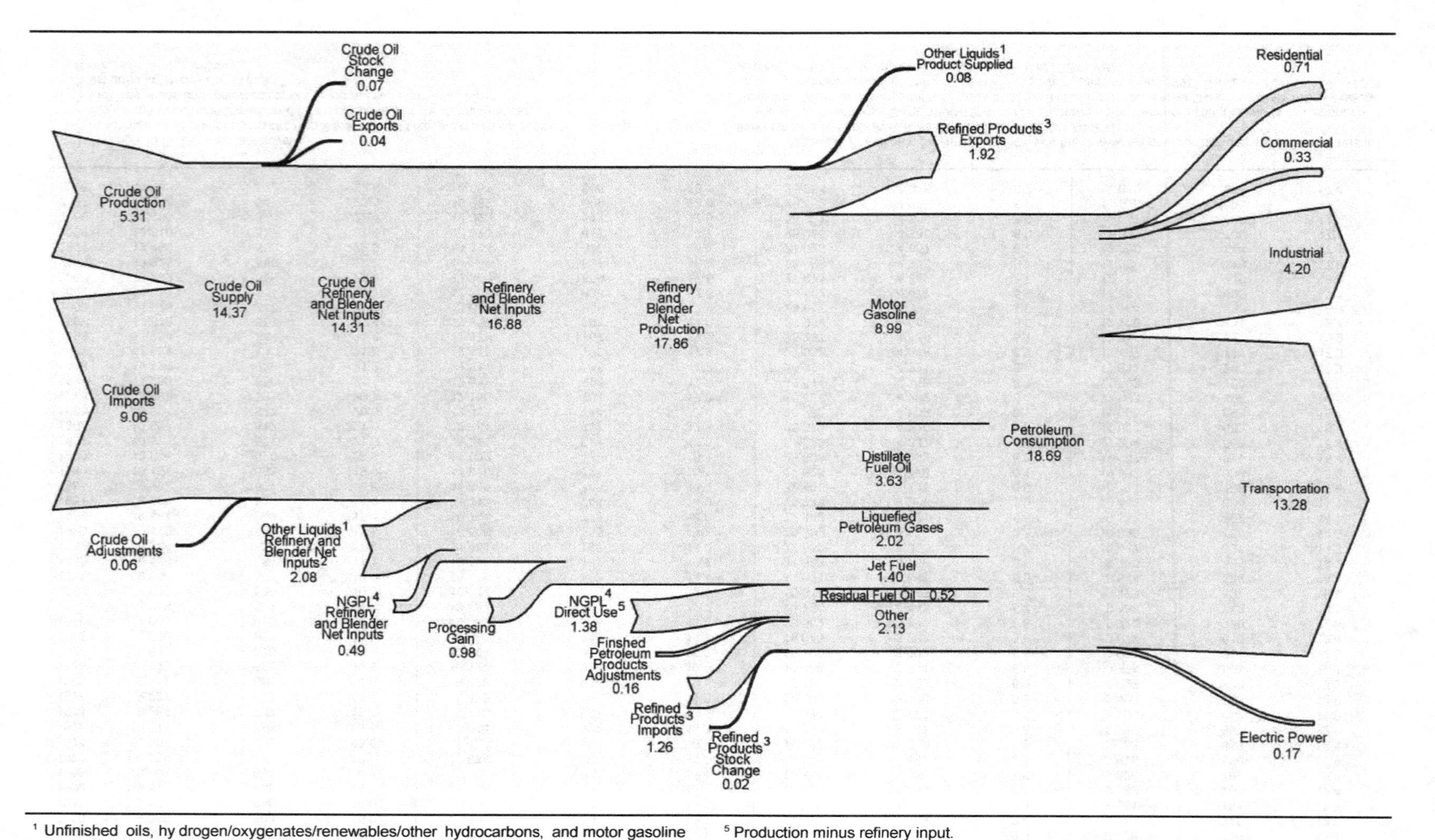


[1] Unfinished oils, hy drogen/oxygenates/renewables/other hydrocarbons, and motor gasoline and aviation gasoline blending components.
[2] Renewable fuels and oxygenate plant net production (0.75), net imports (1.34) and adjustments (-0.03) minus stock change (0.06) and product supplied (-0.08).
[3] Finished petroleum products, liquefied petroleum gases, and pentanes plus.
[4] Natural gas plant liquids.
[5] Production minus refinery input.

Notes: • Data are preliminary. • Values are derived from source data prior to rounding for publication. • Totals may not equal sum of components due to independent rounding.

Sources: Tables 5.1, 5.3, 5.5, 5.8, 5.11, 5.13a-5.13d, 5.16, and *Petroleum Supply Monthly*, February 2010, Table 4.

U.S. Energy Information Administration / Annual Energy Review 2009

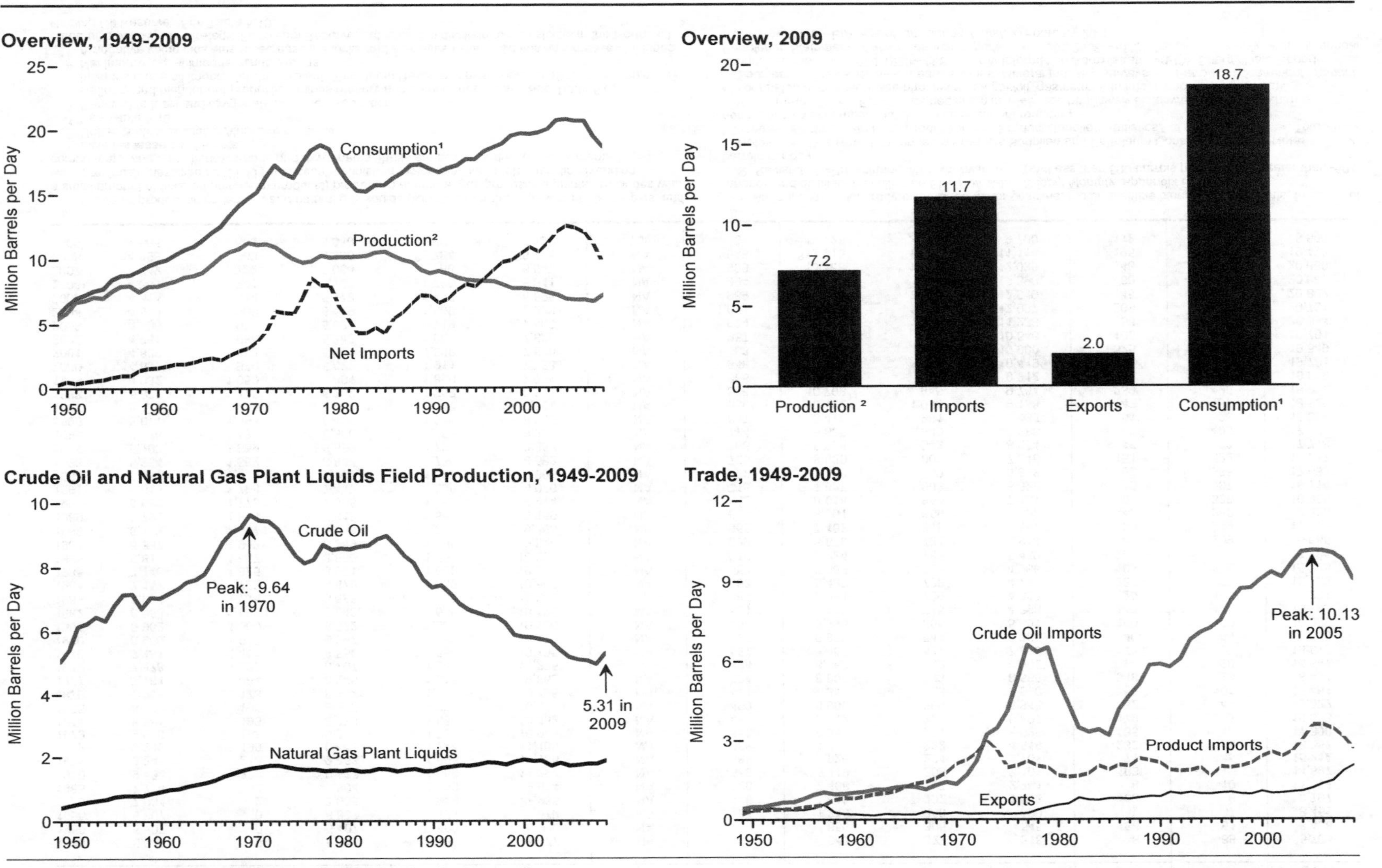


¹ Petroleum products supplied is used as an approximation for consumption.
² Crude oil and natural gas plant liquids production.

U.S. Energy Information Administration / Annual Energy Review 2009

## Petroleum Overview, Selected Years, 1949-2009

(Thousand Barrels per Day)

| | Field Production [1] | | | | | Renewable Fuels and Oxygenates [5] | Processing Gain [6] | Trade | | | Stock Change [9] | Adjust-ments [10] | Petroleum Products Supplied |
|---|---|---|---|---|---|---|---|---|---|---|---|---|---|
| | Crude Oil [2] | | | Natural Gas Plant Liquids [4] | Total | | | Imports [7] | Exports | Net Imports [8] | | | |
| Year | 48 States [3] | Alaska | Total | | | | | | | | | | |
| 1949 | 5,046 | 0 | 5,046 | 430 | 5,477 | NA | -2 | 645 | 327 | 318 | -8 | -38 | 5,763 |
| 1950 | 5,407 | 0 | 5,407 | 499 | 5,906 | NA | 2 | 850 | 305 | 545 | -56 | -51 | 6,458 |
| 1955 | 6,807 | 0 | 6,807 | 771 | 7,578 | NA | 34 | 1,248 | 368 | 880 | (s) | -37 | 8,455 |
| 1960 | 7,034 | 2 | 7,035 | 929 | 7,965 | NA | 146 | 1,815 | 202 | 1,613 | -83 | -8 | 9,797 |
| 1965 | 7,774 | 30 | 7,804 | 1,210 | 9,014 | NA | 220 | 2,468 | 187 | 2,281 | -8 | -10 | 11,512 |
| 1970 | 9,408 | 229 | 9,637 | 1,660 | 11,297 | NA | 359 | 3,419 | 259 | 3,161 | 103 | -16 | 14,697 |
| 1971 | 9,245 | 218 | 9,463 | 1,693 | 11,155 | NA | 382 | 3,926 | 224 | 3,701 | 71 | 45 | 15,212 |
| 1972 | 9,242 | 199 | 9,441 | 1,744 | 11,185 | NA | 388 | 4,741 | 222 | 4,519 | -232 | 43 | 16,367 |
| 1973 | 9,010 | 198 | 9,208 | 1,738 | 10,946 | NA | 453 | 6,256 | 231 | 6,025 | 135 | 18 | 17,308 |
| 1974 | 8,581 | 193 | 8,774 | 1,688 | 10,462 | NA | 480 | 6,112 | 221 | 5,892 | 179 | -2 | 16,653 |
| 1975 | 8,183 | 191 | 8,375 | 1,633 | 10,007 | NA | 460 | 6,056 | 209 | 5,846 | 32 | 41 | 16,322 |
| 1976 | 7,958 | 173 | 8,132 | 1,604 | 9,736 | NA | 477 | 7,313 | 223 | 7,090 | -58 | 101 | 17,461 |
| 1977 | 7,781 | 464 | 8,245 | 1,618 | 9,862 | NA | 524 | 8,807 | 243 | 8,565 | 548 | 28 | 18,431 |
| 1978 | 7,478 | 1,229 | 8,707 | 1,567 | 10,275 | NA | 496 | 8,363 | 362 | 8,002 | -94 | -20 | 18,847 |
| 1979 | 7,151 | 1,401 | 8,552 | 1,584 | 10,135 | NA | 527 | 8,456 | 471 | 7,985 | 173 | 38 | 18,513 |
| 1980 | 6,980 | 1,617 | 8,597 | 1,573 | 10,170 | NA | 597 | 6,909 | 544 | 6,365 | 140 | 64 | 17,056 |
| 1981 | 6,962 | 1,609 | 8,572 | 1,609 | 10,180 | NA | 508 | 5,996 | 595 | 5,401 | 160 | 129 | 16,058 |
| 1982 | 6,953 | 1,696 | 8,649 | 1,550 | 10,199 | NA | 531 | 5,113 | 815 | 4,298 | -147 | 121 | 15,296 |
| 1983 | 6,974 | 1,714 | 8,688 | 1,559 | 10,246 | NA | 488 | 5,051 | 739 | 4,312 | -20 | 165 | 15,231 |
| 1984 | 7,157 | 1,722 | 8,879 | 1,630 | 10,509 | NA | 553 | 5,437 | 722 | 4,715 | 280 | 228 | 15,726 |
| 1985 | 7,146 | 1,825 | 8,971 | 1,609 | 10,581 | NA | 557 | 5,067 | 781 | 4,286 | -103 | 200 | 15,726 |
| 1986 | 6,814 | 1,867 | 8,680 | 1,551 | 10,231 | NA | 616 | 6,224 | 785 | 5,439 | 202 | 197 | 16,281 |
| 1987 | 6,387 | 1,962 | 8,349 | 1,595 | 9,944 | NA | 639 | 6,678 | 764 | 5,914 | 41 | 209 | 16,665 |
| 1988 | 6,123 | 2,017 | 8,140 | 1,625 | 9,765 | NA | 655 | 7,402 | 815 | 6,587 | -28 | 249 | 17,283 |
| 1989 | 5,739 | 1,874 | 7,613 | 1,546 | 9,159 | NA | 661 | 8,061 | 859 | 7,202 | -43 | 260 | 17,325 |
| 1990 | 5,582 | 1,773 | 7,355 | 1,559 | 8,914 | NA | 683 | 8,018 | 857 | 7,161 | 107 | 338 | 16,988 |
| 1991 | 5,618 | 1,798 | 7,417 | 1,659 | 9,076 | NA | 715 | 7,627 | 1,001 | 6,626 | -10 | 287 | 16,714 |
| 1992 | 5,457 | 1,714 | 7,171 | 1,697 | 8,868 | NA | 772 | 7,888 | 950 | 6,938 | -68 | 386 | 17,033 |
| 1993 | 5,264 | 1,582 | 6,847 | 1,736 | 8,582 | NA | 766 | 8,620 | 1,003 | 7,618 | 151 | 422 | 17,237 |
| 1994 | 5,103 | 1,559 | 6,662 | 1,727 | 8,388 | NA | 768 | 8,996 | 942 | 8,054 | 15 | 523 | 17,718 |
| 1995 | 5,076 | 1,484 | 6,560 | 1,762 | 8,322 | NA | 774 | 8,835 | 949 | 7,886 | -246 | 496 | 17,725 |
| 1996 | 5,071 | 1,393 | 6,465 | 1,830 | 8,295 | NA | 837 | 9,478 | 981 | 8,498 | -151 | 528 | 18,309 |
| 1997 | 5,156 | 1,296 | 6,452 | 1,817 | 8,269 | NA | 850 | 10,162 | 1,003 | 9,158 | 143 | 487 | 18,620 |
| 1998 | 5,077 | 1,175 | 6,252 | 1,759 | 8,011 | NA | 886 | 10,708 | 945 | 9,764 | 239 | 495 | 18,917 |
| 1999 | 4,832 | 1,050 | 5,881 | 1,850 | 7,731 | NA | 886 | 10,852 | 940 | 9,912 | -422 | 567 | 19,519 |
| 2000 | 4,851 | 970 | 5,822 | 1,911 | 7,733 | NA | 948 | 11,459 | 1,040 | 10,419 | -69 | 532 | 19,701 |
| 2001 | 4,839 | 963 | 5,801 | 1,868 | 7,670 | NA | 903 | 11,871 | 971 | 10,900 | 325 | 501 | 19,649 |
| 2002 | 4,761 | 984 | 5,746 | 1,880 | 7,626 | NA | 957 | 11,530 | 984 | 10,546 | -105 | 527 | 19,761 |
| 2003 | 4,706 | 974 | 5,681 | 1,719 | 7,400 | NA | 974 | 12,264 | 1,027 | 11,238 | 56 | 478 | 20,034 |
| 2004 | 4,510 | 908 | 5,419 | 1,809 | 7,228 | NA | 1,051 | 13,145 | 1,048 | 12,097 | 209 | 564 | 20,731 |
| 2005 | 4,314 | 864 | 5,178 | 1,717 | 6,895 | NA | 989 | 13,714 | 1,165 | 12,549 | 145 | 513 | 20,802 |
| 2006 | 4,361 | 741 | 5,102 | 1,739 | 6,841 | NA | 994 | 13,707 | 1,317 | 12,390 | 60 | 522 | 20,687 |
| 2007 | 4,342 | 722 | 5,064 | 1,783 | 6,847 | NA | 996 | 13,468 | 1,433 | 12,036 | -148 | 653 | 20,680 |
| 2008 | R4,268 | 683 | R4,950 | R1,784 | R6,734 | NA | R993 | R12,915 | R1,802 | R11,114 | R195 | R852 | R19,498 |
| 2009P | 4,665 | 645 | 5,310 | 1,886 | 7,196 | 735 | 981 | 11,726 | 2,026 | 9,700 | 112 | 186 | 18,686 |

[1] Crude oil production on leases, and natural gas liquids (liquefied petroleum gases, pentanes plus, and a small amount of finished petroleum products) production at natural gas processing plants. Excludes what was previously classified as "Field Production" of finished motor gasoline, motor gasoline blending components, and other hydrocarbons and oxygenates; these are now included in "Adjustments."
[2] Includes lease condensate.
[3] United States excluding Alaska and Hawaii.
[4] See Table 5.10.
[5] Renewable fuels and oxygenate plant net production.
[6] Refinery and blender net production minus refinery and blender net inputs. See Table 5.8.
[7] Includes crude oil imports for the Strategic Petroleum Reserve, which began in 1977. See Table 5.17.
[8] Net imports equal imports minus exports.
[9] A negative value indicates a decrease in stocks and a positive value indicates an increase. Includes crude oil stocks in the Strategic Petroleum Reserve, but excludes distillate fuel oil stocks in the Northeast Heating Oil Reserve. See Table 5.16.
[10] An adjustment for crude oil, finished motor gasoline, motor gasoline blending components, fuel ethanol, and distillate fuel oil. See EIA, *Petroleum Supply Monthly,* Appendix B, Note 3.

R=Revised. P=Preliminary. NA=Not available. (s)=Less than 500 barrels per day and greater than -500 barrels per day.

Notes: • See Note 1, "Petroleum Products Supplied and Petroleum Consumption," and Note 2, "Changes Affecting Petroleum Production and Product Supplied Statistics," at end of section. • Totals may not equal sum of components due to independent rounding.

Web Pages: • For all data beginning in 1949, see http://www.eia.gov/emeu/aer/petro.html. • For related information, see http://www.eia.gov/oil_gas/petroleum/info_glance/petroleum.html.

Sources: • 1949-1975—Bureau of Mines, Mineral Industry Surveys, *Petroleum Statement, Annual,* annual reports. • 1976-1980—U.S. Energy Information Administration (EIA), Energy Data Reports, *Petroleum Statement, Annual,* annual reports. • 1981-2008—EIA, *Petroleum Supply Annual,* annual reports. • 2009—EIA, *Petroleum Supply Monthly* (February 2010).

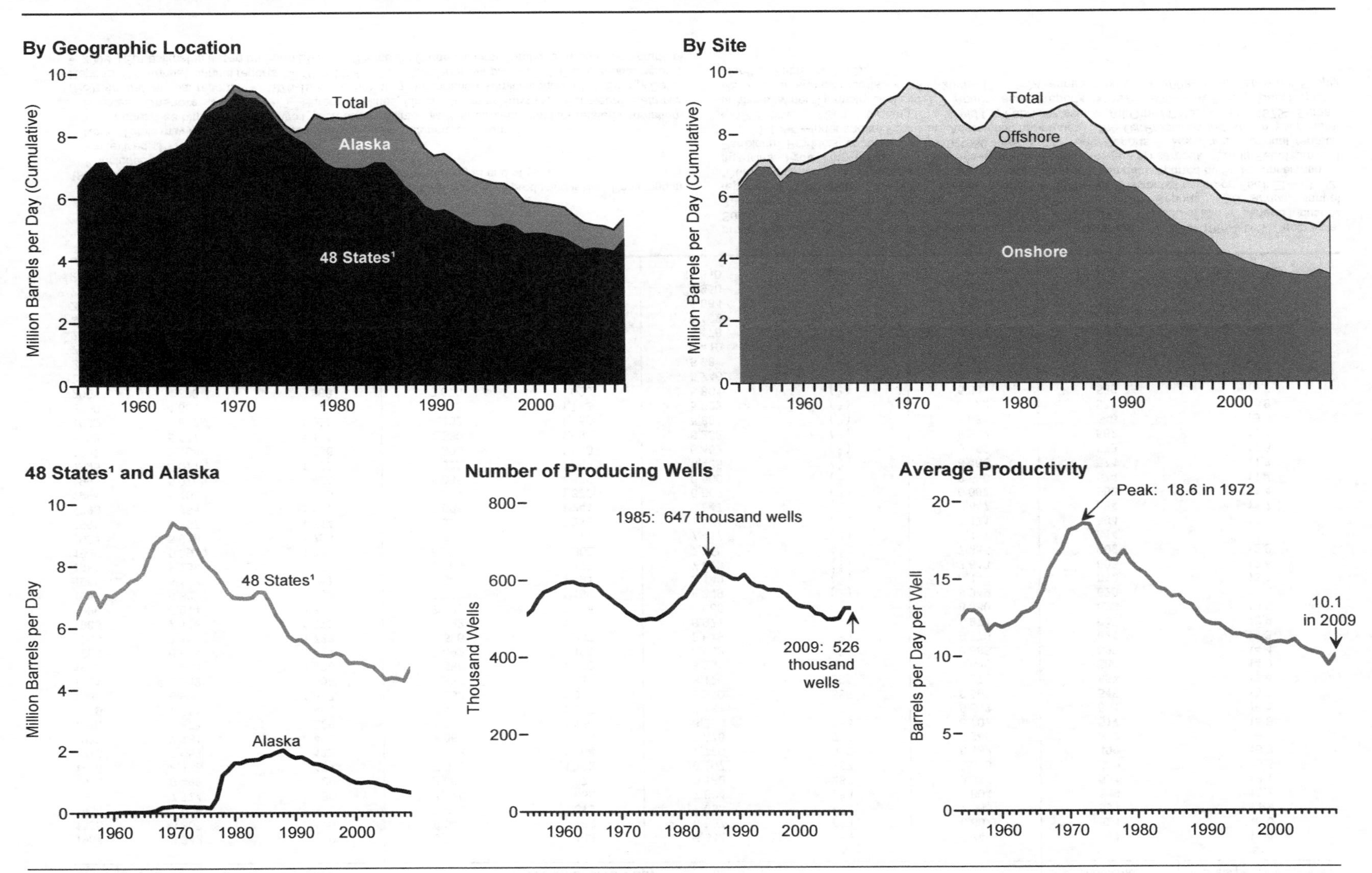


U.S. Energy Information Administration / Annual Energy Review 2009

## Crude Oil Production and Crude Oil Well Productivity, Selected Years, 1954-2009

| Year | Geographic Location | | Site | | Type | | Total Production | Crude Oil Well [1] Productivity | |
|---|---|---|---|---|---|---|---|---|---|
| | 48 States [2] | Alaska | Onshore | Offshore | Crude Oil | Lease Condensate | | Producing Wells [3] | Average Productivity [4] |
| | Thousand Barrels per Day | | | | | | | Thousands | Barrels per Day per Well |
| 1954 | 6,342 | 0 | 6,209 | 133 | 6,342 | (5) | 6,342 | 511 | 12.4 |
| 1955 | 6,807 | 0 | 6,645 | 162 | 6,807 | (5) | 6,807 | 524 | 13.0 |
| 1960 | 7,034 | 2 | 6,716 | 319 | 7,035 | (5) | 7,035 | 591 | 11.9 |
| 1965 | 7,774 | 30 | 7,140 | 665 | 7,804 | (5) | 7,804 | 589 | 13.2 |
| 1970 | 9,408 | 229 | 8,060 | 1,577 | 9,180 | 457 | 9,637 | 531 | 18.1 |
| 1975 | 8,183 | 191 | 7,012 | 1,362 | 8,007 | 367 | 8,375 | 500 | 16.8 |
| 1976 | 7,958 | 173 | 6,868 | 1,264 | 7,776 | 356 | 8,132 | 499 | 16.3 |
| 1977 | 7,781 | 464 | 7,069 | 1,176 | 7,875 | 370 | 8,245 | 507 | 16.3 |
| 1978 | 7,478 | 1,229 | 7,571 | 1,136 | 8,353 | 355 | 8,707 | 517 | 16.8 |
| 1979 | 7,151 | 1,401 | 7,485 | 1,067 | 8,181 | 371 | 8,552 | 531 | 16.1 |
| 1980 | 6,980 | 1,617 | 7,562 | 1,034 | 8,210 | 386 | 8,597 | 548 | 15.7 |
| 1981 | 6,962 | 1,609 | 7,537 | 1,034 | 8,176 | 395 | 8,572 | 557 | 15.4 |
| 1982 | 6,953 | 1,696 | 7,538 | 1,110 | 8,261 | 387 | 8,649 | 580 | 14.9 |
| 1983 | 6,974 | 1,714 | 7,492 | 1,196 | 8,688 | (5) | 8,688 | 603 | 14.4 |
| 1984 | 7,157 | 1,722 | 7,596 | 1,283 | 8,879 | (5) | 8,879 | 621 | 14.3 |
| 1985 | 7,146 | 1,825 | 7,722 | 1,250 | 8,971 | (5) | 8,971 | 647 | 13.9 |
| 1986 | 6,814 | 1,867 | 7,426 | 1,254 | 8,680 | (5) | 8,680 | 623 | 13.9 |
| 1987 | 6,387 | 1,962 | 7,153 | 1,196 | 8,349 | (5) | 8,349 | 620 | 13.5 |
| 1988 | 6,123 | 2,017 | 6,949 | 1,191 | 8,140 | (5) | 8,140 | 612 | 13.3 |
| 1989 | 5,739 | 1,874 | 6,486 | 1,127 | 7,613 | (5) | 7,613 | 603 | 12.6 |
| 1990 | 5,582 | 1,773 | 6,273 | 1,082 | 7,355 | (5) | 7,355 | 602 | 12.2 |
| 1991 | 5,618 | 1,798 | 6,245 | 1,172 | 7,417 | (5) | 7,417 | 614 | 12.1 |
| 1992 | 5,457 | 1,714 | 5,953 | 1,218 | 7,171 | (5) | 7,171 | 594 | 12.1 |
| 1993 | 5,264 | 1,582 | 5,606 | 1,241 | 6,847 | (5) | 6,847 | 584 | 11.7 |
| 1994 | 5,103 | 1,559 | 5,291 | 1,370 | 6,662 | (5) | 6,662 | 582 | 11.4 |
| 1995 | 5,076 | 1,484 | 5,035 | 1,525 | 6,560 | (5) | 6,560 | 574 | 11.4 |
| 1996 | 5,071 | 1,393 | 4,902 | 1,562 | 6,465 | (5) | 6,465 | 574 | 11.3 |
| 1997 | 5,156 | 1,296 | 4,803 | 1,648 | 6,452 | (5) | 6,452 | 573 | 11.3 |
| 1998 | 5,077 | 1,175 | 4,560 | 1,692 | 6,252 | (5) | 6,252 | 562 | 11.1 |
| 1999 | 4,832 | 1,050 | 4,132 | 1,750 | 5,881 | (5) | 5,881 | 546 | 10.8 |
| 2000 | 4,851 | 970 | 4,049 | 1,773 | 5,822 | (5) | 5,822 | 534 | 10.9 |
| 2001 | 4,839 | 963 | 3,879 | 1,923 | 5,801 | (5) | 5,801 | 530 | 10.9 |
| 2002 | 4,761 | 984 | 3,743 | 2,003 | 5,746 | (5) | 5,746 | 529 | 10.9 |
| 2003 | 4,706 | 974 | 3,668 | 2,012 | 5,681 | (5) | 5,681 | 513 | 11.1 |
| 2004 | 4,510 | 908 | 3,536 | 1,883 | 5,419 | (5) | 5,419 | 510 | 10.6 |
| 2005 | 4,314 | 864 | 3,466 | 1,712 | 5,178 | (5) | 5,178 | 498 | 10.4 |
| 2006 | 4,361 | 741 | 3,401 | 1,701 | 5,102 | (5) | 5,102 | 497 | 10.3 |
| 2007 | 4,342 | 722 | 3,407 | 1,657 | 5,064 | (5) | 5,064 | 500 | 10.1 |
| 2008 | R4,268 | 683 | R3,580 | 1,371 | R4,950 | (5) | R4,950 | R526 | 9.4 |
| 2009 | P4,665 | P645 | E3,442 | E1,868 | P5,310 | (5) | P5,310 | 526 | 10.1 |

[1] See "Crude Oil Well" in Glossary.
[2] United States excluding Alaska and Hawaii.
[3] As of December 31.
[4] Through 1976, average productivity is based on the average number of producing wells. Beginning in 1977, average productivity is based on the number of wells producing at end of year.
[5] Included in "Crude Oil."
R=Revised. P=Preliminary. E=Estimate.
Note: Totals may not equal sum of components due to independent rounding.
Web Page: See http://www.eia.gov/oil_gas/petroleum/info_glance/petroleum.html for related infomation.
Sources: **Onshore:** • 1954-1975—Bureau of Mines, Mineral Industry Surveys, *Petroleum Statement (PS), Annual,* annual reports. • 1976-1980—U.S. Energy Information Administration (EIA), Energy Data Reports, *PS, Annual,* annual reports. • 1981-2008—EIA, *Petroleum Supply Annual (PSA),* annual reports. • 2009—EIA estimates based on Form EIA-182, "Domestic Crude Oil First Purchase Report," and crude oil production data reported by State conservation agencies. **Offshore:** • 1954-1969—U.S. Geologic al Survey, *Outer Continental Shelf Statistics* (June 1979). • 1970-1975—Bureau of Mines, Mineral Indust ry Surveys, *PS, Annual,* annual reports. • 1976-1980—EIA, Energy Data Reports, *PS, Annual,* annual reports. • 1981-2008—EIA, *PSA,* annual reports. • 2009—EIA estimates based on Form EIA-18 2, "Domestic Crude Oil First Purchase Report," and crude oil production data reported by State conservatio n agencies. **Producing Wells:** • 1954-1975—Bureau of Mines, *Minerals Yearbook,* "Crude Petroleum and Petroleum Products" chapter. • 1976-1980—EIA, Energy Data Reports, *PS, Annual,* annual reports. • 1981-1994—Independent Petroleum Association of America, *The Oil Producing Industry in Your State.* • 1995 forward—Gulf Publishing Co., *World Oil,* February issues. **All Other Data:** • 1954-1975—Bureau of Mines, Mineral Industry Surveys, *PS, Annual,* annual reports. • 1976-1980—EIA, Energy Data Reports, *PS, Annual,* annual reports. • 1981-2008—EIA, *PSA,* annual reports. • 2009—EIA, *Petroleum Supply Monthly* (February 2010).

## Petroleum Imports by Type

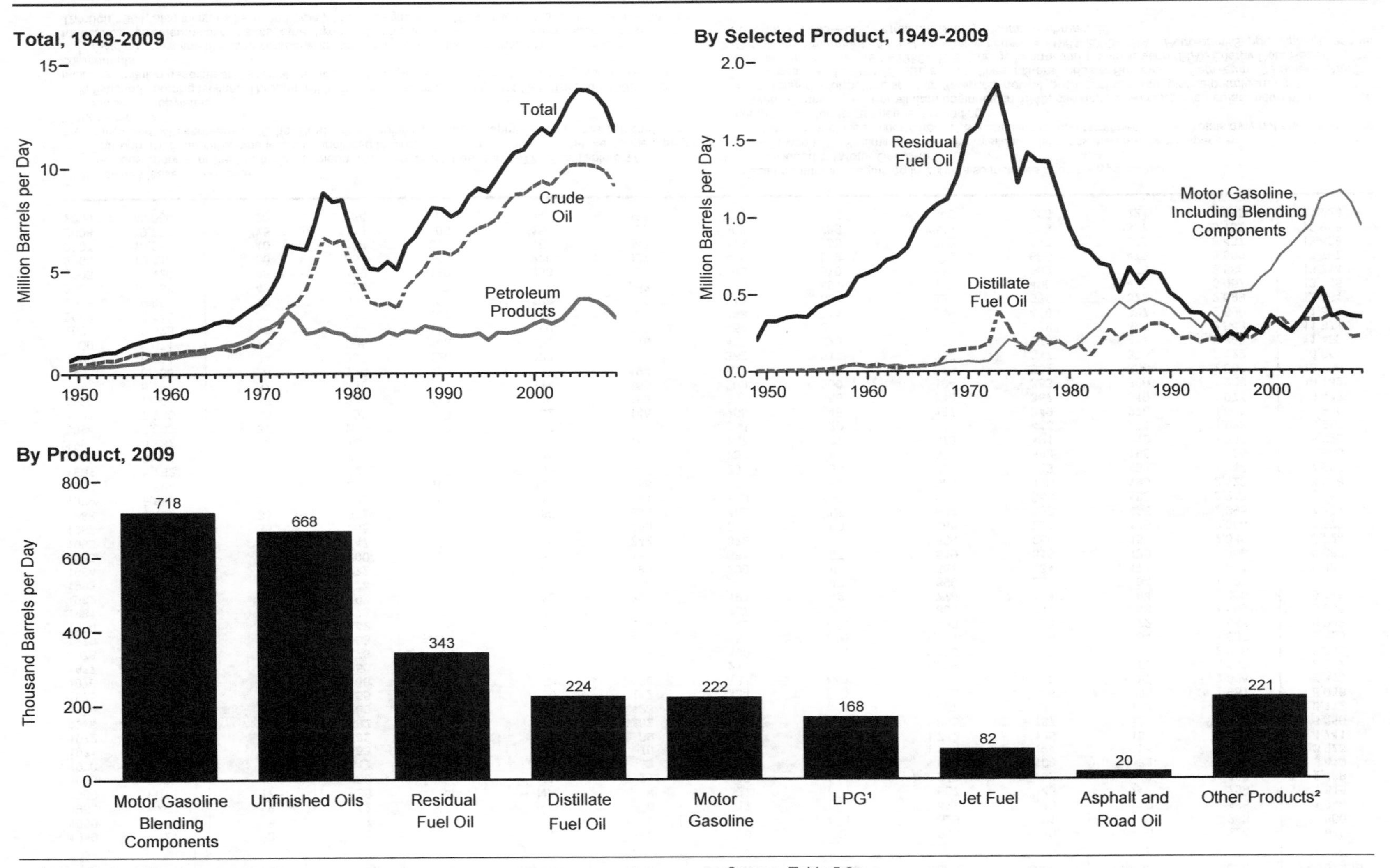


[1] Liquefied petroleum gases.
[2] Aviation gasoline and blending components, kerosene, lubricants, pentanes plus, petrochemical feedstocks, petroleum coke, special naphthas, waxes, other hydrocarbons and oxygenates, and miscellaneous products.

Source: Table 5.3.

## Petroleum Imports by Type, Selected Years, 1949-2009

(Thousand Barrels per Day)

| Year | Crude Oil [1,2] | Petroleum Products | | | | | | | | | | | Total Petroleum |
|---|---|---|---|---|---|---|---|---|---|---|---|---|---|
| | | Asphalt and Road Oil | Distillate Fuel Oil | Jet Fuel [3] | Liquefied Petroleum Gases | | Motor Gasoline [5] | Motor Gasoline Blending Components | Residual Fuel Oil | Unfinished Oils | Other Products [6] | Total | |
| | | | | | Propane [4] | Total | | | | | | | |
| 1949 | 421 | 3 | 5 | (3) | 0 | 0 | 0 | 0 | 206 | 10 | 0 | 224 | 645 |
| 1950 | 487 | 5 | 7 | (3) | 0 | 0 | (s) | (7) | 329 | 21 | 1 | 363 | 850 |
| 1955 | 782 | 9 | 12 | (3) | 0 | 0 | 13 | (7) | 417 | 15 | 0 | 466 | 1,248 |
| 1960 | 1,015 | 17 | 35 | 34 | NA | 4 | 27 | (7) | 637 | 45 | (s) | 799 | 1,815 |
| 1965 | 1,238 | 17 | 36 | 81 | NA | 21 | 28 | (7) | 946 | 92 | 10 | 1,229 | 2,468 |
| 1970 | 1,324 | 17 | 147 | 144 | 26 | 52 | 67 | (7) | 1,528 | 108 | 32 | 2,095 | 3,419 |
| 1971 | 1,681 | 20 | 153 | 180 | 32 | 70 | 59 | (7) | 1,583 | 124 | 56 | 2,245 | 3,926 |
| 1972 | 2,216 | 25 | 182 | 194 | 43 | 89 | 68 | (7) | 1,742 | 125 | 101 | 2,525 | 4,741 |
| 1973 | 3,244 | 23 | 392 | 212 | 71 | 132 | 134 | (7) | 1,853 | 137 | 129 | 3,012 | 6,256 |
| 1974 | 3,477 | 31 | 289 | 163 | 59 | 123 | 204 | (7) | 1,587 | 121 | 117 | 2,635 | 6,112 |
| 1975 | 4,105 | 14 | 155 | 133 | 60 | 112 | 184 | (7) | 1,223 | 36 | 95 | 1,951 | 6,056 |
| 1976 | 5,287 | 11 | 146 | 76 | 68 | 130 | 131 | (7) | 1,413 | 32 | 87 | 2,026 | 7,313 |
| 1977 | 6,615 | 4 | 250 | 75 | 86 | 161 | 217 | (7) | 1,359 | 31 | 95 | 2,193 | 8,807 |
| 1978 | 6,356 | 2 | 173 | 86 | 57 | 123 | 190 | (7) | 1,355 | 27 | 50 | 2,008 | 8,363 |
| 1979 | 6,519 | 4 | 193 | 78 | 88 | 217 | 181 | (7) | 1,151 | 59 | 54 | 1,937 | 8,456 |
| 1980 | 5,263 | 4 | 142 | 80 | 69 | 216 | 140 | (7) | 939 | 55 | 72 | 1,646 | 6,909 |
| 1981 | 4,396 | 4 | 173 | 38 | 70 | 244 | 157 | 24 | 800 | 112 | 48 | 1,599 | 5,996 |
| 1982 | 3,488 | 5 | 93 | 29 | 63 | 226 | 197 | 42 | 776 | 174 | 84 | 1,625 | 5,113 |
| 1983 | 3,329 | 7 | 174 | 29 | 44 | 190 | 247 | 47 | 699 | 234 | 94 | 1,722 | 5,051 |
| 1984 | 3,426 | 18 | 272 | 62 | 67 | 195 | 299 | 83 | 681 | 231 | 171 | 2,011 | 5,437 |
| 1985 | 3,201 | 35 | 200 | 39 | 67 | 187 | 381 | 67 | 510 | 318 | 130 | 1,866 | 5,067 |
| 1986 | 4,178 | 29 | 247 | 57 | 110 | 242 | 326 | 72 | 669 | 250 | 153 | 2,045 | 6,224 |
| 1987 | 4,674 | 36 | 255 | 67 | 88 | 190 | 384 | 60 | 565 | 299 | 146 | 2,004 | 6,678 |
| 1988 | 5,107 | 31 | 302 | 90 | 106 | 209 | 405 | 57 | 644 | 360 | 196 | 2,295 | 7,402 |
| 1989 | 5,843 | 31 | 306 | 106 | 111 | 181 | 369 | 66 | 629 | 348 | 183 | 2,217 | 8,061 |
| 1990 | 5,894 | 32 | 278 | 108 | 115 | 188 | 342 | 62 | 504 | 413 | 198 | 2,123 | 8,018 |
| 1991 | 5,782 | 28 | 205 | 67 | 91 | 147 | 297 | 36 | 453 | 413 | 198 | 1,844 | 7,627 |
| 1992 | 6,083 | 27 | 216 | 82 | 85 | 131 | 294 | 41 | 375 | 443 | 195 | 1,805 | 7,888 |
| 1993 | 6,787 | 32 | 184 | 100 | 103 | 160 | 247 | 27 | 373 | 491 | 219 | 1,833 | 8,620 |
| 1994 | 7,063 | 37 | 203 | 117 | 124 | 183 | 356 | 20 | 314 | 413 | 291 | 1,933 | 8,996 |
| 1995 | 7,230 | 36 | 193 | 106 | 102 | 146 | 265 | 48 | 187 | 349 | 276 | 1,605 | 8,835 |
| 1996 | 7,508 | 27 | 230 | 111 | 119 | 166 | 336 | 166 | 248 | 367 | 319 | 1,971 | 9,478 |
| 1997 | 8,225 | 32 | 228 | 91 | 113 | 169 | 309 | 200 | 194 | 353 | 360 | 1,936 | 10,162 |
| 1998 | 8,706 | 28 | 210 | 124 | 137 | 194 | 311 | 209 | 275 | 302 | 350 | 2,002 | 10,708 |
| 1999 | 8,731 | 34 | 250 | 128 | 122 | 182 | 382 | 217 | 237 | 317 | 375 | 2,122 | 10,852 |
| 2000 | 9,071 | 28 | 295 | 162 | 161 | 215 | 427 | 223 | 352 | 274 | 414 | 2,389 | 11,459 |
| 2001 | 9,328 | 26 | 344 | 148 | 140 | 206 | 454 | 298 | 295 | 378 | 393 | 2,543 | 11,871 |
| 2002 | 9,140 | 27 | 267 | 107 | 145 | 183 | 498 | 311 | 249 | 410 | 337 | 2,390 | 11,530 |
| 2003 | 9,665 | 12 | 333 | 109 | 168 | 225 | 518 | 367 | 327 | 335 | 373 | 2,599 | 12,264 |
| 2004 | 10,088 | 43 | 325 | 127 | 209 | 263 | 496 | 451 | 426 | 490 | 436 | 3,057 | 13,145 |
| 2005 | 10,126 | 43 | 329 | 190 | 233 | 328 | 603 | 510 | 530 | 582 | 473 | 3,588 | 13,714 |
| 2006 | 10,118 | 50 | 365 | 186 | 228 | 332 | 475 | 669 | 350 | 689 | 473 | 3,589 | 13,707 |
| 2007 | 10,031 | 40 | 304 | 217 | 182 | 247 | 413 | 753 | 372 | 717 | 375 | 3,437 | 13,468 |
| 2008 | R9,783 | R25 | R213 | R103 | R185 | R253 | R302 | R789 | R349 | 763 | R337 | R3,132 | R12,915 |
| 2009P | 9,060 | 20 | 224 | 82 | 142 | 168 | 222 | 718 | 343 | 668 | 221 | 2,665 | 11,726 |

[1] Includes lease condensate.

[2] Includes imports for the Strategic Petroleum Reserve, which began in 1977. See Table 5.17.

[3] Through 1955, naphtha-type jet fuel is included in "Motor Gasoline." Through 1964, kerosene-type jet fuel is included with kerosene in "Other Products." Beginning in 2005, naphtha-type jet fuel is included in "Other Products."

[4] Includes propylene.

[5] Finished motor gasoline. Through 1955, also includes naphtha-type jet fuel. Through 1963, also includes aviation gasoline and special naphthas. Through 1980, also includes motor gasoline blending components.

[6] Aviation gasoline blending components, kerosene, lubricants, pentanes plus, petrochemical feedstocks, petroleum coke, waxes, other hydrocarbons and oxygenates, and miscellaneous products. Through 1964, also includes kerosene-type jet fuel. Beginning in 1964, also includes aviation gasoline and special naphthas. Beginning in 2005, also includes naphtha-type jet fuel.

[7] Included in "Motor Gasoline."

R=Revised. P=Preliminary. NA=Not available. (s)=Less than 500 barrels per day.

Notes: • Includes imports from U.S. possessions and territories. • Totals may not equal sum of components due to independent rounding.

Web Pages: • For all data beginning in 1949, see http://www.eia.gov/emeu/aer/petro.htm l. • For related information, see http://www.eia.gov/oil_gas/petroleum/info_glance/petroleum.html.

Sources: • 1949-1975—Bureau of Mines, Mineral Industry Surveys, *Petroleum Statement, Annual,* annual reports. • 1976-1980—U.S. Energy Information Administration (EIA), Energy Data Report s, *Petroleum Statement, Annual,* annual reports. • 1981-2008—EIA, *Petroleum Supply Annual,* annual reports. • 2009—EIA, *Petroleum Supply Monthly* (February 2010).

## Petroleum Imports by Country of Origin

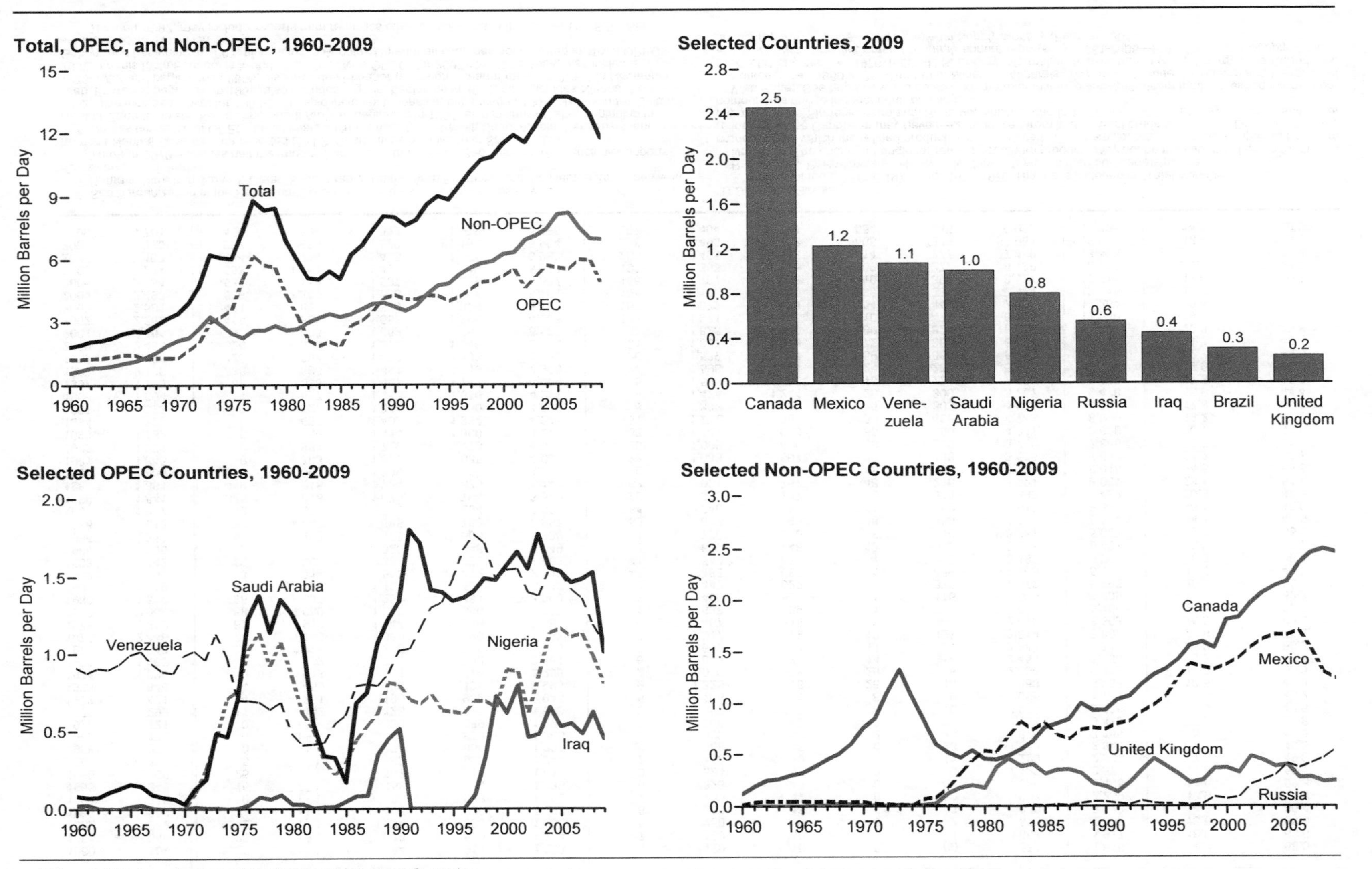


Note: OPEC=Organization of the Petroleum Exporting Countries.

U.S. Energy Information Administration / Annual Energy Review 2009

## Petroleum Imports by Country of Origin, 1960-2009

| Year | Persian Gulf [2] | Selected OPEC [1] Countries: Iraq | Nigeria | Saudi Arabia [3] | Venezuela | Total OPEC [4] | Selected Non-OPEC [1] Countries: Brazil | Canada | Mexico | Russia [5] | United Kingdom | Total Non-OPEC [4] | Total Imports | Imports From Persian Gulf [2] as Share of Total Imports | Imports From OPEC [1] as Share of Total Imports |
|---|---|---|---|---|---|---|---|---|---|---|---|---|---|---|---|
| | Thousand Barrels per Day | | | | | | | | | | | | | Percent | |
| 1960 | NA | 22 | ([6]) | 84 | 911 | 1,233 | 1 | 120 | 16 | 0 | (s) | 581 | 1,815 | NA | 68.0 |
| 1961 | 346 | 25 | ([6]) | 73 | 879 | 1,224 | 7 | 190 | 40 | 0 | 1 | 693 | 1,917 | 18.0 | 63.8 |
| 1962 | 272 | 2 | ([6]) | 74 | 906 | 1,265 | 5 | 250 | 49 | 0 | 2 | 816 | 2,082 | 13.0 | 60.8 |
| 1963 | 303 | 1 | ([6]) | 108 | 900 | 1,282 | 6 | 265 | 48 | 0 | 3 | 840 | 2,123 | 14.3 | 60.4 |
| 1964 | 315 | 0 | ([6]) | 131 | 933 | 1,352 | 1 | 299 | 47 | 0 | (s) | 907 | 2,259 | 13.9 | 59.8 |
| 1965 | 345 | 16 | ([6]) | 158 | 994 | 1,439 | 0 | 323 | 48 | 0 | (s) | 1,029 | 2,468 | 14.0 | 58.3 |
| 1966 | 306 | 26 | ([6]) | 147 | 1,018 | 1,444 | 0 | 384 | 45 | 0 | 6 | 1,129 | 2,573 | 11.9 | 56.1 |
| 1967 | 198 | 5 | ([6]) | 92 | 938 | 1,247 | 2 | 450 | 49 | 0 | 11 | 1,290 | 2,537 | 7.8 | 49.2 |
| 1968 | 202 | 0 | ([6]) | 74 | 886 | 1,287 | (s) | 506 | 45 | 0 | 28 | 1,553 | 2,840 | 7.1 | 45.3 |
| 1969 | 179 | 0 | ([6]) | 65 | 875 | 1,286 | 0 | 608 | 43 | 2 | 20 | 1,879 | 3,166 | 5.7 | 40.6 |
| 1970 | 121 | 0 | ([6]) | 30 | 989 | 1,294 | 2 | 766 | 42 | 3 | 11 | 2,126 | 3,419 | 3.5 | 37.8 |
| 1971 | 299 | 11 | 102 | 128 | 1,020 | 1,673 | 3 | 857 | 27 | 0 | 10 | 2,253 | 3,926 | 7.6 | 42.6 |
| 1972 | 471 | 4 | 251 | 190 | 959 | 2,046 | 5 | 1,108 | 21 | 8 | 9 | 2,695 | 4,741 | 9.9 | 43.2 |
| 1973 | 848 | 4 | 459 | 486 | 1,135 | 2,993 | 9 | 1,325 | 16 | 26 | 15 | 3,263 | 6,256 | 13.6 | 47.8 |
| 1974 | 1,039 | 0 | 713 | 461 | 979 | 3,256 | 2 | 1,070 | 8 | 20 | 8 | 2,856 | 6,112 | 17.0 | 53.3 |
| 1975 | 1,165 | 2 | 762 | 715 | 702 | 3,601 | 5 | 846 | 71 | 14 | 14 | 2,454 | 6,056 | 19.2 | 59.5 |
| 1976 | 1,840 | 26 | 1,025 | 1,230 | 700 | 5,066 | 0 | 599 | 87 | 11 | 31 | 2,247 | 7,313 | 25.2 | 69.3 |
| 1977 | 2,448 | 74 | 1,143 | 1,380 | 690 | 6,193 | 0 | 517 | 179 | 12 | 126 | 2,614 | 8,807 | 27.8 | 70.3 |
| 1978 | 2,219 | 62 | 919 | 1,144 | 646 | 5,751 | 0 | 467 | 318 | 8 | 180 | 2,612 | 8,363 | 26.5 | 68.8 |
| 1979 | 2,069 | 88 | 1,080 | 1,356 | 690 | 5,637 | 1 | 538 | 439 | 1 | 202 | 2,819 | 8,456 | 24.5 | 66.7 |
| 1980 | 1,519 | 28 | 857 | 1,261 | 481 | 4,300 | 3 | 455 | 533 | 1 | 176 | 2,609 | 6,909 | 22.0 | 62.2 |
| 1981 | 1,219 | (s) | 620 | 1,129 | 406 | 3,323 | 23 | 447 | 522 | 5 | 375 | 2,672 | 5,996 | 20.3 | 55.4 |
| 1982 | 696 | 3 | 514 | 552 | 412 | 2,146 | 47 | 482 | 685 | 1 | 456 | 2,968 | 5,113 | 13.6 | 42.0 |
| 1983 | 442 | 10 | 302 | 337 | 422 | 1,862 | 41 | 547 | 826 | 1 | 382 | 3,189 | 5,051 | 8.8 | 36.9 |
| 1984 | 506 | 12 | 216 | 325 | 548 | 2,049 | 60 | 630 | 748 | 13 | 402 | 3,388 | 5,437 | 9.3 | 37.7 |
| 1985 | 311 | 46 | 293 | 168 | 605 | 1,830 | 61 | 770 | 816 | 8 | 310 | 3,237 | 5,067 | 6.1 | 36.1 |
| 1986 | 912 | 81 | 440 | 685 | 793 | 2,837 | 50 | 807 | 699 | 18 | 350 | 3,387 | 6,224 | 14.7 | 45.6 |
| 1987 | 1,077 | 83 | 535 | 751 | 804 | 3,060 | 84 | 848 | 655 | 11 | 352 | 3,617 | 6,678 | 16.1 | 45.8 |
| 1988 | 1,541 | 345 | 618 | 1,073 | 794 | 3,520 | 98 | 999 | 747 | 29 | 315 | 3,882 | 7,402 | 20.8 | 47.6 |
| 1989 | 1,861 | 449 | 815 | 1,224 | 873 | 4,140 | 82 | 931 | 767 | 48 | 215 | 3,921 | 8,061 | 23.1 | 51.4 |
| 1990 | 1,966 | 518 | 800 | 1,339 | 1,025 | 4,296 | 49 | 934 | 755 | 45 | 189 | 3,721 | 8,018 | 24.5 | 53.6 |
| 1991 | 1,845 | 0 | 703 | 1,802 | 1,035 | 4,092 | 22 | 1,033 | 807 | 29 | 138 | 3,535 | 7,627 | 24.2 | 53.7 |
| 1992 | 1,778 | 0 | 681 | 1,720 | 1,170 | 4,092 | 20 | 1,069 | 830 | 18 | 230 | 3,796 | 7,888 | 22.5 | 51.9 |
| 1993 | 1,782 | 0 | 740 | 1,414 | 1,300 | 4,273 | 33 | 1,181 | 919 | 55 | 350 | 4,347 | 8,620 | 20.7 | 49.6 |
| 1994 | 1,728 | 0 | 637 | 1,402 | 1,334 | 4,247 | 31 | 1,272 | 984 | 30 | 458 | 4,749 | 8,996 | 19.2 | 47.2 |
| 1995 | 1,573 | 0 | 627 | 1,344 | 1,480 | 4,002 | 8 | 1,332 | 1,068 | 25 | 383 | 4,833 | 8,835 | 17.8 | 45.3 |
| 1996 | 1,604 | 1 | 617 | 1,363 | 1,676 | 4,211 | 9 | 1,424 | 1,244 | 25 | 308 | 5,267 | 9,478 | 16.9 | 44.4 |
| 1997 | 1,755 | 89 | 698 | 1,407 | 1,773 | 4,569 | 5 | 1,563 | 1,385 | 13 | 226 | 5,593 | 10,162 | 17.3 | 45.0 |
| 1998 | 2,136 | 336 | 696 | 1,491 | 1,719 | 4,905 | 26 | 1,598 | 1,351 | 24 | 250 | 5,803 | 10,708 | 19.9 | 45.8 |
| 1999 | 2,464 | 725 | 657 | 1,478 | 1,493 | 4,953 | 26 | 1,539 | 1,324 | 89 | 365 | 5,899 | 10,852 | 22.7 | 45.6 |
| 2000 | 2,488 | 620 | 896 | 1,572 | 1,546 | 5,203 | 51 | 1,807 | 1,373 | 72 | 366 | 6,257 | 11,459 | 21.7 | 45.4 |
| 2001 | 2,761 | 795 | 885 | 1,662 | 1,553 | 5,528 | 82 | 1,828 | 1,440 | 90 | 324 | 6,343 | 11,871 | 23.3 | 46.6 |
| 2002 | 2,269 | 459 | 621 | 1,552 | 1,398 | 4,605 | 116 | 1,971 | 1,547 | 210 | 478 | 6,925 | 11,530 | 19.7 | 39.9 |
| 2003 | 2,501 | 481 | 867 | 1,774 | 1,376 | 5,162 | 108 | 2,072 | 1,623 | 254 | 440 | 7,103 | 12,264 | 20.4 | 42.1 |
| 2004 | 2,493 | 656 | 1,140 | 1,558 | 1,554 | 5,701 | 104 | 2,138 | 1,665 | 298 | 380 | 7,444 | 13,145 | 19.0 | 43.4 |
| 2005 | 2,334 | 531 | 1,166 | 1,537 | 1,529 | 5,587 | 156 | 2,181 | 1,662 | 410 | 396 | 8,127 | 13,714 | 17.0 | 40.7 |
| 2006 | 2,211 | 553 | 1,114 | 1,463 | 1,419 | 5,517 | 193 | 2,353 | 1,705 | 369 | 272 | 8,190 | 13,707 | 16.1 | 40.2 |
| 2007 | 2,163 | 484 | 1,134 | 1,485 | 1,361 | 5,980 | 200 | 2,455 | 1,532 | 414 | 277 | 7,489 | 13,468 | 16.1 | 44.4 |
| 2008 | R2,370 | 627 | R988 | R1,529 | R1,189 | R5,954 | 258 | R2,493 | R1,302 | R465 | R236 | R6,961 | R12,915 | 18.4 | R46.1 |
| 2009P | 1,701 | 450 | 804 | 1,012 | 1,078 | 4,786 | 307 | 2,464 | 1,234 | 554 | 245 | 6,939 | 11,726 | 14.5 | 40.8 |

[1] See "Organization of the Petroleum Exporting Countries (OPEC)" in Glossary.

[2] Bahrain, Iran, Iraq, Kuwait, Qatar, Saudi Arabia, United Arab Emirates, and the Neutral Zone (between Kuwait and Saudi Arabia).

[3] Through 1970, includes half the imports from the Neutral Zone. Beginning in 1971, includes imports from the Neutral Zone that are reported to U.S. Customs as originating in Saudi Arabia.

[4] On this table, "Total OPEC" for all years includes Iran, Iraq, Kuwait, Saudi Arabia, Venezuela, and the Neutral Zone (between Kuwait and Saudi Arabia); beginning in 1961, also includes Qatar; beginning in 1962, also includes Libya; for 1962-2008, also includes Indonesia; beginning in 1967, also includes United Arab Emirates; beginning in 1969, also includes Algeria; beginning in 1971, also includes Nigeria; for 1973-1992 and beginning in 2008, also includes Ecuador (although Ecuador rejoined OPEC in November 2007, on this table Ecuador is included in "Total Non-OPEC" for 2007); for 1975-1994, also includes Gabon; and beginning in 2007, also includes Angola. Data for all countries not included in "Total OPEC" are included in "Total Non-OPEC."

[5] Through 1992, may include imports from republics other than Russia in the former U.S.S.R. See "U.S.S.R." in Glossary.

[6] Nigeria joined OPEC in 1971. For 1960-1970, Nigeria is included in "Total Non-OPEC."

R=Revised. P=Preliminary. NA=Not available. (s)=Less than 500 barrels per day.

Notes: • The country of origin for refined petroleum products may not be the country of origin for the crude oil from which the refined products were produced. For example, refined products imported from refineries in the Caribbean may have been produced from Middle East crude oil. • Data include any imports for the Strategic Petroleum Reserve, which began in 1977. • Totals may not equal sum of components due to independent rounding.

Web Page: See http://www.eia.gov/oil_gas/petroleum/info_glance/petroleum.html for related information.

Sources: • 1960-1975—Bureau of Mines, *Minerals Yearbook,* "Crude Petroleum and Petroleum Products" chapter. • 1976-1980—U.S. Energy Information Administration (EIA), Energy Data Reports, *P.A.D. Districts Supply/Demand, Annual,* annual reports. • 1981-2008—EIA, *Petroleum Supply Annual,* annual reports. • 2009—EIA, *Petroleum Supply Monthly* (February 2010).

# Petroleum Exports by Type

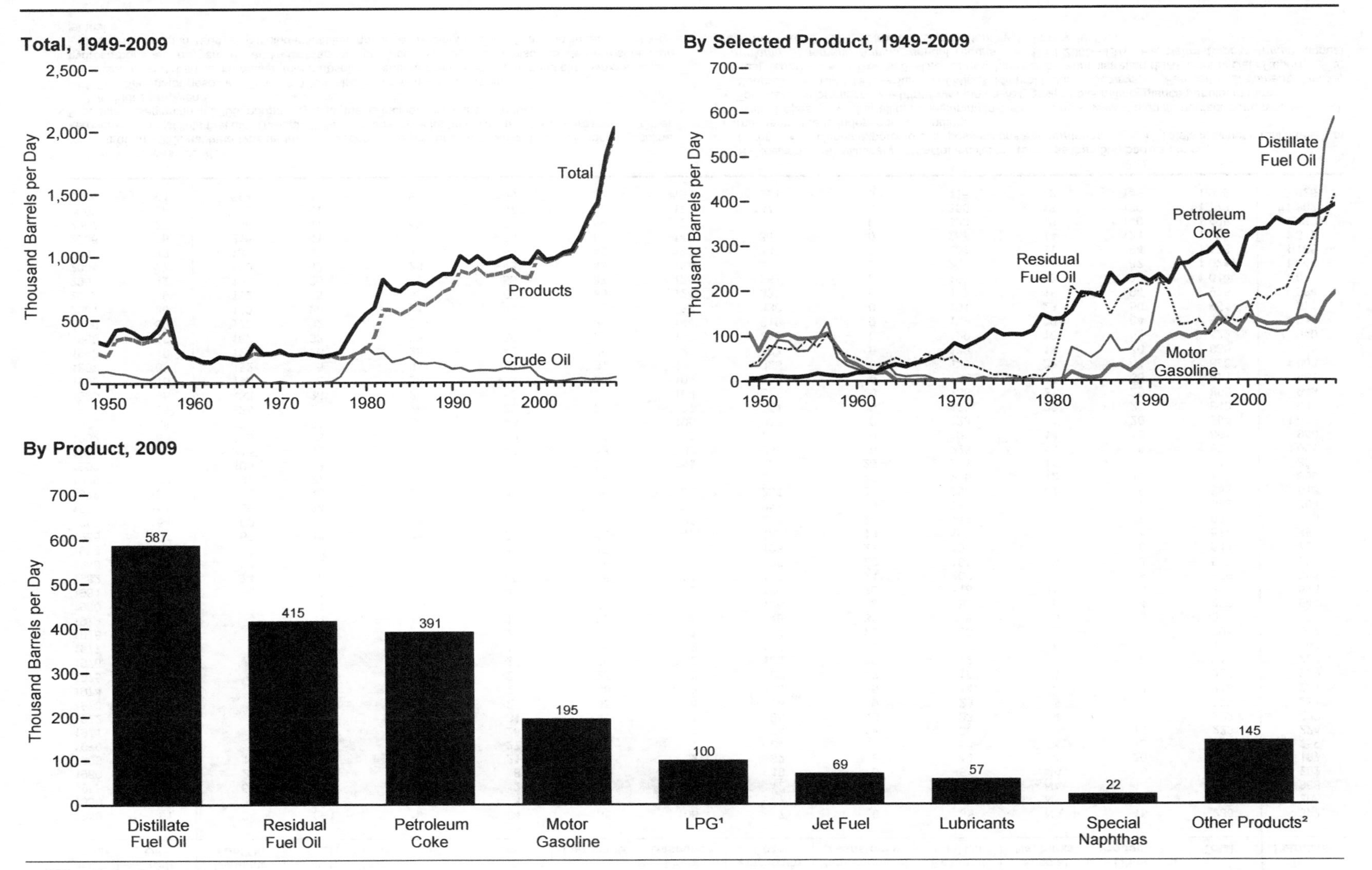


[1] Liquefied petroleum gases.

[2] Asphalt and road oil, aviation gasoline, kerosene, motor gasoline blending components, pentanes plus, waxes, other hydrocarbons and oxygenates, and miscellaneous products.

U.S. Energy Information Administration / Annual Energy Review 2009

## Petroleum Exports by Type, Selected Years, 1949-2009

(Thousand Barrels per Day)

| Year | Crude Oil [1] | Petroleum Products | | | | | | | | | | | | Total Petroleum |
|---|---|---|---|---|---|---|---|---|---|---|---|---|---|---|
| | | Distillate Fuel Oil | Jet Fuel [2] | Liquefied Petroleum Gases | | Lubricants | Motor Gasoline [4] | Petroleum Coke | Petrochemical Feedstocks | Residual Fuel Oil | Special Naphthas | Other Products [5] | Total | |
| | | | | Propane [3] | Total | | | | | | | | | |
| 1949 | 91 | 34 | ([2]) | NA | 4 | 35 | 108 | 7 | 0 | 35 | NA | 15 | 236 | 327 |
| 1950 | 95 | 35 | ([2]) | NA | 4 | 39 | 68 | 7 | 0 | 44 | NA | 12 | 210 | 305 |
| 1955 | 32 | 67 | (s) | NA | 12 | 39 | 95 | 12 | 0 | 93 | NA | 18 | 336 | 368 |
| 1960 | 8 | 27 | (s) | NA | 8 | 43 | 37 | 19 | 0 | 51 | NA | 9 | 193 | 202 |
| 1965 | 3 | 10 | 3 | NA | 21 | 45 | 2 | 32 | 5 | 41 | 4 | 20 | 184 | 187 |
| 1970 | 14 | 2 | 6 | 6 | 27 | 44 | 1 | 84 | 10 | 54 | 4 | 12 | 245 | 259 |
| 1971 | 1 | 8 | 4 | 13 | 26 | 43 | 1 | 74 | 14 | 36 | 4 | 12 | 223 | 224 |
| 1972 | 1 | 3 | 3 | 18 | 31 | 41 | 1 | 85 | 13 | 33 | 4 | 8 | 222 | 222 |
| 1973 | 2 | 9 | 4 | 15 | 27 | 35 | 4 | 96 | 19 | 23 | 5 | 8 | 229 | 231 |
| 1974 | 3 | 2 | 3 | 14 | 25 | 33 | 2 | 113 | 15 | 14 | 4 | 7 | 218 | 221 |
| 1975 | 6 | 1 | 2 | 13 | 26 | 25 | 2 | 102 | 22 | 15 | 3 | 6 | 204 | 209 |
| 1976 | 8 | 1 | 2 | 13 | 25 | 26 | 3 | 103 | 30 | 12 | 7 | 6 | 215 | 223 |
| 1977 | 50 | 1 | 2 | 10 | 18 | 26 | 2 | 102 | 24 | 6 | 4 | 7 | 193 | 243 |
| 1978 | 158 | 3 | 1 | 9 | 20 | 27 | 1 | 111 | 23 | 13 | 2 | 2 | 204 | 362 |
| 1979 | 235 | 3 | 1 | 8 | 15 | 23 | (s) | 146 | 31 | 9 | 5 | 3 | 236 | 471 |
| 1980 | 287 | 3 | 1 | 10 | 21 | 23 | 1 | 136 | 29 | 33 | 5 | 4 | 258 | 544 |
| 1981 | 228 | 5 | 2 | 18 | 42 | 19 | 2 | 138 | 26 | 118 | 11 | 4 | 367 | 595 |
| 1982 | 236 | 74 | 6 | 31 | 65 | 16 | 20 | 156 | 24 | 209 | 5 | 4 | 579 | 815 |
| 1983 | 164 | 64 | 6 | 43 | 73 | 16 | 10 | 195 | 20 | 185 | 3 | 3 | 575 | 739 |
| 1984 | 181 | 51 | 9 | 30 | 48 | 15 | 6 | 193 | 21 | 190 | 2 | 6 | 541 | 722 |
| 1985 | 204 | 67 | 13 | 48 | 62 | 15 | 10 | 187 | 19 | 197 | 1 | 4 | 577 | 781 |
| 1986 | 154 | 100 | 18 | 28 | 42 | 23 | 33 | 238 | 22 | 147 | 1 | 8 | 631 | 785 |
| 1987 | 151 | 66 | 24 | 24 | 38 | 23 | 35 | 213 | 20 | 186 | 2 | 7 | 613 | 764 |
| 1988 | 155 | 69 | 28 | 31 | 49 | 26 | 22 | 231 | 23 | 200 | 7 | 6 | 661 | 815 |
| 1989 | 142 | 97 | 27 | 24 | 35 | 19 | 39 | 233 | 26 | 215 | 12 | 15 | 717 | 859 |
| 1990 | 109 | 109 | 43 | 28 | 40 | 20 | 55 | 220 | 26 | 211 | 11 | 13 | 748 | 857 |
| 1991 | 116 | 215 | 43 | 28 | 41 | 18 | 82 | 235 | 0 | 226 | 15 | 9 | 885 | 1,001 |
| 1992 | 89 | 219 | 43 | 33 | 49 | 16 | 96 | 216 | 0 | 193 | 14 | 16 | 861 | 950 |
| 1993 | 98 | 274 | 59 | 26 | 43 | 19 | 105 | 258 | 0 | 123 | 4 | 20 | 904 | 1,003 |
| 1994 | 99 | 234 | 20 | 24 | 38 | 22 | 97 | 261 | 0 | 125 | 20 | 26 | 843 | 942 |
| 1995 | 95 | 183 | 26 | 38 | 58 | 25 | 104 | 277 | 0 | 136 | 21 | 25 | 855 | 949 |
| 1996 | 110 | 190 | 48 | 28 | 51 | 34 | 104 | 285 | 0 | 102 | 21 | 36 | 871 | 981 |
| 1997 | 108 | 152 | 35 | 32 | 50 | 31 | 137 | 306 | 0 | 120 | 22 | 44 | 896 | 1,003 |
| 1998 | 110 | 124 | 26 | 25 | 42 | 25 | 125 | 267 | 0 | 138 | 18 | 70 | 835 | 945 |
| 1999 | 118 | 162 | 32 | 33 | 50 | 28 | 111 | 242 | 0 | 129 | 16 | 52 | 822 | 940 |
| 2000 | 50 | 173 | 32 | 53 | 74 | 26 | 144 | 319 | 0 | 139 | 20 | 64 | 990 | 1,040 |
| 2001 | 20 | 119 | 29 | 31 | 44 | 26 | 133 | 336 | 0 | 191 | 23 | 50 | 951 | 971 |
| 2002 | 9 | 112 | 15 | 55 | 67 | 33 | 124 | 337 | 0 | 177 | 15 | 94 | 975 | 984 |
| 2003 | 12 | 107 | 20 | 37 | 56 | 37 | 125 | 361 | 0 | 197 | 22 | 89 | 1,014 | 1,027 |
| 2004 | 27 | 110 | 40 | 28 | 43 | 41 | 124 | 350 | 0 | 205 | 27 | 82 | 1,021 | 1,048 |
| 2005 | 32 | 138 | 53 | 37 | 53 | 40 | 136 | 347 | 0 | 251 | 21 | 94 | 1,133 | 1,165 |
| 2006 | 25 | 215 | 41 | 45 | 56 | 55 | 142 | 366 | 0 | 283 | 14 | 121 | 1,292 | 1,317 |
| 2007 | 27 | 268 | 41 | 42 | 57 | 59 | 127 | 366 | 0 | 330 | 18 | 140 | 1,405 | 1,433 |
| 2008 | 29 | R528 | 61 | 53 | 67 | 60 | 172 | R377 | 0 | 355 | 13 | R139 | R1,773 | R1,802 |
| 2009P | 44 | 587 | 69 | 85 | 100 | 57 | 195 | 391 | 0 | 415 | 22 | 145 | 1,982 | 2,026 |

[1] Includes lease condensate.

[2] Through 1952, naphtha-type jet fuel is included in the products from which it was blended: gasoline, kerosene, and distillate fuel oil. Through 1964, kerosene-type jet fuel is included with kerosene in "Other Products." Beginning in 2005, naphtha-type jet fuel is included in "Other Products."

[3] Includes propylene.

[4] Finished motor gasoline. Through 1963, also includes aviation gasoline.

[5] Asphalt and road oil, kerosene, motor gasoline blending components, pentanes plus, waxes, other hydrocarbons and oxygenates, and miscellaneous products. Through 1964, also includes kerosene-type jet fuel. Beginning in 1964, also includes aviation gasoline. Beginning in 2005, also includes naphtha-type jet fuel.

R=Revised. P=Preliminary. NA=Not available. (s)=Less than 500 barrels per day.

Notes: • Includes exports to U.S. possessions and territories. • Totals may not equal sum of components due to independent rounding.

Web Pages: • For all data beginning in 1949, see http://www.eia.gov/emeu/aer/petro.html. • For related information, see http://www.eia.gov/oil_gas/petroleum/info_glance/petroleum.html.

Sources: • 1949-1975—Bureau of Mines, Mineral Industry Surveys, *Petroleum Statement, Annual*, annual reports. • 1976-1980—U.S. Energy Information Administration (EIA), Energy Data Reports, *Petroleum Statement, Annual*, annual reports. • 1981-2008—EIA, *Petroleum Supply Annual*, annual reports. • 2009—EIA, *Petroleum Supply Monthly* (February 2010).

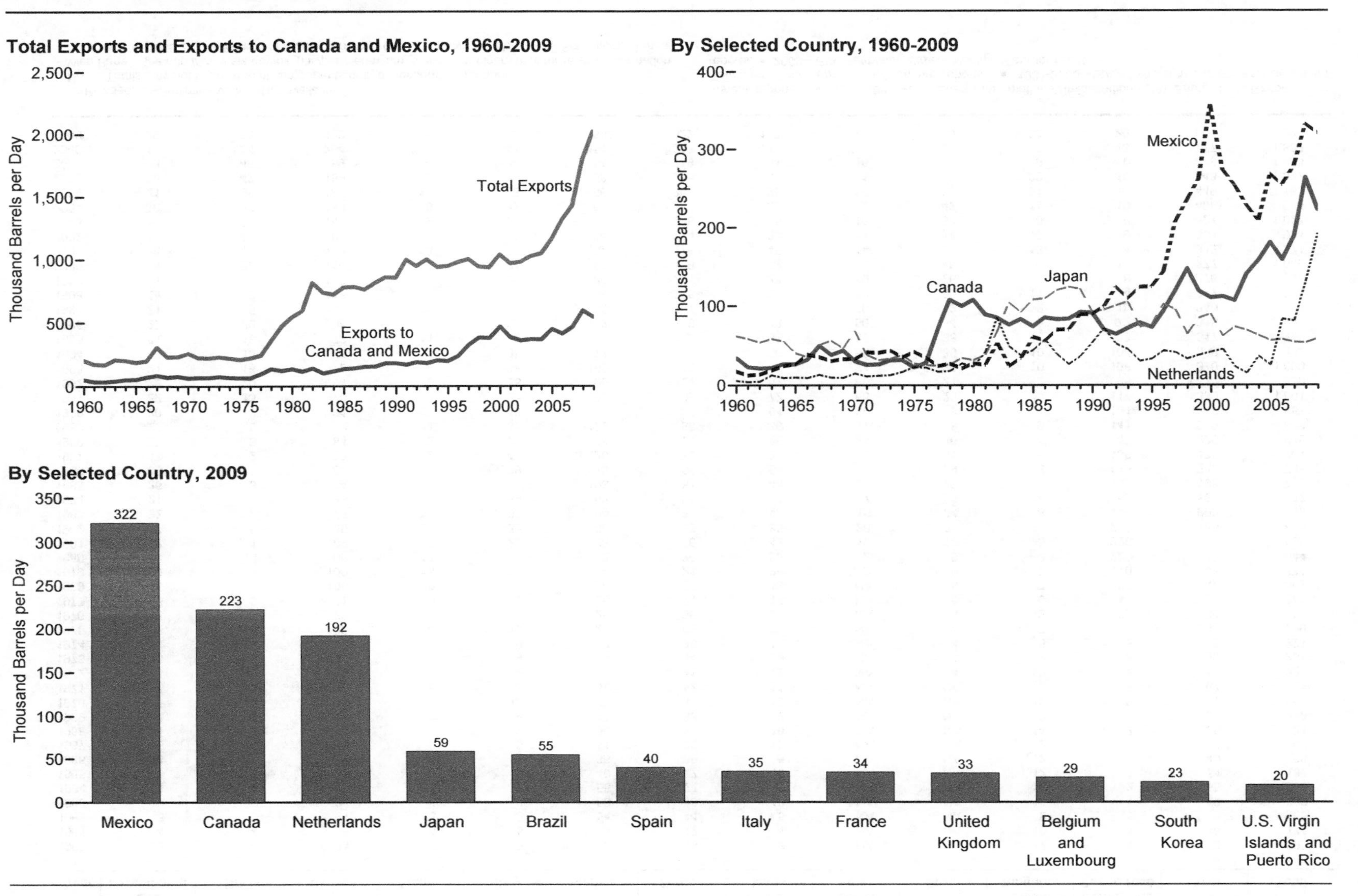
Petroleum Exports by Country of Destination
Total Exports and Exports to Canada and Mexico, 1960-2009
Thousand Barrels per Day
2,500
2,000
1,500
1,000
500
0
1960 1965 1970 1975 1980 1985 1990 1995 2000 2005
Total Exports
Exports to Canada and Mexico
By Selected Country, 1960-2009
Thousand Barrels per Day
400
300
200
100
0
1960 1965 1970 1975 1980 1985 1990 1995 2000 2005
Mexico
Canada
Japan
Netherlands
By Selected Country, 2009
Thousand Barrels per Day
350
300
250
200
150
100
50
0
322
223
192
59
55
40
35
34
33
29
23
20
Mexico
Canada
Netherlands
Japan
Brazil
Spain
Italy
France
United Kingdom
Belgium and Luxembourg
South Korea
U.S. Virgin Islands and Puerto Rico
U.S. Energy Information Administration / Annual Energy Review 2009

## Petroleum Exports by Country of Destination, 1960-2009

(Thousand Barrels per Day)

| Year | Belgium and Luxembourg | Brazil | Canada | France | Italy | Japan | Mexico | Nether-lands | South Korea | Spain | United Kingdom | U.S. Virgin Islands and Puerto Rico | Other | Total |
|---|---|---|---|---|---|---|---|---|---|---|---|---|---|---|
| 1960 | 3 | 4 | 34 | 4 | 6 | 62 | 18 | 6 | NA | NA | 12 | 1 | 52 | 202 |
| 1961 | 4 | 4 | 23 | 4 | 5 | 59 | 12 | 4 | NA | NA | 10 | 1 | 48 | 174 |
| 1962 | 3 | 5 | 21 | 3 | 5 | 54 | 14 | 5 | NA | NA | 8 | 1 | 50 | 168 |
| 1963 | 9 | 4 | 22 | 4 | 8 | 58 | 19 | 13 | NA | NA | 11 | 1 | 59 | 208 |
| 1964 | 4 | 4 | 27 | 4 | 8 | 56 | 24 | 9 | NA | NA | 10 | 2 | 55 | 202 |
| 1965 | 3 | 3 | 26 | 3 | 7 | 40 | 27 | 10 | NA | NA | 12 | 1 | 54 | 187 |
| 1966 | 3 | 4 | 32 | 4 | 7 | 36 | 39 | 9 | NA | NA | 12 | 3 | 49 | 198 |
| 1967 | 5 | 6 | 50 | 3 | 9 | 51 | 36 | 13 | NA | NA | 62 | 7 | 65 | 307 |
| 1968 | 4 | 8 | 39 | 4 | 8 | 56 | 31 | 10 | NA | NA | 14 | 2 | 55 | 231 |
| 1969 | 4 | 7 | 44 | 4 | 9 | 47 | 33 | 9 | NA | NA | 13 | 2 | 59 | 233 |
| 1970 | 5 | 7 | 31 | 5 | 10 | 69 | 33 | 15 | NA | NA | 12 | 2 | 71 | 259 |
| 1971 | 7 | 9 | 26 | 5 | 8 | 39 | 42 | 11 | NA | NA | 9 | 3 | 67 | 224 |
| 1972 | 13 | 9 | 26 | 5 | 9 | 32 | 41 | 12 | NA | 4 | 10 | 4 | 59 | 222 |
| 1973 | 15 | 8 | 31 | 5 | 9 | 34 | 44 | 13 | NA | 4 | 9 | 3 | 56 | 231 |
| 1974 | 13 | 9 | 32 | 4 | 9 | 38 | 35 | 17 | NA | 4 | 6 | 6 | 48 | 221 |
| 1975 | 9 | 6 | 22 | 6 | 10 | 27 | 42 | 23 | NA | 4 | 7 | 12 | 40 | 209 |
| 1976 | 12 | 7 | 28 | 6 | 10 | 25 | 35 | 22 | NA | 4 | 13 | 22 | 39 | 223 |
| 1977 | 16 | 6 | 71 | 9 | 10 | 25 | 24 | 17 | NA | 5 | 9 | 11 | 39 | 243 |
| 1978 | 15 | 8 | 108 | 9 | 10 | 26 | 27 | 18 | NA | 5 | 7 | 86 | 42 | 362 |
| 1979 | 19 | 7 | 100 | 13 | 15 | 34 | 21 | 28 | 2 | 9 | 7 | 170 | 45 | 471 |
| 1980 | 20 | 4 | 108 | 11 | 14 | 32 | 28 | 23 | 2 | 8 | 7 | 220 | 70 | 544 |
| 1981 | 12 | 1 | 89 | 15 | 22 | 38 | 26 | 42 | 10 | 18 | 5 | 220 | 97 | 595 |
| 1982 | 17 | 8 | 85 | 24 | 32 | 68 | 53 | 85 | 28 | 24 | 14 | 212 | 165 | 815 |
| 1983 | 22 | 2 | 76 | 23 | 35 | 104 | 24 | 49 | 15 | 34 | 8 | 144 | 202 | 739 |
| 1984 | 21 | 1 | 83 | 18 | 39 | 92 | 35 | 37 | 17 | 29 | 14 | 152 | 182 | 722 |
| 1985 | 26 | 3 | 74 | 11 | 30 | 108 | 61 | 44 | 27 | 28 | 14 | 162 | 193 | 781 |
| 1986 | 30 | 3 | 85 | 11 | 39 | 110 | 56 | 58 | 12 | 39 | 8 | 113 | 222 | 785 |
| 1987 | 17 | 2 | 83 | 12 | 42 | 120 | 70 | 39 | 25 | 31 | 6 | 136 | 179 | 764 |
| 1988 | 25 | 3 | 84 | 12 | 29 | 124 | 70 | 26 | 24 | 36 | 9 | 147 | 226 | 815 |
| 1989 | 23 | 5 | 92 | 11 | 37 | 122 | 89 | 36 | 17 | 28 | 9 | 141 | 249 | 859 |
| 1990 | 20 | 2 | 91 | 17 | 48 | 92 | 89 | 54 | 60 | 33 | 11 | 101 | 240 | 857 |
| 1991 | 22 | 13 | 70 | 27 | 55 | 95 | 99 | 72 | 66 | 23 | 13 | 117 | 330 | 1,001 |
| 1992 | 22 | 20 | 64 | 9 | 38 | 100 | 124 | 52 | 80 | 21 | 12 | 95 | 315 | 950 |
| 1993 | 21 | 16 | 72 | 8 | 34 | 105 | 110 | 45 | 74 | 30 | 10 | 108 | 370 | 1,003 |
| 1994 | 26 | 15 | 78 | 11 | 35 | 74 | 124 | 30 | 66 | 30 | 10 | 104 | 338 | 942 |
| 1995 | 21 | 16 | 73 | 11 | 46 | 76 | 125 | 33 | 57 | 38 | 14 | 123 | 317 | 949 |
| 1996 | 27 | 29 | 94 | 18 | 32 | 102 | 143 | 43 | 60 | 34 | 9 | 72 | 318 | 981 |
| 1997 | 21 | 15 | 119 | 11 | 30 | 95 | 207 | 41 | 50 | 42 | 12 | 18 | 340 | 1,003 |
| 1998 | 14 | 18 | 148 | 8 | 30 | 64 | 235 | 33 | 33 | 30 | 11 | 4 | 317 | 945 |
| 1999 | 11 | 27 | 119 | 7 | 25 | 84 | 261 | 38 | 49 | 26 | 9 | 8 | 276 | 940 |
| 2000 | 14 | 28 | 110 | 10 | 34 | 90 | 358 | 42 | 20 | 40 | 10 | 10 | 277 | 1,040 |
| 2001 | 16 | 23 | 112 | 13 | 33 | 62 | 274 | 45 | 14 | 51 | 13 | 4 | 312 | 971 |
| 2002 | 19 | 26 | 106 | 12 | 29 | 74 | 254 | 23 | 11 | 54 | 12 | 9 | 354 | 984 |
| 2003 | 13 | 27 | 141 | 9 | 39 | 69 | 228 | 15 | 10 | 39 | 6 | 9 | 421 | 1,027 |
| 2004 | 20 | 27 | 158 | 18 | 32 | 63 | 209 | 36 | 12 | 42 | 14 | 10 | 408 | 1,048 |
| 2005 | 21 | 39 | 181 | 14 | 28 | 56 | 268 | 25 | 16 | 35 | 21 | 11 | 449 | 1,165 |
| 2006 | 23 | 42 | 159 | 13 | 39 | 58 | 255 | 83 | 21 | 42 | 28 | 10 | 543 | 1,317 |
| 2007 | 13 | 46 | 189 | 24 | 34 | 54 | 279 | 81 | 16 | 48 | 9 | 10 | 629 | 1,433 |
| 2008 | 18 | [R]54 | 264 | [R]27 | 41 | [R]54 | [R]333 | 131 | [R]18 | [R]54 | 17 | 13 | [R]777 | [R]1,802 |
| 2009[P] | 29 | 55 | 223 | 34 | 35 | 59 | 322 | 192 | 23 | 40 | 33 | 20 | 961 | 2,026 |

R=Revised. P=Preliminary. NA=Not available.
Note: Totals may not equal sum of components due to independent rounding.
Web Page: See http://www.eia.gov/oil_gas/petroleum/info_glance/petroleum.html for related information.
Sources: • 1960-1975—Bureau of Mines, Mineral Industry Surveys, *Petroleum Statement, Annual,* annual reports. • 1976-1980—U.S. Energy Information Administration (EIA), Energy Data Reports, *Petroleum Statement, Annual,* annual reports. • 1981-2008—EIA, *Petroleum Supply Annual,* annual reports. • 2009—EIA, *Petroleum Supply Monthly* (February 2010).

## Natural Gas Flow, 2009 (Trillion Cubic Feet)

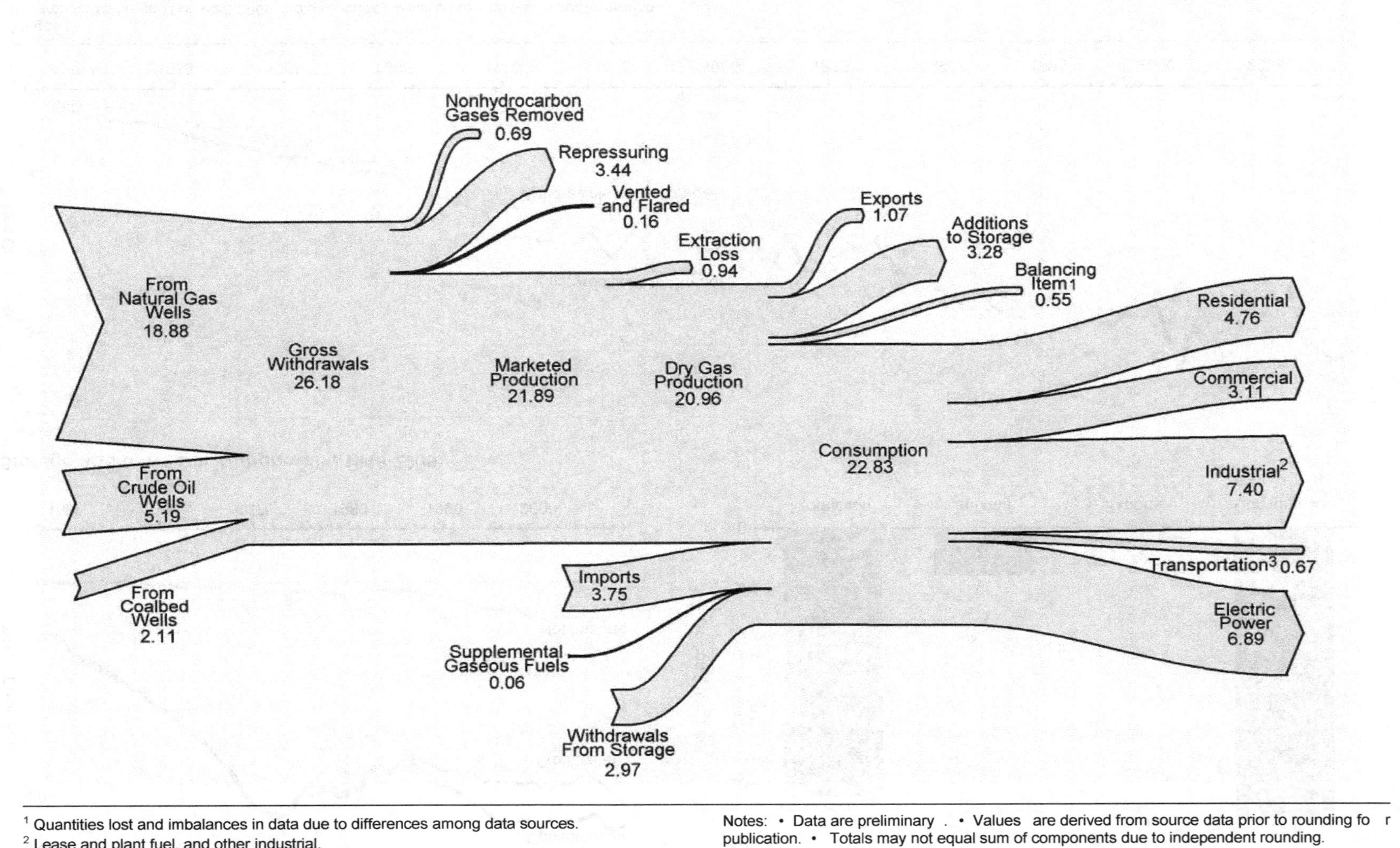


[1] Quantities lost and imbalances in data due to differences among data sources.

[2] Lease and plant fuel, and other industrial.

[3] Natural gas consumed in the operation of pipeli nes (primarily in compressors), and as fuel in the delivery of natural gas to consumers; plus a small quantity used as vehicle fuel.

Notes: • Data are preliminary . • Values are derived from source data prior to rounding fo r publication. • Totals may not equal sum of components due to independent rounding.

## Natural Gas Overview

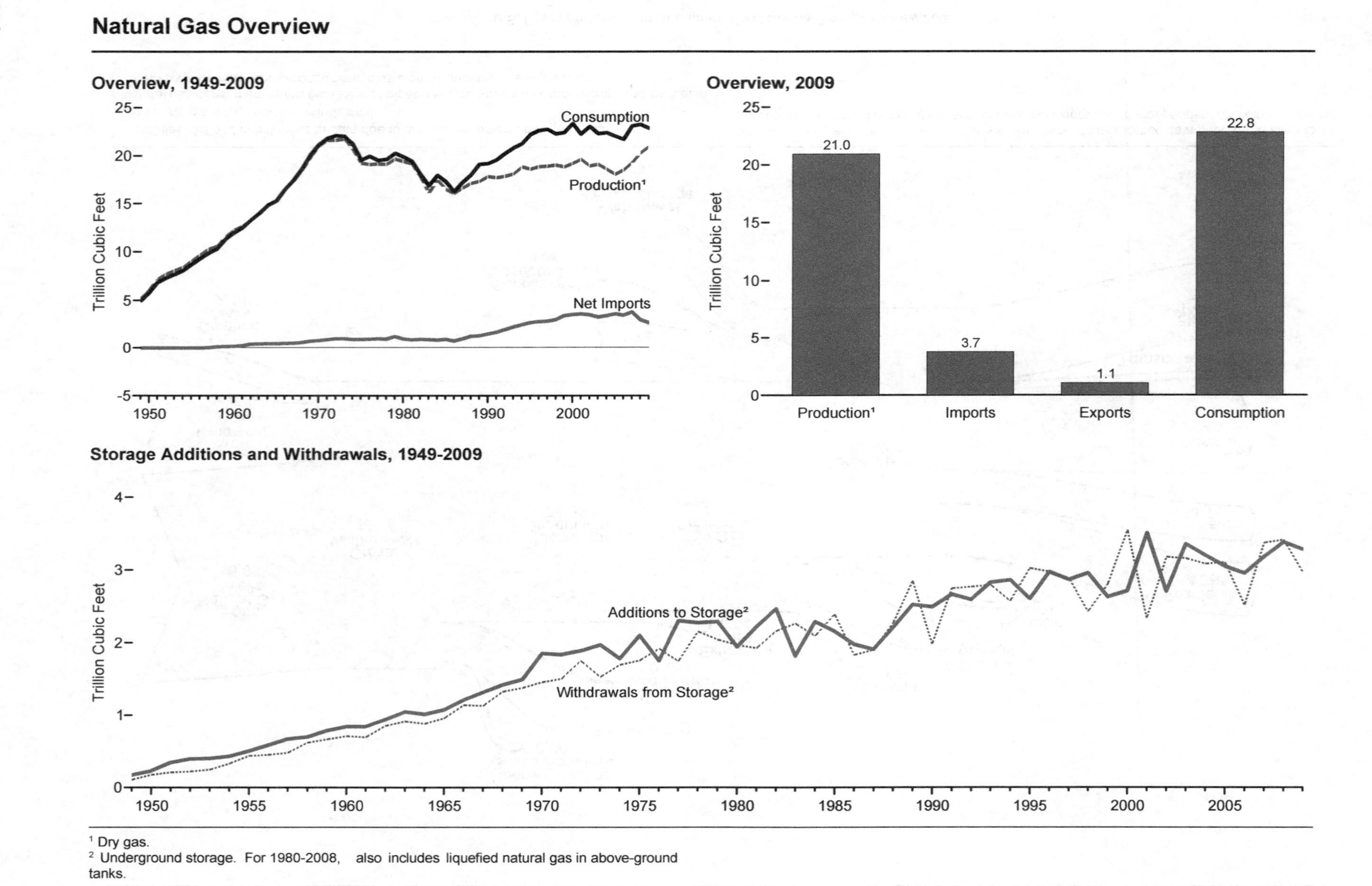


[1] Dry gas.
[2] Underground storage. For 1980-2008, also includes liquefied natural gas in above-ground tanks.

U.S. Energy Information Administration / Annual Energy Review 2009

## Natural Gas Overview, Selected Years, 1949-2009

(Billion Cubic Feet)

| Year | Dry Gas Production | Supplemental Gaseous Fuels [2] | Trade | | | Storage [1] Activity | | | Balancing Item [5] | Consumption [6] |
|---|---|---|---|---|---|---|---|---|---|---|
| | | | Imports | Exports | Net Imports [3] | Withdrawals | Additions | Net Withdrawals [4] | | |
| 1949 | 5,195 | NA | 0 | 20 | -20 | 106 | 172 | -66 | -139 | 4,971 |
| 1950 | 6,022 | NA | 0 | 26 | -26 | 175 | 230 | -54 | -175 | 5,767 |
| 1955 | 9,029 | NA | 11 | 31 | -20 | 437 | 505 | -68 | -247 | 8,694 |
| 1960 | 12,228 | NA | 156 | 11 | 144 | 713 | 844 | -132 | -274 | 11,967 |
| 1965 | 15,286 | NA | 456 | 26 | 430 | 960 | 1,078 | -118 | -319 | 15,280 |
| 1970 | 21,014 | NA | 821 | 70 | 751 | 1,459 | 1,857 | -398 | -228 | 21,139 |
| 1971 | 21,610 | NA | 935 | 80 | 854 | 1,508 | 1,839 | -332 | -339 | 21,793 |
| 1972 | 21,624 | NA | 1,019 | 78 | 941 | 1,757 | 1,893 | -136 | -328 | 22,101 |
| 1973 | 21,731 | NA | 1,033 | 77 | 956 | 1,533 | 1,974 | -442 | -196 | 22,049 |
| 1974 | 20,713 | NA | 959 | 77 | 882 | 1,701 | 1,784 | -84 | -289 | 21,223 |
| 1975 | 19,236 | NA | 953 | 73 | 880 | 1,760 | 2,104 | -344 | -235 | 19,538 |
| 1976 | 19,098 | NA | 964 | 65 | 899 | 1,921 | 1,756 | 165 | -216 | 19,946 |
| 1977 | 19,163 | NA | 1,011 | 56 | 955 | 1,750 | 2,307 | -557 | -41 | 19,521 |
| 1978 | 19,122 | NA | 966 | 53 | 913 | 2,158 | 2,278 | -120 | -287 | 19,627 |
| 1979 | 19,663 | NA | 1,253 | 56 | 1,198 | 2,047 | 2,295 | -248 | -372 | 20,241 |
| 1980 | 19,403 | 155 | 985 | 49 | 936 | 1,972 | 1,949 | 23 | -640 | 19,877 |
| 1981 | 19,181 | 176 | 904 | 59 | 845 | 1,930 | 2,228 | -297 | -500 | 19,404 |
| 1982 | 17,820 | 145 | 933 | 52 | 882 | 2,164 | 2,472 | -308 | -537 | 18,001 |
| 1983 | 16,094 | 132 | 918 | 55 | 864 | 2,270 | 1,822 | 447 | -703 | 16,835 |
| 1984 | 17,466 | 110 | 843 | 55 | 788 | 2,098 | 2,295 | -197 | -217 | 17,951 |
| 1985 | 16,454 | 126 | 950 | 55 | 894 | 2,397 | 2,163 | 235 | -428 | 17,281 |
| 1986 | 16,059 | 113 | 750 | 61 | 689 | 1,837 | 1,984 | -147 | -493 | 16,221 |
| 1987 | 16,621 | 101 | 993 | 54 | 939 | 1,905 | 1,911 | -6 | -444 | 17,211 |
| 1988 | 17,103 | 101 | 1,294 | 74 | 1,220 | 2,270 | 2,211 | 59 | -453 | 18,030 |
| 1989 | 17,311 | 107 | 1,382 | 107 | 1,275 | 2,854 | 2,528 | 326 | 101 | [7]19,119 |
| 1990 | 17,810 | 123 | 1,532 | 86 | 1,447 | 1,986 | 2,499 | -513 | 307 | [7]19,174 |
| 1991 | 17,698 | 113 | 1,773 | 129 | 1,644 | 2,752 | 2,672 | 80 | 27 | [7]19,562 |
| 1992 | 17,840 | 118 | 2,138 | 216 | 1,921 | 2,772 | 2,599 | 173 | 176 | [7]20,228 |
| 1993 | 18,095 | 119 | 2,350 | 140 | 2,210 | 2,799 | 2,835 | -36 | 401 | 20,790 |
| 1994 | 18,821 | 111 | 2,624 | 162 | 2,462 | 2,579 | 2,865 | -286 | 139 | 21,247 |
| 1995 | 18,599 | 110 | 2,841 | 154 | 2,687 | 3,025 | 2,610 | 415 | 396 | 22,207 |
| 1996 | 18,854 | 109 | 2,937 | 153 | 2,784 | 2,981 | 2,979 | 2 | 860 | 22,609 |
| 1997 | 18,902 | 103 | 2,994 | 157 | 2,837 | 2,894 | 2,870 | 24 | 871 | 22,737 |
| 1998 | 19,024 | 102 | 3,152 | 159 | 2,993 | 2,432 | 2,961 | -530 | 657 | 22,246 |
| 1999 | 18,832 | 98 | 3,586 | 163 | 3,422 | 2,808 | 2,636 | 172 | -119 | 22,405 |
| 2000 | 19,182 | 90 | 3,782 | 244 | 3,538 | 3,550 | 2,721 | 829 | [R]-305 | 23,333 |
| 2001 | 19,616 | 86 | 3,977 | 373 | 3,604 | 2,344 | 3,510 | -1,166 | 99 | 22,239 |
| 2002 | 18,928 | 68 | 4,015 | 516 | 3,499 | 3,180 | 2,713 | 467 | 45 | 23,007 |
| 2003 | 19,099 | 68 | 3,944 | 680 | 3,264 | 3,161 | 3,358 | -197 | 44 | 22,277 |
| 2004 | 18,591 | 60 | 4,259 | 854 | 3,404 | 3,088 | 3,202 | -114 | 448 | 22,389 |
| 2005 | 18,051 | 64 | 4,341 | 729 | 3,612 | 3,107 | 3,055 | 52 | 232 | 22,011 |
| 2006 | 18,504 | 66 | 4,186 | 724 | 3,462 | 2,527 | 2,963 | -436 | 89 | 21,685 |
| 2007 | [R]19,266 | 63 | 4,608 | 822 | 3,785 | [R]3,375 | [R]3,183 | [R]192 | [R]-209 | [R]23,097 |
| 2008 | [R]20,286 | [R]61 | [R]3,984 | [R]1,006 | [R]2,979 | [R]3,417 | [R]3,383 | [R]34 | [R]-133 | [R]23,227 |
| 2009 | [E]20,955 | [P]64 | [P]3,748 | [P]1,071 | [P]2,677 | [P]2,968 | [P]3,281 | [P]-313 | [P]-549 | [P]22,834 |

[1] Underground storage. For 1980-2008, also includes liquefied natural gas in above-ground tanks.
[2] See Note 1, "Supplemental Gaseous Fuels," at end of section.
[3] Net imports equal imports minus exports. Minus sign indicates exports are greater than imports.
[4] Net withdrawals equal withdrawals minus additions. Minus sign indicates additions are greater than withdrawals.
[5] Quantities lost and imbalances in data due to differences among data sources. Since 1980, excludes intransit shipments that cross the U.S.-Canada border (i.e., natural gas delivered to its destination via the other country).
[6] See Note 2, "Natural Gas Consumption," at end of section.
[7] For 1989-1992, a small amount of consumption at independent power producers may be counted in both "Other Industrial" and "Electric Power Sector" on Table 6.5. See Note 3, "Natural Gas Consumption, 1989-1992," at end of section.

R=Revised. P=Preliminary. E=Estimate. NA=Not available.

Notes: • Beginning with 1965, all volumes are shown on a pressure base of 14.73 p.s.i.a. at 60 °F. For prior years, the pressure base was 14.65 p.s.i.a. at 60 °F. • Totals may not equal sum of components due to independent rounding.

Web Pages: • For all data beginning in 1949, see http://www.eia.gov/emeu/aer/natgas.html. • For related information, see http://www.eia.gov/oil_gas/natural_gas/info_glance/natural_gas.html.

Sources: **Dry Gas Production:** Table 6.2. **Supplemental Gaseous Fuels:** • 1980-2004—U.S. Energy Information Administration (EIA), *Natural Gas Annual (NGA)*, annual reports. • 2005 forward—EIA, *Natural Gas Monthly (NGM)* (April 2010), Table 1. **Trade:** Table 6.3. **Storage Activity:** • 1949-2008—EIA, *NGA*, annual reports. • 2009—EIA, *NGM* (April 2010), Table 6. **Balancing Item:** Calculated as consumption minus dry gas production, supplemental gaseous fuels, net imports, and net withdrawals. **Consumption:** Table 6.5.

## Coal Flow, 2009 (Million Short Tons)

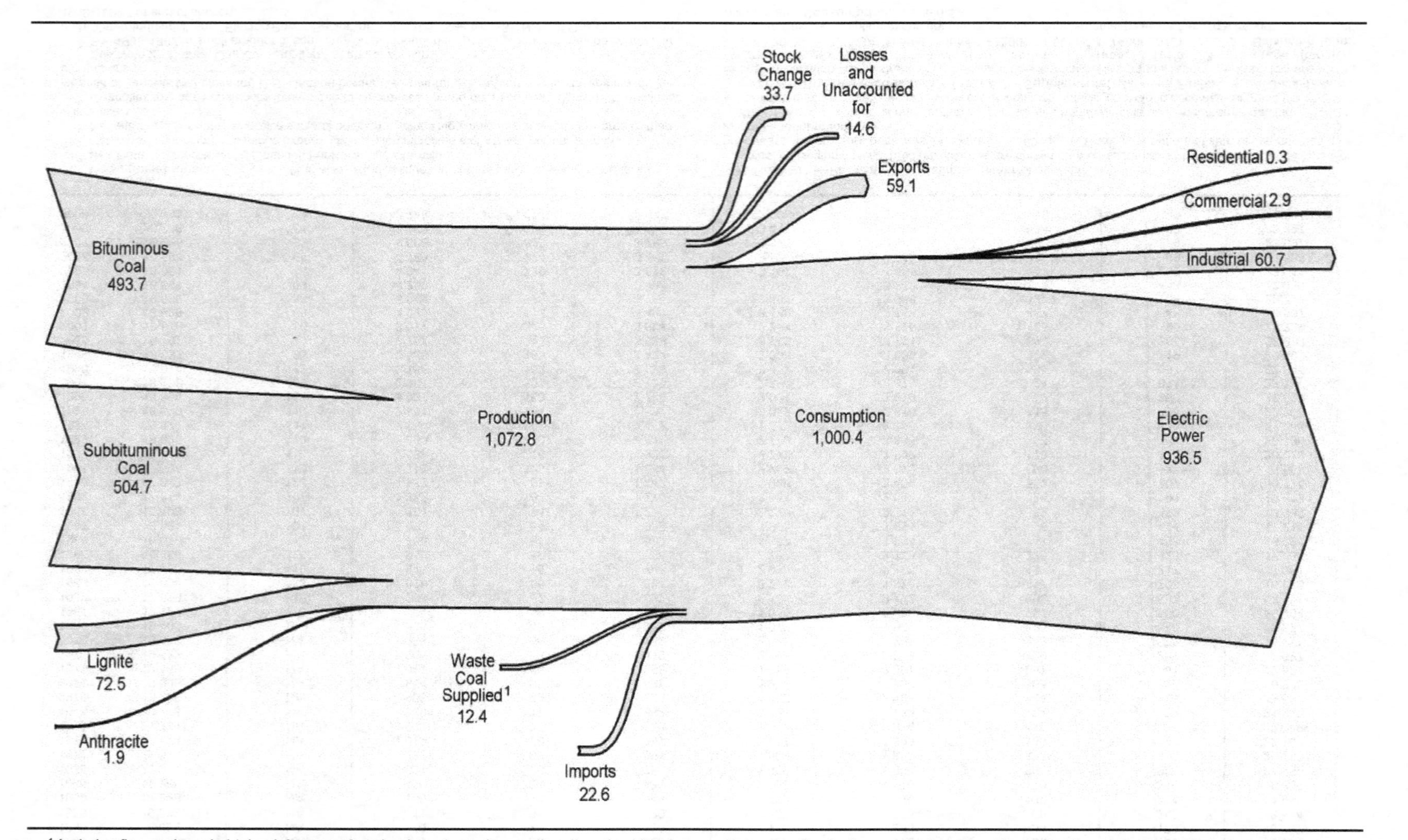


[1] Includes fine coal, coal obtained from a refuse bank or slurry dam, anthracite culm, bituminous gob, and lignite waste that are consumed by the electric power and industrial sectors.

Notes: • Production categories are estimated; other data are preliminary. • Values are derived from source data prior to rounding for publication. • Totals may not equal sum of components due to independent rounding.

U.S. Energy Information Administration / Annual Energy Review 2009

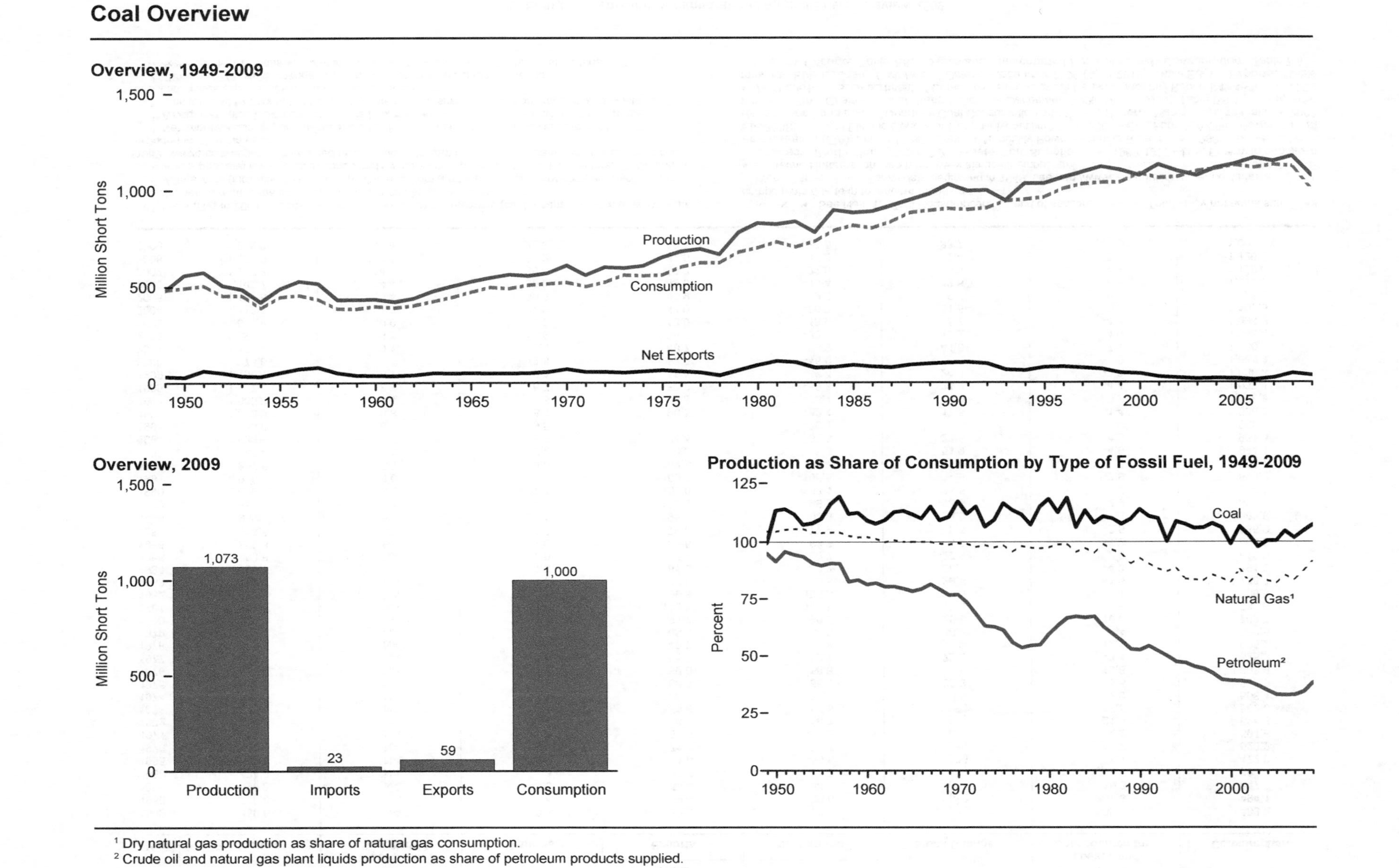
Coal Overview
Overview, 1949-2009
Million Short Tons
1,500
1,000
500
0
Production
Consumption
Net Exports
1950
1955
1960
1965
1970
1975
1980
1985
1990
1995
2000
2005
Overview, 2009
1,500
1,000
500
0
1,073
23
59
1,000
Production
Imports
Exports
Consumption
Production as Share of Consumption by Type of Fossil Fuel, 1949-2009
Percent
125
100
75
50
25
0
Coal
Natural Gas¹
Petroleum²
1950
1960
1970
1980
1990
2000
¹ Dry natural gas production as share of natural gas consumption.
² Crude oil and natural gas plant liquids production as share of petroleum products supplied.
U.S. Energy Information Administration / Annual Energy Review 2009

## Coal Overview, Selected Years, 1949-2009

(Million Short Tons)

| Year | Production [1] | Waste Coal Supplied [2] | Trade | | | Stock Change [4] | Losses and Unaccounted for [5] | Consumption |
|---|---|---|---|---|---|---|---|---|
| | | | Imports | Exports | Net Imports [3] | | | |
| 1949 | 480.6 | NA | 0.3 | 32.8 | -32.5 | ([6]) | [6]-35.1 | 483.2 |
| 1950 | 560.4 | NA | .4 | 29.4 | -29.0 | ([6]) | [6]37.3 | 494.1 |
| 1955 | 490.8 | NA | .3 | 54.4 | -54.1 | ([6]) | [6]-10.3 | 447.0 |
| 1960 | 434.3 | NA | .3 | 38.0 | -37.7 | ([6]) | [6]-1.5 | 398.1 |
| 1965 | 527.0 | NA | .2 | 51.0 | -50.8 | ([6]) | [6]4.1 | 472.0 |
| 1970 | 612.7 | NA | (s) | 71.7 | -71.7 | ([6]) | [6]17.7 | 523.2 |
| 1971 | 560.9 | NA | .1 | 57.3 | -57.2 | ([6]) | [6]2.2 | 501.6 |
| 1972 | 602.5 | NA | (s) | 56.7 | -56.7 | ([6]) | [6]21.5 | 524.3 |
| 1973 | 598.6 | NA | .1 | 53.6 | -53.5 | ([6]) | [6]-17.5 | 562.6 |
| 1974 | 610.0 | NA | 2.1 | 60.7 | -58.6 | -8.9 | 2.0 | 558.4 |
| 1975 | 654.6 | NA | .9 | 66.3 | -65.4 | 32.2 | -5.5 | 562.6 |
| 1976 | 684.9 | NA | 1.2 | 60.0 | -58.8 | 8.5 | 13.8 | 603.8 |
| 1977 | 697.2 | NA | 1.6 | 54.3 | -52.7 | 22.6 | -3.4 | 625.3 |
| 1978 | 670.2 | NA | 3.0 | 40.7 | -37.8 | -4.9 | 12.1 | 625.2 |
| 1979 | 781.1 | NA | 2.1 | 66.0 | -64.0 | 36.2 | .4 | 680.5 |
| 1980 | 829.7 | NA | 1.2 | 91.7 | -90.5 | 25.6 | 10.8 | 702.7 |
| 1981 | 823.8 | NA | 1.0 | 112.5 | -111.5 | -19.0 | -1.4 | 732.6 |
| 1982 | 838.1 | NA | .7 | 106.3 | -105.5 | 22.6 | 3.1 | 706.9 |
| 1983 | 782.1 | NA | 1.3 | 77.8 | -76.5 | -29.5 | -1.6 | 736.7 |
| 1984 | 895.9 | NA | 1.3 | 81.5 | -80.2 | 28.7 | -4.3 | 791.3 |
| 1985 | 883.6 | NA | 2.0 | 92.7 | -90.7 | -27.9 | 2.8 | 818.0 |
| 1986 | 890.3 | NA | 2.2 | 85.5 | -83.3 | 4.0 | -1.2 | 804.2 |
| 1987 | 918.8 | NA | 1.7 | 79.6 | -77.9 | 6.5 | -2.5 | 836.9 |
| 1988 | 950.3 | NA | 2.1 | 95.0 | -92.9 | -24.9 | -1.3 | 883.6 |
| 1989 | 980.7 | 1.4 | 2.9 | 100.8 | -98.0 | -13.7 | 2.9 | 895.0 |
| 1990 | 1,029.1 | 3.3 | 2.7 | 105.8 | -103.1 | 26.5 | -1.7 | 904.5 |
| 1991 | 996.0 | 4.0 | 3.4 | 109.0 | -105.6 | -.9 | -3.9 | 899.2 |
| 1992 | 997.5 | 6.3 | 3.8 | 102.5 | -98.7 | -3.0 | .5 | 907.7 |
| 1993 | 945.4 | 8.1 | 8.2 | 74.5 | -66.3 | -51.9 | -4.9 | 944.1 |
| 1994 | 1,033.5 | 8.2 | 8.9 | 71.4 | -62.5 | 23.6 | 4.3 | 951.3 |
| 1995 | 1,033.0 | 8.6 | 9.5 | 88.5 | -79.1 | -.3 | .6 | 962.1 |
| 1996 | 1,063.9 | 8.8 | 8.1 | 90.5 | -82.4 | -17.5 | 1.4 | 1,006.3 |
| 1997 | 1,089.9 | 8.1 | 7.5 | 83.5 | -76.1 | -11.3 | 3.7 | 1,029.5 |
| 1998 | 1,117.5 | 8.7 | 8.7 | 78.0 | -69.3 | 24.2 | -4.4 | 1,037.1 |
| 1999 | 1,100.4 | 8.7 | 9.1 | 58.5 | -49.4 | 24.0 | -2.9 | 1,038.6 |
| 2000 | 1,073.6 | 9.1 | 12.5 | 58.5 | -46.0 | -48.3 | .9 | 1,084.1 |
| 2001 | [1]1,127.7 | 10.1 | 19.8 | 48.7 | -28.9 | 41.6 | 7.1 | 1,060.1 |
| 2002 | 1,094.3 | 9.1 | 16.9 | 39.6 | -22.7 | 10.2 | 4.0 | 1,066.4 |
| 2003 | 1,071.8 | 10.0 | 25.0 | 43.0 | -18.0 | -26.7 | -4.4 | 1,094.9 |
| 2004 | 1,112.1 | 11.3 | 27.3 | 48.0 | -20.7 | -11.5 | 6.9 | 1,107.3 |
| 2005 | 1,131.5 | 13.4 | 30.5 | 49.9 | -19.5 | -9.7 | 9.1 | 1,126.0 |
| 2006 | 1,162.7 | 14.4 | 36.2 | 49.6 | -13.4 | 42.6 | 8.8 | 1,112.3 |
| 2007 | 1,146.6 | 14.1 | 36.3 | 59.2 | -22.8 | 5.8 | 4.1 | 1,128.0 |
| 2008 | R1,171.8 | R14.1 | 34.2 | 81.5 | -47.3 | R12.4 | R5.7 | R1,120.5 |
| 2009P | 1,072.8 | 12.4 | 22.6 | 59.1 | -36.5 | 33.7 | 14.6 | 1,000.4 |

[1] Beginning in 2001, includes a small amount of refuse recovery (coal recaptured from a refuse mine, and cleaned to reduce the concentration of noncombustible materials).

[2] Waste coal (including fine coal, coal obtained from a refuse bank or slurry dam, anthracite culm, bituminous gob, and lignite waste) consumed by the electric power and industrial sectors. Beginning in 1989, waste coal supplied is counted as a supply-side item to balance the same amount of waste coal included in "Consumption."

[3] Net imports equal imports minus exports. Minus sign indicates exports are greater than imports.

[4] A negative value indicates a decrease in stocks and a positive value indicates an increase.

[5] "Losses and Unaccounted for" is calculated as the sum of production, imports, and waste coal supplied, minus exports, stock change, and consumption.

[6] Through 1973, stock change is included in "Losses and Unaccounted for."

R=Revised. P=Preliminary. NA=Not available. (s)=Less than 0.05 million short tons.

Notes: • See Note 1, "Coal Consumption," at end of section. • Totals may not equal sum of components due to independent rounding.

Web Pages: • For all data beginning in 1949, see http://www.eia.gov/emeu/aer/coal.htm l. • For related information, see http://www.eia.gov/fuelcoal.html.

Sources: **Production:** Table 7.2. **Waste Coal Supplied:** • 1989-1997—U.S. Energy Informatio n Administration (EIA), Form EIA-867, "Annual Nonutility Power Producer Report." • 1998-2000—EIA, Form EIA-860B, "Annual Electric Generator Report—Nonutility." • 2001—EIA, Form EIA-906, "Power Pla nt Report," and Form EIA-3, "Quarterly Coal Consumption and Quality Report—Manufacturing Plants. • 2002 forward—EIA, *Quarterly Coal Report October-December 2009* (April 2010), Table ES-1. **Imports:** • 1949-2001—U.S. Department of Commerce, Bureau of the Census, "Monthly Report IM 145." • 2002 forward—EIA, *Quarterly Coal Report October-December 2009* (April 2010), Table ES-1. **Exports:** Table 7.4. **Stock Change:** Table 7.5. **Losses and Unaccounted for:** Calculated. **Consumption:** Table 7.3.

## Electricity Flow, 2009 (Quadrillion Btu)

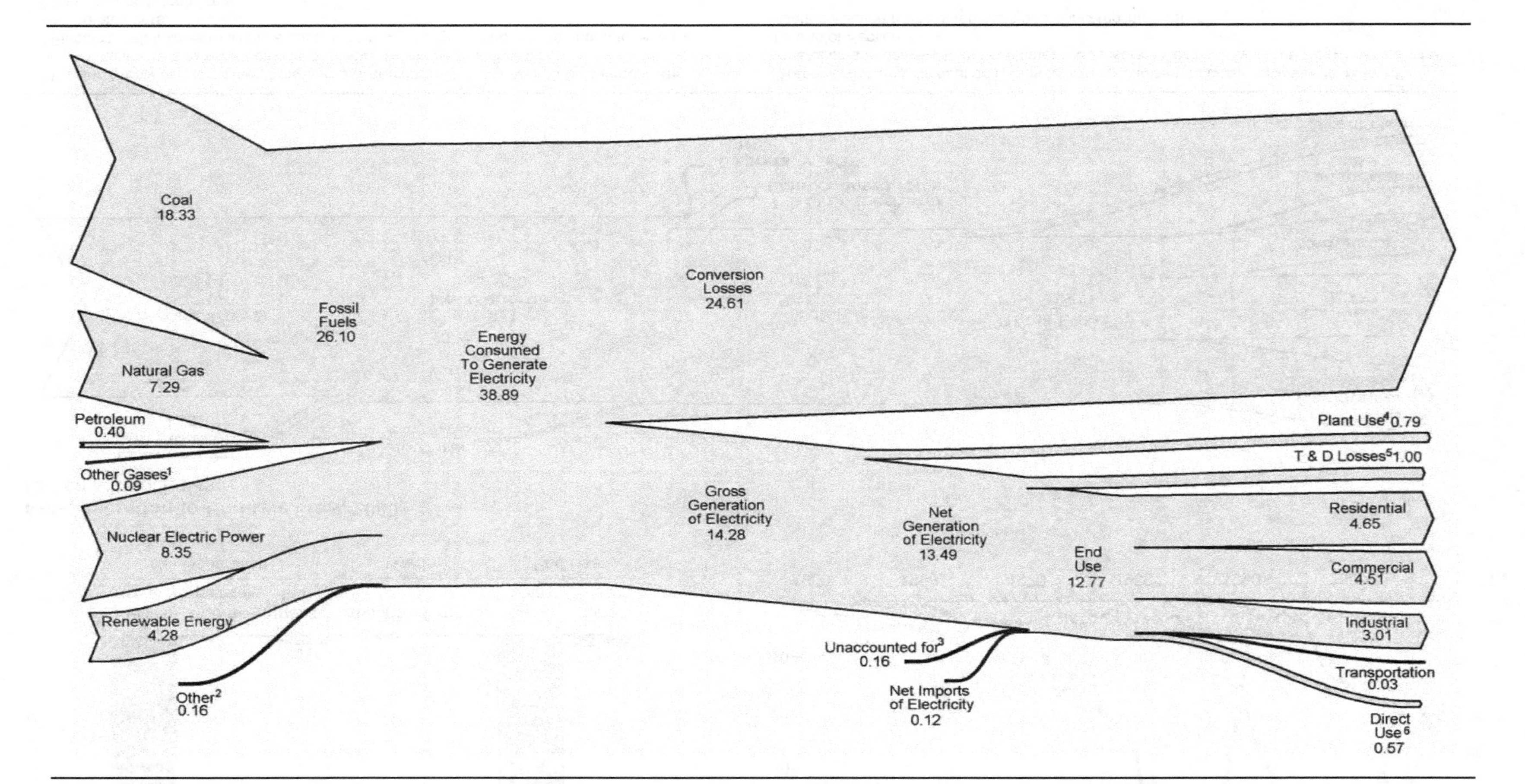


[1] Blast furnace gas, propane gas, and other manufactured and waste gases derived from fossil fuels.

[2] Batteries, chemicals, hydrogen, pitch, purchased steam, sulfur, mi scellaneous technologies, and non-renewable waste (municipal solid w aste from non-biogenic sources, and tire-derived fuels).

[3] Data collection frame differences and nonsam pling error. Derived for the diagram by subtracting the "T & D Losses" estimate from "T & D Losses and Unaccounted for" derived from Table 8.1.

[4] Electric energy used in the operation of power plants.

[5] Transmission and distribution losses (electricity losses that occur between the point of generation and delivery to the customer) are estimated as 7 percent of gross generation.

[6] Use of electricity that is 1) self-generated, 2) produced by either the same entity that consumes the power or an affiliate, and 3) used in direct support of a service or industrial process located within the same facility or group of facilities that house the generating equipment. Direct use is exclusive of station use.

Notes: • Data are preliminary. • See Note, "Electrical System Energy Losses," at the end of Section 2. • Net generation of electricity includes pumped storage facility production minus energy used for pumping. • Values are derived from source data prior to rounding for publication. • Totals may not equal sum of components due to independent rounding.

Sources: T ables 8.1, 8.4a, 8.9, A6 (column 4), and U.S. Energy Information Administration, Form EIA-923, "Power Plant Operations Report."

# Electricity Overview

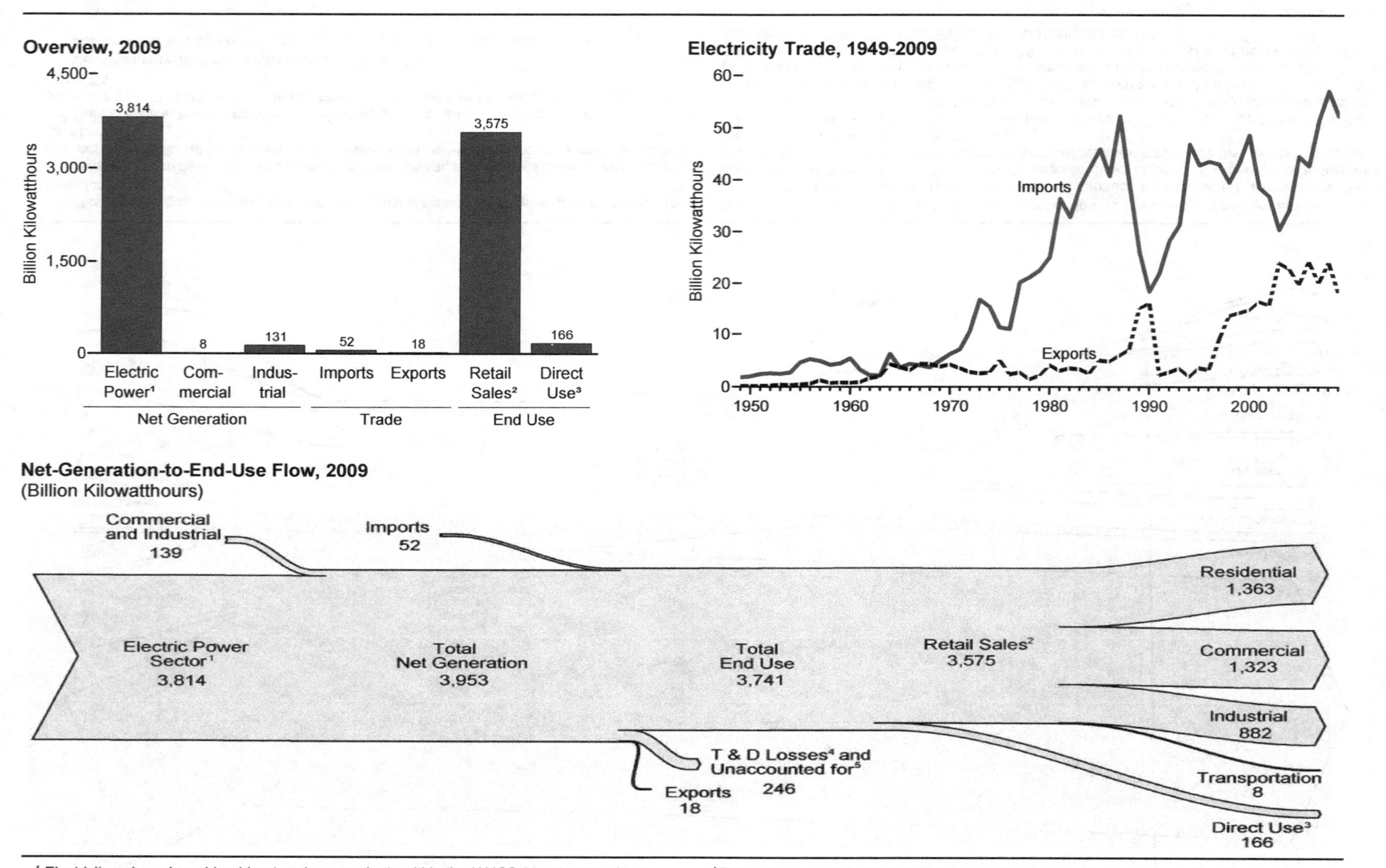


[1] Electricity-only and combined-heat-and-power plants within the NAICS 22 category whose primary business is to sell electricity, or electricity and heat, to the public.
[2] Electricity retail sales to ultimate customers reported by electric utilities and other energy service providers.
[3] See Table 8.1, footnote 8.
[4] Transmission and distribution losses (electricity losses that occur between the point of generation and delivery to the customer). See Note, "Electrical System Energy Losses," at the end of Section 2.
[5] Data collection frame differences and nonsampling error.

## Electricity Overview, Selected Years, 1949-2009

(Billion Kilowatthours)

| Year | Net Generation | | | | Trade | | | | | T & D Losses [5] and Unaccounted for [6] | End Use | | |
|---|---|---|---|---|---|---|---|---|---|---|---|---|---|
| | Electric Power Sector [2] | Commercial Sector [3] | Industrial Sector [4] | Total | Imports [1] From Canada | Imports [1] Total | Exports [1] To Canada | Exports [1] Total | Net Imports [1] Total | | Retail Sales [7] | Direct Use [8] | Total |
| 1949 | 291 | NA | 5 | 296 | NA | 2 | NA | (s) | 2 | 43 | 255 | NA | 255 |
| 1950 | 329 | NA | 5 | 334 | NA | 2 | NA | (s) | 2 | 44 | 291 | NA | 291 |
| 1955 | 547 | NA | 3 | 550 | NA | 5 | NA | (s) | 4 | 58 | 497 | NA | 497 |
| 1960 | 756 | NA | 4 | 759 | NA | 5 | NA | 1 | 5 | 76 | 688 | NA | 688 |
| 1965 | 1,055 | NA | 3 | 1,058 | NA | 4 | NA | 4 | (s) | 104 | 954 | NA | 954 |
| 1970 | 1,532 | NA | 3 | 1,535 | NA | 6 | NA | 4 | 2 | 145 | 1,392 | NA | 1,392 |
| 1971 | 1,613 | NA | 3 | 1,616 | NA | 7 | NA | 4 | 4 | 150 | 1,470 | NA | 1,470 |
| 1972 | 1,750 | NA | 3 | 1,753 | NA | 10 | NA | 3 | 8 | 166 | 1,595 | NA | 1,595 |
| 1973 | 1,861 | NA | 3 | 1,864 | NA | 17 | NA | 3 | 14 | 165 | 1,713 | NA | 1,713 |
| 1974 | 1,867 | NA | 3 | 1,870 | NA | 15 | NA | 3 | 13 | 177 | 1,706 | NA | 1,706 |
| 1975 | 1,918 | NA | 3 | 1,921 | NA | 11 | NA | 5 | 6 | 180 | 1,747 | NA | 1,747 |
| 1976 | 2,038 | NA | 3 | 2,041 | NA | 11 | NA | 2 | 9 | 194 | 1,855 | NA | 1,855 |
| 1977 | 2,124 | NA | 3 | 2,127 | NA | 20 | NA | 3 | 17 | 197 | 1,948 | NA | 1,948 |
| 1978 | 2,206 | NA | 3 | 2,209 | NA | 21 | NA | 1 | 20 | 211 | 2,018 | NA | 2,018 |
| 1979 | 2,247 | NA | 3 | 2,251 | NA | 23 | NA | 2 | 20 | 200 | 2,071 | NA | 2,071 |
| 1980 | 2,286 | NA | 3 | 2,290 | NA | 25 | NA | 4 | 21 | 216 | 2,094 | NA | 2,094 |
| 1981 | 2,295 | NA | 3 | 2,298 | NA | 36 | NA | 3 | 33 | 184 | 2,147 | NA | 2,147 |
| 1982 | 2,241 | NA | 3 | 2,244 | NA | 33 | NA | 4 | 29 | 187 | 2,086 | NA | 2,086 |
| 1983 | 2,310 | NA | 3 | 2,313 | NA | 39 | NA | 3 | 35 | 198 | 2,151 | NA | 2,151 |
| 1984 | 2,416 | NA | 3 | 2,419 | NA | 42 | NA | 3 | 40 | 173 | 2,286 | NA | 2,286 |
| 1985 | 2,470 | NA | 3 | 2,473 | NA | 46 | NA | 5 | 41 | 190 | 2,324 | NA | 2,324 |
| 1986 | 2,487 | NA | 3 | 2,490 | NA | 41 | NA | 5 | 36 | 158 | 2,369 | NA | 2,369 |
| 1987 | 2,572 | NA | 3 | 2,575 | NA | 52 | NA | 6 | 46 | 164 | 2,457 | NA | 2,457 |
| 1988 | 2,704 | NA | 3 | 2,707 | NA | 39 | NA | 7 | 32 | 161 | 2,578 | NA | 2,578 |
| 1989 | [2]2,848 | 4 | [4]115 | 2,967 | NA | 26 | NA | 15 | 11 | 222 | 2,647 | 109 | 2,756 |
| 1990 | 2,901 | 6 | 131 | 3,038 | 16 | 18 | 16 | 16 | 2 | 203 | 2,713 | 125 | 2,837 |
| 1991 | 2,936 | 6 | 133 | 3,074 | 20 | 22 | 2 | 2 | 20 | 207 | 2,762 | 124 | 2,886 |
| 1992 | 2,934 | 6 | 143 | 3,084 | 26 | 28 | 2 | 3 | 25 | 212 | 2,763 | 134 | 2,897 |
| 1993 | 3,044 | 7 | 146 | 3,197 | 29 | 31 | 3 | 4 | 28 | 224 | 2,861 | 139 | 3,001 |
| 1994 | 3,089 | 8 | 151 | 3,248 | 45 | 47 | 1 | 2 | 45 | 211 | 2,935 | 146 | 3,081 |
| 1995 | 3,194 | 8 | 151 | 3,353 | 41 | 43 | 2 | 4 | 39 | 229 | 3,013 | 151 | 3,164 |
| 1996 | 3,284 | 9 | 151 | 3,444 | 42 | 43 | 2 | 3 | 40 | 231 | 3,101 | 153 | 3,254 |
| 1997 | 3,329 | 9 | 154 | 3,492 | 43 | 43 | 7 | 9 | 34 | 224 | 3,146 | 156 | 3,302 |
| 1998 | 3,457 | 9 | 154 | 3,620 | 40 | 40 | 12 | 14 | 26 | 221 | 3,264 | 161 | 3,425 |
| 1999 | 3,530 | 9 | 156 | 3,695 | 43 | 43 | 13 | 14 | 29 | 240 | 3,312 | 172 | 3,484 |
| 2000 | 3,638 | 8 | 157 | 3,802 | 49 | 49 | 13 | 15 | 34 | 244 | 3,421 | 171 | 3,592 |
| 2001 | 3,580 | 7 | 149 | 3,737 | 38 | 39 | 16 | 16 | 22 | 202 | 3,394 | 163 | 3,557 |
| 2002 | 3,698 | 7 | 153 | 3,858 | 37 | 37 | 15 | 16 | 21 | 248 | 3,465 | 166 | 3,632 |
| 2003 | 3,721 | 7 | 155 | 3,883 | 29 | 30 | 24 | 24 | 6 | 228 | 3,494 | 168 | 3,662 |
| 2004 | 3,808 | 8 | 154 | 3,971 | 33 | 34 | 22 | 23 | 11 | 266 | 3,547 | 168 | 3,716 |
| 2005 | 3,902 | 8 | 145 | 4,055 | 43 | 45 | 19 | 20 | 25 | 269 | 3,661 | 150 | 3,811 |
| 2006 | 3,908 | 8 | 148 | 4,065 | 42 | 43 | 23 | 24 | 18 | 266 | 3,670 | 147 | 3,817 |
| 2007 | 4,005 | 8 | 143 | 4,157 | 50 | 51 | 20 | 20 | 31 | 264 | 3,765 | 159 | 3,924 |
| 2008 | [R]3,974 | 8 | [R]137 | [R]4,119 | 56 | 57 | 23 | 24 | 33 | [R]246 | [R]3,733 | [R]173 | [R]3,906 |
| 2009[P] | 3,814 | 8 | 131 | 3,953 | 51 | 52 | 17 | 18 | 34 | 246 | 3,575 | [E]166 | 3,741 |

[1] Electricity transmitted across U.S. borders. Net imports equal imports minus exports.

[2] Electricity-only and combined-heat-and-power (CHP) plants within the NAICS 22 category whose primary business is to sell electricity, or electricity and heat, to the public. Through 1988, data are for electric utilities only; beginning in 1989, data are for electric utilities and independent power producers.

[3] Commercial combined-heat-and-power (CHP) and commercial electricity-only plants.

[4] Industrial combined-heat-and-power (CHP) and industrial electricity-only plants. Through 1988, data are for industrial hydroelectric power only.

[5] Transmission and distribution losses (electricity losses that occur between the point of generation and delivery to the customer). See Note, "Electrical System Energy Losses," at end of Section 2.

[6] Data collection frame differences and nonsampling error.

[7] Electricity retail sales to ultimate customers by electric utilities and, beginning in 1996, other energy service providers.

[8] Use of electricity that is 1) self-generated, 2) produced by either the same entity that consumes the power or an affiliate, and 3) used in direct support of a service or industrial process located within the same facility or group of facilities that house the generating equipment. Direct use is exclusive of station use.

R=Revised. P=Preliminary. E=Estimate. NA=Not available. (s)=Less than 0.5 billion kilowatthours.

Notes: • See Note 1, "Coverage of Electricity Statistics," and Note 2, "Classification of Power Plants Into Energy-Use Sectors," at end of section. • Totals may not equal sum of components due to independent rounding.

Web Pages: • For all data beginning in 1949, see http://www.eia.gov/emeu/aer/elect.html.
• For related information, see http://www.eia.gov/fuelelectric.html.

U.S. Energy Information Administration / Annual Energy Review 2009

# Nuclear Generating Units

## Operable Units,[1] 1957-2009

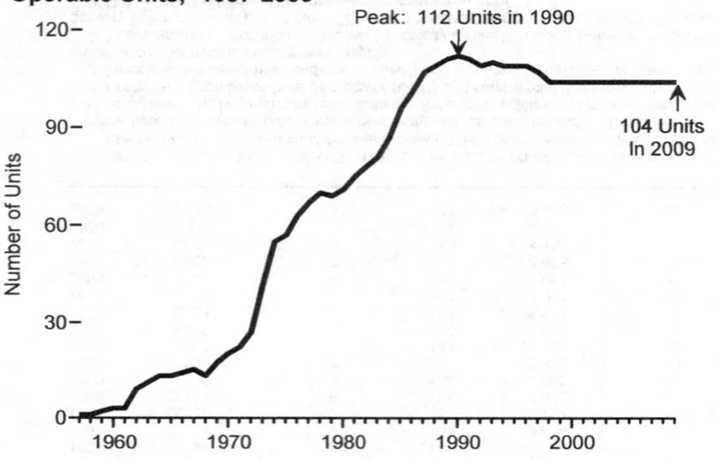

Peak: 112 Units in 1990

104 Units In 2009

Number of Units

## Nuclear Net Summer Capacity Change, 1950-2009

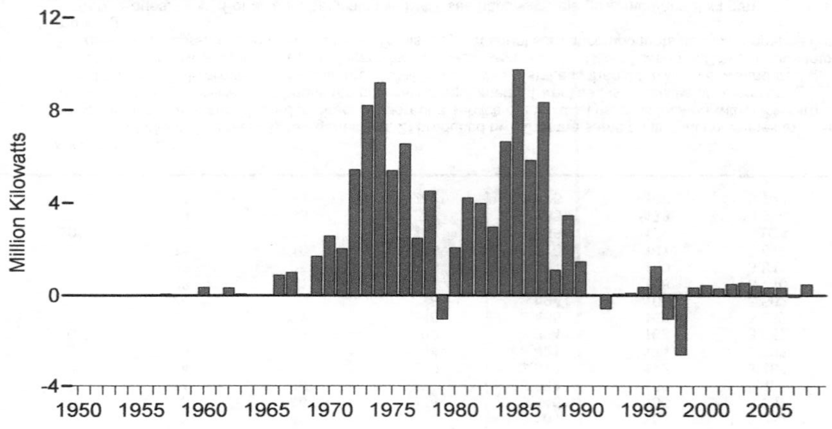

Million Kilowatts

## Status of All Nuclear Generating Units, 2009

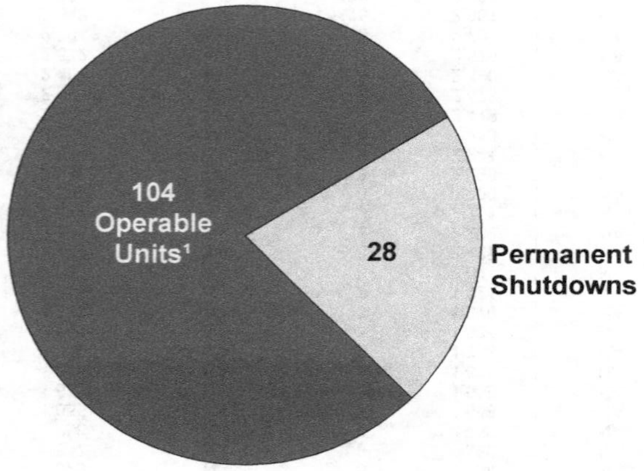

104 Operable Units[1]

28 Permanent Shutdowns

## Permanent Shutdowns by Year, 1955-2009

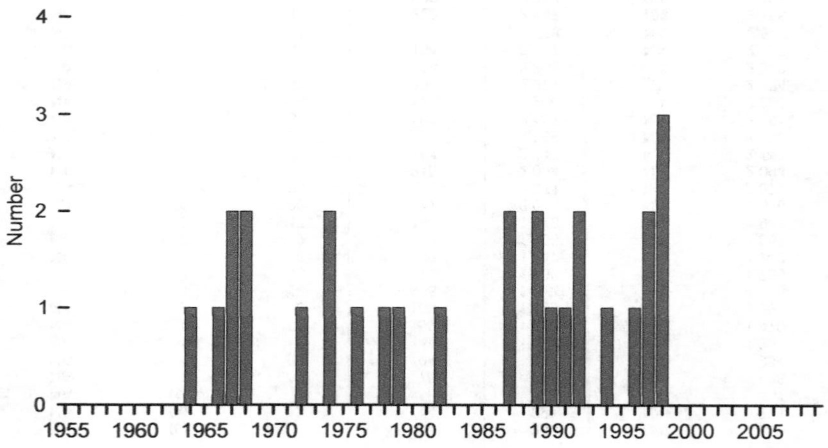

Number

[1] Units holding full-power operating licenses, or equivalent permission to operate, at the end of the year.

Note: Data are at end of year.

U.S. Energy Information Administration / Annual Energy Review 2009

## Nuclear Generating Units, 1955-2009

| Year | Original Licensing Regulations (10 CFR Part 50) [1] | | | Current Licensing Regulations (10 CFR Part 52) [1] | | | Permanent Shutdowns | Operable Units [6] |
|---|---|---|---|---|---|---|---|---|
| | Construction Permits Issued [2,3] | Low-Power Operating Licenses Issued [3,4] | Full-Power Operating Licenses Issued [3,5] | Early Site Permits Issued [3] | Combined License Applications Received | Combined Licenses Issued [3] | | |
| 1955 | 1 | 0 | 0 | – – | – – | – – | 0 | 0 |
| 1956 | 3 | 0 | 0 | – – | – – | – – | 0 | 0 |
| 1957 | 1 | 1 | 1 | – – | – – | – – | 0 | 1 |
| 1958 | 0 | 0 | 0 | – – | – – | – – | 0 | 1 |
| 1959 | 3 | 1 | 1 | – – | – – | – – | 0 | 2 |
| 1960 | 7 | 1 | 1 | – – | – – | – – | 0 | 3 |
| 1961 | 0 | 0 | 0 | – – | – – | – – | 0 | 3 |
| 1962 | 1 | 7 | 6 | – – | – – | – – | 0 | 9 |
| 1963 | 1 | 3 | 2 | – – | – – | – – | 0 | 11 |
| 1964 | 3 | 2 | 3 | – – | – – | – – | 1 | 13 |
| 1965 | 1 | 0 | 0 | – – | – – | – – | 0 | 13 |
| 1966 | 5 | 1 | 2 | – – | – – | – – | 1 | 14 |
| 1967 | 14 | 3 | 3 | – – | – – | – – | 2 | 15 |
| 1968 | 23 | 0 | 0 | – – | – – | – – | 2 | 13 |
| 1969 | 7 | 4 | 4 | – – | – – | – – | 0 | 17 |
| 1970 | 10 | 4 | 3 | – – | – – | – – | 0 | 20 |
| 1971 | 4 | 5 | 2 | – – | – – | – – | 0 | 22 |
| 1972 | 8 | 6 | 6 | – – | – – | – – | 1 | 27 |
| 1973 | 14 | 12 | 15 | – – | – – | – – | 0 | 42 |
| 1974 | 23 | 14 | 15 | – – | – – | – – | 2 | 55 |
| 1975 | 9 | 3 | 2 | – – | – – | – – | 0 | 57 |
| 1976 | 9 | 7 | 7 | – – | – – | – – | 1 | 63 |
| 1977 | 15 | 4 | 4 | – – | – – | – – | 0 | 67 |
| 1978 | 13 | 3 | 4 | – – | – – | – – | 1 | 70 |
| 1979 | 2 | 0 | 0 | – – | – – | – – | 1 | 69 |
| 1980 | 0 | 5 | 2 | – – | – – | – – | 0 | 71 |
| 1981 | 0 | 3 | 4 | – – | – – | – – | 0 | 75 |
| 1982 | 0 | 6 | 4 | – – | – – | – – | 1 | 78 |
| 1983 | 0 | 3 | 3 | – – | – – | – – | 0 | 81 |
| 1984 | 0 | 7 | 6 | – – | – – | – – | 0 | 87 |
| 1985 | 0 | 7 | 9 | – – | – – | – – | 0 | 96 |
| 1986 | 0 | 7 | 5 | – – | – – | – – | 0 | 101 |
| 1987 | 0 | 6 | 8 | – – | – – | – – | 2 | 107 |
| 1988 | 0 | 1 | 2 | – – | – – | – – | 0 | 109 |
| 1989 | 0 | 3 | 4 | – – | – – | – – | 2 | 111 |
| 1990 | 0 | 1 | 2 | – – | – – | – – | 1 | 112 |
| 1991 | 0 | 0 | 0 | – – | – – | – – | 1 | 111 |
| 1992 | 0 | 0 | 0 | – – | – – | – – | 2 | 109 |
| 1993 | 0 | 1 | 1 | – – | – – | – – | 0 | 110 |
| 1994 | 0 | 0 | 0 | – – | – – | – – | 1 | 109 |
| 1995 | 0 | 1 | 0 | – – | – – | – – | 0 | 109 |
| 1996 | 0 | 0 | 1 | – – | – – | – – | 1 | 109 |
| 1997 | 0 | 0 | 0 | 0 | 0 | 0 | 2 | 107 |
| 1998 | 0 | 0 | 0 | 0 | 0 | 0 | 3 | 104 |
| 1999-2006 | 0 | 0 | 0 | 0 | 0 | 0 | 0 | 104 |
| 2007 | 0 | 0 | 0 | 3 | 5 | 0 | 0 | 104 |
| 2008 | 0 | 0 | 0 | 0 | 11 | 0 | 0 | 104 |
| 2009 | 0 | 0 | 0 | 1 | 1 | 0 | 0 | 104 |
| Total | 177 | 132 | 132 | 4 | 17 | 0 | 28 | – – |

[1] Data in columns 1-3 are based on the U.S. Nuclear Regulatory Commission (NRC) regulation 10 CFR Part 50. Data in columns 4-6 are based on the NRC regulation 10 CFR Part 52. See Note 1, "Pending Actions on Nuclear Generating Units," at end of section.

[2] Issuance by regulatory authority of a permit, or equivalent permission, to begin construction. Under current licensing regulations, the construction permit is no longer issued separately from the operating license.

[3] Numbers reflect permits or licenses issued in a given year, not extant permits or licenses.

[4] Issuance by regulatory authority of license, or equivalent permission, to conduct testing but not to operate at full power.

[5] Issuance by regulatory authority of full-power operating license, or equivalent permission (note that some units receive full-power licenses the same year they receive low-power licenses). Units initially undergo low-power testing prior to commercial operation.

[6] Total of nuclear generating units holding full-power licenses, or equivalent permission to operate, at the end of the year (the number of operable units equals the cumulative number of units holding full-power licenses minus the cumulative number of permanent shutdowns).

– – = Not applicable.

Note: See Note 2, "Coverage of Nuclear Energy Statistics," at end of section.

Web Page: For related information, see http://www.eia.gov/fuelnuclear.html.

Nuclear Power Plant Operations

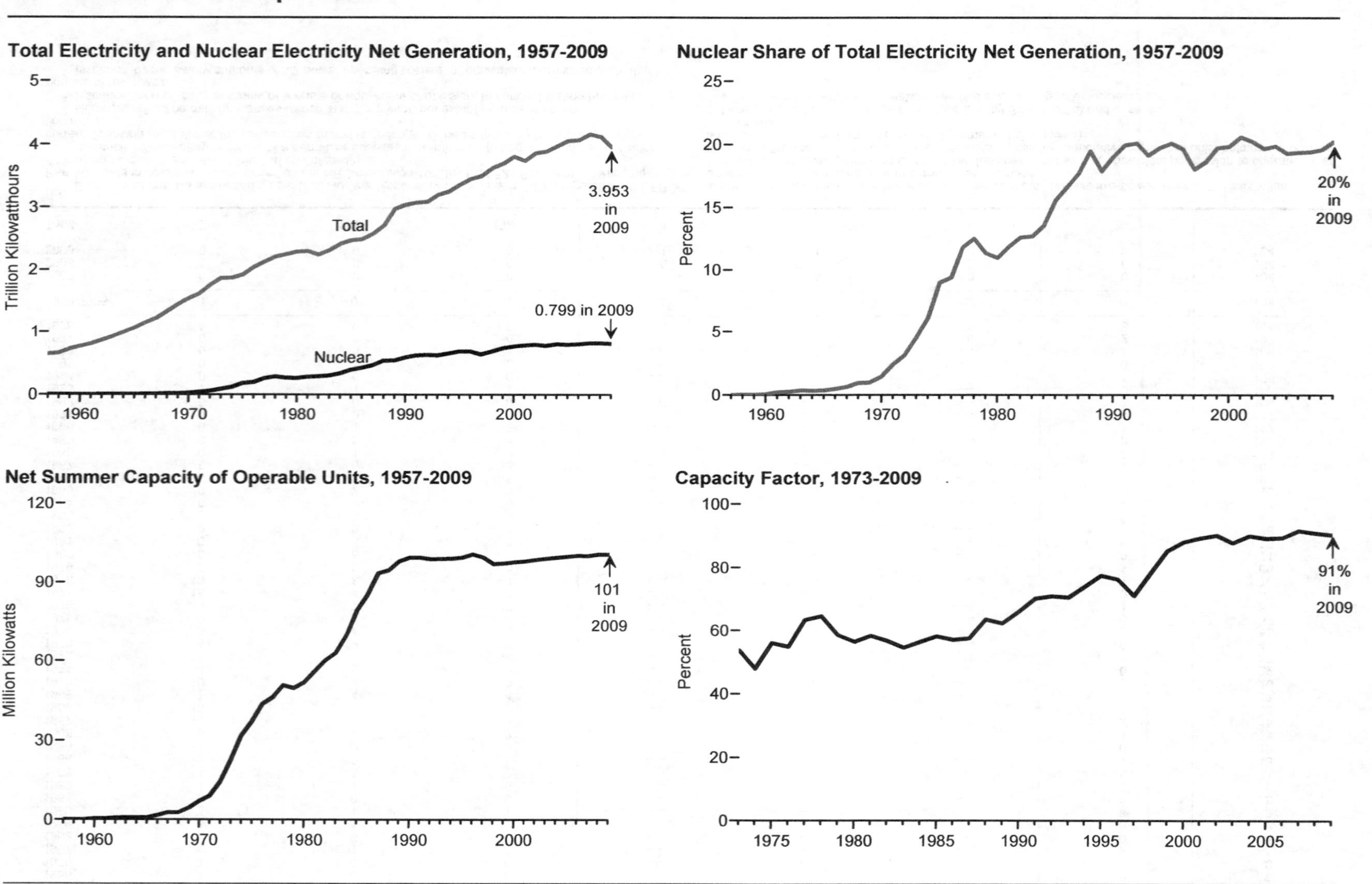
Total Electricity and Nuclear Electricity Net Generation, 1957-2009
Trillion Kilowatthours
Total
Nuclear
3.953 in 2009
0.799 in 2009
Nuclear Share of Total Electricity Net Generation, 1957-2009
Percent
20% in 2009
Net Summer Capacity of Operable Units, 1957-2009
Million Kilowatts
101 in 2009
Capacity Factor, 1973-2009
Percent
91% in 2009
U.S. Energy Information Administration / Annual Energy Review 2009

## Nuclear Power Plant Operations, 1957-2009

| Year | Nuclear Electricity Net Generation | Nuclear Share of Total Electricity Net Generation | Net Summer Capacity of Operable Units [1] | Capacity Factor [2] |
|---|---|---|---|---|
| | Billion Kilowatthours | Percent | Million Kilowatts | Percent |
| 1957 | (s) | (s) | 0.1 | NA |
| 1958 | .2 | (s) | .1 | NA |
| 1959 | .2 | (s) | .1 | NA |
| 1960 | .5 | .1 | .4 | NA |
| 1961 | 1.7 | .2 | .4 | NA |
| 1962 | 2.3 | .3 | .7 | NA |
| 1963 | 3.2 | .3 | .8 | NA |
| 1964 | 3.3 | .3 | .8 | NA |
| 1965 | 3.7 | .3 | .8 | NA |
| 1966 | 5.5 | .5 | 1.7 | NA |
| 1967 | 7.7 | .6 | 2.7 | NA |
| 1968 | 12.5 | .9 | 2.7 | NA |
| 1969 | 13.9 | 1.0 | 4.4 | NA |
| 1970 | 21.8 | 1.4 | 7.0 | NA |
| 1971 | 38.1 | 2.4 | 9.0 | NA |
| 1972 | 54.1 | 3.1 | 14.5 | NA |
| 1973 | 83.5 | 4.5 | 22.7 | 53.5 |
| 1974 | 114.0 | 6.1 | 31.9 | 47.8 |
| 1975 | 172.5 | 9.0 | 37.3 | 55.9 |
| 1976 | 191.1 | 9.4 | 43.8 | 54.7 |
| 1977 | 250.9 | 11.8 | 46.3 | 63.3 |
| 1978 | 276.4 | 12.5 | 50.8 | 64.5 |
| 1979 | 255.2 | 11.3 | 49.7 | 58.4 |
| 1980 | 251.1 | 11.0 | 51.8 | 56.3 |
| 1981 | 272.7 | 11.9 | 56.0 | 58.2 |
| 1982 | 282.8 | 12.6 | 60.0 | 56.6 |
| 1983 | 293.7 | 12.7 | 63.0 | 54.4 |
| 1984 | 327.6 | 13.5 | 69.7 | 56.3 |
| 1985 | 383.7 | 15.5 | 79.4 | 58.0 |
| 1986 | 414.0 | 16.6 | 85.2 | 56.9 |
| 1987 | 455.3 | 17.7 | 93.6 | 57.4 |
| 1988 | 527.0 | 19.5 | 94.7 | 63.5 |
| 1989 | 529.4 | 17.8 | 98.2 | 62.2 |
| 1990 | 576.9 | 19.0 | 99.6 | 66.0 |
| 1991 | 612.6 | 19.9 | 99.6 | 70.2 |
| 1992 | 618.8 | 20.1 | 99.0 | 70.9 |
| 1993 | 610.3 | 19.1 | 99.0 | 70.5 |
| 1994 | 640.4 | 19.7 | 99.1 | 73.8 |
| 1995 | 673.4 | 20.1 | 99.5 | 77.4 |
| 1996 | 674.7 | 19.6 | 100.8 | 76.2 |
| 1997 | 628.6 | 18.0 | 99.7 | 71.1 |
| 1998 | 673.7 | 18.6 | 97.1 | 78.2 |
| 1999 | 728.3 | 19.7 | 97.4 | 85.3 |
| 2000 | 753.9 | 19.8 | 97.9 | 88.1 |
| 2001 | 768.8 | 20.6 | 98.2 | 89.4 |
| 2002 | 780.1 | 20.2 | 98.7 | 90.3 |
| 2003 | 763.7 | 19.7 | 99.2 | 87.9 |
| 2004 | 788.5 | 19.9 | 99.6 | 90.1 |
| 2005 | 782.0 | 19.3 | 100.0 | 89.3 |
| 2006 | 787.2 | 19.4 | 100.3 | 89.6 |
| 2007 | 806.4 | 19.4 | 100.3 | 91.8 |
| 2008 | 806.2 | 19.6 | R100.8 | R91.1 |
| 2009P | 798.7 | 20.2 | 100.8 | 90.5 |

[1] At end of year. See "Generator Net Summer Capacity" in Glossary.
[2] See "Generator Capacity Factor" in Glossary.
R=Revised. P=Preliminary. NA=Not available. (s)=Less than 0.05.
Note: See Note 2, "Coverage of Nuclear Energy Statistics," at end of section.
Web Page: For related information, see http://www.eia.gov/fuelnuclear.html.

Sources: **Nuclear Electricity Net Generation** and **Nuclear Share of Electricity Net Generation**: Table 8.2a. **Net Summer Capacity of Operable Units:** • 1949-2008: Table 8.11a. • 2009—U.S. Energy Information Administration (EIA), *Monthly Energy Review (MER)* (April 2010), Table 8.1. **Capacity Factor:** EIA, *MER* (April 2010), Table 8.1. Annual capacity factors are weighted averages of monthly capacity factors.

## Renewable Energy Consumption by Major Source

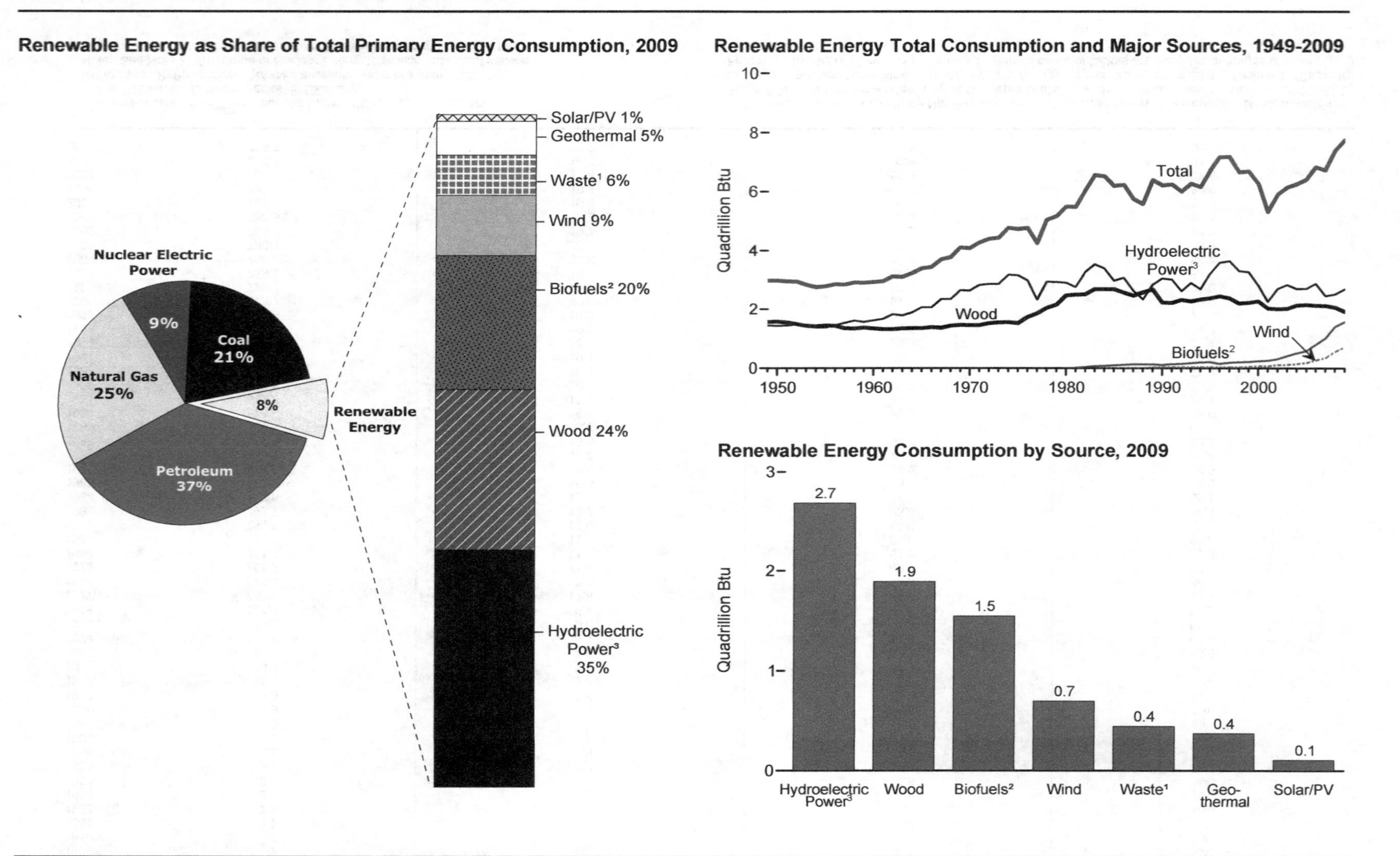


[1] Municipal solid waste from biogenic sources, landfill gas, sludge waste, agricultural byproducts, and other biomass.

[2] Fuel ethanol (minus denaturant) and biodiesel consumption, plus losses and co-products from the production of fuel ethanol and biodiesel.

[3] Conventional hydroelectric power.

U.S. Energy Information Administration / Annual Energy Review 2009

## Renewable Energy Production and Consumption by Primary Energy Source, Selected Years, 1949-2009

(Trillion Btu)

| | Production [1] | | | Consumption | | | | | | | | |
|---|---|---|---|---|---|---|---|---|---|---|---|---|
| | Biomass | | Total Renewable Energy [4] | Hydro-electric Power [5] | Geo-thermal [6] | Solar/PV [7] | Wind [8] | Biomass | | | | Total Renewable Energy |
| Year | Biofuels [2] | Total [3] | | | | | | Wood [9] | Waste [10] | Biofuels [11] | Total | |
| 1949 | NA | 1,549 | 2,974 | 1,425 | NA | NA | NA | 1,549 | NA | NA | 1,549 | 2,974 |
| 1950 | NA | 1,562 | 2,978 | 1,415 | NA | NA | NA | 1,562 | NA | NA | 1,562 | 2,978 |
| 1955 | NA | 1,424 | 2,784 | 1,360 | NA | NA | NA | 1,424 | NA | NA | 1,424 | 2,784 |
| 1960 | NA | 1,320 | 2,929 | 1,608 | 1 | NA | NA | 1,320 | NA | NA | 1,320 | 2,929 |
| 1965 | NA | 1,335 | 3,398 | 2,059 | 4 | NA | NA | 1,335 | NA | NA | 1,335 | 3,398 |
| 1970 | NA | 1,431 | 4,076 | 2,634 | 11 | NA | NA | 1,429 | 2 | NA | 1,431 | 4,076 |
| 1971 | NA | 1,432 | 4,268 | 2,824 | 12 | NA | NA | 1,430 | 2 | NA | 1,432 | 4,268 |
| 1972 | NA | 1,503 | 4,398 | 2,864 | 31 | NA | NA | 1,501 | 2 | NA | 1,503 | 4,398 |
| 1973 | NA | 1,529 | 4,433 | 2,861 | 43 | NA | NA | 1,527 | 2 | NA | 1,529 | 4,433 |
| 1974 | NA | 1,540 | 4,769 | 3,177 | 53 | NA | NA | 1,538 | 2 | NA | 1,540 | 4,769 |
| 1975 | NA | 1,499 | 4,723 | 3,155 | 70 | NA | NA | 1,497 | 2 | NA | 1,499 | 4,723 |
| 1976 | NA | 1,713 | 4,768 | 2,976 | 78 | NA | NA | 1,711 | 2 | NA | 1,713 | 4,768 |
| 1977 | NA | 1,838 | 4,249 | 2,333 | 77 | NA | NA | 1,837 | 2 | NA | 1,838 | 4,249 |
| 1978 | NA | 2,038 | 5,039 | 2,937 | 64 | NA | NA | 2,036 | 1 | NA | 2,038 | 5,039 |
| 1979 | NA | 2,152 | 5,166 | 2,931 | 84 | NA | NA | 2,150 | 2 | NA | 2,152 | 5,166 |
| 1980 | NA | 2,476 | 5,485 | 2,900 | 110 | NA | NA | 2,474 | 2 | NA | 2,476 | 5,485 |
| 1981 | 13 | R2,596 | R5,477 | 2,758 | 123 | NA | NA | 2,496 | 88 | 13 | R2,596 | R5,477 |
| 1982 | R34 | R2,663 | 6,034 | 3,266 | 105 | NA | NA | 2,510 | 119 | R34 | R2,663 | 6,034 |
| 1983 | R63 | R2,904 | R6,561 | 3,527 | 129 | NA | (s) | 2,684 | 157 | R63 | R2,904 | R6,561 |
| 1984 | R77 | R2,971 | R6,522 | 3,386 | 165 | (s) | (s) | 2,686 | 208 | R77 | R2,971 | R6,522 |
| 1985 | R93 | R3,016 | R6,185 | 2,970 | 198 | (s) | (s) | 2,687 | 236 | R93 | R3,016 | R6,185 |
| 1986 | R107 | R2,932 | R6,223 | 3,071 | 219 | (s) | (s) | 2,562 | 263 | R107 | R2,932 | R6,223 |
| 1987 | R123 | R2,875 | R5,739 | 2,635 | 229 | (s) | (s) | 2,463 | 289 | R123 | R2,875 | R5,739 |
| 1988 | R124 | R3,016 | R5,568 | 2,334 | 217 | (s) | (s) | 2,577 | 315 | R124 | R3,016 | R5,568 |
| 1989 | R125 | R3,159 | R6,391 | 2,837 | 317 | 55 | 22 | 2,680 | 354 | R125 | R3,159 | R6,391 |
| 1990 | R111 | R2,735 | R6,206 | 3,046 | 336 | 60 | 29 | 2,216 | 408 | R111 | R2,735 | R6,206 |
| 1991 | R128 | R2,782 | R6,237 | 3,016 | 346 | 63 | 31 | 2,214 | 440 | R128 | R2,782 | R6,238 |
| 1992 | R145 | R2,932 | R5,992 | 2,617 | 349 | 64 | 30 | 2,313 | 473 | R145 | R2,932 | R5,992 |
| 1993 | R169 | R2,908 | R6,261 | 2,892 | 364 | 66 | 31 | 2,260 | 479 | R169 | R2,908 | R6,261 |
| 1994 | R188 | R3,028 | R6,153 | 2,683 | 338 | 69 | 36 | 2,324 | 515 | R188 | R3,028 | R6,153 |
| 1995 | R198 | R3,099 | R6,701 | 3,205 | 294 | 70 | 33 | 2,370 | 531 | R200 | R3,101 | R6,703 |
| 1996 | R141 | R3,155 | R7,165 | 3,590 | 316 | 71 | 33 | 2,437 | 577 | R143 | R3,157 | R7,166 |
| 1997 | R186 | R3,108 | R7,177 | 3,640 | 325 | 70 | 34 | 2,371 | 551 | R184 | R3,105 | R7,175 |
| 1998 | R202 | R2,929 | R6,655 | 3,297 | 328 | 70 | 31 | 2,184 | 542 | R201 | R2,928 | R6,654 |
| 1999 | R211 | R2,965 | R6,678 | 3,268 | 331 | 69 | 46 | 2,214 | 540 | R209 | R2,963 | R6,677 |
| 2000 | R233 | R3,006 | R6,257 | 2,811 | 317 | 66 | 57 | 2,262 | 511 | R236 | R3,008 | R6,260 |
| 2001 | R254 | R2,624 | R5,312 | 2,242 | 311 | 65 | 70 | 2,006 | 364 | R253 | R2,622 | R5,311 |
| 2002 | R308 | R2,705 | R5,892 | 2,689 | 328 | 64 | 105 | 1,995 | 402 | R303 | R2,701 | R5,888 |
| 2003 | R402 | R2,805 | R6,139 | 2,825 | 331 | 64 | 115 | 2,002 | 401 | R404 | R2,807 | R6,141 |
| 2004 | R487 | R2,998 | R6,235 | 2,690 | 341 | 65 | 142 | 2,121 | 389 | R500 | R3,010 | R6,247 |
| 2005 | R564 | R3,104 | R6,393 | 2,703 | 343 | 66 | 178 | 2,136 | 403 | R577 | R3,117 | R6,406 |
| 2006 | R720 | R3,226 | R6,774 | 2,869 | 343 | 72 | 264 | R2,109 | R397 | R771 | R3,277 | R6,824 |
| 2007 | R978 | R3,489 | R6,706 | 2,446 | 349 | 81 | 341 | R2,098 | R413 | R991 | R3,503 | R6,719 |
| 2008 | R1,387 | R3,867 | R7,381 | R2,511 | R360 | R97 | R546 | R2,044 | R436 | R1,372 | R3,852 | R7,366 |
| 2009P | 1,562 | 3,900 | 7,761 | 2,682 | 373 | 109 | 697 | 1,891 | 447 | 1,545 | 3,883 | 7,744 |

[1] Production equals consumption for all renewable energy sources except biofuels.
[2] Total biomass inputs to the production of fuel ethanol and biodiesel.
[3] Wood and wood-derived fuels, biomass waste, and total biomass inputs to the production of fuel ethanol and biodiesel.
[4] Hydroelectric power, geothermal, solar thermal/photovoltaic, wind, and biomass.
[5] Conventional hydroelectricity net generation (converted to Btu using the fossil-fueled plants heat rate).
[6] Geothermal electricity net generation (converted to Btu using the geothermal energy plants heat rate), and geothermal heat pump and direct use energy.
[7] Solar thermal and photovoltaic (PV) electricity net generation (converted to Btu using the fossil-fueled plants heat rate), and solar thermal direct use energy.
[8] Wind electricity net generation (converted to Btu using the fossil-fueled plants heat rate).
[9] Wood and wood-derived fuels.
[10] Municipal solid waste from biogenic sources, landfill gas, sludge waste, agricultural byproducts, and other biomass. Through 2000, also includes non-renewable waste (municipal solid waste from non-biogenic sources, and tire-derived fuels).
[11] Fuel ethanol (minus denaturant) and biodiesel consumption, plus losses and co-products from the production of fuel ethanol and biodiesel.

R=Revised. P=Preliminary. NA=Not available. (s)=Less than 0.5 trillion Btu.

Notes: • Most data for the residential, commercial, industrial, and transportation sectors are estimates. See notes and sources for Tables 10.2a and 10.2b. • See Section 8, Tables 8.2a-d and 8.3a-c, for electricity net generation and useful thermal output from renewable energy sources; Tables 8.4a-c, 8.5a-d, 8.6a-c, and 8.7a-c for renewable energy consumption for electricity generation and useful thermal output; and Tables 8.11a-d for renewable energy electric net summer capacity. • See Note, "Renewable Energy Production and Consumption," at end of section. • See Table E1 for estimated renewable energy consumption for 1635-1945. • Totals may not equal sum of components due to independent rounding.

Web Pages: • For all data beginning in 1949, see http://www.eia.gov/emeu/aer/renew.html. • For related information, see http://www.eia.gov/fuelrenewable.html.

# Renewable Energy Consumption: End-Use Sectors, 1989-2009

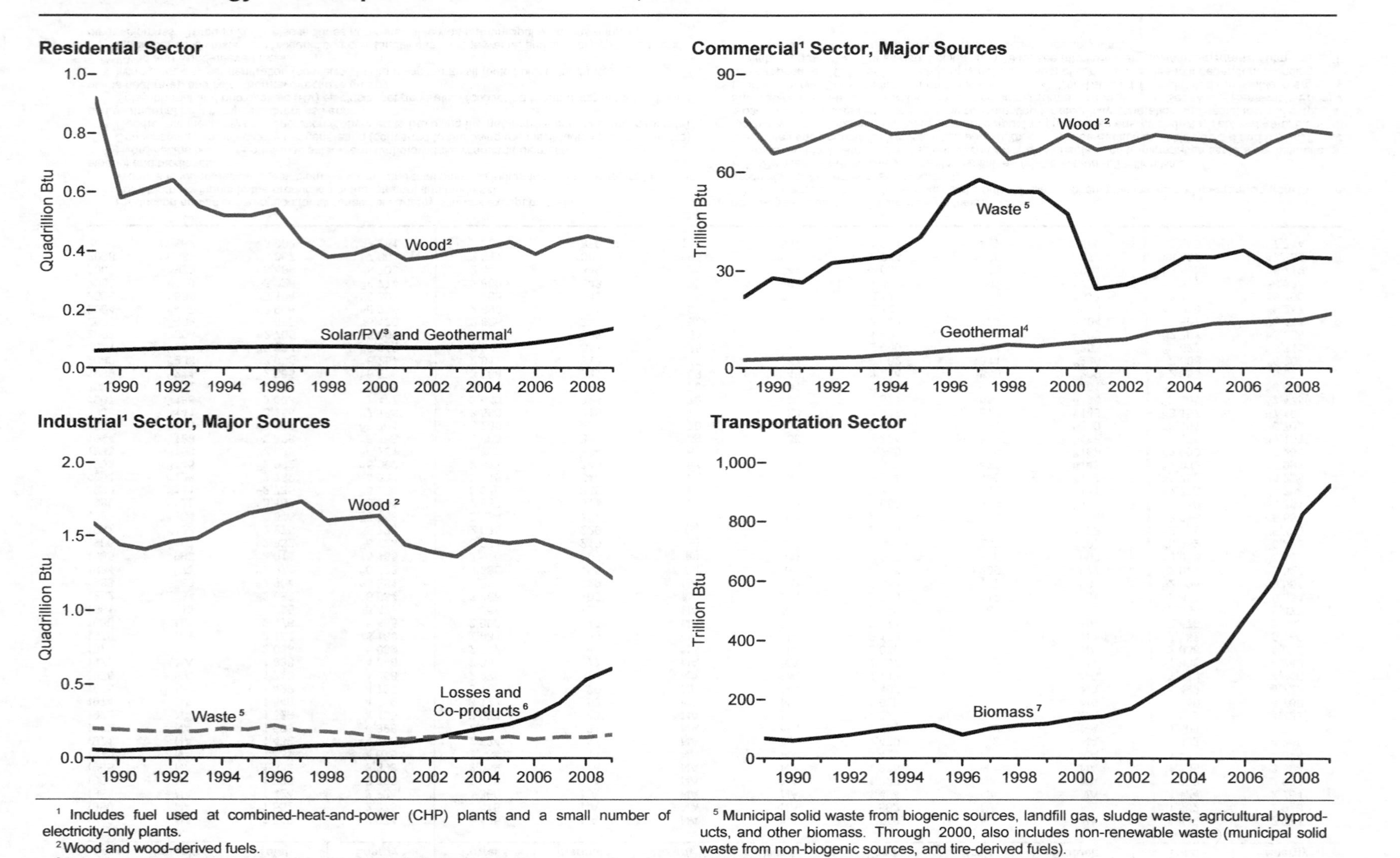


[1] Includes fuel used at combined-heat-and-power (CHP) plants and a small number of electricity-only plants.

[2] Wood and wood-derived fuels.

[3] Solar thermal direct use energy , and photovoltaic (PV) electricity net generation. Includes small amounts of distributed solar thermal and PV energy used in the commercial, industrial, and electric power sectors.

[4] Geothermal heat pump and direct use energy.

[5] Municipal solid waste from biogenic sources, landfill gas, sludge waste, agricultural byproducts, and other biomass. Through 2000, also includes non-renewable waste (municipal solid waste from non-biogenic sources, and tire-derived fuels).

[6] From the production of fuel ethanol and biodiesel.

[7] The fuel ethanol (minus denaturant) portion of motor fuels (such as E10 and E85), and biofuels used as diesel fuel substitutes, additives or extenders.

U.S. Energy Information Administration / Annual Energy Review 2009

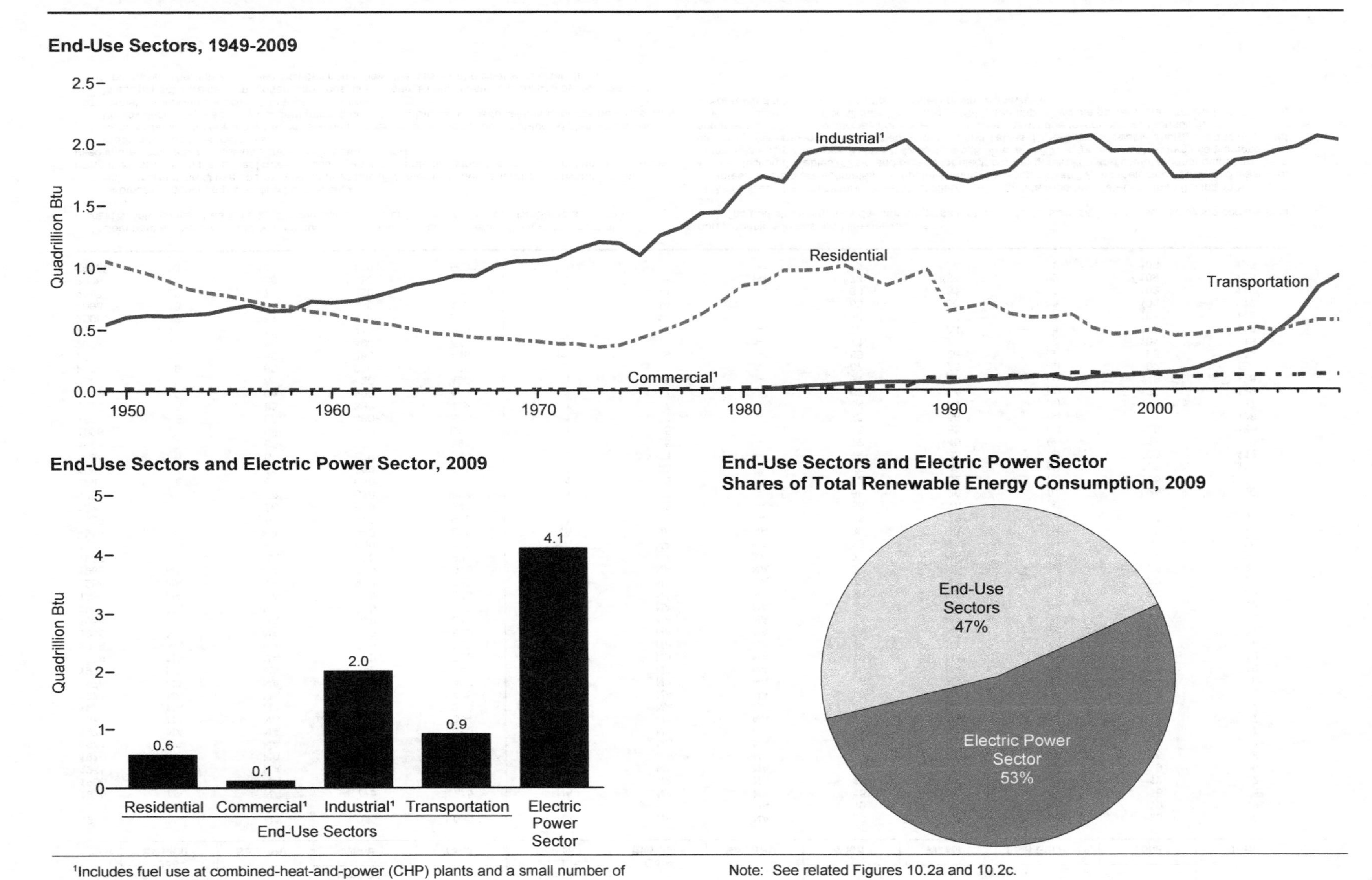
Renewable Energy Consumption: End-Use Sectors and Electric Power Sector
End-Use Sectors, 1949-2009
Quadrillion Btu
2.5
2.0
1.5
1.0
0.5
0.0
1950
1960
1970
1980
1990
2000
Industrial¹
Residential
Transportation
Commercial¹
End-Use Sectors and Electric Power Sector, 2009
Quadrillion Btu
5
4
3
2
1
0
0.6
0.1
2.0
0.9
4.1
Residential
Commercial¹
Industrial¹
Transportation
End-Use Sectors
Electric Power Sector
End-Use Sectors and Electric Power Sector Shares of Total Renewable Energy Consumption, 2009
End-Use Sectors 47%
Electric Power Sector 53%
¹Includes fuel use at combined-heat-and-power (CHP) plants and a small number of electricity-only plants.
Note: See related Figures 10.2a and 10.2c.
U.S. Energy Information Administration / Annual Energy Review 2009

## Renewable Energy Consumption: Residential and Commercial Sectors, Selected Years, 1949-2009

(Trillion Btu)

| Year | Residential Sector | | | | Commercial Sector [1] | | | | | | | |
|---|---|---|---|---|---|---|---|---|---|---|---|---|
| | Geo-thermal [2] | Solar/PV [3] | Biomass | Total | Hydro-electric Power [5] | Geo-thermal [2] | Solar/PV [6] | Biomass | | | | Total |
| | | | Wood [4] | | | | | Wood [4] | Waste [7] | Fuel Ethanol [8] | Total | |
| 1949 | NA | NA | 1,055 | 1,055 | NA | NA | NA | 20 | NA | NA | 20 | 20 |
| 1950 | NA | NA | 1,006 | 1,006 | NA | NA | NA | 19 | NA | NA | 19 | 19 |
| 1955 | NA | NA | 775 | 775 | NA | NA | NA | 15 | NA | NA | 15 | 15 |
| 1960 | NA | NA | 627 | 627 | NA | NA | NA | 12 | NA | NA | 12 | 12 |
| 1965 | NA | NA | 468 | 468 | NA | NA | NA | 9 | NA | NA | 9 | 9 |
| 1970 | NA | NA | 401 | 401 | NA | NA | NA | 8 | NA | NA | 8 | 8 |
| 1971 | NA | NA | 382 | 382 | NA | NA | NA | 7 | NA | NA | 7 | 7 |
| 1972 | NA | NA | 380 | 380 | NA | NA | NA | 7 | NA | NA | 7 | 7 |
| 1973 | NA | NA | 354 | 354 | NA | NA | NA | 7 | NA | NA | 7 | 7 |
| 1974 | NA | NA | 371 | 371 | NA | NA | NA | 7 | NA | NA | 7 | 7 |
| 1975 | NA | NA | 425 | 425 | NA | NA | NA | 8 | NA | NA | 8 | 8 |
| 1976 | NA | NA | 482 | 482 | NA | NA | NA | 9 | NA | NA | 9 | 9 |
| 1977 | NA | NA | 542 | 542 | NA | NA | NA | 10 | NA | NA | 10 | 10 |
| 1978 | NA | NA | 622 | 622 | NA | NA | NA | 12 | NA | NA | 12 | 12 |
| 1979 | NA | NA | 728 | 728 | NA | NA | NA | 14 | NA | NA | 14 | 14 |
| 1980 | NA | NA | 850 | 850 | NA | NA | NA | 21 | NA | NA | 21 | 21 |
| 1981 | NA | NA | 870 | 870 | NA | NA | NA | 21 | NA | (s) | 21 | 21 |
| 1982 | NA | NA | 970 | 970 | NA | NA | NA | 22 | NA | (s) | 22 | 22 |
| 1983 | NA | NA | 970 | 970 | NA | NA | NA | 22 | NA | (s) | 22 | 22 |
| 1984 | NA | NA | 980 | 980 | NA | NA | NA | 22 | NA | (s) | 22 | 22 |
| 1985 | NA | NA | 1,010 | 1,010 | NA | NA | NA | 24 | NA | (s) | 24 | 24 |
| 1986 | NA | NA | 920 | 920 | NA | NA | NA | 27 | NA | (s) | 27 | 27 |
| 1987 | NA | NA | 850 | 850 | NA | NA | NA | 29 | NA | 1 | 30 | 30 |
| 1988 | NA | NA | 910 | 910 | NA | NA | NA | 32 | NA | 1 | 33 | 33 |
| 1989 | 5 | 53 | 920 | 978 | 1 | 3 | – | 76 | 22 | 1 | 99 | 102 |
| 1990 | 6 | 56 | 580 | 641 | 1 | 3 | – | 66 | 28 | R (s) | 94 | 98 |
| 1991 | 6 | 58 | 610 | 674 | 1 | 3 | – | 68 | 26 | (s) | 95 | 100 |
| 1992 | 6 | 60 | 640 | 706 | 1 | 3 | – | 72 | 32 | (s) | 105 | 109 |
| 1993 | 7 | 62 | 550 | 618 | 1 | 3 | – | 76 | 33 | (s) | 109 | 114 |
| 1994 | 6 | 64 | 520 | 590 | 1 | 4 | – | 72 | 35 | (s) | 106 | 112 |
| 1995 | 7 | 65 | 520 | 591 | 1 | 5 | – | 72 | 40 | (s) | 113 | 118 |
| 1996 | 7 | 65 | 540 | 612 | 1 | 5 | – | 76 | 53 | (s) | 129 | 135 |
| 1997 | 8 | 65 | 430 | 503 | 1 | 6 | – | 73 | 58 | (s) | 131 | 138 |
| 1998 | 8 | 65 | 380 | 452 | 1 | 7 | – | 64 | 54 | (s) | 118 | 127 |
| 1999 | 9 | 64 | 390 | 462 | 1 | 7 | – | 67 | 54 | (s) | 121 | 129 |
| 2000 | 9 | 61 | 420 | 490 | 1 | 8 | – | 71 | 47 | (s) | 119 | 128 |
| 2001 | 9 | 60 | 370 | 439 | 1 | 8 | – | 67 | 25 | (s) | 92 | 101 |
| 2002 | 10 | 59 | 380 | 449 | (s) | 9 | – | 69 | 26 | (s) | 95 | 104 |
| 2003 | 13 | 58 | 400 | 471 | 1 | 11 | – | 71 | 29 | 1 | 101 | 113 |
| 2004 | 14 | 59 | 410 | 483 | 1 | 12 | – | 70 | 34 | 1 | 105 | 118 |
| 2005 | 16 | 61 | 430 | 507 | 1 | 14 | – | 70 | 34 | 1 | 105 | 119 |
| 2006 | 18 | 67 | 390 | 475 | 1 | 14 | – | 65 | 36 | 1 | 102 | 117 |
| 2007 | 22 | 75 | 430 | 527 | 1 | 14 | – | 69 | 31 | 2 | 102 | 118 |
| 2008 | 26 | R88 | R450 | R565 | 1 | 15 | (s) | R73 | R34 | R2 | R109 | R125 |
| 2009P | 33 | 101 | 430 | 563 | 1 | 17 | (s) | 72 | 34 | 2 | 108 | 125 |

[1] Commercial sector, including commercial combined-heat-and-power (CHP) and commercial electricity-only plants. See Note 2, "Classification of Power Plants Into Energy-Use Sectors," at end of Section 8.

[2] Geothermal heat pump and direct use energy.

[3] Solar thermal direct use energy, and photovoltaic (PV) electricity net generation (converted to Btu using the fossil-fueled plants heat rate). Includes small amounts of distributed solar thermal and PV energy used in the commercial, industrial, and electric power sectors.

[4] Wood and wood-derived fuels.

[5] Conventional hydroelectricity net generation (converted to Btu using the fossil-fueled plants heat rate).

[6] Photovoltaic (PV) electricity net generation (converted to Btu using the fossil-fueled plants heat rate) at commercial plants with capacity of 1 megawatt or greater.

[7] Municipal solid waste from biogenic sources, landfill gas, sludge waste, agricultural byproducts, and other biomass. Through 2000, also includes non-renewable waste (municipal solid waste from non-biogenic sources, and tire-derived fuels).

[8] The fuel ethanol (minus denaturant) portion of motor fuels, such as E10, consumed by the commercial sector.

R=Revised. P=Preliminary. NA=Not available. – = No data reported. (s)=Less than 0.5 trillion Btu.

Notes: • Data are estimates, except for commercial sector solar/PV, hydroelectric power, and waste. • See Section 8, Tables 8.2a-d and 8.3a-c, for electricity net generation and useful thermal output from renewable energy sources; Tables 8.4a-c, 8.5a-d, 8.6a-c, and 8.7a-c for renewable energy consumption for electricity generation and useful thermal output; and Tables 8.11a-d for renewable energy electric net summer capacity. • Totals may not equal sum of components due to independent rounding.

Web Pages: • For all data beginning in 1949, see http://www.eia.gov/emeu/aer/renew.html. • For related information, see http://www.eia.gov/fuelrenewable.html.

## Renewable Energy Consumption: Industrial and Transportation Sectors, Selected Years, 1949-2009

(Trillion Btu)

| | Industrial Sector [1] | | | | | | | | Transportation Sector | | |
|---|---|---|---|---|---|---|---|---|---|---|---|
| | | | Biomass | | | | | | Biomass | | |
| Year | Hydro-electric Power [2] | Geo-thermal [3] | Wood [4] | Waste [5] | Fuel Ethanol [6] | Losses and Co-products [7] | Total | Total | Fuel Ethanol [8] | Biodiesel [9] | Total |
| 1949 | 76 | NA | 468 | NA | NA | NA | 468 | 544 | NA | NA | NA |
| 1950 | 69 | NA | 532 | NA | NA | NA | 532 | 602 | NA | NA | NA |
| 1955 | 38 | NA | 631 | NA | NA | NA | 631 | 669 | NA | NA | NA |
| 1960 | 39 | NA | 680 | NA | NA | NA | 680 | 719 | NA | NA | NA |
| 1965 | 33 | NA | 855 | NA | NA | NA | 855 | 888 | NA | NA | NA |
| 1970 | 34 | NA | 1,019 | NA | NA | NA | 1,019 | 1,053 | NA | NA | NA |
| 1971 | 34 | NA | 1,040 | NA | NA | NA | 1,040 | 1,074 | NA | NA | NA |
| 1972 | 34 | NA | 1,113 | NA | NA | NA | 1,113 | 1,147 | NA | NA | NA |
| 1973 | 35 | NA | 1,165 | NA | NA | NA | 1,165 | 1,200 | NA | NA | NA |
| 1974 | 33 | NA | 1,159 | NA | NA | NA | 1,159 | 1,192 | NA | NA | NA |
| 1975 | 32 | NA | 1,063 | NA | NA | NA | 1,063 | 1,096 | NA | NA | NA |
| 1976 | 33 | NA | 1,220 | NA | NA | NA | 1,220 | 1,253 | NA | NA | NA |
| 1977 | 33 | NA | 1,281 | NA | NA | NA | 1,281 | 1,314 | NA | NA | NA |
| 1978 | 32 | NA | 1,400 | NA | NA | NA | 1,400 | 1,432 | NA | NA | NA |
| 1979 | 34 | NA | 1,405 | NA | NA | NA | 1,405 | 1,439 | NA | NA | NA |
| 1980 | 33 | NA | 1,600 | NA | NA | NA | 1,600 | 1,633 | NA | NA | NA |
| 1981 | 33 | NA | 1,602 | 87 | (s) | 6 | 1,695 | 1,728 | 7 | NA | 7 |
| 1982 | 33 | NA | 1,516 | 118 | (s) | 16 | 1,650 | 1,683 | R18 | NA | R18 |
| 1983 | 33 | NA | 1,690 | 155 | (s) | 29 | R1,874 | 1,908 | 34 | NA | 34 |
| 1984 | 33 | NA | 1,679 | 204 | 1 | R35 | R1,918 | R1,951 | R41 | NA | R41 |
| 1985 | 33 | NA | 1,645 | 230 | 1 | R42 | R1,918 | R1,951 | R50 | NA | R50 |
| 1986 | 33 | NA | 1,610 | 256 | 1 | R48 | R1,915 | R1,948 | R57 | NA | R57 |
| 1987 | 33 | NA | 1,576 | 282 | 1 | R55 | R1,914 | R1,947 | R66 | NA | R66 |
| 1988 | 33 | NA | 1,625 | 308 | 1 | R55 | R1,989 | R2,022 | R67 | NA | R67 |
| 1989 | 28 | 2 | 1,584 | 200 | 1 | R56 | R1,841 | R1,871 | R68 | NA | R68 |
| 1990 | 31 | 2 | 1,442 | 192 | 1 | R49 | R1,684 | R1,717 | R60 | NA | R60 |
| 1991 | 30 | 2 | 1,410 | 185 | 1 | R56 | R1,652 | R1,684 | R70 | NA | R70 |
| 1992 | 31 | 2 | 1,461 | 179 | 1 | R64 | R1,705 | R1,737 | R80 | NA | R80 |
| 1993 | 30 | 2 | 1,484 | 181 | 1 | R74 | R1,741 | R1,773 | R94 | NA | R94 |
| 1994 | 62 | 3 | 1,580 | 199 | 1 | R82 | R1,862 | R1,927 | R105 | NA | R105 |
| 1995 | 55 | 3 | 1,652 | 195 | 2 | R86 | R1,934 | R1,992 | R113 | NA | R113 |
| 1996 | 61 | 3 | 1,683 | 224 | 1 | R61 | R1,969 | R2,033 | R81 | NA | R81 |
| 1997 | 58 | 3 | 1,731 | 184 | 1 | R80 | R1,996 | R2,057 | R102 | NA | R102 |
| 1998 | 55 | 3 | 1,603 | 180 | 1 | R86 | R1,872 | R1,929 | R113 | NA | R113 |
| 1999 | 49 | 4 | 1,620 | 171 | 1 | R90 | R1,882 | R1,934 | R118 | NA | R118 |
| 2000 | 42 | 4 | 1,636 | 145 | 1 | R99 | R1,881 | R1,928 | R135 | NA | R135 |
| 2001 | 33 | 5 | 1,443 | 129 | 3 | R108 | R1,681 | R1,719 | R141 | 1 | R142 |
| 2002 | 39 | 5 | 1,396 | 146 | 3 | R130 | R1,676 | R1,720 | R168 | 2 | R170 |
| 2003 | 43 | 3 | 1,363 | 142 | R4 | R169 | R1,679 | R1,726 | R228 | 2 | R230 |
| 2004 | 33 | 4 | 1,476 | 132 | 6 | R203 | R1,817 | R1,853 | R286 | 3 | R290 |
| 2005 | 32 | 4 | 1,452 | 148 | 7 | R230 | R1,837 | R1,873 | R328 | 12 | R339 |
| 2006 | 29 | 4 | R1,472 | R130 | 10 | R285 | R1,897 | R1,930 | R442 | 33 | R475 |
| 2007 | 16 | 5 | R1,413 | R144 | 10 | R377 | R1,944 | R1,964 | R557 | 46 | R603 |
| 2008 | R17 | 5 | R1,344 | R144 | R12 | R532 | R2,031 | R2,053 | R786 | R40 | R827 |
| 2009P | 18 | 4 | 1,217 | 160 | 13 | 607 | 1,997 | 2,019 | 879 | 43 | 922 |

[1] Industrial sector, including industrial combined-heat-and-power (CHP) and industrial electricity-only plants. See Note 2, "Classification of Power Plants Into Energy-Use Sectors," at end of Section 8.
[2] Conventional hydroelectricity net generation (converted to Btu using the fossil-fueled plants heat rate).
[3] Geothermal heat pump and direct use energy.
[4] Wood and wood-derived fuels.
[5] Municipal solid waste from biogenic sources, landfill gas, sludge waste, agricultural byproducts, and other biomass. Through 2000, also includes non-renewable waste (municipal solid waste from non-biogenic sources, and tire-derived fuels).
[6] The fuel ethanol (minus denaturant) portion of motor fuels, such as E10, consumed by the industrial sector.
[7] Losses and co-products from the production of fuel ethanol and biodiesel. Does not include natural gas, electricity, and other non-biomass energy used in the production of fuel ethanol and biodiesel—these are included in the industrial sector consumption statistics for the appropriate energy source.
[8] The fuel ethanol (minus denaturant) portion of motor fuels, such as E10 and E85, consumed by the transportation sector.
[9] "Biodiesel" is any liquid biofuel suitable as a diesel fuel substitute, additive, or extender. See "Biodiesel" in Glossary.

R=Revised. P=Preliminary. NA=Not available. (s)=Less than 0.5 trillion Btu.

Notes: • Data are estimates, except for industrial sector hydroelectric power in 1949-1978 and 1989 forward. • See Section 8, Tables 8.2a-d and 8.3a-c, for electricity net generation and useful thermal output from renewable energy sources; Tables 8.4a-c, 8.5a-d, 8.6a-c, and 8.7a-c for renewable energy consumption for electricity generation and useful thermal output; and Tables 8.11a-d for renewable energy electric net summer capacity. • Totals may not equal sum of components due to independent rounding.

Web Pages: • For all data beginning in 1949, see http://www.eia.gov/emeu/aer/renew.html. • For related information, see http://www.eia.gov/fuelrenewable.html.

U.S. Energy Information Administration / Annual Energy Review 2009

# Renewable Energy Consumption: Electric Power Sector

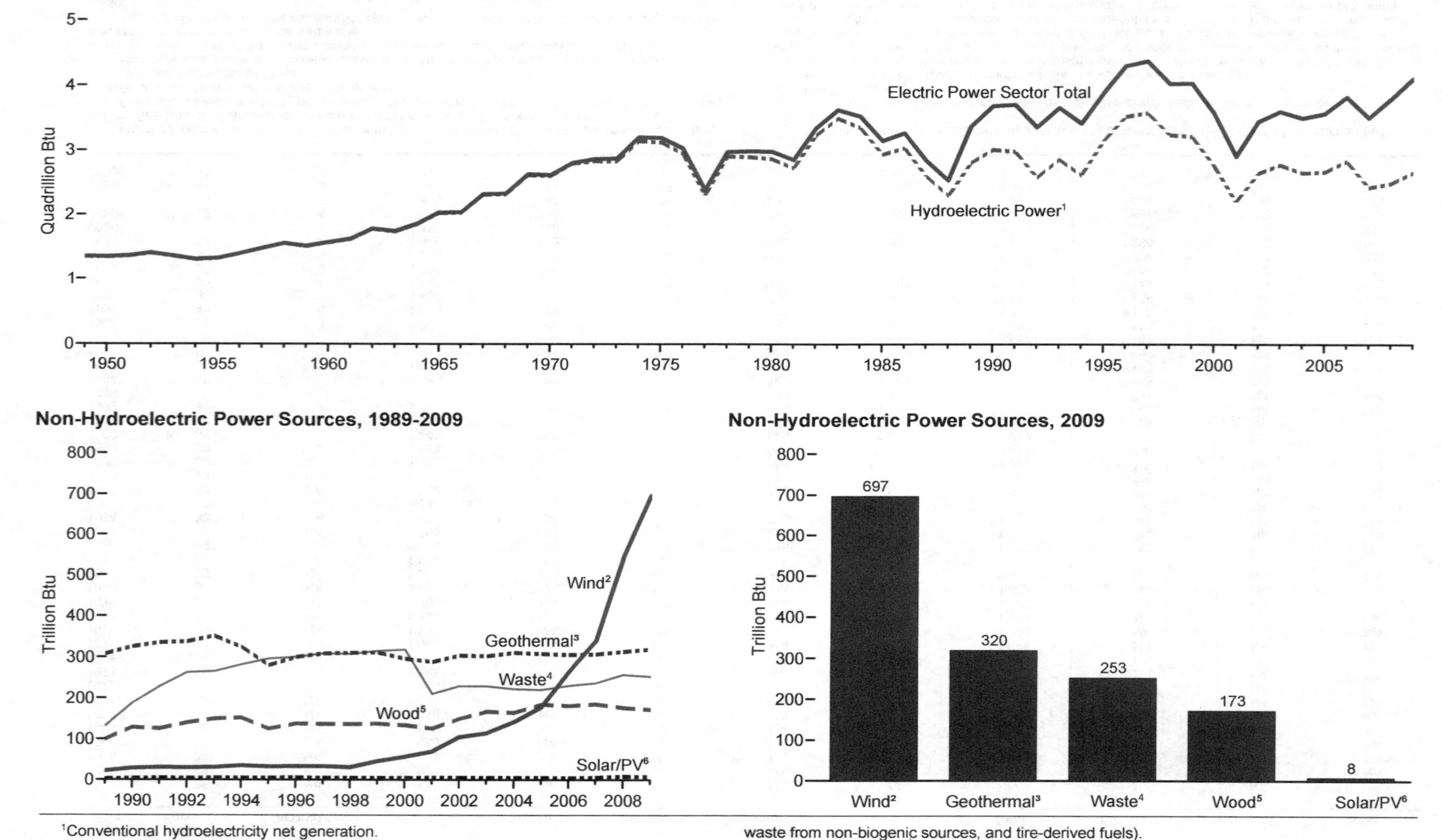


[1]Conventional hydroelectricity net generation.
[2]Wind electricity net generation.
[3]Geothermal electricity net generation.
[4]Municipal solid waste from biogenic sources, landfill gas, sludge waste, agricultural byproducts, and other biomass. Through 2000, also includes non-renewable waste (municipal solid waste from non-biogenic sources, and tire-derived fuels).
[5]Wood and wood-derived fuels.
[6]Solar thermal and photovoltaic (PV) electricity net generation.
Note: See related Figures 10.2a and 10.2b on the end-use sectors.

U.S. Energy Information Administration / Annual Energy Review 2009

## Renewable Energy Consumption: Electric Power Sector, Selected Years, 1949-2009

(Trillion Btu)

| Year | Hydroelectric Power [1] | Geothermal [2] | Solar/PV [3] | Wind [4] | Biomass: Wood [5] | Biomass: Waste [6] | Biomass: Total | Total |
|---|---|---|---|---|---|---|---|---|
| 1949 | 1,349 | NA | NA | NA | 6 | NA | 6 | 1,355 |
| 1950 | 1,346 | NA | NA | NA | 5 | NA | 5 | 1,351 |
| 1955 | 1,322 | NA | NA | NA | 3 | NA | 3 | 1,325 |
| 1960 | 1,569 | 1 | NA | NA | 2 | NA | 2 | 1,571 |
| 1965 | 2,026 | 4 | NA | NA | 3 | NA | 3 | 2,033 |
| 1970 | 2,600 | 11 | NA | NA | 1 | 2 | 4 | 2,615 |
| 1971 | 2,790 | 12 | NA | NA | 1 | 2 | 3 | 2,806 |
| 1972 | 2,829 | 31 | NA | NA | 1 | 2 | 3 | 2,864 |
| 1973 | 2,827 | 43 | NA | NA | 1 | 2 | 3 | 2,873 |
| 1974 | 3,143 | 53 | NA | NA | 1 | 2 | 3 | 3,199 |
| 1975 | 3,122 | 70 | NA | NA | (s) | 2 | 2 | 3,194 |
| 1976 | 2,943 | 78 | NA | NA | 1 | 2 | 3 | 3,024 |
| 1977 | 2,301 | 77 | NA | NA | 3 | 2 | 5 | 2,383 |
| 1978 | 2,905 | 64 | NA | NA | 2 | 1 | 3 | 2,973 |
| 1979 | 2,897 | 84 | NA | NA | 3 | 2 | 5 | 2,986 |
| 1980 | 2,867 | 110 | NA | NA | 3 | 2 | 5 | 2,982 |
| 1981 | 2,725 | 123 | NA | NA | 3 | 1 | 4 | 2,852 |
| 1982 | 3,233 | 105 | NA | NA | 2 | 1 | 3 | 3,341 |
| 1983 | 3,494 | 129 | NA | (s) | 2 | 2 | 4 | 3,627 |
| 1984 | 3,353 | 165 | (s) | (s) | 5 | 4 | 9 | 3,527 |
| 1985 | 2,937 | 198 | (s) | (s) | 8 | 7 | 14 | 3,150 |
| 1986 | 3,038 | 219 | (s) | (s) | 5 | 7 | 12 | 3,270 |
| 1987 | 2,602 | 229 | (s) | (s) | 8 | 7 | 15 | 2,846 |
| 1988 | 2,302 | 217 | (s) | (s) | 10 | 8 | 17 | 2,536 |
| 1989[7] | 2,808 | 308 | 3 | 22 | 100 | 132 | 232 | 3,372 |
| 1990 | 3,014 | 326 | 4 | 29 | 129 | 188 | 317 | 3,689 |
| 1991 | 2,985 | 335 | 5 | 31 | 126 | 229 | 354 | 3,710 |
| 1992 | 2,586 | 338 | 4 | 30 | 140 | 262 | 402 | 3,360 |
| 1993 | 2,861 | 351 | 5 | 31 | 150 | 265 | 415 | 3,662 |
| 1994 | 2,620 | 325 | 5 | 36 | 152 | 282 | 434 | 3,420 |
| 1995 | 3,149 | 280 | 5 | 33 | 125 | 296 | 422 | 3,889 |
| 1996 | 3,528 | 300 | 5 | 33 | 138 | 300 | 438 | 4,305 |
| 1997 | 3,581 | 309 | 5 | 34 | 137 | 309 | 446 | 4,375 |
| 1998 | 3,241 | 311 | 5 | 31 | 137 | 308 | 444 | 4,032 |
| 1999 | 3,218 | 312 | 5 | 46 | 138 | 315 | 453 | 4,034 |
| 2000 | 2,768 | 296 | 5 | 57 | 134 | 318 | 453 | 3,579 |
| 2001 | 2,209 | 289 | 6 | 70 | 126 | 211 | 337 | 2,910 |
| 2002 | 2,650 | 305 | 6 | 105 | 150 | 230 | 380 | 3,445 |
| 2003 | 2,781 | 303 | 5 | 115 | 167 | 230 | 397 | 3,601 |
| 2004 | 2,656 | 311 | 6 | 142 | 165 | 223 | 388 | 3,503 |
| 2005 | 2,670 | 309 | 6 | 178 | 185 | 221 | 406 | 3,568 |
| 2006 | 2,839 | 306 | 5 | 264 | 182 | 231 | 412 | 3,827 |
| 2007 | 2,430 | 308 | 6 | 341 | 186 | 237 | 423 | 3,508 |
| 2008 | R2,494 | R314 | R9 | R546 | R177 | R258 | R435 | R3,798 |
| 2009P | 2,663 | 320 | 8 | 697 | 173 | 253 | 426 | 4,113 |

[1] Conventional hydroelectricity net generation (converted to Btu using the fossil-fueled plants heat rate).
[2] Geothermal electricity net generation (converted to Btu using the geothermal energy plants heat rate).
[3] Solar thermal and photovoltaic (PV) electricity net generation (converted to Btu using the fossil-fueled plants heat rate).
[4] Wind electricity net generation (converted to Btu using the fossil-fueled plants heat rate).
[5] Wood and wood-derived fuels.
[6] Municipal solid waste from biogenic sources, landfill gas, sludge waste, agricultural byproducts, and other biomass. Through 2000, also includes non-renewable waste (municipal solid waste from non-biogenic sources, and tire-derived fuels).
[7] Through 1988, data are for electric utilities only. Beginning in 1989, data are for electric utilities and independent power producers.

R=Revised. P=Preliminary. NA=Not available. (s)=Less than 0.5 trillion Btu.
Notes: • The electric power sector comprises electricity-only and combined-heat-and-power (CHP) plants within the NAICS 22 category whose primary business is to sell electricity, or electricity and heat, to the public. • See Section 8, Tables 8.2a-d and 8.3a-c, for electricity net generation and useful thermal output from renewable energy sources; Tables 8.4a-c, 8.5a-d, 8.6a-c, and 8.7a-c for renewable energy consumption for electricity generation and useful thermal output; and Tables 8.11a-d for renewable energy electric net summer capacity. • See Note 3, "Electricity Imports and Exports," at end of Section 8.
• Totals may not equal sum of components due to independent rounding.
Web Pages: • For all data beginning in 1949, see http://www.eia.gov/emeu/aer/renew.html.
• For related information, see http://www.eia.gov/fuelrenewable.html.

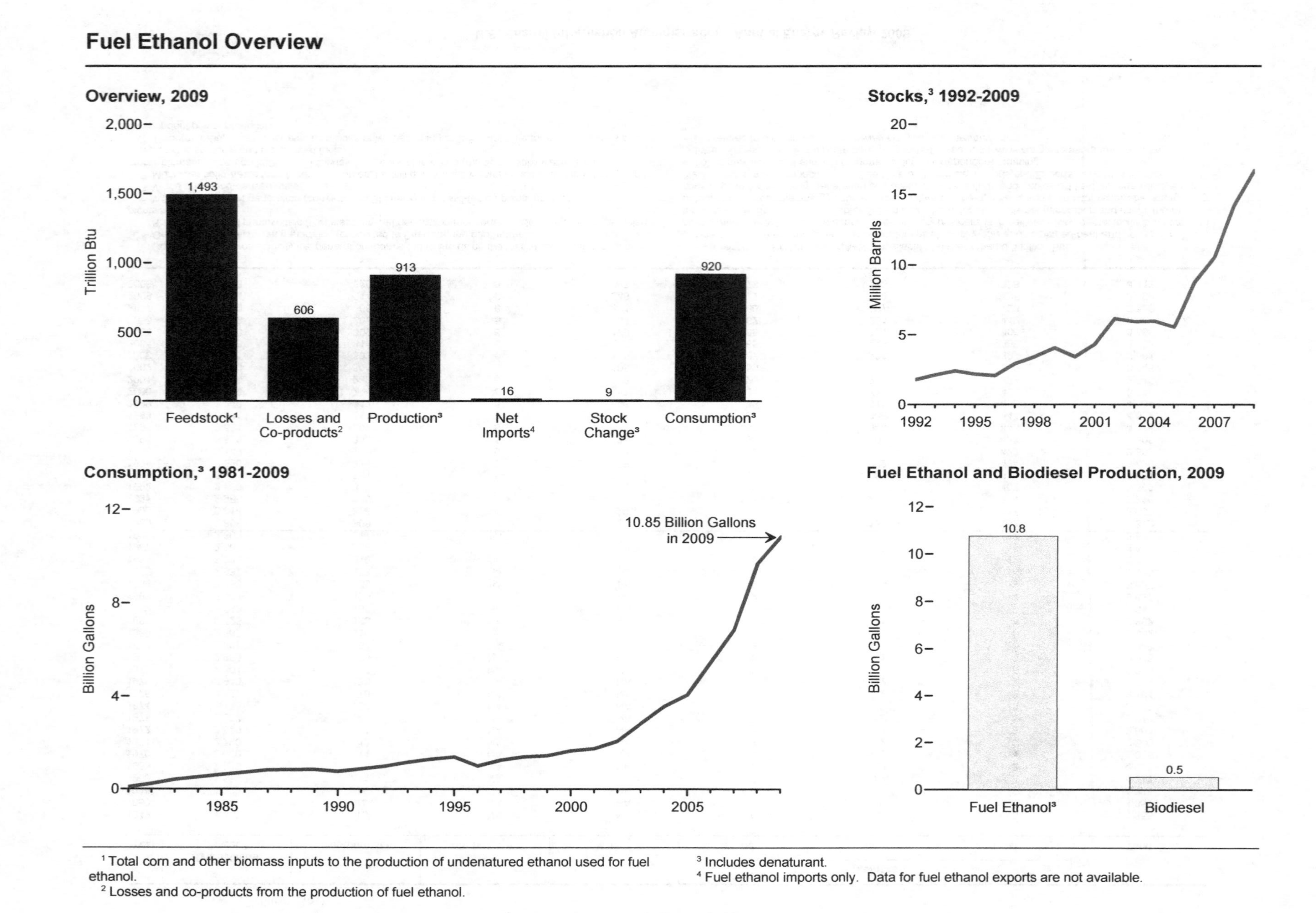


[1] Total corn and other biomass inputs to the production of undenatured ethanol used for fuel ethanol.
[2] Losses and co-products from the production of fuel ethanol.
[3] Includes denaturant.
[4] Fuel ethanol imports only. Data for fuel ethanol exports are not available.

U.S. Energy Information Administration / Annual Energy Review 2009

## Fuel Ethanol Overview, 1981-2009

| Year | Feed-stock [1] | Losses and Co-products [2] | Dena-turant [3] | Production [4] | | | Trade [4] Imports | Trade [4] Exports | Trade [4] Net Imports [5] | Stocks [4,6] | Stock Change [4,7] | Consumption [4] | | | *Consumption Minus Denaturant* [8] |
|---|---|---|---|---|---|---|---|---|---|---|---|---|---|---|---|
| | Trillion Btu | Trillion Btu | Thousand Barrels | Thousand Barrels | Million Gallons | Trillion Btu | Thousand Barrels | Thousand Barrels | Thousand Barrels | Thousand Barrels | Thousand Barrels | Thousand Barrels | Million Gallons | Trillion Btu | *Trillion Btu* |
| 1981 | 13 | 6 | 40 | 1,978 | 83 | 7 | NA | NA | NA | NA | NA | 1,978 | 83 | 7 | *7* |
| 1982 | R34 | 16 | 107 | 5,369 | 225 | 19 | NA | NA | NA | NA | NA | 5,369 | 225 | 19 | *19* |
| 1983 | R63 | 29 | 198 | 9,890 | 415 | 35 | NA | NA | NA | NA | NA | 9,890 | 415 | 35 | *34* |
| 1984 | R77 | R35 | 243 | 12,150 | 510 | 43 | NA | NA | NA | NA | NA | 12,150 | 510 | 43 | *42* |
| 1985 | R93 | R42 | 294 | 14,693 | 617 | 52 | NA | NA | NA | NA | NA | 14,693 | 617 | 52 | *51* |
| 1986 | R107 | R48 | 339 | 16,954 | 712 | 60 | NA | NA | NA | NA | NA | 16,954 | 712 | 60 | *59* |
| 1987 | R123 | R55 | 390 | 19,497 | 819 | 69 | NA | NA | NA | NA | NA | 19,497 | 819 | 69 | *68* |
| 1988 | R124 | R55 | 396 | 19,780 | 831 | 70 | NA | NA | NA | NA | NA | 19,780 | 831 | 70 | *69* |
| 1989 | R125 | R56 | 401 | 20,062 | 843 | 71 | NA | NA | NA | NA | NA | 20,062 | 843 | 71 | *70* |
| 1990 | R111 | R49 | 356 | 17,802 | 748 | 63 | NA | NA | NA | NA | NA | 17,802 | 748 | 63 | *62* |
| 1991 | R128 | R56 | 413 | 20,627 | 866 | 73 | NA | NA | NA | NA | NA | 20,627 | 866 | 73 | *72* |
| 1992 | R145 | R64 | 469 | 23,453 | 985 | R84 | NA | NA | NA | 1,791 | NA | 23,453 | 985 | R84 | *81* |
| 1993 | R169 | R74 | 550 | 27,484 | 1,154 | R98 | 244 | NA | 244 | 2,114 | 323 | 27,405 | 1,151 | R98 | *95* |
| 1994 | R188 | R82 | 614 | 30,689 | 1,289 | 109 | 279 | NA | 279 | 2,393 | 279 | 30,689 | 1,289 | 109 | *106* |
| 1995 | R198 | R86 | 647 | 32,325 | 1,358 | R115 | 387 | NA | 387 | 2,186 | -207 | 32,919 | 1,383 | 117 | *114* |
| 1996 | R141 | R61 | 464 | 23,178 | 973 | R83 | 313 | NA | 313 | 2,065 | -121 | 23,612 | 992 | 84 | *82* |
| 1997 | R186 | R80 | 613 | 30,674 | 1,288 | 109 | 85 | NA | 85 | 2,925 | 860 | 29,899 | 1,256 | R107 | *104* |
| 1998 | R202 | R86 | 669 | 33,453 | 1,405 | R119 | 66 | NA | 66 | 3,406 | 481 | 33,038 | 1,388 | R118 | *115* |
| 1999 | R211 | R90 | 698 | 34,881 | 1,465 | R124 | 87 | NA | 87 | 4,024 | 618 | 34,350 | 1,443 | 122 | *119* |
| 2000 | R233 | R99 | 773 | 38,627 | 1,622 | R138 | 116 | NA | 116 | 3,400 | -624 | 39,367 | 1,653 | R140 | *137* |
| 2001 | R253 | R108 | 841 | 42,028 | 1,765 | R150 | 315 | NA | 315 | 4,298 | 898 | 41,445 | 1,741 | R148 | *144* |
| 2002 | R307 | R130 | 1,019 | 50,956 | 2,140 | R182 | 306 | NA | 306 | 6,200 | 1,902 | 49,360 | 2,073 | R176 | *171* |
| 2003 | R400 | R169 | 1,335 | 66,772 | 2,804 | R238 | 292 | NA | 292 | 5,978 | -222 | 67,286 | 2,826 | R240 | *233* |
| 2004 | R484 | R203 | 1,621 | 81,058 | 3,404 | R289 | 3,542 | NA | 3,542 | 6,002 | 24 | 84,576 | 3,552 | R301 | *293* |
| 2005 | R552 | R230 | 1,859 | 92,961 | 3,904 | R331 | 3,234 | NA | 3,234 | 5,563 | -439 | 96,634 | 4,059 | R344 | *335* |
| 2006 | R688 | R285 | 2,326 | 116,294 | 4,884 | R414 | 17,408 | NA | 17,408 | 8,760 | 3,197 | 130,505 | 5,481 | R465 | *453* |
| 2007 | R914 | R376 | 3,105 | 155,263 | 6,521 | R553 | 10,457 | NA | 10,457 | 10,535 | 1,775 | 163,945 | 6,886 | R584 | *569* |
| 2008 | R1,300 | R531 | 4,433 | R221,637 | R9,309 | R790 | R12,610 | NA | R12,610 | R14,226 | R3,691 | R230,556 | R9,683 | R821 | *800* |
| 2009P | 1,493 | 606 | 5,507 | 256,149 | 10,758 | 913 | 4,614 | – | 4,614 | 16,711 | [9]2,492 | 258,271 | 10,847 | 920 | *894* |

[1] Total corn and other biomass inputs to the production of undenatured ethanol used for fuel ethanol.
[2] Losses and co-products from the production of fuel ethanol. Does not include natural gas, electricity, and other non-biomass energy used in the production of fuel ethanol—these are included in the industrial sector consumption statistics for the appropriate energy source.
[3] The amount of denaturant in fuel ethanol produced.
[4] Includes denaturant.
[5] Net imports equal imports minus exports.
[6] Stocks are at end of year.
[7] A negative value indicates a decrease in stocks and a positive value indicates an increase.
[8] Consumption of fuel ethanol minus denaturant. Data for fuel ethanol minus denaturant are used to develop data for "Renewable Energy/Biomass" in Tables 10.1-10.2b, as well as in Sections 1 and 2.
[9] Derived from the preliminary 2008 stocks value, not the final 2008 value shown in this table.
R=Revised. P=Preliminary. NA=Not available. – = No data reported.
Notes: • Fuel ethanol data in thousand barrels are converted to million gallons by multiplying by 0.042, and are converted to Btu by multiplying by the approximate heat content of fuel ethanol—see Table A3. • Through 1980, data are not available. For 1981-1992, data are estimates. For 1993-2008, only data for feedstock, losses and co-products, and denaturant are estimates. For 2009, only data for feedstock, and losses and co-products, are estimates. • See "Denaturant," "Ethanol," "Fuel Ethanol," and "Fuel Ethanol Minus Denaturant" in Glossary. • Totals may not equal sum of components due to independent rounding.
Web Page: http://www.eia.gov/oil_gas/petroleum/data_publications/petroleum_supply_monthly/psm.html.
Sources: **Feedstock:** Calculated as fuel ethanol production (in thousand barrels) minus denaturant, and then multiplied by the fuel ethanol feedstock factor—see Table A3. **Losses and Co-products:** Calculated as fuel ethanol feedstock plus denaturant minus fuel ethanol production. **Denaturant:** • 1981-2008—Data in thousand barrels for petroleum denaturant in fuel ethanol produced are estimated as 2 percent of fuel ethanol production; these data are converted to Btu by multiplying by 4.641 million Btu per barrel (the estimated quantity-weighted factor of pentanes plus and conventional motor gasoline used as denaturant). • 2009—U.S. Energy Information Administration (EIA), *Petroleum Supply Monthly (PSM)*, monthly reports, Table 1. Data in thousand barrels for net production of pentanes plus at renewable fuels and oxygenate plants are multiplied by -1; these data are converted to Btu by multiplying by 4.620 million Btu per barrel (the approximate heat content of pentanes plus). Data in thousand barrels for net production of conventional motor gasoline at renewable fuels and oxygenate plants are multiplied by -1; these data are converted to Btu by multiplying by 5.253 million Btu per barrel (the approximate heat content of conventional motor gasoline). Total denaturant is the sum of the values for pentanes plus and conventional motor gasoline. **Production:** • 1981-1992—Fuel ethanol production is assumed to equal fuel ethanol consumption—see sources for "Consumption." • 1993-2004—Calculated as fuel ethanol consumption plus fuel ethanol stock change minus fuel ethanol net imports. These data differ slightly from the original production data from EIA, Form EIA-819, "Monthly Oxygenate Report," and predecessor form, which were not reconciled and updated to be consistent with the final balance. • 2005-2008—EIA, Form EIA-819, "Monthly Oxygenate Report." • 2009—EIA, *PSM*, monthly reports. **Trade, Stocks, and Stock Change:** • 1992-2008—EIA, *Petroleum Supply Annual (PSA)*, annual reports. • 2009—EIA, *PSM*, monthly reports. **Consumption:** • 1981-1989—EIA, *Estimates of U.S. Biofuels Consumption 1990*, Table 10; and EIA, Office of Coal, Nuclear, Electric and Alternate Fuels (CNEAF), estimates. • 1990-1992—EIA, *Estimates of U.S. Biomass Energy Consumption 1992*, Table D2; and EIA, CNEAF, estimates. • 1993-2004—EIA, *PSA*, annual reports, Tables 2 and 16. Calculated as 10 percent of oxygenated finished motor gasoline field production (Table 2), plus fuel ethanol refinery input (Table 16). • 2005-2008—EIA, *PSA*, annual reports, Tables 1 and 15. Calculated as motor gasoline blending components adjustments (Table 1), plus finished motor gasoline adjustments (Table 1), plus fuel ethanol refinery and blender net inputs (Table 15). • 2009—EIA, *PSM*, monthly reports, Table 1. Calculated as fuel ethanol refinery and blender net inputs minus fuel ethanol adjustments. **Consumption Minus Denaturant:** Calculated as fuel ethanol consumption minus the amount of denaturant in fuel ethanol consumed. Denaturant in fuel ethanol consumed is estimated by multiplying denaturant in fuel ethanol produced by the fuel ethanol consumption-to-production ratio.

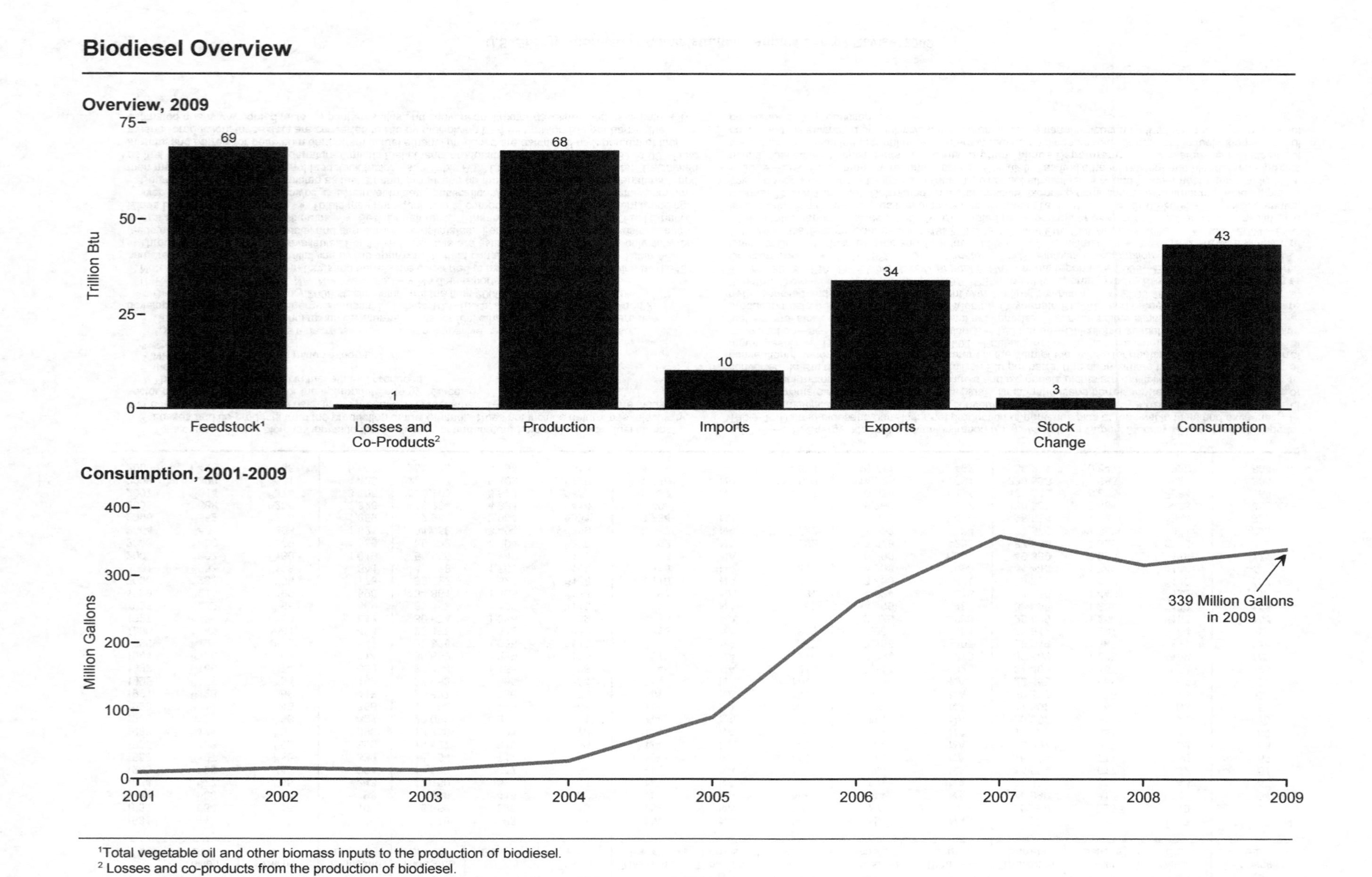


[1]Total vegetable oil and other biomass inputs to the production of biodiesel.
[2] Losses and co-products from the production of biodiesel.

U.S. Energy Information Administration / Annual Energy Review 2009

# Biodiesel Overview, 2001-2009

| | Feedstock [1] | Losses and Co-products [2] | Production | | | Trade | | | Stocks [4] | Stock Change [5] | Balancing Item [6] | Consumption | | |
|---|---|---|---|---|---|---|---|---|---|---|---|---|---|---|
| | | | | | | Imports | Exports | Net Imports [3] | | | | | | |
| Year | Trillion Btu | Trillion Btu | Thousand Barrels | Million Gallons | Trillion Btu | Thousand Barrels | Thousand Barrels | Thousand Barrels | Thousand Barrels | Thousand Barrels | Thousand Barrels | Thousand Barrels | Million Gallons | Trillion Btu |
| 2001 | 1 | (s) | 204 | 9 | 1 | 78 | 39 | 39 | NA | NA | NA | 243 | 10 | 1 |
| 2002 | 1 | (s) | 250 | 10 | 1 | 191 | 56 | 135 | NA | NA | NA | 385 | 16 | 2 |
| 2003 | 2 | (s) | 338 | 14 | 2 | 94 | 110 | -16 | NA | NA | NA | 322 | 14 | 2 |
| 2004 | 4 | (s) | 666 | 28 | 4 | 97 | 124 | -26 | NA | NA | NA | 640 | 27 | 3 |
| 2005 | 12 | (s) | 2,162 | 91 | 12 | 207 | 206 | 1 | NA | NA | NA | 2,163 | 91 | 12 |
| 2006 | 32 | (s) | 5,963 | 250 | 32 | 1,069 | 828 | 242 | NA | NA | NA | 6,204 | 261 | 33 |
| 2007 | 63 | 1 | 11,662 | 490 | 62 | 3,342 | 6,477 | -3,135 | NA | NA | NA | 8,528 | 358 | 46 |
| 2008 | 88 | 1 | R16,145 | R678 | 87 | 7,502 | 16,128 | -8,626 | NA | NA | NA | R7,519 | R316 | R40 |
| 2009P | 69 | 1 | 12,657 | 532 | 68 | 1,844 | 6,332 | -4,489 | 506 | 506 | 419 | 8,082 | 339 | 43 |

[1] Total vegetable oil and other biomass inputs to the production of biodiesel.

[2] Losses and co-products from the production of biodiesel. Does not include natural gas, electricity, and other non-biomass energy used in the production of biodiesel—these are included in the industrial sector consumption statistics for the appropriate energy source.

[3] Net imports equal imports minus exports.

[4] Stocks are at end of year.

[5] A negative value indicates a decrease in stocks and a positive value indicates an increase.

[6] Beginning in 2009, because of incomplete data coverage and different data sources, "Balancing Item" is used to balance biodiesel supply and disposition.

R=Revised. P=Preliminary. NA=Not available. (s)=Less than 0.5 trillion Btu.

Notes: • Biodiesel data in thousand barrels are converted to million gallons by multiplying by 0.042, and are converted to Btu by multiplying by 5.359 million Btu per barrel (the approximate heat content of biodiesel—see Table A3). • Through 2000, data are not available. Beginning in 2001, data not from U.S. Energy Information Administration (EIA) surveys are estimates. • Totals may not equal sum of components due to independent rounding.

Web Page: See http://www.census.gov/manufacturing/cir/historical_data/m311k/index.html for related information.

Sources: **Feedstock:** Calculated as biodiesel production in thousand barrels multiplied by 5.433 million Btu per barrel (the biodiesel feedstock factor—see Table A3). **Losses and Co-products:** Calculated as biodiesel feedstock minus biodiesel production. **Production:** • 2001-2005—U.S. Department of Agriculture, Commodity Credit Corporation, Bioenergy Program records. Annual data are derived from quarterly data. • 2006—U.S. Department of Commerce, Bureau of the Census, "M311K - Fats and Oils: Production, Consumption, and Stocks," data for soybean oil consumed in methyl esters (biodiesel). In addition, the EIA, Office of Integrated Analysis and Forecasting, estimates that 14.4 million gallons of yellow grease were consumed in methyl esters (biodiesel). • 2007—U.S. Department of Commerce, Bureau of the Census, "M311K - Fats and Oils: Production, Consumption, and Stocks," data for all fats and oils consumed in methyl esters (biodiesel). • 2008—EIA, *Monthly Biodiesel Production Report, March 2009* (release date April 2010), Table 10. • 2009—The 2009 annual value is the sum of January-March 2009 values from EIA, *Monthly Biodiesel Production Report, March 2009* (release date April 2010), Table 10; and April-December 2009 values from U.S. Department of Commerce, Bureau of the Census, "M311K - Fats and Oils: Production, Consumption, and Stocks," data for all fats and oils consumed in methyl esters (biodiesel). **Trade:** U.S. Department of Agriculture, imports data for Harmonized Tariff Schedule code 3824.90.40.20 (Fatty Esters Animal/Vegetable/Mixture), and exports data for Schedule B code 3824.90.40.00 (Fatty Substances Animal/Vegetable/Mixture). Although these categories include products other than biodiesel (such as those destined for soaps, cosmetics, and other items), biodiesel is the largest component. In the absence of other reliable data for biodiesel trade, EIA sees these data as good estimates. **Stocks** and **Stock Change:** • 2009—EIA *Petroleum Supply Monthly (PSM),* monthly reports, Table 1, data for renewable fuels except fuel ethanol. **Balancing Item:** • 2009—Calculated as biodiesel consumption and biodiesel stock change minus biodiesel production and biodiesel net imports. **Consumption:** • 2001-2008—Calculated as biodiesel production plus biodiesel net imports. • 2009—Calculated as the sum of the monthly consumption data. Data for January and February 2009 are from EIA, *PSM,* monthly reports, Table 1, refinery and blender net inputs of renewable fuels except fuel ethanol. Data for March-December 2009 are calculated as biodiesel production plus biodiesel net imports minus biodiesel stock change.

# Estimated Number of Alternative-Fueled Vehicles in Use and Alternative Fuel Consumption

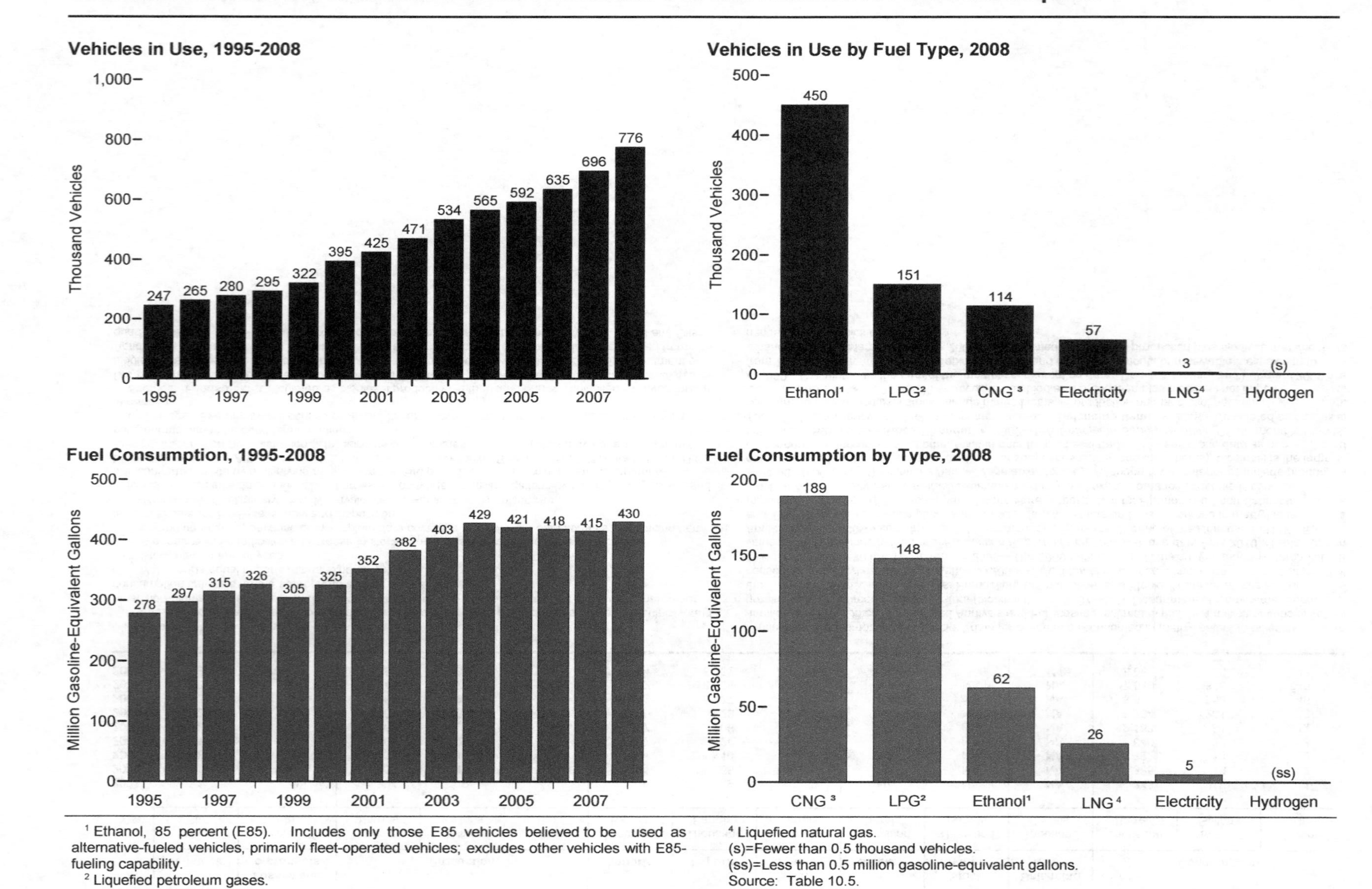


[1] Ethanol, 85 percent (E85). Includes only those E85 vehicles believed to be used as alternative-fueled vehicles, primarily fleet-operated vehicles; excludes other vehicles with E85-fueling capability.
[2] Liquefied petroleum gases.
[3] Compressed natural gas.
[4] Liquefied natural gas.
(s)=Fewer than 0.5 thousand vehicles.
(ss)=Less than 0.5 million gasoline-equivalent gallons.
Source: Table 10.5.

U.S. Energy Information Administration / Annual Energy Review 2009

## Estimated Number of Alternative-Fueled Vehicles in Use and Fuel Consumption, 1992-2008

| Year | Alternative and Replacement Fuels [1] | | | | | | | | | | | | | | | |
|---|---|---|---|---|---|---|---|---|---|---|---|---|---|---|---|---|
| | Liquefied Petroleum Gases | Compressed Natural Gas | Liquefied Natural Gas | Methanol, 85 Percent (M85) [3] | Methanol, Neat (M100) [4] | Ethanol, 85 Percent (E85) [3,5] | Ethanol, 95 Percent (E95) [3] | Electricity [6] | Hydrogen | Other Fuels [7] | Subtotal | Oxygenates [2]: Methyl Tertiary Butyl Ether [8] | Oxygenates [2]: Ethanol in Gasohol [9] | Oxygenates [2]: Total | Biodiesel [10] | Total |
| | **Alternative-Fueled Vehicles in Use [11] (number)** | | | | | | | | | | | | | | | |
| 1992 | NA | 23,191 | 90 | 4,850 | 404 | 172 | 38 | 1,607 | NA | NA | NA | NA | NA | NA | NA | NA |
| 1993 | NA | 32,714 | 299 | 10,263 | 414 | 441 | 27 | 1,690 | NA | NA | NA | NA | NA | NA | NA | NA |
| 1994 | NA | 41,227 | 484 | 15,484 | 415 | 605 | 33 | 2,224 | NA | NA | NA | NA | NA | NA | NA | NA |
| 1995 | 172,806 | 50,218 | 603 | 18,319 | 386 | 1,527 | 136 | 2,860 | 0 | 0 | 246,855 | NA | NA | NA | NA | NA |
| 1996 | 175,585 | 60,144 | 663 | 20,265 | 172 | 4,536 | 361 | 3,280 | 0 | 0 | 265,006 | NA | NA | NA | NA | NA |
| 1997 | 175,679 | 68,571 | 813 | 21,040 | 172 | 9,130 | 347 | 4,453 | 0 | 0 | 280,205 | NA | NA | NA | NA | NA |
| 1998 | 177,183 | 78,782 | 1,172 | 19,648 | 200 | 12,788 | 14 | 5,243 | 0 | 0 | 295,030 | NA | NA | NA | NA | NA |
| 1999 | 178,610 | 91,267 | 1,681 | 18,964 | 198 | 24,604 | 14 | 6,964 | 0 | 0 | 322,302 | NA | NA | NA | NA | NA |
| 2000 | 181,994 | 100,750 | 2,090 | 10,426 | 0 | 87,570 | 4 | 11,830 | 0 | 0 | 394,664 | NA | NA | NA | NA | NA |
| 2001 | 185,053 | 111,851 | 2,576 | 7,827 | 0 | 100,303 | 0 | 17,847 | 0 | 0 | 425,457 | NA | NA | NA | NA | NA |
| 2002 | 187,680 | 120,839 | 2,708 | 5,873 | 0 | 120,951 | 0 | 33,047 | 0 | 0 | 471,098 | NA | NA | NA | NA | NA |
| 2003 | 190,369 | 114,406 | 2,640 | 0 | 0 | 179,090 | 0 | 47,485 | 9 | 0 | 533,999 | NA | NA | NA | NA | NA |
| 2004 | 182,864 | 118,532 | 2,717 | 0 | 0 | 211,800 | 0 | 49,536 | 43 | 0 | 565,492 | NA | NA | NA | NA | NA |
| 2005 | 173,795 | 117,699 | 2,748 | 0 | 0 | 246,363 | 0 | 51,398 | 119 | 3 | 592,125 | NA | NA | NA | NA | NA |
| 2006 | 164,846 | 116,131 | 2,798 | 0 | 0 | 297,099 | 0 | 53,526 | 159 | 3 | 634,562 | NA | NA | NA | NA | NA |
| 2007 | 158,254 | 114,391 | 2,781 | 0 | 0 | 364,384 | 0 | 55,730 | 223 | 3 | 695,766 | NA | NA | NA | NA | NA |
| 2008P | 151,049 | 113,973 | 3,101 | 0 | 0 | 450,327 | 0 | 56,901 | 313 | 3 | 775,667 | NA | NA | NA | NA | NA |
| | **Fuel Consumption [12] (thousand gasoline-equivalent gallons)** | | | | | | | | | | | | | | | |
| 1992 | NA | 17,159 | 598 | 1,121 | 2,672 | 22 | 87 | 359 | NA | NA | NA | 1,175,964 | 719,408 | 1,895,372 | NA | NA |
| 1993 | NA | 22,035 | 1,944 | 1,671 | 3,321 | 49 | 82 | 288 | NA | NA | NA | 2,070,897 | 779,958 | 2,850,854 | NA | NA |
| 1994 | NA | 24,643 | 2,398 | 2,455 | 3,347 | 82 | 144 | 430 | NA | NA | NA | 2,020,455 | 868,113 | 2,888,569 | NA | NA |
| 1995 | 233,178 | 35,865 | 2,821 | 2,122 | 2,255 | 195 | 1,021 | 663 | 0 | 0 | 278,121 | 2,693,407 | 934,615 | 3,628,022 | NA | 3,906,142 |
| 1996 | 239,648 | 47,861 | 3,320 | 1,862 | 364 | 712 | 2,770 | 773 | 0 | 0 | 297,310 | 2,751,955 | 677,537 | 3,429,492 | NA | 3,726,802 |
| 1997 | 238,845 | 66,495 | 3,798 | 1,630 | 364 | 1,314 | 1,166 | 1,010 | 0 | 0 | 314,621 | 3,106,745 | 852,514 | 3,959,260 | NA | 4,273,880 |
| 1998 | 241,881 | 73,859 | 5,463 | 1,271 | 471 | 1,772 | 61 | 1,202 | 0 | 0 | 325,980 | 2,905,781 | 912,858 | 3,818,639 | NA | 4,144,620 |
| 1999 | 210,247 | 81,211 | 5,959 | 1,126 | 469 | 4,019 | 64 | 1,524 | 0 | 0 | 304,618 | 3,405,390 | 975,255 | 4,380,645 | NA | 4,685,263 |
| 2000 | 213,012 | 88,478 | 7,423 | 614 | 0 | 12,388 | 13 | 3,058 | 0 | 0 | 324,986 | 3,298,803 | 1,114,313 | 4,413,116 | 6,828 | 4,744,930 |
| 2001 | 216,319 | 106,584 | 9,122 | 461 | 0 | 15,007 | 0 | 4,066 | 0 | 0 | 351,558 | 3,354,949 | 1,173,323 | 4,528,272 | 10,627 | 4,890,457 |
| 2002 | 223,600 | 123,081 | 9,593 | 354 | 0 | 18,250 | 0 | 7,274 | 0 | 0 | 382,152 | 3,122,859 | 1,450,721 | 4,573,580 | 16,824 | 4,972,556 |
| 2003 | 224,697 | 133,222 | 13,503 | 0 | 0 | 26,376 | 0 | 5,141 | 2 | 0 | 402,941 | 2,368,400 | 1,919,572 | 4,287,972 | 14,082 | 4,704,995 |
| 2004 | 211,883 | 158,903 | 20,888 | 0 | 0 | 31,581 | 0 | 5,269 | 8 | 0 | 428,532 | 1,877,300 | 2,414,167 | 4,291,467 | R27,616 | R4,747,615 |
| 2005 | 188,171 | 166,878 | 22,409 | 0 | 0 | 38,074 | 0 | 5,219 | 25 | 2 | 420,778 | 1,654,500 | 2,756,663 | 4,411,163 | R93,281 | R4,925,222 |
| 2006 | 173,130 | 172,011 | 23,474 | 0 | 0 | 44,041 | 0 | 5,104 | 41 | 2 | 417,803 | 435,000 | 3,729,168 | 4,164,168 | R267,623 | R4,849,594 |
| 2007 | 152,360 | 178,565 | 24,594 | 0 | 0 | 54,091 | 0 | 5,037 | 66 | 2 | 414,715 | R0 | 4,694,304 | R4,694,304 | R367,764 | R5,476,783 |
| 2008P | 147,784 | 189,358 | 25,554 | 0 | 0 | 62,464 | 0 | 5,050 | 117 | 2 | 430,329 | 0 | 6,442,781 | 6,442,781 | 324,329 | 7,197,439 |

[1] See "Alternative Fuel" and "Replacement Fuel" in Glossary.
[2] See "Oxygenates" in Glossary.
[3] Remaining portion is motor gasoline. Consumption data include the motor gasoline portion of the fuel.
[4] One hundred percent methanol.
[5] Includes only those E85 vehicles believed to be used as alternative-fuels vehicles (AFVs), primarily fleet-operated vechicles; excludes other vehicles with E85-fueling capability. In 1997, some vehicle manufacturers began including E85-fueling capability in certain model lines of vehicles. For 2008, the U.S. Energy Information Administration (EIA) estimates that the number of E85 vehicles that are capable of operating on E85, motor gasoline, or both, is about 7.1 million. Many of these AFVs are sold and used as traditional gasoline-powered vehicles.
[6] Excludes gasoline-electric hybrids.
[7] May include P-Series fuel or any other fuel designated by the Secretary of Energy as an alternative fuel in acordance with the Energy Policy Act of 1995.
[8] In addition to methyl tertiary butyl ether (MTBE), includes a very small amount of other ethers, primarily tertiary amyl methyl ether (TAME) and ethyl tertiary butyl ether (ETBE).
[9] Data do not include the motor gasoline portion of the fuel.
[10] "Biodiesel" may be used as a diesel fuel substitute or diesel fuel additive or extender. See "Biodiesel" in Glossary.
[11] "Vehicles in Use" data represent accumulated acquisitions, less retirements, as of the end of each calendar year; data do not include concept and demonstration vehicles that are not ready for delivery to end users. See "Alternative-Fueled Vehicle" in Glossary.
[12] Fuel consumption quantities are expressed in a common base unit of gasoline-equivalent gallons to allow comparisons of different fuel types. Gasoline-equivalent gallons do not represent gasoline displacement. Gasoline equivalent is computed by dividing the gross heat content of the replacement fuel by the gross heat content of gasoline (using an approximate heat content of 122,619 Btu per gallon) and multiplying the result by the replacement fuel consumption value. See "Heat Content" in Glossary.

R=Revised. P=Preliminary. NA=Not available.

Note: Totals may not equal sum of components due to independent rounding.

Web Page: For related information, see http://www.eia.gov/fuelrenewable.html.

Sources: • 1992-1994—Science Applications International Corporation, "Alternative Transportation Fuels and Vehicles Data Development," unpublished final report prepared for the EIA, (McLean, VA, July 1996), and U.S. Department of Energy, Office of Energy Efficiency and Renewable Energy. Data were revised by using gross instead of net heat contents. For a table of gross and net heat contents, see EIA, *Alternatives to Traditional Transportation Fuels: An Overview* (June 1994), Table 22. • 1995-2002—EIA, "Alternatives to Traditional Transportation Fuels 2003 Estimated Data" (February 2004), Tables 1 and 10, and unpublished revisions. Data were revised by using gross instead of net heat contents. • 2003 forward—EIA, "Alternatives to Traditional Transportation Fuels," annual reports, Tables V1 and C1, and unpublished revisions.

## Solar Thermal Collector Shipments by Type, Price, and Trade

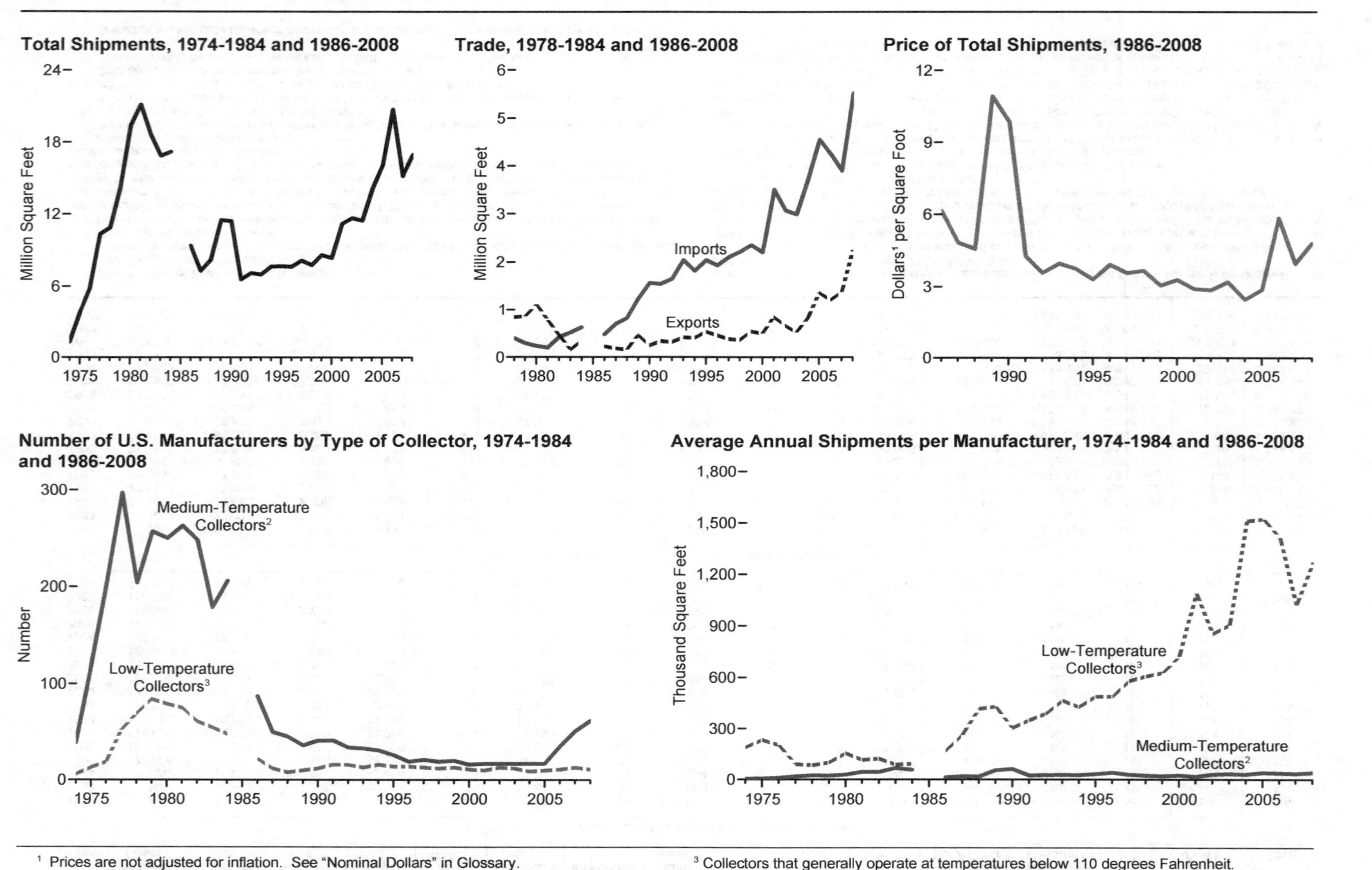


[1] Prices are not adjusted for inflation. See "Nominal Dollars" in Glossary.
[2] Collectors that generally operate in the temperature range of 140 degrees Fahrenheit to 180 degrees Fahrenheit but can also operate at temperatures as low as 110 degrees Fahrenheit. Special collectors–evacuated tube collectors or concentrating (focusing) collectors–are included in the medium-temperature category.
[3] Collectors that generally operate at temperatures below 110 degrees Fahrenheit.

Notes: • Shipments are for domestic and export shipments, and may include imports that subsequently were shipped to domestic or foreign customers. • Data were not collected for 1985.

U.S. Energy Information Administration / Annual Energy Review 2009

## Solar Thermal Collector Shipments by Type, Price, and Trade, 1974-2008

(Thousand Square Feet, Except as Noted)

| | Low-Temperature Collectors [1] | | | | Medium-Temperature Collectors [2] | | | | High-Temperature Collectors [3] | | Total Shipments | | Trade | |
|---|---|---|---|---|---|---|---|---|---|---|---|---|---|---|
| Year | Number of U.S. Manu-facturers | Quantity Shipped | Shipments per Manu-facturer | Price [4] (dollars [5] per square foot) | Number of U.S. Manu-facturers | Quantity Shipped | Shipments per Manu-facturer | Price [4] (dollars [5] per square foot) | Quantity Shipped | Price [4] (dollars [5] per square foot) | Quantity Shipped | Price [4] (dollars [5] per square foot) | Imports | Exports |
| 1974 | 6 | 1,137 | 190 | NA | 39 | 137 | 4 | NA | NA | NA | 1,274 | NA | NA | NA |
| 1975 | 13 | 3,026 | 233 | NA | 118 | 717 | 6 | NA | NA | NA | 3,743 | NA | NA | NA |
| 1976 | 19 | 3,876 | 204 | NA | 203 | 1,925 | 10 | NA | NA | NA | 5,801 | NA | NA | NA |
| 1977 | 52 | 4,743 | 91 | NA | 297 | 5,569 | 19 | NA | NA | NA | 10,312 | NA | NA | NA |
| 1978 | 69 | 5,872 | 85 | NA | 204 | 4,988 | 25 | NA | NA | NA | 10,860 | NA | 396 | 840 |
| 1979 | 84 | 8,394 | 100 | NA | 257 | 5,856 | 23 | NA | NA | NA | 14,251 | NA | 290 | 855 |
| 1980 | 79 | 12,233 | 155 | NA | 250 | 7,165 | 29 | NA | NA | NA | 19,398 | NA | 235 | 1,115 |
| 1981 | 75 | 8,677 | 116 | NA | 263 | 11,456 | 44 | NA | NA | NA | 21,133 | NA | 196 | 771 |
| 1982 | 61 | 7,476 | 123 | NA | 248 | 11,145 | 45 | NA | NA | NA | 18,621 | NA | 418 | 455 |
| 1983 | 55 | 4,853 | 88 | NA | 179 | 11,975 | 67 | NA | NA | NA | 16,828 | NA | 511 | 159 |
| 1984 | 48 | 4,479 | 93 | NA | 206 | 11,939 | 58 | NA | 773 | NA | 17,191 | NA | 621 | 348 |
| 1985[6] | NA | NA | NA | NA | NA | NA | NA | NA | NA | NA | NA | NA | NA | NA |
| 1986 | 22 | 3,751 | 171 | 2.30 | 87 | 1,111 | 13 | 18.30 | 4,498 | NA | 9,360 | 6.14 | 473 | 224 |
| 1987 | 12 | 3,157 | 263 | 2.18 | 50 | 957 | 19 | 13.50 | 3,155 | NA | 7,269 | 4.82 | 691 | 182 |
| 1988 | 8 | 3,326 | 416 | 2.24 | 45 | 732 | 16 | 14.88 | 4,116 | NA | 8,174 | 4.56 | 814 | 158 |
| 1989 | 10 | 4,283 | 428 | 2.60 | 36 | 1,989 | 55 | 11.74 | 5,209 | 17.76 | 11,482 | 10.92 | 1,233 | 461 |
| 1990 | 12 | 3,645 | 304 | 2.90 | 41 | 2,527 | 62 | 7.68 | 5,237 | 15.74 | 11,409 | 9.86 | 1,562 | 245 |
| 1991 | 16 | 5,585 | 349 | 2.90 | 41 | 989 | 24 | 11.94 | 1 | 31.94 | 6,574 | 4.26 | 1,543 | 332 |
| 1992 | 16 | 6,187 | 387 | 2.50 | 34 | 897 | 26 | 10.96 | 2 | 75.66 | 7,086 | 3.58 | 1,650 | 316 |
| 1993 | 13 | 6,025 | 464 | 2.80 | 33 | 931 | 28 | 11.74 | 12 | 22.12 | 6,968 | 3.96 | 2,039 | 411 |
| 1994 | 16 | 6,823 | 426 | 2.54 | 31 | 803 | 26 | 13.54 | 2 | 177.00 | 7,627 | 3.74 | 1,815 | 405 |
| 1995 | 14 | 6,813 | 487 | 2.32 | 26 | 840 | 32 | 10.48 | 13 | 53.26 | 7,666 | 3.30 | 2,037 | 530 |
| 1996 | 14 | 6,821 | 487 | 2.67 | 19 | 785 | 41 | 14.48 | 10 | 18.75 | 7,616 | 3.91 | 1,930 | 454 |
| 1997 | 13 | 7,524 | 579 | 2.60 | 21 | 606 | 29 | 15.17 | 7 | 25.00 | 8,138 | 3.56 | 2,102 | 379 |
| 1998 | 12 | 7,292 | 607 | 2.83 | 19 | 443 | 23 | 15.17 | 21 | 53.21 | 7,756 | 3.66 | 2,206 | 360 |
| 1999 | 13 | 8,152 | 627 | 2.08 | 20 | 427 | 21 | 19.12 | 4 | 286.49 | 8,583 | 3.05 | 2,352 | 537 |
| 2000 | 11 | 7,948 | 723 | 2.09 | 16 | 400 | 25 | W | 5 | W | 8,354 | 3.28 | 2,201 | 496 |
| 2001 | 10 | 10,919 | 1,092 | 2.15 | 17 | 268 | 16 | W | 2 | W | 11,189 | 2.90 | 3,502 | 840 |
| 2002 | 13 | 11,126 | 856 | 1.97 | 17 | 535 | 31 | W | 2 | W | 11,663 | 2.85 | 3,068 | 659 |
| 2003 | 12 | 10,877 | 906 | 2.08 | 17 | 560 | 33 | W | 7 | W | 11,444 | 3.19 | 2,986 | 518 |
| 2004 | 9 | 13,608 | 1,512 | 1.80 | 17 | 506 | 30 | 19.30 | R – | – – | 14,114 | 2.43 | 3,723 | 813 |
| 2005 | 10 | 15,224 | 1,522 | 2.00 | 17 | 702 | 41 | W | 115 | W | 16,041 | 2.86 | 4,546 | 1,361 |
| 2006 | 11 | 15,546 | 1,413 | 1.95 | 35 | 1,346 | 38 | W | 3,852 | W | 20,744 | 5.84 | 4,244 | 1,211 |
| 2007 | 13 | 13,323 | 1,025 | 1.97 | 51 | 1,797 | 35 | W | 33 | W | 15,153 | 3.95 | 3,891 | 1,376 |
| 2008 | 11 | 14,015 | 1,274 | 1.89 | 62 | 2,560 | 41 | W | 388 | 11.96 | 16,963 | 4.80 | 5,517 | 2,247 |

[1] Low-temperature collectors are solar thermal collectors that generally operate at temperatures below 110° F.

[2] Medium-temperature collectors are solar thermal collectors that generally operate in the temperature range of 140° F to 180° F but can also operate at temperatures as low as 110 ° F. Special collectors are included in this category. Special collectors are evacuated tube collectors or concentrating (focusing) collectors. They operate in the temperature range from just above ambient temperature (low concentration for pool heating) to several hundred degrees Fahrenheit (high concentration for air conditioning and specialized industrial processes).

[3] High-temperature collectors are solar thermal collectors that generally operate at temperatures above 180° F. High-temperature collector shipments are dominated by one manufacturer, and the collectors are used by the electric power sector to build new central station solar thermal power plants and generate electricity. Year-to-year fluctations depend on how much new capacity is brought online.

[4] Prices equal shipment value divided by quantity shipped. Value includes charges for advertising and warranties. Excluded are excise taxes and the cost of freight or transportation for the shipments.

[5] Prices are not adjusted for inflation. See "Nominal Dollars" in Glossary.

[6] No data are available for 1985.

R=Revised. NA=Not available. – = No data reported. – – = Not applicable. W=Value withheld to avoid disclosure of proprietary company data.

Notes: • Shipments data are for domestic and export shipments, and may include imports that subsequently were shipped to domestic or foreign customers. • Manufacturers producing more than one type of collector are accounted for in both groups.

Web Page: For related information, see http://www.eia.gov/fuelrenewable.html.

Sources: • 1974-1992—U.S. Energy Information Administration (EIA), *Solar Collector Manufacturing Activity,* annual reports, and Form CE-63A, "Annual Solar Thermal Collector Manufacturers Survey," and predecessor forms. • 1993-2002—EIA, *Renewable Energy Annual,* annual reports, and Form EIA-63A, "Annual Solar Thermal Collector Manufacturers Survey," and predecessor form. • 2003 forward—EIA, *Solar Thermal Collector Manufacturing Activities* (and predecessor reports), annual reports, and Form EIA-63A, "Annual Solar Thermal Collector Manufacturers Survey."

## Solar Thermal Collector Domestic Shipments by Market Sector, End-Use, and Type, 2008

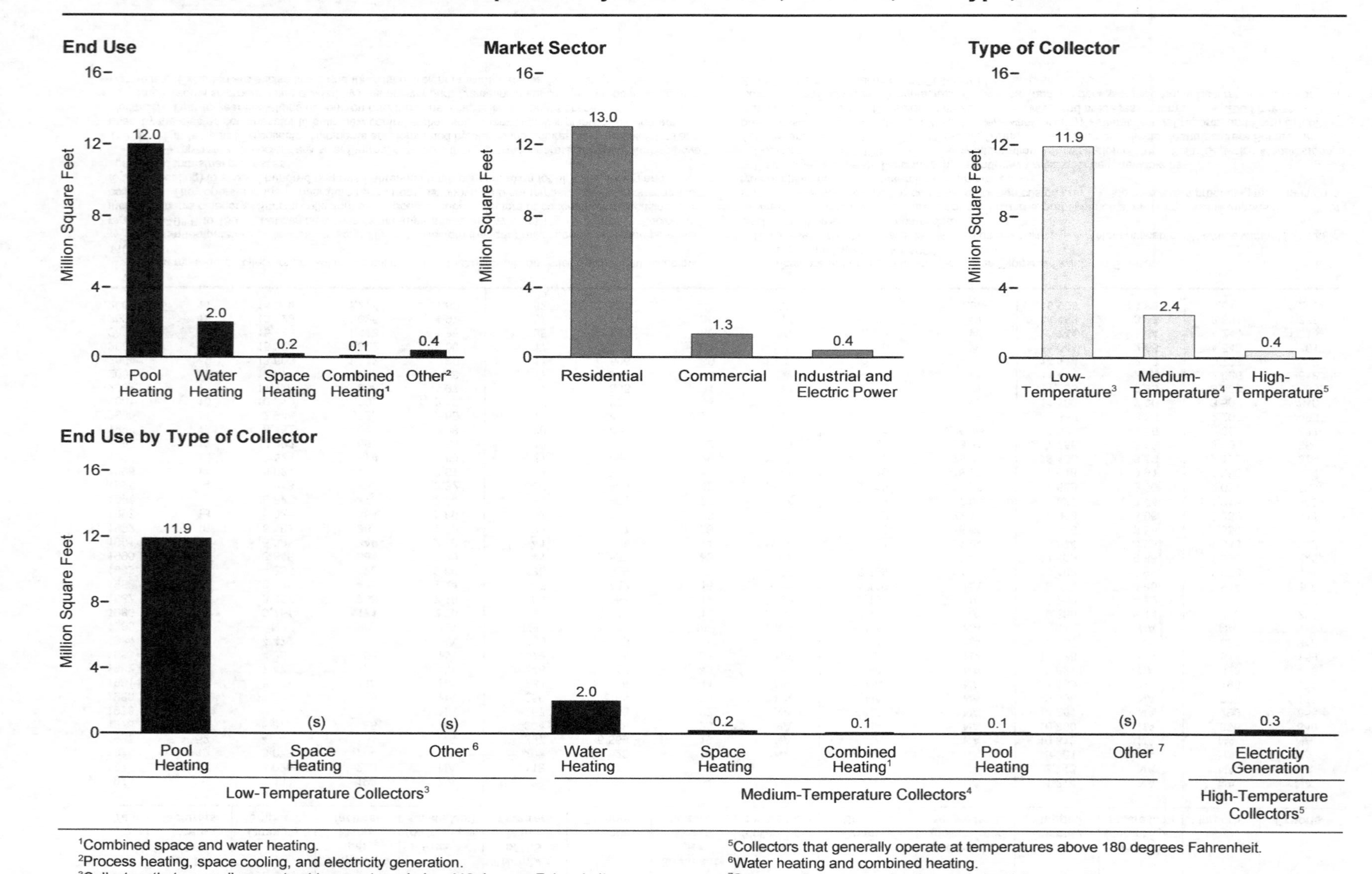


[1]Combined space and water heating.
[2]Process heating, space cooling, and electricity generation.
[3]Collectors that generally operate at temperatures below 110 degrees Fahrenheit.
[4]Collectors that generally operate in the temperature range of 140 degrees Fahrenheit to 180 degrees Fahrenheit but can also operate at temperatures as low as 110 degrees Fahrenheit.
[5]Collectors that generally operate at temperatures above 180 degrees Fahrenheit.
[6]Water heating and combined heating.
[7]Space cooling, process heating, and electricity generation.
(s)=Less than 0.05 million square feet.

U.S. Energy Information Administration / Annual Energy Review 2009

## Solar Thermal Collector Shipments by Market Sector, End Use, and Type, 2000-2008

(Thousand Square Feet)

| Year and Type | By Market Sector | | | | | By End Use | | | | | | | Total |
|---|---|---|---|---|---|---|---|---|---|---|---|---|---|
| | Residential | Commercial [1] | Industrial [2] | Electric Power [3] | Other [4] | Pool Heating | Water Heating | Space Heating | Space Cooling | Combined Heating [5] | Process Heating | Electricity Generation | |
| **Total Shipments [6]** | | | | | | | | | | | | | |
| **2000 Total** | 7,473 | 810 | 57 | 5 | 10 | 7,863 | 367 | 99 | 0 | 2 | 20 | 3 | 8,355 |
| Low [7] | 7,102 | 786 | 52 | 0 | 9 | 7,836 | 0 | 92 | 0 | 0 | 20 | 0 | 7,949 |
| Medium [8] | 370 | 23 | 5 | (s) | 1 | 26 | 365 | 7 | 0 | 2 | 0 | 0 | 400 |
| High [9] | 1 | 1 | 0 | 3 | 0 | 0 | 2 | 0 | 0 | (s) | 0 | 3 | 5 |
| **2001 Total** | 10,125 | 1,012 | 17 | 1 | 35 | 10,797 | 274 | 70 | 0 | 12 | 34 | 2 | 11,189 |
| Low [7] | 9,885 | 987 | 12 | 0 | 34 | 10,782 | 42 | 61 | 0 | 0 | 34 | 0 | 10,919 |
| Medium [8] | 240 | 24 | 5 | 0 | 1 | 16 | 232 | 9 | 0 | 12 | 0 | 0 | 268 |
| High [9] | 0 | 1 | 0 | 1 | 0 | 0 | 0 | 0 | 0 | 0 | 0 | 2 | 2 |
| **2002 Total** | 11,000 | 595 | 62 | 4 | 1 | 11,073 | 423 | 146 | (s) | 17 | 4 | 0 | 11,663 |
| Low [7] | 10,519 | 524 | 2 | 0 | 0 | 11,045 | 1 | 0 | 0 | 0 | 0 | 0 | 11,046 |
| Medium [8] | 481 | 69 | 60 | 4 | 1 | 28 | 422 | 146 | (s) | 15 | 4 | 0 | 615 |
| High [9] | 0 | 2 | 0 | 0 | 0 | 0 | 0 | 0 | 0 | 2 | 0 | 0 | 2 |
| **2003 Total** | 10,506 | 864 | 71 | 0 | 2 | 10,800 | 511 | 76 | (s) | 23 | 34 | 0 | 11,444 |
| Low [7] | 9,993 | 813 | 71 | 0 | 0 | 10,778 | 0 | 65 | 0 | 0 | 34 | 0 | 10,877 |
| Medium [8] | 513 | 44 | 0 | 0 | 2 | 22 | 511 | 11 | (s) | 16 | 0 | 0 | 560 |
| High [9] | 0 | 7 | 0 | 0 | 0 | 0 | 0 | 0 | 0 | 7 | 0 | 0 | 7 |
| **2004 Total** | 12,864 | 1,178 | 70 | 0 | 3 | 13,634 | 452 | 13 | 0 | 16 | 0 | 0 | 14,115 |
| Low [7] | 12,386 | 1,178 | 44 | 0 | 0 | 13,600 | 0 | 8 | 0 | 0 | 0 | 0 | 13,608 |
| Medium [8] | 478 | 0 | 26 | 0 | 3 | 33 | 452 | 5 | 0 | 16 | 0 | 0 | 506 |
| High [9] | 0 | 0 | 0 | 0 | 0 | 0 | 0 | 0 | 0 | 0 | 0 | 0 | 0 |
| **2005 Total** | 14,681 | 1,160 | 31 | 114 | 56 | 15,041 | 640 | 228 | 2 | 16 | 0 | 114 | 16,041 |
| Low [7] | 14,045 | 1,099 | 30 | 0 | 50 | 15,022 | 12 | 190 | 0 | 0 | 0 | 0 | 15,224 |
| Medium [8] | 636 | 58 | 1 | 0 | 6 | 20 | 628 | 38 | 0 | 16 | 0 | 0 | 702 |
| High [9] | 0 | 2 | 0 | 114 | 0 | 0 | 0 | 0 | 2 | 0 | 0 | 114 | 115 |
| **2006 Total** | 15,123 | 1,626 | 42 | 3,845 | 107 | 15,362 | 1,136 | 330 | 3 | 66 | 0 | 3,847 | 20,744 |
| Low [7] | 13,906 | 1,500 | 40 | 0 | 100 | 15,225 | 10 | 290 | 0 | 21 | 0 | 0 | 15,546 |
| Medium [8] | 1,217 | 120 | 2 | 0 | 7 | 137 | 1,126 | 40 | 3 | 38 | 0 | 2 | 1,346 |
| High [9] | 0 | 7 | 0 | 3,845 | 0 | 0 | 0 | 0 | 0 | 7 | 0 | 3,845 | 3,852 |
| **Domestic Shipments [6]** | | | | | | | | | | | | | |
| **2007 Total** | 12,799 | 931 | 46 | 1 | – | 12,076 | 1,393 | 189 | 13 | 73 | 27 | 6 | 13,777 |
| Low [7] | 11,352 | 633 | – | 1 | – | 11,917 | 4 | 63 | – | – | – | 1 | 11,986 |
| Medium [8] | 1,447 | 298 | 18 | – | – | 158 | 1,389 | 126 | 13 | 73 | – | 5 | 1,764 |
| High [9] | – | (s) | 27 | – | – | – | (s) | – | – | – | 27 | – | 27 |
| **2008 Total** | 13,000 | 1,294 | 128 | 294 | – | 11,973 | 1,978 | 186 | 18 | 148 | 50 | 361 | 14,716 |
| Low [7] | 10,983 | 918 | – | – | – | 11,880 | 8 | 10 | – | 2 | – | – | 11,900 |
| Medium [8] | 2,017 | 376 | 33 | 6 | – | 93 | 1,971 | 176 | 18 | 141 | 21 | 12 | 2,432 |
| High [9] | – | – | 95 | 289 | – | – | – | – | – | 5 | 29 | 349 | 383 |

[1] Through 2006, data are for the commercial sector, excluding government, which is included in "Other." Beginning in 2007, data are for the commercial sector, including government.
[2] Through 2006, data are for the industrial sector and independent power producers. Beginning in 2007, data are for the industrial sector only; independent power producers are included in "Electric Power."
[3] Through 2006, data are for electric utilities only; independent power producers are included in "Industrial." Beginning in 2007, data are for electric utilities and independent power producers.
[4] Through 2006, data are for other sectors such as government, including the military, but excluding space applications. Beginning in 2007, data are for the transportation sector.
[5] Combined space and water heating.
[6] Through 2006, data are for domestic and export shipments, and may include imports that subsequently were shipped to domestic or foreign customers. Beginning in 2007, data are for domestic shipments only.
[7] Low-temperature collectors are solar thermal collectors that generally operate at temperatures below 110° F.
[8] Medium-temperature collectors are solar thermal collectors that generally operate in the temperature range of 140° F to 180° F, but can also operate at temperatures as low as 110 ° F. Special collectors are included in this category. Special collectors are evacuated tube collectors or concentrating (focusing) collectors. They operate in the temperature range from just above ambient temperature (low concentration for pool heating) to several hundred degrees Fahrenheit (high concentration for air conditioning and specialized industrial processes).
[9] High-temperature collectors are solar thermal collectors that generally operate at temperatures above 180° F. These are parabolic dish/trough collectors used primarily by the electric power sector to generate electricity for the electric grid.
– = No data reported. (s)=Less than 0.5 thousand square feet.
Note: Totals may not equal sum of components due to independent rounding.
Web Page: For related information, see http://www.eia.gov/fuelrenewable.html.
Sources: • 2000-2002—U.S. Energy Information Administration (EIA), *Renewable Energy Annual*, annual reports, and Form EIA-63A, "Annual Solar Thermal Collector Manufacturers Survey." • 2003 forward—EIA, *Solar Thermal Collector Manufacturing Activities* (and predecessor reports), annual reports, and Form EIA-63A, "Annual Solar Thermal Collector Manufacturers Survey."

U.S. Energy Information Administration / Annual Energy Review 2009

## Photovoltaic Cell and Module Shipments, Trade, and Prices

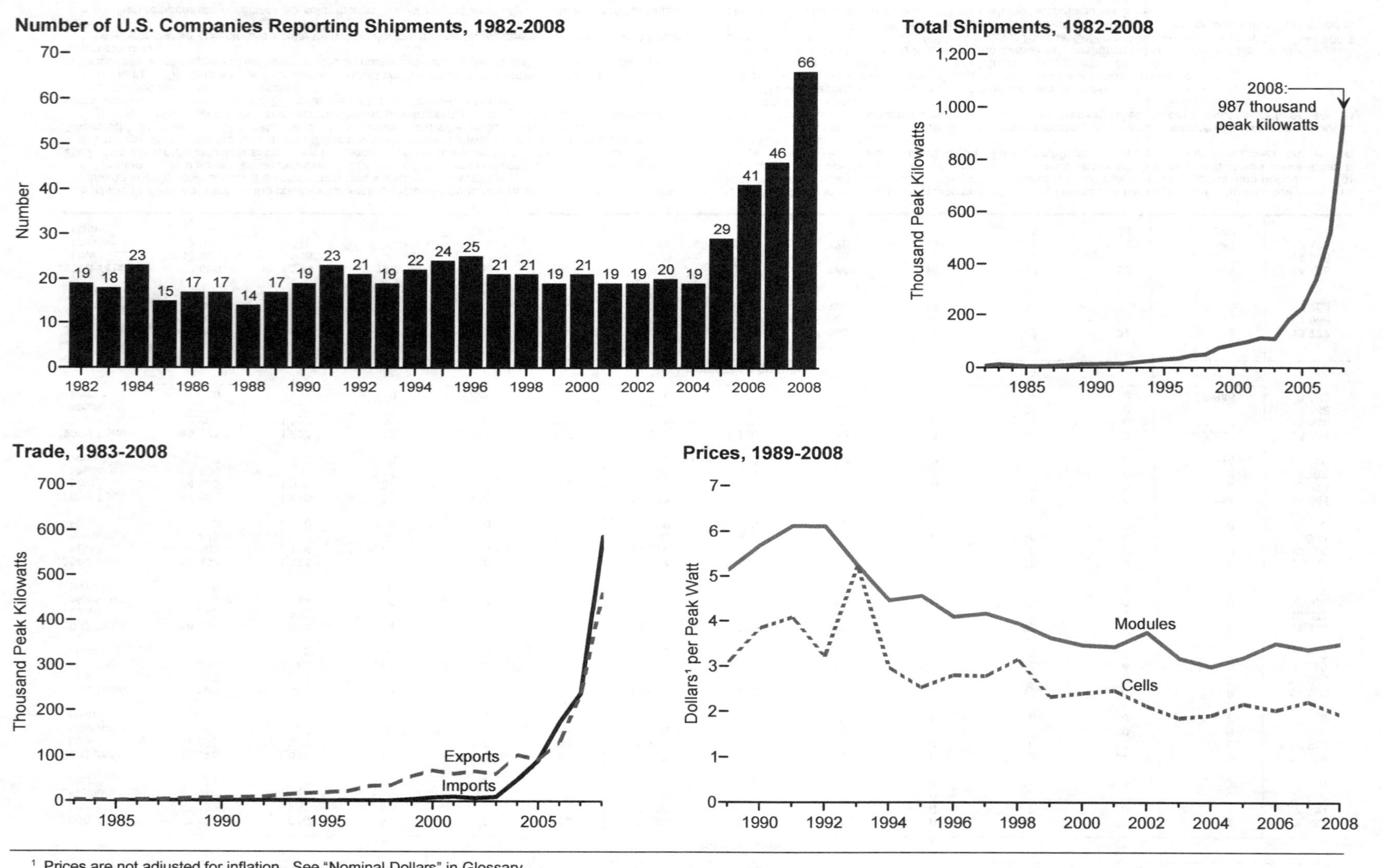


[1] Prices are not adjusted for inflation. See "Nominal Dollars" in Glossary.

Note: Shipments are for domestic and export shipments, and may include imports that subsequently were shipped to domestic and foreign customers.

U.S. Energy Information Administration / Annual Energy Review 2009

## Photovoltaic Cell and Module Shipments by Type, Trade, and Prices, 1982-2008

| Year | U.S. Companies Reporting Shipments | Shipments | | | Trade | | Prices [1] | |
|---|---|---|---|---|---|---|---|---|
| | | Crystalline Silicon | Thin-Film Silicon | Total [2] | Imports | Exports | Modules | Cells |
| | Number | Peak Kilowatts [3] | | | | | Dollars [4] per Peak Watt [3] | |
| 1982 | 19 | NA | NA | 6,897 | NA | NA | NA | NA |
| 1983 | 18 | NA | NA | 12,620 | NA | 1,903 | NA | NA |
| 1984 | 23 | NA | NA | 9,912 | NA | 2,153 | NA | NA |
| 1985 | 15 | 5,461 | 303 | 5,769 | 285 | 1,670 | NA | NA |
| 1986 | 17 | 5,806 | 516 | 6,333 | 678 | 3,109 | NA | NA |
| 1987 | 17 | 5,613 | 1,230 | 6,850 | 921 | 3,821 | NA | NA |
| 1988 | 14 | 7,364 | 1,895 | 9,676 | 1,453 | 5,358 | NA | NA |
| 1989 | 17 | 10,747 | 1,628 | 12,825 | 826 | 7,363 | 5.14 | 3.08 |
| 1990 | [5]19 | 12,492 | 1,321 | [5]13,837 | 1,398 | 7,544 | 5.69 | 3.84 |
| 1991 | 23 | 14,205 | 723 | 14,939 | 2,059 | 8,905 | 6.12 | 4.08 |
| 1992 | 21 | 14,457 | 1,075 | 15,583 | 1,602 | 9,823 | 6.11 | 3.21 |
| 1993 | 19 | 20,146 | 782 | 20,951 | 1,767 | 14,814 | 5.24 | 5.23 |
| 1994 | 22 | 24,785 | 1,061 | 26,077 | 1,960 | 17,714 | 4.46 | 2.97 |
| 1995 | 24 | 29,740 | 1,266 | 31,059 | 1,337 | 19,871 | 4.56 | 2.53 |
| 1996 | 25 | 33,996 | 1,445 | 35,464 | 1,864 | 22,448 | 4.09 | 2.80 |
| 1997 | 21 | 44,314 | 1,886 | 46,354 | 1,853 | 33,793 | 4.16 | 2.78 |
| 1998 | 21 | 47,186 | 3,318 | 50,562 | 1,931 | 35,493 | 3.94 | 3.15 |
| 1999 | 19 | 73,461 | 3,269 | 76,787 | 4,784 | R55,585 | 3.62 | 2.32 |
| 2000 | 21 | 85,155 | 2,736 | 88,221 | 8,821 | 68,382 | 3.46 | 2.40 |
| 2001 | 19 | 84,651 | 12,541 | 97,666 | 10,204 | 61,356 | 3.42 | 2.46 |
| 2002 | 19 | 104,123 | 7,396 | 112,090 | 7,297 | 66,778 | 3.74 | 2.12 |
| 2003 | 20 | 97,940 | 10,966 | 109,357 | 9,731 | 60,693 | 3.17 | 1.86 |
| 2004 | 19 | 159,138 | 21,978 | 181,116 | 47,703 | 102,770 | 2.99 | 1.92 |
| 2005 | 29 | 172,965 | 53,826 | 226,916 | 90,981 | 92,451 | 3.19 | 2.17 |
| 2006 | 41 | 233,518 | 101,766 | 337,268 | 173,977 | 130,757 | 3.50 | 2.03 |
| 2007 | 46 | 310,330 | 202,519 | 517,684 | 238,018 | 237,209 | 3.37 | 2.22 |
| 2008 | 66 | 665,795 | 293,182 | 986,504 | 586,558 | 462,252 | 3.49 | 1.94 |

[1] Prices equal shipment value divided by quantity shipped. Value includes charges for advertising and warranties. Excluded are excise taxes and the cost of freight or transportation for the shipments.
[2] Includes all types of photovoltaic cells and modules (single-crystal silicon, cast silicon, ribbon silicon, thin-film silicon, and concentrator silicon). Excludes cells and modules for space and satellite applications.
[3] See "Peak Kilowatt" and "Peak Watt" in Glossary.
[4] Prices are not adjusted for inflation. See "Nominal Dollars" in Glossary.
[5] Data were imputed for one nonrespondent who exited the industry during 1990.
R=Revised. NA=Not available.

Note: Shipments data are for domestic and export shipments, and may include imports th at subsequently were shipped to domestic or foreign customers.
Web Page: For related information, see http://www.eia.gov/fuelrenewable.html.
Sources: • 1982-1992—U.S. Energy Information Administration (EIA), *Solar Collector Manufacturing Activity*, annual reports. • 1993-2002—EIA, *Renewable Energy Annual*, annual reports. • 2003 forward—EIA, *Solar Photovoltaic Cell/Module Manufacturing Activities* (and predecessor reports), annual reports.

## Photovoltaic Cell and Module Domestic Shipments by Market Sector and End Use, 2008

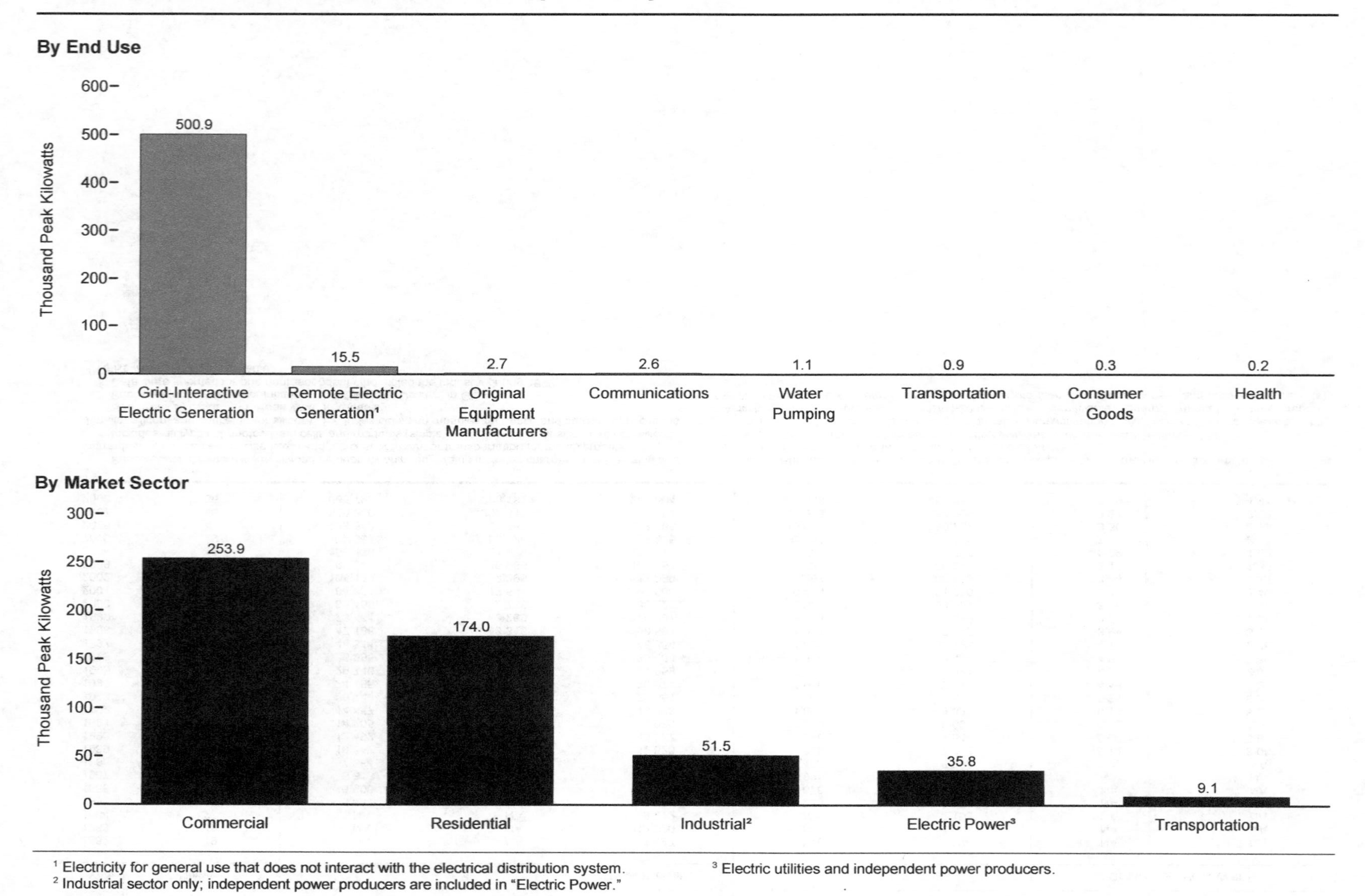


[1] Electricity for general use that does not interact with the electrical distribution system.
[2] Industrial sector only; independent power producers are included in "Electric Power."
[3] Electric utilities and independent power producers.

U.S. Energy Information Administration / Annual Energy Review 2009

## Photovoltaic Cell and Module Shipments by Market Sector and End Use, 1989-2008

(Peak Kilowatts [1])

| | By Market Sector | | | | | | | By End Use | | | | | | | | | |
|---|---|---|---|---|---|---|---|---|---|---|---|---|---|---|---|---|---|
| | | | | | | | | | | Electricity Generation [2] | | | | | | | |
| Year | Residential | Commercial [3] | Government | Industrial [4] | Transportation | Electric Power [5] | Other [6] | Communications | Consumer Goods | Grid-Interactive | Remote | Health | Original Equipment Manufacturers [7] | Transportation | Water Pumping | Other [8] | Total |
| Total Shipments [9] | | | | | | | | | | | | | | | | | |
| 1989 | 1,439 | 3,850 | 1,077 | 3,993 | 1,130 | 785 | 551 | 2,590 | 2,788 | 1,251 | 2,620 | 5 | 1,595 | 1,196 | 711 | 69 | 12,825 |
| 1990 | 1,701 | 6,086 | 1,002 | 2,817 | 974 | 826 | 432 | 4,340 | 2,484 | 469 | 3,097 | 5 | 1,119 | 1,069 | 1,014 | 240 | 13,837 |
| 1991 | 3,624 | 3,345 | 815 | 3,947 | 1,555 | 1,275 | 377 | 3,538 | 3,312 | 856 | 3,594 | 61 | 1,315 | 1,523 | 729 | 13 | 14,939 |
| 1992 | 4,154 | 2,386 | 1,063 | 4,279 | 1,673 | 1,553 | 477 | 3,717 | 2,566 | 1,227 | 4,238 | 67 | 828 | 1,602 | 809 | 530 | 15,583 |
| 1993 | 5,237 | 4,115 | 1,325 | 5,352 | 2,564 | 1,503 | 856 | 3,846 | 946 | 1,096 | 5,761 | 674 | 2,023 | 4,238 | 2,294 | 74 | 20,951 |
| 1994 | 6,632 | 5,429 | 2,114 | 6,855 | 2,174 | 2,364 | 510 | 5,570 | 3,239 | 2,296 | 9,253 | 79 | 1,849 | 2,128 | 1,410 | 254 | 26,077 |
| 1995 | 6,272 | 8,100 | 2,000 | 7,198 | 2,383 | 3,759 | 1,347 | 5,154 | 1,025 | 4,585 | 8,233 | 776 | 3,188 | 4,203 | 2,727 | 1,170 | 31,059 |
| 1996 | 8,475 | 5,176 | 3,126 | 8,300 | 3,995 | 4,753 | 1,639 | 6,041 | 1,063 | 4,844 | 10,884 | 977 | 2,410 | 5,196 | 3,261 | 789 | 35,464 |
| 1997 | 10,993 | 8,111 | 3,909 | 11,748 | 3,574 | 5,651 | 2,367 | 7,383 | 347 | 8,273 | 8,630 | 1,303 | 5,245 | 6,705 | 3,783 | 4,684 | 46,354 |
| 1998 | 15,936 | 8,460 | 2,808 | 13,232 | 3,440 | 3,965 | 2,720 | 8,280 | 1,198 | 14,193 | 8,634 | 1,061 | 5,044 | 6,356 | 4,306 | 1,491 | 50,562 |
| 1999 | 19,817 | 17,283 | 3,107 | 24,972 | 4,341 | 5,876 | 1,392 | 12,147 | 2,292 | 24,782 | 10,829 | 1,466 | 12,400 | 8,486 | 4,063 | 322 | 76,787 |
| 2000 | 24,814 | 13,692 | 4,417 | 28,808 | 5,502 | 6,298 | 4,690 | 12,269 | 2,870 | 21,713 | 14,997 | 2,742 | 12,153 | 12,804 | 5,644 | 3,028 | 88,221 |
| 2001 | 33,262 | 15,710 | 5,728 | 28,063 | 8,486 | 5,846 | 571 | 14,743 | 4,059 | 27,226 | 21,447 | 3,203 | 6,268 | 12,636 | 7,444 | 641 | 97,666 |
| 2002 | 29,315 | 20,578 | 8,565 | 32,218 | 12,932 | 7,640 | 841 | 17,290 | 3,400 | 33,983 | 21,693 | 4,202 | 7,869 | 16,028 | 7,532 | 93 | 112,090 |
| 2003 | 23,389 | 32,604 | 5,538 | 27,951 | 11,089 | 8,474 | 313 | 14,185 | 2,995 | 42,485 | 15,025 | 2,924 | 11,334 | 14,143 | 6,073 | 194 | 109,357 |
| 2004 | 53,928 | 74,509 | 3,257 | 30,493 | 1,380 | 3,233 | 14,316 | 11,348 | 6,444 | 129,265 | 18,371 | 341 | 6,452 | 1,380 | 1,322 | 6,193 | 181,116 |
| 2005 | 75,040 | 89,459 | 28,683 | 22,199 | 1,621 | 143 | 9,772 | 8,666 | 5,787 | 168,474 | 24,958 | 0 | 11,677 | 2,159 | 1,343 | 3,853 | 226,916 |
| 2006 | 95,815 | 180,852 | 7,688 | 28,618 | 2,458 | 3,981 | 17,857 | 6,888 | 4,030 | 274,197 | 18,003 | 0 | 6,132 | 2,438 | 2,093 | 23,487 | 337,268 |
| Domestic Shipments [9] | | | | | | | | | | | | | | | | | |
| 2007 | 68,417 | [10]140,434 | ([10]) | 32,702 | 3,627 | 35,294 | – – | 2,836 | 589 | 253,101 | 10,867 | 410 | 4,802 | 4,018 | 3,852 | – – | 280,475 |
| 2008 | 173,989 | [10]253,852 | ([10]) | 51,493 | 9,100 | 35,819 | – – | 2,622 | 312 | 500,854 | 15,527 | 217 | 2,659 | 916 | 1,145 | – – | 524,252 |

[1] See "Peak Killowatt" in Glossary.

[2] Grid-interactive means connection to the electrical distribution system; remote means electricity for general use that does not interact with the electrical distribution system, such as at an isolated residential site or mobile home. The other end uses in this table also include electricity generation, but only for the specific use cited.

[3] Through 2006, data are for the commercial sector, excluding government, which is included in "Government." Beginning in 2007, data are for the commercial sector, including government.

[4] Through 2006, data are for the industrial sector and independent power producers. Beginning in 2007, data are for the industrial sector only; independent power producers are included in "Electric Power."

[5] Through 2006, data are for electric utilities only; independent power producers are included in "Industrial." Beginning in 2007, data are for electric utilities and independent power producers.

[6] Through 2006, data are for shipments for specialty purposes such as research.

[7] "Original Equipment Manufacturers" are non-photovoltaic manufacturers that combine photovoltaic technology into existing or newly developed product lines.

[8] Through 2006, includes applications such as cooking food, desalinization, and distilling.

[9] Through 2006, data are for domestic and export shipments, and may include imports that subsequently were shipped to domestic or foreign customers. Beginning in 2007, data are for domestic shipments only.

[10] Beginning in 2007, the government sector is included in "Commercial."

– – = Not applicable.

Note: Totals may not equal sum of components due to independent rounding.

Web Page: For related information, see http://www.eia.gov/fuelrenewable.html.

Sources: • 1989-1992—U.S. Energy Information Administration (EIA), *Solar Collector Manufacturing Activity*, annual reports. • 1993-2002—EIA, *Renewable Energy Annual*, annual reports. • 2003 forward—EIA, *Solar Photovoltaic Cell/Module Manufacturing Activities* (and predecessor reports), annual reports.

## Environmental Revenue and Expenditures, by State

| State | Revenue ($/per capita) | | | | | Current Operational Expenses ($/per capita) | | | | | |
|---|---|---|---|---|---|---|---|---|---|---|---|
| | NRF[2] | NRO[3] | PR[4] | S[5] | SWM[6] | NRFG[1] | NRF[2] | NRO[3] | PR[4] | S[5] | SWM[6] |
| Alabama | 0.93 | 0.81 | 4.77 | 0.00 | 0.00 | 4.96 | 5.51 | 45.86 | 4.12 | 0.00 | 1.44 |
| Alaska | 3.37 | 33.58 | 3.91 | 0.00 | 0.00 | 215.84 | 71.69 | 123.37 | 11.56 | 0.00 | 0.00 |
| Arizona | 0.03 | 4.31 | 2.58 | 0.00 | 0.03 | 9.58 | 3.84 | 33.69 | 9.82 | 0.00 | 5.57 |
| Arkansas | 0.86 | 5.06 | 9.57 | 0.00 | 0.00 | 17.11 | 7.90 | 50.48 | 30.21 | 0.00 | 9.02 |
| California | 0.16 | 29.35 | 2.63 | 0.00 | 0.00 | 6.75 | 11.73 | 68.28 | 8.06 | 3.95 | 35.92 |
| Colorado | 0.00 | 4.69 | 0.54 | 0.00 | 0.82 | 18.12 | 0.00 | 26.11 | 12.11 | 0.42 | 4.61 |
| Connecticut | 0.11 | 0.83 | 1.32 | 0.00 | 52.07 | 7.24 | 2.74 | 13.57 | 4.75 | 0.01 | 49.02 |
| Delaware | 0.89 | 6.31 | 8.45 | 0.00 | 75.05 | 15.71 | 4.56 | 67.12 | 43.13 | 0.00 | 45.38 |
| Florida | 0.64 | 2.52 | 1.77 | 0.00 | 1.77 | 10.33 | 3.29 | 59.72 | 8.13 | 0.00 | 12.96 |
| Georgia | 0.42 | 2.53 | 18.90 | 0.00 | 0.00 | 4.26 | 5.31 | 36.97 | 19.52 | 0.56 | 3.93 |
| Hawaii | 0.15 | 12.08 | 7.26 | 0.00 | 0.50 | 3.59 | 8.71 | 76.59 | 42.57 | 0.20 | 0.15 |
| Idaho | 43.71 | 2.10 | 0.64 | 0.00 | 0.00 | 45.56 | 26.80 | 59.65 | 14.90 | 0.00 | 0.10 |
| Illinois | 0.00 | 0.75 | 0.89 | 0.00 | 0.00 | 3.39 | 0.15 | 9.58 | 8.15 | 1.30 | 8.77 |
| Indiana | 0.45 | 5.80 | 4.07 | 0.00 | 1.47 | 5.36 | 2.16 | 35.93 | 7.96 | 0.72 | 2.35 |
| Iowa | 0.05 | 10.58 | 1.43 | 0.24 | 1.41 | 7.79 | 1.51 | 61.04 | 6.44 | 0.98 | 1.22 |
| Kansas | 0.00 | 6.72 | 1.47 | 0.00 | 2.16 | 9.42 | 0.00 | 56.81 | 7.41 | 0.00 | 1.72 |
| Kentucky | 0.20 | 10.88 | 15.85 | 0.03 | 0.00 | 7.92 | 3.92 | 67.15 | 26.87 | 4.41 | 18.11 |
| Louisiana | 0.49 | 1.16 | 10.60 | 0.00 | 0.00 | 11.61 | 3.88 | 63.38 | 24.85 | 0.00 | 2.12 |
| Maine | 0.04 | 12.19 | 2.95 | 0.00 | 14.43 | 39.32 | 1.15 | 82.81 | 8.24 | 1.44 | 7.46 |
| Maryland | 0.75 | 1.80 | 4.07 | 0.02 | 0.70 | 19.87 | 2.15 | 42.50 | 15.08 | 14.28 | 0.00 |
| Massachusetts | 0.00 | 5.87 | 7.02 | 0.00 | 0.64 | 4.23 | 0.18 | 26.89 | 20.59 | 23.64 | 6.61 |
| Michigan | 2.71 | 0.36 | 0.45 | 0.00 | 0.00 | 6.62 | 3.14 | 20.14 | 7.52 | 0.00 | 0.93 |
| Minnesota | 5.66 | 6.75 | 4.36 | 0.00 | 0.00 | 16.96 | 9.18 | 41.74 | 25.47 | 0.00 | 7.19 |
| Mississippi | 0.87 | 5.09 | 3.35 | 0.00 | 0.00 | 13.07 | 8.67 | 49.67 | 10.22 | 0.00 | 0.00 |
| Missouri | 0.66 | 1.21 | 0.61 | 0.00 | 0.26 | 2.66 | 1.54 | 46.65 | 6.39 | 0.00 | 4.86 |
| Montana | 18.25 | 15.44 | 1.76 | 0.00 | 0.07 | 65.28 | 36.92 | 90.16 | 13.53 | 0.22 | 2.09 |
| Nebraska | 0.00 | 12.01 | 1.86 | 0.00 | 0.00 | 16.72 | 0.00 | 63.07 | 16.18 | 0.00 | 0.00 |
| Nevada | 1.24 | 1.61 | 2.10 | 0.00 | 1.30 | 9.17 | 7.03 | 35.04 | 10.34 | 0.00 | 8.30 |
| New Hampshire | 0.12 | 2.45 | 9.97 | 0.11 | 0.09 | 13.38 | 1.97 | 26.62 | 6.94 | 2.29 | 8.70 |
| New Jersey | 0.00 | 2.83 | 37.36 | 1.65 | 6.07 | 2.50 | 1.12 | 31.46 | 48.63 | 2.39 | 12.36 |
| New Mexico | 0.10 | 18.81 | 3.89 | 0.00 | 0.00 | 13.31 | 4.61 | 96.76 | 28.25 | 0.00 | 2.35 |
| New York | 0.00 | 1.81 | 11.85 | 0.00 | 1.41 | 2.17 | 1.80 | 13.77 | 19.06 | 0.00 | 4.14 |
| North Carolina | 1.15 | 2.87 | 1.06 | 0.00 | 0.06 | 8.09 | 6.20 | 26.84 | 12.48 | 0.13 | 4.07 |
| North Dakota | 0.58 | 18.93 | 3.08 | 0.00 | 0.00 | 13.46 | 2.40 | 166.05 | 22.20 | 0.00 | 0.00 |
| Ohio | 0.35 | 1.66 | 1.30 | 0.29 | 0.00 | 3.84 | 1.89 | 18.68 | 5.47 | 0.87 | 2.33 |
| Oklahoma | 0.07 | 1.19 | 7.08 | 0.00 | 0.29 | 7.18 | 2.27 | 43.56 | 20.78 | 0.37 | 3.34 |
| Oregon | 30.92 | 4.95 | 5.87 | 0.00 | 2.33 | 26.35 | 32.11 | 40.65 | 15.58 | 0.00 | 9.90 |
| Pennsylvania | 3.77 | 2.56 | 1.15 | 0.00 | 0.00 | 7.32 | 4.78 | 26.18 | 9.70 | 0.05 | 5.10 |
| Rhode Island | 0.06 | 3.76 | 4.13 | 1.19 | 8.17 | 5.28 | 2.94 | 27.35 | 13.29 | 31.35 | 50.95 |
| South Carolina | 1.30 | 3.93 | 4.88 | 0.00 | 1.91 | 13.64 | 4.93 | 23.81 | 16.49 | 0.00 | 0.94 |
| South Dakota | 2.71 | 6.86 | 11.39 | 0.00 | 0.00 | 39.98 | 10.07 | 89.42 | 27.77 | 0.00 | 0.00 |
| Tennessee | 0.50 | 1.23 | 5.87 | 0.01 | 1.00 | 10.98 | 5.48 | 21.67 | 18.30 | 2.78 | 5.96 |
| Texas | 0.00 | 1.76 | 1.66 | 0.51 | 0.78 | 6.83 | 0.80 | 28.51 | 2.92 | 0.12 | 7.00 |
| Utah | 0.21 | 6.80 | 0.41 | 0.02 | 1.62 | 19.25 | 7.60 | 39.45 | 14.10 | 2.83 | 4.92 |
| Vermont | 1.33 | 0.93 | 11.36 | 0.00 | 0.99 | 22.12 | 7.84 | 114.55 | 21.13 | 0.00 | 16.56 |
| Virginia | 0.24 | 0.02 | 1.91 | 0.00 | 0.00 | 7.15 | 3.69 | 15.42 | 12.85 | 10.14 | 5.87 |
| Washington | 29.08 | 4.04 | 0.82 | 0.00 | 1.30 | 21.49 | 12.34 | 45.85 | 9.55 | 0.00 | 8.34 |
| West Virginia | 0.41 | 14.11 | 9.26 | 10.20 | 5.32 | 9.19 | 3.36 | 71.43 | 27.72 | 15.85 | 6.45 |
| Wisconsin | 1.25 | 8.81 | 2.82 | 0.00 | 0.21 | 13.74 | 11.53 | 39.19 | 4.16 | 0.00 | 1.95 |
| Wyoming | 2.22 | 12.47 | 2.24 | 0.00 | 0.01 | 80.58 | 14.55 | 160.84 | 49.67 | 0.00 | 4.82 |

***Notes:*** *(1) Natural Resources, Fish & Game; (2) Natural Resources, Forestry; (3) Natural Resources, Other; (4) Parks & Recreation; (5) Sewerage; (6) Solid Waste Management*

***Source:*** *U.S. Census Bureau, State and Local Government Finances, 2006*

## Environmental Revenue and Expenditures for U.S Counties with Populations of 100,000+

| County | State | Revenue ($/per capita) | | | Current Operational Expenses ($/per capita) | | | |
|---|---|---|---|---|---|---|---|---|
| | | Parks & Recreation | Sewerage | Solid Waste Management | Parks & Recreation | Sewerage | Solid Waste Management | Natural Resources, Other |
| Ada | ID | 1.74 | 0.00 | 20.78 | 1.30 | 0.00 | 11.12 | 10.10 |
| Adams | CO | 7.62 | 0.00 | 0.51 | 10.63 | 0.00 | 1.90 | 9.83 |
| Adams | PA | 0.00 | 0.00 | 0.00 | 3.96 | 0.00 | 2.90 | 15.25 |
| Aiken | SC | 7.53 | 34.76 | 12.69 | 8.56 | 37.20 | 32.25 | 0.24 |
| Alachua | FL | 0.00 | 0.00 | 30.80 | 6.03 | 0.76 | 61.85 | 14.82 |
| Alamance | NC | 0.39 | 0.00 | 20.97 | 8.99 | 0.00 | 10.75 | 2.15 |
| Alameda | CA | 0.00 | 0.00 | 0.00 | 0.36 | 0.00 | 0.00 | 24.04 |
| Albany | NY | 20.22 | 3.39 | 0.00 | 20.20 | 28.33 | 0.00 | 0.00 |
| Allegan | MI | 0.00 | 0.00 | 0.00 | 0.00 | 0.00 | 0.00 | 0.00 |
| Allegheny | PA | 2.99 | 0.00 | 0.00 | 31.86 | 0.00 | 0.00 | 0.16 |
| Allen | IN | 18.55 | 0.00 | 0.00 | 29.41 | 0.00 | 0.00 | 3.69 |
| Allen | OH | 0.00 | 41.41 | 0.00 | 0.00 | 41.60 | 0.00 | 16.31 |
| Anderson | SC | 0.00 | 23.20 | 20.77 | 12.71 | 16.42 | 18.72 | 0.15 |
| Androscoggin | ME | 0.00 | 0.00 | 0.00 | 0.00 | 0.00 | 0.00 | 0.00 |
| Anne Arundel | MD | 18.26 | 123.75 | 88.21 | 85.03 | 40.95 | 82.22 | 2.36 |
| Anoka | MN | 7.10 | 0.00 | 18.52 | 17.51 | 0.00 | 22.91 | 1.48 |
| Arapahoe | CO | 0.07 | 0.00 | 0.00 | 12.88 | 0.16 | 0.00 | 0.00 |
| Arlington | VA | 22.42 | 23.48 | 47.65 | 153.97 | 33.52 | 67.17 | 37.69 |
| Ashtabula | OH | 3.08 | 17.37 | 4.14 | 0.03 | 14.10 | 7.20 | 3.53 |
| Atlantic | NJ | 2.12 | 0.00 | 0.00 | 7.34 | 0.00 | 0.19 | 3.13 |
| Baldwin | AL | 0.00 | 0.00 | 114.32 | 2.83 | 0.00 | 51.14 | 0.57 |
| Baltimore | MD | 12.45 | 208.72 | 1.44 | 26.52 | 66.23 | 55.34 | 8.37 |
| Barnstable | MA | 0.00 | 0.00 | 0.00 | 0.00 | 0.00 | 0.00 | 10.03 |
| Bay | FL | 0.31 | 44.33 | 70.22 | 18.78 | 66.02 | 81.99 | 1.19 |
| Bay | MI | 5.12 | 24.37 | 0.00 | 21.48 | 23.99 | 0.00 | 7.71 |
| Beaufort | SC | 0.00 | 0.00 | 0.00 | 26.86 | 0.00 | 0.00 | 0.00 |
| Beaver | PA | 0.00 | 0.00 | 0.00 | 7.48 | 0.00 | 2.33 | 0.03 |
| Bell | TX | 10.66 | 0.00 | 0.00 | 15.04 | 0.00 | 0.00 | 1.55 |
| Benton | AR | 0.00 | 0.00 | 0.00 | 0.04 | 0.00 | 0.49 | 0.00 |
| Benton | WA | 0.08 | 0.00 | 0.64 | 4.36 | 0.00 | 0.58 | 3.27 |
| Bergen | NJ | 6.69 | 0.00 | 0.00 | 10.74 | 0.00 | 0.07 | 0.00 |
| Berkeley | SC | 3.99 | 0.00 | 60.01 | 9.09 | 0.00 | 35.50 | 0.00 |
| Berks | PA | 0.28 | 0.00 | 11.82 | 6.01 | 0.23 | 0.50 | 8.32 |
| Bernalillo | NM | 1.91 | 0.00 | 6.54 | 19.09 | 0.12 | 6.92 | 3.24 |
| Berrien | MI | 2.04 | 0.00 | 11.68 | 5.24 | 0.00 | 11.24 | 6.51 |
| Bexar | TX | 0.00 | 0.00 | 0.00 | 1.39 | 0.00 | 0.00 | 0.43 |
| Bibb | GA | 3.53 | 0.00 | 15.66 | 13.02 | 0.00 | 17.13 | 1.38 |
| Black Hawk | IA | 2.16 | 1.43 | 0.00 | 14.25 | 0.34 | 0.00 | 0.41 |
| Blair | PA | 0.00 | 0.00 | 0.00 | 0.89 | 0.00 | 1.52 | 1.45 |
| Blount | TN | 0.00 | 0.00 | 0.00 | 0.00 | 0.00 | 0.83 | 6.96 |
| Boone | KY | 1.50 | 0.00 | 1.20 | 13.28 | 0.00 | 0.02 | 0.00 |
| Boone | MO | 0.00 | 0.00 | 0.00 | 0.27 | 0.72 | 0.14 | 0.44 |
| Bossier Parish | LA | 0.00 | 0.00 | 0.00 | 1.99 | 0.00 | 0.01 | 8.14 |
| Boulder | CO | 0.65 | 0.00 | 13.88 | 38.82 | 0.00 | 13.28 | 1.63 |
| Brazoria | TX | 0.11 | 0.00 | 0.00 | 7.62 | 0.00 | 0.20 | 1.84 |
| Brazos | TX | 0.00 | 0.00 | 0.19 | 0.00 | 0.00 | 0.00 | 1.47 |
| Brevard | FL | 11.05 | 0.00 | 62.47 | 54.61 | 0.00 | 87.06 | 32.44 |
| Bristol | MA | 0.00 | 0.00 | 0.00 | 0.00 | 0.00 | 0.00 | 0.00 |
| Broome | NY | 3.57 | 0.00 | 39.16 | 25.64 | 0.00 | 28.34 | 2.69 |
| Broward | FL | 8.75 | 28.80 | 67.65 | 36.96 | 15.98 | 64.67 | 10.44 |
| Brown | WI | 9.66 | 0.00 | 20.38 | 36.28 | 0.00 | 22.59 | 14.38 |
| Bucks | PA | 0.00 | 0.00 | 0.00 | 6.55 | 0.00 | 0.26 | 0.61 |
| Buncombe | NC | 0.29 | 0.00 | 27.83 | 28.15 | 0.00 | 29.37 | 1.20 |
| Burlington | NJ | 0.00 | 0.00 | 56.12 | 0.24 | 0.00 | 32.19 | 2.07 |

| County | State | Revenue ($/per capita) | | | Current Operational Expenses ($/per capita) | | | |
|---|---|---|---|---|---|---|---|---|
| | | Parks & Recreation | Sewerage | Solid Waste Management | Parks & Recreation | Sewerage | Solid Waste Management | Natural Resources, Other |
| Butler | OH | 0.00 | 51.47 | 0.00 | 0.00 | 30.08 | 0.04 | 1.26 |
| Butler | PA | 0.00 | 0.00 | 1.04 | 2.38 | 0.00 | 0.89 | 5.11 |
| Butte | CA | 0.00 | 0.00 | 29.03 | 1.50 | 0.00 | 27.48 | 5.29 |
| Cabarrus | NC | 5.09 | 0.00 | 18.34 | 14.35 | 0.00 | 5.83 | 4.42 |
| Caddo Parish | LA | 0.06 | 3.28 | 0.00 | 3.22 | 1.68 | 7.12 | 2.22 |
| Calcasieu Parish | LA | 9.60 | 0.69 | 0.00 | 45.99 | 0.67 | 19.36 | 25.45 |
| Calhoun | AL | 0.85 | 0.00 | 15.09 | 2.52 | 0.00 | 13.92 | 0.00 |
| Calhoun | MI | 0.72 | 0.00 | 0.00 | 0.94 | 0.07 | 2.95 | 3.57 |
| Cambria | PA | 0.19 | 0.00 | 0.00 | 0.05 | 0.00 | 0.00 | 12.59 |
| Camden | NJ | 0.38 | 0.00 | 73.37 | 9.43 | 0.00 | 59.17 | 0.29 |
| Cameron | TX | 10.18 | 0.00 | 0.64 | 8.37 | 0.00 | 0.19 | 1.23 |
| Canadian | OK | 0.00 | 0.00 | 0.00 | 0.00 | 0.00 | 0.00 | 2.28 |
| Canyon | ID | 0.00 | 0.00 | 17.05 | 0.73 | 0.00 | 5.29 | 0.05 |
| Carroll | GA | 2.38 | 0.00 | 26.96 | 14.64 | 0.00 | 42.40 | 0.99 |
| Carroll | MD | 8.95 | 24.32 | 37.92 | 0.97 | 27.49 | 42.13 | 4.11 |
| Cass | ND | 0.00 | 0.00 | 0.00 | 5.21 | 0.00 | 0.00 | 5.63 |
| Catawba | NC | 0.00 | 0.00 | 38.24 | 2.36 | 0.32 | 24.70 | 2.50 |
| Centre | PA | 0.00 | 0.00 | 0.00 | 0.33 | 0.00 | 1.16 | 10.00 |
| Champaign | IL | 0.00 | 0.00 | 0.00 | 0.00 | 0.00 | 0.16 | 12.94 |
| Charles | MD | 22.59 | 115.87 | 40.12 | 18.57 | 80.60 | 43.10 | 4.30 |
| Charleston | SC | 27.89 | 10.57 | 102.31 | 13.83 | 11.45 | 94.57 | 0.00 |
| Charlotte | FL | 6.93 | 108.52 | 182.47 | 60.19 | 84.14 | 143.90 | 23.23 |
| Chatham | GA | 5.43 | 2.75 | 4.23 | 15.59 | 17.28 | 12.48 | 2.58 |
| Chautauqua | NY | 0.00 | 25.84 | 59.69 | 5.62 | 24.21 | 31.15 | 0.20 |
| Cherokee | GA | 0.00 | 0.00 | 15.89 | 9.05 | 0.00 | 2.66 | 0.91 |
| Chester | PA | 0.54 | 0.00 | 0.00 | 14.53 | 0.00 | 0.00 | 4.08 |
| Chesterfield | VA | 1.60 | 86.56 | 8.26 | 29.96 | 57.01 | 9.66 | 14.91 |
| Chittenden | VT | 0.00 | 0.00 | 0.00 | 0.00 | 0.00 | 0.00 | 0.00 |
| Citrus | FL | 0.23 | 15.90 | 42.64 | 12.59 | 4.30 | 21.80 | 61.21 |
| Clackamas | OR | 11.21 | 50.54 | 0.00 | 28.53 | 40.09 | 4.62 | 3.51 |
| Clark | IN | 0.00 | 0.00 | 9.33 | 0.00 | 0.00 | 0.00 | 2.68 |
| Clark | NV | 37.09 | 54.26 | 0.00 | 134.59 | 23.62 | 0.00 | 6.18 |
| Clark | OH | 0.00 | 31.57 | 6.20 | 0.00 | 15.68 | 7.39 | 4.37 |
| Clark | WA | 5.89 | 15.99 | 2.92 | 10.00 | 4.91 | 4.04 | 2.25 |
| Clay | FL | 0.00 | 0.00 | 31.16 | 5.30 | 0.00 | 58.28 | 3.49 |
| Clay | MO | 13.19 | 0.00 | 0.00 | 18.87 | 0.00 | 0.00 | 0.00 |
| Clayton | GA | 8.20 | 0.00 | 11.28 | 24.48 | 0.00 | 16.75 | 1.30 |
| Clermont | OH | 0.00 | 72.07 | 1.48 | 0.00 | 37.83 | 3.47 | 0.00 |
| Cleveland | OK | 0.00 | 0.00 | 0.00 | 0.00 | 0.00 | 0.00 | 2.72 |
| Cobb | GA | 7.62 | 84.30 | 8.79 | 26.75 | 52.24 | 15.35 | 0.04 |
| Cochise | AZ | 0.00 | 26.68 | 0.00 | 7.35 | 24.94 | 6.26 | 0.00 |
| Coconino | AZ | 0.00 | 0.00 | 0.00 | 0.00 | 0.00 | 0.00 | 0.00 |
| Collier | FL | 14.15 | 92.74 | 74.81 | 68.13 | 78.08 | 77.70 | 32.10 |
| Collin | TX | 0.15 | 0.00 | 0.00 | 0.87 | 0.00 | 0.00 | 0.48 |
| Columbia | GA | 4.50 | 83.63 | 19.29 | 16.24 | 32.74 | 8.83 | 1.25 |
| Columbiana | OH | 0.00 | 16.39 | 0.00 | 0.00 | 17.03 | 0.00 | 0.00 |
| Comanche | OK | 0.00 | 0.00 | 0.00 | 0.00 | 0.00 | 0.00 | 3.00 |
| Contra Costa | CA | 0.00 | 22.24 | 1.53 | 0.00 | 16.59 | 0.00 | 27.27 |
| Cook | IL | 8.96 | 0.00 | 0.00 | 22.36 | 0.00 | 0.00 | 0.00 |
| Coweta | GA | 1.90 | 0.00 | 4.61 | 12.35 | 0.00 | 8.18 | 1.01 |
| Cumberland | ME | 0.00 | 0.00 | 0.00 | 0.00 | 0.00 | 0.00 | 0.50 |
| Cumberland | NJ | 0.00 | 0.00 | 0.00 | 1.95 | 0.00 | 0.00 | 2.99 |
| Cumberland | NC | 5.06 | 0.41 | 10.52 | 26.64 | 0.66 | 17.04 | 1.64 |
| Cumberland | PA | 0.00 | 0.00 | 0.34 | 0.25 | 0.00 | 2.30 | 2.52 |
| Cuyahoga | OH | 0.00 | 2.42 | 1.50 | 0.00 | 7.28 | 1.60 | 5.59 |
| Dakota | MN | 1.70 | 0.00 | 9.64 | 13.81 | 0.00 | 18.29 | 13.82 |

| County | State | Revenue ($/per capita) | | | Current Operational Expenses ($/per capita) | | | |
|---|---|---|---|---|---|---|---|---|
| | | Parks & Recreation | Sewerage | Solid Waste Management | Parks & Recreation | Sewerage | Solid Waste Management | Natural Resources, Other |
| Dallas | TX | 0.00 | 0.00 | 0.00 | 1.15 | 0.00 | 0.05 | 0.04 |
| Dane | WI | 20.38 | 0.00 | 17.71 | 26.82 | 0.00 | 18.43 | 6.33 |
| Dauphin | PA | 0.15 | 0.00 | 2.75 | 5.90 | 0.00 | 2.88 | 9.87 |
| Davidson | NC | 0.13 | 0.57 | 23.94 | 3.55 | 0.54 | 23.77 | 3.07 |
| Davis | UT | 0.00 | 0.00 | 57.62 | 9.95 | 0.00 | 54.13 | 5.87 |
| Dekalb | GA | 4.85 | 78.30 | 91.28 | 14.95 | 39.71 | 160.24 | 1.77 |
| Dekalb | IL | 0.00 | 0.00 | 0.00 | 6.07 | 0.00 | 0.00 | 4.25 |
| Delaware | IN | 0.00 | 0.00 | 0.00 | 0.50 | 0.00 | 0.00 | 8.00 |
| Delaware | OH | 0.00 | 129.26 | 0.74 | 0.00 | 36.46 | 0.63 | 1.80 |
| Delaware | PA | 0.00 | 0.00 | 0.00 | 2.29 | 0.00 | 20.38 | 0.40 |
| Denton | TX | 0.00 | 0.00 | 0.00 | 0.16 | 0.00 | 0.00 | 0.58 |
| Deschutes | OR | 13.53 | 0.00 | 51.81 | 14.60 | 0.00 | 29.96 | 1.94 |
| Desoto | MS | 0.00 | 0.00 | 7.65 | 7.56 | 0.00 | 8.37 | 3.24 |
| Dona Ana | NM | 0.00 | 0.00 | 0.00 | 0.00 | 0.00 | 0.00 | 0.00 |
| Dorchester | SC | 0.00 | 42.06 | 35.77 | 31.53 | 32.04 | 36.62 | 0.00 |
| Douglas | CO | 2.06 | 0.00 | 0.02 | 14.35 | 0.00 | 0.30 | 3.57 |
| Douglas | GA | 1.68 | 0.00 | 14.96 | 20.03 | 0.00 | 15.20 | 0.03 |
| Douglas | KS | 0.00 | 0.00 | 0.00 | 1.22 | 0.00 | 0.00 | 0.59 |
| Douglas | NE | 0.00 | 0.00 | 27.85 | 4.29 | 0.00 | 22.94 | 23.74 |
| Douglas | OR | 34.72 | 3.53 | 3.40 | 44.49 | 3.50 | 21.60 | 78.86 |
| Dupage | IL | 5.48 | 9.58 | 0.00 | 2.44 | 16.04 | 0.00 | 37.47 |
| Durham | NC | 0.00 | 21.93 | 4.60 | 0.00 | 8.67 | 6.94 | 7.11 |
| Dutchess | NY | 0.28 | 0.00 | 0.01 | 8.43 | 0.00 | 4.38 | 2.70 |
| Eaton | MI | 0.44 | 0.00 | 2.25 | 4.27 | 0.00 | 1.87 | 5.60 |
| Ector | TX | 5.60 | 0.00 | 0.00 | 9.42 | 0.00 | 0.00 | 0.76 |
| El Dorado | CA | 0.19 | 0.00 | 0.93 | 4.88 | 0.00 | 0.00 | 57.61 |
| El Paso | CO | 0.68 | 0.00 | 1.33 | 6.98 | 0.00 | 1.06 | 2.17 |
| El Paso | TX | 2.97 | 0.07 | 0.00 | 6.27 | 0.00 | 0.26 | 1.08 |
| Elkhart | IN | 0.60 | 0.00 | 21.31 | 7.84 | 0.00 | 14.38 | 3.87 |
| Ellis | TX | 0.00 | 0.00 | 0.00 | 0.00 | 0.00 | 0.00 | 0.00 |
| Erie | NY | 3.43 | 5.83 | 0.00 | 18.84 | 27.62 | 0.00 | 0.00 |
| Erie | PA | 0.00 | 0.00 | 0.00 | 0.00 | 0.00 | 0.00 | 0.00 |
| Escambia | FL | 14.98 | 0.00 | 40.61 | 26.66 | 0.00 | 43.64 | 8.09 |
| Essex | NJ | 6.75 | 0.00 | 0.00 | 19.82 | 0.00 | 0.12 | 0.67 |
| Etowah | AL | 0.00 | 0.00 | 0.00 | 0.13 | 0.00 | 0.34 | 0.00 |
| Fairfax | VA | 44.09 | 122.08 | 107.97 | 99.88 | 71.95 | 117.39 | 6.02 |
| Fairfield | OH | 0.00 | 20.82 | 0.00 | 0.00 | 11.88 | 2.42 | 2.92 |
| Faulkner | AR | 0.00 | 0.00 | 0.00 | 0.06 | 0.00 | 0.00 | 1.86 |
| Fayette | GA | 1.44 | 49.24 | 1.13 | 9.06 | 19.74 | 1.13 | 0.93 |
| Fayette | PA | 0.00 | 0.00 | 0.00 | 0.00 | 0.00 | 0.00 | 0.00 |
| Florence | SC | 3.59 | 0.00 | 10.62 | 30.82 | 0.00 | 26.07 | 0.00 |
| Forsyth | GA | 7.85 | 171.92 | 4.53 | 43.62 | 69.86 | 0.00 | 2.66 |
| Forsyth | NC | 15.42 | 0.00 | 0.00 | 25.14 | 0.00 | 0.00 | 1.84 |
| Fort Bend | TX | 0.37 | 0.00 | 0.00 | 3.28 | 0.00 | 0.00 | 11.72 |
| Franklin | MO | 0.00 | 0.00 | 0.00 | 0.00 | 0.00 | 0.00 | 1.66 |
| Franklin | OH | 3.74 | 2.97 | 0.00 | 4.64 | 1.25 | 0.00 | 0.00 |
| Franklin | PA | 0.00 | 0.00 | 0.00 | 0.00 | 0.00 | 0.00 | 0.00 |
| Frederick | MD | 2.49 | 52.31 | 67.25 | 19.40 | 44.77 | 65.39 | 10.99 |
| Fresno | CA | 0.30 | 8.32 | 17.15 | 3.39 | 6.08 | 8.97 | 8.86 |
| Fulton | GA | 0.47 | 46.87 | 0.66 | 3.77 | 31.15 | 2.63 | 0.92 |
| Galveston | TX | 0.01 | 0.00 | 0.00 | 8.41 | 0.00 | 0.00 | 5.45 |
| Gaston | NC | 0.39 | 0.00 | 17.26 | 4.72 | 0.00 | 14.83 | 2.55 |
| Genesee | MI | 3.94 | 0.00 | 0.00 | 14.08 | 0.00 | 0.00 | 7.36 |
| Gloucester | NJ | 4.12 | 0.00 | 0.00 | 11.83 | 0.00 | 0.00 | 1.48 |
| Grayson | TX | 0.00 | 0.00 | 0.00 | 0.00 | 0.00 | 0.00 | 0.00 |

| County | State | Revenue ($/per capita) | | | Current Operational Expenses ($/per capita) | | | |
|---|---|---|---|---|---|---|---|---|
| | | Parks & Recreation | Sewerage | Solid Waste Management | Parks & Recreation | Sewerage | Solid Waste Management | Natural Resources, Other |
| Greene | MO | 0.00 | 0.00 | 0.00 | 0.06 | 0.03 | 0.00 | 0.60 |
| Greene | OH | 0.48 | 111.10 | 8.46 | 14.06 | 60.68 | 5.54 | 3.91 |
| Greenville | SC | 0.00 | 0.00 | 12.82 | 0.00 | 0.00 | 34.64 | 0.00 |
| Gregg | TX | 0.00 | 0.00 | 0.00 | 0.00 | 0.00 | 0.00 | 1.23 |
| Guilford | NC | 0.00 | 1.05 | 0.00 | 3.03 | 0.00 | 1.82 | 1.14 |
| Gwinnett | GA | 4.00 | 101.87 | 0.99 | 37.19 | 79.05 | 0.57 | 0.00 |
| Hall | GA | 0.53 | 0.07 | 28.62 | 15.62 | 1.12 | 25.10 | 2.18 |
| Hamilton | IN | 1.43 | 0.00 | 0.00 | 9.98 | 0.00 | 0.00 | 8.06 |
| Hamilton | OH | 0.00 | 0.00 | 0.00 | 0.00 | 0.00 | 0.00 | 0.00 |
| Hamilton | TN | 1.58 | 0.64 | 1.21 | 25.86 | 0.00 | 0.00 | 1.55 |
| Harford | MD | 3.12 | 47.90 | 54.66 | 33.40 | 22.56 | 48.77 | 17.98 |
| Harnett | NC | 0.00 | 93.03 | 30.37 | 2.85 | 55.63 | 27.91 | 0.00 |
| Harris | TX | 0.66 | 0.00 | 0.00 | 5.54 | 0.00 | 0.01 | 21.56 |
| Harrison | MS | 0.00 | 0.00 | 0.00 | 23.20 | 0.00 | 15.08 | 2.36 |
| Hawaii | HI | 8.40 | 38.27 | 46.32 | 88.88 | 38.88 | 135.52 | 2.72 |
| Hays | TX | 0.00 | 0.00 | 0.00 | 9.45 | 0.00 | 0.00 | 0.00 |
| Hendricks | IN | 0.00 | 0.00 | 6.09 | 8.09 | 6.14 | 3.45 | 4.21 |
| Hennepin | MN | 0.01 | 0.00 | 62.46 | 0.99 | 0.00 | 43.79 | 0.00 |
| Henrico | VA | 5.49 | 143.02 | 61.39 | 48.02 | 76.87 | 34.74 | 1.16 |
| Henry | GA | 2.09 | 0.00 | 0.00 | 16.86 | 0.00 | 0.00 | 1.69 |
| Hernando | FL | 2.26 | 59.05 | 41.51 | 33.53 | 50.96 | 58.21 | 8.39 |
| Hidalgo | TX | 0.11 | 0.00 | 0.26 | 3.22 | 0.00 | 5.28 | 10.92 |
| Hillsborough | FL | 1.39 | 0.00 | 66.99 | 44.21 | 0.00 | 64.72 | 30.08 |
| Hillsborough | NH | 0.00 | 0.00 | 0.00 | 0.00 | 0.00 | 0.00 | 1.08 |
| Hinds | MS | 0.00 | 0.00 | 0.00 | 6.88 | 0.00 | 0.00 | 3.15 |
| Horry | SC | 25.91 | 0.00 | 3.46 | 21.12 | 0.00 | 17.72 | 9.26 |
| Houston | GA | 0.00 | 17.14 | 50.08 | 0.02 | 13.79 | 49.52 | 0.65 |
| Howard | MD | 47.01 | 69.64 | 58.16 | 41.10 | 107.81 | 46.49 | 4.94 |
| Hudson | NJ | 0.00 | 0.00 | 0.00 | 8.11 | 0.00 | 0.00 | 0.11 |
| Humboldt | CA | 2.03 | 0.00 | 4.29 | 4.35 | 0.00 | 2.46 | 4.89 |
| Hunterdon | NJ | 10.81 | 0.00 | 0.00 | 19.48 | 0.00 | 0.85 | 3.18 |
| Imperial | CA | 0.36 | 0.00 | 14.92 | 3.54 | 0.15 | 11.49 | 16.66 |
| Indian River | FL | 28.97 | 87.79 | 70.49 | 67.82 | 91.80 | 145.96 | 8.35 |
| Ingham | MI | 1.15 | 0.00 | 0.00 | 19.57 | 0.00 | 0.00 | 4.04 |
| Iredell | NC | 1.75 | 0.00 | 44.13 | 6.05 | 0.00 | 27.81 | 2.62 |
| Jackson | MI | 5.57 | 0.00 | 55.33 | 10.14 | 0.00 | 40.41 | 19.72 |
| Jackson | MS | 4.11 | 0.00 | 0.00 | 21.91 | 0.00 | 23.90 | 2.68 |
| Jackson | MO | 7.37 | 23.51 | 0.00 | 20.25 | 13.64 | 0.00 | 0.04 |
| Jackson | OR | 10.43 | 0.00 | 0.00 | 17.32 | 0.00 | 0.96 | 11.32 |
| Jasper | MO | 0.00 | 0.00 | 0.00 | 0.00 | 0.00 | 0.00 | 0.00 |
| Jefferson | AL | 0.00 | 0.00 | 0.00 | 0.00 | 0.00 | 0.00 | 0.00 |
| Jefferson | CO | 0.00 | 0.00 | 0.00 | 17.97 | 0.35 | 0.41 | 1.40 |
| Jefferson | MO | 0.00 | 0.00 | 0.00 | 0.00 | 0.00 | 0.00 | 0.00 |
| Jefferson | NY | 0.00 | 0.00 | 16.18 | 3.65 | 0.00 | 19.34 | 2.76 |
| Jefferson | TX | 14.29 | 0.00 | 0.00 | 20.70 | 0.00 | 3.62 | 0.00 |
| Jefferson Parish | LA | 8.78 | 39.77 | 32.85 | 45.85 | 66.45 | 49.42 | 183.22 |
| Johnson | IN | 0.62 | 0.00 | 0.03 | 6.83 | 1.65 | 0.00 | 2.06 |
| Johnson | IA | 0.00 | 0.00 | 0.00 | 9.78 | 0.00 | 4.23 | 1.21 |
| Johnson | KS | 29.66 | 118.45 | 0.00 | 47.55 | 62.62 | 0.00 | 2.32 |
| Johnson | TX | 0.00 | 0.00 | 0.00 | 0.00 | 0.00 | 0.00 | 0.83 |
| Johnston | NC | 0.00 | 95.59 | 38.81 | 0.98 | 39.61 | 32.96 | 3.23 |
| Kalamazoo | MI | 2.49 | 0.00 | 0.00 | 11.69 | 0.00 | 0.00 | 1.83 |
| Kanawha | WV | 0.00 | 0.00 | 0.00 | 0.00 | 0.00 | 0.00 | 0.00 |
| Kane | IL | 0.00 | 0.00 | 9.37 | 0.00 | 0.00 | 7.79 | 8.89 |
| Kankakee | IL | 0.00 | 0.00 | 0.00 | 0.00 | 0.00 | 0.00 | 0.97 |
| Kennebec | ME | 0.00 | 0.00 | 0.00 | 0.00 | 0.00 | 0.00 | 0.00 |

| County | State | Revenue ($/per capita) | | | Current Operational Expenses ($/per capita) | | | |
|---|---|---|---|---|---|---|---|---|
| | | Parks & Recreation | Sewerage | Solid Waste Management | Parks & Recreation | Sewerage | Solid Waste Management | Natural Resources, Other |
| Kenosha | WI | 19.27 | 0.00 | 3.92 | 28.20 | 0.00 | 0.00 | 12.49 |
| Kent | DE | 0.00 | 90.21 | 12.41 | 12.36 | 57.43 | 12.86 | 0.00 |
| Kent | MI | 2.31 | 0.00 | 81.45 | 13.85 | 3.54 | 72.38 | 1.73 |
| Kenton | KY | 31.02 | 211.41 | 0.00 | 28.26 | 115.46 | 0.44 | 0.32 |
| Kern | CA | 8.69 | 1.34 | 54.01 | 20.02 | 3.87 | 51.62 | 6.45 |
| King | WA | 3.81 | 133.38 | 48.75 | 20.72 | 55.54 | 42.60 | 4.20 |
| Kings | CA | 0.27 | 0.00 | 0.00 | 9.44 | 0.00 | 0.00 | 12.88 |
| Kitsap | WA | 2.46 | 55.23 | 57.87 | 20.56 | 37.14 | 50.23 | 1.78 |
| Knox | TN | 0.71 | 0.00 | 0.00 | 22.53 | 0.00 | 12.04 | 1.43 |
| Kootenai | ID | 0.11 | 0.11 | 69.23 | 2.50 | 0.00 | 44.85 | 4.11 |
| La Crosse | WI | 3.05 | 0.00 | 82.08 | 5.98 | 0.00 | 106.03 | 13.84 |
| La Porte | IN | 0.00 | 0.00 | 0.00 | 0.00 | 0.00 | 0.00 | 0.00 |
| La Salle | IL | 0.00 | 0.00 | 1.99 | 1.18 | 0.00 | 0.00 | 0.00 |
| Lackawanna | PA | 0.00 | 0.00 | 0.00 | 0.00 | 0.00 | 0.00 | 0.00 |
| Lake | FL | 0.79 | 0.00 | 51.89 | 3.73 | 0.00 | 89.36 | 9.23 |
| Lake | IL | 7.79 | 29.47 | 1.82 | 5.65 | 24.57 | 2.03 | 33.48 |
| Lake | IN | 10.57 | 0.40 | 0.00 | 21.35 | 0.26 | 0.00 | 4.37 |
| Lake | OH | 0.00 | 65.74 | 29.99 | 0.00 | 40.03 | 33.66 | 0.97 |
| Lancaster | NE | 0.00 | 0.00 | 0.00 | 3.72 | 0.00 | 0.00 | 0.00 |
| Lancaster | PA | 0.55 | 0.00 | 0.00 | 5.30 | 0.00 | 0.00 | 7.93 |
| Lane | OR | 2.01 | 0.00 | 34.99 | 5.97 | 0.00 | 31.37 | 14.68 |
| Larimer | CO | 0.45 | 0.00 | 17.05 | 11.88 | 0.00 | 10.31 | 24.59 |
| Lebanon | PA | 0.00 | 0.00 | 0.00 | 0.00 | 0.00 | 0.00 | 0.00 |
| Lee | AL | 0.00 | 0.00 | 18.61 | 0.44 | 0.00 | 15.83 | 0.64 |
| Lee | FL | 5.16 | 55.71 | 93.04 | 56.73 | 33.85 | 67.34 | 28.24 |
| Lehigh | PA | 2.87 | 10.49 | 1.70 | 8.16 | 9.78 | 2.16 | 1.72 |
| Leon | FL | 0.12 | 0.00 | 26.50 | 11.05 | 11.23 | 38.93 | 21.68 |
| Lexington | SC | 4.08 | 0.00 | 6.55 | 15.54 | 0.00 | 26.74 | 0.26 |
| Licking | OH | 0.00 | 10.75 | 0.00 | 3.19 | 14.81 | 14.11 | 3.50 |
| Linn | IA | 2.29 | 0.00 | 0.00 | 18.15 | 0.00 | 0.15 | 2.83 |
| Linn | OR | 13.39 | 0.00 | 0.00 | 14.00 | 0.00 | 0.00 | 2.83 |
| Livingston | MI | 0.00 | 0.00 | 0.00 | 0.00 | 0.00 | 0.00 | 0.00 |
| Livingston Parish | LA | 5.92 | 12.05 | 0.00 | 8.14 | 7.88 | 0.00 | 21.87 |
| Lorain | OH | 0.00 | 3.53 | 10.86 | 0.00 | 2.87 | 11.17 | 0.32 |
| Los Angeles | CA | 9.50 | 4.79 | 3.56 | 24.68 | 4.38 | 2.29 | 41.19 |
| Loudoun | VA | 37.73 | 0.00 | 6.39 | 94.41 | 0.00 | 13.10 | 6.22 |
| Lubbock | TX | 0.00 | 0.00 | 0.00 | 1.18 | 0.00 | 0.00 | 0.69 |
| Lucas | OH | 10.05 | 16.87 | 4.44 | 21.71 | 8.89 | 5.40 | 1.04 |
| Luzerne | PA | 0.00 | 0.00 | 0.00 | 0.00 | 0.00 | 0.00 | 0.00 |
| Lycoming | PA | 0.00 | 0.00 | 104.16 | 0.31 | 0.00 | 79.20 | 8.64 |
| Macomb | MI | 0.32 | 0.00 | 0.00 | 1.36 | 4.22 | 0.00 | 2.18 |
| Macon | IL | 0.00 | 0.00 | 0.00 | 0.00 | 0.00 | 1.96 | 0.00 |
| Madera | CA | 0.07 | 0.00 | 29.86 | 0.14 | 0.00 | 23.56 | 9.61 |
| Madison | AL | 0.00 | 0.00 | 24.17 | 1.78 | 13.33 | 22.28 | 0.00 |
| Madison | IL | 0.00 | 5.99 | 4.28 | 3.17 | 7.32 | 2.96 | 0.00 |
| Madison | IN | 0.00 | 0.00 | 0.00 | 0.00 | 3.82 | 0.64 | 8.33 |
| Mahoning | OH | 0.00 | 75.00 | 12.41 | 0.00 | 60.17 | 16.15 | 0.00 |
| Manatee | FL | 20.17 | 152.38 | 109.75 | 55.56 | 116.77 | 101.06 | 22.16 |
| Marathon | WI | 4.70 | 0.00 | 13.91 | 22.34 | 0.00 | 19.39 | 9.89 |
| Maricopa | AZ | 1.09 | 0.00 | 0.01 | 2.00 | 0.00 | 1.65 | 9.98 |
| Marin | CA | 8.83 | 25.52 | 0.00 | 51.29 | 20.41 | 0.00 | 24.43 |
| Marion | FL | 1.84 | 26.55 | 10.39 | 8.47 | 9.11 | 119.62 | 4.29 |
| Marion | OR | 0.36 | 1.02 | 52.22 | 1.37 | 2.25 | 56.83 | 1.55 |
| Martin | FL | 3.97 | 76.65 | 149.20 | 67.08 | 42.93 | 174.97 | 33.70 |
| Maui | HI | 12.60 | 199.09 | 68.24 | 172.76 | 112.10 | 91.33 | 0.00 |
| Mchenry | IL | 0.00 | 0.00 | 0.00 | 0.00 | 0.00 | 0.00 | 0.00 |

| County | State | Revenue ($/per capita) | | | Current Operational Expenses ($/per capita) | | | |
|---|---|---|---|---|---|---|---|---|
| | | Parks & Recreation | Sewerage | Solid Waste Management | Parks & Recreation | Sewerage | Solid Waste Management | Natural Resources, Other |
| Mclean | IL | 0.00 | 0.00 | 0.00 | 5.65 | 0.00 | 0.00 | 0.00 |
| Mclennan | TX | 0.00 | 0.00 | 0.00 | 0.07 | 0.00 | 0.30 | 0.88 |
| Mecklenburg | NC | 3.21 | 0.00 | 16.67 | 37.92 | 0.00 | 15.29 | 0.00 |
| Medina | OH | 0.00 | 87.83 | 49.68 | 0.41 | 64.55 | 45.94 | 0.63 |
| Merced | CA | 1.32 | 0.00 | 42.68 | 5.86 | 0.00 | 32.52 | 10.96 |
| Mercer | NJ | 15.99 | 0.00 | 0.00 | 32.89 | 0.00 | 0.00 | 0.00 |
| Mercer | PA | 0.00 | 0.00 | 0.00 | 0.28 | 0.00 | 0.04 | 6.26 |
| Merrimack | NH | 0.00 | 0.00 | 0.00 | 0.00 | 0.00 | 0.00 | 2.72 |
| Mesa | CO | 0.00 | 0.25 | 26.23 | 6.69 | 1.14 | 22.09 | 3.13 |
| Miami | OH | 0.00 | 12.83 | 46.05 | 0.48 | 14.23 | 43.44 | 5.33 |
| Miami-Dade | FL | 14.96 | 108.39 | 103.12 | 61.83 | 57.97 | 126.97 | 15.24 |
| Middlesex | NJ | 0.50 | 0.00 | 0.29 | 9.37 | 0.00 | 2.92 | 0.38 |
| Midland | TX | 0.00 | 0.00 | 0.00 | 0.47 | 0.00 | 0.02 | 1.25 |
| Milwaukee | WI | 34.23 | 0.00 | 0.05 | 105.23 | 0.00 | 0.00 | 0.00 |
| Minnehaha | SD | 0.10 | 0.00 | 0.00 | 7.15 | 0.00 | 0.00 | 1.88 |
| Missoula | MT | 9.80 | 0.00 | 0.39 | 13.36 | 0.00 | 2.99 | 15.88 |
| Mobile | AL | 0.73 | 0.00 | 0.00 | 6.94 | 0.00 | 1.40 | 0.45 |
| Mohave | AZ | 7.23 | 0.00 | 7.19 | 7.44 | 8.93 | 0.00 | 9.89 |
| Monmouth | NJ | 9.93 | 0.00 | 48.42 | 30.08 | 0.00 | 49.54 | 0.65 |
| Monroe | IN | 1.44 | 0.00 | 0.00 | 5.19 | 0.00 | 0.00 | 1.44 |
| Monroe | MI | 0.04 | 0.00 | 0.00 | 4.13 | 18.86 | 0.00 | 16.09 |
| Monroe | NY | 2.25 | 27.54 | 9.17 | 17.99 | 55.32 | 13.75 | 0.00 |
| Monroe | PA | 2.68 | 0.00 | 0.00 | 4.70 | 0.00 | 0.00 | 26.53 |
| Monterey | CA | 11.43 | 3.59 | 0.00 | 14.14 | 4.12 | 0.27 | 59.59 |
| Montgomery | AL | 0.00 | 0.00 | 0.00 | 5.69 | 0.00 | 0.43 | 0.00 |
| Montgomery | MD | 28.55 | 123.61 | 103.10 | 115.88 | 70.55 | 103.19 | 4.31 |
| Montgomery | OH | 0.00 | 75.43 | 41.25 | 1.57 | 50.91 | 25.35 | 0.00 |
| Montgomery | PA | 0.00 | 0.00 | 0.16 | 7.22 | 0.00 | 0.80 | 0.00 |
| Montgomery | TN | 0.02 | 0.00 | 0.00 | 0.33 | 0.00 | 0.72 | 2.09 |
| Montgomery | TX | 0.78 | 0.00 | 0.00 | 2.34 | 0.00 | 0.00 | 1.78 |
| Morgan | AL | 0.00 | 0.00 | 20.62 | 8.45 | 0.00 | 19.26 | 0.00 |
| Morris | NJ | 23.25 | 0.00 | 0.00 | 26.30 | 0.00 | 0.00 | 0.51 |
| Multnomah | OR | 0.00 | 0.01 | 0.00 | 0.00 | 0.49 | 0.00 | 0.04 |
| Muskegon | MI | 3.44 | 71.60 | 11.70 | 3.25 | 57.51 | 3.72 | 0.00 |
| Napa | CA | 0.00 | 33.55 | 0.00 | 8.22 | 78.82 | 0.00 | 214.38 |
| Nassau | NY | 17.30 | 1.14 | 0.00 | 38.73 | 77.96 | 0.00 | 0.26 |
| Navajo | AZ | 0.00 | 0.00 | 0.00 | 0.00 | 0.00 | 0.00 | 0.00 |
| New Castle | DE | 2.95 | 112.46 | 0.00 | 8.99 | 101.91 | 0.00 | 0.80 |
| New Hanover | NC | 0.45 | 68.93 | 71.49 | 21.70 | 29.06 | 55.47 | 3.37 |
| Niagara | NY | 2.64 | 1.17 | 2.35 | 15.64 | 13.94 | 1.42 | 2.70 |
| Norfolk | MA | 1.52 | 0.00 | 0.00 | 2.22 | 0.00 | 0.00 | 0.00 |
| Northampton | PA | 0.04 | 0.00 | 0.00 | 3.63 | 0.00 | 0.00 | 3.33 |
| Nueces | TX | 0.00 | 0.00 | 0.00 | 7.81 | 0.00 | 0.00 | 0.64 |
| Oakland | MI | 7.47 | 14.74 | 0.00 | 7.50 | 77.77 | 0.00 | 6.97 |
| Ocean | NJ | 2.61 | 0.00 | 0.00 | 9.33 | 0.00 | 5.06 | 2.43 |
| Okaloosa | FL | 4.79 | 58.21 | 44.99 | 24.52 | 51.11 | 49.81 | 3.24 |
| Oklahoma | OK | 0.00 | 0.00 | 0.00 | 0.00 | 0.00 | 0.00 | 0.00 |
| Olmsted | MN | 0.00 | 0.17 | 97.08 | 12.74 | 0.06 | 65.47 | 4.14 |
| Oneida | NY | 0.00 | 31.58 | 0.00 | 2.52 | 30.51 | 0.00 | 13.22 |
| Onondaga | NY | 5.65 | 109.62 | 0.00 | 33.50 | 90.33 | 0.00 | 4.46 |
| Onslow | NC | 0.51 | 0.00 | 38.58 | 8.68 | 0.00 | 14.34 | 0.65 |
| Ontario | NY | 0.25 | 24.15 | 0.00 | 3.46 | 16.31 | 11.04 | 5.11 |
| Orange | CA | 13.82 | 0.00 | 37.89 | 0.04 | 0.00 | 26.30 | 15.46 |
| Orange | FL | 44.74 | 80.12 | 60.21 | 27.85 | 108.34 | 42.75 | 14.02 |
| Orange | NY | 4.45 | 12.54 | 35.34 | 16.06 | 14.39 | 35.17 | 4.39 |
| Orange | NC | 10.64 | 0.33 | 57.65 | 23.49 | 0.84 | 67.13 | 3.76 |

| County | State | Revenue ($/per capita) | | | Current Operational Expenses ($/per capita) | | | |
|---|---|---|---|---|---|---|---|---|
| | | Parks & Recreation | Sewerage | Solid Waste Management | Parks & Recreation | Sewerage | Solid Waste Management | Natural Resources, Other |
| Osceola | FL | 14.36 | 0.00 | 21.81 | 38.87 | 0.00 | 99.35 | 5.90 |
| Oswego | NY | 0.10 | 0.00 | 29.50 | 7.96 | 0.00 | 35.98 | 1.02 |
| Ottawa | MI | 0.00 | 0.00 | 0.00 | 0.00 | 0.00 | 0.00 | 0.00 |
| Ouachita Parish | LA | 0.87 | 16.87 | 0.00 | 6.16 | 11.91 | 0.00 | 0.12 |
| Outagamie | WI | 0.45 | 0.00 | 29.02 | 5.93 | 0.00 | 50.48 | 11.17 |
| Palm Beach | FL | 9.84 | 29.69 | 126.28 | 45.18 | 28.28 | 104.53 | 16.21 |
| Pasco | FL | 1.52 | 67.52 | 34.87 | 18.16 | 78.82 | 51.14 | 1.63 |
| Passaic | NJ | 2.97 | 0.00 | 0.55 | 5.38 | 0.00 | 0.62 | 0.27 |
| Paulding | GA | 2.95 | 58.55 | 4.36 | 17.61 | 45.02 | 5.06 | 0.07 |
| Penobscot | ME | 0.00 | 0.00 | 0.00 | 0.00 | 0.00 | 0.00 | 0.00 |
| Peoria | IL | 0.00 | 0.00 | 1.07 | 0.00 | 0.00 | 1.45 | 0.00 |
| Pickens | SC | 1.87 | 0.00 | 0.00 | 7.36 | 0.00 | 31.88 | 0.77 |
| Pierce | WA | 3.56 | 34.42 | 3.87 | 11.24 | 21.14 | 5.96 | 8.27 |
| Pima | AZ | 2.15 | 109.85 | 4.79 | 19.28 | 80.13 | 7.21 | 2.33 |
| Pinal | AZ | 0.00 | 0.00 | 0.00 | 3.04 | 0.00 | 2.21 | 3.99 |
| Pinellas | FL | 3.79 | 59.75 | 83.21 | 27.14 | 51.19 | 61.30 | 14.54 |
| Pitt | NC | 0.00 | 0.00 | 48.27 | 0.43 | 0.00 | 46.40 | 3.89 |
| Placer | CA | 8.55 | 29.64 | 18.24 | 12.48 | 23.57 | 9.11 | 6.12 |
| Plymouth | MA | 0.00 | 0.00 | 0.00 | 0.00 | 0.00 | 0.00 | 0.00 |
| Polk | FL | 0.45 | 0.00 | 75.10 | 8.99 | 0.00 | 57.60 | 16.09 |
| Polk | IA | 25.41 | 1.73 | 0.00 | 26.20 | 1.93 | 1.85 | 10.23 |
| Portage | OH | 0.00 | 56.60 | 20.19 | 0.00 | 38.62 | 20.24 | 0.00 |
| Porter | IN | 0.60 | 0.00 | 0.00 | 2.63 | 0.00 | 0.00 | 4.90 |
| Potter | TX | 0.00 | 0.00 | 0.00 | 0.00 | 0.00 | 0.00 | 1.26 |
| Prince Georges | MD | 22.94 | 118.50 | 92.87 | 133.51 | 87.96 | 83.73 | 0.84 |
| Prince William | VA | 35.37 | 106.65 | 37.19 | 66.90 | 45.11 | 32.08 | 0.00 |
| Pueblo | CO | 4.76 | 0.00 | 0.41 | 8.58 | 0.00 | 0.41 | 3.78 |
| Pulaski | AR | 0.00 | 0.00 | 6.78 | 0.56 | 0.00 | 6.47 | 0.91 |
| Putnam | NY | 26.06 | 0.00 | 0.00 | 45.95 | 0.00 | 3.65 | 65.95 |
| Racine | WI | 3.27 | 0.00 | 0.95 | 10.60 | 0.00 | 0.00 | 6.78 |
| Ramsey | MN | 9.93 | 0.00 | 1.70 | 19.33 | 0.00 | 38.10 | 2.00 |
| Randall | TX | 0.00 | 0.00 | 0.00 | 0.00 | 0.00 | 0.00 | 2.31 |
| Randolph | NC | 0.00 | 0.00 | 19.31 | 0.00 | 0.00 | 21.64 | 2.84 |
| Rankin | MS | 0.00 | 0.00 | 18.10 | 9.58 | 0.00 | 20.95 | 1.76 |
| Rapides Parish | LA | 1.66 | 4.06 | 0.00 | 5.11 | 4.12 | 0.00 | 4.00 |
| Rensselaer | NY | 0.00 | 25.42 | 0.00 | 2.85 | 27.29 | 0.00 | 0.45 |
| Richland | OH | 0.00 | 21.35 | 0.00 | 0.99 | 7.71 | 0.00 | 2.75 |
| Richland | SC | 0.00 | 0.00 | 37.80 | 0.00 | 0.00 | 45.57 | 0.83 |
| Riverside | CA | 3.70 | 0.00 | 35.15 | 4.66 | 0.00 | 30.35 | 32.68 |
| Robeson | NC | 0.16 | 47.47 | 32.55 | 6.01 | 28.95 | 18.45 | 2.62 |
| Rock | WI | 0.57 | 0.00 | 0.00 | 4.88 | 0.00 | 0.00 | 2.62 |
| Rock Island | IL | 7.83 | 0.00 | 0.00 | 17.55 | 0.00 | 0.00 | 17.66 |
| Rockingham | NH | 0.00 | 0.00 | 0.00 | 0.00 | 0.00 | 0.00 | 1.74 |
| Rockland | NY | 0.02 | 0.03 | 0.00 | 6.86 | 46.38 | 0.00 | 0.00 |
| Rowan | NC | 7.11 | 0.04 | 27.05 | 15.25 | 0.00 | 13.86 | 1.96 |
| Rutherford | TN | 0.00 | 0.00 | 0.00 | 0.00 | 0.00 | 0.00 | 0.00 |
| Sacramento | CA | 9.74 | 12.53 | 47.51 | 20.54 | 16.28 | 44.80 | 2.84 |
| Saginaw | MI | 0.29 | 0.00 | 1.78 | 26.54 | 0.00 | 2.22 | 3.04 |
| Saint Charles | MO | 17.21 | 0.00 | 0.00 | 23.24 | 0.00 | 0.00 | 1.53 |
| Saint Clair | IL | 0.00 | 0.00 | 0.00 | 0.00 | 0.00 | 0.00 | 0.00 |
| Saint Clair | MI | 0.00 | 4.67 | 30.97 | 8.77 | 4.45 | 21.80 | 5.92 |
| Saint Johns | FL | 11.00 | 38.11 | 65.71 | 50.08 | 24.10 | 76.08 | 6.02 |
| Saint Joseph | IN | 0.00 | 0.00 | 0.00 | 0.00 | 0.00 | 0.00 | 0.00 |
| Saint Lawrence | NY | 0.02 | 0.00 | 31.16 | 2.56 | 0.00 | 29.41 | 4.97 |
| Saint Louis | MN | 0.00 | 0.00 | 28.31 | 3.67 | 0.00 | 28.89 | 34.21 |
| Saint Louis | MO | 1.01 | 2.98 | 1.39 | 29.29 | 2.65 | 1.48 | 0.00 |

| County | State | Revenue ($/per capita) | | | Current Operational Expenses ($/per capita) | | | |
|---|---|---|---|---|---|---|---|---|
| | | Parks & Recreation | Sewerage | Solid Waste Management | Parks & Recreation | Sewerage | Solid Waste Management | Natural Resources, Other |
| Saint Lucie | FL | 9.52 | 8.17 | 69.81 | 76.31 | 8.25 | 46.80 | 39.14 |
| Saint Tammany Parish | LA | 3.74 | 6.12 | 1.63 | 13.97 | 7.93 | 2.03 | 9.86 |
| Salt Lake | UT | 7.35 | 0.00 | 11.79 | 76.35 | 0.00 | 14.71 | 4.91 |
| San Bernardino | CA | 3.46 | 0.97 | 35.39 | 7.67 | 1.33 | 33.98 | 18.44 |
| San Diego | CA | 0.98 | 5.40 | 2.88 | 6.96 | 4.20 | 5.73 | 6.67 |
| San Joaquin | CA | 1.91 | 0.00 | 30.64 | 5.56 | 1.38 | 23.79 | 12.56 |
| San Juan | NM | 0.55 | 0.00 | 0.00 | 22.04 | 0.00 | 17.55 | 4.25 |
| San Luis Obispo | CA | 27.29 | 10.18 | 6.01 | 26.64 | 10.14 | 2.35 | 42.55 |
| San Mateo | CA | 1.94 | 8.33 | 6.48 | 10.22 | 11.84 | 11.32 | 1.73 |
| Sandoval | NM | 0.00 | 0.00 | 13.76 | 0.00 | 0.00 | 0.00 | 0.00 |
| Sangamon | IL | 0.00 | 0.00 | 0.00 | 0.00 | 0.00 | 0.00 | 0.00 |
| Santa Barbara | CA | 8.19 | 14.03 | 57.58 | 20.58 | 9.04 | 47.83 | 28.26 |
| Santa Clara | CA | 2.10 | 1.08 | 0.43 | 14.14 | 1.11 | 0.64 | 2.28 |
| Santa Cruz | CA | 5.57 | 76.26 | 47.99 | 24.35 | 50.11 | 46.18 | 10.80 |
| Santa Fe | NM | 0.00 | 0.00 | 0.00 | 5.26 | 0.00 | 9.30 | 0.04 |
| Santa Rosa | FL | 0.10 | 0.00 | 26.47 | 3.87 | 0.00 | 13.27 | 14.89 |
| Sarasota | FL | 3.56 | 96.89 | 68.38 | 60.03 | 81.38 | 90.24 | 28.15 |
| Saratoga | NY | 0.00 | 47.63 | 0.00 | 4.01 | 32.76 | 0.00 | 5.37 |
| Sarpy | NE | 0.00 | 19.05 | 27.61 | 0.00 | 14.64 | 17.64 | 0.70 |
| Schenectady | NY | 1.52 | 0.00 | 1.12 | 4.57 | 0.00 | 1.93 | 3.71 |
| Schuylkill | PA | 0.00 | 0.00 | 0.00 | 3.47 | 0.00 | 2.54 | 28.70 |
| Scott | IA | 6.27 | 0.00 | 0.00 | 12.03 | 0.00 | 3.71 | 7.52 |
| Scott | MN | 0.00 | 0.00 | 0.00 | 12.07 | 0.00 | 0.00 | 7.45 |
| Sebastian | AR | 7.11 | 0.00 | 0.00 | 8.89 | 0.00 | 0.94 | 0.00 |
| Sedgwick | KS | 4.53 | 0.00 | 1.94 | 19.00 | 0.00 | 1.94 | 3.09 |
| Seminole | FL | 1.78 | 42.31 | 34.75 | 9.38 | 44.54 | 72.26 | 2.67 |
| Shasta | CA | 0.00 | 0.00 | 21.12 | 0.73 | 0.00 | 9.03 | 0.00 |
| Shawnee | KS | 4.37 | 0.00 | 37.76 | 37.80 | 0.00 | 34.45 | 3.10 |
| Sheboygan | WI | 0.00 | 0.00 | 0.00 | 3.71 | 0.00 | 0.99 | 14.60 |
| Shelby | AL | 0.00 | 14.34 | 25.64 | 4.73 | 9.27 | 10.21 | 0.57 |
| Shelby | TN | 0.58 | 0.72 | 0.03 | 1.24 | 0.00 | 0.22 | 0.38 |
| Skagit | WA | 6.17 | 0.00 | 74.00 | 16.02 | 4.88 | 61.88 | 24.75 |
| Smith | TX | 0.00 | 0.00 | 0.00 | 0.00 | 0.00 | 0.00 | 0.00 |
| Snohomish | WA | 4.30 | 0.00 | 63.93 | 10.66 | 9.59 | 50.92 | 2.62 |
| Solano | CA | 0.88 | 0.00 | 0.00 | 3.13 | 0.00 | 0.00 | 0.00 |
| Somerset | NJ | 4.49 | 0.00 | 1.54 | 26.01 | 0.00 | 10.79 | 3.18 |
| Sonoma | CA | 4.43 | 34.34 | 69.02 | 33.05 | 20.05 | 244.57 | 0.00 |
| Spartanburg | SC | 2.48 | 0.00 | 26.28 | 17.18 | 0.00 | 16.57 | 0.00 |
| Spokane | WA | 7.56 | 42.75 | 0.00 | 15.92 | 15.62 | 2.31 | 0.82 |
| Spotsylvania | VA | 5.79 | 115.95 | 3.57 | 18.52 | 74.13 | 33.62 | 1.01 |
| Stafford | VA | 12.20 | 133.81 | 0.00 | 39.13 | 66.68 | 1.82 | 1.35 |
| Stanislaus | CA | 3.16 | 0.00 | 11.60 | 12.75 | 0.00 | 7.79 | 0.01 |
| Stark | OH | 0.00 | 51.60 | 0.00 | 0.00 | 33.12 | 0.36 | 0.00 |
| Stearns | MN | 0.00 | 0.00 | 2.16 | 5.85 | 0.00 | 3.47 | 7.05 |
| Strafford | NH | 0.00 | 0.00 | 0.00 | 0.00 | 0.00 | 0.00 | 2.01 |
| Suffolk | NY | 6.97 | 16.00 | 0.00 | 11.83 | 49.52 | 0.00 | 4.39 |
| Sullivan | TN | 1.76 | 0.00 | 5.38 | 3.82 | 0.00 | 10.25 | 0.00 |
| Summit | OH | 0.00 | 46.78 | 0.00 | 0.00 | 41.61 | 0.25 | 0.43 |
| Sumner | TN | 0.00 | 0.00 | 0.00 | 0.00 | 0.00 | 0.00 | 1.64 |
| Sumter | SC | 0.00 | 0.00 | 0.00 | 0.00 | 0.00 | 0.00 | 0.00 |
| Sussex | DE | 0.00 | 189.45 | 0.00 | 0.29 | 54.33 | 0.00 | 2.46 |
| Sussex | NJ | 0.00 | 0.00 | 0.00 | 1.00 | 0.00 | 0.27 | 1.17 |
| Tangipahoa Parish | LA | 0.00 | 5.39 | 18.84 | 6.40 | 6.03 | 33.30 | 11.41 |
| Tarrant | TX | 0.02 | 0.00 | 0.00 | 0.01 | 0.00 | 0.00 | 0.00 |
| Taylor | TX | 0.00 | 0.00 | 0.00 | 0.00 | 0.00 | 0.00 | 3.23 |
| Tazewell | IL | 0.00 | 0.00 | 0.00 | 0.00 | 0.00 | 3.48 | 0.00 |

| County | State | Revenue ($/per capita) | | | Current Operational Expenses ($/per capita) | | | |
|---|---|---|---|---|---|---|---|---|
| | | Parks & Recreation | Sewerage | Solid Waste Management | Parks & Recreation | Sewerage | Solid Waste Management | Natural Resources, Other |
| Thurston | WA | 2.82 | 2.23 | 63.23 | 6.18 | 1.10 | 50.01 | 5.42 |
| Tippecanoe | IN | 0.00 | 0.00 | 0.00 | 0.00 | 0.00 | 0.00 | 0.00 |
| Tom Green | TX | 0.07 | 0.00 | 0.00 | 1.29 | 0.00 | 0.00 | 1.52 |
| Tompkins | NY | 0.00 | 0.00 | 53.38 | 16.70 | 0.00 | 40.29 | 7.52 |
| Travis | TX | 1.59 | 0.00 | 0.00 | 6.40 | 0.00 | 0.68 | 3.24 |
| Trumbull | OH | 0.00 | 37.15 | 0.00 | 0.00 | 28.84 | 0.00 | 0.00 |
| Tulare | CA | 0.46 | 0.33 | 23.37 | 3.84 | 0.30 | 28.74 | 0.51 |
| Tulsa | OK | 4.03 | 0.00 | 0.00 | 12.66 | 0.00 | 0.00 | 1.59 |
| Tuscaloosa | AL | 0.00 | 0.00 | 0.00 | 0.00 | 0.00 | 0.00 | 0.00 |
| Ulster | NY | 0.54 | 0.00 | 86.00 | 4.70 | 0.00 | 88.26 | 2.75 |
| Union | NJ | 8.35 | 0.00 | 0.00 | 20.18 | 0.00 | 0.63 | 0.90 |
| Union | NC | 2.76 | 146.28 | 21.45 | 7.86 | 44.50 | 19.03 | 1.69 |
| Utah | UT | 2.10 | 13.94 | 13.72 | 5.48 | 16.72 | 14.09 | 0.66 |
| Vanderburgh | IN | 35.47 | 0.00 | 0.00 | 67.39 | 0.52 | 0.00 | 2.30 |
| Ventura | CA | 5.02 | 26.71 | 1.58 | 6.05 | 22.60 | 3.78 | 72.27 |
| Vigo | IN | 0.00 | 0.28 | 6.31 | 44.66 | 2.71 | 0.00 | 4.76 |
| Volusia | FL | 19.26 | 7.94 | 52.21 | 48.30 | 5.60 | 38.23 | 27.62 |
| Wake | NC | 0.17 | 0.29 | 25.32 | 15.33 | 0.10 | 21.39 | 0.00 |
| Walworth | WI | 0.01 | 0.00 | 2.18 | 0.87 | 0.00 | 3.36 | 16.00 |
| Warren | KY | 2.09 | 0.00 | 0.00 | 16.55 | 0.00 | 0.00 | 3.72 |
| Warren | NJ | 0.00 | 0.00 | 139.19 | 2.34 | 0.00 | 54.43 | 3.08 |
| Warren | OH | 0.00 | 59.66 | 0.32 | 0.00 | 34.45 | 1.17 | 0.00 |
| Washington | AR | 0.00 | 0.00 | 0.00 | 0.00 | 0.00 | 0.00 | 0.00 |
| Washington | MD | 10.80 | 58.06 | 68.10 | 11.78 | 37.51 | 63.05 | 3.06 |
| Washington | MN | 0.00 | 0.00 | 0.00 | 12.58 | 0.00 | 0.00 | 0.97 |
| Washington | OR | 0.00 | 150.93 | 0.00 | 1.36 | 77.20 | 0.00 | 3.02 |
| Washington | PA | 0.33 | 0.00 | 0.00 | 3.07 | 0.00 | 0.00 | 0.00 |
| Washington | TN | 0.00 | 0.00 | 0.00 | 0.00 | 0.00 | 0.00 | 0.00 |
| Washington | UT | 0.00 | 17.28 | 51.85 | 6.25 | 17.11 | 51.33 | 5.20 |
| Washington | WI | 13.53 | 0.00 | 0.00 | 41.25 | 0.00 | 0.51 | 13.80 |
| Washoe | NV | 39.40 | 51.17 | 0.00 | 107.97 | 7.94 | 0.00 | 0.00 |
| Washtenaw | MI | 0.00 | 0.00 | 0.00 | 0.00 | 0.00 | 0.00 | 0.00 |
| Waukesha | WI | 14.64 | 0.00 | 3.11 | 33.62 | 0.00 | 5.87 | 1.34 |
| Wayne | MI | 1.82 | 38.04 | 0.00 | 33.79 | 33.79 | 0.00 | 7.28 |
| Wayne | NC | 0.00 | 4.05 | 30.81 | 0.63 | 0.68 | 20.35 | 5.28 |
| Wayne | OH | 0.00 | 14.00 | 0.22 | 0.00 | 2.46 | 3.86 | 4.98 |
| Webb | TX | 0.00 | 0.00 | 0.00 | 1.24 | 0.00 | 0.00 | 0.62 |
| Weber | UT | 0.00 | 0.00 | 0.00 | 28.70 | 0.00 | 0.00 | 0.89 |
| Weld | CO | 0.00 | 0.00 | 6.01 | 1.41 | 0.00 | 2.77 | 4.00 |
| Westchester | NY | 32.22 | 3.35 | 28.41 | 45.25 | 67.20 | 61.93 | 0.92 |
| Westmoreland | PA | 0.00 | 0.00 | 0.00 | 7.12 | 0.00 | 0.00 | 6.79 |
| Whatcom | WA | 1.68 | 0.19 | 4.50 | 16.99 | 0.00 | 5.69 | 12.43 |
| Wichita | TX | 0.00 | 0.00 | 0.00 | 0.00 | 0.00 | 0.00 | 0.82 |
| Will | IL | 0.00 | 0.00 | 1.12 | 0.00 | 0.00 | 0.77 | 10.36 |
| Williamson | TN | 16.54 | 0.00 | 9.34 | 41.56 | 0.00 | 31.00 | 2.05 |
| Williamson | TX | 0.00 | 0.00 | 0.00 | 2.05 | 0.00 | 0.00 | 0.79 |
| Wilson | TN | 0.00 | 0.00 | 3.42 | 0.00 | 0.00 | 15.25 | 6.42 |
| Winnebago | IL | 7.86 | 0.00 | 0.00 | 11.89 | 0.00 | 0.00 | 12.50 |
| Winnebago | WI | 1.28 | 0.00 | 20.89 | 11.48 | 0.00 | 47.80 | 11.08 |
| Wood | OH | 0.00 | 0.00 | 23.51 | 1.93 | 0.00 | 21.41 | 6.01 |
| Woodbury | IA | 0.00 | 0.00 | 0.00 | 9.41 | 0.00 | 0.67 | 8.45 |
| Wright | MN | 0.00 | 0.00 | 1.29 | 7.95 | 0.00 | 3.31 | 2.47 |
| Yakima | WA | 0.40 | 0.51 | 27.57 | 1.23 | 0.35 | 17.92 | 7.76 |
| Yavapai | AZ | 0.00 | 0.00 | 0.00 | 0.00 | 7.60 | 9.09 | 0.60 |
| Yellowstone | MT | 26.34 | 0.00 | 2.11 | 35.89 | 0.00 | 3.74 | 1.96 |

| County | State | Revenue ($/per capita) | | | Current Operational Expenses ($/per capita) | | | |
|---|---|---|---|---|---|---|---|---|
| | | Parks & Recreation | Sewerage | Solid Waste Management | Parks & Recreation | Sewerage | Solid Waste Management | Natural Resources, Other |
| Yolo | CA | 0.52 | 0.00 | 45.42 | 3.44 | 0.07 | 39.85 | 0.37 |
| York | ME | 0.00 | 0.00 | 0.00 | 0.00 | 0.00 | 0.00 | 0.00 |
| York | PA | 1.80 | 0.00 | 0.00 | 8.20 | 0.00 | 0.00 | 4.83 |
| York | SC | 1.64 | 16.80 | 23.62 | 22.31 | 3.58 | 35.78 | 2.62 |
| Yuma | AZ | 0.00 | 0.00 | 0.39 | 0.71 | 0.00 | 4.43 | 0.00 |

***Source:*** *U.S. Census Bureau, State and Local Governments Finances, 2006*

## Environmental Revenue and Expenditures for U.S. Cities/Townships with Populations of 50,000+

| City | State | Revenue ($/per capita) | | | Current Operational Expenses ($/per capita) | | | |
|---|---|---|---|---|---|---|---|---|
| | | Park & Recreation | Sewerage | Solid Waste Management | Parks & Recreation | Sewerage | Solid Waste Management | Natural Resources, Other |
| Abilene (city) | TX | 5.43 | 59.44 | 89.59 | 46.09 | 41.16 | 69.56 | 0.00 |
| Abington (township) | PA | 12.45 | 118.28 | 80.35 | 107.77 | 106.08 | 98.03 | 0.00 |
| Addison (township) | IL | 0.00 | 0.00 | 0.00 | 0.00 | 0.00 | 0.00 | 0.00 |
| Akron (city) | OH | 4.65 | 178.82 | 35.58 | 43.14 | 121.44 | 42.65 | 0.00 |
| Albany (city) | GA | 16.84 | 118.46 | 101.85 | 94.59 | 99.41 | 93.83 | 0.00 |
| Albany (city) | NY | 11.17 | 0.00 | 114.00 | 55.37 | 0.00 | 75.24 | 1.17 |
| Albuquerque (city) | NM | 28.72 | 99.02 | 95.06 | 145.41 | 45.73 | 89.28 | 0.00 |
| Alexandria (city) | VA | 16.55 | 141.94 | 31.65 | 157.49 | 161.61 | 39.90 | 0.00 |
| Alhambra (city) | CA | 31.53 | 12.88 | 64.08 | 70.52 | 10.03 | 79.86 | 0.00 |
| Allentown (city) | PA | 11.98 | 164.25 | 83.79 | 24.09 | 78.76 | 84.89 | 0.00 |
| Amarillo (city) | TX | 20.28 | 61.00 | 77.32 | 59.93 | 49.91 | 47.24 | 0.00 |
| Ames (city) | IA | 21.22 | 88.06 | 75.02 | 49.77 | 64.84 | 64.36 | 0.00 |
| Amherst (town) | NY | 19.53 | 2.00 | 1.73 | 74.44 | 119.51 | 57.50 | 28.39 |
| Anaheim (city) | CA | 104.21 | 0.00 | 148.00 | 144.10 | 0.00 | 140.12 | 0.00 |
| Anchorage (municipality) | AK | 22.45 | 103.15 | 78.69 | 83.20 | 73.49 | 57.86 | 0.00 |
| Anderson (city) | IN | 6.85 | 253.34 | 0.00 | 58.42 | 165.05 | 54.21 | 0.00 |
| Ann Arbor (city) | MI | 8.73 | 96.44 | 7.28 | 90.00 | 84.99 | 75.58 | 0.00 |
| Antioch (city) | CA | 23.13 | 45.01 | 0.00 | 55.73 | 21.07 | 0.00 | 0.00 |
| Apple Valley (city) | MN | 48.77 | 70.29 | 0.00 | 107.92 | 68.05 | 0.00 | 0.00 |
| Appleton (city) | WI | 15.46 | 110.50 | 7.07 | 67.39 | 144.16 | 53.53 | 0.00 |
| Arcadia (city) | CA | 12.30 | 14.82 | 0.00 | 39.37 | 8.14 | 0.00 | 0.00 |
| Arlington (city) | TX | 29.25 | 107.85 | 12.21 | 73.13 | 82.09 | 11.58 | 0.00 |
| Arlington Heights (village) | IL | 0.00 | 0.00 | 24.60 | 3.53 | 0.00 | 20.96 | 0.00 |
| Arvada (city) | CO | 89.81 | 57.04 | 0.10 | 178.19 | 65.70 | 0.16 | 0.00 |
| Asheville (city) | NC | 38.47 | 262.83 | 12.01 | 188.00 | 144.43 | 51.59 | 0.00 |
| Athens-Clarke Co. (cons. city) | GA | 7.43 | 97.93 | 55.31 | 61.44 | 73.69 | 47.17 | 0.99 |
| Atlanta (city) | GA | 63.79 | 199.48 | 103.97 | 139.98 | 130.82 | 92.09 | 0.00 |
| Auburn (city) | AL | 6.11 | 103.98 | 47.03 | 72.65 | 66.91 | 49.05 | 0.00 |
| Augusta-Richmond (cons. city/co.) | GA | 5.15 | 131.10 | 112.71 | 50.51 | 86.95 | 85.58 | 1.44 |
| Aurora (city) | CO | 43.68 | 95.72 | 0.00 | 113.86 | 81.11 | 0.00 | 0.00 |
| Aurora (city) | IL | 13.03 | 18.09 | 0.05 | 63.53 | 14.80 | 0.00 | 0.00 |
| Aurora (township) | IL | 0.00 | 0.00 | 0.00 | 0.00 | 0.00 | 0.00 | 0.00 |
| Austin (city) | TX | 0.00 | 0.00 | 0.00 | 0.00 | 0.00 | 0.00 | 0.00 |
| Avondale (city) | AZ | 0.00 | 92.12 | 45.14 | 16.34 | 39.92 | 38.66 | 0.00 |
| Babylon (town) | NY | 5.92 | 0.00 | 116.84 | 42.98 | 0.29 | 209.34 | 2.98 |
| Bakersfield (city) | CA | 2.44 | 104.60 | 108.24 | 41.47 | 31.89 | 103.73 | 0.00 |
| Baldwin Park (city) | CA | 4.56 | 0.00 | 4.34 | 31.65 | 0.00 | 4.09 | 0.00 |
| Baltimore (city) | MD | 15.86 | 216.05 | 20.64 | 85.94 | 220.61 | 107.43 | 0.00 |
| Baton Rouge (cons. city/parish) | LA | 15.09 | 253.94 | 93.77 | 24.97 | 149.15 | 168.08 | 0.00 |
| Battle Creek (city) | MI | 0.00 | 0.00 | 0.00 | 0.00 | 0.00 | 0.00 | 0.00 |
| Bayonne (city) | NJ | 0.00 | 105.26 | 0.00 | 67.84 | 96.59 | 67.35 | 0.00 |
| Baytown (city) | TX | 0.49 | 140.41 | 45.75 | 45.43 | 70.45 | 40.21 | 0.01 |
| Beaumont (city) | TX | 10.26 | 101.73 | 64.31 | 32.21 | 43.58 | 50.71 | 0.00 |
| Beaverton (city) | OR | 0.00 | 30.83 | 0.00 | 1.43 | 25.59 | 0.00 | 0.00 |
| Bellevue (city) | WA | 29.56 | 264.66 | 8.25 | 140.03 | 228.62 | 11.61 | 0.97 |
| Bellingham (city) | WA | 22.78 | 185.12 | 2.61 | 115.05 | 64.03 | 21.61 | 3.81 |
| Bend (city) | OR | 0.00 | 121.27 | 0.00 | 0.00 | 56.01 | 0.00 | 2.04 |
| Bensalem (township) | PA | 8.39 | 0.00 | 0.00 | 17.29 | 0.00 | 0.00 | 0.00 |
| Berkeley (city) | CA | 65.97 | 140.72 | 266.94 | 168.94 | 119.62 | 253.34 | 0.00 |
| Berwyn (city) | IL | 6.43 | 38.15 | 57.99 | 18.30 | 31.37 | 59.72 | 0.00 |
| Bethlehem (city) | PA | 22.17 | 123.12 | 15.74 | 46.45 | 69.90 | 21.81 | 0.00 |
| Billings (city) | MT | 4.25 | 96.70 | 74.45 | 34.85 | 62.52 | 60.41 | 0.00 |
| Birmingham (city) | AL | 4.26 | 24.53 | 2.07 | 90.25 | 14.26 | 213.69 | 0.00 |
| Bismarck (city) | ND | 56.25 | 96.98 | 68.81 | 70.25 | 68.54 | 65.02 | 0.00 |

| City | State | Revenue ($/per capita) | | | Current Operational Expenses ($/per capita) | | | |
|---|---|---|---|---|---|---|---|---|
| | | Park & Recreation | Sewerage | Solid Waste Management | Parks & Recreation | Sewerage | Solid Waste Management | Natural Resources, Other |
| Blaine (city) | MN | 3.97 | 73.21 | 41.22 | 37.47 | 10.90 | 36.25 | 0.00 |
| Bloom (township) | IL | 0.78 | 0.00 | 0.00 | 1.71 | 0.00 | 0.00 | 0.00 |
| Bloomington (city) | IN | 28.49 | 189.44 | 16.17 | 98.70 | 234.12 | 22.25 | 0.00 |
| Bloomington (city) | MN | 43.00 | 139.34 | 6.44 | 125.64 | 27.53 | 8.90 | 0.00 |
| Bloomington City (township) | IL | 0.00 | 0.00 | 0.00 | 0.00 | 0.00 | 0.00 | 0.00 |
| Blue Springs (city) | MO | 25.70 | 103.11 | 0.00 | 84.31 | 76.55 | 0.00 | 0.00 |
| Boca Raton (city) | FL | 74.91 | 156.10 | 2.57 | 302.21 | 165.31 | 50.31 | 29.91 |
| Boise (city) | ID | 28.50 | 116.12 | 78.65 | 73.80 | 88.93 | 80.67 | 0.26 |
| Bolingbrook (village) | IL | 102.47 | 44.06 | 0.00 | 149.57 | 40.77 | 59.11 | 0.00 |
| Bossier City (city) | LA | 49.33 | 105.60 | 50.96 | 85.64 | 63.04 | 41.86 | 0.00 |
| Boston (city) | MA | 0.00 | 205.78 | 0.00 | 25.25 | 40.74 | 96.50 | 1.98 |
| Boulder (city) | CO | 70.31 | 109.09 | 0.00 | 225.03 | 73.13 | 0.00 | 0.00 |
| Bowling Green (city) | KY | 54.12 | 115.30 | 0.00 | 169.51 | 62.87 | 0.47 | 0.00 |
| Boynton Beach (city) | FL | 46.44 | 219.50 | 114.46 | 128.58 | 356.76 | 86.61 | 5.89 |
| Bremen (township) | IL | 0.11 | 0.00 | 0.00 | 0.00 | 0.00 | 0.00 | 0.00 |
| Brick (township) | NJ | 4.50 | 242.36 | 0.00 | 24.76 | 157.97 | 88.34 | 0.03 |
| Bridgeport (city) | CT | 12.12 | 159.04 | 0.99 | 42.50 | 118.83 | 66.38 | 0.00 |
| Bristol (city) | CT | 6.41 | 65.66 | 25.28 | 36.20 | 53.71 | 99.57 | 0.03 |
| Bristol (township) | PA | 0.00 | 9.49 | 0.00 | 12.45 | 0.00 | 34.56 | 0.00 |
| Brockton (city) | MA | 9.24 | 105.55 | 75.68 | 13.79 | 67.32 | 66.63 | 0.57 |
| Broken Arrow (city) | OK | 20.06 | 56.99 | 46.23 | 38.82 | 43.95 | 34.25 | 0.00 |
| Brookhaven (town) | NY | 7.27 | 0.00 | 189.09 | 36.21 | 0.11 | 172.94 | 0.33 |
| Brookline (town) | MA | 28.22 | 198.13 | 40.84 | 57.77 | 12.31 | 48.41 | 2.05 |
| Brooklyn Park (city) | MN | 55.60 | 69.19 | 15.03 | 97.74 | 57.10 | 16.90 | 0.00 |
| Brownsville (city) | TX | 11.73 | 109.67 | 86.80 | 50.13 | 67.86 | 17.98 | 0.00 |
| Bryan (city) | TX | 19.76 | 157.64 | 98.36 | 50.92 | 85.70 | 65.86 | 3.82 |
| Buena Park (city) | CA | 14.90 | 10.91 | 30.40 | 44.84 | 9.39 | 29.66 | 0.00 |
| Buffalo (city) | NY | 5.39 | 202.56 | 66.00 | 18.09 | 125.66 | 69.54 | 0.00 |
| Burbank (city) | CA | 43.54 | 125.70 | 100.71 | 82.21 | 76.27 | 91.58 | 0.00 |
| Burnsville (city) | MN | 27.04 | 48.11 | 0.00 | 69.25 | 11.73 | 0.00 | 11.97 |
| Calumet (township) | IN | 0.01 | 0.00 | 0.00 | 3.25 | 0.00 | 0.00 | 0.00 |
| Camarillo (city) | CA | 0.00 | 134.87 | 81.52 | 0.00 | 97.35 | 80.41 | 0.00 |
| Cambridge (city) | MA | 22.01 | 333.78 | 0.16 | 130.65 | 13.91 | 53.17 | 0.63 |
| Camden (city) | NJ | 0.00 | 73.94 | 0.00 | 29.41 | 39.28 | 57.99 | 0.00 |
| Canton (city) | OH | 0.00 | 135.98 | 52.58 | 30.66 | 96.97 | 54.88 | 0.00 |
| Canton Charter (township) | MI | 42.56 | 169.57 | 0.00 | 93.46 | 49.87 | 35.83 | 0.00 |
| Cape Coral (city) | FL | 43.05 | 99.92 | 0.00 | 88.65 | 259.97 | 0.00 | 7.23 |
| Capital (township) | IL | 0.00 | 0.00 | 0.00 | 0.00 | 0.00 | 0.00 | 0.00 |
| Carlsbad (city) | CA | 17.26 | 91.39 | 20.82 | 135.18 | 41.17 | 18.28 | 0.00 |
| Carmel (city) | IN | 0.00 | 78.44 | 0.00 | 43.37 | 81.29 | 0.00 | 0.00 |
| Carrollton (city) | TX | 24.19 | 90.96 | 36.82 | 69.45 | 6.16 | 30.77 | 5.99 |
| Carson (city) | CA | 25.57 | 0.00 | 0.01 | 157.76 | 0.00 | 1.82 | 0.00 |
| Carson City (city) | NV | 23.31 | 88.66 | 47.95 | 205.01 | 79.94 | 30.62 | 1.07 |
| Cary (town) | NC | 21.37 | 242.43 | 41.00 | 67.89 | 132.73 | 57.92 | 93.77 |
| Casper (city) | WY | 71.72 | 109.81 | 63.01 | 159.98 | 113.73 | 60.91 | 0.00 |
| Cedar Park (city) | TX | 5.71 | 144.72 | 0.00 | 31.98 | 84.44 | 0.00 | 0.00 |
| Cedar Rapids (city) | IA | 42.28 | 207.87 | 95.96 | 52.75 | 138.39 | 109.78 | 0.02 |
| Centennial (city) | CO | 0.00 | 0.00 | 0.00 | 1.37 | 0.00 | 0.00 | 0.00 |
| Center (township) | IN | 0.12 | 0.00 | 0.00 | 3.63 | 0.00 | 0.00 | 0.00 |
| Center (township) | IN | 0.00 | 0.00 | 0.00 | 0.00 | 0.00 | 0.00 | 0.00 |
| Cerritos (city) | CA | 149.05 | 1.24 | 42.25 | 329.91 | 0.00 | 91.55 | 0.00 |
| Champaign (city) | IL | 0.77 | 28.17 | 0.00 | 0.00 | 14.63 | 0.00 | 0.00 |
| Chandler (city) | AZ | 0.81 | 85.82 | 50.62 | 3.91 | 59.09 | 42.62 | 0.00 |
| Charleston (city) | SC | 23.29 | 297.58 | 0.00 | 199.48 | 165.28 | 99.04 | 0.00 |
| Charleston (city) | WV | 32.14 | 269.89 | 78.85 | 168.02 | 238.54 | 84.79 | 0.00 |
| Charlotte (city) | NC | 0.24 | 298.33 | 16.99 | 17.62 | 105.20 | 56.90 | 7.33 |

| City | State | Revenue ($/per capita) | | | Current Operational Expenses ($/per capita) | | | |
|---|---|---|---|---|---|---|---|---|
| | | Park & Recreation | Sewerage | Solid Waste Management | Parks & Recreation | Sewerage | Solid Waste Management | Natural Resources, Other |
| Chattanooga (city) | TN | 18.75 | 272.18 | 35.78 | 88.36 | 215.81 | 106.66 | 16.83 |
| Cheektowaga (town) | NY | 7.97 | 0.04 | 1.16 | 49.36 | 110.25 | 65.94 | 4.14 |
| Cherry Hill (township) | NJ | 0.00 | 54.55 | 0.00 | 8.56 | 53.70 | 77.70 | 0.00 |
| Chesapeake (city) | VA | 2.86 | 42.33 | 0.00 | 50.48 | 39.63 | 42.27 | 0.00 |
| Cheyenne (city) | WY | 28.40 | 126.98 | 173.14 | 101.60 | 88.33 | 75.03 | 0.00 |
| Chicago (city) | IL | 0.00 | 50.66 | 0.00 | 25.56 | 21.88 | 64.49 | 0.00 |
| Chicopee (city) | MA | 12.35 | 115.60 | 17.80 | 30.55 | 68.66 | 24.33 | 0.00 |
| Chino (city) | CA | 10.13 | 67.78 | 149.33 | 101.28 | 70.83 | 140.93 | 0.00 |
| Chino Hills (city) | CA | 10.51 | 58.27 | 46.66 | 40.12 | 43.52 | 46.66 | 0.00 |
| Chula Vista (city) | CA | 6.05 | 100.67 | 2.15 | 95.95 | 36.73 | 3.90 | 0.00 |
| Cicero (town) | IL | 0.00 | 0.00 | 0.00 | 0.00 | 0.00 | 0.00 | 0.00 |
| Cincinnati (city) | OH | 34.94 | 0.00 | 0.99 | 113.87 | 265.21 | 48.38 | 1.42 |
| Citrus Heights (city) | CA | 0.00 | 0.00 | 3.39 | 3.96 | 0.00 | 3.39 | 0.00 |
| Clarks (town) | NY | 24.83 | 0.00 | 127.09 | 67.14 | 7.43 | 236.94 | 16.77 |
| Clarksville (city) | TN | 10.55 | 133.07 | 0.00 | 46.36 | 82.52 | 0.00 | 0.00 |
| Clay (township) | IN | 0.02 | 0.00 | 0.00 | 0.00 | 0.00 | 0.00 | 0.00 |
| Clay (town) | NY | 4.10 | 0.00 | 0.00 | 14.28 | 8.36 | 35.39 | 12.82 |
| Clearwater (city) | FL | 50.39 | 167.81 | 177.74 | 233.27 | 119.70 | 162.15 | 0.00 |
| Cleveland (city) | OH | 11.76 | 45.14 | 5.44 | 99.60 | 38.07 | 59.45 | 0.00 |
| Clifton (city) | NJ | 1.44 | 0.00 | 3.32 | 13.69 | 0.08 | 77.88 | 0.01 |
| Clinton Charter (township) | MI | 4.16 | 123.01 | 37.65 | 25.93 | 49.26 | 40.73 | 0.00 |
| Clovis (city) | CA | 42.03 | 184.49 | 135.98 | 38.85 | 62.18 | 116.51 | 0.00 |
| Colerain (township) | OH | 0.00 | 0.00 | 0.00 | 19.80 | 0.00 | 0.00 | 0.00 |
| College Station (city) | TX | 14.76 | 130.71 | 69.49 | 91.99 | 52.78 | 66.60 | 0.00 |
| Colonie (town) | NY | 24.25 | 72.48 | 115.66 | 55.49 | 47.72 | 48.39 | 0.31 |
| Colorado Springs (city) | CO | 14.74 | 106.89 | 0.00 | 56.76 | 80.66 | 0.00 | 0.31 |
| Columbia (city) | MO | 41.35 | 91.03 | 119.53 | 104.20 | 61.86 | 102.31 | 0.00 |
| Columbia (city) | SC | 0.00 | 0.00 | 0.00 | 0.00 | 0.00 | 0.00 | 0.00 |
| Columbus (consolidated city) | GA | 10.61 | 116.07 | 50.67 | 58.09 | 72.96 | 43.53 | 0.64 |
| Columbus (city) | OH | 14.35 | 234.24 | 0.02 | 73.11 | 106.70 | 43.27 | 0.00 |
| Compton (city) | CA | 0.94 | 11.02 | 89.78 | 14.84 | 2.07 | 86.34 | 0.00 |
| Concord (city) | CA | 33.89 | 140.18 | 0.00 | 91.91 | 117.25 | 0.00 | 0.00 |
| Concord (city) | NC | 28.76 | 365.48 | 1.07 | 55.75 | 225.85 | 64.42 | 0.00 |
| Conway (city) | AR | 9.61 | 83.46 | 100.97 | 15.54 | 50.58 | 80.96 | 0.00 |
| Coon Rapids (city) | MN | 37.13 | 75.84 | 0.00 | 59.66 | 66.33 | 2.04 | 0.00 |
| Coral Springs (city) | FL | 26.86 | 62.28 | 0.00 | 95.96 | 40.18 | 0.00 | 0.00 |
| Corona (city) | CA | 8.39 | 134.51 | 40.34 | 52.86 | 108.41 | 42.34 | 0.00 |
| Corpus Christi (city) | TX | 40.12 | 127.45 | 80.48 | 78.39 | 78.22 | 51.80 | 0.00 |
| Costa Mesa (city) | CA | 11.86 | 0.00 | 2.61 | 32.05 | 3.60 | 0.05 | 0.00 |
| Council Bluffs (city) | IA | 19.38 | 88.98 | 48.03 | 59.40 | 68.99 | 56.91 | 0.32 |
| Cranston (city) | RI | 0.00 | 207.06 | 0.00 | 28.82 | 180.17 | 0.00 | 0.00 |
| Cuyahoga Falls (city) | OH | 100.28 | 143.16 | 77.48 | 143.78 | 96.83 | 72.64 | 0.00 |
| Dallas (city) | TX | 28.11 | 141.04 | 53.30 | 74.57 | 66.93 | 42.67 | 1.25 |
| Daly City (city) | CA | 38.95 | 142.69 | 16.57 | 106.08 | 107.74 | 14.56 | 0.00 |
| Danbury (city) | CT | 23.14 | 132.31 | 0.00 | 55.12 | 63.80 | 3.71 | 0.00 |
| Davenport (city) | IA | 42.09 | 105.83 | 37.96 | 95.44 | 86.76 | 40.60 | 0.00 |
| Davie (town) | FL | 0.00 | 0.00 | 0.00 | 56.10 | 0.00 | 0.00 | 0.00 |
| Dayton (city) | OH | 18.70 | 177.31 | 0.00 | 76.98 | 126.50 | 0.00 | 0.00 |
| Daytona Beach (city) | FL | 104.87 | 238.33 | 131.44 | 135.32 | 114.08 | 114.03 | 42.00 |
| Dearborn (city) | MI | 33.43 | 223.04 | 0.00 | 161.54 | 32.24 | 51.47 | 0.00 |
| Dearborn Heights (city) | MI | 5.30 | 135.46 | 0.00 | 18.56 | 70.61 | 64.24 | 0.00 |
| Decatur (city) | AL | 53.48 | 135.25 | 68.86 | 163.33 | 88.37 | 106.21 | 0.00 |
| Decatur (city) | IL | 0.00 | 25.22 | 6.09 | 0.99 | 10.10 | 0.00 | 0.00 |
| Decatur (township) | IL | 0.00 | 0.00 | 0.00 | 0.00 | 0.00 | 0.00 | 0.00 |
| Denton (city) | TX | 23.86 | 175.49 | 124.14 | 87.34 | 114.13 | 115.63 | 25.69 |

| City | State | Revenue ($/per capita) | | | Current Operational Expenses ($/per capita) | | | |
|---|---|---|---|---|---|---|---|---|
| | | Park & Recreation | Sewerage | Solid Waste Management | Parks & Recreation | Sewerage | Solid Waste Management | Natural Resources, Other |
| Denver (city) | CO | 79.02 | 123.85 | 0.00 | 250.89 | 45.18 | 0.00 | 0.00 |
| Des Moines (city) | IA | 10.56 | 129.61 | 51.56 | 66.05 | 83.50 | 54.95 | 0.23 |
| Des Plaines (city) | IL | 0.61 | 33.03 | 44.85 | 0.00 | 17.55 | 44.41 | 0.00 |
| Detroit (city) | MI | 4.91 | 402.45 | 0.17 | 76.92 | 226.84 | 31.26 | 2.56 |
| Dothan (city) | AL | 23.04 | 53.96 | 17.47 | 123.38 | 86.68 | 74.63 | 0.00 |
| Dover (township) | NJ | 24.77 | 158.67 | 0.00 | 37.29 | 40.16 | 96.20 | 0.55 |
| Downey (city) | CA | 34.94 | 1.24 | 4.30 | 59.93 | 9.25 | 3.85 | 0.00 |
| Dubuque (city) | IA | 21.58 | 85.55 | 41.56 | 90.77 | 57.21 | 42.55 | 0.87 |
| Duluth (city) | MN | 134.03 | 205.51 | 0.00 | 180.90 | 172.55 | 0.00 | 0.00 |
| Dundee (township) | IL | 0.00 | 0.00 | 0.00 | 0.00 | 0.00 | 0.00 | 0.00 |
| Durham (city) | NC | 14.24 | 223.55 | 45.86 | 57.09 | 109.87 | 95.67 | 17.21 |
| Eagan (city) | MN | 66.29 | 73.96 | 0.00 | 99.66 | 16.08 | 0.20 | 0.00 |
| East Orange (city) | NJ | 10.29 | 82.56 | 0.00 | 43.82 | 76.18 | 82.03 | 0.00 |
| Eau Claire (city) | WI | 18.74 | 97.26 | 2.07 | 88.08 | 80.62 | 2.07 | 4.88 |
| Eden Prairie (city) | MN | 0.00 | 65.89 | 0.00 | 70.06 | 56.49 | 33.58 | 0.00 |
| Edison (township) | NJ | 1.09 | 115.11 | 0.00 | 28.78 | 95.76 | 91.08 | 0.00 |
| Edmond (city) | OK | 36.01 | 62.51 | 75.44 | 53.77 | 40.72 | 68.43 | 0.00 |
| El Cajon (city) | CA | 4.58 | 126.06 | 0.00 | 58.86 | 125.11 | 0.00 | 0.00 |
| El Monte (city) | CA | 4.05 | 0.00 | 0.76 | 41.52 | 0.00 | 0.00 | 0.00 |
| El Paso (city) | TX | 4.30 | 70.19 | 46.95 | 41.27 | 35.52 | 11.14 | 0.00 |
| Elgin (city) | IL | 50.15 | 69.67 | 2.67 | 138.68 | 12.91 | 41.73 | 0.00 |
| Elizabeth (city) | NJ | 0.85 | 125.14 | 9.98 | 67.60 | 87.55 | 66.62 | 0.00 |
| Elk Grove (city) | CA | 0.00 | 0.00 | 66.98 | 0.95 | 0.00 | 64.51 | 0.00 |
| Elk Grove (township) | IL | 0.00 | 1.31 | 0.00 | 0.00 | 0.23 | 0.00 | 0.00 |
| Elkhart (city) | IN | 0.00 | 0.00 | 0.00 | 0.00 | 0.00 | 0.00 | 0.00 |
| Elyria (city) | OH | 5.69 | 175.32 | 54.14 | 40.60 | 125.61 | 55.40 | 0.00 |
| Encinitas (city) | CA | 41.66 | 131.17 | 6.82 | 90.62 | 58.39 | 8.47 | 0.00 |
| Erie (city) | PA | 6.00 | 171.62 | 38.99 | 27.95 | 116.58 | 25.26 | 0.00 |
| Escondido (city) | CA | 26.13 | 152.56 | 0.00 | 52.83 | 118.66 | 0.00 | 0.00 |
| Eugene (city) | OR | 33.08 | 273.59 | 0.00 | 149.73 | 131.12 | 12.21 | 1.74 |
| Euless (city) | TX | 109.25 | 88.82 | 4.63 | 136.84 | 28.32 | 0.00 | 0.00 |
| Evanston (city) | IL | 57.05 | 212.82 | 0.00 | 156.57 | 28.20 | 57.86 | 0.00 |
| Evanston (township) | IL | 0.00 | 0.00 | 0.00 | 0.00 | 0.00 | 0.00 | 0.00 |
| Evansville (city) | IN | 43.29 | 214.09 | 4.11 | 130.79 | 214.12 | 4.05 | 10.06 |
| Everett (city) | WA | 42.55 | 243.97 | 10.31 | 112.45 | 127.75 | 4.48 | 2.28 |
| Fairfield (city) | CA | 62.50 | 0.00 | 0.00 | 109.29 | 0.00 | 0.00 | 0.00 |
| Fairfield (town) | CT | 14.13 | 61.01 | 50.55 | 65.45 | 58.74 | 67.23 | 0.00 |
| Fall River (city) | MA | 0.08 | 155.21 | 0.00 | 20.17 | 104.59 | 12.46 | 0.52 |
| Fargo (city) | ND | 44.13 | 99.03 | 87.96 | 66.13 | 35.12 | 62.85 | 10.08 |
| Farmington Hills (city) | MI | 54.29 | 112.18 | 0.00 | 93.93 | 31.63 | 41.56 | 0.00 |
| Fayetteville (city) | AR | 7.19 | 171.09 | 113.03 | 46.49 | 114.54 | 98.93 | 0.00 |
| Fayetteville (city) | NC | 4.01 | 227.69 | 0.00 | 52.75 | 137.15 | 45.89 | 0.62 |
| Federal Way (city) | WA | 8.73 | 0.00 | 1.97 | 43.87 | 20.00 | 3.30 | 0.00 |
| Flagstaff (city) | AZ | 0.00 | 111.25 | 166.20 | 0.00 | 84.77 | 144.09 | 0.00 |
| Flint (city) | MI | 8.85 | 160.44 | 0.47 | 44.60 | 133.91 | 42.40 | 0.00 |
| Fontana (city) | CA | 13.62 | 55.04 | 0.00 | 26.42 | 54.22 | 0.00 | 0.00 |
| Fort Collins (city) | CO | 77.34 | 101.72 | 0.00 | 137.29 | 73.28 | 0.00 | 30.80 |
| Fort Lauderdale (city) | FL | 35.46 | 0.00 | 99.12 | 150.82 | 8.27 | 101.62 | 0.09 |
| Fort Myers (city) | FL | 150.86 | 408.30 | 180.55 | 220.60 | 152.62 | 125.08 | 0.00 |
| Fort Smith (city) | AR | 0.00 | 119.27 | 133.34 | 37.65 | 66.65 | 93.17 | 0.00 |
| Fort Wayne (city) | IN | 22.23 | 152.57 | 36.92 | 72.58 | 131.15 | 36.99 | 1.32 |
| Fort Worth (city) | TX | 7.68 | 120.85 | 62.54 | 60.73 | 57.12 | 51.75 | 0.00 |
| Fountain Valley (city) | CA | 10.22 | 21.52 | 51.90 | 52.65 | 10.26 | 49.43 | 0.00 |
| Framingham (town) | MA | 6.02 | 222.83 | 2.52 | 32.36 | 35.89 | 55.93 | 1.67 |
| Frankfort (township) | IL | 1.94 | 0.00 | 0.00 | 2.03 | 0.00 | 0.00 | 0.00 |
| Franklin (township) | NJ | 0.00 | 176.51 | 0.00 | 8.64 | 98.42 | 5.86 | 0.00 |

| City | State | Revenue ($/per capita) | | | Current Operational Expenses ($/per capita) | | | |
|---|---|---|---|---|---|---|---|---|
| | | Park & Recreation | Sewerage | Solid Waste Management | Parks & Recreation | Sewerage | Solid Waste Management | Natural Resources, Other |
| Franklin (city) | TN | 0.00 | 0.00 | 0.00 | 0.00 | 0.00 | 0.00 | 0.00 |
| Frederick (city) | MD | 46.64 | 90.28 | 0.51 | 79.18 | 89.79 | 53.50 | 0.00 |
| Fremont (city) | CA | 14.78 | 0.00 | 28.09 | 54.21 | 0.00 | 16.84 | 0.00 |
| Fresno (city) | CA | 14.83 | 93.85 | 100.98 | 57.65 | 56.48 | 91.11 | 0.00 |
| Frisco (city) | TX | 0.00 | 0.00 | 0.00 | 0.00 | 0.00 | 0.00 | 0.00 |
| Fullerton (city) | CA | 11.00 | 38.32 | 67.01 | 36.04 | 8.46 | 67.11 | 0.00 |
| Gainesville (city) | FL | 15.94 | 275.20 | 67.10 | 85.86 | 137.57 | 65.60 | 0.67 |
| Gaithersburg (city) | MD | 51.92 | 0.00 | 0.00 | 124.85 | 0.00 | 34.11 | 0.00 |
| Galveston (city) | TX | 76.72 | 195.14 | 68.36 | 197.85 | 102.01 | 49.93 | 18.25 |
| Garden Grove (city) | CA | 7.63 | 0.00 | 0.00 | 30.74 | 0.00 | 0.00 | 0.00 |
| Garland (city) | TX | 17.65 | 123.96 | 60.71 | 50.12 | 60.26 | 57.55 | 12.56 |
| Gary (city) | IN | 7.59 | 209.15 | 0.00 | 64.54 | 171.86 | 69.27 | 3.93 |
| Gastonia (city) | NC | 22.17 | 308.34 | 15.91 | 54.96 | 192.68 | 76.35 | 0.00 |
| Gilbert (town) | AZ | 10.98 | 72.89 | 60.65 | 44.46 | 40.24 | 54.89 | 0.00 |
| Glendale (city) | AZ | 0.00 | 75.39 | 90.55 | 49.47 | 40.32 | 77.58 | 0.00 |
| Glendale (city) | CA | 10.77 | 83.56 | 81.74 | 67.04 | 24.85 | 67.44 | 0.00 |
| Gloucester (township) | NJ | 3.58 | 63.33 | 0.00 | 21.07 | 55.41 | 74.96 | 0.00 |
| Grand Forks (city) | ND | 66.03 | 145.80 | 147.60 | 100.75 | 75.64 | 105.34 | 0.00 |
| Grand Prairie (city) | TX | 35.85 | 79.83 | 51.52 | 69.84 | 58.01 | 36.47 | 0.00 |
| Grand Rapids (city) | MI | 25.03 | 191.03 | 27.85 | 72.93 | 91.77 | 58.81 | 0.00 |
| Great Falls (city) | MT | 37.27 | 115.91 | 48.67 | 61.91 | 52.92 | 36.82 | 0.00 |
| Greece (town) | NY | 5.16 | 1.51 | 0.00 | 13.72 | 10.06 | 4.37 | 2.01 |
| Greeley (city) | CO | 0.00 | 0.00 | 0.00 | 0.00 | 0.00 | 0.00 | 0.00 |
| Green (township) | OH | 0.00 | 0.00 | 0.00 | 6.34 | 0.00 | 0.00 | 0.00 |
| Green Bay (city) | WI | 32.95 | 113.07 | 1.76 | 67.57 | 110.18 | 55.25 | 30.41 |
| Greenburgh (town) | NY | 12.98 | 0.00 | 0.32 | 80.63 | 4.61 | 42.37 | 0.00 |
| Greensboro (city) | NC | 61.91 | 258.61 | 49.34 | 152.48 | 162.99 | 92.69 | 11.33 |
| Greenville (city) | NC | 25.12 | 270.89 | 53.88 | 74.95 | 236.86 | 48.92 | 2.57 |
| Greenville (city) | SC | 120.06 | 45.41 | 7.80 | 273.72 | 24.22 | 0.00 | 0.00 |
| Greenwich (town) | CT | 39.95 | 0.63 | 11.87 | 145.27 | 57.99 | 116.02 | 7.09 |
| Gresham (city) | OR | 0.09 | 138.18 | 3.60 | 5.38 | 62.90 | 4.74 | 0.00 |
| Gulfport (city) | MS | 5.24 | 168.15 | 66.08 | 73.40 | 136.00 | 53.44 | 0.36 |
| Hamburg (town) | NY | 18.22 | 0.77 | 0.00 | 65.80 | 56.11 | 3.15 | 0.25 |
| Hamden (town) | CT | 12.31 | 6.81 | 0.00 | 35.34 | 3.98 | 0.00 | 0.00 |
| Hamilton (township) | NJ | 0.00 | 149.20 | 0.00 | 33.38 | 117.92 | 94.91 | 0.01 |
| Hamilton (city) | OH | 14.04 | 184.44 | 36.62 | 39.47 | 124.67 | 38.58 | 5.17 |
| Hammond (city) | IN | 32.37 | 145.62 | 21.25 | 92.84 | 152.10 | 58.33 | 3.44 |
| Hampton (city) | VA | 129.16 | 65.56 | 69.49 | 210.25 | 46.90 | 62.19 | 0.55 |
| Harlingen (city) | TX | 26.99 | 98.70 | 83.72 | 73.22 | 108.84 | 74.69 | 0.00 |
| Hartford (city) | CT | 0.08 | 0.00 | 4.97 | 20.79 | 0.00 | 118.01 | 0.00 |
| Haverhill (city) | MA | 0.00 | 105.17 | 0.00 | 0.66 | 71.74 | 42.92 | 2.33 |
| Hawthorne (city) | CA | 6.27 | 12.23 | 0.00 | 41.30 | 6.12 | 0.00 | 0.00 |
| Hayward (city) | CA | 4.77 | 120.04 | 0.00 | 7.37 | 84.48 | 3.86 | 0.00 |
| Hempstead (town) | NY | 8.20 | 0.00 | 13.55 | 89.37 | 0.00 | 148.75 | 9.72 |
| Henderson (city) | NV | 0.00 | 0.00 | 0.00 | 0.00 | 0.00 | 0.00 | 0.00 |
| Hialeah (city) | FL | 6.66 | 116.10 | 57.75 | 50.00 | 135.56 | 70.92 | 0.00 |
| High Point (city) | NC | 33.53 | 260.72 | 57.42 | 59.49 | 119.09 | 38.69 | 25.70 |
| Hillsboro (city) | OR | 0.00 | 0.00 | 0.00 | 0.00 | 0.00 | 0.00 | 0.00 |
| Hoffman Estates (village) | IL | 0.00 | 21.08 | 0.00 | 4.25 | 25.57 | 17.72 | 0.00 |
| Hollywood (city) | FL | 33.98 | 201.61 | 115.73 | 61.62 | 137.89 | 107.45 | 0.00 |
| Homestead (city) | FL | 3.66 | 160.92 | 125.17 | 54.85 | 86.22 | 111.05 | 0.00 |
| Honolulu (city) | HI | 25.51 | 160.00 | 104.68 | 77.93 | 73.67 | 143.40 | 0.00 |
| Hoover (city) | AL | 14.05 | 64.27 | 0.09 | 78.99 | 19.85 | 66.99 | 0.00 |
| Houston (city) | TX | 12.05 | 153.33 | 0.52 | 47.90 | 83.80 | 30.46 | 0.19 |
| Howell (township) | NJ | 0.00 | 96.82 | 0.00 | 22.20 | 94.54 | 16.30 | 0.00 |
| Huntington (town) | NY | 34.08 | 3.69 | 118.39 | 62.41 | 14.68 | 220.95 | 5.17 |

| City | State | Revenue ($/per capita) | | | Current Operational Expenses ($/per capita) | | | |
|---|---|---|---|---|---|---|---|---|
| | | Park & Recreation | Sewerage | Solid Waste Management | Parks & Recreation | Sewerage | Solid Waste Management | Natural Resources, Other |
| Huntington Beach (city) | CA | 23.47 | 35.12 | 53.80 | 85.11 | 16.34 | 56.54 | 0.00 |
| Huntsville (city) | AL | 45.74 | 142.21 | 54.72 | 155.07 | 63.72 | 63.03 | 3.10 |
| Idaho Falls (city) | ID | 11.01 | 131.30 | 45.88 | 115.67 | 87.73 | 40.90 | 0.00 |
| Independence (city) | MO | 2.82 | 127.26 | 0.00 | 43.89 | 107.65 | 0.00 | 0.00 |
| Indianapolis (city) | IN | 32.60 | 83.31 | 11.47 | 72.28 | 122.19 | 46.78 | 1.01 |
| Inglewood (city) | CA | 1.95 | 24.28 | 94.24 | 67.30 | 14.81 | 92.56 | 0.00 |
| Iowa City (city) | IA | 12.66 | 188.43 | 113.43 | 75.61 | 66.59 | 94.29 | 0.00 |
| Irondequoit (town) | NY | 7.07 | 0.20 | 0.00 | 38.83 | 40.85 | 3.83 | 8.30 |
| Irvine (city) | CA | 387.05 | 0.00 | 0.00 | 155.78 | 0.00 | 0.00 | 0.00 |
| Irving (city) | TX | 15.61 | 104.88 | 38.48 | 82.81 | 86.49 | 37.80 | 11.41 |
| Irvington (township) | NJ | 0.00 | 53.99 | 0.00 | 15.61 | 44.05 | 46.57 | 0.00 |
| Islip (town) | NY | 19.02 | 0.00 | 0.01 | 35.26 | 0.00 | 124.82 | 3.76 |
| Jackson (city) | MS | 3.70 | 135.19 | 50.94 | 45.02 | 81.73 | 50.10 | 0.00 |
| Jackson (township) | NJ | 0.00 | 0.00 | 0.00 | 13.00 | 0.00 | 23.48 | 0.00 |
| Jackson (city) | TN | 6.76 | 0.00 | 154.93 | 60.68 | 0.00 | 148.36 | 0.00 |
| Jacksonville (city) | FL | 19.57 | 153.87 | 44.59 | 65.56 | 64.97 | 108.54 | 15.41 |
| Jacksonville (city) | NC | 4.16 | 117.62 | 34.31 | 31.24 | 118.70 | 69.54 | 6.49 |
| Janesville (city) | WI | 13.48 | 90.59 | 57.37 | 44.16 | 88.92 | 64.38 | 7.64 |
| Jersey City (city) | NJ | 0.00 | 137.45 | 8.28 | 38.28 | 54.73 | 96.60 | 0.00 |
| Johnson City (city) | TN | 42.56 | 145.47 | 171.35 | 94.73 | 84.96 | 131.18 | 0.00 |
| Joliet (city) | IL | 29.49 | 85.74 | 27.73 | 51.39 | 52.99 | 53.31 | 0.00 |
| Joliet (township) | IL | 0.00 | 0.00 | 0.00 | 1.22 | 0.00 | 0.00 | 0.00 |
| Jonesboro (city) | AR | 0.00 | 62.18 | 13.46 | 55.33 | 43.26 | 78.73 | 0.00 |
| Kalamazoo (city) | MI | 27.38 | 267.15 | 0.00 | 56.75 | 227.23 | 29.23 | 9.73 |
| Kansas City (city) | KS | 10.83 | 108.92 | 34.80 | 44.08 | 116.12 | 21.89 | 6.81 |
| Kansas City (city) | MO | 25.16 | 153.86 | 0.00 | 133.13 | 102.63 | 0.00 | 0.00 |
| Kenner (city) | LA | 3.26 | 54.89 | 33.94 | 76.11 | 72.73 | 407.86 | 0.00 |
| Kennewick (city) | WA | 32.93 | 91.08 | 0.79 | 93.52 | 27.94 | 2.47 | 0.00 |
| Kenosha (city) | WI | 6.57 | 95.59 | 1.92 | 69.21 | 97.75 | 44.13 | 7.81 |
| Kent (city) | WA | 39.98 | 182.01 | 0.00 | 103.56 | 193.77 | 0.17 | 2.34 |
| Kettering (city) | OH | 98.42 | 0.00 | 0.00 | 192.37 | 0.00 | 0.00 | 0.00 |
| Killeen (city) | TX | 10.06 | 138.62 | 101.42 | 32.14 | 64.49 | 63.86 | 3.95 |
| Kissimmee (city) | FL | 16.73 | 20.17 | 52.62 | 87.94 | 0.00 | 50.76 | 0.00 |
| Knoxville (city) | TN | 14.41 | 226.23 | 4.16 | 63.46 | 129.81 | 52.49 | 0.00 |
| La Crosse (city) | WI | 75.40 | 106.67 | 1.03 | 124.20 | 118.25 | 41.46 | 8.69 |
| Lafayette (city) | IN | 0.00 | 0.00 | 0.00 | 0.00 | 0.00 | 0.00 | 0.00 |
| Lafayette (cons. city/parish) | LA | 89.67 | 135.37 | 71.45 | 176.18 | 103.02 | 66.58 | 45.18 |
| Lake Charles (city) | LA | 29.72 | 78.39 | 0.00 | 86.02 | 70.76 | 43.46 | 0.00 |
| Lakeland (city) | FL | 83.50 | 168.47 | 124.19 | 224.48 | 162.86 | 106.44 | 11.76 |
| Lakeville (city) | MN | 48.95 | 120.45 | 0.23 | 46.09 | 50.91 | 0.77 | 0.00 |
| Lakewood (city) | CA | 12.58 | 0.00 | 50.70 | 118.51 | 0.00 | 31.34 | 0.00 |
| Lakewood (city) | CO | 54.98 | 17.35 | 0.00 | 100.40 | 18.86 | 6.36 | 5.71 |
| Lakewood (township) | NJ | 0.00 | 40.76 | 0.69 | 19.61 | 10.19 | 24.47 | 0.03 |
| Lakewood (city) | OH | 14.20 | 82.73 | 0.00 | 43.99 | 69.24 | 80.03 | 7.53 |
| Lakewood (city) | WA | 0.85 | 0.49 | 0.00 | 15.96 | 12.26 | 0.10 | 0.00 |
| Lancaster (city) | CA | 11.41 | 0.94 | 0.00 | 90.34 | 0.00 | 0.00 | 0.00 |
| Lancaster (city) | PA | 4.87 | 170.76 | 0.00 | 94.27 | 90.89 | 7.17 | 0.00 |
| Lansing (city) | MI | 59.28 | 234.39 | 35.97 | 114.74 | 112.60 | 38.26 | 0.00 |
| Laredo (city) | TX | 2.69 | 62.13 | 67.70 | 32.17 | 31.61 | 60.09 | 0.00 |
| Las Cruces (city) | NM | 0.00 | 137.12 | 114.23 | 91.23 | 78.20 | 103.02 | 0.00 |
| Las Vegas (city) | NV | 19.15 | 140.79 | 0.00 | 79.58 | 72.29 | 8.81 | 3.62 |
| Lawrence (township) | IN | 0.00 | 0.00 | 0.00 | 0.00 | 0.00 | 0.00 | 0.00 |
| Lawrence (city) | KS | 25.40 | 91.87 | 94.24 | 82.69 | 46.09 | 88.57 | 0.00 |
| Lawrence (city) | MA | 0.25 | 59.83 | 0.00 | 10.37 | 18.21 | 49.49 | 0.00 |
| Lawton (city) | OK | 11.01 | 89.17 | 59.45 | 78.34 | 57.94 | 20.00 | 0.00 |
| Layton (city) | UT | 10.71 | 77.36 | 47.47 | 58.06 | 47.07 | 15.35 | 0.00 |

| City | State | Revenue ($/per capita) | | | Current Operational Expenses ($/per capita) | | | |
|---|---|---|---|---|---|---|---|---|
| | | Park & Recreation | Sewerage | Solid Waste Management | Parks & Recreation | Sewerage | Solid Waste Management | Natural Resources, Other |
| League City (city) | TX | 3.06 | 113.97 | 0.00 | 32.84 | 54.87 | 0.00 | 0.00 |
| Lees Summit (city) | MO | 25.23 | 148.88 | 33.35 | 59.97 | 109.02 | 32.13 | 0.00 |
| Lewisville (city) | TX | 15.98 | 93.19 | 28.08 | 47.67 | 31.73 | 3.75 | 0.00 |
| Lexington-Fayette (cons. city/co.) | KY | 15.77 | 105.14 | 29.54 | 58.13 | 85.13 | 67.95 | 0.00 |
| Leyden (township) | IL | 0.95 | 0.00 | 0.00 | 0.24 | 0.00 | 4.64 | 0.00 |
| Libertyville (township) | IL | 0.00 | 0.00 | 0.00 | 1.69 | 0.00 | 0.00 | 0.00 |
| Lincoln (city) | NE | 28.83 | 70.43 | 16.60 | 70.79 | 43.47 | 21.69 | 0.00 |
| Little Rock (city) | AR | 29.24 | 165.98 | 80.99 | 80.40 | 154.43 | 57.98 | 0.00 |
| Livermore (city) | CA | 27.72 | 256.06 | 9.37 | 31.32 | 139.98 | 8.36 | 0.00 |
| Livonia (city) | MI | 58.05 | 147.70 | 0.85 | 90.33 | 127.41 | 115.10 | 0.00 |
| Lockport (township) | IL | 0.00 | 7.06 | 0.00 | 0.02 | 7.16 | 0.00 | 0.00 |
| Long Beach (city) | CA | 58.21 | 20.57 | 148.42 | 192.22 | 12.50 | 123.61 | 0.00 |
| Longmont (city) | CO | 67.72 | 87.05 | 65.84 | 128.14 | 73.06 | 51.27 | 2.86 |
| Longview (city) | TX | 7.88 | 136.21 | 49.84 | 65.05 | 69.43 | 42.44 | 0.00 |
| Lorain (city) | OH | 0.00 | 136.69 | 0.00 | 13.63 | 81.50 | 0.65 | 0.00 |
| Los Angeles (city) | CA | 27.25 | 126.14 | 29.42 | 77.88 | 74.75 | 55.58 | 0.00 |
| Louisville-Jefferson (cons. city/co.) | KY | 15.96 | 0.00 | 0.65 | 53.81 | 0.00 | 25.92 | 0.00 |
| Loveland (city) | CO | 92.36 | 92.32 | 64.25 | 95.81 | 61.47 | 58.75 | 0.00 |
| Lowell (city) | MA | 10.99 | 124.46 | 17.96 | 25.41 | 89.26 | 60.63 | 0.00 |
| Lower Merion (township) | PA | 1.57 | 107.48 | 84.51 | 38.12 | 79.71 | 105.06 | 0.00 |
| Lubbock (city) | TX | 0.00 | 0.00 | 0.00 | 0.00 | 0.00 | 0.00 | 0.00 |
| Lynchburg (city) | VA | 4.16 | 227.48 | 106.60 | 122.78 | 96.56 | 59.05 | 0.00 |
| Lynn (city) | MA | 8.10 | 150.37 | 0.00 | 10.54 | 105.15 | 55.32 | 0.01 |
| Lyons (township) | IL | 0.00 | 0.00 | 0.00 | 0.00 | 0.00 | 0.00 | 0.00 |
| Macomb (township) | MI | 23.72 | 90.11 | 0.00 | 35.45 | 59.15 | 0.00 | 0.00 |
| Macon (city) | GA | 4.02 | 0.00 | 61.12 | 53.94 | 4.08 | 42.29 | 0.00 |
| Madison (city) | WI | 76.79 | 81.12 | 12.03 | 194.32 | 81.38 | 65.15 | 55.54 |
| Maine (township) | IL | 0.00 | 0.00 | 0.00 | 0.00 | 0.00 | 0.00 | 0.00 |
| Malden (city) | MA | 0.00 | 217.38 | 0.00 | 11.33 | 0.07 | 59.68 | 0.09 |
| Manchester (town) | CT | 4.01 | 100.31 | 115.27 | 43.54 | 73.62 | 90.69 | 0.00 |
| Manchester (city) | NH | 18.04 | 109.96 | 0.28 | 70.49 | 67.12 | 21.37 | 0.13 |
| Manhattan (city) | KS | 16.79 | 90.17 | 0.00 | 73.75 | 41.21 | 0.00 | 0.00 |
| Mansfield (city) | OH | 0.00 | 167.29 | 0.00 | 15.20 | 124.03 | 0.00 | 0.00 |
| Maple Grove (city) | MN | 101.46 | 67.38 | 7.61 | 120.64 | 61.45 | 0.00 | 4.54 |
| Marietta (city) | GA | 31.42 | 184.27 | 51.94 | 40.08 | 142.12 | 48.87 | 0.00 |
| Mcallen (city) | TX | 11.26 | 83.11 | 88.37 | 87.14 | 44.13 | 42.00 | 0.00 |
| Medford (city) | MA | 0.25 | 178.00 | 0.00 | 12.18 | 24.50 | 87.57 | 0.11 |
| Medford (city) | OR | 0.00 | 158.48 | 0.00 | 53.44 | 88.51 | 0.00 | 0.00 |
| Melbourne (city) | FL | 38.72 | 175.32 | 0.00 | 108.38 | 165.38 | 0.01 | 0.00 |
| Memphis (city) | TN | 16.37 | 82.96 | 68.71 | 82.85 | 48.17 | 62.48 | 1.22 |
| Mentor (city) | OH | 31.19 | 0.00 | 0.00 | 159.44 | 0.00 | 0.00 | 0.00 |
| Meriden (city) | CT | 45.59 | 103.79 | 2.19 | 13.17 | 79.54 | 31.65 | 0.02 |
| Meridian (city) | ID | 2.56 | 111.13 | 4.63 | 14.06 | 87.49 | 6.05 | 0.00 |
| Mesa (city) | AZ | 26.74 | 118.79 | 87.33 | 102.94 | 51.04 | 52.68 | 0.00 |
| Mesquite (city) | TX | 10.95 | 95.00 | 41.36 | 45.67 | 63.96 | 36.30 | 0.00 |
| Miami (city) | FL | 14.85 | 0.00 | 77.27 | 66.56 | 4.33 | 50.86 | 3.70 |
| Miami Beach (city) | FL | 63.65 | 322.94 | 56.47 | 556.77 | 307.70 | 108.59 | 9.93 |
| Miami Gardens (city) | FL | 6.02 | 0.00 | 0.00 | 30.69 | 0.00 | 0.00 | 5.54 |
| Middletown (township) | NJ | 6.85 | 118.04 | 0.00 | 34.17 | 84.41 | 62.58 | 0.04 |
| Middletown (city) | OH | 33.63 | 108.66 | 47.85 | 52.08 | 94.13 | 43.36 | 0.00 |
| Midland (city) | TX | 15.22 | 79.36 | 61.40 | 93.23 | 57.80 | 52.93 | 0.00 |
| Midwest City (city) | OK | 30.86 | 82.41 | 57.58 | 38.78 | 64.39 | 42.28 | 0.00 |
| Milford (city) | CT | 1.92 | 0.00 | 3.17 | 15.26 | 84.98 | 84.21 | 2.05 |
| Millcreek (township) | PA | 3.22 | 136.07 | 0.00 | 14.79 | 122.69 | 0.00 | 4.49 |
| Milpitas (city) | CA | 21.20 | 152.29 | 9.55 | 66.26 | 109.05 | 6.95 | 0.00 |
| Milton (township) | IL | 0.00 | 0.00 | 0.00 | 0.00 | 0.00 | 0.00 | 0.00 |

| City | State | Revenue ($/per capita) Park & Recreation | Sewerage | Solid Waste Management | Current Operational Expenses ($/per capita) Parks & Recreation | Sewerage | Solid Waste Management | Natural Resources, Other |
|---|---|---|---|---|---|---|---|---|
| Milwaukee (city) | WI | 6.21 | 106.92 | 83.59 | 4.22 | 33.80 | 132.01 | 0.00 |
| Minneapolis (city) | MN | 60.30 | 182.38 | 79.14 | 297.26 | 30.93 | 68.80 | 0.00 |
| Mission Viejo (city) | CA | 18.78 | 0.00 | 0.00 | 116.91 | 0.00 | 0.00 | 0.00 |
| Missoula (city) | MT | 0.66 | 103.35 | 0.00 | 46.99 | 50.94 | 0.00 | 0.00 |
| Mobile (city) | AL | 26.44 | 221.76 | 1.72 | 141.00 | 168.21 | 56.15 | 0.00 |
| Modesto (city) | CA | 28.60 | 140.37 | 6.40 | 65.97 | 83.61 | 4.30 | 0.00 |
| Monroe (city) | LA | 39.47 | 86.39 | 51.96 | 146.31 | 60.44 | 83.72 | 0.00 |
| Montebello (city) | CA | 61.16 | 0.00 | 43.86 | 143.36 | 0.00 | 43.86 | 0.00 |
| Montgomery (city) | AL | 20.07 | 7.21 | 49.98 | 123.60 | 7.91 | 75.56 | 0.00 |
| Moreno Valley (city) | CA | 6.71 | 0.00 | 0.52 | 57.98 | 0.00 | 0.58 | 0.00 |
| Mount Pleasant (town) | SC | 27.88 | 154.92 | 0.00 | 64.98 | 154.33 | 90.22 | 0.00 |
| Mount Prospect (village) | IL | 0.00 | 17.90 | 18.95 | 5.25 | 9.90 | 65.29 | 0.00 |
| Mountain View (city) | CA | 75.52 | 165.96 | 116.76 | 314.48 | 161.92 | 120.46 | 0.00 |
| Muncie (city) | IN | 6.79 | 128.63 | 0.00 | 21.52 | 139.19 | 119.79 | 0.00 |
| Murfreesboro (city) | TN | 30.09 | 107.62 | 0.08 | 96.78 | 62.40 | 40.45 | 0.00 |
| Murrieta (city) | CA | 8.57 | 0.00 | 0.00 | 84.17 | 0.00 | 0.00 | 0.00 |
| Nampa (city) | ID | 66.75 | 116.12 | 58.59 | 156.49 | 43.82 | 58.59 | 0.00 |
| Napa (city) | CA | 30.61 | 0.00 | 181.60 | 95.88 | 0.00 | 162.91 | 0.00 |
| Naperville (city) | IL | 5.23 | 52.23 | 0.00 | 43.93 | 43.70 | 36.93 | 0.00 |
| Naperville (township) | IL | 0.00 | 0.00 | 0.00 | 0.00 | 0.00 | 0.00 | 0.00 |
| Nashua (city) | NH | 1.85 | 104.80 | 49.00 | 34.75 | 76.43 | 45.31 | 0.06 |
| Nashville-Davidson (cons. city/co.) | TN | 16.64 | 141.13 | 7.18 | 129.00 | 84.01 | 38.57 | 8.81 |
| New Bedford (city) | MA | 5.07 | 169.48 | 0.08 | 13.29 | 121.11 | 46.31 | 3.38 |
| New Britain (city) | CT | 34.36 | 80.65 | 0.00 | 135.29 | 62.63 | 80.97 | 0.00 |
| New Brunswick (city) | NJ | 0.00 | 146.20 | 0.00 | 35.92 | 20.17 | 45.54 | 0.00 |
| New Haven (city) | CT | 8.87 | 21.39 | 5.13 | 71.03 | 15.32 | 5.15 | 0.00 |
| New Orleans (city) | LA | 135.19 | 262.32 | 78.35 | 240.54 | 298.14 | 121.81 | 0.00 |
| New Trier (township) | IL | 0.00 | 0.00 | 0.00 | 0.00 | 0.00 | 0.00 | 0.00 |
| New York (city) | NY | 7.64 | 139.20 | 0.97 | 62.79 | 35.06 | 122.01 | 0.00 |
| Newark (city) | NJ | 0.00 | 134.16 | 0.00 | 51.50 | 28.61 | 91.42 | 0.00 |
| Newport Beach (city) | CA | 32.89 | 39.47 | 0.00 | 346.47 | 43.17 | 0.00 | 0.00 |
| Newport News (city) | VA | 34.35 | 87.42 | 58.23 | 127.09 | 25.11 | 52.69 | 0.00 |
| Newton (city) | MA | 1.73 | 226.26 | 0.00 | 55.40 | 34.55 | 73.44 | 0.76 |
| Niles (township) | IL | 0.00 | 0.00 | 0.00 | 0.88 | 0.00 | 0.00 | 0.00 |
| Norfolk (city) | VA | 26.24 | 130.07 | 44.02 | 154.26 | 74.62 | 68.08 | 0.00 |
| Norman (city) | OK | 0.00 | 0.00 | 0.00 | 0.00 | 0.00 | 0.00 | 0.00 |
| North (township) | IN | 4.70 | 0.00 | 0.00 | 7.85 | 0.00 | 0.00 | 0.00 |
| North Bergen (township) | NJ | 0.00 | 203.36 | 0.00 | 23.69 | 193.46 | 0.94 | 0.26 |
| North Charleston (city) | SC | 12.91 | 0.00 | 0.00 | 124.99 | 0.00 | 52.91 | 0.00 |
| North Hempstead (town) | NY | 40.88 | 4.38 | 70.02 | 80.17 | 40.92 | 113.30 | 0.00 |
| North Las Vegas (city) | NV | 26.02 | 222.85 | 0.00 | 72.39 | 74.35 | 0.00 | 0.00 |
| North Little Rock (city) | AR | 16.18 | 160.67 | 8.69 | 78.44 | 103.37 | 104.17 | 0.00 |
| North Miami (city) | FL | 6.87 | 166.01 | 104.42 | 77.09 | 286.61 | 101.94 | 105.81 |
| North Richland Hills (city) | TX | 99.61 | 117.37 | 0.00 | 130.76 | 79.64 | 0.00 | 0.00 |
| Northfield (township) | IL | 0.00 | 2.47 | 0.00 | 0.00 | 1.98 | 0.00 | 0.00 |
| Norwalk (city) | CA | 4.59 | 0.00 | 0.00 | 66.78 | 0.00 | 0.00 | 0.00 |
| Norwalk (city) | CT | 106.25 | 121.97 | 0.00 | 140.37 | 68.63 | 0.00 | 0.00 |
| Novi (city) | MI | 48.78 | 177.12 | 0.00 | 94.46 | 118.75 | 0.00 | 0.00 |
| O'fallon (city) | MO | 29.26 | 73.54 | 50.79 | 70.73 | 44.14 | 38.63 | 0.00 |
| Oak Lawn (village) | IL | 0.00 | 37.56 | 42.90 | 6.06 | 36.58 | 46.97 | 0.00 |
| Oak Park (village) | IL | 0.00 | 42.21 | 41.32 | 0.00 | 34.63 | 44.72 | 0.00 |
| Oak Park (township) | IL | 0.00 | 0.00 | 0.00 | 0.00 | 0.00 | 0.00 | 0.00 |
| Oakland (city) | CA | 5.47 | 62.15 | 0.00 | 118.40 | 47.23 | 0.00 | 0.00 |
| Ocala (city) | FL | 46.96 | 359.11 | 204.52 | 141.10 | 202.96 | 194.46 | 0.00 |
| Oceanside (city) | CA | 8.31 | 124.27 | 119.04 | 47.19 | 126.69 | 113.22 | 0.00 |
| Odessa (city) | TX | 0.51 | 108.73 | 73.46 | 33.70 | 40.53 | 65.54 | 0.00 |

| City | State | Revenue ($/per capita) | | | Current Operational Expenses ($/per capita) | | | |
|---|---|---|---|---|---|---|---|---|
| | | Park & Recreation | Sewerage | Solid Waste Management | Parks & Recreation | Sewerage | Solid Waste Management | Natural Resources, Other |
| Ogden (city) | UT | 15.19 | 74.64 | 49.87 | 86.06 | 64.34 | 50.14 | 0.00 |
| Oklahoma City (city) | OK | 21.74 | 97.97 | 57.32 | 88.58 | 61.24 | 52.33 | 0.00 |
| Olathe (city) | KS | 8.02 | 81.24 | 71.32 | 52.77 | 48.46 | 71.49 | 0.00 |
| Old Bridge (township) | NJ | 11.30 | 162.53 | 0.14 | 31.85 | 65.61 | 10.93 | 0.00 |
| Omaha (city) | NE | 13.71 | 79.30 | 1.36 | 46.13 | 56.05 | 40.72 | 28.39 |
| Ontario (city) | CA | 4.98 | 68.89 | 161.65 | 55.65 | 74.16 | 136.30 | 0.00 |
| Orange (city) | CA | 5.12 | 0.00 | 143.86 | 38.55 | 0.00 | 144.78 | 0.00 |
| Orem (city) | UT | 18.13 | 92.96 | 31.92 | 43.96 | 82.23 | 30.53 | 0.00 |
| Orlando (city) | FL | 78.48 | 242.21 | 88.03 | 199.04 | 225.39 | 91.97 | 99.10 |
| Oshkosh (city) | WI | 17.07 | 118.14 | 1.84 | 72.30 | 106.28 | 39.40 | 3.81 |
| Overland Park (city) | KS | 27.78 | 18.20 | 0.00 | 46.92 | 18.40 | 0.00 | 0.00 |
| Owensboro (city) | KY | 18.95 | 54.84 | 90.59 | 62.08 | 30.56 | 71.57 | 0.00 |
| Oxnard (city) | CA | 24.04 | 139.73 | 197.80 | 106.97 | 87.75 | 197.02 | 0.00 |
| Oyster Bay (town) | NY | 14.08 | 1.32 | 42.46 | 77.16 | 8.75 | 177.38 | 7.30 |
| Palatine (village) | IL | 0.00 | 32.18 | 59.99 | 0.00 | 18.47 | 61.40 | 0.00 |
| Palm Bay (city) | FL | 5.68 | 63.00 | 1.28 | 39.50 | 41.65 | 0.00 | 0.00 |
| Palmdale (city) | CA | 32.14 | 0.86 | 0.00 | 78.46 | 0.00 | 0.00 | 0.00 |
| Palo Alto (city) | CA | 116.31 | 562.82 | 428.91 | 300.01 | 414.68 | 456.31 | 0.00 |
| Palos (township) | IL | 0.00 | 0.00 | 0.00 | 0.00 | 0.00 | 0.00 | 0.00 |
| Parma (city) | OH | 18.25 | 1.47 | 0.00 | 46.17 | 11.10 | 48.12 | 0.21 |
| Parsippany-Troy Hills (township) | NJ | 76.85 | 238.84 | 33.39 | 57.93 | 145.03 | 36.54 | 0.00 |
| Pasadena (city) | CA | 80.66 | 37.17 | 74.90 | 140.60 | 21.11 | 69.18 | 0.00 |
| Pasadena (city) | TX | 8.38 | 56.23 | 35.55 | 59.75 | 42.45 | 45.26 | 0.00 |
| Passaic (city) | NJ | 0.28 | 62.72 | 0.00 | 13.70 | 68.19 | 70.70 | 0.00 |
| Paterson (city) | NJ | 0.00 | 36.54 | 4.59 | 18.81 | 0.00 | 70.39 | 0.00 |
| Pawtucket (city) | RI | 4.08 | 0.00 | 3.40 | 36.07 | 8.81 | 38.54 | 0.00 |
| Peabody (city) | MA | 37.40 | 179.51 | 16.22 | 57.91 | 17.38 | 56.81 | 0.23 |
| Pearland (city) | TX | 6.31 | 64.26 | 48.15 | 38.04 | 69.25 | 57.58 | 0.00 |
| Pembroke Pines (city) | FL | 58.67 | 97.51 | 0.00 | 103.34 | 216.69 | 0.00 | 23.34 |
| Penn (township) | IN | 0.00 | 0.00 | 0.00 | 1.49 | 0.00 | 0.00 | 0.00 |
| Pensacola (city) | FL | 14.91 | 0.00 | 100.74 | 220.63 | 0.00 | 97.58 | 0.00 |
| Peoria (city) | AZ | 41.85 | 112.42 | 66.55 | 122.58 | 50.30 | 42.37 | 0.00 |
| Peoria (city) | IL | 0.00 | 0.00 | 2.62 | 22.68 | 0.00 | 0.72 | 0.00 |
| Peoria City (township) | IL | 0.00 | 0.00 | 0.00 | 0.00 | 0.00 | 0.00 | 0.00 |
| Perris (city) | CA | 2.59 | 27.01 | 10.99 | 40.72 | 22.55 | 11.34 | 0.00 |
| Perry (township) | IN | 0.00 | 0.00 | 0.00 | 0.00 | 0.00 | 0.00 | 0.00 |
| Pharr (city) | TX | 0.00 | 0.00 | 0.00 | 0.00 | 0.00 | 0.00 | 0.00 |
| Philadelphia (city) | PA | 0.61 | 164.73 | 0.94 | 47.37 | 86.29 | 66.61 | 0.00 |
| Phoenix (city) | AZ | 17.43 | 75.72 | 77.38 | 81.10 | 59.82 | 56.60 | 0.00 |
| Pike (township) | IN | 0.00 | 0.00 | 0.00 | 0.00 | 0.00 | 0.00 | 0.00 |
| Pine Bluff (city) | AR | 19.65 | 109.61 | 47.72 | 74.06 | 103.73 | 48.61 | 0.00 |
| Piscataway (township) | NJ | 0.00 | 114.53 | 0.00 | 25.35 | 47.91 | 14.00 | 0.00 |
| Pittsburg (city) | CA | 38.40 | 57.00 | 0.00 | 57.95 | 17.88 | 0.00 | 0.00 |
| Pittsburgh (city) | PA | 0.43 | 0.00 | 0.00 | 20.38 | 0.00 | 0.00 | 0.00 |
| Plain (township) | OH | 2.98 | 0.00 | 0.00 | 3.99 | 0.00 | 0.60 | 0.00 |
| Plano (city) | TX | 31.20 | 131.13 | 45.78 | 96.55 | 24.54 | 62.25 | 7.56 |
| Plantation (city) | FL | 32.15 | 58.38 | 2.98 | 104.11 | 37.56 | 6.03 | 0.00 |
| Pleasanton (city) | CA | 88.69 | 166.00 | 0.00 | 236.94 | 149.71 | 0.00 | 0.00 |
| Plymouth (town) | MA | 0.04 | 63.77 | 18.63 | 9.31 | 49.93 | 21.71 | 0.11 |
| Plymouth (city) | MN | 36.15 | 87.89 | 15.31 | 102.89 | 29.59 | 13.05 | 0.00 |
| Pocatello (city) | ID | 18.21 | 132.33 | 92.80 | 56.72 | 60.47 | 66.01 | 0.00 |
| Pomona (city) | CA | 2.46 | 18.58 | 48.04 | 28.14 | 13.79 | 55.03 | 0.00 |
| Pompano Beach (city) | FL | 39.93 | 144.26 | 36.75 | 106.32 | 132.58 | 26.28 | 104.75 |
| Pontiac (city) | MI | 54.84 | 159.82 | 13.63 | 112.67 | 126.81 | 47.12 | 0.00 |
| Port Arthur (city) | TX | 0.00 | 71.72 | 89.59 | 33.37 | 51.83 | 96.78 | 0.00 |

| City | State | Revenue ($/per capita) | | | Current Operational Expenses ($/per capita) | | | |
|---|---|---|---|---|---|---|---|---|
| | | Park & Recreation | Sewerage | Solid Waste Management | Parks & Recreation | Sewerage | Solid Waste Management | Natural Resources, Other |
| Port St Lucie (city) | FL | 12.09 | 154.02 | 0.00 | 47.97 | 28.71 | 0.00 | 1.35 |
| Portage (township) | IN | 0.00 | 0.00 | 0.00 | 0.00 | 0.00 | 0.00 | 0.00 |
| Portland (city) | ME | 22.66 | 266.10 | 20.92 | 100.08 | 59.59 | 57.53 | 0.00 |
| Portland (city) | OR | 45.24 | 379.75 | 2.32 | 126.00 | 156.66 | 4.99 | 0.00 |
| Portsmouth (city) | VA | 31.81 | 144.73 | 80.19 | 104.39 | 56.27 | 80.54 | 0.00 |
| Providence (city) | RI | 36.37 | 0.00 | 0.00 | 58.78 | 2.16 | 44.94 | 0.00 |
| Provo (city) | UT | 29.50 | 42.96 | 14.95 | 54.50 | 48.57 | 22.13 | 0.00 |
| Pueblo (city) | CO | 23.79 | 69.09 | 1.98 | 52.76 | 53.38 | 9.78 | 0.00 |
| Quincy (city) | MA | 0.13 | 159.07 | 0.00 | 28.11 | 28.54 | 67.64 | 0.05 |
| Racine (city) | WI | 13.20 | 148.13 | 7.38 | 104.81 | 136.47 | 57.53 | 4.64 |
| Raleigh (city) | NC | 57.22 | 227.37 | 35.12 | 128.00 | 127.61 | 48.47 | 0.00 |
| Ramapo (town) | NY | 24.99 | 0.40 | 3.93 | 57.28 | 12.81 | 22.78 | 0.00 |
| Rancho Cordova (city) | CA | 0.00 | 0.00 | 0.00 | 0.00 | 0.00 | 0.00 | 0.00 |
| Rancho Cucamonga (city) | CA | 22.30 | 27.50 | 6.17 | 25.45 | 0.00 | 4.32 | 0.00 |
| Rapid City (city) | SD | 91.14 | 99.27 | 105.99 | 180.74 | 56.70 | 70.67 | 0.00 |
| Reading (city) | PA | 15.95 | 248.04 | 40.19 | 21.58 | 150.43 | 33.79 | 0.00 |
| Redding (city) | CA | 8.70 | 141.06 | 184.45 | 127.92 | 77.60 | 156.10 | 0.00 |
| Redlands (city) | CA | 3.89 | 86.74 | 128.93 | 50.87 | 66.18 | 109.53 | 0.00 |
| Redwood (city) | CA | 22.88 | 170.09 | 0.00 | 104.35 | 167.89 | 0.00 | 0.00 |
| Reno (city) | NV | 0.00 | 0.00 | 0.00 | 0.00 | 0.00 | 0.00 | 0.00 |
| Renton (city) | WA | 56.27 | 239.30 | 160.47 | 151.91 | 181.36 | 132.90 | 4.82 |
| Rialto (city) | CA | 12.41 | 120.89 | 12.61 | 39.54 | 77.73 | 5.57 | 0.00 |
| Rich (township) | IL | 0.00 | 0.00 | 0.00 | 0.00 | 0.00 | 0.00 | 0.00 |
| Richardson (city) | TX | 45.23 | 118.62 | 102.73 | 114.63 | 121.43 | 94.51 | 0.00 |
| Richmond (city) | CA | 15.32 | 122.95 | 0.00 | 67.53 | 65.48 | 0.00 | 0.00 |
| Richmond (city) | VA | 14.92 | 263.27 | 58.60 | 97.24 | 129.38 | 80.12 | 10.97 |
| Rio Rancho (city) | NM | 3.09 | 142.65 | 0.00 | 65.93 | 134.15 | 0.13 | 0.00 |
| Riverside (city) | CA | 2.50 | 87.06 | 52.20 | 77.62 | 60.84 | 43.11 | 0.00 |
| Roanoke (city) | VA | 0.00 | 0.00 | 0.00 | 0.00 | 0.00 | 0.00 | 0.00 |
| Rochester (city) | MN | 54.60 | 105.26 | 0.00 | 113.30 | 61.01 | 0.00 | 4.42 |
| Rochester (city) | NY | 10.61 | 0.00 | 112.61 | 71.15 | 3.22 | 105.87 | 0.00 |
| Rochester Hills (city) | MI | 36.44 | 117.51 | 0.00 | 141.62 | 40.12 | 0.00 | 0.00 |
| Rock Hill (city) | SC | 18.42 | 257.47 | 82.57 | 71.71 | 175.24 | 72.23 | 30.83 |
| Rockford (city) | IL | 0.00 | 0.00 | 43.25 | 0.00 | 4.09 | 41.77 | 0.00 |
| Rockford (township) | IL | 0.00 | 0.00 | 0.00 | 0.00 | 0.00 | 0.00 | 0.00 |
| Rockville (city) | MD | 91.16 | 90.27 | 81.64 | 280.34 | 81.33 | 79.64 | 0.00 |
| Rocky Mount (city) | NC | 7.71 | 290.10 | 96.38 | 87.32 | 201.32 | 86.98 | 0.00 |
| Rogers (city) | AR | 2.95 | 144.13 | 0.00 | 48.43 | 84.78 | 0.00 | 0.00 |
| Roseville (city) | CA | 62.96 | 230.87 | 163.85 | 183.56 | 134.73 | 131.74 | 0.00 |
| Roswell (city) | GA | 39.40 | 15.05 | 109.61 | 99.61 | 13.46 | 85.00 | 6.56 |
| Royal Oak (city) | MI | 53.57 | 197.81 | 0.00 | 58.29 | 146.01 | 96.89 | 0.00 |
| Sacramento (city) | CA | 53.76 | 38.02 | 94.66 | 154.24 | 5.24 | 92.59 | 0.00 |
| Saginaw (city) | MI | 0.00 | 0.00 | 0.00 | 0.00 | 0.00 | 0.00 | 0.00 |
| Salem (city) | OR | 0.00 | 213.60 | 0.00 | 67.16 | 143.70 | 0.00 | 0.00 |
| Salinas (city) | CA | 8.98 | 31.00 | 0.00 | 48.38 | 21.09 | 0.00 | 0.00 |
| Salt Lake City (city) | UT | 46.17 | 123.41 | 13.44 | 96.38 | 65.42 | 32.86 | 0.00 |
| San Angelo (city) | TX | 8.71 | 84.07 | 9.18 | 34.14 | 48.22 | 6.49 | 0.00 |
| San Antonio (city) | TX | 18.54 | 170.73 | 45.58 | 93.13 | 86.97 | 43.43 | 0.84 |
| San Bernardino (city) | CA | 7.71 | 102.33 | 119.35 | 35.30 | 77.27 | 99.67 | 0.00 |
| San Buenaventura (city) | CA | 48.57 | 135.89 | 0.00 | 127.18 | 83.67 | 0.00 | 0.00 |
| San Clemente (city) | CA | 56.20 | 129.73 | 2.69 | 144.44 | 98.17 | 2.44 | 0.00 |
| San Diego (city) | CA | 50.32 | 265.95 | 48.87 | 108.00 | 167.47 | 41.18 | 9.04 |
| San Francisco (city) | CA | 33.43 | 221.36 | 0.00 | 259.28 | 139.82 | 0.00 | 0.00 |
| San Jose (city) | CA | 16.51 | 117.81 | 80.72 | 91.51 | 92.54 | 85.17 | 0.00 |
| San Leandro (city) | CA | 30.13 | 145.11 | 0.00 | 107.75 | 86.54 | 0.00 | 0.00 |
| San Marcos (city) | CA | 16.39 | 0.00 | 8.04 | 85.72 | 0.00 | 0.00 | 0.00 |

| City | State | Revenue ($/per capita) | | | Current Operational Expenses ($/per capita) | | | |
|---|---|---|---|---|---|---|---|---|
| | | Park & Recreation | Sewerage | Solid Waste Management | Parks & Recreation | Sewerage | Solid Waste Management | Natural Resources, Other |
| San Mateo (city) | CA | 70.81 | 185.84 | 11.33 | 129.08 | 113.56 | 8.55 | 0.00 |
| Sandy (city) | UT | 26.44 | 37.72 | 38.69 | 63.25 | 10.50 | 35.59 | 0.00 |
| Sandy Springs (city) | GA | 0.00 | 0.00 | 0.00 | 14.57 | 0.00 | 0.00 | 0.00 |
| Santa Ana (city) | CA | 11.29 | 11.16 | 40.80 | 47.93 | 6.80 | 37.31 | 0.00 |
| Santa Barbara (city) | CA | 48.49 | 140.33 | 173.96 | 187.71 | 41.05 | 154.46 | 0.00 |
| Santa Clara (city) | CA | 98.09 | 108.94 | 135.35 | 97.81 | 92.99 | 142.87 | 0.00 |
| Santa Clarita (city) | CA | 28.02 | 0.00 | 13.58 | 51.95 | 0.00 | 6.99 | 0.00 |
| Santa Cruz (city) | CA | 53.23 | 238.78 | 246.92 | 214.48 | 201.19 | 186.52 | 0.00 |
| Santa Fe (city) | NM | 37.65 | 119.64 | 124.08 | 163.36 | 87.56 | 52.97 | 0.00 |
| Santa Maria (city) | CA | 32.53 | 61.74 | 194.01 | 78.75 | 28.99 | 107.21 | 0.00 |
| Santa Monica (city) | CA | 111.52 | 128.06 | 194.19 | 327.23 | 128.85 | 193.66 | 0.00 |
| Santa Rosa (city) | CA | 58.35 | 354.61 | 0.00 | 82.77 | 196.23 | 0.00 | 0.00 |
| Sarasota (city) | FL | 224.85 | 273.02 | 186.85 | 296.93 | 173.64 | 187.64 | 57.16 |
| Savannah (city) | GA | 5.11 | 160.16 | 355.61 | 144.87 | 172.36 | 418.57 | 0.00 |
| Schaumburg (village) | IL | 5.74 | 26.32 | 0.00 | 33.26 | 25.70 | 0.00 | 0.00 |
| Schaumburg (township) | IL | 1.30 | 0.00 | 0.00 | 0.00 | 0.00 | 0.00 | 0.00 |
| Schenectady (city) | NY | 13.73 | 146.07 | 59.05 | 24.40 | 98.85 | 62.90 | 0.00 |
| Scottsdale (city) | AZ | 25.04 | 143.10 | 74.28 | 134.83 | 102.47 | 62.06 | 0.00 |
| Scranton (city) | PA | 0.00 | 0.00 | 0.00 | 0.00 | 0.00 | 0.00 | 0.00 |
| Seattle (city) | WA | 71.48 | 303.00 | 190.97 | 266.50 | 98.97 | 180.12 | 0.00 |
| Shawnee (city) | KS | 7.34 | 0.00 | 0.00 | 59.09 | 0.00 | 0.00 | 0.00 |
| Shelby Charter (township) | MI | 3.89 | 72.50 | 0.00 | 33.08 | 14.78 | 0.00 | 0.00 |
| Shreveport (city) | LA | 10.01 | 123.32 | 32.72 | 67.23 | 62.00 | 84.76 | 0.00 |
| Simi Valley (city) | CA | 0.00 | 106.76 | 0.83 | 4.44 | 85.91 | 10.03 | 0.00 |
| Sioux City (city) | IA | 11.23 | 135.84 | 50.96 | 130.20 | 137.09 | 62.01 | 0.00 |
| Sioux Falls (city) | SD | 2.66 | 60.87 | 32.13 | 93.28 | 38.68 | 22.09 | 0.01 |
| Skokie (village) | IL | 26.84 | 0.00 | 0.00 | 27.84 | 0.00 | 55.76 | 0.00 |
| Smith (town) | NY | 12.92 | 0.00 | 63.87 | 39.47 | 0.00 | 152.56 | 6.26 |
| Somerville (city) | MA | 0.00 | 172.41 | 0.00 | 9.58 | 6.77 | 56.95 | 0.36 |
| South Bend (city) | IN | 36.04 | 187.94 | 37.69 | 132.02 | 174.70 | 39.12 | 0.00 |
| South Gate (city) | CA | 7.87 | 13.70 | 27.37 | 38.89 | 8.70 | 27.58 | 0.00 |
| South San Francisco (city) | CA | 44.54 | 338.27 | 2.82 | 111.55 | 125.26 | 1.26 | 0.00 |
| Southampton (town) | NY | 35.99 | 0.00 | 22.92 | 631.21 | 1.20 | 47.74 | 8.96 |
| Southfield (city) | MI | 35.89 | 0.00 | 37.44 | 116.72 | 0.00 | 39.49 | 0.00 |
| Sparks (city) | NV | 0.00 | 0.00 | 0.00 | 0.00 | 0.00 | 0.00 | 0.00 |
| Spokane (city) | WA | 32.02 | 221.50 | 342.03 | 61.83 | 90.05 | 202.00 | 1.43 |
| Spokane Valley (city) | WA | 0.00 | 0.00 | 0.00 | 0.00 | 0.00 | 0.00 | 0.00 |
| Springdale (city) | AR | 7.61 | 179.42 | 0.00 | 31.15 | 93.72 | 0.00 | 0.00 |
| Springfield (city) | IL | 0.98 | 52.50 | 1.72 | 20.90 | 25.97 | 0.62 | 0.00 |
| Springfield (city) | MA | 15.87 | 195.41 | 0.00 | 37.66 | 117.64 | 59.33 | 0.00 |
| Springfield (city) | MO | 25.84 | 136.32 | 20.41 | 87.59 | 70.36 | 19.66 | 0.00 |
| Springfield (city) | OH | 0.00 | 150.71 | 0.00 | 56.44 | 67.90 | 0.00 | 0.00 |
| Springfield (city) | OR | 0.00 | 478.84 | 0.00 | 0.00 | 339.85 | 0.00 | 0.00 |
| St Charles (city) | MO | 31.65 | 106.71 | 0.00 | 75.12 | 54.28 | 0.00 | 0.00 |
| St Clair Shores (city) | MI | 51.76 | 142.18 | 0.00 | 71.22 | 124.10 | 49.66 | 0.00 |
| St Cloud (city) | MN | 48.68 | 85.31 | 36.83 | 100.23 | 57.44 | 33.14 | 0.00 |
| St George (city) | UT | 92.63 | 136.79 | 41.59 | 169.43 | 78.47 | 41.59 | 0.00 |
| St Joseph (township) | IN | 0.00 | 0.00 | 0.00 | 0.22 | 0.00 | 0.00 | 0.00 |
| St Joseph (city) | MO | 42.48 | 120.09 | 38.55 | 67.75 | 84.13 | 36.48 | 0.00 |
| St Louis (city) | MO | 1.40 | 0.00 | 0.00 | 84.46 | 0.18 | 37.23 | 0.00 |
| St Paul (city) | MN | 61.02 | 150.24 | 0.00 | 135.13 | 27.80 | 9.33 | 0.00 |
| St Peters (city) | MO | 60.19 | 63.79 | 103.54 | 142.91 | 54.45 | 117.74 | 0.00 |
| St Petersburg (city) | FL | 60.46 | 123.19 | 124.78 | 153.99 | 27.34 | 134.55 | 0.00 |
| Stamford (city) | CT | 26.84 | 113.74 | 23.03 | 125.61 | 62.54 | 91.99 | 0.00 |
| Sterling Heights (city) | MI | 3.53 | 106.35 | 0.00 | 18.08 | 1.98 | 31.47 | 0.00 |
| Stockton (city) | CA | 32.08 | 146.78 | 0.00 | 83.41 | 101.29 | 0.13 | 0.00 |

| City | State | Revenue ($/per capita) | | | Current Operational Expenses ($/per capita) | | | |
|---|---|---|---|---|---|---|---|---|
| | | Park & Recreation | Sewerage | Solid Waste Management | Parks & Recreation | Sewerage | Solid Waste Management | Natural Resources, Other |
| Suffolk (city) | VA | 23.31 | 100.75 | 0.00 | 64.25 | 68.29 | 41.22 | 0.89 |
| Sugar Land (city) | TX | 2.54 | 100.82 | 33.52 | 26.88 | 43.18 | 33.86 | 0.00 |
| Sunnyvale (city) | CA | 62.13 | 143.07 | 374.31 | 133.35 | 111.99 | 358.15 | 0.00 |
| Sunrise (city) | FL | 19.45 | 301.48 | 143.44 | 111.98 | 207.00 | 145.71 | 0.00 |
| Surprise (city) | AZ | 53.86 | 138.30 | 45.32 | 134.17 | 85.45 | 34.99 | 0.00 |
| Syracuse (city) | NY | 3.19 | 27.95 | 0.85 | 48.28 | 22.76 | 45.24 | 0.00 |
| Tacoma (city) | WA | 11.05 | 207.18 | 331.01 | 16.71 | 171.98 | 168.88 | 17.84 |
| Tallahassee (city) | FL | 22.88 | 180.73 | 113.63 | 115.02 | 204.02 | 109.98 | 13.09 |
| Tampa (city) | FL | 0.00 | 220.71 | 195.08 | 153.48 | 158.21 | 131.43 | 0.00 |
| Taunton (city) | MA | 0.57 | 81.09 | 19.72 | 19.85 | 57.71 | 26.96 | 0.37 |
| Taylor (city) | MI | 91.03 | 91.27 | 0.00 | 131.86 | 36.02 | 0.00 | 0.00 |
| Taylorsville (city) | UT | 0.00 | 0.00 | 0.00 | 2.91 | 0.00 | 0.00 | 0.00 |
| Tempe (city) | AZ | 40.50 | 112.98 | 76.54 | 115.80 | 73.51 | 63.74 | 0.00 |
| Terre Haute (city) | IN | 18.06 | 189.04 | 0.00 | 43.59 | 180.44 | 184.27 | 0.00 |
| Thornton (city) | CO | 23.32 | 74.71 | 39.78 | 100.98 | 71.85 | 27.87 | 0.00 |
| Thornton (township) | IL | 0.00 | 0.00 | 0.00 | 0.00 | 0.00 | 0.00 | 0.00 |
| Thousand Oaks (city) | CA | 47.90 | 179.23 | 11.20 | 56.44 | 82.60 | 8.83 | 0.00 |
| Toledo (city) | OH | 1.00 | 174.87 | 0.00 | 14.97 | 114.84 | 47.36 | 1.00 |
| Tonawanda (town) | NY | 62.18 | 59.69 | 5.30 | 111.91 | 132.65 | 57.77 | 2.80 |
| Topeka (city) | KS | 23.96 | 191.58 | 0.00 | 90.99 | 132.97 | 0.00 | 0.00 |
| Torrance (city) | CA | 31.32 | 11.58 | 71.82 | 120.15 | 19.01 | 71.63 | 0.00 |
| Tracy (city) | CA | 12.81 | 107.35 | 188.61 | 51.71 | 63.82 | 175.19 | 0.00 |
| Trenton (city) | NJ | 0.00 | 148.27 | 0.00 | 52.99 | 115.17 | 105.70 | 0.30 |
| Troy (city) | MI | 72.81 | 124.76 | 0.00 | 139.03 | 12.30 | 53.84 | 0.00 |
| Tucson (city) | AZ | 39.59 | 0.00 | 74.13 | 130.87 | 0.00 | 70.10 | 4.48 |
| Tulsa (city) | OK | 3.58 | 126.65 | 93.97 | 57.93 | 85.90 | 79.98 | 0.00 |
| Tuscaloosa (city) | AL | 0.00 | 160.77 | 26.92 | 49.35 | 76.93 | 74.07 | 0.10 |
| Tustin (city) | CA | 11.25 | 0.11 | 0.00 | 52.70 | 0.00 | 0.00 | 0.00 |
| Tyler (city) | TX | 3.13 | 72.82 | 91.51 | 44.18 | 51.03 | 69.83 | 0.00 |
| Union (township) | NJ | 0.00 | 83.99 | 0.00 | 8.90 | 66.12 | 79.58 | 0.00 |
| Union (town) | NY | 2.88 | 9.18 | 0.00 | 17.36 | 24.72 | 25.05 | 0.68 |
| Union City (city) | NJ | 0.00 | 0.00 | 1.44 | 21.81 | 0.00 | 106.27 | 0.00 |
| Upper Darby (township) | PA | 6.49 | 91.89 | 39.14 | 25.29 | 77.74 | 26.36 | 0.00 |
| Vacaville (city) | CA | 35.00 | 232.03 | 0.00 | 86.39 | 152.81 | 0.00 | 0.00 |
| Vallejo (city) | CA | 516.41 | 176.80 | 15.47 | 461.99 | 127.51 | 19.98 | 0.00 |
| Vancouver (city) | WA | 19.82 | 155.82 | 3.10 | 71.27 | 85.43 | 8.43 | 0.00 |
| Victoria (city) | TX | 0.00 | 0.00 | 0.00 | 0.00 | 0.00 | 0.00 | 0.00 |
| Vineland (city) | NJ | 0.00 | 127.73 | 0.00 | 10.74 | 116.25 | 3.47 | 0.00 |
| Virginia Beach (city) | VA | 40.34 | 121.01 | 6.31 | 128.49 | 56.09 | 63.45 | 2.43 |
| Visalia (city) | CA | 24.79 | 131.23 | 109.25 | 44.51 | 109.11 | 77.82 | 0.00 |
| Vista (city) | CA | 26.20 | 189.31 | 8.13 | 90.87 | 38.48 | 1.15 | 0.00 |
| Waco (city) | TX | 30.10 | 122.35 | 118.42 | 110.97 | 57.33 | 79.68 | 0.00 |
| Waltham (city) | MA | 6.44 | 199.64 | 0.00 | 19.53 | 21.87 | 86.47 | 1.10 |
| Warner Robins (city) | GA | 6.22 | 85.25 | 110.31 | 30.78 | 68.12 | 102.21 | 0.00 |
| Warren (township) | IL | 0.00 | 0.00 | 0.00 | 28.20 | 0.00 | 0.00 | 0.00 |
| Warren (township) | IN | 0.00 | 0.00 | 0.00 | 0.00 | 0.00 | 0.00 | 0.00 |
| Warren (city) | MI | 14.94 | 107.94 | 0.00 | 51.08 | 94.11 | 58.93 | 0.00 |
| Warwick (city) | RI | 9.19 | 165.75 | 0.00 | 28.84 | 58.23 | 27.88 | 0.00 |
| Washington (township) | IN | 0.00 | 0.00 | 0.00 | 0.00 | 0.00 | 0.00 | 0.00 |
| Washington (township) | NJ | 0.00 | 101.22 | 1.27 | 1.49 | 91.59 | 73.38 | 0.10 |
| Washington (township) | OH | 25.06 | 0.00 | 0.00 | 45.44 | 0.00 | 0.00 | 0.00 |
| Washington Dc (city) | DC | 44.03 | 303.94 | 2.85 | 175.67 | 160.19 | 60.57 | 0.52 |
| Waterbury (city) | CT | 13.61 | 178.25 | 1.09 | 14.81 | 103.47 | 54.08 | 0.00 |
| Waterford Charter (township) | MI | 13.57 | 118.28 | 0.00 | 27.78 | 101.32 | 0.00 | 0.00 |
| Waterloo (city) | IA | 29.08 | 150.07 | 64.46 | 71.23 | 72.46 | 43.97 | 0.00 |
| Waukegan (city) | IL | 0.00 | 0.00 | 0.00 | 0.00 | 0.00 | 0.00 | 0.00 |

| City | State | Revenue ($/per capita) | | | Current Operational Expenses ($/per capita) | | | |
|---|---|---|---|---|---|---|---|---|
| | | Park & Recreation | Sewerage | Solid Waste Management | Parks & Recreation | Sewerage | Solid Waste Management | Natural Resources, Other |
| Waukegan (township) | IL | 0.00 | 0.00 | 0.00 | 0.00 | 0.00 | 0.00 | 0.00 |
| Waukesha (city) | WI | 14.86 | 130.33 | 0.00 | 73.17 | 113.16 | 28.92 | 9.33 |
| Wayne (township) | IL | 0.00 | 0.00 | 0.00 | 0.00 | 0.00 | 0.00 | 0.00 |
| Wayne (township) | IN | 0.00 | 0.00 | 0.00 | 0.00 | 0.00 | 0.00 | 0.00 |
| Wayne (township) | IN | 0.20 | 0.00 | 0.00 | 0.00 | 0.00 | 0.00 | 0.00 |
| Wayne (township) | NJ | 8.13 | 152.49 | 0.00 | 49.97 | 113.18 | 88.44 | 4.19 |
| West Allis (city) | WI | 1.67 | 72.36 | 0.34 | 9.18 | 70.38 | 66.92 | 24.19 |
| West Bloomfield Charter (township) | MI | 6.36 | 133.35 | 0.00 | 36.31 | 104.24 | 0.00 | 0.00 |
| West Chester (township) | OH | 0.00 | 0.00 | 0.00 | 0.00 | 0.00 | 0.00 | 0.00 |
| West Covina (city) | CA | 11.56 | 0.00 | 0.00 | 33.80 | 0.00 | 0.00 | 0.00 |
| West Des Moines (city) | IA | 22.23 | 107.57 | 22.82 | 66.07 | 38.47 | 28.86 | 0.46 |
| West Hartford (town) | CT | 57.61 | 0.00 | 0.00 | 74.91 | 0.00 | 0.00 | 0.00 |
| West Haven (city) | CT | 0.00 | 0.00 | 0.00 | 0.00 | 0.00 | 0.00 | 0.00 |
| West Jordan (city) | UT | 1.00 | 47.08 | 34.83 | 17.69 | 62.73 | 34.02 | 0.00 |
| West Palm Beach (city) | FL | 33.15 | 189.05 | 128.24 | 136.12 | 127.94 | 80.52 | 50.56 |
| West Valley City (city) | UT | 36.21 | 26.37 | 28.80 | 64.58 | 16.71 | 21.03 | 0.00 |
| Westland (city) | MI | 12.24 | 121.60 | 0.00 | 71.65 | 116.61 | 53.12 | 0.00 |
| Westminster (city) | CA | 2.38 | 0.00 | 0.00 | 33.24 | 0.00 | 0.00 | 0.00 |
| Westminster (city) | CO | 106.65 | 106.29 | 0.00 | 116.90 | 49.51 | 0.00 | 0.00 |
| Weston (city) | FL | 11.43 | 0.00 | 51.52 | 73.07 | 0.00 | 44.77 | 0.00 |
| Weymouth (town) | MA | 0.00 | 221.17 | 0.00 | 13.56 | 21.71 | 83.63 | 0.00 |
| Whittier (city) | CA | 28.28 | 15.26 | 106.80 | 89.96 | 12.84 | 94.84 | 0.04 |
| Wichita (city) | KS | 23.69 | 96.66 | 1.20 | 65.79 | 55.13 | 8.27 | 3.82 |
| Wichita Falls (city) | TX | 8.55 | 51.68 | 91.93 | 46.44 | 33.17 | 76.04 | 0.00 |
| Wilmington (city) | DE | 0.00 | 0.00 | 0.00 | 0.00 | 0.00 | 0.00 | 0.00 |
| Wilmington (city) | NC | 15.08 | 304.31 | 68.45 | 64.98 | 165.86 | 71.16 | 12.30 |
| Winston-Salem (city) | NC | 34.04 | 223.58 | 71.32 | 72.13 | 145.70 | 120.87 | 0.00 |
| Woodbridge (township) | NJ | 53.37 | 162.42 | 5.57 | 71.56 | 116.01 | 96.09 | 0.00 |
| Woodbury (city) | MN | 61.91 | 88.24 | 0.00 | 89.78 | 37.01 | 0.64 | 0.00 |
| Woodland (city) | CA | 6.67 | 123.59 | 7.59 | 86.64 | 71.11 | 5.34 | 0.00 |
| Worcester (city) | MA | 6.08 | 123.77 | 17.87 | 20.72 | 103.65 | 25.61 | 0.00 |
| Worth (township) | IL | 0.41 | 0.00 | 0.00 | 0.00 | 0.00 | 0.00 | 0.00 |
| Wyoming (city) | MI | 3.42 | 161.88 | 0.00 | 60.81 | 125.79 | 0.33 | 0.00 |
| Yakima (city) | WA | 12.49 | 166.70 | 41.93 | 44.30 | 62.69 | 26.41 | 12.87 |
| Yonkers (city) | NY | 7.11 | 13.13 | 0.00 | 40.76 | 9.09 | 100.95 | 0.00 |
| Yorba Linda (city) | CA | 118.01 | 0.00 | 53.43 | 152.30 | 0.00 | 61.99 | 0.00 |
| York (township) | IL | 0.00 | 0.00 | 0.00 | 0.00 | 0.00 | 0.00 | 0.00 |
| Youngstown (city) | OH | 0.00 | 209.83 | 0.00 | 37.57 | 166.11 | 0.00 | 0.00 |
| Ypsilanti Charter (township) | MI | 21.42 | 0.00 | 0.00 | 46.11 | 4.26 | 43.34 | 0.00 |
| Yuma (city) | AZ | 37.38 | 173.20 | 21.87 | 130.13 | 121.54 | 49.14 | 0.00 |

***Source:*** *U.S. Census Bureau, State and Local Government Finances, 2006*

## Recent Trends in U.S. Greenhouse Gas Emissions and Sinks (Tg CO2 Eq.)

| Gas/Source | 1990 | 2000 | 2005 | 2006 | 2007 | 2008 | 2009 |
|---|---|---|---|---|---|---|---|
| **$CO_2$** | **5,099.7** | **5,975.0** | **6,113.8** | **6,021.1** | **6,120.0** | **5,921.4** | **5,505.2** |
| Fossil Fuel Combustion | 4,738.4 | 5,594.8 | 5,753.2 | 5,653.1 | 5,756.7 | 5,565.9 | 5,209.0 |
| Electricity Generation | *1,820.8* | *2,296.9* | *2,402.1* | *2,346.4* | *2,412.8* | *2,360.9* | *2,154.0* |
| Transportation | *1,485.9* | *1,809.5* | *1,896.6* | *1,878.1* | *1,894.0* | *1,789.9* | *1,719.7* |
| Industrial | *846.5* | *851.1* | *823.1* | *848.2* | *842.0* | *802.9* | *730.4* |
| Residential | *338.3* | *370.7* | *357.9* | *321.5* | *342.4* | *348.2* | *339.2* |
| Commercial | *219.0* | *230.8* | *223.5* | *208.6* | *219.4* | *224.2* | *224.0* |
| U.S. Territories | *27.9* | *35.9* | *50.0* | *50.3* | *46.1* | *39.8* | *41.7* |
| Non-Energy Use of Fuels | 118.6 | 144.9 | 143.4 | 145.6 | 137.2 | 141.0 | 123.4 |
| Iron and Steel Production & Metallurgical Coke Production | 99.5 | 85.9 | 65.9 | 68.8 | 71.0 | 66.0 | 41.9 |
| Natural Gas Systems | 37.6 | 29.9 | 29.9 | 30.8 | 31.1 | 32.8 | 32.2 |
| Cement Production | 33.3 | 40.4 | 45.2 | 45.8 | 44.5 | 40.5 | 29.0 |
| Incineration of Waste | 8.0 | 11.1 | 12.5 | 12.5 | 12.7 | 12.2 | 12.3 |
| Ammonia Production and Urea Consumption | 16.8 | 16.4 | 12.8 | 12.3 | 14.0 | 11.9 | 11.8 |
| Lime Production | 11.5 | 14.1 | 14.4 | 15.1 | 14.6 | 14.3 | 11.2 |
| Cropland Remaining Cropland | 7.1 | 7.5 | 7.9 | 7.9 | 8.2 | 8.7 | 7.8 |
| Limestone and Dolomite Use | 5.1 | 5.1 | 6.8 | 8.0 | 7.7 | 6.3 | 7.6 |
| Soda Ash Production and Consumption | 4.1 | 4.2 | 4.2 | 4.2 | 4.1 | 4.1 | 4.3 |
| Aluminum Production | 6.8 | 6.1 | 4.1 | 3.8 | 4.3 | 4.5 | 3.0 |
| Petrochemical Production | 3.3 | 4.5 | 4.2 | 3.8 | 3.9 | 3.4 | 2.7 |
| Carbon Dioxide Consumption | 1.4 | 1.4 | 1.3 | 1.7 | 1.9 | 1.8 | 1.8 |
| Titanium Dioxide Production | 1.2 | 1.8 | 1.8 | 1.8 | 1.9 | 1.8 | 1.5 |
| Ferroalloy Production | 2.2 | 1.9 | 1.4 | 1.5 | 1.6 | 1.6 | 1.5 |
| Wetlands Remaining Wetlands | 1.0 | 1.2 | 1.1 | 0.9 | 1.0 | 1.0 | 1.1 |
| Phosphoric Acid Production | 1.5 | 1.4 | 1.4 | 1.2 | 1.2 | 1.2 | 1.0 |
| Zinc Production | 0.7 | 1.0 | 1.1 | 1.1 | 1.1 | 1.2 | 1.0 |
| Lead Production | 0.5 | 0.6 | 0.6 | 0.6 | 0.6 | 0.6 | 0.5 |
| Petroleum Systems | 0.6 | 0.5 | 0.5 | 0.5 | 0.5 | 0.5 | 0.5 |
| Silicon Carbide Production and Consumption | 0.4 | 0.2 | 0.2 | 0.2 | 0.2 | 0.2 | 0.1 |
| *Land Use, Land-Use Change, and Forestry (Sink)*[a] | *(861.5)* | *(576.6)* | *(1,056.5)* | *(1,064.3)* | *(1,060.9)* | *(1,040.5)* | *(1,015.1)* |
| *Biomass—Wood*[b] | 215.2 | 218.1 | 206.9 | 203.8 | 203.3 | 198.4 | 183.8 |
| *International Bunker Fuels*[c] | *111.8* | *98.5* | *109.7* | *128.4* | *127.6* | *133.7* | *123.1* |

| | | | | | | | |
|---|---|---|---|---|---|---|---|
| *Biomass—Ethanol*[b] | 4.2 | 9.4 | 23.0 | 31.0 | 38.9 | 54.8 | 61.2 |
| **$CH_4$** | **674.9** | **659.9** | **631.4** | **672.1** | **664.6** | **676.7** | **686.3** |
| Natural Gas Systems | 189.8 | 209.3 | 190.4 | 217.7 | 205.2 | 211.8 | 221.2 |
| Enteric Fermentation | 132.1 | 136.5 | 136.5 | 138.8 | 141.0 | 140.6 | 139.8 |
| Landfills | 147.4 | 111.7 | 112.5 | 111.7 | 111.3 | 115.9 | 117.5 |
| Coal Mining | 84.1 | 60.4 | 56.9 | 58.2 | 57.9 | 67.1 | 71.0 |
| Manure Management | 31.7 | 42.4 | 46.6 | 46.7 | 50.7 | 49.4 | 49.5 |
| Petroleum Systems | 35.4 | 31.5 | 29.4 | 29.4 | 30.0 | 30.2 | 30.9 |
| Wastewater Treatment | 23.5 | 25.2 | 24.3 | 24.5 | 24.4 | 24.5 | 24.5 |
| Forest Land Remaining Forest Land | 3.2 | 14.3 | 9.8 | 21.6 | 20.0 | 11.9 | 7.8 |
| Rice Cultivation | 7.1 | 7.5 | 6.8 | 5.9 | 6.2 | 7.2 | 7.3 |
| Stationary Combustion | 7.4 | 6.6 | 6.6 | 6.2 | 6.5 | 6.5 | 6.2 |
| Abandoned Underground Coal Mines | 6.0 | 7.4 | 5.5 | 5.5 | 5.6 | 5.9 | 5.5 |
| Mobile Combustion | 4.7 | 3.4 | 2.5 | 2.3 | 2.2 | 2.0 | 2.0 |
| Composting | 0.3 | 1.3 | 1.6 | 1.6 | 1.7 | 1.7 | 1.7 |
| Petrochemical Production | 0.9 | 1.2 | 1.1 | 1.0 | 1.0 | 0.9 | 0.8 |
| Iron and Steel Production & Metallurgical Coke Production | 1.0 | 0.9 | 0.7 | 0.7 | 0.7 | 0.6 | 0.4 |
| Field Burning of Agriculture Residues | 0.3 | 0.3 | 0.2 | 0.2 | 0.2 | 0.3 | 0.2 |
| Ferroalloy Production | + | + | + | + | + | + | + |
| Silicon Carbide Production and Consumption | + | + | + | + | + | + | + |
| Incineration of Waste | + | + | + | + | + | + | + |
| *International Bunker Fuels*[c] | *0.2* | *0.1* | *0.1* | *0.2* | *0.2* | *0.2* | *0.1* |
| **$N_2O$** | **315.2** | **341.0** | **322.9** | **326.4** | **325.1** | **310.8** | **295.6** |
| Agricultural Soil Management | 197.8 | 206.8 | 211.3 | 208.9 | 209.4 | 210.7 | 204.6 |
| Mobile Combustion | 43.9 | 53.2 | 36.9 | 33.6 | 30.3 | 26.1 | 23.9 |
| Manure Management | 14.5 | 17.1 | 17.3 | 18.0 | 18.1 | 17.9 | 17.9 |
| Nitric Acid Production | 17.7 | 19.4 | 16.5 | 16.2 | 19.2 | 16.4 | 14.6 |
| Stationary Combustion | 12.8 | 14.6 | 14.7 | 14.4 | 14.6 | 14.2 | 12.8 |
| Forest Land Remaining Forest Land | 2.7 | 12.1 | 8.4 | 18.0 | 16.7 | 10.1 | 6.7 |
| Wastewater Treatment | 3.7 | 4.5 | 4.8 | 4.8 | 4.9 | 5.0 | 5.0 |
| $N_2O$ from Product Uses | 4.4 | 4.9 | 4.4 | 4.4 | 4.4 | 4.4 | 4.4 |
| Adipic Acid Production | 15.8 | 5.5 | 5.0 | 4.3 | 3.7 | 2.0 | 1.9 |
| Composting | 0.4 | 1.4 | 1.7 | 1.8 | 1.8 | 1.9 | 1.8 |
| Settlements Remaining Settlements | 1.0 | 1.1 | 1.5 | 1.5 | 1.6 | 1.5 | 1.5 |
| Incineration of Waste | 0.5 | 0.4 | 0.4 | 0.4 | 0.4 | 0.4 | 0.4 |
| Field Burning of Agricultural Residues | 0.1 | 0.1 | 0.1 | 0.1 | 0.1 | 0.1 | 0.1 |
| Wetlands Remaining Wetlands | + | + | + | + | + | + | + |
| *International Bunker Fuels*[c] | *1.1* | *0.9* | *1.0* | *1.2* | *1.2* | *1.2* | *1.1* |
| **HFCs** | **36.9** | **103.2** | **120.2** | **123.5** | **129.5** | **129.4** | **125.7** |
| Substitution of Ozone Depleting Substances[d] | 0.3 | 74.3 | 104.2 | 109.4 | 112.3 | 115.5 | 120.0 |
| HCFC-22 Production | 36.4 | 28.6 | 15.8 | 13.8 | 17.0 | 13.6 | 5.4 |
| Semiconductor Manufacture | 0.2 | 0.3 | 0.2 | 0.3 | 0.3 | 0.3 | 0.3 |
| **PFCs** | **20.8** | **13.5** | **6.2** | **6.0** | **7.5** | **6.6** | **5.6** |
| Semiconductor Manufacture | 2.2 | 4.9 | 3.2 | 3.5 | 3.7 | 4.0 | 4.0 |
| Aluminum Production | 18.5 | 8.6 | 3.0 | 2.5 | 3.8 | 2.7 | 1.6 |
| **$SF_6$** | **34.4** | **20.1** | **19.0** | **17.9** | **16.7** | **16.1** | **14.8** |

| | | | | | | | |
|---|---|---|---|---|---|---|---|
| Electrical Transmission and Distribution | 28.4 | 16.0 | 15.1 | 14.1 | 13.2 | 13.3 | 12.8 |
| Magnesium Production and Processing | 5.4 | 3.0 | 2.9 | 2.9 | 2.6 | 1.9 | 1.1 |
| Semiconductor Manufacture | 0.5 | | 1.0 | 1.0 | 0.8 | 0.9 | 1.0 |
| **Total** | **6,181.8** | **7,112.7** | **7,213.5** | **7,166.9** | **7,263.4** | **7,061.1** | **6,633.2** |
| **Net Emissions (Sources and Sinks)** | **5,320.3** | **6,536.1** | **6,157.1** | **6,102.6** | **6,202.5** | **6,020.7** | **5,618.2** |

+ Does not exceed 0.05 Tg $CO_2$ Eq.

[a] The net $CO_2$ flux total includes both emissions and sequestration, and constitutes a sink in the United States. Sinks are only included in net emissions total. Parentheses indicate negative values or sequestration.

[b] Emissions from Wood Biomass and Ethanol Consumption are not included specifically in summing energy sector totals. Net carbon fluxes from changes in biogenic carbon reservoirs are accounted for in the estimates for Land Use, Land-Use Change, and Forestry.

[c] Emissions from International Bunker Fuels are not included in totals.

[d] Small amounts of PFC emissions also result from this source.

Note: Totals may not sum due to independent rounding.

*Source: Inventory of U.S. Greenhouse Gas Emissions and Sinks: 1990-2009*

### Recent Trends in U.S. Greenhouse Gas Emissions and Sinks (Gg)

| Gas/Source | 1990 | 2000 | 2005 | 2006 | 2007 | 2008 | 2009 |
|---|---|---|---|---|---|---|---|
| **$CO_2$** | **5,099,719** | **5,974,991** | **6,113,751** | **6,021,089** | **6,120,009** | **5,921,443** | **5,505,204** |
| Fossil Fuel Combustion | 4,738,422 | 5,594,848 | 5,753,200 | 5,653,116 | 5,756,746 | 5,565,925 | 5,208,981 |
| Electricity Generation | *1,820,818* | *2,296,894* | *2,402,142* | *2,346,406* | *2,412,827* | *2,360,919* | *2,154,025* |
| Transportation | *1,485,937* | *1,809,514* | *1,896,606* | *1,878,125* | *1,893,994* | *1,789,918* | *1,719,685* |
| Industrial | *846,475* | *851,094* | *823,069* | *848,206* | *842,048* | *802,856* | *730,422* |
| Residential | *338,347* | *370,666* | *357,903* | *321,513* | *342,397* | *348,221* | *339,203* |
| Commercial | *218,964* | *230,828* | *223,512* | *208,582* | *219,356* | *224,167* | *223,993* |
| U.S. Territories | *27,882* | *35,853* | *49,968* | *50,284* | *46,123* | *39,845* | *41,652* |
| Non-Energy Use of Fuels | 118,630 | 144,933 | 143,392 | 145,574 | 137,233 | 140,952 | 123,356 |
| Iron and Steel Production & Metallurgical Coke Production | 99,528 | 85,935 | 65,925 | 68,772 | 71,045 | 66,015 | 41,871 |
| Natural Gas Systems | 37,574 | 29,877 | 29,902 | 30,755 | 31,050 | 32,828 | 32,171 |
| Cement Production | 33,278 | 40,405 | 45,197 | 45,792 | 44,538 | 40,531 | 29,018 |
| Incineration of Waste | 7,989 | 11,112 | 12,450 | 12,531 | 12,700 | 12,169 | 12,300 |
| Ammonia Production and Urea Consumption | 16,831 | 16,402 | 12,849 | 12,300 | 14,038 | 11,949 | 11,797 |
| Lime Production | 11,533 | 14,088 | 14,379 | 15,100 | 14,595 | 14,330 | 11,223 |
| Cropland Remaining Cropland | 7,084 | 7,541 | 7,854 | 7,875 | 8,202 | 8,654 | 7,832 |
| Limestone and Dolomite Use | 5,127 | 5,056 | 6,768 | 8,035 | 7,702 | 6,276 | 7,649 |
| Soda Ash Production and Consumption | 4,141 | 4,181 | 4,228 | 4,162 | 4,140 | 4,111 | 4,265 |
| Aluminum Production | 6,831 | 6,086 | 4,142 | 3,801 | 4,251 | 4,477 | 3,009 |
| Petrochemical Production | 3,311 | 4,479 | 4,181 | 3,837 | 3,931 | 3,449 | 2,735 |
| Carbon Dioxide Consumption | 1,416 | 1,421 | 1,321 | 1,709 | 1,867 | 1,780 | 1,763 |
| Titanium Dioxide Production | 1,195 | 1,752 | 1,755 | 1,836 | 1,930 | 1,809 | 1,541 |
| Ferroalloy Production | 2,152 | 1,893 | 1,392 | 1,505 | 1,552 | 1,599 | 1,469 |
| Wetlands Remaining Wetlands | 1,033 | 1,227 | 1,079 | 879 | 1,012 | 992 | 1,090 |
| Phosphoric Acid Production | 1,529 | 1,382 | 1,386 | 1,167 | 1,166 | 1,187 | 1,035 |
| Zinc Production | 667 | 997 | 1,088 | 1,088 | 1,081 | 1,230 | 966 |

| | | | | | | | |
|---|---|---|---|---|---|---|---|
| Lead Production | 516 | 594 | 553 | 560 | 562 | 551 | 525 |
| Petroleum Systems | 555 | 534 | 490 | 488 | 474 | 453 | 463 |
| Silicon Carbide Production and Consumption | 375 | 248 | 219 | 207 | 196 | 175 | 145 |
| *Land Use, Land-Use Change, and Forestry (Sink)*[a] | *(861,535)* | *(576,588)* | *(1,056,459)* | *(1,064,330)* | *(1,060,882)* | *(1,040,461)* | *(1,015,074)* |
| *Biomass - Wood*[b] | *215,186* | *218,088* | *206,865* | *203,846* | *203,316* | *198,361* | *183,777* |
| *International Bunker Fuels*[c] | *111,828* | *98,482* | *109,750* | *128,384* | *127,618* | *133,704* | *123,127* |
| *Biomass - Ethanol*[b] | *4,229* | *9,352* | *22,956* | *31,002* | *38,946* | *54,770* | *61,231* |
| **$CH_4$** | **32,136** | **31,423** | **30,069** | **32,004** | **31,647** | **32,225** | **32,680** |
| Natural Gas Systems | 9,038 | 9,968 | 9,069 | 10,364 | 9,771 | 10,087 | 10,535 |
| Enteric Fermentation | 6,290 | 6,502 | 6,500 | 6,611 | 6,715 | 6,696 | 6,655 |
| Landfills | 7,018 | 5,317 | 5,358 | 5,321 | 5,299 | 5,520 | 5,593 |
| Coal Mining | 4,003 | 2,877 | 2,710 | 2,774 | 2,756 | 3,196 | 3,382 |
| Manure Management | 1,511 | 2,019 | 2,217 | 2,226 | 2,416 | 2,353 | 2,356 |
| Petroleum Systems | 1,685 | 1,501 | 1,398 | 1,398 | 1,427 | 1,439 | 1,473 |
| Wastewater Treatment | 1,118 | 1,199 | 1,159 | 1,167 | 1,163 | 1,168 | 1,167 |
| Forest Land Remaining Forest Land | 152 | 682 | 467 | 1,027 | 953 | 569 | 372 |
| Rice Cultivation | 339 | 357 | 326 | 282 | 295 | 343 | 349 |
| Stationary Combustion | 354 | 315 | 312 | 293 | 308 | 310 | 293 |
| Abandoned Underground Coal Mines | 288 | 350 | 264 | 261 | 267 | 279 | 262 |
| Mobile Combustion | 223 | 160 | 119 | 112 | 105 | 97 | 93 |
| Composting | 15 | 60 | 75 | 75 | 79 | 80 | 79 |
| Petrochemical Production | 41 | 59 | 51 | 48 | 48 | 43 | 40 |
| Iron and Steel Production & Metallurgical Coke Production | 46 | 44 | 34 | 35 | 33 | 31 | 17 |
| Field Burning of Agricultural Residues | 13 | 12 | 9 | 11 | 11 | 13 | 12 |
| Ferroalloy Production | 1 | 1 | + | + | + | + | + |
| Silicon Carbide Production and Consumption | 1 | 1 | + | + | + | + | + |
| Incineration of Waste | + | + | + | + | + | + | + |
| *International Bunker Fuels*[c] | *8* | *6* | *7* | *8* | *8* | *8* | *7* |
| **$N_2O$** | **1,017** | **1,100** | **1,042** | **1,053** | **1,049** | **1,002** | **954** |
| Agricultural Soil Management | 638 | 667 | 682 | 674 | 675 | 680 | 660 |
| Mobile Combustion | 142 | 172 | 119 | 108 | 98 | 84 | 77 |
| Manure Management | 47 | 55 | 56 | 58 | 58 | 58 | 58 |
| Nitric Acid Production | 57 | 63 | 53 | 52 | 62 | 53 | 47 |
| Stationary Combustion | 41 | 47 | 47 | 47 | 47 | 46 | 41 |
| Forest Land Remaining Forest Land | 9 | 39 | 27 | 58 | 54 | 33 | 22 |
| Wastewater Treatment | 12 | 14 | 15 | 16 | 16 | 16 | 16 |
| $N_2O$ from Product Uses | 14 | 16 | 14 | 14 | 14 | 14 | 14 |
| Adipic Acid Production | 51 | 18 | 16 | 14 | 12 | 7 | 6 |
| Composting | 1 | 4 | 6 | 6 | 6 | 6 | 6 |
| Settlements Remaining Settlements | 3 | 4 | 5 | 5 | 5 | 5 | 5 |

| | | | | | | | |
|---|---|---|---|---|---|---|---|
| Incineration of Waste | 2 | 1 | 1 | 1 | 1 | 1 | 1 |
| Field Burning of Agricultural Residues | + | + | + | + | + | + | + |
| Wetlands Remaining Wetlands | + | + | + | + | + | + | + |
| *International Bunker Fuels*[c] | *3* | *3* | *3* | *4* | *4* | *4* | *4* |
| **HFCs** | **M** | **M** | **M** | **M** | **M** | **M** | **M** |
| Substitution of Ozone Depleting Substances[d] | M | M | M | M | M | M | M |
| HCFC-22 Production | 3 | 2 | 1 | 1 | 1 | 1 | + |
| Semiconductor Manufacture | + | + | + | + | + | + | + |
| **PFCs** | **M** | **M** | **M** | **M** | **M** | **M** | **M** |
| Semiconductor Manufacture | M | M | M | M | M | M | M |
| Aluminum Production | M | M | M | M | M | M | M |
| **$SF_6$** | **1** | **1** | **1** | **1** | **1** | **1** | **1** |
| Electrical Transmission and Distribution | 1 | 1 | 1 | 1 | 1 | 1 | 1 |
| Magnesium Production and Processing | + | + | + | + | + | + | + |
| Semiconductor Manufacture | + | + | + | + | + | + | + |

+ Does not exceed 0.5 Gg.
M Mixture of multiple gases
[a] The net $CO_2$ flux total includes both emissions and sequestration, and constitutes a sink in the United States. Sinks are only included in net emissions total. Parentheses indicate negative values or sequestration.
[b] Emissions from Wood Biomass and Ethanol Consumption are not included specifically in summing energy sector totals. Net carbon fluxes from changes in biogenic carbon reservoirs are accounted for in the estimates for Land Use, Land-Use Change, and Forestry
[c] Emissions from International Bunker Fuels are not included in totals.
[d] Small amounts of PFC emissions also result from this source.
Note: Totals may not sum due to independent rounding.
*Source: Inventory of U.S. Greenhouse Gas Emissions and Sinks: 1990-2009*

**Emissions from Energy (Tg $CO_2$ Eq.)**

| Gas/Source | 1990 | 2000 | 2005 | 2006 | 2007 | 2008 | 2009 |
|---|---|---|---|---|---|---|---|
| **$CO_2$** | **4,903.2** | **5,781.3** | **5,939.4** | **5,842.5** | **5,938.2** | **5,752.3** | **5,377.3** |
| Fossil Fuel Combustion | 4,738.4 | 5,594.8 | 5,753.2 | 5,653.1 | 5,756.7 | 5,565.9 | 5,209.0 |
| Electricity Generation | 1,820.8 | 2,296.9 | 2,402.1 | 2,346.4 | 2,412.8 | 2,360.9 | 2,154.0 |
| Transportation | 1,485.9 | 1,809.5 | 1,896.6 | 1,878.1 | 1,894.0 | 1,789.9 | 1,719.7 |
| Industrial | 846.5 | 851.1 | 823.1 | 848.2 | 842.0 | 802.9 | 730.4 |
| Residential | 338.3 | 370.7 | 357.9 | 321.5 | 342.4 | 348.2 | 339.2 |
| Commercial | 219.0 | 230.8 | 223.5 | 208.6 | 219.4 | 224.2 | 224.0 |
| U.S. Territories | 27.9 | 35.9 | 50.0 | 50.3 | 46.1 | 39.8 | 41.7 |
| Non-Energy Use of Fuels | 118.6 | 144.9 | 143.4 | 145.6 | 137.2 | 141.0 | 123.4 |
| Natural Gas Systems | 37.6 | 29.9 | 29.9 | 30.8 | 31.1 | 32.8 | 32.2 |
| Incineration of Waste | 8.0 | 11.1 | 12.5 | 12.5 | 12.7 | 12.2 | 12.3 |
| Petroleum Systems | 0.6 | 0.5 | 0.5 | 0.5 | 0.5 | 0.5 | 0.5 |
| *Biomass - Wood*[a] | *215.2* | *218.1* | *206.9* | *203.8* | *203.3* | *198.4* | *183.8* |
| *International Bunker Fuels*[b] | *111.8* | *98.5* | *109.7* | *128.4* | *127.6* | *133.7* | *123.1* |
| *Biomass - Ethanol*[a] | *4.2* | *9.4* | *23.0* | *31.0* | *38.9* | *54.8* | *61.2* |
| **$CH_4$** | **327.4** | **318.6** | **291.3** | **319.2** | **307.3** | **323.6** | **336.8** |
| Natural Gas Systems | 189.8 | 209.3 | 190.4 | 217.7 | 205.2 | 211.8 | 221.2 |
| Coal Mining | 84.1 | 60.4 | 56.9 | 58.2 | 57.9 | 67.1 | 71.0 |
| Petroleum Systems | 35.4 | 31.5 | 29.4 | 29.4 | 30.0 | 30.2 | 30.9 |
| Stationary Combustion | 7.4 | 6.6 | 6.6 | 6.2 | 6.5 | 6.5 | 6.2 |
| Abandoned Underground Coal Mines | 6.0 | 7.4 | 5.5 | 5.5 | 5.6 | 5.9 | 5.5 |
| Mobile Combustion | 4.7 | 3.4 | 2.5 | 2.3 | 2.2 | 2.0 | 2.0 |
| Incineration of Waste | + | + | + | + | + | + | + |
| *International Bunker Fuels*[b] | *0.2* | *0.1* | *0.1* | *0.2* | *0.2* | *0.2* | *0.1* |
| **$N_2O$** | **57.2** | **68.1** | **52.1** | **48.5** | **45.2** | **40.7** | **37.0** |
| Mobile Combustion | 43.9 | 53.2 | 36.9 | 33.6 | 30.3 | 26.1 | 23.9 |
| Stationary Combustion | 12.8 | 14.6 | 14.7 | 14.4 | 14.6 | 14.2 | 12.8 |
| Incineration of Waste | 0.5 | 0.4 | 0.4 | 0.4 | 0.4 | 0.4 | 0.4 |
| *International Bunker Fuels*[b] | *1.1* | *0.9* | *1.0* | *1.2* | *1.2* | *1.2* | *1.1* |
| **Total** | **5,287.8** | **6,168.0** | **6,282.8** | **6,210.2** | **6,290.7** | **6,116.6** | **5,751.1** |

+ Does not exceed 0.05 Tg $CO_2$ Eq.

[a] Emissions from Wood Biomass and Ethanol Consumption are not included specifically in summing energy sector totals. Net carbon fluxes from changes in biogenic carbon reservoirs are accounted for in the estimates for Land Use, Land-Use Change, and Forestry

[b] Emissions from International Bunker Fuels are not included in totals.

Note: Totals may not sum due to independent rounding.

*Source: Inventory of U.S. Greenhouse Gas Emissions and Sinks: 1990-2009*

## $CO_2$ Emissions from Fossil Fuel Combustion by End-Use Sector (Tg $CO_2$ Eq.)

| End-Use Sector | 1990 | 2000 | 2005 | 2006 | 2007 | 2008 | 2009 |
|---|---|---|---|---|---|---|---|
| **Transportation** | **1,489.0** | **1,813.0** | **1,901.3** | **1,882.6** | **1,899.0** | **1,794.6** | **1,724.1** |
| Combustion | 1,485.9 | 1,809.5 | 1,896.6 | 1,878.1 | 1,894.0 | 1,789.9 | 1,719.7 |
| Electricity | 3.0 | 3.4 | 4.7 | 4.5 | 5.0 | 4.7 | 4.4 |
| **Industrial** | **1,533.2** | **1,640.8** | **1,560.0** | **1,560.2** | **1,572.0** | **1,517.7** | **1,333.7** |
| Combustion | 846.5 | 851.1 | 823.1 | 848.2 | 842.0 | 802.9 | 730.4 |
| Electricity | 686.7 | 789.8 | 737.0 | 712.0 | 730.0 | 714.8 | 603.3 |
| **Residential** | **931.4** | **1,133.1** | **1,214.7** | **1,152.4** | **1,198.5** | **1,182.2** | **1,123.8** |
| Combustion | 338.3 | 370.7 | 357.9 | 321.5 | 342.4 | 348.2 | 339.2 |
| Electricity | 593.0 | 762.4 | 856.7 | 830.8 | 856.1 | 834.0 | 784.6 |
| **Commercial** | **757.0** | **972.1** | **1,027.2** | **1,007.6** | **1,041.1** | **1,031.6** | **985.7** |
| Combustion | 219.0 | 230.8 | 223.5 | 208.6 | 219.4 | 224.2 | 224.0 |
| Electricity | 538.0 | 741.3 | 803.7 | 799.0 | 821.7 | 807.4 | 761.7 |
| **U.S. Territories** | **27.9** | **35.9** | **50.0** | **50.3** | **46.1** | **39.8** | **41.7** |
| **Total** | **4,738.4** | **5,594.8** | **5,753.2** | **5,653.1** | **5,756.7** | **5,565.9** | **5,209.0** |
| **Electricity Generation** | **1,820.8** | **2,296.9** | **2,402.1** | **2,346.4** | **2,412.8** | **2,360.9** | **2,154.0** |

Note: Totals may not sum due to independent rounding. Combustion-related emissions from electricity generation are allocated based on aggregate national electricity consumption by each end-use sector.

*Source: Inventory of U.S. Greenhouse Gas Emissions and Sinks: 1990-2009*

**Emissions from Industrial Processes (Tg $CO_2$ Eq.)**

| Gas/Source | 1990 | 2000 | 2005 | 2006 | 2007 | 2008 | 2009 |
|---|---|---|---|---|---|---|---|
| **$CO_2$** | **188.4** | **184.9** | **165.4** | **169.9** | **172.6** | **159.5** | **119.0** |
| Iron and Steel Production & Metallurgical Coke Production | 99.5 | 85.9 | 65.9 | 68.8 | 71.0 | 66.0 | 41.9 |
| *Iron and Steel Production* | *97.1* | *83.7* | *63.9* | *66.9* | *69.0* | *63.7* | *40.9* |
| *Metallurgical Coke Production* | *2.5* | *2.2* | *2.0* | *1.9* | *2.1* | *2.3* | *1.0* |
| Cement Production | 33.3 | 40.4 | 45.2 | 45.8 | 44.5 | 40.5 | 29.0 |
| Ammonia Production & Urea Consumption | 16.8 | 16.4 | 12.8 | 12.3 | 14.0 | 11.9 | 11.8 |
| Lime Production | 11.5 | 14.1 | 14.4 | 15.1 | 14.6 | 14.3 | 11.2 |
| Limestone and Dolomite Use | 5.1 | 5.1 | 6.8 | 8.0 | 7.7 | 6.3 | 7.6 |
| Soda Ash Production and Consumption | 4.1 | 4.2 | 4.2 | 4.2 | 4.1 | 4.1 | 4.3 |
| Aluminum Production | 6.8 | 6.1 | 4.1 | 3.8 | 4.3 | 4.5 | 3.0 |
| Petrochemical Production | 3.3 | 4.5 | 4.2 | 3.8 | 3.9 | 3.4 | 2.7 |
| Carbon Dioxide Consumption | 1.4 | 1.4 | 1.3 | 1.7 | 1.9 | 1.8 | 1.8 |
| Titanium Dioxide Production | 1.2 | 1.8 | 1.8 | 1.8 | 1.9 | 1.8 | 1.5 |
| Ferroalloy Production | 2.2 | 1.9 | 1.4 | 1.5 | 1.6 | 1.6 | 1.5 |
| Phosphoric Acid Production | 1.5 | 1.4 | 1.4 | 1.2 | 1.2 | 1.2 | 1.0 |
| Zinc Production | 0.7 | 1.0 | 1.1 | 1.1 | 1.1 | 1.2 | 1.0 |
| Lead Production | 0.5 | 0.6 | 0.6 | 0.6 | 0.6 | 0.6 | 0.5 |
| Silicon Carbide Production and Consumption | 0.4 | 0.2 | 0.2 | 0.2 | 0.2 | 0.2 | 0.1 |
| **$CH_4$** | **1.9** | **2.2** | **1.8** | **1.7** | **1.7** | **1.6** | **1.2** |
| Petrochemical Production | 0.9 | 1.2 | 1.1 | 1.0 | 1.0 | 0.9 | 0.8 |
| Iron and Steel Production & Metallurgical Coke Production | 1.0 | 0.9 | 0.7 | 0.7 | 0.7 | 0.6 | 0.4 |
| *Iron and Steel Production* | *1.0* | *0.9* | *0.7* | *0.7* | *0.7* | *0.6* | *0.4* |
| *Metallurgical Coke Production* | + | + | + | + | + | + | + |
| Ferroalloy Production | + | + | + | + | + | + | + |
| Silicon Carbide Production and Consumption | + | + | + | + | + | + | + |
| **$N_2O$** | **33.5** | **24.9** | **21.5** | **20.5** | **22.9** | **18.5** | **16.5** |
| Nitric Acid Production | 17.7 | 19.4 | 16.5 | 16.2 | 19.2 | 16.4 | 14.6 |
| Adipic Acid Production | 15.8 | 5.5 | 5.0 | 4.3 | 3.7 | 2.0 | 1.9 |
| **HFCs** | **36.9** | **103.2** | **120.2** | **123.4** | **129.5** | **129.4** | **125.7** |
| Substitution of Ozone Depleting Substances[a] | 0.3 | 74.3 | 104.2 | 109.4 | 112.3 | 115.5 | 120.0 |
| HCFC-22 Production | 36.4 | 28.6 | 15.8 | 13.8 | 17.0 | 13.6 | 5.4 |
| Semiconductor Manufacture | 0.2 | 0.3 | 0.2 | 0.3 | 0.3 | 0.3 | 0.3 |
| **PFCs** | **20.8** | **13.5** | **6.2** | **6.0** | **7.5** | **6.6** | **5.6** |
| Semiconductor Manufacture | 2.2 | 4.9 | 3.2 | 3.5 | 3.7 | 4.0 | 4.0 |
| Aluminum Production | 18.5 | 8.6 | 3.0 | 2.5 | 3.8 | 2.7 | 1.6 |
| **$SF_6$** | **34.4** | **20.1** | **19.0** | **17.9** | **16.7** | **16.1** | **14.8** |
| Electrical Transmission and Distribution | 28.4 | 16.0 | 15.1 | 14.1 | 13.2 | 13.3 | 12.8 |
| Magnesium Production and Processing | 5.4 | 3.0 | 2.9 | 2.9 | 2.6 | 1.9 | 1.1 |
| Semiconductor Manufacture | 0.5 | 1.1 | 1.0 | 1.0 | 0.8 | 0.9 | 1.0 |
| **Total** | **315.8** | **348.8** | **334.1** | **339.4** | **350.9** | **331.7** | **282.9** |

+ Does not exceed 0.05 Tg $CO_2$ Eq.

[a] Small amounts of PFC emissions also result from this source.

Note: Totals may not sum due to independent rounding.

*Source: Inventory of U.S. Greenhouse Gas Emissions and Sinks: 1990-2009*

**Emissions from Agriculture (Tg $CO_2$ Eq.)**

| Gas/Source | 1990 | 2000 | 2005 | 2006 | 2007 | 2008 | 2009 |
|---|---|---|---|---|---|---|---|
| **$CH_4$** | **171.2** | **186.7** | **190.1** | **191.7** | **198.2** | **197.5** | **196.8** |
| Enteric Fermentation | 132.1 | 136.5 | 136.5 | 138.8 | 141.0 | 140.6 | 139.8 |
| Manure Management | 31.7 | 42.4 | 46.6 | 46.7 | 50.7 | 49.4 | 49.5 |
| Rice Cultivation | 7.1 | 7.5 | 6.8 | 5.9 | 6.2 | 7.2 | 7.3 |
| Field Burning of Agricultural Residues | 0.3 | 0.3 | 0.2 | 0.2 | 0.2 | 0.3 | 0.2 |
| **$N_2O$** | **212.4** | **224.0** | **228.7** | **227.1** | **227.6** | **228.8** | **222.5** |
| Agricultural Soil Management | 197.8 | 206.8 | 211.3 | 208.9 | 209.4 | 210.7 | 204.6 |
| Manure Management | 14.5 | 17.1 | 17.3 | 18.0 | 18.1 | 17.9 | 17.9 |
| Field Burning of Agricultural Residues | 0.1 | 0.1 | 0.1 | 0.1 | 0.1 | 0.1 | 0.1 |
| **Total** | **383.6** | **410.6** | **418.8** | **418.8** | **425.8** | **426.3** | **419.3** |

Note: Totals may not sum due to independent rounding.
*Source: Inventory of U.S. Greenhouse Gas Emissions and Sinks: 1990-2009*

**Emissions from Land Use, Land-Use Change, and Forestry (Tg $CO_2$ Eq.)**

| Source Category | 1990 | 2000 | 2005 | 2006 | 2007 | 2008 | 2009 |
|---|---|---|---|---|---|---|---|
| **$CO_2$** | **8.1** | **8.8** | **8.9** | **8.8** | **9.2** | **9.6** | **8.9** |
| Cropland Remaining Cropland: Liming of Agricultural Soils | 4.7 | 4.3 | 4.3 | 4.2 | 4.5 | 5.0 | 4.2 |
| Cropland Remaining Cropland: Urea Fertilization | 2.4 | 3.2 | 3.5 | 3.7 | 3.7 | 3.6 | 3.6 |
| Wetlands Remaining Wetlands: Peatlands Remaining Peatlands | 1.0 | 1.2 | 1.1 | 0.9 | 1.0 | 1.0 | 1.1 |
| **$CH_4$** | **3.2** | **14.3** | **9.8** | **21.6** | **20.0** | **11.9** | **7.8** |
| Forest Land Remaining Forest Land: Forest Fires | 3.2 | 14.3 | 9.8 | 21.6 | 20.0 | 11.9 | 7.8 |
| **$N_2O$** | **3.7** | **13.2** | **9.8** | **19.5** | **18.3** | **11.6** | **8.3** |
| Forest Land Remaining Forest Land: Forest Fires | 2.6 | 11.7 | 8.0 | 17.6 | 16.3 | 9.8 | 6.4 |
| Forest Land Remaining Forest Land: Forest Soils | 0.1 | 0.4 | 0.4 | 0.4 | 0.4 | 0.4 | 0.4 |
| Settlements Remaining Settlements: Settlement Soils | 1.0 | 1.1 | 1.5 | 1.5 | 1.6 | 1.5 | 1.5 |
| Wetlands Remaining Wetlands: Peatlands Remaining Peatlands | + | + | + | + | + | + | + |
| **Total** | **15.0** | **36.3** | **28.6** | **49.8** | **47.5** | **33.2** | **25.0** |

+ Less than 0.05 Tg $CO_2$ Eq.

Note: Totals may not sum due to independent rounding.

*Source: Inventory of U.S. Greenhouse Gas Emissions and Sinks: 1990-2009*

**Emissions from Waste (Tg $CO_2$ Eq.)**

| Gas/Source | 1990 | 2000 | 2005 | 2006 | 2007 | 2008 | 2009 |
|---|---|---|---|---|---|---|---|
| **$CH_4$** | **171.2** | **138.1** | **138.4** | **137.8** | **137.4** | **142.1** | **143.6** |
| Landfills | 147.4 | 111.7 | 112.5 | 111.7 | 111.3 | 115.9 | 117.5 |
| Wastewater Treatment | 23.5 | 25.2 | 24.3 | 24.5 | 24.4 | 24.5 | 24.5 |
| Composting | 0.3 | 1.3 | 1.6 | 1.6 | 1.7 | 1.7 | 1.7 |
| **$N_2O$** | **4.0** | **5.9** | **6.5** | **6.6** | **6.7** | **6.8** | **6.9** |
| Wastewater Treatment | 3.7 | 4.5 | 4.8 | 4.8 | 4.9 | 5.0 | 5.0 |
| Composting | 0.4 | 1.4 | 1.7 | 1.8 | 1.8 | 1.9 | 1.8 |
| **Total** | **175.2** | **143.9** | **144.9** | **144.4** | **144.1** | **149.0** | **150.5** |

Note: Totals may not sum due to independent rounding.

*Source: Inventory of U.S. Greenhouse Gas Emissions and Sinks: 1990-2009*

**U.S. Greenhouse Gas Emissions Allocated to Economic Sectors (Tg $CO_2$ Eq. & Pct. of Total in 2009)**

| Sector/Source | 1990 | 2000 | 2005 | 2006 | 2007 | 2008 | 2009 | Percent[a] |
|---|---|---|---|---|---|---|---|---|
| **Electric Power Industry** | **1,868.9** | **2,337.6** | **2,444.6** | **2,388.2** | **2,454.0** | **2,400.7** | **2,193.0** | **33.1%** |
| $CO_2$ from Fossil Fuel Combustion | 1,820.8 | 2,296.9 | 2,402.1 | 2,346.4 | 2,412.8 | 2,360.9 | 2,154.0 | 32.5% |
| Electrical Transmission and Distribution | 28.4 | 16.0 | 15.1 | 14.1 | 13.2 | 13.3 | 12.8 | 0.2% |
| Incineration of Waste | 8.5 | 11.5 | 12.9 | 12.9 | 13.1 | 12.5 | 12.7 | 0.2% |
| Stationary Combustion | 8.6 | 10.6 | 11.0 | 10.8 | 11.0 | 10.8 | 9.7 | 0.1% |
| Limestone and Dolomite Use | 2.6 | 2.5 | 3.4 | 4.0 | 3.9 | 3.1 | 3.8 | 0.1% |
| **Transportation** | **1,545.2** | **1,932.3** | **2,017.4** | **1,994.4** | **2,003.8** | **1,890.7** | **1,812.4** | **27.3%** |
| $CO_2$ from Fossil Fuel Combustion | 1,485.9 | 1,809.5 | 1,896.6 | 1,878.1 | 1,894.0 | 1,789.9 | 1,719.7 | 25.9% |
| Substitution of Ozone Depleting Substances | + | 55.7 | 72.9 | 72.2 | 68.8 | 64.9 | 60.2 | 0.9% |
| Mobile Combustion | 47.4 | 55.1 | 37.7 | 34.2 | 30.7 | 26.4 | 24.0 | 0.4% |
| Non-Energy Use of Fuels | 11.8 | 12.1 | 10.2 | 9.9 | 10.2 | 9.5 | 8.5 | 0.1% |
| **Industry** | **1,564.4** | **1,544.0** | **1,441.9** | **1,497.3** | **1,483.0** | **1,446.9** | **1,322.7** | **19.9%** |
| $CO_2$ from Fossil Fuel Combustion | 815.4 | 812.3 | 776.3 | 799.2 | 793.6 | 757.4 | 683.8 | 10.3% |
| Natural Gas Systems | 227.4 | 239.2 | 220.4 | 248.4 | 236.2 | 244.6 | 253.4 | 3.8% |
| Non-Energy Use of Fuels | 101.1 | 122.8 | 125.2 | 126.8 | 119.8 | 123.1 | 111.1 | 1.7% |
| Coal Mining | 84.1 | 60.4 | 56.9 | 58.2 | 57.9 | 67.1 | 71.0 | 1.1% |
| Iron and Steel Production & Metallurgical Coke Production | 100.5 | 86.9 | 66.6 | 69.5 | 71.7 | 66.7 | 42.2 | 0.6% |
| Petroleum Systems | 35.9 | 32.0 | 29.9 | 29.8 | 30.4 | 30.7 | 31.4 | 0.5% |
| Cement Production | 33.3 | 40.4 | 45.2 | 45.8 | 44.5 | 40.5 | 29.0 | 0.4% |
| Nitric Acid Production | 17.7 | 19.4 | 16.5 | 16.2 | 19.2 | 16.4 | 14.6 | 0.2% |
| Ammonia Production and Urea Consumption | 16.8 | 16.4 | 12.8 | 12.3 | 14.0 | 11.9 | 11.8 | 0.2% |
| Lime Production | 11.5 | 14.1 | 14.4 | 15.1 | 14.6 | 14.3 | 11.2 | 0.2% |
| Substitution of Ozone Depleting Substances | + | 3.2 | 6.4 | 7.1 | 7.8 | 8.5 | 10.9 | 0.2% |
| Abandoned Underground Coal Mines | 6.0 | 7.4 | 5.5 | 5.5 | 5.6 | 5.9 | 5.5 | 0.1% |
| HCFC-22 Production | 36.4 | 28.6 | 15.8 | 13.8 | 17.0 | 13.6 | 5.4 | 0.1% |
| Semiconductor Manufacture | 2.9 | 6.2 | 4.4 | 4.7 | 4.8 | 5.1 | 5.3 | 0.1% |
| Aluminum Production | 25.4 | 14.7 | 7.1 | 6.3 | 8.1 | 7.2 | 4.6 | 0.1% |
| $N_2O$ from Product Uses | 4.4 | 4.9 | 4.4 | 4.4 | 4.4 | 4.4 | 4.4 | 0.1% |
| Soda Ash Production and Consumption | 4.1 | 4.2 | 4.2 | 4.2 | 4.1 | 4.1 | 4.3 | 0.1% |
| Limestone and Dolomite Use | 2.6 | 2.5 | 3.4 | 4.0 | 3.9 | 3.1 | 3.8 | 0.1% |
| Stationary Combustion | 4.7 | 4.8 | 4.4 | 4.6 | 4.4 | 4.1 | 3.6 | 0.1% |
| Petrochemical Production | 4.2 | 5.7 | 5.3 | 4.8 | 4.9 | 4.4 | 3.6 | 0.1% |
| Adipic Acid Production | 15.8 | 5.5 | 5.0 | 4.3 | 3.7 | 2.0 | 1.9 | + |
| Carbon Dioxide Consumption | 1.4 | 1.4 | 1.3 | 1.7 | 1.9 | 1.8 | 1.8 | + |
| Titanium Dioxide Production | 1.2 | 1.8 | 1.8 | 1.8 | 1.9 | 1.8 | 1.5 | + |
| Ferroalloy Production | 2.2 | 1.9 | 1.4 | 1.5 | 1.6 | 1.6 | 1.5 | + |
| Mobile Combustion | 0.9 | 1.1 | 1.3 | 1.3 | 1.3 | 1.3 | 1.3 | + |
| Magnesium Production and Processing | 5.4 | 3.0 | 2.9 | 2.9 | 2.6 | 1.9 | 1.1 | + |
| Phosphoric Acid Production | 1.5 | 1.4 | 1.4 | 1.2 | 1.2 | 1.2 | 1.0 | + |
| Zinc Production | 0.7 | 1.0 | 1.1 | 1.1 | 1.1 | 1.2 | 1.0 | + |
| Lead Production | 0.5 | 0.6 | 0.6 | 0.6 | 0.6 | 0.6 | 0.5 | + |
| Silicon Carbide Production and Consumption | 0.4 | 0.3 | 0.2 | 0.2 | 0.2 | 0.2 | 0.2 | + |

| | | | | | | | | |
|---|---|---|---|---|---|---|---|---|
| **Agriculture** | **429.0** | **485.1** | **493.2** | **516.7** | **520.7** | **503.9** | **490.0** | **7.4%** |
| $N_2O$ from Agricultural Soil Management | 197.8 | 206.8 | 211.3 | 208.9 | 209.4 | 210.7 | 204.6 | 3.1% |
| Enteric Fermentation | 132.1 | 136.5 | 136.5 | 138.8 | 141.0 | 140.6 | 139.8 | 2.1% |
| Manure Management | 46.2 | 59.5 | 63.8 | 64.8 | 68.9 | 67.3 | 67.3 | 1.0% |
| $CO_2$ from Fossil Fuel Combustion | 31.04 | 38.79 | 46.81 | 49.04 | 48.44 | 45.44 | 46.66 | 0.7% |
| $CH_4$ and $N_2O$ from Forest Fires | 5.8 | 26.0 | 17.8 | 39.2 | 36.4 | 21.7 | 14.2 | 0.2% |
| Rice Cultivation | 7.1 | 7.5 | 6.8 | 5.9 | 6.2 | 7.2 | 7.3 | 0.1% |
| Liming of Agricultural Soils | 4.7 | 4.3 | 4.3 | 4.2 | 4.5 | 5.0 | 4.2 | 0.1% |
| Urea Fertilization | 2.4 | 3.2 | 3.5 | 3.7 | 3.7 | 3.6 | 3.6 | 0.1% |
| $CO_2$ and $N_2O$ from Managed Peatlands | 1.0 | 1.2 | 1.1 | 0.9 | 1.0 | 1.0 | 1.1 | + |
| Mobile Combustion | 0.3 | 0.4 | 0.5 | 0.5 | 0.5 | 0.5 | 0.5 | + |
| $N_2O$ from Forest Soils | 0.1 | 0.4 | 0.4 | 0.4 | 0.4 | 0.4 | 0.4 | + |
| Field Burning of Agricultural Residues | 0.4 | 0.4 | 0.3 | 0.3 | 0.3 | 0.4 | 0.4 | + |
| Stationary Combustion | + | + | + | + | + | + | + | + |
| **Commercial** | **395.5** | **381.4** | **387.2** | **375.2** | **389.6** | **403.5** | **409.5** | **6.2%** |
| $CO_2$ from Fossil Fuel Combustion | 219.0 | 230.8 | 223.5 | 208.6 | 219.4 | 224.2 | 224.0 | 3.4% |
| Landfills | 147.4 | 111.7 | 112.5 | 111.7 | 111.3 | 115.9 | 117.5 | 1.8% |
| Substitution of Ozone Depleting Substances | + | 5.4 | 17.6 | 21.1 | 24.9 | 29.1 | 33.7 | 0.5% |
| Wastewater Treatment | 23.5 | 25.2 | 24.3 | 24.5 | 24.4 | 24.5 | 24.5 | 0.4% |
| Human Sewage | 3.7 | 4.5 | 4.8 | 4.8 | 4.9 | 5.0 | 5.0 | 0.1% |
| Composting | 0.7 | 2.6 | 3.3 | 3.3 | 3.5 | 3.5 | 3.5 | 0.1% |
| Stationary Combustion | 1.3 | 1.3 | 1.2 | 1.2 | 1.2 | 1.2 | 1.2 | + |
| **Residential** | **345.1** | **386.2** | **371.0** | **335.8** | **358.9** | **367.1** | **360.1** | **5.4%** |
| $CO_2$ from Fossil Fuel Combustion | 338.3 | 370.7 | 357.9 | 321.5 | 342.4 | 348.2 | 339.2 | 5.1% |
| Substitution of Ozone Depleting Substances | 0.3 | 10.1 | 7.3 | 8.9 | 10.7 | 12.9 | 15.1 | 0.2% |
| Stationary Combustion | 5.5 | 4.3 | 4.3 | 3.9 | 4.2 | 4.4 | 4.2 | 0.1% |
| Settlement Soil Fertilization | 1.0 | 1.1 | 1.5 | 1.5 | 1.6 | 1.5 | 1.5 | + |
| **U.S. Territories** | **33.7** | **46.0** | **58.2** | **59.3** | **53.5** | **48.4** | **45.5** | **0.7%** |
| $CO_2$ from Fossil Fuel Combustion | 27.9 | 35.9 | 50.0 | 50.3 | 46.1 | 39.8 | 41.7 | 0.6% |
| Non-Energy Use of Fuels | 5.7 | 10.0 | 8.1 | 8.8 | 7.2 | 8.4 | 3.7 | 0.1% |
| Stationary Combustion | 0.1 | 0.1 | 0.2 | 0.2 | 0.2 | 0.2 | 0.2 | + |
| **Total Emissions** | **6,181.8** | **7,112.7** | **7,213.5** | **7,166.9** | **7,263.4** | **7,061.1** | **6,633.2** | **100.0%** |
| **Sinks** | **(861.5)** | **(576.6)** | **(1,056.5)** | **(1,064.3)** | **(1,060.9)** | **(1,040.5)** | **(1,015.1)** | **-15.3%** |
| $CO_2$ Flux from Forests[b] | (681.1) | (378.3) | (911.5) | (917.5) | (911.9) | (891.0) | (863.1) | -13.0% |
| Urban Trees | (57.1) | (77.5) | (87.8) | (89.8) | (91.9) | (93.9) | (95.9) | -1.4% |
| $CO_2$ Flux from Agricultural Soil Carbon Stocks | (99.2) | (107.6) | (45.6) | (46.1) | (46.3) | (44.4) | (43.4) | -0.7% |
| Landfilled Yard Trimmings and Food Scraps | (24.2) | (13.2) | (11.5) | (11.0) | (10.9) | (11.2) | (12.6) | -0.2% |
| **Net Emissions** | **5,320.3** | **6,536.1** | **6,157.1** | **6,102.6** | **6,202.5** | **6,020.7** | **5,618.2** | **84.7%** |

Note: Includes all emissions of $CO_2$, $CH_4$, $N_2O$, HFCs, PFCs, and $SF_6$. Parentheses indicate negative values or sequestration. Totals may not sum due to independent rounding.
ODS (Ozone Depleting Substances)
+ Does not exceed 0.05 Tg $CO_2$ Eq. or 0.05 percent.
[a] Percent of total emissions for year 2009.
[b] Includes the effects of net additions to stocks of carbon stored in harvested wood products.

*Source: Inventory of U.S. Greenhouse Gas Emissions and Sinks: 1990-2009*

## Electricity Generation-Related Greenhouse Gas Emissions (Tg $CO_2$ Eq.)

| Gas/Fuel Type or Source | 1990 | 2000 | 2005 | 2006 | 2007 | 2008 | 2009 |
|---|---|---|---|---|---|---|---|
| **$CO_2$** | **1,831.4** | **2,310.5** | **2,418.0** | **2,363.0** | **2,429.4** | **2,376.2** | **2,170.1** |
| $CO_2$ from Fossil Fuel Combustion | 1,820.8 | 2,296.9 | 2,402.1 | 2,346.4 | 2,412.8 | 2,360.9 | 2,154.0 |
| *Coal* | *1,547.6* | *1,927.4* | *1,983.8* | *1,953.7* | *1,987.3* | *1,959.4* | *1,747.6* |
| *Natural Gas* | *175.3* | *280.8* | *318.8* | *338.0* | *371.3* | *361.9* | *373.1* |
| *Petroleum* | *97.5* | *88.4* | *99.2* | *54.4* | *53.9* | *39.2* | *32.9* |
| *Geothermal* | *0.4* | *0.4* | *0.4* | *0.4* | *0.4* | *0.4* | *0.4* |
| Incineration of Waste | 8.0 | 11.1 | 12.5 | 12.5 | 12.7 | 12.2 | 12.3 |
| Limestone and Dolomite Use | 2.6 | 2.5 | 3.4 | 4.0 | 3.9 | 3.1 | 3.8 |
| **$CH_4$** | **0.6** | **0.7** | **0.7** | **0.7** | **0.7** | **0.7** | **0.7** |
| Stationary Combustion* | 0.6 | 0.7 | 0.7 | 0.7 | 0.7 | 0.7 | 0.7 |
| Incineration of Waste | + | + | + | + | + | + | + |
| **$N_2O$** | **8.5** | **10.4** | **10.7** | **10.5** | **10.6** | **10.4** | **9.4** |
| Stationary Combustion* | 8.1 | 10.0 | 10.3 | 10.1 | 10.2 | 10.1 | 9.0 |
| Incineration of Waste | 0.5 | 0.4 | 0.4 | 0.4 | 0.4 | 0.4 | 0.4 |
| **$SF_6$** | **28.4** | **16.0** | **15.1** | **14.1** | **13.2** | **13.3** | **12.8** |
| Electrical Transmission and Distribution | 28.4 | 16.0 | 15.1 | 14.1 | 13.2 | 13.3 | 12.8 |
| **Total** | **1,868.9** | **2,337.6** | **2,444.6** | **2,388.2** | **2,454.0** | **2,400.7** | **2,193.0** |

Note: Totals may not sum due to independent rounding.
* Includes only stationary combustion emissions related to the generation of electricity.
+ Does not exceed 0.05 Tg $CO_2$ Eq. or 0.05 percent.
*Source: Inventory of U.S. Greenhouse Gas Emissions and Sinks: 1990-2009*

**Transportation-Related Greenhouse Gas Emissions (Tg $CO_2$ Eq.)**

| Gas/Vehicle Type | 1990 | 2000 | 2005 | 2006 | 2007 | 2008 | 2009 |
|---|---|---|---|---|---|---|---|
| **Passenger Cars** | **657.4** | **695.3** | **709.5** | **682.9** | **672.0** | **632.5** | **627.4** |
| $CO_2$ | 629.3 | 644.2 | 662.3 | 639.1 | 632.8 | 597.9 | 597.2 |
| $CH_4$ | 2.6 | 1.6 | 1.1 | 1.0 | 0.9 | 0.8 | 0.7 |
| $N_2O$ | 25.4 | 25.2 | 17.8 | 15.7 | 13.8 | 11.7 | 10.1 |
| HFCs | + | 24.3 | 28.4 | 27.1 | 24.6 | 22.1 | 19.3 |
| **Light-Duty Trucks** | **336.6** | **512.1** | **551.3** | **564.0** | **570.3** | **553.8** | **551.0** |
| $CO_2$ | 321.1 | 467.0 | 505.9 | 519.5 | 528.4 | 515.1 | 514.5 |
| $CH_4$ | 1.4 | 1.1 | 0.7 | 0.7 | 0.6 | 0.6 | 0.6 |
| $N_2O$ | 14.1 | 22.4 | 13.7 | 12.6 | 11.2 | 9.5 | 9.4 |
| HFCs | + | 21.7 | 31.0 | 31.2 | 30.1 | 28.6 | 26.6 |
| **Medium- and Heavy-Duty Trucks** | **231.1** | **354.6** | **408.4** | **418.6** | **425.2** | **403.1** | **365.6** |
| $CO_2$ | 230.1 | 345.8 | 396.0 | 406.1 | 412.5 | 390.4 | 353.1 |
| $CH_4$ | 0.2 | 0.1 | 0.1 | 0.1 | 0.1 | 0.1 | 0.1 |
| $N_2O$ | 0.8 | 1.2 | 1.1 | 1.1 | 1.1 | 1.0 | 0.8 |
| HFCs | + | 7.4 | 11.1 | 11.4 | 11.5 | 11.6 | 11.6 |
| **Buses** | **8.4** | **11.2** | **12.0** | **12.3** | **12.5** | **12.2** | **11.2** |
| $CO_2$ | 8.4 | 11.1 | 11.8 | 12.0 | 12.1 | 11.8 | 10.8 |
| $CH_4$ | + | + | + | + | + | + | + |
| $N_2O$ | + | + | + | + | + | + | + |
| HFCs | + | 0.1 | 0.2 | 0.3 | 0.3 | 0.4 | 0.4 |
| **Motorcycles** | **1.8** | **1.9** | **1.7** | **1.9** | **2.1** | **2.2** | **2.2** |
| $CO_2$ | 1.7 | 1.8 | 1.6 | 1.9 | 2.1 | 2.1 | 2.1 |
| $CH_4$ | + | + | + | + | + | + | + |
| $N_2O$ | + | + | + | + | + | + | + |
| **Commercial Aircraft**[a] | **136.8** | **170.9** | **162.8** | **138.5** | **139.5** | **123.4** | **112.5** |
| $CO_2$ | 135.4 | 169.2 | 161.2 | 137.1 | 138.1 | 122.2 | 111.4 |
| $CH_4$ | 0.1 | 0.1 | 0.1 | 0.1 | 0.1 | 0.1 | 0.1 |
| $N_2O$ | 1.3 | 1.6 | 1.5 | 1.3 | 1.3 | 1.2 | 1.1 |
| **Other Aircraft**[b] | **44.4** | **33.5** | **35.9** | **35.1** | **33.2** | **35.2** | **29.6** |
| $CO_2$ | 43.9 | 33.1 | 35.5 | 34.7 | 32.8 | 34.8 | 29.3 |
| $CH_4$ | 0.1 | 0.1 | 0.1 | 0.1 | 0.1 | 0.1 | + |
| $N_2O$ | 0.4 | 0.3 | 0.3 | 0.3 | 0.3 | 0.3 | 0.3 |
| **Ships and Boats**[c] | **45.1** | **61.0** | **45.2** | **48.4** | **55.2** | **37.1** | **30.5** |
| $CO_2$ | 44.5 | 60.0 | 44.5 | 47.7 | 54.4 | 36.6 | 30.0 |
| $CH_4$ | + | + | + | + | + | + | + |
| $N_2O$ | 0.6 | 0.9 | 0.6 | 0.7 | 0.8 | 0.5 | 0.4 |
| HFCs | + | 0.1 | + | + | + | + | + |
| **Rail** | **39.0** | **48.1** | **53.0** | **55.1** | **54.3** | **50.6** | **43.3** |
| $CO_2$ | 38.5 | 45.6 | 50.3 | 52.4 | 51.6 | 47.9 | 40.6 |
| $CH_4$ | 0.1 | 0.1 | 0.1 | 0.1 | 0.1 | 0.1 | 0.1 |
| $N_2O$ | 0.3 | 0.3 | 0.4 | 0.4 | 0.4 | 0.4 | 0.3 |
| HFCs | + | 2.0 | 2.2 | 2.2 | 2.2 | 2.3 | 2.3 |
| Other Emissions from Electricity Generation[d] | 0.1 | + | 0.1 | 0.1 | 0.1 | 0.1 | 0.1 |
| **Pipelines**[e] | **36.0** | **35.2** | **32.2** | **32.3** | **34.3** | **35.7** | **35.2** |
| $CO_2$ | 36.0 | 35.2 | 32.2 | 32.3 | 34.3 | 35.7 | 35.2 |
| **Lubricants** | **11.8** | **12.1** | **10.2** | **9.9** | **10.2** | **9.5** | **8.5** |
| $CO_2$ | 11.8 | 12.1 | 10.2 | 9.9 | 10.2 | 9.5 | 8.5 |
| **Total Transportation** | **1,548.3** | **1,935.8** | **2,022.2** | **1,999.0** | **2,008.9** | **1,895.4** | **1,816.9** |
| *International Bunker* | *113.0* | *99.5* | *110.9* | *129.7* | *129.0* | *135.1* | *124.4* |

| *Fuels*[f] | | |
|---|---|---|

Note: Totals may not sum due to independent rounding. Passenger cars and light-duty trucks include vehicles typically used for personal travel and less than 8500 lbs; medium- and heavy-duty trucks include vehicles larger than 8500 lbs. HFC emissions primarily reflect HFC-134a.

+ Does not exceed 0.05 Tg $CO_2$ Eq.

[a] Consists of emissions from jet fuel consumed by domestic operations of commercial aircraft (no bunkers).

[b] Consists of emissions from jet fuel and aviation gasoline consumption by general aviation and military aircraft.

[c] Fluctuations in emission estimates are associated with fluctuations in reported fuel consumption, and may reflect data collection problems.

[d] Other emissions from electricity generation are a result of waste incineration (as the majority of municipal solid waste is combusted in "trash-to-steam" electricity generation plants), electrical transmission and distribution, and a portion of limestone and dolomite use (from pollution control equipment installed in electricity generation plants).

[e] $CO_2$ estimates reflect natural gas used to power pipelines, but not electricity. While the operation of pipelines produces $CH_4$ and $N_2O$, these emissions are not directly attributed to pipelines in the US Inventory.

[f] Emissions from International Bunker Fuels include emissions from both civilian and military activities; these emissions are not included in the transportation totals.

*Source: Inventory of U.S. Greenhouse Gas Emissions and Sinks: 1990-2009*

**Emissions of NOx, CO, NMVOCs, and SO2 (Gg)**

| Gas/Activity | 1990 | 2000 | 2005 | 2006 | 2007 | 2008 | 2009 |
|---|---|---|---|---|---|---|---|
| **$NO_x$** | **21,707** | **19,116** | **15,900** | **15,039** | **14,380** | **13,547** | **11,468** |
| Mobile Fossil Fuel Combustion | 10,862 | 10,199 | 9,012 | 8,488 | 7,965 | 7,441 | 6,206 |
| Stationary Fossil Fuel Combustion | 10,023 | 8,053 | 5,858 | 5,545 | 5,432 | 5,148 | 4,159 |
| Industrial Processes | 591 | 626 | 569 | 553 | 537 | 520 | 568 |
| Oil and Gas Activities | 139 | 111 | 321 | 319 | 318 | 318 | 393 |
| Incineration of Waste | 82 | 114 | 129 | 121 | 114 | 106 | 128 |
| Agricultural Burning | 8 | 8 | 6 | 7 | 8 | 8 | 8 |
| Solvent Use | 1 | 3 | 3 | 4 | 4 | 4 | 3 |
| Waste | 0 | 2 | 2 | 2 | 2 | 2 | 2 |
| **CO** | **130,038** | **92,243** | **70,809** | **67,238** | **63,625** | **60,039** | **51,452** |
| Mobile Fossil Fuel Combustion | 119,360 | 83,559 | 62,692 | 58,972 | 55,253 | 51,533 | 43,355 |
| Stationary Fossil Fuel Combustion | 5,000 | 4,340 | 4,649 | 4,695 | 4,744 | 4,792 | 4,543 |
| Industrial Processes | 4,125 | 2,216 | 1,555 | 1,597 | 1,640 | 1,682 | 1,549 |
| Incineration of Waste | 978 | 1,670 | 1,403 | 1,412 | 1,421 | 1,430 | 1,403 |
| Agricultural Burning | 268 | 259 | 184 | 233 | 237 | 270 | 247 |
| Oil and Gas Activities | 302 | 146 | 318 | 319 | 320 | 322 | 345 |
| Waste | 1 | 8 | 7 | 7 | 7 | 7 | 7 |
| Solvent Use | 5 | 45 | 2 | 2 | 2 | 2 | 2 |
| **NMVOCs** | **20,930** | **15,227** | **13,761** | **13,594** | **13,423** | **13,254** | **9,313** |
| Mobile Fossil Fuel Combustion | 10,932 | 7,229 | 6,330 | 6,037 | 5,742 | 5,447 | 4,151 |
| Solvent Use | 5,216 | 4,384 | 3,851 | 3,846 | 3,839 | 3,834 | 2,583 |
| Industrial Processes | 2,422 | 1,773 | 1,997 | 1,933 | 1,869 | 1,804 | 1,322 |
| Stationary Fossil Fuel Combustion | 912 | 1,077 | 716 | 918 | 1,120 | 1,321 | 424 |
| Oil and Gas Activities | 554 | 388 | 510 | 510 | 509 | 509 | 599 |
| Incineration of Waste | 222 | 257 | 241 | 238 | 234 | 230 | 159 |
| Waste | 673 | 119 | 114 | 113 | 111 | 109 | 76 |
| Agricultural Burning | NA | NA | NA | NA | NA | NA | NA |
| **$SO_2$** | **20,935** | **14,830** | **13,466** | **12,388** | **11,799** | **10,368** | **8,599** |
| Stationary Fossil Fuel Combustion | 18,407 | 12,849 | 11,541 | 10,612 | 10,172 | 8,891 | 7,167 |
| Industrial Processes | 1,307 | 1,031 | 831 | 818 | 807 | 795 | 798 |
| Mobile Fossil Fuel Combustion | 793 | 632 | 889 | 750 | 611 | 472 | 455 |
| Oil and Gas Activities | 390 | 287 | 181 | 182 | 184 | 187 | 154 |
| Incineration of Waste | 38 | 29 | 24 | 24 | 24 | 23 | 24 |
| Waste | 0 | 1 | 1 | 1 | 1 | 1 | 1 |
| Solvent Use | 0 | 1 | 0 | 0 | 0 | 0 | 0 |
| Agricultural Burning | NA | NA | NA | NA | NA | NA | NA |

Source: (EPA 2010, EPA 2009) except for estimates from field burning of agricultural residues.
NA (Not Available)
Note: Totals may not sum due to independent rounding.

*Source: Inventory of U.S. Greenhouse Gas Emissions and Sinks: 1990-2009*

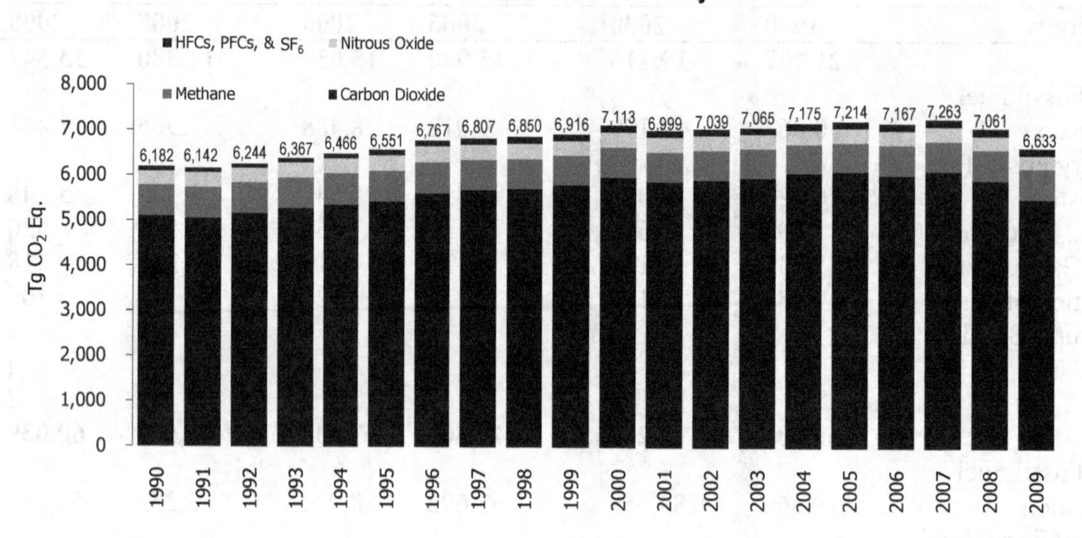


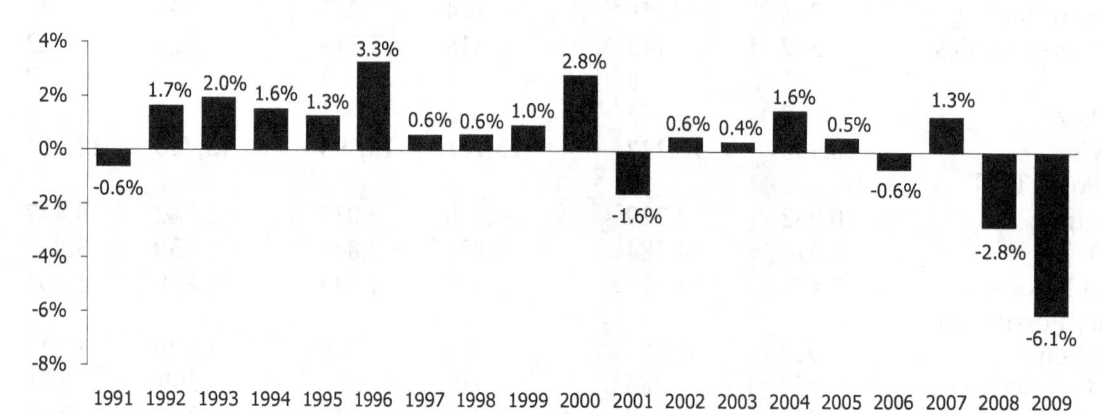


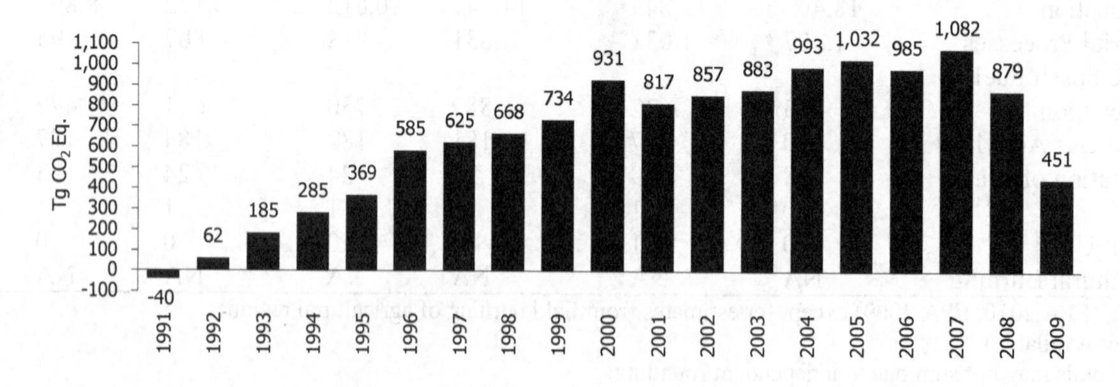


*Source: U.S. EPA, Inventory of U.S. Greenhouse Gas Emissions and Sinks: 1990-2009*

## 2009 Greenhouse Gas Emissions by Gas (percents based on Tg CO2Eq.)

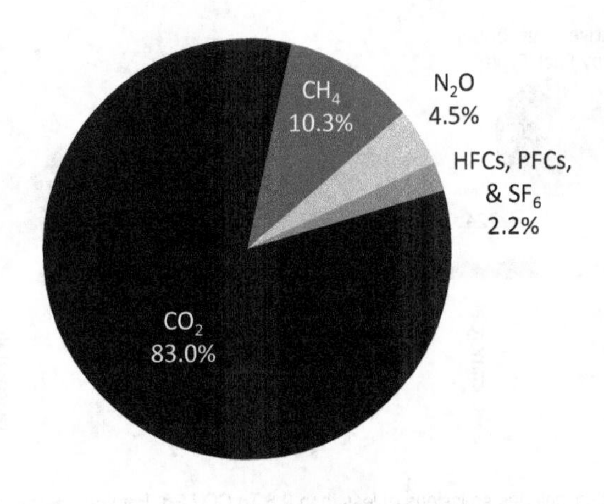


## 2009 Sources of CO2 Emissions

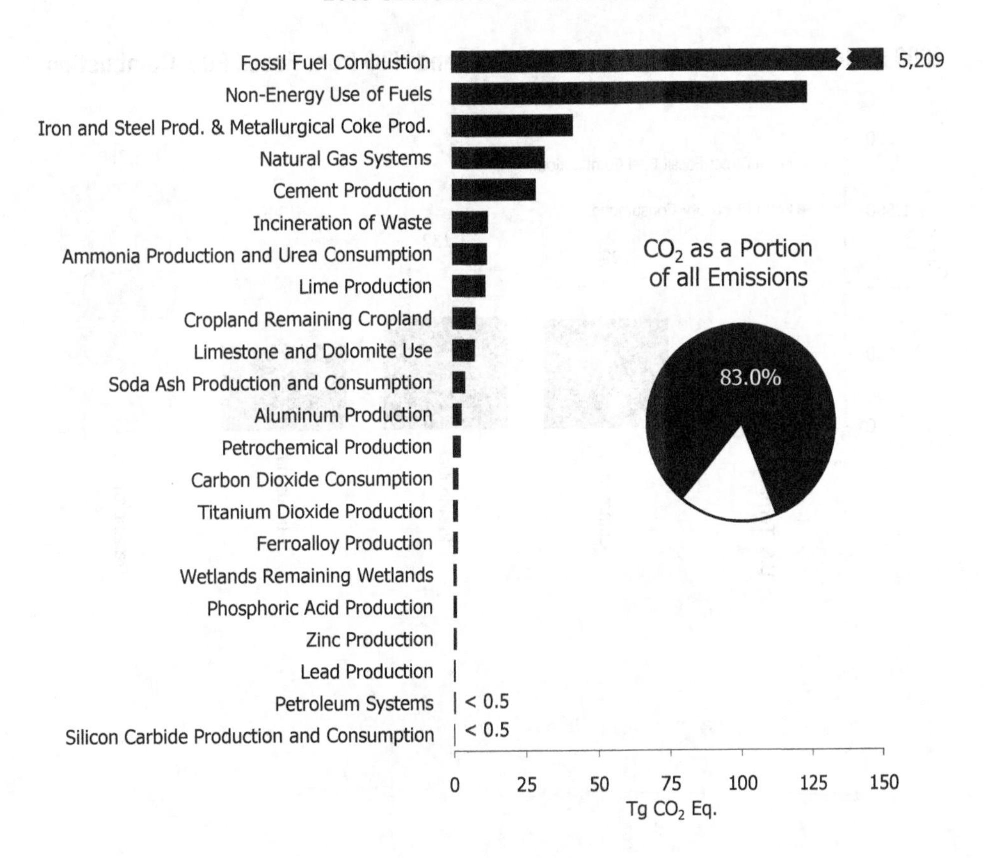


*Source: U.S. EPA, Inventory of U.S. Greenhouse Gas Emissions and Sinks: 1990-2009*

## 2009 CO2 Emissions from Fossil Fuel Combustion by Sector and Fuel Type

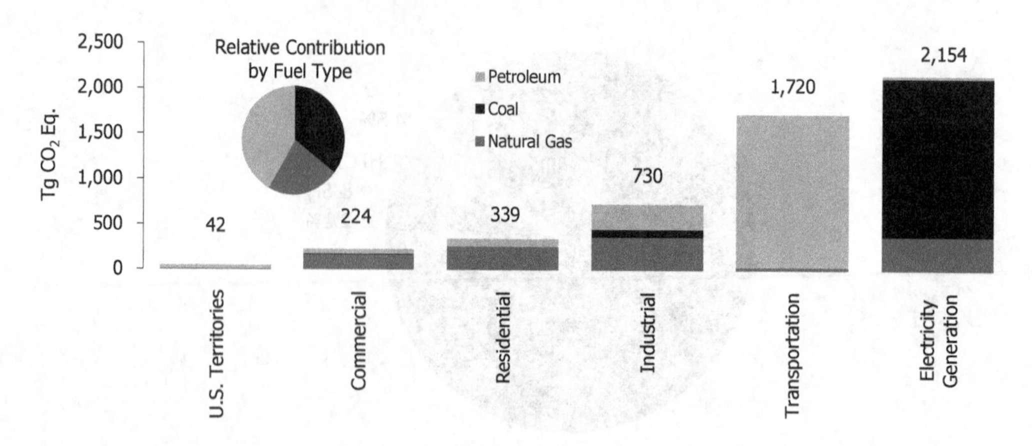


Note: Electricity generation also includes emissions of less than 0.5 Tg CO2 Eq. from geothermal-based electricity generation.

## 2009 End-Use Sector Emissions of CO2, CH4, and N2O from Fossil Fuel Combustion

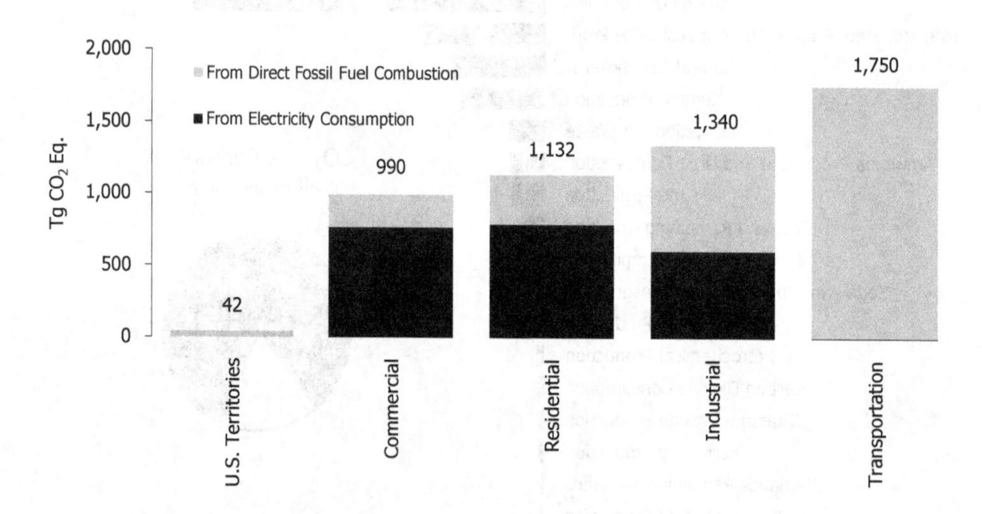


*Source: U.S. EPA, Inventory of U.S. Greenhouse Gas Emissions and Sinks: 1990-2009*

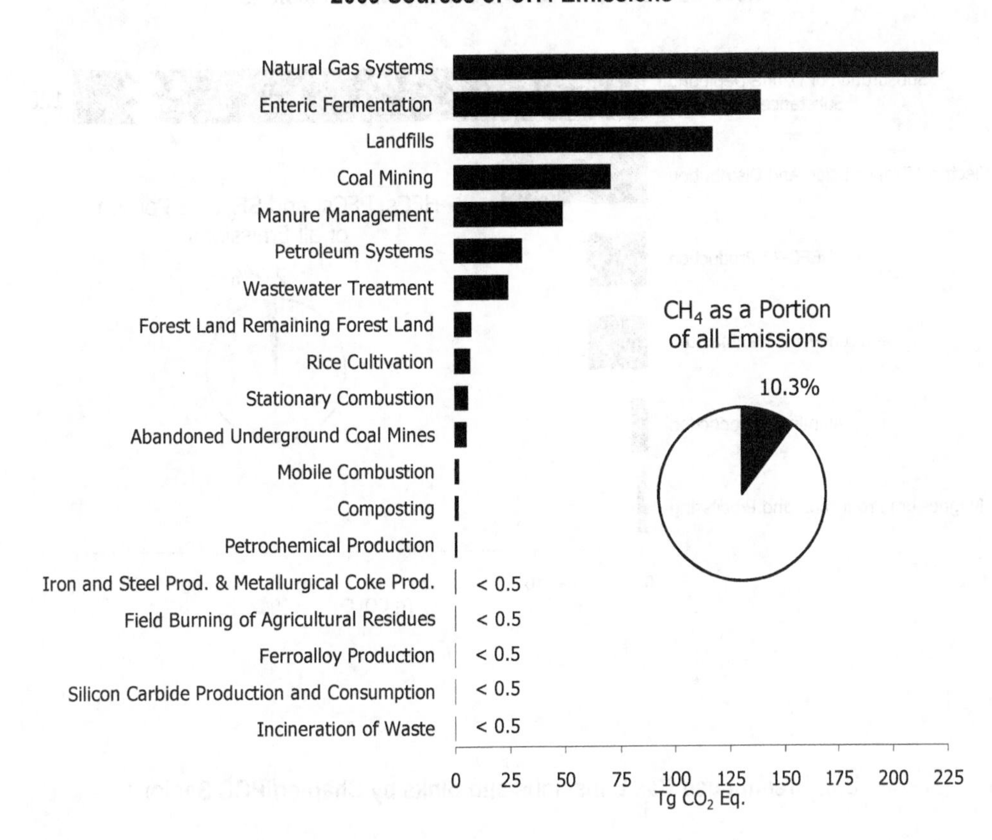
2009 Sources of CH4 Emissions
Natural Gas Systems
Enteric Fermentation
Landfills
Coal Mining
Manure Management
Petroleum Systems
Wastewater Treatment
Forest Land Remaining Forest Land
Rice Cultivation
Stationary Combustion
Abandoned Underground Coal Mines
Mobile Combustion
Composting
Petrochemical Production
Iron and Steel Prod. & Metallurgical Coke Prod. < 0.5
Field Burning of Agricultural Residues < 0.5
Ferroalloy Production < 0.5
Silicon Carbide Production and Consumption < 0.5
Incineration of Waste < 0.5
$CH_4$ as a Portion of all Emissions
10.3%
0 25 50 75 100 125 150 175 200 225
Tg $CO_2$ Eq.

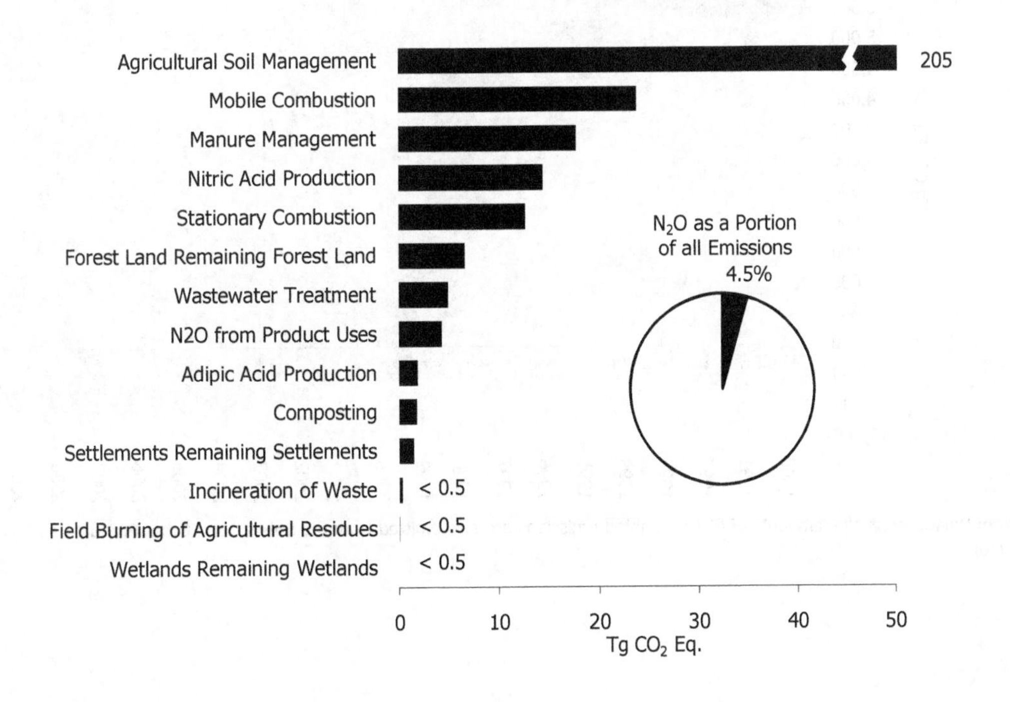
2009 Sources of N2O Emissions
Agricultural Soil Management 205
Mobile Combustion
Manure Management
Nitric Acid Production
Stationary Combustion
Forest Land Remaining Forest Land
Wastewater Treatment
N2O from Product Uses
Adipic Acid Production
Composting
Settlements Remaining Settlements
Incineration of Waste < 0.5
Field Burning of Agricultural Residues < 0.5
Wetlands Remaining Wetlands < 0.5
$N_2O$ as a Portion of all Emissions
4.5%
0 10 20 30 40 50
Tg $CO_2$ Eq.

*Source: U.S. EPA, Inventory of U.S. Greenhouse Gas Emissions and Sinks: 1990-2009*

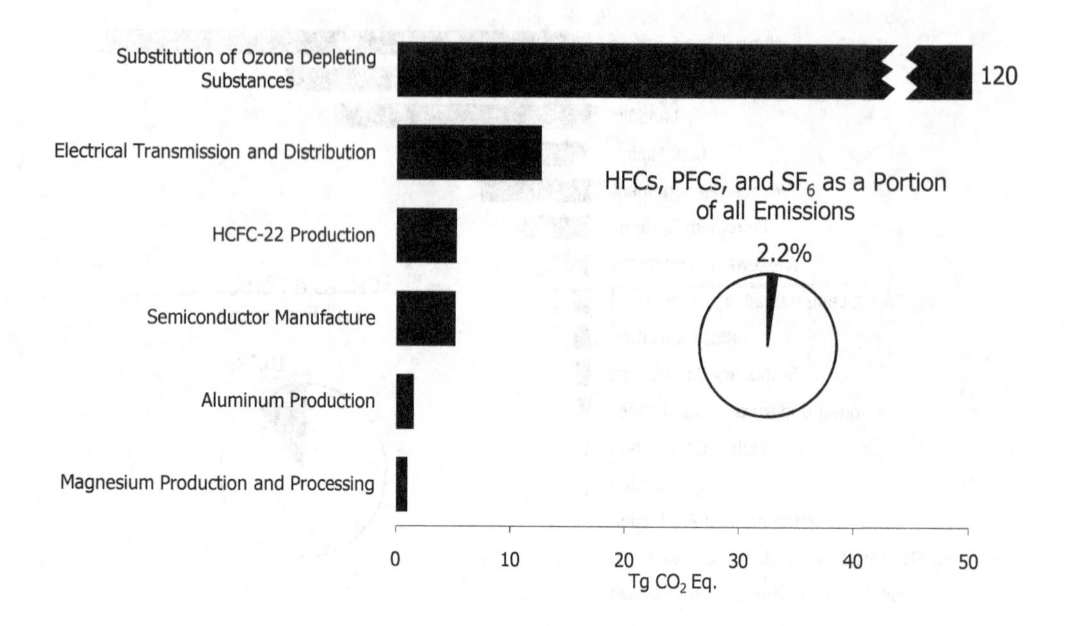


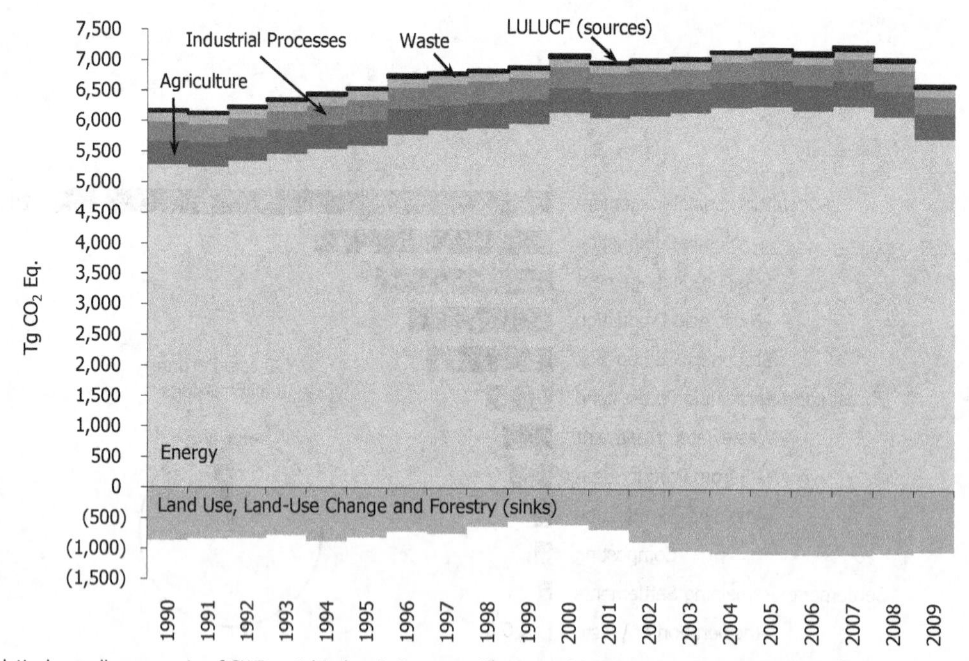


Note: Relatively smaller amounts of GWP-weighted emissions are also emitted from the Solvent and Other Product Use sectors

*Source: U.S. EPA, Inventory of U.S. Greenhouse Gas Emissions and Sinks: 1990-2009*

**2009 U.S. Energy Consumption by Energy Source**

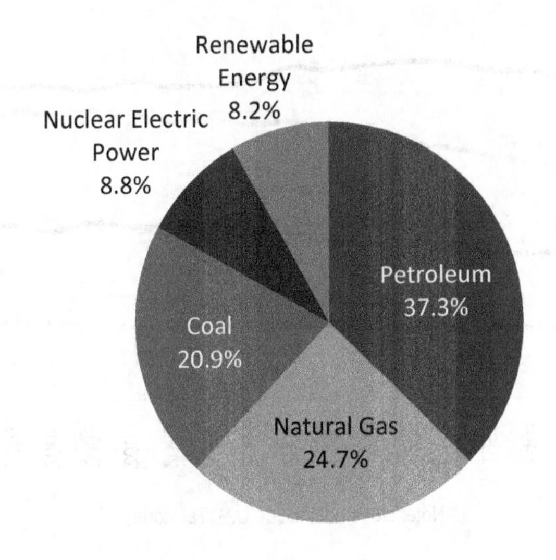


**Emissions Allocated to Economic Sectors**

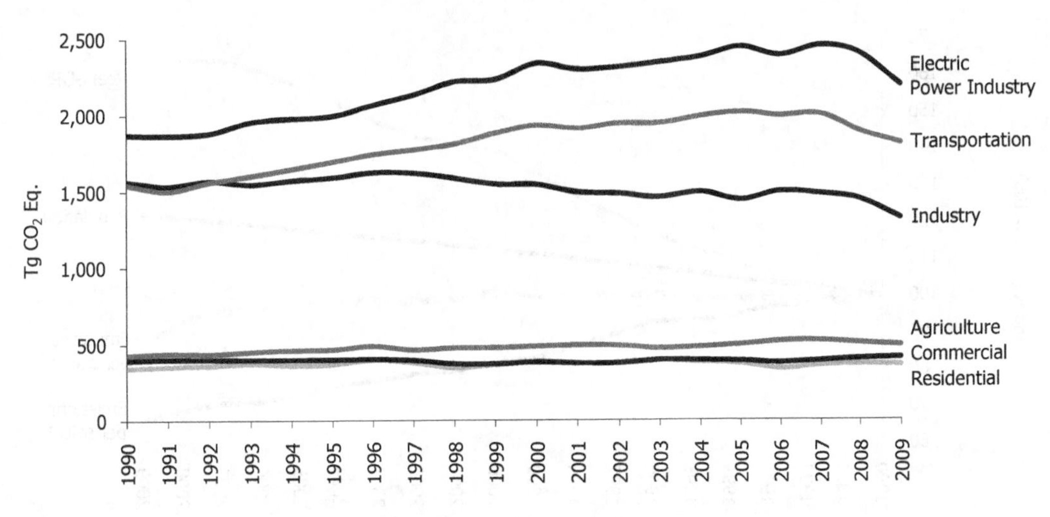


Note: Does not include U.S. Territories.

*Source: U.S. EPA, Inventory of U.S. Greenhouse Gas Emissions and Sinks: 1990-2009*

### Emissions with Electricity Distributed to Economic Sectors

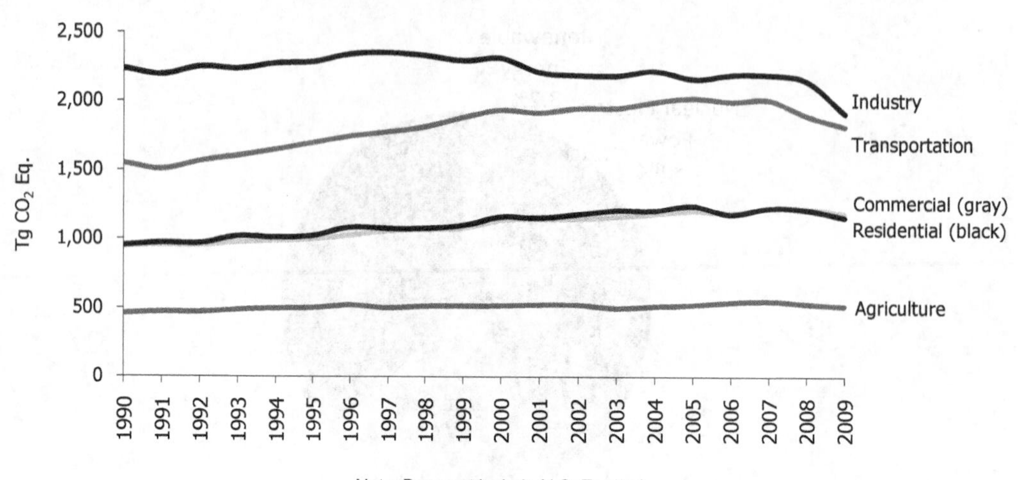


Note: Does not include U.S. Territories.

### U.S. Greenhouse Gas Emissions Per Capita and Per Dollar of Gross Domestic Product

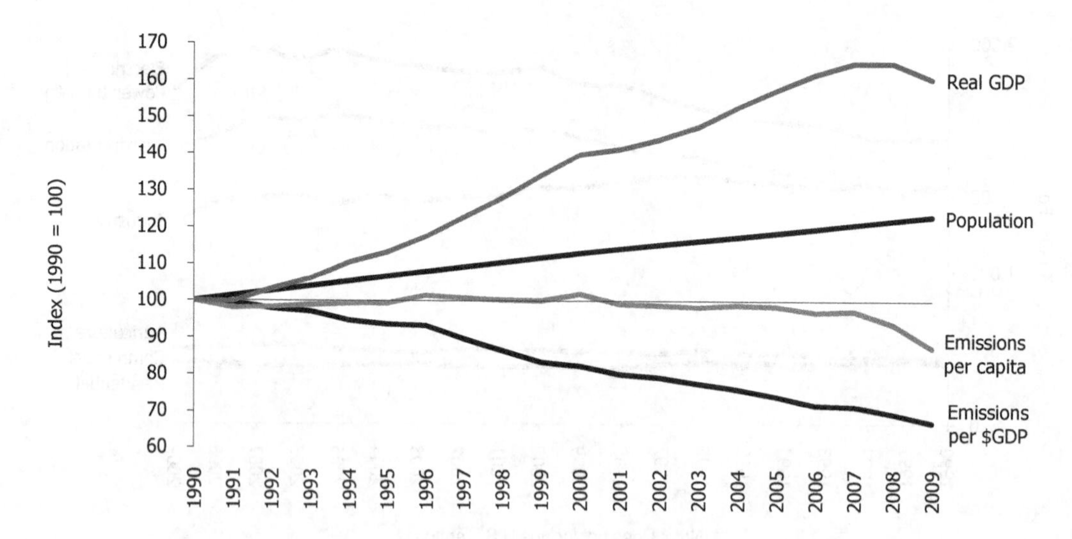


*Source: U.S. EPA, Inventory of U.S. Greenhouse Gas Emissions and Sinks: 1990-2009*

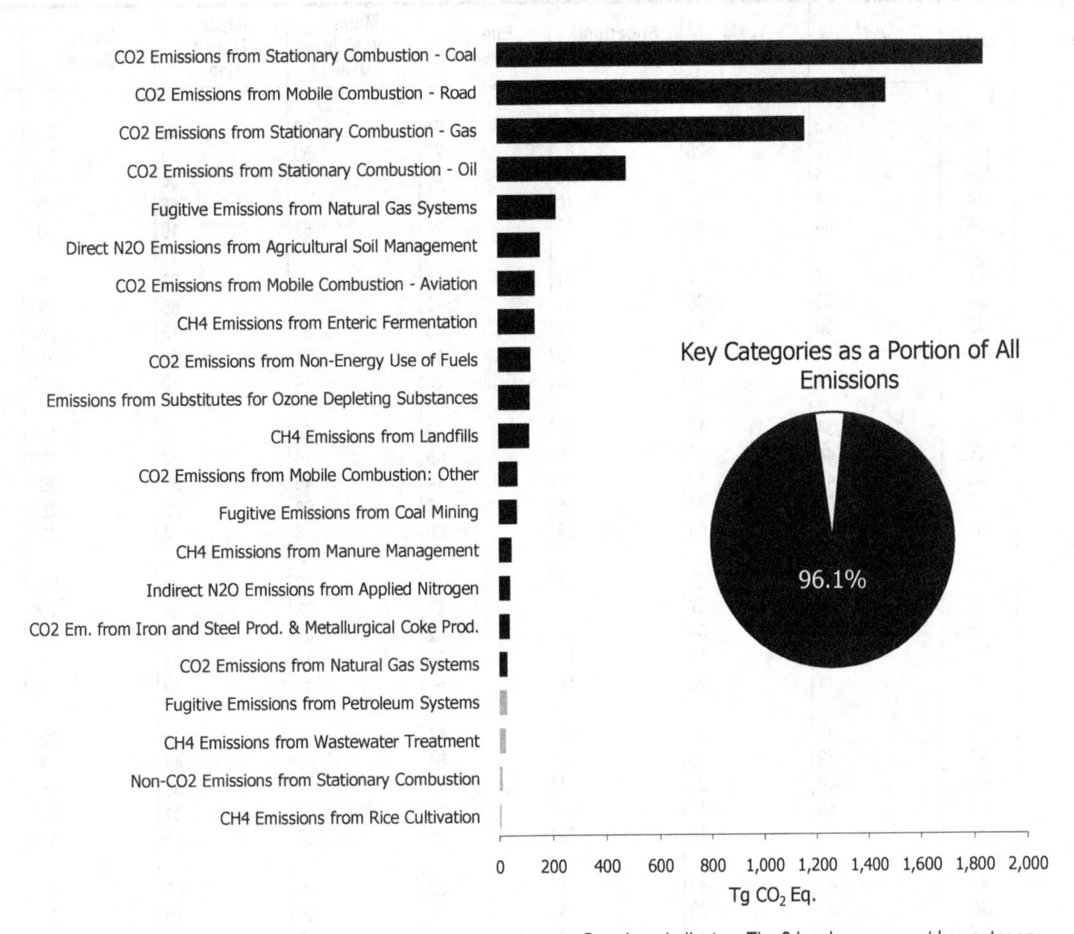


Notes: Black bars indicate a Tier 1 level assessment key category. Gray bars indicate a Tier 2 level assessment key category.

*Source: U.S. EPA, Inventory of U.S. Greenhouse Gas Emissions and Sinks: 1990-2009*

## Green Metro Area Rankings by Category

| Metro Area | State | Air Quality[1] | Toxic Releases[2] | Superfund Sites[3] | Energy Use[4] | Motor Vehicle Use[5] | Mass Transit Use[6] | Overall Score[7] | Overall Rank[8] |
|---|---|---|---|---|---|---|---|---|---|
| Akron | OH | 3 | 33 | 37 | 61 | 28 | 63 | 225 | 35 |
| Albany | NY | 17 | 28 | 62 | 72 | 50 | 33 | 262 | 53 |
| Albuquerque | NM | 74 | 2 | 46 | 40 | 47 | 56 | 265 | 55 |
| Allentown | PA | 24 | 64 | 70 | 53 | 17 | 62 | 290 | 62 |
| Atlanta | GA | 64 | 46 | 1 | 24 | 57 | 10 | 202 | 24 |
| Austin | TX | 7 | 7 | 1 | 23 | 59 | 20 | 117 | 4 |
| Bakersfield | CA | 58 | 38 | 31 | 12 | 3 | 55 | 197 | 21 |
| Baltimore | MD | 50 | 54 | 42 | 45 | 37 | 13 | 241 | 43 |
| Baton Rouge | LA | 18 | 74 | 48 | 13 | 61 | 69 | 283 | 58 |
| Birmingham | AL | 71 | 70 | 22 | 25 | 74 | 70 | 332 | 73 |
| Boston | MA | 34 | 17 | 67 | 49 | 25 | 5 | 197 | 21 |
| Buffalo | NY | 44 | 40 | 35 | 66 | 10 | 37 | 232 | 38 |
| Charleston | SC | 47 | 69 | 54 | 14 | 45 | 67 | 296 | 64 |
| Charlotte | NC | 51 | 42 | 55 | 28 | 68 | 22 | 266 | 56 |
| Chicago | IL | 63 | 53 | 29 | 68 | 6 | 4 | 223 | 33 |
| Cincinnati | OH | 16 | 63 | 47 | 48 | 32 | 34 | 240 | 42 |
| Cleveland | OH | 25 | 57 | 8 | 60 | 23 | 26 | 199 | 23 |
| Columbus | OH | 6 | 34 | 14 | 52 | 41 | 61 | 208 | 26 |
| Dallas-Ft. Worth | TX | 48 | 18 | 10 | 32 | 43 | 31 | 182 | 17 |
| Dayton | OH | 11 | 26 | 74 | 56 | 44 | 49 | 260 | 52 |
| Denver | CO | 43 | 29 | 51 | 57 | 30 | 12 | 222 | 32 |
| Detroit | MI | 72 | 58 | 36 | 64 | 49 | 44 | 323 | 71 |
| El Paso | TX | 68 | 22 | 1 | 27 | 9 | 35 | 162 | 13 |
| Fresno | CA | 60 | 11 | 61 | 15 | 8 | 51 | 206 | 25 |
| Grand Rapids | MI | 59 | 15 | 72 | 73 | 52 | 45 | 316 | 70 |
| Greensboro | NC | 41 | 43 | 1 | 33 | 60 | 50 | 228 | 36 |
| Greenville | SC | 67 | 23 | 64 | 26 | 56 | 75 | 311 | 68 |
| Hartford | CT | 37 | 13 | 52 | 62 | 46 | 41 | 251 | 45 |
| Honolulu | HI | 1 | 35 | 44 | 22 | 12 | 7 | 121 | 6 |
| Houston | TX | 66 | 68 | 45 | 16 | 69 | 23 | 287 | 60 |
| Indianapolis | IN | 52 | 66 | 21 | 54 | 53 | 65 | 311 | 68 |
| Jacksonville | FL | 15 | 60 | 59 | 8 | 64 | 53 | 259 | 51 |
| Kansas City | MO | 57 | 52 | 41 | 58 | 55 | 60 | 323 | 71 |
| Knoxville | TN | 65 | 61 | 38 | 31 | 71 | 68 | 334 | 74 |
| Las Vegas | NV | 32 | 30 | 1 | 38 | 65 | 21 | 187 | 18 |
| Little Rock | AR | 36 | 27 | 19 | 34 | 73 | 66 | 255 | 49 |
| Los Angeles | CA | 55 | 16 | 16 | 2 | 21 | 11 | 121 | 6 |
| Louisville | KY | 73 | 67 | 28 | 44 | 51 | 47 | 310 | 67 |
| Memphis | TN | 56 | 56 | 49 | 35 | 42 | 57 | 295 | 63 |
| Miami | FL | 20 | 4 | 25 | 18 | 36 | 17 | 120 | 5 |
| Milwaukee | WI | 69 | 47 | 50 | 74 | 31 | 28 | 299 | 66 |
| Minneapolis-St. Paul | MN | 22 | 50 | 56 | 75 | 39 | 16 | 258 | 50 |
| Nashville | TN | 31 | 48 | 9 | 36 | 67 | 46 | 237 | 41 |
| New Orleans | LA | 39 | 73 | 33 | 11 | 1 | 64 | 221 | 30 |
| New York-Northern NJ | NY | 54 | 8 | 60 | 46 | 2 | 1 | 171 | 15 |
| Oklahoma City | OK | 5 | 14 | 30 | 41 | 72 | 73 | 235 | 39 |
| Omaha | NE | 35 | 62 | 27 | 70 | 33 | 71 | 298 | 65 |
| Orlando | FL | 30 | 19 | 39 | 9 | 63 | 27 | 187 | 18 |
| Oxnard | CA | 26 | 1 | 32 | 3 | 11 | 43 | 116 | 3 |
| Philadelphia | PA | 45 | 36 | 75 | 47 | 7 | 8 | 218 | 29 |
| Phoenix | AZ | 75 | 6 | 15 | 39 | 24 | 32 | 191 | 20 |
| Pittsburgh | PA | 49 | 71 | 20 | 55 | 16 | 19 | 230 | 37 |
| Portland | OR | 13 | 44 | 53 | 20 | 5 | 9 | 144 | 11 |
| Providence | RI | 19 | 24 | 71 | 51 | 14 | 30 | 209 | 27 |
| Raleigh | NC | 46 | 12 | 34 | 30 | 75 | 54 | 251 | 45 |

| Metro Area | State | Air Quality[1] | Toxic Releases[2] | Superfund Sites[3] | Energy Use[4] | Motor Vehicle Use[5] | Mass Transit Use[6] | Overall Score[7] | Overall Rank[8] |
|---|---|---|---|---|---|---|---|---|---|
| Richmond | VA | 33 | 59 | 43 | 37 | 58 | 58 | 288 | 61 |
| Riverside | CA | 61 | 10 | 24 | 6 | 18 | 39 | 158 | 12 |
| Rochester | NY | 4 | 49 | 40 | 67 | 19 | 42 | 221 | 30 |
| Sacramento | CA | 62 | 5 | 26 | 7 | 4 | 29 | 133 | 9 |
| Saint Louis | MO | 70 | 65 | 65 | 50 | 62 | 25 | 337 | 75 |
| Salt Lake City | UT | 29 | 75 | 69 | 59 | 22 | 14 | 268 | 57 |
| San Antonio | TX | 38 | 31 | 11 | 21 | 48 | 24 | 173 | 16 |
| San Diego | CA | 53 | 9 | 7 | 1 | 26 | 15 | 111 | 2 |
| San Francisco | CA | 10 | 21 | 17 | 4 | 15 | 3 | 70 | 1 |
| San Jose | CA | 14 | 3 | 73 | 5 | 13 | 18 | 126 | 8 |
| Seattle | WA | 21 | 20 | 57 | 17 | 20 | 6 | 141 | 10 |
| Springfield | MA | 23 | 25 | 18 | 71 | 34 | 52 | 223 | 33 |
| Syracuse | NY | 2 | 39 | 63 | 69 | 40 | 38 | 251 | 45 |
| Tampa-St. Petersburg | FL | 42 | 41 | 58 | 10 | 54 | 48 | 253 | 48 |
| Toledo | OH | 9 | 72 | 1 | 65 | 38 | 59 | 244 | 44 |
| Tucson | AZ | 28 | 51 | 12 | 19 | 66 | 36 | 212 | 28 |
| Tulsa | OK | 12 | 55 | 13 | 42 | 70 | 72 | 264 | 54 |
| Virginia Beach | VA | 27 | 37 | 68 | 29 | 35 | 40 | 236 | 40 |
| Washington | DC | 40 | 32 | 23 | 43 | 27 | 2 | 167 | 14 |
| Youngstown | OH | 8 | 45 | 66 | 63 | 29 | 74 | 285 | 59 |

***Note:*** *The Green Metro Index compares 75 major metropolitan areas in the U.S. on measures of environmental quality and performance appropriate to metro areas as a whole. The index is based on federal and private data for six environmental measures including: air quality, toxic releases, superfund sites, energy use, mass transit use and motor vehicle use; The figures above rank how each metro area fared in each category. Lower numbers are better; (1) Based on the percent of days the Air Quality Index (AQI) was in the "Good" range in 2008; (2) Calculated by adding the total toxic releases for each metro area and dividing by the metro area population. Data is from the Environmental Protection Agency's Toxic Release Inventory for 2009; (3) Based on the per capita number of final and proposed Superfund Sites located within each metro area. Data is from the Environmental Protection Agency's Superfund National Priorities list (data extracted 4/1/2011); (4) Based on total heating and cooling degree days per year. Data is from Weather America, A Thirty-Year Summary of Statistical Weather Data and Rankings, 2011; (5) Based on the DVMT (daily vehicle-miles of travel) per capita for each urbanized area. Data is from the Department of Transportation's Urbanized Areas: 2008 Selected Characteristics report; (6) Calculated by dividing the total mass transit passenger miles by the population of the urbanized area. Data is from the Federal Transit Administration's 2009 National Transit Summaries and Trends report; (7) The overall score was calculated by combining the rankings of all six environmental indicators, giving equal weight to each indicator; (8) 1=best, 75=worst*

***Sources:*** *U.S Environmental Protection Agency; U.S. Department of Transportation; Federal Transit Administration; Grey House Publishing, Weather America, A Thirty-Year Summary of Statistical Weather Data and Rankings, 2011*

## Green Metro Area Overall Rankings

| Metro Area | State | Overall Rank[1] | Overall Score[2] |
|---|---|---|---|
| San Francisco | CA | 1 | 70 |
| San Diego | CA | 2 | 111 |
| Oxnard | CA | 3 | 116 |
| Austin | TX | 4 | 117 |
| Miami | FL | 5 | 120 |
| Los Angeles | CA | 6 | 121 |
| Honolulu | HI | 6 | 121 |
| San Jose | CA | 8 | 126 |
| Sacramento | CA | 9 | 133 |
| Seattle | WA | 10 | 141 |
| Portland | OR | 11 | 144 |
| Riverside | CA | 12 | 158 |
| El Paso | TX | 13 | 162 |
| Washington | DC | 14 | 167 |
| New York-Northern NJ | NY | 15 | 171 |
| San Antonio | TX | 16 | 173 |
| Dallas-Ft. Worth | TX | 17 | 182 |
| Las Vegas | NV | 18 | 187 |
| Orlando | FL | 18 | 187 |
| Phoenix | AZ | 20 | 191 |
| Boston | MA | 21 | 197 |
| Bakersfield | CA | 21 | 197 |
| Cleveland | OH | 23 | 199 |
| Atlanta | GA | 24 | 202 |
| Fresno | CA | 25 | 206 |
| Columbus | OH | 26 | 208 |
| Providence | RI | 27 | 209 |
| Tucson | AZ | 28 | 212 |
| Philadelphia | PA | 29 | 218 |
| New Orleans | LA | 30 | 221 |
| Rochester | NY | 30 | 221 |
| Denver | CO | 32 | 222 |
| Chicago | IL | 33 | 223 |
| Springfield | MA | 33 | 223 |
| Akron | OH | 35 | 225 |
| Greensboro | NC | 36 | 228 |
| Pittsburgh | PA | 37 | 230 |
| Buffalo | NY | 38 | 232 |
| Oklahoma City | OK | 39 | 235 |
| Virginia Beach | VA | 40 | 236 |
| Nashville | TN | 41 | 237 |
| Cincinnati | OH | 42 | 240 |
| Baltimore | MD | 43 | 241 |
| Toledo | OH | 44 | 244 |
| Syracuse | NY | 45 | 251 |
| Hartford | CT | 45 | 251 |
| Raleigh | NC | 45 | 251 |
| Tampa-St. Petersburg | FL | 48 | 253 |
| Little Rock | AR | 49 | 255 |
| Minneapolis-St. Paul | MN | 50 | 258 |
| Jacksonville | FL | 51 | 259 |
| Dayton | OH | 52 | 260 |
| Albany | NY | 53 | 262 |
| Tulsa | OK | 54 | 264 |
| Albuquerque | NM | 55 | 265 |

| Metro Area | State | Overall Rank[1] | Overall Score[2] |
|---|---|---|---|
| Charlotte | NC | 56 | 266 |
| Salt Lake City | UT | 57 | 268 |
| Baton Rouge | LA | 58 | 283 |
| Youngstown | OH | 59 | 285 |
| Houston | TX | 60 | 287 |
| Richmond | VA | 61 | 288 |
| Allentown | PA | 62 | 290 |
| Memphis | TN | 63 | 295 |
| Charleston | SC | 64 | 296 |
| Omaha | NE | 65 | 298 |
| Milwaukee | WI | 66 | 299 |
| Louisville | KY | 67 | 310 |
| Greenville | SC | 68 | 311 |
| Indianapolis | IN | 68 | 311 |
| Grand Rapids | MI | 70 | 316 |
| Kansas City | MO | 71 | 323 |
| Detroit | MI | 71 | 323 |
| Birmingham | AL | 73 | 332 |
| Knoxville | TN | 74 | 334 |
| Saint Louis | MO | 75 | 337 |

***Note:*** *The Green Metro Index compares 75 major metropolitan areas in the U.S. on measures of environmental quality and performance appropriate to metro areas as a whole. The index is based on federal data for six environmental measures including: air quality, toxic releases, superfund sites, energy use, mass transit use and motor vehicle use; (1) A lower number indicates better environmental quality or performance; (2) The overall score was calculated by combining the rankings of all six environmental indicators, giving equal weight to each indicator.*

***Sources:*** *U.S Environmental Protection Agency; U.S. Department of Transportation; Federal Transit Administration; Grey House Publishing, Weather America, A Thirty-Year Summary of Statistical Weather Data and Rankings, 2011*

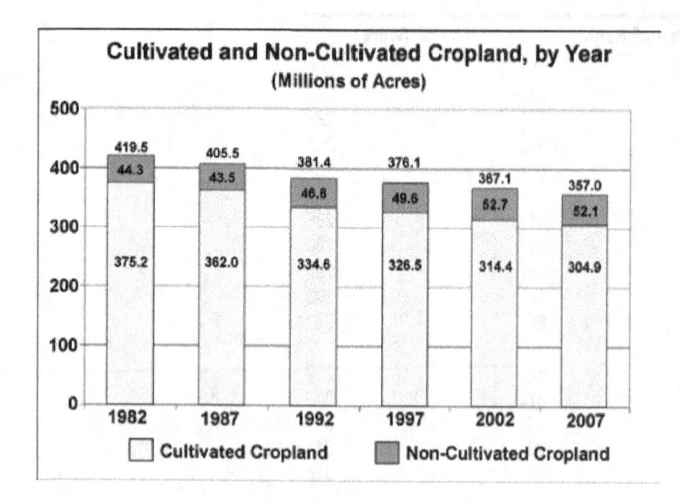


- Soil erosion on cropland decreased 43 percent between 1982 and 2007. Water (sheet and rill) erosion declined from 1.68 billion tons per year to 960 million tons per year, and erosion due to wind decreased from 1.38 billion to 765 million tons per year.

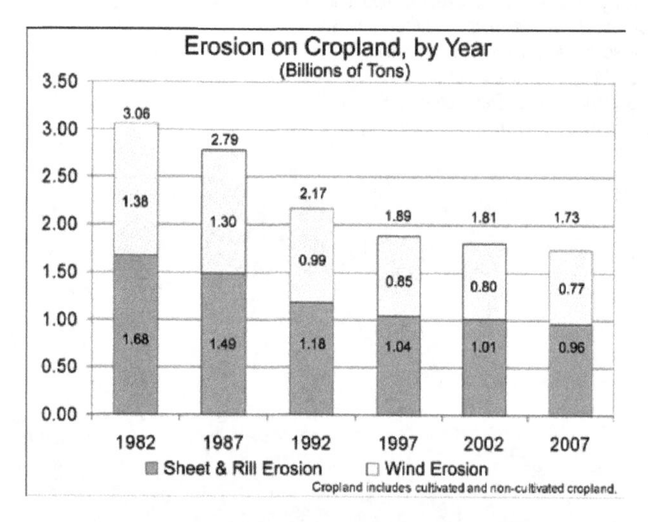


- About 40 million acres of land were newly developed between 1982 and 2007, bringing the total to about 111 million acres; that represents a 56 percent increase. This means that more than one-third of all land that has ever been developed in the lower 48 states was developed during the last quarter century.

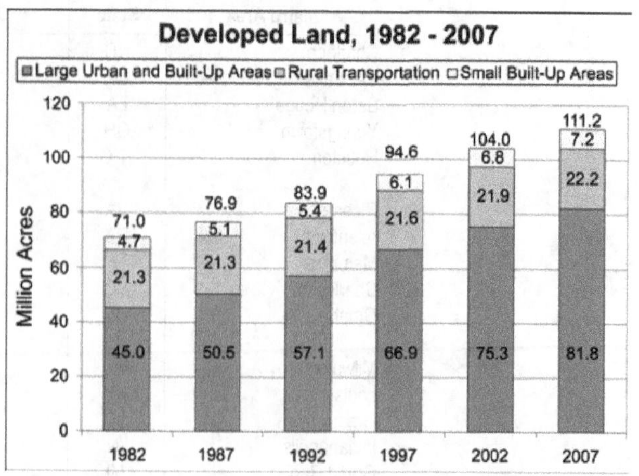


- About 24 percent (or 326 million acres) of the non-Federal rural land base is classified as prime farmland. This represents a 14-million-acre loss since 1982; most of this loss was due to development.

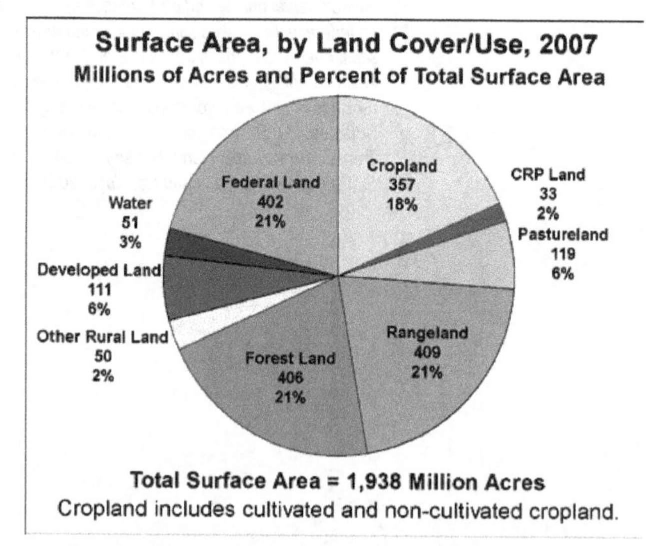


- Cropland acreage declined from 420 million acres in 1982 to 357 million acres in 2007. About half of the reduction in cropland acreage is due to enrollments of environmentally sensitive cropland in the Conservation Reserve Program. The share of noncultivated cropland — permanent hayland and horticultural cropland — increased from 11 percent to 15 percent of all U.S. cropland over the period.

U.S. Department of Agriculture. 2009. *Summary Report: 2007 National Resources Inventory*, Natural Resources Conservation Service, Washington, DC, and Center for Survey Statistics and Methodology, Iowa State University, Ames, Iowa. 123 pages.

**Surface area of non-Federal and Federal land and water areas, by State and year**
In thousands of acres, with margins of error

| State | Year | Federal land | Water areas | Non-Federal land | | | Total surface area |
|---|---|---|---|---|---|---|---|
| | | | | Developed | Rural | Total | |
| Alabama | 1982 | 949.3<br>-- | 1,166.2<br>-- | 1,624.1<br>±113.1 | 29,684.2<br>±109.7 | 31,308.3<br>±27.8 | 33,423.8<br>-- |
| | 1987 | 950.1<br>-- | 1,183.5<br>-- | 1,815.7<br>±120.0 | 29,474.5<br>±115.8 | 31,290.2<br>±27.4 | 33,423.8<br>-- |
| | 1992 | 970.0<br>-- | 1,203.0<br>-- | 1,946.4<br>±129.9 | 29,304.4<br>±125.2 | 31,250.8<br>±27.9 | 33,423.8<br>-- |
| | 1997 | 997.9<br>-- | 1,225.8<br>-- | 2,263.6<br>±140.4 | 28,936.5<br>±139.2 | 31,200.1<br>±28.2 | 33,423.8<br>-- |
| | 2002 | 997.9<br>-- | 1,278.1<br>-- | 2,673.5<br>±207.9 | 28,474.3<br>±213.9 | 31,147.8<br>±35.8 | 33,423.8<br>-- |
| | 2007 | 997.9<br>-- | 1,290.1<br>-- | 2,942.9<br>±203.7 | 28,192.9<br>±214.9 | 31,135.8<br>±146.4 | 33,423.8<br>-- |
| Arizona | 1982 | 31,005.4<br>-- | 182.9<br>-- | 1,018.1<br>±271.0 | 40,758.0<br>±270.9 | 41,776.1<br>±7.5 | 72,964.4<br>-- |
| | 1987 | 30,790.4<br>-- | 184.6<br>-- | 1,184.8<br>±287.3 | 40,804.6<br>±286.7 | 41,989.4<br>±7.6 | 72,964.4<br>-- |
| | 1992 | 30,426.2<br>-- | 187.0<br>-- | 1,281.4<br>±294.8 | 41,069.8<br>±294.2 | 42,351.2<br>±8.7 | 72,964.4<br>-- |
| | 1997 | 30,426.2<br>-- | 188.9<br>-- | 1,388.0<br>±307.1 | 40,961.3<br>±306.7 | 42,349.3<br>±8.9 | 72,964.4<br>-- |
| | 2002 | 30,426.2<br>-- | 189.5<br>-- | 1,753.5<br>±390.9 | 40,595.2<br>±389.7 | 42,348.7<br>±8.6 | 72,964.4<br>-- |
| | 2007 | 30,426.2<br>-- | 189.7<br>-- | 2,006.2<br>±462.3 | 40,342.3<br>±460.4 | 42,348.5<br>±8.5 | 72,964.4<br>-- |
| Arkansas | 1982 | 3,041.7<br>-- | 818.8<br>-- | 1,232.2<br>±98.4 | 28,944.2<br>±104.5 | 30,176.4<br>±26.6 | 34,036.9<br>-- |
| | 1987 | 3,049.1<br>-- | 852.9<br>-- | 1,272.8<br>±101.9 | 28,862.1<br>±108.2 | 30,134.9<br>±26.8 | 34,036.9<br>-- |
| | 1992 | 3,102.5<br>-- | 859.2<br>-- | 1,338.7<br>±106.8 | 28,736.5<br>±112.6 | 30,075.2<br>±27.2 | 34,036.9<br>-- |
| | 1997 | 3,102.8<br>-- | 884.9<br>-- | 1,528.4<br>±110.4 | 28,520.8<br>±116.8 | 30,049.2<br>±30.2 | 34,036.9<br>-- |
| | 2002 | 3,104.2<br>-- | 897.6<br>-- | 1,687.2<br>±125.5 | 28,347.9<br>±132.0 | 30,035.1<br>±30.7 | 34,036.9<br>-- |
| | 2007 | 3,104.2<br>-- | 902.1<br>-- | 1,809.3<br>±142.2 | 28,221.3<br>±142.0 | 30,030.6<br>±59.9 | 34,036.9<br>-- |
| California | 1982 | 46,007.5<br>-- | 1,856.2<br>-- | 4,082.8<br>±342.6 | 49,563.7<br>±341.5 | 53,646.5<br>±28.3 | 101,510.2<br>-- |
| | 1987 | 46,029.0<br>-- | 1,866.5<br>-- | 4,344.6<br>±359.3 | 49,270.1<br>±357.0 | 53,614.7<br>±28.4 | 101,510.2<br>-- |
| | 1992 | 46,633.4<br>-- | 1,869.1<br>-- | 4,836.9<br>±371.7 | 48,170.8<br>±368.0 | 53,007.7<br>±28.2 | 101,510.2<br>-- |
| | 1997 | 46,633.4<br>-- | 1,860.2<br>-- | 5,381.9<br>±370.7 | 47,634.7<br>±368.1 | 53,016.6<br>±28.2 | 101,510.2<br>-- |
| | 2002 | 46,639.0<br>-- | 1,867.6<br>-- | 5,807.2<br>±391.3 | 47,196.4<br>±387.3 | 53,003.6<br>±27.9 | 101,510.2<br>-- |
| | 2007 | 46,639.0<br>-- | 1,874.9<br>-- | 6,173.8<br>±403.8 | 46,822.5<br>±401.9 | 52,996.3<br>±29.2 | 101,510.2<br>-- |

| State | Year | Federal land | Water areas | Non-Federal land | | | Total surface area |
|---|---|---|---|---|---|---|---|
| | | | | Developed | Rural | Total | |
| Colorado | 1982 | 23,606.8<br>-- | 327.3<br>-- | 1,187.4<br>±109.5 | 41,503.0<br>±108.9 | 42,690.4<br>±21.7 | 66,624.5<br>-- |
| | 1987 | 23,741.2<br>-- | 328.9<br>-- | 1,329.4<br>±140.8 | 41,225.0<br>±138.3 | 42,554.4<br>±21.5 | 66,624.5<br>-- |
| | 1992 | 23,802.9<br>-- | 328.4<br>-- | 1,472.1<br>±148.9 | 41,021.1<br>±147.1 | 42,493.2<br>±21.9 | 66,624.5<br>-- |
| | 1997 | 23,793.8<br>-- | 329.4<br>-- | 1,577.6<br>±165.2 | 40,923.7<br>±163.9 | 42,501.3<br>±22.2 | 66,624.5<br>-- |
| | 2002 | 23,796.9<br>-- | 330.9<br>-- | 1,778.7<br>±203.8 | 40,718.0<br>±203.4 | 42,496.7<br>±21.9 | 66,624.5<br>-- |
| | 2007 | 23,796.9<br>-- | 332.1<br>-- | 1,934.3<br>±222.5 | 40,561.2<br>±223.1 | 42,495.5<br>±22.1 | 66,624.5<br>-- |
| Connecticut | 1982 | 9.7<br>-- | 127.0<br>-- | 823.0<br>±33.4 | 2,235.0<br>±34.9 | 3,058.0<br>±5.6 | 3,194.7<br>-- |
| | 1987 | 14.3<br>-- | 126.9<br>-- | 874.3<br>±34.3 | 2,179.2<br>±35.4 | 3,053.5<br>±5.5 | 3,194.7<br>-- |
| | 1992 | 14.5<br>-- | 127.9<br>-- | 916.3<br>±34.9 | 2,136.0<br>±36.0 | 3,052.3<br>±5.3 | 3,194.7<br>-- |
| | 1997 | 14.5<br>-- | 128.2<br>-- | 958.8<br>±33.1 | 2,093.2<br>±34.0 | 3,052.0<br>±5.2 | 3,194.7<br>-- |
| | 2002 | 14.5<br>-- | 128.2<br>-- | 1,016.2<br>±38.5 | 2,035.8<br>±39.4 | 3,052.0<br>±5.3 | 3,194.7<br>-- |
| | 2007 | 14.5<br>-- | 128.6<br>-- | 1,051.6<br>±41.4 | 2,000.0<br>±42.1 | 3,051.6<br>±5.4 | 3,194.7<br>-- |
| Delaware | 1982 | 31.1<br>-- | 288.1<br>-- | 158.7<br>±26.5 | 1,055.6<br>±27.0 | 1,214.3<br>±2.8 | 1,533.5<br>-- |
| | 1987 | 31.0<br>-- | 288.3<br>-- | 175.4<br>±29.0 | 1,038.8<br>±29.5 | 1,214.2<br>±2.8 | 1,533.5<br>-- |
| | 1992 | 31.0<br>-- | 288.4<br>-- | 192.5<br>±29.9 | 1,021.6<br>±30.1 | 1,214.1<br>±3.0 | 1,533.5<br>-- |
| | 1997 | 31.0<br>-- | 288.7<br>-- | 214.4<br>±31.7 | 999.4<br>±31.7 | 1,213.8<br>±3.3 | 1,533.5<br>-- |
| | 2002 | 31.0<br>-- | 289.4<br>-- | 251.0<br>±32.4 | 962.1<br>±32.3 | 1,213.1<br>±3.2 | 1,533.5<br>-- |
| | 2007 | 31.0<br>-- | 290.2<br>-- | 280.1<br>±35.3 | 932.2<br>±35.7 | 1,212.3<br>±3.3 | 1,533.5<br>-- |
| Florida | 1982 | 3,630.9<br>-- | 3,041.4<br>-- | 2,771.8<br>±239.6 | 28,089.6<br>±241.1 | 30,861.4<br>±26.2 | 37,533.7<br>-- |
| | 1987 | 3,656.2<br>-- | 3,048.8<br>-- | 3,081.9<br>±246.1 | 27,746.8<br>±247.3 | 30,828.7<br>±24.8 | 37,533.7<br>-- |
| | 1992 | 3,784.2<br>-- | 3,076.0<br>-- | 3,677.3<br>±272.1 | 26,996.2<br>±273.5 | 30,673.5<br>±26.7 | 37,533.7<br>-- |
| | 1997 | 3,784.2<br>-- | 3,071.4<br>-- | 4,368.2<br>±295.9 | 26,309.9<br>±304.3 | 30,678.1<br>±31.7 | 37,533.7<br>-- |
| | 2002 | 3,784.2<br>-- | 3,096.2<br>-- | 4,945.7<br>±331.0 | 25,707.6<br>±342.9 | 30,653.3<br>±34.5 | 37,533.7<br>-- |
| | 2007 | 3,784.2<br>-- | 3,133.6<br>-- | 5,515.2<br>±405.5 | 25,100.7<br>±422.2 | 30,615.9<br>±37.1 | 37,533.7<br>-- |

| State | Year | Federal land | Water areas | Non-Federal land | | | Total surface area |
|---|---|---|---|---|---|---|---|
| | | | | Developed | Rural | Total | |
| Georgia | 1982 | 2,099.9 -- | 949.8 -- | 2,227.1 ±116.0 | 32,463.7 ±114.1 | 34,690.8 ±24.5 | 37,740.5 -- |
| | 1987 | 2,107.0 -- | 956.4 -- | 2,469.1 ±128.4 | 32,208.0 ±125.9 | 34,677.1 ±24.8 | 37,740.5 -- |
| | 1992 | 2,125.6 -- | 990.6 -- | 2,910.5 ±139.2 | 31,713.8 ±138.8 | 34,624.3 ±24.4 | 37,740.5 -- |
| | 1997 | 2,124.0 -- | 1,012.6 -- | 3,702.2 ±158.6 | 30,901.7 ±159.7 | 34,603.9 ±25.7 | 37,740.5 -- |
| | 2002 | 2,124.0 -- | 1,046.3 -- | 4,233.0 ±205.6 | 30,337.2 ±206.9 | 34,570.2 ±28.1 | 37,740.5 -- |
| | 2007 | 2,124.0 -- | 1,059.3 -- | 4,639.9 ±220.6 | 29,917.3 ±227.4 | 34,557.2 ±35.0 | 37,740.5 -- |
| Idaho | 1982 | 33,601.8 -- | 545.3 -- | 562.2 ±48.3 | 18,778.2 ±50.6 | 19,340.4 ±10.9 | 53,487.5 -- |
| | 1987 | 33,395.5 -- | 548.1 -- | 624.9 ±52.5 | 18,919.0 ±54.9 | 19,543.9 ±10.8 | 53,487.5 -- |
| | 1992 | 33,480.9 -- | 550.8 -- | 680.0 ±59.8 | 18,775.8 ±61.5 | 19,455.8 ±10.4 | 53,487.5 -- |
| | 1997 | 33,563.3 -- | 552.9 -- | 777.2 ±66.8 | 18,594.1 ±67.8 | 19,371.3 ±10.0 | 53,487.5 -- |
| | 2002 | 33,563.3 -- | 554.0 -- | 830.4 ±67.2 | 18,539.8 ±67.6 | 19,370.2 ±9.9 | 53,487.5 -- |
| | 2007 | 33,563.3 -- | 557.6 -- | 907.3 ±75.3 | 18,459.3 ±75.7 | 19,366.6 ±11.2 | 53,487.5 -- |
| Illinois | 1982 | 476.4 -- | 723.6 -- | 2,622.5 ±115.2 | 32,236.2 ±120.6 | 34,858.7 ±24.0 | 36,058.7 -- |
| | 1987 | 477.1 -- | 719.2 -- | 2,759.7 ±123.7 | 32,102.7 ±129.6 | 34,862.4 ±21.9 | 36,058.7 -- |
| | 1992 | 492.1 -- | 715.7 -- | 2,855.2 ±125.3 | 31,995.7 ±128.3 | 34,850.9 ±20.5 | 36,058.7 -- |
| | 1997 | 490.3 -- | 711.0 -- | 3,092.9 ±125.5 | 31,764.5 ±126.2 | 34,857.4 ±20.6 | 36,058.7 -- |
| | 2002 | 491.1 -- | 725.4 -- | 3,226.3 ±136.3 | 31,615.9 ±136.6 | 34,842.2 ±19.2 | 36,058.7 -- |
| | 2007 | 491.1 -- | 732.5 -- | 3,383.3 ±166.3 | 31,451.8 ±166.1 | 34,835.1 ±19.4 | 36,058.7 -- |
| Indiana | 1982 | 473.3 -- | 348.6 -- | 1,778.3 ±92.1 | 20,558.2 ±94.5 | 22,336.5 ±18.1 | 23,158.4 -- |
| | 1987 | 472.3 -- | 357.9 -- | 1,892.7 ±92.0 | 20,435.5 ±95.3 | 22,328.2 ±17.8 | 23,158.4 -- |
| | 1992 | 473.5 -- | 361.6 -- | 1,997.0 ±99.2 | 20,326.3 ±102.0 | 22,323.3 ±17.4 | 23,158.4 -- |
| | 1997 | 472.4 -- | 358.5 -- | 2,186.7 ±110.8 | 20,140.8 ±113.8 | 22,327.5 ±17.3 | 23,158.4 -- |
| | 2002 | 472.4 -- | 366.5 -- | 2,328.1 ±112.5 | 19,991.4 ±115.7 | 22,319.5 ±17.2 | 23,158.4 -- |
| | 2007 | 472.4 -- | 373.1 -- | 2,446.0 ±123.8 | 19,866.9 ±125.1 | 22,312.9 ±18.5 | 23,158.4 -- |

| State | Year | Federal land | Water areas | Non-Federal land | | | Total surface area |
|---|---|---|---|---|---|---|---|
| | | | | Developed | Rural | Total | |
| Iowa | 1982 | 151.1<br>-- | 447.9<br>-- | 1,634.1<br>±95.3 | 33,783.4<br>±92.8 | 35,417.5<br>±17.4 | 36,016.5<br>-- |
| | 1987 | 152.2<br>-- | 451.5<br>-- | 1,661.4<br>±102.3 | 33,751.4<br>±100.9 | 35,412.8<br>±17.5 | 36,016.5<br>-- |
| | 1992 | 150.7<br>-- | 463.6<br>-- | 1,689.8<br>±103.4 | 33,712.4<br>±102.1 | 35,402.2<br>±17.6 | 36,016.5<br>-- |
| | 1997 | 172.0<br>-- | 470.5<br>-- | 1,765.6<br>±106.0 | 33,608.4<br>±103.8 | 35,374.0<br>±18.4 | 36,016.5<br>-- |
| | 2002 | 172.4<br>-- | 479.7<br>-- | 1,838.7<br>±103.5 | 33,525.7<br>±103.5 | 35,364.4<br>±17.7 | 36,016.5<br>-- |
| | 2007 | 172.4<br>-- | 485.9<br>-- | 1,892.3<br>±111.5 | 33,465.9<br>±112.0 | 35,358.2<br>±19.6 | 36,016.5<br>-- |
| Kansas | 1982 | 494.1<br>-- | 518.8<br>-- | 1,732.4<br>±69.3 | 49,915.5<br>±73.3 | 51,647.9<br>±17.8 | 52,660.8<br>-- |
| | 1987 | 494.8<br>-- | 519.3<br>-- | 1,759.6<br>±68.9 | 49,887.1<br>±73.8 | 51,646.7<br>±17.7 | 52,660.8<br>-- |
| | 1992 | 504.0<br>-- | 519.3<br>-- | 1,860.0<br>±70.9 | 49,777.5<br>±75.0 | 51,637.5<br>±18.7 | 52,660.8<br>-- |
| | 1997 | 504.0<br>-- | 530.3<br>-- | 1,959.2<br>±70.8 | 49,667.3<br>±77.0 | 51,626.5<br>±20.0 | 52,660.8<br>-- |
| | 2002 | 504.0<br>-- | 538.6<br>-- | 2,033.5<br>±78.9 | 49,584.7<br>±82.2 | 51,618.2<br>±19.2 | 52,660.8<br>-- |
| | 2007 | 504.0<br>-- | 554.4<br>-- | 2,095.7<br>±94.0 | 49,506.7<br>±96.8 | 51,602.4<br>±22.9 | 52,660.8<br>-- |
| Kentucky | 1982 | 1,107.1<br>-- | 585.4<br>-- | 1,124.0<br>±69.9 | 23,046.9<br>±73.2 | 24,170.9<br>±16.2 | 25,863.4<br>-- |
| | 1987 | 1,148.5<br>-- | 589.6<br>-- | 1,312.5<br>±79.7 | 22,812.8<br>±84.6 | 24,125.3<br>±16.0 | 25,863.4<br>-- |
| | 1992 | 1,187.2<br>-- | 605.4<br>-- | 1,470.1<br>±84.5 | 22,600.7<br>±89.8 | 24,070.8<br>±17.0 | 25,863.4<br>-- |
| | 1997 | 1,187.2<br>-- | 613.1<br>-- | 1,703.0<br>±97.0 | 22,360.1<br>±99.8 | 24,063.1<br>±16.9 | 25,863.4<br>-- |
| | 2002 | 1,295.4<br>-- | 625.8<br>-- | 1,952.4<br>±96.7 | 21,989.8<br>±101.5 | 23,942.2<br>±18.8 | 25,863.4<br>-- |
| | 2007 | 1,295.4<br>-- | 630.9<br>-- | 2,093.1<br>±105.9 | 21,844.0<br>±110.7 | 23,937.1<br>±18.9 | 25,863.4<br>-- |
| Louisiana | 1982 | 1,180.7<br>-- | 3,684.1<br>-- | 1,232.1<br>±57.2 | 25,279.9<br>±70.7 | 26,512.0<br>±36.0 | 31,376.8<br>-- |
| | 1987 | 1,239.7<br>-- | 3,730.5<br>-- | 1,381.3<br>±58.9 | 25,025.3<br>±68.5 | 26,406.6<br>±34.0 | 31,376.8<br>-- |
| | 1992 | 1,308.1<br>-- | 3,769.3<br>-- | 1,454.6<br>±61.9 | 24,844.8<br>±70.7 | 26,299.4<br>±35.0 | 31,376.8<br>-- |
| | 1997 | 1,308.1<br>-- | 3,779.7<br>-- | 1,594.7<br>±65.8 | 24,694.3<br>±75.1 | 26,289.0<br>±39.8 | 31,376.8<br>-- |
| | 2002 | 1,310.0<br>-- | 3,822.4<br>-- | 1,738.7<br>±68.8 | 24,505.7<br>±84.6 | 26,244.4<br>±45.0 | 31,376.8<br>-- |
| | 2007 | 1,310.0<br>-- | 3,924.2<br>-- | 1,862.8<br>±80.0 | 24,279.8<br>±92.0 | 26,142.6<br>±49.1 | 31,376.8<br>-- |

| State | Year | Federal land | Water areas | Non-Federal land | | | Total surface area |
|---|---|---|---|---|---|---|---|
| | | | | Developed | Rural | Total | |
| Maine | 1982 | 168.0<br>– | 1,254.5<br>– | 506.8<br>±68.5 | 19,036.9<br>±74.5 | 19,543.7<br>±25.5 | 20,966.2<br>– |
| | 1987 | 193.4<br>– | 1,255.5<br>– | 552.7<br>±73.1 | 18,964.6<br>±79.7 | 19,517.3<br>±25.6 | 20,966.2<br>– |
| | 1992 | 197.7<br>– | 1,256.1<br>– | 595.6<br>±77.2 | 18,916.8<br>±83.2 | 19,512.4<br>±25.7 | 20,966.2<br>– |
| | 1997 | 207.1<br>– | 1,255.8<br>– | 704.7<br>±85.9 | 18,798.6<br>±90.9 | 19,503.3<br>±26.0 | 20,966.2<br>– |
| | 2002 | 207.2<br>– | 1,256.1<br>– | 797.9<br>±89.6 | 18,705.0<br>±110.3 | 19,502.9<br>±68.1 | 20,966.2<br>– |
| | 2007 | 207.2<br>– | 1,256.6<br>– | 851.1<br>±97.2 | 18,651.3<br>±116.0 | 19,502.4<br>±68.2 | 20,966.2<br>– |
| Maryland | 1982 | 161.5<br>– | 1,651.4<br>– | 963.3<br>±115.6 | 5,093.7<br>±114.8 | 6,057.0<br>±7.8 | 7,869.9<br>– |
| | 1987 | 161.9<br>– | 1,653.9<br>– | 1,047.6<br>±120.6 | 5,006.5<br>±119.9 | 6,054.1<br>±7.6 | 7,869.9<br>– |
| | 1992 | 168.9<br>– | 1,655.6<br>– | 1,116.8<br>±123.8 | 4,928.6<br>±123.2 | 6,045.4<br>±7.9 | 7,869.9<br>– |
| | 1997 | 168.9<br>– | 1,658.0<br>– | 1,306.1<br>±127.5 | 4,736.9<br>±127.2 | 6,043.0<br>±7.9 | 7,869.9<br>– |
| | 2002 | 168.9<br>– | 1,659.5<br>– | 1,408.2<br>±126.4 | 4,633.3<br>±126.6 | 6,041.5<br>±7.7 | 7,869.9<br>– |
| | 2007 | 168.9<br>– | 1,661.6<br>– | 1,496.7<br>±130.0 | 4,542.7<br>±130.2 | 6,039.4<br>±8.1 | 7,869.9<br>– |
| Massachusetts | 1982 | 97.3<br>– | 368.3<br>– | 1,086.1<br>±65.6 | 3,787.3<br>±65.4 | 4,873.4<br>±14.2 | 5,339.0<br>– |
| | 1987 | 97.7<br>– | 368.6<br>– | 1,196.5<br>±68.7 | 3,676.2<br>±68.3 | 4,872.7<br>±13.9 | 5,339.0<br>– |
| | 1992 | 97.7<br>– | 370.3<br>– | 1,332.4<br>±69.0 | 3,538.6<br>±68.6 | 4,871.0<br>±13.7 | 5,339.0<br>– |
| | 1997 | 97.7<br>– | 368.9<br>– | 1,554.7<br>±70.7 | 3,317.7<br>±71.7 | 4,872.4<br>±12.9 | 5,339.0<br>– |
| | 2002 | 97.1<br>– | 368.7<br>– | 1,644.4<br>±74.6 | 3,228.8<br>±75.2 | 4,873.2<br>±13.0 | 5,339.0<br>– |
| | 2007 | 97.1<br>– | 366.6<br>– | 1,716.4<br>±83.3 | 3,158.9<br>±84.3 | 4,875.3<br>±12.8 | 5,339.0<br>– |
| Michigan | 1982 | 3,194.5<br>– | 1,099.7<br>– | 2,839.5<br>±129.0 | 30,215.5<br>±135.9 | 33,055.0<br>±28.4 | 37,349.2<br>– |
| | 1987 | 3,228.2<br>– | 1,097.9<br>– | 3,050.2<br>±140.3 | 29,972.9<br>±147.8 | 33,023.1<br>±28.3 | 37,349.2<br>– |
| | 1992 | 3,274.7<br>– | 1,102.3<br>– | 3,320.1<br>±148.9 | 29,652.1<br>±155.1 | 32,972.2<br>±28.0 | 37,349.2<br>– |
| | 1997 | 3,274.7<br>– | 1,101.5<br>– | 3,700.9<br>±161.3 | 29,272.1<br>±170.3 | 32,973.0<br>±29.0 | 37,349.2<br>– |
| | 2002 | 3,273.6<br>– | 1,112.4<br>– | 4,030.0<br>±162.5 | 28,933.2<br>±170.0 | 32,963.2<br>±24.3 | 37,349.2<br>– |
| | 2007 | 3,273.6<br>– | 1,120.8<br>– | 4,227.6<br>±168.1 | 28,727.2<br>±176.2 | 32,954.8<br>±24.5 | 37,349.2<br>– |

| State | Year | Federal land | Water areas | Non-Federal land | | | Total surface area |
|---|---|---|---|---|---|---|---|
| | | | | Developed | Rural | Total | |
| Minnesota | 1982 | 3,326.3<br>-- | 3,125.8<br>-- | 1,715.6<br>±77.0 | 45,842.2<br>±87.0 | 47,557.8<br>±29.0 | 54,009.9<br>-- |
| | 1987 | 3,342.3<br>-- | 3,133.6<br>-- | 1,839.0<br>±82.1 | 45,695.0<br>±91.1 | 47,534.0<br>±28.6 | 54,009.9<br>-- |
| | 1992 | 3,336.3<br>-- | 3,136.8<br>-- | 1,949.9<br>±88.7 | 45,586.9<br>±97.0 | 47,536.8<br>±28.8 | 54,009.9<br>-- |
| | 1997 | 3,336.3<br>-- | 3,135.0<br>-- | 2,181.3<br>±98.7 | 45,357.3<br>±106.9 | 47,538.6<br>±30.9 | 54,009.9<br>-- |
| | 2002 | 3,336.3<br>-- | 3,142.6<br>-- | 2,305.6<br>±107.4 | 45,225.4<br>±115.8 | 47,531.0<br>±32.1 | 54,009.9<br>-- |
| | 2007 | 3,336.1<br>-- | 3,144.9<br>-- | 2,395.2<br>±116.7 | 45,133.7<br>±124.2 | 47,528.9<br>±33.7 | 54,009.9<br>-- |
| Mississippi | 1982 | 1,634.6<br>-- | 723.8<br>-- | 1,123.3<br>±95.8 | 27,045.6<br>±97.8 | 28,168.9<br>±28.8 | 30,527.3<br>-- |
| | 1987 | 1,673.5<br>-- | 794.2<br>-- | 1,197.0<br>±96.8 | 26,862.6<br>±105.0 | 28,059.6<br>±34.4 | 30,527.3<br>-- |
| | 1992 | 1,751.9<br>-- | 832.6<br>-- | 1,269.3<br>±96.7 | 26,673.5<br>±105.5 | 27,942.8<br>±33.5 | 30,527.3<br>-- |
| | 1997 | 1,769.7<br>-- | 857.4<br>-- | 1,475.4<br>±106.7 | 26,424.8<br>±117.2 | 27,900.2<br>±32.8 | 30,527.3<br>-- |
| | 2002 | 1,794.8<br>-- | 892.3<br>-- | 1,656.7<br>±117.9 | 26,183.5<br>±138.2 | 27,840.2<br>±38.7 | 30,527.3<br>-- |
| | 2007 | 1,794.8<br>-- | 889.9<br>-- | 1,811.9<br>±124.4 | 26,030.7<br>±142.5 | 27,842.6<br>±38.8 | 30,527.3<br>-- |
| Missouri | 1982 | 1,920.0<br>-- | 766.2<br>-- | 2,150.8<br>±93.4 | 39,776.9<br>±95.7 | 41,927.7<br>±17.1 | 44,613.9<br>-- |
| | 1987 | 1,889.9<br>-- | 797.0<br>-- | 2,256.8<br>±96.0 | 39,670.2<br>±96.6 | 41,927.0<br>±17.7 | 44,613.9<br>-- |
| | 1992 | 1,904.5<br>-- | 812.2<br>-- | 2,373.9<br>±101.9 | 39,523.3<br>±103.9 | 41,897.2<br>±18.3 | 44,613.9<br>-- |
| | 1997 | 1,916.1<br>-- | 825.8<br>-- | 2,609.4<br>±115.5 | 39,262.6<br>±115.5 | 41,872.0<br>±18.1 | 44,613.9<br>-- |
| | 2002 | 1,919.4<br>-- | 847.9<br>-- | 2,777.2<br>±118.8 | 39,069.4<br>±115.0 | 41,846.6<br>±21.2 | 44,613.9<br>-- |
| | 2007 | 1,919.4<br>-- | 866.3<br>-- | 2,931.5<br>±204.8 | 38,896.7<br>±198.9 | 41,828.2<br>±21.3 | 44,613.9<br>-- |
| Montana | 1982 | 27,273.5<br>-- | 1,053.0<br>-- | 820.2<br>±91.7 | 64,963.3<br>±98.1 | 65,783.5<br>±38.5 | 94,110.0<br>-- |
| | 1987 | 27,183.6<br>-- | 1,047.6<br>-- | 828.7<br>±91.9 | 65,050.1<br>±99.0 | 65,878.8<br>±38.6 | 94,110.0<br>-- |
| | 1992 | 27,089.7<br>-- | 1,038.5<br>-- | 877.7<br>±96.5 | 65,104.1<br>±106.9 | 65,981.8<br>±35.4 | 94,110.0<br>-- |
| | 1997 | 27,089.7<br>-- | 1,027.6<br>-- | 935.9<br>±101.2 | 65,056.8<br>±111.4 | 65,992.7<br>±32.6 | 94,110.0<br>-- |
| | 2002 | 27,092.0<br>-- | 1,028.5<br>-- | 972.6<br>±104.1 | 65,016.9<br>±117.2 | 65,989.5<br>±35.0 | 94,110.0<br>-- |
| | 2007 | 27,092.0<br>-- | 1,039.4<br>-- | 1,047.0<br>±119.5 | 64,931.6<br>±132.5 | 65,978.6<br>±35.8 | 94,110.0<br>-- |

| State | Year | Federal land | Water areas | Non-Federal land | | | Total surface area |
|---|---|---|---|---|---|---|---|
| | | | | Developed | Rural | Total | |
| Nebraska | 1982 | 575.8 -- | 455.5 -- | 1,030.7 ±104.1 | 47,447.6 ±109.7 | 48,478.3 ±19.7 | 49,509.6 -- |
| | 1987 | 583.2 -- | 463.5 -- | 1,043.0 ±107.3 | 47,419.9 ±113.1 | 48,462.9 ±19.4 | 49,509.6 -- |
| | 1992 | 649.5 -- | 465.5 -- | 1,060.6 ±111.2 | 47,334.0 ±117.0 | 48,394.6 ±19.5 | 49,509.6 -- |
| | 1997 | 647.6 -- | 470.0 -- | 1,107.2 ±112.6 | 47,284.8 ±118.7 | 48,392.0 ±18.9 | 49,509.6 -- |
| | 2002 | 647.6 -- | 474.2 -- | 1,132.2 ±111.1 | 47,255.6 ±117.2 | 48,387.8 ±20.8 | 49,509.6 -- |
| | 2007 | 647.6 -- | 476.2 -- | 1,156.5 ±108.6 | 47,229.3 ±116.1 | 48,385.8 ±21.5 | 49,509.6 -- |
| Nevada | 1982 | 59,871.3 -- | 423.9 -- | 237.8 ±64.1 | 10,230.1 ±65.0 | 10,467.9 ±6.4 | 70,763.1 -- |
| | 1987 | 59,779.8 -- | 430.4 -- | 278.2 ±69.6 | 10,274.7 ±71.6 | 10,552.9 ±6.4 | 70,763.1 -- |
| | 1992 | 59,870.7 -- | 431.4 -- | 307.5 ±69.9 | 10,153.5 ±72.0 | 10,461.0 ±6.4 | 70,763.1 -- |
| | 1997 | 59,870.7 -- | 431.7 -- | 330.7 ±71.8 | 10,130.0 ±74.4 | 10,460.7 ±6.4 | 70,763.1 -- |
| | 2002 | 59,868.9 -- | 430.8 -- | 493.7 ±113.7 | 9,969.7 ±114.7 | 10,463.4 ±5.6 | 70,763.1 -- |
| | 2007 | 59,868.9 -- | 431.1 -- | 582.9 ±138.0 | 9,880.2 ±138.5 | 10,463.1 ±5.4 | 70,763.1 -- |
| New Hampshire | 1982 | 735.1 -- | 234.0 -- | 386.9 ±45.8 | 4,585.0 ±50.0 | 4,971.9 ±12.6 | 5,941.0 -- |
| | 1987 | 735.3 -- | 234.2 -- | 479.7 ±58.3 | 4,491.8 ±62.3 | 4,971.5 ±12.7 | 5,941.0 -- |
| | 1992 | 756.3 -- | 235.0 -- | 538.0 ±59.2 | 4,411.7 ±63.6 | 4,949.7 ±12.7 | 5,941.0 -- |
| | 1997 | 763.2 -- | 236.0 -- | 603.4 ±63.5 | 4,338.4 ±68.0 | 4,941.8 ±12.8 | 5,941.0 -- |
| | 2002 | 763.2 -- | 235.9 -- | 646.6 ±63.1 | 4,295.3 ±67.9 | 4,941.9 ±12.6 | 5,941.0 -- |
| | 2007 | 763.2 -- | 236.2 -- | 695.5 ±63.1 | 4,246.1 ±68.6 | 4,941.6 ±12.8 | 5,941.0 -- |
| New Jersey | 1982 | 135.6 -- | 514.5 -- | 1,178.5 ±65.8 | 3,387.0 ±68.3 | 4,565.5 ±9.6 | 5,215.6 -- |
| | 1987 | 138.4 -- | 516.3 -- | 1,384.7 ±70.3 | 3,176.2 ±74.1 | 4,560.9 ±9.4 | 5,215.6 -- |
| | 1992 | 148.3 -- | 519.5 -- | 1,453.8 ±70.5 | 3,094.0 ±74.2 | 4,547.8 ±9.3 | 5,215.6 -- |
| | 1997 | 148.3 -- | 523.5 -- | 1,653.0 ±75.7 | 2,890.8 ±78.3 | 4,543.8 ±8.7 | 5,215.6 -- |
| | 2002 | 148.3 -- | 525.4 -- | 1,782.4 ±79.2 | 2,759.5 ±83.1 | 4,541.9 ±9.4 | 5,215.6 -- |
| | 2007 | 148.3 -- | 526.0 -- | 1,849.3 ±83.6 | 2,692.0 ±87.3 | 4,541.3 ±10.0 | 5,215.6 -- |

| State | Year | Federal land | Water areas | Non-Federal land | | | Total surface area |
|---|---|---|---|---|---|---|---|
| | | | | Developed | Rural | Total | |
| New Mexico | 1982 | 25,645.6<br>-- | 149.8<br>-- | 707.0<br>±95.9 | 51,320.9<br>±99.1 | 52,027.9<br>±16.2 | 77,823.3<br>-- |
| | 1987 | 26,186.2<br>-- | 150.4<br>-- | 772.1<br>±103.9 | 50,714.6<br>±108.3 | 51,486.7<br>±16.3 | 77,823.3<br>-- |
| | 1992 | 26,448.5<br>-- | 146.6<br>-- | 829.1<br>±116.7 | 50,399.1<br>±120.6 | 51,228.2<br>±16.5 | 77,823.3<br>-- |
| | 1997 | 26,448.5<br>-- | 151.7<br>-- | 995.4<br>±126.7 | 50,227.7<br>±129.5 | 51,223.1<br>±15.8 | 77,823.3<br>-- |
| | 2002 | 26,448.5<br>-- | 152.8<br>-- | 1,190.3<br>±158.8 | 50,031.7<br>±160.0 | 51,222.0<br>±15.8 | 77,823.3<br>-- |
| | 2007 | 26,448.5<br>-- | 153.1<br>-- | 1,261.9<br>±180.9 | 49,959.8<br>±182.5 | 51,221.7<br>±15.2 | 77,823.3<br>-- |
| New York | 1982 | 219.3<br>-- | 1,257.6<br>-- | 2,808.9<br>±123.3 | 27,075.0<br>±126.7 | 29,883.9<br>±20.6 | 31,360.8<br>-- |
| | 1987 | 218.7<br>-- | 1,259.0<br>-- | 2,915.5<br>±129.6 | 26,967.6<br>±133.0 | 29,883.1<br>±20.7 | 31,360.8<br>-- |
| | 1992 | 208.9<br>-- | 1,264.9<br>-- | 3,058.9<br>±132.9 | 26,828.1<br>±135.6 | 29,887.0<br>±21.2 | 31,360.8<br>-- |
| | 1997 | 208.9<br>-- | 1,267.4<br>-- | 3,400.3<br>±142.9 | 26,484.2<br>±145.9 | 29,884.5<br>±20.9 | 31,360.8<br>-- |
| | 2002 | 205.3<br>-- | 1,282.7<br>-- | 3,657.3<br>±155.7 | 26,215.5<br>±160.2 | 29,872.8<br>±23.7 | 31,360.8<br>-- |
| | 2007 | 205.3<br>-- | 1,289.9<br>-- | 3,793.9<br>±160.6 | 26,071.7<br>±165.8 | 29,865.6<br>±24.4 | 31,360.8<br>-- |
| North Carolina | 1982 | 2,181.2<br>-- | 2,724.4<br>-- | 2,317.5<br>±107.6 | 26,486.2<br>±101.2 | 28,803.7<br>±16.8 | 33,709.3<br>-- |
| | 1987 | 2,335.9<br>-- | 2,741.0<br>-- | 2,731.1<br>±116.8 | 25,901.3<br>±110.1 | 28,632.4<br>±17.3 | 33,709.3<br>-- |
| | 1992 | 2,506.6<br>-- | 2,748.8<br>-- | 3,191.3<br>±144.3 | 25,262.6<br>±139.3 | 28,453.9<br>±18.1 | 33,709.3<br>-- |
| | 1997 | 2,507.5<br>-- | 2,757.4<br>-- | 3,673.0<br>±165.6 | 24,771.4<br>±164.7 | 28,444.4<br>±20.7 | 33,709.3<br>-- |
| | 2002 | 2,507.5<br>-- | 2,772.2<br>-- | 4,404.7<br>±209.6 | 24,024.9<br>±206.1 | 28,429.6<br>±20.9 | 33,709.3<br>-- |
| | 2007 | 2,507.5<br>-- | 2,783.3<br>-- | 4,796.7<br>±245.7 | 23,621.8<br>±243.1 | 28,418.5<br>±22.7 | 33,709.3<br>-- |
| North Dakota | 1982 | 1,727.1<br>-- | 971.1<br>-- | 902.1<br>±69.4 | 41,650.4<br>±71.3 | 42,552.5<br>±16.9 | 45,250.7<br>-- |
| | 1987 | 1,743.2<br>-- | 968.4<br>-- | 910.9<br>±69.0 | 41,628.2<br>±71.2 | 42,539.1<br>±16.5 | 45,250.7<br>-- |
| | 1992 | 1,785.0<br>-- | 967.3<br>-- | 924.7<br>±70.4 | 41,573.7<br>±73.0 | 42,498.4<br>±16.8 | 45,250.7<br>-- |
| | 1997 | 1,785.0<br>-- | 1,033.9<br>-- | 954.2<br>±71.7 | 41,477.6<br>±76.5 | 42,431.8<br>±17.0 | 45,250.7<br>-- |
| | 2002 | 1,784.8<br>-- | 1,087.3<br>-- | 967.5<br>±71.3 | 41,411.1<br>±76.2 | 42,378.6<br>±18.4 | 45,250.7<br>-- |
| | 2007 | 1,784.8<br>-- | 1,088.6<br>-- | 973.2<br>±69.5 | 41,404.1<br>±75.2 | 42,377.3<br>±18.8 | 45,250.7<br>-- |

| State | Year | Federal land | Water areas | Non-Federal land | | | Total surface area |
|---|---|---|---|---|---|---|---|
| | | | | Developed | Rural | Total | |
| Ohio | 1982 | 352.5<br>-- | 382.4<br>-- | 2,867.0<br>±116.8 | 22,842.9<br>±119.2 | 25,709.9<br>±12.5 | 26,444.8<br>-- |
| | 1987 | 350.0<br>-- | 386.8<br>-- | 3,073.5<br>±121.6 | 22,634.5<br>±124.6 | 25,708.0<br>±12.2 | 26,444.8<br>-- |
| | 1992 | 373.3<br>-- | 390.9<br>-- | 3,342.9<br>±134.4 | 22,337.7<br>±137.5 | 25,680.6<br>±12.3 | 26,444.8<br>-- |
| | 1997 | 373.3<br>-- | 390.5<br>-- | 3,718.5<br>±147.7 | 21,962.5<br>±151.2 | 25,681.0<br>±12.4 | 26,444.8<br>-- |
| | 2002 | 373.3<br>-- | 400.5<br>-- | 3,938.3<br>±147.4 | 21,732.7<br>±151.3 | 25,671.0<br>±12.9 | 26,444.8<br>-- |
| | 2007 | 373.3<br>-- | 408.5<br>-- | 4,140.3<br>±164.6 | 21,522.7<br>±169.6 | 25,663.0<br>±14.9 | 26,444.8<br>-- |
| Oklahoma | 1982 | 1,161.6<br>-- | 1,003.9<br>-- | 1,448.7<br>±89.7 | 41,123.9<br>±97.5 | 42,572.6<br>±24.7 | 44,738.1<br>-- |
| | 1987 | 1,148.7<br>-- | 1,021.0<br>-- | 1,522.4<br>±88.4 | 41,046.0<br>±95.8 | 42,568.4<br>±23.0 | 44,738.1<br>-- |
| | 1992 | 1,148.3<br>-- | 1,043.2<br>-- | 1,584.2<br>±96.5 | 40,962.4<br>±102.0 | 42,546.6<br>±27.1 | 44,738.1<br>-- |
| | 1997 | 1,148.3<br>-- | 1,054.0<br>-- | 1,734.0<br>±108.9 | 40,801.8<br>±114.2 | 42,535.8<br>±28.7 | 44,738.1<br>-- |
| | 2002 | 1,148.3<br>-- | 1,075.2<br>-- | 1,913.2<br>±121.4 | 40,601.4<br>±127.6 | 42,514.6<br>±35.7 | 44,738.1<br>-- |
| | 2007 | 1,148.3<br>-- | 1,088.7<br>-- | 2,056.8<br>±128.4 | 40,444.3<br>±134.4 | 42,501.1<br>±38.4 | 44,738.1<br>-- |
| Oregon | 1982 | 31,095.9<br>-- | 806.8<br>-- | 967.9<br>±100.3 | 29,290.4<br>±98.8 | 30,258.3<br>±18.4 | 62,161.0<br>-- |
| | 1987 | 31,114.3<br>-- | 923.1<br>-- | 1,057.8<br>±110.6 | 29,065.8<br>±109.9 | 30,123.6<br>±18.5 | 62,161.0<br>-- |
| | 1992 | 31,275.4<br>-- | 658.1<br>-- | 1,134.5<br>±115.5 | 29,093.0<br>±113.9 | 30,227.5<br>±18.6 | 62,161.0<br>-- |
| | 1997 | 31,260.4<br>-- | 819.5<br>-- | 1,241.0<br>±122.3 | 28,840.1<br>±123.3 | 30,081.1<br>±18.2 | 62,161.0<br>-- |
| | 2002 | 31,260.4<br>-- | 824.8<br>-- | 1,323.4<br>±134.1 | 28,752.4<br>±135.7 | 30,075.8<br>±19.6 | 62,161.0<br>-- |
| | 2007 | 31,260.4<br>-- | 826.8<br>-- | 1,389.6<br>±141.1 | 28,684.2<br>±142.8 | 30,073.8<br>±20.1 | 62,161.0<br>-- |
| Pennsylvania | 1982 | 719.9<br>-- | 466.8<br>-- | 2,763.8<br>±103.4 | 25,044.7<br>±109.8 | 27,808.5<br>±19.1 | 28,995.2<br>-- |
| | 1987 | 721.2<br>-- | 468.6<br>-- | 2,944.5<br>±111.6 | 24,860.9<br>±119.5 | 27,805.4<br>±19.4 | 28,995.2<br>-- |
| | 1992 | 723.9<br>-- | 471.5<br>-- | 3,375.3<br>±130.0 | 24,424.5<br>±136.1 | 27,799.8<br>±19.1 | 28,995.2<br>-- |
| | 1997 | 723.9<br>-- | 472.9<br>-- | 3,911.8<br>±145.3 | 23,886.6<br>±150.1 | 27,798.4<br>±18.2 | 28,995.2<br>-- |
| | 2002 | 724.3<br>-- | 477.1<br>-- | 4,173.6<br>±160.0 | 23,620.2<br>±163.8 | 27,793.8<br>±18.3 | 28,995.2<br>-- |
| | 2007 | 724.3<br>-- | 480.8<br>-- | 4,360.7<br>±173.3 | 23,429.4<br>±177.4 | 27,790.1<br>±19.1 | 28,995.2<br>-- |

| State | Year | Federal land | Water areas | Non-Federal land | | | Total surface area |
|---|---|---|---|---|---|---|---|
| | | | | Developed | Rural | Total | |
| Rhode Island | 1982 | 6.0<br>-- | 151.3<br>-- | 171.6<br>±12.4 | 484.4<br>±12.5 | 656.0<br>±2.3 | 813.3<br>-- |
| | 1987 | 6.2<br>-- | 151.1<br>-- | 181.4<br>±13.7 | 474.6<br>±14.0 | 656.0<br>±2.2 | 813.3<br>-- |
| | 1992 | 6.8<br>-- | 151.4<br>-- | 198.5<br>±13.8 | 456.6<br>±14.0 | 655.1<br>±2.2 | 813.3<br>-- |
| | 1997 | 3.5<br>-- | 151.3<br>-- | 205.1<br>±14.2 | 453.4<br>±14.4 | 658.5<br>±2.3 | 813.3<br>-- |
| | 2002 | 3.5<br>-- | 151.3<br>-- | 221.7<br>±15.4 | 436.8<br>±15.6 | 658.5<br>±2.3 | 813.3<br>-- |
| | 2007 | 3.5<br>-- | 151.3<br>-- | 232.2<br>±17.0 | 426.3<br>±17.4 | 658.5<br>±2.3 | 813.3<br>-- |
| South Carolina | 1982 | 1,032.2<br>-- | 773.8<br>-- | 1,359.2<br>±77.8 | 16,774.1<br>±77.1 | 18,133.3<br>±18.5 | 19,939.3<br>-- |
| | 1987 | 1,029.2<br>-- | 784.0<br>-- | 1,525.1<br>±91.3 | 16,601.0<br>±90.5 | 18,126.1<br>±18.3 | 19,939.3<br>-- |
| | 1992 | 1,036.2<br>-- | 787.0<br>-- | 1,749.6<br>±101.6 | 16,366.5<br>±101.0 | 18,116.1<br>±18.8 | 19,939.3<br>-- |
| | 1997 | 1,036.2<br>-- | 791.3<br>-- | 2,117.5<br>±111.5 | 15,994.3<br>±112.6 | 18,111.8<br>±19.6 | 19,939.3<br>-- |
| | 2002 | 1,036.2<br>-- | 803.4<br>-- | 2,447.1<br>±102.0 | 15,652.6<br>±105.1 | 18,099.7<br>±18.6 | 19,939.3<br>-- |
| | 2007 | 1,036.2<br>-- | 811.6<br>-- | 2,672.6<br>±112.9 | 15,418.9<br>±114.6 | 18,091.5<br>±19.3 | 19,939.3<br>-- |
| South Dakota | 1982 | 3,029.5<br>-- | 865.5<br>-- | 811.6<br>±63.2 | 44,651.4<br>±64.1 | 45,463.0<br>±13.7 | 49,358.0<br>-- |
| | 1987 | 3,065.0<br>-- | 872.7<br>-- | 817.6<br>±64.3 | 44,602.7<br>±65.5 | 45,420.3<br>±13.6 | 49,358.0<br>-- |
| | 1992 | 3,107.9<br>-- | 875.0<br>-- | 870.6<br>±89.7 | 44,504.5<br>±90.7 | 45,375.1<br>±14.0 | 49,358.0<br>-- |
| | 1997 | 3,107.9<br>-- | 879.6<br>-- | 921.6<br>±99.3 | 44,448.9<br>±100.7 | 45,370.5<br>±14.0 | 49,358.0<br>-- |
| | 2002 | 3,112.5<br>-- | 879.0<br>-- | 939.2<br>±103.5 | 44,427.3<br>±104.7 | 45,366.5<br>±14.2 | 49,358.0<br>-- |
| | 2007 | 3,112.2<br>-- | 879.4<br>-- | 962.8<br>±110.4 | 44,403.6<br>±112.4 | 45,366.4<br>±14.6 | 49,358.0<br>-- |
| Tennessee | 1982 | 1,212.7<br>-- | 758.9<br>-- | 1,640.0<br>±98.7 | 23,362.0<br>±97.0 | 25,002.0<br>±11.5 | 26,973.6<br>-- |
| | 1987 | 1,233.7<br>-- | 760.8<br>-- | 1,875.1<br>±105.7 | 23,104.0<br>±105.2 | 24,979.1<br>±11.0 | 26,973.6<br>-- |
| | 1992 | 1,232.2<br>-- | 769.0<br>-- | 2,157.9<br>±111.3 | 22,814.5<br>±110.1 | 24,972.4<br>±10.7 | 26,973.6<br>-- |
| | 1997 | 1,232.2<br>-- | 773.9<br>-- | 2,606.3<br>±129.8 | 22,361.2<br>±130.0 | 24,967.5<br>±10.9 | 26,973.6<br>-- |
| | 2002 | 1,302.6<br>-- | 784.3<br>-- | 2,811.6<br>±137.3 | 22,075.1<br>±136.6 | 24,886.7<br>±9.6 | 26,973.6<br>-- |
| | 2007 | 1,302.6<br>-- | 790.8<br>-- | 3,038.3<br>±158.7 | 21,841.9<br>±159.2 | 24,880.2<br>±10.8 | 26,973.6<br>-- |

| State | Year | Federal land | Water areas | Non-Federal land | | | Total surface area |
|---|---|---|---|---|---|---|---|
| | | | | Developed | Rural | Total | |
| Texas | 1982 | 2,769.2<br>-- | 3,691.4<br>-- | 5,073.1<br>±180.7 | 159,518.2<br>±185.5 | 164,591.3<br>±43.3 | 171,051.9<br>-- |
| | 1987 | 2,813.9<br>-- | 3,836.4<br>-- | 5,572.4<br>±196.9 | 158,829.2<br>±200.5 | 164,401.6<br>±46.9 | 171,051.9<br>-- |
| | 1992 | 2,909.9<br>-- | 3,960.9<br>-- | 6,102.0<br>±234.2 | 158,079.1<br>±237.5 | 164,181.1<br>±48.4 | 171,051.9<br>-- |
| | 1997 | 2,909.9<br>-- | 4,041.9<br>-- | 6,770.0<br>±257.1 | 157,330.1<br>±264.6 | 164,100.1<br>±46.2 | 171,051.9<br>-- |
| | 2002 | 2,909.9<br>-- | 4,080.7<br>-- | 7,710.9<br>±291.0 | 156,350.4<br>±294.0 | 164,061.3<br>±47.5 | 171,051.9<br>-- |
| | 2007 | 2,909.9<br>-- | 4,126.1<br>-- | 8,515.7<br>±345.2 | 155,500.2<br>±352.1 | 164,015.9<br>±56.7 | 171,051.9<br>-- |
| Utah | 1982 | 34,508.1<br>-- | 1,767.3<br>-- | 433.7<br>±89.4 | 17,629.8<br>±107.1 | 18,063.5<br>±30.4 | 54,338.9<br>-- |
| | 1987 | 34,153.3<br>-- | 2,356.2<br>-- | 474.6<br>±99.0 | 17,354.8<br>±115.9 | 17,829.4<br>±30.2 | 54,338.9<br>-- |
| | 1992 | 34,278.2<br>-- | 1,786.3<br>-- | 530.5<br>±109.4 | 17,743.9<br>±124.2 | 18,274.4<br>±30.3 | 54,338.9<br>-- |
| | 1997 | 34,278.2<br>-- | 1,800.5<br>-- | 601.1<br>±118.9 | 17,659.1<br>±132.7 | 18,260.2<br>±30.4 | 54,338.9<br>-- |
| | 2002 | 34,278.8<br>-- | 1,800.5<br>-- | 682.7<br>±133.1 | 17,576.9<br>±148.0 | 18,259.6<br>±21.8 | 54,338.9<br>-- |
| | 2007 | 34,278.8<br>-- | 1,800.7<br>-- | 744.6<br>±139.6 | 17,514.8<br>±153.4 | 18,259.4<br>±21.7 | 54,338.9<br>-- |
| Vermont | 1982 | 321.9<br>-- | 261.2<br>-- | 261.9<br>±23.4 | 5,308.6<br>±27.0 | 5,570.5<br>±7.7 | 6,153.6<br>-- |
| | 1987 | 353.0<br>-- | 261.7<br>-- | 303.8<br>±25.0 | 5,235.1<br>±27.8 | 5,538.9<br>±7.7 | 6,153.6<br>-- |
| | 1992 | 370.9<br>-- | 262.1<br>-- | 332.2<br>±26.6 | 5,188.4<br>±28.7 | 5,520.6<br>±7.6 | 6,153.6<br>-- |
| | 1997 | 392.4<br>-- | 261.2<br>-- | 345.4<br>±27.3 | 5,154.6<br>±29.7 | 5,500.0<br>±7.6 | 6,153.6<br>-- |
| | 2002 | 422.6<br>-- | 261.8<br>-- | 372.9<br>±28.8 | 5,096.3<br>±29.8 | 5,469.2<br>±5.1 | 6,153.6<br>-- |
| | 2007 | 422.6<br>-- | 261.9<br>-- | 393.2<br>±30.8 | 5,075.9<br>±32.2 | 5,469.1<br>±5.6 | 6,153.6<br>-- |
| Virginia | 1982 | 2,608.3<br>-- | 1,917.8<br>-- | 1,841.9<br>±108.0 | 20,719.1<br>±107.6 | 22,561.0<br>±13.6 | 27,087.1<br>-- |
| | 1987 | 2,626.2<br>-- | 1,920.2<br>-- | 2,082.7<br>±107.4 | 20,458.0<br>±106.6 | 22,540.7<br>±13.8 | 27,087.1<br>-- |
| | 1992 | 2,646.4<br>-- | 1,927.9<br>-- | 2,285.3<br>±115.5 | 20,227.5<br>±114.5 | 22,512.8<br>±14.8 | 27,087.1<br>-- |
| | 1997 | 2,646.4<br>-- | 1,929.1<br>-- | 2,627.7<br>±111.9 | 19,883.9<br>±111.5 | 22,511.6<br>±15.6 | 27,087.1<br>-- |
| | 2002 | 2,646.4<br>-- | 1,934.4<br>-- | 2,901.3<br>±113.6 | 19,605.0<br>±115.1 | 22,506.3<br>±15.5 | 27,087.1<br>-- |
| | 2007 | 2,646.4<br>-- | 1,943.1<br>-- | 3,101.2<br>±127.5 | 19,396.4<br>±130.0 | 22,497.6<br>±17.4 | 27,087.1<br>-- |

| State | Year | Federal land | Water areas | Non-Federal land | | | Total surface area |
|---|---|---|---|---|---|---|---|
| | | | | Developed | Rural | Total | |
| Washington | 1982 | 11,897.9<br>-- | 1,536.9<br>-- | 1,594.7<br>±147.8 | 29,005.8<br>±154.7 | 30,600.5<br>±21.4 | 44,035.3<br>-- |
| | 1987 | 11,917.9<br>-- | 1,537.7<br>-- | 1,675.3<br>±148.3 | 28,904.4<br>±155.2 | 30,579.7<br>±21.2 | 44,035.3<br>-- |
| | 1992 | 11,921.9<br>-- | 1,537.2<br>-- | 1,897.2<br>±162.6 | 28,679.0<br>±168.7 | 30,576.2<br>±21.5 | 44,035.3<br>-- |
| | 1997 | 11,923.4<br>-- | 1,536.5<br>-- | 2,150.5<br>±176.4 | 28,424.9<br>±181.7 | 30,575.4<br>±20.0 | 44,035.3<br>-- |
| | 2002 | 11,923.5<br>-- | 1,541.7<br>-- | 2,357.7<br>±186.6 | 28,212.4<br>±190.9 | 30,570.1<br>±22.8 | 44,035.3<br>-- |
| | 2007 | 11,923.5<br>-- | 1,544.0<br>-- | 2,464.5<br>±192.7 | 28,103.3<br>±197.8 | 30,567.8<br>±23.0 | 44,035.3<br>-- |
| West Virginia | 1982 | 1,107.3<br>-- | 163.7<br>-- | 633.2<br>±44.9 | 13,604.0<br>±47.6 | 14,237.2<br>±8.5 | 15,508.2<br>-- |
| | 1987 | 1,120.5<br>-- | 164.4<br>-- | 674.8<br>±46.3 | 13,548.5<br>±48.9 | 14,223.3<br>±8.5 | 15,508.2<br>-- |
| | 1992 | 1,210.4<br>-- | 168.1<br>-- | 759.4<br>±51.5 | 13,370.3<br>±53.7 | 14,129.7<br>±8.1 | 15,508.2<br>-- |
| | 1997 | 1,211.4<br>-- | 171.4<br>-- | 958.7<br>±59.6 | 13,166.7<br>±62.3 | 14,125.4<br>±9.1 | 15,508.2<br>-- |
| | 2002 | 1,211.9<br>-- | 178.1<br>-- | 1,079.5<br>±58.3 | 13,038.7<br>±60.6 | 14,118.2<br>±11.4 | 15,508.2<br>-- |
| | 2007 | 1,211.9<br>-- | 180.6<br>-- | 1,151.6<br>±57.3 | 12,964.1<br>±58.5 | 14,115.7<br>±11.9 | 15,508.2<br>-- |
| Wisconsin | 1982 | 1,819.6<br>-- | 1,289.1<br>-- | 1,974.2<br>±103.2 | 30,837.1<br>±112.2 | 32,811.3<br>±28.3 | 35,920.0<br>-- |
| | 1987 | 1,826.5<br>-- | 1,290.2<br>-- | 2,087.4<br>±108.6 | 30,715.9<br>±117.8 | 32,803.3<br>±29.5 | 35,920.0<br>-- |
| | 1992 | 1,845.3<br>-- | 1,288.5<br>-- | 2,213.5<br>±114.2 | 30,572.7<br>±123.1 | 32,786.2<br>±28.1 | 35,920.0<br>-- |
| | 1997 | 1,845.3<br>-- | 1,283.1<br>-- | 2,400.2<br>±122.1 | 30,391.4<br>±130.2 | 32,791.6<br>±29.2 | 35,920.0<br>-- |
| | 2002 | 1,845.3<br>-- | 1,287.7<br>-- | 2,555.8<br>±142.6 | 30,231.2<br>±151.1 | 32,787.0<br>±30.3 | 35,920.0<br>-- |
| | 2007 | 1,845.3<br>-- | 1,292.2<br>-- | 2,724.9<br>±172.3 | 30,057.6<br>±180.7 | 32,782.5<br>±29.9 | 35,920.0<br>-- |
| Wyoming | 1982 | 28,700.7<br>-- | 436.2<br>-- | 535.8<br>±67.8 | 32,930.1<br>±66.7 | 33,465.9<br>±14.7 | 62,602.8<br>-- |
| | 1987 | 28,700.3<br>-- | 438.1<br>-- | 578.8<br>±68.7 | 32,885.6<br>±70.2 | 33,464.4<br>±15.6 | 62,602.8<br>-- |
| | 1992 | 28,748.0<br>-- | 438.3<br>-- | 590.3<br>±71.2 | 32,826.2<br>±73.3 | 33,416.5<br>±16.0 | 62,602.8<br>-- |
| | 1997 | 28,748.0<br>-- | 438.4<br>-- | 621.5<br>±73.7 | 32,794.9<br>±76.8 | 33,416.4<br>±16.0 | 62,602.8<br>-- |
| | 2002 | 28,748.0<br>-- | 440.2<br>-- | 640.5<br>±78.5 | 32,774.1<br>±82.1 | 33,414.6<br>±16.0 | 62,602.8<br>-- |
| | 2007 | 28,748.0<br>-- | 441.1<br>-- | 681.1<br>±92.7 | 32,732.6<br>±98.0 | 33,413.7<br>±16.3 | 62,602.8<br>-- |

| State | Year | Federal land | Water areas | Non-Federal land | | | Total surface area |
|---|---|---|---|---|---|---|---|
| | | | | Developed | Rural | Total | |
| Total | 1982 | 399,076.8<br>-- | 48,657.9<br>-- | 70,964.1<br>±768.3 | 1,418,965.4<br>±805.0 | 1,489,929.5<br>±163.0 | 1,937,664.2<br>-- |
| | 1987 | 399,419.5<br>-- | 49,837.4<br>-- | 76,871.0<br>±837.6 | 1,411,536.3<br>±871.0 | 1,488,407.3<br>±161.8 | 1,937,664.2<br>-- |
| | 1992 | 401,517.0<br>-- | 49,414.1<br>-- | 83,902.3<br>±960.8 | 1,402,830.8<br>±976.9 | 1,486,733.1<br>±154.6 | 1,937,664.2<br>-- |
| | 1997 | 401,685.7<br>-- | 49,902.8<br>-- | 94,578.9<br>±993.5 | 1,391,496.8<br>±1,022.8 | 1,486,075.7<br>±159.0 | 1,937,664.2<br>-- |
| | 2002 | 401,937.4<br>-- | 50,426.2<br>-- | 104,030.8<br>±1,274.8 | 1,381,269.8<br>±1,300.0 | 1,485,300.6<br>±174.5 | 1,937,664.2<br>-- |
| | 2007 | 401,936.9<br>-- | 50,817.3<br>-- | 111,251.2<br>±1,499.4 | 1,373,658.8<br>±1,500.2 | 1,484,910.0<br>±252.4 | 1,937,664.2<br>-- |

Notes:
• Acreages for Federal land, water areas, and total surface area are established through geospatial processes and administrative records; therefore, statistical margins of error are not applicable and shown as a dashed line (--).

U.S. Department of Agriculture. 2009. *Summary Report: 2007 National Resources Inventory*, Natural Resources Conservation Service, Washington, DC, and Center for Survey Statistics and Methodology, Iowa State University, Ames, Iowa. 123 pages.

**Land Cover/use of non-Federal rural land, by State and year**
In thousands of acres, with margins of error

| State | Year | Cropland | CRP land | Pastureland | Rangeland | Forest land | Other rural land | Total rural land |
|---|---|---|---|---|---|---|---|---|
| Alabama | 1982 | 4,472.2 ±176.4 | -- | 3,826.0 ±195.8 | 73.6 ±54.3 | 20,828.4 ±183.0 | 484.0 ±65.6 | 29,684.2 ±109.7 |
| | 1987 | 3,957.9 ±189.1 | 206.1 – | 3,643.9 ±166.9 | 72.6 ±52.2 | 21,115.4 ±178.4 | 478.6 ±60.6 | 29,474.5 ±115.8 |
| | 1992 | 3,122.8 ±188.9 | 534.4 – | 3,746.2 ±160.4 | 72.5 ±52.2 | 21,205.8 ±185.2 | 622.7 ±74.7 | 29,304.4 ±125.2 |
| | 1997 | 2,921.9 ±202.4 | 521.9 – | 3,549.5 ±145.3 | 73.5 ±53.2 | 21,281.5 ±199.5 | 588.2 ±67.7 | 28,936.5 ±139.2 |
| | 2002 | 2,475.2 ±215.0 | 467.8 – | 3,459.1 ±266.0 | *73.3 ±147.7* | 21,546.7 ±267.4 | 452.2 ±111.6 | 28,474.3 ±213.9 |
| | 2007 | 2,221.9 ±219.2 | 456.6 – | 3,464.2 ±297.5 | *73.3 ±73.3* | 21,529.4 ±389.2 | 447.5 ±130.2 | 28,192.9 ±214.9 |
| Arizona | 1982 | 1,247.3 ±130.1 | -- | 86.1 ±48.7 | 32,421.9 ±1,136.7 | 4,405.0 ±963.0 | 2,597.7 ±581.8 | 40,758.0 ±270.9 |
| | 1987 | 1,244.8 ±123.8 | 0.0 – | 75.9 ±38.4 | 32,413.3 ±1,138.7 | 4,383.2 ±966.8 | 2,687.4 ±588.9 | 40,804.6 ±286.7 |
| | 1992 | 1,203.9 ±124.1 | 0.0 – | 82.0 ±34.5 | 32,799.6 ±1,157.1 | 4,230.2 ±985.4 | 2,754.1 ±568.0 | 41,069.8 ±294.2 |
| | 1997 | 1,197.5 ±122.8 | 0.0 – | 72.1 ±33.4 | 32,605.5 ±1,202.6 | 4,159.5 ±998.7 | 2,926.7 ±563.1 | 40,961.3 ±306.7 |
| | 2002 | 846.5 ±196.5 | 0.0 – | 78.3 ±52.7 | 32,608.4 ±1,166.4 | 4,114.7 ±983.2 | 2,947.3 ±558.7 | 40,595.2 ±389.7 |
| | 2007 | 753.4 ±196.3 | 0.0 – | 90.8 ±72.7 | 32,497.2 ±1,206.5 | 4,094.8 ±982.4 | 2,906.1 ±535.3 | 40,342.3 ±460.4 |
| Arkansas | 1982 | 8,063.2 ±442.5 | -- | 5,634.5 ±326.0 | *41.5 ±57.2* | 14,887.0 ±454.7 | 318.0 ±44.7 | 28,944.2 ±104.5 |
| | 1987 | 7,969.5 ±434.3 | 99.4 – | 5,569.8 ±321.8 | *41.5 ±57.2* | 14,851.6 ±458.3 | 330.3 ±42.3 | 28,862.1 ±108.2 |
| | 1992 | 7,724.5 ±436.3 | 233.9 – | 5,505.7 ±324.1 | *37.6 ±52.1* | 14,892.1 ±457.6 | 342.7 ±43.7 | 28,736.5 ±112.6 |
| | 1997 | 7,608.7 ±429.8 | 230.2 – | 5,301.0 ±295.4 | *37.6 ±52.1* | 14,977.3 ±464.3 | 366.0 ±52.4 | 28,520.8 ±116.8 |
| | 2002 | 7,520.9 ±468.8 | 148.1 – | 5,278.1 ±332.3 | *37.6 ±52.2* | 14,985.1 ±462.0 | 378.1 ±59.6 | 28,347.9 ±132.0 |
| | 2007 | 7,379.5 ±489.2 | 156.3 – | 5,167.5 ±420.2 | *37.6 ±37.6* | 15,095.9 ±426.1 | 384.5 ±85.1 | 28,221.3 ±142.0 |
| California | 1982 | 10,430.7 ±548.9 | -- | 1,356.6 ±224.2 | 19,132.1 ±993.6 | 14,853.8 ±791.3 | 3,790.5 ±651.1 | 49,563.7 ±341.5 |
| | 1987 | 10,174.3 ±592.5 | 117.5 – | 1,447.0 ±251.6 | 18,842.6 ±1,039.4 | 14,866.6 ±818.8 | 3,822.1 ±659.1 | 49,270.1 ±357.0 |
| | 1992 | 10,066.0 ±575.8 | 180.7 – | 1,115.1 ±237.9 | 18,274.6 ±976.3 | 14,625.1 ±825.0 | 3,909.3 ±655.1 | 48,170.8 ±368.0 |
| | 1997 | 9,659.1 ±609.8 | 173.0 – | 1,071.7 ±234.0 | 18,284.8 ±1,031.1 | 14,428.1 ±817.3 | 4,018.0 ±657.5 | 47,634.7 ±368.1 |
| | 2002 | 9,572.6 ±678.0 | 174.3 – | 1,156.4 ±252.6 | 17,757.3 ±1,015.0 | 14,401.9 ±692.8 | 4,133.9 ±603.9 | 47,196.4 ±387.3 |
| | 2007 | 9,489.4 ±662.5 | 174.3 – | 1,119.5 ±272.0 | 17,531.9 ±1,129.1 | 14,389.9 ±801.0 | 4,117.5 ±609.1 | 46,822.5 ±401.9 |

| State | Year | Cropland | CRP land | Pastureland | Rangeland | Forest land | Other rural land | Total rural land |
|---|---|---|---|---|---|---|---|---|
| Colorado | 1982 | 10,626.9 ±632.5 | -- | 1,062.3 ±142.0 | 25,368.2 ±791.8 | 3,629.4 ±436.7 | 816.2 ±154.9 | 41,503.0 ±108.9 |
| | 1987 | 9,720.9 ±651.1 | 1,112.5 -- | 1,053.4 ±142.5 | 24,906.0 ±777.5 | 3,575.8 ±427.9 | 856.4 ±168.4 | 41,225.0 ±138.3 |
| | 1992 | 8,877.2 ±606.3 | 1,913.5 -- | 1,098.2 ±152.6 | 24,845.3 ±787.8 | 3,391.3 ±442.3 | 895.6 ±165.3 | 41,021.1 ±147.1 |
| | 1997 | 8,817.7 ±600.0 | 1,890.2 -- | 1,133.3 ±146.7 | 24,833.3 ±772.6 | 3,350.2 ±441.9 | 899.0 ±163.4 | 40,923.7 ±163.9 |
| | 2002 | 8,162.7 ±610.9 | 2,203.9 -- | 947.2 ±191.1 | 25,245.2 ±787.1 | 3,253.3 ±430.1 | 905.7 ±174.5 | 40,718.0 ±203.4 |
| | 2007 | 7,609.4 ±645.2 | 2,446.9 -- | 1,032.9 ±203.9 | 25,275.8 ±788.0 | 3,243.8 ±636.4 | 952.4 ±188.5 | 40,561.2 ±223.1 |
| Connecticut | 1982 | 235.3 ±39.4 | -- | 117.4 ±23.6 | 0.0 -- | 1,776.2 ±57.3 | 106.1 ±22.2 | 2,235.0 ±34.9 |
| | 1987 | 222.7 ±34.9 | 0.0 -- | 116.1 ±26.3 | 0.0 -- | 1,735.1 ±55.0 | 105.3 ±22.1 | 2,179.2 ±35.4 |
| | 1992 | 216.4 ±34.5 | 0.0 -- | 110.7 ±21.6 | 0.0 -- | 1,707.4 ±54.0 | 101.5 ±20.0 | 2,136.0 ±36.0 |
| | 1997 | 198.5 ±32.7 | 0.0 -- | 107.9 ±21.2 | 0.0 -- | 1,689.8 ±52.6 | 97.0 ±18.8 | 2,093.2 ±34.0 |
| | 2002 | 177.3 ±32.4 | 0.0 -- | 109.5 ±23.6 | 0.0 -- | 1,650.9 ±54.4 | 98.1 ±18.4 | 2,035.8 ±39.4 |
| | 2007 | 172.0 ±33.5 | 0.0 -- | 105.2 ±24.9 | 0.0 -- | 1,620.4 ±55.5 | 102.4 ±18.6 | 2,000.0 ±42.1 |
| Delaware | 1982 | 525.6 ±38.3 | -- | 34.5 ±10.5 | 0.0 -- | 372.3 ±43.6 | 123.2 ±25.3 | 1,055.6 ±27.0 |
| | 1987 | 515.3 ±37.8 | 0.0 -- | 32.3 ±10.0 | 0.0 -- | 368.7 ±45.2 | 122.5 ±25.1 | 1,038.8 ±29.5 |
| | 1992 | 506.8 ±38.1 | 0.9 -- | 27.2 ±8.4 | 0.0 -- | 362.3 ±43.8 | 124.4 ±24.6 | 1,021.6 ±30.1 |
| | 1997 | 489.9 ±35.6 | 0.9 -- | 24.2 ±8.1 | 0.0 -- | 356.1 ±41.2 | 128.3 ±24.6 | 999.4 ±31.7 |
| | 2002 | 463.8 ±35.6 | 0.0 -- | 29.7 ±8.4 | 0.0 -- | 345.6 ±40.6 | 123.0 ±26.7 | 962.1 ±32.3 |
| | 2007 | 420.5 ±41.4 | 0.0 -- | 37.4 ±16.4 | 0.0 -- | 339.4 ±40.3 | 134.9 ±29.4 | 932.2 ±35.7 |
| Florida | 1982 | 3,622.1 ±235.5 | -- | 4,373.1 ±252.6 | 4,388.3 ±322.6 | 13,254.2 ±352.2 | 2,451.9 ±296.9 | 28,089.6 ±241.1 |
| | 1987 | 3,236.5 ±247.4 | 92.4 -- | 4,678.2 ±280.6 | 4,060.2 ±333.3 | 13,220.6 ±368.6 | 2,458.9 ±301.6 | 27,746.8 ±247.3 |
| | 1992 | 3,078.9 ±263.5 | 122.5 -- | 4,592.2 ±309.5 | 3,524.2 ±294.8 | 13,205.8 ±375.5 | 2,472.6 ±313.3 | 26,996.2 ±273.5 |
| | 1997 | 2,813.0 ±273.4 | 119.8 -- | 4,391.9 ±331.4 | 3,211.9 ±321.4 | 13,124.2 ±392.3 | 2,649.1 ±340.5 | 26,309.9 ±304.3 |
| | 2002 | 2,961.9 ±333.3 | 88.2 -- | 3,908.8 ±353.5 | 2,881.3 ±333.1 | 13,258.2 ±372.2 | 2,609.2 ±317.7 | 25,707.6 ±342.9 |
| | 2007 | 2,880.4 ±459.0 | 82.8 -- | 3,633.1 ±508.0 | 2,636.0 ±360.5 | 13,169.7 ±454.3 | 2,698.7 ±400.2 | 25,100.7 ±422.2 |

| State | Year | Cropland | CRP land | Pastureland | Rangeland | Forest land | Other rural land | Total rural land |
|---|---|---|---|---|---|---|---|---|
| Georgia | 1982 | 6,580.1 ±210.6 | -- | 2,945.5 ±169.3 | 0.0 -- | 22,027.2 ±380.6 | 910.9 ±118.3 | 32,463.7 ±114.1 |
| | 1987 | 5,973.9 ±207.8 | 307.1 -- | 2,897.7 ±151.0 | 0.0 -- | 22,144.7 ±355.8 | 884.6 ±113.3 | 32,208.0 ±125.9 |
| | 1992 | 5,207.1 ±252.0 | 610.9 -- | 3,040.5 ±162.3 | 0.0 -- | 21,981.9 ±386.4 | 873.4 ±108.0 | 31,713.8 ±138.8 |
| | 1997 | 4,759.9 ±192.8 | 595.3 -- | 2,881.3 ±170.7 | 0.0 -- | 21,805.0 ±373.8 | 860.2 ±122.5 | 30,901.7 ±159.7 |
| | 2002 | 4,412.5 ±224.0 | 316.8 -- | 2,791.8 ±238.3 | 0.0 -- | 21,968.6 ±369.4 | 847.5 ±124.9 | 30,337.2 ±206.9 |
| | 2007 | 3,995.0 ±415.3 | 300.3 -- | 2,809.7 ±339.7 | 0.0 -- | 21,963.9 ±409.2 | 848.4 ±164.9 | 29,917.3 ±227.4 |
| Idaho | 1982 | 6,400.2 ±314.8 | -- | 1,204.2 ±161.4 | 6,691.9 ±363.4 | 3,998.3 ±347.2 | 483.6 ±112.3 | 18,778.2 ±50.6 |
| | 1987 | 6,044.4 ±306.6 | 448.8 -- | 1,210.2 ±144.3 | 6,621.0 ±353.6 | 4,089.6 ±345.3 | 505.0 ±111.0 | 18,919.0 ±54.9 |
| | 1992 | 5,571.1 ±311.3 | 826.7 -- | 1,242.4 ±155.0 | 6,591.8 ±355.0 | 4,028.1 ±349.5 | 515.7 ±113.1 | 18,775.8 ±61.5 |
| | 1997 | 5,488.1 ±315.8 | 784.3 -- | 1,259.4 ±150.6 | 6,556.6 ±363.4 | 3,967.7 ±342.2 | 538.0 ±117.1 | 18,594.1 ±67.8 |
| | 2002 | 5,378.2 ±334.5 | 785.4 -- | 1,276.3 ±168.3 | 6,519.6 ±367.8 | 4,019.8 ±316.0 | 560.5 ±114.4 | 18,539.8 ±67.6 |
| | 2007 | 5,246.1 ±360.1 | 797.2 -- | 1,307.6 ±209.0 | 6,514.3 ±384.1 | 4,015.9 ±333.9 | 578.2 ±122.4 | 18,459.3 ±75.7 |
| Illinois | 1982 | 24,745.6 ±268.1 | -- | 3,203.1 ±223.7 | 0.0 -- | 3,631.8 ±169.5 | 655.7 ±33.7 | 32,236.2 ±120.6 |
| | 1987 | 24,749.8 ±260.8 | 120.4 -- | 2,918.4 ±208.8 | 0.0 -- | 3,653.9 ±170.9 | 660.2 ±37.8 | 32,102.7 ±129.6 |
| | 1992 | 24,138.0 ±251.8 | 710.8 -- | 2,781.1 ±202.8 | 0.0 -- | 3,708.6 ±172.2 | 657.2 ±41.3 | 31,995.7 ±128.3 |
| | 1997 | 24,069.2 ±223.0 | 725.0 -- | 2,522.8 ±191.7 | 0.0 -- | 3,803.5 ±166.5 | 644.0 ±45.7 | 31,764.5 ±126.2 |
| | 2002 | 24,116.6 ±287.6 | 655.9 -- | 2,256.2 ±242.6 | 0.0 -- | 3,924.0 ±164.3 | 663.2 ±66.2 | 31,615.9 ±136.6 |
| | 2007 | 23,910.5 ±285.1 | 664.6 -- | 2,249.5 ±238.6 | 0.0 -- | 3,934.8 ±191.1 | 692.4 ±93.6 | 31,451.8 ±166.1 |
| Indiana | 1982 | 13,834.9 ±189.1 | -- | 2,203.5 ±120.2 | 0.0 -- | 3,798.5 ±139.8 | 721.3 ±69.3 | 20,558.2 ±94.5 |
| | 1987 | 13,899.8 ±197.3 | 144.3 -- | 1,917.8 ±116.1 | 0.0 -- | 3,810.9 ±139.4 | 662.7 ±69.0 | 20,435.5 ±95.3 |
| | 1992 | 13,577.2 ±204.6 | 415.1 -- | 1,844.1 ±112.3 | 0.0 -- | 3,816.3 ±143.2 | 673.6 ±75.7 | 20,326.3 ±102.0 |
| | 1997 | 13,476.8 ±221.7 | 378.6 -- | 1,835.0 ±108.6 | 0.0 -- | 3,802.8 ±144.0 | 647.6 ±65.0 | 20,140.8 ±113.8 |
| | 2002 | 13,380.1 ±226.6 | 239.7 -- | 1,887.7 ±168.3 | 0.0 -- | 3,833.0 ±151.7 | 650.9 ±94.0 | 19,991.4 ±115.7 |
| | 2007 | 13,219.9 ±309.2 | 213.1 -- | 1,926.1 ±197.3 | 0.0 -- | 3,829.2 ±184.7 | 678.6 ±128.8 | 19,866.9 ±125.1 |

| State | Year | Cropland | CRP land | Pastureland | Rangeland | Forest land | Other rural land | Total rural land |
|---|---|---|---|---|---|---|---|---|
| Iowa | 1982 | 26,385.2 ±243.4 | -- | 4,549.9 ±179.4 | 0.0 -- | 1,919.0 ±146.4 | 929.3 ±64.0 | 33,783.4 ±92.8 |
| | 1987 | 25,651.8 ±231.5 | 1,235.2 -- | 3,992.8 ±186.7 | 0.0 -- | 1,987.4 ±145.9 | 884.2 ±63.6 | 33,751.4 ±100.9 |
| | 1992 | 24,923.7 ±225.7 | 2,087.6 -- | 3,711.7 ±193.2 | 0.0 -- | 2,122.7 ±155.3 | 866.7 ±62.5 | 33,712.4 ±102.1 |
| | 1997 | 25,246.3 ±229.0 | 1,738.4 -- | 3,530.9 ±180.4 | 0.0 -- | 2,223.5 ±160.1 | 869.3 ±61.6 | 33,608.4 ±103.8 |
| | 2002 | 25,396.0 ±237.5 | 1,508.1 -- | 3,424.6 ±201.2 | 0.0 -- | 2,329.2 ±148.5 | 867.8 ±76.0 | 33,525.7 ±103.5 |
| | 2007 | 25,446.2 ±250.0 | 1,427.5 -- | 3,304.8 ±237.6 | 0.0 -- | 2,354.7 ±136.8 | 932.7 ±114.2 | 33,465.9 ±112.0 |
| Kansas | 1982 | 29,097.4 ±393.9 | -- | 2,113.5 ±146.9 | 16,528.2 ±340.6 | 1,491.0 ±100.8 | 685.4 ±65.2 | 49,915.5 ±73.3 |
| | 1987 | 28,467.8 ±391.2 | 637.7 -- | 2,162.3 ±142.0 | 16,443.4 ±336.8 | 1,495.0 ±102.2 | 680.9 ±67.0 | 49,887.1 ±73.8 |
| | 1992 | 26,483.5 ±404.8 | 2,875.8 -- | 2,322.2 ±144.1 | 15,831.5 ±337.5 | 1,575.8 ±105.9 | 688.7 ±60.6 | 49,777.5 ±75.0 |
| | 1997 | 26,490.2 ±412.3 | 2,848.8 -- | 2,317.2 ±142.9 | 15,730.6 ±348.9 | 1,584.5 ±102.3 | 696.0 ±63.5 | 49,667.3 ±77.0 |
| | 2002 | 26,397.3 ±432.7 | 2,663.7 -- | 2,383.0 ±209.0 | 15,810.0 ±412.9 | 1,609.6 ±107.3 | 721.1 ±84.2 | 49,584.7 ±82.2 |
| | 2007 | 25,635.6 ±525.1 | 3,164.9 -- | 2,497.6 ±348.9 | 15,787.5 ±553.8 | 1,685.5 ±185.2 | 735.6 ±89.7 | 49,506.7 ±96.8 |
| Kentucky | 1982 | 5,907.9 ±157.2 | -- | 5,954.4 ±160.8 | 0.0 -- | 10,508.9 ±187.5 | 675.7 ±62.1 | 23,046.9 ±73.2 |
| | 1987 | 5,428.3 ±184.8 | 205.2 -- | 5,920.3 ±163.9 | 0.0 -- | 10,558.1 ±203.4 | 700.9 ±66.8 | 22,812.8 ±84.6 |
| | 1992 | 5,094.9 ±165.0 | 421.0 -- | 5,894.7 ±140.2 | 0.0 -- | 10,626.1 ±208.5 | 564.0 ±55.9 | 22,600.7 ±89.8 |
| | 1997 | 5,186.0 ±151.9 | 332.2 -- | 5,679.3 ±124.3 | 0.0 -- | 10,701.1 ±211.0 | 461.5 ±40.8 | 22,360.1 ±99.8 |
| | 2002 | 5,307.5 ±327.3 | 273.5 -- | 5,214.4 ±293.1 | 0.0 -- | 10,657.2 ±208.3 | 537.2 ±73.2 | 21,989.8 ±101.5 |
| | 2007 | 5,173.0 ±394.9 | 285.0 -- | 5,242.1 ±303.1 | 0.0 -- | 10,590.9 ±246.5 | 553.0 ±144.9 | 21,844.0 ±110.7 |
| Louisiana | 1982 | 6,431.2 ±193.5 | -- | 2,343.3 ±160.5 | 214.2 ±42.3 | 13,321.2 ±242.2 | 2,970.0 ±160.8 | 25,279.9 ±70.7 |
| | 1987 | 6,268.4 ±161.5 | 43.3 -- | 2,338.8 ±163.3 | 210.4 ±40.8 | 13,153.9 ±226.3 | 3,010.5 ±156.0 | 25,025.3 ±68.5 |
| | 1992 | 5,958.4 ±170.9 | 147.9 -- | 2,412.3 ±146.0 | 208.7 ±36.7 | 13,120.9 ±230.5 | 2,996.6 ±148.9 | 24,844.8 ±70.7 |
| | 1997 | 5,616.4 ±156.6 | 140.3 -- | 2,516.6 ±140.4 | 202.2 ±33.7 | 13,218.1 ±220.3 | 3,000.7 ±151.5 | 24,694.3 ±75.1 |
| | 2002 | 5,338.2 ±187.1 | 201.8 -- | 2,423.9 ±151.1 | 224.7 ±62.2 | 13,302.5 ±244.9 | 3,014.6 ±150.0 | 24,505.7 ±84.6 |
| | 2007 | 5,107.3 ±250.6 | 226.6 -- | 2,458.1 ±198.8 | 221.2 ±88.4 | 13,306.5 ±337.0 | 2,960.1 ±144.1 | 24,279.8 ±92.0 |

| State | Year | Cropland | CRP land | Pastureland | Rangeland | Forest land | Other rural land | Total rural land |
|---|---|---|---|---|---|---|---|---|
| Maine | 1982 | 518.5 ±102.2 | -- | 283.1 ±47.7 | 0.0 -- | 17,669.6 ±185.5 | 565.7 ±136.9 | 19,036.9 ±74.5 |
| | 1987 | 507.1 ±97.4 | 0.0 -- | 230.3 ±46.9 | 0.0 -- | 17,670.9 ±183.1 | 556.3 ±135.0 | 18,964.6 ±79.7 |
| | 1992 | 447.4 ±85.6 | 35.6 -- | 168.4 ±53.6 | 0.0 -- | 17,683.2 ±186.7 | 582.2 ±119.0 | 18,916.8 ±83.2 |
| | 1997 | 425.0 ±81.4 | 29.7 -- | 139.1 ±38.8 | 0.0 -- | 17,706.0 ±204.1 | 498.8 ±118.3 | 18,798.6 ±90.9 |
| | 2002 | 391.9 ±87.5 | 29.7 -- | 144.8 ±32.1 | 0.0 -- | 17,655.3 ±217.3 | 483.3 ±142.9 | 18,705.0 ±110.3 |
| | 2007 | 372.8 ±92.8 | 29.7 -- | 141.6 ±45.8 | 0.0 -- | 17,632.1 ±211.7 | 475.1 ±150.9 | 18,651.3 ±116.0 |
| Maryland | 1982 | 1,771.8 ±88.3 | -- | 540.7 ±43.6 | 0.0 -- | 2,441.4 ±108.6 | 339.8 ±28.7 | 5,093.7 ±114.8 |
| | 1987 | 1,717.2 ±83.2 | 1.2 -- | 559.9 ±36.3 | 0.0 -- | 2,411.4 ±106.2 | 316.8 ±31.2 | 5,006.5 ±119.9 |
| | 1992 | 1,657.5 ±83.8 | 17.8 -- | 554.5 ±36.2 | 0.0 -- | 2,384.2 ±106.1 | 314.6 ±30.7 | 4,928.6 ±123.2 |
| | 1997 | 1,590.2 ±82.4 | 18.9 -- | 477.7 ±36.5 | 0.0 -- | 2,338.4 ±102.5 | 311.7 ±31.6 | 4,736.9 ±127.2 |
| | 2002 | 1,486.0 ±98.1 | 13.9 -- | 454.7 ±59.2 | 0.0 -- | 2,340.2 ±109.9 | 338.5 ±29.8 | 4,633.3 ±126.6 |
| | 2007 | 1,413.0 ±108.4 | 10.7 -- | 463.4 ±71.4 | 0.0 -- | 2,317.4 ±124.9 | 338.2 ±32.7 | 4,542.7 ±130.2 |
| Massachusetts | 1982 | 285.5 ±50.5 | -- | 187.0 ±38.4 | 0.0 -- | 3,062.0 ±103.8 | 252.8 ±36.4 | 3,787.3 ±65.4 |
| | 1987 | 277.0 ±51.7 | 0.0 -- | 161.7 ±33.7 | 0.0 -- | 2,978.7 ±102.8 | 258.8 ±38.2 | 3,676.2 ±68.3 |
| | 1992 | 258.1 ±52.5 | 0.0 -- | 162.4 ±33.7 | 0.0 -- | 2,866.4 ±101.7 | 251.7 ±37.1 | 3,538.6 ±68.6 |
| | 1997 | 263.0 ±50.8 | 0.0 -- | 129.7 ±27.8 | 0.0 -- | 2,692.0 ±100.5 | 233.0 ±35.9 | 3,317.7 ±71.7 |
| | 2002 | 239.8 ±51.3 | 0.0 -- | 138.7 ±33.5 | 0.0 -- | 2,643.8 ±106.3 | 206.5 ±39.4 | 3,228.8 ±75.2 |
| | 2007 | 237.7 ±63.8 | 0.0 -- | 135.4 ±31.8 | 0.0 -- | 2,589.0 ±110.4 | 196.8 ±50.0 | 3,158.9 ±84.3 |
| Michigan | 1982 | 9,382.1 ±282.3 | -- | 2,938.4 ±182.1 | 0.0 -- | 15,847.7 ±283.3 | 2,047.3 ±132.0 | 30,215.5 ±135.9 |
| | 1987 | 9,246.5 ±253.6 | 56.6 -- | 2,614.6 ±185.2 | 0.0 -- | 16,036.7 ±267.8 | 2,018.5 ±134.3 | 29,972.9 ±147.8 |
| | 1992 | 8,935.1 ±247.7 | 261.4 -- | 2,402.3 ±169.2 | 0.0 -- | 16,051.1 ±260.8 | 2,002.2 ±132.7 | 29,652.1 ±155.1 |
| | 1997 | 8,478.8 ±262.6 | 321.0 -- | 2,054.9 ±137.3 | 0.0 -- | 16,326.5 ±258.6 | 2,090.9 ±137.2 | 29,272.1 ±170.3 |
| | 2002 | 8,118.5 ±256.1 | 258.9 -- | 2,099.1 ±142.2 | 0.0 -- | 16,570.3 ±267.0 | 1,886.4 ±165.6 | 28,933.2 ±170.0 |
| | 2007 | 7,844.1 ±344.5 | 192.0 -- | 2,213.9 ±239.9 | 0.0 -- | 16,568.3 ±284.5 | 1,908.9 ±175.8 | 28,727.2 ±176.2 |

| State | Year | Cropland | CRP land | Pastureland | Rangeland | Forest land | Other rural land | Total rural land |
|---|---|---|---|---|---|---|---|---|
| Minnesota | 1982 | 22,959.2 ±479.0 | -- | 3,822.7 ±255.9 | 0.0 -- | 16,347.2 ±613.2 | 2,713.1 ±207.9 | 45,842.2 ±87.0 |
| | 1987 | 22,342.9 ±475.0 | 778.7 -- | 3,576.4 ±241.4 | 0.0 -- | 16,240.6 ±626.9 | 2,756.4 ±208.3 | 45,695.0 ±91.1 |
| | 1992 | 21,342.7 ±463.4 | 1,810.1 -- | 3,375.4 ±213.4 | 0.0 -- | 16,306.9 ±626.0 | 2,751.8 ±204.2 | 45,586.9 ±97.0 |
| | 1997 | 21,408.8 ±490.0 | 1,543.7 -- | 3,366.9 ±197.1 | 0.0 -- | 16,399.9 ±595.7 | 2,638.0 ±182.3 | 45,357.3 ±106.9 |
| | 2002 | 21,084.1 ±527.9 | 1,436.5 -- | 3,540.5 ±341.4 | 0.0 -- | 16,467.9 ±598.7 | 2,696.4 ±189.4 | 45,225.4 ±115.8 |
| | 2007 | 20,693.9 ±539.3 | 1,453.8 -- | 3,759.8 ±477.1 | 0.0 -- | 16,541.2 ±586.2 | 2,685.0 ±188.3 | 45,133.7 ±124.2 |
| Mississippi | 1982 | 7,379.3 ±264.1 | -- | 3,980.6 ±218.1 | 0.0 -- | 15,351.4 ±309.7 | 334.3 ±38.1 | 27,045.6 ±97.8 |
| | 1987 | 6,638.7 ±252.4 | 292.4 -- | 3,864.5 ±213.6 | 0.0 -- | 15,736.9 ±317.8 | 330.1 ±37.6 | 26,862.6 ±105.0 |
| | 1992 | 5,686.2 ±241.6 | 778.1 -- | 3,937.5 ±212.5 | 0.0 -- | 15,944.0 ±321.3 | 327.7 ±37.0 | 26,673.5 ±105.5 |
| | 1997 | 5,346.7 ±232.9 | 797.6 -- | 3,680.2 ±176.2 | 0.0 -- | 16,227.7 ±301.7 | 372.6 ±47.1 | 26,424.8 ±117.2 |
| | 2002 | 4,925.7 ±266.3 | 805.3 -- | 3,360.1 ±243.4 | 0.0 -- | 16,675.3 ±300.6 | 417.1 ±82.0 | 26,183.5 ±138.2 |
| | 2007 | 4,703.9 ±261.2 | 780.0 -- | 3,249.3 ±305.6 | 0.0 -- | 16,826.8 ±343.7 | 470.7 ±113.4 | 26,030.7 ±142.5 |
| Missouri | 1982 | 14,885.0 ±281.2 | -- | 12,538.4 ±356.4 | 129.0 ±46.3 | 11,557.1 ±298.9 | 667.4 ±49.0 | 39,776.9 ±95.7 |
| | 1987 | 14,321.1 ±281.3 | 567.9 -- | 12,121.8 ±353.4 | 92.2 ±39.5 | 11,916.8 ±303.8 | 650.4 ±45.6 | 39,670.2 ±96.6 |
| | 1992 | 13,212.7 ±307.0 | 1,599.9 -- | 11,842.5 ±310.3 | 87.5 ±39.8 | 12,143.9 ±305.2 | 636.8 ±47.9 | 39,523.3 ±103.9 |
| | 1997 | 13,564.4 ±278.3 | 1,605.9 -- | 10,919.8 ±292.1 | 77.8 ±38.8 | 12,462.9 ±307.7 | 631.8 ±49.1 | 39,262.6 ±115.5 |
| | 2002 | 13,533.6 ±285.7 | 1,486.9 -- | 10,762.0 ±407.2 | 77.7 ±39.3 | 12,548.5 ±368.5 | 660.7 ±80.1 | 39,069.4 ±115.0 |
| | 2007 | 13,285.7 ±339.7 | 1,463.3 -- | 10,950.4 ±474.3 | 83.1 ±39.7 | 12,430.1 ±496.2 | 684.1 ±102.5 | 38,896.7 ±198.9 |
| Montana | 1982 | 17,014.5 ±870.0 | -- | 3,288.5 ±344.9 | 37,851.2 ±981.7 | 5,496.5 ±524.4 | 1,312.6 ±214.1 | 64,963.3 ±98.1 |
| | 1987 | 16,168.1 ±839.5 | 1,488.5 -- | 3,215.2 ±258.2 | 37,368.7 ±1,007.5 | 5,501.9 ±519.7 | 1,307.7 ±211.2 | 65,050.1 ±99.0 |
| | 1992 | 15,027.3 ±870.7 | 2,785.4 -- | 3,403.1 ±325.0 | 37,094.7 ±985.0 | 5,501.9 ±528.5 | 1,291.7 ±212.5 | 65,104.1 ±106.9 |
| | 1997 | 15,139.8 ±837.0 | 2,720.8 -- | 3,523.9 ±363.2 | 36,841.0 ±965.9 | 5,516.6 ±519.0 | 1,314.7 ±213.4 | 65,056.8 ±111.4 |
| | 2002 | 14,480.5 ±852.3 | 3,262.1 -- | 3,659.4 ±440.6 | 36,834.4 ±1,004.4 | 5,496.4 ±518.7 | 1,284.1 ±203.1 | 65,016.9 ±117.2 |
| | 2007 | 13,930.5 ±1,032.0 | 3,315.7 -- | 3,960.1 ±663.5 | 36,953.4 ±991.4 | 5,488.1 ±507.8 | 1,283.8 ±222.0 | 64,931.6 ±132.5 |

| State | Year | Cropland | CRP land | Pastureland | Rangeland | Forest land | Other rural land | Total rural land |
|---|---|---|---|---|---|---|---|---|
| Nebraska | 1982 | 20,232.0 ±446.2 | -- | 1,970.8 ±127.2 | 23,691.1 ±489.9 | 807.1 ±97.4 | 746.6 ±52.0 | 47,447.6 ±109.7 |
| | 1987 | 19,966.3 ±401.6 | 587.4 -- | 1,881.7 ±141.4 | 23,407.7 ±463.9 | 823.8 ±97.9 | 753.0 ±56.2 | 47,419.9 ±113.1 |
| | 1992 | 19,296.3 ±395.6 | 1,365.2 -- | 1,871.7 ±139.3 | 23,213.2 ±463.4 | 826.6 ±99.2 | 761.0 ±58.0 | 47,334.0 ±117.0 |
| | 1997 | 19,499.5 ±405.0 | 1,244.4 -- | 1,785.5 ±122.3 | 23,158.8 ±457.1 | 839.7 ±100.0 | 756.9 ±62.7 | 47,284.8 ±118.7 |
| | 2002 | 19,529.6 ±428.9 | 1,101.6 -- | 1,810.9 ±140.0 | 23,174.6 ±455.9 | 824.7 ±97.6 | 814.2 ±71.4 | 47,255.6 ±117.2 |
| | 2007 | 19,526.2 ±573.0 | 1,198.3 -- | 1,773.5 ±140.4 | 23,107.0 ±557.3 | 823.7 ±100.2 | 800.6 ±88.8 | 47,229.3 ±116.1 |
| Nevada | 1982 | 815.7 ±165.2 | -- | 293.6 ±87.2 | 8,360.9 ±205.8 | 364.6 ±135.6 | 395.3 ±89.8 | 10,230.1 ±65.0 |
| | 1987 | 786.7 ±167.1 | 0.0 -- | 301.9 ±89.5 | 8,404.8 ±209.3 | 373.0 ±135.6 | 408.3 ±96.0 | 10,274.7 ±71.6 |
| | 1992 | 746.8 ±162.8 | 1.4 -- | 288.5 ±88.4 | 8,358.6 ±198.8 | 372.9 ±136.0 | 385.3 ±97.8 | 10,153.5 ±72.0 |
| | 1997 | 686.6 ±156.3 | 2.4 -- | 280.3 ±91.7 | 8,437.5 ±203.6 | 310.6 ±118.8 | 412.6 ±91.3 | 10,130.0 ±74.4 |
| | 2002 | 602.3 ±151.0 | 0.0 -- | 259.5 ±83.7 | 8,357.4 ±217.1 | 312.1 ±119.1 | 438.4 ±100.6 | 9,969.7 ±114.7 |
| | 2007 | 469.8 ±204.3 | 0.0 -- | 248.8 ±94.9 | 8,400.2 ±273.9 | 312.1 ±118.6 | 449.3 ±128.4 | 9,880.2 ±138.5 |
| New Hampshire | 1982 | 160.1 ±26.7 | -- | 130.7 ±29.3 | 0.0 -- | 4,122.7 ±73.7 | 171.5 ±34.4 | 4,585.0 ±50.0 |
| | 1987 | 147.2 ±26.6 | 0.0 -- | 116.7 ±28.5 | 0.0 -- | 4,047.6 ±75.1 | 180.3 ±32.0 | 4,491.8 ±62.3 |
| | 1992 | 140.0 ±25.8 | 0.0 -- | 112.2 ±28.6 | 0.0 -- | 3,976.7 ±78.6 | 182.8 ±34.6 | 4,411.7 ±63.6 |
| | 1997 | 132.6 ±25.5 | 0.0 -- | 107.2 ±29.8 | 0.0 -- | 3,933.1 ±82.9 | 165.5 ±31.5 | 4,338.4 ±68.0 |
| | 2002 | 123.2 ±27.1 | 0.0 -- | 100.1 ±27.8 | 0.0 -- | 3,902.0 ±88.6 | 170.0 ±39.9 | 4,295.3 ±67.9 |
| | 2007 | 109.7 ±27.9 | 0.0 -- | 107.9 ±32.0 | 0.0 -- | 3,879.8 ±85.7 | 148.7 ±37.7 | 4,246.1 ±68.6 |
| New Jersey | 1982 | 818.1 ±59.6 | -- | 225.3 ±36.8 | 0.0 -- | 1,951.5 ±62.1 | 392.1 ±44.0 | 3,387.0 ±68.3 |
| | 1987 | 704.5 ±54.8 | 0.0 -- | 174.8 ±25.3 | 0.0 -- | 1,907.6 ±68.6 | 389.3 ±41.7 | 3,176.2 ±74.1 |
| | 1992 | 665.2 ±52.5 | 0.6 -- | 159.8 ±25.0 | 0.0 -- | 1,883.0 ±72.7 | 385.4 ±43.2 | 3,094.0 ±74.2 |
| | 1997 | 601.6 ±52.7 | 0.6 -- | 129.8 ±20.9 | 0.0 -- | 1,777.8 ±70.7 | 381.0 ±44.0 | 2,890.8 ±78.3 |
| | 2002 | 540.0 ±60.4 | 0.0 -- | 120.6 ±26.0 | 0.0 -- | 1,713.4 ±77.5 | 385.5 ±45.5 | 2,759.5 ±83.1 |
| | 2007 | 487.8 ±67.6 | 0.0 -- | 141.0 ±35.2 | 0.0 -- | 1,677.7 ±87.6 | 385.5 ±47.6 | 2,692.0 ±87.3 |

| State | Year | Cropland | CRP land | Pastureland | Rangeland | Forest land | Other rural land | Total rural land |
|---|---|---|---|---|---|---|---|---|
| New Mexico | 1982 | 2,384.2 ±167.5 | -- | *157.8 ±176.8* | 41,819.5 ±935.6 | 5,237.3 ±775.3 | 1,722.1 ±368.3 | 51,320.9 ±99.1 |
| | 1987 | 1,937.2 ±183.3 | 430.1 – | 193.7 ±177.2 | 41,237.7 ±945.7 | 5,080.4 ±764.8 | 1,835.5 ±372.6 | 50,714.6 ±108.3 |
| | 1992 | 1,894.8 ±188.1 | 482.7 – | 220.8 ±176.5 | 40,627.3 ±877.7 | 5,277.3 ±750.0 | 1,896.2 ±390.1 | 50,399.1 ±120.6 |
| | 1997 | 1,868.3 ±174.0 | 467.0 – | 230.5 ±179.2 | 40,206.6 ±944.1 | 5,565.3 ±794.8 | 1,890.0 ±407.8 | 50,227.7 ±129.5 |
| | 2002 | 1,541.7 ±234.6 | 590.9 – | 243.0 ±182.0 | 40,255.3 ±1,026.9 | 5,457.8 ±868.9 | 1,943.0 ±391.8 | 50,031.7 ±160.0 |
| | 2007 | 1,465.5 ±257.9 | 585.1 – | 212.7 ±196.4 | 40,323.6 ±1,087.8 | 5,444.5 ±930.6 | 1,928.4 ±414.5 | 49,959.8 ±182.5 |
| New York | 1982 | 5,873.5 ±229.4 | -- | 3,888.5 ±160.2 | 0.0 – | 16,662.9 ±219.0 | 650.1 ±66.6 | 27,075.0 ±126.7 |
| | 1987 | 5,738.0 ±223.2 | 17.8 – | 3,426.4 ±163.5 | 0.0 – | 16,986.7 ±229.1 | 798.7 ±67.6 | 26,967.6 ±133.0 |
| | 1992 | 5,579.5 ±207.8 | 57.0 – | 3,077.5 ±143.8 | 0.0 – | 17,327.9 ±244.5 | 786.2 ±66.1 | 26,828.1 ±135.6 |
| | 1997 | 5,357.8 ±218.1 | 53.7 – | 2,737.2 ±131.5 | 0.0 – | 17,570.0 ±251.0 | 765.5 ±73.4 | 26,484.2 ±145.9 |
| | 2002 | 5,233.1 ±216.8 | 51.0 – | 2,632.5 ±184.8 | 0.0 – | 17,504.0 ±279.6 | 794.9 ±82.8 | 26,215.5 ±160.2 |
| | 2007 | 5,000.0 ±240.5 | 49.3 – | 2,693.8 ±230.6 | 0.0 – | 17,518.9 ±310.9 | 809.7 ±95.2 | 26,071.7 ±165.8 |
| North Carolina | 1982 | 6,680.5 ±258.4 | -- | 1,935.8 ±120.5 | 0.0 – | 17,104.7 ±290.3 | 765.2 ±98.7 | 26,486.2 ±101.2 |
| | 1987 | 6,405.2 ±275.5 | 30.2 – | 1,927.1 ±132.6 | 0.0 – | 16,760.1 ±267.3 | 778.7 ±98.9 | 25,901.3 ±110.1 |
| | 1992 | 5,984.3 ±257.7 | 138.2 – | 1,976.5 ±118.3 | 0.0 – | 16,375.4 ±274.9 | 788.2 ±99.8 | 25,262.6 ±139.3 |
| | 1997 | 5,664.4 ±252.9 | 131.4 – | 2,041.0 ±127.2 | 0.0 – | 16,119.9 ±271.3 | 814.7 ±95.5 | 24,771.4 ±164.7 |
| | 2002 | 5,457.1 ±258.6 | 85.5 – | 1,897.3 ±196.3 | 0.0 – | 15,728.2 ±294.1 | 856.8 ±104.8 | 24,024.9 ±206.1 |
| | 2007 | 5,240.0 ±287.9 | 85.9 – | 1,869.7 ±187.8 | 0.0 – | 15,546.7 ±316.8 | 879.5 ±137.2 | 23,621.8 ±243.1 |
| North Dakota | 1982 | 26,985.2 ±403.8 | -- | 1,200.7 ±166.0 | 11,526.2 ±376.3 | 462.2 ±83.9 | 1,476.1 ±127.4 | 41,650.4 ±71.3 |
| | 1987 | 27,021.6 ±384.3 | 525.5 – | 1,192.4 ±130.8 | 10,957.1 ±377.7 | 460.2 ±83.8 | 1,471.4 ±115.3 | 41,628.2 ±71.2 |
| | 1992 | 24,651.6 ±398.8 | 2,906.2 – | 1,169.6 ±142.0 | 10,912.2 ±392.2 | 456.7 ±83.2 | 1,477.4 ±114.0 | 41,573.7 ±73.0 |
| | 1997 | 24,932.0 ±368.1 | 2,802.1 – | 1,040.1 ±139.4 | 10,758.6 ±395.7 | 447.8 ±80.6 | 1,497.0 ±121.4 | 41,477.6 ±76.5 |
| | 2002 | 24,153.1 ±412.8 | 3,205.0 – | 1,059.9 ±181.1 | 11,022.1 ±407.7 | 466.2 ±89.2 | 1,504.8 ±121.4 | 41,411.1 ±76.2 |
| | 2007 | 23,951.6 ±568.2 | 3,211.3 – | 1,194.9 ±221.9 | 11,018.8 ±549.8 | 466.3 ±95.4 | 1,561.2 ±176.2 | 41,404.1 ±75.2 |

| State | Year | Cropland | CRP land | Pastureland | Rangeland | Forest land | Other rural land | Total rural land |
|---|---|---|---|---|---|---|---|---|
| Ohio | 1982 | 12,385.5<br>±161.6 | -- | 2,772.0<br>±135.6 | 0.0<br>-- | 6,671.4<br>±184.5 | 1,014.0<br>±63.7 | 22,842.9<br>±119.2 |
| | 1987 | 12,269.8<br>±173.6 | 60.0<br>-- | 2,461.4<br>±121.0 | 0.0<br>-- | 6,901.6<br>±204.5 | 941.7<br>±71.2 | 22,634.5<br>±124.6 |
| | 1992 | 11,859.3<br>±164.5 | 317.2<br>-- | 2,311.5<br>±124.6 | 0.0<br>-- | 6,930.4<br>±215.8 | 919.3<br>±56.9 | 22,337.7<br>±137.5 |
| | 1997 | 11,593.9<br>±200.2 | 322.8<br>-- | 2,020.9<br>±150.9 | 0.0<br>-- | 7,051.7<br>±206.0 | 973.2<br>±48.3 | 21,962.5<br>±151.2 |
| | 2002 | 11,277.1<br>±206.8 | 257.2<br>-- | 2,196.6<br>±132.0 | 0.0<br>-- | 7,128.2<br>±227.0 | 873.6<br>±101.5 | 21,732.7<br>±151.3 |
| | 2007 | 11,055.1<br>±291.6 | 221.9<br>-- | 2,276.4<br>±184.1 | 0.0<br>-- | 7,087.1<br>±298.1 | 882.2<br>±108.1 | 21,522.7<br>±169.6 |
| Oklahoma | 1982 | 11,570.2<br>±329.1 | -- | 7,149.1<br>±217.8 | 15,118.1<br>±410.5 | 6,869.4<br>±280.5 | 417.1<br>±52.3 | 41,123.9<br>±97.5 |
| | 1987 | 10,963.2<br>±348.0 | 593.8<br>-- | 7,533.9<br>±196.1 | 14,492.4<br>±411.7 | 7,025.0<br>±290.0 | 437.7<br>±54.1 | 41,046.0<br>±95.8 |
| | 1992 | 10,173.2<br>±350.9 | 1,159.3<br>-- | 7,777.1<br>±210.5 | 14,217.7<br>±413.6 | 7,171.2<br>±299.8 | 463.9<br>±54.0 | 40,962.4<br>±102.0 |
| | 1997 | 9,750.5<br>±353.2 | 1,137.6<br>-- | 7,998.3<br>±209.9 | 14,129.2<br>±422.1 | 7,339.6<br>±288.1 | 446.6<br>±51.2 | 40,801.8<br>±114.2 |
| | 2002 | 9,106.1<br>±426.3 | 1,044.1<br>-- | 8,394.8<br>±294.5 | 14,164.4<br>±454.3 | 7,445.7<br>±261.5 | 446.3<br>±61.1 | 40,601.4<br>±127.6 |
| | 2007 | 8,785.3<br>±499.4 | 1,059.6<br>-- | 8,420.8<br>±426.0 | 14,193.4<br>±543.9 | 7,486.7<br>±359.1 | 498.5<br>±103.5 | 40,444.3<br>±134.4 |
| Oregon | 1982 | 4,348.5<br>±290.1 | -- | 2,037.2<br>±255.0 | 9,528.6<br>±632.6 | 12,767.7<br>±466.8 | 608.4<br>±144.5 | 29,290.4<br>±98.8 |
| | 1987 | 3,941.6<br>±312.8 | 400.3<br>-- | 2,001.9<br>±253.9 | 9,351.0<br>±632.4 | 12,760.9<br>±474.0 | 610.1<br>±146.5 | 29,065.8<br>±109.9 |
| | 1992 | 3,746.9<br>±312.7 | 528.2<br>-- | 2,014.6<br>±249.0 | 9,465.5<br>±635.4 | 12,736.3<br>±495.1 | 601.5<br>±134.8 | 29,093.0<br>±113.9 |
| | 1997 | 3,733.3<br>±296.8 | 482.8<br>-- | 1,963.7<br>±233.2 | 9,311.1<br>±645.1 | 12,685.3<br>±500.8 | 663.9<br>±141.9 | 28,840.1<br>±123.3 |
| | 2002 | 3,636.4<br>±336.1 | 474.4<br>-- | 1,812.1<br>±233.7 | 9,381.3<br>±625.7 | 12,743.6<br>±511.7 | 704.6<br>±156.9 | 28,752.4<br>±135.7 |
| | 2007 | 3,601.8<br>±320.9 | 526.0<br>-- | 1,717.2<br>±223.1 | 9,381.1<br>±681.6 | 12,739.4<br>±609.0 | 718.7<br>±189.8 | 28,684.2<br>±142.8 |
| Pennsylvania | 1982 | 5,878.1<br>±251.6 | -- | 2,652.5<br>±150.4 | 0.0<br>-- | 15,583.4<br>±282.7 | 930.7<br>±91.9 | 25,044.7<br>±109.8 |
| | 1987 | 5,747.3<br>±251.2 | 15.7<br>-- | 2,491.1<br>±168.6 | 0.0<br>-- | 15,656.7<br>±291.8 | 950.1<br>±93.3 | 24,860.9<br>±119.5 |
| | 1992 | 5,585.0<br>±252.3 | 92.4<br>-- | 2,305.9<br>±152.7 | 0.0<br>-- | 15,512.9<br>±290.6 | 928.3<br>±88.9 | 24,424.5<br>±136.1 |
| | 1997 | 5,467.7<br>±259.1 | 90.2<br>-- | 1,872.7<br>±147.4 | 0.0<br>-- | 15,559.3<br>±301.1 | 896.7<br>±92.8 | 23,886.6<br>±150.1 |
| | 2002 | 5,101.2<br>±291.5 | 55.1<br>-- | 2,037.4<br>±178.9 | 0.0<br>-- | 15,639.6<br>±316.3 | 786.9<br>±98.7 | 23,620.2<br>±163.8 |
| | 2007 | 4,935.9<br>±325.0 | 51.4<br>-- | 2,050.1<br>±175.3 | 0.0<br>-- | 15,590.1<br>±341.5 | 801.9<br>±121.1 | 23,429.4<br>±177.4 |

| State | Year | Cropland | CRP land | Pastureland | Rangeland | Forest land | Other rural land | Total rural land |
|---|---|---|---|---|---|---|---|---|
| Rhode Island | 1982 | 26.8 ±6.6 | -- | 35.0 ±11.6 | 0.0 -- | 399.3 ±19.4 | 23.3 ±8.6 | 484.4 ±12.5 |
| | 1987 | 23.6 ±6.9 | 0.0 -- | 35.1 ±10.7 | 0.0 -- | 395.4 ±20.8 | 20.5 ±8.5 | 474.6 ±14.0 |
| | 1992 | 22.9 ±7.0 | 0.0 -- | 25.6 ±8.5 | 0.0 -- | 386.3 ±20.3 | 21.8 ±8.4 | 456.6 ±14.0 |
| | 1997 | 19.3 ±6.8 | 0.0 -- | 26.6 ±7.5 | 0.0 -- | 385.3 ±19.9 | 22.2 ±8.6 | 453.4 ±14.4 |
| | 2002 | 18.0 ±7.8 | 0.0 -- | 26.5 ±8.1 | 0.0 -- | 375.6 ±20.1 | 16.7 ±8.6 | 436.8 ±15.6 |
| | 2007 | 17.6 ±6.3 | 0.0 -- | 22.1 ±7.6 | 0.0 -- | 366.7 ±23.2 | 19.9 ±10.7 | 426.3 ±17.4 |
| South Carolina | 1982 | 3,554.7 ±124.2 | -- | 1,173.3 ±119.6 | 0.0 -- | 11,335.7 ±164.6 | 710.4 ±72.0 | 16,774.1 ±77.1 |
| | 1987 | 3,296.2 ±121.2 | 97.6 -- | 1,148.3 ±110.2 | 0.0 -- | 11,339.0 ±158.3 | 719.9 ±78.1 | 16,601.0 ±90.5 |
| | 1992 | 2,953.2 ±109.5 | 264.1 -- | 1,149.9 ±108.1 | 0.0 -- | 11,273.0 ±163.8 | 726.3 ±81.0 | 16,366.5 ±101.0 |
| | 1997 | 2,570.6 ±103.8 | 262.0 -- | 1,161.0 ±99.8 | 0.0 -- | 11,244.4 ±176.8 | 756.3 ±79.4 | 15,994.3 ±112.6 |
| | 2002 | 2,425.2 ±102.2 | 186.2 -- | 1,065.2 ±133.6 | 0.0 -- | 11,208.7 ±169.9 | 767.3 ±85.6 | 15,652.6 ±105.1 |
| | 2007 | 2,230.1 ±176.1 | 170.6 -- | 1,070.3 ±190.9 | 0.0 -- | 11,167.9 ±243.9 | 780.0 ±87.0 | 15,418.9 ±114.6 |
| South Dakota | 1982 | 17,009.8 ±308.2 | -- | 2,645.7 ±237.4 | 22,974.5 ±429.5 | 529.5 ±108.4 | 1,491.9 ±142.5 | 44,651.4 ±64.1 |
| | 1987 | 17,567.2 ±372.7 | 358.1 -- | 2,244.4 ±207.6 | 22,412.7 ±443.5 | 529.5 ±106.1 | 1,490.8 ±140.2 | 44,602.7 ±65.5 |
| | 1992 | 16,428.3 ±366.9 | 1,756.3 -- | 2,164.2 ±195.2 | 22,140.7 ±444.1 | 524.7 ±104.7 | 1,490.3 ±138.9 | 44,504.5 ±90.7 |
| | 1997 | 16,747.6 ±367.2 | 1,685.2 -- | 2,042.5 ±173.0 | 21,944.3 ±456.1 | 512.5 ±105.3 | 1,516.8 ±144.0 | 44,448.9 ±100.7 |
| | 2002 | 17,025.9 ±389.0 | 1,297.0 -- | 2,017.8 ±306.7 | 22,029.7 ±483.9 | 530.7 ±112.1 | 1,526.2 ±153.7 | 44,427.3 ±104.7 |
| | 2007 | 16,764.4 ±528.3 | 1,342.3 -- | 2,089.5 ±313.2 | 22,189.7 ±572.9 | 524.2 ±121.3 | 1,493.5 ±154.9 | 44,403.6 ±112.4 |
| Tennessee | 1982 | 5,525.3 ±156.7 | -- | 5,289.6 ±210.3 | 0.0 -- | 12,060.5 ±276.0 | 486.6 ±62.0 | 23,362.0 ±97.0 |
| | 1987 | 5,296.7 ±146.1 | 173.7 -- | 5,077.3 ±209.1 | 0.0 -- | 12,092.7 ±273.5 | 463.6 ±52.0 | 23,104.0 ±105.2 |
| | 1992 | 4,765.8 ±141.3 | 440.8 -- | 5,099.8 ±195.5 | 0.0 -- | 12,062.4 ±264.6 | 445.7 ±48.3 | 22,814.5 ±110.1 |
| | 1997 | 4,573.7 ±170.6 | 374.2 -- | 4,912.3 ±211.2 | 0.0 -- | 11,978.0 ±264.2 | 523.0 ±52.4 | 22,361.2 ±130.0 |
| | 2002 | 4,505.3 ±238.8 | 240.5 -- | 4,837.4 ±295.3 | 0.0 -- | 11,938.5 ±284.6 | 553.4 ±74.9 | 22,075.1 ±136.6 |
| | 2007 | 4,142.4 ±278.8 | 255.3 -- | 4,977.6 ±352.1 | 0.0 -- | 11,834.6 ±341.5 | 632.0 ±93.8 | 21,841.9 ±159.2 |

| State | Year | Cropland | CRP land | Pastureland | Rangeland | Forest land | Other rural land | Total rural land |
|---|---|---|---|---|---|---|---|---|
| Texas | 1982 | 33,542.9<br>±705.2 | -- | 17,000.6<br>±455.8 | 97,318.3<br>±911.7 | 9,785.7<br>±242.3 | 1,870.7<br>±170.9 | 159,518.2<br>±185.5 |
| | 1987 | 31,389.3<br>±725.5 | 1,587.5<br>-- | 16,764.9<br>±486.0 | 96,817.7<br>±975.4 | 10,268.6<br>±296.6 | 2,001.2<br>±171.6 | 158,829.2<br>±200.5 |
| | 1992 | 28,525.8<br>±713.2 | 3,973.1<br>-- | 16,682.2<br>±513.1 | 96,369.1<br>±1,006.7 | 10,483.1<br>±302.6 | 2,045.8<br>±175.6 | 158,079.1<br>±237.5 |
| | 1997 | 27,227.9<br>±697.5 | 3,905.1<br>-- | 16,016.6<br>±464.4 | 97,041.2<br>±954.1 | 10,960.0<br>±347.6 | 2,179.3<br>±181.8 | 157,330.1<br>±264.6 |
| | 2002 | 25,830.9<br>±748.6 | 4,044.4<br>-- | 16,207.3<br>±533.1 | 97,247.5<br>±949.8 | 10,778.5<br>±375.7 | 2,241.8<br>±153.3 | 156,350.4<br>±294.0 |
| | 2007 | 24,003.6<br>±942.8 | 4,020.7<br>-- | 16,329.8<br>±746.2 | 98,070.3<br>±1,288.3 | 10,650.6<br>±470.8 | 2,425.2<br>±230.2 | 155,500.2<br>±352.1 |
| Utah | 1982 | 2,051.8<br>±284.9 | -- | 533.8<br>±81.0 | 10,854.4<br>±693.8 | 1,818.6<br>±441.4 | 2,371.2<br>±531.9 | 17,629.8<br>±107.1 |
| | 1987 | 1,907.2<br>±309.5 | 152.1<br>-- | 598.5<br>±87.5 | 10,677.7<br>±684.7 | 1,805.1<br>±444.9 | 2,214.2<br>±531.4 | 17,354.8<br>±115.9 |
| | 1992 | 1,830.3<br>±316.1 | 222.6<br>-- | 639.2<br>±95.2 | 10,908.8<br>±680.3 | 1,765.2<br>±440.1 | 2,377.8<br>±529.2 | 17,743.9<br>±124.2 |
| | 1997 | 1,678.1<br>±312.4 | 216.5<br>-- | 695.4<br>±100.5 | 10,775.4<br>±716.3 | 1,894.6<br>±511.2 | 2,399.1<br>±519.9 | 17,659.1<br>±132.7 |
| | 2002 | 1,618.8<br>±273.0 | 200.3<br>-- | 695.9<br>±118.8 | 10,772.6<br>±711.9 | 1,902.8<br>±518.1 | 2,386.5<br>±527.4 | 17,576.9<br>±148.0 |
| | 2007 | 1,421.1<br>±282.9 | 208.2<br>-- | 695.1<br>±147.7 | 10,898.0<br>±715.9 | 1,902.5<br>±500.1 | 2,389.9<br>±527.2 | 17,514.8<br>±153.4 |
| Vermont | 1982 | 643.5<br>±53.7 | -- | 444.9<br>±50.2 | 0.0<br>-- | 4,137.8<br>±72.6 | 82.4<br>±17.4 | 5,308.6<br>±27.0 |
| | 1987 | 641.0<br>±61.9 | 0.0<br>-- | 378.4<br>±49.2 | 0.0<br>-- | 4,136.1<br>±72.3 | 79.6<br>±16.4 | 5,235.1<br>±27.8 |
| | 1992 | 627.2<br>±64.5 | 0.0<br>-- | 347.3<br>±38.0 | 0.0<br>-- | 4,131.8<br>±69.9 | 82.1<br>±19.0 | 5,188.4<br>±28.7 |
| | 1997 | 596.0<br>±61.7 | 0.0<br>-- | 343.0<br>±38.0 | 0.0<br>-- | 4,129.3<br>±71.3 | 86.3<br>±20.8 | 5,154.6<br>±29.7 |
| | 2002 | 558.0<br>±59.4 | 0.0<br>-- | 320.7<br>±46.1 | 0.0<br>-- | 4,128.0<br>±76.0 | 89.6<br>±22.1 | 5,096.3<br>±29.8 |
| | 2007 | 540.6<br>±62.0 | 0.0<br>-- | 308.1<br>±47.4 | 0.0<br>-- | 4,113.4<br>±81.6 | 113.8<br>±24.0 | 5,075.9<br>±32.2 |
| Virginia | 1982 | 3,392.8<br>±172.4 | -- | 3,238.2<br>±158.2 | 0.0<br>-- | 13,473.1<br>±225.8 | 615.0<br>±53.5 | 20,719.1<br>±107.6 |
| | 1987 | 3,071.2<br>±181.2 | 22.7<br>-- | 3,238.2<br>±155.0 | 0.0<br>-- | 13,536.8<br>±224.4 | 589.1<br>±56.0 | 20,458.0<br>±106.6 |
| | 1992 | 2,889.2<br>±176.1 | 73.8<br>-- | 3,208.1<br>±153.8 | 0.0<br>-- | 13,497.1<br>±215.8 | 559.3<br>±49.1 | 20,227.5<br>±114.5 |
| | 1997 | 2,886.7<br>±179.3 | 70.5<br>-- | 2,994.3<br>±161.9 | 0.0<br>-- | 13,362.7<br>±211.7 | 569.7<br>±47.2 | 19,883.9<br>±111.5 |
| | 2002 | 2,849.1<br>±190.0 | 42.1<br>-- | 2,941.1<br>±173.9 | 0.0<br>-- | 13,216.8<br>±214.7 | 555.9<br>±63.6 | 19,605.0<br>±115.1 |
| | 2007 | 2,757.6<br>±250.8 | 42.9<br>-- | 2,948.3<br>±195.2 | 0.0<br>-- | 13,060.0<br>±256.8 | 587.6<br>±96.8 | 19,396.4<br>±130.0 |

| State | Year | Cropland | CRP land | Pastureland | Rangeland | Forest land | Other rural land | Total rural land |
|---|---|---|---|---|---|---|---|---|
| Washington | 1982 | 7,759.3 ±410.2 | -- | 1,287.1 ±168.7 | 6,090.1 ±366.7 | 13,075.0 ±398.4 | 794.3 ±120.7 | 29,005.8 ±154.7 |
| | 1987 | 7,243.7 ±415.4 | 459.9 – | 1,357.0 ±203.3 | 6,073.4 ±358.0 | 12,995.3 ±393.3 | 775.1 ±116.6 | 28,904.4 ±155.2 |
| | 1992 | 6,765.8 ±403.5 | 1,027.4 – | 1,339.9 ±217.5 | 5,874.7 ±350.2 | 12,882.7 ±385.2 | 788.5 ±111.3 | 28,679.0 ±168.7 |
| | 1997 | 6,654.0 ±405.3 | 1,017.0 – | 1,237.6 ±176.5 | 5,903.7 ±368.9 | 12,756.0 ±380.2 | 856.6 ±117.8 | 28,424.9 ±181.7 |
| | 2002 | 6,592.9 ±483.7 | 1,214.2 – | 971.1 ±150.0 | 5,921.1 ±373.9 | 12,654.4 ±396.9 | 858.7 ±126.1 | 28,212.4 ±190.9 |
| | 2007 | 6,465.1 ±398.9 | 1,360.8 – | 897.7 ±199.1 | 5,927.0 ±372.1 | 12,587.0 ±411.9 | 865.7 ±141.9 | 28,103.3 ±197.8 |
| West Virginia | 1982 | 1,084.0 ±85.6 | -- | 1,899.4 ±116.7 | 0.0 – | 10,388.4 ±157.7 | 232.2 ±39.0 | 13,604.0 ±47.6 |
| | 1987 | 987.0 ±80.9 | 0.6 – | 1,753.4 ±112.6 | 0.0 – | 10,562.2 ±156.4 | 245.3 ±42.6 | 13,548.5 ±48.9 |
| | 1992 | 904.1 ±77.8 | 0.6 – | 1,646.5 ±110.1 | 0.0 – | 10,544.5 ±152.4 | 274.6 ±53.7 | 13,370.3 ±53.7 |
| | 1997 | 856.3 ±79.2 | 0.0 – | 1,521.0 ±104.0 | 0.0 – | 10,523.2 ±149.0 | 266.2 ±41.6 | 13,166.7 ±62.3 |
| | 2002 | 785.8 ±100.8 | 0.6 – | 1,495.5 ±128.9 | 0.0 – | 10,516.9 ±150.5 | 239.9 ±113.4 | 13,038.7 ±60.6 |
| | 2007 | 760.4 ±144.5 | 0.6 – | 1,440.3 ±166.8 | 0.0 – | 10,510.3 ±193.8 | 252.5 ±109.7 | 12,964.1 ±58.5 |
| Wisconsin | 1982 | 11,437.5 ±285.5 | -- | 3,520.7 ±178.4 | 0.0 – | 14,262.1 ±238.9 | 1,616.8 ±135.5 | 30,837.1 ±112.2 |
| | 1987 | 11,318.1 ±312.3 | 217.0 – | 3,191.8 ±196.4 | 0.0 – | 14,352.2 ±245.4 | 1,636.8 ±134.4 | 30,715.9 ±117.8 |
| | 1992 | 10,826.5 ±290.1 | 664.7 – | 3,070.2 ±205.7 | 0.0 – | 14,376.3 ±245.3 | 1,635.0 ±132.0 | 30,572.7 ±123.1 |
| | 1997 | 10,591.2 ±305.6 | 661.9 – | 3,009.2 ±208.6 | 0.0 – | 14,478.2 ±267.3 | 1,650.9 ±127.6 | 30,391.4 ±130.2 |
| | 2002 | 10,237.3 ±281.4 | 603.3 – | 3,078.3 ±242.5 | 0.0 – | 14,578.4 ±282.8 | 1,733.9 ±157.4 | 30,231.2 ±151.1 |
| | 2007 | 10,022.9 ±341.4 | 540.9 – | 3,161.4 ±277.3 | 0.0 – | 14,599.3 ±312.2 | 1,733.1 ±175.3 | 30,057.6 ±180.7 |
| Wyoming | 1982 | 2,585.2 ±306.4 | -- | 826.7 ±360.5 | 27,777.7 ±437.5 | 1,033.9 ±269.2 | 706.6 ±237.3 | 32,930.1 ±66.7 |
| | 1987 | 2,430.9 ±331.0 | 129.8 – | 912.4 ±363.6 | 27,670.2 ±461.8 | 1,034.1 ±265.7 | 708.2 ±235.3 | 32,885.6 ±70.2 |
| | 1992 | 2,261.2 ±329.5 | 251.7 – | 987.8 ±378.1 | 27,460.6 ±471.8 | 1,038.4 ±265.5 | 826.5 ±253.5 | 32,826.2 ±73.3 |
| | 1997 | 2,192.6 ±338.8 | 246.6 – | 1,127.7 ±415.1 | 27,421.2 ±478.2 | 1,029.5 ±252.7 | 777.3 ±214.6 | 32,794.9 ±76.8 |
| | 2002 | 2,184.1 ±320.8 | 276.4 – | 777.2 ±562.6 | 27,813.7 ±511.1 | 963.6 ±248.3 | 759.1 ±217.5 | 32,774.1 ±82.1 |
| | 2007 | 2,127.3 ±331.6 | 277.8 – | *648.7* *±938.1* | 27,999.0 ±606.9 | 963.2 ±250.7 | 716.6 ±433.1 | 32,732.6 ±98.0 |

| State | Year | Cropland | CRP land | Pastureland | Rangeland | Forest land | Other rural land | Total rural land |
|---|---|---|---|---|---|---|---|---|
| Total | 1982 | 419,546.9 ±1,988.1 | -- | 130,896.3 ±1,357.3 | 417,899.5 ±3,366.1 | 403,379.6 ±2,664.4 | 47,243.1 ±1,379.4 | 1,418,965.4 ±805.0 |
| | 1987 | 405,545.4 ±1,936.5 | 13,815.0 -- | 126,722.0 ±1,327.5 | 412,574.3 ±3,359.3 | 405,335.0 ±2,833.7 | 47,544.6 ±1,448.8 | 1,411,536.3 ±871.0 |
| | 1992 | 381,440.6 ±1,942.1 | 34,093.5 -- | 125,018.8 ±1,274.6 | 408,916.4 ±3,247.2 | 405,294.8 ±2,845.0 | 48,066.7 ±1,346.6 | 1,402,830.8 ±976.9 |
| | 1997 | 376,138.1 ±2,010.1 | 32,690.5 -- | 119,780.7 ±1,278.1 | 407,542.4 ±3,386.4 | 406,596.7 ±2,893.8 | 48,748.4 ±1,279.7 | 1,391,496.8 ±1,022.8 |
| | 2002 | 367,099.6 ±2,269.2 | 31,990.3 -- | 117,783.0 ±1,629.8 | 408,209.2 ±3,494.5 | 407,256.4 ±2,689.3 | 48,931.3 ±1,225.7 | 1,381,269.8 ±1,300.0 |
| | 2007 | 357,023.5 ±2,688.7 | 32,850.2 -- | 118,615.7 ±2,347.0 | 409,119.4 ±3,992.9 | 406,410.4 ±3,065.4 | 49,639.6 ±1,359.1 | 1,373,658.8 ±1,500.2 |

Notes:
• Acreages for Conservation Reserve Program (CRP) land are established through geospatial processes and administrative records; therefore, statistical margins of error are not applicable and shown as a dashed line (--). CRP was not implemented until 1985.
• Cropland includes cultivated and noncultivated cropland.
• When the estimate is 0.0, margins of error are not applicable and shown as a dashed line (--).
• Instances where the margin of error is greater than or equal to the estimate are displayed in italics indicating that the confidence interval includes zero and that the estimate should not be used.

U.S. Department of Agriculture. 2009. *Summary Report: 2007 National Resources Inventory*, Natural Resources Conservation Service, Washington, DC, and Center for Survey Statistics and Methodology, Iowa State University, Ames, Iowa. 123 pages.

**Changes in land cover/use between 2002 and 2007**
In thousands of acres, with margins of error

| Land cover/use in 2002 | Land cover/use in 2007 | | | | | | | | 2002 total |
|---|---|---|---|---|---|---|---|---|---|
| | Cropland | CRP land | Pastureland | Rangeland | Forest land | Other rural land | Developed land | Water areas & Federal land | |
| **Cropland** | 352,866.1<br>±2,667.8 | 1,729.3<br>-- | 7,974.4<br>±850.0 | 1,279.6<br>±528.3 | 673.7<br>±352.1 | 828.0<br>±211.9 | 1,657.3<br>±138.3 | 91.2<br>-- | 367,099.6<br>±2,242.8 |
| **CRP land** | 365.9<br>-- | 31,008.7<br>-- | 347.5<br>-- | 77.5<br>-- | 182.1<br>-- | 4.7<br>-- | 0.5<br>-- | 3.4<br>-- | 31,990.3<br>-- |
| **Pastureland** | 2,975.4<br>±558.5 | 110.4<br>-- | 109,574.8<br>±2,079.6 | 1,508.4<br>±708.9 | 1,865.0<br>±504.4 | 367.6<br>±138.0 | 1,310.2<br>±182.9 | 71.2<br>-- | 117,783.0<br>±1,770.8 |
| **Rangeland** | 449.4<br>±297.4 | 0.9<br>-- | 69.4<br>±45.2 | 405,891.3<br>±3,805.1 | *307.2*<br>*±374.6* | 326.3<br>±191.6 | 1,112.3<br>±176.5 | 52.4<br>-- | 408,209.2<br>±3,713.4 |
| **Forest land** | *55.6*<br>*±59.1* | 0.9<br>-- | 348.7<br>±185.3 | *105.5*<br>*±149.5* | 402,874.8<br>±3,002.2 | 453.0<br>±210.3 | 3,236.1<br>±131.0 | 181.8<br>-- | 407,256.4<br>±3,020.5 |
| **Other rural land** | 227.0<br>±61.6 | 0.0<br>-- | 231.9<br>±121.2 | 220.4<br>±158.7 | 346.2<br>±255.3 | 47,649.6<br>±1,342.4 | 174.9<br>±28.3 | 81.3<br>-- | 48,931.3<br>±1,332.4 |
| **Developed land** | 58.7<br>±12.0 | 0.0<br>-- | 54.6<br>±9.0 | 31.1<br>±7.6 | 124.9<br>±15.5 | *8.1*<br>*±11.8* | 103,753.0<br>±1,261.9 | 0.4<br>-- | 104,030.8<br>±1,269.0 |
| **Water areas & Federal land** | 25.4<br>-- | 0.0<br>-- | 14.4<br>-- | 5.6<br>-- | 36.5<br>-- | 2.3<br>-- | 6.9<br>-- | 452,272.5<br>-- | 452,363.6<br>-- |
| **2007 total** | 357,023.5<br>±2,688.7 | 32,850.2<br>-- | 118,615.7<br>±2,347.0 | 409,119.4<br>±3,992.9 | 406,410.4<br>±3,065.4 | 49,639.6<br>±1,359.1 | 111,251.2<br>±1,499.4 | 452,754.2<br>-- | 1,937,664.2<br>±163.8 |

Notes:
- Acreages for Conservation Reserve Program (CRP) Land and Water areas and Federal land are established through geospatial processes and administrative records; therefore, statistical margins of error are not applicable and shown as a dashed line (--). CRP was not implemented until 1985.
- Cropland includes cultivated and noncultivated cropland.
- When the estimate is 0.0, margins of error are not applicable and shown as a dashed line (--).
- Instances where the margin of error is greater than or equal to the estimate are displayed in italics indicating that the confidence interval includes zero and that the estimate should not be used.

2002 land cover/use totals are listed in the right hand vertical column, titled 2002 total. 2007 land cover/use totals are listed in the bottom horizontal row, titled 2007 total. The number at the intersection of rows and columns with the same land cover/use designation represents acres that did not change from 2002 to 2007. Reading to the right or left of this number are the acres that were lost to another cover/use by 2007. Reading up or down from this number are the acres that were gained from another cover/use by 2007.

U.S. Department of Agriculture. 2009. *Summary Report: 2007 National Resources Inventory*, Natural Resources Conservation Service, Washington, DC, and Center for Survey Statistics and Methodology, Iowa State University, Ames, Iowa. 123 pages.

**Changes in land cover/use between 1982 and 2007**
In thousands of acres, with margins of error

| Land cover/use in 1982 | Land cover/use in 2007 | | | | | | | | |
|---|---|---|---|---|---|---|---|---|---|
| | Cropland | CRP land | Pastureland | Rangeland | Forest land | Other rural land | Developed land | Water areas & Federal land | 1982 total |
| Cropland | 326,196.4 ±2,675.3 | 30,168.6 -- | 30,344.7 ±1,148.8 | 6,895.4 ±1,025.2 | 8,922.7 ±617.8 | 4,136.4 ±428.5 | 11,117.5 ±403.8 | 1,765.2 -- | 419,546.9 ±2,441.3 |
| CRP land | -- | -- | -- | -- | -- | -- | -- | -- | -- |
| Pastureland | 18,526.6 ±1,055.1 | 1,351.6 -- | 78,372.2 ±2,050.0 | 5,085.3 ±943.5 | 17,760.5 ±1,039.9 | 2,036.1 ±384.5 | 6,845.0 ±338.0 | 919.0 -- | 130,896.3 ±1,493.8 |
| Rangeland | 7,430.8 ±1,292.5 | 1,124.5 -- | 3,369.1 ±719.4 | 391,615.0 ±3,681.9 | 3,379.4 ±802.1 | 2,272.5 ±565.5 | 5,201.0 ±544.1 | 3,507.2 -- | 417,899.5 ±3,741.7 |
| Forest land | 2,121.7 ±328.4 | 144.4 -- | 4,847.6 ±841.7 | 2,175.6 ±970.1 | 371,660.4 ±2,942.2 | 2,229.1 ±464.3 | 17,083.5 ±417.1 | 3,117.3 -- | 403,379.6 ±2,731.8 |
| Other rural land | 1,685.2 ±231.3 | 56.4 -- | 1,159.0 ±265.6 | 915.5 ±422.1 | 3,310.2 ±372.2 | 38,734.9 ±1,262.4 | 1,077.8 ±110.2 | 304.1 -- | 47,243.1 ±1,308.6 |
| Developed land | 264.1 ±22.3 | 0.0 -- | 163.7 ±19.1 | 176.6 ±22.9 | 442.6 ±27.7 | 18.4 ±6.9 | 69,896.9 ±783.7 | 1.8 -- | 70,964.1 ±779.7 |
| Water areas & Federal land | 798.7 -- | 4.7 -- | 359.4 -- | 2,256.0 -- | 934.6 -- | 212.2 -- | 29.5 -- | 443,139.6 -- | 447,734.7 -- |
| 2007 total | 357,023.5 ±2,688.7 | 32,850.2 -- | 118,615.7 ±2,347.0 | 409,119.4 ±3,992.9 | 406,410.4 ±3,065.4 | 49,639.6 ±1,359.1 | 111,251.2 ±1,499.4 | 452,754.2 -- | 1,937,664.2 ±163.8 |

Notes:
- Acreages for Conservation Reserve Program (CRP) Land and Water areas and Federal land are established through geospatial processes and administrative records; therefore, statistical margins of error are not applicable and shown as a dashed line (--). CRP was not implemented until 1985.
- Cropland includes cultivated and noncultivated cropland.
- When the estimate is 0.0, margins of error are not applicable and shown as a dashed line (--).

1982 land cover/use totals are listed in the right hand vertical column, titled 1982 total. 2007 land cover/use totals are listed in the bottom horizontal row, titled 2007 total. The number at the intersection of rows and columns with the same land cover/use designation represents acres that did not change from 1982 to 2007. Reading to the right or left of this number are the acres that were lost to another cover/use by 2007. Reading up or down from this number are the acres that were gained from another cover/use by 2007.

U.S. Department of Agriculture. 2009. *Summary Report: 2007 National Resources Inventory*, Natural Resources Conservation Service, Washington, DC, and Center for Survey Statistics and Methodology, Iowa State University, Ames, Iowa. 123 pages.

**Wetlands and deepwater habitats on water areas and non-Federal land in 2007, by State**
In thousands of acres, with margins of error

| State | Palustrine and Estuarine wetlands | | | Other aquatic habitats | | | Total |
|---|---|---|---|---|---|---|---|
| | Palustrine | Estuarine | Total | Lacustrine | Other (*) | Total | |
| Alabama | 3,669.5 ±206.7 | 0.0 – | 3,669.5 ±206.7 | 514.9 ±17.1 | 601.2 ±22.3 | 1,116.1 ±25.7 | 4,785.6 ±213.4 |
| Arizona | *60.1 ±62.7* | 0.0 – | *60.1 ±62.7* | 139.9 ±2.2 | 204.2 ±154.0 | 344.1 ±153.7 | 404.2 ±154.5 |
| Arkansas | 3,084.0 ±283.2 | 0.0 – | 3,084.0 ±283.2 | 579.5 ±24.5 | 263.1 ±34.6 | 842.6 ±39.5 | 3,926.6 ±292.8 |
| California | 1,205.6 ±259.9 | 96.7 ±59.6 | 1,302.3 ±244.8 | 1,151.8 ±128.4 | 773.1 ±90.3 | 1,924.9 ±103.9 | 3,227.2 ±283.9 |
| Colorado | 552.0 ±129.4 | 0.0 – | 552.0 ±129.4 | 199.5 ±7.7 | 124.4 ±20.8 | 323.9 ±22.3 | 875.9 ±134.2 |
| Connecticut | 363.2 ±39.0 | 22.7 ±19.6 | 385.9 ±34.0 | 60.5 ±3.4 | 47.9 ±4.3 | 108.4 ±4.7 | 494.3 ±33.7 |
| Delaware | 161.5 ±28.9 | 100.8 ±26.2 | 262.3 ±42.6 | 24.4 ±0.0 | 261.2 ±2.5 | 285.6 ±2.5 | 547.9 ±43.3 |
| Florida | 7,885.4 ±326.4 | 478.7 ±190.2 | 8,364.1 ±318.4 | 1,111.6 ±17.5 | 1,852.3 ±26.9 | 2,963.9 ±28.1 | 11,328.0 ±312.1 |
| Georgia | 6,093.6 ±248.4 | 437.6 ±144.5 | 6,531.2 ±231.0 | 565.6 ±28.7 | 350.1 ±19.1 | 915.7 ±34.0 | 7,446.9 ±223.4 |
| Idaho | 665.3 ±162.7 | 0.0 – | 665.3 ±162.7 | 434.2 ±3.3 | 108.1 ±31.2 | 542.3 ±32.2 | 1,207.6 ±176.1 |
| Illinois | 1,182.3 ±109.2 | 0.0 – | 1,182.3 ±109.2 | 346.8 ±9.7 | 298.6 ±16.3 | 645.4 ±16.3 | 1,827.7 ±115.4 |
| Indiana | 733.5 ±108.8 | 0.0 – | 733.5 ±108.8 | 173.2 ±6.7 | 143.9 ±15.1 | 317.1 ±16.0 | 1,050.6 ±114.4 |
| Iowa | 953.0 ±144.0 | 0.0 – | 953.0 ±144.0 | 194.5 ±6.5 | 241.9 ±14.5 | 436.4 ±17.6 | 1,389.4 ±146.2 |
| Kansas | 841.9 ±88.4 | 0.0 – | 841.9 ±88.4 | 220.1 ±9.1 | 207.3 ±28.6 | 427.4 ±28.5 | 1,269.3 ±103.7 |
| Kentucky | 446.3 ±45.3 | 0.0 – | 446.3 ±45.3 | 285.1 ±7.2 | 265.4 ±12.6 | 550.5 ±13.7 | 996.8 ±48.4 |
| Louisiana | 7,743.9 ±255.4 | 2,464.1 ±237.9 | 10,208.0 ±251.6 | 1,145.2 ±34.4 | 2,598.4 ±20.4 | 3,743.6 ±40.8 | 13,951.6 ±256.2 |
| Maine | 5,613.1 ±465.5 | *13.1 ±27.2* | 5,626.2 ±470.3 | 893.3 ±8.3 | 352.2 ±19.7 | 1,245.5 ±23.9 | 6,871.7 ±472.4 |
| Maryland | 724.5 ±71.7 | 231.4 ±35.9 | 955.9 ±76.0 | 36.8 ±2.0 | 1,585.8 ±7.1 | 1,622.6 ±6.7 | 2,578.5 ±77.8 |
| Massachusetts | 528.3 ±56.6 | 36.5 ±25.7 | 564.8 ±63.9 | 136.3 ±8.7 | 221.6 ±6.4 | 357.9 ±11.2 | 922.7 ±66.3 |
| Michigan | 6,003.2 ±322.8 | 0.0 – | 6,003.2 ±322.8 | 841.1 ±17.8 | 217.1 ±12.2 | 1,058.2 ±23.0 | 7,061.4 ±324.7 |
| Minnesota | 10,876.2 ±456.2 | 0.0 – | 10,876.2 ±456.2 | 2,430.3 ±27.0 | 635.9 ±14.4 | 3,066.2 ±28.8 | 13,942.4 ±461.2 |

| State | Palustrine and Estuarine wetlands | | | Other aquatic habitats | | | Total |
|---|---|---|---|---|---|---|---|
| | Palustrine | Estuarine | Total | Lacustrine | Other (*) | Total | |
| Mississippi | 4,546.1 ±259.2 | 50.3 ±42.8 | 4,596.4 ±254.9 | 371.4 ±12.9 | 291.0 ±20.6 | 662.4 ±24.1 | 5,258.8 ±253.6 |
| Missouri | 948.6 ±194.7 | 0.0 -- | 948.6 ±194.7 | 427.8 ±9.4 | 229.0 ±11.7 | 656.8 ±15.5 | 1,605.4 ±194.3 |
| Montana | 1,168.2 ±261.1 | 0.0 -- | 1,168.2 ±261.1 | 699.0 ±23.4 | 323.2 ±63.9 | 1,022.2 ±75.5 | 2,190.4 ±273.1 |
| Nebraska | 1,183.2 ±276.1 | 0.0 -- | 1,183.2 ±276.1 | 221.8 ±10.2 | 205.6 ±24.3 | 427.4 ±27.9 | 1,610.6 ±278.7 |
| Nevada | 386.3 ±154.9 | 0.0 -- | 386.3 ±154.9 | 365.2 ±1.8 | 63.3 ±4.7 | 428.5 ±4.9 | 814.8 ±156.1 |
| New Hampshire | 485.2 ±54.4 | *13.8 ±15.1* | 499.0 ±52.0 | 168.1 ±6.9 | 48.6 ±8.5 | 216.7 ±13.1 | 715.7 ±54.6 |
| New Jersey | 514.7 ±62.4 | 222.7 ±58.1 | 737.4 ±75.7 | 75.1 ±8.1 | 444.6 ±15.9 | 519.7 ±18.1 | 1,257.1 ±73.3 |
| New Mexico | 43.9 ±16.6 | 0.0 -- | 43.9 ±16.6 | 83.4 ±4.8 | 75.9 ±23.1 | 159.3 ±22.6 | 203.2 ±30.1 |
| New York | 3,547.7 ±161.6 | *2.6 ±6.6* | 3,550.3 ±162.1 | 677.4 ±15.5 | 554.9 ±31.4 | 1,232.3 ±35.3 | 4,782.6 ±172.2 |
| North Carolina | 4,536.2 ±195.6 | 181.9 ±90.0 | 4,718.1 ±174.7 | 362.8 ±10.2 | 2,342.6 ±17.0 | 2,705.4 ±20.4 | 7,423.5 ±175.6 |
| North Dakota | 3,493.0 ±187.3 | 0.0 -- | 3,493.0 ±187.3 | 855.7 ±13.4 | 158.7 ±28.0 | 1,014.4 ±29.5 | 4,507.4 ±189.8 |
| Ohio | 898.7 ±77.3 | 0.0 -- | 898.7 ±77.3 | 196.1 ±7.5 | 168.5 ±10.6 | 364.6 ±11.6 | 1,263.3 ±79.5 |
| Oklahoma | 402.9 ±59.1 | 0.0 -- | 402.9 ±59.1 | 629.2 ±20.2 | 295.4 ±55.0 | 924.6 ±60.5 | 1,327.5 ±80.2 |
| Oregon | 1,391.6 ±251.3 | *28.6 ±46.3* | 1,420.2 ±241.9 | 531.8 ±9.3 | 380.7 ±73.7 | 912.5 ±72.9 | 2,332.7 ±273.0 |
| Pennsylvania | 918.2 ±92.7 | 0.0 -- | 918.2 ±92.7 | 223.3 ±8.8 | 228.1 ±14.5 | 451.4 ±18.5 | 1,369.6 ±93.1 |
| Rhode Island | 89.1 ±13.4 | *6.2 ±6.3* | 95.3 ±15.0 | 20.9 ±1.7 | 128.2 ±1.2 | 149.1 ±2.0 | 244.4 ±14.9 |
| South Carolina | 3,276.1 ±137.6 | 422.4 ±84.5 | 3,698.5 ±135.6 | 375.0 ±14.5 | 396.4 ±49.9 | 771.4 ±49.9 | 4,469.9 ±140.1 |
| South Dakota | 2,150.0 ±189.7 | 0.0 -- | 2,150.0 ±189.7 | 632.3 ±8.1 | 151.9 ±10.2 | 784.2 ±13.1 | 2,934.2 ±194.0 |
| Tennessee | 642.1 ±62.7 | 0.0 -- | 642.1 ±62.7 | 472.5 ±6.0 | 234.7 ±9.4 | 707.2 ±9.2 | 1,349.3 ±63.7 |
| Texas | 4,791.8 ±361.9 | 372.5 ±108.7 | 5,164.3 ±339.6 | 1,547.0 ±43.2 | 2,357.2 ±79.6 | 3,904.2 ±87.4 | 9,068.5 ±377.4 |
| Utah | 1,118.4 ±404.0 | 0.0 -- | 1,118.4 ±404.0 | 2,047.7 ±839.6 | *61.0 ±615.8* | 2,108.7 ±373.6 | 3,227.1 ±580.3 |
| Vermont | 560.5 ±53.0 | 0.0 -- | 560.5 ±53.0 | 226.5 ±3.3 | 28.8 ±3.9 | 255.3 ±4.8 | 815.8 ±53.3 |

| State | Palustrine and Estuarine wetlands | | | Other aquatic habitats | | | Total |
|---|---|---|---|---|---|---|---|
| | Palustrine | Estuarine | Total | Lacustrine | Other (*) | Total | |
| Virginia | 1,388.0 ±88.5 | 165.9 ±51.8 | 1,553.9 ±96.8 | 146.9 ±7.2 | 1,739.9 ±17.2 | 1,886.8 ±19.6 | 3,440.7 ±98.9 |
| Washington | 925.8 ±140.4 | *44.6* *±90.5* | 970.4 ±163.7 | 770.4 ±13.6 | 785.3 ±45.6 | 1,555.7 ±46.5 | 2,526.1 ±173.1 |
| West Virginia | 98.2 ±25.4 | 0.0 -- | 98.2 ±25.4 | 71.9 ±7.5 | 102.3 ±6.9 | 174.2 ±11.3 | 272.4 ±30.1 |
| Wisconsin | 5,581.1 ±298.0 | 0.0 -- | 5,581.1 ±298.0 | 919.5 ±19.6 | 293.2 ±22.6 | 1,212.7 ±30.8 | 6,793.8 ±306.6 |
| Wyoming | 806.4 ±223.7 | 0.0 -- | 806.4 ±223.7 | 357.6 ±15.8 | 67.5 ±21.0 | 425.1 ±19.5 | 1,231.5 ±223.9 |
| Total | 105,278.4 ±1,429.8 | 5,393.1 ±337.1 | 110,671.5 ±1,358.7 | 24,960.9 ±870.2 | 23,510.2 ±633.6 | 48,471.1 ±474.7 | 159,142.6 ±1,405.3 |

Notes:
• (*) includes Estuarine deepwater, and all Riverine and Marine systems.
• When the estimate is 0.0, margins of error are not applicable and shown as a dashed line (--).
• Instances where the margin of error is greater than or equal to the estimate are displayed in italics indicating that the confidence interval includes zero and that the estimate should not be used.

U.S. Department of Agriculture. 2009. *Summary Report: 2007 National Resources Inventory*, Natural Resources Conservation Service, Washington, DC, and Center for Survey Statistics and Methodology, Iowa State University, Ames, Iowa. 123 pages.

## Superfund National Priorities List

| St. | City/Area | County | Site Name | Status | Score[1] |
|---|---|---|---|---|---|
| AK | Adak | Aleutians West Census Are | Adak Naval Air Station | Final | 51.37 |
| AK | Anchorage | Anchorage | Standard Steel & Metal Salvage Yard (Usdot) | Deleted | 46.25 |
| AK | Anchorage | Anchorage Borough | Elmendorf Air Force Base | Final | 45.91 |
| AK | Anchorage | Anchorage Borough | Fort Richardson (Usarmy) | Final | 50.00 |
| AK | Fairbanks | Fairbanks North Star | Alaska Battery Enterprises | Deleted | 30.98 |
| AK | Fairbanks | Fairbanks North Star Boro | Arctic Surplus | Deleted | 42.24 |
| AK | Fairbanks | Fairbanks North Star | Eielson Air Force Base | Final | 48.14 |
| AK | Fort Wainwright | Fairbanks North Star Boro | Fort Wainwright | Final | 42.40 |
| AK | Thorne Bay | Outer Ketchikan | Salt Chuck Mine | Final | 50.00 |
| AL | Anniston | Calhoun | Anniston Army Depot (Southeast Industrial Area) | Final | 51.91 |
| AL | Axis | Mobile | Stauffer Chemical Co. (Lemoyne Plant) | Final | 32.34 |
| AL | Bucks | Mobile | Stauffer Chemical Co. (Cold Creek Plant) | Final | 46.77 |
| AL | Childersburg | Talladega | Alabama Army Ammunition Plant | Final | 36.83 |
| AL | Greenville | Butler | Mowbray Engineering Co. | Deleted | 53.67 |
| AL | Headland | Henry | American Brass Inc. | Final | 55.61 |
| AL | Huntsville | Madison | Usarmy/Nasa Redstone Arsenal | Final | 50.00 |
| AL | Leeds | Jefferson | Interstate Lead Co. (Ilco) | Final | 42.86 |
| AL | Limestone/Morgan | Morgan/Limestone/Madison | Triana/Tennessee River | Final | 61.42 |
| AL | Mcintosh | Washington | Ciba-Geigy Corp. (Mcintosh Plant) | Final | 53.42 |
| AL | Mcintosh | Washington | Olin Corp. (Mcintosh Plant) | Final | 39.71 |
| AL | Montgomery | Montgomery | Capitol City Plume | Proposed | 50.00 |
| AL | Montgomery | Montgomery | T.H. Agriculture & Nutrition Co. (Montgomery Plant) | Final | 44.46 |
| AL | Perdido | Baldwin | Perdido Ground Water Contamination | Final | 30.29 |
| AL | Saraland | Mobile | Redwing Carriers, Inc. (Saraland) | Final | 30.83 |
| AL | Vincent | Shelby | Alabama Plating Co Inc | Proposed | 30.21 |
| AR | Edmondsen | Crittenden | Gurley Pit | Deleted | 40.13 |
| AR | El Dorado | Union | Popile, Inc. | Final | 50.03 |
| AR | Fort Smith | Sebastian | Industrial Waste Control | Deleted | 30.31 |
| AR | Jacksonville | Lonoke | Jacksonville Municipal Landfill | Deleted | 29.64 |
| AR | Jacksonville | Pulaski | Rogers Road Municipal Landfill | Deleted | 29.64 |
| AR | Jacksonville | Pulaski | Vertac, Inc. | Final | 65.46 |
| AR | Mena | Polk | Mid-South Wood Products | Final | 45.87 |
| AR | Newport | Jackson | Cecil Lindsey | Deleted | 35.60 |
| AR | Ola/Birta | Yell | Midland Products | Final | 30.77 |
| AR | Omaha | Boone | Arkwood, Inc. | Final | 28.95 |
| AR | Paragould | Greene | Monroe Auto Equipment Co. (Paragould Pit) | Final | 46.01 |
| AR | Plainview | Yell | Mountain Pine Pressure Treating | Final | 41.93 |
| AR | Reader | Ouachita | Ouachita Nevada Wood Treater | Final | 50.00 |
| AR | Walnut Ridge | Lawrence | Frit Industries | Deleted | 39.47 |
| AR | West Memphis | Crittenden | South 8th Street Landfill | Deleted | 50.27 |
| AS | Pago Pago | [Blank County] | Taputimu Farm | Deleted | - |
| AZ | Chandler | Maricopa | Williams Air Force Base | Final | 37.93 |
| AZ | Dewey-Humboldt | Yavapai | Iron King Mine - Humboldt Smelter | Final | 52.69 |
| AZ | Glendale | Maricopa | Luke Air Force Base | Deleted | 37.93 |
| AZ | Globe | Gila | Mountain View Mobile Home Estates | Deleted | - |
| AZ | Goodyear | Maricopa | Phoenix-Goodyear Airport Area | Final | 45.91 |
| AZ | Hassayampa | Maricopa | Hassayampa Landfill | Final | 42.79 |
| AZ | Phoenix | Maricopa | Motorola, Inc. (52nd Street Plant) | Final | 40.83 |
| AZ | Phoenix | Maricopa | Nineteenth Avenue Landfill | Deleted | 54.27 |
| AZ | Saint David | Cochise | Apache Powder Co. | Final | 39.09 |
| AZ | Scottsdale | Maricopa | Indian Bend Wash Area | Final | 42.24 |
| AZ | Tucson | Pima | Tucson International Airport Area | Final | 57.80 |
| AZ | Yuma | Yuma | Yuma Marine Corps Air Station | Final | 32.24 |
| CA | Alameda | Alameda | Alameda Naval Air Station | Final | 50.00 |
| CA | Alhambra | Los Angeles | San Gabriel Valley (Area 3) | Final | 28.90 |
| CA | Alviso | Santa Clara | South Bay Asbestos Area | Final | 44.68 |

| St. | City/Area | County | Site Name | Status | Score[1] |
|---|---|---|---|---|---|
| CA | Arvin | Kern | Brown & Bryant, Inc. (Arvin Plant) | Final | 53.36 |
| CA | Baldwin Park | Los Angeles | San Gabriel Valley (Area 2) | Final | 42.24 |
| CA | Barstow | San Bernardino | Barstow Marine Corps Logistics Base | Final | 37.93 |
| CA | Camp Pendleton | San Diego | Camp Pendleton Marine Corps Base | Final | 33.79 |
| CA | Casmalia | Santa Barbara | Casmalia Resources | Final | 30.00 |
| CA | Clearlake Oaks | Lake | Sulphur Bank Mercury Mine | Final | 44.42 |
| CA | Cloverdale | Sonoma | Mgm Brakes | Final | 34.70 |
| CA | Coalinga | Fresno | Atlas Asbestos Mine | Final | 45.55 |
| CA | Coalinga | Fresno | Coalinga Asbestos Mine | Deleted | 45.55 |
| CA | Concord | Contra Costa | Concord Naval Weapons Station | Final | 50.00 |
| CA | Crescent City | Del Norte | Del Norte Pesticide Storage | Deleted | 35.79 |
| CA | Cupertino | Santa Clara | Intersil Inc./Siemens Components | Final | 28.90 |
| CA | Davis | Solano | Laboratory For Energy-Related Health Research/Old Campus Landfill (Usdoe) | Final | 50.00 |
| CA | Davis | Yolo | Frontier Fertilizer | Final | 35.04 |
| CA | Edwards Afb | Kern | Edwards Air Force Base | Final | 33.62 |
| CA | El Monte | Los Angeles | San Gabriel Valley (Area 1) | Final | 42.24 |
| CA | El Toro | Orange | El Toro Marine Corps Air Station | Final | 37.43 |
| CA | Fillmore | Ventura | Pacific Coast Pipe Lines | Final | 46.01 |
| CA | Fresno | Fresno | Fresno Municipal Sanitary Landfill | Final | 35.57 |
| CA | Fresno | Fresno | Industrial Waste Processing | Final | 51.13 |
| CA | Fresno | Fresno | T.H. Agriculture & Nutrition Co. | Deleted | 42.24 |
| CA | Fullerton | Orange | Mccoll | Final | 41.77 |
| CA | Glendale | Los Angeles | San Fernando Valley (Area 2) | Final | 42.24 |
| CA | Glendale | Los Angeles | San Fernando Valley (Area 3) | Deleted | 42.24 |
| CA | Hoopa | Humboldt | Celtor Chemical Works | Deleted | 30.31 |
| CA | Idria | San Benito | New Idria Mercury Mine | Proposed | 31.66 |
| CA | Imperial | Imperial | Stoker Company | Proposed | 70.94 |
| CA | La Puente | Los Angeles | San Gabriel Valley (Area 4) | Final | 28.90 |
| CA | Lathrop | San Joaquin | Sharpe Army Depot | Final | 42.24 |
| CA | Livermore | Alameda | Lawrence Livermore Natl Lab, Main Site (Usdoe) | Final | 42.24 |
| CA | Los Angeles | Los Angeles | Del Amo | Final | 47.12 |
| CA | Los Angeles | Los Angeles | San Fernando Valley (Area 4) | Final | 35.57 |
| CA | Malaga | Fresno | Purity Oil Sales, Inc. | Final | 43.27 |
| CA | Marina | Monterey | Fort Ord | Final | 42.24 |
| CA | Markleeville | Alpine | Leviathan Mine | Final | 50.00 |
| CA | Mather | Sacramento | Mather Air Force Base (Ac&W Disposal Site) | Final | 28.90 |
| CA | Maywood | Los Angeles | Pemaco Maywood | Final | 45.23 |
| CA | Mcclellan Afb | Sacramento | Mcclellan Air Force Base (Ground Water Contamination) | Final | 57.93 |
| CA | Merced | Merced | Castle Air Force Base (6 Areas) | Final | 37.93 |
| CA | Mira Loma | Riverside | Stringfellow | Final | – |
| CA | Modesto | Stanislaus | Modesto Ground Water Contamination | Final | 28.90 |
| CA | Moffett Field | Santa Clara | Moffett Naval Air Station | Final | 29.49 |
| CA | Monterey Park | Los Angeles | Operating Industries, Inc., Landfill | Final | 57.22 |
| CA | Mountain View | Santa Clara | Cts Printex, Inc. | Final | 33.62 |
| CA | Mountain View | Santa Clara | Fairchild Semiconductor Corp. (Mountain View Plant) | Final | 31.94 |
| CA | Mountain View | Santa Clara | Intel Corp. (Mountain View Plant) | Final | 29.76 |
| CA | Mountain View | Santa Clara | Jasco Chemical Corp. | Final | 35.36 |
| CA | Mountain View | Santa Clara | Raytheon Corp. | Final | 29.76 |
| CA | Mountain View | Santa Clara | Spectra-Physics, Inc. | Final | 37.20 |
| CA | Mountain View | Santa Clara | Teledyne Semiconductor | Final | 35.35 |
| CA | Nevada City | Nevada | Lava Cap Mine | Final | 33.66 |
| CA | North Hollywood | Los Angeles | San Fernando Valley (Area 1) | Final | 42.24 |
| CA | Oakland | Alameda | Amco Chemical | Final | 50.00 |
| CA | Oroville | Butte | Koppers Co., Inc. (Oroville Plant) | Final | 33.73 |
| CA | Oroville | Butte | Louisiana-Pacific Corp. | Deleted | 33.73 |
| CA | Oroville | Butte | Western Pacific Railroad Co. | Deleted | 39.79 |
| CA | Oxnard | Ventura | Halaco Engineering Company | Final | 58.31 |
| CA | Palo Alto | Santa Clara | Hewlett-Packard (620-640 Page Mill Road) | Final | 29.76 |
| CA | Pasadena | Los Angeles | Jet Propulsion Laboratory (Nasa) | Final | 50.00 |

| St. | City/Area | County | Site Name | Status | Score[1] |
|---|---|---|---|---|---|
| CA | Paso Robles | San Luis Obispo | Klau/Buena Vista Mine | Final | 50.00 |
| CA | Petaluma | Sonoma | Sola Optical Usa, Inc. | Final | 33.39 |
| CA | Porterville | Tulare | Beckman Instruments (Porterville Plant) | Final | 34.21 |
| CA | Rancho Cordova | Sacramento | Aerojet General Corp. | Final | 54.63 |
| CA | Redding | Shasta | Iron Mountain Mine | Final | 56.16 |
| CA | Rialto | San Bernardino | B.F. Goodrich | Final | 50.00 |
| CA | Richmond | Contra Costa | Liquid Gold Oil Corp. | Deleted | 43.32 |
| CA | Richmond | Contra Costa | United Heckathorn Co. | Final | 38.49 |
| CA | Riverbank | Stanislaus | Riverbank Army Ammunition Plant | Final | 63.94 |
| CA | Riverside | Riverside | Alark Hard Chrome | Final | 50.50 |
| CA | Riverside | Riverside | March Air Force Base | Final | 31.94 |
| CA | Rogue River-Siskiyou Nf | Siskiyou | Blue Ledge Mine | Proposed | 50.28 |
| CA | Sacramento | Sacramento | Jibboom Junkyard | Deleted | 28.94 |
| CA | Sacramento | Sacramento | Sacramento Army Depot | Final | 44.46 |
| CA | Salinas | Monterey | Crazy Horse Sanitary Landfill | Final | 37.93 |
| CA | Salinas | Monterey | Firestone Tire & Rubber Co. (Salinas Plant) | Deleted | 45.91 |
| CA | San Bernardino | San Bernardino | Newmark Ground Water Contamination | Final | 35.57 |
| CA | San Bernardino | San Bernardino | Norton Air Force Base (Lndfl #2) | Final | 39.65 |
| CA | San Francisco | San Francisco | Treasure Island Naval Station-Hunters Point Annex | Final | 48.77 |
| CA | San Jose | Santa Clara | Fairchild Semiconductor Corp. (South San Jose Plant) | Final | 44.46 |
| CA | San Jose | Santa Clara | Lorentz Barrel & Drum Co. | Final | 33.94 |
| CA | Santa Clara | Santa Clara | Applied Materials | Final | 31.94 |
| CA | Santa Clara | Santa Clara | Intel Corp. (Santa Clara Iii) | Final | 31.94 |
| CA | Santa Clara | Santa Clara | Intel Magnetics | Final | 31.94 |
| CA | Santa Clara | Santa Clara | National Semiconductor Corp. | Final | 35.57 |
| CA | Santa Clara | Santa Clara | Synertek, Inc. (Building 1) | Final | 31.94 |
| CA | Santa Fe Springs | Los Angeles | Waste Disposal, Inc. | Final | 34.60 |
| CA | Scotts Valley | Santa Cruz | Watkins-Johnson Co. (Stewart Division Plant) | Final | 28.90 |
| CA | Selma | Fresno | Selma Treating Co. | Final | 48.83 |
| CA | South Gate | Los Angeles | Cooper Drum Co. | Final | 50.00 |
| CA | Stockton | San Joaquin | Mccormick & Baxter Creosoting Co. | Final | 74.86 |
| CA | Sunnyvale | Santa Clara | Advanced Micro Devices, Inc. | Final | 37.93 |
| CA | Sunnyvale | Santa Clara | Advanced Micro Devices, Inc. (Building 915) | Final | 31.94 |
| CA | Sunnyvale | Santa Clara | Monolithic Memories | Final | 35.57 |
| CA | Sunnyvale | Santa Clara | Trw Microwave, Inc (Building 825) | Final | 31.94 |
| CA | Sunnyvale | Santa Clara | Westinghouse Electric Corp. (Sunnyvale Plant) | Final | 39.93 |
| CA | Torrance | Los Angeles | Montrose Chemical Corp. | Final | 32.10 |
| CA | Tracy | San Joaquin | Lawrence Livermore Natl Lab (Site 300) (Usdoe) | Final | 31.58 |
| CA | Tracy | San Joaquin | Tracy Defense Depot (Usarmy) | Final | 37.16 |
| CA | Travis Afb | Solano | Travis Air Force Base | Final | 29.49 |
| CA | Turlock | Stanislaus | Valley Wood Preserving, Inc. | Final | 32.01 |
| CA | Ukiah | Mendocino | Coast Wood Preserving | Final | 44.73 |
| CA | Victorville | San Bernardino | George Air Force Base | Final | 33.62 |
| CA | Visalia | Tulare | Southern California Edison Co. (Visalia Poleyard) | Deleted | 48.91 |
| CA | Weed | Siskiyou | J.H. Baxter & Co. | Final | 34.78 |
| CA | Westminster | Orange | Ralph Gray Trucking Co. | Deleted | 35.04 |
| CA | Whittier | Los Angeles | Omega Chemical Corporation | Final | 30.94 |
| CO | Adams County | Adams | Rocky Mountain Arsenal (Usarmy) | Final | 58.15 |
| CO | Aspen | Pitkin | Smuggler Mountain | Deleted | 31.31 |
| CO | Aurora | Arapahoe | Lowry Landfill | Final | 48.36 |
| CO | Boulder | Boulder | Marshall Landfill | Final | – |
| CO | Canon City | Fremont | Lincoln Park | Final | 31.31 |
| CO | Commerce City | Adams | Sand Creek Industrial | Deleted | 59.65 |
| CO | Commerce City | Adams | Woodbury Chemical Co. | Deleted | 44.87 |
| CO | Creede | Mineral | Nelson Tunnel/Commodore Waste Rock | Final | 48.03 |
| CO | Denver | Adams | Asarco, Inc. (Globe Plant) | Proposed | 70.71 |
| CO | Denver | Adams | Broderick Wood Products | Final | 35.13 |
| CO | Denver | Denver | Chemical Sales Co. | Final | 37.93 |
| CO | Denver | Denver | Denver Radium Site | Final | 44.11 |

| St. | City/Area | County | Site Name | Status | Score[1] |
|---|---|---|---|---|---|
| CO | Denver | Denver | Vasquez Boulevard And I-70 | Final | 50.00 |
| CO | Golden | Jefferson | Rocky Flats Plant (Usdoe) | Final | 64.32 |
| CO | Gunnison National Forest | Gunnison | Standard Mine | Final | 50.00 |
| CO | Idaho Springs | Clear Creek | Central City, Clear Creek | Final | 51.39 |
| CO | Leadville | Lake | California Gulch | Final | 55.84 |
| CO | Littleton | Jefferson | Air Force Plant Pjks | Final | 42.93 |
| CO | Minturn | Eagle | Eagle Mine | Final | 47.19 |
| CO | Rio Grande County | Rio Grande | Summitville Mine | Final | 50.00 |
| CO | Salida | Chaffee | Smeltertown Site | Proposed | 58.56 |
| CO | Uravan | Montrose | Uravan Uranium Project (Union Carbide Corp.) | Final | 43.53 |
| CO | Ward | Boulder | Captain Jack Mill | Final | 50.56 |
| CT | Barkhamsted | Litchfield | Barkhamsted-New Hartford Landfill | Final | 38.05 |
| CT | Beacon Falls | New Haven | Beacon Heights Landfill | Final | 46.77 |
| CT | Canterbury | Windham | Yaworski Waste Lagoon | Final | 36.72 |
| CT | Cheshire | New Haven | Cheshire Ground Water Contamination | Deleted | 35.57 |
| CT | Durham | Middlesex | Durham Meadows | Final | 33.94 |
| CT | East Windsor | Hartford | Broad Brook Mill | Proposed | 54.35 |
| CT | Naugatuck Borough | New Haven | Laurel Park, Inc. | Final | – |
| CT | New London | New London | New London Submarine Base | Final | 36.53 |
| CT | Norwalk | Fairfield | Kellogg-Deering Well Field | Final | 39.92 |
| CT | Plainfield | Windham | Gallup's Quarry | Final | 46.29 |
| CT | Southington | Hartford | Old Southington Landfill | Final | 54.35 |
| CT | Southington | Hartford | Solvents Recovery Service Of New England | Final | 44.93 |
| CT | Sterling | Windham | Revere Textile Prints Corp. | Deleted | 41.06 |
| CT | Stratford | Fairfield | Raymark Industries, Inc. | Final | – |
| CT | Vernon | Tolland | Precision Plating Corp. | Final | 49.10 |
| CT | Waterbury | New Haven | Scovill Industrial Landfill | Final | 50.00 |
| CT | Wolcott | New Haven | Nutmeg Valley Road | Deleted | 42.69 |
| CT | Woodstock | Windham | Linemaster Switch Corp. | Final | 33.71 |
| DC | Washington | District Of Columbia | Washington Navy Yard | Final | 48.57 |
| DE | Cheswold | Kent | Coker's Sanitation Service Landfills | Final | 52.15 |
| DE | Delaware City | New Castle | Delaware City Pvc Plant | Final | 30.55 |
| DE | Dover | Kent | Chem-Solv, Inc. | Final | 37.93 |
| DE | Dover | Kent | Dover Air Force Base | Final | 35.89 |
| DE | Dover | Kent | Dover Gas Light Co. | Final | 35.57 |
| DE | Dover | Kent | Wildcat Landfill | Deleted | 30.61 |
| DE | Kirkwood | New Castle | Harvey & Knott Drum, Inc. | Final | 30.77 |
| DE | Laurel | Sussex | Sussex County Landfill No. 5 | Deleted | 28.90 |
| DE | Middletown | New Castle | Sealand Limited | Deleted | 33.10 |
| DE | Millsboro | Sussex | Millsboro Tce | Proposed | 50.00 |
| DE | Millsboro | Sussex | Ncr Corp. (Millsboro Plant) | Final | 38.21 |
| DE | New Castle | New Castle | Army Creek Landfill | Final | 69.92 |
| DE | New Castle | New Castle | Delaware Sand & Gravel Landfill | Final | 46.60 |
| DE | New Castle | New Castle | Halby Chemical Co. | Final | 30.90 |
| DE | New Castle | New Castle | New Castle Spill | Deleted | 38.33 |
| DE | New Castle | New Castle | New Castle Steel | Deleted | 30.40 |
| DE | New Castle | New Castle | Standard Chlorine Of Delaware, Inc. | Final | 35.42 |
| DE | New Castle | New Castle | Tybouts Corner Landfill | Final | – |
| DE | Newport | New Castle | E.I. Du Pont De Nemours & Co., Inc. (Newport Pigment Plant Landfill) | Final | 51.91 |
| DE | Newport | New Castle | Koppers Co., Inc. (Newport Plant) | Final | 33.56 |
| DE | Smyrna | Kent | Tyler Refrigeration Pit | Deleted | 33.94 |
| FL | Baldwin | Duval | Yellow Water Road | Deleted | 30.26 |
| FL | Brandon | Hillsborough | Sydney Mine Sludge Ponds | Final | 38.93 |
| FL | Cantonment | Escambia | Dubose Oil Products Co. | Deleted | 34.18 |
| FL | Clermont | Lake | Tower Chemical Co. | Final | 44.03 |
| FL | Cottondale | Jackson | Sapp Battery Salvage | Final | 47.70 |
| FL | Davie | Broward | Davie Landfill | Deleted | 57.86 |
| FL | Deland | Volusia | Sherwood Medical Industries | Final | 39.83 |
| FL | Duval County | Duval | Hipps Road Landfill | Final | 31.94 |

| St. | City/Area | County | Site Name | Status | Score[1] |
|---|---|---|---|---|---|
| FL | Fort Lauderdale | Broward | Florida Petroleum Reprocessors | Final | 50.00 |
| FL | Fort Lauderdale | Broward | Hollingsworth Solderless Terminal | Final | 44.53 |
| FL | Fort Lauderdale | Broward | Wingate Road Municipal Incinerator Dump | Final | 31.72 |
| FL | Gainesville | Alachua | Cabot/Koppers | Final | 36.69 |
| FL | Hialeah | Miami-Dade | B&B Chemical Co., Inc. | Final | 35.35 |
| FL | Hialeah | Miami-Dade | Northwest 58th Street Landfill | Deleted | 49.43 |
| FL | Hialeah | Miami-Dade | Standard Auto Bumper Corp. | Deleted | 42.79 |
| FL | Homestead Air Force Base | Miami-Dade | Homestead Air Force Base | Final | 42.40 |
| FL | Indiantown | Martin | Florida Steel Corp. | Final | 45.92 |
| FL | Jacksonville | Duval | Jacksonville Naval Air Station | Final | 32.08 |
| FL | Jacksonville | Duval | Kerr-Mcgee Chemical Corp - Jacksonville | Final | 70.71 |
| FL | Jacksonville | Duval | Pickettville Road Landfill | Final | 42.94 |
| FL | Jacksonville | Duval | Usn Air Station Cecil Field | Final | 31.99 |
| FL | Lake Alfred | Polk | Callaway & Son Drum Service | Deleted | 46.22 |
| FL | Lake Park | Palm Beach | Bmi-Textron | Deleted | 35.34 |
| FL | Lake Park | Palm Beach | Trans Circuits, Inc. | Final | 50.00 |
| FL | Lakeland | Polk | Alpha Chemical Corp. | Deleted | 43.24 |
| FL | Lakeland | Polk | Landia Chemical Company | Final | 50.00 |
| FL | Live Oak | Suwannee | Brown Wood Preserving | Deleted | 45.51 |
| FL | Longwood | Seminole | General Dynamics Longwood | Final | 50.00 |
| FL | Madison | Madison | Madison County Sanitary Landfill | Final | 37.93 |
| FL | Marianna | Jackson | United Metals, Inc. | Final | 33.73 |
| FL | Medley | Miami-Dade | Pepper Steel & Alloys, Inc. | Final | 31.92 |
| FL | Miami | Miami-Dade | Airco Plating Co. | Final | 42.47 |
| FL | Miami | Miami-Dade | Anaconda Aluminum Co./Milgo Electronics Corp. | Deleted | 31.03 |
| FL | Miami | Miami-Dade | Gold Coast Oil Corp. | Deleted | 57.80 |
| FL | Miami | Miami-Dade | Miami Drum Services | Final | 53.56 |
| FL | Miami | Miami-Dade | Varsol Spill | Deleted | 44.46 |
| FL | Milton | Santa Rosa | Whiting Field Naval Air Station | Final | 50.00 |
| FL | Mount Pleasant | Gadsden | Parramore Surplus | Deleted | 37.61 |
| FL | North Miami | Miami-Dade | Munisport Landfill | Deleted | 32.37 |
| FL | North Miami Beach | Miami-Dade | Anodyne, Inc. | Final | 31.03 |
| FL | Orlando | Orange | Chevron Chemical Co. (Ortho Division) | Final | 50.00 |
| FL | Orlando | Orange | City Industries, Inc. | Final | 32.00 |
| FL | Palm Bay | Brevard | Harris Corp. (Palm Bay Plant) | Final | 35.57 |
| FL | Panama City | Bay | Tyndall Air Force Base | Final | 50.00 |
| FL | Pembroke Park | Broward | Petroleum Products Corp. | Final | 40.11 |
| FL | Pensacola | Escambia | Agrico Chemical Co. | Final | 44.98 |
| FL | Pensacola | Escambia | American Creosote Works, Inc. (Pensacola Plant) | Final | 58.41 |
| FL | Pensacola | Escambia | Beulah Landfill | Deleted | 38.17 |
| FL | Pensacola | Escambia | Escambia Wood - Pensacola | Final | 50.00 |
| FL | Pensacola | Escambia | Pensacola Naval Air Station | Final | 42.40 |
| FL | Plant City | Hillsborough | Schuylkill Metals Corp. | Deleted | 59.16 |
| FL | Pompano Beach | Broward | Chemform, Inc. | Deleted | 37.93 |
| FL | Pompano Beach | Broward | Flash Cleaners | Final | 50.00 |
| FL | Pompano Beach | Broward | Wilson Concepts Of Florida, Inc. | Deleted | 37.93 |
| FL | Port Salerno | Martin | Solitron Microwave | Final | 50.00 |
| FL | Princeton | Miami-Dade | Woodbury Chemical Co. (Princeton Plant) | Deleted | 39.43 |
| FL | Ruskin | Hillsborough | Jj Seifert Machine | Final | 50.00 |
| FL | Sanford | Seminole | Sanford Dry Cleaners | Final | 50.00 |
| FL | Seffner | Hillsborough | Taylor Road Landfill | Final | 51.37 |
| FL | Tampa | Hillsborough | Alaric Area Gw Plume | Final | 41.91 |
| FL | Tampa | Hillsborough | Helena Chemical Co. (Tampa Plant) | Final | 30.19 |
| FL | Tampa | Hillsborough | Kassauf-Kimerling Battery Disposal | Deleted | 53.42 |
| FL | Tampa | Hillsborough | Mri Corp (Tampa) | Final | 37.62 |
| FL | Tampa | Hillsborough | Peak Oil Co./Bay Drum Co. | Final | 58.15 |
| FL | Tampa | Hillsborough | Raleigh Street Dump | Final | 50.00 |
| FL | Tampa | Hillsborough | Reeves Southeastern Galvanizing Corp. | Final | 58.75 |
| FL | Tampa | Hillsborough | Sixty-Second Street Dump | Deleted | 49.09 |

| St. | City/Area | County | Site Name | Status | Score[1] |
|---|---|---|---|---|---|
| FL | Tampa | Hillsborough | Southern Solvents, Inc. | Final | 50.00 |
| FL | Tampa | Hillsborough | Stauffer Chemical Co (Tampa) | Final | 59.81 |
| FL | Tampa | Hillsborough | Tri-City Oil Conservationist, Inc | Deleted | 39.30 |
| FL | Tarpon Springs | Pinellas | Stauffer Chemical Co. (Tarpon Springs) | Final | 50.00 |
| FL | Temple Terrace | Hillsborough | Normandy Park Apartments | Proposed | 49.98 |
| FL | Thonotosassa | Hillsborough | Arkla Terra Property | Final | 50.00 |
| FL | Vero Beach | Indian River | Piper Aircraft Corp./Vero Beach Water & Sewer Department | Final | 31.13 |
| FL | Warrington | Escambia | Pioneer Sand Co. | Deleted | 51.97 |
| FL | Whitehouse | Duval | Coleman-Evans Wood Preserving Co. | Final | 46.18 |
| FL | Whitehouse | Duval | Whitehouse Oil Pits | Final | 52.58 |
| FL | Zellwood | Orange | Zellwood Ground Water Contamination | Final | 51.91 |
| FM | Palau | [Blank County] | Pcb Wastes | Deleted | – |
| GA | Albany | Dougherty | Firestone Tire & Rubber Co. (Albany Plant) | Final | 30.08 |
| GA | Albany | Dougherty | Marine Corps Logistics Base | Final | 44.65 |
| GA | Albany | Dougherty | T.H. Agriculture & Nutrition Co. (Albany Plant) | Final | 40.93 |
| GA | Athens | Clarke | Luminous Processes, Inc. | Deleted | – |
| GA | Augusta | Richmond | Alternate Energy Resources Inc | Final | 50.00 |
| GA | Augusta | Richmond | Monsanto Corp. (Augusta Plant) | Deleted | 35.65 |
| GA | Augusta | Richmond | Peach Orchard Rd Pce Groundwater Plume Site | Final | 50.00 |
| GA | Brunswick | Glynn | Brunswick Wood Preserving | Final | 54.49 |
| GA | Brunswick | Glynn | Hercules 009 Landfill | Final | 52.58 |
| GA | Brunswick | Glynn | Lcp Chemicals Georgia | Final | – |
| GA | Brunswick | Glynn | Terry Creek Dredge Spoil Areas/Hercules Outfall | Proposed | 50.18 |
| GA | Camilla | Mitchell | Camilla Wood Preserving Company | Final | 50.00 |
| GA | Cedartown | Polk | Cedartown Industries, Inc. | Deleted | 42.00 |
| GA | Cedartown | Polk | Cedartown Municipal Landfill | Deleted | 33.62 |
| GA | Cedartown | Polk | Diamond Shamrock Corp. Landfill | Final | 35.60 |
| GA | Fort Valley | Peach | Woolfolk Chemical Works, Inc. | Final | 42.24 |
| GA | Houston County | Houston | Robins Air Force Base (Landfill #4/Sludge Lagoon) | Final | 51.66 |
| GA | Kensington | Walker | Mathis Brothers Landfill (South Marble Top Road) | Final | 30.78 |
| GA | Macon | Bibb | Armstrong World Industries | Proposed | 50.00 |
| GA | Peach County | Peach | Powersville Site | Deleted | 35.53 |
| GA | Tifton | Tift | Marzone Inc./Chevron Chemical Co. | Final | 30.26 |
| GU | Agana | Guam | Ordot Landfill | Final | – |
| GU | Yigo | Guam | Andersen Air Force Base | Final | 50.00 |
| HI | Kunia | Honolulu | Del Monte Corp. (Oahu Plantation) | Final | 50.00 |
| HI | Pearl Harbor | Honolulu | Pearl Harbor Naval Complex | Final | 70.82 |
| HI | Schofield | Honolulu | Schofield Barracks (Usarmy) | Deleted | 28.90 |
| HI | Wahiawa | Honolulu | Naval Computer And Telecommunications Area Master Station Eastern Pacific | Final | 50.00 |
| IA | Camanche | Clinton | Lawrence Todtz Farm | Final | 52.11 |
| IA | Cedar Rapids | Linn | Electro-Coatings, Inc. | Final | 42.24 |
| IA | Charles City | Floyd | Labounty | Deleted | 70.73 |
| IA | Charles City | Floyd | Shaw Avenue Dump | Final | 30.01 |
| IA | Charles City | Floyd | White Farm Equipment Co. Dump | Deleted | 43.40 |
| IA | Des Moines | Polk | Des Moines Tce | Final | 42.28 |
| IA | Dubuque | Dubuque | Peoples Natural Gas Co. | Final | 46.24 |
| IA | Fairfield | Jefferson | Fairfield Coal Gasification Plant | Final | 38.05 |
| IA | Hospers | Sioux | Farmers' Mutual Cooperative | Deleted | 33.74 |
| IA | Kellogg | Jasper | Midwest Manufacturing/North Farm | Final | 32.04 |
| IA | Keokuk | Lee | Sheller-Globe Corp. Disposal | Deleted | 33.66 |
| IA | Mason City | Cerro Gordo | Mason City Coal Gasification Plant | Final | 69.33 |
| IA | Mason City | Cerro Gordo | Northwestern States Portland Cement Co. | Deleted | 57.80 |
| IA | Maurice | Sioux | Vogel Paint & Wax Co. | Final | 31.45 |
| IA | Middletown | Des Moines | Iowa Army Ammunition Plant | Final | 29.73 |
| IA | Mineola | Mills | Aidex Corp. | Deleted | – |
| IA | Ottumwa | Wapello | John Deere (Ottumwa Works Landfills) | Deleted | 42.32 |
| IA | Red Oak | Montgomery | Red Oak City Landfill | Deleted | 34.13 |
| IA | Sergeant Bluff | Woodbury | Mid-America Tanning Co. | Deleted | 47.91 |

| St. | City/Area | County | Site Name | Status | Score[1] |
|---|---|---|---|---|---|
| IA | Waterloo | Black Hawk | Waterloo Coal Gasification Plant | Proposed | 50.00 |
| IA | West Des Moines | Polk | Railroad Avenue Groundwater Contamination | Final | 50.00 |
| IA | West Point | Lee | E.I. Du Pont De Nemours & Co., Inc. (County Road X23) | Deleted | 46.01 |
| ID | Idaho Falls | Butte/Clark/Jefferson | Idaho National Engineering Laboratory (Usdoe) | Final | 51.91 |
| ID | Lemhi County | Lemhi | Blackbird Mine | Proposed | 50.00 |
| ID | Mountain Home | Elmore | Mountain Home Air Force Base | Final | 57.80 |
| ID | Pocatello | Bannock | Pacific Hide & Fur Recycling Co. | Deleted | 42.30 |
| ID | Pocatello | Bannock | Union Pacific Railroad Co. | Deleted | 53.47 |
| ID | Pocatello | Power/Bannock | Eastern Michaud Flats Contamination | Final | 57.80 |
| ID | Rathdrum | Kootenai | Arrcom (Drexler Enterprises) | Deleted | 29.28 |
| ID | Smelterville | Shoshone | Bunker Hill Mining & Metallurgical Complex | Final | 54.76 |
| ID | Soda Springs | Caribou | Kerr-Mcgee Chemical Corp. (Soda Springs Plant) | Final | 51.91 |
| ID | Soda Springs | Caribou | Monsanto Chemical Co. (Soda Springs Plant) | Final | 54.77 |
| ID | St. Maries | Benewah | St. Maries Creosote | Proposed | 50.00 |
| ID | Stibnite | Valley | Stibnite/Yellow Pine Mining Area | Proposed | 50.00 |
| IL | Antioch | Lake | H.O.D. Landfill | Final | 34.68 |
| IL | Beckemeyer | Clinton | Circle Smelting Corp. | Proposed | 70.71 |
| IL | Belvidere | Boone | Belvidere Municipal Landfill | Final | 28.62 |
| IL | Belvidere | Boone | Mig/Dewane Landfill | Final | 49.91 |
| IL | Belvidere | Boone | Parsons Casket Hardware Co. | Final | 55.58 |
| IL | Byron | Ogle | Byron Salvage Yard | Final | 33.93 |
| IL | Carterville | Williamson | Sangamo Electric Dump/Crab Orchard National Wildlife Refuge (Usdoi) | Final | 43.70 |
| IL | Chicago | Cook | Lake Calumet Cluster | Final | 30.00 |
| IL | Danville | Vermilion | Hegeler Zinc | Final | 50.00 |
| IL | Depue | Bureau | Depue/New Jersey Zinc/Mobil Chemical Corp. | Final | 70.71 |
| IL | Dupage County | Dupage | Kerr-Mcgee (Kress Creek/West Branch Of Dupage River) | Final | 39.05 |
| IL | East Cape Girardeau | Alexander | Ilada Energy Co. | Deleted | 34.21 |
| IL | Galesburg | Knox | Galesburg/Koppers Co. | Final | 34.78 |
| IL | Granite City | Madison | Jennison-Wright Corporation | Final | 40.30 |
| IL | Granite City | Madison | Nl Industries/Taracorp Lead Smelter | Final | 38.11 |
| IL | Greenup | Cumberland | A & F Material Reclaiming, Inc. | Final | 55.49 |
| IL | Hartford | Madison | Chemetco | Final | 30.00 |
| IL | Hillsboro | Montgomery | Eagle Zinc Co Div T L Diamond | Final | 50.00 |
| IL | Joliet | Will | Amoco Chemicals (Joliet Landfill) | Final | 39.44 |
| IL | Joliet | Will | Joliet Army Ammunition Plant (Load-Assembly-Packing Area) | Final | 35.23 |
| IL | Joliet | Will | Joliet Army Ammunition Plant (Manufacturing Area) | Final | 32.08 |
| IL | La Salle | La Salle | Lasalle Electric Utilities | Final | 42.06 |
| IL | La Salle | La Salle | Matthiessen And Hegeler Zinc Company | Final | 50.00 |
| IL | Lawrenceville | Lawrence | Indian Refinery-Texaco Lawrenceville | Final | 56.67 |
| IL | Lemont | Dupage | Lenz Oil Service, Inc. | Final | 42.33 |
| IL | Libertyville | Lake | Petersen Sand & Gravel | Deleted | 38.43 |
| IL | Marshall | Clark | Velsicol Chemical Corp. (Marshall Plant) | Final | 48.78 |
| IL | Morristown | Winnebago | Acme Solvent Reclaiming, Inc. (Morristown Plant) | Final | 31.98 |
| IL | Ottawa | La Salle | Ottawa Radiation Areas | Final | 50.00 |
| IL | Pembroke Township | Kankakee | Cross Brothers Pail Recycling (Pembroke) | Final | 42.04 |
| IL | Quincy | Adams | Adams County Quincy Landfills 2&3 | Final | 34.21 |
| IL | Rantoul | Champaign | Chanute Air Force Base | Proposed | 48.30 |
| IL | Rockford | Winnebago | Interstate Pollution Control, Inc. | Final | 46.01 |
| IL | Rockford | Winnebago | Pagel's Pit | Final | 45.91 |
| IL | Rockford | Winnebago | Southeast Rockford Ground Water Contamination | Final | 42.24 |
| IL | Rockton | Winnebago | Beloit Corp. | Final | 52.08 |
| IL | Sandoval | Marion | Sandoval Zinc Company | Proposed | 30.00 |
| IL | Sauget | St. Clair | Sauget Area 1 | Proposed | 50.00 |
| IL | Sauget | St. Clair | Sauget Area 2 | Proposed | 50.00 |
| IL | Savanna | Jo Daviess | Savanna Army Depot Activity | Final | 42.20 |
| IL | South Elgin | Kane | Tri-County Landfill Co./Waste Management Of Illinois, Inc. | Final | 42.76 |
| IL | Taylor Springs | Montgomery | Asarco Taylor Springs | Final | 30.00 |
| IL | Taylorville | Christian | Central Illinois Public Service Co. | Final | 28.95 |
| IL | Warrenville | Dupage | Dupage County Landfill/Blackwell Forest Preserve | Final | 35.57 |

| St. | City/Area | County | Site Name | Status | Score[1] |
|---|---|---|---|---|---|
| IL | Wauconda | Lake | Wauconda Sand & Gravel | Final | 53.42 |
| IL | Waukegan | Lake | Johns-Manville Corp. | Final | 38.20 |
| IL | Waukegan | Lake | Outboard Marine Corp. | Final | – |
| IL | Waukegan | Lake | Yeoman Creek Landfill | Final | 33.23 |
| IL | West Chicago | Dupage | Kerr-Mcgee (Reed-Keppler Park) | Deleted | 39.51 |
| IL | West Chicago | Dupage | Kerr-Mcgee (Residential Areas) | Final | 38.15 |
| IL | West Chicago | Dupage | Kerr-Mcgee (Sewage Treatment Plant) | Final | 35.20 |
| IL | Winnebago County | Winnebago | Evergreen Manor Ground Water Contamination | Proposed | 50.00 |
| IL | Woodstock | Mchenry | Woodstock Municipal Landfill | Final | 50.10 |
| IN | Bloomington | Monroe | Bennett Stone Quarry | Final | 32.55 |
| IN | Bloomington | Monroe | Lemon Lane Landfill | Final | 29.31 |
| IN | Bloomington | Monroe | Neal's Landfill (Bloomington) | Final | 42.93 |
| IN | Claypool | Kosciusko | Lakeland Disposal Service, Inc. | Final | 34.10 |
| IN | Columbia City | Whitley | Wayne Waste Oil | Final | 42.33 |
| IN | Columbus | Bartholomew | Columbus Old Municipal Landfill #1 | Final | 45.31 |
| IN | Columbus | Bartholomew | Tri-State Plating | Deleted | 29.28 |
| IN | East Chicago | Lake | U.S. Smelter And Lead Refinery, Inc. | Final | 58.31 |
| IN | Elkhart | Elkhart | Conrail Rail Yard (Elkhart) | Final | 42.24 |
| IN | Elkhart | Elkhart | Himco Dump | Final | 42.31 |
| IN | Elkhart | Elkhart | Lane Street Ground Water Contamination | Final | 40.53 |
| IN | Elkhart | Elkhart | Lusher Street Ground Water Contamination | Final | 50.00 |
| IN | Elkhart | Elkhart | Main Street Well Field | Final | 42.49 |
| IN | Evansville | Vanderburgh | Jacobsville Neighborhood Soil Contamination | Final | 35.52 |
| IN | Fort Wayne | Allen | Fort Wayne Reduction Dump | Final | 42.47 |
| IN | Gary | Lake | Gary Development Landfill | Proposed | 30.00 |
| IN | Gary | Lake | Lake Sandy Jo (M&M Landfill) | Final | 38.21 |
| IN | Gary | Lake | Midco I | Final | 46.44 |
| IN | Gary | Lake | Midco Ii | Final | 30.16 |
| IN | Gary | Lake | Ninth Avenue Dump | Final | 40.32 |
| IN | Griffith | Lake | American Chemical Service, Inc. | Final | 34.98 |
| IN | Hancock County | Hancock | Poer Farm | Deleted | 37.38 |
| IN | Indianapolis | Marion | Carter Lee Lumber Co. | Deleted | 35.40 |
| IN | Indianapolis | Marion | Reilly Tar & Chemical Corp. (Indianapolis Plant) | Final | 34.03 |
| IN | Indianapolis | Marion | Southside Sanitary Landfill | Deleted | 41.94 |
| IN | Kokomo | Howard | Continental Steel Corp. | Final | 31.85 |
| IN | La Porte | La Porte | Fisher-Calo | Final | 52.05 |
| IN | Lafayette | Tippecanoe | Tippecanoe Sanitary Landfill, Inc. | Final | 42.24 |
| IN | Lebanon | Boone | Wedzeb Enterprises, Inc. | Deleted | 31.27 |
| IN | Marion | Grant | Marion (Bragg) Dump | Final | 35.25 |
| IN | Michigan City | La Porte | Waste, Inc., Landfill | Deleted | 50.63 |
| IN | Mishawaka | St. Joseph | Douglass Road/Uniroyal, Inc., Landfill | Final | 36.61 |
| IN | Osceola | St. Joseph | Galen Myers Dump/Drum Salvage | Final | 42.24 |
| IN | Seymour | Jackson | Seymour Recycling Corp. | Final | – |
| IN | South Bend | St. Joseph | Whiteford Sales & Service Inc./Nationalease | Deleted | 51.87 |
| IN | Spencer | Owen | Neal's Dump (Spencer) | Deleted | 36.55 |
| IN | Terre Haute | Vigo | Elm Street Ground Water Contamination | Final | 50.00 |
| IN | Terre Haute | Vigo | International Minerals (E. Plant) | Deleted | 57.80 |
| IN | Vincennes | Knox | Prestolite Battery Division | Final | 40.63 |
| IN | Westville | La Porte | Cam-Or Inc. | Final | 58.91 |
| IN | Zionsville | Boone | Envirochem Corp. | Final | 46.44 |
| IN | Zionsville | Boone | Northside Sanitary Landfill, Inc | Final | 46.04 |
| KS | Arkansas City | Cowley | Arkansas City Dump | Deleted | – |
| KS | Colby | Thomas | Ace Services | Final | 50.00 |
| KS | Delavan | Morris | Tri-County Public Airport | Proposed | 50.00 |
| KS | El Dorado | Butler | Pester Refinery Co. | Final | 30.16 |
| KS | Galena | Cherokee | Cherokee County | Final | 58.15 |
| KS | Great Bend | Barton | Plating, Inc. | Final | 50.00 |
| KS | Hutchinson | Reno | Obee Road | Final | 33.62 |
| KS | Junction City | Geary | Fort Riley | Final | 33.79 |

| St. | City/Area | County | Site Name | Status | Score[1] |
|---|---|---|---|---|---|
| KS | Olathe | Johnson | Chemical Commodities, Inc. | Final | 50.00 |
| KS | Shawnee Mission | Johnson | Doepke Disposal (Holliday) | Final | 47.46 |
| KS | Topeka | Shawnee | Hydro-Flex Inc. | Deleted | 42.79 |
| KS | Wichita | Sedgwick | 29th & Mead Ground Water Contamination | Deleted | 35.35 |
| KS | Wichita | Sedgwick | 57th And North Broadway Streets Site | Final | 50.00 |
| KS | Wichita | Sedgwick | Big River Sand Co. | Deleted | 32.56 |
| KS | Wichita | Sedgwick | Johns' Sludge Pond | Deleted | 35.94 |
| KS | Winfield | Cowley | Strother Field Industrial Park | Final | 33.62 |
| KS | Wright | Ford | Wright Ground Water Contamination | Final | 50.00 |
| KY | Auburn | Logan | Caldwell Lace Leather Co., Inc. | Final | 34.21 |
| KY | Brooks | Bullitt | A.L. Taylor (Valley Of Drums) | Deleted | – |
| KY | Brooks | Bullitt | Smith's Farm | Final | 32.69 |
| KY | Calvert City | Marshall | Airco | Final | 33.29 |
| KY | Calvert City | Marshall | B.F. Goodrich | Final | 33.01 |
| KY | Dayhoit | Harlan | National Electric Coil Co./Cooper Industries | Final | 50.00 |
| KY | Hawesville | Hancock | National Southwire Aluminum Co. | Final | 50.00 |
| KY | Hillsboro | Fleming | Maxey Flats Nuclear Disposal | Final | 31.71 |
| KY | Howe Valley | Hardin | Howe Valley Landfill | Deleted | 36.73 |
| KY | Island | Mclean | Brantley Landfill | Final | 52.73 |
| KY | Jefferson County | Jefferson | Distler Farm | Final | 34.62 |
| KY | Louisville | Jefferson | Lee's Lane Landfill | Deleted | 39.52 |
| KY | Maceo | Daviess | Green River Disposal, Inc. | Final | 29.12 |
| KY | Mayfield | Graves | General Tire & Rubber Co. (Mayfield Landfill) | Deleted | 32.94 |
| KY | Newport | Campbell | Newport Dump | Deleted | 37.63 |
| KY | Olaton | Ohio | Fort Hartford Coal Co. Stone Quarry | Final | 43.84 |
| KY | Paducah | Mccracken | Paducah Gaseous Diffusion Plant (Usdoe) | Final | 56.95 |
| KY | Peewee Valley | Oldham | Red Penn Sanitation Co. Landfill | Deleted | 38.10 |
| KY | Shepherdsville | Bullitt | Tri-City Disposal Co. | Final | 33.82 |
| KY | West Point | Hardin | Distler Brickyard | Final | 44.77 |
| LA | Abbeville | Vermilion | D.L. Mud, Inc. | Deleted | 32.37 |
| LA | Abbeville | Vermilion Parish | Gulf Coast Vacuum Services | Deleted | 42.78 |
| LA | Abbeville | Vermilion Parish | Pab Oil & Chemical Service, Inc. | Deleted | 38.94 |
| LA | Alexandria | Rapides Parish | Ruston Foundry | Deleted | 43.17 |
| LA | Ascension Parish | Ascension Parish | Dutchtown Treatment Plant | Deleted | 36.41 |
| LA | Bayou Sorrel | Iberville Parish | Bayou Sorrel | Deleted | 34.69 |
| LA | Bossier City | Bossier Parish | Highway 71/72 Refinery | Proposed | 50.00 |
| LA | Darrow | Ascension Parish | Old Inger Oil Refinery | Deleted | – |
| LA | Denham Springs | Livingston Parish | Combustion, Inc. | Final | 33.79 |
| LA | Doyline | Webster Parish | Louisiana Army Ammunition Plant | Final | 30.26 |
| LA | Grand Cheniere | Cameron Parish | Mallard Bay Landing Bulk Plant | Deleted | 48.55 |
| LA | Lake Charles | Calcasieu Parish | Gulf State Utilities-North Ryan Street | Proposed | 50.43 |
| LA | Madisonville | St. Tammany | Madisonville Creosote Works | Final | 48.01 |
| LA | Marion | Union Parish | Marion Pressure Treating | Final | 50.00 |
| LA | New Orleans | Orleans Parish | Agriculture Street Landfill | Final | 50.00 |
| LA | Ponchatoula | Tangipahoa Parish | Delatte Metals | Deleted | 50.00 |
| LA | Scotlandville | East Baton Rouge | Devil's Swamp Lake | Proposed | 50.00 |
| LA | Scotlandville | East Baton Rouge Parish | Petro-Processors Of Louisiana, Inc. | Final | 41.44 |
| LA | Slaughter | East Feliciana Parish | Central Wood Preserving Co. | Deleted | 48.53 |
| LA | Slidell | St. Tammany | Bayou Bonfouca | Final | 29.78 |
| LA | Slidell | St. Tammany Parish | Southern Shipbuilding | Deleted | 50.00 |
| LA | Sorrento | Ascension Parish | Cleve Reber | Deleted | 48.80 |
| LA | Winnfield | Winn Parish | American Creosote Works, Inc. (Winnfield Plant) | Final | 50.70 |
| MA | Acton | Middlesex | W.R. Grace & Co., Inc. (Acton Plant) | Final | 59.31 |
| MA | Ashland | Middlesex | Nyanza Chemical Waste Dump | Final | 69.22 |
| MA | Bedford | Middlesex | Hanscom Field/Hanscom Air Force Base | Final | 50.00 |
| MA | Bedford | Middlesex | Naval Weapons Industrial Reserve Plant | Final | 50.00 |
| MA | Billerica | Middlesex | Iron Horse Park | Final | 42.93 |
| MA | Bridgewater | Plymouth | Cannon Engineering Corp. (Cec) | Final | 39.89 |
| MA | Concord | Middlesex | Nuclear Metals, Inc. | Final | 58.31 |

| St. | City/Area | County | Site Name | Status | Score[1] |
|---|---|---|---|---|---|
| MA | Dartmouth | Bristol | Re-Solve, Inc. | Final | 47.71 |
| MA | Fairhaven | Bristol | Atlas Tack Corp. | Final | 42.60 |
| MA | Falmouth | Barnstable | Otis Air National Guard Base/Camp Edwards | Final | 45.92 |
| MA | Fort Devens | Worcester/Middlesex | Fort Devens | Final | 42.24 |
| MA | Groveland | Essex | Groveland Wells | Final | 40.74 |
| MA | Haverhill | Essex | Haverhill Municipal Landfill | Final | 30.29 |
| MA | Holbrook | Norfolk | Baird & Mcguire | Final | 66.35 |
| MA | Lanesboro | Berkshire | Rose Disposal Pit | Final | 33.03 |
| MA | Lowell | Middlesex | Silresim Chemical Corp. | Final | 42.72 |
| MA | Mansfield | Bristol | Hatheway & Patterson | Final | 56.60 |
| MA | Natick | Middlesex | Natick Laboratory Army Research, Development, And Engineering Center | Final | 50.00 |
| MA | New Bedford | Bristol | New Bedford | Final | – |
| MA | New Bedford | Bristol | Sullivan's Ledge | Final | 32.77 |
| MA | Norton/Attleboro | Bristol | Shpack Landfill | Final | 29.45 |
| MA | Norwood | Norfolk | Norwood Pcbs | Final | 29.43 |
| MA | Palmer | Hampden | Psc Resources | Final | 38.66 |
| MA | Pittsfield | Berkshire | Ge - Housatonic River | Proposed | 70.71 |
| MA | Plymouth | Plymouth | Plymouth Harbor/Cannon Engineering Corp. | Deleted | 54.82 |
| MA | Salem | Essex | Salem Acres | Deleted | 34.94 |
| MA | Sudbury | Middlesex | Fort Devens-Sudbury Training Annex | Deleted | 35.57 |
| MA | Tewksbury | Middlesex | Sutton Brook Disposal Area | Final | 57.12 |
| MA | Tyngsborough | Middlesex | Charles-George Reclamation Trust Landfill | Final | 47.20 |
| MA | Walpole | Norfolk | Blackburn & Union Privileges | Final | 50.00 |
| MA | Watertown | Middlesex | Materials Technology Laboratory (Usarmy) | Deleted | 48.57 |
| MA | Westborough | Worcester | Hocomonco Pond | Final | 44.80 |
| MA | Weymouth | Norfolk | South Weymouth Naval Air Station | Final | 50.00 |
| MA | Wilmington | Middlesex | Olin Chemical | Final | 50.00 |
| MA | Woburn | Middlesex | Industri-Plex | Final | 72.42 |
| MA | Woburn | Middlesex | Wells G&H | Final | 42.71 |
| MD | Aberdeen | Harford | Aberdeen Proving Ground (Michaelsville Landfill) | Final | 31.09 |
| MD | Abingdon | Harford | Bush Valley Landfill | Final | 40.30 |
| MD | Andrews Air Force Base | Prince George's | Andrews Air Force Base | Final | 50.00 |
| MD | Annapolis | Anne Arundel | Middletown Road Dump | Deleted | 29.36 |
| MD | Baltimore | Anne Arundel | Curtis Bay Coast Guard Yard | Final | 50.00 |
| MD | Baltimore | Baltimore City | Chemical Metals Industries, Inc. | Deleted | – |
| MD | Baltimore | Baltimore City | Kane & Lombard Street Drums | Final | 30.15 |
| MD | Beltsville | Prince George's | Beltsville Agricultural Research Center (Usda) | Final | 50.00 |
| MD | Brandywine | Prince George's | Brandywine Drmo | Final | 50.15 |
| MD | Colora | Cecil | Woodlawn County Landfill | Final | 48.13 |
| MD | Cumberland | Allegany | Limestone Road | Final | 30.54 |
| MD | Dundalk | Baltimore | Sauer Dump | Proposed | 50.00 |
| MD | Edgewood | Harford | Aberdeen Proving Ground (Edgewood Area) | Final | 53.57 |
| MD | Elkton | Cecil | Dwyer Property Ground Water Plume | Final | 50.00 |
| MD | Elkton | Cecil | Sand, Gravel And Stone | Final | 41.08 |
| MD | Elkton | Cecil | Spectron, Inc. | Final | 51.42 |
| MD | Fort Detrick | Frederick | Fort Detrick Area B Ground Water | Final | 49.52 |
| MD | Hagerstown | Washington | Central Chemical (Hagerstown) | Final | 50.00 |
| MD | Harmans | Anne Arundel | Mid-Atlantic Wood Preservers, Inc. | Deleted | 42.31 |
| MD | Hollywood | St. Mary's | Southern Maryland Wood Treating | Deleted | 34.21 |
| MD | Indian Head | Charles | Indian Head Naval Surface Warfare Center | Final | 50.00 |
| MD | North East | Cecil | Ordnance Products, Inc. | Final | 32.15 |
| MD | Odenton | Anne Arundel | Fort George G. Meade | Final | 51.44 |
| MD | Patuxent River | St. Mary's | Patuxent River Naval Air Station | Final | 50.00 |
| MD | Rosedale | Baltimore City | 68th Street Dump/Industrial Enterprises | Proposed | 50.00 |
| ME | Augusta | Kennebec | O'connor Co. | Final | 31.86 |
| ME | Brooksville (Cape Rosier) | Hancock | Callahan Mining Corp | Final | 50.00 |
| ME | Brunswick | Cumberland | Brunswick Naval Air Station | Final | 43.38 |
| ME | Corinna | Penobscot | Eastland Woolen Mill | Final | 70.71 |

| St. | City/Area | County | Site Name | Status | Score[1] |
|---|---|---|---|---|---|
| ME | Gray | Cumberland | Mckin Co. | Final | 60.97 |
| ME | Kittery | York | Portsmouth Naval Shipyard | Final | 50.00 |
| ME | Limestone | Aroostook | Loring Air Force Base | Final | 34.49 |
| ME | Meddybemps | Washington | Eastern Surplus | Final | 50.00 |
| ME | Plymouth | Penobscot | West Site/Hows Corners | Final | 50.00 |
| ME | Saco | York | Saco Municipal Landfill | Final | 29.49 |
| ME | Saco | York | Saco Tannery Waste Pits | Deleted | 43.19 |
| ME | South Hope | Knox | Union Chemical Co., Inc. | Final | 32.11 |
| ME | Washburn | Aroostook | Pinette's Salvage Yard | Deleted | 33.98 |
| ME | Winthrop | Kennebec | Winthrop Landfill | Final | 35.62 |
| MI | Adrian | Lenawee | Anderson Development Co. | Deleted | 31.02 |
| MI | Albion | Calhoun | Albion-Sheridan Township Landfill | Final | 33.79 |
| MI | Albion | Calhoun | Mcgraw Edison Corp. | Final | 33.42 |
| MI | Allegan | Allegan | Rockwell International Corp. (Allegan Plant) | Final | 52.15 |
| MI | Battle Creek | Calhoun | Verona Well Field | Final | 46.86 |
| MI | Bay City | Bay | Bay City Middlegrounds | Proposed | 50.00 |
| MI | Belding | Ionia | H & K Sales | Deleted | – |
| MI | Benton Harbor | Berrien | Aircraft Components (D & L Sales) | Final | – |
| MI | Brighton | Livingston | Rasmussen's Dump | Final | 31.80 |
| MI | Bronson | Branch | North Bronson Industrial Area | Final | 33.93 |
| MI | Buchanan | Berrien | Electrovoice | Final | 35.36 |
| MI | Cadillac | Wexford | Kysor Industrial Corp. | Final | 33.94 |
| MI | Cadillac | Wexford | Northernaire Plating | Final | 57.93 |
| MI | Charlevoix | Charlevoix | Charlevoix Municipal Well | Deleted | 37.94 |
| MI | Clare | Clare | Clare Water Supply | Final | 38.43 |
| MI | Dalton Township | Muskegon | Duell & Gardner Landfill | Final | 34.68 |
| MI | Dalton Township | Muskegon | Ott/Story/Cordova Chemical Co. | Final | 53.41 |
| MI | Davisburg | Oakland | Springfield Township Dump | Final | 51.97 |
| MI | Detroit | Wayne | Carter Industrials, Inc. | Deleted | 37.79 |
| MI | Filer City | Manistee | Packaging Corp. Of America | Final | 51.91 |
| MI | Grand Ledge | Eaton | Parsons Chemical Works, Inc. | Final | 31.32 |
| MI | Grand Rapids | Kent | Butterworth #2 Landfill | Final | 50.31 |
| MI | Grand Rapids | Kent | Folkertsma Refuse | Deleted | 33.12 |
| MI | Grand Rapids | Kent | H. Brown Co., Inc. | Final | 39.88 |
| MI | Grand Rapids | Kent | State Disposal Landfill, Inc. | Final | 42.24 |
| MI | Grandville | Kent | Organic Chemicals, Inc. | Final | 32.93 |
| MI | Green Oak Township | Livingston | Spiegelberg Landfill | Final | 53.61 |
| MI | Greilickville | Leelanau | Grand Traverse Overall Supply Co. | Final | 35.53 |
| MI | Hartford | Van Buren | Burrows Sanitation | Final | 30.59 |
| MI | Highland | Oakland | Hi-Mill Manufacturing Co. | Final | 49.54 |
| MI | Holland | Ottawa | Waste Management Of Michigan (Holland Lagoons) | Final | 37.20 |
| MI | Houghton County | Houghton | Torch Lake | Final | 46.72 |
| MI | Howard Township | Cass | U.S. Aviex | Final | 33.66 |
| MI | Howell | Livingston | Shiawassee River | Final | 31.01 |
| MI | Ionia | Ionia | American Anodco, Inc. | Final | 57.99 |
| MI | Ionia | Ionia | Ionia City Landfill | Final | 31.31 |
| MI | Kalamazoo | Kalamazoo | Allied Paper, Inc./Portage Creek/Kalamazoo River | Final | 36.41 |
| MI | Kalamazoo | Kalamazoo | Auto Ion Chemicals, Inc. | Final | 32.07 |
| MI | Kalamazoo | Kalamazoo | Michigan Disposal Service (Cork Street Landfill) | Final | 37.93 |
| MI | Kalamazoo | Kalamazoo | Roto-Finish Co., Inc. | Final | 40.70 |
| MI | Kent City | Kent | Kent City Mobile Home Park | Deleted | 33.62 |
| MI | Kentwood | Kent | Kentwood Landfill | Final | 35.39 |
| MI | Lake Ann | Benzie | Metal Working Shop | Deleted | 28.82 |
| MI | Lansing | Ingham | Adam's Plating | Final | 29.64 |
| MI | Lansing | Ingham | Barrels, Inc. | Final | 42.24 |
| MI | Lansing Township | Ingham | Motor Wheel, Inc. | Final | 48.91 |
| MI | Macomb Township | Macomb | South Macomb Disposal Authority (Landfills #9 And #9a) | Final | 33.67 |
| MI | Mancelona Township | Antrim | Tar Lake | Final | 48.55 |
| MI | Marquette | Marquette | Cliff/Dow Dump | Deleted | 34.50 |
| MI | Metamora | Lapeer | Metamora Landfill | Final | 35.51 |

| St. | City/Area | County | Site Name | Status | Score[1] |
|---|---|---|---|---|---|
| MI | Muskegon | Muskegon | Bofors Nobel, Inc. | Final | 53.42 |
| MI | Muskegon | Muskegon | Kaydon Corp. | Final | 34.21 |
| MI | Muskegon | Muskegon | Peerless Plating Co. | Final | 43.94 |
| MI | Muskegon | Muskegon | Thermo-Chem, Inc. | Final | 53.36 |
| MI | Muskegon Heights | Muskegon | Sca Independent Landfill | Final | 34.75 |
| MI | Oscoda | Iosco | Hedblum Industries | Final | 37.29 |
| MI | Oscoda | Iosco | Wurtsmith Air Force Base | Proposed | 50.00 |
| MI | Oshtemo Township | Kalamazoo | K&L Avenue Landfill | Final | 38.10 |
| MI | Ossineke | Alpena | Ossineke Ground Water Contamination | Deleted | 33.78 |
| MI | Otisville | Genesee | Forest Waste Products | Final | 38.64 |
| MI | Park Township | Ottawa | Southwest Ottawa County Landfill | Final | 39.66 |
| MI | Pere Marquette Twp | Mason | Mason County Landfill | Deleted | 34.18 |
| MI | Petoskey | Emmet | Pmc Groundwater | Final | 42.68 |
| MI | Pleasant Plains Twp | Lake | Wash King Laundry | Final | 40.03 |
| MI | Rochester Hills | Oakland | J & L Landfill | Final | 31.65 |
| MI | Rose Center | Oakland | Cemetery Dump | Deleted | 34.16 |
| MI | Rose Township | Oakland | Rose Township Dump | Final | 50.92 |
| MI | Sault Ste Marie | Chippewa | Cannelton Industries, Inc. | Final | 30.16 |
| MI | Sparta Township | Kent | Sparta Landfill | Final | 32.00 |
| MI | St. Clair Shores | Macomb | Ten-Mile Drain | Final | 48.88 |
| MI | St. Joseph | Berrien | Bendix Corp./Allied Automotive | Final | 37.27 |
| MI | St. Louis | Gratiot | Gratiot County Golf Course | Deleted | 40.22 |
| MI | St. Louis | Gratiot | Gratiot County Landfill | Final | – |
| MI | St. Louis | Gratiot | Velsicol Burn Pit | Final | 29.54 |
| MI | St. Louis | Gratiot | Velsicol Chemical Corp. (Michigan) | Final | 52.29 |
| MI | Sturgis | St. Joseph | Sturgis Municipal Wells | Final | 42.24 |
| MI | Swartz Creek | Genesee | Berlin & Farro | Deleted | 66.74 |
| MI | Temperance | Monroe | Novaco Industries | Deleted | 38.20 |
| MI | Traverse City | Grand Traverse | Avenue E Ground Water Contamination | Deleted | 31.19 |
| MI | Utica | Macomb | G&H Landfill | Final | 49.09 |
| MI | Utica | Macomb | Liquid Disposal, Inc. | Final | 63.28 |
| MI | Whitehall | Muskegon | Muskegon Chemical Co. | Final | 34.19 |
| MI | Whitehall | Muskegon | Whitehall Municipal Wells | Deleted | 35.45 |
| MI | Wyandotte | Wayne | Lower Ecorse Creek Dump | Deleted | – |
| MI | Wyoming | Kent | Spartan Chemical Co. | Final | 41.05 |
| MI | Wyoming Township | Kent | Chem Central | Final | 38.20 |
| MN | Adrian | Nobles | Adrian Municipal Well Field | Deleted | 33.62 |
| MN | Andover | Anoka | South Andover Site | Final | 35.41 |
| MN | Andover | Anoka | Waste Disposal Engineering | Deleted | 50.92 |
| MN | Baytown Township | Washington | Baytown Township Ground Water Plume | Final | 35.62 |
| MN | Bemidji | Beltrami | Kummer Sanitary Landfill | Deleted | 35.57 |
| MN | Brainerd/Baxter | Crow Wing | Burlington Northern (Brainerd/Baxter Plant) | Final | 46.77 |
| MN | Brooklyn Center | Hennepin | Joslyn Manufacturing & Supply Co. | Final | 44.30 |
| MN | Burnsville | Dakota | Freeway Sanitary Landfill | Final | 45.91 |
| MN | Cannon Falls | Dakota | Dakhue Sanitary Landfill | Deleted | 42.24 |
| MN | Cass Lake | Cass | St. Regis Paper Co. | Final | 52.88 |
| MN | Dakota County | Dakota | Pine Bend Sanitary Landfill | Deleted | 52.11 |
| MN | East Bethel Township | Anoka | East Bethel Demolition Landfill | Deleted | 28.75 |
| MN | Fairview Township | Cass | Agate Lake Scrapyard | Deleted | 29.68 |
| MN | Faribault | Rice | Nutting Truck & Caster Co. | Final | 37.87 |
| MN | Fridley | Anoka | Boise Cascade/Onan Corp./Medtronics, Inc. | Deleted | 50.06 |
| MN | Fridley | Anoka | Fmc Corp. (Fridley Plant) | Final | 65.50 |
| MN | Fridley | Anoka | Fridley Commons Park Well Field | Final | 50.00 |
| MN | Fridley | Anoka | Kurt Manufacturing Co. | Final | 31.41 |
| MN | Fridley | Anoka | Naval Industrial Reserve Ordnance Plant | Final | 30.83 |
| MN | Hermantown | St. Louis | Arrowhead Refinery Co. | Final | 43.75 |
| MN | Lagrand Township | Douglas | Lagrand Sanitary Landfill | Deleted | 37.51 |
| MN | Lake Elmo | Washington | Washington County Landfill | Deleted | 42.24 |
| MN | Lehillier | Blue Earth | Lehillier/Mankato | Final | 42.49 |

| St. | City/Area | County | Site Name | Status | Score[1] |
|---|---|---|---|---|---|
| MN | Long Prairie | Todd | Long Prairie Ground Water Contamination | Final | 31.94 |
| MN | Minneapolis | Hennepin | General Mills/Henkel Corp. | Final | 36.28 |
| MN | Minneapolis | Hennepin | South Minneapolis Residential Soil Contamination | Final | 44.58 |
| MN | Minneapolis | Hennepin | Twin Cities Air Force Reserve Base (Small Arms Range Landfill) | Deleted | 33.62 |
| MN | Minneapolis | Hennepin | Union Scrap Iron & Metal Co. | Deleted | 42.63 |
| MN | Minneapolis | Hennepin | Whittaker Corp. | Deleted | 40.03 |
| MN | Morris | Stevens | Morris Arsenic Dump | Deleted | 38.27 |
| MN | New Brighton | Ramsey | Macgillis & Gibbs Co./Bell Lumber & Pole Co. | Final | 48.33 |
| MN | New Brighton | Ramsey | New Brighton/Arden Hills/Tcaap (Usarmy) | Final | 59.16 |
| MN | Oak Grove Township | Anoka | Oak Grove Sanitary Landfill | Deleted | 43.40 |
| MN | Oakdale | Washington | Oakdale Dump | Final | 55.71 |
| MN | Oronoco | Olmsted | Olmsted County Sanitary Landfill | Deleted | 40.70 |
| MN | Perham | Otter Tail | Perham Arsenic Site | Final | 37.98 |
| MN | Pine Bend | Dakota | Koch Refining Co./N-Ren Corp. | Deleted | 31.14 |
| MN | Rosemount | Dakota | University Of Minnesota (Rosemount Research Center) | Deleted | 45.91 |
| MN | Sebeka | Wadena | Ritari Post & Pole | Final | 29.81 |
| MN | St. Augusta Township | Stearns | St. Augusta Sanitary Landfill/Engen Dump | Deleted | 33.85 |
| MN | St. Louis County | St. Louis | St. Louis River Site | Final | 32.08 |
| MN | St. Louis Park | Hennepin | Nl Industries/Taracorp/Golden Auto | Deleted | 39.97 |
| MN | St. Louis Park | Hennepin | Reilly Tar & Chemical Corp. (St. Louis Park Plant) | Final | – |
| MN | St. Paul | Ramsey | Koppers Coke | Final | 55.05 |
| MN | Waite Park | Stearns | Waite Park Wells | Final | 31.94 |
| MN | Windom | Cottonwood | Windom Dump | Deleted | 38.17 |
| MO | Amazonia | Andrew | Wheeling Disposal Service Co., Inc., Landfill | Deleted | 48.58 |
| MO | Annapolis | Iron | Annapolis Lead Mine | Final | 56.67 |
| MO | Bridgeton | St. Louis | Westlake Landfill | Final | 29.85 |
| MO | Caledonia | Washington | Washington County Lead District - Furnace Creek | Final | 50.00 |
| MO | Cape Girardeau | Cape Girardeau | Kem-Pest Laboratories | Deleted | 33.89 |
| MO | Cape Girardeau | Cape Girardeau | Missouri Electric Works | Final | 31.20 |
| MO | Desloge | St. Francois | Big River Mine Tailings/St. Joe Minerals Corp. | Final | 84.91 |
| MO | Ellisville | St. Louis | Ellisville Site | Final | – |
| MO | Fredericktown | Madison | Madison County Mines | Final | 58.41 |
| MO | Granby | Newton | Newton County Mine Tailings | Final | 50.00 |
| MO | Imperial | Jefferson | Minker/Stout/Romaine Creek | Final | 36.78 |
| MO | Independence | Jackson | Lake City Army Ammunition Plant (Northwest Lagoon) | Final | 33.62 |
| MO | Jefferson County | Jefferson | Southwest Jefferson County Mining | Final | 70.71 |
| MO | Joplin | Jasper | Oronogo-Duenweg Mining Belt | Final | 46.20 |
| MO | Joplin | Newton | Newton County Wells | Final | 50.00 |
| MO | Kansas City | Jackson | Conservation Chemical Co. | Final | 29.85 |
| MO | Liberty | Clay | Lee Chemical | Final | 46.81 |
| MO | Malden | Dunklin | Bee Cee Manufacturing Co. | Final | 28.59 |
| MO | Moscow Mills | Lincoln | Shenandoah Stables | Deleted | 30.09 |
| MO | Neosho | Newton | Pools Prairie | Final | 50.00 |
| MO | New Haven | Franklin | Riverfront | Final | 50.00 |
| MO | North Kansas City | Clay | Armour Road | Final | 50.00 |
| MO | Old Mines | Washington | Washington County Lead District - Old Mines | Final | 76.81 |
| MO | Potosi | Washington | Washington County Lead District - Potosi | Final | 50.00 |
| MO | Republic | Greene | Solid State Circuits, Inc. | Final | 37.93 |
| MO | Richwoods | Washington | Washington County Lead District - Richwoods | Final | 76.81 |
| MO | Sikeston | Scott | Quality Plating | Final | 40.70 |
| MO | Springfield | Greene | Fulbright Landfill | Final | 40.60 |
| MO | Springfield | Greene | North-U Drive Well Contamination | Deleted | 28.90 |
| MO | St. Charles | St. Charles | Weldon Spring Former Army Ordnance Works | Final | 30.26 |
| MO | St. Charles | St. Charles | Weldon Spring Quarry/Plant/Pits (Usdoe/Army) | Final | 58.60 |
| MO | St. Louis | St. Louis | St. Louis Airport/Hazelwood Interim Storage/Futura Coatings Co. | Final | 38.31 |
| MO | Sullivan | Franklin | Oak Grove Village Well | Final | 50.00 |
| MO | Times Beach | St. Louis | Times Beach | Deleted | 40.08 |
| MO | Valley Park | St. Louis | Valley Park Tce | Final | 35.57 |
| MO | Verona | Lawrence | Syntex Facility | Final | 43.78 |

| St. | City/Area | County | Site Name | Status | Score[1] |
|---|---|---|---|---|---|
| MO | Vienna | Maries | Vienna Wells | Final | 50.00 |
| MP | Garapan | Saipan | Pcb Warehouse | Deleted | – |
| MS | Clarksdale | Coahoma | Red Panther Chemical Company | Proposed | 39.43 |
| MS | Columbia | Marion | Newsom Brothers/Old Reichhold Chemicals, Inc. | Deleted | 45.70 |
| MS | Columbus | Lowndes | Kerr-Mcgee Chemical Corp - Columbus | Proposed | 52.47 |
| MS | Flowood | Rankin | Flowood Site | Deleted | – |
| MS | Flowood | Rankin | Sonford Products | Final | 31.66 |
| MS | Greenville | Washington | Walcotte Chemical Co. Warehouses | Deleted | – |
| MS | Gulfport | Harrison | Chemfax, Inc. | Proposed | 39.12 |
| MS | Hattiesburg | Lamar | Davis Timber Company | Final | 48.57 |
| MS | Louisville | Winston | American Creosote Works Inc | Final | 62.20 |
| MS | Picayune | Pearl River | Picayune Wood Treating Site | Final | 51.03 |
| MS | Wesson | Copiah | Potter Co. | Proposed | 50.00 |
| MT | Anaconda | Deer Lodge | Anaconda Co. Smelter | Final | 58.71 |
| MT | Basin | Jefferson | Basin Mining Area | Final | 61.15 |
| MT | Billings | Yellowstone | Lockwood Solvent Ground Water Plume | Final | 45.69 |
| MT | Black Eagle | Cascade | Acm Smelter And Refinery | Final | 54.26 |
| MT | Bozeman | Gallatin | Idaho Pole Co. | Final | 38.29 |
| MT | Butte | Silver Bow | Montana Pole And Treating | Final | 33.03 |
| MT | Butte | Silver Bow/Deer Lodge | Silver Bow Creek/Butte Area | Final | 63.76 |
| MT | Columbus | Stillwater | Mouat Industries | Final | 31.66 |
| MT | East Helena | Lewis And Clark | East Helena Site | Final | 61.65 |
| MT | Helena | Lewis And Clark | Upper Tenmile Creek Mining Area | Final | 50.00 |
| MT | Libby | Lincoln | Libby Asbestos Site | Final | – |
| MT | Libby | Lincoln | Libby Ground Water Contamination | Final | 37.67 |
| MT | Livingston | Park | Burlington Northern Livingston Shop Complex | Proposed | 50.00 |
| MT | Milltown | Missoula | Milltown Reservoir Sediments | Final | 43.78 |
| MT | Monarch | Cascade/Judith Basin | Barker Hughesville Mining District | Final | 50.00 |
| MT | Neihart | Cascade | Carpenter Snow Creek Mining District | Final | 50.00 |
| MT | Superior | Mineral | Flat Creek Imm | Final | 51.33 |
| NC | 210 Miles of Roads | Warren | Roadside PCB Spill | Deleted | – |
| NC | Aberdeen | Moore | Aberdeen Contaminated Ground Water | Final | 50.00 |
| NC | Aberdeen | Moore | Aberdeen Pesticide Dumps | Final | 52.70 |
| NC | Aberdeen | Moore | Geigy Chemical Corp. (Aberdeen Plant) | Final | 33.02 |
| NC | Arden | Buncombe | Blue Ridge Plating Company | Final | 38.67 |
| NC | Ashe County | Ashe | Ore Knob Mine | Final | 50.00 |
| NC | Asheville | Buncombe | Cts Of Asheville, Inc. | Proposed | 48.64 |
| NC | Belmont | Gaston | Jadco-Hughes Facility | Final | 42.00 |
| NC | Castle Hayne | New Hanover | Reasor Chemical Company | Final | 32.14 |
| NC | Charlotte | Mecklenburg | Martin-Marietta, Sodyeco, Inc. | Final | 51.93 |
| NC | Charlotte | Mecklenburg | Ram Leather Care Site | Final | 40.43 |
| NC | Concord | Cabarrus | Bypass 601 Ground Water Contamination | Final | 37.93 |
| NC | Cordova | Richmond | Charles Macon Lagoon And Drum Storage | Final | 47.10 |
| NC | East Flat Rock | Henderson | General Electric Co/Shepherd Farm | Final | 70.71 |
| NC | Fayetteville | Cumberland | Cape Fear Wood Preserving | Final | 34.09 |
| NC | Fayetteville | Cumberland | Carolina Transformer Co. | Final | 33.76 |
| NC | Gastonia | Gaston | Davis Park Road Tce | Final | 33.50 |
| NC | Havelock | Craven | Cherry Point Marine Corps Air Station | Final | 70.71 |
| NC | Hazelwood | Haywood | Benfield Industries, Inc. | Final | 31.67 |
| NC | Jacksonville | Onslow | Abc One Hour Cleaners | Final | 29.11 |
| NC | Maco | Brunswick | Potter's Septic Tank Service Pits | Final | 29.14 |
| NC | Morrisville | Wake | Koppers Co., Inc. (Morrisville Plant) | Final | 41.89 |
| NC | Navassa | Brunswick | Kerr-Mcgee Chemical Corp - Navassa | Final | 50.00 |
| NC | North Belmont | Gaston | North Belmont Pce | Final | 50.00 |
| NC | Onslow County | Onslow | Camp Lejeune Military Res. (Usnavy) | Final | 33.13 |
| NC | Oxford | Granville | Jfd Electronics/Channel Master | Final | 39.03 |
| NC | Raleigh | Wake | North Carolina State University (Lot 86, Farm Unit #1) | Final | 48.36 |
| NC | Raleigh | Wake | Ward Transformer | Final | 50.00 |
| NC | Riegelwood | Columbus | Wright Chemical Corporation | Final | 48.03 |

| St. | City/Area | County | Site Name | Status | Score[1] |
|---|---|---|---|---|---|
| NC | Roxboro | Person | Gmh Electronics | Final | 50.00 |
| NC | Salisbury | Rowan | National Starch & Chemical Corp. | Final | 46.51 |
| NC | Shelby | Cleveland | Celanese Corp. (Shelby Fiber Operations) | Final | 48.98 |
| NC | Statesville | Iredell | Fcx, Inc. (Statesville Plant) | Final | 37.93 |
| NC | Statesville | Iredell | Sigmon's Septic Tank Service | Final | 30.03 |
| NC | Swannanoa | Buncombe | Chemtronics, Inc. | Final | 30.16 |
| NC | Washington | Beaufort | Fcx, Inc. (Washington Plant) | Final | 40.39 |
| NC | Waynesville | Haywood | Barber Orchard | Final | 70.71 |
| NC | Wilmington | New Hanover | Horton Iron And Metal | Proposed | 48.03 |
| NC | Wilmington | New Hanover | New Hanover Cnty Airport Burn Pit | Final | 39.39 |
| ND | Lidgerwood/Wyndmere | Richland/Ransom/Sargent | Arsenic Trioxide Site | Deleted | – |
| ND | Minot | Ward | Minot Landfill | Deleted | 33.58 |
| NE | Bruno | Butler | Bruno Co-Op Association/Associated Properties | Final | 50.00 |
| NE | Columbus | Platte | 10th Street Site | Final | 28.90 |
| NE | Grand Island | Hall | Cleburn Street Well | Final | 50.00 |
| NE | Grand Island | Hall | Cornhusker Army Ammunition Plant | Final | 51.13 |
| NE | Grand Island | Hall | Parkview Well | Final | 50.00 |
| NE | Hastings | Adams | Garvey Elevator | Final | 50.00 |
| NE | Hastings | Adams | Hastings Ground Water Contamination | Final | 42.24 |
| NE | Hastings | Adams | West Highway 6 & Highway 281 | Final | 50.00 |
| NE | Lindsay | Platte | Lindsay Manufacturing Co. | Final | 47.91 |
| NE | Mead | Saunders | Nebraska Ordnance Plant (Former) | Final | 31.94 |
| NE | Norfolk | Madison | Sherwood Medical Co. | Final | 50.00 |
| NE | Ogallala | Keith | Ogallala Ground Water Contamination | Final | 50.00 |
| NE | Omaha | Douglas | Omaha Lead | Final | 50.00 |
| NE | Waverly | Lancaster | Waverly Ground Water Contamination | Deleted | 37.93 |
| NH | Barrington | Strafford | Tibbetts Road | Final | 41.09 |
| NH | Berlin | Coos | Chlor-Alkali Facility (Former) | Final | 30.54 |
| NH | Conway | Carroll | Kearsarge Metallurgical Corp. | Final | 38.45 |
| NH | Dover | Strafford | Dover Municipal Landfill | Final | 36.98 |
| NH | Epping | Rockingham | Keefe Environmental Services (Kes) | Final | 65.19 |
| NH | Kingston | Rockingham | Ottati & Goss/Kingston Steel Drum | Final | 53.41 |
| NH | Londonderry | Rockingham | Auburn Road Landfill | Final | 36.30 |
| NH | Londonderry | Rockingham | Tinkham Garage | Final | 43.24 |
| NH | Londonderry | Rockingham | Town Garage/Radio Beacon | Final | 31.94 |
| NH | Merrimack | Hillsborough | New Hampshire Plating Co. | Final | 50.00 |
| NH | Milford | Hillsborough | Fletcher's Paint Works & Storage | Final | 35.39 |
| NH | Milford | Hillsborough | Savage Municipal Water Supply | Final | 37.52 |
| NH | Nashua | Hillsborough | Mohawk Tannery | Proposed | 52.40 |
| NH | Nashua | Hillsborough | Sylvester | Final | – |
| NH | North Hampton | Rockingham | Coakley Landfill | Final | 29.16 |
| NH | Peterborough | Hillsborough | South Municipal Water Supply Well | Final | 35.64 |
| NH | Plaistow | Rockingham | Beede Waste Oil | Final | 70.71 |
| NH | Portsmouth/Newington | Rockingham | Pease Air Force Base | Final | 39.42 |
| NH | Raymond | Rockingham | Mottolo Pig Farm | Final | 40.70 |
| NH | Somersworth | Strafford | Somersworth Sanitary Landfill | Final | 65.56 |
| NH | Troy | Cheshire | Troy Mills Landfill | Final | 50.00 |
| NJ | Alexandria Township | Hunterdon | Crown Vantage Landfill | Final | 50.00 |
| NJ | Asbury Park | Monmouth | M&T Delisa Landfill | Deleted | 32.27 |
| NJ | Atlantic County | Atlantic | Federal Aviation Administration Technical Center (Usdot) | Final | 39.65 |
| NJ | Bayville | Ocean | Denzer & Schafer X-Ray Co. | Deleted | 40.36 |
| NJ | Berkeley Township | Ocean | Beachwood/Berkley Wells | Deleted | 42.24 |
| NJ | Beverly | Burlington | Cosden Chemical Coatings Corp. | Final | 33.86 |
| NJ | Boonton | Morris | Pepe Field | Deleted | 33.83 |
| NJ | Bound Brook | Somerset | American Cyanamid Co | Final | 50.28 |
| NJ | Bound Brook | Somerset | Brook Industrial Park | Final | 58.12 |
| NJ | Brick Township | Ocean | Brick Township Landfill | Final | 58.13 |
| NJ | Bridgeport | Gloucester | Bridgeport Rental & Oil Services | Final | 60.73 |
| NJ | Bridgeport | Gloucester | Chemical Leaman Tank Lines, Inc. | Final | 47.53 |

| St. | City/Area | County | Site Name | Status | Score[1] |
|---|---|---|---|---|---|
| NJ | Byram | Sussex | Mansfield Trail Dump | Final | 50.00 |
| NJ | Camden | Camden | Martin Aaron, Inc. | Final | 50.00 |
| NJ | Camden And Gloucester Cit | Camden | Welsbach & General Gas Mantle (Camden Radiation) | Final | 41.46 |
| NJ | Carlstadt | Bergen | Scientific Chemical Processing | Final | 55.97 |
| NJ | Chester Township | Morris | Combe Fill South Landfill | Final | 45.22 |
| NJ | Cinnaminson Township | Burlington | Cinnaminson Township (Block 702) Ground Water Contamination | Final | 37.93 |
| NJ | Colts Neck | Monmouth | Naval Weapons Station Earle (Site A) | Final | 29.65 |
| NJ | Dover | Morris | Dover Municipal Well 4 | Final | 28.90 |
| NJ | East Brunswick Township | Middlesex | Fried Industries | Final | 33.61 |
| NJ | East Rutherford | Bergen | Universal Oil Products (Chemical Division) | Final | 54.63 |
| NJ | Edgewater | Bergen | Quanta Resources | Final | 50.00 |
| NJ | Edison Township | Middlesex | Chemical Insecticide Corp. | Final | 37.93 |
| NJ | Edison Township | Middlesex | Kin-Buc Landfill | Final | 50.64 |
| NJ | Edison Township | Middlesex | Renora, Inc. | Deleted | 40.44 |
| NJ | Egg Harbor Township | Atlantic | Delilah Road | Deleted | 49.33 |
| NJ | Elizabeth | Union | Chemical Control | Final | 47.13 |
| NJ | Evesham Township | Burlington | Ellis Property | Final | 34.62 |
| NJ | Fair Lawn | Bergen | Fair Lawn Well Field | Final | 42.49 |
| NJ | Fairfield | Essex | Caldwell Trucking Co. | Final | 58.30 |
| NJ | Florence | Burlington | Roebling Steel Co. | Final | 41.02 |
| NJ | Florence Township | Burlington | Florence Land Recontouring, Inc., Landfill | Deleted | 47.39 |
| NJ | Franklin Borough | Sussex | Metaltec/Aerosystems | Final | 48.95 |
| NJ | Franklin Township | Gloucester | Franklin Burn | Final | 40.67 |
| NJ | Franklin Township | Hunterdon | Myers Property | Final | 33.83 |
| NJ | Franklin Township | Somerset | Higgins Farm | Final | 30.47 |
| NJ | Freehold Township | Monmouth | Lone Pine Landfill | Final | 66.33 |
| NJ | Galloway Township | Atlantic | Emmell's Septic Landfill | Final | 50.00 |
| NJ | Galloway Township | Atlantic | Mannheim Avenue Dump | Deleted | 36.56 |
| NJ | Galloway Township | Atlantic | Pomona Oaks Residential Wells | Deleted | 31.94 |
| NJ | Garfield | Bergen | Garfield Ground Water Contamination | Proposed | – |
| NJ | Gibbsboro | Camden | Route 561 Dump | Proposed | 50.00 |
| NJ | Gibbsboro | Camden | Sherwin-Williams/Hilliards Creek | Final | 50.00 |
| NJ | Gibbsboro | Camden | United States Avenue Burn | Final | 50.00 |
| NJ | Gibbstown | Gloucester | Hercules, Inc. (Gibbstown Plant) | Final | 40.36 |
| NJ | Glen Ridge | Essex | Glen Ridge Radium Site | Deleted | 49.14 |
| NJ | Gloucester Township | Camden | Gems Landfill | Final | 68.53 |
| NJ | Green Village | Morris | Rolling Knolls Lf | Final | 58.31 |
| NJ | Hamilton Township | Atlantic | D'imperio Property | Final | 55.79 |
| NJ | Hillsborough Township | Somerset | Krysowaty Farm | Deleted | 55.14 |
| NJ | Hoboken | Hudson | Grand Street Mercury | Deleted | – |
| NJ | Howell Township | Monmouth | Bog Creek Farm | Final | 43.23 |
| NJ | Howell Township | Monmouth | Zschiegner Refining | Final | 50.00 |
| NJ | Jackson Township | Ocean | Jackson Township Landfill | Deleted | 38.11 |
| NJ | Jersey City | Hudson | Pjp Landfill | Final | 28.73 |
| NJ | Kearny | Hudson | Diamond Head Oil Refinery Div. | Final | 30.00 |
| NJ | Kearny | Hudson | Standard Chlorine | Final | 50.00 |
| NJ | Kingston | Somerset | Higgins Disposal | Final | 30.87 |
| NJ | Kingwood Township | Hunterdon | De Rewal Chemical Co. | Final | 35.72 |
| NJ | Lakehurst | Ocean | Naval Air Engineering Center | Final | 50.53 |
| NJ | Linden | Union | Lcp Chemicals Inc. | Final | 50.00 |
| NJ | Lodi | Bergen | Lodi Municipal Well | Deleted | 33.39 |
| NJ | Mantua Township | Gloucester | Helen Kramer Landfill | Final | 72.66 |
| NJ | Manville | Somerset | Federal Creosote | Final | 50.00 |
| NJ | Marlboro Township | Monmouth | Burnt Fly Bog | Final | 59.16 |
| NJ | Maywood/Rochelle Park | Bergen | Maywood Chemical Co. | Final | 51.19 |
| NJ | Middlesex | Middlesex | Middlesex Sampling Plant (Usdoe) | Final | 50.00 |
| NJ | Milford | Hunterdon | Curtis Specialty Papers, Inc | Final | 50.00 |
| NJ | Millington | Morris | Asbestos Dump | Deleted | 39.61 |
| NJ | Millville | Cumberland | Nascolite Corp. | Final | 51.13 |

| St. | City/Area | County | Site Name | Status | Score[1] |
|---|---|---|---|---|---|
| NJ | Minotola | Atlantic | Garden State Cleaners Co. | Final | 28.90 |
| NJ | Minotola | Atlantic | South Jersey Clothing Co. | Final | 42.24 |
| NJ | Monroe Township | Middlesex | Monroe Township Landfill | Deleted | 42.37 |
| NJ | Montclair/West Orange | Essex | Montclair/West Orange Radium Site | Deleted | 49.14 |
| NJ | Montgomery Township | Somerset | Montgomery Township Housing Development | Final | 37.93 |
| NJ | Morganville | Monmouth | Imperial Oil Co., Inc./Champion Chemicals | Final | 33.87 |
| NJ | Mount Holly | Burlington | Landfill & Development Co. | Final | 33.62 |
| NJ | Mount Olive Township | Morris | Combe Fill North Landfill | Deleted | 47.79 |
| NJ | Newark | Essex | Diamond Alkali Co. | Final | 35.40 |
| NJ | Newark | Essex | White Chemical Corp. | Final | – |
| NJ | Newfield Borough | Gloucester | Shieldalloy Corp. | Final | 58.75 |
| NJ | Oakland | Bergen | Witco Chemical Corp. (Oakland Plant) | Deleted | 30.63 |
| NJ | Old Bridge Township | Middlesex | Cps/Madison Industries | Final | 69.73 |
| NJ | Old Bridge Township | Middlesex | Evor Phillips Leasing | Final | 36.64 |
| NJ | Old Bridge Township | Middlesex | Global Sanitary Landfill | Final | 45.92 |
| NJ | Old Bridge Twp/Sayreville | Middlesex | Raritan Bay Slag | Final | 50.00 |
| NJ | Orange | Essex | U.S. Radium Corp. | Final | 37.79 |
| NJ | Parsippany/Troy Hills | Morris | Sharkey Landfill | Final | 48.85 |
| NJ | Pedricktown (Oldmans Town | Salem | Nl Industries | Final | 52.96 |
| NJ | Pemberton Township | Burlington | Fort Dix (Landfill Site) | Final | 37.40 |
| NJ | Pemberton Township | Burlington | Lang Property | Final | 48.89 |
| NJ | Pennsauken Township | Camden | Puchack Well Field | Final | 50.00 |
| NJ | Pennsauken Township | Camden | Swope Oil & Chemical Co. | Final | 35.68 |
| NJ | Piscataway | Middlesex | Chemsol, Inc. | Final | 42.69 |
| NJ | Pitman | Gloucester | Lipari Landfill | Final | 75.60 |
| NJ | Pleasant Plains | Ocean | Reich Farms | Final | 53.48 |
| NJ | Pleasantville | Atlantic | Price Landfill | Final | – |
| NJ | Plumstead Township | Ocean | Goose Farm | Final | 47.71 |
| NJ | Plumstead Township | Ocean | Hopkins Farm | Deleted | 34.09 |
| NJ | Plumstead Township | Ocean | Pijak Farm | Deleted | 43.48 |
| NJ | Plumstead Township | Ocean | Spence Farm | Deleted | 45.87 |
| NJ | Plumstead Township | Ocean/Monmouth | Wilson Farm | Deleted | 33.93 |
| NJ | Ringwood Borough | Passaic | Ringwood Mines/Landfill | Final | 52.58 |
| NJ | Rockaway Township | Morris | Picatinny Arsenal (Usarmy) | Final | 42.92 |
| NJ | Rockaway Township | Morris | Radiation Technology, Inc. | Final | 42.56 |
| NJ | Rockaway Township | Morris | Rockaway Borough Well Field | Final | 42.34 |
| NJ | Rockaway Township | Morris | Rockaway Township Wells | Final | 28.90 |
| NJ | Rocky Hill Borough | Somerset | Rocky Hill Municipal Well | Final | 37.93 |
| NJ | Saddle Brook Twp | Bergen | Curcio Scrap Metal, Inc. | Final | 34.37 |
| NJ | Sayreville | Middlesex | Atlantic Resources | Final | 50.00 |
| NJ | Sayreville | Middlesex | Horseshoe Road | Final | 51.37 |
| NJ | Sayreville | Middlesex | Sayreville Landfill | Final | 37.05 |
| NJ | Shamong Township | Burlington | Ewan Property | Final | 50.19 |
| NJ | South Brunswick | Middlesex | Jis Landfill | Final | 45.14 |
| NJ | South Brunswick | Middlesex | South Brunswick Landfill | Deleted | 53.42 |
| NJ | South Kearny | Hudson | Syncon Resins | Final | 43.43 |
| NJ | South Plainfield | Middlesex | Cornell Dubilier Electronics Inc. | Final | 50.27 |
| NJ | South Plainfield | Middlesex | Woodbrook Road Dump | Final | 50.00 |
| NJ | Sparta Township | Sussex | A. O. Polymer | Final | 28.91 |
| NJ | Springfield Twp(Jobstown) | Burlington | Kauffman & Minteer, Inc. | Final | 28.51 |
| NJ | Swainton Middle | Cape May | Williams Property | Final | 40.45 |
| NJ | Tabernacle Township | Burlington | Tabernacle Drum Dump | Deleted | 36.83 |
| NJ | Thorofare | Gloucester | Matteo & Sons Inc. | Final | 50.00 |
| NJ | Toms River | Ocean | Ciba-Geigy Corp. | Final | 50.33 |
| NJ | Upper Deerfield Township | Cumberland | Upper Deerfield Township Sanitary Landfill | Deleted | 33.62 |
| NJ | Upper Freehold Twp | Monmouth | Friedman Property | Deleted | 33.88 |
| NJ | Vineland | Cumberland | Iceland Coin Laundry Area Gw Plume | Final | 30.30 |
| NJ | Vineland | Cumberland | Vineland Chemical Co., Inc. | Final | 59.16 |
| NJ | Vineland | Cumberland | Vineland State School | Deleted | 40.84 |
| NJ | Voorhees Township | Camden | Cooper Road | Deleted | 36.79 |

| St. | City/Area | County | Site Name | Status | Score[1] |
|---|---|---|---|---|---|
| NJ | Wall Township | Monmouth | Monitor Devices, Inc./Intercircuits, Inc. | Final | 41.93 |
| NJ | Wall Township | Monmouth | Waldick Aerospace Devices, Inc. | Final | 44.86 |
| NJ | Wall Twp | Monmouth | White Swan Laundry And Cleaner Inc. | Final | 41.63 |
| NJ | Wallington Borough | Bergen | Industrial Latex Corp. | Deleted | 32.38 |
| NJ | Warren County | Warren | Pohatcong Valley Ground Water Contamination | Final | 28.90 |
| NJ | Wayne Township | Passaic | W.R. Grace & Co., Inc./Wayne Interim Storage Site (Usdoe) | Final | 47.14 |
| NJ | Wharton Borough | Morris | Dayco Corp./L.E Carpenter Co. | Final | 46.13 |
| NJ | Winslow Township | Camden | King Of Prussia | Final | 47.19 |
| NJ | Winslow Township | Camden | Lightman Drum Company | Final | 42.03 |
| NJ | Wood Ridge Borough | Bergen | Ventron/Velsicol | Final | 51.38 |
| NJ | Woodland Township | Burlington | Woodland Route 532 Dump | Final | 34.98 |
| NJ | Woodland Township | Burlington | Woodland Route 72 Dump | Final | 31.17 |
| NJ | Wrightstown | Burlington | Mcguire Air Force Base #1 | Final | 47.20 |
| NM | Albuquerque | Bernalillo | At&Sf (Albuquerque) | Final | 50.00 |
| NM | Albuquerque | Bernalillo | Fruit Avenue Plume | Final | 50.00 |
| NM | Albuquerque | Bernalillo | South Valley | Final | – |
| NM | Carrizozo | Lincoln | Cimarron Mining Corp. | Final | 38.93 |
| NM | Church Rock | Mckinley | United Nuclear Corp. | Final | 30.36 |
| NM | Clovis | Curry | At & Sf (Clovis) | Deleted | 33.62 |
| NM | Espanola | Rio Arriba | North Railroad Avenue Plume | Final | 50.00 |
| NM | Farmington | San Juan | Lee Acres Landfill (Usdoi) | Final | 39.37 |
| NM | Grants | Cibola | Grants Chlorinated Solvents | Final | 50.00 |
| NM | Las Cruces | Dona Ana | Griggs & Walnut Ground Water Plume | Final | 50.00 |
| NM | Lemitar | Socorro | Cal West Metals (Ussba) | Deleted | 59.37 |
| NM | Los Lunas | Valencia | Pagano Salvage | Deleted | 35.57 |
| NM | Milan | Cibola | Homestake Mining Co. | Final | 34.21 |
| NM | Prewitt | Mckinley | Prewitt Abandoned Refinery | Final | 44.24 |
| NM | Questa | Taos | Molycorp, Inc. | Proposed | 50.00 |
| NM | Roswell | Chaves | Mcgaffey And Main Groundwater Plume | Final | 50.00 |
| NM | Silver City | Grant | Cleveland Mill | Deleted | 40.37 |
| NM | Socorro | Socorro | Eagle Picher Carefree Battery | Final | 50.00 |
| NV | Dayton | Lyon | Carson River Mercury Site | Final | 39.07 |
| NY | Amenia | Dutchess | Sarney Farm | Final | 33.20 |
| NY | Batavia | Genesee | Batavia Landfill | Deleted | 50.18 |
| NY | Bohemia | Suffolk | Bioclinical Laboratories, Inc. | Deleted | 32.91 |
| NY | Brant | Erie | Wide Beach Development | Deleted | 56.58 |
| NY | Brooklyn | Kings | Gowanus Canal | Final | 50.00 |
| NY | Brooklyn/Queens | Kings/Queens | Newtown Creek | Final | 50.00 |
| NY | Byron Township | Genesee | Byron Barrel & Drum | Final | 37.27 |
| NY | Caledonia | Livingston | Jones Chemicals, Inc. | Final | 33.62 |
| NY | Carthage | Jefferson | Crown Cleaners Of Watertown Inc. | Final | 49.00 |
| NY | Central Islip | Suffolk | Mackenzie Chemical Works | Final | 50.00 |
| NY | Cheektowaga | Erie | Pfohl Brothers Landfill | Deleted | 50.11 |
| NY | Clayville | Oneida | Ludlow Sand & Gravel | Final | 36.88 |
| NY | Cold Springs | Putnam | Marathon Battery Corp. | Deleted | 30.27 |
| NY | Colonie | Albany | Mercury Refining, Inc. | Final | 44.58 |
| NY | Conklin | Broome | Conklin Dumps | Deleted | 33.93 |
| NY | Copiague | Suffolk | Action Anodizing, Plating, & Polishing Corp. | Deleted | 34.72 |
| NY | Cortland | Cortland | Rosen Brothers Scrap Yard/Dump | Final | 51.35 |
| NY | Dayton | Cattaraugus | Peter Cooper Corporation (Markhams) | Deleted | 30.00 |
| NY | Deer Park | Suffolk | Sms Instruments, Inc. | Deleted | 37.32 |
| NY | East Farmingdale | Suffolk | Circuitron Corp. | Final | 54.27 |
| NY | East Fishkill | Dutchess | Shenandoah Road Groundwater Contamination | Final | 50.00 |
| NY | Ellenville | Ulster | Ellenville Scrap Iron And Metal | Final | 50.27 |
| NY | Elmira | Chemung | Facet Enterprises, Inc. | Final | 46.67 |
| NY | Farmingdale | Nassau | Liberty Industrial Finishing | Final | 50.65 |
| NY | Farmingdale | Suffolk | Kenmark Textile Corp. | Deleted | 31.72 |
| NY | Farmingdale | Suffolk | Preferred Plating Corp. | Final | 35.06 |
| NY | Farmingdale | Suffolk | Tronic Plating Co., Inc. | Deleted | 45.14 |

| St. | City/Area | County | Site Name | Status | Score[1] |
|---|---|---|---|---|---|
| NY | Franklin Square | Nassau | Genzale Plating Co. | Final | 33.79 |
| NY | Fulton | Oswego | Fulton Terminals | Final | 36.50 |
| NY | Garden City | Nassau | Old Roosevelt Field Contaminated Gw Area | Final | 50.00 |
| NY | Glen Cove | Nassau | Li Tungsten Corp. | Final | 50.00 |
| NY | Glen Cove | Nassau | Mattiace Petrochemical Co., Inc. | Final | 31.90 |
| NY | Glenwood Landing | Nassau | Applied Environmental Services | Final | 41.15 |
| NY | Gowanda | Cattaraugus | Peter Cooper | Final | 50.00 |
| NY | Great Neck | Nassau | Stanton Cleaners Area Ground Water Contamination | Final | 35.76 |
| NY | Hamilton | Madison | C & J Disposal Leasing Co. Dump | Deleted | 35.10 |
| NY | Hauppauge | Suffolk | Computer Circuits | Final | 50.00 |
| NY | Hempstead | Nassau | Pasley Solvents & Chemicals, Inc. | Final | 39.65 |
| NY | Hewlett | Nassau | Peninsula Boulevard Groundwater Plume | Final | 50.00 |
| NY | Hicksville | Nassau | Anchor Chemicals | Deleted | 37.20 |
| NY | Hicksville | Nassau | Hooker Chemical & Plastics Corp./Ruco Polymer Corp. | Final | 41.60 |
| NY | High Falls | Ulster | Mohonk Road Industrial Plant | Final | 50.00 |
| NY | Hillburn | Rockland | Hudson Technologies, Inc. | Proposed | 50.00 |
| NY | Holbrook | Suffolk | Goldisc Recordings, Inc. | Final | 33.39 |
| NY | Holley | Orleans | Diaz Chemical | Final | 50.00 |
| NY | Hopewell Junction | Dutchess | Hopewell Precision | Final | 50.00 |
| NY | Horseheads | Chemung | Kentucky Avenue Well Field | Final | 39.65 |
| NY | Hudson River | Washington | Hudson River Pcbs | Final | 54.66 |
| NY | Hyde Park | Dutchess | Jones Sanitation | Deleted | 52.52 |
| NY | Islip | Suffolk | Islip Municipal Sanitary Landfill | Final | 33.39 |
| NY | Le Roy | Genesee | Lehigh Valley Railroad | Final | 50.00 |
| NY | Lincklaen | Chenango | Solvent Savers | Final | 34.78 |
| NY | Lisbon | St. Lawrence | Sealand Restoration, Inc. | Final | 29.36 |
| NY | Little Valley | Cattaraugus | Little Valley | Final | – |
| NY | Malta | Saratoga | Malta Rocket Fuel Area | Final | 33.62 |
| NY | Massena | St. Lawrence | General Motors (Central Foundry Division) | Final | 40.71 |
| NY | Maybrook | Orange | Nepera Chemical Co., Inc. | Final | 39.87 |
| NY | Mineola/North Hempstead | Nassau | Jackson Steel | Final | 50.00 |
| NY | Moira | Franklin | York Oil Co. | Final | 47.70 |
| NY | Nassau | Rensselaer | Dewey Loeffel Landfill | Final | 50.00 |
| NY | New Cassel/Hicksville | Nassau | New Cassel/Hicksville Ground Water Contamination | Proposed | 50.00 |
| NY | Newburgh | Orange | Consolidated Iron And Metal | Final | 50.00 |
| NY | Niagara Falls | Niagara | Forest Glen Mobile Home Subdivision | Final | – |
| NY | Niagara Falls | Niagara | Hooker (102nd Street) | Deleted | 30.48 |
| NY | Niagara Falls | Niagara | Hooker (Hyde Park) | Final | 34.77 |
| NY | Niagara Falls | Niagara | Hooker (S Area) | Final | 51.62 |
| NY | Niagara Falls | Niagara | Love Canal | Deleted | 52.23 |
| NY | North Hempstead | Nassau | Fulton Avenue | Final | 33.08 |
| NY | North Sea | Suffolk | North Sea Municipal Landfill | Deleted | 33.74 |
| NY | Noyack/Sag Harbor | Suffolk | Rowe Industries Ground Water Contamination | Final | 31.94 |
| NY | Old Bethpage | Nassau | Claremont Polychemical | Final | 31.62 |
| NY | Olean | Cattaraugus | Olean Well Field | Final | 44.46 |
| NY | Oswego | Oswego | Pollution Abatement Services | Final | – |
| NY | Oyster Bay | Nassau | Old Bethpage Landfill | Final | 58.83 |
| NY | Oyster Bay | Nassau | Syosset Landfill | Deleted | 54.27 |
| NY | Plattekill | Ulster | Hertel Landfill | Final | 33.62 |
| NY | Plattsburgh | Clinton | Plattsburgh Air Force Base | Final | 30.34 |
| NY | Port Crane | Broome | Tri-Cities Barrel Co., Inc. | Final | 44.06 |
| NY | Port Jefferson Station | Suffolk | Lawrence Aviation Industries, Inc. | Final | 50.00 |
| NY | Port Jervis | Orange | Carroll & Dubies Sewage Disposal | Final | 33.74 |
| NY | Port Washington | Nassau | Port Washington Landfill | Final | 45.46 |
| NY | Putnam County | Putnam | Brewster Well Field | Final | 37.93 |
| NY | Queens | Queens | Radium Chemical Co., Inc. | Deleted | – |
| NY | Ramapo | Rockland | Ramapo Landfill | Final | 44.73 |
| NY | Rome | Oneida | Griffiss Air Force Base (11 Areas) | Final | 34.20 |
| NY | Romulus | Seneca | Seneca Army Depot | Final | 35.52 |

| St. | City/Area | County | Site Name | Status | Score[1] |
|---|---|---|---|---|---|
| NY | Saratoga Springs | Saratoga | Niagara Mohawk Power Corp. (Saratoga Springs Plant) | Final | 35.48 |
| NY | Sidney | Delaware | Sidney Landfill | Final | 29.36 |
| NY | Sidney Center | Delaware | Richardson Hill Road Landfill/Pond | Final | 34.86 |
| NY | Smithtown | Suffolk | Smithtown Ground Water Contamination | Final | 50.00 |
| NY | South Cairo | Greene | American Thermostat Co. | Final | 33.61 |
| NY | South Glens Falls | Saratoga | Ge Moreau | Final | 58.21 |
| NY | Syracuse | Onondaga | Onondaga Lake | Final | 50.00 |
| NY | Town Of Bedford | Westchester | Katonah Municipal Well | Deleted | 35.35 |
| NY | Town Of Champion | Jefferson | Black River Pcbs | Final | 48.03 |
| NY | Town Of Colesville | Broome | Colesville Municipal Landfill | Final | 30.26 |
| NY | Town Of Granby | Oswego | Clothier Disposal | Deleted | 34.48 |
| NY | Town Of Hyde Park | Dutchess | Haviland Complex | Final | 33.62 |
| NY | Town Of Johnstown | Fulton | Johnstown City Landfill | Final | 48.36 |
| NY | Town Of Shelby | Orleans | Fmc Corp. (Dublin Road Landfill) | Final | 32.90 |
| NY | Town Of Vestal | Broome | Robintech, Inc./National Pipe Co. | Final | 30.75 |
| NY | Town Of Volney | Oswego | Volney Municipal Landfill | Final | 32.89 |
| NY | Union Springs | Cayuga | Cayuga Groundwater Contamination Site | Final | 50.00 |
| NY | Upton | Suffolk | Brookhaven National Laboratory (Usdoe) | Final | 39.92 |
| NY | Vestal | Broome | Bec Trucking | Deleted | 30.75 |
| NY | Vestal | Broome | Vestal Water Supply Well 1-1 | Final | 37.93 |
| NY | Vestal | Broome | Vestal Water Supply Well 4-2 | Deleted | 42.24 |
| NY | Vil Of Narrowsburg | Sullivan | Cortese Landfill | Final | 32.11 |
| NY | Village Of Endicott | Broome | Endicott Village Well Field | Final | 35.57 |
| NY | Village Of Sidney | Delaware | Gcl Tie And Treating Inc. | Final | 48.54 |
| NY | Village Of Suffern | Rockland | Suffern Village Well Field | Deleted | 35.57 |
| NY | Warwick | Orange | Warwick Landfill | Deleted | 29.41 |
| NY | Wellsville | Allegany | Sinclair Refinery | Final | 53.90 |
| NY | West Winfield | Herkimer | Hiteman Leather | Final | 50.00 |
| NY | Wheatfield | Niagara | Niagara County Refuse | Deleted | 39.85 |
| OH | Ashtabula | Ashtabula | Fields Brook | Final | 44.95 |
| OH | Beavercreek | Greene | Lammers Barrel Factory | Final | 69.33 |
| OH | Circleville | Pickaway | Bowers Landfill | Deleted | 50.49 |
| OH | Cleveland | Cuyahoga | Chemical & Minerals Reclamation | Deleted | – |
| OH | Columbus | Franklin | Air Force Plant 85 | Proposed | 50.00 |
| OH | Copley | Summit | Copley Square Plaza | Final | 50.00 |
| OH | Darke County | Darke | Arcanum Iron & Metal | Deleted | 62.26 |
| OH | Dayton | Greene/Montgomery | Wright-Patterson Air Force Base | Final | 57.85 |
| OH | Dayton | Montgomery | Behr Dayton Thermal System Voc Plume | Final | 50.00 |
| OH | Dayton | Montgomery | North Sanitary Landfill | Final | 50.00 |
| OH | Dayton | Montgomery | Powell Road Landfill | Final | 31.62 |
| OH | Dayton | Montgomery | Sanitary Landfill Co. (Industrial Waste Disposal Co., Inc.) | Final | 35.57 |
| OH | Deerfield Township | Portage | Summit National | Final | 52.28 |
| OH | Dover | Tuscarawas | Dover Chemical Corp. | Proposed | 50.00 |
| OH | Dover | Tuscarawas | Reilly Tar & Chemical Corp. (Dover Plant) | Final | 31.38 |
| OH | Elyria | Lorain | Republic Steel Corp. Quarry | Deleted | 29.85 |
| OH | Fernald | Hamilton/Butler | Feed Materials Production Center (Usdoe) | Final | 57.56 |
| OH | Franklin Township | Coshocton | Coshocton Landfill | Deleted | 39.14 |
| OH | Gnadenhutten | Tuscarawas | Alsco Anaconda | Deleted | 42.94 |
| OH | Hamilton | Butler | Armco Incorporation-Hamilton Plant | Proposed | 69.34 |
| OH | Hamilton | Butler | Chem-Dyne | Final | – |
| OH | Hamilton Township | Lawrence | E.H. Schilling Landfill | Final | 34.56 |
| OH | Hannibal | Monroe | Ormet Corp. | Final | 46.44 |
| OH | Ironton | Lawrence | Allied Chemical & Ironton Coke | Final | 47.05 |
| OH | Jackson Township | Guernsey | Fultz Landfill | Final | 39.42 |
| OH | Jefferson Township | Ashtabula | Laskin/Poplar Oil Co. | Deleted | 35.95 |
| OH | Kings Mills | Warren | Peters Cartridge Factory | Proposed | 50.00 |
| OH | Kingsville | Ashtabula | Big D Campground | Final | 30.77 |
| OH | Lockbourne | Franklin | Rickenbacker Air National Guard (Usaf) | Proposed | 50.00 |
| OH | Marietta | Washington | Van Dale Junkyard | Final | 33.03 |

| St. | City/Area | County | Site Name | Status | Score[1] |
|---|---|---|---|---|---|
| OH | Marion County | Marion | Little Scioto River | Final | 48.03 |
| OH | Miamisburg | Montgomery | Mound Plant (Usdoe) | Final | 34.61 |
| OH | Milford | Clermont | Milford Contaminated Aquifer | Final | 50.00 |
| OH | Minerva | Stark | Trw, Inc. (Minerva Plant) | Final | 38.08 |
| OH | Moraine | Montgomery | South Dayton Dump & Landfill | Proposed | 48.63 |
| OH | New Carlisle | Clark | New Carlisle Landfill | Final | 46.40 |
| OH | New Lyme | Ashtabula | New Lyme Landfill | Final | 31.19 |
| OH | Painesville | Lake | Diamond Shamrock Corp. (Painesville Works) | Proposed | 50.00 |
| OH | Reading | Hamilton | Pristine, Inc. | Final | 35.25 |
| OH | Rock Creek | Ashtabula | Old Mill | Final | 35.95 |
| OH | Salem | Mahoning | Nease Chemical | Final | 47.19 |
| OH | South Point | Lawrence | South Point Plant | Final | 46.33 |
| OH | St. Clairsville | Belmont | Buckeye Reclamation | Final | 35.10 |
| OH | Troy | Miami | East Troy Contaminated Aquifer | Final | 50.00 |
| OH | Troy | Miami | Miami County Incinerator | Final | 57.84 |
| OH | Troy | Miami | United Scrap Lead Co., Inc. | Final | 58.15 |
| OH | Uniontown | Stark | Industrial Excess Landfill | Final | 51.13 |
| OH | West Chester | Butler | Skinner Landfill | Final | 30.23 |
| OH | Zanesville | Muskingum | Zanesville Well Field | Final | 35.59 |
| OK | Ardmore | Carter | Imperial Refining Company | Final | 30.00 |
| OK | Bartlesville | Washington | National Zinc Corp. | Proposed | 50.00 |
| OK | Collinsville | Tulsa | Tulsa Fuel And Manufacturing | Final | 50.00 |
| OK | Criner | Mcclain | Hardage/Criner | Final | 51.01 |
| OK | Cushing | Payne | Hudson Refinery | Final | 29.34 |
| OK | Cyril | Caddo | Oklahoma Refining Co. | Final | 46.01 |
| OK | Oklahoma City | Oklahoma | Double Eagle Refinery Co. | Deleted | 30.83 |
| OK | Oklahoma City | Oklahoma | Fourth Street Abandoned Refinery | Deleted | 30.67 |
| OK | Oklahoma City | Oklahoma | Mosley Road Sanitary Landfill | Final | 38.06 |
| OK | Oklahoma City | Oklahoma | Tenth Street Dump/Junkyard | Deleted | 30.98 |
| OK | Oklahoma City | Oklahoma | Tinker Air Force Base (Soldier Creek/Building 3001) | Final | 42.24 |
| OK | Ottawa County | Ottawa | Tar Creek (Ottawa County) | Final | 58.15 |
| OK | Sand Springs | Tulsa | Sand Springs Petrochemical Complex | Deleted | 28.86 |
| OK | Tulsa | Tulsa | Compass Industries (Avery Drive) | Deleted | 36.57 |
| OR | Albany | Linn | Teledyne Wah Chang | Final | 54.27 |
| OR | Astoria | Clatsop | Astoria Marine Construction Company | Proposed | 50.00 |
| OR | Clackamas | Clackamas | Northwest Pipe & Casing/Hall Process Company | Final | 51.09 |
| OR | Corvallis | Benton | United Chrome Products, Inc. | Final | 31.07 |
| OR | Cottage Grove | Lane | Black Butte Mine | Final | 50.00 |
| OR | Hermiston | Morrow | Umatilla Army Depot (Lagoons) | Final | 31.31 |
| OR | Joseph | Wallowa | Joseph Forest Products | Deleted | 32.60 |
| OR | Klamath Falls | Klamath | North Ridge Estates | Proposed | – |
| OR | Lakeview | Lake | Fremont National Forest/White King And Lucky Lass Uranium Mines (Usda) | Final | 50.00 |
| OR | Portland | Multnomah | Allied Plating, Inc. | Deleted | 39.25 |
| OR | Portland | Multnomah | Gould, Inc. | Deleted | 32.12 |
| OR | Portland | Multnomah | Harbor Oil Inc. | Final | 48.00 |
| OR | Portland | Multnomah | Mccormick & Baxter Creosoting Co. (Portland Plant) | Final | 50.00 |
| OR | Portland | Multnomah | Portland Harbor | Final | 50.00 |
| OR | Riddle | Douglas | Formosa Mine | Final | 50.00 |
| OR | Sheridan | Yamhill | Taylor Lumber And Treating | Final | 71.78 |
| OR | The Dalles | Wasco | Martin-Marietta Aluminum Co. | Deleted | 43.70 |
| OR | The Dalles | Wasco | Union Pacific Railroad Co. Tie-Treating Plant | Final | 37.93 |
| OR | Troutdale | Multnomah | Reynolds Metals Company | Final | 70.71 |
| PA | Ambler | Montgomery | Ambler Asbestos Piles | Deleted | 34.47 |
| PA | Ambler | Montgomery | Borit Asbestos | Final | 50.00 |
| PA | Antis/Logan Twps | Blair | Delta Quarries & Disposal, Inc./Stotler Landfill | Final | 41.08 |
| PA | Bally | Berks | Bally Ground Water Contamination | Final | 37.93 |
| PA | Bloomsburg | Columbia | Safety Light Corporation | Final | 70.71 |
| PA | Bridgeton Township | Bucks | Boarhead Farms | Final | 39.92 |

| St. | City/Area | County | Site Name | Status | Score[1] |
|---|---|---|---|---|---|
| PA | Bruin Borough | Butler | Bruin Lagoon | Deleted | 73.11 |
| PA | Buffalo Township | Butler | Hranica Landfill | Deleted | 51.94 |
| PA | Chambersburg | Franklin | Letterkenny Army Depot (Se Area) | Final | 34.21 |
| PA | Chester | Delaware | Wade (Abm) | Deleted | 36.63 |
| PA | Columbia | Lancaster | Ugi Columbia Gas Plant | Final | 50.78 |
| PA | Coraopolis | Allegheny | Breslube-Penn, Inc. | Final | 50.00 |
| PA | Croydon Township | Bucks | Croydon Tce | Final | 31.60 |
| PA | Darby Twp | Delaware | Lower Darby Creek Area | Final | 50.00 |
| PA | Delaware County | Delaware | Austin Avenue Radiation Site | Deleted | – |
| PA | Denver | Lancaster | Berkley Products Co. Dump | Deleted | 30.00 |
| PA | Douglassville | Berks | Douglassville Disposal | Final | 55.18 |
| PA | Doylestown | Bucks | Chem-Fab | Final | 50.00 |
| PA | Dublin Borough | Bucks | Dublin Tce Site | Final | 28.90 |
| PA | E Coventry Twp | Chester | Recticon/Allied Steel Corp. | Final | 32.06 |
| PA | Eagleville | Montgomery | Moyers Landfill | Final | 37.62 |
| PA | East Whiteland Township | Chester | Foote Mineral Co. | Final | 50.00 |
| PA | Elizabethtown | Lancaster | Elizabethtown Landfill | Final | 28.98 |
| PA | Emmaus Borough | Lehigh | Rodale Manufacturing Co., Inc. | Final | 50.00 |
| PA | Erie | Erie | Mill Creek Dump | Final | 49.31 |
| PA | Erie | Erie | Presque Isle | Deleted | 40.59 |
| PA | Exton | Chester | A.I.W. Frank/Mid-County Mustang | Final | 42.40 |
| PA | Falls Creek | Jefferson | Jackson Ceramics, Inc | Final | 30.22 |
| PA | Foster Township | Luzerne | C & D Recycling | Final | 43.92 |
| PA | Frackville | Schuylkill | Metropolitan Mirror And Glass Co., Inc. | Deleted | 34.33 |
| PA | Franklin County | Franklin | Letterkenny Army Depot (Pdo Area) | Final | 37.51 |
| PA | Gettysburg | Adams | Westinghouse Elevator Co. Plant | Final | 36.37 |
| PA | Girard Township | Erie | Lord-Shope Landfill | Final | 38.89 |
| PA | Glen Rock | York | Amp, Inc. (Glen Rock Facility) | Deleted | 39.03 |
| PA | Grove City | Mercer | Osborne Landfill | Final | 54.60 |
| PA | Hamburg | Berks | Brown's Battery Breaking | Final | 37.34 |
| PA | Hamburg | Berks | Price Battery | Final | 37.86 |
| PA | Harrison Township | Allegheny | Lindane Dump | Final | 51.62 |
| PA | Hatboro | Montgomery | Raymark | Final | 53.42 |
| PA | Hatfield | Montgomery | North Penn - Area 2 | Final | 35.57 |
| PA | Haverford | Delaware | Havertown Pcp | Final | 38.34 |
| PA | Heidelberg Twp | Berks | Ryeland Road Arsenic Site | Final | 60.30 |
| PA | Hellertown | Northampton | Hellertown Manufacturing Co. | Final | 51.91 |
| PA | Hereford Township | Berks | Crossley Farm | Final | 29.66 |
| PA | Hermitage | Mercer | River Road Landfill (Waste Management, Inc.) | Deleted | 43.12 |
| PA | Hickory Township | Mercer | Sharon Steel Corp (Farrell Works Disposal Area) | Final | 50.00 |
| PA | Hometown | Schuylkill | Eastern Diversified Metals | Final | 31.02 |
| PA | Honeybrook Township | Chester | Walsh Landfill | Final | 33.64 |
| PA | Hopewell Township | York | York County Solid Waste And Refuse Authority Landfill | Deleted | 44.26 |
| PA | Horsham | Montgomery | Willow Grove Naval Air And Air Reserve Station | Final | 50.00 |
| PA | Jackson Township | Lebanon | Whitmoyer Laboratories | Final | 46.25 |
| PA | Jefferson Borough | Allegheny | Resin Disposal | Deleted | 37.69 |
| PA | Kimberton Borough | Chester | Kimberton | Final | 29.44 |
| PA | King Of Prussia | Montgomery | Stanley Kessler | Final | 33.89 |
| PA | Lansdale | Montgomery | North Penn - Area 6 | Final | 35.57 |
| PA | Lansdowne | Delaware | Lansdowne Radiation Site | Deleted | – |
| PA | Lock Haven | Clinton | Drake Chemical | Final | 38.52 |
| PA | Longswamp Township | Berks | Berks Sand Pit | Final | 32.02 |
| PA | Lower Pottsgrove Township | Montgomery | Occidental Chemical Corp./Firestone Tire & Rubber Co. | Final | 45.91 |
| PA | Lower Providence Township | Montgomery | Commodore Semiconductor Group | Final | 42.35 |
| PA | Lower Salford Township | Montgomery | Salford Quarry | Final | 50.00 |
| PA | Lower Windsor Twp | York | Modern Sanitation Landfill | Final | 33.93 |
| PA | Maitland | Mifflin | Jacks Creek/Sitkin Smelting & Refining, Inc. | Final | 40.37 |
| PA | Malvern | Chester | Malvern Tce | Final | 46.69 |
| PA | Marcus Hook | Delaware | East Tenth Street | Proposed | 67.68 |
| PA | Mcadoo Borough | Schuylkill | Mcadoo Associates | Deleted | – |

| St. | City/Area | County | Site Name | Status | Score[1] |
|---|---|---|---|---|---|
| PA | Mechanicsburg | Cumberland | Navy Ships Parts Control Center | Final | 50.00 |
| PA | Middletown | Dauphin | Middletown Air Field | Deleted | 35.69 |
| PA | Montgomery Township | Montgomery | North Penn - Area 5 | Final | 35.57 |
| PA | Mountain Top | Luzerne | Foster Wheeler Energy Corp./Church Road Tce | Proposed | 50.00 |
| PA | Nesquehoning | Carbon | Tonolli Corp. | Final | 46.58 |
| PA | Neville Island | Allegheny | Ohio River Park | Final | 42.24 |
| PA | Newlin Township | Chester | Strasburg Landfill | Final | 30.71 |
| PA | Nockamixon Township | Bucks | Revere Chemical Co. | Final | 31.31 |
| PA | North Wales | Montgomery | North Penn - Area 7 | Final | 35.57 |
| PA | North Whitehall Twp | Lehigh | Heleva Landfill | Final | 50.23 |
| PA | Old Forge | Lackawanna | Lackawanna Refuse | Deleted | 36.57 |
| PA | Old Forge | Lackawanna | Lehigh Electric & Engineering Co. | Deleted | 30.26 |
| PA | Palmerton | Carbon | Palmerton Zinc Pile | Final | 42.93 |
| PA | Paoli | Chester | Paoli Rail Yard | Final | 32.18 |
| PA | Parker | Armstrong | Craig Farm Drum | Final | 28.72 |
| PA | Philadelphia | Philadelphia | Enterprise Avenue | Deleted | 40.80 |
| PA | Philadelphia | Philadelphia | Franklin Slag Pile (Mdc) | Final | 50.20 |
| PA | Philadelphia | Philadelphia | Metal Banks | Final | 33.23 |
| PA | Philadelphia | Philadelphia | Publicker Industries Inc. | Deleted | 59.06 |
| PA | Pittston Township | Luzerne | Butler Mine Tunnel | Final | 49.51 |
| PA | Pocono Summit | Monroe | Route 940 Drum Dump | Deleted | 44.06 |
| PA | Richland Township | Bucks | Watson Johnson Landfill | Final | 70.71 |
| PA | Sadsburyville | Chester | Old Wilmington Road Gw Contamination | Final | 50.00 |
| PA | Saegertown | Crawford | Saegertown Industrial Area | Final | 33.62 |
| PA | Scott Township | Lackawanna | Aladdin Plating | Deleted | 35.57 |
| PA | Seven Valleys | York | Old City Of York Landfill | Final | 33.93 |
| PA | Sharon | Mercer | Westinghouse Electric Corp. (Sharon Plant) | Final | 41.33 |
| PA | Souderton | Montgomery | North Penn - Area 1 | Final | 35.57 |
| PA | South Montrose | Susquehanna | Bendix Flight Systems Division | Final | 33.74 |
| PA | South Whitehall Township | Lehigh | Novak Sanitary Landfill | Final | 42.31 |
| PA | Spring Township | Berks | Berks Landfill | Deleted | 46.10 |
| PA | Springettsbury Township | York | East Mount Zion | Final | 41.01 |
| PA | State College Borough | Centre | Centre County Kepone | Final | 45.09 |
| PA | Straban Township | Adams | Hunterstown Road | Final | 48.27 |
| PA | Straban Township | Adams | Shriver's Corner | Final | 46.13 |
| PA | Stroudsburg | Monroe | Brodhead Creek | Deleted | 31.09 |
| PA | Stroudsburg | Monroe | Butz Landfill | Final | 32.00 |
| PA | Taylor Borough | Lackawanna | Taylor Borough Dump | Deleted | 30.94 |
| PA | Terry Township | Bradford | Bell Landfill | Final | 34.79 |
| PA | Tobyhanna | Monroe | Tobyhanna Army Depot | Final | 37.93 |
| PA | Union Township | Adams | Keystone Sanitation Landfill | Final | 33.76 |
| PA | Upper Macungie Township | Lehigh | Dorney Road Landfill | Final | 46.10 |
| PA | Upper Macungie Twp | Lehigh | Reeser's Landfill | Deleted | 30.35 |
| PA | Upper Merion Township | Montgomery | Crater Resources, Inc./Keystone Coke Co./Alan Wood Steel Co. | Final | 50.00 |
| PA | Upper Merion Township | Montgomery | Henderson Road | Final | 41.69 |
| PA | Upper Merion Twp | Montgomery | Tysons Dump | Final | 63.10 |
| PA | Upper Saucon Twp | Lehigh | Voortman Farm | Deleted | 28.62 |
| PA | Valley Township | Montour | Mw Manufacturing | Final | 46.44 |
| PA | Warminster | Bucks | Fischer & Porter Co. | Final | 29.07 |
| PA | Warminster Township | Bucks | Naval Air Development Center (8 Waste Areas) | Final | 57.93 |
| PA | Weisenberg Township | Lehigh | Hebelka Auto Salvage Yard | Deleted | 31.94 |
| PA | West Caln Township | Chester | Blosenski Landfill | Final | 30.57 |
| PA | West Caln Township | Chester | William Dick Lagoons | Final | 36.64 |
| PA | West Hazleton | Luzerne | Valmont Tce Site (Former - Valmont Industrial Park) | Final | 43.16 |
| PA | Westline | Mckean | Westline | Deleted | 31.71 |
| PA | Williams Township | Northampton | Industrial Lane | Final | 42.47 |
| PA | Williamsport | Lycoming | Avco Lycoming (Williamsport Division) | Final | 42.24 |
| PA | Worcester | Montgomery | North Penn - Area 12 | Final | 28.90 |
| PA | Worman Township | Berks | Cryochem, Inc. | Final | 28.58 |

| St. | City/Area | County | Site Name | Status | Score[1] |
|---|---|---|---|---|---|
| PR | Almirante Norte Ward | Vega Baja | V&M/Albaladejo | Deleted | 50.00 |
| PR | Arecibo | Arecibo | Pesticide Warehouse I | Final | 50.00 |
| PR | Barceloneta | Barceloneta | Upjohn Facility | Final | 41.92 |
| PR | Barceloneta | Barceloneta Municipality | Rca Del Caribe | Deleted | 31.14 |
| PR | Cabo Rojo | Cabo Rojo | Cabo Rojo Ground Water Contamination | Final | 50.00 |
| PR | Caguas | Caguas | Hormigas Ground Water Plume | Final | 50.00 |
| PR | Candeleria Ward | Toa Baja Municipality | Scorpio Recycling, Inc. | Final | 50.00 |
| PR | Cidra | Cidra | Cidra Groundwater Contamination | Final | 50.00 |
| PR | Florida Afuera | Florida Municipality | Barceloneta Landfill | Final | 41.11 |
| PR | Jobos | Guayama Municipality | Fibers Public Supply Wells | Final | 35.34 |
| PR | Juana Diaz | Juana Diaz Municipality | Ge Wiring Devices | Deleted | 31.24 |
| PR | Juncos | Juncos | Juncos Landfill | Final | 32.57 |
| PR | Manati | Manati | Pesticide Warehouse Iii | Final | 50.00 |
| PR | Maunabo | Maunabo | Maunabo Urbano Public Wells | Final | 50.00 |
| PR | Rio Abajo | Humacao | Frontera Creek | Deleted | 43.07 |
| PR | Rio Abajo Ward | Vega Baja | Vega Baja Solid Waste Disposal | Final | 50.37 |
| PR | Sabana Seca | Toa Baja Municipality | Naval Security Group Activity | Deleted | 34.28 |
| PR | San German | San German | San German Ground Water Contamination | Final | 50.00 |
| PR | Utuado | Utuado Municipality | Papelera Puertorriquena, Inc. | Final | 34.69 |
| PR | Vega Alta | Vega Alta Municipality | Vega Alta Public Supply Wells | Final | 42.24 |
| PR | Vieques | Vieques | Atlantic Fleet Weapons Training Area | Final | – |
| RI | Burrillville | Providence | Western Sand & Gravel | Final | 51.35 |
| RI | Coventry | Kent | Picillo Farm | Final | – |
| RI | Glocester | Providence | Davis (Gsr) Landfill | Deleted | 38.89 |
| RI | Johnston | Providence | Central Landfill | Final | 46.71 |
| RI | Lincoln/Cumberland | Providence | Peterson/Puritan, Inc. | Final | 40.10 |
| RI | Newport | Newport | Newport Naval Education & Training Center | Final | 32.25 |
| RI | North Kingstown | Washington | Davisville Naval Construction Battalion Center | Final | 34.52 |
| RI | North Providence | Providence | Centredale Manor Restoration Project | Final | 70.71 |
| RI | North Smithfield | Providence | Landfill & Resource Recovery, Inc. (L&Rr) | Final | 49.58 |
| RI | North Smithfield | Providence | Stamina Mills, Inc. | Final | 34.07 |
| RI | Smithfield | Providence | Davis Liquid Waste | Final | 47.25 |
| RI | South Kingstown | Washington | Rose Hill Regional Landfill | Final | 38.11 |
| RI | South Kingstown | Washington | West Kingston Town Dump/Uri Disposal Area | Final | 50.00 |
| SC | Aiken | Aiken/Barnwell/Allendal | Savannah River Site (Usdoe) | Final | 47.70 |
| SC | Barnwell | Barnwell | Shuron Inc. | Final | 68.26 |
| SC | Beaufort | Beaufort | Independent Nail Co. | Deleted | 57.90 |
| SC | Beaufort | Beaufort | Kalama Specialty Chemicals | Final | 57.90 |
| SC | Burton | Beaufort | Wamchem, Inc. | Final | 47.70 |
| SC | Cayce | Lexington | Lexington County Landfill Area | Final | 37.93 |
| SC | Cayce | Lexington | Scrdi Dixiana | Final | 40.70 |
| SC | Charleston | Charleston | Koppers Co., Inc. (Charleston Plant) | Final | 50.00 |
| SC | Columbia | Richland | Palmetto Recycling, Inc. | Deleted | 29.46 |
| SC | Columbia | Richland | Scrdi Bluff Road | Final | – |
| SC | Dixiana | Lexington | Palmetto Wood Preserving | Final | 38.43 |
| SC | Fairfax | Allendale | Helena Chemical Co. Landfill | Final | 33.89 |
| SC | Florence | Florence | Koppers Co., Inc. (Florence Plant) | Final | 51.27 |
| SC | Fort Lawn | Chester | Carolawn, Inc. | Final | 32.04 |
| SC | Fountain Inn | Greenville | Beaunit Corp. (Circular Knit & Dyeing Plant) | Final | 32.44 |
| SC | Gaffney | Cherokee | Medley Farm Drum Dump | Final | 31.58 |
| SC | Greenville | Greenville | Us Finishing/Cone Mills | Proposed | 50.00 |
| SC | Greer | Spartanburg | Aqua-Tech Environmental Inc (Groce Labs) | Final | 50.00 |
| SC | Greer | Spartanburg | Elmore Waste Disposal | Final | 31.45 |
| SC | Jefferson | Chesterfield | Brewer Gold Mine | Final | 50.00 |
| SC | Mccormick | Mccormick | Barite Hill/Nevada Goldfields | Final | 50.00 |
| SC | North Charleston | Charleston | Macalloy Corporation | Final | 50.00 |
| SC | Parris Island | Beaufort | Parris Island Marine Corps Recruit Depot | Final | 50.00 |
| SC | Pickens | Pickens | Sangamo Weston, Inc./Twelve-Mile Creek/Lake Hartwell Pcb Contamination | Final | 37.63 |

| St. | City/Area | County | Site Name | Status | Score[1] |
|---|---|---|---|---|---|
| SC | Pontiac | Richland | Townsend Saw Chain Co. | Final | 35.94 |
| SC | Rantoules | Charleston | Geiger (C & M Oil) | Final | 32.25 |
| SC | Rock Hill | York | Leonard Chemical Co., Inc. | Final | 47.10 |
| SC | Rock Hill | York | Rock Hill Chemical Co. | Final | 40.29 |
| SC | Simpsonville | Greenville | Golden Strip Septic Tank Service | Deleted | 40.30 |
| SC | Simpsonville | Greenville | Para-Chem Southern, Inc. | Final | 32.94 |
| SC | Travelers Rest | Greenville | Rochester Property | Deleted | 36.72 |
| SD | Ellsworth Afb | Meade/Pennington | Ellsworth Air Force Base | Final | 33.62 |
| SD | Lead | Lawrence | Gilt Edge Mine | Final | 50.00 |
| SD | Sioux Falls | Minnehaha | Williams Pipe Line Co. Disposal Pit | Deleted | 42.24 |
| SD | Whitewood | Lawrence | Whitewood Creek | Deleted | – |
| TN | Alamo | Crockett | Alamo Contaminated Ground Water | Proposed | 50.00 |
| TN | Arlington | Shelby | Arlington Blending & Packaging | Final | 39.03 |
| TN | Chattanooga | Hamilton | Amnicola Dump | Deleted | 40.91 |
| TN | Chattanooga | Hamilton | Tennessee Products | Final | – |
| TN | Collierville | Shelby | Carrier Air Conditioning Co. | Final | 48.91 |
| TN | Collierville | Shelby | Smalley-Piper | Final | 50.00 |
| TN | Gallaway | Fayette | Gallaway Pits | Deleted | 30.77 |
| TN | Jackson | Madison | American Creosote Works, Inc. (Jackson Plant) | Final | 35.22 |
| TN | Jackson | Madison | Icg Iselin Railroad Yard | Deleted | 50.00 |
| TN | Knoxville | Knox | Smokey Mountain Smelters | Final | 50.00 |
| TN | Lawrenceburg | Lawrence | Murray-Ohio Dump | Final | 46.44 |
| TN | Lewisburg | Marshall | Lewisburg Dump | Deleted | 33.45 |
| TN | Memphis | Shelby | Memphis Defense Depot (Dla) | Final | 58.06 |
| TN | Memphis | Shelby | North Hollywood Dump | Deleted | – |
| TN | Milan | Carroll | Milan Army Ammunition Plant | Final | 58.15 |
| TN | Moscow | Fayette | Chemet Co. | Deleted | 50.00 |
| TN | Oak Ridge | Anderson/Roane | Oak Ridge Reservation (Usdoe) | Final | 51.13 |
| TN | Rossville | Fayette | Ross Metals Inc. | Final | 37.65 |
| TN | Toone | Hardeman | Velsicol Chemical Corp. (Hardeman County) | Final | 47.71 |
| TN | Tullahoma/Manchester | Coffee/Franklin | Arnold Engineering Development Center (Usaf) | Proposed | 50.00 |
| TN | Waynesboro | Wayne | Mallory Capacitor Co. | Final | 29.44 |
| TN | Wrigley | Hickman | Wrigley Charcoal Plant | Final | 36.14 |
| TX | Bell County | Bell | Rockwool Industries Inc. | Final | 48.00 |
| TX | Bridge City | Orange | Bailey Waste Disposal | Deleted | 53.42 |
| TX | Bridge City | Orange | Triangle Chemical Co. | Deleted | 28.75 |
| TX | Channelview | Harris | San Jacinto River Waste Pits | Final | 50.00 |
| TX | Conroe | Montgomery | Conroe Creosoting Co. | Final | 48.00 |
| TX | Conroe | Montgomery | United Creosoting Co. | Final | 37.29 |
| TX | Corpus Christi | Nueces | Brine Service Company | Final | 50.00 |
| TX | Crosby | Harris | French, Ltd. | Final | 63.33 |
| TX | Crosby | Harris | Sikes Disposal Pits | Final | 61.62 |
| TX | Crystal City | Zavala | Crystal City Airport | Deleted | 32.26 |
| TX | Dallas | Dallas | Rsr Corporation | Final | 50.00 |
| TX | Deer Park | Harris | Patrick Bayou | Final | 47.83 |
| TX | Donna | Hidalgo | Donna Reservoir And Canal System | Final | 50.00 |
| TX | Fort Worth | Tarrant | Air Force Plant #4 (General Dynamics) | Final | 39.92 |
| TX | Fort Worth | Tarrant | Pesses Chemical Co. | Deleted | 28.86 |
| TX | Freeport | Brazoria | Gulfco Marine Maintenance | Final | 50.00 |
| TX | Friendswood | Harris | Brio Refining, Inc. | Deleted | 50.38 |
| TX | Friendswood | Harris | Dixie Oil Processors, Inc. | Deleted | 34.21 |
| TX | Grand Prairie | Dallas | Bio-Ecology Systems, Inc. | Deleted | 35.06 |
| TX | Greenville | Hunt | Old Esco Manufacturing | Final | 40.81 |
| TX | Happy | Swisher | North East 2nd Street Site | Final | 32.33 |
| TX | Hempstead | Waller | Sheridan Disposal Services | Final | 30.16 |
| TX | Highlands | Harris | Highlands Acid Pit | Final | 37.77 |
| TX | Houston | Harris | Crystal Chemical Co. | Final | 60.90 |
| TX | Houston | Harris | Geneva Industries/Fuhrmann Energy | Final | 59.46 |
| TX | Houston | Harris | Harris (Farley Street) | Deleted | 33.94 |

| St. | City/Area | County | Site Name | Status | Score[1] |
|---|---|---|---|---|---|
| TX | Houston | Harris | Jones Road Ground Water Plume | Final | 46.50 |
| TX | Houston | Harris | Many Diversified Interests, Inc. | Final | 32.07 |
| TX | Houston | Harris | North Cavalcade Street | Final | 37.08 |
| TX | Houston | Harris | Sol Lynn/Industrial Transformers | Final | 39.65 |
| TX | Houston | Harris | South Cavalcade Street | Final | 38.69 |
| TX | Ingleside | San Patricio | Falcon Refinery | Proposed | 50.00 |
| TX | Jasper | Jasper | Hart Creosoting Company | Final | 48.00 |
| TX | Jasper | Jasper | Jasper Creosoting Company Inc. | Final | 50.00 |
| TX | Jefferson County | Jefferson | State Marine Of Port Arthur | Final | 48.00 |
| TX | Karnack | Harrison | Longhorn Army Ammunition Plant | Final | 39.83 |
| TX | La Marque | Galveston | Motco, Inc. | Final | – |
| TX | Levelland | Hockley | State Road 114 Groundwater Plume | Final | 42.41 |
| TX | Liberty | Liberty | Petro-Chemical Systems, Inc. (Turtle Bayou) | Final | 29.94 |
| TX | Longview | Gregg | Garland Creosoting | Final | 49.10 |
| TX | Midland | Midland | Midessa Ground Water Plume | Final | 50.00 |
| TX | Midland | Midland | West County Road 112 Ground Water | Final | 50.00 |
| TX | Odessa | Ector | East 67th Street Ground Water Plume | Final | 50.00 |
| TX | Odessa | Ector | Odessa Chromium #1 | Final | 42.24 |
| TX | Odessa | Ector | Odessa Chromium #2 (Andrews Highway) | Deleted | 42.24 |
| TX | Odessa | Ector | Sprague Road Ground Water Plume | Final | 43.21 |
| TX | Pantex Village | Carson | Pantex Plant (Usdoe) | Final | 51.22 |
| TX | Pelican Bay | Tarrant | Sandy Beach Road Ground Water Plume | Final | 50.00 |
| TX | Perryton | Ochiltree | City Of Perryton Well No. 2 | Final | 50.00 |
| TX | Point Comfort | Calhoun | Alcoa (Point Comfort)/Lavaca Bay | Final | 50.00 |
| TX | Port Arthur | Jefferson | Palmer Barge Line | Final | 50.00 |
| TX | Port Neches | Jefferson | Star Lake Canal | Final | 50.00 |
| TX | San Antonio | Bexar | Bandera Road Ground Water Plume | Final | 50.00 |
| TX | San Antonio | Bexar | R & H Oil/Tropicana | Proposed | 50.00 |
| TX | Terrell | Kaufman | Van Der Horst Usa Corporation | Final | 48.00 |
| TX | Texarkana | Bowie | Koppers Co., Inc. (Texarkana Plant) | Final | 31.31 |
| TX | Texarkana | Bowie | Lone Star Army Ammunition Plant | Final | 31.85 |
| TX | Texarkana | Bowie | Texarkana Wood Preserving Co. | Final | 40.19 |
| TX | Texas City | Galveston | Malone Service Co - Swan Lake Plant | Final | 50.00 |
| TX | Texas City | Galveston | Tex-Tin Corp. | Final | 50.00 |
| TX | Waskom | Harrison | Stewco, Inc. | Deleted | 48.86 |
| UT | Bountiful | Davis | Bountiful/Woods Cross 5th S. Pce Plume | Final | 50.00 |
| UT | Bountiful | Davis | Intermountain Waste Oil Refinery | Final | 50.00 |
| UT | Eureka | Juab | Eureka Mills | Final | 50.00 |
| UT | Hill Afb | Davis/Weber | Hill Air Force Base | Final | 49.94 |
| UT | Magna | Salt Lake | Kennecott (North Zone) | Proposed | 59.18 |
| UT | Midvale | Salt Lake | Midvale Slag | Final | 42.47 |
| UT | Midvale | Salt Lake | Sharon Steel Corp. (Midvale Tailings) | Deleted | 41.85 |
| UT | Monticello | San Juan | Monticello Mill Tailings (Usdoe) | Final | 35.86 |
| UT | Monticello | San Juan | Monticello Radioactively Contaminated Properties | Deleted | 35.03 |
| UT | Murray City | Salt Lake | Murray Smelter | Proposed | 86.60 |
| UT | Ogden | Weber | Ogden Defense Depot (Dla) | Final | 45.10 |
| UT | Park City | Summit | Richardson Flat Tailings | Proposed | 50.23 |
| UT | Salt Lake City | Salt Lake | Petrochem Recycling Corp./Ekotek Plant | Deleted | 62.81 |
| UT | Salt Lake City | Salt Lake | Portland Cement (Kiln Dust 2 & 3) | Final | 54.40 |
| UT | Salt Lake City | Salt Lake | Rose Park Sludge Pit | Deleted | – |
| UT | Salt Lake City | Salt Lake | Utah Power & Light/American Barrel Co. | Final | 37.93 |
| UT | Salt Lake City | Salt Lake | Wasatch Chemical Co. (Lot 6) | Final | 49.91 |
| UT | Sandy | Salt Lake | Davenport And Flagstaff Smelters | Final | 32.50 |
| UT | Stockton | Tooele | Jacobs Smelter | Final | 50.00 |
| UT | Tooele | Tooele | International Smelting And Refining | Final | 58.31 |
| UT | Tooele | Tooele | Tooele Army Depot (North Area) | Final | 53.95 |
| UT | Tooele County | Tooele | Us Magnesium | Final | 59.18 |
| UT | Woods Cross/Bountiful | Davis | Five Points Pce Plume | Final | 50.00 |
| VA | Buckingham | Buckingham | Buckingham County Landfill | Final | 40.70 |

| St. | City/Area | County | Site Name | Status | Score[1] |
|---|---|---|---|---|---|
| VA | Chesapeake | Chesapeake City | St. Juliens Creek Annex (U.S. Navy) | Final | 50.00 |
| VA | Chesterfield County | Chesterfield | C & R Battery Co., Inc. | Final | 46.44 |
| VA | Chesterfield County | Chesterfield | Defense General Supply Center (Dla) | Final | 33.85 |
| VA | Chuckatuck | Suffolk City | Saunders Supply Co. | Final | 36.88 |
| VA | Culpeper | Culpeper | Culpeper Wood Preservers, Inc. | Final | 45.91 |
| VA | Dahlgren | King George | Naval Surface Warfare Center - Dahlgren | Final | 50.03 |
| VA | Farrington | Hanover | H & H Inc., Burn Pit | Final | 33.71 |
| VA | Frederick County | Frederick | Rhinehart Tire Fire Dump | Deleted | 30.57 |
| VA | Front Royal | Warren | Avtex Fibers, Inc. | Final | 35.39 |
| VA | Hampton | Hampton City | Langley Air Force Base/Nasa Langley Research Center | Final | 50.00 |
| VA | Montross | Westmoreland | Arrowhead Associates, Inc./Scovill Corp. | Final | 37.15 |
| VA | Newport News | Newport News City | Fort Eustis (Us Army) | Final | 50.00 |
| VA | Newtown | Albemarle | Greenwood Chemical Co. | Final | 53.17 |
| VA | Norfolk | Norfolk City | Norfolk Naval Base (Sewells Point Naval Complex) | Final | 50.00 |
| VA | Piney River | Nelson | U.S. Titanium | Final | 34.78 |
| VA | Pittsylvania County | Pittsylvania | First Piedmont Corp. Rock Quarry (Route 719) | Final | 30.16 |
| VA | Portsmouth | Portsmouth City | Abex Corp. | Final | 36.53 |
| VA | Portsmouth | Portsmouth City | Atlantic Wood Industries, Inc. | Final | 37.14 |
| VA | Portsmouth | Portsmouth City | Norfolk Naval Shipyard | Final | 50.00 |
| VA | Portsmouth | Portsmouth City | Peck Iron And Metal | Final | 48.52 |
| VA | Quantico | Prince William | Marine Corps Combat Development Command | Final | 50.00 |
| VA | Richmond | Henrico | Rentokil, Inc. (Virginia Wood Preserving Division) | Final | 30.34 |
| VA | Roanoke | Roanoke City | Matthews Electroplating | Deleted | – |
| VA | Salem | Roanoke | Dixie Caverns County Landfill | Deleted | 35.27 |
| VA | Saltville | Smyth | Saltville Waste Disposal Ponds | Final | 29.52 |
| VA | Selma | Alleghany | Kim-Stan Landfill | Final | 50.00 |
| VA | Spotsylvania | Spotsylvania | L.A. Clarke & Son | Final | 34.24 |
| VA | Sterling | Loudoun | Hidden Lane Landfill | Final | 50.00 |
| VA | Suffolk | Suffolk City | Former Nansemond Ordnance Depot | Final | 70.71 |
| VA | Suffolk | Suffolk City | Suffolk City Landfill | Deleted | 35.76 |
| VA | Virginia Beach | Virginia Beach City | Naval Amphibious Base Little Creek | Final | 50.00 |
| VA | York County | York | Chisman Creek | Final | 47.19 |
| VA | Yorktown | York | Naval Weapons Station - Yorktown | Final | 50.00 |
| VA | Yorktown | York | Nws Yorktown - Cheatham Annex | Final | 49.27 |
| VI | Christiansted | St. Croix | Island Chemical Corp/Virgin Islands Chemical Corp. | Deleted | 50.00 |
| VI | Tutu | St. Thomas | Tutu Wellfield | Final | 50.00 |
| VT | Bennington | Bennington | Bennington Municipal Sanitary Landfill | Final | 49.07 |
| VT | Bennington | Bennington | Tansitor Electronics, Inc. | Deleted | 35.72 |
| VT | Burlington | Chittenden | Pine Street Canal | Final | – |
| VT | Corinth | Orange | Pike Hill Copper Mine | Final | 50.00 |
| VT | Lyndon | Caledonia | Darling Hill Dump | Deleted | 43.92 |
| VT | Lyndon | Caledonia | Parker Sanitary Landfill | Final | 52.29 |
| VT | Pownal | Bennington | Pownal Tannery | Final | 50.00 |
| VT | Rockingham | Windham | Bfi Sanitary Landfill (Rockingham) | Final | 41.92 |
| VT | Springfield | Windsor | Old Springfield Landfill | Final | 34.79 |
| VT | Strafford | Orange | Elizabeth Mine | Final | 50.00 |
| VT | Vershire | Orange | Ely Copper Mine | Final | 50.00 |
| VT | Williston | Chittenden | Commerce Street Plume | Final | 48.48 |
| VT | Woodford | Bennington | Burgess Brothers Landfill | Final | 52.58 |
| WA | Bainbridge Island | Kitsap | Wyckoff Co./Eagle Harbor | Final | 32.55 |
| WA | Bellingham | Whatcom | Oeser Co. | Final | 69.34 |
| WA | Benton County | Benton | Hanford 100-Area (Usdoe) | Final | 46.38 |
| WA | Benton County | Benton | Hanford 1100-Area (Usdoe) | Deleted | 36.34 |
| WA | Benton County | Benton | Hanford 200-Area (Usdoe) | Final | 69.05 |
| WA | Benton County | Benton | Hanford 300-Area (Usdoe) | Final | 65.23 |
| WA | Bremerton | Kitsap | Bangor Ordnance Disposal (Usnavy) | Final | 30.42 |
| WA | Bremerton | Kitsap | Puget Sound Naval Shipyard Complex | Final | 50.00 |
| WA | Brush Prairie | Clark | Toftdahl Drums | Deleted | 40.22 |
| WA | Centralia | Lewis | Centralia Municipal Landfill | Final | 36.36 |
| WA | Chehalis | Lewis | American Crossarm & Conduit Co. | Final | 30.44 |

| St. | City/Area | County | Site Name | Status | Score[1] |
|---|---|---|---|---|---|
| WA | Chehalis | Lewis | Hamilton/Labree Roads Gw Contamination | Final | 37.65 |
| WA | Everson | Whatcom | Northwest Transformer | Deleted | 33.82 |
| WA | Everson | Whatcom | Northwest Transformer (South Harkness Street) | Deleted | 30.56 |
| WA | Indian Island | Jefferson | Port Hadlock Detachment (Usnavy) | Deleted | 50.00 |
| WA | Kent | King | Midway Landfill | Final | 54.27 |
| WA | Kent | King | Seattle Municipal Landfill (Kent Highlands) | Final | 52.19 |
| WA | Kent | King | Western Processing Co., Inc. | Final | 58.63 |
| WA | Keyport | Kitsap | Naval Undersea Warfare Engineering Station (4 Waste Areas) | Final | 33.60 |
| WA | Kitsap County | Kitsap | Jackson Park Housing Complex (Usnavy) | Final | 50.00 |
| WA | Lakewood | Pierce | Lakewood | Final | 42.49 |
| WA | Loomis | Okanogan | Silver Mountain Mine | Deleted | 29.98 |
| WA | Manchester | Kitsap | Old Navy Dump/Manchester Laboratory (Usepa/Noaa) | Final | 50.00 |
| WA | Maple Valley | King | Queen City Farms | Final | 34.38 |
| WA | Marysville | Snohomish | Tulalip Landfill | Deleted | 50.00 |
| WA | Mead | Spokane | Kaiser Aluminum (Mead Works) | Final | 38.07 |
| WA | Mica | Spokane | Mica Landfill | Final | 34.64 |
| WA | Moses Lake | Grant | Moses Lake Wellfield Contamination | Final | 50.00 |
| WA | North Bonneville | Skamania | Hamilton Island Landfill (Usa/Coe) | Deleted | 51.97 |
| WA | Pasco | Franklin | Pasco Sanitary Landfill | Final | 44.46 |
| WA | Pierce County | Pierce | Hidden Valley Landfill (Thun Field) | Final | 37.93 |
| WA | Renton | King | Pacific Car & Foundry Co. | Final | 42.33 |
| WA | Renton | King | Quendall Terminal | Final | 50.00 |
| WA | Seattle | King | Harbor Island (Lead) | Final | 34.60 |
| WA | Seattle | King | Lockheed West Seattle | Final | 50.00 |
| WA | Seattle | King | Lower Duwamish Waterway | Final | 50.00 |
| WA | Seattle | King | Pacific Sound Resources | Final | 70.71 |
| WA | Silverdale | Kitsap | Bangor Naval Submarine Base | Final | 55.91 |
| WA | Spokane | Spokane | Colbert Landfill | Final | 41.59 |
| WA | Spokane | Spokane | Fairchild Air Force Base (4 Waste Areas) | Final | 31.98 |
| WA | Spokane | Spokane | General Electric Co. (Spokane Apparatus Service Shop) | Final | 57.80 |
| WA | Spokane | Spokane | North Market Street | Final | 32.61 |
| WA | Spokane | Spokane | Northside Landfill | Final | 28.90 |
| WA | Spokane | Spokane | Old Inland Pit | Deleted | 29.35 |
| WA | Spokane | Spokane | Spokane Junkyard/Associated Properties | Deleted | 50.00 |
| WA | Spokane County | Spokane | Greenacres Landfill | Final | 28.90 |
| WA | Tacoma | Pierce | American Lake Gardens/Mcchord Afb | Final | 28.90 |
| WA | Tacoma | Pierce | Commencement Bay, Near Shore/Tide Flats | Final | 42.20 |
| WA | Tacoma | Pierce | Commencement Bay, South Tacoma Channel | Final | 54.63 |
| WA | Tacoma | Pierce | Fort Lewis (Landfill No. 5) | Deleted | 33.79 |
| WA | Tacoma | Pierce | Mcchord Air Force Base (Wash Rack/Treatment Area) | Deleted | 42.24 |
| WA | Tillicum | Pierce | Fort Lewis Logistics Center | Final | 35.48 |
| WA | Tumwater | Thurston | Palermo Well Field Ground Water Contamination | Final | 50.00 |
| WA | Vancouver | Clark | Alcoa (Vancouver Smelter) | Deleted | 57.80 |
| WA | Vancouver | Clark | Bonneville Power Administration Ross Complex (Usdoe) | Deleted | 53.67 |
| WA | Vancouver | Clark | Boomsnub/Airco | Final | – |
| WA | Vancouver | Clark | Frontier Hard Chrome, Inc. | Final | 57.93 |
| WA | Vancouver | Clark | Vancouver Water Station #1 Contamination | Final | 50.00 |
| WA | Vancouver | Clark | Vancouver Water Station #4 Contamination | Final | 50.00 |
| WA | Wellpinit | Stevens | Midnite Mine | Final | 50.00 |
| WA | Whidbey Island | Island | Naval Air Station, Whidbey Island (Ault Field) | Final | 47.58 |
| WA | Whidbey Island | Island | Naval Air Station, Whidbey Island (Seaplane Base) | Deleted | 39.64 |
| WA | Yakima | Yakima | Fmc Corp. (Yakima Pit) | Final | 38.80 |
| WA | Yakima | Yakima | Pesticide Lab (Yakima) | Deleted | 29.33 |
| WA | Yakima | Yakima | Yakima Plating Co. | Deleted | 37.93 |
| WI | Algoma | Kewaunee | Algoma Municipal Landfill | Final | 39.99 |
| WI | Appleton | Outagamie | N.W. Mauthe Co., Inc. | Final | – |
| WI | Ashippin | Dodge | Oconomowoc Electroplating Co., Inc. | Final | 31.86 |
| WI | Ashland | Ashland | Ashland/Northern States Power Lakefront | Final | 50.00 |
| WI | Blooming Grove | Dane | Madison Metropolitan Sewerage District Lagoons | Final | 32.65 |

| St. | City/Area | County | Site Name | Status | Score[1] |
|---|---|---|---|---|---|
| WI | Brookfield | Waukesha | Master Disposal Service Landfill | Final | 47.49 |
| WI | Brookfield | Waukesha | Waste Management Of Wisconsin, Inc. (Brookfield Sanitary Landfill) | Final | 28.90 |
| WI | Caledonia | Racine | Hunts Disposal Landfill | Final | 31.02 |
| WI | Cedarburg | Ozaukee | Amcast Industrial Corporation | Final | 46.86 |
| WI | Cleveland Township | Marathon | Mid-State Disposal, Inc. Landfill | Final | 35.23 |
| WI | Daniels | Burnett | Penta Wood Products | Final | 50.00 |
| WI | De Pere | Brown | Better Brite Plating Co. Chrome And Zinc Shops | Final | 48.91 |
| WI | Delavan | Walworth | Delavan Municipal Well #4 | Final | 28.90 |
| WI | Dunn | Dane | City Disposal Corp. Landfill | Final | 36.84 |
| WI | Eau Claire | Chippewa | National Presto Industries, Inc. | Final | 42.39 |
| WI | Eau Claire | Eau Claire | Eau Claire Municipal Well Field | Final | 35.57 |
| WI | Eau Claire | Eau Claire | Waste Research & Reclamation Co. | Deleted | 32.13 |
| WI | Excelsior | Sauk | Sauk County Landfill | Final | 34.21 |
| WI | Fond Du Lac County | Fond Du Lac | Ripon City Landfill | Final | 39.04 |
| WI | Franklin | Milwaukee | Fadrowski Drum Disposal | Deleted | 31.08 |
| WI | Franklin Township | Manitowoc | Lemberger Transport & Recycling | Final | 34.58 |
| WI | Germantown | Washington | Omega Hills North Landfill | Deleted | 58.54 |
| WI | Green Bay | Brown | Fox River Nrda/Pcb Releases | Proposed | 50.00 |
| WI | Harrison | Calumet | Schmalz Dump | Final | 48.92 |
| WI | Janesville | Rock | Janesville Ash Beds | Final | 57.90 |
| WI | Janesville | Rock | Janesville Old Landfill | Final | 57.93 |
| WI | Kohler | Sheboygan | Kohler Co. Landfill | Final | 42.93 |
| WI | La Prairie Township | Rock | Wheeler Pit | Deleted | 57.80 |
| WI | Medford | Taylor | Scrap Processing Co., Inc. | Final | 34.24 |
| WI | Menomonee Falls | Waukesha | Lauer I Sanitary Landfill | Final | 42.69 |
| WI | Middleton | Dane | Refuse Hideaway Landfill | Final | 34.67 |
| WI | Milwaukee | Milwaukee | Moss-American Co., Inc. (Kerr-Mcgee Oil Co.) | Final | 32.14 |
| WI | Muskego | Waukesha | Muskego Sanitary Landfill | Final | 51.91 |
| WI | Onalaska | La Crosse | Onalaska Municipal Landfill | Final | 42.47 |
| WI | Sheboygan | Sheboygan | Sheboygan Harbor & River | Final | 33.79 |
| WI | Sparta | Monroe | Northern Engraving Co. | Deleted | 38.75 |
| WI | Spencer | Marathon | Spickler Landfill | Final | 44.24 |
| WI | Stoughton | Dane | Hagen Farm | Final | 32.06 |
| WI | Stoughton | Dane | Stoughton City Landfill | Final | 35.79 |
| WI | Tomah | Monroe | Tomah Armory | Final | 30.63 |
| WI | Tomah | Monroe | Tomah Fairgrounds | Deleted | 32.87 |
| WI | Tomah | Monroe | Tomah Municipal Sanitary Landfill | Final | 45.91 |
| WI | Wausau | Marathon | Wausau Ground Water Contamination | Final | 28.91 |
| WI | Whitelaw | Manitowoc | Lemberger Landfill, Inc. | Final | 34.07 |
| WI | Williamstown | Dodge | Hechimovich Sanitary Landfill | Final | 47.91 |
| WV | Fairmont | Marion | Big John Salvage - Hoult Road | Final | 48.57 |
| WV | Fairmont | Marion | Sharon Steel Corp (Fairmont Coke Works) | Final | 57.08 |
| WV | Follansbee | Brooke | Follansbee | Deleted | 33.77 |
| WV | Leetown | Jefferson | Leetown Pesticide | Deleted | 36.72 |
| WV | Mineral County | Mineral | Allegany Ballistics Laboratory (Usnavy) | Final | 50.00 |
| WV | Morgantown | Monongalia | Ordnance Works Disposal Areas | Final | 35.62 |
| WV | Moundsville | Marshall | Hanlin-Allied-Olin | Final | 53.98 |
| WV | Nitro | Putnam | Fike Chemical, Inc. | Final | 36.30 |
| WV | Point Pleasant | Mason | West Virginia Ordnance (Usarmy) | Final | – |
| WV | Ravenswood | Jackson | Ravenswood Pce | Final | 50.00 |
| WV | Vienna | Wood | Vienna Tetrachloroethene | Final | 50.00 |
| WY | Cheyenne | Laramie | F.E. Warren Air Force Base | Final | 39.23 |
| WY | Evansville | Natrona | Mystery Bridge Rd/U.S. Highway 20 | Final | 32.10 |
| WY | Laramie | Albany | Baxter/Union Pacific Tie Treating | Deleted | 37.24 |

***Notes:*** *(1) Federal Register Hazard Rankings System (HRS) score. The HRS is a model that is used to evaluate the relative threats to human health and the environment posed by actual or potential releases of hazardous substances, pollutants, and contaminants. The HRS criteria take into account the population at risk, the hazard potential of the substances, as well as the potential for contamination of drinking water supplies, direct human contact, destruction of sensitive ecosystems, damage to natural resources affecting the human food chain, contamination of surface water used for recreation or potable water consumption, and contamination of ambient air. The higher the score, the higher the potential threat to human health or the environment.*

***Source:*** *U.S. Environmental Protection Agency, CERCLIS Hazardous Waste Sites, April 1, 2011*

## 2007 CERCLA List of Priority Hazardous Substances

| 2007 Rank | Substance Name | Totals Points | 2005 Rank | CAS Number[1] |
|---|---|---|---|---|
| 1 | Arsenic | 1672.58 | 1 | 007440-38-2 |
| 2 | Lead | 1534.07 | 2 | 007439-92-1 |
| 3 | Mercury | 1504.69 | 3 | 007439-97-6 |
| 4 | Vinyl Chloride | 1387.75 | 4 | 000075-01-4 |
| 5 | Polychlorinated Biphenyls | 1365.78 | 5 | 001336-36-3 |
| 6 | Benzene | 1355.96 | 6 | 000071-43-2 |
| 7 | Cadmium | 1324.22 | 8 | 007440-43-9 |
| 8 | Polycyclic Aromatic Hydrocarbons | 1316.98 | 7 | 130498-29-2 |
| 9 | Benzo(a)Pyrene | 1312.45 | 9 | 000050-32-8 |
| 10 | Benzo(b)Fluoranthene | 1266.55 | 10 | 000205-99-2 |
| 11 | Chloroform | 1223.03 | 11 | 000067-66-3 |
| 12 | DDT, p,p'- | 1193.36 | 12 | 000050-29-3 |
| 13 | Aroclor 1254 | 1182.63 | 13 | 011097-69-1 |
| 14 | Aroclor 1260 | 1177.77 | 14 | 011096-82-5 |
| 15 | Dibenzo(a,h)Anthracene | 1165.88 | 15 | 000053-70-3 |
| 16 | Trichloroethylene | 1154.73 | 16 | 000079-01-6 |
| 17 | Dieldrin | 1150.91 | 17 | 000060-57-1 |
| 18 | Chromium, Hexavalent | 1149.98 | 18 | 018540-29-9 |
| 19 | Phosphorus, White | 1144.77 | 19 | 007723-14-0 |
| 20 | Chlordane | 1133.21 | 21 | 000057-74-9 |
| 21 | DDE, p,p'- | 1132.49 | 20 | 000072-55-9 |
| 22 | Hexachlorobutadiene | 1129.63 | 22 | 000087-68-3 |
| 23 | Coal Tar Creosote | 1124.32 | 23 | 008001-58-9 |
| 24 | Aldrin | 1117.22 | 25 | 000309-00-2 |
| 25 | DDD, p,p'- | 1114.83 | 24 | 000072-54-8 |
| 26 | Benzidine | 1114.24 | 26 | 000092-87-5 |
| 27 | Aroclor 1248 | 1112.20 | 27 | 012672-29-6 |
| 28 | Cyanide | 1099.48 | 28 | 000057-12-5 |
| 29 | Aroclor 1242 | 1093.14 | 29 | 053469-21-9 |
| 30 | Aroclor | 1091.52 | 62 | 012767-79-2 |
| 31 | Toxaphene | 1086.65 | 30 | 008001-35-2 |
| 32 | Hexachlorocyclohexane, Gamma- | 1081.63 | 32 | 000058-89-9 |
| 33 | Tetrachloroethylene | 1080.43 | 31 | 000127-18-4 |
| 34 | Heptachlor | 1072.67 | 33 | 000076-44-8 |
| 35 | 1,2-Dibromoethane | 1064.06 | 34 | 000106-93-4 |
| 36 | Hexachlorocyclohexane, Beta- | 1060.22 | 37 | 000319-85-7 |
| 37 | Acrolein | 1059.07 | 36 | 000107-02-8 |
| 38 | Disulfoton | 1058.85 | 35 | 000298-04-4 |
| 39 | Benzo(a)Anthracene | 1057.96 | 38 | 000056-55-3 |
| 40 | 3,3'-Dichlorobenzidine | 1051.61 | 39 | 000091-94-1 |
| 41 | Endrin | 1048.57 | 41 | 000072-20-8 |
| 42 | Beryllium | 1046.12 | 40 | 007440-41-7 |
| 43 | Hexachlorocyclohexane, Delta- | 1038.27 | 42 | 000319-86-8 |
| 44 | 1,2-Dibromo-3-Chloropropane | 1035.55 | 43 | 000096-12-8 |
| 45 | Pentachlorophenol | 1028.01 | 45 | 000087-86-5 |
| 46 | Heptachlor Epoxide | 1027.12 | 44 | 001024-57-3 |
| 47 | Carbon Tetrachloride | 1023.32 | 46 | 000056-23-5 |
| 48 | Aroclor 1221 | 1018.41 | 47 | 011104-28-2 |
| 49 | Cobalt | 1015.57 | 50 | 007440-48-4 |
| 50 | DDT, o,p'- | 1014.71 | 49 | 000789-02-6 |
| 51 | Aroclor 1016 | 1014.33 | 48 | 012674-11-2 |
| 52 | Di-N-Butyl Phthalate | 1007.49 | 52 | 000084-74-2 |
| 53 | Nickel | 1005.40 | 55 | 007440-02-0 |
| 54 | Endosulfan | 1004.65 | 54 | 000115-29-7 |
| 55 | Endosulfan Sulfate | 1003.56 | 53 | 001031-07-8 |

| 2007 Rank | Substance Name | Totals Points | 2005 Rank | CAS Number[1] |
|---|---|---|---|---|
| 56 | Diazinon | 1002.08 | 57 | 000333-41-5 |
| 57 | Endosulfan, Alpha | 1001.30 | 58 | 000959-98-8 |
| 58 | Xylenes, Total | 996.07 | 59 | 001330-20-7 |
| 59 | Cis-Chlordane | 995.08 | 51 | 005103-71-9 |
| 60 | Dibromochloropropane | 994.87 | 60 | 067708-83-2 |
| 61 | Methoxychlor | 994.47 | 61 | 000072-43-5 |
| 62 | Benzo(k)Fluoranthene | 981.26 | 63 | 000207-08-9 |
| 63 | Endrin Ketone | 978.99 | 64 | 053494-70-5 |
| 64 | Trans-Chlordane | 973.99 | 56 | 005103-74-2 |
| 65 | Chromium(vi) Oxide | 969.58 | 66 | 001333-82-0 |
| 66 | Methane | 959.78 | 67 | 000074-82-8 |
| 67 | Endosulfan, Beta | 959.19 | 65 | 033213-65-9 |
| 68 | Aroclor 1232 | 955.64 | 68 | 011141-16-5 |
| 69 | Endrin Aldehyde | 954.86 | 69 | 007421-93-4 |
| 70 | Benzofluoranthene | 951.48 | 70 | 056832-73-6 |
| 71 | Toluene | 947.50 | 71 | 000108-88-3 |
| 72 | 2-Hexanone | 942.02 | 72 | 000591-78-6 |
| 73 | 2,3,7,8-Tetrachlorodibenzo-P-Dioxin | 938.11 | 73 | 001746-01-6 |
| 74 | Zinc | 932.89 | 74 | 007440-66-6 |
| 75 | Dimethylarsinic Acid | 922.06 | 75 | 000075-60-5 |
| 76 | Di(2-Ethylhexyl)Phthalate | 919.02 | 76 | 000117-81-7 |
| 77 | Chromium | 908.52 | 77 | 007440-47-3 |
| 78 | Naphthalene | 896.67 | 78 | 000091-20-3 |
| 79 | 1,1-Dichloroethene | 891.19 | 79 | 000075-35-4 |
| 80 | Methylene Chloride | 888.96 | 81 | 000075-09-2 |
| 81 | Aroclor 1240 | 888.11 | 80 | 071328-89-7 |
| 82 | 2,4,6-Trinitrotoluene | 883.59 | 82 | 000118-96-7 |
| 83 | Bromodichloroethane | 870.00 | 83 | 000683-53-4 |
| 84 | Hydrazine | 864.41 | 85 | 000302-01-2 |
| 85 | 1,2-Dichloroethane | 863.99 | 84 | 000107-06-2 |
| 86 | 2,4,6-Trichlorophenol | 863.71 | 86 | 000088-06-2 |
| 87 | 2,4-Dinitrophenol | 860.45 | 87 | 000051-28-5 |
| 88 | Bis(2-Chloroethyl) Ether | 859.88 | 88 | 000111-44-4 |
| 89 | Thiocyanate | 849.21 | 89 | 000302-04-5 |
| 90 | Asbestos | 841.54 | 90 | 001332-21-4 |
| 91 | Chlorine | 840.37 | 92 | 007782-50-5 |
| 92 | Cyclotrimethylenetrinitramine (RDX) | 840.28 | 91 | 000121-82-4 |
| 93 | Hexachlorobenzene | 838.34 | 93 | 000118-74-1 |
| 94 | 2,4-Dinitrotoluene | 837.88 | 96 | 000121-14-2 |
| 95 | Radium-226 | 835.93 | 94 | 013982-63-3 |
| 96 | Ethion | 834.03 | 97 | 000563-12-2 |
| 97 | 1,1,1-Trichloroethane | 833.81 | 95 | 000071-55-6 |
| 98 | Uranium | 833.41 | 98 | 007440-61-1 |
| 99 | Ethylbenzene | 832.13 | 99 | 000100-41-4 |
| 100 | Radium | 828.07 | 100 | 007440-14-4 |
| 101 | Thorium | 825.17 | 101 | 007440-29-1 |
| 102 | 4,6-Dinitro-O-Cresol | 822.78 | 102 | 000534-52-1 |
| 103 | 1,3,5-Trinitrobenzene | 820.17 | 103 | 000099-35-4 |
| 104 | Chlorobenzene | 819.69 | 105 | 000108-90-7 |
| 105 | Radon | 817.89 | 104 | 010043-92-2 |
| 106 | Radium-228 | 816.76 | 106 | 015262-20-1 |
| 107 | Thorium-230 | 814.72 | 107 | 014269-63-7 |
| 107 | Uranium-235 | 814.72 | 107 | 015117-96-1 |
| 109 | Barium | 813.46 | 109 | 007440-39-3 |
| 110 | Fluoranthene | 812.40 | 113 | 000206-44-0 |
| 111 | Uranium-234 | 812.11 | 110 | 013966-29-5 |
| 112 | N-Nitrosodi-N-Propylamine | 811.05 | 111 | 000621-64-7 |
| 113 | Thorium-228 | 810.36 | 112 | 014274-82-9 |
| 114 | Radon-222 | 809.78 | 114 | 014859-67-7 |

| 2007 Rank | Substance Name | Totals Points | 2005 Rank | CAS Number[1] |
|---|---|---|---|---|
| 115 | Hexachlorocyclohexane, Alpha- | 809.56 | 116 | 000319-84-6 |
| 116 | 1,2,3-Trichlorobenzene | 808.41 | 143 | 000087-61-6 |
| 117 | Manganese | 807.90 | 115 | 007439-96-5 |
| 118 | Coal Tars | 807.07 | 117 | 008007-45-2 |
| 119 | Chrysotile Asbestos | 806.68 | 119 | 012001-29-5 |
| 119 | Strontium-90 | 806.68 | 119 | 010098-97-2 |
| 121 | Plutonium-239 | 806.67 | 118 | 015117-48-3 |
| 122 | Polonium-210 | 806.39 | 122 | 013981-52-7 |
| 123 | Methylmercury | 806.39 | 121 | 022967-92-6 |
| 124 | Plutonium-238 | 806.01 | 123 | 013981-16-3 |
| 125 | Lead-210 | 805.90 | 124 | 014255-04-0 |
| 126 | Plutonium | 805.23 | 125 | 007440-07-5 |
| 127 | Chlorpyrifos | 804.93 | 125 | 002921-88-2 |
| 128 | Copper | 804.86 | 133 | 007440-50-8 |
| 129 | Americium-241 | 804.55 | 128 | 086954-36-1 |
| 130 | Radon-220 | 804.54 | 127 | 022481-48-7 |
| 131 | Amosite Asbestos | 804.07 | 129 | 012172-73-5 |
| 132 | Iodine-131 | 803.48 | 130 | 010043-66-0 |
| 133 | Hydrogen Cyanide | 803.08 | 132 | 000074-90-8 |
| 134 | Tributyltin | 802.61 | 131 | 000688-73-3 |
| 135 | Guthion | 802.32 | 134 | 000086-50-0 |
| 136 | Neptunium-237 | 802.13 | 135 | 013994-20-2 |
| 137 | Chrysene | 802.10 | 139 | 000218-01-9 |
| 138 | Chlordecone | 801.64 | 136 | 000143-50-0 |
| 138 | Iodine-129 | 801.64 | 136 | 015046-84-1 |
| 138 | Plutonium-240 | 801.64 | 136 | 014119-33-6 |
| 141 | S,S,S-Tributyl Phosphorotrithioate | 797.88 | 140 | 000078-48-8 |
| 142 | Bromine | 789.15 | 142 | 007726-95-6 |
| 143 | Polybrominated Biphenyls | 789.11 | 141 | 067774-32-7 |
| 144 | Dicofol | 787.56 | 144 | 000115-32-2 |
| 145 | Parathion | 784.14 | 145 | 000056-38-2 |
| 146 | 1,1,2,2-Tetrachloroethane | 782.15 | 146 | 000079-34-5 |
| 147 | Selenium | 778.98 | 147 | 007782-49-2 |
| 148 | Hexachlorocyclohexane, Technical Grade | 774.91 | 148 | 000608-73-1 |
| 149 | Trichlorofluoroethane | 770.74 | 149 | 027154-33-2 |
| 150 | Trifluralin | 770.12 | 150 | 001582-09-8 |
| 151 | DDD, o,p'- | 768.73 | 151 | 000053-19-0 |
| 152 | 4,4'-Methylenebis(2-Chloroaniline) | 766.66 | 152 | 000101-14-4 |
| 153 | Hexachlorodibenzo-P-Dioxin | 760.42 | 153 | 034465-46-8 |
| 154 | Heptachlorodibenzo-P-Dioxin | 754.47 | 154 | 037871-00-4 |
| 155 | Pentachlorobenzene | 753.58 | 155 | 000608-93-5 |
| 156 | 1,3-Butadiene | 747.31 | 201 | 000106-99-0 |
| 157 | Ammonia | 745.55 | 156 | 007664-41-7 |
| 158 | 2-Methylnaphthalene | 743.24 | 157 | 000091-57-6 |
| 159 | 1,4-Dichlorobenzene | 737.32 | 159 | 000106-46-7 |
| 160 | 1,1-Dichloroethane | 736.23 | 158 | 000075-34-3 |
| 161 | Acenaphthene | 731.25 | 160 | 000083-32-9 |
| 162 | 1,2,3,4,6,7,8,9-Octachlorodibenzofuran | 726.14 | 161 | 039001-02-0 |
| 163 | 1,1,2-Trichloroethane | 724.96 | 162 | 000079-00-5 |
| 164 | Trichloroethane | 723.32 | 163 | 025323-89-1 |
| 165 | Hexachlorocyclopentadiene | 719.01 | 164 | 000077-47-4 |
| 166 | Heptachlorodibenzofuran | 718.58 | 165 | 038998-75-3 |
| 167 | 1,2-Diphenylhydrazine | 713.90 | 166 | 000122-66-7 |
| 168 | 2,3,4,7,8-Pentachlorodibenzofuran | 710.71 | 167 | 057117-31-4 |
| 169 | Tetrachlorobiphenyl | 709.21 | 168 | 026914-33-0 |
| 170 | Cresol, Para- | 707.83 | 169 | 000106-44-5 |
| 171 | Oxychlordane | 706.32 | 170 | 027304-13-8 |
| 172 | 1,2-Dichlorobenzene | 704.91 | 171 | 000095-50-1 |

| 2007 Rank | Substance Name | Totals Points | 2005 Rank | CAS Number[1] |
|---|---|---|---|---|
| 173 | 1,2-Dichloroethene, Trans- | 704.04 | 178 | 000156-60-5 |
| 174 | Indeno(1,2,3-Cd)Pyrene | 703.30 | 180 | 000193-39-5 |
| 175 | Gamma-Chlordene | 702.59 | 172 | 056641-38-4 |
| 176 | Carbon Disulfide | 702.55 | 174 | 000075-15-0 |
| 177 | Tetrachlorophenol | 702.54 | 173 | 025167-83-3 |
| 178 | Americium | 701.62 | 175 | 007440-35-9 |
| 178 | Uranium-233 | 701.62 | 175 | 013968-55-3 |
| 180 | Palladium | 700.66 | 177 | 007440-05-3 |
| 181 | Hexachlorodibenzofuran | 700.56 | 179 | 055684-94-1 |
| 182 | Phenol | 696.96 | 183 | 000108-95-2 |
| 183 | Chloroethane | 693.90 | 182 | 000075-00-3 |
| 184 | Acetone | 693.31 | 181 | 000067-64-1 |
| 185 | P-Xylene | 690.20 | 185 | 000106-42-3 |
| 186 | Dibenzofuran | 689.19 | 187 | 000132-64-9 |
| 187 | Aluminum | 688.13 | 186 | 007429-90-5 |
| 188 | 2,4-Dimethylphenol | 685.76 | 189 | 000105-67-9 |
| 189 | Carbon Monoxide | 684.49 | 188 | 000630-08-0 |
| 190 | Tetrachloroethane | 677.97 | 190 | 025322-20-7 |
| 191 | Hydrogen Sulfide | 676.51 | 193 | 007783-06-4 |
| 192 | Pentachlorodibenzofuran | 673.21 | 192 | 030402-15-4 |
| 193 | Chloromethane | 670.19 | 191 | 000074-87-3 |
| 194 | Bis(2-Methoxyethyl) Phthalate | 666.08 | 194 | 034006-76-3 |
| 195 | Butyl Benzyl Phthalate | 659.38 | 195 | 000085-68-7 |
| 196 | Cresol, Ortho- | 658.66 | 196 | 000095-48-7 |
| 197 | Hexachloroethane | 653.10 | 199 | 000067-72-1 |
| 198 | Vanadium | 651.70 | 198 | 007440-62-2 |
| 199 | N-Nitrosodimethylamine | 650.71 | 200 | 000062-75-9 |
| 200 | 1,2,4-Trichlorobenzene | 647.30 | 203 | 000120-82-1 |
| 201 | Bromoform | 643.53 | 202 | 000075-25-2 |
| 202 | Tetrachlorodibenzo-P-Dioxin | 635.74 | 204 | 041903-57-5 |
| 203 | 1,3-Dichlorobenzene | 631.41 | 205 | 000541-73-1 |
| 204 | Pentachlorodibenzo-P-Dioxin | 625.12 | 207 | 036088-22-9 |
| 205 | N-Nitrosodiphenylamine | 624.79 | 208 | 000086-30-6 |
| 206 | 1,2-Dichloroethylene | 622.49 | 206 | 000540-59-0 |
| 207 | 2,3,7,8-Tetrachlorodibenzofuran | 622.15 | 210 | 051207-31-9 |
| 208 | 2-Butanone | 620.01 | 209 | 000078-93-3 |
| 209 | 2,4-Dichlorophenol | 616.45 | 212 | 000120-83-2 |
| 210 | 1,4-Dioxane | 616.29 | 215 | 000123-91-1 |
| 211 | Fluorine | 613.28 | 214 | 007782-41-4 |
| 212 | Nitrite | 612.64 | 216 | 014797-65-0 |
| 213 | Cesium-137 | 612.50 | 217 | 010045-97-3 |
| 214 | Silver | 612.19 | 213 | 007440-22-4 |
| 215 | Chromium Trioxide | 610.85 | 218 | 007738-94-5 |
| 216 | Nitrate | 610.66 | 219 | 014797-55-8 |
| 217 | Potassium-40 | 608.91 | 220 | 013966-00-2 |
| 218 | Dinitrotoluene | 607.65 | 221 | 025321-14-6 |
| 219 | Antimony | 605.37 | 222 | 007440-36-0 |
| 220 | Coal Tar Pitch | 605.33 | 224 | 065996-93-2 |
| 221 | Thorium-227 | 605.32 | 223 | 015623-47-9 |
| 222 | 2,4,5-Trichlorophenol | 604.83 | 225 | 000095-95-4 |
| 223 | Arsenic Acid | 604.45 | 226 | 007778-39-4 |
| 224 | Arsenic Trioxide | 604.36 | 227 | 001327-53-3 |
| 225 | Phorate | 603.10 | 228 | 000298-02-2 |
| 226 | Benzopyrene | 603.00 | 230 | 073467-76-2 |
| 227 | Cresols | 602.74 | 229 | 001319-77-3 |
| 228 | Chlordane, Technical | 602.62 | 231 | 012789-03-6 |
| 229 | Dimethoate | 602.61 | 232 | 000060-51-5 |
| 230 | Actinium-227 | 602.57 | 233 | 014952-40-0 |

| 2007 Rank | Substance Name | Totals Points | 2005 Rank | CAS Number[1] |
|---|---|---|---|---|
| 230 | Strobane | 602.57 | 233 | 008001-50-1 |
| 232 | 4-Aminobiphenyl | 602.51 | 235 | 000092-67-1 |
| 232 | Pyrethrum | 602.51 | 235 | 008003-34-7 |
| 234 | Arsine | 602.42 | 237 | 007784-42-1 |
| 235 | Naled | 602.32 | 238 | 000300-76-5 |
| 236 | Dibenzofurans, Chlorinated | 602.13 | 239 | 042934-53-2 |
| 236 | Ethoprop | 602.13 | 239 | 013194-48-4 |
| 238 | Alpha-Chlordene | 601.94 | 241 | 056534-02-2 |
| 238 | Carbophenothion | 601.94 | 241 | 000786-19-6 |
| 240 | Dichlorvos | 601.64 | 243 | 000062-73-7 |
| 241 | Calcium Arsenate | 601.45 | 244 | 007778-44-1 |
| 241 | Mercuric Chloride | 601.45 | 244 | 007487-94-7 |
| 241 | Sodium Arsenite | 601.45 | 244 | 007784-46-5 |
| 244 | Formaldehyde | 599.64 | 247 | 000050-00-0 |
| 245 | 2-Chlorophenol | 599.62 | 248 | 000095-57-8 |
| 246 | Phenanthrene | 597.68 | 249 | 000085-01-8 |
| 247 | Hydrogen Fluoride | 588.03 | 250 | 007664-39-3 |
| 248 | 2,4-D Acid | 584.47 | 251 | 000094-75-7 |
| 249 | Dibromochloromethane | 580.59 | 252 | 000124-48-1 |
| 250 | Diuron | 579.16 | 253 | 000330-54-1 |
| 251 | Butylate | 578.43 | 254 | 002008-41-5 |
| 252 | Dimethyl Formamide | 578.23 | 255 | 000068-12-2 |
| 253 | Pyrene | 577.95 | 256 | 000129-00-0 |
| 254 | Dichlorobenzene | 577.70 | 211 | 025321-22-6 |
| 255 | Ethyl Ether | 572.47 | 257 | 000060-29-7 |
| 256 | Dichloroethane | 570.46 | 258 | 001300-21-6 |
| 257 | 4-Nitrophenol | 567.79 | 259 | 000100-02-7 |
| 258 | 1,3-Dichloropropene, Cis- | 561.82 | 184 | 010061-01-5 |
| 259 | Phosphine | 559.74 | 260 | 007803-51-2 |
| 260 | Trichlorobenzene | 557.96 | 261 | 012002-48-1 |
| 261 | 2,6-Dinitrotoluene | 555.20 | 262 | 000606-20-2 |
| 262 | Fluoride Ion | 549.64 | 263 | 016984-48-8 |
| 263 | 1,2,3,4,6,7,8-Heptachlorodibenzo-P-Dioxin | 547.90 | 264 | 035822-46-9 |
| 264 | Methyl Parathion | 545.83 | 265 | 000298-00-0 |
| 265 | Pentaerythritol Tetranitrate | 545.59 | 266 | 000078-11-5 |
| 266 | 1,3-Dichloropropene, Trans- | 543.37 | 267 | 010061-02-6 |
| 267 | Bis(2-Ethylhexyl)Adipate | 540.20 | 268 | 000103-23-1 |
| 268 | Carbazole | 534.52 | 269 | 000086-74-8 |
| 269 | Methyl Isobutyl Ketone | 533.24 | 271 | 000108-10-1 |
| 270 | 1,2-Dichloroethene, Cis- | 533.15 | 270 | 000156-59-2 |
| 271 | Styrene | 532.70 | 272 | 000100-42-5 |
| 272 | Carbaryl | 530.98 | 273 | 000063-25-2 |
| 273 | 1,2,3,4,6,7,8-Heptachlorodibenzofuran | 529.45 | 274 | 067562-39-4 |
| 274 | Acrylonitrile | 528.28 | 275 | 000107-13-1 |
| 275 | 1-Methylnaphthalene | 526.51 | NEW | - |

***Notes:*** *Substances were assigned the same rank when two (or more) substances received equivalent total scores; (1) CAS Number = Chemical Abstracts Service registry number*

*The Comprehensive Environmental Response, Compensation, and Liability Act (CERCLA) section 104 (i), as amended by the Superfund Amendments and Reauthorization Act (SARA), requires ATSDR and the EPA to prepare a list, in order of priority, of substances that are most commonly found at facilities on the National Priorities List (NPL) and which are determined to pose the most significant potential threat to human health due to their known or suspected toxicity and potential for human exposure at these NPL sites. CERCLA also requires this list to be revised periodically to reflect additional information on hazardous substances.*

*This CERCLA priority list is revised and published on a 2-year basis, with a yearly informal review and revision. Each substance on the CERCLA List of Priority Hazardous Substances is a candidate to become the subject of a toxicological profile prepared by ATSDR and subsequently a candidate for the identification of priority data needs. This priority list is based on an algorithm that utilizes the following three components: frequency of occurrence at NPL sites, toxicity, and potential for human exposure to the substances found at NPL sites. This algorithm utilizes data from ATSDR's HazDat database, which contains information from ATSDR's public health assessments and health consultations.*

*It should be noted that this priority list is not a list of "most toxic" substances, but rather a prioritization of substances based on a combination of their frequency, toxicity, and potential for human exposure at NPL sites. Thus, it is possible for substances with low toxicity but high NPL frequency of occurrence and exposure to be on this priority list. The objective of this priority list is to rank substances across all NPL hazardous waste sites to provide guidance in selecting which substances will be the subject of toxicological profiles prepared by ATSDR.*

***Source:*** *Center for Disease Control, Agency for Toxic Substances and Disease Registry, 2007 CERCLA List of Priority Hazardous Substances*

## TRI On-site and Off-site Reported Disposed of or Otherwise Released (in pounds), for Facilities in All Industries, for All Chemicals, by State, U.S., 2009

| State | Total On-site Disposal or Other Releases[1] | Total Off-site Disposal or Other Releases[2] | Total On- and Off-site Disposal or Other Releases |
|---|---|---|---|
| Alabama | 77,457,607 | 13,618,043 | 91,075,650 |
| Alaska | 695,742,084 | 185,771 | 695,927,855 |
| Arizona | 59,992,157 | 907,142 | 60,899,299 |
| Arkansas | 30,181,159 | 3,828,363 | 34,009,523 |
| California | 32,591,583 | 4,126,743 | 36,718,326 |
| Colorado | 17,063,236 | 3,105,777 | 20,169,013 |
| Connecticut | 2,213,713 | 1,101,200 | 3,314,913 |
| Delaware | 5,299,999 | 2,785,314 | 8,085,313 |
| District of Columbia | 21,410 | 2,399 | 23,809 |
| Florida | 80,366,218 | 4,575,163 | 84,941,381 |
| Georgia | 77,958,102 | 1,866,647 | 79,824,749 |
| Hawaii | 2,603,536 | 343,728 | 2,947,264 |
| Idaho | 45,485,652 | 2,385,585 | 47,871,236 |
| Illinois | 52,535,739 | 42,532,834 | 95,068,572 |
| Indiana | 91,528,145 | 40,546,216 | 132,074,361 |
| Iowa | 30,168,175 | 16,308,881 | 46,477,055 |
| Kansas | 18,214,819 | 2,908,586 | 21,123,404 |
| Kentucky | 133,043,180 | 9,564,367 | 142,607,547 |
| Louisiana | 112,512,546 | 7,014,860 | 119,527,406 |
| Maine | 7,652,685 | 810,996 | 8,463,681 |
| Maryland | 32,848,719 | 2,910,592 | 35,759,311 |
| Massachusetts | 3,281,355 | 2,093,654 | 5,375,009 |
| Michigan | 53,511,055 | 17,235,270 | 70,746,325 |
| Minnesota | 19,723,245 | 2,506,495 | 22,229,740 |
| Mississippi | 49,748,397 | 4,368,400 | 54,116,796 |
| Missouri | 73,713,150 | 2,014,235 | 75,727,384 |
| Montana | 39,828,747 | 1,340,346 | 41,169,093 |
| Nebraska | 26,815,775 | 2,746,980 | 29,562,754 |
| Nevada | 181,169,941 | 2,201,222 | 183,371,163 |
| New Hampshire | 2,686,647 | 212,627 | 2,899,274 |
| New Jersey | 10,537,890 | 2,406,707 | 12,944,597 |
| New Mexico | 15,147,092 | 150,808 | 15,297,901 |
| New York | 18,326,871 | 4,952,652 | 23,279,524 |
| North Carolina | 57,250,465 | 6,332,180 | 63,582,646 |
| North Dakota | 13,232,034 | 7,972,526 | 21,204,560 |
| Ohio | 130,176,178 | 28,332,380 | 158,508,558 |
| Oklahoma | 28,161,342 | 1,392,603 | 29,553,945 |
| Oregon | 12,219,887 | 5,043,683 | 17,263,571 |
| Pennsylvania | 71,701,398 | 48,698,910 | 120,400,308 |
| Rhode Island | 174,700 | 220,531 | 395,231 |
| South Carolina | 43,741,271 | 5,660,580 | 49,401,851 |
| South Dakota | 4,250,837 | 338,805 | 4,589,642 |
| Tennessee | 79,078,441 | 10,023,312 | 89,101,753 |
| Texas | 166,711,267 | 23,068,126 | 189,779,393 |
| Utah | 145,082,313 | 2,291,065 | 147,373,378 |
| Vermont | 157,912 | 104,118 | 262,030 |
| Virginia | 50,885,261 | 5,149,931 | 56,035,192 |
| Washington | 13,848,098 | 1,781,489 | 15,629,587 |
| West Virginia | 37,759,318 | 5,183,387 | 42,942,705 |
| Wisconsin | 19,525,783 | 13,408,815 | 32,934,598 |
| Wyoming | 23,614,749 | 1,345,980 | 24,960,729 |
| American Samoa | 6 | 0 | 6 |
| Guam | 224,459 | 49 | 224,508 |
| Northern Mariana Isl | 5,780 | 328 | 6,108 |
| Puerto Rico | 5,121,483 | 1,334,268 | 6,455,751 |

| State | Total On-site Disposal or Other Releases[1] | Total Off-site Disposal or Other Releases[2] | Total On- and Off-site Disposal or Other Releases |
|---|---|---|---|
| Virgin Islands | 641,358 | 26,669 | 668,027 |
| Total | 3,003,534,967 | 371,368,337 | 3,374,903,304 |

***Notes:*** *TRI = Toxic Release Inventory; Reporting year (RY) 2009 is the most recent TRI data available. Facilities reporting to TRI were required to submit RY 2009 data to EPA by July 1, 2010. Facilities may submit revisions at any time. This is the National Analysis dataset released to the public in December 2010 and includes updates for the years 1988 to 2009.Revisions submitted to EPA after this time are not reflected in TRI Explorer reports. TRI data may also be obtained through EPA Envirofacts.*

*(1) On-site Disposal or Other Releases include Underground Injection to Class I Wells (Section 5.4.1), RCRA Subtitle C Landfills (5.5.1A), Other Landfills (5.5.1B), Fugitive or Non-point Air Emissions (5.1), Stack or Point Air Emissions (5.2), Surface Water Discharges (5.3), Underground Injection to Class II-V Wells (5.4.2), Land Treatment/Application Farming (5.5.2), RCRA Subtitle C Surface Impoundments (5.5.3A), Other Surface Impoundments (5.5.3B), and Other Land Disposal (5.5.4). Off-site Disposal or Other Releases include from Section 6.2 Class I Underground Injection Wells (M81), Class II-V Underground Injection Wells (M82, M71), RCRA Subtitle C Landfills (M65), Other Landfills (M64, M72), Storage Only (M10), Solidification/Stabilization - Metals and Metal Category Compounds only (M41 or M40), Wastewater Treatment (excluding POTWs) - Metals and Metal Category Compounds only (M62 or M61), RCRA Subtitle C Surface Impoundments (M66), Other Surface Impoundments (M67, M63), Land Treatment (M73), Other Land Disposal (M79), Other Off-site Management (M90), Transfers to Waste Broker - Disposal (M94, M91), and Unknown (M99) and, from Section 6.1 Transfers to POTWs (metals and metal category compounds only).*

*(2) Off-site disposal or other releases show only net off-site disposal or other releases, that is, off-site disposal or other releases transferred to other TRI facilities reporting such transfers as on-site disposal or other releases are not included to avoid double counting.*

*This report may not include all states in the US. A state may not be included in this report for two reasons: 1) there are no facilities reporting to TRI in the particular state; or 2) the facilities reporting to TRI in the particular state did not report to TRI for the user-specified selection criteria.*

*Users of TRI information should be aware that TRI data reflect releases and other waste management activities of chemicals, not whether (or to what degree) the public has been exposed to those chemicals. Release estimates alone are not sufficient to determine exposure or to calculate potential adverse effects on human health and the environment. TRI data, in conjunction with other information, can be used as a starting point in evaluating exposures that may result from releases and other waste management activities which involve toxic chemicals. The determination of potential risk depends upon many factors, including the toxicity of the chemical, the fate of the chemical, and the amount and duration of human or other exposure to the chemical after it is released.*

***Source:*** *U.S. Environmental Protection Agency, TRI Explorer, May 16, 2011*

**TRI On-site and Off-site Reported Disposed of or Otherwise Released (in grams), for Facilities in All Industries, Dioxin and Dioxin-like Compounds, by State, U.S., 2009**

| State | Total On-site Disposal or Other Releases[1] | Total Off-site Disposal or Other Releases[2] | Total On- and Off-site Disposal or Other Releases |
|---|---|---|---|
| Alabama | 116.17 | 63.23 | 179.41 |
| Alaska | 0.69 | 0.00 | 0.69 |
| Arizona | 12.89 | 0.00 | 12.89 |
| Arkansas | 42.48 | 51.05 | 93.53 |
| California | 55.58 | 38.46 | 94.03 |
| Colorado | 6.64 | 0.61 | 7.25 |
| Connecticut | 2.12 | 0.09 | 2.20 |
| Delaware | 10.38 | 972.28 | 982.65 |
| Florida | 69.59 | 1.85 | 71.45 |
| Georgia | 113.95 | 0.40 | 114.35 |
| Hawaii | 3.09 | 1.00 | 4.08 |
| Idaho | 2.23 | 8.34 | 10.57 |
| Illinois | 29.02 | 72.50 | 101.52 |
| Indiana | 122.62 | 67.17 | 189.79 |
| Iowa | 20.65 | 0.00 | 20.65 |
| Kansas | 9.82 | 0.00 | 9.82 |
| Kentucky | 306.84 | 2,136.80 | 2,443.64 |
| Louisiana | 457.53 | 762.50 | 1,220.03 |
| Maine | 13.41 | 1.31 | 14.72 |
| Maryland | 64.45 | 0.00 | 64.45 |
| Massachusetts | 6.26 | 0.00 | 6.26 |
| Michigan | 299.64 | 5.31 | 304.95 |
| Minnesota | 38.15 | 578.06 | 616.21 |
| Mississippi | 1,740.55 | 141.50 | 1,882.05 |
| Missouri | 98.21 | 0.00 | 98.21 |
| Montana | 5.24 | 1.37 | 6.61 |
| Nebraska | 2.82 | 0.00 | 2.82 |
| Nevada | 7.21 | 0.00 | 7.21 |
| New Hampshire | 0.46 | 0.00 | 0.46 |
| New Jersey | 5.53 | 22.37 | 27.90 |
| New Mexico | 3.96 | 0.00 | 3.96 |
| New York | 21.82 | 2.79 | 24.61 |
| North Carolina | 72.57 | 1.22 | 73.80 |
| North Dakota | 18.56 | 0.00 | 18.56 |
| Ohio | 1,964.70 | 103.69 | 2,068.38 |
| Oklahoma | 23.84 | 36.64 | 60.48 |
| Oregon | 542.99 | 20.97 | 563.96 |
| Pennsylvania | 52.63 | 5.50 | 58.13 |
| South Carolina | 48.10 | 2.25 | 50.36 |
| South Dakota | 8.18 | 0.00 | 8.18 |
| Tennessee | 2,064.44 | 117.00 | 2,181.45 |
| Texas | 5,598.70 | 7,507.42 | 13,106.11 |
| Utah | 4,371.83 | 29.86 | 4,401.69 |
| Virginia | 15.66 | 2.61 | 18.27 |
| Washington | 46.98 | 249.56 | 296.54 |
| West Virginia | 13.07 | 35.09 | 48.16 |
| Wisconsin | 22.36 | 565.85 | 588.22 |
| Wyoming | 16.89 | 0.00 | 16.89 |
| Guam | 0.16 | 0.00 | 0.16 |
| Puerto Rico | 14.05 | 0.08 | 14.13 |
| Virgin Islands | 2.84 | 0.00 | 2.84 |
| Total | 18,588.55 | 13,606.74 | 32,195.29 |

***Notes:*** *TRI = Toxic Release Inventory; Reporting year (RY) 2009 is the most recent TRI data available. Facilities reporting to TRI were required to submit RY 2009 data to EPA by July 1, 2010. Facilities may submit revisions at any time. This is the National Analysis dataset released to the public in December 2010 and includes updates for the years 1988 to*

*2009.Revisions submitted to EPA after this time are not reflected in TRI Explorer reports. TRI data may also be obtained through EPA Envirofacts.*

*(1) On-site Disposal or Other Releases include Underground Injection to Class I Wells (Section 5.4.1), RCRA Subtitle C Landfills (5.5.1A), Other Landfills (5.5.1B), Fugitive or Non-point Air Emissions (5.1), Stack or Point Air Emissions (5.2), Surface Water Discharges (5.3), Underground Injection to Class II-V Wells (5.4.2), Land Treatment/Application Farming (5.5.2), RCRA Subtitle C Surface Impoundments (5.5.3A), Other Surface Impoundments (5.5.3B), and Other Land Disposal (5.5.4). Off-site Disposal or Other Releases include from Section 6.2 Class I Underground Injection Wells (M81), Class II-V Underground Injection Wells (M82, M71), RCRA Subtitle C Landfills (M65), Other Landfills (M64, M72), Storage Only (M10), Solidification/Stabilization - Metals and Metal Category Compounds only (M41 or M40), Wastewater Treatment (excluding POTWs) - Metals and Metal Category Compounds only (M62 or M61), RCRA Subtitle C Surface Impoundments (M66), Other Surface Impoundments (M67, M63), Land Treatment (M73), Other Land Disposal (M79), Other Off-site Management (M90), Transfers to Waste Broker - Disposal (M94, M91), and Unknown (M99) and, from Section 6.1 Transfers to POTWs (metals and metal category compounds only).*

*(2) Off-site disposal or other releases show only net off-site disposal or other releases, that is, off-site disposal or other releases transferred to other TRI facilities reporting such transfers as on-site disposal or other releases are not included to avoid double counting.*

*This report may not include all states in the US. A state may not be included in this report for two reasons: 1) there are no facilities reporting to TRI in the particular state; or 2) the facilities reporting to TRI in the particular state did not report to TRI for the user-specified selection criteria.*

*Users of TRI information should be aware that TRI data reflect releases and other waste management activities of chemicals, not whether (or to what degree) the public has been exposed to those chemicals. Release estimates alone are not sufficient to determine exposure or to calculate potential adverse effects on human health and the environment. TRI data, in conjunction with other information, can be used as a starting point in evaluating exposures that may result from releases and other waste management activities which involve toxic chemicals. The determination of potential risk depends upon many factors, including the toxicity of the chemical, the fate of the chemical, and the amount and duration of human or other exposure to the chemical after it is released.*

***Source:*** *U.S. Environmental Protection Agency, TRI Explorer, May 16, 2011*

## TRI On-site and Off-site Reported Disposed of or Otherwise Released (in pounds), for Facilities in All Industries, for All Chemicals, Top 200 Counties, 2009

| Rank | County | State | Total On-site Disposal or Other Releases[1] | Total Off-site Disposal or Other Releases[2] | Total On- and Off-site Disposal or Other Releases |
|---|---|---|---|---|---|
| 1 | Northwest Arctic Borough | Alaska | 637,528,125 | 482 | 637,528,607 |
| 2 | Salt Lake | Utah | 125,226,068 | 515,726 | 125,741,794 |
| 3 | Humboldt | Nevada | 50,585,462 | 901 | 50,586,363 |
| 4 | Juneau Borough | Alaska | 47,244,157 | 0 | 47,244,157 |
| 5 | Elko | Nevada | 43,142,969 | 612 | 43,143,581 |
| 6 | Gila | Arizona | 41,046,753 | 657,495 | 41,704,248 |
| 7 | Lander | Nevada | 40,140,837 | 1,792 | 40,142,629 |
| 8 | Harris | Texas | 23,184,991 | 14,713,404 | 37,898,395 |
| 9 | Henderson | Kentucky | 35,293,139 | 18,921 | 35,312,060 |
| 10 | Escambia | Florida | 30,985,180 | 302,459 | 31,287,639 |
| 11 | Eureka | Nevada | 28,999,551 | 1,156 | 29,000,707 |
| 12 | Shoshone | Idaho | 25,920,743 | 0 | 25,920,743 |
| 13 | Lake | Indiana | 15,169,947 | 9,928,563 | 25,098,511 |
| 14 | Beaver | Pennsylvania | 1,871,740 | 22,546,498 | 24,418,238 |
| 15 | Hancock | Kentucky | 21,333,758 | 112,225 | 21,445,983 |
| 16 | Montgomery | Tennessee | 21,154,583 | 113,037 | 21,267,621 |
| 17 | Wayne | Michigan | 7,840,563 | 13,141,881 | 20,982,445 |
| 18 | Spencer | Indiana | 20,004,075 | 903,788 | 20,907,863 |
| 19 | Humphreys | Tennessee | 20,635,232 | 56,309 | 20,691,541 |
| 20 | Brazoria | Texas | 20,105,816 | 507,433 | 20,613,248 |
| 21 | Iron | Missouri | 19,867,696 | 99,910 | 19,967,606 |
| 22 | Silver Bow | Montana | 19,633,199 | 32,481 | 19,665,680 |
| 23 | Reynolds | Missouri | 19,567,652 | 0 | 19,567,652 |
| 24 | Calhoun | Texas | 18,673,692 | 10,795 | 18,684,487 |
| 25 | Jefferson | Louisiana | 14,134,334 | 4,257,297 | 18,391,631 |
| 26 | Armstrong | Pennsylvania | 16,840,691 | 80 | 16,840,771 |
| 27 | Jefferson | Texas | 15,448,619 | 1,286,201 | 16,734,820 |
| 28 | Jefferson | Ohio | 12,212,138 | 4,386,296 | 16,598,433 |
| 29 | Washington | Ohio | 14,526,604 | 1,989,806 | 16,516,410 |
| 30 | Harrison | Mississippi | 16,335,210 | 2,649 | 16,337,859 |
| 31 | Monroe | Michigan | 15,722,887 | 58,744 | 15,781,631 |
| 32 | Peoria | Illinois | 1,424,001 | 13,668,286 | 15,092,287 |
| 33 | Lucas | Ohio | 14,238,379 | 637,682 | 14,876,061 |
| 34 | Kings | California | 14,797,262 | 24,543 | 14,821,805 |
| 35 | Ohio | Kentucky | 14,448,418 | 412 | 14,448,831 |
| 36 | St. Charles | Louisiana | 13,910,963 | 229,560 | 14,140,523 |
| 37 | Ascension | Louisiana | 13,471,973 | 489,768 | 13,961,741 |
| 38 | Anne Arundel | Maryland | 13,689,759 | 185,488 | 13,875,247 |
| 39 | Allegheny | Pennsylvania | 5,052,897 | 8,294,075 | 13,346,972 |
| 40 | Allen | Ohio | 12,850,425 | 490,174 | 13,340,599 |
| 41 | Calcasieu | Louisiana | 12,596,502 | 538,003 | 13,134,505 |
| 42 | Montgomery | Virginia | 12,443,677 | 239,101 | 12,682,778 |
| 43 | Carroll | Kentucky | 9,474,339 | 3,085,296 | 12,559,636 |
| 44 | Jefferson | Alabama | 9,586,192 | 2,552,351 | 12,138,543 |
| 45 | East Baton Rouge | Louisiana | 11,929,573 | 124,098 | 12,053,671 |
| 46 | White Pine | Nevada | 11,949,150 | 0 | 11,949,150 |
| 47 | San Juan | New Mexico | 8,196,635 | 3,570,745 | 11,767,380 |
| 48 | De Kalb | Indiana | 428,287 | 11,240,927 | 11,669,214 |
| 49 | De Soto | Louisiana | 11,287,705 | 0 | 11,287,705 |
| 50 | Fairbanks North Star Borough | Alaska | 10,879,024 | 8,690 | 10,887,714 |
| 51 | Richmond | Georgia | 10,833,289 | 47,258 | 10,880,547 |
| 52 | Marion | Indiana | 2,211,642 | 8,663,372 | 10,875,015 |
| 53 | Sumter | Alabama | 10,839,750 | 30,857 | 10,870,607 |
| 54 | Ashtabula | Ohio | 10,604,690 | 214,416 | 10,819,106 |
| 55 | Jefferson | Missouri | 10,770,730 | 33,404 | 10,804,134 |

| Rank | County | State | Total On-site Disposal or Other Releases[1] | Total Off-site Disposal or Other Releases[2] | Total On- and Off-site Disposal or Other Releases |
|---|---|---|---|---|---|
| 56 | Berkeley | South Carolina | 8,165,237 | 2,576,404 | 10,741,641 |
| 57 | Rosebud | Montana | 10,376,895 | 116,202 | 10,493,097 |
| 58 | Citrus | Florida | 10,350,195 | 8,591 | 10,358,786 |
| 59 | Monroe | Mississippi | 10,162,343 | 71,774 | 10,234,116 |
| 60 | Indiana | Pennsylvania | 10,154,453 | 861 | 10,155,314 |
| 61 | Jefferson | Kentucky | 9,467,015 | 610,621 | 10,077,637 |
| 62 | Muhlenberg | Kentucky | 9,772,917 | 77,633 | 9,850,550 |
| 63 | Laramie | Wyoming | 9,755,062 | 2 | 9,755,064 |
| 64 | Tooele | Utah | 8,741,643 | 1,008,908 | 9,750,550 |
| 65 | Muscatine | Iowa | 1,264,732 | 7,807,824 | 9,072,556 |
| 66 | Adams | Ohio | 8,822,495 | 306 | 8,822,801 |
| 67 | Putnam | West Virginia | 7,225,792 | 1,535,762 | 8,761,554 |
| 68 | Mercer | North Dakota | 2,346,318 | 6,162,388 | 8,508,706 |
| 69 | St. Clair | Michigan | 8,393,567 | 46,946 | 8,440,513 |
| 70 | Bartow | Georgia | 8,336,691 | 31,891 | 8,368,582 |
| 71 | Ouachita | Louisiana | 8,271,157 | 43,081 | 8,314,238 |
| 72 | Titus | Texas | 8,184,592 | 10,675 | 8,195,267 |
| 73 | Hillsborough | Florida | 7,681,914 | 442,222 | 8,124,136 |
| 74 | Gallia | Ohio | 8,078,576 | 10,192 | 8,088,768 |
| 75 | Charles | Maryland | 7,098,638 | 971,804 | 8,070,442 |
| 76 | Cook | Illinois | 2,636,041 | 5,402,124 | 8,038,166 |
| 77 | Richland | South Carolina | 7,732,972 | 65,644 | 7,798,616 |
| 78 | Major | Oklahoma | 7,514,080 | 0 | 7,514,080 |
| 79 | Sherburne | Minnesota | 7,491,691 | 111 | 7,491,802 |
| 80 | Mobile | Alabama | 4,338,918 | 3,148,106 | 7,487,024 |
| 81 | Shelby | Tennessee | 4,263,636 | 3,108,172 | 7,371,808 |
| 82 | Chatham | Georgia | 7,161,280 | 204,715 | 7,365,995 |
| 83 | Tuscaloosa | Alabama | 272,270 | 7,051,134 | 7,323,404 |
| 84 | Jefferson | Montana | 7,267,218 | 0 | 7,267,218 |
| 85 | Caribou | Idaho | 7,074,084 | 3,423 | 7,077,507 |
| 86 | Duval | Florida | 5,468,053 | 1,597,736 | 7,065,789 |
| 87 | Monroe | Georgia | 6,935,057 | 1,534 | 6,936,591 |
| 88 | Los Angeles | California | 3,417,667 | 3,496,831 | 6,914,499 |
| 89 | Gibson | Indiana | 6,903,566 | 10,064 | 6,913,630 |
| 90 | Madison | Illinois | 4,667,213 | 2,154,630 | 6,821,842 |
| 91 | Putnam | Georgia | 6,792,635 | 207 | 6,792,842 |
| 92 | Clermont | Ohio | 6,736,323 | 20,316 | 6,756,640 |
| 93 | Iberville | Louisiana | 4,980,547 | 1,669,147 | 6,649,695 |
| 94 | Nueces | Texas | 5,983,949 | 610,517 | 6,594,466 |
| 95 | Garfield | Oklahoma | 6,443,370 | 0 | 6,443,370 |
| 96 | Rusk | Texas | 6,431,135 | 4,403 | 6,435,537 |
| 97 | Pike | Indiana | 5,347,283 | 1,043,009 | 6,390,292 |
| 98 | Lawrence | Kentucky | 6,351,006 | 2,120 | 6,353,125 |
| 99 | Bay | Florida | 6,185,741 | 80,314 | 6,266,055 |
| 100 | Coshocton | Ohio | 5,190,372 | 995,260 | 6,185,633 |
| 101 | Orange | Texas | 5,135,752 | 1,029,540 | 6,165,291 |
| 102 | Shelby | Alabama | 6,155,169 | 7,048 | 6,162,217 |
| 103 | Galveston | Texas | 4,995,902 | 1,021,667 | 6,017,569 |
| 104 | Sullivan | Tennessee | 5,331,700 | 610,780 | 5,942,480 |
| 105 | Decatur | Georgia | 5,777,442 | 2 | 5,777,444 |
| 106 | Sandusky | Ohio | 5,679,774 | 86,145 | 5,765,919 |
| 107 | Will | Illinois | 3,051,761 | 2,676,149 | 5,727,910 |
| 108 | Chesterfield | Virginia | 5,580,758 | 2,917 | 5,583,675 |
| 109 | Lee | Iowa | 4,940,313 | 548,219 | 5,488,533 |
| 110 | Harrison | West Virginia | 5,425,247 | 1,077 | 5,426,324 |
| 111 | Victoria | Texas | 5,110,799 | 310,857 | 5,421,656 |
| 112 | Stewart | Tennessee | 5,382,110 | 49 | 5,382,159 |
| 113 | New Castle | Delaware | 2,591,124 | 2,779,911 | 5,371,035 |

| Rank | County | State | Total On-site Disposal or Other Releases[1] | Total Off-site Disposal or Other Releases[2] | Total On- and Off-site Disposal or Other Releases |
|---|---|---|---|---|---|
| 114 | Woodbury | Iowa | 5,071,385 | 192,049 | 5,263,434 |
| 115 | Northampton | Pennsylvania | 3,637,835 | 1,606,755 | 5,244,590 |
| 116 | Sweetwater | Wyoming | 5,213,932 | 2,108 | 5,216,041 |
| 117 | McLean | North Dakota | 5,181,206 | 34,370 | 5,215,576 |
| 118 | Dawson | Nebraska | 5,014,605 | 109,197 | 5,123,802 |
| 119 | Rock Island | Illinois | 4,835,206 | 264,606 | 5,099,812 |
| 120 | York | Pennsylvania | 4,079,587 | 1,001,162 | 5,080,749 |
| 121 | Polk | Florida | 4,394,462 | 646,503 | 5,040,965 |
| 122 | Hamilton | Ohio | 3,997,578 | 990,152 | 4,987,730 |
| 123 | St. James | Louisiana | 4,901,615 | 35,151 | 4,936,766 |
| 124 | Pottawattamie | Iowa | 4,205,375 | 696,800 | 4,902,175 |
| 125 | New Hanover | North Carolina | 4,238,825 | 633,839 | 4,872,665 |
| 126 | Jefferson | Arkansas | 4,814,164 | 47,108 | 4,861,272 |
| 127 | Fort Bend | Texas | 4,745,405 | 109,641 | 4,855,047 |
| 128 | Warrick | Indiana | 4,199,097 | 587,324 | 4,786,421 |
| 129 | Prince Georges | Maryland | 3,966,815 | 793,317 | 4,760,132 |
| 130 | Apache | Arizona | 4,750,828 | 0 | 4,750,828 |
| 131 | Mason | West Virginia | 4,729,082 | 5,332 | 4,734,414 |
| 132 | Yamhill | Oregon | 56,283 | 4,660,425 | 4,716,708 |
| 133 | Lake | Ohio | 3,307,276 | 1,339,619 | 4,646,895 |
| 134 | Harrison | Texas | 4,449,381 | 130,234 | 4,579,614 |
| 135 | Cass | Texas | 4,574,517 | 0 | 4,574,517 |
| 136 | Person | North Carolina | 4,374,669 | 178,100 | 4,552,769 |
| 137 | Georgetown | South Carolina | 4,551,003 | 31 | 4,551,034 |
| 138 | Whiteside | Illinois | 6,350 | 4,520,917 | 4,527,267 |
| 139 | Vermilion | Illinois | 4,438,908 | 59,778 | 4,498,686 |
| 140 | Pointe Coupee | Louisiana | 2,847,554 | 1,619,044 | 4,466,598 |
| 141 | Independence | Arkansas | 4,462,732 | 207 | 4,462,939 |
| 142 | Scott | Iowa | 411,034 | 4,034,240 | 4,445,274 |
| 143 | Dakota | Nebraska | 4,346,771 | 89,657 | 4,436,428 |
| 144 | Pima | Arizona | 4,325,509 | 10,624 | 4,336,133 |
| 145 | Cuyahoga | Ohio | 1,445,967 | 2,884,981 | 4,330,947 |
| 146 | Ottawa | Michigan | 4,263,265 | 53,817 | 4,317,082 |
| 147 | Wood | Wisconsin | 4,044,686 | 269,172 | 4,313,858 |
| 148 | Jackson | Alabama | 4,289,764 | 24 | 4,289,787 |
| 149 | Greene | Pennsylvania | 4,288,224 | 3 | 4,288,227 |
| 150 | Jasper | Indiana | 4,287,719 | 48 | 4,287,766 |
| 151 | Nye | Nevada | 4,242,102 | 6,107 | 4,248,209 |
| 152 | Hutchinson | Texas | 4,123,148 | 108,572 | 4,231,720 |
| 153 | Putnam | Florida | 4,132,830 | 93,532 | 4,226,362 |
| 154 | Oliver | North Dakota | 2,469,317 | 1,721,567 | 4,190,884 |
| 155 | Jefferson | Indiana | 4,164,329 | 792 | 4,165,121 |
| 156 | Macon | Illinois | 2,291,024 | 1,859,392 | 4,150,416 |
| 157 | Columbus | North Carolina | 4,053,799 | 53,297 | 4,107,096 |
| 158 | Marshall | West Virginia | 4,008,302 | 13,838 | 4,022,141 |
| 159 | Wayne | Georgia | 4,014,435 | 851 | 4,015,286 |
| 160 | Colfax | Nebraska | 4,008,599 | 2,574 | 4,011,173 |
| 161 | Rutherford | North Carolina | 3,979,185 | 16,917 | 3,996,102 |
| 162 | Berks | Pennsylvania | 2,394,352 | 1,586,077 | 3,980,428 |
| 163 | Bladen | North Carolina | 3,943,326 | 7,287 | 3,950,614 |
| 164 | Lorain | Ohio | 2,532,714 | 1,378,285 | 3,910,999 |
| 165 | Cass | Illinois | 3,906,361 | 0 | 3,906,361 |
| 166 | Kanawha | West Virginia | 2,346,199 | 1,550,295 | 3,896,494 |
| 167 | Warren | New York | 3,797,849 | 60,247 | 3,858,097 |
| 168 | Bradford | Pennsylvania | 3,272,043 | 563,045 | 3,835,088 |
| 169 | Salem | New Jersey | 3,740,118 | 49,153 | 3,789,271 |
| 170 | Hopewell City | Virginia | 3,545,207 | 212,222 | 3,757,430 |
| 171 | Christian | Illinois | 1,608,225 | 2,114,150 | 3,722,375 |

| Rank | County | State | Total On-site Disposal or Other Releases[1] | Total Off-site Disposal or Other Releases[2] | Total On- and Off-site Disposal or Other Releases |
|---|---|---|---|---|---|
| 172 | Coweta | Georgia | 3,530,608 | 179,528 | 3,710,136 |
| 173 | Lincoln | Nebraska | 3,702,595 | 0 | 3,702,595 |
| 174 | Pleasants | West Virginia | 3,681,364 | 197 | 3,681,561 |
| 175 | Monroe | New York | 2,967,473 | 693,336 | 3,660,809 |
| 176 | Gilliam | Oregon | 3,645,955 | 1,783 | 3,647,738 |
| 177 | Floyd | Georgia | 3,416,739 | 219,515 | 3,636,255 |
| 178 | Rogers | Oklahoma | 3,454,754 | 169,654 | 3,624,408 |
| 179 | Jackson | Mississippi | 3,444,209 | 161,747 | 3,605,956 |
| 180 | Carroll | Georgia | 3,563,703 | 28,544 | 3,592,247 |
| 181 | Hendricks | Indiana | 40,717 | 3,479,380 | 3,520,097 |
| 182 | Campbell | Wyoming | 2,174,313 | 1,323,340 | 3,497,654 |
| 183 | Grant | New Mexico | 3,494,459 | 1,000 | 3,495,459 |
| 184 | Marion | Ohio | 203,044 | 3,267,637 | 3,470,681 |
| 185 | St. Charles | Missouri | 3,446,448 | 3,357 | 3,449,805 |
| 186 | McCurtain | Oklahoma | 3,389,015 | 28,948 | 3,417,963 |
| 187 | Bertie | North Carolina | 3,385,427 | 0 | 3,385,427 |
| 188 | McCracken | Kentucky | 3,362,578 | 813 | 3,363,391 |
| 189 | Covington City | Virginia | 3,333,482 | 0 | 3,333,482 |
| 190 | Lawrence | Alabama | 3,331,262 | 0 | 3,331,262 |
| 191 | Linn | Oregon | 3,089,862 | 230,652 | 3,320,514 |
| 192 | Catawba | North Carolina | 3,261,499 | 57,737 | 3,319,236 |
| 193 | Linn | Kansas | 3,299,681 | 10 | 3,299,691 |
| 194 | Freestone | Texas | 3,295,189 | 0 | 3,295,189 |
| 195 | Teller | Colorado | 3,294,736 | 0 | 3,294,736 |
| 196 | Coles | Illinois | 561,617 | 2,726,999 | 3,288,616 |
| 197 | Sullivan | Indiana | 3,287,296 | 1 | 3,287,296 |
| 198 | St. John the Baptist | Louisiana | 1,320,277 | 1,964,911 | 3,285,188 |
| 199 | Warren | Mississippi | 3,223,952 | 33,421 | 3,257,373 |
| 200 | Morgan | Alabama | 2,710,523 | 531,753 | 3,242,276 |

***Notes:*** *TRI = Toxic Release Inventory; Reporting year (RY) 2009 is the most recent TRI data available. Facilities reporting to TRI were required to submit RY 2009 data to EPA by July 1, 2010. Facilities may submit revisions at any time. This is the National Analysis dataset released to the public in December 2010 and includes updates for the years 1988 to 2009.Revisions submitted to EPA after this time are not reflected in TRI Explorer reports. TRI data may also be obtained through EPA Envirofacts.*

*(1) On-site Disposal or Other Releases include Underground Injection to Class I Wells (Section 5.4.1), RCRA Subtitle C Landfills (5.5.1A), Other Landfills (5.5.1B), Fugitive or Non-point Air Emissions (5.1), Stack or Point Air Emissions (5.2), Surface Water Discharges (5.3), Underground Injection to Class II-V Wells (5.4.2), Land Treatment/Application Farming (5.5.2), RCRA Subtitle C Surface Impoundments (5.5.3A), Other Surface Impoundments (5.5.3B), and Other Land Disposal (5.5.4). Off-site Disposal or Other Releases include from Section 6.2 Class I Underground Injection Wells (M81), Class II-V Underground Injection Wells (M82, M71), RCRA Subtitle C Landfills (M65), Other Landfills (M64, M72), Storage Only (M10), Solidification/Stabilization - Metals and Metal Category Compounds only (M41 or M40), Wastewater Treatment (excluding POTWs) - Metals and Metal Category Compounds only (M62 or M61), RCRA Subtitle C Surface Impoundments (M66), Other Surface Impoundments (M67, M63), Land Treatment (M73), Other Land Disposal (M79), Other Off-site Management (M90), Transfers to Waste Broker - Disposal (M94, M91), and Unknown (M99) and, from Section 6.1 Transfers to POTWs (metals and metal category compounds only).*

*(2) Off-site disposal or other releases show only net off-site disposal or other releases, that is, off-site disposal or other releases transferred to other TRI facilities reporting such transfers as on-site disposal or other releases are not included to avoid double counting.*

*This report may not include all states in the US. A state may not be included in this report for two reasons: 1) there are no facilities reporting to TRI in the particular state; or 2) the facilities reporting to TRI in the particular state did not report to TRI for the user-specified selection criteria.*

*Users of TRI information should be aware that TRI data reflect releases and other waste management activities of chemicals, not whether (or to what degree) the public has been exposed to those chemicals. Release estimates alone are not sufficient to determine exposure or to calculate potential adverse effects on human health and the environment. TRI data, in conjunction with other information, can be used as a starting point in evaluating exposures that may result from releases and other waste management activities which involve toxic chemicals. The determination of potential risk depends upon many factors, including the toxicity of the chemical, the fate of the chemical, and the amount and duration of human or other exposure to the chemical after it is released.*

***Source:*** *U.S. Environmental Protection Agency, TRI Explorer, May 16, 2011*

## TRI On-site and Off-site Reported Disposed of or Otherwise Released (in grams), for Facilities in All Industries, Dioxin and Dioxin-like Compounds, Top 200 Counties, 2009

| Rank | County | State | Total On-site Disposal or Other Releases[1] | Total Off-site Disposal or Other Releases[2] | Total On- and Off-site Disposal or Other Releases |
|---|---|---|---|---|---|
| 1 | Harris | Texas | 12.45 | 7,121.29 | 7,133.74 |
| 2 | Brazoria | Texas | 5,499.87 | 0.00 | 5,499.87 |
| 3 | Tooele | Utah | 4,356.21 | 0.00 | 4,356.21 |
| 4 | Marshall | Kentucky | 238.93 | 2,082.70 | 2,321.63 |
| 5 | Humphreys | Tennessee | 2,036.81 | 0.01 | 2,036.82 |
| 6 | Ashtabula | Ohio | 1,849.33 | 2.01 | 1,851.34 |
| 7 | Harrison | Mississippi | 1,343.06 | 0.30 | 1,343.36 |
| 8 | New Castle | Delaware | 10.05 | 972.28 | 982.33 |
| 9 | Iberville | Louisiana | 368.37 | 349.77 | 718.14 |
| 10 | Dakota | Minnesota | 19.98 | 577.83 | 597.82 |
| 11 | Gilliam | Oregon | 533.48 | 0.00 | 533.48 |
| 12 | Sheboygan | Wisconsin | 6.15 | 526.26 | 532.41 |
| 13 | Monroe | Mississippi | 299.30 | 0.00 | 299.30 |
| 14 | Midland | Michigan | 278.94 | 0.00 | 278.94 |
| 15 | Rapides | Louisiana | 5.47 | 263.42 | 268.89 |
| 16 | San Patricio | Texas | 3.46 | 227.18 | 230.64 |
| 17 | Pierce | Washington | 3.24 | 192.12 | 195.36 |
| 18 | Kemper | Mississippi | 59.02 | 135.24 | 194.26 |
| 19 | Calcasieu | Louisiana | 18.39 | 118.73 | 137.12 |
| 20 | Calhoun | Texas | 6.00 | 96.00 | 102.00 |
| 21 | Wabash | Indiana | 87.79 | 0.96 | 88.75 |
| 22 | Tuscarawas | Ohio | 4.27 | 81.08 | 85.35 |
| 23 | Paulding | Ohio | 82.96 | 0.00 | 82.96 |
| 24 | Orange | Texas | 23.60 | 58.52 | 82.12 |
| 25 | Porter | Indiana | 6.34 | 63.60 | 69.94 |
| 26 | Pike | Missouri | 68.97 | 0.00 | 68.97 |
| 27 | Chatham | Georgia | 66.30 | 0.02 | 66.32 |
| 28 | Peoria | Illinois | 2.14 | 53.90 | 56.04 |
| 29 | Union | Arkansas | 0.12 | 54.65 | 54.78 |
| 30 | Cowlitz | Washington | 4.12 | 44.14 | 48.26 |
| 31 | Barren | Kentucky | 0.47 | 46.74 | 47.20 |
| 32 | Baltimore | Maryland | 45.20 | 0.00 | 45.20 |
| 33 | Etowah | Alabama | 0.52 | 43.76 | 44.28 |
| 34 | Mecklenburg | North Carolina | 40.93 | 0.00 | 40.93 |
| 35 | Greene | Tennessee | 0.40 | 39.21 | 39.61 |
| 36 | Creek | Oklahoma | 1.92 | 36.64 | 38.56 |
| 37 | Kern | California | 36.22 | 0.00 | 36.22 |
| 38 | St. Charles | Louisiana | 10.89 | 25.26 | 36.15 |
| 39 | Eau Claire | Wisconsin | 0.00 | 36.13 | 36.13 |
| 40 | Maury | Tennessee | 2.60 | 33.02 | 35.63 |
| 41 | Los Angeles | California | 2.29 | 30.65 | 32.94 |
| 42 | East Baton Rouge | Louisiana | 32.09 | 0.00 | 32.09 |
| 43 | Tyler | West Virginia | 1.58 | 30.12 | 31.70 |
| 44 | Weber | Utah | 0.14 | 29.86 | 30.00 |
| 45 | Bibb | Alabama | 27.06 | 0.00 | 27.06 |
| 46 | Escambia | Alabama | 25.74 | 0.00 | 25.74 |
| 47 | DeKalb | Tennessee | 0.08 | 25.00 | 25.08 |
| 48 | Butler | Kentucky | 23.83 | 0.00 | 23.83 |
| 49 | Spokane | Washington | 17.11 | 5.54 | 22.64 |
| 50 | Lane | Oregon | 2.16 | 20.25 | 22.41 |
| 51 | Cook | Illinois | 2.09 | 18.59 | 20.68 |
| 52 | Jefferson | Alabama | 5.47 | 13.67 | 19.14 |
| 53 | Camden | New Jersey | 1.55 | 16.88 | 18.43 |
| 54 | York | South Carolina | 17.45 | 0.00 | 17.45 |
| 55 | San Bernardino | California | 9.79 | 6.34 | 16.13 |

| Rank | County | State | Total On-site Disposal or Other Releases[1] | Total Off-site Disposal or Other Releases[2] | Total On- and Off-site Disposal or Other Releases |
|---|---|---|---|---|---|
| 56 | Grenada | Mississippi | 11.52 | 3.72 | 15.24 |
| 57 | Cass | Texas | 14.92 | 0.00 | 14.92 |
| 58 | Mercer | North Dakota | 14.84 | 0.00 | 14.84 |
| 59 | Stone | Mississippi | 13.55 | 0.00 | 13.55 |
| 60 | Columbiana | Ohio | 0.03 | 13.03 | 13.06 |
| 61 | Schuylkill | Pennsylvania | 12.03 | 0.00 | 12.03 |
| 62 | Loudon | Tennessee | 0.79 | 11.14 | 11.93 |
| 63 | Nassau | Florida | 11.86 | 0.04 | 11.90 |
| 64 | Henderson | Kentucky | 11.80 | 0.00 | 11.80 |
| 65 | Hillsborough | Florida | 11.67 | 0.00 | 11.67 |
| 66 | Ashley | Arkansas | 11.29 | 0.00 | 11.29 |
| 67 | Putnam | Florida | 11.17 | 0.02 | 11.20 |
| 68 | Platte | Wyoming | 11.06 | 0.00 | 11.06 |
| 69 | McCurtain | Oklahoma | 10.84 | 0.00 | 10.84 |
| 70 | Sullivan | Tennessee | 6.39 | 4.20 | 10.59 |
| 71 | Fulton | Georgia | 10.12 | . | 10.12 |
| 72 | Little River | Arkansas | 9.95 | 0.00 | 9.95 |
| 73 | Escambia | Florida | 7.75 | 1.79 | 9.53 |
| 74 | Wilcox | Alabama | 9.50 | 0.00 | 9.50 |
| 75 | Millard | Utah | 9.39 | 0.00 | 9.39 |
| 76 | Ascension | Louisiana | 6.87 | 2.46 | 9.33 |
| 77 | Mobile | Alabama | 9.32 | 0.00 | 9.32 |
| 78 | Penobscot | Maine | 8.00 | 1.17 | 9.16 |
| 79 | Georgetown | South Carolina | 8.76 | 0.00 | 8.76 |
| 80 | Warren | Kentucky | 2.60 | 6.06 | 8.66 |
| 81 | Jefferson | Arkansas | 7.94 | 0.00 | 7.94 |
| 82 | Skagit | Washington | 7.78 | 0.00 | 7.78 |
| 83 | Kootenai | Idaho | 0.39 | 7.34 | 7.73 |
| 84 | Lorain | Ohio | 1.76 | 5.88 | 7.64 |
| 85 | Monroe | New York | 7.41 | 0.19 | 7.60 |
| 86 | Duval | Florida | 7.58 | 0.00 | 7.58 |
| 87 | Apache | Arizona | 7.26 | 0.00 | 7.26 |
| 88 | Lake | Indiana | 7.00 | 0.23 | 7.23 |
| 89 | St. Louis | Minnesota | 7.12 | 0.00 | 7.12 |
| 90 | Wayne | Georgia | 7.02 | 0.00 | 7.02 |
| 91 | Frederick | Maryland | 6.94 | 0.00 | 6.94 |
| 92 | Monroe | Michigan | 6.92 | 0.00 | 6.92 |
| 93 | York | Pennsylvania | 6.54 | 0.24 | 6.77 |
| 94 | Snohomish | Washington | 0.57 | 6.04 | 6.60 |
| 95 | Walla Walla | Washington | 6.55 | 0.00 | 6.55 |
| 96 | Scott | Iowa | 6.29 | 0.00 | 6.29 |
| 97 | Massac | Illinois | 6.20 | 0.00 | 6.20 |
| 98 | Marshall | West Virginia | 4.81 | 1.33 | 6.14 |
| 99 | Choctaw | Alabama | 6.14 | 0.00 | 6.14 |
| 100 | Lawrence | Pennsylvania | 5.92 | 0.00 | 5.92 |
| 101 | Hancock | Kentucky | 5.92 | 0.00 | 5.92 |
| 102 | Yankton | South Dakota | 5.89 | 0.00 | 5.89 |
| 103 | Richland | South Carolina | 5.82 | 0.00 | 5.82 |
| 104 | Choctaw | Mississippi | 5.80 | 0.00 | 5.80 |
| 105 | San Joaquin | California | 5.73 | . | 5.73 |
| 106 | Columbus | North Carolina | 5.71 | 0.00 | 5.71 |
| 107 | Richmond | Georgia | 5.61 | 0.09 | 5.70 |
| 108 | Martin | North Carolina | 5.55 | 0.00 | 5.55 |
| 109 | Sebastian | Arkansas | 5.54 | 0.00 | 5.54 |
| 110 | Union | New Jersey | 0.04 | 5.49 | 5.53 |
| 111 | Lancaster | Pennsylvania | 2.21 | 3.29 | 5.50 |
| 112 | Branch | Michigan | 0.26 | 5.09 | 5.35 |
| 113 | St. Clair | Alabama | 0.26 | 5.03 | 5.29 |

| Rank | County | State | Total On-site Disposal or Other Releases[1] | Total Off-site Disposal or Other Releases[2] | Total On- and Off-site Disposal or Other Releases |
|---|---|---|---|---|---|
| 114 | Atascosa | Texas | 5.21 | 0.00 | 5.21 |
| 115 | Niagara | New York | 4.82 | 0.37 | 5.19 |
| 116 | Baltimore City | Maryland | 5.13 | 0.00 | 5.13 |
| 117 | Blount | Tennessee | 1.21 | 3.90 | 5.11 |
| 118 | Dallas | Alabama | 5.03 | 0.00 | 5.03 |
| 119 | Randolph | Missouri | 4.97 | 0.00 | 4.97 |
| 120 | Iron | Missouri | 4.93 | 0.00 | 4.93 |
| 121 | Bristol | Massachusetts | 4.92 | 0.00 | 4.92 |
| 122 | Beaver | Pennsylvania | 4.79 | 0.00 | 4.79 |
| 123 | Carlton | Minnesota | 4.79 | . | 4.79 |
| 124 | Isle of Wight | Virginia | 2.83 | 1.96 | 4.79 |
| 125 | Florence | South Carolina | 4.75 | 0.00 | 4.75 |
| 126 | New Madrid | Missouri | 4.74 | 0.00 | 4.74 |
| 127 | Ouachita | Louisiana | 4.73 | 0.00 | 4.73 |
| 128 | Pike | Alabama | 4.54 | 0.00 | 4.54 |
| 129 | Person | North Carolina | 4.26 | 0.00 | 4.26 |
| 130 | Bowie | Texas | 0.44 | 3.68 | 4.12 |
| 131 | Berkeley | South Carolina | 4.11 | 0.00 | 4.11 |
| 132 | Salinas | Puerto Rico | 4.08 | 0.00 | 4.08 |
| 133 | Carroll | Kentucky | 2.58 | 1.30 | 3.88 |
| 134 | Macon | Illinois | 3.75 | 0.00 | 3.75 |
| 135 | Independence | Arkansas | 3.72 | 0.00 | 3.72 |
| 136 | Orange | Florida | 3.69 | 0.00 | 3.69 |
| 137 | Major | Oklahoma | 3.67 | 0.00 | 3.67 |
| 138 | Braxton | West Virginia | 0.01 | 3.64 | 3.65 |
| 139 | Honolulu | Hawaii | 2.62 | 1.00 | 3.61 |
| 140 | Marathon | Wisconsin | 1.57 | 2.02 | 3.59 |
| 141 | Guayanilla | Puerto Rico | 3.58 | 0.00 | 3.58 |
| 142 | Indiana | Pennsylvania | 3.54 | 0.00 | 3.54 |
| 143 | Will | Illinois | 3.52 | 0.00 | 3.52 |
| 144 | Forsyth | North Carolina | 3.50 | 0.00 | 3.50 |
| 145 | Knox | Tennessee | 3.39 | 0.00 | 3.39 |
| 146 | Sandusky | Ohio | 3.39 | 0.00 | 3.39 |
| 147 | Rosebud | Montana | 3.38 | 0.00 | 3.38 |
| 148 | Middlesex | New Jersey | 3.37 | 0.00 | 3.37 |
| 149 | Jefferson | Kentucky | 3.37 | 0.00 | 3.37 |
| 150 | Haywood | North Carolina | 2.71 | 0.65 | 3.36 |
| 151 | Sherburne | Minnesota | 3.35 | 0.00 | 3.35 |
| 152 | Marengo | Alabama | 3.33 | 0.00 | 3.33 |
| 153 | Talladega | Alabama | 3.31 | 0.00 | 3.31 |
| 154 | Stewart | Tennessee | 3.30 | 0.00 | 3.30 |
| 155 | San Juan | New Mexico | 3.24 | 0.00 | 3.24 |
| 156 | Delta | Michigan | 3.22 | 0.00 | 3.22 |
| 157 | Glynn | Georgia | 3.22 | 0.00 | 3.22 |
| 158 | Northampton | Pennsylvania | 2.07 | 1.10 | 3.17 |
| 159 | Gallia | Ohio | 3.16 | 0.00 | 3.16 |
| 160 | Taylor | Florida | 3.15 | 0.00 | 3.15 |
| 161 | Muhlenberg | Kentucky | 3.15 | 0.00 | 3.15 |
| 162 | Clark | Nevada | 3.15 | 0.00 | 3.15 |
| 163 | Pottawatomie | Kansas | 3.03 | 0.00 | 3.03 |
| 164 | Jefferson | Ohio | 3.03 | 0.00 | 3.03 |
| 165 | Monroe | Alabama | 3.03 | 0.00 | 3.03 |
| 166 | El Paso | Colorado | 2.98 | 0.00 | 2.98 |
| 167 | St. James | Louisiana | 0.80 | 2.10 | 2.90 |
| 168 | Morrow | Oregon | 2.89 | 0.00 | 2.89 |
| 169 | Wood | Wisconsin | 2.34 | 0.53 | 2.88 |
| 170 | Grays Harbor | Washington | 2.12 | 0.69 | 2.81 |
| 171 | Charleston | South Carolina | 0.82 | 1.99 | 2.81 |

| Rank | County | State | Total On-site Disposal or Other Releases[1] | Total Off-site Disposal or Other Releases[2] | Total On- and Off-site Disposal or Other Releases |
|---|---|---|---|---|---|
| 172 | Clarke | Alabama | 2.68 | 0.12 | 2.80 |
| 173 | Uintah | Utah | 2.80 | 0.00 | 2.80 |
| 174 | Pike | Indiana | 2.75 | 0.00 | 2.75 |
| 175 | Lawrence | Alabama | 2.75 | 0.00 | 2.75 |
| 176 | Covington City | Virginia | 2.73 | 0.00 | 2.73 |
| 177 | Jefferson | Texas | 2.72 | 0.00 | 2.72 |
| 178 | Robertson | Texas | 2.69 | 0.00 | 2.69 |
| 179 | Humboldt | Nevada | 2.66 | 0.00 | 2.66 |
| 180 | Randolph | Illinois | 2.63 | 0.00 | 2.63 |
| 181 | Jackson | Mississippi | 0.45 | 2.13 | 2.58 |
| 182 | Contra Costa | California | 1.12 | 1.46 | 2.58 |
| 183 | Muscatine | Iowa | 2.48 | 0.04 | 2.52 |
| 184 | Berks | Pennsylvania | 2.28 | 0.24 | 2.51 |
| 185 | Navajo | Arizona | 2.51 | 0.00 | 2.51 |
| 186 | Wyandotte | Kansas | 2.51 | 0.00 | 2.51 |
| 187 | Lowndes | Georgia | 2.48 | 0.00 | 2.48 |
| 188 | Lewis | Washington | 2.47 | 0.00 | 2.47 |
| 189 | Cambria | Pennsylvania | 2.46 | 0.00 | 2.46 |
| 190 | Warren | Mississippi | 2.46 | 0.00 | 2.46 |
| 191 | Warrick | Indiana | 2.42 | 0.00 | 2.42 |
| 192 | Nueces | Texas | 2.41 | 0.00 | 2.41 |
| 193 | Guayama | Puerto Rico | 2.40 | 0.00 | 2.40 |
| 194 | Carroll | Georgia | 2.01 | 0.35 | 2.36 |
| 195 | Knox | Indiana | 0.71 | 1.65 | 2.36 |
| 196 | Pottawattamie | Iowa | 2.35 | 0.00 | 2.35 |
| 197 | Mayes | Oklahoma | 2.34 | 0.00 | 2.34 |
| 198 | Citrus | Florida | 2.30 | 0.00 | 2.30 |
| 199 | Nez Perce | Idaho | 1.30 | 1.00 | 2.30 |
| 200 | Floyd | Georgia | 2.29 | 0.00 | 2.29 |

***Notes:*** *TRI = Toxic Release Inventory; Reporting year (RY) 2009 is the most recent TRI data available. Facilities reporting to TRI were required to submit RY 2009 data to EPA by July 1, 2010. Facilities may submit revisions at any time. This is the National Analysis dataset released to the public in December 2010 and includes updates for the years 1988 to 2009.Revisions submitted to EPA after this time are not reflected in TRI Explorer reports. TRI data may also be obtained through EPA Envirofacts.*

*(1) On-site Disposal or Other Releases include Underground Injection to Class I Wells (Section 5.4.1), RCRA Subtitle C Landfills (5.5.1A), Other Landfills (5.5.1B), Fugitive or Non-point Air Emissions (5.1), Stack or Point Air Emissions (5.2), Surface Water Discharges (5.3), Underground Injection to Class II-V Wells (5.4.2), Land Treatment/Application Farming (5.5.2), RCRA Subtitle C Surface Impoundments (5.5.3A), Other Surface Impoundments (5.5.3B), and Other Land Disposal (5.5.4). Off-site Disposal or Other Releases include from Section 6.2 Class I Underground Injection Wells (M81), Class II-V Underground Injection Wells (M82, M71), RCRA Subtitle C Landfills (M65), Other Landfills (M64, M72), Storage Only (M10), Solidification/Stabilization - Metals and Metal Category Compounds only (M41 or M40), Wastewater Treatment (excluding POTWs) - Metals and Metal Category Compounds only (M62 or M61), RCRA Subtitle C Surface Impoundments (M66), Other Surface Impoundments (M67, M63), Land Treatment (M73), Other Land Disposal (M79), Other Off-site Management (M90), Transfers to Waste Broker - Disposal (M94, M91), and Unknown (M99) and, from Section 6.1 Transfers to POTWs (metals and metal category compounds only).*

*(2) Off-site disposal or other releases show only net off-site disposal or other releases, that is, off-site disposal or other releases transferred to other TRI facilities reporting such transfers as on-site disposal or other releases are not included to avoid double counting.*

*This report may not include all states in the US. A state may not be included in this report for two reasons: 1) there are no facilities reporting to TRI in the particular state; or 2) the facilities reporting to TRI in the particular state did not report to TRI for the user-specified selection criteria.*

*Users of TRI information should be aware that TRI data reflect releases and other waste management activities of chemicals, not whether (or to what degree) the public has been exposed to those chemicals. Release estimates alone are not sufficient to determine exposure or to calculate potential adverse effects on human health and the environment. TRI data, in conjunction with other information, can be used as a starting point in evaluating exposures that may result from releases and other waste management activities which involve toxic chemicals. The determination of potential risk depends upon many factors, including the toxicity of the chemical, the fate of the chemical, and the amount and duration of human or other exposure to the chemical after it is released.*

***Source:*** *U.S. Environmental Protection Agency, TRI Explorer, May 16, 2011*

## TRI On-site and Off-site Reported Disposed of or Otherwise Released (in pounds), for Facilities in All Industries, for All Chemicals, Top 75 Zipcodes, 2009

| Rank | Zip | City | State | Total On-site Disposal or Other Releases[1] | Total Off-site Disposal or Other Releases[2] | Total On- and Off-site Disposal or Other Releases |
|---|---|---|---|---|---|---|
| 1 | 99752 | Kotzebue | Alaska | 637,528,125 | 482 | 637,528,607 |
| 2 | 84006 | Bingham Canyon | Utah | 101,613,933 | 16,870 | 101,630,804 |
| 3 | 89414 | Golconda | Nevada | 49,094,111 | 859 | 49,094,971 |
| 4 | 99801 | Juneau | Alaska | 47,244,157 | 0 | 47,244,157 |
| 5 | 89803 | Elko | Nevada | 42,299,724 | 612 | 42,300,336 |
| 6 | 42452 | Robards | Kentucky | 35,282,641 | 5 | 35,282,646 |
| 7 | 89822 | Carlin | Nevada | 28,569,238 | 1,107 | 28,570,344 |
| 8 | 84044 | Magna | Utah | 23,036,620 | 10,656 | 23,047,276 |
| 9 | 37040 | Clarksville | Tennessee | 21,154,370 | 112,534 | 21,266,904 |
| 10 | 15061 | Monaca | Pennsylvania | 795,268 | 20,442,642 | 21,237,910 |
| 11 | 42348 | Hawesville | Kentucky | 21,198,968 | 14,785 | 21,213,753 |
| 12 | 85135 | Hayden | Arizona | 20,996,554 | 13 | 20,996,567 |
| 13 | 47635 | Rockport | Indiana | 19,985,335 | 903,788 | 20,889,123 |
| 14 | 89820 | Battle Mountain | Nevada | 20,579,711 | 1,289 | 20,581,001 |
| 15 | 32533 | Cantonment | Florida | 20,153,030 | 302,082 | 20,455,112 |
| 16 | 85532 | Claypool | Arizona | 19,766,319 | 657,371 | 20,423,690 |
| 17 | 65440 | Boss | Missouri | 19,867,637 | 98,737 | 19,966,374 |
| 18 | 59701 | Butte | Montana | 19,633,199 | 32,481 | 19,665,680 |
| 19 | 89821 | Crescent Valley | Nevada | 19,633,504 | 503 | 19,634,007 |
| 20 | 37134 | New Johnsonville | Tennessee | 18,779,785 | 56,309 | 18,836,094 |
| 21 | 83846 | Mullan | Idaho | 18,812,439 | . | 18,812,439 |
| 22 | 70094 | Westwego | Louisiana | 14,072,237 | 4,254,297 | 18,326,534 |
| 23 | 77979 | Port Lavaca | Texas | 16,264,269 | 2,521 | 16,266,790 |
| 24 | 15774 | Shelocta | Pennsylvania | 16,081,607 | 0 | 16,081,607 |
| 25 | 21226 | Curtis Bay | Maryland | 15,207,732 | 203,904 | 15,411,637 |
| 26 | 39571 | Pass Christian | Mississippi | 15,301,471 | 136 | 15,301,607 |
| 27 | 93239 | Kettleman City | California | 14,686,876 | 6,256 | 14,693,132 |
| 28 | 48161 | Monroe | Michigan | 14,627,445 | 57,963 | 14,685,409 |
| 29 | 43616 | Oregon | Ohio | 13,796,015 | 473,171 | 14,269,187 |
| 30 | 42328 | Centertown | Kentucky | 13,763,503 | 0 | 13,763,503 |
| 31 | 77512 | Alvin | Texas | 13,418,784 | 0 | 13,418,784 |
| 32 | 24141 | Radford | Virginia | 12,429,115 | 233,148 | 12,662,263 |
| 33 | 41045 | Ghent | Kentucky | 9,388,021 | 3,025,229 | 12,413,250 |
| 34 | 46402 | Gary | Indiana | 11,481,656 | 525,763 | 12,007,419 |
| 35 | 70070 | Luling | Louisiana | 11,640,323 | 452 | 11,640,775 |
| 36 | 61615 | Peoria | Illinois | 167,019 | 11,257,164 | 11,424,183 |
| 37 | 71052 | Mansfield | Louisiana | 11,287,705 | 0 | 11,287,705 |
| 38 | 46721 | Butler | Indiana | 202,421 | 11,004,303 | 11,206,723 |
| 39 | 89319 | Ruth | Nevada | 11,127,558 | 0 | 11,127,558 |
| 40 | 35459 | Emelle | Alabama | 10,839,481 | 30,857 | 10,870,338 |
| 41 | 32514 | Pensacola | Florida | 10,795,988 | 0 | 10,795,988 |
| 42 | 44004 | Ashtabula | Ohio | 10,477,934 | 110,554 | 10,588,488 |
| 43 | 77503 | Pasadena | Texas | 703,568 | 9,833,999 | 10,537,566 |
| 44 | 59323 | Colstrip | Montana | 10,376,895 | 116,202 | 10,493,097 |
| 45 | 34428 | Crystal River | Florida | 10,345,131 | 8,591 | 10,353,722 |
| 46 | 70805 | Baton Rouge | Louisiana | 9,948,189 | 102,124 | 10,050,313 |
| 47 | 39746 | Hamilton | Mississippi | 10,042,683 | 5,605 | 10,048,288 |
| 48 | 82001 | Cheyenne | Wyoming | 9,670,779 | 0 | 9,670,779 |
| 49 | 46312 | East Chicago | Indiana | 535,274 | 9,132,082 | 9,667,355 |
| 50 | 63629 | Bunker | Missouri | 9,510,303 | 0 | 9,510,303 |
| 51 | 42337 | Drakesboro | Kentucky | 9,317,312 | 8 | 9,317,321 |
| 52 | 45805 | Lima | Ohio | 9,205,858 | 1,293 | 9,207,151 |
| 53 | 52761 | Muscatine | Iowa | 2,176,690 | 6,714,707 | 8,891,397 |
| 54 | 45144 | Manchester | Ohio | 8,822,495 | 306 | 8,822,801 |
| 55 | 25213 | Winfield | West Virginia | 7,223,991 | 1,501,252 | 8,725,243 |

| Rank | Zip | City | State | Total On-site Disposal or Other Releases[1] | Total Off-site Disposal or Other Releases[2] | Total On- and Off-site Disposal or Other Releases |
|---|---|---|---|---|---|---|
| 56 | 45715 | Beverly | Ohio | 8,676,351 | 4,504 | 8,680,855 |
| 57 | 46231 | Indianapolis | Indiana | 10,710 | 8,611,147 | 8,621,857 |
| 58 | 30901 | Augusta | Georgia | 8,396,328 | 41,548 | 8,437,876 |
| 59 | 30120 | Cartersville | Georgia | 8,285,784 | 19,021 | 8,304,805 |
| 60 | 75455 | Mount Pleasant | Texas | 8,184,592 | 10,675 | 8,195,267 |
| 61 | 45620 | Cheshire | Ohio | 8,078,556 | 9,192 | 8,087,748 |
| 62 | 20664 | Newburg | Maryland | 7,069,743 | 971,777 | 8,041,520 |
| 63 | 77627 | Nederland | Texas | 7,937,606 | 7,798 | 7,945,405 |
| 64 | 63048 | Herculaneum | Missouri | 7,802,880 | 1,392 | 7,804,271 |
| 65 | 48054 | East China | Michigan | 7,799,542 | 0 | 7,799,542 |
| 66 | 15748 | Homer City | Pennsylvania | 7,611,302 | 855 | 7,612,158 |
| 67 | 87421 | Waterflow | New Mexico | 4,062,520 | 3,541,344 | 7,603,864 |
| 68 | 43961 | Stratton | Ohio | 5,902,335 | 1,667,533 | 7,569,868 |
| 69 | 73860 | Waynoka | Oklahoma | 7,518,548 | 19,119 | 7,537,667 |
| 70 | 55308 | Becker | Minnesota | 7,482,552 | 111 | 7,482,663 |
| 71 | 63633 | Centerville | Missouri | 7,108,903 | . | 7,108,903 |
| 72 | 70346 | Donaldsonville | Louisiana | 7,044,683 | 34,409 | 7,079,092 |
| 73 | 83276 | Soda Springs | Idaho | 7,074,084 | 3,423 | 7,077,507 |
| 74 | 31046 | Juliette | Georgia | 6,935,057 | 1,534 | 6,936,591 |
| 75 | 35404 | Tuscaloosa | Alabama | 9,194 | 6,911,552 | 6,920,746 |

***Notes:*** *TRI = Toxic Release Inventory; Reporting year (RY) 2009 is the most recent TRI data available. Facilities reporting to TRI were required to submit RY 2009 data to EPA by July 1, 2010. Facilities may submit revisions at any time. This is the National Analysis dataset released to the public in December 2010 and includes updates for the years 1988 to 2009.Revisions submitted to EPA after this time are not reflected in TRI Explorer reports. TRI data may also be obtained through EPA Envirofacts.*

*(1) On-site Disposal or Other Releases include Underground Injection to Class I Wells (Section 5.4.1), RCRA Subtitle C Landfills (5.5.1A), Other Landfills (5.5.1B), Fugitive or Non-point Air Emissions (5.1), Stack or Point Air Emissions (5.2), Surface Water Discharges (5.3), Underground Injection to Class II-V Wells (5.4.2), Land Treatment/Application Farming (5.5.2), RCRA Subtitle C Surface Impoundments (5.5.3A), Other Surface Impoundments (5.5.3B), and Other Land Disposal (5.5.4). Off-site Disposal or Other Releases include from Section 6.2 Class I Underground Injection Wells (M81), Class II-V Underground Injection Wells (M82, M71), RCRA Subtitle C Landfills (M65), Other Landfills (M64, M72), Storage Only (M10), Solidification/Stabilization - Metals and Metal Category Compounds only (M41 or M40), Wastewater Treatment (excluding POTWs) - Metals and Metal Category Compounds only (M62 or M61), RCRA Subtitle C Surface Impoundments (M66), Other Surface Impoundments (M67, M63), Land Treatment (M73), Other Land Disposal (M79), Other Off-site Management (M90), Transfers to Waste Broker - Disposal (M94, M91), and Unknown (M99) and, from Section 6.1 Transfers to POTWs (metals and metal category compounds only).*

*(2) Off-site disposal or other releases show only net off-site disposal or other releases, that is, off-site disposal or other releases transferred to other TRI facilities reporting such transfers as on-site disposal or other releases are not included to avoid double counting.*

*This report may not include all states in the US. A state may not be included in this report for two reasons: 1) there are no facilities reporting to TRI in the particular state; or 2) the facilities reporting to TRI in the particular state did not report to TRI for the user-specified selection criteria.*

*Users of TRI information should be aware that TRI data reflect releases and other waste management activities of chemicals, not whether (or to what degree) the public has been exposed to those chemicals. Release estimates alone are not sufficient to determine exposure or to calculate potential adverse effects on human health and the environment. TRI data, in conjunction with other information, can be used as a starting point in evaluating exposures that may result from releases and other waste management activities which involve toxic chemicals. The determination of potential risk depends upon many factors, including the toxicity of the chemical, the fate of the chemical, and the amount and duration of human or other exposure to the chemical after it is released.*

***Source:*** *U.S. Environmental Protection Agency, TRI Explorer, May 16, 2011*

## TRI On-site and Off-site Reported Disposed of or Otherwise Released (in grams), for Facilities in All Industries, Dioxin and Dioxin-like Compounds, Top 75 Zipcodes, 2009

| Rank | Zip | City | State | Total On-site Disposal or Other Releases[1] | Total Off-site Disposal or Other Releases[2] | Total On- and Off-site Disposal or Other Releases |
|---|---|---|---|---|---|---|
| 1 | 99752 | Kotzebue | Alaska | 637,528,125 | 482 | 637,528,607 |
| 2 | 84006 | Bingham Canyon | Utah | 101,613,933 | 16,870 | 101,630,804 |
| 3 | 89414 | Golconda | Nevada | 49,094,111 | 859 | 49,094,971 |
| 4 | 99801 | Juneau | Alaska | 47,244,157 | 0 | 47,244,157 |
| 5 | 89803 | Elko | Nevada | 42,299,724 | 612 | 42,300,336 |
| 6 | 42452 | Robards | Kentucky | 35,282,641 | 5 | 35,282,646 |
| 7 | 89822 | Carlin | Nevada | 28,569,238 | 1,107 | 28,570,344 |
| 8 | 84044 | Magna | Utah | 23,036,620 | 10,656 | 23,047,276 |
| 9 | 37040 | Clarksville | Tennessee | 21,154,370 | 112,534 | 21,266,904 |
| 10 | 15061 | Monaca | Pennsylvania | 795,268 | 20,442,642 | 21,237,910 |
| 11 | 42348 | Hawesville | Kentucky | 21,198,968 | 14,785 | 21,213,753 |
| 12 | 85135 | Hayden | Arizona | 20,996,554 | 13 | 20,996,567 |
| 13 | 47635 | Rockport | Indiana | 19,985,335 | 903,788 | 20,889,123 |
| 14 | 89820 | Battle Mountain | Nevada | 20,579,711 | 1,289 | 20,581,001 |
| 15 | 32533 | Cantonment | Florida | 20,153,030 | 302,082 | 20,455,112 |
| 16 | 85532 | Claypool | Arizona | 19,766,319 | 657,371 | 20,423,690 |
| 17 | 65440 | Boss | Missouri | 19,867,637 | 98,737 | 19,966,374 |
| 18 | 59701 | Butte | Montana | 19,633,199 | 32,481 | 19,665,680 |
| 19 | 89821 | Crescent Valley | Nevada | 19,633,504 | 503 | 19,634,007 |
| 20 | 37134 | New Johnsonville | Tennessee | 18,779,785 | 56,309 | 18,836,094 |
| 21 | 83846 | Mullan | Idaho | 18,812,439 | . | 18,812,439 |
| 22 | 70094 | Westwego | Louisiana | 14,072,237 | 4,254,297 | 18,326,534 |
| 23 | 77979 | Port Lavaca | Texas | 16,264,269 | 2,521 | 16,266,790 |
| 24 | 15774 | Shelocta | Pennsylvania | 16,081,607 | 0 | 16,081,607 |
| 25 | 21226 | Curtis Bay | Maryland | 15,207,732 | 203,904 | 15,411,637 |
| 26 | 39571 | Pass Christian | Mississippi | 15,301,471 | 136 | 15,301,607 |
| 27 | 93239 | Kettleman City | California | 14,686,876 | 6,256 | 14,693,132 |
| 28 | 48161 | Monroe | Michigan | 14,627,445 | 57,963 | 14,685,409 |
| 29 | 43616 | Oregon | Ohio | 13,796,015 | 473,171 | 14,269,187 |
| 30 | 42328 | Centertown | Kentucky | 13,763,503 | 0 | 13,763,503 |
| 31 | 77512 | Alvin | Texas | 13,418,784 | 0 | 13,418,784 |
| 32 | 24141 | Radford | Virginia | 12,429,115 | 233,148 | 12,662,263 |
| 33 | 41045 | Ghent | Kentucky | 9,388,021 | 3,025,229 | 12,413,250 |
| 34 | 46402 | Gary | Indiana | 11,481,656 | 525,763 | 12,007,419 |
| 35 | 70070 | Luling | Louisiana | 11,640,323 | 452 | 11,640,775 |
| 36 | 61615 | Peoria | Illinois | 167,019 | 11,257,164 | 11,424,183 |
| 37 | 71052 | Mansfield | Louisiana | 11,287,705 | 0 | 11,287,705 |
| 38 | 46721 | Butler | Indiana | 202,421 | 11,004,303 | 11,206,723 |
| 39 | 89319 | Ruth | Nevada | 11,127,558 | 0 | 11,127,558 |
| 40 | 35459 | Emelle | Alabama | 10,839,481 | 30,857 | 10,870,338 |
| 41 | 32514 | Pensacola | Florida | 10,795,988 | 0 | 10,795,988 |
| 42 | 44004 | Ashtabula | Ohio | 10,477,934 | 110,554 | 10,588,488 |
| 43 | 77503 | Pasadena | Texas | 703,568 | 9,833,999 | 10,537,566 |
| 44 | 59323 | Colstrip | Montana | 10,376,895 | 116,202 | 10,493,097 |
| 45 | 34428 | Crystal River | Florida | 10,345,131 | 8,591 | 10,353,722 |
| 46 | 70805 | Baton Rouge | Louisiana | 9,948,189 | 102,124 | 10,050,313 |
| 47 | 39746 | Hamilton | Mississippi | 10,042,683 | 5,605 | 10,048,288 |
| 48 | 82001 | Cheyenne | Wyoming | 9,670,779 | 0 | 9,670,779 |
| 49 | 46312 | East Chicago | Indiana | 535,274 | 9,132,082 | 9,667,355 |
| 50 | 63629 | Bunker | Missouri | 9,510,303 | 0 | 9,510,303 |
| 51 | 42337 | Drakesboro | Kentucky | 9,317,312 | 8 | 9,317,321 |
| 52 | 45805 | Lima | Ohio | 9,205,858 | 1,293 | 9,207,151 |
| 53 | 52761 | Muscatine | Iowa | 2,176,690 | 6,714,707 | 8,891,397 |
| 54 | 45144 | Manchester | Ohio | 8,822,495 | 306 | 8,822,801 |
| 55 | 25213 | Winfield | West Virginia | 7,223,991 | 1,501,252 | 8,725,243 |

| Rank | Zip | City | State | Total On-site Disposal or Other Releases[1] | Total Off-site Disposal or Other Releases[2] | Total On- and Off-site Disposal or Other Releases |
|---|---|---|---|---|---|---|
| 56 | 45715 | Beverly | Ohio | 8,676,351 | 4,504 | 8,680,855 |
| 57 | 46231 | Indianapolis | Indiana | 10,710 | 8,611,147 | 8,621,857 |
| 58 | 30901 | Augusta | Georgia | 8,396,328 | 41,548 | 8,437,876 |
| 59 | 30120 | Cartersville | Georgia | 8,285,784 | 19,021 | 8,304,805 |
| 60 | 75455 | Mount Pleasant | Texas | 8,184,592 | 10,675 | 8,195,267 |
| 61 | 45620 | Cheshire | Ohio | 8,078,556 | 9,192 | 8,087,748 |
| 62 | 20664 | Newburg | Maryland | 7,069,743 | 971,777 | 8,041,520 |
| 63 | 77627 | Nederland | Texas | 7,937,606 | 7,798 | 7,945,405 |
| 64 | 63048 | Herculaneum | Missouri | 7,802,880 | 1,392 | 7,804,271 |
| 65 | 48054 | East China | Michigan | 7,799,542 | 0 | 7,799,542 |
| 66 | 15748 | Homer City | Pennsylvania | 7,611,302 | 855 | 7,612,158 |
| 67 | 87421 | Waterflow | New Mexico | 4,062,520 | 3,541,344 | 7,603,864 |
| 68 | 43961 | Stratton | Ohio | 5,902,335 | 1,667,533 | 7,569,868 |
| 69 | 73860 | Waynoka | Oklahoma | 7,518,548 | 19,119 | 7,537,667 |
| 70 | 55308 | Becker | Minnesota | 7,482,552 | 111 | 7,482,663 |
| 71 | 63633 | Centerville | Missouri | 7,108,903 | . | 7,108,903 |
| 72 | 70346 | Donaldsonville | Louisiana | 7,044,683 | 34,409 | 7,079,092 |
| 73 | 83276 | Soda Springs | Idaho | 7,074,084 | 3,423 | 7,077,507 |
| 74 | 31046 | Juliette | Georgia | 6,935,057 | 1,534 | 6,936,591 |
| 75 | 35404 | Tuscaloosa | Alabama | 9,194 | 6,911,552 | 6,920,746 |

**Notes:** *TRI = Toxic Release Inventory; Reporting year (RY) 2009 is the most recent TRI data available. Facilities reporting to TRI were required to submit RY 2009 data to EPA by July 1, 2010. Facilities may submit revisions at any time. This is the National Analysis dataset released to the public in December 2010 and includes updates for the years 1988 to 2009.Revisions submitted to EPA after this time are not reflected in TRI Explorer reports. TRI data may also be obtained through EPA Envirofacts.*

*(1) On-site Disposal or Other Releases include Underground Injection to Class I Wells (Section 5.4.1), RCRA Subtitle C Landfills (5.5.1A), Other Landfills (5.5.1B), Fugitive or Non-point Air Emissions (5.1), Stack or Point Air Emissions (5.2), Surface Water Discharges (5.3), Underground Injection to Class II-V Wells (5.4.2), Land Treatment/Application Farming (5.5.2), RCRA Subtitle C Surface Impoundments (5.5.3A), Other Surface Impoundments (5.5.3B), and Other Land Disposal (5.5.4). Off-site Disposal or Other Releases include from Section 6.2 Class I Underground Injection Wells (M81), Class II-V Underground Injection Wells (M82, M71), RCRA Subtitle C Landfills (M65), Other Landfills (M64, M72), Storage Only (M10), Solidification/Stabilization - Metals and Metal Category Compounds only (M41 or M40), Wastewater Treatment (excluding POTWs) - Metals and Metal Category Compounds only (M62 or M61), RCRA Subtitle C Surface Impoundments (M66), Other Surface Impoundments (M67, M63), Land Treatment (M73), Other Land Disposal (M79), Other Off-site Management (M90), Transfers to Waste Broker - Disposal (M94, M91), and Unknown (M99) and, from Section 6.1 Transfers to POTWs (metals and metal category compounds only).*

*(2) Off-site disposal or other releases show only net off-site disposal or other releases, that is, off-site disposal or other releases transferred to other TRI facilities reporting such transfers as on-site disposal or other releases are not included to avoid double counting.*

*This report may not include all states in the US. A state may not be included in this report for two reasons: 1) there are no facilities reporting to TRI in the particular state; or 2) the facilities reporting to TRI in the particular state did not report to TRI for the user-specified selection criteria.*

*Users of TRI information should be aware that TRI data reflect releases and other waste management activities of chemicals, not whether (or to what degree) the public has been exposed to those chemicals. Release estimates alone are not sufficient to determine exposure or to calculate potential adverse effects on human health and the environment. TRI data, in conjunction with other information, can be used as a starting point in evaluating exposures that may result from releases and other waste management activities which involve toxic chemicals. The determination of potential risk depends upon many factors, including the toxicity of the chemical, the fate of the chemical, and the amount and duration of human or other exposure to the chemical after it is released.*

**Source:** *U.S. Environmental Protection Agency, TRI Explorer, May 16, 2011*

## Generation, Materials Recovery, Composting, Combustion with Energy Recovery, and Discards of Municipal Solid Waste, 1960 – 2009 (In millions of tons)

| Activity | 1960 | 1970 | 1980 | 1990 | 2000 | 2005 | 2007 | 2008 | 2009 |
|---|---|---|---|---|---|---|---|---|---|
| **Generation** | 88.1 | 121.1 | 151.6 | 208.3 | 242.5 | 252.4 | 255.0 | 251.0 | 243.0 |
| Recovery for recycling | 5.6 | 8.0 | 14.5 | 29.0 | 53.0 | 59.3 | 63.1 | 61.8 | 61.3 |
| Recovery for composting* | Neg. | Neg. | Neg. | 4.2 | 16.5 | 20.6 | 21.7 | 22.1 | 20.8 |
| **Total materials recovery** | 5.6 | 8.0 | 14.5 | 33.2 | 69.5 | 79.9 | 84.8 | 83.9 | 82.0 |
| **Combustion with energy recovery†** | 0.0 | 0.4 | 2.7 | 29.7 | 33.7 | 31.6 | 32.0 | 31.6 | 29.0 |
| **Discards to landfill, other disposal‡** | 82.5 | 112.7 | 134.4 | 145.3 | 139.4 | 140.9 | 138.2 | 135.6 | 131.9 |

* Composting of yard trimmings, food scraps and other MSW organic material. Does not include backyard composting.

† Includes combustion of MSW in mass burn or refuse-derived fuel form, and combustion with energy recovery of source separated materials in MSW (e.g., wood pallets and tire-derived fuel). See Table 29 footnote for more detail.

‡ Discards after recovery minus combustion with energy recovery. Discards include combustion without energy recovery. Details may not add to totals due to rounding.

## Generation, Materials Recovery, Composting, Combustion with Energy Recovery, and Discards of Municipal Solid Waste, 1960 – 2009 (In percent of total generation)

| Activity | 1960 | 1970 | 1980 | 1990 | 2000 | 2005 | 2007 | 2008 | 2009 |
|---|---|---|---|---|---|---|---|---|---|
| **Generation** | 100.0% | 100.0% | 100.0% | 100.0% | 100.0% | 100.0% | 100.0% | 100.0% | 100.0% |
| Recovery for recycling | 6.4% | 6.6% | 9.6% | 14.0% | 21.9% | 23.5% | 24.8% | 24.6% | 25.2% |
| Recovery for composting* | Neg. | Neg. | Neg. | 2.0% | 6.7% | 8.1% | 8.5% | 8.8% | 8.6% |
| **TTotal materials recovery** | 6.4% | 6.6% | 9.6% | 16.0% | 28.6% | 31.6% | 33.3% | 33.4% | 33.8% |
| **Combustion with energy recovery†** | 0.0% | 0.3% | 1.8% | 14.2% | 13.9% | 12.5% | 12.5% | 12.6% | 11.9% |
| **Discards to landfill, other disposal‡** | 93.6% | 93.1% | 88.6% | 69.8% | 57.5% | 55.9% | 54.2% | 54.0% | 54.3% |

* Composting of yard trimmings, food scraps and other MSW organic material. Does not include backyard composting.

† Includes combustion of MSW in mass burn or refuse-derived fuel form, and combustion with energy recovery of source separated materials in MSW (e.g., wood pallets and tire-derived fuel). See Table 29 footnote for more detail.

‡ Discards after recovery minus combustion with energy recovery. Discards include combustion without energy recovery. Details may not add to totals due to rounding.

***SOURCE: U.S. EPA, Municipal Solid Waste in the United States: 2009 Facts and Figures***

## Generation, Materials Recovery, Composting, Combustion with Energy Recovery, and Discards of Municipal Solid Waste, 1960 – 2009 (In pounds per person per day)

| Activity | 1960 | 1970 | 1980 | 1990 | 2000 | 2005 | 2007 | 2008 | 2009 |
|---|---|---|---|---|---|---|---|---|---|
| **Generation** | 2.68 | 3.25 | 3.66 | 4.57 | 4.72 | 4.67 | 4.63 | 4.52 | 4.34 |
| Recovery for recycling | 0.17 | 0.22 | 0.35 | 0.64 | 1.03 | 1.10 | 1.15 | 1.11 | 1.09 |
| Recovery for composting* | Neg. | Neg. | Neg. | 0.09 | 0.32 | 0.38 | 0.39 | 0.40 | 0.37 |
| **Total materials recovery** | 0.17 | 0.22 | 0.35 | 0.73 | 1.35 | 1.48 | 1.54 | 1.51 | 1.46 |
| **Combustion with energy recovery†** | 0.00 | 0.01 | 0.07 | 0.65 | 0.66 | 0.58 | 0.58 | 0.57 | 0.52 |
| **Discards to landfill, other disposal‡** | 2.51 | 3.02 | 3.24 | 3.19 | 2.71 | 2.61 | 2.51 | 2.44 | 2.36 |
| Population (millions) | 179.979 | 203.984 | 227.255 | 249.907 | 281.422 | 296.410 | 301.621 | 304.060 | 307.007 |

* Composting of yard trimmings, food scraps and other MSW organic material. Does not include backyard composting.

† Includes combustion of MSW in mass burn or refuse-derived fuel form, and combustion with energy recovery of source separated materials in MSW (e.g., wood pallets and tire-derived fuel). See Table 29 footnote for more detail.

‡ Discards after recovery minus combustion with energy recovery. Discards include combustion without energy recovery. Details may not add to totals due to rounding.

## MSW Generation Rates, 1960 to 2009

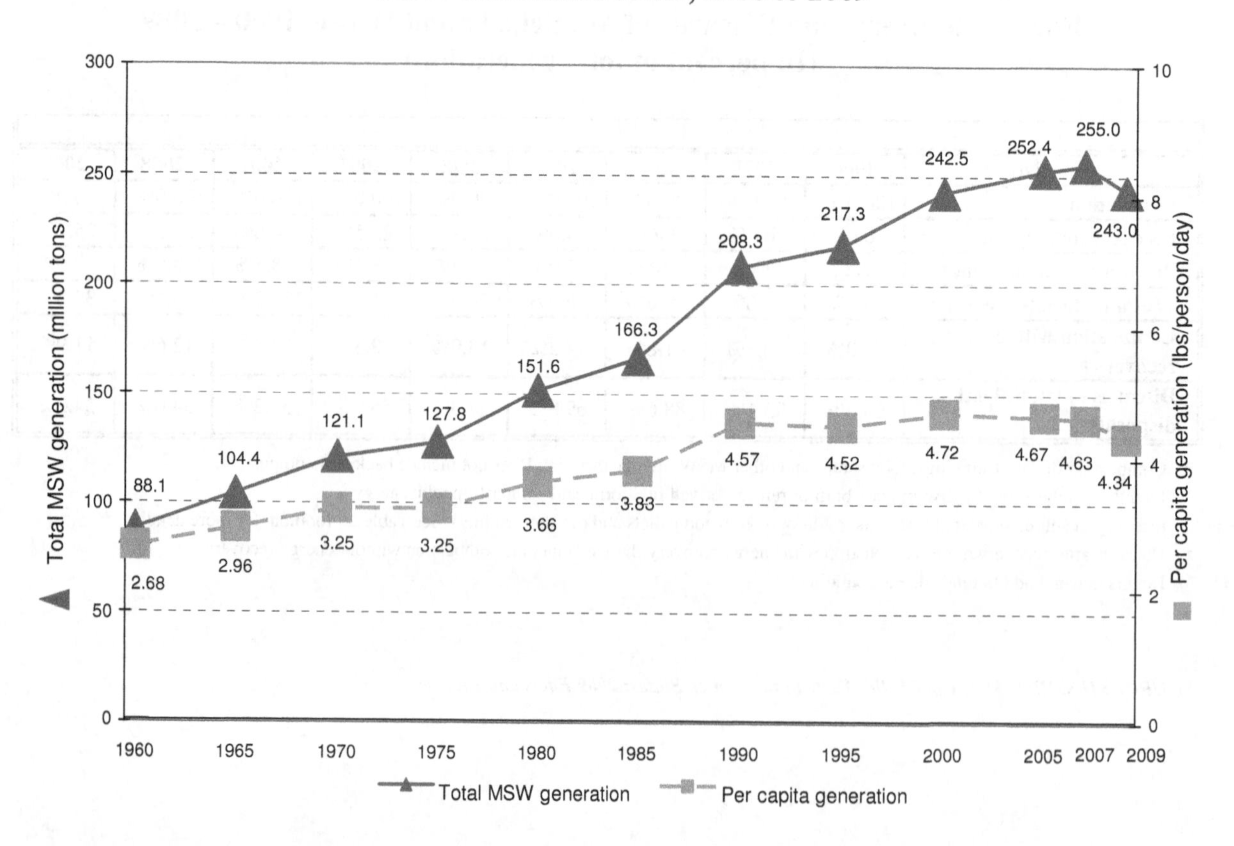


*SOURCE: U.S. EPA, Municipal Solid Waste in the United States: 2009 Facts and Figures*

## MSW Recycling Rates, 1960 to 2009

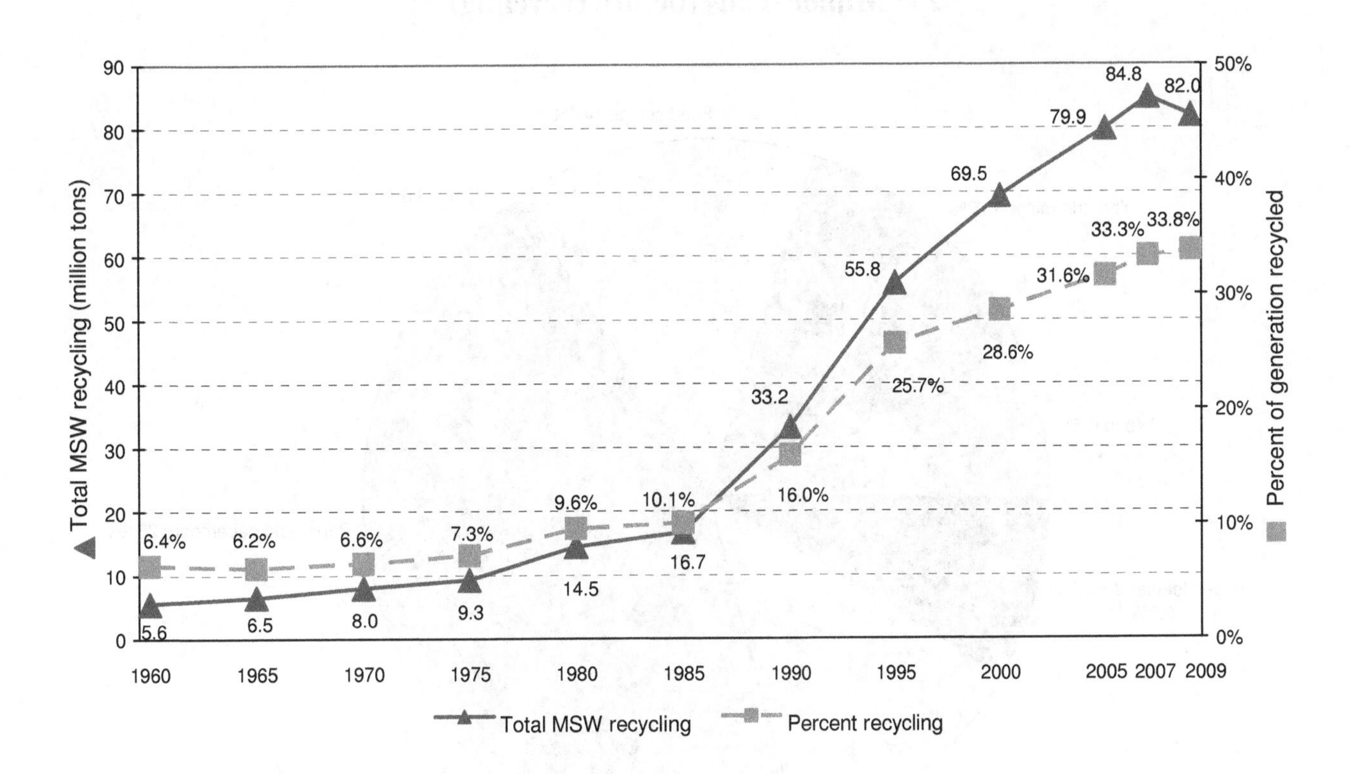


*SOURCE: U.S. EPA, Municipal Solid Waste in the United States: 2009 Facts and Figures*

## Materials Generation in MSW, 2009
## 243 Million Tons (before recycling)

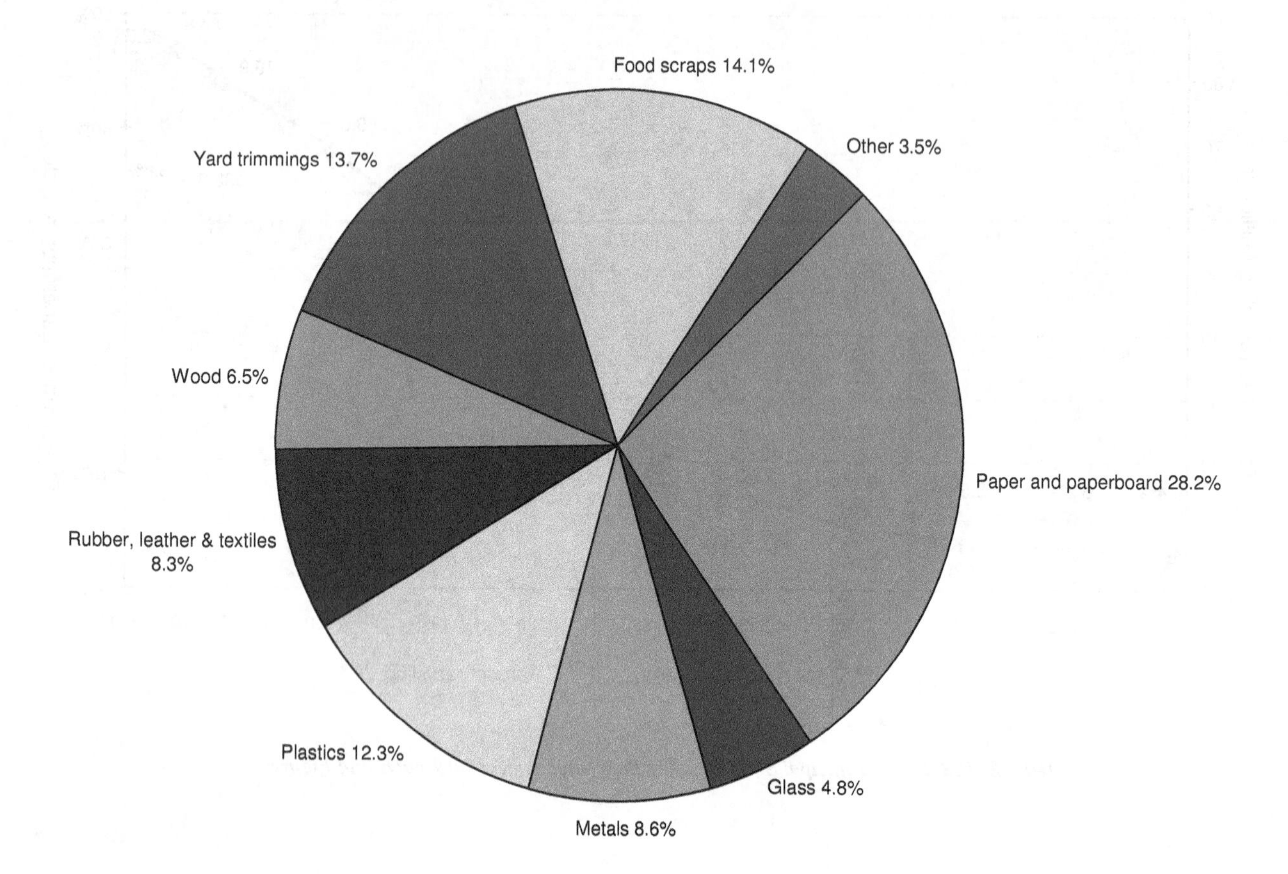


*SOURCE: U.S. EPA, Municipal Solid Waste in the United States: 2009 Facts and Figures*

## Generation and Recovery of Materials in MSW, 2009
## (In millions of tons and percent of generation of each material)

| Material | Weight Generated | Weight Recovered | Recovery As a Percent of Generation |
|---|---|---|---|
| Paper and paperboard | 68.43 | 42.50 | 62.1% |
| Glass | 11.78 | 3.00 | 25.5% |
| Metals | | | |
| Steel | 15.62 | 5.23 | 33.5% |
| Aluminum | 3.40 | 0.69 | 20.3% |
| Other nonferrous metals* | 1.89 | 1.30 | 68.8% |
| *Total metals* | *20.91* | *7.22* | *34.5%* |
| Plastics | 29.83 | 2.12 | 7.1% |
| Rubber and leather | 7.49 | 1.07 | 14.3% |
| Textiles | 12.73 | 1.90 | 14.9% |
| Wood | 15.84 | 2.23 | 14.1% |
| Other materials | 4.64 | 1.23 | 26.5% |
| ***Total Materials in Products*** | *171.65* | *61.27* | *35.7%* |
| Other wastes | | | |
| Food, other** | 34.29 | 0.85 | 2.5% |
| Yard trimmings | 33.20 | 19.90 | 59.9% |
| Miscellaneous inorganic wastes | 3.82 | Neg. | Neg. |
| ***Total Other Wastes*** | *71.31* | *20.75* | *29.1%* |
| ***TOTAL MUNICIPAL SOLID WASTE*** | 242.96 | 82.02 | 33.8% |

Includes waste from residential, commercial, and institutional sources.

* Includes lead from lead-acid batteries.

** Includes recovery of other MSW organics for composting.

Details may not add to totals due to rounding.

Neg. = Less than 5,000 tons or 0.05 percent.

***SOURCE: U.S. EPA, Municipal Solid Waste in the United States: 2009 Facts and Figures***

## Products Generated in MSW, 2009
## 243 Million Tons (before recycling)

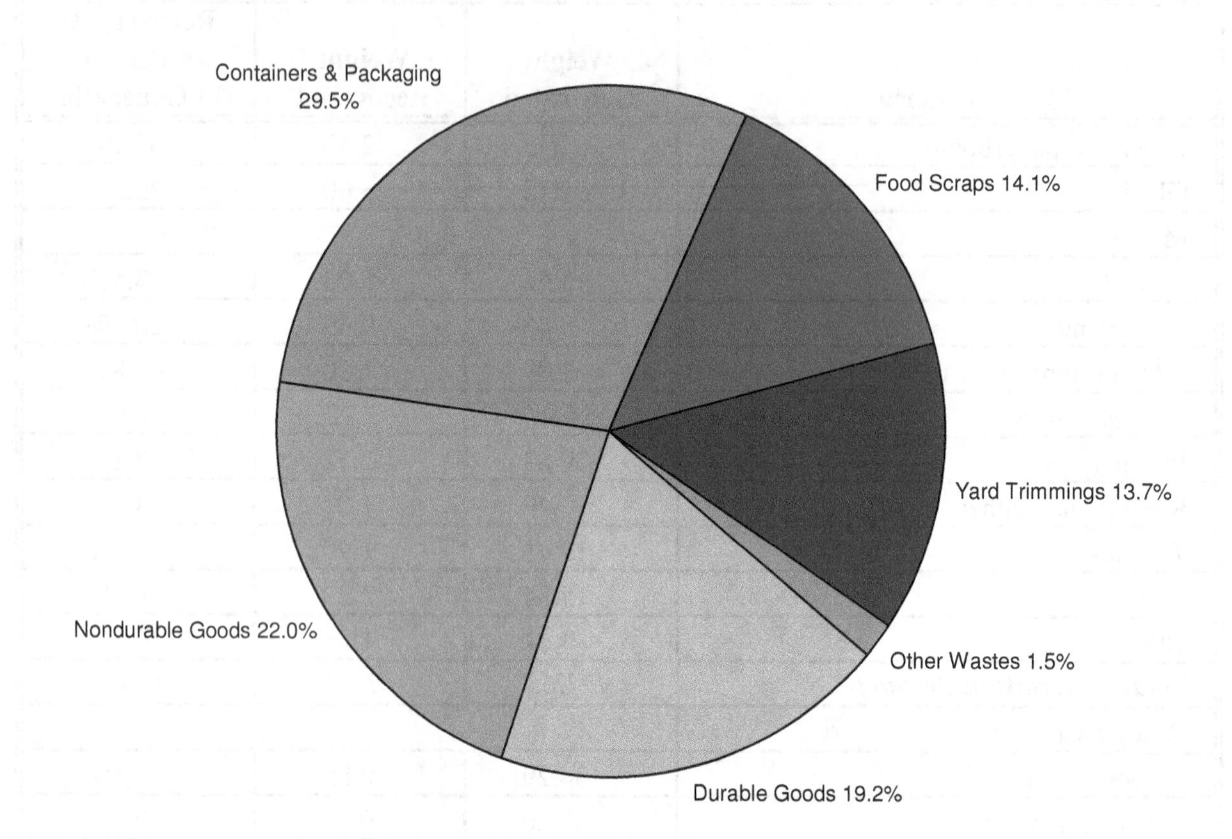


*SOURCE: U.S. EPA, Municipal Solid Waste in the United States: 2009 Facts and Figures*

## Generation and Recovery of Products in MSW by Material, 2009
## (In millions of tons and percent of generation of each product)

| Products | Weight Generated | Weight Recovered | Recovery as a Percent of Generation |
|---|---|---|---|
| **Durable Goods** | | | |
| Steel | 13.34 | 3.72 | 27.9% |
| Aluminum | 1.35 | Neg. | Neg. |
| Other non-ferrous metals* | 1.89 | 1.30 | 68.8% |
| Glass | 2.12 | Neg. | Neg. |
| Plastics | 10.65 | 0.40 | 3.8% |
| Rubber and leather | 6.43 | 1.07 | 16.6% |
| Wood | 5.76 | Neg. | Neg. |
| Textiles | 3.49 | 0.44 | 12.6% |
| Other materials | 1.61 | 1.23 | 76.4% |
| ***Total durable goods*** | 46.64 | 8.16 | 17.5% |
| **Nondurable Goods** | | | |
| Paper and paperboard | 33.48 | 17.43 | 52.1% |
| Plastics | 6.65 | Neg. | Neg. |
| Rubber and leather | 1.06 | Neg. | Neg. |
| Textiles | 9.00 | 1.46 | 16.2% |
| Other materials | 3.25 | Neg. | Neg. |
| ***Total nondurable goods*** | 53.44 | 18.89 | 35.3% |
| **Containers and Packaging** | | | |
| Steel | 2.28 | 1.51 | 66.2% |
| Aluminum | 1.84 | 0.69 | 37.5% |
| Glass | 9.66 | 3.00 | 31.1% |
| Paper and paperboard | 34.94 | 25.07 | 71.8% |
| Plastics | 12.53 | 1.72 | 13.7% |
| Wood | 10.08 | 2.23 | 22.1% |
| Other materials | 0.24 | Neg. | Neg. |
| ***Total containers and packaging*** | 71.57 | 34.22 | 47.8% |
| **Other Wastes** | | | |
| Food, other** | 34.29 | 0.85 | 2.5% |
| Yard trimmings | 33.20 | 19.9 | 59.9% |
| Miscellaneous inorganic wastes | 3.82 | Neg. | Neg. |
| ***Total other wastes*** | 71.31 | 20.75 | 29.1% |
| ***TOTAL MUNICIPAL SOLID WASTE*** | 242.96 | 82.02 | 33.8% |

Includes waste from residential, commercial, and institutional sources.

* Includes lead from lead-acid batteries.

** Includes recovery of other MSW organics for composting.

Details may not add to totals due to rounding.

Neg. = Less than 5,000 tons or 0.05 percent.

***SOURCE: U.S. EPA, Municipal Solid Waste in the United States: 2009 Facts and Figures***

## Number of Landfills in the United States, 1988 - 2009

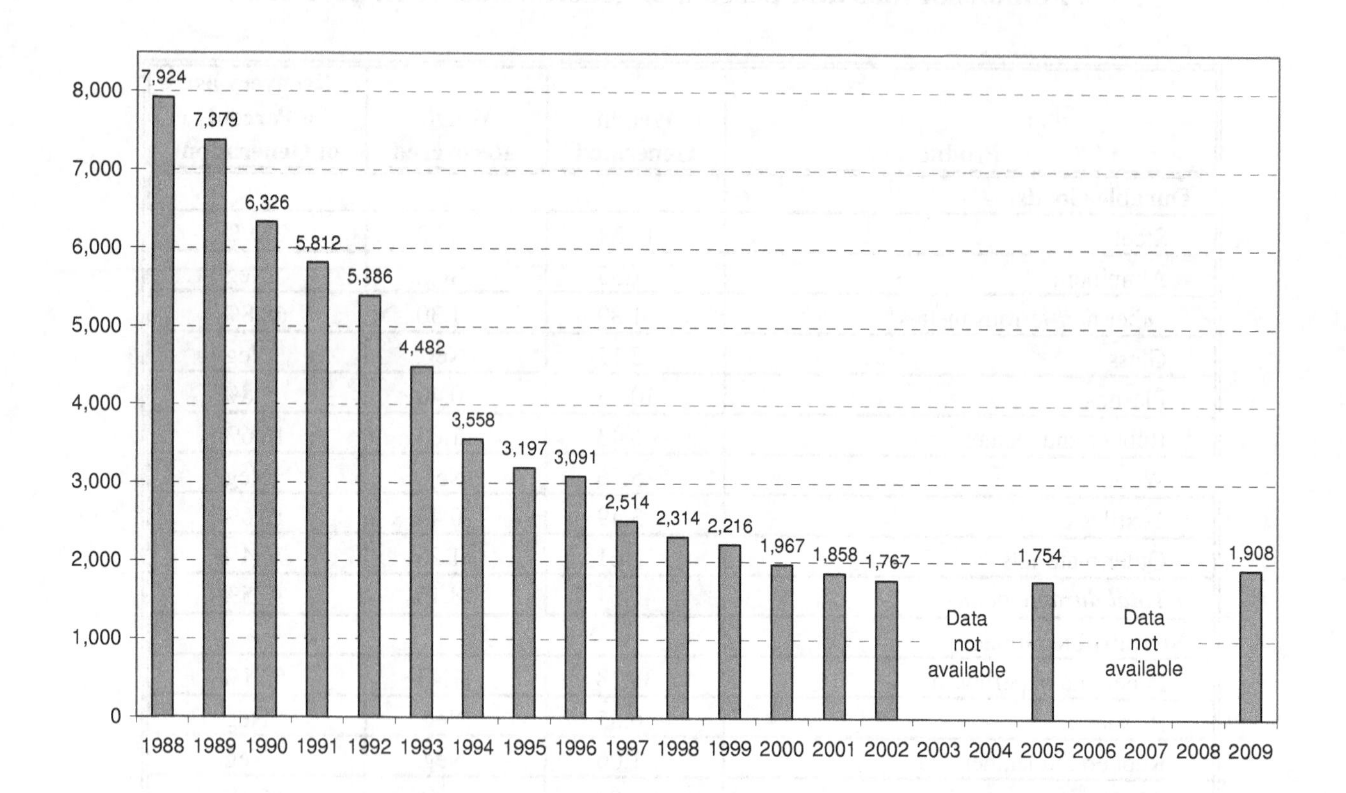


*SOURCE: U.S. EPA, Municipal Solid Waste in the United States: 2009 Facts and Figures*

## Management of MSW in the United States, 2009

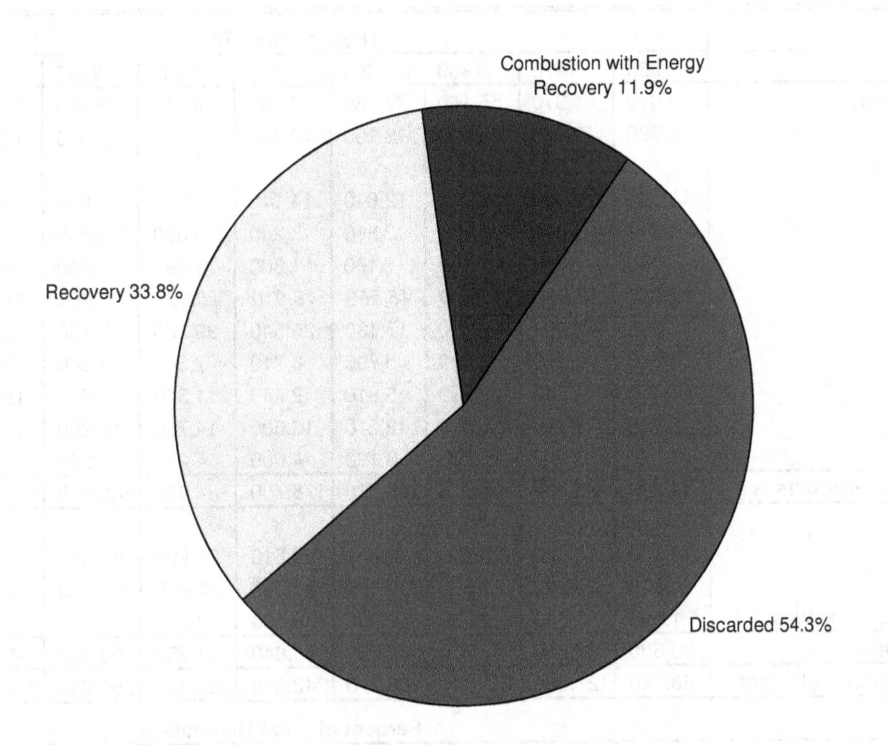


*SOURCE: U.S. EPA, Municipal Solid Waste in the United States: 2009 Facts and Figures*

**MATERIALS GENERATED* IN THE MUNICIPAL WASTE STREAM, 1960 TO 2009**
**(In thousands of tons and percent of total generation)**

| | Thousands of Tons | | | | | | | | |
|---|---|---|---|---|---|---|---|---|---|
| **Materials** | **1960** | **1970** | **1980** | **1990** | **2000** | **2005** | **2007** | **2008** | **2009** |
| Paper and Paperboard | 29,990 | 44,310 | 55,160 | 72,730 | 87,740 | 84,840 | 82,530 | 77,420 | 68,430 |
| Glass | 6,720 | 12,740 | 15,130 | 13,100 | 12,760 | 12,540 | 12,520 | 12,150 | 11,780 |
| Metals | | | | | | | | | |
| Ferrous | 10,300 | 12,360 | 12,620 | 12,640 | 14,110 | 14,990 | 15,640 | 15,730 | 15,620 |
| Aluminum | 340 | 800 | 1,730 | 2,810 | 3,200 | 3,330 | 3,360 | 3,410 | 3,400 |
| Other Nonferrous | 180 | 670 | 1,160 | 1,100 | 1,600 | 1,860 | 1,880 | 1,960 | 1,890 |
| *Total Metals* | *10,820* | *13,830* | *15,510* | *16,550* | *18,910* | *20,180* | *20,880* | *21,100* | *20,910* |
| Plastics | 390 | 2,900 | 6,830 | 17,130 | 25,540 | 29,260 | 30,750 | 30,060 | 29,830 |
| Rubber and Leather | 1,840 | 2,970 | 4,200 | 5,790 | 6,710 | 7,360 | 7,540 | 7,630 | 7,490 |
| Textiles | 1,760 | 2,040 | 2,530 | 5,810 | 9,440 | 11,380 | 11,940 | 12,430 | 12,730 |
| Wood | 3,030 | 3,720 | 7,010 | 12,210 | 13,600 | 14,790 | 15,280 | 15,540 | 15,840 |
| Other ** | 70 | 770 | 2,520 | 3,190 | 4,000 | 4,280 | 4,550 | 4,670 | 4,640 |
| ***Total Materials in Products*** | 54,620 | 83,280 | 108,890 | 146,510 | 178,700 | 184,630 | 185,990 | 181,000 | 171,650 |
| **Other Wastes** | | | | | | | | | |
| Food Scraps | 12,200 | 12,800 | 13,000 | 23,860 | 29,810 | 31,990 | 32,610 | 33,340 | 34,290 |
| Yard Trimmings | 20,000 | 23,200 | 27,500 | 35,000 | 30,530 | 32,070 | 32,630 | 32,900 | 33,200 |
| Miscellaneous Inorganic Wastes | 1,300 | 1,780 | 2,250 | 2,900 | 3,500 | 3,690 | 3,750 | 3,780 | 3,820 |
| ***Total Other Wastes*** | 33,500 | 37,780 | 42,750 | 61,760 | 63,840 | 67,750 | 68,990 | 70,020 | 71,310 |
| ***Total MSW Generated - Weight*** | 88,120 | 121,060 | 151,640 | 208,270 | 242,540 | 252,380 | 254,980 | 251,020 | 242,960 |
| | **Percent of Total Generation** | | | | | | | | |
| **Materials** | **1960** | **1970** | **1980** | **1990** | **2000** | **2005** | **2007** | **2008** | **2009** |
| Paper and Paperboard | 34.0% | 36.6% | 36.4% | 34.9% | 36.2% | 33.6% | 32.4% | 30.8% | 28.2% |
| Glass | 7.6% | 10.5% | 10.0% | 6.3% | 5.3% | 5.0% | 4.9% | 4.8% | 4.8% |
| Metals | | | | | | | | | |
| Ferrous | 11.7% | 10.2% | 8.3% | 6.1% | 5.8% | 5.9% | 6.1% | 6.3% | 6.4% |
| Aluminum | 0.4% | 0.7% | 1.1% | 1.3% | 1.3% | 1.3% | 1.3% | 1.4% | 1.4% |
| Other Nonferrous | 0.2% | 0.6% | 0.8% | 0.5% | 0.7% | 0.7% | 0.7% | 0.8% | 0.8% |
| *Total Metals* | *12.3%* | *11.4%* | *10.2%* | *7.9%* | *7.8%* | *8.0%* | *8.2%* | *8.4%* | *8.6%* |
| Plastics | 0.4% | 2.4% | 4.5% | 8.2% | 10.5% | 11.6% | 12.1% | 12.0% | 12.3% |
| Rubber and Leather | 2.1% | 2.5% | 2.8% | 2.8% | 2.8% | 2.9% | 3.0% | 3.0% | 3.1% |
| Textiles | 2.0% | 1.7% | 1.7% | 2.8% | 3.9% | 4.5% | 4.7% | 5.0% | 5.2% |
| Wood | 3.4% | 3.1% | 4.6% | 5.9% | 5.6% | 5.9% | 6.0% | 6.2% | 6.5% |
| Other ** | 0.1% | 0.6% | 1.7% | 1.5% | 1.6% | 1.7% | 1.8% | 1.9% | 1.9% |
| ***Total Materials in Products*** | 62.0% | 68.8% | 71.8% | 70.3% | 73.7% | 73.2% | 72.9% | 72.1% | 70.6% |
| **Other Wastes** | | | | | | | | | |
| Food Scraps | 13.8% | 10.6% | 8.6% | 11.5% | 12.3% | 12.7% | 12.8% | 13.3% | 14.1% |
| Yard Trimmings | 22.7% | 19.2% | 18.1% | 16.8% | 12.6% | 12.7% | 12.8% | 13.1% | 13.7% |
| Miscellaneous Inorganic Wastes | 1.5% | 1.5% | 1.5% | 1.4% | 1.4% | 1.5% | 1.5% | 1.5% | 1.6% |
| ***Total Other Wastes*** | 38.0% | 31.2% | 28.2% | 29.7% | 26.3% | 26.8% | 27.1% | 27.9% | 29.4% |
| ***Total MSW Generated - %*** | 100.0% | 100.0% | 100.0% | 100.0% | 100.0% | 100.0% | 100.0% | 100.0% | 100.0% |

* Generation before materials recovery or combustion. Does not include construction & demolition debris, industrial process wastes, or certain other wastes.
** Includes electrolytes in batteries and fluff pulp, feces, and urine in disposable diapers.
Details may not add to totals due to rounding.
Source: Franklin Associates, A Division of ERG

**RECOVERY* OF MUNICIPAL SOLID WASTE, 1960 TO 2009**
**(In thousands of tons and percent of generation of each material)**

| | Thousands of Tons | | | | | | | | |
|---|---|---|---|---|---|---|---|---|---|
| **Materials** | **1960** | **1970** | **1980** | **1990** | **2000** | **2005** | **2007** | **2008** | **2009** |
| Paper and Paperboard | 5,080 | 6,770 | 11,740 | 20,230 | 37,560 | 41,960 | 44,480 | 42,940 | 42,500 |
| Glass | 100 | 160 | 750 | 2,630 | 2,880 | 2,590 | 2,880 | 2,810 | 3,000 |
| Metals | | | | | | | | | |
| Ferrous | 50 | 150 | 370 | 2,230 | 4,680 | 5,030 | 5,280 | 5,310 | 5,230 |
| Aluminum | Neg. | 10 | 310 | 1,010 | 860 | 690 | 730 | 720 | 690 |
| Other Nonferrous | Neg. | 320 | 540 | 730 | 1,060 | 1,280 | 1,300 | 1,360 | 1,300 |
| *Total Metals* | *50* | *480* | *1,220* | *3,970* | *6,600* | *7,000* | *7,310* | *7,390* | *7,220* |
| Plastics | Neg. | Neg. | 20 | 370 | 1,480 | 1,770 | 2,100 | 2,130 | 2,120 |
| Rubber and Leather | 330 | 250 | 130 | 370 | 820 | 1,100 | 1,140 | 1,140 | 1,070 |
| Textiles | 50 | 60 | 160 | 660 | 1,320 | 1,850 | 1,920 | 1,910 | 1,900 |
| Wood | Neg. | Neg. | Neg. | 130 | 1,370 | 1,830 | 2,020 | 2,130 | 2,230 |
| Other ** | Neg. | 300 | 500 | 680 | 980 | 1,210 | 1,240 | 1,300 | 1,230 |
| ***Total Materials in Products*** | 5,610 | 8,020 | 14,520 | 29,040 | 53,010 | 59,310 | 63,090 | 61,750 | 61,270 |
| **Other Wastes** | | | | | | | | | |
| Food Scraps | Neg. | Neg. | Neg. | Neg. | 680 | 690 | 810 | 800 | 850 |
| Yard Trimmings | Neg. | Neg. | Neg. | 4,200 | 15,770 | 19,860 | 20,900 | 21,300 | 19,900 |
| Miscellaneous Inorganic Wastes | Neg. | Neg. | Neg. | Neg. | Neg. | Neg. | Neg. | Neg. | Neg. |
| ***Total Other Wastes*** | Neg. | Neg. | Neg. | 4,200 | 16,450 | 20,550 | 21,710 | 22,100 | 20,750 |
| ***Total MSW Recovered - Weight*** | 5,610 | 8,020 | 14,520 | 33,240 | 69,460 | 79,860 | 84,800 | 83,850 | 82,020 |
| | **Percent of Generation of Each Material** | | | | | | | | |
| **Materials** | **1960** | **1970** | **1980** | **1990** | **2000** | **2005** | **2007** | **2008** | **2009** |
| Paper and Paperboard | 16.9% | 15.3% | 21.3% | 27.8% | 42.8% | 49.5% | 53.9% | 55.5% | 62.1% |
| Glass | 1.5% | 1.3% | 5.0% | 20.1% | 22.6% | 20.7% | 23.0% | 23.1% | 25.5% |
| Metals | | | | | | | | | |
| Ferrous | 0.5% | 1.2% | 2.9% | 17.6% | 33.2% | 33.6% | 33.8% | 33.8% | 33.5% |
| Aluminum | Neg. | 1.3% | 17.9% | 35.9% | 26.9% | 20.7% | 21.7% | 21.1% | 20.3% |
| Other Nonferrous | Neg. | 47.8% | 46.6% | 66.4% | 66.3% | 68.8% | 69.1% | 69.4% | 68.8% |
| *Total Metals* | *0.5%* | *3.5%* | *7.9%* | *24.0%* | *34.9%* | *34.7%* | *35.0%* | *35.0%* | *34.5%* |
| Plastics | Neg. | Neg. | 0.3% | 2.2% | 5.8% | 6.0% | 6.8% | 7.1% | 7.1% |
| Rubber and Leather | 17.9% | 8.4% | 3.1% | 6.4% | 12.2% | 14.9% | 15.1% | 14.9% | 14.3% |
| Textiles | 2.8% | 2.9% | 6.3% | 11.4% | 14.0% | 16.3% | 16.1% | 15.4% | 14.9% |
| Wood | Neg. | Neg. | Neg. | 1.1% | 10.1% | 12.4% | 13.2% | 13.7% | 14.1% |
| Other ** | Neg. | 39.0% | 19.8% | 21.3% | 24.5% | 28.3% | 27.3% | 27.8% | 26.5% |
| ***Total Materials in Products*** | 10.3% | 9.6% | 13.3% | 19.8% | 29.7% | 32.1% | 33.9% | 34.1% | 35.7% |
| **Other Wastes** | | | | | | | | | |
| Food, Other^ | Neg. | Neg. | Neg. | Neg. | 2.3% | 2.2% | 2.5% | 2.4% | 2.5% |
| Yard Trimmings | Neg. | Neg. | Neg. | 12.0% | 51.7% | 61.9% | 64.1% | 64.7% | 59.9% |
| Miscellaneous Inorganic Wastes | Neg. | Neg. | Neg. | Neg. | Neg. | Neg. | Neg. | Neg. | Neg. |
| ***Total Other Wastes*** | Neg. | Neg. | Neg. | 6.8% | 25.8% | 30.3% | 31.5% | 31.6% | 29.1% |
| ***Total MSW Recovered - %*** | 6.4% | 6.6% | 9.6% | 16.0% | 28.6% | 31.6% | 33.3% | 33.4% | 33.8% |

* Recovery of postconsumer wastes; does not include converting/fabrication scrap.
** Recovery of electrolytes in batteries; probably not recycled.
Neg. = Less than 5,000 tons or 0.05 percent.
^ Includes recovery of paper and mixed MSW for composting.
Details may not add to totals due to rounding.
Source: Franklin Associates, A Division of ERG

**MATERIALS DISCARDED* IN THE MUNICIPAL WASTE STREAM, 1960 TO 2009**
**(In thousands of tons and percent of total discards)**

| | Thousands of Tons | | | | | | | | |
|---|---|---|---|---|---|---|---|---|---|
| **Materials** | **1960** | **1970** | **1980** | **1990** | **2000** | **2005** | **2007** | **2008** | **2009** |
| Paper and Paperboard | 24,910 | 37,540 | 43,420 | 52,500 | 50,180 | 42,880 | 38,050 | 34,480 | 25,930 |
| Glass | 6,620 | 12,580 | 14,380 | 10,470 | 9,880 | 9,950 | 9,640 | 9,340 | 8,780 |
| Metals | | | | | | | | | |
| Ferrous | 10,250 | 12,210 | 12,250 | 10,410 | 9,430 | 9,960 | 10,360 | 10,420 | 10,390 |
| Aluminum | 340 | 790 | 1,420 | 1,800 | 2,340 | 2,640 | 2,630 | 2,690 | 2,710 |
| Other Nonferrous | 180 | 350 | 620 | 370 | 540 | 580 | 580 | 600 | 590 |
| *Total Metals* | *10,770* | *13,350* | *14,290* | *12,580* | *12,310* | *13,180* | *13,570* | *13,710* | *13,690* |
| Plastics | 390 | 2,900 | 6,810 | 16,760 | 24,060 | 27,490 | 28,650 | 27,930 | 27,710 |
| Rubber and Leather | 1,510 | 2,720 | 4,070 | 5,420 | 5,890 | 6,260 | 6,400 | 6,490 | 6,420 |
| Textiles | 1,710 | 1,980 | 2,370 | 5,150 | 8,120 | 9,530 | 10,020 | 10,520 | 10,830 |
| Wood | 3,030 | 3,720 | 7,010 | 12,080 | 12,230 | 12,960 | 13,260 | 13,410 | 13,610 |
| Other ** | 70 | 470 | 2,020 | 2,510 | 3,020 | 3,070 | 3,310 | 3,370 | 3,410 |
| ***Total Materials in Products*** | 49,010 | 75,260 | 94,370 | 117,470 | 125,690 | 125,320 | 122,900 | 119,250 | 110,380 |
| **Other Wastes** | | | | | | | | | |
| Food Scraps | 12,200 | 12,800 | 13,000 | 23,860 | 29,130 | 31,300 | 31,800 | 32,540 | 33,440 |
| Yard Trimmings | 20,000 | 23,200 | 27,500 | 30,800 | 14,760 | 12,210 | 11,730 | 11,600 | 13,300 |
| Miscellaneous Inorganic Wastes | 1,300 | 1,780 | 2,250 | 2,900 | 3,500 | 3,690 | 3,750 | 3,780 | 3,820 |
| ***Total Other Wastes*** | 33,500 | 37,780 | 42,750 | 57,560 | 47,390 | 47,200 | 47,280 | 47,920 | 50,560 |
| ***Total MSW Discarded - Weight*** | 82,510 | 113,040 | 137,120 | 175,030 | 173,080 | 172,520 | 170,180 | 167,170 | 160,940 |
| | Percent of Total Discards | | | | | | | | |
| **Materials** | **1960** | **1970** | **1980** | **1990** | **2000** | **2005** | **2007** | **2008** | **2009** |
| Paper and Paperboard | 30.2% | 33.2% | 31.7% | 30.0% | 29.0% | 24.9% | 22.4% | 20.6% | 16.1% |
| Glass | 8.0% | 11.1% | 10.5% | 6.0% | 5.7% | 5.8% | 5.7% | 5.6% | 5.5% |
| Metals | | | | | | | | | |
| Ferrous | 12.4% | 10.8% | 8.9% | 5.9% | 5.4% | 5.8% | 6.1% | 6.2% | 6.5% |
| Aluminum | 0.4% | 0.7% | 1.0% | 1.0% | 1.4% | 1.5% | 1.5% | 1.6% | 1.7% |
| Other Nonferrous | 0.2% | 0.3% | 0.5% | 0.2% | 0.3% | 0.3% | 0.3% | 0.4% | 0.4% |
| *Total Metals* | *13.1%* | *11.8%* | *10.4%* | *7.2%* | *7.1%* | *7.6%* | *8.0%* | *8.2%* | *8.5%* |
| Plastics | 0.5% | 2.6% | 5.0% | 9.6% | 13.9% | 15.9% | 16.8% | 16.7% | 17.2% |
| Rubber and Leather | 1.8% | 2.4% | 3.0% | 3.1% | 3.4% | 3.6% | 3.8% | 3.9% | 4.0% |
| Textiles | 2.1% | 1.8% | 1.7% | 2.9% | 4.7% | 5.5% | 5.9% | 6.3% | 6.7% |
| Wood | 3.7% | 3.3% | 5.1% | 6.9% | 7.1% | 7.5% | 7.8% | 8.0% | 8.5% |
| Other ** | 0.1% | 0.4% | 1.5% | 1.4% | 1.7% | 1.8% | 1.9% | 2.0% | 2.1% |
| ***Total Materials in Products*** | 59.4% | 66.6% | 68.8% | 67.1% | 72.6% | 72.6% | 72.2% | 71.3% | 68.6% |
| **Other Wastes** | | | | | | | | | |
| Food Scraps | 14.8% | 11.3% | 9.5% | 13.6% | 16.8% | 18.1% | 18.7% | 19.5% | 20.8% |
| Yard Trimmings | 24.2% | 20.5% | 20.1% | 17.6% | 8.5% | 7.1% | 6.9% | 6.9% | 8.3% |
| Miscellaneous Inorganic Wastes | 1.6% | 1.6% | 1.6% | 1.7% | 2.0% | 2.1% | 2.2% | 2.3% | 2.4% |
| ***Total Other Wastes*** | 40.6% | 33.4% | 31.2% | 32.9% | 27.4% | 27.4% | 27.8% | 28.7% | 31.4% |
| ***Total MSW Discarded - %*** | 100.0% | 100.0% | 100.0% | 100.0% | 100.0% | 100.0% | 100.0% | 100.0% | 100.0% |

* Discards after materials and compost recovery. In this table, discards include combustion with energy recovery. Does not include construction & demolition debris, industrial process wastes, or certain other wastes.

** Includes electrolytes in batteries and fluff pulp, feces, and urine in disposable diapers.
Details may not add to totals due to rounding.
Source: Franklin Associates, A Division of ERG

## PAPER AND PAPERBOARD PRODUCTS IN MSW, 2009
### (In thousands of tons and percent of generation)

| Product Category | Generation (Thousand tons) | Recovery† (Thousand tons) | Recovery† (Percent of generation) | Discards (Thousand tons) |
|---|---|---|---|---|
| **Nondurable Goods** | | | | |
| Newspapers | | | | |
| Newsprint | 5,060 | 4,490 | 88.7% | 570 |
| Groundwood Inserts | 2,700 | 2,350 | 87.0% | 350 |
| ***Total Newspapers*** | 7,760 | 6,840 | 88.1% | 920 |
| Books | 960 | 320 | 33.3% | 640 |
| Magazines | 1,450 | 780 | 53.8% | 670 |
| Office-type Papers* | 5,380 | 3,990 | 74.2% | 1,390 |
| Telephone Directories | 650 | 240 | 36.9% | 410 |
| Standard Mail** | 4,650 | 2,950 | 63.4% | 1,700 |
| Other Commercial Printing | 3,490 | 2,310 | 66.2% | 1,180 |
| Tissue Paper and Towels | 3,490 | Neg. | Neg. | 3,490 |
| Paper Plates and Cups | 1,170 | Neg. | Neg. | 1,170 |
| Other Nonpackaging Paper*** | 4,480 | Neg. | Neg. | 4,480 |
| ***Total Paper and Paperboard Nondurable Goods*** | 33,480 | 17,430 | 52.1% | 16,050 |
| **Containers and Packaging** | | | | |
| Corrugated Boxes | 27,190 | 22,100 | 81.3% | 5,090 |
| Gable Top/Aseptic Cartons‡ | 460 | 30 | 6.5% | 430 |
| Folding Cartons | 4,980 | 2,490 | 50.0% | 2,490 |
| Other Paperboard Packaging | 90 | Neg. | Neg. | 90 |
| Bags and Sacks | 910 | 450 | 49.5% | 460 |
| Other Paper Packaging | 1,310 | Neg. | Neg. | 1,310 |
| ***Total Paper and Paperboard Containers and Packaging*** | 34,940 | 25,070 | 71.8% | 9,870 |
| ***Total Paper and Paperboard^*** | 68,420 | 42,500 | 62.1% | 25,920 |

† Since 2008, recycling rates increased due to generation going down and applying default values to increased single stream recovered mixed paper products.

* High-grade papers such as copy paper and printer paper; both residential and commercial.

** Formerly called Third Class Mail by the U.S. Postal Service.

*** Includes tissue in disposable diapers, paper in games and novelties, cards, etc.

‡ Includes milk, juice, and other products packaged in gable top cartons and liquid food aseptic cartons

^ Table 4 does not include 10,000 tons of paper used in durable goods (Table 1).

Neg. = Less than 5,000 tons or 0.05 percent.

Details may not add to totals due to rounding.

Source: Franklin Associates, A Division of ERG

**GLASS PRODUCTS IN MSW, 2009**
**(In thousands of tons and percent of generation)**

| Product Category | Generation (Thousand tons) | Recovery (Thousand tons) | Recovery (Percent of generation) | Discards (Thousand tons) |
|---|---|---|---|---|
| **Durable Goods*** | 2,120 | Neg. | Neg. | 2,120 |
| **Containers and Packaging** | | | | |
| Beer and Soft Drink Bottles** | 6,000 | 2,340 | 39.0% | 3,660 |
| Wine and Liquor Bottles | 1,710 | 310 | 18.1% | 1,400 |
| Other Bottles and Jars | 1,950 | 350 | 17.9% | 1,600 |
| ***Total Glass Containers*** | 9,660 | 3,000 | 31.1% | 6,660 |
| ***Total Glass*** | 11,780 | 3,000 | 25.5% | 8,780 |

* Glass as a component of appliances, furniture, consumer electronics, etc.

** Includes carbonated drinks and non-carbonated water, teas, flavored drinks, and ready-to-drink alcoholic coolers and cocktails.

Neg. = Less than 5,000 tons or 0.05 percent.

Details may not add to totals due to rounding.

Source: Franklin Associates, A Division of ERG

## METAL PRODUCTS IN MSW, 2009
(In thousands of tons and percent of generation)

| Product Category | Generation (Thousand tons) | Recovery (Thousand tons) | Recovery (Percent of generation) | Discards (Thousand tons) |
|---|---|---|---|---|
| **Durable Goods** | | | | |
| Ferrous Metals* | 13,340 | 3,720 | 27.9% | 9,620 |
| Aluminum** | 1,350 | Neg. | Neg. | 1,350 |
| Lead† | 1,350 | 1,300 | 96.3% | 50 |
| Other Nonferrous Metals‡ | 540 | Neg. | Neg. | 540 |
| ***Total Metals in Durable Goods*** | 16,580 | 5,020 | 30.3% | 11,560 |
| **Nondurable Goods** | | | | |
| Aluminum | 210 | Neg. | Neg. | 210 |
| **Containers and Packaging** | | | | |
| **Steel** | | | | |
| Cans | 1,940 | 1,280 | 66.0% | 660 |
| Other Steel Packaging | 340 | 230 | 67.6% | 110 |
| ***Total Steel Packaging*** | 2,280 | 1,510 | 66.2% | 770 |
| **Aluminum** | | | | |
| Beer and Soft Drink Cans | 1,360 | 690 | 50.7% | 670 |
| Other Cans | 70 | NA | | 70 |
| Foil and Closures | 410 | NA | | 410 |
| ***Total Aluminum Packaging*** | 1,840 | 690 | 37.5% | 1,150 |
| ***Total Metals in Containers and Packaging*** | 4,120 | 2,200 | 53.4% | 1,920 |
| ***Total Metals*** | 20,910 | 7,220 | 34.5% | 13,690 |
| Ferrous | 15,620 | 5,230 | 33.5% | 10,390 |
| Aluminum | 3,400 | 690 | 20.3% | 2,710 |
| Other nonferrous | 1,890 | 1,300 | 68.8% | 590 |

* Ferrous metals (iron and steel) in appliances, furniture, tires, and miscellaneous durables.
** Aluminum in appliances, furniture, and miscellaneous durables.
† Lead in lead-acid batteries.
‡ Other nonferrous metals in appliances and miscellaneous durables.
Neg. = Less than 5,000 tons or 0.05 percent. NA = Not Available
Details may not add to totals due to rounding.
Source: Franklin Associates, A Division of ERG

**PLASTICS IN PRODUCTS IN MSW, 2009**
**(In thousands of tons, and percent of generation by resin)**

| Product Category | Generation (Thousand tons) | Recovery (Thousand tons) | Recovery (Percent of Gen.) | Discards (Thousand tons) |
|---|---|---|---|---|
| **Durable Goods** | | | | |
| PET | 410 | | | |
| HDPE | 1,190 | | | |
| PVC | 360 | | | |
| LDPE/LLDPE | 900 | | | |
| PP | 2,630 | | | |
| PS | 710 | | | |
| Other resins | 4,450 | | | |
| ***Total Plastics in Durable Goods*** | **10,650** | **400** | **3.8%** | **10,250** |
| **Nondurable Goods** | | | | |
| Plastic Plates and Cups | | | | |
| LDPE/LLDPE | 20 | | | 20 |
| PP | 170 | | | 170 |
| PS | 710 | | | 710 |
| ***Subtotal Plastic Plates and Cups*** | 900 | Neg. | | 900 |
| Trash Bags | | | | |
| HDPE | 230 | | | 230 |
| LDPE/LLDPE | 770 | | | 770 |
| ***Subtotal Trash Bags*** | 1,000 | | | 1,000 |
| All other nondurables* | | | | |
| PET | 400 | | | 400 |
| HDPE | 410 | | | 410 |
| PVC | 330 | | | 330 |
| LDPE/LLDPE | 1,390 | | | 1,390 |
| PP | 890 | | | 890 |
| PS | 580 | | | 580 |
| Other resins | 750 | | | 750 |
| ***Subtotal All Other Nondurables*** | 4,750 | | | 4,750 |
| **Total Plastics in Nondurable Goods, by resin** | | | | |
| PET | 400 | | | 400 |
| HDPE | 640 | | | 640 |
| PVC | 330 | | | 330 |
| LDPE/LLDPE | 2,180 | | | 2,180 |
| PP | 1,060 | | | 1,060 |
| PS | 1,290 | | | 1,290 |
| Other resins | 750 | | | 750 |
| ***Total Plastics in Nondurable Goods*** | **6,650** | **Neg.** | **Neg.** | **6,650** |
| **Plastic Containers & Packaging** | | | | |
| Bottles and Jars** | | | | |
| PET | 2,570 | 720 | 28.0% | 1,850 |
| Natural Bottles† | | | | |
| HDPE | 760 | 220 | 28.9% | 540 |

HDPE = High density polyethylene
LDPE = Low density polyethylene
LLDPE = Linear low density polyethylene
PET = Polyethylene terephthalate PS = Polystyrene
PP = Polypropylene PVC = Polyvinyl chloride
Neg. = negligible, less than 5,000 tons or 0.05 percent

* All other nondurables include plastics in disposable diapers, clothing, footwear, etc.

** Injection stretch blow molded PET containers as defined in the 2008 Report on Postconsumer PET Container Recycling Activity Final Report. National Association for PET Container Resources.

† White translucent homopolymer bottles as defined in the 2007 United States National Postconsumer Plastics Bottles Recycling Report. American Chemistry Council and the Association of Postconsumer Plastic Recyclers.

Source: Franklin Associates, A Division of ERG

**PLASTICS IN PRODUCTS IN MSW, 2009**
**(In thousands of tons, and percent of generation by resin)**

| Product Category | Generation (Thousand tons) | Recovery (Thousand tons) | Recovery (Percent of Gen.) | Discards (Thousand tons) |
|---|---|---|---|---|
| **Plastic Containers & Packaging, cont.** | | | | |
| Other plastic containers | | | | |
| HDPE | 1,340 | 270 | 20.1% | 1,070 |
| PVC | 30 | Neg. | | 30 |
| LDPE/LLDPE | 40 | Neg. | | 40 |
| PP | 270 | 20 | 7.4% | 250 |
| PS | 70 | Neg. | | 70 |
| ***Subtotal Other Containers*** | **1,750** | **290** | **16.6%** | **1,460** |
| Bags, sacks, & wraps | | | | |
| HDPE | 660 | 40 | 6.1% | 620 |
| PVC | 60 | | | 60 |
| LDPE/LLDPE | 2,380 | 320 | 13.4% | 2,060 |
| PP | 640 | | | 640 |
| PS | 110 | | | 110 |
| ***Subtotal Bags, Sacks, & Wraps*** | **3,850** | **360** | **9.4%** | **3,490** |
| Other Plastics Packaging‡ | | | | |
| PET | 150 | 10 | 6.7% | 140 |
| HDPE | 620 | 60 | 9.7% | 560 |
| PVC | 340 | Neg. | | 340 |
| LDPE/LLDPE | 800 | Neg. | | 800 |
| PP | 930 | 30 | 3.2% | 900 |
| PS | 290 | 20 | 6.9% | 270 |
| Other resins | 470 | 10 | 2.1% | 460 |
| ***Subtotal Other Packaging*** | **3,600** | **130** | **3.6%** | **3,470** |
| **Total Plastics in Containers & Packaging, by resin** | | | | |
| PET | 2,720 | 730 | 26.8% | 1,990 |
| HDPE | 3,380 | 590 | 17.5% | 2,790 |
| PVC | 430 | | | 430 |
| LDPE/LLDPE | 3,220 | 320 | 9.9% | 2,900 |
| PP | 1,840 | 50 | 2.7% | 1,790 |
| PS | 470 | 20 | 4.3% | 450 |
| Other resins | 470 | 10 | 2.1% | 460 |
| ***Total Plastics in Cont. & Packaging*** | **12,530** | **1,720** | **13.7%** | **10,810** |
| **Total Plastics in MSW, by resin** | | | | |
| PET | 3,530 | 730 | 20.7% | 2,800 |
| HDPE | 5,210 | 590 | 11.3% | 4,620 |
| PVC | 1,120 | | | 1,120 |
| LDPE/LLDPE | 6,300 | 320 | 5.1% | 5,980 |
| PP | 5,530 | 50 | 0.9% | 5,480 |
| PS | 2,470 | 20 | 0.8% | 2,450 |
| Other resins | 5,670 | 410 | 7.2% | 5,260 |
| ***Total Plastics in MSW*** | **29,830** | **2,120** | **7.1%** | **27,710** |

HDPE = High density polyethylene
LDPE = Low density polyethylene
LLDPE = Linear low density polyethylene
PET = Polyethylene terephthalate PS = Polystyrene
PP = Polypropylene PVC = Polyvinyl chloride
NA = Not Available

‡ Other plastic packaging includes coatings, closures, lids, caps, clamshells, egg cartons, produce baskets, trays, shapes, loose fill, etc.

Some detail of recovery by resin omitted due to lack of data.

Source: Franklin Associates, A Division of ERG

## RUBBER AND LEATHER PRODUCTS IN MSW, 2009
### (In thousands of tons and percent of generation)

| Product Category | Generation (Thousand tons) | Recovery (Thousand tons) | Recovery (Percent of generation) | Discards (Thousand tons) |
|---|---|---|---|---|
| **Durable Goods** | | | | |
| Rubber in Tires* | 3,040 | 1,070 | 35.2% | 1,970 |
| Other Durables** | 3,390 | Neg. | Neg. | 3,390 |
| ***Total Rubber & Leather Durable Goods*** | 6,430 | 1,070 | 16.6% | 5,360 |
| **Nondurable Goods** | | | | |
| Clothing and Footwear | 790 | Neg. | Neg. | 790 |
| Other Nondurables | 270 | Neg. | Neg. | 270 |
| ***Total Rubber & Leather Nondurable Goods*** | 1,060 | Neg. | Neg. | 1,060 |
| ***Total Rubber & Leather*** | 7,490 | 1,070 | 14.3% | 6,420 |

* Automobile and truck tires. Does not include other materials in tires.
** Includes carpets and rugs and other miscellaneous durables.
Neg. = Less than 5,000 tons or 0.05 percent.
Details may not add to totals due to rounding.

Source: Franklin Associates, A Division of ERG

*SOURCE: U.S. EPA, Municipal Solid Waste in the United States: 2009 Facts and Figures*

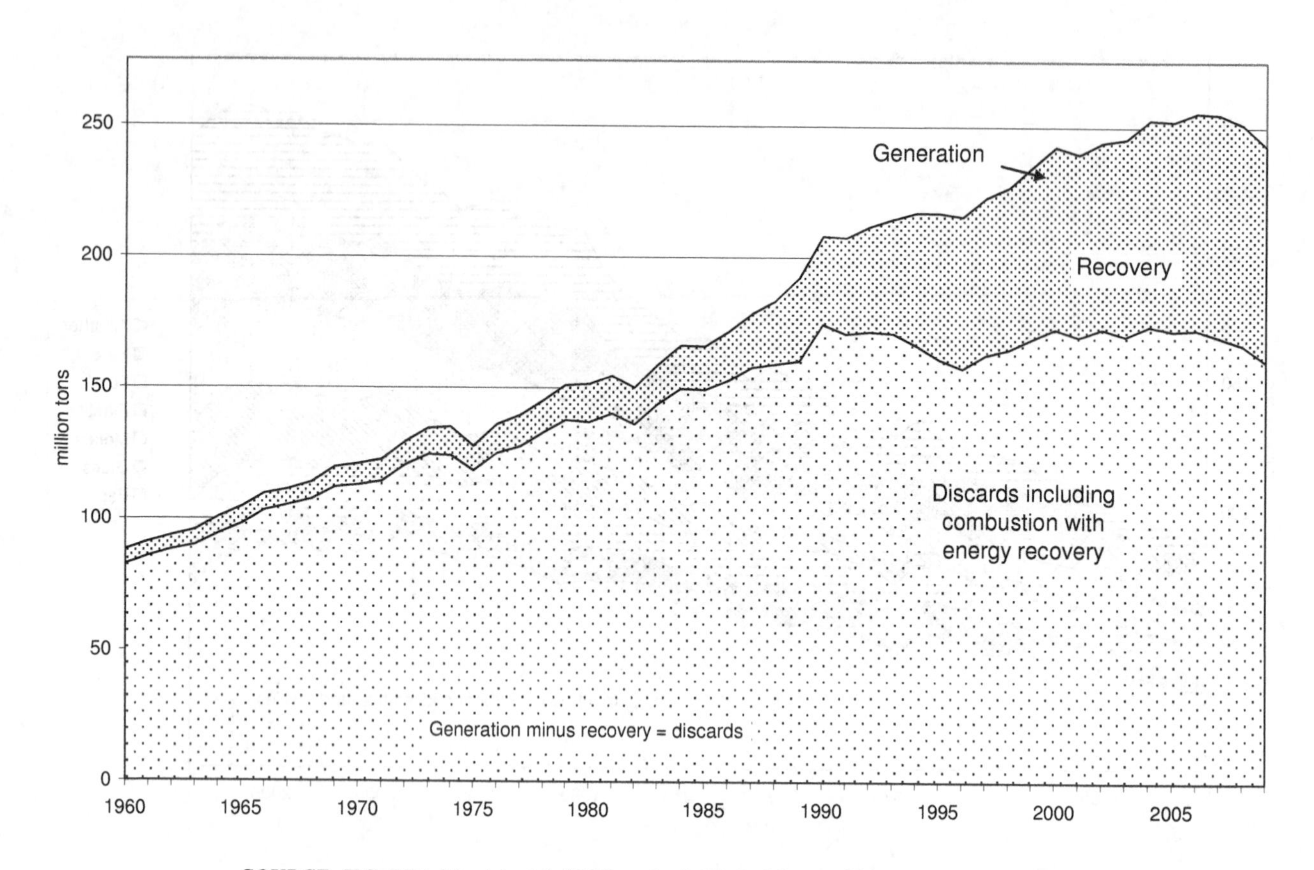


*SOURCE: U.S. EPA, Municipal Solid Waste in the United States: 2009 Facts and Figures*

**PRODUCTS GENERATED* IN THE MUNICIPAL WASTE STREAM, 1960 TO 2009**
**(WITH DETAIL ON DURABLE GOODS)**
**(In thousands of tons and percent of total generation)**

| | Thousands of Tons | | | | | | | | |
|---|---|---|---|---|---|---|---|---|---|
| **Products** | **1960** | **1970** | **1980** | **1990** | **2000** | **2005** | **2007** | **2008** | **2009** |
| **Durable Goods** | | | | | | | | | |
| Major Appliances | 1,630 | 2,170 | 2,950 | 3,310 | 3,640 | 3,610 | 3,620 | 3,690 | 3,760 |
| Small Appliances** | | | | 460 | 1,040 | 1,180 | 1,390 | 1,530 | 1,630 |
| Furniture and Furnishings | 2,150 | 2,830 | 4,760 | 6,790 | 7,990 | 8,870 | 9,340 | 9,610 | 9,870 |
| Carpets and Rugs** | | | | 1,660 | 2,570 | 2,980 | 3,140 | 3,220 | 3,450 |
| Rubber Tires | 1,120 | 1,890 | 2,720 | 3,610 | 4,930 | 4,960 | 5,000 | 5,020 | 4,730 |
| Batteries, Lead-Acid | Neg. | 820 | 1,490 | 1,510 | 2,280 | 2,740 | 2,800 | 2,930 | 2,800 |
| Miscellaneous Durables | | | | | | | | | |
| Selected Consumer Electronics*** | | | | | 1,900 | 2,630 | 3,010 | 3,160 | 3,190 |
| Other Miscellaneous Durables | | | | | 14,500 | 17,680 | 17,470 | 17,240 | 17,210 |
| *Total Miscellaneous Durables* | 5,020 | 6,950 | 9,880 | 12,470 | 16,400 | 20,310 | 20,480 | 20,400 | 20,400 |
| ***Total Durable Goods*** | 9,920 | 14,660 | 21,800 | 29,810 | 38,850 | 44,650 | 45,770 | 46,400 | 46,640 |
| **Nondurable Goods** | 17,330 | 25,060 | 34,420 | 52,170 | 64,010 | 63,650 | 61,760 | 58,690 | 53,440 |
| *(Detail in Table 15)* | | | | | | | | | |
| **Containers and Packaging** | 27,370 | 43,560 | 52,670 | 64,530 | 75,840 | 76,330 | 78,460 | 75,910 | 71,570 |
| *(Detail in Table 18)* | | | | | | | | | |
| ***Total Product Wastes†*** | 54,620 | 83,280 | 108,890 | 146,510 | 178,700 | 184,630 | 185,990 | 181,000 | 171,650 |
| **Other Wastes** | | | | | | | | | |
| Food Scraps | 12,200 | 12,800 | 13,000 | 23,860 | 29,810 | 31,990 | 32,610 | 33,340 | 34,290 |
| Yard Trimmings | 20,000 | 23,200 | 27,500 | 35,000 | 30,530 | 32,070 | 32,630 | 32,900 | 33,200 |
| Miscellaneous Inorganic Wastes | 1,300 | 1,780 | 2,250 | 2,900 | 3,500 | 3,690 | 3,750 | 3,780 | 3,820 |
| ***Total Other Wastes*** | 33,500 | 37,780 | 42,750 | 61,760 | 63,840 | 67,750 | 68,990 | 70,020 | 71,310 |
| ***Total MSW Generated - Weight*** | 88,120 | 121,060 | 151,640 | 208,270 | 242,540 | 252,380 | 254,980 | 251,020 | 242,960 |
| | **Percent of Total Generation** | | | | | | | | |
| **Products** | **1960** | **1970** | **1980** | **1990** | **2000** | **2005** | **2007** | **2008** | **2009** |
| **Durable Goods** | | | | | | | | | |
| Major Appliances | 1.8% | 1.8% | 1.9% | 1.6% | 1.5% | 1.4% | 1.4% | 1.5% | 1.5% |
| Small Appliances** | | | | 0.2% | 0.4% | 0.5% | 0.5% | 0.6% | 0.7% |
| Furniture and Furnishings | 2.4% | 2.3% | 3.1% | 3.3% | 3.3% | 3.5% | 3.7% | 3.8% | 4.1% |
| Carpets and Rugs** | | | | 0.8% | 1.1% | 1.2% | 1.2% | 1.3% | 1.4% |
| Rubber Tires | 1.3% | 1.6% | 1.8% | 1.7% | 2.0% | 2.0% | 2.0% | 2.0% | 1.9% |
| Batteries, Lead-Acid | Neg. | 0.7% | 1.0% | 0.7% | 0.9% | 1.1% | 1.1% | 1.2% | 1.2% |
| Miscellaneous Durables | | | | | | | | | |
| Selected Consumer Electronics*** | | | | | 0.8% | 1.0% | 1.2% | 1.3% | 1.3% |
| Other Miscellaneous Durables | | | | | 6.0% | 7.0% | 6.9% | 6.9% | 7.1% |
| *Total Miscellaneous Durables* | 5.7% | 5.7% | 6.5% | 6.0% | 6.8% | 8.0% | 8.0% | 8.1% | 8.4% |
| ***Total Durable Goods*** | 11.3% | 12.1% | 14.4% | 14.3% | 16.0% | 17.7% | 18.0% | 18.5% | 19.2% |
| **Nondurable Goods** | 19.7% | 20.7% | 22.7% | 25.0% | 26.4% | 25.2% | 24.2% | 23.4% | 22.0% |
| *(Detail in Table 15)* | | | | | | | | | |
| **Containers and Packaging** | 31.1% | 36.0% | 34.7% | 31.0% | 31.3% | 30.2% | 30.8% | 30.2% | 29.5% |
| *(Detail in Table 19)* | | | | | | | | | |
| ***Total Product Wastes†*** | 62.0% | 68.8% | 71.8% | 70.3% | 73.7% | 73.2% | 72.9% | 72.1% | 70.6% |
| **Other Wastes** | | | | | | | | | |
| Food Scraps | 13.8% | 10.6% | 8.6% | 11.5% | 12.3% | 12.7% | 12.8% | 13.3% | 14.1% |
| Yard Trimmings | 22.7% | 19.2% | 18.1% | 16.8% | 12.6% | 12.7% | 12.8% | 13.1% | 13.7% |
| Miscellaneous Inorganic Wastes | 1.5% | 1.5% | 1.5% | 1.4% | 1.4% | 1.5% | 1.5% | 1.5% | 1.6% |
| ***Total Other Wastes*** | 38.0% | 31.2% | 28.2% | 29.7% | 26.3% | 26.8% | 27.1% | 27.9% | 29.4% |
| ***Total MSW Generated - %*** | 100.0% | 100.0% | 100.0% | 100.0% | 100.0% | 100.0% | 100.0% | 100.0% | 100.0% |

* Generation before materials recovery or combustion. Does not include construction & demolition debris, industrial process wastes, or certain other wastes. Details may not add to totals due to rounding.

** Not estimated separately prior to 1990.

*** Not estimated separately prior to 1999. Preliminary data; may undergo revision.

† Other than food products.

Neg. = Less than 5,000 tons or 0.05 percent.

Source: Franklin Associates, A Division of ERG

**RECOVERY* OF PRODUCTS IN MUNICIPAL SOLID WASTE, 1960 TO 2009**
**(WITH DETAIL ON DURABLE GOODS)**
**(In thousands of tons and percent of generation of each product)**

| | Thousands of Tons | | | | | | | | |
|---|---|---|---|---|---|---|---|---|---|
| **Products** | **1960** | **1970** | **1980** | **1990** | **2000** | **2005** | **2007** | **2008** | **2009** |
| **Durable Goods** | | | | | | | | | |
| Major Appliances | 10 | 50 | 130 | 1,070 | 2,000 | 2,420 | 2,430 | 2,470 | 2,510 |
| Small Appliances** | | | | 10 | 20 | 20 | 20 | 110 | 110 |
| Furniture and Furnishings | Neg. | Neg. | Neg. | Neg. | Neg. | Neg. | Neg. | 10 | 10 |
| Carpets and Rugs** | | | | Neg. | 190 | 250 | 280 | 270 | 270 |
| Rubber Tires | 330 | 250 | 150 | 440 | 1,290 | 1,720 | 1,770 | 1,780 | 1,670 |
| Batteries, Lead-Acid | Neg. | 620 | 1,040 | 1,470 | 2,130 | 2,630 | 2,690 | 2,810 | 2,680 |
| Miscellaneous Durables | | | | | | | | | |
| Selected Consumer Electronics*** | | | | | 190 | 360 | 550 | 560 | 600 |
| Other Miscellaneous Durables | | | | | 760 | 640 | 480 | 350 | 310 |
| *Total Miscellaneous Durables* | 10 | 20 | 40 | 470 | 950 | 1,000 | 1,030 | 910 | 910 |
| ***Total Durable Goods*** | 350 | 940 | 1,360 | 3,460 | 6,580 | 8,040 | 8,220 | 8,360 | 8,160 |
| **Nondurable Goods** | 2,390 | 3,730 | 4,670 | 8,800 | 17,560 | 19,770 | 20,970 | 19,310 | 18,890 |
| *(Detail in Table 16)* | | | | | | | | | |
| **Containers and Packaging** | 2,870 | 3,350 | 8,490 | 16,780 | 28,870 | 31,500 | 33,900 | 34,080 | 34,220 |
| *(Detail in Table 20)* | | | | | | | | | |
| ***Total Product Wastes†*** | 5,610 | 8,020 | 14,520 | 29,040 | 53,010 | 59,310 | 63,090 | 61,750 | 61,270 |
| **Other Wastes** | | | | | | | | | |
| Food Scraps | Neg. | Neg. | Neg. | Neg. | 680 | 690 | 810 | 800 | 850 |
| Yard Trimmings | Neg. | Neg. | Neg. | 4,200 | 15,770 | 19,860 | 20,900 | 21,300 | 19,900 |
| Miscellaneous Inorganic Wastes | Neg. | Neg. | Neg. | Neg. | Neg. | Neg. | Neg. | Neg. | Neg. |
| ***Total Other Wastes*** | Neg. | Neg. | Neg. | 4,200 | 16,450 | 20,550 | 21,710 | 22,100 | 20,750 |
| ***Total MSW Recovered - Weight*** | 5,610 | 8,020 | 14,520 | 33,240 | 69,460 | 79,860 | 84,800 | 83,850 | 82,020 |
| | Percent of Generation of Each Product | | | | | | | | |
| **Products** | **1960** | **1970** | **1980** | **1990** | **2000** | **2005** | **2007** | **2008** | **2009** |
| **Durable Goods** | | | | | | | | | |
| Major Appliances | 0.6% | 2.3% | 4.4% | 32.3% | 54.9% | 67.0% | 67.1% | 66.9% | 66.8% |
| Small Appliances** | | | | 2.2% | 1.9% | 1.7% | 1.4% | 7.2% | 6.7% |
| Furniture and Furnishings | Neg. | Neg. | Neg. | Neg. | Neg. | Neg. | Neg. | 0.1% | 0.1% |
| Carpets and Rugs** | | | | Neg. | 7.4% | 8.4% | 8.9% | 8.4% | 7.8% |
| Rubber Tires | 29.5% | 13.2% | 5.5% | 12.2% | 26.2% | 34.7% | 35.4% | 35.5% | 35.3% |
| Batteries, Lead-Acid | Neg. | 75.6% | 69.8% | 97.4% | 93.4% | 96.0% | 96.1% | 95.9% | 95.7% |
| Miscellaneous Durables | | | | | | | | | |
| Selected Consumer Electronics*** | | | | | 10.0% | 13.7% | 18.3% | 17.7% | 18.8% |
| Other Miscellaneous Durables | | | | | 5.2% | 3.6% | 2.7% | 2.0% | 1.8% |
| *Total Miscellaneous Durables* | 0.2% | 0.3% | 0.4% | 3.8% | 5.8% | 4.9% | 5.0% | 4.5% | 4.5% |
| ***Total Durable Goods*** | 3.5% | 6.4% | 6.2% | 11.6% | 16.9% | 18.0% | 18.0% | 18.0% | 17.5% |
| **Nondurable Goods** | 13.8% | 14.9% | 13.6% | 16.9% | 27.4% | 31.1% | 34.0% | 32.9% | 35.3% |
| *(Detail in Table 16)* | | | | | | | | | |
| **Containers and Packaging** | 10.5% | 7.7% | 16.1% | 26.0% | 38.1% | 41.3% | 43.2% | 44.9% | 47.8% |
| *(Detail in Table 21)* | | | | | | | | | |
| ***Total Product Wastes†*** | 10.3% | 9.6% | 13.3% | 19.8% | 29.7% | 32.1% | 33.9% | 34.1% | 35.7% |
| **Other Wastes** | | | | | | | | | |
| Food Scraps | Neg. | Neg. | Neg. | Neg. | 2.3% | 2.2% | 2.5% | 2.4% | 2.5% |
| Yard Trimmings | Neg. | Neg. | Neg. | 12.0% | 51.7% | 61.9% | 64.1% | 64.7% | 59.9% |
| Miscellaneous Inorganic Wastes | Neg. | Neg. | Neg. | Neg. | Neg. | Neg. | Neg. | Neg. | Neg. |
| ***Total Other Wastes*** | Neg. | Neg. | Neg. | 6.8% | 25.8% | 30.3% | 31.5% | 31.6% | 29.1% |
| ***Total MSW Recovered - %*** | 6.4% | 6.6% | 9.6% | 16.0% | 28.6% | 31.6% | 33.3% | 33.4% | 33.8% |

* Recovery of postconsumer wastes; does not include converting/fabrication scrap.
** Not estimated separately prior to 1990. *** Not estimated separately prior to 1999.
† Other than food products.
Neg. = Less than 5,000 tons or 0.05 percent. Details may not add to totals due to rounding.
Source: Franklin Associates, A Division of ERG

**PRODUCTS DISCARDED* IN THE MUNICIPAL WASTE STREAM, 1960 TO 2009**
**(WITH DETAIL ON DURABLE GOODS)**
**(In thousands of tons and percent of total discards)**

| | Thousands of Tons | | | | | | | | |
|---|---|---|---|---|---|---|---|---|---|
| **Products** | **1960** | **1970** | **1980** | **1990** | **2000** | **2005** | **2007** | **2008** | **2009** |
| **Durable Goods** | | | | | | | | | |
| Major Appliances | 1,620 | 2,120 | 2,820 | 2,240 | 1,640 | 1,190 | 1,190 | 1,220 | 1,250 |
| Small Appliances** | | | | 450 | 1,020 | 1,160 | 1,370 | 1,420 | 1,520 |
| Furniture and Furnishings | 2,150 | 2,830 | 4,760 | 6,790 | 7,990 | 8,870 | 9,340 | 9,600 | 9,860 |
| Carpets and Rugs** | | | | 1,660 | 2,380 | 2,730 | 2,860 | 2,950 | 3,180 |
| Rubber Tires | 790 | 1,640 | 2,570 | 3,170 | 3,640 | 3,240 | 3,230 | 3,240 | 3,060 |
| Batteries, Lead-Acid | Neg. | 200 | 450 | 40 | 150 | 110 | 110 | 120 | 120 |
| Miscellaneous Durables | | | | | | | | | |
| Selected Consumer Electronics*** | | | | | 1,710 | 2,270 | 2,460 | 2,600 | 2,590 |
| Other Miscellaneous Durables | | | | | 13,740 | 17,040 | 16,990 | 16,890 | 16,900 |
| *Total Miscellaneous Durables* | 5,010 | 6,930 | 9,840 | 12,000 | 15,450 | 19,310 | 19,450 | 19,490 | 19,490 |
| ***Total Durable Goods*** | 9,570 | 13,720 | 20,440 | 26,350 | 32,270 | 36,610 | 37,550 | 38,040 | 38,480 |
| **Nondurable Goods** | 14,940 | 21,330 | 29,750 | 43,370 | 46,450 | 43,880 | 40,790 | 39,380 | 34,550 |
| *(Detail in Table 17)* | | | | | | | | | |
| **Containers and Packaging** | 24,500 | 40,210 | 44,180 | 47,750 | 46,970 | 44,830 | 44,560 | 41,830 | 37,350 |
| *(Detail in Table 22)* | | | | | | | | | |
| ***Total Product Wastes†*** | 49,010 | 75,260 | 94,370 | 117,470 | 125,690 | 125,320 | 122,900 | 119,250 | 110,380 |
| **Other Wastes** | | | | | | | | | |
| Food Scraps | 12,200 | 12,800 | 13,000 | 23,860 | 29,130 | 31,300 | 31,800 | 32,540 | 33,440 |
| Yard Trimmings | 20,000 | 23,200 | 27,500 | 30,800 | 14,760 | 12,210 | 11,730 | 11,600 | 13,300 |
| Miscellaneous Inorganic Wastes | 1,300 | 1,780 | 2,250 | 2,900 | 3,500 | 3,690 | 3,750 | 3,780 | 3,820 |
| ***Total Other Wastes*** | 33,500 | 37,780 | 42,750 | 57,560 | 47,390 | 47,200 | 47,280 | 47,920 | 50,560 |
| ***Total MSW Discarded - Weight*** | 82,510 | 113,040 | 137,120 | 175,030 | 173,080 | 172,520 | 170,180 | 167,170 | 160,940 |
| | **Percent of Total Discards** | | | | | | | | |
| **Products** | **1960** | **1970** | **1980** | **1990** | **2000** | **2005** | **2007** | **2008** | **2009** |
| **Durable Goods** | | | | | | | | | |
| Major Appliances | 2.0% | 1.9% | 2.1% | 1.3% | 0.9% | 0.7% | 0.7% | 0.7% | 0.8% |
| Small Appliances** | | | | 0.3% | 0.6% | 0.7% | 0.8% | 0.8% | 0.9% |
| Furniture and Furnishings | 2.6% | 2.5% | 3.5% | 3.9% | 4.6% | 5.1% | 5.5% | 5.7% | 6.1% |
| Carpets and Rugs** | | | | 0.9% | 1.4% | 1.6% | 1.7% | 1.8% | 2.0% |
| Rubber Tires | 1.0% | 1.5% | 1.9% | 1.8% | 2.1% | 1.9% | 1.9% | 1.9% | 1.9% |
| Batteries, Lead-Acid | Neg. | 0.2% | 0.3% | 0.0% | 0.1% | 0.1% | 0.1% | 0.1% | 0.1% |
| Miscellaneous Durables | | | | | | | | | |
| Selected Consumer Electronics*** | | | | | 1.0% | 1.3% | 1.4% | 1.6% | 1.6% |
| Other Miscellaneous Durables | | | | | 7.9% | 9.9% | 10.0% | 10.1% | 10.5% |
| *Total Miscellaneous Durables* | 6.1% | 6.1% | 7.2% | 6.9% | 8.9% | 11.2% | 11.4% | 11.7% | 12.1% |
| ***Total Durable Goods*** | 11.6% | 12.1% | 14.9% | 15.1% | 18.6% | 21.2% | 22.1% | 22.8% | 23.9% |
| **Nondurable Goods** | 18.1% | 18.9% | 21.7% | 24.8% | 26.8% | 25.4% | 24.0% | 23.6% | 21.5% |
| *(Detail in Table 17)* | | | | | | | | | |
| **Containers and Packaging** | 29.7% | 35.6% | 32.2% | 27.3% | 27.1% | 26.0% | 26.2% | 25.0% | 23.2% |
| *(Detail in Table 23)* | | | | | | | | | |
| ***Total Product Wastes†*** | 59.4% | 66.6% | 68.8% | 67.1% | 72.6% | 72.6% | 72.2% | 71.3% | 68.6% |
| **Other Wastes** | | | | | | | | | |
| Food Scraps | 14.8% | 11.3% | 9.5% | 13.6% | 16.8% | 18.1% | 18.7% | 19.5% | 20.8% |
| Yard Trimmings | 24.2% | 20.5% | 20.1% | 17.6% | 8.5% | 7.1% | 6.9% | 6.9% | 8.3% |
| Miscellaneous Inorganic Wastes | 1.6% | 1.6% | 1.6% | 1.7% | 2.0% | 2.1% | 2.2% | 2.3% | 2.4% |
| ***Total Other Wastes*** | 40.6% | 33.4% | 31.2% | 32.9% | 27.4% | 27.4% | 27.8% | 28.7% | 31.4% |
| ***Total MSW Discarded - %*** | 100.0% | 100.0% | 100.0% | 100.0% | 100.0% | 100.0% | 100.0% | 100.0% | 100.0% |

* Discards after materials and compost recovery. In this table, discards include combustion with energy recovery.
Does not include construction & demolition debris, industrial process wastes, or certain other wastes.
** Not estimated separately prior to 1990. *** Not estimated separately prior to 1999. Preliminary data; may undergo revision.
† Other than food products.
Neg. = Less than 5,000 tons or 0.05 percent. Details may not add to totals due to rounding.
Source: Franklin Associates, A Division of ERG

**PRODUCTS GENERATED* IN THE MUNICIPAL WASTE STREAM, 1960 TO 2009**
**(WITH DETAIL ON NONDURABLE GOODS)**
**(In thousands of tons and percent of total generation)**

| | Thousands of Tons | | | | | | | | |
|---|---|---|---|---|---|---|---|---|---|
| **Products** | **1960** | **1970** | **1980** | **1990** | **2000** | **2005** | **2007** | **2008** | **2009** |
| **Durable Goods** | 9,920 | 14,660 | 21,800 | 29,810 | 38,850 | 44,650 | 45,770 | 46,400 | 46,640 |
| *(Detail in Table 12)* | | | | | | | | | |
| **Nondurable Goods** | | | | | | | | | |
| Newspapers | 7,110 | 9,510 | 11,050 | 13,430 | 14,790 | 12,790 | 10,780 | 8,800 | 7,760 |
| Books and Magazines | 1,920 | 2,470 | 3,390 | | | | | | |
| Books** | | | | 970 | 1,240 | 1,100 | 1,270 | 1,340 | 960 |
| Magazines** | | | | 2,830 | 2,230 | 2,580 | 2,550 | 2,050 | 1,450 |
| Office-Type Papers | 1,520 | 2,650 | 4,000 | 6,410 | 7,420 | 6,620 | 6,060 | 6,050 | 5,380 |
| Directories** | | | | 610 | 680 | 660 | 760 | 840 | 650 |
| Standard Mail*** | | | | 3,820 | 5,570 | 5,830 | 5,910 | 5,510 | 4,650 |
| Other Commercial Printing | 1,260 | 2,130 | 3,120 | 4,460 | 7,380 | 6,440 | 6,200 | 5,130 | 3,490 |
| Tissue Paper and Towels | 1,090 | 2,080 | 2,300 | 2,960 | 3,220 | 3,460 | 3,500 | 3,460 | 3,490 |
| Paper Plates and Cups | 270 | 420 | 630 | 650 | 960 | 1,160 | 1,230 | 1,250 | 1,170 |
| Plastic Plates and Cups† | | | 190 | 650 | 870 | 930 | 860 | 780 | 900 |
| Trash Bags** | | | | 780 | 850 | 1,060 | 1,070 | 930 | 1,000 |
| Disposable Diapers | Neg. | 350 | 1,930 | 2,700 | 3,230 | 3,410 | 3,730 | 3,770 | 3,810 |
| Other Nonpackaging Paper | 2,700 | 3,630 | 4,230 | 3,840 | 4,250 | 4,490 | 4,260 | 4,630 | 4,420 |
| Clothing and Footwear | 1,360 | 1,620 | 2,170 | 4,010 | 6,470 | 7,890 | 8,320 | 8,820 | 9,080 |
| Towels, Sheets and Pillowcases** | | | | 710 | 820 | 980 | 1,100 | 1,160 | 1,230 |
| Other Miscellaneous Nondurables | 100 | 200 | 1,410 | 3,340 | 4,030 | 4,250 | 4,160 | 4,170 | 4,000 |
| ***Total Nondurable Goods*** | 17,330 | 25,060 | 34,420 | 52,170 | 64,010 | 63,650 | 61,760 | 58,690 | 53,440 |
| **Containers and Packaging** | 27,370 | 43,560 | 52,670 | 64,530 | 75,840 | 76,330 | 78,460 | 75,910 | 71,570 |
| *(Detail in Table 18)* | | | | | | | | | |
| ***Total Product Wastes‡*** | 54,620 | 83,280 | 108,890 | 146,510 | 178,700 | 184,630 | 185,990 | 181,000 | 171,650 |
| **Other Wastes** | 33,500 | 37,780 | 42,750 | 61,760 | 63,840 | 67,750 | 68,990 | 70,020 | 71,310 |
| ***Total MSW Generated - Weight*** | 88,120 | 121,060 | 151,640 | 208,270 | 242,540 | 252,380 | 254,980 | 251,020 | 242,960 |

| | Percent of Total Generation | | | | | | | | |
|---|---|---|---|---|---|---|---|---|---|
| **Products** | **1960** | **1970** | **1980** | **1990** | **2000** | **2005** | **2007** | **2008** | **2009** |
| **Durable Goods** | 11.3% | 12.1% | 14.4% | 14.3% | 16.0% | 17.7% | 18.0% | 18.5% | 19.2% |
| *(Detail in Table 12)* | | | | | | | | | |
| **Nondurable Goods** | | | | | | | | | |
| Newspapers | 8.1% | 7.9% | 7.3% | 6.4% | 6.1% | 5.1% | 4.2% | 3.5% | 3.2% |
| Books and Magazines | 2.2% | 2.0% | 2.2% | | | | | | |
| Books** | | | | 0.5% | 0.5% | 0.4% | 0.5% | 0.5% | 0.4% |
| Magazines** | | | | 1.4% | 0.9% | 1.0% | 1.0% | 0.8% | 0.6% |
| Office-Type Papers*** | 1.7% | 2.2% | 2.6% | 3.1% | 3.1% | 2.6% | 2.4% | 2.4% | 2.2% |
| Directories** | | | | 0.3% | 0.3% | 0.3% | 0.3% | 0.3% | 0.3% |
| Standard Mail§ | | | | 1.8% | 2.3% | 2.3% | 2.3% | 2.2% | 1.9% |
| Other Commercial Printing | 1.4% | 1.8% | 2.1% | 2.1% | 3.0% | 2.6% | 2.4% | 2.0% | 1.4% |
| Tissue Paper and Towels | 1.2% | 1.7% | 1.5% | 1.4% | 1.3% | 1.4% | 1.4% | 1.4% | 1.4% |
| Paper Plates and Cups | 0.3% | 0.3% | 0.4% | 0.3% | 0.4% | 0.5% | 0.5% | 0.5% | 0.5% |
| Plastic Plates and Cups† | | | 0.1% | 0.3% | 0.4% | 0.4% | 0.3% | 0.3% | 0.4% |
| Trash Bags** | | | | 0.4% | 0.4% | 0.4% | 0.4% | 0.4% | 0.4% |
| Disposable Diapers | Neg. | 0.3% | 1.3% | 1.3% | 1.3% | 1.4% | 1.5% | 1.5% | 1.6% |
| Other Nonpackaging Paper | 3.1% | 3.0% | 2.8% | 1.8% | 1.8% | 1.8% | 1.7% | 1.8% | 1.8% |
| Clothing and Footwear | 1.5% | 1.3% | 1.4% | 1.9% | 2.7% | 3.1% | 3.3% | 3.5% | 3.7% |
| Towels, Sheets and Pillowcases** | | | | 0.3% | 0.3% | 0.4% | 0.4% | 0.5% | 0.5% |
| Other Miscellaneous Nondurables | 0.1% | 0.2% | 0.9% | 1.6% | 1.7% | 1.7% | 1.6% | 1.7% | 1.6% |
| ***Total Nondurables*** | 19.7% | 20.7% | 22.7% | 25.0% | 26.4% | 25.2% | 24.2% | 23.4% | 22.0% |
| **Containers and Packaging** | 31.1% | 36.0% | 34.7% | 31.0% | 31.3% | 30.2% | 30.8% | 30.2% | 29.5% |
| *(Detail in Table 19)* | | | | | | | | | |
| ***Total Product Wastes‡*** | 62.0% | 68.8% | 71.8% | 70.3% | 73.7% | 73.2% | 72.9% | 72.1% | 70.6% |
| **Other Wastes** | 38.0% | 31.2% | 28.2% | 29.7% | 26.3% | 26.8% | 27.1% | 27.9% | 29.4% |
| ***Total MSW Generated - %*** | 100.0% | 100.0% | 100.0% | 100.0% | 100.0% | 100.0% | 100.0% | 100.0% | 100.0% |

* Generation before materials recovery or combustion. Does not include construction & demolition debris, industrial process wastes, or certain other wastes. Details may not add to totals due to rounding.
** Not estimated separately prior to 1990.
*** High-grade paper such as printer paper; generated in both commercial and residential sources.
§ Not estimated separately prior to 1990. Formerly called Third Class Mail and Standard (A) Mail by the U.S. Postal Service.
† Not estimated separately prior to 1980.
‡ Other than food products.
Neg. = Less than 5,000 tons or 0.05 percent.
Source: Franklin Associates, A Division of ERG

**RECOVERY* OF PRODUCTS IN MUNICIPAL SOLID WASTE, 1960 TO 2009**
**(WITH DETAIL ON NONDURABLE GOODS)**
**(In thousands of tons and percent of generation of each product)**

| | Thousands of Tons | | | | | | | | |
|---|---|---|---|---|---|---|---|---|---|
| **Products** | **1960** | **1970** | **1980** | **1990** | **2000** | **2005** | **2007** | **2008** | **2009** |
| **Durable Goods** | 350 | 940 | 1,360 | 3,460 | 6,580 | 8,040 | 8,220 | 8,360 | 8,160 |
| *(Detail in Table 13)* | | | | | | | | | |
| **Nondurable Goods** | | | | | | | | | |
| Newspapers | 1,820 | 2,250 | 3,020 | 5,110 | 8,720 | 9,360 | 8,550 | 7,740 | 6,840 |
| Books and Magazines | 100 | 260 | 280 | | | | | | |
| Books** | | | | 100 | 240 | 270 | 360 | 390 | 320 |
| Magazines** | | | | 300 | 710 | 960 | 1,010 | 820 | 780 |
| Office-Type Papers | 250 | 710 | 870 | 1,700 | 4,090 | 4,110 | 4,300 | 4,290 | 3,990 |
| Directories** | | | | 50 | 120 | 120 | 140 | 180 | 240 |
| Standard Mail*** | | | | 200 | 1,830 | 2,090 | 2,380 | 2,240 | 2,950 |
| Other Commercial Printing | 130 | 340 | 350 | 700 | 810 | 1,440 | 2,790 | 2,200 | 2,310 |
| Tissue Paper and Towels | Neg. | Neg. | Neg. | Neg. | Neg. | Neg. | Neg. | Neg. | Neg. |
| Paper Plates and Cups | Neg. | Neg. | Neg. | Neg. | Neg. | Neg. | Neg. | Neg. | Neg. |
| Plastic Plates and Cups† | | | Neg. | Neg. | Neg. | Neg. | Neg. | Neg. | Neg. |
| Trash Bags** | | | | Neg. | Neg. | Neg. | Neg. | Neg. | Neg. |
| Disposable Diapers | | | | Neg. | Neg. | Neg. | Neg. | Neg. | Neg. |
| Other Nonpackaging Paper | 40 | 110 | Neg. | Neg. | Neg. | Neg. | Neg. | Neg. | Neg. |
| Clothing and Footwear | 50 | 60 | 150 | 520 | 900 | 1,250 | 1,250 | 1,250 | 1,250 |
| Towels, Sheets and Pillowcases** | | | | 120 | 140 | 170 | 190 | 200 | 210 |
| Other Miscellaneous Nondurables | Neg. | Neg. | Neg. | Neg. | Neg. | Neg. | Neg. | Neg. | Neg. |
| ***Total Nondurable Goods*** | 2,390 | 3,730 | 4,670 | 8,800 | 17,560 | 19,770 | 20,970 | 19,310 | 18,890 |
| **Containers and Packaging** | 2,870 | 3,350 | 8,490 | 16,780 | 28,870 | 31,500 | 33,900 | 34,080 | 34,220 |
| *(Detail in Table 20)* | | | | | | | | | |
| ***Total Product Wastes‡*** | 5,610 | 8,020 | 14,520 | 29,040 | 53,010 | 59,310 | 63,090 | 61,750 | 61,270 |
| **Other Wastes** | Neg. | Neg. | Neg. | 4,200 | 16,450 | 20,550 | 21,710 | 22,100 | 20,750 |
| ***Total MSW Recovered - Weight*** | 5,610 | 8,020 | 14,520 | 33,240 | 69,460 | 79,860 | 84,800 | 83,850 | 82,020 |
| | Percent of Generation of Each Product | | | | | | | | |
| **Products** | **1960** | **1970** | **1980** | **1990** | **2000** | **2005** | **2007** | **2008** | **2009** |
| **Durable Goods** | 3.5% | 6.4% | 6.2% | 11.6% | 16.9% | 18.0% | 18.0% | 18.0% | 17.5% |
| *(Detail in Table 13)* | | | | | | | | | |
| **Nondurable Goods** | | | | | | | | | |
| Newspapers | 25.6% | 23.7% | 27.3% | 38.0% | 59.0% | 73.2% | 79.3% | 88.0% | 88.1% |
| Books and Magazines | 5.2% | 10.5% | 8.3% | | | | | | |
| Books** | | | | 10.3% | 19.4% | 24.5% | 28.3% | 29.1% | 33.3% |
| Magazines** | | | | 10.6% | 31.8% | 37.2% | 39.6% | 40.0% | 53.8% |
| Office-Type Papers*** | 16.4% | 26.8% | 21.8% | 26.5% | 55.1% | 62.1% | 71.0% | 70.9% | 74.2% |
| Directories** | | | | 8.2% | 17.6% | 18.2% | 18.4% | 21.4% | 36.9% |
| Standard Mail§ | | | | 5.2% | 32.9% | 35.8% | 40.3% | 40.7% | 63.4% |
| Other Commercial Printing | 10.3% | 16.0% | 11.2% | 15.7% | 11.0% | 22.4% | 45.0% | 42.9% | 66.2% |
| Tissue Paper and Towels | Neg. | Neg. | Neg. | Neg. | Neg. | Neg. | Neg. | Neg. | Neg. |
| Paper Plates and Cups | Neg. | Neg. | Neg. | Neg. | Neg. | Neg. | Neg. | Neg. | Neg. |
| Plastic Plates and Cups† | | | Neg. | Neg. | Neg. | Neg. | Neg. | Neg. | Neg. |
| Trash Bags** | | | | Neg. | Neg. | Neg. | Neg. | Neg. | Neg. |
| Disposable Diapers | | | | Neg. | Neg. | Neg. | Neg. | Neg. | Neg. |
| Other Nonpackaging Paper | 1.5% | 3.0% | Neg. | Neg. | Neg. | Neg. | Neg. | Neg. | Neg. |
| Clothing and Footwear | Neg. | Neg. | Neg. | 13.0% | 13.9% | 15.8% | 15.0% | 14.2% | 13.8% |
| Towels, Sheets and Pillowcases** | | | | 16.9% | 17.1% | 17.3% | 17.3% | 17.2% | 17.1% |
| Other Miscellaneous Nondurables | Neg. | Neg. | Neg. | Neg. | Neg. | Neg. | Neg. | Neg. | Neg. |
| ***Total Nondurables*** | 13.8% | 14.9% | 13.6% | 16.9% | 27.4% | 31.1% | 34.0% | 32.9% | 35.3% |
| **Containers and Packaging** | 10.5% | 7.7% | 16.1% | 26.0% | 38.1% | 41.3% | 43.2% | 44.9% | 47.8% |
| *(Detail in Table 21)* | | | | | | | | | |
| ***Total Product Wastes‡*** | 10.3% | 9.6% | 13.3% | 19.8% | 29.7% | 32.1% | 33.9% | 34.1% | 35.7% |
| **Other Wastes** | Neg. | Neg. | Neg. | 6.8% | 25.8% | 30.3% | 31.5% | 31.6% | 29.1% |
| ***Total MSW Recovered - %*** | 6.4% | 6.6% | 9.6% | 16.0% | 28.6% | 31.6% | 33.3% | 33.4% | 33.8% |

* Recovery of postconsumer wastes; does not include converting/fabrication scrap.
Details may not add to totals due to rounding.
** Not estimated separately prior to 1990.
*** High-grade paper such as printer paper; generated in both commercial and residential sources.
§ Not estimated separately prior to 1990. Formerly called Third Class Mail and Standard (A) Mail by the U.S. Postal Service.
† Not estimated separately prior to 1980.
‡ Other than food products.
Neg. = Less than 5,000 tons or 0.05 percent.
Source: Franklin Associates, A Division of ERG

**PRODUCTS DISCARDED* IN THE MUNICIPAL WASTE STREAM, 1960 TO 2009**
**(WITH DETAIL ON NONDURABLE GOODS)**
**(In thousands of tons and percent of total discards)**

| | Thousands of Tons | | | | | | | | |
|---|---|---|---|---|---|---|---|---|---|
| **Products** | **1960** | **1970** | **1980** | **1990** | **2000** | **2005** | **2007** | **2008** | **2009** |
| **Durable Goods** | 9,570 | 13,720 | 20,440 | 26,350 | 32,270 | 36,610 | 37,550 | 38,040 | 38,480 |
| *(Detail in Table 14)* | | | | | | | | | |
| **Nondurable Goods** | | | | | | | | | |
| Newspapers | 5,290 | 7,260 | 8,030 | 8,320 | 6,070 | 3,430 | 2,230 | 1,060 | 920 |
| Books and Magazines | 1,820 | 2,210 | 3,110 | | | | | | |
| Books** | | | | 870 | 1,000 | 830 | 910 | 950 | 640 |
| Magazines** | | | | 2,530 | 1,520 | 1,620 | 1,540 | 1,230 | 670 |
| Office-Type Papers | 1,270 | 1,940 | 3,130 | 4,710 | 3,330 | 2,510 | 1,760 | 1,760 | 1,390 |
| Directories** | | | | 560 | 560 | 540 | 620 | 660 | 410 |
| Standard Mail*** | | | | 3,620 | 3,740 | 3,740 | 3,530 | 3,270 | 1,700 |
| Other Commercial Printing | 1,130 | 1,790 | 2,770 | 3,760 | 6,570 | 5,000 | 3,410 | 2,930 | 1,180 |
| Tissue Paper and Towels | 1,090 | 2,080 | 2,300 | 2,960 | 3,220 | 3,460 | 3,500 | 3,460 | 3,490 |
| Paper Plates and Cups | 270 | 420 | 630 | 650 | 960 | 1,160 | 1,230 | 1,250 | 1,170 |
| Plastic Plates and Cups† | | | 190 | 650 | 870 | 930 | 860 | 780 | 900 |
| Trash Bags** | | | | 780 | 850 | 1,060 | 1,070 | 930 | 1,000 |
| Disposable Diapers | Neg. | 350 | 1,930 | 2,700 | 3,230 | 3,410 | 3,730 | 3,770 | 3,810 |
| Other Nonpackaging Paper | 2,660 | 3,520 | 4,230 | 3,840 | 4,250 | 4,490 | 4,260 | 4,630 | 4,420 |
| Clothing and Footwear | 1,310 | 1,560 | 2,020 | 3,490 | 5,570 | 6,640 | 7,070 | 7,570 | 7,830 |
| Towels, Sheets and Pillowcases** | | | | 590 | 680 | 810 | 910 | 960 | 1,020 |
| Other Miscellaneous Nondurables | 100 | 200 | 1,410 | 3,340 | 4,030 | 4,250 | 4,160 | 4,170 | 4,000 |
| ***Total Nondurable Goods*** | 14,940 | 21,330 | 29,750 | 43,370 | 46,450 | 43,880 | 40,790 | 39,380 | 34,550 |
| **Containers and Packaging** | 24,500 | 40,210 | 44,180 | 47,750 | 46,970 | 44,830 | 44,560 | 41,830 | 37,350 |
| *(Detail in Table 22)* | | | | | | | | | |
| ***Total Product Wastes‡*** | 49,010 | 75,260 | 94,370 | 117,470 | 125,690 | 125,320 | 122,900 | 119,250 | 110,380 |
| **Other Wastes** | 33,500 | 37,780 | 42,750 | 57,560 | 47,390 | 47,200 | 47,280 | 47,920 | 50,560 |
| ***Total MSW Discarded - Weight*** | 82,510 | 113,040 | 137,120 | 175,030 | 173,080 | 172,520 | 170,180 | 167,170 | 160,940 |
| | **Percent of Total Discards** | | | | | | | | |
| **Products** | **1960** | **1970** | **1980** | **1990** | **2000** | **2005** | **2007** | **2008** | **2009** |
| **Durable Goods** | 11.6% | 12.1% | 14.9% | 15.1% | 18.6% | 21.2% | 22.1% | 22.8% | 23.9% |
| *(Detail in Table 14)* | | | | | | | | | |
| **Nondurable Goods** | | | | | | | | | |
| Newspapers | 6.4% | 6.4% | 5.9% | 4.8% | 3.5% | 2.0% | 1.3% | 0.6% | 0.6% |
| Books and Magazines | 2.2% | 2.0% | 2.3% | | | | | | |
| Books** | | | | 0.5% | 0.6% | 0.5% | 0.5% | 0.6% | 0.4% |
| Magazines** | | | | 1.4% | 0.9% | 0.9% | 0.9% | 0.7% | 0.4% |
| Office-Type Papers*** | 1.5% | 1.7% | 2.3% | 2.7% | 1.9% | 1.5% | 1.0% | 1.1% | 0.9% |
| Directories** | | | | 0.3% | 0.3% | 0.3% | 0.4% | 0.4% | 0.3% |
| Standard Mail§ | | | | 2.1% | 2.2% | 2.2% | 2.1% | 2.0% | 1.1% |
| Other Commercial Printing | 1.4% | 1.6% | 2.0% | 2.1% | 3.8% | 2.9% | 2.0% | 1.8% | 0.7% |
| Tissue Paper and Towels | 1.3% | 1.8% | 1.7% | 1.7% | 1.9% | 2.0% | 2.1% | 2.1% | 2.2% |
| Paper Plates and Cups | 0.3% | 0.4% | 0.5% | 0.4% | 0.6% | 0.7% | 0.7% | 0.7% | 0.7% |
| Plastic Plates and Cups† | | | 0.1% | 0.4% | 0.5% | 0.5% | 0.5% | 0.5% | 0.6% |
| Trash Bags** | | | | 0.4% | 0.5% | 0.6% | 0.6% | 0.6% | 0.6% |
| Disposable Diapers | Neg. | 0.3% | 1.4% | 1.5% | 1.9% | 2.0% | 2.2% | 2.3% | 2.4% |
| Other Nonpackaging Paper | 3.2% | 3.1% | 3.1% | 2.2% | 2.5% | 2.6% | 2.5% | 2.8% | 2.7% |
| Clothing and Footwear | 1.6% | 1.4% | 1.5% | 2.0% | 3.2% | 3.8% | 4.2% | 4.5% | 4.9% |
| Towels, Sheets and Pillowcases** | | | | 0.3% | 0.4% | 0.5% | 0.5% | 0.6% | 0.6% |
| Other Miscellaneous Nondurables | 0.1% | 0.2% | 1.7% | 1.9% | 2.3% | 2.5% | 2.4% | 2.5% | 2.5% |
| ***Total Nondurables*** | 18.1% | 18.9% | 21.7% | 24.8% | 26.8% | 25.4% | 24.0% | 23.6% | 21.5% |
| **Containers and Packaging** | 29.7% | 35.6% | 32.2% | 27.3% | 27.1% | 26.0% | 26.2% | 25.0% | 23.2% |
| *(Detail in Table 23)* | | | | | | | | | |
| ***Total Product Wastes‡*** | 59.4% | 66.6% | 68.8% | 67.1% | 72.6% | 72.6% | 72.2% | 71.3% | 68.6% |
| **Other Wastes** | 40.6% | 33.4% | 31.2% | 32.9% | 27.4% | 27.4% | 27.8% | 28.7% | 31.4% |
| ***Total MSW Discarded - %*** | 100.0% | 100.0% | 100.0% | 100.0% | 100.0% | 100.0% | 100.0% | 100.0% | 100.0% |

* Discards after materials and compost recovery. In this table, discards include combustion with energy recovery. Does not include construction & demolition debris, industrial process wastes, or certain other wastes.
** Not estimated separately prior to 1990.
*** High-grade paper such as printer paper; generated in both commercial and residential sources.
§ Not estimated separately prior to 1990. Formerly called Third Class Mail and Standard (A) Mail by the U.S. Postal Service.
† Not estimated separately prior to 1980.
‡ Other than food products.
Neg. = Less than 5,000 tons or 0.05 percent. Details may not add to totals due to rounding.
Source: Franklin Associates, A Division of ERG

**PRODUCTS GENERATED* IN THE MUNICIPAL WASTE STREAM, 1960 TO 2009**
**(WITH DETAIL ON CONTAINERS AND PACKAGING)**
**(In thousands of tons)**

| | Thousands of Tons | | | | | | | | |
|---|---|---|---|---|---|---|---|---|---|
| **Products** | **1960** | **1970** | **1980** | **1990** | **2000** | **2005** | **2007** | **2008** | **2009** |
| **Durable Goods** | 9,920 | 14,660 | 21,800 | 29,810 | 38,850 | 44,650 | 45,770 | 46,400 | 46,640 |
| *(Detail in Table 12)* | | | | | | | | | |
| **Nondurable Goods** | 17,330 | 25,060 | 34,420 | 52,170 | 64,010 | 63,650 | 61,760 | 58,690 | 53,440 |
| *(Detail in Table 15)* | | | | | | | | | |
| **Containers and Packaging** | | | | | | | | | |
| **Glass Packaging** | | | | | | | | | |
| Beer and Soft Drink Bottles** | 1,400 | 5,580 | 6,740 | 5,640 | 5,710 | 6,540 | 6,760 | 6,350 | 6,000 |
| Wine and Liquor Bottles | 1,080 | 1,900 | 2,450 | 2,030 | 1,910 | 1,630 | 1,620 | 1,610 | 1,710 |
| Other Bottles & Jars | 3,710 | 4,440 | 4,780 | 4,160 | 3,420 | 2,290 | 2,030 | 2,090 | 1,950 |
| ***Total Glass Packaging*** | 6,190 | 11,920 | 13,970 | 11,830 | 11,040 | 10,460 | 10,410 | 10,050 | 9,660 |
| **Steel Packaging** | | | | | | | | | |
| Beer and Soft Drink Cans | 640 | 1,570 | 520 | 150 | Neg. | Neg. | Neg. | Neg. | Neg. |
| Cans | 3,760 | 3,540 | 2,850 | 2,540 | 2,630 | 2,130 | 2,430 | 2,310 | 1,940 |
| Other Steel Packaging | 260 | 270 | 240 | 200 | 240 | 240 | 240 | 240 | 340 |
| ***Total Steel Packaging*** | 4,660 | 5,380 | 3,610 | 2,890 | 2,870 | 2,370 | 2,670 | 2,550 | 2,280 |
| **Aluminum Packaging** | | | | | | | | | |
| Beer and Soft Drink Cans | Neg. | 100 | 850 | 1,550 | 1,520 | 1,450 | 1,420 | 1,390 | 1,360 |
| Other Cans | Neg. | 60 | 40 | 20 | 50 | 80 | 30 | 70 | 70 |
| Foil and Closures | 170 | 410 | 380 | 330 | 380 | 400 | 430 | 420 | 410 |
| ***Total Aluminum Packaging*** | 170 | 570 | 1,270 | 1,900 | 1,950 | 1,930 | 1,880 | 1,880 | 1,840 |
| **Paper & Paperboard Pkg** | | | | | | | | | |
| Corrugated Boxes | 7,330 | 12,760 | 17,080 | 24,010 | 30,210 | 30,930 | 31,230 | 29,710 | 27,190 |
| Gable Top/Aseptic Cartons‡ | | | 790 | 510 | 550 | 500 | 500 | 490 | 460 |
| Folding Cartons | | | 3,820 | 4,300 | 5,820 | 5,530 | 5,530 | 5,340 | 4,980 |
| Other Paperboard Packaging | 3,840 | 4,830 | 230 | 290 | 200 | 160 | 150 | 120 | 90 |
| Bags and Sacks | | | 3,380 | 2,440 | 1,490 | 1,120 | 1,140 | 1,170 | 910 |
| Wrapping Papers | | | 200 | 110 | Neg. | Neg. | Neg. | Neg. | Neg. |
| Other Paper Packaging | 2,940 | 3,810 | 850 | 1,020 | 1,670 | 1,400 | 1,390 | 1,460 | 1,310 |
| ***Total Paper & Board Pkg*** | 14,110 | 21,400 | 26,350 | 32,680 | 39,940 | 39,640 | 39,940 | 38,290 | 34,940 |
| **Plastics Packaging** | | | | | | | | | |
| PET Bottles and Jars | | | 260 | 430 | 1,720 | 2,540 | 2,840 | 2,680 | 2,570 |
| HDPE Natural Bottles | | | 230 | 530 | 690 | 800 | 820 | 750 | 760 |
| Other Containers | 60 | 910 | 890 | 1,430 | 1,740 | 1,420 | 1,910 | 1,900 | 1,750 |
| Bags and Sacks | | | 390 | 940 | 1,650 | 1,640 | 1,010 | 940 | 660 |
| Wraps | | | 840 | 1,530 | 2,550 | 2,810 | 3,180 | 3,020 | 3,190 |
| *Subtotal Bags, Sacks, and Wraps* | | | 1,230 | 2,470 | 4,200 | 4,450 | 4,190 | 3,960 | 3,850 |
| Other Plastics Packaging | 60 | 1,180 | 790 | 2,040 | 2,840 | 3,210 | 3,870 | 3,720 | 3,600 |
| ***Total Plastics Packaging*** | 120 | 2,090 | 3,400 | 6,900 | 11,190 | 12,420 | 13,630 | 13,010 | 12,530 |
| Wood Packaging | 2,000 | 2,070 | 3,940 | 8,180 | 8,610 | 9,230 | 9,610 | 9,820 | 10,040 |
| Other Misc. Packaging | 120 | 130 | 130 | 150 | 240 | 280 | 320 | 310 | 280 |
| ***Total Containers & Pkg*** | 27,370 | 43,560 | 52,670 | 64,530 | 75,840 | 76,330 | 78,460 | 75,910 | 71,570 |
| ***Total Product Wastes†*** | 54,620 | 83,280 | 108,890 | 146,510 | 178,700 | 184,630 | 185,990 | 181,000 | 171,650 |
| **Other Wastes** | | | | | | | | | |
| Food Scraps | 12,200 | 12,800 | 13,000 | 23,860 | 29,810 | 31,990 | 32,610 | 33,340 | 34,290 |
| Yard Trimmings | 20,000 | 23,200 | 27,500 | 35,000 | 30,530 | 32,070 | 32,630 | 32,900 | 33,200 |
| Miscellaneous Inorganic Wastes | 1,300 | 1,780 | 2,250 | 2,900 | 3,500 | 3,690 | 3,750 | 3,780 | 3,820 |
| ***Total Other Wastes*** | 33,500 | 37,780 | 42,750 | 61,760 | 63,840 | 67,750 | 68,990 | 70,020 | 71,310 |
| ***Total MSW Generated - Weight*** | 88,120 | 121,060 | 151,640 | 208,270 | 242,540 | 252,380 | 254,980 | 251,020 | 242,960 |

* Generation before materials recovery or combustion.
** Includes carbonated drinks and non-carbonated water, teas, flavored drinks, and ready-to-drink alcoholic coolers and cocktails.
† Other than food products.
‡ Includes milk, juice, and other products packaged in gable top cartons and liquid food aseptic cartons.
Details may not add to totals due to rounding.
Neg. = Less than 5,000 tons or 0.05 percent.
Source: Franklin Associates, A Division of ERG

PRODUCTS GENERATED* IN THE MUNICIPAL WASTE STREAM, 1960 TO 2009
(WITH DETAIL ON CONTAINERS AND PACKAGING)
(In percent of total generation)

| Products | Percent of Total Generation | | | | | | | | |
|---|---|---|---|---|---|---|---|---|---|
| | 1960 | 1970 | 1980 | 1990 | 2000 | 2005 | 2007 | 2008 | 2009 |
| **Durable Goods** | 11.3% | 12.1% | 14.4% | 14.3% | 16.0% | 17.7% | 18.0% | 18.0% | 19.2% |
| *(Detail in Table 12)* | | | | | | | | | |
| **Nondurable Goods** | 19.7% | 20.7% | 22.7% | 25.0% | 26.4% | 25.2% | 24.2% | 24.2% | 22.0% |
| *(Detail in Table 15)* | | | | | | | | | |
| **Containers and Packaging** | | | | | | | | | |
| **Glass Packaging** | | | | | | | | | |
| Beer and Soft Drink Bottles** | 1.6% | 4.6% | 4.4% | 2.7% | 2.4% | 2.6% | 2.7% | 2.7% | 2.5% |
| Wine and Liquor Bottles | 1.2% | 1.6% | 1.6% | 1.0% | 0.8% | 0.6% | 0.6% | 0.6% | 0.7% |
| Other Bottles & Jars | 4.2% | 3.7% | 3.2% | 2.0% | 1.4% | 0.9% | 0.8% | 0.8% | 0.8% |
| ***Total Glass Packaging*** | 7.0% | 9.8% | 9.2% | 5.7% | 4.6% | 4.1% | 4.1% | 4.1% | 4.0% |
| **Steel Packaging** | | | | | | | | | |
| Beer and Soft Drink Cans | 0.7% | 1.3% | 0.3% | 0.1% | Neg. | Neg. | Neg. | Neg. | Neg. |
| Cans | 4.3% | 2.9% | 1.9% | 1.2% | 1.1% | 0.8% | 1.0% | 1.0% | 0.8% |
| Other Steel Packaging | 0.3% | 0.2% | 0.2% | 0.1% | 0.1% | 0.1% | 0.1% | 0.1% | 0.1% |
| ***Total Steel Packaging*** | 5.3% | 4.4% | 2.4% | 1.4% | 1.2% | 0.9% | 1.0% | 1.0% | 0.9% |
| **Aluminum Packaging** | | | | | | | | | |
| Beer and Soft Drink Cans | Neg. | 0.1% | 0.6% | 0.7% | 0.6% | 0.6% | 0.6% | 0.6% | 0.6% |
| Other Cans | Neg. | Neg. | Neg. | Neg. | Neg. | Neg. | 0.01% | 0.01% | 0.03% |
| Foil and Closures | 0.2% | 0.3% | 0.3% | 0.2% | 0.2% | 0.2% | 0.2% | 0.2% | 0.2% |
| ***Total Aluminum Packaging*** | 0.2% | 0.5% | 0.8% | 0.9% | 0.8% | 0.8% | 0.7% | 0.7% | 0.8% |
| **Paper & Paperboard Pkg** | | | | | | | | | |
| Corrugated Boxes | 8.3% | 10.5% | 11.3% | 11.5% | 12.5% | 12.3% | 12.2% | 12.2% | 11.2% |
| Gable Top/Aseptic Cartons‡ | | | 0.5% | 0.2% | 0.2% | 0.2% | 0.2% | 0.2% | 0.2% |
| Folding Cartons | | | 2.5% | 2.1% | 2.4% | 2.2% | 2.2% | 2.2% | 2.0% |
| Other Paperboard Packaging | 4.4% | 4.0% | 0.2% | 0.1% | 0.1% | 0.1% | 0.1% | 0.1% | 0.0% |
| Bags and Sacks | | | 2.2% | 1.2% | 0.6% | 0.4% | 0.4% | 0.4% | 0.4% |
| Wrapping Papers | | | 0.1% | 0.1% | Neg. | Neg. | Neg. | Neg. | Neg. |
| Other Paper Packaging | 3.3% | 3.1% | 0.6% | 0.5% | 0.7% | 0.6% | 0.5% | 0.5% | 0.5% |
| ***Total Paper & Board Pkg*** | 16.0% | 17.7% | 17.4% | 15.7% | 16.5% | 15.7% | 15.7% | 15.7% | 14.4% |
| **Plastics Packaging** | | | | | | | | | |
| PET Bottles and Jars | | | 0.2% | 0.2% | 0.7% | 1.0% | 1.1% | 1.1% | 1.1% |
| HDPE Natural Bottles | | | 0.2% | 0.3% | 0.3% | 0.3% | 0.3% | 0.3% | 0.3% |
| Other Containers | 0.1% | 0.8% | 0.6% | 0.7% | 0.7% | 0.6% | 0.7% | 0.7% | 0.7% |
| Bags and Sacks | | | 0.3% | 0.5% | 0.7% | 0.6% | 0.4% | 0.4% | 0.3% |
| Wraps | | | 0.6% | 0.7% | 1.1% | 1.1% | 1.2% | 1.2% | 1.3% |
| *Subtotal Bags, Sacks, and Wraps* | | | 0.8% | 1.2% | 1.7% | 1.8% | 1.6% | 1.6% | 1.6% |
| Other Plastics Packaging | 0.1% | 1.0% | 0.5% | 1.0% | 1.2% | 1.3% | 1.5% | 1.5% | 1.5% |
| ***Total Plastics Packaging*** | 0.1% | 1.7% | 2.2% | 3.3% | 4.6% | 4.9% | 5.3% | 5.3% | 5.2% |
| Wood Packaging | 2.3% | 1.7% | 2.6% | 3.9% | 3.5% | 3.7% | 3.8% | 3.8% | 4.1% |
| Other Misc. Packaging | 0.1% | 0.1% | 0.1% | 0.1% | 0.1% | 0.1% | 0.1% | 0.1% | 0.1% |
| ***Total Containers & Pkg*** | 31.1% | 36.0% | 34.7% | 31.0% | 31.3% | 30.2% | 30.8% | 30.8% | 29.5% |
| ***Total Product Wastes†*** | 62.0% | 68.8% | 71.8% | 70.3% | 73.7% | 73.2% | 72.9% | 72.9% | 70.6% |
| **Other Wastes** | | | | | | | | | |
| Food Scraps | 13.8% | 10.6% | 8.6% | 11.5% | 12.3% | 12.7% | 12.8% | 12.8% | 14.1% |
| Yard Trimmings | 22.7% | 19.2% | 18.1% | 16.8% | 12.6% | 12.7% | 12.8% | 12.8% | 13.7% |
| Miscellaneous Inorganic Wastes | 1.5% | 1.5% | 1.5% | 1.4% | 1.4% | 1.5% | 1.5% | 1.5% | 1.6% |
| ***Total Other Wastes*** | 38.0% | 31.2% | 28.2% | 29.7% | 26.3% | 26.8% | 27.1% | 27.1% | 29.4% |
| ***Total MSW Generated - %*** | 100.0% | 100.0% | 100.0% | 100.0% | 100.0% | 100.0% | 100.0% | 100.0% | 100.0% |

* Generation before materials recovery or combustion.
** Includes carbonated drinks and non-carbonated water, teas, flavored drinks, and ready-to-drink alcoholic coolers and cocktails.
† Other than food products.
‡ Includes milk, juice, and other products packaged in gable top cartons and liquid food aseptic cartons.
Details may not add to totals due to rounding.
Neg. = Less than 5,000 tons or 0.05 percent.
Source: Franklin Associates, A Division of ERG

**RECOVERY* OF PRODUCTS IN MUNICIPAL SOLID WASTE, 1960 TO 2009**
**(WITH DETAIL ON CONTAINERS AND PACKAGING)**
**(In thousands of tons)**

| | Thousands of Tons | | | | | | | | |
|---|---|---|---|---|---|---|---|---|---|
| **Products** | **1960** | **1970** | **1980** | **1990** | **2000** | **2005** | **2007** | **2008** | **2009** |
| **Durable Goods** | 350 | 940 | 1,360 | 3,460 | 6,580 | 8,040 | 8,220 | 8,360 | 8,160 |
| *(Detail in Table 13)* | | | | | | | | | |
| **Nondurable Goods** | 2,390 | 3,730 | 4,670 | 8,800 | 17,560 | 19,770 | 20,970 | 19,310 | 18,890 |
| *(Detail in Table 16)* | | | | | | | | | |
| **Containers and Packaging** | | | | | | | | | |
| **Glass Packaging** | | | | | | | | | |
| Beer and Soft Drink Bottles** | 90 | 140 | 730 | 1,890 | 1,530 | 2,000 | 2,340 | 2,260 | 2,340 |
| Wine and Liquor Bottles | 10 | 10 | 20 | 210 | 430 | 250 | 240 | 240 | 310 |
| Other Bottles & Jars | Neg. | Neg. | Neg. | 520 | 920 | 340 | 300 | 310 | 350 |
| ***Total Glass Packaging*** | 100 | 150 | 750 | 2,620 | 2,880 | 2,590 | 2,880 | 2,810 | 3,000 |
| **Steel Packaging** | | | | | | | | | |
| Beer and Soft Drink Cans | 10 | 20 | 50 | 40 | Neg. | Neg. | Neg. | Neg. | Neg. |
| Cans | 20 | 60 | 150 | 590 | 1,530 | 1,340 | 1,570 | 1,450 | 1,280 |
| Other Steel Packaging | Neg. | Neg. | Neg. | 60 | 160 | 160 | 160 | 160 | 230 |
| ***Total Steel Packaging*** | 30 | 80 | 200 | 690 | 1,690 | 1,500 | 1,730 | 1,610 | 1,510 |
| **Aluminum Packaging** | | | | | | | | | |
| Beer and Soft Drink Cans | Neg. | 10 | 320 | 990 | 830 | 650 | 690 | 670 | 690 |
| Other Cans | Neg. | Neg. | Neg. | Neg. | Neg. | Neg. | Neg. | 10 | NA |
| Foil and Closures | Neg. | Neg. | Neg. | 20 | 30 | 40 | 40 | 40 | NA |
| ***Total Aluminum Pkg*** | Neg. | 10 | 320 | 1,010 | 860 | 690 | 730 | 720 | 690 |
| **Paper & Paperboard Pkg** | | | | | | | | | |
| Corrugated Boxes | 2,520 | 2,760 | 6,390 | 11,530 | 20,330 | 22,100 | 22,980 | 22,760 | 22,100 |
| Gable Top/Aseptic Cartons‡ | | | Neg. | Neg. | Neg. | Neg. | Neg. | Neg. | 30 |
| Folding Cartons | | | 520 | 340 | 410 | 1,190 | 1,550 | 1,880 | 2,490 |
| Other Paperboard Packaging | | | Neg. | Neg. | Neg. | Neg. | Neg. | Neg. | Neg. |
| Bags and Sacks | | | Neg. | 200 | 300 | 320 | 420 | 440 | 450 |
| Wrapping Papers | | | Neg. | Neg. | Neg. | Neg. | Neg. | Neg. | Neg. |
| Other Paper Packaging | 220 | 350 | 300 | Neg. | Neg. | Neg. | Neg. | Neg. | Neg. |
| ***Total Paper & Board Pkg*** | 2,740 | 3,110 | 7,210 | 12,070 | 21,040 | 23,610 | 24,950 | 25,080 | 25,070 |
| **Plastics Packaging** | | | | | | | | | |
| PET Bottles and Jars | | | 10 | 140 | 380 | 590 | 700 | 730 | 720 |
| HDPE Natural Bottles | | | Neg. | 20 | 210 | 230 | 230 | 220 | 220 |
| Other Containers | Neg. | Neg. | Neg. | 20 | 170 | 140 | 190 | 280 | 290 |
| Bags and Sacks | | | | | | | | | |
| Wraps | | | | | | | | | |
| *Subtotal Bags, Sacks, and Wraps* | | | Neg. | 60 | 180 | 230 | 380 | 370 | 360 |
| Other Plastics Packaging | Neg. | Neg. | Neg. | 20 | 90 | 90 | 90 | 130 | 130 |
| ***Total Plastics Packaging*** | Neg. | Neg. | 10 | 260 | 1,030 | 1,280 | 1,590 | 1,730 | 1,720 |
| Wood Packaging | Neg. | Neg. | Neg. | 130 | 1,370 | 1,830 | 2,020 | 2,130 | 2,230 |
| Other Misc. Packaging | Neg. | Neg. | Neg. | Neg. | Neg. | Neg. | Neg. | Neg. | Neg. |
| ***Total Containers & Pkg*** | 2,870 | 3,350 | 8,490 | 16,780 | 28,870 | 31,500 | 33,900 | 34,080 | 34,220 |
| ***Total Product Wastes†*** | 5,610 | 8,020 | 14,520 | 29,040 | 53,010 | 59,310 | 63,090 | 61,750 | 61,270 |
| **Other Wastes** | | | | | | | | | |
| Food Scraps | Neg. | Neg. | Neg. | Neg. | 680 | 690 | 810 | 800 | 850 |
| Yard Trimmings | Neg. | Neg. | Neg. | 4,200 | 15,770 | 19,860 | 20,900 | 21,300 | 19,900 |
| Miscellaneous Inorganic Wastes | Neg. | Neg. | Neg. | Neg. | Neg. | Neg. | Neg. | Neg. | Neg. |
| ***Total Other Wastes*** | Neg. | Neg. | Neg. | 4,200 | 16,450 | 20,550 | 21,710 | 22,100 | 20,750 |
| ***Total MSW Recovered - Weight*** | 5,610 | 8,020 | 14,520 | 33,240 | 69,460 | 79,860 | 84,800 | 83,850 | 82,020 |

* Recovery of postconsumer wastes; does not include converting/fabrication scrap.
** Includes carbonated drinks and non-carbonated water, teas, flavored drinks, and ready-to-drink alcoholic coolers and cocktails.
† Other than food products.
‡ Includes milk, juice, and other products packaged in gable top cartons and liquid food aseptic cartons.
Details may not add to totals due to rounding.
Neg. = Less than 5,000 tons or 0.05 percent. NA = Not Available
Source: Franklin Associates, A Division of ERG

**RECOVERY* OF PRODUCTS IN MUNICIPAL SOLID WASTE, 1960 TO 2009**
**(WITH DETAIL ON CONTAINERS AND PACKAGING)**
**(In percent of generation of each product)**

| | Percent of Generation of Each Product | | | | | | | | |
|---|---|---|---|---|---|---|---|---|---|
| **Products** | **1960** | **1970** | **1980** | **1990** | **2000** | **2005** | **2007** | **2008** | **2009** |
| **Durable Goods** | 3.5% | 6.4% | 6.2% | 11.6% | 16.9% | 18.0% | 18.0% | 18.0% | 17.5% |
| *(Detail in Table 13)* | | | | | | | | | |
| **Nondurable Goods** | 13.8% | 14.9% | 13.6% | 16.9% | 27.4% | 31.1% | 34.0% | 32.9% | 35.3% |
| *(Detail in Table 16)* | | | | | | | | | |
| **Containers and Packaging** | | | | | | | | | |
| **Glass Packaging** | | | | | | | | | |
| Beer and Soft Drink Bottles** | 6.4% | 2.5% | 10.8% | 33.5% | 26.8% | 30.6% | 34.6% | 35.6% | 39.0% |
| Wine and Liquor Bottles | Neg. | Neg. | Neg. | 10.3% | 22.5% | 15.3% | 14.8% | 14.9% | 18.1% |
| Other Bottles & Jars | Neg. | Neg. | Neg. | 12.5% | 26.9% | 14.8% | 14.8% | 14.8% | 17.9% |
| ***Total Glass Packaging*** | 1.6% | 1.3% | 5.4% | 22.1% | 26.1% | 24.8% | 27.7% | 28.0% | 31.1% |
| **Steel Packaging** | | | | | | | | | |
| Beer and Soft Drink Cans | 1.6% | 1.3% | 9.6% | 26.7% | Neg. | Neg. | Neg. | Neg. | Neg. |
| Cans | Neg. | 1.7% | 5.3% | 23.2% | 58.2% | 62.9% | 64.6% | 62.8% | 66.0% |
| Other Steel Packaging | Neg. | Neg. | Neg. | 30.0% | 66.7% | 66.7% | 66.7% | 66.7% | 67.6% |
| ***Total Steel Packaging*** | Neg. | 1.5% | 5.5% | 23.9% | 58.9% | 63.3% | 64.8% | 63.1% | 66.2% |
| **Aluminum Packaging** | | | | | | | | | |
| Beer and Soft Drink Cans | Neg. | 10.0% | 37.6% | 63.9% | 54.6% | 44.8% | 48.6% | 48.2% | 50.7% |
| Other Cans | Neg. | Neg. | Neg. | Neg. | Neg. | Neg. | Neg. | 14.3% | NA |
| Foil and Closures | Neg. | Neg. | Neg. | 6.1% | 7.9% | 10.0% | 9.3% | 9.5% | NA |
| ***Total Aluminum Pkg*** | Neg. | 1.8% | 25.2% | 53.2% | 44.1% | 35.8% | 38.8% | 38.3% | 37.5% |
| **Paper & Paperboard Pkg** | | | | | | | | | |
| Corrugated Boxes | 34.4% | 21.6% | 37.4% | 48.0% | 67.3% | 71.5% | 73.6% | 76.6% | 81.3% |
| Gable Top/Aseptic Cartons‡ | | | Neg. | Neg. | Neg. | Neg. | Neg. | Neg. | 6.5% |
| Folding Cartons | | | Neg. | Neg. | 7.0% | 21.5% | 28.0% | 35.2% | 50.0% |
| Other Paperboard Packaging | | | Neg. | Neg. | Neg. | Neg. | Neg. | Neg. | Neg. |
| Bags and Sacks | | | Neg. | Neg. | 20.1% | 28.6% | 36.8% | 37.6% | 49.5% |
| Wrapping Papers | | | Neg. | Neg. | Neg. | Neg. | Neg. | Neg. | Neg. |
| Other Paper Packaging | 7.5% | 9.2% | 35.3% | Neg. | Neg. | Neg. | Neg. | Neg. | Neg. |
| ***Total Paper & Board Pkg*** | 19.4% | 14.5% | 27.4% | 36.9% | 52.7% | 59.6% | 62.5% | 65.5% | 71.8% |
| **Plastics Packaging** | | | | | | | | | |
| PET Bottles and Jars | | | 3.8% | 32.6% | 22.1% | 23.2% | 24.6% | 27.2% | 28.0% |
| HDPE Natural Bottles | | | Neg. | 3.8% | 30.4% | 28.8% | 28.0% | 29.3% | 28.9% |
| Other Containers | Neg. | Neg. | Neg. | 1.4% | 9.8% | 9.9% | 9.9% | 14.7% | 16.6% |
| Bags and Sacks | | | | | | | | | |
| Wraps | | | | | | | | | |
| *Subtotal Bags, Sacks, and Wraps* | | | Neg. | 2.4% | 4.3% | 5.2% | 9.1% | 9.3% | 9.4% |
| Other Plastics Packaging | Neg. | Neg. | Neg. | 1.0% | 3.2% | 2.8% | 2.3% | 3.5% | 3.6% |
| ***Total Plastics Packaging*** | Neg. | Neg. | Neg. | 3.8% | 9.2% | 10.3% | 11.7% | 13.3% | 13.7% |
| Wood Packaging | Neg. | Neg. | Neg. | 1.6% | 15.9% | 19.8% | 21.0% | 21.7% | 22.2% |
| Other Misc. Packaging | Neg. | Neg. | Neg. | Neg. | Neg. | Neg. | Neg. | Neg. | Neg. |
| ***Total Containers & Pkg*** | 10.5% | 7.7% | 16.1% | 26.0% | 38.1% | 41.3% | 43.2% | 44.9% | 47.8% |
| ***Total Product Wastes†*** | 10.3% | 9.6% | 13.3% | 19.8% | 29.7% | 32.1% | 33.9% | 34.1% | 35.7% |
| **Other Wastes** | | | | | | | | | |
| Food Scraps | Neg. | Neg. | Neg. | Neg. | 2.3% | 2.2% | 2.5% | 2.4% | 2.5% |
| Yard Trimmings | Neg. | Neg. | Neg. | 12.0% | 51.7% | 61.9% | 64.1% | 64.7% | 59.9% |
| Miscellaneous Inorganic Wastes | Neg. | Neg. | Neg. | Neg. | Neg. | Neg. | Neg. | Neg. | Neg. |
| ***Total Other Wastes*** | Neg. | Neg. | Neg. | 6.8% | 25.8% | 30.3% | 31.5% | 31.6% | 29.1% |
| ***Total MSW Recovered - %*** | 6.4% | 6.6% | 9.6% | 16.0% | 28.6% | 31.6% | 33.3% | 33.4% | 33.8% |

* Recovery of postconsumer wastes; does not include converting/fabrication scrap.
** Includes carbonated drinks and non-carbonated water, teas, flavored drinks, and ready-to-drink alcoholic coolers and cocktails.
† Other than food products.
‡ Includes milk, juice, and other products packaged in gable top cartons and liquid food aseptic cartons.
Details may not add to totals due to rounding.
Neg. = Less than 5,000 tons or 0.05 percent. NA = Not Available
Source: Franklin Associates, A Division of ERG

**PRODUCTS DISCARDED* IN THE MUNICIPAL WASTE STREAM, 1960 TO 2009**
**(WITH DETAIL ON CONTAINERS AND PACKAGING)**
**(In thousands of tons)**

| | Thousands of Tons | | | | | | | | |
|---|---|---|---|---|---|---|---|---|---|
| **Products** | **1960** | **1970** | **1980** | **1990** | **2000** | **2005** | **2007** | **2008** | **2009** |
| **Durable Goods** | 9,570 | 13,720 | 20,440 | 26,350 | 32,270 | 36,610 | 37,550 | 38,040 | 38,480 |
| *(Detail in Table 14)* | | | | | | | | | |
| **Nondurable Goods** | 14,940 | 21,330 | 29,750 | 43,370 | 46,450 | 43,880 | 40,790 | 39,380 | 34,550 |
| *(Detail in Table 17)* | | | | | | | | | |
| **Containers and Packaging** | | | | | | | | | |
| **Glass Packaging** | | | | | | | | | |
| Beer and Soft Drink Bottles** | 1,310 | 5,440 | 6,010 | 3,750 | 4,180 | 4,540 | 4,420 | 4,090 | 3,660 |
| Wine and Liquor Bottles | 1,070 | 1,890 | 2,430 | 1,820 | 1,480 | 1,380 | 1,380 | 1,370 | 1,400 |
| Other Bottles & Jars | 3,710 | 4,440 | 4,780 | 3,640 | 2,500 | 1,950 | 1,730 | 1,780 | 1,600 |
| ***Total Glass Packaging*** | 6,090 | 11,770 | 13,220 | 9,210 | 8,160 | 7,870 | 7,530 | 7,240 | 6,660 |
| **Steel Packaging** | | | | | | | | | |
| Beer and Soft Drink Cans | 630 | 1,550 | 470 | 110 | Neg. | Neg. | Neg. | Neg. | Neg. |
| Cans | 3,740 | 3,480 | 2,700 | 1,950 | 1,100 | 790 | 860 | 860 | 660 |
| Other Steel Packaging | 260 | 270 | 240 | 140 | 80 | 80 | 80 | 80 | 110 |
| ***Total Steel Packaging*** | 4,630 | 5,300 | 3,410 | 2,200 | 1,180 | 870 | 940 | 940 | 770 |
| **Aluminum Packaging** | | | | | | | | | |
| Beer and Soft Drink Cans | Neg. | 90 | 530 | 560 | 690 | 800 | 730 | 720 | 670 |
| Other Cans | Neg. | 60 | 40 | 20 | 50 | 80 | 30 | 60 | 70 |
| Foil and Closures | 170 | 410 | 380 | 310 | 350 | 360 | 390 | 380 | 410 |
| ***Total Aluminum Pkg*** | 170 | 560 | 950 | 890 | 1,090 | 1,240 | 1,150 | 1,160 | 1,150 |
| **Paper & Paperboard Pkg** | | | | | | | | | |
| Corrugated Boxes | 4,810 | 10,000 | 10,690 | 12,480 | 9,880 | 8,830 | 8,250 | 6,950 | 5,090 |
| Gable Top/Aseptic Cartons‡ | | | 790 | 510 | 550 | 500 | 500 | 490 | 430 |
| Folding Cartons | | | 3,300 | 3,960 | 5,410 | 4,340 | 3,980 | 3,460 | 2,490 |
| Other Paperboard Packaging | 3,840 | 4,830 | 230 | 290 | 200 | 160 | 150 | 120 | 90 |
| Bags and Sacks | | | 3,380 | 2,240 | 1,190 | 800 | 720 | 730 | 460 |
| Wrapping Papers | | | 200 | 110 | Neg. | Neg. | Neg. | Neg. | Neg. |
| Other Paper Packaging | 2,720 | 3,460 | 550 | 1,020 | 1,670 | 1,400 | 1,390 | 1,460 | 1,310 |
| ***Total Paper & Board Pkg*** | 11,370 | 18,290 | 19,140 | 20,610 | 18,900 | 16,030 | 14,990 | 13,210 | 9,870 |
| **Plastics Packaging** | | | | | | | | | |
| PET Bottles and Jars | | | 250 | 290 | 1,340 | 1,950 | 2,140 | 1,950 | 1,850 |
| HDPE Natural Bottles | | | 230 | 510 | 480 | 570 | 590 | 530 | 540 |
| Other Containers | 60 | 910 | 890 | 1,410 | 1,570 | 1,280 | 1,720 | 1,620 | 1,460 |
| Bags and Sacks | | | | | | | | | |
| Wraps | | | | | | | | | |
| *Subtotal Bags, Sacks, and Wraps* | | | 1,230 | 2,410 | 4,020 | 4,220 | 3,810 | 3,590 | 3,490 |
| Other Plastics Packaging | 60 | 1,180 | 790 | 2,020 | 2,750 | 3,120 | 3,780 | 3,590 | 3,470 |
| ***Total Plastics Packaging*** | 120 | 2,090 | 3,390 | 6,640 | 10,160 | 11,140 | 12,040 | 11,280 | 10,810 |
| Wood Packaging | 2,000 | 2,070 | 3,940 | 8,050 | 7,240 | 7,400 | 7,590 | 7,690 | 7,810 |
| Other Misc. Packaging | 120 | 130 | 130 | 150 | 240 | 280 | 320 | 310 | 280 |
| ***Total Containers & Pkg*** | 24,500 | 40,210 | 44,180 | 47,750 | 46,970 | 44,830 | 44,560 | 41,830 | 37,350 |
| ***Total Product Wastes†*** | 49,010 | 75,260 | 94,370 | 117,470 | 125,690 | 125,320 | 122,900 | 119,250 | 110,380 |
| **Other Wastes** | | | | | | | | | |
| Food Scraps | 12,200 | 12,800 | 13,000 | 23,860 | 29,130 | 31,300 | 31,800 | 32,540 | 33,440 |
| Yard Trimmings | 20,000 | 23,200 | 27,500 | 30,800 | 14,760 | 12,210 | 11,730 | 11,600 | 13,300 |
| Miscellaneous Inorganic Wastes | 1,300 | 1,780 | 2,250 | 2,900 | 3,500 | 3,690 | 3,750 | 3,780 | 3,820 |
| ***Total Other Wastes*** | 33,500 | 37,780 | 42,750 | 57,560 | 47,390 | 47,200 | 47,280 | 47,920 | 50,560 |
| ***Total MSW Discarded - Weight*** | 82,510 | 113,040 | 137,120 | 175,030 | 173,080 | 172,520 | 170,180 | 167,170 | 160,940 |

* Discards after materials and compost recovery. In this table, discards include combustion with energy recovery. Does not include construction & demolition debris, industrial process wastes, or certain other wastes.
** Includes carbonated drinks and non-carbonated water, teas, flavored drinks, and ready-to-drink alcoholic coolers and cocktails.
† Other than food products.
‡ Includes milk, juice, and other products packaged in gable top cartons and liquid food aseptic cartons.
Neg. = Less than 5,000 tons or 0.05 percent. Details may not add to totals due to rounding.
Source: Franklin Associates, A Division of ERG

**PRODUCTS DISCARDED* IN THE MUNICIPAL WASTE STREAM, 1960 TO 2009**
**(WITH DETAIL ON CONTAINERS AND PACKAGING)**
**(In percent of total discards)**

| | Percent of Total Discards | | | | | | | | |
|---|---|---|---|---|---|---|---|---|---|
| **Products** | **1960** | **1970** | **1980** | **1990** | **2000** | **2005** | **2007** | **2008** | **2009** |
| **Durable Goods** | 11.6% | 12.1% | 14.9% | 15.1% | 18.6% | 21.2% | 22.1% | 22.8% | 23.9% |
| *(Detail in Table 14)* | | | | | | | | | |
| **Nondurable Goods** | 18.1% | 18.9% | 21.7% | 24.8% | 26.8% | 25.4% | 24.0% | 23.6% | 21.5% |
| *(Detail in Table 17)* | | | | | | | | | |
| **Containers and Packaging** | | | | | | | | | |
| **Glass Packaging** | | | | | | | | | |
| Beer and Soft Drink Bottles** | 1.6% | 4.8% | 4.4% | 2.1% | 2.4% | 2.6% | 2.6% | 2.4% | 2.3% |
| Wine and Liquor Bottles | 1.3% | 1.7% | 1.8% | 1.0% | 0.9% | 0.8% | 0.8% | 0.8% | 0.9% |
| Other Bottles & Jars | 4.5% | 3.9% | 3.5% | 2.1% | 1.4% | 1.1% | 1.0% | 1.1% | 1.0% |
| ***Total Glass Packaging*** | 7.4% | 10.4% | 9.6% | 5.3% | 4.7% | 4.6% | 4.4% | 4.3% | 4.1% |
| **Steel Packaging** | | | | | | | | | |
| Beer and Soft Drink Cans | 0.8% | 1.4% | 0.3% | 0.1% | Neg. | Neg. | Neg. | Neg. | Neg. |
| Cans | 4.5% | 3.1% | 2.0% | 1.1% | 0.6% | 0.5% | 0.5% | 0.5% | 0.4% |
| Other Steel Packaging | 0.3% | 0.2% | 0.2% | 0.1% | 0.0% | 0.0% | 0.0% | 0.0% | 0.1% |
| ***Total Steel Packaging*** | 5.6% | 4.7% | 2.5% | 1.3% | 0.7% | 0.5% | 0.6% | 0.6% | 0.5% |
| **Aluminum Packaging** | | | | | | | | | |
| Beer and Soft Drink Cans | Neg. | 0.1% | 0.4% | 0.3% | 0.4% | 0.5% | 0.4% | 0.4% | 0.4% |
| Other Cans | Neg. | Neg. | Neg. | Neg. | Neg. | Neg. | Neg. | Neg. | Neg. |
| Foil and Closures | 0.2% | 0.4% | 0.3% | 0.2% | 0.2% | 0.2% | 0.2% | 0.2% | 0.3% |
| ***Total Aluminum Pkg*** | 0.2% | 0.5% | 0.7% | 0.5% | 0.6% | 0.7% | 0.7% | 0.7% | 0.7% |
| **Paper & Paperboard Pkg** | | | | | | | | | |
| Corrugated Boxes | 5.8% | 8.8% | 7.8% | 7.1% | 5.7% | 5.1% | 4.8% | 4.2% | 3.2% |
| Gable Top/Aseptic Cartons‡ | | | 0.6% | 0.3% | 0.3% | 0.3% | 0.3% | 0.3% | 0.3% |
| Folding Cartons | | | 2.4% | 2.3% | 3.1% | 2.5% | 2.3% | 2.1% | 1.5% |
| Other Paperboard Packaging | 4.7% | 4.3% | 0.2% | 0.2% | 0.1% | 0.1% | 0.1% | 0.1% | 0.1% |
| Bags and Sacks | | | 2.5% | 1.3% | 0.7% | 0.5% | 0.4% | 0.4% | 0.3% |
| Wrapping Papers | | | 0.1% | 0.1% | Neg. | Neg. | Neg. | Neg. | Neg. |
| Other Paper Packaging | 3.3% | 3.1% | 0.4% | 0.6% | 1.0% | 0.8% | 0.8% | 0.9% | 0.8% |
| ***Total Paper & Board Pkg*** | 13.8% | 16.2% | 14.0% | 11.8% | 10.9% | 9.3% | 8.8% | 7.9% | 6.1% |
| **Plastics Packaging** | | | | | | | | | |
| PET Bottles and Jars | | | 0.2% | 0.2% | 0.8% | 1.1% | 1.3% | 1.2% | 1.1% |
| HDPE Natural Bottles | | | 0.2% | 0.3% | 0.3% | 0.3% | 0.3% | 0.3% | 0.3% |
| Other Containers | 0.1% | 0.8% | 0.6% | 0.8% | 0.9% | 0.7% | 1.0% | 1.0% | 0.9% |
| Bags and Sacks | | | | | | | | | |
| Wraps | | | | | | | | | |
| *Subtotal Bags, Sacks, and Wraps* | | | 0.9% | 1.4% | 2.3% | 2.4% | 2.2% | 2.1% | 2.2% |
| Other Plastics Packaging | 0.1% | 1.0% | 0.6% | 1.2% | 1.6% | 1.8% | 2.2% | 2.1% | 2.2% |
| ***Total Plastics Packaging*** | 0.1% | 1.8% | 2.5% | 3.8% | 5.9% | 6.5% | 7.1% | 6.7% | 6.7% |
| Wood Packaging | 2.4% | 1.8% | 2.9% | 4.6% | 4.2% | 4.3% | 4.5% | 4.6% | 4.9% |
| Other Misc. Packaging | 0.1% | 0.1% | 0.1% | 0.1% | 0.1% | 0.2% | 0.2% | 0.2% | 0.2% |
| ***Total Containers & Pkg*** | 29.7% | 35.6% | 32.2% | 27.3% | 27.1% | 26.0% | 26.2% | 25.0% | 23.2% |
| ***Total Product Wastes†*** | 59.4% | 66.6% | 68.8% | 67.1% | 72.6% | 72.6% | 72.2% | 71.3% | 68.6% |
| **Other Wastes** | | | | | | | | | |
| Food Scraps | 14.8% | 11.3% | 9.5% | 13.6% | 16.8% | 18.1% | 18.7% | 19.5% | 20.8% |
| Yard Trimmings | 24.2% | 20.5% | 20.1% | 17.6% | 8.5% | 7.1% | 6.9% | 6.9% | 8.3% |
| Miscellaneous Inorganic Wastes | 1.6% | 1.6% | 1.6% | 1.7% | 2.0% | 2.1% | 2.2% | 2.3% | 2.4% |
| ***Total Other Wastes*** | 40.6% | 33.4% | 31.2% | 32.9% | 27.4% | 27.4% | 27.8% | 28.7% | 31.4% |
| ***Total MSW Discarded - %*** | 100.0% | 100.0% | 100.0% | 100.0% | 100.0% | 100.0% | 100.0% | 100.0% | 100.0% |

* Discards after materials and compost recovery. In this table, discards include combustion with energy recovery.
Does not include construction & demolition debris, industrial process wastes, or certain other wastes.

** Includes carbonated drinks and non-carbonated water, teas, flavored drinks, and ready-to-drink alcoholic coolers and cocktails.

† Other than food products.

‡ Includes milk, juice, and other products packaged in gable top cartons and liquid food aseptic cartons.

Neg. = Less than 5,000 tons or 0.05 percent. Details may not add to totals due to rounding.

Source: Franklin Associates, A Division of ERG

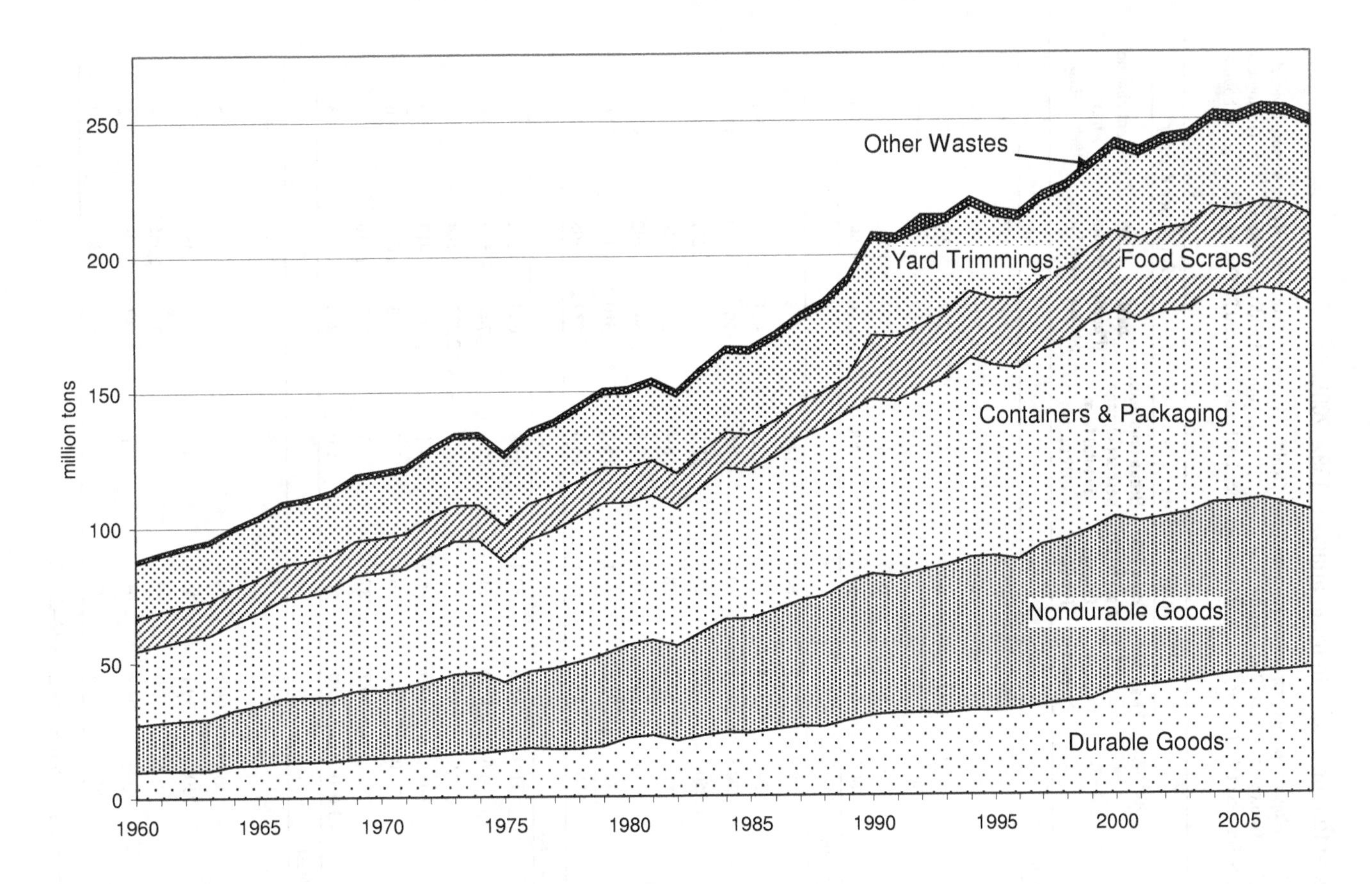


*SOURCE: U.S. EPA, Municipal Solid Waste in the United States: 2009 Facts and Figures*

## Selected information for stream sites dominated by urban land use and selected for pesticide trend analysis, 1992–2008.

[USGS, U.S. Geological Survey; $km^2$, square kilometers; basins were delineated by a number of USGS hydrologists and geographers in National Water-Quality Assessment Program study units using a variety of methods using data sources at multiple scales and were current as of December 2009 (N. Nakagaki, U.S. Geological Survey, 2009, written commun.); NLCD, National Landcover Database, U.S. Geological Survey, 2007, and Fry and others, 2009; population data based on 1990 and 2000 population counts (U.S. Bureau of the Census, 1992, and Geolytics, Inc., 2001) and 1990 and 2000 census block group boundaries (U.S. Bureau of the Census, 2001a and 2001b)]

| Site number (fig. 1) | Site short name | USGS station number | Site name | Drainage area ($km^2$) | Percent land use | | | Population density | | |
|---|---|---|---|---|---|---|---|---|---|---|
| | | | | | Urban 2001 from NLCD Change Product | Urban 1992 from NLCD Change Product | Urban land use, percent change in total from 1992 to 2001 | 2000 (People per $km^2$) | 1990 (People per $km^2$) | Percent change from 1990 to 2000 |
| | | | **Northeast** | | | | | | | |
| 1 | ABERJ | 01102500 | Aberjona River at Winchester, Mass. | 59.8 | 79.3 | 78.0 | 1.7 | 1,141 | 1,126 | 1.3 |
| 2 | CHRLS | 01104615 | Charles River above Watertown Dam at Watertown, Mass. | 695 | 41.2 | 40.1 | 2.7 | 571 | 534 | 6.9 |
| 3 | NRWLK | 01209710 | Norwalk River at Winnipauk, Conn. | 85.1 | 27.6 | 27.1 | 1.7 | 281 | 255 | 10.2 |
| 4 | LISHA | 01356190 | Lisha Kill near Niskayuna, N.Y. | 40.0 | 51.2 | 49.3 | 3.9 | 552 | 524 | 5.3 |
| 5 | BOUND | 01403900 | Bound Brook at Middlesex, N.J. | 126 | 61.2 | 60.3 | 1.6 | 1,391 | 1,292 | 7.7 |
| 6 | ACCOT | 01654000 | Accotink Creek near Annandale, Va. | 60.7 | 61.8 | 59.0 | 4.8 | 1,610 | 1,440 | 11.8 |
| | | | **South** | | | | | | | |
| 7 | SWIFT | 02087580 | Swift Creek near Apex, N.C. | 53.9 | 73.9 | 64.4 | 14.8 | 726 | 489 | 48.5 |
| 8 | GILLS | 02169570 | Gills Creek at Columbia, S.C. | 154 | 51.8 | 50.7 | 2.2 | 481 | 445 | 8.1 |
| 9 | SOPEC | 02335870 | Sope Creek near Marietta, Ga. | 79.5 | 74.4 | 68.3 | 9.0 | 902 | 793 | 13.7 |
| 10 | CHATT | 02338000 | Chattahoochee River near Whitesburg, Ga. | 6,250 | 28.8 | 25.4 | 13.3 | 311 | 231 | 34.6 |
| 11 | CAHAB | 0242354750 | Cahaba Valley Creek at Cross Creek Road at Pelham, Ala. | 66.1 | 30.2 | 28.4 | 6.4 | 275 | 216 | 27.3 |
| 12 | FLTCH | 07031692 | Fletcher Creek at Sycamore View Road at Memphis, Tenn. | 79.0 | 85.0 | 70.7 | 20.3 | 847 | 442 | 91.6 |
| 13 | WHITE | 08057200 | White Rock Creek at Greenville Avenue at Dallas, Tex. | 173 | 91.2 | 77.1 | 18.2 | 1,510 | 987 | 53.0 |
| 14 | SALAD | 08178800 | Salado Creek at Loop 13 at San Antonio, Tex. | 506 | 53.7 | 46.1 | 16.6 | 624 | 508 | 22.8 |
| | | | **Midwest** | | | | | | | |
| 15 | HOLES | 393944084120700 | Holes Creek at Huffman Park at Kettering, Ohio | 51.9 | 85.5 | 82.4 | 3.8 | 650 | 572 | 13.6 |
| 16 | LBUCK | 03353637 | Little Buck Creek near Indianapolis, Ind. | 44.6 | 86.9 | 71.9 | 20.9 | 749 | 572 | 30.9 |
| 17 | LINCO | 040869415 | Lincoln Creek at 47th Street at Milwaukee, Wis. | 26.0 | 91.6 | 91.0 | .7 | 2,222 | 2,184 | 1.7 |
| 18 | CLINT | 04161820 | Clinton River at Sterling Heights, Mich. | 803 | 48.5 | 46.6 | 4.1 | 469 | 413 | 13.6 |
| 19 | SHING | 05288705 | Shingle Creek at Queen Avenue at Minneapolis, Minn. | 73.0 | 78.4 | 76.3 | 2.7 | 1,093 | 1,045 | 4.6 |
| 20 | SALTC | 05531500 | Salt Creek at Western Springs, Ill. | 291 | 90.7 | 87.0 | 4.3 | 1,183 | 1,116 | 6.0 |
| 21 | DPLAI | 05532500 | Des Plaines River at Riverside, Ill. | 1,630 | 63.0 | 60.0 | 5.0 | 868 | 780 | 11.3 |
| | | | **West** | | | | | | | |
| 22 | LCOTT | 10168000 | Little Cottonwood Creek at Jordan River near Salt Lake City, Utah | 117 | 29.0 | 28.1 | 3.3 | 493 | 473 | 4.2 |
| 23 | WARMC | 11060400 | Warm Creek near San Bernardino, Calif. | 30.9 | 96.1 | 95.6 | .5 | 1,900 | 1,887 | .7 |
| 24 | SANTA | 11074000 | Santa Ana River below Prado Dam, Calif. | 3,730 | 39.1 | 37.5 | 4.2 | 539 | 451 | 19.5 |
| 25 | ARCAD | 11447360 | Arcade Creek near Del Paso Heights, Calif. | 81.5 | 100.0 | 97.9 | 2.1 | 2,034 | 1,986 | 2.4 |
| 26 | THORN | 12128000 | Thornton Creek near Seattle, Wash. | 29.2 | 95.8 | 95.7 | .1 | 2,364 | 2,205 | 7.2 |
| 27 | FANNO | 14206950 | Fanno Creek at Durham, Oreg. | 80.7 | 87.0 | 85.2 | 2.1 | 1,502 | 1,202 | 25.0 |

Source: U.S. Geological Survey, Trends in Pesticide Concentrations in Urban Streams in the United States, 1992–2008

Number of uncensored concentrations for pesticides in urban-stream samples, 1992–2008.

[Shaded cells indicate site/pesticide samples with at least 10 uncensored concentrations during the trend assessment period; DCPA, dimethyl tetrachloroterephthalate; --, indicates that fipronil and its degradates were not considered for trend analysis in the first two periods because samples were not analyzed for fipronil and its degradates until 1999]

| Trend assessment period | Site number (fig. 1) | Site short name | Number of uncensored concentrations | | | | | | | | |
|---|---|---|---|---|---|---|---|---|---|---|---|
| | | | Simazine | Prometon | Atrazine | Deethylatrazine | Metolachlor | Trifluralin | Pendimethalin | Tebuthiuron | Dacthal (DCPA) |
| 1992–2000 | 3 | NRWLK | 43 | 97 | 80 | 42 | 32 | 10 | 1 | 0 | 11 |
| | 6 | ACCOT | 76 | 78 | 64 | 48 | 74 | 11 | 29 | 6 | 21 |
| | 9 | SOPEC | 78 | 39 | 75 | 36 | 9 | 8 | 23 | 61 | 5 |
| | 16 | LBUCK | 113 | 118 | 118 | 112 | 118 | 29 | 21 | 46 | 47 |
| 1996–2004 | 1 | ABERJ | 21 | 51 | 40 | 17 | 28 | 16 | 7 | 9 | 0 |
| | 2 | CHRLS | 5 | 14 | 29 | 8 | 3 | 0 | 1 | 0 | 4 |
| | 3 | NRWLK | 28 | 112 | 88 | 58 | 42 | 13 | 0 | 1 | 1 |
| | 5 | BOUND | 45 | 66 | 67 | 58 | 52 | 16 | 13 | 28 | 20 |
| | 6 | ACCOT | 104 | 112 | 88 | 64 | 99 | 25 | 30 | 10 | 12 |
| | 8 | GILLS | 77 | 71 | 76 | 68 | 19 | 0 | 4 | 74 | 1 |
| | 9 | SOPEC | 92 | 57 | 88 | 64 | 5 | 11 | 23 | 90 | 7 |
| | 10 | CHATT | 88 | 67 | 87 | 43 | 13 | 1 | 4 | 60 | 2 |
| | 11 | CAHAB | 86 | 30 | 85 | 82 | 6 | 12 | 10 | 12 | 0 |
| | 12 | FLTCH | 56 | 43 | 54 | 51 | 54 | 13 | 20 | 19 | 2 |
| | 13 | WHITE | 120 | 114 | 120 | 120 | 117 | 11 | 66 | 37 | 16 |
| | 14 | SALAD | 43 | 70 | 70 | 65 | 15 | 0 | 3 | 69 | 2 |
| | 15 | HOLES | 74 | 99 | 104 | 94 | 73 | 27 | 23 | 0 | 0 |
| | 16 | LBUCK | 96 | 116 | 116 | 110 | 112 | 11 | 13 | 8 | 8 |
| | 18 | CLINT | 58 | 60 | 69 | 63 | 63 | 6 | 4 | 1 | 7 |
| | 19 | SHING | 6 | 82 | 73 | 51 | 56 | 8 | 4 | 32 | 19 |
| | 20 | SALTC | 44 | 59 | 66 | 63 | 54 | 11 | 5 | 2 | 5 |
| | 21 | DPLAI | 30 | 50 | 53 | 51 | 48 | 9 | 3 | 25 | 3 |
| | 22 | LCOTT | 11 | 68 | 54 | 45 | 0 | 7 | 17 | 39 | 26 |
| | 23 | WARMC | 42 | 36 | 2 | 2 | 2 | 0 | 0 | 18 | 25 |
| | 25 | ARCAD | 50 | 69 | 16 | 0 | 48 | 18 | 13 | 23 | 41 |
| | 26 | THORN | 30 | 68 | 17 | 0 | 0 | 6 | 0 | 1 | 1 |
| 2000–2008 | 2 | CHRLS | 4 | 13 | 38 | 15 | 7 | 0 | 1 | 0 | 0 |
| | 3 | NRWLK | 10 | 73 | 54 | 37 | 20 | 8 | 0 | 1 | 0 |
| | 4 | LISHA | 8 | 39 | 29 | 16 | 33 | 1 | 3 | 0 | 0 |
| | 5 | BOUND | 20 | 45 | 48 | 37 | 34 | 5 | 4 | 14 | 3 |
| | 6 | ACCOT | 85 | 93 | 70 | 51 | 78 | 17 | 21 | 7 | 4 |
| | 7 | SWIFT | 83 | 74 | 69 | 36 | 44 | 7 | 9 | 8 | 1 |
| | 8 | GILLS | 56 | 53 | 58 | 52 | 23 | 0 | 0 | 53 | 1 |
| | 9 | SOPEC | 109 | 82 | 106 | 85 | 4 | 9 | 24 | 105 | 5 |
| | 10 | CHATT | 106 | 82 | 105 | 69 | 23 | 2 | 5 | 76 | 5 |
| | 11 | CAHAB | 81 | 24 | 81 | 78 | 5 | 9 | 7 | 6 | 0 |
| | 13 | WHITE | 107 | 92 | 107 | 107 | 96 | 9 | 60 | 13 | 8 |
| | 14 | SALAD | 59 | 80 | 80 | 77 | 9 | 3 | 8 | 76 | 7 |
| | 17 | LINCO | 24 | 64 | 70 | 63 | 51 | 1 | 11 | 60 | 3 |
| | 18 | CLINT | 37 | 44 | 53 | 47 | 47 | 3 | 2 | 1 | 1 |
| | 19 | SHING | 2 | 67 | 59 | 44 | 45 | 1 | 3 | 26 | 6 |
| | 20 | SALTC | 37 | 62 | 75 | 72 | 63 | 11 | 6 | 2 | 2 |
| | 22 | LCOTT | 7 | 62 | 49 | 47 | 0 | 7 | 12 | 39 | 24 |
| | 24 | SANTA | 83 | 69 | 59 | 37 | 16 | 1 | 2 | 6 | 48 |
| | 25 | ARCAD | 53 | 76 | 12 | 2 | 60 | 20 | 26 | 17 | 44 |
| | 26 | THORN | 22 | 61 | 7 | 0 | 0 | 11 | 0 | 1 | 6 |
| | 27 | FANNO | 87 | 71 | 83 | 43 | 48 | 24 | 6 | 57 | 2 |

Number of uncensored concentrations for pesticides in urban-stream samples, 1992–2008.—Continued

[Shaded cells indicate site/pesticide samples with at least 10 uncensored concentrations during the trend assessment period; DCPA, dimethyl tetrachloroterephthalate; --, indicates that fipronil and its degradates were not considered for trend analysis in the first two periods because samples were not analyzed for fipronil and its degradates until 1999]

| Trend assessment period | Site number (fig. 1) | Site short name | Number of uncensored concentrations | | | | | | |
|---|---|---|---|---|---|---|---|---|---|
| | | | Chlorpyrifos | Malathion | Diazinon | Fipronil | Fipronil sulfide | Desulfinyl-fipronil | Carbaryl |
| 1992–2000 | 3 | NRWLK | 2 | 2 | 30 | -- | -- | -- | 26 |
| | 6 | ACCOT | 41 | 11 | 75 | -- | -- | -- | 49 |
| | 9 | SOPEC | 39 | 10 | 70 | -- | -- | -- | 33 |
| | 16 | LBUCK | 64 | 26 | 109 | -- | -- | -- | 36 |
| 1996–2004 | 1 | ABERJ | 3 | 0 | 55 | -- | -- | -- | 38 |
| | 2 | CHRLS | 0 | 0 | 21 | -- | -- | -- | 13 |
| | 3 | NRWLK | 3 | 1 | 36 | -- | -- | -- | 28 |
| | 5 | BOUND | 24 | 7 | 59 | -- | -- | -- | 37 |
| | 6 | ACCOT | 25 | 11 | 105 | -- | -- | -- | 69 |
| | 8 | GILLS | 32 | 32 | 63 | -- | -- | -- | 21 |
| | 9 | SOPEC | 20 | 5 | 76 | -- | -- | -- | 30 |
| | 10 | CHATT | 9 | 3 | 81 | -- | -- | -- | 44 |
| | 11 | CAHAB | 22 | 2 | 60 | -- | -- | -- | 15 |
| | 12 | FLTCH | 30 | 25 | 47 | -- | -- | -- | 37 |
| | 13 | WHITE | 54 | 22 | 120 | -- | -- | -- | 55 |
| | 14 | SALAD | 12 | 9 | 52 | -- | -- | -- | 24 |
| | 15 | HOLES | 18 | 8 | 90 | -- | -- | -- | 36 |
| | 16 | LBUCK | 18 | 12 | 111 | -- | -- | -- | 34 |
| | 18 | CLINT | 8 | 1 | 51 | -- | -- | -- | 17 |
| | 19 | SHING | 1 | 4 | 54 | -- | -- | -- | 19 |
| | 20 | SALTC | 2 | 6 | 54 | -- | -- | -- | 26 |
| | 21 | DPLAI | 1 | 3 | 41 | -- | -- | -- | 21 |
| | 22 | LCOTT | 1 | 11 | 64 | -- | -- | -- | 29 |
| | 23 | WARMC | 5 | 3 | 40 | -- | -- | -- | 5 |
| | 25 | ARCAD | 42 | 33 | 71 | -- | -- | -- | 58 |
| | 26 | THORN | 3 | 6 | 55 | -- | -- | -- | 15 |
| 2000–2008 | 2 | CHRLS | 0 | 0 | 12 | 11 | 19 | 16 | 19 |
| | 3 | NRWLK | 1 | 1 | 24 | 0 | 4 | 2 | 18 |
| | 4 | LISHA | 1 | 1 | 31 | 0 | 1 | 1 | 31 |
| | 5 | BOUND | 3 | 0 | 30 | 13 | 16 | 19 | 18 |
| | 6 | ACCOT | 6 | 8 | 69 | 36 | 21 | 27 | 52 |
| | 7 | SWIFT | 4 | 0 | 50 | 52 | 49 | 52 | 39 |
| | 8 | GILLS | 3 | 15 | 31 | 16 | 18 | 19 | 16 |
| | 9 | SOPEC | 15 | 4 | 58 | 43 | 42 | 40 | 35 |
| | 10 | CHATT | 5 | 2 | 62 | 49 | 40 | 41 | 55 |
| | 11 | CAHAB | 17 | 1 | 31 | 31 | 31 | 27 | 20 |
| | 13 | WHITE | 30 | 26 | 87 | 48 | 36 | 46 | 49 |
| | 14 | SALAD | 3 | 7 | 32 | 16 | 20 | 26 | 26 |
| | 17 | LINCO | 0 | 2 | 45 | 0 | 1 | 5 | 16 |
| | 18 | CLINT | 1 | 0 | 28 | 6 | 1 | 1 | 20 |
| | 19 | SHING | 0 | 2 | 36 | 4 | 8 | 17 | 19 |
| | 20 | SALTC | 9 | 5 | 33 | 33 | 24 | 29 | 29 |
| | 22 | LCOTT | 1 | 3 | 46 | 0 | 1 | 1 | 26 |
| | 24 | SANTA | 3 | 9 | 39 | 23 | 31 | 41 | 23 |
| | 25 | ARCAD | 39 | 35 | 77 | 58 | 50 | 56 | 60 |
| | 26 | THORN | 0 | 1 | 21 | 5 | 4 | 16 | 22 |
| | 27 | FANNO | 17 | 2 | 61 | 22 | 26 | 30 | 62 |

Source: U.S. Geological Survey, Trends in Pesticide Concentrations in Urban Streams in the United States, 1992–2008

Trends, in percent per year, for simazine, prometon, atrazine, and deethylatrazine for the 1996–2004 period.

Simazine, 1996–2004
Prometon, 1996–2004
Atrazine, 1996–2004
Deethylatrazine, 1996–2004

SITE NUMBER AND SHORT NAME (TABLE 1)

NORTHEAST
1. ABERJ
2. CHRLS
3. NRWLK
4. LISHA
5. BOUND
6. ACCOT

SOUTH
7. SWIFT
8. GILLS
9. SOPEC
10. CHATT
11. CAHAB
12. FLTCH
13. WHITE
14. SALAD

MIDWEST
15. HOLES
16. LBUCK
17. LINCO
18. CLINT
19. SHING
20. SALTC
21. DPLAI

WEST
22. LCOTT
23. WARMC
24. SANTA
25. ARCAD
26. THORN
27. FANNO

Simazine: 2. CHRLS HC; 4. LISHA NR; 7. SWIFT NR; 17. LINCO NR; 19. SHING HC; 23. WARMC *; 24. SANTA NR; 27. FANNO NR

Prometon: 4. LISHA NR; 7. SWIFT NR; 17. LINCO NR; 24. SANTA NR; 27. FANNO NR

Atrazine: 4. LISHA NR; 7. SWIFT NR; 17. LINCO NR; 23. WARMC HC; 24. SANTA NR; 27. FANNO NR

Deethylatrazine: 2. CHRLS HC; 4. LISHA NR; 7. SWIFT NR; 17. LINCO NR; 23. WARMC HC; 24. SANTA NR; 25. ARCAD HC; 26. THORN HC; 27. FANNO NR

-60 -30 0 30 60

TREND, IN PERCENT PER YEAR

**EXPLANATION**

**95–percent confidence limits**

Nonsignificant trend (p>0.10)
Significant trend (p<0.1): Downward, Upward
Highly significant trend (p<0.01): Downward, Upward

HC Too highly censored to analyze trends
NR Samples not representative of trend assessment period
< Less than
> Greater than
| Estimated trend value

*Confidence limit box outside of graphs represents trends with confid-ence limits beyond the scale of the other trends. Scale was not extended in order to preserve a scale representative of majority of trends.

Source: U.S. Geological Survey, Trends in Pesticide Concentrations in Urban Streams in the United States, 1992–2008

Trends, in percent per year, for simazine, prometon, atrazine, and deethylatrazine for the 2000–2008 period.

Simazine, 2000–2008 | Prometon, 2000–2008 | Atrazine, 2000–2008 | Deethylatrazine, 2000–2008

SITE NUMBER AND SHORT NAME (TABLE 1)

NORTHEAST: 1. ABERJ, 2. CHRLS, 3. NRWLK, 4. LISHA, 5. BOUND, 6. ACCOT

SOUTH: 7. SWIFT, 8. GILLS, 9. SOPEC, 10. CHATT, 11. CAHAB, 12. FLTCH, 13. WHITE, 14. SALAD

MIDWEST: 15. HOLES, 16. LBUCK, 17. LINCO, 18. CLINT, 19. SHING, 20. SALTC, 21. DPLAI

WEST: 22. LCOTT, 23. WARMC, 24. SANTA, 25. ARCAD, 26. THORN, 27. FANNO

–60 –30 0 30 60

TREND, IN PERCENT PER YEAR

**EXPLANATION**

**95-percent confidence limits**

Nonsignificant trend (p>0.10)

Downward / Upward — Significant trend (p<0.1)

Downward / Upward — Highly significant trend (p<0.01)

HC Too highly censored to analyze trends

NR Samples not representative of trend assessment period

< Less than

> Greater than

| Estimated trend value

Source: U.S. Geological Survey, Trends in Pesticide Concentrations in Urban Streams in the United States, 1992–2008

Trends, in percent per year, for chlorpyrifos, diazinon, and carbaryl for the 1996–2004 period.



Source: U.S. Geological Survey, Trends in Pesticide Concentrations in Urban Streams in the United States, 1992–2008

Trends, in percent per year, for chlorpyrifos, diazinon, and carbaryl for the 2000–2008 period.

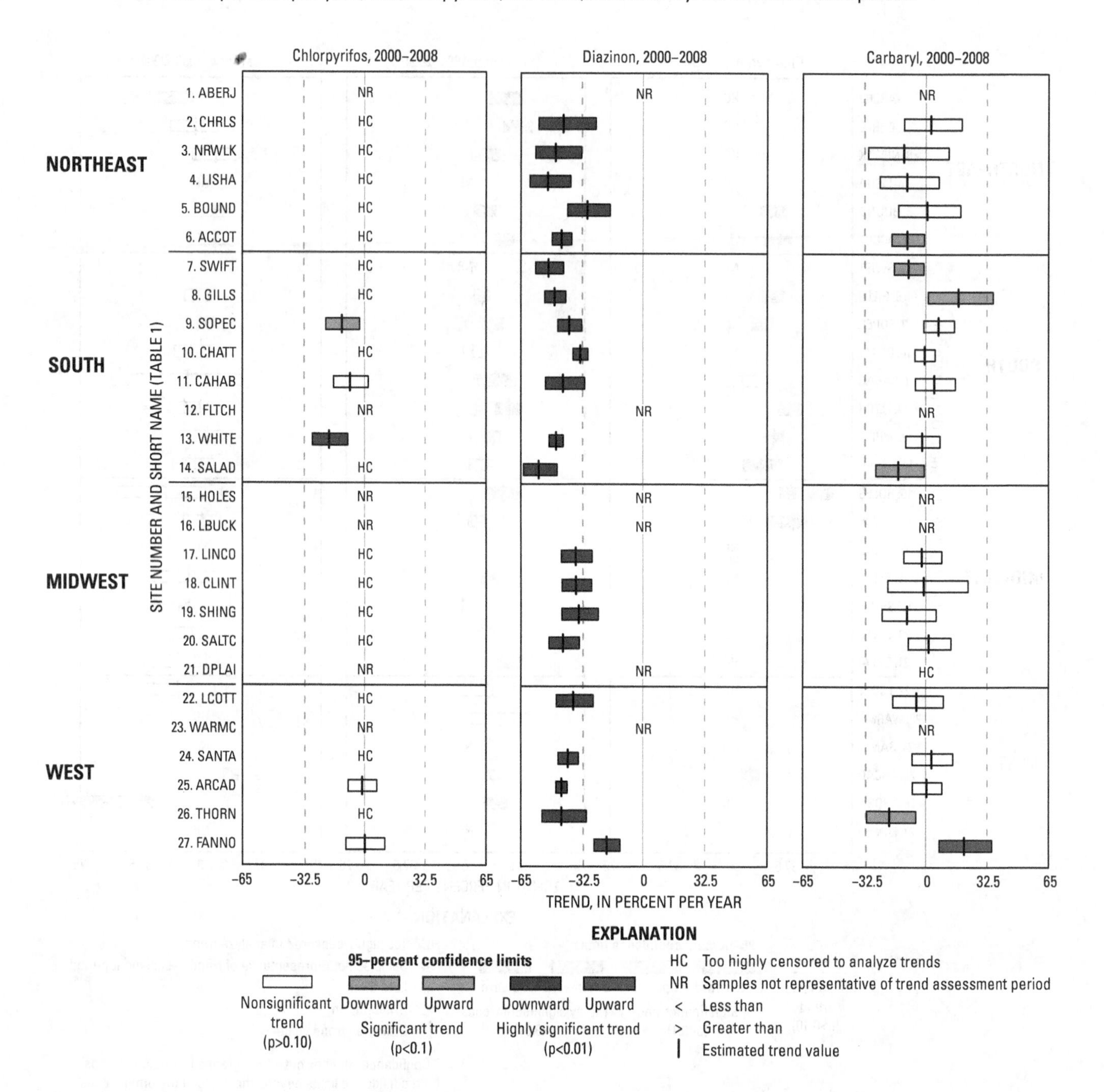


Source: U.S. Geological Survey, Trends in Pesticide Concentrations in Urban Streams in the United States, 1992–2008

Flow-adjusted trends, in percent per year, for fipronil, fipronil sulfide, and desulfinylfipronil for the 2000–2008 period.

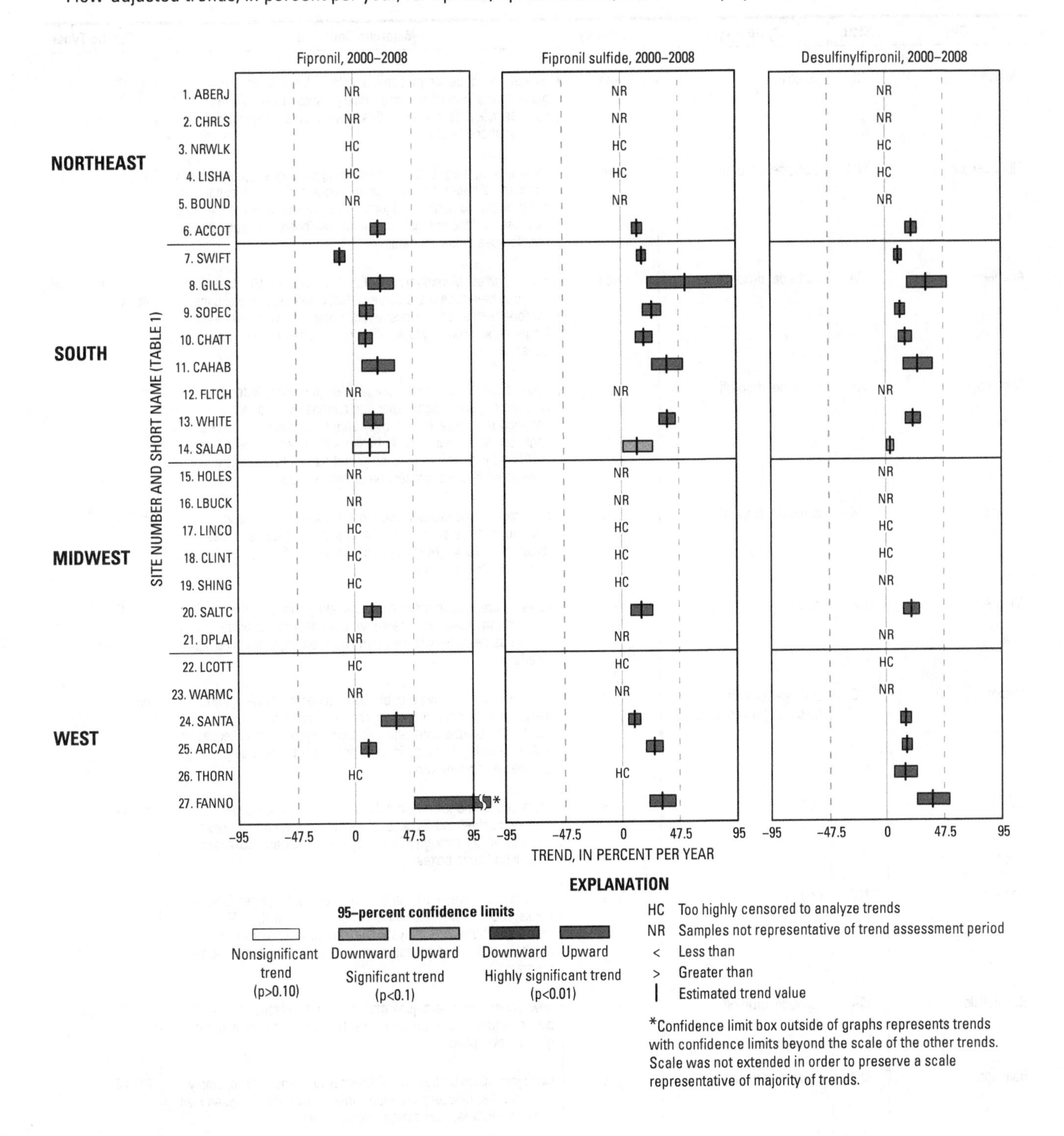


Source: U.S. Geological Survey, Trends in Pesticide Concentrations in Urban Streams in the United States, 1992–2008

## Recycling Profiles of 100 Major U.S. Cities

| City | State | System Type | Voluntary | Materials Collected | Plastic Types |
|---|---|---|---|---|---|
| Akron | OH | curbside | yes | all metal food, beverage cans, aluminum trays, aluminum foil, glass bottles, jars of any color, rigid plastics marked #1 thru #7 such as milk jugs, plastic bottles, newspapers, magazines, junk mail, or phone books | #1, #2 |
| Albuquerque | NM | curbside, drop-off | yes | newspapers, magazines, and shopping catalogues; junk mail and home office paper; tin/steel (small pieces/containers); aluminum cans; all plastic bottles and jugs with a neck or screw top; #1 and #2 plastic bottles or jugs; corrugated cardboard (flattened and bundled-see below for details) | #1, #2 |
| Anaheim | CA | curbside, drop-off | yes | plastic bottles, aluminum cans (CRV and Non-CRV), glass bottles, food and beverage glass bottles, plastic bottles, bi-metal cans, cardboard/kraftpaper, newspapers, computer paper, white/color ledger paper, coated stock, office papers, magazines, mixed paper | #1, #2, #3, #4, #5, #6, #7 |
| Anchorage | AK | curbside, drop-off | yes | mixed paper: newspapers, magazines, junk mail, flattened cardboard, paper bags, paper egg cartons, file folders, phonebooks, cereal boxes, copy paper; aluminum and tin/steel cans; plastic bottles (PET #1 bottles with a neck); plastic jugs (HDPE #2 jugs with a neck, no yogurt tubs); plastic bags and film: newspaper sleeves, grocery bags, shrink wrap | #1, #2 |
| Arlington | TX | curbside, drop-off | yes | newspapers, magazines, telephone books, junk mail and envelopes, office paper, cardboard, paper bags, jars/bottles (clear, brown and green), plastic jugs and bottles, aluminum/steel/tin cans | #1, #2 |
| Atlanta | GA | curbside | yes | newspapers, plastic bottles and jars, aluminum/steel/tin cans, glass (brown, clear and green), yard waste, organic waste, Christmas trees, leaves, tree trimmings, brush, grass clippings, weeds | #1, #2 |
| Aurora | CO | curbside (private - no public trash collection) | yes | The city does not have a public trash collection system, and therefore does not provide recycling services directly. However, most of the private companies that are registered to collect trash in Aurora also provide their customers curbside recycling services. Programs vary. | na |
| Austin | TX | curbside | yes | newspapers, magazines, catalogs, junk mail, office paper; aluminum, steel and tin cans; glass bottles and jars, all colors; rigid plastic (#1 through #7); corrugated cardboard; boxboard (cereal and soda boxes) | #1, #2 |
| Babylon | NY | drop-off | yes | plastics/glass, waste oil, car batteries, computer paper, telephone books, rocks/bricks, household batteries, junk mail, oil filters, metal/aluminum cans, polystyrene, newspapers, cardboard, concrete, tires. Also accepted are bicycles for the adopt-a-bike program. | na |
| Bakersfield | CA | curbside, drop-off | yes | newspaper, cardboard, junk mail, #1 and # 2 plastic bottles, office paper, magazines, aluminum cans, tin cans, clear glass, green glass, brown glass | #1, #2 |
| Baltimore | MD | curbside, drop-off | yes | glass jars, glass bottles, aluminum/steel/tin cans, plastic, empty aerosol cans, newspapers, magazines, telephone books, ad mail, cardboard/boxes, mixed paper, scrap paper | #1, #2 |
| Baton Rouge | LA | curbside, drop-off | yes | newspaper, magazines, scrap paper, cardboard, glass, plastics with a 1 - 7 inside the recycle arrow, milk cartons, juice boxes, detergent refill containers, aluminum cans, tin cans, steel cans, bi-metal cans and metal lids | #1, #2 |
| Birmingham | AL | curbside, drop-off | yes | glass (all types); mixed paper (cereal boxes, box packaged foods, phone books, magazines, office paper, junk mail, etc.); newspaper; corrugated cardboard (broken down); aluminum, steel and tin cans; plastics #1 and #2 (no longer accepting #3-7!); cell phones; printer and ink toner cartridges; batteries (rechargeable and single-use). E-waste: computers, monitors (no TV's) printers, peripherals, other plug-in electronics (processed locally by Technical Knock Out) | #1, #2 |

| City | State | System Type | Voluntary | Materials Collected | Plastic Types |
|---|---|---|---|---|---|
| Boston | MA | curbside, drop-off | yes | newspaper (with inserts); magazines/catalogues; junk mail (remove free samples; plastic envelope window is ok); white and colored paper/brown bags; telephone books; flattened food boxes; paperback books; milk and juice cartons; juice/soy milk boxes; cardboard boxes; pizza boxes; glass bottles/jars; tin and aluminum cans, foil, and pie plates; all plastic containers (no motor oil or chemical containers); cardboard/spiral cans; rigid plastics | #1, #2, #3, #4, #5, #6, #7 |
| Brookhaven | NY | curbside, drop-off | yes | glass bottles (clear and colored), aluminum/bimetallic/ tin cans, aerosol spray cans, plastic, aluminum foil/containers, newspapers, grocery bags, corrugated cardboard cartons, mixed low grade paper, white page phone books | #1, #2 |
| Buffalo | NY | curbside | yes | paper (newspaper, office paper, cardboard, and other paper types); yard trimmings (grass, leaves, shrub and tree clippings are recycled by composting); glass bottles and jars (clear, green, and amber); aluminum and steel cans; batteries - both dry cell (toy/watch/flashlight batteries) and wet cell (vehicle batteries); used motor oil and filters; plastics (soda bottles, milk jugs, bags, detergent containers, etc.); tires | na |
| Charlotte | NC | curbside | yes | all glass containers, magazines, shopping catalogs, milk jugs, newspapers and inserts, cardboard, plastic soft drink and liquor bottles, spiral paper cans, aluminum/steel/tin cans, telephone books, office paper | #1, #2 |
| Chesapeake | VA | curbside, drop-off | yes | aluminum and steel cans; pie plates and foil; clear and brown glass; newspapers; #1 and #2 plastic bottles | #1, #2 |
| Chicago | IL | curbside, dropoff | yes | city-served blue bag program (clean paper, newspaper, magazines, junk mail, cardboard, clean food boxes, phone books, catalogs, brown paper bags, gift wrap, metals, glass and plastics, empty aluminum and steel cans, empty aerosol cans, rinsed aluminum foil and pie plates, milk, juice, soft drink, water, and detergent bottles bearing the #1 or #2 symbol, clear, green, and brown glass bottles and jars, yard waste) | #1, #2 |
| Cincinnati | OH | curbside | yes | newspaper; office paper; junk mail and envelopes; cardboard (broken down 3' X 3'); paperboard (such as cereal boxes); brown paper bags; magazines; plastic bottles and jugs (remove lids); aluminum and steel cans; empty aerosol cans (remove lids and tips); glass bottles and jars (remove lids) | #1, #2 |
| Cleveland | OH | curbside | yes | glass bottles and jars, metal cans, plastic containers, newspapers, cardboard | #1, #2 |
| Colorado Springs | CO | drop-off | yes | motor oil, paint, hazardous chemicals | na |
| Columbus | OH | curbside, drop-off | yes | newspaper, including all inserts; magazines, catalogs and telephone books; mail, scrap paper and envelopes (windows ok); brown paper bags; paperboard (such as cereal or snack boxes); cardboard boxes (please break down and, if needed, cut to 3 feet by 3 feet); bottles and jars; plastic bottles (any); glass bottles and jars; aluminum cans, gutters and siding; steel soup, food and aerosol cans | #1, #2 |
| Corpus Christi | TX | curbside, drop-off | yes | newspapers (with ads and inserts); mixed paper (keep these separate from newspaper) - unwanted mail and envelopes, magazines, catalogs and phone books, paperboard boxes (like cereal boxes), shoe boxes and other similar paper; corrugated cardboard (flattened to less than 2ft square); glass (NOT accepted); aluminum, tin and steel; plastic bottles and containers (soap, shampoo or soft drink bottles, detergent containers); food containers marked with a #1 or #2 recycling symbol | #1, #2 |
| Dallas | TX | curbside, drop-off | yes | newspapers and inserts, mixed paper, magazines, junk mail, home/office paper, chipboard, glass containers, plastic bottles, aluminum cans, steel/tin food cans, empty aerosol cans | #1, #2, #3 |
| Denver | CO | curbside, drop-off | yes | newspapers and inserts, mixed paper, magazines, junk mail, home/office paper, glass containers, aluminum cans, steel/tin food cans, empty aerosol cans | na |

| City | State | System Type | Voluntary | Materials Collected | Plastic Types |
|---|---|---|---|---|---|
| Des Moines | IA | curbside | yes | glass bottles and jars (clear, green and brown), newspapers and inserts, brown paper bags, corrugated cardboard, junk mail, telephone books, magazines, catalogs, paperback books, colored paper, folders and window envelopes, all metal cans, plastic containers | #1, #2 |
| Detroit | MI | drop-off | yes | newspaper; mixed paper (items that CAN NOT be recycled are: tissue paper, receipts, napkins, wrapping paper, and paper towel); glossy paper; books; cardboard and chipboard; glass; all metals; plastic (#1 and #2 plastics together, and #4, #5, #6, and #7 plastics together); styrofoam; aseptic containers; batteries; computers and electronics; # 1, #2, and #4 plastic bags | #1, #2, #4, #5, #6, #7 |
| El Paso | TX | curbside, drop-off | no | corrugated cardboard, brown paper bags, newspapers and inserts, magazines, junk mail, white/colored bond paper, computer paper, plastic, aluminum/steel/tin cans, aluminum foil and pie plates, copper, brass, iron, aluminum | #1, #2 |
| Fort Wayne | IN | curbside | yes | newspapers, magazines, catalogues, cardboard, fiberboard, phonebooks, plastic, glass (brown, clear, green), cans (aluminum, bi-metal, tin, steel), aluminum foil and pie pans, and empty steel paint cans | #1, #2 |
| Fort Worth | TX | curbside | yes | paper, cardboard, catalogs, envelopes, junk mail, magazines, newspapers (all sections), paper bags, telephone books, aluminum cans/baking tins, steel/tin food cans/lids, empty aerosol cans, steel paint cans, glass bottles and jars, plastic bottles/cups | #1, #2, #3, #4, #5, #6, #7 |
| Fremont | CA | curbside, drop-off | yes | newspapers and inserts, most white/colored paper, magazines, junk mail, brown paper bags, catalogs, window envelopes, paper egg cartons, telephone books, flattened cereal/cracker/shoe boxes, glass bottles and jars, plastic containers, aluminum/steel/tin cans, yard waste, food scraps | #1, #2, #3, #4, #5, #6, #7 |
| Fresno | CA | curbside | no | aluminum/tin cans, cardboard, catalogs, chipboard, glass bottles and jars (all colors), junk mail (including envelopes), magazines, newspapers and inserts, plastic bottles (clear/green plastic soda and water bottles, plastic containers, phone books, yard waste | #1, #2, #3 |
| Garland | TX | curbside, drop-off | yes | newspapers and inserts, magazines, aluminum/steel/tin cans, empty aerosol cans, glass bottles and jars (clear and colored), plastic, six pack rings, corrugated cardboard, junk mail, brown paper bags, telephone books, white office/computer paper, chipboard | #1, #2 |
| Glendale | AZ | curbside, drop-off | yes | aluminum cans/foil/foil baking pans, cardboard, cartons, chipboard (without inserts), magazines, junk mail, catalogs, brown paper bags, telephone directories, newspapers and inserts, plastic containers, steel/tin cans | #1, #2 |
| Grand Rapids | MI | curbside | yes | aluminum/steel cans, glass bottles and jars (all colors), plastic bottles, newspapers and inserts, junk mail, corrugated cardboard, magazines, telephone books, white ledger paper, cereal boxes (with liners removed), colored/mixed paper, yard trimmings | #1, #2 |
| Greensboro | NC | curbside, drop-off | yes | plastic bottles and jugs, newspapers, magazines, all aluminum beverage cans, office paper, mail, notebook paper, corrugated cardboard, chipboard, brown/gray egg cartons, all steel beverage and food cans, glass (all colors, shapes and sizes), all aerosol cans | #1, #2 |
| Hempstead | NY | curbside, drop-off | yes | newspapers, cans, plastic, bottles/glass, aluminum foil, corrugated cardboard | #1, #2 |
| Hialeah | FL | curbside | yes | newspapers, aluminum/steel cans, plastics, glass bottles and jars, plastic coated cardboard, milk and juice containers | #1, #2 |
| Honolulu | HI | drop-off, curbside (pilot program) | man | newspaper, corrugated cardboard, glass bottles and jars, aluminum cans and plastic bottles and jugs, grass, tree and hedge trimmings, christmas trees | #1, #2 |
| Houston | TX | curbside, drop-off | no | newspapers, magazines, telephone books, aluminum and tin cans, junk mail, corrugated cardboard, plastic soft drink/milk/water containers, used oil | #1, #2 |

| City | State | System Type | Voluntary | Materials Collected | Plastic Types |
|---|---|---|---|---|---|
| Indianapolis | IN | curbside, drop-off | yes | glass; #1 & #2 plastics; aluminum, tin, and steel beverage and food cans; newspapers; cardboard and magazines | #1, #2 |
| Islip | NY | curbside | yes | newspapers, cans, plastic, bottles/glass, corrugated cardboard | na |
| Jacksonville | FL | curbside | yes | plastic, glass bottles and jars, metal and aluminum cans, newspapers and inserts, magazines, catalogs, phone books, brown paper bags, corrugated cardboard | #1, #2 |
| Jersey City | NJ | curbside, drop-off | no | mixed newspaper, magazines, junk mail, office paper, telephone books, cardboard boxes, corrugated and laundry detergent boxes, glass bottles, aluminum cans, metal cans, milk cartons, drink boxes, water containers, food containers, household cleaner containers, laundry detergent bottles, shampoo bottles, and other related plastic containers | na |
| Kansas City | MO | drop-off | yes | aluminum cans, car batteries, household (dry cell) batteries, corrugated cardboard, clothing (dry), foil/pie pans, glass bottles, magazines, newspapers, mixed office paper, paperboard, plastic bottles, tin cans, telephone books, scrap metal, yard waste | #1, #2 |
| Las Vegas | NV | curbside | yes | newspaper, glass/plastic containers, aluminum/tin cans, corrugated cardboard, phone books | na |
| Lexington-Fayette | KY | curbside | no | boxboard, brown paper bags, catalogs, corrugated cardboard, magazines, newspapers/inserts, office paper, telephone books, junk mail, empty aerosol cans, aluminum/steel cans, plastic bottles/jugs only, glass bottles and jars(blue, brown, clear, and green) | na |
| Lincoln | NE | drop-off | yes | newspaper, glass containers, aluminum/steel cans, old corrugated cardboard, mixed residential paper, plastic | #1, #2 |
| Long Beach | CA | curbside, drop-off | yes | glass, aluminum, steel, tin, plastic, aerosol and empty paint cans, mixed paper, bundled newspapers, corrugated cardboard, used motor oil in recycling container | #1, #2, #3, #4 |
| Los Angeles | CA | curbside, dropoff | yes | paper, all cardboard boxes and chipboard, all aluminum, tin, metal, and bi-metal cans, pie tins, clean aluminum foils; empty paint and aerosol cans with plastic caps removed, and wire hangers, all glass bottles and jars, plastics 1 through 7, polystyrene (styrofoam(r)) | #1, #2, #3, #4, #5, #6, #7 |
| Louisville | KY | curbside, drop-off | yes | junk mail, brown paper bags, telephone books, magazines, catalogs, newspapers and inserts, flattened cardboard, glass bottles and jars (clear, brown, green and blue) aluminum/steel/tin cans, empty aerosol cans, aluminum foil/pie and cake pans, plastic, office paper and envelopes | #1, #2 |
| Lubbock | TX | drop-off | yes | glass (clear and colored), plastic, aluminum and beverage cans, tin food cans, corrugated cardboard, newspaper, computer paper (green bar or shredded), white ledger paper, yard waste | #1, #2 |
| Madison | WI | curbside | no | newspapers, corrugated cardboard, magazines, catalogs, brown paper bags, telephone books, glass, aluminum, tin/steel cans, plastic containers | #1, #2 |
| Memphis | TN | curbside, drop-off | yes | aluminum/steel cans, empty aerosol cans, plastic bottles, glass bottles and jars (clear, brown and green), newspapers and inserts, magazines, telephone books, office paper, junk mail, white/colored paper, envelopes, manila folders, stationery | #1, #2 |
| Mesa | AZ | curbside, drop-off | yes | aluminum, cardboard, chipboard, glass, metal cans, newspapers, magazines, mixed paper, plastic bottles/jugs/jars that have a neck smaller than the base or those with a screw top lid, telephone books | na |
| Miami | FL | curbside | yes | glass/plastic bottles, aluminum cans, magazines, office/mixed paper, corrugated cardboard, white goods, yard trash, tires | na |
| Milwaukee | WI | curbside | yes | plastic, glass jars and bottles (all colors), food and beverage containers only, aluminum beverage cans, steel food cans, empty aerosol cans | #1, #2 |

| City | State | System Type | Voluntary | Materials Collected | Plastic Types |
|---|---|---|---|---|---|
| Minneapolis | MN | curbside, drop-off | yes | dry boxboard, office paper, mail, cans, corrugated cardboard, glass, household batteries, magazines, newspapers, telephone books, plastic bottles | na |
| Mobile | AL | curbside, drop-off | yes | plastic beverage containers, aluminum beverage cans, steel cans, corrugated cardboard, newspapers, magazines, junk mail, telephone books, computer paper, cereal boxes, glass jars (brown/amber, green/blue and clear), styrofoam packing peanuts, pine straw | na |
| Montgomery | AL | curbside, drop-off | yes | newspaper, white paper, computer paper, chipboard, magazines, brown paper bags, aluminum/steel cans, plastic | #1, #2 |
| Nashville | TN | curbside, drop-off (glass must be dropped off) | yes | paper (newspaper, magazines, junk mail, phone books, paperback books, paperboard, cereal boxes, freezer food boxes); cardboard, aluminum and steel cans, plastic containers (plastic bottles, and dairy containers labeled #1 through #7); glass bottles should be taken to one of the drop-off or convenience centers operated by the division of waste management | #1, #2, #3, #4, #5, #6, #7 |
| New Orleans | LA | curbside | yes | newspapers, magazines, telephone books, catalogs, brown paper bags, plastic, aluminum/steel/tin cans, glass (all colors), clean, unsoiled corrugated boxes | #1, #2 |
| New York | NY | curbside, drop-off | no | paper, mail and envelopes, wrapping paper, smooth cardboard, paper bags, cardboard egg cartons and trays, newspapers, magazines, catalogs, phone books, softcover books, corrugated cardboard, metal cans, aluminum foil, household metals, bulk metal (metal furniture, cabinets, large appliances, etc.), glass bottles and jars, plastic bottles and jugs, milk cartons and juice boxes, leaf program (fall) | #1, #2 |
| Newark | NJ | curbside, drop-off | no | glass bottles and jars, cans, bottles, corrugated cardboard, newspapers, magazines, mixed high-grade white paper, aluminum and bimetal cans, used motor oil, leaves | na |
| Norfolk | VA | curbside, drop-off | yes | aluminum cans, glass bottles and jars (clear/brown), corrugated cardboard, household batteries, newspapers, plastic soda/water bottles, plastic milk/water/detergent jugs, steel cans, mixed office paper | #1, #2, #3, #4, #5, #6, #7 |
| North Hempstead | NY | curbside, drop-off | yes | newspapers/inserts, magazines, direct mail, catalogs, construction/wrapping paper, index/greeting cards, paperback books, plastic, food/beverage cans, aluminum foil/pie tins, glass (clear and colored) | #1, #2 |
| Oakland | CA | curbside, drop-off | yes | glass bottles and jars, narrow neck plastic bottles and jugs (where the neck is smaller than the base), tin and aluminum cans/foil/pie plates, milk cartons, empty spray cans, empty and dry metal latex paint containers, newspapers, catalogs, magazines, co | na |
| Oklahoma City | OK | curbside | yes | plastic milk jugs and beverage bottles, aluminum and steel food and beverage cans, glass food and beverage jars and bottles, newspapers and inserts, magazines | na |
| Omaha | NE | curbside, drop-off | yes | newspapers and inserts, magazines, catalogs, telephone books, junk mail, detergent boxes, wrapping paper, paperback books, office/school paper, plastic, glass bottles and jars (clear, green, brown and blue), aluminum/steel/tin cans, corrugated cardboard | #1, #2 |
| Oyster Bay | NY | curbside, drop-off | yes | newspapers, cans, plastics, bottles/glass | na |
| Philadelphia | PA | curbside, drop-off | no | mixed paper (newspapers including inserts, junk mail, envelopes, telephone books, magazines and catalogs, cereal type boxes - no liners, home office paper, stationery and other clean paper); metal cans, aluminum cans, empty aerosol cans (no caps), empty paint cans (air dried), paint can lids (separated from the paint cans), glass bottles and jars, plastic and cardboard - *the city of Philadelphia only picks up plastic or cardboard curbside in some neighborhoods of the city* | #1, #2 |
| Phoenix | AZ | curbside, drop-off | yes | telephone books, plastic, food/glass bottles/jars, office paper, newspapers, magazines, cardboard, chipboard, milk/juice cartons, juice boxes, junk mail, aluminum cans/pie plates/foil, steel cans, metal hangers, scrap metal, aerosol cans | #1, #2, #6 |

| City | State | System Type | Voluntary | Materials Collected | Plastic Types |
|---|---|---|---|---|---|
| Pittsburgh | PA | curbside, drop-off | yes | plastic/metal containers, newspapers and inserts, corrugated cardboard, magazines, catalogs, glossy paper, white office paper, glass (clear and colored), all metal cans, empty aerosol and paint cans, aluminum foil/containers, plastic bottles, jugs, and jars | #1, #2, #3, #4, #5 |
| Plano | TX | curbside, drop-off | yes | newspapers, magazines, catalogs, junk mail, telephone books, brown paper bags, chipboard/ boxboard, corrugated cardboard boxes, aluminum/ steel/tin cans, plastic, aerosol cans, glass jars/ containers/dishes/drinking glasses/vases (any color), office paper | #1, #2 |
| Portland | OR | curbside | no | newspapers, scrap paper, glass bottles and jars, magazines, corrugated cardboard, Kraft paper, plastic bottles (including milk jugs), steel/tin cans | na |
| Raleigh | NC | curbside, drop-off | yes | newspapers and inserts, magazines, catalogs, junk mail, food and beverage cans, plastic drink bottles, glass food and beverage containers | #1, #2 |
| Richmond | VA | curbside, drop-off | yes | newspapers, mixed paper, aluminum cans/foil, steel cans, glass bottles and jars, milk cartons, juice boxes, plastic | #1, #2 |
| Riverside | CA | curbside, drop-off | yes | aluminum/tin cans, ferrous, non-ferrous, car batteries, appliances, electronics, CRV and non-CRV glass, polystyrene, newspapers, white office paper, computer paper, cardboard, magazines, junk mail, gas run equipment, air conditioners, refrigerators | #1, #2 |
| Rochester | NY | curbside | no | newspapers and inserts, magazines, glossy catalogs, corrugated cardboard, glass (clear, green and brown), plastic food and beverage containers, milk/juice cartons, empty aerosol cans, aluminum/tin/bi-metal cans, hi-grade paper | #1, #2 |
| Sacramento | CA | curbside | yes | glass bottles and jars (all colors), newspapers, junk mail, envelopes, telephone books, magazines, catalogs, brown paper bags, paper egg cartons, shoe boxes, computer/colored paper, paper 6-pk containers and boxes, cardboard, aluminum/tin cans, plastic | #1, #2 |
| Saint Louis | MO | curbside, drop-off | yes | aluminum cans, glass bottles and jars (clear, green and brown), newspapers and inserts, magazines, thin catalogs, steel food cans, plastic bottles and jugs, paperboard, office paper, corrugated cardboard | #1, #2 |
| Saint Paul | MN | curbside, drop-off | yes | newspapers/inserts, mail, envelopes, magazines, catalogs, office/school paper, boxboard, corrugated cardboard boxes, glass bottles and jars (clear/colors), metal cans/lids, jar lids, bottle caps, aluminum foil/trays, good clothes/linens | #1, #2 |
| Saint Petersburg | FL | drop-off | yes | mixed paper; newspapers; glass and plastic bottles; aluminum cans; corrugated cardboard boxes; appliances and mixed metal; yard waste | na |
| San Antonio | TX | curbside | yes | paper (ad circulars, catalogs, carbonless paper, dry goods packaging with liners removed, envelopes, file folders, cardboard, junk mail, magazines, newspapers, office paper, paperback books, paper bags, paper towel/toilet paper cores, phone books, non-metallic gift wrap); plastics (#1 through #7); glass (bottles and jars all colors); metal (aluminum, steel and tin beverage and food cans) | #1, #2, #3, #4, #5, #6, #7 |
| San Diego | CA | curbside, dropoff | yes | glass bottles and jars, empty aerosol cans, plastic bottles and jars, cardboard, aluminum cans, paper bags, aluminum foil, foil trays, bagged shredded paper, newspapers, metal cans, phone books, paper or frozen food boxes, mail and magazines, paper and catalogs | na |
| San Francisco | CA | curbside, drop-off | yes | paper, envelopes (windows okay), corrugated cardboard, aluminum cans and foil, cereal boxes (without lining), glass bottles and jars, egg cartons (paper only), plastic bottles (#1 through #7), plastic tubs and lids (#2, #4 and #5), junk mail and brochures, spray cans (must be empty), magazines, steel (tin) cans, newspapers, phone books, wrapping paper | #1, #2, #3, #4, #5, #6, #7 |
| San Jose | CA | curbside | yes | metal cans, milk and juice cartons, glass bottles and jars (brown, clear and green), plastic, carbonless paper, cardboard, catalogs, envelopes, junk mail, magazines, newspapers and inserts, paper bags, polystyrene, scrap metals, textiles | #1, #2, #3, #4, #5, #6, #7 |

| City | State | System Type | Voluntary | Materials Collected | Plastic Types |
|---|---|---|---|---|---|
| Santa Ana | CA | curbside, drop-off | yes | newspaper, mixed/white paper, cardboard, junk mail, magazines, telephone books, cereal/tissue boxes, glass/plastic bottles and jars, aluminum/steel/tin cans, empty aerosol cans, pie tins, plastic milk containers | na |
| Scottsdale | AZ | curbside, drop-off | yes | aseptic boxes, corrugated cardboard/chipboard, glass (clear, green, amber), magazines, telephone books, aluminum/steel/tin cans, empty aerosol cans, newspapers/inserts, junk mail, brown paper bags, plastic jugs/bottles (regardless of recycling #) | na |
| Seattle | WA | curbside, drop-off | yes | cardboard, magazines, junk mail, envelopes, aseptic packaging, telephone books, glass bottles and jars (all colors), steel/tin cans, plastic bottles/jugs/jars, milk and juice cartons, shredded paper and aluminum | na |
| Shreveport | LA | curbside, drop-off | yes | paper; plastic and glass bottles; metal cans; magazines; cardboard; detergent bottles | na |
| Stockton | CA | curbside, drop-off | yes | aluminum beverage containers, steel/tin cans, glass bottles/jars, plastic, newspapers, telephone books | #1, #2 |
| Tampa | FL | curbside | yes | glass bottles and jars (brown, green and clear), newspapers, aluminum cans, milk and juice cartons made from paper, plastic bottles | na |
| Toledo | OH | curbside | yes | junk mail, paper, boxboard, corrugated cardboard, newspapers, magazines, glass bottles, tin and aluminum cans | na |
| Tucson | AZ | curbside, drop-off | yes | white/colored paper, envelopes, junk mail, magazines, catalogs, paperboard/chipboard, phone books, fiberboard, cartons, newspapers, brown paper bags, corrugated cardboard, plastic, steel/tin cans, aluminum cans/foil/baking pans, glass food/beverage containers | #1, #2 |
| Tulsa | OK | curbside, drop-off | yes | glass jars and bottles (green, brown and clear), plastic bottles, aluminum/steel cans, newspapers and inserts, magazines, office paper, household and auto batteries, motor oil and antifreeze | #1, #2 |
| Virginia Beach | VA | curbside, drop-off | yes | newspapers, cardboard, chipboard, junk mail, catalogs, magazines, telephone books, glass bottles and jars (clear, green and brown), plastic bottles with a spout only, aluminum/steel/tin cans | na |
| Washington | DC | curbside | yes | glass (clear, brown, and green glass food and beverage containers); metals (aluminum pie plates, tin, aluminum and steel cans); plastic bottles (all narrow-necked or screw-topped bottles marked with a #1 or #2); paper (white and colored papers, envelopes, forms, file folders, tablets, junk mail, cereal boxes, shoeboxes, wrapping paper, shredded paper/mail, catalogs, magazines, paperback books and phone books); newspaper (including inserts); corrugated cardboard; brown paper bags | #1, #2 |
| Wichita | KS | curbside (private-no public trash collection) | yes | aluminum cans; ammunition; appliances; batteries; books; carpet pad; catalogs; cellular phones; clothes hangers; clothing; compact fluorescent light bulbs; computers; curbside recycling; eyeglasses; fluorescent lights; furniture; glass; hazardous materials; hearing aids; magazines; medication; metal; motor oil; packaging material; paper items; phone directories; plastics; plastic bags; printer cartridges; televisions; tires; unsolicited mail; wood and yard waste | na |

***Source:*** *Independent research by the editors, March 2010*

## Occupational and Environmental Exposures Capable of Causing Illness

| AOEC Code[1] | Primary Name | Synonym | Use[2] | Asthma Inducer? | CAS Number[3] |
|---|---|---|---|---|---|
| 010.00 | Dust, NOS | Dust, NOS | | | |
| 010.00 | Dust, NOS | Inorganic Dust, NOS | | | |
| 010.01 | Abrasives, NOS | Abrasives, NOS | | | |
| 010.01 | Abrasives, NOS | Grinding Dust | | | |
| 010.02 | Asbestos | Amosite | | | 1332-21-4 |
| 010.02 | Asbestos | Asbestos | | | 1332-21-4 |
| 010.02 | Asbestos | Chrysotile | | | 12001-29-5 |
| 010.02 | Asbestos | Crocidolite | | | 12001-28-4 |
| 010.02 | Asbestos | Tremolite | | | 77536-68-6 |
| 010.03 | Cement Dust | Cement Dust | | | 65997-15-1 |
| 010.03 | Cement Dust | Portland Cement | | | 65997-15-1 |
| 010.04 | Carbon Black | Actibon | | | |
| 010.04 | Carbon Black | Activated Carbon | | | 7440-44-0 |
| 010.04 | Carbon Black | Carbon Black | | | 1333-86-4 |
| 010.04 | Carbon Black | Toner, Copier | | | |
| 010.05 | Clay | Clay | | | 1332-58-7 |
| 010.05 | Clay | Terra Cotta | | | |
| 010.06 | Coal | Coal | | | |
| 010.07 | Fly Ash | Fly Ash | | | |
| 010.08 | Graphite | Graphite | | | 7440-44-0 |
| 010.09 | Man-Made Mineral Fibers | Fiberglass | | | |
| 010.09 | Man-Made Mineral Fibers | Fibrous Glass | | | |
| 010.09 | Man-Made Mineral Fibers | Man-Made Mineral Fibers | | | |
| 010.09 | Man-Made Mineral Fibers | Mineral Wool | | | |
| 010.09 | Man-Made Mineral Fibers | MMMF | | | |
| 010.09 | Man-Made Mineral Fibers | Refractory Ceramic Fiber | | | |
| 010.09 | Man-Made Mineral Fibers | Rock Wool | | | |
| 010.10 | Plaster | Plaster | | | |
| 010.11 | Porcelain | Porcelain | | | |
| 010.12 | Silica, Amorphous | Diatomaceous Earth | | | 61790-53-2 |
| 010.12 | Silica, Amorphous | Silica, Amorphous | | | 61790-53-2 |
| 010.13 | Silica, Crystalline | Feldspar Dust | | | |
| 010.13 | Silica, Crystalline | Quartz Dust | | | |
| 010.13 | Silica, Crystalline | Silica Flour | | | |
| 010.13 | Silica, Crystalline | Silica Sand | | | |
| 010.13 | Silica, Crystalline | Silica, Crystalline | | | |
| 010.13 | Silica, Crystalline | Silicon Dioxide | | | 60676-86-0 |
| 010.14 | Vinyl Dust | Plastic Dust | | | |
| 010.14 | Vinyl Dust | PVC Dust | | | |
| 010.14 | Vinyl Dust | Rubber Dust | | | |
| 010.14 | Vinyl Dust | Vinyl Dust | | | 9002-86-2 |
| 010.15 | Perlite | Perlite | | | 93763-70-3 |
| 010.16 | Ash, NOS | Ash, NOS | | | |
| 010.16 | Ash, NOS | Wood Ash | | | |
| 010.17 | Gypsum | Gypsum | | | 13397-24-5 |
| 010.17 | Gypsum | Plaster of Paris | | | 26499-65-0 |
| 010.18 | Silicon Carbide | Carborundum | | | 409-21-2 |
| 010.18 | Silicon Carbide | Silicon Carbide | | | 409-21-2 |
| 010.19 | Tooth Enamel Dust | Tooth Enamel Dust | | | |
| 010.20 | Vermiculite | Vermiculite | | | 1318-00-9 |
| 010.21 | Sulfur, Elemental | Sulfur | | | 7704-34-9 |
| 010.22 | Nylon Flock | Polyamide Fibers | | | |
| 010.23 | Drywall Mud | Drywall Mud | | | |
| 011.00 | Rock, NOS | Rock, NOS | | | |
| 011.01 | Granite | Granite | | | |

| AOEC Code[1] | Primary Name | Synonym | Use[2] | Asthma Inducer? | CAS Number[3] |
|---|---|---|---|---|---|
| 012.00 | Talc | Talc | | | |
| 012.01 | Talc, Fibrous | Talc, Fibrous | | | |
| 012.02 | Talc, Nonasbestiform | Talc, Nonasbestiform | | | 14807-96-6 |
| 020.00 | Metal Fumes, NOS | Metal Fumes, NOS | | | |
| 020.00 | Metal Fumes, NOS | Metallic Oxides, NOS | | | |
| 020.01 | Aluminum | Aluminum | | Yes | 7429-90-5 |
| 020.01 | Aluminum | Aluminum Compounds | | Yes | |
| 020.01 | Aluminum Chloride | Aluminum Chloride | | Yes | 7446-70-0 |
| 020.02 | Aluminum Oxide | Alumina | | Yes | 1344-28-1 |
| 020.02 | Aluminum Oxide | Aluminum Oxide | | Yes | 1344-28-1 |
| 020.02 | Aluminum Oxide | Corundum | | Yes | 1302-74-5 |
| 020.03 | Antimony | Antimony | | | 7440-36-0 |
| 020.03 | Antimony | Antimony Compounds | | | |
| 020.04 | Stibine | Antimony Hydride | | | 7803-52-3 |
| 020.04 | Stibine | Stibine | | | 7803-52-3 |
| 020.05 | Arsenic | Arsenic | | | 7440-38-2 |
| 020.05 | Arsenic | Arsenic Compounds | | | |
| 020.06 | Arsine | Arsine | | | 7784-42-1 |
| 020.07 | Barium | Barium | | | 7440-39-3 |
| 020.07 | Barium | Barium Compounds | | | |
| 020.08 | Beryllium | Beryllium | | | 7440-41-7 |
| 020.08 | Beryllium | Beryllium Compounds | | | |
| 020.09 | Bismuth | Bismuth | | | 7440-69-9 |
| 020.09 | Bismuth | Bismuth Compounds | | | |
| 020.10 | Boron | Boric Acid | | | 1043-35-3 |
| 020.10 | Boron | Boron | | | 7440-42-8 |
| 020.10 | Boron | Boron Compounds | | | |
| 020.11 | Diborane | Boron Hydride | | | 19287-45-7 |
| 020.11 | Diborane | Diborane | | | 19287-45-7 |
| 020.12 | Cadmium | Cadmium | | | 7440-43-9 |
| 020.12 | Cadmium | Cadmium Compounds | | | |
| 020.12 | Cadmium | Cadmium Salts | | | |
| 020.13 | Cerium | Cerium | | | |
| 020.13 | Cerium | Cerium Compounds | | | |
| 020.14 | Chromium, Not Hexavalent | Chromium Metal | | Yes | 7440-47-3 |
| 020.14 | Chromium, Not Hexavalent | Chromium, Not Hexavalent | | Yes | 7440-47-3 |
| 020.15 | Cobalt | Cobalt | | Yes | 7440-48-4 |
| 020.15 | Cobalt | Cobalt Compounds | | Yes | |
| 020.16 | Copper | Copper | | | 7440-50-8 |
| 020.16 | Copper | Copper Compounds | | | |
| 020.17 | Germanium | Germanium | | | 7440-56-4 |
| 020.17 | Germanium | Germanium Compounds | | | |
| 020.18 | Gold | Gold | | | 7440-57-5 |
| 020.18 | Gold | Gold Compounds | | | |
| 020.19 | Indium | Indium | | | 7440-74-6 |
| 020.19 | Indium | Indium Solder | | | |
| 020.20 | Iron | Ferric Chloride | | | 7705-08-0 |
| 020.20 | Iron | Iron | | | 7439-89-6 |
| 020.20 | Iron | Iron Compounds | | | |
| 020.21 | Lead, Inorganic | Inorganic Lead Compounds | | | |
| 020.21 | Lead, Inorganic | Lead, Inorganic | | | 7439-92-1 |
| 020.21 | Lead, Inorganic | Lead-based Paint | | | |
| 020.22 | Lead, Organic | Lead, Organic | | | |
| 020.22 | Lead, Organic | Organic Lead Compounds | | | |
| 020.22 | Lead, Organic | Tetraethyl Lead | | | 78-00-2 |
| 020.23 | Magnesium | Magnesium | | | 7439-95-4 |
| 020.23 | Magnesium | Magnesium Aluminum Silicate | | | 1327-43-1 |
| 020.23 | Magnesium | Magnesium Compounds | | | |
| 020.24 | Manganese | Manganese | | | 7439-96-5 |

| AOEC Code[1] | Primary Name | Synonym | Use[2] | Asthma Inducer? | CAS Number[3] |
|---|---|---|---|---|---|
| 020.24 | Manganese | Manganese Compounds | | | |
| 020.25 | Mercury, Inorganic | Inorganic Mercury Compounds | | | |
| 020.25 | Mercury, Inorganic | Mercuric Chloride | | | 7487-94-7 |
| 020.25 | Mercury, Inorganic | Mercuric Salts | | | |
| 020.25 | Mercury, Inorganic | Mercury, Inorganic | | | |
| 020.26 | Mercury, Organic | Mercury, Organic | | | |
| 020.26 | Mercury, Organic | Organic Mercury Compounds | | | |
| 020.26 | Mercury, Organic | Methyl Mercury due to Fish Consumption | | | |
| 020.27 | Molybdenum | Molybdenum | | | 7439-98-7 |
| 020.27 | Molybdenum | Molybdenum Compounds | | | |
| 020.28 | Nickel | Nickel | | Yes | 7440-02-0 |
| 020.28 | Nickel | Nickel Compounds | | Yes | |
| 020.29 | Osmium | Osmium | | | 4/2/7440 |
| 020.29 | Osmium | Osmium Compounds | | | |
| 020.30 | Platinum | Platinum | | | 6/4/7440 |
| 020.30 | Platinum | Platinum Compounds | | | |
| 020.31 | Selenium | Selenium | | | 7782-49-2 |
| 020.31 | Selenium | Selenium Compounds | | | |
| 020.32 | Silver | Silver | | | 7440-22-4 |
| 020.32 | Silver | Silver Compounds | | | |
| 020.32 | Silver | Silver Sulfate | | | |
| 020.33 | Tellurium | Tellurium | | | 13494-80-9 |
| 020.33 | Tellurium | Tellurium Compounds | | | |
| 020.34 | Thorium | Thorium | | | 7440-29-1 |
| 020.34 | Thorium | Thorium Compounds | | | |
| 020.35 | Tin, Inorganic | Inorganic Tin Compounds | | | |
| 020.35 | Tin, Inorganic | Tin, Inorganic | | | |
| 020.36 | Tin, Organic | Organic Tin Compounds | | | 7440-31-5 |
| 020.36 | Tin, Organic | Tin, Organic | | | 7440-31-5 |
| 020.36 | Tributyl Tin Oxide | Tributyl Tin Oxide | P | Yes | 56-35-9 |
| 020.37 | Tungsten Carbide | Tungsten Carbide/Cobalt | | Yes | 12070-12-1 |
| 020.38 | Vanadium | Vanadium | | | 7440-62-2 |
| 020.38 | Vanadium | Vanadium Compounds | | | |
| 020.39 | Zinc | Zinc | | | 7440-66-6 |
| 020.39 | Zinc | Zinc Compounds | | | |
| 020.39 | Zinc Oxide | Zinc Oxide | | Yes | 1314-13-2 |
| 020.40 | Zinc Chloride | Zinc Chloride | | | 7646-85-7 |
| 020.41 | Zirconium | Zirconium | | | 7440-67-7 |
| 020.41 | Zirconium | Zirconium Compounds | | | |
| 020.43 | Brass | Brass | | | |
| 020.44 | Metal Carbonyls | Metal Carbonyls | | | |
| 020.45 | Hard Metal | Cobalt/Tungsten Carbide | | Yes | |
| 020.45 | Hard Metal | Hard Metal | | Yes | |
| 020.46 | Heavy Metals, NOS | Heavy Metals, NOS | | | |
| 020.47 | Metals, NOS | Metals, NOS | | | |
| 020.48 | Titanium | Titanium | | | 7440-32-6 |
| 020.48 | Titanium | Titanium Compounds | | | |
| 020.49 | Copper Phthalocyanine | Copper Phthalocyanine | | | 147-14-8 |
| 020.49 | Copper Phthalocyanine | Cuprilinic Blue | | | 147-14-8 |
| 020.50 | Strontium | Strontium | | | |
| 020.50 | Strontium | Strontium Compounds | | | |
| 020.51 | Copper Sulfate | Copper Sulfate | | | 7758-98-7 |
| 020.51 | Copper Sulfate | Copper Sulfate (Anhydrous) | | | 7758-98-7 |
| 020.52 | Tungsten | Tungsten Compounds | | | |
| 020.53 | Rhodium | Rhodium | | Yes | 7440-16-6 |
| 021.00 | Metal Dust, NOS | Metal Dust, NOS | | | |
| 022.00 | Chromium, Hexavalent, NOS | Chromium, Hexavalent, NOS | | Yes | |
| 022.01 | Potassium Dichromate | Potassium Dichromate | | | 7778-50-9 |

| AOEC Code[1] | Primary Name | Synonym | Use[2] | Asthma Inducer? | CAS Number[3] |
|---|---|---|---|---|---|
| 022.02 | Ammonium Bichromate | Ammonium Bichromate | | Yes | 9/5/7789 |
| 022.03 | Sodium Dichromate | Sodium Dichromate | | | 10588-01-9 |
| 022.04 | Chromic Acid | Chromic Acid | | Yes | 1333-82-0 |
| 023.00 | Welding, NOS | Brazing, NOS | | | |
| 023.00 | Welding, NOS | Burning Fume | | | |
| 023.00 | Welding, NOS | Welding Fume, NOS | | | |
| 023.00 | Welding, NOS | Welding, NOS | | | |
| 023.01 | Welding Fume, Stainless Steel | Welding Fume, Stainless Steel | | Yes | |
| 023.02 | Welding Fume, Iron or Steel | Welding Fume, Iron or Steel | | | |
| 023.03 | Welding Fume, Copper/Nickel | Welding Fume, Copper/Nickel | | | |
| 023.04 | Soldering Flux, NOS | Soldering Flux, NOS | | | |
| 023.05 | Colophony | Colophony | | Yes | |
| 023.06 | Welding Fume, Galvanized Metal | Welding Fume, Galvanized Metal | | | |
| 023.07 | Gas Metal Arc Welding on Uncoated Mild Steel | Gas Metal Arc Welding on Uncoated Mild Steel | | Yes | |
| 023.08 | Soldering Flux, Zinc Chloride/Ammonium Chloride | Soldering Flux, Zinc Chloride/Ammonium Chloride | | Yes | |
| 023.09 | Stick Welding | Stick Welding | | | |
| 023.10 | Soldering, NOS | Soldering, NOS | | | |
| 024.00 | Soluble Halogenated Platinum Compounds, NOS | Soluble Halogenated Platinum Compounds, NOS | | Yes | |
| 024.01 | Ammonium Hexachloroplatinate (IV) | Ammonium Hexachloroplatinate (IV) | | Yes | 16919-58-7 |
| 030.01 | Bromine | Bromine | | | 7726-95-6 |
| 030.02 | Chlorine | Chlorine | | Yes | 7782-50-5 |
| 030.03 | Fluorine | Fluorine | | Yes | 7782-41-5 |
| 030.04 | Iodine | Iodine | | | 7553-56-2 |
| 030.05 | Chlorine Trifluoride | Chlorine Trifluoride | | | 7790-91-2 |
| 030.06 | Chlorine Dioxide | Chlorine Dioxide | | | 10049-04-4 |
| 040.00 | Inorganic Compounds, NOS | Inorganic Compounds, NOS | | | |
| 040.01 | Argon | Argon | | | 7440-37-1 |
| 040.02 | Calcium Salts, NOS | Calcium Salts, NOS | | | |
| 040.03 | Carbon Dioxide | Carbon Dioxide | | | 124-38-9 |
| 040.04 | Carbon Monoxide | Carbon Monoxide | | | 630-08-0 |
| 040.04 | Carbon Monoxide | CO | | | 630-08-0 |
| 040.05 | Fluxes, NOS | Fluxes, NOS | | | |
| 040.06 | Hydrogen Sulfide | H2S | | | 6/4/7783 |
| 040.06 | Hydrogen Sulfide | Hydrogen Sulfide | | | 6/4/7783 |
| 040.06 | Hydrogen Sulfide | Sewer Gas | | | |
| 040.07 | Nitrogen | Nitrogen | | | 7727-37-9 |
| 040.08 | Nitrogen Oxides | Nitric Oxide | | | 10102-43-9 |
| 040.08 | Nitrogen Oxides | Nitrogen Dioxide | | | 10102-44-0 |
| 040.08 | Nitrogen Oxides | Nitrogen Oxides | | | |
| 040.08 | Nitrogen Oxides | NO | | | 10102-43-9 |
| 040.08 | Nitrogen Oxides | NO2 | | | 10102-44-0 |
| 040.09 | Nitrous Oxide | N2O | | | 10024-97-2 |
| 040.09 | Nitrous Oxide | Nitrous Oxide | | | 10024-97-2 |
| 040.10 | Oxygen, Liquid | Oxygen, Liquid | | | 7782-44-7 |
| 040.11 | Ozone | Ozone | | | 10028-15-6 |
| 040.12 | Phosgene | Phosgene | | | 75-44-5 |
| 040.13 | Phosphine | Phosphine | P | | 7803-51-2 |
| 040.14 | Phosphorus Pentasulfide | Phosphorus Pentasulfide | | | 1314-80-3 |
| 040.14 | Phosphorus Pentasulfide | Phosphorus Sulfide | | | 1314-80-3 |
| 040.14 | Phosphorus Pentasulfide | Thiophosphoric Anhydride | | | |
| 040.15 | Potassium Salts, NOS | Potassium Salts, NOS | | | |
| 040.16 | Potassium Chlorate | Potassium Chlorate | | | 4/9/3811 |
| 040.17 | Sodium Salts, NOS | Sodium Salts, NOS | | | |
| 040.18 | Sodium Bisulfite | Sodium Bisulfite | | | 7631-90-5 |
| 040.18 | Sodium Bisulfite | Sodium Sulfite | | | 7631-90-5 |
| 040.19 | Sodium Sulfide | Sodium Sulfide | | | 16721-80-5 |
| 040.20 | Sulfur Oxides | SO2 | | | 9/5/7446 |
| 040.20 | Sulfur Oxides | Sulfur Dioxide | | | 9/5/7446 |

| AOEC Code[1] | Primary Name | Synonym | Use[2] | Asthma Inducer? | CAS Number[3] |
|---|---|---|---|---|---|
| 040.20 | Sulfur Oxides | Sulfur Oxides | | | |
| 040.21 | Sulfur Gas | Sulfur Gas | | | |
| 040.22 | Sulfur Chloride | Sulfur Chloride | | | 10025-67-9 |
| 040.22 | Sulfur Chloride | Sulfur Monochloride | | | 10025-67-9 |
| 040.23 | Thallium Salts | Thallium Compounds | P | | |
| 040.23 | Thallium Salts | Thallium Salts | P | | |
| 040.24 | Irritant Gases, NOS | Irritant Gases, NOS | | | |
| 040.25 | Asphyxiant Gases, NOS | Asphyxiant Gases, NOS | | | |
| 040.26 | Persulfate Salts | Persulfate Salts | | Yes | |
| 040.27 | Dichlorosilane | Dichlorosilane | | | 4109-96-0 |
| 040.27 | Methyl Dichlorosilane | Methyl Dichlorosilane | | | 20156-50-7 |
| 040.28 | Sodium Metabisulfite | Sodium Metabisulfite | | Yes | 7681-57-4 |
| 040.29 | Sodium Borohydride | Sodium Borohydride | | | |
| 040.30 | Vanadium Hydroxide Oxide Phosphate | Vanadium Hydroxide Oxide Phosphate | | | |
| 040.31 | Aluminum Phosphide | Aluminum Phosphide | P | | 20859-73-8 |
| 040.31 | Aluminum Phosphide | Phostoxin | P | | 20859-73-8 |
| 040.32 | Ammonium Phosphate | Ammonium Phosphate | P | | 7722-76-1 |
| 040.32 | Ammonium Phosphate | Monoammonium Phosphate | P | | 7722-76-1 |
| 040.32 | Ammonium Phosphate | MAP | P | | 7722-76-1 |
| 040.33 | Fluorescent Lamp Phosphor | Fluorescent Lamp Phosphor | | | |
| 041.01 | Calcium Chloride | Calcium Chloride | | | 10043-52-4 |
| 041.02 | Sodium Chloride | Sodium Chloride | | | 7647-14-5 |
| 041.03 | Sodium Chlorite | Sodium Chlorite | | | 7758-19-2 |
| 042.01 | Potassium Nitrate | Potassium Nitrate | | | 7757-79-1 |
| 042.02 | Sodium Nitrate | Sodium Nitrate | | | |
| 050.00 | Acids, Bases, Oxidizers, NOS | Acids, Bases, Oxidizers, NOS | | | |
| 050.00 | Acids, Bases, Oxidizers, NOS | Inorganic Acids, NOS | | | |
| 050.03 | Calcium Bisulfite | Calcium Bisulfite | | | 13780-03-5 |
| 050.05 | Calcium Oxide | Burnt Lime | | | 1305-78-8 |
| 050.05 | Calcium Oxide | Calcium Oxide | | | 1305-78-8 |
| 050.05 | Calcium Oxide | Lime | | | 1305-78-8 |
| 050.05 | Calcium Oxide | Quicklime | | | 1305-78-8 |
| 050.06 | Chloramine, NOS | Chloramine, NOS | | | |
| 050.06 | Nitrogen Mustard Hydrochloride | Nitrogen Mustard Hydrochloride | | | 55-86-7 |
| 050.07 | Citric Acid | Citric Acid | | | 77-92-9 |
| 050.08 | Hydrazoic Acid | Hydrazoic Acid | | | 7782-79-8 |
| 050.09 | Hydrobromic Acid | HBR | | | 10035-10-6 |
| 050.09 | Hydrobromic Acid | Hydrobromic Acid | | | 10035-10-6 |
| 050.09 | Hydrobromic Acid | Hydrogen Bromide | | | 10035-10-6 |
| 050.10 | Hydrochloric Acid | HCL | | Yes | 7647-01-0 |
| 050.10 | Hydrochloric Acid | Hydrochloric Acid | | Yes | 7647-01-0 |
| 050.10 | Hydrochloric Acid | Hydrogen Chloride | | Yes | 7647-01-0 |
| 050.10 | Hydrochloric Acid | Muriatic Acid | | Yes | 7647-01-0 |
| 050.11 | Hydrofluoric Acid | HF | | | 7664-39-3 |
| 050.11 | Hydrofluoric Acid | Hydrofluoric Acid | | | 7664-39-3 |
| 050.11 | Hydrofluoric Acid | Hydrogen Fluoride | | | 7664-39-3 |
| 050.12 | Hydrogen Peroxide | Hydrogen Peroxide | | | |
| 050.13 | Nitric Acid | Nitric Acid | | | 7697-37-2 |
| 050.14 | Peroxides | Peroxides | | | |
| 050.14 | Dicumyl-peroxide | Dicumyl-peroxide | | Yes | 80-43-3 |
| 050.15 | Phosphoric Acid | Phosphoric Acid | | | 7664-38-2 |
| 050.16 | Phosphorus Trichloride | Phosphorus Trichloride | | | 12/2/7719 |
| 050.17 | Potassium Hydroxide | KOH | | | 1310-58-3 |
| 050.17 | Potassium Hydroxide | Potassium Hydroxide | | | 1310-58-3 |
| 050.18 | Sodium Hydroxide | Caustic Soda | | | 1310-73-2 |
| 050.18 | Sodium Hydroxide | Lye | | | 1310-73-2 |
| 050.18 | Sodium Hydroxide | NaOH | | | 1310-73-2 |
| 050.18 | Sodium Hydroxide | Sodium Hydroxide | | | 1310-73-2 |
| 050.19 | Potassium Bicarbonate | Potassium Bicarbonate | | | |

| AOEC Code[1] | Primary Name | Synonym | Use[2] | Asthma Inducer? | CAS Number[3] |
|---|---|---|---|---|---|
| 050.20 | Trisodium Phosphate | Trisodium Phosphate | | | 7601-54-9 |
| 050.21 | Sodium Carbonate | Soda Ash | | | 497-19-8 |
| 050.21 | Sodium Carbonate | Sodium Carbonate | | | |
| 050.22 | Sodium Metasilicate | Sodium Metasilicate | | | 6834-92-0 |
| 050.23 | Sodium Silicate | Sodium Silicate | | | 6834-92-0 |
| 050.24 | Sulfuric Acid | Hydrogen Sulfate | | Yes | 7664-93-9 |
| 050.24 | Sulfuric Acid | Sulfuric Acid | | Yes | 7664-93-9 |
| 050.25 | Acid Stripper | Acid Stripper | | | |
| 050.26 | Acid Solder | Acid Solder | | | |
| 050.27 | Benzoyl Peroxide | Benzoyl Peroxide | | | 94-36-0 |
| 050.28 | Disinfectants, NOS | Disinfectants, NOS | | | |
| 050.30 | Aluminum Hydroxide | Aluminum Hydroxide | | | 21645-51-2 |
| 050.31 | Sodium Benzoate | Sodium Benzoate | | | |
| 050.32 | Phosphorus Bromide | Phosphorus Bromide | | | |
| 050.32 | Phosphorus Bromide | Phosphorus Tribromide | | | |
| 050.33 | Acid & Base Mixture | Acid & Base Mixture | | | |
| 050.34 | Glacial Acetic Acid | Glacial Acetic Acid | | Yes | 64-19-7 |
| 050.34 | Glacial Acetic Acid | Acetic Acid | | Yes | 64-19-7 |
| 050.35 | Calcium Carbonate | Calcium Carbonate | | | 1317-65-3 |
| 050.36 | Calcium Carbide | Calcium Carbide | | | 75-20-7 |
| 050.37 | Trifluoroacetic Acid | Trifluoroacetic Acid | | | 76-05-1 |
| 050.38 | Sodium Tripolyphosphate | Sodium Tripolyphosphate | | | 13573-18-7 |
| 050.39 | Potassium Permanganate | Potassium Permanganate | | | 7722-64-7 |
| 050.40 | Sodium Bisulfate | Sodium Bisulfate | | | 7681-38-1 |
| 050.41 | Potassium Carbonate | Potassium Carbonate | | | 584-08-7 |
| 050.42 | Peroxyacetic Acid | Peroxyacetic Acid | | Yes | 79-21-0 |
| 050.43 | Trichloroacetic Acid | Trichloroacetic Acid | | | 76-03-9 |
| 050.44 | Hydrofluosilicic Acid | Hydrofluosilicic Acid | | | 16961-83-4 |
| 050.44 | Hydrofluosilicic Acid | Fluosilicic Acid | | | 16961-83-4 |
| 050.45 | Sodium Oxide | Sodium Oxide | | | 1313-59-3 |
| 052.01 | New code 322.07 | Ammonia | | Yes | |
| 052.01 | Anhydrous Ammonia | Anhydrous Ammonia | | Yes | 7664-41-7 |
| 052.03 | Ammonium Salts | Ammonium Chloride | | | 12125-02-9 |
| 052.03 | Ammonium Salts | Ammonium Salts | | | |
| 052.04 | Ammonium Persulfate | Ammonium Persulfate | | | 7727-54-0 |
| 060.00 | Aliphatic Hydrocarbons, NOS | Aliphatic Hydrocarbons, NOS | | | |
| 060.00 | Alicyclic Hydrocarbons, NOS | Alicyclic Hydrocarbons, NOS | | | |
| 060.01 | Acetylene | Acetylene | | | 74-86-2 |
| 060.01 | Acetylene | Ethine | | | 74-86-2 |
| 060.01 | Acetylene | Ethyne | | | 74-86-2 |
| 060.02 | Hexane | Hexane | S | | 110-54-3 |
| 060.02 | Hexane | N-Hexane | S | | 110-54-3 |
| 060.03 | Heptane | Heptane | S | | 142-82-5 |
| 060.03 | Heptane | N-Heptane | S | | 142-82-5 |
| 060.04 | Isobutane | 2-Methylpropane | | | 75-28-5 |
| 060.04 | Isobutane | Isobutane | | | 75-28-5 |
| 060.05 | Limonene | Dipentene | S | | 138-86-3 |
| 060.05 | Limonene | Limonene | S | | 138-86-3 |
| 060.05 | Limonene | Methyl-4-Isopropenyl Cyclohexene-1 | S | | |
| 060.05 | Limonene | P-Mentha-1,8-Diene | S | | |
| 060.06 | Methane | Methane | | | 74-82-8 |
| 060.07 | Mineral Oil | Mineral Oil | | | 8012-95-1 |
| 060.07 | Mineral Oil | Mineral Oil Mist | | | 8012-95-1 |
| 060.07 | Mineral Oil | Paraffin Oil | | | 8012-95-1 |
| 060.07 | Mineral Oil | Petrolatum Liquid | | | |
| 060.07 | Mineral Oil | White Oil | | | |
| 060.08 | Natural Gas | Natural Gas | | | 74-82-8 |
| 060.09 | Paraffin | Paraffin | | | 8002-74-2 |

| AOEC Code[1] | Primary Name | Synonym | Use[2] | Asthma Inducer? | CAS Number[3] |
|---|---|---|---|---|---|
| 060.09 | Paraffin | Paraffin Wax | | | |
| 060.10 | Turpentine | Spirits of Turpentine | S | | 8006-64-2 |
| 060.10 | Turpentine | Turpentine | S | | 8006-64-2 |
| 060.11 | 4-Phenylcyclohexene | 4-PC | | | |
| 060.11 | 4-Phenylcyclohexene | New Carpet Odor | | | |
| 060.11 | 4-Phenylcyclohexene | 4-Phenylcyclohexene | | | |
| 060.12 | Isooctane | Isooctane | | | 540-84-1 |
| 060.13 | Cyclopentadiene | Cyclopentadiene | | | 542-92-7 |
| 060.14 | Cyclohexane | Cyclohexane | S | | 110-82-7 |
| 060.15 | Linseed Oil | Linseed Oil | | | 8001-26-1 |
| 060.16 | Tetrahydrofuran | Tetrahydrofuran | | | 109-99-9 |
| 060.17 | Terpene | Terpene | S | | 9003-74-1 |
| 060.18 | D-Limonene | D-Limonene | S | | 5989-27-5 |
| 060.19 | Pentane | Pentane | S | | 109-66-0 |
| 060.19 | Pentane | N-Pentane | S | | 109-66-0 |
| 060.20 | Propane | Dimethyl Methane | | | 74-98-6 |
| 060.20 | Propane | Propyl Hydride | | | 74-98-6 |
| 060.20 | Propane | LPG | | | 74-98-6 |
| 060.20 | Propane | Propane | | | 74-98-6 |
| 060.21 | Carene | Carene | | Yes | 13466-78-9 |
| 060.22 | Rust Inhibitor | Rust Inhibitor | | | |
| 060.23 | Corrosive Preventative | Corrosive Preventative | | | |
| 060.24 | Linseed Oilcake | Linseed Oilcake | | Yes | |
| 060.25 | Cyclohexanediamine_(1,2-diaminocyclohexane) | Cyclohexanediamine_(1,2-diaminocyclohexane) | | | 694-83-7 |
| 061.00 | Petroleum Fractions, NOS | Petroleum Distillates, NOS | S | | |
| 061.00 | Petroleum Fractions, NOS | Petroleum Fractions, NOS | S | | |
| 061.01 | Petroleum Spirits | Mineral Spirits | S | | 64475-85-0 |
| 061.01 | Petroleum Spirits | Mineral Thinner | S | | |
| 061.01 | Petroleum Spirits | Mineral Turpentine | S | | |
| 061.01 | Petroleum Spirits | Petroleum Spirits | S | | 64475-85-0 |
| 061.01 | Petroleum Spirits | Refined Petroleum Solvent | S | | |
| 061.01 | Petroleum Spirits | Solvent Naphtha | S | | |
| 061.01 | Petroleum Spirits | Stoddard Solvent | S | | 8052-41-3 |
| 061.01 | Petroleum Spirits | Varnish Makers' & Painters' Naphtha | S | | |
| 061.01 | Petroleum Spirits | Varsol | S | | |
| 061.01 | Petroleum Spirits | VM & P Naphtha | S | | |
| 061.01 | Petroleum Spirits | White Spirits | S | | |
| 061.02 | Naphtha | Naphtha | S | | 8030-30-6 |
| 061.02 | Naphtha | Petroleum Naphtha | S | | 8030-30-6 |
| 061.03 | Kerosene | Astral Oil | S | | |
| 061.03 | Kerosene | Kerosene | S | | 8008-20-6 |
| 061.03 | Kerosene | Kerosine | S | | 8008-20-6 |
| 061.03 | Kerosene | No. 1 Fuel Oil | S | | |
| 061.04 | Gasoline | Gasoline | S | | |
| 061.04 | Gasoline | Petrol | S | | |
| 061.05 | Jet Fuel | Jet Fuel | S | | |
| 061.06 | Diesel Fuel | Diesel Fuel | S | | |
| 061.06 | Diesel Fuel | Diesel Oil | S | | 68334-30-5 |
| 061.06 | Diesel Fuel | No. 2 Fuel Oil | S | | |
| 061.07 | Asphalt | Asphalt | | | 8052-42-4 |
| 061.07 | Asphalt | Petroleum | | | |
| 061.07 | Asphalt | Road Pitch | | | |
| 061.07 | Asphalt | Road Tar | | | |
| 061.08 | Shale | Shale | | | |
| 070.00 | Alcohols, NOS | Alcohols, NOS | S | | |
| 070.01 | Allyl Alcohol | 2-Propen-1-ol | | | 107-18-6 |
| 070.01 | Allyl Alcohol | Allyl Alcohol | | | 107-18-6 |
| 070.01 | Allyl Alcohol | Vinyl Carbinol | | | 107-18-6 |

| AOEC Code[1] | Primary Name | Synonym | Use[2] | Asthma Inducer? | CAS Number[3] |
|---|---|---|---|---|---|
| 070.02 | Amyl Alcohol | 1-Pentanol | | | 71-41-0 |
| 070.02 | Amyl Alcohol | Amyl Alcohol | | | 71-41-0 |
| 070.02 | Amyl Alcohol | N-Amyl Alcohol | | | 71-41-0 |
| 070.03 | Beta-Chloroethyl Alcohol | 2-Chloroethanol | | | 107-07-3 |
| 070.03 | Beta-Chloroethyl Alcohol | Beta-Chloroethyl Alcohol | | | 107-07-3 |
| 070.03 | Beta-Chloroethyl Alcohol | Ethylene Chlorohydrin | | | 107-07-3 |
| 070.04 | N-Butyl Alcohol | 1-Butanol | S | | 71-36-3 |
| 070.04 | N-Butyl Alcohol | N-Butyl Alcohol | S | | 71-36-3 |
| 070.05 | Ethanol | Alcohol | S | | |
| 070.05 | Ethanol | Ethanol | S | | 64-17-5 |
| 070.05 | Ethanol | Ethyl Alcohol | S | | 64-17-5 |
| 070.05 | Ethanol | Grain Alcohol | S | | 64-17-5 |
| 070.05 | Ethanol | Methyl Carbinol | S | | 64-17-5 |
| 070.06 | Isopropyl Alcohol | 2-Propanol | S | | 67-63-0 |
| 070.06 | Isopropyl Alcohol | Isopropanol | S | | 67-63-0 |
| 070.06 | Isopropyl Alcohol | Isopropyl Alcohol | S | | 67-63-0 |
| 070.06 | Isopropyl Alcohol | Propanol | S | | 67-63-0 |
| 070.06 | Isopropyl Alcohol | Sec-Propyl Alcohol | S | | 67-63-0 |
| 070.07 | Methanol | Carbinol | S | | 67-56-1 |
| 070.07 | Methanol | Methanol | S | | 67-56-1 |
| 070.07 | Methanol | Methyl Alcohol | S | | 67-56-1 |
| 070.07 | Methanol | Wood Alcohol | S | | 67-56-1 |
| 070.07 | Methanol | Wood Spirits | S | | 67-56-1 |
| 070.08 | Propyl Alcohol | 1-Propanol | S | | 71-23-8 |
| 070.08 | Propyl Alcohol | Propyl Alcohol | S | | 71-23-8 |
| 070.09 | Furfuryl Alcohol | 2-Furanmethanol | | Yes | 98-00-0 |
| 070.09 | Furfuryl Alcohol | Furfuryl Alcohol | | Yes | 98-00-0 |
| 070.10 | Alkyl Aryl Polyether Alcohol/Polypropylene Glycol | Alkyl Aryl Polyether Alcohol/Polypropylene Glycol Mixt. | | Yes | |
| 080.01 | Ethylene Glycol | 1,2-Ethanediol | | | 107-21-1 |
| 080.01 | Ethylene Glycol | Antifreeze | | | |
| 080.01 | Ethylene Glycol | Ethylene Glycol | | | 107-21-1 |
| 080.02 | Diethylene Glycol | Diethylene Glycol | | | 111-46-6 |
| 080.03 | Hexylene Glycol | Hexylene Glycol | | | 107-41-5 |
| 080.04 | 1,3-Butylene Glycol | 1,3-Butylene Glycol | | | 107-88-0 |
| 090.00 | Glycol Ethers, NOS | Glycol Ethers, NOS | S | | |
| 090.01 | Propylene Glycol Ethers | Dipropylene Glycol Methyl Ether | S | | 34590-94-8 |
| 090.01 | Propylene Glycol Ethers | Propylene Glycol Ethers | S | | |
| 091.00 | Ethylene Glycol Ethers, NOS | Ethylene Glycol Ethers, NOS | S | | |
| 091.01 | Ethylene Glycol Monomethyl Ether | EGME | S | | 109-86-4 |
| 091.01 | Ethylene Glycol Monomethyl Ether | Ethylene Glycol Monomethyl Ether | S | | 109-86-4 |
| 091.01 | Ethylene Glycol Monomethyl Ether | Methyl Cellosolve | S | | 109-86-4 |
| 091.01 | Ethylene Glycol Monomethyl Ether | Methaoxyethanol | S | | 109-86-4 |
| 091.02 | Ethylene Glycol Monoethyl Ether | 2-Ethoxyethanol | S | | 110-80-5 |
| 091.02 | Ethylene Glycol Monoethyl Ether | Cellosolve | S | | 110-80-5 |
| 091.02 | Ethylene Glycol Monoethyl Ether | EGEE | S | | 110-80-5 |
| 091.02 | Ethylene Glycol Monoethyl Ether | Ethylene Glycol Monoethyl Ether | S | | 110-80-5 |
| 091.03 | 2-Butoxyethanol | 2-Butoxyethanol | S | | 111-76-2 |
| 091.03 | 2-Butoxyethanol | Butyl-Cellosolve | S | | 111-76-2 |
| 091.03 | 2-Butoxyethanol | EGBE | S | | 111-76-2 |
| 091.03 | 2-Butoxyethanol | Ethylene Glycol Monobutyl Ether | S | | 111-76-2 |
| 100.01 | Bis Chloromethyl Ether | BCME | | | 542-88-1 |
| 100.01 | Bis Chloromethyl Ether | Bis Chloromethyl Ether | | | 542-88-1 |
| 100.01 | Bis Chloromethyl Ether | Chloromethyl Ether | | | 542-88-1 |
| 100.01 | Bis Chloromethyl Ether | Dichloromethyl Ether | | | 542-88-1 |
| 100.02 | Chloromethyl Methyl Ether | Chloromethyl Methyl Ether | | | 107-30-2 |
| 100.02 | Chloromethyl Methyl Ether | CMME | | | 107-30-2 |
| 100.03 | Dichloroethyl Ether | Bis 2-Chloroethyl Ether | | | 111-44-4 |
| 100.03 | Dichloroethyl Ether | Chlorex | | | 111-44-4 |

| AOEC Code[1] | Primary Name | Synonym | Use[2] | Asthma Inducer? | CAS Number[3] |
|---|---|---|---|---|---|
| 100.03 | Dichloroethyl Ether | Dichloroethyl Ether | | | 111-44-4 |
| 100.04 | Dioxane | 1,4-Dioxane | S | | 123-91-1 |
| 100.04 | Dioxane | Dioxane | S | | 123-91-1 |
| 100.05 | Ethyl Ether | Diethyl Ether | | | 60-29-7 |
| 100.05 | Ethyl Ether | Ether | | | 60-29-7 |
| 100.05 | Ethyl Ether | Ethoxyethane | | | 60-29-7 |
| 100.05 | Ethyl Ether | Ethyl Ether | | | 60-29-7 |
| 100.05 | Ethyl Ether | Ethyl Oxide | | | |
| 100.06 | Ethyl Methyl Ether | Ethoxymethane | | | |
| 100.06 | Ethyl Methyl Ether | Ethyl Methyl Ether | | | 540-67-0 |
| 100.07 | Eugenol | 1-Allyl-3-Methoxy-4-Hydroxybenzene | | | 97-53-0 |
| 100.07 | Eugenol | Allyl Guaiacol | | | 97-53-0 |
| 100.07 | Eugenol | Eugenol | | | 97-53-0 |
| 100.08 | Methyl Tertiary Butyl Ether | Methyl Tertiary Butyl Ether | | | 1634-04-4 |
| 100.08 | Methyl Tertiary Butyl Ether | MTBE | | | 1634-04-4 |
| 100.09 | Guaiacol | Guaiacol | | | 90-05-1 |
| 100.09 | Guaiacol | o-methoxyphenol | | | 90-05-1 |
| 100.09 | Guaiacol | Methylcatechol | | | 90-05-1 |
| 100.09 | Guaiacol | 1-Hydroxy-2-methoxybenzene | | | 90-05-1 |
| 110.01 | Epichlorohydrin | 1-Chloro-2,3-Epoxypropane | | | 106-89-8 |
| 110.01 | Epichlorohydrin | Epichlorohydrin | | | 106-89-8 |
| 110.02 | Epoxy Resins | Epoxies | | Yes | |
| 110.02 | Epoxy Resins | Epoxy Resin Hardeners | | Yes | |
| 110.02 | Epoxy Resins | Epoxy Resins | | Yes | |
| 110.03 | Ethylene Oxide | Ethylene Oxide | | Yes | 75-21-8 |
| 110.03 | Ethylene Oxide | ETO | | Yes | |
| 110.04 | Tall Oil, Crude | Tall Oil (Crude) | | Yes | 8002-26-4 |
| 110.04 | Tall Oil, Rosin | Tall Oil (Rosin) | | Yes | 10/6/8052 |
| 110.04 | Tall Oil, Fatty Acids | Tall Oil (Fatty Acids) | | | 61790-12-3 |
| 110.05 | Paint, Epoxy | Coating, Epoxy | | | |
| 110.05 | Paint, Epoxy | Paint, Epoxy | | | |
| 110.06 | Adhesive, Epoxy | Adhesive, Epoxy | | Yes | |
| 110.07 | Propylene Oxide | Propylene Oxide | | | 75-56-9 |
| 120.01 | Acetaldehyde | Acetaldehyde | | | 75-07-0 |
| 120.02 | Acrolein | 2-Propenal | | | 107-02-8 |
| 120.02 | Acrolein | Acrolein | | | 107-02-8 |
| 120.02 | Acrolein | Acrylic Aldehyde | | | 107-02-8 |
| 120.03 | Formaldehyde | Formaldehyde | | Yes | 50-00-0 |
| 120.03 | Formaldehyde | Formalin | | Yes | 50-00-0 |
| 120.04 | Furfural | Furfural | | | 98-01-1 |
| 120.05 | Glutaraldehyde | Glutaraldehyde | | Yes | 111-30-8 |
| 120.05 | Glutaraldehyde | Isolyzer | | Yes | |
| 120.06 | Paraformaldehyde | Paraformaldehyde | | Yes | 30525-89-4 |
| 120.07 | Paraldehyde | Paraldehyde | | | 123-63-7 |
| 120.08 | Cinnamic Aldehyde | Cinnamic Aldehyde | | | 104-55-2 |
| 120.08 | Cinnamic Aldehyde | Cinnamon Oil | | | 8007-80-5 |
| 120.09 | Crotonaldehyde | Crotonaldehyde | | | 4170-30-3 |
| 120.10 | Ortho-phthalaldehyde | Cidex OPA | | | 643-79-8 |
| 120.10 | Ortho-phthalaldehyde | 1,2-benzenedicarboxaldehyde | | | 643-79-8 |
| 120.10 | Ortho-phthalaldehyde | Phthalic Aldehyde | | | 643-79-8 |
| 120.10 | Ortho-phthalaldehyde | OPA | | | 643-79-8 |
| 120.10 | Ortho-phthalaldehyde | Ortho-phthalaldehyde | | | 643-79-8 |
| 120.11 | Phenol Formaldehyde | Phenol Formaldehyde | | | |
| 130.00 | Ketones, NOS | Ketones, NOS | S | | |
| 130.01 | Acetone | 2-Propanone | S | | 67-64-1 |
| 130.01 | Acetone | Acetone | S | | 67-64-1 |
| 130.01 | Acetone | Dimethyl Ketone | S | | 67-64-1 |
| 130.02 | MEK Peroxide | MEK Peroxide | | | 1338-23-4 |

| AOEC Code[1] | Primary Name | Synonym | Use[2] | Asthma Inducer? | CAS Number[3] |
|---|---|---|---|---|---|
| 130.02 | MEK Peroxide | Methyl Ethyl Ketone Peroxide | | | 1338-23-4 |
| 130.02 | MEK Peroxide | Peroxide 2-Butanone | | | 1338-23-4 |
| 130.03 | Methyl Ethyl Ketone | 2-Butanone | S | | 78-93-3 |
| 130.03 | Methyl Ethyl Ketone | MEK | S | | 78-93-3 |
| 130.03 | Methyl Ethyl Ketone | Methyl Ethyl Ketone | S | | 78-93-3 |
| 130.04 | Methyl Isobutyl Ketone | 4-Methyl-2-Pentanone | S | | 108-10-1 |
| 130.04 | Methyl Isobutyl Ketone | Hexone | S | | 108-10-1 |
| 130.04 | Methyl Isobutyl Ketone | Methyl Isobutyl Ketone | S | | 108-10-1 |
| 130.04 | Methyl Isobutyl Ketone | MIBK | S | | 108-10-1 |
| 130.05 | Methyl N-Butyl Ketone | Methyl N-Butyl Ketone | S | | 591-78-6 |
| 130.05 | Methyl N-Butyl Ketone | MNBK | S | | |
| 130.06 | Butyl Ketone | 5-Nonanone | S | | 502-56-7 |
| 130.06 | Butyl Ketone | Butyl Ketone | S | | 502-56-7 |
| 130.07 | Cyclohexanone | Cyclohexanone | S | | 108-94-1 |
| 130.08 | Dihydroxyacetone | 1,3-Dihydroxy-2-propanone | | | 96-26-4 |
| 130.08 | Dihydroxyacetone | DHA | | | 96-26-4 |
| 130.08 | Dihydroxyacetone | Dihydroxyacetone | | | 96-26-4 |
| 140.01 | Formates, NOS | Formates, NOS | | | |
| 140.02 | Diethyl Silicate | Diethyl Silicate | | | |
| 140.03 | Dioctylphthalate | Dioctylphthalate | | | |
| 140.04 | Phthalate Ester | Phthalate Ester | | | |
| 140.05 | Methyl Salicylate | Methyl 2-Hydroxybenzoate | | | 119-36-8 |
| 140.05 | Methyl Salicylate | Methyl Salicylate | | | 119-36-8 |
| 140.05 | Methyl Salicylate | Oil of Wintergreen | | | 119-36-8 |
| 140.06 | Phosphate Ester, NOS | Phosphate Ester, NOS | | | |
| 141.00 | Acetates, NOS | Acetates, NOS | | | |
| 141.01 | Amyl Acetate | Amyl Acetate | | | 628-63-7 |
| 141.01 | Amyl Acetate | N-Amyl Acetate | | | 628-63-7 |
| 141.01 | Amyl Acetate | N-Pentyl Acetate | | | 628-63-7 |
| 141.02 | Butyl Acetate | Butyl Acetate | S | | 123-86-4 |
| 141.02 | Butyl Acetate | N-Butyl Acetate | S | | 123-86-4 |
| 141.03 | Ethyl Acetate | Acetic Acid Ethyl Ester | S | | 141-78-6 |
| 141.03 | Ethyl Acetate | Acetic Ether | S | | 141-78-6 |
| 141.03 | Ethyl Acetate | Ethyl Acetate | S | | 141-78-6 |
| 141.04 | Vinyl Acetate | Vinyl Acetate | | | 108-05-4 |
| 141.05 | 2-Ethoxyethyl Acetate | 2-Ethoxyethyl Acetate | | | 111-15-9 |
| 141.05 | 2-Ethoxyethyl Acetate | Glycol Monoethyl Ether Acetate | | | 111-15-9 |
| 142.00 | Acrylates, NOS | Acrylates, NOS | | | |
| 142.01 | Acrylic Monomer | Acrylate | | | |
| 142.01 | Acrylic Monomer | Acrylic Monomer | | | |
| 142.02 | Butyl Acrylate | Acrylic Acid Butyl Ester | | | 141-32-2 |
| 142.02 | Butyl Acrylate | N-Butyl Acrylate | | | 141-32-2 |
| 142.02 | Butyl Acrylate | Butyl Acrylate | | | 141-32-2 |
| 142.03 | Ethyl Methacrylate | Ethyl Methacrylate | | | 97-63-2 |
| 142.04 | Methyl Methacrylate | Methyl Methacrylate | | Yes | 80-62-6 |
| 142.05 | Ethyl Acrylate | Ethyl Acrylate | | | 140-88-5 |
| 142.06 | Diethylaminoethyl Acrylate | Acrylic Acid, 2(Diethylamino)ethyl Ester | | | 2426-54-2 |
| 142.06 | Diethylaminoethyl Acrylate | Diethylaminoethyl Acrylate | | | 2426-54-2 |
| 142.07 | Cyanoacrylates, NOS | Alkylcyanoacrylates, NOS | | Yes | |
| 142.07 | Cyanoacrylates, NOS | Cyanoacrylates, NOS | | Yes | |
| 142.08 | Trimethylolpropane Triacrylate | Acrylic Acid, 1,1,1-(Trihydroxymethyl) Propane Triester | | | 15625-89-5 |
| 142.08 | Trimethylolpropane Triacrylate | Trimethylolpropane Triacrylate | | | 15625-89-5 |
| 142.08 | Trimethylolpropane Triacrylate | TMPTA | | | 15625-89-5 |
| 142.09 | 2-Hydroxypropyl Acrylate | 2-Hydroxypropyl Acrylate | | | 999-61-1 |
| 142.09 | 2-Hydroxypropyl Acrylate | Acrylic Acid, 2-Hydroxypropylester | | | 999-61-1 |
| 142.10 | Trimethylolpropane Triacrylate/2-Hydroxypropyl Acrylate | Trimethylolpropane Triacrylate/2-Hydroxypropyl Acrylate | | Yes | |
| 142.11 | Polymethyl Methacrylate | Polymethyl Methacrylate (PMMA) | | Yes | 9011-14-7 |

| AOEC Code[1] | Primary Name | Synonym | Use[2] | Asthma Inducer? | CAS Number[3] |
|---|---|---|---|---|---|
| 142.12 | N-Hexyl Acrylate | N-Hexyl Acrylate | | | 2499-95-8 |
| 150.01 | Formic Acid | Formic Acid | | | 64-18-6 |
| 150.02 | Benzoic Acid | Benzoic Acid | | | 65-85-0 |
| 150.03 | Oxalic Acid | Ethanedioic Acid | | | 144-62-7 |
| 150.03 | Oxalic Acid | Oxalic Acid | | | 144-62-7 |
| 150.05 | Stearic Acid | Stearic Acid | | | 57-11-4 |
| 150.06 | Tannic Acid | Tannic Acid | | | 1401-55-4 |
| 150.07 | Tartaric Acid | Tartaric Acid | | | 87-69-4 |
| 150.08 | Acrylic Acid | Acrylic Acid | | | 79-10-7 |
| 151.00 | Anhydride, NOS | Anhydride, NOS | | | |
| 151.01 | Phthalic Anhydride | Phthalic Acid Anhydride | | Yes | 85-44-9 |
| 151.01 | Phthalic Anhydride | Phthalic Anhydride | | Yes | 85-44-9 |
| 151.02 | Maleic Anhydride | Maleic Anhydride | | Yes | 108-31-6 |
| 151.03 | Trimellitic Anhydride | 1,2,4-Benzenetricarboxylic Acid 1,2-Anhydride | | Yes | 552-30-7 |
| 151.03 | Trimellitic Anhydride | TMA | | Yes | 552-30-7 |
| 151.03 | Trimellitic Anhydride | Trimellitic Anhydride | | Yes | 552-30-7 |
| 151.04 | Pyromellitic Dianhydride | Pyromellitic Acid Dianhydride | | Yes | 89-32-7 |
| 151.04 | Pyromellitic Dianhydride | Pyromellitic Dianhydride | | Yes | 89-32-7 |
| 151.05 | Methyl Tetrahydrophthalic Anhydride | Methyl Tetrahydrophthalic Anhydride | | Yes | 26590-20-5 |
| 151.06 | Tetrachlorophthalic Anhydride | Tetrachlorophthalic Anhydride | | Yes | 117-08-8 |
| 151.07 | Himic Anhydride | Himic Anhydride | | Yes | |
| 151.08 | Hexahydrophthalic Anhydride | Hexahydrophthalic Anhydride | | Yes | |
| 151.09 | Acetic Anhydride | Acetic Anhydride | | | 108-24-7 |
| 151.09 | Acetic Anhydride | Acetic Oxide | | | 108-24-7 |
| 160.00 | Aromatic Hydrocarbons, NOS | Aromatic Hydrocarbons, NOS | S | | |
| 160.00 | Aromatic Hydrocarbons, NOS | Aromatic Solvents, NOS | S | | |
| 160.01 | Benzene | Benzene | S | | 71-43-2 |
| 160.01 | Benzene | Benzol | S | | 71-43-2 |
| 160.01 | Benzene | Coal Naphtha | S | | 71-43-2 |
| 160.01 | Benzene | Mineral Naphtha | S | | 71-43-2 |
| 160.02 | Toluene | Methylbenzene | S | | 108-88-3 |
| 160.02 | Toluene | Phenylmethane | S | | 108-88-3 |
| 160.02 | Toluene | Toluene | S | | 108-88-3 |
| 160.02 | Toluene | Toluol | S | | 108-88-3 |
| 160.03 | Xylene | Dimethylbenzene | S | | 1330-20-7 |
| 160.03 | Xylene | Xylene | S | | 1330-20-7 |
| 160.03 | Xylene | Xylol | S | | 1330-20-7 |
| 160.04 | Styrene | Cinnamene | S | Yes | 100-42-5 |
| 160.04 | Styrene | Styrene | S | Yes | 100-42-5 |
| 160.04 | Styrene | Styrene Monomer | S | Yes | 100-42-5 |
| 160.04 | Styrene | Vinylbenzene | S | Yes | 100-42-5 |
| 160.05 | Naphthalene | Naphthalene | | | 91-20-3 |
| 160.05 | Naphthalene | Naphthalin | | | 91-20-3 |
| 160.05 | Naphthalene | Tar Camphor | | | 91-20-3 |
| 160.05 | Naphthalene | White Tar | | | 91-20-3 |
| 160.06 | Oil of Clove | Clove Oil | | | |
| 160.06 | Oil of Clove | Oil of Clove | | | |
| 160.07 | Ethyl Benzene | Ethyl Benzene | S | | 100-41-4 |
| 160.08 | Vinyl Toluene | Vinyl Toluene | S | | 25013-15-4 |
| 160.08 | Vinyl Toluene | Methylvinylbenzene | S | | 25013-15-4 |
| 160-08 | Vinyl Toluene | Methylstyrene | S | | 25013-15-4 |
| 161.00 | Polycyclic Aromatic Hydrocarbons, NOS | Coal Tar | | | |
| 161.00 | Polycyclic Aromatic Hydrocarbons, NOS | Coke Oven Emissions | | | |
| 161.00 | Polycyclic Aromatic Hydrocarbons, NOS | PAH | | | |
| 161.00 | Polycyclic Aromatic Hydrocarbons, NOS | PNA | | | |
| 161.00 | Polycyclic Aromatic Hydrocarbons, NOS | Polycyclic Aromatic Hydrocarbons, NOS | | | |
| 161.00 | Polycyclic Aromatic Hydrocarbons, NOS | Polynuclear Aromatics | | | |
| 161.00 | Polycyclic Aromatic Hydrocarbons, NOS | Producer Gas Residue | | | |

| AOEC Code[1] | Primary Name | Synonym | Use[2] | Asthma Inducer? | CAS Number[3] |
|---|---|---|---|---|---|
| 161.01 | Anthracene | Anthracene | | | 120-12-7 |
| 168.00 | Drilling Mud, NOS | Drilling Mud, NOS | | | |
| 168.01 | Drilling mud, water based | Drilling mud, water based | | | |
| 168.02 | Drilling mud, mineral oil based | Drilling mud, mineral oil based | | | |
| 168.03 | Drilling mud, vegetable oil based | Drilling mud, vegetable oil based | | | |
| 168.04 | Drilling mud, oil based NOS | Drilling mud, oil based NOS | | | |
| 168.05 | Drilling mud, synthetic based | Drilling mud, synthetic based | | | |
| 170.00 | Hydrocarbons, NOS | Hydrocarbons, NOS | | | |
| 170.00 | Hydrocarbons, NOS | Organic Chemicals, NOS | | | |
| 170.00 | Hydrocarbons, NOS | Petrochemicals, NOS | | | |
| 170.00 | Hydrocarbons, NOS | VOC, NOS | | | |
| 170.01 | Cutting Oils | Cutting Oils | | Yes | 8012-95-1 |
| 170.01 | Cutting Oils | Metal Working Fluids | | Yes | 8012-95-1 |
| 170.01 | Cutting Oils | Oil Mist | | Yes | 8012-95-1 |
| 170.02 | Inks, NOS | Inks, NOS | | | |
| 170.03 | Oils, NOS | Fats | | | |
| 170.03 | Oils, NOS | Greases | | | |
| 170.03 | Oils, NOS | Oils, NOS | | | |
| 170.04 | Crude oil | Crude oil | | | |
| 170.06 | Waxes, NOS | Waxes, NOS | | | |
| 171.00 | Solvents, NOS | Multiple Solvents | S | | |
| 171.00 | Solvents, NOS | Solvents, NOS | S | | |
| 171.00 | Solvents, NOS | Unspecified Solvents | S | | |
| 171.01 | Paint, NOS | Paint, NOS | S | | |
| 171.02 | Thinner | Enamel Thinner | S | | |
| 171.02 | Thinner | Lacquer Thinner | S | | |
| 171.02 | Thinner | Paint Thinner | S | | |
| 171.02 | Thinner | Thinner | S | | |
| 171.03 | Stripper | Stripper | S | | |
| 171.04 | Degreaser, NOS | Degreaser, NOS | | | |
| 171.05 | Paint, Latex | Paint, Latex | | | |
| 171.06 | Paint, Oil-Based | Paint, Oil-Based | | | |
| 171.07 | Lacquer | Lacquer | | | |
| 171.07 | Lacquer | Varnish | | | |
| 180.00 | Phenols, NOS | Phenols, NOS | | | |
| 180.01 | Creosote | Brick Oil | | | 8001-58-9 |
| 180.01 | Creosote | Coal Tar Oil | | | 8001-58-9 |
| 180.01 | Creosote | Creosote | | | 8001-58-9 |
| 180.02 | Cresol | Cresol | | | 1319-77-3 |
| 180.02 | Cresol | Cresylic Acid | | | 1319-77-3 |
| 180.02 | Cresol | Methylphenol | | | 1319-77-3 |
| 180.03 | Hydroquinone | Hydroquinone | | | 123-31-9 |
| 180.03 | Hydroquinone | P-Dihydroxybenzene | | | 123-31-9 |
| 180.03 | Hydroquinone | Photo Developer, Black & White | | | |
| 180.03 | Hydroquinone | P-Hydroxyphenol | | | 123-31-9 |
| 180.03 | Hydroquinone | X-Ray Developer | | | |
| 180.04 | Phenol | Carbolic Acid | | | 108-95-2 |
| 180.04 | Phenol | Hydroxybenzene | | | 108-95-2 |
| 180.04 | Phenol | Phenol | | | 108-95-2 |
| 180.05 | Quinone | Quinone | | | 106-51-4 |
| 180.06 | 1,2,3-Trihydroxybenzene | 1,2,3-Trihydroxybenzene | | | 87-66-1 |
| 180.06 | 1,2,3-Trihydroxybenzene | Pyrogallic Acid | | | 87-66-1 |
| 180.06 | 1,2,3-Trihydroxybenzene | Pyrogallol | | | 87-66-1 |
| 180.07 | Orthophenylphenol | Orthophenylphenol | P | Yes | 90-43-7 |
| 180.08 | Dodecylphenol | Dodecylphenol | | | 7193-86-8 |
| 181.00 | Chlorinated Phenols, NOS | Chlorinated Phenols, NOS | | | |
| 181.01 | Pentachlorophenol | PCP | | | 87-86-5 |
| 181.01 | Pentachlorophenol | Pentachlorophenol | | | 87-86-5 |
| 181.02 | Hexachlorophene | Hexachlorophene | | Yes | 70-30-4 |

| AOEC Code[1] | Primary Name | Synonym | Use[2] | Asthma Inducer? | CAS Number[3] |
|---|---|---|---|---|---|
| 181.03 | Triclosan | Triclosan | | | 3380-34-5 |
| 181.03 | Triclosan | Irgasan | | | 3380-34-5 |
| 181.03 | Triclosan | 5-chloro-2-(2,4-dichlorohydroxy) phenol | | | 3380-34-5 |
| 190.00 | Chlorinated Hydrocarbons, NOS | Chlorinated Hydrocarbons, NOS | S | | |
| 190.00 | Chlorinated Hydrocarbons, NOS | Chlorinated Solvents, NOS | S | | |
| 190.01 | Carbon Tetrachloride | Carbon Tetrachloride | S | | 56-23-5 |
| 190.01 | Carbon Tetrachloride | Tetrachloromethane | S | | 56-23-5 |
| 190.02 | Chloroethane | Chloroethane | S | | 75-00-3 |
| 190.02 | Chloroethane | Ethyl Chloride | S | | 75-00-3 |
| 190.02 | Chloroethane | Monochloroethane | S | | 75-00-3 |
| 190.03 | Chloroform | Chloroform | | | 67-66-3 |
| 190.03 | Chloroform | Trichloromethane | | | 67-66-3 |
| 190.04 | Chloroprene | Chlorobutadiene | | | 126-99-8 |
| 190.04 | Chloroprene | Chloroprene | | | 126-99-8 |
| 190.05 | Chloropropane | 1-Propyl Chloride | S | | 540-54-5 |
| 190.05 | Chloropropane | Chloropropane | S | | 540-54-5 |
| 190.05 | Chloropropane | N-Chloropropane | S | | 540-54-5 |
| 190.06 | 1,2-Dichloroethylene | 1,2-Dichloroethylene | S | | 540-59-0 |
| 190.06 | 1,2-Dichloroethylene | Acetylene Dichloride | S | | 540-59-0 |
| 190.07 | Ethylene Dichloride | 1,2-Dichloroethane | S | | 107-06-2 |
| 190.07 | Ethylene Dichloride | Ethylene Dichloride | S | | 107-06-2 |
| 190.08 | Methyl Chloroform | 1,1,1-Trichloroethane | S | | 71-55-6 |
| 190.08 | Methyl Chloroform | Methyl Chloroform | S | | 71-55-6 |
| 190.09 | Methylene Chloride | Dichloromethane | S | | 75-09-2 |
| 190.09 | Methylene Chloride | Methylene Chloride | S | | 75-09-2 |
| 190.09 | Methylene Chloride | Methylene Dichloride | S | | 75-09-2 |
| 190.10 | Perchlorethylene | Perc | S | | 127-18-4 |
| 190.10 | Perchlorethylene | Perchlorethylene | S | | 127-18-4 |
| 190.10 | Perchlorethylene | Perchloroethylene | S | | 127-18-4 |
| 190.10 | Perchlorethylene | Tetrachloroethylene | S | | 127-18-4 |
| 190.11 | Propylene Dichloride | 1,2-Dichloropropane | S | | 78-87-5 |
| 190.11 | Propylene Dichloride | Propylene Dichloride | S | | 78-87-5 |
| 190.12 | 1,1,2-Trichloroethane | 1,1,2-Trichloroethane | S | | 79-00-5 |
| 190.12 | 1,1,2-Trichloroethane | Ethane Trichloride | S | | 79-00-5 |
| 190.12 | 1,1,2-Trichloroethane | Vinyl Trichloride | S | | 79-00-5 |
| 190.13 | Trichloroethylene | Acetylene Trichloride | S | | 79-01-6 |
| 190.13 | Trichloroethylene | Trichloroethylene | S | | 79-01-6 |
| 190.14 | Vinyl Chloride Monomer | Chloroethene | | | 75-01-4 |
| 190.14 | Vinyl Chloride Monomer | Monochloroethylene | | | 75-01-4 |
| 190.14 | Vinyl Chloride Monomer | Vinyl Chloride Monomer | | | 75-01-4 |
| 190.14 | Vinyl Chloride Monomer | Vinyl Monomer | | | 75-01-4 |
| 190.15 | Anesthetic Gases, Halogenated | Enflurane | | Yes | 13838-16-9 |
| 190.15 | Anesthetic Gases, Halogenated | Methoxyflurane | | | 76-38-0 |
| 190.15 | Anesthetic Gases, Halogenated | Halothane | | | 151-67-7 |
| 190.15 | Anesthetic Gases, Halogenated | Anesthetic Gases, Halogenated | | | |
| 190.16 | Anesthetic Gases, Halogenated | Anesthetic Ethers, NOS | | | |
| 190.16 | Mustard Gas | Bis-2-Chloroethyl Sulfide | | | 505-60-2 |
| 190.16 | Mustard Gas | Mustard Gas | | | 505-60-2 |
| 190.17 | 1,3-Dichloro-2-Propanol | 1,3-Dichloro-2-Propanol | | | 96-23-1 |
| 191.00 | Brominated Pesticides, NOS | Brominated Pesticides, NOS | P | | |
| 191.01 | Ethylene Dibromide | 1,2-Dibromoethane | P | | 106-93-4 |
| 191.01 | Ethylene Dibromide | EDB | P | | 106-93-4 |
| 191.01 | Ethylene Dibromide | Ethylene Dibromide | P | | 106-93-4 |
| 191.03 | Methyl Bromide | Bromomethane | P | | 74-83-9 |
| 191.03 | Methyl Bromide | Methyl Bromide | P | | 74-83-9 |
| 191.03 | Methyl Bromide | Monobromomethane | P | | 74-83-9 |
| 191.04 | 1-Bromo-3-chloro-5,5-dimethylhydantoin | BCDMH | | | 16079-88-2 |
| 191.04 | 1-Bromo-3-chloro-5,5-dimethylhydantoin | di-Halo | | | 16079-88-2 |

| AOEC Code[1] | Primary Name | Synonym | Use[2] | Asthma Inducer? | CAS Number[3] |
|---|---|---|---|---|---|
| 192.00 | Fluorocarbons, NOS | Fluorocarbons, NOS | | | |
| 192.01 | Freon, NOS | Freon, NOS | | | 75-45-6 |
| 192.02 | Chlorofluorocarbon, NOS | CFC | | | |
| 192.02 | Chlorofluorocarbon, NOS | Chlorofluorocarbon, NOS | | | |
| 192.03 | Brominated Fluorocarbon | Brominated Fluorocarbon | | | |
| 192.04 | Trichlorotrifluoroethane | FC 113 | | | 76-13-1 |
| 192.04 | Trichlorotrifluoroethane | Trichlorotrifluoroethane | | | 76-13-1 |
| 192.04 | Trichlorotrifluoroethane | Trifluorotrichloroethane | | | 76-13-1 |
| 192.05 | Difluoroethane | Difluoroethane | | | 25497-28-3 |
| 192.06 | Freon, Heated | Freon, Heated | | Yes | |
| 192.06 | Freon, Heated | Welding Freon | | Yes | |
| 192.07 | Freon, Unheated | Freon, Unheated | | | 75-45-6 |
| 200.01 | Agent Orange | 2,4,5-T | P | | 93-76-5 |
| 200.01 | Agent Orange | 2,4-D | P | | 94-75-7 |
| 200.01 | Agent Orange | Agent Orange | P | | 39277-47-9 |
| 200.02 | Benzyl Chloride | Alpha-Chlorotoluene | | | 100-44-7 |
| 200.02 | Benzyl Chloride | Benzyl Chloride | | | 100-44-7 |
| 200.03 | Chlorinated Naphthalene | Chlorinated Naphthalene | | | |
| 200.03 | Chlorinated Naphthalene | Halowax | | | |
| 200.04 | Chlorinated Dibenzodioxins | 2,3,7,8-TCDD | P | | 1746-01-6 |
| 200.04 | Chlorinated Dibenzodioxins | Chlorinated Dibenzodioxins | P | | |
| 200.04 | Chlorinated Dibenzodioxins | Dioxin | P | | |
| 200.05 | Isophthaloyl Chloride | Isophthalic Acid Chloride | | | 99-63-8 |
| 200.05 | Isophthaloyl Chloride | Isophthaloyl Chloride | | | 99-63-8 |
| 200.05 | Isophthaloyl Chloride | Phthaloyl Chloride | | | 99-63-8 |
| 200.06 | Pentachloronitrobenzene | Pentachloronitrobenzene | P | | 82-68-8 |
| 200.07 | Polychlorinated Biphenyls | Arochlor | | | 1336-36-3 |
| 200.07 | Polychlorinated Biphenyls | Chlorodiphenyl | | | 1336-36-3 |
| 200.07 | Polychlorinated Biphenyls | Kanechlor | | | 1336-36-3 |
| 200.07 | Polychlorinated Biphenyls | PCBs | | | 1336-36-3 |
| 200.07 | Polychlorinated Biphenyls | Polychlorinated Biphenyls | | | 1336-36-3 |
| 200.08 | Chlorhexidine | Chlorhexidine | | Yes | 55-56-1 |
| 200.09 | Benzoyl Chloride | Benzoyl Chloride | | | 98-88-4 |
| 200.10 | 2,6-Dichlorobenzonitrile | 2,6-Dichlorobenzonitrile | P | | 1194-65-6 |
| 200.10 | 2,6-Dichlorobenzonitrile | Dichlobenil | P | | 1194-65-6 |
| 200.11 | Chlorophenoxy Herbicides, NOS | Chlorophenoxy Herbicides, NOS | P | | |
| 201.00 | Chlorinated Benzenes, NOS | Chlorinated Benzenes, NOS | | | |
| 201.01 | Dichlorobenzene | DCB | P | | |
| 201.01 | Dichlorobenzene | Dichlorobenzene | P | | |
| 201.01 | Dichlorobenzene | Mixed DCB | P | | |
| 201.01 | Dichlorobenzene | P-Dichlorobenzene | P | | 106-46-7 |
| 201.02 | 1,2,4-Trichlorobenzene | 1,2,4-Trichlorobenzene | | | 120-82-1 |
| 201.03 | Hexachlorobenzene | Hexachlorobenzene | P | | 118-74-1 |
| 201.03 | Hexachlorobenzene | Perchlorobenzene Fungicide | P | | |
| 201.04 | Chlorobenzene | Chlorbenzol | S | | 108-90-7 |
| 201.04 | Chlorobenzene | Chlorobenzene | S | | 108-90-7 |
| 201.04 | Chlorobenzene | Phenyl Chloride | S | | 108-90-7 |
| 210.01 | Acetonitrile | Acetonitrile | | | 75-05-8 |
| 210.02 | Acrylonitrile | Acrylonitrile | | | 107-13-1 |
| 210.02 | Acrylonitrile | Propenenitrile | | | 107-13-1 |
| 210.02 | Acrylonitrile | Vinyl Cyanide | | | 107-13-1 |
| 210.03 | Chlorothalonil | Chlorothalonil | P | Yes | 1897-45-6 |
| 210.03 | Chlorothalonil | Tetrachloro-Isophthalonitrile | P | Yes | 1897-45-6 |
| 211.00 | Cyanides, NOS | Cyanides, NOS | | | |
| 211.01 | Hydrogen Cyanide | HCN | | | 74-90-8 |
| 211.01 | Hydrogen Cyanide | Hydrogen Cyanide | | | 74-90-8 |
| 211.02 | Potassium Cyanide | Potassium Cyanide | | | 151-50-8 |
| 211.03 | Sodium Cyanide | Sodium Cyanide | | | 143-33-9 |

| AOEC Code[1] | Primary Name | Synonym | Use[2] | Asthma Inducer? | CAS Number[3] |
|---|---|---|---|---|---|
| 211.04 | Sodium Azide | Sodium Azide | | | 26628-22-8 |
| 220.00 | Isocyanates, NOS | Isocyanates, NOS | | | |
| 220.01 | Methyl Isocyanate | Methyl Ester Isocyanic Acid | | | |
| 220.01 | Methyl Isocyanate | Methyl Isocyanate | | | 624-83-9 |
| 220.01 | Methyl Isocyanate | MIC | | | 624-83-9 |
| 220.02 | Methyl Isothiocyanate | Methyl Isothiocyanate | P | | 556-61-6 |
| 221.00 | Diisocyanates, NOS | Diisocyanates, NOS | | Yes | |
| 221.01 | Toluene Diisocyanate | TDI | | Yes | 584-84-9 |
| 221.01 | Toluene Diisocyanate | Toluene Diisocyanate | | Yes | 584-84-9 |
| 221.01 | Toluene Diisocyanate | Toluene-2,4-Diisocyanate | | Yes | 584-84-9 |
| 221.02 | Methylene Bisphenyl Diisocyanate | Diphenylmethane Diisocyanate | | Yes | 101-68-8 |
| 221.02 | Methylene Bisphenyl Diisocyanate | MDI | | Yes | 101-68-8 |
| 221.02 | Methylene Bisphenyl Diisocyanate | Methylene Bisphenyl Diisocyanate | | Yes | 101-68-8 |
| 221.03 | Naphthalene Diisocyanate | 1,5-Naphthylene Ester Isocyanic Acid | | Yes | |
| 221.03 | Naphthalene Diisocyanate | Naphthalene Diisocyanate | | Yes | |
| 221.03 | Naphthalene Diisocyanate | NDI | | Yes | |
| 221.04 | Hexamethylene Diisocyanate | 1,6-Diisocyanato-Hexane | | Yes | 822-06-0 |
| 221.04 | Hexamethylene Diisocyanate | HDI | | Yes | 822-06-0 |
| 221.04 | Hexamethylene Diisocyanate | Hexamethylene Diisocyanate | | Yes | 822-06-0 |
| 221.05 | Isophorone Diisocyanate | IPDI | | Yes | 4098-71-9 |
| 221.05 | Isophorone Diisocyanate | Isophorone Diisocyanate | | Yes | 4098-71-9 |
| 221.06 | Polymethylene Polyphenylisocyanate | Polymethylene Polyphenylisocyanate | | Yes | 9016-87-9 |
| 221.06 | Polymethylene Polyphenylisocyanate | PPI | | Yes | 9016-87-9 |
| 221.07 | TDI Prepolymers | TDI Prepolymers | | Yes | |
| 221.08 | HDI Prepolymers | HDI Prepolymers | | Yes | |
| 230.01 | N-Butylamine | N-Butylamine | | | 109-73-9 |
| 230.02 | Diethylamine | Diethylamine | | | 109-89-7 |
| 230.03 | Isopropylamine | 2-Aminopropane | | | |
| 230.03 | Isopropylamine | Isopropylamine | | | |
| 230.04 | Dimethylamine | Dimethylamine | | | 124-40-3 |
| 230.05 | Triethylamine | Triethylamine | | | 121-44-8 |
| 230.05 | Triethylamine | N,N-Diethylethanamine | | | 121-44-8 |
| 230.06 | Cyclohexylamine | Cyclohexylamine | | | 108-91-8 |
| 230.07 | Polyethylamine | Polyethylamine | | | |
| 231.00 | Ethanolamines, NOS | Ethanolamines, NOS | | Yes | 141-43-5 |
| 231.01 | Monoethanolamine | 2-Aminoethanol | | Yes | 141-43-5 |
| 231.01 | Monoethanolamine | Monoethanolamine | | Yes | 141-43-5 |
| 231.02 | Triethanolamine | Triethanolamine | | Yes | 102-71-6 |
| 231.03 | Aminoethyl Ethanolamine | Aminoethyl Ethanolamine | | Yes | 111-41-1 |
| 231.03 | Aminoethyl Ethanolamine | Ethanol Ethylene Diamine | | Yes | 111-41-1 |
| 231.04 | Diethanolamine | DEA (Diethanolamine) | | Yes | 111-42-2 |
| 231.04 | Diethanolamine | Diethanolamine | | Yes | 111-42-2 |
| 231.04 | Dimethylethanolamine | Dimethylaminoethanol | | Yes | 108-01-0 |
| 231.04 | Dimethylethanolamine | Dimethylethanolamine | | Yes | 108-01-0 |
| 231.04 | N,N-Dimethylethanolamine | N,N-Dimethylethanolamine | | Yes | 108-01-0 |
| 232.00 | Polyamines, NOS | Polyamines, NOS | | | |
| 232.01 | Ethylenediamine | Ethylenediamine | | Yes | 107-15-3 |
| 232.02 | Hexamethylenetetramine | Hexamethylenetetramine | | | 100-97-0 |
| 232.03 | Triethylenetetramine | Triethylenetetramine | | Yes | 112-24-3 |
| 232.04 | Trimethylhexane-1,6-diamine/Isophorondiamine Mixture | Trimethylhexanediamine/Isophorondiamine Mixture | | Yes | 25620-58-0 |
| 232.04 | Trimethylhexane-1,6-diamine/Isophorondiamine Mixture | Trimethylhexane-1,6-diamine/Isophorondiamine Mixture | | Yes | 25620-58-0 |
| 232.05 | 3-DMAPA | 3-Dimethylamino Propylamine | | Yes | |
| 232.05 | 3-DMAPA | 3-DMAPA | | Yes | |
| 232.06 | EPO 60 | EPO 60 | | Yes | 142443-98- |
| 232.06 | EPO 60 | Polyamine EPO 60 | | Yes | 142443-98- |
| 232.07 | Diethylenetriamine | Diethylenetriamine | | | 111-40-0 |
| 232.08 | Ethylenediamine Tetraacetic Acid | EDTA/Edetic Acid | | | 60-00-4 |

| AOEC Code[1] | Primary Name | Synonym | Use[2] | Asthma Inducer? | CAS Number[3] |
|---|---|---|---|---|---|
| 232.09 | Isophorone diamine | Isophorone diamine | | | 2855-13-2 |
| 232.10 | Hexamethylenediamine | Hexamethylenediamine | | | 124-09-4 |
| 232.10 | Hexamethylenediamine | 1,6-Hexanediamine | | | 124-09-4 |
| 233.00 | Diamines, NOS | Diamines, NOS | | | |
| 240.01 | N-Nitrosodimethylamine | DMN | | | 62-75-9 |
| 240.01 | N-Nitrosodimethylamine | DMNA | | | 62-75-9 |
| 240.01 | N-Nitrosodimethylamine | N,N-Dimethylnitrosamine | | | 62-75-9 |
| 240.01 | N-Nitrosodimethylamine | N-Nitrosodimethylamine | | | 62-75-9 |
| 250.01 | Aniline | Aminobenzene | | | 62-53-3 |
| 250.01 | Aniline | Aminophen | | | 62-53-3 |
| 250.01 | Aniline | Aniline | | | 62-53-3 |
| 250.01 | Aniline | Phenylamine | | | 62-53-3 |
| 250.02 | 8-Hydroxyquinoline | 8-Hydroxyquinoline | | | 148-24-3 |
| 250.02 | 8-Hydroxyquinoline | 8-Quinolinol | | | |
| 250.03 | 4,4'-Methylenebis(2-Chloroaniline) | 4,4'-Methylenebis(2-Chloroaniline) | | | 101-14-4 |
| 250.03 | 4,4'-Methylenebis(2-Chloroaniline) | ME-MDA | | | 101-14-4 |
| 250.03 | 4,4'-Methylenebis(2-Chloroaniline) | MOCA | | | 101-14-4 |
| 250.04 | Nitrobenzene | Nitrobenzene | | | 98-95-3 |
| 250.04 | Nitrobenzene | Nitrobenzol | | | 98-95-3 |
| 250.04 | Nitrobenzene | Oil of Mirbane | | | 98-95-3 |
| 250.05 | Dinitrobenzene | Dinitrobenzene | | | |
| 250.05 | Dinitrobenzene | Dinitrobenzol | | | |
| 250.06 | Dinitro-o-Cresol | Dinitrol | | | |
| 250.06 | Dinitro-o-Cresol | Dinitro-o-Cresol | | | 1335-85-9 |
| 250.06 | Dinitro-o-Cresol | DNOC | | | |
| 250.07 | Dinitrotoluene | Dinitrotoluene | | | |
| 250.07 | Dinitrotoluene | Dinitrotoluol | | | |
| 250.07 | Dinitrotoluene | DNT | | | |
| 250.08 | Trinitrotoluene | TNT | | | |
| 250.08 | Trinitrotoluene | Trinitrotoluene | | | 118-96-7 |
| 250.08 | Trinitrotoluene | Trinitrotoluol | | | |
| 250.09 | Picric Acid | 2,4,6-Trinitrophenol | | | 88-89-1 |
| 250.09 | Picric Acid | Carbazotic Acid | | | 88-89-1 |
| 250.09 | Picric Acid | Picric Acid | | | 88-89-1 |
| 250.10 | Tetryl | Nitramine | | | 479-45-8 |
| 250.10 | Tetryl | Tetryl | | | 479-45-8 |
| 250.11 | Nitrophenol | Nitrophenol | | | |
| 250.12 | Dinitrophenol | Dinitrophenol | | | 25550-58-7 |
| 250.13 | 3,3'-Dichlorobenzidene & Salts | 3,3'-Dichlorobenzidene & Salts | | | 91-94-1 |
| 250.14 | Pyridine | Azabenzene | | | 110-86-1 |
| 250.14 | Pyridine | Azine | | | 110-86-1 |
| 250.14 | Pyridine | Pyridine | | | 110-86-1 |
| 250.15 | Acridine | Acridine | | | 260-94-6 |
| 250.16 | Methylenedianiline | 4,4-Diaminodiphenylmethane | | | 101-77-9 |
| 250.16 | Methylenedianiline | MDA | | | 101-77-9 |
| 250.16 | Methylenedianiline | Methylenedianiline | | | 101-77-9 |
| 250.17 | Dyes, NOS | Dye Intermediates, NOS | | | |
| 250.17 | Dyes, NOS | Dyes, NOS | | | |
| 250.17 | Ethidium Bromide | Ethidium Bromide | | | 1239-45-8 |
| 250.18 | Diethylaniline | DEA (Diethylaniline) | | | 91-66-7 |
| 250.18 | Diethylaniline | Diethylaniline | | | 91-66-7 |
| 250.19 | Amino Acids, NOS | Amino Acids, NOS | | | |
| 250.20 | Pteridine | Pteridine | | | 91-18-9 |
| 251.00 | Aniline Dyes, NOS | Aniline Dyes, NOS | | | |
| 251.01 | Para-Aminophenol | 4-Hydroxyaniline | | | 123-30-8 |
| 251.01 | Para-Aminophenol | Activol | | | 123-30-8 |
| 251.01 | Para-Aminophenol | PAP | | | 123-30-8 |
| 251.01 | Para-Aminophenol | Para-Aminophenol | | | 123-30-8 |

| AOEC Code[1] | Primary Name | Synonym | Use[2] | Asthma Inducer? | CAS Number[3] |
|---|---|---|---|---|---|
| 251.02 | Auramine | Auramine | | | 492-80-8 |
| 251.03 | Benzidine | 4,4-Aminodiphenyl | | | 92-87-5 |
| 251.03 | Benzidine | Benzidine | | | 92-87-5 |
| 251.04 | Alpha-Naphthylamine | Alpha-Naphthylamine | | | 134-32-7 |
| 251.05 | Beta-Naphthylamine | 2-Naphthylamine | | | |
| 251.05 | Beta-Naphthylamine | Beta-Naphthylamine | | | |
| 251.06 | Phenylenediamine-Deleted incorrect entry | Phenylene Diamine-Deleted incorrect entry | | | |
| 251.06 | Benzenediamine | Phenylene Diamine | | | 25265-76-3 |
| 251.06 | 1,4-Benzenediamine | 1,4-Benzenediamine | | | 106-50-3 |
| 251.07 | Magenta | Magenta | | | 632-99-5 |
| 252.00 | Azo Compounds, NOS | Azo Compounds, NOS | | | |
| 252.00 | Azo Compounds, NOS | Azo Dyes, NOS | | | |
| 252.01 | Amaranth | Amaranth | | | 915-67-3 |
| 252.01 | Amaranth | FD&C Red No. 2 | | | 915-67-3 |
| 252.02 | Oil Orange SS | Oil Orange SS | | | 2646-17-5 |
| 252.03 | Diazonium Salt | Diazonium Salt | | Yes | |
| 260.01 | Acetamide | Acetamide | | | 60-35-5 |
| 260.02 | Acrylamide | Acrylamide | | | 79-06-1 |
| 260.02 | Acrylamide | Acrylic Amide | | | 79-06-1 |
| 260.02 | Acrylamide | Propenamide | | | 79-06-1 |
| 260.03 | Benomyl | Benomyl | P | | 17804-35-2 |
| 260.03 | Benomyl | Benlate Fungicide | P | | 17804-35-2 |
| 260.04 | Calcium Cyanamide | Calcium Cyanamide | | | 156-62-7 |
| 260.05 | Captan | Captan | P | | 133-06-2 |
| 260.06 | Chloropicrin | Chloropicrin | P | | 76-06-2 |
| 260.06 | Chloropicrin | Trichloronitromethane | P | | 76-06-2 |
| 260.07 | N,N-Dimethylacetamide | N,N-Dimethylacetamide | | | 127-19-5 |
| 260.07 | N,N-Dimethylacetamide | Dimethylacetamide | | | 127-19-5 |
| 260.07 | N,N-Dimethylacetamide | DMAC | | | 127-19-5 |
| 260.08 | N,N-Dimethylformamide | DMF | S | | 68-12-2 |
| 260.08 | N,N-Dimethylformamide | N,N-Dimethylformamide | S | | 68-12-2 |
| 260.09 | Dimethylhydrazine | Dimethylhydrazine | | | 57-14-7 |
| 260.10 | Ethylenimine | Aziridine | | | 151-56-4 |
| 260.10 | Ethylenimine | Ethylenimine | | | 151-56-4 |
| 260.10 | Polyfunctional aziridine | PFA | | Yes | 64265-57-2 |
| 260.11 | Hydrazine | Hydrazine | | | 302-01-2 |
| 260.11 | Hydrazine | Hydrazine Derivatives | | | |
| 260.12 | Methyl Pyrrolidone | 1-Methyl-2-Pyrrolidinone | | | 872-50-4 |
| 260.12 | Methyl Pyrrolidone | Methyl Pyrrolidone | | | 51013-18-4 |
| 260.13 | Nicotine Sulfate | Nicotine Sulfate | P | | 65-30-5 |
| 260.14 | Nitroparaffins | 2-Nitropropane | | | 79-46-9 |
| 260.14 | Nitroparaffins | Nitroethane | | | 79-24-3 |
| 260.14 | Nitroparaffins | Nitromethane | | | 75-52-5 |
| 260.14 | Nitroparaffins | Nitroparaffins | | | |
| 260.15 | Paraquat | Paraquat | P | | 4685-14-7 |
| 260.16 | Strychnine | Strychnine | P | | 57-24-9 |
| 260.17 | Triazines, NOS | Triazines, NOS | P | | 101-05-3 |
| 260.18 | Azodicarbamide | 1,1-Azobisformamide | | Yes | 123-77-3 |
| 260.18 | Azodicarbamide | Azobisformamide | | Yes | 123-77-3 |
| 260.18 | Azodicarbamide | Azodicarbamide | | Yes | 123-77-3 |
| 260.19 | Piperazine Hydrochloride | Piperazine Hydrochloride | | Yes | 142-64-3 |
| 260.20 | N-Methylmorpholine | N-Methylmorpholine | | Yes | 109-02-4 |
| 260.21 | Chloramine T | Chloramine T | | Yes | 127-65-1 |
| 260.22 | Tetrazene | Tetrazene | | Yes | 70816-59-0 |
| 260.23 | Diazomethane | Diazomethane | | | 334-88-3 |
| 260.24 | Diethyl Formamide | Diethyl Formamide | S | | 617-84-5 |
| 260.25 | Garlon 4 | Garlon 4 | P | | 64700-56-7 |
| 260.26 | N,N-diethyl-m-toluamide | DEET | P | | 134-62-3 |
| 260.26 | N,N-diethyl-m-toluamide | m-Delphene | P | | 134-62-3 |

| AOEC Code[1] | Primary Name | Synonym | Use[2] | Asthma Inducer? | CAS Number[3] |
|---|---|---|---|---|---|
| 260.26 | N,N-diethyl-m-toluamide | Diethyltoluamide | P | | 134-62-3 |
| 260.26 | N,N-diethyl-m-toluamide | Benzamide | P | | 134-62-3 |
| 260.26 | N,N-diethyl-m-toluamide | N,N-diethyl-3-methyl- | P | | 134-62-3 |
| 260.27 | Amyl Nitrite | Amyl Nitrite | | | 110-46-3 |
| 260.27 | Amyl Nitrite | Poppers | | | 110-46-3 |
| 260.27 | Amyl Nitrite | Snappers | | | 110-46-3 |
| 260.27 | Amyl Nitrite | Isoamyl Nitrite | | | 110-46-3 |
| 260.28 | Melamine | Cyanurotramide | | | 108-78-1 |
| 260.28 | Melamine | Melamine | | | 108-78-1 |
| 260.29 | Polyhexamethylene Biguanide | Polyhexamethylene Biguanide | P | | 28757-47-3 |
| 260.29 | Polyhexamethylene Biguanide | Baquacil | P | | 28757-47-3 |
| 260.30 | 3-Amino-5-mercapto-1,2,4-triazole | 3-Amino-5-mercapto-1,2,4-triazole | | Yes | 16691-43-3 |
| 260.30 | 3-Amino-5-mercapto-1,2,4-triazole | AMT | | Yes | 16691-43-3 |
| 260.31 | 2-Ethylhexyl Nitrate | 2-Ethylhexyl Nitrate | | | 27247-96-7 |
| 261.01 | Nitroglycerin | Glyceryl Trinitrate | | | 55-63-0 |
| 261.01 | Nitroglycerin | Nitroglycerin | | | 55-63-0 |
| 261.01 | Nitroglycerin | Trinitroglycerin | | | 55-63-0 |
| 261.02 | Pentaerythritol Tetranitrate | Pentaerythritol Tetranitrate | | | 78-11-5 |
| 261.02 | Pentaerythritol Tetranitrate | PETN | | | 78-11-5 |
| 270.00 | Polymers, NOS | Plastics, NOS | | | |
| 270.00 | Polymers, NOS | Plastics, Pre-Polymer | | | |
| 270.00 | Polymers, NOS | Polymers, NOS | | | |
| 270.01 | Acrylics | Acrylic Acid Polymer | | | |
| 270.01 | Acrylics | Acrylic Resins | | | |
| 270.01 | Acrylics | Acrylics | | | |
| 270.01 | Acrylics | Paint, Acrylic | | | |
| 270.02 | Latex, Natural Rubber | Latex Gloves, NOS | | Yes | |
| 270.02 | Latex, Natural Rubber | Latex, Natural Rubber | | Yes | |
| 270.03 | Latex, Synthetic | Latex, Synthetic | | | |
| 270.03 | Latex, Synthetic | Polydimethyl Siloxane | | | 9016-00-6 |
| 270.04 | Phenolics | Phenolic Resins | | | |
| 270.04 | Phenolics | Phenolics | | | |
| 270.05 | Polyester Resin | Polyester Resin | | | |
| 270.06 | Polyethylene, NOS | PE, NOS | | | |
| 270.06 | Polyethylene, NOS | Polyethylene, NOS | | | |
| 270.07 | Polyurethane | Polycarbamate | | | 64440-88-6 |
| 270.07 | Polyurethane | Polyurethane | | | |
| 270.07 | Polyurethane | Urethane | | | 51-79-6 |
| 270.07 | Polyurethane | Urethane Enamel Paint | | | |
| 270.07 | Polyurethane | Varethane Paint | | | |
| 270.08 | Polyvinyl Alcohol | Polyvinyl Alcohol | | | 9002-89-5 |
| 270.08 | Polyvinyl Alcohol | PVA | | | 9002-89-5 |
| 270.09 | Polyvinyl Chloride (heated) | PVC (heated) | | Yes | |
| 270.09 | Polyvinyl Chloride (heated) | Polyvinyl Chloride (heated) | | Yes | |
| 270.09 | Polyvinyl Chloride | PVC (Non-heated) | | Yes | 9002-86-2 |
| 270.10 | Silicone | Silicone | | | |
| 270.10 | Silicone | Silicone Fluid | | | |
| 270.10 | Silicone | Silicone Rubber | | | |
| 270.10 | Silicone | Siloxanes | | | |
| 270.11 | Sodium Carboxymethyl Cellulose | Sodium Carboxymethyl Cellulose | | | 9004-32-4 |
| 270.12 | Polytetrafluoroethylene | Polytetrafluoroethylene | | | 9002-84-0 |
| 270.12 | Polytetrafluoroethylene | PTFE | | | 9002-84-0 |
| 270.12 | Polytetrafluoroethylene | Teflon | | | 9002-84-0 |
| 270.13 | Plasticizers | Plasticizers | | | |
| 270.14 | Polyvinyl Pyrrolidone | Polyvinyl Pyrrolidone | | | |
| 270.15 | Resin Systems, NOS | Resin Systems, NOS | | | |
| 270.15 | Resins, NOS | Resins, NOS | | | |
| 270.16 | Urea Formaldehyde | Urea Formaldehyde | | Yes | |

| AOEC Code[1] | Primary Name | Synonym | Use[2] | Asthma Inducer? | CAS Number[3] |
|---|---|---|---|---|---|
| 270.16 | Urea Formaldehyde | Urea Formaldehyde Resin | | Yes | |
| 270.17 | Nylon | Nylon | | | 63428-83-1 |
| 270.18 | Plexiglass Dust | Plexiglass Dust | | Yes | |
| 270.19 | Polyimides | Polyimides | | | |
| 270.20 | Polystyrene | Polystyrene | | | 9003-53-6 |
| 270.21 | Alkyd Resins | Alkyd Resins | | | 63148-69-6 |
| 270.22 | Amino Resins | Amino Resins | | | |
| 270.23 | Styrene-Maleic Anhydride Resin | Styrene-Maleic Anhydride Polymer | | | 9011-13-6 |
| 270.23 | Styrene-Maleic Anhydride Resin | Styrene-Maleic Anhydride Resin | | | 9011-13-6 |
| 270.24 | Polypropylene, NOS | Polypropylene, NOS | | | 9003-07-0 |
| 270.25 | Polyethylene Terephthalate/Polybutylene Terephthal | Heated Electrostatic Polyester Paint | | Yes | |
| 270.25 | Polyethylene Terephthalate/Polybutylene Terephthal | Polyethylene Terephthalate/Polybutylene Terephthalate | | Yes | |
| 270.26 | Polyethylene Terephthalate | Mylar | | | 25038-59-9 |
| 270.26 | Polyethylene Terephthalate | Polyethylene Terephthalate | | | 25038-59-9 |
| 270.27 | Vinyl Plastic Wrap | Vinyl Plastic Wrap | | | |
| 270.28 | Polyvinylbutyral | Polyvinyl Butyral Resins | | | 63148-65-2 |
| 270.28 | Polyvinylbutyral | Polyvinylbutyral | | | 63148-65-2 |
| 270.29 | Acrylonitrile-Butadiene-Styrene Copolymer | ABS Copolymer | | | 9003-56-9 |
| 270.29 | Acrylonitrile-Butadiene-Styrene Copolymer | Acrylonitrile-Butadiene-Styrene Copolymer | | | 9003-56-9 |
| 270.30 | Acrylonitrile-Butadiene-Styrene-Polyvinyl Chloride | Acrylonitrile-Butadiene-Styrene-Polyvinyl Chloride | | | |
| 270.31 | Polyethylene, Heated | Polyethylene, Heated | | | 9002-88-4 |
| 270.32 | Polyethylene, Unheated | Polyethylene, Unheated | | | |
| 270.33 | Polypropylene, Heated | Polypropylene, Heated | | Yes | 9003-07-0 |
| 270.34 | Polypropylene, Unheated | Polypropylene, Unheated | | | |
| 270.35 | Poly (P-Phenylenediamine) | Kevlar | | | 26125-61-1 |
| 270.35 | Poly (P-Phenylenediamine) | Poly (P-Phenylenediamine) | | | 26125-61-1 |
| 270.36 | Carbopol, NOS | Carbopol, NOS | | | |
| 270.36 | Carbopol, NOS | Carbopol | | | |
| 270.37 | Polytetrafluoroethylene, Thermal Decomposition Products | Polytetrafluoroethylene, Thermal Decomposition Products | | | |
| 270.37 | Polytetrafluoroethylene, Thermal Decomposition Products | Teflon, Thermal Decomposition Products | | | |
| 270.38 | Polyvinyl Chloride, Thermal Decomposition Products | Polyvinyl Chloride, Thermal Decomposition Products | | Yes | |
| 270.39 | Collodion | Collodium | | | |
| 270.39 | Collodion | Collodion | | | |
| 270.40 | Dodecanedioic Acid | Dodecanedioic Acid | | Yes | 693-23-2 |
| 271.00 | Rubber, NOS | Rubber, NOS | | | |
| 271.01 | Butadiene & Styrene | Butadiene & Styrene | | | |
| 271.01 | Butadiene & Styrene | SBR | | | |
| 271.01 | Butadiene & Styrene | Styrene-Butadiene Copolymer | | | |
| 271.02 | Neoprene | Neoprene | | | 9010-98-4 |
| 271.02 | Neoprene | Polychlorobutadiene | | | |
| 271.02 | Neoprene | Polychloroprene | | | |
| 280.00 | Organochlorine Pesticides, NOS | Organochlorine Insecticides, NOS | P | | |
| 280.00 | Organochlorine Pesticides, NOS | Organochlorine Pesticides, NOS | P | | |
| 280.01 | Aldrin | Aldrin | P | | 309-00-2 |
| 280.02 | Chlordane | Chlordane | P | | 57-74-9 |
| 280.03 | DDT | DDT | P | | 50-29-3 |
| 280.03 | DDT | Dichlorodiphenyltrichloroethane | P | | 50-29-3 |
| 280.04 | Dieldrin | Dieldrin | P | | 60-57-1 |
| 280.05 | Endrin | Endrin | P | | 72-20-8 |
| 280.06 | Lindane | Lindane | P | | 58-89-9 |
| 280.07 | Methoxychlor | Methoxychlor | P | | 72-43-5 |
| 280.08 | Heptachlor | Heptachlor | P | | 76-44-8 |
| 280.08 | Heptachlor | Haptachlorine | P | | 76-44-8 |
| 290.01 | Piperonyl Butoxide | Piperonyl Butoxide | P | | 51-03-6 |

| AOEC Code[1] | Primary Name | Synonym | Use[2] | Asthma Inducer? | CAS Number[3] |
|---|---|---|---|---|---|
| 291.00 | Organophosphate Pesticides, NOS | Organophosphate Pesticides, NOS | P | | |
| 291.01 | Malathion | Malathion | P | Yes | 121-75-5 |
| 291.02 | Parathion | Parathion | P | | 56-38-2 |
| 291.03 | Diazinon | Diazinon | P | Yes | 333-41-5 |
| 291.04 | Nerve Gas | Nerve Gas | | | |
| 291.05 | Chlorpyrifos | Chlorpyrifos | P | | 2921-88-2 |
| 291.05 | Chlorpyrifos | Dursban | P | | 2921-88-2 |
| 291.06 | Nemacur | Nemacur | P | Yes | 22224-92-6 |
| 291.07 | Dimethoate | Dimethoate | P | Yes | 60-51-5 |
| 291.08 | Safrotin | Propetamphos | P | Yes | 31218-83-4 |
| 291.08 | Safrotin | Safrotin | P | Yes | 31218-83-4 |
| 291.09 | Pyrfon | Pyrfon | P | Yes | 25311-71-1 |
| 291.10 | Dichlorvos | Dichlorvos | P | | 62-73-7 |
| 291.11 | Acephate | Acephate | P | Yes | 30560-19-1 |
| 291.11 | Acephate | Orthene | P | Yes | 30560-19-1 |
| 291.12 | Ethephon | Ethephon | P | | 16672-87-0 |
| 291.12 | Ethephon | 2-Chloroethylphosphonic Acid | P | | 16672-87-0 |
| 291.13 | Fenthion | Fenthion | | Yes | 55-38-9 |
| 292.00 | Carbamate Pesticides, NOS | Carbamate Pesticides, NOS | P | | |
| 292.01 | Sodium N-methyldithiocarbamate | Metam Sodium | P | | 137-42-8 |
| 292.01 | Sodium N-methyldithiocarbamate | Metham Sodium | P | | 137-42-8 |
| 292.01 | Sodium N-methyldithiocarbamate | Vapam | P | | 137-42-8 |
| 292.02 | Carbaryl | Carbaryl | P | | 63-25-2 |
| 292.02 | Carbaryl | Sevin | P | | 63-25-2 |
| 292.03 | Ficam | Ficam | P | | 22781-23-3 |
| 292.04 | Propoxur | Baygon | P | | 114-26-1 |
| 292.04 | Propoxur | Propoxur | P | | 114-26-1 |
| 300.00 | Organic Phosphates, Nonpesticide | Organic Phosphates, Nonpesticide | | | |
| 300.01 | Tris | Tris | | | 126-72-7 |
| 300.01 | Tris | Tris 2,3-Dibromopropyl Phosphate | | | 126-72-7 |
| 310.01 | Carbon Disulfide | Carbon Disulfide | | | 75-15-0 |
| 310.02 | Diethyl Sulfate | Diethyl Sulfate | | | 64-67-5 |
| 310.02 | Diethyl Sulfate | Ethyl Sulfate | | | 64-67-5 |
| 310.02 | Diethyl Sulfate | Sulfuric Acid Diethyl Ester | | | 64-67-5 |
| 310.03 | Dimethyl Sulfate | Dimethyl Sulfate | | | 77-78-1 |
| 310.03 | Dimethyl Sulfate | DMS | | | 77-78-1 |
| 310.03 | Dimethyl Sulfate | Sulfuric Acid Methyl Ester | | | |
| 310.04 | Mercaptoethanol | 2-Hydroxyethanethiol | | | 60-24-2 |
| 310.04 | Mercaptoethanol | B-Mercaptoethanol | | | 60-24-2 |
| 310.04 | Mercaptoethanol | Mercaptoethanol | | | 60-24-2 |
| 310.04 | Mercaptoethanol | Thioglycol | | | 60-24-2 |
| 310.06 | Omite | Omite | P | | 2312-35-8 |
| 310.06 | Omite | Omite Cr | P | | |
| 310.07 | Thiourea | 2-Thiocarbamide | | | |
| 310.07 | Thiourea | Thiourea | | | 62-56-6 |
| 310.07 | Thiourea | THU | | | 62-56-6 |
| 310.08 | Thiuram | Thiram | | | 137-26-8 |
| 310.08 | Thiuram | Thiuram | | | 137-26-8 |
| 310.10 | Sulfites, NOS | Sulfites, NOS | | | |
| 310.12 | 3-mercaptopropionic Acid | 3-mercaptopropionic Acid | | | 107-96-0 |
| 310.12 | 3-mercaptopropionic Acid | Mercaptopropionic Acid | | | 107-96-0 |
| 310.12 | 3-mercaptopropionic Acid | Beta-mercaptopropionic Acid | | | 107-96-0 |
| 310.12 | 3-mercaptopropionic Acid | 3-thiopropionic Acid | | | 107-96-0 |
| 310.12 | 3-mercaptopropionic Acid | Beta-thiopropionic Acid | | | 107-96-0 |
| 310.13 | 3-mercaptopropionate | Pentaerythritol Tetrakis | | | 7575-23-7 |
| 310.13 | 3-mercaptopropionate | 3-mercaptopropionate | | | 7575-23-7 |
| 310.14 | Dimethyl Sulfoxide | Methyl Sulfoxide | | | 67-68-5 |
| 310.14 | Dimethyl Sulfoxide | Dimethyl Sulfoxide | | | 67-68-5 |

| AOEC Code[1] | Primary Name | Synonym | Use[2] | Asthma Inducer? | CAS Number[3] |
|---|---|---|---|---|---|
| 310.14 | Dimethyl Sulfoxide | Dermasorb | | | 67-68-5 |
| 310.14 | Dimethyl Sulfoxide | DMSO | | | 67-68-5 |
| 310.15 | Propionic Acid | Propionic Acid | | | 79-09-4 |
| 311.00 | Mercaptans, NOS | Mercaptans, NOS | | | |
| 311.01 | N-Butyl Mercaptan | N-Butyl Mercaptan | | | 109-79-5 |
| 320.01 | Air Pollutants, Indoor | Air Pollutants, Indoor | | | |
| 320.01 | Air Pollutants, Indoor | Sick Building | | | |
| 320.01 | Air Pollutants, Indoor | Ventilation, Inadequate | | | |
| 320.02 | Air Pollutants, Outdoor | Air Pollutants, Outdoor | | | |
| 320.03 | Anesthetic Gases, NOS | Anesthetic Gases, NOS | | | |
| 320.04 | Beta-Propiolactone | 2-Oxetanone | | | 57-57-8 |
| 320.04 | Beta-Propiolactone | Beta-Propiolactone | | | 57-57-8 |
| 320.04 | Beta-Propiolactone | BPL | | | 57-57-8 |
| 320.05 | Carbonless Paper | Carbonless Paper | | | |
| 320.05 | Carbonless Paper | NCR Paper | | | |
| 320.06 | Chemicals, NOS | Chemical Dust, NOS | | | |
| 320.06 | Chemicals, NOS | Chemicals, NOS | | | |
| 320.06 | Chemicals, NOS | Chemicals, Unknown | | | |
| 320.06 | Chemicals, NOS | Multiple Chemicals | | | |
| 320.07 | Electroplating Chemicals, NOS | Electroplating Chemicals, NOS | | | |
| 320.08 | Fungicide, NOS | Fungicide, NOS | P | | |
| 320.09 | Fertilizers, NOS | Fertilizers, NOS | | | |
| 320.10 | Fire Extinguisher Discharge | Fire Extinguisher Discharge | | | |
| 320.10 | Halon | Fire retardent | | | |
| 320.11 | Glues, NOS | Adhesive, NOS | | | |
| 320.11 | Glues, NOS | Glues, NOS | | | |
| 320.12 | Hair Products | Hair Products | | | |
| 320.12 | Hair Products | Hair Solutions | | | |
| 320.12 | Hair Products | Hair Spray | | | |
| 320.13 | Herbicides, NOS | Herbicides, NOS | P | | |
| 320.13 | N-(2,6-difluorophenyl)-5-methyl-(1,2,4) triazolo (1,5-a) pyrimidine-2-sulfo | N-(2,6-difluorophenyl)-5-methyl-(1,2,4) triazolo (1,5-a) pyrimidine-2-sulfo | P | | 98967-40-9 |
| 320.13 | N-(2,6-difluorophenyl)-5-methyl-(1,2,4) triazolo (1,5-a) pyrimidine-2-sulfo | Flumetsulam | P | | 98967-40-9 |
| 320.13 | Glyphosate | Round Up | | | |
| 320.14 | Lubricants, NOS | Coolants | | | |
| 320.14 | Lubricants, NOS | Hydraulics | | | |
| 320.14 | Lubricants, NOS | Lubricants, NOS | | | |
| 320.14 | Lubricants, NOS | Transmission Fluid | | | |
| 320.15 | Odors | Odors | | | |
| 320.16 | Pesticides, NOS | Pesticides, NOS | P | | |
| 320.17 | Photo Developing Chemicals, NOS | Fixer | | | |
| 320.17 | Photo Developing Chemicals, NOS | Photo Developer, Color | | | |
| 320.17 | Photo Developing Chemicals, NOS | Photo Developing Chemicals, NOS | | | |
| 320.18 | Pyrethrins | Pyrethrins | P | Yes | |
| 320.18 | Pyrethroids | Pyrethroids | P | | |
| 320.18 | Tetramethrin | Tetramethrin | | Yes | 7696-12-0 |
| 320.19 | Surfactants, NOS | Polyethylene Glycol Stearates | | | |
| 320.19 | Surfactants, NOS | Surfactants, NOS | | | |
| 320.19 | Surfactant-Specific amines | alkylamine ethoxylate,alkyleneoxy diamine, ethylenediamine | | Yes | 68155-39-5 |
| 320.20 | Theatrical Fog, NOS | Theatrical Fog, NOS | | | |
| 320.21 | Theatrical Fog, Glycol-Based | Theatrical Fog, Glycol-Based | | | |
| 320.22 | Textile Dust, NOS | Textile Dust, NOS | | | |
| 320.23 | Perfume, NOS | Perfume, NOS | | | |
| 320.24 | Carmine | Carmine | | | 1260-17-9 |
| 320.24 | Carmine | Carminic Acid | | | 1260-17-9 |
| 320.25 | Ninhydrin | Ninhydrin | | | 485-47-2 |
| 320.27 | Mace | 2-Chloroacetophenone | | | 532-27-4 |

| AOEC Code[1] | Primary Name | Synonym | Use[2] | Asthma Inducer? | CAS Number[3] |
|---|---|---|---|---|---|
| 320.27 | Mace | Acetophenone | | | 532-27-4 |
| 320.27 | Mace | Mace | | | 532-27-4 |
| 320.27 | Mace | Tear Gas | | | 532-27-4 |
| 320.28 | Cosmetics, NOS | Cosmetology Chemicals, NOS | | | |
| 320.28 | Cosmetics, NOS | Cosmetics, NOS | | | |
| 320.29 | Printing Chemicals, NOS | Printing Chemicals, NOS | | | |
| 320.30 | N-Octyl Bicycloheptene Dicarboximide | MGK 264 | P | | 113-48-4 |
| 320.30 | N-Octyl Bicycloheptene Dicarboximide | N-Octyl Bicycloheptene Dicarboximide | P | | 113-48-4 |
| 320.30 | N-Octyl Bicycloheptene Dicarboximide | Octacide 264 | P | | 113-48-4 |
| 320.31 | Chil-Perm CP-30 | Chil-Perm CP-30 | S | | |
| 320.32 | Radiographic Fixative | Radiographic Fixative | | Yes | |
| 320.33 | Indoor Air Pollutants from Building Renovation | Indoor Air Pollutants from Building Renovation | | | |
| 320.33 | Indoor Air Pollutants from Building Renovation | Air Pollutants, Indoor, from Building Renovation | | | |
| 320.33 | Indoor Air Pollutants from Building Renovation | Building Renovation | | | |
| 320.34 | Fingerprint Powder | Fingerprint Powder | | | |
| 320.34 | Fingerprint Powder | Lightning Powder | | | |
| 320.35 | Carpet Dust | Carpet Dust | | | |
| 320.35 | Carpet Dust | Carpet Fibers | | | |
| 320.36 | Eucalyptus Oil | Eucalyptus Scent | | | 8000-48-4 |
| 320.36 | Eucalyptus Oil | Eucalyptus Oil | | | 8000-48-4 |
| 320.37 | Denatonium Benzoate | Denatonium Benzoate | | | 3734-33-6 |
| 320.37 | Denatonium Benzoate | Bitrex | | | 3734-33-6 |
| 320.38 | Saccharin | 1,2-Benzisothiazol-3(2H)one | | | 81-07-2 |
| 320.38 | Saccharin | Saccharin | | | 81-07-2 |
| 320.39 | World Trade Center Pollution | World Trade Center Pollution | | | |
| 320.39 | World Trade Center Pollution | World Trade Center Dust | | | |
| 320.39 | World Trade Center Pollution | Pollution from Acts of Terrorism/War | | | |
| 320.40 | Methamphetamine Laboratory | Methamphetamine Laboratory | | | |
| 320.40 | Methamphetamine Laboratory | Clandestine Drug laboratory | | | |
| 320.41 | Nail Care Products | Nail Care Products | | | |
| 320.42 | Air Freshener | Air Freshener | | | |
| 320.42 | Air Freshener | Deodorant, Aerosol | | | |
| 320.43 | Habanolide | Habanolide | | | 111879-80- |
| 320.43 | Habanolide | Oxacyclohexedecen-2-one | | | 111879-80- |
| 320.44 | Air Bag Discharge Products | Air Bag Discharge Products | | | |
| 320.45 | Insecticides, NOS | Insecticides, NOS | P | | |
| 320.45 | Imidacloprid | Imidacloprid | P | | 138261-41- |
| 320.46 | Nonoxynol | Nonoxynol | | | |
| 321.00 | Pharmaceuticals, NOS | Pharmaceuticals, NOS | | | |
| 321.01 | Chemotherapeutic Drugs | Chemotherapeutic Drugs | | | |
| 321.02 | Estrogens | Estrogens | | | |
| 321.03 | Coumarin | Coumarin | | | |
| 321.04 | Penicillins | Ampicillin | | Yes | 69-53-4 |
| 321.04 | Penicillins | Penicillins | | Yes | 69-53-4 |
| 321.05 | Penicillamine | Penicillamine | | Yes | |
| 321.06 | Cephalosporins | Cephalosporins | | | |
| 321.06 | Cefadroxil | Cefadroxil | | | 15686-71-2 |
| 321.06 | Ceftazidine | Ceftazidine | | | 72558-82-8 |
| 321.06 | Cephalexin | Cephalexin | | | 66592-87-8 |
| 321.06 | 7-aminocephalosporanic acid | 7 ACA | | | 957-68-6 |
| 321.07 | 7-Amino-3-thiomethyl-3-cephalosporanic acid | 7TACA | | | |
| 321.07 | Tosyalate dihydrate | 7CTD | | | |
| 321.07 | Phenylglycine Acid Chloride | Phenylglycine Acid Chloride | | Yes | |
| 321.08 | Psyllium | Psyllium | | Yes | |
| 321.09 | Methyldopa | Methyldopa | | Yes | 555-30-6 |
| 321.10 | Spiramycin | Spiramycin | | Yes | 8025-81-8 |
| 321.11 | Amprolium | Amprolium | | Yes | |
| 321.12 | Tetracycline | Tetracycline | | Yes | 60-54-8 |
| 321.13 | Isonicotinic Acid Hydrazide | INH | | Yes | |

| AOEC Code[1] | Primary Name | Synonym | Use[2] | Asthma Inducer? | CAS Number[3] |
|---|---|---|---|---|---|
| 321.13 | Isonicotinic Acid Hydrazide | Isoniazid | | Yes | |
| 321.13 | Isonicotinic Acid Hydrazide | Isonicotinic Acid Hydrazide | | Yes | |
| 321.14 | Hydralazine | Hydralazine | | Yes | 86-54-4 |
| 321.15 | Tylosin Tartrate | Tylosin Tartrate | | Yes | 1405-54-5 |
| 321.16 | Ipecacuanha | Ipecac | | Yes | |
| 321.16 | Ipecacuanha | Ipecacuanha | | Yes | |
| 321.17 | Cimetidine | Cimetidine | | Yes | 51481-61-9 |
| 321.18 | Opiate Compounds | Opiate Compounds | | Yes | |
| 321.18 | Morphine | Morphine | | Yes | 57-27-2 |
| 321.18 | Codeine | Codeine | | Yes | 76-57-3 |
| 321.18 | Thebaine | Thebaine | | Yes | 115-37-7 |
| 321.18 | Papaverine | Papaverine | | Yes | 58-74-2710 |
| 321.19 | Diacetyl morphine | Heroin | | Yes | 561-27-3 |
| 321.19 | Oxycodone | Oxycodone | | Yes | 76-46-6 |
| 321.19 | Hydrocodone | Hydrocodone | | Yes | 125-29-1 |
| 321.19 | Senna | Senna | | | 11/4/8013 |
| 321.20 | Imipenem | Imipenem | | | 64221-86-9 |
| 321.21 | Salbutamol Intermediate | Salbutamol Intermediate | | Yes | |
| 321.22 | Trental | Trental | | | 5/6/6493 |
| 321.23 | Pentamidine | Aerosolized Pentamidine | | | 100-33-4 |
| 321.23 | Pentamidine | Pentamidine | | | 100-33-4 |
| 321.24 | Theophylline | Theophylline | | | 58-55-9 |
| 321.25 | Salicylic Acid | Salicylate | | | 63-36-5 |
| 321.25 | Salicylic Acid | Salicylic Acid | | | 63-36-5 |
| 321.26 | Ceclor | Ceclor | | | 53994-73-3 |
| 321.26 | Ceclor | Cefaclor | | | 53994-73-3 |
| 321.27 | Griseofulvin | Griseoufulvin | | | 126-07-8 |
| 321.27 | Griseofulvin | Gris-PEG | | | 126-07-8 |
| 321.27 | Griseofulvin | Fulvin P/G | | | 126-07-8 |
| 321.27 | Griseofulvin | Grisactin-Ultra | | | 126-07-8 |
| 321.28 | Mustargen HCL | Mustargen HCL | | | 55-86-7 |
| 321.28 | Mustargen HCL | Nitrogen Mustard Hydrochloride | | | 55-86-7 |
| 321.28 | Mustargen HCL | Mustargen | | | 55-86-7 |
| 321.29 | Cyclophosphamide | Cyclophosphamide | | | 50-18-0 |
| 321.29 | Cyclophosphamide | Cytoxan | | | 50-18-0 |
| 321.29 | Cyclophosphamide | Neosar | | | 50-18-0 |
| 321.30 | Micotil | Micotil | | | 137330-133 |
| 321.30 | Micotil | Tilmicosin Phosphate | | | 137330-133 |
| 321.31 | Proventil | Proventil | | | 18559-94-9 |
| 321.31 | Proventil | Albuterol Aerosol | | | 18559-94-9 |
| 321.32 | Ribaviran | Virazole | | | 36791-04-5 |
| 321.32 | Ribaviran | 1,2,4-Triazole-3-carboxamide | | | 36791-04-5 |
| 321.32 | Ribaviran | 1-beta-D-ribofuranosyl- | | | 36791-04-5 |
| 321.32 | Ribaviran | Ribaviran | | | 36791-04-5 |
| 321.33 | Vancomycin | Vancomycin | | Yes | 1404-90-6 |
| 321.34 | Colistin | Colistin | | Yes | 1066-17-7 |
| 322.00 | Cleaning Materials, NOS | Cleaning Materials, NOS | | | |
| 322.01 | Soap, excluding Laundry Soap/Detergent | Soap (excluding Laundry Soap/Detergent) | | | |
| 322.02 | Metal Polish, Tarnish Remover, or Preventative | Metal Polish, Tarnish Remover, or Preventative | | | |
| 322.03 | Cleaning Fluids, Photocopier | Cleaning Fluids, Photocopier | | | |
| 322.04 | Cleaners, Household, General Purpose | Cleaners, Household, General Purpose | | | |
| 322.05 | Iodophors | Iodophors | | | 8037-86-3 |
| 322.06 | Cleaners, Solvent-Based | ILA Soap | S | | |
| 322.06 | Cleaners, Solvent-Based | Cleaners, Solvent-Based | S | | |
| 322.07 | Ammonium Hydroxide, NOS | Ammonia Solution, NOS | | Yes | 1336-21-6 |
| 322.07 | Ammonium Hydroxide, NOS | Ammonia | | Yes | 1336-21-6 |
| 322.07 | Ammonium Hydroxide, NOS | Ammonium Hydroxide, NOS | | Yes | 1336-21-6 |
| 322.08 | Ammonia Solution (29%) | Ammonia Solution (29%) | | Yes | 1336-21-6 |

| AOEC Code[1] | Primary Name | Synonym | Use[2] | Asthma Inducer? | CAS Number[3] |
|---|---|---|---|---|---|
| 322.09 | Ammonia Solution (10%) | Ammonia Solution (10%) | | Yes | 1336-21-6 |
| 322.09 | Ammonia Solution (10%) | Ammonia, Household | | Yes | 1336-21-6 |
| 322.10 | Sodium Hypochlorite | Bleach | | | 7681-52-9 |
| 322.10 | Sodium Hypochlorite | NaClO | | | 7681-52-9 |
| 322.10 | Sodium Hypochlorite | Sodium Hypchlorite | | | 7681-52-91 |
| 322.11 | Bleach plus Acid (mixture) | Bleach plus Acid (mixture) | | | |
| 322.12 | Bleach plus Ammonia (mixture) | Bleach plus Ammonia (mixture) | | | 7681-52-9 |
| 322.13 | Calcium Hypochloride | Bleaching Powder | | | 7778-54-3 |
| 322.13 | Calcium Hypochloride | Calcium Hypochloride | | | 7778-54-3 |
| 322.13 | Calcium Hypochloride | Lime Chloride | | | 7778-54-3 |
| 322.14 | Cleaners, Abrasive | Cleaners, Abrasive | | | |
| 322.15 | Cleaners, Acid | Cleaners, Acid | | | |
| 322.16 | Cleaners, Carpet | Cleaners, Carpet | | | |
| 322.17 | Cleaners, Caustic (excluding Lye) | Cleaners, Caustic (excluding Lye) | | | |
| 322.18 | Cleaners, Detergent, NOS | Cleaners, Detergent, NOS | | | |
| 322.19 | Cleaners, Disinfectant, NOS | Cleaners, Disinfectant, NOS | | | |
| 322.20 | Cleaners, Drain | Cleaners, Drain | | | |
| 322.21 | Cleaners, Floor Stripping | Cleaners, Floor Stripping | | | |
| 322.22 | Cleaners, Graffiti Removing | Cleaners, Graffiti Removing | S | | |
| 322.23 | Cleaners, Laundry Soap/Detergent | Cleaners, Laundry Soap/Detergent | | | |
| 322.24 | Cleaners, Lye | Cleaners, Lye | | | |
| 322.25 | Cleaners, Oven | Cleaners, Oven | | | |
| 322.26 | Cleaners, Pine Oil | Cleaners, Pine Oil | | | |
| 322.27 | Cleaners, Tile | Cleaners, Tile | | | |
| 322.28 | Cleaners, Toilet Bowl | Cleaners, Toilet Bowl | | | |
| 322.29 | Cleaners, Wallpaper | Cleaners, Wallpaper | | | |
| 322.30 | Cleaning Fluids/Spot Removers | Cleaning Fluids/Spot Removers | S | | |
| 322.31 | Cleaning Mixtures (excluding Bleach plus Acid or Ammonia) | Cleaning Mixtures (excluding Bleach plus Acid or Ammonia) | | | |
| 322.32 | Quaternary Ammonium Compounds, NOS | Quaternary Ammonium Compounds, NOS | | Yes | |
| 322.32 | Benzalkonium Chloride | Benzalkonium Chloride | | Yes | 8001-54-5 |
| 322.32 | Benzyl-C10-16-alkyldimethyl, chlorides | Benzyl-C10-16-alkyldimethyl, chlorides | | Yes | 68989-00-4 |
| 322.32 | BTC 927 | BTC 927 | | Yes | 8045-22-5 |
| 322.32 | Dialkyl Methyl Benzyl Ammonium Chloride | Dialkyl Methyl Benzyl Ammonium Chloride | | Yes | 73049-75-9 |
| 322.32 | Dimethyl Ethyl Benzyl Ammonium Chloride | Dimethyl Ethyl Benzyl Ammonium Chloride | | Yes | 68956-79-6 |
| 322.32 | Benzyldimethylstearylammonium Chloride | Benzyldimethylstearylammonium Chloride | | Yes | 122-19-0 |
| 322.32 | Dodecyl-dimethyl-benzylammonium Chloride | Dodecyl-dimethyl-benzylammonium Chloride | | Yes | 139-07-1 |
| 322.32 | Dodycyl-dimethyl-benzylammonium Chloride | Lauryl Dimethyl Benzyl Ammonium Chloride | | Yes | 139-07-1 |
| 322.32 | Didecyl Dimethyl Ammonium Chloride | Didecyl Dimethyl Ammonium Chloride | | Yes | 7173-51-5 |
| 322.32 | Cetalkonium Chloride | Cetalkonium Chloride | | Yes | 122-18-9 |
| 322.33 | BTC 776 | BTC 776 | | Yes | 53516-76-0 |
| 322.33 | Alkyl Dimethyl Benzyl Ammonium Chloride | Alkyl Dimethyl Benzyl Ammonium Chloride | | Yes | 61789-71-7 |
| 322.33 | Benzyl-C12-18-alkyldimethyl, chlorides | Benzyl-C12-18-alkyldimethyl, chlorides | | Yes | 68391-01-5 |
| 322.33 | Benzyl-C12-16-alkyldimethyl, chlorides | Benzyl-C12-16-alkyldimethyl, chlorides | | Yes | 68424-85-1 |
| 322.33 | Benzyl-C16-18-alkyldimethyl, chlorides | Benzyl-C16-18-alkyldimethyl, chlorides | | Yes | 68607-20-5 |
| 322.33 | Sulfonates, NOS | Isononanoyl Oxybenzene Sulfonate | | | 123354-92- |
| 322.33 | Sulfonates, NOS | Sulfonates, NOS | | | |
| 322.34 | Dry Cleaning Fluid, NOS | Dry Cleaning Fluid, NOS | S | | |
| 322.35 | Floor Wax | Floor Finish | | | |
| 322.35 | Floor Wax | Floor Wax | | | |
| 322.36 | Cleaner, Citric | Cleaner, Citric | | | |
| 323.00 | Waste, NOS | Waste, NOS | | | |
| 323.01 | Waste, Hazardous | Waste, Hazardous | | | |
| 323.01 | Waste, Hazardous | Waste, Hazardous Acid | | | |
| 323.02 | Leachate | Leachate | | | |
| 323.03 | Sewer Water | Sewage | | | |
| 323.03 | Sewer Water | Sewer Water | | | |
| 323.04 | Waste, Treated Human Sludge | Waste, Treated Human Sludge | | | |

| AOEC Code[1] | Primary Name | Synonym | Use[2] | Asthma Inducer? | CAS Number[3] |
|---|---|---|---|---|---|
| 324.00 | Enzymes, NOS | Catalysts, NOS | | | |
| 324.00 | Enzymes, NOS | Enzymes, NOS | | | |
| 324.00 | Enzymes, NOS | Proteolytic Enzymes, NOS | | | |
| 324.01 | Bacillus Subtilis | Bacillus Subtilis | | Yes | 68038-70-0 |
| 324.02 | Trypsin | Trypsin | | Yes | 7/7/9002 |
| 324.03 | Papain | Papain | | Yes | 9001-73-4 |
| 324.04 | Pepsin | Pepsin | | Yes | 9001-75-6 |
| 324.05 | Pancreatin | Pancreatin | | Yes | 8049-47-6 |
| 324.06 | Flaviastase | Flaviastase | | Yes | |
| 324.07 | Bromelain | Bromelin | | Yes | 9001-00-7 |
| 324.07 | Bromelain | Ananase | | Yes | 9001-00-7 |
| 324.07 | Bromelain | Extranase | | Yes | 9001-00-7 |
| 324.07 | Bromelain | Inflamen | | Yes | 9001-00-7 |
| 324.07 | Bromelain | Traumanase | | Yes | 9001-00-7 |
| 324.07 | Bromelain | Bromelain | | Yes | 9001-00-7 |
| 324.08 | Egg Lysozyme | Egg Lysozyme | | Yes | |
| 324.09 | Fungal Amylase | Alpha Amylase | | Yes | 1/8/9013 |
| 324.09 | Fungal Amylase | Fungal Amylase | | Yes | 1/8/9013 |
| 324.10 | Esperase | Esperase | | Yes | |
| 324.11 | Fungal Amyloglucosidade | Fungal Amyloglucosidade | | Yes | |
| 324.12 | Fungal Hemicellulase | Fungal Hemicellulase | | Yes | |
| 324.13 | Peroxidase Catalyst | Peroxidase Catalyst | | | |
| 324.14 | Lactase | Lactase | | | 11/2/9031 |
| 324.14 | Lactase | B-Galactosidase | | | 11/2/9031 |
| 324.14 | Lactase | Beta-Galactosidase | | | 11/2/9031 |
| 324.14 | Lactase | Beta-Lactosidase | | | 11/2/9031 |
| 324.15 | Zeolite | Zeolite | | | 1318-02-1 |
| 325.00 | Reactive Dyes, NOS | Reactive Dyes, NOS | | Yes | |
| 325.01 | Levafix Brilliant Yellow E36 | Levafix Brilliant Yellow E36 | | Yes | 37300-23-5 |
| 325.02 | Drimaren Brilliant Yellow K-3GL | Drimaren Brilliant Yellow-K-3GL | | Yes | |
| 325.03 | Cibachrome Brilliant Scarlet 32 | Cibachrome Brilliant Scarlet 32 | | Yes | |
| 325.04 | Drimaren Brilliant Blue K-BL | Drimaren Brilliant Blue K-BL | | Yes | 51811-44-0 |
| 325.05 | Rifacion Orange HE 2G | Color Index No. 0-20 | | Yes | |
| 325.05 | Rifacion Orange HE 2G | Rifacion Orange HE 2G | | Yes | |
| 325.06 | Rifafix Yellow 3 RN | Rifafix Yellow 3 RN | | Yes | |
| 325.07 | Rifazol Brilliant Orange 3R | Color Index No. 0-16 | | Yes | |
| 325.07 | Rifazol Brilliant Orange 3R | Rifazol Brilliant Orange 3R | | Yes | |
| 325.08 | Rifazol Black GR | Color Index No. BK-5 | | Yes | |
| 325.08 | Rifazol Black GR | Rifazol Black GR | | Yes | |
| 325.09 | Lanasol Yellow 4G | Lanasol Yellow 4G | | Yes | 70247-70-0 |
| 325.10 | Methyl Blue | Methyl Blue | | Yes | 28983-56-4 |
| 326.01 | Pickle Processing (Unknown Causal Agent) | Pickle Processing (Unknown Causal Agent) | | | |
| 326.02 | Heat Shrink Wrapping | Heat Shrink Wrapping | | | |
| 327.00 | Water Contamination, NOS | Water Contamination, NOS | | | |
| 327.01 | Water Chlorination Byproducts | Water Chlorination Byproducts | | | |
| 327.02 | Water Contamination, Inorganic | Water Contamination, Inorganic | | | |
| 327.03 | Water Contamination, Organic | Water Contamination, Organic | | | |
| 330.01 | Cigarette Smoke | Cigarette Smoke | | | |
| 330.01 | Cigarette Smoke | Environmental Tobacco Smoke | | | |
| 330.01 | Cigarette Smoke | ETS | | | |
| 330.02 | Plastic Smoke | Plastic Smoke | | | |
| 330.02 | Plastic Smoke | Polymer Fume | | | |
| 330.02 | Plastic Smoke | Vinyl Fumes | | | |
| 330.03 | Smoke, NOS | Combustion Products, NOS | | | |
| 330.03 | Smoke, NOS | Fumes, NOS | | | |
| 330.03 | Smoke, NOS | Smoke Inhalation | | | |
| 330.03 | Smoke, NOS | Smoke, NOS | | | |
| 330.04 | Incinerator Fume, NOS | Incinerator Fume, NOS | | | |
| 330.05 | Marijuana Smoke | Marijuana Smoke | | | |

| AOEC Code[1] | Primary Name | Synonym | Use[2] | Asthma Inducer? | CAS Number[3] |
|---|---|---|---|---|---|
| 330.06 | Smoke, Lead-Containing | Smoke, Lead-Containing | | | |
| 330.07 | Incense Smoke | Incense Smoke | | | |
| 331.00 | Exhaust, NOS | Exhaust, NOS | | | |
| 331.01 | Diesel Exhaust | Diesel Exhaust | | | |
| 331.02 | Engine Exhaust | Engine Exhaust | | | |
| 331.02 | Engine Exhaust | Gasoline Exhaust | | | |
| 331.03 | Jet Exhaust | Jet Exhaust | | | |
| 331.04 | Propane Exhaust | Propane Exhaust | | | |
| 350.00 | Physical Factors, NOS | Physical Factors, NOS | | | |
| 350.01 | Noise | Noise | | | |
| 350.02 | Cold | Cold | | | |
| 350.02 | Cold | Low Temperature | | | |
| 350.03 | Heat | Heat | | | |
| 350.03 | Heat | High Temperature | | | |
| 350.03 | Heat | Hot Liquid | | | |
| 350.03 | Heat | Steam | | | |
| 350.03 | Heat | Thermal Energy | | | |
| 350.04 | Humidity, Low | Dry Air | | | |
| 350.04 | Humidity, Low | Humidity, Low | | | |
| 350.05 | Humidity, High | Humidity, High | | | |
| 350.05 | Humidity, High | Moisture | | | |
| 350.05 | Humidity, High | Wet Weather | | | |
| 350.06 | Ultrasound | Ultrasound | | | |
| 350.07 | Air Pressure, Changes | Air Pressure, Changes | | | |
| 350.07 | Air Pressure, Changes | Barometric Pressure, Changes | | | |
| 351.00 | Radiation, Ionizing, NOS | Radiation, Ionizing, NOS | | | |
| 351.01 | Plutonium | Plutonium | | | |
| 351.02 | Nuclear Reactor Release | Nuclear Reactor Release | | | |
| 351.03 | Radon | Radon | | | 10043-92-2 |
| 351.04 | Therapeutic Radiation | Radioisotopes | | | |
| 351.04 | Therapeutic Radiation | Therapeutic Radiation | | | |
| 351.04 | Therapeutic Radiation | Xrays | | | |
| 351.05 | Uranium | Uranium | | | 7440-61-1 |
| 352.00 | Radiation, Nonionizing, NOS | Radiation, Nonionizing, NOS | | | |
| 352.01 | Radiation, Electromagnetic | CRT Radiation | | | |
| 352.01 | Radiation, Electromagnetic | Electromagnetic Fields | | | |
| 352.01 | Radiation, Electromagnetic | ELF | | | |
| 352.01 | Radiation, Electromagnetic | EMF | | | |
| 352.01 | Radiation, Electromagnetic | Extremely Low Frequency Elctromagnetic Radiation | | | |
| 352.01 | Radiation, Electromagnetic | Radiation, Electromagnetic | | | |
| 352.01 | Radiation, Electromagnetic | VDT Radiation | | | |
| 352.02 | Lasers | Lasers | | | |
| 352.03 | Radiation, Microwave | Radiation, Microwave | | | |
| 352.03 | Radiation, Microwave | Radio Frequency Radiation | | | |
| 352.04 | Radiation, Ultraviolet | Radiation, Ultraviolet | | | |
| 352.04 | Radiation, Ultraviolet | UV Light | | | |
| 352.04 | Radiation, Ultraviolet | UV Radiation | | | |
| 352.05 | Infrared Light | Infrared Light | | | |
| 352.05 | Infrared Light | IR Light | | | |
| 352.05 | Infrared Light | IR | | | |
| 353.00 | Trauma, Acute, NOS | Trauma, Acute, NOS | | | |
| 353.01 | Electrical Shock | Electrical Shock | | | |
| 353.01 | Electrical Shock | Electricity | | | |
| 353.02 | Explosion | Explosion | | | |
| 353.03 | Fall, NOS | Fall From Height | | | |
| 353.03 | Fall, NOS | Fall, NOS | | | |
| 353.03 | Fall, NOS | Slip, Trip, or Fall on Same Level | | | |
| 353.03 | Fall, NOS | Struck By/Against Object as Result of Fall | | | |

| AOEC Code[1] | Primary Name | Synonym | Use[2] | Asthma Inducer? | CAS Number[3] |
|---|---|---|---|---|---|
| 353.04 | Hypoxia | Hypoxia | | | |
| 353.04 | Hypoxia | Oxygen Deficiency | | | |
| 353.04 | Hypoxia | Suffocation | | | |
| 353.05 | Motor Vehicle Accident | Auto Accident | | | |
| 353.05 | Motor Vehicle Accident | Car Accident | | | |
| 353.05 | Motor Vehicle Accident | Car Crash | | | |
| 353.05 | Motor Vehicle Accident | Motor Vehicle Accident | | | |
| 353.06 | Struck by Motor Vehicle (Road) | Struck by Motor Vehicle (Road) | | | |
| 353.07 | Struck by Vehicle or Equipment (Non-road) | Struck by Vehicle or Equipment (Non-road) | | | |
| 353.08 | Struck by Falling Object | Struck by Falling Object | | | |
| 353.09 | Struck Against/Struck By Objects or Persons | Struck Against/Struck By Objects or Persons | | | |
| 353.10 | Caught In or Between Objects | Caught In or Between Objects | | | |
| 353.11 | Cutting or Piercing Object, Except Blood-Contam. Sharps | Cutting or Piercing Object, Except Blood-Contam. Sharps | | | |
| 353.12 | Assault, Physical | Assault, Physical | | | |
| 353.12 | Assault, Physical | Violence, Physical Assault | | | |
| 353.13 | Violence, Other than Physical Assault | Violence, Other than Physical Assault | | | |
| 354.00 | Vibration, NOS | Vibration, NOS | | | |
| 354.01 | Vibration, Regional | Hand-Arm Vibration | | | |
| 354.01 | Vibration, Regional | Local Vibration | | | |
| 354.01 | Vibration, Regional | Vibration, Regional | | | |
| 354.02 | Vibration, Whole Body | Vibration, Whole Body | | | |
| 360.00 | Ergonomic Factors, NOS | Ergonomic Factors, NOS | | | |
| 360.01 | Contact Pressure | Contact Pressure | | | |
| 360.01 | Contact Pressure | Mechanical Pressure | | | |
| 360.01 | Contact Pressure | Skin Contact | | | |
| 360.02 | Keyboard Use | Adding Machine | | | |
| 360.02 | Keyboard Use | Calculator | | | |
| 360.02 | Keyboard Use | Computer Keyboard | | | |
| 360.02 | Keyboard Use | Computer Mouse Use | | | |
| 360.02 | Keyboard Use | Key Punching | | | |
| 360.02 | Keyboard Use | Keyboard and Mouse Use | | | |
| 360.02 | Keyboard Use | Keyboard Use | | | |
| 360.02 | Keyboard Use | Typewriter | | | |
| 360.02 | Keyboard Use | Typing | | | |
| 360.02 | Keyboard Use | VDT Keyboard | | | |
| 360.02 | Keyboard Use | VDT Typing | | | |
| 360.02 | Keyboard Use | Cash Register Use | | | |
| 360.02 | Keyboard Use | Word Processing | | | |
| 360.03 | Repetitive Motion | Repetitive Motion | | | |
| 360.03 | Repetitive Motion | Repetitive Trauma | | | |
| 360.04 | Stress | Job Control | | | |
| 360.04 | Stress | Job Demand | | | |
| 360.04 | Stress | Mental Factors | | | |
| 360.04 | Stress | Psychological Factors | | | |
| 360.04 | Stress | Rotating Shifts | | | |
| 360.04 | Stress | Stress | | | |
| 360.05 | VDT Screen/Visual | VDT Screen/Visual | | | |
| 360.06 | Exercise | Exercise | | | |
| 360.07 | Bodily Reaction | Bodily Reaction | | | |
| 360.08 | Walking | Walking | | | |
| 361.01 | Forceful Movements, NOS | Forceful Movements, NOS | | | |
| 361.01 | Forceful Movements, NOS | High Force | | | |
| 361.01 | Forceful Movements, NOS | Pulling | | | |
| 361.01 | Forceful Movements, NOS | Pushing | | | |
| 361.02 | Lifting | Carrying | | | |
| 361.02 | Lifting | Heavy Lifting | | | |
| 361.02 | Lifting | Lifting | | | |
| 361.02 | Lifting | Repetitive Lifting | | | |

| AOEC Code[1] | Primary Name | Synonym | Use[2] | Asthma Inducer? | CAS Number[3] |
|---|---|---|---|---|---|
| 361.03 | Gripping, Forceful | Gripping, Forceful | | | |
| 361.03 | Gripping, Forceful | Handwriting | | | |
| 361.03 | Gripping, Forceful | Pinching | | | |
| 362.00 | Posture, NOS | Posture, NOS | | | |
| 362.01 | Posture, Upper Extremity | Hand-Arm Posture | | | |
| 362.01 | Posture, Upper Extremity | Posture, Upper Extremity | | | |
| 362.01 | Posture, Upper Extremity | Upper Extremity Awkward Positions | | | |
| 362.02 | Posture, Body - Static | Kneeling | | | |
| 362.02 | Posture, Body - Static | Posture, Body - Static | | | |
| 362.02 | Posture, Body - Static | Prolonged Position | | | |
| 362.02 | Posture, Body - Static | Sitting | | | |
| 362.02 | Posture, Body - Static | Standing | | | |
| 362.03 | Posture, Body - Dynamic | Bending | | | |
| 362.03 | Posture, Body - Dynamic | Posture, Body - Dynamic | | | |
| 362.03 | Posture, Body - Dynamic | Stooping | | | |
| 362.03 | Posture, Body - Dynamic | Twisting | | | |
| 370.00 | Herbal Tea, NOS | Herbal Tea, NOS | | | |
| 370.00 | Granola, NOS | Granola, NOS | | | |
| 370.00 | Organic Dusts, NOS | Organic Dusts, NOS | | | |
| 370.00 | Plant Material, NOS | Plant Material, NOS | | | |
| 370.01 | Thymol | Thymol | | Yes | 89-83-8 |
| 370.01 | Paper Dust | Paper Dust | | | |
| 370.02 | Cotton Dust | Cotton Dust | | | |
| 370.03 | Grain Dust | Grain Dust | | Yes | |
| 370.04 | Vegetable Dust | Vegetable Dust | | | |
| 370.05 | Fruit Juices | Fruit Juices | | | |
| 370.05 | Fruit Juices | Raspberries | | | |
| 370.05 | Fruit Juices | Vegetable Juices | | | |
| 370.06 | Poisonous Plants | Poison Ivy | | | |
| 370.06 | Poisonous Plants | Poison Oak | | | |
| 370.06 | Poisonous Plants | Poison Sumac | | | |
| 370.06 | Poisonous Plants | Poisonous Plants | | | |
| 370.07 | Grass Cuttings | Grass Cuttings | | | |
| 370.07 | Grass Cuttings | Plant Waste | | | |
| 370.08 | Hay | Hay | | | |
| 370.10 | Pollen | Pollen | | | |
| 370.11 | Vicia Sativa | Vetch | | Yes | |
| 370.11 | Vicia Sativa | Vicia Sativa | | Yes | |
| 370.12 | Coffee Bean | Coffee Bean | | Yes | |
| 370.12 | Coffee Bean | Green Coffee Bean | | Yes | |
| 370.13 | Castor Bean | Castor Bean | | Yes | |
| 370.14 | Tea | Epigallocatechin gallate | | Yes | |
| 370.14 | Tea | Tea | | Yes | |
| 370.15 | Tobacco Leaf | Tobacco Leaf | | Yes | |
| 370.16 | Hops | Hops | | Yes | |
| 370.17 | Baby's Breath | Baby's Breath | | Yes | |
| 370.17 | Baby's Breath | Gypsophilia Paniculata | | Yes | |
| 370.18 | Chicory | Chicory | | Yes | |
| 370.19 | Rose Hips | Rose Hips | | Yes | |
| 370.20 | Sunflower | Sunflower | | Yes | |
| 370.21 | Garlic Dust | Garlic Dust | | Yes | |
| 370.22 | Nacre Dust | Nacre Dust | | Yes | |
| 370.23 | Pectin | Pectin | | Yes | 9000-69-5 |
| 370.24 | Weeping Fig | Weeping Fig | | Yes | |
| 370.25 | Lycopodium | Lycopodium | | Yes | |
| 370.26 | Sericin | Sericin | | Yes | |
| 370.27 | Henna | Henna | | Yes | 83-72-7 |
| 370.28 | Yeast | Yeast | | | |

| AOEC Code[1] | Primary Name | Synonym | Use[2] | Asthma Inducer? | CAS Number[3] |
|---|---|---|---|---|---|
| 370.29 | Lathyrus Sativus | Lathyrus Sativus | | Yes | |
| 370.30 | Freesia | Freesia | | Yes | |
| 370.31 | Paprika | Paprika | | Yes | |
| 370.32 | Cacoon Seed | Cacoon Seed | | Yes | |
| 370.33 | Thapsigargin | Thapsia Garganica L | | | 67526-95-8 |
| 370.33 | Thapsigargin | Thapsigargin | | | 67526-95-8 |
| 370.34 | Oils, Vegetable | Oils, Vegetable | | | |
| 370.35 | Capsicum | Capsicum | | | |
| 370.35 | Capsicum | Pepper Spray | | | |
| 370.36 | Chorella Algae | Chorella Algae | | Yes | |
| 370.37 | Green Beans | Green Beans | | Yes | |
| 370.38 | Limonium Tataricum | Limonium Tataricum | | Yes | |
| 370.39 | Dioscorea Batatas | Dioscorea Batatas | | Yes | |
| 370.40 | Pinellia Ternata | Pinellia Ternata | | Yes | |
| 370.41 | Soybean Lecithin | Soybean Lecithin | | Yes | 90320-57-3 |
| 370.41 | Soybean Lecithin | Soybean Lectin | | Yes | 90320-57-3 |
| 370.42 | Fenugreek | Fenugreek | | Yes | 68990-15-8 |
| 370.43 | Kapok | Ceiba Pentandra Gaertner | | Yes | |
| 370.43 | Kapok | Kapok | | Yes | |
| 370.44 | Wool Dust | Wool Dust | | | |
| 370.45 | Sesame Seed Dust | Sesame Seed Dust | | | |
| 370.46 | Jalapeno Pepper | Jalapeno Pepper | | | |
| 370.47 | Cocoa Bean | Cocoa Bean | | | |
| 370.47 | Cocoa Bean | Cocoa | | | |
| 370.48 | Rice Dust | Rice Dust | | Yes | |
| 370.49 | Rye Dust | Rye Dust | | | |
| 370.50 | Wheat Dust | Wheat Dust | | | |
| 370.51 | Corn Dust | Corn Dust | | | |
| 370.52 | Arabidopsis Thaliana | Arabidopsis Thaliana | | Yes | |
| 370.53 | Brazil Ginseng | Brazilian Ginseng | | Yes | |
| 370.53 | Brazil Ginseng | Brazil Ginseng | | Yes | |
| 370.54 | Chamomile | Chamomile | | Yes | |
| 370.55 | Tomato | Tomato | | | |
| 370.56 | Malt | Malt | | Yes | |
| 370.60 | Thiamine | Thiamine | | Yes | 59-43-8 |
| 371.00 | Flour, NOS | Flour, NOS | | Yes | |
| 371.01 | Buckwheat | Buckwheat | | Yes | |
| 371.02 | Gluten | Gluten | | Yes | |
| 371.03 | Rye Flour | Rye Flour | | Yes | |
| 371.04 | Wheat Flour | Wheat Flour | | Yes | |
| 371.05 | Soya Flour | Soya Flour | | Yes | |
| 371.06 | Corn Starch | Corn Starch | | | 9005-25-8 |
| 371.07 | Rice Flour | Rice Flour | | | |
| 372.01 | Gum Arabic | Acacia | | Yes | 1/5/9000 |
| 372.01 | Gum Arabic | Gum Acacia | | Yes | 1/5/9000 |
| 372.01 | Gum Arabic | Gum Arabic | | Yes | 1/5/9000 |
| 372.02 | Tragacanth | Tragacanth | | Yes | 9000-65-1 |
| 372.03 | Karaya | Karaya | | Yes | |
| 372.04 | Guar | Guar | | Yes | 9000-30-0 |
| 372.05 | Gum Arabic | Use 372.01 | | Yes | 1/5/9000 |
| 372.06 | Gutta-percha | Gutta-percha | | Yes | |
| 373.00 | Wood Dust, NOS | Wood Bark, NOS | | | |
| 373.00 | Wood Dust, NOS | Wood Dust, NOS | | | |
| 373.01 | Western Red Cedar | Thuja Plicata | | Yes | |
| 373.01 | Western Red Cedar | Western Red Cedar | | Yes | |
| 373.02 | California Redwood | California Redwood | | Yes | |
| 373.02 | California Redwood | Sequoia Sempervirens | | Yes | |
| 373.03 | Cedar of Lebanon | Cedar of Lebanon | | Yes | |

| AOEC Code[1] | Primary Name | Synonym | Use[2] | Asthma Inducer? | CAS Number[3] |
|---|---|---|---|---|---|
| 373.03 | Cedar of Lebanon | Cedrus Libani | | Yes | |
| 373.04 | Cocabolla | Cocabolla | | Yes | |
| 373.04 | Cocabolla | Cocobolo | | Yes | |
| 373.04 | Cocabolla | Dalbergia Retusa | | Yes | |
| 373.05 | Iroko | Chlorophora Excelsa | | Yes | |
| 373.05 | Iroko | Iroko | | Yes | |
| 373.06 | Oak | Oak | | Yes | |
| 373.06 | Oak | Quercus Rubra | | Yes | |
| 373.07 | Mahogany | Mahogany | | Yes | |
| 373.08 | Abiruana | Abiruana | | Yes | |
| 373.08 | Abiruana | Pouteria | | Yes | |
| 373.09 | African Maple | African Maple | | Yes | |
| 373.09 | African Maple | Triplochiton Scleroxylon | | Yes | |
| 373.10 | Tanganyika Aningre | Tanganyika Aningre | | Yes | |
| 373.11 | Central American Walnut | Central American Walnut | | Yes | |
| 373.11 | Central American Walnut | Juglans Olanchana | | Yes | |
| 373.12 | Kejaat | Kejaat | | Yes | |
| 373.12 | Kejaat | Pterocarpus Angolensis | | Yes | |
| 373.13 | African Zebrawood | African Zebrawood | | Yes | |
| 373.13 | African Zebrawood | Microberlinia | | Yes | |
| 373.14 | Ramin | Gonystylus Bancanus | | Yes | |
| 373.14 | Ramin | Ramin | | Yes | |
| 373.15 | Quillaja Bark | Quillaja Bark | | Yes | |
| 373.15 | Quillaja Bark | Soapbark | | Yes | |
| 373.16 | Fernambouc | Caesalpinia Echinata | | Yes | |
| 373.16 | Fernambouc | Fernambouc | | Yes | |
| 373.17 | Ashwood | Ashwood | | Yes | |
| 373.17 | Ashwood | Fraxinus Americana | | Yes | |
| 373.18 | Pau Marfim | Balfourodendron Riedelianum | | Yes | |
| 373.18 | Pau Marfim | Pau Marfim | | Yes | |
| 373.19 | Eastern White Cedar | Eastern White Cedar | | Yes | |
| 373.19 | Eastern White Cedar | Thuja Occidentalis | | Yes | |
| 373.20 | Ebony | Diospyros Crassiflora | | Yes | |
| 373.20 | Ebony | Ebony | | Yes | |
| 373.21 | Kotibe | Kotibe | | Yes | |
| 373.21 | Kotibe | Nesorgordonia Papaverifera | | Yes | |
| 373.22 | Cinnamon | Cinnamomum Zeylanicum | | Yes | |
| 373.22 | Cinnamon | Cinnamon | | Yes | |
| 373.23 | Cabreuva | Cabreuva | | Yes | |
| 373.23 | Cabreuva | Myrocarpus Fastigiatus Fr. All. | | Yes | |
| 373.24 | Hardwood, Tropical, NOS | Hardwood, Tropical, NOS | | | |
| 373.25 | Pine Wood Dust | Pine Wood Dust | | | |
| 380.00 | Animal Material, NOS | Animal Material, NOS | | | |
| 380.00 | Animal Material, NOS | Animal Material, NOS (see specific animals) | | | |
| 380.00 | Animal Material, NOS | Laboratory Animals, NOS (see specific animals) | | | |
| 380.01 | Antigens, Animal | Antigens, Animal | | | |
| 380.01 | Antigens, Animal | Antigens, Animal (see specific animals) | | | |
| 380.02 | Manure | Barn Dust | | | |
| 380.02 | Manure | Manure | | | |
| 380.03 | Leather Dust | Leather Dust | | | |
| 380.04 | Dander, Animal (NOS) | Dander, Animal (NOS) | | | |
| 380.04 | Cow Dander | Cow Dander | | Yes | |
| 380.05 | Venom | Bee Sting | | | |
| 380.05 | Venom | Snake Bite | | | |
| 380.05 | Venom | Venom | | | |
| 380.06 | Egg Protein | Egg Protein | | Yes | |
| 380.07 | Chicken | Chicken | | Yes | |
| 380.08 | Pig | Pig | | Yes | |
| 380.09 | Frog | Frog | | Yes | |

| AOEC Code[1] | Primary Name | Synonym | Use[2] | Asthma Inducer? | CAS Number[3] |
|---|---|---|---|---|---|
| 380.10 | Lactoserum | Lactoserum | | Yes | |
| 380.11 | Casein | Casein | | Yes | |
| 380.12 | Bat Guano | Bat Guano | | Yes | |
| 380.13 | Bovine Serum | Bovine Serum | | | |
| 380.14 | Mice | Mice | | Yes | |
| 380.15 | Rabbit Antigens | Rabbit Antigens | | Yes | |
| 380.16 | Avian Material, NOS | Avian Material, NOS | | | |
| 380.17 | Guinea Pig Antigens | Guinea Pig Antigens | | Yes | |
| 380.18 | Rat Antigens | Rat Antigens | | Yes | |
| 380.20 | Rat Feces | Rat Feces | | | |
| 380.21 | Pigeon Droppings | Pigeon Droppings | | | |
| 380.22 | Cat | Cat | | Yes | |
| 381.01 | Shellfish | Shellfish | | | |
| 381.02 | Crab | Crab | | Yes | |
| 381.03 | Prawn | Prawn | | Yes | |
| 381.04 | Hoya | Hoya | | | |
| 381.05 | Red Soft Coral | Red Soft Coral | | Yes | |
| 381.06 | Cuttlefish | Cuttlefish | | Yes | |
| 381.07 | Shrimp Meal | Shrimp | | Yes | |
| 381.07 | Shrimp Meal | Shrimp Meal | | Yes | |
| 381.08 | Trout | Trout | | Yes | |
| 381.09 | Fish Feed | Echinodorus Larva | | Yes | |
| 381.09 | Fish Feed | Fish Feed, Mosquito Larva | | Yes | |
| 381.09 | Fish Feed | Fish Feed | | Yes | |
| 381.10 | Clam | Clam | | Yes | |
| 381.11 | Shark Cartilage | Shark Cartilage | | | |
| 382.00 | Insect, NOS | Arthropod, NOS | | | |
| 382.00 | Insect, NOS | Insect, NOS | | | |
| 382.01 | Silkworm | Silkworm | | Yes | |
| 382.01 | Silkworm | Silkworm Larva | | Yes | |
| 382.02 | Locust | Locust | | Yes | |
| 382.03 | Screw Worm Fly | Screw Worm Fly | | Yes | |
| 382.04 | Cricket | Cricket | | Yes | |
| 382.05 | Bee Moth | Bee Moth | | Yes | |
| 382.06 | Mexican Bean Weevil | Mexican Bean Weevil | | Yes | |
| 382.07 | Fruit Fly | Fruit Fly | | Yes | |
| 382.08 | Honeybee | Honeybee | | Yes | |
| 382.09 | Lesser Mealworm | Lesser Mealworm | | Yes | |
| 382.10 | Daphnia | Daphnia | | Yes | |
| 382.11 | Acarian | Acarian | | Yes | |
| 382.12 | Sheep Blowfly | Sheep Blowfly | | Yes | |
| 382.13 | Mites, NOS | Mites, NOS | | Yes | |
| 382.14 | Fowl Mite | Fowl Mite | | Yes | |
| 382.15 | Barn Mite | Barn Mite | | Yes | |
| 382.16 | Grain Mite | Grain Mite | | Yes | |
| 382.16 | Grain Mite | Grain Parasite | | Yes | |
| 382.17 | Parasites, NOS | Parasites, NOS | | | |
| 382.18 | Bloodworm | Bloodworm | | | |
| 382.19 | L. Caesar Larva | L. Caesar Larva | | Yes | |
| 382.20 | New Mexico Range Moth Caterpillar | New Mexico Range Moth Caterpillar | | Yes | |
| 382.21 | Insect Bite, NOS | Insect Bite, NOS | | | |
| 382.21 | Insect Bite, NOS | Mosquito Bite | | | |
| 382.21 | Insect Bite, NOS | Spider Bite | | | |
| 382.22 | Chrysoperla Carnea | C. Carnea | | Yes | |
| 382.22 | Chrysoperla Carnea | Chrysoperla Carnea | | Yes | |
| 382.22 | Chrysoperla Carnea | Green Lacewing | | Yes | |
| 382.23 | Leptinotarsa Decemlineata | Colorado Potato Beetle | | Yes | 73468-57-2 |
| 382.23 | Leptinotarsa Decemlineata | L. Decemlineata | | Yes | 73468-57-2 |

| AOEC Code[1] | Primary Name | Synonym | Use[2] | Asthma Inducer? | CAS Number[3] |
|---|---|---|---|---|---|
| 382.23 | Leptinotarsa Decemlineata | Leptinotarsa Decemlineata | | Yes | 73468-57-2 |
| 382.24 | Ostrinia Nubilalis | European Corn Borer | | Yes | |
| 382.24 | Ostrinia Nubilalis | O. Nubilalis | | Yes | |
| 382.24 | Ostrinia Nubilalis | Ostrinia Nubilalis | | Yes | |
| 382.25 | Ephestia Kuehniella | E. Kuehniella | | Yes | |
| 382.25 | Ephestia Kuehniella | Ephestia Kuehniella | | Yes | |
| 390.00 | Microorganisms, NOS | Bioaerosols | | | |
| 390.00 | Microorganisms, NOS | Microorganisms, NOS | | | |
| 390.01 | Mold, NOS (new code 391.01) | Mold, NOS | | | |
| 390.03 | Histoplasma Capsulatum | Histoplasma Capsulatum | | | |
| 390.05 | Thermophilic Actinomyces | Thermophilic Actinomyces | | | |
| 390.07 | Infectious Agents, NOS | Bacteria | | | |
| 390.07 | Infectious Agents, NOS | Infectious Agents, NOS | | | |
| 390.07 | Infectious Agents, NOS | Viruses | | | |
| 390.07 | Staph Aureus | Staph Aureus | | | |
| 390.08 | HIV Exposure | AIDS Exposure | | | |
| 390.08 | HIV Exposure | HIV Exposure | | | |
| 390.08 | HIV Exposure | Sharps | | | |
| 390.09 | Hepatitis B | Hepatitis B | | | |
| 390.10 | Tuberculosis | Tuberculosis | | | |
| 390.11 | Neurospora | Neurospora | | Yes | |
| 390.12 | Hepatitis C | Hepatitis C | | | |
| 390.13 | Protozoa Giardia | Protozoa Giardia | | | |
| 390.14 | Bacillus Thurigiensis | Bacillus Thurigiensis | P | | 23526-02-5 |
| 390.14 | Bacillus Thurigiensis | Bt | P | | 23526-02-5 |
| 390.15 | Borrelium Virus | Borrelium Virus | | | |
| 390.16 | Body Fluid Exposure (Unknown Infection Status) | Body Fluid Exposure (Unknown Exposure Status) | | | |
| 390.16 | Body Fluid Exposure (Unknown Infection Status) | Blood Exposure (Unknown Infection Status) | | | |
| 391.01 | Mold, NOS | Fungi, NOS | | | |
| 391.01 | Mold, NOS | Mold, NOS | | | |
| 391.02 | Aspergillus | Aspergillus | | | |
| 391.02 | Aspergillus | Aspergillus Flavus | | | |
| 391.02 | Aspergillus | Aspergillus Fumigatus | | | |
| 391.02 | Aspergillus | Aspergillus Glaucus | | | |
| 391.02 | Aspergillus | Aspergillus Niger | | | |
| 391.02 | Aspergillus | Aspergillus Versicolor | | | |
| 391.03 | Penicillium | Penicillium | | Yes | |
| 391.05 | Plasmopara | Plasmopara | | | |
| 391.05 | Plasmopara | Plasmopara Viticola | | | |
| 391.06 | Slime Mold | Slime Mold Dictyostelium Discoideum | | | |
| 391.06 | Slime Mold | Slime Mold | | | |
| 391.07 | Stachybotrys | Stachybotrys | | | |
| 391.08 | Trichoderma | Trichoderma | | | |
| 391.08 | Trichoderma | Trichoderma Koningii | | | |
| 391.09 | Mycotoxins | Aflatoxin | | | 1402-68-2 |
| 391.09 | Mycotoxins | Mycotoxins | | | |
| 391.10 | Mushrooms, NOS | Mushrooms, NOS | | | |
| 391.11 | Silage Microorganisms, NOS | Silage Microorganisms, NOS | | | |
| 391.12 | Alternaria | Alternaria | | Yes | |
| 391.12 | Alternaria | Alternaria spp | | Yes | |
| 391.12 | Alternaria | Alternaria Alternata Toxin | | Yes | |
| 391.12 | Alternaria | Alternaria Aleternata | | Yes | |
| 391.13 | Endotoxin | Endotoxin | | | 11034-88-1 |
| 561-27-3 | Diacetyl morphine | Heroin | | Yes | 561-27-3 |

***Notes:** The items above cover the entire range of occupational and environmental exposures capable of causing illness;*
*(1) Association of Occupational & Environmental Clinics code; (2) S=Solvent; P=Pesticide; (3) Chemical Abstracts Service registry number*
***Source:** Association of Occupational & Environmental Clinics, May 18, 2011*

## Cotinine

CAS No. 486-56-6

*Metabolite of nicotine (a component of tobacco smoke)*

### General Information

Tobacco use is the most important preventable cause of premature morbidity and mortality in the United States. The consequences of smoking and of using smokeless tobacco products are well known and include an increased risk for several types of cancer, emphysema, acute respiratory illness, cardiovascular disease, stroke, and various other disorders (U.S. DHHS, 2006). Persons exposed to secondhand tobacco smoke (environmental tobacco smoke [ETS]) may have adverse health effects that include lung cancer and coronary heart disease; maternal exposure during pregnancy can result in lower birth weight. Children exposed to ETS are at increased risk for sudden infant death syndrome, acute respiratory infections, ear problems, and exacerbated asthma (U.S. DHHS, 2004). The smoke produced by burning tobacco contains at least 250 chemicals that are toxic or carcinogenic, and more than 50 compounds present in ETS are known or reasonably anticipated to be human carcinogens (NTP, 2004).

Cigarettes contain about 1.5% nicotine by weight (Kozlowski et al., 1998), producing roughly 1–2 mg of bioavailable nicotine per cigarette (Benowitz and Jacob,

### Serum Cotinine

*Metabolite of nicotine (component of tobacco smoke)*

Geometric mean and selected percentiles of serum concentrations (in ng/mL) for the **non-smoking** U.S. population from the National Health and Nutrition Examination Survey.

| | Survey years | Geometric mean (95% conf. interval) | Selected percentiles (95% confidence interval) 50th | 75th | 90th | 95th | Sample size |
|---|---|---|---|---|---|---|---|
| **Total** | 99-00 | * | **.060** (<LOD-.080) | **.240** (.190-.302) | **1.02** (.770-1.28) | **1.96** (1.60-2.62) | 5999 |
| | 01-02** | **.062** (.050-.077) | < LOD | **.160** (.120-.220) | **.930** (.740-1.17) | **2.20** (1.83-2.44) | 6819 |
| | 03-04 | **.071** (.057-.089) | **.050** (.040-.070) | **.210** (.140-.310) | **.990** (.740-1.30) | **2.17** (1.81-2.54) | 6320 |
| **Age group** | | | | | | | |
| 3-11 years | 99-00 | **.164** (.115-.234) | **.110** (.066-.188) | **.500** (.260-1.16) | **1.88** (.997-3.44) | **3.44** (1.42-4.79) | 1174 |
| | 01-02** | **.110** (.076-.160) | **.070** (<LOD-.130) | **.570** (.310-1.00) | **2.23** (1.63-2.78) | **3.23** (2.53-4.01) | 1415 |
| | 03-04 | **.137** (.088-.213) | **.120** (.060-.220) | **.620** (.310-1.20) | **2.04** (1.38-2.94) | **3.35** (2.12-4.68) | 1252 |
| 12-19 years | 99-00 | **.163** (.142-.187) | **.110** (.080-.163) | **.540** (.428-.660) | **1.66** (1.50-1.95) | **2.62** (2.09-3.39) | 1773 |
| | 01-02** | **.086** (.059-.126) | **.050** (<LOD-.110) | **.350** (.190-.580) | **1.53** (1.09-2.12) | **3.12** (2.47-3.99) | 1902 |
| | 03-04 | **.110** (.087-.139) | **.080** (.060-.120) | **.510** (.350-.670) | **1.55** (1.21-1.93) | **2.68** (1.96-4.02) | 1783 |
| 20 years and older | 99-00 | * | **.050** (<LOD-.061) | **.167** (.140-.193) | **.630** (.533-.820) | **1.50** (1.28-1.66) | 3052 |
| | 01-02** | **.052** (<LOD-.063) | < LOD | **.110** (.090-.150) | **.630** (.470-.790) | **1.42** (1.14-1.89) | 3502 |
| | 03-04 | **.058** (.047-.071) | **.040** (.030-.050) | **.140** (.100-.200) | **.630** (.480-.840) | **1.54** (1.26-1.92) | 3285 |
| **Gender** | | | | | | | |
| Males | 99-00 | **.124** (.106-.145) | **.080** (.060-.110) | **.308** (.220-.410) | **1.20** (.950-1.49) | **2.39** (1.66-3.22) | 2789 |
| | 01-02** | **.075** (.059-.094) | **.050** (<LOD-.070) | **.230** (.160-.320) | **1.17** (.960-1.49) | **2.44** (2.23-2.99) | 3152 |
| | 03-04 | **.087** (.070-.108) | **.060** (.040-.080) | **.280** (.190-.360) | **1.23** (.910-1.68) | **2.63** (2.09-3.19) | 2937 |
| Females | 99-00 | * | < LOD | **.180** (.148-.230) | **.850** (.600-1.14) | **1.85** (1.33-2.45) | 3210 |
| | 01-02** | **.053** (<LOD-.066) | < LOD | **.120** (.090-.180) | **.710** (.540-.990) | **1.77** (1.32-2.20) | 3667 |
| | 03-04 | **.060** (.047-.077) | **.040** (.030-.060) | **.160** (.110-.260) | **.860** (.580-1.15) | **1.76** (1.32-2.22) | 3383 |
| **Race/ethnicity** | | | | | | | |
| Mexican Americans | 99-00 | * | < LOD | **.140** (.110-.180) | **.506** (.370-.726) | **1.21** (.900-1.70) | 2241 |
| | 01-02** | **.060** (<LOD-.084) | < LOD | **.160** (.080-.310) | **.730** (.480-1.19) | **2.12** (1.19-2.96) | 1878 |
| | 03-04 | **.054** (.043-.068) | **.030** (.020-.050) | **.120** (.080-.180) | **.690** (.430-1.00) | **2.65** (1.87-3.57) | 1707 |
| Non-Hispanic blacks | 99-00 | **.175** (.153-.201) | **.131** (.111-.150) | **.505** (.400-.625) | **1.43** (1.21-1.75) | **2.34** (1.84-3.50) | 1333 |
| | 01-02** | **.164** (.137-.197) | **.130** (.110-.160) | **.580** (.450-.770) | **1.77** (1.55-2.05) | **3.15** (2.50-4.30) | 1602 |
| | 03-04 | **.144** (.104-.198) | **.120** (.080-.180) | **.520** (.350-.770) | **1.54** (1.20-2.14) | **2.77** (2.18-3.54) | 1704 |
| Non-Hispanic whites | 99-00 | * | **.050** (<LOD-.073) | **.216** (.154-.312) | **.950** (.621-1.40) | **1.92** (1.48-3.02) | 1950 |
| | 01-02** | **.052** (<LOD-.068) | < LOD | **.120** (.090-.180) | **.800** (.570-1.11) | **1.88** (1.48-2.30) | 2847 |
| | 03-04 | **.066** (.050-.087) | **.040** (.030-.070) | **.180** (.120-.300) | **.920** (.620-1.32) | **2.01** (1.70-2.49) | 2500 |

Limit of detection (LOD, see Data Analysis section) for Survey years 99-00, and 03-04 are 0.05, and 0.015, respectively.

** In the 2001-2002 survey period, 83% of measurements had an LOD of 0.015 ng/mL, and 17% had an LOD of 0.05 ng/mL.

< LOD means less than the limit of detection, which may vary for some chemicals by year and by individual sample.

* Not calculated: proportion of results below limit of detection was too high to provide a valid result.

1994; Hukkanen et al., 2005). Inhaling tobaccosmoke from either active or passive (ETS) smoking is the main source of nicotine exposure for the general population. Up to 90% of the nicotine delivered in tobacco smoke isabsorbed rapidly from the lungs into the blood stream (Armitage et al., 1975; Iwase et al., 1991). Mean air concentrations of nicotine in public spaces where smoking is allowed range from 0.3 to 30 µg/m³, with higher levels measured in restaurants and bars. In homes with one or more smokers, mean air concentrations typically range from 2 to 14 µg/m³ (NTP, 2004). For an adult, the primary sources for ETS exposure are in a workplace where smoking occurs and in a residence shared with one or more smokers. Children are primarily exposed to ETS by parents and caregivers who smoke.

Nicotine can also be absorbed from the gastrointestinal tract and skin by using snuf f, chewing tobacco, or chewing gum, nasal sprays, or skin patches that contain nicotine. Workers who harvest tobacco can be exposed to nicotine and become intoxicated as a result of the transdermal absorption of nicotine contained in the plant. The tobacco plant, *Nicotiana tabacum*, contains nicotine in larger amounts than other nicotine-containing plants, which include potatoes, tomatoes, eggplants, and peppers. Nicotine is also used commercially as an insecticide in its sulfate and alkaloid forms.

Once absorbed, nicotine has a half-life in blood plasma of several hours (Benowitz, 1996). Cotinine, the primary metabolite of nicotine, is currently regarded as the best biomarker in active smokers and in nonsmokers exposed to ETS. Measuring cotinine is preferred over measuring nicotine because cotinine persists longer in the body with a plasma half-life of about 16 hours (Benowitz and Jacob, 1994). However, non-Hispanic blacks metabolize cotinine more slowly than do non-Hispanic whites (Benowitz et al., 1999; Perez-Stable et al., 1998). Cotinine can be measured in serum, urine, saliva, and hair . Nonsmokers exposed to typical levels of ETS have serum cotinine levels of less than 1 ng/mL, with heavy exposure to ETS producing levels in the 1–10 ng/mL range. Active smokers almost always have levels higher than 10 ng/mL and sometimes higher than 500 ng/mL (Hukkanen et al., 2005).

Nicotine stimulates preganglionic choliner gic receptors within peripheral sympathetic autonomic ganglia and at cholinergic sites within the central nervous system. Acute tobacco or nicotine intoxication can produce dizziness, nausea, vomiting, diaphoresis, salivation, diarrhea, variable changes in blood pressure and heart rate, seizures, and death. Nicotine also indirectly causes a release of dopamine in the brain regions that control pleasure and motivation, a process involved in the development of addiction. Symptoms of nicotine withdrawal include irritability, craving, cognitive and sleep disturbances, and increased appetite.

The IARC and the NTP consider tobacco smoke to be a human carcinogen. NIOSH guidelines consider ETS to be a potential occupational carcinogen and recommend that exposure be reduced to the lowest feasible concentration. The Federal Aviation Administration has banned the smoking of tobacco products on both domestic and foreign air carrier flights in the United States. More information about the effects of smoking and nicotine can be found at: http://www.nida.nih.gov/researchreports/nicotine/nicotine.html.

**Biomonitoring Information**

Serum cotinine levels reflect recent exposure to nicotine in tobacco smoke. Nonsmoking is usually defined as a serum cotinine level of less than or equal to 10 ng/mL (Pirkle et al., 1996).

The serum cotinine levels seen in the NHANES 2003-2004 appear approximately similar to levels seen in the previous survey period (NHANES 2001-2002) for the total population estimates. Serum cotinine has been measured in many studies of nonsmoking populations, with levels showing similar or slightly higher results (depending on the degree of ETS exposure) than those reported in the previous NHANES (CDC 2005; NCI, 1999). Over the previous decade, levels of exposure to ETS appeared to decrease since geometric mean cotinine serum concentrations in nonsmokers had fallen by approximately 70% and the rate of detectable cotinine in nonsmokers fell from 88% to 43% when NHANES 1988–1991 was compared to NHANES 1999–2002, (CDC, 2005; Pirkle et al., 2006). The overall decline in population estima tes of serum cotinine likely reflects decreased ETS exposure among nonsmokers in locations with smoke-free laws (Pickett et al., 2006; Soliman et al., 2004). During each previous NHANES survey, the adjusted geometr ic mean serum cotinine was higher in children (aged 4–1 1 years) than in adults among both non-Hispanic blacks and non-Hispanic whites (Pirkle et al., 2006). Non-Hispanic blacks had higher serum cotinine concentrations compared with either non-Hispanic whites or Mexican-Americans. Higher levels of cotinine have previously been reported for non-Hispanic black smokers (Caraballo et al., 1998). Differences in cotinine concentrations among race/ethnicity and age groups may be influenced by pharmacokinetic differences as well as by ETS exposure (Benowitz et al., 1999; Hukkanen et al., 2005; Wilson et al., 2005).

*Source: CDC, Fourth National Report on Human Exposure to Environmental Chemicals, 2009*

## Triclosan

CAS No. 3380-34-5

### General Information

Triclosan is a phenolic diphenyl ether used for over 30 years as a preservative and antiseptic agent. It acts by inhibiting bacterial fatty acid synthesis. Triclosan has been added to soaps, toothpastes, mouthwashes, acne medications, deodorants, and wound disinfection solutions, and has also been impregnated into some kitchen utensils, toys, and medical devices. Triclosan enters the aquatic environment mainly through residential wastewaters. It can be photochemically and biologically degraded, a process that can result in the formation of small amounts of 2,8-dichlorodibenzo-p-dioxin (Aranami et al., 2007; Mezcua et al., 2004). In 1999-2000, triclosan was found in 57.6% of 139 U.S. streams sampled in 30 states (Kolpin et al., 2002). Triclosan has a low bioaccumulation potential in fish. There is some concern that widespread use of triclosan and other biocides can alter antibiotic resistance in bacteria (Aiello et al., 2007).

General population exposure results from dermal and oral use of products containing triclosan. Triclosan can remain present in the oral saliva for several hours after the use of toothpaste containing triclosan (Gilbert et al., 1987). Triclosan can be absorbed across skin into the blood stream. In the body it is conjugated to glucuronides and sulfates (Bodey et al., 1976; Moss et al., 2000). In animal and human studies, it is excreted over several days in the feces and urine as primarily as unchanged triclosan (Kanetoshi et al., 1988; (Sandborgh-Englund et al., 2006).

Human health effects from triclosan at low environmental doses or at biomonitored levels from low environmental exposures are unknown. Triclosan formulations may rarely cause skin irritation. In animal studies, it has low acute toxicity. Some reports show endocrine effects are observed in amphibians and fish (Foran et al., 2000; Matsumura et al., 2005; Veldhoen et al., 2007). Triclosan is not considered teratogenic at maternally toxic doses, and has not been considered mutagenic or carcinogenic (Bhargava and Leonard, 1996; Lyman and Furia, 1969). IARC and NTP do not have ratings with respect to human carcinogenicity.

### Biomonitoring Information

Urinary triclosan levels reflect recent exposure. In a U.S. representative subsample of NHANES 2003-2004, Calafat et al., 2008 has shown higher levels during the third decade of life and among people with the highest household income, but not by race/ethnicity and sex. In a study of 90 U.S. young girls, the median urinary triclosan level of 7.2 μg/L was comparable to the median level (8.2 μg/L) of children 6-11 years of age who participated in NHANES 2003-2004 (Wolff et al., 2007; Calafat et al., 2008).

Finding measurable amounts of triclosan in the urine does not mean that the levels of triclosan cause an adverse health effect. Biomonitoring studies on levels of triclosan provide physicians and public health officials with a reference values so that they can determine whether people have been exposed to higher levels of triclosan than are found in the general population. Biomonitoring data can also help scientists plan and conduct research on exposure and health effects.

## Urinary Triclosan (2,4,4'-Trichloro-2'-hydroxyphenyl ether)

Geometric mean and selected percentiles of urine concentrations (in µg/L) for the U.S. population from the National Health and Nutrition Examination Survey.

| | Survey years | Geometric mean (95% conf. interval) | Selected percentiles (95% confidence interval) 50th | 75th | 90th | 95th | Sample size |
|---|---|---|---|---|---|---|---|
| **Total** | 03-04 | **13.0** (11.6-14.6) | **9.20** (7.90-10.9) | **47.4** (38.2-58.4) | **249** (188-304) | **461** (383-522) | 2517 |
| **Age group** | | | | | | | |
| 6-11 years | 03-04 | **8.16** (6.20-10.8) | **6.00** (4.00-8.50) | **20.7** (14.3-31.6) | **123** (36.4-163) | **157** (113-380) | 314 |
| 12-19 years | 03-04 | **14.5** (11.0-19.1) | **10.3** (8.20-13.1) | **39.0** (26.5-86.4) | **304** (134-566) | **655** (310-890) | 715 |
| 20 years and older | 03-04 | **13.6** (12.0-15.3) | **9.60** (8.20-11.5) | **51.7** (39.6-65.7) | **261** (198-317) | **472** (406-522) | 1488 |
| **Gender** | | | | | | | |
| Males | 03-04 | **16.2** (13.4-19.6) | **11.7** (9.30-14.8) | **84.9** (50.6-111) | **317** (231-433) | **574** (461-716) | 1229 |
| Females | 03-04 | **10.6** (9.29-12.1) | **7.60** (6.10-9.10) | **33.2** (27.1-39.4) | **144** (96.5-250) | **380** (258-430) | 1288 |
| **Race/ethnicity** | | | | | | | |
| Mexican Americans | 03-04 | **14.6** (10.6-20.1) | **8.80** (5.40-17.5) | **65.4** (32.8-127) | **357** (225-456) | **597** (372-992) | 613 |
| Non-Hispanic blacks | 03-04 | **14.4** (11.4-18.2) | **11.1** (8.70-16.1) | **37.6** (30.2-58.0) | **203** (87.5-341) | **450** (254-750) | 652 |
| Non-Hispanic whites | 03-04 | **12.9** (11.2-14.9) | **9.20** (7.40-11.0) | **49.2** (37.8-63.4) | **245** (163-334) | **461** (383-527) | 1092 |

Limit of detection (LOD, see Data Analysis section) for Survey year 03-04 is 2.3.

## Urinary Triclosan (2,4,4'-Trichloro-2'-hydroxyphenyl ether) (creatinine corrected)

Geometric mean and selected percentiles of urine concentrations (in µg/g of creatinine) for the U.S. population from the National Health and Nutrition Examination Survey.

| | Survey years | Geometric mean (95% conf. interval) | Selected percentiles (95% confidence interval) 50th | 75th | 90th | 95th | Sample size |
|---|---|---|---|---|---|---|---|
| **Total** | 03-04 | **12.7** (11.5-14.1) | **9.48** (8.22-10.4) | **43.9** (33.8-60.6) | **212** (172-241) | **368** (294-463) | 2514 |
| **Age group** | | | | | | | |
| 6-11 years | 03-04 | **9.93** (7.43-13.3) | **7.55** (4.72-13.4) | **25.1** (15.3-35.6) | **116** (39.9-236) | **236** (115-336) | 314 |
| 12-19 years | 03-04 | **10.9** (8.32-14.2) | **7.45** (5.48-10.7) | **31.8** (21.9-61.1) | **193** (90.7-318) | **356** (169-580) | 713 |
| 20 years and older | 03-04 | **13.4** (12.0-15.1) | **10.0** (8.89-11.4) | **50.0** (36.0-73.8) | **224** (186-272) | **385** (308-506) | 1487 |
| **Gender** | | | | | | | |
| Males | 03-04 | **13.2** (11.3-15.6) | **9.21** (6.86-12.1) | **73.1** (45.8-85.9) | **237** (175-294) | **384** (294-506) | 1228 |
| Females | 03-04 | **12.2** (10.6-14.2) | **9.54** (8.45-10.4) | **32.3** (26.2-46.6) | **182** (138-217) | **336** (225-480) | 1286 |
| **Race/ethnicity** | | | | | | | |
| Mexican Americans | 03-04 | **13.3** (9.38-18.8) | **9.18** (5.45-13.9) | **66.7** (28.8-112) | **292** (151-432) | **453** (263-1150) | 612 |
| Non-Hispanic blacks | 03-04 | **9.94** (7.92-12.5) | **7.74** (5.50-10.0) | **30.2** (25.6-37.3) | **132** (78.0-213) | **260** (127-513) | 651 |
| Non-Hispanic whites | 03-04 | **13.3** (11.6-15.1) | **9.82** (8.11-11.5) | **47.0** (34.3-67.7) | **213** (160-272) | **358** (276-480) | 1091 |

*Source: CDC, Fourth National Report on Human Exposure to Environmental Chemicals, 2009*

# Chlorpyrifos

CAS No. 2921-88-2

# Chlorpyrifos-methyl

CAS No. 5598-13-0

### General Information

The chemical 3,5,6-trichloro-2-pyridinol (TCPy) is a metabolite of chlorpyrifos and chlorpyrifos-methyl. Chlorpyrifos is a broad spectrum or ganophosphorus insecticide that has been widely used to control insects on food crops such as corn. It also has been applied directly on animals to kill mites, applied to structures to kill termites, and sprayed to kill mosquitoes. Approximately 21-24 million pounds per year were used domestically from 1987-1998. After 2001, chlorpyrifos was no longer registered for indoor residential uses in the United States; pre- and post-construction structural applications for termite control were to be phased out by 2005 (U.S.EPA, 2002). Chlorpyrifos-methyl is an organophosphorus insecticide also used in agriculture and not registered for residential use. Approximately 80,000 pounds are used per year. Chlorpyrifos is degraded in agricultural soils with a half-life of several months, and on plants for days to several weeks. It has low leachability, staying bound to soil particles, and is infrequently detected in ground water (IPCS, 1999; USGS, 2007), but can be detected in streams receiving runoff from application sites. Chlorpyrifos is very toxic to fish and aquatic invertebrates and shows modest degrees of bioconcentration.

The general population may be exposed to chlorpyrifos via oral, dermal, and inhalation routes. Estimated intakes from diet and water have not exceeded recommended intake limits, although some tolerances for specific food crops have been reduced in the past to avoid exceeding recommended intake limits for total dietary intake in special groups (U.S.EPA, 2002). Exposure can also result from contact with contaminated surfaces, air, and dust. For instance, in 142 urban homes and preschools in North Carolina, chlorpyrifos and TCPy were detected in all indoor air and dust samples (Morgan et al., 2005). Chlorpyrifos is not well absorbed through the skin but dermal exposure can be significant when other routes of exposure are low. Inhalational and dermal routes of exposure are important in pesticide formulators and applicators. Chlorpyrifos is

### Urinary 3,5,6-Trichloro-2-pyridinol

*Metabolite of Chlorpyrifos and Chlorpyrifos-methyl*

Geometric mean and selected percentiles of urine concentrations (in µg/L) for the U.S. population from the National Health and Nutrition Examination Survey.

| | Survey years | Geometric mean (95% conf. interval) | Selected percentiles (95% confidence interval) 50th | 75th | 90th | 95th | Sample size |
|---|---|---|---|---|---|---|---|
| **Total** | 99-00 | **1.77** (1.46-2.14) | **1.70** (1.40-2.20) | **3.50** (2.50-5.20) | **7.30** (4.80-10.0) | **10.0** (7.70-15.0) | 1994 |
| | 01-02 | **1.76** (1.52-2.03) | **2.22** (1.90-2.61) | **4.95** (4.55-5.29) | **8.80** (7.74-9.77) | **12.4** (10.4-15.3) | 2509 |
| **Age group** | | | | | | | |
| 6-11 years | 99-00 | **2.88** (1.99-4.16) | **2.80** (1.60-4.90) | **7.09** (3.40-10.0) | **12.0** (7.70-17.0) | **16.0** (10.0-28.0) | 481 |
| | 01-02 | **2.67** (2.13-3.35) | **3.09** (2.50-4.22) | **6.36** (4.97-7.97) | **10.9** (7.98-15.3) | **15.3** (11.5-24.0) | 573 |
| 12-19 years | 99-00 | **2.37** (1.89-2.97) | **2.10** (1.60-3.00) | **4.50** (2.90-7.10) | **8.10** (5.50-14.0) | **12.5** (8.00-24.0) | 681 |
| | 01-02 | **2.71** (2.19-3.35) | **3.57** (2.66-4.34) | **6.60** (5.61-7.59) | **11.3** (8.66-15.1) | **18.0** (13.7-23.7) | 823 |
| 20-59 years | 99-00 | **1.53** (1.29-1.83) | **1.50** (1.30-1.80) | **2.90** (2.20-4.30) | **5.90** (3.90-8.90) | **8.90** (6.70-11.0) | 832 |
| | 01-02 | **1.51** (1.32-1.72) | **1.91** (1.44-2.26) | **4.44** (3.90-4.80) | **7.78** (7.00-8.91) | **10.9** (9.52-12.4) | 1113 |
| **Gender** | | | | | | | |
| Males | 99-00 | **1.92** (1.60-2.32) | **1.90** (1.50-2.40) | **3.60** (2.70-5.60) | **7.40** (5.04-10.0) | **10.0** (7.70-16.0) | 972 |
| | 01-02 | **2.13** (1.81-2.51) | **2.67** (2.20-3.25) | **5.37** (4.87-6.25) | **9.63** (8.20-11.5) | **14.9** (10.9-18.9) | 1183 |
| Females | 99-00 | **1.63** (1.31-2.02) | **1.50** (1.20-2.00) | **3.30** (2.30-5.30) | **7.20** (4.30-12.0) | **11.0** (7.20-16.0) | 1022 |
| | 01-02 | **1.45** (1.24-1.70) | **1.74** (1.39-2.21) | **4.38** (3.72-4.95) | **7.71** (6.30-9.20) | **10.4** (8.47-13.2) | 1326 |
| **Race/ethnicity** | | | | | | | |
| Mexican Americans | 99-00 | **1.61** (1.31-2.00) | **1.67** (1.30-2.20) | **3.20** (2.60-3.80) | **5.10** (3.80-8.40) | **7.40** (5.10-17.0) | 697 |
| | 01-02 | **2.02** (1.79-2.28) | **2.63** (2.24-3.01) | **4.60** (4.05-5.39) | **9.02** (7.04-10.8) | **12.2** (10.8-15.7) | 660 |
| Non-Hispanic blacks | 99-00 | **2.17** (1.59-2.97) | **1.90** (1.43-2.80) | **4.30** (2.50-8.30) | **9.40** (6.40-13.7) | **13.0** (9.40-26.0) | 521 |
| | 01-02 | **2.19** (1.68-2.84) | **2.89** (2.28-3.47) | **5.47** (4.77-6.96) | **9.27** (7.47-11.6) | **12.3** (10.1-16.8) | 701 |
| Non-Hispanic whites | 99-00 | **1.76** (1.51-2.05) | **1.70** (1.50-2.10) | **3.50** (2.50-4.86) | **7.10** (4.30-11.0) | **10.0** (7.20-14.0) | 602 |
| | 01-02 | **1.71** (1.43-2.03) | **2.15** (1.62-2.64) | **4.94** (4.44-5.37) | **8.68** (7.47-9.97) | **12.4** (9.77-15.9) | 947 |

Limit of detection (LOD, see Data Analysis section) for Survey years 99-00 and 01-02 are 0.4 and 0.4.

rapidly absorbed following ingestion. Once absorbed, phosphorothioates such as chlorpyrifos are metabolically activated to the "oxon" form which has greater toxicity than the parent insecticide. Metabolic hydrolysis leads to the formation of TCPy, dialkyl phosphate metabolites (see section titled "Or ganophosphorus Insecticides: Dialkyl Phosphate Metabolites"), and other metabolites. Chlorpyrifos is eliminated from the body primarily in the urine with a half-life of approximately 27 hours (Nolan et al., 1984). In addition to being a metabolite of chlorpyrifos and chlorpyrifos-methyl in the body, TCPy can also occur in the environment from the breakdown of the parent compounds. TCPy is more persistent in the environment than chlorpyrifos itself (U.S.EPA, 2002). Thus, the detection of TCPy in a person's urine may reflect exposure to the environmental degradates.

Human health effects from chlorpyrifos or chlorpyrifos-methyl at low environmenta l doses or at biomonitored levels from low environmental exposures are unknown. Chlorpyrifos and chlorpyrifos-methyl both demonstrate moderate acute toxicity in animal studies. These organophosphorus insecticides share a mechanism of toxicity: inhibition of the activity of acetylcholinesterase enzymes in the nervous system, resulting in excess acetylcholine at nerve terminals, and producing acute symptoms such as nausea, vomiting, choliner gic effects, weakness, paralysis, and seizures. The metabolite TCPy does not inhibit acetylcholinesterase enzymes. Overt cholinergic toxicity from chlorpyrifos has been described following suicidal ingestion and unintentional high level occupational exposure. Based on animal data and human cholinesterase monitoring during occupational exposure, ubiquitous low-level environmental exposures in humans would not be expected to result in inhibiti on of cholinesterase activity. Recent *in vitro* and *in vivo* animal studies suggest that effects on neuronal morphogenesis, neurotransmission, and behavior may occur at systemically nontoxic doses or at doses of chlorpyrifos that do not result in cholinergic signs (Aldridge et al., 2005; Betancourt et al., 2006; Howard et al., 2005; Ricceri et al., 2006; Roy et al., 2005; Slotkin et al., 2006*a*, 2006*b*). In pesticide applicators, chronic exposure to chlorpyrifos may be associated with slight alterations in some components of neurophysiologic testing (Steenland et al., 2000). Two observational studies of pregnant women and their of fspring exposed to chlorpyrifos at environmental levels have found inconsistent relationships with birth outcomes of weight and length (Eskenazi et al.,

## Urinary 3,5,6-Trichloro-2-pyridinol (creatinine corrected)

*Metabolite of Chlorpyrifos and Chlorpyrifos-methyl*

Geometric mean and selected percentiles of urine concentrations (in µg/g of creatinine) for the U.S. population from the National Health and Nutrition Examination Survey.

| | Survey years | Geometric mean (95% conf. interval) | Selected percentiles (95% confidence interval) 50th | 75th | 90th | 95th | Sample size |
|---|---|---|---|---|---|---|---|
| **Total** | 99-00 | **1.58** (1.35-1.85) | **1.47** (1.24-1.74) | **2.85** (2.12-3.59) | **5.43** (4.22-6.68) | **8.42** (6.25-11.6) | 1994 |
| | 01-02 | **1.73** (1.49-2.01) | **1.88** (1.64-2.24) | **3.76** (2.91-4.62) | **6.15** (4.99-8.31) | **9.22** (6.94-12.3) | 2508 |
| **Age group** | | | | | | | |
| 6-11 years | 99-00 | **3.11** (2.31-4.19) | **3.20** (2.05-4.80) | **6.39** (4.14-8.19) | **10.1** (7.26-14.0) | **14.1** (10.1-21.0) | 481 |
| | 01-02 | **3.48** (2.80-4.32) | **3.76** (3.17-4.36) | **6.22** (4.88-8.57) | **12.2** (7.24-24.4) | **16.9** (12.1-38.0) | 573 |
| 12-19 years | 99-00 | **1.60** (1.34-1.91) | **1.45** (1.21-1.81) | **2.58** (1.97-3.92) | **4.82** (3.44-6.16) | **6.16** (4.43-10.6) | 681 |
| | 01-02 | **2.09** (1.72-2.55) | **2.24** (1.92-2.66) | **3.97** (3.30-4.72) | **6.33** (5.62-7.89) | **10.3** (7.65-15.2) | 822 |
| 20-59 years | 99-00 | **1.41** (1.23-1.62) | **1.33** (1.12-1.58) | **2.37** (1.87-3.01) | **4.29** (3.53-5.56) | **6.42** (5.11-9.02) | 832 |
| | 01-02 | **1.49** (1.30-1.71) | **1.64** (1.39-1.88) | **3.11** (2.60-3.91) | **5.50** (4.33-7.23) | **7.44** (5.80-11.0) | 1113 |
| **Gender** | | | | | | | |
| Males | 99-00 | **1.48** (1.27-1.72) | **1.44** (1.19-1.68) | **2.54** (2.05-3.38) | **4.95** (3.84-6.54) | **7.63** (5.65-11.0) | 972 |
| | 01-02 | **1.71** (1.47-2.00) | **1.88** (1.57-2.22) | **3.46** (2.82-4.28) | **5.93** (4.90-9.24) | **10.5** (6.94-14.3) | 1183 |
| Females | 99-00 | **1.69** (1.42-2.01) | **1.51** (1.25-1.85) | **2.97** (2.24-4.01) | **5.63** (4.27-7.39) | **8.44** (5.79-13.3) | 1022 |
| | 01-02 | **1.75** (1.49-2.07) | **1.93** (1.59-2.33) | **3.91** (3.06-4.85) | **6.47** (5.00-8.11) | **8.98** (6.83-11.8) | 1325 |
| **Race/ethnicity** | | | | | | | |
| Mexican Americans | 99-00 | **1.46** (1.20-1.77) | **1.44** (1.05-1.93) | **2.39** (2.09-2.96) | **3.86** (3.24-5.08) | **5.85** (3.88-9.57) | 697 |
| | 01-02 | **1.86** (1.63-2.12) | **2.06** (1.83-2.35) | **3.81** (3.17-4.56) | **6.52** (5.64-7.58) | **9.00** (7.66-11.8) | 660 |
| Non-Hispanic blacks | 99-00 | **1.47** (1.09-1.99) | **1.33** (.940-1.91) | **2.86** (1.58-5.05) | **5.91** (4.05-8.93) | **9.02** (5.91-13.7) | 521 |
| | 01-02 | **1.56** (1.19-2.03) | **1.92** (1.57-2.40) | **3.53** (2.85-4.28) | **5.58** (4.80-6.07) | **7.06** (5.88-8.82) | 700 |
| Non-Hispanic whites | 99-00 | **1.66** (1.45-1.91) | **1.55** (1.31-1.83) | **2.93** (2.09-3.97) | **5.56** (4.21-6.75) | **8.44** (6.25-12.3) | 602 |
| | 01-02 | **1.78** (1.49-2.14) | **1.95** (1.56-2.35) | **3.82** (2.70-4.97) | **6.55** (4.88-10.5) | **9.98** (7.00-13.7) | 947 |

2004; Perera et al., 2003; Whyatt et al., 2004).

Some reproductive and teratogenic effects in animal testing were only observed at high doses of chlorpyrifos that caused overt maternal toxicity. Chlorpyrifos is not considered to be mutagenic or carcinogenic (NTP, 1992; U.S.EPA, 2002). Additional information about external exposure (i.e., environmental levels) and health effects is available from ATSDR at: http://www.atsdr.cdc.gov/toxpro2.html and from U.S. EPA at: http://www.epa.gov/pesticides/.

**Biomonitoring Information**

Urinary TCPy levels reflect recent exposure. Levels of TCPy in the U.S. subsamples of NHANES 1999-2000 and 2001-2002 (CDC, 2005) appear roughly similar to values reported for a nonrandom subsample of NHANES III (1988-1994) participants (Hill et al., 1995) and were similar to levels reported in studies of healthy adults in Germany (Koch et al., 2001) and Italy (Aprea et al., 1999). In a probability-based sample of 102 Minnesota children aged 3-13 years, the weighted population mean of TCPy measurements was approximately three times higher (Adgate, 2001) than the corresponding values reported for the group aged 6-11 years from the NHANES 1999-2000 subsample (CDC, 2005). MacIntosh et al. (1999) reported mean urinary TCPy levels in a sample of Maryland adults that were about three times higher than adults in the U.S. population (CDC, 2005). Of 482 pregnant women living in an agricultural community, 76% had detectable levels of TCPy and levels were similar to those reported for NHANES 1999-2000 (Eskenazi et al., 2004). Other small studies of environmentally-exposed persons have shown a high frequency of detecting low levels of TCPy.

Following crack-and-crevice application of chlorpyrifos in their homes, urinary TCPy levels in children were reported not to have increased (Hore et al., 2005). Chlorpyrifos levels in house dust and hand rinses did not correlate with levels of TCPy in urine (Lioy et al., 2000). Replacing conventional diets with organic diets in 23 children led to about a fourfold decrease in urinary levels of chlorpyrifos; median urinary levels on the conventional diet were several fold higher than those in the NHANES 1999-2000 subsample (Lu et al., 2006). Measurements of urinary TCPy in single spot urine collections show variability over time in environmentally exposed individuals and are poorly correlated between collections, suggesting changing low-level exposure and variance in collection timing with respect to exposure (Meeker et al., 2005). Estimation of dose or intake based on the urinary excretion of TCPy indicates that environmental doses are generally below recommended limits (Hore et al., 2005; Koch et al., 2001).

In Iowa farm families using several different pesticides, but not chlorpyrifos, the geometric mean urinary TCPy levels were similar in parents and children, but levels were roughly four to six times higher than the geometric means in the U.S. representative subsample of NHANES 1999-2000 (CDC, 2005; Curwin et al., 2007). In Minnesota and South Carolina farmers who used chlorpyrifos, urinary TCPy levels averaged about sixfold higher than those in the NHANES 1999-2000 subsample (Mandel et al., 2005; CDC, 2005). Urinary levels of TCPy have been found to be hundredsfold higher for chlorpyrifos manufacturing workers (Burns et al., 2006) and episodically many times higher for pesticide applicators than median levels from NHANES 1999-2000 (CDC, 2005).

Finding a measurable amount of TCPy in urine does not mean that the level will result in an adverse health effect. Biomonitoring studies of TCPy provide physicians and public health officials with reference values so that they can determine whether people have been exposed to higher levels of chlorpyrifos or chlorpyrifos-methyl than are found in the general population. Biomonitoring data can also help scientists plan and conduct research on exposure and health effects.

*Source: CDC, Fourth National Report on Human Exposure to Environmental Chemicals, 2009*

## Lead

CAS No. 7439-92-1

### General Information

Elemental lead is a soft, malleable, dense, blue-gray metal that occurs naturally in soils and rocks. Lead is most often mined from ores or recycled from scrap metal or batteries. Elemental lead can be combined with other elements to form inorganic and organic compounds, such as lead phosphate and tetraethyl lead. Lead has a variety of uses in manufacturing: storage batteries, solders, metal alloys (e.g. brass, bronze), plastics, leaded glass, ceramic glazes, ammunition, antique-molded or cast ornaments, and for radiation shielding. In the past, lead was added to gasoline and residential paints and used in soldering the seams of food cans. Lead was used in plumbing for centuries and may still be present.

Before the 1980's, the main source of lead exposure for the general U.S. population was aerosolized lead emitted from combustion engines that used leaded gasoline. Aerosolized lead is either inhaled or ingested after it is deposited on surfaces and food crops. Since lead has been eliminated from gasoline, adult lead exposures tend to be limited to

### Blood Lead

Geometric mean and selected percentiles of blood concentrations (in µg/dL) for the U.S. population from the National Health and Nutrition Examination Survey.

| | Survey years | Geometric mean (95% conf. interval) | Selected percentiles (95% confidence interval) 50th | 75th | 90th | 95th | Sample size |
|---|---|---|---|---|---|---|---|
| Total | 99-00 | 1.66 (1.60-1.72) | 1.60 (1.60-1.70) | 2.50 (2.40-2.60) | 3.80 (3.60-4.00) | 5.00 (4.70-5.50) | 7970 |
| | 01-02 | 1.45 (1.39-1.51) | 1.40 (1.40-1.50) | 2.20 (2.10-2.30) | 3.40 (3.20-3.60) | 4.50 (4.20-4.70) | 8945 |
| | 03-04 | 1.43 (1.36-1.50) | 1.40 (1.30-1.50) | 2.10 (2.10-2.20) | 3.20 (3.10-3.30) | 4.20 (3.90-4.40) | 8373 |
| **Age group** | | | | | | | |
| 1-5 years | 99-00 | 2.23 (1.96-2.53) | 2.20 (1.90-2.50) | 3.40 (2.80-3.90) | 4.90 (4.00-6.60) | 7.00 (6.10-8.30) | 723 |
| | 01-02 | 1.70 (1.55-1.87) | 1.60 (1.50-1.80) | 2.50 (2.20-2.90) | 4.20 (3.50-5.20) | 5.80 (4.70-6.90) | 898 |
| | 03-04 | 1.77 (1.60-1.95) | 1.70 (1.50-1.90) | 2.50 (2.30-2.80) | 3.90 (3.30-4.60) | 5.10 (4.10-6.60) | 911 |
| 6-11 years | 99-00 | 1.51 (1.36-1.66) | 1.40 (1.30-1.60) | 2.10 (1.80-2.50) | 3.30 (2.80-3.80) | 4.50 (3.40-6.20) | 905 |
| | 01-02 | 1.25 (1.14-1.36) | 1.20 (1.00-1.30) | 1.70 (1.60-2.00) | 2.80 (2.50-3.10) | 3.70 (3.00-4.70) | 1044 |
| | 03-04 | 1.25 (1.12-1.39) | 1.20 (1.10-1.40) | 1.80 (1.50-2.10) | 2.60 (2.10-3.10) | 3.30 (2.50-4.60) | 856 |
| 12-19 years | 99-00 | 1.10 (1.04-1.17) | 1.10 (1.00-1.20) | 1.50 (1.40-1.70) | 2.30 (2.10-2.40) | 2.90 (2.70-3.00) | 2135 |
| | 01-02 | .942 (.899-.986) | .900 (.900-1.00) | 1.30 (1.20-1.40) | 2.00 (1.90-2.10) | 2.70 (2.40-2.90) | 2231 |
| | 03-04 | .946 (.878-1.02) | .900 (.800-1.00) | 1.30 (1.20-1.40) | 1.90 (1.70-2.10) | 2.60 (2.20-3.00) | 2081 |
| 20 years and older | 99-00 | 1.75 (1.68-1.81) | 1.70 (1.60-1.80) | 2.60 (2.50-2.70) | 3.90 (3.70-4.10) | 5.20 (4.80-5.60) | 4207 |
| | 01-02 | 1.56 (1.49-1.62) | 1.60 (1.50-1.60) | 2.30 (2.30-2.40) | 3.60 (3.40-3.70) | 4.60 (4.30-5.00) | 4772 |
| | 03-04 | 1.52 (1.45-1.60) | 1.50 (1.40-1.60) | 2.30 (2.20-2.40) | 3.30 (3.20-3.50) | 4.30 (4.00-4.60) | 4525 |
| **Gender** | | | | | | | |
| Males | 99-00 | 2.01 (1.93-2.09) | 1.90 (1.90-2.00) | 2.90 (2.80-3.00) | 4.50 (4.10-4.80) | 6.00 (5.50-6.50) | 3913 |
| | 01-02 | 1.78 (1.71-1.86) | 1.80 (1.70-1.80) | 2.70 (2.50-2.80) | 3.90 (3.80-4.10) | 5.40 (5.00-5.50) | 4339 |
| | 03-04 | 1.69 (1.62-1.75) | 1.60 (1.50-1.70) | 2.50 (2.40-2.60) | 3.70 (3.40-3.90) | 4.80 (4.50-5.20) | 4132 |
| Females | 99-00 | 1.37 (1.32-1.43) | 1.30 (1.30-1.40) | 2.00 (1.90-2.10) | 3.10 (2.90-3.30) | 4.00 (3.80-4.20) | 4057 |
| | 01-02 | 1.19 (1.14-1.25) | 1.20 (1.10-1.20) | 1.80 (1.70-1.90) | 2.60 (2.50-2.80) | 3.60 (3.10-4.00) | 4606 |
| | 03-04 | 1.22 (1.14-1.31) | 1.20 (1.10-1.30) | 1.80 (1.70-2.00) | 2.70 (2.50-3.00) | 3.50 (3.10-3.80) | 4241 |
| **Race/ethnicity** | | | | | | | |
| Mexican Americans | 99-00 | 1.83 (1.75-1.91) | 1.80 (1.70-1.90) | 2.80 (2.60-2.90) | 4.20 (3.90-4.60) | 5.80 (5.10-6.60) | 2742 |
| | 01-02 | 1.46 (1.34-1.60) | 1.50 (1.30-1.60) | 2.30 (2.10-2.60) | 3.60 (3.40-4.20) | 5.40 (4.40-6.70) | 2268 |
| | 03-04 | 1.55 (1.43-1.69) | 1.50 (1.40-1.60) | 2.30 (2.10-2.50) | 3.50 (2.90-4.20) | 4.90 (3.90-6.40) | 2085 |
| Non-Hispanic blacks | 99-00 | 1.87 (1.75-2.00) | 1.80 (1.70-2.00) | 2.80 (2.60-3.00) | 4.30 (4.00-4.60) | 5.70 (5.20-6.10) | 1842 |
| | 01-02 | 1.65 (1.52-1.80) | 1.60 (1.40-1.70) | 2.60 (2.30-2.90) | 4.20 (3.80-4.70) | 5.80 (5.30-6.50) | 2219 |
| | 03-04 | 1.69 (1.52-1.89) | 1.60 (1.40-1.80) | 2.60 (2.20-3.00) | 4.10 (3.50-4.70) | 5.30 (4.60-6.60) | 2293 |
| Non-Hispanic whites | 99-00 | 1.62 (1.55-1.69) | 1.60 (1.50-1.70) | 2.40 (2.30-2.50) | 3.60 (3.40-3.90) | 5.00 (4.40-5.70) | 2716 |
| | 01-02 | 1.43 (1.37-1.48) | 1.40 (1.30-1.50) | 2.20 (2.10-2.20) | 3.20 (3.10-3.40) | 4.20 (3.90-4.50) | 3806 |
| | 03-04 | 1.37 (1.32-1.43) | 1.30 (1.30-1.40) | 2.10 (2.00-2.10) | 3.00 (2.80-3.20) | 3.90 (3.60-4.30) | 3478 |

Limit of detection (LOD, see Data Analysis section) for Survey years 99-00, 01-02, and 03-04 are 0.3, 0.3, and 0.28, respectively.

occupational (e.g., battery and radiator manufacturing) and recreational sources. However, the primary source of exposure in children is from deteriorated lead-based paint and the resulting dust and soil contamination (Manton et al., 2000). Children may also be exposed to lead brought into the home on the work clothes of adults whose work involves lead. Less common sources of incidental or unique lead exposure are numerous: lead-glazed ceramic pottery; stained glass framing; pewter utensils and drinking vessels; older plumbing systems with leaded pipes or lead soldered connections; lead-based painted surfaces undergoing renovation or demolition; imported children's trinkets and toys; lead-containing folk remedies and cosmetics; bullet fragments retained in human tissue; lead-contaminated dust in indoor firing ranges; and contact with soil, dust, or water contaminated by mining or smelting operations. Small amounts of environmental lead also may result from burning fossil fuels (ATSDR, 2007; CDC, 1991).

Lead is absorbed into the body after fine lead particulates or fumes are inhaled, or after soluble lead compounds are ingested. Absorption of ingested lead can be as much as five times greater in children than adults and even greater when intakes of dietary minerals are deficient. In the blood, absorbed lead is bound to erythrocytes and then is distributed initially to multiple soft tissues and eventually into bone. Approximately half of the absorbed lead may be incorporated into bone, which is the site of approximately

## Urinary Lead

Geometric mean and selected percentiles of urine concentrations (in µg/L) for the U.S. population from the National Health and Nutrition Examination Survey.

| | Survey years | Geometric mean (95% conf. interval) | Selected percentiles (95% confidence interval) 50th | 75th | 90th | 95th | Sample size |
|---|---|---|---|---|---|---|---|
| **Total** | 99-00 | **.766** (.708-.828) | **.800** (.800-.900) | **1.40** (1.30-1.50) | **2.20** (2.00-2.30) | **2.90** (2.60-3.30) | 2465 |
| | 01-02 | **.677** (.637-.718) | **.700** (.700-.800) | **1.20** (1.20-1.30) | **2.00** (1.90-2.20) | **2.70** (2.50-2.80) | 2690 |
| | 03-04 | **.636** (.595-.680) | **.640** (.580-.690) | **1.04** (.960-1.12) | **1.73** (1.52-1.86) | **2.29** (2.03-2.62) | 2558 |
| **Age group** | | | | | | | |
| 6-11 years | 99-00 | **1.07** (.955-1.20) | **1.10** (.900-1.30) | **1.50** (1.40-1.70) | **2.40** (1.80-3.10) | **3.40** (2.40-5.00) | 340 |
| | 01-02 | **.753** (.661-.857) | **.800** (.600-.900) | **1.20** (1.10-1.40) | **2.10** (1.60-2.40) | **2.60** (2.10-3.70) | 368 |
| | 03-04 | **.795** (.671-.941) | **.790** (.640-.900) | **1.35** (.970-1.86) | **2.27** (1.62-4.09) | **3.33** (2.23-4.41) | 290 |
| 12-19 years | 99-00 | **.659** (.579-.749) | **.700** (.600-.800) | **1.10** (.900-1.30) | **1.80** (1.40-2.20) | **2.20** (1.90-2.80) | 719 |
| | 01-02 | **.564** (.526-.605) | **.600** (.500-.600) | **1.00** (.800-1.10) | **1.60** (1.40-1.70) | **2.00** (1.80-2.40) | 762 |
| | 03-04 | **.604** (.553-.660) | **.630** (.570-.680) | **.920** (.810-1.02) | **1.32** (1.14-1.80) | **1.86** (1.44-2.29) | 725 |
| 20 years and older | 99-00 | **.752** (.691-.818) | **.800** (.700-.900) | **1.40** (1.30-1.50) | **2.20** (2.00-2.40) | **2.90** (2.60-3.30) | 1406 |
| | 01-02 | **.688** (.641-.738) | **.700** (.700-.800) | **1.20** (1.20-1.30) | **2.00** (1.90-2.30) | **2.80** (2.50-2.90) | 1560 |
| | 03-04 | **.625** (.579-.674) | **.620** (.560-.700) | **1.04** (.960-1.11) | **1.70** (1.52-1.80) | **2.21** (2.04-2.49) | 1543 |
| **Gender** | | | | | | | |
| Males | 99-00 | **.923** (.822-1.04) | **.900** (.900-1.00) | **1.60** (1.40-1.80) | **2.50** (2.20-2.90) | **3.40** (2.90-3.80) | 1227 |
| | 01-02 | **.808** (.757-.862) | **.800** (.800-.900) | **1.40** (1.30-1.50) | **2.50** (2.20-2.70) | **3.20** (2.90-3.50) | 1335 |
| | 03-04 | **.731** (.680-.785) | **.730** (.680-.800) | **1.17** (1.07-1.27) | **2.03** (1.78-2.22) | **2.66** (2.33-2.91) | 1281 |
| Females | 99-00 | **.642** (.589-.701) | **.700** (.600-.800) | **1.20** (1.10-1.30) | **1.90** (1.60-2.20) | **2.40** (2.10-3.00) | 1238 |
| | 01-02 | **.573** (.535-.613) | **.600** (.600-.600) | **1.10** (1.00-1.10) | **1.60** (1.50-1.80) | **2.20** (1.90-2.40) | 1355 |
| | 03-04 | **.558** (.506-.616) | **.540** (.480-.620) | **.920** (.820-1.04) | **1.49** (1.24-1.75) | **1.82** (1.59-2.30) | 1277 |
| **Race/ethnicity** | | | | | | | |
| Mexican Americans | 99-00 | **1.02** (.915-1.13) | **1.10** (.900-1.20) | **1.80** (1.60-1.90) | **2.90** (2.50-3.40) | **4.30** (3.10-5.40) | 884 |
| | 01-02 | **.833** (.745-.931) | **.900** (.700-1.00) | **1.50** (1.20-1.70) | **2.50** (2.00-2.90) | **3.30** (2.70-3.80) | 683 |
| | 03-04 | **.815** (.710-.935) | **.840** (.700-.990) | **1.31** (1.18-1.59) | **2.19** (1.86-2.50) | **2.66** (2.13-3.97) | 618 |
| Non-Hispanic blacks | 99-00 | **1.11** (1.00-1.23) | **1.10** (1.00-1.20) | **1.90** (1.50-2.10) | **3.00** (2.40-3.50) | **4.20** (3.30-5.70) | 568 |
| | 01-02 | **.940** (.833-1.06) | **.900** (.800-1.00) | **1.60** (1.30-1.80) | **2.70** (2.10-3.40) | **3.70** (2.90-4.80) | 667 |
| | 03-04 | **.848** (.729-.986) | **.850** (.710-1.00) | **1.40** (1.10-1.72) | **2.14** (1.78-2.64) | **2.82** (2.31-3.89) | 723 |
| Non-Hispanic whites | 99-00 | **.695** (.625-.773) | **.700** (.700-.900) | **1.30** (1.10-1.40) | **2.00** (1.80-2.40) | **2.70** (2.30-3.10) | 822 |
| | 01-02 | **.610** (.572-.651) | **.700** (.600-.700) | **1.10** (1.10-1.20) | **1.90** (1.70-2.00) | **2.40** (2.30-2.60) | 1132 |
| | 03-04 | **.591** (.556-.628) | **.590** (.540-.650) | **.960** (.910-.990) | **1.52** (1.40-1.75) | **2.14** (1.78-2.51) | 1074 |

Limit of detection (LOD, see Data Analysis section) for Survey years 99-00, 01-02, and 03-04 are 0.1, 0.1, and 0.33, respectively.

90% of the body lead burden in most adults. The skeleton acts as a storage depot, and approximately 40 to 70% of lead in blood comes from the skeleton in environmentally exposed adults (Smith et al., 1996). Lead can cross the placenta and enter the developing fetal brain. Lead is cleared from the blood and soft tissues with a half-life of 1 to 2 months and more slowly from the skeleton, with a half-life of years to decades. Approximately 70% of lead excretion occurs via the urine, with lesser amounts eliminated via the feces; scant amounts are lost through sweat, hair, and nails (Leggett, 1993; O'Flaherty, 1993).

The toxic effects of lead result from its interference with the physiologic actions of calcium, zinc, and iron, through the inhibition of certain enzymes, and through binding to ion channels and regulatory proteins. Additional mechanisms include generating reactive oxygen species and altering gene expression (ATSDR, 2007). Large amounts of lead in the body can cause anemia, kidney injury, abdominal pain, seizures, encephalopathy, and paralysis. Equilibrated blood lead levels (BLLs) after chronic intake are associated with certain toxic effects. BLLs and associated toxic effects differ in children and adults. For instance, BLLs near 10 µg/dL can affect blood pressure in adults and neurodevelopment in children (Bellinger, 2004; CDC, 1991; Nash et al., 2003; Schwartz, 1995; Staessen et al., 1995). In 1991, based on prospective population studies, the Centers for Disease Control and Prevention (CDC) established a BLL of 10

### Urinary Lead (creatinine corrected)

Geometric mean and selected percentiles of urine concentrations (in µg/g of creatinine) for the U.S. population from the National Health and Nutrition Examination Survey.

| | Survey years | Geometric mean (95% conf. interval) | Selected percentiles (95% confidence interval) 50th | 75th | 90th | 95th | Sample size |
|---|---|---|---|---|---|---|---|
| **Total** | 99-00 | **.721** (.700-.742) | **.701** (.677-.725) | **1.11** (1.05-1.15) | **1.70** (1.62-1.85) | **2.38** (2.22-2.79) | 2465 |
| | 01-02 | **.639** (.603-.677) | **.635** (.588-.676) | **1.03** (.963-1.08) | **1.52** (1.43-1.61) | **2.03** (1.89-2.22) | 2689 |
| | 03-04 | **.632** (.603-.662) | **.622** (.594-.655) | **.979** (.920-1.03) | **1.49** (1.33-1.64) | **1.97** (1.73-2.26) | 2558 |
| **Age group** | | | | | | | |
| 6-11 years | 99-00 | **1.17** (.975-1.41) | **1.06** (.918-1.22) | **1.55** (1.22-1.97) | **2.71** (1.67-4.66) | **4.66** (1.97-18.0) | 340 |
| | 01-02 | **.918** (.841-1.00) | **.870** (.800-.933) | **1.27** (1.12-1.43) | **2.33** (1.59-3.64) | **3.64** (1.89-5.56) | 368 |
| | 03-04 | **.926** (.812-1.06) | **.914** (.781-1.03) | **1.45** (1.17-1.72) | **2.14** (1.62-3.47) | **3.47** (2.19-5.31) | 290 |
| 12-19 years | 99-00 | **.496** (.460-.535) | **.469** (.408-.508) | **.709** (.655-.828) | **1.11** (.981-1.28) | **1.65** (1.15-2.79) | 719 |
| | 01-02 | **.404** (.380-.428) | **.375** (.342-.400) | **.603** (.541-.702) | **.990** (.882-1.18) | **1.41** (1.07-1.63) | 762 |
| | 03-04 | **.432** (.404-.461) | **.404** (.383-.436) | **.623** (.551-.730) | **.938** (.828-1.06) | **1.23** (1.09-1.35) | 725 |
| 20 years and older | 99-00 | **.720** (.683-.758) | **.712** (.667-.739) | **1.10** (1.02-1.18) | **1.69** (1.53-1.87) | **2.31** (2.15-2.62) | 1406 |
| | 01-02 | **.658** (.617-.703) | **.652** (.608-.702) | **1.05** (.992-1.11) | **1.51** (1.40-1.61) | **2.00** (1.85-2.19) | 1559 |
| | 03-04 | **.641** (.606-.679) | **.633** (.605-.670) | **.988** (.917-1.04) | **1.47** (1.28-1.63) | **1.94** (1.72-2.12) | 1543 |
| **Gender** | | | | | | | |
| Males | 99-00 | **.720** (.679-.763) | **.693** (.645-.734) | **1.10** (.992-1.22) | **1.68** (1.50-2.09) | **2.43** (2.15-3.03) | 1227 |
| | 01-02 | **.639** (.607-.673) | **.638** (.586-.686) | **1.01** (.957-1.08) | **1.55** (1.41-1.61) | **2.06** (1.88-2.43) | 1334 |
| | 03-04 | **.615** (.588-.644) | **.593** (.561-.639) | **.914** (.862-.977) | **1.44** (1.25-1.53) | **2.00** (1.71-2.28) | 1281 |
| Females | 99-00 | **.722** (.681-.765) | **.707** (.667-.746) | **1.11** (1.05-1.18) | **1.74** (1.50-2.02) | **2.38** (2.03-2.88) | 1238 |
| | 01-02 | **.639** (.594-.688) | **.625** (.571-.682) | **1.03** (.946-1.11) | **1.50** (1.39-1.61) | **1.98** (1.85-2.15) | 1355 |
| | 03-04 | **.648** (.601-.698) | **.649** (.604-.718) | **1.03** (.938-1.10) | **1.56** (1.34-1.73) | **1.96** (1.72-2.20) | 1277 |
| **Race/ethnicity** | | | | | | | |
| Mexican Americans | 99-00 | **.940** (.876-1.01) | **.887** (.796-1.03) | **1.43** (1.37-1.58) | **2.38** (2.08-2.77) | **3.46** (2.78-4.18) | 884 |
| | 01-02 | **.810** (.731-.898) | **.774** (.702-.893) | **1.29** (1.09-1.44) | **2.05** (1.75-2.50) | **2.78** (2.56-3.33) | 682 |
| | 03-04 | **.755** (.681-.838) | **.708** (.612-.851) | **1.18** (1.09-1.31) | **1.86** (1.50-2.26) | **2.31** (1.98-2.92) | 618 |
| Non-Hispanic blacks | 99-00 | **.722** (.659-.790) | **.671** (.583-.753) | **1.11** (.988-1.20) | **2.00** (1.56-2.51) | **2.83** (2.20-3.88) | 568 |
| | 01-02 | **.644** (.559-.742) | **.608** (.510-.710) | **.962** (.853-1.20) | **1.79** (1.36-2.33) | **2.75** (2.04-3.98) | 667 |
| | 03-04 | **.609** (.529-.701) | **.569** (.492-.698) | **.900** (.793-1.03) | **1.48** (1.11-1.97) | **2.24** (1.65-2.88) | 723 |
| Non-Hispanic whites | 99-00 | **.696** (.668-.725) | **.677** (.645-.718) | **1.07** (.997-1.14) | **1.66** (1.50-1.83) | **2.31** (1.94-2.82) | 822 |
| | 01-02 | **.615** (.579-.654) | **.621** (.571-.667) | **1.00** (.933-1.07) | **1.46** (1.37-1.52) | **1.88** (1.62-2.03) | 1132 |
| | 03-04 | **.623** (.592-.655) | **.618** (.587-.657) | **.971** (.914-1.03) | **1.44** (1.25-1.61) | **1.85** (1.64-2.10) | 1074 |

µg/dL or higher as the level of concern in children. Recent studies have suggested that neurodevelopmental effects may occur at BLLs lower than 10 µg/dL (Canfield et al., 2003; Lanphear et al., 2000). Many animal studies have established the multiple neurotoxic effects of lead (ATSDR, 2007).

In occupationally exposed adults, subtle or nonspecific neurocognitive effects have been reported at BLLs as low as 20-30 µg/dL (Mantere et al., 1984; Schwartz et al., 2001), with overt encephalopathy, seizures, and peripheral neuropathy generally occurring at much higher levels (e.g., higher than 100-200 µg/dL). BLLs higher than 40 µg/dL can result in proximal tubular dysfunction and decreased glomerular filtration rate leading to interstitial and peritubular fibrosis when high body burdens persist. Low level environmental lead exposure may be associated with small decrements in renal function (Kim et al., 1996; Muntner et al., 2003; Payton et al., 1994). Results of studies of adults with either occupational or environmental lead exposure have shown consistent associations between increased BLLs and increased blood pressure (Nash et al., 2003; Schwartz, 1995; Staessen et al., 1995) and associations between increased bone lead concentrations and blood pressure (Hu et al., 1996; Korrick et al., 1999). High dose occupational lead exposure, usually with BLLs greater than 40 µg/dL, may alter sperm morphology, reduce sperm count, and decrease fertility (Alexander et al., 1996; Telisman et al., 2000). At low environmental exposures, lead in women may be associated with hypertension during pregnancy, premature delivery, and spontaneous abortion (Baghurst et al., 1987; Bellinger 2005; Borja-Aburto et al., 1999).

Workplace standards and guidelines for lead exposure and monitoring have been established by OSHA and ACGIH, respectively. Both drinking water and ambient air standards for lead have been established by the U.S. EPA. IARC considers inorganic lead compounds probable human carcinogens, and organic lead compounds not classifiable with respect to human carcinogenicity. NTP considers lead and its compounds reasonably anticipated to be human carcinogens. Information about external exposure (i.e., environmental levels) and health effects is available from ATSDR at: http://www.atsdr.cdc.gov/toxpro2.html.

#### Biomonitoring Information

Blood lead measurement is the preferred method of evaluating lead exposure and its human health effects. BLLs reflect both recent intake and equilibration with stored lead in other tissues, particularly in the skeleton. Urine levels may reflect recently absorbed lead, though there is greater individual variation in urine lead than in blood and greater potential for contamination.

The Adult Blood Lead Epidemiology and Surveillance program has tracked BLLs reported by states for mostly for occupational but also for non-occupational exposure in U.S. adult residents. Overall, the national prevalence rate for adults with BLLs 25 µg/dL or higher was 7.5 per 100,000 adults; the prevalence rate has declined annually since 1994 (CDC, 2006). A decrease in BLLs is evident also in adult NHANES results reported over past decades (CDC, 2005*a*). The U.S. adult population has similar or slightly lower BLLs than adults in other developed nations (CDC, 2005*b*). A general population survey of adults Germany in 1998 reported a geometric mean blood lead concentration of 3.07 µg/dL (Becker et al., 2002), almost double the geometric mean of 1.75 µg/dL in U.S. adults in the 1999-2000 NHANES sample. A general population survey of adults in Italy tested in 2000 found BLLs slightly more than double those reported for U.S. adults in the 1999-2000 NHANES sample (Apostoli et al., 2002a).

In NHANES 1999-2002 in children 1-5 years old, both the geometric mean (1.9 µg/dL) and percentage of children with BLLs greater than 10 µg/dL (1.6%) were lower than those from NHANES 1991-1994, when the geometric mean BLL was 2.7 µg/dL and 4.4% of children had BLLs of 10µg/dL or higher (CDC, 2005*b*; Pirkle et al., 1998). More recently, Jones et al (2009) showed that the prevalence of BLLs of 10 µg/dL or greater decreased from 8.6% in NHANES 1988-1991 to 1.4% in NHANES 1999-2004, which is an 84% decline. Temporal declines in children's BLLs have been found in other developed countries (Wilhelm et al., 2006). Surveillance data reported by U.S. state childhood lead programs also show a decline in the percentage of children younger than 6 years of age who had BLLs of 10 µg/dL or higher. Data submitted through state public health programs from 2006 showed that 1.21% of approximately 3.3 million children tested had BLLs of 10 µg/dL or higher (http://www.cdc.gov/nceh/lead/surv/database/State_Confirmed_byYear_1997_to_2006.xls). However, BLLs greater than 10 µg/dL continue to be more prevalent among children with known risk factors, including minority race or ethnicity; urban residence; residing in housing built before the 1950's; and low family income (CDC, 1991; CDC, 2002; Jones et al., 2009). For example, approximately 11,000 higher-risk children and adolescents who were tested from 2001 to 2002 at an urban medical center had higher BLLs than the NHANES sample; the geometric mean BLL was 3.2 µg/dL in males and 3.0 µg/dL in females (Soldin et al., 2003).

*Source: CDC, Fourth National Report on Human Exposure to Environmental Chemicals, 2009*

## Mercury

CAS No. 7439-97-6

### General Information

Mercury is a naturally occurring metal that has elemental (metallic), inorganic, and organic forms. Elemental mercury is a shiny, silver-white liquid (quicksilver) obtained predominantly from the refining of mercuric sulfide in cinnabar ore. Elemental mercury is used to produce chlorine gas and caustic soda for industrial applications. Other major uses include electrical equipment (e.g., thermostats and switches), electrical lamps, thermometers, sphygmomanometers and barometers, and dental amalgam. Inhalation of elemental mercury volatilized from dental amalgam is a major source of mercury exposure in the general population (Halbach, 1994; Kingman et al., 1998; Woods et al., 2007). Accidental spills of elemental mercury, which create an episodic potential for volatization and inhalation of mercury vapor, have often required public health intervention (Zeitz et al., 2002). Also, elemental mercury is used in rituals practiced in some Latin American and Caribbean communities.

Elemental mercury is released into the air from the combustion of fossil fuels (primarily coal), solid-waste incineration, and mining and smelting. Atmospheric elemental mercury can be deposited on land and water. In addition, water can be contaminated by the direct release of elemental and inorganic mercury from industrial discharges. Metabolism of mercury by microorganisms in aquatic sediments creates methyl mercury, an organic form of mercury, which can bioaccumulate in aquatic and terrestrial food chains. The ingestion of methyl mercury, predominantly from fish and other seafood, constitutes the main source of dietary mercury exposure in the general population. Apart from methyl mercury, synthetic organomercury compounds were once used in pharmaceutical applications, and mercury compounds are still used as preservatives (e.g., thimerosal, phenylmercuric acetate) or topical antiseptics (e.g., merbromin).

Inorganic mercury exists in two oxidative states (mercurous and mercuric) that combine with other elements, such as chlorine (e.g., mercuric chloride), sulfur, or oxygen, to form inorganic mercury compounds or salts. Inorganic mercury compounds such as mercuric oxide are used in producing batteries and pigments and in synthesizing many organic chemicals. Some cosmetic skin creams from countries other than the U.S. may contain inorganic mercury. Imported folk and alternative medicines occasionally are contaminated with inorganic mercury.

The kinetics of the different forms of mercury vary considerably. Poorly absorbed from the gastrointestinal tract, elemental mercury is absorbed mainly by inhaling volatilized vapor, and is distributed to most tissues, with the highest concentrations occurring in the kidneys (Barregard et al., 1999 ; Hursh et al., 1980; IARC, 1993). After elemental mercury is absorbed, it is oxidized in

### Total Blood Mercury—2003-2004

Geometric mean and selected percentiles of blood concentrations (in µg/L) for the U.S. population from the National Health and Nutrition Examination Survey.

| | Survey years | Geometric mean (95% conf. interval) | Selected percentiles (95% confidence interval) 50th | 75th | 90th | 95th | Sample size |
|---|---|---|---|---|---|---|---|
| **Total** | 03-04 | **.797** (.703-.903) | **.800** (.700-.900) | **1.70** (1.50-1.90) | **3.30** (2.90-3.90) | **4.90** (4.30-5.50) | 8373 |
| **Age group** | | | | | | | |
| 1-5 years | 03-04 | **.326** (.285-.372) | **.300** (.300-.300) | **.500** (.500-.700) | **1.00** (.800-1.60) | **1.80** (1.30-2.50) | 911 |
| 6-11 years | 03-04 | **.419** (.363-.484) | **.400** (.400-.500) | **.700** (.700-.900) | **1.30** (1.00-1.60) | **1.90** (1.40-3.50) | 856 |
| 12-19 years | 03-04 | **.490** (.418-.574) | **.500** (.400-.600) | **1.00** (.800-1.20) | **1.80** (1.40-2.30) | **2.60** (2.10-3.30) | 2081 |
| 20 years and older | 03-04 | **.979** (.860-1.12) | **1.00** (.800-1.10) | **2.00** (1.70-2.30) | **3.80** (3.20-4.40) | **5.40** (4.60-6.70) | 4525 |
| **Gender** | | | | | | | |
| Males | 03-04 | **.814** (.714-.927) | **.800** (.700-.900) | **1.80** (1.50-2.00) | **3.70** (3.20-4.30) | **5.40** (4.60-6.50) | 4132 |
| Females | 03-04 | **.781** (.689-.886) | **.800** (.700-.900) | **1.60** (1.40-1.80) | **3.00** (2.50-3.50) | **4.40** (3.60-5.30) | 4241 |
| **Race/ethnicity** | | | | | | | |
| Mexican Americans | 03-04 | **.563** (.472-.672) | **.600** (.500-.700) | **1.00** (.800-1.30) | **1.90** (1.60-2.40) | **3.00** (2.20-3.80) | 2085 |
| Non-Hispanic blacks | 03-04 | **.877** (.753-1.02) | **.900** (.800-1.00) | **1.60** (1.40-1.80) | **3.00** (2.30-4.00) | **4.40** (3.30-6.00) | 2293 |
| Non-Hispanic whites | 03-04 | **.776** (.655-.919) | **.800** (.700-.900) | **1.70** (1.40-2.00) | **3.20** (2.60-3.90) | **4.70** (4.00-5.60) | 3478 |

Limit of detection (LOD, see Data Analysis section) for Survey year 03-04 is 0.2.

the tissues to mercurous and mercuric inorganic forms. Blood concentrations decline initially with a rapid half-life of approximately 1-3 days followed by a slower half-life of approximately 1-3 weeks (Barregard et al., 1992; Sandborgh-Englund et al., 1998). The slow-phase half-life may be several weeks longer in persons with chronic occupational exposure (Sallsten et al., 1993). After exposure to elemental mercury, excretion of mercury occurs predominantly through the kidney (Sandborgh-Englund et al., 1998), and peak urine levels can lag behind peak blood levels by days to a few weeks (Barregard et al., 1992); thereafter, for both acute and chronic exposures, urinary mercury levels decline with a half-life of approximately 1-3 months (Roels et al., 1991; Jonsson et al., 1999).

Less than 15% of inorganic mercury is absorbed from the human gastrointestinal tract (Rahola et al., 1973). Lesser penetration of inorganic mercury occurs through the blood-brain barrier than occurs with either elemental or methyl mercury (Hattula and Rahola, 1975; Vahter et al., 1994). The half-life of inorganic mercury in blood is similar to the slow-phase half-life of mercury after inhalation of elemental mercury. Excretion occurs by renal and fecal routes.

The fraction of methyl mercury absorbed from the gastrointestinal tract is about 95% (Aberg et al., 1969; Miettinen et al., 1971). Methyl mercury enters the brain and other tissues (Vahter et al., 1994) and then undergoes slow dealkylation to inorganic mercury. Human pharmacokinetic studies indicate that methyl mercury declines in blood and the whole body with a half-life of approximately 50 days, with most elimination occurring through in the feces (Sherlock et al., 1984; Smith et al., 1994; Smith and Farris, 1996). Methyl mercury is incorporated into growing hair, a measure of accumulated dose (Cernichiari et al., 1995; Suzuki et al., 1993), and a useful marker of exposure in epidemiologic studies (Grandjean et al., 1992 and 1999; McDowell et al., 2004; Myers et al., 2003).

Transplacental transport of methyl mercury and elemental mercury has been demonstrated in animals (Kajiwara et al., 1996; Vimy et al., 1990). Mercury levels in the cord blood are higher than in the mother's blood (Stern and Smith, 2003), and the newborn's levels decline gradually over several weeks (Bjornberg et al., 2005). Inorganic mercury and methyl mercury are distributed into human breast milk in relatively low concentrations; the transfer

## Total Blood Mercury–1999-2002

Geometric mean and selected percentiles of blood concentrations (in µg/L) for males and females aged 1 to 5 years and females aged 16 to 49 years in the U.S. population, National Health and Nutrition Examination Survey, 1999-2002.

| | Survey years | Geometric mean (95% conf. interval) | Selected percentiles (95% confidence interval) 50th | 75th | 90th | 95th | Sample size |
|---|---|---|---|---|---|---|---|
| **Age Group** | | | | | | | |
| 1-5 years (females and males) | 99-00 | **.343** (.297-.395) | **.300** (.200-.300) | **.500** (.500-.600) | **1.40** (1.00-2.30) | **2.30** (1.20-3.50) | 705 |
| | 01-02 | **.318** (.268-.377) | **.300** (.200-.300) | **.700** (.500-.800) | **1.20** (.900-1.60) | **1.90** (1.40-2.90) | 872 |
| Females | 99-00 | **.377** (.299-.475) | **.200** (.200-.300) | **.800** (.500-1.10) | **1.60** (1.00-2.80) | **2.70** (1.30-5.50) | 318 |
| | 01-02 | **.329** (.265-.407) | **.300** (.200-.300) | **.700** (.500-.800) | **1.30** (1.00-2.10) | **2.60** (1.30-4.90) | 432 |
| Males | 99-00 | **.317** (.269-.374) | **.200** (.200-.300) | **.500** (.500-.600) | **1.10** (.800-1.60) | **2.10** (1.10-3.50) | 387 |
| | 01-02 | **.307** (.256-.369) | **.300** (.200-.300) | **.600** (.400-.700) | **1.30** (.900-1.70) | **1.70** (1.40-2.00) | 440 |
| 16-49 years (females only) | 99-00 | **1.02** (.825-1.27) | **.900** (.800-1.20) | **2.00** (1.50-3.00) | **4.90** (3.70-6.30) | **7.10** (5.30-11.3) | 1709 |
| | 01-02 | **.833** (.738-.940) | **.700** (.700-.800) | **1.70** (1.40-1.90) | **3.00** (2.70-3.50) | **4.60** (3.70-5.90) | 1928 |
| **Race/ethnicity** (females, 16-49 years) | | | | | | | |
| Mexican Americans | 99-00 | **.820** (.664-1.01) | **.900** (.700-1.00) | **1.40** (1.20-2.00) | **2.60** (2.00-3.60) | **4.00** (2.70-5.50) | 579 |
| | 01-02 | **.667** (.541-.824) | **.700** (.500-.800) | **1.10** (1.00-1.40) | **2.10** (1.70-3.00) | **3.50** (2.30-4.40) | 527 |
| Non-hispanic blacks | 99-00 | **1.35** (1.06-1.73) | **1.30** (1.10-1.70) | **2.60** (1.80-3.40) | **4.80** (3.30-6.60) | **5.90** (4.20-11.7) | 370 |
| | 01-02 | **1.06** (.871-1.29) | **1.10** (.800-1.20) | **1.80** (1.50-2.20) | **3.20** (2.20-3.90) | **4.10** (3.30-6.00) | 436 |
| Non-hispanic whites | 99-00 | **.944** (.726-1.23) | **.900** (.700-1.10) | **1.90** (1.30-3.30) | **5.00** (3.00-6.90) | **6.90** (4.50-12.0) | 588 |
| | 01-02 | **.800** (.697-.919) | **.800** (.700-.800) | **1.50** (1.30-2.00) | **3.00** (2.20-3.70) | **4.60** (3.30-6.80) | 806 |

Limit of detection (LOD, see Data Analysis section) for Survey years 99-00 and 01-02 are 0.14 and 0.14.

may be more efficient for inorganic mercury (Grandjean et al., 1995; Oskarsson et al., 1996). Mercury levels in breast milk also decline in the weeks after birth (Bjornberg et al., 2005; Drexler and Schaller, 1998; Sakamoto et al., 2002; Sakamoto et al., 2004).

The health effects of mercury are diverse and can depend on the form of the mercury to which a person is exposed and the dose and the duration of exposure. Acute, high-dose exposure to elemental mercury vapor may cause severe pneumonitis. At levels below those that cause acute lung injury, overt signs and symptoms of chronic inhalation may include tremor, gingivitis, and neurocognitive and behavioral disturbances, particularly irritability, depression, short-term memory loss, fatigue, anorexia, and sleep disturbance (Bidstrup et al., 1951; Smith et al., 1970; Smith et al., 1983). Low-level exposure from dental amalgams has not been associated with neurologic effects in children or adults (Bates et al., 2004; Bellinger et al., 2006; DeRouen et al., 2006; Factor-Litvak et al., 2003). Occupational exposure to elemental mercury vapor has been associated with subclinical effects on biomarkers of renal dysfunction (Cardenas et al., 1993).

Inorganic mercury exposure usually occurs by ingestion. Large amounts may cause irritant or corrosive effects on the gastrointestinal tract (Sanchez-Sicilia et al., 1963). Once absorbed, the most prominent effect is on the kidneys where mercury accumulates and may lead to renal tubular necrosis. Acrodynia is a sporadic and predominantly pediatric syndrome historically associated with calomel (mercuric oxide) in teething powders and occasionally other inorganic forms of mercury. The constellation of findings may include anorexia, insomnia, irritability, hypertension, maculopapular rash, pain in the extremities, and pinkish discoloration of the hands and feet (Tunnessen et al., 1987).

Overt poisoning from methyl mercury primarily affects the central nervous system, causing parasthesias, ataxia, dysarthria, hearing impairment, and progressive constriction of the visual fields, typically after a latent period of weeks to months. High-level prenatal exposure may result in a constellation of developmental deficits that includes mental retardation, cerebellar ataxia, dysarthria, limb deformities, altered physical growth, sensory impairments, and cerebral palsy (NRC, 2000). In recent epidemiologic studies, lower levels of prenatal exposure due to maternal seafood consumption have been associated with an increased risk for abnormal neurocognitive test results in children (NRC, 2000; Rice, 2004). Although recent investigations have suggested a possible link between chronic ingestion of methyl mercury and an increased risk for cardiovascular disease, the existence of a causal relation is unresolved (Chan and Egeland, 2004; Rissanen et al., 2000; Salonen et al., 1995; Stern 2005; Vupputuri et al., 2005).

Workplace standards for inorganic mercury exposure have been established by OSHA and ACGIH, and a drinking water

### Inorganic Blood Mercury

Geometric mean and selected percentiles of blood concentrations (in µg/L) for the U.S. population from the National Health and Nutrition Examination Survey.

| | Survey years | Geometric mean (95% conf. interval) | Selected percentiles (95% confidence interval) 50th | 75th | 90th | 95th | Sample size |
|---|---|---|---|---|---|---|---|
| **Total** | 03-04 | * | < LOD | < LOD | .600 (.500-.600) | .700 (.700-.700) | 8147 |
| **Age group** | | | | | | | |
| 1-5 years | 03-04 | * | < LOD | < LOD | < LOD | .500 (<LOD-.600) | 792 |
| 6-11 years | 03-04 | * | < LOD | < LOD | < LOD | .600 (.500-.600) | 842 |
| 12-19 years | 03-04 | * | < LOD | < LOD | .500 (<LOD-.500) | .600 (.500-.600) | 2060 |
| 20 years and older | 03-04 | * | < LOD | < LOD | .600 (.500-.600) | .700 (.700-.800) | 4453 |
| **Gender** | | | | | | | |
| Males | 03-04 | * | < LOD | < LOD | .500 (.500-.600) | .600 (.600-.700) | 4015 |
| Females | 03-04 | * | < LOD | < LOD | .600 (.500-.600) | .700 (.700-.800) | 4132 |
| **Race/ethnicity** | | | | | | | |
| Mexican Americans | 03-04 | * | < LOD | < LOD | .500 (.500-.600) | .700 (.600-.800) | 2007 |
| Non-Hispanic blacks | 03-04 | * | < LOD | < LOD | .600 (.500-.600) | .700 (.600-.800) | 2240 |
| Non-Hispanic whites | 03-04 | * | < LOD | < LOD | .600 (.500-.600) | .700 (.600-.700) | 3406 |

Limit of detection (LOD, see Data Analysis section) for Survey year 03-04 is 0.42.
< LOD means less than the limit of detection, which may vary for some chemicals by year and by individual sample.
* Not calculated: proportion of results below limit of detection was too high to provide a valid result.

standard for inorganic mercury has been established by U.S. EPA. IARC considers methylmercury to be a possible human carcinogen and elemental and inorganic mercury to be unclassifiable with regard to human carcinogenicity. Information about external exposure (i.e., environmental levels) and health effects is available from the U.S. EPA at: http://www.epa.gov/mercury and from ATSDR at: http://www.atsdr.cdc.gov/toxprofiles.

**Biomonitoring Information**

In the general population, the total blood mercury concentration is due mostly to the dietary intake of organic forms, particularly methyl mercury. Urinary mercury consists mostly of inorganic mercury (Cianciola et al., 1997; Kingman et al., 1998). These distinctions can help interpret mercury blood levels in people. Total blood mercury levels increase with greater fish consumption (Dewailly et al., 2001; Grandjean et al., 1995; Mahaffey et al., 2004; Sanzo et al., 2001; Schober et al., 2003). Urine mercury levels increase as more occlusal surfaces of teeth are filled with mercury-containing amalgams (Becker et al., 2003).

In Germany the geometric mean for blood mercury was 0.58 µg/L for 4645 adults, aged 18 to 69 years, who participated in a 1998 representative population survey (Becker et al., 2002). From 1996 through 1998, Benes et al. (2000) studied 1216 blood donors in the Czech Republic (896 men and 320 women, average age 33 years; 758 children, average age 9.9 years); the median concentration of blood mercury was 0.78 µg/L for adults and 0.46 µg/L for children. A cohort of 1127 U.S. military veterans (mean age 52.8 years, range 40 years to 78 years) had an average total blood mercury concentration of 2.55 µg/L. These men had no occupational exposure to mercury but previously had received dental amalgams at military facilities (Kingman et al., 1998).

Over the NHANES 1999-2006 survey periods, total blood mercury geometric mean levels in females aged 16-49 years did not change, although non-Hispanic black females had higher levels than non-Hispanic white or Mexican American females. Among the three racial/ethnic groups, total blood mercury increased with age, and the age-related changes differed across the groups (Caldwell et al., 2009). During the same survey periods, total blood mercury levels declined slightly in non-Hispanic black and Mexican American children, and increased slightly in non-Hispanic white children (Caldwell, et al., 2009). In NHANES 1999-2002, slightly higher total blood mercury levels were found in U.S. adult women in several ethnic subgroups (Hightower et al., 2006).

Clinically observable signs of ataxia and paresthesias may occur when blood mercury levels increase to approximately 100 µg/L after methyl mercury poisoning. However, the developing fetus may be the most susceptible to the effects of ongoing methyl mercury exposure (NRC, 2000). A cord blood mercury level of 85 µg/L (lower 95% confidence bound = 58 µg/L) is associated with a 5% increase in the prevalence of an abnormal Boston Naming Test (NRC,

## Urinary Mercury—2003-2004

Geometric mean and selected percentiles of urine concentrations (in µg/L) for the U.S. population from the National Health and Nutrition Examination Survey.

| | Survey years | Geometric mean (95% conf. interval) | Selected percentiles (95% confidence interval) 50th | 75th | 90th | 95th | Sample size |
|---|---|---|---|---|---|---|---|
| **Total** | 03-04 | **.447** (.406-.492) | **.420** (.360-.480) | **1.00** (.870-1.14) | **2.08** (1.78-2.42) | **3.19** (2.76-3.55) | 2538 |
| **Age group** | | | | | | | |
| 6-11 years | 03-04 | **.254** (.213-.304) | **.200** (.160-.250) | **.440** (.330-.580) | **1.16** (.610-1.61) | **1.96** (1.13-2.97) | 287 |
| 12-19 years | 03-04 | **.358** (.313-.408) | **.330** (.290-.370) | **.700** (.530-.840) | **1.60** (1.14-2.52) | **2.93** (1.88-3.66) | 722 |
| 20 years and older | 03-04 | **.495** (.442-.555) | **.480** (.410-.570) | **1.12** (.930-1.29) | **2.20** (1.85-2.65) | **3.33** (2.76-3.88) | 1529 |
| **Gender** | | | | | | | |
| Males | 03-04 | **.433** (.405-.463) | **.400** (.350-.460) | **.940** (.840-1.05) | **1.88** (1.63-2.18) | **2.68** (2.34-3.05) | 1266 |
| Females | 03-04 | **.460** (.396-.534) | **.430** (.330-.530) | **1.07** (.870-1.28) | **2.26** (1.77-2.90) | **3.54** (2.76-4.31) | 1272 |
| **Race/ethnicity** | | | | | | | |
| Mexican Americans | 03-04 | **.416** (.340-.509) | **.360** (.280-.430) | **.960** (.700-1.23) | **2.19** (1.39-3.24) | **3.16** (1.99-6.30) | 619 |
| Non-Hispanic blacks | 03-04 | **.476** (.413-.549) | **.430** (.360-.530) | **.890** (.770-1.00) | **1.96** (1.60-2.31) | **3.09** (2.03-4.89) | 713 |
| Non-Hispanic whites | 03-04 | **.441** (.382-.509) | **.420** (.330-.520) | **1.01** (.840-1.23) | **2.08** (1.67-2.46) | **3.24** (2.67-3.60) | 1066 |

Limit of detection (LOD, see Data Analysis section) for Survey year 03-04 is 0.14.

2000). Levels in U.S. women of childbearing age have generally been much lower than these levels (CDC, 2005). ACGIH recommends that the blood levels due to inorganic mercury exposure in workers not exceed 15 µg/L. Blood mercury levels of women and children in NHANES 1999-2006 were also below levels established as occupational exposure guidelines (Caldwell, et al., 2009). Information about the biological exposure indices is provided here for comparison, not to imply a safety level for general population exposure.

Urinary mercury levels in recent German (Becker et al., 2003), Czech (Benes et al., 2002), and Italian (Apostoli et al., 2002) adult population surveys were similar to those in a U.S. representative sample of women aged 16-49 years reported in NHANES 1999-2006 (Caldwell, et al., 2009). In the study of U.S. military veterans with dental amalgams, mean urinary mercury was 3.1 µg/L. Urine mercury and the number of dental amalgams were correlated, and on average, the urine mercury increased by approximately 0.1 µg/L for each surface with a dental amalgam (Kingman et al., 1998). Recent studies in children with dental amalgams and urinary levels less than 5 µg/g of creatinine did not have changes in cognitive-behavioral testing when followed for 5-7 years (Bellinger et al., 2006; DeRouen et al., 2006). An expert-panel report recently prepared for the U.S. Department of Health and Human Services noted that several studies have observed a modest, reversible increase in urinary N-acetyl-glucosaminidase, a biomarker of perturbation in renal tubular function, among workers with urinary mercury concentrations of 25-35 µg/L or greater (Barregard et al., 1988; Langworth et al., 1992). The ACGIH (2007) currently recommends that urinary inorganic mercury in workers not exceed 35 µg/g of creatinine.

Finding a measurable amount of mercury in blood or urine does not mean that the level of mercury causes an adverse health effect. Biomonitoring studies provide physicians and public health officials with reference ranges so that they can determine whether people have been exposed to higher levels of mercury than are found in the general population. Biomonitoring data will also help scientists plan and conduct research on exposure and health effects.

## Urinary Mercury (creatinine corrected)—2003-2004

Geometric mean and selected percentiles of urine concentrations (in µg/g of creatinine) for the U.S. population from the National Health and Nutrition Examination Survey.

| | Survey years | Geometric mean (95% conf. interval) | Selected percentiles (95% confidence interval) 50th | 75th | 90th | 95th | Sample size |
|---|---|---|---|---|---|---|---|
| **Total** | 03-04 | **.443** (.404-.486) | **.447** (.392-.498) | **.909** (.785-1.00) | **1.65** (1.40-1.86) | **2.35** (1.88-2.85) | 2537 |
| **Age group** | | | | | | | |
| 6-11 years | 03-04 | **.297** (.246-.358) | **.276** (.208-.347) | **.485** (.391-.630) | **1.25** (.667-1.79) | **1.79** (1.11-2.61) | 286 |
| 12-19 years | 03-04 | **.255** (.225-.289) | **.217** (.196-.275) | **.464** (.376-.535) | **1.06** (.714-1.39) | **1.67** (1.13-2.03) | 722 |
| 20 years and older | 03-04 | **.508** (.455-.566) | **.525** (.447-.616) | **1.00** (.875-1.09) | **1.76** (1.46-2.11) | **2.54** (2.04-3.00) | 1529 |
| **Gender** | | | | | | | |
| Males | 03-04 | **.365** (.333-.400) | **.362** (.309-.417) | **.696** (.620-.784) | **1.31** (1.18-1.44) | **1.87** (1.51-2.30) | 1266 |
| Females | 03-04 | **.532** (.472-.599) | **.545** (.455-.652) | **1.06** (.969-1.21) | **1.88** (1.64-2.30) | **2.77** (2.12-3.56) | 1271 |
| **Race/ethnicity** | | | | | | | |
| Mexican Americans | 03-04 | **.384** (.307-.480) | **.365** (.280-.455) | **.768** (.619-.990) | **1.62** (1.23-2.16) | **2.32** (1.78-4.01) | 618 |
| Non-Hispanic blacks | 03-04 | **.343** (.301-.391) | **.306** (.265-.368) | **.587** (.522-.687) | **1.28** (.964-1.63) | **2.13** (1.41-2.87) | 713 |
| Non-Hispanic whites | 03-04 | **.463** (.400-.537) | **.476** (.385-.588) | **.970** (.800-1.07) | **1.67** (1.32-2.11) | **2.40** (1.88-2.90) | 1066 |

## Urinary Mercury–Females Aged 16-49 Years Old, 1999-2002

Geometric mean and selected percentiles of urine concentrations (in µg/L) for females aged 16 to 49 years in the U.S. population, National Health and Nutrition Examination Survey, 1999-2002.

| | Survey years | Geometric mean (95% conf. interval) | Selected percentiles (95% confidence interval) 50th | 75th | 90th | 95th | Sample size |
|---|---|---|---|---|---|---|---|
| **Age group** (females) | | | | | | | |
| 16-49 years | 99-00 | **.719** (.622-.831) | **.760** (.610-.910) | **1.62** (1.43-1.94) | **3.15** (2.55-3.92) | **5.00** (3.59-5.79) | 1748 |
| | 01-02 | **.606** (.553-.665) | **.580** (.500-.670) | **1.37** (1.23-1.55) | **2.91** (2.53-3.17) | **3.99** (3.50-4.63) | 1960 |
| **Race/ethnicity** (females, 16-49 years) | | | | | | | |
| Mexican Americans | 99-00 | **.724** (.656-.799) | **.650** (.560-.810) | **1.69** (1.45-2.07) | **3.68** (3.10-4.45) | **5.62** (4.91-7.38) | 595 |
| | 01-02 | **.592** (.502-.699) | **.560** (.420-.710) | **1.35** (1.09-1.76) | **2.84** (2.32-3.85) | **4.13** (2.81-6.24) | 531 |
| Non-Hispanic blacks | 99-00 | **1.06** (.832-1.35) | **1.03** (.850-1.51) | **2.30** (1.83-3.03) | **4.81** (3.41-6.18) | **6.98** (5.04-10.3) | 381 |
| | 01-02 | **.772** (.616-.966) | **.740** (.540-.930) | **1.76** (1.30-2.37) | **3.50** (2.57-4.97) | **5.18** (3.61-6.92) | 442 |
| Non-Hispanic whites | 99-00 | **.657** (.557-.774) | **.710** (.520-.870) | **1.50** (1.31-1.77) | **2.84** (2.39-3.32) | **4.05** (3.16-5.52) | 594 |
| | 01-02 | **.565** (.501-.637) | **.540** (.450-.650) | **1.31** (1.09-1.56) | **2.70** (2.22-3.16) | **3.62** (3.13-4.54) | 826 |

Limit of detection (LOD, see Data Analysis section) for Survey years 99-00 and 01-02 are 0.14 and 0.14.

## Urinary Mercury (creatinine corrected)–Females Aged 16-49 Years Old, 1999-2002

Geometric mean and selected percentiles of urine concentrations (in µg/g of creatinine) for females aged 16 to 49 years in the U.S. population, National Health and Nutrition Examination Survey, 1999-2002.

| | Survey years | Geometric mean (95% conf. interval) | Selected percentiles (95% confidence interval) 50th | 75th | 90th | 95th | Sample size |
|---|---|---|---|---|---|---|---|
| **Age group** (females) | | | | | | | |
| 16-49 years | 99-00 | **.710** (.624-.806) | **.723** (.636-.833) | **1.41** (1.24-1.65) | **2.48** (2.10-2.97) | **3.27** (2.85-3.92) | 1748 |
| | 01-02 | **.620** (.579-.664) | **.650** (.582-.709) | **1.27** (1.15-1.42) | **2.30** (2.07-2.45) | **3.00** (2.68-3.39) | 1960 |
| **Race/ethnicity** (females, 16-49 years) | | | | | | | |
| Mexican Americans | 99-00 | **.685** (.580-.809) | **.639** (.508-.790) | **1.45** (1.27-1.61) | **2.89** (2.21-3.42) | **4.51** (3.07-5.68) | 595 |
| | 01-02 | **.600** (.526-.686) | **.596** (.426-.709) | **1.32** (1.04-1.47) | **2.41** (2.14-2.77) | **3.21** (2.65-4.46) | 531 |
| Non-Hispanic blacks | 99-00 | **.658** (.520-.831) | **.615** (.475-.892) | **1.22** (.909-1.79) | **2.56** (1.69-3.99) | **3.99** (2.76-5.14) | 381 |
| | 01-02 | **.522** (.410-.665) | **.516** (.387-.664) | **1.03** (.742-1.47) | **1.97** (1.42-3.25) | **3.21** (1.87-4.44) | 442 |
| Non-Hispanic whites | 99-00 | **.706** (.605-.824) | **.721** (.631-.846) | **1.41** (1.23-1.72) | **2.46** (1.99-2.97) | **3.05** (2.46-4.00) | 594 |
| | 01-02 | **.632** (.578-.691) | **.655** (.569-.744) | **1.28** (1.14-1.45) | **2.30** (2.03-2.56) | **2.95** (2.45-3.53) | 826 |

*Source: CDC, Fourth National Report on Human Exposure to Environmental Chemicals, 2009*

## UV Index for 58 U.S. Cities

| City | State | Clear Sky UV Index | | | | | UV Index Forecast | | | | |
|---|---|---|---|---|---|---|---|---|---|---|---|
| | | Extreme | Very High | High | Moderate | Low | Extreme | Very High | High | Moderate | Low |
| Albuquerque | NM | 102 | 60 | 59 | 98 | 34 | 35 | 92 | 61 | 118 | 47 |
| Anchorage | AK | 0 | 0 | 15 | 110 | 228 | 0 | 0 | 4 | 91 | 258 |
| Atlanta | GA | 43 | 132 | 43 | 102 | 33 | 2 | 101 | 44 | 98 | 108 |
| Atlantic City | NJ | 0 | 120 | 53 | 77 | 103 | 0 | 61 | 41 | 88 | 163 |
| Baltimore | MD | 0 | 119 | 59 | 78 | 97 | 0 | 61 | 43 | 88 | 161 |
| Billings | MT | 0 | 85 | 62 | 64 | 142 | 0 | 54 | 54 | 80 | 165 |
| Bismarck | ND | 0 | 56 | 78 | 70 | 149 | 0 | 24 | 63 | 84 | 182 |
| Boise | ID | 3 | 112 | 54 | 62 | 122 | 0 | 80 | 51 | 73 | 149 |
| Boston | MA | 0 | 97 | 41 | 90 | 125 | 0 | 43 | 52 | 89 | 169 |
| Buffalo | NY | 0 | 93 | 47 | 83 | 130 | 0 | 17 | 71 | 88 | 177 |
| Burlington | VT | 0 | 75 | 61 | 73 | 144 | 0 | 12 | 63 | 86 | 192 |
| Charleston | SC | 32 | 143 | 49 | 115 | 14 | 1 | 122 | 47 | 102 | 81 |
| Charleston | WV | 1 | 118 | 65 | 82 | 87 | 0 | 66 | 54 | 71 | 162 |
| Cheyenne | WY | 27 | 102 | 43 | 71 | 110 | 8 | 82 | 55 | 71 | 137 |
| Chicago | IL | 1 | 94 | 61 | 72 | 125 | 0 | 26 | 74 | 84 | 169 |
| Cleveland | OH | 0 | 98 | 59 | 77 | 119 | 0 | 30 | 66 | 86 | 171 |
| Concord | NH | 0 | 93 | 44 | 84 | 132 | 0 | 27 | 60 | 90 | 176 |
| Dallas | TX | 56 | 112 | 54 | 106 | 25 | 9 | 99 | 50 | 99 | 96 |
| Denver | CO | 39 | 97 | 41 | 85 | 91 | 13 | 93 | 51 | 79 | 117 |
| Des Moines | IA | 3 | 98 | 55 | 78 | 119 | 0 | 52 | 52 | 83 | 166 |
| Detroit | MI | 0 | 93 | 55 | 77 | 128 | 0 | 16 | 70 | 94 | 173 |
| Dover | DE | 0 | 121 | 55 | 81 | 96 | 0 | 61 | 38 | 90 | 164 |
| Hartford | CT | 0 | 99 | 46 | 89 | 119 | 0 | 41 | 55 | 93 | 164 |
| Honolulu | HI | 147 | 98 | 103 | 5 | 0 | 124 | 103 | 94 | 28 | 4 |
| Houston | TX | 84 | 124 | 44 | 101 | 0 | 20 | 127 | 52 | 104 | 50 |
| Indianapolis | IN | 2 | 111 | 56 | 81 | 103 | 0 | 51 | 55 | 78 | 169 |
| Jackson | MS | 57 | 124 | 47 | 113 | 12 | 8 | 110 | 57 | 98 | 80 |
| Jacksonville | FL | 56 | 133 | 58 | 106 | 0 | 19 | 141 | 48 | 104 | 41 |
| Las Vegas | NV | 45 | 101 | 47 | 101 | 59 | 10 | 104 | 58 | 103 | 78 |
| Little Rock | AR | 36 | 117 | 55 | 93 | 52 | 2 | 78 | 59 | 92 | 122 |
| Los Angeles | CA | 62 | 96 | 57 | 115 | 23 | 34 | 107 | 52 | 107 | 53 |
| Louisville | KY | 5 | 114 | 62 | 88 | 84 | 0 | 59 | 51 | 92 | 151 |
| Memphis | TN | 30 | 120 | 59 | 89 | 55 | 0 | 89 | 59 | 85 | 120 |
| Miami | FL | 131 | 101 | 54 | 67 | 0 | 83 | 127 | 51 | 83 | 9 |
| Milwaukee | WI | 0 | 88 | 60 | 75 | 130 | 0 | 17 | 73 | 88 | 175 |
| Minneapolis | MN | 0 | 68 | 72 | 74 | 139 | 0 | 16 | 74 | 84 | 179 |
| Mobile | AL | 74 | 123 | 54 | 102 | 0 | 18 | 126 | 57 | 99 | 53 |
| New Orleans | LA | 80 | 126 | 46 | 101 | 0 | 34 | 131 | 57 | 94 | 37 |
| New York | NY | 0 | 111 | 51 | 79 | 112 | 0 | 53 | 50 | 86 | 164 |
| Norfolk | VA | 0 | 132 | 60 | 89 | 72 | 0 | 83 | 51 | 95 | 124 |
| Oklahoma City | OK | 32 | 112 | 60 | 88 | 61 | 5 | 82 | 51 | 103 | 112 |
| Omaha | NE | 4 | 99 | 54 | 77 | 119 | 0 | 54 | 57 | 76 | 166 |
| Philadelphia | PA | 0 | 118 | 52 | 78 | 105 | 0 | 53 | 48 | 84 | 168 |
| Phoenix | AZ | 75 | 86 | 62 | 107 | 23 | 18 | 118 | 63 | 105 | 49 |
| Pittsburgh | PA | 0 | 110 | 56 | 77 | 110 | 0 | 42 | 49 | 89 | 173 |
| Portland | ME | 0 | 80 | 55 | 77 | 141 | 0 | 21 | 58 | 87 | 187 |
| Portland | OR | 0 | 78 | 62 | 77 | 136 | 0 | 49 | 50 | 64 | 190 |
| Providence | RI | 0 | 100 | 44 | 91 | 118 | 0 | 51 | 43 | 89 | 170 |
| Raleigh | NC | 6 | 131 | 65 | 91 | 60 | 0 | 94 | 46 | 93 | 120 |
| Salt Lake City | UT | 38 | 98 | 41 | 71 | 105 | 15 | 86 | 48 | 85 | 119 |
| San Francisco | CA | 7 | 128 | 43 | 91 | 84 | 0 | 106 | 60 | 75 | 112 |
| San Juan | PR | 199 | 98 | 56 | 0 | 0 | 168 | 111 | 60 | 11 | 3 |
| Seattle | WA | 0 | 49 | 75 | 73 | 156 | 0 | 29 | 53 | 82 | 189 |
| Sioux Falls | SD | 1 | 85 | 64 | 71 | 132 | 0 | 36 | 66 | 78 | 173 |
| St. Louis | MO | 8 | 109 | 59 | 90 | 87 | 0 | 66 | 51 | 86 | 150 |

| City | State | Clear Sky UV Index | | | | | UV Index Forecast | | | | |
|---|---|---|---|---|---|---|---|---|---|---|---|
| | | Extreme | Very High | High | Moderate | Low | Extreme | Very High | High | Moderate | Low |
| Tampa | FL | 83 | 129 | 51 | 90 | 0 | 45 | 136 | 52 | 90 | 30 |
| Washington | DC | 0 | 119 | 63 | 78 | 93 | 0 | 66 | 40 | 85 | 162 |
| Wichita | KS | 20 | 105 | 59 | 94 | 75 | 3 | 77 | 44 | 95 | 134 |

***Notes:*** *Figures are the number of days in each exposure category; The days may not add up to 365 due to missing data.*

*The UV Index is a next day forecast of the amount of skin damaging UV radiation expected to reach the earth's surface at the time when the sun is highest in the sky (solar noon). The amount of UV radiation reaching the surface is primarily related to the elevation of the sun in the sky, the amount of ozone in the stratosphere, and the amount of clouds present. The UV Index can range from 0 (when it is night time) to 15 or 16 (in the tropics at high elevations under clear skies). UV radiation is greatest when the sun is highest in the sky and rapidly decreases as the sun approaches the horizon. The higher the UV Index, the greater the dose rate of skin damaging (and eye damaging) UV radiation. Consequently, the higher the UV Index, the smaller the time it takes before skin damage occurs.*

***Source:*** *NOAA, Climate Prediction Center, UV Index: Annual Time Series, 2009*

# Acronyms & Abbreviations

## A

**A&I:** Alternative and Innovative (Wastewater Treatment System)

**AA:** Accountable Area; Adverse Action; Advices of Allowance; Assistant Administrator; Associate Administrator; Atomic Absorption

**AAEE:** American Academy of Environmental Engineers

**AANWR:** Alaskan Arctic National Wildlife Refuge

**AAP:** Asbestos Action Program

**AAPCO:** American Association of Pesticide Control Officials

**AARC:** Alliance for Acid Rain Control

**ABEL:** EPA's computer model for analyzing a violator's ability to pay a civil penalty.

**ABES:** Alliance for Balanced Environmental Solutions

**AC:** Actual Commitment. Advisory Circular

**A&C:** Abatement and Control

**ACA:** American Conservation Association

**ACBM:** Asbestos-Containing Building Material

**ACE:** Alliance for Clean Energy

**ACE:** Any Credible Evidence

**ACEEE:** American Council for an Energy Efficient Economy

**ACFM:** Actual Cubic Feet Per Minute

**ACL:** Alternate Concentration Limit. Analytical Chemistry Laboratory

**ACM:** Asbestos-Containing Material

**ACP:** Agriculture Control Program (Water Quality Management); ACP: Air Carcinogen Policy

**ACQUIRE:** Aquatic Information Retrieval

**ACQR:** Air Quality Control Region

**ACS:** American Chemical Society

**ACT:** Action

**ACTS:** Asbestos Contractor Tracking System

**ACWA:** American Clean Water Association

**ACWM:** Asbestos-Containing Waste Material

**ADABA:** Acceptable Data Base

**ADB:** Applications Data Base

**ADI:** Acceptable Daily Intake

**ADP:** AHERA Designated Person; Automated Data Processing

**ADQ:** Audits of Data Quality

**ADR:** Alternate Dispute Resolution

**ADSS:** Air Data Screening System

**ADT:** Average Daily Traffic

**AEA:** Atomic Energy Act

**AEC:** Associate Enforcement Counsels

**AEE:** Alliance for Environmental Education

**AEERL:** Air and Energy Engineering Research Laboratory

**AEM:** Acoustic Emission Monitoring

**AERE:** Association of Environmental and Resource Economists

**AES:** Auger Electron Spectrometry

**AFA:** American Forestry Association

**AFCA:** Area Fuel Consumption Allocation

**AFCEE:**Air Force Center for Environmental Excellence

**AFS:** AIRS Facility Subsystem

**AFUG:** AIRS Facility Users Group

**AH:** Allowance Holders

**AHERA:** Asbestos Hazard Emergency Response Act

**AHU:** Air Handling Unit

**AI:** Active Ingredient

**AIC:** Active to Inert Conversion

**AICUZ:** Air Installation Compatible Use Zones

**AID:** Agency for International Development

**AIHC:** American Industrial Health Council

**AIP:** Auto Ignition Point

**AIRMON:** Atmospheric Integrated Research Monitoring Network

**AIRS:** Aerometric Information Retrieval System

**AL:** Acceptable Level

**ALA:** Delta-Aminolevulinic Acid

**ALA-O:** Delta-Aminolevulinic Acid Dehydrates

**ALAPO:** Association of Local Air Pollution Control Officers

**ALARA:** As Low As Reasonably Achievable

**ALC:** Application Limiting Constituent

**ALJ:** Administrative Law Judge

**ALMS:** Atomic Line Molecular Spectroscopy

**ALR:** Action Leakage Rate

**AMBIENS:** Atmospheric Mass Balance of Industrially Emitted and Natural Sulfur

**AMOS:** Air Management Oversight System

**AMPS:** Automatic Mapping and Planning System

**AMSA:** Association of Metropolitan Sewer Agencies

**ANC:** Acid Neutralizing Capacity

**ANPR:** Advance Notice of Proposed Rulemaking

**ANRHRD:** Air, Noise, & Radiation Health Research Division/ORD

**ANSS:** American Nature Study Society

**AOAC:** Association of Official Analytical Chemists

**AOC:** Abnormal Operating Conditions

**AOD:** Argon-Oxygen Decarbonization

**AOML:** Atlantic Oceanographic and Meteorological Laboratory

**AP:** Accounting Point

**APA:** Administrative Procedures Act

**APCA:** Air Pollution Control Association

**APCD:** Air Pollution Control District

**APDS:** Automated Procurement Documentation System

**APHA:** American Public Health Association

**APRAC:** Urban Diffusion Model for Carbon Monoxide from Motor Vehicle Traffic

**APTI:** Air Pollution Training Institute

**APWA:** American Public Works Association

**AQ-7:** Non-reactive Pollutant Modelling

**AQCCT:** Air-Quality Criteria and Control Techniques

**AQCP:** Air Quality Control Program

**AQCR:** Air-Quality Control Region

**AQD:** Air-Quality Digest

**AQDHS:** Air-Quality Data Handling System

**AQDM:** Air-Quality Display Model

**AQMA:** Air-Quality Maintenance Area

**AQMD:** Air Quality Management District

**AQMP:** Air-Quality Maintenance Plan; Air-Quality Management Plan

**AQSM:** Air-Quality Simulation Model

**AQTAD:** Air-Quality Technical Assistance Demonstration

**AR:** Administrative Record

**A&R:** Air and Radiation

**ARA:** Assistant Regional Administrator; Associate Regional Administrator

**ARAC:** Acid Rain Advisory Committee

**ARAR:** Applicable or Relevant and Appropriate Standards, Limitations, Criteria, and Requirements

**ARB:** Air Resources Board

**ARC:** Agency Ranking Committee

**ARCC:** American Rivers Conservation Council

**ARCS:** Alternative Remedial Contract Strategy

**ARG:** American Resources Group

**ARIP:** Accidental Release Information Program

**ARL:** Air Resources Laboratory

**ARM:** Air Resources Management

**ARNEWS:** Acid Rain National Early Warning Systems

**ARO:** Alternate Regulatory Option

**ARRP:** Acid Rain Research Program

**ARRPA:** Air Resources Regional Pollution Assessment Model

**ARS:** Agricultural Research Service

**ARZ:** Auto Restricted Zone

**AS:** Area Source

**ASC:** Area Source Category

**ASDWA:** Association of State Drinking Water Administrators

**ASHAA:** Asbestos in Schools Hazard Abatement Act

**ASHRAE:** American Society of Heating, Refrigerating, and Air-Conditioning Engineers

**ASIWCPA:** Association of State and Interstate Water Pollution Control Administrators

**ASMDHS:** Airshed Model Data Handling System

**ASRL:** Atmospheric Sciences Research Laboratory

**AST:** Advanced Secondary (Wastewater) Treatment

**ASTHO:** Association of State and Territorial Health Officers

**ASTM:** American Society for Testing and Materials

**ASTSWMO:** Association of State and Territorial Solid Waste Management Officials

**AT:** Advanced Treatment. Alpha Track Detection

**ATERIS:** Air Toxics Exposure and Risk Information System

**ATS:** Action Tracking System; Allowance Tracking System

**ATSDR:** Agency for Toxic Substances and Disease Registry

**ATTF:** Air Toxics Task Force

**AUSM:** Advanced Utility Simulation Model

**A/WPR:** Air/Water Pollution Report

**AWRA:** American Water Resources Association

**AWT:** Advanced Wastewater Treatment

**AWWA:** American Water Works Association

**AWWARF:** American Water Works Association Research Foundation.

## B

**BAA:** Board of Assistance Appeals

**BAC:** Bioremediation Action Committee; Biotechnology Advisory Committee

**BACM:** Best Available Control Measures

**BACT:** Best Available Control Technology

**BADT:** Best Available Demonstrated Technology

**BAF:** Bioaccumulation Factor

**BaP:** Benzo(a)Pyrene

**BAP:** Benefits Analysis Program

**BART:** Best Available Retrofit Technology

**BASIS:** Battelle's Automated Search Information System

**BAT:** Best Available Technology

**BATEA:** Best Available Treatment Economically Achievable

**BCT:** Best Control Technology

**BCPCT:** Best Conventional Pollutant Control Technology

**BDAT:** Best Demonstrated Achievable Technology

**BDCT:** Best Demonstrated Control Technology

**BDT:** Best Demonstrated Technology

**BEJ:** Best Engineering Judgement. Best Expert Judgment

**BF:** Bonafide Notice of Intent to Manufacture or Import (IMD/OTS)

**BID:** Background Information Document. Buoyancy Induced Dispersion

**BIOPLUME:** Model to Predict the Maximum Extent of Existing Plumes

**BMP:** Best Management Practice(s)

**BMR:** Baseline Monitoring Report

**BO:** Budget Obligations

**BOA:** Basic Ordering Agreement (Contracts)

**BOD:** Biochemical Oxygen Demand. Biological Oxygen Demand

**BOF:** Basic Oxygen Furnace

**BOP:** Basic Oxygen Process

**BOPF:** Basic Oxygen Process Furnace

**BOYSNC:** Beginning of Year Significant Non-Compliers

**BP:** Boiling Point

**BPJ:** Best Professional Judgment

**BPT:** Best Practicable Technology. Pest Practicable Treatment

**BPWTT:** Best Practical Wastewater Treatment Technology

**BRI:** Building-Related Illness

**BRS:** Bibliographic Retrieval Service

**BSI:** British Standards Institute

**BSO:** Benzene Soluble Organics

**BTZ:** Below the Treatment Zone

**BUN:** Blood Urea Nitrogen

## C

**CA:** Citizen Act. Competition Advocate. Cooperative Agreements. Corrective Action

**CAA:** Clean Air Act; Compliance Assurance Agreement

**CAAA:** Clean Air Act Amendments

**CAER:** Community Awareness and Emergency Response

**CAFE:** Corporate Average Fuel Economy

**CAFO:** Concentrated Animal Feedlot; Consent Agreement/Final Order

**CAG:** Carcinogenic Assessment Group

**CAIR:** Clean Air Interstate Rule: Comprehensive Assessment of Information Rule

**CALINE:** California Line Source Model

**CAM:** Compliance Assurance Monitoring rule; Compliance Assurance Monitoring

**CAMP:** Continuous Air Monitoring Program

**CAN:** Common Account Number

**CAO:** Corrective Action Order

**CAP:** Corrective Action Plan. Cost Allocation Procedure. Criteria Air Pollutant

**CAPMoN:** Canadian Air and Precipitation Monitoring Network

**CAR:** Corrective Action Report

**CAS:** Center for Automotive Safety; Chemical Abstract Service

**CASAC:** Clean Air Scientific Advisory Committee

**CASLP:** Conference on Alternative State and Local Practices

**CASTNet:** Clean Air Status and Trends Network

**CATS:** Corrective Action Tracking System

**CAU:** Carbon Adsorption Unit; Command Arithmetic Unit

**CB:** Continuous Bubbler

**CBA:** Chesapeake Bay Agreement. Cost Benefit Analysis

**CBD:** Central Business District

**CBEP:** Community Based Environmental Project

**CBI:** Compliance Biomonitoring Inspection; Confidential Business Information

**CBOD:** Carbonaceous Biochemical Oxygen Demand

**CBP:** Chesapeake Bay Program; County Business Patterns

**CCA:** Competition in Contracting Act

**CCAA:** Canadian Clean Air Act

**CCAP:** Center for Clean Air Policy; Climate Change Action Plan

**CCEA:** Conventional Combustion Environmental Assessment

**CCHW:** Citizens Clearinghouse for Hazardous Wastes

**CCID:** Confidential Chemicals Identification System

**CCMS/NATO:** Committee on Challenges of a Modern Society/North Atlantic Treaty Organization

**CCP:** Composite Correction Plan

**CC/RTS:**Chemical Collection/ Request Tracking System

**CCTP:** Clean Coal Technology Program

**CD:** Climatological Data

**CDB:** Consolidated Data Base

**CDBA:** Central Data Base Administrator

**CDBG:** Community Development Block Grant

**CDD:** Chlorinated dibenzo-p-dioxin

**CDF:** Chlorinated dibenzofuran

**CDHS:** Comprehensive Data Handling System

**CDI:** Case Development Inspection

**CDM:** Climatological Dispersion Model; Comprehensive Data Management

**CDMQC:** Climatological Dispersion Model with Calibration and Source Contribution

**CDNS:** Climatological Data National Summary

**CDP:** Census Designated Places

**CDS:** Compliance Data System

**CE:** Categorical Exclusion. Conditionally Exempt Generator

**CEA:** Cooperative Enforcement Agreement; Cost and Economic Assessment

**CEAT:** Contractor Evidence Audit Team

**CEARC:** Canadian Environmental Assessment Research Council

**CEB:** Chemical Element Balance

**CEC:** Commission for Environmental Cooperation

**CECATS:** CSB Existing Chemicals Assessment Tracking System

**CEE:** Center for Environmental Education

**CEEM:** Center for Energy and Environmental Management

**CEI:** Compliance Evaluation Inspection

**CELRF:** Canadian Environmental Law Research Foundation

**CEM:** Continuous Emission Monitoring

**CEMS:** Continuous Emission Monitoring System

**CEPA:** Canadian Environmental Protection Act

**CEPP:** Chemical Emergency Preparedness Plan

**CEQ:** Council on Environmental Quality

**CERCLA:** Comprehensive Environmental Response, Compensation, and Liability Act (1980)

**CERCLIS:** Comprehensive Environmental Response, Compensation, and Liability Information System

**CERT:** Certificate of Eligibility

**CESQG:** Conditionally Exempt Small Quantity Generator

**CEST:** Community Environmental Service Teams

**CF:** Conservation Foundation

**CFC:** Chlorofluorocarbons

**CFM:** Chlorofluoromethanes

**CFR:** Code of Federal Regulations

**CHABA:** Committee on Hearing and Bio-Acoustics

**CHAMP:** Community Health Air Monitoring Program

**CHEMNET:** Chemical Industry Emergency Mutual Aid Network

**CHESS:** Community Health and Environmental Surveillance System

**CHIP:** Chemical Hazard Information Profiles

**CI:** Compression Ignition. Confidence Interval

**CIAQ:** Council on Indoor Air Quality

**CIBL:** Convective Internal Boundary Layer

**CICA:** Competition in Contracting Act

**CICIS:** Chemicals in Commerce Information System

**CIDRS:** Cascade Impactor Data Reduction System

**CIMI:** Committee on Integrity and Management Improvement

**CIS:** Chemical Information System. Contracts Information System

**CKD:** Cement Kiln Dust

**CKRC:** Cement Kiln Recycling Coalition

**CLC:** Capacity Limiting Constituents

**CLEANS:** Clinical Laboratory for Evaluation and Assessment of Toxic Substances

**CLEVER:** Clinical Laboratory for Evaluation and Validation of Epidemiologic Research

**CLF:** Conservation Law Foundation

**CLI:** Consumer Labelling Initiative

**CLIPS:** Chemical List Index and Processing System

**CLP:** Contract Laboratory Program

**CM:** Corrective Measure

**CMA:** Chemical Manufacturers Association

**CMB:** Chemical Mass Balance

**CME:** Comprehensive Monitoring Evaluation

**CMEL:** Comprehensive Monitoring Evaluation Log

**CMEP:** Critical Mass Energy Project

**CNG:**Compressedd Natural Gas

**COCO:** Contractor-Owned/ Contractor-Operated

**COD:** Chemical Oxygen Demand

**COH:** Coefficient Of Haze

**CPDA:** Chemical Producers and Distributor Association

**CPF:** Carcinogenic Potency Factor

**CPO:** Certified Project Officer

**CQA:** Construction Quality Assurance

**CR:** Continuous Radon Monitoring

**CROP:** Consolidated Rules of Practice

**CRP:** Child-Resistant Packaging; Conservation Reserve Program

**CRR:** Center for Renewable Resources

**CRSTER:** Single Source Dispersion Model

**CSCT:** Committee for Site Characterization

**CSGWPP:** Comprehensive State Ground Water Protection Program

**CSI:** Common Sense Initiative; Compliance Sampling Inspection

**CSIN:** Chemical Substances Information Network

**CSMA:** Chemical Specialties Manufacturers Association

**CSO:** Combined Sewer Overflow

**CSPA:** Council of State Planning Agencies

**CSRL:** Center for the Study of Responsive Law

**CTARC:** Chemical Testing and Assessment Research Commission

**CTG:** Control Techniques Guidelines

**CTSA:** Cleaner TechnologiesSubstitutess Assessment

**CV:** Chemical Vocabulary

**CVS:** Constant Volume Sampler

**CW:** Continuous working-level monitoring

**CWA:** Clean Water Act (aka FWPCA)

**CWAP:** Clean Water Action Project

**CWTC:** Chemical Waste Transportation Council

**CZMA:** Coastal Zone Management Act

**CZARA:** Coastal Zone Management Act Reauthorization Amendments

## D

**DAPSS:** Document and Personnel Security System (IMD)

**DBP:** Disinfection By-Product

**DCI:** Data Call-In

**DCO:** Delayed Compliance Order

**DCO:** Document Control Officer

**DDT:** DichloroDiphenylTrichloroethane

**DERs:** Data Evaluation Records

**DES:** Diethylstilbesterol

**DfE:** Design for the Environment

**DI:** Diagnostic Inspection

**DMR:** Discharge Monitoring Report

**DNA:** Deoxyribonucleic acid

**DNAPL:** Dense Non-Aqueous Phase Liquid

**DO:** Dissolved Oxygen

**DOW:** Defenders Of Wildlife

**DPA:** Deepwater Ports Act

**DPD:** Method of Measuring Chlorine Residual in Water

**DQO:** Data Quality Objective

**DRE:** Destruction and Removal Efficiency

**DRES:** Dietary Risk Evaluation System

**DRMS:** Defense Reutilization and Marketing Service

**DRR:** Data Review Record

**DS:** Dichotomous Sampler

**DSAP:** Data Self Auditing Program

**DSCF:** Dry Standard Cubic Feet

**DSCM:** Dry Standard Cubic Meter

**DSS:** Decision Support System; Domestic Sewage Study

**DT:** Detectors (radon) damaged or lost; Detention Time

**DU:** Decision Unit. Ducks Unlimited; Dobson Unit

**DUC:** Decision Unit Coordinator

**DWEL:** Drinking Water Equivalent Level

**DWS:** Drinking Water Standard

**DWSRF:** Drinking Water State Revolving Fund

## E

**EA:** Endangerment Assessment; Enforcement Agreement; Environmental Action; Environmental Assessment;. Environmental Audit

**EAF:** Electric Arc Furnaces

**EAG:** Exposure Assessment Group

**EAO:** Emergency Administrative Order

**EAP:** Environmental Action Plan

**EAR:** Environmental Auditing Roundtable

**EASI:** Environmental Alliance for Senior Involvement

**EB:** Emissions Balancing

**EC:** Emulsifiable Concentrate; Environment Canada; Effective Concentration

**ECA:** Economic Community for Africa

**ECAP:** Employee Counselling and Assistance Program

**ECD:** Electron Capture Detector

**ECHH:** Electro-Catalytic Hyper-Heaters

**ECHO:** Enforcement and Compliance History Online

**ECL:** Environmental Chemical Laboratory

**ECOS:** Environmental Council of the States

**ECR:** Enforcement Case Review

**ECRA:** Economic Cleanup Responsibility Act

**ED:** Effective Dose

**EDA:** Emergency Declaration Area

**EDB:** Ethylene Dibromide

**EDC:** Ethylene Dichloride

**EDD:** Enforcement Decision Document

**EDF:** Environmental Defense Fund

**EDRS:** Enforcement Document Retrieval System

**EDS:** Electronic Data System; Energy Data System

**EDTA:** Ethylene Diamine Triacetic Acid

**EDX:** Electronic Data Exchange

**EDZ:** Emission Density Zoning

**EEA:** Energy and Environmental Analysis

**EECs:** Estimated Environmental Concentrations

**EER:** Excess Emission Report

**EERL:** Eastern Environmental Radiation Laboratory

**EERU:** Environmental Emergency Response Unit

**EESI:** Environment and Energy Study Institute

**EESL:** Environmental Ecological and Support Laboratory

**EETFC:** Environmental Effects, Transport, and Fate Committee

**EF:** Emission Factor

**EFO:** Equivalent Field Office

**EFTC:** European Fluorocarbon Technical Committee

**EGR:** Exhaust Gas Recirculation

**EH:** Redox Potential

**EHC:** Environmental Health Committee

**EHS:** Extremely Hazardous Substance

**EI:** Emissions Inventory

**EIA:** Environmental Impact Assessment. Economic Impact Assessment

**EIL:** Environmental Impairment Liability

**EIR:** Endangerment Information Report; Environmental Impact Report

**EIS:** Environmental Impact Statement; Environmental Inventory System

**EIS/AS:** Emissions Inventory System/Area Source

**EIS/PS:** Emissions Inventory System/Point Source

**EJ:** Environmental Justice

**EKMA:** Empirical Kinetic Modeling Approach

**EL:** Exposure Level

**ELI:** Environmental Law Institute

**ELR:** Environmental Law Reporter

**EM:** Electromagnetic Conductivity

**EMAP:** Environmental Mapping and Assessment Program

**EMAS:** Enforcement Management and Accountability System

**EMR:** Environmental Management Report

**EMS:** Enforcement Management System

**EMSL:** Environmental Monitoring Support Systems Laboratory

**EMTS:** Environmental Monitoring Testing Site; Exposure Monitoring Test Site

**EnPA:** Environmental Performance Agreement

**EO:** Ethylene Oxide

**EOC:** Emergency Operating Center

**EOF:** Emergency Operations Facility (RTP)

**EOP:** End Of Pipe

**EOT:** Emergency Operations Team

**EP:** Earth Protectors; Environmental Profile; End-use Product; Experimental Product; Extraction Procedure

**EPAA:** Environmental Programs Assistance Act

**EPAAR:** EPA Acquisition Regulations

**EPCA:** Energy Policy and Conservation Act

**EPACT:** Environmental Policy Act

**EPACASR:** EPA Chemical Activities Status Report

**EPCRA:** Emergency Planning and Community Right to Know Act

**EPD:** Emergency Planning District

**EPI:** Environmental Policy Institute

**EPIC:** Environmental Photographic Interpretation Center

**EPNL:** Effective Perceived Noise Level

**EPRI:** Electric Power Research Institute

**EPTC:** Extraction Procedure Toxicity Characteristic

**EQIP:** Environmental Quality Incentives Program

**ER:** Ecosystem Restoration; Electrical Resistivity

**ERA:** Economic Regulatory Agency

**ERAMS:** Environmental Radiation Ambient Monitoring System

**ERC:** Emergency Response Commission. Emissions Reduction Credit, Environmental Research Center

**ERCS:** Emergency Response Cleanup Services

**ERDA:** Energy Research and Development Administration

**ERD&DAA:** Environmental Research, Development and Demonstration Authorization Act

**ERL:** Environmental Research Laboratory

**ERNS:** Emergency Response Notification System

**ERP:** Enforcement Response Policy

**ERT:** Emergency Response Team

**ERTAQ:** ERT Air Quality Model

**ES:** Enforcement Strategy

**ESA:** Endangered Species Act. Environmentally Sensitive Area

**ESC:** Endangered Species Committee

**ESCA:** Electron Spectroscopy for Chemical Analysis

**ESCAP:** Economic and Social Commission for Asia and the Pacific

**ESECA:** Energy Supply and Environmental Coordination Act

**ESH:** Environmental Safety and Health

**ESP:** Electrostatic Precipitators

**ET:** Emissions Trading

**ETI:** Environmental Technology Initiative

**ETP:** Emissions Trading Policy

**ETS:** Emissions Tracking System; Environmental Tobacco Smoke

**ETV:** Environmental Technology Verification Program

**EUP:** End-Use Product; Experimental Use Permit

**EWCC:** Environmental Workforce Coordinating Committee

**EXAMS:** Exposure Analysis Modeling System

**ExEx:** Expected Exceedance

## F

**FACA:** Federal Advisory Committee Act

**FAN:** Fixed Account Number

**FATES:** FIFRA and TSCA Enforcement System

**FBC:** Fluidized Bed Combustion

**FCC:** Fluid Catalytic Converter

**FCCC:** Framework Convention on Climate Change

**FCCU:** Fluid Catalytic Cracking Unit

**FCO:** Federal Coordinating Officer (in disaster areas); Forms Control Officer

**FDF:** Fundamentally Different Factors

**FDL:** Final Determination Letter

**FDO:** Fee Determination Official

**FE:** Fugitive Emissions

**FEDS:** Federal Energy Data System

**FEFx:** Forced Expiratory Flow

**FEIS:** Fugitive Emissions Information System

**FEL:** Frank Effect Level

**FEPCA:** Federal Environmental Pesticide Control Act; enacted as amendments to FIFRA.

**FERC:** Federal Energy Regulatory Commission

**FES:** Factor Evaluation System

**FEV:** Forced Expiratory Volume

**FEV1:** Forced Expiratory Volume—one second; Front End Volatility Index

**FF:** Federal Facilities

**FFAR:** Fuel and Fuel Additive Registration

**FFDCA:** Federal Food, Drug, and Cosmetic Act

**FFEO:** Federal Facilities Enforcement Office

**FFF:** Firm Financial Facility

**FFFSG:** Fossil-Fuel-Fired Steam Generator

**FFIS:** Federal Facilities Information System

**FFP:** Firm Fixed Price

**FGD:** Flue-Gas Desulfurization

**FID:** Flame Ionization Detector

**FIFRA:** Federal Insecticide, Fungicide, and Rodenticide Act

**FIM:** Friable Insulation Material

**FINDS:** Facility Index System

**FIP:** Final Implementation Plan

**FIPS:** Federal Information Procedures System

**FIT:** Field Investigation Team

**FLETC:** Federal Law Enforcement Training Center

**FLM:** Federal Land Manager

**FLP:** Flash Point

**FLPMA:** Federal Land Policy and Management Act

**FMAP:** Financial Management Assistance Project

**F/M:** Food to Microorganism Ratio

**FML:** Flexible Membrane Liner

**FMP:** Facility Management Plan

**FMP:** Financial Management Plan

**FMS:** Financial Management System

**FMVCP:** Federal Motor Vehicle Control Program

**FOE:** Friends Of the Earth

**FOIA:** Freedom Of Information Act

**FOISD:** Fiber Optic Isolated Spherical Dipole Antenna

**FONSI:** Finding Of No Significant Impact

**FORAST:** Forest Response to Anthropogenic Stress

**FP:** Fine Particulate

**FPA:** Federal Pesticide Act

**FPAS:** Foreign Purchase Acknowledgement Statements

**FPD:** Flame Photometric Detector

**FPEIS:** Fine Particulate Emissions Information System

**FPM:** Federal Personnel Manual

**FPPA:** Federal Pollution Prevention Act

**FPR:** Federal Procurement Regulation

**FPRS:** Federal Program Resources Statement; Formal Planning and Supporting System

**FQPA:** Food Quality Protection Act

**FR:** Federal Register. Final Rulemaking

**FRA:** Federal Register Act

**FREDS:** Flexible Regional Emissions Data System

**FRES:** Forest Range Environmental Study

**FRM:** Federal Reference Methods

**FRN:** Federal Register Notice. Final Rulemaking Notice

**FRS:** Formal Reporting System

**FS:** Feasibility Study

**FSA:** Food Security Act

**FSS:** Facility Status Sheet; Federal Supply Schedule

**FTP:** Federal Test Procedure (for motor vehicles)

**FTS:** File Transfer Service

**FTTS:** FIFRA/TSCA Tracking System

**FUA:** Fuel Use Act

**FURS:** Federal Underground Injection Control Reporting System

**FVMP:** Federal Visibility Monitoring Program

**FWCA:** Fish and Wildlife Coordination Act

**FWPCA:** Federal Water Pollution and Control Act (aka CWA). Federal Water Pollution and Control Administration

**FY:** Fiscal Year

## G

**GAAP:** Generally Accepted Accounting Principles

**GAC:** Granular Activated Carbon

**GACT:** Granular Activated Carbon Treatment

**GAW:** Global Atmospheric Watch

**GCC:** Global Climate Convention

**GC/MS:** Gas Chromatograph/ Mass Spectograph

**GCVTC:** Grand Canyon Visibility Transport Commission

**GCWR:** Gross Combination Weight Rating

**GDE:** Generic Data Exemption

**GEI:** Geographic Enforcement Initiative

**GEMI:** Global Environmental Management Initiative

**GEMS:** Global Environmental Monitoring System; Graphical Exposure Modeling System

**GEP:** Good Engineering Practice

**GFF:** Glass Fiber Filter

**GFO:** Grant Funding Order

**GFP:** Government-Furnished Property

**GICS:** Grant Information and Control System

**GIS:** Geographic Information Systems; Global Indexing System

**GLC:** Gas Liquid Chromatography

**GLERL:** Great Lakes Environmental Research Laboratory

**GLNPO:** Great Lakes National Program Office

**GLP:** Good Laboratory Practices

**GLWQA:** Great Lakes Water Quality Agreement

**GMCC:** Global Monitoring for Climatic Change

**G/MI:** Grams per mile

**GOCO:** Government-Owned/ Contractor-Operated

**GOGO:** Government-Owned/ Government-Operated

**GOP:** General Operating Procedures

**GOPO:** Government-Owned/ Privately-Operated

**GPAD:** Gallons-per-acre per-day

**GPG:** Grams-per-Gallon

**GPR:** Ground-Penetrating Radar

**GPRA:** Government Performance and Results Act

**GPS:** Groundwater Protection Strategy

**GR:** Grab Radon Sampling

**GRAS:** Generally Recognized as Safe

**GRCDA:** Government Refuse Collection and Disposal Association

**GRGL:** Groundwater Residue Guidance Level

**GT:** Gas Turbine

**GTN:** Global Trend Network

**GTR:** Government Transportation Request

**GVP:** Gasoline Vapor Pressure

**GVW:** Gross Vehicle Weight

**GVWR:** Gross Vehicle Weight Rating

**GW:** Grab Working-Level Sampling. Groundwater

**GWDR:** Ground Water Disinfection Rule

**GWM:** Groundwater Monitoring

**GWP:** Global Warming Potential

**GWPC:** Ground Water Protection Council

**GWPS:** Groundwater Protection Standard; Groundwater Protection Strategy

## H

**HA:** Health Advisory

**HAD:** Health Assessment Document

**HAP:** Hazardous Air Pollutant

**HAPEMS:** Hazardous Air Pollutant Enforcement Management System

**HAPPS:** Hazardous Air Pollutant Prioritization System

**HATREMS:** Hazardous and Trace Emissions System

**HAZMAT:** Hazardous Materials

**HAZOP:** Hazard and Operability Study

**HBFC:** Hydrobromofluorocarbon

**HC:** Hazardous Constituents; Hydrocarbon

**HCCPD:** Hexachlorocyclo-pentadiene

**HCFC:** Hydrochlorofluorocarbon

**HCP:** Hypothermal Coal Process

**HDD:** Heavy-Duty Diesel

**HDDT:** Heavy-duty Diesel Truck

**HDDV:** Heavy-Duty Diesel Vehicle

**HDE:** Heavy-Duty Engine

**HDG:** Heavy-Duty Gasoline-Powered Vehicle

**HDGT:** Heavy-Duty Gasoline Truck

**HDGV:** Heavy-Duty Gasoline Vehicle

**HDPE:** High Density Polyethylene

**HDT:** Highest Dose Tested in a study. Heavy-Duty Truck

**HDV:** Heavy-Duty Vehicle

**HEAL:** Human Exposure Assessment Location

**HECC:** House Energy and Commerce Committee

**HEI:** Health Effects Institute

**HEM:** Human Exposure Modeling

**HEPA:** High-Efficiency Particulate Air

**HEPA:** Highly Efficient Particulate Air Filter

**HERS:** Hyperion Energy Recovery System

**HFC:** Hydrofluorocarbon

**HHDDV:** Heavy Heavy-Duty Diesel Vehicle

**HHE:** Human Health and the Environment

**HHV:** Higher Heating Value

**HI:** Hazard Index

**HI-VOL:** High-Volume Sampler

**HIWAY:** A Line Source Model for Gaseous Pollutants

**HLRW:** High Level Radioactive Waste

**HMIS:** Hazardous Materials Information System

**HMS:** Highway Mobile Source

**HMTA:** Hazardous Materials Transportation Act

**HMTR:** Hazardous Materials Transportation Regulations

**HOC:** Halogenated Organic Carbons

**HON:** Hazardous Organic NESHAP

**HOV:** High-Occupancy Vehicle

**HP:** Horse Power

**HPLC:** High-Performance Liquid Chromatography

**HPMS:** Highway Performance Monitoring System

**HPV:** High Priority Violator

**HQCDO:** Headquarters Case Development Officer

**HRS:** Hazardous Ranking System

**HRUP:** High-Risk Urban Problem

**HSDB:** Hazardous Substance Data Base

**HSL:** Hazardous Substance List

**HSWA:** Hazardous and Solid Waste Amendments

**HT:** Hypothermally Treated

**HTP:** High Temperature and Pressure

**HVAC:** Heating, Ventilation, and Air-Conditioning system

**HVIO:** High Volume Industrial Organics

**HW:** Hazardous Waste

**HWDMS:** Hazardous Waste Data Management System

**HWGTF:** Hazardous Waste Groundwater Task Force; Hazardous Waste Groundwater Test Facility

**HWIR:** Hazardous Waste Identification Rule

**HWLT:** Hazardous Waste Land Treatment

**HWM:** Hazardous Waste Management

**HWRTF:** Hazardous Waste Restrictions Task Force

**HWTC:** Hazardous Waste Treatment Council

## I

**I/A:** Innovative/Alternative

**IA:** Interagency Agreement

**IAAC:** Interagency Assessment Advisory Committee

**IADN:** Integrated Atmospheric Deposition Network

**IAG:** Interagency Agreement

**IAP:** Incentive Awards Program. Indoor Air Pollution

**IAQ:** Indoor Air Quality

**IARC:** International Agency for Research on Cancer

**IATDB:** Interim Air Toxics Data Base

**IBSIN:** Innovations in Building Sustainable Industries

**IBT:** Industrial Biotest Laboratory

**IC:** Internal Combustion

**ICAIR:** Interdisciplinary Planning and Information Research

**ICAP:** Inductively Coupled Argon Plasma

**ICB:** Information Collection Budget

**ICBN:** International Commission on the Biological Effects of Noise

**ICCP:** International Climate Change Partnership

**ICE:** Industrial Combustion Emissions Model. Internal Combustion Engine

**ICP:** Inductively Coupled Plasma

**ICR:** Information Collection Request

**ICRE:** Ignitability, Corrosivity, Reactivity, Extraction

**ICRP:** International Commission on Radiological Protection

**ICRU:** International Commission of Radiological Units and Measurements

**ICS:** Incident Command System. Institute for Chemical Studies; Intermittent Control Strategies.; Intermittent Control System

**ICWM:** Institute for Chemical Waste Management

**IDEA:** Integrated Data for Enforcement Analysis

**IDLH:** Immediately Dangerous to Life and Health

**IEB:** International Environment Bureau

**IEMP:** Integrated Environmental Management Project

**IES:** Institute for Environmental Studies

**IFB:** Invitation for Bid

**IFCAM:** Industrial Fuel Choice Analysis Model

**IFCS:** International Forum on Chemical Safety

**IFIS:** Industry File Information System

**IFMS:** Integrated Financial Management System

**IFPP:** Industrial Fugitive Process Particulate

**IGCC:** Integrated Gasification Combined Cycle

**IGCI:** Industrial Gas Cleaning Institute

**IIS:** Inflationary Impact Statement

**IINERT:** In-Place Inactivation and Natural Restoration Technologies

**IJC:** International Joint Commission (on Great Lakes)

**I/M:** Inspection/Maintenance

**IMM:** Intersection Midblock Model

**IMPACT:** Integrated Model of Plumes and Atmosphere in Complex Terrain

**IMPROVE:** Interagency Monitoring of Protected Visual Environment

**INPUFF:** Gaussian Puff Dispersion Model

**INT:** Intermittent

**IOB:** Iron Ore Beneficiation

**IOU:** Input/Output Unit

**IPCS:** International Program on Chemical Safety

**IP:** Inhalable Particles

**IPM:** Inhalable Particulate Matter. Integrated Pest Management

**IPP:** Implementation Planning Program. Integrated Plotting Package; Inter-media Priority Pollutant (document); Independent Power Producer

**IPCC:** Intergovernmental Panel on Climate Change

**IPM:** Integrated Pest Management

**IRG:** Interagency Review Group

**IRLG:** Interagency Regulatory Liaison Group (Composed of EPA, CPSC, FDA, and OSHA)

**IRIS:** Instructional Resources Information System. Integrated Risk Information System

**IRM:** Intermediate Remedial Measures

**IRMC:** Inter-Regulatory Risk Management Council

**IRP:** Installation Restoration Program

**IRPTC:** International Register of Potentially Toxic Chemicals

**IRR:** Institute of Resource Recovery

**IRS:** International Referral Systems

**IS:** Interim Status

**ISAM:** Indexed Sequential File Access Method

**ISC:** Industrial Source Complex

**ISCL:** Interim Status Compliance Letter

**ISCLT:** Industrial Source Complex Long Term Model

**ISCST:** Industrial Source Complex Short Term Model

**ISD:** Interim Status Document

**ISE:** Ion-specific electrode

**ISMAP:** Indirect Source Model for Air Pollution

**ISO:** International Organization for Standardization

**ISPF:** (IBM) Interactive System Productivity Facility

**ISS:** Interim Status Standards

**ITC:**Innovative Technology Council

**ITC:** Interagency Testing Committee

**ITRC:** Interstate Technology Regulatory Coordination

**ITRD:** Innovative Treatment Remediation Demonstration

**IUP:** Intended Use Plan

**IUR:** Inventory Update Rule

**IWC:** In-Stream Waste Concentration

**IWS:** Ionizing Wet Scrubber

## J

**JAPCA:** Journal of Air Pollution Control Association

**JCL:** Job Control Language

**JEC:** Joint Economic Committee

**JECFA:** Joint Expert Committee of Food Additives

**JEIOG:** Joint Emissions Inventory Oversight Group

**JLC:** Justification for Limited Competition

**JMPR:** Joint Meeting on Pesticide Residues

**JNCP:** Justification for Non-Competitive Procurement

**JOFOC:** Justification for Other Than Full and Open Competition

**JPA:** Joint Permitting Agreement

**JSD:** Jackson Structured Design

**JSP:** Jackson Structured Programming

**JTU:** Jackson Turbidity Unit

## L

**LAA:** Lead Agency Attorney

**LADD:** Lifetime Average Daily Dose; Lowest Acceptable Daily Dose

**LAER:** Lowest Achievable Emission Rate

**LAI:** Laboratory Audit Inspection

**LAMP:** Lake Acidification Mitigation Project

**LC:** Lethal Concentration. Liquid Chromatography

**LCA:** Life Cycle Assessment

**LCD:** Local Climatological Data

**LCL:** Lower Control Limit

**LCM:** Life Cycle Management

**LCRS:** Leachate Collection and Removal System

**LD:** Land Disposal. Light Duty

**LD L0:** The lowest dosage of a toxic substance that kills test organisms.

**LDAR:** Leak Detection and Repair

**LDC:** London Dumping Convention

**LDCRS:** Leachate Detection, Collection, and Removal System

**LDD:** Light-Duty Diesel

**LDDT:** Light-Duty Diesel Truck

**LDDV:** Light-Duty Diesel Vehicle

**LDGT:** Light-Duty Gasoline Truck

**LDIP:** Laboratory Data Integrity Program

**LDR:** Land Disposal Restrictions

**LDRTF:** Land Disposal Restrictions Task Force

**LDS:** Leak Detection System

**LDT:** Lowest Dose Tested. Light-Duty Truck

**LDV:** Light-Duty Vehicle

**LEL:** Lowest Effect Level. Lower Explosive Limit

**LEP:** Laboratory Evaluation Program

**LEPC:** Local Emergency Planning Committee

**LERC:** Local Emergency Response Committee

**LEV:** Low Emissions Vehicle

**LFG:** Landfill Gas

**LFL:** Lower Flammability Limit

**LGR:** Local Governments Reimbursement Program

**LHDDV:** Light Heavy-Duty Diesel Vehicle

**LI:** Langelier Index

**LIDAR:** Light Detection and Ranging

**LIMB:** Limestone-Injection Multi-Stage Burner

**LLRW:** Low Level Radioactive Waste

**LMFBR:** Liquid Metal Fast Breeder Reactor

**LMOP:** Landfill Methane Outreach Program

**LNAPL:** Light Non-Aqueous Phase Liquid

**LOAEL:** Lowest-Observed-Adverse-Effect-Level

**LOD:** Limit of Detection

**LQER:** Lesser Quantity Emission Rates

**LQG:** Large Quantity Generator

**LRTAP:** Long Range Transboundary Air Pollution

**LUIS:** Label Use Information System

## M

**MAC:** Mobile Air Conditioner

**MACT:** Maximum Achievable Control Technology

**MAPSIM:** Mesoscale Air Pollution Simulation Model

**MATC:** Maximum Acceptable Toxic Concentration

**MBAS:** Methylene-Blue-Active Substances

**MCL:** Maximum Contaminant Level

**MCLG:** Maximum Contaminant Level Goal

**MCS:** Multiple Chemical Sensitivity

**MDL:** Method Detection Limit

**MEC:** Model Energy Code

**MEI:** Maximally (or most) Exposed Individual

**MEP:** Multiple Extraction Procedure

**MHDDV:** Medium Heavy-Duty Diesel Vehicle

**MOBILE5A:** Mobile Source Emission Factor Model

**MOE:** Margin Of Exposure

**MOS:** Margin of Safety

**MP:** Manufacturing-use Product; Melting Point

**MPCA:** Microbial Pest Control Agent

**MPI:** Maximum Permitted Intake

**MPN:** Maximum Possible Number

**MPWC:** Multiprocess Wet Cleaning

**MRBMA:** Mercury-Containing and Rechargeable Battery Management Act

**MRF:** Materials Recovery Facility

**MRID:** Master Record Identification number

**MRL:** Maximum-Residue Limit (Pesticide Tolerance)

**MSW:** Municipal Solid Waste

**MTBE:** Methyl tertiary butyl ether

**MTD:** Maximum Tolerated Dose

**MUP:** Manufacturing-Use Product

**MUTA:** Mutagenicity

**MWC:** Machine Wet Cleaning

## N

**NAA:** Nonattainment Area

**NAAEC:** North American Agreement on Environmental Cooperation

**NAAQS:** National Ambient Air Quality Standards

**NACA:** National Agricultural Chemicals Association

**NACEPT:** National Advisory Council for Environmental Policy and Technology

**NADP/NTN:** National Atmospheric Deposition Program/National Trends Network

**NAMS:** National Air Monitoring Stations

**NAPAP:** National Acid Precipitation Assessment Program

**NAPL:** Non-Aqueous Phase Liquid

**NAPS:** National Air Pollution Surveillance

**NARA:** National Agrichemical Retailers Association

**NARSTO:** North American Research Strategy for Tropospheric Ozone

**NAS:** National Academy of Sciences

**NASA:** National Aeronautics and Space Administration

**NASDA:** National Association of State Departments of Agriculture

**NCAMP:** National Coalition Against the Misuse of Pesticides

**NCEPI:** National Center for Environmental Publications and Information

**NCWS:** Non-Community Water System

**NEDS:** National Emissions Data System

**NEIC:** National Enforcement Investigations Center

**NEPA:** National Environmental Policy Act

**NEPI:** National Environmental Policy Institute

**NEPPS:** National Environmental Performance Partnership System

**NESHAP:** National Emission Standard for Hazardous Air Pollutants

**NIEHS:** National Institute for Environmental Health Sciences

**NETA:** National Environmental Training Association

**NFRAP:** No Further Remedial Action Planned

**NICT:** National Incident Coordination Team

**NIOSH:** National Institute of Occupational Safety and Health

**NIPDWR:** National Interim Primary Drinking Water Regulations

**NISAC:** National Industrial Security Advisory Committee

**NMHC:** Nonmethane Hydrocarbons

**NMOC:** Non-Methane Organic Component

**NMVOC:** Non-methane Volatile Organic Chemicals

**NO:** Nitric Oxide

**NOý:** Nitrogen Dioxide

**NOA:** Notice of Arrival

**NOAA:** National Oceanographic and Atmospheric Agency

**NOAC:** Nature of Action Code

**NOAEL:** No Observable Adverse Effect Level

**NOEL:** No Observable Effect Level

**NOIC:** Notice of Intent to Cancel

**NOIS:** Notice of Intent to Suspend

**$N_2O$:** Nitrous Oxide

**NOV:** Notice of Violation

**$NO_x$:** Nitrogen Oxides

**NORM:** Naturally Occurring Radioactive Material

**NPCA:** National Pest Control Association

**NPDES:** National Pollutant Discharge Elimination System

**NPHAP:** National Pesticide Hazard Assessment Program

**NPIRS:** National Pesticide Information Retrieval System

**NPMS:** National Performance Measures Strategy

**NPTN:** National Pesticide Telecommunications Network

**NRD:** Natural Resource Damage

**NRDC:** Natural Resources Defense Council

**NSDWR:** National Secondary Drinking Water Regulations

**NSEC:** National System for Emergency Coordination

**NSEP:** National System for Emergency Preparedness

**NSPS:** New Source Performance Standards

**NSR:** New Source Review

**NSR/PSD:** National Source Review/Prevention of Significant Deterioration

**NTI:** National Toxics Inventory

**NTIS:** National Technical Information Service

**NTNCWS:** Non-Transient Non-Community Water System

**NTP:** National Toxicology Program

**NTU:** Nephlometric Turbidity Unit

## O

**$O_3$:** Ozone

**OAQPS:** Office of Air Quality Planning and Standards

**OCD:** Offshore and Coastal Dispersion

**ODP:** Ozone-Depleting Potential

**ODS:** Ozone-Depleting Substances

**OECA:** Office of Enforcement and Compliance Assurance

**OECD:** Organization for Economic Cooperation and Development

**OF:** Optional Form

**OI:** Order for Information

**OLC:** Office of Legal Counsel

**OLTS:** On Line Tracking System

**O&M:** Operations and Maintenance

**ORE:** Office of Regulatory Enforcement

**ORM:** Other Regulated Material

**ORP:** Oxidation-Reduction Potential

**OTAG:** Ozone Transport Assessment Group

**OTC:** Ozone Transport Commission

**OTIS:** Online Tracking Information System

**OTR:** Ozone Transport Region

## P

**P2:** Pollution Prevention

**PAG:** Pesticide Assignment Guidelines

**PAH:** Polynuclear Aromatic Hydrocarbons

**PAI:** Performance Audit Inspection (CWA); Pure Active Ingredient compound

**PAM:** Pesticide Analytical Manual

**PAMS:** Photochemical Assessment Monitoring Stations

**PAT:** Permit Assistance Team (RCRA)

**PATS:** Pesticide Action Tracking System; Pesticides Analytical Transport Solution

**Pb:** Lead

**PBA:** Preliminary Benefit Analysis (BEAD)

**PCA:** Principle Component Analysis

**PCB:** Polychlorinated Biphenyl

**PCE:** Perchloroethylene

**PCM:** Phase Contrast Microscopy

**PCN:** Policy Criteria Notice

**PCO:** Pest Control Operator

**PCSD:** President's Council on Sustainable Development

**PDCI:** Product Data Call-In

**PFC:** Perfluorated Carbon

**PFCRA:** Program Fraud Civil Remedies Act

**PHC:** Principal Hazardous Constituent

**PHI:** Pre-Harvest Interval

**PHSA:** Public Health Service Act

**PI:** Preliminary Injunction. Program Information

**PIC:** Products of Incomplete Combustion

**PIGS:** Pesticides in Groundwater Strategy

**PIMS:** Pesticide Incident Monitoring System

**PIN:** Pesticide Information Network

**PIN:** Procurement Information Notice

**PIP:** Public Involvement Program

**PIPQUIC:** Program Integration Project Queries Used in Interactive Command

**PIRG:** Public Interest Research Group

**PIRT:** Pretreatment Implementation Review Task Force

**PIT:** Permit Improvement Team

**PITS:** Project Information Tracking System

**PLIRRA:** Pollution Liability Insurance and Risk Retention Act

**PLM:** Polarized Light Microscopy

**PLUVUE:** Plume Visibility Model

**PM:** Particulate Matter

**PMAS:** Photochemical Assessment Monitoring Stations

**$PM_{2.5}$:** Particulate Matter Smaller than 2.5 Micrometers in Diameter

**$PM_{10}$:** Particulate Matter (nominally 10m and less)

**$PM_{15}$:** Particulate Matter (nominally 15m and less)

**PMEL:** Pacific Marine Environmental Laboratory

**PMN:** Premanufacture Notification

**PMNF:** Premanufacture Notification Form

**PMR:** Pollutant Mass Rate

**PMR:** Proportionate Mortality Ratio

**PMRS:** Performance Management and Recognition System

**PMS:** Program Management System

**PNA:** Polynuclear Aromatic Hydrocarbons

**PO:** Project Officer

**POC:** Point Of Compliance

**POE:** Point Of Exposure

**POGO:** Privately-Owned/ Government-Operated

**POHC:** Principal Organic Hazardous Constituent

**POI:** Point Of Interception

**POLREP:**Pollution Report

**POM:** Particulate Organic Matter. Polycyclic Organic Matter

**POP:** Persistent Organic Pollutant

**POR:** Program of Requirements

**POTW:** Publicly Owned Treatment Works

**POV:** Privately Owned Vehicle

**PP:** Program Planning

**PPA:** Planned Program Accomplishment

**PPB:** Parts Per Billion

**PPE:** Personal Protective Equipment

**PPG:** Performance Partnership Grant

**PPIC:** Pesticide Programs Information Center

**PPIS:** Pesticide Product Information System; Pollution Prevention Incentives for States

**PPMAP:** Power Planning Modeling Application Procedure

**PPM/PPB:** Parts per million/ parts per billion

**PPSP:** Power Plant Siting Program

**PPT:** Parts Per Trillion

**PPTH:** Parts Per Thousand

**PQUA:** Preliminary Quantitative Usage Analysis

**PR:** Pesticide Regulation Notice; Preliminary Review

**PRA:** Paperwork Reduction Act; Planned Regulatory Action

**PRATS:** Pesticides Regulatory Action Tracking System

**PRC:** Planning Research Corporation

**PRI:** Periodic Reinvestigation

**PRM:** Prevention Reference Manuals

**PRN:** Pesticide Registration Notice

**PRP:** Potentially Responsible Party

**PRZM:** Pesticide Root Zone Model

**PS:** Point Source

**PSAM:** Point Source Ambient Monitoring

**PSC:** Program Site Coordinator

**PSD:** Prevention of Significant Deterioration

**PSES:** Pretreatment Standards for Existing Sources

**PSI:** Pollutant Standards Index; Pounds Per Square Inch; Pressure Per Square Inch

**PSIG:** Pressure Per Square Inch Gauge

**PSM:** Point Source Monitoring

**PSNS:** Pretreatment Standards for New Sources

**PSU:** Primary Sampling Unit

**PTDIS:** Single Stack Meteorological Model in EPA UNAMAP Series

**PTE:** Potential to Emit

**PTFE:** Polytetrafluoroethylene (Teflon)

**PTMAX:** Single Stack Meteorological Model in EPA UNAMAP series

**PTPLU:** Point Source Gaussian Diffusion Model

**PUC:** Public Utility Commission

**PV:** Project Verification

**PVC:** Polyvinyl Chloride

**PWB:** Printed Wiring Board

**PWS:** Public Water Supply/ System

**PWSS:** Public Water Supply System

## Q

**QAC:** Quality Assurance Coordinator

**QA/QC:** Quality Assistance/ Quality Control

**QAMIS:** Quality Assurance Management and Information System

**QAO:** Quality Assurance Officer

**QAPP:** Quality Assurance Program (or Project) Plan

**QAT:** Quality Action Team

**QBTU:** Quadrillion British Thermal Units

**QC:** Quality Control

**QCA:** Quiet Communities Act

**QCI:** Quality Control Index

**QCP:** Quiet Community Program

**QL:** Quantification Limit

**QNCR:** Quarterly Noncompliance Report

**QUA:** Qualitative Use Assessment

**QUIPE:** Quarterly Update for Inspector in Pesticide Enforcement

## R

**RA:** Reasonable Alternative; Regulatory Alternatives; Regulatory Analysis; Remedial Action; Resource Allocation; Risk Analysis; Risk Assessment

**RAATS:** RCRA Administrate Action Tracking System

**RAC:** Radiation Advisory Committee. Raw Agricultural Commodity; Regional Asbestos Coordinator. Response Action Coordinator

**RACM:** Reasonably Available Control Measures

**RACT:** Reasonably Available Control Technology

**RAD:** Radiation Adsorbed Dose (unit of measurement of radiation absorbed by humans)

**RADM:** Random Walk Advection and Dispersion Model; Regional Acid Deposition Model

**RAM:** Urban Air Quality Model for Point and Area Source in EPA UNAMAP Series

**RAMP:** Rural Abandoned Mine Program

**RAMS:** Regional Air Monitoring System

**RAP:** Radon Action Program; Registration Assessment Panel; Remedial Accomplishment Plan; Response Action Plan

**RAPS:** Regional Air Pollution Study

**RARG:** Regulatory Analysis Review Group

**RAS:** Routine Analytical Service

**RAT:** Relative Accuracy Test

**RB:** Request for Bid

**RBAC:** Re-use Business Assistance Center

**RBC:** Red Blood Cell

**RC:** Responsibility Center

**RCC:** Radiation Coordinating Council

**RCDO:** Regional Case Development Officer

**RCO:** Regional Compliance Officer

**RCP:** Research Centers Program

**RCRA:** Resource Conservation and Recovery Act

**RCRIS:** Resource Conservation and Recovery Information System

**RD/RA:** Remedial Design/ Remedial Action

**R&D:** Research and Development

**RD&D:** Research, Development and Demonstration

**RDF:** Refuse-Derived Fuel

**RDNA:** Recombinant DNA

**RDU:** Regional Decision Units

**RDV:** Reference Dose Values

**RE:** Reasonable Efforts; Reportable Event

**REAP:** Regional Enforcement Activities Plan

**REE:** Rare Earth Elements

**REEP:** Review of Environmental Effects of Pollutants

**RECLAIM:** Regional Clean Air Initiatives Marker

**RED:** Reregistration Eligibility Decision Document

**REDA:** Recycling Economic Development Advocate

**ReFIT:** Reinvention for Innovative Technologies

**REI:** Restricted Entry Interval

**REM:** (Roentgen Equivalent Man)

**REM/FIT:** Remedial/Field Investigation Team

**REMS:** RCRA Enforcement Management System

**REP:** Reasonable Efforts Program

**REPS:** Regional Emissions Projection System

**RESOLVE:** Center for Environmental Conflict Resolution

**RF:** Response Factor

**RFA:** Regulatory Flexibility Act

**RFB:** Request for Bid

**RfC:** Reference Concentration

**RFD:** Reference Dose Values

**RFI:** Remedial Field Investigation

**RFP:** Reasonable Further Programs. Request for Proposal

**RHRS:** Revised Hazard Ranking System

**RI:** Reconnaissance Inspection

**RI:** Remedial Investigation

**RIA:** Regulatory Impact Analysis; Regulatory Impact Assessment

**RIC:** Radon Information Center

**RICC:** Retirement Information and Counseling Center

**RICO:** Racketeer Influenced and Corrupt Organizations Act

**RI/FS:** Remedial Investigation/ Feasibility Study

**RIM:** Regulatory Interpretation Memorandum

**RIN:** Regulatory Identifier Number

**RIP:** RCRA Implementation Plan

**RISC:** Regulatory Information Service Center

**RJE:** Remote Job Entry

**RLL:** Rapid and Large Leakage (Rate)

**RMCL:** Recommended Maximum Contaminant Level (this phrase being discontinued in favor of MCLG)

**RMDHS:** Regional Model Data Handling System

**RMIS:** Resources Management Information System

**RMP:** Risk Management Plan

**RNA:** Ribonucleic Acid

**ROADCHEM:** Roadway Version that Includes Chemical Reactions of BI, $NO_2$, and $O_3$

**ROADWAY:** A Model to Predict Pollutant Concentrations Near a Roadway

**ROC:** Record Of Communication

**RODS:** Records Of Decision System

**ROG:** Reactive Organic Gases

**ROLLBACK:** A Proportional Reduction Model

**ROM:** Regional Oxidant Model

**ROMCOE:** Rocky Mountain Center on Environment

**ROP:** Rate of Progress; Regional Oversight Policy

**ROPA:** Record Of Procurement Action

**ROSA:** Regional Ozone Study Area

**RP:** Radon Progeny Integrated Sampling. Respirable Particulates. Responsible Party

**RPAR:** Rebuttable Presumption Against Registration

**RPM:** Reactive Plume Model. Remedial Project Manager

**RQ:** Reportable Quantities

**RRC:** Regional Response Center

**RRT:** Regional Response Team; Requisite Remedial Technology

**RS:** Registration Standard

**RSCC:** Regional Sample Control Center

**RSD:** Risk-Specific Dose

**RSE:** Removal Site Evaluation

**RTCM:** Reasonable Transportation Control Measure

**RTDF:** Remediation Technologies Development Forum

**RTDM:** Rough Terrain Diffusion Model

**RTECS:** Registry of Toxic Effects of Chemical Substances

**RTM:** Regional Transport Model

**RTP:** Research Triangle Park

**RUP:** Restricted Use Pesticide

**RVP:** Reid Vapor Pressure

**RWC:** Residential Wood Combustion

## S

**S&A:** Sampling and Analysis. Surveillance and Analysis

**SAB:** Science Advisory Board

**SAC:** Suspended and Cancelled Pesticides

**SAEWG:** Standing Air Emissions Work Group

**SAIC:** Special-Agents-In-Charge

**SAIP:** Systems Acquisition and Implementation Program

**SAMI:** Southern Appalachian Mountains Initiative

**SAMWG:** Standing Air Monitoring Work Group

**SANE:** Sulfur and Nitrogen Emissions

**SANSS:** Structure and Nomenclature Search System

**SAP:** Scientific Advisory Panel

**SAR:** Start Action Request. Structural Activity Relationship (of a qualitative assessment)

**SARA:** Superfund Amendments and Reauthorization Act of 1986

**SAROAD:** Storage and Retrieval Of Aerometric Data

**SAS:** Special Analytical Service. Statistical Analysis System

**SASS:** Source Assessment Sampling System

**SAV:** Submerged Aquatic Vegetation

**SBC:** Single Breath Cannister

**SC:** Sierra Club

**SCAP:** Superfund Consolidated Accomplishments Plan

**SCBA:** Self-Contained Breathing Apparatus

**SCC:** Source Classification Code

**SCD/SWDC:** Soil or Soil and Water Conservation District

**SCFM:** Standard Cubic Feet Per Minute

**SCLDF:** Sierra Club Legal Defense Fund

**SCR:** Selective Catalytic Reduction

**SCRAM:** State Consolidated RCRA Authorization Manual

**SCRC:** Superfund Community Relations Coordinator

**SCS:** Supplementary Control Strategy/System

**SCSA:** Soil Conservation Society of America

**SCSP:** Storm and Combined Sewer Program

**SCW:** Supercritical Water Oxidation

**SDC:** Systems Decision Plan

**SDWA:** Safe Drinking Water Act

**SDWIS:** Safe Drinking Water Information System

**SBS:** Sick Building Syndrome

**SEA:** State Enforcement Agreement

**SEA:** State/EPA Agreement

**SEAM:** Surface, Environment, and Mining

**SEAS:** Strategic Environmental Assessment System

**SEDS:** State Energy Data System

**SEGIP:** State Environmental Goals and Improvement Project

**SEIA:** Socioeconomic Impact Analysis

**SEM:** Standard Error of the Means

**SEP:** Standard Evaluation Procedures

**SEP:** Supplementary Environmental Project

**SEPWC:** Senate Environment and Public Works Committee

**SERC:** State Emergency Planning Commission

**SES:** Secondary Emissions Standard

**SETAC:** Society for Environmental Toxicology and Chemistry

**SETS:** Site Enforcement Tracking System

**SF:** Standard Form. Superfund

**SFA:** Spectral Flame Analyzers

**SFDS:** Sanitary Facility Data System

**SFFAS:** Superfund Financial Assessment System

**SFIP:** Sector Facility Indexing Project

**SFIREG:** State FIFRA Issues Research and Evaluation Group

**SFS:** State Funding Study

**SHORTZ:** Short Term Terrain Model

**SHWL:** Seasonal High Water Level

**SI:** International System of Units. Site Inspection. Surveillance Index. Spark Ignition

**SIC:** Standard Industrial Classification

**SICEA:** Steel Industry Compliance Extension Act

**SIMS:** Secondary Ion-Mass Spectrometry

**SIP:** State Implementation Plan

**SITE:** Superfund Innovative Technology Evaluation

**SLAMS:** State/Local Air Monitoring Station

**SLN:** Special Local Need

**SLSM:** Simple Line Source Model

**SMART:** Simple Maintenance of ARTS

**SMCL:** Secondary Maximum Contaminant Level

**SMCRA:** Surface Mining Control and Reclamation Act

**SME:** Subject Matter Expert

**SMO:** Sample Management Office

**SMOA:** Superfund Memorandum of Agreement

**SMP:** State Management Plan

**SMR:** Standardized Mortality Ratio

**SMSA:** Standard Metropolitan Statistical Area

**SNA:** System Network Architecture

**SNAAQS:** Secondary National Ambient Air Quality Standards

**SNAP:** Significant New Alternatives Project; Significant Noncompliance Action Program

**SNARL:** Suggested No Adverse Response Level

**SNC:** Significant Noncompliers

**SNUR:** Significant New Use Rule

**$SO_2$:** Sulfur Dioxide

**SOC:** Synthetic Organic Chemicals

**SOCMI:** Synthetic Organic Chemicals Manufacturing Industry

**SOFC:** Solid Oxide Fuel Cell

**SOTDAT:** Source Test Data

**SOW:** Scope Of Work

**SPAR:** Status of Permit Application Report

**SPCC:** Spill Prevention, Containment, and Countermeasure

**SPE:** Secondary Particulate Emissions

**SPF:** Structured Programming Facility

**SPI:** Strategic Planning Initiative

**SPLMD:** Soil-pore Liquid Monitoring Device

**SPMS:** Strategic Planning and Management System; Special Purpose Monitoring Stations

**SPOC:** Single Point Of Contact

**SPS:** State Permit System

**SPSS:** Statistical Package for the Social Sciences

**SPUR:** Software Package for Unique Reports

**SQBE:** Small Quantity Burner Exemption

**SQG:** Small Quantity Generator

**SR:** Special Review

**SRAP:** Superfund Remedial Accomplishment Plan

**SRC:** Solvent-Refined Coal

**SRF:** State Revolving Fund

**SRM:** Standard Reference Method

**SRP:** Special Review Procedure

**SRR:** Second Round Review. Submission Review Record

**SRTS:** Service Request Tracking System

**SS:** Settleable Solids. Superfund Surcharge. Suspended Solids

**SSA:** Sole Source Aquifer

**SSAC:** Soil Site Assimilated Capacity

**SSC:** State Superfund Contracts

**SSD:** Standards Support Document

**SSEIS:** Standard Support and Environmental Impact Statement; Stationary Source Emissions and Inventory System.

**SSI:** Size Selective Inlet

**SSMS:** Spark Source Mass Spectrometry

**SSO:** Sanitary Sewer Overflow; Source Selection Official

**SSRP:** Source Reduction Review Project

**SSTS:** Section Seven Tracking System

**SSURO:** Stop Sale, Use and Removal Order

**STALAPCO:** State and Local Air-Pollution Control Officials

**STAPPA:** State and Territorial Air Pollution

**STAR:** Stability Wind Rose. State Acid Rain Projects

**STARS:** Strategic Targeted Activities for Results System

**STEL:** Short Term Exposure Limit

**STEM:** Scanning Transmission-Electron Microscope

**STN:** Scientific and Technical Information Network

**STORET:** Storage and Retrieval of Water-Related Data

**STP:** Sewage Treatment Plant. Standard Temperature and Pressure

**STTF:** Small Town Task Force (EPA)

**SUP:** Standard Unit of Processing

**SURE:** Sulfate Regional Experiment Program

**SV:** Sampling Visit; Significant Violater

**SW:** Slow Wave

**SWAP:** Source Water Assessment Program

**SWARF:** Waste from Metal Grinding Process

**SWC:** Settlement With Conditions

**SWDA:** Solid Waste Disposal Act

**SWIE:** Southern Waste Information Exchange

**SWMU:** Solid Waste Management Unit

**SWPA:** Source Water Protection Area

**SWQPPP:** Source Water Quality Protection Partnership Petitions

**SWTR:** Surface Water Treatment Rule

**SYSOP:** Systems Operator

## T

**TAD:** Technical Assistance Document

**TAG:** Technical Assistance Grant

**TALMS:** Tunable Atomic Line Molecular Spectroscopy

**TAMS:** Toxic Air Monitoring System

**TAMTAC:** Toxic Air Monitoring System Advisory Committee

**TAP:** Technical Assistance Program

**TAPDS:** Toxic Air Pollutant Data System

**TAS:** Tolerance Assessment System

**TBT:** Tributyltin

**TC:** Target Concentration. Technical Center. Toxicity Characteristics. Toxic Concentration:

**TCDD:** Dioxin (Tetrachlorodibenzo-p-dioxin)

**TCDF:** Tetrachlorodi-benzofurans

**TCE:** Trichloroethylene

**TCF:** Total Chlorine Free

**TCLP:** Total Concentrate Leachate Procedure. Toxicity Characteristic Leachate Procedure

**TCM:** Transportation Control Measure

**TCP:** Transportation Control Plan; Trichloropropane;

**TCRI:** Toxic Chemical Release Inventory

**TD:** Toxic Dose

**TDS:** Total Dissolved Solids

**TEAM:** Total Exposure Assessment Model

**TEC:** Technical Evaluation Committee

**TED:** Turtle Excluder Devices

**TEG:** Tetraethylene Glycol

**TEGD:** Technical Enforcement Guidance Document

**TEL:** Tetraethyl Lead

**TEM:** Texas Episodic Model

**TEP:** Typical End-use Product. Technical Evaluation Panel

**TERA:** TSCA Environmental Release Application

**TES:** Technical Enforcement Support

**TEXIN:** Texas Intersection Air Quality Model

**TGO:** Total Gross Output

**TGAI:** Technical Grade of the Active Ingredient

**TGP:** Technical Grade Product

**THC:** Total Hydrocarbons

**THM:** Trihalomethane

**TI:** Temporary Intermittent

**TI:** Therapeutic Index

**TIBL:** Thermal Internal Boundary Layer

**TIC:** Technical Information Coordinator. Tentatively Identified Compounds

**TIM:** Technical Information Manager

**TIP:** Technical Information Package

**TIP:** Transportation Improvement Program

**TIS:** Tolerance Index System

**TISE:** Take It Somewhere Else

**TITC:** Toxic Substance Control Act Interagency Testing Committee

**TLV:** Threshold Limit Value

**TLV-C:** TLV-Ceiling

**TLV-STEL:** TLV-Short Term Exposure Limit

**TLV-TWA:** TLV-Time Weighted Average

**TMDL:** Total Maximum Daily Limit; Total Maximum Daily Load

**TMRC:** Theoretical Maximum Residue Contribution

**TNCWS:** Transient Non-Community Water System

**TNT:** Trinitrotoluene

**TO:** Task Order

**TOA:** Trace Organic Analysis

**TOC:** Total Organic Carbon/ Compound

**TOX:** Tetradichloroxylene

**TP:** Technical Product; Total Particulates

**TPC:** Testing Priorities Committee

**TPI:** Technical Proposal Instructions

**TPQ:** Threshold Planning Quantity

**TPSIS:** Transportation Planning Support Information System

**TPTH:** Triphenyltinhydroxide

**TPY:** Tons Per Year

**TQM:** Total Quality Management

**T-R:** Transformer-Rectifier

**TRC:** Technical Review Committee

**TRD:** Technical Review Document

**TRI:** Toxic Release Inventory

**TRIP:** Toxic Release Inventory Program

**TRIS:** Toxic Chemical Release Inventory System

**TRLN:** Triangle Research Library Network

**TRO:** Temporary Restraining Order

**TSA:** Technical Systems Audit

**TSCA:** Toxic Substances Control Act

**TSCATS:** TSCA Test Submissions Database

**TSCC:** Toxic Substances Coordinating Committee

**TSD:** Technical Support Document

**TSDF:** Treatment, Storage, and Disposal Facility

**TSDG:** Toxic Substances Dialogue Group

**TSI:** Thermal System Insulation

**TSM:** Transportation System Management

**TSO:** Time Sharing Option

**TSP:** Total Suspended Particulates

**TSS:** Total Suspended (non-filterable) Solids

**TTFA:** Target Transformation Factor Analysis

**TTHM:** Total Trihalomethane

**TTN:** Technology Transfer Network

**TTO:** Total Toxic Organics

**TTY:** Teletypewriter

**TVA:** Tennessee Valley Authority

**TVOC:** Total Volatile Organic Compounds

**TWA:** Time Weighted Average

**TWS:** Transient Water System

**TZ:** Treatment Zone

## U

**UAC:** User Advisory Committee

**UAM:** Urban Airshed Model

**UAO:** Unilateral Administrative Order

**UAPSP:** Utility Acid Precipitation Study Program

**UAQI:** Uniform Air Quality Index

**UARG:** Utility Air Regulatory Group

**UCC:** Ultra Clean Coal

**UCCI:** Urea-Formaldehyde Foam Insulation

**UCL:** Upper Control Limit

**UDMH:** Unsymmetrical Dimethyl Hydrazine

**UEL:** Upper Explosive Limit

**UF:** Uncertainty Factor

**UFL:** Upper Flammability Limit

**ug/m$^3$:** Micrograms Per Cubic Meter

**UIC:** Underground Injection Control

**ULEV:** Ultra Low Emission Vehicles

**UMTRCA:** Uranium Mill Tailings Radiation Control Act

**UNAMAP:** Users' Network for Applied Modeling of Air Pollution

**UNECE:** United Nations Economic Commission for Europe

**UNEP:** United Nations Environment Program

**USC:** Unified Soil Classification

**USDA:** United States Department of Agriculture

**USDW:** Underground Sources of Drinking Water

**USFS:** United States Forest Service

**UST:** Underground Storage Tank

**UTM:** Universal Transverse Mercator

**UTP:** Urban Transportation Planning

**UV:** Ultraviolet

**UVA, UVB, UVC:** Ultraviolet Radiation Bands

**UZM:** Unsaturated Zone Monitoring

## V

**VALLEY:** Meteorological Model to Calculate Concentrations on Elevated Terrain

**VCM:** Vinyl Chloride Monomer

**VCP:** Voluntary Cleanup Program

**VE:** Visual Emissions

**VEO:** Visible Emission Observation

**VHS:** Vertical and Horizontal Spread Model

**VHT:** Vehicle-Hours of Travel

**VISTTA:** Visibility Impairment from Sulfur Transformation and Transport in the Atmosphere

**VKT:** Vehicle Kilometers Traveled

**VMT:** Vehicle Miles Traveled

**VOC:** Volatile Organic Compounds

**VOS:** Vehicle Operating Survey

**VOST:** Volatile Organic Sampling Train

**VP:** Vapor Pressure

**VSD:** Virtually Safe Dose

**VSI:** Visual Site Inspection

**VSS:** Volatile Suspended Solids

## W

**WA:** Work Assignment

**WADTF:** Western Atmospheric Deposition Task Force

**WAP:** Waste Analysis Plan

**WAVE:** Water Alliances for Environmental Efficiency

**WB:** Wet Bulb

**WCED:** World Commission on Environment and Development

**WDROP:** Distribution Register of Organic Pollutants in Water

**WENDB:** Water Enforcement National Data Base

**WERL:** Water Engineering Research Laboratory

**WET:** Whole Effluent Toxicity test

**WHO:** World Health Organization

**WHP:** Wellhead Protection Program

**WHPA:** Wellhead Protection Area

**WHWT:** Water and Hazardous Waste Team

**WICEM:** World Industry Conference on Environmental Management

**WL:** Warning Letter; Working Level (radon measurement)

**WLA/TMDL:** Wasteload Allocation/Total Maximum Daily Load

**WLM:** Working Level Months

**WMO:** World Meteorological Organization

**WP:** Wettable Powder

**WPCF:** Water Pollution Control Federation

**WQS:** Water Quality Standard

**WRC:** Water Resources Council

**WRDA:** Water Resources Development Act

**WRI:** World Resources Institute

**WS:** Work Status

**WSF:** Water Soluble Fraction

**WSRA:** Wild and Scenic Rivers Act

**WSTB:** Water Sciences and Technology Board

**WSTP:** Wastewater Sewage Treatment Plant

**WWEMA:** Waste and Wastewater Equipment Manufacturers Association

**WWF:** World Wildlife Fund

**WWTP:** Wastewater Treatment Plant

**WWTU:** Wastewater Treatment Unit

## Z

**ZEV:** Zero Emissions Vehicle

**ZHE:** Zero Headspace Extractor

**ZOI:** Zone Of Incorporation

**ZRL:** Zero Risk Level

***Note:*** *Some acronyms have more than one meaning. Multiple meanings are listed, separated by semi-colons.*

***Source:*** *U.S. Environmental Protection Agency, "Terms of Environment"*

# Glossary of Environmental Terms

## A

**Abandoned Well:** A well whose use has been permanently discontinued or which is in a state of such disrepair that it cannot be used for its intended purpose.

**Abatement:** Reducing the degree or intensity of, or eliminating, pollution.

**Abatement Debris:** Waste from remediation activities.

**Absorbed Dose:** In exposure assessment, the amount of a substance that penetrates an exposed organism's absorption barriers (e.g. skin, lung tissue, gastrointestinal tract) through physical or biological processes. The term is synonymous with internal dose.

**Absorption:** The uptake of water , other fluids, or dissolved chemicals by a cell or an organism (as tree roots absorb dissolved nutrients in soil.)

**Absorption Barrier:** Any of the exchange sites of the body that permit uptake of various substances at different rates (e.g. skin, lung tissue, and gastrointestinal-tract wall)

**Accident Site:** The location of an unexpected occurrence, failure or loss, either at a plant or along a transportation route, resulting in a release of hazardous materials.

**Acclimatization:** The physiological and behavioral adjustments of an organism to changes in its environment.

**Acid:** A corrosive solution with a pH less than 7.

**Acid Aerosol:** Acidic liquid or solid particles small enough to become airborne. High concentrations can irritate the lungs and have been associated with respiratory diseases like asthma.

**Acid Deposition:** A complex chemical and atmospheric phenomenon that occurs when emissions of sulfur and nitrogen compounds and other substances are transformed by chemical processes in the atmosphere, often far from the original sources, and then deposited on earth in either wet or dry form. The wet forms, popularly called "acid rain," can fall to earth as rain, snow, or fog. The dry forms are acidic gases or particulates.

**Acid Mine Drainage:** Drainage of water from areas that have been mined for coal or other mineral ores. The water has a low pH because of its contact with sulfur-bearing material and is harmful to aquatic organisms.

**Acid Neutralizing Capacity:** Measure of ability of a base (e.g. water or soil) to resist changes in pH.

**Acid Rain:** (See: acid deposition.)

**Acidic:** The condition of water or soil that contains a sufficient amount of acid substances to lower the pH below 7.0.

**Action Levels:** 1. Regulatory levels recommended by EPA for enforcement by FDA and USDA when pesticide residues occur in food or feed commodities for reasons other than the direct application of the pesticide. As opposed to "tolerances" which are established for residues occurring as a direct result of proper usage, action levels are set for inadvertent residues resulting from previous legal use or accidental contamination. 2. In the Superfund program, the existence of a contaminant concentration in the environment high enough to warrant action or trigger a response under SARA and the National Oil and Hazardous Substances Contingency Plan. The term is also used in other regulatory programs. (See: tolerances.)

**Activated Carbon:** A highly adsorbent form of carbon used to remove odors and toxic substances from liquid or gaseous emissions. In waste treatment, it is used to remove dissolved organic matter from waste drinking water. It is also used in motor vehicle evaporative control systems.

**Activated Sludge:** Product that results when primary effluent is mixed with bacteria-laden sludge and then agitated and aerated to promote biological treatment, speeding the breakdown of organic matter in raw sewage undergoing secondary waste treatment.

**Activator:** A chemical added to a pesticide to increase its activity.

**Active Ingredient:** In any pesticide product, the component that kills, or otherwise controls, target pests. Pesticides are regulated primarily on the basis of active ingredients.

**Activity Plans:** Written procedures in a school's asbestos-management plan that detail the steps a Local Education Agency (LEA) will follow in performing the initial and additional cleaning, operation and maintenance-program tasks; periodic surveillance; and reinspection required by the Asbestos Hazard Emergency Response Act (AHERA).

**Acute Effect:** An adverse effect on any living organism which results in severe symptoms that develop rapidly; symptoms often subside after the exposure stops.

**Acute Exposure:** A single exposure to a toxic substance which may result in severe biological harm or death. Acute exposures are usually characterized as lasting no longer than a day, as compared to longer, continuing exposure over a period of time.

**Acute Toxicity:** The ability of a substance to cause severe biological harm or death soon after a single exposure or dose. Also, any poisonous effect resulting from a single short-term exposure to a toxic substance. (See: chronic toxicity, toxicity.)

**Adaptation:** Changes in an organism's physiological structure or function or habits that allow it to survive in new surroundings.

**Add-on Control Device:** An air pollution control device such as carbon absorber or incinerator that reduces the pollution in an exhaust gas. The control device usually does not affect the process being controlled and thus is "add-on" technology, as opposed to a scheme to control pollution through altering the basic process itself.

**Adequately Wet:** Asbestos containing material that is sufficiently mixed or penetrated with liquid to prevent the release of particulates.

**Administered Dose:** In exposure assessment, the amount of a substance given to a test subject (human or animal) to determine dose-response relationships. Since exposure to chemicals is usually inadvertent, this quantity is often called potential dose.

**Administrative Order:** A legal document signed by EPA directing an individual, business, or other entity to take corrective action or refrain from an activity. It describes the violations and actions to be taken, and can be enforced in court. Such orders may be issued, for example, as a result of an administrative complaint whereby the respondent is ordered to pay a penalty for violations of a statute.

**Administrative Order On Consent:** A legal agreement signed by EPA and an individual, business, or other entity through which the violator agrees to pay for correction of violations, take the required corrective or cleanup actions, or refrain from an activity. It describes the actions to be taken, may be subject to a comment period, applies to civil actions, and can be enforced in court.

**Administrative Procedures Act:** A law that spells out procedures and requirements related to the promulgation of regulations.

**Administrative Record:** All documents which EPA considered or relied on in selecting the response action at a Superfund site, culminating in the record of decision for remedial action or, an action memorandum for removal actions.

**Adsorption:** Removal of a pollutant from air or water by collecting the pollutant on the surface of a solid material; e.g., an advanced method of treating waste in which activated carbon removes organic matter from waste-water.

**Adulterants:** Chemical impurities or substances that by law do not belong in a food, or pesticide.

**Adulterated:** 1. Any pesticide whose strength or purity falls below the quality stated on its label. 2. A food, feed, or product that contains illegal pesticide residues.

**Advanced Treatment:** A level of wastewater treatment more stringent than secondary treatment; requires an 85-percent reduction in conventional pollutant concentration or a significant reduction in non-conventional pollutants. Sometimes called tertiary treatment.

**Advanced Wastewater Treatment:** Any treatment of sewage that goes beyond the secondary or biological water treatment stage and includes the removal of nutrients such as phosphorus and nitrogen and a high percentage of suspended solids. (See primary, secondary treatment.)

**Adverse Effects Data:** FIFRA requires a pesticide registrant to submit data to EPA on any studies or other information regarding unreasonable adverse effects of a pesticide at any time after its registration.

**Advisory:** A non-regulatory document that communicates risk information to those who may have to make risk management decisions.

**Aerated Lagoon:** A holding and/or treatment pond that speeds up the natural process of biological decomposition of organic waste by stimulating the growth and activity of bacteria that degrade organic waste.

**Aeration:** A process which promotes biological degradation of organic matter in water. The process may be passive (as when waste is exposed to air), or active (as when a mixing or bubbling device introduces the air).

**Aeration Tank:** A chamber used to inject air into water.

**Aerobic:** Life or processes that require, or are not destroyed by, the presence of oxygen. (See: anaerobic.)

**Aerobic Treatment:** Process by which microbes decompose complex organic compounds in the presence of oxygen and use the liberated energy for reproduction and growth. (Such processes include extended aeration, trickling filtration, and rotating biological contactors.)

**Aerosol:** 1. Small droplets or particles suspended in the atmosphere, typically containing sulfur. They are usually emitted naturally (e.g. in volcanic eruptions) and as the result of anthropogenic (human) activities such as burning fossil fuels. 2. The pressurized gas used to propel substances out of a container.

**Aerosol:** A finely divided material suspended in air or other gaseous environment.

**Affected Landfill:** Under the Clean Air Act, landfills that meet criteria for capacity, age, and emissions rates set by the EPA. They are required to collect and combust their gas emissions.

**Affected Public:** 1.The people who live and/or work near a hazardous waste site. 2. The human population adversely impacted following exposure to a toxic pollutant in food, water, air, or soil.

**Afterburner:** In incinerator technology, a burner located so that the combustion gases are made to pass through its flame in order to remove smoke and odors. It may be attached to or be separated from the incinerator proper.

**Age Tank:** A tank used to store a chemical solution of known concentration for feed to a chemical feeder. Also called a day tank.

**Agent:** Any physical, chemical, or biological entity that can be harmful to an organism (synonymous with stressors.)

**Agent Orange:** A toxic herbicide and defoliant used in the Vietnam conflict, containing 2,4,5-trichlorophen-oxyacetic acid (2,4,5-T) and 2-4 dichlorophenoxyacetic acid (2,4-D) with trace amounts of dioxin.

**Agricultural Pollution:** Farming wastes, including runoff and leaching of pesticides and fertilizers; erosion and dust from plowing; improper disposal of animal manure and carcasses; crop residues, and debris.

**Agricultural Waste:** Poultry and livestock manure, and residual materials in liquid or solid form generated from the production and marketing of poultry, livestock or fur-bearing animals; also includes grain, vegetable, and fruit harvest residue.

**Agroecosystem:** Land used for crops, pasture, and livestock; the adjacent uncultivated land that supports other vegetation and wildlife; and the associated atmosphere, the underlying soils, groundwater, and drainage networks.

**AHERA Designated Person (ADP):** A person designated by a Local Education Agency to ensure that the AHERA requirements for asbestos management and abatement are properly implemented.

**Air Binding:** Situation where air enters the filter media and harms both the filtration and backwash processes.

**Air Changes Per Hour (ACH):** The movement of a volume of air in a given period of time; if a house has one air change per hour, it means that the air in the house will be replaced in a one-hour period.

**Air Cleaning:** Indoor-air quality-control strategy to remove various airborne particulates and/or gases from the air. Most common methods are particulate filtration, electrostatic precipitation, and gas sorption.

**Air Contaminant:** Any particulate matter, gas, or combination thereof, other than water vapor. (See: air pollutant.)

**Air Curtain:** A method of containing oil spills. Air bubbling through a perforated pipe causes an upward water flow that slows the spread of oil. It can also be used to stop fish from entering polluted water.

**Air Exchange Rate:** The rate at which outside air replaces indoor air in a given space.

**Air Gap:** Open vertical gap or empty space that separates drinking water supply to be protected from another water system in a treatment plant or other location. The open gap protects the drinking water from contamination by backflow or back siphonage.

**Air Handling Unit:** Equipment that includes a fan or blower, heating and/or cooling coils, regulator controls, condensate drain pans, and air filters.

**Air Mass:** A large volume of air with certain meteorological or polluted characteristics—e.g., a heat inversion or smogginess—while in one location. The characteristics can change as the air mass moves away.

**Air Monitoring:** (See: monitoring.)

**Air/Oil Table:** The surface between the vadose zone and ambient oil; the pressure of oil in the porous medium is equal to atmospheric pressure.

**Air Padding:** Pumping dry air into a container to assist with the withdrawal of liquid or to force a liquefied gas such as chlorine out of the container.

**Air Permeability:** Permeability of soil with respect to air. Important to the design of soil-gas surveys. Measured in darcys or centimeters-per-second.

**Air Plenum:** Any space used to convey air in a building, furnace, or structure. The space above a suspended ceiling is often used as an air plenum.

**Air Pollutant:** Any substance in air that could, in high enough concentration, harm man, other animals, vegetation, or material. Pollutants may include almost any natural or artificial composition of airborne matter capable of being airborne. They may be in the form of solid particles, liquid droplets, gases, or in combination thereof. Generally, they fall into two main groups: (1) those emitted directly from identifiable sources and (2) those produced in the air by interaction between two or more primary pollutants, or by reaction with normal atmospheric constituents, with or without photoactivation. Exclusive of pollen, fog, and dust, which are of natural origin, about 100 contaminants have been identified. Air pollutants are often grouped in categories for ease in classification; some of he categories are: solids, sulfur compounds, volatile organic chemicals, particulate matter, nitrogen compounds, oxygen compounds, halogen compounds, radioactive compound, and odors.

**Air Pollution:** The presence of contaminants or pollutant substances in the air that interfere with human health or welfare, or produce other harmful environmental effects.

**Air Pollution Control Device:** Mechanism or equipment that cleans emissions generated by a source (e.g. an incinerator, industrial smokestack, or an automobile exhaust system) by removing pollutants that would otherwise be released to the atmosphere.

**Air Pollution Episode:** A period of abnormally high concentration of air pollutants, often due to low winds and temperature inversion, that can cause illness and death. (See: episode, pollution.)

**Air Quality Control Region:**

**Air Quality Criteria:** The levels of pollution and lengths of exposure above which adverse health and welfare effects may occur.

**Air Quality Standards:** The level of pollutants prescribed by regulations that are not be exceeded during a given time in a defined area.

**Air Sparging:** Injecting air or oxygen into an aquifer to strip or flush volatile contaminants as air bubbles up through The ground water and is captured by a vapor extraction system.

**Air Stripping:** A treatment system that removes volatile organic compounds (VOCs) from contaminated ground water or surface water by forcing an airstream through the water and causing the compounds to evaporate.

**Air Toxics:** Any air pollutant for which a national ambient air quality standard (NAAQS) does not exist (i.e. excluding ozone, carbon monoxide, PM-10, sulfur dioxide, nitrogen oxide) that may reasonably be anticipated to cause cancer; respiratory, cardiovascular, or developmental effects; reproductive dysfunctions, neurological disorders, heritable gene mutations, or other serious or irreversible chronic or acute health effects in humans.

**Airborne Particulates:** Total suspended particulate matter found in the atmosphere as solid particles or liquid droplets. Chemical composition of particulates varies widely, depending on location and time of year. Sources of airborne particulates include: dust, emissions from industrial processes, combustion products from the burning of wood and coal, combustion products associated with motor vehicle or non-road engine exhausts, and reactions to gases in the atmosphere.

**Airborne Release:** Release of any pollutant into the air.

**Alachlor:** A herbicide, marketed under the trade name Lasso, used mainly to control weeds in corn and soybean fields.

**Alar:** Trade name for daminozide, a pesticide that makes apples redder, firmer, and less likely to drop off trees before growers are ready to pick them. It is also used to a lesser extent on peanuts, tart cherries, concord grapes, and other fruits.

**Aldicarb:** An insecticide sold under the trade name Temik. It is made from ethyl isocyanate.

**Algae:** Simple rootless plants that grow in sunlit waters in proportion to the amount of available nutrients. They can affect water quality adversely by lowering the dissolved oxygen in the water. They are food for fish and small aquatic animals.

**Algal Blooms:** Sudden spurts of algal growth, which can affect water quality adversely and indicate potentially hazardous changes in local water chemistry.

**Algicide:** Substance or chemical used specifically to kill or control algae.

**Aliquot:** A measured portion of a sample taken for analysis. One or more aliquots make up a sample. (See: duplicate.)

**Alkaline:** The condition of water or soil which contains a sufficient amount of alkali substance to raise the pH above 7.0.

**Alkalinity:** The capacity of bases to neutralize acids. An example is lime added to lakes to decrease acidity.

**Allergen:** A substance that causes an allergic reaction in individuals sensitive to it.

**Alluvial:** Relating to and/or sand deposited by flowing water.

**Alternate Method:** Any method of sampling and analyzing for an air or water pollutant that is not a reference or equivalent method but that has been demonstrated in specific cases-to EPA's satisfaction-to produce results adequate for compliance monitoring.

**Alternative Compliance:** A policy that allows facilities to choose among methods for achieving emission-reduction or risk-reduction instead of command-and control regulations that specify standards and how to meet them. Use of a theoretical emissions bubble over a facility to cap the amount of pollution emitted while allowing the company to choose where and how (within the facility) it complies.(See: bubble, emissions trading.)

**Alternative Fuels:** Substitutes for traditional liquid, oil-derived motor vehicle fuels like gasoline and diesel. Includes mixtures of alcohol-based fuels with gasoline, methanol, ethanol, compressed natural gas, and others.

**Alternative Remedial Contract Strategy Contractors:** Government contractors who provide project management and technical services to support remedial response activities at National Priorities List sites.

**Ambient Air:** Any unconfined portion of the atmosphere: open air, surrounding air.

**Ambient Air Quality Standards:** (See: Criteria Pollutants and National Ambient Air Quality Standards.)

**Ambient Measurement:** A measurement of the concentration of a substance or pollutant within the immediate environs of an organism; taken to relate it to the amount of possible exposure.

**Ambient Medium:** Material surrounding or contacting an organism (e.g. outdoor air, indoor air, water, or soil, through which chemicals or pollutants can reach the organism. (See: biological medium, environmental medium.)

**Ambient Temperature:** Temperature of the surrounding air or other medium.

**Amprometric Titration:** A way of measuring concentrations of certain substances in water using an electric current that flows during a chemical reaction.

**Anaerobic:** A life or process that occurs in, or is not destroyed by, the absence of oxygen.

**Anaerobic Decomposition:** Reduction of the net energy level and change in chemical composition of organic matter caused by microorganisms in an oxygen-free environment.

**Animal Dander:** Tiny scales of animal skin, a common indoor air pollutant.

**Animal Studies:** Investigations using animals as surrogates for humans with the expectation that the results are pertinent to humans.

**Anisotropy:** In hydrology, the conditions under which one or more hydraulic properties of an aquifer vary from a reference point.

**Annular Space, Annulus:** The space between two concentric tubes or casings, or between the casing and the borehole wall.

**Antagonism:** Interference or inhibition of the effect of one chemical by the action of another.

**Antarctic "Ozone Hole":** Refers to the seasonal depletion of ozone in the upper atmosphere above a large area of Antarctica. (See: Ozone Hole.)

**Anti-Degradation Clause:** Part of federal air quality and water quality requirements prohibiting deterioration where pollution levels are above the legal limit.

**Anti-Microbial:** An agent that kills microbes.

**Applicable or Relevant and Appropriate Requirements (ARARs):** Any state or federal statute that pertains to protection of human life and the environment in addressing specific conditions or use of a particular cleanup technology at a Superfund site,

**Applied Dose:** In exposure assessment, the amount of a substance in contact with the primary absorption boundaries of an organism (e.g. skin, lung tissue, gastrointestinal track) and available for absorption.

**Aqueous:** Something made up of water.

**Aqueous Solubility:** The maximum concentration of a chemical that will dissolve in pure water at a reference temperature.

**Aquifer:** An underground geological formation, or group of formations, containing water. Are sources of groundwater for wells and springs.

**Aquifer Test:** A test to determine hydraulic properties of an aquifer.

**Aquitard:** Geological formation that may contain groundwater but is not capable of transmitting significant quantities of it under normal hydraulic gradients. May function as confining bed.

**Architectural Coatings:** Coverings such as paint and roof tar that are used on exteriors of buildings.

**Area of Review:** In the UIC program, the area surrounding an injection well that is reviewed during the permitting process to determine if flow between aquifers will be induced by the injection operation.

**Area Source:** Any source of air pollution that is released over a relatively small area but which cannot be classified as a point source. Such sources may include vehicles and other small engines, small businesses and household activities, or biogenic sources such as a forest that releases hydrocarbons.

**Aromatics:** A type of hydrocarbon, such as benzene or toluene, with a specific type of ring structure. Aromatics are sometimes added to gasoline in order to increase octane. Some aromatics are toxic.

**Arsenicals:** Pesticides containing arsenic.

**Artesian (Aquifer or Well):** Water held under pressure in porous rock or soil confined by impermeable geological formations.

**Asbestos:** A mineral fiber that can pollute air or water and cause cancer or asbestosis when inhaled. EPA has banned or severely restricted its use in manufacturing and construction.

**Asbestos Abatement:** Procedures to control fiber release from asbestos-containing materials in a building or to remove them entirely, including removal, encapsulation, repair, enclosure, encasement, and operations and maintenance programs.

**Asbestos Assessment:** In the asbestos-in-schools program, the evaluation of the physical condition and potential for damage of all friable asbestos containing materials and thermal insulation systems.

**Asbestos Program Manager:** A building owner or designated representative who supervises all aspects of the facility asbestos management and control program.

**Asbestos-Containing Waste Materials (ACWM):** Mill tailings or any waste that contains commercial asbestos and is generated by a source covered by the Clean Air Act Asbestos NESHAPS.

**Asbestosis:** A disease associated with inhalation of asbestos fibers. The disease makes breathing progressively more difficult and can be fatal.

**Ash:** The mineral content of a product remaining after complete combustion.

**Assay:** A test for a specific chemical, microbe, or effect.

**Assessment Endpoint:** In ecological risk assessment, an explicit expression of the environmental value to be protected; includes both an ecological entity and specific attributed thereof. entity (e.g. salmon are a valued ecological entity; reproduction and population maintenance—the attribute—form an assessment endpoint.)

**Assimilation:** The ability of a body of water to purify itself of pollutants.

**Assimilative Capacity:** The capacity of a natural body of water to receive wastewaters or toxic materials without deleterious effects and without damage to aquatic life or humans who consume the water.

**Association of Boards of Certification:** An international organization representing boards which certify the operators of waterworks and wastewater facilities.

**Attainment Area:** An area considered to have air quality as good as or better than the national ambient air quality standards as defined in the Clean Air Act. An area may be an attainment area for one pollutant and a non-attainment area for others.

**Attenuation:** The process by which a compound is reduced in concentration over time, through absorption, adsorption, degradation, dilution, and/or transformation. an also be the decrease with distance of sight caused by attenuation of light by particulate pollution.

**Attractant:** A chemical or agent that lures insects or other pests by stimulating their sense of smell.

**Attrition:** Wearing or grinding down of a substance by friction. Dust from such processes contributes to air pollution.

**Availability Session:** Informal meeting at a public location where interested citizens can talk with EPA and state officials on a one-to-one basis.

**Available Chlorine:** A measure of the amount of chlorine available in chlorinated lime, hypochlorite compounds, and other materials used as a source of chlorine when compared with that of liquid or gaseous chlorines.

**Avoided Cost:** The cost a utility would incur to generate the next increment of electric capacity using its own resources; many landfill gas projects' buy back rates are based on avoided costs.

**A-Scale Sound Level:** A measurement of sound approximating the sensitivity of the human ear, used to note the intensity or annoyance level of sounds.

## B

**Back Pressure:** A pressure that can cause water to backflow into the water supply when a user's waste water system is at a higher pressure than the public system.

**Backflow/Back Siphonage:** A reverse flow condition created by a difference in water pressures that causes water to flow back into the distribution pipes of a drinking water supply from any source other than the intended one.

**Background Level:** 1. The concentration of a substance in an environmental media (air, water, or soil) that occurs naturally or is not the result of human activities. 2. In exposure assessment the concentration of a substance in a defined control area, during a fixed period of time before, during, or after a data-gathering operation..

**Backwashing:** Reversing the flow of water back through the filter media to remove entrapped solids.

**Backyard Composting:** Diversion of organic food waste and yard trimmings from the municipal waste stream by composting hem in one's yard through controlled decomposition of organic matter by bacteria and fungi into a humus-like product. It is considered source reduction, not recycling, because the composted materials never enter the municipal waste stream.

**Barrel Sampler:** Open-ended steel tube used to collect soil samples.

**BACT - Best Available Control Technology:** An emission limitation based on the maximum degree of emission reduction (considering energy, environmental, and economic impacts) achievable through application of production processes and available methods, systems, and techniques. BACT does not permit emissions in excess of those allowed under any applicable Clean Air Act provisions. Use of the BACT concept is allowable on a case by case basis for major new or modified emissions sources in attainment areas and applies to each regulated pollutant.

**Bacteria:** (Singular: bacterium) Microscopic living organisms that can aid in pollution control by metabolizing organic matter in sewage, oil spills or other pollutants. However, bacteria in soil, water or air can also cause human, animal and plant health problems.

**Bactericide:** A pesticide used to control or destroy bacteria, typically in the home, schools, or hospitals.

**Baffle:** A flat board or plate, deflector, guide, or similar device constructed or placed in flowing water or slurry systems to cause more uniform flow velocities to absorb energy and to divert, guide, or agitate liquids.

**Baffle Chamber:** In incinerator design, a chamber designed to promote the settling of fly ash and coarse particulate matter by changing the direction and/or reducing the velocity of the gases produced by the combustion of the refuse or sludge.

**Baghouse Filter:** Large fabric bag, usually made of glass fibers, used to eliminate intermediate and large (greater than 20 PM in diameter) particles. This device operates like the bag of an electric vacuum cleaner, passing the air and smaller particles while entrapping the larger ones.

**Bailer:** A pipe with a valve at the lower end, used to remove slurry from the bottom or side of a well as it is being drilled, or to collect groundwater samples from wells or open boreholes. 2. A tube of varying length.

**Baling:** Compacting solid waste into blocks to reduce volume and simplify handling.

**Ballistic Separator:** A machine that sorts organic from inorganic matter for composting.

**Band Application:** The spreading of chemicals over, or next to, each row of plants in a field.

**Banking:** A system for recording qualified air emission reductions for later use in bubble, offset, or netting transactions. (See: emissions trading.)

**Bar Screen:** In wastewater treatment, a device used to remove large solids.

**Barrier Coating(s):** A layer of a material that obstructs or prevents passage of something through a surface that is to be protected; e.g., grout, caulk, or various sealing compounds; sometimes used with polyurethane membranes to prevent corrosion or oxidation of metal surfaces, chemical impacts on various materials, or, for example, to prevent radon infiltration through walls, cracks, or joints in a house.

**Basal Application:** In pesticides, the application of a chemical on plant stems or tree trunks just above the soil line.

**Basalt:** Consistent year-round energy use of a facility; also refers to the minimum amount of electricity supplied continually to a facility.

**Bean Sheet:** Common term for a pesticide data package record.

**Bed Load:** Sediment particles resting on or near the channel bottom that are pushed or rolled along by the flow of water.

**BEN:** EPA's computer model for analyzing a violator's economic gain from not complying with the law.

**Bench-scale Tests:** Laboratory testing of potential cleanup technologies (See: treatability studies.)

**Benefit-Cost Analysis:** An economic method for assessing the benefits and costs of achieving alternative health-based standards at given levels of health protection.

**Benthic/Benthos:** An organism that feeds on the sediment at the bottom of a water body such as an ocean, lake, or river.

**Bentonite:** A colloidal clay, expansible when moist, commonly used to provide a tight seal around a well casing.

**Beryllium:** An metal hazardous to human health when inhaled as an airborne pollutant. It is discharged by machine shops, ceramic and propellant plants, and foundries.

**Best Available Control Measures (BACM):** A term used to refer to the most effective measures (according to EPA guidance) for controlling small or dispersed particulates and other emissions from sources such as roadway dust, soot and ash from woodstoves and open burning of rush, timber, grasslands, or trash.

**Best Available Control Technology (BACT):** For any specific source, the currently available technology producing the greatest reduction of air pollutant emissions,taking into account energy, environmental, economic, and other costs.

**Best Available Control Technology (BACT):** The most stringent technology available for controlling emissions; major sources are required to use BACT, unless it can be demonstrated that it is not feasible for energy, environmental, or economic reasons.

**Best Demonstrated Available Technology (BDAT):** As identified by EPA, the most effective commercially available means of treating specific types of hazardous waste. The BDATs may change with advances in treatment technologies.

**Best Management Practice (BMP):** Methods that have been determined to be the most effective, practical means of preventing or reducing pollution from non-point sources.

**Bimetal:** Beverage containers with steel bodies and aluminum tops; handled differently from pure aluminum in recycling.

**Bioaccumulants:** Substances that increase in concentration in living organisms as they take in contaminated air, water, or food because the substances are very slowly metabolized or excreted. (See: biological magnification.)

**Bioassay:** A test to determine te relative strength of a substance by comparing its effect on a test organism with that of a standard preparation.

**Bioavailability:** Degree of ability to be absorbed and ready to interact in organism metabolism.

**Biochemical Oxygen Demand (BOD):** A measure of the amount of oxygen consumed in the biological processes that break down organic matter in water. The greater the BOD, the greater the degree of pollution.

**Bioconcentration:** The accumulation of a chemical in tissues of a fish or other organism to levels greater than in the surrounding medium.

**Biodegradable:** Capable of decomposing under natural conditions.

**Biodiversity:** Refers to the variety and variability among living organisms and the ecological complexes in which they occur. Diversity can be defined as the number of different items and their relative frequencies. For biological diversity, these items are organized at many levels, ranging from complete ecosystems to the biochemical structures that are the molecular basis of heredity. Thus, the term encompasses different ecosystems, species, and genes.

**Biological Contaminants:** Living organisms or derivates (e.g. viruses, bacteria, fungi, and mammal and bird antigens) that can cause harmful health effects when inhaled, swallowed, or otherwise taken into the body.

**Biological Control:** In pest control, the use of animals and organisms that eat or otherwise kill or out-compete pests.

**Biological Integrity:** The ability to support and maintain balanced, integrated, functionality in the natural habitat of a given region. Concept is applied primarily in drinking water management.

**Biological Magnification:** Refers to the process whereby certain substances such as pesticides or heavy metals move up the food chain, work their way into rivers or lakes, and are eaten by aquatic organisms such as fish, which in turn are eaten by large birds, animals or humans. The substances become concentrated in tissues or internal organs as they move up the chain. (See: bioaccumulants.)

**Biological Measurement:** A measurement taken in a biological medium. For exposure assessment, it is related to the measurement is taken to related it to the established internal dose of a compound.

**Biological Medium:** One of the major component of an organism; e.g. blood, fatty tissue, lymph nodes or breath, in which chemicals can be stored or transformed. (See: ambient medium, environmental medium.)

**Biological Oxidation:** Decomposition of complex organic materials by microorganisms. Occurs in self-purification of water bodies and in activated sludge wastewater treatment.

**Biological Oxygen Demand (BOD):** An indirect measure of the concentration of biologically degradable material present in organic wastes. It usually reflects the amount of oxygen consumed in five days by biological processes breaking down organic waste.

**Biological pesticides:** Certain microorganism, including bacteria, fungi, viruses, and protozoa that are effective in controlling pests. These agents usually do not have toxic effects on animals and people and do not leave toxic or persistent chemical residues in the environment.

**Biological Stressors:** Organisms accidentally or intentionally dropped into habitats in which they do not evolve naturally; e.g. gypsy moths, Dutch elm disease, certain types of algae, and bacteria.

**Biological Treatment:** A treatment technology that uses bacteria to consume organic waste.

**Biologically Effective Dose:** The amount of a deposited or absorbed compound reaching the cells or target sites where adverse effect occur, or where the chemical interacts with a membrane.

**Biologicals:** Vaccines, cultures and other preparations made from living organisms and their products, intended for use in diagnosing, immunizing, or treating humans or animals, or in related research.

**Biomass:** All of the living material in a given area; often refers to vegetation.

**Biome:** Entire community of living organisms in a single major ecological area. (See: biotic community.)

**Biomonitoring:** 1. The use of living organisms to test the suitability of effluents for discharge into receiving waters and to test the quality of such waters downstream from the discharge. 2. Analysis of blood, urine, tissues, etc. to measure chemical exposure in humans.

**Bioremediation:** Use of living organisms to clean up oil spills or remove other pollutants from soil, water, or wastewater; use of organisms such as non-harmful insects to remove agricultural pests or counteract diseases of trees, plants, and garden soil.

**Biosensor:** Analytical device comprising a biological recognition element (e.g. enzyme, receptor, DNA, antibody, or microorganism) in intimate contact with an electrochemical, optical, thermal, or acoustic signal transducer that together permit analyses of chemical properties or quantities. Shows potential development in some areas, including environmental monitoring.

**Biosphere:** The portion of Earth and its atmosphere that can support life.

**Biostabilizer:** A machine that converts solid waste into compost by grinding and aeration.

**Biota:** The animal and plant life of a given region.

**Biotechnology:** Techniques that use living organisms or parts of organisms to produce a variety of products (from medicines to industrial enzymes) to improve plants or animals or to develop microorganisms to remove toxics from bodies of water, or act as pesticides.

**Biotic Community:** A naturally occurring assemblage of plants and animals that live in the same environment and are mutually sustaining and interdependent. (See: biome.)

**Biotransformation:** Conversion of a substance into other compounds by organisms; includes biodegredation.

**Blackwater:** Water that contains animal, human, or food waste.

**Blood Products:** Any product derived from human blood, including but not limited to blood plasma, platelets, red or white corpuscles, and derived licensed products such as interferon.

**Bloom:** A proliferation of algae and/or higher aquatic plants in a body of water; often related to pollution, especially when pollutants accelerate growth.

**BOD5:** The amount of dissolved oxygen consumed in five days by biological processes breaking down organic matter.

**Body Burden:** The amount of a chemical stored in the body at a given time, especially a potential toxin in the body as the result of exposure.

**Bog:** A type of wetland that accumulates appreciable peat deposits. Bogs depend primarily on precipitation for their water source, and are usually acidic and rich in plant residue with a conspicuous mat of living green moss.

**Boiler:** A vessel designed to transfer heat produced by combustion or electric resistance to water. Boilers may provide hot water or steam.

**Boom:** 1. A floating device used to contain oil on a body of water. 2. A piece of equipment used to apply pesticides from a tractor or truck.

**Borehole:** Hole made with drilling equipment.

**Botanical Pesticide:** A pesticide whose active ingredient is a plant-produced chemical such as nicotine or strychnine. Also called a plant-derived pesticide.

**Bottle Bill:** Proposed or enacted legislation which requires a returnable deposit on beer or soda containers and provides for retail store or other redemption. Such legislation is designed to discourage use of throw-away containers.

**Bottom Ash:** The non-airborne combustion residue from burning pulverized coal in a boiler; the material which falls to the bottom of the boiler and is removed mechanically; a concentration of non-combustible materials, which may include toxics.

**Bottom Land Hardwoods:** Forested freshwater wetlands adjacent to rivers in the southeastern United States, especially valuable for wildlife breeding, nesting and habitat.

**Bounding Estimate:** An estimate of exposure, dose, or risk that is higher than that incurred by the person in the population with the currently highest exposure, dose, or risk. Bounding estimates are useful in developing statements that exposures, doses, or risks are not greater than an estimated value.

**Brackish:** Mixed fresh and salt water.

**Breakpoint Chlorination:** Addition of chlorine to water until the chlorine demand has been satisfied.

**Breakthrough:** A crack or break in a filter bed that allows the passage of floc or particulate matter through a filter; will cause an increase in filter effluent turbidity.

**Breathing Zone:** Area of air in which an organism inhales.

**Brine Mud:** Waste material, often associated with well-drilling or mining, composed of mineral salts or other inorganic compounds.

**British Thermal Unit:** Unit of heat energy equal to the amount of heat required to raise the temperature of one pound of water by one degree Fahrenheit at sea level.

**Broadcast Application:** The spreading of pesticides over an entire area.

**Brownfields:** Abandoned, idled, or under used industrial and commercial facilities/sites where expansion or redevelopment is complicated by real or perceived environmental contamination. They can be in urban, suburban, or rural areas. EPA's Brownfields initiative helps communities mitigate potential health risks and restore the economic viability of such areas or properties.

**Bubble:** A system under which existing emissions sources can propose alternate means to comply with a set of emissions limitations; under the bubble concept, sources can control more than required at one emission point where control costs are relatively low in return for a comparable relaxation of controls at a second emission point where costs are higher.

**Bubble Policy:** (See: emissions trading.)

**Buffer:** A solution or liquid whose chemical makeup is such that it minimizes changes in pH when acids or bases are added to it.

**Buffer Strips:** Strips of grass or other erosion-resisting vegetation between or below cultivated strips or fields.

**Building Cooling Load:** The hourly amount of heat that must be removed from a building to maintain indoor comfort (measured in British thermal units (Btus).

**Building Envelope:** The exterior surface of a building's construction—the walls, windows, floors, roof, and floor. Also called building shell.

**Building Related Illness:** Diagnosable illness whose cause and symptoms can be directly attributed to a specific pollutant source within a building (e.g. Legionnaire's disease, hypersensitivity, pneumonitis.) (See: sick building syndrome.)

**Bulk Sample:** A small portion (usually thumbnail size) of a suspect asbestos-containing building material collected by an asbestos inspector for laboratory analysis to determine asbestos content.

**Bulky Waste:** Large items of waste materials, such as appliances, furniture, large auto parts, trees, stumps.

**Burial Ground (Graveyard):** A disposal site for radioactive waste materials that uses earth or water as a shield.

**Buy-Back Center:** Facility where individuals or groups bring reyclables in return for payment.

**By-product:** Material, other than the principal product, generated as a consequence of an industrial process or as a breakdown product in a living system.

## C

**Cadmium (Cd):** A heavy metal that accumulates in the environment.

**Cancellation:** Refers to Section 6 (b) of the Federal Insecticide, Fungicide and Rodenticide Act (FIFRA) which authorizes cancellation of a pesticide registration if unreasonable adverse effects to the environment and public health develop when a product is used according to widespread and commonly recognized practice, or if its labeling or other material required to be submitted does not comply with FIFRA provisions.

**Cap:** A layer of clay, or other impermeable material installed over the top of a closed landfill to prevent entry of rainwater and minimize leachate.

**Capacity Assurance Plan:** A statewide plan which supports a state's ability to manage the hazardous waste generated within its boundaries over a twenty year period.

**Capillary Action:** Movement of water through very small spaces due to molecular forces called capillary forces.

**Capillary Fringe:** The porous material just above the water table which may hold water by capillarity (a property of surface tension that draws water upwards) in the smaller void spaces.

**Capillary Fringe:** The zone above he water table within which the porous medium is saturated by water under less than atmospheric pressure.

**Capture Efficiency:** The fraction of organic vapors generated by a process that are directed to an abatement or recovery device.

**Carbon Absorber:** An add-on control device that uses activated carbon to absorb volatile organic compounds from a gas stream. (The VOCs are later recovered from the carbon.)

**Carbon Adsorption:** A treatment system that removes contaminants from ground water or surface water by forcing it through tanks containing activated carbon treated to attract the contaminants.

**Carbon Monoxide (CO):** A colorless, odorless, poisonous gas produced by incomplete fossil fuel combustion.

**Carbon Tetrachloride (CC14):** Compound consisting of one carbon atom ad four chlorine atoms, once widely used as a industrial raw material, as a solvent, and in the production of

CFCs. Use as a solvent ended when it was discovered to be carcinogenic.

**Carboxyhemoglobin:** Hemoglobin in which the iron is bound to carbon monoxide(CO) instead of oxygen.

**Carcinogen:** Any substance that can cause or aggravate cancer.

**Carrier:** 1.The inert liquid or solid material in a pesticide product that serves as a delivery vehicle for the active ingredient. Carriers do not have toxic properties of their own. 2. Any material or system that can facilitate the movement of a pollutant into the body or cells.

**Carrying Capacity:** 1. In recreation management, the amount of use a recreation area can sustain without loss of quality. 2. In wildlife management, the maximum number of animals an area can support during a given period.

**CAS Registration Number:** A number assigned by the Chemical Abstract Service to identify a chemical.

**Case Study:** A brief fact sheet providing risk, cost, and performance information on alternative methods and other pollution prevention ideas, compliance initiatives, voluntary efforts, etc.

**Cask:** A thick-walled container (usually lead) used to transport radioactive material. Also called a coffin.

**Catalyst:** A substance that changes the speed or yield of a chemical reaction without being consumed or chemically changed by the chemical reaction.

**Catalytic Converter:** An air pollution abatement device that removes pollutants from motor vehicle exhaust, either by oxidizing them into carbon dioxide and water or reducing them to nitrogen.

**Catalytic Incinerator:** A control device that oxidizes volatile organic compounds (VOCs) by using a catalyst to promote the combustion process. Catalytic incinerators require lower temperatures than conventional thermal incinerators, thus saving fuel and other costs.

**Categorical Exclusion:** A class of actions which either individually or cumulatively would not have a significant effect on the human environment and therefore would not require preparation of an environmental assessment or environmental impact statement under the National Environmental Policy Act (NEPA).

**Categorical Pretreatment Standard:** A technology-based effluent limitation for an industrial facility discharging into a municipal sewer system. Analogous in stringency to Best Availability Technology (BAT) for direct dischargers.

**Cathodic Protection:** A technique to prevent corrosion of a metal surface by making it the cathode of an electrochemical cell.

**Cavitation:** The formation and collapse of gas pockets or bubbles on the blade of an impeller or the gate of a valve; collapse of these pockets or bubbles drives water with such force that it can cause pitting of the gate or valve surface.

**Cells:** 1. In solid waste disposal, holes where waste is dumped, compacted, and covered with layers of dirt on a daily basis. 2. The smallest structural part of living matter capable of functioning as an independent unit.

**Cementitious:** Densely packed and nonfibrous friable materials.

**Central Collection Point:** Location were a generator of regulated medical waste consolidates wastes originally generated at various locations in his facility. The wastes are gathered together for treatment on-site or for transportation elsewhere for treatment and/or disposal. This term could also apply to community hazardous waste collections, industrial and other waste management systems.

**Centrifugal Collector:** A mechanical system using centrifugal force to remove aerosols from a gas stream or to remove water from sludge.

**CERCLIS:** The federal Comprehensive Environmental Response, Compensation, and Liability Information System is a database that includes all sites which have been nominated for investigation by the Superfund program.

**Channelization:** Straightening and deepening streams so water will move faster, a marsh-drainage tactic that can interfere with waste assimilation capacity, disturb fish and wildlife habitats, and aggravate flooding.

**Characteristic:** Any one of the four categories used in defining hazardous waste: ignitability, corrosivity, reactivity, and toxicity.

**Characterization of Ecological Effects:** Part of ecological risk assessment that evaluates ability of a stressor to cause adverse effects under given circumstances.

**Characterization of Exposure:** Portion of an ecological risk assessment that evaluates interaction of a stressor with one or more ecological entities.

**Check-Valve Tubing Pump:** Water sampling tool also referred to as a water Pump.

**Chemical Case:** For purposes of review and regulation, the grouping of chemically similar pesticide active ingredients (e.g. salts and esters of the same chemical) into chemical cases.

**Chemical Compound:** A distinct and pure substance formed by the union or two or more elements in definite proportion by weight.

**Chemical Element:** A fundamental substance comprising one kind of atom; the simplest form of matter.

**Chemical Oxygen Demand (COD):** A measure of the oxygen required to oxidize all compounds, both organic and inorganic, in water.

**Chemical Stressors:** Chemicals released to the environment through industrial waste, auto emissions, pesticides, and other human activity that can cause illnesses and even death in plants and animals.

**Chemical Treatment:** Any one of a variety of technologies that use chemicals or a variety of chemical processes to treat waste.

**Chemnet:** Mutual aid network of chemical shippers and contractors that assigns a contracted emergency response company to provide technical support if a representative of the firm whose chemicals are involved in an incident is not readily available.

**Chemosterilant:** A chemical that controls pests by preventing reproduction.

**Chemtrec:** The industry-sponsored Chemical Transportation Emergency Center; provides information and/or emergency assistance to emergency responders.

**Child Resistant Packaging (CRP):** Packaging that protects children or adults from injury or illness resulting from accidental contact with or ingestion of residential pesticides that meet or exceed specific toxicity levels. Required by FIFRA regulations. Term is also used for protective packaging of medicines.

**Chiller:** A device that generates a cold liquid that is circulated through an air-handling unit's cooling coil to cool the air supplied to the building.

**Chilling Effect:** The lowering of the Earth's temperature because of increased particles in the air blocking the sun's rays. (See: greenhouse effect.)

**Chisel Plowing:** Preparing croplands by using a special implement that avoids complete inversion of the soil as in conventional plowing. Chisel plowing can leave a protective cover or crops residues on the soil surface to help prevent erosion and improve filtration.

**Chlorinated Hydrocarbons:** 1. Chemicals containing only chlorine, carbon, and hydrogen. These include a class of persistent, broad-spectrum insecticides that linger in the environment and accumulate in the food chain. Among them are DDT, aldrin, dieldrin, heptachlor, chlordane, lindane, endrin, Mirex, hexachloride, and toxaphene. Other examples include TCE, used as an industrial solvent. 2. Any chlorinated organic compounds including chlorinated solvents such as dichloromethane, trichloromethylene, chloroform.

**Chlorinated Solvent:** An organic solvent containing chlorine atoms(e.g. methylene chloride and 1,1,1-trichloromethane). Uses of chlorinated solvents are include aerosol spray containers, in highway paint, and dry cleaning fluids.

**Chlorination:** The application of chlorine to drinking water, sewage, or industrial waste to disinfect or to oxidize undesirable compounds.

**Chlorinator:** A device that adds chlorine, in gas or liquid form, to water or sewage to kill infectious bacteria.

**Chlorine-Contact Chamber:** That part of a water treatment plant where effluent is disinfected by chlorine.

**Chlorofluorocarbons (CFCs):** A family of inert, nontoxic, and easily liquefied chemicals used in refrigeration, air conditioning, packaging, insulation, or as solvents and aerosol propellants. Because CFCs are not destroyed in the lower atmosphere they drift into the upper atmosphere where their chlorine components destroy ozone. (See: fluorocarbons.)

**Chlorophenoxy:** A class of herbicides that may be found in domestic water supplies and cause adverse health effects.

**Chlorosis:** Discoloration of normally green plant parts caused by disease, lack of nutrients, or various air pollutants.

**Cholinesterase:** An enzyme found in animals that regulates nerve impulses by the inhibition of acetylcholine. Cholinesterase inhibition is associated with a variety of acute symptoms such as nausea, vomiting, blurred vision, stomach cramps, and rapid heart rate.

**Chromium:** (See: heavy metals.)

**Chronic Effect:** An adverse effect on a human or animal in which symptoms recur frequently or develop slowly over a long period of time.

**Chronic Exposure:** Multiple exposures occurring over an extended period of time or over a significant fraction of an animal's or human's lifetime (Usually seven years to a lifetime.)

**Chronic Toxicity:** The capacity of a substance to cause long-term poisonous health effects in humans, animals, fish, and other organisms. (See: acute toxicity.)

**Circle of Influence:** The circular outer edge of a depression produced in the water table by the pumping of water from a well. (See: cone of depression.)

**Cistern:** Small tank or storage facility used to store water for a home or farm; often used to store rain water.

**Clarification:** Clearing action that occurs during wastewater treatment when solids settle out. This is often aided by centrifugal action and chemically induced coagulation in wastewater.

**Clarifier:** A tank in which solids settle to the bottom and are subsequently removed as sludge.

**Class I Area:** Under the Clean Air Act. a Class I area is one in which visibility is protected more stringently than under the national ambient air quality standards; includes national parks, wilderness areas, monuments, and other areas of special national and cultural significance.

**Class I Substance:** One of several groups of chemicals with an ozone depletion potential of 0.2 or higher, including CFCS, Halons, Carbon Tetrachloride, and Methyl Chloroform (listed in the Clean Air Act), and HBFCs and Ethyl Bromide (added by EPA regulations). (See: Global warming potential.)

**Class II Substance:** A substance with an ozone depletion potential of less than 0.2. All HCFCs are currently included in this classification. (See: Global warming potential.)

**Clay Soil:** Soil material containing more than 40 percent clay, less than 45 percent sand, and less than 40 percent silt.

**Clean Coal Technology:** Any technology not in widespread use prior to the Clean Air Act Amendments of 1990. This Act will achieve significant reductions in pollutants associated with the burning of coal.

**Clean Fuels:** Blends or substitutes for gasoline fuels, including compressed natural gas, methanol, ethanol, and liquified petroleum gas.

**Cleaner Technologies Substitutes Assessment:** A document that systematically evaluates the relative risk, performance, and cost trade-offs of technological alternatives; serves as a repository for all the technical data (including methodology and results) developed by a DfE or other pollution prevention or education project.

**Cleanup:** Actions taken to deal with a release or threat of release of a hazardous substance that could affect humans and/or the environment. The term “cleanup” is sometimes used interchangeably with the terms remedial action, removal action, response action, or corrective action.

**Clear Cut:** Harvesting all the trees in one area at one time, a practice that can encourage fast rainfall or snowmelt runoff, erosion, sedimentation of streams and lakes, and flooding, and destroys vital habitat.

**Clear Well:** A reservoir for storing filtered water of sufficient quantity to prevent the need to vary the filtration rate with variations in demand. Also used to provide chlorine contact time for disinfection.

**Climate Change (also referred to as ‘global climate change’):** The term ‘climate change’ is sometimes used to refer to all forms of climatic inconsistency, but because the Earth's climate is never static, the term is more properly used to imply a significant change from one climatic condition to another. In some cases, ‘climate change’ has been used synonymously with the term, ‘global warming’; scientists however, tend to use the term in the wider sense to also include natural changes in climate. (See: global warming.)

**Cloning:** In biotechnology, obtaining a group of genetically identical cells from a single cell; making identical copies of a gene.

**Closed-Loop Recycling:** Reclaiming or reusing wastewater for non-potable purposes in an enclosed process.

**Closure:** The procedure a landfill operator must follow when a landfill reaches its legal capacity for solid ceasing acceptance of solid waste and placing a cap on the landfill site.

**Co-fire:** Burning of two fuels in the same combustion unit; e.g., coal and natural gas, or oil and coal.

**Coagulation:** Clumping of particles in wastewater to settle out impurities, often induced by chemicals such as lime, alum, and iron salts.

**Coal Cleaning Technology:** A precombustion process by which coal is physically or chemically treated to remove some of its sulfur so as to reduce sulfur dioxide emissions.

**Coal Gasification:** Conversion of coal to a gaseous product by one of several available technologies.

**Coastal Zone:** Lands and waters adjacent to the coast that exert an influence on the uses of the sea and its ecology, or whose uses and ecology are affected by the sea.

**Code of Federal Regulations (CFR):** Document that codifies all rules of the executive departments and agencies of the federal government. It is divided into fifty volumes, known as titles. Title 40 of the CFR (referenced as 40 CFR) lists all environmental regulations.

**Coefficient of Haze (COH):** A measurement of visibility interference in the atmosphere.

**Cogeneration:** The consecutive generation of useful thermal and electric energy from the same fuel source.

**Coke Oven:** An industrial process which converts coal into coke, one of the basic materials used in blast furnaces for the conversion of iron ore into iron.

**Cold Temperature CO:** A standard for automobile emissions of carbon monoxide (CO) emissions to be met at a low temperature (i.e. 20 degrees Fahrenheit). Conventional automobile catalytic converters are not efficient in cold weather until they warm up.

**Coliform Index:** A rating of the purity of water based on a count of fecal bacteria.

**Coliform Organism:** Microorganisms found in the intestinal tract of humans and animals. Their presence in water indicates fecal pollution and potentially adverse contamination by pathogens.

**Collector:** Public or private hauler that collects nonhazardous waste and recyclable materials from residential, commercial, institutional and industrial sources. (See: hauler.)

**Collector Sewers:** Pipes used to collect and carry wastewater from individual sources to an interceptor sewer that will carry it to a treatment facility.

**Colloids:** Very small, finely divided solids (that do not dissolve) that remain dispersed in a liquid for a long time due to their small size and electrical charge.

**Combined Sewer Overflows:** Discharge of a mixture of storm water and domestic waste when the flow capacity of a sewer system is exceeded during rainstorms.

**Combined Sewers:** A sewer system that carries both sewage and storm-water runoff. Normally, its entire flow goes to a waste treatment plant, but during a heavy storm, the volume of water may be so great as to cause overflows of untreated mixtures of storm water and sewage into receiving waters. Storm-water runoff may also carry toxic chemicals from industrial areas or streets into the sewer system.

**Combustion:** 1. Burning, or rapid oxidation, accompanied by release of energy in the form of heat and light. 2. Refers to controlled burning of waste, in which heat chemically alters organic compounds, converting into stable inorganics such as carbon dioxide and water.

**Combustion Chamber:** The actual compartment where waste is burned in an incinerator.

**Combustion Product:** Substance produced during the burning or oxidation of a material.

**Command Post:** Facility located at a safe distance upwind from an accident site, where the on-scene coordinator, responders, and technical representatives make response decisions, deploy manpower and equipment, maintain liaison with news media, and handle communications.

**Command-and-Control Regulations:** Specific requirements prescribing how to comply with specific standards defining acceptable levels of pollution.

**Comment Period:** Time provided for the public to review and comment on a proposed EPA action or rulemaking after publication in the Federal Register.

**Commercial Waste:** All solid waste emanating from business establishments such as stores, markets, office buildings, restaurants, shopping centers, and theaters.

**Commercial Waste Management Facility:** A treatment, storage, disposal, or transfer facility which accepts waste from a variety of sources, as compared to a private facility which normally manages a limited waste stream generated by its own operations.

**Commingled Recyclables:** Mixed recyclables that are collected together.

**Comminuter:** A machine that shreds or pulverizes solids to make waste treatment easier.

**Comminution:** Mechanical shredding or pulverizing of waste. Used in both solid waste management and wastewater treatment.

**Common Sense Initiative:** Voluntary program to simplify environmental regulation to achieve cleaner, cheaper, smarter results, starting with six major industry sectors.

**Community:** In ecology, an assemblage of populations of different species within a specified location in space and time. Sometimes, a particular subgrouping may be specified, such as the fish community in a lake or the soil arthropod community in a forest.

**Community Relations:** The EPA effort to establish two-way communication with the public to create understanding of EPA programs and related actions, to ensure public input into decision-making processes related to affected communities, and to make certain that the Agency is aware of and responsive to public concerns. Specific community relations activities are required in relation to Superfund remedial actions.

**Community Water System:** A public water system which serves at least 15 service connections used by year-round residents or regularly serves at least 25 year-round residents.

**Compact Fluorescent Lamp (CFL):** Small fluorescent lamps used as more efficient alternatives to incandescent lighting. Also called PL, CFL, Twin-Tube, or BIAX lamps.

**Compaction:** Reduction of the bulk of solid waste by rolling and tamping.

**Comparative Risk Assessment:** Process that generally uses the judgement of experts to predict effects and set priorities among a wide range of environmental problems.

**Complete Treatment:** A method of treating water that consists of the addition of coagulant chemicals, flash mixing, coagulation-flocculation, sedimentation, and filtration. Also called conventional filtration.

**Compliance Coal:** Any coal that emits less than 1.2 pounds of sulfur dioxide per million Btu when burned. Also known as low sulfur coal.

**Compliance Coating:** A coating whose volatile organic compound content does not exceed that allowed by regulation.

**Compliance Cycle:** The 9-year calendar year cycle, beginning January 1, 1993, during which public water systems must monitor. Each cycle consists of three 3-year compliance periods.

**Compliance Monitoring:** Collection and evaluation of data, including self-monitoring reports, and verification to show whether pollutant concentrations and loads contained in permitted discharges are in compliance with the limits and conditions specified in the permit.

**Compliance Schedule:** A negotiated agreement between a pollution source and a government agency that specifies dates and procedures by which a source will reduce emissions and, thereby, comply with a regulation.

**Composite Sample:** A series of water samples taken over a given period of time and weighted by flow rate.

**Compost:** The relatively stable humus material that is produced from a composting process in which bacteria in soil mixed with garbage and degradable trash break down the mixture into organic fertilizer.

**Composting:** The controlled biological decomposition of organic material in the presence of air to form a humus-like material. Controlled methods of composting include mechanical mixing and aerating, ventilating the materials by dropping them through a vertical series of aerated chambers, or placing the compost in piles out in the open air and mixing it or turning it periodically.

**Composting Facilities:** 1. An offsite facility where the organic component of municipal solid waste is decomposed under controlled conditions; 2.an aerobic process in which organic materials are ground or shredded and then decomposed to humus in windrow piles or in mechanical digesters, drums, or similar enclosures.

**Compressed Natural Gas (CNG):** An alternative fuel for motor vehicles; considered one of the cleanest because of low hydrocarbon emissions and its vapors are relatively non-ozone producing. However, vehicles fueled with CNG do emit a significant quantity of nitrogen oxides.

**Concentration:** The relative amount of a substance mixed with another substance. An example is five ppm of carbon monoxide in air or 1 mg/l of iron in water.

**Condensate:** 1.Liquid formed when warm landfill gas cools as it travels through a collection system. 2. Water created by cooling steam or water vapor.

**Condensate Return System:** System that returns the heated water condensing within steam piping to the boiler and thus saves energy.

**Conditional Registration:** Under special circumstances, the Federal Insecticide, Fungicide, and Rodenticide Act (FIFRA) permits registration of pesticide products that is "conditional" upon the submission of additional data. These special circumstances include a finding by the EPA Administrator that a new product or use of an existing pesticide will not significantly increase the risk of unreasonable adverse effects. A product containing a new (previously unregistered) active ingredient may be conditionally registered only if the Administrator finds that such conditional registration is in the public interest, that a reasonable time for conducting the additional studies has not elapsed, and the use of the pesticide for the period of conditional registration will not present an unreasonable risk.

**Conditionally Exempt Generators (CE):** Persons or enterprises which produce less than 220 pounds of hazardous waste per month. Exempt from most regulation, they are required merely to determine whether their waste is hazardous, notify appropriate state or local agencies, and ship it by an authorized transporter to a permitted facility for proper disposal. (See : small quantity generator.)

**Conductance:** A rapid method of estimating the dissolved solids content of water supply by determining the capacity of a water sample to carry an electrical current. Conductivity is a measure of the ability of a solution to carry and electrical current.

**Conductivity:** A measure of the ability of a solution to carry an electrical current.

**Cone of Depression:** A depression in the water table that develops around a pumped well.

**Cone of Influence:** The depression, roughly conical in shape, produced in a water table by the pumping of water from a well.

**Cone Penterometer Testing (CPT):** A direct push system used to measure lithology based on soil penetration resistance. Sensors in the tip of the cone of the DP rod measure tip resistance and side-wall friction, transmitting electrical signals to digital processing equipment on the ground surface. (See: direct push.)

**Confidential Business Information (CBI):** Material that contains trade secrets or commercial or financial information that has been claimed as confidential by its source (e.g. a pesticide or new chemical formulation registrant). EPA has special procedures for handling such information.

**Confidential Statement of Formula (CSF):** A list of the ingredients in a new pesticide or chemical formulation. The list is submitted at the time for application for registration or change in formulation.

**Confined Aquifer:** An aquifer in which ground water is confined under pressure which is significantly greater than atmospheric pressure.

**Confluent Growth:** A continuous bacterial growth covering all or part of the filtration area of a membrane filter in which the bacteria colonies are not discrete.

**Consent Decree:** A legal document, approved by a judge, that formalizes an agreement reached between EPA and potentially responsible parties (PRPs) through which PRPs will conduct all or part of a cleanup action at a Superfund site; cease or correct actions or processes that are polluting the environment; or otherwise comply with EPA initiated regulatory enforcement actions to resolve the contamination at the Superfund site involved. The consent decree describes the actions PRPs will take and may be subject to a public comment period.

**Conservation:** Preserving and renewing, when possible, human and natural resources. The use, protection, and improvement of natural resources according to principles that will ensure their highest economic or social benefits.

**Conservation Easement:** Easement restricting a landowner to land uses that that are compatible with long-term conservation and environmental values.

**Constituent(s) of Concern:** Specific chemicals that are identified for evaluation in the site assessment process

**Construction and Demolition Waste:** Waste building materials, dredging materials, tree stumps, and rubble resulting from construction, remodeling, repair, and demolition of homes, commercial buildings and other structures and pavements. May contain lead, asbestos, or other hazardous substances.

**Construction Ban:** If, under the Clean Air Act, EPA disapproves an area's planning requirements for correcting nonattainment, EPA can ban the construction or modification of any major stationary source of the pollutant for which the area is in nonattainment.

**Consumptive Water Use:** Water removed from available supplies without return to a water resources system, e.g. water used in manufacturing, agriculture, and food preparation.

**Contact Pesticide:** A chemical that kills pests when it touches them, instead of by ingestion. Also, soil that contains the minute skeletons of certain algae that scratch and dehydrate waxy-coated insects.

**Contaminant:** Any physical, chemical, biological, or radiological substance or matter that has an adverse effect on air, water, or soil.

**Contamination:** Introduction into water, air, and soil of microorganisms, chemicals, toxic substances, wastes, or wastewater in a concentration that makes the medium unfit for its next intended use. Also applies to surfaces of objects, buildings, and various household and agricultural use products.

**Contamination Source Inventory:** An inventory of contaminant sources within delineated State Water-Protection Areas. Targets likely sources for further investigation.

**Contingency Plan:** A document setting out an organized, planned, and coordinated course of action to be followed in case of a fire, explosion, or other accident that releases toxic chemicals, hazardous waste, or radioactive materials that threaten human health or the environment. (See: National Oil and Hazardous Substances Contingency Plan.)

**Continuous Discharge:** A routine release to the environment that occurs without interruption, except for infrequent shutdowns for maintenance, process changes, etc.

**Continuous Sample:** A flow of water, waste or other material from a particular place in a plant to the location where samples are collected for testing. May be used to obtain grab or composite samples.

**Contour Plowing:** Soil tilling method that follows the shape of the land to discourage erosion.

**Contour Strip Farming:** A kind of contour farming in which row crops are planted in strips, between alternating strips of close-growing, erosion-resistant forage crops.

**Contract Labs:** Laboratories under contract to EPA, which analyze samples taken from waste, soil, air, and water or carry out research projects.

**Control Technique Guidelines (CTG):** EPA documents designed to assist state and local pollution authorities to achieve and maintain air quality standards for certain sources (e.g. organic emissions from solvent metal cleaning known as degreasing) through reasonably available control technologies (RACT).

**Controlled Reaction:** A chemical reaction under temperature and pressure conditions maintained within safe limits to produce a desired product or process.

**Conventional Filtration:** (See: complete treatment.)

**Conventional Pollutants:** Statutorily listed pollutants understood well by scientists. These may be in the form of organic waste, sediment, acid, bacteria, viruses, nutrients, oil and grease, or heat.

**Conventional Site Assessment:** Assessment in which most of the sample analysis and interpretation of data is completed off-site; process usually requires repeated mobilization of equipment and staff in order to fully determine the extent of contamination.

**Conventional Systems:** Systems that have been traditionally used to collect municipal wastewater in gravity sewers and convey it to a central primary or secondary treatment plant prior to discharge to surface waters.

**Conventional Tilling:** Tillage operations considered standard for a specific location and crop and that tend to bury the crop residues; usually considered as a base for determining the cost effectiveness of control practices.

**Conveyance Loss:** Water loss in pipes, channels, conduits, ditches by leakage or evaporation.

**Cooling Electricity Use:** Amount of electricity used to meet the building cooling load. (See: building cooling load.)

**Cooling Tower:** A structure that helps remove heat from water used as a coolant; e.g., in electric power generating plants.

**Cooling Tower:** Device which dissipates the heat from water-cooled systems by spraying the water through streams of rapidly moving air.

**Cooperative Agreement:** An assistance agreement whereby EPA transfers money, property, services or anything of value to a state, university, non-profit, or not-for-profit organization for the accomplishment of authorized activities or tasks.

**Core:** The uranium-containing heart of a nuclear reactor, where energy is released.

**Core Program Cooperative Agreement:** An assistance agreement whereby EPA supports states or tribal governments with funds to help defray the cost of non-item-specific administrative and training activities.

**Corrective Action:** EPA can require treatment, storage and disposal (TSDF) facilities handling hazardous waste to undertake corrective actions to clean up spills resulting from failure to follow hazardous waste management procedures or other mistakes. The process includes cleanup procedures designed to guide TSDFs toward in spills.

**Corrosion:** The dissolution and wearing away of metal caused by a chemical reaction such as between water and the pipes, chemicals touching a metal surface, or contact between two metals.

**Corrosive:** A chemical agent that reacts with the surface of a material causing it to deteriorate or wear away.

**Cost/Benefit Analysis:** A quantitative evaluation of the costs which would have incurred by implementing an environmental regulation versus the overall benefits to society of the proposed action.

**Cost Recovery:** A legal process by which potentially responsible parties who contributed to contamination at a Superfund site can be required to reimburse the Trust Fund for money spent during any cleanup actions by the federal government.

**Cost Sharing:** A publicly financed program through which society, as a beneficiary of environmental protection, shares part of the cost of pollution control with those who must actually install the controls. In Superfund, for example, the government may pay part of the cost of a cleanup action with those responsible for the pollution paying the major share.

**Cost-Effective Alternative:** An alternative control or corrective method identified after analysis as being the best available in terms of reliability, performance, and cost. Although costs are one important consideration, regulatory and compliance analysis does not require EPA to choose the least expensive alternative. For example, when selecting or approving a method for cleaning up a Superfund site, the Agency balances costs with the long-term effectiveness of the methods proposed and the potential danger posed by the site.

**Cover Crop:** A crop that provides temporary protection for delicate seedlings and/or provides a cover canopy for seasonal soil protection and improvement between normal crop production periods.

**Cover Material:** Soil used to cover compacted solid waste in a sanitary landfill.

**Cradle-to-Grave or Manifest System:** A procedure in which hazardous materials are identified and followed as they are produced, treated, transported, and disposed of by a series of permanent, linkable, descriptive documents (e.g. manifests). Commonly referred to as the cradle-to-grave system.

**Criteria:** Descriptive factors taken into account by EPA in setting standards for various pollutants. These factors are used to determine limits on allowable concentration levels, and to limit the number of violations per year. When issued by EPA, the criteria provide guidance to the states on how to establish their standards.

**Criteria Pollutants:** The 1970 amendments to the Clean Air Act required EPA to set National Ambient Air Quality Standards for certain pollutants known to be hazardous to human health. EPA has identified and set standards to protect human health and welfare for six pollutants: ozone, carbon monoxide, total suspended particulates, sulfur dioxide, lead, and nitrogen oxide. The term, "criteria pollutants" derives from the requirement that EPA must describe the characteristics and potential health and welfare effects of these pollutants. It is on the basis of these criteria that standards are set or revised.

**Critical Effect:** The first adverse effect, or its known precursor, that occurs as a dose rate increases. Designation is based on evaluation of overall database.

**Crop Consumptive Use:** The amount of water transpired during plant growth plus what evaporated from the soil surface and foliage in the crop area.

**Crop Rotation:** Planting a succession of different crops on the same land rea as opposed to planting the same crop time after time.

**Cross Contamination:** The movement of underground contaminants from one level or area to another due to invasive subsurface activities.

**Cross-Connection:** Any actual or potential connection between a drinking water system and an unapproved water supply or other source of contamination.

**Crumb Rubber:** Ground rubber fragments the size of sand or silt used in rubber or plastic products, or processed further into reclaimed rubber or asphalt products.

**Cryptosporidium:** A protozoan microbe associated with the disease cryptosporidiosis in man. The disease can be transmitted through ingestion of drinking water, person-to-person contact, or other pathways, and can cause acute diarrhea, abdominal pain, vomiting, fever, and can be fatal as it was in the Milwaukee episode.

**Cubic Feet Per Minute (CFM):** A measure of the volume of a substance flowing through air within a fixed period of time. With regard to indoor air, refers to the amount of air, in cubic feet, that is exchanged with outdoor air in a minute's time; i.e. the air exchange rate.

**Cullet:** Crushed glass.

**Cultural Eutrophication:** Increasing rate at which water bodies "die" by pollution from human activities.

**Cultures and Stocks:** Infectious agents and associated biologicals including cultures from medical and pathological laboratories; cultures and stocks of infectious agents from research and industrial laboratories; waste from the production of biologicals; discarded live and attenuated vaccines; and culture dishes and devices used to transfer, inoculate, and mix cultures. (See: regulated medical waste.)

**Cumulative Ecological Risk Assessment:** Consideration of the total ecological risk from multiple stressors to a given eco-zone.

**Cumulative Exposure:** The sum of exposures of an organism to a pollutant over a period of time.

**Cumulative Working Level Months (CWLM):** The sum of lifetime exposure to radon working levels expressed in total working level months.

**Curb Stop:** A water service shutoff valve located in a water service pipe near the curb and between the water main and the building.

**Curbside Collection:** Method of collecting recyclable materials at homes, community districts or businesses.

**Cutie-Pie:** An instrument used to measure radiation levels.

**Cuttings:** Spoils left by conventional drilling with hollow stem auger or rotary drilling equipment.

**Cyclone Collector:** A device that uses centrifugal force to remove large particles from polluted air.

## D

**Data Call-In:** A part of the Office of Pesticide Programs (OPP) process of developing key required test data, especially on the long-term, chronic effects of existing pesticides, in advance of scheduled Registration Standard reviews. Data Call-In from manufacturers is an adjunct of the Registration Standards program intended to expedite re-registration.

**Data Quality Objectives (DQOs):** Qualitative and quantitative statements of the overall level of uncertainty that a decision-maker will accept in results or decisions based on environmental data. They provide the statistical framework for planning and managing environmental data operations consistent with user's needs.

**Day Tank:** Another name for deaerating tank. (See: age tank.)

**DDT:** The first chlorinated hydrocarbon insecticide chemical name: Dichloro-Diphenyl-Trichloroethane. It has a half-life of 15 years and can collect in fatty tissues of certain animals. EPA banned registration and interstate sale of DDT for virtually all but emergency uses in the United States in 1972 because of its persistence in the environment and accumulation in the food chain.

**Dead End:** The end of a water main which is not connected to other parts of the distribution system.

**Deadmen:** Anchors drilled or cemented into the ground to provide additional reactive mass for DP sampling rigs.

**Decant:** To draw off the upper layer of liquid after the heaviest material (a solid or another liquid) has settled.

**Decay Products:** Degraded radioactive materials, often referred to as "daughters" or "progeny"; radon decay products of most concern from a public health standpoint are polonium-214 and polonium-218.

**Dechlorination:** Removal of chlorine from a substance.

**Decomposition:** The breakdown of matter by bacteria and fungi, changing the chemical makeup and physical appearance of materials.

**Decontamination:** Removal of harmful substances such as noxious chemicals, harmful bacteria or other organisms, or radioactive material from exposed individuals, rooms and furnishings in buildings, or the exterior environment.

**Deep-Well Injection:** Deposition of raw or treated, filtered hazardous waste by pumping it into deep wells, where it is contained in the pores of permeable subsurface rock.

**Deflocculating Agent:** A material added to a suspension to prevent settling.

**Defluoridation:** The removal of excess flouride in drinking water to prevent the staining of teeth.

**Defoliant:** An herbicide that removes leaves from trees and growing plants.

**Degasification:** A water treatment that removes dissolved gases from the water.

**Degree-Day:** A rough measure used to estimate the amount of heating required in a given area; is defined as the difference between the mean daily temperature and 65 degrees Fahrenheit. Degree-days are also calculated to estimate cooling requirements.

**Delegated State:** A state (or other governmental entity such as a tribal government) that has received authority to administer an environmental regulatory program in lieu of a federal counterpart. As used in connection with NPDES, UIC, and PWS programs, the term does not connote any transfer of federal authority to a state.

**Delist:** Use of the petition process to have a facility's toxic designation rescinded.

**Demand-side Waste Management:** Prices whereby consumers use purchasing decisions to communicate to product manufacturers that they prefer environmentally sound products packaged with the least amount of waste, made from recycled or recyclable materials, and containing no hazardous substances.

**Demineralization:** A treatment process that removes dissolved minerals from water.

**Denitrification:** The biological reduction of nitrate to nitrogen gas by denitrifying bacteria in soil.

**Dense Non-Aqueous Phase Liquid (DNAPL):** Non-aqueous phase liquids such as chlorinated hydrocarbon solvents or petroleum fractions with a specific gravity greater than 1.0 that sink through the water column until they reach a confining layer. Because they are at the bottom of aquifers instead of floating on the water table, typical monitoring wells do not indicate their presence.

**Density:** A measure of how heavy a specific volume of a solid, liquid, or gas is in comparison to water. depending on the chemical.

**Depletion Curve:** In hydraulics, a graphical representation of water depletion from storage-stream channels, surface soil, and groundwater. A depletion curve can be drawn for base flow, direct runoff, or total flow.

**Depressurization:** A condition that occurs when the air pressure inside a structure is lower that the air pressure outdoors. Depressurization can occur when household appliances such as fireplaces or furnaces, that consume or exhaust house air, are not supplied with enough makeup air. Radon may be drawn into a house more rapidly under depressurized conditions.

**Dermal Absorption/Penetration:** Process by which a chemical penetrates the skin and enters the body as an internal dose.

**Dermal Exposure:** Contact between a chemical and the skin.

**Dermal Toxicity:** The ability of a pesticide or toxic chemical to poison people or animals by contact with the skin. (See: contact pesticide.)

**DES:** A synthetic estrogen, diethylstilbestrol is used as a growth stimulant in food animals. Residues in meat are thought to be carcinogenic.

**Desalination:** [Desalinization] (1) Removing salts from ocean or brackish water by using various technologies. (2) Removal of salts from soil by artificial means, usually leaching.

**Desiccant:** A chemical agent that absorbs moisture; some desiccants are capable of drying out plants or insects, causing death.

**Design Capacity:** The average daily flow that a treatment plant or other facility is designed to accommodate.

**Design Value:** The monitored reading used by EPA to determine an area's air quality status; e.g., for ozone, the fourth highest reading measured over the most recent three years is the design value.

**Designated Pollutant:** An air pollutant which is neither a criteria nor hazardous pollutant, as described in the Clean Air Act, but for which new source performance standards exist. The Clean Air Act does require states to control these pollutants, which include acid mist, total reduced sulfur (TRS), and fluorides.

**Designated Uses:** Those water uses identified in state water quality standards that must be achieved and maintained as required under the Clean Water Act. Uses can include cold water fisheries, public water supply, and irrigation.

**Designer Bugs:** Popular term for microbes developed through biotechnology that can degrade specific toxic chemicals at their source in toxic waste dumps or in ground water.

**Destination Facility:** The facility to which regulated medical waste is shipped for treatment and destruction, incineration, and/or disposal.

**Destratification:** Vertical mixing within a lake or reservoir to totally or partially eliminate separate layers of temperature, plant, or animal life.

**Destroyed Medical Waste:** Regulated medical waste that has been ruined, torn apart, or mutilated through thermal treatment, melting, shredding, grinding, tearing, or breaking, so that it is no longer generally recognized as medical waste, but has not yet been treated (excludes compacted regulated medical waste).

**Destruction and Removal Efficiency (DRE):** A percentage that represents the number of molecules of a compound removed or destroyed in an incinerator relative to the number of molecules entering the system (e.g. a DRE of 99.99 percent means that 9,999 molecules are destroyed for every 10,000 that enter; 99.99 percent is known as "four nines." For some pollutants, the RCRA removal requirement may be as stringent as "six nines").

**Destruction Facility:** A facility that destroys regulated medical waste.

**Desulfurization:** Removal of sulfur from fossil fuels to reduce pollution.

**Detectable Leak Rate:** The smallest leak (from a storage tank), expressed in terms of gallons- or liters-per-hour, that a test can reliably discern with a certain probability of detection or false alarm.

**Detection Criterion:** A predetermined rule to ascertain whether a tank is leaking or not. Most volumetric tests use a threshold value as the detection criterion. (See: volumetric tank tests.)

**Detection Limit:** The lowest concentration of a chemical that can reliably be distinguished from a zero concentration.

**Detention Time:** 1. The theoretical calculated time required for a small amount of water to pass through a tank at a given rate of flow. 2. The actual time that a small amount of water is in a settling basin, flocculating basin, or rapid-mix chamber. 3. In storage reservoirs, the length of time water will be held before being used.

**Detergent:** Synthetic washing agent that helps to remove dirt and oil. Some contain compounds which kill useful bacteria and encourage algae growth when they are in wastewater that reaches receiving waters.

**Development Effects:** Adverse effects such as altered growth, structural abnormality, functional deficiency, or death observed in a developing organism.

**Dewater:** 1. Remove or separate a portion of the water in a sludge or slurry to dry the sludge so it can be handled and disposed of. 2. Remove or drain the water from a tank or trench.

**Diatomaceous Earth (Diatomite):** A chalk-like material (fossilized diatoms) used to filter out solid waste in wastewater treatment plants; also used as an active ingredient in some powdered pesticides.

**Diazinon:** An insecticide. In 1986, EPA banned its use on open areas such as sod farms and golf courses because it posed a danger to migratory birds. The ban did not apply to agricultural, home lawn or commercial establishment uses.

**Dibenzofurans:** A group of organic compounds, some of which are toxic.

**Dicofol:** A pesticide used on citrus fruits.

**Diffused Air:** A type of aeration that forces oxygen into sewage by pumping air through perforated pipes inside a holding tank.

**Diffusion:** The movement of suspended or dissolved particles (or molecules) from a more concentrated to a less concentrated area. The process tends to distribute the particles or molecules more uniformly.

**Digester:** In wastewater treatment, a closed tank; in solid-waste conversion, a unit in which bacterial action is induced and accelerated in order to break down organic matter and establish the proper carbon to nitrogen ratio.

**Digestion:** The biochemical decomposition of organic matter, resulting in partial gasification, liquefaction, and mineralization of pollutants.

**Dike:** A low wall that can act as a barrier to prevent a spill from spreading.

**Diluent:** Any liquid or solid material used to dilute or carry an active ingredient.

**Dilution Ratio:** The relationship between the volume of water in a stream and the volume of incoming water. It affects the ability of the stream to assimilate waste.

**Dimictic:** Lakes and reservoirs that freeze over and normally go through two stratifications and two mixing cycles a year.

**Dinocap:** A fungicide used primarily by apple growers to control summer diseases. EPA proposed restrictions on its use in 1986 when laboratory tests found it caused birth defects in rabbits.

**Dinoseb:** A herbicide that is also used as a fungicide and insecticide. It was banned by EPA in 1986 because it posed the risk of birth defects and sterility.

**Dioxin:** Any of a family of compounds known chemically as dibenzo-p-dioxins. Concern about them arises from their potential toxicity as contaminants in commercial products. Tests on laboratory animals indicate that it is one of the more toxic anthropogenic (man-made) compounds.

**Direct Discharger:** A municipal or industrial facility which introduces pollution through a defined conveyance or system such as outlet pipes; a point source.

**Direct Filtration:** A method of treating water which consists of the addition of coagulent chemicals, flash mixing, coagulation, minimal flocculation, and filtration. Sedimentation is not uses.

**Direct Push:** Technology used for performing subsurface investigations by driving, pushing, and/or vibrating small-diameter hollow steel rods into the ground/ Also known as direct drive, drive point, or push technology.

**Direct Runoff:** Water that flows over the ground surface or through the ground directly into streams, rivers, and lakes.

**Discharge:** Flow of surface water in a stream or canal or the outflow of ground water from a flowing artesian well, ditch, or spring. Can also apply tp discharge of liquid effluent from a facility or to chemical emissions into the air through designated venting mechanisms.

**Disinfectant:** A chemical or physical process that kills pathogenic organisms in water, air, or on surfaces. Chlorine is often used to disinfect sewage treatment effluent, water supplies, wells, and swimming pools.

**Disinfectant By-Product:** A compound formed by the reaction of a disinfenctant such as chlorine with organic material in the water supply; a chemical byproduct of the disinfection process..

**Disinfectant Time:** The time it takes water to move from the point of disinfectant application (or the previous point of residual disinfectant measurement) to a point before or at the point where the residual disinfectant is measured. In pipelines, the time is calculated by dividing the internal volume of the pipe by he maximum hourly flow rate; within mixing basins and storage reservoirs it is determined by tracer studies of an equivalent demonstration.

**Dispersant:** A chemical agent used to break up concentrations of organic material such as spilled oil.

**Displacement Savings:** Saving realized by displacing purchases of natural gas or electricity from a local utility by using landfill gas for power and heat.

**Disposables:** Consumer products, other items, and packaging used once or a few times and discarded.

**Disposal:** Final placement or destruction of toxic, radioactive, or other wastes; surplus or banned pesticides or other chemicals; polluted soils; and drums containing hazardous materials from removal actions or accidental releases. Disposal may be accomplished through use of approved secure landfills, surface impoundments, land farming, deep-well injection, ocean dumping, or incineration.

**Disposal Facilities:** Repositories for solid waste, including landfills and combustors intended for permanent containment or destruction of waste materials. Excludes transfer stations and composting facilities.

**Dissolved Oxygen (DO):** The oxygen freely available in water, vital to fish and other aquatic life and for the prevention of odors. DO levels are considered a most important indicator of a water body's ability to support desirable aquatic life. Secondary and advanced waste treatment are generally designed to ensure adequate DO in waste-receiving waters.

**Dissolved Solids:** Disintegrated organic and inorganic material in water. Excessive amounts make water unfit to drink or use in industrial processes.

**Distillation:** The act of purifying liquids through boiling, so that the steam or gaseous vapors condense to a pure liquid. Pollutants and contaminants may remain in a concentrated residue.

**Disturbance:** Any event or series of events that disrupt ecosystem, community, or population structure and alters the physical environment.

**Diversion:** 1. Use of part of a stream flow as water supply. 2. A channel with a supporting ridge on the lower side constructed across a slope to divert water at a non-erosive velocity to sites where it can be used and disposed of.

**Diversion Rate:** The percentage of waste materials diverted from traditional disposal such as landfilling or incineration to be recycled, composted, or re-used.

**DNA Hybridization:** Use of a segment of DNA, called a DNA probe, to identify its complementary DNA; used to detect specific genes.

**Dobson Unit (DU):** Units of ozone level measurement. measurement of ozone levels. If, for example, 100 DU of ozone were brought to the earth's surface they would form a layer one millimeter thick. Ozone levels vary geographically, even in the absence of ozone depletion.

**Domestic Application:** Pesticide application in and around houses, office buildings, motels, and other living or working areas.(See: residential use.)

**Dosage/Dose:** 1. The actual quantity of a chemical administered to an organism or to which it is exposed. 2. The amount of a substance that reaches a specific tissue (e.g. the liver). 3. The amount of a substance available for interaction with metabolic processes after crossing the outer boundary of an organism. (See: absorbed dose, administered dose, applied dose, potential dose.)

**Dose Equivalent:** The product of the absorbed dose from ionizing radiation and such factors as account for biological differences due to the type of radiation and its distribution in the body in the body.

**Dose Rate:** In exposure assessment, dose per time unit (e.g. mg/day), sometimes also called dosage.

**Dose Response:** Shifts in toxicological responses of an individual (such as alterations in severity) or populations (such as alterations in incidence) that are related to changes in the dose of any given substance.

**Dose Response Curve:** Graphical representation of the relationship between the dose of a stressor and the biological response thereto.

**Dose-Response Assessment:** 1. Estimating the potency of a chemical. 2. In exposure assessment, the process of determining the relationship between the dose of a stressor and a specific biological response. 3. Evaluating the quantitative relationship between dose and toxicological responses.

**Dose-Response Relationship:** The quantitative relationship between the amount of exposure to a substance and the extent of toxic injury or disease produced.

**Dosimeter:** An instrument to measure dosage; many so-called dosimeters actually measure exposure rather than dosage. Dosimetry is the process or technology of measuring and/or estimating dosage.

**DOT Reportable Quantity:** The quantity of a substance specified in a U.S. Department of Transportation regulation that triggers labeling, packaging and other requirements related to shipping such substances.

**Downgradient:** The direction that groundwater flows; similar to "downstream" for surface water.

**Downstream Processors:** Industries dependent on crop production (e.g. canneries and food processors).

**DP Hole:** Hole in the ground made with DP equipment. (See: direct push.)

**Draft:** 1. The act of drawing or removing water from a tank or reservoir. 2. The water which is drawn or removed.

**Draft Permit:** A preliminary permit drafted and published by EPA; subject to public review and comment before final action on the application.

**Drainage:** Improving the productivity of agricultural land by removing excess water from the soil by such means as ditches or subsurface drainage tiles.

**Drainage Basin:** The area of land that drains water, sediment, and dissolved materials to a common outlet at some point along a stream channel.

**Drainage Well:** A well drilled to carry excess water off agricultural fields. Because they act as a funnel from the surface to the groundwater below. Drainage wells can contribute to groundwater pollution.

**Drawdown:** 1. The drop in the water table or level of water in the ground when water is being pumped from a well. 2. The amount of water used from a tank or reservoir. 3. The drop in the water level of a tank or reservoir.

**Dredging:** Removal of mud from the bottom of water bodies. This can disturb the ecosystem and causes silting that kills aquatic life. Dredging of contaminated muds can expose biota to heavy metals and other toxics. Dredging activities may be subject to regulation under Section 404 of the Clean Water Act.

**Drilling Fluid:** Fluid used to lubricate the bit and convey drill cuttings to the surface with rotary drilling equipment. Usually composed of bentonite slurry or muddy water. Can become contaminated, leading to cross contamination, and may require special disposal. Not used with DP methods

**Drinking Water Equivalent Level:** Protective level of exposure related to potentially non-carcinogenic effects of chemicals that are also known to cause cancer.

**Drinking Water State Revolving Fund:** The Fund provides capitalization grants to states to develop drinking water revolving loan funds to help finance system infrastructure improvements, assure source-water protection, enhance operation and management of drinking-water systems, and otherwise promote local water-system compliance and protection of public health.

**Drive Casing:** Heavy duty steel casing driven along with the sampling tool in cased DP systems. Keeps the hole open between sampling runs and is not removed until last sample has been collected.

**Drive Point Profiler:** An exposed groundwater DP system used to collect multiple depth-discrete groundwater samples. Ports in the tip of the probe connect to an internal stainless steel or teflon tube that extends to the surface. Samples are collected via suction or airlift methods. Deionized water is pumped down through the ports to prevent plugging while driving the tool to the next sampling depth.

**Drop-off:** Recyclable materials collection method in which individuals bring them to a designated collection site.

**Dual-Phase Extraction:** Active withdrawal of both liquid and gas phases from a well usually involving the use of a vacuum pump.

**Dump:** A site used to dispose of solid waste without environmental controls.

**Duplicate:** A second aliquot or sample that is treated the same as the original sample in order to determine the precision of the analytical method. (See: aliquot.)

**Dustfall Jar:** An open container used to collect large particles from the air for measurement and analysis.

**Dynamometer. A device used to place a load on an engine and measure its performance.**

**Dystrophic Lakes:** Acidic, shallow bodies of water that contain much humus and/or other organic matter; contain many plants but few fish.

## E

**Ecological Entity:** In ecological risk assessment, a general term referring to a species, a group of species, an ecosystem function or characteristic, or a specific habitat or biome.

**Ecological/Environmental Sustainability:** Maintenance of ecosystem components and functions for future generations.

**Ecological Exposure:** Exposure of a non-human organism to a stressor.

**Ecological Impact:** The effect that a man-caused or natural activity has on living organisms and their non-living (abiotic) environment.

**Ecological Indicator:** A characteristic of an ecosystem that is related to, or derived from, a measure of biotic or abiotic variable, that can provide quantitative information on ecological structure and function. An indicator can contribute to a measure of integrity and sustainability.

**Ecological Integrity:** A living system exhibits integrity if, when subjected to disturbance, it sustains and organizes self-correcting ability to recover toward a biomass end-state that is normal for that system. End-states other than the pristine or naturally whole may be accepted as normal and good.

**Ecological Risk Assessment:** The application of a formal framework, analytical process, or model to estimate the effects of human actions(s) on a natural resource and to interpret the significance of those effects in light of the uncertainties identified in each component of the assessment process. Such analysis includes initial hazard identification, exposure and dose-response assessments, and risk characterization.

**Ecology:** The relationship of living things to one another and their environment, or the study of such relationships.

**Economic Poisons:** Chemicals used to control pests and to defoliate cash crops such as cotton.

**Ecosphere:** The "bio-bubble" that contains life on earth, in surface waters, and in the air. (See: biosphere.)

**Ecosystem:** The interacting system of a biological community and its non-living environmental surroundings.

**Ecosystem Structure:** Attributes related to the instantaneous physical state of an ecosystem; examples include species population density, species richness or evenness, and standing crop biomass.

**Ecotone:** A habitat created by the juxtaposition of distinctly different habitats; an edge habitat; or an ecological zone or boundary where two or more ecosystems meet.

**Effluent:** Wastewater—treated or untreated—that flows out of a treatment plant, sewer, or industrial outfall. Generally refers to wastes discharged into surface waters.

**Effluent Guidelines:** Technical EPA documents which set effluent limitations for given industries and pollutants.

**Effluent Limitation:** Restrictions established by a state or EPA on quantities, rates, and concentrations in wastewater discharges.

**Effluent Standard:** (See: effluent limitation.)

**Ejector:** A device used to disperse a chemical solution into water being treated.

**Electrodialysis:** A process that uses electrical current applied to permeable membranes to remove minerals from water. Often used to desalinize salty or brackish water.

**Electromagnetic Geophysical Methods:** Ways to measure subsurface conductivity via low-frequency electromagnetic induction.

**Electrostatic Precipitator (ESP):** A device that removes particles from a gas stream (smoke) after combustion occurs. The ESP imparts an electrical charge to the particles, causing them to adhere to metal plates inside the precipitator. Rapping on the plates causes the particles to fall into a hopper for disposal.

**Eligible Costs:** The construction costs for wastewater treatment works upon which EPA grants are based.

**EMAP Data:** Environmental monitoring data collected under the auspices of the Environmental Monitoring and Assessment Program. All EMAP data share the common attribute of being of known quality, having been collected in the context of explicit data quality objectives (DQOs) and a consistent quality assurance program.

**Emergency and Hazardous Chemical Inventory:** An annual report by facilities having one or more extremely hazardous substances or hazardous chemicals above certain weight limits.

**Emergency (Chemical):** A situation created by an accidental release or spill of hazardous chemicals that poses a threat to the safety of workers, residents, the environment, or property.

**Emergency Episode:** (See: air pollution episode.)

**Emergency Exemption:** Provision in FIFRA under which EPA can grant temporary exemption to a state or another federal agency to allow the use of a pesticide product not registered for that particular use. Such actions involve unanticipated and/or severe pest problems where there is not time or interest by a manufacturer to register the product for that use. (Registrants cannot apply for such exemptions.)

**Emergency Removal Action:** 1. Steps take to remove contaminated materials that pose imminent threats to local residents (e.g. removal of leaking drums or the excavation of explosive waste.) 2. The state record of such removals.

**Emergency Response Values:** Concentrations of chemicals, published by various groups, defining acceptable levels for short-term exposures in emergencies.

**Emergency Suspension:** Suspension of a pesticide product registration due to an imminent hazard. The action immediately halts distribution, sale, and sometimes actual use of the pesticide involved.

**Emission:** Pollution discharged into the atmosphere from smokestacks, other vents, and surface areas of commercial or industrial facilities; from residential chimneys; and from motor vehicle, locomotive, or aircraft exhausts.

**Emission Cap:** A limit designed to prevent projected growth in emissions from existing and future stationary sources from eroding any mandated reductions. Generally, such provisions require that any emission growth from facilities under the restrictions be offset by equivalent reductions at other facilities under the same cap. (See: emissions trading.)

**Emission Factor:** The relationship between the amount of pollution produced and the amount of raw material processed. For example, an emission factor for a blast furnace making iron would be the number of pounds of particulates per ton of raw materials.

**Emission Inventory:** A listing, by source, of the amount of air pollutants discharged into the atmosphere of a community; used to establish emission standards.

**Emission Standard:** The maximum amount of air polluting discharge legally allowed from a single source, mobile or stationary.

**Emissions Trading:** The creation of surplus emission reductions at certain stacks, vents or similar emissions sources and the use of this surplus to meet or redefine pollution requirements applicable to other emissions sources. This allows one source to increase emissions when another source reduces them, maintaining an overall constant emission level. Facilities that reduce emissions substantially may "bank" their "credits" or sell them to other facilities or industries.

**Emulsifier:** A chemical that aids in suspending one liquid in another. Usually an organic chemical in an aqueous solution.

**Encapsulation:** The treatment of asbestos-containing material with a liquid that covers the surface with a protective coating or embeds fibers in an adhesive matrix to prevent their release into the air.

**Enclosure:** Putting an airtight, impermeable, permanent barrier around asbestos-containing materials to prevent the release of asbestos fibers into the air.

**End User:** Consumer of products for the purpose of recycling. Excludes products for re-use or combustion for energy recovery.

**End-of-the-pipe:** Technologies such as scrubbers on smokestacks and catalytic convertors on automobile tailpipes that reduce emissions of pollutants after they have formed.

**End-use Product:** A pesticide formulation for field or other end use. The label has instructions for use or application to control pests or regulate plant growth. The term excludes products used to formulate other pesticide products.

**Endangered Species:** Animals, birds, fish, plants, or other living organisms threatened with extinction by anthropogenic (man-caused) or other natural changes in their environment. Requirements for declaring a species endangered are contained in the Endangered Species Act.

**Endangerment Assessment:** A study to determine the nature and extent of contamination at a site on the National Priorities List and the risks posed to public health or the environment. EPA or the state conducts the study when a legal action is to be taken to direct potentially responsible parties to clean up a site or pay for it. An endangerment assessment supplements a remedial investigation.

**Endrin:** A pesticide toxic to freshwater and marine aquatic life that produces adverse health effects in domestic water supplies.

**Energy Management System:** A control system capable of monitoring environmental and system loads and adjusting HVAC operations accordingly in order to conserve energy while maintaining comfort.

**Energy Recovery:** Obtaining energy from waste through a variety of processes (e.g. combustion).

**Enforceable Requirements:** Conditions or limitations in permits issued under the Clean Water Act Section 402 or 404 that, if violated, could result in the issuance of a compliance order or initiation of a civil or criminal action under federal or applicable state laws. If a permit has not been issued, the term includes any requirement which, in the Regional Administrator's judgement, would be included in the permit when issued. Where no permit applies, the term includes any requirement which the RA determines is necessary for the best practical waste treatment technology to meet applicable criteria.

**Enforcement:** EPA, state, or local legal actions to obtain compliance with environmental laws, rules, regulations, or agreements and/or obtain penalties or criminal sanctions for violations. Enforcement procedures may vary, depending on the requirements of different environmental laws and related implementing regulations. Under CERCLA, for example, EPA will seek to require potentially responsible parties to clean up a Superfund site, or pay for the cleanup, whereas under the Clean Air Act the Agency may invoke sanctions against cities failing to meet ambient air quality standards that could prevent certain types of construction or federal funding. In other situations, if investigations by EPA and state agencies uncover willful violations, criminal trials and penalties are sought.

**Enforcement Decision Document (EDD):** A document that provides an explanation to the public of EPA's selection of the cleanup alternative at enforcement sites on the National Priorities List. Similar to a Record of Decision.

**Engineered Controls:** Method of managing environmental and health risks by placing a barrier between the contamination and the rest of the site, thus limiting exposure pathways.

**Enhanced Inspection and Maintenance (I&M):** An improved automobile inspection and maintenance program—aimed at reducing automobile emissions—that contains, at a minimum, more vehicle types and model years, tighter inspection, and better management practices. It may also include annual computerized or centralized inspections, under-the-hood inspection—for signs of tampering with pollution control equipment—and increased repair waiver cost.

**Enrichment:** The addition of nutrients (e.g. nitrogen, phosphorus, carbon compounds) from sewage effluent or agricultural runoff to surface water, greatly increases the growth potential for algae and other aquatic plants.

**Entrain:** To trap bubbles in water either mechanically through turbulence or chemically through a reaction.

**Environment:** The sum of all external conditions affecting the life, development and survival of an organism.

**Environmental Assessment:** An environmental analysis prepared pursuant to the National Environmental Policy Act to determine whether a federal action would significantly affect the environment and thus require a more detailed environmental impact statement.

**Environmental Audit:** An independent assessment of the current status of a party's compliance with applicable environmental requirements or of a party's environmental compliance policies, practices, and controls.

**Environmental/Ecological Risk:** The potential for adverse effects on living organisms associated with pollution of the environment by effluents, emissions, wastes, or accidental chemical releases; energy use; or the depletion of natural resources.

**Environmental Equity/Justice:** Equal protection from environmental hazards for individuals, groups, or communities regardless of race, ethnicity, or economic status. This applies to the development, implementation, and enforcement of environmental laws, regulations, and policies, and implies that no population of people should be forced to shoulder a disproportionate share of negative environmental impacts of pollution or environmental hazard due to a lack of political or economic strength levels.

**Environmental Exposure:** Human exposure to pollutants originating from facility emissions. Threshold levels are not necessarily surpassed, but low-level chronic pollutant exposure is one of the most common forms of environmental exposure (See: threshold level).

**Environmental Fate:** The destiny of a chemical or biological pollutant after release into the environment.

**Environmental Fate Data:** Data that characterize a pesticide's fate in the ecosystem, considering factors that foster its degradation (light, water, microbes), pathways and resultant products.

**Environmental Impact Statement:** A document required of federal agencies by the National Environmental Policy Act for major projects or legislative proposals significantly affecting the environment. A tool for decision making, it describes the positive and negative effects of the undertaking and cites alternative actions.

**Environmental Indicator:** A measurement, statistic or value that provides a proximate gauge or evidence of the effects of environmental management programs or of the state or condition of the environment.

**Environmental Justice:** The fair treatment of people of all races, cultures, incomes, and educational levels with respect to the development and enforcement of environmental laws, regulations, and policies.

**Environmental Lien:** A charge, security, or encumbrance on a property's title to secure payment of cost or debt arising from response actions, cleanup, or other remediation of hazardous substances or petroleum products.

**Environmental Medium:** A major environmental category that surrounds or contacts humans, animals, plants, and other organisms (e.g. surface water, ground water, soil or air) and through which chemicals or pollutants move. (See: ambient medium, biological medium.)

**Environmental Monitoring for Public Access and Community Tracking:** Joint EPA, NOAA, and USGS program to provide timely and effective communication of environmental data and information through improved and updated technology solutions that support timely environmental monitoring reporting, interpreting, and use of the information for the benefit of the public. (See: real-time monitoring.)

**Environmental Response Team:** EPA experts located in Edison, N.J., and Cincinnati, OH, who can provide around-the-clock technical assistance to EPA regional offices and states during all types of hazardous waste site emergencies and spills of hazardous substances.

**Environmental Site Assessment:** The process of determining whether contamination is present on a parcel of real property.

**Environmental Sustainability:** Long-term maintenance of ecosystem components and functions for future generations.

**Environmental Tobacco Smoke:** Mixture of smoke from the burning end of a cigarette, pipe, or cigar and smoke exhaled by the smoker. (See: passive smoking/secondhand smoke.)

**Epidemiology:** Study of the distribution of disease, or other health-related states and events in human populations, as related to age, sex, occupation, ethnicity, and economic status in order to identify and alleviate health problems and promote better health.

**Epilimnion:** Upper waters of a thermally stratified lake subject to wind action.

**Episode (Pollution):** An air pollution incident in a given area caused by a concentration of atmospheric pollutants under meteorological conditions that may result in a significant increase in illnesses or deaths. May also describe water pollution events or hazardous material spills.

**Equilibrium:** In relation to radiation, the state at which the radioactivity of consecutive elements within a radioactive series is neither increasing nor decreasing.

**Equivalent Method:** Any method of sampling and analyzing for air pollution which has been demonstrated to the EPA Administrator's satisfaction to be, under specific conditions, an acceptable alternative to normally used reference methods.

**Erosion:** The wearing away of land surface by wind or water, intensified by land-clearing practices related to farming, residential or industrial development, road building, or logging.

**Established Treatment Technologies:** Technologies for which cost and performance data are readily available. (See: Innovative treatment technologies.)

**Estimated Environmental Concentration:** The estimated pesticide concentration in an ecosystem.

**Estuary:** Region of interaction between rivers and near-shore ocean waters, where tidal action and river flow mix fresh and salt water. Such areas include bays, mouths of rivers, salt marshes, and lagoons. These brackish water ecosystems shelter and feed marine life, birds, and wildlife. (See: wetlands.)

**Ethanol:** An alternative automotive fuel derived from grain and corn; usually blended with gasoline to form gasohol.

**Ethylene Dibromide (EDB):** A chemical used as an agricultural fumigant and in certain industrial processes. Extremely toxic and found to be a carcinogen in laboratory animals, EDB has been banned for most agricultural uses in the United States.

**Eutrophic Lakes:** Shallow, murky bodies of water with concentrations of plant nutrients causing excessive production of algae. (See: dystrophic lakes.)

**Eutrophication:** The slow aging process during which a lake, estuary, or bay evolves into a bog or marsh and eventually disappears. During the later stages of eutrophication the water body is choked by abundant plant life due to higher levels of nutritive compounds such as nitrogen and phosphorus. Human activities can accelerate the process.

**Evaporation Ponds:** Areas where sewage sludge is dumped and dried.

**Evapotranspiration:** The loss of water from the soil both by evaporation and by transpiration from the plants growing in the soil.

**Exceedance:** Violation of the pollutant levels permitted by environmental protection standards.

**Exclusion:** In the asbestos program, one of several situations that permit a Local Education Agency (LEA) to delete one or more of the items required by the Asbestos Hazard Emergency Response Act (AHERA); e.g. records of previous asbestos sample collection and analysis may be used by the accredited inspector in lieu of AHERA bulk sampling.

**Exclusionary Ordinance:** Zoning that excludes classes of persons or businesses from a particular neighborhood or area.

**Exempt Solvent:** Specific organic compounds not subject to requirements of regulation because they are deemed by EPA to be of negligible photochemical reactivity.

**Exempted Aquifer:** Underground bodies of water defined in the Underground Injection Control program as aquifers that are potential sources of drinking water though not being used as such, and thus exempted from regulations barring underground injection activities.

**Exemption:** A state (with primacy) may exempt a public water system from a requirement involving a Maximum Contaminant Level (MCL), treatment technique, or both, if the system cannot comply due to compelling economic or other factors, or because the system was in operation before the requirement or MCL was instituted; and the exemption will not create a public health risk. (See: variance.)

**Exotic Species:** A species that is not indigenous to a region.

**Experimental Use Permit:** Obtained by manufacturers for testing new pesticides or uses thereof whenever they conduct experimental field studies to support registration on 10 acres or more of land or one acre or more of water.

**Experimental Use Permit:** A permit granted by EPA that allows a producer to conduct tests of a new pesticide, product and/or use outside the laboratory. The testing is usually done on ten or more acres of land or water surface.

**Explosive Limits:** The amounts of vapor in the air that form explosive mixtures; limits are expressed as lower and upper limits and give the range of vapor concentrations in air that will explode if an ignition source is present.

**Exports :** In solid waste program, municipal solid waste and recyclables transported outside the state or locality where they originated.

**Exposure:** The amount of radiation or pollutant present in a given environment that represents a potential health threat to living organisms.

**Exposure Assessment:** Identifying the pathways by which toxicants may reach individuals, estimating how much of a chemical an individual is likely to be exposed to, and estimating the number likely to be exposed.

**Exposure Concentration:** The concentration of a chemical or other pollutant representing a health threat in a given environment.

**Exposure Indicator:** A characteristic of the environment measured to provide evidence of the occurrence or magnitude of a response indicator's exposure to a chemical or biological stress.

**Exposure Level:** The amount (concentration) of a chemical at the absorptive surfaces of an organism.

**Exposure Pathway:** The path from sources of pollutants via, soil, water, or food to man and other species or settings.

**Exposure Route:** The way a chemical or pollutant enters an organism after contact; i.e. by ingestion, inhalation, or dermal absorption.

**Exposure-Response Relationship:** The relationship between exposure level and the incidence of adverse effects.

**Extraction Procedure (EP Toxic):** Determining toxicity by a procedure which simulates leaching; if a certain concentration of a toxic substance can be leached from a waste, that waste is considered hazardous, i.e."EP Toxic."

**Extraction Well:** A discharge well used to remove groundwater or air.

**Extremely Hazardous Substances:** Any of 406 chemicals identified by EPA as toxic, and listed under SARA Title III. The list is subject to periodic revision.

## F

**Fabric Filter:** A cloth device that catches dust particles from industrial emissions.

**Facilities Plans:** Plans and studies related to the construction of treatment works necessary to comply with the Clean Water Act or RCRA. A facilities plan investigates needs and provides information on the cost-effectiveness of alternatives, a recommended plan, an environmental assessment of the recommendations, and descriptions of the treatment works, costs, and a completion schedule.

**Facility Emergency Coordinator:** Representative of a facility covered by environmental law (e.g, a chemical plant) who participates in the emergency reporting process with the Local Emergency Planning Committee (LEPC).

**Facultative Bacteria:** Bacteria that can live under aerobic or anaerobic conditions.

**Feasibility Study:** 1. Analysis of the practicability of a proposal; e.g., a description and analysis of potential cleanup alternatives for a site such as one on the National Priorities List. The feasibility study usually recommends selection of a cost-effective alternative. It usually starts as soon as the remedial investigation is underway; together, they are commonly referred to as the "RI/FS". 2. A small-scale investigation of a problem to ascertain whether a proposed research approach is likely to provide useful data.

**Fecal Coliform Bacteria:** Bacteria found in the intestinal tracts of mammals. Their presence in water or sludge is an indicator of pollution and possible contamination by pathogens.

**Federal Implementation Plan:** Under current law, a federally implemented plan to achieve attainment of air quality standards, used when a state is unable to develop an adequate plan.

**Federal Motor Vehicle Control Program:** All federal actions aimed at controlling pollution from motor vehicles by such efforts as establishing and enforcing tailpipe and evaporative emission standards for new vehicles, testing methods development, and guidance to states operating inspection and maintenance programs. Federally designated area that is required to meet and maintain federal ambient air quality standards. May include nearby locations in the same state or nearby states that share common air pollution problems.

**Feedlot:** A confined area for the controlled feeding of animals. Tends to concentrate large amounts of animal waste that cannot be absorbed by the soil and, hence, may be carried to nearby streams or lakes by rainfall runoff.

**Fen:** A type of wetland that accumulates peat deposits. Fens are less acidic than bogs, deriving most of their water from groundwater rich in calcium and magnesium. (See: wetlands.)

**Ferrous Metals:** Magnetic metals derived from iron or steel; products made from ferrous metals include appliances, furniture, containers, and packaging like steel drums and barrels. Recycled products include processing tin/steel cans, strapping, and metals from appliances into new products.

**FIFRA Pesticide Ingredient:** An ingredient of a pesticide that must be registered with EPA under the Federal Insecticide, Fungicide, and Rodenticide Act. Products making pesticide claims must register under FIFRA and may be subject to labeling and use requirements.

**Fill:** Man-made deposits of natural soils or rock products and waste materials.

**Filling:** Depositing dirt, mud or other materials into aquatic areas to create more dry land, usually for agricultural or commercial development purposes, often with ruinous ecological consequences.

**Filter Strip:** Strip or area of vegetation used for removing sediment, organic matter, and other pollutants from runoff and wastewater.

**Filtration:** A treatment process, under the control of qualified operators, for removing solid (particulate) matter from water by means of porous media such as sand or a man-made filter; often used to remove particles that contain pathogens.

**Financial Assurance for Closure:** Documentation or proof that an owner or operator of a facility such as a landfill or other waste repository is capable of paying the projected costs of closing the facility and monitoring it afterwards as provided in RCRA regulations.

**Finding of No Significant Impact:** A document prepared by a federal agency showing why a proposed action would not have a significant impact on the environment and thus would not require preparation of an Environmental Impact Statement. An FNSI is based on the results of an environmental assessment.

**Finished Water:** Water is "finished" when it has passed through all the processes in a water treatment plant and is ready to be delivered to consumers.

**First Draw:** The water that comes out when a tap is first opened, likely to have the highest level of lead contamination from plumbing materials.

**Fix a Sample:** A sample is "fixed" in the field by adding chemicals that prevent water quality indicators of interest in the sample from changing before laboratory measurements are made.

**Fixed-Location Monitoring:** Sampling of an environmental or ambient medium for pollutant concentration at one location continuously or repeatedly.

**Flammable:** Any material that ignites easily and will burn rapidly.

**Flare:** A control device that burns hazardous materials to prevent their release into the environment; may operate continuously or intermittently, usually on top of a stack.

**Flash Point:** The lowest temperature at which evaporation of a substance produces sufficient vapor to form an ignitable mixture with air.

**Floc:** A clump of solids formed in sewage by biological or chemical action.

**Flocculation:** Process by which clumps of solids in water or sewage aggregate through biological or chemical action so they can be separated from water or sewage.

**Floodplain:** The flat or nearly flat land along a river or stream or in a tidal area that is covered by water during a flood.

**Floor Sweep:** Capture of heavier-than-air gases that collect at floor level.

**Flow Rate:** The rate, expressed in gallons -or liters-per-hour, at which a fluid escapes from a hole or fissure in a tank. Such measurements are also made of liquid waste, effluent, and surface water movement.

**Flowable:** Pesticide and other formulations in which the active ingredients are finely ground insoluble solids suspended in a liquid. They are mixed with water for application.

**Flowmeter:** A gauge indicating the velocity of wastewater moving through a treatment plant or of any liquid moving through various industrial processes.

**Flue Gas:** The air coming out of a chimney after combustion in the burner it is venting. It can include nitrogen oxides, carbon oxides, water vapor, sulfur oxides, particles and many chemical pollutants.

**Flue Gas Desulfurization:** A technology that employs a sorbent, usually lime or limestone, to remove sulfur dioxide from the gases produced by burning fossil fuels. Flue gas desulfurization is current state-of-the art technology for major $SO_2$ emitters, like power plants.

**Fluidized:** A mass of solid particles that is made to flow like a liquid by injection of water or gas is said to have been fluidized. In water treatment, a bed of filter media is fluidized by backwashing water through the filter.

**Fluidized Bed Incinerator:** An incinerator that uses a bed of hot sand or other granular material to transfer heat directly to waste. Used mainly for destroying municipal sludge.

**Flume:** A natural or man-made channel that diverts water.

**Fluoridation:** The addition of a chemical to increase the concentration of fluoride ions in drinking water to reduce the incidence of tooth decay.

**Fluorides:** Gaseous, solid, or dissolved compounds containing fluorine that result from industrial processes. Excessive amounts in food can lead to fluorosis.

**Fluorocarbons (FCs):** Any of a number of organic compounds analogous to hydrocarbons in which one or more hydrogen atoms are replaced by fluorine. Once used in the United States as a propellant for domestic aerosols, they are now found mainly in coolants and some industrial processes. FCs containing chlorine are called chlorofluorocarbons (CFCs). They are believed to be modifying the ozone layer in the stratosphere, thereby allowing more harmful solar radiation to reach the Earth's surface.

**Flush:** 1. To open a cold-water tap to clear out all the water which may have been sitting for a long time in the pipes. In new homes, to flush a system means to send large volumes of water gushing through the unused pipes to remove loose particles of solder and flux. 2. To force large amounts of water through a system to clean out piping or tubing, and storage or process tanks.

**Flux:** 1. A flowing or flow. 2. A substance used to help metals fuse together.

**Fly Ash:** Non-combustible residual particles expelled by flue gas.

**Fogging:** Applying a pesticide by rapidly heating the liquid chemical so that it forms very fine droplets that resemble smoke or fog. Used to destroy mosquitoes, black flies, and similar pests.

**Food Chain:** A sequence of organisms, each of which uses the next, lower member of the sequence as a food source.

**Food Processing Waste:** Food residues produced during agricultural and industrial operations.

**Food Waste:** Uneaten food and food preparation wastes from residences and commercial establishments such as grocery stores, restaurants, and produce stands, institutional cafeterias and kitchens, and industrial sources like employee lunchrooms.

**Food Web:** The feeding relationships by which energy and nutrients are transferred from one species to another.

**Formaldehyde:** A colorless, pungent, and irritating gas, CH20, used chiefly as a disinfectant and preservative and in synthesizing other compounds like resins.

**Formulation:** The substances comprising all active and inert ingredients in a pesticide.

**Fossil Fuel:** Fuel derived from ancient organic remains; e.g. peat, coal, crude oil, and natural gas.

**Fracture:** A break in a rock formation due to structural stresses; e.g. faults, shears, joints, and planes of fracture cleavage.

**Free Product:** A petroleum hydrocarbon in the liquid free or non aqueous phase. (See: non-aqueous phase liquid.)

**Freeboard:** 1. Vertical distance from the normal water surface to the top of a confining wall. 2. Vertical distance from the sand surface to the underside of a trough in a sand filter.

**Fresh Water:** Water that generally contains less than 1,000 milligrams-per-liter of dissolved solids.

**Friable:** Capable of being crumbled, pulverized, or reduced to powder by hand pressure.

**Friable Asbestos:** Any material containing more than one-percent asbestos, and that can be crumbled or reduced to powder by hand pressure. (May include previously non-friable material which becomes broken or damaged by mechanical force.)

**Fuel Economy Standard:** The Corporate Average Fuel Economy Standard (CAFE) effective in 1978. It enhanced the national fuel conservation effort imposing a miles-per-gallon floor for motor vehicles.

**Fuel Efficiency:** The proportion of energy released by fuel combustion that is converted into useful energy.

**Fuel Switching:** 1. A precombustion process whereby a low-sulfur coal is used in place of a higher sulfur coal in a power plant to reduce sulfur dioxide emissions. 2. Illegally using leaded gasoline in a motor vehicle designed to use only unleaded.

**Fugitive Emissions:** Emissions not caught by a capture system.

**Fume:** Tiny particles trapped in vapor in a gas stream.

**Fumigant:** A pesticide vaporized to kill pests. Used in buildings and greenhouses.

**Functional Equivalent:** Term used to describe EPA's decision-making process and its relationship to the environmental review conducted under the National Environmental Policy Act (NEPA). A review is considered functionally equivalent when it addresses the substantive components of a NEPA review.

**Fungicide:** Pesticides which are used to control, deter, or destroy fungi.

**Fungistat:** A chemical that keeps fungi from growing.

**Fungus (Fungi):** Molds, mildews, yeasts, mushrooms, and puffballs, a group of organisms lacking in chlorophyll (i.e. are not photosynthetic) and which are usually non-mobile, filamentous, and multicellular. Some grow in soil, others attach themselves to decaying trees and other plants whence they obtain nutrients. Some are pathogens, others stabilize sewage and digest composted waste.

**Furrow Irrigation:** Irrigation method in which water travels through the field by means of small channels between each groups of rows.

**Future Liability:** Refers to potentially responsible parties' obligations to pay for additional response activities beyond those specified in the Record of Decision or Consent Decree.

## G

**Game Fish:** Species like trout, salmon, or bass, caught for sport. Many of them show more sensitivity to environmental change than "rough" fish.

**Garbage:** Animal and vegetable waste resulting from the handling, storage, sale, preparation, cooking, and serving of foods.

**Gas Chromatograph/Mass Spectrometer:** Instrument that identifies the molecular composition and concentrations of various chemicals in water and soil samples.

**Gasahol:** Mixture of gasoline and ethanol derived from fermented agricultural products containing at least nine percent ethanol. Gasohol emissions contain less carbon monoxide than those from gasoline.

**Gasification:** Conversion of solid material such as coal into a gas for use as a fuel.

**Gasoline Volatility:** The property of gasoline whereby it evaporates into a vapor. Gasoline vapor is a mixture of volatile organic compounds.

**General Permit:** A permit applicable to a class or category of dischargers.

**General Reporting Facility:** A facility having one or more hazardous chemicals above the 10,000 pound threshold for planning quantities. Such facilities must file MSDS and emergency inventory information with the SERC, LEPC, and local fire departments.

**Generally Recognized as Safe (GRAS):** Designation by the FDA that a chemical or substance (including certain pesticides) added to food is considered safe by experts, and so is exempted from the usual FFDCA food additive tolerance requirements.

**Generator:** 1. A facility or mobile source that emits pollutants into the air or releases hazardous waste into water or soil. 2. Any person, by site, whose act or process produces regulated medical waste or whose act first causes such waste to become subject to regulation. Where more than one person (e.g. doctors with separate medical practices) are located in the same building, each business entity is a separate generator.

**Genetic Engineering:** A process of inserting new genetic information into existing cells in order to modify a specific organism for the purpose of changing one of its characteristics.

**Genotoxic:** Damaging to DNA; pertaining to agents known to damage DNA.

**Geographic Information System (GIS):** A computer system designed for storing, manipulating, analyzing, and displaying data in a geographic context.

**Geological Log:** A detailed description of all underground features (depth, thickness, type of formation) discovered during the drilling of a well.

**Geophysical Log:** A record of the structure and composition of the earth encountered when drilling a well or similar type of test hold or boring.

**Geothermal/Ground Source Heat Pump:** These heat pumps are underground coils to transfer heat from the ground to the inside of a building. (See: heat pump; water source heat pump)

**Germicide:** Any compound that kills disease-causing microorganisms.

**Giardia Lamblia:** Protozoan in the feces of humans and animals that can cause severe gastrointestinal ailments. It is a common contaminant of surface waters.

**Glass Containers:** For recycling purposes, containers like bottles and jars for drinks, food, cosmetics and other products. When being recycled, container glass is generally separated into color categories for conversion into new containers, construction materials or fiberglass insulation.

**Global Warming:** An increase in the near surface temperature of the Earth. Global warming has occurred in the distant past as the result of natural influences, but the term is most often used to refer to the warming predicted to occur as a result of increased emissions of greenhouse gases. Scientists generally agree that the Earth's surface has warmed by about 1 degree Fahrenheit in the past 140 years. The Intergovernmental Panel on Climate Change (IPCC) recently concluded that increased concentrations of greenhouse gases are causing an increase in the Earth's surface temperature and that increased concentrations of sulfate aerosols have led to relative cooling in some regions, generally over and downwind of heavily industrialized areas. (See: climate change)

**Global Warming Potential:** The ratio of the warming caused by a substance to the warming caused by a similar mass of carbon dioxide. CFC-12, for example, has a GWP of 8,500, while water has a GWP of zero. (See: Class I Substance and Class II Substance.)

**Glovebag:** A polyethylene or polyvinyl chloride bag-like enclosure affixed around an asbestos-containing source (most often thermal system insulation) permitting the material to be removed while minimizing release of airborne fibers to the surrounding atmosphere.

**Gooseneck:** A portion of a water service connection between the distribution system water main and a meter. Sometimes called a pigtail.

**Grab Sample:** A single sample collected at a particular time and place that represents the composition of the water, air, or soil only at that time and place.

**Grain Loading:** The rate at which particles are emitted from a pollution source. Measurement is made by the number of grains per cubic foot of gas emitted.

**Granular Activated Carbon Treatment:** A filtering system often used in small water systems and individual homes to remove organics. Also used by municipal water treatment plantsd. GAC can be highly effective in lowering elevated levels of radon in water.

**Grasscycling:** Source reduction activities in which grass clippings are left on the lawn after mowing.

**Grassed Waterway:** Natural or constructed watercourse or outlet that is shaped or graded and established in suitable vegetation for the disposal of runoff water without erosion.

**Gray Water:** Domestic wastewater composed of wash water from kitchen, bathroom, and laundry sinks, tubs, and washers.

**Greenhouse Effect:** The warming of the Earth's atmosphere attributed to a buildup of carbon dioxide or other gases; some scientists think that this build-up allows the sun's rays to heat the Earth, while making the infra-red radiation atmosphere opaque to infra-red radiation, thereby preventing a counterbalancing loss of heat.

**Greenhouse Gas:** A gas, such as carbon dioxide or methane, which contributes to potential climate change.

**Grinder Pump:** A mechanical device that shreds solids and raises sewage to a higher elevation through pressure sewers.

**Gross Alpha/Beta Particle Activity:** The total radioactivity due to alpha or beta particle emissions as inferred from measurements on a dry sample.

**Gross Power-Generation Potential:** The installed power generation capacity that landfill gas can support.

**Ground Cover:** Plants grown to keep soil from eroding.

**Ground Water:** The supply of fresh water found beneath the Earth's surface, usually in aquifers, which supply wells and springs. Because ground water is a major source of drinking water, there is growing concern over contamination from leaching agricultural or industrial pollutants or leaking underground storage tanks.

**Ground Water Under the Direct Influence (UDI) of Surface Water:** Any water beneath the surface of the ground with: 1. significant occurence of insects or other microorganisms, algae, or large-diameter pathogens; 2. significant and relatively rapid shifts in water characteristics such as turbidity, temperature, conductivity, or pH which closely correlate to climatological or surface water conditions. Direct influence is determined for individual sources in accordance with criteria established by a state.

**Ground-Penetrating Radar:** A geophysical method that uses high frequency electromagnetic waves to obtain subsurface information.

**Ground-Water Discharge:** Ground water entering near coastal waters which has been contaminated by landfill leachate, deep well injection of hazardous wastes, septic tanks, etc.

**Ground-Water Disinfection Rule:** A 1996 amendment of the Safe Drinking Water Act requiring EPA to promulgate national primary drinking water regulations requiring disinfection as for all public water systems, including surface waters and ground water systems.

**Gully Erosion:** Severe erosion in which trenches are cut to a depth greater than 30 centimeters (a foot). Generally, ditches deep enough to cross with farm equipment are considered gullies.

## H

**Habitat:** The place where a population (e.g. human, animal, plant, microorganism) lives and its surroundings, both living and non-living.

**Habitat Indicator:** A physical attribute of the environment measured to characterize conditions necessary to support an organism, population, or community in the absence of pollutants; e.g. salinity of estuarine waters or substrate type in streams or lakes.

**Half-Life:** 1. The time required for a pollutant to lose one-half of its original coconcentrationor example, the biochemical half-life of DDT in the environment is 15 years. 2. The time required for half of the atoms of a radioactive element to undergo self-transmutation or decay (half-life of radium is 1620 years). 3. The time required for the elimination of half a total dose from the body.

**Halogen:** A type of incandescent lamp with higher energy-efficiency that standard ones.

**Halon:** Bromine-containing compounds with long atmospheric lifetimes whose breakdown in the stratosphere causes depletion of ozone. Halons are used in firefighting.

**Hammer Mill:** A high-speed machine that uses hammers and cutters to crush, grind, chip, or shred solid waste.

**Hard Water:** Alkaline water containing dissolved salts that interfere with some industrial processes and prevent soap from sudsing.

**Hauler:** Garbage collection company that offers complete refuse removal service; many will also collect recyclables.

**Hazard:** 1. Potential for radiation, a chemical or other pollutant to cause human illness or injury. 2. In the pesticide program, the inherent toxicity of a compound. Hazard identification of a given substances is an informed judgment based on verifiable toxicity data from animal models or human studies.

**Hazard Assessment:** Evaluating the effects of a stressor or determining a margin of safety for an organism by comparing the concentration which causes toxic effects with an estimate of exposure to the organism.

**Hazard Communication Standard:** An OSHA regulation that requires chemical manufacturers, suppliers, and importers to assess the hazards of the chemicals that they make, supply, or import, and to inform employers, customers, and workers of these hazards through MSDS information.

**Hazard Evaluation:** A component of risk evaluation that involves gathering and evaluating data on the types of health injuries or diseases that may be produced by a chemical and on the conditions of exposure under which such health effects are produced.

**Hazard Identification:** Determining if a chemical or a microbe can cause adverse health effects in humans and what those effects might be.

**Hazard Quotient:** The ratio of estimated site-specific exposure to a single chemical from a site over a specified period to the estimated daily exposure level, at which no adverse health effects are likely to occur.

**Hazard Ratio:** A term used to compare an animal's daily dietary intake of a pesticide to its LD 50 value. A ratio greater than 1.0 indicates that the animal is likely to consume an a dose amount which would kill 50 percent of animals of the same species. (See: LD 50 /Lethal Dose.)

**Hazardous Air Pollutants:** Air pollutants which are not covered by ambient air quality standards but which, as defined in the Clean Air Act, may present a threat of adverse human health effects or adverse environmental effects.Such pollutants include asbestos, beryllium, mercury, benzene, coke oven emissions, radionuclides, and vinyl chloride.

**Hazardous Chemical:** An EPA designation for any hazardous material requiring an MSDS under OSHA's Hazard Communication Standard. Such substances are capable of producing fires and explosions or adverse health effects like cancer and dermatitis. Hazardous chemicals are distinct from hazardous waste.(See: Hazardous Waste.)

**Hazardous Ranking System:** The principal screening tool used by EPA to evaluate risks to public health and the environment associated with abandoned or uncontrolled hazardous waste sites. The HRS calculates a score based on the potential of hazardous substances spreading from the site through the air, surface water, or ground water, and on other factors such as density and proximity of human population. This score is the primary factor in deciding if the site should be on the National Priorities List and, if so, what ranking it should have compared to other sites on the list.

**Hazardous Substance:** 1. Any material that poses a threat to human health and/or the environment. Typical hazardous substances are toxic, corrosive, ignitable, explosive, or chemically reactive. 2. Any substance designated by EPA to be reported if a designated quantity of the substance is spilled in the waters of the United States or is otherwise released into the environment.

**Hazardous Waste:** By-products of society that can pose a substantial or potential hazard to human health or the environment when improperly managed. Possesses at least one of four characteristics (ignitability, corrosivity, reactivity, or toxicity), or appears on special EPA lists.

**Hazardous Waste Landfill:** An excavated or engineered site where hazardous waste is deposited and covered.

**Hazardous Waste Minimization:** Reducing the amount of toxicity or waste produced by a facility via source reduction or environmentally sound recycling.

**Hazards Analysis:** Procedures used to (1) identify potential sources of release of hazardous materials from fixed facilities or transportation accidents; (2) determine the vulnerability of a geographical area to a release of hazardous materials; and (3) compare hazards to determine which present greater or lesser risks to a community.

**Hazards Identification:** Providing information on which facilities have extremely hazardous substances, what those chemicals are, how much there is at each facility, how the chemicals are stored, and whether they are used at high temperatures.

**Headspace:** The vapor mixture trapped above a solid or liquid in a sealed vessel.

**Health Advisory Level:** A non-regulatory health-based reference level of chemical traces (usually in ppm) in drinking water at which there are no adverse health risks when ingested over various periods of time. Such levels are established for one day, 10 days, long-term and life-time exposure periods. They contain a wide margin of safety.

**Health Assessment:** An evaluation of available data on existing or potential risks to human health posed by a Superfund site. The Agency for Toxic Substances and Disease Registry (ATSDR) of the Department of Health and Human Services (DHHS) is required to perform such an assessment at every site on the National Priorities List.

**Heat Island Effect:** A "dome" of elevated temperatures over an urban area caused by structural and pavement heat fluxes, and pollutant emissions.

**Heat Pump:** An electric device with both heating and cooling capabilities. It extracts heat from one medium at a lower (the heat source) temperature and transfers it to another at a higher temperature (the heat sink), thereby cooling the first and warming the second. (See: geothermal, water source heat pump.)

**Heavy Metals:** Metallic elements with high atomic weights; (e.g. mercury, chromium, cadmium, arsenic, and lead); can damage living things at low concentrations and tend to accumulate in the food chain.

**Heptachlor:** An insecticide that was banned on some food products in 1975 and in all of them 1978. It was allowed for use in seed treatment until 1983. More recently it was found in milk and other dairy products in Arkansas and Missouri where dairy cattle were illegally fed treated seed.

**Herbicide:** A chemical pesticide designed to control or destroy plants, weeds, or grasses.

**Herbivore:** An animal that feeds on plants.

**Heterotrophic Organisms:** Species that are dependent on organic matter for food.

**High End Exposure (dose) Estimate:** An estimate of exposure, or dose level received anyone in a defined population that is greater than the 90th percentile of all individuals in that population, but less than the exposure at the highest percentile in that population. A high end risk descriptor is an estimate of the risk level for such individuals. Note that risk is based on a combination of exposure and susceptibility to the stressor.

**High Intensity Discharge:** A generic term for mercury vapor, metal halide, and high pressure sodium lamps and fixtures.

**High-Density Polyethylene:** A material used to make plastic bottles and other products that produces toxic fumes when burned.

**High-Level Nuclear Waste Facility:** Plant designed to handle disposal of used nuclear fuel, high-level radioactive waste, and plutonium waste.

**High-Level Radioactive Waste (HLRW):** Waste generated in core fuel of a nuclear reactor, found at nuclear reactors or by nuclear fuel reprocessing; is a serious threat to anyone who comes near the waste without shielding. (See: low-level radioactive waste.)

**High-Line Jumpers:** Pipes or hoses connected to fire hydrants and laid on top of the ground to provide emergency water service for an isolated portion of a distribution system.

**High-Risk Community:** A community located within the vicinity of numerous sites of facilities or other potential sources of envienvironmental exposure/health hazards which may result in high levels of exposure to contaminants or pollutants.

**High-to-Low-Dose Extrapolation:** The process of prediction of low exposure risk to humans and animals from the measured high-exposure-high-risk data involving laboratory animals.

**Highest Dose Tested:** The highest dose of a chemical or substance tested in a study.

**Holding Pond:** A pond or reservoir, usually made of earth, built to store polluted runoff.

**Holding Time:** The maximum amount of time a sample may be stored before analysis.

**Hollow Stem Auger Drilling:** Conventional drilling method that uses augurs to penetrate the soil. As the augers are rotated, soil cuttings are conveyed to the ground surface via augur spirals. DP tools can be used inside the hollow augers.

**Homeowner Water System:** Any water system which supplies piped water to a single residence.

**Homogeneous Area:** In accordance with Asbestos Hazard and Emergency Response Act (AHERA) definitions, an area of surfacing materials, thermal surface insulation, or miscellaneous material that is uniform in color and texture.

**Hood Capture Efficiency:** Ratio of the emissions captured by a hood and directed into a control or disposal device, expressed as a percent of all emissions.

**Host:** 1. In genetics, the organism, typically a bacterium, into which a gene from another organism is transplanted. 2. In medicine, an animal infected or parasitized by another organism.

**Household Hazardous Waste:** Hazardous products used and disposed of by residential as opposed to industrial consumers. Includes paints, stains, varnishes, solvents, pesticides, and other materials or products containing volatile chemicals that can catch fire, react or explode, or that are corrosive or toxic.

**Household Waste (Domestic Waste):** Solid waste, composed of garbage and rubbish, which normally originates in a private home or apartment house. Domestic waste may contain a significant amount of toxic or hazardous waste.

**Human Equivalent Dose:** A dose which, when administered to humans, produces an effect equal to that produced by a dose in animals.

**Human Exposure Evaluation:** Describing the nature and size of the population exposed to a substance and the magnitude and duration of their exposure.

**Human Health Risk:** The likelihood that a given exposure or series of exposures may have damaged or will damage the health of individuals.

**Hydraulic Conductivity:** The rate at which water can move through a permeable medium. (i.e. the coefficient of permeability.)

**Hydraulic Gradient:** In general, the direction of groundwater flow due to changes in the depth of the water table.

**Hydrocarbons (HC):** Chemical compounds that consist entirely of carbon and hydrogen.

**Hydrogen Sulfide (H2S):** Gas emitted during organic decomposition. Also a by-product of oil refining and burning. Smells like rotten eggs and, in heavy concentration, can kill or cause illness.

**Hydrogeological Cycle:** The natural process recycling water from the atmosphere down to (and through) the earth and back to the atmosphere again.

**Hydrogeology:** The geology of ground water, with particular emphasis on the chemistry and movement of water.

**Hydrologic Cycle:** Movement or exchange of water between the atmosphere and earth.

**Hydrology:** The science dealing with the properties, distribution, and circulation of water.

**Hydrolysis:** The decomposition of organic compounds by interaction with water.

**Hydronic:** A ventilation system using heated or cooled water pumped through a building.

**Hydrophilic:** Having a strong affinity for water.

**Hydrophobic:** Having a strong aversion for water.

**Hydropneumatic:** A water system, usually small, in which a water pump is automatically controlled by the pressure in a compressed air tank.

**Hypersensitivity Diseases:** Diseases characterized by allergic responses to pollutants; diseases most clearly associated with indoor air quality are asthma, rhinitis, and pneumonic hypersensitivity.

**Hypolimnion:** Bottom waters of a thermally stratified lake. The hypolimnion of a eutrophic lake is usually low or lacking in oxygen.

**Hypoxia/Hypoxic Waters:** Waters with dissolved oxygen concentrations of less than 2 ppm, the level generally accepted as the minimum required for most marine life to survive and reproduce.

## I

**Identification Code or EPA I.D. Number:** The unique code assigned to each generator, transporter, and treatment, storage, or disposal facility by regulating agencies to facilitate identification and tracking of chemicals or hazardous waste.

**Ignitable:** Capable of burning or causing a fire.

**IM240:** A high-tech, transient dynamometer automobile emissions test that takes up to 240 seconds.

**Imhoff Cone:** A clear, cone-shaped container used to measure the volume of settleable solids in a specific volume of water.

**Immediately Dangerous to Life and Health (IDLH):** The maximum level to which a healthy individual can be exposed to a chemical for 30 minutes and escape without suffering irreversible health effects or impairing symptoms. Used as a "level of concern." (See: level of concern.)

**Imminent Hazard:** One that would likely result in unreasonable adverse effects on humans or the environment or risk unreasonable hazard to an endangered species during the time required for a pesticide registration cancellation proceeding.

**Imminent Threat:** A high probability that exposure is occurring.

**Immiscibility:** The inability of two or more substances or liquids to readily dissolve into one another, such as soil and water. Immiscibility The inability of two or more substances or liquids to readily dissolve into one another, such as soil and water.

**Impermeable:** Not easily penetrated. The property of a material or soil that does not allow, or allows only with great difficulty, the movement or passage of water.

**Imports:** Municipal solid waste and recyclables that have been transported to a state or locality for processing or final disposition (but that did not originate in that state or locality).

**Impoundment:** A body of water or sludge confined by a dam, dike, floodgate, or other barrier.

**In Situ:** In its original place; unmoved unexcavated; remaining at the site or in the subsurface.

**In-Line Filtration:** Pre-treatment method in which chemicals are mixed by the flowing water; commonly used in pressure filtration installations. Eliminates need for flocculation and sedimentation.

**In-Situ Flushing:** Introduction of large volumes of water, at times supplemented with cleaning compounds, into soil, waste, or ground water to flush hazardous contaminants from a site.

**In-Situ Oxidation:** Technology that oxidizes contaminants dissolved in ground water, converting them into insoluble compounds.

**In-Situ Stripping:** Treatment system that removes or "strips" volatile organic compounds from contaminated ground or surface water by forcing an airstream through the water and causing the compounds to evaporate.

**In-Situ Vitrification:** Technology that treats contaminated soil in place at extremely high temperatures, at or more than 3000 degrees Fahrenheit.

**In Vitro:** Testing or action outside an organism (e.g. inside a test tube or culture dish.)

**In Vivo:** Testing or action inside an organism.

**Incident Command Post:** A facility located at a safe distance from an emergency site, where the incident commander, key staff, and technical representatives can make decisions and deploy emergency manpower and equipment.

**Incident Command System (ICS):** The organizational arrangement wherein one person, normally the Fire Chief of the impacted district, is in charge of an integrated, comprehensive emergency response organization and the emergency incident site, backed by an Emergency Operations Center staff with resources, information, and advice.

**Incineration:** A treatment technology involving destruction of waste by controlled burning at high temperatures; e.g., burning sludge to remove the water and reduce the remaining residues to a safe, non-burnable ash that can be disposed of safely on land, in some waters, or in underground locations.

**Incineration at Sea:** Disposal of waste by burning at sea on specially-designed incinerator ships.

**Incinerator:** A furnace for burning waste under controlled conditions.

**Incompatible Waste:** A waste unsuitable for mixing with another waste or material because it may react to form a hazard.

**Indemnification:** In the pesticide program, legal requirement that EPA pay certain end-users, dealers, and distributors for the cost of stock on hand at the time a pesticide registration is suspended.

**Indicator:** In biology, any biological entity or processes, or community whose characteristics show the presence of specific environmental conditions. 2. In chemistry, a substance that shows a visible change, usually of color, at a desired point in a chemical reaction. 3.A device that indicates the result of a measurement; e.g. a pressure gauge or a moveable scale.

**Indirect Discharge:** Introduction of pollutants from a non-domestic source into a publicly owned waste-treatment system. Indirect dischargers can be commercial or industrial facilities whose wastes enter local sewers.

**Indirect Source:** Any facility or building, property, road or parking area that attracts motor vehicle traffic and, indirectly, causes pollution.

**Indoor Air:** The breathable air inside a habitable structure or conveyance.

**Indoor Air Pollution:** Chemical, physical, or biological contaminants in indoor air.

**Indoor Climate:** Temperature, humidity, lighting, air flow and noise levels in a habitable structure or conveyance. Indoor climate can affect indoor air pollution.

**Industrial Pollution Prevention:** Combination of industrial source reduction and toxic chemical use substitution.

**Industrial Process Waste:** Residues produced during manufacturing operations.

**Industrial Sludge:** Semi-liquid residue or slurry remaining from treatment of industrial water and wastewater.

**Industrial Source Reduction:** Practices that reduce the amount of any hazardous substance, pollutant, or contaminant entering any waste stream or otherwise released into the environment. Also reduces the threat to public health and the environment associated with such releases. Term includes equipment or technology modifications, substitution of raw materials, and improvements in housekeeping, maintenance, training or inventory control.

**Industrial Waste:** Unwanted materials from an industrial operation; may be liquid, sludge, solid, or hazardous waste.

**Inert Ingredient:** Pesticide components such as solvents, carriers, dispersants, and surfactants that are not active against target pests. Not all inert ingredients are innocuous.

**Inertial Separator:** A device that uses centrifugal force to separate waste particles.

**Infectious Agent:** Any organism, such as a pathogenic virus, parasite, or or bacterium, that is capable of invading body tissues, multiplying, and causing disease.

**Infectious Waste:** Hazardous waste capable of causing infections in humans, including: contaminated animal waste; human blood and blood products; isolation waste, pathological waste; and discarded sharps (needles, scalpels or broken medical instruments).

**Infiltration:** 1. The penetration of water through the ground surface into sub-surface soil or the penetration of water from the soil into sewer or other pipes through defective joints, connections, or manhole walls. 2. The technique of applying large volumes of waste water to land to penetrate the surface and percolate through the underlying soil. (See: percolation.)

**Infiltration Gallery:** A sub-surface groundwater collection system, typically shallow in depth, constructed with open-jointed or perforated pipes that discharge collected water into a watertight chamber from which the water is pumped to treatment facilities and into the distribution system. Usually located close to streams or ponds.

**Infiltration Rate:** The quantity of water that can enter the soil in a specified time interval.

**Inflow:** Entry of extraneous rain water into a sewer system from sources other than infiltration, such as basement drains, manholes, storm drains, and street washing.

**Influent:** Water, wastewater, or other liquid flowing into a reservoir, basin, or treatment plant.

**Information Collection Request (ICR):** A description of information to be gathered in connection with rules, proposed rules, surveys, and guidance documents that contain information-gathering requirements. The ICR describes what information is needed, why it is needed, how it will be collected, and how much collecting it will cost. The ICR is submitted by the EPA to the Office of Management and Budget (OMB) for approval.

**Information File:** In the Superfund program, a file that contains accurate, up-to-date documents on a Superfund site. The file is usually located in a public building (school, library, or city hall) convenient for local residents.

**Inhalable Particles:** All dust capable of entering the human respiratory tract.

**Initial Compliance Period (Water):** The first full three-year compliance period which begins at least 18 months after promulgation.

**Injection Well:** A well into which fluids are injected for purposes such as waste disposal, improving the recovery of crude oil, or solution mining.

**Injection Zone:** A geological formation receiving fluids through a well.

**Innovative Technologies:** New or inventive methods to treat effectively hazardous waste and reduce risks to human health and the environment.

**Innovative Treatment Technologies:** Technologies whose routine use is inhibited by lack of data on performance and cost. (See: Established treatment technologies.)

**Inoculum:** 1. Bacteria or fungi injected into compost to start biological action. 2. A medium containing organisms, usually bacteria or a virus, that is introduced into cultures or living organisms.

**Inorganic Chemicals:** Chemical substances of mineral origin, not of basically carbon structure.

**Insecticide:** A pesticide compound specifically used to kill or prevent the growth of insects.

**Inspection and Maintenance (I/M):** 1. Activities to ensure that vehicles' emission controls work properly. 2. Also applies to wastewater treatment plants and other anti-pollution facilities and processes.

**Institutional Waste:** Waste generated at institutions such as schools, libraries, hospitals, prisons, etc.

**Instream Use:** Water use taking place within a stream channel; e.g., hydro-electric power generation, navigation, water quality improvement, fish propagation, recreation.

**Integrated Exposure Assessment:** Cumulative summation (over time) of the magnitude of exposure to a toxic chemical in all media.

**Integrated Pest Management (IPM):** A mixture of chemical and other, non-pesticide, methods to control pests.

**Integrated Waste Management:** Using a variety of practices to handle municipal solid waste; can include source reduction, recycling, incineration, and landfilling.

**Interceptor Sewers:** Large sewer lines that, in a combined system, control the flow of sewage to the treatment plant. In a storm, they allow some of the sewage to flow directly into a receiving stream, thus keeping it from overflowing onto the streets. Also used in separate systems to collect the flows from main and trunk sewers and carry them to treatment points.

**Interface:** The common boundary between two substances such as a water and a solid, water and a gas, or two liquids such as water and oil.

**Interfacial Tension:** The strength of the film separating two immiscible fluids (e.g. oil and water) measured in dynes per, or millidynes per centimeter.

**Interim (Permit) Status:** Period during which treatment, storage and disposal facilities coming under RCRA in 1980 are temporarily permitted to operate while awaiting a permanent permit. Permits issued under these circumstances are usually called "Part A" or "Part B" permits.

**Internal Dose:** In exposure assessment, the amount of a substance penetrating the absorption barriers (e.g. skin, lung tissue, gastrointestinal tract) of an organism through either physical or biological processes. (See: absorbed dose)

**Interstate Carrier Water Supply:** A source of water for drinking and sanitary use on planes, buses, trains, and ships operating in more than one state. These sources are federally regulated.

**Interstate Commerce Clause:** A clause of the U.S. Constitution which reserves to the federal government the right to regulate the conduct of business across state lines. Under this clause, for example, the U.S. Supreme Court has ruled that states may not inequitably restrict the disposal of out-of-state wastes in their jurisdictions.

**Interstate Waters:** Waters that flow across or form part of state or international boundaries; e.g. the Great Lakes, the Mississippi River, or coastal waters.

**Interstitial Monitoring:** The continuous surveillance of the space between the walls of an underground storage tank.

**Intrastate Product:** Pesticide products once registered by states for sale and use only in the state. All intrastate products have been converted to full federal registration or canceled.

**Inventory (TSCA):** Inventory of chemicals produced pursuant to Section 8 (b) of the Toxic Substances Control Act.

**Inversion:** A layer of warm air that prevents the rise of cooling air and traps pollutants beneath it; can cause an air pollution episode.

**Ion:** An electrically charged atom or group of atoms.

**Ion Exchange Treatment:** A common water-softening method often found on a large scale at water purification plants that remove some organics and radium by adding calcium oxide or calcium hydroxide to increase the pH to a level where the metals will precipitate out.

**Ionization Chamber:** A device that measures the intensity of ionizing radiation.

**Ionizing Radiation:** Radiation that can strip electrons from atoms; e.g. alpha, beta, and gamma radiation.

**IRIS:** EPA's Integrated Risk Information System, an electronic data base containing the Agency's latest descriptive and quantitative regulatory information on chemical constituents.

**Irradiated Food:** Food subject to brief radioactivity, usually gamma rays, to kill insects, bacteria, and mold, and to permit storage without refrigeration.

**Irradiation:** Exposure to radiation of wavelengths shorter than those of visible light (gamma, x-ray, or ultra- violet), for medical purposes, to sterilize milk or other foodstuffs, or to induce polymerization of monomers or vulcanization of rubber.

**Irreversible Effect:** Effect characterized by the inability of the body to partially or fully repair injury caused by a toxic agent.

**Irrigation:** Applying water or wastewater to land areas to supply the water and nutrient needs of plants.

**Irrigation Efficiency:** The amount of water stored in the crop root zone compared to the amount of irrigation water applied.

**Irrigation Return Flow:** Surface and subsurface water which leaves the field following application of irrigation water.

**Irritant:** A substance that can cause irritation of the skin, eyes, or respiratory system. Effects may be acute from a single high level exposure, or chronic from repeated low-level exposures to such compounds as chlorine, nitrogen dioxide, and nitric acid.

**Isoconcentration:** More than one sample point exhibiting the same isolate concentration.

**Isopleth:** The line or area represented by an isoconcentration.

**Isotope:** A variation of an element that has the same atomic number of protons but a different weight because of the number of neutrons. Various isotopes of the same element may have different radioactive behaviors, some are highly unstable..

**Isotropy:** The condition in which the hydraulic or other properties of an aquifer are the same in all directions.

## J

**Jar Test:** A laboratory procedure that simulates a water treatment plant's coagulation/flocculation units with differing chemical doses, mix speeds, and settling times to estimate the minimum or ideal coagulant dose required to achieve certain water quality goals.

**Joint and Several Liability:** Under CERCLA, this legal concept relates to the liability for Superfund site cleanup and other costs on the part of more than one potentially responsible party (i.e. if there were several owners or users of a site that became contaminated over the years, they could all be considered potentially liable for cleaning up the site.)

## K

**Karst:** A geologic formation of irregular limestone deposits with sinks, underground streams, and caverns.

**Kinetic Energy:** Energy possessed by a moving object or water body.

**Kinetic Rate Coefficient:** A number that describes the rate at which a water constituent such as a biochemical oxygen demand or dissolved oxygen rises or falls, or at which an air pollutant reacts.

## L

**Laboratory Animal Studies:** Investigations using animals as surrogates for humans.

**Lagoon:** 1. A shallow pond where sunlight, bacterial action, and oxygen work to purify wastewater; also used for storage of wastewater or spent nuclear fuel rods. 2. Shallow body of water, often separated from the sea by coral reefs or sandbars.

**Land Application:** Discharge of wastewater onto the ground for treatment or reuse. (See: irrigation.)

**Land Ban:** Phasing out of land disposal of most untreated hazardous wastes, as mandated by the 1984 RCRA amendments.

**Land Disposal Restrictions:** Rules that require hazardous wastes to be treated before disposal on land to destroy or immobilize hazardous constituents that might migrate into soil and ground water.

**Land Farming (of Waste):** A disposal process in which hazardous waste deposited on or in the soil is degraded naturally by microbes.

**Landfills:** 1. Sanitary landfills are disposal sites for non-hazardous solid wastes spread in layers, compacted to the smallest practical volume, and covered by material applied at the end of each operating day. 2. Secure chemical landfills are disposal sites for hazardous waste, selected and designed to minimize the chance of release of hazardous substances into the environment.

**Landscape:** The traits, patterns, and structure of a specific geographic area, including its biological composition, its physical environment, and its anthropogenic or social patterns. An area where interacting ecosystems are grouped and repeated in similar form.

**Landscape Characterization:** Documentation of the traits and patterns of the essential elements of the landscape.

**Landscape Ecology:** The study of the distribution patterns of communities and ecosystems, the ecological processes that affect those patterns, and changes in pattern and process over time.

**Landscape Indicator:** A measurement of the landscape, calculated from mapped or remotely sensed data, used to describe spatial patterns of land use and land cover across a geographic area. Landscape indicators may be useful as measures of certain kinds of environmental degradation such as forest fragmentation.

**Langelier Index (LI):** An index reflecting the equilibrium pH of a water with respect to calcium and alkalinity; used in stabilizing water to control both corrosion and scale deposition.

**Large Quantity Generator:** Person or facility generating more than 2200 pounds of hazardous waste per month. Such generators produce about 90 percent of the nation's hazardous waste, and are subject to all RCRA requirements.

**Large Water System:** A water system that services more than 50,000 customers.

**Laser Induced Fluorescence:** A method for measuring the relative amount of soil and/or groundwater with an in-situ sensor.

**Latency:** Time from the first exposure of a chemical until the appearance of a toxic effect.

**Lateral Sewers:** Pipes that run under city streets and receive the sewage from homes and businesses, as opposed to domestic feeders and main trunk lines.

**Laundering Weir:** Sedimention basin overflow weir.

**LC 50/Lethal Concentration:** Median level concentration, a standard measure of toxicity. It tells how much of a substance is needed to kill half of a group of experimental organisms in a given time. (See: LD 50.)

**LD 50/ Lethal Dose:** The dose of a toxicant or microbe that will kill 50 percent of the test organisms within a designated period. The lower the LD 50, the more toxic the compound.

**Ldlo:** Lethal dose low; the lowest dose in an animal study at which lethality occurs.

**Leachate:** Water that collects contaminants as it trickles through wastes, pesticides or fertilizers. Leaching may occur in farming areas, feedlots, and landfills, and may result in hazardous substances entering surface water, ground water, or soil.

**Leachate Collection System:** A system that gathers leachate and pumps it to the surface for treatment.

**Leaching:** The process by which soluble constituents are dissolved and filtered through the soil by a percolating fluid. (See: leachate.)

**Lead (Pb):** A heavy metal that is hazardous to health if breathed or swallowed. Its use in gasoline, paints, and plumbing compounds has been sharply restricted or eliminated by federal laws and regulations. (See: heavy metals.)

**Lead Service Line:** A service line made of lead which connects the water to the building inlet and any lead fitting connected to it.

**Legionella:** A genus of bacteria, some species of which have caused a type of pneumonia called Legionaires Disease.

**Lethal Concentration 50:** Also referred to as LC50, a concentration of a pollutant or effluent at which 50 percent of the test organisms die; a common measure of acute toxicity.

**Lethal Dose 50:** Also referred to as LD50, the dose of a toxicant that will kill 50 percent of test organisms within a designated period of time; the lower the LD 50, the more toxic the compound.

**Level of Concern (LOC):** The concentration in air of an extremely hazardous substance above which there may be serious immediate health effects to anyone exposed to it for short periods

**Life Cycle of a Product:** All stages of a product's development, from extraction of fuel for power to production, marketing, use, and disposal.

**Lifetime Average Daily Dose:** Figure for estimating excess lifetime cancer risk.

**Lifetime Exposure:** Total amount of exposure to a substance that a human would receive in a lifetime (usually assumed to be 70 years).

**Lift:** In a sanitary landfill, a compacted layer of solid waste and the top layer of cover material.

**Lifting Station:** (See: pumping station.)

**Light Non-Aqueous Phase Liquid (LNAPL):** A non-aqueous phase liquid with a specific gravity less than 1.0. Because the specific gravity of water is 1.0, most LNAPLs float on top of the water table. Most common petroleum hydrocarbon fuels and lubricating oils are LNAPLs.

**Light-Emitting Diode:** A long-lasting illumination technology used for exit signs which requires very little power

**Limestone Scrubbing:** Use of a limestone and water solution to remove gaseous stack-pipe sulfur before it reaches the atmosphere.

**Limit of Detection (LOD):** The minimum concentration of a substance being analyzed test that has a 99 percent probability of being identified.

**Limited Degradation:** An environmental policy permitting some degradation of natural systems but terminating at a level well beneath an established health standard.

**Limiting Factor:** A condition whose absence or excessive concentration, is incompatible with the needs or tolerance of a species or population and which may have a negative influence on their ability to thrive.

**Limnology:** The study of the physical, chemical, hydrological, and biological aspects of fresh water bodies.

**Lindane:** A pesticide that causes adverse health effects in domestic water supplies and is toxic to freshwater fish and aquatic life.

**Liner:** 1. A relatively impermeable barrier designed to keep leachate inside a landfill. Liner materials include plastic and dense clay. 2. An insert or sleeve for sewer pipes to prevent leakage or infiltration.

**Lipid Solubility:** The maximum concentration of a chemical that will dissolve in fatty substances. Lipid soluble substances are insoluble in water. They will very selectively disperse through the environment via uptake in living tissue.

**Liquefaction:** Changing a solid into a liquid.

**Liquid Injection Incinerator:** Commonly used system that relies on high pressure to prepare liquid wastes for incineration by breaking them up into tiny droplets to allow easier combustion.

**List:** Shorthand term for EPA list of violating facilities or firms debarred from obtaining government contracts because they violated certain sections of the Clean Air or Clean Water Acts. The list is maintained by The Office of Enforcement and Compliance Monitoring.

**Listed Waste:** Wastes listed as hazardous under RCRA but which have not been subjected to the Toxic Characteristics Listing Process because the dangers they present are considered self-evident.

**Lithology:** Mineralogy, grain size, texture, and other physical properties of granular soil, sediment, or rock.

**Litter:** 1. The highly visible portion of solid waste carelessly discarded outside the regular garbage and trash collection and disposal system. 2. leaves and twigs fallen from forest trees.

**Littoral Zone:** 1. That portion of a body of fresh water extending from the shoreline lakeward to the limit of occupancy of rooted plants. 2. A strip of land along the shoreline between the high and low water levels.

**Local Education Agency (LEA):** In the asbestos program, an educational agency at the local level that exists primarily to operate schools or to contract for educational services, including primary and secondary public and private schools. A single, unaffiliated school can be considered an LEA for AHERA purposes.

**Local Emergency Planning Committee (LEPC):** A committee appointed by the state emergency response commission, as required by SARA Title III, to formulate a comprehensive emergency plan for its jurisdiction.

**Low Density Polyethylene (LOPE):** Plastic material used for both rigid containers and plastic film applications.

**Low Emissivity (low-E) Windows:** New window technology that lowers the amount of energy loss through windows by inhibiting the transmission of radiant heat while still allowing sufficient light to pass through.

**Low $NO_x$ Burners:** One of several combustion technologies used to reduce emissions of Nitrogen Oxides ($NO_x$.)

**Low-Level Radioactive Waste (LLRW):** Wastes less hazardous than most of those associated with a nuclear reactor; generated by hospitals, research laboratories, and certain industries. The Department of Energy, Nuclear Regulatory Commission, and EPA share responsibilities for managing them. (See: high-level radioactive wastes.)

**Lower Detection Limit:** The smallest signal above background noise an instrument can reliably detect.

**Lower Explosive Limit (LEL):** The concentration of a compound in air below which the mixture will not catch on fire.

**Lowest Acceptable Daily Dose:** The largest quantity of a chemical that will not cause a toxic effect, as determined by animal studies.

**Lowest Achievable Emission Rate:** Under the Clean Air Act, the rate of emissions that reflects (1) the most stringent emission limitation in the implementation plan of any state for such source unless the owner or operator demonstrates such limitations are not achievable; or (2) the most stringent emissions limitation achieved in practice, whichever is more stringent. A proposed new or modified source may not emit pollutants in excess of existing new source standards.

**Lowest Observed Adverse Effect Level (LOAEL):** The lowest level of a stressor that causes statistically and biologically significant differences in test samples as compared to other samples subjected to no stressor.

## M

**Macropores:** Secondary soil features such as root holes or desiccation cracks that can create significant conduits for movement of NAPL and dissolved contaminants, or vapor-phase contaminants.

**Magnetic Separation:** Use of magnets to separate ferrous materials from mixed municipal waste stream.

**Major Modification:** This term is used to define modifications of major stationary sources of emissions with respect to Prevention of Significant Deterioration and New Source Review under the Clean Air Act.

**Major Stationary Sources:** Term used to determine the applicability of Prevention of Significant Deterioration and new source regulations. In a nonattainment area, any stationary pollutant source with potential to emit more than 100 tons per year is considered a major stationary source. In PSD areas the cutoff level may be either 100 or 250 tons, depending upon the source.

**Majors:** Larger publicly owned treatment works (POTWs) with flows equal to at least one million gallons per day (mgd) or servicing a population equivalent to 10,000 persons; certain other POTWs having significant water quality impacts. (See: minors.)

**Man-Made (Anthropogenic) Beta Particle and Photon Emitters:** All radionuclides emitting beta particles and/or photons listed in Maximum Permissible Body Burdens and Maximum Permissible Concentrations of Radonuclides in Air and Water for Occupational Exposure.

**Management Plan:** Under the Asbestos Hazard Emergency Response Act (AHERA), a document that each Local Education Agency is required to prepare, describing all activities planned and undertaken by a school to comply with AHERA regulations, including building inspections to identify asbestos-containing materials, response actions, and operations and maintenance programs to minimize the risk of exposure.

**Managerial Controls:** Methods of nonpoint source pollution control based on decisions about managing agricultural wastes or application times or rates for agrochemicals.

**Mandatory Recycling:** Programs which by law require consumers to separate trash so that some or all recyclable materials are recovered for recycling rather than going to landfills.

**Manifest:** A one-page form used by haulers transporting waste that lists EPA identification numbers, type and quantity of waste, the generator it originated from, the transporter that shipped it, and the storage or disposal facility to which it is being shipped. It includes copies for all participants in the shipping process.

**Manifest System:** Tracking of hazardous waste from "cradle-to-grave" (generation through disposal) with accompanying documents known as manifests.(See: cradle to grave.)

**Manual Separation:** Hand sorting of recyclable or compostable materials in waste.

**Manufacturer's Formulation:** A list of substances or component parts as described by the maker of a coating, pesticide, or other product containing chemicals or other substances.

**Manufacturing Use Product:** Any product intended (labeled) for formulation or repackaging into other pesticide products.

**Margin of Safety:** Maximum amount of exposure producing no measurable effect in animals (or studied humans) divided by the actual amount of human exposure in a population.

**Margin of Exposure (MOE):** The ratio of the no-observed adverse-effect-level to the estimated exposure dose.

**Marine Sanitation Device:** Any equipment or process installed on board a vessel to receive, retain, treat, or discharge sewage.

**Marsh:** A type of wetland that does not accumulate appreciable peat deposits and is dominated by herbaceous vegetation. Marshes may be either fresh or saltwater, tidal or non-tidal. (See: wetlands.)

**Material Category:** In the asbestos program, broad classification of materials into thermal surfacing insulation, surfacing material, and miscellaneous material.

**Material Safety Data Sheet (MSDS):** A compilation of information required under the OSHA Communication Standard on the identity of hazardous chemicals, health, and physical hazards, exposure limits, and precautions. Section 311 of SARA requires facilities to submit MSDSs under certain circumstances.

**Material Type:** Classification of suspect material by its specific use or application; e.g., pipe insulation, fireproofing, and floor tile.

**Materials Recovery Facility (MRF):** A facility that processes residentially collected mixed recyclables into new products available for market.

**Maximally (or Most) Exposed Individual:** The person with the highest exposure in a given population.

**Maximum Acceptable Toxic Concentration:** For a given ecological effects test, the range (or geometric mean) between the No Observable Adverse Effect Level and the Lowest Observable Adverse Effects Level.

**Maximum Available Control Technology (MACT):** The emission standard for sources of air pollution requiring the maximum reduction of hazardous emissions, taking cost and feasibility into account. Under the Clean Air Act Amendments of 1990, the MACT must not be less than the average emission level achieved by controls on the best performing 12 percent of existing sources, by category of industrial and utility sources.

**Maximum Contaminant Level:** The maximum permissible level of a contaminant in water delivered to any user of a public system. MCLs are enforceable standards.

**Maximum Contaminant Level Goal (MCLG):** Under the Safe Drinking Water Act, a non-enforceable concentration of a drinking water contaminant, set at the level at which no known or anticipated adverse effects on human health occur and which allows an adequate safety margin. The MCLG is usually the starting point for determining the regulated Maximum Contaminant Level. (See: maximum contaminant level.)

**Maximum Exposure Range:** Estimate of exposure or dose level received by an individual in a defined population that is greater than the 98th percentile dose for all individuals in that population, but less than the exposure level received by the person receiving the highest exposure level.

**Maximum Residue Level:** Comparable to a U.S. tolerance level, the Maximum Residue Level the enforceable limit on food pesticide levels in some countries. Levels are set by the Codex Alimentarius Commission, a United Nations agency managed and funded jointly by the World Health Organization and the Food and Agriculture Organization.

**Maximum Tolerated Dose:** The maximum dose that an animal species can tolerate for a major portion of its lifetime without significant impairment or toxic effect other than carcinogenicity.

**Measure of Effect/ Measurement Endpoint:** A measurable characteristic of ecological entity that can be related to an assessment endpoint; e.g. a laboratory test for eight species meeting certain requirements may serve as a measure of effect for an assessment endpoint, such as survival of fish, aquatic, invertebrate or algal species under acute exposure.

**Measure of Exposure:** A measurable characteristic of a stressor (such as the specific amount of mercury in a body of water) used to help quantify the exposure of an ecological entity or individual organism.

**Mechanical Aeration:** Use of mechanical energy to inject air into water to cause a waste stream to absorb oxygen.

**Mechanical Separation:** Using mechanical means to separate waste into various components.

**Mechanical Turbulence:** Random irregularities of fluid motion in air caused by buildings or other nonthermal, processes.

**Media:** Specific environments—air, water, soil—which are the subject of regulatory concern and activities.

**Medical Surveillance:** A periodic comprehensive review of a worker's health status; acceptable elements of such surveillance program are listed in the Occupational Safety and Health Administration standards for asbestos.

**Medical Waste:** Any solid waste generated in the diagnosis, treatment, or immunization of human beings or animals, in research pertaining thereto, or in the production or testing of biologicals, excluding hazardous waste identified or listed under 40 CFR Part 261 or any household waste as defined in 40 CFR Sub-section 261.4 (b)(1).

**Medium-size Water System:** A water system that serves 3,300 to 50,000 customers.

**Meniscus:** The curved top of a column of liquid in a small tube.

**Mercury (Hg):** Heavy metal that can accumulate in the environment and is highly toxic if breathed or swallowed. (See:heavy metals.)

**Mesotrophic:** Reservoirs and lakes which contain moderate quantities of nutrients and are moderately productive in terms of aquatic animal and plant life.

**Metabolites:** Any substances produced by biological processes, such as those from pesticides.

**Metalimnion:** The middle layer of a thermally stratified lake or reservoir. In this layer there is a rapid decrease in temperature with depth. Also called thermocline.

**Methane:** A colorless, nonpoisonous, flammable gas created by anaerobic decomposition of organic compounds. A major component of natural gas used in the home.

**Methanol:** An alcohol that can be used as an alternative fuel or as a gasoline additive. It is less volatile than gasoline; when blended with gasoline it lowers the carbon monoxide emissions but increases hydrocarbon emissions. Used as pure fuel, its emissions are less ozone-forming than those from gasoline. Poisonous to humans and animals if ingested.

**Method 18:** An EPA test method which uses gas chromatographic techniques to measure the concentration of volatile organic compounds in a gas stream.

**Method 24:** An EPA reference method to determine density, water content and total volatile content (water and VOC) of coatings.

**Method 25:** An EPA reference method to determine the VOC concentration in a gas stream.

**Method Detection Limit (MDL):** See limit of detection.

**Methoxychlor:** Pesticide that causes adverse health effects in domestic water supplies and is toxic to freshwater and marine aquatic life.

**Methyl Orange Alkalinity:** A measure of the total alkalinity in a water sample in which the color of methyl orange reflects the change in level.

**Microbial Growth:** The amplification or multiplication of microorganisms such as bacteria, algae, diatoms, plankton, and fungi.

**Microbial Pesticide:** A microorganism that is used to kill a pest, but is of minimum toxicity to humans.

**Microclimate:** 1. Localized climate conditions within an urban area or neighborhood. 2. The climate around a tree or shrub or a stand of trees.

**Microenvironmental Method:** A method for sequentially assessing exposure for a series of microenvironments that can be approximated by constant concentrations of a stressor.

**Microenvironments:** Well-defined surroundings such as the home, office, or kitchen that can be treated as uniform in terms of stressor concentration.

**Million-Gallons Per Day (MGD):** A measure of water flow.

**Minimization:** A comprehensive program to minimize or eliminate wastes, usually applied to wastes at their point of origin. (See: waste minimization.)

**Mining of an Aquifer:** Withdrawal over a period of time of ground water that exceeds the rate of recharge of the aquifer.

**Mining Waste:** Residues resulting from the extraction of raw materials from the earth.

**Minor Source:** New emissions sources or modifications to existing emissions sources that do not exceed NAAQS emission levels.

**Minors:** Publicly owned treatment works with flows less than 1 million gallons per day. (See: majors.)

**Miscellaneous ACM:** Interior asbestos-containing building material or structural components, members or fixtures, such as floor and ceiling tiles; does not include surfacing materials or thermal system insulation.

**Miscellaneous Materials:** Interior building materials on structural components, such as floor or ceiling tiles.

**Miscible Liquids:** Two or more liquids that can be mixed and will remain mixed under normal conditions.

**Missed Detection:** The situation that occurs when a test indicates that a tank is "tight" when in fact it is leaking.

**Mist:** Liquid particles measuring 40 to 500 micrometers (pm), are formed by condensation of vapor. By comparison, fog particles are smaller than 40 micrometers (pm).

**Mitigation:** Measures taken to reduce adverse impacts on the environment.

**Mixed Funding:** Settlements in which potentially responsible parties and EPA share the cost of a response action.

**Mixed Glass:** Recovered container glass not sorted into categories (e.g. color, grade).

**Mixed Liquor:** A mixture of activated sludge and water containing organic matter undergoing activated sludge treatment in an aeration tank.

**Mixed Metals:** Recovered metals not sorted into categories such as aluminum, tin, or steel cans or ferrous or non-ferrous metals.

**Mixed Municipal Waste:** Solid waste that has not been sorted into specific categories (such as plastic, glass, yard trimmings, etc.)

**Mixed Paper:** Recovered paper not sorted into categories such as old magazines, old newspapers, old corrugated boxes, etc.

**Mixed Plastic:** Recovered plastic unsorted by category.

**Mobile Incinerator Systems:** Hazardous waste incinerators that can be transported from one site to another.

**Mobile Source:** Any non-stationary source of air pollution such as cars, trucks, motorcycles, buses, airplanes, and locomotives.

**Model Plant:** A hypothetical plant design used for developing economic, environmental, and energy impact analyses as support for regulations or regulatory guidelines; first step in exploring the economic impact of a potential NSPS.

**Modified Bin Method:** Way of calculating the required heating or cooling for a building based on determining how much energy the system would use if outdoor temperatures were within a certain temperature interval and then multiplying the energy use by the time the temperature interval typically occurs.

**Modified Source:** The enlargement of a major stationary pollutant sources is often referred to as modification, implying that more emissions will occur.

**Moisture Content:** 1.The amount of water lost from soil upon drying to a constant weight, expressed as the weight per unit of dry soil or as the volume of water per unit bulk volume of the soil. For a fully saturated medium, moisture content indicates the porosity. 2. Water equivalent of snow on the ground; an indicator of snowmelt flood potential.

**Molecule:** The smallest division of a compound that still retains or exhibits all the properties of the substance.

**Molten Salt Reactor:** A thermal treatment unit that rapidly heats waste in a heat-conducting fluid bath of carbonate salt.

**Monitoring:** Periodic or continuous surveillance or testing to determine the level of compliance with statutory requirements and/or pollutant levels in various media or in humans, plants, and animals.

**Monitoring Well:** 1. A well used to obtain water quality samples or measure groundwater levels. 2. A well drilled at a hazardous waste management facility or Superfund site to collect ground-water samples for the purpose of physical, chemical, or biological analysis to determine the amounts, types, and distribution of contaminants in the groundwater beneath the site.

**Monoclonal Antibodies (Also called MABs and MCAs):** 1. Man-made (anthropogenic) clones of a molecule, produced in quantity for medical or research purposes. 2. Molecules of living organisms that selectively find and attach to other molecules to which their structure conforms exactly. This could also apply to equivalent activity by chemical molecules.

**Monomictic:** Lakes and reservoirs which are relatively deep, do not freeze over during winter, and undergo a single stratification and mixing cycle during the year (usually in the fall).

**Montreal Protocol:** Treaty, signed in 1987, governs stratospheric ozone protection and research, and the production and use of ozone-depleting substances. It provides for the end of production of ozone-depleting substances such as CFCS. Under the Protocol, various research groups continue to assess the ozone layer. The Multilateral Fund provides resources to developing nations to promote the transition to ozone-safe technologies.

**Moratorium:** During the negotiation process, a period of 60 to 90 days during which EPA and potentially responsible parties may reach settlement but no site response activities can be conducted.

**Morbidity:** Rate of disease incidence.

**Mortality:** Death rate.

**Most Probable Number:** An estimate of microbial density per unit volume of water sample, based on probability theory.

**Muck Soils:** Earth made from decaying plant materials.

**Mudballs:** Round material that forms in filters and gradually increases in size when not removed by backwashing.

**Mulch:** A layer of material (wood chips, straw, leaves, etc.) placed around plants to hold moisture, prevent weed growth, and enrich or sterilize the soil.

**Multi-Media Approach:** Joint approach to several environmental media, such as air, water, and land.

**Multiple Chemical Sensitivity:** A diagnostic label for people who suffer multi-system illnesses as a result of contact with, or proximity to, a variety of airborne agents and other substances.

**Multiple Use:** Use of land for more than one purpose; e.g., grazing of livestock, watershed and wildlife protection, recreation, and timber production. Also applies to use of bodies of water for recreational purposes, fishing, and water supply.

**Multistage Remote Sensing:** A strategy for landscape characterization that involves gathering and analyzing information at several geographic scales, ranging from generalized levels of detail at the national level through high levels of detail at the local scale.

**Municipal Discharge:** Discharge of effluent from waste water treatment plants which receive waste water from households, commercial establishments, and industries in the coastal drainage basin. Combined sewer/separate storm overflows are included in this category.

**Municipal Sewage:** Wastes (mostly liquid) orginating from a community; may be composed of domestic wastewaters and/or industrial discharges.

**Municipal Sludge:** Semi-liquid residue remaining from the treatment of municipal water and wastewater.

**Municipal Solid Waste:** Common garbage or trash generated by industries, businesses, institutions, and homes.

**Mutagen/Mutagenicity:** An agent that causes a permanent genetic change in a cell other than that which occurs during normal growth. Mutagenicity is the capacity of a chemical or physical agent to cause such permanent changes.

## N

**National Ambient Air Quality Standards (NAAQS):** Standards established by EPA that apply for outdoor air throughout the country. (See: criteria pollutants, state implementation plans, emissions trading.)

**National Emissions Standards for Hazardous Air Pollutants (NESHAPS):** Emissions standards set by EPA for an air pollutant not covered by NAAQS that may cause an increase in fatalities or in serious, irreversible, or incapacitating illness. Primary standards are designed to protect human health, secondary standards to protect public welfare (e.g. building facades, visibility, crops, and domestic animals).

**National Environmental Performance Partnership Agreements:** System that allows states to assume greater responsibility for environmental programs based on their relative ability to execute them.

**National Estuary Program:** A program established under the Clean Water Act Amendments of 1987 to develop and implement conservation and management plans for protecting estuaries and restoring and maintaining their chemical, physical, and biological integrity, as well as controlling point and nonpoint pollution sources.

**National Municipal Plan:** A policy created in 1984 by EPA and the states in 1984 to bring all publicly owned treatment works (POTWs) into compliance with Clean Water Act requirements.

**National Oil and Hazardous Substances Contingency Plan (NOHSCP/NCP):** The federal regulation that guides determination of the sites to be corrected under both the Superfund program and the program to prevent or control spills into surface waters or elsewhere.

**National Pollutant Discharge Elimination System (NPDES):** A provision of the Clean Water Act which prohibits discharge of pollutants into waters of the United States unless a special permit is issued by EPA, a state, or, where delegated, a tribal government on an Indian reservation.

**National Priorities List (NPL):** EPA's list of the most serious uncontrolled or abandoned hazardous waste sites identified for possible long-term remedial action under Superfund. The list is based primarily on the score a site receives from the Hazard Ranking System. EPA is required to update the NPL at least once a year. A site must be on the NPL to receive money from the Trust Fund for remedial action.

**National Response Center:** The federal operations center that receives notifications of all releases of oil and hazardous substances into the environment; open 24 hours a day, is operated by the U.S. Coast Guard, which evaluates all reports and notifies the appropriate agency.

**National Response Team (NRT):** Representatives of 13 federal agencies that, as a team, coordinate federal responses to nationally significant incidents of pollution—an oil spill, a major chemical release, or a - superfund response action—and provide advice and technical assistance to the responding agency(ies) before and during a response action.

**National Secondary Drinking Water Regulations:** Commonly referred to as NSDWRs.

**Navigable Waters:** Traditionally, waters sufficiently deep and wide for navigation by all, or specified vessels; such waters in the United States come under federal jurisdiction and are protected by certain provisions of the Clean Water Act.

**Necrosis:** Death of plant or animal cells or tissues. In plants, necrosis can discolor stems or leaves or kill a plant entirely.

**Negotiations (Under Superfund):** After potentially responsible parties are identified for a site, EPA coordinates with them to reach a settlement that will result in the PRP paying for or conducting the cleanup under EPA supervision. If negotiations fail, EPA can order the PRP to conduct the cleanup or EPA can pay for the cleanup using Superfund monies and then sue to recover the costs.

**Nematocide:** A chemical agent which is destructive to nematodes.

**Nephelometric:** Method of of measuring turbidity in a water sample by passing light through the sample and measuring the amount of the light that is deflected.

**Netting:** A concept in which all emissions sources in the same area that owned or controlled by a single company are treated as one large source, thereby allowing flexibility in controlling individual sources in order to meet a single emissions standard. (See: bubble.)

**Neutralization:** Decreasing the acidity or alkalinity of a substance by adding alkaline or acidic materials, respectively.

**New Source:** Any stationary source built or modified after publication of final or proposed regulations that prescribe a given standard of performance.

**New Source Performance Standards (NSPS):** Uniform national EPA air emission and water effluent standards which limit the amount of pollution allowed from new sources or from modified existing sources.

**New Source Review (NSR):** A Clean Air Act requirement that State Implementation Plans must include a permit review that applies to the construction and operation of new and modified stationary sources in nonattainment areas to ensure attainment of national ambient air quality standards.

**Nitrate:** A compound containing nitrogen that can exist in the atmosphere or as a dissolved gas in water and which can have harmful effects on humans and animals. Nitrates in water can cause severe illness in infants and domestic animals. A plant nutrient and inorganic fertilizer, nitrate is found in septic systems, animal feed lots, agricultural fertilizers, manure, industrial waste waters, sanitary landfills, and garbage dumps.

**Nitric Oxide (NO):** A gas formed by combustion under high temperature and high pressure in an internal combustion engine; it is converted by sunlight and photochemical processes in ambient air to nitrogen oxide. NO is a precursor of ground-level ozone pollution, or smog..

**Nitrification:** The process whereby ammonia in wastewater is oxidized to nitrite and then to nitrate by bacterial or chemical reactions.

**Nitrilotriacetic Acid (NTA):** A compound now replacing phosphates in detergents.

**Nitrite:** 1. An intermediate in the process of nitrification. 2. Nitrous oxide salts used in food preservation.

**Nitrogen Dioxide ($NO_2$):** The result of nitric oxide combining with oxygen in the atmosphere; major component of photochemical smog.

**Nitrogen Oxide ($NO_x$):** The result of photochemical reactions of nitric oxide in ambient air; major component of photochemical smog. Product of combustion from transportation and stationary sources and a major contributor to the formation of ozone in the troposphere and to acid deposition.

**Nitrogenous Wastes:** Animal or vegetable residues that contain significant amounts of nitrogen.

**Nitrophenols:** Synthetic organopesticides containing carbon, hydrogen, nitrogen, and oxygen.

**No Further Remedial Action Planned:** Determination made by EPA following a preliminary assessment that a site does not pose a significant risk and so requires no further activity under CERCLA.

**No Observable Adverse Effect Level (NOAEL):** An exposure level at which there are no statistically or biologically significant increases in the frequency or severity of adverse effects between the exposed population and its appropriate control; some effects may be produced at this level, but they are not considered as adverse, or as precurors to adverse effects. In an experiment with several NOAELs, the regulatory focus is primarily on the highest one, leading to the common usage of the term NOAEL as the highest exposure without adverse effects.

**No Till:** Planting crops without prior seedbed preparation, into an existing cover crop, sod, or crop residues, and eliminating subsequent tillage operations.

**No-Observed-Effect-Level (NOEL):** Exposure level at which there are no statistically or biological significant differences in the frequency or severity of any effect in the exposed or control populations.

**Noble Metal:** Chemically inactive metal such as gold; does not corrode easily.

**Noise:** Product-level or product-volume changes occurring during a test that are not related to a leak but may be mistaken for one.

**Non-Aqueous Phase Liquid (NAPL):** Contaminants that remain undiluted as the original bulk liquid in the subsurface, e.g. spilled oil. (See: fee product.)

**Non-Attainment Area:** Area that does not meet one or more of the National Ambient Air Quality Standards for the criteria pollutants designated in the Clean Air Act.

**Non-Binding Allocations of Responsibility (NBAR):** A process for EPA to propose a way for potentially responsible parties to allocate costs among themselves.

**Non-Community Water System:** A public water system that is not a community water system; e.g. the water supply at a camp site or national park.

**Non-Compliance Coal:** Any coal that emits greater than 3.0 pounds of sulfur dioxide per million BTU when burned. Also known as high-sulfur coal.

**Non-Contact Cooling Water:** Water used for cooling which does not come into direct contact with any raw material, product, byproduct, or waste.

**Non-Conventional Pollutant:** Any pollutant not statutorily listed or which is poorly understood by the scientific community.

**Non-Degradation:** An environmental policy which disallows any lowering of naturally occurring quality regardless of preestablished health standards.

**Non-Ferrous Metals:** Nonmagnetic metals such as aluminum, lead, and copper. Products made all or in part from such metals include containers, packaging, appliances, furniture, electronic equipment and aluminum foil.

**Non-ionizing Electromagnetic Radiation:** 1. Radiation that does not change the structure of atoms but does heat tissue and may cause harmful biological effects. 2. Microwaves, radio waves, and low-frequency electromagnetic fields from high-voltage transmission lines.

**Non-Methane Hydrocarbon (NMHC):** The sum of all hydrocarbon air pollutants except methane; significant precursors to ozone formation.

**Non-Methane Organic Gases (NMOG):** The sum of all organic air pollutants. Excluding methane; they account for aldehydes, ketones, alcohols, and other pollutants that are not hydrocarbons but are precursors of ozone.

**Non-Point Sources:** Diffuse pollution sources (i.e. without a single point of origin or not introduced into a receiving stream from a specific outlet). The pollutants are generally carried off the land by storm water. Common non-point sources are agriculture, forestry, urban, mining, construction, dams, channels, land disposal, saltwater intrusion, and city streets.

**Non-potable:** Water that is unsafe or unpalatable to drink because it contains pollutants, contaminants, minerals, or infective agents.

**Non-Road Emissions:** Pollutants emitted by combustion engines on farm and construction equipment, gasoline-powered lawn and garden equipment, and power boats and outboard motors.

**Non-Transient Non-Community Water System:** A public water system that regularly serves at least 25 of the same non-resident persons per day for more than six months per year.

**Nondischarging Treatment Plant:** A treatment plant that does not discharge treated wastewater into any stream or river. Most are pond systems that dispose of the total flow they receive by means of evaporation or percolation to groundwater, or facilities that dispose of their effluent by recycling or reuse (e.g. spray irrigation or groundwater discharge).

**Nonfriable Asbestos-Containing Materials:** Any material containing more than one percent asbestos (as determined by Polarized Light Microscopy) that, when dry, cannot be crumbled, pulverized, or reduced to powder by hand pressure.

**Nonhazardous Industrial Waste:** Industrial process waste in wastewater not considered municipal solid waste or hazardous waste under RARA.

**Notice of Deficiency:** An EPA request to a facility owner or operator requesting additional information before a preliminary decision on a permit application can be made.

**Notice of Intent to Cancel:** Notification sent to registrants when EPA decides to cancel registration of a product containing a pesticide.

**Notice of Intent to Deny:** Notification by EPA of its preliminary intent to deny a permit application.

**Notice of Intent to Suspend:** Notification sent to a pesticide registrant when EPA decides to suspend product sale and distribution because of failure to submit requested data in a timely and/or acceptable manner, or because of imminent hazard. (See: emergency suspension.)

**Nuclear Reactors and Support Facilities:** Uranium mills, commercial power reactors, fuel reprocessing plants, and uranium enrichment facilities.

**Nuclear Winter:** Prediction by some scientists that smoke and debris rising from massive fires of a nuclear war could block sunlight for weeks or months, cooling the earth's surface and producing climate changes that could, for example, negatively affect world agricultural and weather patterns.

**Nuclide:** An atom characterized by the number of protons, neturons, and energy in the nucleus.

**Nutrient:** Any substance assimilated by living things that promotes growth. The term is generally applied to nitrogen and phosphorus in wastewater, but is also applied to other essential and trace elements.

**Nutrient Pollution:** Contamination of water resources by excessive inputs of nutrients. In surface waters, excess algal production is a major concern.

## O

**Ocean Discharge Waiver:** A variance from Clean Water Act requirements for discharges into marine waters.

**Odor Threshold:** The minimum odor of a water or air sample that can just be detected after successive dilutions with odorless water. Also called threshold odor.

**OECD Guidelines:** Testing guidelines prepared by the Organization of Economic and Cooperative Development of the United Nations. They assist in preparation of protocols for studies of toxicology, environmental fate, etc.

**Off-Site Facility:** A hazardous waste treatment, storage or disposal area that is located away from the generating site.

**Office Paper:** High grade papers such as copier paper, computer printout, and stationary almost entirely made of uncoated chemical pulp, although some ground wood is used. Such waste is also generated in homes, schools, and elsewhere.

**Offsets:** A concept whereby emissions from proposed new or modified stationary sources are balanced by reductions from existing sources to stabilize total emissions. (See: bubble, emissions trading, netting)

**Offstream Use:** Water withdrawn from surface or groundwater sources for use at another place.

**Oil and Gas Waste:** Gas and oil drilling muds, oil production brines, and other waste associated with exploration for, development and production of crude oil or natural gas.

**Oil Desulfurization:** Widely used precombustion method for reducing sulfur dioxide emissions from oil-burning power plants. The oil is treated with hydrogen, which removes some of the sulfur by forming hydrogen sulfide gas.

**Oil Fingerprinting:** A method that identifies sources of oil and allows spills to be traced to their source.

**Oil Spill:** An accidental or intentional discharge of oil which reaches bodies of water. Can be controlled by chemical dispersion, combustion, mechanical containment, and/or adsorption. Spills from tanks and pipelines can also occur away from water bodies, contaminating the soil, getting into sewer systems and threatening underground water sources.

**Oligotrophic Lakes:** Deep clear lakes with few nutrients, little organic matter and a high dissolved-oxygen level.

**On-Scene Coordinator (OSC):** The predesignated EPA, Coast Guard, or Department of Defense official who coordinates and directs Superfund removal actions or Clean Water Act oil- or hazardous-spill response actions.

**On-Site Facility:** A hazardous waste treatment, storage or disposal area that is located on the generating site.

**Onboard Controls:** Devices placed on vehicles to capture gasoline vapor during refueling and route it to the engines when the vehicle is starting so that it can be efficiently burned.

**Onconogenicity:** The capacity to induce cancer.

**One-hit Model:** A mathematical model based on the biological theory that a single "hit" of some minimum critical amount of a carcinogen at a cellular target such as DNA can start an irreversible series events leading to a tumor.

**Opacity:** The amount of light obscured by particulate pollution in the air; clear window glass has zero opacity, a brick wall is 100 percent opaque. Opacity is an indicator of changes in performance of particulate control systems.

**Open Burning:** Uncontrolled fires in an open dump.

**Open Dump:** An uncovered site used for disposal of waste without environmental controls. (See: dump.)

**Operable Unit:** Term for each of a number of separate activities undertaken as part of a Superfund site cleanup. A typical operable unit would be removal of drums and tanks from the surface of a site.

**Operating Conditions:** Conditions specified in a RCRA permit that dictate how an incinerator must operate as it burns different waste types. A trial burn is used to identify operating conditions needed to meet specified performance standards.

**Operation and Maintenance:** 1. Activities conducted after a Superfund site action is completed to ensure that the action is effective. 2. Actions taken after construction to ensure that facilities constructed to treat waste water will be properly operated and maintained to achieve normative efficiency levels and prescribed effluent limitations in an optimum manner. 3. On-going asbestos management plan in a school or other public building, including regular inspections, various methods of maintaining asbestos in place, and removal when necessary.

**Operator Certification:** Certification of operators of community and nontransient noncommunity water systems, asbestos specialists, pesticide applicators, hazardous waste transporter, and other such specialists as required by the EPA or a state agency implementing an EPA-approved environmental regulatory program.

**Optimal Corrosion Control Treatment:** An erosion control treatment that minimizes the lead and copper concentrations at users' taps while also ensuring that the treatment does not cause the water system to violate any national primary drinking water regulations.

**Oral Toxicity:** Ability of a pesticide to cause injury when ingested.

**Organic:** 1. Referring to or derived from living organisms. 2. In chemistry, any compound containing carbon.

**Organic Chemicals/Compounds:** Naturally occuring (animal or plant-produced or synthetic) substances containing mainly carbon, hydrogen, nitrogen, and oxygen.

**Organic Matter:** Carbonaceous waste contained in plant or animal matter and originating from domestic or industrial sources.

**Organism:** Any form of animal or plant life.

**Organophosphates:** Pesticides that contain phosphorus; short-lived, but some can be toxic when first applied.

**Organophyllic:** A substance that easily combines with organic compounds.

**Organotins:** Chemical compounds used in anti-foulant paints to protect the hulls of boats and ships, buoys, and pilings from marine organisms such as barnacles.

**Original AHERA Inspection/Original Inspection/Inspection:** Examination of school buildings arranged by Local Education Agencies to identify asbestos-containing-materials, evaluate their condition, and take samples of materials suspected to contain asbestos; performed by EPA-accredited inspectors.

**Original Generation Point:** Where regulated medical or other material first becomes waste.

**Osmosis:** The passage of a liquid from a weak solution to a more concentrated solution across a semipermeable membrane that allows passage of the solvent (water) but not the dissolved solids.

**Other Ferrous Metals:** Recyclable metals from strapping, furniture, and metal found in tires and consumer electronics but does not include metals found in construction materials or cars, locomotives, and ships. (See: ferrous metals.)

**Other Glass:** Recyclable glass from furniture, appliances, and consumer electronics. Does not include glass from transportation products (cars trucks or shipping containers) and construction or demolition debris. (See: glass.)

**Other Nonferrous Metals:** Recyclable nonferrous metals such as lead, copper, and zinc from appliances, consumer electronics, and nonpackaging aluminum products. Does not include nonferrous metals from industrial applications and construction and demolition debris. (See: nonferrous metals.)

**Other Paper:** For Recyclable paper from books, third-class mail, commercial printing, paper towels, plates and cups; and other nonpackaging paper such as posters, photographic papers, cards and games, milk cartons, folding boxes, bags, wrapping paper, and paperboard. Does not include wrapping paper or shipping cartons.

**Other Plastics:** Recyclable plastic from appliances, eating utensils, plates, containers, toys, and various kinds of equipment. Does not include heavy-duty plastics such as yielding materials.

**Other Solid Waste:** Recyclable nonhazardous solid wastes, other than municipal solid waste, covered under Subtitle D of RARA. (See: solid waste.)

**Other Wood:** Recyclable wood from furniture, consumer electronics cabinets, and other nonpackaging wood products. Does not include lumber and tree stumps recovered from construction and demolition activities, and industrial process waste such as shavings and sawdust.

**Outdoor Air Supply:** Air brought into a building from outside.

**Outfall:** The place where effluent is discharged into receiving waters.

**Overburden:** Rock and soil cleared away before mining.

**Overdraft:** The pumping of water from a groundwater basin or aquifer in excess of the supply flowing into the basin; results in a depletion or "mining" of the groundwater in the basin. (See: groundwater mining)

**Overfire Air:** Air forced into the top of an incinerator or boiler to fan the flames.

**Overflow Rate:** One of the guidelines for design of the settling tanks and clarifers in a treatment plant; used by plant operators to determine if tanks and clarifiers are over or under-used.

**Overland Flow:** A land application technique that cleanses waste water by allowing it to flow over a sloped surface. As the water flows over the surface, contaminants are absorbed and the water is collected at the bottom of the slope for reuse.

**Oversized Regulated Medical Waste:** Medical waste that is too large for plastic bags or standard containers.

**Overturn:** One complete cycle of top to bottom mixing of previously stratified water masses. This phenomenon may occur in spring or fall, or after storms, and results in uniformity of chemical and physical properties of water at all depths.

**Oxidant:** A collective term for some of the primary constituents of photochemical smog.

**Oxidation Pond:** A man-made (anthropogenic) body of water in which waste is consumed by bacteria, used most frequently with other waste-treatment processes; a sewage lagoon.

**Oxidation:** The chemical addition of oxygen to break down pollutants or organizac waste; e.g., destruction of chemicals such as cyanides, phenols, and organic sulfur compounds in sewage by bacterial and chemical means.

**Oxidation-Reduction Potential:** The electric potential required to transfer electrons from one compound or element (the oxidant) to another compound (the reductant); used as a qualitative measure of the state of oxidation in water treatment systems.

**Oxygenated Fuels:** Gasoline which has been blended with alcohols or ethers that contain oxygen in order to reduce carbon monoxide and other emissions.

**Oxygenated Solvent:** An organic solvent containing oxygen as part of the molecular structure. Alcohols and ketones are oxygenated compounds often used as paint solvents.

**Ozonation/Ozonator:** Application of ozone to water for disinfection or for taste and odor control. The ozonator is the device that does this.

**Ozone ($O_3$):** Found in two layers of the atmosphere, the stratosphere and the troposphere. In the stratosphere (the atmospheric layer 7 to 10 miles or more above the earth's surface) ozone is a natural form of oxygen that provides a protective layer shielding the earth from ultraviolet radiation.In the troposphere (the layer extending up 7 to 10 miles from the earth's surface), ozone is a chemical oxidant and major component of photochemical smog. It can seriously impair the respiratory system and is one of the most wide- spread of all the criteria pollutants for which the Clean Air Act required EPA to set standards. Ozone in the troposphere is produced through complex chemical reactions of nitrogen oxides, which are among the primary pollutants emitted by combustion sources; hydrocarbons, released into the atmosphere through the combustion, handling and processing of petroleum products; and sunlight.

**Ozone Depletion:** Destruction of the stratospheric ozone layer which shields the earth from ultraviolet radiation harmful to life. This destruction of ozone is caused by the breakdown of certain chlorine and/or bromine containing compounds (chlorofluorocarbons or halons), which break down when they reach the stratosphere and then catalytically destroy ozone molecules.

**Ozone Hole:** A thinning break in the stratospheric ozone layer. Designation of amount of such depletion as an "ozone hole" is made when the detected amount of depletion exceeds fifty percent. Seasonal ozone holes have been observed over both the Antarctic and Arctic regions, part of Canada, and the extreme northeastern United States.

**Ozone Layer:** The protective layer in the atmosphere, about 15 miles above the ground, that absorbs some of the sun's ultraviolet rays, thereby reducing the amount of potentially harmful radiation that reaches the earth's surface.

## P

**Packaging:** The assembly of one or more containers and any other components necessary to ensure minimum compliance with a program's storage and shipment packaging requirements. Also, the containers, etc. involved.

**Packed Bed Scrubber:** An air pollution control device in which emissions pass through alkaline water to neutralize hydrogen chloride gas.

**Packed Tower:** A pollution control device that forces dirty air through a tower packed with crushed rock or wood chips while liquid is sprayed over the packing material. The pollutants in the air stream either dissolve or chemically react with the liquid.

**Packer:** An inflatable gland, or balloon, used to create a temporary seal in a borehole, probe hole, well, or drive casing. It is made of rubber or non-reactive materials.

**Palatable Water:** Water, at a desirable temperature, that is free from objectionable tastes, odors, colors, and turbidity.

**Pandemic:** A widespread epidemic throughout an area, nation or the world.

**Paper:** In the recycling business, refers to products and materials, including newspapers, magazines, office papers, corrugated containers, bags and some paperboard packaging that can be recycled into new paper products.

**Paper Processor/Plastics Processor:** Intermediate facility where recovered paper or plastic products and materials are sorted, decontaminated, and prepared for final recycling.

**Parameter:** A variable, measurable property whose value is a determinant of the characteristics of a system; e.g. temperature, pressure, and density are parameters of the atmosphere.

**Paraquat:** A standard herbicide used to kill various types of crops, including marijuana. Causes lung damage if smoke from the crop is inhaled..

**Parshall Flume:** Device used to measure the flow of water in an open channel.

**Part A Permit, Part B Permit:** (See: Interim Permit Status.)

**Participation Rate:** Portion of population participating in a recycling program.

**Particle Count:** Results of a microscopic examination of treated water with a special "particle counter" that classifies suspended particles by number and size.

**Particulate Loading:** The mass of particulates per unit volume of air or water.

**Particulates:** 1. Fine liquid or solid particles such as dust, smoke, mist, fumes, or smog, found in air or emissions. 2. Very small solids suspended in water; they can vary in size, shape, density and electrical charge and can be gathered together by coagulation and flocculation.

**Partition Coefficient:** Measure of the sorption phenomenon, whereby a pesticide is divided between the soil and water phase; also referred to as adsorption partition coefficient.

**Parts Per Billion (ppb)/Parts Per Million (ppm):** Units commonly used to express contamination ratios, as in establishing the maximum permissible amount of a contaminant in water, land, or air.

**Passive Smoking/Secondhand Smoke:** Inhalation of others' tobacco smoke.

**Passive Treatment Walls:** Technology in which a chemical reaction takes place when contaminated ground water comes in contact with a barrier such as limestone or a wall containing iron filings.

**Pathogens:** Microorganisms (e.g., bacteria, viruses, or parasites) that can cause disease in humans, animals and plants.

**Pathway:** The physical course a chemical or pollutant takes from its source to the exposed organism.

**Pay-As-You-Throw/Unit-Based Pricing:** Systems under which residents pay for municipal waste management and disposal services by weight or volume collected, not a fixed fee.

**Peak Electricity Demand:** The maximum electricity used to meet the cooling load of a building or buildings in a given area.

**Peak Levels:** Levels of airborne pollutant contaminants much higher than average or occurring for short periods of time in response to sudden releases.

**Percent Saturatiuon:** The amount of a substance that is dissolved in a solution compared to the amount that could be dissolved in it.

**Perched Water:** Zone of unpressurized water held above the water table by impermeable rock or sediment.

**Percolating Water:** Water that passes through rocks or soil under the force of gravity.

**Percolation:** 1. The movement of water downward and radially through subsurface soil layers, usually continuing downward to ground water. Can also involve upward movement of water. 2. Slow seepage of water through a filter.

**Performance Bond:** Cash or securities deposited before a landfill operating permit is issued, which are held to ensure that all requirements for operating ad subsequently closing the landfill are faithful performed. The money is returned to the owner after proper closure of the landfill is completed. If contamination or other problems appear at any time during operation, or upon closure, and are not addressed, the owner must forfeit all or part of the bond which is then used to cover clean-up costs.

**Performance Data (For Incinerators):** Information collected, during a trial burn, on concentrations of designated organic compounds and pollutants found in incinerator emissions. Data analysis must show that the incinerator meets performance standards under operating conditions specified in the RCRA permit. (See: trial burn; performance standards.)

**Performance Standards:** 1. Regulatory requirements limiting the concentrations of designated organic compounds, particulate matter, and hydrogen chloride in emissions from incinerators. 2. Operating standards established by EPA for various permitted pollution control systems, asbestos inspections, and various program operations and maintenance requirements.

**Periphyton:** Microscopic underwater plants and animals that are firmly attached to solid surfaces such as rocks, logs, and pilings.

**Permeability:** The rate at which liquids pass through soil or other materials in a specified direction.

**Permissible Dose:** The dose of a chemical that may be received by an individual without the expectation of a significantly harmful result.

**Permissible Exposure Limit:** Also referred to as PEL, federal limits for workplace exposure to contaminants as established by OSHA.

**Permit:** An authorization, license, or equivalent control document issued by EPA or an approved state agency to implement the requirements of an environmental regulation; e.g. a permit to operate a wastewater treatment plant or to operate a facility that may generate harmful emissions.

**Persistence:** Refers to the length of time a compound stays in the environment, once introduced. A compound may persist for less than a second or indefinitely.

**Persistent Pesticides:** Pesticides that do not break down chemically or break down very slowly and remain in the environment after a growing season.

**Personal Air Samples:** Air samples taken with a pump that is directly attached to the worker with the collecting filter and cassette placed in the worker's breathing zone (required under OSHA asbestos standards and EPA worker protection rule).

**Personal Measurement:** A measurement collected from an individual's immediate environment.

**Personal Protective Equipment:** Clothing and equipment worn by pesticide mixers, loaders and applicators and re-entry workers, hazmat emergency responders, workers cleaning up Superfund sites, et. al., which is worn to reduce their exposure to potentially hazardous chemicals and other pollutants.

**Pest:** An insect, rodent, nematode, fungus, weed or other form of terrestrial or aquatic plant or animal life that is injurious to health or the environment.

**Pest Control Operator:** Person or company that applies pesticides as a business (e.g. exterminator); usually describes household services, not agricultural applications.

**Pesticide:** Substances or mixture there of intended for preventing, destroying, repelling, or mitigating any pest. Also, any substance or mixture intended for use as a plant regulator, defoliant, or desiccant.

**Pesticide Regulation Notice:** Formal notice to pesticide registrants about important changes in regulatory policy, procedures, regulations.

**Pesticide Tolerance:** The amount of pesticide residue allowed by law to remain in or on a harvested crop. EPA sets these levels well below the point where the compounds might be harmful to consumers.

**PETE (Polyethylene Terepthalate):** Thermoplastic material used in plastic soft drink and rigid containers.

**Petroleum:** Crude oil or any fraction thereof that is liquid under normal conditions of temperature and pressure. The term includes petroleum-based substances comprising a complex blend of hydrocarbons derived from crude oil through the process of separation, conversion, upgrading, and finishing, such as motor fuel, jet oil, lubricants, petroleum solvents, and used oil.

**Petroleum Derivatives:** Chemicals formed when gasoline breaks down in contact with ground water.

**pH:** An expression of the intensity of the basic or acid condition of a liquid; may range from 0 to 14, where 0 is the most acid and 7 is neutral. Natural waters usually have a pH between 6.5 and 8.5.

**Pharmacokinetics:** The study of the way that drugs move through the body after they are swallowed or injected.

**Phenolphthalein Alkalinity:** The alkalinity in a water sample measured by the amount of standard acid needed to lower the pH to a level of 8.3 as indicated by the change of color of the phenolphthalein from pink to clear.

**Phenols:** Organic compounds that are byproducts of petroleum refining, tanning, and textile, dye, and resin manufacturing. Low concentrations cause taste and odor problems in water; higher concentrations can kill aquatic life and humans.

**Phosphates:** Certain chemical compounds containing phosphorus.

**Phosphogypsum Piles (Stacks):** Principal byproduct generated in production of phosphoric acid from phosphate rock. These piles may generate radioactive radon gas.

**Phosphorus:** An essential chemical food element that can contribute to the eutrophication of lakes and other water bodies. Increased phosphorus levels result from discharge of phosphorus-containing materials into surface waters.

**Phosphorus Plants:** Facilities using electric furnaces to produce elemental phosphorous for commercial use, such as high grade phosphoric acid, phosphate-based detergent, and organic chemicals use.

**Photochemical Oxidants:** Air pollutants formed by the action of sunlight on oxides of nitrogen and hydrocarbons.

**Photochemical Smog:** Air pollution caused by chemical reactions of various pollutants emitted from different sources. (See: photochemical oxidants.)

**Photosynthesis:** The manufacture by plants of carbohydrates and oxygen from carbon dioxide mediated by chlorophyll in the presence of sunlight.

**Physical and Chemical Treatment:** Processes generally used in large-scale wastewater treatment facilities. Physical processes may include air-stripping or filtration. Chemical treatment includes coagulation, chlorination, or ozonation. The term can also refer to treatment of toxic materials in surface and ground waters, oil spills, and some methods of dealing with hazardous materials on or in the ground.

**Phytoplankton:** That portion of the plankton community comprised of tiny plants; e.g. algae, diatoms.

**Phytoremediation:** Low-cost remediation option for sites with widely dispersed contamination at low concentrations.

**Phytotoxic:** Harmful to plants.

**Phytotreatment:** The cultivation of specialized plants that absorb specific contaminants from the soil through their roots or foliage. This reduces the concentration of contaminants in the soil, but incorporates them into biomasses that may be released back into the environment when the plant dies or is harvested.

**Picocuries Per Liter pCi/L):** A unit of measure for levels of radon gas; becquerels per cubic meter is metric equivalent.

**Piezometer:** A nonpumping well, generally of small diameter, for measuring the elevation of a water table.

**Pilot Tests:** Testing a cleanup technology under actual site conditions to identify potential problems prior to full-scale implementation.

**Plankton:** Tiny plants and animals that live in water.

**Plasma Arc Reactors:** devices that use an electric arc to thermally decompose organic and inorganic materials at ultra-high temperatures into gases and a vitrified slag residue. A plasma arc reactor can operate as any of the following:

- integral component of chemical, fuel, or electricty production systems, processing high or medium value organic compounds into a synthetic gas used as a fuel

- materials recovery device, processing scrap to recover metal from the slag
- destruction or incineration system, processing waste materials into slag and gases ignited inside of a secondary combustion chamber that follows the reactor

**Plasmid:** A circular piece of DNA that exists apart from the chromosome and replicates independently of it. Bacterial plasmids carry information that renders the bacteria resistant to antibiotics. Plasmids are often used in genetic engineering to carry desired genes into organisms.

**Plastics:** Non-metallic chemoreactive compounds molded into rigid or pliable construction materials, fabrics, etc.

**Plate Tower Scrubber:** An air pollution control device that neutralizes hydrogen chloride gas by bubbling alkaline water through holes in a series of metal plates.

**Plug Flow:** Type of flow the occurs in tanks, basins, or reactors when a slug of water moves through without ever dispersing or mixing with the rest of the water flowing through.

**Plugging:** Act or process of stopping the flow of water, oil, or gas into or out of a formation through a borehole or well penetrating that formation.

**Plume:** 1. A visible or measurable discharge of a contaminant from a given point of origin. Can be visible or thermal in water, or visible in the air as, for example, a plume of smoke. 2 The area of radiation leaking from a damaged reactor. 3. Area downwind within which a release could be dangerous for those exposed to leaking fumes.

**Plutonium:** A radioactive metallic element chemically similar to uranium.

**PM-10/PM-2.5:** PM 10 is measure of particles in the atmosphere with a diameter of less than ten or equal to a nominal 10 micrometers. PM-2.5 is a measure of smaller particles in the air. PM-10 has been the pollutant particulate level standard against which EPA has been measuring Clean Air Act compliance. On the basis of newer scientific findings, the Agency is considering regulations that will make PM-2.5 the new "standard".

**Pneumoconiosis:** Health conditions characterized by permanent deposition of substantial amounts of particulate matter in the lungs and by the tissue reaction to its presence; can range from relatively harmless forms of sclerosis to the destructive fibrotic effect of silicosis.

**Point Source:** A stationary location or fixed facility from which pollutants are discharged; any single identifiable source of pollution; e.g. a pipe, ditch, ship, ore pit, factory smokestack.

**Point-of-Contact Measurement of Exposure:** Estimating exposure by measuring concentrations over time (while the exposure is taking place) at or near the place where it is occurring.

**Point-of-Disinfectant Application:** The point where disinfectant is applied and water downstream of that point is not subject to recontamination by surface water runoff.

**Point-of-Use Treatment Device:** Treatment device applied to a single tap to reduce contaminants in the drinking water at the one faucet.

**Pollen:** The fertilizing element of flowering plants; background air pollutant.

**Pollutant:** Generally, any substance introduced into the environment that adversely affects the usefulness of a resource or the health of humans, animals, or ecosystems..

**Pollutant Pathways:** Avenues for distribution of pollutants. In most buildings, for example, HVAC systems are the primary pathways although all building components can interact to affect how air movement distributes pollutants.

**Pollutant Standard Index (PSI):** Indicator of one or more pollutants that may be used to inform the public about the potential for adverse health effects from air pollution in major cities.

**Pollution:** Generally, the presence of a substance in the environment that because of its chemical composition or quantity prevents the functioning of natural processes and produces undesirable environmental and health effects.Under the Clean Water Act, for example, the term has been defined as the man-made or man-induced alteration of the physical, biological, chemical, and radiological integrity of water and other media.

**Pollution Prevention:** 1. Identifying areas, processes, and activities which create excessive waste products or pollutants in order to reduce or prevent them through, alteration, or eliminating a process. Such activities, consistent with the Pollution Prevention Act of 1990, are conducted across all EPA programs and can involve cooperative efforts with such agencies as the Departments of Agriculture and Energy. 2. EPA has initiated a number of voluntary programs in which industrial, or commercial or "partners" join with EPA in promoting activities that conserve energy, conserve and protect water supply, reduce emissions or find ways of utilizing them as energy resources, and reduce the waste stream. Among these are: Agstar, to reduce methane emissions through manure management. Climate Wise, to lower industrial greenhouse-gas emissions and energy costs. Coalbed Methane Outreach, to boost methane recovery at coal mines. Design for the Environment, to foster including environmental considerations in product design and processes. Energy Star programs, to promote energy efficiency in commercial and residential buildings, office equipment, transformers, computers, office equipment, and home appliances. Environmental Accounting, to help businesses identify environmental costs and factor them into management decision making. Green Chemistry, to promote and recognize cost-effective breakthroughs in chemistry that prevent pollution. Green Lights, to spread the use of energy-efficient lighting technologies. Indoor Environments, to reduce risks from indoor-air pollution. Landfill Methane Outreach, to develop landfill gas-to-energy projects. Natural Gas Star, to reduce methane emissions from the natural gas industry. Ruminant Livestock Methane, to reduce methane emissions from ruminant livestock. Transportation Partners, to reduce carbon dioxide emissions from the transportation sector. Voluntary Aluminum Industrial Partnership, to reduce perfluorocarbon emissions from the primary aluminum industry. WAVE, to promote efficient water use in the lodging industry. Wastewi$e, to reduce business-generated solid waste through prevention, reuse, and recycling. (See: Common Sense Initiative and Project XL.)

**Polychlorinated Biphenyls:** A group of toxic, persistent chemicals used in electrical transformers and capacitors for insulating purposes, and in gas pipeline systems as lubricant. The sale and new use of these chemicals, also known as PCBs, were banned by law in 1979.

**Portal-of-Entry Effect:** A local effect produced in the tissue or organ of first contact between a toxicant and the biological system.

**Polonium:** A radioactive element that occurs in pitchblende and other uranium-containing ores.

**Polyelectrolytes:** Synthetic chemicals that help solids to clump during sewage treatment.

**Polymer:** A natural or synthetic chemical structure where two or more like molecules are joined to form a more complex molecular structure (e.g. polyethylene in plastic).

**Polyvinyl Chloride (PVC):** A tough, environmentally indestructible plastic that releases hydrochloric acid when burned.

**Population:** A group of interbreeding organisms occupying a particular space; the number of humans or other living creatures in a designated area.

**Population at Risk:** A population subgroup that is more likely to be exposed to a chemical, or is more sensitive to the chemical, than is the general population.

**Porosity:** Degree to which soil, gravel, sediment, or rock is permeated with pores or cavities through which water or air can move.

**Post-Chlorination:** Addition of chlorine to plant effluent for disinfectant purposes after the effluent has been treated.

**Post-Closure:** The time period following the shutdown of a waste management or manufacturing facility; for monitoring purposes, often considered to be 30 years.

**Post-Consumer Materials/Waste:** Recovered materials that are diverted from municipal solid waste for the purpose of collection, recycling, and disposition.

**Post-Consumer Recycling:** Use of materials generated from residential and consumer waste for new or similar purposes; e.g. converting wastepaper from offices into corrugated boxes or newsprint.

**Potable Water:** Water that is safe for drinking and cooking.

**Potential Dose:** The amount of a compound contained in material swallowed, breathed, or applied to the skin.

**Potentially Responsible Party (PRP):** Any individual or company—including owners, operators, transporters or generators—potentially responsible for, or contributing to a spill or other contamination at a Superfund site. Whenever possible, through administrative and legal actions, EPA requires PRPs to clean up hazardous sites they have contaminated.

**Potentiation:** The ability of one chemical to increase the effect of another chemical.

**Potentiometric Surface:** The surface to which water in an aquifer can rise by hydrostatic pressure.

**Precautionary Principle:** When information about potential risks is incomplete, basing decisions about the best ways to manage or reduce risks on a preference for avoiding unnecessary health risks instead of on unnecessary economic expenditures.

**Pre-Consumer Materials/Waste:** Materials generated in manufacturing and converting processes such as manufacturing scrap and trimmings and cuttings. Includes print overruns, overissue publications, and obsolete inventories.

**Pre-Harvest Interval:** The time between the last pesticide application and harvest of the treated crops.

**Prechlorination:** The addition of chlorine at the headworks of a treatment plant prior to other treatment processes. Done mainly for disinfection and control of tastes, odors, and aquatic growths, and to aid in coagulation and settling,

**Precipitate:** A substance separated from a solution or suspension by chemical or physical change.

**Precipitation:** Removal of hazardous solids from liquid waste to permit safe disposal; removal of particles from airborne emissions as in rain (e.g. acid precipitation).

**Precipitator:** Pollution control device that collects particles from an air stream.

**Precursor:** In photochemistry, a compound antecedent to a pollutant. For example, volatile organic compounds (VOCs) and nitric oxides of nitrogen react in sunlight to form ozone or other photochemical oxidants. As such, VOCs and oxides of nitrogen are precursors.

**Preliminary Assessment:** The process of collecting and reviewing available information about a known or suspected waste site or release.

**Prescriptive:** Water rights which are acquired by diverting water and putting it to use in accordance with specified procedures; e.g. filing a request with a state agency to use unused water in a stream, river, or lake.

**Pressed Wood Products:** Materials used in building and furniture construction that are made from wood veneers, particles, or fibers bonded together with an adhesive under heat and pressure.

**Pressure Sewers:** A system of pipes in which water, wastewater, or other liquid is pumped to a higher elevation.

**Pressure, Static:** In flowing air, the total pressure minus velocity pressure, pushing equally in all directions.

**Pressure, Total:** In flowing air, the sum of the static and velocity pressures.

**Pressure, Velocity:** In flowing air, the pressure due to velocity and density of air.

**Pretreatment:** Processes used to reduce, eliminate, or alter the nature of wastewater pollutants from non-domestic sources before they are discharged into publicly owned treatment works (POTWs).

**Prevalent Level Samples:** Air samples taken under normal conditions (also known as ambient background samples).

**Prevalent Levels:** Levels of airborne contaminant occurring under normal conditions.

**Prevention of Significant Deterioration (PSD):** EPA program in which state and/or federal permits are required in order to restrict emissions from new or modified sources in places where air quality already meets or exceeds primary and secondary ambient air quality standards.

**Primacy:** Having the primary responsibility for administering and enforcing regulations.

**Primary Drinking Water Regulation:** Applies to public water systems and specifies a contaminant level, which, in the judgment of the EPA Administrator, will not adversely affect human health.

**Primary Effect:** An effect where the stressor acts directly on the ecological component of interest, not on other parts of the ecosystem. (See: secondary effect.)

**Primary Standards:** National ambient air quality standards designed to protect human health with an adequate margin for safety. (See: National Ambient Air Quality Standards, secondary standards.)

**Primary Treatment:** First stage of wastewater treatment in which solids are removed by screening and settling.

**Primary Waste Treatment:** First steps in wastewater treatment; screens and sedimentation tanks are used to remove most materials that float or will settle. Primary treatment removes about 30 percent of carbonaceous biochemical oxygen demand from domestic sewage.

**Principal Organic Hazardous Constituents (POHCs):** Hazardous compounds monitored during an incinerator's trial burn, selected for high concentration in the waste feed and difficulty of combustion.

**Prions:** Microscopic particles made of protein that can cause disease.

**Prior Appropriation:** A doctrine of water law that allocates the rights to use water on a first-come, first-served basis.

**Probability of Detection :** The likelihood, expressed as a percentage, that a test method will correctly identify a leaking tank.

**Process Variable:** A physical or chemical quantity which is usually measured and controlled in the operation of a water treatment plant or industrial plant.

**Process Verification:** Verifying that process raw materials, water usage, waste treatment processes, production rate and other facts relative to quantity and quality of pollutants contained in discharges are substantially described in the permit application and the issued permit.

**Process Wastewater:** Any water that comes into contact with any raw material, product, byproduct, or waste.

**Process Weight:** Total weight of all materials, including fuel, used in a manufacturing process; used to calculate the allowable particulate emission rate.

**Producers:** Plants that perform photosynthesis and provide food to consumers.

**Product Level:** The level of a product in a storage tank.

**Product Water:** Water that has passed through a water treatment plant and is ready to be delivered to consumers.

**Products of Incomplete Combustion (PICs):** Organic compounds formed by combustion. Usually generated in small amounts and sometimes toxic, PICs are heat-altered versions of the original material fed into the incinerator (e.g. charcoal is a P.I.C. from burning wood).

**Project XL:** An EPA initiative to give states and the regulated community the flexibility to develop comprehensive strategies as alternatives to multiple current regulatory requirements in order to exceed compliance and increase overall environmental benefits.

**Propellant:** Liquid in a self-pressurized pesticide product that expels the active ingredient from its container.

**Proportionate Mortality Ratio (PMR):** The number of deaths from a specific cause in a specific period of time per 100 deaths from all causes in the same time period.

**Proposed Plan:** A plan for a site cleanup that is available to the public for comment.

**Proteins:** Complex nitrogenous organic compounds of high molecular weight made of amino acids; essential for growth and repair of animal tissue. Many, but not all, proteins are enzymes.

**Protocol:** A series of formal steps for conducting a test.

**Protoplast:** A membrane-bound cell from which the outer wall has been partially or completely removed. The term often is applied to plant cells.

**Protozoa:** One-celled animals that are larger and more complex than bacteria. May cause disease.

**Public Comment Period:** The time allowed for the public to express its views and concerns regarding an action by EPA (e.g. a Federal Register Notice of proposed rule-making, a public notice of a draft permit, or a Notice of Intent to Deny).

**Public Health Approach:** Regulatory and voluntary focus on effective and feasible risk management actions at the national and community level to reduce human exposures and risks, with priority given to reducing exposures with the biggest impacts in terms of the number affected and severity of effect.

**Public Health Context:** The incidence, prevalence, and severity of diseases in communities or populations and the factors that account for them, including infections, exposure to pollutants, and other exposures or activities.

**Public Hearing:** A formal meeting wherein EPA officials hear the public's views and concerns about an EPA action or proposal. EPA is required to consider such comments when evaluating its actions. Public hearings must be held upon request during the public comment period.

**Public Notice:** 1. Notification by EPA informing the public of Agency actions such as the issuance of a draft permit or scheduling of a hearing. EPA is required to ensure proper public notice, including publication in newspapers and broadcast over radio and television stations. 2. In the safe drinking water program, water suppliers are required to publish and broadcast notices when pollution problems are discovered.

**Public Water System:** A system that provides piped water for human consumption to at least 15 service connections or regularly serves 25 individuals.

**Publicly Owned Treatment Works (POTWs):** A waste-treatment works owned by a state, unit of local government, or Indian tribe, usually designed to treat domestic wastewaters.

**Pumping Station:** Mechanical device installed in sewer or water system or other liquid-carrying pipelines to move the liquids to a higher level.

**Pumping Test:** A test conducted to determine aquifer or well characteristics.

**Purging:** Removing stagnant air or water from sampling zone or equipment prior to sample collection.

**Putrefaction:** Biological decomposition of organic matter; associated with anaerobic conditions.

**Putrescible:** Able to rot quickly enough to cause odors and attract flies.

**Pyrolysis:** Decomposition of a chemical by extreme heat.

## Q

**Qualitative Use Assessment:** Report summarizing the major uses of a pesticide including percentage of crop treated, and amount of pesticide used on a site.

**Quality Assurance/Quality Control:** A system of procedures, checks, audits, and corrective actions to ensure that all EPA research design and performance, environmental monitoring and sampling, and other technical and reporting activities are of the highest achievable quality.

**Quench Tank:** A water-filled tank used to cool incinerator residues or hot materials during industrial processes.

## R

**Radiation:** Transmission of energy though space or any medium. Also known as radiant energy.

**Radiation Standards:** Regulations that set maximum exposure limits for protection of the public from radioactive materials.

**Radio Frequency Radiation:** (See non-ionizing electromagnetic radiation.)

**Radioactive Decay:** Spontaneous change in an atom by emission of of charged particles and/or gamma rays; also known as radioactive disintegration and radioactivity.

**Radioactive Substances:** Substances that emit ionizing radiation.

**Radioactive Waste:** Any waste that emits energy as rays, waves, streams or energetic particles. Radioactive materials are often mixed with hazardous waste, from nuclear reactors, research institutions, or hospitals.

**Radioisotopes:** Chemical variants of radioactive elements with potentially oncogenic, teratogenic, and mutagenic effects on the human body.

**Radionuclide:** Radioactive particle, man-made (anthropogenic) or natural, with a distinct atomic weight number. Can have a long life as soil or water pollutant.

**Radius of Vulnerability Zone:** The maximum distance from the point of release of a hazardous substance in which the airborne concentration could reach the level of concern under specified weather conditions.

**Radius of Influence:** 1. The radial distance from the center of a wellbore to the point where there is no lowering of the water table or potentiometric surface (the edge of the cone of depression); 2. the radial distance from an extraction well that has adequate air flow for effective removal of contaminants when a vacuum is applied to the extraction well.

**Radon:** A colorless naturally occurring, radioactive, inert gas formed by radioactive decay of radium atoms in soil or rocks.

**Radon Daughters/Radon Progeny:** Short-lived radioactive decay products of radon that decay into longer-lived lead isotopes that can attach themselves to airborne dust and other particles and, if inhaled, damage the linings of the lungs.

**Radon Decay Products:** A term used to refer collectively to the immediate products of the radon decay chain. These include Po-218, Pb-214, Bi-214, and Po-214, which have an average combined half-life of about 30 minutes.

**Rainbow Report:** Comprehensive document giving the status of all pesticides now or ever in registration or special reviews. Known as the "rainbow report" because chapters are printed on different colors of paper.

**Rasp:** A machine that grinds waste into a manageable material and helps prevent odor.

**Raw Agricultural Commodity:** An unprocessed human food or animal feed crop (e.g., raw carrots, apples, corn, or eggs.)

**Raw Sewage:** Untreated wastewater and its contents.

**Raw Water:** Intake water prior to any treatment or use.

**Re-entry:** (In indoor air program) Refers to air exhausted from a building that is immediately brought back into the system through the air intake and other openings.

**Reactivity:** Refers to those hazardous wastes that are normally unstable and readily undergo violent chemical change but do not explode.

**Reaeration:** Introduction of air into the lower layers of a reservoir. As the air bubbles form and rise through the water, the oxygen dissolves into the water and replenishes the dissolved oxygen. The rising bubbles also cause the lower waters to rise to the surface where they take on oxygen from the atmosphere.

**Real-Time Monitoring:** Monitoring and measuring environmental developments with technology and communications systems that provide time-relevant information to the public in an easily understood format people can use in day-to-day decision-making about their health and the environment.

**Reasonable Further Progress:** Annual incremental reductions in air pollutant emissions as reflected in a State Implementation Plan that EPA deems sufficient to provide for the attainment of the applicable national ambient air quality standards by the statutory deadline.

**Reasonable Maximum Exposure:** The maximum exposure reasonably expected to occur in a population.

**Reasonable Worst Case:** An estimate of the individual dose, exposure, or risk level received by an individual in a defined population that is greater than the 90th percentile but less than that received by anyone in the 98th percentile in the same population.

**Reasonably Available Control Measures (RACM):** A broadly defined term referring to technological and other measures for pollution control.

**Reasonably Available Control Technology (RACT):** Control technology that is reasonably available, and both technologically and economically feasible. Usually applied to existing sources in nonattainment areas; in most cases is less stringent than new source performance standards.

**Recarbonization:** Process in which carbon dioxide is bubbled into water being treated to lower the pH.

**Receiving Waters:** A river, lake, ocean, stream or other watercourse into which wastewater or treated effluent is discharged.

**Receptor:** Ecological entity exposed to a stressor.

**Recharge:** The process by which water is added to a zone of saturation, usually by percolation from the soil surface; e.g., the recharge of an aquifer.

**Recharge Area:** A land area in which water reaches the zone of saturation from surface infiltration, e.g., where rainwater soaks through the earth to reach an aquifer.

**Recharge Rate:** The quantity of water per unit of time that replenishes or refills an aquifer.

**Reclamation:** (In recycling) Restoration of materials found in the waste stream to a beneficial use which may be for purposes other than the original use.

**Recombinant Bacteria:** A microorganism whose genetic makeup has been altered by deliberate introduction of new genetic elements. The offspring of these altered bacteria also contain these new genetic elements; i.e. they "breed true."

**Recombinant DNA:** The new DNA that is formed by combining pieces of DNA from different organisms or cells.

**Recommended Maximum Contaminant Level (RMCL):** The maximum level of a contaminant in drinking water at which no known or anticipated adverse effect on human health would occur, and that includes an adequate margin of safety. Recommended levels are nonenforceable health goals. (See: maximum contaminant level.)

**Reconstructed Source:** Facility in which components are replaced to such an extent that the fixed capital cost of the new components exceeds 50 percent of the capital cost of constructing a comparable brand-new facility. New-source performance standards may be applied to sources reconstructed after the proposal of the standard if it is technologically and economically feasible to meet the standards.

**Reconstruction of Dose:** Estimating exposure after it has occurred by using evidence within an organism such as chemical levels in tissue or fluids.

**Record of Decision (ROD):** A public document that explains which cleanup alternative(s) will be used at National Priorities List sites where, under CERCLA, Trust Funds pay for the cleanup.

**Recovery Rate:** Percentage of usable recycled materials that have been removed from the total amount of municipal solid waste generated in a specific area or by a specific business.

**Recycle/Reuse:** Minimizing waste generation by recovering and reprocessing usable products that might otherwise become waste (.i.e. recycling of aluminum cans, paper, and bottles, etc.).

**Recycling and Reuse Business Assistance Centers:** Located in state solid-waste or economic-development agencies, these centers provide recycling businesses with customized and targeted assistance.

**Recycling Economic Development Advocates:** Individuals hired by state or tribal economic development offices to focus financial, marketing, and permitting resources on creating recycling businesses.

**Recycling Mill:** Facility where recovered materials are remanufactured into new products.

**Recycling Technical Assistance Partnership National Network:** A national information-sharing resource designed to help businesses and manufacturers increase their use of recovered materials.

**Red Bag Waste:** (See: infectious waste.)

**Red Border:** An EPA document undergoing review before being submitted for final management decision-making.

**Red Tide:** A proliferation of a marine plankton toxic and often fatal to fish, perhaps stimulated by the addition of nutrients. A tide can be red, green, or brown, depending on the coloration of the plankton.

**Redemption Program:** Program in which consumers are monetarily compensated for the collection of recyclable materials, generally through prepaid deposits or taxes on beverage containers. In some states or localities legislation has enacted redemption programs to help prevent roadside litter. (See: bottle bill.)

**Reduction:** The addition of hydrogen, removal of oxygen, or addition of electrons to an element or compound.

**Reentry Interval:** The period of time immediately following the application of a pesticide during which unprotected workers should not enter a field.

**Reference Dose (RfD):** The RfD is a numerical estimate of a daily oral exposure to the human population, including sensitive subgroups such as children, that is not likely to cause harmful effects during a lifetime. RfDs are generally used for health effects that are thought to have a threshold or low dose limit for producing effects.

**Reformulated Gasoline:** Gasoline with a different composition from conventional gasoline (e.g., lower aromatics content) that cuts air pollutants.

**Refueling Emissions:** Emissions released during vehicle re-fueling.

**Refuse:** (See: solid waste.)

**Refuse Reclamation:** Conversion of solid waste into useful products; e.g., composting organic wastes to make soil conditioners or separating aluminum and other metals for recycling.

**Regeneration:** Manipulation of cells to cause them to develop into whole plants.

**Regional Response Team (RRT):** Representatives of federal, local, and state agencies who may assist in coordination of activities at the request of the On-Scene Coordinator before and during a significant pollution incident such as an oil spill, major chemical release, or Superfund response.

**Registrant:** Any manufacturer or formulator who obtains registration for a pesticide active ingredient or product.

**Registration:** Formal listing with EPA of a new pesticide before it can be sold or distributed. Under the Federal Insecticide, Fungicide, and Rodenticide Act, EPA is responsible for registration (pre-market licensing) of pesticides on the basis of data demonstrating no unreasonable adverse effects on human health or the environment when applied according to approved label directions.

**Registration Standards:** Published documents which include summary reviews of the data available on a pesticide's active ingredient, data gaps, and the Agency's existing regulatory position on the pesticide.

**Regulated Asbestos-Containing Material (RACM):** Friable asbestos material or nonfriable ACM that will be or has been subjected to sanding, grinding, cutting, or abrading or has crumbled, or been pulverized or reduced to powder in the course of demolition or renovation operations.

**Regulated Medical Waste:** Under the Medical Waste Tracking Act of 1988, any solid waste generated in the diagnosis, treatment, or immunization of human beings or animals, in research pertaining thereto, or in the production or testing of biologicals. Included are cultures and stocks of infectious agents; human blood and blood products; human pathological body wastes from surgery and autopsy; contaminated animal carcasses from medical research; waste from patients with communicable diseases; and all used sharp implements, such as needles and scalpels, and certain unused sharps. (See: treated medical waste; untreated medical waste; destroyed medical waste.)

**Relative Ecological Sustainability:** Ability of an ecosystem to maintain relative ecological integrity indefinitely.

**Relative Permeability:** The permeability of a rock to gas, NAIL, or water, when any two or more are present.

**Relative Risk Assessment:** Estimating the risks associated with different stressors or management actions.

**Release:** Any spilling, leaking, pumping, pouring, emitting, emptying, discharging, injecting, escaping, leaching, dumping, or disposing into the environment of a hazardous or toxic chemical or extremely hazardous substance.

**Remedial Action (RA):** The actual construction or implementation phase of a Superfund site cleanup that follows remedial design.

**Remedial Design:** A phase of remedial action that follows the remedial investigation/feasibility study and includes development of engineering drawings and specifications for a site cleanup.

**Remedial Investigation:** An in-depth study designed to gather data needed to determine the nature and extent of contamination at a Superfund site; establish site cleanup criteria; identify preliminary alternatives for remedial action; and support technical and cost analyses of alternatives. The remedial investigation is usually done with the feasibility study. Together they are usually referred to as the "RI/FS".

**Remedial Project Manager (RPM):** The EPA or state official responsible for overseeing on-site remedial action.

**Remedial Response:** Long-term action that stops or substantially reduces a release or threat of a release of hazardous substances that is serious but not an immediate threat to public health.

**Remediation:** 1. Cleanup or other methods used to remove or contain a toxic spill or hazardous materials from a Superfund site; 2. for the Asbestos Hazard Emergency Response program, abatement methods including evaluation, repair, enclosure, encapsulation, or removal of greater than 3 linear feet or square feet of asbestos-containing materials from a building.

**Remote Sensing:** The collection and interpretation of information about an object without physical contact with the object; e.g., satellite imaging, aerial photography, and open path measurements.

**Removal Action:** Short-term immediate actions taken to address releases of hazardous substances that require expedited response. (See: cleanup.)

**Renewable Energy Production Incentive (REPI):** Incentive established by the Energy Policy Act available to renewable energy power projects owned by a state or local government or nonprofit electric cooperative.

**Repeat Compliance Period:** Any subsequent compliance period after the initial one.

**Reportable Quantity (RQ):** Quantity of a hazardous substance that triggers reports under CERCLA. If a substance exceeds its RQ, the release must be reported to the National Response Center, the SERC, and community emergency coordinators for areas likely to be affected.

**Repowering:** Rebuilding and replacing major components of a power plant instead of building a new one.

**Representative Sample:** A portion of material or water that is as nearly identical in content and consistency as possible to that in the larger body of material or water being sampled.

**Reregistration:** The reevaluation and relicensing of existing pesticides originally registered prior to current scientific and regulatory standards. EPA reregisters pesticides through its Registration Standards Program.

**Reserve Capacity:** Extra treatment capacity built into solid waste and wastewater treatment plants and interceptor sewers to accommodate flow increases due to future population growth.

**Reservoir:** Any natural or artificial holding area used to store, regulate, or control water.

**Residential Use:** Pesticide application in and around houses, office buildings, apartment buildings, motels, and other living or working areas.

**Residential Waste:** Waste generated in single and multi-family homes, including newspapers, clothing, disposable tableware, food packaging, cans, bottles, food scraps, and yard trimmings other than those that are diverted to backyard composting. (See: Household hazardous waste.)

**Residual:** Amount of a pollutant remaining in the environment after a natural or technological process has taken place; e.g., the sludge remaining after initial wastewater treatment, or particulates remaining in air after it passes through a scrubbing or other process.

**Residual Risk:** The extent of health risk from air pollutants remaining after application of the Maximum Achievable Control Technology (MACT).

**Residual Saturation:** Saturation level below which fluid drainage will not occur.

**Residue:** The dry solids remaining after the evaporation of a sample of water or sludge.

**Resistance:** For plants and animals, the ability to withstand poor environmental conditions or attacks by chemicals or disease. May be inborn or acquired.

**Resource Recovery:** The process of obtaining matter or energy from materials formerly discarded.

**Response Action:** 1. Generic term for actions taken in response to actual or potential health-threatening environmental events such as spills, sudden releases, and asbestos abatement/management problems. 2. A CERCLA-authorized action involving either a short-term removal action or a long-term removal response. This may include but is not limited to: removing hazardous materials from a site to an EPA-approved hazardous waste facility for treatment, containment or treating the waste on-site, identifying and removing the sources of ground-water contamination and halting further migration of contaminants. 3. Any of the following actions taken in school buildings in response to AHERA to reduce the risk of exposure to asbestos: removal, encapsulation, enclosure, repair, and operations and maintenance. (See: cleanup.)

**Responsiveness Summary:** A summary of oral and/or written public comments received by EPA during a comment period on key EPA documents, and EPA's response to those comments.

**Restoration:** Measures taken to return a site to pre-violation conditions.

**Restricted Entry Interval:** The time after a pesticide application during which entry into the treated area is restricted.

**Restricted Use:** A pesticide may be classified (under FIFRA regulations) for restricted use if it requires special handling because of its toxicity, and, if so, it may be applied only by trained, certified applicators or those under their direct supervision.

**Restriction Enzymes:** Enzymes that recognize specific regions of a long DNA molecule and cut it at those points.

**Retrofit:** Addition of a pollution control device on an existing facility without making major changes to the generating plant. Also called backfit.

**Reuse:** Using a product or component of municipal solid waste in its original form more than once; e.g., refilling a glass bottle that has been returned or using a coffee can to hold nuts and bolts.

**Reverse Osmosis:** A treatment process used in water systems by adding pressure to force water through a semi-permeable membrane. Reverse osmosis removes most drinking water contaminants. Also used in wastewater treatment. Large-scale reverse osmosis plants are being developed.

**Reversible Effect:** An effect which is not permanent; especially adverse effects which diminish when exposure to a toxic chemical stops.

**Ribonucleic Acid (RNA):** A molecule that carries the genetic message from DNA to a cellular protein-producing mechanism.

**Rill:** A small channel eroded into the soil by surface runoff; can be easily smoothed out or obliterated by normal tillage.

**Ringlemann Chart:** A series of shaded illustrations used to measure the opacity of air pollution emissions, ranging from light grey through black; used to set and enforce emissions standards.

**Riparian Habitat:** Areas adjacent to rivers and streams with a differing density, diversity, and productivity of plant and animal species relative to nearby uplands.

**Riparian Rights:** Entitlement of a land owner to certain uses of water on or bordering the property, including the right to prevent diversion or misuse of upstream waters. Generally a matter of state law.

**Risk:** A measure of the probability that damage to life, health, property, and/or the environment will occur as a result of a given hazard.

**Risk (Adverse) for Endangered Species:** Risk to aquatic species if anticipated pesticide residue levels equal one-fifth of LD10 or one-tenth of LC50; risk to terrestrial species if anticipated pesticide residue levels equal one-fifth of LC10 or one-tenth of LC50.

**Risk Assessment:** Qualitative and quantitative evaluation of the risk posed to human health and/or the environment by the actual or potential presence and/or use of specific pollutants.

**Risk Characterization:** The last phase of the risk assessment process that estimates the potential for adverse health or ecological effects to occur from exposure to a stressor and evaluates the uncertainty involved.

**Risk Communication:** The exchange of information about health or environmental risks among risk assessors and managers, the general public, news media, interest groups, etc.

**Risk Estimate:** A description of the probability that organisms exposed to a specific dose of a chemical or other pollutant will develop an adverse response, e.g., cancer.

**Risk Factor:** Characteristics (e.g., race, sex, age, obesity) or variables (e.g., smoking, occupational exposure level) associated with increased probability of a toxic effect.

**Risk for Non-Endangered Species:** Risk to species if anticipated pesticide residue levels are equal to or greater than LC50.

**Risk Management:** The process of evaluating and selecting alternative regulatory and non-regulatory responses to risk. The selection process necessarily requires the consideration of legal, economic, and behavioral factors.

**Risk-based Targeting:** The direction of resources to those areas that have been identified as having the highest potential or actual adverse effect on human health and/or the environment.

**Risk-Specific Dose:** The dose associated with a specified risk level.

**River Basin:** The land area drained by a river and its tributaries.

**Rodenticide:** A chemical or agent used to destroy rats or other rodent pests, or to prevent them from damaging food, crops, etc.

**Rotary Kiln Incinerator:** An incinerator with a rotating combustion chamber that keeps waste moving, thereby allowing it to vaporize for easier burning.

**Rough Fish:** Fish not prized for sport or eating, such as gar and suckers. Most are more tolerant of changing environmental conditions than are game or food species.

**Route of Exposure:** The avenue by which a chemical comes into contact with an organism, e.g., inhalation, ingestion, dermal contact, injection.

**Rubbish:** Solid waste, excluding food waste and ashes, from homes, institutions, and workplaces.

**Run-Off:** That part of precipitation, snow melt, or irrigation water that runs off the land into streams or other surface-water. It can carry pollutants from the air and land into receiving waters.

**Running Losses:** Evaporation of motor vehicle fuel from the fuel tank while the vehicle is in use.

## S

**Sacrifical Anode:** An easily corroded material deliberately installed in a pipe or intake to give it up (sacrifice it) to corrosion while the rest of the water supply facility remains relatively corrosion-free.

**Safe:** Condition of exposure under which there is a practical certainty that no harm will result to exposed individuals.

**Safe Water:** Water that does not contain harmful bacteria, toxic materials, or chemicals, and is considered safe for drinking even if it may have taste, odor, color, and certain mineral problems.

**Safe Yield:** The annual amount of water that can be taken from a source of supply over a period of years without depleting that source beyond its ability to be replenished naturally in "wet years."

**Safener:** A chemical added to a pesticide to keep it from injuring plants.

**Salinity:** The percentage of salt in water.

**Salt Water Intrusion:** The invasion of fresh surface or ground water by salt water. If it comes from the ocean it may be called sea water intrusion.

**Salts:** Minerals that water picks up as it passes through the air, over and under the ground, or from households and industry.

**Salvage:** The utilization of waste materials.

**Sampling Frequency:** The interval between the collection of successive samples.

**Sanctions:** Actions taken by the federal government for failure to provide or implement a State Implementation Plan (SIP). Such action may include withholding of highway funds and a ban on construction of new sources of potential pollution.

**Sand Filters:** Devices that remove some suspended solids from sewage. Air and bacteria decompose additional wastes filtering through the sand so that cleaner water drains from the bed.

**Sanitary Landfill:** (See: landfills.)

**Sanitary Sewers:** Underground pipes that carry off only domestic or industrial waste, not storm water.

**Sanitary Survey:** An on-site review of the water sources, facilities, equipment, operation and maintenance of a public water system to evaluate the adequacy of those elements for producing and distributing safe drinking water.

**Sanitary Water (Also known as gray water):** Water discharged from sinks, showers, kitchens, or other non-industrial operations, but not from commodes.

**Sanitation:** Control of physical factors in the human environment that could harm development, health, or survival.

**Saprolite:** A soft, clay-rich, thoroughly decomposed rock formed in place by chemical weathering of igneous or metamorphic rock. Forms in humid, tropical, or subtropical climates.

**Saprophytes:** Organisms living on dead or decaying organic matter that help natural decomposition of organic matter in water.

**Saturated Zone:** The area below the water table where all open spaces are filled with water under pressure equal to or greater than that of the atmosphere.

**Saturation:** The condition of a liquid when it has taken into solution the maximum possible quantity of a given substance at a given temperature and pressure.

**Science Advisory Board (SAB):** A group of external scientists who advise EPA on science and policy.

**Scrap:** Materials discarded from manufacturing operations that may be suitable for reprocessing.

**Scrap Metal Processor:** Intermediate operating facility where recovered metal is sorted, cleaned of contaminants, and prepared for recycling.

**Screening:** Use of screens to remove coarse floating and suspended solids from sewage.

**Screening Risk Assessment:** A risk assessment performed with few data and many assumptions to identify exposures that should be evaluated more carefully for potential risk.

**Scrubber:** An air pollution device that uses a spray of water or reactant or a dry process to trap pollutants in emissions.

**Secondary Drinking Water Regulations:** Non-enforceable regulations applying to public water systems and specifying the maximum contamination levels that, in the judgment of EPA, are required to protect the public welfare. These regulations apply to any contaminants that may adversely affect the odor or appearance of such water and consequently may cause people served by the system to discontinue its use.

**Secondary Effect:** Action of a stressor on supporting components of the ecosystem, which in turn impact the ecological component of concern. (See: primary effect.)

**Secondary Materials:** Materials that have been manufactured and used at least once and are to be used again.

**Secondary Standards:** National ambient air quality standards designed to protect welfare, including effects on soils, water, crops, vegetation, man-made (anthropogenic) materials, animals, wildlife, weather, visibility, and climate; damage to property; transportation hazards; economic values, and personal comfort and well-being.

**Secondary Treatment:** The second step in most publicly owned waste treatment systems in which bacteria consume the organic parts of the waste. It is accomplished by bringing together waste, bacteria, and oxygen in trickling filters or in the activated sludge process. This treatment removes floating and settleable solids and about 90 percent of the oxygen-demanding substances and suspended solids. Disinfection is the final stage of secondary treatment. (See: primary, tertiary treatment.)

**Secure Chemical Landfill:** (See:landfills.)

**Secure Maximum Contaminant Level:** Maximum permissible level of a contaminant in water delivered to the free flowing outlet of the ultimate user, or of contamination resulting from corrosion of piping and plumbing caused by water quality.

**Sediment:** Topsoil, sand, and minerals washed from the land into water, usually after rain or snow melt.

**Sediment Yield:** The quantity of sediment arriving at a specific location.

**Sedimentation:** Letting solids settle out of wastewater by gravity during treatment.

**Sedimentation Tanks:** Wastewater tanks in which floating wastes are skimmed off and settled solids are removed for disposal.

**Sediments:** Soil, sand, and minerals washed from land into water, usually after rain. They pile up in reservoirs, rivers and harbors, destroying fish and wildlife habitat, and clouding the water so that sunlight cannot reach aquatic plants. Careless farming, mining, and building activities will expose sediment materials, allowing them to wash off the land after rainfall.

**Seed Protectant:** A chemical applied before planting to protect seeds and seedlings from disease or insects.

**Seepage:** Percolation of water through the soil from unlined canals, ditches, laterals, watercourses, or water storage facilities.

**Selective Pesticide:** A chemical designed to affect only certain types of pests, leaving other plants and animals unharmed.

**Semi-Confined Aquifer:** An aquifer partially confined by soil layers of low permeability through which recharge and discharge can still occur.

**Semivolatile Organic Compounds:** Organic compounds that volatilize slowly at standard temperature (20 degrees C and 1 atm pressure).

**Senescence:** The aging process. Sometimes used to describe lakes or other bodies of water in advanced stages of eutrophication. Also used to describe plants and animals.

**Septic System:** An on-site system designed to treat and dispose of domestic sewage. A typical septic system consists of tank that receives waste from a residence or business and a system of tile lines or a pit for disposal of the liquid effluent (sludge) that remains after decomposition of the solids by bacteria in the tank and must be pumped out periodically.

**Septic Tank:** An underground storage tank for wastes from homes not connected to a sewer line. Waste goes directly from the home to the tank. (See: septic system.)

**Service Connector:** The pipe that carries tap water from a public water main to a building.

**Service Line Sample:** A one-liter sample of water that has been standing for at least 6 hours in a service pipeline and is collected according to federal regulations.

**Service Pipe:** The pipeline extending from the water main to the building served or to the consumer's system.

**Set-Back:** Setting a thermometer to a lower temperature when the building is unoccupied to reduce consumption of heating energy. Also refers to setting the thermometer to a higher temperature during unoccupied periods in the cooling season.

**Settleable Solids:** Material heavy enough to sink to the bottom of a wastewater treatment tank.

**Settling Chamber:** A series of screens placed in the way of flue gases to slow the stream of air, thus helping gravity to pull particles into a collection device.

**Settling Tank:** A holding area for wastewater, where heavier particles sink to the bottom for removal and disposal.

**7Q10:** Seven-day, consecutive low flow with a ten year return frequency; the lowest stream flow for seven consecutive days that would be expected to occur once in ten years.

**Sewage:** The waste and wastewater produced by residential and commercial sources and discharged into sewers.

**Sewage Lagoon:** (See: lagoon.)

**Sewage Sludge:** Sludge produced at a Publicly Owned Treatment Works, the disposal of which is regulated under the Clean Water Act.

**Sewer:** A channel or conduit that carries wastewater and storm-water runoff from the source to a treatment plant or receiving stream. "Sanitary" sewers carry household, industrial, and commercial waste. "Storm" sewers carry runoff from rain or snow. "Combined" sewers handle both.

**Sewerage:** The entire system of sewage collection, treatment, and disposal.

**Shading Coefficient:** The amount of the sun's heat transmitted through a given window compared with that of a standard 1/8- inch-thick single pane of glass under the same conditions.

**Sharps:** Hypodermic needles, syringes (with or without the attached needle), Pasteur pipettes, scalpel blades, blood vials, needles with attached tubing, and culture dishes used in animal or human patient care or treatment, or in medical, research or industrial laboratories. Also included are other types of broken or unbroken glassware that were in contact with infectious agents, such as used slides and cover slips, and unused hypodermic and suture needles, syringes, and scalpel blades.

**Shock Load:** The arrival at a water treatment plant of raw water containing unusual amounts of algae, colloidal matter. color, suspended solids, turbidity, or other pollutants.

**Short-Circuiting:** When some of the water in tanks or basins flows faster than the rest; may result in shorter contact, reaction, or settling times than calculated or presumed.

**Sick Building Syndrome:** Building whose occupants experience acute health and/or comfort effects that appear to be linked to time spent therein, but where no specific illness or cause can be identified. Complaints may be localized in a particular room or zone, or may spread throughout the building. (See: building-related illness.)

**Signal:** The volume or product-level change produced by a leak in a tank.

**Signal Words:** The words used on a pesticide label—Danger, Warning, Caution—to indicate level of toxicity.

**Significant Deterioration:** Pollution resulting from a new source in previously "clean" areas. (See: prevention of significant deterioration.)

**Significant Municipal Facilities:** Those publicly owned sewage treatment plants that discharge a million gallons per day or more and are therefore considered by states to have the potential to substantially affect the quality of receiving waters.

**Significant Non-Compliance:** (See significant violations.)

**Significant Potential Source of Contamination:** A facility or activity that stores, uses, or produces compounds with potential for significant contaminating impact if released into the source water of a public water supply.

**Significant Violations:** Violations by point source dischargers of sufficient magnitude or duration to be a regulatory priority.

**Silt:** Sedimentary materials composed of fine or intermediate-sized mineral particles.

**Silviculture:** Management of forest land for timber.

**Single-Breath Canister:** Small one-liter canister designed to capture a single breath. Used in air pollutant ingestion research.

**Sink:** Place in the environment where a compound or material collects.

**Sinking:** Controlling oil spills by using an agent to trap the oil and sink it to the bottom of the body of water where the agent and the oil are biodegraded.

**SIP Call:** EPA action requiring a state to resubmit all or part of its State Implementation Plan to demonstrate attainment of the require national ambient air quality standards within the statutory deadline. A SIP Revision is a revision of a SIP altered at the request of EPA or on a state's initiative. (See: State Implementation Plan.)

**Site:** An area or place within the jurisdiction of the EPA and/or a state.

**Site Assessment Program:** A means of evaluating hazardous waste sites through preliminary assessments and site inspections to develop a Hazard Ranking System score.

**Site Inspection:** The collection of information from a Superfund site to determine the extent and severity of hazards posed by the site. It follows and is more extensive than a preliminary assessment. The purpose is to gather information necessary to score the site, using the Hazard Ranking System, and to determine if it presents an immediate threat requiring prompt removal.

**Site Safety Plan:** A crucial element in all removal actions, it includes information on equipment being used, precautions to be taken, and steps to take in the event of an on-site emergency.

**Siting:** The process of choosing a location for a facility.

**Skimming:** Using a machine to remove oil or scum from the surface of the water.

**Slow Sand Filtration:** Passage of raw water through a bed of sand at low velocity, resulting in substantial removal of chemical and biological contaminants.

**Sludge:** A semi-solid residue from any of a number of air or water treatment processes; can be a hazardous waste.

**Sludge Digester:** Tank in which complex organic substances like sewage sludges are biologically dredged. During these reactions, energy is released and much of the sewage is converted to methane, carbon dioxide, and water.

**Slurry:** A watery mixture of insoluble matter resulting from some pollution control techniques.

**Small Quantity Generator (SQG-sometimes referred to as "Squeegee"):** Persons or enterprises that produce 220-2200 pounds per month of hazardous waste; they are required to keep more records than conditionally exempt generators. The largest category of hazardous waste generators, SQGs, include automotive shops, dry cleaners, photographic developers, and many other small businesses. (See: conditionally exempt generators.)

**Smelter:** A facility that melts or fuses ore, often with an accompanying chemical change, to separate its metal content. Emissions cause pollution. "Smelting" is the process involved.

**Smog:** Air pollution typically associated with oxidants. (See: photochemical smog.)

**Smoke:** Particles suspended in air after incomplete combustion.

**Soft Detergents:** Cleaning agents that break down in nature.

**Soft Water:** Any water that does not contain a significant amount of dissolved minerals such as salts of calcium or magnesium.

**Soil Adsorption Field:** A sub-surface area containing a trench or bed with clean stones and a system of piping through which treated sewage may seep into the surrounding soil for further treatment and disposal.

**Soil and Water Conservation Practices:** Control measures consisting of managerial, vegetative, and structural practices to reduce the loss of soil and water.

**Soil Conditioner:** An organic material like humus or compost that helps soil absorb water, build a bacterial community, and take up mineral nutrients.

**Soil Erodibility:** An indicator of a soil's susceptibility to raindrop impact, runoff, and other erosive processes.

**Soil Gas:** Gaseous elements and compounds in the small spaces between particles of the earth and soil. Such gases can be moved or driven out under pressure.

**Soil Moisture:** The water contained in the pore space of the unsaturated zone.

**Soil Sterilant:** A chemical that temporarily or permanently prevents the growth of all plants and animals,

**Solder:** Metallic compound used to seal joints between pipes. Until recently, most solder contained 50 percent lead. Use of solder containing more than 0.2 percent lead in pipes carrying drinking water is now prohibited.

**Sole-Source Aquifer:** An aquifer that supplies 50-percent or more of the drinking water of an area.

**Solid Waste:** Non-liquid, non-soluble materials ranging from municipal garbage to industrial wastes that contain complex and sometimes hazardous substances. Solid wastes also include sewage sludge, agricultural refuse, demolition wastes, and mining residues. Technically, solid waste also refers to liquids and gases in containers.

**Solid Waste Disposal:** The final placement of refuse that is not salvaged or recycled.

**Solid Waste Management:** Supervised handling of waste materials from their source through recovery processes to disposal.

**Solidification and Stabilization:** Removal of wastewater from a waste or changing it chemically to make it less permeable and susceptible to transport by water.

**Solubility:** The amount of mass of a compound that will dissolve in a unit volume of solution. Aqueous Solubility is the maximum concentration of a chemical that will dissolve in pure water at a reference temperature.

**Soot:** Carbon dust formed by incomplete combustion.

**Sorption:** The action of soaking up or attracting substances; process used in many pollution control systems.

**Source Area:** The location of liquid hydrocarbons or the zone of highest soil or groundwater concentrations, or both, of the chemical of concern.

**Source Characterization Measurements:** Measurements made to estimate the rate of release of pollutants into the environment from a source such as an incinerator, landfill, etc.

**Source Reduction:** Reducing the amount of materials entering the waste stream from a specific source by redesigning products or patterns of production or consumption (e.g., using returnable beverage containers). Synonymous with waste reduction.

**Source Separation:** Segregating various wastes at the point of generation (e.g., separation of paper, metal and glass from other wastes to make recycling simpler and more efficient).

**Source-Water Protection Area:** The area delineated by a state for a Public Water Supply or including numerous such suppliers, whether the source is ground water or surface water or both.

**Sparge or Sparging:** Injection of air below the water table to strip dissolved volatile organic compounds and/or oxygenate ground water to facilitate aerobic biodegradation of organic compounds.

**Special Local-Needs Registration:** Registration of a pesticide product by a state agency for a specific use that is not federally registered. However, the active ingredient must be federally registered for other uses. The special use is specific to that state and is often minor, thus may not warrant the additional cost of a full federal registration process. SLN registration cannot be issued for new active ingredients, food-use active ingredients without tolerances, or for a canceled registration. The products cannot be shipped across state lines.

**Special Review:** Formerly known as Rebuttable Presumption Against Registration (RPAR), this is the regulatory process through which existing pesticides suspected of posing unreasonable risks to human health, non-target organisms, or the environment are referred for review by EPA. Such review requires an intensive risk/benefit analysis with opportunity for public comment. If risk is found to outweigh social and economic benefits, regulatory actions can be initiated, ranging from label revisions and use-restriction to cancellation or suspended registration.

**Special Waste:** Items such as household hazardous waste, bulky wastes (refrigerators, pieces of furniture, etc.) tires, and used oil.

**Species:** 1. A reproductively isolated aggregate of interbreeding organisms having common attributes and usually designated by a common name.2. An organism belonging to belonging to such a category.

**Specific Conductance:** Rapid method of estimating the dissolved solid content of a water supply by testing its capacity to carry an electrical current.

**Specific Yield:** The amount of water a unit volume of saturated permeable rock will yield when drained by gravity.

**Spill Prevention, Containment, and Countermeasures Plan (SPCP):** Plan covering the release of hazardous substances as defined in the Clean Water Act.

**Spoil:** Dirt or rock removed from its original location—destroying the composition of the soil in the process—as in strip-mining, dredging, or construction.

**Sprawl:** Unplanned development of open land.

**Spray Tower Scrubber:** A device that sprays alkaline water into a chamber where acid gases are present to aid in neutralizing the gas.

**Spring:** Ground water seeping out of the earth where the water table intersects the ground surface.

**Spring Melt/Thaw:** The process whereby warm temperatures melt winter snow and ice. Because various forms of acid deposition may have been stored in the frozen water, the melt can result in abnormally large amounts of acidity entering streams and rivers, sometimes causing fish kills.

**Stabilization:** Conversion of the active organic matter in sludge into inert, harmless material.

**Stabilization Ponds:** (See: lagoon.)

**Stable Air:** A motionless mass of air that holds, instead of dispersing, pollutants.

**Stack:** A chimney, smokestack, or vertical pipe that discharges used air.

**Stack Effect:** Air, as in a chimney, that moves upward because it is warmer than the ambient atmosphere.

**Stack Effect:** Flow of air resulting from warm air rising, creating a positive pressure area at the top of a building and negative pressure area at the bottom. This effect can overpower the mechanical system and disrupt building ventilation and air circulation.

**Stack Gas:** (See: flue gas.)

**Stage II Controls:** Systems placed on service station gasoline pumps to control and capture gasoline vapors during refuelling.

**Stagnation:** Lack of motion in a mass of air or water that holds pollutants in place.

**Stakeholder:** Any organization, governmental entity, or individual that has a stake in or may be impacted by a given approach to environmental regulation, pollution prevention, energy conservation, etc.

**Standard Industrial Classification Code:** Also known as SIC Codes, a method of grouping industries with similar products or services and assigning codes to these groups.

**Standard Sample:** The part of finished drinking water that is examined for the presence of coliform bacteria.

**Standards:** Norms that impose limits on the amount of pollutants or emissions produced. EPA establishes minimum standards, but states are allowed to be stricter.

**Start of a Response Action:** The point in time when there is a guarantee or set-aside of funding by EPA, other federal agencies, states or Principal Responsible Parties in order to begin response actions at a Superfund site.

**State Emergency Response Commission (SERC):** Commission appointed by each state governor according to the requirements of SARA Title III. The SERCs designate emergency planning districts, appoint local emergency planning committees, and supervise and coordinate their activities.

**State Environmental Goals and Indication Project:** Program to assist state environmental agencies by providing technical and financial assistance in the development of environmental goals and indicators.

**State Implementation Plans (SIP):** EPA approved state plans for the establishment, regulation, and enforcement of air pollution standards.

**State Management Plan:** Under FIFRA, a state management plan required by EPA to allow states, tribes, and U.S. territories the flexibility to design and implement ways to protect ground water from the use of certain pesticides.

**Static Water Depth:** The vertical distance from the centerline of the pump discharge down to the surface level of the free pool while no water is being drawn from the pool or water table.

**Static Water Level:** 1. Elevation or level of the water table in a well when the pump is not operating. 2. The level or elevation to which water would rise in a tube connected to an artesian aquifer or basin in a conduit under pressure.

**Stationary Source:** A fixed-site producer of pollution, mainly power plants and other facilities using industrial combustion processes. (See: point source.)

**Sterilization:** The removal or destruction of all microorganisms, including pathogenic and other bacteria, vegetative forms, and spores.

**Sterilizer:** One of three groups of anti-microbials registered by EPA for public health uses. EPA considers an antimicrobial to be a sterilizer when it destroys or eliminates all forms of bacteria, viruses, and fungi and their spores. Because spores are considered the most difficult form of microorganism to destroy, EPA considers the term sporicide to be synonymous with sterilizer.

**Storage:** Temporary holding of waste pending treatment or disposal, as in containers, tanks, waste piles, and surface impoundments.

**Storm Sewer:** A system of pipes (separate from sanitary sewers) that carries water runoff from buildings and land surfaces.

**Stratification:** Separating into layers.

**Stratigraphy:** Study of the formation, composition, and sequence of sediments, whether consolidated or not.

**Stratosphere:** The portion of the atmosphere 10-to-25 miles above the earth's surface.

**Stressors:** Physical, chemical, or biological entities that can induce adverse effects on ecosystems or human health.

**Strip-Cropping:** Growing crops in a systematic arrangement of strips or bands that serve as barriers to wind and water erosion.

**Strip-Mining:** A process that uses machines to scrape soil or rock away from mineral deposits just under the earth's surface.

**Structural Deformation:** Distortion in walls of a tank after liquid has been added or removed.

**Subchronic:** Of intermediate duration, usually used to describe studies or periods of exposure lasting between 5 and 90 days.

**Subchronic Exposure:** Multiple or continuous exposures lasting for approximately ten percent of an experimental species lifetime, usually over a three-month period.

**Submerged Aquatic Vegetation:** Vegetation that lives at or below the water surface; an important habitat for young fish and other aquatic organisms.

**Subwatershed:** Topographic perimeter of the catchment area of a stream tributary.

**Sulfur Dioxide ($SO_2$):** A pungent, colorless, gasformed primarily by the combustion of fossil fuels; becomes a pollutant when present in large amounts.

**Sump:** A pit or tank that catches liquid runoff for drainage or disposal.

**Superchlorination:** Chlorination with doses that are deliberately selected to produce water free of combined residuals so large as to require dechlorination.

**Supercritical Water:** A type of thermal treatment using moderate temperatures and high pressures to enhance the ability of water to break down large organic molecules into smaller, less toxic ones. Oxygen injected during this process combines with simple organic compounds to form carbon dioxide and water.

**Superfund:** The program operated under the legislative authority of CERCLA and SARA that funds and carries out EPA solid waste emergency and long-term removal and remedial activities. These activities include establishing the National Priorities List, investigating sites for inclusion on the list, determining their priority, and conducting and/or supervising cleanup and other remedial actions.

**Superfund Innovative Technology Evaluation (SITE) Program:** EPA program to promote development and use of innovative treatment and site characterization technologies in Superfund site cleanups.

**Supplemental Registration:** An arrangement whereby a registrant licenses another company to market its pesticide product under the second company's registration.

**Supplier of Water:** Any person who owns or operates a public water supply.

**Surface Impoundment:** Treatment, storage, or disposal of liquid hazardous wastes in ponds.

**Surface Runoff:** Precipitation, snow melt, or irrigation water in excess of what can infiltrate the soil surface and be stored in small surface depressions; a major transporter of non-point source pollutants in rivers, streams, and lakes..

**Surface Uranium Mines:** Strip mining operations for removal of uranium-bearing ore.

**Surface Water:** All water naturally open to the atmosphere (rivers, lakes, reservoirs, ponds, streams, impoundments, seas, estuaries, etc.)

**Surface-Water Treatment Rule:** Rule that specifies maximum contaminant level goals for Giardia lamblia, viruses, and Legionella and promulgates filtration and disinfection requirements for public water systems using surface-water or ground-water sources under the direct influence of surface water. The regulations also specify water quality, treatment, and watershed protection criteria under which filtration may be avoided.

**Surfacing ACM:** Asbestos-containing material that is sprayed or troweled on or otherwise applied to surfaces, such as acoustical plaster on ceilings and fireproofing materials on structural members.

**Surfacing Material:** Material sprayed or troweled onto structural members (beams, columns, or decking) for fire protection; or on ceilings or walls for fireproofing, acoustical or decorative purposes. Includes textured plaster, and other textured wall and ceiling surfaces.

**Surfactant:** A detergent compound that promotes lathering.

**Surrogate Data:** Data from studies of test organisms or a test substance that are used to estimate the characteristics or effects on another organism or substance.

**Surveillance System:** A series of monitoring devices designed to check on environmental conditions.

**Susceptibility Analysis:** An analysis to determine whether a Public Water Supply is subject to significant pollution from known potential sources.

**Suspect Material:** Building material suspected of containing asbestos; e.g., surfacing material, floor tile, ceiling tile, thermal system insulation.

**Suspended Loads:** Specific sediment particles maintained in the water column by turbulence and carried with the flow of water.

**Suspended Solids:** Small particles of solid pollutants that float on the surface of, or are suspended in, sewage or other liquids. They resist removal by conventional means.

**Suspension:** Suspending the use of a pesticide when EPA deems it necessary to prevent an imminent hazard resulting from its continued use. An emergency suspension takes effect immediately; under an ordinary suspension a registrant can request a hearing before the suspension goes into effect. Such a hearing process might take six months.

**Suspension Culture:** Cells growing in a liquid nutrient medium.

**Swamp:** A type of wetland dominated by woody vegetation but without appreciable peat deposits. Swamps may be fresh or salt water and tidal or non-tidal. (See: wetlands.)

**Synergism:** An interaction of two or more chemicals that results in an effect greater than the sum of their separate effects.

**Synthetic Organic Chemicals (SOCs):** Man-made (anthropogenic) organic chemicals. Some SOCs are volatile; others tend to stay dissolved in water instead of evaporating.

**System With a Single Service Connection:** A system that supplies drinking water to consumers via a single service line.

**Systemic Pesticide:** A chemical absorbed by an organism that interacts with the organism and makes the organism toxic to pests.

## T

**Tail Water:** The runoff of irrigation water from the lower end of an irrigated field.

**Tailings:** Residue of raw material or waste separated out during the processing of crops or mineral ores.

**Tailpipe Standards:** Emissions limitations applicable to mobile source engine exhausts.

**Tampering:** Adjusting, negating, or removing pollution control equipment on a motor vehicle.

**Technical Assistance Grant (TAG):** As part of the Superfund program, Technical Assistance Grants of up to $50,000 are provided to citizens' groups to obtain assistance in interpreting information related to clean-ups at Superfund sites or those proposed for the National Priorities List. Grants are used by such groups to hire technical advisors to help them understand the site-related technical information for the duration of response activities.

**Technical-Grade Active Ingredient (TGA):** A pesticide chemical in pure form as it is manufactured prior to being formulated into an end-use product (e.g. wettable powders, granules, emulsifiable concentrates). Registered manufactured products composed of such chemicals are known as Technical Grade Products.

**Technology-Based Limitations:** Industry-specific effluent limitations based on best available preventive technology applied to a discharge when it will not cause a violation of water quality standards at low stream flows. Usually applied to discharges into large rivers.

**Technology-Based Standards:** Industry-specific effluent limitations applicable to direct and indirect sources which are developed on a category-by-category basis using statutory factors, not including water-quality effects.

**Teratogen:** A substance capable of causing birth defects.

**Teratogenesis:** The introduction of nonhereditary birth defects in a developing fetus by exogenous factors such as physical or chemical agents acting in the womb to interfere with normal embryonic development.

**Terracing:** Dikes built along the contour of sloping farm land that hold runoff and sediment to reduce erosion.

**Tertiary Treatment:** Advanced cleaning of wastewater that goes beyond the secondary or biological stage, removing nutrients such as phosphorus, nitrogen, and most BOD and suspended solids.

**Theoretical Maximum Residue Contribution:** The theoretical maximum amount of a pesticide in the daily diet of an average person. It assumes that the diet is composed of all food items for which there are tolerance-level residues of the pesticide. The TMRC is expressed as milligrams of pesticide/kilograms of body weight/day.

**Therapeutic Index:** The ratio of the dose required to produce toxic or lethal effects to the dose required to produce nonadverse or therapeutic response.

**Thermal Pollution:** Discharge of heated water from industrial processes that can kill or injure aquatic organisms.

**Thermal Stratification:** The formation of layers of different temperatures in a lake or reservoir.

**Thermal System Insulation (TSI):** Asbestos-containing material applied to pipes, fittings, boilers, breeching, tanks, ducts, or other interior structural components to prevent heat loss or gain or water condensation.

**Thermal Treatment:** Use of elevated temperatures to treat hazardous wastes. (See: incineration; pyrolysis.)

**Thermocline:** The middle layer of a thermally stratified lake or reservoir. In this layer, there is a rapid decrease in temperatures in a lake or reservoir.

**Threshold:** The lowest dose of a chemical at which a specified measurable effect is observed and below which it is not observed.

**Threshold:** The dose or exposure level below which a significant adverse effect is not expected.

**Threshold Level:** Time-weighted average pollutant concentration values, exposure beyond which is likely to adversely affect human health. (See: environmental exposure)

**Threshold Limit Value (TLV):** The concentration of an airborne substance to which an average person can be repeatedly exposed without adverse effects. TLVs may be expressed in three ways: (1) TLV-TWA—Time weighted average, based on an allowable exposure averaged over a normal 8-hour workday or 40-hour work- week; (2) TLV-STEL—Short-term exposure limit or maximum concentration for a brief specified period of time, depending on a specific chemical (TWA must still be met); and (3) TLV-C—Ceiling Exposure Limit or maximum exposure concentration not to be exceeded under any circumstances. (TWA must still be met.)

**Threshold Odor:** (See: Odor threshold)

**Threshold Planning Quantity:** A quantity designated for each chemical on the list of extremely hazardous substances that triggers notification by facilities to the State Emergency Response Commission that such facilities are subject to emergency planning requirements under SARA Title III.

**Thropic Levels:** A functional classification of species that is based on feeding relationships (e.g. generally aquatic and terrestrial green plants comprise the first thropic level, and herbivores comprise the second.)

**Tidal Marsh:** Low, flat marshlands traversed by channels and tidal hollows, subject to tidal inundation; normally, the only vegetation present is salt-tolerant bushes and grasses. (See: wetlands.)

**Tillage:** Plowing, seedbed preparation, and cultivation practices.

**Time-weighted Average (TWA):** In air sampling, the average air concentration of contaminants during a given period.

**Tire Processor:** Intermediate operating facility where recovered tires are processed in preparation for recycling.

**Tires:** As used in recycling, passenger car and truck tires (excludes airplane, bus, motorcycle and special service military, agricultural, off-the-road and-slow speed industrial tires). Car and truck tires are recycled into rubber products such as trash cans, storage containers, rubberized asphalt or used whole for playground and reef construction.

**Tolerance Petition:** A formal request to establish a new tolerance or modify an existing one.

**Tolerances:** Permissible residue levels for pesticides in raw agricultural produce and processed foods. Whenever a pesticide is registered for use on a food or a feed crop, a tolerance (or exemption from the tolerance requirement) must be established. EPA establishes the tolerance levels, which are enforced by the Food and Drug Administration and the Department of Agriculture.

**Tonnage:** The amount of waste that a landfill accepts, usually expressed in tons per month. The rate at which a landfill accepts waste is limited by the landfill's permit.

**Topography:** The physical features of a surface area including relative elevations and the position of natural and man-made (anthropogenic) features.

**Total Dissolved Phosphorous:** The total phosphorous content of all material that will pass through a filter, which is determined as orthophosphate without prior digestion or hydrolysis. Also called soluble P. or ortho P.

**Total Dissolved Solids (TDS):** All material that passes the standard glass river filter; now called total filtrable residue. Term is used to reflect salinity.

**Total Petroleum Hydrocarbons (TPH):** Measure of the concentration or mass of petroleum hydrocarbon constituents present in a given amount of soil or water. The word "total" is a misnomer—few, if any, of the procedures for quantifying hydrocarbons can measure all of them in a given sample. Volatile ones are usually lost in the process and not quantified and non-petroleum hydrocarbons sometimes appear in the analysis.

**Total Recovered Petroleum Hydrocarbon:** A method for measuring petroleum hydrocarbons in samples of soil or water.

**Total Suspended Particles (TSP):** A method of monitoring airborne particulate matter by total weight.

**Total Suspended Solids (TSS):** A measure of the suspended solids in wastewater, effluent, or water bodies, determined by tests for "total suspended non-filterable solids." (See: suspended solids.)

**Toxaphene:** Chemical that causes adverse health effects in domestic water supplies and is toxic to fresh water and marine aquatic life.

**Toxic Chemical:** Any chemical listed in EPA rules as "Toxic Chemicals Subject to Section 313 of the Emergency Planning and Community Right-to-Know Act of 1986."

**Toxic Chemical Release Form:** Information form required of facilities that manufacture, process, or use (in quantities above a specific amount) chemicals listed under SARA Title III.

**Toxic Chemical Use Substitution:** Replacing toxic chemicals with less harmful chemicals in industrial processes.

**Toxic Cloud:** Airborne plume of gases, vapors, fumes, or aerosols containing toxic materials.

**Toxic Concentration:** The concentration at which a substance produces a toxic effect.

**Toxic Dose:** The dose level at which a substance produces a toxic effect.

**Toxic Pollutants:** Materials that cause death, disease, or birth defects in organisms that ingest or absorb them. The quantities and exposures necessary to cause these effects can vary widely.

**Toxic Release Inventory:** Database of toxic releases in the United States compiled from SARA Title III Section 313 reports.

**Toxic Substance:** A chemical or mixture that may present an unreasonable risk of injury to health or the environment.

**Toxic Waste:** A waste that can produce injury if inhaled, swallowed, or absorbed through the skin.

**Toxicant:** A harmful substance or agent that may injure an exposed organism.

**Toxicity:** The degree to which a substance or mixture of substances can harm humans or animals. Acute toxicity involves harmful effects in an organism through a single or short-term exposure. Chronic toxicity is the ability of a substance or mixture of substances to cause harmful effects over an extended period, usually upon repeated or continuous exposure sometimes lasting for the entire life of the exposed organism. Subchronic toxicity is the ability of the substance to cause effects for more than one year but less than the lifetime of the exposed organism.

**Toxicity Assessment:** Characterization of the toxicological properties and effects of a chemical, with special emphasis on establishment of dose-response characteristics.

**Toxicity Testing:** Biological testing (usually with an invertebrate, fish, or small mammal) to determine the adverse effects of a compound or effluent.

**Toxicological Profile:** An examination, summary, and interpretation of a hazardous substance to determine levels of exposure and associated health effects.

**Transboundary Pollutants:** Air pollution that travels from one jurisdiction to another, often crossing state or international boundaries. Also applies to water pollution.

**Transfer Station:** Facility where solid waste is transferred from collection vehicles to larger trucks or rail cars for longer distance transport.

**Transient Water System:** A non-community water system that does not serve 25 of the same nonresidents per day for more than six months per year.

**Transmission Lines:** Pipelines that transport raw water from its source to a water treatment plant, then to the distribution grid system.

**Transmissivity:** The ability of an aquifer to transmit water.

**Transpiration:** The process by which water vapor is lost to the atmosphere from living plants. The term can also be applied to the quantity of water thus dissipated.

**Transportation Control Measures (TCMs):** Steps taken by a locality to reduce vehicular emission and improve air quality by reducing or changing the flow of traffic; e.g. bus and HOV lanes, carpooling and other forms of ride-shairing, public transit, bicycle lanes.

**Transporter:** Hauling firm that picks up properly packaged and labeled hazardous waste from generators and transports it to designated facilities for treatment, storage, or disposal. Transporters are subject to EPA and DOT hazardous waste regulations.

**Trash:** Material considered worthless or offensive that is thrown away. Generally defined as dry waste material, but in common usage it is a synonym for garbage, rubbish, or refuse.

**Trash-to-Energy Plan:** Burning trash to produce energy.

**Treatability Studies:** Tests of potential cleanup technologies conducted in a laboratory (See: bench-scale tests.)

**Treated Regulated Medical Waste:** Medical waste treated to substantially reduce or eliminate its pathogenicity, but that has not yet been destroyed.

**Treated Wastewater:** Wastewater that has been subjected to one or more physical, chemical, and biological processes to reduce its potential of being health hazard.

**Treatment:** (1) Any method, technique, or process designed to remove solids and/or pollutants from solid waste, waste-streams, effluents, and air emissions. (2) Methods used to change the biological character or composition of any regulated medical waste so as to substantially reduce or eliminate its potential for causing disease.

**Treatment Plant:** A structure built to treat wastewater before discharging it into the environment.

**Treatment, Storage, and Disposal Facility:** Site where a hazardous substance is treated, stored, or disposed of. TSD facilities are regulated by EPA and states under RCRA.

**Tremie:** Device used to place concrete or grout under water.

**Trial Burn:** An incinerator test in which emissions are monitored for the presence of specific organic compounds, particulates, and hydrogen chloride.

**Trichloroethylene (TCE):** A stable, low boiling-point colorless liquid, toxic if inhaled. Used as a solvent or metal degreasing agent, and in other industrial applications.

**Trickle Irrigation:** Method in which water drips to the soil from perforated tubes or emitters.

**Trickling Filter:** A coarse treatment system in which wastewater is trickled over a bed of stones or other material covered with bacteria that break down the organic waste and produce clean water.

**Trihalomethane (THM):** One of a family of organic compounds named as derivative of methane. THMs are generally by-products of chlorination of drinking water that contains organic material.

**Troposhpere:** The layer of the atmosphere closest to the earth's surface.

**Trust Fund (CERCLA):** A fund set up under the Comprehensive Environmental Response, Compensation and Liability Act (CERCLA) to help pay for cleanup of hazardous waste sites and for legal action to force those responsible for the sites to clean them up.

**Tube Settler:** Device using bundles of tubes to let solids in water settle to the bottom for removal by conventional sludge collection means; sometimes used in sedimentation basins and clarifiers to improve particle removal.

**Tuberculation:** Development or formation of small mounds of corrosion products on the inside of iron pipe. These tubercules roughen the inside of the pipe, increasing its resistance to water flow.

**Tundra:** A type of treeless ecosystem dominated by lichens, mosses, grasses, and woody plants. Tundra is found at high latitudes (arctic tundra) and high altitudes (alpine tundra). Arctic tundra is underlain by permafrost and is usually water saturated. (See: wetlands.)

**Turbidimeter:** A device that measures the cloudiness of suspended solids in a liquid; a measure of the quantity of suspended solids.

**Turbidity:** 1. Haziness in air caused by the presence of particles and pollutants. 2. A cloudy condition in water due to suspended silt or organic matter.

## U

**Ultra Clean Coal (UCC):** Coal that is washed, ground into fine particles, then chemically treated to remove sulfur, ash, silicone, and other substances; usually briquetted and coated with a sealant made from coal.

**Ultraviolet Rays:** Radiation from the sun that can be useful or potentially harmful. UV rays from one part of the spectrum (UV-A) enhance plant life. UV rays from other parts of the spectrum (UV-B) can cause skin cancer or other tissue damage. The ozone layer in the atmosphere partly shields us from ultraviolet rays reaching the earth's surface.

**Uncertainty Factor:** One of several factors used in calculating the reference dose from experimental data. UFs are intended to account for (1) the variation in sensitivity among humans; (2) the uncertainty in extrapolating animal data to humans; (3) the uncertainty in extrapolating data obtained in a study that covers less than the full life of the exposed animal or human; and (4) the uncertainty in using LOAEL data rather than NOAEL data.

**Unconfined Aquifer:** An aquifer containing water that is not under pressure; the water level in a well is the same as the water table outside the well.

**Underground Injection Control (UIC):** The program under the Safe Drinking Water Act that regulates the use of wells to pump fluids into the ground.

**Underground Injection Wells:** Steel- and concrete-encased shafts into which hazardous waste is deposited by force and under pressure.

**Underground Sources of Drinking Water:** Aquifers currently being used as a source of drinking water or those capable of supplying a public water system. They have a total dissolved solids content of 10,000 milligrams per liter or less, and are not "exempted aquifers." (See: exempted aquifer.)

**Underground Storage Tank (UST):** A tank located at least partially underground and designed to hold gasoline or other petroleum products or chemicals.

**Unreasonable Risk:** Under the Federal Insecticide, Fungicide, and Rodenticide Act (FIFRA), "unreasonable adverse effects" means any unreasonable risk to man or the environment, taking into account the medical, economic, social, and environmental costs and benefits of any pesticide.

**Unsaturated Zone:** The area above the water table where soil pores are not fully saturated, although some water may be present.

**Upper Detection Limit:** The largest concentration that an instrument can reliably detect.

**Uranium Mill Tailings Piles:** Former uranium ore processing sites that contain leftover radioactive materials (wastes), including radium and unrecovered uranium.

**Uranium Mill-Tailings Waste Piles:** Licensed active mills with tailings piles and evaporation ponds created by acid or alkaline leaching processes.

**Urban Runoff:** Storm water from city streets and adjacent domestic or commercial properties that carries pollutants of various kinds into the sewer systems and receiving waters.

**Urea-Formaldehyde Foam Insulation:** A material once used to conserve energy by sealing crawl spaces, attics, etc.; no longer used because emissions were found to be a health hazard.

**Use Cluster:** A set of competing chemicals, processes, and/or technologies that can substitute for one another in performing a particular function.

**Used Oil:** Spent motor oil from passenger cars and trucks collected at specified locations for recycling (not included in the category of municipal solid waste).

**User Fee:** Fee collected from only those persons who use a particular service, as compared to one collected from the public in general.

**Utility Load:** The total electricity demand for a utility district.

## V

**Vadose Zone:** The zone between land surface and the water table within which the moisture content is less than saturation (except in the capillary fringe) and pressure is less than atmospheric. Soil pore space also typically contains air or other gases. The capillary fringe is included in the vadose zone. (See: Unsaturated Zone.)

**Valued Environmental Attributes/Components:** Those aspects(components/processes/functions) of ecosystems, human health, and environmental welfare considered to be important and potentially at risk from human activity or natural hazards. Similar to the term "valued environmental components" used in environmental impact assessment.

**Vapor:** The gas given off by substances that are solids or liquids at ordinary atmospheric pressure and temperatures.

**Vapor Capture System:** Any combination of hoods and ventilation system that captures or contains organic vapors so they may be directed to an abatement or recovery device.

**Vapor Dispersion:** The movement of vapor clouds in air due to wind, thermal action, gravity spreading, and mixing.

**Vapor Plumes:** Flue gases visible because they contain water droplets.

**Vapor Pressure:** A measure of a substance's propensity to evaporate, vapor pressure is the force per unit area exerted by vapor in an equilibrium state with surroundings at a given pressure. It increases exponentially with an increase in temperature. A relative measure of chemical volatility, vapor pressure is used to calculate water partition coefficients and volatilization rate constants.

**Vapor Recovery System:** A system by which the volatile gases from gasoline are captured instead of being released into the atmosphere.

**Variance:** Government permission for a delay or exception in the application of a given law, ordinance, or regulation.

**Vector:** 1. An organism, often an insect or rodent, that carries disease. 2. Plasmids, viruses, or bacteria used to transport genes into a host cell. A gene is placed in the vector; the vector then "infects" the bacterium.

**Vegetative Controls:** Non-point source pollution control practices that involve vegetative cover to reduce erosion and minimize loss of pollutants.

**Vehicle Miles Travelled (VMT):** A measure of the extent of motor vehicle operation; the total number of vehicle miles travelled within a specific geographic area over a given period of time.

**Ventilation Rate:** The rate at which indoor air enters and leaves a building. Expressed as the number of changes of outdoor air per unit of time (air changes per hour (ACH), or the rate at which a volume of outdoor air enters in cubic feet per minute (CFM).

**Ventilation/Suction:** The act of admitting fresh air into a space in order to replace stale or contaminated air; achieved by blowing air into the space. Similarly, suction represents the admission of fresh air into an interior space by lowering the pressure outside of the space, thereby drawing the contaminated air outward.

**Venturi Scrubbers:** Air pollution control devices that use water to remove particulate matter from emissions.

**Vinyl Chloride:** A chemical compound, used in producing some plastics, that is believed to be oncogenic.

**Virgin Materials:** Resources extracted from nature in their raw form, such as timber or metal ore.

**Viscosity:** The molecular friction within a fluid that produces flow resistance.

**Volatile:** Any substance that evaporates readily.

**Volatile Liquids:** Liquids which easily vaporize or evaporate at room temperature.

**Volatile Organic Compound (VOC):** Any organic compound that participates in atmospheric photochemical reactions except those designated by EPA as having negligible photochemical reactivity.

**Volatile Solids:** Those solids in water or other liquids that are lost on ignition of the dry solids at 550ø centigrade.

**Volatile Synthetic Organic Chemicals:** Chemicals that tend to volatilize or evaporate.

**Volume Reduction:** Processing waste materials to decrease the amount of space they occupy, usually by compacting, shredding, incineration, or composting.

**Volumetric Tank Test:** One of several tests to determine the physical integrity of a storage tank; the volume of fluid in the tank is measured directly or calculated from product-level changes. A marked drop in volume indicates a leak.

**Vulnerability Analysis:** Assessment of elements in the community that are susceptible to damage if hazardous materials are released.

**Vulnerable Zone:** An area over which the airborne concentration of a chemical accidentally released could reach the level of concern.

## W

**Waste:** 1. Unwanted materials left over from a manufacturing process. 2. Refuse from places of human or animal habitation.

**Waste Characterization:** Identification of chemical and microbiological constituents of a waste material.

**Waste Exchange:** Arrangement in which companies exchange their wastes for the benefit of both parties.

**Waste Feed:** The continuous or intermittent flow of wastes into an incinerator.

**Waste Generation:** The weight or volume of materials and products that enter the waste stream before recycling, composting, landfilling, or combustion takes place. Also can represent the amount of waste generated by a given source or category of sources.

**Waste Load Allocation:** 1. The maximum load of pollutants each discharger of waste is allowed to release into a particular waterway. Discharge limits are usually required for each specific water quality criterion being, or expected to be, violated. 2. The portion of a stream's total assimilative capacity assigned to an individual discharge.

**Waste Minimization:** Measures or techniques that reduce the amount of wastes generated during industrial production processes; term is also applied to recycling and other efforts to reduce the amount of waste going into the waste stream.

**Waste Piles:** Non-containerized, lined or unlined accumulations of solid, nonflowing waste.

**Waste Reduction:** Using source reduction, recycling, or composting to prevent or reduce waste generation.

**Waste Stream:** The total flow of solid waste from homes, businesses, institutions, and manufacturing plants that is recycled, burned, or disposed of in landfills, or segments thereof such as the "residential waste stream" or the "recyclable waste stream."

**Waste Treatment Lagoon:** Impoundment made by excavation or earth fill for biological treatment of wastewater.

**Waste Treatment Plant:** A facility containing a series of tanks, screens, filters and other processes by which pollutants are removed from water.

**Waste Treatment Stream:** The continuous movement of waste from generator to treater and disposer.

**Waste-Heat Recovery:** Recovering heat discharged as a byproduct of one process to provide heat needed by a second process.

**Waste-to-Energy Facility/Municipal-Waste Combustor:** Facility where recovered municipal solid waste is converted into a usable form of energy, usually via combustion.

**Wastewater:** The spent or used water from a home, community, farm, or industry that contains dissolved or suspended matter.Water Pollution: The presence in water of enough harmful or objectionable material to damage the water's quality.

**Wastewater Infrastructure:** The plan or network for the collection, treatment, and disposal of sewage in a community. The level of treatment will depend on the size of the community, the type of discharge, and/or the designated use of the receiving water.

**Wastewater Operations and Maintenance:** Actions taken after construction to ensure that facilities constructed to treat wastewater will be operated, maintained, and managed to reach prescribed effluent levels in an optimum manner.

**Wastewater Treatment Plan:** A facility containing a series of tanks, screens, filters, and other processes by which pollutants are removed from water. Most treatments include chlorination to attain safe drinking water standards.

**Water Purveyor:** A public utility, mutual water company, county water district, or municipality that delivers drinking water to customers.

**Water Quality Criteria:** Levels of water quality expected to render a body of water suitable for its designated use. Criteria are based on specific levels of pollutants that would make the water harmful if used for drinking, swimming, farming, fish production, or industrial processes.

**Water Quality Standards:** State-adopted and EPA-approved ambient standards for water bodies. The standards prescribe the use of the water body and establish the water quality criteria that must be met to protect designated uses.

**Water Quality-Based Limitations:** Effluent limitations applied to dischargers when mere technology-based limitations would cause violations of water quality standards. Usually applied to discharges into small streams.

**Water Quality-Based Permit:** A permit with an effluent limit more stringent than one based on technology performance. Such limits may be necessary to protect the designated use of receiving waters (e.g. recreation, irrigation, industry or water supply).

**Water Solubility:** The maximum possible concentration of a chemical compound dissolved in water. If a substance is water soluble it can very readily disperse through the environment.

**Water Storage Pond:** An impound for liquid wastes designed to accomplish some degree of biochemical treatment.

**Water Supplier:** One who owns or operates a public water system.

**Water Supply System:** The collection, treatment, storage, and distribution of potable water from source to consumer.

**Water Table:** The level of groundwater.

**Water Treatment Lagoon:** An impound for liquid wastes designed to accomplish some degree of biochemical treatment.

**Water Well:** An excavation where the intended use is for location, acquisition, development, or artificial recharge of ground water.

**Water-Soluble Packaging:** Packaging that dissolves in water; used to reduce exposure risks to pesticide mixers and loaders.

**Water-Source Heat Pump:** Heat pump that uses wells or heat exchangers to transfer heat from water to the inside of a building. Most such units use ground water. (See: groundsource heat pump; heat pump.)

**Waterborne Disease Outbreak:** The significant occurence of acute illness associated with drinking water from a public water system that is deficient in treatment, as determined by appropriate local or state agencies.

**Watershed:** The land area that drains into a stream; the watershed for a major river may encompass a number of smaller watersheds that ultimately combine at a common point.

**Watershed Approach:** A coordinated framework for environmental management that focuses public and private efforts on the highest priority problems within hydrologically-defined geographic areas taking into consideration both ground and surface water flow.

**Watershed Area:** A topographic area within a line drawn connecting the highest points uphill of a drinking waterintake into which overland flow drains.

**Weight of Scientific Evidence:** Considerations in assessing the interpretation of published information about toxicity—quality of testing methods, size and power of study design, consistency of results across studies, and biological plausibility of exposure-response relationships and statistical associations.

**Weir:** 1. A wall or plate placed in an open channel to measure the flow of water. 2. A wall or obstruction used to control flow from settling tanks and clarifiers to ensure a uniform flow rate and avoid short-circuiting. (See: short-circuiting.)

**Well:** A bored, drilled, or driven shaft, or a dug hole whose depth is greater than the largest surface dimension and whose purpose is to reach underground water supplies or oil, or to store or bury fluids below ground.

**Well Field:** Area containing one or more wells that produce usable amounts of water or oil.

**Well Injection:** The subsurface emplacement of fluids into a well.

**Well Monitoring:** Measurement by on-site instruments or laboratory methods of well water quality.

**Well Plug:** A watertight, gastight seal installed in a bore hole or well to prevent movement of fluids.

**Well Point:** A hollow vertical tube, rod, or pipe terminating in a perforated pointed shoe and fitted with a fine-mesh screen.

**Wellhead Protection Area:** A protected surface and subsurface zone surrounding a well or well field supplying a public water system to keep contaminants from reaching the well water.

**Wetlands:** An area that is saturated by surface or ground water with vegetation adapted for life under those soil conditions, as swamps, bogs, fens, marshes, and estuaries.

**Wettability:** The relative degree to which a fluid will spread into or coat a solid surface in the presence of other immiscible fluids.

**Wettable Powder:** Dry formulation that must be mixed with water or other liquid before it is applied.

**Wheeling:** The transmission of electricity owned by one entity through the facilities owned by another (usually a utility).

**Whole-Effluent-Toxicity Tests:** Tests to determine the toxicity levels of the total effluent from a single source as opposed to a series of tests for individual contaminants.

**Wildlife Refuge:** An area designated for the protection of wild animals, within which hunting and fishing are either prohibited or strictly controlled.

**Wire-to-Wire Efficiency:** The efficiency of a pump and motor together.

**Wood Packaging:** Wood products such as pallets, crates, and barrels.

**Wood Treatment Facility:** An industrial facility that treats lumber and other wood products for outdoor use. The process employs chromated copper arsenate, which is regulated as a hazardous material.

**Wood-Burning-Stove Pollution:** Air pollution caused by emissions of particulate matter, carbon monoxide, total suspended particulates, and polycyclic organic matter from wood-burning stoves.

**Working Level (WL):** A unit of measure for documenting exposure to radon decay products, the so-called "daughters." One working level is equal to approximately 200 picocuries per liter.

**Working Level Month (WLM):** A unit of measure used to determine cumulative exposure to radon.

## X

**Xenobiota:** Any biotum displaced from its normal habitat; a chemical foreign to a biological system.

## Y

**Yard Waste:** The part of solid waste composed of grass clippings, leaves, twigs, branches, and other garden refuse.

**Yellow-Boy:** Iron oxide flocculant (clumps of solids in waste or water); usually observed as orange-yellow deposits in surface streams with excess iron content. (See: floc, flocculation.)

**Yield:** The quantity of water (expressed as a rate of flow or total quantity per year) that can be collected for a given use from surface or groundwater sources.

## Z

**Zero Air:** Atmospheric air purified to contain less than 0.1 ppm total hydrocarbons.

**Zooplankton:** Small (often microscopic) free-floating aquatic plants or animals.

**Zone of Saturation:** The layer beneath the surface of the land containing openings that may fill with water.

***Source:** U.S. Environmental Protection Agency, "Terms of Environment"*

## A

# Entry Name Index

## B

# Entry Name Index

## C

# Entry Name Index

## D

### E

## F

## G

## J

## K



## L

## M

## N

## O

## P

## Q



## R

## S

## T

## U

## V

## X

## Y



## Z

## Alabama



## Alaska



## Arizona

## Arkansas



## California

## Colorado

## Connecticut

## Delaware



## District of Columbia

## Florida

## Georgia

## Hawaii



## Idaho

## Illinois



## Indiana

## Iowa



## Kansas



## Kentucky

## Louisiana



## Maine

## Maryland

## Massachusetts

## Michigan



## Minnesota

## Mississippi



## Missouri



## Montana

## Nebraska



## Nevada



## New Hampshire



## New Jersey

## New Mexico



## New York

## North Carolina



## North Dakota

## Ohio



## Oklahoma



## Oregon

## Pennsylvania

## Rhode Island



## South Carolina

## South Dakota



## Tennessee



## Texas

## Utah



## Vermont

## Virginia

## Washington

## West Virginia



## Wisconsin



## Wyoming

## Canada

## Coral reef ecology



## Deforestation



## Desertification



## Dolphins



## Drinking water

*(See charts starting on page 504)*

## Ecological preserves



## Ecology

*(See also: Conservation of natural resources)*



## Ecology, Tropical



## Electricity

*(See charts starting on page 518)*

## Endangered species, Animals

*(See also: charts starting on page 515)*



## Endangered species, Plants

*(See also: charts starting on page 515)*



## Energy conservation



## Energy consumption

*(See charts starting on page 518)*

## Energy conversion



## Energy economics



## Energy management



## Energy policy



## Energy resources

*(See also: Biomass energy)*

## Energy technology



## Environmental chemistry

*(See also: Air pollution)*



## Environmental compliance



## Environmental design



## Environmental economics

*(See also: Energy economics)*



## Environmental education

### Environmental engineering

*(See also: Environmental design)*

### Environmental ethics



### Environmental finances

*(See charts starting on page 604)*

### Environmental health

*(See also: Air pollution)*

## Environmental impact analysis



## Environmental monitoring



## Environmental policy

### Environmental protection



### Environmental quality



### Environmental sciences



### Estuaries



### Fertilizers

### Fire ecology



### Fish feeding and feeds



### Fish habitat



### Fish reproduction and growth



### Fisheries



### Fishery management



### Fishery policy



### Fishes



### Food science and technology



### Forest ecology



### Forest management

*(See also: Fire ecology)*



### Forests and forestry

*(See also: National forests)*



### Fuels

*(See also: Biomass energy, and charts starting on page 518)*



### Geographic information systems



### Geothermal resources



### Glaciology



### Global warming

*(See also: charts starting on page 628)*

### Groundwater



### Habitat, Ecology

*(See also: charts starting on page 658)*



### Harbors



### Hazardous substances



### Hazardous waste

*(See also: charts starting on page 690)*



### Hazardous waste management



### Human ecology

*(See also: Conservation of natural resources)*



### Hydrocarbons

### Municipal waste

*(See charts starting on page 741)*



### National forests

*(See also: Forests and forestry)*



### National parks and reserves

*(See also: Parks)*



### Natural resources



### Natural resources development



### Noise pollution



### Nuclear energy

*(See also: charts starting on page 518)*



### Nuclear engineering



### Nuclear safety



### Occupational diseases



### Ocean

### Ocean engineering



### Oceanography



### Organic gardening



### Ornithology



### Ozone



### Parks

*(See also: National parks and reserves)*



### Pest control, Integrated



### Pesticides

*(See also: charts starting on page 774)*



### Pesticides and the environment



### Petroleum

*(See also: charts starting on page 518)*



### Plant communities



### Plant conservation



### Plants, Protection of

*(See also: Plant conservation)*



### Plastics



### Polar regions

*(See also: Arctic ecology)*



### Pollution

*(See also: Air pollution)*



### Pollution control



### Pollution, Environmental effects of



### Population



### Prairie ecology



### Public awareness/information programs

### Pulp and paper technology



### Radiation effects



### Radioactive pollution

*(See also: Radioactive wastes)*



### Radioactive wastes

*(See also: Hazardous waste management)*



### Radon



### Rain forest



### Recreation areas



### Recycling

*(See also: Energy conservation, and charts starting on page 782)*



### Renewable energy sources

*(See also: Solar energy, Wind energy)*



### Renewable natural resources



### Reservoirs

*(See also: Water resources)*



### Risk

### Water treatment

*(See also: Water reclamation)*



### Water use

*(See also: Water reclamation)*



### Waterfowl



### Watersheds



### Wetlands

*(See also: Estuaries)*



### Whales



### Wilderness area



### Wildlife

*(See also: Animal ecology; Wildlife management)*

### Wildlife management



### Wind energy

# Grey House Publishing
# 2011 Title List

Visit **www.greyhouse.com** for Product Information, Table of Contents and Sample Pages

## General Reference

American Environmental Leaders: From Colonial Times to the Present
An African Biographical Dictionary
Encyclopedia of African-American Writing
Encyclopedia of American Industries
Encyclopedia of Emerging Industries
Encyclopedia of Global Industries
Encyclopedia of Gun Control & Gun Rights
Encyclopedia of Invasions & Conquests
Encyclopedia of Prisoners of War & Internment
Encyclopedia of Religion & Law in America
Encyclopedia of Rural America
Encyclopedia of the United States Cabinet, 1789-2010
Encyclopedia of Warrior Peoples & Fighting Groups
Environmental Resource Handbook
From Suffrage to the Senate: America's Political Women
Global Terror & Political Risk Assessment
Historical Dictionary of War Journalism
Human Rights in the United States
Nations of the World
Political Corruption in America
Speakers of the House of Representatives, 1789-2009
The Environmental Debate: A Documentary History
The Evolution Wars: A Guide to the Debates
The Religious Right: A Reference Handbook
The Value of a Dollar: 1860-2009
The Value of a Dollar: Colonial Era
University & College Museums, Galleries & Related Facilities
Weather America
World Cultural Leaders of the 20th & 21st Centuries
Working Americans 1880-1999 Vol. I: The Working Class
Working Americans 1880-1999 Vol. II: The Middle Class
Working Americans 1880-1999 Vol. III: The Upper Class
Working Americans 1880-1999 Vol. IV: Their Children
Working Americans 1880-2003 Vol. V: At War
Working Americans 1880-2005 Vol. VI: Women at Work
Working Americans 1880-2006 Vol. VII: Social Movements
Working Americans 1880-2007 Vol. VIII: Immigrants
Working Americans 1770-1869 Vol. IX: Revol. War to the Civil War
Working Americans 1880-2009 Vol. X: Sports & Recreation
Working Americans 1880-2010 Vol. XI: Entrepreneurs & Inventors

## Bowker's Books In Print® Titles

Books In Print®
Books In Print® Supplement
American Book Publishing Record® Annual
American Book Publishing Record® Monthly
Books Out Loud™
Bowker's Complete Video Directory™
Children's Books In Print®
Complete Directory of Large Print Books & Serials™
El-Hi Textbooks & Serials In Print®
Forthcoming Books®
Law Books & Serials In Print™
Medical & Health Care Books In Print™
Publishers, Distributors & Wholesalers of the US™
Subject Guide to Books In Print®
Subject Guide to Children's Books In Print®

## Business Information

Directory of Business Information Resources
Directory of Mail Order Catalogs
Directory of Venture Capital & Private Equity Firms
Food & Beverage Market Place
Grey House Homeland Security Directory
Grey House Performing Arts Directory
Hudson's Washington News Media Contacts Directory
New York State Directory
Sports Market Place Directory
The Rauch Guides – Industry Market Research Reports

## Statistics & Demographics

America's Top-Rated Cities
America's Top-Rated Small Towns & Cities
America's Top-Rated Smaller Cities
Comparative Guide to American Suburbs
Comparative Guide to Health in America
Profiles of... Series – State Handbooks

## Health Information

Comparative Guide to American Hospitals
Comparative Guide to Health in America
Complete Directory for Pediatric Disorders
Complete Directory for People with Chronic Illness
Complete Directory for People with Disabilities
Complete Mental Health Directory
Directory of Health Care Group Purchasing Organizations
Directory of Hospital Personnel
HMO/PPO Directory
Medical Device Register
Older Americans Information Directory

## Education Information

Charter School Movement
Comparative Guide to American Elementary & Secondary Schools
Complete Learning Disabilities Directory
Educators Resource Directory
Special Education

## TheStreet.com Ratings Guides

TheStreet.com Ratings Consumer Box Set
TheStreet.com Ratings Guide to Bank Fees & Service Charges
TheStreet.com Ratings Guide to Banks & Thrifts
TheStreet.com Ratings Guide to Bond & Money Market Mutual Funds
TheStreet.com Ratings Guide to Common Stocks
TheStreet.com Ratings Guide to Credit Unions
TheStreet.com Ratings Guide to Exchange-Traded Funds
TheStreet.com Ratings Guide to Health Insurers
TheStreet.com Ratings Guide to Life & Annuity Insurers
TheStreet.com Ratings Guide to Property & Casualty Insurers
TheStreet.com Ratings Guide to Stock Mutual Funds
TheStreet.com Ratings Ultimate Guided Tour of Stock Investing

## Canadian General Reference

Associations Canada
Canadian Almanac & Directory
Canadian Environmental Resource Guide
Canadian Parliamentary Guide
Financial Services Canada
History of Canada
Libraries Canada

**Grey House Publishing**
4919 Route 22, PO Box 56, Amenia NY 12501-0056 | (800) 562-2139 | www.greyhouse.com | books@greyhouse.com